$\operatorname{sgn}(\sigma)$	signature of a permutation (Theorem a
Span	span (Definition 2.4.3)
σ	standard deviation (Definition 3.8.6). A
$\sum$	sum (Section 0.1)
sup	supremum; least upper bound (Definitic
$\operatorname{Supp}(f)$	support of a function f (Definition 4.1.2)
τ	(tau) torsion (Definition 3.9.14)
$T_{\mathbf{x}}X$	tangent space to manifold (Definition 3.2.1)
tr	trace of a matrix (Definition 1.4.13)
$\vec{\mathbf{v}} \cdot \vec{\mathbf{w}}$	dot product of two vectors, (Definition 1.4.1)
$\vec{\mathbf{v}} \times \vec{\mathbf{w}}$	cross product of two vectors (Definition 1.4.17)
$\lvert\vec{\mathbf{v}}\rvert$	length of vector $\vec{\mathbf{v}}$ (Definition 1.4.2)
$(\vec{\mathbf{v}})^{\perp}$	orthogonal complement to subspace spanned by $\vec{\mathbf{v}}$ (proof of Theorem 3.7.15)
$\operatorname{Var} f$	variance (Definitions 3.8.6)
$[x]_k$	k-truncation (Definition A1.2)

Notation particular to this book

$[0]$	matrix with all entries 0 (equation 1.7.48)
$\underset{L}{=}$	equal in the sense of Lebesgue (Definition 4.11.6)
$\widehat{\ }$ (e.g., $\widehat{\vec{\mathbf{v}}_i}$)	"hat" indicating omitted factor (equation 6.5.25)
$\widetilde{A}$	result of row reducing A to echelon form (Theorem 2.1.7)
$[A\lvert\vec{\mathbf{b}}]$	matrix formed from columns of A and $\vec{\mathbf{b}}$ (sentence after equation 2.1.7)
$B_r(\mathbf{x})$	ball of radius r around $\mathbf{x}$ (Definition 1.5.1)
β_n	volume of unit ball (Example 4.5.7)
$\mathcal{D}_N(\mathbb{R}^n)$	dyadic paving (Definition 4.1.7)
$[\mathbf{Df}(\mathbf{a})]$	derivative of $\mathbf{f}$ at $\mathbf{a}$ (Proposition and Definition 1.7.9)
$\mathbb{D}$	set of finite decimals (Definition A1.4)
$D_I f$	higher partial derivatives (equation 3.3.11)
$\lvert d^n\mathbf{x}\rvert$	integrand for multiple integral (Section 4.1); see also Section 5.3
$\partial^s_M X$	smooth part of the boundary of X (Definition 6.6.2)
$\Gamma(f)$	graph of f (Definition 3.1.1)
$[h]_R$	R-truncation of h (equation 4.11.26)
$\Phi_{\vec{F}}$	flux form (Definition 6.5.2)
$\Phi_{\{\mathbf{v}\}}$	concrete to abstract function (Definition 2.6.12)
$\mathcal{I}^k_n$	set of multi-exponents (Notation 3.3.5)
$[\mathbf{Jf}(\mathbf{a})]$	Jacobian matrix (Definition 1.7.7)
$L(f)$	lower integral (Definition 4.1.10)
$m_A(f)$	infimum of $f(\mathbf{x})$ for $\mathbf{x} \in A$ (Definition 4.1.3)
M_f	mass form (Definition 6.5.4)
$M_A(f)$	supremum of $f(\mathbf{x})$ for $\mathbf{x} \in A$ (Definition 4.1.3)
$\operatorname{Mat}(n, m)$	space of $n \times m$ matrices (Discussion before Proposition 1.5.38)
$\operatorname{osc}_A(f)$	oscillation of f over A (Definition 4.1.4)
Ω	orientation (Definition 6.3.1)
$\Omega^{\{\mathbf{v}\}}$	orientation specified by $\{\mathbf{v}\}$ (paragraph after Definition 6.3.1)
Ω^{st}	standard orientation (Section 6.3)
$P^k_{f,\mathbf{a}}$	Taylor polynomial (Definition 3.3.13)
$P(\vec{\mathbf{v}}_1, \dots \vec{\mathbf{v}}_k)$	k-parallelogram (Definition 4.9.3)
$P_{\mathbf{x}}(\vec{\mathbf{v}}_1, \dots \vec{\mathbf{v}}_k)$	anchored k-parallelogram (Section 5.1)
$[P_{\mathbf{v}'\to\mathbf{v}}]$	change of basis matrix (Proposition and Definition 2.6.17)
Q_n, Q	unit n-dimensional cube (Definition 4.9.4)
U^{OK}	(equation 5.2.18)
$U(f)$	upper integral (Definition 4.1.10)
$\vec{\mathbf{v}}$ and $\mathbf{v}$	column vector and point (Definition 1.1.2)
vol_n	n-dimensional volume (Definition 4.1.17)
$W_{\vec{F}}$	work form (Definition 6.5.1)
$\dot{x}, \dot{y}$	used in denoting tangent space (paragraph before Example 3.2.2)

Vector Calculus, Linear Algebra, and Differential Forms

A Unified Approach

5th Edition

John Hamal Hubbard

Cornell University
Université Aix-Marseille

Barbara Burke Hubbard

Matrix Editions Ithaca, NY 14850 MatrixEditions.com

The Library of Congress has cataloged the 4th edition as follows:

Hubbard, John H.
Vector calculus, linear algebra, and differential forms : a unified approach / John Hamal Hubbard, Barbara Burke Hubbard. – 4th ed.
p. cm.
Includes bibliographical references and index.
ISBN 978-0-9715766-5-0 (alk. paper)
1. Calculus. 2. Algebras, Linear. I. Hubbard, Barbara Burke, 1948- II. Title.
QA303.2.H83 2009
515′.63–dc22

2009016333

214 University Ave. Ithaca, NY 14850
www.MatrixEditions.com

Printed in the United States of America

10 9 8 7 6 5 4 3 2 1

ISBN 978-0-9715766-8-1

Cover image: *The Wave* by the American painter Albert Bierstadt (1830–1902). A breaking wave is just the kind of thing to which a physicist would want to apply Stokes's theorem: the balance between surface tension (a surface integral) and gravity and momentum (a volume integral) is the key to the cohesion of the water. The crashing of the wave is poorly understood; we speculate that it corresponds to the loss of the conditions to be a piece-with-boundary (see Section 6.6). The crest carries no surface tension; when it acquires positive area it breaks the balance. In this picture, this occurs rather suddenly as you travel from left to right along the wave, but a careful look at real waves will show that the picture is remarkably accurate.

A Adrien et Régine Douady, pour l'inspiration d'une vie entière

Contents

Preface

Joseph Fourier (1768–1830)

Fourier was arrested during the French Revolution and threatened with the guillotine, but survived and later accompanied Napoleon to Egypt; in his day he was as well known for his studies of Egypt as for his contributions to mathematics and physics. He found a way to solve linear partial differential equations while studying heat diffusion. An emphasis on computationally effective algorithms is one theme of this book.

... The numerical interpretation ... is however necessary. ... So long as it is not obtained, the solutions may be said to remain incomplete and useless, and the truth which it is proposed to discover is no less hidden in the formulae of analysis than it was in the physical problem itself.

–Joseph Fourier, *The Analytic Theory of Heat*

Chapters 1 through 6 of this book cover the standard topics in multivariate calculus and a first course in linear algebra. The book can also be used for a course in analysis, using the proofs in the Appendix.

The organization and selection of material differs from the standard approach in three ways, reflecting the following principles.

First, we believe that at this level linear algebra should be more a convenient setting and language for multivariate calculus than a subject in its own right. The guiding principle of this unified approach is that locally, a nonlinear function behaves like its derivative.

When we have a question about a nonlinear function we answer it by looking carefully at a linear transformation: its derivative. In this approach, everything learned about linear algebra pays off twice: first for understanding linear equations, then as a tool for understanding nonlinear equations. We discuss abstract vector spaces in Section 2.6, but the emphasis is on $\mathbb{R}^n$, as we believe that most students find it easiest to move from the concrete to the abstract.

Second, we emphasize computationally effective algorithms, and we prove theorems by showing that these algorithms work.

We feel this better reflects the way mathematics is used today, in applied and pure mathematics. Moreover, it can be done with no loss of rigor. For linear equations, row reduction is the central tool; we use it to prove all the standard results about dimension and rank. For nonlinear equations, the cornerstone is Newton's method, the best and most widely used method for solving nonlinear equations; we use it both as a computational tool and in proving the inverse and implicit function theorems. We include a section on numerical methods of integration, and we encourage the use of computers both to reduce tedious calculations and as an aid in visualizing curves and surfaces.

Third, we use differential forms to generalize the fundamental theorem of calculus to higher dimensions.

The great conceptual simplification gained by doing electromagnetism in the language of forms is a central motivation for using forms. We apply the language of forms to electromagnetism and potentials in Sections 6.12 and 6.13.

In our experience, differential forms can be taught to freshmen and sophomores if forms are presented geometrically, as integrands that take an oriented piece of a curve, surface, or manifold, and return a number. We are aware that students taking courses in other fields need to master the language of vector calculus, and we devote three sections of Chapter 6 to integrating the standard vector calculus into the language of forms.

Other significant ways this book differs from standard texts include

◇ Applications involving big matrices
◇ The treatment of eigenvectors and eigenvalues
◇ Lebesgue integration
◇ Rules for computing Taylor polynomials

A few minutes spent on the Internet finds a huge range of applications of principal component analysis.

Big data Example 2.7.12 discussing the Google PageRank algorithm shows the power of the Perron-Frobenius theorem. Example 3.8.10 illustrates an application of principal component analysis, which is built on the singular value decomposition.

Eigenvectors and eigenvalues In keeping with our prejudice in favor of computationally effective algorithms, we provide in Section 2.7 a theory of eigenvectors and eigenvalues that bypasses determinants, which are more or less uncomputable for large matrices. This treatment is also stronger theoretically: Theorem 2.7.9 gives an "if and only if" statement for the existence of eigenbases. In addition, our emphasis on defining an eigenvector $\mathbf{v}$ as satisfying $A\mathbf{v} = \lambda\mathbf{v}$ has the advantage of working when A is a linear transformation between infinite-dimensional vector spaces, whereas the definition in terms of roots of the characteristic polynomial does not. However, in Section 4.8 we define the characteristic polynomial of a matrix, connecting eigenvalues and eigenvectors to determinants.

In our experience, undergraduates, even freshmen, are quite prepared to approach the Lebesgue integral via the Riemann integral, but the approach via measurable sets and σ-algebras of measurable sets is inconceivable.

Lebesgue integration We give a new approach to Lebesgue integration, tying it much more closely to Riemann integrals. We had two motivations. First, integrals over unbounded domains and integrals of unbounded functions are really important, for instance in physics and probability, and students will need to know about such integrals before they take a course in analysis. Second, there simply does not appear to be a successful theory of improper multiple integrals.

Rules for computing Taylor polynomials Even good graduate students are often unaware of the rules that make computing Taylor polynomials in higher dimensions palatable. We give these in Section 3.4.

How the book has evolved: the first four editions

The first edition of this book, published by Prentice Hall in 1999, was a mere 687 pages. The basic framework of our guiding principles was there, but we had no Lebesgue integration and no treatment of electromagnetism.

The second edition, published by Prentice Hall in 2002, grew to 800 pages. The biggest change was replacing improper integrals by Lebesgue integrals. We also added approximately 270 new exercises and 50 new examples and reworked the treatment of orientation. This edition first saw the inclusion of photos of mathematicians.

In September 2006 we received an email from Paul Bamberg, senior lecturer on mathematics at Harvard University, saying that Prentice Hall had declared the book out of print (something Prentice Hall had neglected to mention to us). We obtained the copyright and set to work on the third edition. We put exercises in small type, freeing up space for quite a bit of new material, including a section on electromagnetism; a discussion of eigenvectors, eigenvalues, and diagonalization; a discussion of the determinant and eigenvalues; and a section on integration and curvature.

The major impetus for the fourth edition (2009) was that we finally hit on what we consider the right way to define orientation of manifolds. We also expanded the page count to 818, which made it possible to add a proof of Gauss's *theorem egregium*, a discussion of Faraday's experiments, a trick for finding Lipschitz ratios for polynomial functions, a way to classify constrained critical points using the augmented Hessian matrix, and a proof of Poincaré's lemma for arbitrary forms, using the cone operator.

What's new in the fifth edition

The initial impetus for producing a new edition rather than reprinting the fourth edition was to reconcile differences in numbering (propositions, examples, etc.) in the first and second printings of the fourth edition.

An additional impetus came from discussions John Hubbard had at an AMS meeting in Washington, DC, in October 2014. Mathematicians and computer scientists there told him that existing textbooks lack examples of "big matrices". This led to two new examples illustrating the power of linear algebra and calculus: Example 2.7.12, showing how Google uses the Perron-Frobenius theorem to rank web pages, and Example 3.8.10, showing how the singular value decomposition (Theorem 3.8.1) can be used for computer face recognition.

The more we worked on the new edition, the more we wanted to change. An inclusive list is impossible; here are some additional highlights.

- ◇ In several places in the fourth edition (for instance, the proof of Proposition 6.4.8 on orientation-preserving parametrizations) we noted that "in Chapter 3 we failed to define the differentiability of functions defined on manifolds, and now we pay the price". For this edition we "paid the price" (Proposition and Definition 3.2.9) and the effort paid off handsomely, allowing us to shorten and simplify a number of proofs.
- ◇ We rewrote the discussion of multi-index notation in Section 3.3.

- ◇ We rewrote the proof of Stokes's theorem and moved most of it out of the Appendix and into the main text.
- ◇ We added Example 2.4.17 on Fourier series.
- ◇ We rewrote the discussion of classifying constrained critical points.
- ◇ We use differentiation under the integral sign to compute the Fourier transform of the Gaussian, and discuss its relation to the Heisenberg uncertainty principle.
- ◇ We added a new proposition (2.4.18) about orthonormal bases.
- ◇ We greatly expanded the discussion of orthogonal matrices.
- ◇ We added a new section in Chapter 3 on finite probability, showing the connection between probability and geometry. The new section also includes the statement and proof of the singular value decomposition.
- ◇ We added about 40 new exercises.

A student solution manual, with solutions to odd-numbered exercises, is available from Matrix Editions. Instructors who wish to acquire the instructors' solution manual should write hubbard@matrixeditions.com.

Practical information

Chapter 0 and back cover Chapter 0 is intended as a resource. We recommend that students skim through it to see if any material is unfamiliar. The inside back cover lists some useful formulas.

Jean Dieudonné (1906–1992)

Dieudonné, one of the founding members of "Bourbaki", a group of young mathematicians who published collectively under the pseudonym Nicolas Bourbaki, and whose goal was to put modern mathematics on a solid footing, was the personification of rigor in mathematics. Yet in his book *Infinitesimal Calculus* he put the harder proofs in small type, saying "a beginner will do well to accept plausible results without taxing his mind with subtle proofs."

Errata Errata will be posted at

http://www.MatrixEditions.com

Exercises Exercises are given at the end of each section; chapter review exercises are given at the end of each chapter, except Chapter 0 and the Appendix. Exercises range from very easy exercises intended to make students familiar with vocabulary, to quite difficult ones. The hardest exercises are marked with an asterisk (in rare cases, two asterisks).

Notation Mathematical notation is not always uniform. For example, $|A|$ can mean the length of a matrix A (the usage in this book) or the determinant of A (which we denote by $\det A$). Different notations for partial derivatives also exist. This should not pose a problem for readers who begin at the beginning and end at the end, but for those who are using only selected chapters, it could be confusing. Notations used in the book are listed on the front inside cover, along with an indication of where they are first introduced.

In this edition, we have changed the notation for sequences: we now denote sequences by "$i \mapsto x_i$" rather than "x_i" or "$x_1, x_2, \dots$". We are also more careful to distinguish between equalities that are true by definition, denoted $\stackrel{\text{def}}{=}$, and those true by reasoning, denoted $=$. But, to avoid too heavy notation, we write $=$ in expressions like "set $\mathbf{x} = \gamma(\mathbf{u})$".

Numbering Theorems, lemmas, propositions, corollaries, and examples share the same numbering system: Proposition 2.3.6 is not the sixth proposition of Section 2.3; it is the sixth numbered item of that section.

Figures and tables share their own numbering system; Figure 4.5.2 is the second figure or table of Section 4.5. Virtually all displayed equations and inequalities are numbered, with the numbers given at right; equation 4.2.3 is the third equation of Section 4.2.

We often refer back to theorems, examples, and so on, and believe this numbering makes them easier to find.

Readers are welcome to propose additional programs (or translations of these programs into other programming languages); if interested, please write John Hubbard at jhh8@cornell.edu.

Programs The NEWTON.M program used in Section 2.8 works in MATLAB; it is posted at http://MatrixEditions.com/Programs.html. (Two other programs available there, MONTE CARLO and DETERMINANT, are written in PASCAL and probably no longer usable.)

Symbols We use $\triangle$ to mark the end of an example or remark, and $\square$ to mark the end of a proof. Sometimes we specify what proof is being ended: $\square$ Corollary 1.6.16 means "end of the proof of Corollary 1.6.16".

Using this book as a calculus text or as an analysis text

This book can be used at different levels of rigor. Chapters 1 through 6 contain material appropriate for a course in linear algebra and multivariate calculus. Appendix A contains the technical, rigorous underpinnings appropriate for a course in analysis. It includes proofs of those statements not proved in the main text, and a painstaking justification of arithmetic.

In deciding what to include in this appendix, and what to put in the main text, we used the analogy that learning calculus is like learning to drive a car with standard transmission – acquiring the understanding and intuition to shift gears smoothly when negotiating hills, curves, and the stops and starts of city streets. Analysis is like designing and building a car. To use this book to "learn how to drive", Appendix A should be omitted.

The SAT test used to have a section of analogies; the "right" answer sometimes seemed contestable. In that spirit,

Calculus is to analysis as playing a sonata is to composing one.

Calculus is to analysis as performing in a ballet is to choreographing it.

Analysis involves more painstaking technical work, which at times may seem like drudgery, but it provides a level of mastery that calculus alone cannot give.

Most of the proofs included in this appendix are more difficult than the proofs in the main text, but difficulty was not the only criterion; many students find the proof of the fundamental theorem of algebra (Section 1.6) quite difficult. But we find this proof qualitatively different from the proof of the Kantorovich theorem, for example. A professional mathematician who has understood the proof of the fundamental theorem of algebra should be able to reproduce it. A professional mathematician who has read through the proof of the Kantorovich theorem, and who agrees that each step is justified, might well want to refer to notes in order to reproduce it. In this sense, the first proof is more conceptual, the second more technical.

One-year courses

At Cornell University this book is used for the honors courses Math 2230 and 2240. Students are expected to have a 5 on the Advanced Placement BC Calculus exam, or the equivalent. When John Hubbard teaches the course, he typically gets to the middle of Chapter 4 in the first semester, sometimes skipping Section 3.9 on the geometry of curves and surfaces, and going through Sections 4.2–4.4 rather rapidly, in order to get to Section 4.5 on Fubini's theorem and begin to compute integrals. In the second semester he gets to the end of Chapter 6 and goes on to teach some of the material that will appear in a sequel volume, in particular differential equations.[1]

One could also spend a year on Chapters 1–6. Some students might need to review Chapter 0; others may be able to include some proofs from the appendix.

Semester courses

1. A semester course for students who have had a solid course in linear algebra

We used an earlier version of this text with students who had taken a course in linear algebra, and feel they gained a great deal from seeing how linear algebra and multivariate calculus mesh. Such students could be expected to cover chapters 1–6, possibly omitting some material. For a less fast-paced course, the book could also be covered in a year, possibly including some proofs from the appendix.

2. A semester course in analysis for students who have studied multivariable calculus

In one semester one could hope to cover all six chapters and some or most of the proofs in the appendix. This could be done at varying levels of difficulty; students might be expected to follow the proofs, for example, or they might be expected to understand them well enough to construct similar proofs.

Use by graduate students

Many graduate students have told us that they found the book very useful in preparing for their qualifying exams.

John H. Hubbard

Barbara Burke Hubbard

jhh8@cornell.edu, hubbard@matrixeditions.com

John H. Hubbard (BA Harvard University, PhD University of Paris) is professor of mathematics at Cornell University and professor emeritus at the Université Aix-Marseille; he is the author of several books on differential equations (with Beverly West), a book on Teichmüller theory, and a two-volume book in French on scientific computing (with Florence Hubert). His research mainly concerns

[1]Eventually, he would like to take three semesters to cover chapters 1–6 of the current book and material in the forthcoming sequel, including differential equations, inner products (with Fourier analysis and wavelets), and advanced topics in differential forms.

complex analysis, differential equations, and dynamical systems. He believes that mathematics research and teaching are activities that enrich each other and should not be separated.

Barbara Burke Hubbard (BA Harvard University) is the author of *The World According to Wavelets*, which was awarded the prix d'Alembert by the French Mathematical Society in 1996. She founded Matrix Editions in 2002.

ACKNOWLEDGMENTS

For changes in this edition, we'd like in particular to thank Matthew Ando, Paul Bamberg, Alexandre Bardet, Xiaodong Cao, Calvin Chong, Kevin Huang, Jon Kleinberg, Tan Lei, and Leonidas Nguyen.

Many people – colleagues, students, readers, friends – contributed to the previous editions and thus also contributed to this one. We are grateful to them all: Nikolas Akerblom, Travis Allison, Omar Anjum, Ron Avitzur, Allen Back, Adam Barth, Nils Barth, Brian Beckman, Barbara Beeton, David Besser, Daniel Bettendorf, Joshua Bowman, Robert Boyer, Adrian Brown, Ryan Budney, Xavier Buff, Der-Chen Chang, Walter Chang, Robin Chapman, Gregory Clinton, Adrien Douady, Régine Douady, Paul DuChateau, Bill Dunbar, David Easley, David Ebin, Robert Ghrist, Manuel Heras Gilsanz, Jay Gopalakrishnan, Robert Gross, Jean Guex, Dion Harmon, Skipper Hartley, Matt Holland, Tara Holm, Chris Hruska, Ashwani Kapila, Jason Kaufman, Todd Kemp, Ehssan Khanmohammadi, Hyun Kyu Kim, Sarah Koch, Krystyna Kuperberg, Daniel Kupriyenko, Margo Levine, Anselm Levskaya, Brian Lukoff, Adam Lutoborski, Thomas Madden, Francisco Martin, Manuel López Mateos, Jim McBride, Mark Millard.

We also thank John Milnor, Colm Mulcahy, Ralph Oberste-Vorth, Richard Palas, Karl Papadantonakis, Peter Papadopol, David Park, Robert Piche, David Quinonez, Jeffrey Rabin, Ravi Ramakrishna, Daniel Alexander Ramras, Oswald Riemenschneider, Lewis Robinson, Jon Rosenberger, Bernard Rothman, Johannes Rueckert, Ben Salzberg, Ana Moura Santos, Dierk Schleicher, Johan De Schrijver, George Sclavos, Scott Selikoff, John Shaw, Ted Shifrin, Leonard Smiley, Birgit Speh, Jed Stasch, Mike Stevens, Ernest Stitzinger, Chan-Ho Suh, Shai Szulanski, Robert Terrell, Eric Thurschwell, Stephen Treharne, Leo Trottier, Vladimir Veselov, Hans van den Berg, Charles Yu, and Peng Zhao.

We thank Philippe Boulanger of Pour la Science for many pictures of mathematicians. The MacTutor History of Mathematics archive at www-groups.dcs.st-and.ac.uk/ history/ was helpful in providing historical information.

We also wish to thank our children, Alexander Hubbard, Eleanor Hubbard (creator of the goat picture in Section 3.9), Judith Hubbard, and Diana Hubbard.

We apologize to anyone whose name has been inadvertently omitted.

Producing this book and the previous editions would have been a great deal more difficult without the mathematical typesetting program Textures, created by Barry Smith. With Textures, an 800-page book can be typeset in seconds. Mr. Smith died in 2012; he is sorely missed. For updates on efforts keep Textures available for Macs, see www.blueskytex.com.

We also wish to thank Cornell undergraduates in Math 2230 and 2240 (formerly, Math 223 and 224).

0

Preliminaries

Allez en avant, et la foi vous viendra
(Keep going; faith will come.)—Jean d'Alembert (1717–1783), to those questioning calculus

0.0 Introduction

This chapter is intended as a resource. You may be familiar with its contents, or there may be topics you never learned or that you need to review. You should not feel that you need to master Chapter 0 before beginning Chapter 1; just refer back to it as needed. (A possible exception is Section 0.7 on complex numbers.)

We have included reminders in the main text; for example, in Section 1.5 we write, "You may wish to review the discussion of quantifiers in Section 0.2."

In Section 0.1 we share some guidelines that in our experience make reading mathematics easier, and discuss specific issues like sum notation.

Section 0.2 analyzes the rather tricky business of negating mathematical statements. (To a mathematician, the statement "All eleven-legged alligators are orange with blue spots" is an obviously true statement, not an obviously meaningless one.) We first use this material in Section 1.5.

Set theory notation is discussed in Section 0.3. The "eight words" of set theory are used beginning in Section 1.1. The discussion of Russell's paradox is not necessary; we include it because it is fun and not hard.

Section 0.4 defines the word "function" and discusses the relationship between a function being "onto" or "one to one" and the existence and uniqueness of solutions. This material is first needed in Section 1.3.

Most of this text concerns real numbers, but we think that anyone beginning a course in multivariate calculus should know what complex numbers are and be able to compute with them.

Real numbers are discussed in Section 0.5, in particular, least upper bounds, convergence of sequences and series, and the intermediate value theorem. This material is first used in Sections 1.5 and 1.6.

The discussion of countable and uncountable sets in Section 0.6 is fun and not hard. These notions are fundamental to Lebesgue integration.

In our experience, most students studying vector calculus for the first time are comfortable with complex numbers, but a sizable minority have either never heard of complex numbers or have forgotten everything they once knew. If you are among them, we suggest reading at least the first few pages of Section 0.7 and doing some of the exercises.

0.1 Reading mathematics

The most efficient logical order for a subject is usually different from the best psychological order in which to learn it. Much mathematical writing is based too closely on the logical order of deduction in a subject, with too many definitions without, or before, the examples which motivate them, and too many answers before, or without, the questions they address.—William Thurston

Many students do well in high school mathematics courses without reading their texts. At the college level you are expected to read the book. Better yet, read ahead. If you read a section before listening to a lecture on it, the lecture will be more comprehensible, and if there is something in the text you don't understand, you will be able to listen more actively and ask questions.

Reading mathematics is different from other reading. We think the following guidelines can make it easier. There are two parts to understanding a theorem: understanding the statement, and understanding the proof. *The first is more important than the second.*

What if you don't understand the statement? If there's a symbol in the formula you don't understand, perhaps a δ, look to see whether the next line continues, "where δ is such and such." In other words, read the whole sentence before you decide you can't understand it.

If you're still having trouble, *skip ahead to examples.* This may contradict what you have been told – that mathematics is sequential, and that you must understand each sentence before going on to the next. In reality, although mathematical writing is necessarily sequential, mathematical understanding is not: you (and the experts) never understand perfectly up to some point and not at all beyond. The "beyond", where understanding is only partial, is an essential part of the motivation and the conceptual background of the "here and now". You may often find that when you return to something you left half-understood, it will have become clear in the light of the further things you have studied, even though the further things are themselves obscure.

Many students are uncomfortable in this state of partial understanding, like a beginning rock climber who wants to be in stable equilibrium at all times. To learn effectively one must be willing to leave the cocoon of equilibrium. *If you don't understand something perfectly, go on ahead and then circle back.*

In particular, an example will often be easier to follow than a general statement; you can then go back and reconstitute the meaning of the statement in light of the example. Even if you still have trouble with the general statement, you will be ahead of the game if you understand the examples. We feel so strongly about this that we have sometimes flouted mathematical tradition and given examples before the proper definition.

Read with pencil and paper in hand, making up little examples for yourself as you go on.

Some of the difficulty in reading mathematics is notational. A pianist who has to stop and think whether a given note on the staff is A or F will not be able to sight-read a Bach prelude or Schubert sonata. The temptation, when faced with a long, involved equation, may be to give up. You need to take the time to identify the "notes".

The Greek Alphabet

Greek letters that look like Roman letters are not used as mathematical symbols; for example, A is capital a, not capital α. The letter χ is pronounced "kye" to rhyme with "sky"; φ, ψ, and ξ may rhyme with either "sky" or "tea".

α	A	alpha
β	B	beta
γ	Γ	gamma
δ	Δ	delta
ϵ	E	epsilon
ζ	Z	zeta
η	H	eta
θ	Θ	theta
ι	I	iota
κ	K	kappa
λ	Λ	lambda
μ	M	mu
ν	N	nu
ξ	Ξ	xi
o	O	omicron
π	Π	pi
ρ	P	rho
σ	Σ	sigma
τ	T	tau
υ	Υ	upsilon
φ, ϕ	Φ	phi
χ	X	chi
ψ	Ψ	psi
ω	Ω	omega

Learn the names of Greek letters – not just the obvious ones like alpha, beta, and pi, but the more obscure psi, xi, tau, omega. The authors know a mathematician who calls all Greek letters "xi" (ξ), except for omega (ω), which he calls "w". This leads to confusion. Learn not just to recognize these letters, but how to pronounce them. Even if you are not reading mathematics out loud, it is hard to think about formulas if ξ, ψ, τ, ω, φ are all "squiggles" to you.

Sum and product notation

In equation 0.1.3, the symbol $\sum_{k=1}^{n}$ says that the sum will have n terms. Since the expression being summed is $a_{i,k}b_{k,j}$, each of those n terms will have the form ab.

Usually the quantity being summed has an index matching the index of the sum (for instance, k in formula 0.1.1). If not, it is understood that you add one term for every "whatever" that you are summing over. For example, $\sum_{i}^{10} 1 = 10$.

Sum notation can be confusing at first; we are accustomed to reading in one dimension, from left to right, but something like

$$\sum_{k=1}^{n} a_{i,k}b_{k,j} \qquad 0.1.1$$

requires what we might call two-dimensional (or even three-dimensional) thinking. It may help at first to translate a sum into a linear expression:

$$\sum_{i=0}^{\infty} 2^i = 2^0 + 2^1 + 2^2 \dots \qquad 0.1.2$$

$$\text{or} \quad c_{i,j} = \sum_{k=1}^{n} a_{i,k}b_{k,j} = a_{i,1}b_{1,j} + a_{i,2}b_{2,j} + \cdots + a_{i,n}b_{n,j}. \qquad 0.1.3$$

Two $\sum$ placed side by side do not denote the product of two sums; one sum is used to talk about one index, the other about another. The same thing could be written with one $\sum$, with information about both indices underneath. For example,

$$\begin{aligned} \sum_{i=1}^{3}\sum_{j=2}^{4}(i+j) &= \sum_{\substack{i \text{ from 1 to 3,} \\ j \text{ from 2 to 4}}} (i+j) \\ &= \left(\sum_{j=2}^{4} 1+j\right) + \left(\sum_{j=2}^{4} 2+j\right) + \left(\sum_{j=2}^{4} 3+j\right) \qquad 0.1.4 \\ &= \big((1+2)+(1+3)+(1+4)\big) \\ &\quad + \big((2+2)+(2+3)+(2+4)\big) \\ &\quad + \big((3+2)+(3+3)+(3+4)\big); \end{aligned}$$

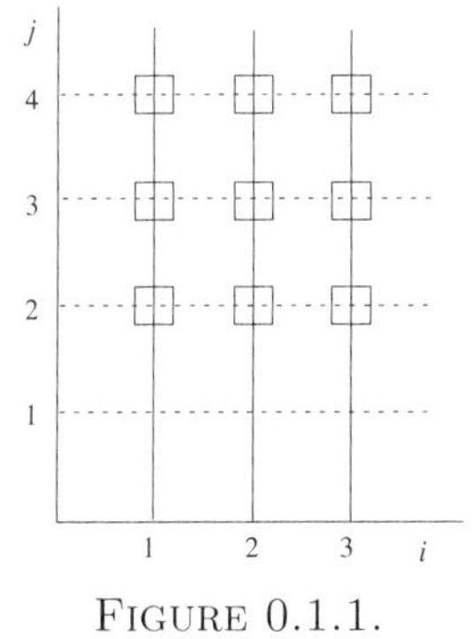

FIGURE 0.1.1.

In the double sum of equation 0.1.4, each sum has three terms, so the double sum has nine terms.

this double sum is illustrated in Figure 0.1.1.

Rules for product notation $\prod$ are analogous to those for sum notation:

$$\prod_{i=1}^{n} a_i = a_1 \cdot a_2 \cdots a_n; \quad \text{for example,} \quad \prod_{i=1}^{n} i = n!. \qquad 0.1.5$$

Proofs

We said earlier that it is more important to understand a mathematical statement than to understand its proof. We have put some of the harder proofs in Appendix A; these can safely be skipped by a student studying multivariate calculus for the first time. We urge you, however, to read the proofs in the main text. By reading many proofs you will learn what a proof is, so that (for one thing) you will know when you have proved something and when you have not.

In addition, a good proof doesn't just convince you that something is true; it tells you why it is true. You presumably don't lie awake at night worrying about the truth of the statements in this or any other math textbook. (This is known as "proof by eminent authority": you assume the authors know what they are talking about.) But reading the proofs will help you understand the material.

If you get discouraged, keep in mind that the contents of this book represent a cleaned-up version of many false starts. For example, John Hubbard started by trying to prove Fubini's theorem in the form presented in equation 4.5.1. When he failed, he realized (something he had known and forgotten) that the statement was in fact false. He then went through a stack of scrap paper before coming up with a correct proof. Other statements in the book represent the efforts of some of the world's best mathematicians over many years.

When Jacobi complained that Gauss's proofs appeared unmotivated, Gauss is said to have answered, *You build the building and remove the scaffolding.* Our sympathy is with Jacobi's reply: he likened Gauss to *the fox who erases his tracks in the sand with his tail.*

FIGURE 0.1.2.
Nathaniel Bowditch (1773–1838)

0.2 Quantifiers and negation

Interesting mathematical statements are seldom like "2 + 2 = 4"; more typical is the statement "every prime number such that if you divide it by 4 you have a remainder of 1 is the sum of two squares." In other words, most interesting mathematical statements are about infinitely many cases; in the case above, it is about all those prime numbers such that if you divide them by 4 you have a remainder of 1 (there are infinitely many such numbers).

In a mathematical statement, every variable has a corresponding quantifier, either implicit or explicitly stated. There are two such quantifiers: "for all" (the *universal quantifier*), written symbolically $\forall$, and "there exists" (the *existential quantifier*), written $\exists$. Above we have a single quantifier, "every". More complicated statements have several quantifiers, for example, the statement, "For all $x \in \mathbb{R}$ and for all $\epsilon > 0$, there exists $\delta > 0$ such that for all $y \in \mathbb{R}$, if $|y - x| < \delta$, then $|y^2 - x^2| < \epsilon$." This true statement says that the squaring function is continuous.

The order in which these quantifiers appears matters. If we change the order of quantifiers in the preceding statement about the squaring function to "For all $\epsilon > 0$, there exists $\delta > 0$ such that for all $x, y \in \mathbb{R}$, if $|y - x| < \delta$, then $|y^2 - x^2| < \epsilon$," we have a meaningful mathematical sentence but it is false. (It claims that the squaring function is uniformly continuous, which it is not.)

According to a contemporary, the French mathematician Laplace (1749–1827) wrote *il est aisé à voir* ("it's easy to see") whenever he couldn't remember the details of a proof.

"I never come across one of Laplace's '*Thus it plainly appears*' without feeling sure that I have hours of hard work before me to fill up the chasm and find out and show *how* it plainly appears," wrote Bowditch.

Forced to leave school at age 10 to help support his family, the American Bowditch taught himself Latin in order to read Newton, and French in order to read French mathematics. He made use of a scientific library captured by a privateer and taken to Salem. In 1806 he was offered a professorship at Harvard but turned it down.

Even professional mathematicians have to be careful when negating a mathematical statement with several quantifiers. The rules are:

1. The opposite of

$$\begin{aligned} &[\text{For all } x,\ P(x) \text{ is true}] \\ \text{is}\quad &[\text{There exists } x \text{ for which } P(x) \text{ is not true}]. \end{aligned} \qquad 0.2.1$$

Above, P stands for "property." Symbolically the sentence is written

$$\text{The opposite of} \quad (\forall x)P(x) \quad \text{is} \quad (\exists x) \text{ not } P(x). \qquad 0.2.2$$

Another standard notation for $(\exists x)$ not $P(x)$ is $(\exists x)|$ not $P(x)$, where the bar | means "such that."

2. The opposite of

$$\begin{aligned} &[\text{There exists } x \text{ for which } P(x) \text{ is true}] \\ \text{is}\quad &[\text{For all } x, P(x) \text{ is not true}]. \end{aligned} \qquad 0.2.3$$

Symbolically the same sentence is written

$$\text{The opposite of} \quad (\exists x)P(x) \quad \text{is} \quad (\forall x) \text{ not } P(x). \qquad 0.2.4$$

These rules may seem reasonable and simple. Clearly the opposite of the (false) statement "All rational numbers equal 1" is the statement "There exists a rational number that does not equal 1."

However, by the same rules, the statement, "All eleven-legged alligators are orange with blue spots" is true, since if it were false, then there would exist an eleven-legged alligator that is not orange with blue spots. The statement, "All eleven-legged alligators are black with white stripes" is equally true.

In addition, mathematical statements are rarely as simple as "All rational numbers equal 1." Often there are many quantifiers, and even the experts have to watch out. At a lecture attended by one of the authors, it was not clear to the audience in what order the lecturer was taking the quantifiers; when he was forced to write down a precise statement, he discovered that he didn't know what he meant and the lecture fell apart.

Example 0.2.1 (Order of quantifiers). The statement

$$(\forall n \text{ integer}, n \geq 2)(\exists p \text{ prime})\ n/p \text{ is an integer} \quad \text{is true,}$$

$$(\exists p \text{ prime})(\forall n \text{ integer}, n \geq 2)\ n/p \text{ is an integer} \quad \text{is false.}$$

Example 0.2.2 (Order of quantifiers in defining continuity). In the definitions of continuity and uniform continuity, the order of quantifiers really counts. A function f is *continuous* if for all x, and for all $\epsilon > 0$, there exists $\delta > 0$ such that for all y, if $|x - y| < \delta$, then $|f(x) - f(y)| < \epsilon$. That is, f is continuous if

$$(\forall x)(\forall \epsilon > 0)(\exists \delta > 0)(\forall y)\ |x - y| < \delta \text{ implies } |f(x) - f(y)| < \epsilon, \qquad 0.2.5$$

Note that in ordinary English, the word "any" can be used to mean either "for all" or "there exists". The sentence "any execution of an innocent person invalidates the death penalty" means one single execution; the sentence "any fool knows that" means "every fool knows that". Usually in mathematical writing the meaning is clear from context, but not always. The solution is to use language that sounds stilted, but is at least unambiguous.

Most mathematicians avoid the symbolic notation, instead writing out quantifiers in full, as in formula 0.2.1. But when there is a complicated string of quantifiers, they often use the symbolic notation to avoid ambiguity.

Statements that to the ordinary mortal are false or meaningless are thus accepted as true by mathematicians; if you object, the mathematician will retort, "find me a counterexample."

Notice that when we have a "for all" followed by "there exists", the thing that exists is allowed to depend on the preceding variable. For instance, when we write "for all ϵ there exists δ", there can be a different δ for each ϵ. But if we write "there exists δ such that for all ϵ" , the single δ has to work for all ϵ.

which can also be written

$$(\forall x)(\forall \epsilon > 0)(\exists \delta > 0)(\forall y)\ \Big(|x - y| < \delta \implies |f(x) - f(y)| < \epsilon\Big). \qquad 0.2.6$$

A function f is *uniformly continuous* if for all $\epsilon > 0$, there exists $\delta > 0$ such that, for all x and all y, if $|x - y| < \delta$, then $|f(x) - f(y)| < \epsilon$. That is, f is uniformly continuous if

$$(\forall \epsilon > 0)(\exists \delta > 0)(\forall x)(\forall y)\ \big(|x - y| < \delta \implies |f(x) - f(y)| < \epsilon\big). \qquad 0.2.7$$

For the continuous function, we can choose *different* δ for different x; for the uniformly continuous function, we start with ϵ and have to find a *single* δ that works for all x.

For example, the function $f(x) = x^2$ is continuous but not uniformly continuous: as you choose bigger and bigger x, you will need a smaller δ if you want the statement $|x - y| < \delta$ to imply $|f(x) - f(y)| < \epsilon$, because the function keeps climbing more and more steeply. But $\sin x$ is uniformly continuous; you can find one δ that works for all x and all y. △

It is often easiest to negate a complicated mathematical sentence using symbolic notation: replace every ∀ by ∃ and vice versa, and then negate the conclusion. For example, to negate formula 0.2.5, write

$$(\exists x)(\exists \epsilon > 0)(\forall \delta > 0)(\exists y)$$

such that

$$|x - y| < \delta \text{ and } |f(x) - f(y)| \geq \epsilon.$$

Of course one could also negate formula 0.2.5, or any mathematical statement, by putting "not" in the very front, but that is not very useful when you are trying to determine whether a complicated statement is true.

You can also reverse some leading quantifiers, then insert a "not" and leave the remainder as it was. Usually getting the not at the end is most useful: you finally come down to a statement that you can check.

Exercise for Section 0.2

0.2.1 Negate the following statements:

a. Every prime number such that if you divide it by 4 you have a remainder of 1 is the sum of two squares.

b. For all $x \in \mathbb{R}$ and for all $\epsilon > 0$, there exists $\delta > 0$ such that for all $y \in \mathbb{R}$, if $|y - x| < \delta$, then $|y^2 - x^2| < \epsilon$.

c. For all $\epsilon > 0$, there exists $\delta > 0$ such that for all $x, y \in \mathbb{R}$, if $|y - x| < \delta$, then $|y^2 - x^2| < \epsilon$.

0.2.2 Explain why one of these statements is true and the other false:

$$(\forall \text{ man } M)(\exists \text{ woman } W) \mid W \text{ is the mother of } M$$
$$(\exists \text{ woman } W)(\forall \text{ man } M) \mid W \text{ is the mother of } M$$

0.3 Set theory

FIGURE 0.3.1.
An artist's image of Euclid.

The Latin word *locus* means "place"; its plural is *loci*.

There is nothing new about the concept of a "set" composed of elements such that some property is true. Euclid spoke of geometric *loci*, a locus being the set of points defined by some property. But historically, mathematicians apparently did not think in terms of sets, and the introduction of set theory was part of a revolution at the end of the nineteenth century that included topology and measure theory; central to this revolution was Cantor's discovery (discussed in Section 0.6) that some infinities are bigger than others.

At the level at which we are working, set theory is a language, with a vocabulary consisting of eight words:

$\in$	"is an element of"
$\{a \mid p(a)\}$	"the set of a such that $p(a)$ is true"
$=$	"equality"; $A = B$ if A and B have the same elements.
$\subset$	"is a subset of": $A \subset B$ means that every element of A is an element of B. Note that with this definition, every set is a subset of itself: $A \subset A$, and the empty set $\emptyset$ is a subset of every set.
$\cap$	"intersect": $A \cap B$ is the set of elements of both A and B.
$\cup$	"union": $A \cup B$ is the set of elements of either A or B or both.
$\times$	"cross": $A \times B$ is the set of ordered pairs (a, b) with $a \in A$ and $b \in B$.
$-$	"complement": $A - B$ is the set of elements in A that are not in B.

> In spoken mathematics, the symbols $\in$ and $\subset$ often become "in": $\mathbf{x} \in \mathbb{R}^n$ becomes "x in Rn" and $U \subset \mathbb{R}^n$ becomes "U in Rn". Make sure you know whether "in" means element of or subset of.
>
> The symbol $\notin$ ("not in") means "not an element of"; similarly, $\not\subset$ means "not a subset of" and $\neq$ means "not equal".
>
> The expression "$x, y \in V$" means that x and y are both elements of V.
>
> In mathematics, the word "or" means one or the other or both.

You should think that *set, subset, intersection, union,* and *complement* mean precisely what they mean in English. However, this suggests that any property can be used to define a set; we will see, when we discuss Russell's paradox, that this is too naive. But for our purposes, naive set theory is sufficient.

The symbol $\emptyset$ denotes the empty set, which has no elements, and is a subset of every set. There are also sets of numbers with standard names; they are written in *blackboard bold*, a font we use only for these sets.

$\mathbb{N}$	the *natural numbers* $\{0, 1, 2, \dots\}$
$\mathbb{Z}$	the *integers*, i.e., signed whole numbers $\{\dots, -1, 0, 1, \dots\}$
$\mathbb{Q}$	the *rational numbers* p/q, with $p, q \in \mathbb{Z}$, $q \neq 0$
$\mathbb{R}$	the *real numbers*, which we will think of as infinite decimals
$\mathbb{C}$	the *complex numbers* $\{a + ib \mid a, b \in \mathbb{R}\}$

> $\mathbb{N}$ is for "natural", $\mathbb{Z}$ is for "Zahl", the German word for number, $\mathbb{Q}$ is for "quotient", $\mathbb{R}$ is for "real", and $\mathbb{C}$ is for "complex".
>
> When writing with chalk on a blackboard, it's hard to distinguish between normal letters and bold letters. Blackboard bold font is characterized by double lines, as in $\mathbb{N}$ and $\mathbb{R}$.
>
> Some authors do not include 0 in $\mathbb{N}$.

Often we use slight variants of the notation above: $\{3, 5, 7\}$ is the set consisting of $3, 5,$ and 7; more generally, the set consisting of some list of elements is denoted by that list, enclosed in curly brackets, as in

$$\{\, n \mid n \in \mathbb{N} \text{ and } n \text{ is even}\,\} = \{0, 2, 4, \dots\}, \qquad 0.3.1$$

where again the vertical line $\mid$ means "such that".

The symbols are sometimes used backwards; for example, $A \supset B$ means $B \subset A$, as you probably guessed. Expressions are sometimes condensed:

$$\{\, x \in \mathbb{R} \mid x \text{ is a square}\,\} \quad \text{means} \quad \{\, x \mid x \in \mathbb{R} \text{ and } x \text{ is a square}\,\} \qquad 0.3.2$$

(i.e., the set of nonnegative real numbers).

A slightly more elaborate variation is *indexed unions and intersections*: if S_α is a collection of sets indexed by $\alpha \in A$, then $\bigcap_{\alpha \in A} S_\alpha$ denotes the intersection of all the S_α, and $\bigcup_{\alpha \in A} S_\alpha$ denotes their union. For instance, if $l_n \subset \mathbb{R}^2$ is the line of equation $y = n$, then $\bigcup_{n \in \mathbb{Z}} l_n$ is the set of points in the plane whose y-coordinate is an integer.

Although it may seem a bit pedantic, you should notice that

$$\bigcup_{n \in \mathbb{Z}} l_n \quad \text{and} \quad \{l_n |\ n \in \mathbb{Z}\}$$

are not the same thing: the first is a subset of the plane; an element of it is a point on one of the lines. The second is a set of lines, not a set of points. This is similar to one of the molehills which became mountains in the new-math days: telling the difference between ϕ and $\{\phi\}$, the set whose only element is the empty set.

We will use exponents to denote multiple products of sets; $A \times A \times \cdots \times A$ with n terms is denoted A^n: the set of n-tuples of elements of A. (The set of n-tuples of real numbers, $\mathbb{R}^n$, is central to this book; to a lesser extent, we will be interested in the set of n-tuples of complex numbers, $\mathbb{C}^n$.)

Finally, note that the order in which elements of a set are listed (assuming they are listed) does not matter, and that duplicating does not affect the set; $\{1, 2, 3\} = \{1, 2, 3, 3\} = \{3, 1, 2\}$.

FIGURE 0.3.2.

Russell's paradox has a long history. The Greeks knew it as the paradox of a barber living on the island of Milos, who decided to shave all the men of the island who did not shave themselves. Does the barber shave himself? Here the barber is Bertrand Russell. (Picture by Roger Hayward, provided by Pour la Science.)

Russell's paradox

In 1902, Bertrand Russell (1872–1970) wrote the logician Gottlob Frege a letter containing the following argument: Consider the set X of all sets that do not contain themselves. If $X \in X$, then X does contain itself, so $X \notin X$. But if $X \notin X$, then X is a set which does not contain itself, so $X \in X$.

"Your discovery of the contradiction caused me the greatest surprise and, I would almost say, consternation," Frege replied, "since it has shaken the basis on which I intended to build arithmetic ... your discovery is very remarkable and will perhaps result in a great advance in logic, unwelcome as it may seem at first glance."[1]

As Figure 0.3.2 suggests, Russell's paradox was (and remains) extremely perplexing. The "solution", such as it is, is to say that the naive idea that any property defines a set is untenable, and that sets must be built up, allowing you to take subsets, unions, products, ... of sets already defined; moreover, to make the theory interesting, you must assume the existence of an infinite set. Set theory (still an active subject of research) consists of describing exactly the allowed construction procedures, and seeing what consequences can be derived.

EXERCISE FOR SECTION 0.3

0.3.1 Let E be a set, with subsets $A \subset E$ and $B \subset E$, and let $*$ be the operation

[1]These letters by Russell and Frege are published in *From Frege to Gödel: A Source Book in Mathematical Logic, 1879–1931* (Harvard University Press, Cambridge, 1967), by Jean van Heijenoort, who in his youth was bodyguard to Leon Trotsky.

$A * B = (E - A) \cap (E - B)$. Express the sets

   a. $A \cup B$ b. $A \cap B$ c. $E - A$

using A, B, and $*$.

0.4 FUNCTIONS

In the eighteenth century, when mathematicians spoke of functions they generally meant functions such as $f(x) = x^2$ or $f(x) = \sin x$. Such a function f associates to a number x another number y according to a precise, computational rule.

Such a restrictive definition is inadequate when mathematics is used to study and describe physical phenomena. If we are interested in pressure of a gas as a function of temperature, or life expectancy as a function of education, or changes in climate over time, there is no reason to expect that such a relationship could be expressed by a simple formula. Nor does there exist a computational rule (for example, a finite computer program) that can, on the basis of sheer logic and arithmetic, describe changes in climate over time.

Yet the notion of rule has not been abandoned. In Definitions 0.4.1 and 0.4.3 we give two definitions of function, one that uses the word "rule" and one that does not. The two definitions are compatible if one is sufficiently elastic in defining "rule".

When we write $f(x) = y$, the function is f, the element x is the *argument* of f and $y = f(x)$ is the *value* of f at x. Out loud, $f : X \to Y$ is read "f from X to Y". Such a function is said to be "on" X, or "defined on" X.

You are familiar with the notation $f(x) = x^2$ to denote the "rule" for a function, in this case, the squaring function. Another notation uses $\mapsto$ (the "maps to" symbol):

$$f : \mathbb{R} \to \mathbb{R}, \quad f : x \mapsto x^2.$$

Do not confuse $\mapsto$ with $\to$ ("to").

Definition 0.4.1 (Function as rule). A *function* consists of three things: two sets, called the *domain* and the *codomain*, and a rule that associates to any element in the domain exactly one element in the codomain.

Definition 0.4.1: For any set X, there is an especially simple function $\mathrm{id}_X : X \to X$, whose domain and codomain are X, and whose rule is

$$\mathrm{id}_X(x) = x.$$

This identity function is often denoted simply id.

Typically, we will say "let $f : X \to Y$ be a function" to indicate that the domain is X, the codomain is Y, and the rule is f. For instance, the function that associates to every real number x the largest integer $n \leq x$ is a function $f : \mathbb{R} \to \mathbb{Z}$, often called "floor". (Evaluated on 4.3, that function returns 4; evaluated on -5.3 it returns -6.)

It must be possible to evaluate the function on every element of the domain, and every output (value of the function) must be in the codomain. But it is not necessary that every element of the codomain be a value of the function. We use the word "image" to denote the set of elements in the codomain that are actually reached.

We can think of the domain as the "space of departure" and of the codomain as the "target space".

Some authors use "range" to denote what we mean by image. Others either have no word for what we call the codomain, or use the word "range" interchangeably to mean both codomain and image.

Definition 0.4.2 (Image). The set of all values of f is called its *image*: y is an element of the image of a function $f : X \to Y$ if there exists an $x \in X$ such that $f(x) = y$.

For example, the image of the squaring function $f : \mathbb{R} \to \mathbb{R}$ given by $f(x) = x^2$ is the nonnegative real numbers; the codomain is $\mathbb{R}$.

The codomain of a function may be considerably larger than its image. Moreover, you cannot think of a function without knowing its codomain, i.e., without having some idea of what kind of object the function produces. (This is especially important when dealing with vector-valued functions.) Knowing the image, on the other hand, may be difficult.

The words *function*, *mapping*, and *map* are synonyms, generally used in different contexts. A *function* normally returns a number. *Mapping* is a more recent word; it was first used in topology and geometry and has spread to all parts of mathematics. In higher dimensions, we tend to use the word mapping rather than function.

What do we mean by rule?

The rule used to define a function may be a computational scheme specifiable in finitely many words, but it need not be. If we are measuring the conductivity of copper as a function of temperature, then an element in the domain is a temperature and an element in the codomain is a measure of conductivity; all other variables held constant, each temperature is associated to one and only one measure of conductivity. The "rule" here is "for each temperature, measure the conductivity and write down the result".

In English it is more natural to say, "John's father" rather than "the father of John". A school of algebraists exists that uses this notation: they write $(x)f$ rather than $f(x)$. The notation $f(x)$ was established by the Swiss mathematician Leonhard Euler (1707–1783). He set the notation we use from high school on: sin, cos, and tan for the trigonometric functions are also due to him.

We can also devise functions where the "rule" is "because I said so". For example, we can devise the function $M : [0,1] \to \mathbb{R}$ that takes every number in the interval $[0,1]$ that can be written in base 3 without using 1, changes every 2 to a 1, and then considers the result as a number in base 2. If the number written in base 3 must contain a 1, the function M changes every digit after the first 1 to 0, then changes every 2 to 1, and considers the result as a number in base 2. Cantor proposed this function to point out the need for greater precision in a number of theorems, in particular the fundamental theorem of calculus.

When Cantor proposed this function, it was viewed as pathological, but it turns out to be important for understanding Newton's method for complex cubic polynomials. A surprising discovery of the early 1980s was that functions just like it occur everywhere in complex dynamics.

In other cases the rule may be simply "look it up". Thus to define a function associating to each student in a class his or her final grade, all you need is the final list of grades; you do not need to know how the professor graded and weighted various exams, homeworks, and papers (although you could define such a function, which, if you were given access to the student's work for the year, would allow you to compute his or her final grade).

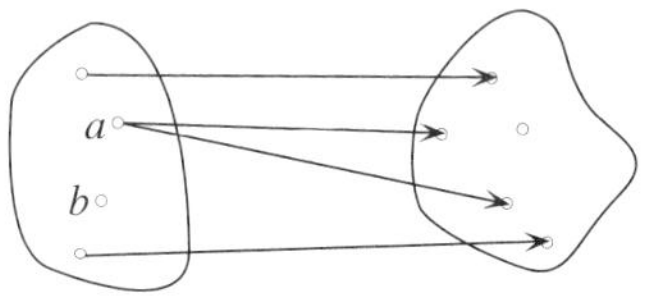

FIGURE 0.4.1.

Not a function: Not well defined at a, not defined at b.

Moreover, if the rule is "look it up in the table", the table need not be finite. One of the fundamental differences between mathematics and virtually everything else is that mathematics routinely deals with the infinite. We are going to be interested in things like the set of all continuous functions f that take an element of $\mathbb{R}$ (i.e., any real number) and return an element of $\mathbb{R}$. If we restrict ourselves to functions that are finitely specifiable, then much of what we might want to say about such sets is not true or has quite a different meaning. For instance, any time we want to take the maximum of some infinite set of numbers, we would have to specify a way of finding the maximum.

Thus the "rule" in Definition 0.4.1 can be virtually anything at all, just so long as *every* element of the domain (which in most cases contains infinitely many elements) can be associated to *one and only one* element of the codomain (which in most cases also contains infinitely many elements).

Definition 0.4.3 emphasizes that it is this result that is crucial; any procedure that arrives at it is acceptable.

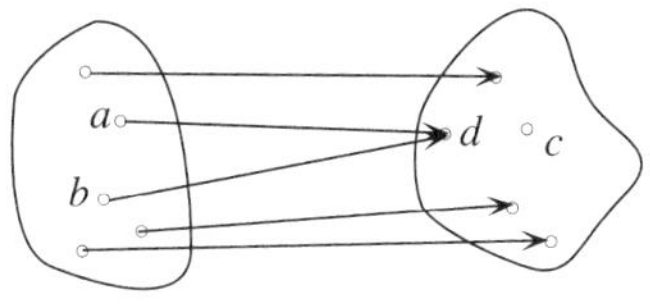

FIGURE 0.4.2.

A function: Every point on the left goes to only one point on the right. The fact that a function takes you unambiguously from any point in the domain to a single point in the codomain does not mean that you can go unambiguously, or at all, in the reverse direction; here, going backward from d in the codomain takes you to either a or b in the domain, and there is no path from c in the codomain to any point in the domain.

Definition 0.4.3 (Set theoretic definition of function). A *function* $f : X \to Y$ is a subset $\Gamma_f \subset X \times Y$ having the property that for every $x \in X$, there exists a unique $y \in Y$ such that $(x, y) \in \Gamma_f$.

Is arcsin a function? Natural domains and other ambiguities

We use functions from early childhood, typically with the word "of" or its equivalent: "the price of a book" associates a price to a book; "the father of" associates a man to a person. Yet not all such expressions are true functions in the mathematical sense. Nor are all expressions of the form $f(x) = y$ true functions. As both Definitions 0.4.1 and 0.4.3 express in different words, a function must be defined at every point of the domain (*everywhere defined*), and for each, it must return a unique element of the codomain (it must be *well defined*). This is illustrated by Figures 0.4.1 and 0.4.2.

"The daughter of", as a "rule" from people to girls and women, is not everywhere defined, because not everyone has a daughter; it is not well defined because some people have more than one daughter. It is not a mathematical function. But "the number of daughters of" is a function from women to numbers: it is everywhere defined and well defined, at a particular time. So is "the biological father of" as a rule from people to men; every person has a biological father, and only one.

The mathematical definition of function then seems straightforward and unambiguous. Yet what are we to make of the arcsin "function" key on your calculator? Figure 0.4.3 shows the "graph" of "arcsin". Clearly the argument 1/2 does not return one and only one value in the codomain; $\arcsin(1/2) = \pi/6$ but we also have $\arcsin(1/2) = 5\pi/6$ and so on. But if you ask your calculator to compute $\arcsin(1/2)$ it returns only the answer $.523599 \approx \pi/6$. The people who programmed the calculator declared "arcsin" to be the function $\arcsin : [-1, 1] \to \mathbb{R}$ whose rule is "$\arcsin(x)$ is the unique angle θ satisfying $-\pi/2 \le \theta \le \pi/2$ and $\sin\theta = x$."

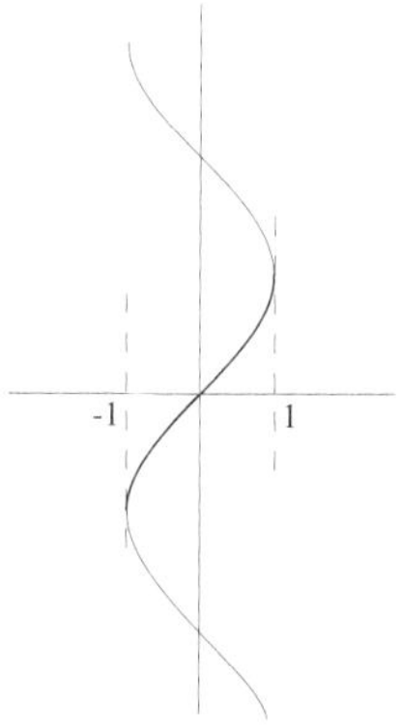

FIGURE 0.4.3.

The graph of arcsin. The part in bold is the graph of the "function" arcsin as defined by calculators and computers: the function $\arcsin : [-1, 1] \to \mathbb{R}$ whose rule is "$\arcsin(x)$ is the unique angle θ satisfying $-\pi/2 \le \theta \le \pi/2$ and $\sin\theta = x$."

Remark. In the past, some textbooks spoke of "multi-valued functions" that assign different values to the same argument; such a "definition" would allow arcsin to be a function. In his book *Calcul Infinitésimal*, published in 1980, the French mathematician Jean Dieudonné pointed out that such definitions are meaningless, "for the authors of such texts refrain from giving the least rule for how to perform calculations using these new mathematical objects that they claim to define, which makes the so-called 'definition' unusable."

Computers have shown just how right he was. Computers do not tolerate ambiguity. If the "function" assigns more than one value to a single argument, the computer will choose one without telling you that it is making a choice. Computers are in effect redefining certain expressions to be functions. When the authors were in school, $\sqrt{4}$ was two numbers, $+2$ and -2. Increasingly, "square root" is taken to mean "positive square root", because a computer cannot compute if each time it lands on a square root it must consider both positive and negative square roots. $\triangle$

Natural domain

Often people refer to functions without specifying the domain or the codomain; they speak of something like "the function $\ln(1+x)$". When the word "function" is used in this way, there is an implicit domain consisting of all numbers for which the formula makes sense. In the case of $\ln(1+x)$ it is the set of numbers $x > -1$. This default domain is called the formula's *natural domain*.

Discovering the natural domain of a formula can be complicated, and the answer may depend on context. In this book we have tried to be scrupulous about specifying a function's domain and codomain.

When working with complex numbers, choosing a "natural domain" is more difficult. The natural domain is usually ambiguous: a choice of a domain for a formula f is referred to a "choosing a branch of f". For instance, one speaks of "the branch of $\sqrt{z}$ defined in $\operatorname{Re} z > 0$, taking positive values on the positive real axis". Historically, the notion of *Riemann surface* grew out of trying to find natural domains.

Example 0.4.4 (Natural domain). The natural domain of the formula $f(x) = \ln x$ is the positive real numbers; the natural domain of the formula $f(x) = \sqrt{x}$ is the nonnegative real numbers. The notion of natural domain may depend on context: for both $\ln x$ and $\sqrt{x}$ we are assuming that the domain and codomain are restricted to be real numbers, not complex numbers.

What is the natural domain of the formula

$$f(x) = \sqrt{x^2 - 3x + 2}\,? \qquad 0.4.1$$

This can be evaluated only if $x^2 - 3x + 2 \geq 0$, which happens if $x \leq 1$ or $x \geq 2$. So the natural domain is $(-\infty, 1] \cup [2, \infty)$. $\triangle$

Parentheses denote an open interval and brackets denote a closed one; (a, b) is open, $[a, b]$ is closed:

$$(a,b) = \{x \in \mathbb{R} \mid a < x < b\}$$
$$[a,b] = \{x \in \mathbb{R} \mid a \leq x \leq b\}.$$

We discuss open and closed sets in Section 1.5.

We could "define" the natural domain of a formula to consist of those arguments for which a computer does not return an error message.

Most often, a computer discovers that a number is not in the natural domain of a formula when it lands on an illegal procedure like dividing by 0. When working with computers, failure to be clear about a function's domain can be dangerous. One does not wish a computer to shut down an airplane's engines or the cooling system of a nuclear power plant because an input has been entered that is not in a formula's natural domain.

Obviously it would be desirable to know before feeding a formula a number whether the result will be an error message; an active field of computer science research consists of trying to figure out ways to guarantee that a set is in a formula's natural domain.

Often a mathematical function modeling a real system has a codomain considerably larger than the realistic values. We may say that the codomain of the function assigning height in centimeters to children is $\mathbb{R}$, but clearly many real numbers do not correspond to the height of any child.

Latitude in choosing a codomain

A function consists of three things: a rule, a domain, and a codomain. A rule comes with a natural domain, but there is no similar notion of natural codomain. The codomain must be at least as big as the image, but it can be a little bigger, or a lot bigger; if $\mathbb{R}$ will do, then so will $\mathbb{C}$, for example. In this sense we can speak of the "choice" of a codomain. Since the codomain is part of what defines a particular function, then, strictly speaking, choosing one allowable codomain as opposed to another means creating a different function. But generally, two functions with the same rule and same domain but different codomains behave the same.

When we work with computers, the situation is more complicated. We mentioned earlier the floor function $f : \mathbb{R} \to \mathbb{Z}$ that associates to every real number x the largest integer $n \le x$. We could also consider the floor function as a function $\mathbb{R} \to \mathbb{R}$, since an integer is a real number. If you are working with pen and paper, these two (strictly speaking different) functions will behave the same. But a computer will treat them differently: if you write a computer program to compute them, in Pascal for instance, one will be introduced by the line

```
function floor(x:real):integer;
```

whereas the other will be introduced by

```
function floor(x:real):real.
```

These functions are indeed different: in the computer, reals and integers are not stored the same way and cannot be used interchangeably. For instance, you cannot perform a division with remainder unless the divisor is an integer, and if you attempt such a division using the output of the second "floor" function above, you will get a TYPE MISMATCH error.

Of course, computers cannot actually compute with real numbers; they compute with approximations to real numbers.

The computer language C is if anything more emphatic about specifying the codomain of a function. In C, the first word of a function declaration describes the codomain. The functions at right would be introduced by the lines

```
integer floor(double x);
```

and

```
double floor(double x).
```

The first word indicates the type of output (the word "double" is C's name for a particular encoding of the reals); the second word is the name of the function; and the expression in parentheses describes the type of input.

Existence and uniqueness of solutions

Given a function f, is there a solution to the equation $f(x) = b$, for every b in the codomain? If so, the function is said to be *onto*, or *surjective*. "Onto" is thus a way to talk about the *existence of solutions*. The function "the father of" as a function from people to men is not onto, because not all men are fathers. There is no solution to the equation "The father of x is Mr. Childless". An onto function is shown in Figure 0.4.4.

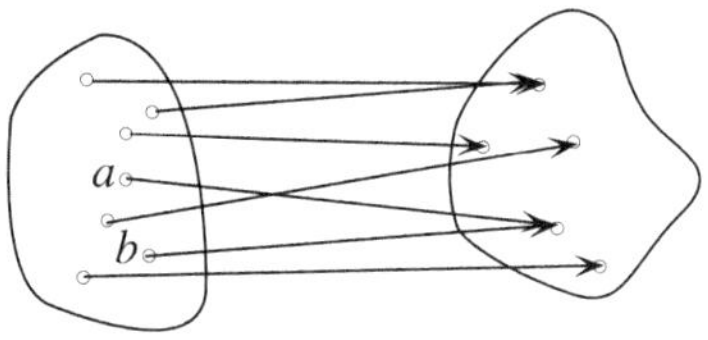

FIGURE 0.4.4.

An onto function, not 1–1: a and b go to the same point.

A second question of interest concerns uniqueness of solutions. Given any particular b in the codomain, is there at most *one* value of x that solves the equation $T(x) = b$, or might there be many? If for each b there is at most one solution to the equation $T(x) = b$, the function T is said to be *one to one*, or *injective*. The mapping "the father of" is not one to one. There are, in fact, four solutions to the equation "The father of x is John Hubbard". But the function "the twin sibling of", as a function from twins to twins, is one to one: the equation "the twin sibling of $x = y$" has a unique solution for each y. "One to one" is thus a way to talk about the *uniqueness of solutions*. A one to one function is shown in Figure 0.4.5.

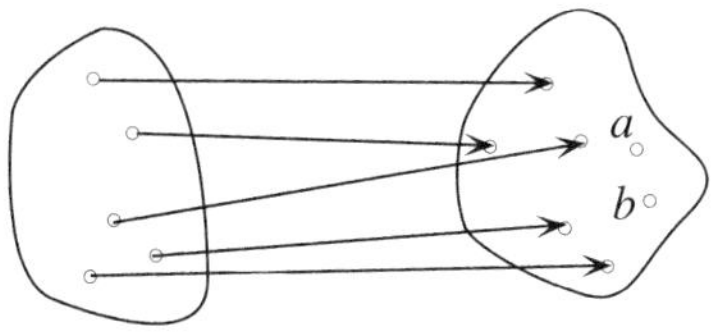

FIGURE 0.4.5.

A function: 1–1, not onto, no points go to a or to b.

A function T that is both *onto* and *one to one* has an inverse function T^{-1} that undoes it. Because T is onto, T^{-1} is everywhere defined; because T is one to one, T^{-1} is well defined. So T^{-1} qualifies as a function. To summarize:

Definition 0.4.5 (Onto). A function $f : X \to Y$ is *onto* (or *surjective*) if for every $y \in Y$ there exists $x \in X$ such that $f(x) = y$.

Thus f is onto if every element of the set of arrival (the codomain Y) corresponds to *at least one* element of the set of departure (the domain X).

The inverse function of f is usually called simply "f inverse".

Definition 0.4.6 (One to one). A function $f : X \to Y$ is *one to one* (or *injective*) if for every $y \in Y$ there is at most one $x \in X$ such that $f(x) = y$.

Thus f is one to one if every element of the set of arrival corresponds to *at most one* element of the set of departure. The *horizontal line test* to see whether a function is one to one is shown in Figure 0.4.6.

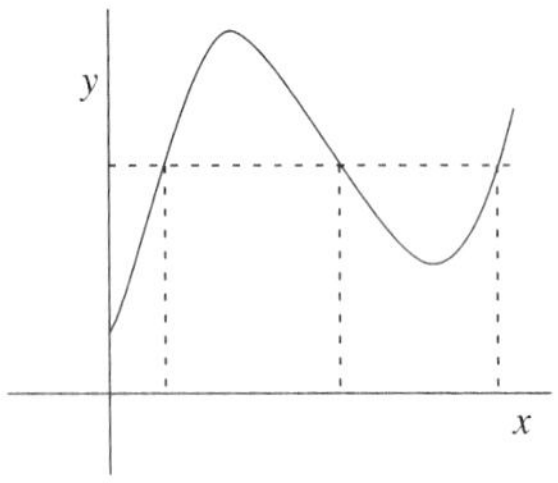

FIGURE 0.4.6.

The function graphed above is not one to one. It fails the "horizontal line test": the horizontal dotted line cuts it in three places, showing that three different values of x give the same value of y.

Definition 0.4.7 (Invertible). A mapping f is *invertible* (or *bijective*) if it is both onto and one to one. The inverse function of f is denoted f^{-1}.

An invertible function can be undone; if $f(a) = b$, then $f^{-1}(b) = a$. The words "invertible" and "inverse" are particularly appropriate for multiplication; to undo multiplication by a, we multiply by its inverse, $1/a$. (But the inverse function of the function $f(x) = x$ is $f^{-1}(x) = x$, so that in this case the inverse of x is x, not $1/x$, which is the multiplicative inverse. Usually it is clear from context whether "inverse" means "inverse mapping", as in Definition 0.4.7, or "multiplicative inverse", but sometimes there can be ambiguity.)

Example 0.4.8 (One to one; onto). The mapping "the Social Security number of" as a mapping from United States citizens to numbers is not onto because there exist numbers that aren't Social Security numbers. But it is one to one: no two U.S. citizens have the same Social Security number.

The mapping $f(x) = x^2$ from real numbers to real nonnegative numbers is onto because every real nonnegative number has a real square root, but it is not one to one because every real positive number has both a positive and a negative square root. $\triangle$

If a function is not invertible, we can still speak of the *inverse image* of a set under f.

For the map of Example 0.4.10, $f^{-1}(\{-1\}) = \phi$; it is a well-defined set. If we had defined $g(x) = x^2$ as a map from $\mathbb{R}$ to the nonnegative reals, then $g^{-1}(\{-1\})$ and $g^{-1}(\{-1, 4, 9, 16\})$ would not exist.

Definition 0.4.9 (Inverse image). Let $f : X \to Y$ be a function, and let $C \subset Y$ be a subset of the codomain of f. Then the *inverse image* of C under f, denoted $f^{-1}(C)$, consists of those elements $\mathbf{x} \in X$ such that $f(\mathbf{x}) \in C$.

Example 0.4.10 (Inverse image). Let $f : \mathbb{R} \to \mathbb{R}$ be the (noninvertible) mapping $f(x) = x^2$. The inverse image of $\{-1, 4, 9, 16\}$ under f is $\{-4, -3, -2, 2, 3, 4\}$:

$$f^{-1}(\{-1, 4, 9, 16\}) = \{-4, -3, -2, 2, 3, 4\}. \quad \triangle \qquad 0.4.2$$

You are asked to prove Proposition 0.4.11 in Exercise 0.4.6.

Proposition 0.4.11 (Inverse image of intersection, union).

1. *The inverse image of an intersection equals the intersection of the inverse images:*

$$f^{-1}(A \cap B) = f^{-1}(A) \cap f^{-1}(B). \qquad 0.4.3$$

2. *The inverse image of a union equals the union of the inverse images:*

$$f^{-1}(A \cup B) = f^{-1}(A) \cup f^{-1}(B). \qquad 0.4.4$$

It is often easier to understand a composition if one writes it in diagram form; $(f \circ g) : A \to D$ can be written

$$A \underset{g}{\to} B \subset C \underset{f}{\to} D.$$

Composition of mappings

Often one wishes to apply, consecutively, more than one mapping. This is known as *composition*.

A composition is written from left to right but computed from right to left: you apply the mapping g to the argument x and then apply the mapping f to the result. Exercise 0.4.7 provides some practice.

Definition 0.4.12 (Composition). If $f : C \to D$ and $g : A \to B$ are two mappings with $B \subset C$, then the *composition* $(f \circ g) : A \to D$ is the mapping given by

$$(f \circ g)(x) = f\big(g(x)\big). \qquad 0.4.5$$

Note that for the composition $f \circ g$ to make sense, the codomain of g must be contained in the domain of f.

When computers do compositions, it is not quite true that composition is associative. One way of doing the calculation may be more computationally effective than another; because of round-off errors, the computer may even come up with different answers, depending on where the parentheses are placed.

Example 0.4.13 (Composition of "the father of" and "the mother of"). Consider the following two mappings from the set of persons to the set of persons (alive or dead): F, "the father of", and M, "the mother of". Composing these gives:

$F \circ M$ (the father of the mother of = maternal grandfather of)

$M \circ F$ (the mother of the father of = paternal grandmother of).

It is clear in this case that composition is associative:

$$F \circ (F \circ M) = (F \circ F) \circ M. \qquad 0.4.6$$

The father of David's maternal grandfather is the same person as the paternal grandfather of David's mother. Of course, it is not commutative: the "father of the mother" is not the "mother of the father".) △

Example 0.4.14 (Composition of two functions). If $f(x) = x - 1$, and $g(x) = x^2$, then

$$(f \circ g)(x) = f\big(g(x)\big) = x^2 - 1. \quad \triangle \qquad 0.4.7$$

Proposition 0.4.15 (Composition is associative). *Composition is associative:*

$$(f \circ g) \circ h = f \circ (g \circ h). \qquad 0.4.8$$

Proof. This is simply the computation

$$\big((f \circ g) \circ h\big)(x) = (f \circ g)\big(h(x)\big) = f\big(g(h(x))\big) \quad \text{whereas}$$
$$\big(f \circ (g \circ h)\big)(x) = f\big((g \circ h)(x)\big) = f\big(g(h(x))\big). \quad \square \qquad 0.4.9$$

Although composition is associative, in many settings,

$$\Big((f \circ g) \circ h\Big) \quad \text{and} \quad \Big(f \circ (g \circ h)\Big)$$

correspond to different ways of thinking. The author of a biography might use "the father of the maternal grandfather" when focusing on the relationship between the subject's grandfather and the grandfather's father, and use "the paternal grandfather of the mother" when focusing on the relationship between the subject's mother and her grandfather.

You may find this proof devoid of content. Composition of mappings is part of our basic thought processes: you use a composition any time you speak of "the this of the that of the other". So the statement that composition is associative may seem too obvious to need proving, and the proof may seem too simple to be a proof.

Proposition 0.4.16 (Composition of onto functions). *Let the functions $f : B \to C$ and $g : A \to B$ be onto. Then the composition $(f \circ g)$ is onto.*

Proposition 0.4.17 (Composition of one to one functions). *Let $f : B \to C$ and $g : A \to B$ be one to one. Then the composition $(f \circ g)$ is one to one.*

You are asked to prove Propositions 0.4.16 and 0.4.17 in Exercise 0.4.8

Exercises for Section 0.4

0.4.1 Are the following true functions? That is, are they both everywhere defined and well defined?

a. "The aunt of", from people to people.

b. $f(x) = \frac{1}{x}$, from real numbers to real numbers.

c. "The capital of", from countries to cities (careful – at least one country, Bolivia, has two capitals.)

0.4.2 a. Make up a nonmathematical function that is onto but not one to one.
b. Make up a mathematical function that is onto but not one to one.

0.4.3 a. Make up a nonmathematical function that is bijective (onto and one to one).
b. Make up a mathematical function that is bijective.

0.4.4 a. Make up a nonmathematical function that is one to one but not onto.
b. Make up a mathematical function that is one to one but not onto.

0.4.5 Given the functions $f : A \to B$, $g : B \to C$, $h : A \to C$, and $k : C \to A$, which of the following compositions are well defined? For those that are, give the domain and codomain of each.

a. $f \circ g$	b. $g \circ f$	c. $h \circ f$	d. $g \circ h$	e. $k \circ h$	f. $h \circ g$
g. $h \circ k$	h. $k \circ g$	i. $k \circ f$	j. $f \circ h$	k. $f \circ k$	l. $g \circ k$

0.4.6 Prove Proposition 0.4.11.

0.4.7 Evaluate $(f \circ g \circ h)(x)$ at $x = a$ for the following.

a. $f(x) = x^2 - 1, \quad g(x) = 3x, \quad h(x) = -x + 2$, for $a = 3$.

b. $f(x) = x^2, \quad g(x) = x - 3, \quad h(x) = x - 3$, for $a = 1$.

0.4.8 a. Prove Proposition 0.4.16. b. Prove Proposition 0.4.17.

0.4.9 What is the natural domain of $\sqrt{\frac{x}{y}}$?

0.4.10 What is the natural domain of

a. $\ln \circ \ln$? b. $\ln \circ \ln \circ \ln$? c. ln composed with itself n times?

0.4.11 What subset of $\mathbb{R}$ is the natural domain of the function $(1 + x)^{1/x}$?

0.4.12 The function $f(x) = x^2$ from real numbers to real nonnegative numbers is onto but not one to one.

a. Can you make it one to one by changing its domain? By changing its codomain?

b. Can you make it not onto by changing its domain? Its codomain?

0.5 Real numbers

Calculus is about limits, continuity, and approximation. These concepts involve real numbers and complex numbers, as opposed to integers and rationals. In this Section (and in Appendix A1), we present real numbers and establish some of their most useful properties. Our approach privileges the writing of numbers in base 10; as such it is a bit unnatural, but we hope you will like our real numbers being exactly the numbers you are used to.

Showing that all such constructions lead to the same numbers is a fastidious exercise, which we will not pursue.

Numbers and their ordering

By definition, the set of real numbers is the set of infinite decimals: expressions like $2.957653920457\ldots$, preceded by a plus or a minus sign (often the $+$ is omitted). The number that you think of as 3 is the infinite decimal $3.0000\ldots$, ending in all 0's. The following identification is vital: a number ending in all 9's is equal to the "rounded up" number ending in all 0's:

$$0.34999999\ldots = 0.350000\ldots. \tag{0.5.1}$$

Also, $+.0000\ldots = -.0000\ldots$. Other than these exceptions, there is only one way of writing a real number.

Real numbers are actually bi-infinite decimals with 0's to the left: a number like $3.0000\ldots$ is actually

$$\ldots 00003.0000\ldots.$$

By convention, leading 0's are usually omitted. One exception is credit card expiration dates: the month March is 03, not 3.

Numbers starting with a $+$ sign, except $+0.000\ldots$, are positive; those starting with a $-$ sign, except $-0.00\ldots$, are negative. If x is a real number, then $-x$ has the same string of digits, but with the opposite sign in front.

When a number is written in base 10, the digit in the 10^kth position refers to the number of 10^ks. For instance, $217.4\ldots$ has 2 hundreds, 1 ten, 7 ones, and 4 tenths, corresponding to 2 in the 10^2 position, 1 in the 10^1

position, and so on. We denote by $[x]_k$ the number formed by keeping all digits to the left of and including the 10^kth position, setting all others to 0. Thus for $a = 5\,129.359\ldots$, we have $[a]_2 = 5\,100.00\ldots$, $[a]_0 = 5\,129.00\ldots$, $[a]_{-2} = 5\,129.3500\ldots$. If x has two decimal expressions, we define $[x]_k$ to be the finite decimal built from the infinite decimal ending in 0's; for the number in formula 0.5.1, $[x]_{-3} = 0.350$; it is not 0.349.

Real numbers can be defined in more elegant ways: Dedekind cuts, for instance (see, for example, M. Spivak, *Calculus*, second edition, Publish or Perish, 1980, pp. 554–572), or Cauchy sequences of rational numbers. One could also mirror the present approach, writing numbers in any base, for instance 2. Since this section is partially motivated by the treatment of floating-point numbers on computers, base 2 would seem very natural.

Given two different finite numbers x and y, one is always bigger than the other, as follows. If x is positive and y is nonpositive, then $x > y$. If both are positive, then in their decimal expansions there is a left-most digit in which they differ; whichever has the larger digit in that position is larger. If both x and y are negative, then $x > y$ if $-y > -x$.

Least upper bound

Definition 0.5.1 (Upper bound; least upper bound). A number a is an *upper bound* for a subset $X \subset \mathbb{R}$ if for every $x \in X$ we have $x \leq a$. A *least upper bound*, also known as the *supremum*, is an upper bound b such that for any other upper bound a, we have $b \leq a$. It is denoted $\sup X$. If X is unbounded above, $\sup X$ is defined to be $+\infty$.

Definition 0.5.2 (Lower bound; greatest lower bound). A number a is a *lower bound* for a subset $X \subset \mathbb{R}$ if for every $x \in X$ we have $x \geq a$. A *greatest lower bound* is a lower bound b such that for any other lower bound a, we have $b \geq a$. The greatest lower bound, or *infimum*, is denoted $\inf X$. If X is unbounded below, $\inf X$ is defined to be $-\infty$.

The least upper bound property of the reals is often taken as an axiom; indeed, it characterizes the real numbers, and it **lies at the foundation of every theorem in calculus**. However, at least with the preceding description of the reals, it is a theorem, not an axiom.

The least upper bound $\sup X$ is sometimes denoted l.u.b.X; the notation $\max X$ is also used, but it suggests to some people that $\max X \in X$, which may not be the case.

Theorem 0.5.3 (The real numbers are complete). *Every nonempty subset $X \subset \mathbb{R}$ that has an upper bound has a least upper bound $\sup X$. Every nonempty subset $X \subset \mathbb{R}$ that has a lower bound has a greatest lower bound $\inf X$.*

Proof. We will construct successive decimals of $\sup X$. Suppose that $x \in X$ is an element (which we know exists, since $X \neq \phi$) and that a is an upper bound. We will assume that $x > 0$ (the case $x \leq 0$ is slightly different). If $x = a$, we are done: the least upper bound is a.

If $a = 29.86\ldots$, $b = 29.73\ldots$, then

$$[b]_{-2} < [a]_{-2}, \quad [b]_{-1} < [a]_{-1};$$

$j = -1$ is the largest j such that $[b]_j < [a]_j$.

If $x \neq a$, there is then a largest j such that $[x]_j < [a]_j$. There are 10 numbers that have the same kth digit as x for $k > j$ and that have 0 as the kth digit for $k < j$; consider those that are in $[\,[x]_j, a]$. This set is not empty, since $[x]_j$ is one of them. Let b_j be the largest of these ten numbers such that $X \cap [b_j, a] \neq \phi$; such a b_j exists, since $x \in X \cap [[x]_j, a]$.

Consider the set of numbers in $[b_j, a]$ that have the same kth digit as b_j for $k > j-1$, and 0 for $k < j-1$. This is a nonempty set with at most 10 elements, and b_j is one of them (the smallest). Call b_{j-1} the largest such that $X \cap [b_{j-1}, a] \neq \phi$. Such a b_{j-1} exists, since if necessary we can choose b_j. Keep going this way, defining b_{j-2}, b_{j-3}, and so on, and let b be the number whose nth decimal digit (for all n) is the same as the nth decimal digit of b_n.

We claim that $b = \sup X$. Indeed, if there exists $y \in X$ with $y > b$, then there is a first k such that the kth digit of y differs from the kth digit of b. This contradicts our assumption that b_k was the largest number (out of 10) such that $X \cap [b_k, a] \neq \phi$, since using the kth digit of y would give a bigger one. So b is an upper bound. Now suppose that $b' < b$. If b' is an upper bound for X, then $(b', a] \cap X = \phi$. Again there is a first k such that the kth digit of b differs from the kth digit of b'. Then $(b', a] \cap X \supset [b_k, a] \cap X \neq \phi$. Thus b' is not an upper bound for X. □

The procedure we give for proving the existence of $b = \sup X$ gives no recipe for finding it. Like the proof of Theorem 1.6.3, this proof is *non-constructive*. Example 1.6.4 illustrates the kind of difficulty one might encounter when trying to construct b.

Sequences and series

A sequence is an infinite list $a_1, a_2, \dots$ (of numbers or vectors or matrices ...). We denote such a list by $n \mapsto a_n$, where n is assumed to be a positive (or sometimes nonnegative) integer.

The symbol $\mapsto$ ("maps to") describes what a function does to an input; see the margin note about function notation, page 9. Using this notation for sequences is reasonable, since a sequence really is a map from the positive integers to whatever space the sequence lives in.

A sequence $i \mapsto a_i$ can also be written as (a_i) or as $(a_i)_{i \in \mathbb{N}}$ or even as a_i. We used the notation a_i ourselves in previous editions, but we have become convinced that $i \mapsto a_i$ is best.

Definition 0.5.4 (Convergent sequence). A sequence $n \mapsto a_n$ of real numbers *converges* to the limit a if for all $\epsilon > 0$, there exists N such that for all $n > N$, we have $|a - a_n| < \epsilon$.

Many important sequences appear as partial sums of series. A *series* is a sequence whose terms are to be added. If we consider the sequence $a_1, a_2, \dots$ as a series, then the associated sequence of partial sums is the sequence $s_1, s_2, \dots$, where

$$s_N = \sum_{n=1}^{N} a_n. \qquad 0.5.2$$

For example, if $a_1 = 1$, $a_2 = 2$, $a_3 = 3$, and so on, then $s_4 = 1 + 2 + 3 + 4$.

If a series converges, then the same list of numbers viewed as a sequence must converge to 0. The converse is not true. For example, the harmonic series

$$1 + \frac{1}{2} + \frac{1}{3} + \cdots$$

does not converge, although the terms tend to 0.

Definition 0.5.5 (Convergent series). If the sequence of partial sums of a series has a limit S, the series *converges*, and its limit is

$$S \stackrel{\text{def}}{=} \sum_{n=1}^{\infty} a_n. \qquad 0.5.3$$

Example 0.5.6 (Convergent geometric series). If $|r| < 1$, then

$$\sum_{n=0}^{\infty} ar^n = \frac{a}{1-r}. \qquad 0.5.4$$

For example, $2.020202\ldots = 2 + 2(.01) + 2(.01)^2 + \cdots = \dfrac{2}{1-(.01)} = \dfrac{200}{99}$.

Indeed, the following subtraction shows that $S_n(1-r) = a - ar^{n+1}$:

$$\begin{aligned} S_n &\stackrel{\text{def}}{=} a + ar + ar^2 + ar^3 + \cdots + ar^n \\ S_n r &= \quad\;\; ar + ar^2 + ar^3 + \cdots + ar^n + ar^{n+1} \\ \hline S_n(1-r) &= a \qquad\qquad\qquad\qquad\qquad\qquad\quad - ar^{n+1}. \end{aligned} \qquad 0.5.5$$

In practice, the index set for a series may vary; for instance, in Example 0.5.6, n goes from 0 to ∞, not from 1 to ∞. For Fourier series, n goes from $-\infty$ to ∞. But series are usually written with the sum running from 1 to ∞.

But $\lim_{n\to\infty} ar^{n+1} = 0$ when $|r| < 1$, so we can forget about the $-ar^{n+1}$: as $n \to \infty$, we have $S_n \to a/(1-r)$. △

Proving convergence

The weakness of the definition of a convergent sequence or series is that it involves the limit value; it is hard to see how you will ever be able to prove that a sequence has a limit if you don't know the limit ahead of time. The first result along these lines is Theorem 0.5.7. It and its corollaries underlie all of calculus.

Theorem 0.5.7. *A nondecreasing sequence $n \mapsto a_n$ of real numbers converges if and only if it is bounded.*

Proof. If a sequence $n \mapsto a_n$ of real numbers converges, it is clearly bounded. If it is bounded, then (by Theorem 0.5.3) it has a least upper bound A. We claim that A is the limit. This means that for any $\epsilon > 0$, there exists N such that if $n > N$, then $|a_n - A| < \epsilon$. Choose $\epsilon > 0$; if $A - a_n > \epsilon$ for all n, then $A - \epsilon$ is an upper bound for the sequence, contradicting the definition of A. So there is a first N with $A - a_N < \epsilon$, and it will do, since when $n > N$, we must have $A - a_n \leq A - a_N < \epsilon$. □

Theorem 0.5.7 has the following consequence:

Theorem 0.5.8 (Absolute convergence implies convergence). *If the series of absolute values*

$$\sum_{n=1}^{\infty} |a_n| \quad \text{converges, then so does the series} \quad \sum_{n=1}^{\infty} a_n.$$

Proof. The series $\sum_{n=1}^{\infty}(a_n + |a_n|)$ is a series of nonnegative numbers, so the partial sums $b_m = \sum_{n=1}^{m}(a_n + |a_n|)$ are nondecreasing. They are also bounded:

$$b_m = \sum_{n=1}^{m}(a_n + |a_n|) \leq \sum_{n=1}^{m} 2|a_n| = 2\sum_{n=1}^{m}|a_n| \leq 2\sum_{n=1}^{\infty}|a_n|. \qquad 0.5.6$$

So (by Theorem 0.5.7) $m \mapsto b_m$ is a convergent sequence, and $\sum_{n=1}^{\infty} a_n$ can be represented as the sum of two numbers, each the sum of a convergent series:

$$\sum_{n=1}^{\infty} a_n = \sum_{n=1}^{\infty}\Big(a_n + |a_n|\Big) + \Big(-\sum_{n=1}^{\infty}|a_n|\Big). \qquad □ \qquad 0.5.7$$

The intermediate value theorem

The intermediate value theorem appears to be obviously true, and is often useful. It follows easily from Theorem 0.5.3 and the definition of continuity.

Theorem 0.5.7: Of course it is also true that a nonincreasing sequence converges if and only if it is bounded. Most sequences are neither nondecreasing nor nonincreasing.

In mathematical analysis, problems are usually solved by exhibiting a sequence that converges to the solution. Since we don't know the solution, it is essential to guarantee convergence without knowing the limit. Coming to terms with this was a watershed in the history of mathematics, associated first with a rigorous construction of the real numbers, and later with the definition of the Lebesgue integral, which allows the construction of Banach spaces and Hilbert spaces where "absolute convergence implies convergence", again giving convergence without knowing the limit. The use of these notions is also a watershed in mathematical education: elementary calculus gives solutions exactly, more advanced calculus constructs them as limits.

In contrast to the real numbers and the complex numbers, it is impossible to prove that a sequence of rational numbers or algebraic numbers has a rational or algebraic limit without exhibiting it specifically.

Inequality 0.5.6: the sum

$$\sum_{n=1}^{\infty} |a_n|$$

is bounded by the hypothesis of the theorem.

Theorem 0.5.9 (Intermediate value theorem). *If $f : [a, b] \to \mathbb{R}$ is a continuous function and c is a number such that $f(a) \le c$ and $f(b) \ge c$, then there exists $x_0 \in [a, b]$ such that $f(x_0) = c$.*

Proof. Let X be the set of $x \in [a, b]$ such that $f(x) \le c$. Note that X is nonempty (a is in it) and it has an upper bound, namely b, so that it has a least upper bound, which we call x_0. We claim $f(x_0) = c$.

Since f is continuous, for any $\epsilon > 0$, there exists $\delta > 0$ such that when $|x_0 - x| < \delta$, then $|f(x_0) - f(x)| < \epsilon$. If $f(x_0) > c$, we can set $\epsilon = f(x_0) - c$, and find a corresponding δ. Since x_0 is a *least* upper bound for X, there exists $x \in X$ such that $x_0 - x < \delta$, so

$$f(x) = f(x_0) + f(x) - f(x_0) \ge f(x_0) - |f(x) - f(x_0)| > c + \epsilon - \epsilon = c,$$

contradicting that x is in X; see Figure 0.5.1.

If $f(x_0) < c$, a similar argument shows that there exists $\delta > 0$ such that $f(x_0 + \delta/2) < c$, contradicting the assumption that x_0 is an upper bound for X. The only choice left is $f(x_0) = c$. $\square$

One unsuccessful nineteenth-century definition of continuity stated that a function f is continuous if it satisfies the intermediate value theorem. You are asked in Exercise 0.5.2 to show that this does not coincide with the usual definition (and presumably not with anyone's intuition of what continuity should mean).

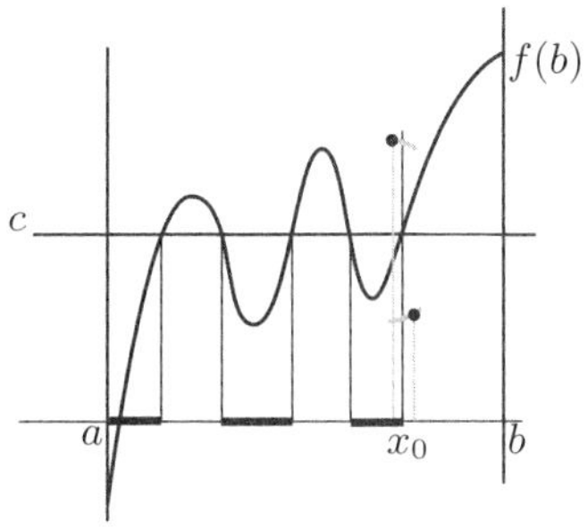

FIGURE 0.5.1.

The bold intervals in $[a, b]$ are the set X of x such that $f(x) \le c$. The point $x \in X$ slightly to the left of x_0 gives rise to $f(x) > c$, contradicting the definition of X. The point x slightly to the right gives rise to $f(x) < c$, contradicting x_0 being an upper bound.

Exercises for Section 0.5

The exercises for Section 0.5 are fairly difficult.

0.5.1 Show that if p is a polynomial of odd degree with real coefficients, then there is a real number c such that $p(c) = 0$.

0.5.2 a. Show that the function

$$f(x) = \begin{cases} \sin \frac{1}{x} & \text{if } x \ne 0 \\ 0 & \text{if } x = 0 \end{cases} \quad \text{is not continuous.}$$

b. Show that f satisfies the conclusion of the intermediate value theorem: if $f(x_1) = a_1$ and $f(x_2) = a_2$, then for any number a between a_1 and a_2, there exists a number x between x_1 and x_2 such that $f(x) = a$.

0.5.3 Suppose $a \le b$. Show that if $f : [a, b] \to [a, b]$ is continuous, there exists $c \in [a, b]$ with $f(c) = c$.

0.5.4 Let

$$a_n = \frac{(-1)^{n+1}}{n}, \quad \text{for } n = 1, 2, \dots.$$

a. Show that the series $\sum a_n$ is convergent.

*b. Show that $\sum_{n=1}^{\infty} a_n = \ln 2$.

c. Explain how to rearrange the terms of the series so it converges to 5.

d. Explain how to rearrange the terms of the series so that it diverges.

0.5.5 Show that if a series $\sum_{k=1}^{\infty} a_n$ is absolutely convergent, then any rearrangement of the series is still convergent and converges to the same limit. *Hint*: For any $\epsilon > 0$, there exists N such that $\sum_{n=N+1}^{\infty} |a_n| < \epsilon$. For any rearrangement $\sum_{k=1}^{\infty} b_n$ of the series, there exists M such that all of $a_1, \dots, a_N$ appear among $b_1, \dots, b_M$. Show that $|\sum_{n=1}^{N} a_n - \sum_{n=1}^{M} b_n| < \epsilon$.

Exercise 0.5.1: Exercise 1.6.11 repeats this exercise, with hints.

Exercise 0.5.3: By convention, $[a, b]$ implies $a \le b$. Exercise 0.5.3 is the one-dimensional case of the celebrated *Brouwer fixed point theorem*, to be discussed in a subsequent volume. In dimension one it is an easy consequence of the intermediate value theorem, but in higher dimensions (even two) it is quite a delicate result.

Exercise 0.5.4 illustrates how complicated convergence can be when a series is not absolutely convergent. Exercise 0.5.5 shows that these problem do not arise for absolutely convergent series.

0.6 Infinite sets

FIGURE 0.6.1.
Georg Cantor (1845–1918)
After thousands of years of philosophical speculation about the infinite, Cantor found a fundamental notion that had been completely overlooked.

Recall (Section 0.3) that $\mathbb{N}$ is the "natural numbers" $0, 1, 2, \dots$; $\mathbb{Z}$ is the integers; $\mathbb{R}$ is the real numbers.

It would seem likely that $\mathbb{R}$ and $\mathbb{R}^2$ have different infinities of elements, but that is not the case (see Exercise A1.5).

FIGURE 0.6.2.
Joseph Liouville (1809–1882)

One reason set theory is accorded so much importance is that Georg Cantor (1845–1918) discovered that two infinite sets need not have the same "number" of elements; there isn't just one infinity. You might think this is just obvious; for example, that clearly there are more whole numbers than even whole numbers. But with the definition Cantor gave, two sets A and B have the same number of elements (the same *cardinality*) if you can set up a bijective correspondence between them (i.e., a mapping that is one to one and onto). For instance,

$$\begin{matrix} 0 & 1 & 2 & 3 & 4 & 5 & 6 & \dots \\ 0 & 2 & 4 & 6 & 8 & 10 & 12 & \dots \end{matrix} \qquad 0.6.1$$

establishes a bijective correspondence between the natural numbers and the even natural numbers. More generally, any set whose elements you can list has the same cardinality as $\mathbb{N}$: for instance,

$$0,\ 1,\ 1/2,\ 1/3,\ 2/3,\ 1/4,\ 3/4,\ 1/5,\ 2/5,\ 3/5,\ 4/5, \dots \qquad 0.6.2$$

is the beginning of a list of the rational numbers in $[0, 1]$.

But in 1873 Cantor discovered that $\mathbb{R}$ does not have the same cardinality as $\mathbb{N}$: it has a *bigger infinity of elements*. Indeed, imagine making any infinite list of real numbers, say between 0 and 1, so that written as decimals, your list might look like

$$\begin{array}{l} .\mathbf{1}54362786453429823763490652367347548757\dots \\ .9\mathbf{8}7354621943756598673562940657349327658\dots \\ .22\mathbf{9}573521903564355423035465523390080742\dots \\ .104\mathbf{7}52018746267653209365723689076565787\dots \\ .0263\mathbf{2}8560082356835654432879897652377327\dots \\ \dots . \end{array} \qquad 0.6.3$$

Now consider the decimal $.18972\dots$ formed by the diagonal digits (in bold in formula 0.6.3), and modify it (almost any way you want) so that every digit is changed, for instance according to the rule "change 7's to 5's and change anything that is not a 7 to a 7": in this case, your number becomes $.77757\dots$. Clearly this last number does not appear in your list: it is not the nth element of the list, because it doesn't have the same nth decimal digit.

Infinite sets that can be put in one-to-one correspondence with the natural numbers are called *countable* or *countably infinite*. Those that cannot are called *uncountable*; the set $\mathbb{R}$ of real numbers is uncountable.

Existence of transcendental numbers

An *algebraic number* is a root of a polynomial equation with integer coefficients: the rational number p/q is algebraic, since it is a solution of

$qx - p = 0$, and so is $\sqrt{2}$, since it is a root of $x^2 - 2 = 0$. A number that is not algebraic is called *transcendental.* In 1851 Joseph Liouville came up with the transcendental number (now called the *Liouvillian number*)

$$\sum_{n=1}^{\infty} \frac{1}{10^{n!}} = 0.110001000000000000000000100\ldots, \qquad 0.6.4$$

the number with 1 in every position corresponding to $n!$ and 0's elsewhere. In 1873 Charles Hermite proved a much harder result, that e is transcendental. But Cantor's work on cardinality made it obvious that there must exist uncountably many transcendental numbers: all those real numbers left over when one tries to put the real numbers in one-to-one correspondence with the algebraic numbers.

FIGURE 0.6.3.
Charles Hermite (1822–1901)
For Hermite, there was something scandalous about Cantor's proof of the existence of infinitely many transcendental numbers, which required no computations and virtually no effort and failed to come up with a single example.

Here is one way to show that the algebraic numbers are countable. First list the polynomials a_1x+a_0 of degree ≤ 1 with integer coefficients satisfying $|a_i| \leq 1$, then the polynomials $a_2x^2 + a_1x + a_0$ of degree ≤ 2 with $|a_i| \leq 2$, etc. The list starts

$$\begin{aligned} &-x-1, -x+0, -x+1, -1, 0, 1, x-1, x, x+1, -2x^2-2x-2, \\ &-2x^2-2x-1, -2x^2-2x, -2x^2-2x+1, -2x^2-2x+2, \ldots. \end{aligned} \qquad 0.6.5$$

(The polynomial -1 in formula 0.6.5 is $0 \cdot x - 1$.) Then we go over the list, crossing out repetitions.

Next we write a second list, putting first the roots of the first polynomial in formula 0.6.5, then the roots of the second polynomial, etc.; again, go through the list and cross out repetitions. This lists all algebraic numbers, showing that they form a countable set.

Other consequences of different cardinalities

Two sets A and B have the same cardinality (denoted $A \asymp B$) if there exists an invertible mapping $A \to B$. A set A is countable if $A \asymp \mathbb{N}$, and it has *the cardinality of the continuum* if $A \asymp \mathbb{R}$. We will say that the cardinality of a set A is at most that of B (denoted $A \preceq B$) if there exists a one-to-one map from A to B. The Schröder–Bernstein theorem, sketched in Exercise 0.6.5, shows that if $A \preceq B$ and $B \preceq A$, then $A \asymp B$.

The fact that $\mathbb{R}$ and $\mathbb{N}$ have different cardinalities raises all sorts of questions. Are there other infinities besides those of $\mathbb{N}$ and $\mathbb{R}$? We will see in Proposition 0.6.1 that there are infinitely many.

For any set E, we denote by $\mathcal{P}(E)$ the set of all subsets of E, called the *power set* of E. Clearly for any set E there exists a one-to-one map $f: E \to \mathcal{P}(E)$; for instance, the map $f(a) = \{a\}$. So the cardinality of E is at most that of $\mathcal{P}(E)$. In fact, it is strictly less. If E is finite and has n elements, then $\mathcal{P}(E)$ has 2^n elements, clearly more than E (see Exercise 0.6.2). Proposition 0.6.1 says that this is still true if E is infinite.

Proposition 0.6.1. *A mapping $f : E \to \mathcal{P}(E)$ is never onto.*

Proof. Let $A \subset E$ be the set of x such that $x \notin f(x)$. We will show that A is not in the image of f, and thus that f is not onto. Suppose A is in the image of f. Then A must be of the form $f(y)$ for an appropriate $y \in E$. But if y is in $A = f(y)$, this implies that $y \notin A$, and if y is not in $A = f(y)$, this implies that $y \in A$. Thus A is not an element of the image of f, and f is not onto. $\square$

Thus $\mathbb{R}$, $\mathcal{P}(\mathbb{R})$, $\mathcal{P}(\mathcal{P}(\mathbb{R})), \dots$ all have different cardinalities, each bigger than the last. Exercise 0.6.8 asks you to show that $\mathcal{P}(\mathbb{N})$ has the same cardinality as $\mathbb{R}$.

FIGURE 0.6.4.
Paul Cohen (1934–2007)
Proving the continuum hypothesis was the first in the famous list of 23 problems Hilbert presented to the Second International Congress of Mathematicians in 1900. Cohen and Gödel, working some 30 years apart, proved that the problem is unsolvable.

The continuum hypothesis

Another natural question is: are there infinite subsets of $\mathbb{R}$ that cannot be put into one-to-one correspondence with either $\mathbb{R}$ or $\mathbb{N}$?

Cantor's *continuum hypothesis* says that the answer is no. This statement has been shown to be unsolvable: it is consistent with the other axioms of set theory to assume it is true (Gödel, 1938) or false (Paul Cohen, 1965). This means that if there is a contradiction in set theory assuming the continuum hypothesis, then there is a contradiction without assuming it, and if there is a contradiction in set theory assuming that the continuum hypothesis is false, then again there is a contradiction without assuming it is false.

FIGURE 0.6.5.
Kurt Gödel (1906–1978) is considered the greatest logician of all time. His brother, a doctor, wrote that Gödel " ... believed all his life that he was always right not only in mathematics but also in medicine, so he was a very difficult patient for doctors."

Many of the exercises for Section 0.6 are difficult, requiring a lot of imagination.

Exercises for Section 0.6

0.6.1 a. Show that the set of rational numbers is countable (i.e., that you can list all rational numbers).

b. Show that the set $\mathbb{D}$ of finite decimals is countable.

0.6.2 a. Show that if E is finite and has n elements, then the power set $\mathcal{P}(E)$ has 2^n elements.

b. Choose a map $f : \{a, b, c\} \to \mathcal{P}(\{a, b, c\})$, and compute for that map the set $\{x | x \notin f(x)\}$. Verify that this set is not in the image of f.

0.6.3 Show that $(-1, 1)$ has the same infinity of points as the reals. *Hint*: Consider the function $g(x) = \tan(\pi x/2)$.

0.6.4 a. How many polynomials of degree d are there with coefficients with absolute value $\leq d$?

b. Give an upper bound for the position in the list of formula 0.6.5 of the polynomial $x^4 - x^3 + 5$.

c. Give an upper bound for the position of the real cube root of 2 in the list of numbers made from the list of formula 0.6.5.

0.6.5 Let $f : A \to B$ and $g : B \to A$ be one to one. We will sketch how to construct a mapping $h : A \to B$ that is one to one and onto.

Let an (f,g)-chain be a sequence consisting s alternately of elements of A and B, with the element following an element $a \in A$ being $f(a) \in B$, and the element following an element $b \in B$ being $g(b) \in A$.

Exercise 0.6.5: The existence of such a mapping h is the *Schröder–Bernstein theorem*; we owe the proof sketched here to Thierry Gallouet at Aix–Marseille University.

a. Show that every (f,g)-chain can be uniquely continued forever to the right, and to the left can

1. either be uniquely continued forever, or
2. can be continued to an element of A that is not in the image of g, or
3. can be continued to an element of B that is not in the image of f.

b. Show that every element of A and every element of B is an element of a unique such maximal (f,g)-chain.

c. Construct $h : A \to B$ by setting

$$h(a) = \begin{cases} f(a) & \text{if } a \text{ belongs to a maximal chain of type 1 or 2 above,} \\ g^{-1}(a) & \text{if } a \text{ belongs to a maximal chain of type 3.} \end{cases}$$

Show that $h : A \to B$ is well defined, one to one, and onto.

d. Take $A = [-1,1]$ and $B = (-1,1)$. It is surprisingly difficult to write a mapping $h : A \to B$. Take $f : A \to B$ defined by $f(x) = x/2$, and $g : B \to A$ defined by $g(x) = x$.

What elements of $[-1,1]$ belong to chains of type 1, 2, 3?

What map $h : [-1,1] \to (-1,1)$ does the construction in part c give?

Exercise 0.6.7: See Section 1.1 for a discussion of the notation $\mathbb{R}^2$ and $\mathbb{R}^n$. Hint for part a: Consider the "mapping" that takes $(0.a_1a_2\dots, 0.b_1b_2\dots)$ and returns the real number $0.a_1b_1a_2b_2\dots$. There are difficulties with numbers that can be written either ending in 0's or 9's; make choices so as to make the mapping one to one, and apply Bernstein's theorem.

Part c: In 1877 Cantor wrote to Dedekind, proving that points on the interval $[0,1]$ can be put in one-to-one correspondence with points in $\mathbb{R}^n$. "I see it, but I don't believe it!" he wrote.

0.6.6 Show that the points of the circle $\left\{ \begin{pmatrix} x \\ y \end{pmatrix} \in \mathbb{R}^2 \mid x^2 + y^2 = 1 \right\}$ have the same infinity of elements as $\mathbb{R}$. *Hint*: This is easy if you use Bernstein's theorem (Exercise 0.6.5.)

***0.6.7** a. Show that $[0,1) \times [0,1)$ has the same cardinality as $[0,1)$.

b. Show that $\mathbb{R}^2$ has the same infinity of elements as $\mathbb{R}$.

c. Show that $\mathbb{R}^n$ has the same infinity of elements as $\mathbb{R}$.

***0.6.8** Show that the power set $\mathcal{P}(\mathbb{N})$ has the same cardinality as $\mathbb{R}$.

***0.6.9** Is it possible to list (all) the rationals in $[0,1]$, written as decimals, so that the entries on the diagonal also give a rational number?

0.7 Complex numbers

Complex numbers were introduced about 1550 by several Italian mathematicians trying to solve cubic equations. Their work represented the rebirth of mathematics in Europe after a sleep of more than 15 centuries.

A complex number is written $a + bi$, where a and b are real numbers, and addition and multiplication are defined in equations 0.7.1 and 0.7.2. It follows from those rules that $i^2 = -1$.

FIGURE 0.7.1.
The Italian Girolamo Cardano (1501–1576) used complex numbers (considered "impossible") as a crutch that made it possible to find *real* roots of *real* cubic polynomials. But complex numbers turned out to have immense significance in mathematics, and even physics. In quantum mechanics a state of a physical system is given by a "wave function", which is a *complex-valued function*; real-valued functions will not do.

Equation 0.7.2 is not the only possible definition of multiplication. For instance, we could define

$$(a_1 + ib_1)*(a_2 + ib_2) = (a_1a_2) + i(b_1b_2).$$

But then there would be lots of elements by which one could not divide, since the product of any purely real number and any purely imaginary number would be 0:

$$(a_1 + i0) * (0 + ib_2) = 0.$$

The four properties concerning addition don't depend on the special nature of complex numbers; we can similarly define addition for n-tuples of real numbers, and these rules will still be true.

The complex number $a+ib$ is often plotted as the point $\begin{pmatrix} a \\ b \end{pmatrix} \in \mathbb{R}^2$. The real number a is called the *real part* of $a+ib$, denoted $\text{Re}\,(a+ib)$; the real number b is called the *imaginary part*, denoted $\text{Im}\,(a+ib)$. The reals $\mathbb{R}$ can be considered as a subset of the complex numbers $\mathbb{C}$, by identifying $a \in \mathbb{R}$ with $a + i0 \in \mathbb{C}$; such complex numbers are called *real*, as you might imagine. Real numbers are systematically identified with the real complex numbers, and $a + i0$ is usually denoted simply a.

Numbers of the form $0 + ib$ are called *purely imaginary*. What complex numbers, if any, are both real and purely imaginary?[2] If we plot $a + ib$ as the point $\begin{pmatrix} a \\ b \end{pmatrix} \in \mathbb{R}^2$, what do the purely real numbers correspond to? The purely imaginary numbers?[3]

Arithmetic in $\mathbb{C}$

Complex numbers are added in the obvious way:

$$(a_1 + ib_1) + (a_2 + ib_2) = (a_1 + a_2) + i(b_1 + b_2). \qquad 0.7.1$$

Thus the identification with $\mathbb{R}^2$ preserves the operation of addition.

What makes $\mathbb{C}$ interesting is that complex numbers can also be multiplied:

$$(a_1 + ib_1)(a_2 + ib_2) = (a_1a_2 - b_1b_2) + i(a_1b_2 + a_2b_1). \qquad 0.7.2$$

This formula consists of multiplying $a_1 + ib_1$ and $a_2 + ib_2$ (treating i like the variable x of a polynomial) to find

$$(a_1 + ib_1)(a_2 + ib_2) = a_1a_2 + i(a_1b_2 + a_2b_1) + i^2(b_1b_2) \qquad 0.7.3$$

and then setting $i^2 = -1$.

Example 0.7.1 (Multiplying complex numbers).

$$(2 + i)(1 - 3i) = (2 + 3) + i(1 - 6) = 5 - 5i; \qquad (1 + i)^2 = 2i. \quad \triangle \qquad 0.7.4$$

Addition and multiplication of reals viewed as complex numbers coincides with ordinary addition and multiplication:

$$(a + i0) + (b + i0) = (a + b) + i0 \qquad (a + i0)(b + i0) = (ab) + i0. \qquad 0.7.5$$

Exercise 0.7.5 asks you to check the following nine rules, for $z_1, z_2 \in \mathbb{C}$:

1. $(z_1+z_2)+z_3 = z_1+(z_2+z_3)$ — Addition is associative.
2. $z_1 + z_2 = z_2 + z_1$ — Addition is commutative.
3. $z + 0 = z$ — 0 (i.e., the complex number $0 + 0i$) is an additive identity.
4. $(a + ib) + (-a - ib) = 0$ — $(-a - ib)$ is the additive inverse of $a + ib$.

[2]The only complex number that is both real and purely imaginary is $0 = 0+0i$.

[3]The purely real numbers are all found on the x-axis, the purely imaginary numbers on the y-axis.

The multiplication here is, of course, the special multiplication of complex numbers, defined in equation 0.7.2. We can think of the multiplication of two complex numbers as the multiplication of two pairs of real numbers. If we were to define a new kind of number as the 3-tuple (a, b, c), there would be no way to multiply two such 3-tuples that satisfies these five requirements.

There is a way to define multiplication for 4-tuples that satisfies all but commutativity; it is called *Hamilton's quaternions*. It is even possible to define multiplication of 8-tuples, but this multiplication is neither commutative nor associative. That is the end of the list: for any other n, there is no multiplication $\mathbb{R}^n \times \mathbb{R}^n \to \mathbb{R}^n$ that allows for division.

Immense psychological difficulties had to be overcome before complex numbers were accepted as an integral part of mathematics; when Gauss gave his proof of the fundamental theorem of algebra, complex numbers were still not sufficiently respectable that he could use them in his statement of the theorem (although the proof depends on them).

5. $(z_1 z_2)z_3 = z_1(z_2 z_3)$ — Multiplication is associative.
6. $z_1 z_2 = z_2 z_1$ — Multiplication is commutative.
7. $1z = z$ — 1 (i.e., the complex number $1 + 0i$) is a multiplicative identity.
8. $(a + ib)\left(\frac{a}{a^2+b^2} - i\frac{b}{a^2+b^2}\right) = 1$ — If $z \neq 0$, then z has a multiplicative inverse.
9. $z_1(z_2 + z_3) = z_1 z_2 + z_1 z_3$ — Multiplication is distributive over addition.

The complex conjugate

Definition 0.7.2 (Complex conjugate). The *complex conjugate* of the complex number $z = a + ib$ is the number $\overline{z} \stackrel{\text{def}}{=} a - ib$.

Complex conjugation preserves all of arithmetic:

$$\overline{z + w} = \overline{z} + \overline{w} \quad \text{and} \quad \overline{zw} = \overline{z}\,\overline{w}. \tag{0.7.6}$$

The real numbers are the complex numbers z that are equal to their complex conjugates: $\overline{z} = z$, and the purely imaginary complex numbers are those that are the opposites of their complex conjugates: $\overline{z} = -z$.

Note that for any complex number

$$\operatorname{Im} z = \frac{z - \overline{z}}{2i} \quad \text{and} \quad \operatorname{Re} z = \frac{z + \overline{z}}{2}, \tag{0.7.7}$$

since

$$\frac{z - \overline{z}}{i} = \frac{a + ib - (a - ib)}{i} = 2b = 2\operatorname{Im} z$$

$$z + \overline{z} = (a + ib) + (a - ib) = 2a = 2\operatorname{Re} z.$$

The *modulus* of a complex number, also known as its *absolute value*, can be written in terms of complex conjugates.

Definition 0.7.3 (Modulus of complex number). The *modulus* of a complex number $z = a + ib$ is the number

$$|z| \stackrel{\text{def}}{=} |a + ib| = \sqrt{z\overline{z}} = \sqrt{a^2 + b^2}. \tag{0.7.8}$$

Clearly, $|z| = |a + ib|$ is the distance from the origin to $\begin{pmatrix} a \\ b \end{pmatrix}$.

Complex numbers in polar coordinates

Let $z = a + ib \neq 0$ be a complex number. Then the point $\begin{pmatrix} a \\ b \end{pmatrix}$ can be represented in polar coordinates as $\begin{pmatrix} r\cos\theta \\ r\sin\theta \end{pmatrix}$, where

$$r = \sqrt{a^2 + b^2} = |z|, \tag{0.7.9}$$

and the "polar angle" θ is an angle such that $\cos\theta = \dfrac{a}{r}$ and $\sin\theta = \dfrac{b}{r}$, so that

$$z = r(\cos\theta + i\sin\theta). \qquad 0.7.10$$

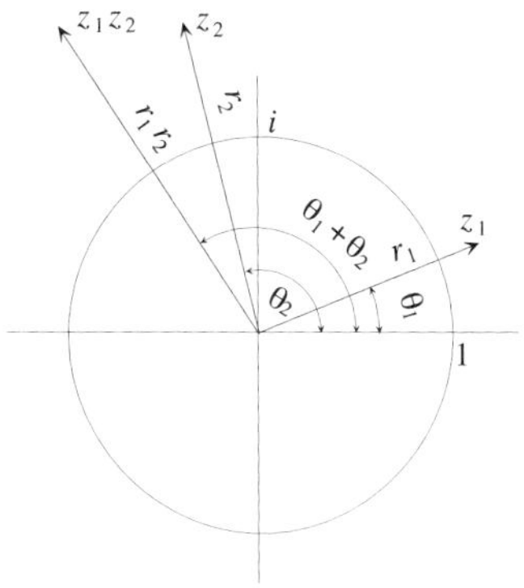

FIGURE 0.7.2.
When multiplying two complex numbers, the moduli (absolute values) are multiplied and the polar angles (arguments) are added.

Definition 0.7.4 (Argument of a complex number). The polar angle θ is called an *argument* of z. It is determined by equation 0.7.10 up to addition of a multiple of 2π.

The marvelous thing about this polar representation is that it gives a geometric representation of multiplication, as shown in Figure 0.7.2.

Proposition 0.7.5 (Geometrical representation of multiplication of complex numbers). *The modulus of the product is the product of the moduli:*

$$|z_1z_2| = |z_1|\,|z_2|. \qquad 0.7.11$$

The polar angle of the product is the sum of the polar angles θ_1, θ_2:

$$\Big(r_1(\cos\theta_1+i\sin\theta_1)\Big)\Big(r_2(\cos\theta_2+i\sin\theta_2)\Big) = r_1r_2\Big(\cos(\theta_1+\theta_2)+i\sin(\theta_1+\theta_2)\Big).$$

Proof. Multiply out and apply the addition rules of trigonometry:

$$\begin{aligned}\cos(\theta_1+\theta_2) &= \cos\theta_1\cos\theta_2 - \sin\theta_1\sin\theta_2\\ \sin(\theta_1+\theta_2) &= \sin\theta_1\cos\theta_2 + \cos\theta_1\sin\theta_2. \quad \square\end{aligned} \qquad 0.7.12$$

The following formula follows immediately:

FIGURE 0.7.3.
Abraham de Moivre (1667–1754)
De Moivre, a Protestant, fled France in 1685 when King Louis XIV revoked the Edict of Nantes, which had given Protestants freedom to practice their religion. In 1697 he became a fellow of the Royal Society in England.

Corollary 0.7.6 (De Moivre's formula). *If $z = r(\cos\theta + i\sin\theta)$, then*

$$z^n = r^n(\cos n\theta + i\sin n\theta). \qquad 0.7.13$$

De Moivre's formula itself has an important consequence. While it is not true that any real number has n real nth roots, every complex number (and thus every real number) has n complex nth roots. Figure 0.7.4 illustrates Proposition 0.7.7 for $n = 5$. Recall that $r^{1/n}$ denotes the positive real nth root of the positive number r.

Proposition 0.7.7 (Roots of a complex number). *Every complex number $z = r(\cos\theta + i\sin\theta)$ with $r \neq 0$ has n distinct complex nth roots, the numbers*

$$r^{1/n}\left(\cos\frac{\theta+2k\pi}{n} + i\sin\frac{\theta+2k\pi}{n}\right), \quad k = 0,\dots,n-1. \qquad 0.7.14.$$

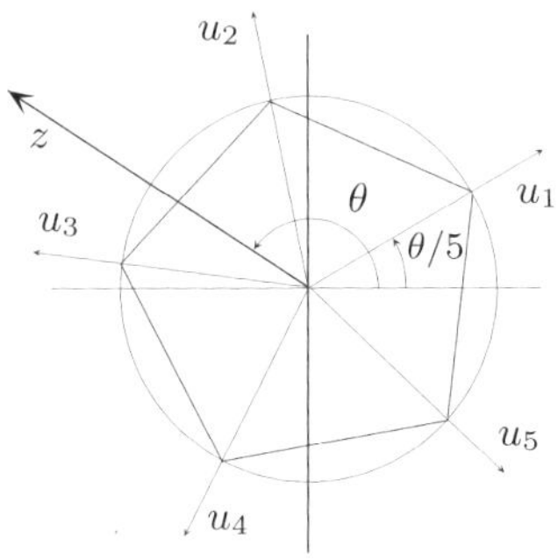

FIGURE 0.7.4.
The fifth roots of z form a regular pentagon, with one vertex at polar angle $\theta/5$ and the others rotated from that one by multiples of $2\pi/5$.

FIGURE 0.7.5.
Niccolo Fontana Tartaglia
Tartaglia's beard hid scars from sabre wounds he suffered as a 12-year-old when the French sacked his town of Brescia in 1512. He was left for dead; wounds to his jaw and palate resulted in a stammer that earned him the nickname Tartaglia, "the stammerer".

In 1535, Tartaglia won a contest with a man who had boasted of his ability to solve cubic equations; his success inspired Cardano to work on cubics and quartics.

For instance, the number $8 = 8 + i0$ has $r = 8$ and $\theta = 0$, so the three cube roots of 8 are

1. $2(\cos 0 + i \sin 0) = 2$; 2. $2\left(\cos \frac{2\pi}{3} + i \sin \frac{2\pi}{3}\right) = -1 + i\sqrt{3}$
3. $2\left(\cos \frac{4\pi}{3} + i \sin \frac{4\pi}{3}\right) = -1 - i\sqrt{3}$.

Proof. All that needs to be checked is that

1. $\left(r^{1/n}\right)^n = r$, which is true by definition;
2. $$\cos n\frac{\theta + 2k\pi}{n} = \cos\theta \quad \text{and} \quad \sin n\frac{\theta + 2k\pi}{n} = \sin\theta, \qquad 0.7.15$$
   which is true since $n\frac{\theta+2k\pi}{n} = \theta + 2k\pi$, and sin and cos are periodic with period 2π;
3. the numbers in formula 0.7.14 are distinct, which is true since the polar angles do not differ by an integer multiple of 2π; they differ by multiples $2\pi k/n$, with $0 < k < n$. $\square$

The fundamental theorem of algebra

We saw in Proposition 0.7.7 that all complex numbers have n nth roots. In other words, the equation $z^n = a$ has n roots. A great deal more is true: *all polynomial equations with complex coefficients have all the roots one might hope for*. This is the content of the fundamental theorem of algebra, Theorem 1.6.14, proved by d'Alembert in 1746 and by Gauss around 1799. This milestone of mathematics followed by some 200 years the first introduction of complex numbers, by several Italian mathematicians trying to solve cubic equations, notably Niccolo Tartaglia and Girolamo Cardano. Cubic and quartic equations are discussed in Appendix A2.

"Graphing" complex-valued functions

In one-variable calculus, graphing functions is tremendously helpful in understanding their properties. Please note that the graph of a function $f : \mathbb{R} \to \mathbb{R}$ is a subset of $\mathbb{R}^2$; it is the set of pairs $\begin{pmatrix} x \\ y \end{pmatrix}$ such that $y = f(x)$.

Graphing maps $f : \mathbb{R}^n \to \mathbb{R}^m$ would no doubt be tremendously helpful also, but there is a serious problem: the graph is a subset of $\mathbb{R}^{n+m}$, and most people are unable to visualize such graphs when $n + m > 3$. Thus the only cases other than $n = m = 1$ where we can really use our ability to visualize functions are $n = 2$, $m = 1$, where the graph is a surface in space, and $n = 1$, $m = 2$, where the graph is a curve in space.

Even these can be difficult, because usually what we have at hand is two dimensional: the page of a book, a blackboard, a computer screen. There is no very satisfactory way out of this difficulty: understanding the behavior of mappings $\mathbb{R}^n \to \mathbb{R}^m$ when n or m (or both) are even moderately large is really difficult.

Computer programs that animate two-dimensional representations of three-dimensional objects are very helpful for visualization; for example, when we created Figure 5.2.2 we spent some time turning the surface by computer before choosing the angle we found most suggestive of the actual surface.

How then can we visualize a mapping $f : \mathbb{C} \to \mathbb{C}$, which is equivalent to a function $f : \mathbb{R}^2 \to \mathbb{R}^2$, whose graph is a subset of $\mathbb{R}^4$? One method commonly used is to draw two planes, one corresponding to the domain and one to the codomain, and to try to understand how the point $f(z)$ moves in the codomain as the point z moves in the domain. Figure 0.7.6 is an example. It shows the mapping $f(z) = z^2$, where z is complex. We have drawn a fencer in the domain; the "fencer squared" has stabbed himself in the stomach and is about to kick himself in the nose.

Since every complex number ($\neq 0$) has two square roots, the "fencer squared" on the right in Figure 0.7.6 should have two square roots; these are represented in Figure 0.7.7: each is stabbing the other in the stomach and kicking him in the head. There is in fact no mapping "$\pm\sqrt{z}$", and in the complex case there is no precise analogue of the positive square root and the negative square root.

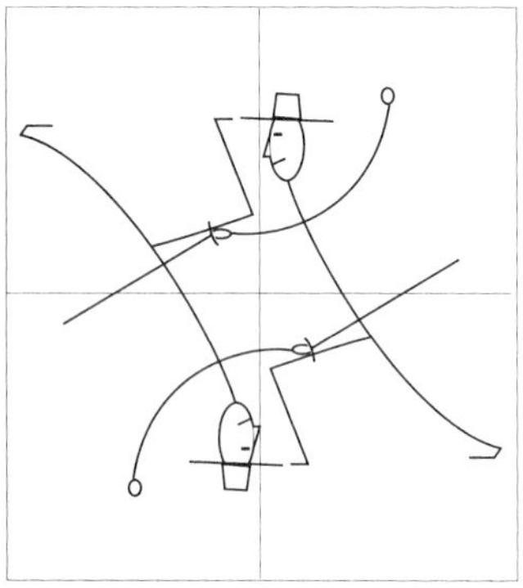

FIGURE 0.7.7.
The "fencer squared" of Figure 0.7.6 has two square roots.

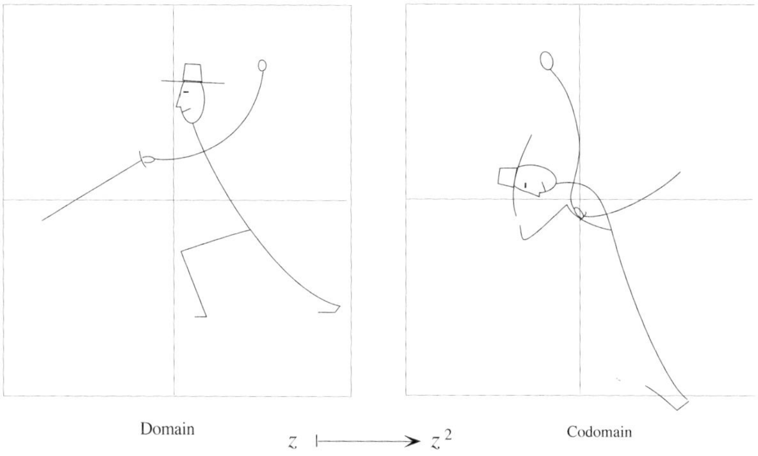

FIGURE 0.7.6. "Graph" of the complex-valued squaring function $(z) = z^2$.

EXERCISES FOR SECTION 0.7

0.7.1 If z is the complex number $a + ib$, which of the following are synonyms?

modulus of z	$\lvert z\rvert$	$\bar{z}$	$\operatorname{Re} z$
absolute value of z	$\operatorname{Im} z$	complex conjugate of z	a
b	real part of z	imaginary part of z	

0.7.2 If $z = a + ib$, which of the following are real numbers?

$\lvert z\rvert$	$\operatorname{Re} z$	$\sqrt{z\bar{z}}$	$\bar{z}$
a	$\operatorname{Im} z$	b	$z + \bar{z}$
modulus of z	real part of z	imaginary part of z	$z - \bar{z}$

0.7.3 Find the absolute value and argument of each of the following complex numbers:

a. $2 + 4i$ b. $(3 + 4i)^{-1}$ c. $(1 + i)^5$ d. $1 + 4i$.

0.7.4 Find the modulus and polar angle of the following complex numbers:

a. $3 + 2i$ b. $(1 - i)^4$ c. $2 + i$ d. $\sqrt[7]{3 + 4i}$

0.7.5 Verify the nine rules for addition and multiplication of complex numbers. Statements 5 and 9 are the only ones that are not immediate.

0.7.6 a. Solve $x^2 + ix = -1$. b. Solve $x^4 + x^2 + 1 = 0$.

0.7.7 Describe the set of all complex number $z = x + iy$ such that

a. $|z - u| + |z - v| = c$ for $u, v \in \mathbb{C}$ and $c \in \mathbb{R}$ b. $|z| < 1 - \operatorname{Re} z$

Exercise 0.7.8, part b: There is a fairly obvious real root x_1. Find it by trial and error. Then divide the equation by $x - x_1$ and solve the resulting quadratic equation.

0.7.8 a. Solve $x^2 + 2ix - 1 = 0$. b. Solve $x^3 - x^2 - x = 2$.

0.7.9 Solve the following equations where x, y are complex numbers:

a. $x^2 + ix + 2 = 0$ b. $x^4 + x^2 + 2 = 0$ c. $\begin{cases} ix - (2 + i)y = 3 \\ (1 + i)x + iy = 4 \end{cases}$

0.7.10 Describe the set of all complex numbers $z = x + iy$ such that

a. $\bar{z} = -z$ b. $|z - a| = |z - b|$ for given $a, b \in \mathbb{C}$ c. $\bar{z} = z^{-1}$

In Exercises 0.7.11 and 0.7.12, the symbol $\mapsto$ ("maps to") describes what a mapping does to an input. One could give the name f to the map $z \mapsto z^2$ and rewrite it as $f(z) = z^2$.

0.7.11 a. Describe the loci in $\mathbb{C}$ given by the following equations:

i. $\operatorname{Re} z = 1$, ii. $|z| = 3$.

b. What is the image of each locus under the mapping $z \mapsto z^2$?

c. What is the inverse image of each locus under the mapping $z \mapsto z^2$?

0.7.12 a. Describe the loci in $\mathbb{C}$ given by the following equations:

i. $\operatorname{Im} z = -|z|$, ii. $\frac{1}{2}|z - i| = z$.

b. What is the image of each locus under the mapping $z \mapsto z^2$?

c. What is the inverse image of each locus under the mapping $z \mapsto z^2$?

0.7.13 a. Find all the cubic roots of 1.

b. Find all the 4th roots of 1.

c. Find all the 6th roots of 1.

0.7.14 Show that for any complex numbers z_1, z_2, we have $\operatorname{Im}(\bar{z}_1 z_2) \le |z_1||z_2|$.

0.7.15 a. Find all the fifth roots of 1, using a formula that only includes square roots (no trigonometric functions).

*b. Use your formula, ruler, and compass to construct a regular pentagon.

1

Vectors, matrices, and derivatives

It is sometimes said that the great discovery of the nineteenth century was that the equations of nature were linear, and the great discovery of the twentieth century is that they are not.—Tom Körner, *Fourier Analysis*

1.0 Introduction

In this chapter, we introduce the principal actors of linear algebra and multivariable calculus.

By and large, first year calculus deals with functions f that associate *one* number $f(x)$ to *one* number x. In most realistic situations, this is inadequate: the description of most systems depends on many functions of many variables.

In physics, a gas might be described by pressure and temperature as a function of position and time, two functions of four variables. In biology, one might be interested in numbers of sharks and sardines as functions of position and time; a famous study of sharks and sardines in the Adriatic, described in *The Mathematics of the Struggle for Life* by Vito Volterra, founded the subject of mathematical ecology.

In microeconomics, a company might be interested in production as a function of input, where that function has as many coordinates as the number of products the company makes, each depending on as many inputs as the company uses. Even thinking of the variables needed to describe a macroeconomic model is daunting (although economists and the government base many decisions on such models). Countless examples are found in every branch of science and social science.

Mathematically, all such things are represented by functions $\mathbf{f}$ that take n numbers and return m numbers; such functions are denoted $\mathbf{f} : \mathbb{R}^n \to \mathbb{R}^m$. In that generality, there isn't much to say; we must impose restrictions on the functions we will consider before any theory can be elaborated.

The strongest requirement we can make is that $\mathbf{f}$ should be *linear*; roughly speaking, a function is linear if when we double the input, we double the output. Such linear functions are fairly easy to describe completely, and a thorough understanding of their behavior is the foundation for everything else.

The first four sections of this chapter lay the foundations of linear algebra. In the first three sections we introduce the main actors – vectors and matrices – and relate them to the notion of a linear function.

One problem of great interest at present is protein folding. The human genome is now known; in particular, we know the sequence of amino acids of all the proteins in the body. But proteins are only active when they are curled up in just the right way; the world's biggest computers are busy trying to derive from the sequence of amino acids just how they will fold.

Specifying the position and orientation of all N amino acid requires $6N$ numbers; each such configuration has a potential energy, and the preferred folding corresponds to a minimum of this potential energy. Thus the problem of protein folding is essentially finding the minima of a function of $6N$ variables, where N might be 1000.

Although many approaches of this problem are actively being pursued, there is no very satisfactory solution so far. Understanding this function of $6N$ variables is one of the main challenges of the age.

One object of linear algebra is to extend to higher dimensions the geometric language and intuition we have concerning the plane and space, familiar to us all from everyday experience.

When we discuss solving systems of equations in Chapter 2, we will also want to be able to interpret algebraic statements geometrically. You learned in school that saying that two equations in two unknowns have a unique solution is the same as saying that the two lines given by those equations intersect in a single point. In higher dimensions we will want, for example, to be able to speak of the "space of solutions" of a particular system of equations as being a four-dimensional subspace of $\mathbb{R}^7$.

Next we develop the geometrical language that we will need in multivariable calculus. The realization by René Decartes (1596–1650) that an equation can denote a curve or surface was a crucial moment in the history of mathematics, integrating two fields, algebra and geometry, that had previously seemed unrelated. We do not want to abandon this double perspective when we move to higher dimensions. Just as the equation $x^2 + y^2 = 1$ and the circle of radius 1 centered at the origin are one and the same object, we will want to speak both of the 9-dimensional unit sphere in $\mathbb{R}^{10}$ and of the equation denoting that sphere. In Section 1.4 we define such notions as the length of a vector in $\mathbb{R}^n$, the length of a matrix, and the angle between two vectors. This will enable us to think and speak in geometric terms about higher-dimensional objects.

In Section 1.5 we discuss sequences, subsequences, limits, and convergence. In Section 1.6 we will expand on that discussion, developing the topology needed for a rigorous treatment of calculus.

Most functions are not linear, but very often they are well approximated by linear functions, at least for some values of the variables. For instance, as long as there are few hares, their number may well quadruple every three or four months, but as soon as they become numerous, they will compete with each other, and their rate of increase (or decrease) will become more complex. In the last three sections of this chapter we begin exploring how to approximate a nonlinear function by a linear function – specifically, by its higher-dimensional derivative.

1.1 Introducing the actors: Points and vectors

Much of linear algebra and multivariate calculus takes place within $\mathbb{R}^n$. This is the space of ordered lists of n real numbers.

The notion that one can think about and manipulate higher-dimensional spaces by considering a point in n-dimensional space as a list of its n "coordinates" did not always appear as obvious to mathematicians as it does today. In 1846, the English mathematician Arthur Cayley pointed out that a point with four coordinates can be interpreted geometrically without recourse to "any metaphysical notion concerning the possibility of four-dimensional space."

You are probably used to thinking of a point in the plane in terms of its two coordinates: the familiar Cartesian plane with its x, y axes is $\mathbb{R}^2$. A point in space (after choosing axes) is specified by its three coordinates: Cartesian space is $\mathbb{R}^3$. Similarly, a point in $\mathbb{R}^n$ is specified by its n coordinates; it is a list of n real numbers. Such ordered lists occur everywhere, from grades on a transcript to prices on the stock exchange. Seen this way, higher dimensions are no more complicated than $\mathbb{R}^2$ and $\mathbb{R}^3$; the lists of coordinates just get longer.

Example 1.1.1 (The stock market). The following data is from the *Ithaca Journal*, Dec. 14, 1996.

Local NYSE Stocks

	Vol	**High**	**Low**	**Close**	**Chg**
Airgas	193	24 1/2	23 1/8	23 5/8	-3/8
AT&T	36606	39 1/4	38 3/8	39	3/8
Borg Warner	74	38 3/8	38	38	-3/8
Corning	4575	44 3/4	43	44 1/4	1/2
Dow Jones	1606	33 1/4	32 1/2	33 1/4	1/8
Eastman Kodak	7774	80 5/8	79 1/4	79 3/8	-3/4
Emerson Elec.	3335	97 3/8	95 5/8	95 5/8	-1 1/8
Federal Express	5828	42 1/2	41	41 5/8	1 1/2

We can think of this table as five columns, each an element of $\mathbb{R}^8$:

"Vol" denotes the number of shares traded, "High" and "Low" the highest and lowest price paid per share, "Close" the price when trading stopped at the end of the day, and "Chg" the difference between the closing price and the closing price on the previous day.

$$\underbrace{\begin{bmatrix} 193 \\ 36606 \\ 74 \\ 4575 \\ 1606 \\ 7774 \\ 3335 \\ 5828 \end{bmatrix}}_{\text{Vol}} \quad \underbrace{\begin{pmatrix} 24\,^1\!/_2 \\ 39\,^1\!/_4 \\ 38\,^3\!/_8 \\ 44\,^3\!/_4 \\ 33\,^1\!/_4 \\ 80\,^5\!/_8 \\ 97\,^3\!/_8 \\ 42\,^1\!/_2 \end{pmatrix}}_{\text{High}} \quad \underbrace{\begin{pmatrix} 23\,^1\!/_8 \\ 38\,^3\!/_8 \\ 38 \\ 43 \\ 32\,^1\!/_2 \\ 79\,^1\!/_4 \\ 95\,^5\!/_8 \\ 41 \end{pmatrix}}_{\text{Low}} \underbrace{\begin{pmatrix} 23\,^5\!/_8 \\ 39 \\ 38 \\ 44\,^1\!/_4 \\ 33\,^1\!/_4 \\ 79\,^3\!/_8 \\ 95\,^5\!/_8 \\ 41\,^5\!/_8 \end{pmatrix}}_{\text{Close}} \quad \underbrace{\begin{bmatrix} -\,^3\!/_8 \\ ^3\!/_8 \\ -\,^3\!/_8 \\ ^1\!/_2 \\ ^1\!/_8 \\ -\,^3\!/_4 \\ -1\,^1\!/_8 \\ 1\,^1\!/_2 \end{bmatrix}}_{\text{Chg}}$$

Note that we use parentheses for "positional" data (for example, highest price paid per share), and brackets for "incremental" data (for example, change in price). Note also that we write elements of $\mathbb{R}^n$ as *columns*, not rows. The reason for preferring columns will become clear later: we want the order of terms in matrix multiplication to be consistent with the notation $f(x)$, where the function is placed before the variable.

Remark. Time is sometimes referred to as "the fourth dimension". This is misleading. A point in $\mathbb{R}^4$ is simply four numbers. If the first three numbers give the x, y, z coordinates, the fourth number might give time. But the fourth number could also give temperature, or density, or some other information. In addition, as shown in the above example, there is no need for any of the numbers to denote a position in physical space; in higher dimensions, it can be more helpful to think of a point as a "state" of a system. If 3356 stocks are listed on the New York Stock Exchange, the list of closing prices for those stocks is an element of $\mathbb{R}^{3356}$, and every element of $\mathbb{R}^{3356}$ is one theoretically possible state of closing prices on the stock market. (Of course, some such states will never occur; for instance, stock prices are positive.) $\triangle$

Points and vectors: Positional data versus incremental data

An element of $\mathbb{R}^n$ is simply an ordered list of n numbers, but such a list can be interpreted in two ways: as a *point* representing a position or as a *vector* representing a displacement or increment.

Definition 1.1.2 (Point, vector, and coordinates). The element of $\mathbb{R}^n$ with coordinates $x_1, x_2, \cdots, x_n$ can be interpreted as the *point* $\mathbf{x} = \begin{pmatrix} x_1 \\ \vdots \\ x_n \end{pmatrix}$ or as the *vector* $\vec{\mathbf{x}} = \begin{bmatrix} x_1 \\ \vdots \\ x_n \end{bmatrix}$, which represents an increment.

Example 1.1.3 (An element of $\mathbb{R}^2$ as a point and as a vector). The element of $\mathbb{R}^2$ with coordinates $x = 2$, $y = 3$ can be interpreted as the point

$\begin{pmatrix} 2 \\ 3 \end{pmatrix}$ in the plane, as shown in Figure 1.1.1. But it can also be interpreted as the instructions "start anywhere and go two units right and three units up," rather like instructions for a treasure hunt: "take two giant steps to the east, and three to the north"; this is shown in Figure 1.1.2. Here we are interested in the displacement: if we start at *any* point and travel $\begin{bmatrix} 2 \\ 3 \end{bmatrix}$, how far will we have gone, in what direction? When we interpret an element of $\mathbb{R}^n$ as a position, we call it a *point*; when we interpret it as a displacement or increment, we call it a *vector*. △

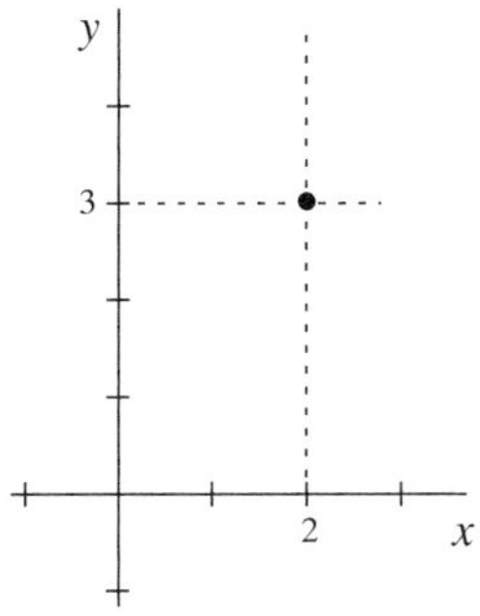

FIGURE 1.1.1.
The point $\begin{pmatrix} 2 \\ 3 \end{pmatrix}$.

Example 1.1.4 (Points and vectors in $\mathbb{R}^{3356}$). If 3356 stocks are listed on the New York Stock Exchange, the list of closing prices for those stocks is a point in $\mathbb{R}^{3356}$. The list telling how much each stock gained or lost compared with the previous day is also an element of $\mathbb{R}^{3356}$, but this corresponds to thinking of the element as a vector, with direction and magnitude: did the price of each stock go up or down? How much? △

In the plane and in three-dimensional space a vector can be depicted as an arrow pointing in the direction of the displacement. The amount of displacement is the length of the arrow. This does not extend well to higher dimensions. How are we to picture the "arrow" in $\mathbb{R}^{3356}$ representing the change in prices on the stock market? How long is it, and in what "direction" does it point? We will show how to compute magnitudes and directions for vectors in $\mathbb{R}^n$ in Section 1.4.

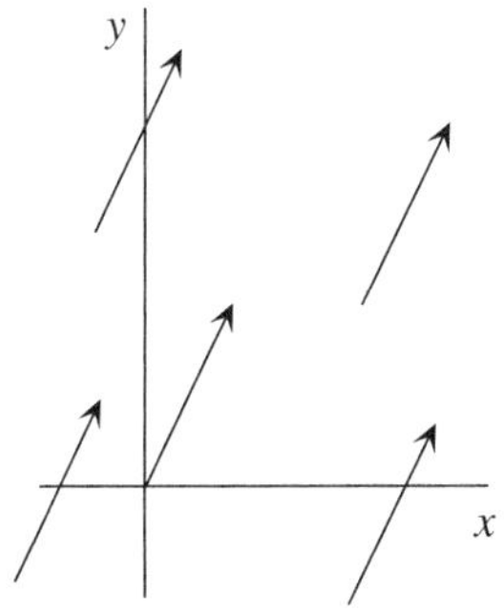

FIGURE 1.1.2.
All the arrows represent the same vector, $\begin{bmatrix} 2 \\ 3 \end{bmatrix}$.

Remark. In physics textbooks and some first year calculus books, vectors are often said to represent quantities (velocity, forces) that have both "magnitude" and "direction," while other quantities (length, mass, volume, temperature) have only "magnitude" and are represented by numbers (scalars). We think this focuses on the wrong distinction, suggesting that some quantities are always represented by vectors while others never are, and that it takes more information to specify a quantity with direction than one without.

The volume of a balloon is a single number, but so is the vector expressing the difference in volume between an inflated balloon and one that has popped. The first is a number in $\mathbb{R}$, while the second is a vector in $\mathbb{R}$. The height of a child is a single number, but so is the vector expressing how much she has grown since her last birthday. A temperature can be a "magnitude," as in "It got down to -20 last night," but it can also have "magnitude and direction," as in "It is 10 degrees colder today than yesterday." Nor can "static" information always be expressed by a single number: the state of the stock market at a given instant requires one number for each stock listed – as does the vector describing the change in the stock market from one day to the next. △

Philosophically the point zero and the zero vector are quite different. The zero vector (i.e., the "zero increment") has a universal meaning, the same regardless of the frame of reference. The point zero is arbitrary, just as "zero degrees" has a different meaning in the Celsius system and in Fahrenheit.

Sometimes, often at a key point in a proof, we will suddenly start thinking of points as vectors, or vice versa; this happens in the proof of Kantorovich's theorem in Appendix A5, for example.

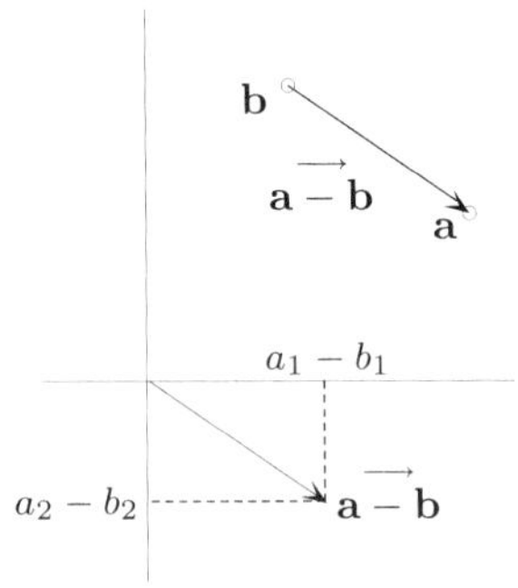

FIGURE 1.1.3.
The difference $\overrightarrow{\mathbf{a} - \mathbf{b}}$ between point **a** and point **b** is the vector joining them. The difference can be computed by subtracting the coordinates of **b** from those of **a**.

Points can't be added; vectors can

As a rule, it doesn't make sense to add points, any more than it makes sense to "add" the positions "Boston" and "New York" or the temperatures 50 degrees Fahrenheit and 70 degrees Fahrenheit. (If you opened a door between two rooms at those temperatures, the result would not be two rooms at 120 degrees!) But it does make sense to measure the difference between points (i.e., to subtract them): you can talk about the distance between Boston and New York, or about the difference in temperature between two rooms. The result of subtracting one point from another is thus a vector specifying the increment you need to add to get from one point to another.

You can also add increments (vectors) together, giving another increment. For instance, the vectors "advance five meters east then take two giant steps south" and "take three giant steps north and go seven meters west" can be added, to get "advance two meters west and one giant step north."

Similarly, in the NYSE table in Example 1.1.1, adding the *Close* columns on two successive days does not produce a meaningful answer. But adding the *Chg* columns for each day of a week produces a perfectly meaningful increment: the change in the market over that week. It is also meaningful to add increments to points (giving a point): adding a *Chg* column to the previous day's *Close* column produces the current day's *Close* – the new state of the system.

To help distinguish these two kinds of elements of $\mathbb{R}^n$, we denote them differently: points are denoted by boldface lowercase letters, and vectors by boldface lowercase letters with arrows above them. Thus **x** is a point in $\mathbb{R}^2$, while $\vec{\mathbf{x}}$ is a vector in $\mathbb{R}^2$. We do not distinguish between entries of points and entries of vectors; they are all written in plain type, with subscripts. However, when we write elements of $\mathbb{R}^n$ as columns, we use parentheses for a point **x** and square brackets for a vector $\vec{\mathbf{x}}$: in $\mathbb{R}^2$, $\mathbf{x} = \begin{pmatrix} x_1 \\ x_2 \end{pmatrix}$ and $\vec{\mathbf{x}} = \begin{bmatrix} x_1 \\ x_2 \end{bmatrix}$.

Remark. An element of $\mathbb{R}^n$ is an ordered list of numbers whether it is interpreted as a point or as a vector. But we have very different images of points and vectors, and we hope that sharing them with you explicitly will help you build a sound intuition. In linear algebra, you should think of elements of $\mathbb{R}^n$ as vectors. However, differential calculus is all about increments to points. It is because the increments are vectors that linear algebra is a prerequisite for multivariate calculus: it provides the right language and tools for discussing these increments.

Subtraction and addition of vectors and points

The difference between point **a** and point **b** is the vector $\overrightarrow{\mathbf{a} - \mathbf{b}}$, as shown in Figure 1.1.3.

Vectors are added by adding their coordinates:

$$\underbrace{\begin{bmatrix} v_1 \\ v_2 \\ \vdots \\ v_n \end{bmatrix}}_{\vec{\mathbf{v}}} + \underbrace{\begin{bmatrix} w_1 \\ w_2 \\ \vdots \\ w_n \end{bmatrix}}_{\vec{\mathbf{w}}} = \underbrace{\begin{bmatrix} v_1 + w_1 \\ v_2 + w_2 \\ \vdots \\ v_n + w_n \end{bmatrix}}_{\vec{\mathbf{v}} + \vec{\mathbf{w}}}; \qquad 1.1.1$$

the result is a vector. Similarly, vectors are subtracted by subtracting the corresponding coordinates to get a new vector. A point and a vector are added by adding the corresponding coordinates; the result is a point.

In the plane, the sum $\vec{\mathbf{v}} + \vec{\mathbf{w}}$ is the diagonal of the parallelogram of which two adjacent sides are $\vec{\mathbf{v}}$ and $\vec{\mathbf{w}}$, as shown in Figure 1.1.4 (left). We can also add vectors by placing the beginning of one vector at the end of the other, as shown in Figure 1.1.4 (right).

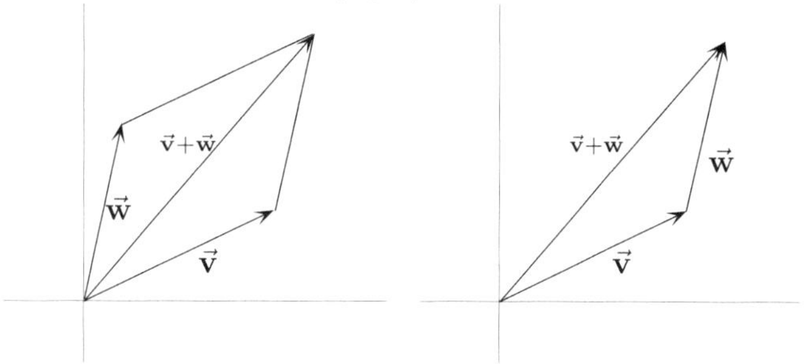

FIGURE 1.1.4. In the plane, the sum $\vec{\mathbf{v}} + \vec{\mathbf{w}}$ is the diagonal of the parallelogram at left. We can also add them by putting them head to tail.

Multiplying vectors by scalars

Multiplication of a vector by a *scalar* is straightforward:

$$a \begin{bmatrix} x_1 \\ \vdots \\ x_n \end{bmatrix} = \begin{bmatrix} ax_1 \\ \vdots \\ ax_n \end{bmatrix}; \quad \text{for example,} \quad \sqrt{3} \begin{bmatrix} 3 \\ -1 \\ 2 \end{bmatrix} = \begin{bmatrix} 3\sqrt{3} \\ -\sqrt{3} \\ 2\sqrt{3} \end{bmatrix}. \qquad 1.1.2$$

We use the word "scalar" rather than "real number" because most theorems in linear algebra are just as true for complex vector spaces or rational vector spaces as for real ones, and we don't want to restrict the validity of the statements unnecessarily.

In this book, our vectors will most often be columns of real numbers, so that our scalars – the kinds of numbers we are allowed to multiply vectors by – are real numbers. If the entries of our vectors were complex numbers, our scalars would be complex numbers; in number theory, scalars might be rational numbers; in coding theory, they might be elements of a finite field. (You may have run into such things under the name "clock arithmetic".)

Subsets and subspaces of $\mathbb{R}^n$

A subset of $\mathbb{R}^n$ is a collection of elements of $\mathbb{R}^n$. The symbol $\subset$ denotes this relationship: $X \subset \mathbb{R}^n$ means that X is a subset of $\mathbb{R}^n$.

Do not confuse $\subset$ ("subset of") with $\in$ ("element of"). Out loud, both are often rendered as "in". If you are unfamiliar with the notation of set theory, see the discussion in Section 0.3.

Subsets can be as simple as the unit circle in the plane, which is a subset of $\mathbb{R}^2$, or as wild as the set of rational numbers, which is a subset of $\mathbb{R}$. (Each rational number is arbitrarily close to another rational number but never next to one, since an irrational number always gets in the way.)

Note that a subset of $\mathbb{R}^n$ need not be "n dimensional". (We use quotes because haven't defined "dimension" yet.) Consider the unit circle. The set of points $\left\{ \begin{pmatrix} x \\ y \end{pmatrix} \middle| x^2 + y^2 = 1 \right\}$ is a subset of $\mathbb{R}^2$; the set of points

$$\left\{ \begin{pmatrix} x \\ y \\ z \end{pmatrix} \middle| x^2 + y^2 = 1, \ z = 0 \right\} \qquad 1.1.3$$

is a subset of $\mathbb{R}^3$. Both these unit circles are one dimensional.

Only a very few, very special subsets of $\mathbb{R}^n$ are *subspaces* (or *vector subspaces*) of $\mathbb{R}^n$. Roughly, a subspace is a flat subset that goes through the origin.

To be closed under multiplication, a subspace must contain the zero vector, so that if $\vec{\mathbf{v}}$ is in V,

$$0 \cdot \vec{\mathbf{v}} = \vec{\mathbf{0}} \quad \text{is in } V.$$

Definition 1.1.5 (Subspace of $\mathbb{R}^n$). A nonempty subset $V \subset \mathbb{R}^n$ is a *vector subspace* if it is closed under addition and closed under multiplication by scalars; that is, V is a vector subspace if when

$$\vec{\mathbf{x}}, \vec{\mathbf{y}} \in V \text{ and } a \in \mathbb{R}, \qquad \text{then} \qquad \vec{\mathbf{x}} + \vec{\mathbf{y}} \in V \text{ and } a\vec{\mathbf{x}} \in V.$$

(This $\mathbb{R}^n$ should be thought of as made up of vectors, not points.)

In Section 2.6 we will discuss "abstract" vector spaces whose elements, still called vectors, cannot be considered as an ordered list of numbers denoted in column form. For example, a polynomial can be an element of a vector space; so can a function. In this book our vector spaces will most often be subspaces of $\mathbb{R}^n$, but we will sometimes use subspaces of $\mathbb{C}^n$ (whose elements are column vectors with complex numbers as entries), and occasionally we will use abstract vector spaces.

Example 1.1.6 (Subspaces of $\mathbb{R}^2$). Let us first give some examples of subsets that are *not* subspaces. The unit circle $S^1 \subset \mathbb{R}^2$ of equation $x^2 + y^2 = 1$ is not a subspace of $\mathbb{R}^2$. It is not closed under addition: for instance, $\begin{bmatrix} 1 \\ 0 \end{bmatrix} \in S^1$ and $\begin{bmatrix} 0 \\ 1 \end{bmatrix} \in S^1$, but $\begin{bmatrix} 1 \\ 0 \end{bmatrix} + \begin{bmatrix} 0 \\ 1 \end{bmatrix} = \begin{bmatrix} 1 \\ 1 \end{bmatrix} \notin S^1$. It is not closed under multiplication by scalars either: $\begin{bmatrix} 1 \\ 0 \end{bmatrix}$ is an element of S^1, but $2 \begin{bmatrix} 1 \\ 0 \end{bmatrix} = \begin{bmatrix} 2 \\ 0 \end{bmatrix}$ is not.

The subset

$$X = \left\{ \begin{bmatrix} x \\ y \end{bmatrix} \in \mathbb{R}^2 \mid x \text{ is an integer} \right\} \qquad 1.1.4$$

is not a subspace of $\mathbb{R}^2$. It is closed under addition, but not under multiplication by real numbers. For instance, $\begin{bmatrix} 1 \\ 0 \end{bmatrix} \in X$, but $.5 \begin{bmatrix} 1 \\ 0 \end{bmatrix} = \begin{bmatrix} .5 \\ 0 \end{bmatrix} \notin X$.

The union A of the two axes in $\mathbb{R}^2$ (defined by the equation $xy = 0$) is not a subspace of $\mathbb{R}^2$. It is closed under multiplication by scalars, but it is not closed under addition:

$$\begin{bmatrix} 1 \\ 0 \end{bmatrix} \in A \quad \text{and} \quad \begin{bmatrix} 0 \\ 1 \end{bmatrix} \in A \quad \text{but} \quad \begin{bmatrix} 1 \\ 0 \end{bmatrix} + \begin{bmatrix} 0 \\ 1 \end{bmatrix} = \begin{bmatrix} 1 \\ 1 \end{bmatrix} \notin A. \qquad 1.1.5$$

Straight lines are usually not subspaces. For instance, the line L of equation $x + y = 1$ is not a subspace: both $\begin{bmatrix}1\\0\end{bmatrix}$ and $\begin{bmatrix}0\\1\end{bmatrix}$ are in L, but $\begin{bmatrix}1\\0\end{bmatrix} + \begin{bmatrix}0\\1\end{bmatrix} = \begin{bmatrix}1\\1\end{bmatrix}$ is not in L.

The trivial subspace $\{\vec{\mathbf{0}}\}$ is not the empty set; it contains $\vec{\mathbf{0}}$.

So what *is* a subspace of $\mathbb{R}^2$? There are exactly three kinds: lines through the origin (the main kind), and the two *trivial subspaces* $\{\vec{\mathbf{0}}\}$ and $\mathbb{R}^2$ itself. Consider, for instance, the line through the origin of equation $y = 3x$. If $\begin{bmatrix}x_1\\y_1\end{bmatrix}$ and $\begin{bmatrix}x_2\\y_2\end{bmatrix}$ are on the line, so that $y_1 = 3x_1$ and $y_2 = 3x_2$, then

$$\begin{bmatrix}x_1\\y_1\end{bmatrix} + \begin{bmatrix}x_2\\y_2\end{bmatrix} = \begin{bmatrix}x_1+x_2\\y_1+y_2\end{bmatrix} \tag{1.1.6}$$

is also on the line:

$$y_1 + y_2 = 3x_1 + 3x_2 = 3(x_1 + x_2). \tag{1.1.7}$$

If $\begin{bmatrix}x\\y\end{bmatrix}$ is on the line, then $a\begin{bmatrix}x\\y\end{bmatrix} = \begin{bmatrix}ax\\ay\end{bmatrix}$ is on the line for any real number a: indeed, $ay = a(3x) = 3(ax)$. $\triangle$

Many students find it less obvious that the sphere of equation

$$x^2 + y^2 + z^2 = 1$$

(which of course is not a subspace) is two dimensional, but it also takes only two numbers (for example, latitude and longitude) to locate a point on a sphere (by which we mean just the shell, not the inside).

Intuitively, it is clear that a line through the origin is a subspace of dimension 1, and a plane through the origin is a subspace of dimension 2: it takes one number to locate a point on a line, and two numbers to locate a point on a plane. Being precise about what this means requires some "machinery" (mainly the notions of linear independence and span), introduced in Section 2.4.

The standard basis vectors

We will meet one particular family of vectors in $\mathbb{R}^n$ often: the *standard basis vectors*. In $\mathbb{R}^2$ there are two standard basis vectors; in $\mathbb{R}^3$, there are three:

The notation for the standard basis vectors is ambiguous; at right we have three different vectors denoted $\vec{\mathbf{e}}_1$. The subscript tells us which entry is 1 but does not say how many entries the vector has; the vector $\vec{\mathbf{e}}_1$ could be in $\mathbb{R}^2$ or $\mathbb{R}^{27}$ ....

$$\text{in } \mathbb{R}^2 : \vec{\mathbf{e}}_1 = \begin{bmatrix}1\\0\end{bmatrix}, \vec{\mathbf{e}}_2 = \begin{bmatrix}0\\1\end{bmatrix}; \quad \text{in } \mathbb{R}^3 : \vec{\mathbf{e}}_1 = \begin{bmatrix}1\\0\\0\end{bmatrix}, \vec{\mathbf{e}}_2 = \begin{bmatrix}0\\1\\0\end{bmatrix}, \vec{\mathbf{e}}_3 = \begin{bmatrix}0\\0\\1\end{bmatrix}.$$

In $\mathbb{R}^5$ there are five standard basis vectors:

$$\vec{\mathbf{e}}_1 = \begin{bmatrix}1\\0\\0\\0\\0\end{bmatrix}, \quad \vec{\mathbf{e}}_2 = \begin{bmatrix}0\\1\\0\\0\\0\end{bmatrix}, \ldots, \quad \vec{\mathbf{e}}_5 = \begin{bmatrix}0\\0\\0\\0\\1\end{bmatrix}. \tag{1.1.8}$$

Definition 1.1.7 (Standard basis vectors). The *standard basis vectors* in $\mathbb{R}^n$ are the vectors $\vec{\mathbf{e}}_j$ with n entries, the jth entry 1 and the others 0.

The standard basis vectors in $\mathbb{R}^2$ and $\mathbb{R}^3$ are sometimes denoted $\vec{i}$, $\vec{j}$, and $\vec{k}$:

$$\vec{i} = \vec{\mathbf{e}}_1 = \begin{bmatrix} 1 \\ 0 \end{bmatrix} \text{ or } \begin{bmatrix} 1 \\ 0 \\ 0 \end{bmatrix}$$

$$\vec{j} = \vec{\mathbf{e}}_2 = \begin{bmatrix} 0 \\ 1 \end{bmatrix} \text{ or } \begin{bmatrix} 0 \\ 1 \\ 0 \end{bmatrix}$$

$$\vec{k} = \vec{\mathbf{e}}_3 = \begin{bmatrix} 0 \\ 0 \\ 1 \end{bmatrix}$$

We don't use this notation.

Geometrically, there is a close connection between the standard basis vectors in $\mathbb{R}^2$ and a choice of axes in the Euclidean plane. When in school you drew an x-axis and y-axis on a piece of paper and marked off units so that you could plot a point, you were identifying the plane with $\mathbb{R}^2$: each point on the plane corresponds to a pair of real numbers – its coordinates with respect to those axes. A set of axes providing such an identification must have an origin, and each axis must have a direction (so you know what is positive and what is negative) and it must have units (so you know, for example, where $x = 3$ is).

Such axes need not be at right angles, and the units on one axis need not be the same as those on the other, as shown in Figure 1.1.5. But it is often convenient to choose the axes at right angles and the units equal; the plane with such axes is known as the *Cartesian plane*. We can think that $\vec{\mathbf{e}}_1$ measures one unit along the horizontal axis, going to the right, and $\vec{\mathbf{e}}_2$ measures one unit along the vertical axis, going "up". This is the standard way to draw $\mathbb{R}^2$.

Vector fields

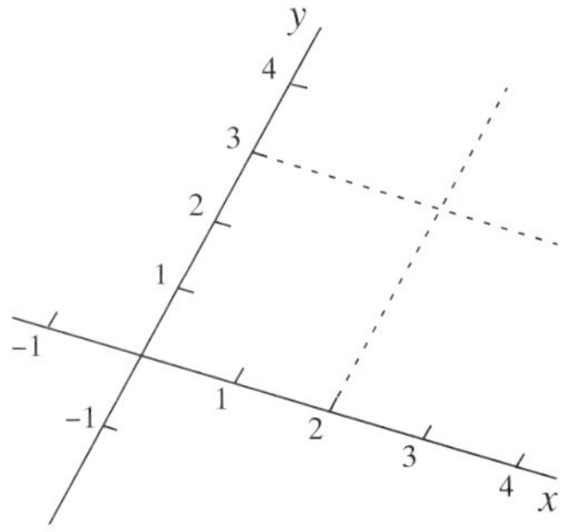

FIGURE 1.1.5.

The intersection of the dotted lines is the point $\begin{pmatrix} 2 \\ 3 \end{pmatrix}$ in this coordinate system, where the axes are not at right angles.

Virtually all of physics deals with fields. The electric and magnetic fields of electromagnetism, the gravitational and other force fields of mechanics, the velocity fields of fluid flow, and the wave function of quantum mechanics are all "fields". Fields are also used in epidemiology and population studies.

By "field" we mean data that varies from point to point. Some fields, like temperature or pressure distribution, are scalar fields: they associate a number to every point. Some fields, like the Newtonian gravitation field, are best modeled by vector fields, which associate a vector to every point. Others, like the electromagnetic field and charge distributions, are best modeled by *form fields*, discussed in Chapter 6. Still others, like the Einstein field of general relativity (a field of pseudo inner products), are none of the above.

Definition 1.1.8 (Vector field). A *vector field* on $\mathbb{R}^n$ is a function whose input is a point in $\mathbb{R}^n$ and whose output is a vector in $\mathbb{R}^n$ emanating from that point.

(Actually, a vector field simply associates to each point a vector; how you imagine that vector is up to you. But it is helpful to imagine each vector anchored at, or emanating from, the corresponding point.)

We will distinguish between functions and vector fields by putting arrows on vector fields, as in $\vec{F}$ in Example 1.1.9.

Example 1.1.9 (Vector fields in $\mathbb{R}^2$). The identity function in $\mathbb{R}^2$ $f\begin{pmatrix} x \\ y \end{pmatrix} = \begin{pmatrix} x \\ y \end{pmatrix}$ takes a point in $\mathbb{R}^2$ and returns the same point. The radial vector field $\vec{F}\begin{pmatrix} x \\ y \end{pmatrix} = \begin{bmatrix} x \\ y \end{bmatrix}$ in Figure 1.1.6 assigns to each point in $\mathbb{R}^2$ the corresponding vector. $\triangle$

Vector fields are often used to describe the flow of fluids or gases: the vector assigned to each point gives the velocity and direction of the flow. For flows that don't change over time (steady-state flows), such a vector field gives a complete description. In more realistic cases where the flow is constantly changing, the vector field gives a snapshot of the flow at a given instant.

Drawing vector fields is an art. In Figure 1.1.7 we drew a vector at each point on a grid. Sometimes it is better to draw more vectors where the vector field is changing fast, and fewer where less is happening.

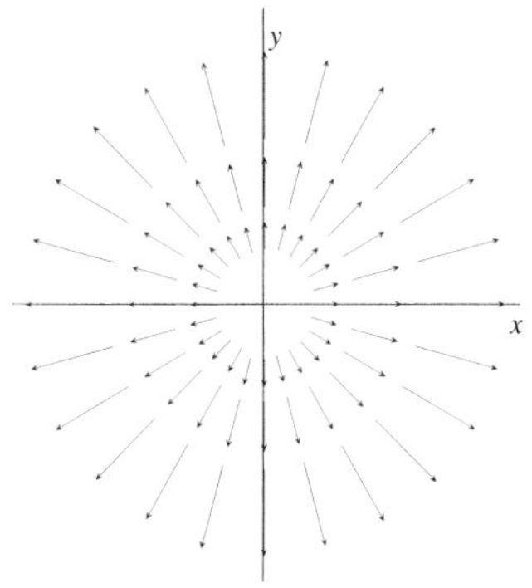

FIGURE 1.1.6.

Vector fields are often easier to depict when the vectors are scaled down, as they are for the vector field above, the radial vector field

$$\vec{F}\begin{pmatrix} x \\ y \end{pmatrix} = \begin{bmatrix} x \\ y \end{bmatrix}.$$

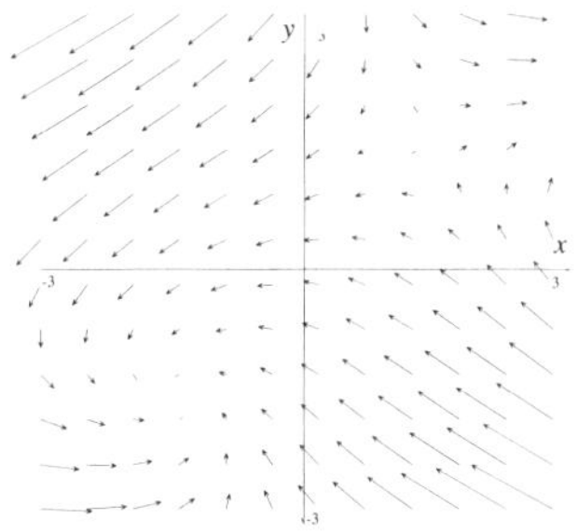

FIGURE 1.1.7.
The vector field

$$\vec{F}\begin{pmatrix} x \\ y \end{pmatrix} = \begin{bmatrix} xy-2 \\ x-y \end{bmatrix}.$$

$$\begin{bmatrix} 1 \\ 1 \\ \vdots \\ 1 \\ 1 \end{bmatrix}, \quad \begin{bmatrix} 1 \\ 2 \\ \vdots \\ n-1 \\ n \end{bmatrix}, \quad \begin{bmatrix} 0 \\ 0 \\ 3 \\ 4 \\ \vdots \\ n-1 \\ n \end{bmatrix}$$

Vectors in $\mathbb{R}^n$ for Exercise 1.1.5

Exercises for Section 1.1

1.1.1 Compute the following vectors by coordinates and sketch what you did:

$$\text{a. } \begin{bmatrix} 1 \\ 3 \end{bmatrix} + \begin{bmatrix} 2 \\ 1 \end{bmatrix} \quad \text{b. } 2\begin{bmatrix} 2 \\ 4 \end{bmatrix} \quad \text{c. } \begin{bmatrix} 1 \\ 3 \end{bmatrix} - \begin{bmatrix} 2 \\ 1 \end{bmatrix} \quad \text{d. } \begin{bmatrix} 3 \\ 2 \end{bmatrix} + \vec{\mathbf{e}}_1$$

1.1.2 Compute the following vectors:

$$\text{a. } \begin{bmatrix} 3 \\ \pi \\ 1 \end{bmatrix} + \begin{bmatrix} 1 \\ -1 \\ \sqrt{2} \end{bmatrix} \quad \text{b. } \begin{bmatrix} 1 \\ 4 \\ c \\ 2 \end{bmatrix} + \vec{\mathbf{e}}_2 \quad \text{c. } \begin{bmatrix} 1 \\ 4 \\ c \\ 2 \end{bmatrix} - \vec{\mathbf{e}}_4$$

1.1.3 In the following, translate the word "in" by the appropriate symbol: $\in$ (element of) or $\subset$ (subset of).

a. A vector $\vec{\mathbf{v}}$ in $\mathbb{R}^3$ b. A line L in $\mathbb{R}^2$
c. A curve C in $\mathbb{R}^3$ d. A point $\mathbf{x}$ in $\mathbb{C}^2$
e. A sequence of nested balls B_0 in B_1 in $B_2 \dots$

1.1.4 a. Name the two trivial subspaces of $\mathbb{R}^n$.

b. Let $S^1 \subset \mathbb{R}^2$ be the unit circle, of equation $x^2 + y^2 = 1$. Do there exist *any* two elements of S^1 whose sum is an element of the set?

1.1.5 Using sum notation (discussed in Section 0.1), write the vectors in the margin at left as a sum of multiples of standard basis vectors.

1.1.6 Sketch the following vector fields:

$$\text{a. } \vec{F}\begin{pmatrix} x \\ y \end{pmatrix} = \begin{bmatrix} 0 \\ 1 \end{bmatrix} \quad \text{b. } \vec{F}\begin{pmatrix} x \\ y \end{pmatrix} = \begin{bmatrix} x \\ 0 \end{bmatrix} \quad \text{c. } \vec{F}\begin{pmatrix} x \\ y \end{pmatrix} = \begin{bmatrix} x \\ y \end{bmatrix}$$

$$\text{d. } \vec{F}\begin{pmatrix} x \\ y \end{pmatrix} = \begin{bmatrix} x \\ -y \end{bmatrix} \quad \text{e. } \vec{F}\begin{pmatrix} x \\ y \end{pmatrix} = \begin{bmatrix} y \\ x \end{bmatrix} \quad \text{f. } \vec{F}\begin{pmatrix} x \\ y \end{pmatrix} = \begin{bmatrix} -y \\ x \end{bmatrix}$$

$$\text{g. } \vec{F}\begin{pmatrix} x \\ y \end{pmatrix} = \begin{bmatrix} y \\ x-y \end{bmatrix} \quad \text{h. } \vec{F}\begin{pmatrix} x \\ y \end{pmatrix} = \begin{bmatrix} x-y \\ x+y \end{bmatrix} \quad \text{i. } \vec{F}\begin{pmatrix} x \\ y \end{pmatrix} = \begin{bmatrix} x^2-y-1 \\ x-y \end{bmatrix}$$

1.1.7 Sketch the following vector fields:

$$\text{a. } \vec{F}\begin{pmatrix} x \\ y \\ z \end{pmatrix} = \begin{bmatrix} 0 \\ 0 \\ x^2+y^2 \end{bmatrix} \quad \text{b. } \vec{F}\begin{pmatrix} x \\ y \\ z \end{pmatrix} = \begin{bmatrix} y \\ -x \\ -z \end{bmatrix} \quad \text{c. } \vec{F}\begin{pmatrix} x \\ y \\ z \end{pmatrix} = \begin{bmatrix} x-y \\ x+y \\ -z \end{bmatrix}$$

1.1.8 Suppose that water is flowing through a pipe of radius r, with speed $r^2 - a^2$, where a is the distance to the axis of the pipe.

a. Write the vector field describing the flow if the pipe is in the direction of the z-axis.

b. Write the vector field describing the flow if the axis of the pipe is the unit circle in the (x, y)-plane.

Exercise 1.1.9: If you are not familiar with complex numbers, see Section 0.7.

1.1.9 Show that the set of complex numbers $\{z | \operatorname{Re}(wz) = 0\}$ with fixed $w \in \mathbb{C}$ is a subspace of $\mathbb{R}^2 = \mathbb{C}$. Describe this subspace.

1.2 Introducing the actors: Matrices

Probably no other area of mathematics has been applied in such numerous and diverse contexts as the theory of matrices. In mechanics, electromagnetics, statistics, economics, operations research, the social sciences, and so on, the list of applications seems endless. By and large this is due to the utility of matrix structure and methodology in conceptualizing sometimes complicated relationships and in the orderly processing of otherwise tedious algebraic calculations and numerical manipulations.—James Cochran, *Applied Mathematics: Principles, Techniques, and Applications*

When a matrix is described, height is given first, then width: an $m \times n$ matrix is m high and n wide; an $n \times m$ matrix is n high and m wide. After struggling for years to remember which goes first, we hit on a mnemonic: first take the elevator, then walk down the hall.

The other central actor in linear algebra is the *matrix*.

Definition 1.2.1 (Matrix). An $m \times n$ *matrix* is a rectangular array of entries, m high and n wide. We denote by $\operatorname{Mat}(m, n)$ the set of $m \times n$ matrices.

We use capital letters to denote matrices. Usually our matrices will be arrays of numbers, real or complex, but matrices can be arrays of polynomials, or of more general functions; a matrix can even be an array of other matrices. A vector $\vec{\mathbf{v}} \in \mathbb{R}^m$ is an $m \times 1$ matrix; a number is a 1×1 matrix.

Addition of matrices and multiplication by a scalar work in the obvious way:

How would you add the matrices

$$\begin{bmatrix} 1 & 2 & 5 \\ 0 & 2 & 3 \end{bmatrix} \quad \text{and} \quad \begin{bmatrix} 1 & 2 \\ 0 & 2 \end{bmatrix}?$$

You can't: matrices can be added only if they have the same height and same width.

Example 1.2.2 (Addition of matrices; multiplication by scalars).

$$\begin{bmatrix} 1 & 0 \\ 2 & -1 \\ 4 & 2 \end{bmatrix} + \begin{bmatrix} 0 & -3 \\ 1 & -2 \\ 3 & 1 \end{bmatrix} = \begin{bmatrix} 1 & -3 \\ 3 & -3 \\ 7 & 3 \end{bmatrix} \quad \text{and} \quad 2\begin{bmatrix} 1 & 4 \\ -2 & 3 \end{bmatrix} = \begin{bmatrix} 2 & 8 \\ -4 & 6 \end{bmatrix} \qquad \triangle$$

So far, it's not clear why we should bother with matrices. What do we gain by talking about the 2×2 matrix $\begin{bmatrix} a & b \\ c & d \end{bmatrix}$ rather than the point $\begin{pmatrix} a \\ b \\ c \\ d \end{pmatrix} \in \mathbb{R}^4$? The answer is that the matrix format allows us to perform another operation: *matrix multiplication.* We will see in Section 1.3 that every linear transformation corresponds to multiplication by a matrix, and (Theorem 1.3.10) that composition of linear transformations corresponds to multiplying together the corresponding matrices. This is one reason matrix multiplication is a natural and important operation; other important applications of matrix multiplication are found in probability theory and graph theory.

"When Werner Heisenberg discovered 'matrix' mechanics in 1925, he didn't know what a matrix was (Max Born had to tell him), and neither Heisenberg nor Born knew what to make of the appearance of matrices in the context of the atom."—Manfred R. Schroeder, "Number Theory and the Real World," *Mathematical Intelligencer*, Vol. 7, No. 4.

The entry $a_{i,j}$ of a matrix A is the entry at the intersection of the ith row and jth column; it is the entry you find by taking the elevator to the ith floor (row) and walking down the corridor to the jth room.

One student objected that here we speak of the ith row and jth column, but in Example 1.2.6 we write that "the ith column of A is $A\vec{\mathbf{e}}_i$." Do not assume that i has to be a row and j a column. "The ith column of A is $A\vec{\mathbf{e}}_i$" is just a convenient way to say "The first column of A is $A\vec{\mathbf{e}}_1$, the second column is $A\vec{\mathbf{e}}_2$, etc." The thing to remember is that the first index of $a_{i,j}$ refers to the row, and the second to the column; thus $a_{3,2}$ corresponds to the entry of the third row, second column.

Another reader pointed out that sometimes we write "$n \times m$ matrix" and sometimes "$m \times n$". Again, which is n and which is m doesn't matter, and we could just as well write $s \times v$. The point is that the first letter refers to height and the second to width.

Matrix multiplication is best learned by example. The simplest way to compute AB is to write B above and to the right of A. Then the product AB fits in the space to the right of A and below B, the (i,j)th entry of AB being the intersection of the ith row of A and the jth column of B, as shown in Example 1.2.3. Note that for AB to exist, *the width of A must equal the height of B.* The resulting matrix then has the height of A and the width of B.

Example 1.2.3 (Matrix multiplication). The first entry of the first row of AB is obtained by multiplying, one by one, the entries of the first *row* of A by those of the first *column* of B, and adding these products together: in equation 1.2.1, $(2 \times 1) + (-1 \times 3) = -1$. The second entry of the first row is obtained by multiplying the first row of A by the second column of B: $(2 \times 4) + (-1 \times 0) = 8$. After multiplying the first row of A by all the columns of B, the process is repeated with the second row of A: $(3 \times 1) + (2 \times 3) = 9$, and so on:

$$\begin{matrix} & \begin{bmatrix} B \end{bmatrix} \\ \begin{bmatrix} A \end{bmatrix} & \begin{bmatrix} AB \end{bmatrix} \end{matrix} \qquad \begin{matrix} & \overbrace{\begin{bmatrix} 1 & 4 & -2 \\ 3 & 0 & 2 \end{bmatrix}}^{B} \\ \underbrace{\begin{bmatrix} 2 & -1 \\ 3 & 2 \end{bmatrix}}_{A} & \underbrace{\begin{bmatrix} -1 & 8 & -6 \\ 9 & 12 & -2 \end{bmatrix}}_{AB} \end{matrix} \qquad \triangle \qquad 1.2.1$$

Given the matrices

$$A = \begin{bmatrix} 1 & 0 \\ 2 & 3 \end{bmatrix} \quad B = \begin{bmatrix} 0 & 1 \\ 0 & 1 \end{bmatrix} \quad C = \begin{bmatrix} 1 & -1 & 1 \\ 1 & 0 & -1 \end{bmatrix} \quad D = \begin{bmatrix} 1 & 0 \\ 2 & 2 \\ 1 & 1 \end{bmatrix},$$

what are the products AB, AC and CD? Check your answers in the footnote below.[1] Now compute BA. What do you notice? What if you try to compute CA?[2]

Next we state the formal definition of the process we've just described. If the indices bother you, refer to Figure 1.2.1.

---

[1] $AB = \begin{bmatrix} 0 & 1 \\ 0 & 5 \end{bmatrix}$; $AC = \begin{bmatrix} 1 & -1 & 1 \\ 5 & -2 & -1 \end{bmatrix}$; $CD = \begin{bmatrix} 0 & -1 \\ 0 & -1 \end{bmatrix}$.

[2] Matrix multiplication is *not* commutative; $BA = \begin{bmatrix} 2 & 3 \\ 2 & 3 \end{bmatrix} \neq AB = \begin{bmatrix} 0 & 1 \\ 0 & 5 \end{bmatrix}$. Although the product AC exists, CA does not.

Definition 1.2.4 (Matrix multiplication). If A is an $m \times n$ matrix whose (i,j)th entry is $a_{i,j}$, and B is an $n \times p$ matrix whose (i,j)th entry is $b_{i,j}$, then $C = AB$ is the $m \times p$ matrix with entries

$$\begin{aligned} c_{i,j} &= \sum_{k=1}^{n} a_{i,k} b_{k,j} \\ &= a_{i,1}b_{1,j} + a_{i,2}b_{2,j} + \cdots + a_{i,n}b_{n,j}. \end{aligned} \qquad 1.2.2$$

Definition 1.2.4: Note that the summation is over the inner index k of $a_{i,k}b_{k,j}$.

Definition 1.2.4 says nothing new, but it provides some practice moving between the concrete (multiplying two particular matrices) and the symbolic (expressing this operation so that it applies to any two matrices of appropriate dimensions, even if the entries are complex numbers or functions, not real numbers.) In linear algebra one is constantly moving from one form of representation to another. For example, as we have seen, a point $\mathbf{b}$ in $\mathbb{R}^n$ can be considered as a single entity, $\mathbf{b}$, or as the ordered list of its coordinates; a matrix A can be thought of as a single entity, A, or as the rectangular array of its entries.

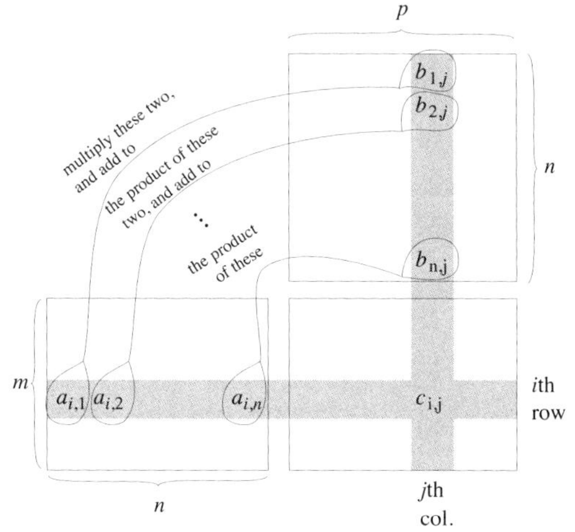

FIGURE 1.2.1. The entry $c_{i,j}$ of the matrix $C = AB$ is the sum of the products of the entries $a_{i,k}$ of the matrix A and the corresponding entry $b_{k,j}$ of the matrix B. The entries $a_{i,k}$ are all in the ith row of A; the first index i is constant, and the second index k varies. The entries $b_{k,j}$ are all in the jth column of B; the first index k varies, and the second index j is constant. Since the width of A equals the height of B, the entries of A and those of B can be paired up exactly.

Remark. Often matrix multiplication is written in a row: $[A][B] = [AB]$. The format shown in Example 1.2.3 avoids confusion: the product of the ith row of A and the jth column of B lies at the intersection of that row and column. It also avoids recopying matrices when doing repeated multiplications. For example, to multiply A times B times C times D we write

$$\begin{array}{cccc} & \begin{bmatrix} B \end{bmatrix} & \begin{bmatrix} C \end{bmatrix} & \begin{bmatrix} D \end{bmatrix} \\ \begin{bmatrix} A \end{bmatrix} & \begin{bmatrix} (AB) \end{bmatrix} & \begin{bmatrix} (AB)C \end{bmatrix} & \begin{bmatrix} (ABC)D \end{bmatrix} \end{array} \quad \triangle \qquad 1.2.3$$

Noncommutativity of matrix multiplication

As we saw earlier, matrix multiplication is *not commutative*. It may well be possible to multiply A by B but not B by A. Even if both matrices have the same number of rows and columns, AB will usually not equal BA, as shown in Example 1.2.5.

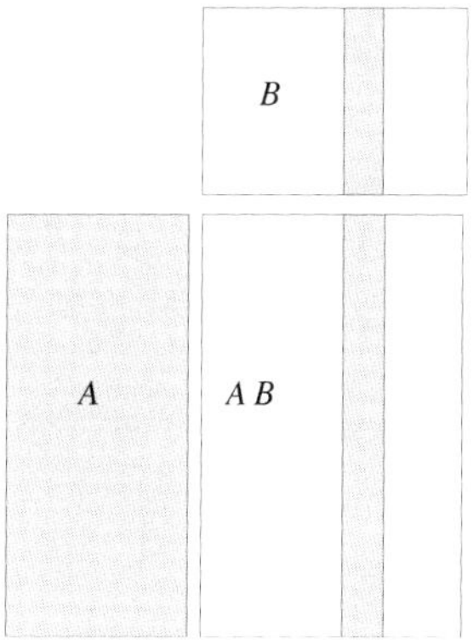

FIGURE 1.2.2.
The ith column of the product AB depends on all the entries of A but only the ith column of B.

Example 1.2.5 (Matrix multiplication is not commutative).

$$\begin{matrix} & \begin{bmatrix} 0 & 1 \\ 1 & 0 \end{bmatrix} \\ \begin{bmatrix} 0 & 1 \\ 1 & 1 \end{bmatrix} & \begin{bmatrix} 1 & 0 \\ 1 & 1 \end{bmatrix} \end{matrix} \quad \text{is not equal to} \quad \begin{matrix} & \begin{bmatrix} 0 & 1 \\ 1 & 1 \end{bmatrix} \\ \begin{bmatrix} 0 & 1 \\ 1 & 0 \end{bmatrix} & \begin{bmatrix} 1 & 1 \\ 0 & 1 \end{bmatrix} \end{matrix} \quad \triangle \qquad 1.2.4$$

Multiplying a matrix by a standard basis vector

Multiplying a matrix A by the standard basis vector $\vec{\mathbf{e}}_i$ selects out the ith column of A, as shown in the following example. We will use this fact often.

Example 1.2.6 (The ith column of A is $A\vec{\mathbf{e}}_i$). Below, we show that the second column of A is $A\vec{\mathbf{e}}_2$:

$$\begin{matrix} & \overbrace{\begin{bmatrix} 0 \\ 1 \\ 0 \end{bmatrix}}^{\vec{\mathbf{e}}_2} \\ \underbrace{\begin{bmatrix} 3 & -2 & 0 \\ 2 & 1 & 2 \\ 0 & 4 & 3 \\ 1 & 0 & 2 \end{bmatrix}}_{A} & \underbrace{\begin{bmatrix} -2 \\ 1 \\ 4 \\ 0 \end{bmatrix}}_{A\vec{\mathbf{e}}_2} \end{matrix} \quad \text{corresponds to} \quad \begin{bmatrix} -2 \\ 1 \\ 4 \\ 0 \end{bmatrix} \times 1 = \begin{bmatrix} -2 \\ 1 \\ 4 \\ 0 \end{bmatrix}. \quad \triangle$$

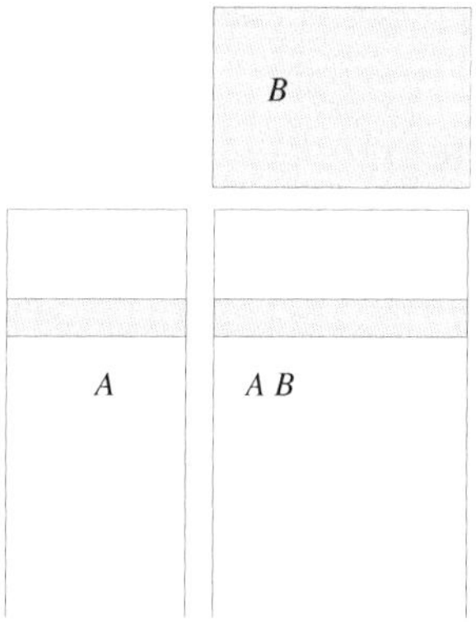

FIGURE 1.2.3.
The jth row of the product AB depends on all the entries of B but only the jth row of A.

Similarly, the ith column of AB is $A\vec{\mathbf{b}}_i$, where $\vec{\mathbf{b}}_i$ is the ith column of B, as shown in Example 1.2.7 and represented in Figure 1.2.2. The jth *row* of AB is the product of the jth row of A and the matrix B, as shown in Example 1.2.8 and Figure 1.2.3.

Example 1.2.7 (The ith column of AB is $A\vec{\mathbf{b}}_i$). The second column of the product AB is the product of A and the second column of B:

$$\begin{matrix} & \overbrace{\begin{bmatrix} 1 & 4 & -2 \\ 3 & 0 & 2 \end{bmatrix}}^{B} \\ \underbrace{\begin{bmatrix} 2 & -1 \\ 3 & 2 \end{bmatrix}}_{A} & \underbrace{\begin{bmatrix} -1 & 8 & -6 \\ 9 & 12 & -2 \end{bmatrix}}_{AB} \end{matrix} \qquad \begin{matrix} & \overbrace{\begin{bmatrix} 4 \\ 0 \end{bmatrix}}^{\vec{\mathbf{b}}_2} \\ \underbrace{\begin{bmatrix} 2 & -1 \\ 3 & 2 \end{bmatrix}}_{A} & \underbrace{\begin{bmatrix} 8 \\ 12 \end{bmatrix}}_{A\vec{\mathbf{b}}_2} \end{matrix} \qquad 1.2.5$$

Example 1.2.8. The second row of the product AB is the product of the second row of A and the matrix B:

$$\begin{array}{cc} & \overbrace{\begin{bmatrix} 1 & 4 & -2 \\ 3 & 0 & 2 \end{bmatrix}}^{B} \\ \underbrace{\begin{bmatrix} 2 & -1 \\ 3 & 2 \end{bmatrix}}_{A} & \underbrace{\begin{bmatrix} -1 & 8 & -6 \\ 9 & 12 & -2 \end{bmatrix}}_{AB} \end{array} \qquad \begin{array}{cc} & \overbrace{\begin{bmatrix} 1 & 4 & -2 \\ 3 & 0 & 2 \end{bmatrix}}^{B} \\ \begin{bmatrix} 3 & 2 \end{bmatrix} & \begin{bmatrix} 9 & 12 & -2 \end{bmatrix} \end{array} \qquad 1.2.6$$

FIGURE 1.2.4.

Arthur Cayley (1821–1895) introduced matrices in 1858.

Cayley worked as a lawyer until 1863, when he was appointed professor at Cambridge. As professor, he "had to manage on a salary only a fraction of that which he had earned as a skilled lawyer. However, Cayley was very happy to have the chance to devote himself entirely to mathematics."—From a biographical sketch by J. J. O'Connor and E. F. Robertson. For more , see the MacTutor History of Mathematics archive at http://www-history .mcs.st-and.ac.uk/history/

Matrix multiplication is associative

When multiplying the matrices A, B, and C, we could set up the repeated multiplication as we did in equation 1.2.3, which corresponds to the product $(AB)C$. We can use another format to get the product $A(BC)$:

$$\begin{array}{ccc} & \begin{bmatrix} B \end{bmatrix} & \begin{bmatrix} C \end{bmatrix} \\ \begin{bmatrix} A \end{bmatrix} & \begin{bmatrix} AB \end{bmatrix} & \begin{bmatrix} (AB)C \end{bmatrix} \end{array} \quad \text{or} \quad \begin{array}{ccc} & & \begin{bmatrix} C \end{bmatrix} \\ & \begin{bmatrix} B \end{bmatrix} & \begin{bmatrix} (BC) \end{bmatrix} \\ & \begin{bmatrix} A \end{bmatrix} & \begin{bmatrix} A(BC) \end{bmatrix} \end{array}. \qquad 1.2.7$$

Is $(AB)C$ the same as $A(BC)$? In Section 1.3 we see that associativity of matrix multiplication follows from Theorem 1.3.10. Here we give a computational proof.

Proposition 1.2.9 (Matrix multiplication is associative). *If A is an $n \times m$ matrix, B an $m \times p$ matrix, and C a $p \times q$ matrix, so that $(AB)C$ and $A(BC)$ are both defined, then they are equal:*

$$(AB)C = A(BC). \qquad 1.2.8$$

In his 1858 article on matrices, Cayley stated that matrix multiplication is associative but gave no proof. The impression one gets is that he played around with matrices (mostly 2×2 and 3×3) to get some feeling for how they behave, without worrying about rigor. Concerning another matrix result, Theorem 4.8.27 (the Cayley-Hamilton theorem) he verified it for 2×2 and 3×3 matrices, and stopped there.

Proof. Figure 1.2.5 shows that the (i,j)th entry of both $A(BC)$ and $(AB)C$ depend only on the ith row of A and the jth column of C (but on all the entries of B). Without loss of generality we can assume that A is a line matrix and C is a column matrix ($n = q = 1$), so that both $(AB)C$ and $A(BC)$ are numbers. Now apply associativity of multiplication of numbers:

$$\begin{aligned} (AB)C &= \sum_{l=1}^{p} \underbrace{\left(\sum_{k=1}^{m} a_k b_{k,l} \right)}_{l\text{th entry of } AB} c_l \\ &= \sum_{l=1}^{p} \sum_{k=1}^{m} a_k b_{k,l} c_l = \sum_{k=1}^{m} a_k \underbrace{\left(\sum_{l=1}^{p} b_{k,l} c_l \right)}_{k\text{th entry of } BC} = A(BC). \qquad \square \end{aligned} \qquad 1.2.9$$

Exercise 1.2.24 asks you to show that matrix multiplication is distributive over addition:

$$A(B+C) = AB + AC$$
$$(B+C)A = BA + CA.$$

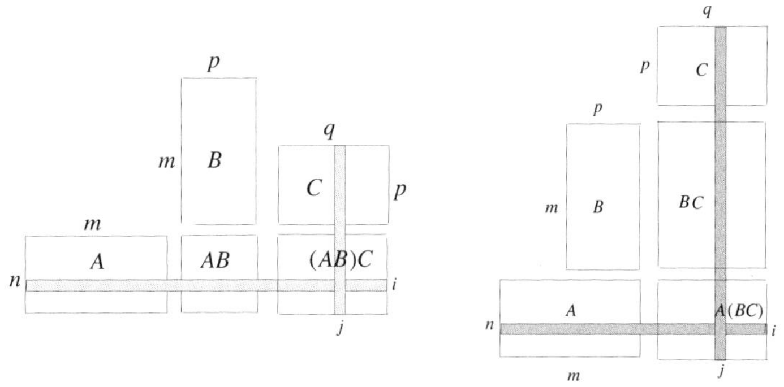

FIGURE 1.2.5. LEFT: This way of writing the matrices corresponds to calculating $(AB)C$. The ith row of AB depends on the ith row of A and the entire matrix B. RIGHT: This format corresponds to calculating $A(BC)$. The jth column of BC depends on the jth column of C and the entire matrix B.

Not all operations are associative. For example, the operation that takes two matrices A, B and gives $AB - BA$ is not associative. The cross product, discussed in Section 1.4, is also not associative.

The identity matrix

The identity matrix I plays the same role in matrix multiplication as the number 1 in multiplication of numbers: $IA = A = AI$.

Definition 1.2.10 (Identity matrix). The *identity matrix* I_n is the $n \times n$-matrix with 1's along the main diagonal (the diagonal from top left to bottom right) and 0's elsewhere.

The main diagonal is also called the *diagonal*. The diagonal from bottom left to top right is the *antidiagonal*.

For example, $I_2 = \begin{bmatrix} 1 & 0 \\ 0 & 1 \end{bmatrix}$ and $I_3 = \begin{bmatrix} 1 & 0 & 0 \\ 0 & 1 & 0 \\ 0 & 0 & 1 \end{bmatrix}$.

If A is an $n \times m$ matrix, then

$$IA = AI = A, \qquad \text{or, more precisely,} \qquad I_n A = A I_m = A, \qquad 1.2.10$$

since if $n \neq m$ one must change the size of the identity matrix to match the size of A. When the context is clear, we will omit the index.

The columns of the identity matrix I_n are, of course, the standard basis vectors $\vec{\mathbf{e}}_1, \dots, \vec{\mathbf{e}}_n$:

$$I_4 = \begin{bmatrix} 1 & 0 & 0 & 0 \\ 0 & 1 & 0 & 0 \\ 0 & 0 & 1 & 0 \\ 0 & 0 & 0 & 1 \end{bmatrix}.$$
$$\;\;\vec{\mathbf{e}}_1\;\vec{\mathbf{e}}_2\;\vec{\mathbf{e}}_3\;\vec{\mathbf{e}}_4$$

Matrix inverses

The inverse A^{-1} of a matrix A plays the same role in matrix multiplication as the inverse $1/a$ for the number a.

The only number with no inverse is 0, but many nonzero matrices do not have inverses. In addition, the noncommutativity of matrix multiplication makes the definition more complicated.

Definition 1.2.11 (Left and right inverses of matrices). Let A be a matrix. If there is a matrix B such that $BA = I$, then B is a *left inverse* of A. If there is a matrix C such that $AC = I$, then C is a *right inverse* of A.

Example 1.2.12 (A matrix with neither right nor left inverse). The matrix $\begin{bmatrix} 1 & 0 \\ 0 & 0 \end{bmatrix}$ does not have a right or a left inverse. To see this, assume it has a right inverse. Then there exists a matrix $\begin{bmatrix} a & b \\ c & d \end{bmatrix}$ such that

$$\begin{bmatrix} 1 & 0 \\ 0 & 0 \end{bmatrix} \begin{bmatrix} a & b \\ c & d \end{bmatrix} = \begin{bmatrix} 1 & 0 \\ 0 & 1 \end{bmatrix}. \qquad 1.2.11$$

But that product is $\begin{bmatrix} a & b \\ 0 & 0 \end{bmatrix}$, with 0 in the bottom right corner, not the required 1. A similar computation shows that there is no left inverse. △

It is possible for a nonsquare matrix to have lots of left inverses and no right inverse, or lots of right inverses and no left inverse, as explored in Exercise 1.2.23.

Definition 1.2.13 (Invertible matrix). An *invertible matrix* is a matrix that has both a left inverse and a right inverse.

We will see in Section 2.2 (discussion following Corollary 2.2.7) that only square matrices can have a two-sided inverse (i.e., an inverse). Furthermore, if a *square* matrix has a left inverse, then that left inverse is necessarily also a right inverse; if it has a right inverse, that right inverse is a left inverse.

Associativity of matrix multiplication gives the following result:

Proposition and Definition 1.2.14 (Matrix inverse). *If a matrix A has both a left and a right inverse, then it has only one left inverse and one right inverse, and they are identical; such a matrix is called the inverse of A and is denoted A^{-1}.*

Proof. If a matrix A has a right inverse B, then $AB = I$. If it has a left inverse C, then $CA = I$. So

$$C(AB) = CI = C \quad \text{and} \quad (CA)B = IB = B, \quad \text{so} \quad C = B. \quad \square \qquad 1.2.12$$

While we can write the inverse of a number x either as x^{-1} or as $1/x$, giving $x\,x^{-1} = x\,(1/x) = 1$, the inverse of a matrix A is only written A^{-1}. We *cannot divide by a matrix.* If for two matrices A and B you were to write A/B, it would be unclear whether this meant

$$B^{-1}A \quad \text{or} \quad AB^{-1}.$$

We discuss how to find inverses of matrices in Section 2.3. A formula exists for 2×2 matrices: the inverse of

$$A = \begin{bmatrix} a & b \\ c & d \end{bmatrix} \quad \text{is} \quad A^{-1} = \frac{1}{ad-bc}\begin{bmatrix} d & -b \\ -c & a \end{bmatrix}, \qquad 1.2.13$$

as Exercise 1.2.12 asks you to confirm. The formula for the inverse of a 3×3 matrix is given in Exercise 1.4.20.

Notice that a 2×2 matrix is invertible if $ad - bc \neq 0$. The converse is also true: if $ad - bc = 0$, the matrix is not invertible, as you are asked to show in Exercise 1.2.13.

Associativity of matrix multiplication is also used to prove the following:

Proposition 1.2.15: We are indebted to Robert Terrell for the mnemonic, "socks on, shoes on; shoes off, socks off". To undo a process, you undo first the last thing you did:

$$(f \circ g)^{-1} = g^{-1} \circ f^{-1}.$$

Proposition 1.2.15 (Inverse of product of matrices). *If A and B are invertible matrices, then AB is invertible, and the inverse is given by*

$$(AB)^{-1} = B^{-1}A^{-1}. \qquad 1.2.14$$

Proof. The computation

$$(AB)(B^{-1}A^{-1}) = A(BB^{-1})A^{-1} = AA^{-1} = I \qquad 1.2.15$$

and a similar one for $(B^{-1}A^{-1})(AB)$ prove the result. □

Where does this use associativity? Check the footnote below.[3]

The transpose

Do not confuse a matrix with its transpose, and *never write a vector horizontally.* A vector written horizontally is its transpose; confusion between a vector (or matrix) and its transpose leads to endless difficulties with the order in which things should be multiplied, as you can see from Theorem 1.2.17.

If $\vec{\mathbf{v}} = \begin{bmatrix} 1 \\ 0 \\ 2 \end{bmatrix}$, its transpose is $\vec{\mathbf{v}}^\top = [1\ 0\ 2]$.

Definition 1.2.16 (Transpose). The *transpose* $A^\top$ of a matrix A is formed by interchanging the rows and columns of A, reading the rows from left to right, and columns from top to bottom.

For example, if $A = \begin{bmatrix} 1 & 4 & -2 \\ 3 & 0 & 2 \end{bmatrix}$, then $A^\top = \begin{bmatrix} 1 & 3 \\ 4 & 0 \\ -2 & 2 \end{bmatrix}$.

The transpose of a single row of a matrix is a vector.

$$\begin{bmatrix} 2 & 1 & 3 \\ 1 & 5 & 4 \\ 3 & 4 & 0 \end{bmatrix}$$

A symmetric matrix

If A is any matrix, then $A^\top A$ is symmetric, as Exercise 1.2.16 asks you to show.

Theorem 1.2.17 (Transpose of product). *The transpose of a product is the product of the transposes in reverse order:*

$$(AB)^\top = B^\top A^\top. \tag{1.2.16}$$

The proof is straightforward and is left as Exercise 1.2.14.

$$\begin{bmatrix} 0 & 1 & 2 \\ -1 & 0 & 3 \\ -2 & -3 & 0 \end{bmatrix}$$

An antisymmetric matrix

Symmetric and antisymmetric matrices are necessarily square.

Special kinds of matrices

Definition 1.2.18 (Symmetric and antisymmetric matrices). A symmetric matrix is equal to its transpose. An antisymmetric matrix is equal to minus its transpose.

$$\begin{bmatrix} 1 & 1 & 0 & 3 \\ \mathbf{0} & 2 & 0 & 0 \\ \mathbf{0} & \mathbf{0} & 1 & 0 \\ \mathbf{0} & \mathbf{0} & \mathbf{0} & 0 \end{bmatrix}$$

An upper triangular matrix

Definition 1.2.19 (Triangular matrix). An *upper triangular matrix* is a square matrix with nonzero entries only on or above the main diagonal. A *lower triangular matrix* is a square matrix with nonzero entries only on or below the main diagonal.

[3]Associativity is used for the first two equalities below:

$$\underbrace{(AB)}_{(AB)}\underbrace{(B^{-1}A^{-1})}_{C} = \underbrace{A}_{A}\underbrace{(\overbrace{B}^{D}\overbrace{(B^{-1}A^{-1})}^{(EF)})}_{(BC)} = A\Big(\overbrace{(BB^{-1})}^{(DE)}\overbrace{A^{-1}}^{F}\Big) = A(IA^{-1}) = I.$$

Exercise 1.3 asks you to show that if A and B are upper triangular $n \times n$ matrices, then so is AB. Exercise 2.4 asks you to show that a triangular matrix is invertible if and only if its diagonal entries are all nonzero.

$$\begin{bmatrix} \mathbf{2} & 0 & 0 & 0 \\ 0 & \mathbf{2} & 0 & 0 \\ 0 & 0 & \mathbf{0} & 0 \\ 0 & 0 & 0 & \mathbf{1} \end{bmatrix}$$

A diagonal matrix

Definition 1.2.20 (Diagonal matrix). A *diagonal matrix* is a square matrix with nonzero entries (if any) only on the main diagonal.

What happens if you square the matrix $\begin{bmatrix} a & 0 \\ 0 & b \end{bmatrix}$? If you cube it?[4]

Applications: probabilities and graphs

From the perspective of this book, matrices are most important because they represent linear transformations, discussed in the next section. But matrix multiplication has other important applications. Two good examples are probability theory and graph theory.

A situation like this, with each outcome depending only on the one just before it, is called a *Markov chain*.

This kind of approach is useful in determining efficient storage. How should a lumber yard store wood, to minimize time lost digging out a particular plank from under others? How should the operating system of a computer store data most efficiently?

Sometimes easy access isn't the goal. In Zola's novel *Au Bonheur des Dames*, the story of the growth of the first big department store in Paris, the hero places goods in the most inconvenient arrangement possible, forcing customers to pass through parts of the store where they otherwise wouldn't set foot, and which are mined with temptations for impulse shopping.

Example 1.2.21 (Matrices and probabilities). Suppose you have three reference books on a shelf: a thesaurus, a French dictionary, and an English dictionary. Each time you consult one of these books, you put it back on the shelf at the far left. When you need a reference, we denote by P_1 the probability that it will be the thesaurus, by P_2 the probability that it will be the French dictionary, by P_3 the probability it will be the English dictionary.

There are six possible arrangements on the shelf: $1\,2\,3$, $1\,3\,2$, and so on. We can write the following 6×6 transition matrix T, indicating the probability of going from one arrangement to another:

	(1 2 3)	(1 3 2)	(2 1 3)	(2 3 1)	(3 1 2)	(3 2 1)
(1 2 3)	P_1	0	P_2	0	P_3	0
(1 3 2)	0	P_1	P_2	0	P_3	0
(2 1 3)	P_1	0	P_2	0	0	P_3
(2 3 1)	P_1	0	0	P_2	0	P_3
(3 1 2)	0	P_1	0	P_2	P_3	0
(3 2 1)	0	P_1	0	P_2	0	P_3

The move from $(2\,1\,3)$ to $(3\,2\,1)$ has probability P_3, since if you start with the order $(2\,1\,3)$ (French dictionary, thesaurus, English dictionary), consult the English dictionary, and put it back to the far left, you will then have the order $(3\,2\,1)$. So the entry at the 3rd row, 6th column is P_3. The move from $(2\,1\,3)$ to $(3\,1\,2)$ has probability 0, since moving the English dictionary from third to first position won't change the position of the other books. So the entry at the 3rd row, 5th column is 0.

Now say you start with the fourth arrangement, $(2\,3\,1)$. Multiplying the line matrix $[0\,0\,0\,1\,0\,0]$ (probability 1 for the fourth choice, 0 for the others)

---

[4] $\begin{bmatrix} a & 0 \\ 0 & b \end{bmatrix}^2 = \begin{bmatrix} a^2 & 0 \\ 0 & b^2 \end{bmatrix}$; $\begin{bmatrix} a & 0 \\ 0 & b \end{bmatrix}^3 = \begin{bmatrix} a^3 & 0 \\ 0 & b^3 \end{bmatrix}$. We will see in Section 2.7 how supremely important this observation is.

by the transition matrix T gives the probabilities $P_1, 0, 0, P_2, 0, P_3$. This is, of course, just the 4th row of the matrix.

The interesting point here is to explore the long-term probabilities. At the second step, we would multiply the line matrix $[P_1\,0\,0\,P_2\,0\,P_3]$ by T; at the third we would multiply that product by T, ... If we know actual values for P_1, P_2, and P_3, we can compute the probabilities for the various configurations after a great many iterations. If we don't know the probabilities, we can use this system to deduce them from the configuration of the bookshelf after different numbers of iterations. △

Example 1.2.22: An adjacency matrix can also represent a graph whose vertices are web pages and whose edges are links between pages; such a matrix is used to construct the "Google matrix" used by Google to rank web pages. To construct such an adjacency matrix A, list all web pages and use that list to label the rows and columns of a matrix, each web page corresponding to some row i (and also to column i). If web page j links to web page i, make the entry $a_{i,j}$ a 1; otherwise make it 0. In an elaboration of this scheme, if web page j has k outgoing links, assign $1/k$ rather than 1 to each $a_{i,j} \neq 0$; then the total "value" of all links from a given web page will always be 1. We discuss this further in Example 2.7.12.

Example 1.2.22 provides an entertaining setting for practice at matrix multiplication, while showing some of its power.

Example 1.2.22 (Matrices and graphs). We are going to take walks on the edges of a unit cube; if in going from a vertex V_i to another vertex V_k we walk along n edges, we will say that our walk is of length n. For example, in Figure 1.2.6, if we go from vertex V_1 to V_6, passing by V_4 and V_5, the length of our walk is 3. We will stipulate that each segment of the walk has to take us from one vertex to a different vertex; the shortest possible walk from a vertex to itself is of length 2.

How many walks of length n are there that go from a vertex to itself, or, more generally, from a given vertex to a different vertex? As we will see in Proposition 1.2.23, we answer that question by raising to the nth power the *adjacency matrix* of the graph. The adjacency matrix for our cube is the 8×8 matrix A whose rows and columns are labeled by the vertices $V_1, \dots, V_8$, and such that the (i,j)th entry is 1 if there is an edge joining V_i to V_j, and 0 if not, as shown in Figure 1.2.6. For example, $a_{4,1} = 1$ because an edge joins V_4 to V_1; the entry $a_{4,6}$ is 0 because no edge joins V_4 to V_6.

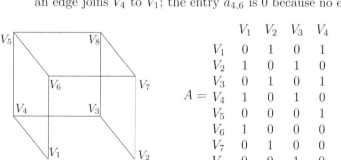

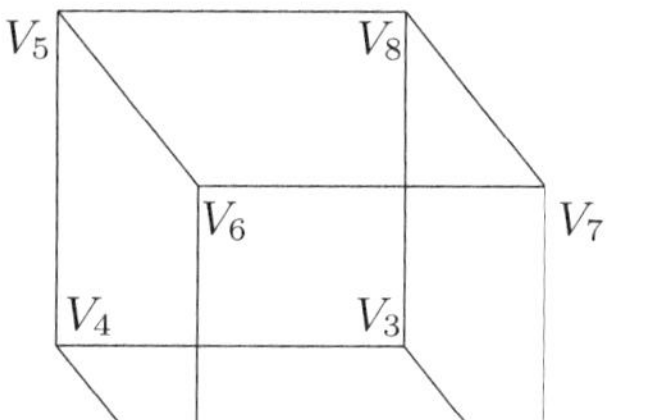

$$A = \begin{array}{c|cccccccc} & V_1 & V_2 & V_3 & V_4 & V_5 & V_6 & V_7 & V_8 \\ \hline V_1 & 0 & 1 & 0 & 1 & 0 & 1 & 0 & 0 \\ V_2 & 1 & 0 & 1 & 0 & 0 & 0 & 1 & 0 \\ V_3 & 0 & 1 & 0 & 1 & 0 & 0 & 0 & 1 \\ V_4 & 1 & 0 & 1 & 0 & 1 & 0 & 0 & 0 \\ V_5 & 0 & 0 & 0 & 1 & 0 & 1 & 0 & 1 \\ V_6 & 1 & 0 & 0 & 0 & 1 & 0 & 1 & 0 \\ V_7 & 0 & 1 & 0 & 0 & 0 & 1 & 0 & 1 \\ V_8 & 0 & 0 & 1 & 0 & 1 & 0 & 1 & 0 \end{array}$$

FIGURE 1.2.6. LEFT: The graph of a cube. RIGHT: Its adjacency matrix A. If vertices V_i and V_j are joined by a single edge, the (i,j)th and (j,i)th entries of the matrix are 1; otherwise they are 0.

The reason this matrix is important is the following. You may appreciate Proposition 1.2.23 more if you try to make a rough estimate of the number of walks of length 4 from a vertex to itself. The authors did and discovered later that they had missed quite a few possible walks.

Exercise 1.2.17 asks you to construct the adjacency matrix for a triangle and for a square. We can also make a matrix that allows for one-way streets (one-way edges); this is explored in Exercise 1.5, one of the review exercises at the end of the chapter.

As you would expect, all the 1's in the adjacency matrix A have turned into 0's in A^4; if two vertices are connected by a single edge, then when n is even there will be no walks of length n between them.

Of course we used a computer to compute this matrix. For all but simple problems involving matrix multiplication, use MATLAB, MAPLE, or an equivalent.

Like the transition matrices of probability theory, matrices representing the length of walks from one vertex of a graph to another have important applications for computers and multiprocessing.

Earlier we denoted the (i,j)th entry of a matrix A by $a_{i,j}$. Here we denote it by $(A)_{i,j}$. Both notations are standard.

Proposition 1.2.23. *For any graph formed of vertices connected by edges, the number of possible walks of length n from vertex V_i to vertex V_j is given by the (i,j)th entry of the matrix A^n formed by taking the nth power of the graph's adjacency matrix A.*

For example, there are 20 different walks of length 4 from V_5 to V_7 (or vice versa), but no walks of length 4 from V_4 to V_3 because when A is the matrix of the cube shown in Figure 1.2.6,

$$A^4 = \begin{bmatrix} 21 & 0 & 20 & 0 & 20 & 0 & 20 & 0 \\ 0 & 21 & 0 & 20 & 0 & 20 & 0 & 20 \\ 20 & 0 & 21 & 0 & 20 & 0 & 20 & 0 \\ 0 & 20 & 0 & 21 & 0 & 20 & 0 & 20 \\ 20 & 0 & 20 & 0 & 21 & 0 & 20 & 0 \\ 0 & 20 & 0 & 20 & 0 & 21 & 0 & 20 \\ 20 & 0 & 20 & 0 & 20 & 0 & 21 & 0 \\ 0 & 20 & 0 & 20 & 0 & 20 & 0 & 21 \end{bmatrix}. \qquad 1.2.17$$

Proof. The proof is by induction, in the context of the graph above; the general case is the same. Let B_n be the 8×8 matrix whose (i,j)th entry is the number of walks from V_i to V_j of length n, for a graph with eight vertices; we must prove $B_n = A^n$. First notice that $B_1 = A^1 = A$: the entry $(A)_{i,j}$ of the matrix A is exactly the number of walks of length 1 from v_i to v_j.

Next, suppose it is true for n; let us see that it is true for $n+1$. A walk of length $n+1$ from V_i to V_j must be at some vertex V_k at time n. The number of such walks is the sum, over all vertices V_k, of the number of ways of getting from V_i to V_k in n steps, times the number of ways of getting from V_k to V_j in one step (which will be 1 if V_k is next to V_j, and 0 otherwise). In symbols, this is

$$\underbrace{(B_{n+1})_{i,j}}_{\substack{\text{no. of ways} \\ i \text{ to } j \text{ in } n+1 \text{ steps}}} = \sum_{\substack{k=1 \\ \text{for all} \\ \text{vertices } k}}^{8} \underbrace{(B_n)_{i,k}}_{\substack{\text{no. ways } i \text{ to} \\ k \text{ in } n \text{ steps}}} \underbrace{(B_1)_{k,j}}_{\substack{\text{no. ways } k \text{ to} \\ j \text{ in } 1 \text{ step}}}$$

$$= \sum_{k=1}^{8} \underbrace{(A^n)_{i,k}}_{\substack{\text{inductive} \\ \text{hypothesis}}} \underbrace{(A)_{k,j}}_{\substack{\text{def.} \\ \text{of } A}} \underbrace{=}_{\substack{\text{def.} \\ 1.2.4}} (A^{n+1})_{i,j}, \qquad 1.2.18$$

which is precisely the definition of A^{n+1}. $\square$

Remark. In Proposition 1.2.23 and equation 1.2.18, what do we mean by A^n? If you look at the proof, you will see that what we used was

$$A^n = \underbrace{\big(((\ldots)A)A\big)A}_{n \text{ factors}}. \qquad 1.2.19$$

Matrix multiplication is associative, so you can also put the parentheses any way you want; for example,

$$A^n = A\big(A(A(\ldots))\big). \qquad 1.2.20$$

In this case, we can see that it is true, and simultaneously make the associativity less abstract: with the definition given by equation 1.2.18, $B_n B_m = B_{n+m}$. Indeed, a walk of length $n+m$ from V_i to V_j is a walk of length n from V_i to some V_k, followed by a walk of length m from V_k to V_j. In formulas, this gives

$$(B_{n+m})_{i,j} = \sum_{k=1}^{8} (B_n)_{i,k}(B_m)_{k,j}. \qquad \triangle \tag{1.2.21}$$

Exercises for Section 1.2

1.2.1 a. What are the dimensions of the following matrices?

$$\text{i. } \begin{bmatrix} a & b & c \\ d & e & f \end{bmatrix} \quad \text{ii. } \begin{bmatrix} 4 & 1 \\ 0 & 2 \end{bmatrix} \quad \text{iii. } \begin{bmatrix} \pi & 1 \\ 0 & 1 \\ 1 & 0 \end{bmatrix} \quad \text{iv. } \begin{bmatrix} 1 & 0 & 0 & 1 \\ 0 & 1 & 0 & 1 \\ 1 & 0 & 1 & 0 \end{bmatrix} \quad \text{v. } \begin{bmatrix} 1 & 0 & 0 \\ 0 & 1 & 0 \\ 0 & 0 & 1 \end{bmatrix}$$

b. Which of the preceding matrices can be multiplied together?

In Exercise 1.2.2, remember to use the format

$$\begin{array}{cc} & \begin{bmatrix} 7 & 8 \\ 9 & 0 \\ 1 & 2 \end{bmatrix} \\ \begin{bmatrix} 1 & 2 & 3 \\ 4 & 5 & 6 \end{bmatrix} & \begin{bmatrix} .. & .. \\ .. & .. \end{bmatrix} \end{array}.$$

1.2.2 Perform the following matrix multiplications when possible.

$$\text{a. } \begin{bmatrix} 1 & 2 & 3 \\ 4 & 5 & 6 \end{bmatrix} \begin{bmatrix} 7 & 8 \\ 9 & 0 \\ 1 & 2 \end{bmatrix}; \quad \text{b. } \begin{bmatrix} 1 & 2 \\ 0 & 3 \end{bmatrix} \begin{bmatrix} 1 & 4 \\ -1 & 3 \\ -2 & 2 \end{bmatrix}$$

$$\text{c. } \begin{bmatrix} 1 & -1 & 1 \\ -1 & 0 & 2 \\ -1 & 1 & 1 \end{bmatrix} \begin{bmatrix} 0 & 1 & -1 \\ -1 & 1 & 2 \\ 2 & 0 & -2 \end{bmatrix}; \quad \text{d. } \begin{bmatrix} 7 & 1 \\ -1 & 0 \\ 2 & 3 \end{bmatrix} \begin{bmatrix} 5 \\ -4 \end{bmatrix}$$

$$\text{e. } \begin{bmatrix} 1 & 2 \\ 0 & 3 \end{bmatrix} \begin{bmatrix} 1 & 4 \\ -1 & 3 \end{bmatrix} \begin{bmatrix} 0 & 1 \\ -1 & 3 \end{bmatrix}; \quad \text{f. } \begin{bmatrix} 0 & 2 & 1 \\ 1 & 3 & 2 \end{bmatrix} \begin{bmatrix} 0 & 1 \\ 3 & 5 \end{bmatrix}$$

$$A = \begin{bmatrix} 1 & 2 & 0 \\ 3 & 1 & -1 \end{bmatrix} \qquad B = \begin{bmatrix} 2 & 5 & 1 \\ 1 & 4 & 2 \\ 1 & 3 & 3 \end{bmatrix}$$

Matrices for Exercise 1.2.3

1.2.3 Given the matrices A and B in the margin at left,

a. compute the third column of AB *without* computing the entire matrix AB.

b. compute the second *row* of AB, again *without* computing the entire matrix AB.

1.2.4 Compute the following without doing any arithmetic.

$$\text{a. } \begin{bmatrix} 7 & 2 & \sqrt{3} & 4 \\ 6 & 8 & a^2 & 2 \\ 3 & \sqrt{5} & a & 7 \end{bmatrix} \begin{bmatrix} 0 \\ 1 \\ 0 \\ 0 \end{bmatrix} \quad \text{b. } \begin{bmatrix} 6a & 2 & 3a^2 \\ 4 & 2\sqrt{a} & 2 \\ 5 & 12 & 3 \end{bmatrix} \vec{\mathbf{e}}_2 \quad \text{c. } \begin{bmatrix} 2 & 1 & 8 & 6 \\ 3 & 2 & \sqrt{3} & 4 \end{bmatrix} \vec{\mathbf{e}}_3$$

1.2.5 Let A and B be $n \times n$ matrices, with A symmetric. Are the following true or false?

a. $(AB)^\top = B^\top A$ b. $(A^\top B)^\top = B^\top A^\top$

c. $(A^\top B)^\top = BA$ d. $(AB)^\top = A^\top B^\top$

1.2.6 Which of the following matrices are diagonal? symmetric? triangular? antisymmetric? Do the matrix multiplication by hand.

$$\text{a. } \begin{bmatrix} a & o \\ o & a \end{bmatrix} \quad \text{b. } \begin{bmatrix} a & 0 \\ 0 & a \end{bmatrix}^2 \quad \text{c. } \begin{bmatrix} a & 0 \\ 0 & a \end{bmatrix} \begin{bmatrix} 0 & 0 \\ b & b \end{bmatrix} \quad \text{d. } \begin{bmatrix} a & 0 \\ 0 & b \end{bmatrix}^2$$

e. $\begin{bmatrix} 0 & 0 \\ a & a \end{bmatrix}^2$ f. $\begin{bmatrix} 0 & 0 \\ a & a \end{bmatrix}^3$ g. $\begin{bmatrix} 1 & 0 \\ 1 & 1 \end{bmatrix}\begin{bmatrix} 1 & 0 \\ -1 & 1 \end{bmatrix}$ h. $\begin{bmatrix} 1 & 1 & -1 \\ 1 & 0 & 1 \\ -1 & 1 & 0 \end{bmatrix}$

i. $\begin{bmatrix} 1 & 0 \\ -1 & 1 \end{bmatrix}^3$ j. $\begin{bmatrix} 1 & 0 & 1 \\ 0 & 1 & 0 \\ 1 & 0 & 1 \end{bmatrix}^2$ k. $\begin{bmatrix} 1 & 0 & -1 \\ 0 & 1 & 0 \\ 1 & 0 & 1 \end{bmatrix}^2$ l. $\begin{bmatrix} 1 & 0 \\ -1 & 1 \end{bmatrix}^4$

1.2.7 Which of the following matrices are transposes of each other?

a. $\begin{bmatrix} 1 & 2 & 3 \\ x & 0 & \sqrt{3} \\ 1 & x^2 & 2 \end{bmatrix}$ b. $\begin{bmatrix} 1 & x & 1 \\ 2 & 0 & \sqrt{3} \\ 3 & x^2 & 2 \end{bmatrix}$ c. $\begin{bmatrix} 1 & x^2 & 2 \\ x & 0 & \sqrt{3} \\ 1 & 2 & 3 \end{bmatrix}$

d. $\begin{bmatrix} 3 & \sqrt{3} & 2 \\ 2 & 0 & x^2 \\ 1 & x & 1 \end{bmatrix}$ e. $\begin{bmatrix} 1 & x & 1 \\ x^2 & 0 & 2 \\ 2 & \sqrt{3} & 3 \end{bmatrix}$ f. $\begin{bmatrix} 1 & 2 & 3 \\ x & 0 & x^2 \\ 1 & \sqrt{3} & 2 \end{bmatrix}$

1.2.8 For what values of a do the matrices

$$A = \begin{bmatrix} 1 & 1 \\ 1 & 0 \end{bmatrix} \quad \text{and} \quad B = \begin{bmatrix} 1 & 0 \\ a & 1 \end{bmatrix} \quad \text{satisfy } AB = BA?$$

1.2.9 Given the two matrices $A = \begin{bmatrix} 1 & 0 \\ 1 & 0 \end{bmatrix}$ and $B = \begin{bmatrix} 1 & 0 & 1 \\ 2 & 1 & 0 \end{bmatrix}$,

a. What are their transposes?

b. Without computing AB, what is $(AB)^\top$?

c. Confirm your result by computing AB.

d. What happens if you do part b using the *incorrect* formula $(AB)^\top = A^\top B^\top$?

1.2.10 What is the inverse of the matrix $A = \begin{bmatrix} a & b \\ 0 & a \end{bmatrix}$ for $a \neq 0$?

$$A = \begin{bmatrix} 1 & 0 \\ 1 & 0 \end{bmatrix}$$

$$B = \begin{bmatrix} 1 & 0 & 1 \\ 1 & 0 & 1 \end{bmatrix}$$

$$C = \begin{bmatrix} 1 & 1 & 0 \\ 1 & 0 & 1 \\ 1 & 1 & 0 \end{bmatrix}$$

Matrices for Exercise 1.2.11

1.2.11 Given the matrices A, B, and C at left, which of the following expressions make no sense (are undefined)?

a. AB b. BA c. $A + B$ d. AC

e. BC f. CB g. $\frac{B}{C}$ h. $B^\top A$ i. $B^\top C$

1.2.12 Confirm by matrix multiplication that the inverse of

$$A = \begin{bmatrix} a & b \\ c & d \end{bmatrix} \quad \text{is} \quad A^{-1} = \frac{1}{ad - bc}\begin{bmatrix} d & -b \\ -c & a \end{bmatrix}.$$

1.2.13 Prove that a matrix $\begin{bmatrix} a & b \\ c & d \end{bmatrix}$ is not invertible if $ad - bc = 0$.

1.2.14 Prove Theorem 1.2.17: that the transpose of a product is the product of the transposes in reverse order: $(AB)^\top = B^\top A^\top$.

1.2.15 Show that

$$\begin{bmatrix} 1 & a & b \\ 0 & 1 & c \\ 0 & 0 & 1 \end{bmatrix} \text{ has an inverse of the form } \begin{bmatrix} 1 & x & y \\ 0 & 1 & z \\ 0 & 0 & 1 \end{bmatrix}. \text{ Find it.}$$

1.2.16 Show that if A is any matrix, then $A^\top A$ and $AA^\top$ are symmetric.

1.2.17 a. Compute the adjacency matrix A_T for a triangle and A_S for a square.

b. For each of these, compute the powers up to 5, and explain the meaning of the diagonal entries.

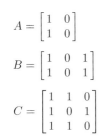

c. For the triangle, observe that the diagonal terms differ by 1 from the off-diagonal terms. Can you prove that this will be true for all powers of A_T?

d. For the square, you should observe that half the terms are 0 for even powers, and the other half are 0 for odd powers. Can you prove that this will be true for all powers of A_S?

Asterisks indicate difficult exercises.

*e. Show that you can color the vertices of a connected graph (one on which you can walk from any vertex to any other) in two colors, such that no two adjacent vertices have the same color, if and only if for all sufficiently high powers n of the adjacency matrix, those entries that are 0 for A^n are nonzero for A^{n+1}, and those that are nonzero for A^n are 0 for A^{n+1}.

1.2.18 a. For the adjacency matrix A corresponding to the cube (shown in Figure 1.2.6), compute A^2, A^3, and A^4. Check directly that $(A^2)(A^2) = (A^3)A$.

b. The diagonal entries of A^4 should all be 21; count the number of walks of length 4 from a vertex to itself directly.

c. For this same matrix A, some entries of A^n are always 0 when n is even, and others (the diagonal entries for instance) are always 0 when n is odd. Can you explain why? Think of coloring the vertices of the cube in two colors, so that each edge connects vertices of opposite colors.

d. Is this phenomenon true for A_T (the adjacency matrix for a triangle) or for A_S (the adjacency matrix for a square)? Explain why, or why not.

1.2.19 Suppose we redefined a walk on the cube of Example 1.2.22 to allow stops: in one time unit you may either go to an adjacent vertex, or stay where you are.

Exercise 1.2.20 says that the matrices M_z provide a model for the complex numbers within the 2×2 real matrices.

a. Find a matrix B such that $B^n_{i,j}$ counts the walks of length n from V_i to V_j.

b. Compute B^2 and B^3. Explain the diagonal entries of B^3.

1.2.20 Associate to $z = x + iy \in \mathbb{C}$ the matrix $M_z = \begin{bmatrix} x & y \\ -y & x \end{bmatrix}$; let $z_1, z_2 \in \mathbb{C}$. Show that $M_{z_1+z_2} = M_{z_1} + M_{z_2}$ and $M_{z_1 z_2} = M_{z_1} M_{z_2}$.

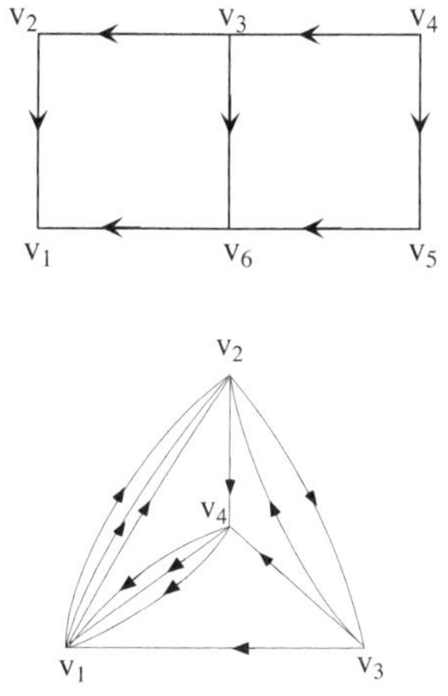

Graphs for Exercise 1.2.22

1.2.21 An oriented walk of length n on an oriented graph consists of a sequence of vertices $V_0, V_1, \dots, V_n$ such that V_i, V_{i+1} are, respectively, the beginning and the end of an oriented edge.

a. Show that if A is the oriented adjacency matrix of an oriented graph, then the (i,j)th entry of A^n is the number of oriented walks of length n going from vertex i to vertex j.

b. What does it mean for the oriented adjacency matrix of an oriented graph to be upper triangular? lower triangular? diagonal?

1.2.22 Suppose all the edges of a graph are oriented by an arrow on them. Define the oriented adjacency matrix to be the square matrix with both rows and columns labeled by the vertices, where the (i,j)th entry is m if there are m oriented edges leading from vertex i to vertex j.

What are the oriented adjacency matrices of the two graphs in the margin, (where some edges of the graph are one-way superhighways, with more than one lane)?

1.2.23 a. Let $A = \begin{bmatrix} 0 & 0 \\ 1 & 0 \\ 0 & 1 \end{bmatrix}$. Show that $\begin{bmatrix} a & 1 & 0 \\ b & 0 & 1 \end{bmatrix}$ is a left inverse of A.

b. Show that the matrix A of part a has no right inverse.

c. Find a matrix that has infinitely many right inverses. (Try transposing.)

1.2.24 Show that matrix multiplication is distributive over addition.

1.3 What the actors do: Matrix multiplication as a linear transformation

The central notion of linear algebra is that multiplication of a vector by a matrix is a *linear transformation.*

A map $\mathbf{f} : \mathbb{R}^n \to \mathbb{R}^m$ is a black box that takes as input a list of n numbers and gives as output a list of m numbers. The innards of the box (the rule determining the output) may be very complicated. In this section we discuss the simplest maps from $\mathbb{R}^n$ to $\mathbb{R}^m$, where the black box multiplies the input by a matrix. (Thus it will be convenient to think of the input and output as vectors, not points.)

In Section 2.2 we will see how matrices are used to solve systems of linear equations.

The words *map*, *mapping*, *function*, and *transformation* are synonyms.

Transformation is more common in linear algebra.

Recall from Section 0.4 that the *codomain* of a function is not the same as its *image*. Knowing that the codomain of a function is $\mathbb{R}^m$ tells you that the output is a list of m numbers. It doesn't say which lists. The image consists of the lists that can actually come out of the "black box."

For example, the matrix $A = \begin{bmatrix} 1 & 1 & 2 \\ 1 & 0 & 1 \end{bmatrix}$ is a transformation whose domain is $\mathbb{R}^3$ and whose codomain is $\mathbb{R}^2$; evaluated on $\vec{\mathbf{v}} = \begin{bmatrix} 1 \\ 2 \\ 1 \end{bmatrix}$ it returns $\vec{\mathbf{w}} = \begin{bmatrix} 5 \\ 2 \end{bmatrix}$: $A\vec{\mathbf{v}} = \vec{\mathbf{w}}$. To go from $\mathbb{R}^3$, where vectors have three entries, to $\mathbb{R}^4$, where vectors have four entries, you could multiply $\vec{\mathbf{v}} \in \mathbb{R}^3$ on the left by a 4×3 matrix:

$$\begin{matrix} & \begin{bmatrix} v_1 \\ v_2 \\ v_3 \end{bmatrix} \\ \begin{bmatrix} \dots & \dots & \dots \\ \dots & \dots & \dots \\ \dots & \dots & \dots \\ \dots & \dots & \dots \end{bmatrix} & \begin{bmatrix} w_1 \\ w_2 \\ w_3 \\ w_4 \end{bmatrix} \end{matrix}. \qquad 1.3.1$$

Here is a more concrete example.

Example 1.3.1 (Food processing plant). In a food processing plant making three types of frozen dinners, one might associate the number of different dinners produced to the total ingredients needed (beef, chicken, noodles, cream, salt, ...). As shown in Figure 1.3.1, this mapping is given by multiplication (on the left) by the matrix A, giving the amount of each ingredient needed for each dinner: A tells how to go from $\vec{\mathbf{b}}$ (how many dinners of each kind are produced) to the product $\vec{\mathbf{c}}$ (the total ingredients needed to produce those dinners). For example, to produce 60 beef Stroganoff dinners, 30 ravioli dinners, and 40 fried chicken dinners, 21 pounds of beef are needed, because

$$(.25 \times 60) + (.20 \times 30) + (0 \times 40) = 21. \qquad 1.3.2$$

Figure 1.3.1: In real applications – designing buildings, modeling the economy, modeling airflow over the wing of an airplane – vectors of input data contain tens of thousands of entries or more, and the matrix giving the transformation has millions of entries.

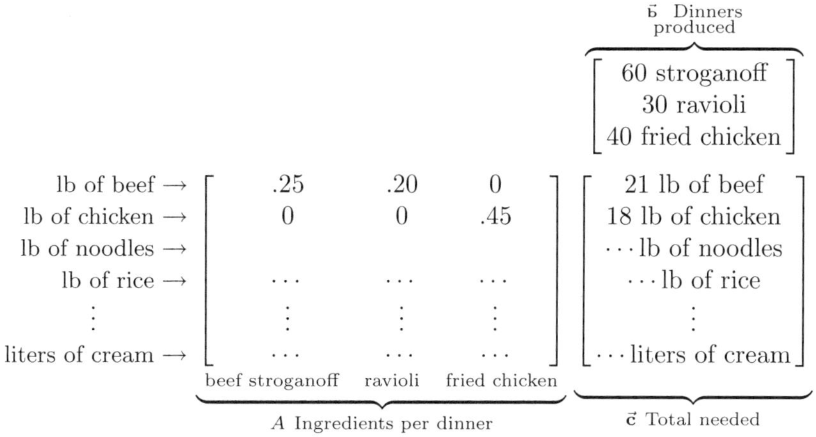

FIGURE 1.3.1. The matrix A is the transformation associating the number of dinners of various sorts produced to the total ingredients needed. △

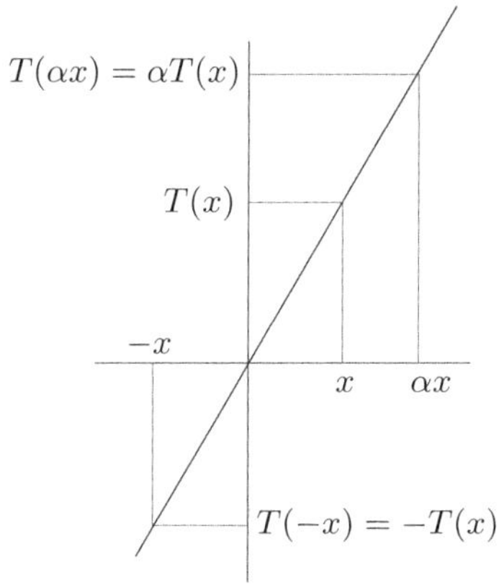

FIGURE 1.3.2.
For any linear transformation T and any scalar α,

$$T(\alpha \mathbf{x}) = \alpha T(\mathbf{x}).$$

It follows from the rules of matrix multiplication that any transformation represented by a matrix has the property "if you double the input, you double the output", or, more generally, if you add inputs, you add outputs. This characterizes "linear response", which is the simplest sort of response one can imagine.

Definition 1.3.2 (Linear transformation from $\mathbb{R}^n$ to $\mathbb{R}^m$). A *linear transformation* $T : \mathbb{R}^n \to \mathbb{R}^m$ is a mapping such that for all scalars a and all $\vec{\mathbf{v}}, \vec{\mathbf{w}} \in \mathbb{R}^n$,

$$T(\vec{\mathbf{v}} + \vec{\mathbf{w}}) = T(\vec{\mathbf{v}}) + T(\vec{\mathbf{w}}) \quad \text{and} \quad T(a\vec{\mathbf{v}}) = aT(\vec{\mathbf{v}}). \qquad 1.3.3$$

The second formula is illustrated by Figure 1.3.2. The two formulas can be combined into one (where b is also a scalar):

$$T(a\vec{\mathbf{v}} + b\vec{\mathbf{w}}) = aT(\vec{\mathbf{v}}) + bT(\vec{\mathbf{w}}). \qquad 1.3.4$$

What does this formula imply in the case where the input is $\mathbf{0}$?[5]

Example 1.3.3 (Linearity at the checkout counter). At a supermarket checkout counter, the transformation T performed by the scanner takes a shopping cart of groceries and gives a total price. (The supermarket has n different products on its shelves; the ith entry of the vector $\vec{\mathbf{v}} \in \mathbb{R}^n$ corresponds to the quantity you are buying of the ith item, and T is a $1 \times n$ row matrix giving the price of each item.) The first of equations 1.3.3 says that if you distribute your groceries in two shopping carts and pay for them separately, you will pay the same amount as you will if you go through the checkout line once, with everything in one cart. The second equation says that if you buy $2\vec{\mathbf{v}}$ items (two gallons of milk, no bananas, four boxes of

[5]The output must be $\mathbf{0}$: a linear transformation takes the origin to the origin.

FIGURE 1.3.3.
The Italian mathematician Salvatore Pincherle (1853–1936) one of the pioneers of linear algebra, called a linear transformation a *distributive transformation*. This name is perhaps more suggestive of the formulas than is "linear".

One of the great discoveries at the end of the 19th century was that the natural way to do mathematics is to look at sets with structure, such as $\mathbb{R}^n$, with addition and multiplication by scalars, and to consider the mappings that preserve that structure. Linear transformations "preserve structure" in the sense that one can first add, then map, or first map, then add, and get the same answer; similarly, first multiplying by a scalar and then mapping gives the same result as first mapping and then multiplying by a scalar.

oatmeal ...), you will pay the same price if you buy them all at once or if you go through the checkout line twice, taking $\vec{\mathbf{v}}$ items each time. △

Remark. The assumption that a transformation is linear is the main simplifying assumption that scientists and social scientists (especially economists) make when modeling the world. The simplicity of linear transformations is both a great strength and a great weakness. Linear transformations can be mastered: completely understood. They contain no surprises. But when linear transformations are used to model real life or physical systems, they are rarely accurate. Even at the supermarket, is the price transformation always linear? If you buy one box of cereal with a $1 coupon, and a second box at full price, the price of the two boxes is not twice the price of one box. Nor does a linear model of the "price transformation" allow for the possibility that if you buy more you will get a discount for quantity, or that if you buy even more you might create scarcity and drive prices up. The failure to take such effects into consideration is a fundamental weakness of all models that linearize mappings and interactions.

But linear transformations are not just a powerful tool for making imperfect models of reality. They are the essential first step in understanding nonlinear mappings. We will begin the study of calculus and nonlinear mappings in Section 1.7; at every step we will call upon our knowledge of linear transformations. Whenever we have a question about a nonlinear function, we will answer it by looking carefully at a linear transformation: its derivative. When we wish to say something about solutions to a system of nonlinear equations, our tool will be Newton's method, which involves solving linear equations. When in Section 3.7 we wish to determine the maximum value a function can achieve on a certain ellipse, we will use two linear transformations: the derivative of the function to be maximized and the derivative of the function describing the ellipse. When in Section 4.10 we use the change of variables formula to integrate over subsets of $\mathbb{R}^n$, we will again need the derivative. Everything you learn about linear algebra in the coming pages will pay off twice. This is why we feel linear algebra and multivariate calculus should be taught as one integrated course. △

Now we will relate matrices and linear transformations more precisely.

The reason we write vectors as columns, not rows, is so that $T(\vec{\mathbf{v}}) = [T]\vec{\mathbf{v}}$, where $[T]$ is the matrix corresponding to the linear transformation T. If we wrote vectors as rows (i.e., as the transpose of the column), then we would have to write

$$T(\vec{\mathbf{v}}) = \vec{\mathbf{v}}[T]^\top.$$

Theorem 1.3.4 (Matrices and linear transformations).

1. *Any $m \times n$ matrix A defines a linear transformation $T : \mathbb{R}^n \to \mathbb{R}^m$ by matrix multiplication:*

$$T(\vec{\mathbf{v}}) = A\vec{\mathbf{v}}. \tag{1.3.5}$$

2. *Every linear transformation $T : \mathbb{R}^n \to \mathbb{R}^m$ is given by multiplication by the $m \times n$ matrix $[T]$:*

$$T(\vec{\mathbf{v}}) = [T]\vec{\mathbf{v}}, \tag{1.3.6}$$

where the ith column of $[T]$ is $T(\vec{\mathbf{e}}_i)$.

Part 2 of Theorem 1.3.4 is powerful and surprising. It says not just that every linear transformation from $\mathbb{R}^n$ to $\mathbb{R}^m$ is given by a matrix; it also says that one can construct the matrix by seeing how the transformation acts on the standard basis vectors. This is rather remarkable. A priori the notion of a transformation from $\mathbb{R}^n$ to $\mathbb{R}^m$ is vague and abstract; one might not think that merely by imposing the condition of linearity one could say something so precise about this shapeless set of mappings as saying that each is given by a matrix, which we know how to construct.

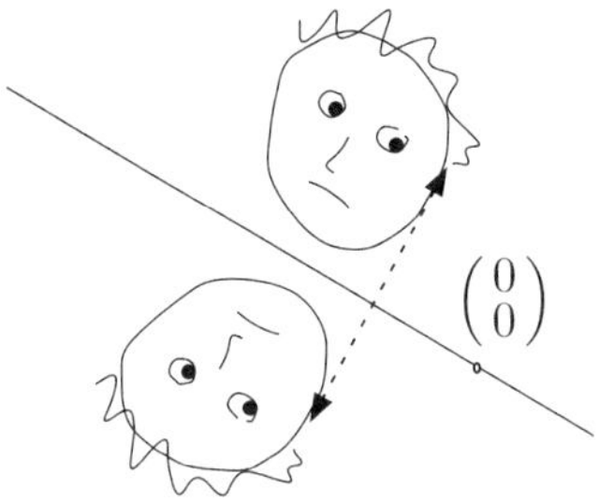

FIGURE 1.3.4.

Example 1.3.5: Every point on one face is reflected to the corresponding point of the other. The straight line between the two faces goes through the origin.

Proof of Theorem 1.3.4. 1. This follows immediately from the rules of matrix multiplication; you are asked to spell it out in Exercise 1.3.15.

2. Start with a linear transformation $T : \mathbb{R}^n \to \mathbb{R}^m$, and manufacture the matrix $[T] = [T\vec{\mathbf{e}}_1, \dots, T\vec{\mathbf{e}}_n]$. We may write any vector $\vec{\mathbf{v}} \in \mathbb{R}^n$ in terms of the standard basis vectors:

$$\vec{\mathbf{v}} = v_1\vec{\mathbf{e}}_1 + v_2\vec{\mathbf{e}}_2 + \cdots + v_n\vec{\mathbf{e}}_n, \quad \text{or, with sum notation,} \quad \vec{\mathbf{v}} = \sum_{i=1}^{n} v_i\vec{\mathbf{e}}_i. \qquad 1.3.7$$

Then, by linearity,

$$T(\vec{\mathbf{v}}) = T\sum_{i=1}^{n} v_i\vec{\mathbf{e}}_i = \sum_{i=1}^{n} v_i T(\vec{\mathbf{e}}_i), \qquad 1.3.8$$

which is precisely the column vector $[T]\vec{\mathbf{v}}$. If this isn't apparent, try translating it out of sum notation:

$$T(\vec{\mathbf{v}}) = \sum_{i=1}^{n} v_i T(\vec{\mathbf{e}}_i) = v_1 \underbrace{\overset{T(\vec{\mathbf{e}}_1)}{\begin{bmatrix} \\ \\ \end{bmatrix}}}_{\substack{\text{1st col.}\\ \text{of } [T]}} + \cdots + v_n \underbrace{\overset{T(\vec{\mathbf{e}}_n)}{\begin{bmatrix} \\ \\ \end{bmatrix}}}_{\substack{n\text{th col.}\\ \text{of } [T]}} = \begin{bmatrix} & T & \end{bmatrix} \begin{bmatrix} v_1 \\ v_2 \\ \vdots \\ v_n \end{bmatrix} = [T]\vec{\mathbf{v}}. \qquad \square \qquad 1.3.9$$

Example 1.3.5 (Finding the matrix of a linear transformation). Consider the transformation T that reflects with respect to a line through the origin; one such transformation is shown in Figure 1.3.4.

The "reflection transformation" is indeed linear: as shown in Figure 1.3.5 (where we are reflecting with respect to a different line than in Figure 1.3.4), given two vectors $\vec{\mathbf{v}}$ and $\vec{\mathbf{w}}$, we have $T(\vec{\mathbf{v}} + \vec{\mathbf{w}}) = T(\vec{\mathbf{v}}) + T(\vec{\mathbf{w}})$. It is also apparent from the figure that $T(c\vec{\mathbf{v}}) = cT(\vec{\mathbf{v}})$. To find the matrix of this transformation, all we have to do is figure out what T does to $\vec{\mathbf{e}}_1$ and $\vec{\mathbf{e}}_2$. To obtain the first column of our matrix we thus consider where $\vec{\mathbf{e}}_1$ is mapped to. Suppose that our line makes an angle θ (theta) with the x-axis, as shown in Figure 1.3.6. Then $\vec{\mathbf{e}}_1$ is mapped to $\begin{bmatrix} \cos 2\theta \\ \sin 2\theta \end{bmatrix}$.

To get the second column, we see that $\vec{\mathbf{e}}_2$ is mapped to

$$\begin{bmatrix} \cos(2\theta - 90^\circ) \\ \sin(2\theta - 90^\circ) \end{bmatrix} = \begin{bmatrix} \sin 2\theta \\ -\cos 2\theta \end{bmatrix}. \qquad 1.3.10$$

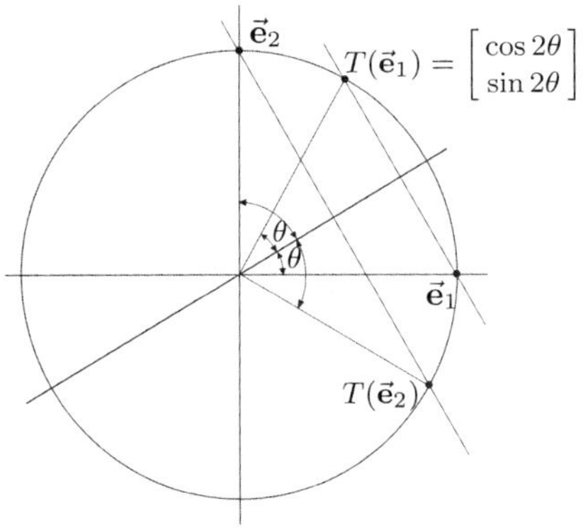

FIGURE 1.3.6.
The reflection map T takes

$$\vec{\mathbf{e}}_1 = \begin{bmatrix} 1 \\ 0 \end{bmatrix} \text{ to } \begin{bmatrix} \cos 2\theta \\ \sin 2\theta \end{bmatrix};$$

$$\vec{\mathbf{e}}_2 = \begin{bmatrix} 0 \\ 1 \end{bmatrix} \text{ to } \begin{bmatrix} \sin 2\theta \\ -\cos 2\theta \end{bmatrix}.$$

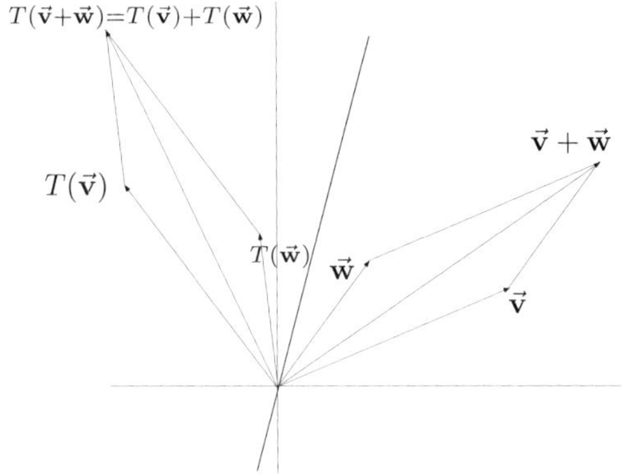

FIGURE 1.3.5. Reflection is linear.

So the "reflection" matrix is

$$\begin{bmatrix} \cos 2\theta & \sin 2\theta \\ \sin 2\theta & -\cos 2\theta \end{bmatrix}; \qquad 1.3.11$$

we can apply T to any point, by multiplying it by this matrix. For example, the point $\begin{pmatrix} 2 \\ 1 \end{pmatrix}$ reflects to the point $\begin{pmatrix} 2\cos 2\theta + \sin 2\theta \\ 2\sin 2\theta - \cos 2\theta \end{pmatrix}$, since

$$\begin{bmatrix} \cos 2\theta & \sin 2\theta \\ \sin 2\theta & -\cos 2\theta \end{bmatrix} \begin{bmatrix} 2 \\ 1 \end{bmatrix} = \begin{bmatrix} 2\cos 2\theta + \sin 2\theta \\ 2\sin 2\theta - \cos 2\theta \end{bmatrix}. \quad \triangle \qquad 1.3.12$$

To take the *projection* of a point onto a line, we draw a line from the point, perpendicular to the line, and see where on the line it arrives.

What transformation takes a point in $\mathbb{R}^2$ and gives its projection on the x-axis, as shown in Figure 1.3.7? What is the projection on the line of equation $x = y$ of the point $\begin{pmatrix} 3 \\ -1 \end{pmatrix}$? Assume both transformations are linear. Check your answers in the footnote below.[6]

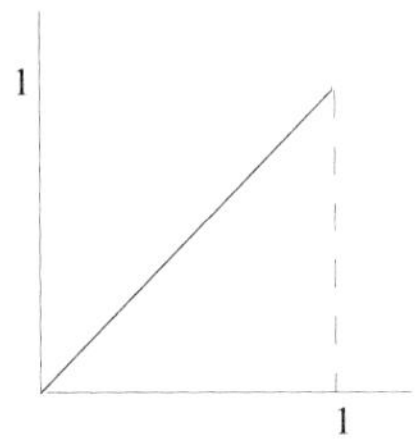

FIGURE 1.3.7.
The projection of the point $\begin{pmatrix} 1 \\ 1 \end{pmatrix}$ onto the x-axis is $\begin{pmatrix} 1 \\ 0 \end{pmatrix}$.

Geometric interpretation of linear transformations

As was suggested by Figures 1.3.4 and 1.3.5, a linear transformation can be applied to a subset, not just individual vectors: if X is a subset of $\mathbb{R}^n$,

---

[6]The first transformation is $\begin{bmatrix} 1 & 0 \\ 0 & 0 \end{bmatrix}$. To get the first column, ask, what does the transformation do to $\vec{\mathbf{e}}_1$? Since $\vec{\mathbf{e}}_1$ lies on the x-axis, it is projected onto itself. The second standard basis vector is projected onto the origin, so the second column of the matrix is $\begin{bmatrix} 0 \\ 0 \end{bmatrix}$. The second transformation is $\begin{bmatrix} 1/2 & 1/2 \\ 1/2 & 1/2 \end{bmatrix}$, since the perpendicular from $\begin{bmatrix} 1 \\ 0 \end{bmatrix}$ to the line of equation $x = y$ intersects that line at $\begin{pmatrix} 1/2 \\ 1/2 \end{pmatrix}$, as does the perpendicular line from $\begin{bmatrix} 0 \\ 1 \end{bmatrix}$. To find the projection of $\begin{pmatrix} 3 \\ -1 \end{pmatrix}$, we multiply $\begin{bmatrix} 1/2 & 1/2 \\ 1/2 & 1/2 \end{bmatrix} \begin{bmatrix} 3 \\ -1 \end{bmatrix} = \begin{bmatrix} 1 \\ 1 \end{bmatrix}$. Note that we have to consider the point $\begin{pmatrix} 3 \\ -1 \end{pmatrix}$ as a vector; we can't multiply a matrix and a point.

$T(X)$ corresponds to multiplying each point of X by $[T]$. (To do this we write the points as vectors.)

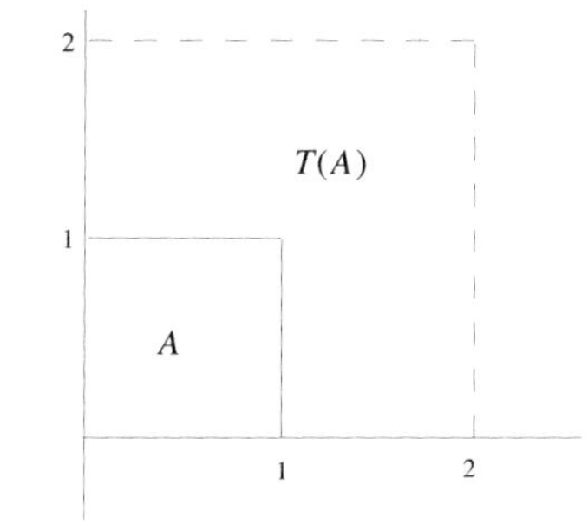

FIGURE 1.3.8.
The scaling transformation given by $\begin{bmatrix} 2 & 0 \\ 0 & 2 \end{bmatrix}$ turns the unit square into the square with sidelength 2.

Example 1.3.6 (Identity transformation). The identity transformation id : $\mathbb{R}^n \to \mathbb{R}^n$ is linear and is given by the matrix I_n. Applying this transformation to a subset of $\mathbb{R}^n$ leaves it unchanged. $\triangle$

Example 1.3.7 (Scaling transformation). The transformation T that enlarges everything in $\mathbb{R}^2$ by a factor of a is given by the matrix $\begin{bmatrix} a & 0 \\ 0 & a \end{bmatrix}$, since $T\vec{\mathbf{e}}_1 = \begin{bmatrix} a \\ 0 \end{bmatrix}$ and $T\vec{\mathbf{e}}_2 = \begin{bmatrix} 0 \\ a \end{bmatrix}$. Thus the matrix $\begin{bmatrix} 2 & 0 \\ 0 & 2 \end{bmatrix}$ stretches vectors by a factor of 2, whereas $\begin{bmatrix} 1/2 & 0 \\ 0 & 1/2 \end{bmatrix}$ shrinks them by the same factor. Figure 1.3.8 shows the result of applying $[T] = \begin{bmatrix} 2 & 0 \\ 0 & 2 \end{bmatrix}$ to the unit square A. More generally, aI_n, the $n \times n$ matrix where each diagonal entry is a and every other entry is 0, scales every vector $\vec{\mathbf{x}} \in \mathbb{R}^n$ by a factor of a. If $a = 1$, we have the identity transformation, and if $a = -1$, we have reflection with respect to the origin. $\triangle$

A scaling transformation is a special case of a stretching transformation.

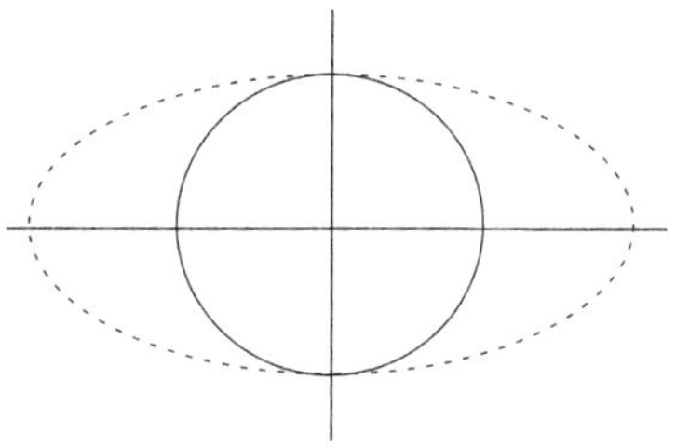

FIGURE 1.3.9.
The linear transformation given by the matrix $\begin{bmatrix} 2 & 0 \\ 0 & 1 \end{bmatrix}$ stretches the unit circle into an ellipse.

Example 1.3.8 (Stretching transformation). The linear transformation $\begin{bmatrix} 2 & 0 \\ 0 & 1 \end{bmatrix}$ stretches the unit square into a rectangle 2 wide and 1 high; it turns the unit circle into the ellipse shown in Figure 1.3.9. More generally, any diagonal matrix with positive diagonal entries $x_1, \dots, x_n$ stretches by x_i in the direction of the axis corresponding to $\vec{\mathbf{e}}_i$. $\triangle$

Example 1.3.9 (Rotation by an angle θ). Figure 1.3.10 shows that the transformation R giving rotation by θ counterclockwise around the origin is linear, and that its matrix is

$$[R(\vec{\mathbf{e}}_1), R(\vec{\mathbf{e}}_2)] = \begin{bmatrix} \cos\theta & -\sin\theta \\ \sin\theta & \cos\theta \end{bmatrix}. \qquad 1.3.13$$

Exercise 1.3.16 asks you to derive the fundamental theorems of trigonometry, using composition of the transformation in Example 1.3.9. $\triangle$

We discuss compositions in Section 0.4.

Now we will see that composition corresponds to matrix multiplication.

Theorem 1.3.10 (Composition corresponds to matrix multiplication). *Suppose $S : \mathbb{R}^n \to \mathbb{R}^m$ and $T : \mathbb{R}^m \to \mathbb{R}^l$ are linear transformations given by the matrices $[S]$ and $[T]$ respectively. Then the composition $T \circ S$ is linear and*

$$[T \circ S] = [T][S]. \qquad 1.3.14$$

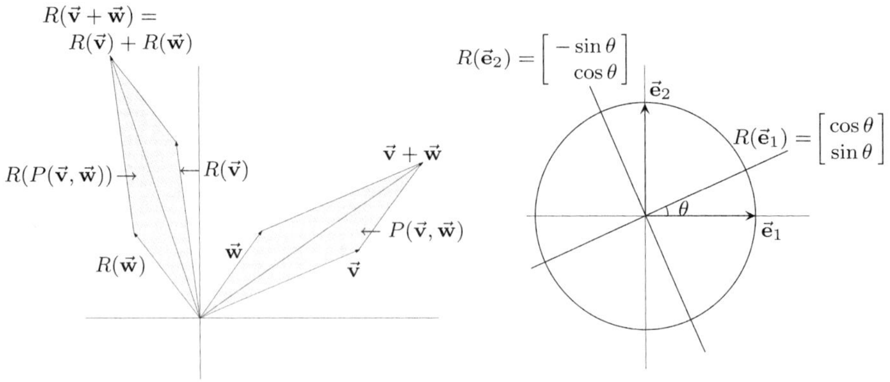

FIGURE 1.3.10. Figure for Example 1.3.9. LEFT: The sum of the rotations is the rotation of the sum. Rather than rotating the vectors $\vec{\mathbf{v}}$ and $\vec{\mathbf{w}}$, we rotate the whole parallelogram $P(\vec{\mathbf{v}},\vec{\mathbf{w}})$ that they span (the parallelogram with sides $\vec{\mathbf{v}}$ and $\vec{\mathbf{w}}$). Then $R(P(\vec{\mathbf{v}},\vec{\mathbf{w}}))$ is the parallelogram spanned by $R(\vec{\mathbf{v}}), R(\vec{\mathbf{w}})$, and the diagonal of $R(P(\vec{\mathbf{v}},\vec{\mathbf{w}}))$ is $R(\vec{\mathbf{v}}+\vec{\mathbf{w}})$. It is also true that $R(c\vec{\mathbf{v}}) = cR(\vec{\mathbf{v}})$, so rotation is linear. RIGHT: We see that the matrix of the transformation R giving rotation by θ counterclockwise around the origin is $\begin{bmatrix} \cos\theta & -\sin\theta \\ \sin\theta & \cos\theta \end{bmatrix}$.

We will sometimes, but not always, distinguish between a linear transformation A and its associated matrix $[A]$.

To a mathematician, matrices are merely a convenient way to encode linear transformations from $\mathbb{R}^n$ to $\mathbb{R}^m$, and matrix multiplication then conveniently encodes compositions of such mappings. Linear transformations and their compositions are the central issue; some courses on linear algebra barely mention matrices.

Students, however, tend to be much more comfortable with matrices than with the abstract notion of linear transformation; this is why we have based our treatment of linear algebra on matrices.

Proof of Theorem 1.3.10. The following computation shows that $T \circ S$ is linear:

$$\begin{aligned}(T\circ S)(a\vec{\mathbf{v}}+b\vec{\mathbf{w}}) &= T\big(S(a\vec{\mathbf{v}}+b\vec{\mathbf{w}})\big) = T\big(aS(\vec{\mathbf{v}})+bS(\vec{\mathbf{w}})\big)\\ &= aT\big(S(\vec{\mathbf{v}})\big)+bT\big(S(\vec{\mathbf{w}})\big) = a(T\circ S)(\vec{\mathbf{v}})+b(T\circ S)(\vec{\mathbf{w}}).\end{aligned}$$

Equation 1.3.14 is a statement about matrix multiplication. We will use the following facts:

1. $A\vec{\mathbf{e}}_i$ is the ith column of A (see Example 1.2.6).
2. The ith column of AB is $A\vec{\mathbf{b}}_i$, where $\vec{\mathbf{b}}_i$ is the ith column of B (see Example 1.2.7).

Since $T \circ S$ is linear, it is given by the matrix $[T\circ S]$. The definition of composition gives the second equality of equation 1.3.15. Next we replace S by its matrix $[S]$, and T by its matrix:

$$[T\circ S]\vec{\mathbf{e}}_i = (T\circ S)(\vec{\mathbf{e}}_i) = T\big(S(\vec{\mathbf{e}}_i)\big) = T([S]\vec{\mathbf{e}}_i) = [T]\big([S]\vec{\mathbf{e}}_i\big). \qquad 1.3.15$$

So $[T\circ S]\vec{\mathbf{e}}_i$, which is the ith column of $[T\circ S]$ by fact (1), is equal to

$$[T] \text{ times the } i\text{th column of } [S], \qquad 1.3.16$$

which is the ith column of $[T][S]$ by fact (2).

Each column of $[T\circ S]$ is equal to the corresponding column of $[T][S]$, so the two matrices are equal. □

We will see (Theorem 3.8.1 and Exercise 4.8.23) that any linear transformation can be thought of as some combination of rotation, reflection, and stretching in the direction of the axes. In the following example we stretch and rotate the unit square.

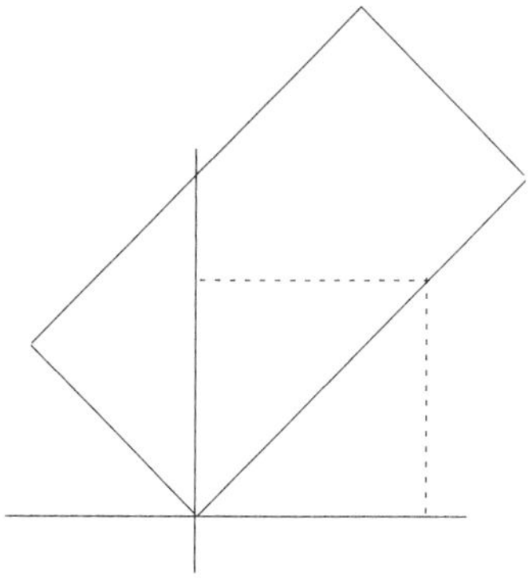

FIGURE 1.3.11.
The result of first stretching, then rotating, the unit square. Exercise 1.3.18 asks you to rotate the square first, then stretch it.

GRAPHING CALCULATOR, available at www.PacificT.com, allows you to apply linear (and nonlinear) transformations to subsets of $\mathbb{R}^2$ and see what happens.

Example 1.3.11 (Composition: stretching and rotating a square). If we apply the transformation $\begin{bmatrix} 2 & 0 \\ 0 & 1 \end{bmatrix}$ to the unit square, then rotate the result by 45 degrees counterclockwise using the matrix $\begin{bmatrix} \cos\pi/4 & -\sin\pi/4 \\ \sin\pi/4 & \cos\pi/4 \end{bmatrix}$, we get the rectangle shown in Figure 1.3.11. We could get the same result using the composition

$$\begin{bmatrix} \cos\pi/4 & -\sin\pi/4 \\ \sin\pi/4 & \cos\pi/4 \end{bmatrix} \begin{bmatrix} 2 & 0 \\ 0 & 1 \end{bmatrix} = \begin{bmatrix} 2\cos\pi/4 & -\sin\pi/4 \\ 2\sin\pi/4 & \cos\pi/4 \end{bmatrix}. \quad \triangle \qquad 1.3.17$$

We gave a computational proof of the associativity of matrix multiplication in Proposition 1.2.9; this associativity is also an immediate consequence of Theorem 1.3.10, since composition of mappings is associative.

Example 1.3.12 (Associativity of matrix multiplication). For the food plant of Example 1.3.1, one might make a line matrix D, n wide, giving the price, per pound or liter, of each of the n ingredients. Since A is an $n \times 3$ matrix, the product DA is a line matrix 3 wide; A tells how much of each ingredient is in each dinner, so DA gives the cost of the ingredients in each type of dinner. The product $(DA)\vec{\mathbf{b}}$ would give the total cost of the ingredients for all $\vec{\mathbf{b}}$ dinners. We could compose these transformations in a different order. We could first compute how much of each ingredient we need for all $\vec{\mathbf{b}}$ dinners: the product $A\vec{\mathbf{b}}$. Then the total cost is $D(A\vec{\mathbf{b}})$. Clearly, $(DA)\vec{\mathbf{b}} = D(A\vec{\mathbf{b}})$. $\triangle$

Invertibility of matrices and linear transformations

Recall (Definition 0.4.7) that a mapping $f : X \to Y$ is invertible if it is one to one and onto.

Proposition 1.3.13. *A linear transformation $T : \mathbb{R}^n \to \mathbb{R}^m$ is invertible if and only if the $m \times n$ matrix $[T]$ is invertible. If it is invertible, then*

$$[T^{-1}] = [T]^{-1}. \qquad 1.3.18$$

Proof. Suppose that $[T]$ is invertible, and let $S : \mathbb{R}^m \to \mathbb{R}^n$ be the linear transformation such that $[S] = [T]^{-1}$. Then for any $\vec{\mathbf{y}} \in \mathbb{R}^m$, we have

$$\vec{\mathbf{y}} = ([T][S])\vec{\mathbf{y}} = T\big(S(\vec{\mathbf{y}})\big), \qquad 1.3.19$$

showing that T is onto: for any $\vec{\mathbf{y}} \in \mathbb{R}^m$, there is a solution, $S(\vec{\mathbf{y}})$, to the equation $T(\vec{\mathbf{x}}) = \vec{\mathbf{y}}$. If $\vec{\mathbf{x}}_1, \vec{\mathbf{x}}_2 \in \mathbb{R}^n$, and $T(\vec{\mathbf{x}}_1) = T(\vec{\mathbf{x}}_2)$, then

$$\vec{\mathbf{x}}_1 = ([S][T])\vec{\mathbf{x}}_1 = S\big(T(\vec{\mathbf{x}}_1)\big) = S\big(T(\vec{\mathbf{x}}_2)\big) = ([S][T])\vec{\mathbf{x}}_2 = \vec{\mathbf{x}}_2, \qquad 1.3.20$$

showing that T is one to one. Thus if the matrix $[T]$ is invertible, then $T : \mathbb{R}^n \to \mathbb{R}^m$ is invertible, and its inverse mapping is S.

Exercise 1.3.17 asks you to confirm by matrix multiplication that reflecting a point across the line, and then back again, lands you back at the original point.

The converse is more delicate. First we show that if $T : \mathbb{R}^n \to \mathbb{R}^m$ is an invertible linear transformation, then $T^{-1} : \mathbb{R}^m \to \mathbb{R}^n$ is also linear. Let $\vec{\mathbf{y}}_1$ and $\vec{\mathbf{y}}_2$ be two elements of $\mathbb{R}^m$. Then

$$T\Big(aT^{-1}(\vec{\mathbf{y}}_1) + bT^{-1}(\vec{\mathbf{y}}_2)\Big) = aT\Big(T^{-1}(\vec{\mathbf{y}}_1)\Big) + bT\Big(T^{-1}(\vec{\mathbf{y}}_2)\Big) = a\vec{\mathbf{y}}_1 + b\vec{\mathbf{y}}_2$$
$$= T \circ T^{-1}(a\vec{\mathbf{y}}_1 + b\vec{\mathbf{y}}_2). \qquad 1.3.21$$

Since T is one to one, we deduce from this equation that

In equation 1.3.23, the first id is the identity mapping from $\mathbb{R}^m$ to itself, and the second is the identity mapping from $\mathbb{R}^n$ to itself.

$$aT^{-1}(\vec{\mathbf{y}}_1) + bT^{-1}(\vec{\mathbf{y}}_2) = T^{-1}(a\vec{\mathbf{y}}_1 + b\vec{\mathbf{y}}_2). \qquad 1.3.22$$

Thus if $T : \mathbb{R}^n \to \mathbb{R}^m$ is invertible as a mapping, its inverse mapping T^{-1} is linear and has a matrix $[T^{-1}]$. We now see that

$$[T][T^{-1}] = [TT^{-1}] = [\mathrm{id}] = I_m, \qquad [T^{-1}][T] = [T^{-1}T] = [\mathrm{id}] = I_n. \qquad 1.3.23$$

Thus the matrix $[T]$ is invertible, and its inverse is $[T]^{-1} = [T^{-1}]$. □

We will see in Corollary 2.2.7 that only square matrices can have inverses; thus for a linear transformation $T : \mathbb{R}^n \to \mathbb{R}^m$ to be invertible, we must have $m = n$.[7]

Affine maps

Both affine and linear transformations give as their output first-degree polynomials in their inputs.

An affine function is not linear unless the constant term is 0 (but linear functions are affine).

You probably met $y = ax + b$ in middle school; it expresses the line of slope a and y-intercept b as the graph of an affine function.

Do not confuse linear maps with maps of degree 1. For instance, lines are graphs of functions of the form $f(x) = ax + b$, and planes are graphs of functions of the form $f\begin{pmatrix} x \\ y \end{pmatrix} = ax + by + c$. Such maps are called *affine*.

Definition 1.3.14 (Affine function, affine subspace). A function $\vec{\mathbf{f}} : \mathbb{R}^n \to \mathbb{R}^m$ is *affine* if the function $\vec{\mathbf{x}} \mapsto \vec{\mathbf{f}}(\vec{\mathbf{x}}) - \vec{\mathbf{f}}(\vec{\mathbf{0}})$ is linear. The image of an affine function is called an *affine subspace*.

A transformation that would be linear if the image of the origin were translated back to the origin is affine. For example, the map

$$\begin{pmatrix} x \\ y \end{pmatrix} \mapsto \begin{pmatrix} x - y + 2 \\ 2x + y + 1 \end{pmatrix} \quad \text{is affine.}$$

[7] It might appear obvious that for a map $f : \mathbb{R}^n \to \mathbb{R}^m$ to be invertible, we must have $m = n$. For example, there appear to be more points in the plane than in a line, so establishing a one-to-one correspondence would appear to be impossible. This is false; Exercise 0.6.7 asked you to establish such a set-theoretic correspondence. Thus it is possible to have an invertible mapping $\mathbb{R} \to \mathbb{R}^2$. (This mapping is wildly nonlinear, even wildly discontinuous.) Surprisingly enough, especially in light of Peano curves (Exercise A1.5), we must have $m = n$ if $f : \mathbb{R}^n \to \mathbb{R}^m$ is continuous and invertible; in that case, the inverse is automatically continuous. This is a famous result, called the *invariance of domain theorem*; it is due to Brouwer. The proof requires algebraic topology.

Exercises for Section 1.3

$$A = \begin{bmatrix} 1 & 3 & 0 & 1 \\ 0 & 3 & 1 & 5 \\ 1 & 2 & 0 & 1 \end{bmatrix}$$

$$B = \begin{bmatrix} a_1 & b_1 \\ a_2 & b_2 \\ a_3 & b_3 \\ a_4 & b_4 \\ a_5 & b_5 \end{bmatrix}$$

$$C = \begin{bmatrix} \pi & 1 & 0 & \sqrt{2} \\ 0 & -1 & 2 & 1 \end{bmatrix}$$

$$D = \begin{bmatrix} 1 & 0 & -2 & 5 \end{bmatrix}$$

Matrices for Exercise 1.3.3

1.3.1 a. Give an example of a linear transformation $T : \mathbb{R}^4 \to \mathbb{R}^2$.

b. Give an example of a linear transformation $T : \mathbb{R}^3 \to \mathbb{R}$.

1.3.2 For the following linear transformations, what must be the dimensions of the corresponding matrix? Remember that a 3×2 matrix is 3 tall and 2 wide.

a. $T : \mathbb{R}^2 \to \mathbb{R}^3$ b. $T : \mathbb{R}^3 \to \mathbb{R}^3$ c. $T : \mathbb{R}^4 \to \mathbb{R}^2$ d. $T : \mathbb{R}^4 \to \mathbb{R}$

1.3.3 For the matrices A–D at left, what are the domain and codomain of the corresponding transformation?

1.3.4 a. Let T be a linear transformation such that $T\begin{bmatrix} v_1 \\ v_2 \\ v_3 \end{bmatrix} = \begin{bmatrix} 2v_1 \\ v_2 \\ v_3 \end{bmatrix}$. What is its matrix?

b. Repeat part a for $T\begin{bmatrix} v_1 \\ v_2 \\ v_3 \end{bmatrix} = \begin{bmatrix} v_2 \\ v_1 + 2v_2 \\ v_3 + v_1 \end{bmatrix}$.

1.3.5 For a class of 150 students, grades on a midterm exam, 10 homework assignments, and the final were entered as a matrix A, each row corresponding to a student, the first column giving the grade on the mid-term, the next 10 columns giving grades on the homeworks, and the last column corresponding to the grade on the final. The final counts for 50%, the midterm for 25%, and each homework for 2.5% of the final grade. What matrix operation should one perform to assign to each student his or her final grade?

a. The increase in height of a child from birth to age 18.

b. "You get what you pay for."

c. The value of a bank account at 5% interest, compounded daily, as a function of time.

d. "Two can live as cheaply as one."

e. "Cheaper by the dozen"

For Exercise 1.3.6

1.3.6 Which of the expressions a–e in the margin are linear?

1.3.7 Let $T : \mathbb{R}^3 \to \mathbb{R}^4$ be a linear transformation such that

$$T\begin{bmatrix} 0 \\ 1 \\ 1 \end{bmatrix} = \begin{bmatrix} -1 \\ 3 \\ 4 \\ 1 \end{bmatrix}, \quad T\begin{bmatrix} 1 \\ 0 \\ 0 \end{bmatrix} = \begin{bmatrix} 3 \\ 1 \\ 2 \\ 1 \end{bmatrix}, \quad T\begin{bmatrix} 1 \\ 1 \\ 1 \end{bmatrix} = \begin{bmatrix} 2 \\ 4 \\ 6 \\ 2 \end{bmatrix} \quad T\begin{bmatrix} 1 \\ 1 \\ 0 \end{bmatrix} = \begin{bmatrix} 2 \\ 2 \\ 5 \\ 1 \end{bmatrix}$$

$$T\begin{bmatrix} 1 \\ 0 \\ 1 \end{bmatrix} = \begin{bmatrix} 3 \\ 3 \\ 3 \\ 2 \end{bmatrix}, \quad T\begin{bmatrix} 0 \\ 1 \\ 0 \end{bmatrix} = \begin{bmatrix} -1 \\ 1 \\ 3 \\ 0 \end{bmatrix} \quad T\begin{bmatrix} -1 \\ 1 \\ 0 \end{bmatrix} = \begin{bmatrix} -4 \\ 0 \\ 1 \\ -1 \end{bmatrix} \quad T\begin{bmatrix} 0 \\ 0 \\ 1 \end{bmatrix} = \begin{bmatrix} 0 \\ 2 \\ 1 \\ 1 \end{bmatrix}.$$

How much of this information do you need to determine the matrix of T? What is that matrix?

1.3.8 a. Suppose you have a partner who is given the matrix of a linear transformation $T : \mathbb{R}^5 \to \mathbb{R}^6$. He is not allowed to tell you what the matrix is, but he can answer questions about what it does. In order to reconstitute the matrix, how many questions do you need answered? What are they?

b. Repeat, for a linear transformation $T : \mathbb{R}^6 \to \mathbb{R}^5$.

c. In each case, is there only one right answer to the question, "what are they?"

1.3.9 Evaluated on the five vectors below, the transformation T gives

$$T\begin{bmatrix} 1 \\ 0 \\ 0 \end{bmatrix} = \begin{bmatrix} 2 \\ 1 \\ 1 \end{bmatrix}, \; T\begin{bmatrix} 0 \\ 1 \\ 0 \end{bmatrix} = \begin{bmatrix} 1 \\ 2 \\ 1 \end{bmatrix}, \; T\begin{bmatrix} 0 \\ 0 \\ 1 \end{bmatrix} = \begin{bmatrix} 1 \\ 0 \\ 1 \end{bmatrix}, \; T\begin{bmatrix} 1 \\ 1 \\ 1 \end{bmatrix} = \begin{bmatrix} 4 \\ 3 \\ 3 \end{bmatrix}, \; T\begin{bmatrix} 2 \\ -1 \\ 4 \end{bmatrix} = \begin{bmatrix} 7 \\ 0 \\ 4 \end{bmatrix}.$$

Can T be linear? Justify your answer.

1.3.10 Is there a linear transformation $T : \mathbb{R}^3 \to \mathbb{R}^3$ such that

$$T\begin{bmatrix}1\\0\\0\end{bmatrix} = \begin{bmatrix}3\\0\\1\end{bmatrix}, \quad T\begin{bmatrix}1\\1\\0\end{bmatrix} = \begin{bmatrix}4\\2\\4\end{bmatrix}, \quad T\begin{bmatrix}1\\1\\1\end{bmatrix} = \begin{bmatrix}2\\3\\3\end{bmatrix}?$$

If so, what is its matrix?

1.3.11 The transformation in the plane that rotates by θ clockwise around the origin is a linear transformation. What is its matrix?

1.3.12 a. What is the matrix of the linear transformation $S : \mathbb{R}^3 \to \mathbb{R}^3$ that corresponds to reflection in the plane of equation $x = y$? What is the matrix corresponding to reflection $T : \mathbb{R}^3 \to \mathbb{R}^3$ in the plane $y = z$? What is the matrix of $S \circ T$?

b. What is the relationship between $[S \circ T]$ and $[T \circ S]$?

1.3.13 Let $A : \mathbb{R}^n \to \mathbb{R}^m$, $B : \mathbb{R}^m \to \mathbb{R}^k$, and $C : \mathbb{R}^k \to \mathbb{R}^n$ be linear transformations, with k, m, n all different. Which of the following make sense? For those that make sense, give the domain and codomain of the composition.

a. $A \circ B$ b. $C \circ B$ c. $A \circ C$ d. $B \circ A \circ C$ e. $C \circ A$

f. $B \circ C$ g. $B \circ A$ h. $A \circ C \circ B$ i. $C \circ B \circ A$ j. $B \circ C \circ A$

1.3.14 The unit square is the square in $\mathbb{R}^2$ with length 1 and bottom left corner at $\begin{pmatrix}0\\0\end{pmatrix}$. Let T be the translation that takes every point of this square and moves it to the corresponding point of the unit square with bottom left corner at $\begin{pmatrix}1\\1\end{pmatrix}$. Since it is a translation, it maps the bottom right corner to $\begin{pmatrix}2\\1\end{pmatrix}$. Give a formula for T. Is T linear? If so, find its matrix. If not, what goes wrong?

1.3.15 Prove part 1 of Theorem 1.3.4: show that the mapping from $\mathbb{R}^n$ to $\mathbb{R}^m$ described by the product $A\vec{\mathbf{v}}$ is indeed linear.

1.3.16 Use composition of transformations to derive from the transformation in Example 1.3.9 the fundamental theorems of trigonometry:

$$\cos(\theta_1 + \theta_2) = \cos\theta_1 \cos\theta_2 - \sin\theta_1 \sin\theta_2$$

$$\sin(\theta_1 + \theta_2) = \sin\theta_1 \cos\theta_2 + \cos\theta_1 \sin\theta_2.$$

1.3.17 Confirm (Example 1.3.5) by matrix multiplication that reflecting a point across the line, and then back again, lands you back at the original point.

1.3.18 Rotate the unit square by 45 degrees counterclockwise, then stretch it using the linear transformation $\begin{bmatrix}2 & 0\\0 & 1\end{bmatrix}$. Sketch the result.

1.3.19 If A and B are $n \times n$ matrices, their *Jordan product* is $\dfrac{AB + BA}{2}$. Show that this product is commutative but not associative.

1.3.20 Identify $\mathbb{R}^2$ to $\mathbb{C}$ by identifying $\begin{pmatrix}a\\b\end{pmatrix}$ to $z = a + ib$.

Show that the following mappings $\mathbb{C} \to \mathbb{C}$ are linear transformations, and give their matrices:

Exercise 1.3.20: Note that the symbol $\mapsto$ ("maps to") is different from the symbol $\to$ ("to"). While $\to$ describes the relationship between the domain and codomain of a mapping, as in $T : U \to V$, the symbol $\mapsto$ describes what a mapping does to a particular input. One could write $f(x) = x^2$ as $f : x \mapsto x^2$ (or just as $x \mapsto x^2$, but then we wouldn't have a name for the function).

a. $\text{Re} : z \mapsto \text{Re}\,(z)$ (the real part of z);

b. $\text{Im} : z \mapsto \text{Im}\,(z)$ (the imaginary part of z);

c. $c : z \mapsto \overline{z}$ (the complex conjugate of z, i.e., $\overline{z} = a - ib$ if $z = a + ib$);

d. $m_w : z \mapsto wz$, where $w = u + iv$ is a fixed complex number.

1.4 The geometry of $\mathbb{R}^n$

To acquire the feeling for calculus that is indispensable even in the most abstract speculations, one must have learned to distinguish that which is "big" from that which is "little," that which is "preponderant" and that which is "negligible."—Jean Dieudonné, *Calcul infinitésimal*

For a long time, the difficulty of visualizing higher dimensions hobbled mathematicians. In 1827, August Möbius, for whom the Möbius band is named, wrote that if two flat figures can be moved in space so that they coincide at every point, they are "equal and similar". (The two faces of Figure 1.3.4 are "equal and similar"; you could cut one out and place it on top of the other. But to achieve this you must move them out of the plane into $\mathbb{R}^3$.)

To speak of equal and similar objects in three dimensions, he wrote, one would have to be able to move them in four-dimensional space. "But since such space cannot be imagined, coincidence in this case is impossible."

By the end of the nineteenth century, mathematicians no longer felt such motions were impossible. You can turn a right-handed glove into a left-handed glove by moving it in $\mathbb{R}^4$.

Algebra is all about equalities. Calculus is about inequalities: about things being arbitrarily small or large, about some terms being dominant or negligible compared with others. Rather than saying that things are exactly true, we need to be able to say that they are almost true, so that they "become true in the limit". For example, $(5 + h)^3 = 125 + 75h + \cdots$, so if $h = .01$, we could use the approximation

$$(5.01)^3 \approx 125 + (75 \cdot .01) = 125.75. \qquad 1.4.1$$

The issue then is to quantify the error.

Such notions cannot be discussed in the language about $\mathbb{R}^n$ that has been developed so far: we need lengths of vectors to say that vectors are small, or that points are close to each other. We will also need lengths of matrices to say that linear transformations are "close" to each other. Having a notion of distance between transformations will be crucial in proving that under appropriate circumstances Newton's method converges to a solution (Section 2.8).

In this section we introduce these notions. The formulas are all more or less immediate generalizations of the Pythagorean theorem and the cosine law, but they acquire a whole new meaning in higher dimensions (and more yet in infinitely many dimensions).

The dot product

The dot product is the basic construct giving rise to all the geometric notions of lengths and angles. It is also known as the *standard inner product.*

Definition 1.4.1 (Dot product). The *dot product* $\vec{\mathbf{x}} \cdot \vec{\mathbf{y}}$ of two vectors $\vec{\mathbf{x}}, \vec{\mathbf{y}} \in \mathbb{R}^n$ is

$$\vec{\mathbf{x}} \cdot \vec{\mathbf{y}} = \begin{bmatrix} x_1 \\ x_2 \\ \vdots \\ x_n \end{bmatrix} \cdot \begin{bmatrix} y_1 \\ y_2 \\ \vdots \\ y_n \end{bmatrix} \stackrel{\text{def}}{=} x_1 y_1 + x_2 y_2 + \cdots + x_n y_n. \qquad 1.4.2$$

The symbol $\stackrel{\text{def}}{=}$ means "equal by definition". In contrast, the equalities in equation 1.4.3 can be justified by computations.

Example of the dot product:

$$\begin{bmatrix} 1 \\ 2 \\ 3 \end{bmatrix} \cdot \begin{bmatrix} 1 \\ 0 \\ 1 \end{bmatrix} = (1 \times 1) + (2 \times 0) + (3 \times 1) = 4.$$

The dot product of a vector $\vec{\mathbf{x}}$ with a standard basis vector $\vec{\mathbf{e}}_i$ selects the ith coordinate of $\vec{\mathbf{x}}$:

$$\vec{\mathbf{x}} \cdot \vec{\mathbf{e}}_i = x_i.$$

If A is an $n \times n$ matrix and $\vec{\mathbf{v}}$ and $\vec{\mathbf{w}}$ are vectors in $\mathbb{R}^n$, then

$$\vec{\mathbf{v}} \cdot A\vec{\mathbf{w}} = \sum_{i=1}^{n} \sum_{j=1}^{n} a_{i,j} v_i w_j.$$

The dot product is obviously commutative: $\vec{\mathbf{x}} \cdot \vec{\mathbf{y}} = \vec{\mathbf{y}} \cdot \vec{\mathbf{x}}$, and it is not much harder to check that it is distributive, that is,

$$\begin{aligned} \vec{\mathbf{x}} \cdot (\vec{\mathbf{y}}_1 + \vec{\mathbf{y}}_2) &= (\vec{\mathbf{x}} \cdot \vec{\mathbf{y}}_1) + (\vec{\mathbf{x}} \cdot \vec{\mathbf{y}}_2), \text{ and} \\ (\vec{\mathbf{x}}_1 + \vec{\mathbf{x}}_2) \cdot \vec{\mathbf{y}} &= (\vec{\mathbf{x}}_1 \cdot \vec{\mathbf{y}}) + (\vec{\mathbf{x}}_2 \cdot \vec{\mathbf{y}}). \end{aligned} \qquad 1.4.3$$

The dot product of two vectors can be written as the matrix product of the transpose of one vector by the other: $\vec{\mathbf{x}} \cdot \vec{\mathbf{y}} = \vec{\mathbf{x}}^\top \vec{\mathbf{y}} = \vec{\mathbf{y}}^\top \vec{\mathbf{x}}$;

$$\underbrace{\begin{bmatrix} x_1 \\ x_2 \\ \vdots \\ x_n \end{bmatrix} \cdot \begin{bmatrix} y_1 \\ y_2 \\ \vdots \\ y_n \end{bmatrix}}_{\vec{\mathbf{x}} \cdot \vec{\mathbf{y}}} \text{ is the same as } \begin{matrix} & \begin{bmatrix} y_1 \\ y_2 \\ \vdots \\ y_n \end{bmatrix} \\ \underbrace{\begin{bmatrix} x_1 & x_2 & \cdots & x_n \end{bmatrix}}_{\text{transpose } \vec{\mathbf{x}}^\top} & \underbrace{\begin{bmatrix} x_1 y_1 + x_2 y_2 + \cdots + x_n y_n \end{bmatrix}}_{\vec{\mathbf{x}}^\top \vec{\mathbf{y}}} \end{matrix}.$$

Conversely, the (i, j)th entry of the matrix product AB is the dot product of the jth column of B and the transpose of the ith row of A. For example, the second entry of the first row of AB below is 5:

$$\begin{matrix} & \overbrace{\begin{bmatrix} 1 & 3 \\ 1 & 1 \end{bmatrix}}^{B} \\ \underbrace{\begin{bmatrix} 1 & 2 \\ 3 & 4 \end{bmatrix}}_{A} & \underbrace{\begin{bmatrix} 3 & 5 \\ 7 & 13 \end{bmatrix}}_{AB} \end{matrix}; \qquad 5 = \underbrace{\begin{bmatrix} 1 \\ 2 \end{bmatrix}}_{\substack{\text{transpose,} \\ \text{1st row of } A}} \cdot \underbrace{\begin{bmatrix} 3 \\ 1 \end{bmatrix}}_{\substack{\text{2nd col.} \\ \text{of } B}}. \qquad 1.4.4$$

What we call the length of a vector is often called the *Euclidean norm*.

Some texts use double lines to denote the length of a vector: $\|\vec{\mathbf{v}}\|$ rather than $|\vec{\mathbf{v}}|$. We reserve double lines to denote the norm of a matrix, defined in Section 2.9.

Note that in one dimension, the length of a "vector" (i.e., number) coincides with its absolute value; the "length" of $\vec{\mathbf{v}} = [-2]$ is $\sqrt{2^2} = 2$.

Definition 1.4.2 (Length of a vector). The *length* $|\vec{\mathbf{x}}|$ of a vector $\vec{\mathbf{x}} \in \mathbb{R}^n$ is

$$|\vec{\mathbf{x}}| \stackrel{\text{def}}{=} \sqrt{\vec{\mathbf{x}} \cdot \vec{\mathbf{x}}} = \sqrt{x_1^2 + x_2^2 + \cdots + x_n^2}. \qquad 1.4.5$$

For example, the length $|\vec{\mathbf{v}}|$ of $\vec{\mathbf{v}} = \begin{bmatrix} 1 \\ 1 \\ 1 \end{bmatrix}$ is $\sqrt{1^2 + 1^2 + 1^2} = \sqrt{3}$.

Dividing a vector by its length to give it unit length is called *normalizing* the vector.

A vector of length 1 is called a *unit vector*. Given a vector $\vec{\mathbf{v}}$, it is easy to create a unit vector $\vec{\mathbf{w}}$ pointing in the same direction: simply divide $\vec{\mathbf{v}}$ by its length:

$$|\vec{\mathbf{w}}| = \left| \frac{\vec{\mathbf{v}}}{|\vec{\mathbf{v}}|} \right| = \frac{|\vec{\mathbf{v}}|}{|\vec{\mathbf{v}}|} = 1. \qquad 1.4.6$$

Length and dot product: geometric interpretation

In the plane and in space, $|\vec{\mathbf{x}}|$ is the ordinary distance between $\mathbf{0}$ and $\mathbf{x}$. As shown by Figure 1.4.1, this follows from the Pythagorean theorem.

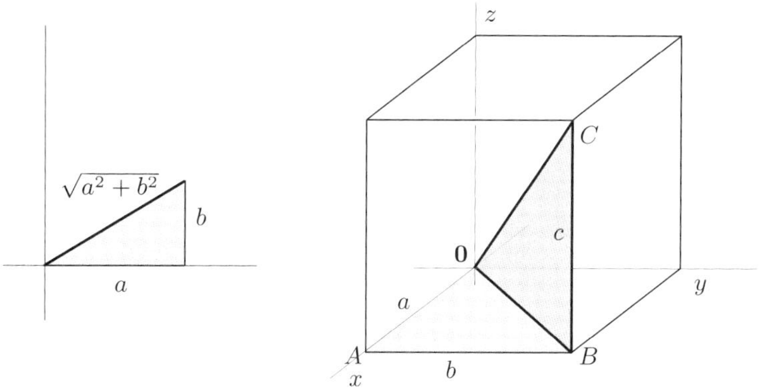

FIGURE 1.4.1. In the plane, the length of the vector with coordinates (a, b) is the ordinary distance from $\mathbf{0}$ to the point $\begin{pmatrix} a \\ b \end{pmatrix}$. In space, the length of the vector from $\mathbf{0}$ to C is the ordinary distance between $\mathbf{0}$ and C: $\mathbf{0}BC$ is a right triangle, the distance from $\mathbf{0}$ to B is $\sqrt{a^2+b^2}$ (applying the Pythagorean theorem to the pale triangle), and the distance from B to C is c, so (applying the Pythagorean theorem to the darker triangle) the distance from $\mathbf{0}$ to C is $\sqrt{a^2+b^2+c^2}$.

The dot product also has an interpretation in ordinary geometry:

Proposition 1.4.3 (Geometric interpretation of dot product). *Let $\vec{\mathbf{x}}, \vec{\mathbf{y}}$ be vectors in $\mathbb{R}^2$ or in $\mathbb{R}^3$, and let α be the angle between them. Then*

$$\vec{\mathbf{x}} \cdot \vec{\mathbf{y}} = |\vec{\mathbf{x}}|\,|\vec{\mathbf{y}}|\cos\alpha. \tag{1.4.7}$$

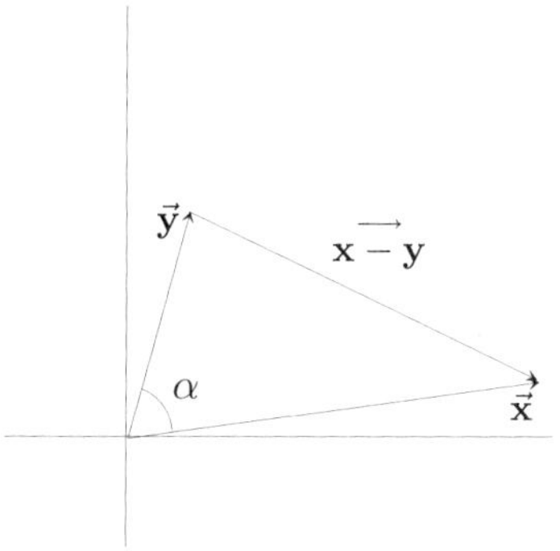

FIGURE 1.4.2.
The cosine law gives
$|\vec{\mathbf{x}}-\vec{\mathbf{y}}|^2 = |\vec{\mathbf{x}}|^2+|\vec{\mathbf{y}}|^2-2|\vec{\mathbf{x}}||\vec{\mathbf{y}}|\cos\alpha$.

Thus the dot product is independent of the coordinate system used: we can rotate a pair of vectors in the plane or in space without changing the dot product, since rotations don't change the lengths of vectors or the angle between them.

Proof. This is an application of the cosine law from trigonometry, which says that if you know all of a triangle's sides, or two sides and the angle between them, or two angles and a side, then you can determine the others. Let a triangle have sides of length a, b, c, and let γ be the angle opposite the side with length c. Then

$$c^2 = a^2 + b^2 - 2ab\cos\gamma. \qquad \text{(cosine law)} \tag{1.4.8}$$

Consider the triangle formed by the three vectors $\vec{\mathbf{x}}$, $\vec{\mathbf{y}}$, $\vec{\mathbf{x}}-\vec{\mathbf{y}}$; let α be the angle between $\vec{\mathbf{x}}$ and $\vec{\mathbf{y}}$ (see Figure 1.4.2). By the cosine law,

$$|\vec{\mathbf{x}}-\vec{\mathbf{y}}|^2 = |\vec{\mathbf{x}}|^2 + |\vec{\mathbf{y}}|^2 - 2|\vec{\mathbf{x}}||\vec{\mathbf{y}}|\cos\alpha. \tag{1.4.9}$$

But we can also write (remembering that the dot product is distributive)

$$\begin{aligned}|\vec{\mathbf{x}}-\vec{\mathbf{y}}|^2 &= (\vec{\mathbf{x}}-\vec{\mathbf{y}})\cdot(\vec{\mathbf{x}}-\vec{\mathbf{y}}) = \big((\vec{\mathbf{x}}-\vec{\mathbf{y}})\cdot\vec{\mathbf{x}}\big) - \big((\vec{\mathbf{x}}-\vec{\mathbf{y}})\cdot\vec{\mathbf{y}}\big)\\ &= (\vec{\mathbf{x}}\cdot\vec{\mathbf{x}}) - (\vec{\mathbf{y}}\cdot\vec{\mathbf{x}}) - (\vec{\mathbf{x}}\cdot\vec{\mathbf{y}}) + (\vec{\mathbf{y}}\cdot\vec{\mathbf{y}}) \qquad 1.4.10\\ &= (\vec{\mathbf{x}}\cdot\vec{\mathbf{x}}) + (\vec{\mathbf{y}}\cdot\vec{\mathbf{y}}) - 2\vec{\mathbf{x}}\cdot\vec{\mathbf{y}} = |\vec{\mathbf{x}}|^2 + |\vec{\mathbf{y}}|^2 - 2\vec{\mathbf{x}}\cdot\vec{\mathbf{y}}.\end{aligned}$$

Comparing these two equations gives

$$\vec{\mathbf{x}}\cdot\vec{\mathbf{y}} = |\vec{\mathbf{x}}||\vec{\mathbf{y}}|\cos\alpha. \qquad \square \qquad 1.4.11$$

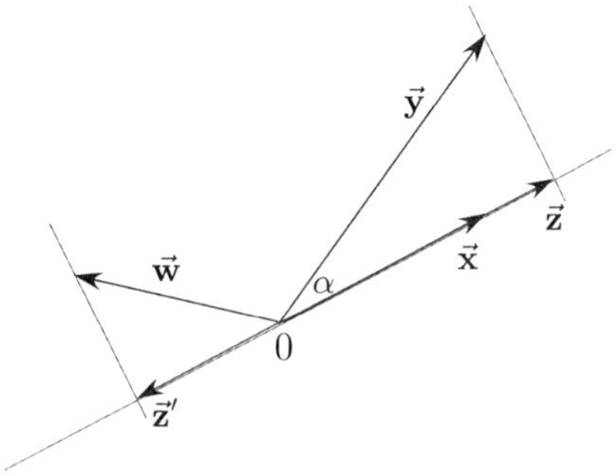

FIGURE 1.4.3.
The projection of $\vec{\mathbf{y}}$ onto the line spanned by $\vec{\mathbf{x}}$ is $\vec{\mathbf{z}}$, giving

$$\begin{aligned}|\vec{\mathbf{x}}\cdot\vec{\mathbf{y}}| &= |\vec{\mathbf{x}}||\vec{\mathbf{y}}|\,|\cos\alpha|\\ &= |\vec{\mathbf{x}}||\vec{\mathbf{y}}|\frac{|\vec{\mathbf{z}}|}{|\vec{\mathbf{y}}|} = |\vec{\mathbf{x}}||\vec{\mathbf{z}}|.\end{aligned}$$

The signed length of $\vec{\mathbf{z}}$ is positive, since $\vec{\mathbf{z}}$ points in the direction of $\vec{\mathbf{x}}$. The projection of $\vec{\mathbf{w}}$ onto the lined spanned by $\vec{\mathbf{x}}$ is $\vec{\mathbf{z}}'$; its signed length is negative.

Corollary 1.4.4 restates Proposition 1.4.3 in terms of projections; it is illustrated by Figure 1.4.3. By "line spanned by $\vec{\mathbf{x}}$" we mean the line formed of all multiples of $\vec{\mathbf{x}}$.

Corollary 1.4.4 (The dot product in terms of projections). *If $\vec{\mathbf{x}}$ and $\vec{\mathbf{y}}$ are two vectors in $\mathbb{R}^2$ or $\mathbb{R}^3$, then $\vec{\mathbf{x}}\cdot\vec{\mathbf{y}}$ is the product of $|\vec{\mathbf{x}}|$ and the signed length of the projection of $\vec{\mathbf{y}}$ onto the line spanned by $\vec{\mathbf{x}}$. The signed length of the projection is positive if it points in the direction of $\vec{\mathbf{x}}$; it is negative if it points in the opposite direction.*

Defining angles between vectors in $\mathbb{R}^n$

We want to use equation 1.4.7 to provide a definition of angles in $\mathbb{R}^n$, where we can't invoke elementary geometry when $n > 3$. Thus, we want to define

$$\alpha = \arccos\frac{\vec{\mathbf{v}}\cdot\vec{\mathbf{w}}}{|\vec{\mathbf{v}}||\vec{\mathbf{w}}|}, \quad \text{i.e., define } \alpha \text{ so that} \quad \cos\alpha = \frac{\vec{\mathbf{v}}\cdot\vec{\mathbf{w}}}{|\vec{\mathbf{v}}||\vec{\mathbf{w}}|}. \qquad 1.4.12$$

But there's a problem: how do we know that

$$-1 \le \frac{\vec{\mathbf{v}}\cdot\vec{\mathbf{w}}}{|\vec{\mathbf{v}}||\vec{\mathbf{w}}|} \le 1, \qquad 1.4.13$$

so that the arccosine exists? *Schwarz's inequality* provides the answer.[8] It is an absolutely fundamental result about dot products.

FIGURE 1.4.4.
Hermann Schwarz (1843–1921) often translated geometrical ideas into the language of analysis. In addition to his work as professor at the University of Berlin, he was captain of the local volunteer fire brigade.

Theorem 1.4.5 (Schwarz's inequality). *For any vectors $\vec{\mathbf{v}},\vec{\mathbf{w}}\in\mathbb{R}^n$,*

$$|\vec{\mathbf{v}}\cdot\vec{\mathbf{w}}| \le |\vec{\mathbf{v}}|\,|\vec{\mathbf{w}}|. \qquad 1.4.14$$

The two sides are equal if and only if $\vec{\mathbf{v}}$ or $\vec{\mathbf{w}}$ is a multiple of the other by a scalar.

Proof. If either $\vec{\mathbf{v}}$ or $\vec{\mathbf{w}}$ is $\vec{\mathbf{0}}$, the statement is obvious, so suppose both are nonzero. Consider the function $|\vec{\mathbf{v}}+t\vec{\mathbf{w}}|^2$ as a function of t. It is a second-degree polynomial of the form at^2+bt+c:

$$|t\vec{\mathbf{w}}+\vec{\mathbf{v}}|^2 = |\vec{\mathbf{w}}|^2t^2 + 2(\vec{\mathbf{v}}\cdot\vec{\mathbf{w}})t + |\vec{\mathbf{v}}|^2. \qquad 1.4.15$$

[8] A more abstract form of Schwarz's inequality concerns inner products of vectors in possibly infinite-dimensional vector spaces, not just the standard dot product in $\mathbb{R}^n$. The general case is no more difficult to prove than this version involving the dot product.

FIGURE 1.4.5.
Viktor Bunyakovsky (1804–1889)

The Schwarz inequality is also known as the Cauchy-Bunyakovsky-Schwarz and the Cauchy-Schwarz inequality. Bunyakovsky published the result in 1859, 25 years before Schwarz.

All its values are ≥ 0, since the left side is a square; therefore, the graph of the polynomial must not cross the t-axis. But remember from high school the quadratic formula: for an equation of the form $at^2 + bt + c = 0$,

$$t = \frac{-b \pm \sqrt{b^2 - 4ac}}{2\,a}. \qquad 1.4.16$$

If the *discriminant* (the quantity $b^2 - 4ac$ under the square root sign) is positive, the equation will have two distinct solutions, and its graph will cross the t-axis twice, as shown in the leftmost graph in Figure 1.4.6.

The discriminant of equation 1.4.15 (substituting $|\vec{\mathbf{w}}|^2$ for a, $2\vec{\mathbf{v}} \cdot \vec{\mathbf{w}}$ for b, and $|\vec{\mathbf{v}}|^2$ for c) is

$$4(\vec{\mathbf{v}} \cdot \vec{\mathbf{w}})^2 - 4|\vec{\mathbf{v}}|^2|\vec{\mathbf{w}}|^2. \qquad 1.4.17$$

Since the graph of equation 1.4.15 does not cross the t-axis, the discriminant cannot be positive, so

$$4(\vec{\mathbf{v}} \cdot \vec{\mathbf{w}})^2 - 4|\vec{\mathbf{v}}|^2|\vec{\mathbf{w}}|^2 \leq 0, \quad \text{and therefore} \quad |\vec{\mathbf{v}} \cdot \vec{\mathbf{w}}| \leq |\vec{\mathbf{v}}||\vec{\mathbf{w}}|, \qquad 1.4.18$$

which is what we wanted to show.

The second part of Schwarz's inequality, that $|\vec{\mathbf{v}} \cdot \vec{\mathbf{w}}| = |\vec{\mathbf{v}}|\,|\vec{\mathbf{w}}|$ if and only if $\vec{\mathbf{v}}$ or $\vec{\mathbf{w}}$ is a multiple of the other by a scalar, has two directions. If $\vec{\mathbf{w}}$ is a multiple of $\vec{\mathbf{v}}$, say $\vec{\mathbf{w}} = t\vec{\mathbf{v}}$, then

$$|\vec{\mathbf{v}} \cdot \vec{\mathbf{w}}| = |t||\vec{\mathbf{v}}|^2 = (|\vec{\mathbf{v}}|)(|t||\vec{\mathbf{v}}|) = |\vec{\mathbf{v}}||\vec{\mathbf{w}}|, \qquad 1.4.19$$

so that Schwarz's inequality is satisfied as an equality.

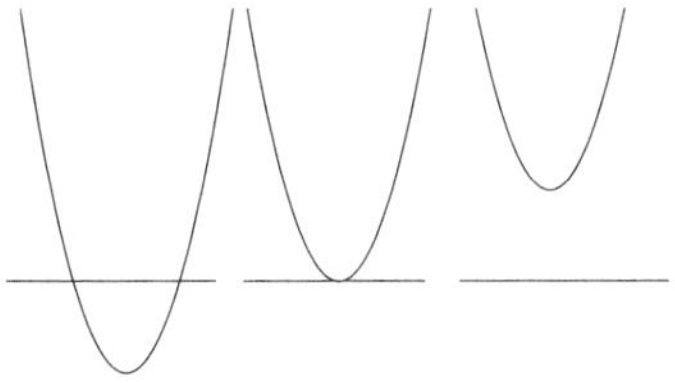

FIGURE 1.4.6.
LEFT TO RIGHT: A positive discriminant gives two roots; a zero discriminant gives one root; a negative discriminant gives no roots.

Conversely, if $|\vec{\mathbf{v}} \cdot \vec{\mathbf{w}}| = |\vec{\mathbf{v}}||\vec{\mathbf{w}}|$, then the discriminant in formula 1.4.17 is zero, so the polynomial has a single root t_0:

$$|\vec{\mathbf{v}} + t_0\vec{\mathbf{w}}|^2 = 0, \quad (\text{i.e.,} \quad \vec{\mathbf{v}} = -t_0\vec{\mathbf{w}}) \qquad 1.4.20$$

and $\vec{\mathbf{v}}$ is a multiple of $\vec{\mathbf{w}}$. □

The proof of Schwarz's inequality is clever; you can follow it line by line, but you won't find it by simply following your nose.

Schwarz's inequality allows us to define the angle between two vectors, since we are now assured that

$$-1 \leq \frac{\vec{\mathbf{a}} \cdot \vec{\mathbf{b}}}{|\vec{\mathbf{a}}||\vec{\mathbf{b}}|} \leq 1. \qquad 1.4.21$$

Definition 1.4.6 (The angle between two vectors). The *angle* between two vectors $\vec{\mathbf{v}}$ and $\vec{\mathbf{w}}$ in $\mathbb{R}^n$ is that angle α satisfying $0 \leq \alpha \leq \pi$ such that

$$\cos\alpha = \frac{\vec{\mathbf{v}} \cdot \vec{\mathbf{w}}}{|\vec{\mathbf{v}}|\,|\vec{\mathbf{w}}|}. \qquad 1.4.22$$

The dot product of two vectors is positive if the angle between them is less than $\pi/2$, and negative if it is bigger than $\pi/2$.

Example 1.4.7 (Finding an angle). What is the angle between the diagonal of a cube and any edge? Let us assume that our cube is the unit cube $0 \le x, y, z \le 1$, so that the standard basis vectors $\vec{\mathbf{e}}_1, \vec{\mathbf{e}}_2, \vec{\mathbf{e}}_3$ are edges, and the vector $\vec{\mathbf{d}} = \begin{bmatrix} 1 \\ 1 \\ 1 \end{bmatrix}$ is a diagonal. The length of the diagonal is $|\vec{\mathbf{d}}| = \sqrt{3}$, so the required angle α satisfies

$$\cos \alpha = \frac{\vec{\mathbf{d}} \cdot \vec{\mathbf{e}}_i}{|\vec{\mathbf{d}}||\vec{\mathbf{e}}_i|} = \frac{1}{\sqrt{3}}. \qquad 1.4.23$$

Thus $\alpha = \arccos \sqrt{3}/3 \approx 54.7°$. △

One student asked, "When $n > 3$, how can n vectors be orthogonal, or is that some weird math thing you just can't visualize?" We can't visualize four (or 17, or 98) vectors all perpendicular to each other. But we still find the mental image of a right angle helpful. To determine that two vectors are orthogonal, we check that their dot product is 0. But we think of orthogonal vectors $\vec{\mathbf{v}}, \vec{\mathbf{w}}$ as being at right angles to each other; we don't think $v_1w_1 + \cdots + v_nw_n = 0$.

We prefer the word "orthogonal" to its synonym "perpendicular". Orthogonal comes from the Greek for "right angle," while perpendicular comes from the Latin for "plumb line," which suggests a vertical line. The word "normal" is also used, both as noun and as adjective, to express a right angle.

Corollary 1.4.8. *Two vectors are orthogonal (form a right angle) if their dot product is zero.*

Schwarz's inequality also gives us the *triangle inequality*: to travel from London to Paris, it is shorter to go across the English Channel than by way of Moscow. We give two variants of the inequality on pages 109 and 113.

Theorem 1.4.9 (The triangle inequality). *For vectors* $\vec{\mathbf{x}}, \vec{\mathbf{y}} \in \mathbb{R}^n$,

$$|\vec{\mathbf{x}} + \vec{\mathbf{y}}| \le |\vec{\mathbf{x}}| + |\vec{\mathbf{y}}|. \qquad 1.4.24$$

Proof. This inequality is proved by the following computation:

$$|\vec{\mathbf{x}}+\vec{\mathbf{y}}|^2 = |\vec{\mathbf{x}}|^2+2\vec{\mathbf{x}}\cdot\vec{\mathbf{y}}+|\vec{\mathbf{y}}|^2 \underset{\text{Schwarz}}{\le} |\vec{\mathbf{x}}|^2+2|\vec{\mathbf{x}}||\vec{\mathbf{y}}|+|\vec{\mathbf{y}}|^2 = \big(|\vec{\mathbf{x}}|+|\vec{\mathbf{y}}|\big)^2, \qquad 1.4.25$$

so that $|\vec{\mathbf{x}} + \vec{\mathbf{y}}| \le |\vec{\mathbf{x}}| + |\vec{\mathbf{y}}|$. □

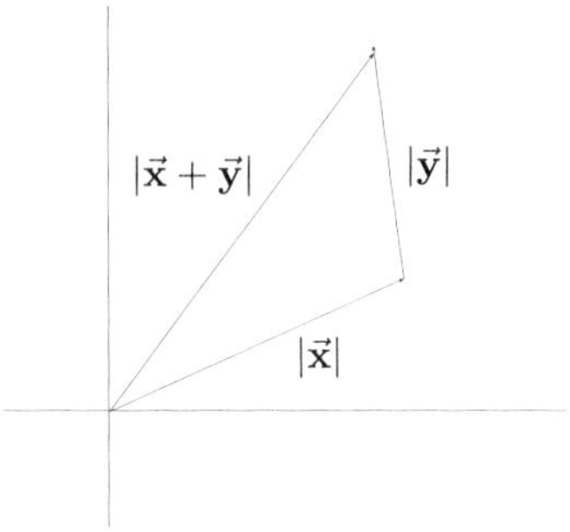

FIGURE 1.4.7. The triangle inequality: $|\vec{\mathbf{x}} + \vec{\mathbf{y}}| < |\vec{\mathbf{x}}| + |\vec{\mathbf{y}}|$.

The triangle inequality can be interpreted (in the case of strict inequality, not $\le$) as the statement that the length of one side of a triangle is less than the sum of the lengths of the other two sides, as shown in Figure 1.4.7.

Measuring matrices

The dot product gives a way to measure the length of a vector. We will also need a way to measure the "length" of a matrix (not to be confused with either its height or its width). There is an obvious way to do this: consider an $m \times n$ matrix as a point in $\mathbb{R}^{nm}$, and use the ordinary dot product.

Definition 1.4.10 (The length of a matrix). If A is an $n \times m$ matrix, its length $|A|$ is the square root of the sum of the squares of all its entries:

$$|A|^2 \stackrel{\text{def}}{=} \sum_{i=1}^{n}\sum_{j=1}^{m} a_{i,j}^2. \qquad 1.4.26$$

For example, the length $|A|$ of the matrix $A = \begin{bmatrix} 1 & 2 \\ 0 & 1 \end{bmatrix}$ is $\sqrt{6}$, since $1+4+0+1=6$. What is the length of the matrix $B = \begin{bmatrix} 1 & 2 & 1 \\ 1 & 0 & 3 \end{bmatrix}$?[9]

If you find double sum notation confusing, equation 1.4.26 can be rewritten as a single sum:

$$|A|^2 = \sum_{\substack{i=1,\dots,n \\ j=1,\dots,m}} a_{i,j}^2. \text{ (We sum all } a_{i,j}^2 \text{ for } i \text{ from 1 to } n \text{ and } j \text{ from 1 to } m.)$$

In some texts, $|A|$ denotes the determinant of the matrix A. We use $\det A$ to denote the determinant.

The length $|A|$ is also called the *Frobenius norm* (and the *Schur norm* and *Hilbert-Schmidt norm*). We find it simpler to call it the length, generalizing the notion of length of a vector. Indeed, the length of an $n \times 1$ matrix is identical to the length of the vector in $\mathbb{R}^n$ with the same entries. (And the length of the 1×1 matrix consisting of the single entry $[n]$ is the absolute value of n.)

You shouldn't take the word "length" too literally; it's just a name for one way to measure matrices. (A more sophisticated measure, considerably harder to compute, is discussed in Section 2.9.)

If we think of an $n \times m$ matrix as a point in $\mathbb{R}^{nm}$, we see that two matrices A and B (and the corresponding linear transformations) are close if the length of their difference is small (i.e., if $|A - B|$ is small).

Length and matrix multiplication

We said at the beginning of Section 1.2 that the point of writing the entries of $\mathbb{R}^{mn}$ as matrices is to allow matrix multiplication, yet it isn't clear that this notion of length, in which a matrix is considered as a list of numbers, is related to matrix multiplication. The following proposition says that it is. This will soon become an old friend; it is a very useful tool in a number of proofs. It is really a corollary of Schwarz's lemma, Theorem 1.4.5.

Proposition 1.4.11 (Lengths of products). *Let A be an $n \times m$ matrix, B an $m \times k$ matrix, and $\vec{\mathbf{a}}$ a vector in $\mathbb{R}^m$. Then*

$$|A\vec{\mathbf{b}}| \le |A||\vec{\mathbf{b}}|. \qquad 1.4.27$$

$$|AB| \le |A|\,|B| \qquad 1.4.28$$

Inequality 1.4.27 is the special case of inequality 1.4.28 where $k = 1$, but the intuitive content is sufficiently different that we state the two parts separately. The proof of the second part follows from the first.

Example 1.4.12. Let $A = \begin{bmatrix} 2 & 1 & 3 \\ 1 & 0 & 2 \\ 1 & 2 & 1 \end{bmatrix}, \vec{\mathbf{b}} = \begin{bmatrix} 1 \\ 1 \\ 1 \end{bmatrix}, \ B = \begin{bmatrix} 1 & 0 & 1 \\ 1 & 2 & 0 \\ 1 & 1 & 0 \end{bmatrix}$.

Then $|A| = 5$, $|\vec{\mathbf{b}}| = \sqrt{3}$, and $|B| = 3$, giving

$$|A\vec{\mathbf{b}}| = \sqrt{61} \approx 7.8 < |A||\vec{\mathbf{b}}| = 5\sqrt{3} \approx 8.7$$

$$|AB| = 11 < |A||B| = 15. \qquad \triangle$$

Proof. In inequality 1.4.27, note that if A consists of a single row (i.e., if $A = \vec{\mathbf{a}}^\top$ is the transpose of a vector $\vec{\mathbf{a}}$), the theorem is Schwarz's inequality:

$$\underbrace{|A\vec{\mathbf{b}}|}_{|\vec{\mathbf{a}}^\top \vec{\mathbf{b}}|} = |\vec{\mathbf{a}} \cdot \vec{\mathbf{b}}| \le |\vec{\mathbf{a}}|\,|\vec{\mathbf{b}}| = \underbrace{|A||\vec{\mathbf{b}}|}_{|\vec{\mathbf{a}}^\top||\vec{\mathbf{b}}|}. \qquad 1.4.29$$

[9] $|B| = 4$, since $1+4+0+1+1+9 = 16$.

We will consider the rows of A as transposes of vectors $\vec{\mathbf{a}}_1, \dots, \vec{\mathbf{a}}_n$, and we will apply the argument above to each row separately; see Figure 1.4.8.

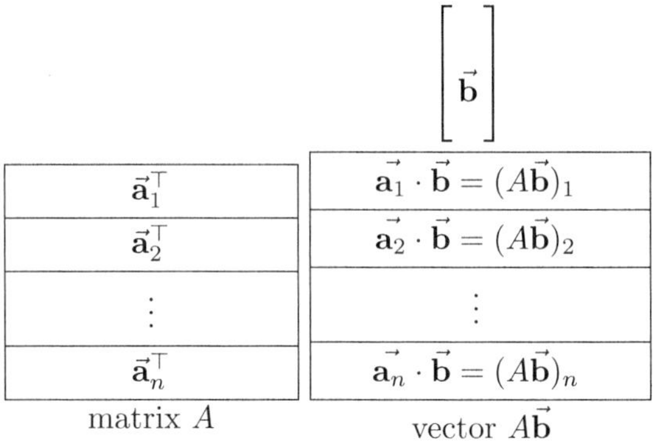

FIGURE 1.4.8. The product $\vec{\mathbf{a}}_i^{\top}\vec{\mathbf{b}}$ is the same as the dot product $\vec{\mathbf{a}}_i \cdot \vec{\mathbf{b}}$. Note that $A\vec{\mathbf{b}}$ is a vector.

Since the ith row of A is $\vec{\mathbf{a}}_i^\top$, the ith entry $(A\vec{\mathbf{b}})_i$ of $A\vec{\mathbf{b}}$ is the dot product $\vec{\mathbf{a}}_i \cdot \vec{\mathbf{b}}$. This accounts for the equality marked (1) in equation 1.4.30. This leads to

$$|A\vec{\mathbf{b}}|^2 = \sum_{i=1}^{n} (A\vec{\mathbf{b}})_i^2 \underset{(1)}{=} \sum_{i=1}^{n} (\vec{\mathbf{a}}_i \cdot \vec{\mathbf{b}})^2. \qquad 1.4.30$$

Now use Schwarz's inequality, marked (2) in formula 1.4.31; factor out $|\vec{\mathbf{b}}|^2$ (step 3), and consider (step 4) the length squared of A to be the sum of the squares of the lengths of $\vec{\mathbf{a}}_i$. (Of course, $|\vec{\mathbf{a}}_i|^2 = |\vec{\mathbf{a}}_i^\top|^2$.) Thus,

$$\sum_{i=1}^{n} (\vec{\mathbf{a}}_i \cdot \vec{\mathbf{b}})^2 \underset{(2)}{\le} \sum_{i=1}^{n} |\vec{\mathbf{a}}_i|^2 \, |\vec{\mathbf{b}}|^2 \underset{(3)}{=} \left(\sum_{i=1}^{n} |\vec{\mathbf{a}}_i|^2 \right) |\vec{\mathbf{b}}|^2 \underset{(4)}{=} |A|^2 |\vec{\mathbf{b}}|^2. \qquad 1.4.31$$

This gives us the result we wanted:

$$|A\vec{\mathbf{b}}|^2 \le |A|^2 |\vec{\mathbf{b}}|^2. \qquad 1.4.32$$

For formula 1.4.28, we decompose the matrix B into its columns and proceed as above. Let $\vec{\mathbf{b}}_1, \dots, \vec{\mathbf{b}}_k$ be the columns of B. Then

$$|AB|^2 = \sum_{j=1}^{k} |A\vec{\mathbf{b}}_j|^2 \le \sum_{j=1}^{k} |A|^2 |\vec{\mathbf{b}}_j|^2 = |A|^2 \sum_{j=1}^{k} |\vec{\mathbf{b}}_j|^2 = |A|^2 |B|^2. \quad \square \qquad 1.4.33$$

The determinant and trace in $\mathbb{R}^2$

The *determinant* and *trace* are functions of square matrices: they take a square matrix as input and return a number. We define the determinant in $\mathbb{R}^3$ in Definition 1.4.15, but defining it in $\mathbb{R}^n$ is rather delicate; we discuss determinants in higher dimensions in Section 4.8. In contrast, the trace is easy to define in all dimensions: it is the sum of the diagonal entries.

Recall that the inverse of a 2×2 matrix $A = \begin{bmatrix} a & b \\ c & d \end{bmatrix}$ is

$$A^{-1} = \frac{1}{ad-bc}\begin{bmatrix} d & -b \\ -c & a \end{bmatrix}.$$

So a 2×2 matrix A is invertible if and only if $\det A \neq 0$.

Definition 1.4.13 (Determinant and trace in $\mathbb{R}^2$). The *determinant* det and *trace* tr of a 2×2 matrix $\begin{bmatrix} a_1 & b_1 \\ a_2 & b_2 \end{bmatrix}$ are given by

$$\det \begin{bmatrix} a_1 & b_1 \\ a_2 & b_2 \end{bmatrix} \overset{\text{def}}{=} a_1b_2 - a_2b_1 \quad \text{and} \quad \operatorname{tr} \begin{bmatrix} a_1 & b_1 \\ a_2 & b_2 \end{bmatrix} \overset{\text{def}}{=} a_1 + b_2. \qquad 1.4.34$$

If we think of the determinant as a function of the vectors $\vec{\mathbf{a}}$ and $\vec{\mathbf{b}}$ in $\mathbb{R}^2$, it has a geometric interpretation, illustrated by Figure 1.4.9. (The trace is harder to understand geometrically; it reappears in Theorem 4.8.15.)

Proposition 1.4.14 (Geometric interpretation of the determinant in $\mathbb{R}^2$).

1. *The area of the parallelogram spanned by the vectors $\vec{\mathbf{a}}$ and $\vec{\mathbf{b}}$ is $|\det[\vec{\mathbf{a}}, \vec{\mathbf{b}}]|$.*
2. *If $\mathbb{R}^2$ is drawn in the standard way, with $\vec{\mathbf{e}}_1$ clockwise from $\vec{\mathbf{e}}_2$, then $\det[\vec{\mathbf{a}}, \vec{\mathbf{b}}]$ is positive if and only if $\vec{\mathbf{a}}$ lies clockwise from $\vec{\mathbf{b}}$; it is negative if and only if $\vec{\mathbf{a}}$ lies counterclockwise from $\vec{\mathbf{b}}$.*

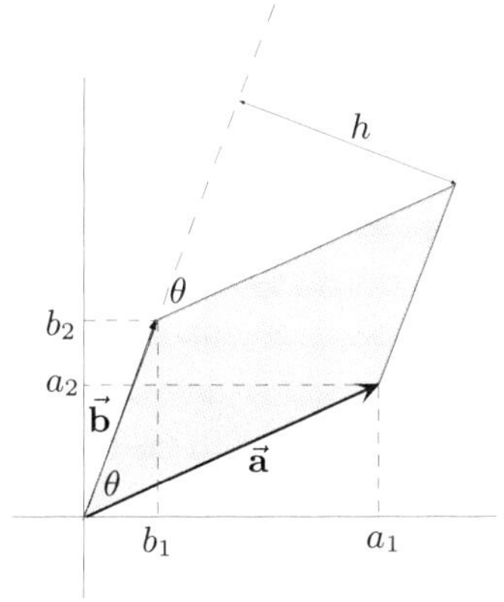

FIGURE 1.4.9.
We say that the shaded parallelogram above is "spanned" by the vectors

$$\vec{\mathbf{a}} = \begin{bmatrix} a_1 \\ a_2 \end{bmatrix} \quad \text{and} \quad \vec{\mathbf{b}} = \begin{bmatrix} b_1 \\ b_2 \end{bmatrix}.$$

Its area is $|\det[\vec{\mathbf{a}}, \vec{\mathbf{b}}]|$.

Exercise 1.4.18 gives a more geometric proof of part 1 of Proposition 1.4.14.

Proof. 1. The area of the parallelogram is height times base. We will choose as base $|\vec{\mathbf{b}}| = \sqrt{b_1^2 + b_2^2}$. If θ is the angle between $\vec{\mathbf{a}}$ and $\vec{\mathbf{b}}$, the height h of the parallelogram is

$$h = \sin\theta |\vec{\mathbf{a}}| = \sin\theta\sqrt{a_1^2 + a_2^2}. \qquad 1.4.35$$

To compute $\sin\theta$ we first compute $\cos\theta$ using equation 1.4.7:

$$\cos\theta = \frac{\vec{\mathbf{a}} \cdot \vec{\mathbf{b}}}{|\vec{\mathbf{a}}||\vec{\mathbf{b}}|} = \frac{a_1b_1 + a_2b_2}{\sqrt{a_1^2 + a_2^2}\sqrt{b_1^2 + b_2^2}}. \qquad 1.4.36$$

We then get $\sin\theta$ as follows:

$$\begin{aligned} \sin\theta &= \sqrt{1 - \cos^2\theta} = \sqrt{\frac{(a_1^2 + a_2^2)(b_1^2 + b_2^2) - (a_1b_1 + a_2b_2)^2}{(a_1^2 + a_2^2)(b_1^2 + b_2^2)}} \\ &= \sqrt{\frac{a_1^2b_1^2 + a_1^2b_2^2 + a_2^2b_1^2 + a_2^2b_2^2 - a_1^2b_1^2 - 2a_1b_1a_2b_2 - a_2^2b_2^2}{(a_1^2 + a_2^2)(b_1^2 + b_2^2)}} \qquad 1.4.37 \\ &= \sqrt{\frac{(a_1b_2 - a_2b_1)^2}{(a_1^2 + a_2^2)(b_1^2 + b_2^2)}}. \end{aligned}$$

When solving big systems of linear questions was out of the question, determinants were a reasonable approach to the theory of linear equations. With the advent of computers they lost importance: systems of linear equations can be solved far more effectively with row reduction, discussed in Chapter 2. But determinants have an interesting geometric interpretation; in Chapters 4–6, we use them constantly.

Using this value for $\sin\theta$ in the equation for the area of a parallelogram gives

$$\text{Area} = \underbrace{|\vec{\mathbf{b}}|}_{\text{base}}\ \underbrace{|\vec{\mathbf{a}}|\sin\theta}_{\text{height}} \qquad 1.4.38$$

$$= \underbrace{\sqrt{b_1^2+b_2^2}}_{\text{base}}\ \underbrace{\sqrt{a_1^2+a_2^2}\sqrt{\frac{(a_1b_2-a_2b_1)^2}{(a_1^2+a_2^2)(b_1^2+b_2^2)}}}_{\text{height}} = \underbrace{|a_1b_2-a_2b_1|}_{\text{determinant}}.$$

2. When we rotate the vector $\vec{\mathbf{a}}$ counterclockwise by $\pi/2$, it becomes $\vec{\mathbf{c}} = \begin{bmatrix}-a_2\\ a_1\end{bmatrix}$, and we see that $\vec{\mathbf{c}}\cdot\vec{\mathbf{b}} = \det[\vec{\mathbf{a}},\vec{\mathbf{b}}]$:

$$\begin{bmatrix}-a_2\\ a_1\end{bmatrix}\cdot\begin{bmatrix}b_1\\ b_2\end{bmatrix} = -a_2b_1 + a_1b_2 = \det\begin{bmatrix}a_1 & b_1\\ a_2 & b_2\end{bmatrix}. \qquad 1.4.39$$

Since (Proposition 1.4.3) the dot product of two vectors is positive if the angle between them is less than $\pi/2$, the determinant is positive if the angle between $\vec{\mathbf{b}}$ and $\vec{\mathbf{c}}$ is less than $\pi/2$. This corresponds to $\vec{\mathbf{b}}$ being counterclockwise from $\vec{\mathbf{a}}$, as shown in Figure 1.4.10. □

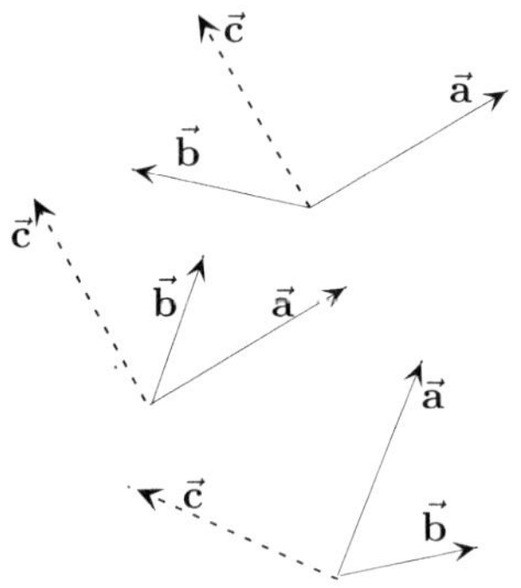

FIGURE 1.4.10.

In all three cases above, the angle between $\vec{\mathbf{a}}$ and $\vec{\mathbf{c}}$ is $\pi/2$. In the two cases at the top, the angle between $\vec{\mathbf{b}}$ and $\vec{\mathbf{c}}$ is less than $\pi/2$, so $\det(\vec{\mathbf{a}},\vec{\mathbf{b}}) > 0$; $\vec{\mathbf{b}}$ is counterclockwise from $\vec{\mathbf{a}}$. At the bottom, the angle between $\vec{\mathbf{b}}$ and $\vec{\mathbf{c}}$ is more than $\pi/2$; $\vec{\mathbf{b}}$ is clockwise from $\vec{\mathbf{a}}$, and $\det(\vec{\mathbf{a}},\vec{\mathbf{b}})$ is negative.

Determinants and cross products in $\mathbb{R}^3$

In Definition 1.4.15 we give the formula for the determinant of a 3×3 matrix. For larger matrices, the formulas rapidly get out of hand; we will see in Section 4.8 that such determinants can be computed much more reasonably by row (or column) reduction.

Definition 1.4.15 (Determinant in $\mathbb{R}^3$). The *determinant of a* 3×3 *matrix* is

$$\det\begin{bmatrix}a_1 & b_1 & c_1\\ a_2 & b_2 & c_2\\ a_3 & b_3 & c_3\end{bmatrix} \overset{\text{def}}{=} a_1\det\begin{bmatrix}b_2 & c_2\\ b_3 & c_3\end{bmatrix} - a_2\det\begin{bmatrix}b_1 & c_1\\ b_3 & c_3\end{bmatrix} + a_3\det\begin{bmatrix}b_1 & c_1\\ b_2 & c_2\end{bmatrix}$$

$$= a_1(b_2c_3 - b_3c_2) - a_2(b_1c_3 - b_3c_1) + a_3(b_1c_2 - b_2c_1).$$

The determinant can also be computed using entries of the first row as coefficients, rather than entries of the first column.

Each entry of the first column of the original matrix serves as the coefficient for the determinant of a 2×2 matrix; the first and third (a_1 and a_3) are positive, the middle one is negative. To remember which 2×2 matrix goes with which coefficient, cross out the row and column the coefficient is in; what is left is the matrix you want. To get the 2×2 matrix for the coefficient a_2,

$$\begin{bmatrix}a_1 & b_1 & c_1\\ \not{a_2} & \not{b_2} & \not{c_2}\\ a_3 & b_3 & c_3\end{bmatrix} = \begin{bmatrix}b_1 & c_1\\ b_3 & c_3\end{bmatrix}. \qquad 1.4.40$$

Example 1.4.16 (Determinant of a 3×3 matrix).

$$\det \begin{bmatrix} 3 & 1 & -2 \\ 1 & 2 & 4 \\ 2 & 0 & 1 \end{bmatrix} = 3 \det \begin{bmatrix} 2 & 4 \\ 0 & 1 \end{bmatrix} - 1 \det \begin{bmatrix} 1 & -2 \\ 0 & 1 \end{bmatrix} + 2 \det \begin{bmatrix} 1 & -2 \\ 2 & 4 \end{bmatrix}$$
$$= 3\,(2-0) - (1+0) + 2\,(4+4) = 21. \quad \triangle \qquad 1.4.41$$

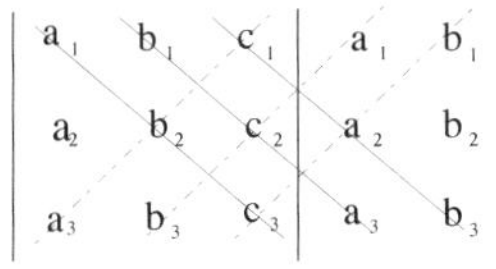

FIGURE 1.4.11.

Another way to compute the determinant of a 3×3 matrix is to write the matrix followed by the first two columns of the matrix, as shown above. Add the products of the entries on each solid diagonal lines:

$$a_1b_2c_3 + b_1c_2a_3 + c_1a_2b_3$$

and subtract from that total the sum of the products of the entries on each dotted line:

$$-a_3b_2c_1 - b_3c_2a_1 - c_3a_2b_1.$$

Exercise 1.4.20 shows that a 3×3 matrix is invertible if and only if its determinant is not 0 (Theorem 4.8.3 generalizes this to $n \times n$ matrices). Figure 1.4.11 shows another way to compute the determinant of a 3×3 matrix.

The cross product of two vectors

The cross product exists only in $\mathbb{R}^3$ (and to some extent in $\mathbb{R}^7$). The determinant is a number, as is the dot product, but the cross product is a *vector*. Exercise 1.4.14 asks you to show that the cross product is not associative, but that it is distributive.

Definition 1.4.17 (Cross product in $\mathbb{R}^3$). The *cross product* $\vec{\mathbf{a}} \times \vec{\mathbf{b}}$ in $\mathbb{R}^3$ is

$$\begin{bmatrix} a_1 \\ a_2 \\ a_3 \end{bmatrix} \times \begin{bmatrix} b_1 \\ b_2 \\ b_3 \end{bmatrix} \overset{\text{def}}{=} \begin{bmatrix} \det \begin{bmatrix} a_2 & b_2 \\ a_3 & b_3 \end{bmatrix} \\ -\det \begin{bmatrix} a_1 & b_1 \\ a_3 & b_3 \end{bmatrix} \\ \det \begin{bmatrix} a_1 & b_1 \\ a_2 & b_2 \end{bmatrix} \end{bmatrix} = \begin{bmatrix} a_2b_3 - a_3b_2 \\ -a_1b_3 + a_3b_1 \\ a_1b_2 - a_2b_1 \end{bmatrix}. \qquad 1.4.42$$

A trick for remembering how to compute the cross product is to write

$$\vec{\mathbf{a}} \times \vec{\mathbf{b}} = \det \begin{bmatrix} \vec{\mathbf{e}}_1 & a_1 & b_1 \\ \vec{\mathbf{e}}_2 & a_2 & b_2 \\ \vec{\mathbf{e}}_3 & a_3 & b_3 \end{bmatrix}.$$

It isn't clear what meaning to attach to a matrix with some entries vectors and some scalars, but if you just treat each entry as something to be multiplied, it gives the right answer.

Think of your vectors as a 3×2 matrix; first cover up the first row and take the determinant of what's left. That gives the first entry of the cross product. Then cover up the second row and take *minus* the determinant of what's left, giving the second entry. The third entry is obtained by covering up the third row and taking the determinant of what's left.

Example 1.4.18 (Cross product of two vectors in $\mathbb{R}^3$).

$$\begin{bmatrix} 3 \\ 0 \\ 1 \end{bmatrix} \times \begin{bmatrix} 2 \\ 1 \\ 4 \end{bmatrix} = \begin{bmatrix} \det \begin{bmatrix} 0 & 1 \\ 1 & 4 \end{bmatrix} \\ -\det \begin{bmatrix} 3 & 2 \\ 1 & 4 \end{bmatrix} \\ \det \begin{bmatrix} 3 & 2 \\ 0 & 1 \end{bmatrix} \end{bmatrix} = \begin{bmatrix} -1 \\ -10 \\ 3 \end{bmatrix}. \quad \triangle \qquad 1.4.43$$

It follows immediately from Definition 1.4.17 that changing the order of the vectors changes the sign of the cross product: $\vec{\mathbf{a}} \times \vec{\mathbf{b}} = -\vec{\mathbf{b}} \times \vec{\mathbf{a}}$.

Proposition 1.4.19. *The cross product and the determinant satisfy*

$$\det[\vec{\mathbf{a}}, \vec{\mathbf{b}}, \vec{\mathbf{c}}] = \vec{\mathbf{a}} \cdot (\vec{\mathbf{b}} \times \vec{\mathbf{c}}).$$

Proof. This follows immediately from Definitions 1.4.15 and 1.4.17:

$$\overbrace{\begin{bmatrix} a_1 \\ a_2 \\ a_3 \end{bmatrix}}^{\vec{\mathbf{a}}} \cdot \overbrace{\begin{bmatrix} \det \begin{bmatrix} b_2 & c_2 \\ b_3 & c_3 \end{bmatrix} \\ -\det \begin{bmatrix} b_1 & c_1 \\ b_3 & c_3 \end{bmatrix} \\ \det \begin{bmatrix} b_1 & c_1 \\ b_2 & c_2 \end{bmatrix} \end{bmatrix}}^{\vec{\mathbf{b}}\times\vec{\mathbf{c}}} = \underbrace{a_1 \det \begin{bmatrix} b_2 & c_2 \\ b_3 & c_3 \end{bmatrix} - a_2 \det \begin{bmatrix} b_2 & c_2 \\ b_3 & c_3 \end{bmatrix} + a_3 \det \begin{bmatrix} b_1 & c_1 \\ b_2 & c_2 \end{bmatrix}}_{\det[\vec{\mathbf{a}},\vec{\mathbf{b}},\vec{\mathbf{c}}]}. \quad \square$$

Engineers and physicists sometimes call $\vec{\mathbf{a}}\cdot(\vec{\mathbf{b}}\times\vec{\mathbf{c}})$ the *scalar triple product*.

Proposition 1.4.20, part 3: $\vec{\mathbf{a}}$ and $\vec{\mathbf{b}}$ "not collinear" means that they are not multiples of each other; they do not lie on a single line. In Section 2.4 we will see that saying that two vectors in $\mathbb{R}^n$ are "not collinear" is saying that they are *linearly independent*, or alternately, that they *span* a plane.

It follows from the definition of the determinant that

$$\det[\vec{\mathbf{a}}, \vec{\mathbf{b}}, \vec{\mathbf{c}}] = \det[\vec{\mathbf{c}}, \vec{\mathbf{a}}, \vec{\mathbf{b}}] = \det[\vec{\mathbf{b}}, \vec{\mathbf{c}}, \vec{\mathbf{a}}]$$

so we could replace

$$\det[\vec{\mathbf{a}}, \vec{\mathbf{b}}, \vec{\mathbf{a}} \times \vec{\mathbf{b}}] > 0$$

in part 3 of Proposition 1.4.20 by

$$\det[\vec{\mathbf{a}} \times \vec{\mathbf{b}}, \vec{\mathbf{a}}, \vec{\mathbf{b}}] > 0.$$

Proposition 1.4.20 (Geometric interpretation of cross product). *The cross product $\vec{\mathbf{a}} \times \vec{\mathbf{b}}$ is the vector satisfying three properties:*

1. *It is orthogonal to the plane spanned by $\vec{\mathbf{a}}$ and $\vec{\mathbf{b}}$:*
$$\vec{\mathbf{a}} \cdot (\vec{\mathbf{a}} \times \vec{\mathbf{b}}) = 0 \quad \text{and} \quad \vec{\mathbf{b}} \cdot (\vec{\mathbf{a}} \times \vec{\mathbf{b}}) = 0. \qquad 1.4.44$$
2. *Its length $|\vec{\mathbf{a}} \times \vec{\mathbf{b}}|$ is the area of the parallelogram spanned by $\vec{\mathbf{a}}$ and $\vec{\mathbf{b}}$.*
3. *If $\vec{\mathbf{a}}$ and $\vec{\mathbf{b}}$ are not collinear, then* $\det[\vec{\mathbf{a}}, \vec{\mathbf{b}}, \vec{\mathbf{a}} \times \vec{\mathbf{b}}] > 0$.

Proof. 1. To check that the cross product $\vec{\mathbf{a}} \times \vec{\mathbf{b}}$ is orthogonal to $\vec{\mathbf{a}}$ and $\vec{\mathbf{b}}$, we check that the dot product in each case is zero (Corollary 1.4.8). For instance, $\vec{\mathbf{a}} \times \vec{\mathbf{b}}$ is orthogonal to $\vec{\mathbf{a}}$ because

$$\vec{\mathbf{a}} \cdot (\vec{\mathbf{a}} \times \vec{\mathbf{b}}) = \begin{bmatrix} a_1 \\ a_2 \\ a_3 \end{bmatrix} \cdot \overbrace{\begin{bmatrix} a_2b_3 - a_3b_2 \\ -a_1b_3 + a_3b_1 \\ a_1b_2 - a_2b_1 \end{bmatrix}}^{\text{Definition 1.4.17 of } \vec{\mathbf{a}}\times\vec{\mathbf{b}}} \qquad 1.4.45$$
$$= a_1a_2b_3 - a_1a_3b_2 - a_1a_2b_3 + a_2a_3b_1 + a_1a_3b_2 - a_2a_3b_1 = 0.$$

2. The area of the parallelogram spanned by $\vec{\mathbf{a}}$ and $\vec{\mathbf{b}}$ is $|\vec{\mathbf{a}}||\vec{\mathbf{b}}|\sin\theta$, where θ is the angle between $\vec{\mathbf{a}}$ and $\vec{\mathbf{b}}$. We know (equation 1.4.7) that

$$\cos\theta = \frac{\vec{\mathbf{a}} \cdot \vec{\mathbf{b}}}{|\vec{\mathbf{a}}||\vec{\mathbf{b}}|} = \frac{a_1b_1 + a_2b_2 + a_3b_3}{\sqrt{a_1^2 + a_2^2 + a_3^2}\sqrt{b_1^2 + b_2^2 + b_3^2}}, \qquad 1.4.46$$

so we have

$$\sin\theta = \sqrt{1 - \cos^2\theta} = \sqrt{1 - \frac{(a_1b_1 + a_2b_2 + a_3b_3)^2}{(a_1^2 + a_2^2 + a_3^2)(b_1^2 + b_2^2 + b_3^2)}}$$
$$= \sqrt{\frac{(a_1^2 + a_2^2 + a_3^2)(b_1^2 + b_2^2 + b_3^2) - (a_1b_1 + a_2b_2 + a_3b_3)^2}{(a_1^2 + a_2^2 + a_3^2)(b_1^2 + b_2^2 + b_3^2)}}, \qquad 1.4.47$$

so that

$$|\vec{\mathbf{a}}||\vec{\mathbf{b}}|\sin\theta = \sqrt{(a_1^2+a_2^2+a_3^2)(b_1^2+b_2^2+b_3^2)-(a_1b_1+a_2b_2+a_3b_3)^2}. \quad 1.4.48$$

Carrying out the multiplication results in a formula for the area that looks worse than it is: a long string of terms too big to fit on this page under one square root sign. That's a good excuse for omitting it here. But after cancellations we have the following on the right side:

$$\sqrt{\underbrace{a_1^2b_2^2+a_2^2b_1^2-2a_1b_1a_2b_2}_{(a_1b_2-a_2b_1)^2}+\underbrace{a_1^2b_3^2+a_3^2b_1^2-2a_1b_1a_3b_3}_{(a_1b_3-a_3b_1)^2}+\underbrace{a_2^2b_3^2+a_3^2b_2^2-2a_2b_2a_3b_3}_{(a_2b_3-a_3b_2)^2}}, \quad 1.4.49$$

which conveniently gives us

$$\begin{aligned}\text{Area } &= |\vec{\mathbf{a}}||\vec{\mathbf{b}}|\sin\theta = \sqrt{(a_1b_2-a_2b_1)^2+(a_1b_3-a_3b_1)^2+(a_2b_3-a_3b_2)^2}\\ &= |\vec{\mathbf{a}}\times\vec{\mathbf{b}}|. \end{aligned} \quad 1.4.50$$

3. By Proposition 1.4.19,

$$\det[\vec{\mathbf{a}}\times\vec{\mathbf{b}},\vec{\mathbf{a}},\vec{\mathbf{b}}] = (\vec{\mathbf{a}}\times\vec{\mathbf{b}})\cdot(\vec{\mathbf{a}}\times\vec{\mathbf{b}}) = |\vec{\mathbf{a}}\times\vec{\mathbf{b}}|^2. \quad 1.4.51$$

Thus $\det[\vec{\mathbf{a}}\times\vec{\mathbf{b}},\vec{\mathbf{a}},\vec{\mathbf{b}}] \geq 0$, but by part 2 if $\vec{\mathbf{a}},\vec{\mathbf{b}}$ are not collinear then $\vec{\mathbf{a}}\times\vec{\mathbf{b}} \neq 0$, so $\det[\vec{\mathbf{a}}\times\vec{\mathbf{b}},\vec{\mathbf{a}},\vec{\mathbf{b}}] > 0$. □

Proposition 1.4.21: The word "parallelepiped" seems to have fallen into disuse. It is simply a possibly slanted box: a box with six faces, each of which is a parallelogram; opposite faces are equal.

Proposition 1.4.21 (Determinant in $\mathbb{R}^3$).

1. *If P is the parallelepiped spanned by three vectors $\vec{\mathbf{a}},\vec{\mathbf{b}},\vec{\mathbf{c}}$, then*

$$\textit{volume of } P = |\vec{\mathbf{a}}\cdot(\vec{\mathbf{b}}\times\vec{\mathbf{c}})| = |\det[\vec{\mathbf{a}},\vec{\mathbf{b}},\vec{\mathbf{c}}]|.$$

2. *If $\vec{\mathbf{a}},\vec{\mathbf{b}},\vec{\mathbf{c}}$ are coplanar, then $\det[\vec{\mathbf{a}},\vec{\mathbf{b}},\vec{\mathbf{c}}]=0$.*
3. *If $\vec{\mathbf{e}}_1,\vec{\mathbf{e}}_2,\vec{\mathbf{e}}_3$ are drawn the standard way, so they satisfy the right-hand rule, then $\det[\vec{\mathbf{a}},\vec{\mathbf{b}},\vec{\mathbf{c}}]>0$ if $\vec{\mathbf{a}}$, $\vec{\mathbf{b}}$, $\vec{\mathbf{c}}$ (in that order) satisfy the right-hand rule.*

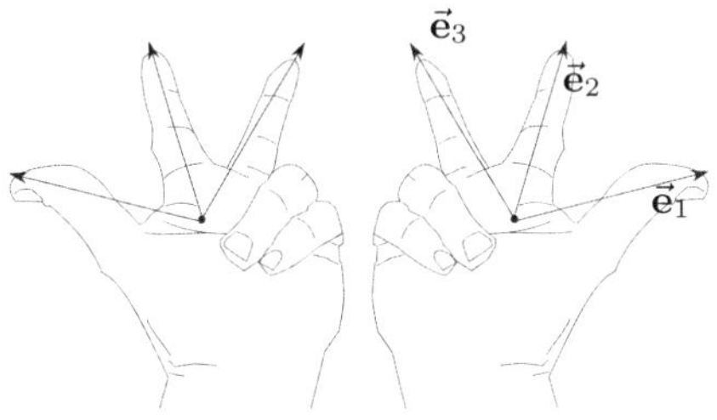

FIGURE 1.4.12.

In $\mathbb{R}^3$, the standard basis vectors, drawn the usual way, "fit" the right hand but not the left hand. Any three vectors that are not coplanar and "fit" the right hand like this are said to satisfy the right-hand rule.

In part 3, note that three different permutations of the vectors give the same result: $\vec{\mathbf{a}},\vec{\mathbf{b}},\vec{\mathbf{c}}$ satisfy the right-hand rule if and only if $\vec{\mathbf{b}},\vec{\mathbf{c}},\vec{\mathbf{a}}$ and $\vec{\mathbf{c}},\vec{\mathbf{a}},\vec{\mathbf{b}}$ satisfy the right-hand rule. (These are two instances of *even permutations*, which we will explore in Section 4.8.)

Proof of Proposition 1.4.21. 1. The proof is illustrated by Figure 1.4.13. The volume is height × area of the base. The base is the parallelogram spanned by $\vec{\mathbf{b}}$ and $\vec{\mathbf{c}}$; its area is $|\vec{\mathbf{b}}\times\vec{\mathbf{c}}|$. The height is the distance from $\vec{\mathbf{a}}$ to the plane spanned by $\vec{\mathbf{b}}$ and $\vec{\mathbf{c}}$. Let θ be the angle between $\vec{\mathbf{a}}$ and the line through the origin perpendicular to the plane spanned by $\vec{\mathbf{b}}$ and $\vec{\mathbf{c}}$; the height is then $|\vec{\mathbf{a}}|\cos\theta$. This gives

$$\text{Volume of parallelepiped } = |\vec{\mathbf{b}}\times\vec{\mathbf{c}}||\vec{\mathbf{a}}|\cos\theta = \vec{\mathbf{a}}\cdot(\vec{\mathbf{b}}\times\vec{\mathbf{c}})|. \quad 1.4.52$$

2. If $\vec{\mathbf{a}},\vec{\mathbf{b}},\vec{\mathbf{c}}$ are coplanar, then the volume of the parallelepiped they span is 0, so by part 1 $\det[\vec{\mathbf{a}},\vec{\mathbf{b}},\vec{\mathbf{c}}]=0$.

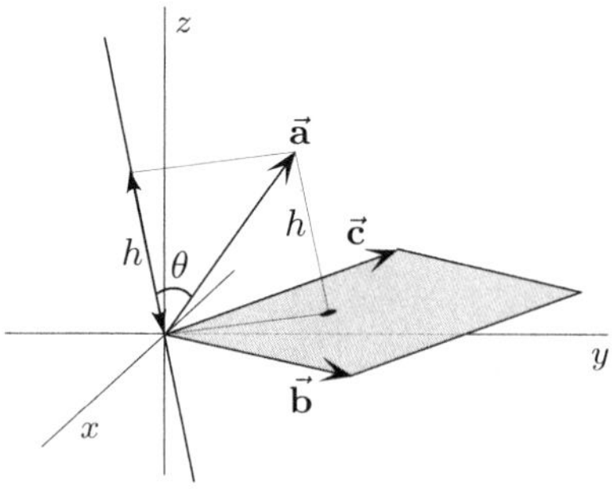

FIGURE 1.4.13.
The volume of the parallelepiped spanned by the vectors $\vec{\mathbf{a}}, \vec{\mathbf{b}}, \vec{\mathbf{c}}$ is $|\det[\vec{\mathbf{a}}, \vec{\mathbf{b}}, \vec{\mathbf{c}}]|$. The vector $\vec{\mathbf{b}} \times \vec{\mathbf{c}}$ is colinear with the line marked h through the origin.

3. If $\vec{\mathbf{a}}, \vec{\mathbf{b}}, \vec{\mathbf{c}}$ are not coplanar, you can rotate $\mathbb{R}^3$ to place them at $\vec{\mathbf{a}}', \vec{\mathbf{b}}', \vec{\mathbf{c}}'$, where $\vec{\mathbf{a}}'$ is a positive multiple of $\vec{\mathbf{e}}_1$, the vector $\vec{\mathbf{b}}'$ is in the (x,y)-plane with positive y-coordinate (i.e., counterclockwise from $\vec{\mathbf{e}}_1$), and the z-coordinate of $\vec{\mathbf{c}}'$ is nonzero. While being rotated, the vectors never become coplanar so the determinant never vanishes, and $\det[\vec{\mathbf{a}}, \vec{\mathbf{b}}, \vec{\mathbf{c}}] = \det[\vec{\mathbf{a}}', \vec{\mathbf{b}}', \vec{\mathbf{c}}']$. Then

$$\det[\vec{\mathbf{a}}, \vec{\mathbf{b}}, \vec{\mathbf{c}}] = \det \begin{bmatrix} a_1' & b_1' & c_1' \\ 0 & b_2' & c_2' \\ 0 & 0 & c_3' \end{bmatrix} = a_1' b_2' c_3'$$

has the same sign as c_3'. The three vector will fit the same hand as $\vec{\mathbf{e}}_1, \vec{\mathbf{e}}_2, \vec{\mathbf{e}}_3$ if $c_3' > 0$, and the other hand if $c_3' < 0$. See Figure 1.4.12. △

The relationships between algebra and geometry form a constant theme of mathematics. Table 1.4.14 summarizes those discussed in this section.

Correspondence between Algebra and Geometry		
Operation	Algebra	Geometry
dot product	$\vec{\mathbf{v}} \cdot \vec{\mathbf{w}} = \sum v_i w_i$	$\vec{\mathbf{v}} \cdot \vec{\mathbf{w}} = \lvert\vec{\mathbf{v}}\rvert\lvert\vec{\mathbf{w}}\rvert \cos\theta$
determinant of 2 × 2 matrix	$\det \begin{bmatrix} a_1 & b_1 \\ a_2 & b_2 \end{bmatrix} = a_1 b_2 - a_2 b_1$	$\left\lvert \det \begin{bmatrix} a_1 & b_1 \\ a_2 & b_2 \end{bmatrix} \right\rvert$ = area of parallelogram
cross product	$\begin{bmatrix} a_1 \\ a_2 \\ a_3 \end{bmatrix} \times \begin{bmatrix} b_1 \\ b_2 \\ b_3 \end{bmatrix} = \begin{bmatrix} a_2 b_3 - b_2 a_3 \\ b_1 a_3 - a_1 b_3 \\ a_1 b_2 - a_2 b_1 \end{bmatrix}$	$(\vec{\mathbf{a}} \times \vec{\mathbf{b}}) \perp \vec{\mathbf{a}}$, $(\vec{\mathbf{a}} \times \vec{\mathbf{b}}) \perp \vec{\mathbf{b}}$ length = area of parallelogram right-hand rule
determinant of 3 × 3 matrix	$\det \begin{bmatrix} a_1 & b_1 & c_1 \\ a_2 & b_2 & c_2 \\ a_3 & b_3 & c_3 \end{bmatrix} = \det[\vec{\mathbf{a}}, \vec{\mathbf{b}}, \vec{\mathbf{c}}]$ $= \vec{\mathbf{a}} \cdot (\vec{\mathbf{b}} \times \vec{\mathbf{c}})$	$\lvert\det[\vec{\mathbf{a}}, \vec{\mathbf{b}}, \vec{\mathbf{c}}]\rvert$ = volume of parallelepiped

TABLE 1.4.14. Mathematical "objects" often have two interpretations: algebraic and geometric. The symbol ⊥ means orthogonal; it is pronounced "perp".

EXERCISES FOR SECTION 1.4

1.4.1 a. If $\vec{\mathbf{v}}$ and $\vec{\mathbf{w}}$ are vectors and A is a matrix, which of the following are numbers? Which are vectors?

$$\vec{\mathbf{v}} \times \vec{\mathbf{w}}; \quad \vec{\mathbf{v}} \cdot \vec{\mathbf{w}}; \quad |\vec{\mathbf{v}}|; \quad |A|; \quad \det A; \quad A\vec{\mathbf{v}}$$

b. What do the dimensions of the vectors and matrices need to be for the expressions in part a to be defined?

a. $\begin{bmatrix} 1 \\ 2 \end{bmatrix}$ b. $\begin{bmatrix} \sqrt{2} \\ \sqrt{7} \end{bmatrix}$

c. $\begin{bmatrix} 1 \\ -1 \\ 1 \end{bmatrix}$ d. $\begin{bmatrix} 1 \\ -2 \\ 2 \end{bmatrix}$

Vectors for Exercise 1.4.2

1.4.2 What are the lengths of the vectors in the margin?

1.4.3 Normalize the following vectors: a. $\begin{bmatrix} 0 \\ 1 \\ 4 \end{bmatrix}$ b. $\begin{bmatrix} -3 \\ 7 \end{bmatrix}$ c. $\begin{bmatrix} \sqrt{2} \\ -2 \\ -5 \end{bmatrix}$

1.4.4 a. What is the angle between the vectors in Exercise 1.4.2, a and b?
b. What is the angle between the vectors in Exercise 1.4.2, c and d?

1.4.5 Calculate the angles between the following pairs of vectors:

a. $\begin{bmatrix} 1 \\ 0 \\ 0 \end{bmatrix}, \begin{bmatrix} 1 \\ 1 \\ 1 \end{bmatrix}$ b. $\begin{bmatrix} 1 \\ 0 \\ -1 \\ 0 \end{bmatrix}, \begin{bmatrix} 1 \\ 1 \\ 1 \\ 1 \end{bmatrix}$

a. $\begin{bmatrix} 3 & 1 \\ 0 & 2 \end{bmatrix}$ b. $\begin{bmatrix} 1 & 0 & 2 \\ 2 & 4 & 1 \\ 0 & 1 & 3 \end{bmatrix}$

c. $\begin{bmatrix} -2 & 5 & 3 \\ -1 & 3 & 4 \\ -2 & 3 & 7 \end{bmatrix}$

d. $\begin{bmatrix} 1 & 2 & -6 \\ 0 & 1 & -3 \\ 1 & 0 & -2 \end{bmatrix}$

Matrices for Exercise 1.4.6

1.4.6 Compute the determinants of the matrices at left.

1.4.7 For each of the following matrices, compute its inverse if it exists, and its determinant.

a. $\begin{bmatrix} 2 & -1 \\ 1 & 0 \end{bmatrix}$ b. $\begin{bmatrix} 1 & 1 \\ 1 & 1 \end{bmatrix}$ c. $\begin{bmatrix} a & b \\ 0 & d \end{bmatrix}$ d. $\begin{bmatrix} 1 & -1 \\ 1 & -1 \end{bmatrix}$

1.4.8 Compute the determinants of the matrices in the margin at left.

1.4.9 Compute the following cross products:

a. $\begin{bmatrix} 2x \\ -y \\ 3z \end{bmatrix} \times \begin{bmatrix} x \\ 2y \\ 0 \end{bmatrix}$ b. $\begin{bmatrix} 1 \\ 2 \\ 5 \end{bmatrix} \times \begin{bmatrix} 2 \\ 0 \\ 3 \end{bmatrix}$ c. $\begin{bmatrix} 2 \\ -1 \\ -6 \end{bmatrix} \times \begin{bmatrix} 3 \\ 0 \\ 2 \end{bmatrix}$

1.4.10 a. Let A be a matrix. Show that $|A^k| \le |A|^k$. Compute $|A^3|$ and $|A|^3$ for $A = \begin{bmatrix} 1 & 2 \\ 1 & 1 \end{bmatrix}$.

b. Let $\vec{\mathbf{u}} = \begin{bmatrix} 1 \\ 2 \\ 3 \end{bmatrix}$, $\vec{\mathbf{v}} = \begin{bmatrix} -2 \\ -4 \\ -6 \end{bmatrix}$, $\vec{\mathbf{w}} = \begin{bmatrix} 2 \\ 0 \\ 6 \end{bmatrix}$. Without computations, explain why the following are true or false:

i. $|\vec{\mathbf{u}} \cdot \vec{\mathbf{v}}| = |\vec{\mathbf{u}}||\vec{\mathbf{v}}|$ ii. $|\vec{\mathbf{u}} \cdot \vec{\mathbf{w}}| = |\vec{\mathbf{u}}||\vec{\mathbf{w}}|$.

Confirm by computations.

c. Let $\vec{\mathbf{v}} = \begin{bmatrix} v_1 \\ v_2 \end{bmatrix}$ and $\vec{\mathbf{w}} = \begin{bmatrix} w_1 \\ w_2 \end{bmatrix}$ be two vectors such that $v_1 w_2 < v_2 w_1$. Does $\vec{\mathbf{w}}$ lie clockwise or counterclockwise from $\vec{\mathbf{v}}$?

d. Let $\vec{\mathbf{v}} = \begin{bmatrix} 1 \\ 2 \\ 3 \end{bmatrix}$, and let $\vec{\mathbf{w}}$ be a vector such that $\vec{\mathbf{v}} \cdot \vec{\mathbf{w}} = 42$. What is the shortest $\vec{\mathbf{w}}$ can be? The longest it can be?

a. $\begin{bmatrix} 1 & 2 & 3 \\ -1 & 1 & 1 \\ 2 & 2 & 2 \end{bmatrix}$

b. $\begin{bmatrix} a & b & c \\ 0 & d & e \\ 0 & 0 & f \end{bmatrix}$

c. $\begin{bmatrix} a & b & 0 \\ c & d & 0 \\ e & f & g \end{bmatrix}$

Matrices for Exercise 1.4.8

1.4.11 Given the vectors $\vec{\mathbf{u}} = \begin{bmatrix} bc \\ -ac \\ 2ab \end{bmatrix}$, $\vec{\mathbf{v}} = \begin{bmatrix} a \\ -b \\ -c \end{bmatrix}$, $\vec{\mathbf{w}} = \begin{bmatrix} -2a \\ 2b \\ 2c \end{bmatrix}$, which of the following statements are true?

a. $|\vec{\mathbf{v}} \cdot \vec{\mathbf{w}}| = |\vec{\mathbf{v}}||\vec{\mathbf{w}}|$ b. $\vec{\mathbf{u}} \cdot (\vec{\mathbf{v}} \times \vec{\mathbf{w}}) = |\vec{\mathbf{u}}|(\vec{\mathbf{v}} \times \vec{\mathbf{w}})$

c. $\det[\vec{\mathbf{u}}, \vec{\mathbf{v}}, \vec{\mathbf{w}}] = \det[\vec{\mathbf{u}}, \vec{\mathbf{w}}, \vec{\mathbf{v}}]$ d. $|\vec{\mathbf{u}} \cdot \vec{\mathbf{w}}| = |\vec{\mathbf{u}}||\vec{\mathbf{w}}|$

e. $\det[\vec{\mathbf{u}}, \vec{\mathbf{v}}, \vec{\mathbf{w}}] = \vec{\mathbf{u}} \cdot (\vec{\mathbf{v}} \times \vec{\mathbf{w}})$ f. $\vec{\mathbf{u}} \cdot \vec{\mathbf{w}} = \vec{\mathbf{w}} \cdot \vec{\mathbf{u}}$

1.4.12 a. Show that $|\vec{\mathbf{v}}+\vec{\mathbf{w}}| \geq |\vec{\mathbf{v}}| - |\vec{\mathbf{w}}|$.

b. True or false? $|\det[\vec{\mathbf{a}}, \vec{\mathbf{b}}, \vec{\mathbf{c}}]| \leq |\vec{\mathbf{a}}|\,|\vec{\mathbf{b}}\times\vec{\mathbf{c}}|$. Explain your answer. What does it mean geometrically?

1.4.13 Show that the cross product of two vectors pointing in the same direction is zero.

$$\underset{\vec{\mathbf{u}}}{\begin{bmatrix}1\\2\\1\end{bmatrix}}, \quad \underset{\vec{\mathbf{v}}}{\begin{bmatrix}2\\0\\1\end{bmatrix}}, \quad \underset{\vec{\mathbf{w}}}{\begin{bmatrix}1\\0\\-1\end{bmatrix}}$$

Vectors for Exercise 1.4.14

1.4.14 Given the vectors $\vec{\mathbf{u}}$, $\vec{\mathbf{v}}$, $\vec{\mathbf{w}}$ shown in the margin,

a. Compute $\vec{\mathbf{u}}\times(\vec{\mathbf{v}}\times\vec{\mathbf{w}})$ and $(\vec{\mathbf{u}}\times\vec{\mathbf{v}})\times\vec{\mathbf{w}}$.

b. Confirm that $\vec{\mathbf{v}}\cdot(\vec{\mathbf{v}}\times\vec{\mathbf{w}}) = 0$. What is the geometrical relationship of $\vec{\mathbf{v}}$ and $\vec{\mathbf{v}}\times\vec{\mathbf{w}}$?

c. Show that the cross product is distributive: given vectors $\vec{\mathbf{a}}, \vec{\mathbf{b}}, \vec{\mathbf{c}} \in \mathbb{R}^3$, show that $\vec{\mathbf{a}}\times(\vec{\mathbf{b}}+\vec{\mathbf{c}}) = (\vec{\mathbf{a}}\times\vec{\mathbf{b}}) + (\vec{\mathbf{a}}\times\vec{\mathbf{c}})$.

1.4.15 Given two vectors $\vec{\mathbf{v}}$ and $\vec{\mathbf{w}}$ in $\mathbb{R}^3$, show that $(\vec{\mathbf{v}}\times\vec{\mathbf{w}}) = -(\vec{\mathbf{w}}\times\vec{\mathbf{v}})$.

1.4.16 a. What is the area of the parallelogram with vertices at

$$\begin{pmatrix}0\\0\end{pmatrix}, \begin{pmatrix}1\\2\end{pmatrix}, \begin{pmatrix}5\\1\end{pmatrix}, \begin{pmatrix}6\\3\end{pmatrix}?$$

b. What is the area of the parallelogram with vertices at

$$\begin{pmatrix}0\\0\end{pmatrix}, \begin{pmatrix}1\\2\end{pmatrix}, \begin{pmatrix}5\\-1\end{pmatrix}, \begin{pmatrix}6\\1\end{pmatrix}?$$

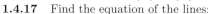

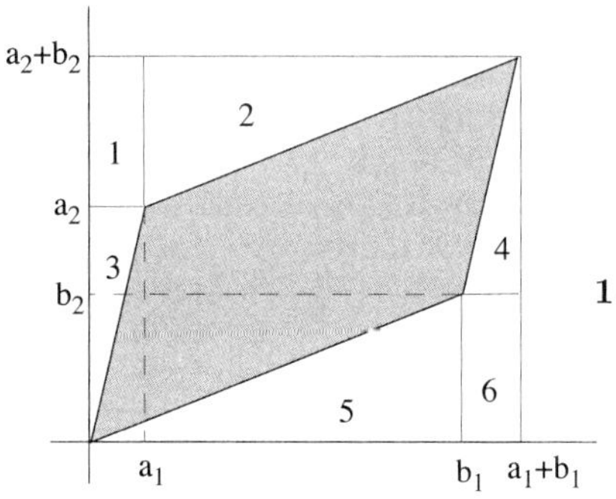

Figure for Exercise 1.4.18

1.4.17 Find the equation of the lines:

a. in the plane through the origin and perpendicular to the vector $\begin{bmatrix}2\\-1\end{bmatrix}$.

b. in the plane through $\begin{pmatrix}2\\3\end{pmatrix}$ and perpendicular to $\begin{bmatrix}2\\-4\end{bmatrix}$.

1.4.18 a. Prove part 1 of Proposition 1.4.14 in the case where the coordinates of $\mathbf{a}$ and $\mathbf{b}$ are ≥ 0 by subtracting pieces 1–6 from $(a_1+b_1)(a_2+b_2)$, as suggested by the figure in the margin.

b. Repeat for the case where b_1 is negative.

1.4.19 a. What is the length of $\vec{\mathbf{v}}_n = \vec{\mathbf{e}}_1 + \cdots + \vec{\mathbf{e}}_n \in \mathbb{R}^n$?

b. What is the angle α_n between $\vec{\mathbf{v}}_n$ and $\vec{\mathbf{e}}_1$? What is $\lim_{n\to\infty}\alpha_n$?

1.4.20 Confirm the following formula for the inverse of a 3×3 matrix:

$$A^{-1} = \begin{bmatrix} a_1 & b_1 & c_1\\ a_2 & b_2 & c_2\\ a_3 & b_3 & c_3\end{bmatrix}^{-1} = \frac{1}{\det A}\begin{bmatrix} b_2c_3 - b_3c_2 & b_3c_1 - b_1c_3 & b_1c_2 - b_2c_1\\ a_3c_2 - a_2c_3 & a_1c_3 - a_3c_1 & a_2c_1 - a_1c_2\\ a_2b_3 - a_3b_2 & a_3b_1 - a_1b_3 & a_1b_2 - a_2b_1\end{bmatrix}.$$

1.4.21 For the two matrices and the vector

$$A = \begin{bmatrix}1 & 0\\1 & 2\end{bmatrix}, \quad B = \begin{bmatrix}2 & 0\\0 & 1\end{bmatrix}, \quad \vec{\mathbf{c}} = \begin{bmatrix}1\\3\end{bmatrix},$$

a. compute $|A|, |B|, |\vec{\mathbf{c}}|$;

b. confirm that $|AB| \leq |A||B|$, $\quad |A\vec{\mathbf{c}}| \leq |A||\vec{\mathbf{c}}|$, $\quad$ and $|B\vec{\mathbf{c}}| \leq |B||\vec{\mathbf{c}}|$.

Hint for Exercise 1.4.23: You may find it helpful to use the formulas

$$1+2+\cdots+n = \frac{n(n+1)}{2}$$

and

$$1+4+\cdots+n^2 = \frac{n(n+1)(2n+1)}{6}.$$

1.4.22 a. When is inequality 1.4.27 (Proposition 1.4.11) an equality?

b. When is inequality 1.4.28 an equality?

1.4.23 a. What is the length of $\vec{\mathbf{w}}_n = \vec{\mathbf{e}}_1 + 2\vec{\mathbf{e}}_2 + \cdots + n\vec{\mathbf{e}}_n = \sum_{i=1}^n i\vec{\mathbf{e}}_i$?

b. What is the angle $\alpha_{n,k}$ between $\vec{\mathbf{w}}_n$ and $\vec{\mathbf{e}}_k$?

*c. What are the limits

$$\lim_{n\to\infty} \alpha_{n,k} \quad , \quad \lim_{n\to\infty} \alpha_{n,n} \quad , \quad \lim_{n\to\infty} \alpha_{n,[n/2]}$$

where $[n/2]$ stands for the largest integer not greater than $n/2$?

Exercise 1.4.24: the symbol $\perp$ means" orthogonal to"; it is pronounced "perp". The space $\vec{\mathbf{v}}^\perp$ is called the *orthogonal complement* of $\vec{\mathbf{v}}$.

1.4.24 Let $\vec{\mathbf{v}} \in \mathbb{R}^n$ be a nonzero vector, and denote by $\vec{\mathbf{v}}^\perp \subset \mathbb{R}^n$ the set of vectors $\vec{\mathbf{w}} \in \mathbb{R}^n$ such that $\vec{\mathbf{v}} \cdot \vec{\mathbf{w}} = 0$.

a. Show that $\vec{\mathbf{v}}^\perp$ is a subspace of $\mathbb{R}^n$.

b. Given any vector $\vec{\mathbf{a}} \in \mathbb{R}^n$, show that $\vec{\mathbf{a}} - \dfrac{\vec{\mathbf{a}} \cdot \vec{\mathbf{v}}}{|\vec{\mathbf{v}}|^2}\vec{\mathbf{v}}$ is an element of $\vec{\mathbf{v}}^\perp$.

c. Define the projection of $\vec{\mathbf{a}}$ onto $\vec{\mathbf{v}}^\perp$ by the formula

$$P_{\vec{\mathbf{v}}^\perp}(\vec{\mathbf{a}}) = \vec{\mathbf{a}} - \frac{\vec{\mathbf{a}} \cdot \vec{\mathbf{v}}}{|\vec{\mathbf{v}}|^2}\vec{\mathbf{v}}.$$

Show that there is a unique number $t(\vec{\mathbf{a}})$ such that $\left(\vec{\mathbf{a}} + t(\vec{\mathbf{a}})\vec{\mathbf{v}}\right) \in \vec{\mathbf{v}}^\perp$. Show that

$$\vec{\mathbf{a}} + t(\vec{\mathbf{a}})\vec{\mathbf{v}} = P_{\vec{\mathbf{v}}^\perp}(\vec{\mathbf{a}}).$$

1.4.25 Use direct computation to prove Schwarz's inequality (Theorem 1.4.5) in $\mathbb{R}^2$ for the standard inner product (dot product); that is, show that for any numbers x_1, x_2, y_1, y_2, we have

$$|x_1y_1 + x_2y_2| \le \sqrt{x_1^2 + x_2^2}\,\sqrt{y_1^2 + y_2^2}.$$

1.4.26 Let $A = \begin{bmatrix} 1 & -2 \\ 3 & 4 \end{bmatrix}$.

a. What is the angle $\alpha\begin{pmatrix} x \\ y \end{pmatrix}$ between $\begin{bmatrix} x \\ y \end{bmatrix}$ and $A\begin{bmatrix} x \\ y \end{bmatrix}$?

b. Is there any nonzero vector that is rotated by $\pi/2$?

1.4.27 Let $\vec{\mathbf{a}} = \begin{bmatrix} a \\ b \\ c \end{bmatrix}$ be a unit vector in $\mathbb{R}^3$, so that $a^2 + b^2 + c^2 = 1$.

a. Show that the transformation $T_{\vec{\mathbf{a}}}$ defined by $T_{\vec{\mathbf{a}}}(\vec{\mathbf{v}}) = \vec{\mathbf{v}} - 2(\vec{\mathbf{a}} \cdot \vec{\mathbf{v}})\vec{\mathbf{a}}$ is a linear transformation $\mathbb{R}^3 \to \mathbb{R}^3$.

b. What is $T_{\vec{\mathbf{a}}}(\vec{\mathbf{a}})$? If $\vec{\mathbf{v}}$ is orthogonal to $\vec{\mathbf{a}}$, what is $T_{\vec{\mathbf{a}}}(\vec{\mathbf{v}})$? Can you give a name to $T_{\vec{\mathbf{a}}}$?

c. Write the matrix M of $T_{\vec{\mathbf{a}}}$ (in terms of a, b, c, of course). What can you say of M^2 ?

1.4.28 Let M be an $n \times m$ matrix. Show that $|M|^2 = \operatorname{tr}(M^\top M)$, where tr is the trace (sum of diagonal entries; see Definition 4.8.14).

1.5 Limits and continuity

Integrals, derivatives, series, approximations: calculus is all about convergence and limits. It could easily be argued that these notions are the hardest and deepest of all of mathematics; mathematicians struggled for two hundred years to come up with correct definitions. More students have foundered on these definitions than on anything else in calculus: the combination of Greek letters, precise order of quantifiers, and inequalities is a hefty obstacle. Fortunately, these notions do not become more difficult in several variables than they are in one variable. Working through a few examples will help you understand what the definitions mean, but a proper appreciation can probably only come from use; we hope you have already started on this path in one-variable calculus.

The inventors of calculus in the seventeenth century did not have rigorous definitions of limits and continuity; these were achieved only in the 1870s. Rigor is ultimately necessary in mathematics, but it does not always come first, as Archimedes acknowledged in a manuscript discovered in 1906.

In it Archimedes reveals that his deepest results were found using dubious arguments, and only later proved rigorously, because "it is of course easier to supply the proof when we have previously acquired some knowledge of the questions by the method, than it is to find it without any previous knowledge." (We found this story in John Stillwell's *Mathematics and Its History*, Springer Verlag, 1997.)

Open and closed sets

In mathematics we often need to speak of an *open set* U; whenever we want to approach points of a set U from every side, U must be open.

Think of a set or subset as your property, surrounded by a fence. The set is open if the entire fence belongs to your neighbor. As long as you stay on your property, you can get closer and closer to the fence, but you can never reach it; no matter how close you are to your neighbor's property, there is an epsilon-thin buffer zone of your property between you and it.

The set is closed if you own the fence. Now, if you sit on your fence, there is nothing between you and your neighbor's property. If you move even an epsilon further, you will be trespassing.

What if some of the fence belongs to you and some belongs to your neighbor? Then the set is neither open nor closed.

To state this in proper mathematical language, we first need to define an open ball. Imagine a balloon of radius r, centered around a point $\mathbf{x}$. The open ball of radius r around $\mathbf{x}$ consists of all points $\mathbf{y}$ inside the balloon, but *not* the skin of the balloon itself. We use a subscript to indicate the radius of a ball B; the argument gives the center of the ball: $B_2(\mathbf{a})$ is a ball of radius 2 centered at the point $\mathbf{a}$.

FIGURE 1.5.1.

An open set includes none of the fence; however close a point in the open set is to the fence, you can always surround it with a ball of other points in the open set.

Definition 1.5.1 (Open ball). For any $\mathbf{x} \in \mathbb{R}^n$ and any $r > 0$, the *open ball* of radius r around $\mathbf{x}$ is the subset

$$B_r(\mathbf{x}) \stackrel{\text{def}}{=} \{\mathbf{y} \in \mathbb{R}^n \text{ such that } |\mathbf{x} - \mathbf{y}| < r\}. \qquad 1.5.1$$

Note that $|\mathbf{x} - \mathbf{y}|$ must be *less than* r for the ball to be open; it *cannot* be $= r$. In dimension 1, a ball is an interval.

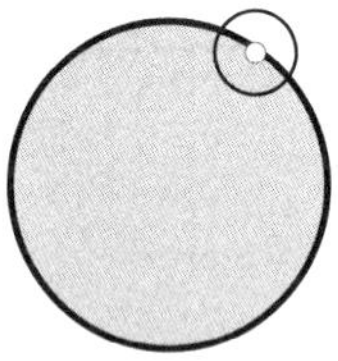

FIGURE 1.5.2.

A closed set includes its fence.

Definition 1.5.2 (Open set of $\mathbb{R}^n$). A subset $U \subset \mathbb{R}^n$ is *open* if for every point $\mathbf{x} \in U$, there exists $r > 0$ such that the open ball $B_r(\mathbf{x})$ is contained in U.

An open set is shown in Figure 1.5.1, a closed set in Figure 1.5.2.

Example 1.5.3 (Open sets).

1. If $a < b$, then the interval $(a, b) = \{\, x \in \mathbb{R} \mid a < x < b \,\}$ is open. Indeed, if $x \in (a, b)$, set $r = \min\{x - a, b - x\}$. Both these numbers

are strictly positive, since $a < x < b$, and so is their minimum. Then the ball $\{ y \mid |y - x| < r \}$ is a subset of (a, b).

2. The rectangle

$$(a,b) \times (c,d) = \left\{ \begin{pmatrix} x \\ y \end{pmatrix} \in \mathbb{R}^2 \mid a < x < b \ , \ c < y < d \right\} \text{ is open.} \qquad 1.5.2$$

3. The infinite intervals (a, ∞), $(-\infty, b)$ are also open, but the intervals

$$(a,b] = \{ x \in \mathbb{R} \mid a < x \leq b \} \quad \text{and } [a,b] = \{ x \in \mathbb{R} \mid a \leq x \leq b \} \qquad 1.5.3$$

are not. △

Parentheses denote an open interval and brackets denote a closed one: (a, b) is open, $[a, b]$ is closed.

Sometimes backwards brackets are used to denote an open interval: $]a, b[= (a, b)$.

Open and closed subsets of $\mathbb{R}^n$ are special; you shouldn't expect a subset to be either. For instance, the set of rationals $\mathbb{Q} \subset \mathbb{R}$ is neither open nor closed: every neighborhood of a rational number contains irrational numbers, and every neighborhood of an irrational number contains rational numbers.

A door that is not open is closed, but a set that is not open is not necessarily closed. An open set owns *none* of its fence. A closed set owns *all* of its fence:

Definition 1.5.4 (Closed set of $\mathbb{R}^n$). A subset $C \subset \mathbb{R}^n$ is *closed* if its complement $\mathbb{R}^n - C$ is open.

We discussed the *natural domain* (the "domain of definition") of a formula in Section 0.4. We will often be interested in whether this domain is open or closed or neither.

Example 1.5.5 (Natural domain: open or closed?). Is the natural domain of the formula

$$f\begin{pmatrix} x \\ y \end{pmatrix} = \sqrt{\frac{x}{y}} \quad \text{open or closed?} \qquad 1.5.4$$

If the argument of the square root is nonnegative, the square root can be evaluated, so the first and the third quadrants are in the natural domain. The x-axis is not (since $y = 0$ there), but the y-axis with the origin removed is in the natural domain, since x/y is zero there. So the natural domain is the region drawn in Figure 1.5.3; it is neither open nor closed. △

FIGURE 1.5.3.
The natural domain of

$$f\begin{pmatrix} x \\ y \end{pmatrix} = \sqrt{\frac{x}{y}}$$

is neither open nor closed. It includes the y-axis (with the origin removed) but not the x-axis.

Remark 1.5.6 (Why we specify open set). Even very good students often don't see the point of specifying that a set is open. But it is absolutely essential, for example in computing derivatives. If a function f is defined on a set that is not open, and thus contains at least one point x that is part of the fence, then talking of the derivative of f at x is meaningless. To compute $f'(x)$ we need to compute

$$f'(x) = \lim_{h \to 0} \frac{1}{h}\big(f(x+h) - f(x)\big), \qquad 1.5.5$$

but $f(x+h)$ won't necessarily exist even for h arbitrarily small, since $x+h$ may be outside the fence and thus not in the domain of f. This situation gets much worse in $\mathbb{R}^n$.[10] △

[10]For simple closed sets, fairly obvious ad hoc definitions of derivatives exist. For arbitrary closed sets, one can also make sense of the notion of derivatives, but these results, due to the great American mathematician Hassler Whitney, are extremely difficult, well beyond the scope of this book.

Neighborhood, closure, interior, and boundary

Before discussing the crucial topics of this section – limits and continuity – we need to introduce some more vocabulary.

A set is open if it contains a neighborhood of every one of its points.

We will use the word *neighborhood* often; it is handy when we want to describe a region around a point without requiring it to be open; a neighborhood contains an open ball but need not be open itself.

An open subset $U \subset \mathbb{R}^n$ is "thick" or "chunky". No matter where you are in U, there is always a little room: the ball $B_r(\mathbf{x})$ guarantees that you can stretch out at least $r > 0$ in any direction possible in your particular $\mathbb{R}^n$, without leaving U. Thus an open subset $U \subset \mathbb{R}^n$ is necessarily n-dimensional. In contrast, a plane in $\mathbb{R}^3$ cannot be open; a flatworm living in the plane cannot lift its head out of the plane. A line in a plane, or in space, cannot be open; a "lineworm" living in a line that is a subset of a plane or of space can neither wiggle from side to side nor lift its head.

A closed n-dimensional subset of $\mathbb{R}^n$ is "thick" in its interior (if the interior isn't empty) but not on its fence. If an n-dimensional subset of $\mathbb{R}^n$ is neither closed nor open, then it is "thick" everywhere except on the part of fence that it owns.

Definition 1.5.7 (Neighborhood). A *neighborhood* of a point $\mathbf{x} \in \mathbb{R}^n$ is a subset $X \subset \mathbb{R}^n$ such that there exists $\epsilon > 0$ with $B_\epsilon(\mathbf{x}) \subset X$.

Most often, we deal with sets that are neither open nor closed. But every set is contained in a smallest closed set, called its *closure*, and every set contains a biggest open set, called its *interior*. (Exercise 1.5.4 asks you to show that these characterizations of closure and interior are equivalent to the following definitions.)

Definition 1.5.8 (Closure). If A is a subset of $\mathbb{R}^n$, the *closure* of A, denoted $\overline{A}$, is the set of $\mathbf{x} \in \mathbb{R}^n$ such that for all $r > 0$,

$$B_r(\mathbf{x}) \cap A \neq \phi. \tag{1.5.6}$$

Definition 1.5.9 (Interior). If A is a subset of $\mathbb{R}^n$, the *interior* of A, denoted $\mathring{A}$, is the set of $\mathbf{x} \in \mathbb{R}^n$ such that there exists $r > 0$ such that

$$B_r(\mathbf{x}) \subset A. \tag{1.5.7}$$

The closure of a closed set is itself; the interior of an open set is itself.

We spoke of the "fence" of a set when we defined closed and open sets informally. The technical term is *boundary*. A closed set contains its boundary; an open set contains none of its boundary. We used the word "fence" because we think it is easier to think of owning a fence than owning a boundary. But "boundary" is generally more appropriate. The boundary of the rational numbers is all of $\mathbb{R}$, which would be difficult to imagine as a picket fence.

Definition 1.5.10 (Boundary of subset). The *boundary* of a subset $A \subset \mathbb{R}^n$, denoted ∂A, consists of those points $\mathbf{x} \in \mathbb{R}^n$ such that every neighborhood of $\mathbf{x}$ intersects both A and the complement of A.

Equation 1.5.8: Remember that $\cup$ denotes "union": $A \cup B$ is the set of elements of either A or B or both. The notation of set theory is discussed in Section 0.3.

The closure of A is thus A plus its boundary; its interior is A minus its boundary:

$$\overline{A} = A \cup \partial A \quad \text{and} \quad \mathring{A} = A - \partial A. \tag{1.5.8}$$

The boundary is the closure minus the interior:

$$\partial A = \overline{A} - \mathring{A}. \tag{1.5.9}$$

Exercise 1.18 asks you to show that the closure of U is the subset of $\mathbb{R}^n$ made up of all limits of sequences in U that converge in $\mathbb{R}^n$.

Exercise 1.5.6 asks you to show that ∂A is the intersection of the closure of A and the closure of the complement of A.

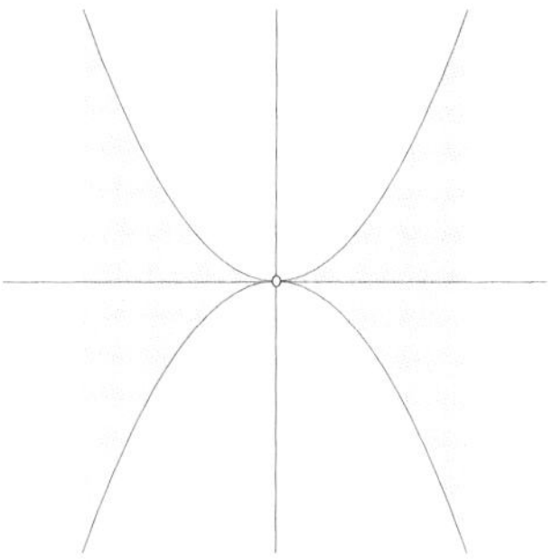

FIGURE 1.5.4.

The shaded region is U in Example 1.5.11, part 5. You can approach the origin from this region, but only in rather special ways.

FIGURE 1.5.5.
Karl Weierstrass (1815–1897)

At his father's insistence, Weierstrass studied law, finance, and economics, but he refused to take his final exams. Instead he became a teacher, teaching mathematics, history, physical education and handwriting. In 1854 an article on Abelian functions brought him to the notice of the mathematical world.

Exercise 1.5.4 asks you to justify equations 1.5.8 and 1.5.9.

Examples 1.5.11 (Closure, interior, and boundary).

1. The sets $(0,1)$, $[0,1]$, $[0,1)$, and $(0,1]$ all have the same closure, interior, and boundary: the closure is $[0,1]$, the interior is $(0,1)$, and the boundary consists of the two points 0 and 1.
2. The sets
$$\left\{ \begin{pmatrix} x \\ y \end{pmatrix} \in \mathbb{R}^2 \mid x^2+y^2<1 \right\} \quad \text{and} \quad \left\{ \begin{pmatrix} x \\ y \end{pmatrix} \in \mathbb{R}^2 \mid x^2+y^2\le 1 \right\}$$
both have the same closure, interior, and boundary: the closure is the disc of equation $x^2+y^2 \le 1$, the interior is the disc of equation $x^2+y^2<1$, and the boundary is the circle of equation $x^2+y^2=1$.
3. The closure of the rational numbers $\mathbb{Q} \subset \mathbb{R}$ is all of $\mathbb{R}$: the intersection of every neighborhood of every real number with $\mathbb{Q}$ is not empty. The interior is empty, and the boundary is $\mathbb{R}$: every neighborhood of every real number contains both rationals and irrationals.
4. The closure of the open unit disc with the origin removed:
$$U = \left\{ \begin{pmatrix} x \\ y \end{pmatrix} \in \mathbb{R}^2 \mid 0 < x^2+y^2<1 \right\} \qquad 1.5.10$$
is the closed unit disc; the interior is itself, since it is open; and its boundary consists of the unit circle and the origin.
5. The closure of the region U between two parabolas touching at the origin, shown in Figure 1.5.4:
$$U = \left\{ \begin{pmatrix} x \\ y \end{pmatrix} \in \mathbb{R}^2 \mid |y| < x^2 \right\} \qquad 1.5.11$$
is the region given by
$$\overline{U} = \left\{ \begin{pmatrix} x \\ y \end{pmatrix} \in \mathbb{R}^2 \mid |y| \le x^2 \right\}; \qquad 1.5.12$$
in particular, it contains the origin. The set is open, so $\mathring{U} = U$. The boundary is the union of the two parabolas. $\triangle$

Convergence and limits

Limits of sequences and limits of functions are the fundamental construct of calculus, as you will already have seen in first year calculus: both derivatives and integrals are defined as limits. Nothing can be done in calculus without limits.

The notion of limit is complicated; historically, coming up with a correct definition took 200 years. It wasn't until Karl Weierstrass (1815–1897) wrote an incontestably correct definition that calculus became a rigorous subject.[11] A major stumbling block to writing an unambiguous definition

[11]Many great mathematicians wrote correct definitions of limits before Weierstrass: Newton, Euler, Cauchy, among others. However, Weierstrass was the first to show that his definition, which is the modern definition, provides an adequate foundation for analysis.

was understanding in what order quantifiers should be taken. (You may wish to review the discussion of quantifiers in Section 0.2.)

We will define two notions of limits: limits of sequences and limits of functions. Limits of sequences are simpler; for limits of functions one needs to think carefully about domains.

Unless we state explicitly that a sequence is finite, it is infinite.

Definition 1.5.12: Recall from Section 0.2 that $\exists$ means "there exists"; $\forall$ means "for all". In words, the sequence $i \mapsto \mathbf{a}_i$ converges to $\mathbf{a}$ if for all $\epsilon > 0$ there exists M such that when $m > M$, then $|\mathbf{a}_m - \mathbf{a}| < \epsilon$.

Saying that $i \mapsto \mathbf{a}_i$ converges to $\mathbf{0}$ is the same as saying that $\lim_{i\to\infty} |\mathbf{a}_i| = 0$; we can show that a sequence of vectors converges to $\mathbf{0}$ by showing that their lengths converge to 0. Note that this is not true if the sequence converges to $\mathbf{a} \neq \mathbf{0}$.

Definition 1.5.12 (Convergent sequence; limit of sequence). A sequence $i \mapsto \mathbf{a}_i$ of points in $\mathbb{R}^n$ *converges* to $\mathbf{a} \in \mathbb{R}^n$ if

$$\forall \epsilon > 0,\ \exists M \mid m > M \implies |\mathbf{a}_m - \mathbf{a}| < \epsilon. \qquad 1.5.13$$

We then call $\mathbf{a}$ the *limit* of the sequence.

Exactly the same definition applies to a sequence of vectors: just replace $\mathbf{a}$ in Definition 1.5.12 by $\vec{\mathbf{a}}$, and substitute the word "vector" for "point".

Convergence in $\mathbb{R}^n$ is just n separate convergences in $\mathbb{R}$:

Proposition 1.5.13 (Convergence in terms of coordinates). *A sequence $m \mapsto \mathbf{a}_m$ with $\mathbf{a}_m \in \mathbb{R}^n$ converges to $\mathbf{a}$ if and only if each coordinate converges; i.e., if for all j with $1 \le j \le n$ the jth coordinate $(a_i)_j$ of $\mathbf{a}_i$ converges to a_j, the jth coordinate of the limit $\mathbf{a}$.*

Infinite decimals are limits of convergent sequences. If

$$a_0 = 3,$$
$$a_1 = 3.1,$$
$$a_2 = 3.14,$$
$$\vdots$$
$$a_n = \pi \text{ to } n \text{ decimal places},$$

how large must M be so that if $n \ge M$, then $|a_n - \pi| < 10^{-3}$? The answer is $M = 3$:

$$\pi - 3.141 = .0005926\ldots.$$

The same argument holds for any real number.

The proof is a good setting for understanding how the (ϵ, M)-game is played (where M is the M of Definition 1.5.12). You should imagine that your opponent gives you an epsilon and challenges you to find an M that works: an M such that when $m > M$, then $|(a_m)_j - a_j| < \epsilon$. You get Brownie points for style for finding a small M, but it is not necessary in order to win the game.

Proof. Let us first see the easy direction: that $m \mapsto \mathbf{a}_m$ converges to $\mathbf{a}$ implies that for each $j = 1, \ldots, n$, the jth coordinate of $\mathbf{a}_m$ converges to a_j. The challenger hands you an epsilon. Write $\mathbf{a}_m$ as $\begin{pmatrix} (a_m)_1 \\ \vdots \\ (a_m)_n \end{pmatrix}$. Fortunately you have a teammate who knows how to play the game for the sequence $m \mapsto \mathbf{a}_m$, and you hand her the epsilon you just got. She promptly hands you back an M with the guarantee that when $m > M$, then $|\mathbf{a}_m - \mathbf{a}| < \epsilon$ (since the sequence $m \mapsto \mathbf{a}_m$ is convergent). The length of the vector $\mathbf{a}_m - \mathbf{a}$ is

$$|\mathbf{a}_m - \mathbf{a}| = \sqrt{\big((a_m)_1 - a_1\big)^2 + \cdots + \big((a_m)_n - a_n\big)^2}, \qquad 1.5.14$$

so you give that M to your challenger, with the argument that

$$|(a_m)_j - a_j| \le |\mathbf{a}_m - \mathbf{a}| < \epsilon. \qquad 1.5.15$$

He promptly concedes defeat.

Now we have to show that convergence of the coordinate sequences implies convergence of the sequence $m \mapsto \mathbf{a}_m$. Again the challenger hands you an $\epsilon > 0$. This time you have n teammates; each knows how the play the game for a single convergent coordinate sequence $m \mapsto (a_m)_j$. After a bit of thought and scribbling on a piece of paper, you pass along $\epsilon/\sqrt{n}$ to each of them. They dutifully return to you cards containing numbers $M_1, \dots, M_n$, with the guarantee that

$$|(a_m)_j - a_j| < \frac{\epsilon}{\sqrt{n}} \quad \text{when } m > M_j. \qquad 1.5.16$$

You sort through the cards and choose the one with the largest number,

$$M = \max\{M_1, \dots, M_n\}, \qquad 1.5.17$$

which you pass on to the challenger with the following message:

if $m > M$, then $m > M_j$ for each $j = 1 \cdots = n$, so $|(a_m)_j - a_j| < \epsilon/\sqrt{n}$, so

$$\begin{aligned} |\mathbf{a}_m - \mathbf{a}| &= \sqrt{\big((a_m)_1 - a_1\big)^2 + \cdots + \big((a_m)_n - a_n\big)^2} \\ &< \sqrt{\left(\frac{\epsilon}{\sqrt{n}}\right)^2 + \cdots + \left(\frac{\epsilon}{\sqrt{n}}\right)^2} = \sqrt{\frac{n\epsilon^2}{n}} = \epsilon. \quad \square \end{aligned} \qquad 1.5.18$$

This is typical of all proofs involving convergence and limits: you are given an ϵ and challenged to come up with a δ (or M or whatever) such that a certain quantity is less than ϵ.

Your "challenger" can give you any $\epsilon > 0$ he likes; statements concerning limits and continuity are of the form "for *all* epsilon, there exists ... ".

The scribbling you did was to figure out that handing $\epsilon/\sqrt{n}$ to your teammates would work. What if you can't figure out how to "slice up" ϵ so that the final answer will be precisely ϵ? Just work directly with ϵ and see where it takes you. If you use ϵ instead of $\epsilon/\sqrt{n}$ in inequalities 1.5.16 and 1.5.18, you will end up with

$$|\mathbf{a}_m - \mathbf{a}| < \epsilon\sqrt{n}. \qquad 1.5.19$$

You then see that to get the exact answer, you should have chosen $\epsilon/\sqrt{n}$.

In fact, the answer in inequality 1.5.19 is good enough and you don't need to go back and fiddle. "Less than epsilon" for any $\epsilon > 0$ and "less than some quantity that goes to 0 when epsilon goes to 0" achieve the same goal: showing that you can make some quantity arbitrarily small. The following theorem states this precisely; you are asked to prove it in Exercise 1.5.11. The first statement is the one mathematicians use most often.

Proposition 1.5.14 (Elegance is not required). *Let u be a function of $\epsilon > 0$ such that $u(\epsilon) \to 0$ as $\epsilon \to 0$. Then a sequence $i \mapsto \mathbf{a}_i$ converges to $\mathbf{a}$ if either of the two equivalent statements are true:*

1. *For all $\epsilon > 0$ there exists M such that when $m > M$, then* $|\mathbf{a}_m - \mathbf{a}| < u(\epsilon)$.
2. *For all $\epsilon > 0$ there exists M such that when $m > M$, then* $|\mathbf{a}_m - \mathbf{a}| < \epsilon$.

The following result is of great importance. It says that the notion of limit is well defined: if the limit is something, then it isn't something else.

Proposition 1.5.15 (Limit of sequence is unique). *If the sequence $i \mapsto \mathbf{a}_i$ of points in $\mathbb{R}^n$ converges to $\mathbf{a}$ and to $\mathbf{b}$, then $\mathbf{a} = \mathbf{b}$.*

Proof. This could be reduced to the one-dimensional case, but we will use it as an opportunity to play the (ϵ, M)-game in more sober fashion. Suppose $\mathbf{a} \neq \mathbf{b}$, and set $\epsilon_0 = (|\mathbf{a} - \mathbf{b}|)/4$; our assumption $\mathbf{a} \neq \mathbf{b}$ implies that $\epsilon_0 > 0$. By the definition of the limit, there exists M_1 such that $|\mathbf{a}_m - \mathbf{a}| < \epsilon_0$ when $m > M_1$, and M_2 such that $|\mathbf{a}_m - \mathbf{b}| < \epsilon_0$ when $m > M_2$. Set $M = \max\{M_1, M_2\}$. If $m > M$, then by the triangle inequality,

$$|\mathbf{a} - \mathbf{b}| = |(\mathbf{a} - \mathbf{a}_m) + (\mathbf{a}_m - \mathbf{b})| \leq \underbrace{|\mathbf{a} - \mathbf{a}_m|}_{<\epsilon_0} + \underbrace{|\mathbf{a}_m - \mathbf{b}|}_{<\epsilon_0} < 2\epsilon_0 = \frac{|\mathbf{a} - \mathbf{b}|}{2}.$$

This is a contradiction, so $\mathbf{a} = \mathbf{b}$. □

Theorem 1.5.16 (The arithmetic of limits of sequences). *Let $i \mapsto \mathbf{a}_i$ and $i \mapsto \mathbf{b}_i$ be two sequences of points in $\mathbb{R}^n$, and let $i \mapsto c_i$ be a sequence of numbers. Then*

1. *If $i \mapsto \mathbf{a}_i$ and $i \mapsto \mathbf{b}_i$ both converge, then so does $i \mapsto \mathbf{a}_i + \mathbf{b}_i$, and*
$$\lim_{i\to\infty} (\mathbf{a}_i + \mathbf{b}_i) = \lim_{i\to\infty} \mathbf{a}_i + \lim_{i\to\infty} \mathbf{b}_i. \qquad 1.5.20$$
2. *If $i \mapsto \mathbf{a}_i$ and $i \mapsto c_i$ both converge, then so does $i \mapsto c_i\mathbf{a}_i$, and*
$$\lim_{i\to\infty} c_i\mathbf{a}_i = \left(\lim_{i\to\infty} c_i\right)\left(\lim_{i\to\infty} \mathbf{a}_i\right). \qquad 1.5.21$$
3. *If the sequences $i \mapsto \mathbf{a}_i$ and $i \mapsto \mathbf{b}_i$ converge, then so do the sequences of vectors $i \mapsto \vec{\mathbf{a}}_i$ and $i \mapsto \vec{\mathbf{b}}_i$, and the limit of the dot products is the dot product of the limits:*
$$\lim_{i\to\infty} (\vec{\mathbf{a}}_i \cdot \vec{\mathbf{b}}_i) = \left(\lim_{i\to\infty} \vec{\mathbf{a}}_i\right) \cdot \left(\lim_{i\to\infty} \vec{\mathbf{b}}_i\right). \qquad 1.5.22$$
4. *If $i \mapsto \mathbf{a}_i$ is bounded and $i \mapsto c_i$ converges to 0, then*
$$\lim_{i\to\infty} c_i\mathbf{a}_i = \mathbf{0}. \qquad 1.5.23$$

Theorem 1.5.16, part 4: "If $\mathbf{a}_m$ is bounded" means "if there exists $R < \infty$ such that $|\mathbf{a}_m| \leq R$ for all m". To see that this requirement is necessary, consider

$$c_m = 1/m \quad \text{and} \quad \mathbf{a}_m = \begin{pmatrix} 2 \\ m \end{pmatrix}.$$

Then $m \mapsto c_m$ converges to 0, but

$$\lim_{m\to\infty} (c_m\mathbf{a}_m) \neq \mathbf{0}.$$

This does not contradict part 4, because $\mathbf{a}_m$ is not bounded.

Proof. Proposition 1.5.13 reduces Theorem 1.5.16 to the one-dimensional case; the details are left as Exercise 1.5.17. □

Proposition 1.5.17 says that there is an intimate relationship between limits of sequences and closed sets: closed sets are "closed under limits".

The second part of Proposition 1.5.17 is one possible definition of a closed set.

Proposition 1.5.17 (Sequence in closed set).

1. *Let $i \mapsto \mathbf{x}_i$ be a sequence in a closed set $C \subset \mathbb{R}^n$ converging to $\mathbf{x}_0 \in \mathbb{R}^n$. Then $\mathbf{x}_0 \in C$.*
2. *Conversely, if every convergent sequence in a set $C \subset \mathbb{R}^n$ converges to a point in C, then C is closed.*

Intuitively, this is not hard to see: a convergent sequence in a closed set can't approach a point outside the set; in the other direction, if every convergent sequence in a set C converges to a point in C, then C must own its fence. (But a sequence in a set that is not closed can converge to a point of the fence that is not in the set. For instance, the sequence $1/2, 1/3, 1/4, \ldots$ in the open set $(0, 1)$ converges to 0, which is not in the set. Similarly, the sequence $1, 1.4, 1.41, 1.414, \ldots$ in $\mathbb{Q}$ formed by adding successive digits of $\sqrt{2}$ has the limit $\sqrt{2}$, which is not rational.)

Proof. 1. If $\mathbf{x}_0 \notin C$, then $\mathbf{x}_0 \in (\mathbb{R}^n - C)$, which is open, so there exists $r > 0$ such that $B_r(\mathbf{x}_0) \subset (\mathbb{R}^n - C)$. Then for all m we have $|\mathbf{x}_m - \mathbf{x}_0| \geq r$. But by the definition of convergence, for any $\epsilon > 0$ we have $|\mathbf{x}_m - \mathbf{x}_0| < \epsilon$ for m sufficiently large. Taking $\epsilon = r$, we see that this is a contradiction.

2. You are asked to prove the converse in Exercise 1.5.13. □

Subsequences

Subsequences are a useful tool, as we will see in Section 1.6. They are not particularly difficult, but they require somewhat complicated indices, which are tedious to type and can be difficult to read.

Definition 1.5.18: When we write $j \mapsto a_{i(j)}$, the index i is the function that associates to the position in the subsequence the position of the same entry in the original sequence. For example, if the original sequence is
$$\underset{a_1}{\frac{1}{1}}, \underset{a_2}{\frac{1}{2}}, \underset{a_3}{\frac{1}{3}}, \underset{a_4}{\frac{1}{4}}, \underset{a_5}{\frac{1}{5}}, \underset{a_6}{\frac{1}{6}}, \ldots$$
and the subsequence is
$$\underset{a_{i(1)}}{\frac{1}{2}}, \underset{a_{i(2)}}{\frac{1}{4}}, \underset{a_{i(3)}}{\frac{1}{6}}, \ldots$$
we see that $i(1) = 2$, since $1/2$ is the second entry of the original sequence. Similarly,
$$i(2) = 4, \quad i(3) = 6, \ldots.$$

The proof of Proposition 1.5.19 is left as Exercise 1.5.18, largely to provide practice with the language.

Definition 1.5.18 (Subsequence). A *subsequence* of a sequence $i \mapsto a_i$ is a sequence formed by taking first some element of the original sequence, then another element further on, then another, yet further on .... It is denoted $j \mapsto a_{i(j)}$ where $i(k) > i(j)$ when $k > j$.

You might take all the even terms, or all the odd terms, or all those whose index is a prime, etc. Of course, any sequence is a subsequence of itself. Sometimes the subsequence $j \mapsto a_{i(j)}$ is denoted $a_{i(1)}, a_{i(2)}, \ldots$ or $a_{i_1}, a_{i_2}, \ldots$.

Proposition 1.5.19 (Subsequence of convergent sequence converges). *If a sequence $k \mapsto \mathbf{a}_k$ converges to $\mathbf{a}$, then any subsequence of the sequence converges to the same limit $\mathbf{a}$.*

Limits of functions

Limits like $\lim_{\mathbf{x} \to \mathbf{x}_0} f(\mathbf{x})$ can only be defined if you can approach $\mathbf{x}_0$ by points where f can be evaluated. Thus when $\mathbf{x}_0$ is in the *closure* of the domain of f, it makes sense to ask whether the limit $\lim_{\mathbf{x} \to \mathbf{x}_0} f(\mathbf{x})$ exists. Of course, this includes the case when $\mathbf{x}_0$ is in the domain of f; in that case for the limit to exist we must have

$$\lim_{\mathbf{x} \to \mathbf{x}_0} f(\mathbf{x}) = f(\mathbf{x}_0). \qquad 1.5.24$$

But the interesting case is when $\mathbf{x}_0$ is in the closure of the domain but *not* in the domain. For example, does it make sense to speak of

$$\lim_{x \to 0} (1 + x)^{1/x} ? \qquad 1.5.25$$

Yes; although we cannot evaluate the function at 0, the natural domain of the function includes $(-1,0) \cup (0,\infty)$, and 0 is in the closure of that set.[12]

Definition 1.5.20: Recall (Definition 1.5.8) that if $X \subset \mathbb{R}^n$, its closure $\overline{X}$ is the set of $\mathbf{x} \in \mathbb{R}^n$ such that for all $r > 0$,

$$B_r(\mathbf{x}) \cap X \neq \phi.$$

Definition 1.5.20 is not standard in the United States but is common in France. The standard U.S. version does not allow for the case $\mathbf{x} = \mathbf{x}_0$: it substitutes

$$0 < |\mathbf{x} - \mathbf{x}_0| < \delta$$

for our $|\mathbf{x} - \mathbf{x}_0| < \delta$. The definition we have adopted makes limits better behaved under composition; see Example 1.5.23.

Formula 1.5.27: Again, elegance is not necessary; by Proposition 1.5.14 it is not necessary to show that $|f(\mathbf{x}) - \mathbf{a}| < \epsilon$; it is sufficient to show that

$$|f(\mathbf{x}) - \mathbf{a}| < u(\epsilon),$$

where u is a function of $\epsilon > 0$, such that $u(\epsilon) \to 0$ as $\epsilon \to 0$.

Definition 1.5.20 (Limit of a function). Let X be a subset of $\mathbb{R}^n$, and $\mathbf{x}_0$ a point in $\overline{X}$. A function $\mathbf{f}: X \to \mathbb{R}^m$ has the *limit* $\mathbf{a}$ at $\mathbf{x}_0$:

$$\lim_{\mathbf{x} \to \mathbf{x}_0} \mathbf{f}(\mathbf{x}) = \mathbf{a} \tag{1.5.26}$$

if for all $\epsilon > 0$ there exists $\delta > 0$ such that for all $\mathbf{x} \in X$,

$$|\mathbf{x} - \mathbf{x}_0| < \delta \implies |\mathbf{f}(\mathbf{x}) - \mathbf{a}| < \epsilon. \tag{1.5.27}$$

Note that we must have $|\mathbf{f}(\mathbf{x}) - \mathbf{a}| < \epsilon$ for *all* $\mathbf{x} \in X$ sufficiently close to $\mathbf{x}_0$. For example, there are x arbitrarily close to 0 such that $\sin 1/x = 1/2$, but $\lim_{x \to 0} \sin 1/x \neq 1/2$; in fact the limit does not exist, because there are other x arbitrarily close to 0 such that $\sin 1/x = -1/2$ (or 0 or -1, or any other number in $[-1,1]$).

The definition also requires that there *exist* $\mathbf{x} \in X$ arbitrarily close to $\mathbf{x}_0$; that is what $\mathbf{x}_0 \in \overline{X}$ means. For example, if we think of the domain of the square root function as the nonnegative reals, then $\lim_{x \to -2} \sqrt{x}$ does not exist, because -2 is not in the closure of the domain (which in this case is equal to the domain).

It is important that limits are well defined:

Proposition 1.5.21 (Limit of function is unique). *If a function has a limit, it is unique.*

Proof. Suppose $\mathbf{f}: X \to \mathbb{R}^m$ has two distinct limits, $\mathbf{a}$ and $\mathbf{b}$, and set $\epsilon_0 = (|\mathbf{a} - \mathbf{b}|)/4$; the assumption that $\mathbf{a} \neq \mathbf{b}$ implies that $\epsilon_0 > 0$. By the definition of limit, there exist $\delta_1 > 0$ such that when $|\mathbf{x} - \mathbf{x}_0| < \delta_1$ and $\mathbf{x} \in X$, then $|\mathbf{f}(\mathbf{x}) - \mathbf{a}| < \epsilon_0$, and $\delta_2 > 0$ such that when $|\mathbf{x} - \mathbf{x}_0| < \delta_2$ and $\mathbf{x} \in X$, then $|\mathbf{f}(\mathbf{x}) - \mathbf{b}| < \epsilon_0$. Let δ be the smaller of δ_1, δ_2. Then (since $\mathbf{x}_0 \in \overline{X}$) there exists $\mathbf{x} \in X$ satisfying $|\mathbf{x} - \mathbf{x}_0| < \delta$, and for that $\mathbf{x}$ we have $|\mathbf{f}(\mathbf{x}) - \mathbf{a}| < \epsilon_0$ and $|\mathbf{f}(\mathbf{x}) - \mathbf{b}| < \epsilon_0$. The triangle inequality then gives

$$|\mathbf{a}-\mathbf{b}| = |\mathbf{f}(\mathbf{x})-\mathbf{a}+\mathbf{b}-\mathbf{f}(\mathbf{x})| \leq |\mathbf{f}(\mathbf{x})-\mathbf{a}|+|\mathbf{f}(\mathbf{x})-\mathbf{b}| < 2\epsilon_0 = \frac{|\mathbf{a}-\mathbf{b}|}{2}, \tag{1.5.28}$$

which is a contradiction. □

Limits behave well with respect to compositions.

[12] We say that the natural domain "includes" $(-1,0) \cup (0,\infty)$ instead of "is" $(-1,0) \cup (0,\infty)$ because some people might argue that -3 is also in the natural domain, since every real number, whether positive or negative, has a unique real cube root. Others might argue that there is no reason to prefer the real root over the complex roots. In any case, any time you want to use the function $(1+x)^{1/x}$ for x complex, you will have to specify which roots you are talking about.

Theorem 1.5.22 (Limit of a composition). *Let $U \subset \mathbb{R}^n$, $V \subset \mathbb{R}^m$ be subsets, and $\mathbf{f} : U \to V$ and $\mathbf{g} : V \to \mathbb{R}^k$ be mappings, so that $\mathbf{g} \circ \mathbf{f}$ is defined in U. If $\mathbf{x}_0$ is a point of U and*

$$\mathbf{y}_0 \stackrel{\text{def}}{=} \lim_{\mathbf{x} \to \mathbf{x}_0} \mathbf{f}(\mathbf{x}) \quad \textit{and} \quad \mathbf{z}_0 \stackrel{\text{def}}{=} \lim_{\mathbf{y} \to \mathbf{y}_0} \mathbf{g}(\mathbf{y}) \qquad 1.5.29$$

both exist, then $\lim_{\mathbf{x} \to \mathbf{x}_0} \mathbf{g} \circ \mathbf{f}(\mathbf{x})$ exists, and

$$\lim_{\mathbf{x} \to \mathbf{x}_0} (\mathbf{g} \circ \mathbf{f})(\mathbf{x}) = \mathbf{z}_0. \qquad 1.5.30$$

There is no natural condition that will guarantee that

$$\mathbf{f}(\mathbf{x}) \neq \mathbf{f}(\mathbf{x}_0);$$

if we had required $\mathbf{x} \neq \mathbf{x}_0$ in our definition of limit, this proof would not work.

Proof. For all $\epsilon > 0$ there exists $\delta_1 > 0$ such that if $|\mathbf{y} - \mathbf{y}_0| < \delta_1$, then $|\mathbf{g}(\mathbf{y}) - \mathbf{z}_0| < \epsilon$. Next, there exists $\delta > 0$ such that if $|\mathbf{x} - \mathbf{x}_0| < \delta$, then $|\mathbf{f}(\mathbf{x}) - \mathbf{y}_0| < \delta_1$. Hence

$$\big|\mathbf{g}\big(\mathbf{f}(\mathbf{x})\big) - \mathbf{z}_0\big| < \epsilon \quad \text{when} \quad |\mathbf{x} - \mathbf{x}_0| < \delta. \quad \square \qquad 1.5.31$$

Example 1.5.23. Consider the functions $f, g : \mathbb{R} \to \mathbb{R}$

$$f(x) = \begin{cases} x \sin \frac{1}{x} & \text{if } x \neq 0 \\ 0 & \text{if } x = 0 \end{cases} \quad \text{and} \quad g(y) = \begin{cases} 1 & \text{if } y \neq 0 \\ 0 & \text{if } y = 0 \end{cases}$$

The standard definition of limit requires $0 < |y - 0| < \delta$, which gives

$$\lim_{y \to 0} g(y) = 1 \quad \text{and} \quad \lim_{x \to 0} f(x) = 0,$$

but $\lim_{x \to 0} (g \circ f)(x)$ does not exist. With Definition 1.5.20, $\lim_{y \to 0} g(y)$ doesn't exist, but Theorem 1.5.22 is true (vacuously: the hypotheses aren't satisfied). $\triangle$

Much of the subtlety of limits in higher dimensions is that there are many different ways to approach a point in the domain, and different approaches may yield different limits, in which case the limit does not exist.

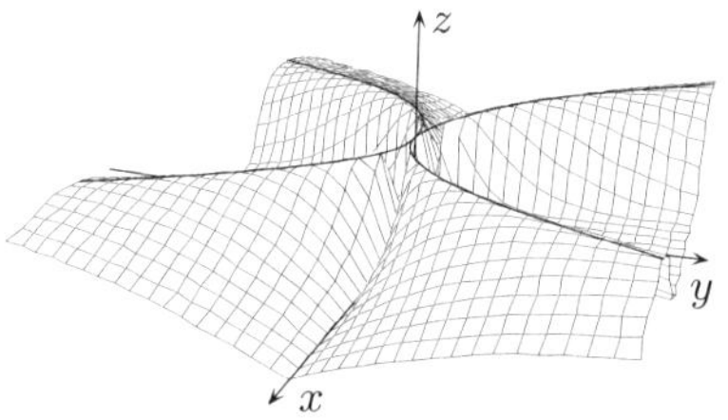

FIGURE 1.5.6.

As you are asked to prove in Exercise 1.7, the function of Example 1.5.24 is continuous (except at the origin, where it is not defined). Its value is $1/e$ along the "crest line" $y = \pm x^2$, but the function vanishes on both axes, forming a very deep canyon along the x-axis. Any straight line $y = mx$ with $m \neq 0$ will enter the broad valley along the y-axis before it reaches the origin; along any such path the limit of f exists and is 0.

Example 1.5.24 (Different approaches that give different limits). Consider the function $f : \mathbb{R}^2 - \begin{pmatrix} 0 \\ 0 \end{pmatrix} \to \mathbb{R}$

$$f\begin{pmatrix} x \\ y \end{pmatrix} = \begin{cases} \dfrac{|y| e^{-|y|/x^2}}{x^2} & \text{if } x \neq 0 \\ 0 & \text{if } x = 0,\ y \neq 0 \end{cases} \qquad 1.5.32$$

shown in Figure 1.5.6. Does $\lim_{\begin{pmatrix} x \\ y \end{pmatrix} \to \begin{pmatrix} 0 \\ 0 \end{pmatrix}} f\begin{pmatrix} x \\ y \end{pmatrix}$ exist?

A first idea is to approach the origin along straight lines. Set $y = mx$. When $m \neq 0$, the limit becomes

$$\lim_{x \to 0} \left| \frac{m}{x} \right| e^{-\left| \frac{m}{x} \right|}; \qquad 1.5.33$$

this limit exists and is always 0, for all values of m. Indeed,

$$\lim_{t \to 0} \frac{1}{t} e^{-1/t} = \lim_{s \to \infty} \frac{s}{e^s} = 0. \qquad 1.5.34$$

So no matter how you approach the origin along straight lines, the limit exists and is always 0. But if you approach the origin along the parabola $y = kx^2$ (i.e., set $y = kx^2$ and let $x \to 0$), you find something quite different:

$$\lim_{x\to 0} |k|e^{-|k|} = |k|e^{-|k|}, \qquad 1.5.35$$

which is some number that varies between 0 and $1/e$, and which depends on k (i.e., on the particular parabola along which you are approaching the origin; see Exercise 1.6.6). Thus if you approach the origin in different ways, the limits may be different. $\triangle$

A mapping $\mathbf{f} : \mathbb{R}^n \to \mathbb{R}^m$ is an "$\mathbb{R}^m$-valued" mapping; its argument is in $\mathbb{R}^n$ and its values are in $\mathbb{R}^m$. Such a mapping is sometimes written in terms of the coordinate functions. For example, the mapping $\mathbf{f} : \mathbb{R}^2 \to \mathbb{R}^3$,

Often $\mathbb{R}^m$-valued mappings are called "vector-valued" mappings, but usually we think of their values as points rather than vectors.

We denote a mapping whose values are points in $\mathbb{R}^m$ with a boldface letter without arrow: $\mathbf{f}$. Sometimes we do want to think of the values of a mapping $\mathbb{R}^n \to \mathbb{R}^m$ as vectors: when we are thinking of vector fields. We denote a vector field with an arrow: $\vec{F}$ or $\vec{\mathbf{f}}$.

$$\mathbf{f}\begin{pmatrix} x \\ y \end{pmatrix} = \begin{pmatrix} xy \\ x^2y \\ x-y \end{pmatrix} \quad \text{can be written} \quad \mathbf{f} = \begin{pmatrix} f_1 \\ f_2 \\ f_3 \end{pmatrix}, \qquad 1.5.36$$

where $f_1\begin{pmatrix} x \\ y \end{pmatrix} = xy$, $f_2\begin{pmatrix} x \\ y \end{pmatrix} = x^2y$, and $f_3\begin{pmatrix} x \\ y \end{pmatrix} = x - y$.

Proposition 1.5.25 (Convergence by coordinates). *Suppose*

$$U \subset \mathbb{R}^n, \quad \mathbf{f} = \begin{pmatrix} f_1 \\ \vdots \\ f_m \end{pmatrix} : U \to \mathbb{R}^m, \quad \textit{and} \quad \mathbf{x}_0 \in \overline{U}.$$

Then

$$\lim_{\mathbf{x}\to\mathbf{x}_0} \mathbf{f}(\mathbf{x}) = \begin{pmatrix} a_1 \\ \vdots \\ a_m \end{pmatrix} \quad \textit{if and only if} \quad \lim_{\mathbf{x}\to\mathbf{x}_0} f_i(\mathbf{x}) = a_i,\ i = 1, \dots, m.$$

Theorem 1.5.26 (Limits of functions). *Let U be a subset of $\mathbb{R}^n$ and let $\mathbf{f}$ and $\mathbf{g}$ be functions $U \to \mathbb{R}^m$ and h a function $U \to \mathbb{R}$.*

In Theorem 1.5.26, when we specify that $\lim_{\mathbf{x}\to\mathbf{x}_0} \mathbf{f}(\mathbf{x})$ exists, this means that $\mathbf{x}_0$ is in the closure of U. It does not have to be in U.

1. *If $\lim_{\mathbf{x}\to\mathbf{x}_0} \mathbf{f}(\mathbf{x})$ and $\lim_{\mathbf{x}\to\mathbf{x}_0} \mathbf{g}(\mathbf{x})$ exist, then $\lim_{\mathbf{x}\to\mathbf{x}_0} (\mathbf{f}+\mathbf{g})(\mathbf{x})$ exists, and*

$$\lim_{\mathbf{x}\to\mathbf{x}_0} (\mathbf{f}+\mathbf{g})(\mathbf{x}) = \lim_{\mathbf{x}\to\mathbf{x}_0} \mathbf{f}(\mathbf{x}) + \lim_{\mathbf{x}\to\mathbf{x}_0} \mathbf{g}(\mathbf{x}). \qquad 1.5.37$$

2. *If $\lim_{\mathbf{x}\to\mathbf{x}_0} \mathbf{f}(\mathbf{x})$ and $\lim_{\mathbf{x}\to\mathbf{x}_0} h(\mathbf{x})$ exist, then $\lim_{\mathbf{x}\to\mathbf{x}_0} (h\mathbf{f})(\mathbf{x})$ exists, and*

$$\lim_{\mathbf{x}\to\mathbf{x}_0} (h\mathbf{f})(\mathbf{x}) = \lim_{\mathbf{x}\to\mathbf{x}_0} h(\mathbf{x}) \lim_{\mathbf{x}\to\mathbf{x}_0} \mathbf{f}(\mathbf{x}). \qquad 1.5.38$$

3. *If $\lim_{\mathbf{x}\to\mathbf{x}_0} \mathbf{f}(\mathbf{x})$ exists, and $\lim_{\mathbf{x}\to\mathbf{x}_0} h(\mathbf{x})$ exists and is different from 0, then $\lim_{\mathbf{x}\to x_0} (\frac{\mathbf{f}}{h})(\mathbf{x})$ exists, and*

$$\lim_{\mathbf{x}\to\mathbf{x}_0} \left(\frac{\mathbf{f}}{h}\right)(\mathbf{x}) = \frac{\lim_{\mathbf{x}\to\mathbf{x}_0} \mathbf{f}(\mathbf{x})}{\lim_{\mathbf{x}\to\mathbf{x}_0} h(\mathbf{x})}. \qquad 1.5.39$$

4. *Set* $(\mathbf{f}\cdot\mathbf{g})(\mathbf{x}) \stackrel{\text{def}}{=} \mathbf{f}(\mathbf{x})\cdot\mathbf{g}(\mathbf{x})$. *If* $\lim_{\mathbf{x}\to\mathbf{x}_0}\mathbf{f}(\mathbf{x})$ *and* $\lim_{\mathbf{x}\to\mathbf{x}_0}\mathbf{g}(\mathbf{x})$ *exist, then* $\lim_{\mathbf{x}\to\mathbf{x}_0}(\mathbf{f}\cdot\mathbf{g})(\mathbf{x})$ *exists, and*

$$\lim_{\mathbf{x}\to\mathbf{x}_0}(\mathbf{f}\cdot\mathbf{g})(\mathbf{x}) = \lim_{\mathbf{x}\to\mathbf{x}_0}\mathbf{f}(\mathbf{x})\cdot\lim_{\mathbf{x}\to\mathbf{x}_0}\mathbf{g}(\mathbf{x}). \qquad 1.5.40$$

5. *If* $\mathbf{f}$ *is bounded on* U *(i.e., there exists* R *such that* $|\mathbf{f}(\mathbf{x})|\le R$ *for all* $\mathbf{x}\in U$*), and if* $\lim_{\mathbf{x}\to\mathbf{x}_0}h(\mathbf{x})=0$, *then*

$$\lim_{\mathbf{x}\to\mathbf{x}_0}(h\mathbf{f})(\mathbf{x}) = \mathbf{0}. \qquad 1.5.41$$

6. *If* $\lim_{\mathbf{x}\to\mathbf{x}_0}\mathbf{f}(\mathbf{x}) = \mathbf{0}$ *and* h *is bounded, then*

$$\lim_{\mathbf{x}\to\mathbf{x}_0}(h\mathbf{f})(\mathbf{x}) = \mathbf{0}. \qquad 1.5.42$$

Proof of Proposition 1.5.25. Let's go through the picturesque description again. For the implication $\Rightarrow$ ("only if"), the challenger hands you an $\epsilon>0$. You give it to a teammate who returns a δ with the guarantee that when $|\mathbf{x}-\mathbf{x}_0|<\delta$ and $\mathbf{f}(\mathbf{x})$ is defined, then $|\mathbf{f}(\mathbf{x})-\mathbf{a}|<\epsilon$. You pass on the same δ, and a_i, to the challenger, with the explanation

$$|f_i(\mathbf{x})-a_i| \le |\mathbf{f}(\mathbf{x})-\mathbf{a}| < \epsilon. \qquad 1.5.43$$

For the implication $\Leftarrow$ ("if"), the challenger hands you an $\epsilon>0$. You pass this ϵ to your teammates, who know how to deal with the coordinate functions. They hand you back $\delta_1,\dots,\delta_m$, with the guarantee that when $|\mathbf{x}-\mathbf{x}_0|<\delta_i$, then $|f_i(\mathbf{x})-a_i|<\epsilon$. You look through these and select the smallest one, which you call δ and pass on to the challenger, with the message

Inequality 1.5.44: If you gave $\epsilon/\sqrt{m}$ to your teammates, as in the proof of Proposition 1.5.13, you would end up with

$$|\mathbf{f}(\mathbf{x})-\mathbf{a}|<\epsilon,$$

rather than $|\mathbf{f}(\mathbf{x})-\mathbf{a}|<\epsilon\sqrt{m}$. In some sense this is more "elegant." But Proposition 1.5.14 says that it is mathematically *just as good* to arrive at anything that goes to 0 when ϵ goes to 0.

"If $|\mathbf{x}-\mathbf{x}_0|<\delta$, then $|\mathbf{x}-\mathbf{x}_0|<\delta_i$, so that $|f_i(\mathbf{x})-a_i|<\epsilon$, so that

$$\begin{aligned}|f(\mathbf{x})-\mathbf{a}| &= \sqrt{\big(f_1(\mathbf{x})-a_1\big)^2+\dots+\big(f_m(\mathbf{x})-a_m\big)^2}\\ &< \underbrace{\sqrt{\epsilon^2+\dots+\epsilon^2}}_{m\text{ terms}} = \epsilon\sqrt{m},\end{aligned} \qquad 1.5.44$$

which goes to 0 as ϵ goes to 0." You win! □

Proof of Theorem 1.5.26. The proofs of all these statements are very similar; we will do only part 4, which is the hardest.

4. Choose ϵ (think of the challenger giving it to you). To lighten notation, set $\mathbf{a}=\lim_{\mathbf{x}\to\mathbf{x}_0}\mathbf{f}(\mathbf{x})$ and $\mathbf{b}=\lim_{\mathbf{x}\to\mathbf{x}_0}\mathbf{g}(\mathbf{x})$. Then

- Find a δ_1 such that when $|\mathbf{x}-\mathbf{x}_0|<\delta_1$, then

$$|\mathbf{g}(\mathbf{x})-\mathbf{b}|<\epsilon. \qquad 1.5.45$$

- Next find a δ_2 such that when $|\mathbf{x}-\mathbf{x}_0|<\delta_2$, then

$$|\mathbf{f}(\mathbf{x})-\mathbf{a}|<\epsilon. \qquad 1.5.46$$

Equation 1.5.47: The first inequality is the triangle inequality. The equality in the 3rd line uses the distributivity of the dot product. The inequality in the 4th line is Schwarz's inequality.

Now set δ to be the smallest of δ_1 and δ_2, and consider the sequence of inequalities

$$\begin{aligned} |\mathbf{f}(\mathbf{x}) \cdot \mathbf{g}(\mathbf{x}) - \mathbf{a} \cdot \mathbf{b}| &= |\mathbf{f}(\mathbf{x}) \cdot \mathbf{g}(\mathbf{x}) \underbrace{- \mathbf{a} \cdot \mathbf{g}(\mathbf{x}) + \mathbf{a} \cdot \mathbf{g}(\mathbf{x})}_{=0} - \mathbf{a} \cdot \mathbf{b}| \\ &\le |\mathbf{f}(\mathbf{x}) \cdot \mathbf{g}(\mathbf{x}) - \mathbf{a} \cdot \mathbf{g}(\mathbf{x})| + |\mathbf{a} \cdot \mathbf{g}(\mathbf{x}) - \mathbf{a} \cdot \mathbf{b}| \\ &= \big|(\mathbf{f}(\mathbf{x}) - \mathbf{a}) \cdot \mathbf{g}(\mathbf{x})\big| + \big|\mathbf{a} \cdot (\mathbf{g}(\mathbf{x}) - \mathbf{b})\big| \\ &\le \big|(\mathbf{f}(\mathbf{x}) - \mathbf{a})\big|\, |\mathbf{g}(\mathbf{x})| + |\mathbf{a}|\, \big|(\mathbf{g}(\mathbf{x}) - \mathbf{b})\big| \\ &\le \epsilon |\mathbf{g}(\mathbf{x})| + \epsilon |\mathbf{a}| = \epsilon\big(|\mathbf{g}(\mathbf{x})| + |\mathbf{a}|\big). \end{aligned} \qquad 1.5.47$$

Since $\mathbf{g}(\mathbf{x})$ is a function, not a point, we might worry that it could get big faster than ϵ gets small. But we know that when $|\mathbf{x} - \mathbf{x}_0| < \delta$, then $|\mathbf{g}(\mathbf{x}) - \mathbf{b}| < \epsilon$, which gives

$$|\mathbf{g}(\mathbf{x})| < \epsilon + |\mathbf{b}|. \qquad 1.5.48$$

So continuing inequality 1.5.47, we get

$$\epsilon\big(|\mathbf{g}(\mathbf{x})| + |\mathbf{a}|\big) < \epsilon\big(\epsilon + |\mathbf{b}|\big) + \epsilon|\mathbf{a}|, \qquad 1.5.49$$

which goes to 0 as ϵ goes to 0. □

Continuous functions

Continuity is *the* fundamental notion of topology, and it arises throughout calculus also. It took mathematicians 200 years to arrive at a correct definition. (Historically, we have our presentation out of order: it was the search for a usable definition of continuity that led to the correct definition of limits.)

Definition 1.5.27: A map $\mathbf{f}$ is continuous at $\mathbf{x}_0$ if you can make the difference between $\mathbf{f}(\mathbf{x})$ and $\mathbf{f}(\mathbf{x}_0)$ arbitrarily small by choosing $\mathbf{x}$ sufficiently close to $\mathbf{x}_0$. Note that $|\mathbf{f}(\mathbf{x}) - \mathbf{f}(\mathbf{x}_0)|$ must be small for *all* $\mathbf{x}$ "sufficiently close" to $\mathbf{x}_0$. It is not enough to find a δ such that for one particular value of $\mathbf{x}$ the statement is true.

However, for different values of $\mathbf{x}_0$, the "sufficiently close" (i.e., the choice of δ) can be *different*.

We started by trying to write the above in one simple sentence, and found it was impossible to do so and avoid mistakes. If definitions of continuity sound stilted, it is because any attempt to stray from the "for all this, there exists that..." inevitably leads to ambiguity.

Definition 1.5.27 (Continuous function). Let $X \subset \mathbb{R}^n$. A mapping $\mathbf{f} : X \to \mathbb{R}^m$ is *continuous at* $\mathbf{x}_0 \in X$ if

$$\lim_{\mathbf{x} \to \mathbf{x}_0} \mathbf{f}(\mathbf{x}) = \mathbf{f}(\mathbf{x}_0); \qquad 1.5.50$$

$\mathbf{f}$ is *continuous on* X if it is continuous at every point of X. Equivalently, $\mathbf{f} : X \to \mathbb{R}^m$ is continuous at $\mathbf{x}_0 \in X$ if and only if for every $\epsilon > 0$, there exists $\delta > 0$ such that when $|\mathbf{x} - \mathbf{x}_0| < \delta$, then $|\mathbf{f}(\mathbf{x}) - \mathbf{f}(\mathbf{x}_0)| < \epsilon$.

The following criterion shows that it is enough to consider $\mathbf{f}$ on sequences converging to $\mathbf{x}_0$.

Proposition 1.5.28 (Criterion for continuity). *Let $X \subset \mathbb{R}^n$. A function $\mathbf{f} : X \to \mathbb{R}^m$ is continuous at $\mathbf{x}_0 \in X$ if and only if for every sequence $\mathbf{x}_i \in X$ converging to $\mathbf{x}_0$,*

$$\lim_{i \to \infty} \mathbf{f}(\mathbf{x}_i) = \mathbf{f}(\mathbf{x}_0). \qquad 1.5.51$$

Topology used to be called "analysis situs" or "geometria situs." Johann Listing disliked the terms and introduced the word "topology" in a letter in 1836; in his book *Vorstudien zur Topologie* he wrote that "by topology we mean ... the laws of connection, of relative position and of succession of points, lines, surfaces, bodies and their parts, or aggregates in space, always without regard to matters of measure or quantity."

The new word became widely adopted after Solomon Lefschetz used it around 1930. According to Lefschetz, "If it's just turning the crank it's algebra, but if it's got an idea in it, it's topology."

Harvard mathematician Barry Mazur once said that for him, topology meant explanation.

Theorems 1.5.29 and 1.5.30 are reformulations of Theorems 1.5.26 and 1.5.22. You are asked to prove them in Exercise 1.22.

When $\mathbf{f}$ and $\mathbf{g}$ are scalar-valued functions f and g, the dot product of part 4 is the same as the ordinary product.

Proof. Suppose the ϵ, δ condition is satisfied, and let $\mathbf{x}_i$, $i = 1, 2, \dots$ be a sequence in X that converges to $\mathbf{x}_0 \in X$. We must show that the sequence $i \mapsto \mathbf{f}(\mathbf{x}_i)$ converges to $\mathbf{f}(\mathbf{x}_0)$: that for any $\epsilon > 0$, there exists N such that when $n > N$ we have $|\mathbf{f}(\mathbf{x}_n) - \mathbf{f}(\mathbf{x}_0)| < \epsilon$. Find δ such that $|\mathbf{x} - \mathbf{x}_0| < \delta$ implies $|\mathbf{f}(\mathbf{x}) - \mathbf{f}(\mathbf{x}_0)| < \epsilon$. Since the sequence $\mathbf{x}_i$ is convergent, there exists N such that if $n > N$, then $|\mathbf{x}_n - \mathbf{x}_0| < \delta$. Clearly this N works.

For the converse, remember how to negate sequences of quantifiers (Section 0.2). Suppose the ϵ, δ condition is not satisfied; then there exists $\epsilon_0 > 0$ such that for all δ, there exists $\mathbf{x} \in X$ such that $|\mathbf{x} - \mathbf{x}_0| < \delta$ but $|\mathbf{f}(\mathbf{x}) - \mathbf{f}(\mathbf{x}_0)| \geq \epsilon_0$. Let $\delta_n = 1/n$, and let $\mathbf{x}_n \in X$ be such a point, satisfying

$$|\mathbf{x}_n - \mathbf{x}_0| < \frac{1}{n} \quad \text{and} \quad |\mathbf{f}(\mathbf{x}_n) - \mathbf{f}(\mathbf{x}_0)| \geq \epsilon_0. \qquad 1.5.52$$

The first part shows that the sequence $\mathbf{x}_n$ converges to $\mathbf{x}_0$, and the second part shows that $\mathbf{f}(\mathbf{x}_n)$ does not converge to $\mathbf{f}(\mathbf{x}_0)$. □

Theorem 1.5.29 (Combining continuous mappings). *Let U be a subset of $\mathbb{R}^n$, $\mathbf{f}$ and $\mathbf{g}$ mappings $U \to \mathbb{R}^m$, and h a function $U \to \mathbb{R}$.*

1. *If $\mathbf{f}$ and $\mathbf{g}$ are continuous at $\mathbf{x}_0 \in U$, so is $\mathbf{f} + \mathbf{g}$.*
2. *If $\mathbf{f}$ and h are continuous at $\mathbf{x}_0 \in U$, so is $h\mathbf{f}$.*
3. *If $\mathbf{f}$ and h are continuous at $\mathbf{x}_0 \in U$, and $h(\mathbf{x}_0) \neq 0$, so is $\frac{\mathbf{f}}{h}$.*
4. *If $\mathbf{f}$ and $\mathbf{g}$ are continuous at $\mathbf{x}_0 \in U$, so is $\mathbf{f} \cdot \mathbf{g}$*
5. *If h is continuous at $\mathbf{x}_0 \in \overline{U}$ with $h(\mathbf{x}_0) = 0$, and there exist $C, \delta > 0$ such that $|\mathbf{f}(\mathbf{x})| \leq C$ for $\mathbf{x} \in U$, $|\mathbf{x} - \mathbf{x}_0| < \delta$, then the map*

$$\mathbf{x} \mapsto \begin{cases} h(\mathbf{x})\mathbf{f}(\mathbf{x}) & \text{for } \mathbf{x} \in U \\ \mathbf{0} & \text{if } \mathbf{x} = \mathbf{x}_0 \end{cases} \quad \text{is continuous at } \mathbf{x}_0.$$

Theorem 1.5.30 (Composition of continuous functions). *Let $U \subset \mathbb{R}^n$, $V \subset \mathbb{R}^m$ be subsets, and $\mathbf{f} : U \to V$ and $\mathbf{g} : V \to \mathbb{R}^k$ be mappings, so that $\mathbf{g} \circ \mathbf{f}$ is defined. If $\mathbf{f}$ is continuous at $\mathbf{x}_0$ and $\mathbf{g}$ is continuous at $\mathbf{f}(\mathbf{x}_0)$, then $\mathbf{g} \circ \mathbf{f}$ is continuous at $\mathbf{x}_0$.*

Theorems 1.5.29 and 1.5.30 show that if $f : \mathbb{R}^n \to \mathbb{R}$ is given by a formula involving addition, multiplication, division, and composition of continuous functions, and f is defined at a point $\mathbf{x}_0$, then f is continuous at $\mathbf{x}_0$. But if there is a division we need to worry: are we dividing by 0? We also need to worry when we see tan: what happens if the argument of tan is $\pi/2 + k\pi$? Similarly, $\ln, \cot, \sec, \csc$ all introduce complications. In one dimension, these problems are often addressed using l'Hôpital's rule (although Taylor expansions often work better).

We can now write down a fairly large collection of continuous functions on $\mathbb{R}^n$: polynomials and rational functions.

A *monomial function* $\mathbb{R}^n \to \mathbb{R}$ is an expression of the form $x_1^{k_1} \cdots x_n^{k_n}$ with integer exponents $k_1, \ldots, k_n \geq 0$. For instance, x^2yz^5 is a monomial on $\mathbb{R}^3$, and $x_1x_2x_4^2$ is a monomial on $\mathbb{R}^4$ (or perhaps on $\mathbb{R}^n$ with $n > 4$). A polynomial function is a finite sum of monomials with real coefficients, like $x^2y + 3yz$. A *rational function* is a ratio of two polynomials, like $\frac{x+y}{xy+z^2}$.

> "Vanishes" means "equals 0".

Corollary 1.5.31 (Continuity of polynomials and rational functions).

1. *Any polynomial function $\mathbb{R}^n \to \mathbb{R}$ is continuous on all of $\mathbb{R}^n$.*
2. *Any rational function is continuous on the subset of $\mathbb{R}^n$ where the denominator does not vanish.*

Uniform continuity

When a function f is continuous at $\mathbf{x}$, how δ depends on ϵ tells us "how nicely f is continuous" at $\mathbf{x}$. When the same δ works for all $\mathbf{x}$, the function is continuous "the same way" everywhere. Such a function is said to be *uniformly continuous*.

> Often, proving that a single continuous function is uniformly continuous is a major result. For example, as of this writing, a major outstanding problem in dynamical systems is proving that the Mandelbrot set is locally connected. If a certain function could be shown to be uniformly continuous, the local connectivity of the Mandelbrot set could be proved tomorrow.

Definition 1.5.32 (Uniformly continuous function). Let X be a subset of $\mathbb{R}^n$. A mapping $\mathbf{f} : X \to \mathbb{R}^m$ is *uniformly continuous* on X if for all $\epsilon > 0$ there exists $\delta > 0$ such that for all $\mathbf{x}, \mathbf{y} \in X$, if $|\mathbf{x} - \mathbf{y}| < \delta$, then $|\mathbf{f}(\mathbf{x}) - \mathbf{f}(\mathbf{y})| < \epsilon$.

Theorem 1.5.33 (Linear functions are uniformly continuous). *Every linear transformation $T : \mathbb{R}^n \to \mathbb{R}^m$ is uniformly continuous.*

Proof. We have to show that for all $\epsilon > 0$ there exists $\delta > 0$ such that for any $\vec{\mathbf{x}}_0, \vec{\mathbf{x}}_1 \in \mathbb{R}^n$ with $|\vec{\mathbf{x}}_1 - \vec{\mathbf{x}}_0| < \delta$, then $|T\vec{\mathbf{x}}_1 - T\vec{\mathbf{x}}_0| < \epsilon$. Take $\delta = \frac{\epsilon}{|T|+1}$, where $|T|$ is the length of the matrix representing T. Then

$$|T\vec{\mathbf{x}}_1 - T\vec{\mathbf{x}}_0| = |T(\vec{\mathbf{x}}_1 - \vec{\mathbf{x}}_0)| \leq |T||\vec{\mathbf{x}}_1 - \vec{\mathbf{x}}_0|$$
$$< (|T| + 1)\delta = \epsilon. \quad \square$$

1.5.53

> Inequality 1.5.53: the $+1$ in $(|T|+1)$ is there to deal with the case where T is the zero linear transformation. Otherwise we could take $\delta = \epsilon/|T|$.
>
> For any C such that for all $\vec{\mathbf{x}}$ we have $|T\vec{\mathbf{x}}| \leq C|\vec{\mathbf{x}}|$, we can take $\delta = \epsilon/C$. The smallest such C is
>
> $$\|T\| \overset{\text{def}}{=} \sup_{\vec{\mathbf{x}} \neq \vec{\mathbf{0}}} \frac{|T\vec{\mathbf{x}}|}{|\vec{\mathbf{x}}|} = \sup_{|\vec{\mathbf{x}}|=1} |T\vec{\mathbf{x}}|.$$
>
> This number $\|T\|$, called the *norm* of T, is very important but hard to compute; it is discussed in Section 2.9.

Series of vectors

Many of the most interesting sequences arise as partial sums of series.

Definition 1.5.34 (Convergent series of vectors). A series $\sum_{i=1}^{\infty} \vec{\mathbf{a}}_i$ is *convergent* if the sequence $n \mapsto \vec{\mathbf{s}}_n$ of partial sums $\vec{\mathbf{s}}_n \overset{\text{def}}{=} \sum_{i=1}^{n} \vec{\mathbf{a}}_i$ is a convergent sequence of vectors. In that case the infinite sum is

$$\sum_{i=1}^{\infty} \vec{\mathbf{a}}_i \overset{\text{def}}{=} \lim_{n \to \infty} \vec{\mathbf{s}}_n.$$

1.5.54

> Figure 1.5.7 shows a convergent series of vectors.

Below, *absolute convergence* means that the absolute values converge.

Proposition 1.5.35 (Absolute convergence implies convergence).
If $\sum_{i=1}^{\infty} |\vec{a}_i|$ converges, then $\sum_{i=1}^{\infty} \vec{a}_i$ converges.

Proof. Set $\vec{a}_i = \begin{bmatrix} a_{1,i} \\ \vdots \\ a_{n,i} \end{bmatrix}$. Then $|a_{k,i}| \leq |\vec{a}_i|$, so $\sum_{i=1}^{\infty} |a_{k,i}|$ converges, so by Theorem 0.5.8, $\sum_{i=1}^{\infty} a_{k,i}$ converges. Proposition 1.5.35 then follows from Proposition 1.5.13. $\square$

Proposition 1.5.35 is important. We use it in Section 2.8 to prove that Newton's method converges. We use it here to prove an extraordinary result named after Euler (equation 1.5.55) and to show that geometric series of matrices can be treated like geometric series of numbers.

Complex exponentials and trigonometric functions

The formula
$$e^{it} = \cos t + i \sin t, \qquad 1.5.55$$
known as *Euler's formula*, was actually found by Roger Cotes in 1714 before Euler rediscovered it in 1748. It is quite remarkable that both exponentials (in the context of compound interest), and sines and cosines (in geometry and especially astronomy) had been in use for 2000 years, but that no one had seen that they were intimately related. Of course, this requires defining e^{it}, or more generally, e^z for z complex.

Proposition 1.5.36 (Complex exponentials). For any complex number z, the series
$$e^z \stackrel{\text{def}}{=} 1 + z + \frac{z^2}{2!} + \cdots = \sum_{k=0}^{\infty} \frac{z^k}{k!} \quad \text{converges.} \qquad 1.5.56$$

(Recall that $0! = 1$; when $k = 0$ we are not dividing by 0.)

Proof. In one-variable calculus, you learn (using the ratio test, for instance) that the series $\sum_{k=0}^{\infty} \frac{x^k}{k!}$ converges to e^x for all $x \in \mathbb{R}$. Since
$$\left| \frac{z^k}{k!} \right| = \frac{|z|^k}{k!}, \qquad 1.5.57$$
the series defining e^z converges absolutely, hence converges by Proposition 1.5.35. (We need Proposition 1.5.35, not Proposition 0.5.8, because z is complex.) $\square$

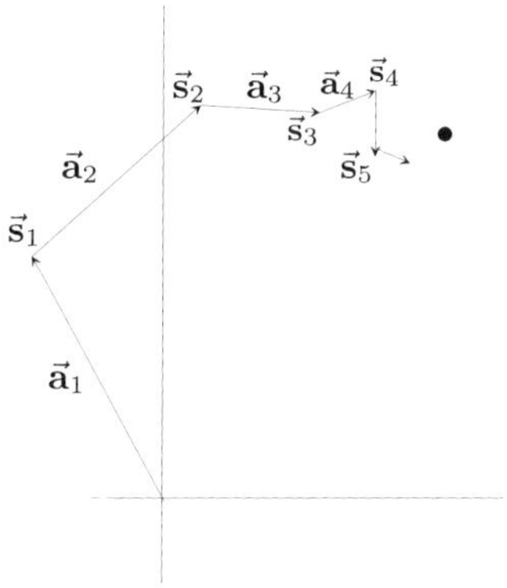

FIGURE 1.5.7.
A convergent series of vectors. The kth partial sum is gotten by putting the first k vectors nose to tail.

When z is an integer (or more generally, a rational number), the definition of e^z given in Proposition 1.5.36 agrees with the usual notion of the power of a number. You can confirm this by showing by direct computation (justified by Exercise 0.5.5) that for e^z as defined in 1.5.56,
$e^{a+b} = e^a e^b$, hence $e^2 = e^1 e^1$.

In 1614, John Napier introduced logarithms, which use the formula $e^{a+b} = e^a e^b$ to turn multiplication into addition, a great labor-saving trick in pre-calculator days. Indeed Laplace (1749–1827) wrote that logarithms " ... by shortening the labours, doubled the life of the astronomer".

Since Euler's formula relates the exponential to trigonometric functions, one might expect that trigonometric functions could also be used to turn multiplication into addition. Surprisingly, more than a century before the discovery of Euler's formula, astronomers in the laboratory of Tycho Brahe (1546–1601) did just that, using the addition formulas for sines and cosines to avoid having to multiply big numbers.

Theorem 1.5.37. *For any real number t we have $e^{it} = \cos t + i \sin t$*

Proof. In terms of power series we have

$$\sin t = t - \frac{t^3}{3!} + \frac{t^5}{5!} - \frac{t^7}{7!} + \cdots \qquad 1.5.58$$

$$\cos t = 1 - \frac{t^2}{2!} + \frac{t^4}{4!} - \frac{t^6}{6!} + \cdots \qquad 1.5.59$$

$$\begin{aligned} e^{it} &= 1 + (it) + \frac{(it)^2}{2!} + \frac{(it)^3}{3!} + \frac{(it)^4}{4!} + \cdots \\ &= 1 + it - \frac{t^2}{2!} - i\frac{t^3}{3!} + \frac{t^4}{4!} \cdots . \end{aligned} \qquad 1.5.60$$

The terms of $\cos t$ are exactly the real terms of e^{it}, and the terms of $\sin t$ are exactly the imaginary terms. We can rearrange the terms by Exercise 0.5.5, appropriately modified to deal with complex numbers. □

Geometric series of matrices

When he introduced matrices, Cayley remarked that square matrices "comport themselves as single quantities." In many ways, one can think of a square matrix as a generalized number. Here we will see that a standard result about the sum of a geometric series applies to square matrices; we will need this result when we discuss Newton's method in Section 2.8. In the exercises we will explore other series of matrices.

Definitions 1.5.12 and 1.5.34 of convergent sequences and series apply just as well to matrices as to vectors. Denote by $\text{Mat}\,(n, m)$ the set of $n \times m$ matrices; where distances are concerned, $\text{Mat}\,(n, m)$ is the same as $\mathbb{R}^{nm}$. In particular, Proposition 1.5.13 applies: a series $\sum_{k=1}^{\infty} A_k$ of $n \times m$ matrices converges if and only if for each position (i, j), the series of the entries $(A_n)_{(i,j)}$ converges.

When we say that $\text{Mat}\,(n, m)$ is "the same" as $\mathbb{R}^{nm}$, we mean that we can identify an $n \times m$ matrix with an element of $\mathbb{R}^{nm}$; for example, we can think of the matrix

$$\begin{bmatrix} a & b \\ c & d \\ e & f \end{bmatrix} \text{ as the point } \begin{pmatrix} a \\ b \\ c \\ d \\ e \\ f \end{pmatrix} \in \mathbb{R}^6.$$

We will explore such "identifications" in Section 2.6.

Recall (Example 0.5.6) that the geometric series $S = a + ar + ar^2 + \cdots$ converges if $|r| < 1$, and that the sum is $a/(1 - r)$.

Proposition 1.5.38. *Let A be a square matrix. If $|A| < 1$, the series*

$$S \stackrel{\text{def}}{=} I + A + A^2 + \cdots \quad \textit{converges to} \quad (I - A)^{-1}. \qquad 1.5.61$$

Example of Proposition 1.5.38: Let

$$A = \begin{bmatrix} 0 & 1/4 \\ 0 & 0 \end{bmatrix}.$$

Then $A^2 = \begin{bmatrix} 0 & 0 \\ 0 & 0 \end{bmatrix}$ (surprise!), so the infinite series of equation 1.5.61 becomes a finite sum:

$$(I - A)^{-1} = I + A,$$

and

$$\begin{bmatrix} 1 & -1/4 \\ 0 & 1 \end{bmatrix}^{-1} = \begin{bmatrix} 1 & 1/4 \\ 0 & 1 \end{bmatrix}.$$

Proof. We use the same trick used in the scalar case of Example 0.5.6:

$$\begin{aligned} S_k &\stackrel{\text{def}}{=} I + A + A^2 + \cdots + A^k \\ S_k A &= \phantom{I +{}} A + A^2 + \cdots + A^k + A^{k+1} \\ \hline S_k(I - A) = S_k - S_k A &= I - A^{k+1}. \end{aligned} \qquad 1.5.62$$

We know (Proposition 1.4.11) that

$$|A^{k+1}| \le |A|^k |A| = |A|^{k+1}. \qquad 1.5.63$$

This tells us that the series of numbers $\sum_{k=1}^{\infty} |A|^k$ converges when $|A| < 1$; hence, by Proposition 1.5.35, the series 1.5.61 defining S converges. Thus

$$S(I - A) = \lim_{k\to\infty} S_k(I - A) = \lim_{k\to\infty} \left(I - A^{k+1}\right) \\ = I - \underbrace{\lim_{k\to\infty} A^{k+1}}_{[0]} = I. \qquad 1.5.64$$

We will see in Section 2.2 (discussion after Corollary 2.2.7) that if a square matrix has either a right or a left inverse, that inverse is necessarily a true inverse; checking both directions is not actually necessary.

Equation 1.5.64 says that S is a left inverse of $(I - A)$. If in equation 1.5.62 we had written AS_k instead of S_kA, the same computation would have given us $(I - A)S = I$, showing that S is a right inverse. So by Proposition 1.2.14, S is the inverse of $(1 - A)$. □

Corollary 1.5.39. *If $|A| < 1$, then $(I - A)$ is invertible.*

Corollary 1.5.40. *The set of invertible $n \times n$ matrices is open.*

Proof. Suppose B is invertible, and $|H| < 1/|B^{-1}|$. Then $|-B^{-1}H| < 1$, so $I + B^{-1}H$ is invertible (by Corollary 1.5.39), and

$$(I + B^{-1}H)^{-1}B^{-1} \underset{\text{Prop. 1.2.15}}{=} \left(B(I + B^{-1}H)\right)^{-1} = (B + H)^{-1}. \qquad 1.5.65$$

Thus if $|H| < 1/|B^{-1}|$, the matrix $B + H$ is invertible, giving an explicit neighborhood of B made up of invertible matrices. □

Exercises for Section 1.5

1.5.1 For each of the following subsets, state whether it is open or closed (or both or neither), and say why.

a. $\{x \in \mathbb{R} \,|\, 0 < x \le 1\}$ as a subset of $\mathbb{R}$

b. $\left\{\begin{pmatrix} x \\ y \end{pmatrix} \in \mathbb{R}^2 \,\middle|\, \sqrt{x^2 + y^2} < 1\right\}$ as a subset of $\mathbb{R}^2$

c. the interval $(0, 1]$ as a subset of $\mathbb{R}$

d. $\left\{\begin{pmatrix} x \\ y \end{pmatrix} \in \mathbb{R}^2 \,\middle|\, \sqrt{x^2 + y^2} \le 1\right\}$ as a subset of $\mathbb{R}^2$

e. $\{x \in \mathbb{R} \,|\, 0 \le x \le 1\}$ as a subset of $\mathbb{R}$

f. $\left\{\begin{pmatrix} x \\ y \\ z \end{pmatrix} \in \mathbb{R}^3 \,\middle|\, \sqrt{x^2 + y^2 + z^2} \le 1, \text{ and } x, y, z \ne 0\right\}$ as a subset of $\mathbb{R}^3$

g. the empty set as a subset of $\mathbb{R}$

Exercise 1.5.2: The unit sphere in $\mathbb{R}^3$ is the set of equation

$$x^2 + y^2 + z^2 = 1.$$

A "solid sphere" is referred to as a ball, and a "solid circle" is referred to as a disc.

1.5.2 For each of the following subsets, state whether it is open or closed (or both or neither), and say why.

a. (x, y)-plane in $\mathbb{R}^3$ b. $\mathbb{R} \subset \mathbb{C}$ c. the line $x = 5$ in the (x, y)-plane

d. $(0, 1) \subset \mathbb{C}$ e. $\mathbb{R}^n \subset \mathbb{R}^n$ f. the unit sphere in $\mathbb{R}^3$

1.5.3 Prove the following statements for open subsets of $\mathbb{R}^n$:

a. Any union of open sets is open.

b. A finite intersection of open sets is open.

c. An infinite intersection of open sets is not necessarily open.

Exercise 1.5.3: In this text our open and closed sets are open and closed sets of $\mathbb{R}^n$, as defined in Definitions 1.5.2 and 1.5.4. More general open and closed sets can exist. A set M of subsets of a set X such that the subsets satisfy properties a and b of Exercise 1.5.3, and such that M contains the empty set and the set X is called a *topology* on X. The field called *general topology* studies the consequences of these axioms.

Hint for Exercise 1.5.3, part a: The union may well be a union of infinitely many open sets. Hint for part c: Find a counterexample.

1.5.4 a. Show that the interior of A is the biggest open set contained in A.

b. Show that the closure of A is the smallest closed set that contains A.

c. Show that the closure of a set A is A plus its boundary: $\overline{A} = A \cup \partial A$.

d. Show that the boundary is the closure minus the interior: $\partial A = \overline{A} - \mathring{A}$.

1.5.5 For each of the following subsets of $\mathbb{R}$ and $\mathbb{R}^2$, state whether it is open or closed (or both or neither), and prove it.

a. $\left\{ \begin{pmatrix} x \\ y \end{pmatrix} \in \mathbb{R}^2 \,\middle|\, 1 < x^2 + y^2 < 2 \right\}$ b. $\left\{ \begin{pmatrix} x \\ y \end{pmatrix} \in \mathbb{R}^2 \,\middle|\, xy \neq 0 \right\}$

c. $\left\{ \begin{pmatrix} x \\ y \end{pmatrix} \in \mathbb{R}^2 \mid y = 0 \right\}$ d. $\{\mathbb{Q} \subset \mathbb{R}\}$ (the rational numbers)

1.5.6 Let A be a subset of $\mathbb{R}^n$. Show that the boundary of A is also the intersection of the closure of A and the closure of the complement of A.

1.5.7 For each of the following formulas, find its natural domain, and show whether the natural domain is open, closed or neither.

a. $\sin \frac{1}{xy}$ b. $\ln \sqrt{x^2 - y}$ c. $\ln \ln x$

d. $\arcsin \frac{3}{x^2+y^2}$ e. $\sqrt{e^{\cos xy}}$ f. $\frac{1}{xyz}$

Exercise 1.5.8 is easier than you might fear!

$$B = \begin{bmatrix} 1 & \epsilon & \epsilon \\ 0 & 1 & \epsilon \\ 0 & 0 & 1 \end{bmatrix}$$

Matrix for Exercise 1.5.8, part a

$$C = \begin{bmatrix} 1 & -\epsilon \\ +\epsilon & 1 \end{bmatrix}$$

Matrix for Exercise 1.5.8, part b

1.5.8 a. Find the inverse of the matrix B at left, by finding the matrix A such that $B = I - A$ and computing the value of the series $S = I + A + A^2 + A^3 + \cdots$.

b. Compute the inverse of the matrix C in the margin, where $|\epsilon| < 1$.

1.5.9 Suppose $\sum_{i=1}^{\infty} \mathbf{x}_i$ is a convergent series in $\mathbb{R}^n$. Show that the triangle inequality applies:

$$\left| \sum_{i=1}^{\infty} \mathbf{x}_i \right| \le \sum_{i=1}^{\infty} |\mathbf{x}_i|.$$

1.5.10 Let A be an $n \times n$ matrix, and define

$$e^A = \sum_{k=0}^{\infty} \frac{1}{k!} A^k = I + A + \frac{1}{2} A^2 + \frac{1}{3!} A^3 + \cdots.$$

a. Show that the series converges for all A, and find a bound for $|e^A|$ in terms of $|A|$ and n.

b. Compute explicitly e^A for values of A given in the margin.

c. Prove the following or find counterexamples:

1. $e^{A+B} = e^A e^B$ for all A and B
2. $e^{A+B} = e^A e^B$ for A and B satisfying $AB = BA$
3. $e^{2A} = \left(e^A\right)^2$ for all A

i. $\begin{bmatrix} a & 0 \\ 0 & b \end{bmatrix}$ ii. $\begin{bmatrix} 0 & a \\ 0 & 0 \end{bmatrix}$

iii. $\begin{bmatrix} 0 & a \\ -a & 0 \end{bmatrix}$

Matrices A for Exercise 1.5.10, part b. For the third matrix, you might look up the power series for $\sin x$ and $\cos x$.

1.5.11 Let $\varphi : (0, \infty) \to (0, \infty)$ be a function such that $\lim_{\epsilon \to 0} \varphi(\epsilon) = 0$.

a. Show that the sequence $i \mapsto \mathbf{a}_i$ in $\mathbb{R}^n$ converges to $\mathbf{a}$ if and only if for any $\epsilon > 0$, there exists N such that for $n > N$, we have $|\mathbf{a}_n - \mathbf{a}| < \varphi(\epsilon)$.

b. Find an analogous statement for limits of functions.

1.5.12 Let u be a strictly positive function of $\epsilon > 0$, such that $u(\epsilon) \to 0$ as $\epsilon \to 0$, and let U be a subset of $\mathbb{R}^n$. Prove that the following statements are equivalent:

1. A function $f : U \to \mathbb{R}$ has the limit a at $\mathbf{x}_0$ if $\mathbf{x}_0 \in \overline{U}$ and if for all $\epsilon > 0$, there exists $\delta > 0$ such that when $|\mathbf{x} - \mathbf{x}_0| < \delta$ and $\mathbf{x} \in U$, then $|f(\mathbf{x}) - a| < \epsilon$.

2. A function $f : U \to \mathbb{R}$ has the limit a at $\mathbf{x}_0$ if $\mathbf{x}_0 \in \overline{U}$ and if for all $\epsilon > 0$, there exists $\delta > 0$ such that when $|\mathbf{x} - \mathbf{x}_0| < \delta$ and $\mathbf{x} \in U$, then $|f(\mathbf{x}) - a| < u(\epsilon)$.

1.5.13 Prove the converse of Proposition 1.5.17 (i.e., prove that if every convergent sequence in a set $C \subset \mathbb{R}^n$ converges to a point in C, then C is closed).

1.5.14 State whether the following limits exist, and prove it.

a. $\displaystyle\lim_{\binom{x}{y} \to \binom{1}{2}} \frac{x^2}{x+y}$ b. $\displaystyle\lim_{\binom{x}{y} \to \binom{0}{0}} \frac{\sqrt{|x|}\,y}{x^2+y^2}$ c. $\displaystyle\lim_{\binom{x}{y} \to \binom{0}{0}} \frac{\sqrt{|xy|}}{\sqrt{x^2+y^2}}$ d. $\displaystyle\lim_{\binom{x}{y} \to \binom{1}{2}} x^2+y^3-3$

1.5.15 Suppose that in Definition 1.5.20 we had omitted the requirement that $\mathbf{x}_0$ be in $\overline{X}$. Would it be true that

Exercise 1.5.15: Think of alligators, Section 0.2.

a. $\lim_{x \to -2} \sqrt{x} = 5$? b. $\lim_{x \to -2} \sqrt{x} = 3$?

1.5.16 a. Let $D^* \subset \mathbb{R}^2$ be the region $0 < x^2 + y^2 < 1$, and let $f : D^* \to \mathbb{R}$ be a function. What does the following assertion mean?

$$\lim_{\binom{x}{y} \to \binom{0}{0}} f\begin{pmatrix} x \\ y \end{pmatrix} = a$$

b. For the following two functions, defined on $\mathbb{R}^2 - \{\mathbf{0}\}$, either show that the limit exists at $\mathbf{0}$ and find it, or show that it does not exist:

$$f\begin{pmatrix} x \\ y \end{pmatrix} = \frac{\sin(x+y)}{\sqrt{x^2+y^2}} \qquad g\begin{pmatrix} x \\ y \end{pmatrix} = \big(|x| + |y|\big) \ln(x^2 + y^4)$$

1.5.17 Prove Theorem 1.5.16.

1.5.18 Prove Proposition 1.5.19: If a sequence $i \mapsto \vec{\mathbf{a}}_i$ converges to $\vec{\mathbf{a}}$, then any subsequence converges to the same limit.

1.5.19 Let $U \subset \operatorname{Mat}(2,2)$ be the set of matrices A such that $I - A$ is invertible.

a. Show that U is open, and find a sequence in U that converges to I.

b. Consider the mapping $f : U \to \operatorname{Mat}(2,2)$ given by $f(A) = (A^2 - I)(A - I)^{-1}$. Does $\lim_{A \to I} f(A)$ exist? If so, what is the limit?

*c. Let $B = \begin{bmatrix} 1 & 0 \\ 0 & -1 \end{bmatrix}$, and let $V \subset \operatorname{Mat}(2,2)$ be the set of matrices A such that $A - B$ is invertible. Again, show that V is open, and that B can be approximated by elements of V.

*(d) Consider the mapping $g : V \to \operatorname{Mat}(2,2)$ given by

$$g(A) = (A^2 - B^2)(A - B)^{-1}.$$

Does $\lim_{A \to B} g(A)$ exist? If so, what is the limit?

1.5.20 Set $A = \begin{bmatrix} a & a \\ a & a \end{bmatrix}$. For what numbers $a \in \mathbb{R}$ does the sequence $k \mapsto A^k$ of matrices in $\operatorname{Mat}(2,2)$ converge as $k \to \infty$? What is the limit?

b. What about 3×3 or $n \times n$ matrices, where every entry is a?

1.5.21 For the following functions, can you choose a value for f at $\begin{pmatrix} 0 \\ 0 \end{pmatrix}$ to make the function continuous at the origin?

a. $f\begin{pmatrix} x \\ y \end{pmatrix} = \dfrac{1}{x^2+y^2+1}$ b. $f\begin{pmatrix} x \\ y \end{pmatrix} = \dfrac{\sqrt{x^2+y^2}}{|x|+|y|^{1/3}}$

c. $f\begin{pmatrix} x \\ y \end{pmatrix} = (x^2+y^2)\ln(x^2+2y^2)$ d. $f\begin{pmatrix} x \\ y \end{pmatrix} = (x^2+y^2)\ln|x+y|$

1.5.22 The matrix $A = \begin{bmatrix} 2 & 2 \\ 0 & 1 \end{bmatrix}$ represents a uniformly continuous mapping $\mathbb{R}^2 \to \mathbb{R}^2$ (see Theorem 1.5.33).

a. Find an explicit δ in terms of ϵ.

*b. Find the largest δ in terms of ϵ that works. *Hint*: Look at the margin note next to the proof of Theorem 1.5.33.

Exercise 1.5.23: If x is a real number, then

$$\lim_{y \to x} \frac{y^2 - x^2}{y - x} = 2x;$$

in particular, the limit exists. This exercise investigates whether this is still true for matrices.

1.5.23 Let A be an $n \times n$ matrix.

a. What does it mean to say that

$$\lim_{B \to A} (A-B)^{-1}(A^2 - B^2) \quad \text{exists?}$$

b. Does the limit exist when $A = \begin{bmatrix} 1 & 0 \\ 0 & 1 \end{bmatrix}$? *c. When $A = \begin{bmatrix} 0 & 1 \\ 1 & 0 \end{bmatrix}$?

***1.5.24** Let a, b, c, d be nonnegative integers. For what values of a, b, c, d does $\displaystyle\lim_{\begin{pmatrix} x \\ y \end{pmatrix} \to \begin{pmatrix} 0 \\ 0 \end{pmatrix}} \frac{x^a y^b}{x^{2c} + y^{2d}}$ exist? For those values, what is the limit?

1.6 Five big theorems

When the examples of Peano and Cantor were discovered, they were thought of as aberrations. In 1899, the French mathematician Henri Poincaré lamented the rise of "a rabble of functions ... whose only job, it seems, is to look as little as possible like decent and useful functions."

"What will the poor student think?" Poincaré worried. "He will think that mathematical science is just an arbitrary heap of useless subtleties; either he will turn from it in aversion, or he will treat it like an amusing game."

In this section we prove five major theorems: a sequence in a compact set has a convergent subsequence, a continuous function on a compact set achieves its minimum and maximum, the mean value theorem, a continuous function on a compact set is uniformly continuous, and the fundamental theorem of algebra.

The first four are only about 130 years old or so. They were recognized as fundamental after various mathematicians (Peano, Weierstrass, Cantor) found that many statements thought to be obvious were false. For example, the statement *a curve in the plane has area 0* may seem obvious. Yet it is possible to construct a continuous curve that completely fills up a triangle, visiting every point at least once! (Such a curve is known as a *Peano curve*; see Exercise A1.5 in Appendix A1.) These discoveries forced mathematicians to rethink their definitions and statements, putting calculus on a rigorous basis.

These results are usually avoided in first and second year calculus, but they are not so hard to prove when one knows a bit of topology: notions like open and closed sets, and maxima and minima of functions, for example.

The existence of a convergent subsequence in a compact set

Ironically, Poincaré was ultimately responsible for showing that such seemingly "pathological" functions are essential in describing nature, leading to such fields as chaotic dynamics, where fractal objects are everywhere.

In Section 1.5 we introduced some basic notions of topology. Now we will use them to prove Theorem 1.6.3, a remarkable result that will enable us to prove the existence of a convergent subsequence without knowing where it is. We will use this theorem to prove two key statements: the mean value theorem (Theorem 1.6.13) and the integrability of continuous functions with bounded support (Theorem 4.3.6). These statements are used – indeed, they are absolutely central – but often they are not proved.[13] We will also use Theorem 1.6.3 to prove the fundamental theorem of algebra and the spectral theorem for symmetric matrices (Theorem 3.7.15).

First we need two definitions.

Exercise 1.6.1 asks you to show that a set is bounded if it is contained in a ball centered anywhere – the ball doesn't have to be centered at the origin.

Definition 1.6.1 (Bounded set). A subset $X \subset \mathbb{R}^n$ is *bounded* if it is contained in a ball in $\mathbb{R}^n$ centered at the origin:

$$X \subset B_R(\mathbf{0}) \quad \text{for some } R < \infty. \qquad 1.6.1$$

Note that a closed set need not be bounded. For instance, $[0, \infty)$ is a closed subset of $\mathbb{R}$; its only fence is the point 0, which is in it. (Alternatively, we can say that its complement $(-\infty, 0)$ is open.)

Definition 1.6.2 (Compact set). A nonempty subset $C \subset \mathbb{R}^n$ is *compact* if it is closed and bounded.

Definition 1.6.2 is amazingly important, invading whole chapters of mathematics; it is the basic "finiteness criterion" for spaces. Something like half of mathematics consists of showing that some space is compact. The following theorem, known as the *Bolzano-Weierstrass theorem*, is as important as the definition, if not more so.

Theorem 1.6.3: Equivalently, a bounded sequence in $\mathbb{R}^n$ has a convergent subsequence.

Two more characterizations of compact sets are proved in Appendix A3: the Heine-Borel theorem and the fact that decreasing intersections of compact subsets are nonempty.

Theorem 1.6.3 (Convergent subsequence in a compact set). *If a compact set $C \subset \mathbb{R}^n$ contains a sequence $i \mapsto \mathbf{x}_i$, then that sequence has a convergent subsequence $j \mapsto \mathbf{x}_{i(j)}$ whose limit is in C.*

Note that Theorem 1.6.3 says nothing about what the convergent subsequence converges to; it just says that a convergent subsequence exists.

Proof. The set C is contained in a box $-10^N \le x_i < 10^N$ for some N. Decompose this box into boxes of sidelength 1 in the obvious way. Then at least one of these boxes, which we'll call B_0, must contain infinitely many terms of the sequence, since the sequence is infinite and we have a finite number of boxes. Choose some term $\mathbf{x}_{i(0)}$ in B_0, and cut up B_0 into 10^n boxes of sidelength $1/10$ (in the plane, 100 boxes; in $\mathbb{R}^3$, 1,000 boxes). At least one of these smaller boxes must contain infinitely many terms of the

[13] One exception is Michael Spivak's *Calculus*.

sequence. Call this box B_1, choose $\mathbf{x}_{i(1)} \in B_1$ with $i(1) > i(0)$. Now keep going: cut up B_1 into 10^n boxes of sidelength $1/10^2$; again, one of these boxes must contain infinitely many terms of the sequence; call one such box B_2 and choose an element $\mathbf{x}_{i(2)} \in B_2$ with $i(2) > i(1) \dots$.

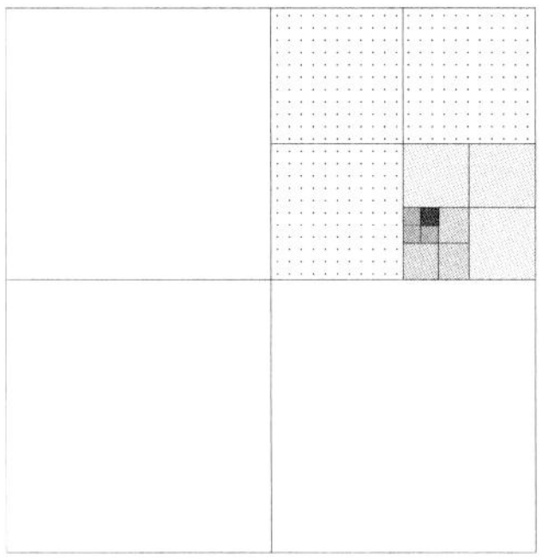

FIGURE 1.6.1.

If the large box contains an infinite sequence, then one of the four quadrants must contain a convergent subsequence. If that quadrant is divided into four smaller boxes, one of those small boxes must contain a convergent subsequence, and so on.

Think of the first box B_0 as giving the coordinates, up to the decimal point, of all the points in B_0. (Because it is hard to illustrate many levels for a decimal system, Figure 1.6.1 illustrates the process for a binary system.) The next box, B_1, gives the first digit after the decimal point.[14] Suppose, for example, that B_0 has vertices $\begin{pmatrix}1\\2\end{pmatrix}$, $\begin{pmatrix}2\\2\end{pmatrix}$, $\begin{pmatrix}1\\3\end{pmatrix}$, and $\begin{pmatrix}2\\3\end{pmatrix}$. Suppose further that B_1 is the small square at the top right-hand corner. Then *all* the points in B_1 are written $\begin{pmatrix}1.9\dots\\2.9\dots\end{pmatrix}$, where the dots indicate arbitrary digits. In other words, all points in B_1 have the same decimal expansion through the first digit after the decimal. When you divide B_1 into 10^2 smaller boxes, the choice of B_2 will determine the next digit; if B_2 is at the bottom right-hand corner, then all points in B_2 are written $\begin{pmatrix}1.99\dots\\2.90\dots\end{pmatrix}$, and so on.

Of course you don't actually know what the coordinates of your points are, because you don't know that B_1 is the small square at the top right corner, or that B_2 is at the bottom right corner. All you know is that there exists a first box B_0 of side 1 that contains infinitely many terms of the original sequence, a second box $B_1 \subset B_0$ of side $1/10$ that also contains infinitely many terms of the original sequence, and so on.

Construct in this way a sequence of nested boxes

$$B_0 \supset B_1 \supset B_2 \supset \cdots \qquad 1.6.2$$

with B_m of sidelength 10^{-m}, and each containing infinitely many terms of the sequence; further choose $\mathbf{x}_{i(m)} \in B_m$ with $i(m+1) > i(m)$.

Consider $\mathbf{a} \in \mathbb{R}^n$ such that the mth decimal digit of each coordinate a_k agrees with the mth decimal digit of the kth coordinate of $\mathbf{x}_{i(m)}$. Then $|\mathbf{x}_{i(m)} - \mathbf{a}| < n10^{-m}$ for all m. Thus $\mathbf{x}_{i(m)}$ converges to $\mathbf{a}$. Since C is closed, $\mathbf{a}$ is in C. □

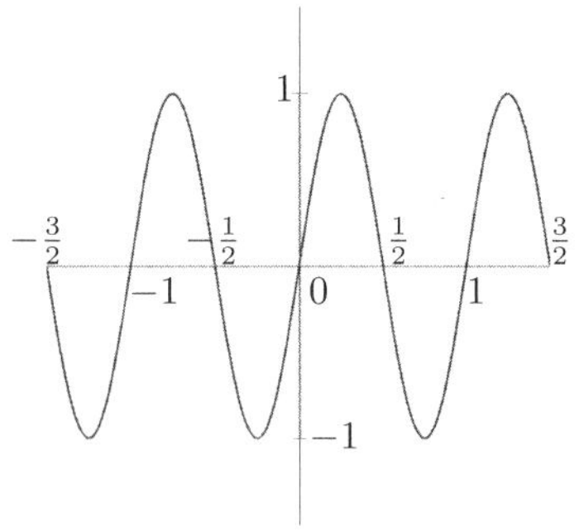

FIGURE 1.6.2.

Graph of $\sin 2\pi\alpha$. If the fractional part of a number α is between 0 and 1/2, then $\sin 2\pi\alpha \geq 0$; if it is between 1/2 and 1, then $\sin 2\pi\alpha \leq 0$.

You may think "what's the big deal?" To see the troubling implications of the proof, consider Example 1.6.4.

Example 1.6.4 (Convergent subsequence). Consider the sequence

$$m \mapsto x_m \stackrel{\text{def}}{=} \sin 10^m \qquad 1.6.3$$

This is a sequence in the compact set $C = [-1, 1]$, so it contains a convergent subsequence. But how do you find it? The first step of the construction is to divide the interval $[-1, 1]$ into three subintervals (our "boxes"), writing

$$[-1, 1] = [-1, 0) \cup [0, 1) \cup \{1\}. \qquad 1.6.4$$

[14] To ensure that all points in the same box have the same decimal expansion up to the appropriate digit, we should say that our boxes are all open on the top and on the right.

How do we choose which of the three "boxes" in equation 1.6.4 should be the first box B_0? We know that x_m will never be in the box $\{1\}$, since $\sin\theta = 1$ if and only if $\theta = \pi/2 + 2k\pi$ for some integer k and (since π is irrational) 10^m cannot be $\pi/2 + 2k\pi$. But how do we choose between $[-1, 0)$ and $[0, 1)$? If we want to choose $[0, 1)$, we must be sure that we have infinitely many positive x_m. So when is x_m positive?

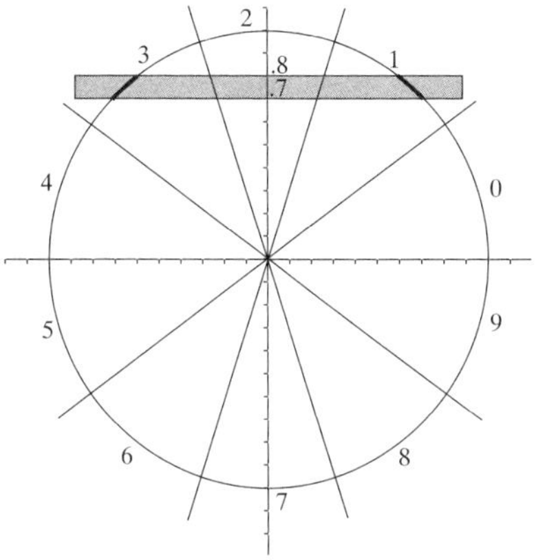

FIGURE 1.6.3.

For $\sin 10^m$ to have 7 as its first digit, i.e, $.7 \le \sin 10^m < .8$, the angle 10^m *radians* (or equivalently $10^m/2\pi$ *turns*) must be in the dark arcs. Whole turns don't count, so only the fractional part of $10^m/2\pi$ matters; in particular, the first digit after the decimal point of $10^m/2\pi$ must be 1 or 3. Thus the $(n+1)$st digit after the decimal point of $1/2\pi$ must be 1 or 3. For $\sin 10^m$ to have a convergent subsequence with limit $.7\ldots$, this must occur for infinitely many m, so there must be infinitely many 1s or 3s in the decimal expansion of $1/2\pi$.

Although $m \mapsto \sin 10^m$ is a sequence in a compact set, and therefore (by Theorem 1.6.3) contains a convergent subsequence, we can't begin to "locate" that subsequence. We can't even say whether it is in $[-1, 0)$ or $[0, 1)$.

Since $\sin\theta$ is positive for $0 < \theta < \pi$, then $x_m = \sin 10^m$ is positive when the fractional part of $10^m/(2\pi)$ is greater than 0 and less than 1/2. (By "fractional part" we mean the part after the decimal; for example, the fractional part of 5/3 is $.666\ldots$, since $5/3 = 1 + 2/3 = 1.666\ldots$.) If you don't see this, consider that (as shown in Figure 1.6.2) $\sin 2\pi\alpha$ depends only on the fractional part of α:

$$\sin 2\pi\alpha \begin{cases} = 0 & \text{if } \alpha \text{ is an integer or half-integer} \\ > 0 & \text{if the fractional part of } \alpha \text{ is } < 1/2 \\ < 0 & \text{if the fractional part of } \alpha \text{ is } > 1/2. \end{cases} \qquad 1.6.5$$

If instead of writing $x_m = \sin 10^m$ we write

$$x_m = \sin 2\pi \frac{10^m}{2\pi}, \quad \text{i.e.,} \quad \alpha = \frac{10^m}{2\pi}, \qquad 1.6.6$$

we see, as stated above, that x_m is positive when the fractional part of $10^m/(2\pi)$ is less than 1/2.

So if a convergent subsequence of the sequence $m \mapsto \sin 10^m$ is contained in the box $[0, 1)$, an infinite number of $10^m/(2\pi)$ must have a fractional part that is less than 1/2. This will ensure that we have infinitely many $x_m = \sin 10^m$ in $[0, 1)$.

For any single x_m, it is enough to know that the first digit of the fractional part of $10^m/(2\pi)$ is 0, 1, 2, 3 or 4: knowing the first digit after the decimal point tells you whether the fractional part is less than or greater than 1/2. Since multiplying by 10^m just moves the decimal point to the right by m, asking whether the fractional part of $10^m/(2\pi)$ is in $[0, 1/2)$ is really asking whether the $(m+1)$st digit of $\frac{1}{2\pi} = .1591549\ldots$ is 0, 1, 2, 3, or 4; saying that the x_m have a limit in $[0, 1/2)$ is the same as saying that at least one of the digits 0, 1, 2, 3, 4 appears infinitely many times in the decimal expansion of $1/2\pi$. (Figure 1.6.3 illustrates the more refined question, whether the fractional part of $10^m/(2\pi)$ is in $[0.7, 0.8)$.)

Note that we are not saying that all the $10^m/(2\pi)$ must have the decimal point followed by 0, 1, 2, 3, or 4! Clearly they don't. We are not interested in all the x_m; we just want to know that we can find a subsequence of the sequence $m \mapsto x_m$ that converges to something inside the box $[0, 1)$. For example, x_1 is not in $[0, 1)$, since $10 \times .1591549\ldots = 1.591549\ldots$; the fractional part starts with 5. Nor is x_2, since the fractional part starts with 9: $10^2 \times .1591549\ldots = 15.91549\ldots$. But x_3 is in $[0, 1)$, since the fractional part of $10^3 \times .1591549\ldots = 159.1549\ldots$ starts with 1.

Everyone believes that the digits 0, 1, 2, 3, 4 appear infinitely many times in the decimal expansion of $\frac{1}{2\pi}$. Indeed, it is widely believed that π

is a *normal number* (every digit appears roughly 1/10th of the time, every pair of digits appears roughly 1/100th of the time, etc.). The first 4 billion digits of π have been computed and appear to bear out this conjecture.

It is also widely believed that $\frac{1}{2\pi}$ is normal. Still, no one knows how to prove it; as far as we know it is conceivable that all the digits after the 10 billionth are 6, 7, and 8. Thus, even choosing the first box B_0 requires some God-like ability to "see" this whole infinite sequence, when there is no obvious way to do it. $\triangle$

Theorem 1.6.3 is *nonconstructive*: it proves that something exists but gives not the slightest hint of how to find it. Many mathematicians of the end of the nineteenth century were deeply disturbed by this type of proof; even today, a school of mathematicians called *constructivists* reject this sort of thinking. They demand that in order for a number to be determined, one give a rule for computing the successive decimals. Constructivists are pretty scarce these days; we have never met one. But we have a certain sympathy with their views and much prefer proofs that involve effectively computable algorithms, at least implicitly.

Definition 1.6.5: The least upper bound is defined in Definition 0.5.1; of course it only exists if the set of $f(c)$ has an upper bound; in that case the function is said to be *bounded above*.

The word "maximum" is sometimes used to denote a point in the domain and sometimes to denote a point in the codomain. We will use "maximum value" for the point M in the codomain and "maximum" for a point b in the domain such that $f(b) = M$.

Some people use "maximum" to denote a least upper bound, rather than a least upper bound that is achieved.

The greatest lower bound is defined in Definition 0.5.2: a number a is a lower bound for a subset $X \subset \mathbb{R}$ if for every $x \in X$ we have $x \geq a$; a greatest lower bound is a lower bound b such that for any other lower bound a, we have $b \geq a$. Some people use the words "greatest lower bound" and "minimum" interchangeably.

On the open set $(0, 1)$ the greatest lower bound of $f(x) = x^2$ is 0, and f has no minimum value. On the closed set $[0, 1]$, the number 0 is both the infimum of f and its minimum value.

Continuous functions on compact sets

We can now explore some consequences of Theorem 1.6.3. One is that a continuous function defined on a compact subset has both a point at which it has a maximum value and a point at which it has a minimum value.

In the definitions below, C is a subset of $\mathbb{R}^n$.

Definition 1.6.5 (Supremum). A number M is the *supremum* of a function $f : C \to \mathbb{R}$ if M is the least upper bound of the values of f (i.e., of the set of numbers $f(\mathbf{c})$ for all $\mathbf{c} \in C$). It is denoted $M = \sup_{\mathbf{c} \in C} f(\mathbf{c})$.

Definition 1.6.6 (Maximum value, maximum). A number M is the *maximum value* of a function $f : C \to \mathbb{R}$ if it is the supremum of f and there exists $\mathbf{b} \in C$ such that $f(\mathbf{b}) = M$. The point $\mathbf{b}$ is called a *maximum* of f.

For example, on the open set $(0, 1)$ the least upper bound of $f(x) = x^2$ is 1, and f has no maximum. On the closed set $[0, 2]$, 4 is both the supremum and the maximum value of f; the maximum is 2.

Definition 1.6.7 (Infimum). A number m is the *infimum* of a function $f : C \to \mathbb{R}$ if m is the greatest lower bound of the set of values of f. It is denoted $m = \inf_{\mathbf{c} \in C} f$.

Definition 1.6.8 (Minimum value; minimum). A number m is the *minimum value* of a function f defined on a set $C \subset \mathbb{R}^n$ if it is the infimum and if there exists $\mathbf{c} \in C$ such that $f(\mathbf{c}) = m$. The point $\mathbf{c}$ is then a *minimum* of the function.

Theorem 1.6.9 enables you to go from *local* knowledge – that the function is continuous – to *global* knowledge: existence of a maximum and a minimum for the entire function. Note that C is compact.

Recall that "compact" means "closed and bounded".

Theorem 1.6.9 (Existence of minima and maxima). *Let $C \subset \mathbb{R}^n$ be a compact subset, and let $f : C \to \mathbb{R}$ be a continuous function. Then there exists a point $\mathbf{a} \in C$ such that $f(\mathbf{a}) \geq f(\mathbf{x})$ for all $\mathbf{x} \in C$, and a point $\mathbf{b} \in C$ such that $f(\mathbf{b}) \leq f(\mathbf{x})$ for all $\mathbf{x} \in C$.*

Theorem 1.6.9: One student objected, "what is there to prove?" It might seem obvious that the graph of a function on $[0, 1]$ that you can draw must have a maximum. But Theorem 1.6.9 also talks about functions defined on compact subsets of $\mathbb{R}^{17}$, about which we have no intuition. Moreover, it isn't quite so clear as you might think, even for functions on $[0, 1]$.

Examples 1.6.10 (Functions with no maxima). Consider the function

$$f(x) = \begin{cases} 0 & \text{when } x = 0 \\ \frac{1}{x} & \text{otherwise,} \end{cases} \qquad 1.6.7$$

defined on the compact set $[0, 1]$. As $x \to 0$, we see that $f(x)$ blows up to infinity; the function does not have a maximum value (or even a supremum); it is not bounded. This function is not continuous, so Theorem 1.6.9 does not apply to it.

The function $f(x) = 1/x$ defined on $(0, 1]$ is continuous but has no maximum; this time the problem is that $(0, 1]$ is not closed, hence not compact. The function $f(x) = x$ defined on all of $\mathbb{R}$ has no maximum (or minimum); this time the problem is that $\mathbb{R}$ is not bounded, hence not compact. Now consider the function $f(x) = \frac{1}{|x-\sqrt{2}|}$, defined on the rational numbers $\mathbb{Q}$ in the closed interval $[0, 2]$. Restricting the domain this way avoids the potential trouble spot at $\sqrt{2}$, where $\frac{1}{|x-\sqrt{2}|}$ would blow up to infinity, but the function is not guaranteed to have a maximum because the rational numbers are not closed in $\mathbb{R}$. In some sense, Theorem 1.6.9 says that $\mathbb{R}$ (and more generally $\mathbb{R}^n$) has no holes, unlike $\mathbb{Q}$, which does (the irrational numbers). △

Exercise 1.6.2 asks you to show that if $A \subset \mathbb{R}^n$ is any noncompact subset, then there always is a continuous unbounded function on A.

Proof of Theorem 1.6.9. We will prove the statement for the maximum. The proof is by contradiction. Assume f is unbounded. Then for any integer N, no matter how large, there exists a point $\mathbf{x}_N \in C$ such that $|f(\mathbf{x}_N)| > N$. By Theorem 1.6.3, the sequence $N \mapsto \mathbf{x}_N$ must contain a convergent subsequence $j \mapsto \mathbf{x}_{N(j)}$ which converges to some point $\mathbf{b} \in C$. Since f is continuous at $\mathbf{b}$, for any ϵ, there exists a $\delta > 0$ such that when $|\mathbf{x} - \mathbf{b}| < \delta$, then $|f(\mathbf{x}) - f(\mathbf{b})| < \epsilon$; that is, $|f(\mathbf{x})| < |f(\mathbf{b})| + \epsilon$.

By the triangle inequality, if $|f(\mathbf{x}) - f(\mathbf{b})| < \epsilon$, then $|f(\mathbf{x})| < |f(\mathbf{b})| + \epsilon$:

$$\begin{aligned} |f(\mathbf{x})| &= |f(\mathbf{x}) - f(\mathbf{b}) + f(\mathbf{b})| \\ &\leq |f(\mathbf{x}) - f(\mathbf{b})| + |f(\mathbf{b})| \\ &< \epsilon + |f(\mathbf{b})|. \end{aligned}$$

Since $j \mapsto \mathbf{x}_{N(j)}$ converges to $\mathbf{b}$, we have $|\mathbf{x}_{N(j)} - \mathbf{b}| < \delta$ for j sufficiently large. But as soon as $N(j) > |f(\mathbf{b})| + \epsilon$, we have

$$|f(\mathbf{x}_{N(j)})| > N(j) > |f(\mathbf{b})| + \epsilon, \qquad 1.6.8$$

a contradiction. Therefore, the set of values of f is bounded, which means that f has a supremum M.

Next we want to show is that f has a maximum: that there exists a point $\mathbf{a} \in C$ such that $f(\mathbf{a}) = M$. There is a sequence $i \mapsto \mathbf{x}_i$ such that

$$\lim_{i \to \infty} f(\mathbf{x}_i) = M. \qquad 1.6.9$$

We can again extract a convergent subsequence $m \mapsto \mathbf{x}_{i(m)}$ that converges to some point $\mathbf{a} \in C$. Then, since $\mathbf{a} = \lim_{m\to\infty} \mathbf{x}_{i(m)}$,

$$f(\mathbf{a}) = \lim_{m\to\infty} f(\mathbf{x}_{i(m)}) = M. \qquad 1.6.10$$

The proof for the minimum works the same way. □

Theorem 1.6.11 is another result that allows us to get global information from local information using compactness.

Theorem 1.6.11 will be essential in Chapter 4.

Theorem 1.6.11. *Let $X \subset \mathbb{R}^n$ be compact. A continuous function $f : X \to \mathbb{R}$ is uniformly continuous.*

Proof. Uniform continuity says:

$$(\forall \epsilon > 0)(\exists \delta > 0)\Big(|\mathbf{x} - \mathbf{y}| < \delta \implies |f(\mathbf{x}) - f(\mathbf{y})| < \epsilon\Big). \qquad 1.6.11$$

By contradiction, suppose f is not uniformly continuous. Then there exist $\epsilon > 0$ and sequences $i \mapsto \mathbf{x}_i$, $i \mapsto y_i$ such that

$$\lim_{i\to\infty} |\mathbf{x}_i - \mathbf{y}_i| = 0, \quad \text{but for all } i \text{ we have} \quad |f(\mathbf{x}_i) - f(\mathbf{y}_i)| \geq \epsilon. \qquad 1.6.12$$

Since X is compact, we can extract a subsequence $j \mapsto \mathbf{x}_{i_j}$ that converges to some point $\mathbf{a} \in X$. Since $\lim_{i\to\infty} |\mathbf{x}_i - \mathbf{y}_i| = 0$, the sequence $j \mapsto \mathbf{y}_{i_j}$ also converges to $\mathbf{a}$.

By hypothesis, f is continuous *at* $\mathbf{a}$, so there exists $\delta > 0$ such that

$$|\mathbf{x} - \mathbf{a}| < \delta \implies |f(\mathbf{x}) - f(\mathbf{a})| < \frac{\epsilon}{3}. \qquad 1.6.13$$

Further, there exists J such that

$$j \geq J \implies |\mathbf{x}_{i_j} - \mathbf{a}| < \delta \quad \text{and} \quad |\mathbf{x}_{i_j} - \mathbf{a}| < \delta. \qquad 1.6.14$$

Thus for $j \geq J$ we have

$$|f(\mathbf{x}_{i_j}) - f(\mathbf{y}_{i_j})| \leq |f(\mathbf{x}_{i_j}) - f(\mathbf{a})| + |f(\mathbf{a}) - f(\mathbf{y}_{i_j})| \leq \frac{\epsilon}{3} + \frac{\epsilon}{3} < \epsilon. \quad 1.6.15$$

This is a contradiction. □

You no doubt saw the following proposition in first year calculus.

Proposition 1.6.12 (Derivative 0 at maximum or minimum). *Let $g : (a,b) \to \mathbb{R}$ be a differentiable function on the open interval (a,b). If $c \in (a,b)$ is a maximum or minimum of g, then $g'(c) = 0$.*

Proof. We will prove it only for the maximum. If c is a maximum, then $g(c) - g(c+h) \geq 0$, so

$$\frac{g(c) - g(c+h)}{h} \begin{cases} \geq 0 & \text{if } h > 0 \\ \leq 0 & \text{if } h < 0; \end{cases} \quad \text{i.e.,} \quad g'(c) = \lim_{h\to 0} \frac{g(c+h) - g(c)}{h} \qquad 1.6.16$$

is simultaneously ≤ 0 and ≥ 0, so it is 0. □

Obviously a car cannot jump from 59 mph to 61 mph without passing through 60 mph, but note that the mean value theorem does not require that the derivative be continuous. Example 1.9.4 describes a function with a discontinuous derivative that oscillates infinitely many times between -1 and 1 (see also Exercise 0.5.2). It follows from the mean value theorem that for functions of one variable, either the derivative is good (continuous) or it is very, very bad (oscillating wildly). It can't simply jump from one value to another, skipping a value in between.

More precisely, the derivative of a differentiable function of one variable satisfies the intermediate value property; this is the object of Exercise 1.6.7.

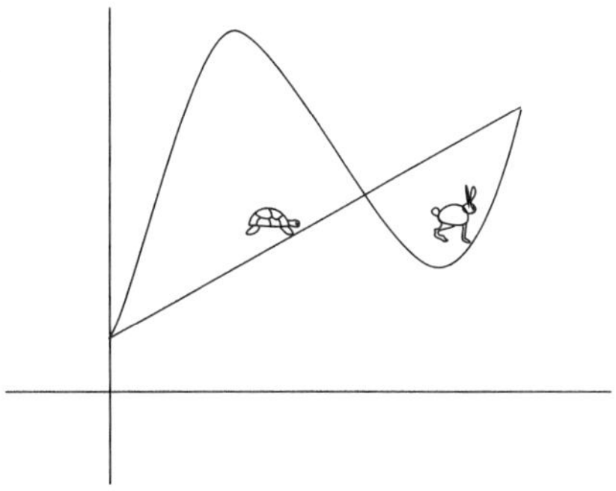

FIGURE 1.6.4.

A race between hare and tortoise ends in a dead heat. The function f represents the progress of the hare, starting at time a and ending at time b. He speeds ahead, overshoots the mark, and returns. Slow-and-steady tortoise is represented by $g(x) = f(a) + m(x - a)$

Our next result is the mean value theorem, without which practically nothing in differential calculus can be proved. The mean value theorem says that you can't drive 60 miles in an hour without going exactly 60 mph at one instant at least: the average change in f over the interval (a, b) is the derivative of f at some point $c \in (a, b)$. This is pretty obvious (but still not so easy to prove) if the velocity is continuous; but if the velocity is not assumed to be continuous, it isn't so clear.

Theorem 1.6.13 (Mean value theorem). *If $f : [a, b] \to \mathbb{R}$ is continuous, and f is differentiable on (a, b), then there exists $c \in (a, b)$ such that*

$$f'(c) = \frac{f(b) - f(a)}{b - a}. \qquad 1.6.17$$

The special case where $f(a) = f(b) = 0$ is called *Rolle's theorem.*

Note that f is defined on the closed and bounded interval $[a, b]$, but we must specify the open interval (a, b) when we talk about where f is differentiable.[15] If we think that f measures position as a function of time, then the right side of equation 1.6.17 measures average speed over the time interval $b - a$.

Proof. Think of f as representing distance traveled (by a car or, as in Figure 1.6.4, by a hare). The distance the hare travels in the time interval $b - a$ is $f(b) - f(a)$, so its average speed is

$$m = \frac{f(b) - f(a)}{b - a}. \qquad 1.6.18$$

The function g represents the steady progress of a tortoise starting at $f(a)$ and constantly maintaining that average speed (alternatively, a car set on cruise control):

$$g(x) = f(a) + m(x - a). \qquad 1.6.19$$

The function h measures the distance between f and g:

$$h(x) = f(x) - g(x) = f(x) - \big(f(a) + m(x - a)\big). \qquad 1.6.20$$

It is a continuous function on $[a, b]$, and $h(a) = h(b) = 0$. (The hare and the tortoise start together and finish in a dead heat.)

If h is 0 everywhere, then $f(x) = g(x) = f(a) + m(x - a)$ has derivative m everywhere, so the theorem is true. If h is not 0 everywhere, then it must take on positive values or negative values somewhere, so (by Theorem 1.6.9) it must have a point where it achieves a positive maximum value or a point where it achieves a negative minimum value, or both. Let c be such a point; then $c \in (a, b)$, so h is differentiable at c, and by Proposition 1.6.12, $h'(c) = 0$.

[15] One could have a left and right derivative at the endpoints, but we are not assuming that such one-sided derivatives exist.

Since $h'(x) = f'(x) - m$ (the terms $f(a)$ and ma in equation 1.6.20 are constant), this gives $0 = h'(c) = f'(c) - m$. □

The fundamental theorem of algebra

The fundamental theorem of algebra is one of the most important results of all mathematics, with a history going back to the Greeks and Babylonians. It was not proved satisfactorily until about 1830. The theorem asserts that every complex polynomial of degree k has at least one complex root. It follows that every polynomial of degree k has k complex roots, counted with multiplicity; this is discussed at the end of this section (Corollary 1.6.15). We will also see that every real polynomial can be factored as a product of real polynomials of degree 1 or 2.

In equation 1.6.21, the coefficient of z^k is 1. A polynomial whose highest-degree term has coefficient 1 is said to be *monic*. Of course, any polynomial can be put into monic form by dividing through by the appropriate number. Using monic polynomials simplifies statements and proofs.

Even if the coefficients a_n are real, the fundamental theorem of algebra does not guarantee that the polynomial has any real roots. The roots may be complex. For instance, $x^2 + 1 = 0$ has no real roots.

Theorem 1.6.14 (Fundamental theorem of algebra). *Let*

$$p(z) = z^k + a_{k-1}z^{k-1} + \cdots + a_0 \qquad 1.6.21$$

be a polynomial of degree $k > 0$ with complex coefficients. Then p has a root: there exists a complex number z_0 such that $p(z_0) = 0$.

When $k = 1$, this is clear: the unique root is $z_0 = -a_0$.

When $k = 2$, the famous *quadratic formula* tells you that the roots are

$$\frac{-a_1 \pm \sqrt{a_1^2 - 4a_0}}{2}. \qquad 1.6.22$$

This was known to the Greeks and Babylonians.

The cases $k = 3$ and $k = 4$ were solved in the sixteenth century by Cardano and others; their solutions are presented in Appendix A2. For the next two centuries, an intense search failed to find anything analogous for equations of higher degree. Finally, around 1830, two young mathematicians with tragic personal histories, the Norwegian Niels Henrik Abel and the Frenchman Evariste Galois, proved that no analogous formulas exist in degrees 5 and higher.[16] Again, these discoveries opened new fields in mathematics.

FIGURE 1.6.5.

Niels Henrik Abel, born in 1802, assumed responsibility for his younger brother and sister after the death of their alcoholic father in 1820. For years he struggled against poverty and illness, trying to obtain a position that would allow him to marry his fiancée; he died from tuberculosis at the age of 26, without learning that he had been appointed professor in Berlin.

Several mathematicians (Laplace, d'Alembert, Gauss) had earlier come to suspect that the fundamental theorem was true, and tried their hands at proving it; d'Alembert's argument was published in 1746, while Gauss gave five "proofs", the first in 1799 and the last in about 1810. In the absence of topological tools, their proofs were necessarily short on rigor, and the criticism each heaped on his competitors does not reflect well on any of them. Our proof illustrates the kind of thing we meant when we said, in the beginning of Section 1.4, that calculus is about "some terms being dominant or negligible compared with other terms".

[16] Although there is no formula for solving higher-degree polynomials, there is an enormous literature on how to search for solutions.

Proof of Theorem 1.6.14. We want to show that there exists a number z such that $p(z) = 0$. The proof requires thinking about complex numbers geometrically, as discussed in Section 0.7, in the subsection "complex numbers in polar coordinates".

The polynomial p is a complex-valued function on $\mathbb{C}$, and $|p|$ is a real-valued function on $\mathbb{C}$. The strategy of the proof is first to establish that $|p|$ has a minimum on $\mathbb{C}$, i.e., that there exists $z_0 \in \mathbb{C}$ such that for all $z \in \mathbb{C}$, we have $|p(z_0)| \leq |p(z)|$, and next to establish that $|p(z_0)| = 0$. The existence of the minimum is the main conceptual difficulty of the argument, though showing that the minimum value is 0 is technically longer.

Note that the function $1/(1+x^2)$ does not have a minimum on $\mathbb{R}$, and the function $|e^z|$ does not have a minimum on $\mathbb{C}$. In both cases, the infimum of the values is 0, but there is no $x_0 \in \mathbb{R}$ such that $1/(1+x_0^2) = 0$, and there no complex number z_0 such that $e^{z_0} = 0$. We must use something about polynomials that doesn't hold for $x \mapsto 1/(1+x^2)$ or for $z \mapsto e^z$.

Our only criterion (very nearly *the* only criterion) for the existence of a minimum is Theorem 1.6.9, which requires that the domain be compact.[17] The domain of $z \mapsto |p(z)|$ is $\mathbb{C}$, which is not compact. To overcome this difficulty, we will show that there exists $R > 0$ such that if $|z| > R$, then $|p(z)| \geq |p(0)| = |a_0|$. This is illustrated by Figure 1.6.8.

The disc $\{\, z \in \mathbb{C} \mid z \leq R \,\}$ is compact, so by Theorem 1.6.9, there is a point z_0 with $|z_0| \leq R$ such that for all z with $|z| \leq R$ we have $|p(z_0)| \leq |p(z)|$. In particular $|p(z_0)| \leq |p(0)| = |a_0|$. Thus if $z \in \mathbb{C}$ satisfies $|z| \leq R$ we have $|p(z_0)| \leq |p(z)|$, and if $|z| > R$ we have $|p(z_0)| \leq |p(z)|$ also, and z_0 is the required minimum. Thus for this half of the proof, the only thing we need to show is that there exists $R > 0$ such that if $|z| > R$, then $|p(z)| \geq |a_0|$.

We said that we must use something about polynomials that doesn't hold for functions in general. The property we use is that for *large* z, the leading term z^k of a polynomial

$$p(z) = z^k + a_{k-1}z^{k-1} + \cdots + a_0 \qquad 1.6.23$$

dominates the sum of all the other terms, $a_{k-1}z^{k-1} + \cdots + a_0$.

What does "large" mean? For one thing, we must require $|z| > 1$, otherwise $|z^k|$ is small rather than large. We must also require that $|z|$ be large as compared to the coefficients $a_{k-1}, \dots, a_0$. After a bit of fiddling, we find that setting $A = \max\{|a_{k-1}|, \dots, |a_0|\}$ and $R = \max\{(k+1)A, 1\}$ works: if $|z| \geq R$, then

$$\begin{aligned} |p(z)| &\geq |z|^k - \big(|a_{k-1}||z|^{k-1} + \cdots + |a_0|\big) \geq |z|^k - kA|z|^{k-1} \\ &= |z|^{k-1}(|z| - kA) \geq |z| - kA \geq R - kA \geq A \geq |a_0|. \end{aligned} \qquad 1.6.24$$

In the first inequality in the second line, we can drop $|z|^{k-1}$ because by hypothesis $|z| \geq R \geq 1$, and because we know that $|z| - kA \geq 0$, since

[17] Using Theorem 1.6.9 to prove existence of a minimum is what d'Alembert and Gauss did not know how to do; that is why their proofs were not rigorous. Topology was invented in large part to make these kinds of arguments possible.

The absolute value of a complex number $z = x + iy$ is (Definition 0.7.3) $|z| = \sqrt{x^2 + y^2}$. It wouldn't make sense to look for a minimum of p, since complex numbers aren't ordered.

Unlike the quadratic formula and Cardano's formulas, this proof *does not provide a recipe to find a root.* This is a serious problem: one very often needs to solve polynomials, and to this day there is no really satisfactory way to do it.

FIGURE 1.6.6.

Born illegitimate in 1717, Jean Le Rond d'Alembert was abandoned by his mother, a former nun; his father sought him out and placed him with a foster mother.

Inequality 1.6.24: The first inequality in the first line is the triangle inequality, which can be stated as

$$|a + b| \geq |a| - |b|,$$

since

$$\begin{aligned} |a| &= |a + b - b| \\ &\leq |a + b| + |-b| \\ &= |a + b| + |b|. \end{aligned}$$

Here, $a + b$ is $p(z)$.

$|z| - kA \geq R - kA \geq (k+1)A - kA = A \geq 0$. This last computation also justifies the remaining inequalities.

Thus any z outside the disc of radius R gives $|p(z)| \geq |a_0|$. Therefore p has a global minimum on $\mathbb{C}$, at some z_0 with $|z_0| \leq R$.

Remark. The argument above works just as well with real numbers; if p is a real polynomial, then $|p|$ achieves its minimum value at some x_0 with $|x_0| < R$, where R is constructed as above. But $p(x_0)$ is not necessarily 0 (i.e, x_0 is not necessarily a root); for example, for the polynomial x^2+1, the minimum value is 1, achieved at 0. Real polynomials do not necessarily have real roots. $\triangle$

FIGURE 1.6.7.

Evariste Galois, born in 1811, twice failed to win admittance to Ecole Polytechnique in Paris, the second time shortly after his father's suicide. In 1831 he was imprisoned for making an implied threat against the king at a republican banquet. He was 20 years old when he died from wounds received in a duel.

Now we need to show that z_0 is a root of the polynomial (i.e., that $p(z_0) = 0$). We will argue by contradiction: we will assume that $p(z_0) \neq 0$, and show that in that case there exists a point z such that $|p(z)| < |p(z_0)|$. Since $|p(z_0)|$ is the minimum value, this will prove that our assumption that $p(z_0) \neq 0$ is false.

More specifically, our strategy will be to show that as z travels on a little circle around z_0, the number $p(z)$ travels around $p(z_0)$; in doing so, it will come between $p(z_0)$ and the origin, and thus $|p(z)|$ will be smaller than the proven minimum value $|p(z_0)|$, which is impossible.

We start with a change of variables; it will be easier to consider numbers in a circle around z_0 if we treat z_0 as the origin. So set $z = z_0 + u$, and consider the function

$$\begin{aligned} p(z) = z^k + a_{k-1}z^{k-1} + \cdots + a_0 &= (z_0+u)^k + a_{k-1}(z_0+u)^{k-1} + \cdots + a_0 \\ &= u^k + b_{k-1}u^{k-1} + \cdots + b_0 = q(u), \end{aligned} \qquad 1.6.25$$

$$\text{where} \quad b_0 = z_0^k + a_{k-1}z_0^{k-1} + \cdots + a_0 = p(z_0). \qquad 1.6.26$$

Equations 1.6.25 and 1.6.26: You might object, what happens to the middle terms, for example, the $2a_2z_0u$ in

$$a_2(z_0+u)^2 = a_2z_0^2 + 2a_2z_0u + a_2u^2?$$

But that is a term in u with coefficient $2a_2z_0$, so the coefficient $2a_2z_0$ just gets absorbed in b_1, the coefficient of u.

Calculating the b_i may involve quite a lot of arithmetic if the degree of p is large.

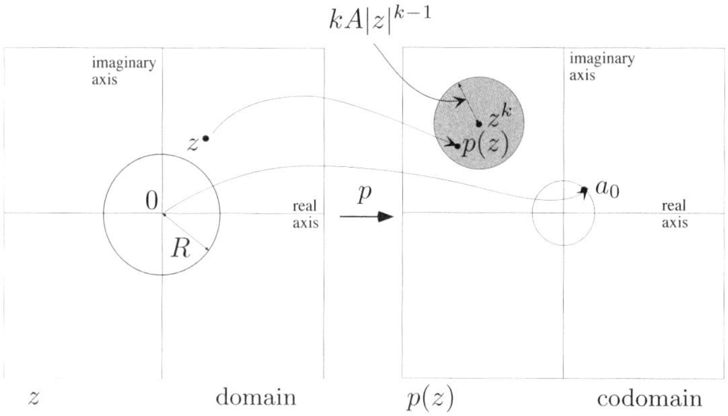

FIGURE 1.6.8. If $|z| \geq R$, the complex number $p(z)$ is in the shaded disc of radius $kA|z|^{k-1}$ around z^k, since

$$|p(z) - z^k| = |a_{k-1}z^{k-1} + \cdots + a_0| \leq kA|z|^{k-1}.$$

Equation 1.6.24 shows that the shaded disc lies outside the disc of radius $|a_0|$ centered at the origin, so $|p(z)| \geq |a_0|$. We discuss "graphing" complex-valued functions in Section 0.7.

Thus $q(u) \stackrel{\text{def}}{=} u^k + b_{k-1}u^{k-1} + \cdots + b_0$ is a polynomial of degree k in u, with constant term $b_0 = p(z_0)$. Our assumption, which we will contradict, is that $b_0 \neq 0$.

This business of numbers "traveling on circles" uses the fact that we are dealing with complex numbers. When a complex number w is written in terms of length ρ and polar angle θ,

$$w = \rho(\cos\theta + i\sin\theta),$$

w turns in a circle of radius ρ around 0 as the angle θ goes from 0 to 2π.

We choose the term of q with the smallest power $j > 0$ that has a nonzero coefficient.[18] We rewrite q as follows:

$$q(u) = b_0 + b_j u^j + (b_{j+1}u^{j+1} + \cdots + u^k). \qquad 1.6.27$$

We will see what happens to $q(u)$ as u goes around the circle $\{\, u \mid |u| = \rho \,\}$. Figure 1.6.9 illustrates the construction: $p(z_0) = b_0$ is the base of a flagpole, and the origin is the doghouse. The complex number $b_0 + b_j u^j$ is a man walking on a circle of radius $|b_j|\rho^j$ around the flagpole. He is walking a dog that is running circles around him, restrained by a leash of length $|b_{j+1}u^{j+1} + \cdots + u^k|$:

De Moivre's formula (Corollary 0.7.6):

$$\Big(r(\cos\theta + i\sin\,\theta)\Big)^k = r^k(\cos k\theta + i\,\sin k\theta).$$

$$q(u) = \overbrace{\underbrace{b_0}_{\text{flagpole}} + b_j u^j}^{\text{position of man}} + \underbrace{b_{j+1}u^{j+1} + \cdots + u^k}_{\text{leash}} = p(z) = \text{ dog}. \qquad 1.6.28$$

We will show – this is the crux of the proof – that for small ρ, the leash is shorter than the distance from man to flagpole, so at some point the dog is closer to the origin (the doghouse) than is the flagpole: $|p(z)| < |p(z_0)|$. But this is impossible, since $|p(z_0)|$ is the minimum value of p.

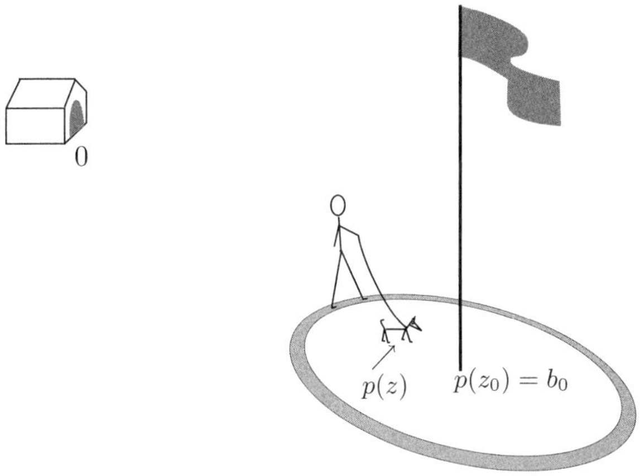

FIGURE 1.6.9. The flagpole is $p(z_0)$; the dog is $p(z)$. We have shown that there is a point z_0 at which $|p|$ has a minimum. We assume that $p(z_0) \neq 0$ and show a contradiction: if $|u|$ is small, the leash is shorter than the radius $|b_j|\rho^j$ of the circle on which the man is walking, so at some point the dog will be closer to the doghouse than the distance from flagpole to doghouse: $|p(z)| < |p(z_0)|$. Since $|p|$ has a minimum at z_0, this is impossible, so our assumption $p(z_0) \neq 0$ is false.

[18] Usually the smallest such power will be $j = 1$; for example, if we had $q(u) = u^4 + 2u^2 + 3u + 10$, that term, which we call $b_j u^j$, would be $3u$. If we had $q(u) = u^5 + 2u^4 + 5u^3 + 1$, that term would be $5u^3$.

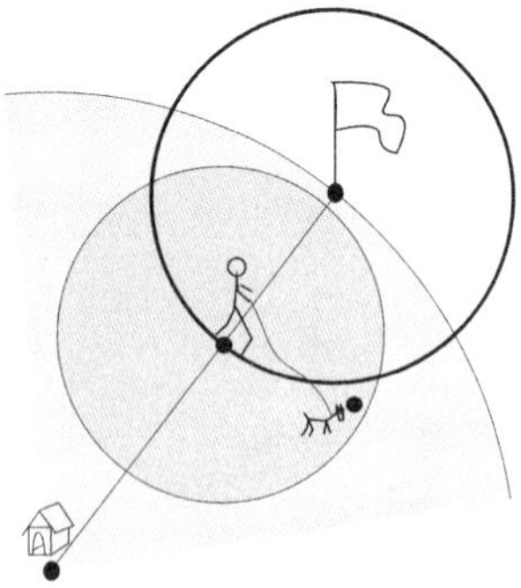

FIGURE 1.6.10.

When the man is between the doghouse and flagpole, the dog is constrained to the dark shaded disc. If inequality 1.6.30 is satisfied, the dog is then always closer to the doghouse than the flagpole is to the doghouse.

Inequality 1.6.31: Note that $|b_j|$ is working in our favor (it is in the numerator, so the bigger it is, the easier it is to satisfy the inequality), and that $|b_{j+1}|, \dots, |b_k|$ are working against us (via B, in the denominator). This is reasonable; if any of these coefficients is very big, we would expect to have to take ρ very small to compensate.

Long division of polynomials reduces finding all the roots of a polynomial to finding one. Unfortunately, this division is quite unstable numerically. If you know a polynomial with a certain precision, you know the result of the division with considerably less precision, and if you need to do this many times to find all the roots, these roots will be less and less accurate. So other schemes are needed to find roots of polynomials of large degree ($k > 10$).

Let us justify this description. If we write u in terms of length ρ and polar angle θ, the number $z = z_0 + u$ in the domain turns in a circle of radius ρ around z_0 as θ goes from 0 to 2π. What is happening in the codomain? De Moivre's formula says that the man representing the number $b_0 + b_j u^j$ travels in a circle of radius $|b_j|\rho^j$ around the flagpole (i.e., around b_0). Thus if $|b_j|\rho^j < |b_0|$ (which is true as long as we choose $\rho < |b_0/b_j|^{1/j}$), then for some values of θ, the man will be *between* b_0 and 0: on the line segment joining 0 to b_0. (In fact, there are precisely j such values of θ; Exercise 1.6.9 asks you to find them.)

But it is not enough to know where the man is; we need to know whether the dog representing $p(z)$ is ever closer to the origin than is the flagpole (see Figure 1.6.10). This will happen if the leash is shorter than the distance $|b_j|\rho^j$ from the man to the flagpole:

$$\underbrace{|b_{j+1}u^{j+1} + \cdots + u^k|}_{\text{length of leash}} \quad < \quad \underbrace{|b_j|\rho^j}_{\text{distance man to pole}}. \qquad 1.6.29$$

which will happen if we take $\rho > 0$ sufficiently small and set $|u| = \rho$.

To see this, set $B = \max\{|b_{j+1}|, \dots, |b_k| = 1\}$ and choose ρ satisfying

$$\rho < \min\left\{ \frac{|b_j|}{(k-j)B}, \left|\frac{b_0}{b_j}\right|^{1/j}, 1 \right\}. \qquad 1.6.30$$

If we set $|u| = \rho$ so that $|u|^i < |u|^{j+1}$ for $i > j+1$, we have

$$\begin{aligned} \underbrace{|b_{j+1}u^{j+1} + \cdots + u^k|}_{\text{length of leash}} &\le (k-j)B\rho^{j+1} = (k-j)B\rho\rho^j \\ &< (k-j)B\frac{|b_j|}{(k-j)B}\rho^j = |b_j|\rho^j. \end{aligned} \qquad 1.6.31$$

Thus when the man is between 0 and the flagpole, the dog, which represents the point $p(z)$, is closer to 0 than is the flagpole. That is, $|p(z)| < |b_0| = |p(z_0)|$. This is impossible, because we proved that $|p(z_0)|$ is the minimum value of our function. Therefore, our assumption that $p(z_0) \neq 0$ is false. □

Corollary 1.6.15 (Polynomial of degree k has k roots). *Every complex polynomial $p(z) = z^k + a_{k-1}z^{k-1} + \cdots + a_0$ with $k > 0$ can be written*

$$p(z) = (z - c_1)^{k_1} \cdots (z - c_m)^{k_m} \quad \textit{with} \quad k_1 + \cdots + k_m = k, \qquad 1.6.32$$

with the c_j all distinct. This expression is unique up to permutation of the factors.

The number k_j is called the *multiplicity* of the root c_j. Thus Corollary 1.6.15 asserts that every polynomial of degree k has exactly k complex roots, counted with multiplicity.

Proof. Let $\widetilde{k}$ be the largest degree of a monic polynomial $\widetilde{p}$ that divides p and is a product of polynomials of degree 1, so that we can write $p(z) = \widetilde{p}(z)q(z)$, where $q(z)$ is a polynomial of degree $k - \widetilde{k}$. (Conceivably, $\widetilde{k}$ might be 0, and $\widetilde{p}$ might be the constant polynomial 1. The proof shows that this is not the case.)

If $\widetilde{k} < k$, then by Theorem 1.6.14, there exists a number c such that $q(c) = 0$, and we can write $q(z) = (z - c)\widetilde{q}(z)$, where $\widetilde{q}(z)$ is a monic polynomial of degree $k - \widetilde{k} - 1$. Now

$$p(z) = \widetilde{p}(z)q(z) = \big(\widetilde{p}(z)(z - c)\big)\widetilde{q}(z), \qquad 1.6.33$$

and we see that $\widetilde{p}(z)(z-c)$ is a polynomial of degree $\widetilde{k}+1$ that factors as a product of factors of degree 1 and divides p. So $\widetilde{k}$ is not the largest degree of such a polynomial: our assumption $\widetilde{k} < k$ is wrong. Therefore $\widetilde{k} = k$ (i.e., p is the product of factors of degree 1). We can collect the factors corresponding to the same root, and write p as

$$p(z) = (z - c_1)^{k_1} \cdot \ldots \cdot (z - c_m)^{k_m}, \quad \text{with the } c_j \text{ distinct.} \qquad 1.6.34$$

Now we show that this expression is unique. Suppose we can write

$$p(z) = (z - c_1)^{k_1} \cdots (z - c_m)^{k_m} = (z - c'_1)^{k'_1} \cdots (z - c'_{m'})^{k'_{m'}}. \qquad 1.6.35$$

Cancel all the common factors on both sides of the equation above. If the two factorizations are different, a factor $z - c_i$ will appear on the left and not on the right. But that means that the polynomial on the left vanishes at c_i and the one on the right doesn't; since the polynomials are equal this is a contradiction, and the two factorizations are really the same. □

When Gauss gave his first proof of the fundamental theorem of algebra, in 1799, he stated the theorem in terms of real polynomials, in the form of Corollary 1.6.16: complex numbers were sufficiently disreputable that stating the theorem in its present form might have cast doubt on the respectability of the result. When he gave his fifth proof, some 12 years later, complex numbers were completely acceptable. The change was probably due to the realization that a complex number can be depicted geometrically as a point in the plane.

Corollary 1.6.16 (Factoring real polynomials). *Every real polynomial of degree $k > 0$ can be factored as a product of real polynomials of degree 1 or 2.*

Proof. We will start as above: let $\widetilde{k}$ be the largest degree of a real monic polynomial $\widetilde{p}$ that divides p and which is a product of real polynomials of degree 1 or 2, so that we can write $p(z) = \widetilde{p}(z)q(z)$, where $q(z)$ is a real polynomial of degree $k - \widetilde{k}$. If $\widetilde{k} < k$, then by Theorem 1.6.14, there exists a number c such that $q(c) = 0$. There are two cases to consider.

If c is real, we can write $q(z) = (z - c)\widetilde{q}(z)$, where $\widetilde{q}(z)$ is a real monic polynomial of degree $k - \widetilde{k} - 1$. Now

$$p(z) = \widetilde{p}(z)q(z) = \big(\widetilde{p}(z)(z - c)\big)\widetilde{q}(z), \qquad 1.6.36$$

and we see that $\widetilde{p}(z)(z-c)$ is a real polynomial of degree $\widetilde{k}+1$ that factors as a product of real polynomials of degree 1 and 2 and divides p. This is a contradiction and shows that $\widetilde{k} = k$ (i.e., that p is the product of real polynomials of degree 1 or 2).

The case where c is not real requires a separate lemma.

Long division of polynomials gives the following result: for any polynomials p_1, p_2, there exist unique polynomials q and r such that

$$p_1 = p_2 q + r,$$

with $\deg r < \deg p_2$. In particular, if $p(a) = 0$, then we can factor p as

$$p(z) = (z - a)q(z).$$

Indeed, there exists r of degree 0 (i.e., a constant) such that

$$p(z) = (z - a)q(z) + r;$$

evaluating at $z = a$, we find

$$0 = 0 \cdot q(a) + r, \quad \text{i.e.,} \quad r = 0.$$

Lemma 1.6.17 is an application of equations 0.7.6.

Lemma 1.6.17. *If q is a real polynomial, and $c \in \mathbb{C}$ is a root of q, then $\bar{c}$ is also a root of q.*

Proof. In essence, the proof is the following sequence of equalities:

$$q(\overline{c}) = \overline{q(c)} = \overline{0} = 0. \qquad 1.6.37$$

We need to justify the first equality. The polynomial is a sum of monomials of the form az^j with a real. Since $\overline{b_1 + b_2} = \overline{b}_1 + \overline{b}_2$, it is enough to prove the result for each monomial. This follows from

$$a(\overline{z})^j = \overline{a}(\overline{z^j}) = \overline{az^j}, \qquad 1.6.38$$

where we use the fact that a is real in the form $a = \overline{a}$. □ Lemma 1.6.17

Thus if c is not real, then $q(\overline{c}) = 0$, and

$$(z-c)(z-\overline{c}) = z^2 - (c+\overline{c})z + c\overline{c} = z^2 - 2\text{Re}\,(c)z + |c|^2 \qquad 1.6.39$$

is a real polynomial that divides q. So we can write $q(z) = (z-c)(z-\overline{c})\widetilde{q}(z)$, where $\widetilde{q}(z)$ is a real monic polynomial of degree $\widetilde{k} - 2$. Since

$$p(z) = \widetilde{p}(z)q(z) = \big(\widetilde{p}(z)(z^2 - 2\text{Re}\,(c)z + |c|^2)\big)\widetilde{q}(z), \qquad 1.6.40$$

we see that $\widetilde{p}(z)(z^2 - 2\text{Re}\,(c)z + |c|^2)$ is a real polynomial of degree $\widetilde{k} + 2$ that factors as a product of real polynomials of degree 1 and 2 and divides p. Again this is a contradiction and shows that $\widetilde{k} = k$. □ Corollary 1.6.16

Exercises for Section 1.6

1.6.1 Show that a set is bounded if it is contained in a ball centered anywhere; it does not have to be centered at the origin.

1.6.2 Let $A \subset \mathbb{R}^n$ be a subset that is not compact. Show that there exists a continuous unbounded function on A.

1.6.3 Set $z = x + iy$, where $x, y \in \mathbb{R}$. Show that the polynomial $p(z) = 1 + x^2y^2$ has no roots. Why doesn't this contradict Theorem 1.6.14?

1.6.4 Find, with justification, a number R such that there is a root of $p(z) = z^5 + 4z^3 + 3iz - 3$ in the disc $|z| \leq R$. (You may use that a minimum of $|p|$ is a root of p.)

1.6.5 Consider the polynomial $p(z) = z^6 + 4z^4 + z + 2 = z^6 + q(z)$.

a. Find R such that $|z^6| > |q(z)|$ when $|z| > R$.

b. Find a number R_1 such that you are sure that the minimum of $|p|$ occurs at a point z such that $|z| < R_1$.

1.6.6 a. Let $f : \mathbb{R} \to \mathbb{R}$ be the function $f(x) = |x|e^{-|x|}$. Show that it has an absolute maximum (i.e., not merely local) at some $x > 0$.

b. Where is the maximum of the function? What is the maximum value?

c. Show that the image of f is $[0, 1/e]$.

1.6.7 Show that if f is differentiable on a neighborhood of $[a, b]$, and we have $f'(a) < m < f'(b)$, then there exists $c \in (a, b)$ such that $f'(c) = m$.

1.6.8 Consider Example 1.6.4. What properties do the digits of $1/(2\pi)$ have to satisfy in order for the x_m to have a limit between 0.7 and 0.8?

Exercise 1.6.9 concerns the proof of Theorem 1.6.14. You will need Proposition 0.7.7.

1.6.9 If $a, b \in \mathbb{C}$, $a, b \neq 0$ and $j \geq 1$, find $p_0 > 0$ such that if $0 < p < p_0$ and $u = p(\cos\theta + i\sin\theta)$, then there exist j values of θ such that $a + bu^j$ is between 0 and a.

1.6.10 Consider the polynomial $p(z) = z^8 + z^4 + z^2 + 1$, where z is complex. Use the construction in the proof of the fundamental theorem of algebra, equation 1.6.28, to find a point z_0 such that $|p(z_0)| < |p(0)| = 1$.

1.6.11 Use the intermediate value theorem and the first part of the proof of the fundamental theorem of algebra (construction of R), to show that every real polynomial of odd degree has a real root.

1.7 Derivatives in several variables as linear transformations

Born: *I should like to put to Herr Einstein a question, namely, how quickly the action of gravitation is propagated in your theory ....*

Einstein: *It is extremely simple to write down the equations for the case when the perturbations that one introduces in the field are infinitely small. ... The perturbations then propagate with the same velocity as light.*

Born: *But for great perturbations things are surely very complicated?*

Einstein: *Yes, it is a mathematically complicated problem. It is especially difficult to find exact solutions of the equations, as the equations are nonlinear.* – Discussion after lecture by Einstein in 1913

Nonlinearities complicate both life and mathematics, but the associated instabilities can be useful. In his book *Fourier Analysis*, T. W. Körner relates that early airplanes were designed for stability until dog fights in World War I showed that stability came at the price of maneuverability. See also "The Forced Damped Pendulum: Chaos, Complication and Control," J. Hubbard, American Mathematical Monthly 106, No. 8, October 1999, available in pdf at MatrixEditions.com.

As mentioned in Section 1.3, a great many problems of interest are not linear. As a rule of thumb, on a small enough scale, things behave linearly; on a larger scale, nonlinearities often appear. If one company cuts costs by firing workers, it will probably increase profits; if all its competitors do the same, no one company will gain a competitive advantage; if enough workers lose jobs, who will buy the company's products?

The federal income tax rate is essentially linear in seven segments; for a single person in 2014, all taxable income up to \$9,075 was taxed at 10 percent, additional income up to \$36,900 was taxed at 15 percent, additional income up to \$89,350 at 25 percent, and so, up to a maximum of 39.6 percent. (The tax table is based on \$50 increments up to \$100,000, so tax owed is actually a step function.)

Water being chilled can be understood as a linear phenomenon – until it reaches the "phase transition" in which it turns to ice.

The object of differential calculus is to study nonlinear mappings by replacing them with linear transformations: we replace nonlinear equations with linear equations, curved surfaces by their tangent planes, and so on. Of course, this linearization is useful only if you understand linear objects reasonably well. Also, this replacement is only more or less justified. Locally, near the point of tangency, a curved surface may be very similar to its tangent plane, but further away it isn't. The hardest part of differential calculus is determining when replacing a nonlinear object by a linear one is justified.

In Section 1.3 we studied linear transformations in $\mathbb{R}^n$. Now we will see what this study contributes to the study of nonlinear transformations, more commonly called maps or mappings.

But nonlinear is much too broad a class to consider. Dividing mappings into linear and nonlinear is like dividing people into left-handed cello players and everyone else. We will study a limited subset of nonlinear mappings: those that are, in a sense we will study with care, "well approximated by linear transformations".

Derivatives and linear approximation in one dimension

In one dimension, the derivative is the main tool used to linearize a function. Recall from one-variable calculus that the derivative of a function $f : \mathbb{R} \to \mathbb{R}$, evaluated at a, is

$$f'(a) = \lim_{h \to 0} \frac{1}{h}\big(f(a+h) - f(a)\big). \qquad 1.7.1$$

Although it sounds less friendly, we really should say the following:

Definition 1.7.1 (Derivative). Let U be an open subset of $\mathbb{R}$, and let $f : U \to \mathbb{R}$ be a function. Then f is *differentiable* at $a \in U$ with *derivative* $f'(a)$ if the limit

$$f'(a) \stackrel{\text{def}}{=} \lim_{h \to 0} \frac{1}{h}\big(f(a+h) - f(a)\big) \quad \text{exists.} \qquad 1.7.2$$

Definition 1.7.1: We may sound finicky when we define the domain of f to be an open subset $U \subset \mathbb{R}$. Why not define the domain to be $\mathbb{R}$ and leave it at that? The problem with that approach is that many interesting functions are not defined on all of $\mathbb{R}$, but they are defined on an appropriate open subset $U \subset \mathbb{R}$. Such functions as $\ln x, \tan x$, and $1/x$ are not defined on all of $\mathbb{R}$; for example, $1/x$ is not defined at 0. If we used equation 1.7.1 as our definition, $\tan x$, $\ln x$, and $1/x$ would not be differentiable. So we are being more tolerant and broadminded, not less, when we allow functions to be defined only on an open subset of $\mathbb{R}$, rather than requiring them to be defined on all of $\mathbb{R}$.

We discussed in Remark 1.5.6 why it is necessary to specify an *open* set when talking about derivatives.

Students often find talk about open sets $U \subset \mathbb{R}$ and domains of definition pointless; what does it mean when we talk about a function $f : U \to \mathbb{R}$? This is the same as saying $f : \mathbb{R} \to \mathbb{R}$, except that $f(x)$ is only defined if the point x is in U.

Example 1.7.2 (Derivative of a function from $\mathbb{R} \to \mathbb{R}$). If $f(x) = x^2$, then $f'(x) = 2x$. This is proved by writing

$$f'(x) = \lim_{h \to 0} \frac{1}{h}\big((x+h)^2 - x^2\big) = \lim_{h \to 0} \frac{1}{h}(2xh + h^2) = 2x + \lim_{h \to 0} h = 2x. \qquad 1.7.3$$

The derivative $2x$ of the function $f(x) = x^2$ is the slope of the line tangent to f at x; we also say that $2x$ is the *slope of the graph of f at*

x. In higher dimensions, this idea of the slope of the tangent to a function still holds, although already picturing a plane tangent to a surface is considerably more difficult than picturing a line tangent to a curve. △

Partial derivatives

One kind of derivative of a function of several variables works just like a derivative of a function of one variable: take the derivative *with respect to one variable*, treating all the others as constants.

The partial derivative $D_i f$ answers the question, how fast does the function change when you vary the ith variable, keeping the other variables constant?

We can state this in terms of standard basis vectors:

$$D_i f(\mathbf{a}) = \lim_{h \to 0} \frac{1}{h} f(\mathbf{a} + h\vec{\mathbf{e}}_i) - f(\mathbf{a});$$

$D_i f$ measures the rate at which $f(\mathbf{x})$ changes as the variable $\mathbf{x}$ moves from $\mathbf{a}$ in the direction $\vec{\mathbf{e}}_i$.

The partial derivatives of Definition 1.7.3 are first partial derivatives. In Section 3.3 we discuss partial derivatives of higher order.

Definition 1.7.3 (Partial derivative). Let U be an open subset of $\mathbb{R}^n$ and $f : U \to \mathbb{R}$ a function. The *partial derivative* of f with respect to the ith variable, and evaluated at $\mathbf{a}$, is the limit

$$D_i f(\mathbf{a}) \stackrel{\text{def}}{=} \lim_{h \to 0} \frac{1}{h} \left(f \begin{pmatrix} a_1 \\ \vdots \\ a_i + h \\ \vdots \\ a_n \end{pmatrix} - f \begin{pmatrix} a_1 \\ \vdots \\ a_i \\ \vdots \\ a_n \end{pmatrix} \right), \qquad 1.7.4$$

if the limit exists, of course.

Partial derivatives are computed the same way derivatives are computed in first year calculus. The partial derivative of $f\begin{pmatrix} x \\ y \end{pmatrix} = x^2 + x \sin y$ with respect to the first variable is $D_1 f = 2x + \sin y$; the partial derivative with respect to the second variable is $D_2 f = x \cos y$. At the point $\mathbf{a} = \begin{pmatrix} a \\ b \end{pmatrix}$, we have $D_1 f(\mathbf{a}) = 2a + \sin b$ and $D_2 f(\mathbf{a}) = a \cos b$. For $f\begin{pmatrix} x \\ y \end{pmatrix} = x^3 + x^2 y + y^2$, what is $D_1 f$? What is $D_2 f$? Check your answers in the footnote below.[19]

Remark. The most common notation for partial derivatives is

$$\frac{\partial f}{\partial x_1}, \dots, \frac{\partial f}{\partial x_n}, \quad \text{i.e.,} \quad \frac{\partial f}{\partial x_i} = D_i f. \qquad 1.7.5$$

We prefer $D_i f$: it is simpler to write, looks better in matrices, and focuses on the important information: with respect to *which variable* the partial derivative is being taken. (In economics, for example, one might write $D_w f$ for the "wages" variable, $D_p f$ for "prime rate".) But we will occasionally use the other notation, so that you will be familiar with it. △

A notation for partial derivatives often used in partial differential equations is

$$f_{x_i} = D_i f.$$

Partial derivatives of $\mathbb{R}^m$-valued functions

The definition of a partial derivative makes just as good sense for an $\mathbb{R}^m$-valued (or "vector-valued") function: a function $\mathbf{f}$ from $\mathbb{R}^n$ to $\mathbb{R}^m$. In such a case, we evaluate the limit for each component of $\mathbf{f}$, defining

[19] $D_1 f = 3x^2 + 2xy$ and $D_2 f = x^2 + 2y$.

Note that the partial derivative of a vector-valued function is a vector.

$$\overrightarrow{D_i}\mathbf{f}(\mathbf{a}) \overset{\text{def}}{=} \lim_{h\to 0}\frac{1}{h}\left(\mathbf{f}\begin{pmatrix} a_1 \\ \vdots \\ a_i+h \\ \vdots \\ a_n\end{pmatrix} - \mathbf{f}\begin{pmatrix} a_1 \\ \vdots \\ a_i \\ \vdots \\ a_n\end{pmatrix}\right) = \begin{bmatrix} D_i f_1(\mathbf{a}) \\ \vdots \\ D_i f_m(\mathbf{a})\end{bmatrix}. \qquad 1.7.6$$

Example 1.7.4 (Partial derivatives). Let $\mathbf{f} : \mathbb{R}^2 \to \mathbb{R}^3$ be given by

We give two versions of equation 1.7.8 to illustrate the two notations and to emphasize the fact that although we used x and y to define the function, we can evaluate it at variables that look different.

$$\mathbf{f}\begin{pmatrix} x \\ y\end{pmatrix} = \begin{pmatrix} xy \\ \sin(x+y) \\ x^2-y^2\end{pmatrix}. \qquad 1.7.7$$

The partial derivative of $\mathbf{f}$ with respect to the first variable is

$$\overrightarrow{D_1}\mathbf{f}\begin{pmatrix} x \\ y\end{pmatrix} = \begin{bmatrix} y \\ \cos(x+y) \\ 2x\end{bmatrix} \quad \text{or} \quad \frac{\partial \mathbf{f}}{\partial x_1}\begin{pmatrix} a \\ b\end{pmatrix} = \begin{bmatrix} b \\ \cos(a+b) \\ 2a\end{bmatrix}. \qquad 1.7.8$$

What are the partial derivatives at $\begin{pmatrix} a \\ b\end{pmatrix}$ of $\mathbf{f}\begin{pmatrix} x \\ y\end{pmatrix} = \begin{pmatrix} x^2y \\ \cos y\end{pmatrix}$? How would you rewrite the answer using the notation of formula 1.7.5?[20]

Pitfalls of partial derivatives

The eminent French mathematician Adrien Douady complained that the notation for the partial derivative omits the most important information: which variables are being kept constant. For instance, does increasing the minimum wage increase or decrease the number of minimum wage jobs? This is a question about the sign of $D_{\text{minimum wage}} f$, where x is the economy and $f(x) =$ number of minimum wage jobs.

But this partial derivative is meaningless until you state what is being held constant, and it isn't easy to see what this means. Is public investment to be held constant, or the discount rate, or is the discount rate to be adjusted to keep total unemployment constant? There are many other variables to consider, who knows how many. You can see here why economists disagree about the sign of this partial derivative: it is hard if not impossible to say what the partial derivative is, never mind evaluating it.

Similarly, if you are studying pressure of a gas as a function of temperature, it makes a big difference whether the volume of gas is kept constant or whether the gas is allowed to expand (for instance, because it fills a balloon).

The derivative in several variables

Often we will want to see how a system changes when *all* the components of its argument are allowed to vary: how $\mathbf{f} : \mathbb{R}^n \to \mathbb{R}^m$ changes when its

---

[20] $\overrightarrow{D_1}\mathbf{f}\begin{pmatrix} a \\ b\end{pmatrix} = \frac{\partial \mathbf{f}}{\partial x_1}\begin{pmatrix} a \\ b\end{pmatrix} = \begin{bmatrix} 2ab \\ 0\end{bmatrix}$; $\overrightarrow{D_2}\mathbf{f}\begin{pmatrix} a \\ b\end{pmatrix} = \frac{\partial \mathbf{f}}{\partial x_2}\begin{pmatrix} a \\ b\end{pmatrix} = \begin{bmatrix} a^2 \\ -\sin b\end{bmatrix}$.

"When its variable $\mathbf{x}$ moves in any direction $\vec{\mathbf{v}}$" means that x_1 becomes $x_1 + hv_1$ and x_2 becomes $x_2 + hv_2$ (for the same h), and so on.

argument $\mathbf{x}$ moves in any direction $\vec{\mathbf{v}}$. We will see that if $\mathbf{f}$ is differentiable, its derivative consists of a matrix, called the *Jacobian matrix*, whose entries are the partial derivatives of the function. Later we will see that this matrix indeed answers the question, given any direction, how does $\mathbf{f}$ change when $\mathbf{x}$ moves in that direction?

Definition 1.7.1 from first year calculus defines the derivative as the limit

$$\frac{\text{change in } f}{\text{change in } x}, \quad \text{i.e.,} \quad \frac{f(a+h)-f(a)}{h}, \qquad 1.7.9$$

as h (the increment to the variable x) approaches 0. This does not generalize well to higher dimensions. When f is a function of several variables, then an increment to the variable will be a vector, and we can't divide by vectors.

It is tempting just to divide by $|\vec{\mathbf{h}}|$, the length of $\vec{\mathbf{h}}$:

$$f'(\mathbf{a}) = \lim_{\vec{\mathbf{h}}\to\vec{\mathbf{0}}} \frac{1}{|\vec{\mathbf{h}}|}\big(f(\mathbf{a}+\vec{\mathbf{h}}) - f(\mathbf{a})\big). \qquad 1.7.10$$

This would allow us to rewrite Definition 1.7.1 in higher dimensions, since we can divide by the length of a vector, which is a number. But this wouldn't work even in dimension 1, because the limit changes sign when h approaches 0 from the left and from the right. In higher dimensions it's much worse. All the different directions from which $\vec{\mathbf{h}}$ could approach $\vec{\mathbf{0}}$ give different limits. By dividing by $|\vec{\mathbf{h}}|$ in equation 1.7.10 we are canceling the magnitude but not the direction.

The Greek letter Δ is the uppercase "delta"; the lowercase delta is δ.

We will rewrite it in a form that does generalize well. This definition emphasizes the idea that a function f is differentiable at a point a if the increment Δf to the function is well approximated by a *linear function of the increment h to the variable.* This linear function is $f'(a)h$.

When we call $f'(a)h$ a linear function, we mean the linear function that takes the variable h and *multiplies* it by $f'(a)$ (i.e., the function $h \mapsto f'(a)\,h$, to be read, "h maps to $f'(a)\,h$"). Usually the derivative of f at a is *not* a linear function of a. If $f(x) = \sin x$ or $f(x) = x^3$, or just about anything except $f(x) = x^2$, then $f'(a)$ is not a linear function of a. But

$$h \mapsto f'(a)\,h$$

is a linear function of h. For example, $h \mapsto (\sin x)h$ is a linear function of h, since

$$(\sin x)(h_1 + h_2) = (\sin x)h_1 + (\sin x)h_2.$$

Note the difference between $\mapsto$ ("maps to") and $\to$ ("to"). The first has a "pusher".

Definition 1.7.5 (Alternative definition of the derivative). Let U be an open subset of $\mathbb{R}$, and $f : U \to \mathbb{R}$ a function. Then f is *differentiable* at a, with *derivative* m, if and only if

$$\lim_{h\to 0} \frac{1}{h}\Bigg(\overbrace{\big(f(a+h)-f(a)\big)}^{\Delta f} - \overbrace{(mh)}^{\text{linear function of } \Delta x}\Bigg) = 0. \qquad 1.7.11$$

The letter Δ denotes "change in"; Δf is the change in the function; $\Delta x = h$ is the change in the variable x. The function mh that multiplies h by the derivative m is thus a linear function of the change in x.

Remark 1.7.6. We are taking the limit as $h \to 0$, so h is small, and dividing by it makes things big; the numerator – the difference between the increment to the function and the approximation of that increment – must be *very small* when h is near 0 for the limit to be zero anyway (see Exercise 1.7.13).

Specifically, the increment to f, i.e., $h \mapsto f(a+h) - f(a)$, and its linear approximation $h \mapsto mh$ must differ by something *smaller than linear* when

h is near 0. If, for example, we have a linear function L such that the difference is linear, say $3h$, then equation 1.7.11 is not satisfied:

$$\lim_{h\to 0}\frac{1}{h}\big(f(a+h)-f(a)-L(h)\big)=\lim_{h\to 0}\frac{1}{h}3h=3\neq 0. \qquad 1.7.12$$

Such a function L does not qualify as the derivative of f. But if we have another linear function, D, such that the difference is smaller than linear, say $3h^2$, then

$$\lim_{h\to 0}\frac{1}{h}\big(f(a+h)-f(a)-D(h)\big)=\lim_{h\to 0}\frac{1}{h}3h^2=\lim_{h\to 0}3h=0. \quad \triangle$$

The following computation shows that Definition 1.7.5 is just a way of restating Definition 1.7.1:

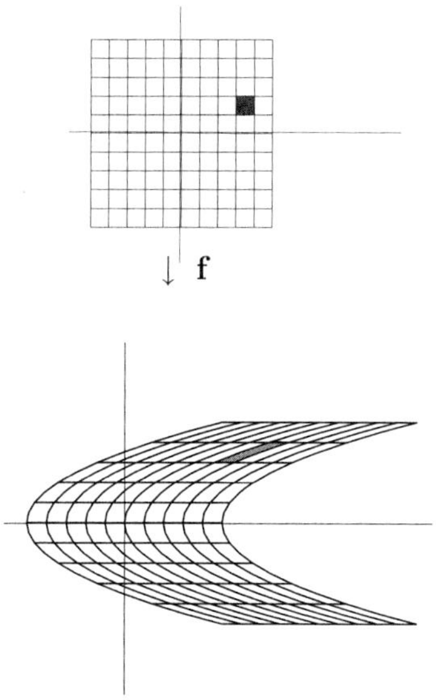

FIGURE 1.7.1.
The mapping

$$\mathbf{f}\begin{pmatrix}x\\y\end{pmatrix}=\begin{pmatrix}x^2+y\\x\end{pmatrix}$$

takes the shaded square in the square at top to the shaded area at bottom.

$$\lim_{h\to 0}\frac{1}{h}\Big(\big(f(a+h)-f(a)\big)-[f'(a)]h\Big)=\overbrace{\lim_{h\to 0}\Big(\frac{f(a+h)-f(a)}{h}}^{f'(a)\text{ by Def. 1.7.1}}-\frac{f'(a)h}{h}\Big)$$
$$=f'(a)-f'(a)=0. \qquad 1.7.13$$

Moreover, the linear function $h\mapsto f'(a)h$ is the *only* linear function satisfying equation 1.7.11. Indeed, any linear function of one variable can be written $h\mapsto mh$, and

$$\begin{aligned}0&=\lim_{h\to 0}\frac{1}{h}\Big(\big(f(a+h)-f(a)\big)-mh\Big)\\&=\lim_{h\to 0}\Big(\frac{f(a+h)-f(a)}{h}-\frac{mh}{h}\Big)=f'(a)-m,\end{aligned} \qquad 1.7.14$$

so we have $f'(a)=m$.

The point of rewriting the definition of the derivative is that with Definition 1.7.5, we can divide by $|h|$ rather than h; the number $m=f'(a)$ is also the unique number such that

$$\lim_{h\to 0}\frac{1}{|h|}\Big(\overbrace{\big(f(a+h)-f(a)\big)}^{\Delta f}-\overbrace{\big(f'(a)h\big)}^{\text{linear function of }h}\Big)=0. \qquad 1.7.15$$

It doesn't matter if the limit changes sign, since the limit is 0; a number close to 0 is close to 0 whether it is positive or negative.

Therefore, we can generalize equation 1.7.15 to mappings in higher dimensions, like the one in Figure 1.7.1. As in the case of functions of one variable, the key to understanding derivatives of functions of several variables is to think that the increment to the function (the output) is *approximately a linear function* of the increment to the variable (the input): the increment

$$\Delta\mathbf{f}=\mathbf{f}(\mathbf{a}+\vec{\mathbf{h}})-\mathbf{f}(\mathbf{a}) \qquad 1.7.16$$

is approximately a linear function of the increment $\vec{\mathbf{h}}$.

In the one-dimensional case, Δf is well approximated by the linear function $h \mapsto f'(a)h$. We saw in Section 1.3 that every linear transformation is given by a matrix; the linear transformation $h \mapsto f'(a)h$ is given by multiplication by the 1×1 matrix $[f'(a)]$.

For a mapping $\mathbf{f} : U \to \mathbb{R}^m$, where $U \subset \mathbb{R}^n$ is open, the role of this 1×1 matrix is played by an $m \times n$ matrix composed of the n partial derivatives of the mapping at $\mathbf{a}$. This matrix is called the *Jacobian matrix* of the mapping $\mathbf{f}$; we denote it $[\mathbf{Jf}(\mathbf{a})]$.

Note that in the Jacobian matrix we write the components of $\mathbf{f}$ from top to bottom, and the variables from left to right. The first column gives the partial derivatives with respect to the first variable; the second column gives the partial derivatives with respect to the second variable, and so on.

Definition 1.7.7 (Jacobian matrix). Let U be an open subset of $\mathbb{R}^n$. The *Jacobian matrix* of a function $\mathbf{f} : U \to \mathbb{R}^m$ is the $m \times n$ matrix composed of the n partial derivatives of $\mathbf{f}$ evaluated at $\mathbf{a}$:

$$[\mathbf{Jf}(\mathbf{a})] \stackrel{\text{def}}{=} \begin{bmatrix} D_1 f_1(\mathbf{a}) & \dots & D_n f_1(\mathbf{a}) \\ \vdots & & \vdots \\ D_1 f_m(\mathbf{a}) & \dots & D_n f_m(\mathbf{a}) \end{bmatrix}. \qquad 1.7.17$$

FIGURE 1.7.2.

Carl Jacobi (1804–1851) was 12 years old when he finished the last year of secondary school, but had to wait until 1821 before entering the University of Berlin, which did not accept students younger than 16. He died after contracting flu and smallpox.-

Example 1.7.8. The Jacobian matrix of $\mathbf{f}\begin{pmatrix} x \\ y \end{pmatrix} = \begin{pmatrix} xy \\ \sin(x+y) \\ x^2 - y^2 \end{pmatrix}$ is

$$\left[\mathbf{Jf}\begin{pmatrix} x \\ y \end{pmatrix}\right] = \begin{bmatrix} y & x \\ \cos(x+y) & \cos(x+y) \\ 2x & -2y \end{bmatrix}. \quad \triangle \qquad 1.7.18$$

What is the Jacobian matrix of $\mathbf{f}\begin{pmatrix} x \\ y \end{pmatrix} = \begin{pmatrix} x^3 y \\ 2x^2y^2 \\ xy \end{pmatrix}$? Check your answer.[21]

When is the Jacobian matrix of a function its derivative?

We would like to say that if the Jacobian matrix exists, then it is the derivative of $\mathbf{f}$: we would like to say that the increment $\mathbf{f}(\mathbf{a} + \vec{\mathbf{h}}) - \mathbf{f}(\mathbf{a})$ is approximately $[\mathbf{Jf}(\mathbf{a})]\vec{\mathbf{h}}$, in the sense that

$$\lim_{\vec{\mathbf{h}} \to \vec{\mathbf{0}}} \frac{1}{|\vec{\mathbf{h}}|}\Big(\big(\mathbf{f}(\mathbf{a} + \vec{\mathbf{h}}) - \mathbf{f}(\mathbf{a})\big) - [\mathbf{Jf}(\mathbf{a})]\vec{\mathbf{h}} \Big) = \vec{\mathbf{0}}. \qquad 1.7.19$$

This is the higher-dimensional analogue of equation 1.7.11. Usually it is true in higher dimensions: you can calculate the derivative of a function with several variables by computing its partial derivatives, and putting them in a matrix.

---

[21] $\left[\mathbf{Jf}\begin{pmatrix} x \\ y \end{pmatrix}\right] = \begin{bmatrix} 3x^2y & x^3 \\ 4xy^2 & 4x^2y \\ y & x \end{bmatrix}$. The first column is $\overrightarrow{D_1\mathbf{f}}$; the second is $\overrightarrow{D_2\mathbf{f}}$. The first row gives the partial derivatives for x^3y; the second the partial derivatives for $2x^2y^2$; the third the partial derivatives for xy.

Unfortunately, it is possible for all partial derivatives of a function $\mathbf{f}$ to exist and yet for equation 1.7.19 not to be true! (We will examine the issue of such pathological functions in Section 1.9; see Example 1.9.3.) The best we can do without extra hypotheses is the following.

Proposition and Definition 1.7.9 (Derivative). *Let $U \subset \mathbb{R}^n$ be an open subset and let $\mathbf{f}: U \to \mathbb{R}^m$ be a mapping; let $\mathbf{a}$ be a point in U. If there exists a linear transformation $L: \mathbb{R}^n \to \mathbb{R}^m$ such that*

$$\lim_{\vec{\mathbf{h}} \to \vec{\mathbf{0}}} \frac{1}{|\vec{\mathbf{h}}|}\Big((\mathbf{f}(\mathbf{a}+\vec{\mathbf{h}}) - \mathbf{f}(\mathbf{a})) - (L(\vec{\mathbf{h}}))\Big) = \vec{\mathbf{0}}, \qquad 1.7.20$$

then $\mathbf{f}$ is differentiable at $\mathbf{a}$, and L is unique and is the derivative of $\mathbf{f}$ at $\mathbf{a}$, denoted $[\mathbf{Df}(\mathbf{a})]$.

Note in equation 1.7.20 that $\vec{\mathbf{h}}$ can approach $\vec{\mathbf{0}}$ from any direction.

The fact that the derivative of a function in several variables is represented by the Jacobian matrix is the reason why linear algebra is a prerequisite to multivariable calculus.

Theorem 1.7.10 concerns only first partial derivatives. In Section 3.3 we discuss partial derivatives of higher order.

Theorem 1.7.10 (The Jacobian matrix and the derivative). *If $\mathbf{f}$ is differentiable at $\mathbf{a}$, then all partial derivatives of $\mathbf{f}$ at $\mathbf{a}$ exist, and the matrix representing $[\mathbf{Df}(\mathbf{a})]$ is $[\mathbf{Jf}(\mathbf{a})]$.*

Remark. We use $[\mathbf{Df}(\mathbf{a})]$ to denote both the derivative of $\mathbf{f}$ at $\mathbf{a}$ and the matrix representing it. In the case of a function $f: \mathbb{R} \to \mathbb{R}$, it is a 1×1 matrix, i.e., a number. It is convenient to write $[\mathbf{Df}(\mathbf{a})]$ rather than writing the Jacobian matrix in full:

$$[\mathbf{Df}(\mathbf{a})] = [\mathbf{Jf}(\mathbf{a})] = \begin{bmatrix} D_1 f_1(\mathbf{a}) & \dots & D_n f_1(\mathbf{a}) \\ \vdots & & \vdots \\ D_1 f_m(\mathbf{a}) & \dots & D_n f_m(\mathbf{a}) \end{bmatrix}. \qquad 1.7.21$$

But when you see $[\mathbf{Df}(\mathbf{a})]$, you should always be aware that it is a linear transformation, and you should know its domain and codomain. Given a function $\mathbf{f}: \mathbb{R}^3 \to \mathbb{R}^2$, what are the dimensions of $[\mathbf{Df}(\mathbf{a})]$?[22] $\triangle$

The symbol ∇ is pronounced "nabla".

We discuss the gradient further in Section 6.8 and in Exercise 6.6.3.

The gradient of a function f is a vector that points in the direction in which f increases fastest; Exercise 1.7.18 asks you to justify this. The length of the gradient is proportional to how fast f increases in that direction.

To evaluate the gradient on $\vec{\mathbf{v}}$ we need to write $\operatorname{grad} f(\mathbf{a}) \cdot \vec{\mathbf{v}}$; we need the dot product, but often there is no natural dot product. The number $[\mathbf{Df}(\mathbf{a})]\vec{\mathbf{v}}$ always makes sense.

Remark. The *gradient* of a function $f: \mathbb{R}^n \to \mathbb{R}$ is

$$\operatorname{grad} f(\mathbf{a}) = \vec{\nabla} f(\mathbf{a}) \stackrel{\text{def}}{=} \begin{bmatrix} D_1 f(\mathbf{a}) \\ \vdots \\ D_n f(\mathbf{a}) \end{bmatrix}. \qquad 1.7.22$$

In other words, $\operatorname{grad} f(\mathbf{a}) = [\mathbf{D}f(\mathbf{a})]^\top = [D_1 f(\mathbf{a}) \dots D_n f(\mathbf{a})]^\top$.

Many people think of the gradient as the derivative. This is unwise, encouraging confusion of row matrices and column vectors. The transpose is not as innocent as it looks; geometrically, it is only defined if we are in a space with an inner product (the generalization of the dot product). When there is no natural inner product, the *vector* $\vec{\nabla} f(\mathbf{a})$ is not natural, unlike the derivative, which as a linear transformation depends on no such choices. $\triangle$

---

[22] Since $\mathbf{f}: \mathbb{R}^3 \to \mathbb{R}^2$ takes a point in $\mathbb{R}^3$ and gives a point in $\mathbb{R}^2$, similarly, $[\mathbf{Df}(\mathbf{a})]$ takes a vector in $\mathbb{R}^3$ and gives a vector in $\mathbb{R}^2$. Therefore, $[\mathbf{Df}(\mathbf{a})]$ is a 2×3 matrix.

Proof of Proposition and Definition 1.7.9 and Theorem 1.7.10. The only part of Proposition 1.7.9 that needs to be proved is uniqueness.

This proof proves that if there is a derivative, it is unique, since a linear transformation has just one matrix.

We know (Theorem 1.3.4) that the linear transformation L is represented by the matrix whose ith column is $L(\vec{\mathbf{e}}_i)$, so we need to show that

$$L(\vec{\mathbf{e}}_i) = \overrightarrow{D_i\mathbf{f}}(\mathbf{a}), \qquad 1.7.23$$

where $\overrightarrow{D_i\mathbf{f}}(\mathbf{a})$ is by definition the ith column of the Jacobian matrix $[\mathbf{Jf}(\mathbf{a})]$.

Equation 1.7.20 says nothing about the direction in which $\vec{\mathbf{h}}$ approaches $\vec{\mathbf{0}}$; it can approach $\vec{\mathbf{0}}$ from any direction. In particular, we can set $\vec{\mathbf{h}} = t\vec{\mathbf{e}}_i$, and let the number t tend to 0, giving

$$\lim_{t\vec{\mathbf{e}}_i \to \vec{\mathbf{0}}} \frac{\big(\mathbf{f}(\mathbf{a}+t\vec{\mathbf{e}}_i) - \mathbf{f}(\mathbf{a})\big) - L(t\vec{\mathbf{e}}_i)}{|t\vec{\mathbf{e}}_i|} = \vec{\mathbf{0}}. \qquad 1.7.24$$

We want to get rid of the absolute value signs in the denominator. Since $|t\vec{\mathbf{e}}_i| = |t||\vec{\mathbf{e}}_i|$ (remember t is a number) and $|\vec{\mathbf{e}}_i| = 1$, we can replace $|t\vec{\mathbf{e}}_i|$ by $|t|$. The limit in equation 1.7.24 is $\vec{\mathbf{0}}$, whether t is positive or negative, so we can replace $|t|$ by t:

$$\lim_{t\vec{\mathbf{e}}_i \to \vec{\mathbf{0}}} \frac{\mathbf{f}(\mathbf{a}+t\vec{\mathbf{e}}_i) - \mathbf{f}(\mathbf{a}) - L(t\vec{\mathbf{e}}_i)}{t} = \vec{\mathbf{0}}. \qquad 1.7.25$$

Using the linearity of the derivative, we see that

$$L(t\vec{\mathbf{e}}_i) = tL(\vec{\mathbf{e}}_i), \qquad 1.7.26$$

so we can rewrite equation 1.7.25 as

$$\lim_{t\vec{\mathbf{e}}_i \to \vec{\mathbf{0}}} \frac{\mathbf{f}(\mathbf{a}+t\vec{\mathbf{e}}_i) - \mathbf{f}(\mathbf{a})}{t} - L(\vec{\mathbf{e}}_i) = \vec{\mathbf{0}}. \qquad 1.7.27$$

The first term is precisely equation 1.7.6 defining the partial derivative. So $L(\vec{\mathbf{e}}_i) = \overrightarrow{D_i\mathbf{f}}(\mathbf{a})$: the ith column of the matrix corresponding to the linear transformation L is indeed $\overrightarrow{D_i\mathbf{f}}(\mathbf{a})$. In other words, the matrix corresponding to L is the Jacobian matrix. □

The converse of Proposition 1.7.11 is of course false: it is easy to find continuous functions that are not differentiable (i.e., not differentiable at some point). It is also possible to find continuous functions that are *nowhere differentiable*, but finding them is considerably harder.

Proposition 1.7.11 (Differentiable implies continuous). *Let U be an open subset of $\mathbb{R}^n$ and let $\mathbf{f}: U \to \mathbb{R}^m$ be a mapping; let $\mathbf{a}$ be a point in U. If $\mathbf{f}$ is differentiable at $\mathbf{a}$, then $\mathbf{f}$ is continuous at $\mathbf{a}$.*

Proof. Since $\mathbf{f}$ is differentiable at $\mathbf{a}$, the limit in equation 1.7.20 exists, which implies the much weaker statement

$$\lim_{\vec{\mathbf{h}} \to \vec{\mathbf{0}}} \Big(\big(\mathbf{f}(\mathbf{a}+\vec{\mathbf{h}}) - \mathbf{f}(\mathbf{a})\big) - L(\vec{\mathbf{h}})\Big) = \vec{\mathbf{0}}. \qquad 1.7.28$$

Since L is linear, $\lim_{\vec{\mathbf{h}} \to \vec{\mathbf{0}}} L(\vec{\mathbf{h}}) = \vec{\mathbf{0}}$. Thus $\lim_{\vec{\mathbf{h}} \to \vec{\mathbf{0}}} (\mathbf{f}(\mathbf{a}+\vec{\mathbf{h}}) - \mathbf{f}(\mathbf{a}))$ must also be $\vec{\mathbf{0}}$, so $\mathbf{f}$ is continuous (see Definition 1.5.27). □

Example 1.7.12 (The Jacobian matrix of a function $\mathbf{f}: \mathbb{R}^2 \to \mathbb{R}^2$). Let's see, for a fairly simple nonlinear mapping from $\mathbb{R}^2$ to $\mathbb{R}^2$, that the Jacobian matrix does indeed provide the desired approximation of the change in the mapping. The Jacobian matrix of the mapping

$$\mathbf{f}\begin{pmatrix} x \\ y \end{pmatrix} = \begin{pmatrix} xy \\ x^2 - y^2 \end{pmatrix} \quad \text{is} \quad \left[\mathbf{Jf}\begin{pmatrix} x \\ y \end{pmatrix}\right] = \begin{bmatrix} y & x \\ 2x & -2y \end{bmatrix}, \qquad 1.7.29$$

since $D_1 xy = y$, $D_2 xy = x$, etc. Our increment vector will be $\vec{\mathbf{h}} = \begin{bmatrix} h \\ k \end{bmatrix}$.

Equation 1.7.19 then becomes

$$\lim_{\begin{bmatrix} h \\ k \end{bmatrix} \to \begin{bmatrix} 0 \\ 0 \end{bmatrix}} \frac{1}{\sqrt{h^2+k^2}} \left(\left(\underbrace{\mathbf{f}\begin{pmatrix} a_1 + h \\ a_2 + k \end{pmatrix}}_{\mathbf{f}(\mathbf{a}+\vec{\mathbf{h}})} - \underbrace{\mathbf{f}\begin{pmatrix} a_1 \\ a_2 \end{pmatrix}}_{\mathbf{f}(\mathbf{a})} \right) - \left(\underbrace{\begin{bmatrix} a_2 & a_1 \\ 2a_1 & -2a_2 \end{bmatrix}}_{\text{Jacobian matrix}} \underbrace{\begin{bmatrix} h \\ k \end{bmatrix}}_{\vec{\mathbf{h}}} \right) \right) \overset{?}{=} \vec{\mathbf{0}}. \qquad 1.7.30$$

Equation 1.7.30: the $\sqrt{h^2+k^2}$ in the denominator is $|\vec{\mathbf{h}}|$, the length of the increment vector.

When we evaluate $\mathbf{f}$ at $\begin{pmatrix} a_1 + h \\ a_2 + k \end{pmatrix}$ and at $\begin{pmatrix} a_1 \\ a_2 \end{pmatrix}$, the left side becomes

$$\lim_{\begin{bmatrix} h \\ k \end{bmatrix} \to \begin{bmatrix} 0 \\ 0 \end{bmatrix}} \frac{1}{\sqrt{h^2+k^2}} \left(\left(\begin{pmatrix} (a_1+h)(a_2+k) \\ (a_1+h)^2 - (a_2+k)^2 \end{pmatrix} - \begin{pmatrix} a_1 a_2 \\ a_1^2 - a_2^2 \end{pmatrix} \right) - \begin{bmatrix} a_2 h + a_1 k \\ 2a_1 h - 2a_2 k \end{bmatrix} \right)$$

$$= \lim_{\begin{bmatrix} h \\ k \end{bmatrix} \to \begin{bmatrix} 0 \\ 0 \end{bmatrix}} \frac{1}{\sqrt{h^2+k^2}} \begin{bmatrix} hk \\ h^2 - k^2 \end{bmatrix}. \qquad 1.7.31$$

Example 1.7.12: For instance, at $\begin{pmatrix} a \\ b \end{pmatrix} = \begin{pmatrix} 1 \\ 1 \end{pmatrix}$, the function $\mathbf{f}$ of equation 1.7.29 gives

$$\mathbf{f}\begin{pmatrix} 1 \\ 1 \end{pmatrix} = \begin{pmatrix} 1 \\ 0 \end{pmatrix},$$

and we are asking whether

$$\mathbf{f}\begin{pmatrix} 1 \\ 1 \end{pmatrix} + \begin{bmatrix} 1 & 1 \\ 2 & -2 \end{bmatrix}\begin{bmatrix} h \\ k \end{bmatrix} = \begin{pmatrix} 1 + h + k \\ 2h - 2k \end{pmatrix}$$

is a good approximation to

$$f\begin{pmatrix} 1+h \\ 1+k \end{pmatrix} = \begin{pmatrix} 1 + h + k + \underline{hk} \\ 2h - 2k + \underline{h^2 - k^2} \end{pmatrix}.$$

That is, we are asking whether the difference between the two is smaller than linear (see Remark 1.7.6). It clearly is: the difference is $\begin{pmatrix} hk \\ h^2 - k^2 \end{pmatrix}$, and both hk and $h^2 - k^2$ are quadratic.

We want to show that this limit is $\vec{\mathbf{0}}$. First we consider the entry $\dfrac{hk}{\sqrt{h^2+k^2}}$. Indeed, $0 \le |h| \le \sqrt{h^2+k^2}$ and $0 \le |k| \le \sqrt{h^2+k^2}$, so

$$0 \le \left| \frac{hk}{\sqrt{h^2+k^2}} \right| = \underbrace{\left| \frac{h}{\sqrt{h^2+k^2}} \right|}_{\le 1} |k| \le |k|, \qquad 1.7.32$$

and we have

$$0 \le \overbrace{\lim_{\begin{bmatrix} h \\ k \end{bmatrix} \to \begin{bmatrix} 0 \\ 0 \end{bmatrix}} \left| \frac{hk}{\sqrt{h^2+k^2}} \right|}^{\text{squeezed between 0 and 0}} \le \lim_{\begin{bmatrix} h \\ k \end{bmatrix} \to \begin{bmatrix} 0 \\ 0 \end{bmatrix}} |k| = 0. \qquad 1.7.33$$

Next we consider $\dfrac{h^2 - k^2}{\sqrt{h^2+k^2}}$:

$$0 \le \left| \frac{h^2 - k^2}{\sqrt{h^2+k^2}} \right| \le |h| \left| \frac{h}{\sqrt{h^2+k^2}} \right| + |k| \left| \frac{k}{\sqrt{h^2+k^2}} \right| \le |h| + |k|, \qquad 1.7.34$$

so

$$0 \le \lim_{\begin{bmatrix} h \\ k \end{bmatrix} \to \begin{bmatrix} 0 \\ 0 \end{bmatrix}} \left| \frac{h^2 - k^2}{\sqrt{h^2+k^2}} \right| \le \lim_{\begin{bmatrix} h \\ k \end{bmatrix} \to \begin{bmatrix} 0 \\ 0 \end{bmatrix}} (|h| + |k|) = 0. \qquad 1.7.35$$

Thus $\mathbf{f}$ is differentiable, and its derivative is its Jacobian matrix. △

Directional derivatives

Directional derivatives are a generalization of partial derivatives. The partial derivative $\overrightarrow{D_i}\mathbf{f}(\mathbf{a})$ describes how $\mathbf{f}(\mathbf{x})$ varies as the variable $\mathbf{x}$ moves from $\mathbf{a}$ in the direction of the standard basis vector $\vec{\mathbf{e}}_i$:

$$\overrightarrow{D_i}\mathbf{f}(\mathbf{a}) = \lim_{h\to 0} \frac{\mathbf{f}(\mathbf{a}+h\vec{\mathbf{e}}_i) - \mathbf{f}(\mathbf{a})}{h}. \qquad 1.7.36$$

If $\mathbf{f}$ is differentiable, then once we know the partial derivatives of $\mathbf{f}$, which measure rate of change in the direction of the standard basis vectors, we can compute the derivatives in *any* direction. This should not come as a surprise. The matrix for any linear transformation is formed by seeing what the transformation does to the standard basis vectors: the ith column of $[T]$ is $T(\vec{\mathbf{e}}_i)$. The Jacobian matrix is the matrix for the "rate of change" transformation, formed by seeing what that transformation does to the standard basis vectors.

If $\mathbf{f}$ is not differentiable, the extrapolation does not work, as we will see in Example 1.9.3.

The *directional derivative* describes how $\mathbf{f}(\mathbf{x})$ varies when the variable moves in any direction $\vec{\mathbf{v}}$, i.e., when the variable $\mathbf{x}$ moves at constant speed from $\mathbf{a}$ to $\mathbf{a}+\vec{\mathbf{v}}$ in time 1. See Figure 1.7.3.

Definition 1.7.13 (Directional derivative). Let $U \subset \mathbb{R}^n$ be open, and let $\mathbf{f} : U \to \mathbb{R}^m$ be a function. The *directional derivative* of $\mathbf{f}$ at $\mathbf{a}$ in the direction $\vec{\mathbf{v}}$ is

$$\lim_{h\to 0} \frac{\mathbf{f}(\mathbf{a}+h\vec{\mathbf{v}}) - \mathbf{f}(\mathbf{a})}{h}. \qquad 1.7.37$$

If a function is differentiable, we can extrapolate all its directional derivatives from its partial derivatives – that is, from its derivative.

Proposition 1.7.14 (Computing directional derivatives from the derivative). *If $U \subset \mathbb{R}^n$ is open, and $\mathbf{f} : U \to \mathbb{R}^m$ is differentiable at $\mathbf{a} \in U$, then all directional derivatives of $\mathbf{f}$ at $\mathbf{a}$ exist, and the directional derivative in the direction $\vec{\mathbf{v}}$ is given by the formula*

$$\lim_{h\to 0} \frac{\mathbf{f}(\mathbf{a}+h\vec{\mathbf{v}}) - \mathbf{f}(\mathbf{a})}{h} = [\mathbf{Df}(\mathbf{a})]\vec{\mathbf{v}}. \qquad 1.7.38$$

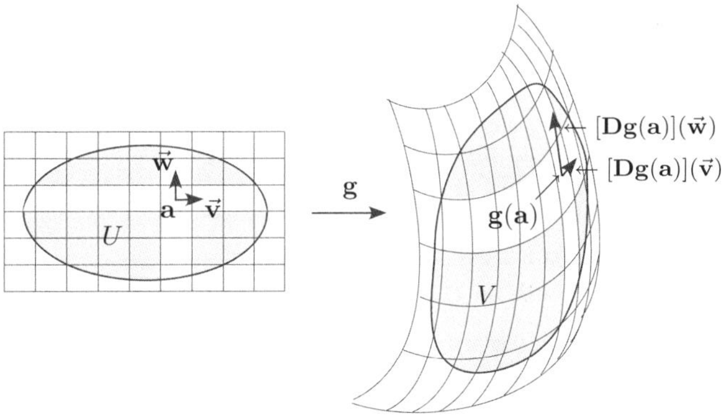

FIGURE 1.7.3. The function $\mathbf{g} : \mathbb{R}^2 \to \mathbb{R}^2$ maps a point $\mathbf{a} \in U$ to a point $\mathbf{g}(\mathbf{a}) \in V$; in particular, it maps the white cross in U to the deformed cross in V. The derivative of $\mathbf{g}$ at $\mathbf{a}$ maps the vector $\vec{\mathbf{v}}$ anchored at $\mathbf{a}$ to $[\mathbf{Dg}(\mathbf{a})](\vec{\mathbf{v}})$, which is anchored at $\mathbf{g}(\mathbf{a})$; it maps the vector $\vec{\mathbf{w}}$ anchored at $\mathbf{a}$ to $[\mathbf{Dg}(\mathbf{a})](\vec{\mathbf{w}})$, which is anchored at $\mathbf{g}(\mathbf{a})$.

Some authors allow only vectors of unit length to be used to define directional derivatives; others define the directional derivative in the direction $\vec{\mathbf{v}}$ to be what we call the directional derivative in the direction $\vec{\mathbf{v}}/|\vec{\mathbf{v}}|$, which has length 1. We feel that these restrictions are undesirable, as they lose the linear character of the directional derivative as a function of $\vec{\mathbf{v}}$.

Moreover, the directional derivative is meaningful even in settings where there is no common unit of length in different directions. Consider a model representing unemployment as a function of taxes, inflation, the minimum wage, and the prime interest rate. An increment is a small change in all these variables. It is perfectly meaningful to speak of the directional derivative in such a model, but the length $|\vec{\mathbf{v}}|$ makes no sense.

Still, when there is a common unit of length in different directions, you should normalize the increment vectors before drawing any conclusions about the direction in which a function varies the fastest. For instance, if f measures temperature on earth, measuring increments in degrees of latitude and longitude, as you might be tempted to do if you study sattelite data, would give different results except at the equator.

Equation 1.7.43 uses the formula
$$\sin(a+b) = \sin a \cos b + \cos a \sin b.$$

Proof of Proposition 1.7.14. The expression
$$\mathbf{r}(\vec{\mathbf{h}}) = \bigl(\mathbf{f}(\mathbf{a}+\vec{\mathbf{h}}) - \mathbf{f}(\mathbf{a})\bigr) - [\mathbf{Df}(\mathbf{a})]\vec{\mathbf{h}} \qquad 1.7.39$$
defines the "remainder" $\mathbf{r}(\vec{\mathbf{h}})$ – the difference between the increment to the function and its linear approximation – as a function of $\vec{\mathbf{h}}$. The hypothesis that $\mathbf{f}$ is differentiable at $\mathbf{a}$ says that
$$\lim_{\vec{\mathbf{h}}\to\vec{\mathbf{0}}} \frac{\mathbf{r}(\vec{\mathbf{h}})}{|\vec{\mathbf{h}}|} = \vec{\mathbf{0}}. \qquad 1.7.40$$
Substituting $h\vec{\mathbf{v}}$ for $\vec{\mathbf{h}}$ in equation 1.7.39, we find
$$\mathbf{r}(h\vec{\mathbf{v}}) = \mathbf{f}(\mathbf{a}+h\vec{\mathbf{v}}) - \mathbf{f}(\mathbf{a}) - [\mathbf{Df}(\mathbf{a})]h\vec{\mathbf{v}}, \qquad 1.7.41$$
and dividing by h (using the linearity of the derivative) gives
$$|\vec{\mathbf{v}}|\frac{\mathbf{r}(h\vec{\mathbf{v}})}{h|\vec{\mathbf{v}}|} = \frac{\mathbf{f}(\mathbf{a}+h\vec{\mathbf{v}})-\mathbf{f}(\mathbf{a})}{h} - \frac{[\mathbf{Df}(\mathbf{a})]h\vec{\mathbf{v}}}{h} = \frac{\mathbf{f}(\mathbf{a}+h\vec{\mathbf{v}})-\mathbf{f}(\mathbf{a})}{h} - [\mathbf{Df}(\mathbf{a})]\vec{\mathbf{v}}.$$
The term $\dfrac{\mathbf{r}(h\vec{\mathbf{v}})}{h|\vec{\mathbf{v}}|}$ on the left side has limit $\vec{\mathbf{0}}$ as $h \to 0$ by equation 1.7.40, so
$$\lim_{h\to 0}\frac{\mathbf{f}(\mathbf{a}+h\vec{\mathbf{v}})-\mathbf{f}(\mathbf{a})}{h} - [\mathbf{Df}(\mathbf{a})]\vec{\mathbf{v}} = \vec{\mathbf{0}}. \quad \square \qquad 1.7.42$$

Example 1.7.15 (Computing a directional derivative from the Jacobian matrix). Let us compute the derivative in the direction $\vec{\mathbf{v}} = \begin{bmatrix} 1 \\ 2 \\ 1 \end{bmatrix}$ of the function $f\begin{pmatrix} x \\ y \\ z \end{pmatrix} = xy\sin z$, evaluated at $\mathbf{a} = \begin{pmatrix} 1 \\ 1 \\ \pi/2 \end{pmatrix}$. By Theorem 1.8.1, this function is differentiable. Its derivative is the 1×3 matrix $[y\sin z \;\; x\sin z \;\; xy\cos z]$; evaluated at $\mathbf{a}$, it is $[1 \;\; 1 \;\; 0]$. So the directional derivative is $[1 \;\; 1 \;\; 0]\vec{\mathbf{v}} = 3$.

Now let's compute this directional derivative from Definition 1.7.13. Since
$h\vec{\mathbf{v}} = h\begin{bmatrix} 1 \\ 2 \\ 1 \end{bmatrix} = \begin{bmatrix} h \\ 2h \\ h \end{bmatrix}$, formula 1.7.37 becomes
$$\lim_{h\to 0}\frac{1}{h}\left(\underbrace{(1+h)(1+2h)\sin\left(\frac{\pi}{2}+h\right)}_{\mathbf{f}(\mathbf{a}+h\vec{\mathbf{v}})} - \underbrace{\left(1\cdot 1\cdot \sin\frac{\pi}{2}\right)}_{\mathbf{f}(\mathbf{a})}\right). \qquad 1.7.43$$
$$= \lim_{h\to 0}\frac{1}{h}\left(\bigl((1+h)(1+2h)(\underbrace{\sin\tfrac{\pi}{2}}_{=1}\cos h + \underbrace{\cos\tfrac{\pi}{2}}_{=0}\sin h)\bigr) - \underbrace{\sin\tfrac{\pi}{2}}_{=1}\right)$$
$$= \lim_{h\to 0}\frac{1}{h}\Bigl(\bigl((1+3h+2h^2)(\cos h)\bigr)-1\Bigr)$$
$$= \lim_{h\to 0}\frac{1}{h}(\cos h - 1) + \lim_{h\to 0}\frac{1}{h}3h\cos h + \lim_{h\to 0}\frac{1}{h}2h^2\cos h = 0+3+0 = 3. \quad \triangle$$

The directional derivative in Example 1.7.15 is a number, since the codomain of f is $\mathbb{R}$. Directional derivatives of a function $\mathbf{f} : \mathbb{R}^n \to \mathbb{R}^m$ are vectors in $\mathbb{R}^m$: the output of such a function is a point in $\mathbb{R}^m$, so the output of its derivative is a vector in $\mathbb{R}^m$.

Example 1.7.16: You may notice that the directional derivatives of this function at the origin are all 0. This does not mean that if you move from the origin in any direction the function remains unchanged. But since both f_1 and f_2 are quadratic, when you move a small distance from the origin, the change in the function is much smaller than the increment to the variable. Recall (Remark 1.7.6) that the difference between the increment to f and the linear function that is its derivative must be *smaller than linear* when the increment to the variable is near 0.

Example 1.7.16 (Directional derivative of an $\mathbb{R}^m$-valued function). Let $\mathbf{f} : \mathbb{R}^2 \to \mathbb{R}^2$ be the function $\mathbf{f}\begin{pmatrix} x \\ y \end{pmatrix} = \begin{pmatrix} xy \\ x^2 - y^2 \end{pmatrix}$. At the point $\begin{pmatrix} 1 \\ 1 \end{pmatrix}$, at what rate is $\mathbf{f}$ varying in the direction $\begin{bmatrix} 2 \\ 1 \end{bmatrix}$? The derivative of $\mathbf{f}$ is

$$\left[\mathbf{Df}\begin{pmatrix} x \\ y \end{pmatrix}\right] = \begin{bmatrix} y & x \\ 2x & -2y \end{bmatrix}, \quad \text{so} \quad \left[\mathbf{Df}\begin{pmatrix} 1 \\ 1 \end{pmatrix}\right] = \begin{bmatrix} 1 & 1 \\ 2 & -2 \end{bmatrix};$$

$\begin{bmatrix} 1 & 1 \\ 2 & -2 \end{bmatrix}\begin{bmatrix} 2 \\ 1 \end{bmatrix} = \begin{bmatrix} 3 \\ 2 \end{bmatrix}$ is the directional derivative in the direction $\begin{bmatrix} 2 \\ 1 \end{bmatrix}$.

The Jacobian matrix: not always the right approach

Computing the derivative of a function $\mathbf{f} : \mathbb{R}^n \to \mathbb{R}^m$ by computing its Jacobian matrix is usually a lot quicker than computing it from the definition, just as in one-variable calculus it is quicker to compute derivatives using handy formulas rather than computing a limit. But it is important to keep in mind the meaning of the derivative: it is the best linear approximation to a function, the linear function L such that

$$\lim_{\vec{\mathbf{h}} \to \vec{\mathbf{0}}} \frac{1}{|\vec{\mathbf{h}}|}\Big((\mathbf{f}(\mathbf{a} + \vec{\mathbf{h}}) - \mathbf{f}(\mathbf{a})) - (L(\vec{\mathbf{h}})) \Big) = \vec{\mathbf{0}}. \qquad 1.7.44$$

This is the definition that can always be used to compute a derivative, even when the domain and codomain of a function are abstract vector spaces, not $\mathbb{R}^n$. Although it may be possible to identify such a function with a function from $\mathbb{R}^n$ to $\mathbb{R}^m$, and thus compute a Jacobian matrix, doing so may be quite difficult.

Example 1.7.17: As mentioned in Section 1.5, we can identify an $n \times n$ matrix with an element of $\mathbb{R}^{n^2}$; for example, we can think of $\begin{bmatrix} a & b \\ c & d \end{bmatrix}$ as the point $\begin{pmatrix} a \\ b \\ c \\ d \end{pmatrix} \in \mathbb{R}^4$. (We could use a different identification, for example, identifying $\begin{bmatrix} a & b \\ c & d \end{bmatrix}$ with $\begin{pmatrix} a \\ c \\ b \\ d \end{pmatrix}$. We shall consider the first to be the "standard" identification.)

In Section 2.6 we will see that Mat (n, m) is an example of an abstract vector space, and we will be more precise about what it means to identify such a space with an appropriate $\mathbb{R}^N$.

Example 1.7.17 (Derivative of the squaring function for matrices). In first year calculus, you learned that if $f(x) = x^2$, then $f'(x) = 2x$. Let us compute the same thing when a matrix is being squared. Since the square of an $n \times n$ matrix is another $n \times n$ matrix, and such a matrix can be "identified" with $\mathbb{R}^{n^2}$, this could be written as a function $\mathbb{R}^{n^2} \to \mathbb{R}^{n^2}$. You are asked to spell this out for $n = 2$ and $n = 3$ in Exercise 1.7.16. But the expression that you get is very unwieldy as soon as $n > 2$, as you will see if you try to solve the exercise.

This is one time when a linear transformation is easier to deal with than the corresponding matrix. Recall that we denote by Mat (n, n) the set of $n \times n$ matrices, and consider the squaring map

$$S : \text{Mat}\,(n, n) \to \text{Mat}\,(n, n) \quad \text{given by} \quad S(A) = A^2. \qquad 1.7.45$$

In this case we can compute the derivative without computing the Jacobian matrix. We shall see that S is differentiable and that its derivative $[\mathbf{D}S(A)]$ is the linear transformation that maps H to $AH + HA$:[23]

$$[\mathbf{D}S(A)]H = AH + HA, \text{ also written } [\mathbf{D}S(A)] : H \mapsto AH + HA. \qquad 1.7.46$$

(Note the parallel with writing the derivative of the function $f(x) = x^2$ as $f'(x) : h \mapsto 2xh$.) Since the increment is a matrix, we denote it H. Note that if matrix multiplication were commutative, we could denote this derivative $2AH$ or $2HA$ – very much like the derivative $f' = 2x$ for the function $f(x) = x^2$.

To make sense of equation 1.7.46, the first thing to realize is that the map

$$[\mathbf{D}S(A)] : \text{Mat}\,(n,n) \to \text{Mat}\,(n,n)\,, \quad H \mapsto AH + HA \qquad 1.7.47$$

is a linear transformation. Exercise 2.20 asks you to check this.

How do we prove equation 1.7.46?

The assertion is that

$$\lim_{H\to[0]} \frac{1}{|H|}\Big(\underbrace{(S(A+H) - S(A))}_{\substack{\text{increment}\\ \text{to mapping}}} - \underbrace{(AH + HA)}_{\substack{\text{linear function of}\\ \text{increment to variable}}} \Big) = [0]. \qquad 1.7.48$$

Equation 1.7.48: Recall that we denote by $[0]$ a matrix with all entries 0.

We write equation 1.7.49 using absolute values because we need them in inequality 1.7.50, so that we can cancel the $|H|$ in the denominator. Recall from the margin note next to Definition 1.5.12 that we can show that a sequence of vectors converges to $\mathbf{0}$ by showing that their lengths converge to 0. The same is true of matrices.

Since $S(A) = A^2$,

$$\begin{aligned} |S(A+H) - S(A) - (AH+HA)| &= |(A+H)^2 - A^2 - AH - HA| \\ &= |\, A^2 + AH + HA + H^2 - A^2 - AH - HA \,| \\ &= |H^2|. \end{aligned} \qquad 1.7.49$$

This gives

$$\lim_{H\to[0]} \frac{|H^2|}{|H|} \le \lim_{H\to[0]} \frac{|H||H|}{|H|} = 0, \qquad 1.7.50$$

so $AH + HA$ is indeed the derivative. △

Exercise 1.7.16 asks you to prove, in the case of 2×2 matrices, that the derivative $AH + HA$ is the "same" as the Jacobian matrix computed with partial derivatives. Much of the difficulty is in understanding S as a mapping from $\mathbb{R}^4 \to \mathbb{R}^4$.

Remark. Equation 1.7.49 shows that subtracting $AH + HA$ leaves only higher degree terms. Thus $AH + HA$ is the derivative; see Remark 1.7.6. This approach can always be used to find derivatives in higher dimensions; it is particularly useful in settings where the Jacobian matrix is not available: compute $F(A+H) - F(A)$ and discard all quadratic and higher terms in the increment H. The result will be the linear function of the increment H that best approximates the function $H \mapsto F(A+H) - F(A)$. △

[23] In equation 1.7.46 we denote this derivative by saying what it does to a vector H. When a derivative is a Jacobian matrix, we don't need to write it this way because we know what function a matrix represents. We could avoid the format of equation 1.7.46 by (for example) denoting by A_L multiplication on the left, and by A_R multiplication on the right. Then we could write $[\mathbf{D}S(A)] = A_L + A_R$. Or we could write the derivative as an $n^2 \times n^2$ matrix. This has obvious drawbacks.

Here is another example of this kind of thing. Recall that if $f(x) = 1/x$, then $f'(a) = -1/a^2$. Proposition 1.7.18 generalizes this to matrices.

Exercise 1.7.20 asks you to compute the Jacobian matrix and verify Proposition 1.7.18 in the case of 2×2 matrices. It should be clear from the exercise that using this approach even for 3×3 matrices would be extremely unpleasant.

Proposition 1.7.18 (Derivative of inverse function for matrices). *If f is the function*

$$f(A) = A^{-1}, \qquad 1.7.51$$

defined on the set of invertible matrices in Mat (n, n), *then f is differentiable, and*

$$[\mathbf{D}f(A)]H = -A^{-1}HA^{-1}. \qquad 1.7.52$$

Note the interesting way in which this reduces to $f'(a)h = -h/a^2$ in one dimension.

Proof. For f to be differentiable, it must be defined on an open subset of Mat (n, n) (see Definition 1.7.1). We proved that the set of invertible matrices is open in Corollary 1.5.40. Now we need to show that

$$\lim_{H \to [0]} \frac{1}{|H|}\Big(\underbrace{(A+H)^{-1} - A^{-1}}_{\text{increment to mapping}} \quad - \quad \underbrace{-A^{-1}HA^{-1}}_{\text{linear function of } H}\Big) = [0]. \qquad 1.7.53$$

Our strategy will be first to fiddle with the quantity $(A+H)^{-1}$. We will use Proposition 1.5.38, which says that if B is a square matrix such that $|B| < 1$, then the series $I + B + B^2 + \cdots$ converges to $(I - B)^{-1}$; here, $-A^{-1}H$ plays the role of B. (Since $H \to [0]$ in equation 1.7.53, we can assume that $|A^{-1}H| < 1$, so treating $(I + A^{-1}H)^{-1}$ as the sum of the series is justified.) This gives the following computation:

Proposition 1.2.15: if A and B are invertible, then

$$(AB)^{-1} = B^{-1}A^{-1}.$$

$$\begin{aligned}
(A+H)^{-1} &= \big(A(I + A^{-1}H)\big)^{-1} \overbrace{=}^{\text{Prop. 1.2.15 (see margin)}} (I + A^{-1}H)^{-1}A^{-1} \\
&= \underbrace{\big(I - (-A^{-1}H)\big)^{-1}}_{\text{sum of series in line below}} A^{-1} \\
&= \underbrace{\Big(I + (-A^{-1}H) + (-A^{-1}H)^2 + \cdots\Big)A^{-1}}_{\text{series } I+B+B^2+\ldots,\text{ where } B=-A^{-1}H} \qquad 1.7.54
\end{aligned}$$

(Now we consider the first term, second terms, and remaining terms:)

$$= \underbrace{A^{-1}}_{\text{1st}} \underbrace{- A^{-1}HA^{-1}}_{\text{2nd}} + \underbrace{\Big((-A^{-1}H)^2 + (-A^{-1}H)^3 + \ldots\Big)A^{-1}}_{\text{others}}$$

Now we can subtract $(A^{-1} - A^{-1}HA^{-1})$ from both sides of equation 1.7.54; on the left this gives the expression that really interests us: the difference between the increment of the function and its approximation:

$$\underbrace{\left((A+H)^{-1}-A^{-1}\right)}_{\text{increment to function}} - \underbrace{\left(-A^{-1}HA^{-1}\right)}_{\substack{\text{linear function of}\\ \text{increment } H \text{ to variable}}}$$

$$= \Big((-A^{-1}H)^2 + (-A^{-1}H)^3 + \cdots\Big)A^{-1} \qquad 1.7.55$$

$$= (-A^{-1}H)\Big((-A^{-1}H) + (-A^{-1}H)^2 + (-A^{-1}H)^3 + \cdots\Big)A^{-1}.$$

Inequality 1.7.56: Switching from matrices to lengths of matrices allows us, via Proposition 1.4.11, to establish an inequality. It also allows us to take advantage of the commutativity of multiplication of numbers, so that we can write

$$|A^{-1}|^2|H|\,|A^{-1}|\,|H| = |A^{-1}|^3|H|^2.$$

Second inequality in inequality 1.7.56: The triangle inequality applies to convergent infinite sums (see Exercise 1.5.9).

Proposition 1.4.11 and the triangle inequality (Theorem 1.4.9) give

$$\begin{aligned} &\left|(A+H)^{-1} - A^{-1} + A^{-1}HA^{-1}\right| \\ &\underbrace{\leq}_{\text{Prop. 1.4.11}} |A^{-1}H|\,\Big|(-A^{-1}H) + (-A^{-1}H)^2 + (-A^{-1}H)^3 + \cdots\Big|\,|A^{-1}| \\ &\underbrace{\leq}_{\text{triangle ineq.}} |A^{-1}H|\,|A^{-1}|\,\Big(\overbrace{|A^{-1}H| + |A^{-1}H|^2 + \cdots}^{\text{convergent geometric series of numbers}}\Big) \qquad 1.7.56 \\ &\leq |A^{-1}|^2|H|\,\frac{|A^{-1}|\,|H|}{1-|A^{-1}H|} = \frac{|A^{-1}|^3|H|^2}{1-|A^{-1}H|} \end{aligned}$$

Now suppose H so small that $|A^{-1}H| < 1/2$, so that

$$\frac{1}{1-|A^{-1}H|} \leq 2. \qquad 1.7.57$$

We see that

$$\lim_{H\to[0]} \frac{\left|(A+H)^{-1} - A^{-1} + A^{-1}HA^{-1}\right|}{|H|} \leq \lim_{H\to[0]} \frac{2|H|^2|A^{-1}|^3}{|H|} = [0]. \quad \square$$

Exercises for Section 1.7

Exercises 1.7.1–1.7.4 provide some review of tangents and derivatives.

1.7.1 Find the equation of the line tangent to the graph of $f(x)$ at $\begin{pmatrix} a \\ f(a) \end{pmatrix}$ for the following functions:

a. $f(x) = \sin x\ ,\quad a = 0$ b. $f(x) = \cos x\ ,\quad a = \pi/3$

c. $f(x) = \cos x\ ,\quad a = 0$ d. $f(x) = 1/x\ ,\quad a = 1/2$

1.7.2 For what a is the tangent to the graph of $f(x) = e^{-x}$ at $\begin{pmatrix} a \\ e^{-a} \end{pmatrix}$ a line of the form $y = mx$?

1.7.3 Find $f'(x)$ for the following functions f:

a. $f(x) = \sin^3(x^2 + \cos x)$ b. $f(x) = \cos^2\Big((x + \sin x)^2\Big)$ c. $f(x) = (\cos x)^4 \sin x$

d. $f(x) = (x + \sin^4 x)^3$ e. $f(x) = \dfrac{\sin x^2 \sin^3 x}{2 + \sin x}$ f. $f(x) = \sin\left(\dfrac{x^3}{\sin x^2}\right)$

1.7.4 Example 1.7.2 may lead you to expect that if $f : \mathbb{R} \to \mathbb{R}$ is differentiable at a, then $f(a+h) - f(a) - f'(a)h$ will be comparable to h^2. It is not true that once you get rid of the linear term you always have a term that includes h^2.

Using the definition, check whether the following functions are differentiable at 0.

a. $f(x) = |x|^{3/2}$ b. $f(x) = \begin{cases} x \ln|x| & \text{if } x \neq 0 \\ 0 & \text{if } x = 0 \end{cases}$ c. $f(x) = \begin{cases} x/\ln|x| & \text{if } x \neq 0 \\ 0 & \text{if } x = 0 \end{cases}$

In each case, if f is differentiable at 0, is $f(0+h) - f(0) - f'(0)h$ comparable to h^2?

a. $f\begin{pmatrix} x \\ y \end{pmatrix} = \sqrt{x^2 + y}$

b. $f\begin{pmatrix} x \\ y \end{pmatrix} = x^2y + y^4$

c. $f\begin{pmatrix} x \\ y \end{pmatrix} = \cos xy + y \cos y$

d. $f\begin{pmatrix} x \\ y \end{pmatrix} = \dfrac{xy^2}{\sqrt{x + y^2}}$

Functions for Exercise 1.7.5.

1.7.5 What are the partial derivatives $D_1 f$ and $D_2 f$ of the functions in the margin, at the points $\begin{pmatrix} 2 \\ 1 \end{pmatrix}$ and $\begin{pmatrix} 1 \\ -2 \end{pmatrix}$?

1.7.6 Calculate the partial derivatives $\dfrac{\partial \mathbf{f}}{\partial x}$ and $\dfrac{\partial \mathbf{f}}{\partial y}$ for the $\mathbb{R}^m$-valued functions

$$\text{a. } \mathbf{f}\begin{pmatrix} x \\ y \end{pmatrix} = \begin{pmatrix} \cos x \\ x^2y + y^2 \\ \sin(x^2 - y) \end{pmatrix} \qquad \text{b. } \mathbf{f}\begin{pmatrix} x \\ y \end{pmatrix} = \begin{pmatrix} \sqrt{x^2 + y^2} \\ xy \\ \sin^2 xy \end{pmatrix}$$

Exercise 1.7.6: You must learn the notation for partial derivatives used in Exercise 1.7.6, as it is used practically everywhere, but we much prefer $\overrightarrow{D_i}\mathbf{f}$.

1.7.7 Write the answers to Exercise 1.7.6 in the form of the Jacobian matrix.

1.7.8 a. Given a function $\mathbf{f}\begin{pmatrix} x \\ y \end{pmatrix} = \begin{pmatrix} f_1 \\ f_2 \end{pmatrix}$ with Jacobian matrix

$$\begin{bmatrix} 2x\cos(x^2 + y) & \cos(x^2 + y) \\ ye^{xy} & xe^{xy} \end{bmatrix},$$

what is D_1 of the function f_1? D_2 of the function f_1? D_2 of f_2?

b. What are the dimensions of the Jacobian matrix of $\mathbf{f}\begin{pmatrix} x \\ y \end{pmatrix} = \begin{pmatrix} f_1 \\ f_2 \\ f_3 \end{pmatrix}$?

a. $\mathbf{f} : \mathbb{R}^n \to \mathbb{R}^m$

b. $f : \mathbb{R}^3 \to \mathbb{R}$

c. $\mathbf{f} : \mathbb{R} \to \mathbb{R}^4$

Functions for Exercise 1.7.9.

1.7.9 Assume the functions in the margin are differentiable. What can you say about their derivatives? (What form do they take?)

1.7.10 Let $\mathbf{f} : \mathbb{R}^2 \to \mathbb{R}^2$ be a function.

a. Prove that if $\mathbf{f}$ is affine, then for any $\mathbf{a}, \mathbf{v} \in \mathbb{R}^2$,

$$\mathbf{f}\begin{pmatrix} a_1 + v_1 \\ a_2 + v_2 \end{pmatrix} = \mathbf{f}\begin{pmatrix} a_1 \\ a_2 \end{pmatrix} + [\mathbf{Df}(\mathbf{a})]\vec{\mathbf{v}}.$$

b. Prove that if $\mathbf{f}$ is not affine, this is not true.

1.7.11 Find the Jacobian matrices of the following mappings:

a. $f\begin{pmatrix} x \\ y \end{pmatrix} = \sin(xy)$ b. $f\begin{pmatrix} x \\ y \end{pmatrix} = e^{(x^2+y^3)}$

c. $\mathbf{f}\begin{pmatrix} x \\ y \end{pmatrix} = \begin{pmatrix} xy \\ x + y \end{pmatrix}$ d. $\mathbf{f}\begin{pmatrix} r \\ \theta \end{pmatrix} = \begin{pmatrix} r\cos\theta \\ r\sin\theta \end{pmatrix}$

1.7.12 Let $f\begin{pmatrix} x \\ y \end{pmatrix} = \begin{pmatrix} xy \\ x^2 - y^2 \end{pmatrix}$ and let $\mathbf{p} = \begin{pmatrix} 1 \\ 1 \end{pmatrix}$, $\vec{\mathbf{v}} = \begin{pmatrix} 2 \\ 1 \end{pmatrix}$. Compute

$$f(\mathbf{p} + t\vec{\mathbf{v}}) - f(\mathbf{p}) - t[\mathbf{D}f(\mathbf{p})]\vec{\mathbf{v}}$$

for $t = 1, \frac{1}{10}, \frac{1}{100}, \frac{1}{1000}$. Does the difference scale like t^k for some k?

1.7.13 Show that if $f(x) = |x|$, then for any number m,

$$\lim_{h\to 0}\Big(f(0+h) - f(0) - mh\Big) = 0, \quad \text{but}$$

$$\lim_{h\to 0}\frac{1}{h}\Big(f(0+h) - f(0) - mh\Big) = 0$$

is never true: there is no number m such that mh is a "good approximation" to $f(h) - f(0)$ in the sense of Definition 1.7.5.

1.7.14 Let U be an open subset of $\mathbb{R}^n$; let $\mathbf{a}$ be a point in U; let $\mathbf{g} : U \to \mathbb{R}$ be a function that is differentiable at $\mathbf{a}$. Prove that

$$\frac{|\mathbf{g}(\mathbf{a} + \vec{\mathbf{h}}) - \mathbf{g}(\mathbf{a})|}{|\vec{\mathbf{h}}|} \quad \text{is bounded when } \vec{\mathbf{h}} \to \vec{\mathbf{0}}.$$

1.7.15 a. Define what it means for a mapping $F : \operatorname{Mat}(n,m) \to \operatorname{Mat}(k,l)$ to be differentiable at a point $A \in \operatorname{Mat}(n,m)$.

b. Consider the function $F : \operatorname{Mat}(n,m) \to \operatorname{Mat}(n,n)$ given by $F(A) = AA^\top$. Show that F is differentiable, and compute the derivative $[\mathbf{D}F(A)]$.

1.7.16 Let $A = \begin{bmatrix} a & b \\ c & d \end{bmatrix}$ and $A^2 = \begin{bmatrix} a_1 & b_1 \\ c_1 & d_1 \end{bmatrix}$.

a. Write the formula for the function $S : \mathbb{R}^4 \to \mathbb{R}^4$ defined by

$$S\begin{pmatrix} a \\ b \\ c \\ d \end{pmatrix} = \begin{pmatrix} a_1 \\ b_1 \\ c_1 \\ d_1 \end{pmatrix}.$$

b. Find the Jacobian matrix of S.

c. Check that your answer agrees with Example 1.7.17.

d. (For the courageous): Do the same for 3×3 matrices.

1.7.17 Let $A = \begin{bmatrix} 1 & 1 \\ 0 & 1 \end{bmatrix}$. If $H = \begin{bmatrix} 0 & 0 \\ \epsilon & 0 \end{bmatrix}$ represents an "instantaneous velocity" at A, what 2×2 matrix is the corresponding velocity vector in the codomain of the squaring function S given by $S(A) = A^2$? First use the derivative of S (Example 1.7.17); then compute both A^2 and $(A+H)^2$.

1.7.18 Let U be an open subset of $\mathbb{R}^n$ and let $f : U \to \mathbb{R}$ be differentiable at $\mathbf{a} \in U$. Show that if $\mathbf{v}$ is a *unit* vector making an angle θ with the gradient $\vec{\nabla} f(\mathbf{a})$, then

$$[\mathbf{D}f(\mathbf{a})]\mathbf{v} = |\vec{\nabla} f(\mathbf{a})| \cos\theta.$$

Why does this justify saying that $\vec{\nabla} f(\mathbf{a})$ points in the direction in which f increases the fastest, and that $|\vec{\nabla} f(\mathbf{a})|$ is this fastest rate of increase?

1.7.19 Is the mapping $\mathbf{f} : \mathbb{R}^n \to \mathbb{R}^n$ given by $\mathbf{f}(\vec{\mathbf{x}}) = |\vec{\mathbf{x}}|\vec{\mathbf{x}}$ differentiable at the origin? If so, what is its derivative?

1.7.20 a. Let A be a 2×2 matrix. Compute the derivative (Jacobian matrix) for the function $f(A) = A^{-1}$ described in Proposition 1.7.18.

b. Show that your result agrees with the result of Proposition 1.7.18.

1.7.21 Considering the determinant as a function only of 2×2 matrices (i.e., $\det : \text{Mat}\,(2,2) \mapsto \mathbb{R}$), show that

$$[\mathbf{D}\det(I)]H = h_{1,1} + h_{2,2},$$

where I is the identity and H is the increment matrix $H = \begin{bmatrix} h_{1,1} & h_{1,2} \\ h_{2,1} & h_{2,2} \end{bmatrix}$. Note that $h_{1,1} + h_{2,2}$ is the trace of H.

Exercise 1.7.20, part a : Think of $A = \begin{bmatrix} a & b \\ c & d \end{bmatrix}$ as the element $\begin{pmatrix} a \\ b \\ c \\ d \end{pmatrix}$ of $\mathbb{R}^4$. Use the formula for computing the inverse of a 2×2 matrix (equation 1.2.13).

1.8 Rules for computing derivatives

This section gives rules that allow you to differentiate any function that is given s by a formula. Some are grouped in Theorem 1.8.1; the chain rule is discussed in Theorem 1.8.3.

Theorem 1.8.1 (Rules for computing derivatives). *Let $U \subset \mathbb{R}^n$ be open.*

1. *If $\mathbf{f} : U \to \mathbb{R}^m$ is a constant function, then $\mathbf{f}$ is differentiable, and its derivative is $[0]$ (it is the zero linear transformation $\mathbb{R}^n \to \mathbb{R}^m$, represented by the $m \times n$ matrix filled with 0's).*
2. *If $\mathbf{f} : \mathbb{R}^n \to \mathbb{R}^m$ is linear, then it is differentiable everywhere, and its derivative at all points $\mathbf{a}$ is $\mathbf{f}$, i.e., $[\mathbf{D}\mathbf{f}(\mathbf{a})]\vec{\mathbf{v}} = \mathbf{f}(\vec{\mathbf{v}})$.*
3. *If $f_1, \dots, f_m : U \to \mathbb{R}$ are m scalar-valued functions differentiable at $\mathbf{a}$, then the vector-valued mapping $\mathbf{f} = \begin{pmatrix} f_1 \\ \vdots \\ f_m \end{pmatrix} : U \to \mathbb{R}^m$ is differentiable at $\mathbf{a}$, with derivative*

$$[\mathbf{D}\mathbf{f}(\mathbf{a})]\vec{\mathbf{v}} = \begin{bmatrix} [\mathbf{D}f_1(\mathbf{a})]\vec{\mathbf{v}} \\ \vdots \\ [\mathbf{D}f_m(\mathbf{a})]\vec{\mathbf{v}} \end{bmatrix}. \qquad 1.8.1$$

Conversely, if $\mathbf{f}$ is differentiable at $\mathbf{a}$, each f_i is differentiable at $\mathbf{a}$, and $[\mathbf{D}f_i(\mathbf{a})] = [D_1 f_i(\mathbf{a}), \dots, D_n f_i(\mathbf{a})]$.

4. *If $\mathbf{f}, \mathbf{g} : U \to \mathbb{R}^m$ are differentiable at $\mathbf{a}$, then so is $\mathbf{f} + \mathbf{g}$, and*

$$[\mathbf{D}(\mathbf{f} + \mathbf{g})(\mathbf{a})] = [\mathbf{D}\mathbf{f}(\mathbf{a})] + [\mathbf{D}\mathbf{g}(\mathbf{a})]. \qquad 1.8.2$$

In Theorem 1.8.1, we considered writing $\mathbf{f}$ and $\mathbf{g}$ as $\vec{\mathbf{f}}$ and $\vec{\mathbf{g}}$, since some of the computations only make sense for vectors: the dot product $\mathbf{f} \cdot \mathbf{g}$ in part 7, for example. We did not do so partly to avoid heavy notation. In practice, you can compute without worrying about the distinction.

We discussed in Remark 1.5.6 the importance of limiting the domain to an open subset.

In part 4 (and in part 7, on the next page) the expression

$$\mathbf{f}, \mathbf{g} : U \to \mathbb{R}^m$$

is shorthand for $\mathbf{f} : U \to \mathbb{R}^m$ and $\mathbf{g} : U \to \mathbb{R}^m$.

5. *If $f : U \to \mathbb{R}$ and $\mathbf{g} : U \to \mathbb{R}^m$ are differentiable at $\mathbf{a}$, then so is $f\mathbf{g}$, and the derivative is given by*

$$[\mathbf{D}(f\mathbf{g})(\mathbf{a})]\vec{\mathbf{v}} = \underbrace{f(\mathbf{a})}_{\mathbb{R}}\underbrace{[\mathbf{D}\mathbf{g}(\mathbf{a})]\vec{\mathbf{v}}}_{\mathbb{R}^m} + \underbrace{([\mathbf{D}f(\mathbf{a})]\vec{\mathbf{v}})}_{\mathbb{R}}\underbrace{\mathbf{g}(\mathbf{a})}_{\mathbb{R}^m}. \qquad 1.8.3$$

6. *If $f : U \to \mathbb{R}$ and $\mathbf{g} : U \to \mathbb{R}^m$ are differentiable at $\mathbf{a}$, and $f(\mathbf{a}) \neq 0$, then so is $\mathbf{g}/f$, and the derivative is given by*

$$\left[\mathbf{D}\left(\frac{\mathbf{g}}{f}\right)(\mathbf{a})\right]\vec{\mathbf{v}} = \frac{[\mathbf{D}\mathbf{g}(\mathbf{a})]\vec{\mathbf{v}}}{f(\mathbf{a})} - \frac{([\mathbf{D}f(\mathbf{a})]\vec{\mathbf{v}})(\mathbf{g}(\mathbf{a}))}{(f(\mathbf{a}))^2}. \qquad 1.8.4$$

7. *If $\mathbf{f}, \mathbf{g} : U \to \mathbb{R}^m$ are both differentiable at $\mathbf{a}$, then so is the dot product $\mathbf{f}\cdot\mathbf{g} : U \to \mathbb{R}$, and (as in one dimension)*

$$[\mathbf{D}(\mathbf{f}\cdot\mathbf{g})(\mathbf{a})]\vec{\mathbf{v}} = \underbrace{[\mathbf{D}\mathbf{f}(\mathbf{a})]\vec{\mathbf{v}}}_{\mathbb{R}^m}\cdot\underbrace{\mathbf{g}(\mathbf{a})}_{\mathbb{R}^m} + \underbrace{\mathbf{f}(\mathbf{a})}_{\mathbb{R}^m}\cdot\underbrace{[\mathbf{D}\mathbf{g}(\mathbf{a})]\vec{\mathbf{v}}}_{\mathbb{R}^m}. \qquad 1.8.5$$

FIGURE 1.8.1.
Gottfried Leibniz (1646–1716)
Parts 5 and 7 of Theorem 1.8.1 are versions of Leibniz's rule.

Leibniz (or Leibnitz) was both philosopher and mathematician. He wrote the first book on calculus and invented the notation for derivatives and integrals still used today. Contention between Leibniz and Newton for who should get credit for calculus contributed to a schism between English and continental mathematics that lasted into the 20th century.

We saw in Corollary 1.5.31 that polynomials are continuous, and rational functions are continuous wherever the denominator does not vanish. It follows from Theorem 1.8.1 that they are also differentiable.

Corollary 1.8.2 (Differentiability of polynomials and rational functions).

1. *Any polynomial function $\mathbb{R}^n \to \mathbb{R}$ is differentiable on all of $\mathbb{R}^n$.*
2. *Any rational function is differentiable on the subset of $\mathbb{R}^n$ where the denominator does not vanish.*

Proof of Theorem 1.8.1. Proving most parts of Theorem 1.8.1 is straightforward; parts 5 and 6 are a bit tricky.

1. If $\mathbf{f}$ is a constant function, then $\mathbf{f}(\mathbf{a}+\vec{\mathbf{h}}) = \mathbf{f}(\mathbf{a})$, so the derivative $[\mathbf{D}\mathbf{f}(\mathbf{a})]$ is the zero linear transformation:

$$\lim_{\vec{\mathbf{h}}\to\vec{\mathbf{0}}} \frac{1}{|\vec{\mathbf{h}}|}\Big(\mathbf{f}(\mathbf{a}+\vec{\mathbf{h}}) - \mathbf{f}(\mathbf{a}) - \underbrace{[0]\vec{\mathbf{h}}}_{[\mathbf{D}\mathbf{f}(\mathbf{a})]\vec{\mathbf{h}}}\Big) = \lim_{\vec{\mathbf{h}}\to 0}\frac{1}{|\vec{\mathbf{h}}|}\vec{\mathbf{0}} = \vec{\mathbf{0}}, \qquad 1.8.6$$

where $[0]$ is a zero matrix.

2. Since $\mathbf{f}$ is linear, $\mathbf{f}(\mathbf{a}+\vec{\mathbf{h}}) = \mathbf{f}(\mathbf{a}) + \mathbf{f}(\vec{\mathbf{h}})$, so

$$\lim_{\vec{\mathbf{h}}\to\vec{\mathbf{0}}} \frac{1}{|\vec{\mathbf{h}}|}\big(\mathbf{f}(\mathbf{a}+\vec{\mathbf{h}}) - \mathbf{f}(\mathbf{a}) - \mathbf{f}(\vec{\mathbf{h}})\big) = \vec{\mathbf{0}}. \qquad 1.8.7$$

It follows that $[\mathbf{D}\mathbf{f}(\mathbf{a})] = \mathbf{f}$, i.e., for every $\vec{\mathbf{h}} \in \mathbb{R}^n$ we have $[\mathbf{D}\mathbf{f}(\mathbf{a})]\vec{\mathbf{h}} = \mathbf{f}(\vec{\mathbf{h}})$.

3. The assumption that $\mathbf{f}$ is differentiable can be written

$$\lim_{\vec{\mathbf{h}}\to\vec{\mathbf{0}}} \frac{1}{|\vec{\mathbf{h}}|}\left(\begin{pmatrix} f_1(\mathbf{a}+\vec{\mathbf{h}}) \\ \vdots \\ f_m(\mathbf{a}+\vec{\mathbf{h}}) \end{pmatrix} - \begin{pmatrix} f_1(\mathbf{a}) \\ \vdots \\ f_m(\mathbf{a}) \end{pmatrix} - \begin{bmatrix} [\mathbf{D}f_1(\mathbf{a})]\vec{\mathbf{h}} \\ \vdots \\ [\mathbf{D}f_m(\mathbf{a})]\vec{\mathbf{h}} \end{bmatrix}\right) = \begin{bmatrix} 0 \\ \vdots \\ 0 \end{bmatrix}. \qquad 1.8.8$$

Equation 1.8.3: Note that the terms on the right belong to the indicated spaces, and therefore the whole expression makes sense; it is the sum of two vectors in $\mathbb{R}^m$, each of which is the product of a vector in $\mathbb{R}^m$ and a number. Note that $[\mathbf{D}f(\mathbf{a})]\vec{\mathbf{v}}$ is the product of a line matrix and a vector, hence it is a number.

Proof, part 2 of Theorem 1.8.1: We used the variable $\vec{\mathbf{v}}$ in the statement of the theorem, because we were thinking of it as a random vector. In the proof the variable naturally is an increment, so we use $\vec{\mathbf{h}}$, which we tend to use for increments.

The assumption that $f_1, \dots, f_m$ are differentiable can be written

$$\begin{bmatrix} \lim_{\vec{\mathbf{h}}\to\vec{\mathbf{0}}} \frac{1}{|\vec{\mathbf{h}}|}\left(f_1(\mathbf{a}+\vec{\mathbf{h}}) - f_1(\mathbf{a}) - [\mathbf{D}f_1(\mathbf{a})]\vec{\mathbf{h}}\right) \\ \vdots \\ \lim_{\vec{\mathbf{h}}\to\vec{\mathbf{0}}} \frac{1}{|\vec{\mathbf{h}}|}\left(f_m(\mathbf{a}+\vec{\mathbf{h}}) - f_m(\mathbf{a}) - [\mathbf{D}f_m(\mathbf{a})]\vec{\mathbf{h}}\right) \end{bmatrix} = \begin{bmatrix} 0 \\ \vdots \\ 0 \end{bmatrix}. \qquad 1.8.9$$

The expressions on the left sides are equal by Proposition 1.5.25.

4. Functions are added point by point, so we can separate out $\mathbf{f}$ and $\mathbf{g}$:

$$\begin{aligned} &(\mathbf{f}+\mathbf{g})(\mathbf{a}+\vec{\mathbf{h}}) - (\mathbf{f}+\mathbf{g})(\mathbf{a}) - \big([\mathbf{Df}(\mathbf{a})] + [\mathbf{Dg}(\mathbf{a})]\big)\vec{\mathbf{h}} \qquad 1.8.10 \\ &\quad = \big(\mathbf{f}(\mathbf{a}+\vec{\mathbf{h}}) - \mathbf{f}(\mathbf{a}) - [\mathbf{Df}(\mathbf{a})]\vec{\mathbf{h}}\big) + \big(\mathbf{g}(\mathbf{a}+\vec{\mathbf{h}}) - \mathbf{g}(\mathbf{a}) - [\mathbf{Dg}(\mathbf{a})]\vec{\mathbf{h}}\big). \end{aligned}$$

Now divide by $|\vec{\mathbf{h}}|$, and take the limit as $|\vec{\mathbf{h}}| \to \vec{\mathbf{0}}$. The right side gives $\vec{\mathbf{0}} + \vec{\mathbf{0}} = \vec{\mathbf{0}}$, so the left side does too.

In the second line of equation 1.8.11 we added and subtracted $f(\mathbf{a})g(\mathbf{a}+\vec{\mathbf{h}})$; to go from the second equality to the third, we added and subtracted

$$g(\mathbf{a}+\vec{\mathbf{h}})[\mathbf{D}f(\mathbf{a})]\vec{\mathbf{h}}.$$

How did we know to do this? We began with the fact that f and g are differentiable; that's really all we had to work with, so clearly we needed to use that information. Thus we wanted to end up with something of the form

$$f(\mathbf{a}+\vec{\mathbf{h}}) - f(\mathbf{a}) - [\mathbf{D}f(\mathbf{a})]\vec{\mathbf{h}}$$

and

$$g(\mathbf{a}+\vec{\mathbf{h}}) - g(\mathbf{a}) - [\mathbf{D}g(\mathbf{a})]\vec{\mathbf{h}}.$$

So we looked for quantities we could add and subtract in order to arrive at something that includes those quantities. (The ones we chose to add and subtract are not the only possibilities.)

5. We need to show both that $f\mathbf{g}$ is differentiable and that its derivative is given by equation 1.8.3. By part 3, we may assume that $m = 1$ (i.e., that $\mathbf{g} = g$ is scalar valued). Then

$$\begin{aligned} &\lim_{\vec{\mathbf{h}}\to\vec{\mathbf{0}}} \frac{1}{|\vec{\mathbf{h}}|}\Big((fg)(\mathbf{a}+\vec{\mathbf{h}}) - (fg)(\mathbf{a}) \overbrace{- f(\mathbf{a})([\mathbf{D}g(\mathbf{a})]\vec{\mathbf{h}}) - ([\mathbf{D}f(\mathbf{a})]\vec{\mathbf{h}})g(\mathbf{a})}^{-[\mathbf{D}(fg)(\mathbf{a})]\vec{\mathbf{h}}, \text{ according to Thm. 1.8.1}}\Big) \\ &= \lim_{\vec{\mathbf{h}}\to\vec{\mathbf{0}}} \frac{1}{|\vec{\mathbf{h}}|}\Bigg(f(\mathbf{a}+\vec{\mathbf{h}})g(\mathbf{a}+\vec{\mathbf{h}}) \underbrace{- f(\mathbf{a})g(\mathbf{a}+\vec{\mathbf{h}}) + f(\mathbf{a})g(\mathbf{a}+\vec{\mathbf{h}})}_{0} - f(\mathbf{a})g(\mathbf{a}) \\ &\qquad - \Big([\mathbf{D}f(\mathbf{a})]\vec{\mathbf{h}}\Big)g(\mathbf{a}) - \Big([\mathbf{D}g(\mathbf{a})]\vec{\mathbf{h}}\Big)f(\mathbf{a})\Bigg) \\ &= \lim_{\vec{\mathbf{h}}\to\vec{\mathbf{0}}} \frac{1}{|\vec{\mathbf{h}}|}\Bigg(\Big(f(\mathbf{a}+\vec{\mathbf{h}}) - f(\mathbf{a})\Big)\big(g(\mathbf{a}+\vec{\mathbf{h}})\big) + f(\mathbf{a})\Big(g(\mathbf{a}+\vec{\mathbf{h}}) - g(\mathbf{a})\Big) \\ &\qquad - \Big([\mathbf{D}f(\mathbf{a})]\vec{\mathbf{h}}\Big)g(\mathbf{a}) - \Big([\mathbf{D}g(\mathbf{a})]\vec{\mathbf{h}}\Big)f(\mathbf{a})\Bigg) \qquad 1.8.11 \\ &= \lim_{\vec{\mathbf{h}}\to\vec{\mathbf{0}}} \left(\frac{f(\mathbf{a}+\vec{\mathbf{h}}) - f(\mathbf{a}) - [\mathbf{D}f(\mathbf{a})]\vec{\mathbf{h}}}{|\vec{\mathbf{h}}|}\right) g(\mathbf{a}+\vec{\mathbf{h}}) \\ &\qquad + \lim_{\vec{\mathbf{h}}\to\vec{\mathbf{0}}} \Big(g(\mathbf{a}+\vec{\mathbf{h}}) - g(\mathbf{a})\Big)\frac{[\mathbf{D}f(\mathbf{a})]\vec{\mathbf{h}}}{|\vec{\mathbf{h}}|} \\ &\qquad + \lim_{\vec{\mathbf{h}}\to\vec{\mathbf{0}}} f(\mathbf{a})\frac{\Big(g(\mathbf{a}+\vec{\mathbf{h}}) - g(\mathbf{a}) - [\mathbf{D}g(\mathbf{a})]\vec{\mathbf{h}}\Big)}{|\vec{\mathbf{h}}|} = \vec{\mathbf{0}} + \vec{\mathbf{0}} + \vec{\mathbf{0}} = \vec{\mathbf{0}}. \end{aligned}$$

By Theorem 1.5.29, part 5, the first limit is $\vec{\mathbf{0}}$: by the definition of the derivative of f, the first factor has limit $\vec{\mathbf{0}}$, and the second factor $g(\mathbf{a}+\vec{\mathbf{h}})$ is bounded in a neighborhood of $\vec{\mathbf{h}} = \vec{\mathbf{0}}$; see the margin.

We again use Theorem 1.5.29, part 5, to show that the second limit is $\vec{\mathbf{0}}$: $g(\mathbf{a}+\vec{\mathbf{h}}) - g(\mathbf{a})$ goes to $\vec{\mathbf{0}}$ as $\vec{\mathbf{h}} \to \vec{\mathbf{0}}$ (since g is differentiable, hence continuous, at $\mathbf{a}$) and the second factor $\frac{[\mathbf{D}f(\mathbf{a})]\vec{\mathbf{h}}}{|\vec{\mathbf{h}}|}$ is bounded, since

$$\left|\frac{[\mathbf{D}f(\mathbf{a})]\vec{\mathbf{h}}}{|\vec{\mathbf{h}}|}\right| \overbrace{\leq}^{\text{Prop. 1.4.11}} \left|[\mathbf{D}f(\mathbf{a})]\right|\frac{|\vec{\mathbf{h}}|}{|\vec{\mathbf{h}}|} = \left|[\mathbf{D}f(\mathbf{a})]\right|. \qquad 1.8.12$$

The third limit is $\vec{\mathbf{0}}$ by the definition of the derivative of g (and the fact that $f(\mathbf{a})$ is constant).

6. Applying part 5 to the function $1/f$, we see that it is enough to show that

$$\left[\mathbf{D}\left(\frac{1}{f}\right)(\mathbf{a})\right]\vec{\mathbf{v}} = -\frac{[\mathbf{D}f(\mathbf{a})]\vec{\mathbf{v}}}{(f(\mathbf{a}))^2}. \qquad 1.8.13$$

This is seen as follows:

$$\begin{aligned}
&\frac{1}{|\vec{\mathbf{h}}|}\left(\frac{1}{f(\mathbf{a}+\vec{\mathbf{h}})} - \frac{1}{f(\mathbf{a})} + \frac{[\mathbf{D}f(\mathbf{a})]\vec{\mathbf{h}}}{(f(\mathbf{a}))^2}\right) = \frac{1}{|\vec{\mathbf{h}}|}\left(\frac{f(\mathbf{a}) - f(\mathbf{a}+\vec{\mathbf{h}})}{f(\mathbf{a}+\vec{\mathbf{h}})f(\mathbf{a})} + \frac{[\mathbf{D}f(\mathbf{a})]\vec{\mathbf{h}}}{(f(\mathbf{a}))^2}\right)\\
&= \frac{1}{|\vec{\mathbf{h}}|}\left(\frac{f(\mathbf{a}) - f(\mathbf{a}+\vec{\mathbf{h}}) + [\mathbf{D}f(\mathbf{a})]\vec{\mathbf{h}}}{(f(\mathbf{a}))^2}\right) - \frac{1}{|\vec{\mathbf{h}}|}\left(\frac{f(\mathbf{a}) - f(\mathbf{a}+\vec{\mathbf{h}})}{(f(\mathbf{a}))^2} - \frac{f(\mathbf{a}) - f(\mathbf{a}+\vec{\mathbf{h}})}{f(\mathbf{a}+\vec{\mathbf{h}})(f(\mathbf{a}))}\right)\\
&= \frac{1}{(f(\mathbf{a}))^2}\underbrace{\left(\frac{f(\mathbf{a}) - f(\mathbf{a}+\vec{\mathbf{h}}) + [\mathbf{D}f(\mathbf{a})]\vec{\mathbf{h}}}{|\vec{\mathbf{h}}|}\right)}_{\text{limit as } \vec{\mathbf{h}}\to\vec{\mathbf{0}} \text{ is } \vec{\mathbf{0}} \text{ by def. of deriv.}} - \frac{1}{f(\mathbf{a})}\underbrace{\frac{f(\mathbf{a}) - f(\mathbf{a}+\vec{\mathbf{h}})}{|\vec{\mathbf{h}}|}}_{\text{bounded}}\underbrace{\left(\frac{1}{f(\mathbf{a})} - \frac{1}{f(\mathbf{a}+\vec{\mathbf{h}})}\right)}_{\text{limit as } \vec{\mathbf{h}}\to\vec{\mathbf{0}} \text{ is } \vec{\mathbf{0}}}.
\end{aligned}$$

To see why we label one term in the last line as bounded, note that since $\mathbf{f}$ is differentiable,

$$\lim_{\vec{\mathbf{h}}\to\vec{\mathbf{0}}}\left(\frac{\mathbf{f}(\vec{\mathbf{a}}+\vec{\mathbf{h}}) - \mathbf{f}(\vec{\mathbf{a}})}{|\vec{\mathbf{h}}|} - \underbrace{\frac{[\mathbf{D}f(\mathbf{a})](\vec{\mathbf{h}})}{|\vec{\mathbf{h}}|}}_{\text{bounded; see eq. 1.8.12}}\right) = \vec{\mathbf{0}}. \qquad 1.8.14$$

If the first term in the parentheses were unbounded, then the sum would be unbounded, so the limit could not be zero.

7. We do not need to prove that $\mathbf{f}\cdot\mathbf{g}$ is differentiable, since it is the sum of products of differentiable functions, thus differentiable by parts 4 and 5. So we just need to compute the Jacobian matrix:

Equation 1.8.15: The second equality uses part 4: $\mathbf{f}\cdot\mathbf{g}$ is the sum of the $f_i g_i$, so the derivative of the sum is the sum of the derivatives. The third equality uses part 5.

Exercise 1.8.6 sketches a more conceptual proof of parts 5 and 7.

$$\begin{aligned}
[\mathbf{D}(\mathbf{f}\cdot\mathbf{g})(\mathbf{a})]\vec{\mathbf{h}} &\overset{\text{def. of dot prod.}}{=} \left[\mathbf{D}\left(\sum_{i=1}^n f_i g_i\right)(\mathbf{a})\right]\vec{\mathbf{h}} \overset{(4)}{=} \sum_{i=1}^n [\mathbf{D}(f_i g_i)(\mathbf{a})]\vec{\mathbf{h}}\\
&\overset{(5)}{=} \sum_{i=1}^n \left([\mathbf{D}f_i(\mathbf{a})]\vec{\mathbf{h}}\right)g_i(\mathbf{a}) + f_i(\mathbf{a})\left([\mathbf{D}g_i(\mathbf{a})]\vec{\mathbf{h}}\right) \qquad 1.8.15\\
&\overset{\text{def. of dot prod.}}{=} \left([\mathbf{Df}(\mathbf{a})]\vec{\mathbf{h}}\right)\cdot\mathbf{g}(\mathbf{a}) + \mathbf{f}(\mathbf{a})\cdot\left([\mathbf{Dg}(\mathbf{a})]\vec{\mathbf{h}}\right). \qquad \square
\end{aligned}$$

The chain rule

One rule for differentiation is so fundamental that it deserves a subsection of its own: the *chain rule*, which states that *the derivative of a composition is the composition of the derivatives.* It is proved in Appendix A4.

Some physicists claim that the chain rule is the most important theorem in all mathematics.

For the composition $\mathbf{f} \circ \mathbf{g}$ to be well defined, the codomain of $\mathbf{g}$ and the domain of $\mathbf{f}$ must be the same (i.e., V in the theorem). In terms of matrix multiplication, this is equivalent to saying that the width of $[\mathbf{Df}(\mathbf{g}(\mathbf{a}))]$ must equal the height of $[\mathbf{Dg}(\mathbf{a})]$ for the multiplication to be possible.

Theorem 1.8.3 (Chain rule). *Let $U \subset \mathbb{R}^n$, $V \subset \mathbb{R}^m$ be open sets, let $\mathbf{g}: U \to V$ and $\mathbf{f}: V \to \mathbb{R}^p$ be mappings, and let $\mathbf{a}$ be a point of U. If $\mathbf{g}$ is differentiable at $\mathbf{a}$ and $\mathbf{f}$ is differentiable at $\mathbf{g}(\mathbf{a})$, then the composition $\mathbf{f} \circ \mathbf{g}$ is differentiable at $\mathbf{a}$, and its derivative is given by*

$$[\mathbf{D}(\mathbf{f} \circ \mathbf{g})(\mathbf{a})] = [\mathbf{Df}\,(\mathbf{g}(\mathbf{a}))] \circ [\mathbf{Dg}(\mathbf{a})]. \qquad 1.8.16$$

The chain rule is illustrated by Figure 1.8.2.

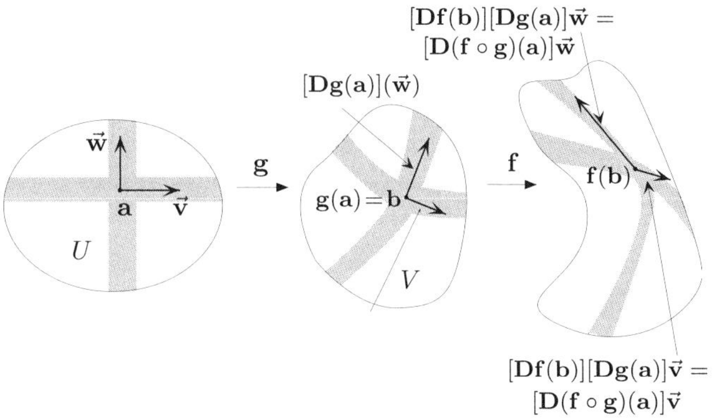

FIGURE 1.8.2. The function $\mathbf{g}$ maps a point $\mathbf{a} \in U$ to a point $\mathbf{g}(\mathbf{a}) \in V$. The function $\mathbf{f}$ maps the point $\mathbf{g}(\mathbf{a}) = \mathbf{b}$ to the point $\mathbf{f}(\mathbf{b})$. The derivative of $\mathbf{g}$ maps the vector $\vec{\mathbf{v}}$ to $[\mathbf{Dg}(\mathbf{a})](\vec{\mathbf{v}})$. The derivative of $\mathbf{f} \circ \mathbf{g}$ maps $\vec{\mathbf{v}}$ to $[\mathbf{Df}(\mathbf{b})][\mathbf{Dg}(\mathbf{a})]\vec{\mathbf{v}}$, which, of course, is identical to $[\mathbf{Df}\,(\mathbf{g}(\mathbf{a}))][\mathbf{Dg}(\mathbf{a})]\vec{\mathbf{v}}$.

One motivation for discussing the relationship between matrix multiplication and composition of linear transformations at the beginning of this chapter was to have these tools available now. In coordinates, and using matrix multiplication, the chain rule states that

$$D_j(\mathbf{f} \circ \mathbf{g})_i(\mathbf{a}) = \sum_{k=1}^{m} D_k f_i\big(\mathbf{g}(\mathbf{a})\big) D_j g_k(\mathbf{a}).$$

As a statement, this form of the chain rule is a disaster: it turns a fundamental, transparent statement into a messy formula, the proof of which seems to be a computational miracle.

In practice, when using the chain rule, most often we represent linear transformations by their matrices, and we compute the right side of equation 1.8.16 by multiplying the matrices together:

$$[\mathbf{D}(\mathbf{f} \circ \mathbf{g})(\mathbf{a})] = [\mathbf{Df}\,(\mathbf{g}(\mathbf{a}))][\mathbf{Dg}(\mathbf{a})]. \qquad 1.8.17$$

Remark. Note that the equality $(\mathbf{f} \circ \mathbf{g})(\mathbf{a}) = \mathbf{f}(\mathbf{g}(\mathbf{a}))$ does *not* mean you can stick a $\mathbf{D}$ in front of each and claim that $[\mathbf{D}(\mathbf{f} \circ \mathbf{g})(\mathbf{a})]$ equals $[\mathbf{Df}(\mathbf{g}(\mathbf{a}))]$; this is *wrong*. Remember that $[\mathbf{D}(\mathbf{f} \circ \mathbf{g})(\mathbf{a})]$ is the derivative of $\mathbf{f} \circ \mathbf{g}$ evaluated at the point $\mathbf{a}$; $[\mathbf{Df}(\mathbf{g}(\mathbf{a}))]$ is the derivative of $\mathbf{f}$ evaluated at the point $\mathbf{g}(\mathbf{a})$. △

Example 1.8.4 (The derivative of a composition). Define $\mathbf{g}: \mathbb{R} \to \mathbb{R}^3$ and $f: \mathbb{R}^3 \to \mathbb{R}$ by

$$f\begin{pmatrix} x \\ y \\ z \end{pmatrix} = x^2 + y^2 + z^2; \qquad \mathbf{g}(t) = \begin{pmatrix} t \\ t^2 \\ t^3 \end{pmatrix}. \qquad 1.8.18$$

The derivative of f is the 1×3 matrix $\left[\mathbf{D}f\begin{pmatrix}x\\y\\z\end{pmatrix}\right] = [2x, 2y, 2z]$; evaluated at $\mathbf{g}(t)$ it is $[2t, 2t^2, 2t^3]$. The derivative of $\mathbf{g}$ at t is $[\mathbf{D}\mathbf{g}(t)] = \begin{bmatrix}1\\2t\\3t^2\end{bmatrix}$, so

$$[\mathbf{D}(f\circ \mathbf{g})(t)] \;=\; [\mathbf{D}f(\mathbf{g}(t))][\mathbf{D}\mathbf{g}(t)] = [2t, 2t^2, 2t^3]\begin{bmatrix}1\\2t\\3t^2\end{bmatrix} = 2t+4t^3+6t^5.$$

Note that the dimensions are right: the composition $f\circ\mathbf{g}$ goes from $\mathbb{R}$ to $\mathbb{R}$, so its derivative is a number.

In this case it is also easy to compute the derivative of the composition directly: since $(f\circ g)(t) = t^2+t^4+t^6$, we have

$$[\mathbf{D}(f\circ \mathbf{g})(t)] \;=(t^2+t^4+t^6)' = 2t+4t^3+6t^5. \quad \triangle \qquad 1.8.19$$

Example 1.8.5 (Composition of a function with itself). Let $\mathbf{f}:\mathbb{R}^3\to\mathbb{R}^3$ be given by $\mathbf{f}\begin{pmatrix}x\\y\\z\end{pmatrix} = \begin{pmatrix}z^2\\xy-3\\2y\end{pmatrix}$. What is the derivative of $\mathbf{f}\circ\mathbf{f}$, evaluated at $\begin{pmatrix}1\\1\\1\end{pmatrix}$? We need to compute

$$\left[\mathbf{D}(\mathbf{f}\circ\mathbf{f})\begin{pmatrix}1\\1\\1\end{pmatrix}\right] = \left[\mathbf{D}\mathbf{f}\left(\mathbf{f}\begin{pmatrix}1\\1\\1\end{pmatrix}\right)\right]\left[\mathbf{D}\mathbf{f}\begin{pmatrix}1\\1\\1\end{pmatrix}\right]. \qquad 1.8.20$$

We use $[\mathbf{D}\mathbf{f}]$ twice, evaluating it at $\begin{pmatrix}1\\1\\1\end{pmatrix}$ and at $\mathbf{f}\begin{pmatrix}1\\1\\1\end{pmatrix} = \begin{pmatrix}1\\-2\\2\end{pmatrix}$.

Since $\left[\mathbf{D}\mathbf{f}\begin{pmatrix}x\\y\\z\end{pmatrix}\right] = \begin{bmatrix}0&0&2z\\y&x&0\\0&2&0\end{bmatrix}$, equation 1.8.20 gives

$$\left[\mathbf{D}(\mathbf{f}\circ\mathbf{f})\begin{pmatrix}1\\1\\1\end{pmatrix}\right] = \begin{bmatrix}0&0&4\\-2&1&0\\0&2&0\end{bmatrix}\begin{bmatrix}0&0&2\\1&1&0\\0&2&0\end{bmatrix} = \begin{bmatrix}0&8&0\\1&1&-4\\2&2&0\end{bmatrix}. \qquad 1.8.21$$

Now compute the derivative of the composition directly and check in the footnote below.[24] $\triangle$

Example 1.8.6 (Composition of linear transformations). Here is a case where it is easier to think of the derivative as a linear transformation

[24]Since $\mathbf{f}\circ\mathbf{f}\begin{pmatrix}x\\y\\z\end{pmatrix} = \mathbf{f}\begin{pmatrix}z^2\\xy-3\\2y\end{pmatrix} = \begin{pmatrix}4y^2\\xyz^2-3z^2-3\\2xy-6\end{pmatrix}$, the derivative of the composition at $\begin{pmatrix}x\\y\\z\end{pmatrix}$ is $\begin{bmatrix}0&8y&0\\yz^2&xz^2&2xyz-6z\\2y&2x&0\end{bmatrix}$; at $\begin{pmatrix}1\\1\\1\end{pmatrix}$, it is $\begin{bmatrix}0&8&0\\1&1&-4\\2&2&0\end{bmatrix}$.

than as a matrix, and it is easier to think of the chain rule as concerning a composition of linear transformations rather than a product of matrices. If A and H are $n \times n$ matrices, and $f(A) = A^2$ and $g(A) = A^{-1}$, then $(f \circ g)(A) = A^{-2}$. Below we compute the derivative of $f \circ g$:

$$[\mathbf{D}(f\circ g)(A)]H \underbrace{=}_{\text{chain rule}} [\mathbf{D}f(g(A))]\underbrace{[\mathbf{D}g(A)]H}_{\substack{\text{new increment}\\ H \text{ for } \mathbf{D}f}} = [\mathbf{D}f(A^{-1})]\underbrace{(-A^{-1}HA^{-1})}_{\text{eq. 1.7.52}}$$

$$\underbrace{=}_{\text{eq. 1.7.46}} A^{-1}(-A^{-1}HA^{-1}) + (-A^{-1}HA^{-1})A^{-1}$$

$$= -\left(A^{-2}HA^{-1} + A^{-1}HA^{-2}\right). \quad \triangle \qquad 1.8.22$$

Example 1.8.6: Equation 1.7.46 says that the derivative of the "squaring function" f is

$$[\mathbf{D}f(A)]H = AH + HA.$$

Equation 1.7.52 says that the derivative of the inverse function for matrices is

$$[\mathbf{D}f(A)]H = -A^{-1}HA^{-1}.$$

In the second line of equation 1.8.22, $g(A) = A^{-1}$ plays the role of A, and $-A^{-1}HA^{-1}$ plays the role of H.

Notice the interesting way this result is related to the one-variable computation: if $f(x) = x^{-2}$, then $f'(x) = -2x^{-3}$. Notice also how much easier this computation is, using the chain rule, than the proof of Proposition 1.7.18, without the chain rule.

Exercise 1.32 asks you to compute the derivatives of the maps $A \mapsto A^{-3}$ and $A \mapsto A^{-n}$.

Exercises for Section 1.8

1.8.1 Let $f : \mathbb{R}^3 \to \mathbb{R}$, $\mathbf{f} : \mathbb{R}^2 \to \mathbb{R}^3$, $g : \mathbb{R}^2 \to \mathbb{R}$, $\mathbf{g} : \mathbb{R}^3 \to \mathbb{R}^2$ be differentiable functions. Which of the following compositions make sense? For each that does, what are the dimensions of its derivative?

a. $f \circ g$ b. $g \circ f$ c. $\mathbf{g} \circ \mathbf{f}$

d. $\mathbf{f} \circ \mathbf{g}$ e. $f \circ \mathbf{f}$ f. $\mathbf{f} \circ f$

1.8.2 a. Given the functions $f\begin{pmatrix} x \\ y \\ z \end{pmatrix} = x^2 + y^2 + 2z^2$ and $\mathbf{g}(t) = \begin{pmatrix} t \\ t^2 \\ t^3 \end{pmatrix}$, what is $\left[\mathbf{D}(\mathbf{g} \circ f)\begin{pmatrix} a \\ b \\ c \end{pmatrix}\right]$?

b. Let $\mathbf{f}\begin{pmatrix} x \\ y \\ z \end{pmatrix} = \begin{pmatrix} x^2 + z \\ yz \end{pmatrix}$ and $g\begin{pmatrix} a \\ b \end{pmatrix} = a^2 + b^2$. What is the derivative of $g \circ \mathbf{f}$ at $\begin{pmatrix} x \\ y \\ z \end{pmatrix}$?

1.8.3 Is $f\begin{pmatrix} x \\ y \end{pmatrix} = \sin(e^{xy})$ differentiable at $\begin{pmatrix} 0 \\ 0 \end{pmatrix}$?

1.8.4 a. What compositions can you form using the following functions?

i. $f\begin{pmatrix} x \\ y \\ z \end{pmatrix} = x^2+y^2$ ii. $g\begin{pmatrix} a \\ b \end{pmatrix} = 2a+b^2$ iii. $\mathbf{f}(t) = \begin{pmatrix} t \\ 2t \\ t^2 \end{pmatrix}$ iv. $\mathbf{g}\begin{pmatrix} x \\ y \end{pmatrix} = \begin{pmatrix} \cos x \\ x+y \\ \sin y \end{pmatrix}$

b. Compute these compositions.

c. Compute their derivatives, both by using the chain rule and directly from the compositions.

1.8.5 The following "proof" of part 5 of Theorem 1.8.1 is correct as far as it goes, but it is not a complete proof. Why not?

"Proof": By part 3 of Theorem 1.8.1, we may assume that $m = 1$ (i.e., that $\mathbf{g} = g$ is scalar valued). Then

$$\begin{aligned}
[\mathbf{D}fg(\mathbf{a})]\vec{\mathbf{h}} &= \overbrace{[(D_1fg)(\mathbf{a}), \dots, (D_nfg)(\mathbf{a})]}^{\text{Jacobian matrix of } fg}\vec{\mathbf{h}} \\
&= \underbrace{[f(\mathbf{a})(D_1g)(\mathbf{a}) + (D_1f)(\mathbf{a})g(\mathbf{a}), \dots, f(\mathbf{a})(D_ng)(\mathbf{a}) + (D_nf)(\mathbf{a})g(\mathbf{a})]}_{\text{in one variable, } (fg)'=fg'+f'g}\vec{\mathbf{h}} \\
&= f(\mathbf{a})\underbrace{[(D_1g)(\mathbf{a}), \dots, (D_ng)(\mathbf{a})]}_{\text{Jacobian matrix of } g}\vec{\mathbf{h}} + \underbrace{[(D_1f)(\mathbf{a}), \dots, (D_nf)(\mathbf{a})]}_{\text{Jacobian matrix of } f}g(\mathbf{a})\vec{\mathbf{h}} \\
&= f(\mathbf{a})\Big([\mathbf{D}g(\mathbf{a})]\vec{\mathbf{h}}\Big) + \Big([\mathbf{D}f(\mathbf{a})]\vec{\mathbf{h}}\Big)g(\mathbf{a}).
\end{aligned}$$

1.8.6 a. Prove the rule for differentiating dot products (part 7 of Theorem 1.8.1) directly from the definition of the derivative.

b. Let $U \in \mathbb{R}^3$ be open. Show by a similar argument that if $\mathbf{f}, \mathbf{g} : U \to \mathbb{R}^3$ are both differentiable at $\mathbf{a}$, then so is the cross product $\mathbf{f} \times \mathbf{g} : U \to \mathbb{R}^3$. Find the formula for this derivative.

1.8.7 Consider the function

$$f\begin{pmatrix} x_1 \\ \vdots \\ x_n \end{pmatrix} = \sum_{i=1}^{n-1} x_i x_{i+1} \quad \text{and the curve } \gamma : \mathbb{R} \to \mathbb{R}^n \text{ given by } \quad \gamma(t) = \begin{pmatrix} t \\ t^2 \\ \vdots \\ t^n \end{pmatrix}.$$

What is the derivative of the function $t \mapsto f\big(\gamma(t)\big)$?

1.8.8 True or false? Justify your answer. If $\mathbf{f} : \mathbb{R}^2 \to \mathbb{R}^2$ is a differentiable function with

$$\mathbf{f}\begin{pmatrix} 0 \\ 0 \end{pmatrix} = \begin{pmatrix} 1 \\ 1 \end{pmatrix} \quad \text{and} \quad \left[\mathbf{Df}\begin{pmatrix} 0 \\ 0 \end{pmatrix}\right] = \begin{bmatrix} 1 & 1 \\ 1 & 1 \end{bmatrix},$$

there is no differentiable mapping $\mathbf{g} : \mathbb{R}^2 \to \mathbb{R}^2$ with

$$\mathbf{g}\begin{pmatrix} 1 \\ 1 \end{pmatrix} = \begin{pmatrix} 0 \\ 0 \end{pmatrix} \quad \text{and} \quad \mathbf{f} \circ \mathbf{g}\begin{pmatrix} x \\ y \end{pmatrix} = \begin{pmatrix} y \\ x \end{pmatrix}.$$

1.8.9 Let $\varphi : \mathbb{R} \to \mathbb{R}$ be any differentiable function. Show that the function

$$f\begin{pmatrix} x \\ y \end{pmatrix} = y\varphi(x^2 - y^2)$$

satisfies the equation

$$\frac{1}{x}D_1f\begin{pmatrix} x \\ y \end{pmatrix} + \frac{1}{y}D_2f\begin{pmatrix} x \\ y \end{pmatrix} = \frac{1}{y^2}f\begin{pmatrix} x \\ y \end{pmatrix}.$$

1.8.10 a. Show that if a function $f : \mathbb{R}^2 \to \mathbb{R}$ can be written $\varphi(x^2 + y^2)$ for some differentiable function $\varphi : \mathbb{R} \to \mathbb{R}$, then it satisfies

$$x\overrightarrow{D_2f} - y\overrightarrow{D_1f} = 0.$$

Exercise 1.8.10, hint for part b: What is the "partial derivative of f with respect to the polar angle θ"?

*b. Show the converse: every function satisfying $x\overrightarrow{D_2f} - y\overrightarrow{D_1f} = 0$ can be written $\varphi(x^2 + y^2)$ for some function $\varphi : \mathbb{R} \to \mathbb{R}$.

1.8.11 Show that if $f\begin{pmatrix} x \\ y \end{pmatrix} = \varphi\left(\frac{x+y}{x-y}\right)$ for some differentiable function $\varphi : \mathbb{R} \to \mathbb{R}$, then

$$xD_1 f + yD_2 f = 0.$$

1.8.12 True or false? Explain your answers.

a. If $\mathbf{f} : \mathbb{R}^2 \to \mathbb{R}^2$ is differentiable and $[\mathbf{Df}(\mathbf{0})]$ is not invertible, then there is no differentiable function $\mathbf{g} : \mathbb{R}^2 \to \mathbb{R}^2$ such that

$$(\mathbf{g} \circ \mathbf{f})(\mathbf{x}) = \mathbf{x}.$$

b. Differentiable functions have continuous partial derivatives.

1.8.13 Let $U \subset \text{Mat}\,(n, n)$ be the set of matrices A such that $A + A^2$ is invertible. Compute the derivative of the map $F : U \to \text{Mat}\,(n, n)$ given by $F(A) = (A + A^2)^{-1}$.

1.9 The mean value theorem and criteria for differentiability

I turn with terror and horror from this lamentable scourge of continuous functions with no derivatives.—Charles Hermite, in a letter to Thomas Stieltjes, 1893

In this section we discuss two applications of the mean value theorem. The first extends that theorem to functions of several variables, and the second gives a criterion for determining when a function is differentiable.

The mean value theorem for functions of several variables

The derivative measures the difference of the values of functions at different points. For functions of one variable, the mean value theorem (Theorem 1.6.13) says that if $f : [a, b] \to \mathbb{R}$ is continuous, and f is differentiable on (a, b), then there exists $c \in (a, b)$ such that

$$f(b) - f(a) = f'(c)(b - a). \qquad 1.9.1$$

Theorem 1.9.1: The segment $[\mathbf{a}, \mathbf{b}]$ is the image of the map

$$t \mapsto (1 - t)\mathbf{a} + t\mathbf{b},$$

for $0 \le t \le 1$.

The analogous statement in several variables is the following.

Theorem 1.9.1 (Mean value theorem for functions of several variables). *Let $U \subset \mathbb{R}^n$ be open, let $f : U \to \mathbb{R}$ be differentiable, and let the segment $[\mathbf{a}, \mathbf{b}]$ joining $\mathbf{a}$ to $\mathbf{b}$ be contained in U. Then there exists $\mathbf{c}_0 \in [\mathbf{a}, \mathbf{b}]$ such that*

$$f(\mathbf{b}) - f(\mathbf{a}) = [\mathbf{D}f(\mathbf{c}_0)](\overrightarrow{\mathbf{b} - \mathbf{a}}). \qquad 1.9.2$$

Corollary 1.9.2. *If f is a function as defined in Theorem 1.9.1, then*

$$|f(\mathbf{b}) - f(\mathbf{a})| \leq \left(\sup_{\mathbf{c}\in[\mathbf{a},\mathbf{b}]} \left| [\mathbf{D}f(\mathbf{c})] \right| \right) |\overrightarrow{\mathbf{b}-\mathbf{a}}|. \quad 1.9.3$$

Why do we write inequality 1.9.3 with the sup, rather than

$$|f(\mathbf{b}) - f(\mathbf{a})| \leq |[\mathbf{D}f(\mathbf{c})]|\ |\overrightarrow{\mathbf{b}-\mathbf{a}}|,$$

which of course is also true? Using the sup means that we do not need to know the value of $\mathbf{c}$ in order to relate how fast f is changing to its derivative; we can run through all $\mathbf{c} \in [\mathbf{a},\mathbf{b}]$ and choose the one where the derivative is greatest. This will be useful in Section 2.8 when we discuss Lipschitz ratios.

Proof of Corollary 1.9.2. This follows immediately from Theorem 1.9.1 and Proposition 1.4.11. □

Proof of Theorem 1.9.1. As t varies from 0 to 1, the point $(1-t)\mathbf{a}+t\mathbf{b}$ moves from $\mathbf{a}$ to $\mathbf{b}$. Consider the mapping $g(t) = f((1-t)\mathbf{a}+t\mathbf{b})$. By the chain rule, g is differentiable, and by the one-variable mean value theorem, there exists t_0 such that

$$g(1) - g(0) = g'(t_0)(1-0) = g'(t_0). \quad 1.9.4$$

Set $\mathbf{c}_0 = (1-t_0)\mathbf{a} + t_0\mathbf{b}$. By Proposition 1.7.14, we can express $g'(t_0)$ in terms of the derivative of f:

$$\begin{aligned} g'(t_0) &= \lim_{s\to 0} \frac{g(t_0+s)-g(t_0)}{s} \\ &= \lim_{s\to 0} \frac{f\big(\mathbf{c}_0 + s(\overrightarrow{\mathbf{b}-\mathbf{a}})\big) - f(\mathbf{c}_0)}{s} = [\mathbf{D}f(\mathbf{c}_0)](\overrightarrow{\mathbf{b}-\mathbf{a}}). \end{aligned} \quad 1.9.5$$

So equation 1.9.4 reads

$$f(\mathbf{b}) - f(\mathbf{a}) = [\mathbf{D}f(\mathbf{c}_0)](\overrightarrow{\mathbf{b}-\mathbf{a}}). \quad □ \quad 1.9.6$$

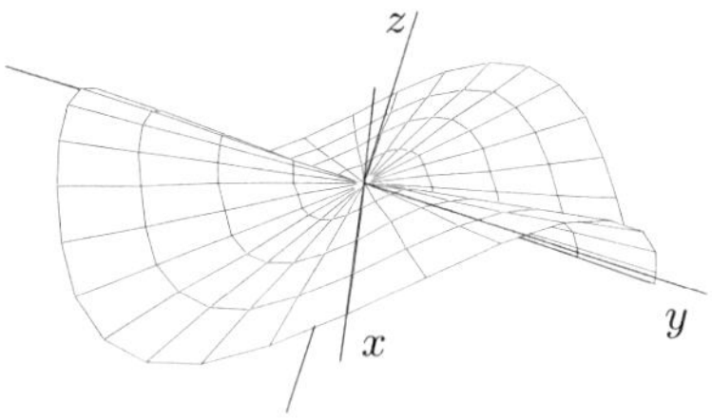

FIGURE 1.9.1.

The graph of the function f defined in equation 1.9.7 is made up of straight lines through the origin, so if you leave the origin in any direction, the directional derivative in that direction certainly exists. Both axes are among the lines making up the graph, so the directional derivatives in those directions are 0. But clearly there is no tangent plane to the graph at the origin.

Differentiability and pathological functions

Most often the Jacobian matrix of a function is its derivative. But as we mentioned in Section 1.7, there are exceptions. It is possible for all partial derivatives of f to exist, and even all directional derivatives, and yet for f not to be differentiable! In such a case the Jacobian matrix exists but does not represent the derivative.

Example 1.9.3 (Nondifferentiable function with Jacobian matrix). This happens even for the innocent-looking function

$$f\begin{pmatrix} x \\ y \end{pmatrix} = \frac{x^2 y}{x^2+y^2} \quad 1.9.7$$

shown in Figure 1.9.1. Actually, we should write this function as

$$f\begin{pmatrix} x \\ y \end{pmatrix} = \begin{cases} \dfrac{x^2y}{x^2+y^2} & \text{if } \begin{pmatrix} x \\ y \end{pmatrix} \neq \begin{pmatrix} 0 \\ 0 \end{pmatrix} \\ 0 & \text{if } \begin{pmatrix} x \\ y \end{pmatrix} = \begin{pmatrix} 0 \\ 0 \end{pmatrix}. \end{cases} \quad 1.9.8$$

You have probably learned to be suspicious of functions that are defined by different formulas for different values of the variable. In this case, the value at $\begin{pmatrix}0\\0\end{pmatrix}$ is really natural, in the sense that as $\begin{pmatrix}x\\y\end{pmatrix}$ approaches $\begin{pmatrix}0\\0\end{pmatrix}$, the function f approaches 0. This is not one of those functions whose value takes a sudden jump; indeed, f is continuous everywhere. Away from the origin, this is obvious by Corollary 1.5.31: away from the origin, f is a rational function whose denominator does not vanish. So we can compute both its partial derivatives at any point $\begin{pmatrix}x\\y\end{pmatrix} \neq \begin{pmatrix}0\\0\end{pmatrix}$.

That f is continuous at the origin requires a little checking, as follows. If $x^2+y^2=r^2$, then $|x|\le r$ and $|y|\le r$ so $|x^2y|\le r^3$. Therefore,

$$\left|f\begin{pmatrix}x\\y\end{pmatrix}\right| \le \frac{r^3}{r^2} = r, \quad \text{and} \quad \lim_{\begin{pmatrix}x\\y\end{pmatrix}\to\begin{pmatrix}0\\0\end{pmatrix}} f\begin{pmatrix}x\\y\end{pmatrix} = 0. \qquad 1.9.9$$

So f is continuous at the origin. Moreover, f vanishes identically on both axes, so both partial derivatives of f vanish at the origin.

"Identically" means "at every point."

So far, f looks perfectly civilized: it is continuous, and both partial derivatives exist everywhere. But consider the derivative in the direction of the vector $\begin{bmatrix}1\\1\end{bmatrix}$:

$$\lim_{t\to 0}\frac{f\left(\begin{pmatrix}0\\0\end{pmatrix}+t\begin{bmatrix}1\\1\end{bmatrix}\right)-f\begin{pmatrix}0\\0\end{pmatrix}}{t} = \lim_{t\to 0}\frac{t^3}{2t^3} = \frac{1}{2}. \qquad 1.9.10$$

This is *not* what we get when we compute the same directional derivative by multiplying the Jacobian matrix of f by the vector $\begin{bmatrix}1\\1\end{bmatrix}$, as on the right side of equation 1.7.38:

$$\underbrace{\left[D_1 f\begin{pmatrix}0\\0\end{pmatrix}, D_2 f\begin{pmatrix}0\\0\end{pmatrix}\right]}_{\text{Jacobian matrix } [\mathbf{J}f(\mathbf{0})]}\begin{bmatrix}1\\1\end{bmatrix} = [0,0]\begin{bmatrix}1\\1\end{bmatrix} = 0. \qquad 1.9.11$$

Thus, by Proposition 1.7.14, f is not differentiable. △

If we change the function of Example 1.9.3, replacing the x^2y in the numerator of

$$\frac{x^2y}{x^2+y^2}$$

by xy, then the resulting function, which we'll call g, will *not* be continuous at the origin. If $x=y$, then $g=1/2$ no matter how close $\begin{pmatrix}x\\y\end{pmatrix}$ gets to the origin: we then have

$$g\begin{pmatrix}x\\x\end{pmatrix} = \frac{x^2}{2x^2} = \frac{1}{2}.$$

Things can get worse. The function we just discussed is continuous, but it is possible for all directional derivatives of a function to exist, and yet for the function not to be continuous, or even bounded in a neighborhood of $\mathbf{0}$. For instance, the function discussed in Example 1.5.24 is not continuous in a neighborhood of the origin; if we redefine it to be 0 at the origin, then all directional derivatives would exist everywhere, but the function would not be continuous. Exercise 1.9.2 provides another example. Thus knowing that a function has partial derivatives or directional derivatives does not tell you either that the function is differentiable or even that it is continuous.

Even knowing that a function is differentiable tells you less than you might think. The function in Example 1.9.4 has a positive derivative at x although it does not increase in a neighborhood of x!

Example 1.9.4 (A differentiable yet pathological function). Consider the function $f:\mathbb{R}\to\mathbb{R}$ defined by

$$f(x) = \frac{x}{2} + x^2\sin\frac{1}{x}, \qquad 1.9.12$$

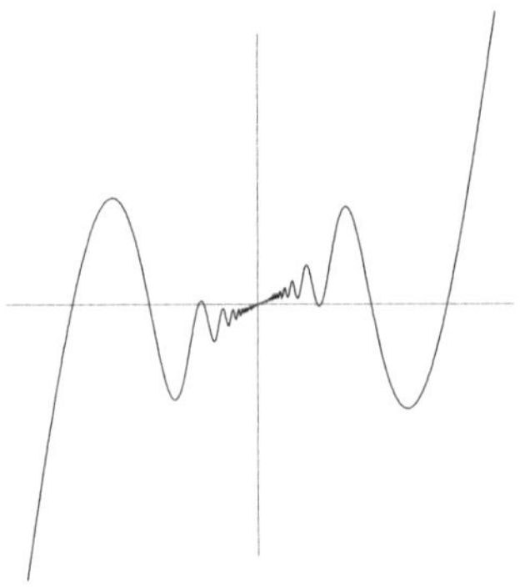

FIGURE 1.9.2.
Graph of the function

$$f(x) = \frac{x}{2} + 6x^2 \sin\frac{1}{x}.$$

The derivative of f does not have a limit at the origin, but the curve still has slope 1/2 there.

a variant of which is shown in Figure 1.9.2. To be precise, one should add $f(0) = 0$, since $\sin 1/x$ is not defined there, but this was the only reasonable value, since

$$\lim_{x\to 0} x^2 \sin\frac{1}{x} = 0. \tag{1.9.13}$$

Moreover, we will see that the function f is differentiable at the origin. This is one case where you *must* use the definition of the derivative as a limit; you *cannot* use the rules for computing derivatives blindly. In fact, let's try. We find

$$f'(x) = \frac{1}{2} + 2x\sin\frac{1}{x} + x^2\left(\cos\frac{1}{x}\right)\left(-\frac{1}{x^2}\right) = \frac{1}{2} + 2x\sin\frac{1}{x} - \cos\frac{1}{x}. \tag{1.9.14}$$

This formula is certainly correct for $x \neq 0$, but $f'(x)$ doesn't have a limit when $x \to 0$. Indeed,

$$\lim_{x\to 0}\left(\frac{1}{2} + 2x\sin\frac{1}{x}\right) = \frac{1}{2} \tag{1.9.15}$$

does exist, but $\cos 1/x$ oscillates infinitely many times between -1 and 1. So f' will oscillate from a value near $-1/2$ to a value near $3/2$. *This does not mean that f isn't differentiable at* 0. We can compute the derivative at 0 using the definition of the derivative:

$$\begin{aligned} f'(0) &= \lim_{h\to 0}\frac{1}{h}\left(\frac{0+h}{2} + (0+h)^2\sin\frac{1}{0+h}\right) = \lim_{h\to 0}\frac{1}{h}\left(\frac{h}{2} + h^2\sin\frac{1}{h}\right) \\ &= \frac{1}{2} + \lim_{h\to 0} h\sin\frac{1}{h} = \frac{1}{2}, \end{aligned} \tag{1.9.16}$$

since (Theorem 1.5.26, part 5) $\lim_{h\to 0} h \sin\frac{1}{h}$ exists, and indeed vanishes.

Finally, we can see that although the derivative *at* 0 is positive, the function is not increasing *in any neighborhood* of 0, since in any interval arbitrarily close to 0 the derivative takes negative values; as we saw above, it oscillates from a value near $-1/2$ to a value near $3/2$. △

This is *very bad.* Our whole point is that locally, the function should behave like its best linear approximation, and in this case it emphatically does not. We could easily make up examples in several variables where the same thing occurs: where the function is differentiable, so that the Jacobian matrix represents the derivative, but where that derivative doesn't tell you much. Of course we don't claim that derivatives are worthless. The problem in these pathological cases is that the partial derivatives of the function are not continuous.

Example 1.9.5 (Discontinuous derivatives). Let us go back to the function of Example 1.9.3:

$$f\begin{pmatrix}x\\y\end{pmatrix} = \begin{cases} \dfrac{x^2y}{x^2+y^2} & \text{if } \begin{pmatrix}x\\y\end{pmatrix} \neq \begin{pmatrix}0\\0\end{pmatrix} \\ 0 & \text{if } \begin{pmatrix}x\\y\end{pmatrix} = \begin{pmatrix}0\\0\end{pmatrix}, \end{cases} \tag{1.9.17}$$

which we saw is not differentiable although its Jacobian matrix exists.

Both partial derivatives are 0 at the origin. Away from the origin – that is, if $\begin{pmatrix} x \\ y \end{pmatrix} \neq \begin{pmatrix} 0 \\ 0 \end{pmatrix}$ – then

$$\begin{aligned} D_1 f\begin{pmatrix} x \\ y \end{pmatrix} &= \frac{(x^2+y^2)(2xy) - x^2y(2x)}{(x^2+y^2)^2} = \frac{2xy^3}{(x^2+y^2)^2} \\ D_2 f\begin{pmatrix} x \\ y \end{pmatrix} &= \frac{(x^2+y^2)(x^2) - x^2y(2y)}{(x^2+y^2)^2} = \frac{x^4 - x^2y^2}{(x^2+y^2)^2}. \end{aligned} \qquad 1.9.18$$

These partial derivatives are not continuous at the origin, as you will see if you approach the origin from any direction other than one of the axes. For example, if you compute the first partial derivative at the point $\begin{pmatrix} t \\ t \end{pmatrix}$ of the diagonal, you find the limit

$$\lim_{t \to 0} D_1 f\begin{pmatrix} t \\ t \end{pmatrix} = \frac{2t^4}{(2t^2)^2} = \frac{1}{2}, \qquad 1.9.19$$

which is not the value of

$$D_1 f\begin{pmatrix} 0 \\ 0 \end{pmatrix} = 0. \qquad 1.9.20$$

In the case of the differentiable but pathological function of Example 1.9.4, the discontinuities are worse. This is a function of one variable, so (as discussed in Section 1.6) the only kind of discontinuities its derivative can have are wild oscillations. Indeed, $f'(x) = \frac{1}{2} + 2x \sin \frac{1}{x} - \cos \frac{1}{x}$, and $\cos 1/x$ oscillates infinitely many times between -1 and 1. △

The moral of the story: Only study *continuously differentiable* functions.

Definition 1.9.6 (Continuously differentiable function). A function is *continuously differentiable* on $U \subset \mathbb{R}^n$ if all its partial derivatives exist and are continuous on U. Such a function is known as a C^1 function.

This definition can be generalized; in Section 3.3 we will need functions that are "smoother" than C^1.

Definition 1.9.7 (C^p function). A *C^p function* on $U \subset \mathbb{R}^n$ is a function that is p times continuously differentiable: all of its partial derivatives up to order p exist and are continuous on U.

If you come across a function that is not continuously differentiable, you should be aware that none of the usual tools of calculus can be relied upon. Each such function is an outlaw, obeying none of the standard theorems.

Theorem 1.9.8 guarantees that a "continuously differentiable" function is indeed differentiable: from the continuity of its partial derivatives, one can infer the existence of its derivative. Most often, the criterion of Theorem 1.9.8 is the tool used to determine whether a function is differentiable.

Theorem 1.9.8 (Criterion for differentiability). *If U is an open subset of $\mathbb{R}^n$, and $\mathbf{f} : U \to \mathbb{R}^m$ is a C^1 mapping, then $\mathbf{f}$ is differentiable on U, and its derivative is given by its Jacobian matrix.*

A function that meets this criterion is not only differentiable; it is also guaranteed not to be pathological. By "not pathological" we mean that locally, its derivative is a reliable guide to its behavior.

Note that the last part of Theorem 1.9.8 – "and its derivative is given by its Jacobian matrix" – is obvious; if a function is differentiable, Theorem 1.7.10 tells us that its derivative is given by its Jacobian matrix. So the point is to prove that the function is differentiable.

In equation 1.9.21 we use the interval $(a, a+h)$, rather than (a, b), so instead of the equation

$$f'(c) = \frac{f(b)-f(a)}{b-a}$$

of the mean value theorem we have statement

$$f'(c) = \frac{f(a+h)-f(a)}{h},$$

or

$$hf'(c) = f(a+h) - f(a).$$

Proof of Theorem 1.9.8. This is an application of Theorem 1.9.1, the mean value theorem. What we need to show is that

$$\lim_{\vec{\mathbf{h}}\to\vec{\mathbf{0}}} \frac{1}{|\vec{\mathbf{h}}|}\Big(\mathbf{f}(\mathbf{a}+\vec{\mathbf{h}}) - \mathbf{f}(\mathbf{a}) - [\mathbf{Jf}(\mathbf{a})]\vec{\mathbf{h}}\Big) = \vec{\mathbf{0}}. \tag{1.9.21}$$

First, note that by part 3 of Theorem 1.8.1, it is enough to prove it when $m = 1$ (i.e., for $f : U \to \mathbb{R}$). Next write

$$f(\mathbf{a}+\vec{\mathbf{h}}) - f(\mathbf{a}) = f\begin{pmatrix} a_1+h_1 \\ a_2+h_2 \\ \vdots \\ a_n+h_n \end{pmatrix} - f\begin{pmatrix} a_1 \\ a_2 \\ \vdots \\ a_n \end{pmatrix} \tag{1.9.22}$$

in expanded form, subtracting and adding inner terms:

$$f(\mathbf{a}+\vec{\mathbf{h}}) - f(\mathbf{a}) =$$

$$f\begin{pmatrix} a_1+h_1 \\ a_2+h_2 \\ \vdots \\ a_n+h_n \end{pmatrix} \underbrace{- f\begin{pmatrix} a_1 \\ a_2+h_2 \\ \vdots \\ a_n+h_n \end{pmatrix}}_{\text{subtracted}} \underbrace{+ f\begin{pmatrix} a_1 \\ a_2+h_2 \\ a_3+h_3 \\ \vdots \\ a_n+h_n \end{pmatrix}}_{\text{added}} - f\begin{pmatrix} a_1 \\ a_2 \\ a_3+h_3 \\ \vdots \\ a_n+h_n \end{pmatrix}$$

$$+\cdots\pm\cdots + f\begin{pmatrix} a_1 \\ a_2 \\ \vdots \\ a_{n-1} \\ a_n+h_n \end{pmatrix} - f\begin{pmatrix} a_1 \\ a_2 \\ \vdots \\ a_{n-1} \\ a_n \end{pmatrix}.$$

By the mean value theorem, the ith term is

$$\underbrace{f\begin{pmatrix} a_1 \\ a_2 \\ \vdots \\ a_i+h_i \\ a_{i+1}+h_{i+1} \\ \vdots \\ a_n+h_n \end{pmatrix} - f\begin{pmatrix} a_1 \\ a_2 \\ \vdots \\ a_i \\ a_{i+1}+h_{i+1} \\ \vdots \\ a_n+h_n \end{pmatrix}}_{i\text{th term}} = h_i D_i f\begin{pmatrix} a_1 \\ a_2 \\ \vdots \\ b_i \\ a_{i+1}+h_{i+1} \\ \vdots \\ a_n+h_n \end{pmatrix} \tag{1.9.23}$$

for some $b_i \in [a_i,\, a_i+h_i]$: there is some point b_i in the interval $[a_i,\, a_i+h_i]$ such that the partial derivative $D_i f$ at b_i gives the average change of the function f over that interval, when all variables except the ith are kept constant.

Since f has n variables, we need to find such a point for every i from 1 to n. We will call these points $\mathbf{c}_i$:

$$\mathbf{c}_i = \begin{pmatrix} a_1 \\ a_2 \\ \vdots \\ b_i \\ a_{i+1}+h_{i+1} \\ \vdots \\ a_n+h_n \end{pmatrix}; \qquad \text{this gives} \qquad f(\mathbf{a}+\vec{\mathbf{h}}) - f(\mathbf{a}) = \sum_{i=1}^{n} h_i D_i f(\mathbf{c}_i).$$

Thus we find that

$$\underbrace{f(\mathbf{a}+\vec{\mathbf{h}}) - f(\mathbf{a})}_{=\sum_{i=1}^{n} D_i f(\mathbf{c}_i) h_i} - \sum_{i=1}^{n} D_i f(\mathbf{a}) h_i = \sum h_i \big(D_i f(\mathbf{c}_i) - D_i f(\mathbf{a})\big). \qquad 1.9.24$$

So far we haven't used the hypothesis that the partial derivatives $D_i f$ are continuous. Now we do. Since $D_i f$ is continuous, and since $\mathbf{c}_i$ tends to $\mathbf{a}$ as $\vec{\mathbf{h}} \to \vec{\mathbf{0}}$, we see that the theorem is true:

The inequality in the second line of inequality 1.9.25 comes from the fact that $|h_i|/|\vec{\mathbf{h}}| \le 1$.

$$\begin{aligned} \lim_{\vec{\mathbf{h}}\to\vec{\mathbf{0}}} \frac{\left| f(\mathbf{a}+\vec{\mathbf{h}}) - f(\mathbf{a}) - \overbrace{[\mathbf{J}f(\mathbf{a})]\vec{\mathbf{h}}}^{\sum_{i=1}^{n} D_i f(\mathbf{a}) h_i} \right|}{|\vec{\mathbf{h}}|} &\le \lim_{\vec{\mathbf{h}}\to\vec{\mathbf{0}}} \sum_{i=1}^{n} \frac{|h_i|}{|\vec{\mathbf{h}}|} |D_i f(\mathbf{c}_i) - D_i f(\mathbf{a})| \\ &\le \lim_{\vec{\mathbf{h}}\to\vec{\mathbf{0}}} \sum_{i=1}^{n} \big|D_i f(\mathbf{c}_i) - D_i f(\mathbf{a})\big| = 0. \qquad \square \end{aligned} \qquad 1.9.25$$

Example 1.9.9. Here we work out the preceding computation when f is a scalar-valued function on $\mathbb{R}^2$:

$$\begin{aligned} & f\begin{pmatrix} a_1+h_1 \\ a_2+h_2 \end{pmatrix} - f\begin{pmatrix} a_1 \\ a_2 \end{pmatrix} \qquad 1.9.26 \\ &= f\begin{pmatrix} a_1+h_1 \\ a_2+h_2 \end{pmatrix} \overbrace{- f\begin{pmatrix} a_1 \\ a_2+h_2 \end{pmatrix} + f\begin{pmatrix} a_1 \\ a_2+h_2 \end{pmatrix}}^{0} - f\begin{pmatrix} a_1 \\ a_2 \end{pmatrix} \\ &= h_1 D_1 f\begin{pmatrix} b_1 \\ a_2+h_2 \end{pmatrix} + h_2 D_2 f\begin{pmatrix} a_1 \\ b_2 \end{pmatrix} = h_1 D_1 f(\mathbf{c}_1) + h_2 D_2 f(\mathbf{c}_2). \quad \triangle \end{aligned}$$

Exercises for Section 1.9

1.9.1 Show that the function $f : \mathbb{R}^2 \to \mathbb{R}$ given by

$$f\begin{pmatrix} x \\ y \end{pmatrix} = \begin{cases} \dfrac{x^4+y^4}{x^2+y^2} & \text{if } \begin{pmatrix} x \\ y \end{pmatrix} \neq \begin{pmatrix} 0 \\ 0 \end{pmatrix} \\ 0 & \text{if } \begin{pmatrix} x \\ y \end{pmatrix} = \begin{pmatrix} 0 \\ 0 \end{pmatrix} \end{cases}$$

is differentiable at every point of $\mathbb{R}^2$.

1.9.2 a. Show that for

$$f\begin{pmatrix} x \\ y \end{pmatrix} = \begin{cases} \dfrac{3x^2y-y^3}{x^2+y^2} & \text{if } \begin{pmatrix} x \\ y \end{pmatrix} \neq \begin{pmatrix} 0 \\ 0 \end{pmatrix} \\ 0 & \text{if } \begin{pmatrix} x \\ y \end{pmatrix} = \begin{pmatrix} 0 \\ 0 \end{pmatrix}, \end{cases}$$

all directional derivatives exist, but that f is not differentiable at the origin.

*b. Show that the function g defined in the margin has directional derivatives at every point but is not continuous.

*c. Show that the function h defined in the margin has directional derivatives at every point but is not bounded in a neighborhood of $\mathbf{0}$.

Exercise 1.9.2: Remember, sometimes you have to use the definition of the derivative, rather than the rules for computing derivatives.

The functions g and h for Exercise 1.9.2, parts b and c:

$$g\begin{pmatrix} x \\ y \end{pmatrix} = \begin{cases} \dfrac{x^2y}{x^4+y^2} & \text{if } \begin{pmatrix} x \\ y \end{pmatrix} \neq \mathbf{0} \\ 0 & \text{if } \begin{pmatrix} x \\ y \end{pmatrix} = \mathbf{0}. \end{cases}$$

$$h\begin{pmatrix} x \\ y \end{pmatrix} = \begin{cases} \dfrac{x^2y}{x^6+y^2} & \text{if } \begin{pmatrix} x \\ y \end{pmatrix} \neq \mathbf{0} \\ 0 & \text{if } \begin{pmatrix} x \\ y \end{pmatrix} = \mathbf{0}. \end{cases}$$

1.9.3 Consider the function $f : \mathbb{R}^2 \to \mathbb{R}$ given by

$$f\begin{pmatrix} x \\ y \end{pmatrix} = \begin{cases} \dfrac{\sin(x^2y^2)}{x^2+y^2} & \text{if } \begin{pmatrix} x \\ y \end{pmatrix} \neq \begin{pmatrix} 0 \\ 0 \end{pmatrix} \\ 0 & \text{if } \begin{pmatrix} x \\ y \end{pmatrix} = \begin{pmatrix} 0 \\ 0 \end{pmatrix}. \end{cases}$$

a. What does it mean to say that f is differentiable at $\begin{pmatrix} 0 \\ 0 \end{pmatrix}$?

b. Show that both partial derivatives $D_1 f\begin{pmatrix} 0 \\ 0 \end{pmatrix}$ and $D_2 f\begin{pmatrix} 0 \\ 0 \end{pmatrix}$ exist, and compute them.

c. Is f differentiable at $\begin{pmatrix} 0 \\ 0 \end{pmatrix}$?

Exercise 1.9.3, part c: You may find the following fact useful:

$$|\sin x| \leq |x| \text{ for all } x \in \mathbb{R}.$$

This follows from the mean value theorem:

$$|\sin x| = \left| \int_0^x \cos t \, dt \right| \leq \left| \int_0^x 1 \, dt \right| = |x|.$$

1.10 Review exercises for Chapter 1

1.1 Which of the following lines are subspaces of $\mathbb{R}^2$ (or $\mathbb{R}^n$)? For any that are not, why not?

a. $y = -2x - 5$ b. $y = 2x + 1$ c. $y = \dfrac{5x}{2}$

1.2 For what values of a and b do the matrices

$$A = \begin{bmatrix} 1 & a \\ a & 0 \end{bmatrix} \quad \text{and} \quad B = \begin{bmatrix} 1 & 0 \\ b & 1 \end{bmatrix}$$

satisfy $AB = BA$?

1.3 Show that if A and B are upper triangular $n \times n$ matrices, so is AB.

Exercise 1.4: If you are not comfortable with complex numbers, please read Section 0.7.

1.4 a. Show that the rule associating a complex number $z = \alpha + i\beta$ to the 2×2 matrix $T_z = \begin{bmatrix} \alpha & \beta \\ -\beta & \alpha \end{bmatrix}$ satisfies

$$T_{z_1+z_2} = T_{z_1} + T_{z_2} \quad \text{and} \quad T_{z_1 z_2} = T_{z_1} T_{z_2}.$$

b. What is the inverse of the matrix T_z? How is it related to $1/z$?

c. Find a 2×2 matrix whose square is minus the identity.

1.5 Suppose all the edges of a graph are oriented by an arrow on them. We allow for two-way streets. Define the oriented adjacency matrix to be the square matrix with both rows and columns labeled by the vertices, where the (i, j)th entry is m if there are m oriented edges leading from vertex i to vertex j.

What are the oriented adjacency matrices of the graphs below?

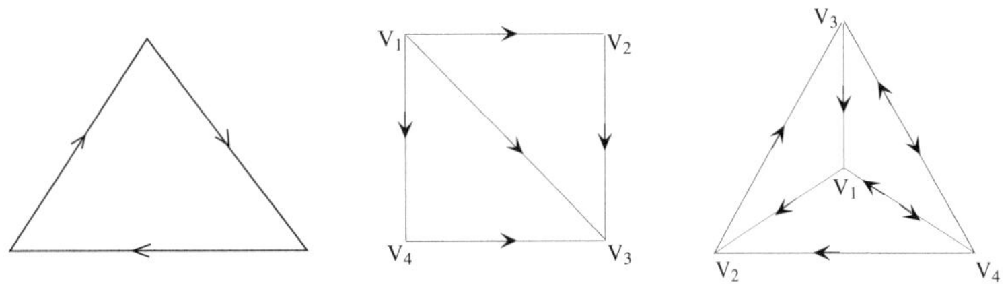

1.6 Are the following maps linear? If so, give their matrices.

$$\text{a.} \quad \begin{bmatrix} x_1 \\ x_2 \\ x_3 \\ x_4 \end{bmatrix} \mapsto \begin{bmatrix} x_2 \\ x_4 \end{bmatrix} \qquad \text{b.} \quad \begin{bmatrix} x_1 \\ x_2 \\ x_3 \\ x_4 \end{bmatrix} \mapsto \begin{bmatrix} x_2 x_4 \\ x_1 + x_3 \end{bmatrix}$$

1.7 a. Show that the function defined in Example 1.5.24 is continuous.

b. Extend f to be 0 at the origin. Show that then all directional derivatives exist at the origin (although the function is not continuous there).

1.8 Let B be a $k \times n$ matrix, A an $n \times n$ matrix, and C a $n \times m$ matrix, so that the product BAC is defined. Show that if $|A| < 1$, the series

$$BC + BAC + BA^2C + BA^3C \cdots$$

converges in $\operatorname{Mat}(k, m)$ to $B(I - A)^{-1}C$.

1.9 a. Is there a linear transformation $T : \mathbb{R}^4 \to \mathbb{R}^3$ such that all of the following are satisfied?

$$T\begin{bmatrix} 1 \\ 0 \\ 1 \\ 0 \end{bmatrix} = \begin{bmatrix} 1 \\ 1 \\ 1 \end{bmatrix}, \quad T\begin{bmatrix} 0 \\ 1 \\ 1 \\ 0 \end{bmatrix} = \begin{bmatrix} 0 \\ 2 \\ 3 \end{bmatrix}, \quad T\begin{bmatrix} 0 \\ 1 \\ 1 \\ 1 \end{bmatrix} = \begin{bmatrix} 2 \\ 1 \\ 2 \end{bmatrix}, \quad T\begin{bmatrix} 1 \\ 0 \\ 0 \\ 1 \end{bmatrix} = \begin{bmatrix} 3 \\ -1 \\ -1 \end{bmatrix}$$

If so, what is its matrix?

b. Let S be a transformation such that the equations of part a are satisfied, and, in addition, $S\begin{bmatrix} 1 \\ 1 \\ 1 \\ 1 \end{bmatrix} = \begin{bmatrix} 0 \\ 3 \\ 2 \end{bmatrix}$. Is S linear?

1.10 Find the matrix for the transformation from $\mathbb{R}^3 \to \mathbb{R}^3$ that rotates by $30°$ around the y-axis.

1.11 a. What are the matrices of the linear transformations $S, T : \mathbb{R}^3 \to \mathbb{R}^3$ corresponding to reflection in the planes of equation $x = y$ and $y = z$?

b. What are the matrices of the compositions $S \circ T$ and $T \circ S$?

c. What relation is there between the matrices in part b?

d. Can you name the linear transformations $S \circ T$ and $T \circ S$?

1.12 Let A be a 2×2 matrix. If we identify the set of 2×2 matrices with $\mathbb{R}^4$ by identifying $\begin{bmatrix} a & b \\ c & d \end{bmatrix}$ with $\begin{bmatrix} a \\ b \\ c \\ d \end{bmatrix}$, what is the angle between A and A^{-1}? Under what condition are A and A^{-1} orthogonal?

1.13 Let P be the parallelepiped $0 \le x \le a,\ 0 \le y \le b,\ 0 \le z \le c$.

a. What angle does a diagonal make with the sides? What relation is there between the length of a side and the corresponding angle?

b. What are the angles between the diagonal and the faces of the parallelepiped? What relation is there between the area of a face and the corresponding angle?

1.14 Let A be a 3×3 matrix with columns $\vec{\mathbf{a}}, \vec{\mathbf{b}}, \vec{\mathbf{c}}$, and let Q_A be the 3×3 matrix with rows $(\vec{\mathbf{b}} \times \vec{\mathbf{c}})^\top, (\vec{\mathbf{c}} \times \vec{\mathbf{a}})^\top, (\vec{\mathbf{a}} \times \vec{\mathbf{b}})^\top$.

Exercise 1.14, part c: Think of the geometric definition of the cross product, and the definition of the determinant of a 3×3 matrix in terms of cross products.

a. Compute Q_A when $A = \begin{bmatrix} 1 & 2 & 0 \\ 0 & -1 & 1 \\ 1 & 1 & -1 \end{bmatrix}$.

b. What is the product $Q_A\, A$ when A is the matrix of part a?

c. What is $Q_A\, A$ for any 3×3 matrix A?

d. Can you relate this problem to Exercise 1.4.20?

1.15 a. Normalize the following vectors:

$$\text{(i)} \begin{bmatrix} 2 \\ 1 \\ 3 \end{bmatrix}; \quad \text{(ii)} \begin{bmatrix} -2 \\ 3 \end{bmatrix}; \quad \text{(iii)} \begin{bmatrix} \sqrt{3} \\ 0 \\ 2 \end{bmatrix}.$$

b. What is the angle between the vectors (i) and (iii)?

Hint for Exercise 1.16: You may find it helpful to use the formulas

$$1 + 2 + \cdots + n = \frac{n(n+1)}{2}$$

and

$$1 + 4 + \cdots + n^2 = \frac{n(n+1)(2n+1)}{6}.$$

1.16 a. What is the angle θ_n between the vectors $\vec{\mathbf{v}}, \vec{\mathbf{w}} \in \mathbb{R}^n$ given by

$$\vec{\mathbf{v}} = \sum_{i=1}^n \vec{\mathbf{e}}_i \quad \text{and} \quad \vec{\mathbf{w}} = \sum_{i=1}^n i\vec{\mathbf{e}}_i\,?$$

b. What is $\lim_{n\to\infty} \theta_n$?

1.17 Prove the following statements for closed subsets of $\mathbb{R}^n$:

a. Any intersection of closed sets is closed.

b. A finite union of closed sets is closed.

c. An infinite union of closed sets is not necessarily closed.

1.18 Show that $\overline{U}$ (the closure of U) is the subset of $\mathbb{R}^n$ made up of all limits of sequences in U that converge in $\mathbb{R}^n$.

1.19 Consider the function

$$\mathbf{f}\begin{pmatrix} x \\ y \\ z \end{pmatrix} = \begin{pmatrix} zy^2 \\ 2x^2 - y^2 \\ x + z \end{pmatrix}, \quad \text{evaluated at } \begin{pmatrix} 1 \\ 1 \\ 1 \end{pmatrix}.$$

You are given the choice of five directions:

$$\vec{\mathbf{e}}_1,\ \vec{\mathbf{e}}_2,\ \vec{\mathbf{e}}_3,\ \vec{\mathbf{v}}_1 = \begin{bmatrix} 1/\sqrt{2} \\ 0 \\ 1/\sqrt{2} \end{bmatrix},\ \vec{\mathbf{v}}_2 = \begin{bmatrix} 0 \\ 1/\sqrt{2} \\ 1/\sqrt{2} \end{bmatrix}.$$

(Note that these vectors all have length 1.) You are to move in one of those directions. In which direction should you start out

a. if you want zy^2 to be increasing as slowly as possible?

b. if you want $2x^2 - y^2$ to be increasing as fast as possible?

c. if you want $2x^2 - y^2$ to be decreasing as fast as possible?

1.20 Let $h : \mathbb{R} \to \mathbb{R}$ be a C^1 function, periodic of period 2π, and define the function $f : \mathbb{R}^2 \to \mathbb{R}$ by

$$f\begin{pmatrix} r\cos\theta \\ r\sin\theta \end{pmatrix} = rh(\theta).$$

a. Show that f is a continuous real-valued function on $\mathbb{R}^2$.

b. Show that f is differentiable on $\mathbb{R}^2 - \{\mathbf{0}\}$.

c. Show that all directional derivatives of f exist at $\mathbf{0}$ if and only if

$$h(\theta) = -h(\theta + \pi) \quad \text{for all } \theta.$$

d. Show that f is differentiable at $\mathbf{0}$ if and only if $h(\theta) = a\cos\theta + b\sin\theta$ for some numbers a and b.

1.21 State whether the following limits exist, and prove it.

a. $\displaystyle\lim_{\binom{x}{y}\to\binom{0}{0}} \frac{x+y}{x^2-y^2}$ b. $\displaystyle\lim_{\binom{x}{y}\to\binom{0}{0}} \frac{(x^2+y^2)^2}{x+y}$ c. $\displaystyle\lim_{\binom{x}{y}\to\binom{0}{0}} (x^2+y^2)\ln(x^2+y^2)$

*d. $\displaystyle\lim_{\binom{x}{y}\to\binom{0}{0}} (x^2+y^2)(\ln|xy|)$, defined when $xy \neq 0$

1.22 Prove Theorems 1.5.29 and 1.5.30.

1.23 Let $\mathbf{a}_n \in \mathbb{R}^n$ be the vector whose entries are π and e, to n places: $\mathbf{a}_1 = \begin{bmatrix} 3.1 \\ 2.7 \end{bmatrix}$, $\mathbf{a}_2 = \begin{bmatrix} 3.14 \\ 2.71 \end{bmatrix}$, and so on. How large does n have to be so that

$$\left|\mathbf{a}_n - \begin{bmatrix} \pi \\ e \end{bmatrix}\right| < 10^{-3}? \quad \text{so that} \quad \left|\mathbf{a}_n - \begin{bmatrix} \pi \\ e \end{bmatrix}\right| < 10^{-4}?$$

1.24 Set $A_m = \begin{bmatrix} \cos m\theta & \sin m\theta \\ -\sin m\theta & \cos m\theta \end{bmatrix}$. For what numbers θ does the sequence of matrices $m \mapsto A_m$ converge? When does it have a convergent subsequence?

1.25 Find a number R for which you can prove that the polynomial

$$p(z) = z^{10} + 2z^9 + 3z^8 + \cdots + 10z + 11$$

has a root for $|z| < R$. Explain your reasoning.

1.26 What is the derivative of the function $f : \mathbb{R}^n \to \mathbb{R}^n$ given by the formula $f(\mathbf{x}) = |\mathbf{x}|^2\mathbf{x}$?

1.27 Using Definition 1.7.1, show that $\sqrt{x^2}$ and $\sqrt[3]{x^2}$ are not differentiable at 0, but that $\sqrt{x^4}$ is.

1.28 a. Show that the mapping $\operatorname{Mat}(n,n) \to \operatorname{Mat}(n,n)$, $A \mapsto A^3$ is differentiable. Compute its derivative.

b. Compute the derivative of the mapping

$$\operatorname{Mat}(n,n) \to \operatorname{Mat}(n,n),\ A \mapsto A^k, \quad \text{for any integer } k \geq 1.$$

1.29 Which of the following functions are differentiable at $\begin{pmatrix} 0 \\ 0 \end{pmatrix}$?

a. $f\begin{pmatrix} x \\ y \end{pmatrix} = \begin{cases} \dfrac{x^2 y}{x^2+y^2} & \text{if } \begin{pmatrix} x \\ y \end{pmatrix} \neq \begin{pmatrix} 0 \\ 0 \end{pmatrix} \\ 0 & \text{if } \begin{pmatrix} x \\ y \end{pmatrix} = \begin{pmatrix} 0 \\ 0 \end{pmatrix} \end{cases}$ b. $f\begin{pmatrix} x \\ y \end{pmatrix} = |x+y|$

c. $f\begin{pmatrix} x \\ y \end{pmatrix} = \begin{cases} \dfrac{\sin(xy)}{x^2+y^2} & \text{if } \begin{pmatrix} x \\ y \end{pmatrix} \neq \begin{pmatrix} 0 \\ 0 \end{pmatrix} \\ 0 & \text{if } \begin{pmatrix} x \\ y \end{pmatrix} = \begin{pmatrix} 0 \\ 0 \end{pmatrix} \end{cases}$

1.30 a. Compute the formula for the mapping $A : \mathbb{R}^2 \to \mathbb{R}$ that gives the area of the parallelogram in $\mathbb{R}^3$ spanned by $\begin{bmatrix} u \\ 0 \\ u^2 \end{bmatrix}$ and $\begin{bmatrix} 0 \\ v^2 \\ v \end{bmatrix}$.

b. What is $\left[\mathbf{D}A\begin{pmatrix} 1 \\ -1 \end{pmatrix}\right]\begin{bmatrix} 1 \\ 2 \end{bmatrix}$?

c. For what unit vector $\vec{\mathbf{v}} \in \mathbb{R}^2$ is $\left[\mathbf{D}A\begin{pmatrix} 1 \\ -1 \end{pmatrix}\right]\vec{\mathbf{v}}$ maximal; that is, in what direction should you begin moving $\begin{pmatrix} u \\ v \end{pmatrix}$ from $\begin{pmatrix} 1 \\ -1 \end{pmatrix}$ so that as you start out, the area is increasing the fastest?

d. What is $\left[\mathbf{D}A\begin{pmatrix} 1 \\ 1 \end{pmatrix}\right]\begin{bmatrix} 1 \\ 2 \end{bmatrix}$?

e. In what direction should you move $\begin{pmatrix} u \\ v \end{pmatrix}$ from $\begin{pmatrix} 1 \\ 1 \end{pmatrix}$ so that as you start out, the area is increasing the fastest?

f. Find a point at which A is not differentiable.

Hint for Exercise 1.31: Think of the composition of

$$t \mapsto \begin{pmatrix} t \\ t^2 \end{pmatrix} \quad \text{and}$$

$$\begin{pmatrix} x \\ y \end{pmatrix} \mapsto \int_x^y \frac{ds}{s+\sin s},$$

both of which you should know how to differentiate.

1.31 a. What is the derivative of the function $f(t) = \displaystyle\int_t^{t^2} \frac{ds}{s+\sin s}$, defined for $t > 1$?

b. When is f increasing or decreasing?

Exercise 1.32: It's a lot easier to think of this as the composition of $A \mapsto A^3$ and $A \mapsto A^{-1}$ and to apply the chain rule than to compute the derivative directly.

1.32 Let A be an $n \times n$ matrix, as in Example 1.8.6.

a. Compute the derivative of the map $A \mapsto A^{-3}$.

b. Compute the derivative of the map $A \mapsto A^{-n}$.

1.33 Let $U \subset \text{Mat}\,(n,n)$ be the set of matrices A such that the matrix $AA^\top + A^\top A$ is invertible. Compute the derivative of the map $F : U \to \text{Mat}\,(n,n)$ given by

$$F(A) = (AA^\top + A^\top A)^{-1}.$$

1.34 Consider the function defined on $\mathbb{R}^2$ and given by the formula in the margin.

a. Show that both partial derivatives exist everywhere.

b. Where is f differentiable?

$$f\begin{pmatrix} x \\ y \end{pmatrix} = \begin{cases} \dfrac{xy}{x^2+y^2} & \text{if } \begin{pmatrix} x \\ y \end{pmatrix} \neq \begin{pmatrix} 0 \\ 0 \end{pmatrix} \\ 0 & \text{if } \begin{pmatrix} x \\ y \end{pmatrix} = \begin{pmatrix} 0 \\ 0 \end{pmatrix} \end{cases}$$

Function for Exercise 1.34

1.35 Consider the function on $\mathbb{R}^3$ defined by

$$f\begin{pmatrix} x \\ y \\ z \end{pmatrix} = \begin{cases} \dfrac{xyz}{x^4+y^4+z^4} & \text{if } \begin{pmatrix} x \\ y \\ z \end{pmatrix} \neq \begin{pmatrix} 0 \\ 0 \\ 0 \end{pmatrix} \\ 0 & \text{if } \begin{pmatrix} x \\ y \\ z \end{pmatrix} = \begin{pmatrix} 0 \\ 0 \\ 0 \end{pmatrix}. \end{cases}$$

a. Show that all partial derivatives exist everywhere.

b. Where is f differentiable?

1.36 a. Show that if an $n \times n$ matrix A is strictly upper triangular or strictly lower triangular, then $A^n = [0]$.

b. Show that $(I - A)$ is invertible.

c. Show that if all the entries of A are nonnegative, then each entry of the matrix $(I - A)^{-1}$ is nonnegative.

Starred exercises are difficult; exercises with two stars are particularly challenging.

***1.37** What 2×2 matrices A satisfy

$$\text{a. } A^2 = 0, \qquad \text{b. } A^2 = I, \qquad \text{c. } A^2 = -I\,?$$

****1.38** (This is very hard.) In the Singapore public garden, there is a statue consisting of a spherical stone ball, with diameter perhaps 1.3 m, weighing at least a ton. This ball is placed in a semispherical stone cup, which it fits almost exactly; moreover, there is a water jet at the bottom of the cup, so the stone is suspended on a film of water, making the friction of the ball with the cup almost 0; it is easy to set it in motion, and it keeps rotating in whatever way you start it for a long time.

Suppose now you are given access to this ball only near the top, so that you can push it to make it rotate around any horizontal axis, but you don't have enough of a grip to make it turn around the vertical axis. Can you make it rotate around the vertical axis anyway?

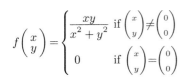

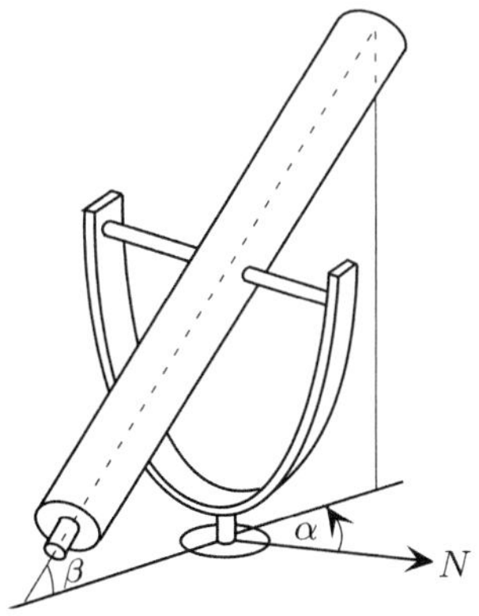

The telescope of Exercise 1.39. The angle α is the azimuth; β is the elevation. North is denoted N.

***1.39** Suppose a telescope is mounted on an equatorial mount, as shown in the margin. This means that mounted on a vertical axis that can pivot there is a U-shaped holder with a bar through its ends, which can rotate also, and the telescope is mounted on the bar. The angle which the horizontal direction perpendicular to the plane of the U makes with north (the angle labeled α in the picture in the margin) is called the *azimuth*, and the angle which the telescope makes with the horizontal (labeled β) is called the *elevation*.

Center your system of coordinates at the center of the bar holding the telescope (which doesn't move either when you change either the azimuth or the elevation), and suppose that the x-axis points north, the y-axis points west, and the z-axis points up. Suppose, moreover, that the telescope is in position azimuth θ_0 and elevation φ_0, where φ_0 is neither 0 nor $\pi/2$.

a. What is the matrix of the linear transformation consisting of raising the elevation of the telescope by φ?

b. Suppose you can rotate the telescope on its own axis. What is the matrix of rotation of the telescope on its axis by ω (measured counterclockwise by an observer sitting at the lower end of the telescope)?

2

Solving equations

In 1985, John Hubbard was asked to testify before the Committee on Science and Technology of the U.S. House of Representatives. He was preceded by a chemist from DuPont, who spoke of modeling molecules, and by an official from the geophysics institute of California, who spoke of exploring for oil and attempting to predict tsunamis.

When it was his turn, he explained that when chemists model molecules, they are solving Schrödinger's equation, that exploring for oil requires solving the Gelfand-Levitan equation, and that predicting tsunamis means solving the Navier-Stokes equation. Astounded, the chairman of the committee interrupted him and turned to the previous speakers. "Is that true, what Professor Hubbard says?" he demanded. "Is it true that what you do is solve equations?"

2.0 Introduction

In every subject, language is intimately related to understanding.

"It is impossible to dissociate language from science or science from language, because every natural science always involves three things: the sequence of phenomena on which the science is based; the abstract concepts which call these phenomena to mind; and the words in which the concepts are expressed. To call forth a concept, a word is needed; to portray a phenomenon, a concept is needed. All three mirror one and the same reality."—Antoine Lavoisier, 1789.

"Professor Hubbard, you always underestimate the difficulty of vocabulary."—Helen Chigirinskaya, Cornell University, 1997.

All readers of this book will have solved systems of simultaneous *linear* equations. Such problems arise throughout mathematics and its applications, so a thorough understanding of the problem is essential.

What most students encounter in high school is systems of n equations in n unknowns, where n might be general or might be restricted to $n = 2$ and $n = 3$. Such a system usually has a unique solution, but sometimes something goes wrong: some equations are "consequences of others" and have infinitely many solutions; other systems of equations are "incompatible" and have no solutions. The first half of this chapter is largely concerned with making these notions systematic.

A language has evolved to deal with these concepts: "linear transformation", "linear combination", "linear independence", "kernel", "span", "basis", and "dimension". These words may sound unfriendly, but they are actually quite transparent if thought of in terms of linear equations. They are needed to answer questions like, "How many equations are consequences of the others?" The relationship of these words to linear equations goes further. Theorems in linear algebra can be proved with abstract induction proofs, but students generally prefer the following method, which we discuss in this chapter:

> *Reduce the statement to a statement about linear equations, row reduce the resulting matrix, and see whether the statement becomes obvious.*

If so, the statement is true; otherwise it is likely to be false.

In Section 2.6 we discuss abstract vector spaces and the change of basis. In Section 2.7 we explore the advantages of expressing a linear transformation in an *eigenbasis*, when such a basis exists, and how to find it when it does. The Perron-Frobenium theorem (Theorem 2.7.10) introduces the notion of a *leading eigenvalue*; Example 2.7.12 shows how Google uses this notion in its PageRank algorithm.

We added Example 2.7.12 on the PageRank algorithm and Example 3.8.10 on the use of *eigenfaces* to identify photographs in response to a request from computer scientists for more "big matrices" reflecting current applications of linear algebra.

Solving nonlinear equations is much harder than solving linear equations. In the days before computers, finding solutions was virtually impossible, and mathematicians had to be satisfied with proving that solutions existed. Today, knowing that a system of equations has solutions is no longer enough; we want a practical algorithm that will enable us to find them. The algorithm most often used is *Newton's method*. In Section 2.8 we show Newton's method in action and state Kantorovich's theorem, which guarantees that under appropriate circumstances Newton's method converges to a solution; in Section 2.9 we see when Newton's method superconverges and state a stronger version of Kantorovich's theorem.

In Section 2.10 we use Kantorovich's theorem to see when a function $\mathbf{f} : \mathbb{R}^n \to \mathbb{R}^n$ has a local inverse function, and when a function $\mathbf{f} : \mathbb{R}^n \to \mathbb{R}^m$, with $n > m$, gives rise to an *implicit function* locally expressing some variables in terms of others.

2.1 The main algorithm: row reduction

Suppose we want to solve the system of linear equations

$$\begin{aligned} 2x + y + 3z &= 1 \\ x - y &= 1 \\ 2x + z &= 1. \end{aligned} \tag{2.1.1}$$

We could add together the first and second equations to get $3x + 3z = 2$. Substituting $(2-3z)/3$ for x in the third equation gives $z = 1/3$, so $x = 1/3$; putting this value for x into the second equation then gives $y = -2/3$.

The big advantage of row reduction is that it requires no cleverness.

In this section we show how to make this approach systematic, using *row reduction*. The first step is to write the system of equations in matrix form:

$$\underbrace{\begin{bmatrix} 2 & 1 & 3 \\ 1 & -1 & 0 \\ 2 & 0 & 1 \end{bmatrix}}_{\text{coefficient matrix } A} \qquad \underbrace{\begin{bmatrix} x \\ y \\ z \end{bmatrix}}_{\text{vector of unknowns } \vec{\mathbf{x}}} \qquad = \qquad \underbrace{\begin{bmatrix} 1 \\ 1 \\ 1 \end{bmatrix}}_{\text{constants } \vec{\mathbf{b}}}, \tag{2.1.2}$$

which can be written as the matrix multiplication $A\vec{\mathbf{x}} = \vec{\mathbf{b}}$:

$$\begin{array}{cc} & \overbrace{\begin{bmatrix} x \\ y \\ z \end{bmatrix}}^{\vec{\mathbf{x}}} \\ \underbrace{\begin{bmatrix} 2 & 1 & 3 \\ 1 & -1 & 0 \\ 2 & 0 & 1 \end{bmatrix}}_{A} & \underbrace{\begin{bmatrix} 1 \\ 1 \\ 1 \end{bmatrix}}_{\vec{\mathbf{b}}} \end{array}. \tag{2.1.3}$$

We now use a shorthand notation, omitting the vector $\vec{\mathbf{x}}$ and writing A and $\vec{\mathbf{b}}$ as a single matrix, with $\vec{\mathbf{b}}$ the last column of the new matrix:

$$\underbrace{\begin{bmatrix} 2 & 1 & 3 \\ 1 & -1 & 0 \\ 2 & 0 & 1 \end{bmatrix}}_{A}\underbrace{\begin{matrix} 1 \\ 1 \\ 1 \end{matrix}}_{\vec{\mathbf{b}}} . \qquad 2.1.4$$

More generally, the system of equations

$$\begin{matrix} a_{1,1}x_1 & + \cdots + & a_{1,n}x_n & = b_1 \\ \vdots & \cdots & \vdots & \vdots \\ a_{m,1}x_1 & + \cdots + & a_{m,n}x_n & = b_m \end{matrix} \qquad 2.1.5$$

is the same as $A\vec{\mathbf{x}} = \vec{\mathbf{b}}$:

$$\underbrace{\begin{bmatrix} a_{1,1} & \cdots & a_{1,n} \\ \vdots & & \vdots \\ a_{m,1} & \cdots & a_{m,n} \end{bmatrix}}_{A} \underbrace{\begin{bmatrix} x_1 \\ \vdots \\ x_n \end{bmatrix}}_{\vec{\mathbf{x}}} = \underbrace{\begin{bmatrix} b_1 \\ \vdots \\ b_m \end{bmatrix}}_{\vec{\mathbf{b}}}, \qquad 2.1.6$$

represented by

$$\underbrace{\begin{bmatrix} a_{1,1} & \cdots & a_{1,n} & b_1 \\ \vdots & \cdots & \vdots & \vdots \\ a_{m,1} & \cdots & a_{m,n} & b_m \end{bmatrix}}_{[A|\vec{\mathbf{b}}]} . \qquad 2.1.7$$

We denote by $[A\,|\,\vec{\mathbf{b}}]$ the matrix obtained by putting $\vec{\mathbf{b}}$ next to the columns of A. The ith column of the matrix A contains the coefficients of x_i; the rows of $[A\,|\,\vec{\mathbf{b}}]$ represent equations. The vertical line in $[A\,|\,\vec{\mathbf{b}}]$ is intended to avoid confusion with multiplication; we are not multiplying A and $\vec{\mathbf{b}}$.

Note that writing the system of equations 2.1.5 in the form 2.1.4 uses position to impart information. In equations 2.1.5 we could write the terms of

$$a_{i,1}x_1 + \cdots + a_{i,n}x_n$$

in any order; in 2.1.4 the coefficient of x_1 must be in the first column, the coefficient of x_2 in the second column, and so on.

Using position to impart information allows for concision; in Roman numerals, 4 084 is

MMMMLXXXIIII.

(When we write IV = 4 and VI = 6 we are using position, but the Romans themselves were quite happy writing their numbers in any order, MMXXM for 3 020, for example.)

Recall that the first subscript in a pair of subscripts refers to vertical position, and the second to horizontal position: $a_{i,j}$ is the entry in the ith row, jth column: *first take the elevator, then walk down the hall.*

Row operations

We can solve a system of linear equations $A\vec{\mathbf{x}} = \vec{\mathbf{b}}$ by *row reducing* the matrix $[A\,|\,\vec{\mathbf{b}}]$, using *row operations.*

Definition 2.1.1 (Row operations). A *row operation* on a matrix is one of three operations:

1. Multiplying a row by a nonzero number
2. Adding a multiple of a row onto another row
3. Exchanging two rows

Exercise 2.1.4 asks you to show that the third operation is not necessary; one can exchange rows using operations 1 and 2.

Column operations are defined by replacing *row* by *column*. We will use column operations in Section 4.8.

Row operations are important for two reasons. First, they require only arithmetic: addition, subtraction, multiplication, and division. This is what computers do well; in some sense it is all they can do. They spend a lot of time doing it: row operations are fundamental to most other mathematical algorithms. The other reason is that row operations enable us to solve systems of linear equations:

Theorem 2.1.2 (Solutions of $A\vec{\mathbf{x}} = \vec{\mathbf{b}}$ unchanged by row operations). *If the matrix $[A|\vec{\mathbf{b}}]$ representing a system of linear equations $A\vec{\mathbf{x}} = \vec{\mathbf{b}}$ can be turned into $[A'|\vec{\mathbf{b}}']$ by a sequence of row operations, then the set of solutions of $A\vec{\mathbf{x}} = \vec{\mathbf{b}}$ and the set of solutions of $A'\vec{\mathbf{x}} = \vec{\mathbf{b}}'$ coincide.*

Formula 2.1.8: We said not to worry about how we did this row reduction. But if you do worry, here are the steps: To get (1), divide row 1 of the original matrix by 2, and add $-1/2$ row 1 to row 2, and subtract row 1 from row 3. To get from (1) to (2), multiply row 2 of the matrix (1) by $-2/3$, and then add that result to row 3. From (2) to (3), subtract half of row 2 from row 1. For (4), subtract row 3 from row 1. For (5), subtract row 3 from row 2.

1. $\begin{bmatrix} 1 & 1/2 & 3/2 & 1/2 \\ 0 & -3/2 & -3/2 & 1/2 \\ 0 & -1 & -2 & 0 \end{bmatrix}$

2. $\begin{bmatrix} 1 & 1/2 & 3/2 & 1/2 \\ 0 & 1 & 1 & -1/3 \\ 0 & 0 & -1 & -1/3 \end{bmatrix}$

3. $\begin{bmatrix} 1 & 0 & 1 & 2/3 \\ 0 & 1 & 1 & -1/3 \\ 0 & 0 & 1 & 1/3 \end{bmatrix}$

4. $\begin{bmatrix} 1 & 0 & 0 & 1/3 \\ 0 & 1 & 1 & -1/3 \\ 0 & 0 & 1 & 1/3 \end{bmatrix}$

5. $\begin{bmatrix} 1 & 0 & 0 & 1/3 \\ 0 & 1 & 0 & -2/3 \\ 0 & 0 & 1 & 1/3 \end{bmatrix}$

Proof. Row operations consist of multiplying one equation by a nonzero number, adding a multiple of one equation to another, and exchanging two equations. Any solution of $A\vec{\mathbf{x}} = \vec{\mathbf{b}}$ is thus a solution of $A'\vec{\mathbf{x}} = \vec{\mathbf{b}}'$. In the other direction, any row operation can be undone by another row operation (Exercise 2.1.5), so any solution $A'\vec{\mathbf{x}} = \vec{\mathbf{b}}'$ is also a solution of $A\vec{\mathbf{x}} = \vec{\mathbf{b}}$. □

Theorem 2.1.2 suggests that we solve $A\vec{\mathbf{x}} = \vec{\mathbf{b}}$ by using row operations to bring the system of equations to the most convenient form. In Example 2.1.3 we apply this technique to equation 2.1.1. For now, don't worry about how the row reduction was achieved. Concentrate instead on what the row-reduced matrix tells us about solutions to the system of equations.

Example 2.1.3 (Solving a system of equations with row operations). To solve the system of equations 2.1.1 we can use row operations to bring the matrix

$$\begin{bmatrix} 2 & 1 & 3 & 1 \\ 1 & -1 & 0 & 1 \\ 2 & 0 & 1 & 1 \end{bmatrix} \quad \text{to the form} \quad \begin{bmatrix} \underbrace{\begin{matrix} 1 & 0 & 0 \\ 0 & 1 & 0 \\ 0 & 0 & 1 \end{matrix}}_{\widetilde{A}} & \underbrace{\begin{matrix} 1/3 \\ -2/3 \\ 1/3 \end{matrix}}_{\widetilde{\mathbf{b}}} \end{bmatrix}. \qquad 2.1.8$$

(To distinguish the new A and $\vec{\mathbf{b}}$ from the old, we put a "tilde" on top: $\widetilde{A}, \widetilde{\mathbf{b}}$; to lighten notation, we drop the arrow on the $\mathbf{b}$.) In this case, the solution can just be read off the matrix. If we put the unknowns back in the matrix, we get

$$\begin{bmatrix} x & 0 & 0 & 1/3 \\ 0 & y & 0 & -2/3 \\ 0 & 0 & z & 1/3 \end{bmatrix} \quad \text{or} \quad \begin{aligned} x &= 1/3 \\ y &= -2/3 \\ z &= 1/3 \end{aligned} \qquad 2.1.9$$

△

Echelon form

Some systems of linear equations may have no solutions, and others may have infinitely many. But if a system has solutions, they can be found by an appropriate sequence of row operations, called *row reduction*, bringing the matrix to *echelon form*, as in the second matrix of formula 2.1.8.

Definition 2.1.4 (Echelon form). A matrix is in *echelon form* if

1. In every row, the first nonzero entry is 1, called a *pivotal 1*.
2. The pivotal 1 of a lower row is always to the right of the pivotal 1 of a higher row.
3. In every column that contains a pivotal 1, all other entries are 0.
4. Any rows consisting entirely of 0's are at the bottom.

Echelon form is not the fastest method for solving systems of linear equations; see Exercise 2.2.11, which describes a faster algorithm, partial row reduction with back substitution. We use echelon form because part 2 of Theorem 2.1.7 is true, and there is no analogous statement for partial row reduction. Thus echelon form is better adapted to proving theorems in linear algebra.

Example 2.1.5 (Matrices in echelon form). Clearly, the identity matrix is in echelon form. So are the following matrices, in which the pivotal 1's are underlined:

$$\begin{bmatrix} \underline{1} & 0 & 0 & 3 \\ 0 & \underline{1} & 0 & -2 \\ 0 & 0 & \underline{1} & 1 \end{bmatrix}, \quad \begin{bmatrix} \underline{1} & 1 & 0 & 0 \\ 0 & 0 & \underline{1} & 0 \\ 0 & 0 & 0 & \underline{1} \end{bmatrix}, \quad \begin{bmatrix} 0 & \underline{1} & 3 & 0 & 0 & 3 & 0 & -4 \\ 0 & 0 & 0 & \underline{1} & -2 & 1 & 0 & 1 \\ 0 & 0 & 0 & 0 & 0 & 0 & \underline{1} & 2 \end{bmatrix}.$$

Example 2.1.6 (Matrices not in echelon form). The following matrices are *not* in echelon form. Can you say why not?[1]

$$\begin{bmatrix} 1 & 0 & 0 & 2 \\ 0 & 0 & 1 & -1 \\ 0 & 1 & 0 & 1 \end{bmatrix}, \quad \begin{bmatrix} 1 & 1 & 0 & 1 \\ 0 & 0 & 2 & 0 \\ 0 & 0 & 0 & 1 \end{bmatrix}, \quad \begin{bmatrix} 0 & 0 & 0 \\ 1 & 0 & 0 \\ 0 & 1 & 0 \end{bmatrix}, \quad \begin{bmatrix} 0 & 1 & 0 & 3 & 0 & -3 \\ 0 & 0 & -1 & 1 & 1 & 1 \\ 0 & 0 & 0 & 0 & 1 & 2 \end{bmatrix}.$$

Row reduction to echelon form is really a systematic form of elimination of variables. The goal is to arrive, if possible, at a situation where each row of the row-reduced matrix corresponds to just one variable. Then, as in formula 2.1.9, the solution can just be read off the matrix.

Exercise 2.1.7 asks you to bring them to echelon form.

How to row reduce a matrix

The following result and its proof are fundamental; essentially every result in the first six sections of this chapter is an elaboration of Theorem 2.1.7.

In MATLAB, the command rref (for "row reduce echelon form") brings a matrix to echelon form.

Theorem 2.1.7. *1. Given any matrix A, there exists a matrix $\widetilde{A}$ in echelon form that can be obtained from A by row operations.*

2. The matrix $\widetilde{A}$ is unique.

Proof. 1. The proof of part 1 is an explicit algorithm for computing $\widetilde{A}$. Called *row reduction* or *Gaussian elimination* (or several other names), it is the main tool for solving linear equations.

[1] The first matrix violates rule 2, the second and fourth violate rules 1 and 3, and the third violates rule 4.

Once you've gotten the hang of row reduction, you'll see that it is perfectly simple (although we find it astonishingly easy to make mistakes). Just as you should know how to add and multiply, you should know how to row reduce, but the goal is not to compete with a computer; that's a losing proposition.

You may want to take shortcuts; for example, if the first row of your matrix starts with a 3 and the third row starts with a 1, you might want to make the third row the first one, rather than dividing through by 3.

Exercise 2.1.3 provides practice in row reduction.

Row reduction: the algorithm. To bring a matrix to echelon form,

1. Find the first column that is not all 0's; call this the first pivotal column and call its first nonzero entry a pivot. If the pivot is not in the first row, move the row containing it to first row position.
2. Divide the first row by the pivot, so that the first entry of the first pivotal column is 1.
3. Add appropriate multiples of the first row to the other rows to make all other entries of the first pivotal column 0. The 1 in the first column is now a pivotal 1.
4. Choose the next column that contains at least one nonzero entry beneath the first row, and put the row containing the new pivot in second row position. Make the pivot a pivotal 1: divide by the pivot, and add appropriate multiples of this row to the other rows, to make all other entries of this column 0.
5. Repeat until the matrix is in echelon form. Each time choose the first column that has a nonzero entry in a lower row than the lowest row containing a pivotal 1, and put the row containing that entry directly below the lowest row containing a pivotal 1.

2. Uniqueness, which is more subtle, is proved in Section 2.2. □

Example 2.1.8 (Row reduction). Here we row reduce a matrix. The R's refer in each case to the rows of the immediately preceding matrix. For example, the second row of the second matrix is labeled $R_1 + R_2$, because that row is obtained by adding the first and second rows of the preceding matrix. Again, we underline the pivotal 1's.

$$\begin{bmatrix} 1 & 2 & 3 & 1 \\ -1 & 1 & 0 & 2 \\ 1 & 0 & 1 & 2 \end{bmatrix} \rightarrow \begin{matrix} \\ R_1 + R_2 \\ R_3 - R_1 \end{matrix} \begin{bmatrix} \underline{1} & 2 & 3 & 1 \\ 0 & 3 & 3 & 3 \\ 0 & -2 & -2 & 1 \end{bmatrix} \rightarrow R_2/3 \begin{bmatrix} \underline{1} & 2 & 3 & 1 \\ 0 & \underline{1} & 1 & 1 \\ 0 & -2 & -2 & 1 \end{bmatrix}$$

$$\rightarrow \begin{matrix} R_1 - 2R_2 \\ \\ R_3 + 2R_2 \end{matrix} \begin{bmatrix} \underline{1} & 0 & 1 & -1 \\ 0 & \underline{1} & 1 & 1 \\ 0 & 0 & 0 & 3 \end{bmatrix} \rightarrow \begin{matrix} \\ \\ R_3/3 \end{matrix} \begin{bmatrix} \underline{1} & 0 & 1 & -1 \\ 0 & \underline{1} & 1 & 1 \\ 0 & 0 & 0 & \underline{1} \end{bmatrix} \rightarrow \begin{matrix} R_1 + R_3 \\ R_2 - R_3 \\ \\ \end{matrix} \begin{bmatrix} \underline{1} & 0 & 1 & 0 \\ 0 & \underline{1} & 1 & 0 \\ 0 & 0 & 0 & \underline{1} \end{bmatrix} \quad \triangle$$

When computers row reduce: avoiding loss of precision

Matrices generated by computer operations often have entries that are really zero but are made nonzero by round-off error: for example, a number may be subtracted from a number that in theory is the same but in practice is off by, say, 10^{-50}, because it has been rounded off. Such an entry is a poor choice for a pivot, because you will need to divide its row through by it, and the row will then contain very large entries. When you then add multiples of that row onto another row, you will be committing the *cardinal sin of computation*: adding numbers of very different sizes, which leads to loss of precision. So what do you do? You skip over that almost-zero entry and choose another pivot. There is, in fact, no reason to choose the first nonzero entry in a given column; in practice, when computers row reduce matrices, they always choose the largest.

Remark. This is not a small issue. Computers spend most of their time solving linear equations by row reduction; keeping loss of precision due to round-off errors from getting out of hand is critical. Entire professional journals are devoted to this topic; at a university like Cornell perhaps half a dozen mathematicians and computer scientists spend their lives trying to understand it. $\triangle$

Example 2.1.9 (Thresholding to minimize round-off errors). If you are computing to 10 significant digits, then $1 + 10^{-10} = 1.0000000001 = 1$. So consider the system of equations

$$\begin{aligned} 10^{-10}x + 2y &= 1 \\ x + y &= 1, \end{aligned} \qquad 2.1.10$$

the solution of which is

$$x = \frac{1}{2 - 10^{-10}}, \quad y = \frac{1 - 10^{-10}}{2 - 10^{-10}}. \qquad 2.1.11$$

If you are computing to 10 significant digits, this is $x = y = .5$. If you use 10^{-10} as a pivot, the row reduction, to 10 significant digits, goes as follows:

Exercise 2.1.9 asks you to analyze precisely where the errors occur in formula 2.1.12.

$$\begin{bmatrix} 10^{-10} & 2 & 1 \\ 1 & 1 & 1 \end{bmatrix} \to \begin{bmatrix} \underline{1} & 2\cdot 10^{10} & 10^{10} \\ 1 & 1 & 1 \end{bmatrix} \to \begin{bmatrix} \underline{1} & 2\cdot 10^{10} & 10^{10} \\ 0 & -2\cdot 10^{10} & -10^{10} \end{bmatrix}$$
$$\to \begin{bmatrix} \underline{1} & 0 & 0 \\ 0 & \underline{1} & .5 \end{bmatrix}. \qquad 2.1.12$$

The "solution" shown by the last matrix reads $x = 0$, but x is supposed to be .5. Now do the row reduction treating 10^{-10} as zero; what do you get? If you have trouble, check the answer in the footnote.[2] $\triangle$

$$\begin{aligned} 3x + y - 4z &= 0 \\ 2y + z &= 4 \\ x - 3y &= 1 \end{aligned}$$

System of equations for Exercise 2.1.1, part a

Exercises for Section 2.1

2.1.1 a. Write the system of linear equations in the margin as the multiplication of a matrix by a vector, using the format of Exercise 1.2.2.

b. Write this system as a single matrix, using the notation of equation 2.1.4.

---

[2]Remember to put the second row in the first row position, as we do in the third step below:

$$\begin{bmatrix} 10^{-10} & 2 & 1 \\ 1 & 1 & 1 \end{bmatrix} \to \begin{bmatrix} 0 & 2 & 1 \\ 1 & 1 & 1 \end{bmatrix} \to \begin{bmatrix} \underline{1} & 1 & 1 \\ 0 & 2 & 1 \end{bmatrix} \to \begin{bmatrix} \underline{1} & 1 & 1 \\ 0 & 1 & .5 \end{bmatrix} \to \begin{bmatrix} \underline{1} & 0 & .5 \\ 0 & \underline{1} & .5 \end{bmatrix}.$$

a. $\begin{bmatrix} 1 & 2 & 3 \\ 4 & 5 & 6 \end{bmatrix}$

b. $\begin{bmatrix} 1 & -1 & 1 \\ -1 & 0 & 2 \\ -1 & 1 & 1 \end{bmatrix}$

c. $\begin{bmatrix} 1 & 2 & 3 & 5 \\ 2 & 3 & 0 & -1 \\ 0 & 1 & 2 & 3 \end{bmatrix}$

d. $\begin{bmatrix} 1 & 3 & -1 & 4 \\ 1 & 2 & 1 & 2 \\ 3 & 7 & 1 & 9 \end{bmatrix}$

e. $\begin{bmatrix} 1 & 1 & 1 & 1 \\ 2 & -3 & 3 & 3 \\ 1 & -4 & 2 & 2 \end{bmatrix}$

Matrices for Exercise 2.1.3

Exercise 2.1.6: This exercise shows that while row operations do not change the set of solutions to a system of linear equations represented by $[A \,|\, \vec{\mathbf{b}}]$, column operations usually do change the solution set. This is not surprising, since column operations change the unknowns.

c. Write the following system of equations as a single matrix:

$$\begin{aligned} x_1 - 7x_2 + 2x_3 &= 1 \\ x_1 - 3x_2 &= 2 \\ 2x_1 - 2x_2 &= -1. \end{aligned}$$

2.1.2 Write each of the following systems of equations as a single matrix, and row reduce the matrix to echelon form.

$$\text{a.}\quad \begin{aligned} 3y - z &= 0 \\ -2x + y + 2z &= 0 \\ x - 5z &= 0 \end{aligned} \qquad \text{b.}\quad \begin{aligned} 2x_1 + 3x_2 - x_3 &= 1 \\ -2x_2 + x_3 &= 2 \\ x_1 - 2x_3 &= -1 \end{aligned}$$

2.1.3 Bring the matrices in the margin to echelon form, using row operations.

2.1.4 Show that the row operation that consists of exchanging two rows is not necessary; one can exchange rows using the other two row operations: (1) multiplying a row by a nonzero number, and (2) adding a multiple of a row onto another row.

2.1.5 Show that any row operation can be undone by another row operation. Note the importance of the word "nonzero" in Definition 2.1.1 of row operations.

2.1.6 a. Perform the following row operations on the matrix $\begin{bmatrix} 2 & 1 & 3 & 1 \\ 1 & -1 & 0 & 1 \\ 2 & 0 & 1 & 1 \end{bmatrix}$ in Example 2.1.3:

i. Multiply the second row by 2. What system of equations does this matrix represent? Confirm that the solution obtained by row reduction is not changed.

ii. Repeat, this time exchanging the first and second rows.

iii. Repeat, this time adding -2 times the second row to the third row.

b. Now use column operations: multiply the second column of the matrix by 2, exchange the first and second columns, and add -2 times the second column to the third column. In each case, what system of equations does this matrix represent? What is its solution?

2.1.7 For each of the four matrices in Example 2.1.6, find (and label) row operations that will bring them to echelon form.

2.1.8 Show that if A is square, and $\tilde{A}$ is A row reduced to echelon form, then either $\tilde{A}$ is the identity, or the last row is a row of zeros.

2.1.9 For Example 2.1.9, analyze where the troublesome errors occur.

2.2 Solving equations with row reduction

In this section we will see what a row-reduced matrix representing a system of linear equations tells us about its solutions. To solve the system of linear equations $A\vec{\mathbf{x}} = \vec{\mathbf{b}}$, form the matrix $[A|\vec{\mathbf{b}}]$ and row reduce it to echelon form. If the system has a unique solution, Theorem 2.2.1 says that the solution can be read off the matrix, as in Example 2.1.3. If it does not, the matrix will tell you whether there is no solution or infinitely many solutions. Although Theorem 2.2.1 is practically obvious, it is the backbone of the entire part of linear algebra that deals with linear equations, dimension, bases, rank, and so forth (but not eigenvectors and eigenvalues, or quadratic forms, or the determinant).

Recall that $A\vec{\mathbf{x}} = \vec{\mathbf{b}}$ represents a system of equations, the matrix A giving the coefficients, the vector $\vec{\mathbf{x}}$ giving the unknowns. The augmented matrix $[A \,|\, \vec{\mathbf{b}}]$ is shorthand for $A\vec{\mathbf{x}} = \vec{\mathbf{b}}$.

The ith column of A corresponds to the ith unknown of the system $A\vec{\mathbf{x}} = \vec{\mathbf{b}}$.

Remark. In Theorem 2.1.7 we used a tilde to denote echelon form: $\widetilde{A}$ is the row-reduced echelon form of A. Here, $[\widetilde{A} \,|\, \widetilde{\mathbf{b}}]$ denotes the echelon form of the entire matrix $[A \,|\, \vec{\mathbf{b}}]$ (i.e., it is $\widetilde{[A \,|\, \vec{\mathbf{b}}]}$). We use two tildes because we need to talk about $\widetilde{\mathbf{b}}$ independently of $\widetilde{A}$. △

Theorem 2.2.1 (Solutions to linear equations). *Represent the system $A\vec{\mathbf{x}} = \vec{\mathbf{b}}$, involving m linear equations in n unknowns, by the $m \times (n+1)$ matrix $[A \,|\, \vec{\mathbf{b}}]$, which row reduces to $[\widetilde{A} \,|\, \widetilde{\mathbf{b}}]$. Then*

1. *If the row-reduced vector $\widetilde{\mathbf{b}}$ contains a pivotal 1, the system has no solutions.*
2. *If $\widetilde{\mathbf{b}}$ does not contain a pivotal 1, then solutions are uniquely determined by the values of the nonpivotal variables:*
   a. *If each column of $\widetilde{A}$ contains a pivotal 1 (so there are no nonpivotal variables), the system has a unique solution.*
   b. *If at least one column of $\widetilde{A}$ is nonpivotal, there are infinitely many solutions: exactly one for each value of the nonpivotal variables.*

The nonlinear versions of Theorems 2.2.1 and 2.2.6 are the inverse function theorem and the implicit function theorem, which are discussed in Section 2.10. As in the linear case, some unknowns are *implicit functions* of the others. But those implicit functions will be defined only in a small region, and which variables determine the others depends on where we compute our linearization.

We will prove Theorem 2.2.1 after looking at some examples. Let us consider the case where the results are most intuitive, where $n = m$. The case where the system of equations has a unique solution is illustrated by Example 2.1.3.

Example 2.2.2 (A system with no solutions). Let us solve

$$\begin{aligned} 2x + y + 3z &= 1 \\ x - y &= 1 \\ x + y + 2z &= 1. \end{aligned} \qquad 2.2.1$$

The matrix

$$\begin{bmatrix} 2 & 1 & 3 & 1 \\ 1 & -1 & 0 & 1 \\ 1 & 1 & 2 & 1 \end{bmatrix} \quad \text{row reduces to} \quad \begin{bmatrix} \underline{1} & 0 & 1 & 0 \\ 0 & \underline{1} & 1 & 0 \\ 0 & 0 & 0 & \underline{1} \end{bmatrix}, \qquad 2.2.2$$

so the equations are incompatible and there are no solutions; the last row says that $0 = 1$. △

Example 2.2.3 (A system with infinitely many solutions). Let us solve

$$\begin{aligned} 2x + y + 3z &= 1 \\ x - y \qquad &= 1 \\ x + y + 2z &= 1/3. \end{aligned} \qquad 2.2.3$$

The matrix

$$\begin{bmatrix} 2 & 1 & 3 & 1 \\ 1 & -1 & 0 & 1 \\ 1 & 1 & 2 & 1/3 \end{bmatrix} \quad \text{row reduces to} \quad \begin{bmatrix} \underline{1} & 0 & 1 & 2/3 \\ 0 & \underline{1} & 1 & -1/3 \\ 0 & 0 & 0 & 0 \end{bmatrix}. \qquad 2.2.4$$

Example 2.2.3: In this case, the solutions form a family that depends on the single nonpivotal variable, z; the matrix $\widetilde{A}$ has one column that does not contain a pivotal 1.

The first row says that $x + z = 2/3$; the second that $y + z = -1/3$. We can choose z arbitrarily, giving the solutions

$$\begin{bmatrix} 2/3 - z \\ -1/3 - z \\ z \end{bmatrix}; \qquad 2.2.5$$

there are as many solutions as there are possible values of z: infinitely many. In this system of equations, the third equation provides no new information; it is a consequence of the first two. If we denote the three equations R_1, R_2, and R_3 respectively, then $R_3 = 1/3\,(2R_1 - R_2)$:

$$\begin{array}{lr} 2R_1 & 4x + 2y + 6z = \;\; 2 \\ -R_2 & -x \;+\; y \qquad = -1 \\ \hline 2R_1 - R_2 = 3R_3 & 3x + 3y + 6z = \;\; 1. \end{array} \quad \triangle \qquad 2.2.6$$

In the examples we have seen so far, $\vec{\mathbf{b}}$ was a vector with numbers as entries. What if its entries are symbolic? Depending on the values of the symbols, different cases of Theorem 2.2.1 may apply.

Example 2.2.4 (Equations with symbolic coefficients). Suppose we want to know what solutions, if any, exist for the system of equations

$$\begin{aligned} x_1 + x_2 &= a_1 \\ x_2 + x_3 &= a_2 \\ x_3 + x_4 &= a_3 \\ x_4 + x_1 &= a_4. \end{aligned} \qquad 2.2.7$$

Row operations bring the matrix

$$\begin{bmatrix} 1 & 1 & 0 & 0 & a_1 \\ 0 & 1 & 1 & 0 & a_2 \\ 0 & 0 & 1 & 1 & a_3 \\ 1 & 0 & 0 & 1 & a_4 \end{bmatrix} \quad \text{to} \quad \begin{bmatrix} 1 & 0 & 0 & 1 & a_1 + a_3 - a_2 \\ 0 & 1 & 0 & -1 & a_2 - a_3 \\ 0 & 0 & 1 & 1 & a_3 \\ 0 & 0 & 0 & 0 & a_2 + a_4 - a_1 - a_3 \end{bmatrix}, \qquad 2.2.8$$

so a first thing to notice is that there are no solutions if $a_2 + a_4 - a_1 - a_3 \neq 0$: we are then in case 1 of Theorem 2.2.1. If $a_2 + a_4 - a_1 - a_3 = 0$, we are in case 2b of Theorem 2.2.1: there is no pivotal 1 in the last column, so the system has infinitely many solutions, depending on the value of the variable x_4, corresponding to the fourth column, the only column of the row-reduced matrix that does not contain a pivotal 1. $\triangle$

$$[\widetilde{A} \,|\, \widetilde{\mathbf{b}}] = \begin{bmatrix} \underline{1} & 0 & 1 & 0 \\ 0 & \underline{1} & 1 & 0 \\ 0 & 0 & 0 & \underline{1} \end{bmatrix}$$
$\widetilde{\mathbf{b}}$

FIGURE 2.2.1.

Theorem 2.2.1, case 1: No solution. The row-reduced column $\widetilde{\mathbf{b}}$ contains a pivotal 1; the third line reads $0 = 1$. The left side of that line must contain all 0's; if the third entry were not 0, it would be a pivotal 1, and then $\widetilde{\mathbf{b}}$ would contain no pivotal 1.

$$[\widetilde{A} \,|\, \widetilde{\mathbf{b}}] = \begin{bmatrix} \underline{1} & 0 & 0 & \widetilde{b}_1 \\ 0 & \underline{1} & 0 & \widetilde{b}_2 \\ 0 & 0 & \underline{1} & \widetilde{b}_3 \\ 0 & 0 & 0 & 0 \end{bmatrix}$$

FIGURE 2.2.2.

Case 2a: Unique solution. Here we have four equations in three unknowns. Each column of the matrix $\widetilde{A}$ contains a pivotal 1, giving

$$x_1 = \widetilde{b}_1; \quad x_2 = \widetilde{b}_2; \quad x_3 = \widetilde{b}_3.$$

$$[\widetilde{A} \,|\, \widetilde{\mathbf{b}}] = \begin{bmatrix} \underline{1} & 0 & -1 & \widetilde{b}_1 \\ 0 & \underline{1} & 2 & \widetilde{b}_2 \\ 0 & 0 & 0 & 0 \end{bmatrix}$$

$$[\widetilde{B} \,|\, \widetilde{\mathbf{b}}] = \begin{bmatrix} \underline{1} & 0 & 3 & 0 & 2 & \widetilde{b}_1 \\ 0 & \underline{1} & 1 & 0 & 0 & \widetilde{b}_2 \\ 0 & 0 & 0 & \underline{1} & 1 & \widetilde{b}_3 \end{bmatrix}$$

FIGURE 2.2.3.

Case 2b: Infinitely many solutions (one for each value of nonpivotal variables).

Proof of Theorem 2.2.1. 1. The set of solutions of $A\vec{\mathbf{x}} = \vec{\mathbf{b}}$ is the same as that of $\widetilde{A}\vec{\mathbf{x}} = \widetilde{\mathbf{b}}$ by Theorem 2.1.2. If the entry $\widetilde{b}_j$ is a pivotal 1, then the jth equation of $\widetilde{A}\vec{\mathbf{x}} = \widetilde{\mathbf{b}}$ reads $0 = 1$ (as illustrated by Figure 2.2.1), so the system is inconsistent.

2a. This occurs only if there are at least as many equations as unknowns (there may be more, as shown in Figure 2.2.2). If each column of $\widetilde{A}$ contains a pivotal 1, and $\widetilde{\mathbf{b}}$ has no pivotal 1, then for each variable x_i there is a unique solution $x_i = \widetilde{b}_i$; all other entries in the ith row will be 0, by the rules of row reduction. If there are more equations than unknowns, the extra equations do not make the system incompatible, since by the rules of row reduction, the corresponding rows will contain all 0's, giving the correct if uninformative equation $0 = 0$.

2b. Assume that the ith column contains a pivotal 1 (this pivotal 1 corresponds to the variable x_i). Suppose the row containing this pivotal 1 is the jth row. By the definition of echelon form, there is only one pivotal 1 per row, so the other nonzero entries of the jth row are in nonpivotal columns of $\widetilde{A}$. Denote these nonzero entries by $\tilde{a}_1, \dots, \tilde{a}_k$ and the corresponding nonpivotal columns of $\widetilde{A}$ by $p_1, \dots p_k$. Then

$$x_i = \widetilde{b}_j - \sum_{l=1}^{k} \tilde{a}_l x_{p_l}. \qquad 2.2.9$$

Thus we have infinitely many solutions, each defined uniquely by a choice of values for the variables corresponding to nonpivotal columns of $\widetilde{A}$. $\square$

For instance, for the matrix $\widetilde{A}$ of Figure 2.2.3 we get $x_1 = \widetilde{b}_1 - (-1)x_3$ and $x_2 = \widetilde{b}_2 - 2x_3$; we can make x_3 anything we like; our choice will determine the values of x_1 and x_2, which both correspond to pivotal columns of $\widetilde{A}$. What are the analogous equations for the matrix $\widetilde{B}$ in Figure 2.2.3?[3]

Uniqueness of matrix in echelon form

Now we can prove part 2 of Theorem 2.1.7: that the matrix $\widetilde{A}$ in echelon form is unique.

Proof of Theorem 2.1.7, part 2. Write $A = [\vec{\mathbf{a}}_1, \dots, \vec{\mathbf{a}}_m]$. For $k \le m$, let $A[k]$ consist of the first k columns of A, i.e., $A[k] = [\vec{\mathbf{a}}_1, \dots, \vec{\mathbf{a}}_k]$. If $\widetilde{A}$ is in echelon form, obtained from A by row operations, then the matrix $\widetilde{A}[k]$ given by the first k columns of $\widetilde{A}$ is also obtained from $A[k]$ by the same row operations.

[3]The first, second, and fourth columns contain pivotal 1's, corresponding to variables x_1, x_2, x_4. These variables depend on our choice of values for x_3, x_5:

$$x_1 = \widetilde{b}_1 - 3x_3 - 2x_5$$
$$x_2 = \widetilde{b}_2 - x_3$$
$$x_4 = \widetilde{b}_3 - x_5.$$

Think of $A[k]$ as the augmented matrix corresponding to the system of equations

$$x_1\vec{\mathbf{a}}_1 + \cdots + x_{k-1}\vec{\mathbf{a}}_{k-1} = \vec{\mathbf{a}}_k. \qquad 2.2.10$$

Note that equation 2.2.10 depends only on A, not on $\widetilde{A}$. Theorem 2.2.1 says that the system 2.2.10 has no solutions if and only if the last column of $\widetilde{A}[k]$, i.e., the kth column of $\widetilde{A}$, is pivotal. So A determines the pivotal columns of $\widetilde{A}$.

If for some k the system of equations 2.2.10 has any solutions, then by Theorem 2.2.1 it has a unique solution for every value of the variables corresponding to nonpivotal columns among the first $k-1$ columns of $\widetilde{A}[k]$. Set these variables to be 0. The solution corresponding to that choice uniquely determines the (nonpivotal) kth column of $\widetilde{A}$.

Thus each column of $\widetilde{A}$ is uniquely specified by the system of equations 2.2.10 (one for every k), hence by A. □

A column is *nonpivotal* if the corresponding column of $\widetilde{A}$ contains no pivotal 1. The corresponding variable is called a *nonpivotal variable*.

When a row of $\widetilde{A}$ contains a pivotal 1, we say that the corresponding row of A is a *pivotal row*.

Earlier we spoke of pivotal and nonpivotal columns of a matrix $\widetilde{A}$ in echelon form. Now we can say that these definitions apply also to A.

Definition 2.2.5 (Pivotal column, pivotal variable). A column of A is *pivotal* if the corresponding column of $\widetilde{A}$ contains a pivotal 1. The corresponding variable in the domain is called a *pivotal variable*.

Thus we can rephrase part 2b of Theorem 2.2.1: *If $\widetilde{\mathbf{b}}$ contains no pivotal 1 and at least one variable is nonpivotal, you can choose freely the values of the nonpivotal variables, and these values uniquely determine the values of the pivotal variables.*

"Pivotal" and "nonpivotal" do not describe some intrinsic quality of a particular unknown. If a system of equations has both pivotal and nonpivotal unknowns, which are pivotal and which are not usually depends on the order in which you list the unknowns, as illustrated by Exercise 2.2.1.

Remark. Since in case 2b the pivotal variables depend on the freely chosen nonpivotal variables, we will occasionally call pivotal variables "passive" and nonpivotal variables "active". (Mnemonic: Both "passive" and "pivotal" start with "p".) We will in particular use this terminology when speaking of systems of nonlinear equations, as "pivotal" and "nonpivotal" are really restricted to linear equations. △

Since all sequences of row operations that turn a matrix A into a matrix in echelon form lead to the same matrix $\widetilde{A}$, we can now speak of A row reducing to the identity. This allows us to rephrase Theorem 2.2.1, part 2a:

Theorem 2.2.6: Since (Theorem 2.1.2) the solutions to $A\vec{\mathbf{x}} = \vec{\mathbf{b}}$ and $\widetilde{A}\vec{\mathbf{x}} = \widetilde{\mathbf{b}}$ coincide, if $\widetilde{A} = I$, then $I\vec{\mathbf{x}} = \widetilde{\mathbf{b}}$ says that $\vec{\mathbf{x}} = \widetilde{\mathbf{b}}$ is the unique solution to $A\vec{\mathbf{x}} = \vec{\mathbf{b}}$.

If $\widetilde{A} \neq I$, then Theorem 2.2.1 says that there are either infinitely many solutions or no solutions.

Theorem 2.2.6. *A system $A\vec{\mathbf{x}} = \vec{\mathbf{b}}$ has a unique solution for every $\vec{\mathbf{b}}$ if and only if A row reduces to the identity.*

"Has a unique solution" means that the linear transformation represented by A is one to one; existence of a "solution for every $\vec{\mathbf{b}}$" means that A is onto. Thus Theorem 2.2.6 can be restated as follows:

Corollary 2.2.7. *A matrix A is invertible if and only if it row reduces to the identity.*

In particular, to be invertible a matrix must be square. Moreover, if a square matrix has a left inverse, the left inverse is also a right inverse, and if it has a right inverse, the right inverse is also a left inverse. To see this, note first that a square matrix A is one to one if and only if it is onto:

The first and third implications in 2.2.11 are true for any matrix A, but the second implication is true only if A is square. A 3×2 matrix may row reduce to $\begin{bmatrix} 1 & 0 \\ 0 & 1 \\ 0 & 0 \end{bmatrix}$, which has no nonpivotal columns but one nonpivotal row.

$$\begin{aligned} A \text{ one to one } &\iff A \text{ has no nonpivotal columns} \\ &\iff A \text{ has no nonpivotal rows} \\ &\iff A \text{ is onto.} \end{aligned} \qquad 2.2.11$$

Next note that if B is a right inverse of A, i.e., $AB\vec{\mathbf{x}} = \vec{\mathbf{x}}$ for every $\vec{\mathbf{x}}$, then A is onto, since it has a solution for every $B\vec{\mathbf{x}}$. If B is a left inverse of A, i.e., $BA\vec{\mathbf{x}} = \vec{\mathbf{x}}$ for every $\vec{\mathbf{x}}$, then A must be one to one. Otherwise, there could be two vectors, $\vec{\mathbf{x}}$ and $\vec{\mathbf{x}}'$, such that $A\vec{\mathbf{x}}' = \vec{\mathbf{x}}$ and $A\vec{\mathbf{x}} = \vec{\mathbf{x}}$, which implies $BA\vec{\mathbf{x}}' = \vec{\mathbf{x}}$, contradicting "$BA\vec{\mathbf{x}} = \vec{\mathbf{x}}$ for every $\vec{\mathbf{x}}$".

Being onto corresponds to having a right inverse; being one to one corresponds to having a left inverse.

Remark 2.2.8. In order to check that a matrix is invertible, it is enough to show that the matrix can be put in upper triangular form with all diagonal entries nonzero. For instance, for the following matrix A, row reduction gives

$$A = \begin{bmatrix} 1 & -2 & -1 \\ 2 & 1 & 1 \\ 3 & 2 & -1 \end{bmatrix} \to \begin{bmatrix} 1 & -2 & -1 \\ 0 & 5 & 3 \\ 0 & 8 & 2 \end{bmatrix} \to \begin{bmatrix} 1 & -2 & -1 \\ 0 & 1 & 3/5 \\ 0 & 8 & 2 \end{bmatrix} \to \begin{bmatrix} 1 & -2 & -1 \\ 0 & 1 & 3/5 \\ 0 & 0 & -14/5 \end{bmatrix}.$$

At this point it is clearly possible to row reduce further until we get the identity. Exercise 2.4 asks you to show that a triangular matrix is invertible if and only if all the diagonal terms are nonzero. △

How many equations in how many unknowns?

In most cases, the outcomes given by Theorem 2.2.1 can be predicted by considering how many equations we have for how many unknowns. If we have n equations for n unknowns, most often there will be a unique solution. In terms of row reduction, A will be square, and most often row reduction will result in every row of A having a pivotal 1 (i.e., $\widetilde{A}$ will be the identity). This is not always the case, however, as we saw in Examples 2.2.2 and 2.2.3.

If we have more equations than unknowns, as in Exercise 2.1.3(b), we would expect to have no solutions; only in very special cases can $n-1$ unknowns satisfy n equations. In terms of row reduction, in this case A has more rows than columns, and at least one row of $\widetilde{A}$ will not have a pivotal 1. A row of $\widetilde{A}$ without a pivotal 1 will consist of 0's; if the adjacent entry of $\widetilde{\mathbf{b}}$ is nonzero (as is likely), then the set of equations will have no solutions.

If we have fewer equations than unknowns, as in Exercise 2.2.2, part e, we would expect infinitely many solutions. In terms of row reduction, A will have fewer rows than columns, so at least one column of $\widetilde{A}$ will contain no pivotal 1: there will be at least one nonpivotal unknown. In most cases, $\widetilde{\mathbf{b}}$ will not contain a pivotal 1. (If it does, then that pivotal 1 is preceded by a row of 0's.)

Geometric interpretation of solutions

Theorem 2.2.1 has a geometric interpretation. The top graph in Figure 2.2.4 shows the (usual) case where two equations in two unknowns have a unique solution.

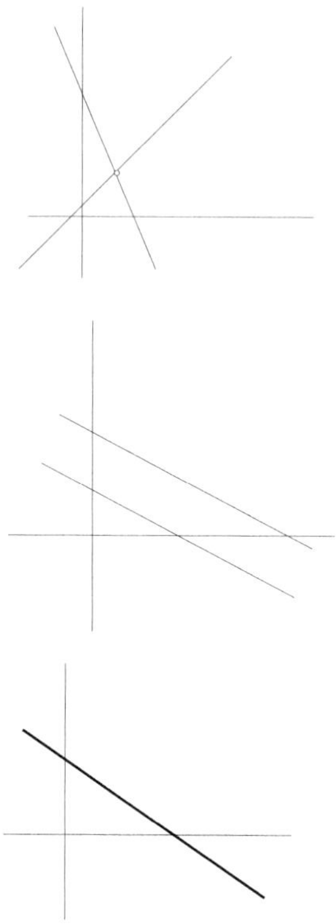

FIGURE 2.2.4.
TOP: Two lines meet in a single point, representing the unique solution to two equations in two unknowns. MIDDLE: A case where two equations in two unknowns have no solution. BOTTOM: Two collinear lines, representing two equations in two unknowns with infinitely many solutions.

As you surely know, two equations in two unknowns,

$$\begin{aligned} a_1x + b_1y &= c_1 \\ a_2x + b_2y &= c_2, \end{aligned} \qquad 2.2.12$$

are incompatible if and only if the lines ℓ_1 and ℓ_2 in $\mathbb{R}^2$ with equations $a_1x + b_1y = c_1$ and $a_2x + b_2y = c_2$ are parallel (middle, Figure 2.2.4). The equations have infinitely many solutions if and only if $\ell_1 = \ell_2$ (bottom, Figure 2.2.4).

When we have three linear equations in three unknowns, each equation describes a plane in $\mathbb{R}^3$. In the top graph of Figure 2.2.5, three planes meet in a single point, the case where three equations in three unknowns have a unique solution.

When three linear equations in three unknowns are incompatible, the planes represented by the equations do not intersect. This happens when two of the planes are parallel, but this is not the only, or even the usual way: the equations will also be incompatible if no two planes are parallel, but the line of intersection of any two is parallel to the third, as shown by the middle graph of Figure 2.2.5. This occurs in Example 2.2.2.

There are also two ways for three linear equations in three unknowns to have infinitely many solutions. The three planes may coincide, but again this is not necessary or usual. The equations will also have infinitely many solutions if the planes intersect in a common line, as shown by the bottom graph of Figure 2.2.5. This second possibility occurs in Example 2.2.3.

Solving several systems of linear equations with one matrix

Theorem 2.2.1 has an additional spinoff. If you want to solve several systems of n linear equations in n unknowns that have the same matrix of coefficients, you can deal with them all at once, using row reduction.

Corollary 2.2.9 (Solving several systems of equations simultaneously). *Several systems of n linear equations in n unknowns, with the same coefficients (e.g. $A\vec{\mathbf{x}} = \vec{\mathbf{b}}_1, \dots, A\vec{\mathbf{x}} = \vec{\mathbf{b}}_k$) can be solved at once with row reduction. Form the matrix*

$$[A\,|\,\vec{\mathbf{b}}_1, \dots, \vec{\mathbf{b}}_k] \quad \textit{and row reduce it to get} \quad [\widetilde{A}\,|\,\widetilde{\mathbf{b}}_1, \dots, \widetilde{\mathbf{b}}_k].$$

If $\widetilde{A}$ is the identity, then $\widetilde{\mathbf{b}}_i$ is the solution to the ith equation $A\vec{\mathbf{x}} = \vec{\mathbf{b}}_i$.

Proof. If A row reduces to the identity, the row reduction is completed by the time one has dealt with the last row of A. The row operations needed to turn A into $\widetilde{A}$ affect each $\vec{\mathbf{b}}_i$, but the $\vec{\mathbf{b}}_i$ do not affect each other. $\square$

Example 2.2.10 (Several systems of equations solved simultaneously). To solve the three systems

$$1.\ \begin{aligned} 2x+y+3z&=1\\ x-y+\ z&=1\\ x+y+2z&=1 \end{aligned}\qquad 2.\ \begin{aligned} 2x+y+3z&=2\\ x-y+\ z&=0\\ x+y+2z&=1 \end{aligned}\qquad 3.\ \begin{aligned} 2x+y+3z&=0\\ x-y+\ z&=1\\ x+y+2z&=1, \end{aligned}$$

we form the matrix

$$\underbrace{\begin{bmatrix} 2 & 1 & 3 & 1 & 2 & 0\\ 1 & -1 & 1 & 1 & 0 & 1\\ 1 & 1 & 2 & 1 & 1 & 1 \end{bmatrix}}_{A \quad \vec{\mathbf{b}}_1\ \vec{\mathbf{b}}_2\ \vec{\mathbf{b}}_3}, \text{ which row reduces to } \underbrace{\begin{bmatrix} 1 & 0 & 0 & -2 & 2 & -5\\ 0 & 1 & 0 & -1 & 1 & -2\\ 0 & 0 & 1 & 2 & -1 & 4 \end{bmatrix}}_{I \quad \tilde{\mathbf{b}}_1\ \tilde{\mathbf{b}}_2\ \tilde{\mathbf{b}}_3}.$$

The solution to the first system of equations is $\begin{bmatrix} -2\\ -1\\ 2 \end{bmatrix}$ (i.e., $x=-2$, $y=-1$, and $z=2$); the solution to the second is $\begin{bmatrix} 2\\ 1\\ -1 \end{bmatrix}$; the third is $\begin{bmatrix} -5\\ -2\\ 4 \end{bmatrix}$. △

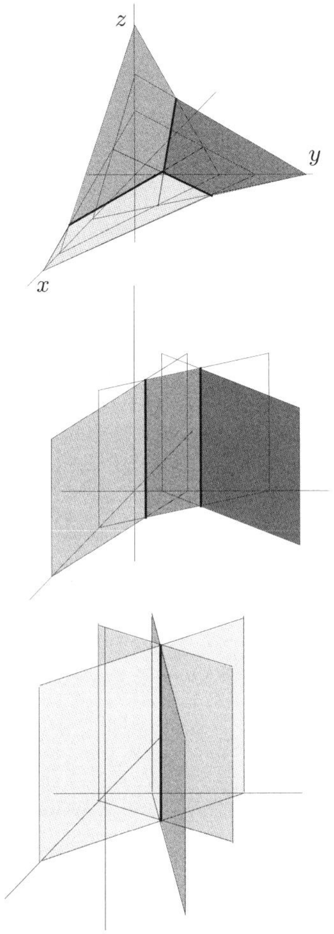

FIGURE 2.2.5.

TOP: Three planes meet in a single point, representing the case where three equations in three unknowns have a unique solution. (Each thick line represents the intersection of two planes.)

MIDDLE: No solution.

BOTTOM: Infinitely many solutions.

EXERCISES FOR SECTION 2.2

2.2.1 a. What system of equations corresponds to $[A|\vec{\mathbf{b}}] = \begin{bmatrix} 2 & 1 & 3 & 1\\ 1 & -1 & 0 & 1\\ 1 & 1 & 2 & 1 \end{bmatrix}$? What are the pivotal and nonpivotal unknowns?

b. Rewrite the system of equations in part a so that y is the first variable, z the second. Now what are the pivotal unknowns?

2.2.2 Predict whether each of the following systems of equations will have a unique solution, no solution, or infinitely many solutions. Solve, using row operations. If your results do not confirm your predictions, can you suggest an explanation for the discrepancy?

$$\text{a.}\ \begin{aligned} 2x+13y-3z&=-7\\ x+y&=\ 1\\ x+7z&=\ 22 \end{aligned}\qquad \text{b.}\ \begin{aligned} x-2y-12z&=12\\ 2x+2y+2z&=\ 4\\ 2x+3y+4z&=\ 3 \end{aligned}\qquad \text{c.}\ \begin{aligned} x+y+z&=\ 5\\ x-y-z&=\ 4\\ 2x+6y+6z&=12 \end{aligned}$$

$$\text{d.}\ \begin{aligned} x+3y+z&=\ 4\\ -x-y+z&=-1\\ 2x+4y&=\ 0 \end{aligned}\qquad \text{e.}\ \begin{aligned} x+2y+z-4w+v&=0\\ x+2y-z+2w-v&=0\\ 2x+4y+z-5w+v&=0\\ x+2y+3z-10w+2v&=0 \end{aligned}$$

2.2.3 a. Confirm the solution for Exercise 2.2.2 e, without using row reduction.

b. On how many parameters does the family of solutions for part e of Exercise 2.2.2 depend?

2.2.4 Compose a system of $(n-1)$ equations in n unknowns, in which $\tilde{\mathbf{b}}$ contains a pivotal 1.

2.2.5 a. For what values of a does the system of equations

$$\begin{aligned} ax + y &= 2 \\ z + ay &= 3 \end{aligned} \quad \text{have a solution?}$$

b. For what values of a does it have a unique solution?

System of equations for Exercise 2.2.6:

$$\begin{aligned} 2x + ay &= 1 \\ x - 3y &= a. \end{aligned}$$

2.2.6 Repeat Exercise 2.2.5, for the system of equations in the margin.

System of equations for Exercise 2.2.7:

$$\begin{aligned} x + y + 2z &= 1 \\ x - y + az &= b \\ 2x - bz &= 0. \end{aligned}$$

2.2.7 Symbolically row reduce the system of linear equations in the margin.

a. For what values of a, b does the system have a unique solution? Infinitely many solutions? No solutions?

b. Which of the preceding possibilities correspond to open subsets of the (a, b)-plane? Closed subsets? Neither?

System of equations for Exercise 2.2.8:

$$\begin{aligned} x + y + az &= 1 \\ x + ay + z &= 1 \\ ax + y + z &= a. \end{aligned}$$

2.2.8 The same as Exercise 2.2.5, for the system of equations in the margin.

***2.2.9** a. For what values of β are there solutions for the system

$$\begin{aligned} x_1 - x_2 - x_3 - 3x_4 + x_5 &= 1 \\ x_1 + x_2 - 5x_3 - x_4 + 7x_5 &= 2 \\ -x_1 + 2x_2 + 2x_3 + 2x_4 + x_5 &= 0 \\ -2x_1 + 5x_2 - 4x_3 + 9x_4 + 7x_5 &= \beta\,? \end{aligned}$$

b. When solutions exist, give values of the pivotal variables in terms of the nonpivotal variables.

2.2.10 Show that it follows from Theorem 2.2.6 and the chain rule that if a function $f : \mathbb{R}^n \to \mathbb{R}^m$ is differentiable with differentiable inverse, then $m = n$.

In Exercise 2.2.11 we use the following rules: a single addition, multiplication, or division has unit cost; administration (i.e., relabeling entries when switching rows, and comparisons) is free.

Hint for part b: There will be $n - k + 1$ divisions,

$$(n-1)(n-k+1)$$

multiplications, $(n-1)(n-k+1)$ additions.

2.2.11 In this exercise, we will estimate how expensive it is to solve a system $A\vec{\mathbf{x}} = \vec{\mathbf{b}}$ of n equations in n unknowns, assuming that there is a unique solution. In particular, we will see that *partial row reduction and back substitution* (defined below) is roughly a third cheaper than full row reduction. First, we will show that the number of operations required to row reduce the augmented matrix $[A|\vec{\mathbf{b}}]$ is

$$R(n) = n^3 + n^2/2 - n/2. \tag{1}$$

a. Compute $R(1), R(2)$; show that formula (1) is correct for $n = 1$ and $n = 2$.

b. Suppose that columns $1, \dots, k-1$ each contain a pivotal 1, and that all other entries in those columns are 0. Show that you will require another $(2n-1)(n-k+1)$ operations for the same to be true of k.

c. Show that $\displaystyle\sum_{k=1}^{n}(2n-1)(n-k+1) = n^3 + \frac{n^2}{2} - \frac{n}{2}$.

Now we will consider a different approach, in which we do all the steps of row reduction, except that we do not make the entries above pivotal 1's be 0. We end up with a matrix of the form in the margin, where $*$ stands for terms that are whatever they are, usually nonzero. Putting the variables back in, when $n = 3$, our system of equations might be

$$\begin{bmatrix} 1 & * & * & \dots & * & \tilde{b}_1 \\ 0 & 1 & * & \dots & * & \tilde{b}_2 \\ \vdots & \vdots & \vdots & \vdots & \vdots & \vdots \\ 0 & 0 & 0 & \dots & 1 & \tilde{b}_n \end{bmatrix}$$

Matrix for Exercise 2.2.11, part c

$$\begin{aligned} x + 2y - z &= 2 \\ y - 3z &= -1 \\ z &= 5, \end{aligned}$$

which can be solved by *back substitution* as follows:

$$z = 5, \qquad y = -1 + 3z = 14, \qquad x = 2 - 2y + z = 2 - 28 + 5 = -21.$$

We will show that partial row reduction and back substitution takes

$$Q(n) = \frac{2}{3}n^3 + \frac{3}{2}n^2 - \frac{7}{6}n \qquad \text{operations.}$$

d. Compute $Q(1), Q(2), Q(3)$. Show that $Q(n) < R(n)$ when $n \geq 3$.

e. By induction on the columns, show that $(n-k+1)(2n-2k+1)$ operations are needed to go from the $(k-1)$th step to the kth step of partial row reduction.

f. Show that $\sum_{k=1}^{n}(n-k+1)(2n-2k+1) = \frac{2}{3}n^3 + \frac{1}{2}n^2 - \frac{1}{6}n.$

g. Show that back substitution requires $n^2 - n$ operations.

h. Compute $Q(n)$.

2.3 Matrix inverses and elementary matrices

In this section we see that matrix inverses give another way to solve equations. We will also introduce the modern view of row reduction: that a row operation is equivalent to multiplying a matrix by an *elementary matrix*.

Proposition 2.3.1 (Solving equations with matrix inverse). *If A has an inverse A^{-1}, then for any $\vec{\mathbf{b}}$ the equation $A\vec{\mathbf{x}} = \vec{\mathbf{b}}$ has a unique solution, namely $\vec{\mathbf{x}} = A^{-1}\vec{\mathbf{b}}$.*

In practice, matrix inverses are rarely a good way to solve linear equations. Row reduction is much less expensive computationally; see Exercise 2.41 in the review exercises for Chapter 2. Even when solving several equations $A\vec{\mathbf{x}}_i = \vec{\mathbf{b}}_i$ with the same A and different $\vec{\mathbf{b}}_i$, the procedure given by Corollary 2.2.9 is usually more effective than computing first A^{-1} and then all the products $A^{-1}\vec{\mathbf{b}}_i$.

The following verifies that $A^{-1}\vec{\mathbf{b}}$ is a solution:

$$A(A^{-1}\vec{\mathbf{b}}) = (AA^{-1})\vec{\mathbf{b}} = I\vec{\mathbf{b}} = \vec{\mathbf{b}}. \qquad 2.3.1$$

This makes use of the associativity of matrix multiplication.

The following computation proves uniqueness:

$$A\vec{\mathbf{x}} = \vec{\mathbf{b}}, \quad \text{so} \quad A^{-1}A\vec{\mathbf{x}} = A^{-1}\vec{\mathbf{b}}; \qquad 2.3.2$$

$$\text{since} \quad A^{-1}A\vec{\mathbf{x}} = \vec{\mathbf{x}}, \quad \text{we have} \quad \vec{\mathbf{x}} = A^{-1}\vec{\mathbf{b}}. \qquad 2.3.3$$

Again we use the associativity of matrix multiplication. Note that in equation 2.3.1, the inverse of A is on the right; in equation 2.3.2, it is on the left.

Computing matrix inverses

Equation 1.2.13 shows how to compute the inverse of a 2×2 matrix. Analogous formulas exist for larger matrices, but they rapidly get out of hand. The effective way to compute matrix inverses for larger matrices is by row reduction, using Theorem 2.3.2.

Theorem 2.3.2 (Computing a matrix inverse). *If A is a $n \times n$ matrix, and you construct the $n \times 2n$ augmented matrix $[A|I]$ and row reduce it, then either:*

1. *The first n columns row reduce to the identity, in which case the last n columns of the row-reduced matrix are the inverse of A, or*
2. *The first n columns do not row reduce to the identity, in which case A does not have an inverse.*

To construct the matrix $[A|I]$ of Theorem 2.3.2, you put A to the left of the corresponding identity matrix I. By "corresponding" we mean that if A is $n \times n$, then I must be $n \times n$ also.

Example 2.3.3 (Computing a matrix inverse).

$$\text{The matrix } A = \begin{bmatrix} 2 & 1 & 3 \\ 1 & -1 & 1 \\ 1 & 1 & 2 \end{bmatrix} \quad \text{has inverse} \quad A^{-1} = \begin{bmatrix} 3 & -1 & -4 \\ 1 & -1 & -1 \\ -2 & 1 & 3 \end{bmatrix},$$

because

$$\begin{bmatrix} 2 & 1 & 3 & 1 & 0 & 0 \\ 1 & -1 & 1 & 0 & 1 & 0 \\ 1 & 1 & 2 & 0 & 0 & 1 \end{bmatrix} \text{ row reduces to } \begin{bmatrix} \underline{1} & 0 & 0 & 3 & -1 & -4 \\ 0 & \underline{1} & 0 & 1 & -1 & -1 \\ 0 & 0 & \underline{1} & -2 & 1 & 3 \end{bmatrix}. \ \triangle$$

Example 2.3.4 (A matrix with no inverse). Consider the matrix of Examples 2.2.2 and 2.2.3, for two systems of linear equations, neither of which has a unique solution:

We haven't row reduced the matrix to echelon form. As soon we see that the first three columns are not the identity matrix, there's no point in continuing; we already know that A has no inverse.

$$A = \begin{bmatrix} 2 & 1 & 3 \\ 1 & -1 & 0 \\ 1 & 1 & 2 \end{bmatrix}. \tag{2.3.4}$$

This matrix has no inverse because

$$\begin{bmatrix} 2 & 1 & 3 & 1 & 0 & 0 \\ 1 & -1 & 0 & 0 & 1 & 0 \\ 1 & 1 & 2 & 0 & 0 & 1 \end{bmatrix} \text{ row reduces to } \begin{bmatrix} \underline{1} & 0 & 1 & 1 & 0 & -1 \\ 0 & \underline{1} & 1 & -1 & 0 & 2 \\ 0 & 0 & 0 & -2 & 1 & 3 \end{bmatrix}. \ \triangle$$

Proof of Theorem 2.3.2. 1. Suppose $[A|I]$ row reduces to $[I|B]$. This is a special case of Corollary 2.2.9; by that corollary, $\vec{\mathbf{b}}_i$ is the solution to $A\vec{\mathbf{x}}_i = \vec{\mathbf{e}}_i$ (i.e., $A\vec{\mathbf{b}}_i = \vec{\mathbf{e}}_i$). Thus $AB = I$ and B is a right inverse of A. Corollary 2.2.7 says that if A row reduces to the identity it is invertible, so by Proposition 1.2.14, B is also a left inverse, hence the inverse of A.

2. This follows from Corollary 2.2.7. $\square$

$$\begin{array}{c} i \\ \begin{bmatrix} 1 & 0 & \dots & 0 & \dots & 0 \\ 0 & 1 & \dots & 0 & \dots & 0 \\ \vdots & \vdots & \ddots & \vdots & & \vdots \\ 0 & 0 & \dots & x & \dots & 0 \\ \vdots & \vdots & & \vdots & \ddots & \vdots \\ 0 & 0 & \dots & 0 & \dots & 1 \end{bmatrix} \begin{matrix} \\ \\ \\ i \\ \\ \\ \end{matrix} \end{array}$$

Type 1: $E_1(i, x)$

Elementary matrices

The modern view of row reduction is that any row operation can be performed on a matrix by multiplying A *on the left* by an *elementary matrix*. Elementary matrices will simplify a number of arguments later in the book.

$$\begin{bmatrix}1&0&0&0\\0&1&0&0\\0&0&2&0\\0&0&0&1\end{bmatrix}$$

Example type 1: $E_1(3,2)$

There are three types of elementary matrices, *all square*, corresponding to the three kinds of row operations. They are defined in terms of the main diagonal, from top left to bottom right. We refer to them as "type 1", "type 2", and "type 3", but there is no standard numbering; we have listed them in the same order that we listed the corresponding row operations in Definition 2.1.1. They are obtained by performing row operations on the identity matrix.

$$\begin{array}{c@{}c} \begin{array}{ccccccc} & & i & & j & & \end{array} & \\ \left[\begin{array}{ccccccc} 1 & \cdots & 0 & \cdots & 0 & \cdots & 0\\ \vdots & \ddots & \vdots & & \vdots & & \vdots\\ 0 & \cdots & 1 & \cdots & x & \cdots & 0\\ \vdots & & \vdots & \ddots & \vdots & & \vdots\\ 0 & \cdots & 0 & \cdots & 1 & \cdots & 0\\ \vdots & & \vdots & & \vdots & \ddots & \vdots\\ 0 & \cdots & 0 & \cdots & 0 & \cdots & 1 \end{array}\right] & \begin{array}{c} \\ \\ i\\ \\ j\\ \\ \\ \end{array} \end{array}$$

Type 2: $E_2(i,j,x)$

The (i,j)th entry is x; the (j,i)th entry is 0.

Definition 2.3.5 (Elementary matrices).

1. The *type 1 elementary matrix* $E_1(i,x)$ is the square matrix where all nondiagonal entries are 0, and every entry on the diagonal is 1 except for the (i,i)th entry, which is $x \neq 0$.
2. The *type 2 elementary matrix* $E_2(i,j,x)$, for $i \neq j$, is the square matrix with all diagonal entries 1, and all other entries 0 except for the (i,j)th, which is x. Remember that the first index, i, refers to the row, and the second, j, refers to the column.
3. The *type 3 elementary matrix* $E_3(i,j)$, $i \neq j$, is the square matrix where the entries i,j and j,i are 1, as are all entries on the diagonal except i,i and j,j, which are 0. All others are 0.

$$\begin{bmatrix}1&0&-3\\0&1&0\\0&0&1\end{bmatrix}$$

Example type 2: $E_2(1,3,-3)$

You are asked in Exercise 2.3.10 to confirm that multiplying a matrix A on the left by an elementary matrix performs the following row operations:

- Multiplying A on the left by $E_1(i,x)$ multiplies the ith row of A by x.
- Multiplying A on the left by $E_2(i,j,x)$ adds (x times the jth row) to the ith row.
- Multiplying A on the left by $E_3(i,j)$ exchanges the ith and the jth rows of A.

$$\begin{array}{cc} & \overbrace{\begin{bmatrix}1&0\\0&2\\2&1\\1&1\end{bmatrix}}^{A}\\ \underbrace{\begin{bmatrix}1&0&0&0\\0&1&0&0\\0&0&2&0\\0&0&0&1\end{bmatrix}}_{E_1(3,2)} & \begin{bmatrix}1&0\\0&2\\4&2\\1&1\end{bmatrix} \end{array}$$

We can multiply by 2 the third row of the matrix A by multiplying it on the left by the type 1 elementary matrix $E_1(3,2)$

$$\underbrace{\begin{array}{c@{}c} \begin{array}{cccccccc} & & i & & j & & & \end{array} & \\ \left[\begin{array}{cccccccc} 1 & \cdots & 0 & \cdots & 0 & \cdots & \cdots & 0\\ \vdots & \ddots & \vdots & \vdots & \vdots & \vdots & \vdots & \vdots\\ 0 & \cdots & 0 & \cdots & 1 & \cdots & \cdots & 0\\ 0 & \cdots & 0 & 1 & 0 & \cdots & \cdots & 0\\ 0 & \cdots & 1 & \cdots & 0 & \cdots & \cdots & 0\\ 0 & \cdots & 0 & 0 & 0 & 1 & \cdots & 0\\ \vdots & \vdots & \vdots & \vdots & \vdots & \vdots & \ddots & \vdots\\ 0 & \cdots & 0 & \cdots & 0 & \cdots & \cdots & 1 \end{array}\right] & \begin{array}{c} \\ \\ i\\ \\ j\\ \\ \\ \\ \end{array} \end{array}}_{\text{Type 3: } E_3(i,j)} \qquad \underbrace{\begin{bmatrix}1&0&0&0&0\\0&0&1&0&0\\0&1&0&0&0\\0&0&0&1&0\\0&0&0&0&1\end{bmatrix}}_{\substack{\text{Example type 3}\\ E_3(2,3)}}$$

It follows from Exercise 2.3.10 and Corollary 2.2.7 that any invertible matrix is a product of invertible matrices: if A is invertible, there exist elementary matrices $E_1, \dots, E_k$ such that

$$I = E_k \cdots E_2 E_1 A, \quad \text{so that} \quad A^{-1} = E_k \cdots E_2 E_1. \qquad 2.3.5$$

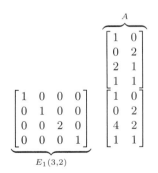

Since any row operation can be undone by another row operation (Exercise 2.1.5), elementary matrices are invertible, and their inverses are elementary matrices.

The proof of Proposition 2.3.6 is left to the reader.

Proposition 2.3.6 (Inverses of elementary matrices).

1. $\big(E_1(i,x)\big)^{-1} = E_1(i, \frac{1}{x})$
2. $\big(E_2(i,j,x)\big)^{-1} = E_2(i,j,-x)$
3. $(E_3(i,j))^{-1} = E_3(i,j)$.

Proposition 2.3.7 (Square matrix approximated by invertible matrices). *Any square matrix A can be approximated by a sequence of invertible matrices.*

Proof. Set $\widetilde{A} = E_k \cdots E_1 A$, with $\widetilde{A}$ upper triangular (i.e., row reduce A until it is either the identity or has rows of 0's at the bottom). If any diagonal entries of $\widetilde{A}$ are 0, change them to $1/n$; denote the resulting invertible matrix $\widetilde{A}_n$. Then the matrices

$$A_n \overset{\text{def}}{=} E_1^{-1} \cdots E_k^{-1} \widetilde{A}_n \qquad 2.3.6$$

are invertible, and converge to A as $n \to \infty$. □

Exercises for Section 2.3

2.3.1 Use the matrix A^{-1} of Example 2.3.3 to solve the system of Example 2.2.10.

$$A = \begin{bmatrix} 2 & 1 & 3 & a \\ 1 & -1 & 1 & b \\ 1 & 1 & 2 & c \end{bmatrix}$$

$$B = \begin{bmatrix} 2 & 1 & 3 \\ 1 & -1 & 1 \\ 1 & 1 & 2 \end{bmatrix}$$

Matrices for Exercise 2.3.4

2.3.2 Find the inverse, or show it does not exist, for each of the following matrices.

a. $\begin{bmatrix} 1 & -5 \\ 9 & 9 \end{bmatrix}$ b. $\begin{bmatrix} 1 & 3 \\ 3 & 9 \end{bmatrix}$ c. $\begin{bmatrix} 1 & 2 & 3 \\ 2 & 3 & 0 \\ 0 & 1 & 2 \end{bmatrix}$ d. $\begin{bmatrix} 1 & 2 \\ 0 & 3 \\ 1 & 0 \end{bmatrix}$

e. $\begin{bmatrix} 3 & 2 & -1 \\ 0 & 1 & 1 \\ 8 & 3 & 9 \end{bmatrix}$ f. $\begin{bmatrix} 1 & 0 & 1 \\ 2 & 1 & -1 \\ 1 & 1 & -1 \end{bmatrix}$ g. $\begin{bmatrix} 1 & 1 & 1 & 1 \\ 1 & 2 & 3 & 4 \\ 1 & 3 & 6 & 10 \\ 1 & 4 & 10 & 20 \end{bmatrix}$

2.3.3 Find matrices A, B where $AB = I$, but $BA \neq I$.

2.3.4 a. Symbolically row reduce the matrix A in the margin.

b. Compute the inverse of the matrix B in the margin.

c. What is the relation between the answers in parts a and b?

2.3.5 Working by hand, solve the system of linear equations

$$\begin{aligned} 3x - y + 3z &= 1 \\ 2x + y - 2z &= 1 \\ x + y + z &= 1 \end{aligned}$$

a. By row reduction

b. By computing and using a matrix inverse.

2.3.6 a. For what values of a and b is the matrix $C = \begin{bmatrix} 1 & -2 & 4 \\ 0 & 5 & -5 \\ 3 & a & b \end{bmatrix}$ invertible?

b. For those values, compute the inverse.

2.3.7 a. Let $A = \begin{bmatrix} 1 & -6 & 3 \\ 2 & -7 & 3 \\ 4 & -12 & 5 \end{bmatrix}$. Compute the matrix product AA.

b. Use your result to solve the system of equations

$$\begin{aligned} x \; - 6y + 3z &= 5 \\ 2x \; - 7y + 3z &= 7 \\ 4x - 12y + 5z &= 11. \end{aligned}$$

2.3.8 a. Predict the effect of multiplying the matrix $\begin{bmatrix} 1 & 0 & -1 \\ 2 & 1 & 1 \\ 0 & 1 & 2 \end{bmatrix}$ by each of the elementary matrices below, with the elementary matrix on the left.

$$\text{i. } \begin{bmatrix} 1 & 0 & 0 \\ 0 & 3 & 0 \\ 0 & 0 & 1 \end{bmatrix} \quad \text{ii. } \begin{bmatrix} 1 & 0 & 0 \\ 0 & 0 & 1 \\ 0 & 1 & 0 \end{bmatrix} \quad \text{iii. } \begin{bmatrix} 1 & 0 & 0 \\ 0 & 1 & 0 \\ 2 & 0 & 1 \end{bmatrix}$$

b. Confirm your answer by carrying out the multiplication.

c. Redo part a and part b placing the elementary matrix on the right.

$$\begin{bmatrix} 1 & 3 & -2 \\ 0 & 2 & 3 \\ 1 & 0 & 4 \end{bmatrix}$$

The matrix B of Exercise 2.3.9

2.3.9 a. Predict the product AB, where B is the matrix in the margin and A is given below.

$$\text{i. } A = \begin{bmatrix} 1 & 0 & -3 \\ 0 & 1 & 0 \\ 0 & 0 & 1 \end{bmatrix} \quad \text{ii. } A = \begin{bmatrix} 1 & 0 & 0 \\ 0 & 2 & 0 \\ 0 & 0 & 1 \end{bmatrix} \quad \text{iii. } A = \begin{bmatrix} 1 & 0 & 0 \\ 0 & 0 & 1 \\ 0 & 1 & 0 \end{bmatrix}$$

b. Verify your prediction by carrying out the multiplication.

Exercise 2.3.10: Remember to put the elementary matrix on the left of the matrix to be multiplied.

2.3.10 a. Confirm that multiplying a matrix by a type 2 elementary matrix as described in Definition 2.3.5 is equivalent to adding rows or multiples of rows.

b. Confirm that multiplying a matrix by a type 3 elementary matrix is equivalent to switching rows.

2.3.11 Show that column operations (see Exercise 2.1.6, part b) can be achieved by multiplication on the right by elementary matrices.

2.3.12 Prove Proposition 2.3.6.

2.3.13 Show that it is possible to switch rows using multiplication by only the first two types of elementary matrices, as described in Definition 2.3.5.

2.3.14 Row reduce the matrices in Exercise 2.1.3, using elementary matrices.

2.3.15 Prove Theorem 2.3.2 using elementary matrices.

2.4 Linear combinations, span, and linear independence

In 1750, questioning the general assumption that every system of n linear equations in n unknowns has a unique solution, the great mathematician Leonhard Euler pointed out the case of the two equations $3x - 2y = 5$ and $4y = 6x - 10$. "We will see that it is not possible to determine the two unknowns x and y," he wrote, "since when one is eliminated, the other disappears by itself, and we are left with an identity from which we can determine nothing. The reason for this accident is immediately obvious, since the second equation can be changed to $6x - 4y = 10$, which, being just the double of the first, is in no way different from it."

FIGURE 2.4.1.
The Swiss mathematician Leonhard Euler (1707–1783) touched on all aspects of the mathematics and physics of his time. His complete works fill 85 large volumes; some were written after he became completely blind in 1771. He was immensely influential both in his research and in teaching: all elementary calculus and algebra textbooks are in some sense rewrites of Euler's books.

Euler spent much of his professional life in St. Petersburg. He and his wife had thirteen children, five of whom survived to adulthood. According to the Mathematics Genealogy Project, he also had (as of May 1, 2015) 87 941 mathematical descendants.

Euler concluded that when claiming that n equations are sufficient to determine n unknowns, "one must add the restriction that all the equations be different from each other, and that none of them is included in the others." Euler's "descriptive and qualitative approach" represented the beginning of a new way of thinking. At the time, mathematicians were interested in solving individual systems of equations, not in analyzing them. Even Euler began his argument by pointing out that attempts to solve the system fail; only then did he explain this failure by the obvious fact that $3x - 2y = 5$ and $4y = 6x - 10$ are really the same equation.

Today, linear algebra provides a systematic approach to both analyzing and solving systems of linear equations, which was unknown in Euler's time. We have already seen something of its power. Writing a system of equations as a matrix and row reducing it to echelon form makes it easy to analyze: Theorem 2.2.1 tells us how to read the row-reduced matrix to find out whether the system has no solution, infinitely many solutions, or a unique solution (and, in the latter case, what it is).

Now we introduce vocabulary that describes concepts implicit in what we have done so far. *Linear combinations*, *span*, and *linear independence* give a precise way to answer the questions, given a collection of linear equations, how many genuinely different equations do we have? How many can be derived from the others?

These ideas apply in all linear settings, such as function spaces and integral and differential equations. Any time the notion of linear combination makes sense we can talk about span, kernels, linear independence, and so forth.

Definition 2.4.1 (Linear combination). If $\vec{\mathbf{v}}_1, \dots, \vec{\mathbf{v}}_k$ is a collection of vectors in $\mathbb{R}^n$, then a *linear combination* of the $\vec{\mathbf{v}}_i$ is a vector $\vec{\mathbf{w}}$ of the form

$$\vec{\mathbf{w}} = \sum_{i=1}^{k} a_i \vec{\mathbf{v}}_i \quad \text{for any scalars } a_i. \qquad 2.4.1$$

The vector $\begin{bmatrix} 3 \\ 4 \end{bmatrix}$ is a linear combination of $\vec{\mathbf{e}}_1$ and $\vec{\mathbf{e}}_2$:

$$\begin{bmatrix} 3 \\ 4 \end{bmatrix} = 3 \begin{bmatrix} 1 \\ 0 \end{bmatrix} + 4 \begin{bmatrix} 0 \\ 1 \end{bmatrix}.$$

The vectors $\vec{\mathbf{e}}_1, \vec{\mathbf{e}}_2 \in \mathbb{R}^2$ are linearly independent. There is only one way to write $\begin{bmatrix} 3 \\ 4 \end{bmatrix}$ in terms of $\vec{\mathbf{e}}_1$ and $\vec{\mathbf{e}}_2$:

$$3\begin{bmatrix} 1 \\ 0 \end{bmatrix} + 4\begin{bmatrix} 0 \\ 1 \end{bmatrix} = \begin{bmatrix} 3 \\ 4 \end{bmatrix}.$$

But the vectors $\begin{bmatrix} 1 \\ 0 \end{bmatrix}, \begin{bmatrix} 0 \\ 1 \end{bmatrix}, \begin{bmatrix} 3 \\ 2 \end{bmatrix}$ are not linearly independent since we can also write

$$\begin{bmatrix} 3 \\ 4 \end{bmatrix} = \begin{bmatrix} 3 \\ 2 \end{bmatrix} + 2\begin{bmatrix} 0 \\ 1 \end{bmatrix}.$$

If you drive a car that can only move parallel to the $\vec{\mathbf{v}}_i$ vectors, the span of the $\vec{\mathbf{v}}_i$ is the set of all accessible destinations.

Saying that the columns of $[T]$ are linearly independent is equivalent to saying that a linear transformation T is one to one.

Saying that the columns of $T : \mathbb{R}^n \to \mathbb{R}^m$ span $\mathbb{R}^m$ is equivalent to saying that T is onto; both statements mean that for every $\vec{\mathbf{b}} \in \mathbb{R}^m$, there is at least one solution to the equation $T(\mathbf{x}) = \vec{\mathbf{b}}$.

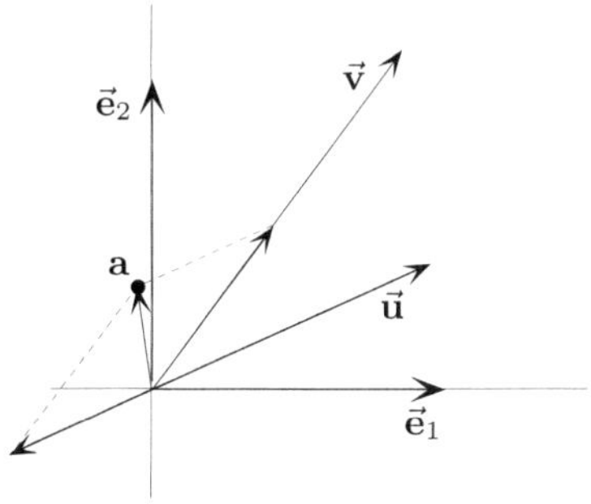

FIGURE 2.4.2.

The vectors $\vec{\mathbf{u}}$ and $\vec{\mathbf{v}}$ span the plane: any vector, such as $\vec{\mathbf{a}}$, can be expressed as the sum of components in the directions $\vec{\mathbf{u}}$ and $\vec{\mathbf{v}}$ (i.e., multiples of $\vec{\mathbf{u}}$ and $\vec{\mathbf{v}}$).

But $\begin{bmatrix} 3 \\ 4 \\ 1 \end{bmatrix}$ is not a linear combination of $\vec{\mathbf{e}}_1 = \begin{bmatrix} 1 \\ 0 \\ 0 \end{bmatrix}$ and $\vec{\mathbf{e}}_2 = \begin{bmatrix} 0 \\ 1 \\ 0 \end{bmatrix}$.

Linear independence and span

Linear independence is a way to talk about uniqueness of solutions to linear equations; *span* is a way to talk about the existence of solutions.

Definition 2.4.2 (Linear independence). A set of vectors $\vec{\mathbf{v}}_1, \dots, \vec{\mathbf{v}}_k$ is *linearly independent* if the *only* solution to

$$a_1\vec{\mathbf{v}}_1 + a_2\vec{\mathbf{v}}_2 + \cdots + a_k\vec{\mathbf{v}}_k = \vec{\mathbf{0}} \quad \text{is} \quad a_1 = a_2 = \cdots = a_k = 0. \qquad 2.4.2$$

An equivalent definition (as you are asked to confirm in Exercise 2.4.15) is that vectors $\vec{\mathbf{v}}_1, \dots, \vec{\mathbf{v}}_k \in \mathbb{R}^n$ are linearly independent if and only if a vector $\vec{\mathbf{w}} \in \mathbb{R}^n$ can be written as a linear combination of those vectors in at most one way:

$$\sum_{i=1}^{k} x_i\vec{\mathbf{v}}_i = \sum_{i=1}^{k} y_i\vec{\mathbf{v}}_i \quad \text{implies} \quad x_1 = y_1,\ x_2 = y_2, \dots, x_k = y_k. \qquad 2.4.3$$

Yet another equivalent statement (as you are asked to show in Exercise 2.8) is to say that $\vec{\mathbf{v}}_i$ are linearly independent if none of the $\vec{\mathbf{v}}_i$ is a linear combination of the others.

Definition 2.4.3 (Span). The *span* of $\vec{\mathbf{v}}_1, \dots, \vec{\mathbf{v}}_k$ is the set of linear combinations $a_1\vec{\mathbf{v}}_1 + \cdots + a_k\vec{\mathbf{v}}_k$. It is denoted $\operatorname{Span}(\vec{\mathbf{v}}_1, \dots, \vec{\mathbf{v}}_k)$.

The word "span" is also used as a verb. For instance, the standard basis vectors $\vec{\mathbf{e}}_1$ and $\vec{\mathbf{e}}_2$ span $\mathbb{R}^2$ but not $\mathbb{R}^3$. They span the plane, because any vector in the plane is a linear combination $a_1\,\vec{\mathbf{e}}_1 + a_2\,\vec{\mathbf{e}}_2$: any point in the (x, y)-plane can be written in terms of its x and y coordinates. The vectors $\vec{\mathbf{u}}$ and $\vec{\mathbf{v}}$ shown in Figure 2.4.2 also span the plane. (We used the word informally in this sense in Section 1.4 when we defined the "line spanned by $\vec{\mathbf{x}}$" as the line formed of all multiples of $\vec{\mathbf{x}}$.)

Exercise 2.4.5 asks you to show that $\operatorname{Span}(\vec{\mathbf{v}}_1, \dots, \vec{\mathbf{v}}_k)$ is a subspace of $\mathbb{R}^n$ and is the smallest subspace containing $\vec{\mathbf{v}}_1, \dots, \vec{\mathbf{v}}_k$. (Recall Definition 1.1.5 of a subspace.)

Examples 2.4.4 (Span: two easy cases). In simple cases it is possible to see immediately whether a given vector is in the span of a set of vectors.

1. Is the vector $\vec{\mathbf{u}} = \begin{bmatrix} 2 \\ 1 \\ 1 \end{bmatrix}$ in the span of $\vec{\mathbf{w}} = \begin{bmatrix} 2 \\ 0 \\ 1 \end{bmatrix}$? Clearly not; no multiple of 0 will give the 1 in the second position of $\vec{\mathbf{u}}$.

2. Given the vectors

$$\vec{\mathbf{v}}_1 = \begin{bmatrix} 1 \\ 0 \\ 0 \\ -1 \end{bmatrix}, \quad \vec{\mathbf{v}}_2 = \begin{bmatrix} -2 \\ -1 \\ 1 \\ 0 \end{bmatrix}, \quad \vec{\mathbf{v}}_3 = \begin{bmatrix} 1 \\ 1 \\ -1 \\ 1 \end{bmatrix}, \quad \vec{\mathbf{v}}_4 = \begin{bmatrix} 0 \\ 0 \\ 1 \\ 0 \end{bmatrix}, \qquad 2.4.4$$

is $\vec{\mathbf{v}}_4$ in Span $\{\vec{\mathbf{v}}_1, \vec{\mathbf{v}}_2, \vec{\mathbf{v}}_3\}$? Check your answer in the footnote below.[4] △

Theorem 2.4.5 translates Theorem 2.2.1 on solutions to linear equations into the language of linear independence and span.

Theorem 2.4.5 (Linear independence and span). *Let $\vec{\mathbf{v}}_1, \dots, \vec{\mathbf{v}}_k$ be vectors in $\mathbb{R}^n$; let A be the $n \times k$ matrix $[\vec{\mathbf{v}}_1, \dots \vec{\mathbf{v}}_k]$. Then*

1. *$\vec{\mathbf{v}}_1, \dots, \vec{\mathbf{v}}_k$ are linearly independent if and only if the row-reduced matrix $\widetilde{A}$ has a pivotal 1 in every column*
2. *$\vec{\mathbf{v}}_1, \dots, \vec{\mathbf{v}}_k$ span $\mathbb{R}^n$ if and only if $\widetilde{A}$ has a pivotal 1 in every row*

Proof. 1. The vectors $\vec{\mathbf{v}}_1, \dots, \vec{\mathbf{v}}_k$ are linearly independent if and only if the only solution to $A\vec{\mathbf{x}} = \vec{\mathbf{0}}$ is $\vec{\mathbf{x}} = \vec{\mathbf{0}}$. Thus part 1 follows immediately from part 2a of Theorem 2.2.1.

2. The vectors $\vec{\mathbf{v}}_1, \dots, \vec{\mathbf{v}}_k$ span $\mathbb{R}^n$ if and only if for any $\vec{\mathbf{b}} \in \mathbb{R}^n$, the equation $A\vec{\mathbf{x}} = \vec{\mathbf{b}}$ has a solution. The row reduction of $[A|\vec{\mathbf{b}}]$ is $[\widetilde{A}, \widetilde{\mathbf{b}}]$. There exists $\widetilde{\mathbf{b}}$ containing a pivotal 1 if and only if $\widetilde{A}$ contains a row of 0's, (i.e., if and only if $\widetilde{A}$ does not have a pivotal 1 in every row). Thus the equation $A\vec{\mathbf{x}} = \vec{\mathbf{b}}$ has a solution for every $\vec{\mathbf{b}}$ if and only if $\widetilde{A}$ has a pivotal 1 in every row. □

Linear independence is not restricted to vectors in $\mathbb{R}^n$: it also applies to functions and matrices (and more generally, to elements of arbitrary vector spaces, discussed in Section 2.6). For example, the matrices A, B and C are linearly independent if the only solution to

$$\alpha_1 A + \alpha_2 B + \alpha_3 C = [0] \quad \text{is}$$
$$\alpha_1 = \alpha_2 = \alpha_3 = 0.$$

(Recall that we denote by $[0]$ the matrix all of whose entries are 0.)

It is clear from Theorem 2.4.5 that n linearly independent vectors in $\mathbb{R}^n$ span $\mathbb{R}^n$: the matrix A formed using those vectors as columns row reduces to the identity, so there is a pivotal 1 in every column and every row.

Example 2.4.6 (Row reducing to check span). Given the vectors

$$\vec{\mathbf{w}}_1 = \begin{bmatrix} 2 \\ 1 \\ 1 \end{bmatrix}, \quad \vec{\mathbf{w}}_2 = \begin{bmatrix} 1 \\ -1 \\ 1 \end{bmatrix}, \quad \vec{\mathbf{w}}_3 = \begin{bmatrix} 3 \\ 0 \\ 2 \end{bmatrix}, \quad \vec{\mathbf{v}} = \begin{bmatrix} 3 \\ 3 \\ 1 \end{bmatrix}, \qquad 2.4.5$$

is $\vec{\mathbf{v}}$ in the span of the other three? We row reduce

[4] No. Since the second and third entries of $\vec{\mathbf{v}}_1$ are 0, if $\vec{\mathbf{v}}_4$ were in the span of $\{\vec{\mathbf{v}}_1, \vec{\mathbf{v}}_2, \vec{\mathbf{v}}_3\}$, its second and third entries would depend only on $\vec{\mathbf{v}}_2$ and $\vec{\mathbf{v}}_3$. To achieve the 0 of the second position, we must give equal weights to $\vec{\mathbf{v}}_2$ and $\vec{\mathbf{v}}_3$, but then we would also have 0 in the third position, whereas we need 1.

$$[A|\vec{\mathbf{v}}] = [\vec{\mathbf{w}}_1, \vec{\mathbf{w}}_2, \vec{\mathbf{w}}_3|\vec{\mathbf{v}}] = \begin{bmatrix} 2 & 1 & 3 & 3 \\ 1 & -1 & 0 & 3 \\ 1 & 1 & 2 & 1 \end{bmatrix} \quad \text{to} \quad \begin{bmatrix} 1 & 0 & 1 & 2 \\ 0 & 1 & 1 & -1 \\ 0 & 0 & 0 & 0 \end{bmatrix} = [\widetilde{A}|\widetilde{\mathbf{v}}]. \qquad 2.4.6$$

We used MATLAB to row reduce the matrix in equation 2.4.6, as we don't enjoy row reduction and tend to make mistakes.

Since $\widetilde{A}$ contains a row of 0's, Theorem 2.4.5 says that $\vec{\mathbf{w}}_1, \vec{\mathbf{w}}_2, \vec{\mathbf{w}}_3$ do not span $\mathbb{R}^3$. However, $\vec{\mathbf{v}}$ is still in the span of those vectors: $\widetilde{\mathbf{v}}$ contains no pivotal 1, so Theorem 2.2.1 says there is a solution to $A\vec{\mathbf{x}} = \vec{\mathbf{v}}$. But the solution is not unique: $\widetilde{A}$ has a column with no pivotal 1, so there are infinitely many ways to express $\vec{\mathbf{v}}$ as a linear combination of $\{\vec{\mathbf{w}}_1, \vec{\mathbf{w}}_2, \vec{\mathbf{w}}_3\}$. For example,

$$\vec{\mathbf{v}} = 2\vec{\mathbf{w}}_1 - \vec{\mathbf{w}}_2 = \mathbf{w}_1 - 2\vec{\mathbf{w}}_2 + \vec{\mathbf{w}}_3. \qquad \triangle$$

Example 2.4.6: If when you row reduce $[A|\vec{\mathbf{b}}]$ you see that $\widetilde{A}$ has a row of 0's, you cannot conclude that $\vec{\mathbf{b}}$ is not in the span of the vectors that make up A. You can only conclude that A is not onto its codomain: that there exist $\vec{\mathbf{b}}$ such that the equation $A\mathbf{x} = \vec{\mathbf{b}}$ has no solution. To determine whether there is a solution for some particular $\vec{\mathbf{b}}$, you must see whether $\widetilde{\mathbf{b}}$ has a pivotal 1. If it does, the equation has no solution.

Is $\vec{\mathbf{v}} = \begin{bmatrix} 0 \\ 1 \\ 1 \end{bmatrix}$ in the span of $\vec{\mathbf{w}}_1 = \begin{bmatrix} 1 \\ 0 \\ 1 \end{bmatrix}, \vec{\mathbf{w}}_2 = \begin{bmatrix} 2 \\ 1 \\ 1 \end{bmatrix}, \vec{\mathbf{w}}_3 = \begin{bmatrix} 1 \\ 3 \\ 0 \end{bmatrix}$?
Is $\vec{\mathbf{b}} = \begin{bmatrix} 0 \\ 1 \\ 1 \end{bmatrix}$ in the span of $\vec{\mathbf{a}}_1 = \begin{bmatrix} 2 \\ 2 \\ 0 \end{bmatrix}, \vec{\mathbf{a}}_2 = \begin{bmatrix} -2 \\ -1 \\ 2 \end{bmatrix}, \vec{\mathbf{a}}_3 = \begin{bmatrix} 1 \\ 1 \\ 0 \end{bmatrix}$? Check your answers in the footnote below.[5]

Example 2.4.7 (Linearly independent vectors). Are the vectors

$$\vec{\mathbf{w}}_1 = \begin{bmatrix} 1 \\ 2 \\ 3 \end{bmatrix}, \quad \vec{\mathbf{w}}_2 = \begin{bmatrix} -2 \\ 1 \\ 2 \end{bmatrix}, \quad \vec{\mathbf{w}}_3 = \begin{bmatrix} -1 \\ 1 \\ -1 \end{bmatrix} \qquad 2.4.7$$

linearly independent? Do they span $\mathbb{R}^3$? The matrix

$$\underset{\vec{\mathbf{w}}_1 \quad \vec{\mathbf{w}}_2 \quad \vec{\mathbf{w}}_3}{\begin{bmatrix} 1 & -2 & -1 \\ 2 & 1 & 1 \\ 3 & 2 & -1 \end{bmatrix}} \quad \text{row reduces to} \quad \begin{bmatrix} 1 & 0 & 0 \\ 0 & 1 & 0 \\ 0 & 0 & 1 \end{bmatrix}, \qquad 2.4.8$$

so by Theorem 2.4.5, part 1, the vectors are linearly independent. By part 2, these three vectors also span $\mathbb{R}^3$. $\triangle$

Vectors that are not linearly independent are *linearly dependent*.

Example 2.4.8. We can make the collection of vectors in Example 2.4.7 linearly dependent, by adding a vector that is a linear combination of some of them, say $\vec{\mathbf{w}}_4 = 2\vec{\mathbf{w}}_2 + \vec{\mathbf{w}}_3$:

$$\vec{\mathbf{w}}_1 = \begin{bmatrix} 1 \\ 2 \\ 3 \end{bmatrix}, \vec{\mathbf{w}}_2 = \begin{bmatrix} -2 \\ 1 \\ 2 \end{bmatrix}, \vec{\mathbf{w}}_3 = \begin{bmatrix} -1 \\ 1 \\ -1 \end{bmatrix}, \vec{\mathbf{w}}_4 = \begin{bmatrix} -5 \\ 3 \\ 3 \end{bmatrix}. \qquad 2.4.9$$

---

[5] Yes, $\vec{\mathbf{v}}$ is in the span of $\vec{\mathbf{w}}_1, \vec{\mathbf{w}}_2, \vec{\mathbf{w}}_3$, since $\begin{bmatrix} 1 & 2 & 1 \\ 0 & 1 & 3 \\ 1 & 1 & 0 \end{bmatrix} \mapsto \begin{bmatrix} 1 & 0 & 0 \\ 0 & 1 & 0 \\ 0 & 0 & 1 \end{bmatrix}$.

No, $\vec{\mathbf{b}}$ is not in the span of the others (as you might have suspected, since $\vec{\mathbf{a}}_1$ is a multiple of $\vec{\mathbf{a}}_3$). If we row reduce the appropriate matrix we get $\begin{bmatrix} 1 & 0 & 1/2 & 0 \\ 0 & 1 & 0 & 0 \\ 0 & 0 & 0 & 1 \end{bmatrix}$; the system of equations has no solution.

How many ways can we write some arbitrary vector[6] in $\mathbb{R}^3$, say $\vec{\mathbf{v}} = \begin{bmatrix} -7 \\ -2 \\ 1 \end{bmatrix}$,

as a linear combination of these four vectors? The matrix

$$\begin{bmatrix} 1 & -2 & -1 & -5 & -7 \\ 2 & 1 & 1 & 3 & -2 \\ 3 & 2 & -1 & 3 & 1 \end{bmatrix} \text{ row reduces to } \begin{bmatrix} 1 & 0 & 0 & 0 & -2 \\ 0 & 1 & 0 & 2 & 3 \\ 0 & 0 & 1 & 1 & -1 \end{bmatrix}. \qquad 2.4.10$$

Example 2.4.8: Saying that $\vec{\mathbf{v}}$ is in the span of the vectors

$$\vec{\mathbf{w}}_1, \vec{\mathbf{w}}_2, \vec{\mathbf{w}}_3, \vec{\mathbf{w}}_4$$

means that the system of equations

$$x_1\vec{\mathbf{w}}_1 + x_2\vec{\mathbf{w}}_2 + x_3\vec{\mathbf{w}}_3 + x_4\vec{\mathbf{w}}_4 = \vec{\mathbf{v}}$$

has a solution; the first matrix in formula 2.4.10 is the augmented matrix representing that system of equations. Since the four vectors $\vec{\mathbf{w}}_1, \vec{\mathbf{w}}_2, \vec{\mathbf{w}}_3, \vec{\mathbf{w}}_4$ are not linearly independent, the solution is not unique.

Since the fourth column is nonpivotal and the last column has no pivotal 1, the system has infinitely many solutions: there are infinitely many ways to write $\vec{\mathbf{v}}$ as a linear combination of the vectors $\vec{\mathbf{w}}_1, \vec{\mathbf{w}}_2, \vec{\mathbf{w}}_3, \vec{\mathbf{w}}_4$. The vector $\vec{\mathbf{v}}$ is in the span of those vectors, but they are not linearly independent. △

Example 2.4.9 (Geometric interpretation of linear independence).

1. One vector is linearly independent if it isn't the zero vector.
2. Two vectors are linearly independent if they do not both lie on the same line.
3. Three vectors are linearly independent if they do not all lie in the same plane.

Calling a single vector linearly independent may seem bizarre; the word "independence" seems to imply that there is something to be independent from. But the case of one vector is simply Definition 2.4.2 with $k = 1$; excluding that case from the definition would create all sorts of difficulties. △

These are not separate definitions; they are all examples of Definition 2.4.2. In Proposition and Definition 2.4.11, we will see that n vectors in $\mathbb{R}^n$ are linearly independent if they do not all lie in a subspace of dimension less than n.

The following theorem is basic to the entire theory.

Theorem 2.4.10. *In $\mathbb{R}^n$, $n+1$ vectors are never linearly independent, and $n-1$ vectors in $\mathbb{R}^n$ never span $\mathbb{R}^n$.*

Proof. This follows immediately from Theorem 2.4.5. When we use $n+1$ vectors in $\mathbb{R}^n$ to form a matrix A, the matrix is $n \times (n+1)$; when we row reduce A to $\widetilde{A}$, at least one column of $\widetilde{A}$ contains no pivotal 1, since there can be at most n pivotal 1's but the matrix has $n+1$ columns. So by part 1 of Theorem 2.4.5, the $n+1$ vectors are not linearly independent.

When we form a matrix A from $n-1$ vectors in $\mathbb{R}^n$, the matrix is $n \times (n-1)$. When we row reduce A, the resulting matrix must have at

---

[6] Actually, not quite arbitrary. Our first choice was $\begin{bmatrix} 1 \\ 1 \\ 1 \end{bmatrix}$, but that resulted in messy fractions, so we looked for a vector that gave a neater answer.

least one row with at all 0's, since we have n rows but at most $n-1$ pivotal 1's. So by part 2 of Theorem 2.4.5, $n-1$ vectors in $\mathbb{R}^n$ cannot span $\mathbb{R}^n$. □

A set of vectors as a basis

Choosing a basis for a subspace of $\mathbb{R}^n$, or for $\mathbb{R}^n$ itself, is like choosing axes (with units marked) in the plane or in space. The direction of basis vectors gives the direction of the axes; the length of the basis vectors provides units for those axes. This allows us to pass from noncoordinate geometry (synthetic geometry) to coordinate geometry (analytic geometry). Bases provide a "frame of reference" for vectors in a subspace. We will prove that the conditions below are equivalent after Proposition 2.4.16.

The zero vector space (whose only element is the vector $\vec{\mathbf{0}}$) has only one basis, the empty set.

Proposition and Definition 2.4.11 (Basis). *Let $V \subset \mathbb{R}^n$ be a subspace. An ordered set of vectors $\vec{\mathbf{v}}_1, \dots, \vec{\mathbf{v}}_k \in V$ is called a basis of V if it satisfies any of the three equivalent conditions.*

1. *The set is a maximal linearly independent set: it is linearly independent, and if you add one more vector, the set will no longer be linearly independent.*
2. *The set is a minimal spanning set: it spans V, and if you drop one vector, it will no longer span V.*
3. *The set is a linearly independent set spanning V.*

Example 2.4.12 (Standard basis). The fundamental example of a basis is the *standard basis* of $\mathbb{R}^n$; vectors in $\mathbb{R}^n$ are lists of numbers written with respect to the "standard basis" of standard basis vectors, $\vec{\mathbf{e}}_1, \dots, \vec{\mathbf{e}}_n$.

Clearly every vector in $\mathbb{R}^n$ is in the span of $\vec{\mathbf{e}}_1, \dots, \vec{\mathbf{e}}_n$:

$$\begin{bmatrix} a_1 \\ \vdots \\ a_n \end{bmatrix} = a_1\vec{\mathbf{e}}_1 + \cdots + a_n\vec{\mathbf{e}}_n. \qquad 2.4.11$$

It is equally clear that $\vec{\mathbf{e}}_1, \dots, \vec{\mathbf{e}}_n$ are linearly independent (Exercise 2.4.1).

Example 2.4.13 (Basis formed of n vectors in $\mathbb{R}^n$). The standard basis is not the only basis of $\mathbb{R}^n$. For instance, $\begin{bmatrix} 1 \\ 1 \end{bmatrix}, \begin{bmatrix} 1 \\ -1 \end{bmatrix}$ form a basis of $\mathbb{R}^2$, as do $\begin{bmatrix} 2 \\ 0 \end{bmatrix}, \begin{bmatrix} 0.5 \\ -3 \end{bmatrix}$, but not $\begin{bmatrix} 2 \\ 0 \end{bmatrix}, \begin{bmatrix} 0.5 \\ 0 \end{bmatrix}$. △

In general, if you randomly choose n vectors in $\mathbb{R}^n$, they will form a basis. In $\mathbb{R}^2$, the odds are against picking two vectors on the same line; in $\mathbb{R}^3$ the odds are against picking three vectors in the same plane.[7]

[7] In fact, the probability of picking at random two vectors on the same line or three vectors in the same plane is 0.

You might think that the standard basis should be enough. But there are times when a problem becomes much more straightforward in a different basis. We will see a striking example of this in Section 2.7, when we discuss Fibonacci numbers (Example 2.7.1); a simple case is illustrated by Figure 2.4.3. (Think also of decimals and fractions. It is a great deal simpler to write 1/7 than 0.142857142857..., yet at other times computing with decimals is easier.)

For a subspace $V \subset \mathbb{R}^n$ it is usually inefficient to describe vectors using all n numbers.

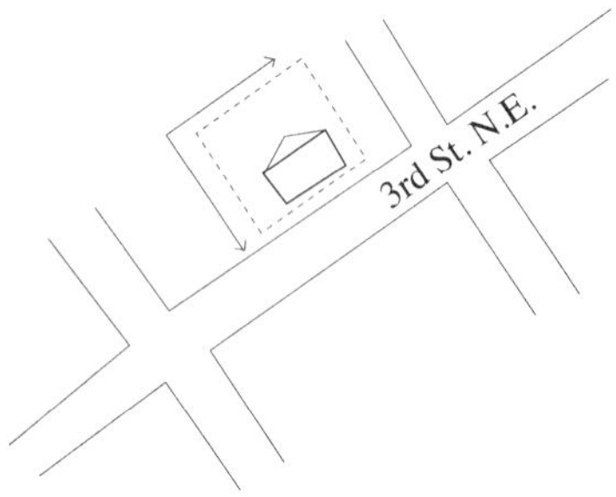

FIGURE 2.4.3.

The standard basis would not be convenient when surveying this yard. Use a basis suited to the job.

Example 2.4.14 (Using two basis vectors in a subspace of $\mathbb{R}^3$). Let $V \subset \mathbb{R}^3$ be the subspace of equation $x + y + z = 0$. Then rather than writing a vector by giving its three entries, we could write them using only two coefficients, a and b, and the vectors $\vec{\mathbf{w}}_1 = \begin{bmatrix} 1 \\ -1 \\ 0 \end{bmatrix}$ and $\vec{\mathbf{w}}_2 = \begin{bmatrix} 1 \\ 0 \\ -1 \end{bmatrix}$. For instance,

$$\vec{\mathbf{v}} = a\vec{\mathbf{w}}_1 + b\vec{\mathbf{w}}_2. \qquad 2.4.12$$

What other two vectors might you choose as a basis for V?[8]

Orthonormal bases

When doing geometry, it is almost always best to work with an *orthonormal basis*. Recall that two vectors are orthogonal if their dot product is zero (Corollary 1.4.80). The standard basis is of course orthonormal.

Definition 2.4.15 (Orthonormal set, orthonormal basis). A set of vectors $\vec{\mathbf{v}}_1, \dots, \vec{\mathbf{v}}_k$ is *orthonormal* if each vector in the set is orthogonal to every other vector in the set, and all vectors have length 1:

$$\vec{\mathbf{v}}_i \cdot \vec{\mathbf{v}}_j = 0 \quad \text{for } i \neq j \quad \text{and} \quad |\vec{\mathbf{v}}_i| = 1 \quad \text{for all } i \leq k.$$

An orthonormal set is an *orthonormal basis* of the subspace $V \subset \mathbb{R}^n$ that it spans.

If all vectors in a set (or basis) are orthogonal to each other, but they don't all have length 1, then the set (or basis) is *orthogonal*. Is either of the two bases of Example 2.4.13 orthogonal? orthonormal?[9]

---

[8]The vectors $\begin{bmatrix} -1 \\ 0 \\ 1 \end{bmatrix}, \begin{bmatrix} 0 \\ 1 \\ -1 \end{bmatrix}$ are a basis for V, as are $\begin{bmatrix} -1/2 \\ -1/2 \\ 1 \end{bmatrix}, \begin{bmatrix} 1 \\ -1 \\ 0 \end{bmatrix}$; the vectors just need to be linearly independent and the sum of the entries of each vector must be 0 (satisfying $x+y+z=0$). Part of the "structure" of the subspace V is thus built into the basis vectors.

[9]The first is orthogonal, since $\begin{bmatrix} 1 \\ 1 \end{bmatrix} \cdot \begin{bmatrix} 1 \\ -1 \end{bmatrix} = 1 - 1 = 0$; the second is not, since $\begin{bmatrix} 2 \\ 0 \end{bmatrix} \cdot \begin{bmatrix} 0.5 \\ -3 \end{bmatrix} = 1 + 0 = 1$. Neither is orthonormal; the vectors of the first basis each have length $\sqrt{2}$, and those of the second have lengths 2 and $\sqrt{9.25}$.

To check that orthogonal vectors form a basis we only need to show that they span, since by Proposition 2.4.16 orthogonal vectors are automatically linearly independent:

Proposition 2.4.16. *An orthogonal set of nonzero vectors $\vec{\mathbf{v}}_1, \dots, \vec{\mathbf{v}}_k$ is linearly independent.*

The surprising thing about Proposition 2.4.16 is that it allows us to assert that a set of vectors is linearly independent looking only at pairs of vectors. Of course, it is not true that if you have a set of vectors and every pair is linearly independent, the whole set is linearly independent; consider, for instance, the three vectors $\begin{bmatrix}1\\0\end{bmatrix}$, $\begin{bmatrix}0\\1\end{bmatrix}$ and $\begin{bmatrix}1\\1\end{bmatrix}$.

Proof. Suppose $a_1\vec{\mathbf{v}}_1 + \dots + a_k\vec{\mathbf{v}}_k = \vec{\mathbf{0}}$. Take the dot product of both sides with $\vec{\mathbf{v}}_i$:

$$(a_1\vec{\mathbf{v}}_1 + \dots + a_k\vec{\mathbf{v}}_k)\cdot\vec{\mathbf{v}}_i = \vec{\mathbf{0}}\cdot\vec{\mathbf{v}}_i = 0, \quad \text{so} \qquad 2.4.13$$

$$a_1(\vec{\mathbf{v}}_1\cdot\vec{\mathbf{v}}_i) + \dots + a_i(\vec{\mathbf{v}}_i\cdot\vec{\mathbf{v}}_i) + \dots + a_k(\vec{\mathbf{v}}_k\cdot\vec{\mathbf{v}}_i) = 0. \qquad 2.4.14$$

Since the $\vec{\mathbf{v}}_j$ form an orthogonal set, all the dot products on the left are zero except for the ith, so $a_i(\vec{\mathbf{v}}_i\cdot\vec{\mathbf{v}}_i) = 0$. Since the vectors are assumed to be nonzero, this says that $a_i = 0$. □

Proposition 2.4.16 is crucially important in infinite-dimensional spaces. In $\mathbb{R}^n$ we can determine whether vectors are linearly independent by row reducing a matrix and using Theorem 2.4.5. If n is big, the task may be long, but it can be done. In infinite-dimensional spaces, this is not an option. For some important classes of functions, Proposition 2.4.16 provides a solution.

A vector space is by definition infinite dimensional if no finite subset spans it. The notion of *basis* for infinite-dimensional spaces is a delicate issue.

Example 2.4.17. Consider the infinite-dimensional space $C[0,\pi]$ of continuous functions on $[0,\pi]$ with the "dot product"

$$\langle f, g\rangle \overset{\text{def}}{=} \int_0^\pi f(x)g(x)\,dx. \qquad 2.4.15$$

We invite the reader to show that this "dot product" has the properties

$$\begin{aligned} \langle af_1 + bf_2, g\rangle &= a\langle f_1, g\rangle + b\langle f_2, g\rangle \\ \langle f, g\rangle &= \langle g, f\rangle \\ \langle f, f\rangle &= 0 \quad \text{if and only if } f = 0. \end{aligned} \qquad 2.4.16$$

The proof of Proposition 2.4.16 used no other properties, so it is true in $C[0,\pi]$.

Example 2.4.17 is the foundation of Fourier series. You are asked in Exercise 2.4.14 to show that the functions $1 = \cos 0x$, $\cos x$, $\cos 2x, \dots$ form an orthogonal set, and to say when they are orthogonal to the functions $\sin nx$.

We claim that the elements $\sin nx, n = 1, 2, \dots$ of $C[0,\pi]$ form an orthogonal set: if $n \ne m$ we have

$$\begin{aligned} \int_0^\pi \sin nx \sin mx\,dx &= \frac{1}{2}\int_0^\pi \Big((\cos(n+m)x + \cos(n-m)x)\Big)\,dx \\ &= \frac{1}{2}\left(\left[\frac{\sin(n+m)x}{n+m}\right]_0^\pi + \left[\frac{\sin(n-m)x}{n-m}\right]_0^\pi\right) = 0 \end{aligned}$$

Therefore these functions are linearly independent: if the n_i are distinct,

$$a_1 \sin n_1 x + a_2 \sin n_2 x + \dots a_k \sin n_k x = 0 \implies a_1 = a_2 = \dots = a_k = 0.$$

It isn't clear how one could prove this directly. △

Proposition 2.4.18 (Properties of orthonormal bases). *Let $\vec{\mathbf{v}}_1, \dots, \vec{\mathbf{v}}_n$ be an orthonormal basis of $\mathbb{R}^n$.*

1. *Any vector $\vec{\mathbf{x}} \in \mathbb{R}^n$ can be written $\vec{\mathbf{x}} = \sum_{i=1}^{n} (\vec{\mathbf{x}} \cdot \vec{\mathbf{v}}_i)\vec{\mathbf{v}}_i$.*
2. *If $\vec{\mathbf{x}} = a_1\vec{\mathbf{v}}_1 + \dots + a_n\vec{\mathbf{v}}_n$, then $|\vec{\mathbf{x}}|^2 = a_1^2 + \dots + a_n^2$.*

Proposition 2.4.18: If $\vec{\mathbf{v}}_1, \dots, \vec{\mathbf{v}}_n$ is an orthonormal basis of $\mathbb{R}^n$, and

$$\vec{\mathbf{w}} = a_1\vec{\mathbf{v}}_1 + \dots + a_n\vec{\mathbf{v}}_n,$$

then

$$a_i = \vec{\mathbf{w}} \cdot \vec{\mathbf{v}}_i.$$

In the context of Fourier series, this simple statement becomes the formula

$$c_n = \int_0^1 f(t)e^{-2\pi int}\,dt$$

for computing Fourier coefficients.

The proof is left as Exercise 2.4.7.

Orthogonal matrices

A matrix whose columns form an orthonormal basis is called an *orthogonal matrix*. In $\mathbb{R}^2$, an orthogonal matrix is either a rotation or a reflection (Exercise 2.15). In higher dimensions, multiplication by orthogonal matrices is the analogue of rotations and reflections; see Exercise 4.8.23.

Proposition and Definition 2.4.19 (Orthogonal matrix). *An $n \times n$ matrix is called orthogonal if it satisfies one of the following equivalent conditions:*

1. *$AA^\top = A^\top A = I$, i.e., $A^\top = A^{-1}$*
2. *The columns of A form an orthonormal basis of $\mathbb{R}^n$*
3. *The matrix A preserves dot products: for any $\vec{\mathbf{v}}, \vec{\mathbf{w}} \in \mathbb{R}^n$, we have*

$$A\vec{\mathbf{v}} \cdot A\vec{\mathbf{w}} = \vec{\mathbf{v}} \cdot \vec{\mathbf{w}}.$$

It would make more sense to call an orthogonal matrix an *orthonormal* matrix.

It follows from part 3 that orthogonal matrices preserve lengths and angles. Conversely, a linear transformation that preserves lengths and angles is orthogonal, by the geometric description of the dot product (Proposition 1.4.3).

Proof. 1. We have seen (discussion after Corollary 2.2.7) that a left inverse of a square matrix is also a right inverse, so $A^\top A = I$ implies $AA^\top = I$.

2. Let $A = [\vec{\mathbf{v}}_1, \dots, \vec{\mathbf{v}}_n]$. Then

$$(A^\top A)_{i,j} = \vec{\mathbf{v}}_i \cdot \vec{\mathbf{v}}_j = \begin{cases} 1 & \text{if } i = j \\ 0 & \text{if } i \neq j \end{cases} \qquad 2.4.17$$

if and only if the $\vec{\mathbf{v}}_i$ form an orthonormal set.

3. If $A^\top A = I$, then for any $\vec{\mathbf{v}}, \vec{\mathbf{w}} \in \mathbb{R}^n$ we have

$$(A\vec{\mathbf{v}}) \cdot (A\vec{\mathbf{w}}) = (A\vec{\mathbf{v}})^\top (A\vec{\mathbf{w}}) = \vec{\mathbf{v}}^\top A^\top A\vec{\mathbf{w}} = \mathbf{v}^\top \vec{\mathbf{w}} = \vec{\mathbf{v}} \cdot \vec{\mathbf{w}}.$$

Conversely, if $A = [\vec{\mathbf{v}}_1, \dots, \vec{\mathbf{v}}_n]$ preserves dot products, then

$$\mathbf{v}_i \cdot \mathbf{v}_j = A\vec{\mathbf{e}}_i \cdot A\vec{\mathbf{e}}_j = \vec{\mathbf{e}}_i \cdot \vec{\mathbf{e}}_j = \begin{cases} 1 & \text{if } i = j \\ 0 & \text{if } i \neq j \end{cases} \qquad 2.4.18$$

Equivalence of the three conditions for a basis

We need to show that the three conditions for a basis given in Definition 2.4.11 are indeed equivalent.

We will show 1 $\implies$ 2: that if a set of vectors is a maximal linearly independent set, it is a minimal spanning set. Let $V \subset \mathbb{R}^n$ be a subspace. If an ordered set of vectors $\vec{\mathbf{v}}_1, \dots, \vec{\mathbf{v}}_k \in V$ is a maximal linearly independent set, then for any other vector $\vec{\mathbf{w}} \in V$, the set $\{\vec{\mathbf{v}}_1, \dots, \vec{\mathbf{v}}_k, \vec{\mathbf{w}}\}$ is linearly *dependent*, and (by Definition 2.4.2) there exists a nontrivial relation

By "nontrivial", we mean a solution other than

$$a_1 = a_2 = \cdots = a_n = b = 0.$$

$$a_1\vec{\mathbf{v}}_1 + \cdots + a_k\vec{\mathbf{v}}_k + b\vec{\mathbf{w}} = \vec{\mathbf{0}}. \qquad 2.4.19$$

The coefficient b is not zero, because if it were, the relation would then involve only $\vec{\mathbf{v}}_1, \dots, \vec{\mathbf{v}}_k$, which are linearly independent by hypothesis. Therefore, we can divide through by b, expressing $\vec{\mathbf{w}}$ as a linear combination of $\vec{\mathbf{v}}_1, \dots, \vec{\mathbf{v}}_k$:

$$\frac{a_1}{b}\vec{\mathbf{v}}_1 + \cdots + \frac{a_k}{b}\vec{\mathbf{v}}_k = -\vec{\mathbf{w}}. \qquad 2.4.20$$

Since $\vec{\mathbf{w}} \in V$ can be any vector in V, we see that the $\vec{\mathbf{v}}$'s do span V.

Moreover, $\vec{\mathbf{v}}_1, \dots, \vec{\mathbf{v}}_k$ is a *minimal* spanning set: if one of the $\vec{\mathbf{v}}_i$ is omitted, the set no longer spans, since the omitted vector is linearly independent of the others and hence cannot be in the span of the others.

The other implications are similar and left as Exercise 2.4.9. □

Now we can restate Theorem 2.4.10:

Corollary 2.4.20. *Every basis of $\mathbb{R}^n$ has exactly n elements.*

The notion of the dimension of a subspace will allow us to talk about such things as the size of the space of solutions to a set of equations or the number of genuinely different equations.

Proposition and Definition 2.4.21 (Dimension). *Every subspace $E \subset \mathbb{R}^n$ has a basis, and any two bases of E have the same number of elements, called the dimension of E, and denoted $\dim E$.*

Proof. First let us see that E has a basis. If $E = \{\vec{\mathbf{0}}\}$, the empty set is a basis of E. Otherwise, choose a sequence of vectors in E as follows: choose $\vec{\mathbf{v}}_1 \neq \vec{\mathbf{0}}$, then $\vec{\mathbf{v}}_2 \notin \text{Span}(\vec{\mathbf{v}}_1)$, then $\vec{\mathbf{v}}_3 \notin \text{Span}(\vec{\mathbf{v}}_1, \vec{\mathbf{v}}_2)$, etc. Vectors chosen this way are clearly linearly independent. Therefore, we can choose at most n such vectors, and for some $k \leq n$, $\vec{\mathbf{v}}_1, \dots, \vec{\mathbf{v}}_k$ span E. (If they don't span, we can choose another.) Since these vectors are linearly independent, they form a basis of E.

We can express any $\vec{\mathbf{w}}_j$ as a linear combination of the $\vec{\mathbf{v}}_i$ because the $\vec{\mathbf{v}}_i$ span E.

To see that any two bases have the same number of elements, suppose $\vec{\mathbf{v}}_1, \dots, \vec{\mathbf{v}}_m$ and $\vec{\mathbf{w}}_1, \dots, \vec{\mathbf{w}}_p$ are two bases of E. Then $\vec{\mathbf{w}}_j$ can be expressed as a linear combination of the $\vec{\mathbf{v}}_i$, so there exists an $m \times p$ matrix A with entries $a_{i,j}$ such that

$$\vec{\mathbf{w}}_j = \sum_{i=1}^{m} a_{i,j}\vec{\mathbf{v}}_i. \qquad 2.4.21$$

We can write this as the matrix multiplication $VA = W$, where V is the $n \times m$ matrix with columns $\mathbf{v}_i$ and W is the $n \times p$ matrix with columns $\mathbf{w}_j$:

$$\begin{array}{cc} & \begin{bmatrix} a_{1,1}, & \dots, & a_{1,p} \\ \vdots & \dots & \vdots \\ a_{m,1}, & \dots, & a_{m,p} \end{bmatrix} \\ [\vec{\mathbf{v}}_1, \dots, \vec{\mathbf{v}}_m] & [\vec{\mathbf{w}}_1, \ \dots, \ \vec{\mathbf{w}}_p] \end{array} \qquad 2.4.22$$

The matrices A and B are examples of the *change of basis matrix*, to be discussed in Proposition and Definition 2.6.17.

There also exists a $p \times m$ matrix B with entries $b_{l,i}$ such that

$$\vec{\mathbf{v}}_i = \sum_{l=1}^{p} b_{l,i} \vec{\mathbf{w}}_l. \qquad 2.4.23$$

We can write this as the matrix multiplication $WB = V$. So

$$VAB = WB = V \quad \text{and} \quad WBA = VA = W. \qquad 2.4.24$$

Equation 2.4.24: We cannot conclude from $VAB = V$ that $AB = I$ or from $WBA = W$ that $BA = I$. Consider the multiplication

$$[1\ 2]\begin{bmatrix} 3 & 1 \\ -1 & 1/2 \end{bmatrix} = [1\ 2].$$

Multiplying a matrix V on the right by AB produces a matrix whose columns are linear combinations of the columns of V.

The equation $VAB = V$ expresses each column of V as a linear combination of the columns of V. Since the columns of V are linearly independent, there is only one such expression:

$$\vec{\mathbf{v}}_i = 0\vec{\mathbf{v}}_1 + \dots + 1\vec{\mathbf{v}}_i + \dots + 0\vec{\mathbf{v}}_n. \qquad 2.4.25$$

Thus AB must be the identity. The equation $WBA = W$ expresses each column of W as a linear combination of the columns of W; by the same argument, BA is the identity. Thus A is invertible, hence square, so $m = p$. □

Corollary 2.4.22. *The only n-dimensional subspace of $\mathbb{R}^n$ is $\mathbb{R}^n$ itself.*

Remark. We said earlier that the terms "linear combinations", "span", and "linear independence" give a precise way to answer the questions, given a collection of linear equations, how many genuinely different equations do we have? We have seen that row reduction provides a systematic way to determine how many of the columns of a matrix are linearly independent. But the equations correspond to the *rows* of a matrix, not to its columns. In the next section we will see that the number of linearly independent equations in the system $A\vec{\mathbf{x}} = \vec{\mathbf{b}}$ is the same as the number of linearly independent columns of A.

Exercises for Section 2.4

$$\underbrace{\begin{bmatrix} 1 \\ 2 \\ 3 \end{bmatrix}}_{\vec{\mathbf{w}}_1}, \underbrace{\begin{bmatrix} -2 \\ 1 \\ 2 \end{bmatrix}}_{\vec{\mathbf{w}}_2}, \underbrace{\begin{bmatrix} -1 \\ 1 \\ -1 \end{bmatrix}}_{\vec{\mathbf{w}}_3}$$

Vectors for Exercise 2.4.2

2.4.1 Show that the standard basis vectors are linearly independent.

2.4.2 a. Do the vectors in the margin form a basis of $\mathbb{R}^3$? If so, is this basis orthogonal?

b. Is $\begin{bmatrix} 4 \\ 1 \\ 2 \end{bmatrix}$ in Span $\left(\begin{bmatrix} 4 \\ 2 \\ 1 \end{bmatrix}, \begin{bmatrix} 3 \\ 0 \\ 4 \end{bmatrix}, \begin{bmatrix} 2 \\ 1 \\ 4 \end{bmatrix}\right)$? In Span $\left(\begin{bmatrix} 4 \\ 2 \\ 1 \end{bmatrix}, \begin{bmatrix} 3 \\ 0 \\ 4 \end{bmatrix}, \begin{bmatrix} 5 \\ 1 \\ 4.5 \end{bmatrix}\right)$?

2.4.3 The vectors $\begin{bmatrix}1\\1\end{bmatrix}, \begin{bmatrix}1\\-1\end{bmatrix}$ form an orthogonal basis of $\mathbb{R}^2$. Use these vectors to create an orthonormal basis for $\mathbb{R}^2$.

$$\begin{bmatrix}1\\1\\0\end{bmatrix}, \begin{bmatrix}1\\2\\1\end{bmatrix}, \begin{bmatrix}0\\1\\\alpha\end{bmatrix}$$

Vectors for Exercise 2.4.4

2.4.4 a. For what values of α are the three vectors in the margin linearly dependent?

b. Show that for each such α the three vectors lie in the same plane, and give an equation of the plane.

2.4.5 Show that if $\vec{\mathbf{v}}_1, \dots, \vec{\mathbf{v}}_k$ are vectors in $\mathbb{R}^n$, then Span $(\vec{\mathbf{v}}_1, \dots, \vec{\mathbf{v}}_k)$ is a subspace of $\mathbb{R}^n$ and is the smallest subspace containing $\vec{\mathbf{v}}_1, \dots, \vec{\mathbf{v}}_k$.

2.4.6 Show that if $\vec{\mathbf{v}}$ and $\vec{\mathbf{w}}$ are two vectors of an orthonormal basis of $\mathbb{R}^3$, the third vector is $\pm\vec{\mathbf{v}} \times \vec{\mathbf{w}}$.

2.4.7 Prove Proposition 2.4.18.

2.4.8 Let A, B, C be three matrices such that $AB = C$. Show that if A is $n \times n$, and C is an $n \times m$ matrix with n linearly independent columns, then A is invertible.

2.4.9 Finish the proof that the three conditions in Definition 2.4.11 of a basis are equivalent.

2.4.10 Let $\vec{\mathbf{v}}_1 = \begin{bmatrix}1\\1\end{bmatrix}$ and $\vec{\mathbf{v}}_2 = \begin{bmatrix}1\\3\end{bmatrix}$. Let x and y be the coordinates with respect to the standard basis $\{\vec{\mathbf{e}}_1, \vec{\mathbf{e}}_2\}$ and let u and v be the coordinates with respect to $\{\vec{\mathbf{v}}_1, \vec{\mathbf{v}}_2\}$. Write the equations to translate from (x, y) to (u, v) and back. Use these equations to write the vector $\begin{bmatrix}3\\-5\end{bmatrix}$ in terms of $\vec{\mathbf{v}}_1$ and $\vec{\mathbf{v}}_2$.

Exercise 2.4.11: In this scheme, the "weights" $a_{i,n}$ are chosen once and for all; they are independent of f. Of course, you can't make the approximation exact, but you can make it exact for all polynomials up to degree n by choosing the $a_{i,n}$ appropriately.

2.4.11 Suppose we want to approximate

$$\int_0^1 f(x)\,dx \approx \sum_{i=0}^{n} a_{i,n} f\left(\frac{i}{n}\right). \qquad (1)$$

a. For $n = 1$, $n = 2$, $n = 3$, write the system of linear equations which the $a_{0,n}, \dots, a_{n,n}$ must satisfy so that formula (1) is an equality for each of the functions

$$f(x) = 1,\ f(x) = x,\ f(x) = x^2,\ \dots,\ f(x) = x^n.$$

b. Solve these equations, and use them to give three different approximations of $\int_0^1 \frac{dx}{x+1}$.

c. Use MATLAB or the equivalent to compute the numbers $a_{i,8}$. Is there anything very different from the earlier cases?

2.4.12 Let $A_t = \begin{bmatrix}2 & t\\0 & 2\end{bmatrix}$.

a. Are the elements I, A_t, A_t^2, A_t^3 linearly independent in Mat $(2,2)$? What is the dimension of the subspace $V_t \subset$ Mat $(2,2)$ which they span? (The answer depends on t.)

b. Show that the set W_t of matrices $B \in$ Mat $(2,2)$ that satisfy $A_t B = B A_t$ is a subspace of Mat $(2,2)$. What is its dimension? (Again, it depends on t.)

c. Show that $V_t \subset W_t$. For what values of t are they equal?

$$\begin{bmatrix}1 & a & a & a\\1 & 1 & a & a\\1 & 1 & 1 & a\\1 & 1 & 1 & 1\end{bmatrix}$$

Matrix A for Exercise 2.4.13

2.4.13 For what values of the parameter a is the matrix A in the margin invertible?

2.4.14 a. Show that the functions $1 = \cos 0x, \cos x, \cos 2x, \dots$ form an orthogonal set in $C[0, \pi]$.

b. When are vectors in this orthogonal set orthogonal to the vectors $\sin nx$ of Example 2.4.17?

2.4.15 Show that vectors $\vec{\mathbf{v}}_1, \dots, \vec{\mathbf{v}}_k \in \mathbb{R}^n$ are linearly independent if and only if a vector $\vec{\mathbf{w}} \in \mathbb{R}^n$ can be written as a linear combination of those vectors in at most one way.

2.5 Kernels, images, and the dimension formula

The kernel is sometimes called the "null space."

Definition 2.5.1, part 1, example: The vector $\begin{bmatrix} -2 \\ -1 \\ 3 \end{bmatrix}$ is in the kernel of $\begin{bmatrix} 1 & 1 & 1 \\ 2 & -1 & 1 \end{bmatrix}$, because

$$\begin{bmatrix} 1 & 1 & 1 \\ 2 & -1 & 1 \end{bmatrix} \begin{bmatrix} -2 \\ -1 \\ 3 \end{bmatrix} = \begin{bmatrix} 0 \\ 0 \end{bmatrix}.$$

The image of T is sometimes denoted $\operatorname{im} T$, but we use $\operatorname{img} T$ to avoid confusion with "the imaginary part," which is also denoted im. For a complex matrix, both "image" and "imaginary part" make sense.

Proposition 2.5.2 says that the kernel and the image are closed under addition and under multiplication by scalars; if you add two elements of the kernel you get an element of the kernel, and so on. You are asked to prove it in Exercise 2.5.5.

The kernel of T is the inverse image of $\{\mathbf{0}\}$ under T:

$$\ker T = T^{-1}\{\vec{\mathbf{0}}\}.$$

(The inverse image is defined in Definition 0.4.9.)

The words *kernel* and *image* give a geometric language in which to discuss existence and uniqueness of solutions to systems of linear equations. We will see that the kernel of a linear transformation is a subspace of its domain and the image is a subspace of its codomain. Their dimensions are related: the *dimension formula* is a kind of conservation law, saying that as the dimension of one gets bigger, the dimension of the other gets smaller. This establishes a relationship between the existence and the uniqueness of solutions to linear equations, which is the real power of linear algebra.

Definition 2.5.1 (Kernel and image). Let $T : \mathbb{R}^n \to \mathbb{R}^m$ be a linear transformation.

1. The *kernel* of T, denoted $\ker T$, is the set of vectors $\vec{\mathbf{x}} \in \mathbb{R}^n$ such that $T(\vec{\mathbf{x}}) = \vec{\mathbf{0}}$.
2. The *image* of T, denoted $\operatorname{img} T$, is the set of vectors $\vec{\mathbf{w}} \in \mathbb{R}^m$ such that there is a vector $\vec{\mathbf{v}} \in \mathbb{R}^n$ with $T(\vec{\mathbf{v}}) = \vec{\mathbf{w}}$.

This definition of image is the same as the set theoretic definition given in Section 0.3 (Definition 0.4.2).

Proposition 2.5.2. *If $T : \mathbb{R}^n \to \mathbb{R}^m$ is a linear transformation, then $\ker T$ is a vector subspace of $\mathbb{R}^n$ and $\operatorname{img} T$ is a vector subspace of $\mathbb{R}^m$.*

Proposition 2.5.3 relates existence and uniqueness of solutions of linear equations to the image and kernel: uniqueness is equivalent to kernel 0, and existence is equivalent to image everything.

Proposition 2.5.3. *Let $T : \mathbb{R}^n \to \mathbb{R}^m$ be a linear transformation. The system of linear equations $T(\vec{\mathbf{x}}) = \vec{\mathbf{b}}$ has*

1. *at most one solution for every $\vec{\mathbf{b}} \in \mathbb{R}^m$ if and only if $\ker T = \{\vec{\mathbf{0}}\}$.*
2. *at least one solution for every $\vec{\mathbf{b}} \in \mathbb{R}^m$ if and only if $\operatorname{img} T = \mathbb{R}^m$.*

Part 1 of Proposition 2.5.3 should become second nature; to check that a *linear* transformation is injective, just check that its kernel is $\{\vec{0}\}$. Nothing like this is true for nonlinear mappings, which don't have kernels (although they do of course have inverse images of $\mathbf{0}$). To check that a nonlinear function is injective, it is not enough to check that this inverse image consists of a single point.

Proof. 1. If the kernel of T is not $\{\vec{\mathbf{0}}\}$, then there is more than one solution to $T(\vec{\mathbf{x}}) = \vec{\mathbf{0}}$. (Of course, one solution is $\vec{\mathbf{x}} = \vec{\mathbf{0}}$).

In the other direction, if there exists a $\vec{\mathbf{b}}$ for which $T(\vec{\mathbf{x}}) = \vec{\mathbf{b}}$ has more than one solution, i.e.,

$$\begin{aligned} T(\vec{\mathbf{x}}_1) = T(\vec{\mathbf{x}}_2) = \vec{\mathbf{b}} \quad \text{and} \quad \vec{\mathbf{x}}_1 \neq \vec{\mathbf{x}}_2, \quad \text{then} \\ T(\vec{\mathbf{x}}_1 - \vec{\mathbf{x}}_2) = T(\vec{\mathbf{x}}_1) - T(\vec{\mathbf{x}}_2) = \vec{\mathbf{b}} - \vec{\mathbf{b}} = \vec{\mathbf{0}}. \end{aligned} \qquad 2.5.1$$

So $(\vec{\mathbf{x}}_1 - \vec{\mathbf{x}}_2)$ is a nonzero element of the kernel, so $\ker T \neq \{\vec{\mathbf{0}}\}$.

2. Saying that $\operatorname{img} T = \mathbb{R}^m$ is exactly saying that T is onto. □

Finding bases for the image and kernel

Let T be a linear transformation. If we row reduce the corresponding matrix $[T]$ to echelon form, we can find a basis for the image, using the following theorem. We prove it after giving some examples.

Theorem 2.5.4 (A basis for the image). *The pivotal columns of* $[T]$ *form a basis for* $\operatorname{img} T$.

Example 2.5.5 (Finding a basis for the image). Consider the matrix A below, which describes a linear transformation from $\mathbb{R}^5$ to $\mathbb{R}^4$:

$$A = \begin{bmatrix} 1 & 2 & 4 & -1 & 2 \\ -1 & 0 & -2 & -1 & 1 \\ 2 & 0 & 4 & 2 & 1 \\ 1 & 1 & 3 & 0 & 2 \end{bmatrix}, \quad \text{which row reduces to } \widetilde{A} = \begin{bmatrix} \underline{1} & 0 & 2 & 1 & 0 \\ 0 & \underline{1} & 1 & -1 & 0 \\ 0 & 0 & 0 & 0 & \underline{1} \\ 0 & 0 & 0 & 0 & 0 \end{bmatrix}.$$

Example 2.5.5: The basis we find using Theorem 2.5.4 is not the only basis for the image.

Note that while the pivotal columns of the original matrix A form a basis for the image, it is usually not the case that the columns of the row-reduced matrix $\widetilde{A}$ containing pivotal 1's form such a basis.

The pivotal 1's of the row-reduced matrix $\widetilde{A}$ are in columns 1, 2 and 5, so columns 1, 2, and 5 of the original matrix A are a basis for the image. We can express any vector in the image of A uniquely as a linear combination of those three vectors. For instance, $\vec{\mathbf{w}} = 2\vec{\mathbf{a}}_1 + \vec{\mathbf{a}}_2 - \vec{\mathbf{a}}_3 + 2\vec{\mathbf{a}}_4 - 3\vec{\mathbf{a}}_5$ can be written

$$\vec{\mathbf{w}} = \begin{bmatrix} -8 \\ -5 \\ 1 \\ -6 \end{bmatrix} = 2\begin{bmatrix} 1 \\ -1 \\ 2 \\ 1 \end{bmatrix} - 2\begin{bmatrix} 2 \\ 0 \\ 0 \\ 1 \end{bmatrix} - 3\begin{bmatrix} 2 \\ 1 \\ 1 \\ 2 \end{bmatrix}. \qquad 2.5.2$$

Note that each vector in the basis for the image has four entries, as it must, since the image is a subspace of $\mathbb{R}^4$. But the image is not of course $\mathbb{R}^4$; a basis for $\mathbb{R}^4$ must have four elements. △

A basis for the kernel

Finding a basis for the kernel is more complicated. A basis for the kernel is a set of vectors such that any vector $\vec{\mathbf{w}}$ satisfying $A\vec{\mathbf{w}} = \vec{\mathbf{0}}$ can be expressed as a linear combination of those basis vectors. The basis vectors must be in the kernel, and they must be linearly independent.

An equation $A\vec{\mathbf{x}} = \vec{\mathbf{0}}$ is called *a homogeneous system of linear equations.*

The dimension of the image of A is the number of pivotal columns, while the dimension of the kernel of A is the number of nonpivotal columns.

Theorem 2.2.1 says that if a system of linear equations has a solution, then it has a unique solution for any value you choose of the nonpivotal unknowns. Clearly $A\vec{\mathbf{w}} = \vec{\mathbf{0}}$ has a solution, namely $\vec{\mathbf{w}} = \vec{\mathbf{0}}$. So the tactic is to choose the values of the nonpivotal (active) unknowns in a convenient way. We take our inspiration from the standard basis vectors: each has one entry equal to 1, and the others 0. We construct a vector $\vec{\mathbf{v}}_i$ for each nonpivotal column, by setting the entry corresponding to that nonpivotal unknown to be 1, and the entries corresponding to the other nonpivotal unknowns to be 0. The entries corresponding to the pivotal (passive) unknowns will be whatever they have to be to satisfy the equation $A\vec{\mathbf{v}}_i = \vec{\mathbf{0}}$.

Theorem 2.5.6 (A basis for the kernel). *Let p be the number of nonpivotal columns of A, and $k_1, \dots, k_p$ their positions. For each nonpivotal column, form the vector $\vec{\mathbf{v}}_i$ satisfying $A\vec{\mathbf{v}}_i = \vec{\mathbf{0}}$ and such that its k_ith entry is 1, and its k_jth entries are all 0, for $j \neq i$. The vectors $\vec{\mathbf{v}}_1, \dots, \vec{\mathbf{v}}_p$ form a basis of ker A.*

The two vectors of equation 2.5.3 are clearly linearly independent; no "linear combination" of $\vec{\mathbf{v}}_1$ could produce the 1 in the fourth entry of $\vec{\mathbf{v}}_2$, and no "linear combination" of $\vec{\mathbf{v}}_2$ could produce the 1 in the third entry of $\vec{\mathbf{v}}_1$. Basis vectors found using the technique given in Theorem 2.5.6 will always be linearly independent, since for each entry corresponding to a nonpivotal unknown, one basis vector will have 1, and all the others will have 0.

Note that each vector in the basis for the kernel has five entries, as it must, since the domain of the transformation is $\mathbb{R}^5$.

Example 2.5.7 (Finding a basis for the kernel). The third and fourth columns of A in Example 2.5.5 are nonpivotal, so $k_1 = 3$ and $k_2 = 4$. The system has a unique solution for any values we choose of the third and fourth unknowns. In particular, there is a unique vector $\vec{\mathbf{v}}_1$ whose third entry is 1 and fourth entry is 0, such that $A\vec{\mathbf{v}}_1 = \vec{\mathbf{0}}$. There is another, $\vec{\mathbf{v}}_2$, whose fourth entry is 1 and third entry is 0, such that $A\vec{\mathbf{v}}_2 = \vec{\mathbf{0}}$:

$$\vec{\mathbf{v}}_1 = \begin{bmatrix} - \\ - \\ 1 \\ 0 \\ - \end{bmatrix}, \quad \vec{\mathbf{v}}_2 = \begin{bmatrix} - \\ - \\ 0 \\ 1 \\ - \end{bmatrix}. \qquad 2.5.3$$

The first, second, and fifth entries of $\vec{\mathbf{v}}_1$ and $\vec{\mathbf{v}}_2$ correspond to the pivotal unknowns. We read their values from the first three rows of $[\widetilde{A}\,|\,\widetilde{\mathbf{0}}]$ (remembering that a solution for $\widetilde{A}\mathbf{x} = \widetilde{\mathbf{0}}$ is also a solution for $A\mathbf{x} = \vec{\mathbf{0}}$):

$$[\widetilde{A}\,|\,\widetilde{\mathbf{0}}] = \begin{bmatrix} \underline{1} & 0 & 2 & 1 & 0 & 0 \\ 0 & \underline{1} & 1 & -1 & 0 & 0 \\ 0 & 0 & 0 & 0 & \underline{1} & 0 \\ 0 & 0 & 0 & 0 & 0 & 0 \end{bmatrix}, \quad \text{that is,} \quad \begin{aligned} x_1 + 2x_3 + x_4 &= 0 \\ x_2 + x_3 - x_4 &= 0 \\ x_5 &= 0, \end{aligned} \qquad 2.5.4$$

which gives

$$\begin{aligned} x_1 &= -2x_3 - x_4 \\ x_2 &= x_4 - x_3 \\ x_5 &= 0. \end{aligned} \qquad 2.5.5$$

Equations 2.5.6: The vector

$$\vec{\mathbf{w}} = \begin{bmatrix} -5 \\ -4 \\ 3 \\ -1 \\ 0 \end{bmatrix}$$

is in the kernel of A, since $A\vec{\mathbf{w}} = \vec{\mathbf{0}}$, so it should be possible to express $\vec{\mathbf{w}}$ as a linear combination of $\vec{\mathbf{v}}_1$ and $\vec{\mathbf{v}}_2$. Indeed it is: $\vec{\mathbf{w}} = 3\vec{\mathbf{v}}_1 - \vec{\mathbf{v}}_2$.

So for $\vec{\mathbf{v}}_1$, where $x_3 = 1$ and $x_4 = 0$, the first entry is $x_1 = -1$, the second is -1 and the fifth is 0; the corresponding entries for $\vec{\mathbf{v}}_2$ are -3, -2, and 0:

$$\vec{\mathbf{v}}_1 = \begin{bmatrix} -2 \\ -1 \\ 1 \\ 0 \\ 0 \end{bmatrix}; \quad \vec{\mathbf{v}}_2 = \begin{bmatrix} -1 \\ 1 \\ 0 \\ 1 \\ 0 \end{bmatrix}. \qquad 2.5.6$$

These two vectors form a basis of the kernel of A. $\triangle$

Now find a basis for the image and kernel of

$$A = \begin{bmatrix} 2 & 1 & 3 & 1 \\ 1 & -1 & 0 & 1 \\ 1 & 1 & 2 & 1 \end{bmatrix}, \text{ which row reduces to } \begin{bmatrix} \underline{1} & 0 & 1 & 0 \\ 0 & \underline{1} & 1 & 0 \\ 0 & 0 & 0 & \underline{1} \end{bmatrix}, \qquad 2.5.7$$

checking your answer in the footnote below.[10]

Proof of Theorem 2.5.4 (A basis for the image). Let $A = [\vec{\mathbf{a}}_1 \dots \vec{\mathbf{a}}_m]$.

1. *The pivotal columns of A (in fact, all columns of A) are in the image*, since $A\vec{\mathbf{e}}_i = \vec{\mathbf{a}}_i$.
2. *The pivotal columns are linearly independent*, by Theorem 2.4.5.
3. *The pivotal columns span the image,* since each nonpivotal column is a linear combination of the preceding pivotal ones. Suppose the kth column of A is nonpivotal. View the first k columns of A as an augmented matrix, i.e., try to express the kth column as a linear combination of the earlier ones. Row reduce the submatrix of A consisting of the first k columns, which is the same thing as considering the first k columns of $\widetilde{A}$. Since the kth column is nonpivotal, there is no pivotal 1 in the last column, so it is possible to express the kth column of A as a linear combination of the earlier ones, and in fact the entries of the kth column of $\widetilde{A}$ tell us how to express it as a linear combination of the earlier pivotal columns. $\square$

Proof of Theorem 2.5.6 (A basis for the kernel).

1. By definition, $A\vec{\mathbf{v}}_i = \vec{\mathbf{0}}$, so $\vec{\mathbf{v}}_i \in \ker A$.
2. *The $\vec{\mathbf{v}}_i$ are linearly independent*, since exactly one has a nonzero number in each position corresponding to nonpivotal unknown.
3. *The $\vec{\mathbf{v}}_i$ span the kernel* means that any $\vec{\mathbf{x}}$ satisfying $A\vec{\mathbf{x}} = \vec{\mathbf{0}}$ can be written as a linear combination of the $\vec{\mathbf{v}}_i$. Suppose $A\vec{\mathbf{x}} = \vec{\mathbf{0}}$. We can construct a vector $\vec{\mathbf{w}} = x_{k_1}\vec{\mathbf{v}}_1 + \dots + x_{k_p}\vec{\mathbf{v}}_p$ that has the same entry x_{k_i} in the nonpivotal column k_i as does $\vec{\mathbf{x}}$. Since $A\vec{\mathbf{v}}_i = \vec{\mathbf{0}}$, we have $A\vec{\mathbf{w}} = \vec{\mathbf{0}}$. But for each value of the nonpivotal variables, there is a unique vector $\vec{\mathbf{x}}$ such that $A\vec{\mathbf{x}} = \vec{\mathbf{0}}$. Therefore, $\vec{\mathbf{x}} = \vec{\mathbf{w}}$. $\square$

---

[10] The vectors $\begin{bmatrix} 2 \\ 1 \\ 1 \end{bmatrix}, \begin{bmatrix} 1 \\ -1 \\ 1 \end{bmatrix}, \begin{bmatrix} 1 \\ 1 \\ 1 \end{bmatrix}$ form a basis for the image; $\begin{bmatrix} -1 \\ -1 \\ 1 \\ 0 \end{bmatrix}$ is a basis for the kernel. The third column of A is nonpivotal, so for the vector of the basis of the kernel we set $x_3 = 1$. The row-reduced matrix $[\widetilde{A}|\widetilde{\mathbf{0}}]$ is $\begin{bmatrix} \underline{1} & 0 & 1 & 0 & 0 \\ 0 & \underline{1} & 1 & 0 & 0 \\ 0 & 0 & 0 & \underline{1} & 0 \end{bmatrix}$, i.e, $x_1 + x_3 = 0$, $x_2 + x_3 = 0$, and $x_4 = 0$. This gives $x_1 = -1$, $x_2 = -1$.

Uniqueness and existence: the dimension formula

Recall (Proposition and Definition 2.4.21) that the dimension of a subspace of $\mathbb{R}^n$ is the number of elements of any basis of the subspace. It is denoted dim.

The dimension of the kernel of a linear transformation is called its *nullity*, but we rarely use that word.

Theorems 2.5.4 and 2.5.6 tell us that if T is a linear transformation given by a matrix $[T]$, then the dimension of its image is the number of pivotal columns of $[T]$, while the dimension of its kernel is the number of nonpivotal columns. Since the total number of columns of $[T]$ is the dimension of the domain of T (a matrix n wide takes as input a vector in $\mathbb{R}^n$), the sum of the dimension of the image and the dimension of the kernel equals the dimension of the domain. This statement, known as the *dimension formula*, is deceptively simple; it provides much of the power of linear algebra. Because it is so important, we state it formally here.

Theorem 2.5.8 (Dimension formula). *Let $T: \mathbb{R}^n \to \mathbb{R}^m$ be a linear transformation. Then*

$$\dim(\ker T) + \dim(\operatorname{img} T) = n, \quad \textit{the dimension of the domain.} \qquad 2.5.8$$

The dimension formula says there is a conservation law concerning the kernel and the image. Since kernels are a way to talk about uniqueness of solutions to linear equations, and images are a way to talk about existence of solutions, this means that saying something about uniqueness says something about existence. We will state this precisely in the case of square matrices in Corollary 2.5.10.

The rank of a matrix (i.e., the number of linearly independent columns) is the most important number to associate to it.

The *power of linear algebra* comes from Corollary 2.5.10. We give examples concerning interpolation and partial fractions later in this section. See also Exercises 2.5.17 and 2.37, which deduce major mathematical results from this corollary.

Definition 2.5.9 (Rank). The dimension of the image of a linear transformation is called its *rank*.

If T is a linear transformation represented by a 3×4 matrix $[T]$ with rank 2, what is its domain and codomain? What is the dimension of its kernel? Is it onto? Check your answers in the footnote below.[11]

The most important case of the dimension formula is when the domain and codomain have the same dimension: then existence of solutions can be deduced from uniqueness. It is remarkable that knowing that $T(\vec{\mathbf{x}}) = \mathbf{0}$ has a unique solution guarantees existence of solutions for all $T(\vec{\mathbf{x}}) = \vec{\mathbf{b}}$!

This is, of course, an elaboration of Theorem 2.2.1 (and of Theorem 2.4.5). But those theorems depend on knowing a matrix. Corollary 2.5.10 can be applied when there is no matrix to write down, as we will see in Example 2.5.14 and Exercises 2.5.12, 2.5.17, and 2.37.

Corollary 2.5.10 (Deducing existence from uniqueness). *Let $T: \mathbb{R}^n \to \mathbb{R}^n$ be a linear transformation. Then the equation $T(\vec{\mathbf{x}}) = \vec{\mathbf{b}}$ has a solution for every $\vec{\mathbf{b}} \in \mathbb{R}^n$ if and only if the only solution to $T(\vec{\mathbf{x}}) = \vec{\mathbf{0}}$ is $\vec{\mathbf{x}} = \vec{\mathbf{0}}$, (i.e., if the kernel has dimension 0).*

Since Corollary 2.5.10 is an "if and only if" statement, it can also be used to deduce uniqueness from existence, when the domain and codomain of a transformation have the same dimension; in practice this is not quite so useful. It is often easier to prove that $T(\vec{\mathbf{x}}) = \vec{\mathbf{0}}$ has a unique solution than it is to construct a solution of $T(\vec{\mathbf{x}}) = \vec{\mathbf{b}}$.

[11] The domain of T is $\mathbb{R}^4$; the codomain is $\mathbb{R}^3$. The dimension of its kernel is 2, since here $\dim(\ker T) + \dim(\operatorname{img} T) = n$ becomes $\dim(\ker T) + 2 = 4$. The transformation is not onto, since a basis for $\mathbb{R}^3$ must have three basis elements.

Proof. Saying that $T(\vec{\mathbf{x}}) = \vec{\mathbf{b}}$ has a solution for every $\vec{\mathbf{b}} \in \mathbb{R}^n$ means that $\mathbb{R}^n$ is the image of T, so dim img$\,T = n$, which is equivalent to dim ker $(T) = 0$. □

The following result is really quite surprising.

Proposition 2.5.11: A matrix A and its transpose $A^\top$ have the same rank.

Proposition 2.5.11. *Let A be an $m \times n$ matrix. Then the number of linearly independent columns of A equals the number of linearly independent rows.*

One way to understand this result is to think of *constraints on the kernel of A*. Think of A as the $m \times n$ matrix made up of its rows:

$$A = \begin{bmatrix} --\ A_1\ -- \\ --\ A_2\ -- \\ \vdots \\ --\ A_m\ -- \end{bmatrix}. \qquad 2.5.9$$

Then the kernel of A is a subspace of $\mathbb{R}^n$; it is made up of the vectors $\vec{\mathbf{x}}$ satisfying the linear constraints $A_1\vec{\mathbf{x}} = 0, \dots, A_m\vec{\mathbf{x}} = 0$. Think of adding in these constraints one at a time. Each time you add one constraint (i.e., one row A_i), you cut down the dimension of the kernel by 1. But this is only true if the new constraint is genuinely new, not a consequence of the previous ones (i.e., if A_i is linearly independent from $A_1, \dots, A_{i-1}$).

Let us call the number of linearly independent rows A_i the *row rank* of A. The argument above leads to the formula

$$\dim\ker A = n - \text{row rank}(A). \qquad 2.5.10$$

The dimension formula says exactly that

$$\dim\ker A = n - \text{rank}(A), \qquad 2.5.11$$

so the rank of A and the row rank of A should be equal.

This argument isn't quite rigorous: it used the intuitively plausible but unjustified "Each time you add one constraint, you cut down the dimension of the kernel by 1." This is true and not hard to prove, but the following argument is shorter (and interesting too).

We defined linear combinations in terms of linear combinations of vectors, but (as we will see in Section 2.6) the same definition can apply to linear combinations of other objects, such as matrices, functions, etc. In this proof we are applying it to row matrices.

Proof. Call the span of the columns of a matrix A the *column space* of A and the span of the rows its *row space*. The rows of $\widetilde{A}$ are linear combinations of the rows of A, and vice versa since row operations are reversible, so if A row reduces to $\widetilde{A}$, the row space of A and of $\widetilde{A}$ coincide.

The rows of $\widetilde{A}$ that contain pivotal 1's are a basis of the row space of $\widetilde{A}$: the other rows are zero so they definitely don't contribute to the row space, and the pivotal rows of A are linearly independent, since all the other entries in a column containing a pivotal 1 are 0. So the dimension of the row space of A is the number of pivotal 1's of $\widetilde{A}$, which we have seen is the dimension of the column space of A. □

FIGURE 2.5.1.
Carl Friedrich Gauss (1777–1855) Often thought of as the greatest mathematician of all times, Gauss was also deeply involved in computations; he invented row reduction (also called *Gaussian elimination*), the Fast Fourier transform, Gaussian quadrature, and much more.

Remark. Proposition 2.5.11 gives us the statement we wanted in Section 2.4: the number of linearly independent equations in a system of linear equations $A\vec{\mathbf{x}} = \vec{\mathbf{b}}$ equals the number of pivotal columns of A. Basing linear algebra on row reduction can be seen as a return to Euler's way of thinking. It is, as Euler said, immediately apparent why you can't determine x and y from the two equations $3x - 2y = 5$ and $4y = 6x - 10$. (In the original, "La raison de cet accident saute d'abord aux yeux": *the reason for this accident leaps to the eyes.*)

When the linear dependence of a system of linear equations no longer *leaps to the eyes*, row reduction can make it obvious. Unfortunately for the history of mathematics, in the same year 1750 that Euler wrote his analysis, Gabriel Cramer published a treatment of linear equations based on determinants, which rapidly took hold, overshadowing the approach begun by Euler. Today, the rise of the computer, with emphasis on computationally effective schemes, has refocused attention on row reduction as an approach to linear algebra. △

We explore Cramer's formulas in Exercise 4.8.18.

In linear algebra there are many different ways of saying the same thing; we summarize some below in Table 2.5.2.

In group 1, statements 2 and 5 are just the definition of "onto", for any map. In group 2, statement 2 is just the definition of "one to one", for any map. The other equivalences depend on A being a *linear* transformation.

Statement 4 in group 1: If an $m \times n$ matrix spans $\mathbb{R}^m$ it has m linearly independent columns, so it follows from Proposition 2.5.11 that all m rows must be linearly independent.

Within each group, if one statement is true, all are; if one statement is false, all are. But if A is a square matrix, then if any statement in either group is true, then all 16 statements are true; if any statement is false, then all 16 statements are false.

Equivalent statements about a linear transformation $A : \mathbb{R}^n \to \mathbb{R}^m$

1. A is onto (also known as surjective).
2. The image of A is $\mathbb{R}^m$.
3. The columns of A span $\mathbb{R}^m$.
4. The rows of A are linearly independent.
5. For every $\mathbf{b} \in \mathbb{R}^m$ there exists a solution to the equation $A\vec{\mathbf{x}} = \vec{\mathbf{b}}$.
6. The dimension of img(A) (also known as the rank of A) is m.
7. The row-reduced matrix $\tilde{A}$ has no row containing all zeros.
8. The row-reduced matrix $\tilde{A}$ has a pivotal 1 in every row.

Equivalent statements about a linear transformation $A : \mathbb{R}^n \to \mathbb{R}^m$

1. A is one to one (also called injective).
2. If the equation $A\vec{\mathbf{x}} = \vec{\mathbf{b}}$ has a solution, it is unique.
3. The columns of A are linearly independent.
4. $\ker A = \{\vec{\mathbf{0}}\}$.
5. The only solution to the equation $A\vec{\mathbf{x}} = \vec{\mathbf{0}}$ is $\vec{\mathbf{x}} = \vec{\mathbf{0}}$.
6. The dimension of $\ker A$ is 0.
7. The row-reduced matrix $\tilde{A}$ has no nonpivotal column.
8. The row-reduced matrix $\tilde{A}$ has a pivotal 1 in every column.

TABLE 2.5.2.

Now let us see two examples of the power of Corollary 2.5.10 (the dimension formula in the case where the domain and codomain of T have the same dimension). One is related to signal processing, the other to partial fractions.

The problem of reproducing a polynomial from its samples long engaged mathematicians; around 1805 Gauss hit on the algorithm now known as the *fast Fourier transform* while trying to determine the position of an asteroid from sampled positions, when the orbits are assumed to be given by a trigonometric polynomial.

Interpolation and the dimension formula

In the mid-eighteenth century, the Italian Joseph Lagrange (1736–1813) developed what is now known as the *Lagrange interpolation formula*: an explicit formula for reconstituting a polynomial of degree at most k from $k+1$ samples (Figure 2.5.3 shows that this is not an effective way to approximate a function). Exercise 2.5.18 asks you to find Lagrange's formula. Here we will use the dimension formula to show that such a formula must exist.

Let P_k be the space of polynomials of degree $\leq k$. Given $k+1$ numbers $c_0, \dots, c_k$, can we find a polynomial $p \in P_k$ such that

$$\begin{aligned} p(0) &= c_0 \\ &\vdots \\ p(k) &= c_k\,? \end{aligned} \qquad 2.5.12$$

Interpolation is fundamental in the communications and computer industry. Usually one does not want to reconstitute a polynomial from sampled values. Rather one wants to find a *function* that fits a data set; or, in the other direction, to encode a function as n sampled values (for example, to store sampled values of music on a CD and then use these values to reproduce the original sound waves).

In 1885 Weierstrass proved that a continuous function on a finite interval of $\mathbb{R}$ can be uniformly approximated as closely as one wants by a polynomial. But if such a function is sampled n times for n even moderately big (more than 10, for example), Lagrange interpolation is a dreadful way to attempt such an approximation. *Bernstein polynomials* (introduced in 1911 by Sergei Bernstein to give a constructive proof of Weierstrass's theorem) are far superior; they are used when a computer draws Bézier curves.

Our tactic will be to consider the linear transformation that takes the polynomial p and returns the sampled values $p(0), \dots, p(k)$; we will then use the dimension formula to show that this transformation is invertible.

Consider the linear transformation $T_k : P_k \to \mathbb{R}^{k+1}$ given by

$$T_k(p) = \begin{bmatrix} p(0) \\ \vdots \\ p(k) \end{bmatrix}. \qquad 2.5.13$$

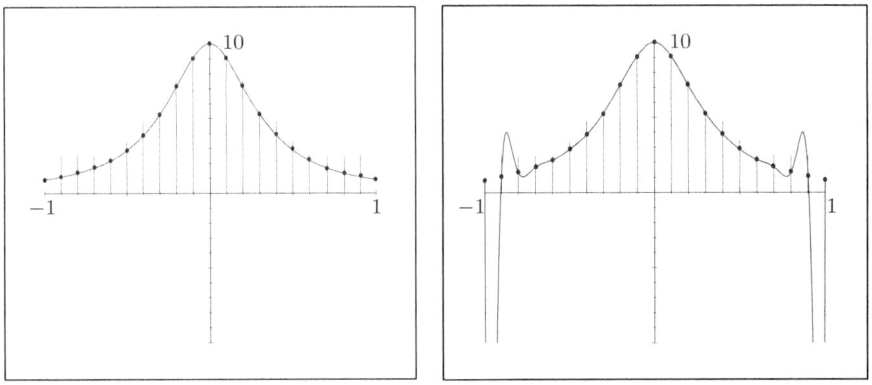

FIGURE 2.5.3. LEFT: The function $\dfrac{1}{x^2 + 1/10}$, between $x = -1$ and $x = 1$. RIGHT: The unique polynomial of degree 20 reconstituted from the values of the function at the 21 points $-1, -.9, \dots, .9, 1$ (the marked dots), using Lagrange interpolation. Near the boundary of the domain of interpolation, Lagrange interpolation is a terrible way to approximate a function.

The kernel of T_k is the space of polynomials $p \in P_k$ that vanish at $0, 1, \dots, k$. But a polynomial of degree $\le k$ cannot have $k+1$ roots, unless it is the zero polynomial. So the kernel of T_k is $\{0\}$. The dimension formula then tells us that its image is all of $\mathbb{R}^{k+1}$:

$$\dim \operatorname{img} T_k + \dim \ker T_k = \underbrace{\dim P_k}_{\text{dim. of domain of } T_k} = k+1. \qquad 2.5.14$$

In particular, T_k is invertible.

This gives the result we want: there is a transformation $T_k^{-1} : \mathbb{R}^{k+1} \to P_k$ such that for the sampled values $c_0, \dots, c_k$ we have

$$T_k^{-1} \begin{bmatrix} c_0 \\ \vdots \\ c_k \end{bmatrix} = \underbrace{a_0 + a_1 x + \cdots + a_k x^k}_{p(x)}, \qquad 2.5.15$$

where p is the unique polynomial with $p(0) = c_0$, $p(1) = c_1$, $\dots$, $p(k) = c_k$.

Since $\ker T_k = \{0\}$, the transformation is one to one; since the domain and codomain of T_k both are of dimension $k+1$, Corollary 2.5.10 says that since the kernel is zero, T_k is onto. Thus it is invertible.

Example 2.5.12 (Interpolation: polynomials of degree at most 2). The matrix of $T_2 : p \mapsto \begin{bmatrix} p(0) \\ p(1) \\ p(2) \end{bmatrix}$, where P_2 is identified to $\mathbb{R}^3$ by identifying $a + bx + cx^2$ to $\begin{pmatrix} a \\ b \\ c \end{pmatrix}$, is $T_2 = \begin{bmatrix} 1 & 0 & 0 \\ 1 & 1 & 1 \\ 1 & 2 & 4 \end{bmatrix}$; multiplying T_2 by $\begin{bmatrix} a \\ b \\ c \end{bmatrix}$ gives

$$\begin{bmatrix} 1 & 0 & 0 \\ 1 & 1 & 1 \\ 1 & 2 & 4 \end{bmatrix} \begin{bmatrix} a \\ b \\ c \end{bmatrix} = \begin{bmatrix} a \\ a+b+c \\ a+2b+4c \end{bmatrix}, \qquad 2.5.16$$

equivalent to the polynomial $p(x) = a + bx + cx^2$ evaluated at $x = 0$, $x = 1$, and $x = 2$. The inverse of T_2 is

$$T_2^{-1} = \begin{bmatrix} 1 & 0 & 0 \\ -3/2 & 2 & -1/2 \\ 1/2 & -1 & 1/2 \end{bmatrix}, \qquad 2.5.17$$

so if we choose (for instance) $c_0 = 1, c_1 = 4, c_2 = 9$, we get $1 + 2x + x^2$:

$$\begin{bmatrix} 1 & 0 & 0 \\ -3/2 & 2 & -1/2 \\ 1/2 & -1 & 1/2 \end{bmatrix} \begin{bmatrix} 1 \\ 4 \\ 9 \end{bmatrix} = \begin{bmatrix} 1 \\ 2 \\ 1 \end{bmatrix}, \qquad 2.5.18$$

which corresponds to the polynomial $p(x) = 1 + 2x + x^2$. Indeed,

$$\underbrace{1 + 2 \cdot 0 + 0^2}_{p(0)} = 1, \quad \underbrace{1 + 2 \cdot 1 + 1^2}_{p(1)} = 4, \quad \underbrace{1 + 2 \cdot 2 + 2^2}_{p(2)} = 9. \quad \triangle$$

Constructing T_2 in Example 2.5.12: The ith column of a matrix T is $T\vec{\mathbf{e}}_i$; here, the standard basis vectors $\vec{\mathbf{e}}_1, \vec{\mathbf{e}}_2, \vec{\mathbf{e}}_3$ are replaced by the polynomials 1, x, and x^2, since $\vec{\mathbf{e}}_1$ is identified to the polynomial $1+0x+0x^2$, $\vec{\mathbf{e}}_2$ is identified to $0+x+0x^2$, and $\vec{\mathbf{e}}_3$ is identified to $0+0+x^2$. Thus the first column of T_2 is

$$T_2(1) = \begin{bmatrix} 1 \\ 1 \\ 1 \end{bmatrix};$$

the constant polynomial 1 is always 1. The second is

$$T_2(x) = \begin{bmatrix} 0 \\ 1 \\ 2 \end{bmatrix},$$

since the polynomial x is 0 when $x = 0$, 1 when $x = 1$, and 2 when $x = 2$. For the third, the polynomial x^2 is 0 at 0, 1 at 1, and 4 at 2.

Exercise 2.5.10 asks you to show that there exist numbers $c_0, \dots, c_k$ such that

$$\int_0^k p(t)\, dt = \sum_{i=0}^k c_i p(i)$$

for all polynomials $p \in P_k$. Thus "sampling" (evaluating) a polynomial $p \in P_k$ at $k+1$ points, giving appropriate weights c_i to each sample, and adding the results, gives the same information as using an integral to sum the values of the polynomial at all points in a given interval.

Remark. Our discussion of interpolation deals with the special case where the $k+1$ numbers $c_0, \dots, c_k$ are the values of a polynomial evaluated at the specific numbers $0, 1, \dots, k$. The result is true when the polynomial is evaluated at *any* distinct $k+1$ points $x_0, \dots, x_k$: there is a unique polynomial such that $p(x_0) = c_0$, $p(x_1) = c_1$, $\dots, p(x_k) = c_k$. $\triangle$

Partial fractions

When d'Alembert proved the fundamental theorem of algebra, his motivation was to be able to decompose rational functions into partial fractions so that they could be explicitly integrated; that is why he called his paper *Recherches sur le calcul intégral* ("Enquiries into integral calculus"). You will presumably have run into partial fractions when studying methods of integration. Recall that decomposition into partial fractions is done by writing undetermined coefficients for the numerators; multiplying out then leads to a system of linear equations for these coefficients. Proposition 2.5.13 proves that such a system of equations always has a unique solution. As in the case of interpolation, the dimension formula, in its special case Corollary 2.5.10, will be essential.

The claim of partial fractions is the following:

In equation 2.5.19, we are requiring that the a_i be distinct. For example, although

$$p(x) = x(x-1)(x-1)(x-1)$$

can be written

$$p(x) = x(x-1)(x-1)^2,$$

this is not an allowable decomposition for use in Proposition 2.5.13, while $p(x) = x(x-1)^3$ is.

Proposition 2.5.13 (Partial fractions). *Let*

$$p(x) = (x-a_1)^{n_1} \cdots (x-a_k)^{n_k} \qquad 2.5.19$$

be a polynomial of degree $n = n_1 + \cdots + n_k$ with the a_i distinct and let q be any polynomial of degree $< n$. Then the rational function q/p can be written uniquely as a sum of simpler terms, called partial fractions:

$$\frac{q(x)}{p(x)} = \frac{q_1(x)}{(x-a_1)^{n_1}} + \cdots + \frac{q_k(x)}{(x-a_k)^{n_k}}, \qquad 2.5.20$$

with each q_i a polynomial of degree $< n_i$.

It follows from Proposition 2.5.13 that you can integrate all rational functions, i.e., ratios of two polynomials. You can integrate each term of the form

$$\frac{q(x)}{(x-a)^n}$$

with $\deg q < n$ in equation 2.5.20, as follows.

Substitute $x = (x-a) + a$ in q and multiply out to see that there exist numbers $b_{n-1}, \dots, b_0$ such that

$$q(x) = b_{n-1}(x-a)^{n-1} + \cdots + b_0.$$

Then

$$\frac{q(x)}{(x-a)^n} = \frac{b_{n-1}(x-a)^{n-1} + \cdots + b_0}{(x-a)^n} = \frac{b_{n-1}}{(x-a)} + \cdots + \frac{b_0}{(x-a)^n}.$$

If $n \neq 1$, then

$$\int \frac{dx}{(x-a)^n} = \frac{-1}{n-1}\frac{1}{(x-a)^{n-1}} + c;$$

if n=1, then

$$\int \frac{dx}{(x-a)} = \ln|x-a| + c$$

Note that by the fundamental theorem of algebra, every polynomial can be written as a product of powers of degree 1 polynomials with the a_i distinct, as in equation 2.5.19. For example,

$$x^2 - 1 = (x+1)(x-1), \text{ with } a_1 = -1, a_2 = 1; n_1 = n_2 = 1, \text{ so } n = 2$$

$$x^3 - 2x^2 + x = x(x-1)^2, \text{ with } a_1 = 0, a_2 = 1; n_1 = 1, n_2 = 2, \text{ so } n = 3.$$

Of course, finding the a_i means finding the roots of the polynomial, which may be very difficult.

Example 2.5.14 (Partial fractions). When

$$q(x) = 2x + 3 \quad \text{and} \quad p(x) = x^2 - 1, \qquad 2.5.21$$

Proposition 2.5.13 says that there exist polynomials q_1 and q_2 of degree less than 1 (i.e., numbers, which we will call A_0 and B_0, the subscript indicating that they are coefficients of the term of degree 0) such that

$$\frac{2x+3}{x^2-1} = \frac{A_0}{x+1} + \frac{B_0}{x-1}. \qquad 2.5.22$$

If $q(x) = x^3 - 1$ and $p(x) = (x+1)^2(x-1)^2$, then the proposition says that there exist two polynomials of degree 1, $q_1 = A_1 x + A_0$ and $q_2 = B_1 x + B_0$, such that

$$\frac{x^3-1}{(x+1)^2(x-1)^2} = \frac{A_1x+A_0}{(x+1)^2} + \frac{B_1x+B_0}{(x-1)^2}. \qquad 2.5.23$$

In simple cases, it's clear how to proceed. In equation 2.5.22, to find A_0 and B_0, we multiply out to get a common denominator:

$$\frac{2x+3}{x^2-1} = \frac{A_0}{x+1} + \frac{B_0}{x-1} = \frac{A_0(x-1)+B_0(x+1)}{x^2-1} = \frac{(A_0+B_0)x+(B_0-A_0)}{x^2-1},$$

so that we get two linear equations in two unknowns:

$$\begin{aligned} -A_0+B_0 &= 3 \\ A_0+B_0 &= 2, \end{aligned} \quad \text{that is, the constants} \quad B_0 = \frac{5}{2},\ A_0 = -\frac{1}{2}. \qquad 2.5.24$$

We can think of the system of linear equations on the left side of equation 2.5.24 as the matrix multiplication

$$\begin{bmatrix} -1 & 1 \\ 1 & 1 \end{bmatrix} \begin{bmatrix} A_0 \\ B_0 \end{bmatrix} = \begin{bmatrix} 3 \\ 2 \end{bmatrix}. \qquad 2.5.25$$

What is the analogous matrix multiplication for equation 2.5.23?[12] △

What about the general case? If we put the right side of equation 2.5.20 on a common denominator, we see that $q(x)/p(x)$ is equal to

$$\frac{q_1(x)(x-a_2)^{n_2}\cdots(x-a_k)^{n_k}+q_2(x)(x-a_1)^{n_1}(x-a_3)^{n_3}\cdots(x-a_k)^{n_k}+\cdots+q_k(x)(x-a_1)^{n_1}\cdots(x-a_{k-1})^{n_{k-1}}}{(x-a_1)^{n_1}(x-a_2)^{n_2}\cdots(x-a_k)^{n_k}}. \qquad 2.5.26$$

As we did in our simpler cases, we could write this as a system of linear equations for the coefficients of the q_i and solve by row reduction. But except in the simplest cases, computing the matrix would be a big job. Worse, how do we know that the resulting system of equations *has* solutions? Might we invest a lot of work only to discover that the equations are inconsistent?

---

[12] Multiplying out, we get

$$\frac{x^3(A_1+B_1)+x^2(-2A_1+A_0+2B_1+B_0)+x(A_1-2A_0+B_1+2B_0)+A_0+B_0}{(x+1)^2(x-1)^2},$$

so

$$\begin{aligned} A_0+B_0 &= -1 && \text{(coefficient of term of degree 0)} \\ A_1-2A_0+B_1+2B_0 &= 0 && \text{(coefficient of term of degree 1)} \\ -2A_1+A_0+2B_1+B_0 &= 0 && \text{(coefficient of term of degree 2)} \\ A_1+B_1 &= 1 && \text{(coefficient of term of degree 3);} \end{aligned}$$

that is,

$$\begin{bmatrix} 0 & 1 & 0 & 1 \\ 1 & -2 & 1 & 2 \\ -2 & 1 & 2 & 1 \\ 1 & 0 & 1 & 0 \end{bmatrix} \begin{bmatrix} A_1 \\ A_0 \\ B_1 \\ B_0 \end{bmatrix} = \begin{bmatrix} -1 \\ 0 \\ 0 \\ 1 \end{bmatrix}.$$

Proposition 2.5.13 assures us that there will always be a solution, and Corollary 2.5.10 provides the key.

Corollary 2.5.10: If

$$T : \mathbb{R}^n \to \mathbb{R}^n$$

is a linear transformation, the equation $T(\vec{\mathbf{x}}) = \vec{\mathbf{b}}$ has a solution for any $\vec{\mathbf{b}} \in \mathbb{R}^n$ if and only if the only solution to the equation $T(\vec{\mathbf{x}}) = \vec{\mathbf{0}}$ is $\vec{\mathbf{x}} = \vec{\mathbf{0}}$.

We are thinking of the transformation T both as the matrix that takes the coefficients of the q_i and returns the coefficients of q, and as the linear function that takes $q_i, \dots, q_k$ and returns the polynomial q.

Proof of Proposition 2.5.13 (Partial fractions). The matrix we would get following the above procedure would necessarily be $n \times n$. This matrix gives a linear transformation that has as its input a vector whose entries are the coefficients of $q_1, \dots, q_k$. There are n such coefficients in all: each polynomial q_i has degree $< n_i$, so it is specified by its coefficients for terms of degree 0 through $(n_i - 1)$ and the sum of the n_i equals n. (Note that some coefficients may be 0, for instance if q_j has degree $< n_j - 1$.) It has as its output a vector giving the n coefficients of q (since q is of degree $< n$, it is specified by its coefficients for the terms of degree $0, \dots, n-1$.)

Thus we can think of the matrix as a linear transformation $T : \mathbb{R}^n \to \mathbb{R}^n$, and by Corollary 2.5.10, Proposition 2.5.13 is true *if and only if the only solution of* $T(q_1, \dots, q_k) = \mathbf{0}$ *is* $q_1 = \dots = q_k = 0$. This will follow from Lemma 2.5.15.

Lemma 2.5.15. *If $q_i \neq 0$ is a polynomial of degree $< n_i$, then*

$$\lim_{x \to a_i} \left| \frac{q_i(x)}{(x - a_i)^{n_i}} \right| = \infty. \qquad 2.5.27$$

That is, if $q_i \neq 0$, then $q_i(x)/(x - a_i)^{n_i}$ blows up to infinity.

Proof of Lemma 2.5.15. For values of x very close to a_i, the denominator $(x - a_i)^{n_i}$ gets very small; if all goes well the entire term then gets very big. But we have to make sure that the numerator does not get small equally fast. Let us make the change of variables $u = x - a_i$, so that

$$q_i(x) = q_i(u + a_i), \quad \text{which we will denote} \quad \widetilde{q}_i(u). \qquad 2.5.28$$

Then we have

$$\lim_{u \to 0} \left| \frac{\widetilde{q}_i(u)}{u^{n_i}} \right| = \infty \quad \text{if } q_i \neq 0. \qquad 2.5.29$$

The numerator in equation 2.5.29 is of degree $< n_i$, while the denominator is of degree n_i.

Indeed, if $q_i \neq 0$, then $\widetilde{q}_i \neq 0$, and there exists a number $m < n_i$ such that

$$\widetilde{q}_i(u) = a_m u^m + \dots + a_{n_i - 1} u^{n_i - 1} \qquad 2.5.30$$

with $a_m \neq 0$. (This a_m is the first nonzero coefficient; as $u \to 0$, the term $a_m u^m$ is bigger than all the other terms.) Dividing by u^{n_i}, we can write

$$\frac{\widetilde{q}_i(u)}{u^{n_i}} = \frac{1}{u^{n_i - m}} (a_m + \cdots), \qquad 2.5.31$$

where the dots ... represent terms containing u to a positive power, since $m < n_i$. In particular,

$$\text{as} \quad u \to 0, \quad \left| \frac{1}{u^{n_i - m}} \right| \to \infty \quad \text{and } (a_m + \cdots) \to a_m. \qquad 2.5.32$$

Thus as $x \to a_i$, the term $q_i(x)/(x - a_i)^{n_i}$ blows up to infinity: the denominator gets smaller and smaller while the numerator tends to $a_m \neq 0$.

□ Lemma 2.5.15

We also have to make sure that as x tends to a_i and $(x-a_i)^{n_i}$ gets small, the other terms in equation 2.5.20 don't compensate. Suppose $q_i \neq 0$. For all the other terms $q_j, j \neq i$, the rational functions

$$\frac{q_j(x)}{(x-a_j)^{n_j}}, \tag{2.5.33}$$

The proof of Proposition 2.5.13 really put linear algebra to work. Even after translating the problem into linear algebra, via the linear transformation T, the answer was not clear, but the dimension formula made the result apparent.

have the finite limits $q_j(a_i)/(a_i - a_j)^{n_j}$ as $x \to a_i$, and therefore the sum

$$\frac{q(x)}{p(x)} = \frac{q_1(x)}{(x-a_1)^{n_1}} + \cdots + \frac{q_k(x)}{(x-a_k)^{n_k}} \tag{2.5.34}$$

which includes the term $\frac{q_i(x)}{(x-a_i)^{n_i}}$ with infinite limit as $x \to a_i$, has infinite limit as $x \to a_i$ and q cannot vanish identically. Thus $T(q_1, \dots, q_k) \neq \mathbf{0}$ if some q_i is nonzero, and we can conclude – without computing matrices or solving systems of equations – that Proposition 2.5.13 is correct: for any polynomial p of degree n and q of degree $< n$, the rational function q/p can be written uniquely as a sum of partial fractions. □

Table 2.5.4 summarizes four different ways to think about and speak of linear transformations.

Understanding a linear transformation $T : \mathbb{R}^n \to \mathbb{R}^m$			
Property of T	Solutions to $T(\vec{\mathbf{x}}) = \vec{\mathbf{b}}$	Matrix of T	Geometry
One to one (injective)	Unique if exist	All columns linearly independent	$\ker T = \{\vec{\mathbf{0}}\}$
Onto (surjective)	Exist for every $\vec{\mathbf{b}} \in \mathbb{R}^m$	All rows linearly independent Columns span $\mathbb{R}^m$	image = codomain
Invertible (1–1 and onto)	All exist and are unique	T is square ($n = m$) and all columns are linearly independent	$\ker T = \{\vec{\mathbf{0}}\}$ and image = codomain

TABLE 2.5.4. Row reduction applied to the matrix of T is the key to understanding the properties of T and solutions to the set of linear equations $T(\vec{\mathbf{x}}) = \vec{\mathbf{b}}$.

$$\underbrace{\begin{bmatrix} 0 \\ 2 \\ 1 \\ -1 \end{bmatrix}}_{\vec{\mathbf{v}}_1}, \underbrace{\begin{bmatrix} 1 \\ 0 \\ 1 \\ 0 \end{bmatrix}}_{\vec{\mathbf{v}}_2}, \underbrace{\begin{bmatrix} 0 \\ 4 \\ 2 \\ -2 \end{bmatrix}}_{\vec{\mathbf{v}}_3}$$

Vectors for Exercise 2.5.1, part a

EXERCISES FOR SECTION 2.5

2.5.1 a. For the matrix $A = \begin{bmatrix} 1 & 0 & 1 & 1 \\ 2 & 1 & 1 & 3 \\ 1 & 0 & 2 & 2 \end{bmatrix}$, which of the vectors $\vec{\mathbf{v}}_1, \vec{\mathbf{v}}_2, \vec{\mathbf{v}}_3$ in the margin (if any) are in the kernel of A?

b. Which vectors have the right height to be in the kernel of T? To be in its image? Can you find a nonzero element of its kernel?

$$T = \begin{bmatrix} 2 & -1 & 3 & 2 & 1 \\ 1 & 0 & 1 & 3 & 0 \\ 2 & -1 & 1 & 0 & 1 \end{bmatrix}, \quad \vec{\mathbf{w}}_1 = \begin{bmatrix} 1 \\ 2 \\ 3 \end{bmatrix}, \quad \vec{\mathbf{w}}_2 = \begin{bmatrix} 0 \\ 1 \\ 1 \\ 2 \end{bmatrix}, \quad \vec{\mathbf{w}}_3 = \begin{bmatrix} 1 \\ 0 \\ 1 \end{bmatrix}, \quad \vec{\mathbf{w}}_4 = \begin{bmatrix} 2 \\ 1 \\ 2 \\ 0 \\ 0 \end{bmatrix}$$

2.5.2 Let the linear transformation $T : \mathbb{R}^n \to \mathbb{R}^m$ be onto. Are the following statements true or false?

a. The columns of T span $\mathbb{R}^n$.

b. T has rank m.

c. For any vector $\vec{\mathbf{v}} \in \mathbb{R}^m$, there exists a solution to $T\vec{\mathbf{x}} = \vec{\mathbf{v}}$.

d. For any vector $\vec{\mathbf{v}} \in \mathbb{R}^n$, there exists a solution to $T\vec{\mathbf{x}} = \vec{\mathbf{v}}$.

e. T has nullity n.

f. The kernel of T is $\{\vec{\mathbf{0}}\}$.

g. For any vector $\vec{\mathbf{v}} \in \mathbb{R}^m$, there exists a unique solution to $T\vec{\mathbf{x}} = \vec{\mathbf{v}}$.

2.5.3 Let T be a linear transformation. Connect each item in the column at left with all synonymous items in the column at right:

	dim ker T
	dim domain T
rank T	dim codomain T
nullity T	dim image T
rank T+ nullity T	no. of linearly independent columns of T
	no. of pivotal columns of T
	no. of nonpivotal columns of T

2.5.4 Let T be an $n \times m$ matrix such that the row-reduced matrix $\widetilde{T}$ has at least one row containing all 0's. What can you deduce about the rank of T? What if $\widetilde{T}$ has precisely one row containing all 0's? Justify your statements.

$$a.\ \begin{bmatrix} 1 & 1 & 3 \\ 2 & 2 & 6 \end{bmatrix}, \quad b.\ \begin{bmatrix} 1 & 2 & 3 \\ -1 & 1 & 1 \\ -1 & 4 & 5 \end{bmatrix}$$

$$c.\ \begin{bmatrix} 1 & 1 & 1 \\ 1 & 2 & 3 \\ 2 & 3 & 4 \end{bmatrix}$$

Matrices for Exercise 2.5.6

$$A = \begin{bmatrix} 2 & 1 & 3 & 1 & 0 & 0 \\ 1 & -1 & 0 & 0 & 1 & 0 \\ 1 & 1 & 1 & 0 & 0 & 1 \end{bmatrix}$$

Matrix for Exercise 2.5.7

2.5.5 Prove Proposition 2.5.2. (This is a "turn the crank" proof.)

2.5.6 For each of the matrices in the margin, find a basis for the kernel and a basis for the image.

2.5.7 Let n be the rank of the matrix A shown in the margin.

a. What is n?

b. Is there more than one way to choose n linearly independent columns of A?

c. Is it possible to choose three columns of A that are not linearly independent?

d. What does this say about solutions to the system of equations

$$\begin{aligned} 2x_1 + x_2 + 3x_3 + x_4 &= 1 \\ x_1 - x_2 + x_5 &= 1 \\ x_1 + x_2 + x_3 + x_6 &= 1\,? \end{aligned}$$

2.5.8 True or false? (Justify your answer.) Let $f : \mathbb{R}^m \to \mathbb{R}^k$ and $g : \mathbb{R}^n \to \mathbb{R}^m$ be linear transformations. Then

$$f \circ g = 0 \quad \text{implies} \quad \operatorname{img} g = \ker f.$$

2.5.9 Let P_2 be the space of polynomials of degree ≤ 2, identified with $\mathbb{R}^3$ by identifying $a + bx + cx^2$ to $\begin{pmatrix} a \\ b \\ c \end{pmatrix}$.

a. Write the matrix of the linear transformation $T : P_2 \to P_2$ given by

$$(T(p))(x) = xp'(x) + x^2p''(x).$$

b. Find a basis for the image and the kernel of T.

2.5.10 Show that for any $a < b$, there exist numbers $c_0, \dots, c_k$ such that

$$\int_a^b p(t)\, dt = \sum_{i=0}^{k} c_i p(i) \quad \text{for all polynomials } p \in P_k.$$

Exercise 2.5.10 is very realistic. When integrating a function numerically you are always going to be taking a weighted average of the values of the function at some points. Here we explain how to choose the weights so that the formula for approximating integrals is exact for polynomials of degree $\leq k$.

2.5.11 Make a sketch, in the (a, b)-plane, of the sets where the kernels of the matrices at left have dimension $0, 1, 2, 3$. Indicate on the same sketch the dimensions of the images.

$$A = \begin{bmatrix} 1 & b \\ a & 2 \end{bmatrix}$$

$$B = \begin{bmatrix} 1 & 2 & a \\ a & b & a \\ b & b & a \end{bmatrix}$$

Matrices for Exercise 2.5.11

2.5.12 Decompose the following into partial fractions, as requested, being explicit in each case about the system of linear equations involved and showing that its matrix is invertible:

a. Write $\dfrac{x + x^2}{(x+1)(x+2)(x+3)}$ as $\dfrac{A}{x+1} + \dfrac{B}{x+2} + \dfrac{C}{x+3}$.

b. Write $\dfrac{x + x^3}{(x+1)^2(x-1)^3}$ as $\dfrac{Ax + B}{(x+1)^2} + \dfrac{Cx^2 + Dx + F}{(x-1)^3}$.

2.5.13 a. For what value of a can you *not* write

$$\frac{x - 1}{(x+1)(x^2 + ax + 5)} = \frac{A_0}{x+1} + \frac{B_1 x + B_0}{x^2 + ax + 5}?$$

b. Why does this not contradict Proposition 2.5.13?

2.5.14 a. Let $f(x) = x + Ax^2 + Bx^3$. Find a polynomial $g(x) = x + \alpha x^2 + \beta x^3$ such that $g(f(x)) - x$ is a polynomial starting with terms of degree 4.

b. Show that if $f(x) = x + \sum_{i=2}^{k} a_i x^i$ is a polynomial, then there exists a unique polynomial

$$g(x) = x + \sum_{i=2}^{k} b_i x^i \text{ with } g \circ f(x) = x + x^{k+1} p(x) \text{ for some polynomial } p.$$

2.5.15 Show that if A and B are $n \times n$ matrices, and AB is invertible, then A and B are invertible.

2.5.16 Let $S\colon \mathbb{R}^n \to \mathbb{R}^m$ and $T\colon \mathbb{R}^m \to \mathbb{R}^p$ be linear transformations. Show that $\operatorname{rank}(T \circ S) \leq \min(\operatorname{rank} T, \operatorname{rank} S)$.

Exercise 2.5.17: The polynomial p constructed in this exercise is called the *Lagrange interpolation polynomial*; it "interpolates" between the assigned values. Consider the map from the space of P_n of polynomials of degree n to $\mathbb{R}^{n+1}$ given by

$$p \mapsto \begin{bmatrix} p(x_0) \\ \vdots \\ p(x_n) \end{bmatrix}.$$

You need to show that this map is onto; by Corollary 2.5.10 it is enough to show that its kernel is $\{0\}$.

***2.5.17** a. Find a polynomial $p(x) = a + bx + cx^2$ of degree 2 such that

$$p(0) = 1, \quad p(1) = 4, \quad \text{and} \quad p(3) = -2.$$

b. Show that if $x_0, \dots, x_n$ are $n + 1$ distinct points in $\mathbb{R}$, and $a_0, \dots, a_n$ are any numbers, there exists a unique polynomial of degree n such that $p(x_i) = a_i$ for each $i = 0, \dots, n$.

c. Let the x_i and a_i be as in part b, and let $b_0, \dots, b_n$ be any numbers. Find a number k such that there exists a unique polynomial of degree k with

$$p(x_i) = a_i \quad \text{and} \quad p'(x_i) = b_i \quad \text{for all } i = 0, \dots, n.$$

2.5.18 a. Prove that given any distinct points $x_0, \dots, x_k \in \mathbb{R}$ and any numbers $c_0, \dots, c_k \in \mathbb{R}$, the polynomial

$$p(x) = \sum_{i=0}^{k} c_i \frac{\prod_{j \neq i}(x - x_j)}{\prod_{j \neq i}(x_i - x_j)} \qquad \text{(Lagrange interpolation formula)}$$

is a polynomial of degree at most k such that $p(x_i) = c_i$.

b. For $k = 1$, $k = 2$, and $k = 3$, write the matrix of the transformation $\begin{pmatrix} c_0 \\ \vdots \\ c_k \end{pmatrix} \mapsto p$, where $p = a_0 + a_1 x + \cdots + a_k x^k \in P_k$ is identified to $\begin{pmatrix} a_0 \\ \vdots \\ a_k \end{pmatrix} \in \mathbb{R}^{k+1}$.

Exercise 2.5.19, part a: Remember, the ith column of the matrix is the image of the ith basis vector.

Part c: If you multiply out $f_{\vec{\mathbf{a}}}$, you get something of the form

$$\frac{p_{\vec{\mathbf{a}}}(x)}{x^n}.$$

What is the degree of $p_{\vec{\mathbf{a}}}$? What must $p_{\vec{\mathbf{a}}}$ satisfy if $\vec{\mathbf{a}} \in \ker H_n$?

2.5.19 For any $\vec{\mathbf{a}} \in \mathbb{R}^n$, set $f_{\vec{\mathbf{a}}}(x) = \frac{a_1}{x} + \frac{a_2}{x^2} + \cdots + \frac{a_n}{x^n}$.

Let $H : \mathbb{R}^n \to \mathbb{R}^n$ be the linear transformation $H_n(\vec{\mathbf{a}}) = \begin{bmatrix} f_{\vec{\mathbf{a}}}(1) \\ \vdots \\ f_{\vec{\mathbf{a}}}(n) \end{bmatrix}$.

a. Write the matrix of H_n when $n = 2$ and $n = 3$.

b. Write the matrix of H_n in general.

c. Show that H_n is invertible.

2.5.20 For any $\vec{\mathbf{a}} \in \mathbb{R}^n$, set $f_{\vec{\mathbf{a}}}(x) = \frac{a_1}{x} + \frac{a_2}{x+1} + \cdots + \frac{a_n}{x+n-1}$.

Let $H : \mathbb{R}^n \to \mathbb{R}^n$ be the linear transformation $H_n(\vec{\mathbf{a}}) = \begin{bmatrix} f_{\vec{\mathbf{a}}}(1) \\ \vdots \\ f_{\vec{\mathbf{a}}}(n) \end{bmatrix}$.

a. Write the matrix of H_n when $n = 2$ and $n = 3$.

b. Write the matrix of H_n in general.

c. For any vector $\vec{\mathbf{b}} \in \mathbb{R}^n$, does there exist $\vec{\mathbf{a}} \in \mathbb{R}^n$ such that

$$f_{\vec{\mathbf{a}}}(1) = b_1;\ f_{\vec{\mathbf{a}}}(2) = b_2\ ;\ \dots\ ;\ f_{\vec{\mathbf{a}}}(n) = b_n?$$

***2.5.21** Let $T_1, T_2 : \mathbb{R}^n \to \mathbb{R}^n$ be linear transformations.

a. Show that there exists a linear transformation $S : \mathbb{R}^n \to \mathbb{R}^n$ such that $T_1 = S \circ T_2$ if and only if $\ker T_2 \subset \ker T_1$.

b. Show that there exists a linear transformation $S : \mathbb{R}^n \to \mathbb{R}^n$ such that $T_1 = T_2 \circ S$ if and only if $\operatorname{img} T_1 \subset \operatorname{img} T_2$.

Section 2.6: Why should we study abstract vector spaces? Why not stick to $\mathbb{R}^n$? One reason is that $\mathbb{R}^n$ comes with the standard basis, which may not be the best basis for the problem at hand.

Another reason is that when you prove something about $\mathbb{R}^n$, you then need to check that your proof was "basis independent" before you can extend it to an arbitrary vector space.

Applied mathematicians generally prefer working in $\mathbb{R}^n$, translating when necessary to abstract vector spaces; pure mathematicians generally prefer working directly in the world of abstract vector spaces. We have sympathy with both views.

2.6 Abstract vector spaces

In this section we briefly discuss *abstract vector spaces*. A vector space is a set in which elements can be added and multiplied by numbers (scalars). We denote a vector in an abstract vector space by a bold letter without arrow, to distinguish it from a vector in $\mathbb{R}^n$: $\mathbf{v} \in V$ as opposed to $\vec{\mathbf{v}} \in \mathbb{R}^n$.

The archetypal example of a vector space is $\mathbb{R}^n$. More generally, a subset of $\mathbb{R}^n$ (endowed with the same addition and multiplication by scalars as $\mathbb{R}^n$ itself) is a vector space if and only if it is a subspace of $\mathbb{R}^n$ (Definition

1.1.5). Other examples that are fairly easy to understand are the space Mat (n, m) of $n \times m$ matrices, with addition and multiplication by scalars defined in Section 1.2, and the space P_k of polynomials of degree at most k. In Section 1.4 we considered an $m \times n$ matrix as a point in $\mathbb{R}^{nm}$, and in Example 1.7.17 we "identified" P_k with $\mathbb{R}^{k+1}$. But other vector spaces, like that in Example 2.6.2, have a different flavor: they are somehow much too big, and just what "identifying with $\mathbb{R}^n$" means is not quite clear.

In general, one cannot multiply together elements of a vector space, although sometimes one can (for instance, matrix multiplication in Mat (n, n) or the cross product in $\mathbb{R}^3$).

What scalars? Two choices are important for this book: real numbers and complex numbers. When the scalars are real numbers, the vector space is said to be a real vector space; when they are complex, it is a complex vector space.

But most of this section would work just as well if we used as scalars elements of any field.

A *field of scalars* is a set in which one can add and multiply; these operations must satisfy certain conditions (for instance, addition and multiplication must be associative and commutative, and multiplication is distributive over addition). A key property is that one can divide by nonzero elements. In number theory vector spaces over the field $\mathbb{Q}$ of rational numbers are important, and in cryptography vector spaces over finite fields play the central role.

The word "field" is unfortunate; fields of scalars are unrelated to vector fields or form fields.

Definition 2.6.1 (Vector space). A *vector space* V is a set of vectors such that two vectors can be added to form another vector in V, and a vector can be multiplied by a scalar to form another vector in V. This addition and multiplication must satisfy these rules:

1. *Additive identity.* There exists a vector $\mathbf{0} \in V$ such that for any $\mathbf{v} \in V$, we have $\mathbf{0} + \mathbf{v} = \mathbf{v}$.
2. *Additive inverse.* For any $\mathbf{v} \in V$, there exists a vector $-\mathbf{v} \in V$ such that $\mathbf{v} + (-\mathbf{v}) = \mathbf{0}$.
3. *Commutative law for addition.* For all $\mathbf{v}, \mathbf{w} \in V$, we have
$$\mathbf{v} + \mathbf{w} = \mathbf{w} + \mathbf{v}.$$
4. *Associative law for addition.* For all $\mathbf{v}_1, \mathbf{v}_2, \mathbf{v}_3 \in V$,
$$\mathbf{v}_1 + (\mathbf{v}_2 + \mathbf{v}_3) = (\mathbf{v}_1 + \mathbf{v}_2) + \mathbf{v}_3.$$
5. *Multiplicative identity.* For all $\mathbf{v} \in V$, we have $1\mathbf{v} = \mathbf{v}$.
6. *Associative law for multiplication.* For all scalars a, b and all $\mathbf{v} \in V$,
$$a(b\mathbf{v}) = (ab)\mathbf{v}.$$
7. *Distributive law for scalar addition.* For all scalars a, b and all $\mathbf{v} \in V$, we have $(a + b)\mathbf{v} = a\mathbf{v} + b\mathbf{v}$.
8. *Distributive law for vector addition.* For all scalars a and all $\mathbf{v}, \mathbf{w} \in V$, we have $a(\mathbf{v} + \mathbf{w}) = a\mathbf{v} + a\mathbf{w}$.

Note that the empty set is not a vector space, because it does not satisfy condition 1. But $\{\mathbf{0}\}$ is a vector space – the trivial vector space.

Example 2.6.2 (An infinite-dimensional vector space). Consider the space $\mathcal{C}(0, 1)$ of continuous real-valued functions defined on $(0, 1)$. The "vectors" of this space are functions $f : (0, 1) \to \mathbb{R}$, with addition defined as usual by $(f + g)(x) = f(x) + g(x)$ and multiplication by scalars by $(\alpha f)(x) = \alpha\, f(x)$. Exercise 2.6.2 asks you to show that this is a vector space. △

In Example 2.6.2 our assumption that addition is well defined in $\mathcal{C}[0, 1]$ uses the fact that the sum of two continuous functions is continuous (Theorem 1.5.29). Multiplication by scalars is well defined, because the product of a continuous function by a constant is continuous.

The vector space $\mathcal{C}(0, 1)$ cannot be identified with $\mathbb{R}^n$; there is no linear transformation from any $\mathbb{R}^n$ to this space that is onto, as we will see in Example 2.6.23. But it has subspaces that can be identified with appropriate $\mathbb{R}^n$'s, as seen in Example 2.6.3.

Example 2.6.3 (A finite-dimensional subspace of $\mathcal{C}(0, 1)$). Consider the space of twice differentiable functions $f : \mathbb{R} \to \mathbb{R}$ such that $D^2 f = 0$ (i.e., functions of one variable whose second derivatives are 0; we could also write this $f'' = 0$). This is a subspace of the vector space of Example 2.6.2 and is also a vector space itself. But since a function has a vanishing second derivative if and only if it is a polynomial of degree at most 1, we see that this space is the set of functions

$$f_{a,b}(x) = a + bx. \tag{2.6.1}$$

Precisely two numbers are needed to specify each element of this vector space; we could choose as our basis the constant function 1 and the function x. Thus in some sense, this space "is" $\mathbb{R}^2$, since $f_{a,b}$ can be identified with $\begin{pmatrix} a \\ b \end{pmatrix} \in \mathbb{R}^2$; this was not obvious from the definition.

But the subspace $\mathcal{C}^1(0,1) \subset \mathcal{C}(0,1)$ of once continuously differentiable functions on $(0,1)$ cannot be identified with any $\mathbb{R}^n$; the elements are more restricted than those of $\mathcal{C}(0,1)$, but not enough so that an element can be specified by finitely many numbers. $\triangle$

Linear transformations

Equation 2.6.2 is a shorter way of writing

$$T(\mathbf{v}_1 + \mathbf{v}_2) = T(\mathbf{v}_1) + T(\mathbf{v}_2)$$
$$T(\alpha\mathbf{v}_1) = \alpha T(\mathbf{v}_1).$$

A bijective linear transformation is often called an *isomorphism* of vector spaces. The word isomorphism depends on context; it means "preserve whatever structure is relevant". Here, the relevant structure is the vector space structure of V and W. In other cases, it might be a group structure, or a ring structure.

Other maps that preserve structure have different names: a *homeomorphism* preserves topological structure; a *diffeomorphism* is a differentiable mapping with differentiable inverse, and thus preserves differential structure.

In Section 1.3 we investigated linear transformations $\mathbb{R}^n \to \mathbb{R}^m$. The same definition can be used for linear transformations of abstract vector spaces.

Definition 2.6.4 (Linear transformation). If V and W are vector spaces, a *linear transformation* $T : V \to W$ is a mapping satisfying

$$T(\alpha\mathbf{v}_1 + \beta\mathbf{v}_2) = \alpha T(\mathbf{v}_1) + \beta T(\mathbf{v}_2) \tag{2.6.2}$$

for all scalars $\alpha, \beta \in \mathbb{R}$ and all $\mathbf{v}_1, \mathbf{v}_2 \in V$.

Theorem 1.3.4 says that every linear transformation $T : \mathbb{R}^m \to \mathbb{R}^n$ is given by the $n \times m$ matrix whose ith column is $T(\vec{\mathbf{e}}_i)$. This provides a complete understanding of linear transformations from $\mathbb{R}^m$ to $\mathbb{R}^n$. Linear transformations between abstract vector spaces don't have this wonderful concreteness. In finite-dimensional vector spaces, it is possible to understand a linear transformation as a matrix but you have to work at it, choosing a basis for the domain and a basis for the codomain; we discuss this in Proposition and Definition 2.6.16.

Even when it is possible to write a linear transformation as a matrix, it may not be the easiest way to deal with things, as shown in Example 2.6.5.

Example 2.6.5 (A linear transformation difficult to write as a matrix). If $A \in \operatorname{Mat}(n,n)$, then the transformation $\operatorname{Mat}(n,n) \to \operatorname{Mat}(n,n)$ given by $H \mapsto AH + HA$ is a linear transformation. We encountered it in Example 1.7.17 as the derivative of the mapping $S : A \mapsto A^2$:

$$[\mathbf{D}S(A)]H = AH + HA. \tag{2.6.3}$$

Even in the case $n = 3$ it would be tedious to identify each 3×3 matrix with a vector in $\mathbb{R}^9$ and write this transformation as a 9×9 matrix; the language of abstract linear transformations is more appropriate. $\triangle$

Example 2.6.6: If we divide $[0,1]$ into n equal parts, so that $[0,1] \times [0,1]$ becomes a grid with n^2 little squares, then we could sample g at each point in the grid, thinking of each sample $g\begin{pmatrix} x \\ y \end{pmatrix}$ as the entry $g_{x,y}$ of an $(n+1)\times(n+1)$ matrix A. Similarly we can sample f at the gridpoints, giving a vector $\vec{\mathbf{b}} \in \mathbb{R}^{n+1}$. Then the matrix product $A\vec{\mathbf{b}}$ is a Riemann sum for the integral in equation 2.6.4. Thus, when n is large, equation 2.6.4 looks a lot like the matrix multiplication $A\vec{\mathbf{b}}$.

This is the kind of thing we meant when we referred to "analogues" of matrices; it is as much like a matrix as you can hope to get in this particular infinite-dimensional setting. It is not true that all transformations from $\mathcal{C}[0,1]$ to $\mathcal{C}[0,1]$ are of this sort; even the identity cannot be written in the form T_g.

Example 2.6.6 (Showing that a transformation is linear). Let $\mathcal{C}[0,1]$ denote the space of continuous real-valued functions defined for $0 \le x \le 1$. Let $g : [0,1] \times [0,1] \to \mathbb{R}$ be a continuous function, and define the mapping $T_g : \mathcal{C}[0,1] \to \mathcal{C}[0,1]$ by

$$\big(T_g(f)\big)(x) = \int_0^1 g\begin{pmatrix} x \\ y \end{pmatrix} f(y)\, dy. \qquad 2.6.4$$

For example, if $g\begin{pmatrix} x \\ y \end{pmatrix} = |x-y|$, then $\big(T_g(f)\big)(x) = \int_0^1 |x-y| f(y)\, dy$.

Let us show that T_g is a linear transformation. We first show that

$$T_g(f_1 + f_2) = T_g(f_1) + T_g(f_2), \qquad 2.6.5$$

which we do as follows:

$$\begin{aligned}
\big(T_g(f_1+f_2)\big)(x) &= \int_0^1 g\begin{pmatrix} x \\ y \end{pmatrix} (f_1+f_2)(y)\, dy = \int_0^1 g\begin{pmatrix} x \\ y \end{pmatrix} \overbrace{\big(f_1(y)+f_2(y)\big)}^{\text{definition of addition in vector space}} dy \\
&= \int_0^1 \left(g\begin{pmatrix} x \\ y \end{pmatrix} f_1(y) + g\begin{pmatrix} x \\ y \end{pmatrix} f_2(y) \right) dy \\
&= \int_0^1 g\begin{pmatrix} x \\ y \end{pmatrix} f_1(y)\, dy + \int_0^1 g\begin{pmatrix} x \\ y \end{pmatrix} f_2(y)\, dy \qquad 2.6.6 \\
&= \big(T_g(f_1)\big)(x) + \big(T_g(f_2)\big)(x) = \big(T_g(f_1) + T_g(f_2)\big)(x).
\end{aligned}$$

Next we show that $T_g(\alpha f)(x) = \alpha T_g(f)(x)$:

$$T_g(\alpha f)(x) = \int_0^1 g\begin{pmatrix} x \\ y \end{pmatrix} (\alpha f)(y)\, dy = \alpha \int_0^1 g\begin{pmatrix} x \\ y \end{pmatrix} f(y)\, dy = \alpha T_g(f)(x). \quad \square \qquad 2.6.7$$

The linear transformation in our next example is a special kind of linear transformation, called a *linear differential operator*. Solving a differential equation is same as looking for the kernel of such a linear transformation.

Example 2.6.7: Recall (Definition 1.9.7) that a function is C^2 if its second partial derivatives exist and are continuous.

We could replace the coefficients in equation 2.6.8 by any continuous functions of x, so this is an example of an important class.

Example 2.6.7 (A linear differential operator). Let $\mathcal{C}^2$ be the space of C^2 functions. The transformation $T : \mathcal{C}^2(\mathbb{R}) \to \mathcal{C}(\mathbb{R})$ given by the formula

$$\big(T(f)\big)(x) = (x^2+1)f''(x) - xf'(x) + 2f(x) \qquad 2.6.8$$

is a linear transformation, as Exercise 2.19 asks you to show. $\triangle$

We will now extend the notions of linear independence, span, and basis to arbitrary real vector spaces.

Definition 2.6.8 (Linear combination). Let V be a vector space and let $\{\mathbf{v}\} \overset{\text{def}}{=} \mathbf{v}_1, \dots, \mathbf{v}_m$ be a finite ordered collection of vectors in V. A *linear combination* of the vectors $\mathbf{v}_1, \dots, \mathbf{v}_m$ is a vector $\mathbf{v}$ of the form

$$\mathbf{v} = \sum_{i=1}^{m} a_i \mathbf{v}_i, \quad \text{with} \quad a_1, \dots, a_m \in \mathbb{R}. \qquad 2.6.9$$

Definition 2.6.9 (Span). The vectors $\mathbf{v}_1, \dots, \mathbf{v}_m$ *span* V if and only if every vector in V is a linear combination of $\mathbf{v}_1, \dots, \mathbf{v}_m$.

Definition 2.6.10: If any one of these conditions is met, all are met.

By Definition 2.6.11, a basis is necessarily finite, but we could have allowed infinite bases. We stick to finite bases because bases tend to be useless in infinite-dimensional vector spaces. The interesting notion for infinite-dimensional vector spaces is not expressing an element of the space as a linear combination of a finite number of basis elements, but expressing it as a linear combination of *infinitely* many basis vectors, that is, as an infinite series $\sum_{i=0}^{\infty} a_i \, \mathbf{v}_i$ (for example, power series or Fourier series). This introduces questions of convergence, which are interesting indeed but a bit foreign to the spirit of linear algebra.

Definition 2.6.10 (Linear independence). The vectors $\mathbf{v}_1, \dots, \mathbf{v}_m$ are *linearly independent* if and only if any of the following three equivalent conditions is met:

1. There is only one way of writing a given linear combination:

$$\sum_{i=1}^{m} a_i \, \mathbf{v}_i = \sum_{i=1}^{m} b_i \, \mathbf{v}_i \quad \text{implies} \quad a_1 = b_1, \; a_2 = b_2, \dots, \; a_m = b_m. \qquad 2.6.10$$

2. The only solution to

$$a_1 \mathbf{v}_1 + a_2 \mathbf{v}_2 + \cdots + a_m \, \mathbf{v}_m = \mathbf{0} \quad \text{is} \quad a_1 = a_2 = \cdots = a_m = 0. \qquad 2.6.11$$

3. None of the $\mathbf{v}_i$ is a linear combination of the others.

Definition 2.6.11 (Basis). An ordered set of vectors $\mathbf{v}_1, \dots, \mathbf{v}_m \in V$ is a *basis* of V if and only if it is linearly independent and spans V.

The following definition is central: the map $\Phi_{\{\mathbf{v}\}}$ enables us to move from the concrete world of $\mathbb{R}^n$ to the abstract world of a vector space V.

Equation 2.6.12:

$$\Phi_{\{\mathbf{v}\}}(\vec{\mathbf{e}}_i) = \mathbf{v}_i.$$

This uses Theorem 1.3.4: the ith column of $[T]$ is $T(\vec{\mathbf{e}}_i)$. For instance,

$$\Phi_{\{\mathbf{v}\}}\left(\begin{bmatrix} 1 \\ 0 \end{bmatrix}\right) = 1\mathbf{v}_1 + 0\mathbf{v}_2 = \mathbf{v}_1$$

$$\Phi_{\{\mathbf{v}\}}\left(\begin{bmatrix} 0 \\ 1 \end{bmatrix}\right) = 0\mathbf{v}_1 + 1\mathbf{v}_2 = \mathbf{v}_2.$$

Definition 2.6.12 ("Concrete to abstract" function $\Phi_{\{\mathbf{v}\}}$). Let $\{\mathbf{v}\} = \mathbf{v}_1, \dots, \mathbf{v}_n$ be a finite, ordered collection of n vectors in a vector space V. The *concrete to abstract* function $\Phi_{\{\mathbf{v}\}}$ is the linear transformation $\Phi_{\{\mathbf{v}\}} : \mathbb{R}^n \to V$ that translates from $\mathbb{R}^n$ to V:

$$\Phi_{\{\mathbf{v}\}}(\vec{\mathbf{a}}) = \Phi_{\{\mathbf{v}\}}\left(\begin{bmatrix} a_1 \\ \vdots \\ a_n \end{bmatrix}\right) \stackrel{\text{def}}{=} a_1\mathbf{v}_1 + \cdots + a_n\mathbf{v}_n. \qquad 2.6.12$$

Example 2.6.13 (Concrete to abstract function). Let P_2 be the space of polynomials of degree at most 2, with basis $\mathbf{v}_1 = 1$, $\mathbf{v}_2 = x$, $\mathbf{v}_3 = x^2$. Then $\Phi_{\{\mathbf{v}\}}\left(\begin{bmatrix} a_1 \\ a_2 \\ a_3 \end{bmatrix}\right) = a_1 + a_2 x + a_3 x^2$ identifies $\mathbb{R}^3$ with P_2. △

Example 2.6.14. In practice, the "abstract" vector space V is often $\mathbb{R}^n$, with a new basis. If $V = \mathbb{R}^2$ and the set $\{\mathbf{v}\}$ consists of $\vec{\mathbf{v}}_1 = \begin{bmatrix} 1 \\ 1 \end{bmatrix}$ and $\vec{\mathbf{v}}_2 = \begin{bmatrix} 1 \\ -1 \end{bmatrix}$, then

$$\Phi_{\{\mathbf{v}\}}\left(\begin{bmatrix} a \\ b \end{bmatrix}\right) = a\vec{\mathbf{v}}_1 + b\vec{\mathbf{v}}_2 = \begin{bmatrix} a+b \\ a-b \end{bmatrix}; \qquad 2.6.13$$

the point with coordinates (a,b) in the basis $\vec{\mathbf{v}}_1, \vec{\mathbf{v}}_2$, equals $\begin{bmatrix} a+b \\ a-b \end{bmatrix}$ in the standard basis:

$$a\vec{\mathbf{v}}_1 + b\vec{\mathbf{v}}_2 = (a+b)\vec{\mathbf{e}}_1 + (a-b)\vec{\mathbf{e}}_2. \qquad 2.6.14$$

For instance, the point with coordinates $(2,3)$ in the new basis $\{\mathbf{v}\}$ is the point with coordinates $(5,-1)$ in the standard basis:

$$\underbrace{\begin{bmatrix} 1 & 1 \\ 1 & -1 \end{bmatrix}}_{\Phi_{\{\mathbf{v}\}}} \begin{bmatrix} 2 \\ 3 \end{bmatrix} = \begin{bmatrix} 5 \\ -1 \end{bmatrix} \text{ means } \Phi_{\{\mathbf{v}\}}\left(\begin{bmatrix} 2 \\ 3 \end{bmatrix}\right) = 2\vec{\mathbf{v}}_1 + 3\vec{\mathbf{v}}_2 = \begin{bmatrix} 2+3 \\ 2-3 \end{bmatrix}. \triangle$$

In Example 2.6.14,

$$\Phi_{\{\mathbf{v}\}} = \begin{bmatrix} 1 & 1 \\ 1 & -1 \end{bmatrix}$$

because

$$\Phi_{\{\mathbf{v}\}}(\vec{\mathbf{e}}_1) = \vec{\mathbf{v}}_1 = \begin{bmatrix} 1 \\ 1 \end{bmatrix}$$
$$\Phi_{\{\mathbf{v}\}}(\vec{\mathbf{e}}_2) = \vec{\mathbf{v}}_2 = \begin{bmatrix} 1 \\ -1 \end{bmatrix}.$$

Proposition 2.6.15 says that any vector space with a basis is "just like" $\mathbb{R}^n$: if $\{\mathbf{v}\}$ is a basis of V, then the linear transformation $\Phi_{\{\mathbf{v}\}} : \mathbb{R}^n \to V$ is invertible, and its inverse allows us to identify V with $\mathbb{R}^n$:

$$\Phi^{-1}_{\{\mathbf{v}\}}(a_1\mathbf{v}_1 + \cdots + a_n\mathbf{v}_n) = \begin{bmatrix} a_1 \\ \vdots \\ a_n \end{bmatrix}.$$

This allows us to replace questions about V with questions about the coordinates in $\mathbb{R}^n$.

Proposition 2.6.15 restates the notions of linear independence, span, and basis in terms of $\Phi_{\{\mathbf{v}\}}$.

Proposition 2.6.15 (Linear independence, span, and basis). *Let $\{\mathbf{v}\} = \mathbf{v}_1, \dots, \mathbf{v}_n$ be vectors in a vector space V, and let $\Phi_{\{\mathbf{v}\}} : \mathbb{R}^n \to V$ be the associated concrete-to-abstract transformation. Then*

1. *The set $\{\mathbf{v}\}$ is linearly independent if and only if $\Phi_{\{\mathbf{v}\}}$ is one to one.*
2. *The set $\{\mathbf{v}\}$ spans V if and only if $\Phi_{\{\mathbf{v}\}}$ is onto.*
3. *The set $\{\mathbf{v}\}$ is a basis of V if and only if $\Phi_{\{\mathbf{v}\}}$ is invertible.*

Proof. 1. Definition 2.6.10 says that $\mathbf{v}_1, \dots, \mathbf{v}_n$ are linearly independent if and only if

$$\sum_{i=1}^n a_i\,\mathbf{v}_i = \sum_{i=1}^n b_i\,\mathbf{v}_i \quad \text{implies } a_1 = b_1,\ a_2 = b_2, \dots,\ a_n = b_n, \qquad 2.6.15$$

that is, if and only if $\Phi_{\{\mathbf{v}\}}$ is injective (one to one).

2. Definition 2.6.9 says that

$$\{\mathbf{v}\} = \mathbf{v}_1, \dots, \mathbf{v}_n \qquad 2.6.16$$

span V if and only if any vector $\mathbf{v} \in V$ is a linear combination of $\mathbf{v}_1, \dots, \mathbf{v}_n$:

$$\mathbf{v} = a_1\mathbf{v}_1 + \cdots + a_n\mathbf{v}_n = \Phi_{\{\mathbf{v}\}}(\vec{\mathbf{a}}). \qquad 2.6.17$$

In other words, $\Phi_{\{\mathbf{v}\}}$ is surjective (onto).

3. Putting these together, $\mathbf{v}_1, \dots, \mathbf{v}_n$ is a basis if and only if it is linearly independent and spans V (i.e., if $\Phi_{\{\mathbf{v}\}}$ is invertible). $\square$

Matrix with respect to a basis and change of basis

Proposition and Definition 2.6.16 says we can use matrices to describe linear transformations between finite-dimensional vector spaces, *once bases have been chosen.* The change of basis matrix described in Proposition and

Definition 2.6.17 then allows us to express a vector in one basis in terms of its expression in another basis. The change of basis formula (Theorem 2.6.20) allows us to relate a linear transformation $T : V \to W$ written in one pair of bases to the same linear transformation written in another pair of bases.

Proposition and Definition 2.6.16 (Matrix with respect to bases) *Let V and W be finite-dimensional vector spaces, with $\{\mathbf{v}\} = \mathbf{v}_1, \dots, \mathbf{v}_n$ a basis of V and $\{\mathbf{w}\} = \mathbf{w}_1, \dots, \mathbf{w}_m$ a basis of W. Let $T : V \to W$ be a linear transformation. Then the matrix $[T]_{\{\mathbf{v}\},\{\mathbf{w}\}}$ of T with respect to the bases $\{\mathbf{v}\}$ and $\{\mathbf{w}\}$ is the $m \times n$ matrix with entries $t_{i,j}$ where*

$$T\mathbf{v}_k = \sum_{\ell=1}^{m} t_{\ell,k}\mathbf{w}_\ell. \qquad 2.6.18$$

Equivalently, if

$$\Phi_{\{\mathbf{v}\}} : \mathbb{R}^n \to V \quad \text{and} \quad \Phi_{\{\mathbf{w}\}} : \mathbb{R}^m \to W \qquad 2.6.19$$

are the associated "concrete to abstract" linear transformations, then the matrix of T with respect to the bases $\{\mathbf{v}\}$ and $\{\mathbf{w}\}$ is

$$[T]_{\{\mathbf{v}\},\{\mathbf{w}\}} = \Phi_{\mathbf{w}}^{-1} \circ T \circ \Phi_{\mathbf{v}}. \qquad 2.6.20$$

Note that equation 2.6.18 is not a matrix multiplication; the summation is over the first index of the p, not the second (see Definition 1.2.4). It is equivalent to the "matrix multiplication"

$$[T\mathbf{v}_1 \dots T\mathbf{v}_n] = [\mathbf{w}_1 \dots \mathbf{w}_m][T]_{\{\mathbf{v}\},\{\mathbf{w}\}}.$$

If V is a subset of $\mathbb{R}^n$ and W a subset of $\mathbb{R}^m$, this is a true matrix multiplication. If V and W are abstract vector spaces, then $[T\mathbf{v}_1 \dots T\mathbf{v}_n]$ and $[\mathbf{w}_1 \dots \mathbf{w}_m]$ are not matrices, but the multiplication makes sense anyway; each $T\mathbf{v}_k$ is a linear combination of the vectors $\mathbf{w}_1, \dots, \mathbf{w}_m$.

Note that the map $\Phi_{\mathbf{w}}^{-1} \circ T \circ \Phi_{\mathbf{v}}$ in equation 2.6.20 is a linear transformation $\mathbb{R}^n \to \mathbb{R}^m$, and hence has a matrix. In particular, the matrix of a linear transformation $\mathbb{R}^n \to \mathbb{R}^m$ as defined in Theorem 1.3.4 is simply the matrix with respect to the standard basis $\{\mathbf{e}_n\}$ in the domain and the standard basis $\{\mathbf{e}_m\}$ in the codomain. In this case, $\Phi_{\{\mathbf{e}_n\}}$ is the $n \times n$ identity matrix, and $\Phi_{\{\mathbf{e}_m\}}^{-1}$ the $m \times m$ identity matrix.

Proof. The kth column of $\Phi_{\mathbf{w}}^{-1} \circ T \circ \Phi_{\mathbf{v}}$ is $(\Phi_{\mathbf{w}}^{-1} \circ T \circ \Phi_{\mathbf{v}})\vec{\mathbf{e}}_k$, so compute

$$\Phi_{\mathbf{w}}^{-1} \circ T \circ \Phi_{\mathbf{v}}\vec{\mathbf{e}}_k = \Phi_{\mathbf{w}}^{-1} \circ T\vec{\mathbf{v}}_k = \Phi_{\mathbf{w}}^{-1} \sum_{\ell=1}^{m} t_{\ell,k}\mathbf{w}_\ell \qquad 2.6.21$$

$$= \sum_{\ell=1}^{m} t_{\ell,k}\Phi_{\mathbf{w}}^{-1}\mathbf{w}_\ell = \sum_{\ell=1}^{m} t_{\ell,k}\vec{\mathbf{e}}_\ell = \begin{bmatrix} t_{1,k} \\ \vdots \\ t_{m,k} \end{bmatrix}. \qquad \square$$

Charles V (1500–1558) was reputed to have said: "I speak Spanish to God, Italian to women, French to men, and German to my horse."

Jean-Pierre Kahane relates that in his youth he was taught that both French and German were suited to mathematics, but that English was not. Today, French mathematicians writing for publication probably write in English as often as they do in French.

A basis gives a name to a vector living in an abstract vector space, but a vector has many embodiments, depending on the choice of basis for V, just as *book*, *livre*, *Buch* all mean the same thing. The *change of basis matrix* $[P_{\mathbf{v}' \to \mathbf{v}}]$ is the dictionary that allows you to translate any vector written in the basis $\{\mathbf{v}'\}$ to the same vector written in the basis $\{\mathbf{v}\}$. We will see how useful it can be to change bases when we discuss eigenvectors, eigenvalues, and diagonalization in Section 2.7.

Proposition and Definition 2.6.17 (Change of basis matrix). *Let V be an n-dimensional vector space. Given two bases $\{\mathbf{v}\}, \{\mathbf{v}'\}$ of V, we can express each vector in $\{\mathbf{v}'\}$ in terms of the vectors in $\{\mathbf{v}\}$:*

$$\mathbf{v}'_i = p_{1,i}\mathbf{v}_1 + p_{2,i}\mathbf{v}_2 + \cdots + p_{n,i}\mathbf{v}_n, \quad \text{i.e.,} \quad \mathbf{v}'_i = \sum_{j=1}^{n} p_{j,i}\mathbf{v}_j. \qquad 2.6.22$$

The change of basis matrix $[P_{\mathbf{v}'\to\mathbf{v}}]$ is then

$$[P_{\mathbf{v}'\to\mathbf{v}}] \stackrel{\text{def}}{=} \begin{bmatrix} p_{1,1} & \cdots & p_{1,n} \\ \vdots & \cdots & \vdots \\ p_{n,1} & \cdots & p_{n,n} \end{bmatrix}; \qquad 2.6.23$$

the ith column of this matrix consists of the coefficients of the ith basis vector of $\{\mathbf{v}'\}$, written in terms of $\mathbf{v}_1, \dots, \mathbf{v}_n$. If

$$[P_{\mathbf{v}'\to\mathbf{v}}]\vec{\mathbf{a}} = \vec{\mathbf{b}}, \quad \textit{then} \quad \sum_{i=1}^{n} a_i\mathbf{v}'_i = \sum_{j=1}^{n} b_j\mathbf{v}_j\,. \qquad 2.6.24$$

The change of basis matrix $[P_{\mathbf{v}'\to\mathbf{v}}]$ is the matrix of the linear transformation $\Phi^{-1}_{\{\mathbf{v}\}} \circ \Phi_{\{\mathbf{v}'\}}$:

$$[P_{\mathbf{v}'\to\mathbf{v}}] = \Phi^{-1}_{\{\mathbf{v}\}} \circ \Phi_{\{\mathbf{v}'\}} : \mathbb{R}^n \to \mathbb{R}^n. \qquad 2.6.25$$

As in equation 2.6.18, equation 2.6.22 is not a matrix multiplication. But it can be written as

$$[\mathbf{v}'_1 \dots \mathbf{v}'_n] = [\mathbf{v}_1 \dots \mathbf{v}_n][P_{\mathbf{v}'\to\mathbf{v}}];$$

the multiplication makes sense.

We already saw the change of basis matrix in the proof of Proposition and Definition 2.4.21, when we showed that every basis of a subset $E \subset \mathbb{R}^n$ has the same number of elements; the matrices A and B there are change of basis matrices.

Thus to translate from basis $\{\mathbf{v}'\}$ to basis $\{\mathbf{v}\}$, we write each basis vector of $\{\mathbf{v}'\}$ as a linear combination of the vectors of $\{\mathbf{v}\}$. These coefficients are the entries of the change of basis matrix $[P_{\mathbf{v}'\to\mathbf{v}}]$. When writing the matrix, make sure you write the coefficients for each basis vector as a column, not a row: the coefficients used to write $\mathbf{v}'_i$ in terms of the vectors $\{\mathbf{v}\}$ form the ith *column* of $[P_{\mathbf{v}'\to\mathbf{v}}]$.

Proof. The following computation shows that $[P_{\mathbf{v}'\to\mathbf{v}}]\vec{\mathbf{a}} = \vec{\mathbf{b}}$ implies $\sum_{i=1}^{n} a_i\mathbf{v}'_i = \sum_{j=1}^{n} b_j\mathbf{v}_j$:

$$\sum_{i=1}^{n} a_i\mathbf{v}'_i = \sum_{i=1}^{n} a_i \sum_{j=1}^{n} p_{j,i}\mathbf{v}_j = \sum_{j=1}^{n}\left(\sum_{i=1}^{n} p_{j,i}a_i\right)\mathbf{v}_j = \sum_{j=1}^{n}\left([P_{\mathbf{v}'\to\mathbf{v}}]\vec{\mathbf{a}}\right)_j\mathbf{v}_j = \sum_{j=1}^{n} b_j\mathbf{v}_j.$$

To see that $[P_{\mathbf{v}'\to\mathbf{v}}]$ is the matrix of $\Phi^{-1}_{\{\mathbf{v}\}} \circ \Phi_{\{\mathbf{v}'\}}$, just check what the image of $\vec{\mathbf{e}}_i$ is under $\Phi^{-1}_{\{\mathbf{v}\}} \circ \Phi_{\{\mathbf{v}'\}}$. We have $\Phi_{\{\mathbf{v}'\}}(\vec{\mathbf{e}}_i) = \mathbf{v}'_i$, and $\Phi^{-1}_{\{\mathbf{v}\}}$ then returns the coordinates of $\mathbf{v}'_i$ with respect to the (old) basis vectors $\mathbf{v}_j$. □

In diagram form this gives

$$\begin{array}{ccc} \mathbb{R}^n & \stackrel{[P_{\mathbf{v}'\to\mathbf{v}}]}{\longrightarrow} & \mathbb{R}^n \\ {\scriptstyle \Phi_{\{\mathbf{v}'\}}}\searrow & & \nearrow{\scriptstyle \Phi^{-1}_{\{\mathbf{v}\}}} \\ & V & \end{array} . \qquad 2.6.26$$

Think of the change of basis matrix as a foreign language dictionary; the existence of such a dictionary implies the existence of a world of objects and ideas that can be assigned different names in different languages:

$$\begin{array}{ccc} \text{Names} & \overset{\text{Fr.-Eng. Dictionary}}{\longrightarrow} & \text{Names} \\ \text{in French} & & \text{in English} \\ \text{“Name to obj.” function.} \searrow & & \nearrow \text{“Obj. to name” function} \\ & \text{Objects \& Ideas} & \end{array} \qquad 2.6.27$$

Note that unlike $\mathbb{R}^3$, for which the "obvious" basis is the standard basis vectors, the subspace $V \subset \mathbb{R}^3$ in Example 2.6.18 does not come with a distinguished basis. We found these bases by randomly choosing two linearly independent vectors with three entries that sum to 0; any two vectors that meet those requirements form a basis for V.

The requirement "entries that sum to 0" is why we can get away with only two basis vectors, although V is a subset of $\mathbb{R}^3$; the 2-dimensional structure of V is built into the choice of basis vectors.

Example 2.6.18 (Change of basis matrix). Suppose V is the plane of equation $x+y+z=0$ in $\mathbb{R}^3$. A basis in V consists of two basis vectors (since V is a plane) each with three entries (since V is a subspace of $\mathbb{R}^3$). Suppose our bases consist of

$$\underbrace{\vec{\mathbf{v}}_1 = \begin{bmatrix} 1 \\ -1 \\ 0 \end{bmatrix}, \ \vec{\mathbf{v}}_2 = \begin{bmatrix} 1 \\ 0 \\ -1 \end{bmatrix}}_{\text{basis } \{\mathbf{v}\}} \quad \text{and} \quad \underbrace{\vec{\mathbf{v}}_1' = \begin{bmatrix} 0 \\ 1 \\ -1 \end{bmatrix}, \ \vec{\mathbf{v}}_2' = \begin{bmatrix} 1 \\ 1 \\ -2 \end{bmatrix}}_{\text{basis } \{\mathbf{v}'\}}. \qquad 2.6.28$$

What is the change of basis matrix $[P_{\mathbf{v}' \to \mathbf{v}}]$? Since $\vec{\mathbf{v}}_1' = -\vec{\mathbf{v}}_1 + \vec{\mathbf{v}}_2$, the coordinates of $\vec{\mathbf{v}}_1'$ expressed in terms of $\vec{\mathbf{v}}_1$ and $\vec{\mathbf{v}}_2$ are $p_{1,1} = -1$ and $p_{2,1} = 1$, giving the first column of the matrix. Similarly, $\vec{\mathbf{v}}_2' = -\vec{\mathbf{v}}_1 + 2\vec{\mathbf{v}}_2$, giving coordinates $p_{1,2} = -1$ and $p_{2,2} = 2$. So the change of basis matrix is

$$[P_{\mathbf{v}' \to \mathbf{v}}] = \begin{bmatrix} p_{1,1} & p_{1,2} \\ p_{2,1} & p_{2,2} \end{bmatrix} = \begin{bmatrix} -1 & -1 \\ 1 & 2 \end{bmatrix}. \qquad 2.6.29$$

The tricky point when changing bases is not getting confused about what direction you are going. We adopted the notation $[P_{\mathbf{v}' \to \mathbf{v}}]$ in hopes that it will help you remember that this matrix translates a vector written in $\{\mathbf{v}'\}$ to the corresponding vector written in $\{\mathbf{v}\}$.

If, for example, $\vec{\mathbf{w}} = \begin{bmatrix} 1 \\ 3 \\ -4 \end{bmatrix}$ in the standard basis, then

$$\vec{\mathbf{w}} = 2\vec{\mathbf{v}}_1' + \vec{\mathbf{v}}_2', \quad \text{i.e., } \begin{bmatrix} 2 \\ 1 \end{bmatrix} \text{ in the basis } \{\mathbf{v}'\}. \qquad 2.6.30$$

In the basis $\{\mathbf{v}\}$, the same vector would be $\vec{\mathbf{w}} = -3\vec{\mathbf{v}}_1 + 4\vec{\mathbf{v}}_2$, since

$$\overbrace{\begin{bmatrix} -1 & -1 \\ 1 & 2 \end{bmatrix} \begin{bmatrix} 2 \\ 1 \end{bmatrix} = \begin{bmatrix} -3 \\ 4 \end{bmatrix}}^{[P_{\mathbf{v}' \to \mathbf{v}}]\vec{\mathbf{a}} = \vec{\mathbf{b}}, \text{ equation 2.6.24}}. \quad \text{Indeed,} \quad -3 \overbrace{\begin{bmatrix} 1 \\ -1 \\ 0 \end{bmatrix}}^{\vec{\mathbf{v}}_1} + 4 \overbrace{\begin{bmatrix} 1 \\ 0 \\ -1 \end{bmatrix}}^{\vec{\mathbf{v}}_2} = \overbrace{\begin{bmatrix} 1 \\ 3 \\ -4 \end{bmatrix}}^{\vec{\mathbf{w}}}. \quad \triangle$$

When translating from a basis $\{\vec{\mathbf{v}}'\}$ of $\mathbb{R}^n$ into the standard basis, the columns of the change of basis matrix are simply the elements of $\{\vec{\mathbf{v}}'\}$:

$$[P_{\vec{\mathbf{v}}' \to \vec{\mathbf{e}}}] = [\vec{\mathbf{v}}_1', \dots, \vec{\mathbf{v}}_n'], \qquad 2.6.31$$

since the entries of $\vec{\mathbf{v}}_i'$ and its coordinates when it is expressed as a linear combination of the standard basis vectors are the same:

$$\vec{\mathbf{v}}_i' = \begin{bmatrix} v_1' \\ \vdots \\ v_n' \end{bmatrix} = v_1'\vec{\mathbf{e}}_1 + \cdots + v_n'\vec{\mathbf{e}}_n. \qquad 2.6.32$$

(The coefficients $v'_1, \dots, v'_n$ are the coefficients $p_{1,i}, \dots, p_{n,i}$ of equation 2.6.22, and thus the entries of the ith column of the change of basis matrix.) But remember (Example 2.6.18) that this is not the case in general!

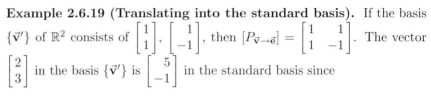

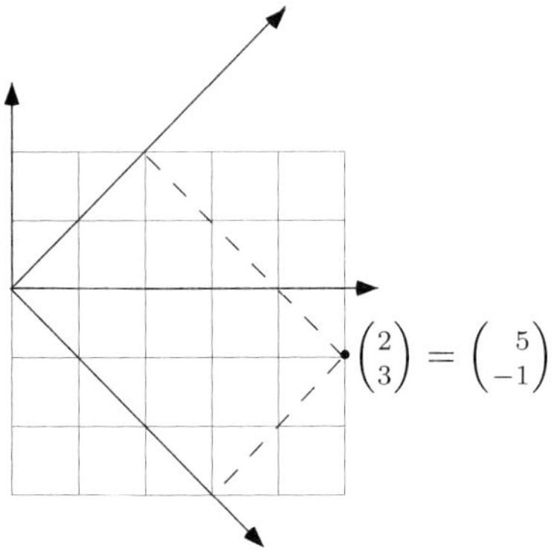

FIGURE 2.6.1.

The point $\begin{pmatrix} 5 \\ -1 \end{pmatrix}$ with respect to the standard basis is the point $\begin{pmatrix} 2 \\ 3 \end{pmatrix}$ with respect to the basis $\{\vec{\mathbf{v}}'\}$ of Example 2.6.19.

Example 2.6.19 (Translating into the standard basis). If the basis $\{\vec{\mathbf{v}}'\}$ of $\mathbb{R}^2$ consists of $\begin{bmatrix} 1 \\ 1 \end{bmatrix}, \begin{bmatrix} 1 \\ -1 \end{bmatrix}$, then $[P_{\vec{\mathbf{v}}'\to\vec{\mathbf{e}}}] = \begin{bmatrix} 1 & 1 \\ 1 & -1 \end{bmatrix}$. The vector $\begin{bmatrix} 2 \\ 3 \end{bmatrix}$ in the basis $\{\vec{\mathbf{v}}'\}$ is $\begin{bmatrix} 5 \\ -1 \end{bmatrix}$ in the standard basis since

$$[P_{\vec{\mathbf{v}}'\to\vec{\mathbf{e}}}]\begin{bmatrix} 2 \\ 3 \end{bmatrix} = \begin{bmatrix} 5 \\ -1 \end{bmatrix}. \qquad 2.6.33$$

This is illustrated by Figure 2.6.1.

Given the basis vectors $\begin{bmatrix} 1 \\ 2 \end{bmatrix}$ and $\begin{bmatrix} 2 \\ -1 \end{bmatrix}$, and the vector $\vec{\mathbf{v}} = \begin{bmatrix} 1 \\ -1 \end{bmatrix}$ written in that basis, what is $\vec{\mathbf{v}}$ written in the standard basis?[13] △

The change of basis formula

Suppose that V is a vector space of dimension n, and W is a vector space of dimension m. Let $\{\mathbf{v}\}$, $\{\mathbf{v}'\}$ be two bases of V, let $\{\mathbf{w}\}$, $\{\mathbf{w}'\}$ be two bases of W, and let $T: V \to W$ be a linear transformation. Then the change of basis formula relates the matrix of T with respect to the bases $\{\mathbf{v}\}$, $\{\mathbf{w}\}$ to the matrix of T with respect to the bases $\{\mathbf{v}'\}$, $\{\mathbf{w}'\}$.

In the first line of equation 2.6.35, we add the identity twice, once in the form $\Phi_{\{\mathbf{w}\}} \circ \Phi^{-1}_{\{\mathbf{w}\}}$, and once in the form $\Phi^{-1}_{\{\mathbf{v}\}} \circ \Phi_{\{\mathbf{v}\}}$.

Theorem 2.6.20 (Change of basis formula).

$$[T]_{\{\mathbf{v}'\},\{\mathbf{w}'\}} = [P_{\mathbf{w}'\to\mathbf{w}}]^{-1}[T]_{\{\mathbf{v}\},\{\mathbf{w}\}}[P_{\mathbf{v}'\to\mathbf{v}}]. \qquad 2.6.34$$

Proof. Write everything using the "concrete to abstract" transformations:

$$[T]_{\{\mathbf{v}'\},\{\mathbf{w}'\}} \underbrace{=}_{\text{eq. 2.6.20}} \Phi^{-1}_{\{\mathbf{w}'\}} \circ T \circ \Phi_{\{\mathbf{v}'\}} = \underbrace{\Phi^{-1}_{\{\mathbf{w}'\}} \circ \Phi_{\{\mathbf{w}\}}}_{[P_{\mathbf{w}'\to\mathbf{w}}]^{-1}} \circ \underbrace{\Phi^{-1}_{\{\mathbf{w}\}} \circ T \circ \Phi_{\{\mathbf{v}\}}}_{[T]_{\{\mathbf{v}\},\{\mathbf{w}\}}} \circ \underbrace{\Phi^{-1}_{\{\mathbf{v}\}} \circ \Phi_{\{\mathbf{v}'\}}}_{[P_{\mathbf{v}'\to\mathbf{v}}]}$$

$$= [P_{\mathbf{w}'\to\mathbf{w}}]^{-1}\,[T]_{\{\mathbf{v}\},\{\mathbf{w}\}}\,[P_{\mathbf{v}'\to\mathbf{v}}]. \qquad 2.6.35$$

This computation is illustrated by the diagram below.

$$\begin{array}{ccccccc}
 & \mathbb{R}^n & & \xrightarrow{[T]_{\{\mathbf{v}'\},\{\mathbf{w}'\}}} & & \mathbb{R}^m & \\
 & & \searrow \Phi_{\{\mathbf{v}'\}} & & \Phi_{\{\mathbf{w}'\}} \swarrow & & \\
[P_{\mathbf{v}'\to\mathbf{v}}] \downarrow & & V & \xrightarrow{T} & W & & \downarrow [P_{\mathbf{w}'\to\mathbf{w}}] \\
 & & \nearrow \Phi_{\{\mathbf{v}\}} & & \Phi_{\{\mathbf{w}\}} \nwarrow & & \\
 & \mathbb{R}^n & & \xrightarrow[{[T]_{\{\mathbf{v}\},\{\mathbf{w}\}}}]{} & & \mathbb{R}^m &
\end{array} \qquad 2.6.36$$

[13] In the standard basis, $\vec{\mathbf{v}} = \begin{bmatrix} -1 \\ 3 \end{bmatrix}$.

In it, we use (twice) the definition of a matrix with respect to bases, and (also twice) the definition of the change of basis matrix. □

Exercise 2.21 asks you to show that in a vector space of dimension n, more than n vectors are never linearly independent, and fewer than n vectors never span.

The realization that the dimension of a vector space needed to be well defined was a turning point in the development of linear algebra.

The dimension of a vector space

In Section 2.4 we proved (Proposition and Definition 2.4.21) that any two bases of a subspace $E \subset \mathbb{R}^n$ have the same number of elements. We can now show the same thing for an abstract vector space.

Proposition and Definition 2.6.21 (Dimension of a vector space). *If a vector space V has a finite basis, then all its bases are finite and contain the same number of elements, called the dimension of V.*

Proof. The proof is essentially the same as the proof of Proposition and Definition 2.4.21. It is shorter because we have developed the appropriate machinery, mainly, the change of basis matrix. Let $\{\mathbf{v}\} = \mathbf{v}_1, \dots, \mathbf{v}_k$ and $\{\mathbf{w}\} = \mathbf{w}_1, \dots, \mathbf{w}_p$ be two bases of a vector space V. The linear transformation

$$\underbrace{\Phi^{-1}_{\{\mathbf{w}\}} \circ \Phi_{\{\mathbf{v}\}} : \mathbb{R}^k \to \mathbb{R}^p}_{\text{change of basis matrix}} \quad \text{(i.e. } \mathbb{R}^k \underset{\Phi_{\{\mathbf{v}\}}}{\longrightarrow} V \underset{\Phi^{-1}_{\{\mathbf{w}\}}}{\longrightarrow} \mathbb{R}^p), \qquad 2.6.37$$

How do we know that $\Phi^{-1}_{\{\mathbf{w}\}}$ and $\Phi^{-1}_{\{\mathbf{v}\}}$ exist? By Proposition 2.6.15, the fact that $\{\mathbf{v}\}$ and $\{\mathbf{w}\}$ are bases means that $\Phi_{\{\mathbf{v}\}}$ and $\Phi_{\{\mathbf{w}\}}$ are invertible.

is given by a $p \times k$ matrix, which is invertible, since we can undo it using $\Phi^{-1}_{\{\mathbf{v}\}} \circ \Phi_{\{\mathbf{w}\}}$. But only square matrices can be invertible. Thus $k = p$. □

Remark. There is something a bit miraculous about the proof of this theorem; we prove an important result about abstract vector spaces using a matrix that seems to drop out of the sky. Without the material developed earlier in this chapter, this result would be quite difficult to prove. △

Example 2.6.22 (Dimension of vector spaces). The space $\text{Mat}\,(n, m)$ is a vector space of dimension nm. The space P_k of polynomials of degree at most k is a vector space of dimension $k + 1$. △

Earlier we talked loosely of "finite-dimensional" and "infinite-dimensional" vector spaces. Now we can be precise: a vector space is *finite dimensional* if it has a finite basis; it is infinite dimensional if it does not.

Example 2.6.23 (An infinite-dimensional vector space). The vector space $\mathcal{C}[0, 1]$ of continuous functions on $[0, 1]$, which we saw in Example 2.6.2, is infinite dimensional. Intuitively it is not hard to see that there are too many such functions to be expressed with any finite number of basis vectors. We can pin it down as follows.

Assume functions $f_1, \dots, f_n$ are a basis, and pick $n + 1$ distinct points $0 = x_1 < x_2 \cdots < x_{n+1} = 1$ in $[0, 1]$. Then given any values $c_1, \dots, c_{n+1}$, there certainly exists a continuous function $f(x)$ with $f(x_i) = c_i$, for instance, the piecewise linear one whose graph consists of the line segments joining up the points $\begin{pmatrix} x_i \\ c_i \end{pmatrix}$.

If we can write $f = \sum_{k=1}^{n} a_k f_k$, then evaluating at the x_i, we get

$$f(x_i) = c_i = \sum_{k=1}^{n} a_k f_k(x_i), \qquad i = 1, \dots, n+1. \qquad 2.6.38$$

For given c_i, this is a system of $n+1$ equations for the n unknowns $a_1, \dots, a_n$; we know by Theorem 2.2.1 that for appropriate c_i the equations will be incompatible. Therefore, functions exist that are not linear combinations of $f_1, \dots, f_n$, so $f_1, \dots, f_n$ do not span $\mathcal{C}[0,1]$. △

Exercises for Section 2.6

Exercise 2.6.1: In Example 1.7.17 we identified $\text{Mat}(n,n)$ with $\mathbb{R}^{n^2}$ by setting

$$\begin{bmatrix} a & b \\ c & d \end{bmatrix} = \begin{bmatrix} a \\ b \\ c \\ d \end{bmatrix}.$$

2.6.1 a. In Example 1.7.17 we identified $\text{Mat}(n,n)$ with $\mathbb{R}^{n^2}$. What basis of $\text{Mat}(2,2)$ does this correspond to? Write $\begin{bmatrix} 2 & 1 \\ 5 & 4 \end{bmatrix}$ as a linear combination of those basis vectors.

b. What basis of $\text{Mat}(2,2)$ corresponds to the identification $\begin{bmatrix} a & b \\ c & d \end{bmatrix} = \begin{bmatrix} a \\ c \\ b \\ d \end{bmatrix}$?

Write $\begin{bmatrix} 2 & 1 \\ 5 & 4 \end{bmatrix}$ as a linear combination of those basis vectors.

2.6.2 Show that the space $\mathcal{C}(0,1)$ of continuous real-valued functions defined for $0 < x < 1$ (Example 2.6.2) satisfies all eight requirements for a vector space.

2.6.3 Let $\{\mathbf{v}\}$ be the basis of $\text{Mat}(2,2)$ consisting of

$$\mathbf{v}_1 = \begin{bmatrix} 1 & 0 \\ 0 & 1 \end{bmatrix}, \quad \mathbf{v}_2 = \begin{bmatrix} 1 & 0 \\ 0 & -1 \end{bmatrix}, \quad \mathbf{v}_3 = \begin{bmatrix} 0 & 1 \\ 1 & 0 \end{bmatrix}, \quad \mathbf{v}_4 = \begin{bmatrix} 0 & -1 \\ 1 & 0 \end{bmatrix}$$

What is $\Phi_{\{\mathbf{v}\}}\left(\begin{bmatrix} a \\ b \\ c \\ d \end{bmatrix}\right)$?

2.6.4 Let V and W be subspaces of $\mathbb{R}^n$ such that $\dim V + \dim W \geq n$. Show that $\dim V \cap W \geq \dim V + \dim W - n$.

2.6.5 Set $A = \begin{bmatrix} a & b \\ c & d \end{bmatrix}$. Let $L_A : \text{Mat}(2,2) \to \text{Mat}(2,2)$ be the linear transformation "multiplication on the left by A", i.e., $L_A : B \mapsto AB$. Similarly, define $R_A : \text{Mat}(2,2) \to \text{Mat}(2,2)$ by $R_A : B \mapsto BA$. Identify $\text{Mat}(2,2)$ with $\mathbb{R}^4$, choosing as the basis $\{\mathbf{v}\}$ of $\text{Mat}(2,2)$

$$\begin{bmatrix} 1 & 0 \\ 0 & 0 \end{bmatrix}, \begin{bmatrix} 0 & 1 \\ 0 & 0 \end{bmatrix}, \begin{bmatrix} 0 & 0 \\ 1 & 0 \end{bmatrix}, \begin{bmatrix} 0 & 0 \\ 0 & 1 \end{bmatrix}, \text{ so } \Phi_{\{\mathbf{v}\}}\begin{pmatrix} a \\ b \\ c \\ d \end{pmatrix} = \begin{bmatrix} a & b \\ c & d \end{bmatrix}$$

Exercise 2.6.5, part a: Here we use the notation of Proposition and Definition 2.6.16. The matrix $[R_A]_{\{\mathbf{v}\},\{\mathbf{v}\}}$ is the matrix of R_A with respect to the basis $\{\mathbf{v}\}$ in both domain and codomain; the matrix $[L_A]_{\{\mathbf{v}\},\{\mathbf{v}\}}$ is the matrix of L_A with respect to the same bases.

a. Show that

$$[R_A]_{\{\mathbf{v}\},\{\mathbf{v}\}} = \begin{bmatrix} a & c & 0 & 0 \\ b & d & 0 & 0 \\ 0 & 0 & a & c \\ 0 & 0 & b & d \end{bmatrix} \quad \text{and} \quad [L_A]_{\{\mathbf{v}\},\{\mathbf{v}\}} = \begin{bmatrix} a & 0 & b & 0 \\ 0 & a & 0 & b \\ c & 0 & d & 0 \\ 0 & c & 0 & d \end{bmatrix}.$$

b. Compute $|R_A|$ and $|L_A|$ in terms of $|A|$.

2.6.6 a. As in Exercise 2.6.5, find the matrices for the linear transformations $L_A : \text{Mat}\,(3,3) \to \text{Mat}\,(3,3)$ and $R_A : \text{Mat}\,(3,3) \to \text{Mat}\,(3,3)$ when A is a 3×3 matrix.

b. Repeat when A is an $n \times n$ matrix.

c. In the $n \times n$ case, compute $|R_A|$ and $|L_A|$ in terms of $|A|$.

2.6.7 Let V be the vector space of C^1 functions on $(0, 1)$. Which of the following are subspaces of V?

a. $\{ f \in V \mid f(x) = f'(x) + 1 \}$ b. $\{ f \in V \mid f(x) = xf'(x) \}$

c. $\{ f \in V \mid f(x) = (f'(x))^2 \}$

Exercise 2.6.8: By "identified to $\mathbb{R}^3$ via the coefficients" we mean that

$$p(x) = a + bx + cx^2 \in P_2$$

is identified to

$$\begin{pmatrix} a \\ b \\ c \end{pmatrix}.$$

Exercise 2.6.8, part c: The pattern should become clear after the first three.

2.6.8 Let P_2 be the space of polynomials of degree at most two, identified to $\mathbb{R}^3$ via the coefficients. Consider the mapping $T : P_2 \to P_2$ given by

$$T(p)(x) = (x^2 + 1)p''(x) - xp'(x) + 2p(x).$$

a. Verify that T is linear, i.e., that $T(ap_1 + bp_2) = aT(p_1) + bT(p_2)$.

b. Choose the basis of P_2 consisting of the polynomials $p_1(x) = 1$, $p_2(x) = x$, $p_3(x) = x^2$. Denote by $\Phi_{\{p\}} : \mathbb{R}^3 \to P_2$ the corresponding concrete-to-abstract linear transformation. Show that the matrix of $\Phi^{-1}_{\{p\}} \circ T \circ \Phi_{\{p\}}$ is $\begin{bmatrix} 2 & 0 & 2 \\ 0 & 1 & 0 \\ 0 & 0 & 2 \end{bmatrix}$.

c. Using the basis $1, x, x^2, \dots, x^n$, compute the matrices of the same differential operator T, viewed first as an operator from P_3 to P_3, then from P_4 to P_4, $\dots$, P_n to P_n (polynomials of degree at most $3, 4, \dots, n$).

Exercise 2.6.9 says that any linearly independent set can be extended to form a basis. In French treatments of linear algebra, this is called the *theorem of the incomplete basis*; it plus induction can be used to prove all the theorems of linear algebra in Chapter 2.

2.6.9 a. Let V be a finite-dimensional vector space, and let $\mathbf{v}_1, \dots, \mathbf{v}_k \in V$ be linearly independent vectors. Show that there exist $\mathbf{v}_{k+1}, \dots, \mathbf{v}_n \in V$ such that $\mathbf{v}_1, \dots, \mathbf{v}_n$ is a basis of V.

b. Let V be a finite-dimensional vector space, and let $\mathbf{v}_1, \dots, \mathbf{v}_k \in V$ be a set of vectors that span V. Show that there exists a subset $i_1, i_2, \dots, i_m$ of $\{1, 2, \dots, k\}$ such that $\mathbf{v}_{i_1}, \dots, \mathbf{v}_{i_m}$ is a basis of V.

2.6.10 Let $A = \begin{bmatrix} 1 & a \\ 0 & 1 \end{bmatrix}$ and $B = \begin{bmatrix} 1 & 0 \\ b & 1 \end{bmatrix}$. Using the "standard" identification of $\text{Mat}\,(2,2)$ with $\mathbb{R}^4$, what is the dimension of

$$\text{Span}\,(A, B, AB, BA) \quad \text{in terms of } a \text{ and } b?$$

2.7 Eigenvectors and eigenvalues

When Werner Heisenberg discovered 'matrix' mechanics in 1925, he didn't know what a matrix was (Max Born had to tell him), and neither Heisenberg nor Born knew what to make of the appearance of matrices in the context of the atom. (David Hilbert is reported to have told them to go look for a differential equation with the same eigenvalues, if that would make them happier. They did not follow Hilbert's well-meant advice and thereby may have missed discovering the Schrödinger wave equation.)—M. R. Schroeder, *Mathematical Intelligencer*, Vol. 7, No. 4

In Section 2.6 we discussed the change of basis matrix. We change bases when a problem is easier in a different basis. Most often this comes down to some problem being easier in an *eigenbasis*: a basis of *eigenvectors*. Before defining the terms, let's give an example.

Example 2.7.1 (Fibonacci numbers). *Fibonacci numbers* are the numbers $1, 1, 2, 3, 5, 8, 13, \ldots$ defined by $a_0 = a_1 = 1$ and $a_{n+1} = a_n + a_{n-1}$ for $n \geq 1$. We propose to prove the formula

$$a_n = \frac{5 + \sqrt{5}}{10}\left(\frac{1 + \sqrt{5}}{2}\right)^n + \frac{5 - \sqrt{5}}{10}\left(\frac{1 - \sqrt{5}}{2}\right)^n. \qquad 2.7.1$$

Equation 2.7.1 is quite amazing: it isn't even obvious that the right side is an integer! The key to understanding it is the matrix equation

$$\begin{bmatrix} a_n \\ a_{n+1} \end{bmatrix} = \begin{bmatrix} 0 & 1 \\ 1 & 1 \end{bmatrix} \begin{bmatrix} a_{n-1} \\ a_n \end{bmatrix}. \qquad 2.7.2$$

The first equation says $a_n = a_n$, and the second says $a_{n+1} = a_n + a_{n-1}$. What have we gained? We see that

$$\begin{bmatrix} a_n \\ a_{n+1} \end{bmatrix} = \begin{bmatrix} 0 & 1 \\ 1 & 1 \end{bmatrix} \begin{bmatrix} a_{n-1} \\ a_n \end{bmatrix} = \begin{bmatrix} 0 & 1 \\ 1 & 1 \end{bmatrix}^2 \begin{bmatrix} a_{n-2} \\ a_{n-1} \end{bmatrix} = \cdots = \begin{bmatrix} 0 & 1 \\ 1 & 1 \end{bmatrix}^n \begin{bmatrix} 1 \\ 1 \end{bmatrix}. \qquad 2.7.3$$

This looks useful, until you start computing the powers of the matrix, and discover that you are just computing Fibonacci numbers the old way. Is there a more effective way to compute the powers of a matrix?

Certainly there is an easy way to compute the powers of a *diagonal* matrix; you just raise all the diagonal entries to that power:

$$\begin{bmatrix} c_1 & 0 & \cdots & 0 \\ 0 & c_2 & \cdots & 0 \\ \vdots & \vdots & \ddots & \vdots \\ 0 & 0 & \cdots & c_m \end{bmatrix}^n = \begin{bmatrix} c_1^n & 0 & \cdots & 0 \\ 0 & c_2^n & \cdots & 0 \\ \vdots & \vdots & \ddots & \vdots \\ 0 & 0 & \cdots & c_m^n \end{bmatrix}. \qquad 2.7.4$$

We will see that we can turn this to our advantage. Let

$$P = \begin{bmatrix} 2 & 2 \\ 1 + \sqrt{5} & 1 - \sqrt{5} \end{bmatrix}, \text{ so } P^{-1} = \frac{1}{4\sqrt{5}}\begin{bmatrix} \sqrt{5} - 1 & 2 \\ \sqrt{5} + 1 & -2 \end{bmatrix} \qquad 2.7.5$$

and "observe" that if we set $A = \begin{bmatrix} 0 & 1 \\ 1 & 1 \end{bmatrix}$, then

$$P^{-1}AP = \begin{bmatrix} \frac{1+\sqrt{5}}{2} & 0 \\ 0 & \frac{1-\sqrt{5}}{2} \end{bmatrix} \text{ is diagonal.} \qquad 2.7.6$$

This has the following remarkable consequence:

$$(P^{-1}AP)^n = (P^{-1}A\underbrace{P)(P^{-1}}_{I}AP)\ldots(P^{-1}A\underbrace{P)(P^{-1}}_{I}AP) = P^{-1}A^nP, \qquad 2.7.7$$

Equation 2.7.1: Note that for large n, the second term in equation 2.7.1 is negligible, since

$$\frac{1 - \sqrt{5}}{2} \approx -0.618.$$

For instance, $\left(\frac{1-\sqrt{5}}{2}\right)^{1000}$ starts with at least 200 zeros after the decimal point. But the first term grows exponentially, since

$$\frac{1 + \sqrt{5}}{2} \approx 1.618.$$

Assume $n = 1000$. Using logarithms base 10 to evaluate the first term, we see that

$$\log_{10} a_{1000} \approx \log_{10} \frac{5 + \sqrt{5}}{10}$$
$$+ \left(1000 \times \log_{10} \frac{1 + \sqrt{5}}{2}\right)$$
$$\approx -.1405 + 1000 \times .20899$$
$$\approx 208.85,$$

so a_{1000} has 209 digits.

Equation 2.7.6: The matrices A and $P^{-1}AP$ represent the same linear transformation, in different bases. Such matrices are said to be *conjugate*. In this case, it is much easier to compute with $P^{-1}AP$.

which we can rewrite as

$$A^n = P(P^{-1}AP)^n P^{-1}. \qquad 2.7.8$$

So applying equations 2.7.4, 2.7.6, and 2.7.8 leads to

$$\overbrace{\begin{bmatrix} 0 & 1 \\ 1 & 1 \end{bmatrix}^n}^{A^n} = \overbrace{\begin{bmatrix} 2 & 2 \\ 1+\sqrt{5} & 1-\sqrt{5} \end{bmatrix}}^{P} \overbrace{\begin{bmatrix} \left(\frac{1+\sqrt{5}}{2}\right)^n & 0 \\ 0 & \left(\frac{1-\sqrt{5}}{2}\right)^n \end{bmatrix}}^{(P^{-1}AP)^n} \overbrace{\begin{bmatrix} \frac{\sqrt{5}-1}{4\sqrt{5}} & \frac{2}{4\sqrt{5}} \\ \frac{\sqrt{5}+1}{4\sqrt{5}} & \frac{-2}{4\sqrt{5}} \end{bmatrix}}^{P^{-1}}$$

$$= \frac{1}{2\sqrt{5}} \begin{bmatrix} \left(\frac{1+\sqrt{5}}{2}\right)^n (\sqrt{5}-1) + \left(\frac{1-\sqrt{5}}{2}\right)^n (1+\sqrt{5}) & 2\left(\frac{1+\sqrt{5}}{2}\right)^n - 2\left(\frac{1-\sqrt{5}}{2}\right)^n \\ \left(\frac{1+\sqrt{5}}{2}\right)^{n+1} (\sqrt{5}-1) + \left(\frac{1-\sqrt{5}}{2}\right)^{n+1} (\sqrt{5}+1) & 2\left(\frac{1+\sqrt{5}}{2}\right)^{n+1} - 2\left(\frac{1-\sqrt{5}}{2}\right)^{n+1} \end{bmatrix}$$

This confirms that our mysterious and miraculous equation 2.7.1 for a_n is correct: if we multiply this value of $\begin{bmatrix} 0 & 1 \\ 1 & 1 \end{bmatrix}^n$ by $\begin{bmatrix} 1 \\ 1 \end{bmatrix}$, the matrix equation 2.7.3 says that the top line of the product is a_n and indeed we get

$$a_n = \frac{5+\sqrt{5}}{10}\left(\frac{1+\sqrt{5}}{2}\right)^n + \frac{5-\sqrt{5}}{10}\left(\frac{1-\sqrt{5}}{2}\right)^n. \qquad 2.7.9$$

So instead of computing $\begin{bmatrix} 0 & 1 \\ 1 & 1 \end{bmatrix}^n$ to determine a_n, we only need to compute the numbers $\left(\frac{1+\sqrt{5}}{2}\right)^n$ and $\left(\frac{1-\sqrt{5}}{2}\right)^n$; the problem has been *decoupled* (see Remark 2.7.4). △

How did we hit on the matrix P (equation 2.7.5) for the Fibonacci numbers? Clearly that choice of matrix was not random. Indeed, the columns of P are *eigenvectors* for $A = \begin{bmatrix} 0 & 1 \\ 1 & 1 \end{bmatrix}$, and the diagonal entries of the handy matrix $P^{-1}AP$ are their corresponding *eigenvalues*.

Definition 2.7.2: In order to have an eigenvector, a matrix must map a vector space *to itself*: $T\mathbf{v}$ is in the codomain, and $\lambda\mathbf{v}$ is in the domain. So *only square matrices can have eigenvectors.*

Often an eigenvalue is defined as a root of the characteristic polynomial (see Definition 4.8.18). Our approach to eigenvectors and eigenvalues was partially inspired by the paper "Down with determinants" by Sheldon Axler. We are not in favor of jettisoning determinants, which we use heavily in Chapters 4, 5, and 6 as a way to measure volume. But determinants are more or less uncomputable for large matrices. The procedure given later in this section for finding eigenvectors and eigenvalues is more amenable to computation. However, it is still not a practical algorithm even for medium-sized matrices; the algorithms that actually work are the QR algorithm or Jacobi's method for symmetric matrices; see Exercise 3.29.

Recall that the Greek letter λ is pronounced "lambda".

An eigenvector must be nonzero, but 0 can be an eigenvalue: this happens if and only if $\mathbf{v} \neq \mathbf{0}$ is in the kernel of T.

Definition 2.7.2 (Eigenvector, eigenvalue, multiplicity). Let V be a complex vector space and $T: V \to V$ a linear transformation. A nonzero vector $\mathbf{v}$ such that

$$T\mathbf{v} = \lambda\mathbf{v} \qquad 2.7.10$$

for some number λ is called an *eigenvector* of T. The number λ is the corresponding *eigenvalue.* The *multiplicity* of an eigenvalue λ is the dimension of the *eigenspace* $\{\, \mathbf{v} \mid T\mathbf{v} = \lambda\mathbf{v} \,\}$.

It is obvious but of great importance that if $T\mathbf{v} = \lambda\mathbf{v}$, then $T^k\mathbf{v} = \lambda^k\mathbf{v}$ (for instance, $TT\mathbf{v} = T\lambda\mathbf{v} = \lambda TV = \lambda^2\mathbf{v}$).

Definition 2.7.3 (Eigenbasis). A basis for a complex vector space V is an *eigenbasis* of V for a linear transformation T if each element of the basis is an eigenvector of T.

The columns of P form an eigenbasis of $\mathbb{R}^2$ for $A = \begin{bmatrix} 0 & 1 \\ 1 & 1 \end{bmatrix}$:

$$\begin{bmatrix} 0 & 1 \\ 1 & 1 \end{bmatrix} \underbrace{\begin{bmatrix} 2 \\ 1+\sqrt{5} \end{bmatrix}}_{\text{1st column of } P} = \underbrace{\frac{1+\sqrt{5}}{2}}_{\text{eigenvalue}} \begin{bmatrix} 2 \\ 1+\sqrt{5} \end{bmatrix}. \qquad 2.7.11$$

$$\begin{bmatrix} 0 & 1 \\ 1 & 1 \end{bmatrix} \underbrace{\begin{bmatrix} 2 \\ 1-\sqrt{5} \end{bmatrix}}_{\text{2nd column of } P} = \underbrace{\frac{1-\sqrt{5}}{2}}_{\text{eigenvalue}} \begin{bmatrix} 2 \\ 1-\sqrt{5} \end{bmatrix}. \qquad 2.7.12$$

Remark 2.7.4. Using eigenvectors, we can analyze again the decoupling of the Fibonacci example, where one term grows exponentially, the other shrinks exponentially. Write $\begin{bmatrix} a_n \\ a_{n+1} \end{bmatrix}$ as a linear combination of the eigenvectors:

$$\begin{bmatrix} a_n \\ a_{n+1} \end{bmatrix} = c_n \begin{bmatrix} 2 \\ 1+\sqrt{5} \end{bmatrix} + d_n \begin{bmatrix} 2 \\ 1-\sqrt{5} \end{bmatrix}. \qquad 2.7.13$$

The following computation, setting as before $A = \begin{bmatrix} 0 & 1 \\ 1 & 1 \end{bmatrix}$, shows that $c_{n+1} = c_n \frac{1+\sqrt{5}}{2}$ and $d_{n+1} = d_n \frac{1-\sqrt{5}}{2}$:

$$c_{n+1}\begin{bmatrix} 2 \\ 1+\sqrt{5} \end{bmatrix} + d_{n+1}\begin{bmatrix} 2 \\ 1-\sqrt{5} \end{bmatrix} = \begin{bmatrix} a_{n+1} \\ a_{n+2} \end{bmatrix} = A\begin{bmatrix} a_n \\ a_{n+1} \end{bmatrix} = A\left(c_n\begin{bmatrix} 2 \\ 1+\sqrt{5} \end{bmatrix} + d_n\begin{bmatrix} 2 \\ 1-\sqrt{5} \end{bmatrix}\right)$$

$$= c_n A\begin{bmatrix} 2 \\ 1+\sqrt{5} \end{bmatrix} + d_n A\begin{bmatrix} 2 \\ 1-\sqrt{5} \end{bmatrix} \underbrace{=}_{\text{Def. 2.7.2}} c_n\frac{1+\sqrt{5}}{2}\begin{bmatrix} 2 \\ 1+\sqrt{5} \end{bmatrix} + d_n\frac{1-\sqrt{5}}{2}\begin{bmatrix} 2 \\ 1-\sqrt{5} \end{bmatrix}.$$

So

$$c_n = \left(\frac{1+\sqrt{5}}{2}\right)^n c_0 \quad \text{and} \quad d_n = \left(\frac{1-\sqrt{5}}{2}\right)^n d_0; \qquad 2.7.14$$

the c_n grow exponentially, but the d_n tend to 0 exponentially.

Figure 2.7.1 illustrates this geometrically. If you were to treat the $\begin{bmatrix} a_n \\ a_{n+1} \end{bmatrix}$ as stepping stones, stepping from $\begin{bmatrix} 1 \\ 1 \end{bmatrix}$ to $\begin{bmatrix} 1 \\ 2 \end{bmatrix}$ to $\begin{bmatrix} 2 \\ 3 \end{bmatrix}$ to $\begin{bmatrix} 3 \\ 5 \end{bmatrix}$ to $\begin{bmatrix} 5 \\ 8 \end{bmatrix}$ to ..., you would start with baby steps but would quickly switch to giant steps; soon you would find it impossible to bridge the ever-growing distance from one stepping stone to the next. And since the component of $\begin{bmatrix} a_n \\ a_{n+1} \end{bmatrix}$ in the direction of the eigenvector $\begin{bmatrix} 2 \\ 1-\sqrt{5} \end{bmatrix}$ shrinks as n gets large, you would soon be moving almost exclusively in the direction of the eigenvector $\begin{bmatrix} 2 \\ 1+\sqrt{5} \end{bmatrix}$. △

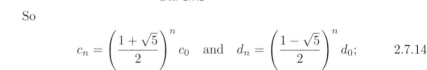

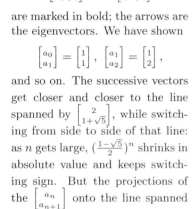

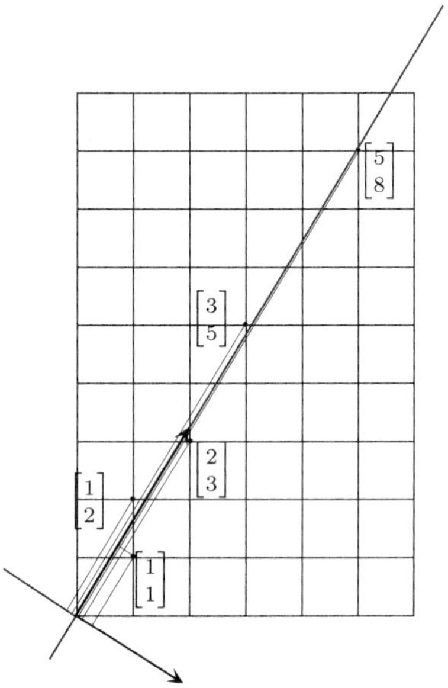

FIGURE 2.7.1.
The axes spanned by the eigenvectors

$$\begin{bmatrix} 2 \\ 1+\sqrt{5} \end{bmatrix} \text{ and } \begin{bmatrix} 2 \\ 1-\sqrt{5} \end{bmatrix}$$

are marked in bold; the arrows are the eigenvectors. We have shown

$$\begin{bmatrix} a_0 \\ a_1 \end{bmatrix} = \begin{bmatrix} 1 \\ 1 \end{bmatrix}, \begin{bmatrix} a_1 \\ a_2 \end{bmatrix} = \begin{bmatrix} 1 \\ 2 \end{bmatrix},$$

and so on. The successive vectors get closer and closer to the line spanned by $\begin{bmatrix} 2 \\ 1+\sqrt{5} \end{bmatrix}$, while switching from side to side of that line: as n gets large, $(\frac{1-\sqrt{5}}{2})^n$ shrinks in absolute value and keeps switching sign. But the projections of the $\begin{bmatrix} a_n \\ a_{n+1} \end{bmatrix}$ onto the line spanned by $\begin{bmatrix} 2 \\ 1+\sqrt{5} \end{bmatrix}$ get farther and farther apart: as n gets large, $(\frac{1+\sqrt{5}}{2})^n$ grows exponentially.

Diagonalization and eigenvectors

The theory of eigenvectors often goes under the name "diagonalization". Proposition 2.7.5 explains why.

Proposition 2.7.5 (Diagonalization and eigenvectors). *Let A be an $n \times n$ matrix and $P = [\vec{\mathbf{v}}_1, \dots, \vec{\mathbf{v}}_n]$ an invertible $n \times n$ matrix.*

1. *The eigenvalues of A and the eigenvalues of $P^{-1}AP$ coincide.*
2. *If $(\vec{\mathbf{v}}_1, \dots, \vec{\mathbf{v}}_n)$ is an eigenbasis of $\mathbb{C}^n$ for A, with $A\vec{\mathbf{v}}_i = \lambda_i \vec{\mathbf{v}}_i$, then $P^{-1}AP$ is a diagonal matrix with diagonal entries $\lambda_1, \dots, \lambda_n$.*
3. *Conversely, if $P^{-1}AP$ is diagonal with diagonal entries λ_i, the columns of P are eigenvectors of A, with eigenvalues λ_i.*

$$\begin{bmatrix} a & 0 & 0 \\ 0 & b & 0 \\ 0 & 0 & c \end{bmatrix} \begin{bmatrix} x \\ y \\ z \end{bmatrix} = \begin{bmatrix} ax \\ by \\ cz \end{bmatrix}.$$

If you multiply a diagonal matrix by a vector, each entry of the vector multiplies a single and distinct entry of the matrix; the various entries do not interact. An $n \times n$ diagonal matrix behaves like n one-dimensional matrices, i.e., like n numbers. It is hard to overstate the importance of this *decoupling*, in both pure and applied mathematics.

A matrix A is *diagonalizable* if there exists a matrix P such that $P^{-1}AP$ is diagonal. We already encountered this in equation 2.7.6. Here is another example.

Example 2.7.6. If $A = \begin{bmatrix} 1 & -1 & 0 \\ -1 & 2 & -1 \\ 0 & -1 & 1 \end{bmatrix}$ and $P = \begin{bmatrix} 1 & -1 & 1 \\ 1 & 0 & -2 \\ 1 & 1 & 1 \end{bmatrix}$, then

$$P^{-1}AP = \begin{bmatrix} 0 & 0 & 0 \\ 0 & 1 & 0 \\ 0 & 0 & 3 \end{bmatrix} \quad \text{is diagonal, so} \quad \begin{bmatrix} 1 \\ 1 \\ 1 \end{bmatrix}, \begin{bmatrix} -1 \\ 0 \\ 1 \end{bmatrix}, \begin{bmatrix} 1 \\ -2 \\ 1 \end{bmatrix}$$

are eigenvectors for A, with eigenvalues 0, 1, and 3. In the eigenbasis $\begin{bmatrix} 1 \\ 1 \\ 1 \end{bmatrix}, \begin{bmatrix} -1 \\ 0 \\ 1 \end{bmatrix}, \begin{bmatrix} 1 \\ -2 \\ 1 \end{bmatrix}$ of $\mathbb{R}^3$, the matrix A becomes $\begin{bmatrix} 0 & 0 & 0 \\ 0 & 1 & 0 \\ 0 & 0 & 3 \end{bmatrix}$. △

Remark. The formula $P^{-1}AP$ should remind you of the change of basis formula (Theorem 2.6.20):

$$[T]_{\{\mathbf{v}'\},\{\mathbf{w}'\}} = [P_{\mathbf{w}' \to \mathbf{w}}]^{-1} [T]_{\{\mathbf{v}\},\{\mathbf{w}\}} [P_{\mathbf{v}' \to \mathbf{v}}].$$

Indeed, $P^{-1}AP$ is a special case of that formula, where the domain is the same as the codomain. Using the notation of the change of basis formula, we would say that if $\{\mathbf{v}\}$ is an eigenbasis of $\mathbb{R}^n$ for A and $\{\mathbf{e}\}$ is the standard basis, then the matrix of A with respect to the eigenbasis is

$$[A]_{\{\mathbf{v}\},\{\mathbf{v}\}} = [P_{\{\mathbf{v}\} \to \{\mathbf{e}\}}]^{-1} [A]_{\{\mathbf{e}\},\{\mathbf{e}\}} [P_{\{\mathbf{v}\} \to \{\mathbf{e}\}}]. \quad \triangle$$

Proof of Proposition 2.7.5. 1. If $A\vec{\mathbf{v}} = \lambda \vec{\mathbf{v}}$, then $P^{-1}\vec{\mathbf{v}}$ is an eigenvector of $P^{-1}AP$ with eigenvalue λ:

$$P^{-1}AP(P^{-1}\vec{\mathbf{v}}) = P^{-1}A\vec{\mathbf{v}} = P^{-1}\lambda\vec{\mathbf{v}} = \lambda(P^{-1}\vec{\mathbf{v}}). \qquad 2.7.15$$

2. If for each i we have $A\vec{\mathbf{v}}_i = \lambda_i \vec{\mathbf{v}}_i$, then

$$P^{-1}AP\vec{\mathbf{e}}_i = P^{-1}A\vec{\mathbf{v}}_i = P^{-1}\lambda_i\vec{\mathbf{v}}_i = P^{-1}\lambda_i P\vec{\mathbf{e}}_i = \lambda_i P^{-1}P\vec{\mathbf{e}}_i = \lambda_i\vec{\mathbf{e}}_i, \qquad 2.7.16$$

$$\text{so we have} \quad P^{-1}AP = \begin{bmatrix} \lambda_1 & 0 & \dots & 0 \\ 0 & \lambda_2 & \dots & 0 \\ \vdots & \vdots & \ddots & \vdots \\ 0 & 0 & \dots & \lambda_n \end{bmatrix}. \qquad 2.7.17$$

3. Recall that $\vec{\mathbf{v}}_i = P\vec{\mathbf{e}}_i$. If $P^{-1}AP$ is a diagonal matrix with diagonal entries λ_i, then the columns of P are eigenvectors of A and the λ_i are the corresponding eigenvalues:

$$A\vec{\mathbf{v}}_i = AP\vec{\mathbf{e}}_i = PP^{-1}AP\vec{\mathbf{e}}_i = P\begin{bmatrix} \lambda_1 & 0 & \dots & 0 \\ 0 & \lambda_2 & \dots & 0 \\ \vdots & \vdots & \ddots & \vdots \\ 0 & 0 & \dots & \lambda_n \end{bmatrix}\vec{\mathbf{e}}_i = P\lambda_i\vec{\mathbf{e}}_i = \lambda_i\vec{\mathbf{v}}_i. \quad \square$$

When do eigenbases exist?

We say "essentially unique" because if you multiply each eigenvector of an eigenbasis by some number the result will still be an eigenbasis.

The converse of Theorem 2.7.7 is not true; k eigenvectors can be linearly independent even if some of the eigenvectors share the same eigenvalue. For instance, every basis of $\mathbb{C}^n$ is an eigenbasis for the identity, but the identity has only one eigenvalue: the number 1.

Since the λ are all distinct, $\lambda_j - \lambda_i \neq 0$, and the a_i cannot all be zero because if they were, equation 2.7.18 would give $\mathbf{v}_j = \mathbf{0}$, but $\mathbf{v}_j$ cannot be zero, since it is an eigenvector.

When do eigenbases exist and how do we find them? Might a linear transformation have lots of eigenvectors and eigenvalues? Theorem 2.7.7 says that if an eigenbasis exists with distinct eigenvalues, it is essentially unique.

Theorem 2.7.7 (Eigenvectors with distinct eigenvalues are linearly independent). *If $A: V \to V$ is a linear transformation, and $\mathbf{v}_1, \dots, \mathbf{v}_k$ are eigenvectors of A with distinct eigenvalues $\lambda_1, \dots, \lambda_k$, then $\mathbf{v}_1, \dots, \mathbf{v}_k$ are linearly independent.*

In particular, if $\dim V = n$, there are at most n eigenvalues.

Proof. We will prove this by contradiction. If $\mathbf{v}_1, \dots, \mathbf{v}_k$ are not linearly independent, then there is a first vector $\mathbf{v}_j$ that is a linear combination of the earlier ones. Thus we can write

$$\mathbf{v}_j = a_1\mathbf{v}_1 + \dots + a_{j-1}\mathbf{v}_{j-1}, \tag{2.7.18}$$

where at least one coefficient is not zero, say a_i. Apply $\lambda_j I - A$ to both sides to get

$$\begin{aligned} \mathbf{0} = \underbrace{(\lambda_j I - A)\mathbf{v}_j}_{\lambda_j\mathbf{v}_j - \lambda_j\mathbf{v}_j} &= (\lambda_j I - A)\big(a_1\mathbf{v}_1 + \dots + a_{j-1}\mathbf{v}_{j-1}\big) \\ &= a_1(\lambda_j I - A)\mathbf{v}_1 + \dots + a_{j-1}(\lambda_j I - A)\mathbf{v}_{j-1} \\ &= a_1(\lambda_j - \lambda_1)\mathbf{v}_1 + \dots + a_{j-1}(\lambda_j - \lambda_{j-1})\mathbf{v}_{j-1}. \end{aligned} \tag{2.7.19}$$

By hypothesis, the λ_i are all different, so the coefficients $a_k(\lambda_j - \lambda_k)$ can't all be zero. Suppose the ith term $a_i(\lambda_j - \lambda_i)\vec{\mathbf{v}}_i$ is the last nonzero term; move it to the other side of the equality and divide out by the (nonzero) coefficient $-a_i(\lambda_j - \lambda_i)$. This expresses some $\mathbf{v}_i$ with $i < j$ as a linear combination of the $\mathbf{v}_1, \dots, \mathbf{v}_{j-1}$, which contradicts the assumption that $\mathbf{v}_j$ is the first vector that is a linear combination of the earlier linearly independent ones. $\square$

Procedure for finding eigenvectors

Our procedure for finding eigenvectors will require knowing how to evaluate polynomials on matrices. Let $p(t) = a_0 + a_1 t + \dots + a_{m-1}t^{m-1} + t^m$ be a polynomial in one variable. Then we can write the matrix

$$p(A) = a_0 I + a_1 A + \cdots + a_{m-1}A^{m-1} + A^m. \qquad 2.7.20$$

Procedure for finding eigenvectors: The matrix A can have real entries, and $\vec{\mathbf{w}}$ can be real, but to use the fundamental theorem of algebra we must allow the root λ to be complex.

We might worry that there might be lots of such polynomials for different choices of $\vec{\mathbf{w}}$, and that these polynomials might be unrelated. This is not the case. For instance, if for some $\vec{\mathbf{w}}$ we find a polynomial p of degree n with distinct roots, then by Theorem 2.7.7 these roots are *all* the eigenvalues, so starting with any $\vec{\mathbf{w}}_1$ will lead to a polynomial p_1 whose roots are among those of p.

Equation 2.7.23: The polynomial q is obtained by dividing p by $t-\lambda$, so $\deg q = m-1$. Thus

$$q(A) = b_0 I + b_1 A + \cdots + A^{m-1}$$

for appropriate $b_0, \dots, b_{m-1}$, and the coefficient of A^{m-1} is 1. Thus

$$q(A)\vec{\mathbf{w}} = b_0\vec{\mathbf{w}} + b_1 A\vec{\mathbf{w}} + \cdots + A^{m-1}\vec{\mathbf{w}},$$

is a linear combination of the m vectors $\vec{\mathbf{w}}, A\vec{\mathbf{w}}, \dots, A^{m-1}\vec{\mathbf{w}}$.

This is a new and powerful tool: it establishes a link between row reduction (linear dependence, kernels, etc.) and polynomials (factoring, roots, long division, the fundamental theorem of algebra).

We now show how to find at least one eigenvector for any $n \times n$ complex matrix A. For any nonzero $\vec{\mathbf{w}} \in \mathbb{C}^n$, there is a smallest m such that $A^m\vec{\mathbf{w}}$ is a linear combination of $\vec{\mathbf{w}}, A\vec{\mathbf{w}}, \dots, A^{m-1}\vec{\mathbf{w}}$: there exist numbers $a_0, \dots, a_{m-1}$ such that

$$a_0\vec{\mathbf{w}} + a_1 A\vec{\mathbf{w}} + \cdots + a_{m-1}A^{m-1}\vec{\mathbf{w}} + A^m\vec{\mathbf{w}} = \vec{\mathbf{0}}. \qquad 2.7.21$$

Then we can use the coefficients to define a polynomial:

$$p(t) \overset{\text{def}}{=} a_0 + a_1 t + \cdots + a_{m-1}t^{m-1} + t^m \qquad 2.7.22$$

satisfies $p(A)\vec{\mathbf{w}} = \vec{\mathbf{0}}$, and it is the lowest degree nonzero polynomial with this property.

By the fundamental theorem of algebra (Theorem 1.6.14), p has at least one root λ, so we can write

$$p(t) = (t-\lambda)q(t) \qquad 2.7.23$$

for some polynomial q of degree $m-1$. Define $\vec{\mathbf{v}} = q(A)\vec{\mathbf{w}}$. Then

$$(A-\lambda I)\vec{\mathbf{v}} = (A-\lambda I)\big(q(A)\vec{\mathbf{w}}\big) = \Big((A-\lambda I)(q(A))\Big)\vec{\mathbf{w}} = p(A)\vec{\mathbf{w}} = \vec{\mathbf{0}}, \qquad 2.7.24$$

so $A\vec{\mathbf{v}} = \lambda\vec{\mathbf{v}}$. To see that $\vec{\mathbf{v}}$ is an eigenvector with eigenvalue λ, we still need to check that $\vec{\mathbf{v}} \neq \vec{\mathbf{0}}$. Since $m > 0$ is the smallest integer such that $A^m\vec{\mathbf{w}}$ is a linear combination of $\vec{\mathbf{w}}, A\vec{\mathbf{w}}, \dots, A^{m-1}\vec{\mathbf{w}}$, the vectors $\vec{\mathbf{w}}, \dots, A^{m-1}\vec{\mathbf{w}}$ are linearly independent. Since the coefficient of $A^{m-1}\vec{\mathbf{w}}$ is not 0 (see the margin note for equation 2.7.23), $\vec{\mathbf{v}} \neq \vec{\mathbf{0}}$.

This procedure will yield one eigenvector for each distinct root of p.

How can we find the polynomial p? That is exactly the kind of problem row reduction handles well. Consider the matrix

$$M = [\vec{\mathbf{w}}, A\vec{\mathbf{w}}, A^2\vec{\mathbf{w}}, \dots, A^n\vec{\mathbf{w}}]. \qquad 2.7.25$$

This matrix has $n+1$ columns and n rows, so if you row reduce it to $\widetilde{M}$, then $\widetilde{M}$ will certainly have a nonpivotal column. Let the $(m+1)$st column be the first nonpivotal column, and consider the first $m+1$ columns of M as an augmented matrix to solve the system of linear equations

$$x_0\vec{\mathbf{w}} + x_1 A\vec{\mathbf{w}} + \cdots + x_{m-1}A^{m-1}\vec{\mathbf{w}} = A^m\vec{\mathbf{w}}. \qquad 2.7.26$$

The matrix row reduces to

$$\begin{bmatrix} 1 & 0 & \dots & 0 & b_0 \\ 0 & 1 & \dots & 0 & b_1 \\ \vdots & \vdots & \ddots & \vdots & \vdots \\ 0 & 0 & \dots & 1 & b_{m-1} \\ 0 & 0 & \dots & 0 & 0 \\ \vdots & \vdots & & \vdots & \vdots \end{bmatrix}, \qquad 2.7.27$$

and by Theorem 2.2.1, the b_i are the solution of our system of equations:

$$b_0\vec{\mathbf{w}} + b_1 A\vec{\mathbf{w}} + \cdots + b_{m-1}A^{m-1}\vec{\mathbf{w}} = A^m\vec{\mathbf{w}}. \qquad 2.7.28$$

Thus p is given by $p(t) = -b_0 - b_1 t - \cdots - b_{m-1}t^{m-1} + t^m$.

Example 2.7.8 (Finding an eigenbasis). Let $A : \mathbb{R}^3 \to \mathbb{R}^3$ be the linear transformation $A = \begin{bmatrix} 1 & -1 & 0 \\ -1 & 2 & -1 \\ 0 & -1 & 1 \end{bmatrix}$, and for $\vec{\mathbf{w}}$ use $\vec{\mathbf{e}}_1$. We want to find the first vector in the list $\vec{\mathbf{e}}_1, A\vec{\mathbf{e}}_1, A^2\vec{\mathbf{e}}_1, A^3\vec{\mathbf{e}}_1$ that is a linear combination of the preceding vectors. Because of the zero entries in $\vec{\mathbf{e}}_1$ and $A\vec{\mathbf{e}}_1$, the first three vectors are clearly linearly independent, but the fourth must then be a linear combination of the first three. We row reduce

$$\begin{bmatrix} 1 & 1 & 2 & 5 \\ 0 & -1 & -3 & -9 \\ 0 & 0 & 1 & 4 \end{bmatrix} \quad \text{to get} \quad \begin{bmatrix} 1 & 0 & 0 & 0 \\ 0 & 1 & 0 & -3 \\ 0 & 0 & 1 & 4 \end{bmatrix}, \qquad 2.7.29$$

which tells us that $0\vec{\mathbf{e}}_1 - 3A\vec{\mathbf{e}}_1 + 4A^2\vec{\mathbf{e}}_1 = A^3\vec{\mathbf{e}}_1$. Thus p is the polynomial $p(t) = t^3 - 4t^2 + 3t$, with roots 0, 1, and 3.

1. $(\lambda = 0)$. Write $p(t) = t(t^2 - 4t + 3) \overset{\text{def}}{=} tq_1(t)$ to find the eigenvector

$$q_1(A)\vec{\mathbf{e}}_1 = \begin{bmatrix} 2 \\ -3 \\ 1 \end{bmatrix} - 4\begin{bmatrix} 1 \\ -1 \\ 0 \end{bmatrix} + 3\begin{bmatrix} 1 \\ 0 \\ 0 \end{bmatrix} = \begin{bmatrix} 1 \\ 1 \\ 1 \end{bmatrix}. \qquad 2.7.30$$

2. $(\lambda = 1)$. Write $p(t) = (t-1)(t^2 - 3t) \overset{\text{def}}{=} (t-1)q_2(t)$ to find the eigenvector

$$q_2(A)\vec{\mathbf{e}}_1 = \begin{bmatrix} 2 \\ -3 \\ 1 \end{bmatrix} - 3\begin{bmatrix} 1 \\ -1 \\ 0 \end{bmatrix} = \begin{bmatrix} -1 \\ 0 \\ 1 \end{bmatrix}. \qquad 2.7.31$$

3. $(\lambda = 3)$. Write $p(t) = (t-3)(t^2 - t) \overset{\text{def}}{=} (t-3)q_3(t)$ to find the eigenvector

$$q_3(A)\vec{\mathbf{e}}_1 = \begin{bmatrix} 2 \\ -3 \\ 1 \end{bmatrix} - \begin{bmatrix} 1 \\ -1 \\ 0 \end{bmatrix} = \begin{bmatrix} 1 \\ -2 \\ 1 \end{bmatrix}. \qquad 2.7.32$$

It follows from equation 2.7.24 that the first eigenvector has eigenvalue 0, the second has eigenvalue 1, and the third eigenvalue 3. (It is also easy to check this directly.) △

Remark. The procedure we describe for finding eigenvectors is not practical even for matrices as small as 5 × 5. For one thing, it requires finding roots of polynomials without saying how to do it; for another, row reduction is too unstable for the polynomial defined in equation 2.7.22 to be accurately known. In practice, the *QR algorithm* is used. It iterates directly in the space of matrices. Although not foolproof, it is the method of choice. △

Example 2.7.8: Below we compute $A\vec{\mathbf{e}}_1$, $A^2\vec{\mathbf{e}}_1$, and $A^3\vec{\mathbf{e}}_1$:

$$\begin{bmatrix} 1 \\ 0 \\ 0 \end{bmatrix} \} \; \vec{\mathbf{e}}_1$$

$$\begin{bmatrix} 1 & -1 & 0 \\ -1 & 2 & -1 \\ 0 & -1 & 1 \end{bmatrix} \begin{bmatrix} 1 \\ -1 \\ 0 \end{bmatrix} \} \; A\vec{\mathbf{e}}_1$$

$$\begin{bmatrix} 1 & -1 & 0 \\ -1 & 2 & -1 \\ 0 & -1 & 1 \end{bmatrix} \begin{bmatrix} 2 \\ -3 \\ 1 \end{bmatrix} \} A^2\vec{\mathbf{e}}_1$$

$$\begin{bmatrix} 1 & -1 & 0 \\ -1 & 2 & -1 \\ 0 & -1 & 1 \end{bmatrix} \begin{bmatrix} 5 \\ -9 \\ 4 \end{bmatrix} \} A^3\vec{\mathbf{e}}_1$$

Equation 2.7.30: Since

$$q_1(t) = t^2 - 4t + 3,$$

we have

$$q_1(A)\vec{\mathbf{e}}_1 = A^2\vec{\mathbf{e}}_1 - 4A\vec{\mathbf{e}}_1 + 3\vec{\mathbf{e}}_1$$
$$= \begin{bmatrix} 2 \\ -3 \\ 1 \end{bmatrix} - 4\begin{bmatrix} 1 \\ -1 \\ 0 \end{bmatrix} + 3\begin{bmatrix} 1 \\ 0 \\ 0 \end{bmatrix}.$$

In the eigenbasis

$$\begin{bmatrix} 1 \\ 1 \\ 1 \end{bmatrix}, \begin{bmatrix} -1 \\ 0 \\ 1 \end{bmatrix}, \begin{bmatrix} 1 \\ -2 \\ 1 \end{bmatrix}$$

the linear transformation

$$A = \begin{bmatrix} 1 & -1 & 0 \\ -1 & 2 & -1 \\ 0 & -1 & 1 \end{bmatrix}$$

is written $\begin{bmatrix} 0 & 0 & 0 \\ 0 & 1 & 0 \\ 0 & 0 & 3 \end{bmatrix}$, i.e., if we set $P = \begin{bmatrix} 1 & -1 & 1 \\ 1 & 0 & -2 \\ 1 & 1 & 1 \end{bmatrix}$, then

$$P^{-1}AP = \begin{bmatrix} 0 & 0 & 0 \\ 0 & 1 & 0 \\ 0 & 0 & 3 \end{bmatrix}.$$

Criterion for the existence of an eigenbasis

When A is an $n \times n$ complex matrix and the polynomial p has degree n and the roots are simple, the above procedure (using a single standard basis vector) will find an eigenbasis. This is the usual case. But exceptionally, there can be complications. If $\deg p < n$, then we will have to repeat the procedure with a different vector.

But what if the roots aren't simple? Theorem 2.7.9 says that in that case, there is no eigenbasis.

Let p_i denote the polynomial constructed as in equations 2.7.21 and 2.7.22, where $\vec{\mathbf{w}} = \vec{\mathbf{e}}_i$, for $i = 1, \dots, n$. Thus p_i is the lowest degree nonzero polynomial satisfying $p_i(A)\vec{\mathbf{e}}_i = \vec{\mathbf{0}}$.

Note that it is possible for the p_i not to have simple roots. Exercise 2.7.1 asks you to show that for the matrix $\begin{bmatrix} 1 & 0 \\ 1 & 1 \end{bmatrix}$, the polynomial p_1 is $p_1(t) = (t-1)^2$.

Theorem 2.7.9. *Let A be an $n \times n$ complex matrix. There exists an eigenbasis of $\mathbb{C}^n$ for A if and only if all the roots of all the p_i are simple.*

We construct p_i by finding the smallest m_i such that $A^{m_i}\vec{\mathbf{e}}_i$ is a linear combination of

$$\vec{\mathbf{e}}_i, A\vec{\mathbf{e}}_i, \dots, A^{m_i-1}(\vec{\mathbf{e}}_i),$$

so Span (E_i) contains m_i linearly independent vectors. See equations 2.7.21 and 2.7.22.

Proof. First assume that all the roots of all the p_i are simple, and denote by m_i the degree of p_i. Let E_i be the span of $\vec{\mathbf{e}}_i, A\vec{\mathbf{e}}_i, A^2\vec{\mathbf{e}}_i, \dots$. It should be clear that this is a vector subspace of $\mathbb{C}^n$ of dimension m_i (see the margin note). Since $E_1 \cup \dots \cup E_n$ contains the standard basis vectors, we have Span $(E_1 \cup \dots \cup E_n) = \mathbb{C}^n$.

The procedure described above yields for each root $\lambda_{i,j}$ of p_i an eigenvector $\vec{\mathbf{v}}_{i,j}$, and since $\vec{\mathbf{v}}_{i,1}, \dots, \vec{\mathbf{v}}_{i,m_i}$ are m_i linearly independent vectors in the space E_i of dimension m_i, they span E_i. Thus the set

$$\bigcup_{i=1}^{n} \bigcup_{j=1}^{m_i} \{\vec{\mathbf{v}}_{i,j}\} \qquad 2.7.33$$

spans all of $\mathbb{C}^n$, and we can select a basis of $\mathbb{C}^n$ from it. This proves the direction (all roots simple $\Longrightarrow$ there exists a basis of eigenvectors).

If A s a real matrix, you might want a real basis. If any eigenvalues are complex, there is no real eigenbasis, but often the real and imaginary parts of eigenvectors will serve the same purposes as eigenvectors. If $A\mathbf{u}_j = \mu\mathbf{u}_j$ with $\mu \notin \mathbb{R}$, then $A\bar{\mathbf{u}}_j = \bar{\mu}\bar{\mathbf{u}}_j$, and the vectors $\mathbf{u}_j$ and $\bar{\mathbf{u}}_j$ are linearly independent in $\mathbb{C}^n$, by Theorem 2.7.7. Moreover, the vectors

$$\mathbf{u}_j + \bar{\mathbf{u}}_j, \quad \frac{\mathbf{u}_j - \bar{\mathbf{u}}_j}{i}$$

are two linearly independent elements of $\mathbb{R}^n$ in Span $(\mathbf{u}_j, \bar{\mathbf{u}}_j)$. See Exercise 2.14.

For the converse, let $\lambda_1, \dots, \lambda_k$ be the distinct eigenvalues of A. Set

$$p(t) = \prod_{j=1}^{k} (t - \lambda_j), \qquad 2.7.34$$

so that p has simple roots. We will show that for all i,

$$p(A)(\vec{\mathbf{e}}_i) = \vec{\mathbf{0}}. \qquad 2.7.35$$

This will imply that all p_i divide p, hence all p_i have simple roots.

Set $V_j = \ker(A - \lambda_j I)$, so that the elements of V_j are precisely the eigenvectors of A with eigenvalue λ_j. Since $V_1 \cup \dots \cup V_k$ contains all the eigenvectors, our hypothesis that there is a basis of eigenvectors says that $V_1 \cup \dots \cup V_k$ spans $\mathbb{C}^n$. Thus each $\vec{\mathbf{e}}_i$ is a linear combination of elements in the V_j: we can write

$$\vec{\mathbf{e}}_i = a_{i,1}\vec{\mathbf{w}}_{i,1} + \dots + a_{i,k}\vec{\mathbf{w}}_{i,k}, \quad \text{with } \vec{\mathbf{w}}_{i,j} \in V_j. \qquad 2.7.36$$

By Definition 2.7.2, $\dim V_j$ is the multiplicity of λ_j as an eigenvalue of A. If $k < n$, then at least one V_j will satisfy $\dim V_j > 1$, and will contain $\dim V_j > 1$ elements of the eigenbasis. The corresponding $\vec{\mathbf{w}}_{i,j}$ will then be a linear combination of these eigenvectors.

Equation 2.7.37: Note that when A is a square matrix,

$$p_1(A)p_2(A) = p_2(A)p_1(A),$$

since A commutes with itself and all its powers. If this were not the case, then in equation 2.7.37 we would not be able to change the order of matrices in the product when we go from the first line to the second, putting the term $(A - \lambda_l I)$ at the end.

Since by definition $V_j = \ker\,(A - \lambda_j I)$, we have $(A - \lambda_\ell I)\vec{\mathbf{w}}_{i,\ell} = \vec{\mathbf{0}}$ for all ℓ from 1 to k, so $p(A)(\vec{\mathbf{e}}_i) = \vec{\mathbf{0}}$:

$$p(A)(\vec{\mathbf{e}}_i) = \underbrace{\left(\prod_{j=1}^{k}(A-\lambda_j I)\right)}_{p(A)}\underbrace{\left(\sum_{\ell=1}^{k} a_{i,\ell}\vec{\mathbf{w}}_{i,\ell}\right)}_{\vec{\mathbf{e}}_i} \qquad 2.7.37$$

$$= \sum_{\ell=1}^{k} a_{i,\ell}\left(\overbrace{\left(\prod_{j\neq\ell}(A-\lambda_j I)\right)\underbrace{(A-\lambda_\ell I)\vec{\mathbf{w}}_{i,\ell}}_{\vec{0}}}^{p(A)\vec{\mathbf{w}}_{i,\ell},\ \text{factored}}\right) = \vec{\mathbf{0}}.$$

It follows that p_i divides p with remainder 0. Indeed, using long division we can write $p = p_i q_i + r_i$, where degree $r_i <$ degree p_i. By definition, $p_i(A)\vec{\mathbf{e}}_i = \vec{\mathbf{0}}$, so by equation 2.7.37,

$$\vec{\mathbf{0}} = p(A)\vec{\mathbf{e}}_i = p_i(A)q_i(A)\vec{\mathbf{e}}_i + r_i(A)\vec{\mathbf{e}}_i = \vec{\mathbf{0}} + r_i(A)\vec{\mathbf{e}}_i. \qquad 2.7.38$$

Thus $r_i(A)\vec{\mathbf{e}}_i = \vec{\mathbf{0}}$. Since degree $r_i <$ degree p_i and p_i is the nonzero polynomial of lowest degree such that $p_i(A)\vec{\mathbf{e}}_i = \vec{\mathbf{0}}$, it follows that $r_i = 0$. Since p has simple roots, so does p_i. □ Theorem 2.7.9

The Perron-Frobenius theorem

Square real matrices with all entries nonnegative are important in many fields; examples include adjacency matrices of graphs (combinatorics) and transition matrices of Markov processes (probability).

Let A be a real matrix (perhaps a column matrix, i.e., a vector). We will write

- $A \geq \mathbf{0}$ if all entries $a_{i,j}$ satisfy $a_{i,j} \geq 0$
- $A > \mathbf{0}$ if all entries $a_{i,j}$ satisfy $a_{i,j} > 0$.

Exercise 2.7.7 extends the Perron-Frobenius theorem, weakening the hypothesis $A > \mathbf{0}$ to $A \geq \mathbf{0}$ and adding the hypothesis that there exists k such that $A^k > \mathbf{0}$.

Theorem 2.7.10: The vector $\vec{\mathbf{v}}$ (and any multiple of $\vec{\mathbf{v}}$, thus with the same eigenvalue) is called the *leading eigenvector*.

Theorem 2.7.10 (Perron-Frobenius theorem). *If A is a real $n \times n$ matrix such that $A > \mathbf{0}$, there exists a unique real eigenvector $\vec{\mathbf{v}} > \mathbf{0}$ with $|\vec{\mathbf{v}}| = 1$. This eigenvector has a simple real eigenvalue $\lambda > 0$, called the leading or dominant eigenvalue. Any other eigenvalue $\mu \in \mathbb{C}$ of A satisfies $|\mu| < \lambda$.*

Further, $\vec{\mathbf{v}}$ can be found by iteration: for any $\vec{\mathbf{w}}$ with $\vec{\mathbf{w}} \geq \mathbf{0}$, we have $\vec{\mathbf{v}} = \lim_{k\to\infty} A^k(\vec{\mathbf{w}})/|A^k(\vec{\mathbf{w}})|$.

Lemma 2.7.11 is key to the proof.

Lemma 2.7.11: We write

$$\vec{\mathbf{v}} \leq \vec{\mathbf{w}} \quad \text{with} \quad \vec{\mathbf{v}} \neq \vec{\mathbf{w}}$$

rather than $\vec{\mathbf{v}} < \vec{\mathbf{w}}$ because some entries of $\vec{\mathbf{v}}$ and $\vec{\mathbf{w}}$ may coincide.

Lemma 2.7.11. *If $A > \mathbf{0}$ and $\mathbf{0} \leq \vec{\mathbf{v}} \leq \vec{\mathbf{w}}$ with $\vec{\mathbf{v}} \neq \vec{\mathbf{w}}$, then $A\vec{\mathbf{v}} < A\vec{\mathbf{w}}$.*

Proof. In order to have $\vec{\mathbf{v}} \neq \vec{\mathbf{w}}$, at least one entry of $\vec{\mathbf{w}}$ must be larger than the corresponding entry of $\vec{\mathbf{v}}$. That entry will be multiplied by some strictly positive entry of $A > \mathbf{0}$, leading to the strict inequality in the product. □

Formula 2.7.39: For instance, set $A = \begin{bmatrix} 1 & 2 \\ 3 & 4 \end{bmatrix}$ and $\vec{\mathbf{w}} = \begin{bmatrix} \frac{1}{\sqrt{2}} \\ \frac{1}{\sqrt{2}} \end{bmatrix}$.

Then $A\vec{\mathbf{w}}' = \begin{bmatrix} \frac{3}{\sqrt{2}} \\ \frac{7}{\sqrt{2}} \end{bmatrix}$, so

$$g(\vec{\mathbf{w}}) = \inf\{3, 7\} = 3$$

and

$$g : (\vec{\mathbf{w}})\vec{\mathbf{w}} = \begin{bmatrix} \frac{3}{\sqrt{2}} \\ \frac{3}{\sqrt{2}} \end{bmatrix} \le \begin{bmatrix} \frac{3}{\sqrt{2}} \\ \frac{7}{\sqrt{2}} \end{bmatrix}.$$

The "finitely many" in the second paragraph of the proof refers, in a neighborhood of any $\mathbf{w} \in \Delta$, to those $A\mathbf{w}/w_i$ for which $w_i \neq 0$. The terms where $w_i = 0$ are all arbitrarily large in a sufficiently small neighborhood of $\mathbf{w}$, so that they cannot contribute to the infimum.

If $A > \mathbf{0}$, then $A^\top > \mathbf{0}$.

We have $\tilde{\mathbf{v}} > \mathbf{0}$ since $A\tilde{\mathbf{v}} = \mu\tilde{\mathbf{v}}$ and $A\tilde{\mathbf{v}} > \mathbf{0}$; we have $\vec{\mathbf{v}}_1^\top > \mathbf{0}$ since $A^\top\vec{\mathbf{v}}_1 = \lambda_1\vec{\mathbf{v}}_1$ and $A^\top\vec{\mathbf{v}}_1 > \mathbf{0}$.

Earlier we showed that $\vec{\mathbf{v}}$ exists, using a nonconstructive proof. Here we show how to construct it.

Proof of Theorem 2.7.10. Let $Q \subset \mathbb{R}^n$ be the "quadrant" $\vec{\mathbf{w}} \ge \mathbf{0}$, and let Δ be the set of unit vectors in Q. If $\vec{\mathbf{w}} \in \Delta$, then $\vec{\mathbf{w}} \ge \mathbf{0}$ and $\vec{\mathbf{w}} \neq \mathbf{0}$, so (by Lemma 2.7.11) $A\vec{\mathbf{w}} > \mathbf{0}$.

Consider the function $g : \Delta \to \mathbb{R}$ given by

$$g : \vec{\mathbf{w}} \mapsto \inf\left\{ \frac{(A\vec{\mathbf{w}})_1}{w_1}, \frac{(A\vec{\mathbf{w}})_2}{w_2}, \dots, \frac{(A\vec{\mathbf{w}})_n}{w_n} \right\}; \qquad 2.7.39$$

then $g(\vec{\mathbf{w}})\vec{\mathbf{w}} \le A\vec{\mathbf{w}}$ for all $\vec{\mathbf{w}} \in \Delta$, and $g(\vec{\mathbf{w}})$ is the largest number for which this is true.

Since g is an infimum of finitely many continuous functions $\Delta \to \mathbb{R}$, the function g is continuous; since Δ is compact, g achieves its maximum at some $\vec{\mathbf{v}} \in \Delta$. Let us see that $\vec{\mathbf{v}}$ is an eigenvector of A with eigenvalue $\lambda \stackrel{\text{def}}{=} g(\vec{\mathbf{v}})$. By contradiction, suppose that $g(\vec{\mathbf{v}})\vec{\mathbf{v}} \neq A\vec{\mathbf{v}}$. By Lemma 2.7.11, $g(\vec{\mathbf{v}})\mathbf{v} \le A\vec{\mathbf{v}}$ and $g(\vec{\mathbf{v}})\vec{\mathbf{v}} \neq A\vec{\mathbf{v}}$ imply

$$g(\vec{\mathbf{v}})A\vec{\mathbf{v}} = Ag(\vec{\mathbf{v}})\vec{\mathbf{v}} < AA\vec{\mathbf{v}}. \qquad 2.7.40$$

Since the inequality $g(\vec{\mathbf{v}})A\vec{\mathbf{v}} < AA\vec{\mathbf{v}}$ is strict, this contradicts the hypothesis that $\vec{\mathbf{v}}$ is an element of Δ at which g achieves its maximum: $g(A\vec{\mathbf{v}})$ is the largest number such that $g(A\vec{\mathbf{v}})A\vec{\mathbf{v}} \le AA\vec{\mathbf{v}}$, so $g(A\vec{\mathbf{v}}) > g(\vec{\mathbf{v}})$.

Thus $A\vec{\mathbf{v}} = g(\vec{\mathbf{v}})\vec{\mathbf{v}}$: we have found an eigenvector $\vec{\mathbf{v}} > \mathbf{0}$, with real eigenvalue $\lambda = g(\vec{\mathbf{v}}) > 0$. Exactly the same argument shows that there exists a vector $\vec{\mathbf{v}}_1 \in \Delta$ such that $A^\top\vec{\mathbf{v}}_1 = \lambda_1\vec{\mathbf{v}}_1$, i.e., $\vec{\mathbf{v}}_1^\top A = \lambda_1\vec{\mathbf{v}}_1^\top$. Let $\tilde{\mathbf{v}} \in \Delta$ be an eigenvector of A with eigenvalue μ. Then

$$\lambda_1\vec{\mathbf{v}}_1^\top\tilde{\mathbf{v}} = \vec{\mathbf{v}}_1^\top A\tilde{\mathbf{v}} = \vec{\mathbf{v}}_1^\top\mu\tilde{\mathbf{v}} = \mu\vec{\mathbf{v}}_1^\top\tilde{\mathbf{v}}. \qquad 2.7.41$$

Since (see the margin) $\vec{\mathbf{v}}_1^\top > \mathbf{0}$ and $\tilde{\mathbf{v}} > \mathbf{0}$, we have $\vec{\mathbf{v}}_1^\top\tilde{\mathbf{v}} \neq 0$, so $\mu = \lambda_1$. Since $\tilde{\mathbf{v}}$ is any eigenvector of A in Δ, it could be $\vec{\mathbf{v}}$, so $\mu = \lambda$. Further, $\vec{\mathbf{v}}$ is the unique eigenvector of A in Δ: if $\vec{\mathbf{v}}' \neq \vec{\mathbf{v}}$ were another such eigenvector, then the restriction of A to the subspace spanned by $\vec{\mathbf{v}}$ and $\vec{\mathbf{v}}'$ would be λid, contradicting $A(\Delta) \subset \overset{\circ}{Q}$ (which is true since $A > \mathbf{0}$).

Now let $\vec{\mathbf{w}} \in \Delta$ be any vector in Δ satisfying $\vec{\mathbf{w}} \neq \vec{\mathbf{v}}$. Then we have $g(\vec{\mathbf{w}})\vec{\mathbf{w}} \le A\vec{\mathbf{w}}$ and $g(\vec{\mathbf{w}})\vec{\mathbf{w}} \neq A\vec{\mathbf{w}}$ (since $\vec{\mathbf{v}}$ is the unique eigenvector of A in Δ), so (again using Lemma 2.7.11),

$$g(\vec{\mathbf{w}})A\vec{\mathbf{w}} = A(g(\vec{\mathbf{w}}))\vec{\mathbf{w}} < A(A\vec{\mathbf{w}}), \quad \text{so} \quad g(A\vec{\mathbf{w}}) > g(\vec{\mathbf{w}}). \qquad 2.7.42$$

Thus for any $\vec{\mathbf{w}} \in \Delta$ with $\vec{\mathbf{w}} \neq \vec{\mathbf{v}}$, the sequence $k \mapsto g(A^k\vec{\mathbf{w}})$ is strictly increasing, and so (since Δ is compact) the sequence $k \mapsto A^k\vec{\mathbf{w}}/|A^k\vec{\mathbf{w}}|$ converges to $\vec{\mathbf{v}}$.

We will now show that any other eigenvalue $\eta \in \mathbb{C}$ of A satisfies $|\eta| < \lambda$.

Let $\vec{\mathbf{u}}_1, \dots, \vec{\mathbf{u}}_m \in \mathbb{C}^n$ be the eigenvectors of A linearly independent from $\vec{\mathbf{v}}$ with eigenvalues η_i satisfying $|\eta_i| \ge \lambda$. Then $\text{Span}\,(\vec{\mathbf{v}}, \vec{\mathbf{u}}_1, \dots, \vec{\mathbf{u}}_m) \subset \mathbb{C}^n$ intersects $\mathbb{R}^n$ in a real subspace of dimension $m + 1$, since the nonreal eigenvectors come in conjugate pairs. If $m > 0$, there are elements $\vec{\mathbf{w}} \neq \vec{\mathbf{v}}$ in Δ with $\vec{\mathbf{w}} \in \text{Span}\,(\vec{\mathbf{v}}, \vec{\mathbf{u}}_1, \dots, \vec{\mathbf{u}}_m)$: they can be written

$$\vec{\mathbf{w}} = a\vec{\mathbf{v}} + \sum_i a_i\vec{\mathbf{u}}_i, \quad \text{which gives} \quad A^k\vec{\mathbf{w}} = a\lambda^k\vec{\mathbf{v}} + \sum_{i=1}^m a_i\eta_i^k\vec{\mathbf{u}}_i, \qquad 2.7.43$$

and our assumption $|\eta_i| \geq \lambda$ means that the sequence $k \mapsto A^k\vec{\mathbf{w}}/|A^k\vec{\mathbf{w}}|$ cannot converge to $\vec{\mathbf{v}}$. Since the sequence does converge to $\vec{\mathbf{v}}$, this shows that all the other eigenvalues $\eta \neq \lambda$ of A satisfy $|\eta| < \lambda$. Exercise 2.7.8 asks you to show that λ is a simple eigenvalue; this completes the proof. □

The patent for the PageRank algorithm was awarded to Stanford University, which granted Google a long-term license for 1.8 million shares of Google stock. Stanford sold the stock in 2005 for \$336 million.

A striking application of the Perron-Frobenius theorem is the PageRank algorithm Google uses to rank the importance of web pages, in order to return relevant search results.

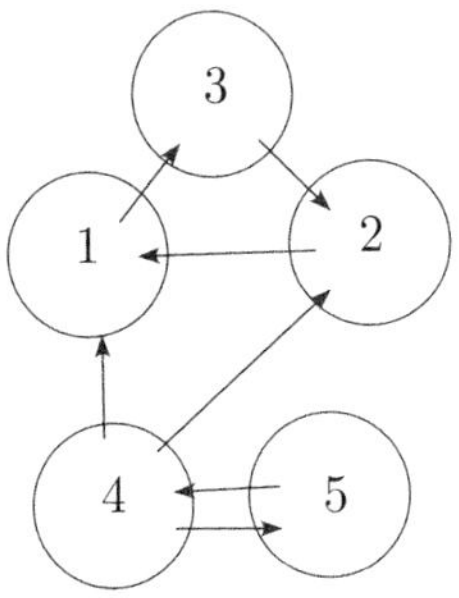

FIGURE 2.7.2.

A directed graph representing five web pages. The corresponding matrix A is

$$A = \begin{bmatrix} 0 & 1 & 0 & 1/3 & 0 \\ 0 & 0 & 1 & 1/3 & 0 \\ 1 & 0 & 0 & 0 & 0 \\ 0 & 0 & 0 & 0 & 1 \\ 0 & 0 & 0 & 1/3 & 0 \end{bmatrix}$$

The first column of matrix G is

$$\begin{bmatrix} \frac{m}{5} \\ \frac{m}{5} \\ 1 - m + \frac{m}{5} \\ \frac{m}{5} \\ \frac{m}{5} \end{bmatrix}.$$

Example 2.7.12 (Google's PageRank). Google's PageRank algorithm uses the Perron-Frobenius theorem to find a vector $\vec{\mathbf{x}}$ whose entries are the "importance" of each web page: x_i gives the importance of the ith page. Importance is defined as having lots of links from other pages ("backlinks"), with a backlink from an important page weighted more heavily than one from a less important page. Each page is given one "vote", split evenly between all the pages to which it links: if the jth page links to eight pages, the contribution of a link to a recipient page is $x_j/8$; if it links to only one page, the contribution to the recipient page is x_j.

Suppose there are n webpages (some huge number) and construct the "weighted adjacency matrix" A of the corresponding graph: $a_{i,j} = 0$ if web page j does not link to web page i, and $a_{i,j} = 1/k_j$ if j does link to web page i, and there are in all k_j links from web page j; see Figure 2.7.2, where $n = 5$. Putting $1/k_j$ rather than 1 reflects "one page, one vote". Then the multiplication $A\vec{\mathbf{x}} = \vec{\mathbf{x}}$ says that x_i (the importance of web page i) is the sum of the weighted links from all other web pages.

In order to guarantee that such a vector exists with eigenvalue 1, we need to modify the matrix A. At least in the original version of PageRank, this was done roughly as follows.

Those columns that contain any nonzero entries now sum to 1, but there may be columns of 0's, corresponding to web pages that do not link out; we replace all entries of these columns by $1/n$; now all columns sum to 1.

Next, A will still have entries that are 0, but the Perron-Frobenius theorem requires $A > \mathbf{0}$.[14] To remedy this, we define $G = (1-m)A + mB$, where B is the matrix all of whose entries are $1/n$, and m is some number satisfying $0 < m < 1$; apparently the original value of m was .15. An entry of A that was 0 becomes an entry m/n of G, and an entry of A that was $1/k_j$ becomes the entry $(1-m)/k_j + m/n$ of G.

Now we have $G > \mathbf{0}$, and the entries of each column of G sum to 1. It follows that the leading eigenvalue of $G^\top$ is 1, since the vector $\operatorname{Re}\vec{\mathbf{w}}$ that is n high with all 1's satisfies $\vec{\mathbf{w}} > \mathbf{0}$ and is an eigenvector of $G^\top$, since $G^\top\vec{\mathbf{w}} = \vec{\mathbf{w}}$. Our proof of Perron-Frobenius shows that the leading eigenvalue of G is also 1.

Now the Perron-Frobenius theorem tells us that G has an eigenvector $\vec{\mathbf{x}}$ (unique up to multiples) with eigenvalue 1: $G\vec{\mathbf{x}} = \vec{\mathbf{x}}$. Moreover, $\vec{\mathbf{x}}$ can be found by iteration. △

[14] Alternatively, we could use $A \geq \mathbf{0}$ as long as there exists k with $A^k > \mathbf{0}$.

Exercises for Section 2.7

2.7.1 Show that for $A = \begin{bmatrix} 1 & 0 \\ 1 & 1 \end{bmatrix}$, the polynomial p_1 is $p_1(t) = (t-1)^2$, so A admits no eigenbasis.

2.7.2 Let $\mathcal{C}^\infty(\mathbb{R})$ denote the vector space of infinitely differentiable functions on $\mathbb{R}$. Consider the linear transformation $D : \mathcal{C}^\infty(\mathbb{R}) \to \mathcal{C}^\infty(\mathbb{R})$ given by $D(f) = f'$.

a. Show that e^{ax} is an eigenvector of D, with eigenvalue a.

b. Show that for any integer $k > 0$,

$$b_0 + b_1 e^x + \cdots + b_k e^{kx} = 0 \implies b_0 = b_1 = \cdots = b_k = 0.$$

Part a of Exercise 2.7.2 shows that eigenvalues are not so special in infinite-dimensional vector spaces: it is possible to find linear transformations such that every number is an eigenvalue.

In part b,

$$b_0 + b_1 e^x + \cdots + b_k e^{kx} = 0$$

means that $b_0 + b_1 e^x + \cdots + b_k e^{kx}$ is the zero function: the zero element of the vector space $\mathcal{C}^\infty(\mathbb{R})$.

2.7.3 a. Let P_3 be the space of polynomials of degree 3. Find the matrix of the linear transformation $T(p) = p + xp''$ with respect to the basis

$$p_1(x) = 1,\ p_2(x) = 1 + x,\ p_3(x) = 1 + x + x^2,\ p_4(x) = 1 + x + x^2 + x^3.$$

b. Let P_k be the space of polynomials of degree k. Find the matrix of the linear transformation $T(p) = p + xp'$ with respect to the basis

$$p_1(x) = 1,\ p_2(x) = 1 + x,\ \ldots,\ p_i(x) = 1 + x + \cdots + x^{i-1}.$$

2.7.4 a. Show that if $P^{-1}AP$ is an $n \times n$ diagonal matrix with $\lambda_1, \ldots, \lambda_n$ on the diagonal, then $e^{tA} = P \begin{bmatrix} e^{t\lambda_1} & \ldots & 0 \\ \vdots & \ddots & \vdots \\ 0 & \ldots & e^{t\lambda_n} \end{bmatrix} P^{-1}$.

$$X = \begin{bmatrix} 1 & -1 & 0 \\ -1 & 2 & -1 \\ 0 & -1 & 1 \end{bmatrix}$$

Matrix for Exercise 2.7.4, part b.

b. For an $n \times n$ matrix A, let e^A be defined as in Exercise 1.5.10. Compute e^X, where X is the matrix in the margin.

c. Let $\vec{\mathbf{x}}$ be a vector in $\mathbb{R}^3$ and consider the function $f : \mathbb{R} \to \mathbb{R}$ defined by $f(t) = \left| e^{tX} \vec{\mathbf{x}} \right|$. For what values of $\vec{\mathbf{x}}$ does $f(t)$ tend to ∞ as $t \to \infty$? Remain bounded as $t \to \infty$? Tend to 0 as $t \to \infty$?

2.7.5 Let $n \mapsto b_n$ be the "Fibonacci-like" sequence defined by $b_0 = b_1 = 1$ and $b_{n+1} = 2b_n + b_{n-1}$ for $n \geq 2$.

a. Find a formula for b_n analogous to equation 2.7.1.

b. How many digits does b_{1000} have? What are the leading digits?

c. Do the same for the sequence $n \mapsto c_n$ defined by $c_0 = c_1 = c_2 = 1$ and $c_{n+1} = c_n + c_{n-1} + c_{n-2}$, for $n \geq 3$.

$$\underbrace{\begin{bmatrix} 0 & -3 & -1 \\ 0 & 2 & 0 \\ 2 & 3 & 3 \end{bmatrix}}_{A}$$

$$\underbrace{\begin{bmatrix} 22 & 23 & 10 & -98 \\ 12 & 18 & 16 & -38 \\ -15 & -19 & -13 & 58 \\ 6 & 7 & 4 & -25 \end{bmatrix}}_{B}$$

Matrices for Exercise 2.7.6

2.7.6 For each matrix A and B in the margin, tell whether it admits an eigenbasis and if so, find it.

2.7.7 a. Let A be a square matrix. Show that if $A \geq \mathbf{0}$ (if every entry of A is ≥ 0) , then A has an eigenvector $\mathbf{v}$ satisfying $\mathbf{v} \geq \mathbf{0}$ (every entry of $\mathbf{v}$ is ≥ 0) .

b. Prove the "improved" Perron-Frobenius theorem: that if $A \geq \mathbf{0}$ and $A^n > \mathbf{0}$ for some $n \geq 1$, there exists an eigenvector $\mathbf{v} > \mathbf{0}$ with simple eigenvalue $\lambda > 0$, and any other eigenvalue $\mu \in \mathbb{C}$ of A satisfies $|\mu| < \lambda$.

2.7.8 Complete the proof of Theorem 2.7.10 by showing that the eigenvalue λ is simple.

2.8 Newton's method

When John Hubbard was teaching first year calculus in France in 1976, he wanted to include some numerical content. Computers for undergraduates did not exist, but Newton's method to solve cubic polynomials just about fit into the 50 steps of program and eight memory registers available with programmable calculators, so he used that as his main example. But what should the initial guess be? He assumed that although he didn't know where to start, the experts surely did. It took some time to discover that no one knew anything about the global behavior of Newton's method.

A natural thing to do was to color each point of the complex plane according to what root (if any) starting at that point led to. (This was before color screens and printers, so he printed some character at every point of some grid: x and 0, for example.) The resulting printouts were the first pictures of fractals arising from complex dynamical systems, with its archetype the Mandelbrot set.

FIGURE 2.8.1.
Isaac Newton (1643–1727)

Theorem 2.2.1 gives a quite complete understanding of linear equations. In practice, one often wants to solve nonlinear equations. This is a genuinely hard problem; the usual response is to apply *Newton's method* and hope for the best. This hope is sometimes justified on theoretical grounds and actually works much more often than any theory explains.

Newton's method requires an initial guess $\mathbf{a}_0$. How do you choose it? You might have a good reason to think that nearby there is a solution, for instance, because $|\vec{\mathbf{f}}(\mathbf{a}_0)|$ is small. In good cases you can then prove that the scheme works. Or it might be wishful thinking: you know roughly what solution you want. Or you might pull your guess out of thin air and start with a collection of initial guesses $\mathbf{a}_0$, hoping that at least one will converge. In some cases, this is just a hope.

Newton's method:

$$\mathbf{a}_1 = \mathbf{a}_0 - [\mathbf{D}\vec{\mathbf{f}}(\mathbf{a}_0)]^{-1}\vec{\mathbf{f}}(\mathbf{a}_0),$$
$$\mathbf{a}_2 = \mathbf{a}_1 - [\mathbf{D}\vec{\mathbf{f}}(\mathbf{a}_1)]^{-1}\vec{\mathbf{f}}(\mathbf{a}_1)$$
$$\mathbf{a}_3 = \mathbf{a}_2 - [\mathbf{D}\vec{\mathbf{f}}(\mathbf{a}_2)]^{-1}\vec{\mathbf{f}}(\mathbf{a}_2)$$
$$\vdots$$

Definition 2.8.1 (Newton's method). Let $\vec{\mathbf{f}}$ be a differentiable map from U to $\mathbb{R}^n$, where U is an open subset of $\mathbb{R}^n$. *Newton's method* consists of starting with some guess $\mathbf{a}_0$ for a solution of $\vec{\mathbf{f}}(\mathbf{x}) = \vec{\mathbf{0}}$. Then linearize the equation at $\mathbf{a}_0$: replace the increment to the function, $\vec{\mathbf{f}}(\mathbf{x}) - \vec{\mathbf{f}}(\mathbf{a}_0)$, by a linear function of the increment, $[\mathbf{D}\vec{\mathbf{f}}(\mathbf{a}_0)](\mathbf{x} - \mathbf{a}_0)$. Now solve the corresponding *linear equation*:

$$\vec{\mathbf{f}}(\mathbf{a}_0) + [\mathbf{D}\vec{\mathbf{f}}(\mathbf{a}_0)](\mathbf{x} - \mathbf{a}_0) = \vec{\mathbf{0}}. \qquad 2.8.1$$

This is a system of n linear equations in n unknowns. We can write it as

$$\underbrace{[\mathbf{D}\vec{\mathbf{f}}(\mathbf{a}_0)]}_{A}\underbrace{(\mathbf{x} - \mathbf{a}_0)}_{\vec{\mathbf{x}}} = \underbrace{-\vec{\mathbf{f}}(\mathbf{a}_0)}_{\vec{\mathbf{b}}}. \qquad 2.8.2$$

If $[\mathbf{D}\vec{\mathbf{f}}(\mathbf{a}_0)]$ is invertible, which will usually be the case, then

$$\mathbf{x} = \mathbf{a}_0 - [\mathbf{D}\vec{\mathbf{f}}(\mathbf{a}_0)]^{-1}\vec{\mathbf{f}}(\mathbf{a}_0). \qquad 2.8.3$$

Call this solution $\mathbf{a}_1$, use it as your new "guess", and solve

$$[\mathbf{D}\vec{\mathbf{f}}(\mathbf{a}_1)](\mathbf{x} - \mathbf{a}_1) = -\vec{\mathbf{f}}(\mathbf{a}_1), \qquad 2.8.4$$

calling the solution $\mathbf{a}_2$, and so on. The hope is that $\mathbf{a}_1$ is a better approximation to a root than $\mathbf{a}_0$, and that the sequence $i \mapsto \mathbf{a}_i$ converges to a root of the equation.

Newton's method is illustrated by Figure 2.8.2.

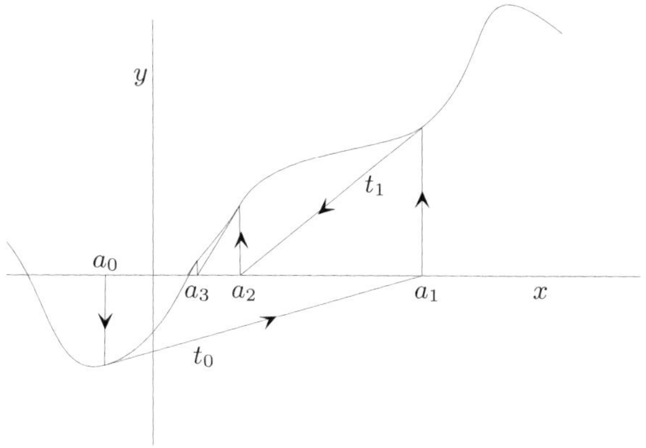

FIGURE 2.8.2. Newton's method: We start with a_0, and draw the tangent to the curve at the point with x-coordinate a_0. The point where that tangent t_0 intersects the x-axis is a_1. Now we draw the tangent to the curve at the point with x-coordinate a_1. That tangent t_1 intersects the x-axis at a_2 .... Each time we calculate a_{n+1} from a_n we are calculating the intersection with the x-axis of the line tangent to the curve at the point with x-coordinate a_n.

Definition 2.8.1: The arrow over $\mathbf{f}$ indicates that elements of the codomain are vectors; $\mathbf{a}_0$ is a point and $\vec{\mathbf{f}}(\mathbf{a}_0)$ is a vector. In equation 2.8.1, the derivative $[\mathbf{D}\vec{\mathbf{f}}(\mathbf{a}_0)]$ is a matrix, and the increment to the variable, $\mathbf{x} - \mathbf{a}_0$, is a vector. Thus on the left side we have the addition of two vectors (elements of the codomain).

In equation 2.8.3 (setting $\mathbf{a}_1 = \mathbf{x}$),

$$\mathbf{a}_1 = \underbrace{\overbrace{\mathbf{a}_0}^{\text{point in domain}} - \overbrace{[\mathbf{D}\vec{\mathbf{f}}(\mathbf{a}_0)]^{-1} \underbrace{\vec{\mathbf{f}}(\mathbf{a}_0)}_{\text{vector in codomain}}}^{\text{increment in domain}}}_{\text{point minus vector equals point}}.$$

Equation 2.8.3 explains the theory behind Newton's method but it is not the way Newton's method is used to solve equations, since finding $[\mathbf{D}\vec{\mathbf{f}}(\mathbf{a}_0)]^{-1}$ is computationally expensive. Instead row reduction is used to solve equation 2.8.2 for $\mathbf{x} - \mathbf{a}_0$; adding $\mathbf{a}_0$ to the solution gives the new $\mathbf{a}_1$. (Partial row reduction and back substitution, discussed in Exercise 2.2.11, are more effective yet.)

When Newton's method is used, most of the computational time is spent doing row operations.

Example 2.8.2: Exercise 2.8.4 asks you to find the corresponding formula for nth roots.

Remark. The domain and codomain of $\vec{\mathbf{f}}$ are usually different spaces, with different units. For example, the units of U might be temperature and the units of the codomain might be volume, with $\vec{\mathbf{f}}$ measuring volume as a function of temperature. In a processing plant $\vec{\mathbf{f}}$ might take n inputs (wheat, oats, fruit, sugar, hours of labor, kilowatt hours of electricity) and produce n different kinds of cereal.

The only requirement is that there must be as many equations as unknowns: the dimensions of the two spaces must be equal. $\triangle$

Example 2.8.2 (Finding a square root). How do calculators compute the square root of a positive number b? They apply Newton's method to the equation $f(x) = x^2 - b = 0$. In this case, this means the following: choose a_0 and plug it into equation 2.8.3. Our equation is in one variable, so we can replace $[\mathbf{D}\vec{\mathbf{f}}(\mathbf{a}_0)]$ by $f'(a_0) = 2a_0$, as shown in equation 2.8.5:

$$a_1 = \underbrace{a_0 - \frac{1}{2a_0}(a_0^2 - b)}_{\text{Newton's method}} = \underbrace{\frac{1}{2}\left(a_0 + \frac{b}{a_0}\right)}_{\text{divide and average}}. \qquad 2.8.5$$

This method is sometimes taught in middle school under the name *divide and average*.

The motivation for *divide and average* is the following: let a be a first guess at $\sqrt{b}$. If your guess is too big (i.e., if $a > \sqrt{b}$), then b/a will be too small, and the average of the two will be better than the original guess. This seemingly naive explanation is quite solid and can easily be turned into a proof that Newton's method works in this case.

Suppose first that $a_0 > \sqrt{b}$; then we want to show that $\sqrt{b} < a_1 < a_0$. Since $a_1 = \frac{1}{2}(a_0 + b/a_0)$, this comes down to showing

$$b < \Bigg(\underbrace{\frac{1}{2}\left(a_0 + \frac{b}{a_0}\right)}_{a_1}\Bigg)^2 < a_0^2, \qquad 2.8.6$$

or, if you develop, $4b < a_0^2 + 2b + \frac{b^2}{a_0^2} < 4a_0^2$. To see the left inequality, subtract $4b$ from each side:

$$a_0^2 - 2b + \frac{b^2}{a_0^2} = \left(a_0 - \frac{b}{a_0}\right)^2 > 0. \qquad 2.8.7$$

The right inequality follows immediately from the inequality $b < a_0^2$, hence $b^2/a_0^2 < a_0^2$:

$$a_0^2 + \underbrace{2b}_{<2a_0^2} + \underbrace{\frac{b^2}{a_0^2}}_{<a_0^2} < 4a_0^2. \qquad 2.8.8$$

FIGURE 2.8.3.

Augustin Louis Cauchy (1789–1857) proved the convergence of Newton's method in one dimension. Cauchy is known both for his great accomplishments – he wrote 789 papers on a wide range of subjects – and for his difficult relations with his colleagues, in particular his mistreatment of the young and impoverished Abel. "Cauchy is mad and there is nothing that can be done about him, although, right now, he is the only one who knows how mathematics should be done," Abel wrote in 1826.

Recall from first year calculus (or from Theorem 0.5.7) that if a decreasing sequence is bounded below, it converges. Hence the a_i converge. The limit a must satisfy

$$a = \lim_{i\to\infty} a_{i+1} = \lim_{i\to\infty} \frac{1}{2}\left(a_i + \frac{b}{a_i}\right) = \frac{1}{2}\left(a + \frac{b}{a}\right), \quad \text{i.e.,} \quad a = \sqrt{b}. \qquad 2.8.9$$

What if you choose $0 < a_0 < \sqrt{b}$? In this case as well, $a_1 > \sqrt{b}$:

$$4a_0^2 < 4b < \underbrace{a_0^2 + 2b + \frac{b^2}{a_0^2}}_{4a_1^2}. \qquad 2.8.10$$

We get the right inequality using the same argument used in equation 2.8.7: $2b < a_0^2 + \frac{b^2}{a_0^2}$, since subtracting $2b$ from both sides gives $0 < \left(a_0 - \frac{b}{a_0}\right)^2$. Then the same argument as before shows that $a_2 < a_1$.

This "divide and average" method can be interpreted geometrically in terms of Newton's method, as was shown in Figure 2.8.2. Each time we calculate a_{n+1} from a_n we are calculating the intersection with the x-axis of the line tangent to the parabola $y = x^2 - b$ at $\begin{pmatrix} a_n \\ a_n^2 - b \end{pmatrix}$. △

There aren't many cases where Newton's method is really well understood far away from the roots; Example 2.8.3 shows one of the problems that can arise. (There are many others.)

Example 2.8.3 (A case where Newton's method doesn't work). Let's apply Newton's method to the equation

$$x^3 - x + \frac{\sqrt{2}}{2} = 0, \qquad 2.8.11$$

starting at $x = 0$ (i.e., our "guess" a_0 is 0). The derivative is $f'(x) = 3x^2 - 1$, so $f'(0) = -1$ and $f(0) = \sqrt{2}/2$, giving

$$a_1 = a_0 - \frac{1}{f'(a_0)} f(a_0) = a_0 - \frac{a_0^3 - a_0 + \frac{\sqrt{2}}{2}}{3a_0^2 - 1} = 0 + \frac{\frac{\sqrt{2}}{2}}{1} = \frac{\sqrt{2}}{2}. \qquad 2.8.12$$

Since $a_1 = \sqrt{2}/2$, we have $f'(a_1) = 1/2$, and

$$a_2 = \frac{\sqrt{2}}{2} - 2\left(\frac{\sqrt{2}}{4} - \frac{\sqrt{2}}{2} + \frac{\sqrt{2}}{2}\right) = 0. \qquad 2.8.13$$

Don't be too discouraged by this example. Most of the time Newton's method does work. It is the best method available for solving nonlinear equations.

We're back to where we started, at $a_0 = 0$. If we continue, we'll bounce back and forth between $\frac{\sqrt{2}}{2}$ and 0, never converging to any root:

$$0 \to \frac{\sqrt{2}}{2} \to 0 \dots . \qquad 2.8.14$$

This uses equation 0.5.4 concerning the sum of a geometric series: If $|r| < 1$, then

$$\sum_{n=0}^{\infty} ar^n = \frac{a}{1-r}.$$

We can substitute $3\epsilon^2$ for r in that equation because ϵ^2 is small.

Now let's try starting at some small $\epsilon > 0$. We have $f'(\epsilon) = 3\epsilon^2 - 1$, and $f(\epsilon) = \epsilon^3 - \epsilon + \sqrt{2}/2$, giving

$$a_1 = \epsilon - \frac{1}{3\epsilon^2 - 1}\left(\epsilon^3 - \epsilon + \frac{\sqrt{2}}{2}\right) = \epsilon + \frac{1}{1 - 3\epsilon^2}\left(\epsilon^3 - \epsilon + \frac{\sqrt{2}}{2}\right). \qquad 2.8.15$$

Now we can treat the factor $\dfrac{1}{1-3\epsilon^2}$ as the sum of the geometric series $(1 + 3\epsilon^2 + 9\epsilon^4 + \cdots)$. This gives

$$a_1 = \epsilon + \left(\epsilon^3 - \epsilon + \frac{\sqrt{2}}{2}\right)(1 + 3\epsilon^2 + 9\epsilon^4 + \cdots). \qquad 2.8.16$$

Does it seem like cheating to *ignore terms that are smaller than* ϵ^2, *ignore the remainder*, or *throw out all the terms with* ϵ^2? Remember (introduction to Section 1.4) that calculus is about "some terms being dominant or negligible compared to other terms." Ignoring these negligible terms will be justified when we get to Section 3.4 on Taylor rules.

Now we ignore terms that are smaller than ϵ^2, getting

$$\begin{aligned} a_1 &= \epsilon + \left(\frac{\sqrt{2}}{2} - \epsilon\right)(1 + 3\epsilon^2) + \text{remainder} \\ &= \frac{\sqrt{2}}{2} + \frac{3\sqrt{2}\epsilon^2}{2} + \text{remainder}. \end{aligned} \qquad 2.8.17$$

Ignoring the remainder, and repeating the process, we get

$$a_2 \approx \frac{\sqrt{2}}{2} + \frac{3\sqrt{2}\epsilon^2}{2} - \frac{\left(\frac{\sqrt{2}}{2} + \frac{3\sqrt{2}}{2}\epsilon^2\right)^3 - \left(\frac{\sqrt{2}}{2} + \frac{3\sqrt{2}}{2}\epsilon^2\right) + \frac{\sqrt{2}}{2}}{3\left(\frac{\sqrt{2}}{2} + \frac{3\sqrt{2}}{2}\epsilon^2\right)^2 - 1}. \qquad 2.8.18$$

This looks unpleasant; let's throw out all the terms with ϵ^2. We get

If we continue, we'll bounce between a region around $\frac{\sqrt{2}}{2}$ and a region around 0, getting closer and closer to these points each time.

$$a_2 \approx \frac{\sqrt{2}}{2} - \frac{\left(\frac{\sqrt{2}}{2}\right)^3 - \frac{\sqrt{2}}{2} + \frac{\sqrt{2}}{2}}{3\left(\frac{\sqrt{2}}{2}\right)^2 - 1} = \frac{\sqrt{2}}{2} - \frac{\frac{1}{2}\left(\frac{\sqrt{2}}{2}\right)}{\frac{1}{2}} = \frac{\sqrt{2}}{2} - \frac{\sqrt{2}}{2} = 0, \qquad 2.8.19$$

so that $a_2 = 0 + c\epsilon^2$, where c is a constant.

We started at $0 + \epsilon$ and we've been sent back to $0 + c\epsilon^2$!

History of the Kantorovich theorem: Cauchy proved the special case of dimension 1. As far as we know, no further work on the subject was done until 1932, when Alexander Ostrowski published a proof in the 2-dimensional case. Ostrowski claimed a student proved the general case in his thesis in 1939, but that student was killed in World War II, and we know of no one who has seen this thesis.

Kantorovich approached the problem from a different point of view, that of nonlinear problems in Banach spaces, important in economics (and in many other fields). He proved the general case in a paper in the 1940s; it was included in a book on functional analysis published in 1959.

The proof we give would work in the infinite-dimensional setting of Banach spaces.

FIGURE 2.8.4.
Rudolf Lipschitz (1832–1903)

Lipschitz's interpretation of Riemann's differential geometry in terms of mechanical laws has been credited with contributing to the development of Einstein's special theory of relativity.

We're not getting anywhere; does that mean there are no roots? Not at all.[15] Let's try once more, with $a_0 = -1$. We have

$$a_1 = a_0 - \frac{a_0^3 - a_0 + \frac{\sqrt{2}}{2}}{3a_0^2 - 1} = \frac{2a_0^3 - \frac{\sqrt{2}}{2}}{3a_0^2 - 1}. \qquad 2.8.20$$

A computer or programmable calculator can be programmed to keep iterating this formula. It's slightly more tedious with a simple scientific calculator; with the one the authors have at hand, we enter "1 +/− Min" to put -1 in the memory ("MR") and then

$$(2 \times MR \times MR \times MR - 2\sqrt{\ }\,\mathrm{div}\,2)\,\mathrm{div}(3 \times MR \times MR - 1).$$

We get $a_1 = -1.35355\dots$; entering that in memory by pushing on the "Min" (or "memory in") key, we repeat the process to get

$$\begin{aligned} a_2 &= -1.26032\dots & a_4 &= -1.25107\dots \\ a_3 &= -1.25116\dots & a_5 &= -1.25107\dots. \end{aligned} \qquad 2.8.21$$

It's then simple to confirm that a_5 is indeed a root, to the limits of precision of the calculator or computer. △

Does Newton's method depend on starting with a lucky guess? Luck sometimes enters into it; with a fast computer one can afford to try out several guesses and see if one converges. But how do we really *know* that solutions are converging? Checking by plugging a root into the equation isn't entirely convincing, because of round-off errors.

Any statement that guarantees that you can find solutions to nonlinear equations in any generality at all is bound to be tremendously important. Kantorovich's theorem (Theorem 2.8.13) *guarantees* that under appropriate circumstances Newton's method converges. Even stating the theorem is difficult, but the effort will pay off. In addition, Newton's method gives a practical algorithm for finding implicit and inverse functions. Kantorovich's theorem proves that these algorithms work. We now lay the groundwork for stating the theorem.

Lipschitz conditions

Imagine an airplane beginning its approach to its destination, its altitude represented by f. If it loses altitude gradually, the derivative f' allows one to approximate the function very well; if you know how high the airplane is at the moment t and what its derivative is at t, you can get a good idea of how high the airplane will be at the moment $t + h$:

$$f(t + h) \approx f(t) + f'(t)h. \qquad 2.8.22$$

But if the airplane suddenly loses power and starts plummeting to earth, the derivative changes abruptly: the derivative of f at t will no longer be a reliable gauge of the airplane's altitude a few seconds later.

[15] Of course not. All odd-degree polynomials have real roots by the intermediate value theorem, Theorem 0.5.9.

The natural way to limit how fast the derivative can change is to bound the second derivative; you probably ran into this when studying Taylor's theorem with remainder. In one variable this is a good idea. If you put an appropriate limit to f'' at t, then the airplane will not suddenly change altitude. Bounding the second derivative of an airplane's altitude function is indeed a pilot's primary goal, except in rare emergencies.

To guarantee that Newton's method starting at a certain point will converge to a root, we need an explicit bound on how good an approximation

$$[\mathbf{D}\vec{\mathbf{f}}(\mathbf{x}_0)]\vec{\mathbf{h}} \quad \text{is to} \quad \vec{\mathbf{f}}(\mathbf{x}_0+\vec{\mathbf{h}})-\vec{\mathbf{f}}(\mathbf{x}_0). \qquad 2.8.23$$

In Definition 2.8.4, note that the domain and the codomain of $\mathbf{f}$ need not have the same dimension. But when we use this definition in the Kantorovich theorem, the dimensions will have to be the same.

A Lipschitz ratio M is often called a *Lipschitz constant*. But M is not a true constant; it depends on the problem at hand. In addition, a mapping will almost always have different M at different points or on different regions. When there is a single Lipschitz ratio that works on all of $\mathbb{R}^n$, we will call it a *global Lipschitz ratio.*

As in the case of the airplane, we will need some assumption on how fast the derivative of $\vec{\mathbf{f}}$ changes. In several variables there are lots of second derivatives, so bounding the second derivatives is complicated; see Proposition 2.8.9. Here we adopt a different approach: demanding that the derivative of $\vec{\mathbf{f}}$ satisfy a *Lipschitz condition.*[16]

Definition 2.8.4 (Lipschitz condition for a derivative). Let $U \subset \mathbb{R}^n$ be open and let $\mathbf{f}: U \to \mathbb{R}^m$ be a differentiable mapping. The derivative $[\mathbf{Df}(\mathbf{x})]$ satisfies a *Lipschitz condition* on a subset $V \subset U$ with *Lipschitz ratio* M if for all $\mathbf{x}, \mathbf{y} \in V$

$$\underbrace{\big|[\mathbf{Df}(\mathbf{x})]-[\mathbf{Df}(\mathbf{y})]\big|}_{\text{distance between derivatives}} \le M \underbrace{|\mathbf{x}-\mathbf{y}|}_{\text{distance between points}}. \qquad 2.8.24$$

Note that a function whose derivative satisfies a Lipschitz condition is certainly *continuously differentiable*. Indeed, requiring that the derivative of a function be Lipschitz is close to demanding that the function be twice continuously differentiable; see Proposition 2.8.9.

Example 2.8.5 (Lipschitz ratio: a simple case). Consider the mapping $\mathbf{f}: \mathbb{R}^2 \to \mathbb{R}^2$

$$\mathbf{f}\begin{pmatrix} x_1 \\ x_2 \end{pmatrix} = \begin{pmatrix} x_1 - x_2^2 \\ x_1^2 + x_2 \end{pmatrix} \text{ with derivative } \left[\mathbf{Df}\begin{pmatrix} x_1 \\ x_2 \end{pmatrix}\right] = \begin{bmatrix} 1 & -2x_2 \\ 2x_1 & 1 \end{bmatrix}. \qquad 2.8.25$$

Given two points $\mathbf{x}$ and $\mathbf{y}$,

$$\left[\mathbf{Df}\begin{pmatrix} x_1 \\ x_2 \end{pmatrix}\right] - \left[\mathbf{Df}\begin{pmatrix} y_1 \\ y_2 \end{pmatrix}\right] = \begin{bmatrix} 0 & -2(x_2-y_2) \\ 2(x_1-y_1) & 0 \end{bmatrix}. \qquad 2.8.26$$

The length of this matrix is

$$2\sqrt{(x_1-y_1)^2+(x_2-y_2)^2} = 2\left|\begin{bmatrix} x_1-y_1 \\ x_2-y_2 \end{bmatrix}\right| = 2|\mathbf{x}-\mathbf{y}|, \qquad 2.8.27$$

[16]More generally, a function $\mathbf{f}$ is Lipschitz if $|\mathbf{f}(\mathbf{x})-\mathbf{f}(\mathbf{y})| \le M|\mathbf{x}-\mathbf{y}|$. For a function of one variable, being Lipschitz means that the chords joining pairs of points have bounded slope. The function $|x|$ is Lipschitz with Lipschitz ratio 1, but it is not differentiable.

so $M = 2$ is a Lipschitz ratio for $[\mathbf{Df}]$. But this example is misleading: there is usually no Lipschitz ratio valid on the entire space. △

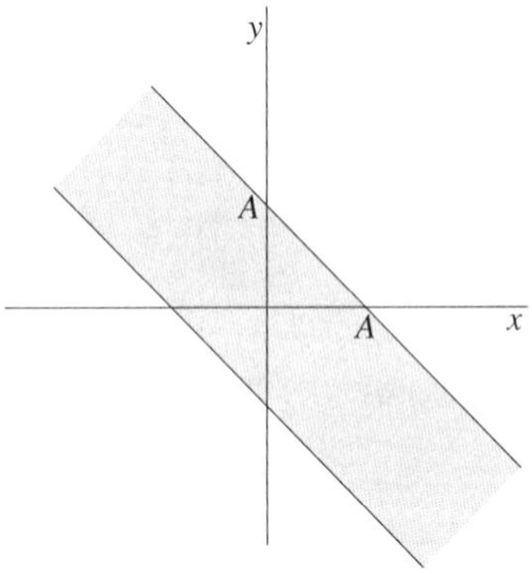

FIGURE 2.8.5.

Inequalities 2.8.30 and 2.8.31 say that when $\begin{pmatrix} x_1 \\ y_1 \end{pmatrix}$ and $\begin{pmatrix} x_2 \\ y_2 \end{pmatrix}$ are in the shaded region (which extends forever on the sides where the boundary is not marked), then $3A$ is a Lipschitz ratio for $[\mathbf{Df}(\mathbf{x})]$. In formula 2.8.34 we translate this statement into a condition on the points

$$\mathbf{x} = \begin{pmatrix} x_1 \\ x_2 \end{pmatrix} \text{ and } \mathbf{y} = \begin{pmatrix} y_1 \\ y_2 \end{pmatrix}.$$

Example 2.8.6 (Lipschitz ratio: a more complicated case). Consider the mapping $\mathbf{f} : \mathbb{R}^2 \to \mathbb{R}^2$ given by

$$\mathbf{f}\begin{pmatrix} x_1 \\ x_2 \end{pmatrix} = \begin{pmatrix} x_1 - x_2^3 \\ x_1^3 + x_2 \end{pmatrix}, \text{ with derivative } \left[\mathbf{Df}\begin{pmatrix} x_1 \\ x_2 \end{pmatrix}\right] = \begin{bmatrix} 1 & -3x_2^2 \\ 3x_1^2 & 1 \end{bmatrix}.$$

Given two points $\mathbf{x}$ and $\mathbf{y}$, we have

$$\left[\mathbf{Df}\begin{pmatrix} x_1 \\ x_2 \end{pmatrix}\right] - \left[\mathbf{Df}\begin{pmatrix} y_1 \\ y_2 \end{pmatrix}\right] = \begin{bmatrix} 0 & -3(x_2^2 - y_2^2) \\ 3(x_1^2 - y_1^2) & 0 \end{bmatrix}, \qquad 2.8.28$$

and taking the length gives

$$\left|\left[\mathbf{Df}\begin{pmatrix} x_1 \\ x_2 \end{pmatrix}\right] - \left[\mathbf{Df}\begin{pmatrix} y_1 \\ y_2 \end{pmatrix}\right]\right| = 3\sqrt{(x_1 - y_1)^2(x_1 + y_1)^2 + (x_2 - y_2)^2(x_2 + y_2)^2}. \qquad 2.8.29$$

Clearly, for this quantity to be bounded we have to put some restrictions on our variables. If we set

$$(x_1 + y_1)^2 \le A^2 \quad \text{and} \quad (x_2 + y_2)^2 \le A^2, \qquad 2.8.30$$

as shown in Figure 2.8.5, we have

$$\left|\left[\mathbf{Df}\begin{pmatrix} x_1 \\ x_2 \end{pmatrix}\right] - \left[\mathbf{Df}\begin{pmatrix} y_1 \\ y_2 \end{pmatrix}\right]\right| \le 3A\left|\begin{pmatrix} x_1 \\ x_2 \end{pmatrix} - \begin{pmatrix} y_1 \\ y_2 \end{pmatrix}\right|; \qquad 2.8.31$$

that is, $3A$ is a Lipschitz ratio for $[\mathbf{Df}(\mathbf{x})]$. When are the inequalities 2.8.30 satisfied? We could just say that it is satisfied when it is satisfied; in what sense can we be more explicit? But the requirement in equation 2.8.30 describes some more or less unimaginable region in $\mathbb{R}^4$. (Keep in mind that inequality 2.8.31 concerns points $\mathbf{x}$ with coordinates x_1, x_2 and $\mathbf{y}$ with coordinates y_1, y_2, *not* the points of Figure 2.8.5, which have coordinates x_1, y_1 and x_2, y_2 respectively.) Moreover, in many settings, what we really want is a ball of radius R such that when two points are in the ball, the Lipschitz condition is satisfied:

$$|[\mathbf{Df}(\mathbf{x})] - [\mathbf{Df}(\mathbf{y})]| \le 3A|\mathbf{x} - \mathbf{y}| \text{ when } |\mathbf{x}| \le R \text{ and } |\mathbf{y}| \le R. \qquad 2.8.32$$

Inequality 2.8.33: We are using the fact that for any two numbers a and b, we always have

$$(a+b)^2 \le 2(a^2 + b^2),$$

since

$$(a+b)^2 \le (a+b)^2 + (a-b)^2 = 2(a^2 + b^2).$$

If we require that $|\mathbf{x}|^2 = x_1^2 + x_2^2 \le A^2/4$ and $|\mathbf{y}|^2 = y_1^2 + y_2^2 \le A^2/4$, then

$$\sup\left\{(x_1 + y_1)^2, (x_2 + y_2)^2\right\} \le 2(x_1^2 + y_1^2 + x_2^2 + y_2^2) = 2(|\mathbf{x}|^2 + |\mathbf{y}|^2) \le A^2. \qquad 2.8.33$$

Thus we can assert that if

$$|\mathbf{x}|, |\mathbf{y}| \le \frac{A}{2}, \quad \text{then} \quad \left|[\mathbf{Df}(\mathbf{x})] - [\mathbf{Df}(\mathbf{y})]\right| \le 3A|\mathbf{x} - \mathbf{y}|. \quad △ \qquad 2.8.34$$

Computing Lipschitz ratios using second partial derivatives

Most students can probably follow the computation in Example 2.8.6 line by line, but even well above average students will probably feel that the tricks used are beyond anything they can come up with on their own. The manipulation of inequalities is a hard skill to acquire, and a hard skill to teach. Here we show that you can compute Lipschitz ratios using second partial derivatives, but it still involves evaluating suprema, and there is no systematic way to do so. (In the case of polynomial functions, there is a trick that requires no evaluation of sups; see Example 2.8.12.)

Higher partial derivatives are so important in scientific applications of mathematics that it seems mildly scandalous to slip them in here, just to solve a computational problem. But in our experience students have such trouble computing Lipschitz ratios – each problem seeming to demand a new trick – that we feel it worthwhile to give a "recipe" for an easier approach.

We discuss higher partial derivatives in Sections 3.3 and 3.4.

Different notations for partial derivatives exist:

$$D_j(D_if)(\mathbf{a}) = \frac{\partial^2 f}{\partial x_j \partial x_i}(\mathbf{a}) = f_{x_i x_j}(\mathbf{a}).$$

As usual, we specify the point $\mathbf{a}$ at which the derivative is evaluated.

Definition 2.8.7 (Second partial derivative). Let $U \subset \mathbb{R}^n$ be open, and let $f : U \to \mathbb{R}$ be a differentiable function. If the function D_if is itself differentiable, then its partial derivative with respect to the jth variable,

$$D_j(D_if),$$

is called a *second partial derivative* of f.

Example 2.8.8 (Second partial derivative). Let f be the function

$$f\begin{pmatrix} x \\ y \\ z \end{pmatrix} = 2x + xy^3 + 2yz^2. \quad \text{Then} \quad D_2(D_1f)\begin{pmatrix} x \\ y \\ z \end{pmatrix} = D_2\underbrace{(2+y^3)}_{D_1f} = 3y^2.$$

$$\text{Similarly, } D_3(D_2f)\begin{pmatrix} x \\ y \\ z \end{pmatrix} = D_3\underbrace{(3xy^2+2z^2)}_{D_2f} = 4z. \quad \triangle$$

We can denote $D_1(D_1f)$ by D_1^2f, we can denote $D_2(D_2f)$ by D_2^2f, and so on. For the function $f\begin{pmatrix} x \\ y \end{pmatrix} = xy^2 + \sin x$, what are D_1^2f, D_2^2f, $D_1(D_2f)$, and $D_2(D_1f)$?[17]

Proposition 2.8.9 says that the derivative of $\mathbf{f}$ is Lipschitz if $\mathbf{f}$ is of class C^2. It shows how to construct a Lipschitz ratio from the supremum of each second partial derivative. We give the proof after a couple of examples.

The proposition would be true for a function going from an open subset of $\mathbb{R}^n$ to $\mathbb{R}^m$, but it would be slightly more complicated to state; the index i would have a different range than j and k.

Proposition 2.8.9 (Derivative of a C^2 mapping is Lipschitz). *Let $U \subset \mathbb{R}^n$ be an open ball, and $\mathbf{f} : U \to \mathbb{R}^n$ a C^2 mapping. If*

$$|D_kD_jf_i(\mathbf{x})| \le c_{i,j,k} \tag{2.8.35}$$

for any $\mathbf{x} \in U$ and for all triples of indices $1 \le i,j,k \le n$, then for $\mathbf{u}, \mathbf{v} \in U$,

$$|[\mathbf{Df}(\mathbf{u})] - [\mathbf{Df}(\mathbf{v})]| \le \left(\sum_{1\le i,j,k \le n} (c_{i,j,k})^2\right)^{1/2} |\mathbf{u}-\mathbf{v}|. \tag{2.8.36}$$

[17] $D_1^2f = D_1(y^2 + \cos x) = -\sin x$, $D_2^2f = D_2(2xy) = 2x$, $D_1(D_2f) = D_1(2xy) = 2y$, and $D_2(D_1f) = D_2(y^2+\cos x) = 2y$.

Example 2.8.10 (Redoing Example 2.8.6). Let's see how much easier it is to find a Lipschitz ratio in Example 2.8.6 using second partial derivatives. First we compute the first and second derivatives, for $f_1 = x_1 - x_2^3$ and $f_2 = x_1^3 + x_2$:

$$D_1 f_1 = 1; \quad D_2 f_1 = -3x_2^2; \quad D_1 f_2 = 3x_1^2; \quad D_2 f_2 = 1. \qquad 2.8.37$$

This gives

$$\begin{aligned} D_1D_1f_1 = 0; \quad D_1D_2f_1 = D_2D_1f_1 = 0; \quad D_2D_2f_1 = -6x_2 \\ D_1D_1f_2 = 6x_1; \quad D_1D_2f_2 = D_2D_1f_2 = 0; \quad D_2D_2f_2 = 0. \end{aligned} \qquad 2.8.38$$

If $|\mathbf{x}| \le A/2$, then $|x_1| \le A/2$, since

$$|x_1| \le \sqrt{x_1^2 + x_2^2}.$$

The same is true of x_2.

If $|\mathbf{x}|, |\mathbf{y}| \le \frac{A}{2}$, we have $|x_1| \le A/2$ and $|x_2| \le A/2$, so

$$|D_2D_2f_1| \le 3A = \underbrace{c_{1,2,2}}_{\substack{\text{bound for}\\ |D_2D_2f_1|}} \quad \text{and} \quad |D_1D_1f_2| \le 3A = \underbrace{c_{2,1,1}}_{\substack{\text{bound for}\\ |D_1D_1f_2|}}$$

with all others 0, so

$$\sqrt{c_{1,2,2}^2 + c_{2,1,1}^2} = 3A\sqrt{2}. \qquad 2.8.39$$

Thus we have

$$\Big|[\mathbf{Df}(\mathbf{x})] - [\mathbf{Df}(\mathbf{y})]\Big| \le 3\sqrt{2}A|\mathbf{x} - \mathbf{y}|. \qquad 2.8.40 \qquad \triangle$$

Inequality 2.8.40: Earlier we got $3A$, a better result. A blunderbuss method guaranteed to work in all cases is unlikely to give as good results as techniques adapted to the problem at hand. But the second partial derivative method gives results that are often good enough.

Using second partial derivatives, recompute the Lipschitz ratio of Example 2.8.5. Do you get the same answer?[18]

Example 2.8.11 (Finding a Lipschitz ratio using second derivatives: a second example). Let us find a Lipschitz ratio for the derivative of the function

$$\mathbf{F}\begin{pmatrix} x \\ y \end{pmatrix} = \begin{pmatrix} \sin(x+y) \\ \cos(xy) \end{pmatrix}, \quad \text{for } |x| < 2,\ |y| < 2. \qquad 2.8.41$$

We compute

$$\begin{aligned} &D_1D_1F_1 = D_2D_2F_1 = D_2D_1F_1 = D_1D_2F_1 = -\sin(x+y), \qquad 2.8.42 \\ &D_1D_1F_2 = -y^2\cos(xy), \quad D_2D_1F_2 = D_1D_2F_2 = -\big(\sin(xy) + yx\cos(xy)\big), \\ &D_2D_2F_2 = -x^2\cos xy. \end{aligned}$$

Since $|\sin|$ and $|\cos|$ are bounded by 1, if we set $|x| < 2, |y| < 2$, this gives

$$\begin{aligned} |D_1D_1F_1| = |D_1D_2F_1| = |D_2D_1F_1| = |D_2D_2F_1| \le 1 \\ |D_1D_1F_2|, |D_2D_2F_2| \le 4, \quad |D_2D_1F_2| = |D_1D_2F_2| \le 5. \end{aligned} \qquad 2.8.43$$

Inequality 2.8.44: By fiddling with the trigonometry, one can get the $\sqrt{86}$ down to $\sqrt{78} \approx 8.8$, but the advantage of Proposition 2.8.9 is that it gives a more or less systematic way to compute Lipschitz ratios.

So for $|x| < 2,\ |y| < 2$, we have a Lipschitz ratio

$$M \le \sqrt{4 + 16 + 16 + 25 + 25} = \sqrt{86} < 9.3; \qquad 2.8.44$$

that is, $\Big|[\mathbf{DF}(\mathbf{u})] - [\mathbf{DF}(\mathbf{v})]\Big| \le 9.3\,|\mathbf{u} - \mathbf{v}|$. □

[18]The only nonzero second partials are $D_1D_1f_2 = 2$ and $D_2D_2f_1 = -2$, bounded by 2 and -2, so the method using second partial derivatives gives $\sqrt{2^2 + (-2)^2} = 2\sqrt{2}$.

Proof of Proposition 2.8.9. Each of the $D_j f_i$ is a scalar-valued function, and Corollary 1.9.2 tells us that

$$|D_j f_i(\mathbf{a}+\vec{\mathbf{h}}) - D_j f_i(\mathbf{a})| \le \left(\sum_{k=1}^{n}(c_{i,j,k})^2\right)^{1/2} |\vec{\mathbf{h}}|. \qquad 2.8.45$$

Inequality 2.8.45 uses the fact that for any function g (in our case, $D_j f_i$),

$$\left|g(\mathbf{a}+\vec{\mathbf{h}}) - g(\mathbf{a})\right| \le \left(\sup_{t\in[0,1]}\left|[\mathbf{D}g(\mathbf{a}+t\vec{\mathbf{h}})]\right|\right)|\vec{\mathbf{h}}|;$$

remember that

$$\left|[\mathbf{D}g(\mathbf{a}+t\vec{\mathbf{h}})]\right| = \sqrt{\sum_{k=1}^{n}\left(D_k g(\mathbf{a}+t\vec{\mathbf{h}})\right)^2}.$$

Using the definition of length of matrices for the first equality below and inequality 2.8.45 for the inequality, we have

$$\begin{aligned}|[\mathbf{Df}(\mathbf{a}+\vec{\mathbf{h}})] - [\mathbf{Df}(\mathbf{a})]| &= \left(\sum_{i,j=1}^{n}\left(D_j f_i(\mathbf{a}+\vec{\mathbf{h}}) - D_j f_i(\mathbf{a})\right)^2\right)^{1/2}\\ &\le \left(\sum_{i,j=1}^{n}\left(\left(\sum_{k=1}^{n}(c_{i,j,k})^2\right)^{1/2}|\vec{\mathbf{h}}|\right)^2\right)^{1/2}\\ &= \left(\sum_{1\le i,j,k\le n}(c_{i,j,k})^2\right)^{1/2}|\vec{\mathbf{h}}|. \qquad 2.8.46\end{aligned}$$

Going from line 2 to line 3 of inequality 2.8.46 may seem daunting, but note that for the inner sum we have a square root squared, and $|\vec{\mathbf{h}}|$ factors out, since it appears in every term as the square root of its square.

The proposition follows by setting $\mathbf{u} = \mathbf{a}+\vec{\mathbf{h}}$, and $\mathbf{v} = \mathbf{a}$. □

Finding a Lipschitz ratio for $[\mathbf{D}f]$ over an appropriate region is usually difficult. But it is easy when f is quadratic, and hence $[\mathbf{D}f]$ has degree 1. All polynomial equations can be reduced to this case, at the expense of adding extra variables.

Example 2.8.12 (Trick for finding Lipschitz ratios). Let $\mathbf{f} : \mathbb{R}^2 \to \mathbb{R}^2$ be given by

$$\mathbf{f}\begin{pmatrix}x\\y\end{pmatrix} = \begin{pmatrix}x^5 - y^2 + xy - a\\ y^4 + x^2y - b\end{pmatrix}. \qquad 2.8.47$$

We will invent new variables, $u_1 = x^2, u_2 = y^2$, $u_3 = x^4 = u_1^2$, so that every term is of degree at most 2, and instead of $\mathbf{f}$ we will consider $\widetilde{\mathbf{f}}$, where

$$\widetilde{\mathbf{f}}\begin{pmatrix}x\\y\\u_1\\u_2\\u_3\end{pmatrix} = \begin{pmatrix}xu_3 - u_2 + xy - a\\ u_2^2 + u_1y - b\\ u_1 - x^2\\ u_2 - y^2\\ u_3 - u_1^2\end{pmatrix}. \qquad 2.8.48$$

In equation 2.8.48, the first two equations are the original system of equations, rewritten in terms of old and new variables. The last three equations of $\widetilde{\mathbf{f}} = \mathbf{0}$ are the definitions of the new variables.

Clearly if $\begin{pmatrix}x\\y\end{pmatrix}$ satisfies $\mathbf{f}\begin{pmatrix}x\\y\end{pmatrix} = \begin{pmatrix}0\\0\end{pmatrix}$, then $\widetilde{\mathbf{f}}\begin{pmatrix}x\\y\\x^2\\y^2\\x^4\end{pmatrix} = \begin{pmatrix}0\\0\\0\\0\\0\end{pmatrix}$.

Thus if we can solve $\widetilde{\mathbf{f}}(\widetilde{\mathbf{x}}) = \widetilde{\mathbf{0}}$, we can also solve $\mathbf{f}(\mathbf{x}) = \mathbf{0}$.

It may seem ridiculous to work in $\mathbb{R}^5$ rather than $\mathbb{R}^2$, but evaluating the Lipschitz ratio of $\mathbf{f}$ is fairly hard, whereas evaluating the Lipschitz ratio of $\widetilde{\mathbf{f}}$ is straightforward.

Indeed,

$$\left[\mathbf{D}\widetilde{\mathbf{f}}\begin{pmatrix} x \\ y \\ u_1 \\ u_2 \\ u_3 \end{pmatrix}\right] = \begin{bmatrix} y+u_3 & x & 0 & -1 & x \\ 0 & u_1 & y & 2u_2 & 0 \\ -2x & 0 & 1 & 0 & 0 \\ 0 & -2y & 0 & 1 & 0 \\ 0 & 0 & -2u_1 & 0 & 1 \end{bmatrix}. \qquad 2.8.49$$

Thus

$$\left|\left[\mathbf{D}\widetilde{\mathbf{f}}\begin{pmatrix} x \\ y \\ u_1 \\ u_2 \\ u_3 \end{pmatrix}\right] - \left[\mathbf{D}\widetilde{\mathbf{f}}\begin{pmatrix} x' \\ y' \\ u_1' \\ u_2' \\ u_3' \end{pmatrix}\right]\right| = \left|\begin{bmatrix} y-y'+u_3-u_3' & x-x' & 0 & 0 & x-x' \\ 0 & u_1-u_1' & y-y' & 2(u_2-u_2') & 0 \\ -2(x-x') & 0 & 0 & 0 & 0 \\ 0 & -2(y-y') & 0 & 0 & 0 \\ 0 & 0 & -2(u_1-u_1') & 0 & 0 \end{bmatrix}\right|$$

$$\leq \Big(6(x-x')^2 + 7(y-y')^2 + 5(u_1-u_1')^2 + 4(u_2-u_2)^2 + 2(u_3-u_3')^2\Big)^{1/2}$$

$$\leq \sqrt{7}\left|\begin{pmatrix} x \\ y \\ u_1 \\ u_2 \\ u_3 \end{pmatrix} - \begin{pmatrix} x' \\ y' \\ u_1' \\ u_2' \\ u_3' \end{pmatrix}\right|. \qquad 2.8.50$$

Going from the first to second line of inequality 2.8.50: Since $(a+b)^2 \leq 2(a^2+b^2)$, we have

$$(y-y'+u_3-u_3')^2 \leq 2(y-y')^2 + 2(u_3-u_3')^2.$$

Thus $\sqrt{7}$ is a Lipschitz ratio for $[\mathbf{D}\widetilde{\mathbf{f}}\,]$. $\triangle$

For all systems of polynomial equations, a similar trick will work and provide a Lipschitz ratio that can be computed, perhaps laboriously, but requiring no invention.

Kantorovich's theorem

Now we are ready to tackle Kantorovich's theorem. It says that if the product of three quantities is $\leq 1/2$, then the equation $\vec{\mathbf{f}}(\mathbf{x}) = \vec{\mathbf{0}}$ has a unique root in a closed ball $\overline{U_0}$, and if you start with an appropriate initial guess $\mathbf{a}_0$, Newton's method will converge to that root.

Note that the domain and codomain of the map $\vec{\mathbf{f}}$ have the *same dimension.* Thus, setting $\vec{\mathbf{f}}(\mathbf{x}) = \vec{\mathbf{0}}$, we get the same number of equations as unknowns. If we had fewer equations than unknowns, we wouldn't expect them to specify a unique solution, and if we had more equations than unknowns, it would be unlikely that there would be any solutions at all.

In addition, if the domain and codomain of the map $\vec{\mathbf{f}}$ had different dimensions, then $[\mathbf{D}\vec{\mathbf{f}}(\mathbf{a}_0)]$ would not be a square matrix, so it would not be invertible.

Theorem 2.8.13 (Kantorovich's theorem). *Let $\mathbf{a}_0$ be a point in $\mathbb{R}^n$, U an open neighborhood of $\mathbf{a}_0$ in $\mathbb{R}^n$, and $\vec{\mathbf{f}} : U \to \mathbb{R}^n$ a differentiable mapping, with its derivative $[\mathbf{D}\vec{\mathbf{f}}(\mathbf{a}_0)]$ invertible. Define*

$$\vec{\mathbf{h}}_0 \stackrel{\text{def}}{=} -[\mathbf{D}\vec{\mathbf{f}}(\mathbf{a}_0)]^{-1}\vec{\mathbf{f}}(\mathbf{a}_0), \quad \mathbf{a}_1 \stackrel{\text{def}}{=} \mathbf{a}_0 + \vec{\mathbf{h}}_0, \quad U_1 \stackrel{\text{def}}{=} B_{|\vec{\mathbf{h}}_0|}(\mathbf{a}_1). \qquad 2.8.51$$

If $\overline{U_1} \subset U$ and the derivative $[\mathbf{D}\vec{\mathbf{f}}(\mathbf{x})]$ satisfies the Lipschitz condition

$$\big|[\mathbf{D}\vec{\mathbf{f}}(\mathbf{u}_1)] - [\mathbf{D}\vec{\mathbf{f}}(\mathbf{u}_2)]\big| \leq M|\mathbf{u}_1 - \mathbf{u}_2| \quad \text{for all points } \mathbf{u}_1, \mathbf{u}_2 \in \overline{U_1}, \qquad 2.8.52$$

and if the inequality

$$\big|\vec{\mathbf{f}}(\mathbf{a}_0)\big|\,\big|[\mathbf{D}\vec{\mathbf{f}}(\mathbf{a}_0)]^{-1}\big|^2 M \leq \frac{1}{2} \qquad 2.8.53$$

is satisfied, the equation $\vec{\mathbf{f}}(\mathbf{x}) = \vec{\mathbf{0}}$ has a unique solution in the closed ball $\overline{U_1}$, and Newton's method with initial guess $\mathbf{a}_0$ converges to it.

The theorem is proved in Appendix A5.

The basic idea is simple. The first of the three quantities that must be small is the value of the function at $\mathbf{a}_0$. If you are in an airplane flying close to the ground, you are more likely to crash (find a root) than if you are several kilometers up.

The second quantity is the square of the length of the *inverse* of the derivative at $\mathbf{a}_0$. In one dimension, we can think that the derivative must be big.[19] If your plane is approaching the ground steeply, it is much more likely to crash than if it is flying almost parallel to the ground.

The third quantity is the Lipschitz ratio M, measuring the change in the derivative (i.e., acceleration). If at the last minute the pilot pulls the plane out of a nose dive, flight attendants may be thrown to the floor as the derivative changes sharply, but a crash will be avoided. (Remember that acceleration need not be a change in speed; it can also be a change in direction.)

But it is not each quantity individually that must be small: the product must be small. If the airplane starts its nose dive too close to the ground, even a sudden change in derivative may not save it. If it starts its nose dive from an altitude of several kilometers, it will still crash if it falls straight down. And if it loses altitude progressively, rather than plummeting to earth, it will still crash (or at least land) if the derivative never changes.

FIGURE 2.8.6.
Leonid Kantorovich (1912–1986)
Kantorovich was among the first to use linear programming in economics, in a paper published in 1939. He was awarded the Nobel Prize in economics in 1975.

When discussing Kantorovich's theorem and Newton's method we write $\vec{\mathbf{f}}(\mathbf{x}) = \vec{\mathbf{0}}$, (with arrows) because we think of the codomain of $\vec{\mathbf{f}}$ as a vector space; the definition of $\vec{\mathbf{h}}_0$ only makes sense if $\vec{\mathbf{f}}(\mathbf{a}_0)$ is a vector. Moreover, $\vec{\mathbf{0}}$ plays a distinguished role in Newton's method (as it does in any vector space): we are trying to solve $\vec{\mathbf{f}}(\mathbf{x}) = \vec{\mathbf{0}}$, not $\mathbf{f}(\mathbf{x}) = \mathbf{a}$ for some random $\mathbf{a}$.

Remarks.

1. To check whether an equation makes sense, first make sure both sides have the same units. In physics and engineering, this is essential. The right side of inequality 2.8.53 is the unitless number 1/2. The left side:
$$\left|\vec{\mathbf{f}}(\mathbf{a}_0)\right| \left|[\mathbf{D}\vec{\mathbf{f}}(\mathbf{a}_0)]^{-1}\right|^2 M \qquad 2.8.54$$
is a complicated mixture of units of domain and codomain, which usually are different. Fortunately, these units cancel. To see this, denote by u the units of the domain U, and by r the units of the codomain $\mathbb{R}^n$. The term $|\vec{\mathbf{f}}(\mathbf{a}_0)|$ has units r. A derivative has units codomain/domain (typically, distance divided by time), so the inverse of the derivative has units domain/codomain $= u/r$, and the term $|[\mathbf{D}\vec{\mathbf{f}}(\mathbf{a}_0)]^{-1}|^2$ has units u^2/r^2. The Lipschitz ratio M is the distance between derivatives divided by a distance in the domain, so its units are r/u divided by u. This gives units $r \times \dfrac{u^2}{r^2} \times \dfrac{r}{u^2}$, which cancel.
2. The Kantorovich theorem does *not* say that if inequality 2.8.53 is not satisfied, the equation has no solutions; it does not even say that

[19] Why the theorem stipulates the *square* of the inverse of the derivative is more subtle. We think of it this way: the theorem should remain true if one changes the scale. Since the "numerator" $\vec{\mathbf{f}}(\mathbf{a}_0)M$ in inequality 2.8.53 contains two terms, scaling up will change it by the scale factor squared. So the "denominator" $|[\mathbf{D}\vec{\mathbf{f}}(\mathbf{a}_0)]^{-1}|^2$ must also contain a square.

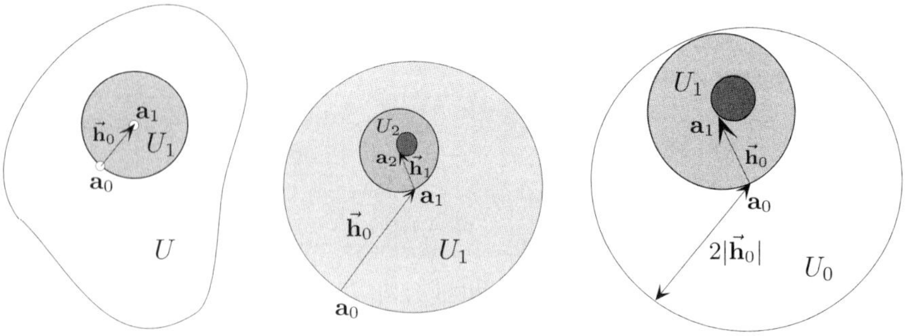

FIGURE 2.8.7. Equation 2.8.51 defines the neighborhood U_1 for which Newton's method is guaranteed to work when inequality 2.8.53 is satisfied. LEFT: the neighborhood U_1 is the ball of radius $|\vec{\mathbf{h}}_0| = |\mathbf{a}_1 - \mathbf{a}_0|$ around $\mathbf{a}_1$, so $\mathbf{a}_0$ is on the border of U_1. MIDDLE: a blow-up of U_1 (shaded), showing the neighborhoods U_2 (dark) and U_3 (darker). RIGHT: The ball U_0 described in Proposition 2.8.14.

if the inequality is not satisfied, there are no solutions in $\overline{U}_1$. In Section 2.9 we will see that if we use a different way to measure $[\mathbf{D}\vec{\mathbf{f}}(\mathbf{a}_0)]$, the *norm*, which is harder to compute, then inequality 2.8.53 is easier to satisfy. That version of Kantorovich's theorem thus guarantees convergence for some equations about which this weaker version of the theorem is silent.

3. To see why it is necessary in the final sentence to specify the closed ball $\overline{U_1}$, not U_1, consider Example 2.9.1, where Newton's method converges to 1, which is not in U_1 but is in its closure. △

If inequality 2.8.53 is satisfied, then at each iteration we create a new ball inside the previous ball: U_2 is in U_1, U_3 is in U_2, ..., as shown in the middle of Figure 2.8.7. As the radius of the balls goes to zero, the sequence $i \mapsto \mathbf{a}_i$ converges to $\mathbf{a}$, which we will see is a root.

In many cases we are interested in having a unique solution in a ball around $\mathbf{a}_0$ (the ball U_0 shown on the right of Figure 2.8.7); and this is easily achieved.

Proposition 2.8.14: The smaller the set in which one can guarantee existence, the better. It is a stronger statement to say that there exists a William Ardvark in the town of Nowhere, New York, population 523, than to say there exists a William Ardvark in New York State.

The larger the set in which one can guarantee uniqueness, the better. It is stronger to say there exists only one John W. Smith in the state of California than to say there exists only one John W. Smith in Tinytown, California.

There are times, such as when proving uniqueness for the inverse function theorem, that one wants the Kantorovich theorem stated for the larger ball U_0. But there are also times when the function is not Lipschitz on U_0, or is not even defined on U_0. In such cases the original Kantorovich theorem is the useful one.

Proposition 2.8.14 (Unique solution in $\overline{U_0}$). *Define*

$$U_0 \stackrel{\text{def}}{=} \{\mathbf{x} \,|\, |\mathbf{x} - \mathbf{a}_0| < 2|\vec{\mathbf{h}}_0|\}. \qquad 2.8.55$$

Then if the Lipschitz condition

$$\left|[\mathbf{D}\vec{\mathbf{f}}(\mathbf{u}_1)] - [\mathbf{D}\vec{\mathbf{f}}(\mathbf{u}_2)]\right| \le M|\mathbf{u}_1 - \mathbf{u}_2| \qquad 2.8.56$$

is satisfied for all points $\mathbf{u}_1, \mathbf{u}_2 \in U_0$, the equation $\vec{\mathbf{f}}(\mathbf{x}) = \vec{\mathbf{0}}$ has a unique solution in $\overline{U_0}$, and Newton's method starting at $\mathbf{a}_0$ converges to it.

The proposition is justified in Appendix A5, when we prove the Kantorovich theorem.

Example 2.8.15 (Using Newton's method). Suppose we want to solve the two equations

$$\begin{aligned}\cos(x-y) &= y\\ \sin(x+y) &= x,\end{aligned} \quad \text{i.e.,} \quad \vec{\mathbf{F}}\begin{pmatrix}x\\y\end{pmatrix} = \begin{bmatrix}\cos(x-y)-y\\ \sin(x+y)-x\end{bmatrix} = \begin{bmatrix}0\\0\end{bmatrix}. \qquad 2.8.57$$

We happen to notice[20] that the equation is close to being satisfied at $\begin{pmatrix}1\\1\end{pmatrix}$:

$$\cos(1-1)-1=0 \quad \text{and} \quad \sin(1+1)-1=-.0907\ldots. \qquad 2.8.58$$

Let us check that starting Newton's method at $\mathbf{a}_0 = \begin{pmatrix}1\\1\end{pmatrix}$ works. To do this we must see that inequality 2.8.53 is satisfied. We just saw that $|\vec{\mathbf{F}}(\mathbf{a}_0)| \sim .0907 < .1$. The derivative at $\mathbf{a}_0$ isn't much worse:

$$[\mathbf{D}\vec{\mathbf{F}}(\mathbf{a}_0)] = \begin{bmatrix}0 & -1\\ (\cos 2)-1 & \cos 2\end{bmatrix}, \quad \text{so} \qquad 2.8.59$$

$$[\mathbf{D}\vec{\mathbf{F}}(\mathbf{a}_0)]^{-1} = \frac{1}{(\cos 2)-1}\begin{bmatrix}\cos 2 & 1\\ 1-\cos 2 & 0\end{bmatrix} \qquad 2.8.60$$

and

$$\left|[\mathbf{D}\vec{\mathbf{F}}(\mathbf{a}_0)]^{-1}\right|^2 = \frac{1}{(\cos 2-1)^2}\Big((\cos 2)^2+1+(1-\cos 2)^2\Big) \sim 1.1727 < 1.2, \qquad 2.8.61$$

as you will see if you put it in your calculator.

Rather than compute the Lipschitz ratio for the derivative using second partial derivatives, we will do it directly, taking advantage of the helpful formulas $|\sin a - \sin b| \le |a-b|$, $|\cos a - \cos b| \le |a-b|$.[21] These make the computation manageable:

$$\begin{aligned}\left|\left[\mathbf{D}\vec{\mathbf{F}}\begin{pmatrix}x_1\\y_1\end{pmatrix}\right] - \left[\mathbf{D}\vec{\mathbf{F}}\begin{pmatrix}x_2\\y_2\end{pmatrix}\right]\right| &= \left|\begin{bmatrix}-\sin(x_1-y_1)+\sin(x_2-y_2) & \sin(x_1-y_1)-\sin(x_2-y_2)\\ \cos(x_1+y_1)-\cos(x_2+y_2) & \cos(x_1+y_1)-\cos(x_2+y_2)\end{bmatrix}\right|\\ &\le \left|\begin{bmatrix}|-(x_1-y_1)+(x_2-y_2)| & |(x_1-y_1)-(x_2-y_2)|\\ |(x_1+y_1)-(x_2+y_2)| & |(x_1+y_1)-(x_2+y_2)|\end{bmatrix}\right|\\ &\le \sqrt{8\big((x_1-x_2)^2+(y_1-y_2)^2\big)} = 2\sqrt{2}\left|\begin{pmatrix}x_1\\y_1\end{pmatrix} - \begin{pmatrix}x_2\\y_2\end{pmatrix}\right|. \qquad 2.8.62\end{aligned}$$

Thus $M = 2\sqrt{2}$ is a Lipschitz ratio for $[\mathbf{D}\vec{\mathbf{F}}]$. Putting these together, we see that

$$|\vec{\mathbf{F}}(\mathbf{a}_0)|\left|[\mathbf{D}\vec{\mathbf{F}}(\mathbf{a}_0)]^{-1}\right|^2 M \le (.1)(1.2)(2\sqrt{2}) \approx .34 < .5 \qquad 2.8.63$$

The Kantorovich theorem does *not* say that the system of equations has a unique solution; it may have many. But *in the ball* $\overline{U}_1$ it has a unique solution, and if you start with the initial guess $\mathbf{a}_0$, Newton's method will find it.

Equation 2.8.60: Recall that the inverse of

$$A = \begin{bmatrix}a & b\\ c & d\end{bmatrix} \quad \text{is}$$

$$A^{-1} = \frac{1}{ad-bc}\begin{bmatrix}d & -b\\ -c & a\end{bmatrix}.$$

Going from the second to the third line of inequality 2.8.62 uses

$$(a+b)^2 \le 2(a^2+b^2).$$

[20] We "happened to notice" that $\sin 2 - 1 = -.0907$ with the help of a calculator. Finding an initial condition for Newton's method is always the delicate part.

[21] By the mean value theorem, there exists a c between a and b such that

$$|\sin a - \sin b| = |\underbrace{\cos}_{\sin' c} c||a-b|;$$

since $|\cos c| \le 1$, we have $|\sin a - \sin b| \le |a-b|$.

so the equation has a solution, and Newton's method starting at $\begin{pmatrix}1\\1\end{pmatrix}$ will converge to it. Moreover,

$$\vec{\mathbf{h}}_0 = \underbrace{\frac{-1}{\cos 2 - 1}\begin{bmatrix}\cos 2 & 1\\ 1-\cos 2 & 0\end{bmatrix}}_{-[\mathbf{D}\vec{\mathbf{F}}(\mathbf{a}_0)]^{-1}}\underbrace{\begin{bmatrix}0\\ \sin 2 - 1\end{bmatrix}}_{\vec{\mathbf{F}}(\mathbf{a}_0)} = \begin{bmatrix}\frac{\sin 2 - 1}{1-\cos 2}\\ 0\end{bmatrix} \approx \begin{bmatrix}-.064\\ 0\end{bmatrix},$$

so Kantorovich's theorem guarantees that the solution is within .064 of the point $\mathbf{a}_1 = \mathbf{a}_0 + \vec{\mathbf{h}}_0 = \begin{pmatrix}.936\\1\end{pmatrix}$. According to our computer, the solution (to three decimal places) is $\begin{pmatrix}.935\\.998\end{pmatrix}$. △

Example 2.8.16: If we wanted to solve equation 2.8.64 without Newton's method, by considering $X = \begin{bmatrix}a & b\\ c & d\end{bmatrix}$ as a point in $\mathbb{R}^4$, we would need to solve four quadratic equations in four unknowns. It's not obvious how to go about it.

Why is $\begin{bmatrix}4 & 0\\ 0 & 4\end{bmatrix}$ an obvious choice? The matrix

$$\begin{bmatrix}4 & 0\\ 0 & 4\end{bmatrix}^2 = \begin{bmatrix}16 & 0\\ 0 & 16\end{bmatrix}$$

is close to $\begin{bmatrix}15 & 0\\ 0 & 17\end{bmatrix}$, and

$$\begin{bmatrix}0 & 1\\ 1 & 0\end{bmatrix}\begin{bmatrix}4 & 0\\ 0 & 4\end{bmatrix} = \begin{bmatrix}0 & 4\\ 4 & 0\end{bmatrix}$$

is close to $\begin{bmatrix}0 & 5\\ 3 & 0\end{bmatrix}$.

Example 2.8.16 (Newton's method and a mapping with matrices). Define $f : \text{Mat}\,(2,2) \to \text{Mat}\,(2,2)$ by

$$f(X) = X^2 + \begin{bmatrix}0 & 1\\ 1 & 0\end{bmatrix}X - \begin{bmatrix}15 & 5\\ 3 & 17\end{bmatrix} \qquad 2.8.64$$

What does Newton's method say about the equation $f(X) = \begin{bmatrix}0 & 0\\ 0 & 0\end{bmatrix}$? There is an obvious initial guess $A_0 = \begin{bmatrix}4 & 0\\ 0 & 4\end{bmatrix}$. Since $f(A_0) = \begin{bmatrix}1 & -1\\ 1 & -1\end{bmatrix}$, we have $|f(A_0)| = 2$. Now we compute the other quantities in the Kantorovich theorem. First, we compute the derivative at A_0 (remembering that since A_0 is a scalar multiple of the identity, $A_0H = HA_0$):

$$[\mathbf{D}f(A_0)]H = \underbrace{A_0H + HA_0}_{\text{see Example 1.7.17}} + \begin{bmatrix}0 & 1\\ 1 & 0\end{bmatrix}H = \begin{bmatrix}8 & 1\\ 1 & 8\end{bmatrix}H, \qquad 2.8.65$$

which gives

$$[\mathbf{D}f(A_0)]^{-1}H = \frac{1}{63}\begin{bmatrix}8 & -1\\ -1 & 8\end{bmatrix}H. \qquad 2.8.66$$

Now we run into complications: we must compute the length of the derivative (to find a Lipschitz ratio) and the length of the inverse of the derivative (for the Kantorovich inequality). Computing $|f(A_0)|$ was no problem, but to compute the length of the derivative we need to figure out what 4×4 matrix that derivative is, when we think of f as a mapping from $\mathbb{R}^4$ to $\mathbb{R}^4$. Write the standard basis vectors in $\mathbb{R}^4$ as the 2×2 matrices

$$\begin{bmatrix}1 & 0\\ 0 & 0\end{bmatrix}, \begin{bmatrix}0 & 1\\ 0 & 0\end{bmatrix}, \begin{bmatrix}0 & 0\\ 1 & 0\end{bmatrix}, \begin{bmatrix}0 & 0\\ 0 & 1\end{bmatrix}. \qquad 2.8.67$$

We get the first column of the 4×4 derivative matrix by applying $[\mathbf{D}f(A_0)]$ to $\begin{bmatrix}1 & 0\\ 0 & 0\end{bmatrix}$:

$$\begin{bmatrix}8 & 1\\ 1 & 8\end{bmatrix}\begin{bmatrix}1 & 0\\ 0 & 0\end{bmatrix} = \begin{bmatrix}8 & 0\\ 1 & 0\end{bmatrix}. \qquad 2.8.68$$

Similarly, the remaining columns are given by the products

$$\begin{bmatrix} 8 & 1 \\ 1 & 8 \end{bmatrix}\begin{bmatrix} 0 & 1 \\ 0 & 0 \end{bmatrix}, \quad \begin{bmatrix} 8 & 1 \\ 1 & 8 \end{bmatrix}\begin{bmatrix} 0 & 0 \\ 1 & 0 \end{bmatrix}, \quad \begin{bmatrix} 8 & 1 \\ 1 & 8 \end{bmatrix}\begin{bmatrix} 0 & 0 \\ 0 & 1 \end{bmatrix}. \qquad 2.8.69$$

Thus, evaluated at A_0, the derivative is

$$[\mathbf{D}f(A_0)] = \begin{bmatrix} 8 & 0 & 1 & 0 \\ 0 & 8 & 0 & 1 \\ 1 & 0 & 8 & 0 \\ 0 & 1 & 0 & 8 \end{bmatrix}. \qquad 2.8.70$$

Using the same procedure applied to $[\mathbf{D}f(A_0)]^{-1}H = \frac{1}{63}\begin{bmatrix} 8 & -1 \\ -1 & 8 \end{bmatrix} H$, we see that the inverse is

$$\frac{1}{63}\begin{bmatrix} 8 & 0 & -1 & 0 \\ 0 & 8 & 0 & -1 \\ -1 & 0 & 8 & 0 \\ 0 & -1 & 0 & 8 \end{bmatrix}, \quad \text{with length} \quad \left|[\mathbf{D}f(A_0)]^{-1}\right| = \frac{2}{63}\sqrt{65} \approx .256.$$

To find a Lipschitz ratio, we evaluate the derivative at $A = \begin{bmatrix} a & b \\ c & d \end{bmatrix}$ and $A_1 = \begin{bmatrix} a' & b' \\ c' & d' \end{bmatrix}$. We get

Inequality 2.8.71: The only tricky part of this computation is the first inequality, which uses

$$(a - a' + d - d')^2 \le 2(a - a')^2 + 2(d - d')^2.$$

Remember that since

$$(u - v)^2 = u^2 + v^2 - 2uv \ge 0,$$

we have $2uv \le u^2 + v^2$, so

$$(u + v)^2 = u^2 + v^2 + 2uv \le 2u^2 + 2v^2.$$

$$\begin{aligned} \left|[\mathbf{D}f(A)] - [\mathbf{D}f(A')]\right|^2 &= 4(a - a')^2 + 4(c - c')^2 + 4(b - b')^2 + 4(d - d')^2 \\ &\quad + 2(a - a' + d - d')^2 \\ &\le 8(a - a')^2 + 4(c - c')^2 + 4(b - b')^2 + 8(d - d')^2 \\ &\le 8\underbrace{\left((a - a')^2 + (c - c')^2 + (b - b')^2 + (d - d')^2\right)}_{|A - A'|^2}. \end{aligned} \qquad 2.8.71$$

So $2\sqrt{2} \approx 2.8$ is a Lipschitz ratio:

$$\left|[\mathbf{D}f(A)] - [\mathbf{D}f(A_1)]\right| \le 2\sqrt{2}\left|A - A_1\right|. \qquad 2.8.72$$

Since

$$\underbrace{2}_{|f(A_0)|}\ \underbrace{(.256)^2}_{|[\mathbf{D}f(A_0)]^{-1}|^2}\ \underbrace{2.8}_{M} = .367 < .5, \qquad 2.8.73$$

the Kantorovich inequality is satisfied, in a ball of radius $|H_0|$ around A_1. Some computations[22] show that $|H_0| = 2/9$ and $A_1 = \frac{1}{9}\begin{bmatrix} 35 & 1 \\ -1 & 37 \end{bmatrix}$. △

---

[22] $f(A_0) = \begin{bmatrix} 1 \\ -1 \\ 1 \\ -1 \end{bmatrix}$ gives $H_0 = -\frac{1}{63}\begin{bmatrix} 8 & 0 & -1 & 0 \\ 0 & 8 & 0 & -1 \\ -1 & 0 & 8 & 0 \\ 0 & -1 & 0 & 8 \end{bmatrix}\begin{bmatrix} 1 \\ -1 \\ 1 \\ -1 \end{bmatrix} = -\frac{1}{9}\begin{bmatrix} 1 \\ -1 \\ 1 \\ -1 \end{bmatrix}$,

so that

$$|H_0| = \frac{1}{9}\sqrt{4 \cdot 1^2} = \frac{2}{9} \quad \text{and} \quad A_1 = A_0 + H_0 = \begin{bmatrix} 4 & 0 \\ 0 & 4 \end{bmatrix} - \frac{1}{9}\begin{bmatrix} 1 & -1 \\ 1 & -1 \end{bmatrix}.$$

Example 2.8.17 (Newton's method, using a computer). Now we will use the MATLAB program NEWTON.M[23] to solve the equations

$$x^2 - y + \sin(x-y) = 2 \quad \text{and} \quad y^2 - x = 3, \tag{2.8.74}$$

starting at $\begin{pmatrix}2\\2\end{pmatrix}$ and at $\begin{pmatrix}-2\\2\end{pmatrix}$.

The equation we are solving is

$$\vec{\mathbf{f}}\begin{pmatrix}x\\y\end{pmatrix} = \begin{bmatrix} x^2 - y + \sin(x-y) - 2 \\ y^2 - x - 3 \end{bmatrix} = \begin{bmatrix}0\\0\end{bmatrix}. \tag{2.8.75}$$

Starting at $\begin{pmatrix}2\\2\end{pmatrix}$, NEWTON.M gives the following values:

$$\begin{aligned} &\mathbf{x}_0 = \begin{pmatrix}2\\2\end{pmatrix}, \quad \mathbf{x}_1 = \begin{pmatrix}2.\overline{1}\\2.2\overline{7}\end{pmatrix}, \quad \mathbf{x}_2 = \begin{pmatrix}2.10131373055664\\2.25868946388913\end{pmatrix}, \\ &\mathbf{x}_3 = \begin{pmatrix}2.10125829441818\\2.25859653392414\end{pmatrix}, \quad \mathbf{x}_4 = \begin{pmatrix}2.10125829294805\\2.25859653168689\end{pmatrix}, \end{aligned} \tag{2.8.76}$$

and the first 14 decimals don't change after that. Newton's method certainly does appear to converge. In fact, it superconverges: the number of correct decimals roughly doubles at each iteration; we see one correct decimal, then three, then eight, then 14. (We discuss superconvergence in Section 2.9.)

But are the conditions of Kantorovich's theorem satisfied? We first computed the Lipschitz ratio for the derivative directly and found it quite tricky. It's considerably easier with second partial derivatives:

$$\begin{aligned} &D_1D_1f_1 = 2 - \sin(x-y); \quad D_1D_2f_1 = \sin(x-y); \quad D_2D_2f_1 = -\sin(x-y) \\ &D_1D_1f_2 = 0; \quad D_1D_2f_2 = 0; \qquad D_2D_2f_2 = 2. \end{aligned} \tag{2.8.77}$$

Since $-1 \le \sin \le 1$,

Recall that the $c_{i,j,k}$ are defined in inequality 2.8.35.

$$\begin{aligned} &\sup|D_1D_1f_1| \le 3 = c_{1,1,1} &&\sup|D_1D_1f_2| = 0 = c_{2,1,1} \\ &\sup|D_1D_2f_1| \le 1 = c_{1,2,1} &&\sup|D_1D_2f_2| = 0 = c_{2,2,1} \\ &\sup|D_2D_1f_1| \le 1 = c_{1,1,2} &&\sup|D_2D_1f_2| = 0 = c_{2,1,2} \\ &\sup|D_2D_2f_1| \le 1 = c_{1,2,2} &&\sup|D_2D_2f_2| = 2 = c_{2,2,2}. \end{aligned}$$

So 4 is a Lipschitz ratio for $\vec{\mathbf{f}}$ on all of $\mathbb{R}^2$:

Equation 2.8.78: The three 1's in the square root correspond to $c_{1,2,1}, c_{2,1,1}, c_{2,2,1}$.

$$\left(\sum_{1\le i,j,k\le n} (c_{i,j,k})^2\right)^{1/2} = \sqrt{9+1+1+1+4} = 4; \tag{2.8.78}$$

We modified the MATLAB program to print out a number, *cond*, at each iteration, equal to $|\vec{\mathbf{f}}(\mathbf{x}_i)|\,\left|[\mathbf{D}\vec{\mathbf{f}}(\mathbf{x}_i)]^{-1}\right|^2$. The Kantorovich theorem says that Newton's method will converge if $cond \cdot 4 \le 1/2$.

[23] found at matrixeditions.com/Programs.html

The first iteration (starting as before at $\begin{pmatrix}2\\2\end{pmatrix}$), gave *cond* $= 0.1419753$ (the exact value is $\sqrt{46}/18$). Since $4 \times 0.1419753 > .5$, Kantorovich's theorem does not assert convergence, but it isn't far off. At the next iteration, we found

$$cond = 0.00874714275069, \qquad 2.8.79$$

which works with a lot to spare.

What happens if we start at $\begin{pmatrix}-2\\2\end{pmatrix}$? The computer gives

$$\mathbf{x}_0 = \begin{pmatrix}-2\\2\end{pmatrix}, \ \mathbf{x}_1 = \begin{pmatrix}-1.785544330702488\\1.30361391732438\end{pmatrix}, \ \mathbf{x}_2 = \begin{pmatrix}-1.82221637692367\\1.10354485721642\end{pmatrix},$$
$$\mathbf{x}_3 = \begin{pmatrix}-1.82152790765992\\1.08572086062422\end{pmatrix}, \ \mathbf{x}_4 = \begin{pmatrix}-1.82151878937233\\1.08557875385529\end{pmatrix},$$
$$\mathbf{x}_5 = \begin{pmatrix}-1.82151878872556\\1.08557874485200\end{pmatrix}, \ \dots, \qquad 2.8.80$$

and again the numbers do not change if we iterate the process further. It certainly converges fast. The *cond* numbers are

$$0.3337, \ 0.1036, \ 0.01045, \ \dots. \qquad 2.8.81$$

We see that the condition of Kantorovich's theorem fails rather badly at the first step (and indeed, the first step is rather large), but it succeeds (just barely) at the second. △

Remark. What if you don't know an initial starting point $\mathbf{a}_0$? *Newton's method is guaranteed to work only when you know something to start out.* If you don't, you have to guess and hope for the best. Actually, this isn't quite true. In the nineteenth century, Cayley showed that for any quadratic equation, Newton's method essentially *always* works. But quadratic equations are the only case where Newton's method does not exhibit chaotic behavior.[24] △

Remark. The hard part of Newton's method is computing the Lipschitz ratio, even using the method of second partial derivatives; bounding second partials is rarely as easy as in Example 2.8.17. In practice, people do not actually compute an upper bound for the second partials (i.e., the $c_{i,j,k}$ of Proposition 2.8.9). For the upper bounds they substitute the actual second partials, evaluated at the point $\mathbf{a}_0$ (and then at $\mathbf{a}_1, \mathbf{a}_2, \dots$); thus in place of the Lipschitz ratio M they use the number obtained by squaring the second partials, adding them together, and taking the square root of the sum.

This calls for some caution. Careful people keep track of how that number changes; it should tend to 0 if the sequence $\mathbf{a}_0, \mathbf{a}_1 \dots$ converges to a

[24]For a precise description of how Newton's method works for quadratic equations and how things can go wrong in other cases, see J. Hubbard and B. West, *Differential Equations, A Dynamical Systems Approach, Part I*, Texts in Applied Mathematics No. 5, Springer-Verlag, New York, 1991, pp. 227–235.

root. If it does not steadily decrease, probably one began in a chaotic region where a small change in initial condition can make a big difference in the ultimate outcome: rather than converging to the nearest root, Newton's method has moved further from it. Most often, it will then start converging, but to a more distant root. If you use this method and have not paid attention, you will then see only the end result: you started at a point $\mathbf{a}_0$ and ended up at a root. It is often useful to know that you are not at the root closest to your starting point. △

Exercises for Section 2.8

2.8.1 In Example 2.8.15 we computed a Lipschitz ratio M directly, getting $M = 2\sqrt{2}$. Compute it using second partial derivatives (Proposition 2.8.9).

2.8.2 a. What happens if you compute $\sqrt{b}$ by Newton's method, setting $a_{n+1} = \frac{1}{2}\left(a_n + \frac{b}{a_n}\right)$, starting with $a_0 < 0$?

b. What happens if you compute $\sqrt[3]{b}$ by Newton's method, with $b > 0$, starting with $a_0 < 0$? *Hint*: Make a careful drawing, focusing on what happens if $a_0 = 0$ or if $a_1 = 0$, etc.

Exercise 2.8.3: Recall that we define a Lipschitz function in the footnote associated with Definition 2.8.4. If $X \subset \mathbb{R}^n$, then a mapping $\mathbf{f} : X \to \mathbb{R}^m$ is Lipschitz if there exists C such that

$$\left|\mathbf{f}(\mathbf{x}) - \mathbf{f}(\mathbf{y})\right| \le C|\mathbf{x} - \mathbf{y}|.$$

(Of course a Lipschitz mapping is continuous; it is better than continuous.)

2.8.3 a. Show that the function $|x|$ is Lipschitz with Lipschitz ratio 1.

b. Show that the function $\sqrt{|x|}$ is not Lipschitz.

2.8.4 a. Find the formula $a_{n+1} = g(a_n)$ to compute the kth root of a number by Newton's method.

b. Interpret this formula as a weighted average.

2.8.5 a. Compute by hand the number $9^{1/3}$ to six decimals, using Newton's method, starting at $a_0 = 2$.

b. Find the relevant quantities h_0, a_1, M of Kantorovich's theorem in this case.

c. Prove that Newton's method does converge. (You are allowed to use Kantorovich's theorem, of course.)

$$\mathbf{f}\begin{pmatrix} x \\ y \end{pmatrix} = \begin{pmatrix} x^2 - y - 12 \\ y^2 - x - 11 \end{pmatrix}$$

Map for Exercise 2.8.6

2.8.6 a. Find a global Lipschitz ratio for the derivative of the map $\mathbf{f} : \mathbb{R}^2 \to \mathbb{R}^2$ in the margin.

b. Do one step of Newton's method to solve $\mathbf{f}\begin{pmatrix} x \\ y \end{pmatrix} = \begin{pmatrix} 0 \\ 0 \end{pmatrix}$, starting at the point $\begin{pmatrix} 4 \\ 4 \end{pmatrix}$.

c. Find a disc that you are sure contains a root.

2.8.7 Consider the system of equations $\begin{cases} \cos x + y = 1.1 \\ x + \cos(x + y) = 0.9 \end{cases}$.

In Exercise 2.8.7 we advocate using a program like MATLAB (NEWTON.M), but it is not too cumbersome for a calculator.

a. Carry out four steps of Newton's method, starting at $\begin{pmatrix} 0 \\ 0 \end{pmatrix}$. How many decimals change between the third and the fourth step?

b. Are the conditions of Kantorovich's theorem satisfied at the first step? At the second step?

2.8.8 Use the MATLAB program NEWTON.M[25] (or the equivalent) to solve the following systems of equations.

Exercise 2.8.8: "Superconvergence" is defined with precision in Section 2.9. In Example 2.8.17, we defined it informally. "Does Newton's method appear to superconverge?" means "does the number of correct decimals appear to double at every step?".

a. $$\begin{aligned} x^2 - y + \sin(x-y) &= 2 \\ y^2 - x &= 3 \end{aligned} \quad \text{starting at } \begin{pmatrix} 2 \\ 2 \end{pmatrix}, \begin{pmatrix} -2 \\ 2 \end{pmatrix}$$

b. $$\begin{aligned} x^3 - y + \sin(x-y) &= 5 \\ y^2 - x &= 3 \end{aligned} \quad \text{starting at } \begin{pmatrix} 2 \\ 2 \end{pmatrix}, \begin{pmatrix} -2 \\ 2 \end{pmatrix}$$

Does Newton's method appear to superconverge?

In all cases, determine the numbers that appear in Kantorovich's theorem, and check whether the theorem guarantees convergence.

2.8.9 Find a number $\epsilon > 0$ such that the set of equations in the margin has a unique solution near $\mathbf{0}$ when $|a|, |b|, |c| < \epsilon$.

$$\begin{cases} x + y^2 = a \\ y + z^2 = b \\ z + x^2 = c \end{cases}$$

Equations for Exercise 2.8.9

2.8.10 Do one step of Newton's method to solve the system of equations

$$\begin{aligned} x + \cos y - 1.1 &= 0 \\ x^2 - \sin y + 0.1 &= 0 \end{aligned} \quad \text{starting at } \mathbf{a}_0 = \begin{pmatrix} 0 \\ 0 \end{pmatrix}.$$

2.8.11 a. Write one step of Newton's method to solve $x^5 - x - 6 = 0$, starting at $x_0 = 2$.

b. Prove that this Newton's method converges.

2.8.12 a. Do one step of Newton's method to solve the equations

$$\begin{aligned} y - x^2 + 8 + \cos x &= 0 \\ x - y^2 + 9 + 2\cos y &= 0 \end{aligned} \quad \text{starting at } \begin{pmatrix} x_0 \\ y_0 \end{pmatrix} = \begin{pmatrix} \pi \\ \pi \end{pmatrix}.$$

b. Does Newton's method converge? You may find it useful to know that $\pi^2 \approx 9.86$ is quite close to 10.

c. Find a number $R < 1$ such that there exists a root of the equations in the disc of radius R around $\begin{pmatrix} \pi \\ \pi \end{pmatrix}$.

Exercise 2.8.13, part a: The blunderbuss (partial derivative) approach works, but a direct calculation gives a better result.

2.8.13 a. Find a global Lipschitz ratio for the derivative of the mapping $\mathbf{F} : \mathbb{R}^2 \to \mathbb{R}^2$ given by

$$\mathbf{F}\begin{pmatrix} x \\ y \end{pmatrix} = \begin{pmatrix} x^2 + y^2 - 2x - 15 \\ xy - x - y \end{pmatrix}.$$

b. Do one step of Newton's method to solve $\mathbf{F}\begin{pmatrix} x \\ y \end{pmatrix} = \begin{pmatrix} 0 \\ 0 \end{pmatrix}$ starting at $\begin{pmatrix} 5 \\ 1 \end{pmatrix}$.

c. Find and sketch a disc in $\mathbb{R}^2$ which you are sure contains a root.

Exercise 2.8.14, part b: *Hint*: Let

$$f(a) = \frac{1}{2}\left(a + \frac{b}{a^{k-1}}\right).$$

Show that $f(b^{1/k}) = b^{1/k}$; compute $f'(b^{1/k})$.

***2.8.14** a. Prove that if you compute $\sqrt[k]{b}$ by Newton's method, as in Exercise 2.8.4, choosing $a_0 > 0$ and $b > 0$, then the sequence $n \mapsto a_n$ converges to the positive kth root.

b. Show that this would not be true if you chose $a_0 > 0$ and $b > 0$ and defined the sequence $n \mapsto a_n$ recursively by the divide and average algorithm:

$$a_{n+1} = \frac{1}{2}\left(a_n + \frac{b}{a_n^{k-1}}\right).$$

[25] found at matrixeditions.com/Programs.html

c. Use Newton's method and "divide and average" (and a calculator or computer) to compute $\sqrt[3]{2}$, starting at $a_0 = 2$. What can you say about the speeds of convergence? (A drawing is recommended, as computing cube roots is considerably harder than computing square roots.)

Exercise 2.8.15: Note that this is a quadratic equation in y, z, λ. Finding eigenvalues always gives rise to quadratic equations.

2.8.15 Find $r > 0$ such that Newton's method to solve

$$\begin{bmatrix} 1+a & a^2 & 0 \\ a & 2 & a \\ -a & 0 & 3 \end{bmatrix} \begin{bmatrix} 1 \\ y \\ z \end{bmatrix} = \lambda \begin{bmatrix} 1 \\ y \\ z \end{bmatrix}, \quad \text{starting at} \begin{pmatrix} 1 \\ y_0 \\ z_0 \end{pmatrix} = \begin{pmatrix} 1 \\ 0 \\ 0 \end{pmatrix}, \quad \lambda_0 = 1,$$

is sure to converge to a solution for $|a| < r$.

Exercise 2.8.16: Try adding variables $u_1 = x^2$, $u_2 = x^4 = u_1^2$, as in Example 2.8.12.

2.8.16 Find $r > 0$ such that if you solve $x^6 + x^5 - x = a$ by Newton's method, starting at 0, you are sure to converge to a solution for $|a| < r$.

2.9 Superconvergence

If Newton's method gave only the linear rate of convergence guaranteed by Kantorovich's theorem, it would be of limited interest. People use it because if we require $< 1/2$ in inequality 2.8.53 rather than $\leq 1/2$, it *superconverges*.

We will also see that if we use a different way to measure matrices: the *norm* rather than the length, then the hypotheses of Kantorovich's theorem are easier to satisfy, so the theorem applies to more systems of equations.

Superconvergence explains why Newton's method is the favorite scheme for solving equations. If at each step Newton's method only halved the distance between guess and root, a number of simpler algorithms (bisection, for example) would work just as well.

When Newton's method works at all, it starts superconverging soon. As a rule of thumb, if Newton's method hasn't converged to a root in seven or eight steps, you've chosen a poor initial condition.

Example 2.9.1 (Slow convergence). Let us solve $f(x) = (x-1)^2 = 0$ by Newton's method, starting at $a_0 = 0$. As Exercise 2.9.1 asks you to show, the best Lipschitz ratio for f' is 2, so

$$|f(a_0)|\left|\left(f'(a_0)\right)^{-1}\right|^2 M = 1 \cdot \left(-\frac{1}{2}\right)^2 \cdot 2 = \frac{1}{2}, \qquad 2.9.1$$

and Theorem 2.8.13 guarantees that Newton's method will converge to the unique root $a = 1$, which is on the boundary of $U_1 = (0, 1)$. The exercise also asks you to check that $h_n = 1/2^{n+1}$, so $a_n = 1 - 1/2^n$. Thus at each step $|h_n|$ is exactly half the preceding $|h_{n-1}|$. The proof of the Kantorovich theorem in Appendix A5 shows that this is exactly the minimal rate of convergence guaranteed. △

Example 2.9.1 is both true and squarely misleading. Newton's method usually converges *much, much faster* than in Example 2.9.1. If the product in inequality 2.8.53 is strictly less than 1/2:

$$|\vec{\mathbf{f}}(\mathbf{a}_0)||[\mathbf{D}\vec{\mathbf{f}}(\mathbf{a}_0)]^{-1}|^2 M = k < \frac{1}{2}, \qquad 2.9.2$$

then Newton's method *superconverges*. How soon it starts superconverging depends on the problem. But once it starts, it is so fast that within four more steps you will have computed your answer to as many digits as a computer can handle.

What do we mean by superconvergence? Our definition is the following.

Definition 2.9.2 (Superconvergence). Set $x_i = |a_{i+1} - a_i|$. The sequence $a_0, a_1, \dots$ *superconverges* if, when the x_i are written in base 2, then each number x_i starts with $2^i - 1 \approx 2^i$ zeros.

In Definition 2.9.2, x_i represents the difference between two successive entries of the sequence.

Example 2.9.3: The 1/2 in $x_0 = 1/2$ is unrelated to the 1/2 of inequality 2.9.2. If we defined superconvergence using digits in base 10, then the same sequence would superconverge starting at $x_3 \le 1/10$. For it to start superconverging at x_0, we would have to have $x_0 \le 1/10$.

Example 2.9.3 (Superconvergence). The sequence $x_{n+1} = x_n^2$, starting with $x_0 = 1/2$ (written .1 in base 2), superconverges to zero, as shown in the left side of Table 2.9.1. The right side shows the convergence achieved in Example 2.9.1, again starting with $x_0 = 1/2$.

$x_0 = .1$	$x_0 = .1$
$x_1 = .01$	$x_1 = .01$
$x_2 = .0001$	$x_2 = .001$
$x_3 = .00000001$	$x_3 = .0001$
$x_4 = .0000000000000001$	$x_4 = .00001$

TABLE 2.9.1. LEFT: Superconvergence. RIGHT: The convergence guaranteed by Kantorovich's theorem. Numbers are written in base 2. △

What goes wrong for Example 2.9.1 is that whenever the product in the Kantorovich inequality equals 1/2, the denominator $1 - 2k$ in equation 2.9.4 is 0 and c is infinite, so the inequality $|\mathbf{h}_n| \le 1/(2c)$ in equation 2.9.5 is meaningless. When the product in the Kantorovich inequality 2.8.53 is strictly less than 1/2, we are guaranteed that superconvergence will occur.

Geometric convergence would replace the 2^m in the exponent in inequality 2.9.5 by m. This would add one digit in base 2 at each iteration, rather than doubling the number of digits.

Even if

$$k = |\vec{\mathbf{f}}(\mathbf{a}_0)|\,|[\mathbf{D}\vec{\mathbf{f}}(\mathbf{a}_0)]^{-1}|^2 M$$

is almost 1/2, so that c is large, the factor $(1/2)^{2^m}$ will soon predominate.

Theorem 2.9.4 (Newton's method superconverges). *Set*

$$k \stackrel{\text{def}}{=} |\vec{\mathbf{f}}(\mathbf{a}_0)|\,\big|[\mathbf{D}\vec{\mathbf{f}}(\mathbf{a}_0)]^{-1}\big|^2 M < \frac{1}{2} \qquad 2.9.3$$

$$c \stackrel{\text{def}}{=} \frac{1-k}{1-2k}\big|[\mathbf{D}\vec{\mathbf{f}}(\mathbf{a}_0)]^{-1}\big|\frac{M}{2}. \qquad 2.9.4$$

$$\textit{If} \quad |\vec{\mathbf{h}}_n| \le \frac{1}{2c}, \qquad \textit{then} \qquad |\vec{\mathbf{h}}_{n+m}| \le \frac{1}{c}\cdot\left(\frac{1}{2}\right)^{2^m}. \qquad 2.9.5$$

Since $\vec{\mathbf{h}}_n = |\mathbf{a}_{n+1} - \mathbf{a}_n|$, starting at step n and using Newton's method for m iterations causes the distance between $\mathbf{a}_n$ and $\mathbf{a}_{n+m}$ to shrink to practically nothing before our eyes; if $m = 10$,

$$|\vec{\mathbf{h}}_{n+m}| \le \frac{1}{c}\cdot\left(\frac{1}{2}\right)^{1024}. \qquad 2.9.6$$

The proof requires the following lemma, proved in Appendix A6.

Lemma 2.9.5. *If the conditions of Theorem 2.9.4 are met, then for all i,*

$$|\vec{\mathbf{h}}_{i+1}| \le c|\vec{\mathbf{h}}_i|^2. \qquad 2.9.7$$

Proof of Theorem 2.9.4. Let $x_i = c|\vec{\mathbf{h}}_i|$. Then, by Lemma 2.9.5,

$$x_{i+1} = c|\vec{\mathbf{h}}_{i+1}| \le c^2|\vec{\mathbf{h}}_i|^2 = x_i^2. \qquad 2.9.8$$

Our assumption that $|\vec{\mathbf{h}}_n| \le \frac{1}{2c}$ tells us that $x_n \le 1/2$. So

$$\begin{aligned} x_{n+1} &\le x_n^2 \le \frac{1}{4} = \left(\frac{1}{2}\right)^{2^1}, \\ x_{n+2} &\le (x_{n+1})^2 \le x_n^4 \le \frac{1}{16} = \left(\frac{1}{2}\right)^{2^2}, \\ &\vdots \\ x_{n+m} &\le x_n^{2^m} \le \left(\frac{1}{2}\right)^{2^m}. \end{aligned} \qquad 2.9.9$$

Since $x_{n+m} = c|\vec{\mathbf{h}}_{n+m}|$ this gives $|\vec{\mathbf{h}}_{n+m}| \le \frac{1}{c} \cdot \left(\frac{1}{2}\right)^{2^m}$. □

Kantorovich's theorem: a stronger version

Now we show that we can state Kantorovich's theorem in such a way that it will apply to a larger class of functions. We do this by using a different way to measure linear mappings: the *norm*.

Definition 2.9.6: We already met the norm in the margin note to the proof of Theorem 1.5.33 and in Exercise 1.5.22.

The norm is also called the *operator norm*. We describe it here for a linear transformation that takes as input a vector in $\mathbb{R}^n$ and gives as output a vector in $\mathbb{R}^m$, but the norm can also be computed for a linear transformation whose inputs and outputs are matrices (see Example 2.9.10) or abstract vectors. The operator norm can be computed for any linear transformation, as long as one has a way to measure the input (to require that it have length 1) and the output (to determine the sup).

Definition 2.9.6 (The norm of a linear transformation). Let $A : \mathbb{R}^n \to \mathbb{R}^m$ be a linear transformation. The *norm* $\|A\|$ of A is

$$\|A\| \overset{\text{def}}{=} \sup|A\vec{\mathbf{x}}|, \text{ when } \vec{\mathbf{x}} \in \mathbb{R}^n \text{ and } |\vec{\mathbf{x}}| = 1. \qquad 2.9.10$$

This means that $\|A\|$ is the maximum amount by which multiplication by A will stretch a vector.

Example 2.9.7: Multiplication by the matrix A can *at most* double the length of a vector; it does not always do so. For instance,

$$\begin{bmatrix} 2 & 0 \\ 0 & 1 \end{bmatrix}\begin{bmatrix} 0 \\ 1 \end{bmatrix} = \begin{bmatrix} 0 \\ 1 \end{bmatrix}.$$

We do not define the norm in full generality here because that would require giving a way to measure the length of an abstract vector, beyond the scope of this book.

Example 2.9.7 (Norm of a matrix). Take $A = \begin{bmatrix} 2 & 0 \\ 0 & 1 \end{bmatrix}$. Then $A\vec{\mathbf{x}} = A\begin{bmatrix} x \\ y \end{bmatrix} = \begin{bmatrix} 2x \\ y \end{bmatrix}$, so

$$\|A\| = \sup_{|\vec{\mathbf{x}}|=1} |A\vec{\mathbf{x}}| = \overbrace{\sup_{\sqrt{x^2+y^2}=1} \sqrt{4x^2+y^2} = 2}^{\text{setting } x=1, y=0}. \quad \triangle \qquad 2.9.11$$

There are many equations for which convergence is guaranteed if one uses the norm, but not if one uses the length.

In Example 2.9.7, note that $\|A\| = 2$, while $|A| = \sqrt{5}$. You are asked in Exercise 2.9.4 to show that it is always true that

$$\|A\| \le |A|. \qquad 2.9.12$$

This is why using the norm rather than the length makes Kantorovich's theorem stronger: the key to that theorem, inequality 2.8.53, is easier to satisfy using the norm.

Theorem 2.9.8 (Kantorovich's theorem: a stronger version). *Kantorovich's theorem 2.8.13 still holds if you replace both lengths of matrices by norms of matrices:* $|[\mathbf{D}\vec{\mathbf{f}}(\mathbf{u}_1)] - [\mathbf{D}\vec{\mathbf{f}}(\mathbf{u}_2)]|$ *is replaced by* $\|[\mathbf{D}\vec{\mathbf{f}}(\mathbf{u}_1)] - [\mathbf{D}\vec{\mathbf{f}}(\mathbf{u}_2)]\|$ *and* $|[\mathbf{D}\vec{\mathbf{f}}(\mathbf{a}_0)]^{-1}|^2$ *by* $\|[\mathbf{D}\vec{\mathbf{f}}(\mathbf{a}_0)]^{-1}\|^2$.

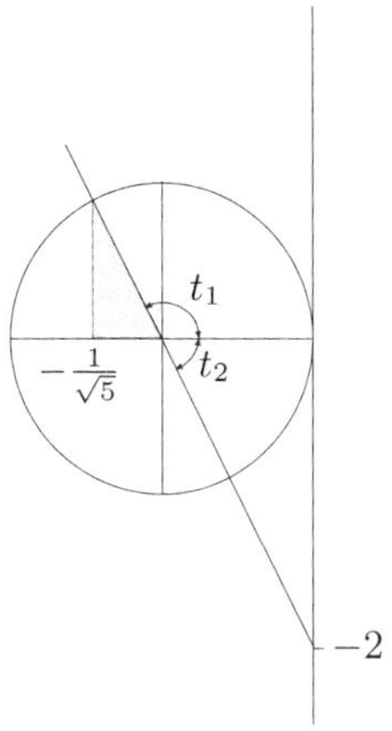

FIGURE 2.9.2.
The diagram for Example 2.9.9.

Proof. In the proof of Theorem 2.8.13 we only used the triangle inequality and Proposition 1.4.11, and these hold for the norm $\|A\|$ of a matrix A as well as for its length $|A|$, as Exercises 2.9.2 and 2.9.3 ask you to show. □

Unfortunately, the norm is usually much harder to compute than the length. In equation 2.9.11, it is not difficult to see that 2 is the largest value of $\sqrt{4x^2+y^2}$ compatible with the requirement that $\sqrt{x^2+y^2}=1$, obtained by setting $x=1$ and $y=0$. Computing the norm is not often that easy.

Example 2.9.9 (Norm is harder to compute). The length of the matrix $A = \begin{bmatrix} 1 & 1 \\ 0 & 1 \end{bmatrix}$ is $\sqrt{1^2+1^2+1^2} = \sqrt{3}$, or about 1.732. The norm is $\frac{1+\sqrt{5}}{2}$, or about 1.618; arriving at that figure takes some work, as follows. A vector $\begin{bmatrix} x \\ y \end{bmatrix}$ with length 1 can be written $\begin{bmatrix} \cos t \\ \sin t \end{bmatrix}$, and the product of A and that vector is $\begin{bmatrix} \cos t + \sin t \\ \sin t \end{bmatrix}$, so the object is to find

$$\sup \sqrt{(\cos t + \sin t)^2 + \sin^2 t}\,. \qquad 2.9.13$$

Proposition 3.8.3 gives a formula for computing the norm of any matrix.

At a minimum or maximum of a function, its derivative is 0, so we need to see where the derivative of $(\cos t + \sin t)^2 + \sin^2 t$ vanishes. That derivative is $2\cos 2t + \sin 2t$, which vanishes for $2t = \arctan(-2)$. We have two possible angles to look for, t_1 and t_2, as shown in Figure 2.9.2; they can be computed with a calculator or with a bit of trigonometry, and we can choose the one that gives the biggest value for equation 2.9.13. Since the entries of the matrix A are all positive, we choose t_1, in $[0,\pi]$, as the best bet.

$$\begin{aligned}&\left((\cos t + \sin t)^2 + \sin^2 t\right)' \\ &= 2(\cos t + \sin t)(\cos t - \sin t) \\ &\quad + 2\sin t \cos t \\ &= 2(\cos^2 t - \sin^2 t) + 2\sin t\cos t \\ &= 2\cos 2t + \sin 2t\end{aligned}$$

By similar triangles, we find that

$$\cos 2t_1 = -\frac{1}{\sqrt{5}} \quad \text{and} \quad \sin 2t_1 = \frac{2}{\sqrt{5}}. \qquad 2.9.14$$

Using the formula $\cos 2t_1 = 2\cos^2 t_1 - 1 = 1 - 2\sin^2 t$, we find that

$$\cos t_1 = \sqrt{\frac{1}{2}\left(1 - \frac{1}{\sqrt{5}}\right)}, \quad \text{and} \quad \sin t_1 = \sqrt{\frac{1}{2}\left(1 + \frac{1}{\sqrt{5}}\right)}, \qquad 2.9.15$$

which, after some computation, gives

$$\left|\begin{bmatrix} \cos t_1 + \sin t_1 \\ \sin t_1 \end{bmatrix}\right|^2 = \frac{3+\sqrt{5}}{2}, \qquad 2.9.16$$

and finally,

$$\|A\| = \left\| \begin{bmatrix} 1 & 1 \\ 0 & 1 \end{bmatrix} \right\| = \sqrt{\frac{3+\sqrt{5}}{2}} = \sqrt{\frac{6+2\sqrt{5}}{4}} = \frac{1+\sqrt{5}}{2}. \quad \triangle \qquad 2.9.17$$

Remark. We could have used the following formula for computing the norm of a 2×2 matrix from its length and its determinant:

$$\|A\| = \sqrt{\frac{|A|^2 + \sqrt{|A|^4 - 4(\det A)^2}}{2}}, \qquad 2.9.18$$

as you are asked to prove in Exercise 2.9.8. In higher dimensions things are much worse. It was to avoid this kind of complication that we used the length rather than the norm when we proved Kantorovich's theorem. $\triangle$

But in some cases the norm is easier to use than the length, as in the following example. In particular, norms of multiples of the identity matrix are easy to compute: such a norm is just the absolute value of the multiple.

Example 2.9.10 (Using the norm in Newton's method). Suppose we want to find a 2×2 matrix A such that $A^2 = \begin{bmatrix} 8 & 1 \\ -1 & 10 \end{bmatrix}$. We define $F : \text{Mat}\,(2,2) \to \text{Mat}\,(2,2)$ by

$$F(A) = A^2 - \begin{bmatrix} 8 & 1 \\ -1 & 10 \end{bmatrix}, \qquad 2.9.19$$

and try to solve it by Newton's method. First we choose an initial point A_0. A logical place to start would seem to be the matrix

$$A_0 = \begin{bmatrix} 3 & 0 \\ 0 & 3 \end{bmatrix}, \quad \text{so that} \quad A_0^2 = \begin{bmatrix} 9 & 0 \\ 0 & 9 \end{bmatrix}. \qquad 2.9.20$$

We want to see whether the Kantorovich inequality 2.8.53 is satisfied: whether

$$|F(A_0)| \cdot M \|[\mathbf{D}F(A_0)]^{-1}\|^2 \le \frac{1}{2}. \qquad 2.9.21$$

First, compute the derivative:

$$[\mathbf{D}F(A)]B = AB + BA. \qquad 2.9.22$$

The following computation shows that $A \mapsto [\mathbf{D}F(A)]$ is Lipschitz with respect to the norm, with Lipschitz ratio 2 on all of $\text{Mat}\,(2,2)$:

$$\begin{aligned} \|[\mathbf{D}F(A_1)] - [\mathbf{D}F(A_2)]\| &= \sup_{|B|=1} \left| \Big([\mathbf{D}F(A_1)] - [\mathbf{D}F(A_2)] \Big) B \right| \\ &= \sup_{|B|=1} |A_1B + BA_1 - A_2B - BA_2| = \sup_{|B|=1} |(A_1 - A_2)B + B(A_1 - A_2)| \\ &\le \sup_{|B|=1} |(A_1 - A_2)B| + |B(A_1 - A_2)| \le \sup_{|B|=1} |A_1 - A_2||B| + |B||A_1 - A_2| \\ &\le \sup_{|B|=1} 2|B||A_1 - A_2| = 2|A_1 - A_2|. \end{aligned} \qquad 2.9.23$$

Example 2.9.10: In a draft of this book we proposed a different example, finding a 2×2 matrix A such that

$$A^2 + A = \begin{bmatrix} 1 & 1 \\ 1 & 1 \end{bmatrix}.$$

A friend pointed out that this problem can be solved explicitly (and more easily) without Newton's method, as Exercise 2.9.5 asks you to do. But trying to solve equation 2.9.19 without Newton's method would be unpleasant.

Equation 2.9.20: You might think that $B = \begin{bmatrix} 3 & 1 \\ -1 & 3 \end{bmatrix}$ would be better, but $B^2 = \begin{bmatrix} 8 & 6 \\ -6 & 8 \end{bmatrix}$. In addition, starting with a diagonal matrix makes our computations easier.

Equation 2.9.21: $F(A_0)$ is a matrix, but it plays the role of the vector $\vec{\mathbf{f}}(\mathbf{a}_0)$ in Theorem 2.8.13, so we use the length, not the norm.

Equation 2.9.22: We first encountered the derivative $AB+BA$ in Example 1.7.17.

Note that when computing the norm of a linear transformation from $\text{Mat}\,(2,2)$ to $\text{Mat}\,(2,2)$, the role of $\vec{\mathbf{x}}$ in Definition 2.9.6 is played by a 2×2 matrix with length 1 (here, the matrix B).

Now we insert A_0 into equation 2.9.19, getting

$$F(A_0) = \begin{bmatrix} 9 & 0 \\ 0 & 9 \end{bmatrix} - \begin{bmatrix} 8 & 1 \\ -1 & 10 \end{bmatrix} = \begin{bmatrix} 1 & -1 \\ 1 & -1 \end{bmatrix}, \qquad 2.9.24$$

so that $|F(A_0)| = \sqrt{4} = 2$.

Now we need to compute $\|[\mathbf{D}F(A_0)]^{-1}\|^2$. Using equation 2.9.22 and the fact that A_0 is three times the identity, we get

$$[\mathbf{D}F(A_0)]B = \underbrace{A_0B + BA_0 = 3B + 3B}_{\substack{\text{multiplication by multiple} \\ \text{of identity is commutative}}} = 6B. \qquad 2.9.25$$

If we used the length rather than the norm in Example 2.9.10 we would have to represent the derivative of F as a 4×4 matrix, as in Example 2.8.16.

So we have

$$[\mathbf{D}F(A_0)]^{-1}B = \frac{B}{6}, \quad \|[\mathbf{D}F(A_0)]^{-1}\| = \sup_{|B|=1} |B/6| = \sup_{|B|=1} \frac{|B|}{6} = 1/6,$$

$$\left\|[\mathbf{D}F(A_0)]^{-1}\right\|^2 = \frac{1}{36}. \qquad 2.9.26$$

The left side of inequality 2.9.21 is $2 \cdot 2 \cdot 1/36 = 1/9$, and we see that the inequality is satisfied with room to spare: if we start at $\begin{bmatrix} 3 & 0 \\ 0 & 3 \end{bmatrix}$ and use Newton's method, we can compute the square root of $\begin{bmatrix} 8 & 1 \\ -1 & 10 \end{bmatrix}$. △

Exercises for Section 2.9

2.9.1 Show (Example 2.9.1) that when solving $f(x) = (x-1)^2 = 0$ by Newton's method, starting at $a_0 = 0$, the best Lipschitz ratio for f' is 2, so inequality 2.8.53 is satisfied as an equality (equation 2.9.1) and Theorem 2.8.13 guarantees that Newton's method will converge to the unique root $a = 1$. Check that $h_n = 1/2^{n+1}$ so $a_n = 1 - 1/2^n$, on the nose the rate of convergence advertised.

2.9.2 Prove that Proposition 1.4.11 is true for the norm $\|A\|$ of a matrix A as well as for its length $|A|$; that is, prove that if A is an $n \times m$ matrix, $\vec{\mathbf{b}}$ a vector in $\mathbb{R}^m$, and B an $m \times k$ matrix, then

$$|A\vec{\mathbf{b}}| \le \|A\|\,|\vec{\mathbf{b}}| \quad \text{and} \quad \|AB\| \le \|A\|\,\|B\|.$$

2.9.3 Prove that the triangle inequality (Theorem 1.4.9) holds for the norm $\|A\|$ of a matrix A, i.e., that for any matrices A and B in $\mathbb{R}^n$,

$$\|A + B\| \le \|A\| + \|B\|.$$

2.9.4 a. Prove (inequality 2.9.12) that the norm of a matrix is at most its length: $\|A\| \le |A|$. *Hint*: This follows from Proposition 1.4.11.

**b. When are they equal?

Exercise 2.9.5: Try a matrix all of whose entries are equal.

2.9.5 a. Find a 2×2 matrix A such that $A^2 + A = \begin{bmatrix} 1 & 1 \\ 1 & 1 \end{bmatrix}$.

b. Show that Newton's method converges when it is used to solve the preceding equation, starting at the identity.

2.9.6 Let X be a 2×2 matrix. For what matrices C can you be sure that the equation $X^2 + X = C$ has a solution that can be found starting at 0? At I?

2.9.7 Repeat the computations of Example 2.8.16, using the norm.

****2.9.8** If $A = \begin{bmatrix} a & b \\ c & d \end{bmatrix}$ is a real matrix, show that

$$\|A\| = \left(\frac{|A|^2 + \sqrt{|A|^4 - 4D^2}}{2} \right)^{1/2}, \quad \text{where } D = ad - bc = \det A.$$

2.10 The inverse and implicit function theorems

We base the inverse and implicit function theorems on Newton's method. This gives more precise statements than the standard approach.

"Implicit" means "implied". The statement $2x - 8 = 0$ implies that $x = 4$; it does not say it explicitly (directly).

In Section 2.2 we completely analyzed systems of linear equations. Given a system of *nonlinear* equations, what solutions do we have? What variables depend on others? Our tools for answering these questions are the implicit function theorem and its special case, the inverse function theorem. These theorems are the backbone of differential calculus, just as their linear analogues, Theorems 2.2.1 and 2.2.6, are the backbone of linear algebra. We will start with inverse functions and then move to the more general case.

An inverse function is a function that "undoes" the original function; by Definition 0.4.7, a function has an inverse if it is one to one and onto. If $f(x) = 2x$, clearly there is a function g such that $g(f(x)) = x$, namely, the function $g(y) = y/2$. Usually finding an inverse isn't so straightforward. But in one variable, there is a simple criterion for a continuous function to have an inverse: the function must be *monotone*. There is also a simple technique for finding the inverse.

Definition 2.10.1 (Monotone function). A function is *strictly monotone* if its graph always goes up or always goes down. If $x < y$ always implies $f(x) < f(y)$, the function is *monotone increasing*; if $x < y$ always implies $f(x) > f(y)$, the function is *monotone decreasing*.

If a continuous function f is monotone, then it has an inverse function g. In addition, you can find g by a series of guesses that converge to the solution, and knowing the derivative of f tells you how to compute the derivative of g. More precisely, we have the following theorem.

Theorem 2.10.2 (Inverse function theorem in one dimension). *Let $f : [a,b] \to [c,d]$ be a continuous function with $f(a) = c$, $f(b) = d$ and with f strictly monotone on $[a,b]$. Then*

1. *There exists a unique continuous function $g : [c,d] \to [a,b]$ such that*

$$f(g(y)) = y, \text{ for all } y \in [c,d], \quad \text{and} \qquad 2.10.1$$

$$g(f(x)) = x, \text{ for all } x \in [a,b]. \qquad 2.10.2$$

2. *You can find $g(y)$ by solving the equation $y - f(x) = 0$ for x by bisection (described below).*
3. *If f is differentiable at $x \in (a,b)$ and $f'(x) \neq 0$, then g is differentiable at $f(x)$ and its derivative satisfies $g'\big(f(x)\big) = 1/f'(x)$.*

You are asked to prove Theorem 2.10.2 in Exercise 2.10.17.

Part 3 of Theorem 2.10.2 justifies the use of implicit differentiation: such statements as

$$\arcsin'(x) = \frac{1}{\sqrt{1-x^2}}.$$

Example 2.10.3 (An inverse function in one dimension). Consider $f(x) = 2x + \sin x$, shown in Figure 2.10.1, and choose $[a,b] = [-k\pi, k\pi]$ for some positive integer k . Then

$$f(a) = f(-k\pi) = -2k\pi + \overbrace{\sin(-k\pi)}^{=0}, \quad f(b) = f(k\pi) = 2k\pi + \overbrace{\sin(k\pi)}^{=0};$$

i.e., $f(a) = 2a$ and $f(b) = 2b$. Since $f'(x) = 2 + \cos x$, which is ≥ 1, we see that f is strictly increasing. Thus Theorem 2.10.2 says that $y = 2x + \sin x$ expresses x implicitly as a function of y for $y \in [-2k\pi, 2k\pi]$: there is a function $g : [-2k\pi, 2k\pi] \to [-k\pi, k\pi]$ such that $g\big(f(x)\big) = g(2x + \sin x) = x$.

But if you take a hardnosed attitude and say "Okay, so what is $g(1)$?", you will see that the question is not so easy to answer. The equation $1 = 2x + \sin x$ is not particularly hard to "solve", but you can't find a formula for the solution using algebra, trigonometry, or even more advanced techniques. Instead you must apply some approximation method. △

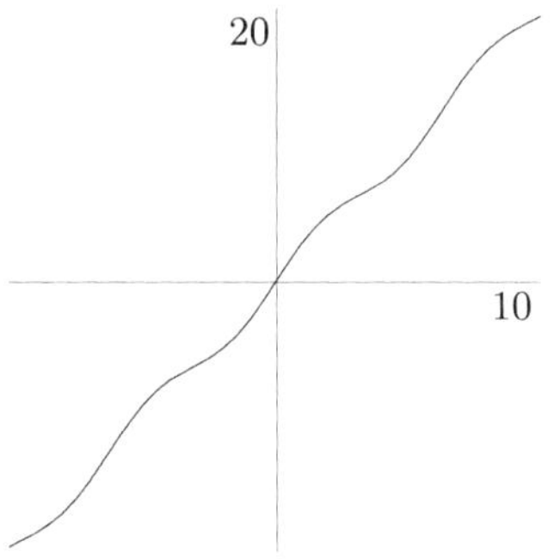

FIGURE 2.10.1.

The function $f(x) = 2x + \sin x$ is monotone increasing; it has an inverse function $g(2x + \sin x) = x$, but finding it requires solving the equation $2x + \sin x = y$, with x the unknown and y known. This can be done, but it requires an approximation technique. You can't find a formula for the solution using algebra, trigonometry, or even more advanced techniques.

In one dimension, we can use *bisection* as our approximation technique. Suppose you want to solve $f(x) = y$, and you know a and b such that $f(a) < y$ and $f(b) > y$. First try the x in the middle of $[a,b]$, computing $f(\frac{a+b}{2})$. If the answer is too small, try the midpoint of the right half-interval; if the answer is too big, try the midpoint of the left half-interval. Next choose the midpoint of the quarter-interval to the right (if your answer was too small) or to the left (if your answer was too big). The sequence of x_n chosen this way will converge to $g(y)$.

If a function is not monotone, we cannot expect to find a global inverse function, but there will usually be monotone stretches of the function for which local inverse functions exist. Consider $f(x) = x^2$. Any inverse function of f can only be defined on the image of f, the positive real numbers. Moreover, there are two such "inverses",

$$g_1(y) = +\sqrt{y} \quad \text{and} \quad g_2(y) = -\sqrt{y}; \qquad 2.10.3$$

they both satisfy $f\big(g(y)\big) = y$, but they do not satisfy $g\big(f(x)\big) = x$. However, g_1 is an inverse if the domain of f is restricted to $x \geq 0$, and g_2 is an inverse if the domain of f is restricted to $x \leq 0$. Figure 2.10.2 shows another example.

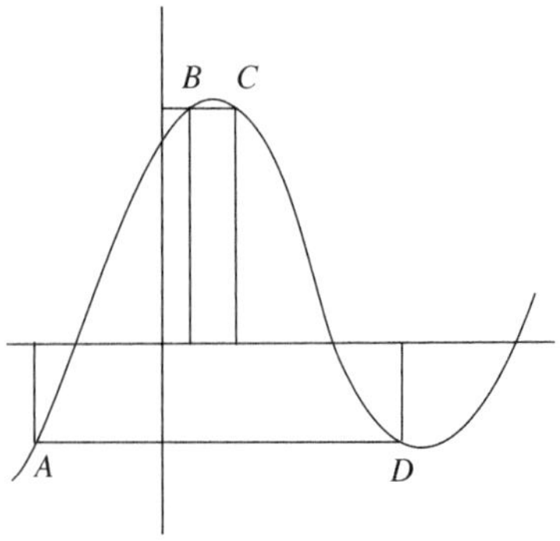

FIGURE 2.10.2.
The function graphed here is not monotone and has no global inverse; the same value of y gives both $x = B$ and $x = C$. Similarly, the same value of y gives $x = A$ and $x = D$. But it has local inverses; the arcs AB and CD both represent x as a function of y.

Inverse functions in higher dimensions

In higher dimensions, we can't speak of a map always increasing or always decreasing, but inverse functions exist in higher dimensions anyway. Monotonicity is replaced by the requirement that the derivative of the mapping be invertible; bisection is replaced by Newton's method. The full statement of the inverse function theorem is given as Theorem 2.10.7; we give a simplified version first, without proof. The key message is:

If the derivative is invertible, the mapping is locally invertible.

More precisely,

Theorem 2.10.4 (Inverse function theorem: short version). *If a mapping* $\mathbf{f}$ *is continuously differentiable, and its derivative is invertible at some point* $\mathbf{x}_0$*, then* $\mathbf{f}$ *is locally invertible, with differentiable inverse, in some neighborhood of the point* $\mathbf{f}(\mathbf{x}_0)$.

Theorem 2.10.4: If $U, V \subset \mathbb{R}^k$ are open subsets and $\mathbf{f} : U \to V$ is a differentiable map with a differentiable inverse, then by the chain rule we see that the linear transformation $[\mathbf{Df}(\mathbf{u})]$ is invertible and the derivative of the inverse is the inverse of the derivative:

$$\begin{aligned} I &= [\mathbf{D}(\mathbf{f}^{-1} \circ \mathbf{f})(\mathbf{u})] \\ &= [\mathbf{Df}^{-1}(\mathbf{f}(\mathbf{u}))][\mathbf{Df}(\mathbf{u})]. \end{aligned}$$

The inverse function theorem is the converse: if the derivative of a function $\mathbf{f}$ is invertible, then $\mathbf{f}$ is locally invertible.

Remark. The condition that the derivative be continuous is really necessary. For instance, the function of Example 1.9.4 has no local inverse near 0, although the derivative does not vanish. Without saying how the derivative is continuous, we can say nothing about the domain of the inverse function. Theorem 2.10.7 requires that the derivative be Lipschitz. Exercise A5.1 explores the use of a weaker continuity condition. $\triangle$

In Section 1.7, we went from a nonlinear function to its linear approximation, the derivative. Here we go the other direction, using information about the (linear) derivative to infer information about the function.

Note that the theorem applies only to functions $\mathbf{f}$ that go from $\mathbb{R}^m$ to $\mathbb{R}^m$, or from a subset of $\mathbb{R}^m$ to a subset of $\mathbb{R}^m$; it deals with the case where we have m equations in m unknowns. This should be obvious: only square matrices can be invertible, and if the derivative of a function is a square matrix, then the dimensions of the domain and codomain of the function must be the same.[26]

The words "locally invertible" are important. We said in Definition 0.4.7 that a transformation has an inverse if it is both onto and one to one: $\mathbf{f} : W \subset \mathbb{R}^m \to \mathbb{R}^m$ is invertible if the equation $\mathbf{f}(\mathbf{x}) = \mathbf{y}$ has a unique solution $\mathbf{x} \in W$ for every $\mathbf{y} \in \mathbb{R}^m$. Such an inverse is global. Most often

[26] As mentioned in Section 1.3, it is possible to have an invertible mapping from $\mathbb{R}^n \to \mathbb{R}^m$, with $n \neq m$, but such a map can never be continuous, much less differentiable. Such maps are wildly discontinuous, very foreign to our present tame world of continuously differentiable functions.

a mapping will not have a global inverse but it will have a local inverse (or several local inverses). Generally, we must be satisfied with asking, if $\mathbf{f}(\mathbf{x}_0) = \mathbf{y}_0$, does there exist a local inverse in a neighborhood of $\mathbf{y}_0$?

Remark. In practice, we answer the question, "is the derivative invertible?" by asking, "is its determinant nonzero?" It follows from the formula for the inverse of a 2 × 2 matrix that the matrix is invertible if and only if its determinant is not 0; Exercise 1.4.20 shows the same is true of 3 × 3 matrices. Theorem 4.8.3 generalizes this to $n \times n$ matrices. △

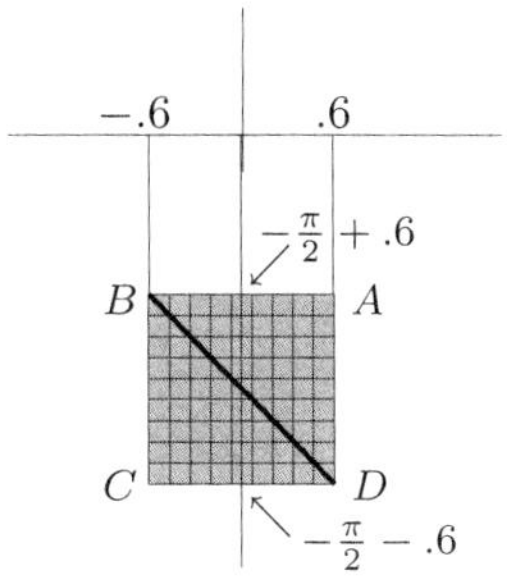

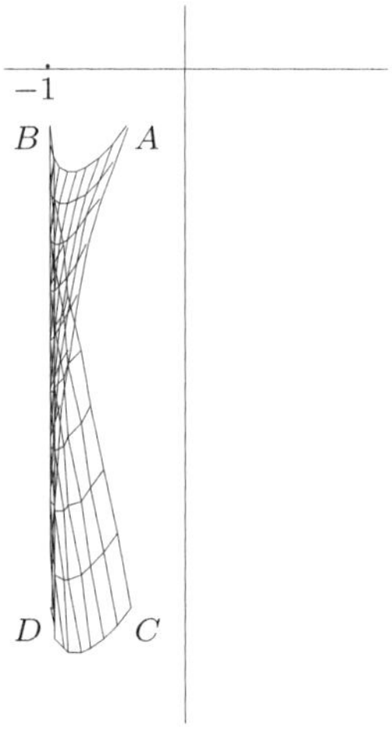

FIGURE 2.10.3.
TOP: The square

$$|x| \le .6, \quad \left|y + \frac{\pi}{2}\right| \le .6$$

BOTTOM: The image of the square under the mapping **f** of Example 2.10.5. We get this figure by folding the square over itself along the line $x+y = -\pi/2$ (the line from B to D). It is evident from the folding that **f** is not invertible on the square.

Example 2.10.5 (Where is f invertible?). Where is the function

$$\mathbf{f}\begin{pmatrix} x \\ y \end{pmatrix} = \begin{pmatrix} \sin(x+y) \\ x^2 - y^2 \end{pmatrix} \tag{2.10.4}$$

locally invertible? The derivative is

$$\left[\mathbf{Df}\begin{pmatrix} x \\ y \end{pmatrix}\right] = \begin{bmatrix} \cos(x+y) & \cos(x+y) \\ 2x & -2y \end{bmatrix}, \tag{2.10.5}$$

which is invertible if $-2y\cos(x+y) - 2x\cos(x+y) \ne 0$. So the function **f** is locally invertible at all points $\mathbf{f}\begin{pmatrix} x_0 \\ y_0 \end{pmatrix}$ that satisfy $-y \ne x$ and $\cos(x+y) \ne 0$ (i.e., $x + y \ne \pi/2 + k\pi$). △

Remark. We recommend using a computer to understand the mapping $\mathbf{f} : \mathbb{R}^2 \to \mathbb{R}^2$ of Example 2.10.5 and, more generally, any mapping from $\mathbb{R}^2$ to $\mathbb{R}^2$. (One thing we can say without a computer's help is that the first coordinate of every point in the image of **f** cannot be bigger than 1 or less than −1, since the sine function oscillates between −1 and 1. So if we graph the image using (x, y)-coordinates, it will be contained in a band between $x = -1$ and $x = 1$.) Figures 2.10.3 and 2.10.4 show just two examples of regions of the domain of **f** and the corresponding region of the image. Figure 2.10.3 shows a region of the image that is folded over; in that region the function has no inverse. Figure 2.10.4 shows a region where it has an inverse. △

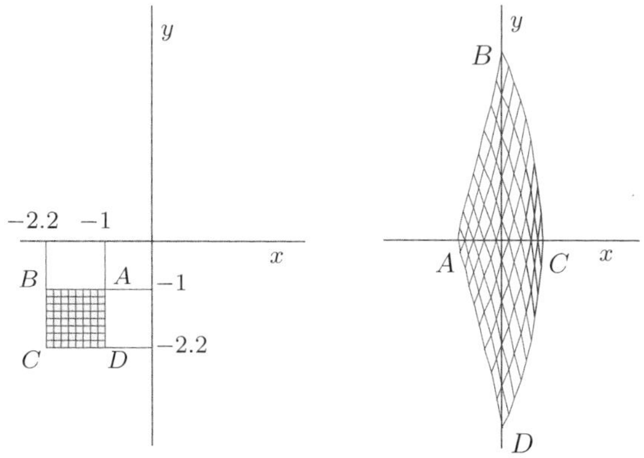

FIGURE 2.10.4. The function **f** of Example 2.10.5 maps the region at left to the region at right. In this region, **f** is invertible.

Example 2.10.6 (An example related to robotics). Let C_1 be the circle of radius 3 centered at the origin in $\mathbb{R}^2$, and let C_2 be the circle of radius 1 centered at $\begin{pmatrix}10\\0\end{pmatrix}$. What is the locus of centers of line segments drawn from a point of C_1 to a point of C_2, as illustrated by Figure 2.10.5?

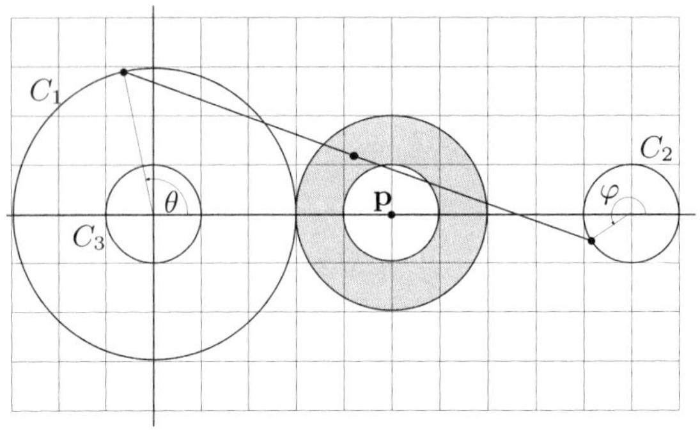

FIGURE 2.10.5. Above, C_1 is the circle of radius 3 centered at the origin, and C_2 is the circle of radius 1 centered at $\begin{pmatrix}10\\0\end{pmatrix}$. The midpoint of every line segment connecting a point in C_1 to a point in C_2 lies in the annular region between the circles of radius 2 and 1 centered at **p**. One such midpoint is shown. Extending the line segment to the far side of C_2 would give a different midpoint.

Does this example seem artificial? It's not. Problems like this come up all the time in robotics; the question of knowing where a robot arm can reach is a question just like this.

The center of the segment joining

$$\begin{pmatrix}3\cos\theta\\3\sin\theta\end{pmatrix} \in C_1 \quad \text{to} \quad \begin{pmatrix}\cos\varphi+10\\\sin\varphi\end{pmatrix} \in C_2 \tag{2.10.6}$$

is the point

$$F\begin{pmatrix}\theta\\\varphi\end{pmatrix} \overset{\text{def}}{=} \frac{1}{2}\begin{pmatrix}3\cos\theta+\cos\varphi+10\\3\sin\theta+\sin\varphi\end{pmatrix}. \tag{2.10.7}$$

We want to find the image of F. If at a point $\begin{pmatrix}\theta\\\varphi\end{pmatrix}$ the derivative of F is invertible, $F\begin{pmatrix}\theta\\\varphi\end{pmatrix}$ will certainly be in the *interior* of the image, since points in the neighborhood of that point are also in the image. To understand the image, we need to know where its boundary is. We will look first for the boundary of the image, then determine what locus can have that boundary.

Candidates for the boundary of the image are those points $F\begin{pmatrix}\theta\\\varphi\end{pmatrix}$ where $\left[\mathbf{D}F\begin{pmatrix}\theta\\\varphi\end{pmatrix}\right]$ is *not* invertible. Since

$$\begin{aligned}\det\left[\mathbf{D}F\begin{pmatrix}\theta\\\varphi\end{pmatrix}\right] &= \frac{1}{4}\det\begin{bmatrix}-3\sin\theta & -\sin\varphi\\3\cos\theta & \cos\varphi\end{bmatrix}\\ &= \frac{-3}{4}(\sin\theta\cos\varphi-\cos\theta\sin\varphi) = \frac{-3}{4}\sin(\theta-\varphi),\end{aligned} \tag{2.10.8}$$

which vanishes when $\theta = \varphi$ and when $\theta = \varphi + \pi$, candidates for the boundary of the image are the points

Equation 2.10.9: To compute

$$F\begin{pmatrix}\theta\\ \theta-\pi\end{pmatrix}$$

we use $\cos(\theta - \pi) = -\cos\theta$ and $\sin(\theta-\pi) = -\sin\theta$.

$$F\begin{pmatrix}\theta\\ \theta\end{pmatrix} = \begin{pmatrix}2\cos\theta+5\\ 2\sin\theta\end{pmatrix} \quad\text{and}\quad F\begin{pmatrix}\theta\\ \theta-\pi\end{pmatrix} = \begin{pmatrix}\cos\theta+5\\ \sin\theta\end{pmatrix}, \qquad 2.10.9$$

i.e., the circles of radius 2 and 1 centered at $\mathbf{p} = \begin{pmatrix}5\\0\end{pmatrix}$.

The only regions whose boundaries are subsets of these sets are the whole disc of radius 2 centered at $\mathbf{p}$ and the annular region between the two circles. We claim that the image of F is the annular region. If it were the whole disc of radius 2, then $\mathbf{p}$ would be in the image, which it cannot be. If $\mathbf{p}$ is the midpoint of a segment with one end on C_2, then the other end is on C_3 (the circle of radius 1 centered at the origin), hence not on C_1. $\triangle$

Remark. If the derivative of a function $\mathbf{f}$ is invertible at some point $\mathbf{x}_0$, then $\mathbf{f}$ is locally invertible in a neighborhood of $\mathbf{f}(\mathbf{x}_0)$; but it is not true that if the derivative is invertible everywhere, the function is invertible. Consider the function $\mathbf{f} : \mathbb{R}^2 \to \mathbb{R}^2$ given by

$$\begin{pmatrix}t\\ \theta\end{pmatrix} \mapsto \begin{pmatrix}e^t\cos\theta\\ e^t\sin\theta\end{pmatrix}; \qquad 2.10.10$$

the derivative is invertible everywhere, since $\det[\mathbf{Df}] = e^t \neq 0$, but $\mathbf{f}$ is not invertible, since it sends $\begin{pmatrix}t\\ \theta\end{pmatrix}$ and $\begin{pmatrix}t\\ \theta+2\pi\end{pmatrix}$ to the same point. $\triangle$

The complete inverse function theorem: quantifying "locally"

So far we have used the short version of the inverse function theorem, which guarantees the existence of a local inverse without specifying a domain for the local inverse. If you ever want to compute an inverse function, you'll need to know in what neighborhood it exists. Theorem 2.10.7 tells how to compute this neighborhood. Unlike the short version, which just requires the derivative to be continuous, Theorem 2.10.7 requires it to be Lipschitz. To compute a neighborhood where an inverse function exists, we must say something about how the derivative is continuous. We chose the Lipschitz condition because we want to use Newton's method to compute the inverse function, and Kantorovich's theorem requires the Lipschitz condition. Exercise A5.1 explores what happens with weaker continuity conditions.

Remark. If the derivative $[\mathbf{Df}(\mathbf{x}_0)]$ is not invertible, then $\mathbf{f}$ has no differentiable inverse in the neighborhood of $\mathbf{f}(\mathbf{x}_0)$; see Exercise 2.10.4. However, it may have a *nondifferentiable* inverse. For example, the derivative at 0 of the function $x \mapsto x^3$ is 0, which is not invertible. But the function is invertible; its inverse is the cube root function $g(y) = y^{1/3}$. This does not contradict the inverse function theorem because g is not differentiable at 0, since

$$g'(y) = \frac{1}{3}\frac{1}{y^{2/3}}. \quad \triangle$$

Theorem 2.10.7 is illustrated by Figure 2.10.6.

Note that we could replace the length of the derivatives by their norm (Definition 2.9.6) to get a slightly stronger theorem.

On first reading, skip the last sentence about the ball with radius R_1. It is a minor point.

The ball V gives a lower bound for the domain of $\mathbf{g}$; the actual domain may be bigger. But there are cases where the largest R satisfying the conditions of Theorem 2.10.7 is optimal. We invite you to check that if $f(x) = (x-1)^2$, with $x_0 = 0$, so that $y_0 = 1$, then the largest R satisfying equation 2.10.11 is $R = 1$. Thus the interval $V = (0, 2)$ is the largest interval centered at 1 on which an inverse can be defined. Indeed, since the function g is $g(y) = 1 - \sqrt{y}$, any value of y smaller than 0 is not in the domain of g.

Theorem 2.10.7 (The inverse function theorem). *Let $W \subset \mathbb{R}^m$ be an open neighborhood of $\mathbf{x}_0$, and let $\mathbf{f} : W \to \mathbb{R}^m$ be a continuously differentiable function. Set $\mathbf{y}_0 = \mathbf{f}(\mathbf{x}_0)$.*

If the derivative $[\mathbf{Df}(\mathbf{x}_0)]$ is invertible, then $\mathbf{f}$ is invertible in some neighborhood of $\mathbf{y}_0$, and the inverse is differentiable.

To quantify this statement, we will specify the radius R of a ball V centered at $\mathbf{y}_0$, in which the inverse function is defined. First simplify notation by setting $L = [\mathbf{Df}(\mathbf{x}_0)]$. Now find $R > 0$ satisfying the following conditions:

1. *The ball W_0 of radius $2R|L^{-1}|$ and centered at $\mathbf{x}_0$ is contained in W.*
2. *In W_0, the derivative of $\mathbf{f}$ satisfies the Lipschitz condition*

$$|[\mathbf{Df}(\mathbf{u})] - [\mathbf{Df}(\mathbf{v})]| \le \frac{1}{2R|L^{-1}|^2} |\mathbf{u} - \mathbf{v}|. \qquad 2.10.11$$

Set $V = B_R(\mathbf{y}_0)$. Then

1. *There exists a unique continuously differentiable mapping $\mathbf{g} : V \to W_0$ such that*

$$\mathbf{g}(\mathbf{y}_0) = \mathbf{x}_0 \quad \text{and} \quad \mathbf{f}(\mathbf{g}(\mathbf{y})) = \mathbf{y} \text{ for all } \mathbf{y} \in V. \qquad 2.10.12$$

   Since the derivative of the identity is the identity, by the chain rule, the derivative of $\mathbf{g}$ is $[\mathbf{Dg}(\mathbf{y})] = [\mathbf{Df}(\mathbf{g}(\mathbf{y}))]^{-1}$.
2. *The image of $\mathbf{g}$ contains the ball of radius R_1 around $\mathbf{x}_0$, where*

$$R_1 = R\left|L^{-1}\right|^2 \left(\sqrt{|L|^2 + \frac{2}{|L^{-1}|^2}} - |L| \right). \qquad 2.10.13$$

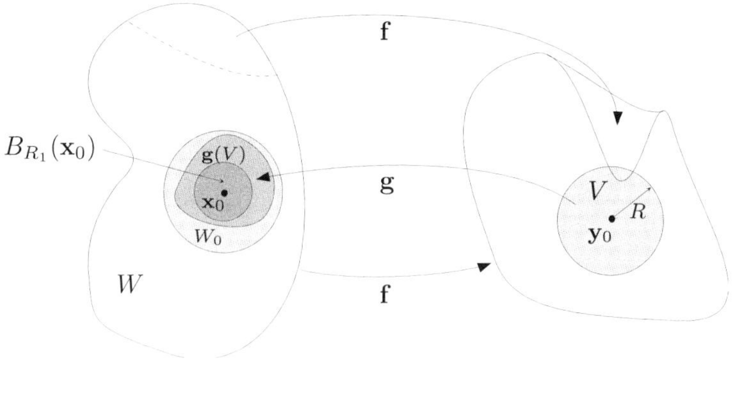

FIGURE 2.10.6. The function $\mathbf{f} : W \to \mathbb{R}^m$ maps every point in $\mathbf{g}(V)$ to a point in V; in particular, it sends $\mathbf{x}_0$ to $\mathbf{f}(\mathbf{x}_0) = \mathbf{y}_0$. Its inverse function $\mathbf{g} : V \to W_0$ undoes that mapping. The image $\mathbf{g}(V)$ of $\mathbf{g}$ may be smaller than its codomain W_0. The ball $B_{R_1}(\mathbf{x}_0)$ is guaranteed to be inside $\mathbf{g}(V)$; it quantifies "locally invertible".

Note that $\mathbf{f}$ may map some points in W to points outside V. This is why we had to write $\mathbf{f}\big(\mathbf{g}(\mathbf{y})\big) = \mathbf{y}$ in equation 2.10.12, rather than $\mathbf{g}\big(\mathbf{f}(\mathbf{x})\big) = \mathbf{x}$. In addition, $\mathbf{f}$ may map more than one point to the same point in V, but only one can come from W_0, and any point from W_0 must come from the subset $\mathbf{g}(V)$. But $\mathbf{g}$ maps a point in V to only one point (since it is a well-defined map) and that point must be in W_0 (since the image of $\mathbf{g}$ must be in the codomain of $\mathbf{g}$).

To come to terms with the details of the theorem, it may help to consider how the size of R affects the statement. In an ideal situation, would we want R to be big or little? We'd like it to be big, because then V will be big, and the inverse function $\mathbf{g}$ will be defined in a bigger neighborhood. What might keep R from being big? First, look at condition 1 of the theorem. We need W_0 (the codomain of $\mathbf{g}$) to be in W (the domain of $\mathbf{f}$). Since the radius of W_0 is $2R|L^{-1}|$, if R is too big, W_0 may no longer fit in W.

Condition 2 is more delicate. Suppose that on W the derivative $[\mathbf{Df}(\mathbf{x})]$ is locally Lipschitz. It will then be Lipschitz on each $W_0 \subset W$, but with a best Lipschitz ratio M_R that starts out at some probably nonzero value when W_0 is just a point (i.e., when $R = 0$), and gets bigger and bigger as R increases (it's harder to satisfy a Lipschitz ratio over a large area than a small one). In contrast, the quantity $1/(2R|L^{-1}|^2)$ starts at infinity when $R = 0$, and decreases as R increases (see Figure 2.10.7). So inequality 2.10.11 is satisfied when R is small, but usually the graphs of M_R and $1/(2R|L^{-1}|^2)$ will cross for some R_0; the inverse function theorem does not guarantee the existence of an inverse in any V with radius larger than R_0.

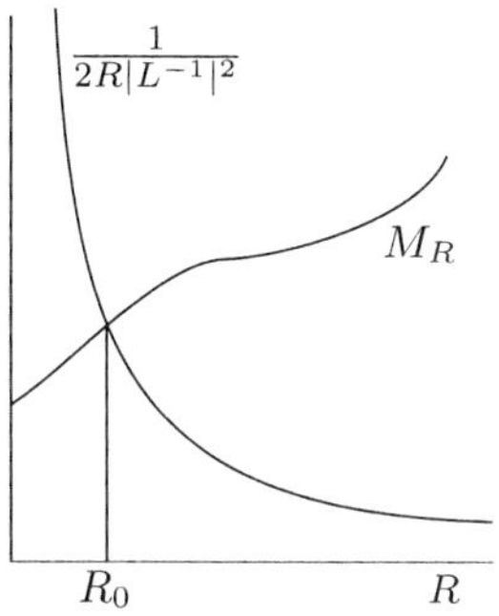

FIGURE 2.10.7.

The graph of the "best Lipschitz ratio" M_R for $[\mathbf{Df}]$ on the ball W_0 of radius $2R|L^{-1}|$ increases with R, and the graph of the function

$$\frac{1}{2R|L^{-1}|^2}$$

decreases. The inverse function theorem only guarantees an inverse on a neighborhood V of radius R when

$$\frac{1}{2R|L^{-1}|^2} > M_R.$$

The main difficulty in applying these principles is that M_R is usually very hard to compute, and $|L^{-1}|$, although usually easier, may be unpleasant too.

Remark 2.10.8. Do we need to worry that maybe no suitable R exists? The answer is no. If $\mathbf{f}$ is differentiable, and the derivative is Lipschitz (with *any* Lipschitz ratio) in some neighborhood of $\mathbf{x}_0$, then the function M_R exists, so the hypotheses on R will be satisfied as soon as $R < R_0$. Thus a differentiable map with Lipschitz derivative has a local inverse near any point where the derivative is invertible: if L^{-1} exists, we can find an R that works. $\triangle$

Proof of the inverse function theorem

We will show that if the conditions of the inverse function theorem are satisfied, then Kantorovich's theorem applies, and Newton's method can be used to find the inverse function.

Given $\mathbf{y} \in V$, we want to find $\mathbf{x}$ such that $\mathbf{f}(\mathbf{x}) = \mathbf{y}$. Since we wish to use Newton's method, we will restate the problem: Define

$$\vec{\mathbf{f}}_{\mathbf{y}}(\mathbf{x}) \overset{\text{def}}{=} \mathbf{f}(\mathbf{x}) - \mathbf{y} = \vec{\mathbf{0}}. \qquad 2.10.14$$

We wish to solve the equation $\vec{\mathbf{f}}_{\mathbf{y}}(\mathbf{x}) = \vec{\mathbf{0}}$ for $\mathbf{y} \in V$, using Newton's method with initial point $\mathbf{x}_0$.

We will use the notation of Theorem 2.8.13, but since the problem depends on $\mathbf{y}$, we will write $\vec{\mathbf{h}}_0(\mathbf{y}), U_1(\mathbf{y})$, etc. Note that

$$[\mathbf{D}\vec{\mathbf{f}}_{\mathbf{y}}(\mathbf{x}_0)] = [\mathbf{Df}(\mathbf{x}_0)] = L, \quad \text{and} \quad \vec{\mathbf{f}}_{\mathbf{y}}(\mathbf{x}_0) = \underbrace{\mathbf{f}(\mathbf{x}_0)}_{=\mathbf{y}_0} - \mathbf{y} = \mathbf{y}_0 - \mathbf{y}, \qquad 2.10.15$$

so that

$$\vec{\mathbf{h}}_0(\mathbf{y}) = -\underbrace{[\mathbf{D}\vec{\mathbf{f}}_{\mathbf{y}}(\mathbf{x}_0)]}_{L}{}^{-1}\vec{\mathbf{f}}_{\mathbf{y}}(\mathbf{x}_0) = -L^{-1}(\mathbf{y}_0 - \mathbf{y}). \qquad 2.10.16$$

We get the first equality in equation 2.10.16 by plugging in appropriate values to the definition of $\vec{\mathbf{h}}_0$ given in the statement of Kantorovich's theorem (equation 2.8.51):

$$\vec{\mathbf{h}}_0 = -[\mathbf{Df}(\mathbf{a}_0)]^{-1}\mathbf{f}(\mathbf{a}_0).$$

This implies that $|\vec{\mathbf{h}}_0(\mathbf{y})| \le |L^{-1}|R$, since $\mathbf{y}_0$ is the center of V, $\mathbf{y}$ is in V, and the radius of V is R, giving $|\mathbf{y}_0 - \mathbf{y}| \le R$. Now we compute $\mathbf{x}_1(\mathbf{y}) = \mathbf{x}_0 + \vec{\mathbf{h}}_0(\mathbf{y})$ (as in equation 2.8.51, where $\mathbf{a}_1 = \mathbf{a}_0 + \vec{\mathbf{h}}_0$). Since $|\vec{\mathbf{h}}_0(\mathbf{y})|$ is at most half the radius of W_0 (i.e., half $2R|L^{-1}|$), we see that $U_1(\mathbf{y})$ (the ball of radius $|\vec{\mathbf{h}}_0(\mathbf{y})|$ centered at $\mathbf{x}_1(\mathbf{y})$) is contained in W_0, as suggested by Figure 2.10.8.

It follows that the Kantorovich inequality 2.8.53 is satisfied:

$$\underbrace{|\vec{\mathbf{f}}_{\mathbf{y}}(\mathbf{x}_0)|}_{|\mathbf{y}_0-\mathbf{y}|\le R}\ \underbrace{|[\mathbf{D}\vec{\mathbf{f}}_{\mathbf{y}}(\mathbf{x}_0)]^{-1}|^2}_{|L^{-1}|^2}\, M \le R|L^{-1}|^2 \underbrace{\frac{1}{2R|L^{-1}|^2}}_{M} = \frac{1}{2}. \qquad 2.10.17$$

Thus Kantorovich's theorem says that Newton's method applied to the equation $\vec{\mathbf{f}}_{\mathbf{y}}(\mathbf{x}) = \vec{\mathbf{0}}$ starting at $\mathbf{x}_0$ converges; denote the limit by $\mathbf{g}(\mathbf{y})$. Since $\mathbf{g}(\mathbf{y})$ is a solution of equation 2.10.14, $\mathbf{f}\big(\mathbf{g}(\mathbf{y})\big) = \mathbf{y}$ for all $\mathbf{y} \in V$. Moreover $\mathbf{g}(\mathbf{y}_0)$ is the limit of the constant sequence $\mathbf{x}_0, \mathbf{x}_0, \dots$ given by Newton's method, so $\mathbf{g}(\mathbf{y}_0) = \mathbf{x}_0$. We now have our inverse function $\mathbf{g}$ satisfying equation 2.10.12.

A complete proof requires showing that $\mathbf{g}$ is continuously differentiable. This is done in Appendix A7, where we also prove equation 2.10.13 about the ball B_{R_1} centered at $\mathbf{x}_0$. □

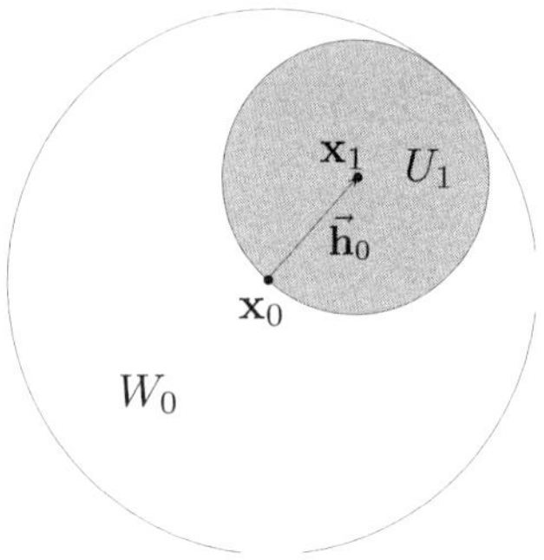

FIGURE 2.10.8.

Newton's method applied to the equation $f_{\mathbf{y}}(\mathbf{x}) = 0$ starting at $\mathbf{x}_0$ converges to a root in $\overline{U_1}$. Kantorovich's theorem tells us this is the unique root in $\overline{U_1}$; the inverse function theorem tells us that it is the unique root in all of W_0.

Example 2.10.9 (Quantifying "locally"). Let's return to the function

$$\mathbf{f}\begin{pmatrix} x \\ y \end{pmatrix} = \begin{pmatrix} \sin(x+y) \\ x^2 - y^2 \end{pmatrix} \qquad 2.10.18$$

of Example 2.10.5. Let's choose a point $\mathbf{x}_0$ where the derivative is invertible and see in how big a neighborhood of $\mathbf{f}(\mathbf{x}_0)$ an inverse function is guaranteed to exist. (We say "guaranteed to exist" because the actual domain of the inverse function may be larger than the ball V.) We know from Example 2.10.5 that the derivative is invertible at $\mathbf{x}_0 = \begin{pmatrix} 0 \\ \pi \end{pmatrix}$. This gives

$$L = \left[\mathbf{Df}\begin{pmatrix} 0 \\ \pi \end{pmatrix}\right] = \begin{bmatrix} -1 & -1 \\ 0 & -2\pi \end{bmatrix}, \qquad 2.10.19$$

so

$$L^{-1} = \frac{1}{2\pi}\begin{bmatrix} -2\pi & 1 \\ 0 & -1 \end{bmatrix}, \quad \text{and } |L^{-1}|^2 = \frac{4\pi^2+2}{4\pi^2}. \qquad 2.10.20$$

Next we compute the Lipschitz ratio M (inequality 2.10.11):

$$
\begin{aligned}
&|[\mathbf{Df}(\mathbf{u})] - [\mathbf{Df}(\mathbf{v})]| \\
&\quad = \left| \begin{bmatrix} \cos(u_1+u_2) - \cos(v_1+v_2) & \cos(u_1+u_2) - \cos(v_1+v_2) \\ 2(u_1 - v_1) & 2(v_2 - u_2) \end{bmatrix} \right| \\
&\quad \le \left| \begin{bmatrix} u_1+u_2-v_1-v_2 & u_1+u_2-v_1-v_2 \\ 2(u_1 - v_1) & 2(v_2 - u_2) \end{bmatrix} \right| \\
&\quad = \sqrt{2\Big((u_1-v_1)+(u_2-v_2)\Big)^2 + 4\Big((u_1-v_1)^2+(u_2-v_2)^2\Big)} \\
&\quad \le \sqrt{4\big((u_1-v_1)^2+(u_2-v_2)^2\big) + 4\big((u_1-v_1)^2+(u_2-v_2)^2\big)} \\
&\quad = \sqrt{8}\,|\mathbf{u} - \mathbf{v}|.
\end{aligned}
\qquad 2.10.21
$$

For the first inequality in inequality 2.10.21, remember that

$$|\cos a - \cos b| \le |a - b|,$$

and set $a = u_1+u_2$ and $b = v_1+v_2$.

In going from the first square root to the second, we use

$$(a+b)^2 \le 2(a^2+b^2),$$

setting $a = u_1 - v_1$ and $b = u_2 - v_2$.

Our Lipschitz ratio M is thus $\sqrt{8} = 2\sqrt{2}$, allowing us to compute R:

$$\frac{1}{2R|L^{-1}|^2} = 2\sqrt{2}, \quad \text{so} \quad R = \frac{4\pi^2}{4\sqrt{2}\,(4\pi^2+2)} \approx 0.16825. \qquad 2.10.22$$

Since the domain W of $\mathbf{f}$ is $\mathbb{R}^2$, the value of R in equation 2.10.22 clearly satisfies the requirement that the ball W_0 with radius $2R|L^{-1}|$ be in W.

The minimum domain V of our inverse function is a ball with radius ≈ 0.17.

What does this say about actually computing an inverse? For example, since

$$\mathbf{f}\begin{pmatrix} 0 \\ \pi \end{pmatrix} = \begin{pmatrix} 0 \\ -\pi^2 \end{pmatrix}, \quad \text{and} \quad \begin{pmatrix} 0.1 \\ -10 \end{pmatrix} \text{ is within .17 of } \begin{pmatrix} 0 \\ -\pi^2 \end{pmatrix}, \qquad 2.10.23$$

the inverse function theorem tells us that by using Newton's method we can solve for $\mathbf{x}$ the equation $\mathbf{f}(\mathbf{x}) = \begin{pmatrix} 0.1 \\ -10 \end{pmatrix}$. △

The implicit function theorem

The inverse function theorem deals with the case where we have n equations in n unknowns. What if we have more unknowns than equations? There is then no inverse function, but often we can express some unknowns in terms of others.

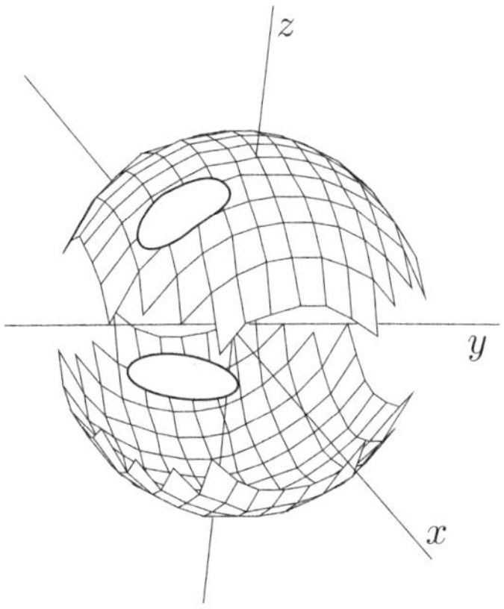

FIGURE 2.10.9.

Example 2.10.10: The entire top hemisphere of the unit sphere expresses z implicitly as a function of x and y. The curvy region at the top is the graph of this function evaluated on the small disc in the (x, y)-plane.

Example 2.10.10 (The unit sphere: three variables, one equation). The equation $x^2+y^2+z^2-1=0$ expresses z as an implicit function of $\begin{pmatrix} x \\ y \end{pmatrix}$ near $\vec{\mathbf{e}}_3$. This implicit function can be made explicit: $z = \sqrt{1-x^2-y^2}$; you can solve for z as a function of x and y (see Figure 2.10.9). △

The implicit function theorem tells under what conditions – and in what neighborhood – an implicit function exists. The idea is to divide the variables into two groups: if we have $n-k$ equations in n variables, we can think of k variables as "known" (variables whose values can be chosen freely). This leaves $n-k$ equations in $n-k$ unknowns. If a solution to these $n-k$ equations exists and is unique, we will have expressed $n-k$ variables in terms of the k known or "active" variables. We then say that the original system of equations expresses the $n-k$ "passive" variables implicitly in terms of the k "active" variables.

If all we want to know is that an implicit function exists on some unspecified neighborhood, then we can streamline the statement of the implicit function theorem; the important question to ask is, "*is the derivative onto*?"

Theorem 2.10.11 (Implicit function theorem: short version). *Let $U \subset \mathbb{R}^n$ be open and $\mathbf{c}$ a point in U. Let $\mathbf{F} : U \to \mathbb{R}^{n-k}$ be a C^1 mapping such that $\mathbf{F}(\mathbf{c}) = \mathbf{0}$ and $[\mathbf{DF}(\mathbf{c})]$ is onto. Then the system of linear equations $[\mathbf{DF}(\mathbf{c})](\vec{\mathbf{x}}) = \vec{\mathbf{0}}$ has $n-k$ pivotal (passive) variables and k nonpivotal (active) variables, and there exists a neighborhood of $\mathbf{c}$ in which $\mathbf{F} = \mathbf{0}$ implicitly defines the $n-k$ passive variables as a function $\mathbf{g}$ of the k active variables.*

Recall (see Theorem 1.9.8) that requiring that $\mathbf{F}$ be C^1 is equivalent to requiring that all its partial derivatives be continuous. Exercise 2.10.3 shows what goes wrong when $[\mathbf{DF}(\mathbf{x})]$ is not continuous with respect to $\mathbf{x}$. But you are unlikely to run into any functions with discontinuous derivatives.

The function $\mathbf{g}$ is the implicit function.

Theorem 2.10.11 is true as stated, but we give no proof; the proof we give of the full statement of the implicit function theorem (Theorem 2.10.14) requires that the derivative be Lipschitz.

Like the inverse function theorem, Theorem 2.10.11 says that *locally, the map behaves like its derivative – i.e., like its linearization.* Since $\mathbf{F}$ goes from a subset of $\mathbb{R}^n$ to $\mathbb{R}^{n-k}$, its derivative goes from $\mathbb{R}^n$ to $\mathbb{R}^{n-k}$. The derivative $[\mathbf{DF}(\mathbf{c})]$ being onto means that its columns span $\mathbb{R}^{n-k}$. Therefore, $[\mathbf{DF}(\mathbf{c})]$ has $n-k$ pivotal columns and k nonpivotal columns. We are in case 2b of Theorem 2.2.1; we can choose freely the values of the k nonpivotal variables; those values determine the values of the $n-k$ pivotal variables.

Example 2.10.12 and Figure 2.10.10 show that we can interpret this geometrically: locally, the graph of the implicit function is the solution set to the equation $\mathbf{F}(\mathbf{x}) = \mathbf{0}$.

Example 2.10.12 (The unit circle and the short implicit function theorem). The unit circle is the set of points $\mathbf{c} = \begin{pmatrix} x \\ y \end{pmatrix}$ such that $F(\mathbf{c}) = 0$ when F is the function $F\begin{pmatrix} x \\ y \end{pmatrix} = x^2 + y^2 - 1$. Here we have one active variable and one passive variable: $n = 2$ and $k = n - k = 1$. The function is differentiable, with derivative

$$\left[\mathbf{D}F\begin{pmatrix} a \\ b \end{pmatrix}\right] = [2a,\ 2b]. \qquad 2.10.24$$

The derivative is onto $\mathbb{R}$ if, when row reduced, it contains one pivotal 1 (i.e., if the derivative contains one linearly independent column).

This derivative will be onto $\mathbb{R}$ as long as it is not $[0, 0]$, which it will never be on the unit circle. If $b \neq 0$, we may consider a to be active; then Theorem 2.10.11 guarantees that in some neighborhood of $\begin{pmatrix} a \\ b \end{pmatrix}$, the equation $x^2 + y^2 - 1 = 0$ implicitly expresses y as a function of x. If $a \neq 0$, we may consider b to be active; then Theorem 2.10.11 guarantees that in some neighborhood of $\begin{pmatrix} a \\ b \end{pmatrix}$, the equation $x^2 + y^2 - 1 = 0$ implicitly expresses x as a function of y. If a and b are both nonzero, we may take either variable to be active.

Locally, the solution set of $x^2 + y^2 - 1 = 0$ is the graph of an implicit function. In this case, the implicit functions can be written explicitly, and we have some choice about what implicit functions to use:

- The upper halfcircle, where $y > 0$, is the graph of $\sqrt{1-x^2}$
- The lower halfcircle is the graph of $-\sqrt{1-x^2}$
- The right halfcircle, where $x > 0$, is the graph of $\sqrt{1-y^2}$
- The left halfcircle is the graph of $-\sqrt{1-y^2}$

Thus near the points where $x, y \neq 0$, the equation $x^2+y^2 = 1$ represents x as a function of y and y as a function of x. Near the two points where $y = 0$ it only represents x as a function of y, as shown in Figure 2.10.10; near $x = 0$ it only represents y as a function of x. △

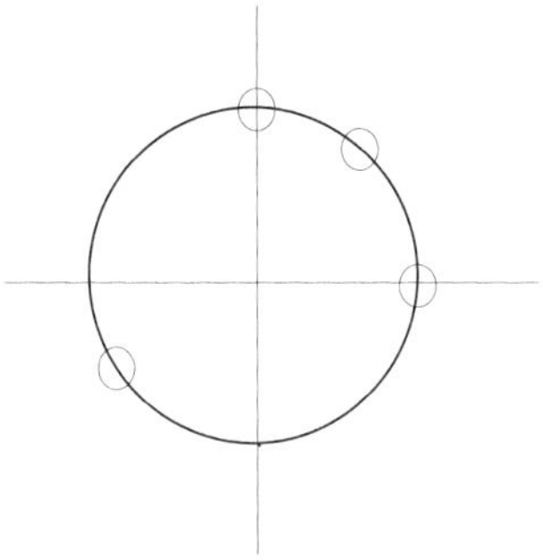

FIGURE 2.10.10.

The unit circle is the solution set to $x^2+y^2-1 = 0$. For $x, y > 0$, the solution set is the graph of both $\sqrt{1-y^2}$ and $\sqrt{1-x^2}$, one expressing x in terms of y, the other y in terms of x. In the third quadrant, it is the graph of both $-\sqrt{1-y^2}$ and $-\sqrt{1-x^2}$. Near $\begin{pmatrix}0\\1\end{pmatrix}$ the solution set is not the graph of $\sqrt{1-y^2}$ because the same value of y gives two different values of x. Near $\begin{pmatrix}1\\0\end{pmatrix}$ it is not the graph of $\sqrt{1-x^2}$ because the same value of x gives two different values of y.

Example 2.10.13 (Implicit function theorem does not guarantee absence of implicit function). Do the equations

$$(x+t)^2 = 0$$
$$y - t = 0$$

express implicitly x and y in terms of t in a neighborhood of $\begin{pmatrix}0\\0\\0\end{pmatrix}$? The implicit function theorem does not guarantee the existence of an implicit function: the derivative of $\mathbf{F}\begin{pmatrix}x\\y\\t\end{pmatrix} = \begin{pmatrix}(x+t)^2\\y-t\end{pmatrix}$ is $\begin{bmatrix}2(x+t) & 0 & 2(x+t)\\0 & 1 & -1\end{bmatrix}$, which at $\begin{pmatrix}0\\0\\0\end{pmatrix}$ is $\begin{bmatrix}0 & 0 & 0\\0 & 1 & -1\end{bmatrix}$. That derivative is not onto $\mathbb{R}^2$.

But there is an implicit function g, defined by $g(t) = \begin{pmatrix}-t\\t\end{pmatrix}$, and such that $\mathbf{F}\begin{pmatrix}g(t)\\t\end{pmatrix} = \begin{pmatrix}0\\0\end{pmatrix}$ in a neighborhood of $\begin{pmatrix}0\\0\\0\end{pmatrix}$, and this implicit function is differentiable. △

The full statement of the implicit function theorem

In Section 3.1 we will see that the short version of the implicit function theorem is enough to tell us when an equation defines a smooth curve, surface, or higher-dimensional analogue. The short version requires only that the mapping be C^1, i.e., that the derivative be continuous. But to get a bound on the domain of the implicit function (which we need to compute implicit functions), we must say *how* the derivative is continuous.

Thus, as was the case for the inverse function theorem, in the long version, we replace the condition that the derivative be continuous by a more demanding condition, requiring that it be Lipschitz.

Figure 2.10.11 illustrates the theorem.

Since $[\mathbf{DF}(\mathbf{c})]$ is onto, it is possible to order the variables in the domain so that the matrix consisting of the first $n-k$ columns of $[\mathbf{DF}(\mathbf{c})]$ row reduces to the identity. (There may be more than one way to achieve this.) Then the variables corresponding to those $n-k$ columns are the pivotal unknowns; the variables corresponding to the remaining k columns are the nonpivotal unknowns.

Equation 2.10.25: The $[0]$ is the $k \times (n-k)$ zero matrix; I_k is the $k \times k$ identity matrix. So L is $n \times n$. If it weren't square, it could not be invertible. Exercise 2.5 justifies the statement that L is invertible.

The implicit function theorem could be stated using a weaker continuity condition than the derivative being Lipschitz; see Exercise A5.1 in Appendix A5. You get what you pay for: with the weaker condition, you get a smaller domain for the implicit function.

The domain $B_R(\mathbf{b})$ of $\mathbf{g}$ has dimension k; its codomain has dimension $n-k$. It takes as input values of the k active variables and gives as output the values of the $n-k$ passive variables.

Equation 2.10.29, which tells us how to compute the derivative of an implicit function, is important; we will use it often.

Theorem 2.10.14 (The implicit function theorem). *Let W be an open neighborhood of $\mathbf{c} \in \mathbb{R}^n$, and let $\mathbf{F} : W \to \mathbb{R}^{n-k}$ be a differentiable function, with $\mathbf{F}(\mathbf{c}) = \mathbf{0}$ and $[\mathbf{DF}(\mathbf{c})]$ onto.*

Order the variables in the domain so that the matrix consisting of the first $n-k$ columns of $[\mathbf{DF}(\mathbf{c})]$ row reduces to the identity. Set $\mathbf{c} = \begin{pmatrix}\mathbf{a}\\ \mathbf{b}\end{pmatrix}$, where the entries of $\mathbf{a}$ correspond to the $n-k$ pivotal unknowns, and the entries of $\mathbf{b}$ correspond to the k nonpivotal (active) unknowns. Then there exists a unique continuously differentiable mapping $\mathbf{g}$ from a neighborhood of $\mathbf{b}$ to a neighborhood of $\mathbf{a}$ such that $\mathbf{F}\begin{pmatrix}\mathbf{x}\\ \mathbf{y}\end{pmatrix} = \mathbf{0}$ expresses the first $n-k$ variables as $\mathbf{g}$ applied to the last k variables: $\mathbf{x} = \mathbf{g}(\mathbf{y})$.

To specify the domain of $\mathbf{g}$, let L be the $n \times n$ matrix

$$L = \begin{bmatrix} [D_1\mathbf{F}(\mathbf{c}), \dots, D_{n-k}\mathbf{F}(\mathbf{c})] & [D_{n-k+1}\mathbf{F}(\mathbf{c}), \dots, D_n\mathbf{F}(\mathbf{c})] \\ [0] & I_k \end{bmatrix}. \qquad 2.10.25$$

Then L is invertible. Now find a number $R > 0$ satisfying the following hypotheses:

1. *The ball W_0 with radius $2R|L^{-1}|$ centered at $\mathbf{c}$ is contained in W.*
2. *For $\mathbf{u}$, $\mathbf{v}$ in W_0 the derivative satisfies the Lipschitz condition*

$$\Big|[\mathbf{DF}(\mathbf{u})] - [\mathbf{DF}(\mathbf{v})]\Big| \le \frac{1}{2R|L^{-1}|^2}|\mathbf{u} - \mathbf{v}|. \qquad 2.10.26$$

Then there exists a unique continuously differentiable mapping

$$\mathbf{g} : B_R(\mathbf{b}) \to B_{2R|L^{-1}|}(\mathbf{a}) \qquad 2.10.27$$

such that

$$\mathbf{g}(\mathbf{b}) = \mathbf{a} \quad \text{and} \quad \mathbf{F}\begin{pmatrix}\mathbf{g}(\mathbf{y})\\ \mathbf{y}\end{pmatrix} = \mathbf{0} \quad \text{for all } \mathbf{y} \in B_R(\mathbf{b}). \qquad 2.10.28$$

By the chain rule, the derivative of this implicit function $\mathbf{g}$ at $\mathbf{b}$ is

$$[\mathbf{Dg}(\mathbf{b})] = -\underbrace{[D_1\mathbf{F}(\mathbf{c}), \dots, D_{n-k}\mathbf{F}(\mathbf{c})]}_{\substack{\text{partial deriv. for the}\\ n-k \text{ pivotal variables}}}{}^{-1}\underbrace{[D_{n-k+1}\mathbf{F}(\mathbf{c}), \dots, D_n\mathbf{F}(\mathbf{c})]}_{\substack{\text{partial deriv. for the}\\ k \text{ nonpivotal variables}}}. \qquad 2.10.29$$

Remarks. 1. The k variables $\mathbf{b}$ determine the $n-k$ variables $\mathbf{a}$, both for the derivative and for the function. This assumption that we express the first $n-k$ variables in terms of the last k is a convenience; in practice the question of what to express in terms of what depends on context.

Usually, if the derivative of $\mathbf{F}$ is onto $\mathbb{R}^{n-k}$ (i.e., has $n-k$ linearly independent columns), there will be more than one way to choose these columns (see Exercise 2.5.7). For any choice of $n-k$ linearly independent columns, the implicit function theorem guarantees that you can express the corresponding variables implicitly in terms of the others.

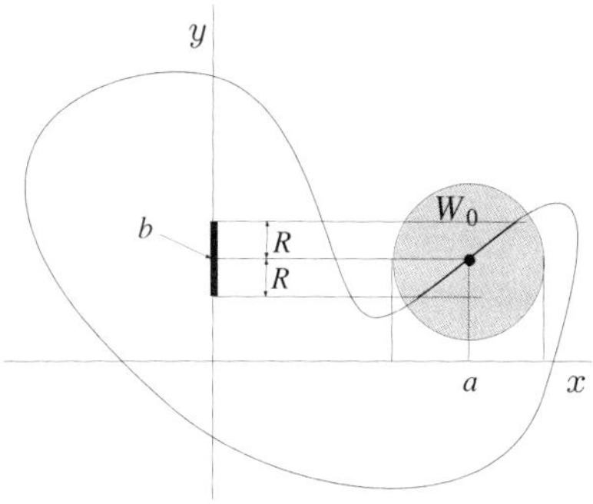

FIGURE 2.10.11. In a neighborhood of $\binom{a}{b}$ the equation $F = 0$ expresses x implicitly as a function of y. That neighborhood is contained in the ball W_0 of radius $r = 2R|L^{-1}|$. But the whole curve is not the graph of a function f expressing x as a function of y; near $y = b$, there are four such functions.

2. The proof of the implicit function theorem given in Appendix A8 shows that every point in W_0 that is a solution to $\mathbf{F}\begin{pmatrix}\mathbf{x}\\ \mathbf{y}\end{pmatrix} = \mathbf{0}$ is of the form $\begin{pmatrix}\mathbf{g}(\mathbf{y})\\ \mathbf{y}\end{pmatrix}$. Thus the intersection of W_0 with the graph of $\mathbf{g}$ is the intersection of W_0 with the solution set of $\mathbf{F}\begin{pmatrix}\mathbf{x}\\ \mathbf{y}\end{pmatrix} = \mathbf{0}$. (The graph of a function $f : \mathbb{R}^k \to \mathbb{R}^{n-k}$ exists in a space whose dimension is n.) △

The inverse function theorem is the special case of the implicit function theorem when we have $n = 2k$ variables, the passive k-dimensional variable $\mathbf{x}$ and the active k-dimensional variable $\mathbf{y}$, and where our original equation is

$$\mathbf{f}(\mathbf{x}) - \mathbf{y} = \vec{\mathbf{0}};$$

i.e., we can separate out $\mathbf{y}$ from $\mathbf{F}\begin{pmatrix}\mathbf{x}\\ \mathbf{y}\end{pmatrix}$.

Example 2.10.15 (The unit circle and the implicit function theorem). In Example 2.10.12 we showed that if $a \neq 0$, then in some neighborhood of $\begin{pmatrix}a\\ b\end{pmatrix}$, the equation $x^2 + y^2 - 1 = 0$ implicitly expresses x as a function of y. Let's see what the strong version of the implicit function theorem says about the domain of this implicit function. The matrix L of equation 2.10.25 is

$$L = \begin{bmatrix} 2a & 2b\\ 0 & 1\end{bmatrix}, \quad \text{and} \quad L^{-1} = \frac{1}{2a}\begin{bmatrix} 1 & -2b\\ 0 & 2a\end{bmatrix}. \qquad 2.10.30$$

Equation 2.10.30: In the lower right corner of L we have the number 1, not the identity matrix I; our function F goes from $\mathbb{R}^2$ to $\mathbb{R}$, so $n = m = 1$, and the 1×1 identity matrix is the number 1.

So we have

$$|L^{-1}| = \frac{1}{2|a|}\sqrt{1 + 4a^2 + 4b^2} = \frac{\sqrt{5}}{2|a|}. \qquad 2.10.31$$

The derivative of $F\begin{pmatrix}x\\ y\end{pmatrix} = x^2 + y^2 - 1$ is Lipschitz with Lipschitz ratio 2:

$$\begin{aligned}\left|\left[\mathbf{D}F\begin{pmatrix}u_1\\ u_2\end{pmatrix}\right] - \left[\mathbf{D}F\begin{pmatrix}v_1\\ v_2\end{pmatrix}\right]\right| &= |[2u_1 - 2v_1,\ 2u_2 - 2v_2]|\\ &= 2|[u_1 - v_1,\ u_2 - v_2]| \leq 2|\mathbf{u} - \mathbf{v}|,\end{aligned} \qquad 2.10.32$$

so (by equation 2.10.26) we can satisfy condition 2 if R satisfies

$$2 = \frac{1}{2R|L^{-1}|^2}; \quad \text{i.e.,} \quad R = \frac{1}{4|L^{-1}|^2} = \frac{a^2}{5}. \qquad 2.10.33$$

Equation 2.10.33: Note the way the radius R of the interval around b shrinks, without ever disappearing, as $a \to 0$. At the points $\begin{pmatrix}0\\ 1\end{pmatrix}$ and $\begin{pmatrix}0\\ -1\end{pmatrix}$, the equation

$$x^2 + y^2 - 1 = 0$$

does not express x in terms of y, but it does express x in terms of y when a is arbitrarily close to 0.

We then see that W_0 is the ball of radius

$$2R|L^{-1}| = \frac{2a^2}{5}\frac{\sqrt{5}}{2|a|} = \frac{|a|}{\sqrt{5}}; \qquad 2.10.34$$

since W is all of $\mathbb{R}^2$, condition 1 is satisfied. Therefore, for all $\begin{pmatrix} a \\ b \end{pmatrix}$ with $a \neq 0$, the equation $x^2 + y^2 - 1 = 0$ expresses x (in the interval of radius $|a|/\sqrt{5}$ around a) as a function of y (in $B_R(b)$, the interval of radius $a^2/5$ around b).

Of course, there are two possible x. One is found by starting Newton's method at a, the other by starting at $-a$.

Of course we don't need the implicit function theorem to understand the unit circle; we know that $x = \pm\sqrt{1-y^2}$. But let's pretend we don't, and go further. The implicit function theorem says that if $\begin{pmatrix} a \\ b \end{pmatrix}$ is a root of the equation $x^2 + y^2 - 1 = 0$, then for any y within $a^2/5$ of b, we can find the corresponding x by starting with the guess $x_0 = a$ and applying Newton's method, iterating

$$x_{n+1} = x_n - \frac{F\begin{pmatrix} x_n \\ y \end{pmatrix}}{D_1F\begin{pmatrix} x_n \\ y \end{pmatrix}} = x_n - \frac{x_n^2 + y^2 - 1}{2x_n}. \qquad 2.10.35$$

(In equation 2.10.35 we write $1/D_1F$ rather than $(D_1F)^{-1}$ because D_1F is a 1×1 matrix, i.e., a number.) $\triangle$

$$\underbrace{\begin{bmatrix} 0 & -1 & 0 & -1 & 0 \\ 0 & 0 & -1 & 0 & -1 \\ -1 & 0 & 0 & 0 & 0 \\ 0 & 0 & 0 & 1 & 0 \\ 0 & 0 & 0 & 0 & 1 \end{bmatrix}}_{L \text{ in Example 2.10.16}}$$

$$\underbrace{\begin{bmatrix} 0 & 0 & -1 & 0 & 0 \\ -1 & 0 & 0 & -1 & 0 \\ 0 & -1 & 0 & 0 & -1 \\ 0 & 0 & 0 & 1 & 0 \\ 0 & 0 & 0 & 0 & 1 \end{bmatrix}}_{L^{-1} \text{ in Example 2.10.16}}$$

Example 2.10.16 (An implicit function in several variables). In what neighborhood of $\mathbf{c} \overset{\text{def}}{=} \begin{pmatrix} 0 \\ 0 \\ 0 \\ 0 \\ 0 \end{pmatrix}$ do the equations $\begin{cases} x^2 - y = a \\ y^2 - z = b \\ z^2 - x = 0 \end{cases}$ determine $\begin{pmatrix} x \\ y \\ z \end{pmatrix}$ implicitly as a function of a and b? Here, $n = 5$, $k = 2$, and

$$\mathbf{F}\begin{pmatrix} x \\ y \\ z \\ a \\ b \end{pmatrix} = \begin{pmatrix} x^2 - y - a \\ y^2 - z - b \\ z^2 - x \end{pmatrix}, \text{ with derivative } \begin{bmatrix} 2x & -1 & 0 & -1 & 0 \\ 0 & 2y & -1 & 0 & -1 \\ -1 & 0 & 2z & 0 & 0 \end{bmatrix}.$$

At $\mathbf{c}$ the derivative is $\begin{bmatrix} 0 & -1 & 0 & -1 & 0 \\ 0 & 0 & -1 & 0 & -1 \\ -1 & 0 & 0 & 0 & 0 \end{bmatrix}$; the first three columns are linearly independent.

Adding the appropriate two bottom lines, we find L and L^{-1} as in the margin, and $|L^{-1}|^2 = 7$.

Since $\mathbf{F}$ is defined on all of $\mathbb{R}^5$, the first restriction on R is vacuous: for any value of R we have $W_0 \subset \mathbb{R}^5$. The second restriction requires that the derivative in W_0 be Lipschitz. Since $M = 2$ is a global Lipschitz ratio for the derivative:

$$\left|\left[\begin{matrix} 2x_1 & -1 & 0 & -1 & 0 \\ 0 & 2y_1 & -1 & 0 & -1 \\ -1 & 0 & 2z_1 & 0 & 0 \end{matrix}\right] - \left[\begin{matrix} 2x_2 & -1 & 0 & -1 & 0 \\ 0 & 2y_2 & -1 & 0 & -1 \\ -1 & 0 & 2z_2 & 0 & 0 \end{matrix}\right]\right|$$

$$= 2\sqrt{(x_1-x_2)^2 + (y_1-y_2)^2 + (z_1-z_2)^2} \le 2\left|\begin{pmatrix} x_1 \\ y_1 \\ z_1 \\ a_1 \\ b_1 \end{pmatrix} - \begin{pmatrix} x_2 \\ y_2 \\ z_2 \\ a_2 \\ b_2 \end{pmatrix}\right|, \qquad 2.10.36$$

this means that we need $\frac{1}{2R|L^{-1}|^2} \ge 2$, i.e., we can take $R = \frac{1}{28}$.

The $\begin{pmatrix} a \\ b \end{pmatrix}$ of this discussion is the $\mathbf{y}$ of equation 2.10.28, and the origin here is the $\mathbf{b}$ of that equation.

Thus we can be sure that for any $\begin{pmatrix} a \\ b \end{pmatrix}$ in the ball of radius 1/28 around the origin there will be a unique solution to the system of equations

$$\begin{cases} x^2 - y = a \\ y^2 - z = b \\ z^2 - x = 0 \end{cases}, \text{ with } \left|\begin{pmatrix} x \\ y \\ z \end{pmatrix}\right| \le \underbrace{2R|L^{-1}|}_{\text{radius of } W_0} = \frac{2}{28}\sqrt{7}. \quad \triangle \qquad 2.10.37$$

In the next example, computing the Lipschitz ratio directly would be harder, so we use a different approach.

When computing implicit functions, the hard part is usually computing the Lipschitz ratio.

Example 2.10.17 (Implicit function theorem: a harder example). In what neighborhood of $\mathbf{0} = \begin{pmatrix} 0 \\ 0 \\ 0 \\ 0 \\ 0 \end{pmatrix}$ do the equations $\begin{cases} x^3 - y = a \\ y^3 - z = b \\ z^3 - x = 0 \end{cases}$ determine $\begin{pmatrix} x \\ y \\ z \end{pmatrix}$ implicitly as a function $\mathbf{g}\begin{pmatrix} a \\ b \end{pmatrix}$ with $\mathbf{g}\begin{pmatrix} 0 \\ 0 \end{pmatrix} = \begin{pmatrix} 0 \\ 0 \\ 0 \end{pmatrix}$? The derivative is

Equation 2.10.38: When we evaluate the first three columns of the derivative at $x = y = z = 0$, we clearly get an invertible matrix, so already we know that in some neighborhood, x, y, z can be expressed implicitly as a function of a and b.

$$\left[\mathbf{DF}\begin{pmatrix} x \\ y \\ x \\ a \\ b \end{pmatrix}\right] = \begin{bmatrix} 3x^2 & -1 & 0 & -1 & 0 \\ 0 & 3y^2 & -1 & 0 & -1 \\ -1 & 0 & 3z^2 & 0 & 0 \end{bmatrix}. \qquad 2.10.38$$

Evaluating the derivative at $\mathbf{0}$, and adding the appropriate rows, we get

$$L = \begin{bmatrix} 0 & -1 & 0 & -1 & 0 \\ 0 & 0 & -1 & 0 & -1 \\ -1 & 0 & 0 & 0 & 0 \\ 0 & 0 & 0 & 1 & 0 \\ 0 & 0 & 0 & 0 & 1 \end{bmatrix}, \quad L^{-1} = \begin{bmatrix} 0 & 0 & -1 & 0 & 0 \\ -1 & 0 & 0 & -1 & 0 \\ 0 & -1 & 0 & 0 & -1 \\ 0 & 0 & 0 & 1 & 0 \\ 0 & 0 & 0 & 0 & 1 \end{bmatrix}, \quad \text{so that } |L^{-1}| = \sqrt{7}.$$

Computing a Lipschitz ratio for the derivative directly would be harder than it was in Example 2.10.16 (although still relatively innocent). Instead we will use partial derivatives, as discussed in Proposition 2.8.9 and

Examples 2.8.10 and 2.8.11. The only nonzero second partials are

$$D_{1,1}F_1 = 6x, \quad D_{2,2}F_2 = 6y, \quad D_{3,3}F_3 = 6z. \qquad 2.10.39$$

Denote by r the radius of W_0 and set

$$x^2+y^2+z^2+a^2+b^2 \le r^2, \quad \text{to get} \quad \sqrt{36(x^2+y^2+z^2)} \le 6r. \qquad 2.10.40$$

(This uses Proposition 2.8.9.) Now the equations

$$\underbrace{2R|L^{-1}|}_{\text{radius of } W_0} = r \quad \text{and} \quad \underbrace{\frac{1}{2R|L^{-1}|^2}}_{\text{Lipsch. eq. 2.10.26}} \ge \underbrace{6r}_{\text{computed Lipsch.}} \qquad 2.10.41$$

give $\dfrac{1}{r|L^{-1}|} \ge 6r$, which gives $r^2 \le \dfrac{1}{6\sqrt{7}}$, so

$$r \approx .2509 \quad \text{and} \quad R = \frac{r}{2|L^{-1}|} \approx \frac{1}{4}\cdot\frac{1}{2\sqrt{7}} \approx .0472. \qquad 2.10.42$$

For any $\begin{pmatrix} a \\ b \end{pmatrix}$ in the ball of radius .0472 around $\begin{pmatrix} 0 \\ 0 \end{pmatrix}$, our system of equations expresses x, y, z in terms of a, b; given such a point $\begin{pmatrix} a \\ b \end{pmatrix}$, the system has a unique solution in the ball of radius $r \approx .2509$. △

Example 2.10.18 doesn't prove that no *non-differentiable* implicit function exists.

Example 2.10.18 (Proving no differentiable implicit function exists). The *trace* tr of a square matrix is the sum of the entries on its diagonal. Consider the set of 2×2 matrices A such that $\det A = 1$ and A^2 is diagonal. In a neighborhood of $A_0 \stackrel{\text{def}}{=} \begin{bmatrix} 0 & 1 \\ -1 & 0 \end{bmatrix}$, are such matrices determined as a differentiable function of $\operatorname{tr}(A^2)$?

Example 2.10.18:

$$\begin{bmatrix} a & b \\ c & d \end{bmatrix}\begin{bmatrix} a & b \\ c & d \end{bmatrix} = \begin{bmatrix} a^2+bc & ab+bd \\ ac+cd & bc+d^2 \end{bmatrix}$$

No they are not. Suppose $A = \begin{bmatrix} a & b \\ c & d \end{bmatrix}$ with A^2 diagonal and $\det A = 1$. Set $t = \operatorname{tr} A^2$. Then

$$\begin{aligned} ad - bc &= 1 \\ ab + bd &= 0 \\ ac + cd &= 0 \\ a^2 + 2bc + d^2 &= t \end{aligned}, \quad \text{since } A^2 = \begin{bmatrix} a^2+bc & ab+bd \\ ac+cd & bc+d^2 \end{bmatrix}. \qquad 2.10.43$$

We are asking (see equation 2.10.28) whether the function $\mathbf{F}$ satisfying

$$\mathbf{F}\begin{pmatrix} a \\ b \\ c \\ d \\ t \end{pmatrix} = \begin{pmatrix} ad-bc-1 \\ ab+bd \\ ac+cd \\ a^2+2bc+d^2-t \end{pmatrix} = \begin{pmatrix} 0 \\ 0 \\ 0 \\ 0 \end{pmatrix} \qquad 2.10.44$$

determines A as a function $t \mapsto G(t)$ with $G(0) = A_0$. Its derivative is

$$\left[\mathbf{DF}\begin{pmatrix} a \\ b \\ c \\ d \\ t \end{pmatrix}\right] = \begin{bmatrix} d & -c & -b & a & 0 \\ b & a+d & 0 & b & 0 \\ c & 0 & a+d & c & 0 \\ 2a & 2c & 2b & 2d & -1 \end{bmatrix}; \text{ at } \begin{pmatrix} A_0 \\ 0 \end{pmatrix} = \begin{pmatrix} 0 \\ 1 \\ -1 \\ 0 \\ 0 \end{pmatrix}, \text{ this is } \begin{bmatrix} 0 & 1 & -1 & 0 & 0 \\ 1 & 0 & 0 & 1 & 0 \\ -1 & 0 & 0 & -1 & 0 \\ 0 & -2 & 2 & 0 & -1 \end{bmatrix}.$$

The first four columns of this matrix, representing our candidate pivotal variables, do not form an invertible matrix. Therefore the implicit function theorem does not guarantee an implicit function G.

We saw in Example 2.10.13 that it is possible for a implicit function to exist even when it is not guaranteed by the implicit function theorem. Could this happen here? No. The chain rule says that

$$\left[\mathbf{D}\left(\mathbf{F}\circ\begin{pmatrix} G \\ \text{id}\end{pmatrix}\right)(t)\right] = \left[\mathbf{DF}\begin{pmatrix} G(t) \\ t\end{pmatrix}\right]\begin{bmatrix}\mathbf{D}G(t) \\ 1\end{bmatrix} = \begin{bmatrix}0\\0\\0\\0\end{bmatrix}, \text{ i.e., for } t=0,$$

$$\begin{bmatrix} 0 & 1 & -1 & 0 & 0 \\ 1 & 0 & 0 & 1 & 0 \\ -1 & 0 & 0 & -1 & 0 \\ 0 & -2 & 2 & 0 & -1 \end{bmatrix}\begin{bmatrix}\mathbf{D}G(t) \\ 1\end{bmatrix} = \begin{bmatrix}0\\0\\0\\0\end{bmatrix} \qquad 2.10.45$$

No $\mathbf{D}G(t)$ satisfying equation 2.10.45 exists. The only way the first entry on the right can be 0 is if the second and third entries of $\mathbf{D}G(t)$ are identical. But then the fourth entry on the right would be -1, not 0. △

Exercises for Section 2.10

2.10.1 Does the inverse function theorem guarantee that the following functions are locally invertible with differentiable inverse?

a. $F\begin{pmatrix} x \\ y \end{pmatrix} = \begin{pmatrix} x^2y \\ -2x \\ y^2 \end{pmatrix}$ at $\begin{pmatrix} 1 \\ 1 \end{pmatrix}$ b. $F\begin{pmatrix} x \\ y \end{pmatrix} = \begin{pmatrix} x^2y \\ -2x \end{pmatrix}$ at $\begin{pmatrix} 1 \\ 1 \end{pmatrix}$

c. $F\begin{pmatrix} x \\ y \\ z \end{pmatrix} = \begin{pmatrix} xyz \\ x^2 \end{pmatrix}$ at $\begin{pmatrix} 0 \\ 0 \\ 0 \end{pmatrix}$ d. $F\begin{pmatrix} x \\ y \\ z \end{pmatrix} = \begin{pmatrix} xyz \\ x^2 \\ z^2 \end{pmatrix}$ at $\begin{pmatrix} 0 \\ 0 \\ 0 \end{pmatrix}$

e. $F\begin{pmatrix} x \\ y \\ z \end{pmatrix} = \begin{pmatrix} xyz \\ x^2 \\ z^2 \end{pmatrix}$ at $\begin{pmatrix} 1 \\ 1 \\ 1 \end{pmatrix}$

2.10.2 Where is the mapping $\mathbf{f}\begin{pmatrix} x \\ y \end{pmatrix} = \begin{pmatrix} xy \\ x^2-y^2 \end{pmatrix}$ guaranteed by the inverse function theorem to be locally invertible?

$$f(x) = \begin{cases} \frac{x}{2} + x^2 \sin\frac{1}{x} & \text{if } x \neq 0 \\ 0 & \text{if } x = 0 \end{cases}$$

Function for Exercise 2.10.3

2.10.3 Let f be the function discussed in Example 1.9.4 (see the margin).

a. Show that f is differentiable at 0 and that the derivative is 1/2.

b. Show that f does not have an inverse on any neighborhood of 0.

c. Why doesn't this contradict the inverse function theorem, Theorem 2.10.2?

2.10.4 Prove that if the derivative $[\mathbf{Df}(\mathbf{x}_0)]$ in Theorem 2.10.7 is not invertible, then $\mathbf{f}$ has no differentiable inverse in the neighborhood of $\mathbf{f}(\mathbf{x}_0)$.

2.10.5 a. Using direct computation, determine where $y^2+y+3x+1=0$ defines y implicitly as a function of x.

b. Check that this result agrees with the implicit function theorem.

c. In what neighborhood of $x = -\frac{1}{2}$ are we guaranteed an implicit function $g(x)$ with $g(-1/2) = \frac{\sqrt{3}-1}{2}$?

d. In what neighborhood of $x = -\frac{13}{4}$ are we guaranteed an implicit function $g(x)$ with $g(-\frac{13}{4}) = \frac{5}{2}$?

$$\mathbf{f}\begin{pmatrix} x \\ y \end{pmatrix} = \begin{pmatrix} \dfrac{(x^2-y^2)}{(x^2+y^2)} \\ \dfrac{xy}{(x^2+y^2)} \end{pmatrix}$$

Function for Exercise 2.10.6

2.10.6 Let $\mathbf{f} : \mathbb{R}^2 - \{\mathbf{0}\} \to \mathbb{R}^2$ be as shown in the margin. Does $\mathbf{f}$ have a local inverse at every point of $\mathbb{R}^2$?

2.10.7 Do the equations $x^2 = 0,\ y = t$ express $\begin{pmatrix} x \\ y \end{pmatrix}$ implicitly in terms of t in a neighborhood of $\mathbf{0}$?

2.10.8 Let $x^2 + y^3 + e^y = 0$ define implicitly $y = f(x)$. Compute $f'(x)$ in terms of x and y.

2.10.9 Does the system of equations $\begin{aligned} x + y + \sin(xy) &= a \\ \sin(x^2 + y) &= 2a \end{aligned}$ have a solution for sufficiently small a?

2.10.10 Does a 2×2 matrix of the form $I + \epsilon B$ have a square root A near

$$\begin{bmatrix} 1 & 0 \\ 0 & -1 \end{bmatrix}?$$

$$A = \begin{bmatrix} a_1 & 0 & \dots & 0 \\ 0 & a_2 & \dots & 0 \\ \vdots & \vdots & \ddots & \vdots \\ 0 & 0 & \dots & a_n \end{bmatrix}$$

Matrix for Exercise 2.10.11

2.10.11 Let A be the matrix in the margin, and consider the linear transformation $T : \operatorname{Mat}(n,n) \to \operatorname{Mat}(n,n)$ given by $T : H \mapsto AH + HA$.

a. If $H_{i,j}$ is the matrix with a 1 in the (i,j)th position and zeros everywhere else, compute $TH_{i,j}$.

b. What condition must the a_i satisfy in order for T to be invertible?

c. True or false? There exists a neighborhood U of the identity in $\operatorname{Mat}(3,3)$ and a differentiable "square root map" $f : U \to \operatorname{Mat}(3,3)$ with $(f(A))^2 = A$ for all $A \in U$, and $f(I) = \begin{bmatrix} -1 & 0 & 0 \\ 0 & 1 & 0 \\ 0 & 0 & 1 \end{bmatrix}$.

2.10.12 True or false? There exists $r > 0$ and a differentiable map

$$g : B_r\left(\begin{bmatrix} -3 & 0 \\ 0 & -3 \end{bmatrix}\right) \to \operatorname{Mat}(2,2) \quad \text{such that}$$

$$g\left(\begin{bmatrix} -3 & 0 \\ 0 & -3 \end{bmatrix}\right) = \begin{bmatrix} 1 & 2 \\ -2 & -1 \end{bmatrix} \text{ and } (g(A))^2 = A \quad \text{for all } A \in B_r\left(\begin{bmatrix} -3 & 0 \\ 0 & -3 \end{bmatrix}\right).$$

2.10.13 True or false? If $f : \mathbb{R}^3 \to \mathbb{R}$ is continuously differentiable, and

$$D_2 f\begin{pmatrix} a \\ b \\ c \end{pmatrix} \neq 0 \;,\quad D_3 f\begin{pmatrix} a \\ b \\ c \end{pmatrix} \neq 0, \quad \text{then there exists}$$

a function h of $\begin{pmatrix} y \\ z \end{pmatrix}$, defined near $\begin{pmatrix} b \\ c \end{pmatrix}$, such that $f\begin{pmatrix} h\begin{pmatrix} y \\ z \end{pmatrix} \\ y \\ z \end{pmatrix} = 0.$

2.10.14 Consider the mapping $S : \text{Mat}\,(2,2) \to \text{Mat}\,(2,2)$ given by $S(A) = A^2$. Observe that $S(-I) = I$. Does there exist an inverse mapping g, i.e., a mapping such that $S(g(A)) = A$, defined in a neighborhood of I, and such that $g(I) = -I$?

2.10.15 a. Show that the mapping $F\begin{pmatrix} x \\ y \end{pmatrix} = \begin{pmatrix} e^x + e^y \\ e^x + e^{-y} \end{pmatrix}$ is locally invertible at every point $\begin{pmatrix} x \\ y \end{pmatrix} \in \mathbb{R}^2$.

b. If $F(\mathbf{a}) = \mathbf{b}$, what is the derivative of F^{-1} at $\mathbf{b}$?

2.10.16 The matrix $A_0 = \begin{bmatrix} 0 & 1 & 0 \\ 0 & 0 & 1 \\ 1 & 0 & 0 \end{bmatrix}$ satisfies $A_0^3 = I$. True or false? There exists a neighborhood $U \subset \text{Mat}\,(3,3)$ of I and a continuously differentiable function $g : U \to \text{Mat}\,(3,3)$ with $g(I) = A_0$ and $(g(A))^3 = A$ for all $A \in U$ (i.e., $g(A)$ is a cube root of A).

2.10.17 Prove Theorem 2.10.2 (the inverse function theorem in one dimension).

2.11 Review exercises for Chapter 2

$$x + y - z = a$$
$$x + 2z = b$$
$$x + ay + z = b$$

Equations for Exercise 2.1

2.1 a. For what values of a and b does the system of linear equations shown in the margin have one solution? No solutions? Infinitely many solutions?

b. For what values of a and b is the matrix of coefficients invertible?

$$\begin{bmatrix} 1 & 2 & 0 & 1 \\ 1 & 1 & 3 & 3 \\ 0 & 1 & 0 & 1 \\ 2 & 1 & 1 & 3 \end{bmatrix}$$

Matrix A of Exercise 2.2

2.2 When A is the matrix at left, multiplication by what elementary matrix corresponds to

a. Exchanging the first and second rows of A?

b. Multiplying the fourth row of A by 3?

c. Adding 2 times the third row of A to the first row of A?

2.3 a. Let $T : \mathbb{R}^n \to \mathbb{R}^m$ be a linear transformation. Are the following statements true or false?

1. If $\ker T = \{\vec{\mathbf{0}}\}$ and $T(\vec{\mathbf{y}}) = \vec{\mathbf{b}}$, then $\vec{\mathbf{y}}$ is the only solution to $T(\vec{\mathbf{x}}) = \vec{\mathbf{b}}$.
2. If $\vec{\mathbf{y}}$ is the only solution to $T(\vec{\mathbf{x}}) = \vec{\mathbf{c}}$, then for any $\vec{\mathbf{b}} \in \mathbb{R}^m$, a solution exists to $T(\vec{\mathbf{x}}) = \vec{\mathbf{b}}$.
3. If $\vec{\mathbf{y}} \in \mathbb{R}^n$ is a solution to $T(\vec{\mathbf{x}}) = \vec{\mathbf{b}}$, it is the only solution.
4. If for any $\vec{\mathbf{b}} \in \mathbb{R}^m$ the equation $T(\vec{\mathbf{x}}) = \vec{\mathbf{b}}$ has a solution, then it is the only solution.

b. For any statements that are false, can one impose conditions on m and n that make them true?

2.4 a. Show that an upper triangular matrix is invertible if and only if its diagonal entries are all nonzero, and that if it is invertible, its inverse is upper triangular.

b. Show the corresponding statement for a lower triangular matrix.

2.5 a. Let A be an $(n-k)\times(n-k)$ matrix, B an $k\times k$ matrix, C an $(n-k)\times k$ matrix, and $[0]$ the $k\times(n-k)$ zero matrix. Show that the $n\times n$ matrix $\begin{bmatrix} A & C \\ [0] & B \end{bmatrix}$ is invertible if and only if A and B are invertible.

b. Find a formula for the inverse.

$$\begin{bmatrix} 1 & -1 & 3 & 0 & -2 \\ -2 & 2 & -6 & 0 & 4 \\ 0 & 2 & 5 & -1 & 0 \\ 2 & -6 & -4 & 2 & -4 \end{bmatrix}$$

Matrix A for Exercise 2.6.

2.6 a. Row reduce the matrix A in the margin.

b. Let $\vec{\mathbf{v}}_m$, $m=1,\dots,5$ be the columns of A. What can you say about the systems of equations

$$[\vec{\mathbf{v}}_1,\dots,\vec{\mathbf{v}}_k]\begin{bmatrix} x_1 \\ \vdots \\ x_k \end{bmatrix} = \vec{\mathbf{v}}_{k+1} \quad \text{for } k=1,2,3,4?$$

Exercise 2.6, part b: For example, for $k=2$ we are asking about the system of equations

$$\begin{bmatrix} 1 & -1 \\ -2 & 2 \\ 0 & 2 \\ 2 & -6 \end{bmatrix}\begin{bmatrix} x_1 \\ x_2 \end{bmatrix} = \begin{bmatrix} 3 \\ -6 \\ 5 \\ -4 \end{bmatrix}.$$

2.7 a. For what values of a is the matrix $\begin{bmatrix} 1 & -1 & -1 \\ 0 & a & 1 \\ 2 & a+2 & a+2 \end{bmatrix}$ invertible?

b. For those values, compute the inverse.

2.8 Show that the following two statements are equivalent to saying that a set of vectors $\vec{\mathbf{v}}_1,\dots,\vec{\mathbf{v}}_k$ is linearly independent:

a. The only way to write the zero vector $\vec{\mathbf{0}}$ as a linear combination of the $\vec{\mathbf{v}}_i$ is to use only zero coefficients.

b. None of the $\vec{\mathbf{v}}_i$ is a linear combination of the others.

2.9 a. Show that $\begin{bmatrix} \cos\theta \\ \sin\theta \end{bmatrix}, \begin{bmatrix} -\sin\theta \\ \cos\theta \end{bmatrix}$ form an orthonormal basis of $\mathbb{R}^2$.

b. Show that $\begin{bmatrix} \cos\theta \\ \sin\theta \end{bmatrix}, \begin{bmatrix} \sin\theta \\ -\cos\theta \end{bmatrix}$ form an orthonormal basis of $\mathbb{R}^2$.

c. Show that any orthogonal 2×2 matrix gives either a reflection or a rotation: a reflection if its determinant is negative, a rotation if its determinant is positive.

Hint for Exercise 2.10: Set

$$z_1 = x^2$$
$$z_2 = y^2$$
$$z_3 = z_1^2$$

2.10 Find a bound on a^2+b^2 such that Newton's method to solve

$$x^3 + x - 3y^2 = a$$
$$x^5 + x^2y^3 - y = b$$

starting at $x=0$, $y=0$, is sure to converge to a solution.

2.11 a. Let $A = \begin{bmatrix} 1 & 2 \\ 2 & 1 \end{bmatrix}$. Are the elements I, A, A^2, A^3 linearly independent in $\operatorname{Mat}(2,2)$? What is the dimension of the subspace $V \subset \operatorname{Mat}(2,2)$ that they span? (Recall that $\operatorname{Mat}(n,m)$ denotes the set of $n\times m$ matrices.)

b. Show that the set W of matrices $B \in \operatorname{Mat}(2,2)$ that satisfy $AB = BA$ is a subspace of $\operatorname{Mat}(2,2)$. What is its dimension?

c. Show that $V \subset W$. Are they equal?

2.12 Let $\vec{\mathbf{v}}_1,\dots,\vec{\mathbf{v}}_k$ be vectors in $\mathbb{R}^n$, and set $V = [\vec{\mathbf{v}}_1,\dots,\vec{\mathbf{v}}_k]$.

a. Show that the set $\vec{\mathbf{v}}_1,\dots,\vec{\mathbf{v}}_k$ is orthogonal if and only if $V^\top V$ is diagonal.

b. Show that the set $\vec{\mathbf{v}}_1,\dots,\vec{\mathbf{v}}_k$ is orthonormal if and only if $V^\top V = I_k$.

2.13 Find a basis for the image and the kernel of the matrices

$$A = \begin{bmatrix} 1 & 1 & 3 & 6 & 2 \\ 2 & -1 & 0 & 4 & 1 \\ 4 & 1 & 6 & 16 & 5 \end{bmatrix} \qquad B = \begin{bmatrix} 2 & 1 & 3 & 6 & 2 \\ 2 & -1 & 0 & 4 & 1 \end{bmatrix},$$

and verify that the dimension formula (equation 2.5.8) is true.

Exercise 2.14: To lighten notation, we omit arrows on the vectors, writing $\overline{\mathbf{v}}$ for the complex conjugate of $\mathbf{v}$.

2.14 Let A be an $n \times n$ matrix with real entries, and let $\mathbf{v} \in \mathbb{C}^n$ be a vector such that $A\mathbf{v} = (a+ib)\mathbf{v}$ with $b \neq 0$. Let $\mathbf{u} = \frac{1}{2}(\mathbf{v} + \overline{\mathbf{v}})$ and $\mathbf{w} = \frac{1}{2i}(\mathbf{v} - \overline{\mathbf{v}})$.

a. Show that $A\overline{\mathbf{v}} = (a-ib)\overline{\mathbf{v}}$, so that $\mathbf{v}$ and $\overline{\mathbf{v}}$ are linearly independent in $\mathbb{C}^n$.

b. Show that $A\mathbf{u} = a\mathbf{u} - b\mathbf{w}$ and $A\mathbf{w} = b\mathbf{u} + a\mathbf{w}$.

Exercise 2.16: For example, the polynomial

$$p = 2x - y + 3xy + 5y^2$$

corresponds to the point $\begin{pmatrix} 0\\2\\-1\\0\\3\\5 \end{pmatrix}$,

so

$$xD_1p = x(2+3y) = 2x + 3xy$$

corresponds to $\begin{pmatrix} 0\\2\\0\\0\\3\\0 \end{pmatrix}$.

2.15 Show that an orthogonal 2×2 matrix is either a rotation or a reflection.

2.16 Let P be the space of polynomials of degree at most 2 in the variables x, y, identifying $a_1 + a_2x + a_3y + a_4x^2 + a_5xy + a_6y^2$ with $\begin{pmatrix} a_1\\ \vdots \\ a_6 \end{pmatrix} \in \mathbb{R}^6$.

a. What are the matrices of the linear transformations $S, T : P \to P$

$$S(p)\begin{pmatrix} x\\y \end{pmatrix} = xD_1p\begin{pmatrix} x\\y \end{pmatrix} \quad \text{and} \quad T(p)\begin{pmatrix} x\\y \end{pmatrix} = yD_2p\begin{pmatrix} x\\y \end{pmatrix}?$$

b. What are the kernel and the image of the linear transformation

$$p \mapsto 2p - S(p) - T(p)?$$

Hint for Exercise 2.17: You should use the fact that a polynomial p of degree d such that $p(n) = p'(n) = 0$ can be written $p(x) = (x-n)^2q(x)$ for some polynomial q of degree $d-2$.

2.17 Let $a_1, \dots, a_k, b_1, \dots, b_k$ be any $2k$ numbers. Show that there exists a unique polynomial p of degree at most $2k-1$ such that $p(n) = a_n, \quad p'(n) = b_n$ for all integers n with $1 \le n \le k$. In other words, show that the values of p and p' at $1, \dots, k$ determine p.

2.18 A square $n \times n$ matrix P such that $P^2 = P$ is called a *projection*.

a. Show that P is a projection if and only if $I - P$ is a projection. Show that if P is invertible, then P is the identity.

Hint for Exercise 2.18, part b: $\vec{\mathbf{v}} = P\vec{\mathbf{v}} + (\vec{\mathbf{v}} - P\vec{\mathbf{v}})$.

b. Let $V_1 = \operatorname{img} P$ and $V_2 = \ker P$. Show that any vector $\vec{\mathbf{v}} \in \mathbb{R}^n$ can be written uniquely $\vec{\mathbf{v}} = \vec{\mathbf{v}}_1 + \vec{\mathbf{v}}_2$ with $\vec{\mathbf{v}}_1 \in V_1$ and $\vec{\mathbf{v}}_2 \in V_2$.

c. Show that there exist a basis $\vec{\mathbf{v}}_1, \dots, \vec{\mathbf{v}}_n$ of $\mathbb{R}^n$ and a number $k \le n$ such that

$$\begin{aligned} &P\vec{\mathbf{v}}_1 = \vec{\mathbf{v}}_1,\ P\vec{\mathbf{v}}_2 = \vec{\mathbf{v}}_2, \dots, P\vec{\mathbf{v}}_k = \vec{\mathbf{v}}_k \quad \text{and} \\ &P\vec{\mathbf{v}}_{k+1} = \mathbf{0},\ P\vec{\mathbf{v}}_{k+2} = \vec{\mathbf{0}}, \dots, P\vec{\mathbf{v}}_n = \vec{\mathbf{0}}. \end{aligned}$$

*d. Show that if P_1 and P_2 are projections such that $P_1P_2 = [0]$, then $Q = P_1 + P_2 - (P_2P_1)$ is a projection, $\ker Q = \ker P_1 \cap \ker P_2$, and the image of Q is the space spanned by the image of P_1 and the image of P_2.

Exercise 2.19: Recall that $\mathcal{C}^2$ is the space of C^2 (twice continuously differentiable) functions.

2.19 Show that the transformation $T : \mathcal{C}^2(\mathbb{R}) \to \mathcal{C}(\mathbb{R})$ given by formula 2.6.8 in Example 2.6.7 is a linear transformation.

2.20 Denote by $\mathcal{L}\Big(\text{Mat}\,(n,n), \text{Mat}\,(n,n)\Big)$ the space of linear transformations from $\text{Mat}\,(n,n)$ to $\text{Mat}\,(n,n)$.

a. Show that $\mathcal{L}\Big(\text{Mat}\,(n,n), \text{Mat}\,(n,n)\Big)$ is a vector space and that it is finite dimensional. What is its dimension?

b. Prove that for any $A \in \text{Mat}\,(n,n)$, the transformations

$$L_A,\ R_A : \text{Mat}\,(n,n) \to \text{Mat}\,(n,n)$$

given by $L_A(B) = AB$ and $R_A(B) = BA$ are linear transformations.

c. Let $\mathcal{M}_L \subset \mathcal{L}\big(\text{Mat}\,(n,n), \text{Mat}\,(n,n)\big)$ be the set of functions of the form L_A. Show that it is a subspace of $\mathcal{L}\big(\text{Mat}\,(n,n), \text{Mat}\,(n,n)\big)$. What is its dimension?

d. Show that there are linear transformations $T : \text{Mat}\,(2,2) \to \text{Mat}\,(2,2)$ that cannot be written as $L_A + R_B$ for any two matrices $A, B \in \text{Mat}\,(2,2)$. Can you find an explicit one?

2.21 Show that in a vector space of dimension n, more than n vectors are never linearly independent, and fewer than n vectors never span.

2.22 Suppose we use the same operator $T : P_2 \to P_2$ as in Exercise 2.6.8, but choose instead to work with the basis

$$q_1(x) = x^2, \quad q_2(x) = x^2 + x, \quad q_3(x) = x^2 + x + 1.$$

Now what is the matrix $\Phi_{\{q\}}^{-1} \circ T \circ \Phi_{\{q\}}$?

We thank Tan Lei and Alexandre Bardet for Exercise 2.23.

2.23 Let A be a $k \times n$ matrix. Show that if you row reduce the augmented matrix $[A^\top | I_n]$ to get $[\widetilde{A}^\top \widetilde{B}^\top]$, the nonzero columns of $\widetilde{A}$ form a basis for the image of A, and the nonzero columns of $\widetilde{B}$ form a basis for the kernel of A.

$$\mathbf{F}\begin{pmatrix} x \\ y \end{pmatrix} = \begin{pmatrix} \sin(x-y) + y^2 \\ \cos(x+y) - x \end{pmatrix}$$

Map for Exercise 2.24

2.24 a. Find a global Lipschitz ratio for the derivative of the map $\mathbf{F}$ defined in the margin.

b. Do one step of Newton's method to solve $\mathbf{F}\begin{pmatrix} x \\ y \end{pmatrix} - \begin{pmatrix} .5 \\ 0 \end{pmatrix} = \begin{pmatrix} 0 \\ 0 \end{pmatrix}$, starting at $\begin{pmatrix} 0 \\ 0 \end{pmatrix}$.

c. Can you be sure that Newton's method converges?

Exercise 2.25: Note that

$$[2\,I]^3 = [8\,I], \quad \text{i.e.,}$$

$$\begin{bmatrix} 2 & 0 & 0 \\ 0 & 2 & 0 \\ 0 & 0 & 2 \end{bmatrix}^3 = \begin{bmatrix} 8 & 0 & 0 \\ 0 & 8 & 0 \\ 0 & 0 & 8 \end{bmatrix}.$$

2.25 Using Newton's method, solve the equation $A^3 = \begin{bmatrix} 9 & 0 & 1 \\ 0 & 7 & 0 \\ 0 & 2 & 8 \end{bmatrix}$.

Exercise 2.26: The computation really does require you to row reduce a 4×4 matrix.

2.26 Consider the map $F : \text{Mat}\,(2,2) \to \text{Mat}\,(2,2)$ given by $F(A) = A^2 + A^{-1}$. Set $A_0 = \begin{bmatrix} 0 & 1 \\ -1 & 0 \end{bmatrix}$ and $B_0 = F(A_0)$, and define

$$U_r = \{\, B \in \text{Mat}\,(2,2) \mid |B - B_0| < r \,\}.$$

Do there exist $r > 0$ and a differentiable mapping $G : U_r \to \text{Mat}\,(2,2)$ such that $F(G(B)) = B$ for every $B \in U_r$?

$$\mathbf{f}\begin{pmatrix} x \\ y \end{pmatrix} = \begin{pmatrix} x^2 - y - 2 \\ y^2 - x - 6 \end{pmatrix}$$

Map for Exercise 2.27

2.27 a. Find a global Lipschitz ratio for the derivative of the mapping $\mathbf{f} : \mathbb{R}^2 \to \mathbb{R}^2$ given in the margin.

b. Do one step of Newton's method to solve $\mathbf{f}\begin{pmatrix} x \\ y \end{pmatrix} = \begin{pmatrix} 0 \\ 0 \end{pmatrix}$ starting at $\begin{pmatrix} 2 \\ 3 \end{pmatrix}$.

c. Find and sketch a disc in $\mathbb{R}^2$ which you are sure contains a root.

2.28 There are other plausible ways to measure matrices other than the length and the norm; for example, we could declare the size $|A|$ of a matrix A to be the largest absolute value of an entry. In this case, $|A + B| \le |A| + |B|$, but the statement $|A\vec{\mathbf{x}}| \le |A||\vec{\mathbf{x}}|$ (where $|\vec{\mathbf{x}}|$ is the ordinary length of a vector) is false. Find an ϵ so that it is false for

$$A = \begin{bmatrix} 1 & 1 & 1+\epsilon \\ 0 & 0 & 0 \\ 0 & 0 & 0 \end{bmatrix}, \quad \text{and} \quad \vec{\mathbf{x}} = \begin{bmatrix} 1 \\ 1 \\ 0 \end{bmatrix}.$$

Exercise 2.29: The norm $\|A\|$ of a matrix A is defined in Section 2.9 (Definition 2.9.6).

2.29 Show that $\|A\| = \|A^\top\|$.

2.30 In Example 2.10.9 we found that $M = 2\sqrt{2}$ is a global Lipschitz ratio for the function $\mathbf{f}\begin{pmatrix} x \\ y \end{pmatrix} = \begin{pmatrix} \sin(x+y) \\ x^2 - y^2 \end{pmatrix}$. What Lipschitz ratio do you get using the method of second partial derivatives? Using that Lipschitz ratio, what minimum domain do you get for the inverse function at $\mathbf{f}\begin{pmatrix} 0 \\ \pi \end{pmatrix}$?

2.31 a. True or false? The equation $\sin(xyz) = z$ expresses x implicitly as a differentiable function of y and z near the point $\begin{pmatrix} x \\ y \\ z \end{pmatrix} = \begin{pmatrix} \pi/2 \\ 1 \\ 1 \end{pmatrix}$.

b. True or false? The equation $\sin(xyz) = z$ expresses z implicitly as a differentiable function of x and y near the same point.

Exercise 2.32: You may use the fact that if

$$S : \text{Mat}\,(2,2) \to \text{Mat}\,(2,2)$$

is the squaring map

$$S(A) = A^2,$$

then

$$[\mathbf{D}S(A)]B = AB + BA.$$

2.32 True or false? There exist a neighborhood $U \subset \text{Mat}\,(2,2)$ of $\begin{bmatrix} 5 & 0 \\ 0 & 5 \end{bmatrix}$ and a C^1 mapping $F : U \to \text{Mat}\,(2,2)$ with

$$F\left(\begin{bmatrix} 5 & 0 \\ 0 & 5 \end{bmatrix}\right) = \begin{bmatrix} 1 & 2 \\ 2 & -1 \end{bmatrix}, \quad \text{and} \quad (F(A))^2 = A.$$

2.33 True or false? There exist $r > 0$ and a differentiable map

$$g : B_r\left(\begin{bmatrix} -3 & 0 \\ 0 & -3 \end{bmatrix}\right) \to \text{Mat}\,(2,2) \quad \text{such that} \quad g\left(\begin{bmatrix} -3 & 0 \\ 0 & -3 \end{bmatrix}\right) = \begin{bmatrix} 1 & 2 \\ -2 & -1 \end{bmatrix}$$

and $(g(A))^2 = A$ for all $A \in B_r\left(\begin{bmatrix} -3 & 0 \\ 0 & -3 \end{bmatrix}\right)$.

Dierk Schleicher contributed Exercise 2.34. Geometrically, the condition given is the condition that there exists a unit cube with a vertex at the origin such that the three sides emanating from the origin are $\vec{\mathbf{a}}$, $\vec{\mathbf{b}}$, and $\vec{\mathbf{c}}$.

2.34 Given three vectors $\begin{bmatrix} a_1 \\ a_2 \end{bmatrix}, \begin{bmatrix} b_1 \\ b_2 \end{bmatrix}, \begin{bmatrix} c_1 \\ c_2 \end{bmatrix}$ in $\mathbb{R}^2$, show that there exist vectors

$\vec{\mathbf{a}} = \begin{bmatrix} a_1 \\ a_2 \\ a_3 \end{bmatrix}, \vec{\mathbf{b}} = \begin{bmatrix} b_1 \\ b_2 \\ b_3 \end{bmatrix}, \vec{\mathbf{c}} = \begin{bmatrix} c_1 \\ c_2 \\ c_3 \end{bmatrix}$ in $\mathbb{R}^3$ such that

$$|\vec{\mathbf{a}}|^2 = |\vec{\mathbf{b}}|^2 = |\vec{\mathbf{c}}|^2 = 1 \quad \text{and} \quad \vec{\mathbf{a}} \cdot \vec{\mathbf{b}} = \vec{\mathbf{a}} \cdot \vec{\mathbf{c}} = \vec{\mathbf{b}} \cdot \vec{\mathbf{c}} = 0$$

if and only if $\vec{\mathbf{v}}_1 = \begin{bmatrix} a_1 \\ b_1 \\ c_1 \end{bmatrix}$ and $\vec{\mathbf{v}}_2 = \begin{bmatrix} a_2 \\ b_2 \\ c_2 \end{bmatrix}$ are of unit length and orthogonal.

Exercise 2.35: There are many "right" answers to this question, so try to think of a few.

2.35 Imagine that, when constructing a Newton sequence

$$\mathbf{x}_{n+1} = \mathbf{x}_n - [\mathbf{Df}(\mathbf{x}_n)]^{-1}\mathbf{f}(\mathbf{x}_n),$$

you happen upon a noninvertible matrix $[\mathbf{Df}(\mathbf{x}_n)]$. What should you do? Suggest ways to deal with the situation.

Exercise 2.36: The answer to part d depends on whether you choose the additive identity to be $\{0\}$ or allow it to be whatever is appropriate for the particular subset you are looking at. In the latter case you might land on *quotient spaces*.

2.36 Let V be a vector space, and denote by $\mathcal{P}^*(V)$ the set of nonempty subsets of V. Define $+ : \mathcal{P}^*(V) \times \mathcal{P}^*(V) \to \mathcal{P}^*(V)$ by

$$A + B \stackrel{\text{def}}{=} \{\, a + b \mid a \in A, b \in B \,\}$$

and scalar multiplication $\mathbb{R} \times \mathcal{P}^*(V) \to \mathcal{P}^*(V)$ by $\alpha A \stackrel{\text{def}}{=} \{\, \alpha a \mid a \in A \,\}$.

a. Show that $+$ is associative: $(A+B)+C = A+(B+C)$ and that $\{0\}$ is a an additive identity.

b. Show that $\alpha(A+B) = \alpha A + \alpha B$, $1A = A$, and $(\alpha\beta)A = \alpha(\beta A)$, for all $\alpha, \beta \in \mathbb{R}$.

c. Is $\mathcal{P}^*(V)$ a vector space with these operations?

*d. Does $\mathcal{P}^*(V)$ have subsets that are vector spaces with these operations?

***2.37** This exercise gives a proof of *Bezout's theorem*. Let p_1 and p_2 be polynomials of degree k_1 and k_2 respectively, and consider the mapping

$$T\colon (q_1, q_2) \mapsto p_1q_1 + p_2q_2,$$

where q_1 is a polynomial of degree at most $k_2 - 1$ and q_2 is a polynomial of degree at most $k_1 - 1$, so that $p_1q_1 + p_2q_2$ is of degree $\le k_1 + k_2 - 1$. Note that the space of such (q_1, q_2) is of dimension $k_1 + k_2$, and the space of polynomials of degree at most $k_1 + k_2 - 1$ is also of dimension $k_1 + k_2$.

Exercise 2.37, part a: It may be easier to work over the complex numbers.

Relatively prime: with no common factors.

a. Show that $\ker T = \{0\}$ if and only if p_1 and p_2 are *relatively prime*.

b. Use Corollary 2.5.10 to prove *Bezout's identity*: if p_1, p_2 are relatively prime, then there exist unique q_1 and q_2 of degree at most $k_2 - 1$ and $k_1 - 1$ such that $p_1q_1 + p_2q_2 = 1$.

****2.38** Let A be an $n \times n$ diagonal matrix: $A = \begin{bmatrix} \lambda_1 & & \\ & \ddots & \\ & & \lambda_n \end{bmatrix}$, and suppose that one of the diagonal entries, say λ_k, satisfies

$$\inf_{k \ne j} |\lambda_k - \lambda_j| \ge m > 0$$

for some number m. Let B be an $n \times n$ matrix. Find a number R, depending on m, such that if $|B| < R$, then Newton's method will converge if it is used to solve

$$(A + B)\mathbf{x} = \mu\mathbf{x}, \quad \text{for } \mathbf{x} \text{ satisfying } |\mathbf{x}|^2 = 1,$$

starting at $\mathbf{x}_0 = \vec{\mathbf{e}}_k$, $\mu_0 = \lambda_k$.

Exercise 2.39: For instance, if $n = 4$, then

$$J_2 = \begin{bmatrix} 0 & 0 & 0 & 0 \\ 0 & 0 & 0 & 0 \\ 0 & 0 & 1 & 0 \\ 0 & 0 & 0 & 1 \end{bmatrix}$$

$$J_3 = \begin{bmatrix} 0 & 0 & 0 & 0 \\ 0 & 1 & 0 & 0 \\ 0 & 0 & 1 & 0 \\ 0 & 0 & 0 & 1 \end{bmatrix}.$$

Of course if an $n \times n$ matrix A has rank n, then $A = QJ_nP^{-1}$ just says that $A = QP^{-1}$, i.e., A is invertible with inverse PQ^{-1}; see Proposition 1.2.15.

2.39 Prove that any $n \times n$ matrix A of rank k can be written $A = QJ_kP^{-1}$, where P and Q are invertible and $J_k = \begin{bmatrix} 0 & 0 \\ 0 & I_k \end{bmatrix}$, where I_k is the $k \times k$ identity matrix.

2.40 Let a sequence of integers $a_0, a_1, a_2, \dots$ be defined inductively by

$$a_0 = 1, \ a_1 = 0, \quad \text{and} \quad a_n = 2a_{n-1} + a_{n-2} \text{ for } n \ge 2.$$

a. Find a matrix M such that $\begin{pmatrix} a_{n+1} \\ a_{n+2} \end{pmatrix} = M \begin{pmatrix} a_n \\ a_{n+1} \end{pmatrix}$. Express $\begin{pmatrix} a_n \\ a_{n+1} \end{pmatrix}$ in terms of powers of M.

b. Find a linear relation between I, M, M^2, and use it to find the eigenvalues of M.

c. Find a matrix P such that $P^{-1}MP$ is diagonal.

d. Compute M^n in terms of powers of numbers. Use the result to find a formula for a_n.

2.41 Exercise 2.2.11 asked you to show that using row reduction to solve n equations in n unknowns takes $n^3 + n^2/2 - n/2$ operations, where a single addition, multiplication, or division counts as one operation. How many operations are needed to compute the inverse of an $n \times n$ matrix A? To perform the matrix multiplication $A^{-1}\vec{\mathbf{b}}$?

3

Manifolds, Taylor polynomials, quadratic forms, and curvature

Thomson [Lord Kelvin] had predicted the problems of the first [transatlantic] cable by mathematics. On the basis of the same mathematics he now promised the company a rate of eight or even 12 words a minute. Half a million pounds was being staked on the correctness of a partial differential equation.—T. W. Körner, *Fourier Analysis*

3.0 Introduction

When a computer calculates sines, it does not look up the answer in some mammoth table of sines; stored in the computer is a polynomial that very well approximates $\sin x$ for x in some particular range. Specifically, it uses a formula very close to equation 3.4.6:

$$\sin x = x + a_3x^3 + a_5x^5 + a_7x^7 + a_9x^9 + a_{11}x^{11} + \epsilon(x),$$

where the coefficients are

$$\begin{aligned} a_3 &= -.1666666664 \\ a_5 &= .0083333315 \\ a_7 &= -.0001984090 \\ a_9 &= .0000027526 \\ a_{11} &= -.0000000239. \end{aligned}$$

When

$$|x| \le \pi/2,$$

the error $\epsilon(x)$ is guaranteed to be less than 2×10^{-9}, good enough for a calculator that computes to eight significant digits. The computer needs only to remember five coefficients and do a bit of arithmetic to replace a huge table of sines.

This chapter is something of a grab bag. The various themes are related, but the relationship is not immediately apparent. We begin with two sections on geometry. In Section 3.1 we use the implicit function theorem to define smooth curves, smooth surfaces, and more general k-dimensional "surfaces" in $\mathbb{R}^n$, called *manifolds*. In Section 3.2 we discuss linear approximations to manifolds: *tangent spaces*.

We switch gears in Section 3.3, where we use higher partial derivatives to construct the Taylor polynomial of a function in several variables. We saw in Section 1.7 how to approximate a nonlinear function by its derivative; here we see that we can better approximate C^k functions using Taylor polynomials when $k \ge 2$. This is useful, since polynomials, unlike sines, cosines, exponentials, square roots, logarithms, ... can actually be computed using arithmetic. Computing Taylor polynomials by calculating higher partial derivatives can be quite unpleasant; in Section 3.4 we show how to compute them by combining Taylor polynomials of simpler functions.

In Sections 3.5 and 3.6 we take a brief detour, introducing quadratic forms and seeing how to classify them according to their "signature": if we consider the second-degree terms of a function's Taylor polynomial as a quadratic form, its signature usually tells us whether at a point where the derivative vanishes the function has a minimum value, a maximum value, or some kind of *saddle*, like a mountain pass. In Section 3.7 we use Lagrange multipliers to find extrema of a function restricted to some manifold $M \subset \mathbb{R}^n$; we use Lagrange multipliers to prove the *spectral theorem*.

In Section 3.8 we introduce finite probability spaces, and show how the singular value decomposition (a consequence of the spectral theorem) gives rise to *principal component analysis*, of immense importance in statistics.

In Section 3.9 we give a brief introduction to the vast and important subject of the geometry of curves and surfaces, using the higher-degree approximations provided by Taylor polynomials: the *curvature* of a curve or surface depends on the quadratic terms of the functions defining it, and the *torsion* of a space curve depends on the cubic terms.

3.1 MANIFOLDS

Everyone knows what a curve is, until he has studied enough mathematics to become confused through the countless number of possible exceptions—Felix Klein

These familiar objects are by no means simple: already, the theory of soap bubbles is a difficult topic, with a complicated partial differential equation controlling the shape of the film.

FIGURE 3.1.1.
Felix Klein (1849–1925)

Klein's work in geometry "has become so much a part of our present mathematical thinking that it is hard for us to realise the novelty of his results."—From a biographical sketch by J. O'Connor and E. F. Robertson. Klein was also instrumental in developing *Mathematische Annalen* into one of the most prestigious mathematical journals.

In this section we introduce one more actor in multivariable calculus. So far, our mappings have been first linear, then nonlinear with good linear approximations. But the domain and codomain of our mappings have been "flat" open subsets of $\mathbb{R}^n$. Now we want to allow "nonlinear" $\mathbb{R}^n$'s, called *smooth manifolds*.

Manifolds are a generalization of the familiar curves and surfaces of every day experience. A one-dimensional manifold is a smooth curve; a two-dimensional manifold is a smooth surface. Smooth curves are idealizations of things like telephone wires or a tangled garden hose. Particularly beautiful smooth surfaces are produced when you blow soap bubbles that wobble and slowly vibrate as they drift through the air. Other examples are shown in Figure 3.1.2.

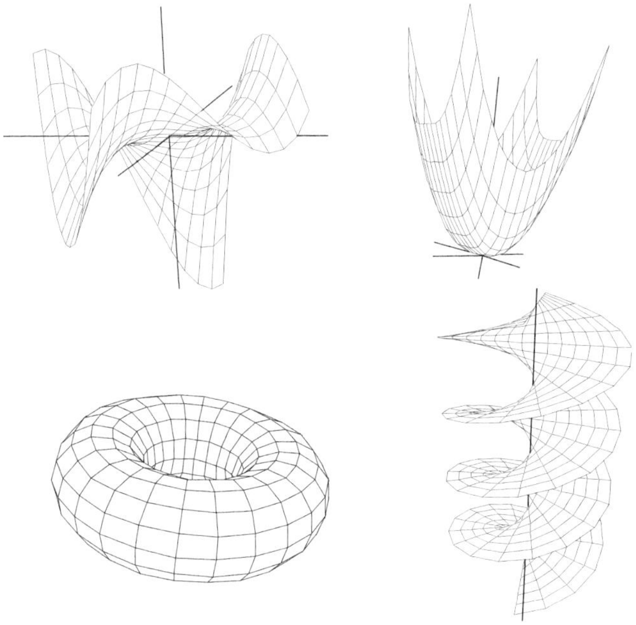

FIGURE 3.1.2. Four surfaces in $\mathbb{R}^3$. The top two are graphs of functions. The bottom two are *locally* graphs of functions. All four qualify as smooth surfaces (two-dimensional manifolds) under Definition 3.1.2.

We will define smooth manifolds mathematically, excluding some objects that we might think of as smooth: a figure eight, for example. We will see how to use the implicit function theorem to tell whether the locus defined by an equation is a smooth manifold. Finally, we will compare knowing a manifold by equations, and knowing it by a parametrization.

Smooth manifolds in $\mathbb{R}^n$

When is a subset $X \subset \mathbb{R}^n$ a smooth manifold? Our definition is based on the notion of a graph.

It is convenient to denote a point in the graph of a function $\mathbf{f}: \mathbb{R}^k \to \mathbb{R}^{n-k}$ as $\begin{pmatrix} \mathbf{x} \\ \mathbf{f}(\mathbf{x}) \end{pmatrix}$, with $\mathbf{x} \in \mathbb{R}^k$ and $\mathbf{f}(\mathbf{x}) \in \mathbb{R}^{n-k}$. But this presupposes that the "active" variables are the first variables, which is a problem, since we usually cannot use the same active variables at all points of the manifold.

How then might we describe a point $\mathbf{z}$ in the graph of a function $\mathbf{f}: \mathbb{R}^k \to \mathbb{R}^{n-k}$, where the variables of the domain of $\mathbf{f}$ have indices $i_1, \dots, i_k$ and the variables of the codomain have indices $j_1, \dots, j_{n-k}$? Here is an accurate if heavy-handed approach, using the "concrete to abstract" linear transformation Φ of Definition 2.6.12. Set

$$\mathbf{x} = \begin{pmatrix} x_1 \\ \vdots \\ x_k \end{pmatrix}, \quad \mathbf{y} = \begin{pmatrix} y_1 \\ \vdots \\ y_{n-k} \end{pmatrix}$$

with $\mathbf{f}(\mathbf{x}) = \mathbf{y}$. Define Φ_{d} (d for the domain of $\mathbf{f}$) and Φ_{c} (c for codomain) by

$$\Phi_{\mathrm{d}}(\mathbf{x}) = x_1\vec{\mathbf{e}}_{i_1} + \cdots + x_k\vec{\mathbf{e}}_{i_k}$$
$$\Phi_{\mathrm{c}}(\mathbf{y}) = y_1\vec{\mathbf{e}}_{j_1} + \cdots + y_{n-k}\vec{\mathbf{e}}_{j_{n-k}},$$

where the $\vec{\mathbf{e}}$'s are standard basis vectors in $\mathbb{R}^n$. Then the graph of $\mathbf{f}$ is the set of $\mathbf{z}$ such that

$$\mathbf{z} = \Phi_{\mathrm{d}}(\mathbf{x}) + \Phi_{\mathrm{c}}(\mathbf{f}(\mathbf{x})).$$

Since the function $\mathbf{f}$ of Definition 3.1.2 is C^1, its domain must be open. If $\mathbf{f}$ is C^p rather than C^1, then the manifold is a C^p manifold.

Definition 3.1.1 (Graph). The *graph* of a function $\mathbf{f}: \mathbb{R}^k \to \mathbb{R}^{n-k}$ is the set of points $\begin{pmatrix} \mathbf{x} \\ \mathbf{y} \end{pmatrix} \in \mathbb{R}^n$ such that $\mathbf{f}(\mathbf{x}) = \mathbf{y}$. It is denoted $\Gamma(\mathbf{f})$.

Thus the graph of a function $\mathbf{f}$ lives in a space whose dimension is the sum of the dimensions of the domain and codomain of $\mathbf{f}$.

Traditionally we graph functions $f: \mathbb{R} \to \mathbb{R}$ with the horizontal x-axis corresponding to the input, and the vertical axis corresponding to the output (values of f). Note that the graph of such a function is a subset of $\mathbb{R}^2$. For example, the graph of $f(x) = x^2$ consists of the points $\begin{pmatrix} x \\ f(x) \end{pmatrix} \in \mathbb{R}^2$, i.e., the points $\begin{pmatrix} x \\ x^2 \end{pmatrix}$.

The top two surfaces shown in Figure 3.1.2 are graphs of functions from $\mathbb{R}^2$ to $\mathbb{R}$: the surface on the left is the graph of $f\begin{pmatrix} x \\ y \end{pmatrix} = x^3 - 2xy^2$; that on the right is the graph of $f\begin{pmatrix} x \\ y \end{pmatrix} = x^2 + y^4$. Although we depict these graphs on a flat piece of paper, they are actually subsets of $\mathbb{R}^3$. The first consists of the points $\begin{pmatrix} x \\ y \\ x^3 - 2xy^2 \end{pmatrix}$, the second of the points $\begin{pmatrix} x \\ y \\ x^2 + y^4 \end{pmatrix}$.

Definition 3.1.2 says that if a function $\mathbf{f}: \mathbb{R}^k \to \mathbb{R}^{n-k}$ is C^1, then its graph is a smooth n-dimensional manifold in $\mathbb{R}^n$. Thus the top two graphs shown in Figure 3.1.2 are two-dimensional manifolds in $\mathbb{R}^3$.

But the torus and helicoid shown in Figure 3.1.2 are also two-dimensional manifolds. Neither is the graph of a single function expressing one variable in terms of the other two. But both are *locally* graphs of functions.

Definition 3.1.2 (Smooth manifold in $\mathbb{R}^n$). A subset $M \subset \mathbb{R}^n$ is a smooth k-dimensional manifold if locally it is the graph of a C^1 mapping $\mathbf{f}$ expressing $n-k$ variables as functions of the other k variables.

With this definition, which depends on chosen coordinates, it isn't obvious that if you rotate a smooth manifold it is still smooth. We will see in Theorem 3.1.16 that it is.

Generally, "smooth" means "as many times differentiable as is relevant to the problem at hand". In this and the next section, it means of class C^1. When speaking of smooth manifolds, we often omit the word smooth.[1]

"Locally" means that every point $\mathbf{x} \in M$ has a neighborhood $U \subset \mathbb{R}^n$ such that $M \cap U$ (the part of M in U) is the graph of a mapping expressing

[1]Some authors use "smooth" to mean C^∞: "infinitely many times differentiable". For our purposes this is overkill.

$n-k$ of the coordinates of each point in $M \cap U$ in terms of the other k. This may sound like an unwelcome complication, *but if we omitted the word "locally" then we would exclude from our definition most interesting manifolds.* We already saw that neither the torus nor the helicoid of Figure 3.1.2 is the graph of a single function expressing one variable as a function of the other two. Even such a simple curve as the unit circle is not the graph of a single function expressing one variable in terms of the other. In Figure 3.1.3 we show another smooth curve that would not qualify as a manifold if we required it to be the graph of a single function expressing one variable in terms of the other; the caption justifies our claim that this curve is a smooth curve.

A manifold M embedded in $\mathbb{R}^n$, denoted $M \subset \mathbb{R}^n$, is sometimes called a *submanifold* of $\mathbb{R}^n$. Strictly speaking, it should not be referred to simply as a "manifold", which can mean an abstract manifold, not embedded in any space. The manifolds in this book are all submanifolds of $\mathbb{R}^n$ for some n.

Especially in higher dimensions, making some kind of global sense of a patchwork of graphs of functions can be quite challenging; a mathematician trying to picture a manifold is rather like a blindfolded person who has never met or seen a picture of an elephant seeking to identify one by patting first an ear, then the trunk or a leg. It is a subject full of open questions, some fully as interesting and demanding as, for example, Fermat's last theorem, whose solution after more than three centuries aroused such passionate interest.

Three-dimensional and four-dimensional manifolds are of particular interest, in part because of applications in representing space-time.

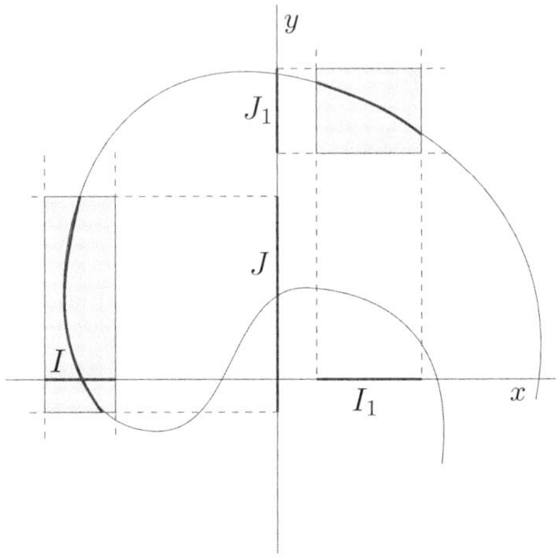

FIGURE 3.1.3. Above, I and I_1 are intervals on the x-axis; J and J_1 are intervals on the y-axis. The darkened part of the curve in the shaded rectangle $I \times J$ is the graph of a function expressing $x \in I$ as a function of $y \in J$, and the darkened part of the curve in $I_1 \times J_1$ is the graph of a function expressing $y \in J_1$ as a function of $x \in I_1$. (By decreasing the size of J_1 a bit, we could also think of the part of the curve in $I_1 \times J_1$ as the graph of a function expressing $x \in I_1$ as a function of $y \in J_1$.) But we cannot think of the darkened part of the curve in $I \times J$ as the graph of a function expressing $y \in J$ as a function of $x \in I$; there are values of x that give two different values of y, and others that give none, so such a "function" is not well defined.

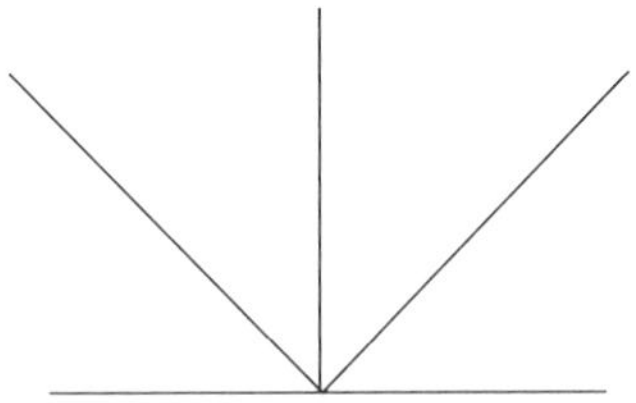

FIGURE 3.1.4. The graph of $f(x) = |x|$ is not a smooth curve.

Example 3.1.3 (Graph of smooth function is smooth manifold). The graph of any smooth function is a smooth manifold. The curve of equation $y = x^2$ is a one-dimensional manifold: the graph of y as the function $f(x) = x^2$. The curve of equation $x = y^2$ is also a one-dimensional manifold: the graph of a function representing x as a function of y. Each surface at the top of Figure 3.1.2 is the graph of a function representing z as a function of x and y. $\triangle$

Example 3.1.4 (Graphs that are not smooth manifolds). The graph of the function $f : \mathbb{R} \to \mathbb{R}$, $f(x) = |x|$, shown in Figure 3.1.4, is not a smooth curve; it is the graph of the function f expressing y as a function of x, of course, but f is not differentiable. Nor is it the graph of a function g expressing x as a function of y, since in a neighborhood of 0 the same value of y gives sometimes two values of x, sometimes none.

In contrast, the graph of the function $f(x) = x^{1/3}$, shown in Figure 3.1.5, is a smooth curve; f is not differentiable at the origin, but the curve is the graph of the function $x = g(y) = y^3$, which is differentiable.

The set $X \subset \mathbb{R}^2$ of equation $xy = 0$ (the union of the two axes), shown on the left in Figure 3.1.6, is not a smooth curve; in any neighborhood of $\begin{pmatrix} 0 \\ 0 \end{pmatrix}$, there are infinitely many y corresponding to $x = 0$, and infinitely many x corresponding to $y = 0$, so it isn't a graph of a function either way. But $X - \left\{ \begin{pmatrix} 0 \\ 0 \end{pmatrix} \right\}$ *is* a smooth curve – even though it consists of four distinct pieces. A manifold need not be connected. △

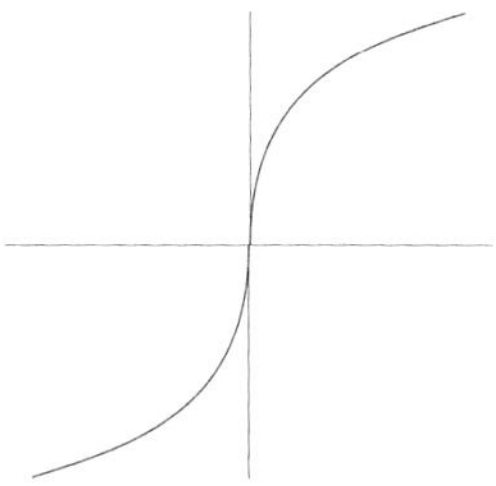

FIGURE 3.1.5.

The graph of $f(x) = x^{1/3}$ is a smooth curve: although f is not differentiable at the origin, the function $g(y) = y^3$ is.

Example 3.1.5 (Unit circle). The unit circle of equation $x^2 + y^2 = 1$ is a smooth curve. Here we need the graphs of four functions to cover the entire circle: the unit circle is only locally the graph of a function. As discussed in Example 2.10.12, the upper half of the circle, made up of points $\begin{pmatrix} x \\ y \end{pmatrix}$ with $y > 0$, is the graph of the function $\sqrt{1 - x^2}$ expressing y in terms of x, the lower half is the graph of $-\sqrt{1 - x^2}$, the right half is the graph of the function $\sqrt{1 - y^2}$ expressing x in terms of y, and the left half is the graph of $-\sqrt{1 - y^2}$. △

Using Definition 3.1.2 to show that a set is a manifold can be quite tedious, even for something as simple (and as obviously smooth) as the unit sphere. In Example 3.1.6 we do it the hard way, working directly from the definition. We will see in Theorem 3.1.10 that there is an easier way.

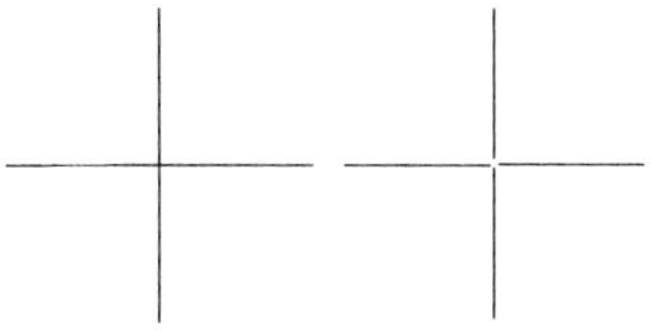

FIGURE 3.1.6.

LEFT: The graph of the x and y axes is not a smooth curve. RIGHT: The graph of the axes minus the origin *is* a smooth curve.

Example 3.1.6: Many students find it hard to call the sphere of equation

$$x^2 + y^2 + z^2 = 1$$

two dimensional. But when we say that Chicago is "at" x latitude and y longitude, we treat the surface of the earth as two dimensional.

Example 3.1.6 (Surface in $\mathbb{R}^3$). By Definition 3.1.2, a subset $S \subset \mathbb{R}^3$ is a smooth surface (two-dimensional manifold) in $\mathbb{R}^3$ if locally it is the graph of a C^1 mapping expressing one variable as a function of the other two variables. That is, S is a smooth surface if for every point $\mathbf{a} = \begin{pmatrix} a \\ b \\ c \end{pmatrix} \in S$, there are neighborhoods I of a, J of b, and K of c, and either a differentiable mapping

- $f : I \times J \to K$, i.e., z as a function of (x, y) or
- $g : I \times K \to J$, i.e., y as a function of (x, z) or
- $h : J \times K \to I$, i.e., x as a function of (y, z),

such that $S \cap (I \times J \times K)$ is the graph of $f, g,$ or h.

For instance, the unit sphere

$$S^2 \stackrel{\text{def}}{=} \left\{ \begin{pmatrix} x \\ y \\ z \end{pmatrix} \text{ such that } x^2 + y^2 + z^2 = 1 \right\} \qquad 3.1.1$$

is a smooth surface. Let

$$D_{x,y} = \left\{ \begin{pmatrix} x \\ y \end{pmatrix} \text{ such that } x^2 + y^2 < 1 \right\} \qquad 3.1.2$$

be the unit disc in the (x, y)-plane and $\mathbb{R}_z^+$ the part of the z-axis where $z > 0$. Then

$$S^2 \cap (D_{x,y} \times \mathbb{R}_z^+) \qquad 3.1.3$$

is the graph of the function $D_{x,y} \to \mathbb{R}_z^+$ expressing z as $\sqrt{1 - x^2 - y^2}$.

This shows that S^2 is a surface near every point where $z > 0$, and considering $z = -\sqrt{1 - x^2 - y^2}$ should convince you that S^2 is also a smooth surface near any point where $z < 0$. Exercise 3.1.4 asks you to consider the case $z = 0$. △

The graph of a function

$$\mathbf{f} : \mathbb{R} \to \mathbb{R}^2$$

lives in $\mathbb{R}^3$. If x determines y, z, we would write this as

$$\mathbf{f}(x) = \begin{pmatrix} f_1(x) \\ f_2(x) \end{pmatrix}; \text{ the graph}$$

consists of points $\begin{pmatrix} x \\ f_1(x) \\ f_2(x) \end{pmatrix}$.

Example 3.1.7 (Smooth curve in $\mathbb{R}^3$). For smooth curves in $\mathbb{R}^2$ and smooth surfaces in $\mathbb{R}^3$, one variable is expressed as a function of the other variable or variables. For curves in space, we have *two* variables expressed as a function of the other variable: a space curve is a one-dimensional manifold in $\mathbb{R}^3$, so it is locally the graph of a C^1 mapping expressing $n - k = 2$ variables as functions of the remaining variable. △

Examples of curves and surfaces in $\mathbb{R}^3$ tend to be misleadingly simple: circles, spheres, tori, all not too hard to draw on paper. But both curves and surfaces can be phenomenally complicated. If you put a ball of yarn in a washing machine you will have a fantastic mess. This is the natural state of a curve in $\mathbb{R}^3$. If you think of the surface of the yarn as a surface in $\mathbb{R}^3$, you will see that surfaces in $\mathbb{R}^3$ can be at least as complicated.

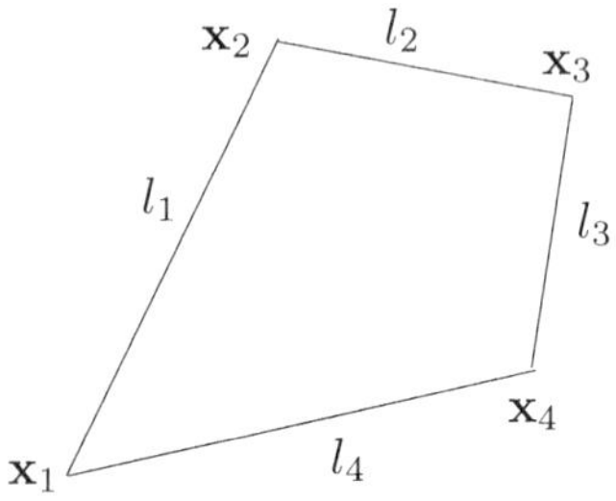

FIGURE 3.1.7.

One possible position of four linked rods, of lengths l_1, l_2, l_3, and l_4, restricted to a plane. Such four-bar mechanisms are used to convert rotational motion into back-and-forth rocking motion (to operate the windshield wipers on a car, for example).

Higher-dimensional manifolds

An open subset $U \subset \mathbb{R}^n$ is the simplest n-dimensional manifold; it is the graph of the zero function taking points in U to the origin, which is the only point of the vector space $\{\vec{0}\}$. (The graph lives in $\mathbb{R}^n \times \{\vec{0}\} = \mathbb{R}^n$.) In particular, $\mathbb{R}^n$ itself is an n-dimensional manifold.

The next example is far more elaborate. The set X_2 it describes is a four-dimensional manifold in $\mathbb{R}^8$; locally, it is the graph of a function expressing four variables in terms of four other variables.

Example 3.1.8 (Linked rods). Linkages of rods are everywhere, in mechanics (consider a railway bridge or the Eiffel tower), in biology (the skeleton), in robotics, in chemistry. One of the simplest examples is formed of four rigid rods, with assigned lengths $l_1, l_2, l_3, l_4 > 0$, connected by universal joints that can achieve any position to form a quadrilateral, as shown in Figure 3.1.7.

What is the set X_2 of positions the linkage can achieve if the points are restricted to a plane? Or the set X_3 of positions the linkage can achieve if the points are allowed to move in space? (In order to guarantee that our sets are not empty, we will require that each rod be shorter than the sum of the lengths of the other three.)

These sets are easy to describe by equations. For X_2 we have

$$X_2 = \text{the set of all } (\mathbf{x}_1, \mathbf{x}_2, \mathbf{x}_3, \mathbf{x}_4) \in (\mathbb{R}^2)^4 \quad \text{such that} \qquad 3.1.4$$
$$|\mathbf{x}_1 - \mathbf{x}_2| = l_1, \quad |\mathbf{x}_2 - \mathbf{x}_3| = l_2, \quad |\mathbf{x}_3 - \mathbf{x}_4| = l_3, \quad |\mathbf{x}_4 - \mathbf{x}_1| = l_4.$$

Thus X_2 is a subset of $\mathbb{R}^8$. Another way of saying this is that X_2 is the subset defined by the equation $\mathbf{f}(\mathbf{x}) = \mathbf{0}$, where $\mathbf{f} : (\mathbb{R}^2)^4 \to \mathbb{R}^4$ is the mapping

$$\mathbf{f}\Bigg(\underbrace{\begin{pmatrix} x_1 \\ y_1 \end{pmatrix}}_{\mathbf{x}_1}, \underbrace{\begin{pmatrix} x_2 \\ y_2 \end{pmatrix}}_{\mathbf{x}_2}, \underbrace{\begin{pmatrix} x_3 \\ y_3 \end{pmatrix}}_{\mathbf{x}_3}, \underbrace{\begin{pmatrix} x_4 \\ y_4 \end{pmatrix}}_{\mathbf{x}_4}\Bigg) = \begin{bmatrix} (x_2 - x_1)^2 + (y_2 - y_1)^2 - l_1^2 \\ (x_3 - x_2)^2 + (y_3 - y_2)^2 - l_2^2 \\ (x_4 - x_3)^2 + (y_4 - y_3)^2 - l_3^2 \\ (x_1 - x_4)^2 + (y_1 - y_4)^2 - l_4^2 \end{bmatrix}. \qquad 3.1.5$$

Equation 3.1.5: This description is remarkably concise and remarkably uninformative. It isn't even clear how many dimensions X_2 and X_3 have; this is typical when you know a set by equations.

(The $\mathbf{x}_i$ have two coordinates, because the points are restricted to a plane.)

Similarly, the set X_3 of positions in space is also described by equation 3.1.4, if we take $\mathbf{x}_i \in \mathbb{R}^3$; X_3 is a subset of $\mathbb{R}^{12}$. Of course, to make equations corresponding to equation 3.1.5 we would have to add a third entry to the $\mathbf{x}_i$, and instead of writing $(x_2 - x_1)^2 + (y_2 - y_1)^2 - l_1^2$ we would need to write $(x_2 - x_1)^2 + (y_2 - y_1)^2 + (z_2 - z_1)^2 - l_1^2$.

If you object that you cannot visualize this manifold, you have our sympathy. Precisely for this reason, it gives a good idea of the kind of problem that comes up: you have a collection of equations defining some set but you have no idea what the set looks like.

Even picturing a two-dimensional manifold based on an equation can be hard; we expect you to be able to visualize the surface given by $x^2 + y^2 + z^2 = 1$, but would you immediately realize that the equation $z = x^3 - 2xy^2$ represents the surface shown in the upper left corner of Figure 3.1.2?

Can we express some of the $\mathbf{x}_i$ as functions of the others? You should feel, on physical grounds, that if the linkage is sitting on the floor, you can move two opposite connectors any way you like and the linkage will follow in a unique way. This is not quite to say that $\mathbf{x}_2$ and $\mathbf{x}_4$ are a function of $\mathbf{x}_1$ and $\mathbf{x}_3$ (or that $\mathbf{x}_1$ and $\mathbf{x}_3$ are a function of $\mathbf{x}_2$ and $\mathbf{x}_4$). This isn't true, as is suggested by Figure 3.1.8.

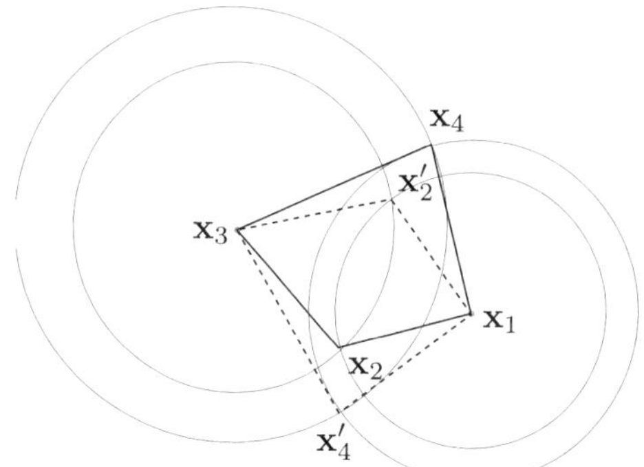

FIGURE 3.1.8. Two possible positions of a linkage with the same $\mathbf{x}_1$ and $\mathbf{x}_3$ are shown in solid and dotted lines. The other two are $\mathbf{x}_1, \mathbf{x}_2, \mathbf{x}_3, \mathbf{x}_4'$ and $\mathbf{x}_1, \mathbf{x}_2', \mathbf{x}_3, \mathbf{x}_4$.

Exercise 3.1.18 asks you to determine how many positions are possible using $\mathbf{x}_1$ and the two angles above – again, except in a few cases. Exercise 3.1.15 asks you to describe X_2 and X_3 when $l_1 = l_2 + l_3 + l_4$.

Usually, $\mathbf{x}_1$ and $\mathbf{x}_3$ determine either no positions of the linkage (if $\mathbf{x}_1$ and $\mathbf{x}_3$ are farther apart than $l_1 + l_2$ or $l_3 + l_4$) or exactly four (if a few other conditions are met; see Exercise 3.1.17). But $\mathbf{x}_2$ and $\mathbf{x}_4$ are *locally* functions of $\mathbf{x}_1, \mathbf{x}_3$. For a given $\mathbf{x}_1$ and $\mathbf{x}_3$ four positions are possible. If the rods are in one of these positions, and you move $\mathbf{x}_1$ and $\mathbf{x}_3$ a *small* amount, then there will be only one possible position of the linkage, and hence of $\mathbf{x}_2$ and $\mathbf{x}_4$, near the starting position.

Even this isn't always true: if any three vertices are aligned (as shown in Figure 3.1.9, for instance, or if one rod is folded back against another), then the two end vertices cannot be used as *parameters* (as the "active" variables that determine the values of the other variables). For instance, if $\mathbf{x}_1, \mathbf{x}_2$, and $\mathbf{x}_3$ are aligned, then you cannot move $\mathbf{x}_1$ and $\mathbf{x}_3$ arbitrarily, since the rods cannot be stretched. But *locally* the position is a function of $\mathbf{x}_2$ and $\mathbf{x}_4$.

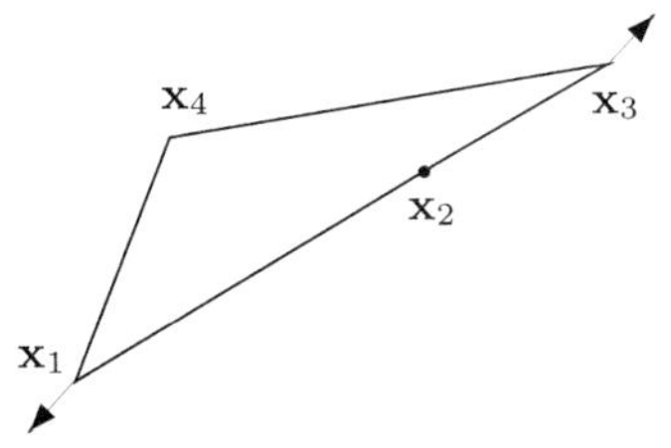

FIGURE 3.1.9.

If three vertices are aligned, the end vertices (here, $\mathbf{x}_1$ and $\mathbf{x}_3$) cannot move freely. For instance, they can't moved in the directions of the arrows without stretching the rods.

There are many other possibilities. For instance, we could choose $\mathbf{x}_2$ and $\mathbf{x}_4$ as the variables that locally determine $\mathbf{x}_1$ and $\mathbf{x}_3$, again making X_2 locally a graph. Or we could use the coordinates of $\mathbf{x}_1$ (two numbers), the polar angle of the first rod with the horizontal line passing through $\mathbf{x}_1$ (one number), and the angle between the first and the second (one number): four numbers in all, the same number we get using the coordinates of $\mathbf{x}_1$ and $\mathbf{x}_3$.[2]

Thus the set X_2 of example 3.1.8 is a 4-manifold in $\mathbb{R}^8$: locally it is the graph of a function expressing four variables (two coordinates each for two points) in terms of four other variables (the coordinates of the other two points or some other choice). It doesn't have to be the same function everywhere. In most neighborhoods, X_2 is the graph of a function of $\mathbf{x}_1$ and $\mathbf{x}_3$, but we saw that this is not true when $\mathbf{x}_1$, $\mathbf{x}_2$, and $\mathbf{x}_3$ are aligned; near such points, X_2 is the graph of a function expressing $\mathbf{x}_1$ and $\mathbf{x}_3$ in terms of $\mathbf{x}_2$ and $\mathbf{x}_4$.[3] $\triangle$

Saying that X_2 is a 4-dimensional manifold doesn't tell us much about what it "looks like". For instance, it doesn't tell us whether X_2 is *connected*. A subset $U \subset \mathbb{R}^n$ is connected if given any two points $\mathbf{x}_0, \mathbf{x}_1 \in U$, there exists a continuous map $\gamma : [0,1] \to U$ with $\gamma(0) = \mathbf{x}_0$ and $\gamma(1) = \mathbf{x}_1$. In other words, U is connected if there is a continuous path joining any two of its points. This very intuitive definition of connectedness is more properly known as *path connectedness*; it is the definition used in this book. (In topology, connectedness means something slightly different, but for open subsets of $\mathbb{R}^n$ and for manifolds, the two notions of connectedness coincide.)

A manifold need not be connected, as we saw already in the case of smooth curves, in Figure 3.1.6. In higher dimensions, determining whether a set is connected may be difficult. For example, as of this writing we don't know for what lengths of bars X_2 is or isn't connected.

The next example shows that when modeling a situation or object as a higher-dimensional manifold, the choice of dimension may depend on the problem to be solved.

Example 3.1.9 (A bicycle as a higher-dimensional manifold). Consider the *configuration space* (set of positions) of a bicycle as a k-dimensional manifold. What number is k? To specify a position, you must say where the center of gravity is (three numbers) and specify the orientation of the bicycle, say the direction in which the horizontal bar is pointing (two numbers) and the angle that the bar supporting the saddle makes with the vertical (one number). Then you need to give the angle that the handlebars make with the body of the bicycle (one number) and the angle that the wheels make with the forks (two numbers). The angle that the pedals make is another number. Presumably the position of the chain is determined by the position of the pedals, so it does not represent an extra free parameter.

[2] Such a system is said to have four *degrees of freedom*.

[3] For some choices of lengths, X_2 is no longer a manifold in a neighborhood of some positions: if all four lengths are equal, then X_2 is not a manifold near the position where it is folded flat.

For most purposes, this should be enough: 10 numbers. Well, maybe the position of the break-pulls is important too, say 12 numbers in all. The space of positions should be a 12-dimensional manifold. But there was a choice made of what to describe and what to ignore. We ignored the positions of the balls in the ball bearings, which might be irrelevant to a bicycle rider but very relevant to an engineer trying to minimize friction. We also ignored the vibrations of the rods, which should be imperceptible to a rider but of central importance to an engineer studying metal fatigue.

The "configuration space" of a system is the set of positions. The "phase space" is the set of positions and velocities.

Although visualizing the 12-dimensional configuration space of a bicycle as a single "12-dimensional surface-like thing" is presumably impossible, learning to ride a bicycle is not. In some sense, knowing how to ride a bicycle is a substitute for visualization; it gives a feel for what "12-dimensionality" really is.

Though there are parts that the rider might miss. For instance, he might think that the positions of the wheels can be ignored: who cares whether the air valves are up or down, except when pumping up the tires? But those variables are actually essential because of their derivatives, which measure how fast the tires are spinning. The angular momentum of the wheels is key to the stability of the motion.

When modeling a real system, this issue always comes up: What variables are relevant? Keep too many, and the model will be awkward and incomprehensible to the point of uselessness; keep too few, and your description may be wrong in essential ways.

Note that if a brake cable snaps, the configuration space suddenly gains a dimension, since the position of the brake calipers and the corresponding brake pull suddenly become uncorrelated. The rider will definitely be aware of this extra dimension. △

Using the implicit function theorem to identify manifolds

Already for something as simple as the unit circle, using Definition 3.1.2 to show that it qualifies as a smooth curve was a bit tedious. Yet that was an easy case, where we know what the curve looks like, and where the functions can be written down explicitly. More typical is the question, "is the locus defined by $x^8 + 2x^3 + y + y^5 = 1$ a smooth curve?"

With the help of a computer, you might be able to visualize such a locus, but using Definition 3.1.2 to determine whether or not it qualifies as a smooth curve would be difficult indeed; you would have to solve for x in terms of y or for y in terms of x, which means solving equations of degree 5 or 8. Using that definition to determine whether a locus is a higher-dimensional manifold is more challenging yet.

Fortunately, the implicit function theorem gives a way to determine that a locus given by an equation or set of equations is a smooth manifold. In Theorem 3.1.10 below, the condition that $[\mathbf{DF}(\mathbf{a})]$ be onto is the crucial condition of the implicit function theorem.

Theorem 3.1.10 is correct as stated, but the proof is based on the implicit function theorem, and our proof of the strong version of the implicit function theorem requires that the derivative of $\mathbf{F}$ be Lipschitz.

We prove Theorem 3.1.10 after Example 3.1.15.

Theorem 3.1.10 (Showing that a locus is a smooth manifold).

1. *Let $U \subset \mathbb{R}^n$ be open, and let $\mathbf{F} : U \to \mathbb{R}^{n-k}$ be a C^1 mapping. Let M be a subset of $\mathbb{R}^n$ such that*

$$M \cap U = \{\, \mathbf{z} \in U \mid \mathbf{F}(\mathbf{z}) = \mathbf{0} \,\}. \tag{3.1.6}$$

If $[\mathbf{DF}(\mathbf{z})]$ is onto for every $\mathbf{z} \in M \cap U$, then $M \cap U$ is a smooth k-dimensional manifold embedded in $\mathbb{R}^n$. If every $\mathbf{z} \in M$ is in such a U, then M is a k-dimensional manifold.

2. *Conversely, if M is a smooth k-dimensional manifold embedded in $\mathbb{R}^n$, then every point $\mathbf{z} \in M$ has a neighborhood $U \subset \mathbb{R}^n$ such that there exists a C^1 mapping $\mathbf{F} : U \to \mathbb{R}^{n-k}$ with $[\mathbf{DF}(\mathbf{z})]$ onto and $M \cap U = \{\, \mathbf{y} \mid \mathbf{F}(\mathbf{y}) = \mathbf{0} \,\}$.*

Remark. We cannot use Theorem 3.1.10 to determine that a locus is *not* a smooth manifold; a locus defined by $\mathbf{F}(\mathbf{z}) = \mathbf{0}$ may be a smooth manifold even though $[\mathbf{DF}(\mathbf{z})]$ is not onto. For example, the set defined by $F\begin{pmatrix} x \\ y \end{pmatrix} = 0$, where $F\begin{pmatrix} x \\ y \end{pmatrix} = (x^2+y^2-1)^2$, is the unit circle "counted twice"; it is a smooth manifold, but the derivative of F is not onto, since

$$\left[\mathbf{D}F\begin{pmatrix} x \\ y \end{pmatrix}\right] = 2(x^2+y^2-1)[2x,\, 2y] = 0. \qquad 3.1.7$$

More generally, if $\mathbf{F}(\mathbf{z}) = \mathbf{0}$ defines a manifold, then $(\mathbf{F}(\mathbf{z}))^2 = \mathbf{0}$ defines the same manifold, but its derivative is always 0. △

Algebraic geometers call the locus of equation

$$F\begin{pmatrix} x \\ y \end{pmatrix} = (x^2+y^2-1)^2$$

a *double circle*. In algebraic geometry, it is a different locus from the unit circle.

Example 3.1.11 (Determining that a locus is a smooth curve). For a one-dimensional manifold in $\mathbb{R}^2$ (i.e., a plane curve), the function of Theorem 3.1.10 goes from an open subset of $\mathbb{R}^2$ to $\mathbb{R}$, and the requirement that $[\mathbf{DF}(\mathbf{z})]$ be onto for every $\mathbf{z} \in M$ is equivalent to requiring that

$$\left[\mathbf{D}F\begin{pmatrix} a \\ b \end{pmatrix}\right] \neq [0\,,\,0] \quad \text{for all} \quad \mathbf{a} = \begin{pmatrix} a \\ b \end{pmatrix} \in M. \qquad 3.1.8$$

For instance, we have no idea what the locus X_c defined by

$$x^8 + 2x^3 + y + y^5 = c \qquad 3.1.9$$

looks like, but we know it is a smooth curve for all c, since the derivative of the function $F\begin{pmatrix} x \\ y \end{pmatrix} = x^8 + 2x^3 + y + y^5 - c$ is

$$\left[\mathbf{D}F\begin{pmatrix} x \\ y \end{pmatrix}\right] = [\underbrace{8x^7 + 6x^2}_{D_1F},\, \underbrace{1 + 5y^4}_{D_2F}], \qquad 3.1.10$$

and the second entry is never 0. △

Example 3.1.11: Here, saying that

$$[\mathbf{D}F(\mathbf{z})] = [D_1F(\mathbf{z}),\, D_2F(\mathbf{z})],$$

is onto means that any real number can be expressed as a linear combination $\alpha D_1F(\mathbf{a}) + \beta D_2F(\mathbf{a})$ for some $\begin{pmatrix} \alpha \\ \beta \end{pmatrix} \in \mathbb{R}^2$. Thus, the following statements mean the same thing: For all $\mathbf{a} \in M$,

1. $[\mathbf{D}F(\mathbf{a})]$ is onto.
2. $[\mathbf{D}F(\mathbf{a})] \neq [0\,,\,0]$.
3. At least one of $D_1F(\mathbf{a})$ or $D_2F(\mathbf{a})$ is not 0.

More generally, for an $(n-1)$-dimensional manifold in any $\mathbb{R}^n$, $\mathbb{R}^{n-k} = \mathbb{R}$, and saying that

$$[\mathbf{D}F(\mathbf{z})] = [D_1F(\mathbf{z}), \dots, D_n(\mathbf{z})]$$

is onto is equivalent to saying that at least one partial derivative is not 0.

Example 3.1.12 (Smooth curve: second example). Consider the function $F\begin{pmatrix} x \\ y \end{pmatrix} = x^4 + y^4 + x^2 - y^2$. We have

$$\left[\mathbf{D}F\begin{pmatrix} x \\ y \end{pmatrix}\right] = [\underbrace{4x^3 + 2x}_{D_1F},\, \underbrace{4y^3 - 2y}_{D_2F}] = \left[2x(2x^2+1),\; 2y(2y^2-1)\right]. \qquad 3.1.11$$

The only places where both partials vanish are $\begin{pmatrix} 0 \\ 0 \end{pmatrix}, \begin{pmatrix} 0 \\ \pm 1/\sqrt{2} \end{pmatrix}$, where F has values 0 and $-1/4$. Thus the locus of equation $x^4 + y^4 + x^2 - y^2 = c$ is a smooth curve for any number $c \neq 0$ and $c \neq -1/4$.

Figure 3.1.10 illustrates this. The locus of equation $F\begin{pmatrix} x \\ y \end{pmatrix} = -1/4$ consists of two points; that of equation $F\begin{pmatrix} x \\ y \end{pmatrix} = 0$ is a figure eight. The others really are things one would want to call smooth curves. △

Remark. Note that the function F in the equation $F\begin{pmatrix} x \\ y \end{pmatrix} = c$ is of a different species than the functions $f(x) = \sqrt{1-x^2}$ and $g(y) = \sqrt{1-y^2}$ used to show that the unit circle is a smooth curve. The function F plays the role of $\mathbf{F}$ in the implicit function theorem; the functions f and g play the role of the implicit function $\mathbf{g}$ in the implicit function theorem. If the conditions of Theorem 3.1.10 are met, then $F(\mathbf{x}) = c$ implicitly expresses x in terms of y or y in terms of x: the locus is the graph of the implicit function g or the implicit function f. $\triangle$

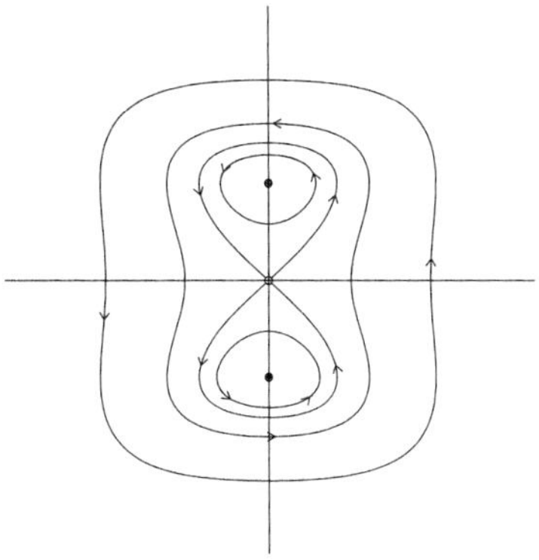

FIGURE 3.1.10.
The locus of equation

$$x^4 + y^4 + x^2 - y^2 = -1/4$$

consists of the two points on the y-axis where $y = \pm 1/\sqrt{2}$; it is not a smooth curve (but it is a zero-dimensional manifold). Nor is the figure eight a smooth curve: near the origin it looks like two intersecting lines; to make it a smooth curve we would have to take out the point where the lines intersect.

The other curves are smooth curves. In addition, by our definition the empty set (the case here if $c < -1/4$) is also a smooth curve! Allowing the empty set to be a smooth curve makes a number of statements simpler.

The arrows on the lines are an artifact of the drawing program.

A locus defined by an equation of the form $F\begin{pmatrix} x \\ y \end{pmatrix} = c$ is called a *level curve*. One way to imagine such a locus is to think of cutting the surface that is the graph of F by the plane $z = c$, as shown in Figure 3.1.11.

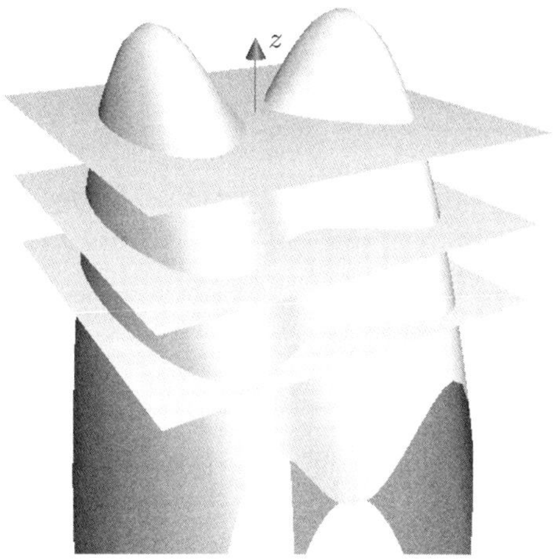

FIGURE 3.1.11. The surface that is the graph of $F\begin{pmatrix} x \\ y \end{pmatrix} = x^2 - 0.2x^4 - y^2$ is sliced horizontally by setting z equal to three different constants. The intersection of the surface and the plane $z = c$ used to slice it is known as a *level set* or *level curve*. This intersection is the same as the locus of equation $F\begin{pmatrix} x \\ y \end{pmatrix} = c$.

If we "sliced" the surface in Figure 3.1.11 by the plane $z = c$ where c is a maximum value of F, we would get a point, not a smooth curve. If we sliced it at a "saddle point" (also a point where the derivative of F is 0, so that the tangent plane is horizontal), we would get a figure eight, not a smooth curve. (Saddle points are discussed in Section 3.6.)

Example 3.1.13 (Smooth surface in $\mathbb{R}^3$). When we use Theorem 3.1.10 to determine whether a locus is a smooth surface in $\mathbb{R}^3$, the function $\mathbf{F}$ goes from an open subset of $\mathbb{R}^3$ to $\mathbb{R}$. In this case the requirement that $[\mathbf{DF}(\mathbf{z})]$ be onto for every $\mathbf{z} \in M$ is equivalent to requiring that for every $\mathbf{z} \in M$, we have $[\mathbf{DF}(\mathbf{z})] \neq [0,\, 0,\, 0]$; at least one partial derivative must not vanish.

For instance, consider the set X defined by the equation

$$F\begin{pmatrix} x \\ y \\ z \end{pmatrix} = \sin(x + yz) = 0. \qquad 3.1.12$$

Is it a smooth surface? The derivative is

$$\left[\mathbf{D}F\begin{pmatrix}a\\b\\c\end{pmatrix}\right] = [\underbrace{\cos(a+bc)}_{D_1F}, \underbrace{c\cos(a+bc)}_{D_2F}, \underbrace{b\cos(a+bc)}_{D_3F}]. \qquad 3.1.13$$

You should be impressed by Example 3.1.13. The implicit function theorem is hard to prove, but the work pays off. Without having any idea what the set defined by equation 3.1.12 might look like, we were able to determine, with hardly any effort, that it is a smooth surface.

Figuring out what the surface looks like – or even whether the set is empty – is another matter. Exercise 3.1.6 asks you to figure out what it looks like in this case, but usually this kind of thing can be quite hard indeed.

On X, by definition, $\sin(a+bc) = 0$, so $\cos(a+bc) \neq 0$, so X is a smooth surface. (In this case, the first partial derivative is never 0, so at all points of the surface, the surface is the graph of a function representing x locally as a function of y and z.) $\triangle$

Example 3.1.14 (Smooth curves in $\mathbb{R}^3$). Theorem 3.1.10 suggests that a natural way to think of a curve C in $\mathbb{R}^3$ is as the intersection of two surfaces. Here $n = 3$, $k = 1$. If surfaces S_1 and S_2 are given by equations $F_1(\mathbf{z}) = 0$ and $F_2(\mathbf{z}) = 0$, each function going from $\mathbb{R}^3$ to $\mathbb{R}$, then $C = S_1 \cap S_2$ is given by the equation $\mathbf{F}(\mathbf{z}) = \mathbf{0}$, where

$$\mathbf{F}(\mathbf{z}) = \begin{pmatrix}F_1(\mathbf{z})\\F_2(\mathbf{z})\end{pmatrix} \text{ is a mapping from } U \to \mathbb{R}^2, \text{ with } U \subset \mathbb{R}^3 \text{ open.}$$

Thus $[\mathbf{DF}(\mathbf{z})]$ is a 2×3 matrix:

$$[\mathbf{DF}(\mathbf{z})] = \begin{bmatrix} D_1F_1(\mathbf{z}) & D_2F_1(\mathbf{z}) & D_3F_1(\mathbf{z}) \\ D_1F_2(\mathbf{z}) & D_2F_2(\mathbf{z}) & D_3F_2(\mathbf{z}) \end{bmatrix} = \begin{bmatrix} [\mathbf{D}F_1(\mathbf{z})] \\ [\mathbf{D}F_2(\mathbf{z})] \end{bmatrix}. \qquad 3.1.14$$

For the 2×3 matrix $[\mathbf{DF}(\mathbf{z})]$ to be onto, the two rows must be linearly independent (see the equivalent statements about onto linear transformations in Section 2.5). Geometrically, this means that at $\mathbf{z}$, the tangent plane to the surface S_1 and the tangent plane to the surface S_2 must be distinct, i.e., the surfaces must not be tangent to each other at $\mathbf{z}$. In that case, the intersection of the two tangent planes is the tangent line to the curve.

The first row of the matrix is the derivative $[\mathbf{D}F_1(\mathbf{z})]$. If it is onto (if any of the three entries of that row is nonzero), then S_1 is a smooth surface. Similarly, if the derivative $[\mathbf{D}F_2(\mathbf{z})]$ is onto, then S_2 is a smooth surface. Thus if only one entry of the matrix is nonzero, the curve is the intersection of two surfaces, at least one of which is smooth. If any entry of the first row is nonzero and any entry of the second row is nonzero, then the curve is the intersection of two smooth surfaces. But the intersection of two smooth surfaces is not necessarily a smooth curve.

For instance, consider the intersection of the surfaces given by $F_1 = 0$ and $F_2 = 0$ where

$$F_1\begin{pmatrix}z_1\\z_2\\z_3\end{pmatrix} = z_3 - z_2^2 \quad \text{and} \quad F_2\begin{pmatrix}z_1\\z_2\\z_3\end{pmatrix} = z_3 - z_1z_2. \qquad 3.1.15$$

Then $[\mathbf{DF}(\mathbf{z})] = \begin{bmatrix} 0 & -2z_2 & 1 \\ -z_2 & -z_1 & 1 \end{bmatrix}$, so the curve given by $\mathbf{F}(\mathbf{z}) = \mathbf{0}$ is the intersection of two smooth surfaces. But at the origin the derivative is not onto $\mathbb{R}^2$, so we cannot conclude that the curve is smooth. $\triangle$

Example 3.1.15 (Checking that the linkage space is a manifold). In Example 3.1.8, X_2 is the set of positions of four rigid rods restricted to the plane. This locus is given by the equation

$$\mathbf{f}(\mathbf{z}) = \mathbf{f}\begin{pmatrix} x_1 \\ y_1 \\ x_2 \\ y_2 \\ x_3 \\ y_3 \\ x_4 \\ y_4 \end{pmatrix} = \begin{bmatrix} (x_2 - x_1)^2 + (y_2 - y_1)^2 - l_1^2 \\ (x_3 - x_2)^2 + (y_3 - y_2)^2 - l_2^2 \\ (x_4 - x_3)^2 + (y_4 - y_3)^2 - l_3^2 \\ (x_1 - x_4)^2 + (y_1 - y_4)^2 - l_4^2 \end{bmatrix} = \begin{bmatrix} 0 \\ 0 \\ 0 \\ 0 \end{bmatrix}. \qquad 3.1.16$$

Each partial derivative at right is a vector with four entries. For example,

$$\vec{D_1}\mathbf{f}(\mathbf{z}) = \begin{bmatrix} D_1 f_1(\mathbf{z}) \\ D_1 f_2(\mathbf{z}) \\ D_1 f_3(\mathbf{z}) \\ D_1 f_4(\mathbf{z}) \end{bmatrix}$$

and so on.

Equation 3.1.17: We had to put the matrix on two lines to make it fit. The second line contains the last four columns of the matrix.

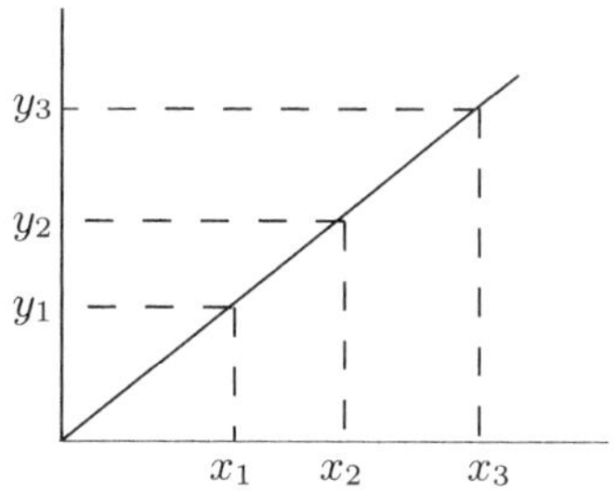

FIGURE 3.1.12.

If the points

$$\begin{pmatrix} x_1 \\ y_1 \end{pmatrix}, \begin{pmatrix} x_2 \\ y_2 \end{pmatrix}, \begin{pmatrix} x_3 \\ y_3 \end{pmatrix}$$

are aligned, then the first two columns of equation 3.1.18 cannot be linearly independent: $y_1 - y_2$ is necessarily a multiple of $x_1 - x_2$, and $y_2 - y_3$ is a multiple of $x_2 - x_3$.

The derivative is composed of the eight partial derivatives (in the second line we label the partial derivatives explicitly by the names of the variables):

$$[\mathbf{Df}(\mathbf{z})] = [\vec{D_1}\mathbf{f}(\mathbf{z}), \vec{D_2}\mathbf{f}(\mathbf{z}), \vec{D_3}\mathbf{f}(\mathbf{z}), \vec{D_4}\mathbf{f}(\mathbf{z}), \vec{D_5}\mathbf{f}(\mathbf{z}), \vec{D_6}\mathbf{f}(\mathbf{z}), \vec{D_7}\mathbf{f}(\mathbf{z}), \vec{D_8}\mathbf{f}(\mathbf{z})]$$

$$= [\vec{D}_{x_1}\mathbf{f}(\mathbf{z}), \vec{D}_{y_1}\mathbf{f}(\mathbf{z}), \vec{D}_{x_2}\mathbf{f}(\mathbf{z}), \vec{D}_{y_2}\mathbf{f}(\mathbf{z}), \vec{D}_{x_3}\mathbf{f}(\mathbf{z}), \vec{D}_{y_3}\mathbf{f}(\mathbf{z}), \vec{D}_{x_4}\mathbf{f}(\mathbf{z}), \vec{D}_{y_4}\mathbf{f}(\mathbf{z})].$$

Computing the partial derivatives gives

$$[\mathbf{Df}(\mathbf{z})] = \left[\begin{matrix} 2(x_1 - x_2) & 2(y_1 - y_2) & -2(x_1 - x_2) & -2(y_1 - y_2) \\ 0 & 0 & 2(x_2 - x_3) & 2(y_2 - y_3) \\ 0 & 0 & 0 & 0 \\ -2(x_4 - x_1) & -2(y_4 - y_1) & 0 & 0 \end{matrix}\right. \qquad 3.1.17$$

$$\left.\begin{matrix} 0 & 0 & 0 & 0 \\ -2(x_2 - x_3) & -2(y_2 - y_3) & 0 & 0 \\ 2(x_3 - x_4) & 2(y_3 - y_4) & -2(x_3 - x_4) & -2(y_3 - y_4) \\ 0 & 0 & 2(x_4 - x_1) & 2(y_4 - y_1) \end{matrix}\right].$$

Since $\mathbf{f}$ is a mapping from $\mathbb{R}^8$ to $\mathbb{R}^4$, the derivative $[\mathbf{Df}(\mathbf{z})]$ is onto if four of its columns are linearly independent. For instance, here you can never use the first four, or the last four, because in both cases there is a row of zeros. How about the third, fourth, seventh, and eighth, i.e., the points $\mathbf{x}_2 = \begin{pmatrix} x_2 \\ y_2 \end{pmatrix}, \mathbf{x}_4 = \begin{pmatrix} x_4 \\ y_4 \end{pmatrix}$? These work as long as the corresponding columns of the matrix

$$\underbrace{\begin{bmatrix} -2(x_1 - x_2) & -2(y_1 - y_2) & 0 & 0 \\ 2(x_2 - x_3) & 2(y_2 - y_3) & 0 & 0 \\ 0 & 0 & -2(x_3 - x_4) & -2(y_3 - y_4) \\ 0 & 0 & 2(x_4 - x_1) & 2(y_4 - y_1) \end{bmatrix}}_{D_{x_2}\mathbf{f}(\mathbf{z}) \quad D_{y_2}\mathbf{f}(\mathbf{z}) \quad D_{x_4}\mathbf{f}(\mathbf{z}) \quad D_{y_4}\mathbf{f}(\mathbf{z})} \qquad 3.1.18$$

are linearly independent. The first two columns are linearly independent precisely when $\mathbf{x}_1$, $\mathbf{x}_2$, and $\mathbf{x}_3$ are not aligned as they are in Figure 3.1.12, and the last two are linearly independent when $\mathbf{x}_3$, $\mathbf{x}_4$, and $\mathbf{x}_1$ are not aligned. The same argument holds for the first, second, fifth, and sixth columns, corresponding to $\mathbf{x}_1$ and $\mathbf{x}_3$. Thus you can use the positions of opposite points to locally parametrize X_2, as long as the other two points are aligned with neither of the two opposite points. The points are never all four in line, unless either one length is the sum of the other three, or

$l_1+l_2=l_3+l_4$, or $l_2+l_3=l_4+l_1$. In all other cases, X_2 is a manifold, and even in these last two cases, it is a manifold except perhaps at the positions where all four rods are aligned. △

Proof of Theorem 3.1.10. 1. This follows immediately from the implicit function theorem.

2. If M is a smooth k-dimensional manifold embedded in $\mathbb{R}^n$, then locally (in a neighborhood of any fixed point $\mathbf{c} \in M$) it is the graph of a C^1 mapping $\mathbf{f}$ expressing $n-k$ variables $z_{j_1}, \dots, z_{j_{n-k}}$ as functions of the other k variables $z_{i_1}, \dots, z_{i_k}$. Set

When the $n-k$ passive variables are the first $n-k$ variables of $\mathbf{z}$, we have $\mathbf{z} = \begin{pmatrix} \mathbf{x} \\ \mathbf{y} \end{pmatrix}$.

$$\mathbf{x} = \underbrace{\begin{pmatrix} z_{j_1} \\ \vdots \\ z_{j_{n-k}} \end{pmatrix}}_{\text{passive variables}} \quad \text{and} \quad \mathbf{y} = \underbrace{\begin{pmatrix} z_{i_1} \\ \vdots \\ z_{i_k} \end{pmatrix}}_{\text{active variables}}. \qquad 3.1.19$$

The manifold is then the locus of equation $\mathbf{F}(\mathbf{z}) = \mathbf{x} - \mathbf{f}(\mathbf{y}) = \vec{\mathbf{0}}$. The derivative is onto $\mathbb{R}^{n-k}$ because it has $n-k$ pivotal (linearly independent) columns, corresponding to the $n-k$ passive variables. □

Manifolds are independent of coordinates

Our definition of a manifold has a serious weakness: it makes explicit reference to a particular coordinate system. Here we remove this restriction: Corollary 3.1.17 says that smooth manifolds can be rotated, translated, and even linearly distorted and remain smooth manifolds. It follows from Theorem 3.1.16, which says a great deal more: the *inverse image* of a smooth manifold by an arbitrary C^1 mappings with surjective derivative is still a smooth manifold.

In Theorem 3.1.16, $\mathbf{f}^{-1}$ is not an inverse mapping; since $\mathbf{f}$ goes from $\mathbb{R}^n$ to $\mathbb{R}^m$, such an inverse map does not exist when $n \neq m$. By $\mathbf{f}^{-1}(M)$ we denote the inverse image: the set of points $\mathbf{x} \in \mathbb{R}^n$ such that $\mathbf{f}(\mathbf{x})$ is in M (see Definition 0.4.9 and Proposition 0.4.11).

Theorem 3.1.16 (Inverse image of manifold by a C^1 mapping). *Let $M \subset \mathbb{R}^m$ be a k-dimensional manifold, U an open subset of $\mathbb{R}^n$, and $\mathbf{f}: U \to \mathbb{R}^m$ a C^1 mapping whose derivative $[\mathbf{Df}(\mathbf{x})]$ is surjective at every $\mathbf{x} \in \mathbf{f}^{-1}(M)$. Then the inverse image $\mathbf{f}^{-1}(M)$ is a submanifold of $\mathbb{R}^n$ of dimension $k+n-m$.*

Proof. Let $\mathbf{x}$ be a point of $\mathbf{f}^{-1}(M)$. By part 2 of Theorem 3.1.10, there exists a neighborhood V of $\mathbf{f}(\mathbf{x})$ such that $M \cap V$ is defined by the equation $\mathbf{F}(\mathbf{y}) = \mathbf{0}$, where $\mathbf{F}: V \to \mathbb{R}^{m-k}$ is a C^1 mapping whose derivative $[\mathbf{DF}(\mathbf{y})]$ is onto for every $\mathbf{y} \in M \cap V$.

Then $W \overset{\text{def}}{=} \mathbf{f}^{-1}(V)$ is an open neighborhood of $\mathbf{x}$, and $\mathbf{f}^{-1}(M) \cap W$ is defined by the equation $\mathbf{F} \circ \mathbf{f} = \mathbf{0}$. By the chain rule,

$$[\mathbf{D}(\mathbf{F} \circ \mathbf{f})(\mathbf{x})] = [\mathbf{DF}(\mathbf{f}(\mathbf{x}))][\mathbf{Df}(\mathbf{x})], \qquad 3.1.20$$

and both $[\mathbf{DF}(\mathbf{f}(\mathbf{x}))]$ and $[\mathbf{Df}(\mathbf{x})]$ are surjective for every $\mathbf{x} \in \mathbf{f}^{-1}(M) \cap W$, so we have verified the condition of part 1 of Theorem 3.1.10. Still by part 1 of Theorem 3.1.10, the dimension of $\mathbf{f}^{-1}(M)$ is the dimension of the domain of $\mathbf{F} \circ \mathbf{f}$ minus the dimension of the codomain, i.e., $n - (m - k)$. □

Direct images are more troublesome. In Corollary 3.1.17 we take direct images by invertible mappings, so that the direct image is the inverse image by the inverse.

Proof of Corollary 3.1.17: Set $\mathbf{f} = \mathbf{g}^{-1}$, so that

$$\mathbf{f}(\mathbf{y}) = A^{-1}(\mathbf{y} - \mathbf{c}).$$

Then $\mathbf{g}(M) = \mathbf{f}^{-1}(M)$, and Theorem 3.1.16 applies.

Corollary 3.1.17 (Manifolds are independent of coordinates). *Let $\mathbf{g} : \mathbb{R}^n \to \mathbb{R}^n$ be a mapping of the form*

$$\mathbf{g}(\mathbf{x}) = A\mathbf{x} + \mathbf{c}, \qquad 3.1.21$$

where A is an invertible matrix. If $M \subset \mathbb{R}^n$ is a smooth k-dimensional manifold, then the direct image $\mathbf{g}(M)$ is also a smooth k-dimensional manifold.

Thus our definition of a manifold, which appeared to be tied to the coordinate system, is in fact independent of the choice of coordinates. In particular, if you rotate or translate a smooth manifold, the result is still a smooth manifold.

Parametrizations of manifolds

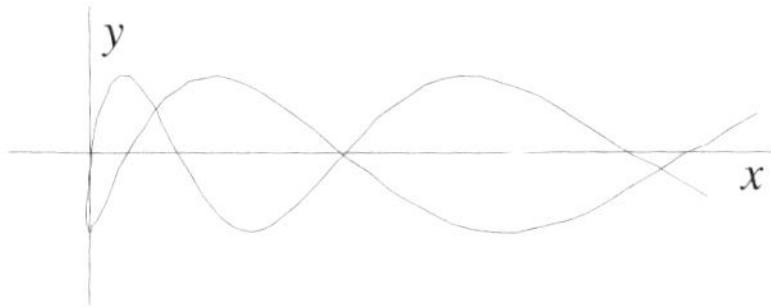

FIGURE 3.1.13.

A curve in the plane, known by the parametrization

$$t \mapsto \begin{pmatrix} x = t^2 - \sin t \\ y = -6 \sin t \cos t \end{pmatrix}.$$

So far we have considered manifolds as loci defined by sets of equations. Technically, such equations describe the manifold completely. In practice (as we saw in equation 3.1.5) such a description is not satisfying; the information is not in a form that can be understood as a picture of the manifold.

There is another way to think of manifolds: *parametrizations*. It is generally far easier to get a picture of a curve or surface if you know it by a parametrization than if you know it by equations. It will generally take a computer milliseconds to compute the coordinates of enough points to give you a good picture of a parametrized curve or surface.

When the word "parametrize" is used loosely, we can say that any map $\mathbf{f} : \mathbb{R}^n \to \mathbb{R}^m$ parametrizes its image by the domain variable in $\mathbb{R}^n$. The map

$$g : t \mapsto \begin{pmatrix} t^2 - \sin t \\ -6 \sin t \cos t \end{pmatrix} \qquad 3.1.22$$

parametrizes the curve shown in Figure 3.1.13. As the parameter t varies, a point in $\mathbb{R}^2$ is chosen; g parametrizes the curve made up of all points in the image of g. Note that often when thinking of functions as parametrizations, the "maps to" notation $\mapsto$ is used, but we could also write this parametrization as $g(t) = \begin{pmatrix} t^2 - \sin t \\ -6 \sin t \cos t \end{pmatrix}$.

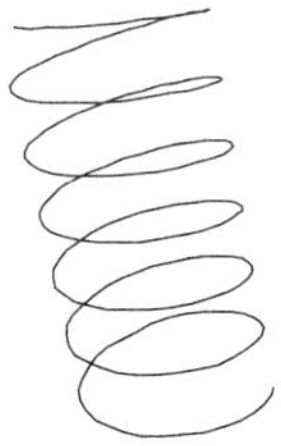

FIGURE 3.1.14.

A curve in space, known by the parametrization $t \mapsto \begin{pmatrix} \cos t \\ \sin t \\ at \end{pmatrix}$.

The mappings $t \mapsto \begin{pmatrix} \cos t \\ \sin t \\ at \end{pmatrix}$ and $\begin{pmatrix} u \\ v \end{pmatrix} \mapsto \begin{pmatrix} u^3 \cos v \\ u^2 + v^2 \\ v^2 \cos u \end{pmatrix}$ parametrize the space curve shown in Figure 3.1.14 and the surface of Figure 3.1.15. The most famous parametrization of surfaces parametrizes the unit sphere in $\mathbb{R}^3$ by latitude u and longitude v:

$$\begin{pmatrix} u \\ v \end{pmatrix} \mapsto \begin{pmatrix} \cos u \cos v \\ \cos u \sin v \\ \sin u \end{pmatrix}. \qquad 3.1.23$$

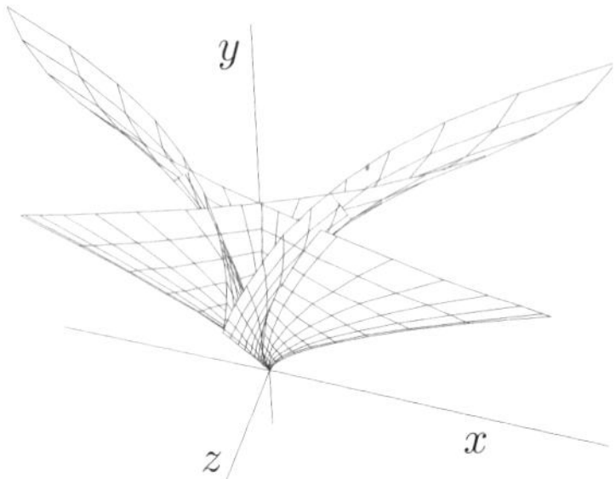

FIGURE 3.1.15.
The surface "parametrized" by

$$\begin{pmatrix} u \\ v \end{pmatrix} \mapsto \begin{pmatrix} u^3 \cos v \\ u^2 + v^2 \\ v^2 \cos u \end{pmatrix}.$$

Note that this is not a smooth surface; it intersects itself.

Using a computer, you can easily produce as many "parametrized" curves and surfaces as you like. If you fill in the blanks of $t \mapsto \begin{pmatrix} - \\ - \end{pmatrix}$, where each blank represents a function of t, and ask a computer to plot it, it will draw a curve in the plane. If you fill in the blanks of $\begin{pmatrix} u \\ v \end{pmatrix} \mapsto \begin{pmatrix} - \\ - \\ - \end{pmatrix}$, where each blank represents a function of u and v, the computer will draw you a surface in $\mathbb{R}^3$.

However, if we want a mapping to parametrize a manifold, we must be much more demanding. In particular, such a mapping must be one to one; we cannot allow our manifolds to intersect themselves, like the surface in Figure 3.1.15.

Definition 3.1.18 (Parametrization of a manifold). A *parametrization* of a k-dimensional manifold $M \subset \mathbb{R}^n$ is a mapping $\gamma : U \subset \mathbb{R}^k \to M$ satisfying the following conditions:

1. U is open.
2. γ is C^1, one to one, and onto M.
3. $[\mathbf{D}\gamma(\mathbf{u})]$ is one to one for every $\mathbf{u} \in U$.

In Chapters 5 and 6 we will need to parametrize our manifolds before we can integrate over them. Fortunately, for the purposes of integration we will be able to "relax" our definition of a parametrization, disregarding such trouble spots as the endpoints of the tubing or the curve going from the North Pole to the South Pole.

We will give some examples, but first, some comments.

1. *Finding a parametrization for a manifold that you know by equations is very hard, and often impossible.* Just as it is rare for a manifold to be the graph of a single function, it is rare to find a map γ that meets the criteria of Definition 3.1.18 and parametrizes the entire manifold. A circle is not like an open interval: if you bend a strip of tubing into a circle, the two endpoints become a single point. A cylinder is not like an open subspace of the plane: if you roll up a piece of paper into a cylinder, two edges become a single line. Neither parametrization is one to one.

   The problem for the sphere is worse. The parametrization by latitude and longitude (equation 3.1.23) satisfies Definition 3.1.18 only if we remove the curve going from the North Pole to the South Pole through Greenwich (for example). There is no general rule for solving such problems.

   The one easy case is when a manifold is the graph of a single function. For example, the surface in the upper left of Figure 3.1.2

Choosing even a local parametrization that is well adapted to the problem at hand is a difficult and important skill and exceedingly difficult to teach. When parametrizing a surface, it sometimes helps to have a computer draw the surface for you; then ask yourself whether there is some way to identify a point on the surface by coordinates (for example, using sine and cosine or latitude and longitude).

is the graph of the function $f\begin{pmatrix} x \\ y \end{pmatrix} = x^3 - 2xy^2$. This surface is parametrized by

$$\mathbf{g}\begin{pmatrix} x \\ y \end{pmatrix} = \begin{pmatrix} x \\ y \\ x^3 - 2xy^2 \end{pmatrix}. \qquad 3.1.24$$

Whenever a manifold is the graph of a function $\mathbf{f}(\mathbf{x}) = \mathbf{y}$, it is parametrized by the mapping $\mathbf{x} \mapsto \begin{pmatrix} \mathbf{x} \\ \mathbf{f}(\mathbf{x}) \end{pmatrix}$.

William Thurston (1946–2012), arguably the best geometer of the twentieth century, said that the right way to know a manifold is from the inside. Imagine yourself inside the manifold, aiming a flashlight first at one spot, then another, through a mist or dust that makes the beam of light visible. If you point the flashlight straight ahead, will you see anything? Will anything be reflected back? Or will you see the light to your side? This approach is illustrated by the video "Not Knot", produced by The Geometry Center, University of Minnesota, and available from CRCPress.com.

2. *Even proving that a candidate parametrization is one to one can be very hard.* An example concerns minimal surfaces (a minimal surface is one that locally minimizes surface area among surfaces with the same boundary; the study of such surfaces was inspired by soap films and water droplets). Weierstrass found a very clever way to "parametrize" all minimal surfaces, but in the more than one hundred years since these mappings were proposed, no one has succeeded in giving a general method for checking that they are indeed one to one. Just showing that one such mapping is a parametrization is quite a feat.
3. *Checking that a derivative is one to one is easy*, and *locally*, the derivative is a good guide to the mapping it approximates. Thus if U is an open subset of $\mathbb{R}^k$ and $\gamma : U \to \mathbb{R}^n$ is a smooth mapping such that $[\mathbf{D}\gamma(\mathbf{u})]$ is one to one, then it is not necessarily true that $\gamma(U)$ is a smooth k-dimensional manifold. But it is true locally: if U is small enough, then the image of the corresponding γ will be a smooth manifold. Exercise 3.1.20 sketches how to prove this.

One program that draws parametrized curves and surfaces is Graphing Calculator by Pacific Tech (www.pacifict.com), version 2.2 or higher. (It also draws surfaces given implicitly by an equation, which is much harder.)

Example 3.1.19 (Parametrization of a curve). In the case of a curve, the open subset $U \subset \mathbb{R}^k$ of Definition 3.1.18 becomes an open interval $I \subset \mathbb{R}$. Think of I as a time interval; as you travel along the curve, the parametrization tells you where you are on the curve at a given time, as shown in Figure 3.1.16. In this interpretation, $\gamma'(t)$ is the *velocity vector*; it is tangent to the curve at $\gamma(t)$ and its length is the speed at which you are traveling at time t.

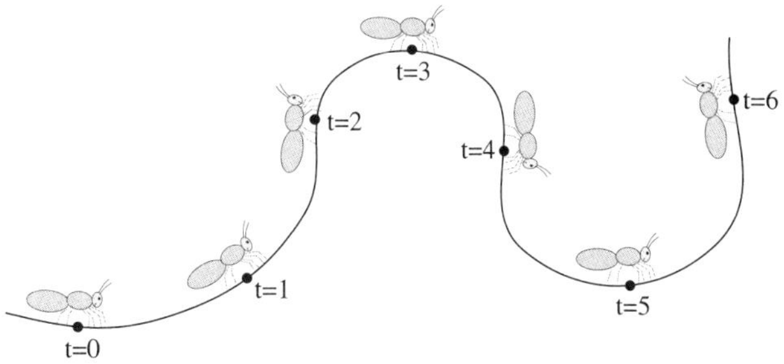

FIGURE 3.1.16. We imagine a parametrized curve as an ant taking a walk in the plane or in space. The parametrization tells where the ant is at any given time.

When you get to Yeniköy look at your mileage reading, then drive on towards Sariyer for exactly four more miles.—Parametrizing the road to Sariyer by arc length (Eric Ambler, *The Light of Day*).

Tangents are discussed in detail in Section 3.2.

When a curve is parametrized, the requirement that $\vec{\gamma}'(t) \neq \vec{\mathbf{0}}$ could also be stated in terms of linear independence: a single vector $\vec{\mathbf{v}}$ is linearly independent if $\vec{\mathbf{v}} \neq \vec{\mathbf{0}}$.

Since the mapping $\gamma : U \to M$ in Example 3.1.20 is one to one and onto, it is invertible. This may seem surprising, since an element of U has two entries and an element of M has three. But both domain and codomain are two dimensional. It takes two numbers, not three, to specify a point in M. The key point is that the codomain of γ is not $\mathbb{R}^3$, or any open subset of $\mathbb{R}^3$; it is a subset of the 2-sphere.

Curves can also be parametrized by arc length (see Definition 3.9.5). The odometer in a car parametrizes curves by arc length.

The requirement that $[\mathbf{D}\gamma(\mathbf{u})]$ be one to one means $\vec{\gamma}'(t) \neq \vec{\mathbf{0}}$: since $\vec{\gamma}'(t) = [\mathbf{D}\gamma(t)]$ is an $n \times 1$ column matrix (i.e., vector), the linear transformation given by this matrix is one to one exactly when $\vec{\gamma}'(t) \neq \vec{\mathbf{0}}$.

The function $\gamma : (0, \pi) \to C$ defined by $\gamma(\theta) = \begin{pmatrix} \cos\theta \\ \sin\theta \end{pmatrix}$ parametrizes the upper half of the unit circle. As defined on the open interval $(0, \pi)$, the function is bijective, and $\vec{\gamma}'(\theta) = \begin{pmatrix} -\sin\theta \\ \cos\theta \end{pmatrix} \neq \begin{pmatrix} 0 \\ 0 \end{pmatrix}$. Note that according to Definition 3.1.18, this mapping is not a parametrization of the entire circle, since in order to make it one to one we must restrict its domain to $(0, 2\pi)$; unfortunately, this restriction misses the point $\begin{pmatrix} 1 \\ 0 \end{pmatrix}$. △

Example 3.1.20 (Parametrization of a surface). In the case of a surface in $\mathbb{R}^3$, the U of Definition 3.1.18 is an open subset of $\mathbb{R}^2$, and $[\mathbf{D}\gamma(\mathbf{u})]$ is a 3×2 matrix. Saying that it is one to one is the same as saying that $\overrightarrow{D_1\gamma}$ and $\overrightarrow{D_2\gamma}$ are linearly independent. For instance, the mapping

$$\gamma : \begin{pmatrix} \theta \\ \varphi \end{pmatrix} \mapsto \begin{pmatrix} \cos\theta\cos\varphi \\ \sin\theta\cos\varphi \\ \sin\varphi \end{pmatrix} \qquad 3.1.25$$

defined on $U = (0, \pi) \times (0, \pi/2)$ parametrizes the quarter of the unit sphere $x^2 + y^2 + z^2 = 1$ where $y, z > 0$, which we will denote by M. The map γ is of class C^1 and the derivative

$$\left[\mathbf{D}\gamma\begin{pmatrix} \theta \\ \varphi \end{pmatrix}\right] = \begin{bmatrix} -\cos\varphi\sin\theta & -\sin\varphi\cos\theta \\ \cos\theta\cos\varphi & -\sin\varphi\sin\theta \\ 0 & \cos\varphi \end{bmatrix} \qquad 3.1.26$$

is one to one because the entries in the third row ensure that the two columns are linearly independent. To check that γ is one to one and onto M, first note that the image of γ is part of the unit sphere, since

$$\cos^2\theta\cos^2\varphi + \sin^2\theta\cos^2\varphi + \sin^2\varphi = 1. \qquad 3.1.27$$

Since $z \in (0, 1)$, there exists a unique $\varphi \in (0, \pi/2)$ such that $\sin\varphi = z$, and since $x \in (-1, 1)$, there then exists a unique $\theta \in (0, \pi)$ such that

$$\cos\theta = \frac{x}{\cos\varphi} = \frac{x}{\sqrt{1 - z^2}} = \frac{x}{\sqrt{x^2 + y^2}}. \qquad 3.1.28$$

For $\theta \in (0, \pi)$, we have $\sin\theta > 0$, so $y = \sin\theta\cos\varphi > 0$. △

Parametrizations versus equations

If you know a curve by a global parametrization, it is easy to find points of the curve but difficult to check whether a given point is on the curve. If you know the curve by an equation, the opposite is true: it may be difficult to find points of the curve, but easy to check whether a point is on the curve.

For example, given the parametrization

$$\gamma : t \mapsto \begin{pmatrix} \cos^3 t - 3\sin t \cos t \\ t^2 - t^5 \end{pmatrix}, \qquad 3.1.29$$

you can find a point by substituting some value of t, like $t = 0$ or $t = 1$. But checking whether a point $\begin{pmatrix} a \\ b \end{pmatrix}$ is on the curve would require showing that there is a solution to the set of nonlinear equations

$$\begin{aligned} a &= \cos^3 t - 3\sin t \cos t \\ b &= t^2 - t^5. \end{aligned} \qquad 3.1.30$$

Now suppose you are given the equation

$$y + \sin xy + \cos(x + y) = 0, \qquad 3.1.31$$

which defines a different curve. You could check whether a given point is on the curve by inserting the values for x and y in the equation, but it's not clear how you would go about finding a point of the curve, i.e., solving one nonlinear equation in two variables. You might try fixing one variable and using Newton's method, but it isn't clear how to choose a starting point where convergence is guaranteed. For instance, you might ask whether the curve crosses the y-axis, i.e., set $x = 0$ and try to solve the equation $y + \cos y = 0$. This has a solution by the intermediate value theorem: $y + \cos y$ is positive when $y > 1$ and negative when $y < -1$. So you might think that using Newton's method starting at $y = 0$ should converge to a root, but inequality 2.8.53 of Kantorovich's theorem is not satisfied. But starting at $y = -\pi/4$ gives $\dfrac{M|f(y_0)|}{\left(f'(y_0)\right)^2} \le 0.027 < \dfrac{1}{2}$.

Relation between differential calculus and linear algebra

Knowing a manifold by an equation $\mathbf{F}(\mathbf{z}) = \mathbf{0}$ is analogous to (and harder than) knowing the kernel of a linear transformation. Knowing a manifold by a parametrization is analogous to (and harder than) knowing the image of a linear transformation. This continues a theme of this book: each construction and technique in linear algebra has a (harder) counterpart in the nonlinear world of differential calculus, as summarized in Table 3.1.17.

	Algorithms	Algebra	Geometry
Linear Algebra	row reduction	Theorem 2.2.6 Theorem 2.2.1	subspaces (flat) kernels images
Differential Calculus	Newton's method	inverse function theorem implicit function theorem	manifolds (curvy) defining manifolds by equations defining manifolds by a parametrizations

TABLE 3.1.17 Correspondences: linear algebra and differential calculus.

Exercises for Section 3.1

Exercise 3.1.1: Unless otherwise stated, the unit sphere is always centered at the origin. It has radius 1.

A number of exercises in this section have counterparts in Section 3.2, which deals with tangent spaces. Exercise 3.1.2 is related to Exercise 3.2.1; Exercise 3.1.5 to 3.2.3.

Exercise 3.1.6: Think that $\sin\alpha = 0$ if and only if $\alpha = k\pi$ for some integer k.

Part a of Exercise 3.1.7 does *not* require the implicit function theorem.

Hint for Exercise 3.1.8, part d: See Figure 3.9.2.

3.1.1 Use Theorem 3.1.10 to show that the unit sphere is a smooth surface.

3.1.2 Show that the set $\left\{ \begin{pmatrix} x \\ y \end{pmatrix} \in \mathbb{R}^2 \mid x + x^2 + y^2 = 2 \right\}$ is a smooth curve.

3.1.3 Show that every straight line in the plane is a smooth curve.

3.1.4 In Example 3.1.6, show that S^2 is a smooth surface, using $D_{x,y}$, $D_{x,z}$, and $D_{y,z}$; the half-axes $\mathbb{R}_z^+$, $\mathbb{R}_x^+$, $\mathbb{R}_y^+$, $\mathbb{R}_z^-$, $\mathbb{R}_x^-$, and $\mathbb{R}_y^-$; and the mappings

$$\pm\sqrt{1-x^2-y^2}, \quad \pm\sqrt{1-x^2-z^2} \quad \text{and} \quad \pm\sqrt{1-y^2-z^2}.$$

3.1.5 a. For what values of c is the set of equation $X_c = x^2 + y^3 = c$ a smooth curve?

b. Sketch these curves for a representative sample of values of c.

3.1.6 What does the surface of equation $F\begin{pmatrix} x \\ y \\ z \end{pmatrix} = \sin(x+yz) = 0$ in Example 3.1.13 look like?

3.1.7 a. Show that for all a and b, the sets X_a and Y_b of equations

$$x^2 + y^3 + z = a \quad \text{and} \quad x + y + z = b,$$

are smooth surfaces in $\mathbb{R}^3$.

b. For what values of a and b does Theorem 3.1.10 guarantee that the intersection $X_a \cap Y_b$ is a smooth curve? For the other values of a and b, what geometric relation is there between the surfaces X_a and Y_b?

3.1.8 a. For what values of a and b are the sets X_a and Y_b of equations $x - y^2 = a$ and $x^2 + y^2 + z^2 = b$ smooth surfaces in $\mathbb{R}^3$?

b. For what values of a and b does Theorem 3.1.10 guarantee that the intersection $X_a \cap Y_b$ is a smooth curve?

c. Sketch the subset of the (a,b)-plane for which $X_a \cap Y_b$ is not guaranteed to be a smooth curve. In each component of the complement, sketch the intersection $X_a \cap Y_b$ in the (x,y)-plane.

d. What geometric relation is there between X_a and Y_b for those values of a and b where $X_a \cap Y_b$ is not guaranteed to be a smooth curve? Sketch the projection onto the (x,y)-plane of representative cases of such values.

3.1.9 a. Find quadratic polynomials p and q for which the function

$$F\begin{pmatrix} x \\ y \end{pmatrix} = x^4 + y^4 + x^2 - y^2 \quad \text{of Example 3.1.12 can be written}$$

$$F\begin{pmatrix} x \\ y \end{pmatrix} = p(x)^2 + q(y)^2 - \frac{1}{2}.$$

b. Sketch the graphs of p, q, p^2, and q^2, and describe the connection between your graph and Figure 3.1.10.

3.1.10 Let $f\begin{pmatrix} x \\ y \end{pmatrix} = 0$ be the equation of a curve $X \subset \mathbb{R}^2$, and suppose that $\left[\mathbf{D}f\begin{pmatrix} x \\ y \end{pmatrix}\right] \neq [0\ 0]$ for all $\begin{pmatrix} x \\ y \end{pmatrix} \in X$.

a. Find an equation for the closure $\overline{CX}$ of the cone $CX \subset \mathbb{R}^3$ over X, when CX is the union of all the lines through the origin and points $\begin{pmatrix} x \\ y \\ 1 \end{pmatrix}$ with $\begin{pmatrix} x \\ y \end{pmatrix} \in X$.

Exercise 3.1.10 has a counterpart in Exercise 3.2.9.

b. If X has the equation $y = x^3$, what is the equation of $\overline{CX}$?

c. When X has the equation $y = x^3$, where is $\overline{CX} - \{\mathbf{0}\}$ guaranteed to be a smooth surface?

3.1.11 a. Find a parametrization for the union X of the lines through the origin and a point of the parametrized curve $t \mapsto \begin{pmatrix} t \\ t^2 \\ t^3 \end{pmatrix}$.

b. Find an equation for the closure $\overline{X}$ of X. Is $\overline{X}$ exactly X?

c. Show that $\overline{X} - \{\mathbf{0}\}$ is a smooth surface.

$$\begin{pmatrix} r \\ \theta \end{pmatrix} \mapsto \begin{pmatrix} r(1+\sin\theta) \\ r\cos\theta \\ r(1-\sin\theta) \end{pmatrix}$$

Map for Exercise 3.1.11, part d

d. Show that the map in the margin is another parametrization of $\overline{X}$. In this form you should have no trouble giving a name to the surface $\overline{X}$.

e. Relate $\overline{X}$ to the set of noninvertible symmetric 2×2 matrices.

3.1.12 Let $X \subset \mathbb{R}^3$ be the set of midpoints of segments joining a point of the curve C_1 of equation $y = x^2$, $z = 0$ to a point of the curve C_2 of equation $z = y^2$, $x = 0$.

a. Parametrize C_1 and C_2.

b. Parametrize X.

c. Find an equation for X.

d. Show that X is a smooth surface.

3.1.13 a. What is the equation of the plane $P \subset \mathbb{R}^3$ that contains the point $\mathbf{a}$ and is perpendicular to a vector $\vec{\mathbf{v}}$?

b. Let $\gamma(t) = \begin{pmatrix} t \\ t^2 \\ t^3 \end{pmatrix}$, and let P_t be the plane through the point $\gamma(t)$ and perpendicular to $\vec{\gamma}'(t)$. What is the equation of P_t?

c. Show that if $t_1 \neq t_2$, the planes P_{t_1} and P_{t_2} always intersect in a line. What are the equations of the line $P_1 \cap P_t$?

d. What is the limiting position of the line $P_1 \cap P_{1+h}$ as h tends to 0?

Exercise 3.1.16 has a counterpart in Exercise 3.2.8.

3.1.14 For what values of the constant c is the locus of equation $\sin(x+y) = c$ a smooth curve? *Hint*: This is not hard , but using Theorem 3.1.10 is not enough.

3.1.15 In Example 3.1.8, describe X_2 and X_3 when $l_1 = l_2 + l_3 + l_4$.

$$\mathbf{p} = \begin{pmatrix} 1+l \\ 1 \\ 0 \\ 0 \end{pmatrix}$$

Point for Exercise 3.1.16, part b

3.1.16 Consider the space X_l of positions of a rod of length l in $\mathbb{R}^3$, where one endpoint is constrained to be on the x-axis, and the other is constrained to be on the unit sphere centered at the origin.

a. Give equations for X_l as a subset of $\mathbb{R}^4$, where the coordinates in $\mathbb{R}^4$ are the x-coordinate of the end of the rod on the x-axis (call it t) and the three coordinates of the other end of the rod.

b. Show that near the point $\mathbf{p}$ in the margin, the set X_l is a manifold.

c. Show that for $l \neq 1$, X_l is a manifold.

d. Show that if $l = 1$, there are positions (i.e., points of X_1) near which X_1 is not a manifold.

3.1.17 In Example 3.1.8, show that knowing $\mathbf{x}_1$ and $\mathbf{x}_3$ determines exactly four positions of the linkage if the distance from $\mathbf{x}_1$ to $\mathbf{x}_3$ is smaller than both $l_1 + l_2$ and $l_3 + l_4$ and greater than $|l_1 - l_2|$ and $|l_3 - l_4|$.

Exercise 3.1.18: When we say "parametrize" by θ_1, θ_2, and the coordinates of $\mathbf{x}_1$, we mean consider the positions of the linkage as being determined by those variables.

3.1.18 a. Parametrize the positions of the linkage of Example 3.1.8 by the coordinates of $\mathbf{x}_1$, the polar angle θ_1 of the first rod with the horizontal line passing through $\mathbf{x}_1$, and the angle θ_2 between the first and the second: four numbers in all. For each value of θ_2 such that

$$(l_3 - l_4)^2 < l_1^2 + l_2^2 - 2l_1 l_2 \cos\theta_2 < (l_3 + l_4)^2, \tag{1}$$

how many positions of the linkage are there?

b. What happens if either of the inequalities in inequality (1) is an equality?

3.1.19 Let $M_k(n, m)$ be the space of $n \times m$ matrices of rank k.

a. Show that the space $M_1(2, 2)$ of 2×2 matrices of rank 1 is a manifold embedded in $\text{Mat}\,(2, 2)$.

b. Show that the space $M_2(3, 3)$ of 3×3 matrices of rank 2 is a manifold embedded in $\text{Mat}\,(3, 3)$. Show (by explicit computation) that $[\mathbf{D}\det(A)] = [0]$ if and only if A has rank < 2.

Exercise 3.1.19, part a: This is the space of matrices $A \neq 0$ such that $\det A = 0$.

Part b: If $A \in M_2(3, 3)$, then $\det A = 0$.

***3.1.20** Let $U \subset \mathbb{R}^2$ be open, $\mathbf{x}_0 \in U$ a point, and $\mathbf{f} : U \to \mathbb{R}^3$ a differentiable mapping with Lipschitz derivative. Suppose that $[\mathbf{Df}(\mathbf{x}_0)]$ is 1–1.

a. Show that there are two standard basis vectors of $\mathbb{R}^3$ spanning a plane E_1 such that if $P : \mathbb{R}^3 \to E_1$ denotes the projection onto the plane spanned by these vectors, then $[\mathbf{D}(P \circ \mathbf{f})(\mathbf{x}_0)]$ is invertible.

b. Show that there exist a neighborhood $V \subset E_1$ of $(P \circ \mathbf{f})(\mathbf{x}_0)$ and a mapping $\mathbf{g} : V \to \mathbb{R}^2$ such that $(P \circ \mathbf{f} \circ \mathbf{g})(\mathbf{y}) = \mathbf{y}$ for all $\mathbf{y} \in V$.

c. Let $W = \mathbf{g}(V)$. Show that $\mathbf{f}(W)$ is the graph of $\mathbf{f} \circ \mathbf{g} : V \to E_2$, where E_2 is the line spanned by the third standard basis vector. Conclude that $\mathbf{f}(W)$ is a smooth surface.

3.1.21 a. Does Theorem 3.1.10 guarantee that the subsets $X_a \subset \mathbb{R}^3$ of equation $e^x + 2e^y + 3e^z = a$ are all smooth surfaces?

b. Does it guarantee that for each (a, b), the subset $X_{a,b}$ given by the equations $e^x + 2e^y + 3e^z = a$ and $x + y + z = b$ is a smooth curve?

3.1.22 a. Is the set of $\begin{pmatrix} \theta \\ \varphi \end{pmatrix} \in \mathbb{R}^2$ such that $\begin{pmatrix} \cos\theta \\ \sin\theta \\ 0 \end{pmatrix}$ and $\begin{pmatrix} \cos\varphi + 1 \\ 0 \\ \sin\varphi \end{pmatrix}$ are distance 2 apart a smooth curve?

b. At what points is this set locally a graph of φ as a function of θ? At what points is it locally a graph of θ as a function of φ?

***3.1.23** Consider Example 3.1.8. If $l_1 + l_2 = l_3 + l_4$, show that X_2 is not a manifold near the position where all four points are aligned, with $\mathbf{x}_2$ and $\mathbf{x}_4$ between $\mathbf{x}_1$ and $\mathbf{x}_3$.

Exercise 3.1.24, part a: It is rather difficult to write equations for $M_1(n, m)$, but it isn't too hard to show that $M_1(n, m)$ is locally the graph of a mapping representing some variables as functions of others. For instance, suppose

$$A = [\mathbf{a}_1, \dots, \mathbf{a}_m] \in M_1(n, m),$$

and that $a_{1,1} \neq 0$. Show that all the entries of

$$\begin{bmatrix} a_{2,2} & \dots & a_{2,m} \\ \vdots & \ddots & \vdots \\ a_{n,2} & \dots & a_{n,m} \end{bmatrix}$$

are functions of the others, for instance, $a_{2,2} = a_{1,2}a_{2,1}/a_{1,1}$.

***3.1.24** Let $M_k(n, m)$ be the space of $n \times m$ matrices of rank k.

a. Show that $M_1(n, m)$ is a manifold embedded in $\text{Mat}\,(n, m)$ for all $n, m \geq 1$.

b. What is the dimension of $M_1(n, m)$?

3.1.25 Show that the mapping $\mathbf{g} : \begin{pmatrix} u \\ v \end{pmatrix} \mapsto \begin{pmatrix} \sin uv + u \\ u + v \\ uv \end{pmatrix}$ is a parametrization of a smooth surface:

a. Show that the image of $\mathbf{g}$ is contained in the locus S of equation

$$z = (x - \sin z)(\sin z - x + y).$$

b. Show that S is a smooth surface.

c. Show that $\mathbf{g}$ maps $\mathbb{R}^2$ onto S.

d. Show that $\mathbf{g}$ is 1–1, and that $\left[\mathbf{Dg}\begin{pmatrix} u \\ v \end{pmatrix}\right]$ is 1–1 for every $\begin{pmatrix} u \\ v \end{pmatrix} \in \mathbb{R}^2$.

3.2 Tangent spaces

The tangent space to a manifold will be essential when we discuss constrained extrema in Section 3.7, differential geometry in Section 3.9, volumes of manifolds in Chapter 5, and orientation of manifolds in Section 6.3.

The guiding principle of differential calculus is to replace nonlinear objects by their linear approximations. We replace a function by its derivative; we replace a curve, surface, or higher-dimensional manifold by its *tangent space*. The main question is always, to what extent does this replacement give an accurate picture of the nonlinear object?

A k-dimensional manifold $M \subset \mathbb{R}^n$ is locally the graph of a C^1 mapping $\mathbf{f}$ expressing $n - k$ variables as functions of the other k variables. Locally, the *tangent space* to a manifold is the graph of the derivative of $\mathbf{f}$. Thus the linear approximation to the graph is the graph of the linear approximation.

Recall (margin note next to Definition 3.1.1) that the notation $\mathbf{z} = \begin{pmatrix} \mathbf{x} \\ \mathbf{y} \end{pmatrix}$ is misleading, since we cannot assume that the active variables comes first. It would be more accurate, but clumsier, to write

$$\mathbf{z} = \Phi_{\mathrm{d}}(\mathbf{x}) + \Phi_{\mathrm{c}}(\mathbf{y}).$$

For a point $\mathbf{z} \in M$, denote by $\mathbf{x}$ the k active variables and by $\mathbf{y}$ the $n-k$ passive variables, so that

$$M = \left\{ \begin{pmatrix} \mathbf{x} \\ \mathbf{y} \end{pmatrix} \in \mathbb{R}^n \mid \mathbf{f}(\mathbf{x}) = \mathbf{y} \right\} \qquad 3.2.1$$

Then the *tangent flat* (tangent line, tangent plane, or higher-dimensional tangent plane) to M at a point $\mathbf{z}_0 = \begin{pmatrix} \mathbf{x}_0 \\ \mathbf{y}_0 \end{pmatrix} \in M$ is given by the equation

$$\underbrace{\mathbf{y} - \mathbf{y}_0}_{\text{change in output}} = [\mathbf{Df}(\mathbf{x}_0)] \underbrace{(\mathbf{x} - \mathbf{x}_0)}_{\text{change in input}}. \qquad 3.2.2$$

To give instead the equation for the *tangent space* at $\mathbf{z}_0$ we denote by $\dot{\mathbf{x}}$ an *increment* to $\mathbf{x}$, and by $\dot{\mathbf{y}}$ an *increment* to $\mathbf{y}$. (This is consistent with the use of dots by physicists to denote increments.) This gives

$$\dot{\mathbf{y}} = [\mathbf{Df}(\mathbf{x}_0)]\dot{\mathbf{x}}. \qquad 3.2.3$$

What is the difference between equation 3.2.2 and equation 3.2.3? Equation 3.2.2 for the tangent flat to M at $\mathbf{z}_0$ expresses $\mathbf{y}$ as an *affine function* of $\mathbf{x}$:

$$\mathbf{y} = \mathbf{y}_0 + [\mathbf{Df}(\mathbf{x}_0)](\mathbf{x} - \mathbf{x}_0) \qquad 3.2.4$$

(see Definition 1.3.14). It describes an affine subspace of $\mathbb{R}^n$.

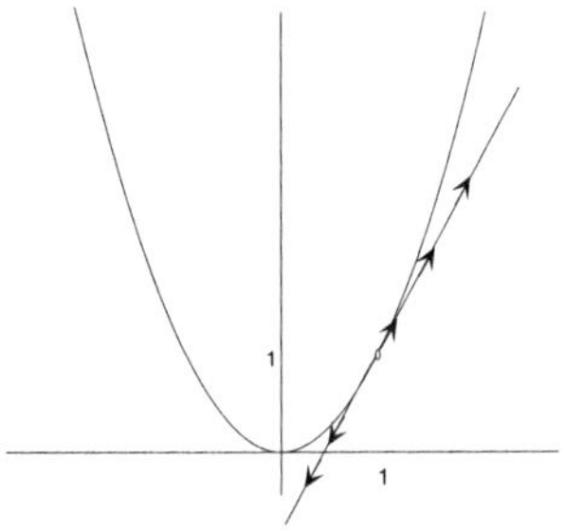

FIGURE 3.2.1.

Elements of the tangent space to the curve given by $f(x) = x^2$ should be thought of as vectors anchored at $\begin{pmatrix}1\\1\end{pmatrix}$ and tangent to the curve; these vectors include the zero vector. The tangent space is a vector space. In contrast, the tangent *line* to the curve at the same point is not a vector space; it does not pass through the origin of the (x, y)-plane.

Equation 3.2.3 expresses $\dot{\mathbf{y}}$ as a *linear function* of $\dot{\mathbf{x}}$: it defines the *vector subspace* $T_{\mathbf{z}_0}M \subset \mathbb{R}^n$. The elements of $T_{\mathbf{z}_0}M$ should be imagined as vectors anchored at $\mathbf{z}_0$; this point of tangency is the "origin" of the tangent space. See Figures 3.2.1 and 3.2.2.

Depending on context, one may be interested in tangent spaces or tangent flats. Tangent spaces are generally more useful: since they are vector spaces, we have all the tools of linear algebra at our disposal.

Definition 3.2.1 (Tangent space to a manifold). Let $M \subset \mathbb{R}^n$ be a k-dimensional manifold. The *tangent space* to M at $\mathbf{z}_0 \stackrel{\text{def}}{=} \begin{pmatrix}\mathbf{x}_0\\ \mathbf{y}_0\end{pmatrix}$, denoted $T_{\mathbf{z}_0}M$, is the graph of the linear transformation $[\mathbf{Df}(\mathbf{x}_0)]$.

Examples 3.2.2 (Tangent line and tangent space to smooth curve). If a smooth plane curve C is the graph of a function g expressing y in terms of x, then the tangent *line* to C at a point $\begin{pmatrix}a\\ g(a)\end{pmatrix}$ is the line of equation

$$y = g(a) + g'(a)(x - a). \tag{3.2.5}$$

(You should recognize this as the equation of the line of slope $g'(a)$ and y-intercept $g(a) - ag'(a)$.) The tangent *space* at $\begin{pmatrix}a\\ g(a)\end{pmatrix}$ is given by

$$\dot{y} = g'(a)\dot{x}. \tag{3.2.6}$$

In Figure 3.2.2, the tangent space to the circle at $\begin{pmatrix}-1\\0\end{pmatrix}$ (where the circle is the graph of a function expressing x in terms of y) has equation $\dot{x} = 0$; it consists of vectors with no increment in the x direction. The tangent space to the circle at $\begin{pmatrix}1\\0\end{pmatrix}$ also has equation $\dot{x} = 0$. But the tangent line to the circle at $\begin{pmatrix}-1\\0\end{pmatrix}$ has equation $x = -1 + g'(0)(y - 0) = -1$, and the tangent line at $\begin{pmatrix}1\\0\end{pmatrix}$ has equation $x = 1$. △

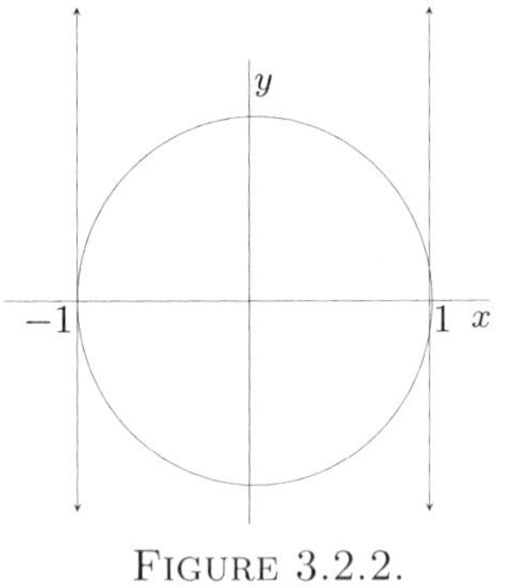

FIGURE 3.2.2.

The unit circle with tangent spaces at $\begin{pmatrix}1\\0\end{pmatrix}$ and at $\begin{pmatrix}-1\\0\end{pmatrix}$. The two tangent spaces consist of vectors such that the increment in the x direction is 0, i.e., $\dot{x} = 0$.

At a point where the curve is neither vertical nor horizontal, it can be thought of locally as either a graph of x as a function of y or as a graph of y as a function of x. Will this give us two different tangent lines or tangent spaces? No. If we have a point

$$\begin{pmatrix}a\\b\end{pmatrix} = \begin{pmatrix}a\\ f(a)\end{pmatrix} = \begin{pmatrix}g(b)\\ b\end{pmatrix} \in C, \tag{3.2.7}$$

where C is a graph of f and a graph of g, then $g \circ f(x) = x$. In particular, $g'(b)f'(a) = 1$ by the chain rule, so the line of equation

$$y - f(a) = f'(a)(x - a) \tag{3.2.8}$$

is also the line of equation $x - g(b) = g'(b)(y - b)$, and our definition of the tangent line is consistent.[4] The equations for the tangent spaces are $\dot{y} = f'(a)\dot{x}$ and $\dot{x} = g'(b)\dot{y}$, which coincide, since $f'(a)g'(b) = 1$.

---

[4]Since $g'(b)f'(a) = 1$, we have $f'(a) = 1/g'(b)$, so $y - f(a) = f'(a)(x - a)$ can be written $y - b = \frac{x-a}{g'(b)} = \frac{x-g(b)}{g'(b)}$; i.e., $x - g(b) = g'(b)(y - b)$.

Example 3.2.3 (Tangent space to a smooth surface). A subset $S \subset \mathbb{R}^3$ is a smooth surface if at every point $\mathbf{a} \in S$, the subset S is the graph of a function expressing one variable in terms of two others. The tangent space to a smooth surface S at $\mathbf{a}$ is the plane composed of the vectors anchored at $\mathbf{a}$, and tangent to the surface.

If at $\mathbf{a} = \begin{pmatrix} a \\ b \\ c \end{pmatrix}$ the surface is the graph of a function f expressing z in terms of x and y, then z is the passive variable: $z = f\begin{pmatrix} x \\ y \end{pmatrix}$. The equation for the tangent plane then becomes

$$z = c + \left[\mathbf{D}f\begin{pmatrix} a \\ b \end{pmatrix}\right]\begin{bmatrix} x-a \\ y-b \end{bmatrix} = c + D_1 f\begin{pmatrix} a \\ b \end{pmatrix}(x-a) + D_2 f\begin{pmatrix} a \\ b \end{pmatrix}(y-b). \quad 3.2.9$$

The equation for the tangent space at the same point is

$$\dot{z} = \left[\mathbf{D}f\begin{pmatrix} a \\ b \end{pmatrix}\right]\begin{bmatrix} \dot{x} \\ \dot{y} \end{bmatrix} = D_1 f\begin{pmatrix} a \\ b \end{pmatrix}\dot{x} + D_2 f\begin{pmatrix} a \\ b \end{pmatrix}\dot{y}. \quad 3.2.10$$

What is the equation for $T_{\mathbf{a}}S$ if S is the graph of a function g expressing y in terms of x and z? If it is the graph of a function h expressing x in terms of y and z?[5]

Tangent lines and tangent spaces to surfaces are part of everyday life. Imagine that a surface is a hill down which you are skiing. At every particular moment, you will be interested in the slope of the tangent *plane* to the surface: how steep is the hill? But you will also be interested in how fast you are going. Your speed at a point $\mathbf{a}$ is represented by a vector in the *tangent space* to the surface at $\mathbf{a}$. The arrow of this vector indicates what direction you are skiing; its length says how fast. If you have come to a halt, the "velocity" vector is the zero vector.

Finding tangent spaces

Usually if we know a manifold by an equation, the equation only expresses some variables implicitly in terms of others. In this case we cannot write the equation for the tangent space using the equations we have been given.

Theorem 3.2.4 gives a different approach. It relates the algebraic notion of ker $[\mathbf{DF}(\mathbf{c})]$ to the geometrical notion of a tangent space.

We use the notation of Theorem 3.1.10: $U \subset \mathbb{R}^n$ is an open subset, and $\mathbf{F} : U \to \mathbb{R}^{n-k}$ is a C^1 mapping.

It should be clear from Theorem 3.2.4 that if the same manifold can be represented as a graph in two different ways, then the tangent spaces should be the same. Indeed, if an equation $\mathbf{F}(\mathbf{z}) = \mathbf{0}$ expresses some variables in terms of others in several different ways, then in all cases, the tangent space is the kernel of the derivative of $\mathbf{F}$ and does not depend on the choice of pivotal variables.

Theorem 3.2.4 (Tangent space to a manifold given by equations). *If $\mathbf{F}(\mathbf{z}) = \mathbf{0}$ describes a manifold M, and $[\mathbf{DF}(\mathbf{z}_0)]$ is onto for some $\mathbf{z}_0 \in M$, then the tangent space $T_{\mathbf{z}_0}M$ is the kernel of $[\mathbf{DF}(\mathbf{z}_0)]$:*

$$T_{\mathbf{z}_0}M = \ker\,[\mathbf{DF}(\mathbf{z}_0)]. \quad 3.2.11$$

[5]The equations are

$$\dot{y} = \left[\mathbf{D}g\begin{pmatrix} a \\ c \end{pmatrix}\right]\begin{bmatrix} \dot{x} \\ \dot{z} \end{bmatrix} = D_1 g\begin{pmatrix} a \\ c \end{pmatrix}\dot{x} + D_2 g\begin{pmatrix} a \\ c \end{pmatrix}\dot{z}$$

$$\dot{x} = \left[\mathbf{D}h\begin{pmatrix} b \\ c \end{pmatrix}\right]\begin{bmatrix} \dot{y} \\ \dot{z} \end{bmatrix} = D_1 h\begin{pmatrix} b \\ c \end{pmatrix}\dot{y} + D_2 h\begin{pmatrix} b \\ c \end{pmatrix}\dot{z}.$$

It follows from equation 3.2.11 that the tangent space at $\mathbf{z}_0$ has equation

$$[\mathbf{DF}(\mathbf{z}_0)]\begin{bmatrix}\dot z_1\\ \vdots \\ \dot z_n\end{bmatrix}=\vec{\mathbf{0}}, \quad \text{i.e.,} \quad \overrightarrow{D_1\mathbf{F}}(\mathbf{z}_0)\,\dot z_1+\cdots+\overrightarrow{D_n\mathbf{F}}(\mathbf{z}_0)\,\dot z_n=\vec{\mathbf{0}}. \qquad 3.2.12$$

The equation for the tangent plane is

$$[\mathbf{DF}(\mathbf{z}_0)]\begin{bmatrix}z_1-z_{0,1}\\ \vdots \\ z_n-z_{0,n}\end{bmatrix}=\vec{\mathbf{0}}, \qquad 3.2.13$$

i.e.,

$$\overrightarrow{D_1\mathbf{F}}(\mathbf{z}_0)\,(z_1-z_{0,1})+\cdots+\overrightarrow{D_n\mathbf{F}}(\mathbf{z}_0)\,(z_n-z_{0,n})=\vec{\mathbf{0}}. \qquad 3.2.14$$

Example 3.2.5: Saying that $\ker\,[\mathbf{D}F(\mathbf{a})]$ is the tangent space to X_c at $\mathbf{a}$ says that every vector $\vec{\mathbf{v}}$ tangent to X_c at $\mathbf{a}$ satisfies the equation

$$[\mathbf{D}F(\mathbf{a})]\vec{\mathbf{v}}=0.$$

This puzzled one student, who argued that for this equation to be true, either $[\mathbf{D}F(\mathbf{a})]$ must be $[0]$ or $\vec{\mathbf{v}}$ must be $\vec{\mathbf{0}}$, yet the requirement that the derivative be onto says that $[\mathbf{D}F(\mathbf{a})]\neq[0\,,\,0]$. This is forgetting that $[\mathbf{D}F(\mathbf{a})]$ is a matrix. For example, if $[\mathbf{D}F(\mathbf{a})]$ is the matrix $[2\,,-2]$, then $[2\,,-2]\begin{bmatrix}1\\1\end{bmatrix}=0$.

Example 3.2.5 (Tangent space and tangent line to smooth curve. The locus X_c defined by $x^9+2x^3+y+y^5=c$ is a smooth curve for all values of c, since the derivative of $F\begin{pmatrix}x\\y\end{pmatrix}=x^9+2x^3+y+y^5$ is

$$\left[\mathbf{D}F\begin{pmatrix}x\\y\end{pmatrix}\right]=[9x^8+6x^2\,,\,1+5y^4], \qquad 3.2.15$$

and $1+5y^4$ is never 0.

What is the equation of the tangent space to the locus defined by $x^9+2x^3+y+y^5=5$, at the point $\begin{pmatrix}1\\1\end{pmatrix}$?

At the point $\begin{pmatrix}1\\1\end{pmatrix}\in X_5$, the derivative is $[15\,,\,6]$, so the tangent space to X_5 at that point is $\ker\,[15\,,\,6]$, and the equation of the tangent space is

$$15\dot x+6\dot y=0. \qquad 3.2.16$$

The equation for the tangent line is

$$15(x-1)+6(y-1)=0, \quad \text{i.e.,} \quad 15x+6y=21. \quad \triangle$$

Example 3.2.6 (Tangent space to smooth surface). We saw in Example 3.1.13 that the set defined by $F\begin{pmatrix}x\\y\\z\end{pmatrix}=\sin(x+yz)=0$ is a smooth surface, since the derivative of F is

$$\left[\mathbf{D}F\begin{pmatrix}a\\b\\c\end{pmatrix}\right]=[\underbrace{\cos(a+bc)}_{D_1F},\underbrace{c\cos(a+bc)}_{D_2F},\underbrace{b\cos(a+bc)}_{D_3F}], \qquad 3.2.17$$

which is onto, since it never vanishes. The equation for its tangent space at the same point is

$$\cos(a+bc)\dot x+c\cos(a+bc)\dot y+b\cos(a+bc)\dot z=0. \quad \triangle \qquad 3.2.18$$

Proof of Theorem 3.2.4. Set $\mathbf{c}=\mathbf{z}_0$. Let $1\le j_1<\cdots\le j_{n-k}\le n$ be indices such that the corresponding columns of $[\mathbf{DF}(\mathbf{c})]$ are linearly independent (pivotal); they exist because $[\mathbf{DF}(\mathbf{c})]$ is onto. Let the other indices be $1 \le i_1 \le \cdots \le i_k \le n$. The implicit function theorem asserts that $\mathbf{F}(\mathbf{z}) = \mathbf{0}$ implicitly defines $z_{j_1}, \dots, z_{j_{n-k}}$ in terms of $z_{i_1}, \dots, z_{i_k}$. Thus locally, near $\mathbf{c}$, the manifold M is defined by the equations

$$\begin{aligned} z_{j_1} &= g_{j_1}(z_{i_1}, \dots, z_{i_k}) \\ &\vdots \\ z_{j_{n-k}} &= g_{j_{n-k}}(z_{i_1}, \dots, z_{i_k}). \end{aligned} \qquad 3.2.19$$

For the variable $\mathbf{z} \in \mathbb{R}^n$ and the point $\mathbf{c} \in M$, set

$$\mathbf{x} = \underbrace{\begin{pmatrix} z_{i_1} \\ \vdots \\ z_{i_k} \end{pmatrix}}_{\text{active variables}}, \quad \mathbf{y} = \underbrace{\begin{pmatrix} z_{j_1} \\ \vdots \\ z_{j_{n-k}} \end{pmatrix}}_{\text{passive variables}}, \quad \mathbf{a} = \begin{pmatrix} c_{i_1} \\ \vdots \\ c_{i_k} \end{pmatrix}, \quad \mathbf{b} = \begin{pmatrix} c_{j_1} \\ \vdots \\ c_{j_{n-k}} \end{pmatrix}$$

The function $\mathbf{g}$ is

$$\mathbf{g} = \begin{pmatrix} g_{j_1} \\ \vdots \\ g_{j_{n-k}} \end{pmatrix}$$

Thus M has equation $\mathbf{y} = \mathbf{g}(\mathbf{x})$, and the tangent space $T_{\mathbf{c}}M$ has equation $\dot{\mathbf{y}} = [\mathbf{Dg}(\mathbf{a})]\dot{\mathbf{x}}$. Equation 2.10.29 of the implicit function theorem tells us that $[\mathbf{Dg}(\mathbf{a})]$ is

$$[\mathbf{Dg}(\mathbf{a})] = -[D_{j_1}\mathbf{F}(\mathbf{c}), \dots, D_{j_{n-k}}\mathbf{F}(\mathbf{c})]^{-1}[D_{i_1}\mathbf{F}(\mathbf{c}), \dots, D_{i_k}\mathbf{F}(\mathbf{c})]. \qquad 3.2.21$$

Substituting this value of $[\mathbf{Dg}(\mathbf{a})]$ in the equation $\dot{\mathbf{y}} = [\mathbf{Dg}(\mathbf{a})]\dot{\mathbf{x}}$, and multiplying each side by $[D_{j_1}\mathbf{F}(\mathbf{c}), \dots, D_{j_{n-k}}\mathbf{F}(\mathbf{c})]$, we get

$$[D_{j_1}\mathbf{F}(\mathbf{c}), \dots, D_{j_{n-k}}\mathbf{F}(\mathbf{c})]\dot{\mathbf{y}} + [D_{i_1}\mathbf{F}(\mathbf{c}), \dots, D_{i_k}\mathbf{F}(\mathbf{c})]\dot{\mathbf{x}} = \vec{\mathbf{0}}, \qquad 3.2.22$$

which is $[\mathbf{DF}(\mathbf{c})]\dot{\mathbf{z}} = \vec{\mathbf{0}}$. □

It follows from Proposition 3.2.7 that the partial derivatives

$$D_1\gamma(\mathbf{u}), \dots, D_k\gamma(\mathbf{u})$$

form a basis of $T_{\gamma(\mathbf{u})}M$.

Proposition 3.2.7 (Tangent space of manifold given by parametrization). *Let $U \subset \mathbb{R}^k$ be open, and let $\gamma : U \to \mathbb{R}^n$ be a parametrization of a manifold M. Then*

$$T_{\gamma(\mathbf{u})}M = \operatorname{img}[\mathbf{D}\gamma(\mathbf{u})]. \qquad 3.2.23$$

Proof. Suppose that in some neighborhood V of $\gamma(\mathbf{u}) \in \mathbb{R}^n$ the manifold M is given by $\mathbf{F}(\mathbf{x}) = \mathbf{0}$, where $\mathbf{F} : V \to \mathbb{R}^{n-k}$ is a C^1 mapping whose derivative is onto, so its kernel has dimension $n - (n-k) = k$.

Let $U_1 = \gamma^{-1}(V)$, so that $U_1 \subset \mathbb{R}^k$ is an open neighborhood of $\mathbf{u}$, and $\mathbf{F} \circ \gamma = \mathbf{0}$ on U_1, since $\gamma(U_1) \subset M$. If we differentiate this equation, we find by the chain rule

To show that the image of $[\mathbf{D}\gamma(\mathbf{u})]$ is the tangent space $T_{\gamma(\mathbf{u})}$, we show that it equals the kernel of $[\mathbf{DF}(\gamma(\mathbf{u}))]$ and then apply Theorem 3.2.4.

$$[\mathbf{D}(\mathbf{F} \circ \gamma)(\mathbf{u})] = [\mathbf{DF}(\gamma(\mathbf{u}))] \circ [\mathbf{D}\gamma(\mathbf{u})] = [0]. \qquad 3.2.24$$

It follows that $\operatorname{img}[\mathbf{D}\gamma(\mathbf{u})] \subset \ker[\mathbf{DF}(\gamma(\mathbf{u}))]$. The spaces are equal because both have dimension k. By Theorem 3.2.4, $T_{\gamma(\mathbf{u})}M = \ker[\mathbf{DF}(\gamma(\mathbf{u}))]$. □

Example 3.2.8. The map $\gamma : \begin{pmatrix} u \\ v \end{pmatrix} \mapsto \begin{pmatrix} u^2 \\ uv \\ v^2 \end{pmatrix}$, for $0 < u < \infty$ and $0 < v < \infty$, parametrizes a surface M of equation $F\begin{pmatrix} x \\ y \\ z \end{pmatrix} = xz - y^2 = 0$. Since

$$\left[\mathbf{D}\gamma\begin{pmatrix} u \\ v \end{pmatrix}\right] = \begin{bmatrix} 2u & 0 \\ v & u \\ 0 & 2v \end{bmatrix}, \qquad 3.2.25$$

the tangent space to M at $\gamma\begin{pmatrix} 1 \\ 1 \end{pmatrix} = \begin{pmatrix} 1 \\ 1 \\ 1 \end{pmatrix}$ is the image of $\begin{bmatrix} 2 & 0 \\ 1 & 1 \\ 0 & 2 \end{bmatrix}$, i.e. (Theorem 2.5.4), the tangent space is spanned by $\begin{bmatrix} 2 \\ 1 \\ 0 \end{bmatrix}, \begin{bmatrix} 0 \\ 1 \\ 2 \end{bmatrix}$.

To see that this tangent space does equal

$$\ker\left[\mathbf{D}F\left(\gamma\begin{pmatrix} 1 \\ 1 \end{pmatrix}\right)\right] = \ker[1 \; -2 \; 1], \qquad 3.2.26$$

as required by Theorem 3.2.4, first we compute

$$[1 \; -2 \; 1]\begin{bmatrix} 2 & 0 \\ 1 & 1 \\ 0 & 2 \end{bmatrix} = [0\,0], \quad \text{showing that} \quad \operatorname{img}\begin{bmatrix} 2 & 0 \\ 1 & 1 \\ 0 & 2 \end{bmatrix} \subset \ker[1 \; -2 \; 1].$$

Then we use the dimension formula to see that both have dimension 2, so they are equal. △

Remark. Theorem 3.2.4 and Proposition 3.2.7 illustrate the golden rule of differential calculus: to find the tangent space to a manifold, *do unto the increment with the derivative whatever you did to points with the function to get your manifold.* For instance,

Note that f' (the derivative of the implicit function) is evaluated at a, not at $\mathbf{a} = \begin{pmatrix} a \\ b \end{pmatrix}$.

- If a curve C is the graph of f, i.e, has equation $y = f(x)$, the tangent space at $\begin{pmatrix} a \\ f(a) \end{pmatrix}$ is the graph of $f'(a)$ and has equation $\dot{y} = f'(a)\dot{x}$.
- If the curve has equation $F\begin{pmatrix} x \\ y \end{pmatrix} = 0$, then the tangent space at $\begin{pmatrix} a \\ b \end{pmatrix}$ has equation $\left[\mathbf{D}F\begin{pmatrix} a \\ b \end{pmatrix}\right]\begin{bmatrix} \dot{x} \\ \dot{y} \end{bmatrix} = 0$.
- If $\gamma : (a, b) \to \mathbb{R}^2$ parametrizes C, then $T_{\gamma(t)}C = \operatorname{img}[\mathbf{D}\gamma(t)]$. △

Differentiable maps on manifolds

To round out our program of manifolds as "curvy objects on which to do calculus" we need to define what it means for maps defined on manifolds to be differentiable, or more precisely to be of class C^p.

Proposition and Definition 3.2.9 (**C^p map defined on manifold**). *Let $M \subset \mathbb{R}^n$ be an m-dimensional manifold, and $\mathbf{f} : M \to \mathbb{R}^k$ a map. Then $\mathbf{f}$ is of class C^p if every $\mathbf{x} \in M$ has a neighborhood $U \subset \mathbb{R}^n$ such that there exists a map $\widetilde{\mathbf{f}} : U \to \mathbb{R}^k$ of class C^p with $\widetilde{\mathbf{f}}|_{U\cap M} = \mathbf{f}|_{U\cap M}$.*

If $p \geq 1$, then

$$[\mathbf{Df}(\mathbf{x})] : T_{\mathbf{x}}M \to \mathbb{R}^k \stackrel{\text{def}}{=} [\mathbf{D}\widetilde{\mathbf{f}}(\mathbf{x})]|_{T_{\mathbf{x}}M} \qquad 3.2.27$$

does not depend on the choice of the extension $\widetilde{\mathbf{f}}$ of $\mathbf{f}$.

Proof. Let $\widetilde{\mathbf{f}}_1$ and $\widetilde{\mathbf{f}}_2$ be two extensions of $\mathbf{f}$ to some neighborhood $U \subset \mathbb{R}^n$ of $\mathbf{x} \in M$. Choose a parametrization $\gamma : V \to M$ of a neighborhood of $\mathbf{x}$ in M, with $V \subset \mathbb{R}^m$ a neighborhood of $\mathbf{0}$ and $\gamma(\mathbf{0}) = \mathbf{x}$.

Since (by Proposition 3.2.7) $[\mathbf{D}\gamma(\mathbf{0})] : \mathbb{R}^m \to T_{\mathbf{x}}M$ is an isomorphism, for any $\vec{\mathbf{v}} \in T_{\mathbf{x}}M$ there exists a unique $\vec{\mathbf{w}} \in \mathbb{R}^m$ with $[\mathbf{D}\gamma(\mathbf{0})]\vec{\mathbf{w}} = \vec{\mathbf{v}}$. Since $\widetilde{\mathbf{f}}_1$ and $\widetilde{\mathbf{f}}_2$ both extend $\mathbf{f}$, we have

$$\widetilde{\mathbf{f}}_1 \circ \gamma = \widetilde{\mathbf{f}}_2 \circ \gamma, \quad \text{so} \quad [\mathbf{D}(\widetilde{\mathbf{f}}_1 \circ \gamma)(\mathbf{0})] = [\mathbf{D}(\widetilde{\mathbf{f}}_2 \circ \gamma)(\mathbf{0})]. \qquad 3.2.28$$

By the chain rule and the identities $\gamma(\mathbf{0}) = \mathbf{x}$ and $[\mathbf{D}\gamma(\mathbf{0})]\vec{\mathbf{w}} = \vec{\mathbf{v}}$,

$$[\mathbf{D}\widetilde{\mathbf{f}}_1(\mathbf{x})]\vec{\mathbf{v}} = \underbrace{[\mathbf{D}(\widetilde{\mathbf{f}}_1 \circ \gamma)(\mathbf{0})]\vec{\mathbf{w}}}_{[\mathbf{D}\widetilde{\mathbf{f}}_1(\gamma(\mathbf{0}))][\mathbf{D}\gamma(\mathbf{0})]\vec{\mathbf{w}}} = [\mathbf{D}(\widetilde{\mathbf{f}}_2 \circ \gamma)(\mathbf{0})]\vec{\mathbf{w}} = [\mathbf{D}\widetilde{\mathbf{f}}_2(\mathbf{x})]\vec{\mathbf{v}}. \quad \square \qquad 3.2.29$$

In Proposition 3.2.10, note that $[\mathbf{Df}(\mathbf{x})]$ goes from $T_{\mathbf{x}}M$ to $\mathbb{R}^k$.

Proposition 3.2.10. *Let $M \subset \mathbb{R}^n$ be an m-dimensional manifold, and $\mathbf{f} : M \to \mathbb{R}^k$ a C^1 map; let $P \subset M$ be the set where $\mathbf{f} = \mathbf{0}$. If $[\mathbf{Df}(\mathbf{x})]$ is onto at every $\mathbf{x} \in P$, then P is an $(m-k)$-dimensional manifold.*

Proof. In a neighborhood U of $\mathbf{x} \in P$ define $M \cap U$ by $\mathbf{g} = \mathbf{0}$, where $\mathbf{g} : U \to \mathbb{R}^{n-m}$ is of class C^1 with $[\mathbf{Dg}(\mathbf{y})]$ onto for every $\mathbf{y} \in M \cap U$. Choose an extension $\widetilde{\mathbf{f}} : U \to \mathbb{R}^k$ as in Proposition and Definition 3.2.9. Then $P \cap U$ is defined by

$$\begin{pmatrix} \mathbf{g} \\ \widetilde{\mathbf{f}} \end{pmatrix} = \begin{pmatrix} \mathbf{0} \\ \mathbf{0} \end{pmatrix}. \qquad 3.2.30$$

To show that P is a $(m-k)$-dimensional manifold we will show that $\left[\mathbf{D}\begin{pmatrix} \mathbf{g} \\ \widetilde{\mathbf{f}} \end{pmatrix}(\mathbf{x})\right]$ is surjective. Choose $\begin{pmatrix} \mathbf{v} \\ \mathbf{w} \end{pmatrix} \in \mathbb{R}^{n-m} \times \mathbb{R}^k$. Since $[\mathbf{Df}(\mathbf{x})]$ is onto, there exists $\vec{\mathbf{w}}_1$ in

$$T_{\mathbf{x}}M \underbrace{=}_{\text{Thm. 3.2.4}} \ker[\mathbf{Dg}(\mathbf{x})] \quad \text{with} \quad [\mathbf{D}\widetilde{\mathbf{f}}(\mathbf{x})]\vec{\mathbf{w}}_1 \underbrace{=}_{\text{eq. 3.2.27}} [\mathbf{Df}(\mathbf{x})]\vec{\mathbf{w}}_1 = \vec{\mathbf{w}}.$$

The linear transformation $[\mathbf{Dg}(\mathbf{x})]$ maps any subspace $V \subset \mathbb{R}^n$ onto $\mathbb{R}^{n-m}$ if V and T_xM together span $\mathbb{R}^n$, and $\ker[\mathbf{D}\widetilde{\mathbf{f}}(\mathbf{x})]$ is such a subspace. So there exists $\vec{\mathbf{v}}_1 \in \ker[\mathbf{D}\widetilde{\mathbf{f}}(\mathbf{x})]$ with $[\mathbf{Dg}(\mathbf{x})]\vec{\mathbf{v}}_1 = \vec{\mathbf{v}}$. So

$$\left[\mathbf{D}\begin{pmatrix} \mathbf{g} \\ \widetilde{\mathbf{f}} \end{pmatrix}(\mathbf{x})\right][\vec{\mathbf{v}}_1 + \vec{\mathbf{w}}_1] = \begin{bmatrix} \vec{\mathbf{v}} \\ \vec{\mathbf{w}} \end{bmatrix}. \qquad 3.2.31$$

Proposition 3.2.11. *Let $M \subset \mathbb{R}^n$ be a manifold, and $\mathbf{f} : M \to \mathbb{R}^k$ a C^1 map. Let $U \subset \mathbb{R}^\ell$ be open and $\mathbf{g} : U \to M$ a C^1 map. Then*

$$[\mathbf{D}(\mathbf{f} \circ \mathbf{g})(\mathbf{x})] = [\mathbf{Df}(\mathbf{g}(\mathbf{x}))][\mathbf{Dg}(\mathbf{x})] \quad \textit{for all } \mathbf{x} \in U.$$

Saying $\mathbf{g} : U \to M$ is C^1 is saying that as a map to $\mathbb{R}^n$, $\mathbf{g}$ is C^1, and its image just happens to lie in M.

Proof. Choose $\mathbf{x} \in \mathbf{g}(U)$, and let $\widetilde{\mathbf{f}}$ be an extension of $\mathbf{f}$ to a neighborhood of $\mathbf{g}(\mathbf{x})$; then by the standard chain rule,

$$[\mathbf{D}(\widetilde{\mathbf{f}} \circ \mathbf{g})(\mathbf{x})] = [\mathbf{D}\widetilde{\mathbf{f}}(\mathbf{g}(\mathbf{x}))][\mathbf{Dg}(\mathbf{x})] \quad \text{for all } \mathbf{x} \in U. \qquad 3.2.32$$

But the image of $\mathbf{g}$ is a subset of M, so $\widetilde{\mathbf{f}} \circ \mathbf{g} = \mathbf{f} \circ \mathbf{g}$, and the image of $[\mathbf{Dg}(\mathbf{x})]$ is a subspace of $T_{\mathbf{g}(\mathbf{x})}M$, so

$$[\mathbf{D}(\mathbf{f} \circ \mathbf{g})(\mathbf{x})] = [\mathbf{D}\widetilde{\mathbf{f}}(\mathbf{g}(\mathbf{x}))][\mathbf{Dg}(\mathbf{x})] = [\mathbf{Df}(\mathbf{g}(\mathbf{x}))][\mathbf{Dg}(\mathbf{x})]. \quad \square$$

The inverse of a parametrization of a manifold M is a map just defined on the manifold, not *a priori* on any neighborhood of M; it is just the kind of map Proposition and Definition 3.2.9 is about.

In Exercise 3.4.12 you are asked to prove the generalization of Proposition 3.2.12: that if γ is C^p, then γ^{-1} is C^p.

Proposition 3.2.12. *If $V \subset \mathbb{R}^m$ is open and $\gamma : V \to \mathbb{R}^n$ is a C^1 parametrization of a manifold $M \subset \mathbb{R}^n$, then $\gamma^{-1} : M \to V$ is of class C^1.*

Note that by the definition of a parametrization, the map γ is injective and onto M, so as a map of sets, $\gamma^{-1} : M \to V$ is well defined.

Proof of Proposition 3.2.12: Recall (margin note for Exercise 1.4.24) that $\perp$ means "orthogonal to". The space $(T_{\mathbf{x}}M)^\perp$ is called the *orthogonal complement* of $T_{\mathbf{x}}M$, also known as the *normal subspace to* $T_{\mathbf{x}}M$.

Proof. To apply Proposition and Definition 3.2.9, we need, for each $\mathbf{x} \in M$, to extend γ^{-1} to a neighborhood $U \subset \mathbb{R}^n$ of $\mathbf{x}$; this requires the inverse function theorem. Let $\mathbf{y} = \gamma^{-1}(\mathbf{x})$. Define $\Psi : V \times (T_{\mathbf{x}}M)^\perp \to \mathbb{R}^n$ by

$$\Psi : (\mathbf{v}, \mathbf{w}) \mapsto \gamma(\mathbf{v}) + \mathbf{w}. \qquad 3.2.33$$

The derivative $[\mathbf{D}\Psi((\mathbf{y}, \mathbf{0}))]$ is the isomorphism $T_{\mathbf{x}}M \times T_{\mathbf{x}}M^\perp \to \mathbb{R}^n$ given by $(\vec{\mathbf{a}}, \vec{\mathbf{b}}) \mapsto \vec{\mathbf{a}} + \vec{\mathbf{b}}$, so there is a neighborhood U of $\mathbf{x}$ in $\mathbb{R}^n$ and a C^1 map $\Phi : U \to V \times T_{\mathbf{x}}M^\perp$ such that $\Phi \circ \Psi = \text{id} : U \to U$. The composition of Φ with the projection onto V is our desired extension. $\square$

Exercises for Section 3.2

3.2.1 Exercise 3.1.2 asked you to show that the equation $x + x^2 + y^2 = 2$ defines a smooth curve.

a. What is an equation for the tangent line to this curve at $\begin{pmatrix} 1 \\ 0 \end{pmatrix}$?

b. What is an equation for the tangent space to the curve at the same point?

3.2.2 Exercise 3.1.14 asked you to find the values of the constant c for which the locus of equation $\sin(x + y) = c$ is a smooth curve. What is the equation for the tangent line to such a curve at a point $\begin{pmatrix} u \\ v \end{pmatrix}$? For the tangent space?

3.2.3 Exercise 3.1.5 asked you to find the values of c for which the set of equation $X_c = x^2 + y^3 = c$ is a smooth curve. Give the equations of the tangent line and tangent space at a point $\begin{pmatrix} u \\ v \end{pmatrix}$ of such a curve X_c.

3.2.4 For each of the following functions f and points $\begin{pmatrix} a \\ b \end{pmatrix}$, state whether there is a tangent plane to the graph of f at the point $\begin{pmatrix} a \\ b \\ f\begin{pmatrix} a \\ b \end{pmatrix} \end{pmatrix}$. If there is such a tangent plane, find its equation, and compute the intersection of the tangent plane with the graph.

Exercise 3.2.4: You are encouraged to use a computer, although it is not absolutely necessary.

a. $f\begin{pmatrix} x \\ y \end{pmatrix} = x^2 - y^2$ at $\begin{pmatrix} 1 \\ 1 \end{pmatrix}$ b. $f\begin{pmatrix} x \\ y \end{pmatrix} = \sqrt{x^2 + y^2}$ at $\begin{pmatrix} 0 \\ 0 \end{pmatrix}$

c. $f\begin{pmatrix} x \\ y \end{pmatrix} = \sqrt{x^2 + y^2}$ at $\begin{pmatrix} 1 \\ -1 \end{pmatrix}$ d. $f\begin{pmatrix} x \\ y \end{pmatrix} = \cos(x^2 + y)$ at $\begin{pmatrix} 0 \\ 0 \end{pmatrix}$

3.2.5 a. Write the equations of the tangent planes P_1, P_2, P_3 to the surface of equation $z = Ax^2 + By^2$ at the points

$$\mathbf{p}_1 = \begin{pmatrix} 0 \\ 0 \\ 0 \end{pmatrix}, \quad \mathbf{p}_2 = \begin{pmatrix} a \\ 0 \\ Aa^2 \end{pmatrix}, \quad \mathbf{p}_3 = \begin{pmatrix} 0 \\ b \\ Bb^2 \end{pmatrix},$$

and find the point $\mathbf{q} = P_1 \cap P_2 \cap P_3$.

b. What is the volume of the tetrahedron with vertices at $\mathbf{p}_1$, $\mathbf{p}_2$, $\mathbf{p}_3$, and $\mathbf{q}$?

3.2.6 a. Show that the subset $X \subset \mathbb{R}^4$ where

$$x_1^2 + x_2^2 - x_3^2 - x_4^2 = 0 \quad \text{and} \quad x_1 + 2x_2 + 3x_3 + 4x_4 = 4$$

is a manifold in $\mathbb{R}^4$ in a neighborhood of the point $\mathbf{p}$ shown in the margin.

$$\mathbf{p} = \begin{pmatrix} 1 \\ 0 \\ 1 \\ 0 \end{pmatrix}$$

Point $\mathbf{p}$ for Exercise 3.2.6

b. What is the tangent space to X at $\mathbf{p}$?

c. What pair of variables do the equations above *not* express as functions of the other two?

d. Is the entire set X a manifold?

3.2.7 Show that if at a particular point $\mathbf{x}_0$ a surface is simultaneously the graph of z as a function of x and y, of y as a function of x and z, and of x as a function of y and z (see Example 3.2.3), then the corresponding equations for the tangent planes to the surface at $\mathbf{x}_0$ denote the same plane.

$$\mathbf{p} = \begin{pmatrix} 1+l \\ 1 \\ 0 \\ 0 \end{pmatrix}.$$

Point for Exercise 3.2.8

3.2.8 Give the equation of the tangent space to the set X_l of Exercise 3.1.16, at the point $\mathbf{p}$ shown in the margin.

3.2.9 For the cone CX of Exercise 3.1.10, what is the equation of the tangent plane to CX at any $\mathbf{x} \in CX - \{\mathbf{0}\}$?

3.2.10 Let C be a helicoid parametrized by $\gamma(t) = \begin{pmatrix} \cos t \\ \sin t \\ t \end{pmatrix}$.

a. Find a parametrization for the union X of all the tangent lines to C. Use a computer program to visualize this surface.

Exercise 3.2.10, part b: A parametrization of this curve is not too hard to find, but a computer will certainly help in describing the curve.

b. What is the intersection of X with the (x, z)-plane?

c. Show that X contains infinitely many curves of double points, where X intersects itself; these curves are helicoids on cylinders $x^2 + y^2 = r_i^2$.

3.2.11 Let $O(n) \subset \text{Mat}\,(n,n)$ be the set of orthogonal matrices. Let $S(n,n)$ be the space of symmetric $n \times n$ matrices. Let A be an $n \times n$ matrix.

a. Show that if $A \in O(n)$ and $B \in O(n)$, then $AB \in O(n)$ and $A^{-1} \in O(n)$.

b. Show that $O(n)$ is compact.

c. Show that for any square matrix A, we have $A^\top A - I \in S(n,n)$.

*d. Define $F : \text{Mat}\,(n,n) \to S(n,n)$ to be $F(A) = A^\top A - I$, so that $O(n) = F^{-1}(0)$. Show that if the matrix A is orthogonal, then the derivative $[\mathbf{D}F(A)] : \text{Mat}\,(n,n) \to S(n,n)$ is onto.

e. Show that $O(n)$ is a manifold embedded in $\text{Mat}\,(n,n)$ and that $T_I O(n)$ (the tangent space at the identity to the orthogonal group) is the space of antisymmetric $n \times n$ matrices.

3.2.12 Let $C \subset \mathbb{R}^3$ be the curve parametrized by $\gamma(t) = \begin{pmatrix} t \\ t^2 \\ t^3 \end{pmatrix}$. Let X be the union of all the lines tangent to C.

a. Find a parametrization of X.

b. Find a function $f : \mathbb{R}^3 \to \mathbb{R}$ such that the equation $f(\mathbf{x}) = 0$ defines X.

*c. Show that $X - C$ is a smooth surface. (Our solution uses material in Exercise 3.1.20.)

d. Find the equation of the curve that is the intersection of X with the plane $x = 0$.

Exercise 3.2.11: Recall (Definition 1.2.18) that a symmetric matrix is a matrix that is equal to its transpose. An antisymmetric matrix is a matrix A such that

$$A = -A^\top.$$

Recall (Proposition and Definition 2.4.19) that $A \in O(n)$ if and only if $A^\top A = I$.

Part a, together with the associativity of matrix multiplication), says exactly that $O(n)$ forms a group, known as the *orthogonal group*.

Exercise 3.2.12, part b: Write that $\mathbf{x} - \gamma(t)$ is a multiple of $\gamma'(t)$, which leads to two equations in x, y, z, and t. Now eliminate t among these equations; it takes a bit of fiddling with the algebra.

Exercise 3.2.12, part c: Show that the only common zeros of f and $[\mathbf{D}f]$ are the points of C; again this requires a bit of fiddling with the algebra.

3.3 Taylor polynomials in several variables

"Although this may seem a paradox, all exact science is dominated by the idea of approximation."—Bertrand Russell

In Sections 3.1 and 3.2 we used first-degree approximations to discuss curves, surfaces, and higher-dimensional manifolds. A function's derivative provides the best linear approximation to the function: at a point $\mathbf{x}$, the function and its derivative differ only by terms that are smaller than linear (usually quadratic and higher).

Now we will discuss higher-degree approximations, using Taylor polynomials and higher partial derivatives. Approximation of functions by polynomials is a central issue in calculus in one and several variables. It is also of great importance in such fields as interpolation and curve fitting, and computer graphics; when a computer graphs a function, most often it is approximating it with cubic piecewise polynomial functions. In Section 3.9 we will apply these notions to the geometry of curves and surfaces.

Higher partial derivatives are essential throughout mathematics and in science. Mathematical physics is essentially the theory of partial differential equations. Electromagnetism is based on Maxwell's equations, general relativity on Einstein's equation, fluid dynamics on the Navier-Stokes equation, and quantum mechanics on Schrödinger's equation. Understanding partial differential equations is a prerequisite for any serious study of these phenomena.

In one variable, you learned that at a point x near a, a function is well approximated by its Taylor polynomial at a. The caveat "at a point x near a" is important. When a differentiable function is defined on an entire interval, it has a Taylor polynomial in a neighborhood of each point of the

Theorem 3.3.1: The polynomial $p^k_{f,a}(a+h)$, called the *Taylor polynomial* of degree k of f at a, is carefully constructed to have the same derivatives at a (up to order k) as f. The theorem is a special case of Theorem 3.3.16, Taylor's theorem in several variables.

Taylor's theorem with remainder provides a way to evaluate the difference between a function and a (finite) Taylor polynomial that approximates it. Taylor's theorem with remainder, in one and several variables, is discussed in Appendix A12.

FIGURE 3.3.1.
Brook Taylor (1685–1731)
Taylor was one of several authors of the result now known as Taylor's theorem. Its importance remained unrecognized until 1772 when Lagrange proclaimed it the basic principle of differential calculus. In addition to his mathematics, Taylor worked on Kepler's second law of planetary motion, capillary action, magnetism, thermometers, and vibrating strings; he also discovered the basic principles of perspective. His personal life was marked by tragedy: his first and second wives died in childbirth; one child survived.

interval; knowing the Taylor polynomial at one point generally tells you nothing about the others.

In Theorem 3.3.1, recall that $f^{(k)}$ denotes the kth derivative of f.

Theorem 3.3.1 (Taylor's theorem without remainder in one variable). *If $U \subset \mathbb{R}$ is an open subset and $f : U \to \mathbb{R}$ is k times continuously differentiable on U, then the polynomial*

$$\underbrace{p^k_{f,a}(a+h)}_{\text{Taylor polynomial of degree } k} \stackrel{\text{def}}{=} f(a) + f'(a)h + \frac{1}{2!}f''(a)h^2 + \cdots + \frac{1}{k!}f^{(k)}(a)h^k \quad 3.3.1$$

is the best approximation to f at a in the sense that it is the unique polynomial of degree $\le k$ such that

$$\lim_{h \to 0} \frac{f(a+h) - p^k_{f,a}(a+h)}{h^k} = 0. \quad 3.3.2$$

One proof, sketched in Exercise 3.3.11, consists of using l'Hôpital's rule k times.

Remark. Note the "k times continuously differentiable" in Theorem 3.3.1. In Sections 3.1 and 3.2 we required that our functions be only C^1: once continuously differentiable. In this section and in Section 3.4 (on rules for computing Taylor polynomials), it really matters exactly how many derivatives exist. △

We will see that there is a polynomial in n variables that in the same sense best approximates functions of n variables.

Multi-exponent notation for polynomials in higher dimensions

First we must introduce some notation. In one variable, it is easy to write the "general polynomial" of degree k as

$$a_0 + a_1x + a_2x^2 + \cdots + a_kx^k = \sum_{i=0}^{k} a_ix^i. \quad 3.3.3$$

It is less obvious how to write polynomials in several variables, like

$$1 + 2x + 3xyz - x^2z + 5xy^5z^2. \quad 3.3.4$$

One effective if cumbersome notation uses *multi-exponents*. A multi-exponent is a way of writing one term (one *monomial*) of a polynomial.

Definition 3.3.2 (Multi-exponent). A *multi-exponent I* is an ordered finite list of nonnegative whole numbers

$$I \stackrel{\text{def}}{=} (i_1, \dots, i_n), \quad 3.3.5$$

which definitely may include 0.

Example 3.3.3 (Multi-exponents). Table 3.3.2 lists all the multi-exponents with nonzero coefficients for $1 + 2x + 3xyz - x^2z + 5xy^5z^2$, the polynomial of formula 3.3.4.

$1 + 2x + 3xyz - x^2z + 5xy^5z^2$					
monomial	1	2x	$3xyz$	$-x^2z$	$5xy^5z^2$
multi-exponent	(0,0,0)	$(1,0,0)$	$(1,1,1)$	$(2,0,1)$	$(1,5,2)$
coefficient	1	2	3	-1	5

TABLE 3.3.2.

Almost the only functions that can be computed are polynomials, or rather piecewise polynomial functions, also known as *splines*: functions formed by stringing together bits of different polynomials. Splines can be computed, since you can put *if* statements in the program that computes your function, allowing you to compute different polynomials for different values of the variables. (Approximation by rational functions, which involves division, is also important in practical applications.)

Polynomials in several variables really are a lot more complicated than in one variable: even the first questions involving factoring, division, etc. lead rapidly to difficult problems in algebraic geometry.

Definition 3.3.4 (Total degree and factorial of multi-exponent). The set of multi-exponents with n entries is denoted $\mathcal{I}_n$:

$$\mathcal{I}_n \stackrel{\text{def}}{=} \{(i_1, \dots, i_n)\}. \qquad 3.3.6$$

For any multi-exponent $I \in \mathcal{I}_n$, the *total degree* of I is $\deg I = i_1 + \cdots + i_n$. The *factorial* of I is $I! \stackrel{\text{def}}{=} i_1! \cdot \cdots \cdot i_n!$.

Definition 3.3.4 and Notation 3.3.6: The total degree of xyz is 3, since $1+1+1 = 3$; the total degree of y^2z is also 3. The set $\mathcal{I}_3^2$ of multi-exponents with three entries and total degree 2 consists of

$$(0,1,1),\ (1,1,0),\ (1,0,1),$$
$$(2,0,0),\ (0,2,0),\ (0,0,2).$$

If $I = (2,0,3)$, then $\deg I = 5$, and $I! = 2!0!3! = 12$. (Recall that $0! = 1$.)

Notation 3.3.5 ($\mathcal{I}_n^k$). We denote by $\mathcal{I}_n^k$ the set of multi-exponents with n entries and of total degree k.

What are the elements of the set $\mathcal{I}_2^3$? Of $\mathcal{I}_3^3$? Check your answers in the footnote below.[6]

Using multi-exponents, we can write a polynomial as a sum of monomials (as we already did in Table 3.3.2).

Notation 3.3.6 ($\mathbf{x}^I$). For any $I \in \mathcal{I}_n$, we denote by $\mathbf{x}^I$ the monomial function $\mathbf{x}^I \stackrel{\text{def}}{=} x_1^{i_1} \dots x_n^{i_n}$ on $\mathbb{R}^n$.

Here i_1 gives the power of x_1, while i_2 gives the power of x_2, and so on. If $I = (2,3,1)$, then $\mathbf{x}^I$ is a monomial of total degree 6:

$$\mathbf{x}^I = \mathbf{x}^{(2,3,1)} = x_1^2 x_2^3 x_3^1. \qquad 3.3.7$$

We can now write the general polynomial p of degree m in n variables as a sum of monomials, each with its own coefficient a_I:

$$p(\mathbf{x}) = \sum_{k=0}^{m} \sum_{I \in \mathcal{I}_n^k} a_I \mathbf{x}^I. \qquad 3.3.8$$

[6] $\mathcal{I}_2^3 = \{(1,2), (2,1), (0,3), (3,0)\}$; $\mathcal{I}_3^3 = \{(1,1,1), (2,1,0), (2,0,1), (1,2,0), (1,0,2), (0,2,1), (0,1,2), (3,0,0), (0,3,0), (0,0,3)\}$.

In formula 3.3.8 (and in the double sum in formula 3.3.9), k is just a placeholder indicating the degree. To write a polynomial with n variables, first we consider the single multi-exponent I of degree $k = 0$ (the constant term) and determine its coefficient. Next we consider the set $\mathcal{I}_n^1$ (multi-exponents of degree $k = 1$) and for each we determine its coefficient. Then we consider the set $\mathcal{I}_n^2$, and so on. We could use the multi-exponent notation without grouping by degree, expressing a polynomial as

$$\sum_{\substack{\deg I \le m \\ I \in \mathcal{I}_n}} a_I \mathbf{x}^I.$$

But it is often useful to group together terms of a polynomial by degree: constant term, linear terms, quadratic terms, cubic terms, etc.

Of course, $D_I f$ is only defined if all partials up to order $\deg I$ exist, and it is also a good idea to assume that they are all continuous, so that the order in which the partials are calculated doesn't matter (Theorem 3.3.8).

Example 3.3.7 (Multi-exponent notation). Using this notation, we can write the polynomial

$$2 + x_1 - x_2x_3 + 4x_1x_2x_3 + 2x_1^2x_2^2 \quad \text{as} \quad \sum_{k=0}^{4} \sum_{I \in \mathcal{I}_3^k} a_I \mathbf{x}^I, \quad \text{where} \qquad 3.3.9$$

$$\begin{aligned} a_{(0,0,0)} &= 2, & a_{(1,0,0)} &= 1, & a_{(0,1,1)} &= -1, \\ a_{(1,1,1)} &= 4, & a_{(2,2,0)} &= 2, \end{aligned} \qquad 3.3.10$$

and all other $a_I = 0$, for $I \in \mathcal{I}_3^k$, $k \le 4$. (There are 35 terms in all.) △

What is the polynomial $\sum_{k=0}^{3} \sum_{I \in \mathcal{I}_2^k} a_I \mathbf{x}^I$, where $a_{(0,0)} = 3$, $a_{(1,0)} = -1$, $a_{(1,2)} = 3$, and $a_{(2,1)} = 2$, and all the other coefficients a_I are 0? Check your answer in the footnote.[7]

Multi-exponent notation and equality of crossed partials

Multi-exponents provide a concise way to write higher partial derivatives in Taylor polynomials in higher dimensions. Recall (Definition 2.8.7) that if the function $D_i f$ is differentiable, then its partial derivative with respect to the jth variable, $D_j(D_i f)$, exists[8]; it is called a *second partial derivative* of f.

To apply multi-exponent notation to higher partial derivatives, set

$$D_I f = D_1^{i_1} D_2^{i_2} \dots D_n^{i_n} f. \qquad 3.3.11$$

For example, for a function f in three variables,

$$D_1\Big(D_1\big(D_2(D_2 f)\big)\Big) = D_1^2(D_2^2 f) \text{ can be written } D_1^2\big(D_2^2(D_3^0 f)\big), \qquad 3.3.12$$

which can be written $D_{(2,2,0)}f$, that is, $D_I f$, where $I = (i_1, i_2, i_3) = (2, 2, 0)$.

What is $D_{(1,0,2)}f$, written in our standard notation for higher partial derivatives? What is $D_{(0,1,1)}f$? Check your answers in the footnote.[9]

Recall that a multi-exponent I is an *ordered* finite list of nonnegative whole numbers. Using multi-exponent notation, how can we distinguish between $D_1(D_3 f)$ and $D_3(D_1 f)$? Both are written $D_{(1,0,1)}$. Similarly, $D_{1,1}$ could denote $D_1(D_2 f)$ or $D_2(D_1 f)$. Is this a problem?

No. If you compute the second partials $D_1(D_3 f)$ and $D_3(D_1 f)$ of the function $x^2 + xy^3 + xz$, you will see that they are equal:

$$D_1(D_3 f)\begin{pmatrix} x \\ y \\ z \end{pmatrix} = D_3(D_1 f)\begin{pmatrix} x \\ y \\ z \end{pmatrix} = 1. \qquad 3.3.13$$

---

[7] It is $3 - x_1 + 3x_1x_2^2 + 2x_1^2x_2$.

[8] This assumes, of course, that $f : U \to \mathbb{R}$ is a differentiable function, and $U \subset \mathbb{R}^n$ is open.

[9] The first is $D_1(D_3^2 f)$, which can also be written $D_1\Big(D_3(D_3 f)\Big)$. The second is $D_2(D_3 f)$.

Similarly, $D_1(D_2 f) = D_2(D_1 f)$, and $D_2(D_3 f) = D_3(D_2 f)$.

Normally, crossed partials are equal.

The equality of crossed partials will simplify the writing of Taylor polynomials in higher dimensions (Definition 3.3.13): for instance, $D_1(D_2 f)$ and $D_2(D_1 f)$ are written $D_{(1,1)}$. This is a benefit of multi-exponent notation.

Theorem 3.3.8 Equality of crossed partials). *Let U be an open subset of $\mathbb{R}^n$ and $f : U \to \mathbb{R}$ a function such that all first partial derivatives $D_i f$ are themselves differentiable at $\mathbf{a} \in U$. Then for every pair of variables x_i, x_j, the crossed partials are equal:*

$$D_j(D_i f)(\mathbf{a}) = D_i(D_j f)(\mathbf{a}). \qquad 3.3.14$$

Theorem 3.3.8 is a surprisingly difficult result, proved in Appendix A9.

We have the following implications:

second partials continuous

⇓

first partials differentiable

⇓

crossed partials equal

(The first implication is Theorem 1.9.8.) In Exercise 4.5.13 we show that Fubini's theorem implies a weaker version of the theorem, where the crossed partials (which are second partials) are required to be continuous. That version is weaker because requiring second partials to be continuous is a stronger requirement than first partials being differentiable.

We thank Xavier Buff for pointing out this stronger result.

The requirement in Theorem 3.3.8 that the first partial derivatives be differentiable is essential, as shown by Example 3.3.9.

Example 3.3.9 (A case where crossed partials aren't equal). Consider the function

$$f\begin{pmatrix} x \\ y \end{pmatrix} = \begin{cases} xy\dfrac{x^2-y^2}{x^2+y^2} & \text{if } \begin{pmatrix} x \\ y \end{pmatrix} \neq \begin{pmatrix} 0 \\ 0 \end{pmatrix} \\ 0 & \text{if } \begin{pmatrix} x \\ y \end{pmatrix} = \begin{pmatrix} 0 \\ 0 \end{pmatrix} \end{cases}. \qquad 3.3.15$$

Then when $\begin{pmatrix} x \\ y \end{pmatrix} \neq \begin{pmatrix} 0 \\ 0 \end{pmatrix}$, we have

$$D_1 f\begin{pmatrix} x \\ y \end{pmatrix} = \frac{4x^2y^3 + x^4y - y^5}{(x^2+y^2)^2} \quad \text{and} \quad D_2 f\begin{pmatrix} x \\ y \end{pmatrix} = \frac{x^5 - 4x^3y^2 - xy^4}{(x^2+y^2)^2}, \qquad 3.3.16$$

and both partials vanish at the origin. So

$$D_1 f\begin{pmatrix} 0 \\ y \end{pmatrix} = \begin{cases} -y & \text{if } y \neq 0 \\ 0 & \text{if } y = 0 \end{cases} = -y, \qquad D_2 f\begin{pmatrix} x \\ 0 \end{pmatrix} = \begin{cases} x & \text{if } x \neq 0 \\ 0 & \text{if } x = 0 \end{cases} = x,$$

giving

$$D_2(D_1 f)\begin{pmatrix} 0 \\ y \end{pmatrix} = D_2(-y) = -1, \qquad D_1(D_2 f)\begin{pmatrix} x \\ 0 \end{pmatrix} = D_1(x) = 1, \qquad 3.3.17$$

the first for any value of y and the second for any value of x; at the origin, the crossed partials $D_2(D_1 f)$ and $D_1(D_2 f)$ are not equal. This does not contradict Theorem 3.3.8 because although the first partial derivatives are continuous, they are not differentiable. △

Example 3.3.9: We invite you to check that $D_1 f$ and $D_2 f$ are not differentiable at the origin. But don't take this example too seriously. The function f here is pathological; such things do not show up unless you go looking for them. Unless you suspect a trap, when computing you should assume that crossed partials are equal.

Corollary 3.3.10. *If $f : U \to \mathbb{R}$ is a function all of whose partial derivatives up to order k are continuous, then the partial derivatives of order up to k do not depend on the order in which they are computed.*

The coefficients of polynomials as derivatives

We can express the coefficients of a polynomial in one variable in terms of the derivatives of the polynomial at 0. If p is a polynomial of degree k

with coefficients $a_0, \dots, a_k$, i.e., $p(x) = a_0 + a_1x + a_2x^2 + \cdots + a_kz^k$, then, denoting by $p^{(i)}$ the ith derivative of p, we have

$$i!a_i = p^{(i)}(0); \qquad \text{i.e.,} \qquad a_i = \frac{1}{i!}p^{(i)}(0). \qquad 3.3.18$$

Evaluating the ith derivative of a polynomial at 0 isolates the coefficient of x^i: the ith derivative of lower terms vanishes, and the ith derivative of higher-degree terms contains positive powers of x, which vanish when evaluated at 0.

For example, take the polynomial $x + 2x^2 + 3x^3$. Then

$$f^{(0)}(0) = f(0) = 0;$$

indeed, $0!a_0 = 0$. Further,

$$f'(x) = 1 + 4x + 9x^2, \text{ so}$$
$$f'(0) = 1; \quad \text{indeed, } 1!\,a_1 = 1.$$
$$f''(x) = 4 + 18x, \quad \text{so}$$
$$f''(0) = 4; \quad \text{indeed, } 2!\,a_2 = 4$$
$$f^{(3)}(x) = 18;$$
$$\text{indeed, } 3!\,a_3 = 6 \cdot 3 = 18.$$

We will want to translate this to the case of several variables. You may wonder why. Our goal is to approximate differentiable functions by polynomials. We will see in Proposition 3.3.17 that if, at a point $\mathbf{a}$, all derivatives up to order k of a function vanish, then the function is small in a neighborhood of that point (small in a sense that depends on k). If we can manufacture a polynomial with the same derivatives up to order k as the function we want to approximate, then the function representing the *difference* between the function and the polynomial will have vanishing derivatives up to order k: hence it will be small.

Proposition 3.3.11 translates equation 3.3.18 to the case of several variables: as in one variable, the coefficients of a polynomial in several variables can be expressed in terms of the partial derivatives of the polynomial at $\mathbf{0}$.

Recall that

$$I \in \mathcal{I} = (i_1, \dots, i_n),$$

$$I! = i_1! \cdots i_n!.$$

Proposition 3.3.11 (Coefficients expressed in terms of partial derivatives at 0). *Let p be the polynomial*

$$p(\mathbf{x}) \stackrel{\text{def}}{=} \sum_{m=0}^{k} \sum_{J \in \mathcal{I}_n^m} a_J \mathbf{x}^J. \qquad 3.3.19$$

Then for any $I \in \mathcal{I}_n$,

$$I!\,a_I = D_I p(\mathbf{0}). \qquad 3.3.20$$

Equation 3.3.23: $D_1^{i_1} x_1^{i_1} = i_1!$, $D_2^{i_2} x_2^{i_2} = i_2!$, and so on. For instance,

$$D^3x^3 = D^2(3x^2) = D(6x)$$
$$= 6 = 3!.$$

As equation 3.3.23 shows, $D_I \mathbf{x}^I$ is a constant, and does not depend on the value of $\mathbf{x}$.

If you find it hard to focus on this proof written in multi-exponent notation, look at Example 3.3.12, which translates multi-exponent notation into more standard (and less concise) notation.

Proof. First, let us see that it is sufficient to show that

$$D_I \mathbf{x}^I(\mathbf{0}) = I! \quad \text{and} \quad D_I \mathbf{x}^J(\mathbf{0}) = 0 \text{ for all } J \neq I. \qquad 3.3.21$$

We can see this by writing

$$D_I p(\mathbf{0}) = D_I \overbrace{\left(\sum_{m=0}^{k} \sum_{J \in \mathcal{I}_n^m} a_J \mathbf{x}^J \right)}^{p \text{ written in multi-exponent form}} (\mathbf{0}) = \sum_{m=0}^{k} \sum_{J \in \mathcal{I}_n^m} a_J D_I \mathbf{x}^J(\mathbf{0}); \qquad 3.3.22$$

if we prove the statements in equation 3.3.21, then all the terms $a_J D_I \mathbf{x}^J(\mathbf{0})$ for $J \neq I$ drop out, leaving $D_I p(\mathbf{0}) = I!\,a_I$.

To prove that $D_I \mathbf{x}^I(\mathbf{0}) = I!$, write

$$\begin{aligned} D_I \mathbf{x}^I &= D_1^{i_1} \cdots D_n^{i_n} x_1^{i_1} x_2^{i_2} \cdots x_n^{i_n} = D_1^{i_1} x_1^{i_1} \cdots D_n^{i_n} x_n^{i_n} \\ &= i_1! \cdots i_n! = I!, \end{aligned} \qquad 3.3.23$$

so in particular, $D_I\mathbf{x}^I(\mathbf{0}) = I!$. To prove $D_I\mathbf{x}^J(\mathbf{0}) = 0$ for all $J \neq I$, write similarly

$$D_I\mathbf{x}^J = D_1^{i_1}\cdots D_n^{i_n} x_1^{j_1}\cdots x_n^{j_n} = D_1^{i_1}x_1^{j_1}\cdots D_n^{i_n}x_n^{j_n}. \qquad 3.3.24$$

At least one j_m must be different from i_m, either bigger or smaller. If it is smaller, then we see a higher derivative than the power, and the derivative is 0. If it is bigger, then $D_I\mathbf{x}^J$ contains a positive power of x_m; when the derivative is evaluated at $\mathbf{0}$, we get 0 again. □

Example 3.3.12 (Coefficients of polynomial in terms of its partial derivatives at 0). Let $p = 3x_1^2x_2^3$. What is $D_1^2D_2^3p$? We have

$$D_2p = 9x_2^2x_1^2, \quad D_2^2p = 18x_2x_1^2, \qquad 3.3.25$$

and so on, ending with $D_1^2D_2^3p = 36$.

In multi-exponent notation, $p = 3x_1^2x_2^3$ is written $3\mathbf{x}^{(2,3)}$, i.e., $a_I\mathbf{x}^I$, where $I = (2,3)$ and $a_{(2,3)} = 3$. The higher partial derivative $D_1^2D_2^3p$ is written $D_{(2,3)}p$. Since $I = (2,3)$, we have $I! = 2!3! = 12$.

Proposition 3.3.11 says

$$a_I = \frac{1}{I!}D_Ip(\mathbf{0}); \quad \text{here, } \frac{1}{I!}D_{(2,3)}p(\mathbf{0}) = \frac{36}{12} = 3, \text{ which is indeed } a_{(2,3)}.$$

What if the multi-exponent I for the higher partial derivatives is different from the multi-exponent J for $\mathbf{x}$? As mentioned in the proof of Proposition 3.3.11, the result is 0. For example, if we take $D_1^2D_2^2$ of the polynomial $p = 3x_1^2x_2^3$, so that $I = (2,2)$ and $J = (2,3)$, we get $36x_2$; evaluated at $\mathbf{0}$, this becomes 0. If any index of I is greater than the corresponding index of J, the result is also 0; for example, what is $D_Ip(\mathbf{0})$ when $I = (2,3)$, $p = a_J\mathbf{x}^J$, $a_J = 3$, and $J = (2,2)$?[10] △

Taylor polynomials in higher dimensions

Although the polynomial in equation 3.3.26 is called the Taylor polynomial of f at $\mathbf{a}$, it is evaluated at $\mathbf{a}+\vec{\mathbf{h}}$, and its value there depends on $\vec{\mathbf{h}}$, the increment to $\mathbf{a}$.

Recall (Definition 1.9.7) that a C^k *function* is k times continuously differentiable: all of its partial derivatives up to order k exist and are continuous on U.

Now we are ready to define Taylor polynomials in higher dimensions and to see in what sense they can be used to approximate functions in n variables.

Definition 3.3.13 (Taylor polynomial in higher dimensions). Let $U \subset \mathbb{R}^n$ be an open subset and let $f : U \to \mathbb{R}$ be a C^k function. Then the polynomial of degree k

$$P^k_{f,\mathbf{a}}(\mathbf{a}+\vec{\mathbf{h}}) \stackrel{\text{def}}{=} \sum_{m=0}^{k}\sum_{I\in\mathcal{I}_n^m}\frac{1}{I!}D_If(\mathbf{a})\vec{\mathbf{h}}^I \qquad 3.3.26$$

is called the *Taylor polynomial* of degree k of f at $\mathbf{a}$.

If $\mathbf{f}: U \to \mathbb{R}^n$ is a C^k function, its Taylor polynomial is the polynomial map $U \to \mathbb{R}^n$ whose coordinate functions are the Taylor polynomials of the coordinate functions of $\mathbf{f}$.

---

[10]This corresponds to $D_1^2D_2^3(3x_1^2x_2^2)$; already, $D_2^3(3x_1^2x_2^2) = 0$.

In equation 3.3.26, remember that I is a multi-exponent; if you want to write the polynomial out in particular cases, it can get complicated, especially if k or n is big.

Example 3.3.14 below illustrates notation; it has no mathematical content. Note that the first term (the term of degree $m = 0$) corresponds to the 0th derivative, i.e., the function f itself.

Example 3.3.14 (Multi-exponent notation for a Taylor polynomial of a function in two variables). Let $f: \mathbb{R}^2 \to \mathbb{R}$ be a function. The formula for the Taylor polynomial of degree 2 of f at $\mathbf{a}$ is then

$$\begin{aligned} P^2_{f,\mathbf{a}}(\mathbf{a}+\vec{\mathbf{h}}) &= \sum_{m=0}^{2} \sum_{I \in \mathcal{I}_2^m} \frac{1}{I!} D_I f(\mathbf{a}) \vec{\mathbf{h}}^I \qquad 3.3.27 \\ &= \underbrace{\frac{1}{0!0!} D_{(0,0)} f(\mathbf{a}) h_1^0 h_2^0}_{m=0;\ \text{i.e., } f(\mathbf{a})} + \underbrace{\frac{1}{1!0!} D_{(1,0)} f(\mathbf{a}) h_1^1 h_2^0 + \frac{1}{0!1!} D_{(0,1)} f(\mathbf{a}) h_1^0 h_2^1}_{\text{terms of degree 1: first derivatives}} \\ &+ \underbrace{\frac{1}{2!0!} D_{(2,0)} f(\mathbf{a}) h_1^2 h_2^0 + \frac{1}{1!1!} D_{(1,1)} f(\mathbf{a}) h_1 h_2 + \frac{1}{0!2!} D_{(0,2)} f(\mathbf{a}) h_1^0 h_2^2}_{\text{terms of degree 2: second derivatives}}, \end{aligned}$$

which we can write more simply as

$$\begin{aligned} P^2_{f,\mathbf{a}}(\mathbf{a}+\vec{\mathbf{h}}) &= f(\mathbf{a}) + D_{(1,0)} f(\mathbf{a}) h_1 + D_{(0,1)} f(\mathbf{a}) h_2 \qquad 3.3.28 \\ &+ \frac{1}{2} D_{(2,0)} f(\mathbf{a}) h_1^2 + D_{(1,1)} f(\mathbf{a}) h_1 h_2 + \frac{1}{2} D_{(0,2)} f(\mathbf{a}) h_2^2. \end{aligned}$$

Remember that $D_{(1,0)} f = D_1 f$, while $D_{(0,1)} f = D_2 f$, and so on. △

Example 3.3.14: Remember (Definition 3.3.6) that

$$\mathbf{x}^I = x_1^{i_1} \dots x_n^{i_n};$$

similarly, $\vec{\mathbf{h}}^I = h_1^{i_1} \dots h_n^{i_n}$. For instance, if $I = (1,1)$ we have

$$\vec{\mathbf{h}}^I = \vec{\mathbf{h}}^{(1,1)} = h_1 h_2;$$

if $I = (2,0,3)$ we have

$$\vec{\mathbf{h}}^I = \vec{\mathbf{h}}^{(2,0,3)} = h_1^2 h_3^3.$$

Since the crossed partials of f are equal,

$$D_{(1,1)} f(\mathbf{a}) h_1 h_2 = \frac{1}{2} D_1 D_2 f(\mathbf{a}) h_1 h_2 + \frac{1}{2} D_2 D_1 f(\mathbf{a}) h_1 h_2.$$

The term $1/I!$ in the formula for the Taylor polynomial weights the various terms to take into account the existence of crossed partials. This advantage of multi-exponent notation is increasingly useful as n gets big.

What are the terms of degree 2 (second derivatives) of the Taylor polynomial at $\mathbf{a}$, of degree 2, of a function with three variables?[11]

Example 3.3.15 (Computing a Taylor polynomial). What is the Taylor polynomial of degree 2 of the function $f\begin{pmatrix} x \\ y \end{pmatrix} = \sin(x+y^2)$, at $\mathbf{0} = \begin{pmatrix} 0 \\ 0 \end{pmatrix}$? The first term, of degree 0, is $f(\mathbf{0}) = \sin 0 = 0$. The terms of degree 1 are $D_{(1,0)} f\begin{pmatrix} x \\ y \end{pmatrix} = \cos(x+y^2)$ and $D_{(0,1)} f\begin{pmatrix} x \\ y \end{pmatrix} = 2y\cos(x+y^2)$,

---

[11] The third term of $P^2_{f,\mathbf{a}}(\mathbf{a}+\vec{\mathbf{h}}) = \sum_{m=0}^{2} \sum_{I \in \mathcal{I}_3^m} \frac{1}{I!} D_I f(\mathbf{a}) \vec{\mathbf{h}}^I$ is

$$\overbrace{D_{(1,1,0)} f(\mathbf{a}) h_1 h_2}^{D_1 D_2} + \overbrace{D_{(1,0,1)} f(\mathbf{a}) h_1 h_3}^{D_1 D_3} + \overbrace{D_{(0,1,1)} f(\mathbf{a}) h_2 h_3}^{D_2 D_3}$$
$$+ \overbrace{\frac{1}{2} D_{(2,0,0)} f(\mathbf{a}) h_1^2}^{D_1^2} + \overbrace{\frac{1}{2} D_{(0,2,0)} f(\mathbf{a}) h_2^2}^{D_2^2} + \overbrace{\frac{1}{2} D_{(0,0,2)} f(\mathbf{a}) h_3^2}^{D_3^2}.$$

so $D_{(1,0)}f\begin{pmatrix}0\\0\end{pmatrix} = 1$ and $D_{(0,1)}f\begin{pmatrix}0\\0\end{pmatrix} = 0$. For the terms of degree 2,

$$\begin{aligned} D_{(2,0)}f\begin{pmatrix}x\\y\end{pmatrix} &= -\sin(x+y^2)\\ D_{(1,1)}f\begin{pmatrix}x\\y\end{pmatrix} &= -2y\sin(x+y^2)\\ D_{(0,2)}f\begin{pmatrix}x\\y\end{pmatrix} &= \ 2\cos(x+y^2) - 4y^2\sin(x+y^2); \end{aligned} \qquad 3.3.29$$

at $\mathbf{0}$, these give 0, 0, and 2. So the Taylor polynomial of degree 2 is

$$P^2_{f,\mathbf{0}}\left(\begin{bmatrix}h_1\\h_2\end{bmatrix}\right) = 0 + h_1 + 0 + 0 + 0 + \frac{2}{2}h_2^2. \qquad 3.3.30$$

For the Taylor polynomial of degree 3, we compute

In Example 3.4.5 we will see how to reduce this computation to two lines, using rules we will give for computing Taylor polynomials.

$$\begin{aligned} D_{(3,0)}f\begin{pmatrix}x\\y\end{pmatrix} &= D_1\big(-\sin(x+y^2)\big) = -\cos(x+y^2)\\ D_{(0,3)}f\begin{pmatrix}x\\y\end{pmatrix} &= -4y\sin(x+y^2) - 8y\sin(x+y^2) - 8y^3\cos(x+y^2)\\ D_{(2,1)}f\begin{pmatrix}x\\y\end{pmatrix} &= D_1\big(-2y\sin(x+y^2)\big) = -2y\cos(x+y^2)\\ D_{(1,2)}f\begin{pmatrix}x\\y\end{pmatrix} &= -2\sin(x+y^2) - 4y^2\cos(x+y^2). \end{aligned} \qquad 3.3.31$$

At $\mathbf{0}$ all are 0 except $D_{(3,0)}$, which is -1. So the term of degree 3 is $(-\frac{1}{3!})h_1^3 = -\frac{1}{6}h_1^3$, and the Taylor polynomial of degree 3 of f at $\mathbf{0}$ is

$$P^3_{f,\mathbf{0}}\left(\begin{bmatrix}h_1\\h_2\end{bmatrix}\right) = h_1 + h_2^2 - \frac{1}{6}h_1^3. \qquad \triangle \quad 3.3.32$$

Theorem 3.3.16 (Taylor's theorem without remainder in higher dimensions). *Let $U \subset \mathbb{R}^n$ be open, $\mathbf{a} \in U$ a point, and $f : U \to \mathbb{R}$ a C^k function.*

1. *The polynomial $P^k_{f,\mathbf{a}}(\mathbf{a}+\vec{\mathbf{h}})$ is the unique polynomial of degree k with the same partial derivatives up to order k at $\mathbf{a}$ as f.*
2. *It best approximates f near $\mathbf{a}$: it is the unique polynomial of degree at most k such that*

$$\lim_{\vec{\mathbf{h}}\to\vec{\mathbf{0}}} \frac{f(\mathbf{a}+\vec{\mathbf{h}}) - P^k_{f,\mathbf{a}}(\mathbf{a}+\vec{\mathbf{h}})}{|\vec{\mathbf{h}}|^k} = 0. \qquad 3.3.33$$

In Theorem 3.3.16, note that since we are dividing by a high power of $|\vec{\mathbf{h}}|$ and $\vec{\mathbf{h}}$ is small, the limit being 0 means that the numerator is very small.

We must require that the partial derivatives be continuous; if they aren't, the statement isn't true even when $k = 1$, as you will see if you look at equation 1.9.10, where f is the function of Example 1.9.5, a function whose partial derivatives are not continuous.

Taylor's theorem with remainder is discussed in Appendix A12.

To prove Theorem 3.3.16 we need the following proposition, which says that if all the partial derivatives of a function g up to some order k equal 0 at a point $\mathbf{a}$, then g is small compared to $|\vec{\mathbf{h}}|^k$ in a neighborhood of $\mathbf{a}$. It is proved in Appendix A10.

Proposition 3.3.17 (Size of a function with many vanishing partial derivatives). *Let U be an open subset of $\mathbb{R}^n$ and let $g : U \to \mathbb{R}$ be a C^k function. If at $\mathbf{a} \in U$ all partial derivatives of g up to order k vanish (including $g(\mathbf{a})$, the 0th partial derivative), then*

$$\lim_{\vec{\mathbf{h}} \to \vec{\mathbf{0}}} \frac{g(\mathbf{a}+\vec{\mathbf{h}})}{|\vec{\mathbf{h}}|^k} = 0. \tag{3.3.34}$$

Equation 3.3.35: The right side follows the general multi-exponent notation for a polynomial given by formula 3.3.8:

$$\sum_{m=0}^{k} \sum_{I \in \mathcal{I}_n^m} a_I \mathbf{x}^I.$$

Here, $\vec{\mathbf{h}}$ corresponds to $\mathbf{x}$, and $\frac{1}{J!} D_J f(\mathbf{a})$ corresponds to a_J.

Proposition 3.3.11: for a polynomial p with coefficient a_I, we have

$$D_I p(\mathbf{0}) = I! a_I.$$

Proof of Theorem 3.3.16. 1. Let $Q^k_{f,\mathbf{a}}$ be the polynomial that, evaluated at $\vec{\mathbf{h}}$, gives the same result as the Taylor polynomial $P^k_{f,\mathbf{a}}$ evaluated at $\mathbf{a}+\vec{\mathbf{h}}$:

$$P^k_{f,\mathbf{a}}(\mathbf{a}+\vec{\mathbf{h}}) = Q^k_{f,\mathbf{a}}(\vec{\mathbf{h}}) = \overbrace{\sum_{m=0}^{k} \sum_{J \in \mathcal{I}_n^m} \underbrace{\frac{1}{J!} D_J f(\mathbf{a})}_{\text{coefficient}} \underbrace{\vec{\mathbf{h}}^J}_{\text{variable}}}^{\text{multi-exponent notation for polynomial } Q^k_{f,\mathbf{a}}}. \tag{3.3.35}$$

By Proposition 3.3.11,

$$\overbrace{D_I Q^k_{f,\mathbf{a}}(\mathbf{0})}^{D_I p(\mathbf{0})} = \overbrace{I! \underbrace{\frac{1}{I!} D_I f(\mathbf{a})}_{\text{coeff. of } \vec{\mathbf{h}}^I \text{ from eq. 3.3.35}}}^{I! \quad a_I} = D_I f(\mathbf{a}). \tag{3.3.36}$$

Since $D_I Q^k_{f,\mathbf{a}}(\mathbf{0}) = D_I P^k_{f,\mathbf{a}}(\mathbf{a})$, we have $D_I P^k_{f,\mathbf{a}}(\mathbf{a}) = D_I f(\mathbf{a})$; the partial derivatives of $P^k_{f,\mathbf{a}}$, up to order k, are the same as the partial derivatives of f, up to order k.

Part 2 then follows from Proposition 3.3.17. To lighten notation, denote by $g(\mathbf{a}+\vec{\mathbf{h}})$ the difference between $f(\mathbf{a}+\vec{\mathbf{h}})$ and the Taylor polynomial of f at $\mathbf{a}$. Since all partials of g up to order k vanish, by Proposition 3.3.17,

$$\lim_{\vec{\mathbf{h}} \to \vec{\mathbf{0}}} \frac{g(\mathbf{a}+\vec{\mathbf{h}})}{|\vec{\mathbf{h}}|^k} = 0. \quad \square \tag{3.3.37}$$

Exercises for Section 3.3

3.3.1 Compute $D_1(D_2 f)$, $D_2(D_3 f)$, $D_3(D_1 f)$, and $D_1\Big(D_2(D_3 f)\Big)$ for the function f defined by $f\begin{pmatrix} x \\ y \\ z \end{pmatrix} = x^2y + xy^2 + yz^2$.

3.3.2 a. Write out the polynomial $\sum_{m=0}^{5} \sum_{I \in \mathcal{I}_3^m} a_I \mathbf{x}^I$, where

$$a_{(0,0,0)} = 4, \quad a_{(0,1,0)} = 3, \quad a_{(1,0,2)} = 4, \quad a_{(1,1,2)} = 2,$$
$$a_{(2,2,0)} = 1, \quad a_{(3,0,2)} = 2, \quad a_{(5,0,0)} = 3,$$

and all other $a_I = 0$, for $I \in \mathcal{I}_3^m$ for $m \le 5$.

b. Write the polynomial $2x_2 + x_1x_2 - x_1x_2x_3 + x_1^2 + 5x_2^2x_3$ using multi-exponent notation.

c. Write $3x_1x_2 - x_2x_3x_4 + 2x_2^2x_3 + x_2^2x_4^4 + x_2^5$ using multi-exponent notation.

3.3.3 Following the format of Example 3.3.14, write the terms of the Taylor polynomial of degree 2 of a function f of three variables x, y, z at a point $\mathbf{a}$.

3.3.4 a. Redo Example 3.3.14, finding the Taylor polynomial of degree 3.

b. Repeat, for degree 4.

3.3.5 Find the cardinality (number of elements) of $\mathcal{I}_n^m$.

3.3.6 a. Let f be a real-valued C^k function defined in a neighborhood of $0 \in \mathbb{R}$. Suppose $f(-x) = -f(x)$. Consider the Taylor polynomial of f at 0; show that all the coefficients of even powers $(a_0, a_2, \dots)$ vanish.

b. Let f be a real-valued C^k function defined in a neighborhood of $\mathbf{0} \in \mathbb{R}^n$. Suppose $f(-\mathbf{x}) = -f(\mathbf{x})$. Consider the Taylor polynomial of f at the origin; show that the coefficients of all terms of even total degree vanish.

3.3.7 Show that if $I \in \mathcal{I}_n^m$, then $(x\vec{\mathbf{h}})^I = x^m \vec{\mathbf{h}}^I$.

3.3.8 For the function f of Example 3.3.9, show that all first and second partial derivatives exist everywhere, that the first partial derivatives are continuous, and that the second partial derivatives are not.

3.3.9 Consider the function

$$f\begin{pmatrix} x \\ y \end{pmatrix} = \begin{cases} \dfrac{x^2y(x-y)}{x^2+y^2} & \text{if } \begin{pmatrix} x \\ y \end{pmatrix} \ne \begin{pmatrix} 0 \\ 0 \end{pmatrix} \\ 0 & \text{if } \begin{pmatrix} x \\ y \end{pmatrix} = \begin{pmatrix} 0 \\ 0 \end{pmatrix}. \end{cases}$$

a. Compute $D_1 f$ and $D_2 f$. Is f of class C^1?

b. Show that all second partial derivatives of f exist everywhere.

c. Show that $D_1\left(D_2 f\begin{pmatrix} 0 \\ 0 \end{pmatrix}\right) \ne D_2\left(D_1 f\begin{pmatrix} 0 \\ 0 \end{pmatrix}\right)$.

d. Why doesn't this contradict Theorem 3.3.8?

The object of Exercise 3.3.10 is to illustrate how long successive derivatives become.

3.3.10 a. Compute the derivatives of $\big(1 + f(x)\big)^m$, up to and including the fourth derivative.

b. Guess how many terms the fifth derivative will have.

**c. Guess how many terms the nth derivative will have.

L'Hôpital's rule (used in Exercise 3.3.11): If $f, g : U \to \mathbb{R}$ are k times differentiable in a neighborhood $U \subset \mathbb{R}$ of 0, and

$$f(0) = \dots = f^{(k-1)}(0) = 0,$$
$$g(0) = \dots = g^{(k-1)}(0) = 0,$$

with $g^{(k)}(0) \ne 0$, then

$$\lim_{t \to 0} \frac{f(t)}{g(t)} = \frac{f^{(k)}(0)}{g^{(k)}(0)}.$$

3.3.11 Prove Theorem 3.3.1. *Hint*: Compute

$$\lim_{h \to 0} \frac{f(a+h) - \left(f(a) + f'(a)h + \dots + \frac{f^{(k)}(a)}{k!}h^k\right)}{h^k}$$

by differentiating, k times, the top and bottom with respect to h, checking each time that the hypotheses of l'Hôpital's rule are satisfied (see the margin note).

Exercise 3.3.12: The techniques of Section 3.4 will make it much easier to solve this exercise. For now, you must do it the hard way.

3.3.12 a. Write to degree 3 the Taylor polynomial of $\sqrt{1 + \dfrac{x+y}{1+xz}}$ at the origin.

b. What is $D_{[1,1,1]} f \begin{pmatrix} 0 \\ 0 \\ 0 \end{pmatrix}$?

3.3.13 Find the Taylor polynomial of degree 2 of the function

$$f\begin{pmatrix} x \\ y \end{pmatrix} = \sqrt{x + y + xy} \quad \text{at the point} \quad \begin{pmatrix} -2 \\ -3 \end{pmatrix}.$$

3.3.14 True or false? Suppose a function $f : \mathbb{R}^2 \to \mathbb{R}$ satisfies Laplace's equation $D_1^2 f + D_2^2 f = 0$. Then the function

$$g\begin{pmatrix} x \\ y \end{pmatrix} = f\begin{pmatrix} x/(x^2+y^2) \\ y/(x^2+y^2) \end{pmatrix} \quad \text{also satisfies Laplace's equation.}$$

Exercise 3.3.15: If S (or, more generally, a k-dimensional manifold M) is of class C^m, $m \geq 2$, then the Taylor polynomial of $\mathbf{f}$ starts with quadratic terms.

These quadratic terms are 1/2 the *second fundamental form* of M at $\mathbf{a}$. We will meet the second fundamental form again in Section 3.9.

3.3.15 Prove the following result: Let $M \subset \mathbb{R}^n$ be a k-dimensional manifold and $\mathbf{a} \in M$ a point. If A is an orthogonal $n \times n$ matrix mapping Span $(\vec{\mathbf{e}}_1, \dots, \vec{\mathbf{e}}_k)$ to $T_{\mathbf{a}}M$, and $\mathbf{F} : \mathbb{R}^n \to \mathbb{R}^n$ is defined by $\mathbf{F}(\mathbf{x}) = \mathbf{a} + A\mathbf{x}$, then $\mathbf{F}^{-1}(M)$ is locally near $\mathbf{0} \in \mathbb{R}^n$ the graph of a map $\mathbf{h}$: Span $(\vec{\mathbf{e}}_1, \dots, \vec{\mathbf{e}}_k) \to$ Span $(\vec{\mathbf{e}}_{k+1}, \dots, \vec{\mathbf{e}}_n)$ such that $\mathbf{h}(\mathbf{0}) = \mathbf{0}$ and $[\mathbf{Dh}(\mathbf{0})] = [0]$.

(In the case of a surface $S \in \mathbb{R}^3$ and a point $\mathbf{a} \in S$, this statement means that we can translate and rotate S to move $\mathbf{a}$ to the origin and the tangent plane to the (x, y)-plane, in which case S becomes locally the graph of a function expressing z as a function of x and y, whose derivative vanishes at the origin.)

3.4 Rules for computing Taylor polynomials

Since the computation of successive derivatives is always painful, we recommend (when it is possible) considering the function as being obtained from simpler functions by elementary operations (sum, product, power, etc.).—Jean Dieudonné, *Calcul Infinitésimal*

FIGURE 3.4.1.

Edmund Landau (1877–1938) taught at Göttingen until forced out by the Nazis in 1933. The mathematician Oswald Teichmüller, a Nazi, organized a boycott of Landau's lectures.

Computing Taylor polynomials is very much like computing derivatives; when the degree is 1, they are essentially the same. Just as we have rules for differentiating sums, products, compositions, etc., there are rules for computing Taylor polynomials of functions obtained by combining simpler functions. Since computing partial derivatives rapidly becomes unpleasant, we strongly recommend using these rules.

To write down the Taylor polynomials of some standard functions, we will use notation invented by Edmund Landau to express the idea that one is computing "up to terms of degree k": the notation o, or "little o" (pronounced "little oh"). In the equations of Proposition 3.4.2 the o term may look like a remainder, but such terms do not give a precise, computable remainder. Little o provides a way to bound one function by another function, in an *unspecified* neighborhood of the point at which you are computing the Taylor polynomial.

Definition 3.4.1 (Little o). *Little o*, denoted o, means "smaller than", in the following sense: Let $U \subset \mathbb{R}^n$ be a neighborhood of $\mathbf{0}$, and let $f, h: U - \{\mathbf{0}\} \to \mathbb{R}$ be two functions, with $h > 0$. Then f is in $o(h)$ if

$$\lim_{\mathbf{x} \to \mathbf{0}} \frac{f(\mathbf{x})}{h(\mathbf{x})} = 0. \qquad 3.4.1$$

Thus as $\mathbf{x} \to \mathbf{0}$, the number $|f(\mathbf{x})|$ becomes infinitely smaller than $h(\mathbf{x})$. For example, anything in $o(|\mathbf{x}|)$ is smaller than $h(\mathbf{x}) = |\mathbf{x}|$ near $\mathbf{x} = \mathbf{0}$; it is smaller than any constant $\epsilon > 0$ times $|\mathbf{x}|$, no matter how small ϵ. But it is not necessarily as small as $|\mathbf{x}|^2$. For instance, $|x|^{3/2}$ is in $o(|x|)$, but it is bigger than $|x|^2$. (That is why, at the beginning of Section 3.3, we said that at a point $\mathbf{x}$, a function and its derivative "usually" differ by terms that are quadratic and higher: terms may exist that are smaller than linear, but bigger than quadratic.)

Using this notation, we can rewrite equation 3.3.33 as

$$f(\mathbf{a} + \vec{\mathbf{h}}) - P^k_{f,\mathbf{a}}(\mathbf{a} + \vec{\mathbf{h}}) \in o(|\vec{\mathbf{h}}|^k). \qquad 3.4.2$$

Very often Taylor polynomials written in terms of bounds with little o are good enough. But in settings where you want to know the error for some particular $\mathbf{x}$, something stronger is required: Taylor's theorem with remainder, discussed in Appendix A12.

Remark. When one is dealing with functions that can be approximated by Taylor polynomials, the only functions h of interest are the functions $|x|^k$ for $k \geq 0$: if a function is C^k it has to look like a nonnegative integer power of x. In other settings, it is interesting to compare nastier functions (not C^k) to a broader class of functions, for instance, one might wish to bound functions by functions such as $\sqrt{|x|}$ or $|x| \ln |x|$ .... (For an example of a "nastier function", see equation 5.3.23.) The art of making such comparisons is called the theory of *asymptotic developments*.

A famous example of an asymptotic development is the *prime number theorem*, which states that if $\pi(x)$ represents the number of prime numbers smaller than x, then, for x near ∞,

$$\pi(x) = \frac{x}{\ln x} + o\left(\frac{x}{\ln x}\right). \qquad 3.4.3$$

(Here π has nothing to do with $\pi \approx 3.1416$.) This was proved independently in 1898 by Jacques Hadamard and Charles de la Vallée-Poussin, after being conjectured a century earlier by Gauss. △

Equation 3.4.3: The stronger statement,

$$\pi(x) = \int_1^x \frac{1}{\ln u}\, du + o\left(|x|^{\frac{1}{2}+\epsilon}\right),$$

for all $\epsilon > 0$, is known as the *Riemann hypothesis*. It is the most famous outstanding problem of mathematics.

In Proposition 3.4.2 we list the functions whose Taylor polynomials we expect you to know from first year calculus. We will write them only near 0, but by translation they can be written at any point where the function is defined. Note that in the equations of Proposition 3.4.2, the Taylor polynomial is the expression on the right side *excluding* the little o term, which indicates how good an approximation the Taylor polynomial is to the corresponding function, without giving any precision.

Strictly speaking, the equations of Proposition 3.4.2 are sloppily written; the "$+o$" term is not a function that makes the equality true but an indication of the kind of function needed to make it true. It would be more correct to use the format of equation 3.4.2, writing, for example,

$$e^x - \left(1 + x + \frac{x^2}{2!} + \cdots + \frac{x^n}{n!}\right) \in o(|x|^n). \qquad 3.4.4$$

An *odd function* is a function f such that $f(-x) = -f(x)$; an *even function* is a function f such that $f(-x) = f(x)$. The Taylor function of an odd function can have only terms of odd degree, and the Taylor function of an even function can have only terms of even degree. Thus the Taylor polynomial for sine contains only odd terms, with alternating signs, while the Taylor polynomial for cosine contains only even terms, again with alternating signs.

The margin note on the first page of this chapter gives the coefficients of a good approximation of $\sin x$ for $|x| \le \pi/2$. They are not exactly equal to the coefficients in equation 3.4.6, because they are optimized to approximate sine over a bigger range.

Equation 3.4.9 is the *binomial formula*.

The proof of Proposition 3.4.2 is left as Exercise 3.4.10.

Note that in the Taylor polynomial of $\ln(1+x)$, there are no factorials in the denominators.

Proposition 3.4.2 (Taylor polynomials of some standard functions). *The following formulas give the Taylor polynomials at 0 of the corresponding functions:*

$$e^x = 1 + x + \frac{x^2}{2!} + \cdots + \frac{x^n}{n!} + o(|x|^n) \qquad 3.4.5$$

$$\sin(x) = x - \frac{x^3}{3!} + \frac{x^5}{5!} - \cdots + (-1)^n \frac{x^{2n+1}}{(2n+1)!} + o(|x|^{2n+1}) \qquad 3.4.6$$

$$\cos(x) = 1 - \frac{x^2}{2!} + \frac{x^4}{4!} - \cdots + (-1)^n \frac{x^{2n}}{(2n)!} + o(|x|^{2n}) \qquad 3.4.7$$

$$\ln(1+x) = x - \frac{x^2}{2} + \cdots + (-1)^{n+1}\frac{x^n}{n} + o(|x|^n) \qquad 3.4.8$$

$$(1+x)^m = 1 + mx + \frac{m(m-1)}{2!}x^2 + \frac{m(m-1)(m-2)}{3!}x^3 + \cdots$$
$$+ \frac{m(m-1)\dots(m-(n-1))}{n!}x^n + o(|x|^n). \qquad 3.4.9$$

Propositions 3.4.3 and 3.4.4, which show how to combine Taylor polynomials, are proved in Appendix A11. We state and prove them for scalar-valued functions, but they are true for vector-valued functions, at least whenever the latter make sense. For instance, the product could be replaced by a dot product, or the product of a scalar with a vector-valued function, or a matrix multiplication. See Exercise 3.22.

Proposition 3.4.3 (Sums and products of Taylor polynomials). *Let $U \subset \mathbb{R}^n$ be open, and let $f, g : U \to \mathbb{R}$ be C^k functions. Then $f+g$ and fg are also of class C^k, and*

1. *The Taylor polynomial of the sum is the sum of the Taylor polynomials:*

$$P^k_{f+g,\mathbf{a}}(\mathbf{a}+\vec{\mathbf{h}}) = P^k_{f,\mathbf{a}}(\mathbf{a}+\vec{\mathbf{h}}) + P^k_{g,\mathbf{a}}(\mathbf{a}+\vec{\mathbf{h}}). \qquad 3.4.10$$

2. *The Taylor polynomial of the product fg is obtained by taking the product*

$$P^k_{f,\mathbf{a}}(\mathbf{a}+\vec{\mathbf{h}}) \cdot P^k_{g,\mathbf{a}}(\mathbf{a}+\vec{\mathbf{h}}) \qquad 3.4.11$$

and discarding the terms of degree $> k$.

Remark. In the proof of Theorem 5.4.6, we will see a striking example of how Proposition 3.4.3 can simplify computations even when the final goal is not to compute a Taylor polynomial. In that proof we compute the derivative of a function by breaking it up into its component parts, computing the Taylor polynomials to degree 1 of those parts, combining them to get the Taylor polynomial to degree 1 of the function, and then discarding the constant terms. Using Taylor polynomials allows us to discard irrelevant higher-degree terms *before* doing our computations. △

Proposition 3.4.4: Why does formula 3.4.12 make sense? The polynomial $P^k_{f,b}(b+u)$ is a good approximation to $f(b+u)$ only when $|u|$ is small. But our requirement that $g(\mathbf{a}) = b$ guarantees precisely that when $\vec{\mathbf{h}}$ is small,

$$P^k_{g,\mathbf{a}}(\mathbf{a}+\vec{\mathbf{h}}) = b + \text{something small}.$$

So it is reasonable to substitute that "something small" for the increment u when evaluating the polynomial $P^k_{f,b}(b+u)$.

Exercise 3.26 asks you to use Proposition 3.4.4 to derive Leibniz's rule.

Proposition 3.4.4 (Chain rule for Taylor polynomials). *Let $U \subset \mathbb{R}^n$ and $V \subset \mathbb{R}$ be open, and $g : U \to V$ and $f : V \to \mathbb{R}$ be of class C^k. Then $f \circ g : U \to \mathbb{R}$ is of class C^k, and if $g(\mathbf{a}) = b$, the Taylor polynomial $P^k_{f\circ g,\mathbf{a}}(\mathbf{a}+\vec{\mathbf{h}})$ is obtained by considering the polynomial*

$$\vec{\mathbf{h}} \mapsto P^k_{f,b}\big(P^k_{g,\mathbf{a}}(\mathbf{a}+\vec{\mathbf{h}})\big) \qquad 3.4.12$$

and discarding the terms of degree $> k$.

Note that the composition of two polynomials is a polynomial. For example, if $f(x) = x^3 + x + 1$ and $g(y) = y^2 - y$, then

$$g \circ f = (x^3+x+1)^2 - (x^3+x+1).$$

Example 3.4.5. Let's use these rules to compute the Taylor polynomial of degree 3 of the function $f\begin{pmatrix} x \\ y \end{pmatrix} = \sin(x+y^2)$ at $\mathbf{0}$, which we already saw in Example 3.3.15. Using Proposition 3.4.4, we simply substitute $x+y^2$ for u in $\sin u = u - u^3/6 + o(u^3)$, omitting all the terms of degree > 3:

$$\begin{aligned} \sin(x+y^2) &= (x+y^2) - \frac{(x+y^2)^3}{6} + \quad \text{terms of degree} > 3 \\ &= \underbrace{x + y^2 - \frac{x^3}{6}}_{\text{Taylor polynomial degree 3}} + \quad \text{terms of degree} > 3. \end{aligned} \qquad 3.4.13$$

Presto: Half a page becomes two lines. △

Equation 3.4.13: the "terms of degree > 3" in the second line includes terms of degree > 3 from

$$\frac{(x+y^2)^3}{6}$$

in the first line.

Example 3.4.6 (A harder example). Let $U \subset \mathbb{R}$ be open, and let $f : U \to \mathbb{R}$ be of class C^2. Let $V \subset U \times U$ be the subset of $\mathbb{R}^2$ where $f(x) + f(y) \neq 0$. Compute the Taylor polynomial of degree 2, at a point $\begin{pmatrix} a \\ b \end{pmatrix} \in V$, of the function $F : V \to \mathbb{R}$ given by

$$F\begin{pmatrix} x \\ y \end{pmatrix} = \frac{1}{f(x)+f(y)}. \qquad 3.4.14$$

Equation 3.4.15, line 2: When computing the Taylor polynomial of a quotient, a good tactic is to factor out the constant terms and apply equation 3.4.9 to what remains. Here, the constant term is $f(a) + f(b)$.

Set $\begin{pmatrix} x \\ y \end{pmatrix} = \begin{pmatrix} a+u \\ b+v \end{pmatrix}$. First we write $f(a+u)$ and $f(b+v)$ in terms of the Taylor polynomial of f evaluated at $a+u$ and $b+v$:

$$\begin{aligned} F\begin{pmatrix} a+u \\ b+v \end{pmatrix} &= \frac{1}{\big(f(a)+f'(a)u+f''(a)u^2/2+o(u^2)\big)+\big(f(b)+f'(b)v+f''(b)v^2/2+o(v^2)\big)} \\ &= \overbrace{\frac{1}{f(a)+f(b)}}^{\text{a constant}} \overbrace{\frac{1}{1+\dfrac{f'(a)u+f''(a)u^2/2+f'(b)v+f''(b)v^2/2}{f(a)+f(b)}}}^{(1+x)^{-1},\text{ where } x \text{ is the fraction in the denominator}} + o(u^2+v^2). \end{aligned} \qquad 3.4.15$$

The second factor is of the form[12] $(1+x)^{-1} = 1 - x + x^2 - \cdots$, leading to

$$F\begin{pmatrix} a+u \\ b+v \end{pmatrix} = \frac{1}{f(a)+f(b)}\left(1 - \frac{f'(a)u + f''(a)u^2/2 + f'(b)v + f''(b)v^2/2}{f(a)+f(b)}\right.$$
$$\left. + \left(\frac{f'(a)u + f''(a)u^2/2 + f'(b)v + f''(b)v^2/2}{f(a)+f(b)}\right)^2 - \cdots\right). \qquad 3.4.16$$

We discard the terms of degree > 2, to find

$$P^2_{F,\begin{pmatrix} a \\ b \end{pmatrix}}\begin{pmatrix} a+u \\ b+v \end{pmatrix} = \frac{1}{f(a)+f(b)} - \frac{f'(a)u+f'(b)v}{\big(f(a)+f(b)\big)^2} - \frac{f''(a)u^2+f''(b)v^2}{2\big(f(a)+f(b)\big)^2} + \frac{\big(f'(a)u+f'(b)v\big)^2}{\big(f(a)+f(b)\big)^3}. \quad \triangle \qquad 3.4.17$$

Taylor polynomials of implicit functions

We are particularly interested in Taylor polynomials of functions given by the inverse and implicit function theorems. Although these functions are only known via some limit process like Newton's method, their Taylor polynomials can be computed algebraically.

It follows from Theorem 3.4.7 that if you write the Taylor polynomial of the implicit function with undetermined coefficients, insert it into the equation specifying the implicit function, and identify like terms, you will be able to determine the coefficients. We do this in the proof of Theorem 3.9.10, in Appendix A15.

Assume we are in the setting of the implicit function theorem, where $\mathbf{F}$ is a function from a neighborhood of $\mathbf{c} = \begin{pmatrix} \mathbf{a} \\ \mathbf{b} \end{pmatrix} \in \mathbb{R}^{n+m}$ to $\mathbb{R}^n$, and $\mathbf{g} : V \to \mathbb{R}^n$ is an implicit function defined on an open subset $V \subset \mathbb{R}^m$ containing $\mathbf{b}$, with $\mathbf{g}(\mathbf{b}) = \mathbf{a}$, and such that $\mathbf{F}\begin{pmatrix} \mathbf{g}(\mathbf{y}) \\ \mathbf{y} \end{pmatrix} = \mathbf{0}$ for all $\mathbf{y} \in V$.

Theorem 3.4.7 (Taylor polynomial of implicit function). *Let $\mathbf{F}$ be a function of class C^k for some $k \geq 1$, such that $\mathbf{F}\begin{pmatrix} \mathbf{a} \\ \mathbf{b} \end{pmatrix} = \mathbf{0}$. Then the implicit function $\mathbf{g}$ is also of class C^k, and its Taylor polynomial of degree k, $P^k_{\mathbf{g},\mathbf{b}} : \mathbb{R}^m \to \mathbb{R}^n$, satisfies*

$$P^k_{\mathbf{F},\begin{pmatrix} \mathbf{a} \\ \mathbf{b} \end{pmatrix}}\begin{pmatrix} P^k_{\mathbf{g},\mathbf{b}}(\mathbf{b}+\vec{\mathbf{h}}) \\ \mathbf{b}+\vec{\mathbf{h}} \end{pmatrix} \in o(|\vec{\mathbf{h}}|^k). \qquad 3.4.18$$

It is the unique polynomial map of degree at most k that does so.

In Example 3.4.8 we use equation 2.10.29 to compute

$$g'(1) = a_1 = -1,$$

but this was not necessary; we could have used undetermined coefficients. If we had written

$$1 + a_1 h + a_2 h^2$$

for x rather than $1 - h + a_2 h^2$ in equation 3.4.22, the vanishing of the linear terms would have imposed $a_1 = -1$.

Proof. By Proposition 3.4.4, it is enough to show that $\mathbf{g}$ is of class C^k. By equation 2.10.29, if $\mathbf{F}$ is of class C^k, then $[\mathbf{Dg}]$ is of class C^{k-1}, hence $\mathbf{g}$ is of class C^k. $\square$

Example 3.4.8 (Taylor polynomial of an implicit function). The equation

$$F\begin{pmatrix} x \\ y \end{pmatrix} = x^3 + xy + y^3 - 3 = 0 \qquad 3.4.19$$

[12]The equation $(1+x)^{-1} = 1 - x + x^2 - \cdots$ is equation 3.4.9 with $m = -1$. We saw this in Example 0.5.6, where we had $\sum_{n=0}^{\infty} ar^n = \frac{a}{1-r}$.

implicitly expresses x in terms of y near $\begin{pmatrix}1\\1\end{pmatrix}$, since at that point the derivative $[3x^2+y\,,\,3y^2+x]$ is $[4\,,\,4]$, and 4 is invertible. So there exists g such that $g(1)=1$, i.e., $\begin{pmatrix}g(y)\\y\end{pmatrix}=\begin{pmatrix}1\\1\end{pmatrix}$. By equation 2.10.29, we have $g'(1)=-1$. We will compute $P^2_{g,1}$, the Taylor polynomial at 1 of this implicit function g to degree 2:

Of course, equation 3.4.19 also implicitly expresses y in terms of x near the same point.

$$g(1+h)=\underbrace{g(1)+g'(1)h+\frac{1}{2}g''(1)h^2}_{P^2_{g,1}}+\cdots=\underbrace{1-h+a_2h^2}_{P^2_{g,1}}+\cdots. \qquad 3.4.20$$

Thus we want to compute a_2 in equation 3.4.20. We will write

$$F\begin{pmatrix}P^2_{g,1}(1+h)\\1+h\end{pmatrix}\in o(h^2), \qquad 3.4.21$$

substituting F for $P^2_{F,\begin{pmatrix}1\\1\end{pmatrix}}$ in equation 3.4.18; this is allowed since they coincide through quadratic terms. This leads to

Equation 3.4.22: It is handy to know that

$$(1+x)^3=1+3x+3x^2+x^3.$$

$$\underbrace{(1-h+a_2h^2)^3}_{x^3}+\underbrace{(1-h+a_2h^2)(h+1)}_{xy}+\underbrace{(1+h)^3}_{y^3}-3\in o(h^2). \qquad 3.4.22$$

If we multiply out, the constant and the linear terms on the left vanish; the quadratic terms are

$$3a_2h^2+3h^2-h^2+a_2h^2+3h^2=(4a_2+5)h^2, \qquad 3.4.23$$

so to satisfy equation 3.4.22 we need $(4a_2+5)h^2=0$ (i.e., $a_2=-5/4$). Thus $g''(1)=-5/2$, and the Taylor polynomial at 1 of the implicit function g is

$$1-h-\frac{5}{4}h^2+\cdots; \qquad 3.4.24$$

its graph is shown in Figure 3.4.2. △

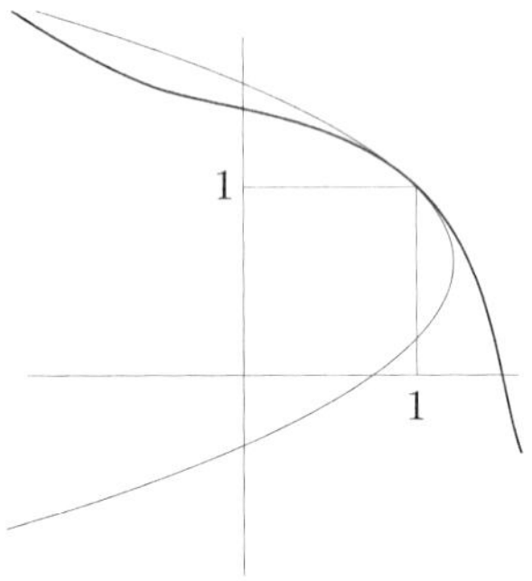

FIGURE 3.4.2.
The darker curve is the curve of equation $F(\mathbf{x})=0$, where

$$F(\mathbf{x})=x^3+xy+y^3-3.$$

The other curve, which is a parabola, is the graph of the Taylor polynomial of degree 2 of the implicit function, which represents x as a function of y. Near $\begin{pmatrix}1\\1\end{pmatrix}$, the two graphs are almost identical.

Example 3.4.9 (Taylor polynomial of implicit function: a harder example.) The equation $F\begin{pmatrix}x\\y\\z\end{pmatrix}=x^2+y^3+xyz^3-3=0$ determines z as a function of x,y in a neighborhood of $\begin{pmatrix}1\\1\\1\end{pmatrix}$, since $D_3F\begin{pmatrix}1\\1\\1\end{pmatrix}=3\neq 0$.

Thus there exists a function g such that $F\begin{pmatrix}x\\y\\g\begin{pmatrix}x\\y\end{pmatrix}\end{pmatrix}=0$. Let us compute the Taylor polynomial of the implicit function g to degree 2. We set

$$\begin{aligned}g\begin{pmatrix}x\\y\end{pmatrix}&=g\begin{pmatrix}1+u\\1+v\end{pmatrix} \qquad 3.4.25\\&=1+a_{1,0}u+a_{0,1}v+\frac{a_{2,0}}{2}u^2+a_{1,1}uv+\frac{a_{0,2}}{2}v^2+o(u^2+v^2).\end{aligned}$$

Inserting this expression for z into $x^2 + y^3 + xyz^3 - 3 = 0$ leads to

$$(1+u)^2+(1+v)^3+\left((1+u)(1+v)\left(1+a_{1,0}u+a_{0,1}v+\frac{a_{2,0}}{2}u^2+a_{1,1}uv+\frac{a_{0,2}}{2}v^2\right)^3\right)-3 \in o(u^2+v^2).$$

Now we multiply out and identify like terms. We get

Constant terms: $3 - 3 = 0$.

The linear terms could have been derived from equation 2.10.29, which gives in this case

$$\left[\mathbf{D}g\begin{pmatrix}1\\1\end{pmatrix}\right] = -[3]^{-1}[3,4]$$
$$= -[1/3][3,4] = [-1,-4/3].$$

Linear terms: $2u + 3v + u + v + 3a_{1,0}u + 3a_{0,1}v = 0$,

i.e., $a_{1,0} = -1$, $a_{0,1} = -4/3$.

Quadratic terms:

$$u^2\left(1+3a_{1,0}+3(a_{1,0})^2+\frac{3}{2}a_{2,0}\right)+v^2\left(3+3a_{0,1}+3(a_{0,1})^2+\frac{3}{2}a_{0,2}\right)$$
$$+\,uv(1+3a_{1,0}+3a_{0,1}+6a_{1,0}a_{0,1}+3a_{1,1})=0. \qquad 3.4.26$$

Identifying the coefficients of u^2, v^2, and uv to 0, and using $a_{1,0} = -1$ and $a_{0,1} = -4/3$ gives

$$a_{2,0} = -2/3,$$
$$a_{0,2} = -26/9, \qquad 3.4.27$$
$$a_{1,1} = -2/3.$$

Finally, this gives the Taylor polynomial of g:

$$P^2_{g,\begin{pmatrix}1\\1\end{pmatrix}} = 1-(x-1)-\frac{4}{3}(y-1)-\frac{1}{3}(x-1)^2-\frac{13}{9}(y-1)^2-\frac{2}{3}(x-1)(y-1). \qquad 3.4.28$$

Therefore,

$$g\begin{pmatrix}x\\y\end{pmatrix} = \overbrace{1-(x-1)-\frac{4}{3}(y-1)-\frac{1}{3}(x-1)^2-\frac{13}{9}(y-1)^2-\frac{2}{3}(x-1)(y-1)}^{P^2_{g,\begin{pmatrix}1\\1\end{pmatrix}}\text{: Taylor polynomial of degree 2 of } g \text{ at } \begin{pmatrix}1\\1\end{pmatrix}}$$
$$+ \underbrace{o\left((x-1)^2+(y-1)^2\right)}_{\text{length squared of } \begin{bmatrix}x-1\\y-1\end{bmatrix}\text{, small near } \begin{pmatrix}1\\1\end{pmatrix}}. \qquad 3.4.29$$

Exercises for Section 3.4

Exercise 3.4.2, part a: The techniques of this section make this easy; see Example 3.4.5.

Hint for part b: Use the Taylor polynomial of $1/(1+u)$.

3.4.1 Write, to degree 2, the Taylor polynomial of $f\begin{pmatrix}x\\y\end{pmatrix} = \sqrt{1+\sin(x+y)}$ at the origin.

3.4.2 a. What is the Taylor polynomial of degree 5 of $\sin(x+y^2)$ at the origin?

b. What is the Taylor polynomial of degree 4 of $1/(1+x^2+y^2)$ at the origin?

Exercise 3.4.3: Notice how much easier this exercise is with the techniques of Section 3.4 than it was without, in Exercise 3.3.13.

3.4.3 Find the Taylor polynomial of degree 2 of the function

$$f\begin{pmatrix} x \\ y \end{pmatrix} = \sqrt{x+y+xy} \quad \text{at the point } \begin{pmatrix} -2 \\ -3 \end{pmatrix}.$$

Exercise 3.4.4: Develop both

$$af(0)+bf(h)+cf(2h)$$

and $\int_0^h f(t)\,dt$

as Taylor polynomials of degree 3 in h, and equate coefficients.

3.4.4 Find numbers a, b, c such that when f is C^3,

$$h\Big(af(0)+bf(h)+cf(2h)\Big) - \int_0^h f(t)\,dt \in o(h^3).$$

3.4.5 Find numbers a, b, c such that when f is C^3,

$$h\Big(af(0)+bf(h)+cf(2h)\Big) - \int_0^{2h} f(t)\,dt \in o(h^3)$$

(same hint as for Exercise 3.4.4.)

3.4.6 Find a Taylor polynomial at the point $x = 0$ for the Fresnel integral $\varphi(x) = \int_0^x \sin t^2\,dt$ that approximates $\varphi(1)$ with an error smaller than 0.001.

3.4.7 Show that the equation $z = \sin(xyz)$ determines z as a function g of x, y near the point $\begin{pmatrix} \pi/2 \\ 1 \\ 1 \end{pmatrix}$. Find the Taylor polynomial of degree 2 of g at $\begin{pmatrix} \pi/2 \\ 1 \end{pmatrix}$. (We owe this exercise, and a number of others in Chapters 3 and 4, to Robert Piche.)

Exercise 3.4.8 generalizes Exercise 3.3.14. The equations

$$D_1h_1 - D_2h_2 = 0$$
$$D_2h_1 + D_1h_2 = 0$$

are known as the *Cauchy-Riemann equations*.

3.4.8 Prove that if $f : \mathbb{R}^2 \to \mathbb{R}$ satisfies Laplace's equation $D_1^2 f + D_2^2 f = 0$ and $\mathbf{h} : \mathbb{R}^2 \to \mathbb{R}^2$ satisfies the equations $D_1h_1 - D_2h_2 = 0$ and $D_2h_1 + D_1h_2 = 0$, then $f \circ \mathbf{h}$ satisfies Laplace's equation.

Hint for Exercise 3.4.9: Use the error estimate for alternating series.

3.4.9 Find a Taylor polynomial at $x = 0$ for the error function

$$\operatorname{erf}(x) = \frac{2}{\sqrt{\pi}} \int_0^x e^{-t^2}\,dt$$

that approximates erf(0.5) with an error smaller than 0.001.

3.4.10 Prove the formulas of Proposition 3.4.2.

Exercise 3.4.11 is much easier with the techniques of Section 3.4 than it was without, in Exercise 3.3.12.

3.4.11 a. Write, to degree 3, the Taylor polynomial of $\sqrt{1 + \dfrac{x+y}{1+xz}}$ at the origin.

b. What is $D_{[1,1,1]}f\begin{pmatrix} 0 \\ 0 \\ 0 \end{pmatrix}$?

3.4.12 Prove the general case of Proposition 3.2.12 (substituting C^p for C^1).

3.5 Quadratic forms

A quadratic form is a polynomial all of whose terms are of degree 2. For instance, $x_1^2 + x_2^2$ and $4x_1^2 + x_1x_2 - x_2^2$ are quadratic forms, as is x_1x_3. But $x_1x_2x_3$ is not a quadratic form; it is a cubic form.

Definition 3.5.1 (Quadratic form). A *quadratic form* $Q : \mathbb{R}^n \to \mathbb{R}$ is a polynomial function in the variables $x_1, \dots, x_n$, all of whose terms are of degree 2.

The margin note next to Exercise 3.5.13 defines a quadratic form on an abstract vector space.

A quadratic form in one dimension is simply an expression of the form ax^2, with $a \in \mathbb{R}$.

Although we will spend much of this section working on quadratic forms that look like $x^2 + y^2$ or $4x^2 + xy - y^2$, the following is a more realistic example. Most often, the quadratic forms one encounters in practice are integrals of functions, often functions in higher dimensions.

Example 3.5.2: The map Q is a function of the coefficients of p. If $p = x^2 + 2x + 1$, then

$$Q(\mathbf{a}) = Q\begin{pmatrix}1\\2\\1\end{pmatrix} = \int_0^1 (x^2 + 2x + 1)^2 dx.$$

Example 3.5.2 (An integral as a quadratic form). Let p be the polynomial $p(t) = a_0 + a_1 t + a_2 t^2$, and let $Q : \mathbb{R}^3 \to \mathbb{R}$ be the function

$$Q(\mathbf{a}) = \int_0^1 \big(p(t)\big)^2 \, dt. \tag{3.5.1}$$

Then Q is a quadratic form, as we can confirm by computing the integral:

$$\begin{aligned} Q(\mathbf{a}) &= \int_0^1 \left(a_0 + a_1 t + a_2 t^2\right)^2 dt \\ &= \int_0^1 (a_0^2 + a_1^2 t^2 + a_2^2 t^4 + 2a_0 a_1 t + 2a_0 a_2 t^2 + 2a_1 a_2 t^3)\, dt \qquad 3.5.2 \\ &= \left[a_0^2 t\right]_0^1 + \left[\frac{a_1^2 t^3}{3}\right]_0^1 + \left[\frac{a_2^2 t^5}{5}\right]_0^1 + \left[\frac{2a_0 a_1 t^2}{2}\right]_0^1 + \left[\frac{2a_0 a_2 t^3}{3}\right]_0^1 + \left[\frac{2a_1 a_2 t^4}{4}\right]_0^1 \\ &= a_0^2 + \frac{a_1^2}{3} + \frac{a_2^2}{5} + a_0 a_1 + \frac{2a_0 a_2}{3} + \frac{a_1 a_2}{2}. \end{aligned}$$

The quadratic form of Example 3.5.2 is fundamental in physics. If a tight string of length 1 (think of a guitar string) is pulled away from its equilibrium position, then its new position is the graph of a function f, and the potential energy of the string is

$$Q(f) = \int_0^1 \big(f(x)\big)^2 \, dx.$$

Similarly, the energy of an electromagnetic field is the integral of the square of the field, so if p is the electromagnetic field, $Q(p)$ gives the amount of energy between 0 and 1.

Above, p is polynomial of degree 2, but the function given by equation 3.5.1 is a quadratic form if p is a polynomial of *any* degree, not just quadratic. This is obvious if p is linear; if $a_2 = 0$, equation 3.5.2 becomes $Q(\mathbf{a}) = a_0^2 + a_1^2/3 + a_0 a_1$. Exercise 3.5.14, part c asks you to show that Q is a quadratic form if p is a cubic polynomial. $\triangle$

In contrast to quadratic forms, no one knows anything about cubic forms. This has ramifications for the understanding of manifolds. The abstract, algebraic view of a four-dimensional manifold is that it is a quadratic form over the integers; because integral quadratic forms are so well understood, a great deal of progress has been made in understanding 4-manifolds. But even the foremost researchers don't know how to approach six-dimensional manifolds; that would require knowing something about cubic forms.

In various guises, quadratic forms have been an important part of mathematics since the ancient Greeks. The quadratic formula, always the centerpiece of high school math, is one aspect.

A much deeper problem is the question, what whole numbers can be written in the form $x^2 + y^2$? Of course, any number a can be written $\sqrt{a}^2 + 0^2$, but suppose you impose that x and y be whole numbers. For instance, 5 can be written as a sum of two squares, $2^2 + 1^2$, but 3 and 7 cannot.

The classification of quadratic forms over the integers is thus a deep and difficult problem, though now reasonably well understood. But the classification over the reals, where we are allowed to extract square roots of positive numbers, is relatively easy. We will be discussing quadratic forms over the reals. In particular, we will be interested in classifying such quadratic forms by associating to each quadratic form two integers, together called its *signature*. In Section 3.6 we will see that signatures of quadratic forms can be used to analyze the behavior of a function at a critical point: a point where the derivative of the function is 0.

Quadratic forms as sums of squares

Essentially everything there is to say about real quadratic forms is summed up by Theorem 3.5.3, which says that a quadratic form can be represented as a *sum of squares of linearly independent linear functions* of the variables. The term "sum of squares" is traditional; it might be more accurate to call it a *combination* of squares, since some squares may be subtracted rather than added.

Theorem 3.5.3 (Quadratic forms as sums of squares).

1. *For any quadratic form $Q : \mathbb{R}^n \to \mathbb{R}$, there exist $m = k+l$ linearly independent linear functions $\alpha_1, \dots, \alpha_m : \mathbb{R}^n \to \mathbb{R}$ such that*

$$Q(\mathbf{x}) = \big(\alpha_1(\mathbf{x})\big)^2 + \dots + \big(\alpha_k(\mathbf{x})\big)^2 - \big(\alpha_{k+1}(\mathbf{x})\big)^2 - \dots - \big(\alpha_{k+l}(\mathbf{x})\big)^2. \quad 3.5.3$$

2. *The number k of plus signs and the number l of minus signs depends only on Q and not on the specific linear functions chosen.*

Theorem 3.5.3: The α_i are row matrices: elements of the vector space $\text{Mat}(1,n)$. Thus the notion of linear independence makes sense. Moreover, $\text{Mat}(1,n)$ has dimension n, so $m \le n$, since there can't be more than n linearly independent elements of a vector space of dimension n.

Definition 3.5.4 (Signature). The *signature* of a quadratic form is the pair of integers (k, l).

More than one quadratic form can have the same signature. The quadratic forms in Examples 3.5.6 and 3.5.7 have signature $(2, 1)$.

The signature of a quadratic form is sometimes called its *type*.

The word "signature" suggests, correctly, that the signature remains unchanged regardless of how the quadratic form is decomposed into a sum of linearly independent linear functions; it suggests, incorrectly, that the signature identifies a quadratic form.

Before giving a proof, or even a precise definition of the terms involved, we want to give some examples of the main technique used in the proof; a careful look at these examples should make the proof almost redundant.

A famous theorem due to Fermat asserts that a prime number $p \neq 2$ can be written as a sum of two squares if and only if the remainder after dividing p by 4 is 1. The proof of this and a world of analogous results (due to Fermat, Euler, Lagrange, Legendre, Gauss, Dirichlet, Kronecker, ...) led to algebraic number theory and the development of abstract algebra.

Completing squares to prove the quadratic formula

The proof is provided by an algorithm for finding the linearly independent functions α_i: *completing squares*. This technique is used in high school to prove the quadratic formula.

Indeed, to solve $ax^2 + bx + c = 0$, write

$$ax^2 + bx + c = ax^2 + bx + \left(\frac{b}{2\sqrt{a}}\right)^2 - \left(\frac{b}{2\sqrt{a}}\right)^2 + c = 0, \qquad 3.5.4$$

which gives

$$\left(\sqrt{a}x + \frac{b}{2\sqrt{a}}\right)^2 = \frac{b^2}{4a} - c. \qquad 3.5.5$$

Taking square roots gives

$$\sqrt{a}x + \frac{b}{2\sqrt{a}} = \pm\sqrt{\frac{b^2-4ac}{4a}}, \quad \text{leading to the famous formula} \tag{3.5.6}$$

$$x = \frac{-b \pm \sqrt{b^2-4ac}}{2a}. \tag{3.5.7}$$

Example 3.5.5 (Quadratic form as a sum of squares). Since

$$x^2 + xy = x^2 + xy + \frac{1}{4}y^2 - \frac{1}{4}y^2 = \left(x + \frac{y}{2}\right)^2 - \left(\frac{y}{2}\right)^2, \tag{3.5.8}$$

the quadratic form $Q(\mathbf{x}) = x^2 + xy$ can be written as $(\alpha_1(\mathbf{x}))^2 - (\alpha_2(\mathbf{x}))^2$, where α_1 and α_2 are the linear functions

$$\alpha_1\begin{pmatrix} x \\ y \end{pmatrix} = x + \frac{y}{2} \quad \text{and} \quad \alpha_2\begin{pmatrix} x \\ y \end{pmatrix} = \frac{y}{2}. \tag{3.5.9}$$

Express the quadratic form $x^2 + xy - y^2$ as a sum of squares, checking your answer in the footnote.[13]

Equations 3.5.9: Clearly the functions

$$\alpha_1\begin{pmatrix} x \\ y \end{pmatrix} = x + \frac{y}{2}, \quad \alpha_2\begin{pmatrix} x \\ y \end{pmatrix} = \frac{y}{2}$$

are linearly independent: no multiple of $y/2$ can give $x + y/2$. If we like, we can be systematic and write these functions as rows of a matrix:

$$\begin{bmatrix} 1 & 1/2 \\ 0 & 1/2 \end{bmatrix}.$$

It is not necessary to row reduce this matrix to see that the rows are linearly independent.

Example 3.5.6 (Completing squares: a more complicated example). Consider the quadratic form

$$Q(\mathbf{x}) = x^2 + 2xy - 4xz + 2yz - 4z^2. \tag{3.5.10}$$

We take all the terms in which x appears, which gives us $x^2 + (2y - 4z)x$. Since $x^2 + (2y-4z)x + (y-2z)^2 = (x+y-2z)^2$, adding and subtracting $(y-2z)^2$ yields

$$\begin{aligned} Q(\mathbf{x}) &= (x+y-2z)^2 - (y^2 - 4yz + 4z^2) + 2yz - 4z^2 \\ &= (x+y-2z)^2 - y^2 + 6yz - 8z^2. \end{aligned} \tag{3.5.11}$$

Collecting all remaining terms in which y appears and completing the square gives

$$Q(\mathbf{x}) = (x+y-2z)^2 - (y-3z)^2 + (z)^2. \tag{3.5.12}$$

In this case, the linear functions are

$$\alpha_1\begin{pmatrix} x \\ y \\ z \end{pmatrix} = x + y - 2z, \quad \alpha_2\begin{pmatrix} x \\ y \\ z \end{pmatrix} = y - 3z, \quad \alpha_3\begin{pmatrix} x \\ y \\ z \end{pmatrix} = z. \tag{3.5.13}$$

These functions are linearly independent, since writing each function as the row of a matrix gives $\begin{bmatrix} 1 & 1 & -2 \\ 0 & 1 & -3 \\ 0 & 0 & 1 \end{bmatrix}$, which is clearly invertible.

13

$$x^2 + xy - y^2 = x^2 + xy + \frac{y^2}{4} - \frac{y^2}{4} - y^2 = \left(x + \frac{y}{2}\right)^2 - \left(\frac{\sqrt{5}y}{2}\right)^2$$

This decomposition of $Q(\mathbf{x})$ is not the only possible one. Exercise 3.5.1 asks you to derive an alternative decomposition. $\triangle$

The algorithm for completing squares should be pretty clear: as long as the square of some coordinate function actually figures in the expression, every appearance of that variable can be incorporated into a perfect square; by subtracting off that perfect square, you are left with a quadratic form in precisely one fewer variable. (The "precisely one fewer variable" guarantees linear independence.) This works when there is at least one square, but what should you do with something like the following?

Example 3.5.7 (Quadratic form with no squares). Consider the quadratic form

$$Q(\mathbf{x}) = xy - xz + yz. \tag{3.5.14}$$

There wasn't anything magical about the choice of u, as Exercise 3.5.2 asks you to show; almost anything would have done.

One possibility is to introduce the new variable $u = x - y$, so that we can trade x for $u + y$, getting

$$\begin{aligned}(u+y)y - (u+y)z + yz = y^2 + uy - uz &= \left(y + \frac{u}{2}\right)^2 - \frac{u^2}{4} - uz - z^2 + z^2 \\ &= \left(y + \frac{u}{2}\right)^2 - \left(\frac{u}{2} + z\right)^2 + z^2 \\ &= \left(\frac{x}{2} + \frac{y}{2}\right)^2 - \left(\frac{x}{2} - \frac{y}{2} + z\right)^2 + z^2.\end{aligned} \tag{3.5.15}$$

Again, to check that the functions

$$\alpha_1\begin{pmatrix}x\\y\\z\end{pmatrix} = \frac{x}{2} + \frac{y}{2}, \quad \alpha_2\begin{pmatrix}x\\y\\z\end{pmatrix} = \frac{x}{2} - \frac{y}{2} + z, \quad \alpha_3\begin{pmatrix}x\\y\\z\end{pmatrix} = z \tag{3.5.16}$$

are linearly independent, we can write them as rows of a matrix:

$$\begin{bmatrix}1/2 & 1/2 & 0\\ 1/2 & -1/2 & 1\\ 0 & 0 & 1\end{bmatrix} \quad \text{row reduces to} \quad \begin{bmatrix}1 & 0 & 0\\ 0 & 1 & 0\\ 0 & 0 & 1\end{bmatrix}. \quad \triangle \tag{3.5.17}$$

Theorem 3.5.3 says that a quadratic form can be expressed as a sum of linearly independent functions of its variables, but it does not say that whenever a quadratic form is expressed as a sum of squares, those squares are necessarily linearly independent.

Example 3.5.8 (Squares that are not linearly independent). We can write $2x^2 + 2y^2 + 2xy$ as

$$x^2 + y^2 + (x+y)^2 \quad \text{or as} \quad \left(\sqrt{2}x + \frac{y}{\sqrt{2}}\right)^2 + \left(\sqrt{\frac{3}{2}}\,y\right)^2. \tag{3.5.18}$$

Only the second decomposition reflects Theorem 3.5.3. In the first, the three functions $\begin{bmatrix}x\\y\end{bmatrix} \mapsto x$, $\begin{bmatrix}x\\y\end{bmatrix} \mapsto y$, and $\begin{bmatrix}x\\y\end{bmatrix} \mapsto x + y$ are linearly dependent. $\triangle$

Proof of Theorem 3.5.3

All the essential ideas for the proof of part 1 are contained in the examples; a formal proof is in Appendix A13.

To prove part 2, which says that the signature (k, l) of a quadratic form does not depend on the specific linear functions chosen for its decomposition, we need to introduce some new vocabulary and a proposition.

Definition 3.5.9 (Positive and negative definite). A quadratic form Q is *positive definite* if $Q(\mathbf{x}) > 0$ when $\mathbf{x} \neq \mathbf{0}$. It is *negative definite* if $Q(\mathbf{x}) < 0$ when $\mathbf{x} \neq \mathbf{0}$.

Definition 3.5.9 is equivalent to saying that a quadratic form on $\mathbb{R}^n$ is positive definite if its signature is $(n, 0)$ (as Exercise 3.5.7 asks you to show) and negative definite if its signature is $(0, n)$.

An example of a positive definite quadratic form is the quadratic form $Q(\mathbf{x}) = |\mathbf{x}|^2$. Another is $Q(p) = \int_0^1 \big(p(t)\big)^2\, dt$, which we saw in Example 3.5.2. Example 3.5.10 gives an important example of a negative definite quadratic form.

The fact that the quadratic form of Example 3.5.10 is negative definite means that the *Laplacian* in one dimension (i.e., the transformation that takes p to p'') is negative. This has important ramifications; for example, it leads to stable equilibria in elasticity.

Example 3.5.10 (Negative definite quadratic form). Let P_k be the space of polynomials of degree $\leq k$, and let $V_{a,b} \subset P_k$ be the space of polynomials p that vanish at a and b for some $a < b$. Consider the quadratic form $Q : V_{a,b} \to \mathbb{R}$ given by

$$Q(p) = \int_a^b p(t)p''(t)\, dt. \qquad 3.5.19$$

Using integration by parts,

$$Q(p) = \int_a^b p(t)p''(t)\, dt = \overbrace{p(b)p'(b) - p(a)p'(a)}^{=0 \text{ by def.}} - \int_a^b \big(p'(t)\big)^2\, dt < 0. \qquad 3.5.20$$

Since $p \in V_{a,b}$, by definition $p(a) = p(b) = 0$; the integral is negative unless $p' = 0$ (i.e., unless p is constant); the only constant in $V_{a,b}$ is 0. △

Recall from Theorem 3.5.3 that when a quadratic form is written as a "sum" of squares of linearly independent functions, k is the number of squares preceded by a plus sign, and l is the number of squares preceded by a minus sign.

Proposition 3.5.11. *The number k is the largest dimension of a subspace of $\mathbb{R}^n$ on which Q is positive definite and the number l is the largest dimension of a subspace on which Q is negative definite.*

Proof. First let us show that Q cannot be positive definite on any subspace of dimension $> k$. Suppose

$$Q(\mathbf{x}) = \Big(\underbrace{\big(\alpha_1(\mathbf{x})\big)^2 + \cdots + \big(\alpha_k(\mathbf{x})\big)^2}_{k \text{ terms}}\Big) - \Big(\underbrace{\big(\alpha_{k+1}(\mathbf{x})\big)^2 + \cdots + \big(\alpha_{k+l}(\mathbf{x})\big)^2}_{l \text{ terms}}\Big) \qquad 3.5.21$$

is a decomposition of Q into squares of linearly independent linear functions, and that $W \subset \mathbb{R}^n$ is a subspace of dimension $k_1 > k$.

Consider the linear transformation $W \to \mathbb{R}^k$ given by

$$\vec{\mathbf{w}} \mapsto \begin{bmatrix} \alpha_1(\vec{\mathbf{w}}) \\ \vdots \\ \alpha_k(\vec{\mathbf{w}}) \end{bmatrix}. \qquad 3.5.22$$

Since the domain has dimension k_1, which is greater than the dimension k of the codomain, this mapping has a nontrivial kernel. Let $\vec{\mathbf{w}} \neq \vec{\mathbf{0}}$ be an element of this kernel. Then, since the terms $(\alpha_1(\vec{\mathbf{w}}))^2 + \cdots + (\alpha_k(\vec{\mathbf{w}}))^2$ vanish, we have

"Nontrivial" kernel means the kernel is not $\{\vec{\mathbf{0}}\}$.

$$Q(\vec{\mathbf{w}}) = -\Big(\big(\alpha_{k+1}(\vec{\mathbf{w}})\big)^2 + \cdots + \big(\alpha_{k+l}(\vec{\mathbf{w}})\big)^2\Big) \leq 0. \qquad 3.5.23$$

So Q cannot be positive definite on any subspace of dimension $> k$.

Now we need to exhibit a subspace of dimension k on which Q is positive definite. We have $k + l$ linearly independent linear functions $\alpha_1, \dots, \alpha_{k+l}$. Add to this set linear functions $\alpha_{k+l+1}, \dots, \alpha_n$ such that $\alpha_1, \dots, \alpha_n$ form a maximal family of linearly independent linear functions; that is, they form a basis of the space of $1 \times n$ row matrices (see Exercise 2.6.9).

Consider the linear transformation $T : \mathbb{R}^n \to \mathbb{R}^{n-k}$ given by

$$T : \vec{\mathbf{x}} \mapsto \begin{bmatrix} \alpha_{k+1}(\vec{\mathbf{x}}) \\ \vdots \\ \alpha_n(\vec{\mathbf{x}}) \end{bmatrix}. \qquad 3.5.24$$

The rows of the matrix corresponding to T are the linearly independent row matrices $\alpha_{k+1}, \dots, \alpha_n$; like Q, they are defined on $\mathbb{R}^n$, so the matrix T is n wide. It is $n-k$ tall. Let us see that $\ker T$ has dimension k, and is thus a subspace of dimension k on which Q is positive definite. By Proposition 2.5.11, the rank of T is equal to $n-k$ (the number of its linearly independent rows). So by the dimension formula,

$$\dim \ker T + \dim \operatorname{img} T = n, \quad \text{i.e.,} \quad \dim \ker T = k. \qquad 3.5.25$$

For any $\vec{\mathbf{v}} \in \ker T$, the terms $\alpha_{k+1}(\vec{\mathbf{v}}), \dots, \alpha_{k+l}(\vec{\mathbf{v}})$ of $Q(\vec{\mathbf{v}})$ vanish, so

$$Q(\vec{\mathbf{v}}) = \big(\alpha_1(\vec{\mathbf{v}})\big)^2 + \cdots + \big(\alpha_k(\vec{\mathbf{v}})\big)^2 \geq 0. \qquad 3.5.26$$

If $Q(\vec{\mathbf{v}}) = 0$, this means that every term is zero, so

$$\alpha_1(\vec{\mathbf{v}}) = \cdots = \alpha_n(\vec{\mathbf{v}}) = 0, \qquad 3.5.27$$

which implies that $\vec{\mathbf{v}} = \vec{\mathbf{0}}$. So if $\vec{\mathbf{v}} \neq \vec{\mathbf{0}}$, then Q is strictly positive. The argument for l is identical. $\square$

Corollary 3.5.12 says that the signature of a quadratic form is independent of coordinates.

Corollary 3.5.12. *If $Q : \mathbb{R}^n \to \mathbb{R}$ is a quadratic form and $A : \mathbb{R}^n \to \mathbb{R}^n$ is invertible, $Q \circ A$ is a quadratic form with the same signature as Q.*

Proof of Theorem 3.5.3, part 2. Since the proof of Proposition 3.5.11 says nothing about any particular choice of decomposition, we see that k and l depend only on the quadratic form, not on the particular linearly independent functions we use to represent it as a sum of squares. $\square$

Classification of quadratic forms

The quadratic form of Example 3.5.5 has rank 2; the quadratic form of Example 3.5.6 has rank 3.

It follows from Exercise 3.5.7 that only nondegenerate forms can be positive definite or negative definite.

Definition 3.5.13 (Rank of a quadratic form). The *rank of a quadratic form* is the number of linearly independent squares that appear when the quadratic form is represented as a sum of linearly independent squares.

Definition 3.5.14 (Degenerate and nondegenerate quadratic forms). A quadratic form on $\mathbb{R}^n$ with rank m is *nondegenerate* if $m = n$. It is *degenerate* if $m < n$.

The examples we have seen so far in this section are all nondegenerate; a degenerate quadratic form is shown in Example 3.5.16.

The following proposition is important; we will use it to prove Theorem 3.6.8 about using quadratic forms to classify critical points of functions.

Proposition 3.5.15 applies just as well to negative definite quadratic forms; just use $-C$ and $\leq$.

Another proof (shorter and less constructive) is sketched in Exercise 3.5.16.

Proposition 3.5.15. *If $Q : \mathbb{R}^n \to \mathbb{R}$ is a positive definite quadratic form, then there exists a constant $C > 0$ such that*

$$Q(\vec{\mathbf{x}}) \geq C|\vec{\mathbf{x}}|^2 \quad \textit{for all } \vec{\mathbf{x}} \in \mathbb{R}^n. \tag{3.5.28}$$

Proof. Since Q has rank n, we can write $Q(\vec{\mathbf{x}})$ as a sum of squares of n linearly independent functions:

$$Q(\vec{\mathbf{x}}) = (\alpha_1(\vec{\mathbf{x}}))^2 + \cdots + (\alpha_n(\vec{\mathbf{x}}))^2. \tag{3.5.29}$$

The linear transformation $T : \mathbb{R}^n \to \mathbb{R}^n$ whose rows are the α_i is invertible.

Since Q is positive definite, all the squares in equation 3.5.29 are preceded by plus signs, and we can consider $Q(\vec{\mathbf{x}})$ as the length squared of the vector $T\vec{\mathbf{x}}$. Since $|\vec{\mathbf{x}}| = |T^{-1}T\vec{\mathbf{x}}| \leq |T^{-1}||T\vec{\mathbf{x}}|$,

$$Q(\vec{\mathbf{x}}) = |T\vec{\mathbf{x}}|^2 \geq \frac{|\vec{\mathbf{x}}|^2}{|T^{-1}|^2}, \tag{3.5.30}$$

so we can take $C = 1/|T^{-1}|^2$. □

Example 3.5.16 (Degenerate quadratic form). The quadratic form

$$Q(p) = \int_0^1 \big(p'(t)\big)^2\, dt \tag{3.5.31}$$

on the space P_k of polynomials of degree at most k is a degenerate quadratic form, because Q vanishes on the constant polynomials. △

Quadratic forms and symmetric matrices

Recall (Definition 1.2.18) that a symmetric matrix is a matrix that is equal to its transpose.

In many treatments of quadratic forms, quadratic polynomials are seldom mentioned; instead quadratic forms are viewed in terms of symmetric matrices. This leads to a treatment of the signature in terms of eigenvalues

Example of equation 3.5.32:

$$\underbrace{\begin{bmatrix} x_1 \\ x_2 \\ x_3 \end{bmatrix}}_{\vec{\mathbf{x}}} \cdot \underbrace{\begin{bmatrix} 1 & 0 & 1 \\ 0 & 1 & 2 \\ 1 & 2 & 9 \end{bmatrix} \begin{bmatrix} x_1 \\ x_2 \\ x_3 \end{bmatrix}}_{A\vec{\mathbf{x}}}$$

$$= \underbrace{x_1^2 + x_2^2 + 2x_1x_3 + 4x_2x_3 + 9x_3^2}_{\text{quadratic form}}$$

Equation 3.5.32 is identical to

$$Q_A(\vec{\mathbf{x}}) = \vec{\mathbf{x}}^\top A\vec{\mathbf{x}}.$$

Note that

$$Q_{T^\top AT} = Q_A \circ T, \quad \text{since}$$

$$\begin{aligned} Q_{T^\top AT}(\vec{\mathbf{x}}) &= \vec{\mathbf{x}}^\top T^\top AT\vec{\mathbf{x}} \\ &= (T\vec{\mathbf{x}})^\top A(T\vec{\mathbf{x}}) \\ &= Q_A(T\vec{\mathbf{x}}) = (Q_A \circ T)\vec{\mathbf{x}} \end{aligned}$$

and eigenvectors rather than in terms of completing squares (we discuss this in Section 3.7; see Theorem 3.7.16). We think this is a mistake: completing squares involves only arithmetic; finding eigenvalues and eigenvectors is much harder. But we will need Proposition 3.5.17 in Section 3.7.

If A is an $n \times n$ matrix, then the function Q_A defined by

$$Q_A(\vec{\mathbf{x}}) \overset{\text{def}}{=} \vec{\mathbf{x}} \cdot A\vec{\mathbf{x}} \qquad 3.5.32$$

is a quadratic form on $\mathbb{R}^n$. A particular quadratic form can be given by many different square matrices, but by only one symmetric matrix. Thus we can identity the set of quadratic forms with the set of symmetric matrices.

Proposition 3.5.17 (Quadratic forms and symmetric matrices). *The mapping $A \mapsto Q_A$ is a bijective map from the space of symmetric $n \times n$ matrices to the space of quadratic forms on $\mathbb{R}^n$.*

The symmetric matrix associated to a quadratic form is constructed as follows: each diagonal entry $a_{i,i}$ is the coefficient of the corresponding variable squared in the quadratic form (i.e., the coefficient of x_i^2) while each entry $a_{i,j}$ is one-half the coefficient of the term x_ix_j. For the quadratic form in the margin, the corresponding matrix has entry $a_{1,1} = 1$ because the coefficient of x_1^2 is 1, while $a_{2,1} = a_{1,2} = 0$ because the coefficient of $x_2x_1 = x_1x_2$ is 0. Exercise 3.16 asks you to turn this into a formal proof.

Exercises for Section 3.5

3.5.1 For the quadratic form $Q(\mathbf{x}) = x^2+2xy-4xz+2yz-4z^2$ of Example 3.5.6, what decomposition into a sum of squares do you find if you start by eliminating the terms in z, then the terms in y, and finally the terms in x?

3.5.2 Consider the quadratic form of Example 3.5.7: $Q(\mathbf{x}) = xy - xz + yz$. Decompose $Q(\mathbf{x})$ with a different choice of u, to support the statement that $u = x - y$ was not a magical choice.

3.5.3 Decompose each of the following quadratic forms by completing squares, and determine its signature.

a. $x^2 + xy - 3y^2$ b. $x^2 + 2xy - y^2$ c. $x^2 + xy + yz$

3.5.4 Are the following quadratic forms degenerate or nondegenerate? Are they positive definite, negative definite, or neither?

a. $x^2 + 4xy + 4y^2$ on $\mathbb{R}^2$

b. $x^2 + 2xy + 2y^2 + 2yz + z^2$ on $\mathbb{R}^3$

c. $2x^2 + 2y^2 + z^2 + w^2 + 4xy + 2xz - 2xw - 2yw$ on $\mathbb{R}^4$

3.5.5 What is the signature of the following quadratic forms?

a. $x^2 + xy$ on $\mathbb{R}^2$ b. $xy + yz$ on $\mathbb{R}^3$

3.5.6 Confirm that the symmetric matrix $A = \begin{bmatrix} 1 & 0 & 1/2 \\ 0 & 0 & -1/2 \\ 1/2 & -1/2 & -1 \end{bmatrix}$

represents the quadratic form $Q = x^2 + xz - yz - z^2$.

3.5.7 Show that a quadratic form on $\mathbb{R}^n$ is positive definite if and only if its signature is $(n, 0)$.

Exercise 3.5.7: The main point is to prove that if the quadratic form Q has signature $(k, 0)$ with $k < n$, there is a vector $\vec{\mathbf{v}} \neq \vec{\mathbf{0}}$ such that $Q(\vec{\mathbf{v}}) = 0$. You can find such a vector using the transformation T of formula 3.5.24.

3.5.8 Identify $\begin{pmatrix} a \\ b \\ d \end{pmatrix} \in \mathbb{R}^3$ with the upper triangular matrix $M = \begin{bmatrix} a & b \\ 0 & d \end{bmatrix}$.

a. What is the signature of the quadratic form $Q(M) = \operatorname{tr}(M^2)$? What kind of surface in $\mathbb{R}^3$ do you get by setting $\operatorname{tr}(M^2) = 1$?

b. What is the signature of the quadratic form $Q(M) = \operatorname{tr}(M^\top M)$? What kind of surface in $\mathbb{R}^3$ do you get by setting $\operatorname{tr}(M^T M) = 1$?

Exercise 3.5.8: Recall from Example 2.10.18 that the trace of a square matrix A, denoted $\operatorname{tr} A$, is the sum of its diagonal elements.

3.5.9 Consider the vector space of 2×2 matrices such that

$$H = \begin{bmatrix} a & b + ic \\ b - ic & d \end{bmatrix},$$

where a, b, c, d are real numbers. (Such a matrix, whose complex conjugate is equal to its transpose, is called a *Hermitian matrix.*) What is the signature of the quadratic form $Q(H) = \det H$?

Exercise 3.5.9: In special relativity, spacetime is a 4-dimensional vector space with a quadratic form of signature $(1, 3)$. This quadratic form assigns length 0 to any two events connected by a light ray. Thus the space of 2×2 Hermitian matrices is a useful model for spacetime.

3.5.10 For each of the following equations, determine the signature of the quadratic form represented by the left side. Where possible, sketch the curve or surface represented by the equation.

a. $x^2 + xy - y^2 = 1$ b. $x^2 + 2xy - y^2 = 1$

c. $x^2 + xy + yz = 1$ d. $xy + yz = 1$

3.5.11 a. Let A be a 2×2 matrix. Compute $\operatorname{tr} A^2$ and $\operatorname{tr} A^\top A$, and show that both are quadratic forms on $\operatorname{Mat}(2, 2)$.

b. What are their signatures?

Exercise 3.5.11: See the note for Exercise 3.5.8.

***3.5.12** What is the signature of $x_1x_2 + x_2x_3 + \cdots + x_{n-1}x_n$ on $\mathbb{R}^n$?

3.5.13 Let V be a vector space. A *symmetric bilinear function* on V is a mapping $B : V \times V \to \mathbb{R}$ such that

1. $B(a\mathbf{v}_1 + b\mathbf{v}_2, \mathbf{w}) = aB(\mathbf{v}_1, \mathbf{w}) + bB(\mathbf{v}_2, \mathbf{w})$ for all $\mathbf{v}_1, \mathbf{v}_2, \mathbf{w} \in V$ and $a, b \in \mathbb{R}$;
2. $B(\mathbf{v}, \mathbf{w}) = B(\mathbf{w}, \mathbf{v})$ for all $\mathbf{v}, \mathbf{w} \in V$.

a. Show that a function $B : \mathbb{R}^n \times \mathbb{R}^n \to \mathbb{R}$ is symmetric and bilinear if and only if there exists a symmetric matrix A such that $B(\vec{\mathbf{v}}, \vec{\mathbf{w}}) = \vec{\mathbf{v}}^\top A \vec{\mathbf{w}}$.

b. Show that every quadratic form on $\mathbb{R}^n$ is of the form $\vec{\mathbf{v}} \mapsto B(\vec{\mathbf{v}}, \vec{\mathbf{v}})$ for some symmetric bilinear function B.

c. Let P_k be the space of polynomials of degree at most k. Show that the function $B : P_k \times P_k \to \mathbb{R}$ given by $B(p, q) = \int_0^1 p(t)q(t)\, dt$ is a symmetric bilinear function.

d. Denote by $p_1(t) = 1$, $p_2(t) = t$, $\ldots$, $p_{k+1}(t) = t^k$ the usual basis of P_k, and by Φ_p the corresponding "concrete to abstract" linear transformation. Show that $B\Big(\Phi_p(\vec{\mathbf{a}}), \Phi_p(\vec{\mathbf{b}})\Big)$ is a symmetric bilinear function on $\mathbb{R}^n$, and find its matrix.

Exercise 3.5.13 tells us what a quadratic form on an abstract vector space should be: it is a map $Q : V \to \mathbb{R}$ of the form

$$Q(\mathbf{v}) = B(\mathbf{v}, \mathbf{v})$$

for some symmetric bilinear function $B : V \times V \to \mathbb{R}$.

3.5.14 a. If $p(t) = a_0 + a_1 t$, show that $Q(\mathbf{a}) = \int_0^1 \Big(p(t)\Big)^2 dt$ is a quadratic form, and write it explicitly.

b. Repeat for $p(t) = a_0 + a_1 t + a_2 t^2$.

c. Repeat for $p(t) = a_0 + a_1 t + a_2 t^2 + a_3 t^3$.

3.5.15 a. Let P_k be the space of polynomials of degree at most k. Show that the function

$$Q(p) = \int_0^1 (p(t))^2 - (p'(t))^2\, dt \quad \text{is a quadratic form on } P_k.$$

b. What is the signature of Q when $k = 2$?

3.5.16 Here is an alternative proof of Proposition 3.5.15. Let $Q : \mathbb{R}^n \to \mathbb{R}$ be a positive definite quadratic form. Show that there exists a constant $C > 0$ such that $Q(\vec{\mathbf{x}}) \geq C|\vec{\mathbf{x}}|^2$ for all $\vec{\mathbf{x}} \in \mathbb{R}^n$, as follows:

a. Let $S^{n-1} \stackrel{\text{def}}{=} \{\vec{\mathbf{x}} \in \mathbb{R}^n \mid |\vec{\mathbf{x}}| = 1\}$. Show that S^{n-1} is compact, so there exists $\vec{\mathbf{x}}_0 \in S^{n-1}$ with $Q(\vec{\mathbf{x}}_0) \leq Q(\vec{\mathbf{x}})$ for all $\vec{\mathbf{x}} \in S^{n-1}$.

b. Show that $Q(\vec{\mathbf{x}}_0) > 0$.

c. Use the formula $Q(\vec{\mathbf{x}}) = |\vec{\mathbf{x}}|^2 Q\left(\frac{\vec{\mathbf{x}}}{|\vec{\mathbf{x}}|}\right)$ to prove Proposition 3.5.15.

3.5.17 Identify (for instance, as an ellipse, hyperbola, ...) and sketch the conic sections and quadratic surfaces of equation $\vec{\mathbf{x}}^\top Q(\vec{\mathbf{x}}) = 1$, where Q is one of the following matrices:

a. $\begin{bmatrix} 2 & 1 \\ 1 & 3 \end{bmatrix}$ b. $\begin{bmatrix} 2 & 1 & 0 \\ 1 & 2 & 1 \\ 0 & 1 & 2 \end{bmatrix}$ c. $\begin{bmatrix} 2 & 0 & 3 \\ 0 & 0 & 0 \\ 3 & 0 & -1 \end{bmatrix}$ d. $\begin{bmatrix} 2 & 4 & -3 \\ 4 & 1 & 3 \\ -3 & 3 & -1 \end{bmatrix}$ e. $\begin{bmatrix} 1 & 2 \\ 2 & 4 \end{bmatrix}$

3.5.18 Let A be a real $n \times m$ matrix, and define $M = A^\top A$. Show that $\vec{\mathbf{x}} \mapsto \vec{\mathbf{x}}^\top M \vec{\mathbf{x}}$ is a positive definite quadratic form if and only if rank $A = m$.

3.6 Classifying critical points of functions

In one-variable calculus we find the maximum or minimum of a function by looking for places where the derivative vanishes.

Minima and maxima of functions (Definitions 1.6.6 and 1.6.8) are both known as *extrema*; the singular is *extremum*.

Part 2: "Strict local minimum" means that there exists a neighborhood $V \subset U$ of x_0 such that $f(x_0) < f(x)$ for all $x \in V - \{x_0\}$.

Part 3: "Strict local maximum" means that there exists a neighborhood $V \subset U$ of x_0 such that $f(x_0) > f(x)$ for all $x \in V - \{x_0\}$.

Theorem 3.6.1 elaborates on Proposition 1.6.12.

Theorem 3.6.1 (Extrema of functions of one variable). *Let $U \subset \mathbb{R}$ be an open interval and $f : U \to \mathbb{R}$ a differentiable function.*

1. *If $x_0 \in U$ is a local minimum or maximum of f, then $f'(x_0) = 0$.*
2. *If f is twice differentiable, and $f'(x_0) = 0$ and $f''(x_0) > 0$, then x_0 is a strict local minimum of f.*
3. *If f is twice differentiable, and $f'(x_0) = 0$ and $f''(x_0) < 0$, then x_0 is a strict local maximum of f.*

Note that Theorem 3.6.1 says nothing about the degenerate case, where $f'(x_0) = 0$ and $f''(x_0) = 0$.

For a function of several variables, classifying *critical points* – points where the derivative vanishes – is more complicated: a nondegenerate critical point may be a local maximum, a local minimum, or a *saddle* (see Figure 3.6.1). In this section we see how to classify nondegenerate critical points of such a function, using the quadratic terms of its Taylor polynomial.

Definition 3.6.2 (Critical point, critical value). Let $U \subset \mathbb{R}^n$ be an open subset and let $f : U \to \mathbb{R}$ be a differentiable function. A *critical point* of f is a point where the derivative vanishes. The value of f at a critical point is a *critical value*.

Part 1 of Theorem 3.6.1 generalizes to functions of several variables in the most obvious way:

Theorem 3.6.3 (Derivative zero at extremum). *Let $U \subset \mathbb{R}^n$ be an open subset and let $f : U \to \mathbb{R}$ be a differentiable function. If $\mathbf{x}_0 \in U$ is a local minimum or maximum of f, then $[\mathbf{D}f(\mathbf{x}_0)] = [0]$.*

The derivative $[\mathbf{D}f(\mathbf{x}_0)]$ in Theorem 3.6.3 is a row matrix n wide.

Proof. Since the derivative is given by the Jacobian matrix, it is enough to show that if $\mathbf{x}_0$ is a local extremum of f, then $D_i f(\mathbf{x}_0) = 0$ for all $i = 1, \dots, n$. But $D_i f(\mathbf{x}_0) = g'(0)$, where g is the function of one variable $g(t) = f(\mathbf{x}_0 + t\vec{\mathbf{e}}_i)$, and our hypothesis also implies that $t = 0$ is an extremum of g, so $g'(0) = 0$ by Theorem 3.6.1. □

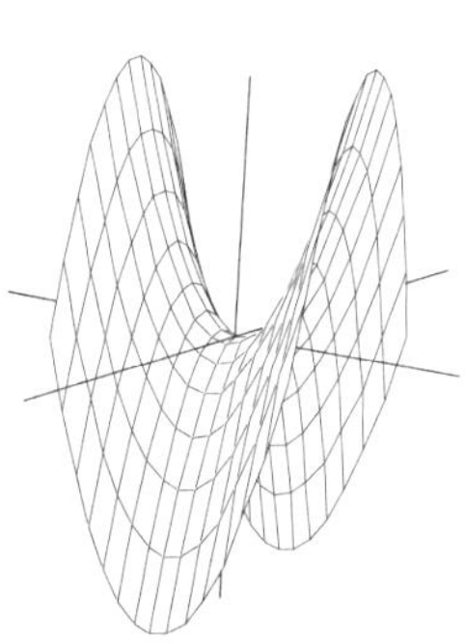

FIGURE 3.6.1.

The graph of $x^2 - y^2$, a typical saddle. Near a saddle point, a surface is like a mountain pass, with mountains to either side, a valley one has left behind, and a different valley in front.

Finding critical points

Theorem 3.6.3 says that when looking for maxima or minima, the first step is to look for critical points, places where the derivative vanishes. For the derivative of a function $f : U \subset \mathbb{R}^n \to \mathbb{R}$ to vanish means that all the partial derivatives vanish. Thus finding critical points means solving $[\mathbf{D}f(\mathbf{x})] = [0]$, a system of n equations in n unknowns. Usually there is no better way to find critical points than applying Newton's method; finding critical points is an important application of Newton's method.

Example 3.6.4 (Finding critical points). What are the critical points of the function $f : \mathbb{R}^2 \to \mathbb{R}$ given by

$$f\begin{pmatrix} x \\ y \end{pmatrix} = x + x^2 + xy + y^3? \qquad 3.6.1$$

The partial derivatives are

$$D_1 f\begin{pmatrix} x \\ y \end{pmatrix} = 1 + 2x + y, \qquad D_2 f\begin{pmatrix} x \\ y \end{pmatrix} = x + 3y^2. \qquad 3.6.2$$

In this case we don't need Newton's method, since the system can be solved explicitly. Substitute $x = -3y^2$ from the second equation into the first, to find

$$1 + y - 6y^2 = 0, \quad \text{which gives} \quad y = \frac{1 \pm \sqrt{1 + 24}}{12} = \frac{1}{2} \text{ or } -\frac{1}{3}. \qquad 3.6.3$$

Substituting this into $x = -(1+y)/2$ (or into $x = -3y^2$) gives the two critical points

$$\mathbf{a}_1 = \begin{pmatrix} -3/4 \\ 1/2 \end{pmatrix} \quad \text{and} \quad \mathbf{a}_2 = \begin{pmatrix} -1/3 \\ -1/3 \end{pmatrix}. \quad \triangle \tag{3.6.4}$$

Remark 3.6.5 (Critical points on closed sets). A major problem in using Theorem 3.6.3 is the hypothesis that U is open. Seeking a minimum or maximum by finding places where the derivative vanishes will usually not work if the critical point is on the boundary.

For instance, the maximum value of x^2 on $[0,2]$ is 4, which occurs at $x = 2$, not a point where the derivative of x^2 vanishes. Especially when we have used Theorem 1.6.9 to assert that a maximum or minimum exists in a compact subset $\overline{U}$, we need to check that the extremum occurs in the interior U before we can say that it is a critical point. $\triangle$

In Section 3.7, we will show how to analyze the behavior of a function restricted to the boundary.

The second derivative criterion

Is either critical point in equation 3.6.4 an extremum? In one variable, we would answer this question by looking at the sign of the second derivative (Theorem 3.6.1). The right generalization of "the second derivative" to higher dimensions is "the quadratic form given by the terms of degree 2 of the Taylor polynomial", and the right generalization of "sign of the second derivative" is "signature of the quadratic form", which we will also call the *signature* of the critical point.

Since a sufficiently differentiable function f is well approximated near a critical point by its Taylor polynomial, it seems reasonable to hope that f should behave like its Taylor polynomial. Evaluating the Taylor polynomial of the function $f\begin{pmatrix} x \\ y \end{pmatrix} = x + x^2 + xy + y^3$ at the critical point $\mathbf{a}_1 = \begin{pmatrix} -3/4 \\ 1/2 \end{pmatrix}$, we get

$$P^2_{f,\mathbf{a}_1}(\mathbf{a}_1 + \vec{\mathbf{h}}) = \underbrace{-\frac{7}{16}}_{f(\mathbf{a}_1)} \; \underbrace{+\frac{1}{2}2h_1^2 + h_1h_2 + \frac{1}{2}3h_2^2}_{\text{terms of degree 2}}. \tag{3.6.5}$$

Equation 3.6.5: In evaluating the second derivative, remember that $D_1^2 f(\mathbf{a})$ is the second partial derivative D_1D_1f, evaluated at $\mathbf{a}$. In this case we have

$$D_1^2 f = 2 \quad \text{and} \quad D_1D_2 f = 1.$$

These are constants, so where we evaluate the derivative doesn't matter. But $D_2^2 f = 6y$; evaluated at $\mathbf{a}_1$ this gives 3.

The terms of degree 2 form a positive definite quadratic form:

$$h_1^2 + h_1h_2 + \frac{3}{2}h_2^2 = \left(h_1 + \frac{h_2}{2}\right)^2 + \frac{5}{4}h_2^2, \quad \text{with signature } (2,0). \tag{3.6.6}$$

How should we interpret this? The quadratic form is positive definite, so (by Proposition 3.5.15) it has a local minimum at $\vec{\mathbf{h}} = \vec{\mathbf{0}}$, and if $P^2_{f,\mathbf{a}_1}$ approximates f sufficiently well near $\mathbf{a}_1$, then f should have a minimum at $\mathbf{a}_1$. (A neighborhood of $\vec{\mathbf{h}} = \vec{\mathbf{0}}$ for the Taylor polynomial corresponds to a neighborhood of $\mathbf{a}_1$ for f.) Similarly, if the signature of a critical point is $(0,n)$, we would expect the critical point to be a maximum.

What happens at the critical point $\mathbf{a}_2 = \begin{pmatrix} -1/3 \\ -1/3 \end{pmatrix}$? Check below.[14]

Definition 3.6.6 (Signature of critical point). Let $U \subset \mathbb{R}^n$ be an open set, let $f : U \to \mathbb{R}$ be of class C^2, and let $\mathbf{a} \in U$ be a critical point of f. The *signature* of the critical point $\mathbf{a}$ is the signature of the quadratic form

$$Q_{f,\mathbf{a}}(\vec{\mathbf{h}}) \stackrel{\text{def}}{=} \sum_{I \in \mathcal{I}_n^2} \frac{1}{I!} (D_I f(\mathbf{a})) \vec{\mathbf{h}}^I. \qquad 3.6.7$$

Definition 3.6.6: In order to define the signature of a critical point, the function f must be at least C^2; we know of no theory of classification of critical points of functions that are only differentiable or of class C^1.

The matrix H of second partials is the *Hessian matrix* of f at $\mathbf{x}$. One can think of it as the "second derivative" of f; it is really only meaningful when $\mathbf{x}$ is a critical point of f. Note that a Hessian matrix is always square and (because cross partials are equal) symmetric.

The quadratic form in equation 3.6.7 is the second-degree term of the Taylor polynomial of f at $\mathbf{a}$ (see Definition 3.3.13).

Recall (Proposition 3.5.17) that every quadratic form is associated to a unique symmetric matrix. The quadratic form $Q_{f,\mathbf{a}}(\vec{\mathbf{h}})$ is associated to the matrix composed of second partial derivatives of f: if we define H by

$$H_{i,j}(\mathbf{x}) = D_i D_j f(\mathbf{x}), \quad \text{then} \quad Q_{f,\mathbf{a}}(\vec{\mathbf{h}}) = \frac{1}{2}\left(\vec{\mathbf{h}} \cdot H\vec{\mathbf{h}}\right). \qquad 3.6.8$$

A critical point $\mathbf{a}$ is called degenerate or nondegenerate precisely when the quadratic form $Q_{f,\mathbf{a}}$ is degenerate or nondegenerate.

Proposition 3.6.7 says that the signature of a critical point is independent of coordinates.

Proposition 3.6.7. *Let $U, V \subset \mathbb{R}^n$ be open, let $\varphi : V \to U$ be a C^2 map and let $f : U \to \mathbb{R}$ be a C^2 function. Let $\mathbf{x}_0 \in U$ be a critical point of f, and $\mathbf{y}_0 \in V$ a point such that $\varphi(\mathbf{y}_0) = \mathbf{x}_0$. Then $\mathbf{y}_0$ is a critical point of $f \circ \varphi$. If the derivative $[\mathbf{D}\varphi(\mathbf{y}_0)]$ is invertible, then $\mathbf{x}_0$ and $\mathbf{y}_0$ have the same signature.*

Proof. If follows from Proposition 3.4.4 that

$$P^2_{f,\mathbf{x}_0} \circ P^1_{\varphi,\mathbf{y}_0} = P^2_{f\circ\varphi,\mathbf{y}_0}. \qquad 3.6.9$$

Why do we write $P^1_{\varphi,\mathbf{y}_0}$ here rather than $P^2_{\varphi,\mathbf{y}_0}$? Since $\mathbf{x}_0$ is a critical point of f, the Taylor polynomial $P^2_{f,\mathbf{x}_0}$ has no linear terms, and quadratic terms of $P^2_{f,\mathbf{x}_0}$ applied to quadratic terms of $P^2_{\varphi,\mathbf{y}_0}$ would give terms of the Taylor polynomial of $f \circ \varphi$ of degree 4. Omitting the constant terms, this leads to

$$Q_{f,\mathbf{x}_0} \circ [\mathbf{D}\varphi(\mathbf{y}_0)] = Q_{f\circ\varphi,\mathbf{y}_0} \qquad 3.6.10$$

and Proposition 3.6.7 follows from Corollary 3.5.12. $\square$

14

$$P^2_{f,\mathbf{a}_2}(\mathbf{a}_2 + \vec{\mathbf{h}}) = -\frac{4}{27} + \frac{1}{2}2h_1^2 + h_1h_2 + \frac{1}{2}(-2)h_2^2, \text{ with quadratic form}$$

$$h_1^2 + h_1h_2 - h_2^2 = \left(h_1 + \frac{h_2}{2}\right)^2 - \frac{5}{4}h_2^2, \quad \text{which has signature } (1,1).$$

Theorem 3.6.8 (Quadratic forms and extrema). *Let $U \subset \mathbb{R}^n$ be an open set, $f : U \to \mathbb{R}$ of class C^2, and $\mathbf{a} \in U$ a critical point of f.*

1. *If the signature of $\mathbf{a}$ is $(n, 0)$, i.e., $Q_{f,\mathbf{a}}$ is positive definite, then $\mathbf{a}$ is a strict local minimum of f. If the signature of $\mathbf{a}$ is (k, l) with $l > 0$, then $\mathbf{a}$ is not a local minimum.*
2. *If the signature of $\mathbf{a}$ is $(0, n)$, i.e., $Q_{f,\mathbf{a}}$ is negative definite, then $\mathbf{a}$ is a strict local maximum of f. If the signature is (k, l) with $k > 0$, then $\mathbf{a}$ is not a local maximum.*

We state Theorem 3.6.8 as we do because if a quadratic form on $\mathbb{R}^n$ is degenerate (i.e., $k + l < n$), then if its signature is $(k, 0)$, it is positive but not positive definite, and the signature does not tell you that there is a local minimum. Similarly, if the signature is $(0, k)$, it does not tell you that there is a local maximum.

Of course the only way the quadratic form of a function in one variable can be degenerate is if it has signature $(0, 0)$.

The first statement in part 1 corresponds to part 2 of Theorem 3.6.1; the first statement in part 2 corresponds to part 3 in that theorem.

What about a critical point that is not positive or negative definite, for instance the critical point $\mathbf{a}_2$ of equation 3.6.4, with signature $(1, 1)$? For functions of one variable, critical points are extrema unless the second derivative vanishes, i.e., unless the quadratic form given by the second derivative is degenerate. This is not the case in higher dimensions: nondegenerate quadratic forms that are neither positive definite nor negative definite are just as "ordinary" as those that are; they correspond to various kinds of *saddles*.

The origin is a saddle for the function $x^2 - y^2$.

Saddle and *saddle point* are synonymous.

Definition 3.6.9 (Saddle). If $\mathbf{a}$ is a critical point of a C^2 function f, and the quadratic form $Q_{f,\mathbf{a}}$ of equation 3.6.7 has signature (k, l) with $k > 0$ and $l > 0$, then $\mathbf{a}$ is a *saddle*.

A saddle can be either degenerate or nondegenerate.

Theorem 3.6.10 (Behavior of a function near a saddle). *Let $U \subset \mathbb{R}^n$ be an open set, and let $f : U \to \mathbb{R}$ be a C^2 function. If f has a saddle at $\mathbf{a} \in U$, then in every neighborhood of $\mathbf{a}$ there are points $\mathbf{b}$ with $f(\mathbf{b}) > f(\mathbf{a})$, and points $\mathbf{c}$ with $f(\mathbf{c}) < f(\mathbf{a})$.*

Proof of Theorem 3.6.8. We will treat part 1 only; part 2 can be derived from it by considering $-f$ rather than f. Since $\mathbf{a}$ is a critical point of f, we can write

$$f(\mathbf{a} + \vec{\mathbf{h}}) = f(\mathbf{a}) + Q_{f,\mathbf{a}}(\vec{\mathbf{h}}) + r(\vec{\mathbf{h}}), \qquad 3.6.11$$

where the remainder satisfies

$$\lim_{\vec{\mathbf{h}} \to \vec{\mathbf{0}}} \frac{r(\vec{\mathbf{h}})}{|\vec{\mathbf{h}}|^2} = 0 \qquad 3.6.12$$

(i.e., the remainder is smaller than quadratic). Thus if $Q_{f,\mathbf{a}}$ is positive definite,

$$\frac{f(\mathbf{a} + \vec{\mathbf{h}}) - f(\mathbf{a})}{|\vec{\mathbf{h}}|^2} = \frac{Q_{f,\mathbf{a}}(\vec{\mathbf{h}})}{|\vec{\mathbf{h}}|^2} + \frac{r(\vec{\mathbf{h}})}{|\vec{\mathbf{h}}|^2} \geq C + \frac{r(\vec{\mathbf{h}})}{|\vec{\mathbf{h}}|^2}, \qquad 3.6.13$$

Equation 3.6.13 uses Proposition 3.5.15.

The constant C depends on $Q_{f,\mathbf{a}}$, not on the vector on which $Q_{f,\mathbf{a}}$ is evaluated, so

$$Q_{f,\mathbf{a}}(\vec{\mathbf{h}}) \geq C|\vec{\mathbf{h}}|^2;$$

i.e.,

$$\frac{Q_{f,\mathbf{a}}(\vec{\mathbf{h}})}{|\vec{\mathbf{h}}|^2} \geq \frac{C|\vec{\mathbf{h}}|^2}{|\vec{\mathbf{h}}|^2} = C.$$

where C is the constant of Proposition 3.5.15 – the constant $C > 0$ such that $Q_{f,\mathbf{a}}(\vec{\mathbf{x}}) \geq C|\mathbf{x}|^2$ for all $\vec{\mathbf{x}} \in \mathbb{R}^n$, when $Q_{f,\mathbf{a}}$ is positive definite.

The right side is positive for $\vec{\mathbf{h}}$ sufficiently small (see equation 3.6.12), so the left side is also, i.e., $f(\mathbf{a}+\vec{\mathbf{h}}) > f(\mathbf{a})$ for $|\vec{\mathbf{h}}|$ sufficiently small. Thus $\mathbf{a}$ is a strict local minimum of f.

If the signature is (k, l) with $l > 0$, there exists $\vec{\mathbf{h}}$ with $Q_{f,\mathbf{a}}(\vec{\mathbf{h}}) < 0$. Then

Equation 3.6.14: Since $Q_{f,\mathbf{a}}$ is a quadratic form,

$$Q_{f,\mathbf{a}}(t\vec{\mathbf{h}}) = t^2 Q_{f,\mathbf{a}}(\vec{\mathbf{h}}).$$

$$f(\mathbf{a}+t\vec{\mathbf{h}}) - f(\mathbf{a}) = Q_{f,\mathbf{a}}(t\vec{\mathbf{h}}) + r(t\vec{\mathbf{h}}) = t^2 Q_{f,\mathbf{a}}(\vec{\mathbf{h}}) + r(t\vec{\mathbf{h}}). \qquad 3.6.14$$

Thus

$$\frac{f(\mathbf{a}+t\vec{\mathbf{h}}) - f(\mathbf{a})}{t^2} = Q_{f,\mathbf{a}}(\vec{\mathbf{h}}) + \frac{r(t\vec{\mathbf{h}})}{t^2}, \qquad 3.6.15$$

and since

$$\lim_{t\to 0} \frac{r(t\vec{\mathbf{h}})}{t^2} = 0, \quad \text{we have} \quad f(\mathbf{a}+t\vec{\mathbf{h}}) < f(\mathbf{a}) \qquad 3.6.16$$

for $|t| > 0$ sufficiently small. Thus $\mathbf{a}$ is not a local minimum. □

Proof of Theorem 3.6.10 (Behavior of functions near saddles). As in equations 3.6.11 and 3.6.12, write

$$f(\mathbf{a}+\vec{\mathbf{h}}) = f(\mathbf{a}) + Q_{f,\mathbf{a}}(\vec{\mathbf{h}}) + r(\vec{\mathbf{h}}) \quad \text{with} \quad \lim_{\vec{\mathbf{h}}\to\vec{\mathbf{0}}} \frac{r(\vec{\mathbf{h}})}{|\vec{\mathbf{h}}|^2} = 0. \qquad 3.6.17$$

A similar argument about W shows that there are also points $\mathbf{c}$ where $f(\mathbf{c}) < f(\mathbf{a})$. Exercise 3.6.3 asks you to spell out this argument.

By Definition 3.6.9, there exist vectors $\vec{\mathbf{h}}$ and $\vec{\mathbf{k}}$ such that $Q_{f,\mathbf{a}}(\vec{\mathbf{h}}) > 0$ and $Q_{f,\mathbf{a}}(\vec{\mathbf{k}}) < 0$. Then

$$\frac{f(\mathbf{a}+t\vec{\mathbf{h}}) - f(\mathbf{a})}{t^2} = \frac{t^2 Q_{f,\mathbf{a}}(\vec{\mathbf{h}}) + r(t\vec{\mathbf{h}})}{t^2} = Q_{f,\mathbf{a}}(\vec{\mathbf{h}}) + \frac{r(t\vec{\mathbf{h}})}{t^2} \qquad 3.6.18$$

is strictly positive for $t > 0$ sufficiently small, and

$$\frac{f(\mathbf{a}+t\vec{\mathbf{k}}) - f(\mathbf{a})}{t^2} = \frac{t^2 Q_{f,\mathbf{a}}(\vec{\mathbf{k}}) + r(t\vec{\mathbf{h}})}{t^2} = Q_{f,\mathbf{a}}(\vec{\mathbf{k}}) + \frac{r(t\vec{\mathbf{k}})}{t^2} \qquad 3.6.19$$

is negative for $t \neq 0$ sufficiently small. □

Theorems 3.6.8 and 3.6.10 say that near a critical point $\mathbf{a}$, the function f "behaves like" the quadratic form $Q_{f,\mathbf{a}}$ (although Theorem 3.6.10 is a weak sense of "behaves like"). The statement can be improved: Morse's lemma[15] says that when $\mathbf{a}$ is nondegenerate, there is a change of variables φ as in Proposition 3.6.7 such that $f \circ \varphi = Q_{f,\mathbf{a}}$.

Degenerate critical points

Degenerate critical points are "exceptional", just as zeros of the first and second derivative of functions of one variable do not usually coincide. We

[15] See lemma 2.2 on page 6 of J. Milnor, *Morse Theory*, Princeton University Press, 1963.

will not attempt to classify them (it is a big job, perhaps impossible), but simply give some examples.

Example 3.6.11 (Degenerate critical points). The three functions $x^2 + y^3$, $x^2 + y^4$, and $x^2 - y^4$ all have the same degenerate quadratic form for the Taylor polynomial of degree 2: x^2. But they behave very differently, as shown in Figure 3.6.2 and the left surface of Figure 3.6.3. The function shown in Figure 3.6.2, bottom, has a minimum; the other three do not. △

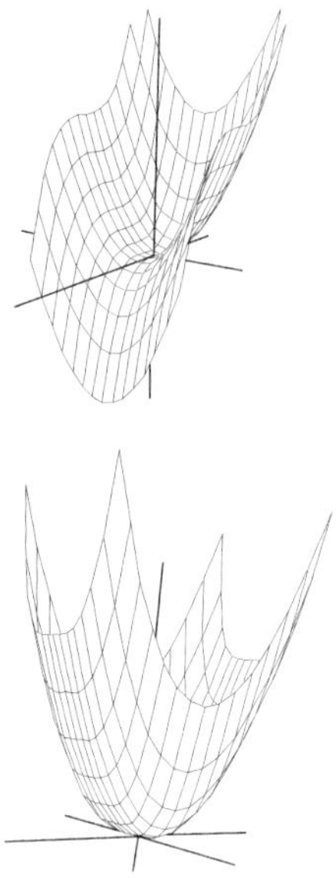

FIGURE 3.6.2. TOP: The surface of equation $z = x^2 + y^3$. BOTTOM: The surface of equation $z = x^2 + y^4$.

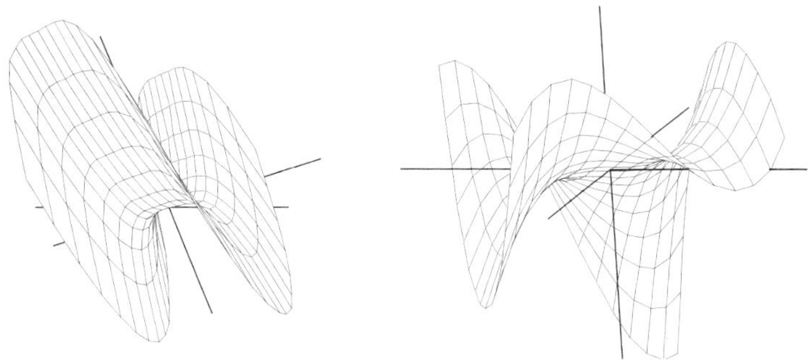

FIGURE 3.6.3. LEFT: The surface of equation $z = x^2 - y^4$. Although its graph look very different from those shown in Figure 3.6.2, all three functions have the same degenerate quadratic form for the Taylor polynomial of degree 2. RIGHT: The monkey saddle; it is the graph of $z = x^3 - 2xy^2$, whose quadratic form is 0. The graph goes up in three directions and down in three also (two for the legs, one for the tail).

EXERCISES FOR SECTION 3.6

3.6.1 a. Show that $f\begin{pmatrix} x \\ y \\ z \end{pmatrix} = x^2 + xy + z^2 - \cos y$ has a critical point at the origin.

b. What kind of critical point does it have?

3.6.2 a. Find the critical points of the function

$$f\begin{pmatrix} x \\ y \end{pmatrix} = x^3 - 12xy + 8y^3.$$

b. Determine the nature of each of the critical points.

3.6.3 Complete the proof of Theorem 3.6.10, showing that if f has a saddle at $\mathbf{a} \in U$, then in every neighborhood of $\mathbf{a}$ there are points $\mathbf{c}$ with $f(\mathbf{c}) < f(\mathbf{a})$.

3.6.4 Use Newton's method (preferably by computer) to find the critical points of $-x^3 + y^3 + xy + 4x - 5y$. Classify them, still using the computer.

3.6.5 a. Find the critical points of the function $f\begin{pmatrix} x \\ y \\ z \end{pmatrix} = xy + yz - xz + xyz$.

b. Determine the nature of each critical point.

3.6.6 Find all the critical points of the following functions:

a. $\sin x \cos y$ b. $xy + \frac{8}{x} + \frac{1}{y}$

*c. $\sin x + \sin y + \sin(x + y)$

3.6.7 a. Find the critical points of the function $f\begin{pmatrix} x \\ y \end{pmatrix} = (x^2 + y^2)e^{x^2 - y^2}$.

b. Determine the nature of each critical point.

3.6.8 a. Write the Taylor polynomial of $f\begin{pmatrix} x \\ y \end{pmatrix} = \sqrt{1 - x + y^2}$ to degree 3 at the origin.

b. Show that $g\begin{pmatrix} x \\ y \end{pmatrix} = \sqrt{1 - x + y^2} + x/2$ has a critical point at the origin. What kind of critical point is it?

3.7 Constrained critical points and Lagrange multipliers

Finding shortest paths goes under the name of the *calculus of variations*. The set of paths from New York to Paris is an infinite-dimensional manifold. We will be restricting ourselves to finite-dimensional problems. But the tools we develop apply quite well to the infinite-dimensional setting.

Another example occurs in Section 2.9: the norm

$$\sup_{|\vec{\mathbf{x}}|=1} |A\vec{\mathbf{x}}|$$

of a matrix A answers the question, what is $\sup |A\vec{\mathbf{x}}|$ when we require that $\vec{\mathbf{x}}$ have length 1?

The shortest path between two points is a straight line. But what is the shortest path if you are restricted to paths that lie on a sphere (for example, because you are flying from New York to Paris)? This example is intuitively clear but quite difficult to address.

Here we will look at easier problems in the same spirit. We will be interested in extrema of a function f when f is either defined on a manifold $X \subset \mathbb{R}^n$ or restricted to it. In the case of the set $X \subset \mathbb{R}^8$ describing the position of four linked rods in the plane (Example 3.1.8), we might imagine that each of the four joints connecting the rods at the vertices $\mathbf{x}_i$ is connected to the origin by a rubber band, and that the vertex $\mathbf{x}_i$ has a "potential" $|\vec{\mathbf{x}}_i|^2$. Then what is the equilibrium position, where the link realizes the minimum of the potential energy? Of course, all four vertices try to be at the origin, but they can't. Where will they go? In this case the function "sum of the $|\vec{\mathbf{x}}_i|^2$" is defined on the ambient space, but there are important functions that are not, such as the curvature of a surface.

In this section we provide tools to answer this sort of question.

Finding constrained critical points using derivatives

Recall that in Section 3.2 we defined the derivative of a function defined on a manifold. Thus we can make the obvious generalization of Definition 3.6.2.

The derivative $[\mathbf{D}f(\mathbf{x})]$ of f is only defined on the tangent space $T_{\mathbf{x}}X$, so saying that it is 0 is saying that it vanishes *on tangent vectors* to X.

Definition 3.7.1 (Critical point of function defined on manifold). Let $X \subset \mathbb{R}^n$ be a manifold, and let $f : X \to \mathbb{R}$ be a C^1 function. Then a critical point of f is a point $\mathbf{x} \in X$ where $[\mathbf{D}f(\mathbf{x})] = 0$.

An important special case is when f is defined not just on X but on an open neighborhood $U \subset \mathbb{R}^n$ of X; in that case we are looking for critical points of the restriction $f|_X$ of f to X.

Analyzing critical points of functions $f : X \to \mathbb{R}$ isn't quite the focus of this section; we are really concerned with functions g defined on a neighborhood of a manifold X and studying critical points of the restriction of g to X.

Traditionally a critical point of g restricted to X is called a *constrained critical point*.

Theorem 3.7.2. *Let $X \subset \mathbb{R}^n$ be a manifold, $f : X \to \mathbb{R}$ a C^1 function, and $\mathbf{c} \in X$ a local extremum of f. Then $\mathbf{c}$ is a critical point of f.*

Proof. Let $\gamma : V \to X$ be a parametrization of a neighborhood of $\mathbf{c} \in X$, with $\gamma(\mathbf{x}_0) = \mathbf{c}$. Then $\mathbf{c}$ is an extremum of f precisely if $\mathbf{x}_0$ is an extremum of $f \circ \gamma$. By Theorem 3.6.3, $[\mathbf{D}(f \circ \gamma)(\mathbf{x}_0)] = [0]$, so, by Proposition 3.2.11,

$$[\mathbf{D}(f \circ \gamma)(\mathbf{x}_0)] = [\mathbf{D}f(\gamma(\mathbf{x}_0))][\mathbf{D}\gamma(\mathbf{x}_0)] = [\mathbf{D}f(\mathbf{c})][\mathbf{D}\gamma(\mathbf{x}_0)] = [0].$$

By Proposition 3.2.7, the image of $[\mathbf{D}\gamma(\mathbf{x}_0)]$ is $T_{\mathbf{c}}X$, so $[\mathbf{D}f(\mathbf{c})]$ vanishes on $T_{\mathbf{c}}X$. The proof is illustrated by Figure 3.7.1.

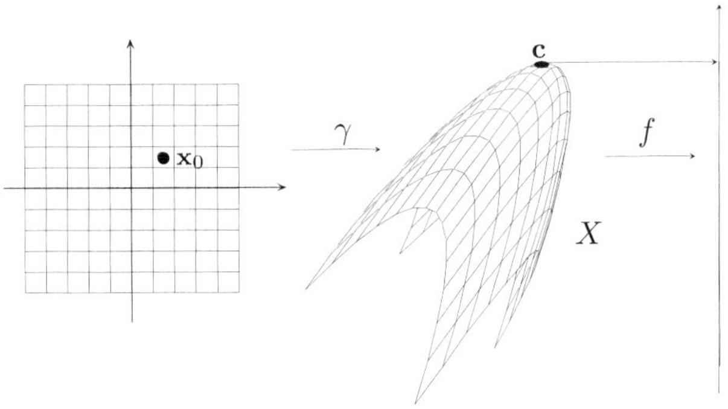

FIGURE 3.7.1 The parametrization γ takes a point in $\mathbb{R}^2$ to the manifold X; f takes it to $\mathbb{R}$. An extremum of the composition $f \circ \gamma$ corresponds to an extremum of f. □

We already used the idea of the proof in Example 2.9.9, where we found the maximum of $|A\mathbf{x}|$ restricted to the unit circle, for $A = \begin{bmatrix} 1 & 1 \\ 0 & 1 \end{bmatrix}$.

In Examples 3.7.3 and 3.7.4 we know a critical point to begin with, and we show that equation 3.7.1 is satisfied:

$$T_{\mathbf{c}}X \subset \ker\,[\mathbf{D}f(\mathbf{c})].$$

Theorem 3.7.5 will show how to find critical points of a function restricted to a manifold (rather than defined on the manifold, as in Definition 3.7.1), when the manifold is known by an equation $\mathbf{F}(\mathbf{z}) = \mathbf{0}$.

Examples 3.7.3 and 3.7.4 illustrate constrained critical points. They show how to check that a maximum or minimum is indeed a critical point satisfying Definition 3.7.1.

Suppose a manifold X is defined by the equation $\mathbf{F}(\mathbf{z}) = \mathbf{0}$, where $\mathbf{F} : U \subset \mathbb{R}^n \to \mathbb{R}^{n-k}$ has onto derivative $[\mathbf{DF}(\mathbf{x})]$ for all $\mathbf{x} \in U \cap X$, and suppose $f : U \to \mathbb{R}$ is a C^1 function. Then Definition 3.7.1 says that $\mathbf{c}$ is a critical point of f restricted to X if

$$T_{\mathbf{c}}X \underbrace{=}_{\text{Thm. 3.2.4}} \ker\,[\mathbf{DF}(\mathbf{c})] \underbrace{\subset}_{\text{Def. 3.7.1}} \ker\,[\mathbf{D}f(\mathbf{c})]. \qquad 3.7.1$$

Note that both derivatives in formula 3.7.1 have the same width, as they must for that equation to make sense; $[\mathbf{DF}(\mathbf{c})]$ is a $(n-k) \times n$ matrix, and $[\mathbf{D}f(\mathbf{c})]$ is a $1 \times n$ matrix, so both can be evaluated on a vector in $\mathbb{R}^n$. It also makes sense that the kernel of the taller matrix should be a subset of the kernel of the shorter matrix. Saying that $\vec{v} \in \ker[\mathbf{DF}(\mathbf{c})]$ means that

$$\begin{bmatrix} D_1 F_1(\mathbf{c}) & \cdots & D_n F_1(\mathbf{c}) \\ \vdots & \vdots & \vdots \\ D_1 F_{n-k}(\mathbf{c}) & \cdots & D_n F_{n-k}(\mathbf{c}) \end{bmatrix} \begin{bmatrix} v_1 \\ \vdots \\ v_n \end{bmatrix} = \begin{bmatrix} 0 \\ \vdots \\ 0 \end{bmatrix}; \qquad 3.7.2$$

$n - k$ equations need to be satisfied. Saying that $\vec{v} \in \ker[\mathbf{D}f(\mathbf{c})]$ means that only one equation needs to be satisfied:

$$[D_1 f(\mathbf{c}) \cdots D_n f(\mathbf{c})] \begin{bmatrix} v_1 \\ \vdots \\ v_n \end{bmatrix} = 0. \qquad 3.7.3$$

Equation 3.7.1 says that at a critical point $\mathbf{c}$, any vector that satisfies equation 3.7.2 also satisfies equation 3.7.3.

Example 3.7.3 (Constrained critical point: a simple example). Suppose we wish to maximize the function $f\begin{pmatrix} x \\ y \end{pmatrix} = xy$ on the first quadrant of the circle $x^2 + y^2 = 1$, which we will denote by X. As shown in Figure 3.7.2, some level sets of that function do not intersect the circle, and some intersect it in two points, but one, $xy = 1/2$, intersects it at the point $\mathbf{c} = \begin{pmatrix} 1/\sqrt{2} \\ 1/\sqrt{2} \end{pmatrix}$. To show that $\mathbf{c}$ is the critical point of f constrained to X, we need to show that $T_\mathbf{c} X \subset \ker[\mathbf{D}f(\mathbf{c})]$.

Since $F\begin{pmatrix} x \\ y \end{pmatrix} = x^2 + y^2 - 1$ is the function defining the circle, we have

$$T_\mathbf{c} X = \ker[\mathbf{DF}(\mathbf{c})] = \ker[2c_1, 2c_2] = \ker\left[\frac{2}{\sqrt{2}}, \frac{2}{\sqrt{2}}\right] \qquad 3.7.4$$

$$\ker[\mathbf{D}f(\mathbf{c})] = \ker[c_2, c_1] = \ker\left[\frac{1}{\sqrt{2}}, \frac{1}{\sqrt{2}}\right]. \qquad 3.7.5$$

Clearly the kernels of both $[\mathbf{DF}(\mathbf{c})]$ and $[\mathbf{D}f(\mathbf{c})]$ consist of vectors $\begin{bmatrix} v_1 \\ v_2 \end{bmatrix}$ such that $v_1 = -v_2$. In particular, $T_\mathbf{c} X \subset \ker[\mathbf{D}f(\mathbf{c})]$, showing that $\mathbf{c}$ is a critical point of f restricted to X. $\triangle$

Example 3.7.4 (Constrained critical point in higher dimensions). Let us find the minimum of the function $f(\mathbf{x}) = x_1^2 + x_2^2 + x_3^2$, when it is constrained to the ellipse X that is the intersection of the cylinder $x_1^2 + x_2^2 = 1$ and the plane of equation $x_1 = x_3$, as shown in Figure 3.7.3.

Since f measures the square of the distance from the origin, we are looking for the points on the ellipse that are closest to the origin. Clearly they are $\mathbf{a} = \begin{pmatrix} 0 \\ 1 \\ 0 \end{pmatrix}$ and $-\mathbf{a} = \begin{pmatrix} 0 \\ -1 \\ 0 \end{pmatrix}$, where $f = 1$. Let us confirm this formally, for $\mathbf{a}$. The ellipse X is given by the equation $\mathbf{F}(\mathbf{x}) = \mathbf{0}$, where

$$\mathbf{F}\begin{pmatrix} x_1 \\ x_2 \\ x_3 \end{pmatrix} = \begin{pmatrix} x_1^2 + x_2^2 - 1 \\ x_1 - x_3 \end{pmatrix}. \qquad 3.7.6$$

A space with more constraints is smaller than a space with fewer constraints: more people belong to the set of musicians than belong to the set of red-headed, left-handed cello players with last name beginning with W.

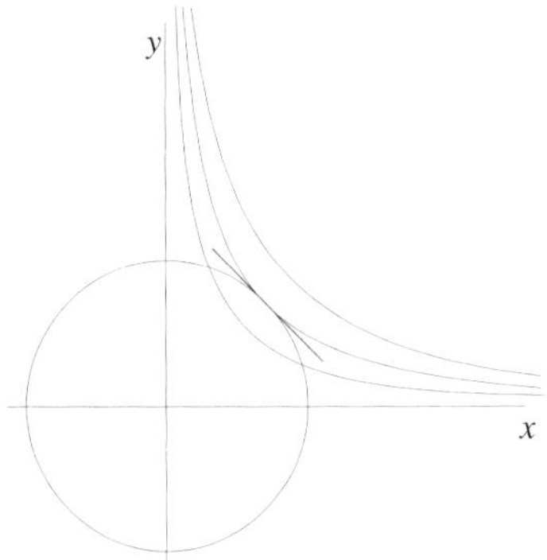

FIGURE 3.7.2.
The unit circle and several level curves of the function xy. The level curve $xy = 1/2$, which realizes the maximum value of xy restricted to the circle, is tangent to the circle at the point $\begin{pmatrix} 1/\sqrt{2} \\ 1/\sqrt{2} \end{pmatrix}$, where the maximum value is realized.

Equation 3.7.6: See Example 3.1.14, which discusses a curve in space as the intersection of two surfaces.

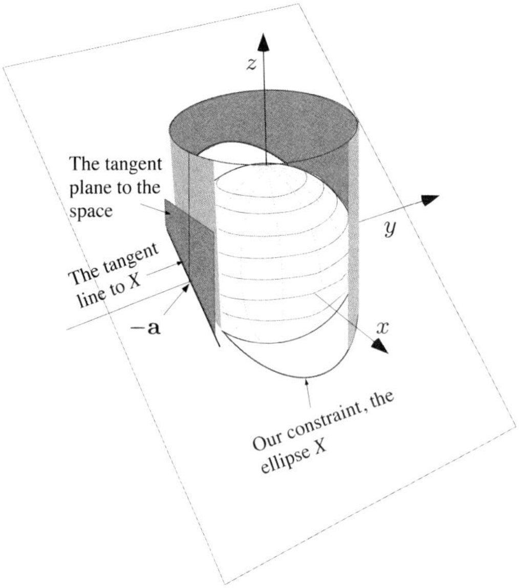

FIGURE 3.7.3 The point $-\mathbf{a}$ is a minimum of the function

$$x^2 + y^2 + z^2$$

restricted to the ellipse; at $-\mathbf{a}$, the one-dimensional tangent space to the ellipse X is a subspace of the two-dimensional tangent space to the sphere given by $x^2 + y^2 + z^2$.

By equation 3.7.1, $\mathbf{a}$ is a critical point of f constrained to X if

$$\ker\,[\mathbf{DF}(\mathbf{a})] \subset \ker\,[\mathbf{D}f(\mathbf{a})]. \qquad 3.7.7$$

We have

$$[\mathbf{DF}(\mathbf{a})] = \begin{bmatrix} 0 & 2 & 0 \\ 1 & 0 & -1 \end{bmatrix} \quad \text{and} \quad [\mathbf{D}f(\mathbf{a})] = [0\ 2\ 0]. \qquad 3.7.8$$

Clearly formula 3.7.7 is satisfied. For a vector $\vec{\mathbf{x}}$ to be in the kernel of $[\mathbf{DF}(\mathbf{a})]$, it must satisfy two equations: $2y = 0$ and $x - z = 0$. To be in the kernel of $[\mathbf{D}f(\mathbf{a})]$, it need only satisfy $2y = 0$. △

Lagrange multipliers

The proof of Theorem 3.7.2 showed how to find equations for critical points of a function defined on (or restricted to) a manifold, *provided* that you know the manifold by a parametrization.

If you know a manifold only by equations, and f is defined on an open neighborhood $U \subset \mathbb{R}^n$ of a manifold X, then *Lagrange multipliers* provide a way to set up the equations for critical points of f constrained to X. (Solving the equations is another matter.)

Theorem and Definition 3.7.5: $F_1, \dots, F_m$ are called *constraint functions* because they define the manifold to which f is restricted.

Note the hypothesis that X be a manifold. Exercise 3.7.15 shows that this is necessary. Here X is an $(n-m)$-dimensional manifold embedded in $\mathbb{R}^n$.

Equation 3.7.9 says that the derivative of f at $\mathbf{a}$ is a linear combination of derivatives of the constraint functions.

Theorem and Definition 3.7.5 (Lagrange multipliers). *Let $U \subset \mathbb{R}^n$ be open, and let $\mathbf{F}: U \to \mathbb{R}^m$ be a C^1 mapping defining a manifold X, with $[\mathbf{DF}(\mathbf{x})]$ onto for every $\mathbf{x} \in X$. Let $f: U \to \mathbb{R}$ be a C^1 mapping. Then $\mathbf{a} \in X$ is a critical point of f restricted to X if and only if there exist numbers $\lambda_1, \dots, \lambda_m$ such that*

$$[\mathbf{D}f(\mathbf{a})] = \lambda_1[\mathbf{D}F_1(\mathbf{a})] + \cdots + \lambda_m[\mathbf{D}F_m(\mathbf{a})]. \qquad 3.7.9$$

The numbers $\lambda_1, \dots, \lambda_m$ are called Lagrange multipliers.

We prove Theorem 3.7.5 after some examples.

Example 3.7.6 (Lagrange multipliers: a simple example). Suppose we want to maximize $f\begin{pmatrix} x \\ y \end{pmatrix} = x + y$ on the ellipse $x^2 + 2y^2 = 1$. We have

$$\underbrace{F\begin{pmatrix} x \\ y \end{pmatrix} = x^2 + 2y^2 - 1}_{\text{constraint function}} \quad \text{and} \quad \left[\mathbf{D}F\begin{pmatrix} x \\ y \end{pmatrix}\right] = [2x,\ 4y], \qquad 3.7.10$$

while $\left[\mathbf{D}f\begin{pmatrix} x \\ y \end{pmatrix}\right] = [1,\ 1]$. So at a critical point, there will exist λ such that

$$[1,\ 1] = \lambda[2x,\ 4y]; \quad \text{i.e.,} \quad x = \frac{1}{2\lambda}; \quad y = \frac{1}{4\lambda}. \qquad 3.7.11$$

Inserting these values into the equation for the ellipse gives

$$\frac{1}{4\lambda^2} + 2\frac{1}{16\lambda^2} = 1, \quad \text{which gives} \quad \lambda = \pm\sqrt{\frac{3}{8}}; \qquad 3.7.12$$

inserting these values for λ in equation 3.7.11 gives

$$x = \pm\frac{1}{2}\sqrt{\frac{8}{3}} \quad \text{and} \quad y = \pm\frac{1}{4}\sqrt{\frac{8}{3}}.$$

So the maximum of the function on the ellipse is

$$\underbrace{\frac{1}{2}\sqrt{\frac{8}{3}}}_{x} + \underbrace{\frac{1}{4}\sqrt{\frac{8}{3}}}_{y} = \frac{3}{4}\sqrt{\frac{8}{3}} = \sqrt{\frac{3}{2}}. \qquad 3.7.13$$

The value $\lambda = -\sqrt{\frac{3}{8}}$ gives the minimum, $x + y = -\sqrt{3/2}$. △

In Example 3.7.6, our constraint manifold is defined by a scalar-valued function F, not by a vector-valued function $\mathbf{F}$.

In Theorem 3.7.5, equation 3.7.9 says that the derivative of f at $\mathbf{a}$ is a linear combination of derivatives of the constraint functions.

The equation $\mathbf{F}(\mathbf{a}) = \mathbf{0}$ defining the manifold is m equations in the n variables $a_1, \dots, a_n$. Equation 3.7.9 is a system of n equations in $n + m$ unknowns: the a_i and $\lambda_1, \dots, \lambda_m$. So together we have $n+m$ equations in $n+m$ unknowns.

FIGURE 3.7.4.
Joseph-Louis Lagrange was born in Italy in 1736, one of 11 children, nine of whom died before reaching adulthood. He was in Paris during the French Revolution; in 1793, the great chemist Antoine-Laurent Lavoisier (who was guillotined the following year) intervened to prevent his arrest as a foreigner. In 1808 Napoleon named him Count of the Empire.

Lagrange's book *Mécanique analytique* set the standard in the subject for a hundred years.

Example 3.7.7 (Critical points of function constrained to ellipse). Let us follow this procedure for the function $f(\mathbf{x}) = x^2 + y^2 + z^2$ of Example 3.7.4, constrained as before to the ellipse given by

$$\mathbf{F}(\mathbf{x}) = \mathbf{F}\begin{pmatrix} x \\ y \\ z \end{pmatrix} = \begin{pmatrix} x^2 + y^2 - 1 \\ x - z \end{pmatrix} = \mathbf{0}. \qquad 3.7.14$$

We have

$$[\mathbf{D}f(\mathbf{x})] = [2x, 2y, 2z], \ [\mathbf{D}F_1(\mathbf{x})] = [2x, 2y, 0], \ [\mathbf{D}F_2(\mathbf{x})] = [1, 0, -1],$$

so Theorem 3.7.5 says that at a critical point there exist numbers λ_1 and λ_2 such that $[2x, 2y, 2z] = \lambda_1[2x, 2y, 0] + \lambda_2[1, 0, -1]$, which gives

$$2x = \lambda_1 2x + \lambda_2, \quad 2y = \lambda_1 2y + 0, \quad 2z = \lambda_1 \cdot 0 - \lambda_2. \qquad 3.7.15$$

If $y \neq 0$, this gives $\lambda_1 = 1$, $\lambda_2 = 0$, and $z = 0$. The equations $F_1 = 0$ and $F_2 = 0$ then say that $x = 0$, $y = \pm 1$, so $\begin{pmatrix} 0 \\ 1 \\ 0 \end{pmatrix}$ and $\begin{pmatrix} 0 \\ -1 \\ 0 \end{pmatrix}$ are critical points. But if $y = 0$, then $F_1 = 0$ and $F_2 = 0$ give $x = z = \pm 1$. So $\begin{pmatrix} 1 \\ 0 \\ 1 \end{pmatrix}$

and $\begin{pmatrix} -1 \\ 0 \\ -1 \end{pmatrix}$ are also critical points. For the first, $\lambda_1 = 2$ and $\lambda_2 = -2$; for the second, $\lambda_1 = 2$ and $\lambda_2 = 2$.

Since the ellipse is compact, the maximum and minimum values of f restricted to $\mathbf{F} = \mathbf{0}$ are attained at constrained critical points of f. Since $f\begin{pmatrix} 0 \\ 1 \\ 0 \end{pmatrix} = f\begin{pmatrix} 0 \\ -1 \\ 0 \end{pmatrix} = 1$ and $f\begin{pmatrix} 1 \\ 0 \\ 1 \end{pmatrix} = f\begin{pmatrix} -1 \\ 0 \\ -1 \end{pmatrix} = 2$, we see that f achieves its maximum value of 2 at $\begin{pmatrix} 1 \\ 0 \\ 1 \end{pmatrix}$ and at $\begin{pmatrix} -1 \\ 0 \\ -1 \end{pmatrix}$ and (as we saw in Example 3.7.4) its minimum value of 1 at $\begin{pmatrix} 0 \\ 1 \\ 0 \end{pmatrix}$ and $\begin{pmatrix} 0 \\ -1 \\ 0 \end{pmatrix}$. △

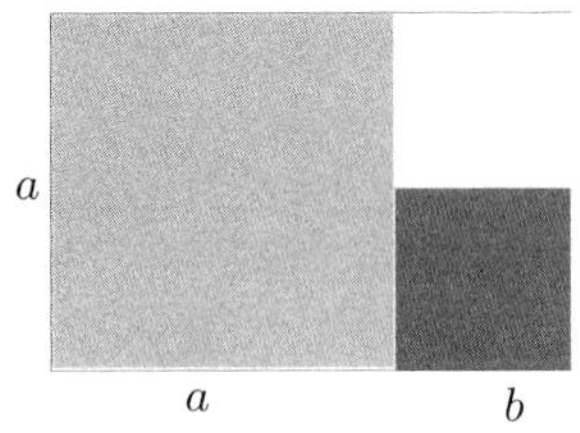

FIGURE 3.7.5.

Example 3.7.8: If $a \geq b$, the smallest rectangle that contains both squares disjointly has sides a and $a + b$.

Example 3.7.8 (Lagrange multipliers: a somewhat harder example). What is the smallest number A such that any two squares S_1, S_2 of total area 1 can be put disjointly into a rectangle of area A? ("Disjointly" means "not overlapping".)

Let us call a and b the lengths of the sides of S_1 and S_2, and we may assume that $a \geq b \geq 0$. Then the smallest rectangle that will contain the two squares disjointly has sides a and $a + b$, and area $a(a + b)$, as shown in Figure 3.7.5. The largest side cannot be less than $a + b$, or the squares would overlap; if it is more than $a + b$, then we are making the rectangle unnecessarily large.

The problem is to maximize the area $a^2 + ab$, subject to the constraints $a^2 + b^2 = 1$, and $a \geq b \geq 0$. Remember that the area A must be big enough so that given *any* two squares with total area 1, we can find a rectangle with area A that can hold them disjointly. If $a = 1$ and $b = 0$, or if $a = b = 1/\sqrt{2}$, then both squares will fit in a rectangle of area 1. If $a = \sqrt{2/3}$ and $b = 1/\sqrt{3}$, a larger rectangle is needed, with area approximately 1.14. To find the smallest A that works, we need to find the largest that $a^2 + ab$ could possibly be, when a and b satisfy $a^2 + b^2 = 1$, and $a \geq b \geq 0$.

We are *not* looking for a single rectangle of area A that will hold any two squares S_1, S_2 of total area 1. Imagine that you have been challenged to find the smallest number of tiny (but very expensive) tiles with which you can make a rectangle that will hold any two squares with total area 1. You buy the tiles you think you need; your opponents challenge you with pairs of squares. You lose if an opponent comes up with a pair of squares that you can't fit into a rectangle made from your tiles. You also lose if someone else succeeds using fewer tiles.

The Lagrange multiplier theorem tells us that at a critical point of the constrained function there exists a number λ such that

$$\underbrace{[2a + b,\ a]}_{\substack{\text{deriv. of} \\ \text{area function}}} = \lambda \underbrace{[2a,\ 2b]}_{\substack{\text{deriv. of} \\ \text{constraint func.}}}. \qquad 3.7.16$$

So we need to solve the system of three nonlinear equations

$$2a + b = 2a\lambda, \quad a = 2b\lambda, \quad a^2 + b^2 = 1. \qquad 3.7.17$$

Substituting the value of a from the second equation into the first, we find

$$4b\lambda^2 - 4b\lambda - b = 0. \qquad 3.7.18$$

One solution is $b = 0$, but this gives $a = 0$, which is incompatible with $a^2 + b^2 = 1$. If $b \neq 0$, then

$$4\lambda^2 - 4\lambda - 1 = 0, \quad \text{which gives} \quad \lambda = \frac{1 \pm \sqrt{2}}{2}. \qquad 3.7.19$$

Our remaining equations are now

$$\frac{a}{b} = 2\lambda = 1 \pm \sqrt{2} \quad \text{and} \quad a^2 + b^2 = 1, \tag{3.7.20}$$

and since we have required $a,\ b \geq 0$ we must take the positive square root. We then have the unique solution

$$a = \frac{\sqrt{2+\sqrt{2}}}{2} \quad \text{and} \quad b = \frac{\sqrt{2-\sqrt{2}}}{2}. \tag{3.7.21}$$

The two endpoints correspond to the two extremes: all the area in one square and none in the other, or both squares with the same area. At $\begin{pmatrix}1\\0\end{pmatrix}$, one square has area 1, and the other has area 0; at $\begin{pmatrix}\sqrt{2}/2\\ \sqrt{2}/2\end{pmatrix}$, the two squares are identical.

This satisfies the constraint $a \geq b \geq 0$, and leads to

$$A = a(a+b) = \frac{\sqrt{2} + \sqrt{6+4\sqrt{2}}}{4} \approx 1.21. \tag{3.7.22}$$

We must check (see Remark 3.6.5) that the maximum is not an endpoint of the constraint region (the point with coordinates $a = 1$, $b = 0$ or the point with coordinates $a = b = \sqrt{2}/2$). It is easy to see that $(a+b)a = 1$ at both endpoints, and since $A > 1$, this is the unique maximum value. △

Example 3.7.9 (Lagrange multipliers: a fourth example). Find the critical points of $F\begin{pmatrix}x\\y\\z\end{pmatrix} = xyz$ on the plane given by the equation $f\begin{pmatrix}x\\y\\z\end{pmatrix} = x + 2y + 3z - 1 = 0$. Theorem 3.7.5 asserts that a critical point is a solution to

In Example 3.7.9, F is the constrained function, and f the constraint function, the reverse of our usual practice. You should not assume that in this type of problem f is necessarily the function with constrained points, and F the constraint function.

Note that equation 3.7.23 is a system of four equations in four unknowns. In general, there is no better way to deal with such a system than using Newton's method or some similar algorithm. This example isn't typical, because there are tricks available for solving equation 3.7.23. You shouldn't expect this to happen in general.

$$\begin{aligned} &1.\ \underbrace{[yz,\ xz,\ xy]}_{\text{deriv. of } F} = \lambda\ \underbrace{[1,\ 2,\ 3]}_{\substack{\text{deriv. of } f\\ \text{(constraint)}}} \\ &2.\ \underbrace{x + 2y + 3z = 1}_{\text{constraint equation}} \end{aligned} \quad \text{or} \quad \begin{aligned} yz &= \lambda \\ xz &= 2\lambda \\ xy &= 3\lambda \\ 1 &= x + 2y + 3z. \end{aligned} \tag{3.7.23}$$

In this case, there are tricks we can use. It is not hard to derive $xz = 2yz$ and $xy = 3yz$, so if $z \neq 0$ and $y \neq 0$, then $y = x/2$ and $z = x/3$. Substituting these values into the last equation gives $x = 1/3$, hence $y = 1/6$ and $z = 1/9$. At this point, F has the value $1/162$.

What if $z = 0$ or $y = 0$? If $z = 0$, the Lagrange multiplier equation is

$$[0,\ 0,\ xy] = \lambda[1,\ 2,\ 3], \tag{3.7.24}$$

which says that $\lambda = 0$, so one of x or y must also vanish. Suppose $y = 0$, then $x = 1$, and the value of F is 0. There are two other similar points. To summarize: there are four critical points,

$$\begin{pmatrix}1\\0\\0\end{pmatrix},\quad \begin{pmatrix}0\\1/2\\0\end{pmatrix},\quad \begin{pmatrix}0\\0\\1/3\end{pmatrix},\quad \begin{pmatrix}1/3\\1/6\\1/9\end{pmatrix}; \tag{3.7.25}$$

at the first three $F = 0$, and at the fourth $F = 1/162$.

Is our fourth point a maximum? The answer is yes (at least, it is a local maximum). The part of the plane of equation $x + 2y + 3z = 1$ that lies in the first octant, where $x, y, z \geq 0$, is compact, since $|x|, |y|, |z| \leq 1$ there; otherwise the equation of the plane cannot be satisfied. So (by Theorem 1.6.9) F has a maximum in that octant. To be sure that this maximum is a critical point, we must check that it isn't on the edge of the octant (i.e., a place where any of x, y, z vanishes). That is straightforward, since the function vanishes on the boundary, while it is positive at the fourth point. So the maximum guaranteed by Theorem 1.6.9 is our fourth point, the critical point that gives the largest value of F. Exercise 3.7.7 asks you to show that the other critical points are saddles. △

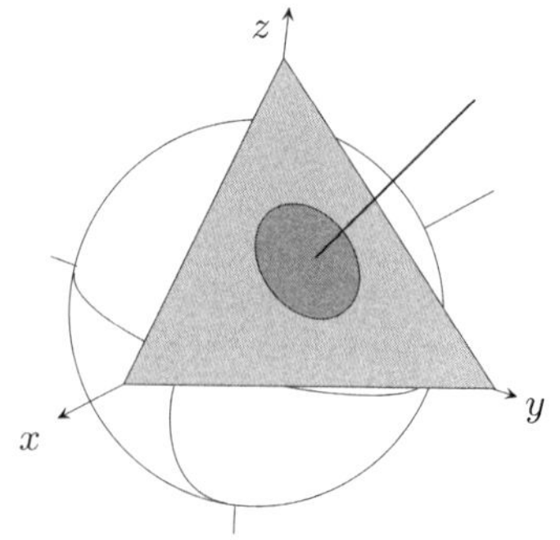

FIGURE 3.7.6.

Example 3.7.10: The lightly shaded region represents the plane $x + y + z = a$, for some a. The dark region is X_a, the part of the unit sphere where $x + y + z \geq a$; you may think of it as a polar cap.

Example 3.7.10 shows that when you are looking for maxima and minima of a function, checking boundary points may be a second Lagrange multiplier problem, perhaps harder than the first. It also points out the value of knowing that the domain of a continuous function is compact.

Example 3.7.10 (Checking boundary values). Let X_a be the part of the sphere of equation $x^2 + y^2 + z^2 = 1$ where $x + y + z \geq a$, as shown in Figure 3.7.6 (the analogous case for the circle, which may be easier to visualize, is shown in Figure 3.7.7). In terms of the parameter a, what are the maxima and minima of the function $f\begin{pmatrix} x \\ y \\ z \end{pmatrix} = xy$ on X_a? Note that the function $x + y + z$ constrained to the sphere achieves its maximum at $x = y = z = 1/\sqrt{3}$, and its minimum at $x = y = z = -1/\sqrt{3}$. Therefore we may restrict to $|a| \leq \sqrt{3}$: if $a > \sqrt{3}$, then X_a is empty, and if $a < -\sqrt{3}$, the set X_a is the whole sphere.

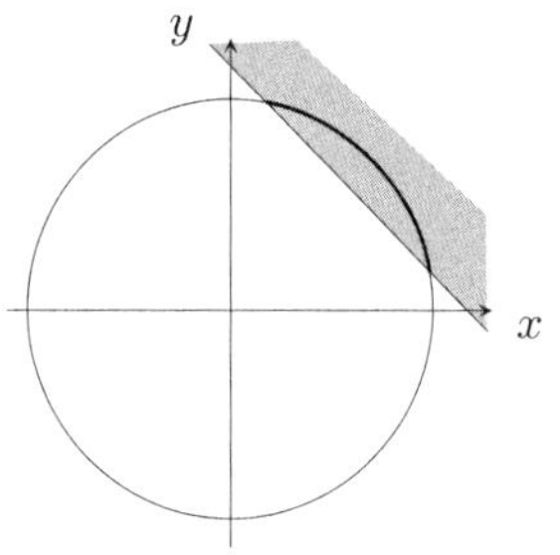

FIGURE 3.7.7.

The shaded region is the part of the plane where $x + y \geq 1.2$; the bold arc of the circle is the part of the unit circle in that region.

Since X_a is compact and f is continuous, there is a maximum and a minimum for every a satisfying $|a| \leq \sqrt{3}$. Each maximum and each minimum occurs either in the interior of X_a or on the boundary.

A first problem is to find the critical points of f on the sphere. Applying Lagrange multipliers, we get the equations

$$\begin{aligned} [y \quad x \quad 0] &= \lambda\, [2x \quad 2y \quad 2z] \\ x^2 + y^2 + z^2 &= 1. \end{aligned} \qquad 3.7.26$$

We leave to you to check that the solutions to these equations are

$$\mathbf{p}_1 = \begin{pmatrix} \sqrt{2}/2 \\ \sqrt{2}/2 \\ 0 \end{pmatrix}, \ \mathbf{p}_2 = \begin{pmatrix} \sqrt{2}/2 \\ -\sqrt{2}/2 \\ 0 \end{pmatrix}, \ \mathbf{p}_3 = \begin{pmatrix} -\sqrt{2}/2 \\ \sqrt{2}/2 \\ 0 \end{pmatrix}, \ \mathbf{p}_4 = \begin{pmatrix} -\sqrt{2}/2 \\ -\sqrt{2}/2 \\ 0 \end{pmatrix}, \mathbf{q}_1 = \begin{pmatrix} 0 \\ 0 \\ 1 \end{pmatrix}, \ \mathbf{q}_2 = \begin{pmatrix} 0 \\ 0 \\ -1 \end{pmatrix}. \qquad 3.7.27$$

Figure 3.7.8 shows which of these points are in X_a, depending on the value of a.

The critical points $\mathbf{p}_1, \dots, \mathbf{p}_4, \mathbf{q}_1, \mathbf{q}_2$ are candidate minima and maxima when they are in the interior of X_a, but we must check the boundary of X_a, where $x + y + z = a$: we must find any critical points of xy restricted to the boundary of X_a, and compare the values of xy there to the values at

We can use Lagrange multipliers to get equation 3.7.28 because the boundary of X_a is a smooth manifold (in fact, a circle) when $|a| < \sqrt{3}$.

On X_a, we have $x + y + z \geq a$.

at $\mathbf{p}_1$: $x + y + z = \sqrt{2}$

at $\mathbf{p}_2$: $x + y + z = 0$

at $\mathbf{p}_3$: $x + y + z = 0$

at $\mathbf{p}_4$: $x + y + z = -\sqrt{2}$

at $\mathbf{q}_1$: $x + y + z = 1$

at $\mathbf{q}_2$: $x + y + z = -1$

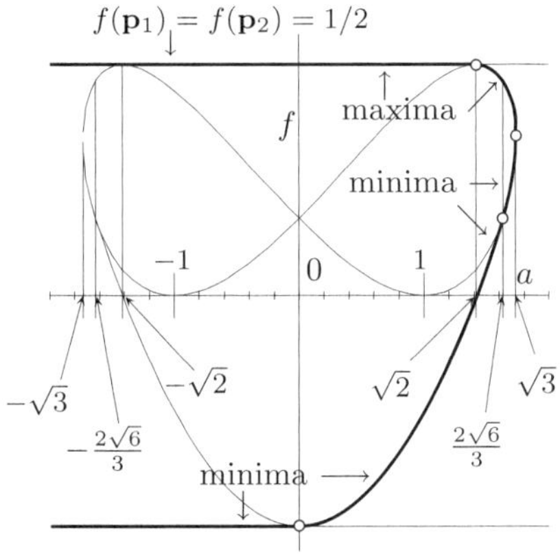

FIGURE 3.7.9.
This picture plots the values of f at its critical points in X_a and on the boundary. The figure eight and the parabola are values for critical points on the boundary; the horizontal lines are critical points in the interior (the values of f don't depend on a; the value of a just determines whether critical points are in X_a).

In bold we have marked the curve corresponding to the maximum and minimum values.

Note that only part of the parabola (the arc where $|a| \leq 2\sqrt{6}/3$) consists of points on the boundary; the parabola goes right on existing but the points they correspond to have nonreal coordinates.

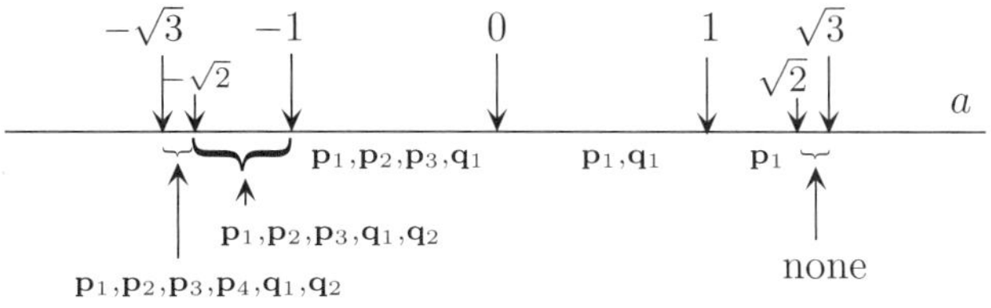

FIGURE 3.7.8. As a decreases from $\sqrt{3}$ to $-\sqrt{3}$, the set X_a grows; when $a \in (\sqrt{2}, \sqrt{3})$, it contains none of the $\mathbf{p}_i, \mathbf{q}_i$, and when $a \in (-\sqrt{2}, -\sqrt{3})$, it contains them all.

the $\mathbf{p}_i$ and $\mathbf{q}_i$ in X_a. Applying Lagrange multipliers, we get five equations in five unknowns:

$$\begin{gathered}[y \quad x \quad 0] = \lambda_1[2x \quad 2y \quad 2z] + \lambda_2[1 \quad 1 \quad 1] \\ x^2 + y^2 + z^2 = 1, \quad x + y + z = a.\end{gathered} \qquad 3.7.28$$

(Remember that a is a fixed parameter for this system.) Solving this system is a bit elaborate. Subtracting one of the first two equations on the first line from the other leads to

$$(x - y)(1 + 2\lambda_1) = 0. \qquad 3.7.29$$

Thus the analysis subdivides into two cases: $x = y$ and $\lambda_1 = -1/2$.

The case $x = y$. Write $z = a - 2x$ and substitute into the equation of the sphere, to find

$$6x^2 - 4ax + a^2 - 1 = 0, \quad \text{with roots} \quad x = \frac{2a \pm \sqrt{6 - 2a^2}}{6}. \qquad 3.7.30$$

The values of f at these roots are

$$\frac{a^2 + 3 + 2a\sqrt{6 - 2a^2}}{18} \quad \text{and} \quad \frac{a^2 + 3 - 2a\sqrt{6 - 2a^2}}{18}. \qquad 3.7.31$$

These values of f as a function of a form the figure eight shown in Figure 3.7.9; every point of the figure eight is a critical point of xy restricted to the boundary.

The case $\lambda_1 = -1/2$. Here the first set of equations in 3.7.28 gives $x + y = \lambda_2$ and $z = \lambda_2$, hence $2\lambda_2 = a$. Substituting into the equation of the sphere gives $4x^2 - 2ax + a^2 - 2 = 0$. Note that this equation has real roots only if $|a| \leq 2\sqrt{6}/3$, and in this case we use $y = \lambda_2 - x$ and $2\lambda_2 = a$ to compute $xy = a^2/4 - 1/2$, whose graph is the parabola shown in Figure 3.7.9.

What does all this mean? If a critical point of xy is in the interior of X_a, it is one of the six points in equation 3.7.27, as specified by Figure 3.7.8. By solving the boundary Lagrange multiplier problem (equation 3.7.28) we have found all the critical points of xy restricted to the boundary of X_a.

To see which critical points are maxima or minima, we need to compare the values of xy at the various points. (Because X_a is compact, we can say that some of the critical points are maxima or minima without any second derivative test.) Plotting the values of f at all critical points in the interior of X_a and the values of f at all critical point on the boundary, we get Figure 3.7.9 (repeated, with abbreviated caption, in the margin at left). We have marked in bold the curve corresponding to the maximum and minimum values as functions of a.

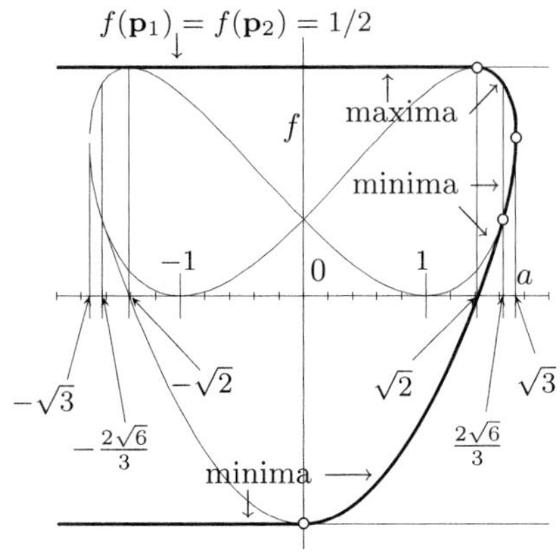

FIGURE 3.7.9.
(Repeated from the previous page)
The figure eight and the parabola are values for critical points on the boundary; the horizontal lines are critical points in the interior (the values of f don't depend on a; the value of a just determines whether critical points are in X_a).

In bold we have marked the curve corresponding to the maximum and minimum values.

In formulas, these maxima and minima are described below:

- $2\sqrt{6}/3 < a < \sqrt{3}$: None of the $\mathbf{p}_i$ or $\mathbf{q}_i$ are in X_a. The max and min are both on the boundary, given by

$$x = y = \frac{2a \pm \sqrt{6-2a^2}}{6}, \quad z = a - 2x \text{ and } f = \frac{a^2 + 3 \pm 2a\sqrt{6-2a^2}}{18}.$$

- $\sqrt{2} < a < 2\sqrt{6}/3$: None of the $\mathbf{p}_i$ or $\mathbf{q}_i$ are in X_a. The max and min are both on the boundary. As above, the maximum is achieved at

$$x = y = \frac{2a + \sqrt{6-2a^2}}{6}, \quad z = a - 2x, \text{ where } f = \frac{a^2 + 3 + 2a\sqrt{6-2a^2}}{18},$$

and the minimum is achieved at two points:

$$x = \frac{a + \sqrt{8-3a^2}}{4}, \quad y = \frac{a - \sqrt{8-3a^2}}{4}, \quad z = \frac{a}{2}, \text{ where } f = \frac{a^2-2}{4} \qquad 3.7.32$$

$$x = \frac{a - \sqrt{8-3a^2}}{4}, \quad y = \frac{a + \sqrt{8-3a^2}}{4}, \quad z = \frac{a}{2}, \text{ where } f = \frac{a^2-2}{4}. \qquad 3.7.33$$

- $0 < a < \sqrt{2}$: The maximum is realized at $\mathbf{p}_1$, with value $1/2$, and the minimum $\frac{a^2-2}{4}$ is still realized at the two points in equations 3.7.32 and 3.7.33.
- $-\sqrt{2} < a < 0$: The maximum $1/2$ is realized at $\mathbf{p}_1$, and the minimum $-1/2$ is realized at the two points $\mathbf{p}_2, \mathbf{p}_3$.
- For $a < -\sqrt{2}$: The maximum value $1/2$ is achieved at $\mathbf{p}_4$ also. $\triangle$

Proof of Theorem 3.7.5 on Lagrange multipliers

This is just an application of Lemma 3.7.11 to equation 3.7.7, which says that $\ker [\mathbf{DF}(\mathbf{a})] \subset \ker [\mathbf{D}f(\mathbf{a})]$.

Lemma 3.7.11. *Let* $A = \begin{bmatrix} \alpha_1 \\ \vdots \\ \alpha_m \end{bmatrix} : \mathbb{R}^n \to \mathbb{R}^m$ *and* $\beta : \mathbb{R}^n \to \mathbb{R}$ *be linear transformations. Then*

$$\ker A \subset \ker \beta \qquad 3.7.34$$

if and only if there exist numbers $\lambda_1, \dots, \lambda_m$ *such that*

$$\beta = \lambda_1 \alpha_1 + \cdots + \lambda_m \alpha_m. \qquad 3.7.35$$

Proof of Lemma 3.7.11. In one direction, if

$$\beta = \lambda_1 \alpha_1 + \cdots + \lambda_m \alpha_m \qquad 3.7.36$$

and $\vec{\mathbf{v}} \in \ker A$, then $\vec{\mathbf{v}} \in \ker \alpha_i$ for $i = 1, \dots, m$, so $\vec{\mathbf{v}} \in \ker \beta$.

For the other direction, if $\ker A \subset \ker \beta$, then

$$\begin{bmatrix} \alpha_1 \\ \vdots \\ \alpha_m \end{bmatrix} \quad \text{and} \quad \begin{bmatrix} \alpha_1 \\ \vdots \\ \alpha_m \\ \beta \end{bmatrix} \qquad 3.7.37$$

have the same kernel, hence (by the dimension formula) the same rank. By Proposition 2.5.11,

$$\text{rank}[\alpha_1^\top \dots \alpha_m^\top] = \text{rank}[\alpha_1^\top \dots \alpha_m^\top \, \beta^\top] \qquad 3.7.38$$

so if you row reduce $[\alpha_1^\top \dots \alpha_m^\top \, \beta^\top]$ there will be no pivotal 1 in the last column. Thus the equation $\beta = \lambda_1\alpha_1 + \cdots + \lambda_m\alpha_m$ has a solution. □

Lemma 3.7.11: A is an $m \times n$ matrix, and β is a row matrix n wide:

$$A = \begin{bmatrix} \alpha_{1,1} & \dots & \alpha_{1,n} \\ \vdots & \dots & \vdots \\ \alpha_{m,1} & \dots & \alpha_{m,n} \end{bmatrix}$$

and $\beta = \begin{bmatrix} \beta_1 & \dots & \beta_n \end{bmatrix}$.

As discussed in connection with equations 3.7.2 and 3.7.3, a space with many constraints is smaller than a space with few constraints. Here, $A\mathbf{x} = \mathbf{0}$ imposes m constraints, and $\beta\mathbf{x} = 0$ imposes only one.

Lemma 3.7.11 is simply saying that the only linear consequences one can draw from a system of linear equations are the linear combinations of those equations.

We don't know anything about the relationship of n and m, but we know that $k \le n$, since $n+1$ vectors in $\mathbb{R}^n$ cannot be linearly independent.

Classifying constrained critical points

In Examples 3.7.6–3.7.10 we classified our constrained critical points, using as our main tool Theorem 1.6.9. But our methods were not systematic and when, in Exercise 3.7.7, we ask you to confirm that three critical points in Example 3.7.9 are saddles, we suggest using a parametrization. This is often impossible; usually, parametrizations cannot be written by formulas. We will now give a straightforward (if lengthy) way to classify critical points restricted to a manifold known by an equation. This is the generalization to constrained critical points of the second derivative criterion.

First we need to define the signature of a constrained critical point.

Proposition and Definition 3.7.12 (Signature of constrained critical point). *Let $U \subset \mathbb{R}^{n+m}$ be open. Let $Z \subset U$ be an n-dimensional manifold, and $f : U \to \mathbb{R}$ a C^2 function. Suppose V is open in $\mathbb{R}^n$ and $\gamma : V \to Z$ is a parametrization of Z near $\mathbf{z}_0 \in Z$; let $\mathbf{v}_0 \in V$ be the point such that $\gamma(\mathbf{v}_0) = \mathbf{z}_0$. Then*

1. *The point $\mathbf{v}_0$ is a critical point of $f \circ \gamma$ if and only if $\mathbf{z}_0$ is a critical point of $f|_Z$.*
2. *The signature (p, q) of $f \circ \gamma$ at $\mathbf{v}_0$ does not depend on the choice of parametrization. It is called the signature of $f|_Z$ at $\mathbf{z}_0$.*

We denote by $f|_Z$ the function f restricted to Z.

Proof. 1. By the chain rule,

$$[\mathbf{D}(f \circ \gamma)(\mathbf{v}_0)] = [\mathbf{D}f(\gamma(\mathbf{v}_0))][\mathbf{D}\gamma(\mathbf{v}_0)] = [\mathbf{D}f(\mathbf{z}_0)][\mathbf{D}\gamma(\mathbf{v}_0)], \qquad 3.7.39$$

so $\mathbf{v}_0$ is a critical point of $f \circ \gamma$ if and only if $[\mathbf{D}f(\mathbf{z}_0)]$ vanishes on $T_{\mathbf{z}_0}X = \text{img}[\mathbf{D}\gamma(\mathbf{v}_0)]$, i.e., if and only if $\mathbf{z}_0$ is a critical point of $f|_Z$.

. The equality

$$T_{\mathbf{z}_0}X = \text{img}[\mathbf{D}\gamma(\mathbf{v}_0)]$$

is Proposition 3.2.7.

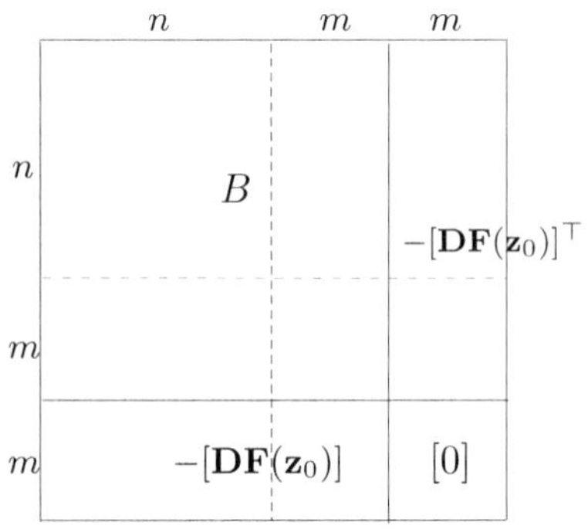

FIGURE 3.7.10.

The augmented Hessian matrix H of equation 3.7.43. Since $\mathbf{F}$ goes from a subset of $\mathbb{R}^{m+n}$ to $\mathbb{R}^m$, the matrix $[\mathbf{DF}(\mathbf{z}_0)]$ is $m \times (m+n)$, and its transpose is $(m+n) \times m$. The matrix B defined in equation 3.7.42 is $(m+n) \times (m+n)$, and $[0]$ is $m \times m$.

Recall from Section 3.6 that an (unaugmented) Hessian matrix is the matrix of second partials at a critical point of f.

2. If V_1, V_2 are open subsets of $\mathbb{R}^n$ and $\gamma_1 : V_1 \to Z$ and $\gamma_2 : V_2 \to Z$ are two parametrizations of class C^2, then the two functions $f \circ \gamma_1$ and $f \circ \gamma_2$ are related by the change of variables $\varphi \stackrel{\text{def}}{=} \gamma_1^{-1} \circ \gamma_2$:

$$f \circ \gamma_2 = f \circ \gamma_1 \circ \gamma_1^{-1} \circ \gamma_2 = (f \circ \gamma_1) \circ \varphi. \qquad 3.7.40$$

Proposition 3.6.7 says that the critical point $\gamma_1^{-1}(\mathbf{z}_0)$ of $f \circ \gamma_1$ and the critical point $\gamma_2^{-1}(\mathbf{z}_0)$ of $f \circ \gamma_2$ have the same signature. □

Usually we know the constraint manifold Z by an equation $\mathbf{F} = \mathbf{0}$, where U is a subset of $\mathbb{R}^{n+m}$ and $\mathbf{F} : U \to \mathbb{R}^m$ is a C^1 map with surjective derivative at every point of Z. We usually don't know any parametrization. By the implicit function theorem, parametrizations exist, but computing them involves some iterative and unwieldy method like Newton's method.

Lagrange multipliers allow us to find constrained critical points without knowing a parametrization; an elaboration of Lagrange multipliers using the *augmented Hessian matrix* allows us to find the signature of such a point, again without knowing a parametrization. Here we will need $\mathbf{F}$ to be C^2.

Suppose that a critical point $\mathbf{z}_0$ of f restricted to Z has been found using Lagrange multipliers, leading to numbers $\lambda_1, \dots, \lambda_m$ such that

$$D_i f(\mathbf{z}_0) = \sum_{j=1}^{m} \lambda_j D_i F_j(\mathbf{z}_0), \qquad 3.7.41$$

where the F_j are coordinate functions of $\mathbf{F}$. Use these Lagrange multipliers λ_k and the second derivatives of f and $\mathbf{F}$ to construct the $(m+n) \times (m+n)$ matrix

$$B = \begin{bmatrix} D_1 D_1 f(\mathbf{z}_0) - \sum_{k=1}^m \lambda_k D_1 D_1 F_k & \cdots & D_1 D_{m+n} f(\mathbf{z}_0) - \sum_{k=1}^m \lambda_k D_1 D_{m+n} F_k \\ \vdots & & \vdots \\ D_{m+n} D_1 f(\mathbf{z}_0) - \sum_{k=1}^m \lambda_k D_{m+n} D_1 F_k & \cdots & D_{m+n} D_{m+n} f(\mathbf{z}_0) - \sum_{k=1}^m \lambda_k D_{m+n} D_{m+n} F_k \end{bmatrix}. \qquad 3.7.42$$

Now construct the $(n+2m) \times (n+2m)$ augmented Hessian matrix

$$H = \begin{bmatrix} B & -[\mathbf{DF}(\mathbf{z}_0)]^\top \\ -[\mathbf{DF}(\mathbf{z}_0)] & [0]. \end{bmatrix}, \qquad 3.7.43$$

where $[0]$ denotes the $m \times m$ matrix with all entries 0; see Figure 3.7.10. Note that if you can find a critical point $\mathbf{z}_0$ of f restricted to Z, this augmented Hessian matrix can be written down; it may be large and unpleasant to write down, but there is no real difficulty.

Since cross partials are equal (Theorem 3.3.8), H is symmetric, and (by Proposition 3.5.17) it defines a quadratic form on $\mathbb{R}^{n+2m}$. Denote its signature by (p_1, q_1); these numbers p_1, q_1 can be calculated by completing squares. Equation 3.7.44, which is quite surprising, says that knowing (p_1, q_1) enables us to compute (p, q), the signature of the constrained critical point.

In Theorem 3.7.13, $f: U \to \mathbb{R}$ is the C^2 function such that the critical points of $f|_Z$ are to be classified, and H is the augmented Hessian matrix defined in equation 3.7.43.

Since Theorem 3.7.13 is local, it is actually enough to say that $[\mathbf{DF}(\mathbf{z}_0)]$ is onto.

Theorem 3.7.13 (Computing the signature of a constrained critical point). *Let $U \subset \mathbb{R}^{n+m}$ be open, $f: U \to \mathbb{R}$ a C^2 function, and $\mathbf{F}: U \to \mathbb{R}^m$ a C^2 map defining a manifold $Z \stackrel{\text{def}}{=} \mathbf{F}^{-1}(\mathbf{0})$, with $[\mathbf{DF}(\mathbf{z})]$ onto at every point $\mathbf{z} \in Z$. Let $\mathbf{z}_0 \in Z$ be a critical point of f restricted to Z of signature (p,q), and let (p_1, q_1) be the signature of the quadratic form defined by the augmented Hessian matrix H. Then*

$$(p,q) = (p_1 - m,\ q_1 - m). \qquad 3.7.44$$

Theorem 3.7.13 is proved in Appendix A14.

Example 3.7.14 (Classifying constrained critical points). We will find and classify the critical points of $f\begin{pmatrix} x \\ y \\ z \end{pmatrix} = ax + by + cz$ on the surface of equation $1/x + 1/y + 1/z = 1$, all in terms of the parameters a, b, c.

In this case the Lagrange multiplier equation becomes

$$[a,\ b,\ c] = \left[-\frac{\lambda}{x^2},\ -\frac{\lambda}{y^2},\ -\frac{\lambda}{z^2}\right], \qquad 3.7.45$$

leading to the four equations

$$x^2 = -\frac{\lambda}{a}, \qquad y^2 = -\frac{\lambda}{b}, \qquad z^2 = -\frac{\lambda}{c}, \qquad \frac{1}{x} + \frac{1}{y} + \frac{1}{z} = 1. \qquad 3.7.46$$

In order for solutions to this system of equations to exist, we must have either $\lambda > 0$ and $a < 0,\ b < 0,\ c < 0$, or $\lambda < 0$ and $a > 0,\ b > 0,\ c > 0$. These two possibilities aren't really different: they correspond to changing f to $-f$, which has the same critical points, with signature (p,q) changed to (q,p). We will assume $\lambda < 0$ and $a > 0,\ b > 0,\ c > 0$.

Without loss of generality, we can uniquely write the solutions

$$x = s_1\frac{\sqrt{-\lambda}}{\sqrt{a}}, \qquad y = s_2\frac{\sqrt{-\lambda}}{\sqrt{b}}, \qquad z = s_3\frac{\sqrt{-\lambda}}{\sqrt{c}}, \qquad 3.7.47$$

with all square roots > 0, and where each s_i is either $+1$ or -1. The constraint becomes

$$s_1\sqrt{a} + s_2\sqrt{b} + s_3\sqrt{c} = \sqrt{-\lambda}, \qquad 3.7.48$$

so a solution with signs s_1, s_2, s_3 exists if and only if $s_1\sqrt{a} + s_2\sqrt{b} + s_3\sqrt{c} > 0$.

If none of the three numbers $\sqrt{a}$, $\sqrt{b}$, $\sqrt{c}$ is larger than the sum of the other two, there are four solutions: all s_i are $+1$, or two are $+1$ and the other is -1. The corresponding solutions have either all coordinates positive, or two positive and one negative.

If one of the three numbers $\sqrt{a}$, $\sqrt{b}$, $\sqrt{c}$ is larger than the sum of the other two, say $\sqrt{a} > \sqrt{b} + \sqrt{c}$, then we must have $s_1 > 0$, and s_2 and s_3 can be anything. Thus there are still four solutions, one with all coordinates positive, two with two coordinates positive and one negative, and one with one coordinate positive and two negative.

Equation 3.7.49: Here $m = 1$ and F is the function
$$\frac{1}{x} + \frac{1}{y} + \frac{1}{z} - 1,$$
so the entry $h_{1,1}$ of H is
$$D_1 D_1 f(\mathbf{a}) - \lambda_1 D_1 D_1 F = 0 - \frac{2\lambda}{x^3},$$
where $\mathbf{a}$ is a critical point of f restricted to $F = 0$.

We discussed how to construct the symmetric matrix associated to a quadratic form immediately after Proposition 3.5.17. Here we go in the other direction. The A, B, C, D in equation 3.7.50 are the coordinates of a vector corresponding to the vector $\vec{x}$ in equation 3.5.32:
$$Q_H \begin{bmatrix} A \\ B \\ C \\ D \end{bmatrix} = \begin{bmatrix} A \\ B \\ C \\ D \end{bmatrix} \cdot H \begin{bmatrix} A \\ B \\ C \\ D \end{bmatrix}.$$

By Theorem 3.7.13, if the signature of the quadratic form associated to the augmented Hessian matrix is (p_1, q_1), then the signature of the quadratic form associated to a constrained critical point is $(p_1 - m, q_1 - m)$. Here, $m = 1$.

In the "degenerate" case where one number equals the sum of the other two, say $\sqrt{a} = \sqrt{b} + \sqrt{c}$, there are only three solutions: the combination $s_1 = +1, s_2 = s_3 = -1$ is impossible.

Now let us analyze the nature of these critical points. Here $n = 2$ and $m = 1$; the augmented Hessian matrix is the 4×4 matrix

$$H = \begin{bmatrix} -\frac{2\lambda}{x^3} & 0 & 0 & \frac{1}{x^2} \\ 0 & -\frac{2\lambda}{y^3} & 0 & \frac{1}{y^2} \\ 0 & 0 & -\frac{2\lambda}{z^3} & \frac{1}{z^2} \\ \frac{1}{x^2} & \frac{1}{y^2} & \frac{1}{z^2} & 0 \end{bmatrix} \quad 3.7.49$$

where x, y, z, λ are the coordinates of one of the critical points found above. This matrix corresponds to the quadratic form

$$-\frac{2\lambda}{x^3} A^2 - \frac{2\lambda}{y^3} B^2 - \frac{2\lambda}{z^3} C^2 + \frac{2}{x^2} AD + \frac{2}{y^2} BD + \frac{2}{z^2} CD$$

$$= -\frac{2}{x^2}\left(\frac{\lambda}{x} A^2 - AD + \frac{x}{4\lambda} D^2\right) - \frac{2}{y^2}\left(\frac{\lambda}{y} B^2 - BD + \frac{y}{4\lambda} D^2\right). \quad 3.7.50$$

$$- \frac{2}{z^2}\left(\frac{\lambda}{z} C^2 - CD + \frac{z}{4\lambda} D^2\right) + \frac{1}{2\lambda}\left(\frac{1}{x} + \frac{1}{y} + \frac{1}{z}\right) D^2$$

The four summands on the right side are all plus or minus perfect squares. The first three terms have the same sign as the corresponding coordinate of the critical point since $\lambda < 0$.[16] Because $\frac{1}{x} + \frac{1}{y} + \frac{1}{z} = 1$ and $\lambda < 0$, the last square comes with a negative sign.

Thus the signature (p_1, q_1) of the quadratic form represented by the augmented Hessian matrix can be

1. $(3, 1)$ in the case where all coordinates of the critical point are positive. In that case the signature of the constrained critical point is $(2, 0)$, so the critical point is a minimum
2. $(2, 2)$ in the case where two coordinates of the critical point are positive and one is negative. In that case the signature of the constrained critical point is $(1, 1)$, so the critical point is a saddle
3. $(1, 3)$. In the case where one coordinate of the critical point is positive and two are negative; in that case the signature of the constrained critical point is $(0, 2)$, so the critical point is a maximum. Note that this occurs only if one of the numbers $\sqrt{a}, \sqrt{b}, \sqrt{c}$ is bigger than the sum of the other two △

The spectral theorem for symmetric matrices

Now we will prove what is probably the most important theorem of linear algebra. It goes under many names: the *spectral theorem*, the *principal axis theorem*, *Sylvester's principle of inertia*.

---

[16]For instance, if $x > 0$, the first summand can be written
$$\frac{2}{x^2}\left(\sqrt{\frac{-\lambda}{x}} A + \frac{1}{2}\sqrt{\frac{x}{-\lambda}} D\right)^2; \text{ if } x < 0, \text{ it can be written } -\frac{2}{x^2}\left(\sqrt{\frac{\lambda}{x}} A - \frac{1}{2}\sqrt{\frac{x}{\lambda}} D\right)^2.$$

We use λ to denote both eigenvalues and the Lagrange multipliers of Theorem 3.7.5. When working on a problem involving constrained critical points, Lagrange used λ to denote what we now call Lagrange multipliers. At that time, linear algebra did not exist, but later (after Hilbert proved the spectral theorem in the harder infinite-dimensional setting!) people realized that Lagrange had proved the finite-dimensional version, Theorem 3.7.15.

Generalizing the spectral theorem to infinitely many dimensions is one of the central topics of functional analysis.

Our proof of the spectral theorem uses Theorem 1.6.9, which guarantees that a continuous function on a compact set has a maximum. That theorem is nonconstructive: it proves existence of a maximum without showing how to find it. Thus Theorem 3.7.15 is also nonconstructive. There is a practical way to find an eigenbasis for a symmetric matrix, called *Jacobi's method*; see Exercise 3.29.

Remember that the maximum of Q_A is an element of its domain; the function achieves a maximal value at a maximum.

Exercise 3.2.11 explores other results concerning orthogonal and symmetric matrices.

Theorem 3.7.15 (Spectral theorem). *Let A be a symmetric $n \times n$ matrix with real entries. Then there exists an orthonormal basis $\vec{\mathbf{v}}_1, \dots, \vec{\mathbf{v}}_n$ of $\mathbb{R}^n$ and numbers $\lambda_1, \dots, \lambda_n \in \mathbb{R}$ such that*

$$A\vec{\mathbf{v}}_i = \lambda_i \vec{\mathbf{v}}_i. \qquad 3.7.51$$

By Proposition 2.7.5, Theorem 3.7.15 is equivalent to the statement that $B^{-1}AB = B^\top AB$ is diagonal, where B is the orthogonal matrix whose columns are $\vec{\mathbf{v}}_1, \dots, \vec{\mathbf{v}}_n$. (Recall that a matrix B is orthogonal if and only if $B^\top B = I$.)

Remark. Recall from Definition 2.7.2 that a vector $\vec{\mathbf{v}}_i$ satisfying equation 3.7.51 is an eigenvector, with eigenvalue λ_i. So the orthonormal basis of Theorem 3.7.15 is an eigenbasis. In Section 2.7 we showed that every square matrix has at least one eigenvector, but since our procedure for finding it used the fundamental theorem of algebra, it had to allow for the possibility that the corresponding eigenvalue might be complex. Square real matrices exist that have no real eigenvectors or eigenvalues: for instance, the matrix $A = \begin{bmatrix} 0 & 1 \\ -1 & 0 \end{bmatrix}$ rotates vectors clockwise by $\pi/2$, so clearly there is no $\lambda \in \mathbb{R}$ and no nonzero vector $\vec{\mathbf{v}} \in \mathbb{R}^2$ such that $A\vec{\mathbf{v}} = \lambda\vec{\mathbf{v}}$; its eigenvalues are $\pm i$. Symmetric real matrices are a very special class of square real matrices, whose eigenvectors are guaranteed not only to exist but also to form an orthonormal basis. $\triangle$

Proof. Our strategy will be to consider the function $Q_A : \mathbb{R}^n \to \mathbb{R}$ given by $Q_A(\vec{\mathbf{x}}) = \vec{\mathbf{x}} \cdot A\vec{\mathbf{x}}$, subject to various constraints. The first constraint ensures that the first basis vector we find has length 1. Adding a second constraint ensures that the second basis vector has length 1 and is orthogonal to the first. Adding a third constraint ensures that the third basis vector has length 1 and is orthogonal to the first two, and so on. We will see that these vectors are eigenvectors.

First we restrict Q_A to the $(n-1)$-dimensional sphere $S \subset \mathbb{R}^n$ of equation $F_1(\vec{\mathbf{x}}) = |\vec{\mathbf{x}}|^2 = 1$; this is the first constraint. Since S is a compact subset of $\mathbb{R}^n$, Q_A restricted to S has a maximum, and the constraint $F_1(\vec{\mathbf{x}}) = |\vec{\mathbf{x}}|^2 = 1$ ensures that this maximum has length 1.

Exercise 3.7.9 asks you to justify that the derivative of Q_A is

$$[\mathbf{D}Q_A(\vec{\mathbf{a}})]\vec{\mathbf{h}} = \vec{\mathbf{a}} \cdot (A\vec{\mathbf{h}}) + \vec{\mathbf{h}} \cdot (A\vec{\mathbf{a}}) = \vec{\mathbf{a}}^\top A\vec{\mathbf{h}} + \vec{\mathbf{h}}^\top A\vec{\mathbf{a}} = 2\vec{\mathbf{a}}^\top A\vec{\mathbf{h}}. \qquad 3.7.52$$

The derivative of the constraint function is

$$[\mathbf{D}F_1(\vec{\mathbf{a}})]\vec{\mathbf{h}} = [2a_1 \dots 2a_n] \begin{bmatrix} h_1 \\ \vdots \\ h_n \end{bmatrix} = 2\vec{\mathbf{a}}^\top \vec{\mathbf{h}}. \qquad 3.7.53$$

Theorem 3.7.5 tells us that if $\mathbf{v}_1$ is a maximum of Q_A restricted to S, then there exists λ_1 such that

$$2\vec{\mathbf{v}}_1^\top A = \lambda_1 2\vec{\mathbf{v}}_1^\top, \text{ so } (\vec{\mathbf{v}}_1^\top A)^\top = (\lambda_1 \vec{\mathbf{v}}_1^\top)^\top, \text{ i.e., } A^\top \vec{\mathbf{v}}_1 = \lambda_1 \vec{\mathbf{v}}_1. \qquad 3.7.54$$

(remember that $(AB)^\top = B^\top A^\top$.)

Since A is symmetric, we can rewrite equation 3.7.54 as $A\vec{\mathbf{v}}_1 = \lambda_1\vec{\mathbf{v}}_1$. Thus $\vec{\mathbf{v}}_1$, the maximum of Q_A restricted to S, satisfies equation 3.7.51 and has length 1; it is the first vector in our orthonormal basis of eigenvectors.

Remark. We can also prove the existence of the first eigenvector by a more geometric argument. By Theorem 3.7.2, at a critical point $\vec{\mathbf{v}}_1$, the derivative $[\mathbf{D}Q_A(\vec{\mathbf{v}}_1)]$ vanishes on $T_{\vec{\mathbf{v}}_1}S$: $[\mathbf{D}Q_A(\vec{\mathbf{v}}_1)]\vec{\mathbf{w}} = 0$ for all $\vec{\mathbf{w}} \in T_{\vec{\mathbf{v}}_1}S$. Since $T_{\vec{\mathbf{v}}_1}S$ consists of vectors tangent to the sphere at $\vec{\mathbf{v}}_1$, hence orthogonal to $\vec{\mathbf{v}}_1$, this means that for all $\vec{\mathbf{w}}$ such that $\vec{\mathbf{w}} \cdot \vec{\mathbf{v}}_1 = 0$,

$$[\mathbf{D}Q_A(\vec{\mathbf{v}}_1)]\vec{\mathbf{w}} = 2(\vec{\mathbf{v}}_1^\top A)\vec{\mathbf{w}} = 0. \qquad 3.7.55$$

Taking transposes of both sides gives $\vec{\mathbf{w}}^\top A\vec{\mathbf{v}}_1 = \vec{\mathbf{w}} \cdot A\vec{\mathbf{v}}_1 = 0$. Since $\vec{\mathbf{w}} \perp \vec{\mathbf{v}}_1$, this says that $A\vec{\mathbf{v}}_1$ is orthogonal to any vector orthogonal to $\vec{\mathbf{v}}_1$. Thus it points in the same direction as $\vec{\mathbf{v}}_1$: $A\vec{\mathbf{v}}_1 = \lambda_1\vec{\mathbf{v}}_1$ for some number λ_1. △

The space $(\vec{\mathbf{v}}_1)^\perp$ is called the *orthogonal complement* to the subspace spanned by $\vec{\mathbf{v}}_1$. More generally, the set of vectors $\vec{\mathbf{x}}$ satisfying

$$\vec{\mathbf{x}} \cdot \vec{\mathbf{v}}_1 = 0, \quad \vec{\mathbf{x}} \cdot \vec{\mathbf{v}}_2 = 0$$

(see equation 3.7.62) is called the *orthogonal complement* of the subspace spanned by $\vec{\mathbf{v}}_1$ and $\vec{\mathbf{v}}_2$.

Equation 3.7.57: The subscript $2,1$ for μ indicates that we are finding our second basis vector; the 1 indicates that we are choosing it orthogonal to the first basis vector.

The first equality of equation 3.7.60 uses the symmetry of A: if A is symmetric,

$$\begin{aligned}\vec{\mathbf{v}} \cdot (A\vec{\mathbf{w}}) &= \vec{\mathbf{v}}^\top (A\vec{\mathbf{w}}) = (\vec{\mathbf{v}}^\top A)\vec{\mathbf{w}} \\ &= (\vec{\mathbf{v}}^\top A^\top)\vec{\mathbf{w}} = (A\vec{\mathbf{v}})^\top \vec{\mathbf{w}} \\ &= (A\vec{\mathbf{v}}) \cdot \vec{\mathbf{w}}.\end{aligned}$$

The third the fact that

$$\vec{\mathbf{v}}_2 \in S \cap (\vec{\mathbf{v}}_1)^\perp.$$

For the second eigenvalue, denote by $\vec{\mathbf{v}}_2$ the maximum of Q_A when Q_A is subject to the two constraints

$$F_1(\vec{\mathbf{x}}) = |\vec{\mathbf{x}}|^2 = 1 \quad \text{and} \quad F_2(\vec{\mathbf{x}}) = \vec{\mathbf{x}} \cdot \vec{\mathbf{v}}_1 = 0. \qquad 3.7.56$$

In other words, we are considering the maximum of Q_A restricted to the space $S \cap (\vec{\mathbf{v}}_1)^\perp$, where $(\vec{\mathbf{v}}_1)^\perp$ is the space of vectors orthogonal to $\vec{\mathbf{v}}_1$.

Since $[\mathbf{D}F_2(\vec{\mathbf{v}}_2)] = \vec{\mathbf{v}}_1^\top$, equations 3.7.52 and 3.7.53 and Theorem 3.7.5 tell us that there exist numbers λ_2 and $\mu_{2,1}$ such that

$$\underbrace{2\vec{\mathbf{v}}_2^\top A}_{[\mathbf{D}Q_A(\vec{\mathbf{v}}_2)]} = \underbrace{\lambda_2 2\vec{\mathbf{v}}_2^\top}_{\lambda_1[\mathbf{D}F_1(\vec{\mathbf{v}}_2)]} + \underbrace{\mu_{2,1}\vec{\mathbf{v}}_1^\top}_{\lambda_2[\mathbf{D}F_2(\vec{\mathbf{v}}_2)]}. \qquad 3.7.57$$

(Note that here λ_2 corresponds to λ_1 in equation 3.7.9, and $\mu_{2,1}$ corresponds to λ_2 in that equation.) Take transposes of both sides (remembering that, since A is symmetric, $A = A^\top$) to get

$$A\vec{\mathbf{v}}_2 = \lambda_2\vec{\mathbf{v}}_2 + \frac{\mu_{2,1}}{2}\vec{\mathbf{v}}_1. \qquad 3.7.58$$

Now take the dot product of each side with $\vec{\mathbf{v}}_1$, to find

$$(A\vec{\mathbf{v}}_2) \cdot \vec{\mathbf{v}}_1 = \lambda_2 \underbrace{\vec{\mathbf{v}}_2 \cdot \vec{\mathbf{v}}_1}_{0} + \frac{\mu_{2,1}}{2} \underbrace{\vec{\mathbf{v}}_1 \cdot \vec{\mathbf{v}}_1}_{1} = \frac{\mu_{2,1}}{2}. \qquad 3.7.59$$

(We have $\vec{\mathbf{v}}_1 \cdot \vec{\mathbf{v}}_1 = 1$ because $\vec{\mathbf{v}}_1$ has length 1; we have $\vec{\mathbf{v}}_2 \cdot \vec{\mathbf{v}}_1 = 0$ because the two vectors are orthogonal to each other.) Using

$$(A\vec{\mathbf{v}}_2) \cdot \vec{\mathbf{v}}_1 = \vec{\mathbf{v}}_2 \cdot (A\vec{\mathbf{v}}_1) = \vec{\mathbf{v}}_2 \cdot (\lambda_1\vec{\mathbf{v}}_1) = 0, \qquad 3.7.60$$

equation 3.7.59 becomes $0 = \mu_{2,1}$, so equation 3.7.58 becomes

$$A\vec{\mathbf{v}}_2 = \lambda_2\vec{\mathbf{v}}_2. \qquad 3.7.61$$

You should be impressed by how easily the existence of eigenvectors drops out of Lagrange multipliers. Of course, we could not have done this without Theorem 1.6.9 guaranteeing that the function Q_A has a maximum and minimum. In addition, we've only proved existence: there is no obvious way to find these constrained maxima of Q_A.

Thus we have found a second eigenvector: it is the maximum of Q_A constrained to the sets given by $F_1 = 0$ and $F_2 = 0$.

It should be clear how to continue, but let us take one further step. Suppose $\vec{\mathbf{v}}_3$ is a maximum of Q_A restricted to $S \cap \vec{\mathbf{v}}_1^\perp \cap \vec{\mathbf{v}}_2^\perp$, that is, maximize Q_A subject to the three constraints

$$F_1(\vec{\mathbf{x}}) = 1, \qquad F_2(\vec{\mathbf{x}}) = \vec{\mathbf{x}} \cdot \vec{\mathbf{v}}_1 = 0, \qquad F_3(\vec{\mathbf{x}}) = \vec{\mathbf{x}} \cdot \vec{\mathbf{v}}_2 = 0. \qquad 3.7.62$$

The argument above says that there exist numbers λ_3, $\mu_{3,1}$, $\mu_{3,2}$ such that

$$A\vec{\mathbf{v}}_3 = \mu_{3,1}\vec{\mathbf{v}}_1 + \mu_{3,2}\vec{\mathbf{v}}_2 + \lambda_3\vec{\mathbf{v}}_3. \qquad 3.7.63$$

Dot this entire equation with $\vec{\mathbf{v}}_1$ (respectively, with $\vec{\mathbf{v}}_2$); you will find $\mu_{3,1} = \mu_{3,2} = 0$. Thus $A\vec{\mathbf{v}}_3 = \lambda_3\vec{\mathbf{v}}_3$. □

The spectral theorem gives another approach to quadratic forms, geometrically more appealing than completing squares.

Theorem 3.7.16. *Let A be a real $n \times n$ symmetric matrix. The quadratic form Q_A has signature (k, l) if and only if there exists an orthonormal basis $\vec{\mathbf{v}}_i, \dots, \vec{\mathbf{v}}_n$ of $\mathbb{R}^n$ with $A\vec{\mathbf{v}}_i = \lambda_i \mathbf{v}_i$, where $\lambda_i, \dots, \lambda_k > 0$, $\lambda_{k+1}, \dots, \lambda_{k+l} < 0$, and all $\lambda_{k+l+1}, \dots, \lambda_n$ are 0 (if $k + l < n$).*

Equation 3.7.65 uses the fact the $\vec{\mathbf{v}}_1, \dots, \vec{\mathbf{v}}_n$ form an orthonormal basis, so $\vec{\mathbf{v}}_i \cdot \vec{\mathbf{v}}_j$ is 0 if $i \neq j$ and 1 if $i = j$.

Proof. By the spectral theorem, there exists an orthonormal eigenbasis $\vec{\mathbf{v}}_1, \dots, \vec{\mathbf{v}}_n$ of $\mathbb{R}^n$ for A; denote by $\lambda_1, \dots, \lambda_n$ the corresponding eigenvalues. Then (Proposition 2.4.18) any vector $\vec{\mathbf{x}} \in \mathbb{R}^n$ can be written

$$\vec{\mathbf{x}} = \sum_{i=1}^{n} (\vec{\mathbf{x}} \cdot \vec{\mathbf{v}}_i)\mathbf{v}_i, \quad \text{so} \quad A\vec{\mathbf{x}} = \sum_{i=1}^{n} \lambda_i (\vec{\mathbf{x}} \cdot \vec{\mathbf{v}}_i)\vec{\mathbf{v}}_i. \qquad 3.7.64$$

The linear functions $\alpha_i(\vec{\mathbf{x}}) \overset{\text{def}}{=} \vec{\mathbf{x}} \cdot \vec{\mathbf{v}}_i$ are linearly independent, and

$$Q_A(\vec{\mathbf{x}}) \underbrace{=}_{\text{eq. 3.5.32}} \vec{\mathbf{x}}^\top A\vec{\mathbf{x}} = \left(\sum_{i=1}^{n} (\vec{\mathbf{x}} \cdot \vec{\mathbf{v}}_i)\vec{\mathbf{v}}_i\right)^\top A \left(\sum_{j=1}^{n} (\vec{\mathbf{x}} \cdot \vec{\mathbf{v}}_j)\vec{\mathbf{v}}_j\right) = \left(\sum_{i=1}^{n} \alpha_i(\vec{\mathbf{x}})\vec{\mathbf{v}}_i\right) \cdot \left(\sum_{j=1}^{n} \alpha_j(\vec{\mathbf{x}})\lambda_j\vec{\mathbf{v}}_j\right)$$

$$= \sum_{i=1}^{n}\sum_{j=1}^{n} \alpha_i(\vec{\mathbf{x}})\alpha_j(\vec{\mathbf{x}})\ \lambda_j\ \vec{\mathbf{v}}_i \cdot \vec{\mathbf{v}}_j = \sum_{i=1}^{n} \lambda_i\ \alpha_i(\vec{\mathbf{x}})^2 \qquad 3.7.65$$

This proves Theorem 3.7.16: if Q_A has signature (k, l), then k of the λ_i are positive, l are negative, and the rest are 0; conversely, if k of the eigenvalues are positive and l are negative, then Q_A has signature (k, l). □

Exercises for Section 3.7

3.7.1 What is the maximum volume of a box of surface area 10, for which one side is exactly twice as long as another?

3.7.2 Consider the "parametrization" $\mathbf{g} : \begin{pmatrix} u \\ v \end{pmatrix} \mapsto \begin{pmatrix} \sin uv \\ u+v \\ uv \end{pmatrix}$ of the surface X of equation $x = \sin z$ (shown in the figure in the margin). Set $f(\mathbf{x}) = x + y + z$.

a. Show that for all $\mathbf{x} \in X$, the derivative $[\mathbf{D}f(\mathbf{x})]$ does not vanish on $T_{\mathbf{x}}X$.

b. Show that $f \circ \mathbf{g}$ does have critical points.

c. By part a, these critical points are not critical points of f constrained to X. What went wrong?

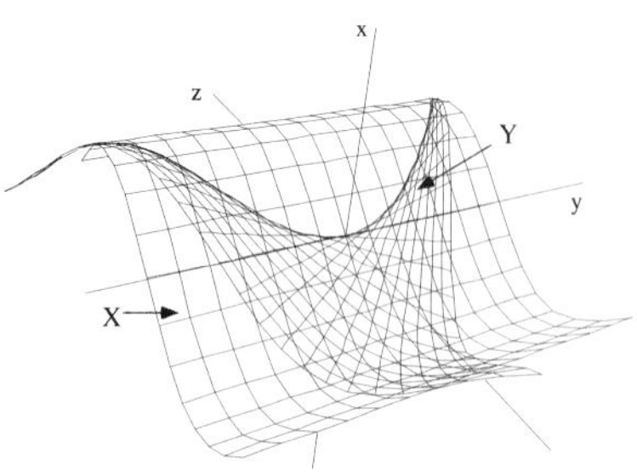

FIGURE FOR EXERCISE 3.7.2.

The "surface" Y is the image of $\mathbf{g} : \begin{pmatrix} u \\ v \end{pmatrix} \mapsto \begin{pmatrix} \sin uv \\ u+v \\ uv \end{pmatrix}$; it is a subset of the surface X of equation $x = \sin z$, which resembles a curved bench. The tangent plane to X at any point is always parallel to the y-axis.

3.7.3 a. Find the critical points of the function $x^3 + y^3 + z^3$ on the intersection of the planes of equation $x + y + z = 2$ and $x + y - z = 3$.

b. Are the critical points maxima, minima, or neither?

3.7.4 a. Show that the set $X \subset \operatorname{Mat}(2,2)$ of 2×2 matrices with determinant 1 is a smooth submanifold. What is its dimension?

b. Find a matrix in X that is closest to the matrix $\begin{bmatrix} 0 & 1 \\ 1 & 0 \end{bmatrix}$.

3.7.5 What is the volume of the largest rectangular parallelepiped aligned with the axes and contained in the ellipsoid $x^2 + 4y^2 + 9z^2 \le 9$?

3.7.6 Find all the critical points of the function

$$f\begin{pmatrix} x \\ y \\ z \end{pmatrix} = 2xy + 2yz - 2x^2 - 2y^2 - 2z^2 \quad \text{on the unit sphere of } \mathbb{R}^3.$$

Exercise 3.7.7: Use the equation of the plane to write z in terms of x and y (i.e., parametrize the plane by x and y).

3.7.7 a. Generalize Example 3.7.9: show that the function xyz has four critical points on the plane of equation $\varphi\begin{pmatrix} x \\ y \\ z \end{pmatrix} = ax + by + cz - 1 = 0$ when $a, b, c > 0$.

b. Show that three of these critical points are saddles and one is a maximum.

Exercise 3.7.8: Think about when the maximum is inside the triangle and when it is on the boundary.

3.7.8 Find the maximum of the function $x^a e^{-x} y^b e^{-y}$ on the triangle $x \ge 0$, $y \ge 0$, $x + y \le 1$, in terms of a and b, for $a, b > 0$.

3.7.9 Justify equation 3.7.51, using the definition of the derivative and the fact that A is symmetric.

***3.7.10** Let D be the closed domain described by the two inequalities $x + y \ge 0$ and $x^2 + y^2 \le 1$, as shown in the margin.

a. Find the maximum and minimum of the function $f\begin{pmatrix} x \\ y \end{pmatrix} = xy$ on D.

b. Try it again with $f\begin{pmatrix} x \\ y \end{pmatrix} = x + 5xy$.

3.7.11 The cone of equation $z^2 = x^2 + y^2$ is cut by the plane $z = 1 + x + y$ in a curve C. Find the points of C closest to and furthest from the origin.

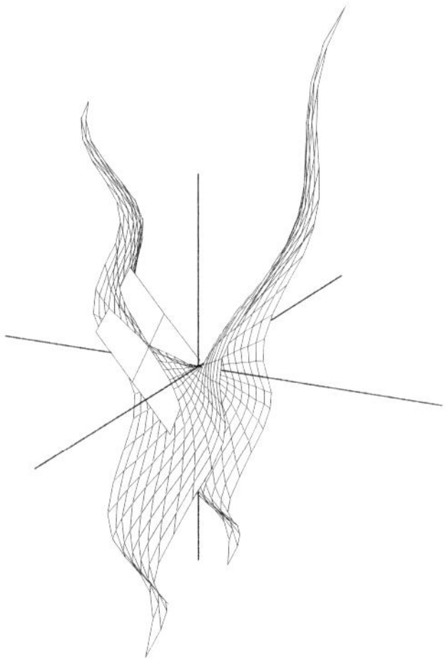

FIGURE FOR EXERCISE 3.7.12.

3.7.12 Find local critical points of the function $f\begin{pmatrix} x \\ y \\ z \end{pmatrix} = x + y + z$ on the surface parametrized by $\mathbf{g} : \begin{pmatrix} u \\ v \end{pmatrix} \mapsto \begin{pmatrix} \sin uv + u \\ u+v \\ uv \end{pmatrix}$, shown in the figure in the margin. (Exercise 3.1.25 asked you to show that $\mathbf{g}$ really is a parametrization.)

3.7.13 Define $f : \mathbb{R}^3 \to \mathbb{R}$ by $f\begin{pmatrix} x \\ y \\ z \end{pmatrix} = xy - x + y + z^2$. Let $\mathbf{x}_0 = \begin{pmatrix} -1 \\ 1 \\ 0 \end{pmatrix}$.

a. Show that f has only one critical point, $\mathbf{x}_0$. What kind of critical point is it?

b. Find all constrained critical points of f on the manifold $S_{\sqrt{2}}(\mathbf{x}_0)$ (the sphere of radius $\sqrt{2}$ centered at $\mathbf{x}_0$), which is given by the equation

$$F\begin{pmatrix} x \\ y \\ z \end{pmatrix} = (x+1)^2 + (y-1)^2 + z^2 - 2 = 0.$$

c. What are the maximum and minimum values of f on the closed ball $\overline{B_{\sqrt{2}}(\mathbf{x}_0)}$?

Hint for Exercise 3.7.13, part c: The maximum and minimum values k occur either at a critical point in the interior $B_{\sqrt{2}}(\mathbf{x}_0)$, or at a constrained critical point on the boundary

$$\partial \overline{B_{\sqrt{2}}(\mathbf{x}_0)} = S_{\sqrt{2}}(\mathbf{x}_0).$$

3.7.14 Analyze the critical points found in Example 3.7.9, using the augmented Hessian matrix.

3.7.15 Let $X \subset \mathbb{R}^2$ be defined by the equation $y^2 = (x-1)^3$.

a. Show that $f\begin{pmatrix} x \\ y \end{pmatrix} = x^2 + y^2$ has a minimum on X.

b. Show that the Lagrange multiplier equation has no solutions.

c. What went wrong?

3.7.16 Find the critical points of $f(x, y, z) = x^3 + y^3 + z^3$ on the surface $x^{-1} + y^{-1} + z^{-1} = 1$.

3.8 Probability and the Singular Value Decomposition

If A is not symmetric, the spectral theorem does not apply. But $AA^\top$ and $A^\top A$ are symmetric (see Exercise 1.2.16), and the spectral theorem applied to these matrices leads to the *singular value decomposition* (SVD). This result has immense importance in statistics, where it is also known as *principal component analysis* (PCA).

We call an $n \times m$ matrix D *rectangular diagonal* if $d_{i,j} = 0$ for $i \neq j$; such a matrix is *nonnegative* if the diagonal entries $d_{i,i}$ are ≥ 0 for all $1 \leq i \leq \min\{n, m\}$.

$$\begin{bmatrix} 2 & 0 & 0 & 0 \\ 0 & 1 & 0 & 0 \\ 0 & 0 & 0 & 0 \end{bmatrix}$$

This rectangular diagonal matrix is nonnegative. The "diagonal" entries are 2, 1, and 0.

In applications of the SVD, it is essential that A be allowed to be nonsquare. The matrices involved are often drastically "lopsided", like $200 \times 1\,000\,000$ in Example 3.8.10.

The matrix D stretches in the direction of the axes; we will see in Exercise 4.8.23 that the orthogonal matrices P and Q are compositions of reflections and rotations.

Theorem 3.8.1 (Singular value decomposition). *Let A be a real $n \times m$ matrix. Then there exist an orthogonal $n \times n$ matrix P, an orthogonal $m \times m$ matrix Q, and a nonnegative rectangular diagonal $n \times m$ matrix D such that $A = PDQ^\top$.*

The nonzero diagonal entries $d_{i,i}$ of D are the square roots of the eigenvalues of $A^\top A$.

The diagonal entries $d_{i,i}$ of D are called the *singular values* of A.

To prove Theorem 3.8.1 we will require the following rather surprising statement.

Lemma 3.8.2: We say that Lemma 3.8.2 is rather surprising because the matrices $A^\top A$ and $AA^\top$ may have drastically different sizes. Note that although they have the same nonzero eigenvalues with the same multiplicities, they may have different numbers of zero eigenvalues.

Lemma 3.8.2. *The symmetric matrices $A^\top A$ and $AA^\top$ have the same nonzero eigenvalues with the same multiplicities: for all $\lambda \neq 0$, the map*

$$\vec{\mathbf{v}} \mapsto \frac{1}{\sqrt{\lambda}} A\vec{\mathbf{v}} \qquad 3.8.1$$

is an isomorphism $\ker(A^\top A - \lambda I) \to \ker(AA^\top - \lambda I)$.

Proof of Lemma 3.8.2. If $\vec{\mathbf{v}}$ is in $\ker(A^\top A - \lambda I)$, i.e., $A^\top A\vec{\mathbf{v}} = \lambda\vec{\mathbf{v}}$, then $A\vec{\mathbf{v}}/\sqrt{\lambda}$ is in $\ker(AA^\top - \lambda I)$:

$$AA^\top\left(\frac{1}{\sqrt{\lambda}}A\vec{\mathbf{v}}\right) = \frac{1}{\sqrt{\lambda}}A(A^\top A\vec{\mathbf{v}}) = \lambda\left(\frac{1}{\sqrt{\lambda}}A\vec{\mathbf{v}}\right). \qquad 3.8.2$$

The map is an isomorphism, since $\vec{\mathbf{w}} \mapsto \frac{1}{\sqrt{\lambda}}A^\top\vec{\mathbf{w}}$ is its inverse. Therefore, $A^\top A$ and $AA^\top$ have the same nonzero eigenvalues, and by Definition 2.7.2, these eigenvalues have the same multiplicities. □

Proof of Theorem 3.8.1. Exercise 3.8.5 asks you to use Lemma 3.8.2 and the spectral theorem to show that $A^\top A$ has an orthonormal eigenbasis $\vec{\mathbf{v}}_1, \dots, \vec{\mathbf{v}}_m$, with eigenvalues $\lambda_1 \geq \cdots \geq \lambda_k > 0$, $\lambda_{k+1} = \cdots = \lambda_m = 0$, and that $AA^\top$ has an orthonormal eigenbasis $\vec{\mathbf{w}}_1, \dots, \vec{\mathbf{w}}_n$, with eigenvalues $\mu_1 = \lambda_1, \dots, \mu_k = \lambda_k$ and $\mu_{k+1} = \cdots = \mu_n = 0$, such that

To find the singular value decomposition of a square matrix A using MATLAB, type

[P,D,Q] = svd(A)

You will get matrices P, D, Q such that $A = PDQ$, where D is diagonal and P and Q are orthogonal. (This Q will be our $Q^\top$.)

$$\vec{\mathbf{w}}_i = \frac{1}{\sqrt{\lambda_i}}A\vec{\mathbf{v}}_i,\ i \leq k \quad \text{and} \quad \vec{\mathbf{v}}_i = \frac{1}{\sqrt{\mu_i}}A^\top\vec{\mathbf{w}}_i,\ i \leq k. \qquad 3.8.3$$

(Note that if A is $n \times m$, then $A^\top A$ is $m \times m$ and $AA^\top$ is $n \times n$; we have $\vec{\mathbf{v}}_i \in \mathbb{R}^m$, $A\vec{\mathbf{v}} \in \mathbb{R}^n$, $\vec{\mathbf{w}} \in \mathbb{R}^n$, and $A^\top\vec{\mathbf{w}} \in \mathbb{R}^m$, so equation 3.8.3 makes sense.) Now set

$$P = [\vec{\mathbf{w}}_1, \dots, \vec{\mathbf{w}}_n] \quad \text{and} \quad Q = [\vec{\mathbf{v}}_1, \dots, \vec{\mathbf{v}}_m] \qquad 3.8.4$$

Set $D = P^\top AQ$. Both P and Q are orthogonal matrices, and the (i,j)th entry of D is

$$\vec{\mathbf{w}}_i^\top A\vec{\mathbf{v}}_j = \begin{cases} 0 & \text{if } i > k \text{ or } j > k \\ \vec{\mathbf{w}}_i \cdot \sqrt{\lambda_j}\vec{\mathbf{w}}_j & \text{if } i, j \leq k \end{cases} \qquad 3.8.5$$

since if $i > k$ then $\vec{\mathbf{w}}_i^\top A = 0$ and if $j > k$ then $A\vec{\mathbf{v}}_j = 0$. Moreover,

Recall that a matrix B is orthogonal if and only if $B^\top B = I$. So

$$\begin{aligned} DD^\top &= P^\top AQQ^\top A^\top P \\ &= P^\top AA^\top P \\ &= P^{-1}AA^\top P. \end{aligned}$$

$$\vec{\mathbf{w}}_i \cdot \sqrt{\lambda_j}\vec{\mathbf{w}}_j = \begin{cases} 0 & \text{if } i \neq j \text{ or } i = j \text{ and } i > k \\ \sqrt{\lambda_j} & \text{if } i = j \text{ and } i, j \leq k. \end{cases} \qquad 3.8.6$$

Thus D is rectangular diagonal with diagonal entries $\sqrt{\lambda_1}, \dots, \sqrt{\lambda_k}$, and since P and Q are orthogonal, $D = P^\top AQ$ implies $A = PDQ^\top$.

The diagonal entries of the diagonal matrix $DD^\top = P^\top AA^\top P$ are the squares of the "diagonal" entries of D. Thus the singular values of A are square roots of the diagonal entries (eigenvalues) of $DD^\top$. Since, by Proposition 2.7.5, the eigenvalues of $DD^\top = P^{-1}AA^\top P$ are also the eigenvalues of $AA^\top$, the singular values of A depend only on A. □

Definition 2.9.6 The *norm* $\|A\|$ of A is

$$\|A\| = \sup |A\vec{\mathbf{x}}|,$$

when $\vec{\mathbf{x}} \in \mathbb{R}^n$ and $|\vec{\mathbf{x}}| = 1$.

FIGURE 3.8.1.
Andrei Kolmogorov (1903–1987), one of the greatest mathematicians of the twentieth century, made probability a rigorous branch of mathematics. Part of his contribution was to say that for any serious treatment of probability, you have to be careful: you need to specify which subsets have a probability (these subsets are called the *measurable subsets*), and you need to replace requirement 3 of Definition 3.8.4 by *countable additivity*: if $i \mapsto A_i$ is a sequence of disjoint measurable subsets of X, then

$$\mathbf{P}\left(\bigcup_{i=1}^{\infty} A_i\right) = \sum_{i=1}^{\infty} \mathbf{P}(A_i);$$

see Theorem A21.6.

Kolmogorov solved the 13th of the 23 problems Hilbert posed in 1900, and made major contributions to dynamical systems and to the theory of complexity and information.

(Photograph by Jürgen Moser)

One interesting consequence is a formula for the norm of a matrix.

Proposition 3.8.3. *For any real matrix A, the norm of A is*

$$\|A\| = \max_{\lambda \text{ eigenvalue of } A^\top A} \sqrt{\lambda}.$$

Proof. By Proposition and Definition 2.4.19, an orthogonal matrix preserves lengths. Thus if P and Q are orthogonal, and B any matrix (not necessarily square) $\|QB\| = \sup_{|\vec{\mathbf{v}}|=1} |QB\vec{\mathbf{v}}| = \sup_{|\vec{\mathbf{v}}|=1} |B\vec{\mathbf{v}}| = \|B\|$ and

$$\|BQ\| = \sup_{|\vec{\mathbf{v}}|=1} |BQ\vec{\mathbf{v}}| = \sup_{|Q\vec{\mathbf{v}}|=1} |BQ\vec{\mathbf{v}}| = \sup_{|\vec{\mathbf{w}}|=1} |B\vec{\mathbf{w}}| = \|B\|,$$

$$\|A\| = \|PDQ^\top\| = \|D\| = \max_{\lambda \text{ eigenvalue of } A^\top A} \sqrt{\lambda}. \qquad \square$$

The language of finite probability

To understand how the singular value decomposition is used in statistics, we need some terminology from probability. In probability we have a *sample space* S, which is the set of all possible outcomes of some experiment, together with a *probability measure* $\mathbf{P}$ that gives the probability of the outcome being in some subset $A \subset S$ (called an *event*). If the experiment consists of throwing a six-sided die, the sample space is $\{1, 2, 3, 4, 5, 6\}$; the event $A = \{3, 4\}$ means "land on either 3 or 4", and $\mathbf{P}(A)$ says how likely it is that you will land on either 3 or 4.

Definition 3.8.4 (Probability measure). The measure $\mathbf{P}$ is required to obey the following rules:

1. $\mathbf{P}(A) \in [0, 1]$
2. $\mathbf{P}(S) = 1$
3. If $\mathbf{P}(A \cap B) = \phi$ for $A, B \subset S$, then

$$\mathbf{P}(A \cup B) = \mathbf{P}(A) + \mathbf{P}(B).$$

In this section we limit ourselves to finite sample spaces; the probability of any subset is then the sum of the probabilities of the outcomes in that subset. In the case of a fair six-sided die, $\mathbf{P}(\{3\}) = 1/6$ and $\mathbf{P}(\{4\}) = 1/6$, so $\mathbf{P}(\{3, 4\}) = 1/3$. In Exercise 3.8.3 we explore some of the difficulties that can arise when a sample space is countably infinite; in Section 4.2 we discuss continuous sample spaces, where individual outcomes have probability 0.

A *random variable* is some feature of an experiment that you can measure. If the experiment consists of throwing two dice, we might choose as our random variable the function that gives the total obtained.

Definition 3.8.5 (Random variable). A random variable is a function $S \to \mathbb{R}$. We denote by $\mathrm{RV}(S)$ the vector space of random variables.

When the sample space is finite: $S = \{s_1, \dots, s_m\}$, the space of random variables on S is simply $\mathbb{R}^m$, with $f \in \mathrm{RV}(S)$ given by the list of numbers $f(s_1), \dots, f(s_m)$. But there is a smarter way of identifying $\mathrm{RV}(S)$ to $\mathbb{R}^m$: the map

The map Φ provides a bridge from probability and statistics on one side and geometry of $\mathbb{R}^m$ on the other.

$$\Phi : \mathrm{RV}(S) \to \mathbb{R}^m : f \mapsto \begin{bmatrix} f(s_1)\sqrt{\mathbf{P}(\{s_1\})} \\ \vdots \\ f(s_m)\sqrt{\mathbf{P}(\{s_m\})} \end{bmatrix}, \qquad 3.8.7$$

which scales the values of f by the square root of the corresponding probability. This is smarter because with this identification, the main constructions of probability and statistics all have natural interpretations in terms of the geometry of $\mathbb{R}^m$. To bring out these relations, we write

$$\langle f, g\rangle_{(S,\mathbf{P})} \stackrel{\text{def}}{=} \Phi(f)\cdot\Phi(g) \qquad 3.8.8$$

$$\|f\|^2_{(S,\mathbf{P})} \stackrel{\text{def}}{=} |\Phi(f)|^2 = \langle f, f\rangle_{(S,\mathbf{P})}. \qquad 3.8.9$$

Equation 3.8.8: $\langle f, g\rangle$ is standard notation for inner products, of which the dot product is the archetype. The geometry of $\mathbb{R}^m$ comes from the dot product; defining a kind of dot product for $\mathrm{RV}(S)$ gives that space a geometry just like the geometry of $\mathbb{R}^m$. Note that the inner product $\langle f, g\rangle_{(S,\mathbf{P})}$ has the same distributive property as the dot product; see equation 1.4.3.

Thus

$$\begin{aligned} \langle f, g\rangle_{(S,\mathbf{P})} &= f(s_1)g(s_1)\,\mathbf{P}(\{s_1\}) + \cdots + f(s_m)g(s_m)\,\mathbf{P}(\{s_m\}) \\ \|f\|^2_{(S,\mathbf{P})} &= \langle f, f\rangle_{(S,\mathbf{P})} = f^2(s_1)\,\mathbf{P}(\{s_1\}) + \cdots + f^2(s_m)\,\mathbf{P}(\{s_m\}). \end{aligned} \qquad 3.8.10$$

Let 1_S be the random variable that is the constant function 1; note that

$$\|1_S\|_{(S,\mathbf{P})} = \sum_{s\in S}\mathbf{P}(\{s\}) = 1. \qquad 3.8.11$$

Definitions 3.8.6: The definition of the expected value says nothing about repeating the experiment, but it follows from the *law of large numbers* that $E(f)$ is the value one would expect to get if one did the experiment a great many times and took the average of the results.

The expected value can be misleading. Suppose the random variable f assigns income to an element of the sample space S, and S consists of 1000 supermarket cashiers and Bill Gates (or, indeed, 1000 school teachers or university professors and Bill Gates); if all you knew was the average income, you might draw very erroneous conclusions.

Definitions 3.8.6. Let $f, g \in \mathrm{RV}(S)$. Then

1. The *expected value* of f is
$$E(f) \stackrel{\text{def}}{=} \sum_{s\in S} f(s)\,\mathbf{P}(\{s\}) = \langle f, 1_S\rangle_{(S,\mathbf{P})}, \qquad 3.8.12$$
which by equations 3.8.10 gives
$$\langle f, g\rangle_{(S,\mathbf{P})} = E(fg) \quad\text{and}\quad \|f\|^2_{(S,\mathbf{P})} = E(f^2). \qquad 3.8.13$$
2. The random variable $\tilde{f} \stackrel{\text{def}}{=} f - E(f)$ is called "*f centered*".
3. The *variance* of f is
$$\mathrm{Var}(f) \stackrel{\text{def}}{=} \|\tilde{f}\|^2_{(S,\mathbf{P})} = E\Big(\big(f - E(f)\big)^2\Big) = E(f^2) - (E(f))^2. \qquad 3.8.14$$
4. The *standard deviation* $\sigma(f)$ of f is
$$\sigma(f) \stackrel{\text{def}}{=} \sqrt{\mathrm{Var}(f)} = \|\tilde{f}\|_{(S,\mathbf{P})}. \qquad 3.8.15$$
5. The *covariance* of f and g is
$$\mathrm{cov}(f, g) \stackrel{\text{def}}{=} \langle\tilde{f}, \tilde{g}\rangle_{(S,\mathbf{P})} = E(fg) - E(f)E(g). \qquad 3.8.16$$

6. The *correlation* of f and g is

$$\operatorname{corr}(f,g) \stackrel{\text{def}}{=} \frac{\operatorname{cov}(f,g)}{\sigma(f)\sigma(g)}. \qquad 3.8.17$$

Each of these words needs an explanation.

1. *Expected value*: The expected value is the average of the values of f, each value weighted by the probability of the corresponding outcome. The words "expected value", "expectation", "mean", and "average" are all synonymous. That $E(f) = \langle f, 1_S\rangle_{(S,\mathbf{P})}$ is a simple computation; for $S = \{s_1, \dots, s_m\}$, we have

$$\langle f, 1_S\rangle_{(S,\mathbf{P})} = \begin{bmatrix} f_1 s_1 \sqrt{\mathbf{P}(\{s_1\})} \\ \vdots \\ f_m s_m \sqrt{\mathbf{P}(\{s_m\})} \end{bmatrix} \cdot \begin{bmatrix} \sqrt{\mathbf{P}(\{s_1\})} \\ \vdots \\ \sqrt{\mathbf{P}(\{s_m\})} \end{bmatrix} = \sum_{i=1}^{m} f_i(s_i)\,\mathbf{P}(\{s_i\}).$$

By Corollary 1.4.4, it follows that the expected value is the signed length of the projection of f onto 1_S, i.e., on the space of constant random variables.

2. *f centered*: The random variable $\widetilde{f}$ tells how far off from average one is. If a random variable f gives weights of 6-month-old babies, then $\widetilde{f}$ tells how much lighter or heavier than average a particular baby is. Clearly $E(\widetilde{f}) = 0$.

Mean absolute deviation: See the margin note next to Example 3.8.9.

3. *Variance*: The variance measures how spread out a random variable is from its expected value. We can't use the average of $f - E(f)$ because f is just as often larger than f than it is smaller, so $E(f - E(f)) = 0$; squaring $f - E(f)$ makes the variance positive. It might seem (it might *be*) more natural to consider the *mean absolute deviation* $E(|f - E(f)|)$, but the absolute deviation is less geometric and therefore much harder to handle. The last equality in the definition of variance is justified in the note below for the covariance.

4. *Standard deviation*: The variance is the length squared of $\widetilde{f}$; the standard deviation turns that into a length.

Geometrically, the covariance has the same relation to the variance as the dot product has to length squared:

$$\operatorname{cov}(f,f) = \operatorname{Var}(f)$$
$$\vec{\mathbf{x}} \cdot \vec{\mathbf{x}} = |\vec{\mathbf{x}}|^2$$

5. *Covariance*: Since (the second equality below uses equation 3.8.13)

$$\operatorname{cov}(f,g) = \langle \widetilde{f}, \widetilde{g}\rangle_{(S,\mathbf{P})} = E(\widetilde{f}\,\widetilde{g}) = E\Big((f - E(f))(g - E(g))\Big) \qquad 3.8.18$$

we see that the covariance measures to what extent f and g vary "together" or "opposite": $(f(s) - E(f))(g(s) - E(g)$ is positive when $f(s)$ and $g(s)$ are both more than or both less than their averages, and negative when one is more than and one less than.

The following computation, which uses the fact that E is a linear function on the vector space of random variables, justifies the second equality in equation 3.8.16 defining the covariance (hence the last equality in equation

3.8.14 defining the variance):

$$\begin{aligned}\operatorname{cov}(f,g) &= \langle \widetilde{f}, \widetilde{g}\rangle_{(S,\mathbf{P})} = E\big((f-E(f))(g-E(g))\big)\\ &= E\Big(fg - fE(g) - gE(f) + E(f)E(g)\Big)\\ &= E(fg) - E(fE(g)) - E(gE(f)) + E(f)E(g)\\ &= E(fg) - 2E(f)E(g) + E(f)E(g)\\ &= E(fg) - E(f)E(g)\\ &= \langle f, g\rangle_{(S,\mathbf{P})} - E(f)E(g).\end{aligned} \qquad 3.8.19$$

Equation 3.8.19, going from the third to the fourth line: Since $E(g)$ is a constant,

$$E(fE(g)) = E(f)E(g);$$

similarly, since $E(f)E(g)$ is a constant,

$$E(E(f)E(g)) = E(f)E(g).$$

6. *Correlation*: Proposition 1.4.3 says that $\vec{\mathbf{v}}\cdot\vec{\mathbf{w}} = |\vec{\mathbf{v}}|\,|\vec{\mathbf{w}}|\cos\theta$. Equation 3.8.17 says that

$$\operatorname{cov}(f,g) = \sigma(f)\sigma(g)\operatorname{corr}(f,g). \qquad 3.8.20$$

So the correlation is the cosine of the angle between two centered random variables.

Recall that the geometric interpretation of the dot product required Schwarz's inequality $|\vec{\mathbf{v}}\cdot\vec{\mathbf{w}}| \le |\vec{\mathbf{v}}|\,|\vec{\mathbf{w}}|$. To interpret the correlation as a cosine we need to check that the analogous result is true here.

Proposition 3.8.7, part 2: We leave it to the reader to check that the correlation is $+1$ if $a > 0$ and -1 if $a < 0$. In particular, when $f = g$,

$$\operatorname{corr}(f,g) = 1$$

and when $f = -g$,

$$\operatorname{corr}(f,g) = -1.$$

Saying that the correlation of two random variables is ± 1 means that one is a multiple of the other. Saying that $\operatorname{corr}(f,g) = 0$ means that f and g are orthogonal; if you know f, you only know $g \in f^\perp$. But $f^\perp \subset \mathbb{R}^n$ is a subspace of dimension $n-1$.

Inequality 3.8.21: The equality marked 1 is equation 3.8.8; the inequality marked 2 is Schwarz; equality 3 is equation 3.8.9.

Proposition 3.8.7. *Let $f, g : (S,\mathbf{P}) \to \mathbb{R}$ be two random variables. Then*

1. $|\operatorname{corr}(f,g)| \le 1$
2. $|\operatorname{corr}(f,g)| = 1$ *if and only if there exist $a, b \in \mathbb{R}$ such that $f = ag + b$.*

Proof. 1. By Schwarz's inequality,

$$\begin{aligned}|\operatorname{cov}(f,g)| = |\langle \widetilde{f}, \widetilde{g}\rangle_{(S,\mathbf{P})}| &\underset{(1)}{=} |\Phi(\widetilde{f})\cdot\Phi(\widetilde{g})|\\ &\underset{(2)}{\le} |\Phi(\widetilde{f})||\Phi(\widetilde{g})| \underset{(3)}{=} \|\widetilde{f}\|_{(S,\mathbf{P})}\|\widetilde{g}\|_{(S,\mathbf{P})} = \sigma(f)\sigma(g).\end{aligned} \qquad 3.8.21$$

So we have

$$|\operatorname{corr}(f,g)| = \left|\frac{\operatorname{cov}(f,g)}{\sigma(f)\sigma(g)}\right| \le 1. \qquad 3.8.22$$

2. Again by Schwarz, we have $|\operatorname{corr}(f,g)| = 1$ if and only if $\Phi(\widetilde{f})$ is a multiple of $\Phi(\widetilde{g})|$ or $\Phi(\widetilde{g})$ is a multiple of $\Phi(\widetilde{f})|$ (likely both). Suppose the former, which occurs if and only if $\widetilde{f}$ is a multiple of $\widetilde{g}$, i.e.,

$$f - E(f) = a(g - E(g)), \quad \text{i.e.,} \quad f = ag + (E(f) - aE(g)) = ag + b \qquad \square$$

Thus we know what $\operatorname{corr}(f,g) = \pm 1$ means: f and g are *affinely related*.

What if $\operatorname{corr}(f,g) = 0$, i.e., f and g are *uncorrelated*? This turns out not to have a clear answer.

Example 3.8.8. Let our sample space be the subset of the plane consisting of the set of points $\begin{pmatrix} \pm x_i \\ x_i^2 \end{pmatrix}$ for some collection of m numbers $x_i \neq 0$, all given the same probability $1/(2m)$. Let f and g be the two random variables

$$f\begin{pmatrix} \pm x_i \\ x_i^2 \end{pmatrix} = \pm x_i, \quad g\begin{pmatrix} \pm x_i \\ x_i^2 \end{pmatrix} = x_i^2, \quad \text{so } g = f^2; \qquad 3.8.23$$

see Figure 3.8.2.

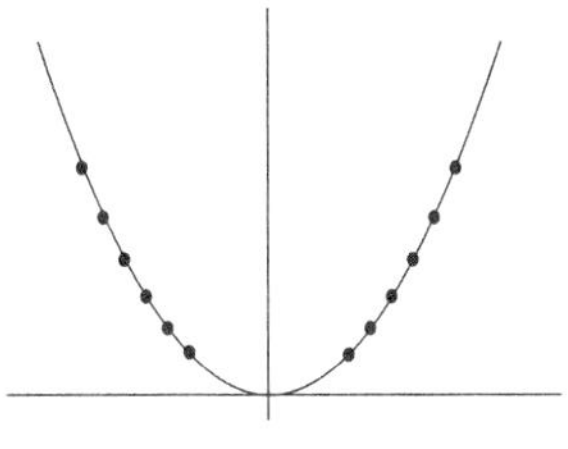

FIGURE 3.8.2 Let S be a collection of points on the parabola $y = x^2$, symmetric with respect to the y-axis. The random variables x and y are uncorrelated, even though $y = x^2$.

In example 3.8.9, the mean absolute deviation for the "total obtained" random variable g is ~ 1.94, significantly less than the standard deviation ~ 2.41. Because of the square in the formula for the variance, the standard deviation weights more heavily values that are far from the expectation than those that are close, whereas the mean absolute deviation treats all deviations equally.

Equation 3.8.26: If you throw two dice 500 times and the average total is not close to 7, you would be justified in suspecting that the dice are loaded. The *central limit theorem* (Theorem 4.2.7) would allow you to quantify your suspicions.

Then, since $E(f) = 0$ and $E(f^3) = 0$, we have

$$\operatorname{cov}(f, f^2) = E(ff^2) - E(f)E(f^2) = 0, \qquad 3.8.24$$

hence corr $(f, g) = 0$. This emphatically does not coincide with our intuitive notion of "uncorrelated": the value of f determines the value of g, and even by a very simple (but nonlinear) function.

The moral is that correlation should really be called linear correlation (or affine correlation); it is not a useful notion when applied to nonlinear functions. This is the main weakness of statistics. △

Example 3.8.9. (Expected value, variance, standard deviation, covariance). Consider throwing two dice, one red and one black. The sample space S is the set consisting of the 36 possible tosses, so RV(S) is $\mathbb{R}^{36}$; an element of this space is a list of 36 numbers, best written as a 6×6 matrix. Each outcome is equally likely: $\mathbf{P}(\{s\}) = \frac{1}{36}$ for any $s \in S$. Below we see three random variables; f_R is "the value of the red die", f_B is "the value of the black die", and g is their sum. The horizontal labels correspond to the red die, vertical labels to the black die. For example, for f_R, if the red die comes up 5 (underlined) and the black comes up 3, the entry $(f_R)_{3,5}$ is 5.

$$f_R = \begin{array}{c} \\ 1\\2\\\underline{3}\\4\\5\\6 \end{array}\begin{array}{c} \begin{array}{cccccc} 1&2&3&4&\underline{5}&6 \end{array}\\ \begin{bmatrix} 1&2&3&4&5&6\\1&2&3&4&5&6\\1&2&3&4&\underline{5}&6\\1&2&3&4&5&6\\1&2&3&4&5&6\\1&2&3&4&5&6 \end{bmatrix}\end{array} \quad f_B = \begin{array}{c} \\ 1\\2\\3\\4\\5\\6 \end{array}\begin{array}{c} \begin{array}{cccccc} 1&2&3&4&5&6 \end{array}\\ \begin{bmatrix} 1&1&1&1&1&1\\2&2&2&2&2&2\\3&3&3&3&3&3\\4&4&4&4&4&4\\5&5&5&5&5&5\\6&6&6&6&6&6 \end{bmatrix}\end{array} \quad g = f_R + f_B = \begin{array}{c} \\ 1\\2\\3\\4\\5\\6 \end{array}\begin{array}{c} \begin{array}{cccccc} 1&2&3&4&5&6 \end{array}\\ \begin{bmatrix} 2&3&4&5&6&7\\3&4&5&6&7&8\\4&5&6&7&8&9\\5&6&7&8&9&10\\6&7&8&9&10&11\\7&8&9&10&11&12 \end{bmatrix}\end{array}$$

The expected value is simply the sum of all 36 entries, divided by 36. Thus

$$E(f_R) = E(f_B) = \frac{6(1+2+3+4+5+6)}{36} = 3.5 \qquad 3.8.25$$

$$E(g) = E(f_R + f_B) = E(f_R) + E(f_B) = 7. \qquad 3.8.26$$

We use the formula $\text{Var}(f) = E(f^2) - (E(f))^2$ (the second equality in equation 3.8.14) to compute the variance and standard deviation of f_R and f_B:

$$\text{Var}(f_R) = \text{Var}(f_B) = \frac{6(1+4+9+16+25+36)}{36} - \left(\frac{7}{2}\right)^2 = \frac{35}{12} \approx 2.92$$

$$\sigma(f_R) = \sigma(f_B) = \sqrt{\frac{35}{12}} \approx 1.71.$$

	1	2	3	4	5	6
1	5^2	4^2	3^2	2^2	1^2	0
2	4^2	3^2	2^2	1^2	0	1^2
3	3^2	2^2	1^2	0	1^2	2^2
4	2^2	1^2	0	1^2	2^2	3^2
5	1^2	0	1^2	2^2	3^2	4^2
6	0	1^2	2^2	3^2	4^2	5^2

The matrix $(g - E(g))^2$

The variance of g can be computed from

$$\text{Var}(g) = E\Big((g - E(g))^2\Big) = \tilde{g}^2(s_1)\,\mathbf{P}(\{s_1\}) + \cdots + \tilde{g}^2(s_m)\,\mathbf{P}(\{s_{36}\}).$$

Adding all the entries of the matrix $(g - E(g))^2$ in the margin and dividing by 36 gives $\text{Var}(g) = \frac{35}{6} \approx 5.83$, so the standard deviation is $\sigma(g) = \sqrt{\frac{35}{6}}$.

	1	2	3	4	5	6
1	1	2	3	4	5	6
2	2	4	6	8	10	12
3	3	6	9	12	15	18
4	4	8	12	16	20	24
5	5	10	15	20	25	30
6	6	12	18	24	30	36

The matrix $f_R f_B$

The second row is the first row times 2, the third row is the first row times 3, and so on. So

$$E(f_R f_B) = \frac{1}{36}(1+2+3+4+5+6)^2 = E(f_R)E(f_B), \quad \text{giving}$$

$$\underbrace{E(f_R f_B) - E(f_R)E(f_B)}_{\text{cov}(f_R, f_B) \text{ by eq. 3.8.16}} = 0.$$

To compute the covariance we will use $\text{cov}(f, g) = E(fg) - E(f)E(g)$ (see equation 3.8.19). The computation in the margin shows that

$$\text{cov}(f_R, f_B) = 0, \quad \text{hence} \quad \text{corr}\,(f_R, f_B) = 0. \qquad 3.8.27$$

The random variables f_R and f_B are *uncorrelated.*

What about the corr (f_R, g)? If we know the throw of the red die, what does it tell us about the throw of two dice? We begin by computing the covariance, this time using the formula $\text{cov}(f, g) = \langle \tilde{f}, \tilde{g} \rangle_{(S,\mathbf{P})}$:

$$\text{cov}(f_R, g) = \langle \widetilde{f_R}, \tilde{g} \rangle_{(S,\mathbf{P})} = \langle \widetilde{f_R}, \widetilde{f_R} + \widetilde{f_B} \rangle_{(S,\mathbf{P})}$$

$$= \overbrace{\langle \widetilde{f_R}, \widetilde{f_R} \rangle_{(S,\mathbf{P})}}^{\|\widetilde{f_R}\|^2_{(S,\mathbf{P})} = \text{Var}(f_R)} + \overbrace{\langle \widetilde{f_R}, \widetilde{f_B} \rangle_{(S,\mathbf{P})}}^{0 \text{ by eq. 3.8.27}} = \text{Var}(f_R) = \frac{35}{12}.$$

The correlation is then

$$\frac{\text{cov}(f_R, g)}{\sigma(f_R)\sigma(g)} = \frac{35/12}{\sqrt{35/12}\sqrt{35/6}} = \frac{1}{\sqrt{2}};$$

the angle between the two random variables f_R and g is 45 degrees. △

The covariance matrix

If $f_1, \dots, f_k$ are random variables, their *covariance matrix* is the $k \times k$ matrix [Cov] whose $(i, j)th$-entry is

$$[\text{Cov}]_{i,j} \stackrel{\text{def}}{=} \text{cov}(f_i, f_j) = \langle \tilde{f}_i, \tilde{f}_j \rangle_{(S,\mathbf{P})} = \Phi(\tilde{f}_i) \cdot \Phi(\tilde{f}_j). \qquad 3.8.28$$

Note that since

$$\vec{\mathbf{v}} \cdot \vec{\mathbf{w}} = \vec{\mathbf{w}} \cdot \vec{\mathbf{v}},$$

the covariance matrix is symmetric. Indeed, for any matrix A, the matrix $A^\top A$ is symmetric; see Exercise 1.2.16. So we can apply the spectral theorem.

We could also write

$$[\text{Cov}] = C^\top C, \quad \text{where } C = [\Phi(\tilde{f}_1), \dots, \Phi(\tilde{f}_k)], \qquad 3.8.29$$

since the (i, j)th entry of $C^\top C$ is the dot product $\Phi(\tilde{f}_i) \cdot \Phi(\tilde{f}_j)$.

Recall that in the proof of the spectral theorem for a symmetric $n \times n$ matrix A, we first found the maximum of the function $Q_A(\mathbf{x}) = \mathbf{x} \cdot A\mathbf{x}$ on the sphere $S = \{\, \mathbf{x} \in \mathbb{R}^n \mid |\mathbf{x}| = 1 \,\}$; the point where the maximum is realized is our first eigenvector $\mathbf{v}_1$ of the matrix A, and the maximum value λ_1 is the first eigenvalue. Our second eigenvector is then the point $\mathbf{v}_2 \in S \cap \mathbf{v}_1^{\perp}$ where Q_A achieves its maximum value λ_2, etc.

In this setting we are interested in $Q_{[\mathrm{Cov}]}(\vec{\mathbf{x}}) = \vec{\mathbf{x}} \cdot [\mathrm{Cov}]\,\vec{\mathbf{x}}$. This function is a quadratic form on $\mathbb{R}^k$ and represents the variance in the sense that

$$\begin{aligned} Q_{[\mathrm{Cov}]}(\vec{\mathbf{x}}) = \vec{\mathbf{x}} \cdot [\mathrm{Cov}]\,\vec{\mathbf{x}} &= \sum_{i=1}^{k}\sum_{j=1}^{k} x_i x_j \langle \widetilde{f}_i, \widetilde{f}_j \rangle_{(S,\mathbf{P})} \\ &= \langle x_1\widetilde{f}_1 + \cdots + x_k\widetilde{f}_k,\; x_1\widetilde{f}_1 + \cdots + x_k\widetilde{f}_k \rangle_{(S,\mathbf{P})} \\ &= \| x_1\widetilde{f}_1 + \cdots + x_k\widetilde{f}_k \|^2_{(S,\mathbf{P})} = \mathrm{Var}(x_1 f_1 + \cdots + x_k f_k). \end{aligned} \qquad 3.8.30$$

In equation 3.8.30,

$$x_1 f_1 + \cdots + x_k f_k$$

is a linear combination of the k random variables. If $\vec{\mathbf{w}}$ is the unit eigenvector of [Cov] with largest eigenvalue, then the variance is maximal at

$$w_1 f_1 + \cdots + w_k f_k$$

among unit vectors. You may find it useful to review the proof of the spectral theorem.

Therefore $\vec{\mathbf{x}} \mapsto Q_{[\mathrm{Cov}]}\vec{\mathbf{x}}$ achieves its maximum among unit vectors at the leading eigenvector of [Cov]. The eigenvectors of [Cov] are called the *principal components*; the subject is called *principal component analysis* (PCA).

PCA and statistics

Statistics is the art of teasing patterns from data. What is data? Usually, it is a cloud of N points in some $\mathbb{R}^k$. We may have many points in a fairly low-dimensional space, if we are measuring a few features of a large population, or few points in a high-dimensional space. Or both dimensions may be large, as in all pictures in the State Department's file of passport pictures and pictures on visa applications: perhaps 100 000 000 pictures, each perhaps a megabyte. Such a cloud of points can be written as a $k \times N$ matrix A: it has as many rows as there are features measured and one column for each data point.[17] In these days of "big data" the matrices may be huge: $1\,000 \times 1\,000\,000$ is nothing unusual, sizes in the billions (Walmart transactions, census data) are common, and even trillions (astronomical objects, weather prediction, geological exploration) are sometimes used.

A few minutes on the Internet reveal that the uses of principal component analysis are myriad: *latent semantic indexing* for querying databases, determining the origin of a gasoline sample taken from a suspected arsonist, analyzing paint in a 16th century altarpiece in Slovakia, determining whether Italian mozzarella comes from Campania or from Apulia.

The object of PCA is to extract useful information from such a huge database A. The idea is to compute an appropriate average $\overline{A}$ to center the data, and apply the SVD to the covariance matrix $[A-\overline{A}]^{\top}[A-\overline{A}]$. The unit eigenvector of this covariance matrix with largest eigenvalue points in the direction in which the data has the most variance; thus it corresponds to the direction that carries the most information. Of the eigenvectors orthogonal to the first eigenvector, we choose the eigenvector with largest eigenvalue, which carries the next greatest amount of information, and so on. The hope is that a relatively small number of these eigenvectors will carry most of the information. If this "most" is actually justified, projecting the data into the space spanned by those directions will lose little information.

Example 3.8.10 (Eigenfaces). Suppose you have a large database of images of faces. Perhaps you are the State Department, with all the faces of visa applicants and passport holders. Or Facebook, with many more

[17] The decision that rows represent features and columns data is arbitrary.

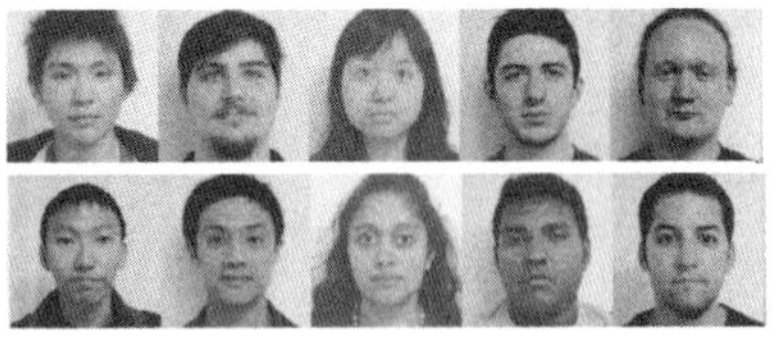

FIGURE 3.8.3.

Kevin Huang used these training faces for a study of eigenfaces when he was an undergraduate at Trinity College (photos reprinted with permission).

FIGURE 3.8.4.

The average of the ten training faces above.

FIGURE 3.8.5.

Five centered faces: test faces with the average $\overline{p}$ subtracted.

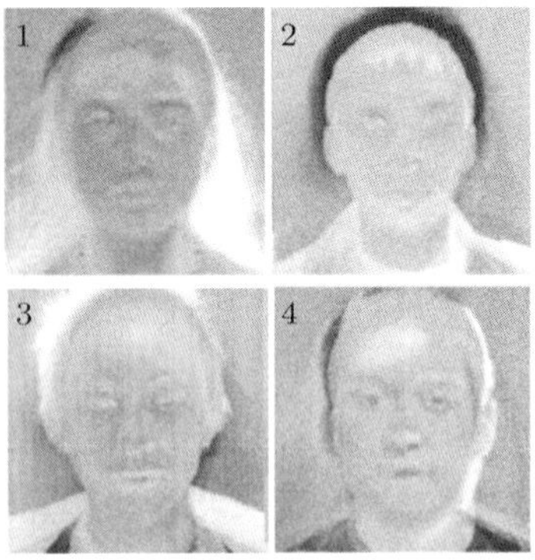

FIGURE 3.8.6.

The four eigenfaces with largest eignvalues created from the faces of Figure 3.8.3. The face marked 1 has the largest egenvalue.

photographs, not so carefully framed. These agencies and companies want to do different things with their databases, but clearly identifying a new photograph as that of someone in a database is one of the basic tasks they will want to accomplish.

To use *eigenfaces*, you first choose a collection of m "training faces". The seminal paper by Matthew Pentland and Alex Turk[18] had just m=16, and to our eyes they all look suspiciously like computer science graduate students, but the paper dates back to 1991, the dinosaur age for technology. The training faces should be chosen randomly within the database for maximum effectiveness. These images might look like Figure 3.8.3.

From the m training faces, form an $m \times n^2$ matrix P, where n might be 128 (for very low resolution pictures), or 1024 (for 1K $\times$ 1K images, still pretty low resolution). The entries of the matrix might be integers from 0 to 255 in the low resolution world, or integers between 0 and 2^{24} for color images. Denote by $p_1, \dots, p_m$ the rows of this matrix; each is a line matrix n^2 wide, representing a picture with $n \times n$ pixels. We think of the shade of each pixel as a random variable.

In any case, n^2 is much larger than m. The space of images is immense: n^2 will be about 10^{12} if the pictures are 1K by 1K. The idea is to locate *face space:* a subspace of this immense space of much lower dimension such that images of faces will lie close to this subspace.

First find the "average picture"

$$\overline{p} = \frac{1}{m}(p_1 + \cdots + p_m); \tag{3.8.31}$$

where the p_i are the training faces. This average $\overline{p}$ is another image but not the picture of anyone; see Figure 3.8.4. Now (see Figure 3.8.5) subtract $\overline{p}$ from each training face p_i to get the "centered" picture $p_i - \overline{p}$, and let $C^\top$ be the $n^2 \times m$ matrix

$$C^\top = [p_1 - \overline{p}, \dots, p_m - \overline{p}]; \tag{3.8.32}$$

i.e., the ith column of $C^\top$ is $p_i - \overline{p}$. Now apply the spectral theorem to the $n^2 \times n^2$ covariance matrix $C^\top C$.

This is a terrifyingly large matrix, but the task is actually quite feasible, because by Lemma 3.8.2 the nonzero eigenvalues of $C^\top C$ are the same as the nonzero eigenvalues of the matrix $CC^\top$, and $CC^\top$ is only $m \times m$, perhaps 500×500 or 1000×1000 (and symmetric, of course). Finding the eigenvalues and eigenvectors such a matrix is routine. Lemma 3.8.2 then tells us how to find the eigenvectors of $C^\top C$. These eigenvectors are called *eigenfaces*: they are vectors n^2 high that encode (usually ghostly) images. See Figure 3.8.6.

The next step involves an element of judgment: which are the "significant" eigenfaces? They should of course be the ones with largest eigenvalues, because they are the directions in which the test pictures show maximal

[18]M. Turk and A. Pentland, Face recognition using eigenfaces.Computer Vision and Pattern Recognition, 1991. Proceedings CVPR '91., IEEE Computer Society Conference on 1991

variance. But how many should you take? There isn't a clean-cut answer to the question: perhaps there is a large gap in the eigenvalues, and the ones beneath the gap can be neglected; perhaps one chooses to cut off at some threshold. In the final analysis, the question is whether the chosen eigenfaces give the user the discrimination needed to accomplish the desired task. Facebook is said to use the largest 125–150 singular values.

Suppose you have chosen k eigenfaces $q_1, \dots, q_k$ as significant; *face space* $\mathcal{F} \subset \mathbb{R}^{n^2}$ is the subspace spanned by these eigenfaces. The eigenfaces form an orthonormal basis for $\mathcal{F}$, and any image p can first be centered by setting $\widetilde{p} = p - \overline{p}$, then be projected to face space. The projection $\pi_{\mathcal{F}}(\widetilde{p})$ can be written as a linear combination

$$\pi_{\mathcal{F}}(\widetilde{p}) = \alpha_1(p)q_1 + \cdots + \alpha_k(p)q_k. \qquad 3.8.33$$

These coefficients should give a lot of information about p, since the chosen eigenfaces are the directions of maximal variance of the sample of faces used. Apply this to all the faces in the database, making a list of elements of $\mathbb{R}^k$ which is supposed to encode the main features of each face of the database. When a new picture p_{new} comes in, project $\widetilde{p}_{new} = p_{new} - \overline{p}$ onto face space to find the corresponding coefficients $\alpha_1, \dots, \alpha_k$.

There are two questions to ask: how large is $|\widetilde{p}_{new} - \pi_{\mathcal{F}}(\widetilde{p}_{new})|$, i.e., how far is $\widetilde{p}_{new}$ from face space. If this distance is too large, the picture is probably not a face at all. Suppose that $\widetilde{p}_{new}$ does pass the "face test"; then how close is $\pi_{\mathcal{F}}(\widetilde{p}_{new})$ to $\widetilde{p}$ for one of the faces p in the database? If this distance is sufficently small, p_{new} is recognized as p, otherwise it is a new face, perhaps to be added to the database. The final judgment of whether the cutoffs were well chosen is whether the algorithm works: does it correctly recognize faces? $\triangle$

Note that what linear algebra tells us are the principal "features" are not features that we can describe. If in Figure 3.8.6 we had not labeled the eigenfaces with largest eigenvalues, you probably could not have even have figured out which had larger eigenvalues.

When we recognize the face of a friend or acquaintance, how do we do it? How do we store and retrieve that data? In the absence of knowledge, we can speculate: perhaps our brains run some form of "eigenfaces". This might explain comments by some whites that "all blacks look the same" or by some Chinese that "all whites look the same": the features we best discern are those of the "training faces" we see as young children. Something similar might explain why we can discern with precision the sounds of our native language, but often we hear the sounds of a foreign language as the same, when to native speakers they sound completely different.

Rather than speak of the largest eigenvalues of $C^\top C$, we could speak of the largest singular values of C, since they are the square roots of the eigenvalues of $C^\top C$ by Theorem 3.8.1 and the definition of singular values.

Exercises for Section 3.8

3.8.1 Suppose an experiment consists of throwing two 6-sided dice, each loaded so that it lands on 4 half the time, while the other outcomes are equally likely. The random variable f gives the total obtained on each throw. What are the probability weights for each outcome?

3.8.2 Repeat Exercise 3.8.1, but this time one die is loaded as before, and the other falls on 3 half the time, with the other outcomes equally likely.

3.8.3 Toss a coin until it comes up heads. The sample space S of this game is the positive integers $S = \{1, 2, \dots\}$; n corresponds to first coming up heads on the nth toss. Suppose the coin is fair, i.e., $\mathbf{P}(\{n\}) = 1/2^n$.

a. Let f be the random variable $f(n) = n$. Show that $E(f) = 2$.

b. Let g be the random variable $g(n) = 2^n$. Show that $E(g) = \infty$.

c. Think of these random variables as payoffs if you play the game: if heads first comes up on throw n, then for the pay-off f you collect n dollars; for g you collect 2^n dollars. How much would you be willing to pay to enter the game with pay-off f? How about g?

3.8.4 Suppose a probability space X consists of n outcomes, $\{1, 2, \dots, n\}$, each with probability $1/n$. Then a random function f on X can be identified with an element $\vec{f} \in \mathbb{R}^n$.

a. Show that $E(f) = \frac{1}{n}(\vec{f} \cdot \vec{1})$, where $\vec{1}$ is the vector with all entries 1.

b. Show that

$$\operatorname{Var}(f) = \frac{1}{n}\left|\vec{f} - E(f)\vec{1}\right|^2 \quad \text{and} \quad \sigma(f) = \frac{1}{\sqrt{n}}\left|\vec{f} - E(f)\vec{1}\right|.$$

c. Show that

$$\operatorname{cov}(f,g) = \frac{1}{n}\left(\vec{f} - E(f)\vec{1}\right) \cdot \left(\vec{g} - E(g)\vec{1}\right);$$

$$\operatorname{corr}(f,g) = \cos\theta,$$

where θ is the angle between the vectors $\vec{f} - E(f)\vec{1}$ and $\vec{g} - E(g)\vec{1}$.

3.8.5 Let A be an $n \times m$ matrix.

a. Show that if $\vec{\mathbf{v}} \in \ker A^\top A$, then $\vec{\mathbf{v}} \in \ker A$ and that if $\vec{\mathbf{w}} \in \ker AA^\top$, then $\vec{\mathbf{w}} \in \ker A^\top$.

b. Using Lemma 3.8.2 and the spectral theorem, show that $A^\top A$ has an orthonormal eigenbasis $\vec{\mathbf{v}}_1, \dots, \vec{\mathbf{v}}_m$ with eigenvalues $\lambda_1 \geq \dots \geq \lambda_k > 0$, $\lambda_{k+1} = \dots = \lambda_m = 0$, and that $AA^\top$ has an orthonormal eigenbasis $\vec{\mathbf{w}}_1, \dots, \vec{\mathbf{w}}_n$, with eigenvalues $\mu_1 = \lambda_1, \dots, \mu_k = \lambda_k$ and $\mu_{k+1} = \dots = \mu_n = 0$, such that

$$\vec{\mathbf{w}}_i = \frac{1}{\sqrt{\lambda_i}} A\vec{\mathbf{v}}_i,\ i \leq k \quad \text{and} \quad \vec{\mathbf{v}}_i = \frac{1}{\sqrt{\mu_i}} A^\top \vec{\mathbf{w}}_i,\ i \leq k.$$

3.9 Geometry of curves and surfaces

In which we apply what we have learned about Taylor polynomials, quadratic forms, and critical points to the geometry of curves and surfaces: in particular, their curvature.

Curvature in geometry manifests itself as gravitation.—C. Misner, K. S. Thorne, J. Wheeler, *Gravitation*

A curve acquires its geometry from the space in which it is embedded. Without that embedding, a curve is boring: geometrically it is a straight line. A one-dimensional worm living inside a smooth curve cannot tell whether the curve is straight or curvy; at most (if allowed to leave a trace behind it) it can tell whether the curve is closed or not.

This is not true of surfaces and higher-dimensional manifolds. Given a long-enough tape measure you could prove that the earth is spherical without recourse to ambient space; Exercise 3.9.7 asks you to compute how long a tape measure you would need.

The central notion used to explore these issues is *curvature*, which comes in several flavors. Its importance cannot be overstated: gravitation is the curvature of spacetime; the electromagnetic field is the curvature of the electromagnetic potential. The geometry of curves and surfaces is an immense field; our treatment cannot be more than the barest overview.[19]

[19]For further reading, we recommend *Riemannian Geometry, A Beginner's Guide*, by Frank Morgan (A K Peters, Ltd., Natick, MA, second edition 1998) and *Differential Geometry of Curves and Surfaces*, by Manfredo P. do Carmo (Prentice Hall, Inc., 1976).

Recall the fuzzy definition of "smooth" as meaning "as many times differentiable as is relevant to the problem at hand". In Sections 3.1 and 3.2, once continuously differentiable was sufficient; here it is not.

We begin with curvature as it applies to curves in the plane and in space. In both cases we write our curve as the graph of a function in the coordinates best adapted to the situation, and read the curvature (and other quantities of interest) from the quadratic terms of the function's Taylor polynomial. Differential geometry only exists for functions that are twice continuously differentiable; without that hypothesis, everything becomes a million times harder. Thus the functions we discuss all have Taylor polynomials of degree at least 2. For curves in space, we will need our functions to be three times continuously differentiable, with Taylor polynomials of degree 3.

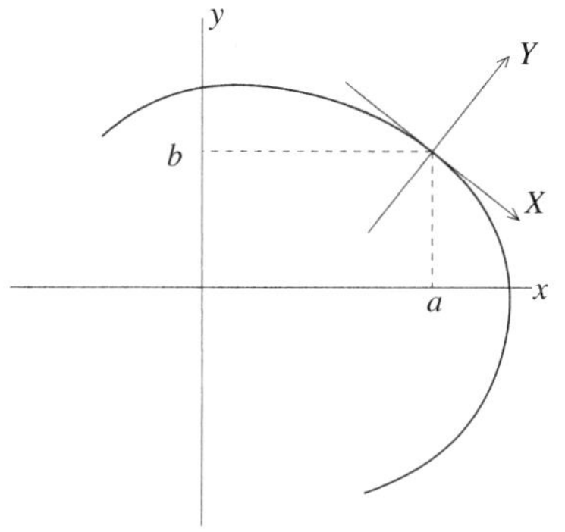

FIGURE 3.9.1.

To study a smooth curve at $\mathbf{a} = \begin{pmatrix} a \\ b \end{pmatrix}$, we make $\mathbf{a}$ the origin of our new coordinates and place the X-axis in the direction of the tangent to the curve at $\mathbf{a}$. Within the shaded region, the curve is the graph of a function $Y = g(X)$ that starts with quadratic terms.

The geometry and curvature of plane curves

For a smooth curve in the plane, the "best coordinate system" X, Y at a point $\mathbf{a} = \begin{pmatrix} a \\ b \end{pmatrix}$ is the system centered at $\mathbf{a}$, with the X-axis in the direction of the tangent line, and the Y-axis orthogonal to the tangent at that point, as shown in Figure 3.9.1. In these (X, Y)-coordinates, the curve is locally the graph of a function $Y = g(X)$, which can be approximated by its Taylor polynomial. This Taylor polynomial contains only quadratic and higher terms[20]:

$$Y = g(X) = \frac{A_2}{2}X^2 + \frac{A_3}{6}X^3 + \cdots, \tag{3.9.1}$$

where A_2 is the second derivative of g (see equation 3.3.1). All the coefficients of this polynomial are *invariants* of the curve: numbers associated to a point of the curve that do not change if you translate or rotate the curve. (Of course, they do depend on the point where you are on the curve.)

In defining the curvature of a plane curve, the coefficient that will interest us is A_2, the second derivative of g.

The Greek letter κ is "kappa".

We could avoid the absolute value in Definition 3.9.1 by defining the *signed* curvature of an oriented curve.

Definition 3.9.1 (Curvature of a curve in $\mathbb{R}^2$). Let a curve in $\mathbb{R}^2$ be locally the graph of a function g, with Taylor polynomial

$$g(X) = \frac{A_2}{2}X^2 + \frac{A_3}{6}X^3 + \cdots. \tag{3.9.2}$$

Then the *curvature* κ of the curve at $\mathbf{0}$ (i.e., at $\mathbf{a}$ in the original (x, y)-coordinates) is

$$\kappa(\mathbf{0}) \stackrel{\text{def}}{=} |A_2|. \tag{3.9.3}$$

The dots in $-\frac{1}{2}X^2 + \cdots$ represent higher-degree terms. We avoided computing the derivatives for $g(X)$ by using the formula for the Taylor series of a binomial (equation 3.4.9). In this case, m is $1/2$ and $a = -X^2$.

Remark. The unit circle has curvature 1: near $\begin{pmatrix} 0 \\ 1 \end{pmatrix}$ the "best coordinates" for the unit circle are $X = x$, $Y = y - 1$, so $y = \sqrt{1 - x^2}$ becomes

$$g(X) = Y = y - 1 = \sqrt{1 - x^2} - 1 = \sqrt{1 - X^2} - 1, \tag{3.9.4}$$

with the Taylor polynomial $g(X) = -\frac{1}{2}X^2 + \cdots$.

[20]The point $\mathbf{a}$ has coordinates $X = 0$, $Y = 0$, so the constant term is 0; the linear term is 0 because the curve is tangent to the X-axis at $\mathbf{a}$.

(When $X = 0$, both $g(X)$ and $g'(0)$ vanish, while $g''(0) = -1$; the quadratic term for the Taylor polynomial is $\frac{1}{2}g''$.) So the unit circle has curvature $|-1| = 1$. $\triangle$

Proposition 3.9.2 tells how to compute the curvature of a smooth plane curve that is locally the graph of a function f (i.e., when we are working in any coordinate system x, y in which the curve is locally the graph of a function, not in the "best" coordinates X, Y). We prove it after giving a couple of examples.

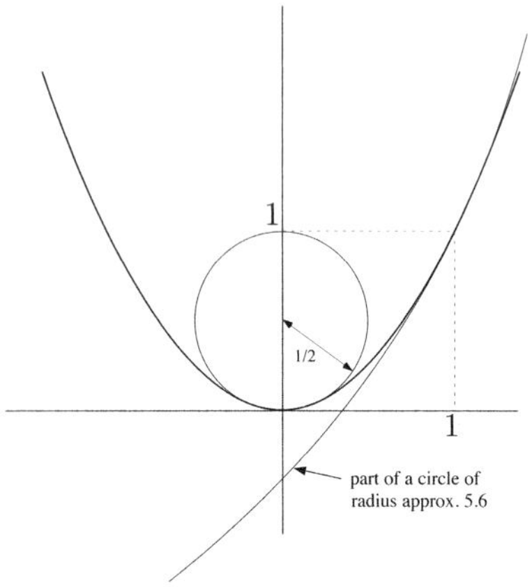

FIGURE 3.9.2.
At $\begin{pmatrix}1\\1\end{pmatrix}$, which corresponds to $a = 1$, the parabola $y = x^2$ looks much flatter than the unit circle. Instead, it resembles a circle of radius $5\sqrt{5}/2 \approx 5.59$. (A portion of such a circle is shown. Note that it crosses the parabola. This is the usual case, occurring when, in adapted coordinates, the cubic terms of the Taylor polynomial of the difference between the circle and the parabola are nonzero.)
At the origin, which corresponds to $a = 0$, the parabola has curvature 2 and resembles the circle of radius 1/2 centered at $\begin{pmatrix}0\\1/2\end{pmatrix}$, which also has curvature 2.

"Resembles" is an understatement. At the origin, the Taylor polynomial of the difference between the circle and the parabola starts with fourth-degree terms.

Proposition 3.9.2 (Computing the curvature of a plane curve known as a graph). *The curvature κ of the curve $y = f(x)$ at $\begin{pmatrix}a\\f(a)\end{pmatrix}$ is*

$$\kappa\begin{pmatrix}a\\f(a)\end{pmatrix} = \frac{|f''(a)|}{\left(1+(f'(a))^2\right)^{3/2}}. \qquad 3.9.5$$

Example 3.9.3 (Curvature of a parabola). The curvature of the curve $y = x^2$ at the point $\begin{pmatrix}a\\a^2\end{pmatrix}$ is

$$\kappa = \frac{2}{(1+4a^2)^{3/2}}. \qquad 3.9.6$$

Note that when $a = 0$, the curvature is 2, twice that of the unit circle. As a increases, the curvature rapidly decreases. Already at $\begin{pmatrix}1\\1\end{pmatrix}$, corresponding to $a = 1$, the curvature is $\frac{2}{5\sqrt{5}} \approx 0.179$. Figure 3.9.2 shows that at this point, the parabola is indeed much flatter than the unit circle. At the origin, it resembles the circle with radius 1/2, centered at $\begin{pmatrix}0\\1/2\end{pmatrix}$. $\triangle$

Example 3.9.4 (Curvature of a curve that is the graph of an implicit function). We saw in Example 3.4.8 that

$$x^3 + xy + y^3 = 3 \qquad 3.9.7$$

implicitly expresses x in terms of y near $\begin{pmatrix}1\\1\end{pmatrix}$ and that the first and second derivatives at 1 of the implicit function g are $g'(1) = -1$, and $g''(1) = -5/2$.

Thus near $\begin{pmatrix}1\\1\end{pmatrix}$, the curve described by equation 3.9.7 has curvature

$$\kappa = \frac{|f''(a)|}{\left(1+(f'(a))^2\right)^{3/2}} = \frac{|-5/2|}{2^{3/2}} = \frac{5}{4\sqrt{2}}. \qquad 3.9.8$$

Proof of Proposition 3.9.2. We express f as its Taylor polynomial, ignoring the constant term, since we can eliminate it by translating the coordinates, without changing any of the derivatives. This gives

$$f(x) = f'(a)x + \frac{f''(a)}{2}x^2 + \cdots. \qquad 3.9.9$$

Now rotate the coordinates x, y by θ to get coordinates X, Y shown in Figure 3.9.3: use the rotation matrix

$$R = \begin{bmatrix} \cos\theta & \sin\theta \\ -\sin\theta & \cos\theta \end{bmatrix} \text{ to get } \begin{bmatrix} X \\ Y \end{bmatrix} = R\begin{bmatrix} x \\ y \end{bmatrix} = \begin{bmatrix} x\cos\theta + y\sin\theta \\ -x\sin\theta + y\cos\theta \end{bmatrix}. \qquad 3.9.10$$

Then express x, y in terms of X, Y by multiplying by R^{-1}:

$$\underbrace{\begin{bmatrix} \cos\theta & -\sin\theta \\ \sin\theta & \cos\theta \end{bmatrix}}_{R^{-1}} \begin{bmatrix} X \\ Y \end{bmatrix} = \underbrace{\begin{bmatrix} \cos\theta & -\sin\theta \\ \sin\theta & \cos\theta \end{bmatrix} \begin{bmatrix} \cos\theta & \sin\theta \\ -\sin\theta & \cos\theta \end{bmatrix}}_{I} \begin{bmatrix} x \\ y \end{bmatrix} = \begin{bmatrix} x \\ y \end{bmatrix}.$$

to get $x = X\cos\theta - Y\sin\theta$ and $y = X\sin\theta + Y\cos\theta$. Substituting these into equation 3.9.9 leads to

$$\underbrace{X\sin\theta + Y\cos\theta}_{y} = f'(a)\underbrace{(X\cos\theta - Y\sin\theta)}_{x} + \frac{f''(a)}{2}\underbrace{(X\cos\theta - Y\sin\theta)^2}_{x^2} + \cdots. \qquad 3.9.11$$

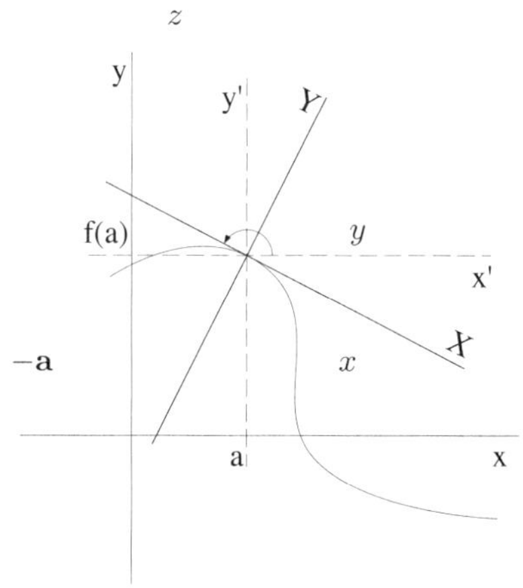

FIGURE 3.9.3 In the coordinates X, Y centered at $\begin{pmatrix} a \\ f(a) \end{pmatrix}$, the curve of equation $y = f(x)$ is the curve of equation $Y = F(X)$,where the Taylor polynomial of F starts with quadratic terms.

Here we are treating Y as the pivotal unknown and X as the nonpivotal unknown, so D_2F corresponds to the invertible matrix $[D_1\mathbf{F}(\mathbf{c}), \dots, D_{n-k}\mathbf{F}(\mathbf{c})]$ in Theorem 2.10.14. Since D_2F is a number, being nonzero and being invertible are the same. (Alternatively, we could say that X is a function of Y if D_1F is invertible.)

Recall (Definition 3.9.1) that curvature is defined for a curve that is locally the graph of a function g whose Taylor polynomial starts with quadratic terms. So we want to choose θ so that equation 3.9.11 expresses Y as a function of X, with derivative 0, so that its Taylor polynomial starts with the quadratic term:

$$Y = g(X) = \frac{A_2}{2}X^2 + \cdots. \qquad 3.9.12$$

If we subtract $X\sin\theta + Y\cos\theta$ from both sides of equation 3.9.11, we can write the equation for the curve in terms of the (X, Y)-coordinates:

$$F\begin{pmatrix} X \\ Y \end{pmatrix} = 0 = -X\sin\theta - Y\cos\theta + f'(a)(X\cos\theta - Y\sin\theta) + \cdots, \qquad 3.9.13$$

with derivative

$$\left[\mathbf{D}F\begin{pmatrix} 0 \\ 0 \end{pmatrix}\right] = [\underbrace{-\sin\theta + f'(a)\cos\theta}_{D_1F}, \underbrace{-\cos\theta - f'(a)\sin\theta}_{D_2F}]. \qquad 3.9.14$$

The implicit function theorem says that there is a function g expressing Y in terms of X if D_2F is invertible (that is, if $-f'(a)\sin\theta - \cos\theta \neq 0$). In that case, equation 2.10.29 for the derivative of an implicit function tells us that in order to have $g'(0) = 0$, so that $g(X)$ starts with quadratic terms, we must have $D_1F = f'(a)\cos\theta - \sin\theta = 0$:

$$g'(0) = 0 = -\underbrace{[D_2F(0)]}_{\neq 0}{}^{-1}\underbrace{(f'(a)\cos\theta - \sin\theta)}_{[D_1F(0)],\text{ must be }0} \tag{3.9.15}$$

Saying that $\tan\theta = f'(a)$ simply says that $f'(a)$ is the slope of the curve.

Thus we must have $\tan\theta = f'(a)$. Indeed, if θ satisfies $\tan\theta = f'(a)$, then

$$D_2F(0) = -f'(a)\sin\theta - \cos\theta = \frac{-1}{\cos\theta} \neq 0, \tag{3.9.16}$$

so the implicit function theorem applies. Now we compute the Taylor polynomial of the implicit function, as we did in Example 3.4.8. We replace Y in equation 3.9.13 by $g(X)$:

$$F\begin{pmatrix} X \\ g(X) \end{pmatrix} = 0 = -X\sin\theta - g(X)\cos\theta + f'(a)(X\cos\theta - g(X)\sin\theta)$$
$$\underbrace{+\frac{f''(a)}{2}(X\cos\theta - g(X)\sin\theta)^2 + \cdots}_{\text{additional term; see eq.3.9.11}}, \tag{3.9.17}$$

Now we rewrite equation 3.9.17, putting the linear terms in $g(X)$ on the left, and everything else on the right. This gives

Equation 3.9.19: $g'(0) = 0$, so $g(X)$ starts with quadratic terms. Moreover, by theorem 3.4.7, the function g is as differentiable as F, hence as f. So the term $Xg(X)$ is of degree 3, and the term $g(X)^2$ is of degree 4.

$$\begin{aligned}(f'(a)\sin\theta + \cos\theta)g(X) &= \overbrace{(f'(a)\cos\theta - \sin\theta)}^{=0}X \\ &\quad + \frac{f''(a)}{2}\big(\cos\theta X - \sin\theta g(X)\big)^2 + \cdots \\ &= \frac{f''(a)}{2}\big(\cos\theta X - \sin\theta g(X)\big)^2 + \cdots .\end{aligned} \tag{3.9.18}$$

We divide by $f'(a)\sin\theta + \cos\theta$ to obtain

$$g(X) = \frac{1}{f'(a)\sin\theta + \cos\theta}\frac{f''(a)}{2}\Big(\cos^2\theta X^2 \underbrace{- 2\cos\theta\sin\theta X g(X) + \sin^2\theta(g(X))^2}_{\text{these are of degree 3 or higher}}\Big) + \cdots . \tag{3.9.19}$$

Now we express the coefficient of X^2 as $A_2/2$ (see Definition 3.9.1), getting

$$A_2 = \frac{f''(a)\cos^2\theta}{f'(a)\sin\theta + \cos\theta}. \tag{3.9.20}$$

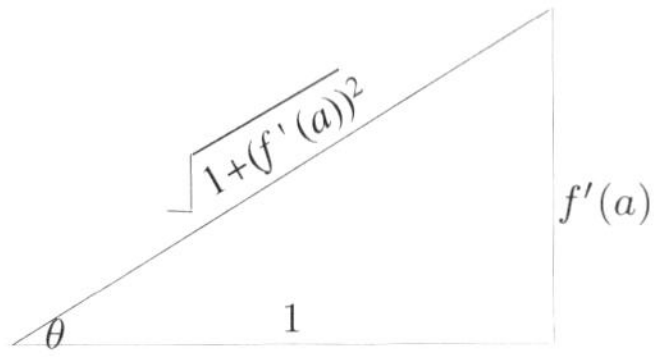

FIGURE 3.9.4.
This justifies equation 3.9.21.

Since $f'(a) = \tan\theta$, we have the right triangle of Figure 3.9.4, so

$$\sin\theta = \frac{f'(a)}{\sqrt{1 + (f'(a))^2}} \quad \text{and} \quad \cos\theta = \frac{1}{\sqrt{1 + (f'(a))^2}}. \tag{3.9.21}$$

Substituting these values in equation 3.9.20, we have

$$A_2 = \frac{f''(a)}{\left(1+\left(f'(a)\right)^2\right)^{3/2}}, \quad \text{so} \quad \kappa = |A_2| = \frac{|f''(a)|}{\left(1+(f'(a))^2\right)^{3/2}}. \quad \square \qquad 3.9.22$$

Geometry of curves parametrized by arc length

There is an alternative approach to the geometry of curves, both in the plane and in space: parametrization by arc length. This method reflects the fact that curves have no interesting intrinsic geometry: if you were a one-dimensional bug living on a curve, you could not make any measurements that would tell whether your universe was a straight line, or all tangled up.

Recall (Definition 3.1.19) that a parametrized curve is a map $\gamma : I \to \mathbb{R}^n$, where I is an interval in $\mathbb{R}$. You can think of I as an interval of time; if you are traveling along the curve, the parametrization tells you where you are on the curve at a given time. The vector $\vec{\gamma}'$ is the velocity vector of the parametrization γ.

FIGURE 3.9.5.

A curve approximated by an inscribed polygon. You may be more familiar with closed polygons, but a polygon need not be closed.

Definition 3.9.5 (Arc length). The *arc length* of the segment $\gamma([a, b])$ of a curve parametrized by γ is given by the integral

$$\int_a^b |\vec{\gamma}'(t)|\, dt. \qquad 3.9.23$$

A more intuitive definition is to consider the lengths of straight line segments ("inscribed polygonal curves") joining points $\gamma(t_0), \gamma(t_1), \dots, \gamma(t_m)$, where $t_0 = a$ and $t_m = b$, as shown in Figure 3.9.5. Then take the limit as the line segments become shorter and shorter. In formulas, this means to consider

$$\sum_{i=0}^{m-1} |\gamma(t_{i+1}) - \gamma(t_i)|, \text{ which is almost } \sum_{i=0}^{m-1} |\vec{\gamma}'(t_i)|(t_{i+1} - t_i). \qquad 3.9.24$$

Equation 3.9.25: If your odometer says you have traveled 50 miles, then you have traveled 50 miles on your curve.

Parametrization by arc length is more attractive in theory than in practice: computing the integral in equation 3.9.25 is painful, and computing the inverse function $t(s)$ is even more so. Later we will see how to compute the curvature of curves known by arbitrary parametrizations, which is much easier (see equation 3.9.61).

(If you have any doubts about the "which is almost", Exercise 3.24 should remove them when γ is twice continuously differentiable.) This last expression is a Riemann sum for $\int_a^b |\vec{\gamma}'(t)|\, dt$.

If you select an origin $\gamma(t_0)$, then you can define $s(t)$ by the formula

$$\underbrace{s(t)}_{\substack{\text{odometer}\\ \text{reading}\\ \text{at time } t}} = \int_{t_0}^{t} \underbrace{|\vec{\gamma}'(u)|}_{\substack{\text{speedometer}\\ \text{reading}\\ \text{at time } u}}\, du; \qquad 3.9.25$$

$s(t)$ gives the odometer reading as a function of time: "how far have you gone since time t_0". It is a monotonically increasing function, so (Theorem 2.10.2) it has an inverse function $t(s)$ (at what time had you gone distance s on the curve?) Composing this function with $\gamma : I \to \mathbb{R}^2$ or $\gamma : I \to \mathbb{R}^3$ now says where you are in the plane, or in space, when you have gone a distance s along the curve (or, if $\gamma : I \to \mathbb{R}^n$, where you are in $\mathbb{R}^n$). The curve

$$\delta(s) = \gamma\big(t(s)\big) \qquad 3.9.26$$

is now parametrized by arc length: distances along the curve are exactly the same as they are in the parameter domain where s lives.

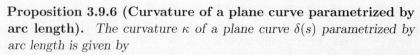

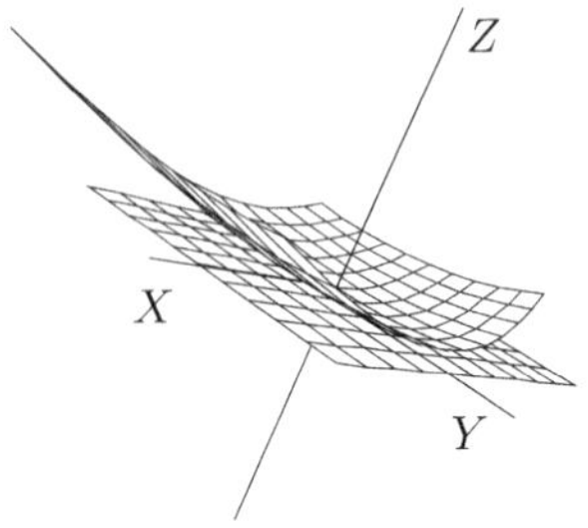

FIGURE 3.9.6.
In an adapted coordinate system, a surface is represented as the graph of a function from the tangent plane to the normal line that starts with quadratic terms. See equation 3.9.28.

Proposition 3.9.6 (Curvature of a plane curve parametrized by arc length). *The curvature κ of a plane curve $\delta(s)$ parametrized by arc length is given by*

$$\kappa\big(\delta(s)\big) = |\vec{\delta}''(s)|. \qquad 3.9.27$$

Exercise 3.9.12 asks you to prove Proposition 3.9.6.

The best coordinates for surfaces

Since surfaces have intrinsic geometry, we cannot generalize the parametrization approach we used for curves. Instead we will return to the "best coordinates" approach, as in equation 3.9.1. Let S be a surface in $\mathbb{R}^3$, and let $\mathbf{a}$ be a point in S.

Coordinates X, Y, Z with respect to an orthonormal basis $\vec{\mathbf{v}}_1, \vec{\mathbf{v}}_2, \vec{\mathbf{v}}_3$ anchored at $\mathbf{a}$ are *adapted to S at* $\mathbf{a}$ if $\vec{\mathbf{v}}_1$ and $\vec{\mathbf{v}}_2$ span $T_{\mathbf{a}}S$, so that $\vec{\mathbf{v}}_3$ is a unit vector orthogonal to S at $\mathbf{a}$, often called a *unit normal* and denoted $\vec{\mathbf{n}}$. In such a coordinate system, the surface S is locally the graph of a function

$$Z = f\begin{pmatrix} X \\ Y \end{pmatrix} = \overbrace{\frac{1}{2}(A_{2,0}X^2 + 2A_{1,1}XY + A_{0,2}Y^2)}^{\text{quadratic terms of Taylor polynomial}} + \text{higher-degree terms}; \qquad 3.9.28$$

see Figure 3.9.6. The quadratic form

$$\begin{pmatrix} X \\ Y \end{pmatrix} \mapsto A_{2,0}X^2 + 2A_{1,1}XY + A_{0,2}Y^2 \qquad 3.9.29$$

is known as the *second fundamental form* of the surface S at the point $\mathbf{a}$; it is a quadratic form on the tangent space $T_{\mathbf{a}}S$. Many interesting things can be read off from the numbers $A_{2,0}$, $A_{1,1}$, and $A_{0,2}$, in particular, the *mean curvature* and the *Gaussian curvature*, both generalizations of the single curvature of smooth curves.

Since $\vec{\mathbf{v}}_1, \vec{\mathbf{v}}_2, \vec{\mathbf{v}}_3$ are an orthonormal basis, the coordinate functions $X, Y, Z : \mathbb{R}^3 \to \mathbb{R}$ are given by

$$X(\mathbf{x}) = (\mathbf{x} - \mathbf{a}) \cdot \vec{\mathbf{v}}_1$$
$$Y(\mathbf{x}) = (\mathbf{x} - \mathbf{a}) \cdot \vec{\mathbf{v}}_2$$
$$Z(\mathbf{x}) = (\mathbf{x} - \mathbf{a}) \cdot \vec{\mathbf{v}}_3.$$

(See Proposition 2.4.18, part 1.)

Since the second fundamental form of S at $\mathbf{a}$ is a quadratic form, it is given by a symmetric matrix, the matrix

$$M = \begin{bmatrix} A_{2,0} & A_{1,1} \\ A_{1,1} & A_{0,2} \end{bmatrix}.$$

The mean curvature is half its trace; the Gaussian curvature (Definition 3.9.8) is its determinant:

$$H(\mathbf{a}) = \frac{1}{2}\operatorname{tr} M$$
$$K(\mathbf{a}) = \det M$$

The eigenvalues of M are the *principal curvatures* of S at $\mathbf{a}$.

Definition 3.9.7 (Mean curvature, mean curvature vector). The *mean curvature* H of a surface at a point $\mathbf{a}$ is

$$H(\mathbf{a}) \stackrel{\text{def}}{=} \frac{1}{2}(A_{2,0} + A_{0,2}). \qquad 3.9.30$$

The *mean curvature vector* is $\vec{H}(\mathbf{a}) \stackrel{\text{def}}{=} H(\mathbf{a})\vec{\mathbf{v}}_3$, where $\vec{\mathbf{v}}_3$ is the chosen unit normal.

Changing the adapted coordinate system by choosing the normal to be $-\vec{\mathbf{v}}_3$ instead of $\vec{\mathbf{v}}_3$ (i.e., changing the sign of Z) changes the sign of the mean curvature $H(\mathbf{a})$, but does not change the mean curvature vector.

The mean curvature vector is fairly easy to understand intuitively. If a surface evolves so as to minimize its area with boundary fixed, like a soap film spanning a loop of wire, the mean curvature points in the direction in which the surface moves (this statement will be justified in Section 5.4; see Theorem 5.4.4 and Corollary 5.4.5). A *minimal surface* is one that locally minimizes surface area among surfaces with the same boundary; according to this description a surface is minimal precisely when the mean curvature vector vanishes.

The flat disc is the minimal surface bounded by a circle; the hemisphere bounded by the same circle is not. Very nice minimal surfaces can be formed by twisting a wire to form a closed curve and dipping it into a soapy solution. The *mean curvature* of the soap film is zero. You may imagine that your wire loop is flat, but it can have many other shapes, and the soap films that span the loop will still be "flat" as far as mean curvature is concerned.

The Gaussian curvature does not depend on the choice of the normal $\vec{\mathbf{v}}_3$. It is the prototype of all the really interesting things in differential geometry. It measures to what extent pieces of a surface can be made flat, without stretching or deformation – as is possible for a cone or cylinder, but not for a sphere.

Definition 3.9.8 (Gaussian curvature of a surface). The *Gaussian curvature* K of a surface at a point $\mathbf{a}$ is

$$K(\mathbf{a}) \overset{\text{def}}{=} A_{2,0}A_{0,2} - A_{1,1}^2. \qquad 3.9.31$$

Exercise 3.9.3 asks you to show that the Gaussian curvature of the unit sphere is 1.

It follows (see Exercise 3.28) that if at some point the quadratic terms of the function f representing S in adapted coordinates form a positive definite or negative definite quadratic form, the Gaussian curvature at that point is positive; otherwise, the Gaussian curvature is nonpositive.

Theorem 3.9.9, proved in Section 5.4, says that the Gaussian curvature measures how big or small a surface is compared to a flat surface. It is a version of Gauss's *Theorema Egregium* (Latin for *remarkable theorem*).

Theorem 3.9.9. *Let $D_r(\mathbf{x})$ be the set of all points $\mathbf{q}$ in a surface $S \subset \mathbb{R}^3$ such that there exists a curve of length $\le r$ in S joining $\mathbf{x}$ to $\mathbf{q}$. Then*

$$\underbrace{\text{Area}(D_r(\mathbf{x}))}_{\text{area of curved disc}} = \underbrace{\pi r^2}_{\text{area of flat disc}} - \frac{K(\mathbf{x})\pi}{12} r^4 + o(r^4). \qquad 3.9.32$$

If the curvature is positive, the curved disc $D_r(\mathbf{x})$ is smaller than the flat disc, and if the curvature is negative, it is larger. The discs have to be measured with a tape measure contained in the surface; in other words, $D_r(\mathbf{x})$ is the set of points that can be connected to $\mathbf{x}$ by a curve *contained in the surface* and of length at most r.

An obvious example of a surface with positive Gaussian curvature is the surface of a ball. Wrap a napkin around a basketball; you will have extra fabric that won't lie smooth. This is why maps of the earth always distort areas: the extra "fabric" won't lie smooth otherwise. An example of a surface with negative Gaussian curvature is a mountain pass. Another is an armpit.

Figure 3.9.7 gives another example of Gaussian curvature.

If you have ever sewed a set-in sleeve on a shirt or dress, you know that when you pin the under part of the sleeve to the main part of the garment, you have extra fabric that doesn't lie flat; sewing the two parts together without puckers or gathers is tricky, and involves distorting the fabric.

Sewing is something of a dying art, but the mathematician Bill Thurston, whose geometric vision was legendary, maintained that it is an excellent way to acquire some feeling for the geometry of surfaces.

FIGURE 3.9.7. The Billy Goat Gruff in the foreground gets just the right amount of grass to eat; he lives on a flat surface, with Gaussian curvature zero. The goat at far left is thin; he lives on the top of a hill, with positive Gaussian curvature; he can reach less grass. The third is fat. His surface has negative Gaussian curvature; with the same length chain, he can get at more grass. This would be true even if the chain were so heavy that it lay on the ground.

Computing curvature of surfaces

Proposition 3.9.11 tells how to compute curvature for a surface in any coordinate system. First we use Proposition 3.9.10 to put the surface into "best" coordinates, if it is not in them already. It is proved in Appendix A15.

Equation 3.9.33: There is no constant term because we translate the surface so that the point under consideration is the origin.

Proposition 3.9.10 (Putting a surface into "best" coordinates). *Let $U \subset \mathbb{R}^2$ be open, $f : U \to \mathbb{R}$ a C^2 function, and S the graph of f. Let the Taylor polynomial of f at the origin be*

$$z = f\begin{pmatrix} x \\ y \end{pmatrix} = a_1 x + a_2 y + \frac{1}{2}\left(a_{2,0}x^2 + 2a_{1,1}xy + a_{0,2}y^2\right) + \cdots . \qquad 3.9.33$$

Set

$$c = \sqrt{a_1^2 + a_2^2}.$$

The number $c = \sqrt{a_1^2 + a_2^2}$ can be thought of as the "length" of the linear term. If $c = 0$, then $a_1 = a_2 = 0$ and there is no linear term; we are already in best coordinates.

The vectors in equation 3.9.34 are the vectors denoted $\vec{\mathbf{v}}_1, \vec{\mathbf{v}}_2, \vec{\mathbf{v}}_3$ at the beginning of the subsection. The first vector is a horizontal unit vector in the tangent plane. The second is a unit vector orthogonal to the first, in the tangent plane. The third is a unit vector orthogonal to the previous two. (It is a downward-pointing normal.) It takes a bit of geometry to find them, but the proof of Proposition 3.9.10 will show that these coordinates are indeed adapted to the surface.

If $c \neq 0$, then with respect to the orthonormal basis

$$\vec{\mathbf{v}}_1 = \begin{bmatrix} -\frac{a_2}{c} \\ \frac{a_1}{c} \\ 0 \end{bmatrix}, \quad \vec{\mathbf{v}}_2 = \begin{bmatrix} \frac{a_1}{c\sqrt{1+c^2}} \\ \frac{a_2}{c\sqrt{1+c^2}} \\ \frac{c}{\sqrt{1+c^2}} \end{bmatrix}, \quad \vec{\mathbf{v}}_3 = \begin{bmatrix} \frac{a_1}{\sqrt{1+c^2}} \\ \frac{a_2}{\sqrt{1+c^2}} \\ \frac{-1}{\sqrt{1+c^2}} \end{bmatrix}, \qquad 3.9.34$$

S is the graph of Z as a function

$$F\begin{pmatrix} X \\ Y \end{pmatrix} = \frac{1}{2}\Big(A_{2,0}X^2 + 2A_{1,1}XY + A_{0,2}Y^2\Big) + \cdots \qquad 3.9.35$$

which starts with quadratic terms, where

$$\begin{aligned} A_{2,0} &= \frac{-1}{c^2\sqrt{1+c^2}}\left(a_{2,0}a_2^2 - 2a_{1,1}a_1a_2 + a_{0,2}a_1^2\right) \\ A_{1,1} &= \frac{1}{c^2(1+c^2)}\left(a_1a_2(a_{2,0} - a_{0,2}) + a_{1,1}(a_2^2 - a_1^2)\right) \\ A_{0,2} &= \frac{-1}{c^2(1+c^2)^{3/2}}\left(a_{2,0}a_1^2 + 2a_{1,1}a_1a_2 + a_{0,2}a_2^2\right). \end{aligned} \qquad 3.9.36$$

Proposition 3.9.11 (Computing curvature of surfaces). *Let S and $\vec{\mathbf{n}}$ be as in Proposition 3.9.10.*

1. *The Gaussian curvature of S at the origin is*

$$K(\mathbf{0}) = \frac{a_{2,0}a_{0,2} - a_{1,1}^2}{(1+c^2)^2}. \qquad 3.9.37$$

2. *The mean curvature of S at the origin with respect to $\vec{\mathbf{n}}$ is*

$$H(\mathbf{0}) = \frac{1}{2(1+c^2)^{3/2}}\left(a_{2,0}(1+a_2^2) - 2a_1a_2a_{1,1} + a_{0,2}(1+a_1^2)\right). \qquad 3.9.38$$

Note the similarities between the equations of Proposition 3.9.11 and equation 3.9.5 for the curvature of a plane curve. In each case the numerator contains second derivatives ($a_{2,0}$, $a_{0,2}$, etc., are coefficients for the second-degree terms of the Taylor polynomial) and the denominator contains something like $1 + |Df|^2$ (the a_1 and a_2 of $c^2 = a_1^2 + a_2^2$ are coefficients of the first-degree term). A more precise relation can be seen if you consider the *surface* of equation $z = f(x)$, y arbitrary, and the plane *curve* $z = f(x)$; see Exercise 3.9.5.

Proposition 3.9.11 gives a reasonably straightforward way to compute the curvature and mean curvature of any surface at any point. It is stated for a surface representing z as a function $f\begin{pmatrix} x \\ y \end{pmatrix}$ such that $f\begin{pmatrix} 0 \\ 0 \end{pmatrix} = 0$ (a surface that passes through the origin); it gives the curvatures at the origin in terms of coefficients of the Taylor polynomial of f at $\begin{pmatrix} 0 \\ 0 \end{pmatrix}$. For any point $\mathbf{a}$ of a surface S, we can translate the coordinates to put $\mathbf{a}$ at the origin and perhaps rename our coordinates, to bring our surface to this situation. Note that even if we cannot find explicitly a function of which S is locally the graph, we can find the Taylor polynomial of that function to degree 2, using Theorem 3.4.7.

Proof of Proposition 3.9.11. For the Gaussian curvature, plug into Definition 3.9.8 the values for $A_{2,0}$, $A_{1,1}$, $A_{0,2}$ given in Proposition 3.9.10:

$$\begin{aligned} K(\mathbf{0}) &= A_{2,0}A_{0,2} - A_{1,1}^2 \\ &= \frac{1}{c^4(1+c^2)^2}\Big(\left(a_{2,0}a_2^2 - 2a_{1,1}a_1a_2 + a_{0,2}a_1^2\right)\left(a_{2,0}a_1^2 + 2a_{1,1}a_1a_2 + a_{0,2}a_2^2\right) \\ &\qquad - \left(a_{2,0}a_1a_2 + a_{1,1}a_2^2 - a_{1,1}a_1^2 - a_{0,2}a_1a_2\right)^2 \Big) \\ &= \frac{a_{2,0}a_{0,2} - a_{1,1}^2}{(1+c^2)^2}. \end{aligned} \qquad 3.9.39$$

This involves some quite miraculous cancellations, as does the computation for the mean curvature, which is left as Exercise A15.1. □

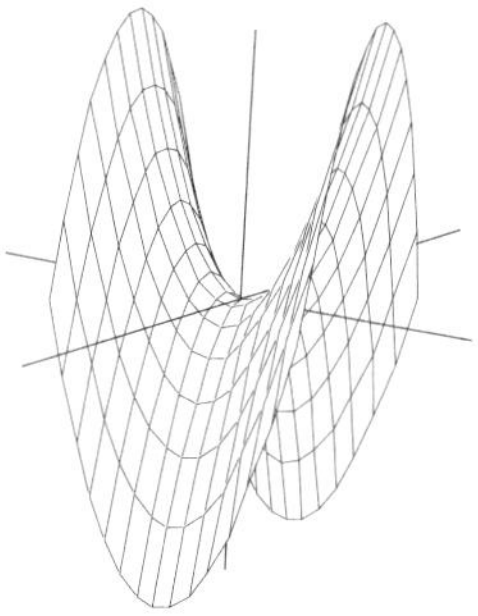

FIGURE 3.9.8.
The graph of $x^2 - y^2$, a typical saddle.

Example 3.9.12 (Computing the Gaussian and mean curvature of a surface given as the graph of a function). Suppose we want to measure the Gaussian curvature at a point $\mathbf{a} \overset{\text{def}}{=} \begin{pmatrix} a \\ b \\ a^2 - b^2 \end{pmatrix}$ of the surface given by the equation $z = x^2 - y^2$ and shown in Figure 3.9.8.

To determine the Gaussian curvature at the point $\mathbf{a}$, we make $\mathbf{a}$ our new origin; i.e., we use new translated coordinates, u, v, w, where

$$\begin{aligned} x &= a + u \\ y &= b + v \\ z &= a^2 - b^2 + w. \end{aligned} \qquad 3.9.40$$

(The u-axis replaces the original x-axis, the v-axis replaces the y-axis, and the w-axis replaces the z-axis.) Now we rewrite $z = x^2 - y^2$ as

$$\begin{aligned} \underbrace{a^2 - b^2 + w}_{z} &= \underbrace{(a+u)^2}_{x^2} - \underbrace{(b+v)^2}_{y^2} \\ &= a^2 + 2au + u^2 - b^2 - 2bv - v^2, \end{aligned} \qquad 3.9.41$$

which gives

$$w = 2au - 2bv + u^2 - v^2 = \underbrace{2a}_{a_1}\, u + \underbrace{-2b}_{a_2}\, v + \frac{1}{2}\Big(\underbrace{2}_{a_{2,0}}\, u^2 + \underbrace{-2}_{a_{0,2}}\, v^2\Big). \qquad 3.9.42$$

Now we have an equation of the form of equation 3.9.33, and (remembering that $c = \sqrt{a_1^2 + a_2^2}$) we can read off the Gaussian curvature, using Proposition 3.9.11:

The Gaussian curvature at $\mathbf{a}$ is

$$K(\mathbf{a}) = \frac{\overbrace{(2 \cdot -2) - 0}^{a_{2,0}a_{0,2} - a_{1,1}^2}}{\underbrace{(1 + 4a^2 + 4b^2)^2}_{(1+c^2)^2}} = \frac{-4}{(1 + 4a^2 + 4b^2)^2}. \qquad 3.9.43$$

Looking at this formula for K, what can you say about the Gaussian curvature of the surface the further one goes from the origin?[21]

Similarly, we can compute the mean curvature:

$$H(\mathbf{a}) = \frac{4(a^2 - b^2)}{(1 + 4a^2 + 4b^2)^{3/2}}. \quad \triangle \tag{3.9.44}$$

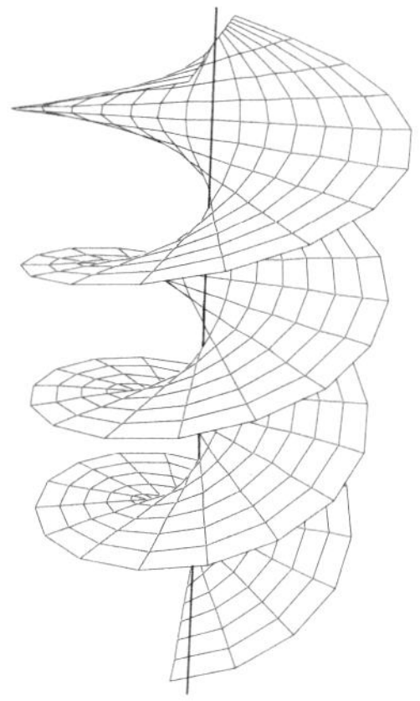

FIGURE 3.9.9.
The helicoid is swept out by a horizontal line, which rotates as it is lifted. "Helicoid" comes from the Greek for "spiral".

Example 3.9.13 (Computing the Gaussian and mean curvature of the helicoid). The helicoid is the surface of equation $y \cos z = x \sin z$. You can imagine it as swept out by a horizontal line going through the z-axis, and which turns steadily as the z-coordinate changes, making an angle z with the parallel to the x-axis through the same point, as shown in Figure 3.9.9.

A first thing to observe is that the mapping

$$\begin{pmatrix} x \\ y \\ z \end{pmatrix} \mapsto \begin{pmatrix} x\cos a + y \sin a \\ -x \sin a + y \cos a \\ z - a \end{pmatrix} \tag{3.9.45}$$

is a rigid motion of $\mathbb{R}^3$ that sends the helicoid to itself. (Can you justify that statement?[22]) In particular, setting $a = z$, this rigid motion sends any point to a point of the form $\begin{pmatrix} r \\ 0 \\ 0 \end{pmatrix}$ (do you see why?[23]), and it is enough to compute the Gaussian curvature $K(r)$ at such a point.

The mapping in Example 3.9.13 simultaneously rotates by a clockwise in the (x, y)-plane and lowers by a along the z-axis. The first two rows of the right side of equation 3.9.45 are the result of multiplying $\begin{bmatrix} x \\ y \end{bmatrix}$ by the rotation matrix we already saw in formula 3.9.10.

We don't know the helicoid as a graph, but, as Exercise 3.11 asks you to show, by the implicit function theorem, the equation of the helicoid determines z as a function $g_r\begin{pmatrix} x \\ y \end{pmatrix}$ near $\begin{pmatrix} r \\ 0 \\ 0 \end{pmatrix}$ when $r \neq 0$. What we need then is the Taylor polynomial of g_r. Introduce the new coordinate u such that $r + u = x$, and write

$$g_r\begin{pmatrix} u \\ y \end{pmatrix} = z = a_2 y + a_{1,1} u y + \frac{1}{2} a_{0,2} y^2 + \cdots . \tag{3.9.46}$$

Part b of Exercise 3.11 asks you to justify our omitting the terms $a_1 u$ and $a_{2,0} u^2$.

Now write $y \cos z$ and $(r + u) \sin z$ using the Taylor polynomials of $\cos z$ and $\sin z$, keeping only quadratic terms, to get

---

[21] The Gaussian curvature of this surface is always negative, but the further you go from the origin, the smaller it is, so the flatter the surface.

[22] If you substitute $x \cos a + y \sin a$ for x, $-x \sin a + y \cos a$ for y, and $z - a$ for z in the equation $y \cos z = x \sin z$, and do some computations, remembering that $\cos(z - a) = \cos z \cos a + \sin z \sin a$ and $\sin(z - a) = \sin z \cos a - \cos z \sin a$, you will land back on the equation $y \cos z = x \sin z$.

[23] If $z = a$, plugging the values of equation 3.9.45 into $y \cos z = x \sin z$ gives $y \cos 0 = 0$, so $y = 0$.

$$y\cos z \approx y \cdot 1 \qquad 3.9.47$$

$$(r+u)\sin z \approx (r+u)\underbrace{\left(a_2 y + a_{1,1}uy + \frac{1}{2}a_{0,2}y^2 + \cdots\right)}_{z \text{ as in equation 3.9.46}} \qquad 3.9.48$$

$$\approx ra_2y + a_2uy + ra_{1,1}uy + \frac{r}{2}a_{0,2}y^2.$$

(In equation 3.9.48, we replace $\sin z$ by the first term of its Taylor polynomial, z; in the second line we keep only linear and quadratic terms.)

So if in $y\cos z = (r+u)\sin z$, we replace the functions by their Taylor polynomials of degree 2, we get

$$y = ra_2y + a_2uy + ra_{1,1}uy + \frac{r}{2}a_{0,2}y^2. \qquad 3.9.49$$

Identifying linear and quadratic terms gives (see the margin)

$$a_1 = 0\ ,\ a_2 = \frac{1}{r}\ ,\ a_{1,1} = -\frac{1}{r^2}\ ,\ a_{0,2} = 0\ ,\ a_{2,0} = 0. \qquad 3.9.50$$

We can now read off the Gaussian and mean curvatures:

$$\underbrace{K(r)}_{\substack{\text{Gaussian}\\ \text{curvature}}} = \frac{-1}{r^4(1+1/r^2)^2} = \frac{-1}{(1+r^2)^2} \quad \text{and} \quad \underbrace{H(r)}_{\substack{\text{mean}\\ \text{curvature}}} = 0. \qquad 3.9.51$$

The first equation shows that the Gaussian curvature is always negative and does not blow up as $r \to 0$: as $r \to 0$, we have $K(r) \to -1$. This is what we should expect, since the helicoid is a smooth surface. The second equation is more interesting yet. It says that the helicoid is a *minimal surface*: every patch of the helicoid minimizes area among surfaces with the same boundary. △

To get equation 3.9.47, we replace $\cos z$ by its Taylor polynomial

$$1 - \frac{z^2}{2!} + \frac{z^4}{4!} - \cdots,$$

keeping only the first term. (The term $z^2/2!$ is quadratic but it isn't kept since $yz^2/2!$ is cubic.)

When identifying linear and quadratic terms, note that the only linear term on the right of equation 3.9.49 is ra_2y, giving $a_2 = 1/r$. The coefficient of uy is $a_2 + ra_{1,1}$, which must be 0.

Equation 3.9.50 says that the coefficients a_2 and $a_{1,1}$ blow up as $r \to 0$, which is what you should expect, since at the origin the helicoid does not represent z as a function of x and y. But the *curvature* does not blow up, since the helicoid is a smooth surface at the origin.

When we first made this computation, we landed on a Gaussian curvature that *did* blow up at the origin. Since this did not make sense, we checked and found that we had a mistake in our statement. We recommend making this kind of internal check (like making sure units in physics come out right), especially if, like us, you are not infallible in your computations.

Coordinates adapted to space curves

Curves in $\mathbb{R}^3$ have considerably simpler local geometry than do surfaces: essentially everything about them is in Propositions 3.9.15 and 3.9.16 below. Their global geometry is quite a different matter: they can tangle, knot, link, etc. in the most fantastic ways.

Knot theory is an active field of research today, with remarkable connections to physics (especially the latest darling of theoretical physicists: string theory).

Suppose $C \subset \mathbb{R}^3$ is a smooth curve, and $\mathbf{a} \in C$ is a point. What new coordinate system X, Y, Z is well adapted to C at $\mathbf{a}$? Of course, we will take $\mathbf{a}$ as the origin of the new system. If we require that the X-axis be tangent to C at $\mathbf{a}$, and call the other two coordinates U and V, then near $\mathbf{a}$ the curve C will have an equation of the form

$$\begin{aligned} U &= f(X) = \frac{1}{2}a_2X^2 + \frac{1}{6}a_3X^3 + \cdots \\ V &= g(X) = \frac{1}{2}b_2X^2 + \frac{1}{6}b_3X^3 + \cdots, \end{aligned} \qquad 3.9.52$$

where both coordinates start with quadratic terms. But we can do better, at least when C is at least three times differentiable, and $a_2^2 + b_2^2 \neq 0$.

Suppose we rotate the coordinate system around the X-axis by an angle θ, and denote by X, Y, Z the new (final) coordinates. Let $c = \cos\theta$ and $s = \sin\theta$ and (as in the proof of Proposition 3.9.2) write

$$U = cY + sZ \quad \text{and} \quad V = -sY + cZ. \tag{3.9.53}$$

Substituting these expressions into equation 3.9.52 leads to

$$\begin{aligned} cY + sZ &= \frac{1}{2}a_2X^2 + \frac{1}{6}a_3X^3 + \cdots \\ -sY + cZ &= \frac{1}{2}b_2X^2 + \frac{1}{6}b_3X^3 + \cdots . \end{aligned} \tag{3.9.54}$$

We solve these equations for Y and Z to get

Remember that $s = \sin\theta$ and $c = \cos\theta$, so $c^2 + s^2 = 1$.

$$\begin{aligned} Y(c^2+s^2) = Y &= \frac{1}{2}(ca_2 - sb_2)X^2 + \frac{1}{6}(ca_3 - sb_3)X^3 \\ Z &= \frac{1}{2}(sa_2 + cb_2)X^2 + \frac{1}{6}(sa_3 + cb_3)X^3. \end{aligned} \tag{3.9.55}$$

The point of all this is that we want to choose θ (the angle by which we rotate the coordinate system around the X-axis) so that the Z-component of the curve begins with cubic terms. We achieve this by setting

If $a_2 = b_2 = 0$ these values of θ do not make sense. At such a point we define the curvature to be 0 (this definition agrees with the general Definition 3.9.14, as the case where $A_2 = 0$).

$$\cos\theta = \frac{a_2}{\sqrt{a_2^2 + b_2^2}} \quad \text{and} \quad \sin\theta = \frac{-\,b_2}{\sqrt{a_2^2 + b_2^2}}, \quad \text{so that } \tan\theta = \frac{-b_2}{a_2};$$

this gives

We choose $\kappa = A_2$ to be the positive square root, $+\sqrt{a_2^2 + b_2^2}$; thus our definition of curvature of a space curve is compatible with Definition 3.9.1 of the curvature of a plane curve.

$$\begin{aligned} Y &= \frac{1}{2}\overbrace{\sqrt{a_2^2 + b_2^2}}^{A_2}\,X^2 + \frac{1}{6}\frac{a_2a_3 + b_2b_3}{\sqrt{a_2^2 + b_2^2}}X^3 = \frac{A_2}{2}X^2 + \frac{A_3}{6}X^3 + \cdots \\ Z &= \frac{1}{6}\frac{-b_2a_3 + a_2b_3}{\sqrt{a_2^2 + b_2^2}}X^3 + \cdots = \frac{B_3}{6}X^3 + \cdots . \end{aligned} \tag{3.9.56}$$

The Z-component measures the distance of the curve from the (X, Y)-plane; since Z is small, the curve stays mainly in that plane (more precisely, it leaves the plane very slowly). The (X, Y)-plane is called the *osculating plane* to C at $\mathbf{a}$.

The word "osculating" comes from the Latin "osculari", meaning "to kiss".

The osculating plane is the plane that the curve is most nearly in, and the torsion measures how fast the curve pulls away from it. It measures the "nonplanarity" of the curve.

This is our best coordinate system for the curve at $\mathbf{a}$, which exists and is unique unless $a_2 = b_2 = 0$. The number $A_2 \geq 0$ is called the *curvature* κ of C at $\mathbf{a}$, and the number B_3/A_2 is called its *torsion* τ.

Definition 3.9.14 (Curvature and torsion of a space curve). Let C be a space curve.

The *curvature* of C at $\mathbf{a}$ is $\kappa(\mathbf{a}) \stackrel{\text{def}}{=} A_2$

The *torsion* of C at $\mathbf{a}$ is $\tau(\mathbf{a}) \stackrel{\text{def}}{=} B_3/A_2$.

Note that torsion is defined only when the curvature is not zero.

Parametrizing space curves by arc length: the Frenet frame

Usually the geometry of space curves is developed using parametrization by arc length rather than adapted coordinates. Above, we emphasized adapted coordinates because they generalize to manifolds of higher dimension, while parametrizations by arc length do not.

A curve in $\mathbb{R}^n$ can be parametrized by arc length because curves have no intrinsic geometry; you could represent the Amazon River as a straight line without distorting its length. Surfaces and other manifolds of higher dimension cannot be parametrized by anything analogous to arc length; any attempt to represent the surface of the globe as a flat map necessarily distorts sizes and shapes of the continents. Gaussian curvature is the obstruction.

When we use parametrization by arc length to study the geometry of curves, the main ingredient is the *Frenet frame*. Imagine driving at unit speed along the curve, perhaps by turning on cruise control. Then (at least if the curve really is curvy, not straight) at each instant you have a distinguished orthonormal basis of $\mathbb{R}^3$, called the Frenet frame.

Imagine that you are riding a motorcycle in the dark and that the first unit vector is the shaft of light produced by your headlight.

The first basis vector is the velocity vector $\vec{\mathbf{t}}$, pointing in the direction of the curve; it is a unit vector since we are traveling at unit speed. The second vector $\vec{\mathbf{n}}$ is the unit vector in the direction of the acceleration vector. It is orthogonal to the curve and points in the direction in which the force is being applied – i.e., in the opposite direction of the centrifugal force you feel. (We know the acceleration must be orthogonal to the curve because the speed is constant; there is no component of acceleration in the direction of motion.[24]) The third is the *binormal* $\vec{\mathbf{b}}$, orthogonal to the other two.

So, if $\delta : \mathbb{R} \to \mathbb{R}^3$ is the parametrization by arc length of a curve C, and $\vec{\delta}''(s) \neq 0$, the three vectors are

$$\underbrace{\vec{\mathbf{t}}(s) = \vec{\delta}'(s),}_{\text{velocity vector}} \quad \underbrace{\vec{\mathbf{n}}(s) = \frac{\vec{\mathbf{t}}'(s)}{|\vec{\mathbf{t}}'(s)|} = \frac{\vec{\delta}''(s)}{|\vec{\delta}''(s)|},}_{\text{normalized acceleration vector}} \quad \underbrace{\vec{\mathbf{b}}(s) = \vec{\mathbf{t}}(s) \times \vec{\mathbf{n}}(s)\,.}_{\text{binormal}} \qquad 3.9.57$$

Proposition 3.9.15: Requiring $\vec{\delta}'' \neq 0$ implies that $\kappa \neq 0$, since $\vec{\mathbf{t}}'(0) = \vec{\delta}''(0)$ (equation 3.9.57) and $\vec{\mathbf{t}}(0) = \kappa\vec{\mathbf{n}}(0)$ (Proposition 3.9.16).

Propositions 3.9.15 and 3.9.16 relate the Frenet frame to adapted coordinates; they provide another description of curvature and torsion, and show that the two approaches coincide. They are proved in Appendix A15.

Proposition 3.9.15 (Frenet frame). *Let $\delta : \mathbb{R} \to \mathbb{R}^3$ parametrize a curve C by arc length, with $\vec{\delta}''(s) \neq 0$. Then the coordinates with respect to the Frenet frame at 0,*

$$\vec{\mathbf{t}}(0),\ \vec{\mathbf{n}}(0),\ \vec{\mathbf{b}}(0), \qquad 3.9.58$$

are adapted to the curve C at $\delta(0)$.

In particular, formula 3.9.59 says that the point $\delta(0)$ in the old coordinates is the point $\begin{pmatrix}0\\0\\0\end{pmatrix}$ in the new coordinates, which is what we want.

Thus the point with coordinates X, Y, Z in the new, adapted coordinates is the point

$$\delta(0) + X\vec{\mathbf{t}}(0) + Y\vec{\mathbf{n}}(0) + Z\vec{\mathbf{b}}(0) \qquad 3.9.59$$

in the old (x, y, z)-coordinates.

[24] Alternatively, you can derive $2\vec{\delta}' \cdot \vec{\delta}'' = 0$ from $|\vec{\delta}'|^2 = 1$.

Proposition 3.9.16 (Frenet frame related to curvature and torsion). *Let $\delta : \mathbb{R} \to \mathbb{R}^3$ parametrize a curve C by arc length. Then the Frenet frame satisfies the following equations, where κ is the curvature of the curve at $\delta(0)$ and τ is its torsion:*

$$\begin{aligned} \vec{\mathbf{t}}'(0) &= \qquad\quad \kappa\vec{\mathbf{n}}(0) \\ \vec{\mathbf{n}}'(0) &= -\kappa\vec{\mathbf{t}}(0) \qquad\qquad + \tau\vec{\mathbf{b}}(0) \\ \vec{\mathbf{b}}'(0) &= \qquad\quad - \tau\vec{\mathbf{n}}(0). \end{aligned} \tag{3.9.60}$$

Proposition 3.9.16: The derivatives are easiest to compute at 0, but the equations $\vec{\mathbf{t}}'(s) = \kappa\vec{\mathbf{n}}(s)$ and so on are true at any point s (any odometer reading).

Equation 3.9.60 corresponds to the antisymmetric matrix

$$A = \begin{bmatrix} 0 & -\kappa & 0 \\ \kappa & 0 & -\tau \\ 0 & \tau & 0 \end{bmatrix},$$

in the sense that

$$[\vec{\mathbf{t}}', \vec{\mathbf{n}}', \vec{\mathbf{b}}'] = [\vec{\mathbf{t}}, \vec{\mathbf{n}}, \vec{\mathbf{b}}]A.$$

Exercise 3.9.11 asks you to explain where this antisymmetry comes from.

The derivatives $\vec{\mathbf{t}}'$, $\vec{\mathbf{n}}'$, and $\vec{\mathbf{b}}'$ are computed with respect to arc length. Think of the point $\delta(0)$ as the place where you set the odometer reading to 0, accounting for $\vec{\mathbf{t}}'(0)$, etc.

Computing curvature and torsion of parametrized curves

We can now give formulas (in Propositions 3.9.17 and 3.9.18) that make the computation of curvature and torsion straightforward for any parametrized curve in $\mathbb{R}^3$.

We already had two equations for computing curvature and torsion of a space curve: equations 3.9.56 and 3.9.60, but they are hard to use. Equation 3.9.56 requires knowing an adapted coordinate system, which leads to cumbersome formulas. Equation 3.9.60 requires a parametrization by arc length; such a parametrization is only known as the inverse of a function which is itself an indefinite integral that can rarely be computed in closed form.

The Frenet formulas can be adapted to any parametrized curve, and give the following propositions.

Propositions 3.9.17 and 3.9.18: Many texts refer to $\kappa(t)$ and $\tau(t)$, not $\kappa(\gamma(t))$ and $\tau(\gamma(t))$. However, curvature and torsion are invariants of a curve: they depend only on the point at which they are evaluated, not on the parametrization used or on time (except in the sense that time can be used to specify a point on the curve). You might think of a curvy mountain road: each point on the road has its intrinsic curvature and torsion. If two drivers both start at time $t = 0$, one driving 30 mph (parametrization γ_1) and another driving 50 mph (parametrization γ_2), they will arrive at a point at different times, but will be confronted by the same curvature and torsion.

Proposition 3.9.17 (Curvature of a parametrized curve). *The curvature κ of a curve parametrized by $\gamma : \mathbb{R} \to \mathbb{R}^3$ is*

$$\kappa\big(\gamma(t)\big) = \frac{|\vec{\gamma}'(t) \times \vec{\gamma}''(t)|}{|\vec{\gamma}'(t)|^3}. \tag{3.9.61}$$

Proposition 3.9.18 (Torsion of a parametrized curve). *The torsion τ of a curve parametrized by $\gamma : \mathbb{R} \to \mathbb{R}^3$ is*

$$\tau\big(\gamma(t)\big) = \frac{(\vec{\gamma}'(t) \times \vec{\gamma}''(t)) \cdot \vec{\gamma}'''(t)}{|\vec{\gamma}'(t) \times \vec{\gamma}''(t)|^2}. \tag{3.9.62}$$

Example 3.9.19. Let $\gamma(t) = \begin{pmatrix} t \\ t^2 \\ t^3 \end{pmatrix}$. Then

$$\vec{\gamma}'(t) = \begin{pmatrix} 1 \\ 2t \\ 3t^2 \end{pmatrix}, \quad \vec{\gamma}''(t) = \begin{pmatrix} 0 \\ 2 \\ 6t \end{pmatrix}, \quad \vec{\gamma}'''(t) = \begin{pmatrix} 0 \\ 0 \\ 6 \end{pmatrix}. \tag{3.9.63}$$

So we find

$$\begin{aligned}\kappa\big(\gamma(t)\big) &= \frac{1}{(1+4t^2+9t^4)^{3/2}}\left|\begin{pmatrix}1\\2t\\3t^2\end{pmatrix}\times\begin{pmatrix}0\\2\\6t\end{pmatrix}\right|\\ &= 2\frac{(1+9t^2+9t^4)^{1/2}}{(1+4t^2+9t^4)^{3/2}}\end{aligned} \qquad 3.9.64$$

and

$$\tau\big(\gamma(t)\big) = \frac{1}{4(1+9t^2+9t^4)}\begin{pmatrix}6t^2\\-6t\\2\end{pmatrix}\cdot\begin{pmatrix}0\\0\\6\end{pmatrix} = \frac{3}{1+9t^2+9t^4}. \qquad 3.9.65$$

Since $\gamma(t) = \begin{pmatrix}t\\t^2\\t^3\end{pmatrix}$, we have $Y = X^2$ and $z = X^3$, so equation 3.9.56 says that $Y = X^2 = \frac{2}{2}X^2$, so $A_2 = 2\ldots.$

At the origin, the standard coordinates are adapted to the curve, so from equation 3.9.56 we find $A_2 = 2$, $B_3 = 6$, giving $\kappa = A_2 = 2$ and $\tau = B_3/A_2 = 3$. This agrees with equations 3.9.64 and 3.9.65 when $t = 0$. $\triangle$

Proof of Proposition 3.9.17 (Curvature of a parametrized curve). We will assume that we have a parametrized curve $\gamma : \mathbb{R} \to \mathbb{R}^3$; you should imagine that you are driving along some winding mountain road, and that $\gamma(t)$ is your position at time t. Since our computation will use equation 3.9.60, we will also use parametrization by arc length; we will denote by $\delta(s)$ the position of the car when the odometer is s, while γ denotes an arbitrary parametrization. These are related by the formula

$$\gamma(t) = \delta\big(s(t)\big), \quad \text{where} \quad s(t) = \int_{t_0}^{t} |\vec{\gamma}'(u)|\, du, \qquad 3.9.66$$

Equation 3.9.68, second line: Note that we are adding vectors to get a vector:

$$\underbrace{\kappa(s(t))(s'(t))^2}_{\text{number}}\,\underbrace{\vec{\mathbf{n}}(s(t))}_{\text{vector}} + \underbrace{s''(t)}_{\text{no.}}\,\underbrace{\vec{\mathbf{t}}(s(t))}_{\text{vec.}}$$

Equation 3.9.69: The derivative κ' is the derivative of κ with respect to arc length. We use equation 3.9.69 in the proof of Proposition 3.9.18, but state it here because it fits nicely.

and t_0 is the time when the odometer was set to 0. The function $t \mapsto s(t)$ gives you the odometer reading as a function of time. The unit vectors $\vec{\mathbf{t}}$, $\vec{\mathbf{n}}$ and $\vec{\mathbf{b}}$ will be considered as functions of s, as will the curvature κ and the torsion τ.[25]

We now use the chain rule to compute three successive derivatives of γ. In equation 3.9.67, recall (equation 3.9.57) that $\vec{\delta}' = \vec{\mathbf{t}}$; in the second line of equation 3.9.68, recall (equation 3.9.60) that $\vec{\mathbf{t}}'(0) = \kappa\vec{\mathbf{n}}(0)$. This gives

$$\vec{\gamma}'(t) = \big(\vec{\delta}'(s(t))s'(t) = s'(t)\vec{\mathbf{t}}\big(s(t)\big), \qquad 3.9.67$$

$$\begin{aligned}\vec{\gamma}''(t) &= \vec{\mathbf{t}}'\big(s(t)\big)\big(s'(t)\big)^2 + \vec{\mathbf{t}}\big(s(t)\big)s''(t)\\ &= \kappa\big(s(t)\big)\big(s'(t)\big)^2\vec{\mathbf{n}}\big(s(t)\big) + s''(t)\vec{\mathbf{t}}\big(s(t)\big),\end{aligned} \qquad 3.9.68$$

[25]Strictly speaking, $\vec{\mathbf{t}}$, $\vec{\mathbf{n}}$, $\vec{\mathbf{b}}$, κ, and τ should be considered as functions of $\delta(s(t))$: the point where the car is at a particular odometer reading, which itself depends on time. However, we hesitate to make the notation any more fearsome than it is.

$$\begin{aligned}\vec{\gamma}'''(t) &= \kappa'\big(s(t)\big)\vec{\mathbf{n}}\big(s(t)\big)\big(s'(t)\big)^3 + \kappa\big(s(t)\big)\vec{\mathbf{n}}'\big(s(t)\big)\big(s'(t)\big)^3\\ &\quad + 2\kappa\big(s(t)\big)\vec{\mathbf{n}}\big(s(t)\big)\big(s'(t)\big)\big(s''(t)\big) + \vec{\mathbf{t}}'\big(s(t)\big)\big(s'(t)\big)\big(s''(t)\big)\\ &\quad + \vec{\mathbf{t}}\big(s(t)\big)\big(s'''(t)\big)\\ &= \Big(-\big(\kappa\big(s(t)\big)\big)^2\big(s'(t)\big)^3 + s'''(t)\Big)\vec{\mathbf{t}}\big(s(t)\big) \qquad 3.9.69\\ &\quad + \Big(\kappa'(s(t)(s'(t))^3 + 3\kappa(s(t))(s'(t))(s''(t))\Big)\vec{\mathbf{n}}\big(s(t)\big)\\ &\quad + \Big(\kappa\big(s(t)\big)\tau\big(s(t)\big)\big(s'(t)\big)^3\Big)\vec{\mathbf{b}}\big(s(t)\big).\end{aligned}$$

Since $\vec{\mathbf{t}}$ has length 1, equation 3.9.67 gives us

$$s'(t) = |\vec{\gamma}'(t)|, \qquad 3.9.70$$

which we already knew from the definition of s. Equations 3.9.67 and 3.9.68 give

$$\vec{\gamma}'(t) \times \vec{\gamma}''(t) = \kappa\big(s(t)\big)\big(s'(t)\big)^3\vec{\mathbf{b}}\big(s(t)\big), \qquad 3.9.71$$

since $\vec{\mathbf{t}} \times \vec{\mathbf{n}} = \vec{\mathbf{b}}$. Since $\vec{\mathbf{b}}$ has length 1,

$$|\vec{\gamma}'(t) \times \vec{\gamma}''(t)| = \kappa\big(s(t)\big)\big(s'(t)\big)^3. \qquad 3.9.72$$

Using equation 3.9.70, this gives the formula for the curvature of Proposition 3.9.17. □

Equation 3.9.71: By equations 3.9.67 and 3.9.68,

$$\begin{aligned}&\vec{\gamma}'(t) \times \vec{\gamma}''(t)\\ &= s'(t)\vec{\mathbf{t}}(s(t))\\ &\times \Big[\kappa(s(t))(s'(t))^2\vec{\mathbf{n}}(s(t))\\ &\qquad + s''(t)\vec{\mathbf{t}}(s(t))\Big].\end{aligned}$$

Since for any vector $\vec{\mathbf{t}} \in \mathbb{R}^3$,

$$\vec{\mathbf{t}} \times \vec{\mathbf{t}} = \vec{\mathbf{0}},$$

since the cross product is distributive, and since (equation 3.9.57)

$$\vec{\mathbf{t}} \times \vec{\mathbf{n}} = \vec{\mathbf{b}}, \quad \text{this gives}$$

$$\begin{aligned}&\vec{\gamma}'(t) \times \vec{\gamma}''(t)\\ &= s'(t)\vec{\mathbf{t}}(s(t))\\ &\times \Big[\kappa(s(t))(s'(t))^2\vec{\mathbf{n}}(s(t))\Big]\\ &= \kappa(s(t))(s'(t))^3\vec{\mathbf{b}}(s(t)).\end{aligned}$$

Proof of Proposition 3.9.18 (Torsion of a parametrized curve). Since $\vec{\gamma}' \times \vec{\gamma}''$ points in the direction of $\vec{\mathbf{b}}$, dotting it with $\vec{\gamma}'''$ picks out the coefficient of $\vec{\mathbf{b}}$ for $\vec{\gamma}'''$. This leads to

$$\big(\vec{\gamma}'(t) \times \vec{\gamma}''(t)\big) \cdot \vec{\gamma}'''(t) = \tau\big(s(t)\big) \underbrace{\Big(\kappa\big(s(t)\big)\Big)^2\big(s'(t)\big)^6}_{\text{square of equation 3.9.72}}. \quad \square \qquad 3.9.73$$

Exercises for Section 3.9

3.9.1 a. What is the curvature of a circle of radius r?

b. What is the curvature of the ellipse of equation $\frac{x^2}{a^2} + \frac{y^2}{b^2} = 1$ at $\begin{pmatrix} u \\ v \end{pmatrix}$?

3.9.2 What is the curvature of the hyperbola of equation $\frac{x^2}{a^2} - \frac{y^2}{b^2} = 1$ at $\begin{pmatrix} u \\ v \end{pmatrix}$?

3.9.3 Show that the absolute value of the mean curvature of the unit sphere is 1 and that the Gaussian curvature of the unit sphere is 1.

3.9.4 a. What is the Gaussian curvature of the sphere of radius r?

b. What are the Gaussian and mean curvature of the ellipsoid of equation

$$\frac{x^2}{a^2} + \frac{y^2}{b^2} + \frac{z^2}{c^2} = 1, \quad \text{at a point } \begin{pmatrix} u \\ v \\ w \end{pmatrix}?$$

3.9.5 Check that if you consider the *surface* of equation $z = f(x)$, y arbitrary, and the plane *curve* $z = f(x)$, the absolute value of the mean curvature of the surface is half the curvature of the plane curve.

Useful fact for Exercise 3.9.7: The arctic circle is those points that are 2 607.5 kilometers south of the north pole. "That radius" means the radius as measured on the surface of the earth from the pole, i.e., 2607.5 kilometers.

3.9.6 What are the Gaussian and mean curvature of the surface of equation

$$\frac{x^2}{a^2} - \frac{y^2}{b^2} + \frac{z^2}{c^2} = 1, \quad \text{at} \begin{pmatrix} u \\ v \\ w \end{pmatrix}?$$

3.9.7 a. How long is the arctic circle? How long would a circle of that radius be if the earth were flat?

b. How big a circle around the pole would you need to measure in order for the difference of its length and the corresponding length in a plane to be one kilometer?

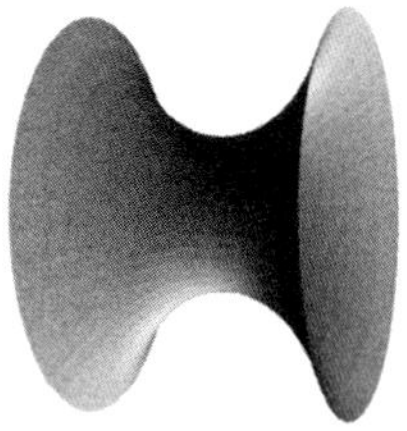

FIGURE FOR EXERCISE 3.9.10.
Part d: The catenoid of equation $y^2 + z^2 = (\cosh x)^2$.

3.9.8 a. Draw the cycloid given parametrically by $\begin{pmatrix} x \\ y \end{pmatrix} = \begin{pmatrix} a(t - \sin t) \\ a(1 - \cos t) \end{pmatrix}$.

b. Can you relate the name "cycloid" to "bicycle"?

c. Find the length of one arc of the cycloid.

3.9.9 Repeat Exercise 3.9.8 for the hypocycloid $\begin{pmatrix} x \\ y \end{pmatrix} = \begin{pmatrix} a\cos^3 t \\ a \sin^3 t \end{pmatrix}$.

3.9.10 a. Let $f : [a, b] \to \mathbb{R}$ be a smooth function satisfying $f(x) > 0$, and consider the surface obtained by rotating its graph around the x-axis. Show that the Gaussian curvature K and the mean curvature H of this surface depend only on the x-coordinate.

b. Show that $K(x) = \dfrac{-f''(x)}{f(x)\left(1 + \left(f'(x)\right)^2\right)^2}$.

c. Show that

$$H(x) = \frac{1}{(1 + (f'(x))^2)^{3/2}} \left(f''(x) - \frac{1 + (f'(x))^2}{f(x)} \right).$$

d. Show that the catenoid of equation $y^2 + z^2 = (\cosh x)^2$ (shown in the margin) has mean curvature 0.

Exercise 3.9.11: The curve

$$F : t \mapsto \left[\vec{\mathbf{t}}(t), \vec{\mathbf{n}}(t), \vec{\mathbf{b}}(t)\right] = T(t)$$

is a mapping $I \mapsto SO(3)$, the space of orthogonal 3×3 matrices with determinant $+1$. So

$$t \mapsto T^{-1}(t_0)T(t)$$

is a curve in $SO(3)$ that passes through the identity at t_0.

***3.9.11** Use Exercise 3.2.11 to explain why the Frenet formulas give an antisymmetric matrix.

3.9.12 Prove Proposition 3.9.6, using Proposition 3.9.16.

3.10 REVIEW EXERCISES FOR CHAPTER 3

3.1 a. Show that the set $X \subset \mathbb{R}^3$ of equation $x^3 + xy^2 + yz^2 + z^3 = 4$ is a smooth surface.

b. Give the equations of the tangent plane and tangent space to X at $\begin{pmatrix} 1 \\ 1 \\ 1 \end{pmatrix}$.

Exercise 3.2: We strongly advocate using MATLAB or similar software.

3.2 a. For what values of c is the set of equation $Y_c = x^2+y^3+z^4 = c$ a smooth surface?

b. Sketch this surface for a representative sample of values of c (for instance, the values $-2,-1,0,1,2$).

c. Give the equations of the tangent plane and tangent space at a point of the surface Y_c.

$$\begin{pmatrix} x_1 \\ y_1 \\ z_1 \\ x_2 \\ y_2 \\ z_2 \end{pmatrix} = \begin{pmatrix} 1 \\ 1 \\ 0 \\ -1 \\ 1 \\ 0 \end{pmatrix}$$

Point for Exercise 3.3, parts b and c.

3.3 Consider the space X of positions of a rod of length 2 in $\mathbb{R}^3$, where one endpoint is constrained to be on the sphere of equation $(x-1)^2+y^2+z^2=1$, and the other on the sphere of equation $(x+1)^2+y^2+z^2=1$.

a. Give equations for X as a subset of $\mathbb{R}^6$, where the coordinates in $\mathbb{R}^6$ are the coordinates $\begin{pmatrix} x_1 \\ y_1 \\ z_1 \end{pmatrix}$ of the end of the rod on the first sphere, and the coordinates $\begin{pmatrix} x_2 \\ y_2 \\ z_2 \end{pmatrix}$ of the other end of the rod.

b. Show that near the point in $\mathbb{R}^6$ shown in the margin, the set X is a manifold. What is the dimension of X near this point?

c. Give the equation of the tangent space to the set X, at the same point as in part b.

Exercise 3.4: The notation $\overline{\mathbf{p},\mathbf{q}}$ means the segment going from $\mathbf{p}$ to $\mathbf{q}$.

3.4 Consider the space X of triples

$$\mathbf{p} = \begin{pmatrix} x \\ 0 \\ 0 \end{pmatrix}, \quad \mathbf{q} = \begin{pmatrix} 0 \\ y \\ 0 \end{pmatrix}, \quad \mathbf{r} = \begin{pmatrix} 0 \\ 0 \\ z \end{pmatrix}$$

such that $y \neq 0$ and the segments $\overline{\mathbf{p},\mathbf{q}}$ and $\overline{\mathbf{q},\mathbf{r}}$ form an angle of $\pi/4$.

a. Write an equation $f\begin{pmatrix} x \\ y \\ z \end{pmatrix} = 0$ which all points of X will satisfy.

b. Show that X is a smooth surface.

c. True or false? Let $\mathbf{a} \stackrel{\text{def}}{=} \begin{pmatrix} 0 \\ 1 \\ 1 \end{pmatrix}$. Then $\mathbf{a} \in X$, and near $\mathbf{a}$ the surface X is locally the graph of a function expressing z as a furnction of x and y.

d. What is the tangent *plane* to X at $\mathbf{a}$? What is the tangent *space* $T_{\mathbf{a}}X$?

3.5 Find the Taylor polynomial of degree 3 of the function

$$f\begin{pmatrix} x \\ y \\ z \end{pmatrix} = \sin(x+y+z) \quad \text{at the point} \quad \begin{pmatrix} \pi/6 \\ \pi/4 \\ \pi/3 \end{pmatrix}.$$

3.6 Show that if $f\begin{pmatrix} x \\ y \end{pmatrix} = \varphi(x-y)$ for some twice continuously differentiable function $\varphi : \mathbb{R} \to \mathbb{R}$, then $D_1^2 f - D_2^2 f = 0$.

3.7 Write, to degree 3, the Taylor polynomial $P^3_{f,\mathbf{0}}$ of

$$f\begin{pmatrix} x \\ y \end{pmatrix} = \cos\Big(1+\sin(x^2+y)\Big) \quad \text{at the origin.}$$

***3.8** a. Let $M_1(m,n) \subset \text{Mat}\,(m,n)$ be the subset of matrices of rank 1. Show that the mapping $\varphi_1 : (\mathbb{R}^m - \{0\})\times \mathbb{R}^{n-1} \to \text{Mat}\,(m,n)$ given by

$$\varphi_1\left(\mathbf{a}, \begin{bmatrix} \lambda_2 \\ \vdots \\ \lambda_n \end{bmatrix}\right) \mapsto [\mathbf{a}, \lambda_2\mathbf{a}, \dots, \lambda_n\mathbf{a}]$$

is a parametrization of the open subset $U_1 \subset M_1(m,n)$ of those matrices whose first column is not $\mathbf{0}$.

b. Show that $M_1(m,n) - U_1$ is a manifold embedded in $M_1(m,n)$. What is its dimension?

c. How many parametrizations like φ_1 do you need to cover every point of $M_1(m,n)$?

***3.9** A homogeneous polynomial in two variables of degree four is an expression of the form $p(x,y) = ax^4 + bx^3y + cx^2y^2 + dxy^3 + ey^4$. Consider the function

A *homogeneous polynomial* is a polynomial in which all terms have the same degree.

$$f\begin{pmatrix} x \\ y \end{pmatrix} = \begin{cases} \dfrac{p(x,y)}{x^2+y^2} & \text{if } \begin{pmatrix} x \\ y \end{pmatrix} \neq \begin{pmatrix} 0 \\ 0 \end{pmatrix} \\ 0 & \text{if } \begin{pmatrix} x \\ y \end{pmatrix} = \begin{pmatrix} 0 \\ 0 \end{pmatrix}, \end{cases}$$

where p is a homogeneous polynomial of degree 4. What condition must the coefficients of p satisfy in order for the crossed partials $D_1(D_2(f))$ and $D_2(D_1(f))$ to be equal at the origin?

3.10 a. Show that $ye^y = x$ implicitly defines y as a function of x, for $x \geq 0$.

b. Find a Taylor polynomial of the implicit function to degree 4.

Exercise 3.11 is relevant to Example 3.9.13. Hint for part b: The x-axis is contained in the surface.

3.11 a. Show that the equation $y\cos z = x\sin z$ expresses z implicitly as a function $z = g_r\begin{pmatrix} x \\ y \end{pmatrix}$ near the point $\begin{pmatrix} r \\ 0 \\ 0 \end{pmatrix}$.

b. Show that $D_1 g_r\begin{pmatrix} r \\ 0 \end{pmatrix} = D_1^2 g_r\begin{pmatrix} r \\ 0 \end{pmatrix} = 0$.

$$Q_1\begin{pmatrix} x \\ y \\ z \end{pmatrix} = \det\begin{bmatrix} 1 & x & y \\ 1 & y & z \\ 1 & z & x \end{bmatrix}$$

$$Q_2\begin{pmatrix} x \\ y \\ z \end{pmatrix} = \det\begin{bmatrix} 0 & x & y \\ x & 0 & z \\ y & z & 0 \end{bmatrix}$$

Functions for Exercise 3.13

3.12 On $\mathbb{R}^4$ as described by $M = \begin{bmatrix} a & c \\ b & d \end{bmatrix}$, consider the quadratic form $Q(M) = \det M$. What is its signature?

3.13 a. Are the functions Q_1 and Q_2 in the margin quadratic forms on $\mathbb{R}^3$?

b. For any that is a quadratic form, what is its signature? Is it degenerate or nondegenerate?

3.14 Let P_k be the space of polynomials p of degree at most k.

a. Show that the function $\delta_a : P_k \to \mathbb{R}$ given by $\delta_a(p) = p(a)$ is a linear function.

b. Show that $\delta_0, \dots, \delta_k$ are linearly independent. First say what it means, being careful with the quantifiers. It may help to think of the polynomial

Exercise 3.14 part c: There is the clever way, and then there is the plodding way.

$$x(x-1)\cdots\Big(x-(j-1)\Big)\Big(x-(j+1)\Big)\cdots(x-k),$$

which vanishes at $0, 1, \dots, j-1, j+1, \dots, k$ but not at j.

c. Show that the function

$$Q(p) = \Big(p(0)\Big)^2 - \Big(p(1)\Big)^2 + \dots + (-1)^k\Big(p(k)\Big)^2$$

is a quadratic form on P_k. When $k = 3$, write it in terms of the coefficients of $p(x) = ax^3 + bx^2 + cx + d$.

d. What is the signature of Q when $k = 3$?

3.15 Show that a 2×2 symmetric matrix $G = \begin{bmatrix} a & b \\ b & d \end{bmatrix}$ represents a positive definite quadratic form if and only if $\det G > 0$, $a + d > 0$.

3.16 Let Q be a quadratic form. Construct a symmetric matrix A as follows: each entry $A_{i,i}$ on the diagonal is the coefficient of x_i^2, while each entry $A_{i,j}$ is half the coefficient of the term $x_i x_j$.

Exercise 3.16, part b: Consider $Q(\vec{\mathbf{e}}_i)$ and $Q(\vec{\mathbf{e}}_i + \vec{\mathbf{e}}_j)$.

a. Show that $Q(\vec{\mathbf{x}}) = \vec{\mathbf{x}} \cdot A\vec{\mathbf{x}}$.

b. Show that A is the unique symmetric matrix with this property.

3.17 a. Find the critical points of the function $f\begin{pmatrix} x \\ y \end{pmatrix} = 3x^2 - 6xy + 2y^3$.

b. What kind of critical points are they?

Exercise 3.18, part a: This is easier if you use

$$\sin(\alpha + \beta) = \sin\alpha\cos\beta + \cos\alpha\sin\beta.$$

3.18 a. What is the Taylor polynomial of degree 2 of the function

$$f\begin{pmatrix} x \\ y \end{pmatrix} = \sin(2x + y) \quad \text{at the point } \begin{pmatrix} \pi/6 \\ \pi/3 \end{pmatrix}?$$

b. Show that $f\begin{pmatrix} x \\ y \end{pmatrix} + \frac{1}{2}\left(2x + y - \frac{2\pi}{3}\right) - \left(x - \frac{\pi}{6}\right)^2$ has a critical point at $\begin{pmatrix} \pi/6 \\ \pi/3 \end{pmatrix}$. What kind of critical point is it?

$$F\begin{pmatrix} x \\ y \\ z \end{pmatrix} = \det \begin{bmatrix} 1 & x & y \\ x & 1 & z \\ y & z & 1 \end{bmatrix}$$

Function of Exercise 3.19

3.19 The function in the margin has exactly five critical points.

a. Find them.

b. For each critical point, what are the quadratic terms of the Taylor polynomial at that point?

c. Say everything you can about the type of critical point each is.

3.20 a. Find the critical points of xyz, if x, y, z belong to the surface S of equation $x + y + z^2 = 16$.

b. Is there a maximum on the whole surface; if so, which critical point is it?

c. Is there a maximum on the part of S where x, y, z are all positive?

3.21 Let A, B, C, D be a convex quadrilateral in the plane, with the vertices free to move but with a the length of AB, b the length of BC, c the length of CD, and d the length of DA, all assigned. Let φ be the angle at A and ψ the angle at C.

Exercise 3.21, part c: You may use the fact that a quadrilateral can be inscribed in a circle if the opposite angles add to π.

a. Show that the angles φ and ψ satisfy the constraint

$$a^2 + d^2 - 2ad\cos\varphi \;=\; b^2 + c^2 - 2bc\cos\psi.$$

b. Find a formula for the area of the quadrilateral in terms of φ, ψ, and a, b, c, d.

c. Show that the area is maximum if the quadrilateral can be inscribed in a circle.

3.22 Let A be an $n \times n$ matrix. What is the Taylor polynomial of degree 2 of $X \mapsto X^3$ at A?

3.23 Compute the Gaussian and mean curvature of the surface of equation $z = \sqrt{x^2+y^2}$ at $\begin{pmatrix} x \\ y \\ z \end{pmatrix} = \begin{pmatrix} a \\ b \\ \sqrt{a^2+b^2} \end{pmatrix}$. Explain your result.

3.24 Suppose $\gamma(t) = \begin{pmatrix} \gamma_1(t) \\ \vdots \\ \gamma_n(t) \end{pmatrix}$ is twice continuously differentiable on a neighborhood of $[a,b]$.

a. Use Taylor's theorem with remainder (or argue directly from the mean value theorem) to show that for any $s_1 < s_2$ in $[a,b]$,

$$|\gamma(s_2) - \gamma(s_1) - \gamma'(s_1)(s_2 - s_1)| \le C|s_2 - s_1|^2, \quad \text{where}$$

$$C = \sqrt{n} \sup_{j=1,\dots,n} \sup_{t\in[a,b]} |\gamma_j''(t)|.$$

b. Use this to show that $\lim \sum_{i=0}^{m-1} |\gamma(t_{i+1} - \gamma(t_i)| = \int_a^b |\gamma'(t)|\, dt$, where $a = t_0 < t_1 \cdots < t_m = b$, and we take the limit as the distances $t_{i+1} - t_i$ tend to 0.

3.25 Analyze the critical points found in Example 3.7.7, this time using the augmented Hessian matrix.

3.26 Use the chain rule for Taylor polynomials (Proposition 3.4.4) to derive Leibniz's rule.

FIGURE FOR EXERCISE 3.27. Scherk's surface, of equation

$$e^z \cos y = \cos x.$$

We thank Francisco Martin for permission to use this picture, and the catenoid picture for Exercise 3.9.10.

3.27 Show that Scherk's surface, of equation $e^z \cos y = \cos x$, is a minimal surface. *Hint*: Write the equation as $z = \ln \cos x - \ln \cos y$.

3.28 Show that the quadratic form $ax^2 + 2bxy + cy^2$

a. is positive definite if and only if $ac - b^2 > 0$ and $a > 0$

b. is negative definite if and only if $ac - b^2 > 0$ and $a < 0$

c. has signature (1,1) if and only if $ac - b^2 < 0$.

3.29 Let $Q_{i,j}(\theta)$ be the matrix of rotation by θ in the (i,j)-plane; note that $(Q_{i,j}(\theta))^{-1} = (Q_{i,j}(\theta))^\top = Q_{i,j}(-\theta)$. Let A be a real $n \times n$ symmetric matrix with $a_{i,j} \ne 0$ for some $i < j$.

a. Find a formula for an angle θ such that if $B = Q_{i,j}(-\theta)AQ_{i,j}(\theta)$, then A' is still symmetric, and $b_{i,j} = 0$.

b. Show that

$$\sum_{1\le k<l\le n} |b_{k,l}|^2 = \sum_{1\le k<l\le n} |a_{k,l}|^2 - |a_{i,j}|^2.$$

c. Use this formula to give a different proof of the spectral theorem; you may need to use the fact that the orthogonal group is compact.

3.30 Show that if $M \subset \mathbb{R}^m$ is a manifold of dimension k, and $P \subset \mathbb{R}^p$ is a manifold of dimension l, then $M \times P \subset \mathbb{R}^m \times \mathbb{R}^p$ is a manifold of dimension $k+l$.

3.31 Find the maximum of the function $x_1x_2\dots x_n$, subject to the constraint

$$x_1^2 + 2x_2^2 + \cdots + nx_n^2 = 1.$$

4

Integration

When you can measure what you are speaking about, and express it in numbers, you know something about it; but when you cannot measure it, when you cannot express it in numbers, your knowledge is of a meager and unsatisfactory kind: it may be the beginning of knowledge, but you have scarcely, in your thoughts, advanced to the stage of science.—William Thomson, Lord Kelvin

4.0 Introduction

FIGURE 4.0.1.
Lord Kelvin (William Thomson, 1824–1907)

An actuary deciding what premium to charge for a life insurance policy needs integrals. So does a banker deciding what to charge for stock options. Black and Scholes received a Nobel Prize for this work, which involves a very fancy stochastic integral.

Chapters 1 and 2 began with algebra, then moved on to calculus. Here, as in Chapter 3, we dive right into calculus. We introduce the relevant linear algebra (determinants) later in the chapter, where we need it.

When students first meet integrals, integrals come in two very different flavors – Riemann sums (the idea) and antiderivatives (the recipe) – rather as derivatives arise as limits, and as something to be computed using Leibnitz's rule, the chain rule, etc.

Since integrals can be systematically computed (by hand) only as antiderivatives, students often take this to be the definition. This is misleading: the definition of an integral is given by a Riemann sum (or by "area under the graph"; Riemann sums are just a way of making the notion of "area" precise). Section 4.1 is devoted to generalizing Riemann sums to functions of several variables. Rather than slice up the domain of a function $f : \mathbb{R} \to \mathbb{R}$ into little intervals and computing the "area under the graph" corresponding to each interval, we will slice up the "n-dimensional domain" of a function $f : \mathbb{R}^n \to \mathbb{R}$ into little n-dimensional cubes.

Computing n-dimensional volume is an important application of multiple integrals. Another is probability theory; probability has become such an important part of integration that integration has almost become part of probability. Even such a mundane problem as quantifying how heavy a child is for his or her height requires multiple integrals. Fancier yet are the uses of probability that arise when physicists study turbulent flows, or engineers try to improve the internal combustion engine. They cannot hope to deal with one molecule at a time; any picture they get of reality at a macroscopic level is necessarily based on a probabilistic picture of what is going on at a microscopic level. We give a brief introduction to this important field in Section 4.2.

Section 4.3 discusses what functions are integrable; in Section 4.4, we use the notion of *measure* to give a sharper criterion for integrability (a criterion that applies to more functions than the criteria of Section 4.3).

FIGURE 4.0.2.

Born in 1826, Bernhard Riemann died of tuberculosis in 1866. The Riemann integral, which he defined when working on trigonometric series, is far from his greatest accomplishment. He made contributions in many fields, including differential geometry and complex analysis. The *Riemann hypothesis* is probably the most famous unsolved problem in mathematics.

In Section 4.5 we discuss Fubini's theorem, which reduces computing the integral of a function of n variables to computing n ordinary integrals. This is an important theoretical tool. Moreover, whenever an integral can be computed in elementary terms, Fubini's theorem is the key tool. Unfortunately, it is usually impossible to compute antiderivatives in elementary terms even for functions of one variable, and this tends to be truer yet of functions of several variables.

In practice, multiple integrals are most often computed using numerical methods, which we discuss in Section 4.6. We will see that although the theory is much the same in $\mathbb{R}^2$ or $\mathbb{R}^{10^{24}}$, the computational issues are quite different. We will encounter some entertaining uses of Newton's method when looking for optimal points at which to evaluate a function, and some fairly deep probability in understanding why Monte Carlo methods work in higher dimensions.

Defining volume using dyadic pavings, as we do in Section 4.1, makes most theorems easiest to prove, but such pavings are rigid; often we will want to have more "paving stones" where the function varies rapidly, and bigger ones elsewhere. Flexibility in choosing pavings is also important for the proof of the *change of variables formula*. Section 4.7 discusses more general pavings.

In Section 4.8 we return to linear algebra to discuss higher-dimensional determinants. In Section 4.9 we show that in all dimensions the determinant measures volume; we use this fact in Section 4.10, where we discuss the change of variables formula.

Many interesting integrals, such as those in Laplace and Fourier transforms, are not integrals of bounded functions over bounded domains. These use a different approach to integration, *Lebesgue integration*, discussed in Section 4.11. Lebesgue integrals cannot be defined as Riemann sums, and require understanding the behavior of integrals under limits. The dominated convergence theorem is the key tool for this.

4.1 Defining the integral

Integration is a summation procedure; it answers the question, how much is there in all? In one dimension, $\rho(x)$ might be the density at point x of a bar parametrized by $[a, b]$; in that case

The Greek letter ρ ("rho") is pronounced "row".

$$\int_a^b \rho(x)\,dx \qquad 4.1.1$$

is the total mass of the bar.

If instead we have a rectangular plate parametrized by $a \le x \le b$ and $c \le y \le d$, and with density $\rho\begin{pmatrix} x \\ y \end{pmatrix}$, then the total mass will be given by the *double integral*

$$\int_{[a,b]\times[c,d]} \rho\begin{pmatrix} x \\ y \end{pmatrix}\,dx\,dy, \qquad 4.1.2$$

where the domain of the entire double integral is $[a,b] \times [c,d]$ (the plate). We will define such multiple integrals in this chapter. But you should always remember that the preceding example is too simple. We might want to understand total rainfall in Britain, whose coastline is a very complicated boundary. (A celebrated article analyzes that coastline as a fractal, with infinite length.) Or we might want to understand the total potential energy stored in the surface tension of a foam; physics tells us that a foam assumes the shape that minimizes this energy.

Thus we want to define integration for rather bizarre domains and functions. Our approach will not work for truly bizarre functions, such as the function that equals 1 at all rational numbers and 0 at all irrational numbers; for that one needs Lebesgue integration (see Section 4.11). But we still have to specify carefully what domains and functions we want to allow.

Our task will be somewhat easier if we keep the domain of integration simple, putting all the complication into the function to be integrated. If we wanted to sum rainfall over Britain, we would use $\mathbb{R}^2$, *not* Britain (with its fractal coastline!) as the domain of integration; we would then define our function to be rainfall over Britain, and 0 elsewhere.

Thus, for a function $f : \mathbb{R}^n \to \mathbb{R}$, we will define the multiple integral

$$\int_{\mathbb{R}^n} f(\mathbf{x}) \, |d^n\mathbf{x}|, \qquad 4.1.3$$

with $\mathbb{R}^n$ the domain of integration.

We emphatically do *not* want to assume that f is continuous, because most often it is not: if, for example, f is defined to be total rainfall for October over Britain, and 0 elsewhere, it will be discontinuous over most of the border of Britain, as shown in Figure 4.1.1. What we actually have is a function g (e.g., rainfall) defined on some subset of $\mathbb{R}^n$ larger than Britain. We then consider that function only over Britain, by setting

$$f(\mathbf{x}) = \begin{cases} g(\mathbf{x}) & \text{if } \mathbf{x} \in \text{Britain} \\ 0 & \text{otherwise.} \end{cases} \qquad 4.1.4$$

We can express this another way, using the *indicator function* **1**.

Definition 4.1.1 (Indicator function). Let $A \subset \mathbb{R}^n$ be a bounded subset. The *indicator function* $\mathbf{1}_A$ is

$$\mathbf{1}_A(\mathbf{x}) \stackrel{\text{def}}{=} \begin{cases} 1 & \text{if } \mathbf{x} \in A \\ 0 & \text{if } \mathbf{x} \notin A. \end{cases} \qquad 4.1.5$$

Equation 4.1.4 can then be rewritten

$$f(\mathbf{x}) = g(\mathbf{x})\mathbf{1}_{\text{Britain}}(\mathbf{x}). \qquad 4.1.6$$

This doesn't get rid of difficulties like the coastline of Britain – indeed, such a function f will usually have discontinuities on the coastline – but putting all the difficulties on the side of the function will make our definitions easier (or at least shorter).

We will see in Section 4.5 that the double integral of formula 4.1.2 can be written

$$\int_c^d \left(\int_a^b \rho \begin{pmatrix} x \\ y \end{pmatrix} dx \right) dy.$$

We are not presupposing this equivalence in this section. One difference worth noting is that $\int_a^b$ specifies a direction: from a to b. (You will recall that direction makes a difference: $\int_a^b = -\int_b^a$.) The integral of formula 4.1.2 specifies a domain but says nothing about direction.

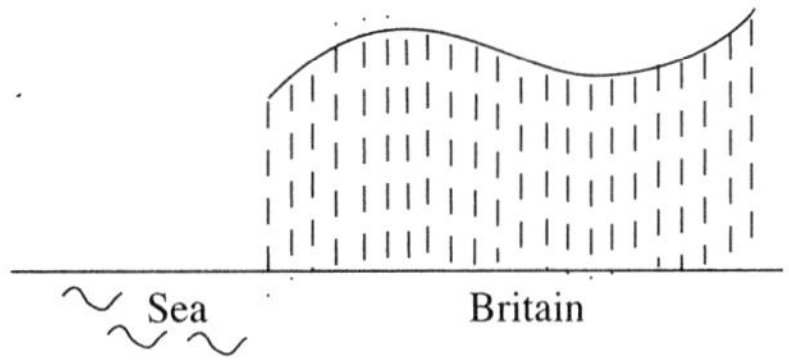

FIGURE 4.1.1.
The function that is rainfall over Britain and 0 elsewhere is discontinuous at the coast.

The indicator function is also known as the *characteristic function*.

So while we really want to integrate g (i.e., rainfall) over Britain, we define that integral in terms of the integral of f over $\mathbb{R}^n$, setting

$$\int_{\text{Britain}} g(\mathbf{x})\,|d^n\mathbf{x}| = \int_{\mathbb{R}^n} g(\mathbf{x})\mathbf{1}_{\text{Britain}}(\mathbf{x})\,|d^n\mathbf{x}| = \int_{\mathbb{R}^n} f(\mathbf{x})\,|d^n\mathbf{x}|. \qquad 4.1.7$$

More generally, when integrating over a subset $A \subset \mathbb{R}^n$,

$$\int_A g(\mathbf{x})\,|d^n\mathbf{x}| = \int_{\mathbb{R}^n} g(\mathbf{x})\mathbf{1}_A(\mathbf{x})\,|d^n\mathbf{x}|. \qquad 4.1.8$$

We tried several notations before choosing $|d^n\mathbf{x}|$. First we used $dx_1 \dots dx_n$. That seemed clumsy, so we switched to dV. But it failed to distinguish between $|d^2\mathbf{x}|$ and $|d^3\mathbf{x}|$, and when changing variables we had to tack on subscripts to keep the variables straight.

But dV had the advantage of suggesting, correctly, that we are not concerned with direction (unlike integration in first year calculus, where $\int_a^b dx \neq \int_b^a dx$). We hesitated at first to convey the same message with absolute value signs, for fear the notation would seem forbidding, but decided that the distinction between oriented and unoriented domains is so important (it is a central theme of Chapter 6) that our notation should reflect that distinction.

The notation Supp (support) should not be confused with sup (supremum). Recall that "supremum" and "least upper bound" are synonymous, as are "infimum" and "greatest lower bound" (Definitions 1.6.5 and 1.6.7).

Some preliminary definitions and notation

Definition 4.1.2 (Support of a function: $\text{Supp}(f)$). The *support* $\text{Supp}(f)$ of a function $f : \mathbb{R}^n \to \mathbb{R}$ is the closure of the set

$$\{\, \mathbf{x} \in \mathbb{R}^n \mid f(\mathbf{x}) \neq 0 \,\}. \qquad 4.1.9$$

Definition 4.1.3 ($M_A(f)$ and $m_A(f)$). If $A \subset \mathbb{R}^n$ is an arbitrary subset, we will denote by $M_A(f)$ the supremum of $f(\mathbf{x})$ for $\mathbf{x} \in A$, and by $m_A(f)$ the infimum of $f(\mathbf{x})$ for $\mathbf{x} \in A$:

$$M_A(f) \stackrel{\text{def}}{=} \sup_{\mathbf{x}\in A} f(\mathbf{x}) \quad \text{and} \quad m_A(f) \stackrel{\text{def}}{=} \inf_{\mathbf{x}\in A} f(\mathbf{x}) \qquad 4.1.10$$

Definition 4.1.4 (Oscillation). The *oscillation* of f over A, denoted $\text{osc}_A(f)$, is the difference between its supremum and its infimum:

$$\text{osc}_A(f) \stackrel{\text{def}}{=} M_A(f) - m_A(f). \qquad 4.1.11$$

In Section 4.11 we will discuss *Lebesgue integration*, which will allow us to define integrals of functions that are not bounded or do not have bounded support, or are locally extremely irregular.

You may have seen *improper integrals* of unbounded functions over unbounded domains. But this only works in dimension 1: improper integrals don't make sense in higher dimensions. In any case, improper integrals that are not absolutely convergent are very tricky, and those that are absolutely convergent also exist as Lebesgue integrals; see Exercise 4.11.19.

Definition of the Riemann integral: dyadic pavings

In Sections 4.1–4.10 we discuss integrals of functions $f : \mathbb{R}^n \to \mathbb{R}$ satisfying

1. $|f|$ is bounded, and
2. f has bounded support (i.e., there exists R such that $f(\mathbf{x}) = 0$ when $|\mathbf{x}| > R$).

With these restrictions on f, and for any subset $A \subset \mathbb{R}^n$, each quantity $M_A(f)$, $m_A(f)$, and $\text{osc}_A(f)$ is a well-defined finite number. This is not true for a function like $f(x) = 1/x$, defined on the open interval $(0, 1)$. In that case $|f|$ is not bounded, and $\sup f(x) = \infty$.

There is quite a bit of choice as to how to define the integral; we will first use the most restrictive definition: *dyadic pavings* of $\mathbb{R}^n$.

To compute an integral in one dimension, we decompose the domain into little intervals, and construct on each the tallest rectangle fitting under the graph and the shortest rectangle containing it, as shown in Figure 4.1.2. If, as we make the rectangles skinnier and skinnier, the sum of the area of the upper rectangles approaches that of the lower rectangles, the function is integrable. We can then approximate the integral by adding areas of rectangles – either the lower rectangles, the upper rectangles, or rectangles constructed some other way. The choice of the point at which to measure the height doesn't matter since the areas of the lower rectangles and the upper rectangles can be made arbitrarily close.

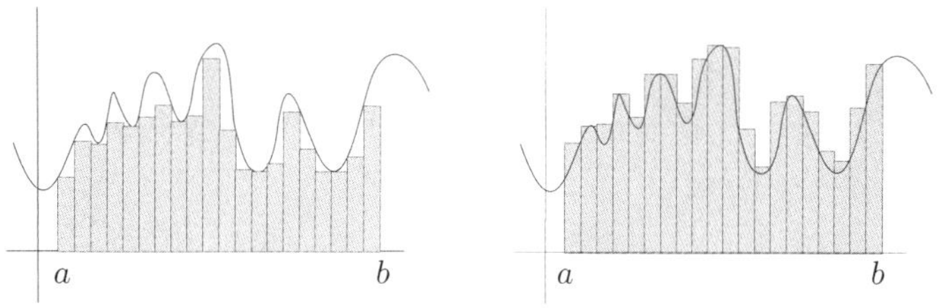

FIGURE 4.1.2. LEFT: Lower Riemann sum for $\int_a^b f(x)\,dx$. RIGHT: Upper Riemann sum. If the two sums converge to a common limit, that limit is the integral of the function.

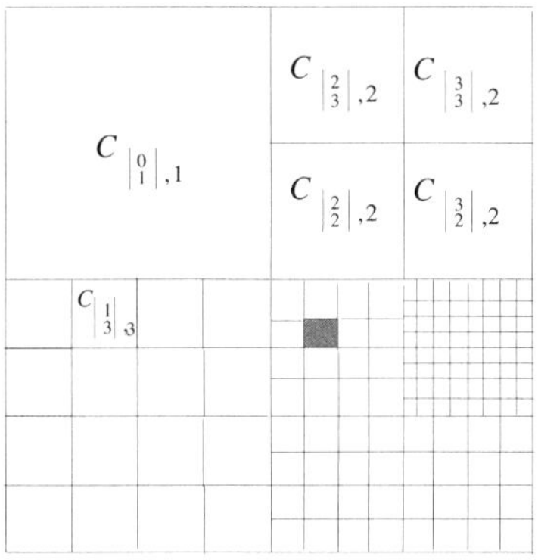

FIGURE 4.1.3.
A dyadic decomposition in $\mathbb{R}^2$. The entire figure is a "cube" in $\mathbb{R}^2$ at level $N = 0$, with sidelength $1/2^0 = 1$. At level 1 (upper left quadrant), cubes have sidelength $1/2^1 = 1/2$; at level 2 (upper right quadrant), they have sidelength $1/2^2 = 1/4$; and so on.
As explained in Example 4.1.6, the small shaded cube in the lower right quadrant is $C_{\begin{pmatrix}9\\6\end{pmatrix},4}$ – the cube where $\mathbf{k} = \begin{pmatrix}9\\6\end{pmatrix}$ and $N = 4$.

In one dimension, dyadic upper and lower sums corresponds to restricting ourselves to decomposing the the domain first at the integers, then the half-integers, then the quarter-integers, etc.

To use dyadic pavings in $\mathbb{R}^n$ we do essentially the same thing. We begin by cutting up $\mathbb{R}^n$ into cubes with sides 1 long, like the big square of Figure 4.1.3. (By "cube" we mean an interval in $\mathbb{R}$, a square in $\mathbb{R}^2$, a cube in $\mathbb{R}^3$, and analogues of cubes in higher dimensions.) Next we cut each side of a cube in half, cutting an interval in half, a square into four equal squares, a cube into eight equal cubes, ... At each step we cut each side in half.

To define dyadic pavings in $\mathbb{R}^n$ precisely, we must first say what we mean by an n-dimensional cube.

Definition 4.1.5 (Dyadic cube). A *dyadic cube* $C_{\mathbf{k},N} \subset \mathbb{R}^n$ is given by

$$C_{\mathbf{k},N} \overset{\text{def}}{=} \left\{ \mathbf{x} \in \mathbb{R}^n \;\middle|\; \frac{k_i}{2^N} \le x_i < \frac{k_i+1}{2^N} \text{ for } 1 \le i \le n \right\}, \qquad 4.1.12$$

where $\mathbf{k} \overset{\text{def}}{=} \begin{pmatrix} k_1 \\ \vdots \\ k_n \end{pmatrix}$ and $k_1, \dots, k_n$ are integers.

In equation 4.1.12, we chose the inequalities $\le$ to the left of x_i and $<$ to the right, so that at every level, every point of $\mathbb{R}^n$ is in exactly one cube. We could just as easily put them in the opposite order; allowing the edges to overlap wouldn't be a problem either.

Each cube C has two indices. The first index, $\mathbf{k}$, locates each cube: it gives the numerators of the coordinates of the cube's lower left corner, when the denominator is 2^N. The second index, N, tells which "level" we are considering, starting with 0; you may think of N as the "fineness" of the cube. The length of a side of a cube is $1/2^N$, so when $N = 0$, each side has length 1; when $N = 1$, each side has length 1/2. The bigger N is, the finer the decomposition and the smaller the cubes.

Example 4.1.6 (Dyadic cubes). The small shaded cube in the lower right quadrant of Figure 4.1.3 is

We call our pavings *dyadic* because each time we divide by a factor of 2; "dyadic" comes from the Greek *dyas*, meaning *two*. We could use decimal pavings instead, cutting each side into ten parts each time, but dyadic pavings are easier to draw. In Section 4.7 we will see that much more general pavings can be used.

$$C_{\begin{pmatrix}9\\6\end{pmatrix},4} = \left\{ \mathbf{x} \in \mathbb{R}^2 \;\middle|\; \underbrace{\frac{9}{16} \le x < \frac{10}{16}}_{\text{width of cube}};\; \underbrace{\frac{6}{16} \le y < \frac{7}{16}}_{\text{height of cube}} \right\}. \qquad 4.1.13$$

If the cube is three dimensional, $\mathbf{k}$ has three entries, and each cube $C_{\mathbf{k},N}$ consists of the $\mathbf{x} = \begin{pmatrix}x\\y\\z\end{pmatrix} \in \mathbb{R}^3$ such that

$$\underbrace{\frac{k_1}{2^N} \le x < \frac{k_1+1}{2^N}}_{\text{width of cube}};\; \underbrace{\frac{k_2}{2^N} \le y < \frac{k_2+1}{2^N}}_{\text{length of cube}};\; \underbrace{\frac{k_3}{2^N} \le z < \frac{k_3+1}{2^N}}_{\text{height of cube}}. \quad \triangle \qquad 4.1.14$$

The collection of all these cubes *paves* $\mathbb{R}^n$:

Definition 4.1.7 (Dyadic paving). The collection of cubes $C_{\mathbf{k},N}$ at a single level N, denoted $\mathcal{D}_N(\mathbb{R}^n)$, is the Nth *dyadic paving* of $\mathbb{R}^n$.

We use vol_n to denote n-dimensional volume.

The n-dimensional volume of a cube C is the product of the lengths of its sides. Since the length of one side is $1/2^N$, the n-dimensional volume is

$$\text{vol}_n\, C = \left(\frac{1}{2^N}\right)^n = \frac{1}{2^{Nn}}. \qquad 4.1.15$$

Note that all $C \in \mathcal{D}_N$ (all cubes at a given resolution N) have the same n-dimensional volume.

You are asked to prove inequality 4.1.16 in Exercise 4.1.8.

The distance between two points $\mathbf{x}, \mathbf{y}$ in a cube $C \in \mathcal{D}_N$ is

$$|\mathbf{x} - \mathbf{y}| \le \frac{\sqrt{n}}{2^N}. \qquad 4.1.16$$

Thus two points in the same cube C are close if N is large.

Upper and lower sums using dyadic pavings

Definition 4.1.8: As in Definition 4.1.3, $M_C(f)$ denotes the least upper bound, and $m_C(f)$ denotes the greatest lower bound.

Since we are assuming that f has bounded support, these sums have only finitely many terms. Each term is finite, since f itself is bounded. We could have allowed our functions to have unbounded support, but required the sums in Definition 4.1.8 to be absolutely convergent series; this complicates the theory.

In one dimension, we say a function is integrable if the sum of areas of upper rectangles and the sum of areas of lower rectangles approach a common limit as the decomposition becomes finer and finer. The common limit is the integral.

We do the same thing here. We define the Nth upper and lower sums:

Definition 4.1.8 (Nth upper sum, Nth lower sum). Let $f : \mathbb{R}^n \to \mathbb{R}$ be bounded with bounded support. The Nth *upper* and *lower sums* of a function f are

$$\underbrace{U_N(f)}_{\substack{N\text{th}\\ \text{upper sum}}} \stackrel{\text{def}}{=} \sum_{C \in \mathcal{D}_N} M_C(f)\, \text{vol}_n\, C, \qquad \underbrace{L_N(f)}_{\substack{N\text{th}\\ \text{lower sum}}} \stackrel{\text{def}}{=} \sum_{C \in \mathcal{D}_N} m_C(f)\, \text{vol}_n\, C.$$

For the Nth upper sum we compute, for each cube C at level N, the product of the least upper bound of the function over the cube and the volume of the cube, and we add the products together. We do the same for the lower sum, using the greatest lower bound. Since for these pavings all the cubes have the same volume $1/2^{nN}$, it can be factored out:

$$U_N(f) = \frac{1}{2^{nN}} \sum_{C \in \mathcal{D}_N} M_C(f), \quad L_N(f) = \frac{1}{2^{nN}} \sum_{C \in \mathcal{D}_N} m_C(f). \qquad 4.1.17$$

Proposition 4.1.9: Think of a two-dimensional function, whose graph is a surface with mountains and valleys. At a coarse level, with big cubes (i.e., squares), a square containing a mountain peak and a valley contributes a lot to the upper sum.

As N increases, the peak is the least upper bound for a much smaller square; other parts of the original big square will have a smaller least upper bound.

The same argument holds, in reverse, for the lower sum; if a large square contains a deep canyon, the entire square will have a low greatest lower bound; as N increases and the squares get smaller, the canyon will have less of an impact, and the lower sum will increase. We invite you to turn this argument into a formal proof.

Proposition 4.1.9. *As N increases, the sequence $N \mapsto U_N(f)$ decreases, and the sequence $N \mapsto L_N(f)$ increases.*

(Strictly speaking, in Proposition 4.1.9, the sequence $N \mapsto U_N(f)$ is nonincreasing, and the sequence $N \mapsto L_N(f)$ is nondecreasing: they may stay constant for some or all steps.)

We are now ready to define the multiple integral.

Definition 4.1.10 (Upper and lower integrals). We call

$$U(f) \stackrel{\text{def}}{=} \lim_{N\to\infty} U_N(f) \quad \text{and} \quad L(f) \stackrel{\text{def}}{=} \lim_{N\to\infty} L_N(f) \qquad 4.1.18$$

the *upper and lower integrals* of f.

We leave the proof of Proposition 4.1.11 to the reader.

Proposition 4.1.11. *If f, g are bounded functions with bounded support and $f \le g$, then*

$$U(f) \le U(g) \quad \textit{and} \quad L(f) \le L(g). \qquad 4.1.19$$

By Definition 4.1.12, an integrable function is real valued, but it is possible to define complex-valued integrable functions, integrating the real and imaginary parts separately, and more generally, vector-valued integrable functions by integrating the coordinate functions separately.

An *integrand* is what comes after an integral sign: in equation 4.1.20, the integrand is $f|d^n\mathbf{x}|$.

Definition 4.1.12 (Integral). A function $f : \mathbb{R}^n \to \mathbb{R}$, bounded with bounded support, is *integrable* if its upper and lower integrals are equal; its integral is then

$$\int_{\mathbb{R}^n} f|d^n\mathbf{x}| \stackrel{\text{def}}{=} U(f) = L(f). \qquad 4.1.20$$

It is rather hard to find integrals that can be computed directly from the definition; here is one.

Example 4.1.13 (Computing an integral). Let

$$f(x) = \begin{cases} x & \text{if } 0 \le x < 1 \\ 0 & \text{otherwise.} \end{cases} \qquad 4.1.21$$

Let us see that the integral

$$\int_0^1 f(x)|dx| = \left[\frac{x^2}{2}\right]_0^1 = \frac{1}{2}, \qquad 4.1.22$$

almost the easiest that calculus provides, can be evaluated by dyadic sums.

Using the indicator function, we can express f as the product

$$f(x) = x\,\mathbf{1}_{[0,1)}(x). \qquad 4.1.23$$

Since we are in dimension 1, our cubes are intervals:

$$C_{k,N} = \left[\frac{k}{2^N}, \frac{k+1}{2^N}\right).$$

Note from equation 4.1.21 that f is bounded with bounded support. In particular, unless $0 \le k/2^N < 1$, we have

$$m_{C_{k,N}}(f) = M_{C_{k,N}}(f) = 0. \qquad 4.1.24$$

If $0 \le k/2^N < 1$, then

$$\underbrace{m_{C_{k,N}}(f) = \frac{k}{2^N}}_{\substack{\text{greatest lower bound of } f \text{ over } C_{k,N}\\ \text{is the beginning of the interval}}} \quad \text{and} \quad \underbrace{M_{C_{k,N}}(f) = \frac{k+1}{2^N}}_{\substack{\text{lowest upper bound of } f \text{ over } C_{k,N}\\ \text{is the beginning of the \textit{next} interval}}}. \qquad 4.1.25$$

Computing multiple integrals by Riemann sums is conceptually no harder than computing one-dimensional integrals, but it is much more time consuming. Even when the dimension is only moderately large (for instance, 3 or 4) this is a serious problem. It becomes much worse when the dimension is 9 or 10; getting a numerical integral correct to six significant digits may be unrealistic.

When the dimension gets really large, like 10^{24}, as happens in quantum field theory and statistical mechanics, even in straightforward cases no one knows how to evaluate such integrals, and their behavior is a central problem in the mathematics of the field. We give an introduction to Riemann sums as they are used in practice in Section 4.6.

Thus equations 4.1.17 for $n = 1$ become

$$L_N(f) = \frac{1}{2^N}\sum_{k=0}^{2^N-1}\frac{k}{2^N}, \qquad U_N(f) = \frac{1}{2^N}\sum_{k=0}^{2^N-1}\frac{k+1}{2^N} = \frac{1}{2^N}\sum_{k=1}^{2^N}\frac{k}{2^N}. \qquad 4.1.26$$

In particular, $U_N(f) - L_N(f) = 2^N/2^{2N} = 1/2^N$, which tends to 0 as N tends to ∞, so f is integrable. Evaluating the integral requires the formula $1 + 2 + \cdots + m = m(m+1)/2$. Using this formula, we find

$$L_N(f) = \frac{1}{2^N}\,\frac{(2^N-1)2^N}{2\cdot 2^N} = \frac{1}{2}\left(1 - \frac{1}{2^N}\right)$$
$$\text{and} \quad U_N(f) = \frac{1}{2^N}\,\frac{2^N(2^N+1)}{2\cdot 2^N} = \frac{1}{2}\left(1 + \frac{1}{2^N}\right). \qquad 4.1.27$$

Clearly both sums converge to 1/2 as N tends to ∞. △

Riemann sums

Computing the upper integral $U(f)$ and the lower integral $L(f)$ may be difficult, as it involves finding maxima and minima. Fortunately, computing upper and lower sums is not really necessary. Suppose we know that f is integrable. Then, just as for Riemann sums in one dimension, we can choose any point $\mathbf{x}_{\mathbf{k},N} \in C_{\mathbf{k},N}$ we like, such as the center of each cube, or the lower left corner, and consider the Riemann sum

Warning: Before doing this you must know that your function is integrable: that the upper and lower sums converge to a *common limit*. It is perfectly possible for a Riemann sum to converge without the function being integrable (see Exercise 4.1.14). In that case, the limit doesn't mean much, and it should be viewed with distrust.

$$R(f,N) = \sum_{\mathbf{k}\in\mathbb{Z}^n} \overbrace{\text{vol}_n(C_{\mathbf{k},N})}^{\text{"width"}}\,\overbrace{f(\mathbf{x}_{\mathbf{k},N})}^{\text{"height"}}. \qquad 4.1.28$$

Since the value of the function at an arbitrary point $\mathbf{x}_{\mathbf{k},N}$ is bounded above by the least upper bound, and below by the greatest lower bound:

$$m_{C_{\mathbf{k},N}} f \le f(\mathbf{x}_{\mathbf{k},N}) \le M_{C_{\mathbf{k},N}} f, \qquad 4.1.29$$

the Riemann sums $R(f,N)$ will converge to the integral. In particular, Definition 4.1.12 justifies the techniques of first-year calculus: if $f : \mathbb{R} \to \mathbb{R}$

is integrable, with support in $[a,b]$ (which necessarily means that $a \le b$), then

$$\int_{\mathbb{R}} f(x)\,|dx| = \int_{-\infty}^{\infty} f(x)\,dx = \int_a^b f(x)\,dx, \qquad 4.1.30$$

and one-dimensional integrals, as defined in Definition 4.1.12, can be computed with the techniques of first-year calculus.[1]

"With support in $[a,b]$" means $f(x)=0$ if $x \notin [a,b]$. If $f(x) \ne 0$ for x in some other interval, say $[c,d]$, we would say "with support in $[a,b] \cup [c,d]$".

In computing a Riemann sum, any point will do, but some are better than others. The sum will converge faster if you use the center point rather than a corner.

Some rules for computing multiple integrals

A certain number of results are more or less obvious:

Proposition 4.1.14 (Rules for computing multiple integrals).

1. *If two functions $f, g : \mathbb{R}^n \to \mathbb{R}$ are both integrable, then $f+g$ is also integrable, and*

$$\int_{\mathbb{R}^n} (f+g)\,|d^n\mathbf{x}| = \int_{\mathbb{R}^n} f\,|d^n\mathbf{x}| + \int_{\mathbb{R}^n} g\,|d^n\mathbf{x}|. \qquad 4.1.31$$

2. *If f is an integrable function, and $a \in \mathbb{R}$, then af is integrable, and*

$$\int_{\mathbb{R}^n} (af)\,|d^n\mathbf{x}| = a\int_{\mathbb{R}^n} f\,|d^n\mathbf{x}|. \qquad 4.1.32$$

3. *If f, g are integrable functions with $f \le g$ (i.e., $f(\mathbf{x}) \le g(\mathbf{x})$ for all $\mathbf{x}$), then*

$$\int_{\mathbb{R}^n} f\,|d^n\mathbf{x}| \le \int_{\mathbb{R}^n} g\,|d^n\mathbf{x}|. \qquad 4.1.33$$

4. *If $f : \mathbb{R}^n \to \mathbb{R}$ is integrable, then $|f|$ is integrable, and*

$$\left|\int_{\mathbb{R}^n} f\,|d^n\mathbf{x}|\right| \le \int_{\mathbb{R}^n} |f|\,|d^n\mathbf{x}|. \qquad 4.1.34$$

Proposition 4.1.14: The converse of part 4 is false. Consider the function

$$f(x) = \begin{cases} 1 & \text{if } x \in [0,1] \cap \mathbb{Q} \\ -1 & \text{if } x \in [0,1] \cap (\mathbb{R}-\mathbb{Q}) \\ 0 & \text{otherwise} \end{cases}$$

The function $|f| = \mathbf{1}_{[0,1]}$ is integrable, but f is not.

Proof. 1. For any subset $A \subset \mathbb{R}^n$, we have

$$M_A(f)+M_A(g) \ge M_A(f+g) \quad \text{and} \quad m_A(f)+m_A(g) \le m_A(f+g). \qquad 4.1.35$$

Applying this to each cube $C \in \mathcal{D}_N(\mathbb{R}^n)$, we get

$$U_N(f)+U_N(g) \ge U_N(f+g) \quad \ge \quad L_N(f+g) \ge L_N(f)+L_N(g). \qquad 4.1.36$$

Since the outer terms have a common limit as $N \to \infty$, the inner ones have the same limit, giving

$$\underbrace{U(f)+U(g)}_{\int_{\mathbb{R}^n}(f)\,|d^n\mathbf{x}|+\int_{\mathbb{R}^n}(g)|d^n\mathbf{x}|} = \underbrace{U(f+g)}_{\int_{\mathbb{R}^n}(f+g)\,|d^n\mathbf{x}|} = L(f+g) = L(f)+L(g). \qquad 4.1.37$$

Inequality 4.1.35: If f and g are functions of census tracts, f assigning to each census tract per capita income for April through September, g assigning per capita income for October through March, then the sum of the maximum value for f and the maximum value for g must be at least the maximum value of $f+g$, and very likely more. A community dependent on the construction industry might have the highest per capita income in the summer, while a ski resort might have the highest per capita income in the winter.

[1] In switching from dx to $|dx|$ we are in some sense throwing away information. There is no way to speak of $\int_b^a f(x)\,dx$ in terms of $|dx|$. The integrand $f(x)dx$ cares about the orientation of the domain; the integrand $f(x)|dx|$ does not. In one dimension, orientation is simple and can be (and often is) swept under the rug. In higher dimensions, orientation is much more subtle and difficult. In Chapter 6 we will learn how to integrate over oriented domains in higher dimensions.

2. If $a > 0$, then $U_N(af) = aU_N(f)$ and $L_N(af) = aL_N(f)$ for any N, so the integral of af is a times the integral of f.

If $a < 0$, then $U_N(af) = aL_N(f)$ and $L_N(af) = aU_N(f)$, so the result is also true: multiplying by a negative number turns the upper limit into a lower limit and vice versa.

3. This is clear: $U_N(f) \le U_N(g)$ for every N.

4. Looking at one cube at a time, we see that for every N we have

$$U_N(|f|) - L_N(|f|) \le U_N(f) - L_N(f) \quad \text{and} \quad |U_N(f)| \le U_N(|f|). \qquad 4.1.38$$

The first inequality shows that $|f|$ is integrable; the second proves inequality 4.1.34. □

Part 1 of Corollary 4.1.15 lets us reduce a problem about an arbitrary function to a problem about nonnegative functions. We will use it to prove Proposition 4.1.16. If $f : \mathbb{R}^n \to \mathbb{R}$ is any function, we define f^+ and f^- by

$$f^+(\mathbf{x}) = \begin{cases} f(\mathbf{x}) & \text{if } f(\mathbf{x}) \ge 0 \\ 0 & \text{if } f(\mathbf{x}) < 0 \end{cases}, \qquad f^-(\mathbf{x}) = \begin{cases} -f(\mathbf{x}) & \text{if } f(\mathbf{x}) \le 0 \\ 0 & \text{if } f(\mathbf{x}) > 0. \end{cases} \qquad 4.1.39$$

Clearly both f^+ and f^- are nonnegative functions, and $f = f^+ - f^-$.

Corollary 4.1.15, part 2:

$$\sup(f,g)(\mathbf{x}) \overset{\text{def}}{=} \sup\{f(\mathbf{x}), g(\mathbf{x})\}$$
$$\inf(f,g)(\mathbf{x}) \overset{\text{def}}{=} \inf\{f(\mathbf{x}), g(\mathbf{x})\}.$$

Corollary 4.1.15. 1. *A bounded function f with bounded support is integrable if and only if both f^+ and f^- are integrable.*

2. *If f and g are integrable, so are $\sup(f,g)$ and $\inf(f,g)$.*

Proof. 1. If f is integrable, then by part 4 of Proposition 4.1.14, $|f|$ is integrable, and so are $f^+ = \frac{1}{2}(|f| + f)$ and $f^- = \frac{1}{2}(|f| - f)$, by parts 1 and 2. Now assume that f^+ and f^- are integrable. Then $f = f^+ - f^-$ is integrable, again by parts 1 and 2.

2. By part 1 and by Proposition 4.1.14, the functions $(f-g)^+$ and $(f-g)^-$ are integrable, so by Proposition 4.1.14, so are the functions

$$\begin{aligned} \inf(f,g) &= \frac{1}{2}\Big(f + g - (f-g)^+ - (f-g)^-\Big), \\ \sup(f,g) &= \frac{1}{2}\Big(f + g + (f-g)^+ + (f-g)^-\Big) \end{aligned} \qquad 4.1.40$$

□

Proposition 4.1.16: You can read equation 4.1.42 to mean "the integral of the product $f_1(\mathbf{x})f_2(\mathbf{y})$ equals the product of the integrals," but please note that we are *not* saying, and it is *not* true, that for two functions with the same variable, the integral of the product is the product of the integrals. There is no formula for $\int f_1(\mathbf{x})f_2(\mathbf{x})$. The two functions f_1 and f_2 of Proposition 4.1.16 have *different* variables.

Proposition 4.1.16. *If $f_1(\mathbf{x})$ is integrable on $\mathbb{R}^n$ and $f_2(\mathbf{y})$ is integrable on $\mathbb{R}^m$, then the function*

$$g(\mathbf{x},\mathbf{y}) = f_1(\mathbf{x})f_2(\mathbf{y}) \quad \text{on } \mathbb{R}^{n+m} \text{ is integrable, and} \qquad 4.1.41$$

$$\int_{\mathbb{R}^{n+m}} g\,|d^n\mathbf{x}||d^m\mathbf{y}| = \left(\int_{\mathbb{R}^n} f_1\,|d^n\mathbf{x}|\right)\left(\int_{\mathbb{R}^m} f_2\,|d^m\mathbf{y}|\right). \qquad 4.1.42$$

Proof. First suppose f_1 and f_2 are nonnegative. For any $A_1 \subset \mathbb{R}^n$ and $A_2 \subset \mathbb{R}^m$, we have

$$M_{A_1 \times A_2}(g) = M_{A_1}(f_1)M_{A_2}(f_2); \quad m_{A_1 \times A_2}(g) = m_{A_1}(f_1)m_{A_2}(f_2). \qquad 4.1.43$$

Since any $C \in \mathcal{D}_N(\mathbb{R}^{n+m})$ is of the form $C_1 \times C_2$ with $C_1 \in \mathcal{D}_N(\mathbb{R}^n)$ and $C_2 \in \mathcal{D}_N(\mathbb{R}^m)$, applying equation 4.1.43 to each cube separately gives

Proof of Proposition 4.1.16: The computation is not difficult, although the resulting equation looks horrendous. We leave the details as Exercise 4.1.7.

$$U_N(g) = U_N(f_1)U_N(f_2) \quad \text{and} \quad L_N(g) = L_N(f_1)L_N(f_2). \qquad 4.1.44$$

For the general case, write $f_1 = f_1^+ - f_1^-$ and $f_2 = f_2^+ - f_2^-$; expand out $f_1 f_2$ and apply the nonnegative case to the individual terms. The result drops out when terms are recollected. □

Volume defined more generally

The computation of volumes, historically the main motivation for integrals, remains an important application. We used the volume of cubes to define the integral; we now use integrals to define volume more generally.

Definition 4.1.17: vol_1 is length of subsets of $\mathbb{R}$, vol_2 is area of subsets of $\mathbb{R}^2$, and so on.

Definition 4.1.17 (n-dimensional volume). When $\mathbf{1}_A$ is integrable, the *n-dimensional volume* of A is

$$\mathrm{vol}_n\, A \overset{\text{def}}{=} \int_{\mathbb{R}^n} \mathbf{1}_A \, |d^n\mathbf{x}|. \qquad 4.1.45$$

We already defined the volume of dyadic cubes in equation 4.1.15. In Proposition 4.1.20 we will see that these definitions are consistent.

Some texts refer to pavable sets as "contented" sets: sets that have content.

Definition 4.1.18 (Pavable set). A set is *pavable* if it has a well-defined volume (i.e., if its indicator function is integrable).

Lemma 4.1.19. *An interval $I = [a, b]$ has length $|b - a|$.*

Proof. Of the cubes (i.e., intervals) $C \in \mathcal{D}_N(\mathbb{R})$, at most two contain an endpoint a or b. The others are either entirely in I or entirely outside; on those

$$M_C(\mathbf{1}_I) = m_C(\mathbf{1}_I) = \begin{cases} 1 & \text{if } C \subset I \\ 0 & \text{if } C \cap I = \emptyset. \end{cases} \qquad 4.1.46$$

The volume of a cube is $\frac{1}{2^{nN}}$, but here $n = 1$.

Thus the difference between upper and lower sums is at most twice the volume of a single cube (i.e., length of a single interval):

$$U_N(\mathbf{1}_I) - L_N(\mathbf{1}_I) \le 2\frac{1}{2^N}. \qquad 4.1.47$$

This quantity tends to 0 as $N \to \infty$, so the upper and lower sums converge to the same limit: $\mathbf{1}_I$ is integrable, and I has volume. We leave its computation as Exercise 4.1.13. □

Similarly, parallelograms with sides parallel to the axes have the volume one expects, namely, the product of the lengths of the sides.

Recall from Section 0.3 that

$$P = I_1 \times \cdots \times I_n \subset \mathbb{R}^n$$

means

$$P = \{ \mathbf{x} \in \mathbb{R}^n \mid x_i \in I_i \};$$

thus P is an interval if $n = 1$, a rectangle if $n = 2$, and a box if $n = 3$.

Equation 4.1.15, which says that the n-dimensional volume of a dyadic cube is

$$\mathrm{vol}_n C = \frac{1}{2^{Nn}},$$

is a special case of Proposition 4.1.20.

"Disjoint" means having no points in common.

A countably infinite union of pavable sets is often not pavable. This is one of the key distinctions between Riemann integration and Lebesgue integration; Theorem A21.6, the Lebesgue integration result corresponding to Theorem 4.1.21, is true for countably infinite unions.

To translate (shift) A by $\vec{\mathbf{v}}$, we add $\vec{\mathbf{v}}$ to each point in A.

Recall (Definition 4.1.10) that L denotes the lower integral and U the upper integral.

Proposition 4.1.20 (Volume of n-dimensional parallelogram). *The n-dimensional parallelogram*

$$P \stackrel{\text{def}}{=} I_1 \times \cdots \times I_n \subset \mathbb{R}^n \qquad 4.1.48$$

formed by the product of intervals $I_i = [a_i, b_i]$ has volume

$$\mathrm{vol}_n(P) = |b_1 - a_1|\,|b_2 - a_2| \,\dots\, |b_n - a_n|. \qquad 4.1.49$$

Proof. This follows immediately from Proposition 4.1.16, applied to

$$\mathbf{1}_P(\mathbf{x}) = \mathbf{1}_{I_1}(x_1)\mathbf{1}_{I_2}(x_2)\dots\mathbf{1}_{I_n}(x_n). \quad \square \qquad 4.1.50$$

The following elementary result has powerful consequences (though these will only become clear later).

Theorem 4.1.21 (Sum of volumes). *If two disjoint sets A, B in $\mathbb{R}^n$ are pavable, then so is their union, and the volume of the union is the sum of the volumes:*

$$\mathrm{vol}_n(A \cup B) = \mathrm{vol}_n A + \mathrm{vol}_n B. \qquad 4.1.51$$

Proof. Since $\mathbf{1}_{A\cup B} = \mathbf{1}_A + \mathbf{1}_B$ if A and B are disjoint, the result follows from part 1 of Proposition 4.1.14. $\square$

We will use Theorem 4.1.21 in proving the following important statement.

Proposition 4.1.22 (Volume invariant under translation). *Let A be any pavable subset of $\mathbb{R}^n$ and $\vec{\mathbf{v}} \in \mathbb{R}^n$ any vector. Denote by $A + \vec{\mathbf{v}}$ the set A translated by $\vec{\mathbf{v}}$. Then $A + \vec{\mathbf{v}}$ is pavable, and*

$$\mathrm{vol}_n(A + \vec{\mathbf{v}}) = \mathrm{vol}_n(A). \qquad 4.1.52$$

In particular, if $C \in \mathcal{D}_N(\mathbb{R})$, then $\mathrm{vol}_n(C + \vec{\mathbf{v}}) = \mathrm{vol}_n(C)$.

Proof. Let K_N be the set of cubes $C \in \mathcal{D}_N(\mathbb{R}^n)$ with $C \subset A$ (i.e., the set of cubes at the Nth level that are entirely inside A), and let H_N be the set of cubes $C \in \mathcal{D}_N(\mathbb{R}^n)$ such that C intersects A (is inside A or straddles the edge, so $K_N \subset H_N$). Then

$$\mathbf{1}_{\bigcup_{C\in K_N}(C+\vec{\mathbf{v}})} \le \mathbf{1}_{(A+\vec{\mathbf{v}})} \le \mathbf{1}_{\bigcup_{C\in H_N}(C+\vec{\mathbf{v}})}, \qquad 4.1.53$$

so

$$L\left(\mathbf{1}_{\bigcup_{C\in K_N}(C+\vec{\mathbf{v}})}\right) \le L\left(\mathbf{1}_{(A+\vec{\mathbf{v}})}\right) \le U\left(\mathbf{1}_{(A+\vec{\mathbf{v}})}\right) \le U\left(\mathbf{1}_{\bigcup_{C\in H_N}(C+\vec{\mathbf{v}})}\right). \qquad 4.1.54$$

If two sets A and B are disjoint, $\mathbf{1}_{A\cup B} = \mathbf{1}_A + \mathbf{1}_B$, so

$$\mathbf{1}_{\bigcup_{C\in K_N}(C+\vec{\mathbf{v}})} = \sum_{C\in K_N} \mathbf{1}_{(C+\vec{\mathbf{v}})}. \qquad 4.1.55$$

To lighten notation, in this proof we denote integrals simply by $\int f$, not by the correct $\int_{\mathbb{R}^n} f|d^n\mathbf{x}|$.

Equation 4.1.57: The sides of $C + \vec{\mathbf{v}}$ have the same lengths as the sides of C.

Proposition 4.1.20 says that any n-parallelogram is pavable, so $C + \vec{\mathbf{v}}$ is pavable; equivalently, $\mathbf{1}_{C+\vec{\mathbf{v}}}$ is integrable. By Proposition 4.1.14, the integral of the sum of integrable functions is the sum of the integrals, so

$$\int \mathbf{1}_{\bigcup_{C\in K_N}(C+\vec{\mathbf{v}})} = \sum_{C\in K_N} \int \mathbf{1}_{(C+\vec{\mathbf{v}})}. \qquad 4.1.56$$

If a function f is integrable, $U(f) = L(f) = \int f$, so we can replace the integral on the left side of equation 4.1.56 by the lower sum:

$$L\left(\mathbf{1}_{\bigcup_{C\in K_N}(C+\vec{\mathbf{v}})}\right) = \sum_{C\in K_N} \int \mathbf{1}_{(C+\vec{\mathbf{v}})} \overbrace{=}^{\text{def. of vol.}} \sum_{C\in K_N} \mathrm{vol}_n(C+\vec{\mathbf{v}}) \overbrace{=}^{\text{Prop. 4.1.20}} \sum_{C\in K_N} \mathrm{vol}_n\, C. \qquad 4.1.57$$

Unfortunately, at the moment there are very few functions we can integrate; we will have to wait until Section 4.5 before we can compute any really interesting examples.

A similar argument gives

$$U\left(\mathbf{1}_{\bigcup_{C\in H_N}(C+\vec{\mathbf{v}})}\right) = \sum_{C\in H_N} \mathrm{vol}_n\, C. \qquad 4.1.58$$

Therefore, inequality 4.1.54 becomes

$$\sum_{C\in K_N} \mathrm{vol}_n\, C \le L\left(\mathbf{1}_{(A+\vec{\mathbf{v}})}\right) \le U\left(\mathbf{1}_{(A+\vec{\mathbf{v}})}\right) \le \sum_{C\in H_N} \mathrm{vol}_n\, C. \qquad 4.1.59$$

We have

$$\lim_{N\to\infty} \sum_{C\in K_N} \mathrm{vol}_n\, C = \lim_{N\to\infty} \sum_{C\in H_N} \mathrm{vol}_n\, C = \mathrm{vol}_n\, A, \qquad 4.1.60$$

which gives

$$L\left(\mathbf{1}_{(A+\vec{\mathbf{v}})}\right) = U\left(\mathbf{1}_{(A+\vec{\mathbf{v}})}\right) = \mathrm{vol}_n(A+\vec{\mathbf{v}}) = \mathrm{vol}_n\, A. \quad \square \qquad 4.1.61$$

You are asked to prove Proposition 4.1.23 in Exercise 4.1.4.

Proposition 4.1.23 (Set with volume 0). *A bounded set $X \subset \mathbb{R}^n$ has volume 0 if and only if for every $\epsilon > 0$ there exists N such that*

$$\sum_{\substack{C\in\mathcal{D}_N(\mathbb{R}^n)\\ C\cap X\neq\emptyset}} \mathrm{vol}_n(C) \le \epsilon. \qquad 4.1.62$$

Proposition 4.1.24 (Scaling volume). *If $A \subset \mathbb{R}^n$ has volume and $t \in \mathbb{R}$, then tA has volume, and*

$$\mathrm{vol}_n(tA) = |t|^n\, \mathrm{vol}_n(A). \qquad 4.1.63$$

Proof. By Proposition 4.1.20, this is true if A is a parallelogram, in particular, a cube $C \in \mathcal{D}_N$. Assume A is a subset of $\mathbb{R}^n$ whose volume is well defined. This means that $\mathbf{1}_A$ is integrable, or, equivalently, that

$$\lim_{N\to\infty} \sum_{C\in\mathcal{D}_N, C\subset A} \mathrm{vol}_n(C) = \lim_{N\to\infty} \sum_{C\in\mathcal{D}_N, C\cap A\neq\emptyset} \mathrm{vol}_n(C), \qquad 4.1.64$$

and that the common limit is $\mathrm{vol}_n(A)$.

Since

$$\bigcup_{C\in\mathcal{D}_N, C\subset A} tC \subset tA \subset \bigcup_{C\in\mathcal{D}_N, C\cap A\neq\emptyset} tC, \qquad 4.1.65$$

and since $\mathrm{vol}_n(tC) = |t|^n \mathrm{vol}_n(C)$ for every cube, this gives

$$\begin{aligned} |t|^n \sum_{C\in\mathcal{D}_N, C\subset A} \mathrm{vol}_n(C) &= \int \sum_{C\in\mathcal{D}_N, C\subset A} \mathbf{1}_{tC} \le L(\mathbf{1}_{tA}) \\ &\le U(\mathbf{1}_{tA}) \le \int \sum_{C\in\mathcal{D}_N, C\cap A\neq\emptyset} \mathbf{1}_{tC} \\ &= |t|^n \sum_{C\in\mathcal{D}_N, C\cap A\neq\emptyset} \mathrm{vol}_n(C). \end{aligned} \qquad 4.1.66$$

Equation 4.1.66: The integral in the first line equals the corresponding lower sum, giving the inequality:

$$\int \sum_{C\in\mathcal{D}_N, C\subset A} \mathbf{1}_{tC} = L\left(\sum_{C\in\mathcal{D}_N, C\subset A} \mathbf{1}_{tC}\right) \le L(\mathbf{1}_{tA}).$$

The outer terms have a common limit as $N \to \infty$, so $L(\mathbf{1}_{tA}) = U(\mathbf{1}_{tA})$, and

$$\mathrm{vol}_n(tA) = \int \mathbf{1}_{tA}(\mathbf{x})|d^n\mathbf{x}| = |t|^n \mathrm{vol}_n(A). \quad \square \qquad 4.1.67$$

Exercises for Section 4.1

4.1.1 a. What is the two-dimensional volume (i.e., area) of a dyadic cube $C \in \mathcal{D}_3(\mathbb{R}^2)$? of $C \in \mathcal{D}_4(\mathbb{R}^2)$? of $C \in \mathcal{D}_5(\mathbb{R}^2)$?

b. What is the volume of a dyadic cube $C \in \mathcal{D}_3(\mathbb{R}^3)$? of $C \in \mathcal{D}_4(\mathbb{R}^3)$? of $C \in \mathcal{D}_5(\mathbb{R}^3)$?

4.1.2 In each group of dyadic cubes below, which has the smallest volume? the largest?

a. $C_{\begin{pmatrix}1\\2\end{pmatrix},4}$; $C_{\begin{pmatrix}1\\2\end{pmatrix},2}$; $C_{\begin{pmatrix}1\\2\end{pmatrix},6}$ b. $C \in \mathcal{D}_2(\mathbb{R}^3)$; $C \in \mathcal{D}_1(\mathbb{R}^3)$; $C \in \mathcal{D}_8(\mathbb{R}^3)$

Dyadic cubes for Exercise 4.1.3:

a. $C_{\begin{pmatrix}1\\2\end{pmatrix},3}$ b. $C_{\begin{pmatrix}0\\1\\3\end{pmatrix},2}$

c. $C_{\begin{pmatrix}0\\1\\1\\1\end{pmatrix},3}$ d. $C_{\begin{pmatrix}0\\1\\4\end{pmatrix},3}$

4.1.3 What is the volume of each of the dyadic cubes in the margin at left? What dimension is the volume? (For example, are the cubes two-dimensional? three-dimensional?) What information is given that you don't need to answer those two questions?

4.1.4 Prove Proposition 4.1.23.

4.1.5 a. Calculate $\sum_{i=0}^{n} i$.

b. Calculate directly from the definition the integrals

$$\int_{\mathbb{R}} x\mathbf{1}_{[0,1)}(x)|dx|, \int_{\mathbb{R}} x\mathbf{1}_{[0,1]}(x)|dx|, \int_{\mathbb{R}} x\mathbf{1}_{(0,1]}(x)|dx|, \int_{\mathbb{R}} x\mathbf{1}_{(0,1)}(x)|dx|.$$

In particular, show that they all exist and that they are equal.

c. Choose $a > 0$, and calculate directly from the definition the integrals

$$\int_{\mathbb{R}} x\mathbf{1}_{[0,a)}(x)|dx|, \int_{\mathbb{R}} x\mathbf{1}_{[0,a]}(x)|dx|, \int_{\mathbb{R}} x\mathbf{1}_{(0,a]}(x)|dx|, \int_{\mathbb{R}} x\mathbf{1}_{(0,a)}(x)|dx|.$$

Exercises 4.1.5 and 4.1.6: In parts c and d, you need to distinguish between the cases where a and b are "dyadic" (i.e., endpoints of dyadic intervals), and the case where they are not.

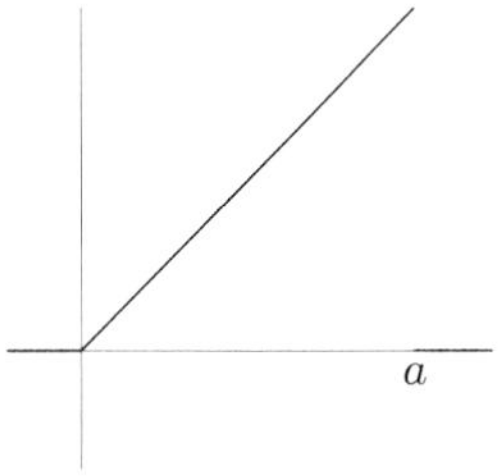

FIGURE FOR EXERCISE 4.1.5.
The dark line is the graph of the function $x\mathbf{1}_{[0,a)}(x)$.

d. If $0 < a < b$, show that $x\mathbf{1}_{[a,b]}$, $x\mathbf{1}_{[a,b)}$, $x\mathbf{1}_{(a,b]}$, $x\mathbf{1}_{(a,b)}$ are all integrable and compute their integrals. (The first is shown in the figure in the margin.) In particular, show that they all exist and that they are equal.

4.1.6 a. Calculate $\sum_{i=0}^{n} i^2$.

b. Choose $a > 0$, and calculate directly from the definition the integrals

$$\int_{\mathbb{R}} x^2\mathbf{1}_{[0,a)}(x)\,|dx|,\ \int_{\mathbb{R}} x^2\mathbf{1}_{[0,a]}(x)\,|dx|,\ \int_{\mathbb{R}} x^2\mathbf{1}_{(0,a]}(x)\,|dx|,\ \int_{\mathbb{R}} x^2\mathbf{1}_{(0,a)}(x)\,|dx|.$$

In particular, show that they all exist and that they are equal.

c. If $0 < a < b$, show that $x^2\mathbf{1}_{[a,b]}$, $x^2\mathbf{1}_{[a,b)}$, $x^2\mathbf{1}_{(a,b]}$, $x^2\mathbf{1}_{(a,b)}$ are all integrable and compute their integrals, which are all equal.

4.1.7 Prove the general case of Proposition 4.1.16.

4.1.8 Prove that the distance $|\mathbf{x}-\mathbf{y}|$ between two points $\mathbf{x}, \mathbf{y}$ in the same cube $C \in \mathcal{D}_N(\mathbb{R}^n)$ is $\le \frac{\sqrt{n}}{2^N}$.

4.1.9 Let $Q \subset \mathbb{R}^2$ be the unit square $0 \le x,\ y \le 1$. Show that the function

$$f\begin{pmatrix} x \\ y \end{pmatrix} = \sin(x-y)\mathbf{1}_Q\begin{pmatrix} x \\ y \end{pmatrix}$$

is integrable by providing an explicit bound for $U_N(f) - L_N(f)$ that tends to 0 as $N \to \infty$.

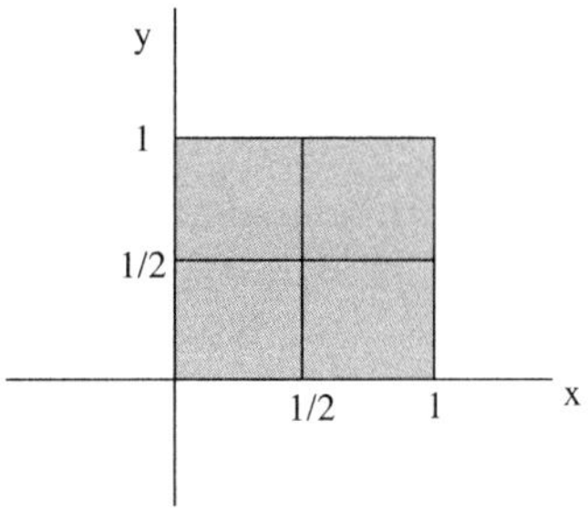

FIGURE FOR EXERCISE 4.1.10.

4.1.10 a. What are the upper and lower sums $U_1(f)$ and $L_1(f)$ for the function

$$f\begin{pmatrix} x \\ y \end{pmatrix} = \begin{cases} x^2+y^2 & \text{if } 0 < x, y < 1 \\ 0 & \text{otherwise,} \end{cases}$$

i.e., the upper and lower sums for the partition $\mathcal{D}_1(\mathbb{R}^2)$, shown in the figure at left?

b. Compute the integral of f and show that it is between the upper and lower sum.

4.1.11 Define the *dilation by a* of a function $f : \mathbb{R}^n \to \mathbb{R}$ by

$$D_a f(\mathbf{x}) = f\left(\frac{\mathbf{x}}{a}\right).$$

Show that if f is integrable, then so is $D_{2^N} f$, and

$$\int_{\mathbb{R}^n} D_{2^N} f(\mathbf{x})\,|d^n\mathbf{x}| = 2^{nN}\int_{\mathbb{R}^n} f(\mathbf{x})\,|d^n\mathbf{x}|.$$

4.1.12 Recall that the dyadic cubes are half open, half closed. Show that the closed cubes also have the same volume. (This result is obvious, but remarkably harder to pin down than you might expect.)

4.1.13 Complete the proof of Lemma 4.1.19.

4.1.14 Consider the function

$$f(x) = \begin{cases} 0 & \text{if } x \notin [0,1], \text{ or } x \text{ is rational} \\ 1 & \text{if } x \in [0,1], \text{ and } x \text{ is irrational.} \end{cases}$$

a. What value do you get for the "left Riemann sum", where for the interval $C_{k,N} = \left\{ x \,\middle|\, \frac{k}{2^N} \le x < \frac{k+1}{2^N} \right\}$ you choose the left endpoint $k/2^N$? For the sum you get when you choose the right endpoint $(k+1)/2^N$? The midpoint Riemann sum?

b. What value do you get for the "geometric mean" Riemann sum, where the point you choose in each $C_{k,N}$ is the geometric mean of the two endpoints,

$$\sqrt{\left(\frac{k}{2^N}\right)\left(\frac{k+1}{2^N}\right)} = \frac{\sqrt{k(k+1)}}{2^N}\ ?$$

4.1.15 a. Show that if X and Y have volume 0, then $X \cap Y$, $X \times Y$, and $X \cup Y$ have volume 0.

b. Show that $\left\{ \begin{pmatrix} x_1 \\ 0 \end{pmatrix} \in \mathbb{R}^2 \middle| 0 \le x_1 \le 1 \right\}$ has volume 0 (i.e., $\text{vol}_2 = 0$).

c. Show that $\left\{ \begin{pmatrix} x_1 \\ x_2 \\ 0 \end{pmatrix} \in \mathbb{R}^3 \middle| 0 \le x_1, x_2 \le 1 \right\}$ has volume 0 (i.e., $\text{vol}_3 = 0$).

Exercise 4.1.16: The unit cube has the lower left corner anchored at the origin.

Exercise 4.1.16 shows that the behavior of an integrable function $f : \mathbb{R}^n \to \mathbb{R}$ on the boundaries of the cubes of $\mathcal{D}_N$ does not affect the integral.

Part c:

$$\overset{\circ}{C} = \left\{ \mathbf{x} \in \mathbb{R}^n \middle| \frac{k_i}{2^N} < x_i < \frac{k_i+1}{2^N} \right\}$$

$$\overline{C} = \left\{ \mathbf{x} \in \mathbb{R}^n \middle| \frac{k_i}{2^N} \le x_i \le \frac{k_i+1}{2^N} \right\}$$

***4.1.16** a. Let Q be the unit cube in $\mathbb{R}^n$, and choose $a \in [0,1)$. Show that the subset $\{\, \mathbf{x} \in S \mid x_i = a \,\}$ has n-dimensional volume 0.

b. Let $X_N = \partial\mathcal{D}_N$ be the set made up of the boundaries of all the cubes $C \in \mathcal{D}_N$. Show that $\text{vol}_n(X_N \cap S) = 0$.

c. For each cube C defined as in equation 4.1.12, consider its closure $\overline{C}$ and its interior $\overset{\circ}{C}$, as defined in Definitions 1.5.8 and 1.5.9 and spelled out in the margin. Show that if $f : \mathbb{R}^n \to \mathbb{R}$ is integrable, then

$$\overset{\circ}{U}(f) = \lim_{N\to\infty} \sum_{C \in \mathcal{D}_N(\mathbb{R}^n)} M_{\overset{\circ}{C}}(f)\, \text{vol}_n(C)$$

$$\overset{\circ}{L}(f) = \lim_{N\to\infty} \sum_{C \in \mathcal{D}_N(\mathbb{R}^n)} m_{\overset{\circ}{C}}(f)\, \text{vol}_n(C)$$

$$\overline{U}(f) = \lim_{N\to\infty} \sum_{C \in \mathcal{D}_N(\mathbb{R}^n)} M_{\overline{C}}(f)\, \text{vol}_n(C)$$

$$\overline{L}(f) = \lim_{N\to\infty} \sum_{C \in \mathcal{D}_N(\mathbb{R}^n)} m_{\overline{C}}(f)\, \text{vol}_n(C)$$

all exist and are all equal to $\int_{\mathbb{R}^n} f(\mathbf{x})|d^n\mathbf{x}|$. *Hint*: You may assume that the support of f is contained in Q, and that $|f| \le 1$. Choose $\epsilon > 0$, then choose N_1 to make

$$U_{N_1}(f) - L_{N_1}(f) < \epsilon/2,$$

then choose $N_2 > N_1$ to make $\text{vol}(X_{\partial\mathcal{D}_{N_2}}) < \epsilon/2$. Now show that for $N > N_2$,

$$\overset{\circ}{U}_N(f) - \overset{\circ}{L}_N(f) < \epsilon \quad \text{and}$$
$$\overline{U}_N(f) - \overline{L}_N(f) < \epsilon.$$

d. Suppose $f : \mathbb{R}^n \to \mathbb{R}$ is integrable, and that $f(-\mathbf{x}) = -f(\mathbf{x})$. Show that $\int_{\mathbb{R}^n} f|d^n\mathbf{x}| = 0$.

4.1.17 An integrand should take a piece of the domain and return a number, in such a way that if we decompose a domain into little pieces, evaluate the integrand on the pieces, and add, the sums should have a limit as the decomposition becomes infinitely fine (and the limit should not depend on how the domain is decomposed). What will happen if we break up $[0,1]$ into intervals $[x_i, x_{i+1}]$, for $i = 0, 1, \dots, n-1$, with $0 = x_0 < x_1 < \dots < x_n = 1$, and assign one of the following numbers to each of the $[x_i, x_{i+1}]$?

a. $|x_{i+1} - x_i|^2$ b. $\sin |x_i - x_{i+1}|$ c. $\sqrt{|x_i - x_{i+1}|}$

d. $|(x_{i+1})^2 - (x_i)^2|$ e. $|(x_{i+1})^3 - (x_i)^3|$

4.1.18 As in Exercise 4.1.17, what will happen if we break up $[0,1]$ into intervals $[x_i, x_{i+1}]$, for $i = 0, 1, \dots, n-1$, with $0 = x_0 < x_1 < \dots < x_n = 1$, and assign one of the following numbers to each of the $[x_i, x_{i+1}]$? In parts a, b, and c, the function f is a C^1 function on a neighborhood of $[0,1]$.

a. $f(x_{i+1}) - f(x_i)$ b. $f\Big((x_{i+1})\Big)^2 - f(x_i^2)$

c. $\Big(f(x_{i+1})\Big)^2 - \Big(f(x_i)\Big)^2$ d. $|x_{i+1} - x_i| \ln |x_{i+1} - x_i|$

4.1.19 Repeat Exercise 4.1.18 but in $\mathbb{R}^2$, the integrand to be integrated over $[0,1]^2$. The integrand takes a rectangle $a < x < b,\quad c < y < d$ and returns the number $|b-a|^2\sqrt{|c-d|}$.

4.1.20 Let $a, b > 0$ be real numbers, and let T be the triangle defined by $x \geq 0$, $y \geq 0$, $x/a + y/b \leq 1$. Using upper and lower sums, compute the integral

$$\int_T x\, dx\, dy = \frac{a^2 b}{6}.$$

Hint: For all $c \in \mathbb{R}$, the *floor* $\lfloor c \rfloor$ is the largest integer $\leq c$. Show that

$$L_N(f) = \frac{1}{2^{2N}} \sum_{k=0}^{\lfloor 2^N a \rfloor} \sum_{j=0}^{\lfloor 2^N b \left(1 - \frac{k+1}{2^N a}\right) \rfloor} \frac{k}{2^N}$$

and find the analogous formula for $U_N(f)$. In evaluating the double sum, note that the term $k/2^N$ does not depend on j, so the inner sum is just a product.

The integral of Exercise 4.1.20 is much easier to compute using Fubini's theorem (Section 4.5). This exercise should convince you that you never want to use upper and lower sums to compute integrals.

To evaluate the outer sum, it is useful to remember that

$$\sum_{k=0}^{m} k = \frac{m(m+1)}{2}$$

$$\sum_{k=0}^{m} k^2 = \frac{m(m+1)(2m+1)}{6}.$$

4.1.21 Let $f, g : \mathbb{R}^n \to \mathbb{R}$ be bounded with bounded support.

a. Show that $U(f+g) \leq U(f) + U(g)$.

b. Find f and g such that $U(f+g) < U(f) + U(g)$.

4.2 Probability and centers of gravity

Computing areas and volumes is one important application of multiple integrals. There are many others, in a wide range of fields: geometry, mechanics, probability, etc. Here we touch on two: computing centers of gravity and computing probabilities when they cannot be computed in terms of individual outcomes. The formulas are so similar that we think each helps in understanding the other.

The *center of gravity* $\mathbf{x}$ of a body A is the point on which the body A could be balanced.

Definition 4.2.1 (Center of gravity of a body).

1. If a body $A \subset \mathbb{R}^n$ (i.e., a pavable set) is made of some homogeneous material, then the *center of gravity* of A is the point $\overline{\mathbf{x}}$ whose ith coordinate is

$$\overline{x}_i \stackrel{\text{def}}{=} \frac{\int_A x_i |d^n\mathbf{x}|}{\int_A |d^n\mathbf{x}|}. \qquad 4.2.1$$

2. If A has variable density given by the function $\mu : A \to \mathbb{R}$, then the mass $M(A)$ of such a body is

$$M(A) \stackrel{\text{def}}{=} \int_A \mu(\mathbf{x})|d^n\mathbf{x}|, \qquad 4.2.2$$

and the *center of gravity* of A is the point $\overline{\mathbf{x}}$ with ith coordinate

$$\overline{x}_i \stackrel{\text{def}}{=} \frac{\int_A x_i \mu(\mathbf{x})|d^n\mathbf{x}|}{M(A)}. \qquad 4.2.3$$

Equation 4.2.1 gives the average value of the ith coordinate over the entire body.

Equation 4.2.2: Integrating density gives mass. In physical situations μ ("mu") is nonnegative. In many problems in probability there is a similar function μ, giving the "density of probability".

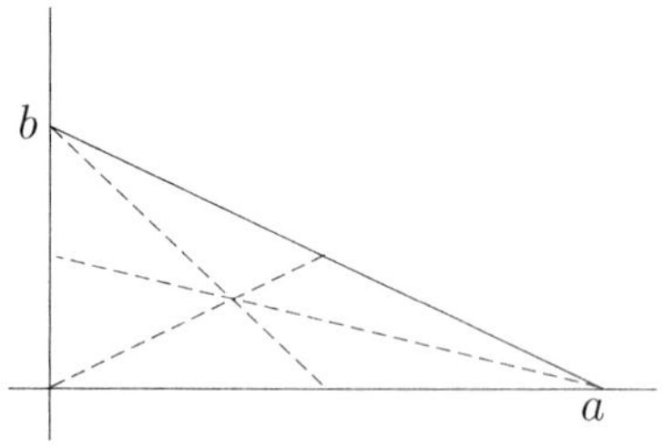

FIGURE 4.2.1.

Each median cuts the triangle into two parts with equal area; the triangle, if made of wood, could be balanced on the median. The intersection of the medians is the center of gravity. The x-coordinate of the center of gravity gives the average value of x over the triangle; the y-coordinate gives the average value of y.

Example 4.2.2 (Center of gravity). What is the center of gravity of the right triangle T shown in Figure 4.2.1? The area of T, i.e., $\int_T dx\,dy$, is $ab/2$. Using the techniques of Section 4.5, we can easily compute

$$\int_T x\,dx\,dy = \frac{a^2b}{6} \quad \text{and} \quad \int_T y\,dx\,dy = \frac{ab^2}{6} \qquad 4.2.4$$

(Exercise 4.1.20 asks you to compute these integrals with upper and lower sums, which is more painful), so by equation 4.2.1, the center of gravity of T is the point $\overline{\mathbf{x}}$ with coordinates

$$\overline{x} = \frac{\frac{a^2b}{6}}{\frac{ab}{2}} = \frac{a}{3} \quad \text{and} \quad \overline{y} = \frac{\frac{ab^2}{6}}{\frac{ab}{2}} = \frac{b}{3} \qquad 4.2.5$$

Note that this agrees with what you probably learned in high school: that the center of gravity of the triangle is the point where the medians (lines from vertex to midpoint of opposite side) intersect, and that the center of gravity cuts every median in the ratio 2:1, as shown in Figure 4.2.1. △

Probability and integration

Recall from Section 3.8 that in probability, a sample space S is the set of all possible outcomes of some experiment, together with a "probability measure" $\mathbf{P}$ assigning numbers to events (i.e., to appropriate subsets of the sample space S). This measure is required to satisfy the conditions of Definition 3.8.4.

When S is finite or countably infinite, we can think of $\mathbf{P}$ in terms of the probability of individual outcomes. But when $S = \mathbb{R}$ or $S = \mathbb{R}^k$, usually the individual outcomes have probability 0. How then can we evaluate $\mathbf{P}$?

In many interesting cases, one can describe the probability measure in terms of a *probability density*.

Definition 4.2.3 (Probability density). Let $S, \mathbf{P}$ be a probability space, with $S \subset \mathbb{R}^n$. If there exists a nonnegative integrable function $\mu : \mathbb{R}^n \to \mathbb{R}$ such that for any event A

$$\mathbf{P}(A) = \int_A \mu(x)\, |d^n\mathbf{x}|, \qquad 4.2.6$$

then μ is called the *probability density* of $S, \mathbf{P}$.

FIGURE 4.2.2.

The French mathematician and natural historian Georges Leclerc Comte de Buffon (1707–1788) hit on this surprising method for computing π. Scientists at the University of Bath report that a species of ants appears to use a Buffon's needle algorithm to measure the area of potential nest sites.

It follows that a probability density must satisfy

$$\mu(\mathbf{x}) \geq 0 \quad \text{and} \quad \int_{\mathbb{R}^k} \mu(\mathbf{x})\, |d^k\mathbf{x}| = 1. \qquad 4.2.7$$

Example 4.2.4 (Computing π using Buffon's needle). Toss a needle of length 1 on a piece of lined paper, with lines parallel to the x-axis spaced 1 apart. The sample space should be the space of positions of the needle. We will be interested in the probability that the needle intersects a line, so we will disregard which line is intersected and where, and will encode a position of needle by its polar angle $\theta \in [0, \pi)$ and the distance $s \in [0, 1/2]$ of the center of the needle to the nearest line; see Figure 4.2.3, left. (You can find simulations of the Buffon needle experiment on the Internet, including one at vis.supstat.com/2013/04/buffons-needle/)

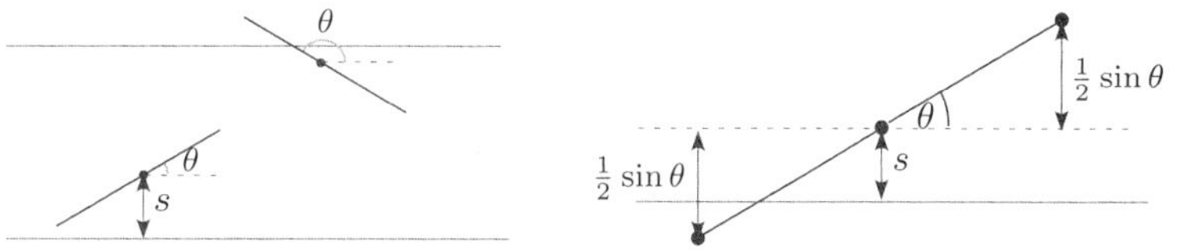

FIGURE 4.2.3. LEFT: The position of a needle is encoded by the distance $s \in [0, 1/2]$ from the center of a needle to the nearest line and the angle $\theta \in [0, \pi)$ that the needle makes with the horizontal. RIGHT: The needle will intersect a line if $s < \frac{1}{2}\sin\theta$.

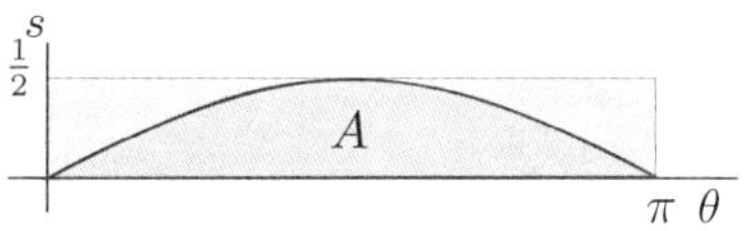

FIGURE 4.2.4.

The shaded rectangle is S; the darker region $A \subset S$ is the subset where a needle intersects a line.

The probability density

$$\mu\begin{pmatrix}\theta\\ s\end{pmatrix} |d\theta\, ds| = \frac{2}{\pi}|d\theta\, ds|, \qquad 4.2.8$$

models an experiment where we throw our needle "randomly", not favoring any position of the needle with respect to the lines of the paper; the $2/\pi$ is needed so that $\int_{[0,1/2]\times[0,\pi]} \mu\begin{pmatrix}\theta\\ s\end{pmatrix} |d\theta\, ds| = 1$.

As shown in Figure 4.2.3, right and Figure 4.2.4, the event "the needle intersects a line" corresponds to the set

$$A = \left\{ s - \frac{1}{2}\sin\theta < 0 \right\}, \qquad 4.2.9$$

So the probability that the needle will intersect a line is

$$\mathbf{P}(A) = \frac{2}{\pi}\int_A |d\theta\, ds| = \frac{2}{\pi}\int_0^\pi \frac{1}{2}\sin\theta\, d\theta = \frac{2}{\pi}. \qquad 4.2.10$$

Thus the probability of intersecting the line is $2/\pi$. This provides a (cumbersome) method for computing π: if as above you toss a needle onto lined paper N times, and for n of the tosses the needle intersects a line, in the long run $2N/n$ should be approximately π. △

Equation 4.2.10: It may seem peculiar that an infinite number of outcomes each with probability 0 can add up to something positive (here, $2/\pi$), but it is the same as the more familiar notion that a line has length, while the points that compose it have length 0.

Example 4.2.5 (Heads, tails, and length). In parlor games, you may be asked to "choose a number, any number". A priori, this is meaningless, but there is a way (there are many ways) to choose $x \in [0,1]$ "at random". Write the number in base 2, and choose the successive binary digits after the binary point by tossing a coin, writing 0 for heads and 1 for tails.

The probability of any individual outcome (number in $[0,1]$) is 0: the probability of any string of length m is $1/2^m$, since there are 2^m such strings and each is equally likely. But it is reasonable to ask about the probability of *events* i.e., subsets of the sample space $[0,1]$. For instance, it is reasonable to ask "what is the probability that the first digit after the binary point is 1", i.e., that the number is between 1/2 and 1. It should be clear that that probability is 1/2: it is precisely the probability of the first coin toss coming up tails, and is not affected by further tosses.

The same argument shows that, for $0 \le k < 2^N$, every dyadic interval $[k/2^N, (k+1)/2^N]$ has probability $1/2^N$, since the numbers in such an interval are precisely those whose first k binary digits are the digits of k, written in base 2. Thus the probability of any dyadic interval is exactly its length. Since our definition of length is in terms of dyadic intervals, this shows that for any $A \subset [0,1]$ that has a length,

$$\mathbf{P}(A) = \int_{\mathbb{R}} \mathbf{1}_A |dx|. \quad △ \qquad 4.2.11$$

Example 4.2.5: It is perhaps surprising that this probabilistic experiment ends up giving something so simple as length. It wouldn't have come out this way if the coin had been biased. Suppose that instead of a coin you toss a die, writing a 1 if it comes up 1, and a 0 otherwise, i.e., if it comes up 2,3,4,5 or 6.

This is also a "way of choosing a number at random". It also gives a probability measure on $[0,1]$, but clearly it does not correspond to a length. For instance

$$\mathbf{P}([0,1/2]) = 5/6,$$
$$\mathbf{P}([1/2,1]) = 1/6,$$

and understanding the function $x \mapsto \mathbf{P}([0,x])$ for $x \in [0,1]$ would be quite difficult.

Definition 4.2.6 (Expected value). Let $S \subset \mathbb{R}^n$ be a sample space with probability of density μ. If f be a random variable such that $f(\mathbf{x})\mu(\mathbf{x})$ is integrable, the *expected value* of f is

$$E(f) \stackrel{\text{def}}{=} \int_S f(\mathbf{x})\mu(\mathbf{x})\, |d^n\mathbf{x}| \qquad 4.2.12$$

The other definitions are identical to those in Definitions 3.8.6, when they are expressed in terms of the expectation. Let $f, g \in \mathrm{RV}(S)$. Then

$$\mathrm{Var}(f) \stackrel{\text{def}}{=} E(f^2) - \big(E(f)\big)^2, \qquad \sigma(f) \stackrel{\text{def}}{=} \sqrt{\mathrm{Var}(f)}$$

$$\mathrm{cov}(f,g) \stackrel{\text{def}}{=} E\Big(f - E(f)\Big)\Big(g - E(g)\Big) = E(fg) - E(f)E(g) \qquad 4.2.13$$

$$\mathrm{corr}\,(f,g) \stackrel{\text{def}}{=} \frac{\mathrm{cov}(f,g)}{\sigma(f)\sigma(g)}.$$

Central limit theorem

One probability density is ubiquitous in probability theory: the *normal distribution* given by

The function μ is called the *Gaussian*; its graph is a bell curve.

$$\mu(x) = \frac{1}{\sqrt{2\pi}} e^{-x^2/2}. \qquad 4.2.14$$

The theorem that makes the normal distribution important is the *central limit theorem*. Suppose you have an experiment and a random variable, with expected value E and standard deviation σ. Suppose that you repeat the experiment n times, with results $x_1, \dots, x_n$. Then the central limit theorem asserts that the average

$$\overline{x} = \frac{1}{n}(x_1 + \cdots + x_n) \qquad 4.2.15$$

is approximately distributed according to the normal distribution with mean E and standard deviation $\sigma/\sqrt{n}$, the approximation getting better and better as $n \to \infty$. Whatever experiment you perform, if you repeat it and average, the normal distribution will describe the results.

As n grows, all the detail of the original experiment gets ironed out, leaving only the normal distribution.

The standard deviation of the new experiment (the "repeat the experiment n times and take the average" experiment) is the standard deviation of the initial experiment divided by $\sqrt{n}$.

Below we will justify this statement in the case of coin tosses. First let us see how to translate the statement into formulas. There are two ways of doing it. One is to say that the probability that $\overline{x}$ is between A and B is approximately

The exponent for e in formula 4.2.16 is so small it may be hard to read; it is

$$-\frac{n}{2}\left(\frac{x-E}{\sigma}\right)^2.$$

$$\frac{\sqrt{n}}{\sqrt{2\pi}\,\sigma} \int_A^B e^{-\frac{n}{2}\left(\frac{x-E}{\sigma}\right)^2}\, dx. \qquad 4.2.16$$

We will use the other in our formal statement of the theorem. For this we make the change of variables $A = E + \sigma a/\sqrt{n}$, $B = E + \sigma b/\sqrt{n}$.

Formula 4.2.16 puts the complication in the exponent; formula 4.2.17 puts it in the domain of integration.

Theorem 4.2.7 (The central limit theorem). *If an experiment and a random variable have expectation E and standard deviation σ, then if the experiment is repeated n times, with average result $\overline{x}$, the probability that $\overline{x}$ is between $E + \frac{\sigma}{\sqrt{n}}a$ and $E + \frac{\sigma}{\sqrt{n}}b$ is approximately*

$$\frac{1}{\sqrt{2\pi}} \int_a^b e^{-y^2/2}\, dy. \qquad 4.2.17$$

There are a great many improvements on and extensions of the central limit theorem; we cannot hope to touch upon them here.

We prove a special case of the central limit theorem in Appendix A16. The proof uses *Stirling's formula*, a very useful result showing how the factorial $n!$ behaves as n becomes large. We recommend reading it if time permits, as it makes interesting use of some of the notions we have studied so far (Taylor polynomials and Riemann sums) as well as some you should remember from high school (logarithms and exponentials).

Example 4.2.8 (Coin toss). As a first example, let us see how the central limit theorem answers the question, what is the probability that a fair coin tossed 1000 times will come up heads between 510 and 520 times? In principle, this is straightforward: just compute the sum

$$\frac{1}{2^{1000}} \sum_{k=510}^{520} \binom{1000}{k}. \qquad 4.2.18$$

Formula 4.2.18: Recall the binomial coefficient:

$$\binom{n}{k} = \frac{n!}{k!(n-k)!}.$$

In practice, computing these numbers would be extremely cumbersome; it is much easier to use the central limit theorem. Our individual experiment consists of throwing a coin, and our random variable returns 1 for heads and 0 for tails. This random variable has expectation $E = .5$ and standard deviation $\sigma = .5$ also, and we are interested in the probability of the average being between .51 and .52. Using the version of the central limit theorem in formula 4.2.16, we see that the probability is approximately

$$\frac{\sqrt{1000}}{\sqrt{2\pi}\,\frac{1}{2}} \int_{.51}^{.52} e^{-\frac{1000}{2}\left(\frac{x-.5}{.5}\right)^2}\,dx. \qquad 4.2.19$$

Now we set

$$1000\left(\frac{x-.5}{.5}\right)^2 = t^2, \quad \text{so that} \quad 2\sqrt{1000}\,dx = dt. \qquad 4.2.20$$

Substituting t^2 and dt in formula 4.2.19, we get

$$\frac{1}{\sqrt{2\pi}} \int_{20/\sqrt{1000}}^{40/\sqrt{1000}} e^{-t^2/2}\,dt \approx 0.1606. \qquad 4.2.21$$

Does this seem large to you? It does to most people. △

We could also write the integral in equation 4.2.21 as

$$\frac{1}{\sqrt{2\pi}} \int_{\sqrt{10}/5}^{2\sqrt{10}/5} e^{-t^2/2}\,dt$$

One way to get the answer in equation 4.2.21 is to look up a table giving values for the "standard normal distribution function".

Another is to use some software. With MATLAB, we use .5 erf to get

```
EDU> a= .5*erf(20/sqrt(2000))
EDU> a = 0.236455371567231
EDU> b= .5*erf(40/sqrt(2000))
EDU> b = 0.397048394633966
EDU> b-a
ans = 0.160593023066735
```

The "error function" erf is related to the "standard normal distribution function" as follows:

$$\frac{1}{\sqrt{2\pi}} \int_0^a e^{-\frac{t^2}{2}}\,dt = \frac{1}{2}\operatorname{erf}\left(\frac{a}{\sqrt{2}}\right).$$

Example 4.2.9 (Political poll). How many people need to be polled to call an election, with a probability of 95% of being within 1% of the "true value"? A mathematical model of this is tossing a biased coin, which falls heads with unknown probability p and tails with probability $1-p$. If we toss this coin n times (i.e., sample n people) and return 1 for heads and 0 for tails (1 for candidate A and 0 for candidate B), the question is: how large does n need to be to achieve 95% probability that the average we get is within 1% of p?

We need to know something about the bell curve to answer this, namely that 95% of the mass is within two standard deviations of the mean, as shown in Figure 4.2.5.

That means that we want $\frac{1}{2}\%$ to be the standard deviation of the experiment of asking n people. The experiment of asking one person has standard deviation $\sigma = \sqrt{p(1-p)}$. Of course, p is what we don't know, but the maximum of $\sqrt{p(1-p)}$ is 1/2 (which occurs for $p = 1/2$). So we will be safe if we choose n so that the standard deviation $\sigma/\sqrt{n}$ is

$$\frac{1}{2\sqrt{n}} = \frac{1}{200}; \quad \text{i.e.,} \quad n = 10000. \qquad 4.2.22$$

How many would you need to ask if you wanted to be 95% sure to be within 2% of the true value? Check in the footnote below.[2] △

Computations like this are used everywhere: for instance, when drug companies figure out how large a population to try out a new drug on, or when industries figure out how long a product can be expected to last.

[2] The number is 2500. Note that gaining 1% quadrupled the price of the poll.

Figure 4.2.5 gives three typical values for the area under the bell curve. For other values, you need to use a table or software, as described in Example 4.2.8.

It is a good idea to memorize Figure 4.2.5; it allows you to distinguish significant results from those that can be explained by chance, so that you can assess results of experiments of all sorts: effectiveness of drugs, election results, etc.

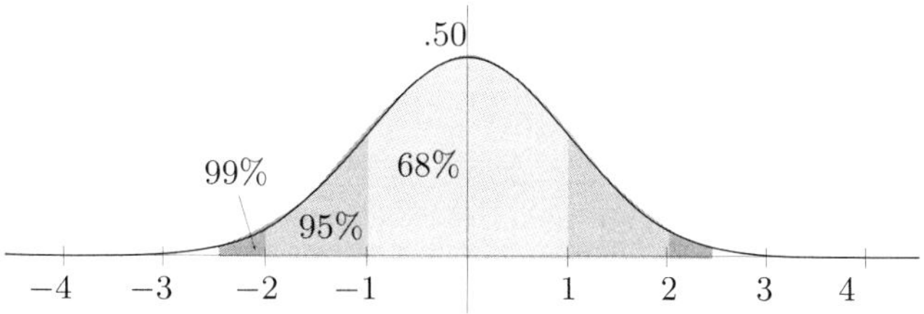

FIGURE 4.2.5. For the normal distribution, 68% of the probability is within one standard deviation; 95% is within two standard deviations; 99% is within 2.5 standard deviations.

Example 4.2.10 (Real data that don't follow normal distribution). The Black-Scholes formula is used to price derivatives; it revolutionized financial mathematics in the 1970s. It is based on the idea that prices oscillate around an expected price, roughly following the normal distribution[3], with a standard deviation called volatility. In this view, price fluctuations are due to a myriad of unknowable causes, which roughly cancel, and should lead to behavior like a random walk. Wild swings are assumed to be virtually impossible. This assumption was proved false on Oct. 19, 1987, when the stock markets lost more than 20 percent of their value. △

EXERCISES FOR SECTION 4.2

4.2.1 Assume you are given the following 15 integers:

$$8,\ 2,\ 9,\ 4,\ 7,\ 7,\ 1,\ 12,\ 6,\ 6,\ 5,\ 10,\ 9,\ 9,\ 1.$$

You are told that each number is the number of heads that came up after tossing a coin 14 times. Is this believable? Do the numbers follow a normal distribution?

4.2.2 What are the expectation, variance, and standard deviation of the function (random variable) $f(x) = x$ for the probability density

$$\mu(x) = \frac{1}{2a}\mathbf{1}_{[-a,a]}(x),$$

where the sample space is all of $\mathbb{R}$?

4.2.3 a. Repeat Exercise 4.2.2 for $f(x) = x^2$.

b. Repeat for $f(x) = x^3$.

4.2.4 Let $A = [a_1, b_1] \times \cdots \times [a_n, b_n]$ be a box in $\mathbb{R}^n$, of constant density $\mu = 1$. Show that the center of gravity $\overline{\mathbf{x}}$ of the box is the center of the box – i.e., the point $\mathbf{c}$ with coordinates $c_i = (a_i + b_i)/2$.

---

[3]More precisely, the *lognormal distribution*, in which prices are equally likely to go up or down by a given percentage, rather than a given number of dollars.

4.3 What functions can be integrated?

What functions are integrable? It would be fairly easy to build up a fair collection by ad hoc arguments, but instead we prove in this section three theorems answering that question. In particular, they will guarantee that all usual functions are integrable.

The first, Theorem 4.3.1, is based on our notion of dyadic pavings. The second, Theorem 4.3.6, states that any continuous function on $\mathbb{R}^n$ with bounded support is integrable. The third, Theorem 4.3.9, is stronger than the second; it tells us that a function with bounded support does not have to be continuous everywhere to be integrable; it is enough that it be continuous except on a set of volume 0.

First, we will discuss Theorem 4.3.1, based on dyadic pavings. Although the index under the sum sign may look unfriendly, the proof is reasonably easy, which doesn't mean that the criterion for integrability that it gives is easy to verify in practice. We don't want to suggest that this theorem is not useful; on the contrary, it is the foundation of the whole subject. But proving that your function satisfies the hypotheses is usually a difficult theorem in its own right. The other theorems state that entire classes of functions satisfy the hypotheses, so that verifying integrability becomes a matter of seeing whether a function belongs to a particular class.

Theorem 4.3.9 is adequate for most functions you will meet. However, it is not the strongest possible statement. In Section 4.4 we prove a harder result: a function $f : \mathbb{R}^n \to \mathbb{R}$, bounded and with bounded support, is integrable if and only if it is continuous except on a set of *measure* 0. The notion of measure 0 is rather subtle and surprising; with this notion, we see that some very strange functions are integrable. Such functions actually arise in statistical mechanics.

In Section 4.11 we discuss Lebesgue integration, which does not require that functions be bounded with bounded support.

Theorem 4.3.1 (Criterion for integrability). *A function $f : \mathbb{R}^n \to \mathbb{R}$ is integrable if and only if it is bounded with bounded support, and for all $\epsilon > 0$, there exists N such that*

$$\overbrace{\sum_{\{C \in \mathcal{D}_N \mid \operatorname{osc}_C(f) > \epsilon\}} \operatorname{vol}_n C}^{\substack{\text{volume of all cubes for which}\\ \text{the oscillation of } f \text{ over the cube is } >\epsilon}} < \epsilon. \qquad 4.3.1$$

Inequality 4.3.1: Recall that $\mathcal{D}_N$ denotes the collection of all dyadic cubes at a single level N, and that $\operatorname{osc}_C(f)$ denotes the oscillation of f over C: the difference between its supremum and infimum over C.

Epsilon has the units of vol_n. If $n = 2$, epsilon is measured in centimeters (or meters ...) squared; if $n = 3$ it is measured in centimeters (or whatever) cubed.

In inequality 4.3.1 we sum the volume of only those cubes for which the oscillation of the function is more than epsilon. If, by making the cubes very small (choosing N sufficiently large) the sum of their volumes is less than epsilon, then the function is integrable: we can make the difference between the upper sum and the lower sum arbitrarily small; the two have a common limit. (The other cubes, with small oscillation, contribute arbitrarily little to the difference between the upper and the lower sum.)

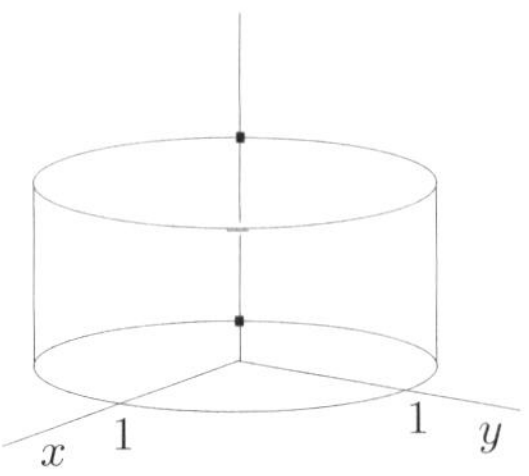

FIGURE 4.3.1.

Graph of the indicator function of the unit disc, $\mathbf{1}_D$.

Examples 4.3.2 (Integrable functions). Consider the indicator function $\mathbf{1}_D$ that is 1 on a disc and 0 outside, as shown in Figure 4.3.1. Cubes C that are completely inside or completely outside the disc have $\operatorname{osc}_C(\mathbf{1}_D) = 0$. Cubes straddling the border have oscillation equal to 1. (Actually, these cubes are squares, since $n = 2$.) By making the squares small enough (choosing N sufficiently large), you can make the area of those that straddle the boundary arbitrarily small. Therefore, $\mathbf{1}_D$ is integrable.

You may object that there will be a whole lot of squares, so how can their volume be less than epsilon? Of course, when we make the squares small, we need more of them to cover the border. But as we divide the original border squares into smaller ones, some of them no longer straddle the border. (Note that this is not quite a proof; it is intended to help you understand the meaning of the statement of Theorem 4.3.1.)

Figure 4.3.2 shows another integrable function, $\sin \frac{1}{x}$. Near 0, we see that a small change in x produces a big change in $f(x)$, leading to a large oscillation. But we can still make the difference between upper and lower sums arbitrarily small by choosing N sufficiently large, and thus the intervals sufficiently small. Theorem 4.3.9 justifies our statement that this function is integrable. $\triangle$

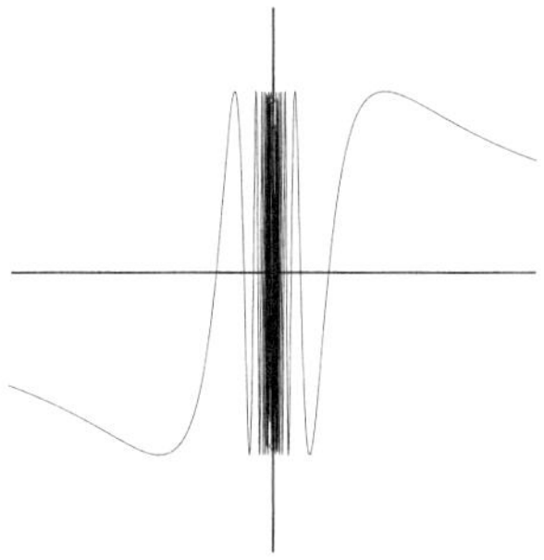

FIGURE 4.3.2.

The function $\sin \frac{1}{x}$ is integrable over any bounded interval. Dyadic intervals sufficiently near 0 always have oscillation 2, but they have small length when the paving is fine. The center region is black because there are infinitely many oscillations in that region.

Example 4.3.3 (A nonintegrable function). The function that is 1 at rational numbers in $[0,1]$ and 0 elsewhere is not integrable. No matter how small you make the cubes (intervals in this case), *each* cube will still contain both rational and irrational numbers and will have osc $= 1$. $\triangle$

Proof of Theorem 4.3.1. It follows from Definition 4.1.12 that to be integrable, a function must be bounded with bounded support, so we are only concerned with inequality 4.3.1. First we will prove that the existence of an N satisfying inequality 4.3.1 implies integrability. Choose any $\epsilon > 0$, and let N satisfy inequality 4.3.1. Then

$$U_N(f) - L_N(f) \leq \overbrace{\sum_{\{C \in \mathcal{D}_N \mid \operatorname{osc}_C(f) > \epsilon\}} 2 \sup|f| \operatorname{vol}_n C}^{\text{contribution from cubes with osc}>\epsilon} + \overbrace{\sum_{\substack{\{C \in \mathcal{D}_N \mid \operatorname{osc}_C(f) \leq \epsilon \\ \text{and } C \cap \operatorname{Supp}(f) \neq \phi\}}} \epsilon \operatorname{vol}_n C}^{\text{contribution from cubes with osc}\leq\epsilon}$$

$$\leq \epsilon\big(2 \sup|f| + \operatorname{vol}_n C_{\operatorname{Supp}}\big), \qquad 4.3.2$$

where $\sup|f|$ is the supremum of $|f|$, and C_{Supp} is a (great big) cube containing the support of f (see Definition 4.1.2).

The first sum on the right side of inequality 4.3.2 concerns only those cubes for which osc $> \epsilon$. Each such cube contributes at most $2 \sup|f| \operatorname{vol}_n C$ to the maximum difference between upper and lower sums. (It is $2 \sup|f|$ rather than $\sup|f|$ because the value of f over a single cube might swing from a positive number to a negative one. We could also express this difference as $\sup f - \inf f$.)

The second sum concerns the cubes for which osc $\leq \epsilon$. We must specify that we count only those cubes for which f has, at least somewhere in the cube, a nonzero value; that is why we say $\{\, C \mid C \cap \operatorname{Supp}(f) \neq \phi \,\}$. Since by definition the oscillation for each of those cubes is at most ϵ, each contributes at most $\epsilon \operatorname{vol}_n C$ to the difference between upper and lower sums.

We have assumed that it is possible to choose N such that the cubes for which osc $> \epsilon$ have total volume less than ϵ, so in the second line of inequality 4.3.2 we replace the first sum by $2\epsilon \sup|f|$. Factoring out ϵ, we see that by choosing N sufficiently large, the difference between upper and lower sums can be made arbitrarily small. Therefore, the function is integrable. This takes care of the "if" part of Theorem 4.3.1.

For the "only if" part we must prove that if f is integrable, then there exists an appropriate N. Suppose not. Then there exists $\epsilon_0 > 0$, such that for *all* N,

$$\sum_{\{\, C\in\mathcal{D}_N \mid \operatorname{osc}_C(f)>\epsilon_0 \,\}} \operatorname{vol}_n C \geq \epsilon_0. \tag{4.3.3}$$

In this case, for any N we have

$$\begin{aligned} U_N(f) - L_N(f) &= \sum_{C\in\mathcal{D}_N} \operatorname{osc}_C(f) \operatorname{vol}_n C \\ &\geq \sum_{\{\, C\in\mathcal{D}_N \mid \operatorname{osc}_C(f)>\epsilon_0 \,\}} \overbrace{\operatorname{osc}_C(f)}^{>\epsilon_0} \ \overbrace{\operatorname{vol}_n C}^{\substack{\text{sum of these is}\\ \geq \epsilon_0}} \geq \epsilon_0^2. \end{aligned} \tag{4.3.4}$$

The sum of $\operatorname{vol}_n C$ is at least ϵ_0, by inequality 4.3.3, so the upper and the lower integrals will differ by at least ϵ_0^2, and will not tend to a common limit. But we started with the assumption that the function is integrable. □

Note in inequality 4.3.2 the surprising but absolutely standard way we prove that something is zero: we prove that it is smaller than an arbitrarily small $\epsilon > 0$. Or we prove that it is smaller than $u(\epsilon)$, when u is a function such that $u(\epsilon) \to 0$ as $\epsilon \to 0$. Theorem 1.5.14 states that these conditions are equivalent.

You might object that in inequality 4.3.4 we argue that the ϵ_0^2 in the last line means the sums *don't* converge; yet the square of a small number is smaller yet. The difference is that inequality 4.3.4 concerns *one particular* $\epsilon_0 > 0$, which is fixed, while inequality 4.3.2 concerns *any* $\epsilon > 0$, which we can choose arbitrarily small.

To review how to negate statements, see Section 0.2.

Proposition 4.3.4 tells us why the indicator function of the disc discussed in Example 4.3.2 is integrable. We argued in that example that we can make the area of cubes straddling the boundary arbitrarily small. Now we justify that argument. The boundary of the disc is the union of two graphs of integrable functions; Proposition 4.3.4 says that any bounded part of the graph of an integrable function has volume 0.

In Proposition 4.3.4, it would be simpler to write

$$\operatorname{vol}_{n+1}\Big(\Gamma(f)\Big) = 0.$$

But our definition for integrability requires that an integrable function have bounded support. Although f is integrable, hence has bounded support, it is *defined* on all of $\mathbb{R}^n$. So although it has value 0 outside some fixed big cube, its graph still exists outside the fixed cube, and the indicator function of its graph does not have bounded support. We fix this problem by speaking of the volume of the intersection of the graph with the $(n+1)$-dimensional bounded region $C_0\times\mathbb{R}$. You should imagine that C_0 is big enough to contain the support of f, though the proof works in any case.

Proposition 4.3.4 (Bounded part of graph of integrable function has volume 0). *Let $f : \mathbb{R}^n \to \mathbb{R}$ be an integrable function with graph $\Gamma(f)$, and let $C_0 \subset \mathbb{R}^n$ be any dyadic cube. Then*

$$\operatorname{vol}_{n+1}\big(\underbrace{\Gamma(f)\cap(C_0\times\mathbb{R})}_{\text{bounded part of graph}}\big) = 0. \tag{4.3.5}$$

As you would expect, a curve in the plane has area 0, a surface in $\mathbb{R}^3$ has volume 0, and so on. Below we must stipulate that such manifolds be compact, since we have defined volume only for bounded subsets of $\mathbb{R}^n$.

Proposition 4.3.5 (Volume of compact submanifold). *If $M \subset \mathbb{R}^n$ is a manifold of dimension $k < n$, then any compact subset $X \subset M$ satisfies $\operatorname{vol}_n(X) = 0$. In particular, any bounded part of a subspace of dimension $k < n$ has n-dimensional volume 0.*

Propositions 4.3.4 and 4.3.5 are proved in Appendix A20.

Integrability of continuous functions with bounded support

What functions satisfy the hypothesis of Theorem 4.3.1? One important class consists of continuous functions with bounded support.

Theorem 4.3.6. *Any continuous function $\mathbb{R}^n \to \mathbb{R}$ with bounded support is integrable.*

The support of a function is closed by definition, so if it is bounded, it is compact.

Proof. Theorem 4.3.6 follows almost immediately from Theorem 1.6.11. Let f be continuous with bounded (hence compact) support. Since $\sup |f|$ is realized on the support, f is bounded. By Theorem 1.6.11, f is uniformly continuous: choose ϵ and find $\delta > 0$ such that

$$|\mathbf{x} - \mathbf{y}| < \delta \implies |f(\mathbf{x}) - f(\mathbf{y})| < \epsilon. \qquad 4.3.6$$

For all N such that $\sqrt{n}/2^N < \delta$, any two points of a cube of $\mathcal{D}_N(\mathbb{R}^n)$ are at most distance δ apart. Thus if $C \in \mathcal{D}_N(\mathbb{R}^n)$, then

$$|f(\mathbf{x}) - f(\mathbf{y})| < \epsilon. \qquad 4.3.7$$

This proves the theorem: f satisfies a much stronger requirement than Theorem 4.3.1 requires. Theorem 4.3.1 only requires that the oscillation be greater than ϵ on a set of cubes of total volume $< \epsilon$, whereas in this case for sufficiently large N there are *no* cubes of $\mathcal{D}_N(\mathbb{R}^n)$ with oscillation $\geq \epsilon$. □

Corollary 4.3.7 is not just a special case of Proposition 4.3.4 because although we could define f on all of $\mathbb{R}^n$ by having it be 0 outside X, we are not requiring that such an extension of f be integrable, and it may not be.

Corollary 4.3.7. *Let $X \subset \mathbb{R}^n$ be compact and let $f : X \to \mathbb{R}$ be continuous. Then the graph $\Gamma_f \subset \mathbb{R}^{n+1}$ has volume 0.*

Proof. Since X is compact, it is bounded, and there is a number A such that for all N, the number of cubes $C \in \mathcal{D}_N(\mathbb{R}^n)$ such that $X \cap C \neq \phi$ is at most $A2^{nN}$. Choose $\epsilon > 0$, and use Theorem 1.6.11 to find $\delta > 0$ such that if $\mathbf{x}_1, \mathbf{x}_2 \in X$,

$$|\mathbf{x}_1 - \mathbf{x}_2| < \delta \implies |f(\mathbf{x}_1) - f(\mathbf{x}_2| < \epsilon. \qquad 4.3.8$$

Further choose N such that $\sqrt{n}/2^N < \delta$, so that for any $C \in \mathcal{D}_N(\mathbb{R}^n)$ and any $\mathbf{x}_1, \mathbf{x}_2 \in C$, we have $|\mathbf{x}_1 - \mathbf{x}_2| < \delta$.

For any $C \in \mathcal{D}_N(\mathbb{R}^n)$ such that $C \cap X \neq \phi$, at most $2^N\epsilon + 2$ cubes of $\mathcal{D}_N(\mathbb{R}^{n+1})$ that project to C intersect Γ_f, hence Γ_f is covered by at most $A2^{nN}(2^N\epsilon + 2)$ cubes with total volume

$$\frac{1}{2^{(n+1)N}} A2^{nN}(2^N\epsilon + 2), \qquad 4.3.9$$

which can be made arbitrarily small by taking ϵ sufficiently small and N sufficiently large. □

Corollary 4.3.8. *Let $U \subset \mathbb{R}^n$ be open and let $f : U \to \mathbb{R}$ be a continuous function. Then any compact part Y of the graph of f has $(n+1)$-dimensional volume 0.*

Corollary 4.3.8: You might expect that any graph of a continuous function $f : U \to \mathbb{R}$ would have $(n+1)$-dimensional volume 0, but this is not true, even if U and f are bounded; see Exercise 4.33.

Proof. The projection of Y into $\mathbb{R}^n$ is compact, and the restriction of f to that projection satisfies the hypotheses of Corollary 4.3.7. □

Integrability of function continuous except on set of volume 0

Our third theorem shows that a function need not be continuous everywhere to be integrable. This theorem is much harder to prove than the first two, but the criterion for integrability is much more useful.

Theorem 4.3.9. *A function $f : \mathbb{R}^n \to \mathbb{R}$, bounded with bounded support, is integrable if it is continuous except on a set of volume 0.*

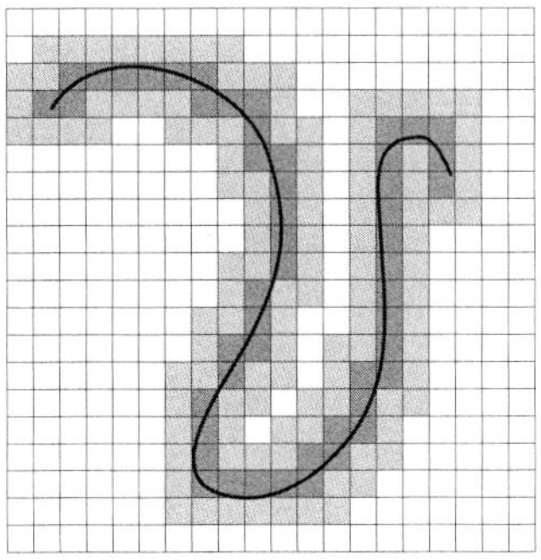

FIGURE 4.3.3.

Proof of Theorem 4.3.9: The black curve is Δ, the set of points where f is discontinuous. The dark cubes intersect Δ. The light cubes are cubes at the same depth that border at least one of the dark cubes. We denote by L the union of all the shaded cubes.

Note that Theorem 4.3.9 is not an "if and only if" statement. As will be seen in Section 4.4, it is possible to find functions that are discontinuous at all the rationals yet still are integrable.

Proof. Denote by Δ ("delta") the set of points where f is discontinuous. Since f is continuous except on a set of volume 0, we have $\operatorname{vol}_n \Delta = 0$. So (by Proposition 4.1.23) for every $\epsilon > 0$, there exist N and some finite union of cubes $C_1, \dots, C_k \in \mathcal{D}_N(\mathbb{R}^n)$ such that

$$\Delta \subset C_1 \cup \dots \cup C_k \quad \text{and} \quad \sum_{i=1}^{k} \operatorname{vol}_n C_i \le \frac{\epsilon}{3^n}. \qquad 4.3.10$$

Now we create a "buffer zone" around the discontinuities: let L be the union of the C_i and all the bordering cubes at level N, as shown in Figure 4.3.3. As illustrated by Figure 4.3.4, we can completely surround each C_i, using $3^n - 1$ cubes. Since the total volume of all the C_i is less than $\epsilon/3^n$,

$$\operatorname{vol}_n(L) \le \epsilon. \qquad 4.3.11$$

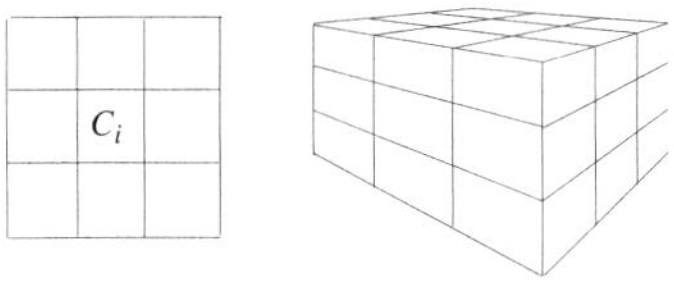

FIGURE 4.3.4.

LEFT: It takes $3^2 - 1 = 8$ cubes to surround a cube C_i in $\mathbb{R}^2$. RIGHT: In $\mathbb{R}^3$, it takes $3^3 - 1 = 26$ cubes. If we include C_i, then 3^2 cubes are enough in $\mathbb{R}^2$, and 3^3 are enough in $\mathbb{R}^3$.

All that remains is to show that there exists $M \ge N$ such that if $C \in \mathcal{D}_M(\mathbb{R}^n)$ and $C \not\subset L$, then $\operatorname{osc}_C(f) \le \epsilon$. If we can do that, we will have shown that a decomposition exists at which the total volume of all cubes over which $\operatorname{osc}(f) > \epsilon$ is less than ϵ, which is the criterion for integrability given by Theorem 4.3.1.

Suppose no such M exists. Then for every $M \ge N$, there is a cube $C \in \mathcal{D}_M$ and points $\mathbf{x}_M, \mathbf{y}_M \in C$ with $|f(\mathbf{x}_M) - f(\mathbf{y}_M)| > \epsilon$.

The sequence $M \mapsto \mathbf{x}_M$ is bounded in $\mathbb{R}^n$, so we can extract a subsequence $i \mapsto \mathbf{x}_{M_i}$ that converges to some point $\mathbf{a}$. Since, by the triangle inequality,

$$|f(\mathbf{x}_{M_i}) - f(\mathbf{a})| + |f(\mathbf{a}) - f(\mathbf{y}_{M_i})| \ge |f(\mathbf{x}_{M_i}) - f(\mathbf{y}_{M_i})| > \epsilon, \qquad 4.3.12$$

we see that at least one of $|f(\mathbf{x}_{M_i}) - f(\mathbf{a})|$ and $|f(\mathbf{y}_{M_i}) - f(\mathbf{a})|$ does not converge to 0, so f is not continuous at $\mathbf{a}$, i.e., $\mathbf{a}$ is in Δ. But this contradicts the fact that $\mathbf{a}$ is a limit of points outside of L. Since the length of a side of a cube is $1/2^N$, all $\mathbf{x}_{M_i}$ are at least $1/2^N$ away from points of Δ, so $\mathbf{a}$ is also at least $1/2^N$ away from points of Δ. $\square$

Corollary 4.3.10. *Let $f:\mathbb{R}^n \to \mathbb{R}$ be integrable, and let $g:\mathbb{R}^n \to \mathbb{R}$ be a bounded function. If $f = g$ except on a set of volume 0, then g is integrable, and*

$$\int_{\mathbb{R}^n} f\,|d^n\mathbf{x}| = \int_{\mathbb{R}^n} g\,|d^n\mathbf{x}|. \tag{4.3.13}$$

Proof. The support of g is bounded, since the support of f is bounded and so is the support of $g - f$. For any N, a cube $C \in \mathcal{D}_N(\mathbb{R}^n)$ where $\text{osc}_C(g) > \epsilon$ is either one where $\text{osc}_C(f) > \epsilon$ or one that intersects the support of $f - g$ (or both). Choose $\epsilon > 0$. The total volume of cubes of first kind is $< \epsilon$ for N sufficiently large by Theorem 4.3.1, and those of the second kind have total volume $< \epsilon$ by Proposition 4.1.23. □

Corollary 4.3.11: For example, the indicator function of the disc is integrable, since the disc is bounded by the union of the graphs of

$$y = +\sqrt{1-x^2}, \quad y = -\sqrt{1-x^2}.$$

Exercise 4.3.3 asks you to give an explicit bound for the number of cubes of $\mathcal{D}_N(\mathbb{R}^2)$ needed to cover the unit circle.

Corollary 4.3.11 says that virtually all examples that occur in vector calculus are integrable. It follows from Theorem 1.6.9, Corollary 4.3.8, and Theorem 4.3.9.

Corollary 4.3.11. *Let $A \subset \mathbb{R}^n$ be a compact region bounded by a finite union of graphs of continuous functions, and let $f : A \to \mathbb{R}$ be continuous. Then the function $\widetilde{f} : \mathbb{R}^n \to \mathbb{R}$ that is $f(\mathbf{x})$ for $\mathbf{x} \in A$ and 0 outside A is integrable.*

Polynomials are of course not integrable, since they do not have bounded support, but they can be integrated over sets of finite volume:

Corollary 4.3.12. *Any polynomial function p can be integrated over any set A of finite volume; that is, $p \cdot \mathbf{1}_A$ is integrable.*

Proof. The function $p \cdot \mathbf{1}_A$ meets the conditions of Theorem 4.3.9: it is bounded with bounded support and is continuous except on the boundary of A, which has volume 0. □

Exercises for Section 4.3

Hint for Exercise 4.3.3: Imitate the proof of Proposition 4.3.4, writing the unit circle as the union of four graphs: the graph of

$$y = \sqrt{1-x^2}$$

for $|x| \le \sqrt{2}/2$ and the three other graphs obtained by rotating that graph around the origin by multiples of $\pi/2$.

4.3.1 a. Prove that if $f:\mathbb{R}^n \to \mathbb{R}$ is integrable, then $|f|$ is integrable and $|\int f| \le \int |f|$.

b. Find an example where $|f|$ is integrable and f is not integrable.

4.3.2 Show that a subset of a set of volume zero has volume zero.

4.3.3 a. Give an explicit upper bound (in terms of N) for the number of squares $C \in \mathcal{D}_N(\mathbb{R}^2)$ needed to cover the unit circle in $\mathbb{R}^2$, so that the area of the squares goes to 0 as N goes to infinity.

b. Use that bound to show that the area of the unit circle is 0.

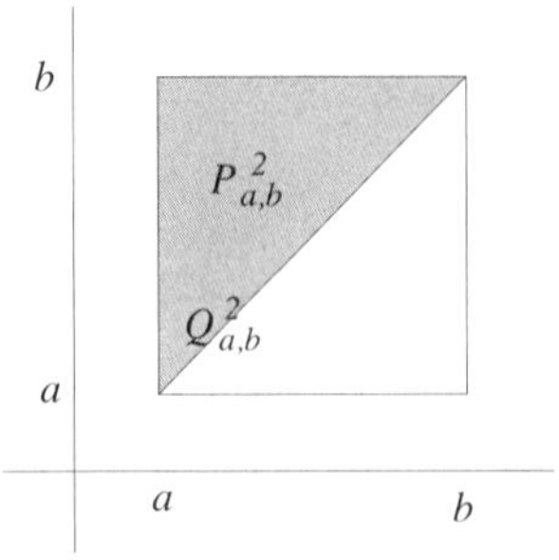

Figure for Exercise 4.3.4

Exercise 4.3.4, part a: Permutations are discussed in Section 4.8. Exercise 4.5.17 gives further applications of Exercise 4.3.4.

4.3.4 For any real numbers $a < b$, let

$$Q^n_{a,b} = \{\, \mathbf{x} \in \mathbb{R}^n \mid a \le x_i \le b \text{ for all } 1 \le i \le n \,\},$$

and let $P^n_{a,b} \subset Q^n_{a,b}$ be the subset where $a \le x_1 \le x_2 \le \cdots \le x_n \le b$. (The case where $n = 2$ is shown in the figure in the margin.)

a. Let $f : \mathbb{R}^n \to \mathbb{R}$ be an integrable function that is symmetric in the sense that

$$f\begin{pmatrix} x_1 \\ \vdots \\ x_n \end{pmatrix} = f\begin{pmatrix} x_{\sigma(1)} \\ \vdots \\ x_{\sigma(n)} \end{pmatrix} \quad \text{for any permutation } \sigma \text{ of the symbols } 1, 2, \dots, n.$$

Show that

$$\int_{Q^n_{a,b}} f(\mathbf{x})\,|d^n\mathbf{x}| = n! \int_{P^n_{a,b}} f(\mathbf{x})\,|d^n\mathbf{x}|.$$

b. Let $f : [a,b] \to \mathbb{R}$ be an integrable function. Show that

$$\int_{P^n_{a,b}} f(x_1)f(x_2)\cdots f(x_n)|d^n\mathbf{x}| = \frac{1}{n!}\left(\int_a^b f(x)|dx|\right)^n.$$

Exercise 4.3.5: You may either apply theorems or prove the result directly. If you use theorems, you must show that they apply.

4.3.5 Let P be the region $x^2 < y < 1$. Prove that $\int_P \sin(y^2)\,|dx\,dy|$ exists.

4.4 Measure zero

There is measure in all things.—Horace

The measure theory approach to integration, *Lebesgue integration*, is discussed in Section 4.11. It is superior to Riemann integration from several points of view. It makes it possible to integrate otherwise nonintegrable functions, and it is better behaved than Riemann integration with respect to limits.

However, Lebesgue integrals are essentially impossible to compute, unless you treat the integral as a Riemann integral or a limit of such integrals.

We mentioned in Section 4.3 that the criterion for integrability given by Theorem 4.3.9 is not sharp. It is not necessary that a function (bounded and with bounded support) be continuous except on a set of volume 0 to be integrable: it is necessary and sufficient that it be continuous except on a set of *measure* 0.

Measure theory is a big topic, discussed in Appendix A21. Fortunately, the notion of *measure* 0 is much more accessible. It is a subtle notion with some bizarre consequences; it gives us a way, for example, of saying that the rational numbers "don't count". Thus it allows us to use Riemann integration to integrate some quite interesting functions, including one we explore in Example 4.4.7 as a reasonable model for space averages in statistical mechanics.

In Definition 4.4.1 of measure 0, a *box* B in $\mathbb{R}^n$ of sidelength $\delta > 0$ is a cube of the form

$$\{\, \mathbf{x} \in \mathbb{R}^n \mid a_i < x_i < a_i + \delta,\ i = 1, \dots, n \,\}. \qquad 4.4.1$$

There is no requirement that the a_i or δ be dyadic.

Definition 4.4.1 (Measure 0). A set $X \subset \mathbb{R}^n$ has *measure* 0 if for every $\epsilon > 0$, there exists an infinite sequence of open boxes B_i such that

$$X \subset \cup B_i \quad \text{and} \quad \sum \text{vol}_n(B_i) \le \epsilon. \qquad 4.4.2$$

You may have noticed that while our dyadic cubes are semi-open, the boxes in Definition 4.4.1 are open. But the theory could be built just as well with closed cubes and boxes (see Exercise 4.4.1).

The theory also applies to boxes with other shapes: Definition 4.4.1 works with the B_i as arbitrary sets with well-defined volume. Exercise 4.4.2 asks you to show that you can use balls; Exercise 4.9 asks you to show that you can use arbitrary pavable sets.

Definition 4.4.2 (Almost everywhere, almost all). The term *almost everywhere*, abbreviated *a.e.*, means *except on a set of measure 0.* A property true of all $\mathbf{x}$ except for $\mathbf{x}$ on a set of measure 0 is said to be true for *almost all* $\mathbf{x}$.

The crucial difference between measure and volume is the word "infinite" in Definition 4.4.1. A set with volume 0 can be contained in a *finite* sequence of cubes whose total volume is arbitrarily small. A set with volume 0 necessarily has measure 0, but it is possible for a set to have measure 0 but not to have a defined volume, as shown in Example 4.4.3. Note that when we speak of measure 0 of a subset $X \subset \mathbb{R}^n$, we mean n-dimensional measure 0.

FIGURE 4.4.1.
The set X, shown as a black line, is covered by boxes that overlap.

Remark. We speak of boxes rather than cubes to avoid confusion with the cubes of our dyadic pavings. In dyadic pavings, we considered "families" of cubes all the same size: the cubes at a particular resolution N, and fitting the dyadic grid. The boxes B_i of Definition 4.4.1 get small as i increases, since their total volume is less than ϵ, but it is not necessarily the case that any particular box is smaller than the one immediately preceding it. The boxes can overlap, as illustrated in Figure 4.4.1, and they are not required to square with any particular grid. △

Exercise A21.1 in the appendix explores some of the bizarre properties of the set U_ϵ of Example 4.4.3. This set is a good one to keep in mind while trying to picture the boxes B_i, because it helps us to see that while the sequence $i \mapsto B_i$ is made of $B_1, B_2, \dots$, in order, these boxes may skip around: if B_1 is centered at 1/2, then B_2 is centered at 1/3, B_3 at 2/3, B_4 at 1/4, and so on. Some boxes may be contained in others: for example, depending on the choice of ϵ, the interval centered at 17/32 may be contained in the interval centered at 1/2.

Example 4.4.3 (A set with measure 0, undefined volume). The set of rational numbers in the interval $[0, 1]$ has measure 0. You can list them in order, for instance, as 1, 1/2, 1/3, 2/3, 1/4, 2/4, 3/4, 1/5, ... (The list is infinite and includes some numbers more than once.) Center an open interval of length $\epsilon/2$ at 1, an open interval of length $\epsilon/4$ at 1/2, an open interval of length $\epsilon/8$ at 1/3, and so on. Call U_ϵ the union of these intervals. The sum of the lengths of these intervals is $\epsilon\,(1/2 + 1/4 + 1/8 + \cdots) = \epsilon$, and the length of the union is $< \epsilon$, since some of the intervals overlap.

You can place all the rationals in $[0, 1]$ in intervals that are infinite in number but whose total length is arbitrarily small! The set U_ϵ thus has measure 0. But it does not have a defined volume because every interval that is not a single point contains both rational and irrational numbers. △

A similar argument proves a stronger result:

Theorem 4.4.4 (A countable union of sets of measure 0 has measure 0). *Let $i \mapsto X_i$ be a sequence of sets of measure 0. Then*

$$X_1 \cup X_2 \cup \dots \quad \textit{is a set of measure 0.}$$

Example 4.4.3 (and, more generally, any countable set) corresponds to the case of Theorem 4.4.4 where each X_i is a single point.

Proof of Theorem 4.4.4. Since there exist infinite sequences $i \mapsto B_{j,i}$ of boxes (one sequence for each j) such that

$$\begin{aligned} X_1 &\subset B_{1,1} \cup B_{1,2} \cup \dots, \quad \text{and} \quad \sum \operatorname{vol} B_{1,i} \le \frac{\epsilon}{2} \\ X_2 &\subset B_{2,1} \cup B_{2,2} \cup \dots, \quad \text{and} \quad \sum \operatorname{vol} B_{2,i} \le \frac{\epsilon}{4} \\ &\vdots \\ X_j &\subset B_{j,1} \cup B_{j,2} \cup \dots, \quad \text{and} \quad \sum \operatorname{vol} B_{j,i} \le \frac{\epsilon}{2^j}, \\ &\vdots \end{aligned} \qquad 4.4.3$$

there exists an infinite sequence of open boxes $B_{j,i}$ such that

$$(X_1 \cup X_2 \cup \dots) \subset \bigcup_{j,i} B_{j,i} \quad \text{and} \quad \sum_{i,j} \operatorname{vol} B_{i,j} \le \epsilon. \qquad \square$$

Theorem 4.4.4 is of course not true for arbitrary unions: any set is the union of its points, which all have measure 0. Thus measure theory depends on distinguishing between countable and uncountable infinities, and could only come after Cantor's work (see Section 0.6). Indeed Riemann integration, which doesn't depend on Cantor's work, came before, but Lebesgue integration, which does, comes after.

Proof of Theorem 4.4.4: We can turn the sequences

$$B_{1,i}, \dots, B_{j,i}, \dots$$

into a single sequence by listing the boxes in some order, just as we did for the rational numbers in Example 4.4.3.

For instance, one could list first the boxes where $i + j = 2$, then those where $i + j = 3$, and so on:

$$B_{1,1}, B_{1,2}, B_{2,1}, B_{1,3}, B_{2,2}, \dots.$$

Corollary 4.4.5. *Let $B_R(\mathbf{0})$ be the ball of radius R centered at $\mathbf{0}$. If for all $R \ge 0$ the subset $X \subset \mathbb{R}^n$ satisfies $\operatorname{vol}_n(X \cap B_R(\mathbf{0})) = 0$, then X has measure 0.*

Proof. Since $X = \cup_{m=1}^{\infty} (X \cap B_m(\mathbf{0}))$, it is a countable union of sets of volume 0, hence measure 0. $\square$

Proposition 4.4.6. *Let X be a subspace of $\mathbb{R}^n$ of dimension $k < n$. Then X has measure 0, and any translate of X has measure 0.*

We could also state Proposition 4.4.6 in terms of affine subspaces: an affine subspace $X \in \mathbb{R}^n$ of dimenion $k < n$ has measure 0.

Proof. The first statement follows from Proposition 4.3.5 and Corollary 4.4.5; the second from Proposition 4.1.22. $\square$

The set described in Example 4.4.3 was already encountered in Example 4.3.3, where we found that we could not integrate the function that is 1 at the rational numbers in $[0, 1]$ and 0 elsewhere. This function is discontinuous everywhere; in every interval, no matter how small, it jumps from 0 to 1 and from 1 to 0. In Example 4.4.7 we see a function that looks similar but is very different. This function *is* continuous except over a set of measure 0 and thus *is* integrable. It arises in real life (statistical mechanics, at least).

Example 4.4.7: "In lowest terms" traditionally includes the requirement that the denominator be positive, so f is well defined for negative numbers: $-\frac{1}{2}$ can be written only as $\frac{-1}{2}$, not as $\frac{1}{-2}$.

Example 4.4.7 (An integrable function with discontinuities on a set not of volume 0). Consider the function

$$f(x) = \begin{cases} \frac{1}{q} & \text{if } x = \frac{p}{q} \text{ is rational, written in lowest terms with } q > 0 \text{ and } |x| \le 1 \\ 0 & \text{if } x \text{ is irrational or } |x| > 1. \end{cases}$$

This function is integrable although it is discontinuous at values of x for which $f(x) \ne 0$. For instance, $f(3/4) = 1/4$, but arbitrarily close to

3/4 we have irrational numbers, giving $f(x) = 0$. But such values form a set of measure 0. The function is continuous at the irrationals: arbitrarily close to any irrational number x you will find rational numbers p/q, but you can choose a neighborhood of x that includes only rational numbers with arbitrarily large denominators q, so that f will be arbitrarily small in such a neighborhood. △

FIGURE 4.4.2.
Ludwig Boltzmann (1844–1906) is credited with the invention of statistical mechanics. A mathematical physicist, he also gave lectures in Vienna on philosophy so popular that no lecture hall large enough could be found.

"The most ordinary things are to philosophy a source of insoluble puzzles," he wrote. "With infinite ingenuity it constructs a concept of space or time and then finds it absolutely impossible that there be objects in this space or that processes occur during this time ... the source of this kind of logic lies in excessive confidence in the so-called laws of thought." In Hegel he found an "unclear thoughtless flow of words"; in Kant, "there were many things that I could grasp so little that ... I almost suspected that he was pulling the reader's leg or was even an imposter."

For the KAM theorem, see "A proof of Kolmogorov's theorem on the conservation of invariant tori," J. Hubbard and Y. Ilyashenko, *Discrete and Continuous Dynamical Systems*, vol. 10, N.1 & 2, Jan. and March 2004, pp. 367–385.

The function of Example 4.4.7 is important because it is a model for functions that show up in an essential way in statistical mechanics (unlike the function of Example 4.3.3, which, as far as we know, is only a pathological example, devised to test the limits of mathematical statements).

In statistical mechanics, one tries to describe a system, typically a gas enclosed in a box, made up of perhaps 10^{25} molecules. Quantities of interest might be temperature, pressure, concentrations of various chemical compounds, etc. A state of the system is a specification of the position and velocity of each molecule (and rotational velocity, vibrational energy, etc., if the molecules have inner structure); to encode this information, one might use a point in some gadzillion-dimensional space.

Mechanics tells us that at the beginning of our experiment, the system is in some state that evolves according to the laws of physics, "exploring" as time proceeds some part of the total state space (and exploring it quite fast relative to our time scale: particles in a gas at room temperature typically travel at several hundred meters per second and undergo millions of collisions per second). The guess underlying thermodynamics is that the quantity one measures, which is really a time average of the quantity as measured along the trajectory of the system, should be nearly equal in the long run to the average over all possible states, called the *space average*. (The "long run" is quite a short run by our clocks.)

This equality of time averages and space averages is called *Boltzmann's ergodic hypothesis*. It is the key hypothesis that connects statistical mechanics to thermodynamics, and physicists believe that it holds in great generality, although there aren't many mechanical systems where it is mathematically proved to be true, and the KAM theorem shows that in some cases where you might expect it to be true, it is not.[4]

What does this have to do with the function f of Example 4.4.7? Even if you believe that a generic time evolution will explore state space fairly evenly, there will always be some trajectories that don't. Consider the (considerably simplified) model of a single particle, moving without friction on a square billiard table, with ordinary bouncing when it hits an edge (the angle of incidence equal to the angle of reflection). Most trajectories (those that start with irrational slope) will evenly fill up the table. But those with

[4]The KAM theorem, due to Andrei Kolmogorov, Vladimir Arnold, and Jürgen Moser, applies to all frictionless systems of classical mechanics. It shows that under some conditions, these systems are inherently stable. The difference between stability and chaos can hinge on a delicate question in number theory: to what extent irrational numbers can be approximated by rational numbers.

rational slopes emphatically will not: they form closed trajectories, which go over and over the same closed path.

Still, as shown in Figure 4.4.3, these closed paths visit more and more of the table as the denominator of the slope becomes large.

Suppose further that the quantity to be observed is some function f on the table with average 0, which is positive near the center and very negative near the corners, as indicated by the lack of shading near the corners in Figure 4.4.3. Suppose we start our particle at the center of the table but don't specify its direction.

Trajectories through the center of the table, and with slope 0, will have positive time averages, as will trajectories with slope ∞; the trajectories will miss the corners. Similarly, we believe that the average, over time, of each trajectory with rational slope will also be positive. But trajectories with irrational slope will have 0 time averages: given enough time, these trajectories will visit each part of the table equally. And trajectories with rational slopes with large denominators will have time averages close to 0.

This billiard table example is a caricature of the kind of thing we might run into in statistical mechanics, for instance, having one gas in half a box and another in another half, and removing the partition between them.

Statistical mechanics is an attempt to apply probability theory to large systems of particles, to estimate average quantities, like temperature, pressure, etc., from the laws of mechanics. Thermodynamics, on the other hand, is a completely macroscopic theory. It tries to relate the same macroscopic quantities (temperature, pressure, etc.) on a phenomenological level. Clearly, one hopes to explain thermodynamics by statistical mechanics.

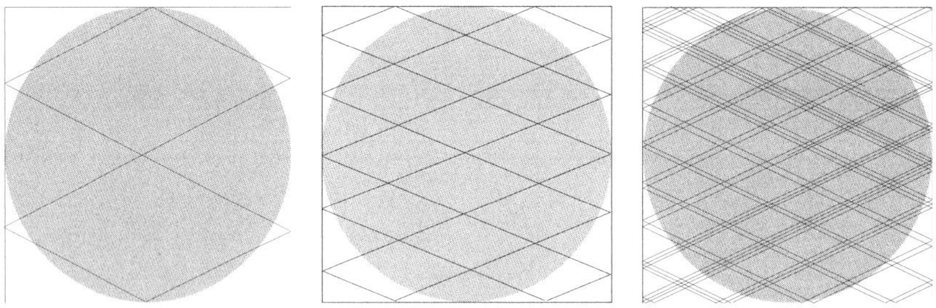

FIGURE 4.4.3. The trajectory with slope $2/5$, at center, visits more of the square than the trajectory with slope $1/2$, at left. The slope of the trajectory at right closely approximates an irrational number; if allowed to continue, this trajectory would visit nearly every part of the square. (The square would soon become black.)

Questions of rational versus irrational numbers are central to understanding the intricate interplay of chaotic and stable behavior exhibited, for example, by the *lakes of Wada*. For more on this topic, see J.H. Hubbard, "What it Means to Understand a Differential Equation," *The College Mathematics Journal*, Vol. 25 (Nov. 5, 1994), 372–384.

Because the rational numbers have measure 0, their contribution to the average does not matter; in this case, at least, Boltzmann's ergodic hypothesis seems correct. △

Integrability of "almost" continuous functions

We are now ready to state and prove Theorem 4.4.8. This theorem makes it clear that whether or not a function is integrable does not depend on where it is placed on some arbitrary grid.

Theorem 4.4.8 (What functions are integrable). *Let $f : \mathbb{R}^n \to \mathbb{R}$ be bounded with bounded support. Then f is integrable if and only if it is continuous except on a set of measure 0.*

Theorem 4.4.8 is stronger than Theorem 4.3.9, since any set of volume 0 also has measure 0, but not vice versa. It is also stronger than Theorem 4.3.6. It is not stronger than Theorem 4.3.1. Theorems 4.4.8 and 4.3.1 both give an "if and only if" condition for integrability; they are exactly equivalent. But it is often easier to verify that a function is integrable using Theorem 4.4.8.

We prune the list of boxes by throwing away any box that is contained in an earlier one. We could prove our result without pruning the list, but it would make the argument more cumbersome.

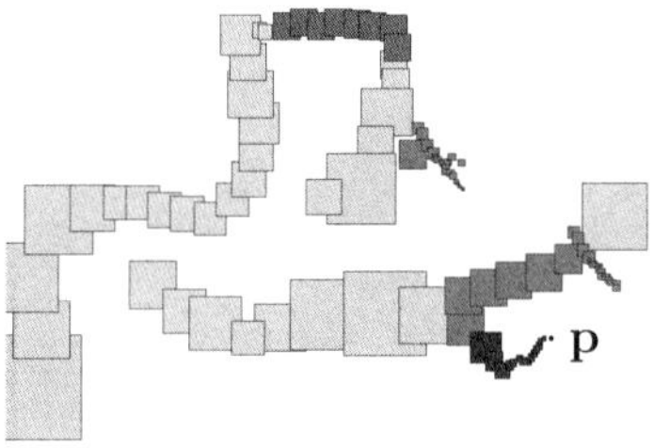

FIGURE 4.4.4.

The collection of boxes B_i covering Δ is lightly shaded. The boxes B_{i_j} with osc $> \epsilon$ are shaded slightly darker. A convergent subsequence of those is shaded darker yet. The key to Lemma 4.4.9 is that the point to which they converge must belong to some box B_i.

Proof. We will start with the harder direction: if $f : \mathbb{R}^n \to \mathbb{R}$, bounded with bounded support, is continuous almost everywhere, then it is integrable. We will use the criterion for integrability given by Theorem 4.3.1; thus we want to prove that for all $\epsilon > 0$ there exists N such that the cubes $C \in \mathcal{D}_N$ over which $\text{osc}_C(f) > \epsilon$ have a combined volume less than ϵ. (Recall from Definition 4.1.4 that $\text{osc}_{B_i}(f)$ is the oscillation of f over B_i.)

Denote by Δ the set of points where f is not continuous, and choose $\epsilon > 0$ (which will remain fixed for the duration of the proof). By Definition 4.4.1 of measure 0, there exists an infinite sequence $i \mapsto B_i$ of open boxes such that

$$\Delta \subset \cup B_i \quad \text{and} \quad \sum \text{vol}_n B_i < \epsilon. \tag{4.4.4}$$

We will choose our boxes so that no box is contained in any other.

The proof is fairly involved. First, we want to get rid of infinity.

Lemma 4.4.9. *Let $i \mapsto B_i$ be a sequence satisfying $\Delta \subset \cup B_i$ and $\sum \text{vol}_n B_i < \epsilon$. Then $\text{osc}_{B_i}(f) > \epsilon$ on only finitely many boxes B_i.*

Denote such boxes by B_{i_j}; denote the union of the B_{i_j} by L. (In Figure 4.4.4, the B_i are the lightly shaded boxes; the B_{i_j} are slightly darker.)

Proof of Lemma 4.4.9. Assume the lemma is false. Then there exist an infinite subsequence of boxes $j \mapsto B_{i_j}$ and two infinite sequences of points, $j \mapsto \mathbf{x}_j, j \mapsto \mathbf{y}_j$ in B_{i_j}, such that $|f(\mathbf{x}_j) - f(\mathbf{y}_j)| > \epsilon$.

The sequence $j \mapsto \mathbf{x}_j$ is bounded, since the support of f is bounded and $\mathbf{x}_j$ is in the support of f. So (by Theorem 1.6.3) it has a convergent subsequence $\mathbf{x}_{j_k}$ converging to some point $\mathbf{p}$. Since $|\mathbf{x}_j - \mathbf{y}_j| \to 0$ as $j \to \infty$, the subsequence $\mathbf{y}_{j_k}$ also converges to $\mathbf{p}$.

The function f is certainly not continuous at $\mathbf{p}$, so $\mathbf{p} \in \Delta$ and by formula 4.4.4, $\mathbf{p}$ has to be in a particular box; we will call it $B_{\mathbf{p}}$. (Since the boxes can overlap, it could be in more than one, but we just need one.) Since $\mathbf{x}_{j_k}$ and $\mathbf{y}_{j_k}$ converge to $\mathbf{p}$, and since the B_{i_j} get small as j gets big (their total volume being less than ϵ), then all $B_{i_{j_k}}$ after a certain point are contained in $B_{\mathbf{p}}$. But this contradicts our assumption that we had pruned our list of B_i so that no one box was contained in any other. Therefore, Lemma 4.4.9 is correct: there are only finitely many B_i on which $\text{osc}_{B_i} f > \epsilon$. □ Lemma 4.4.9

Lemma 4.4.10. *There exists N such that if $C \in \mathcal{D}_N(\mathbb{R}^n)$ and $\text{osc}_C f > \epsilon$, then $C \subset L$.*

If we prove this, we will be finished, because by Theorem 4.3.1, a bounded function with bounded support is integrable if there exists N at which the total volume of cubes with osc $> \epsilon$ is less than ϵ. We know that L is a finite set of B_i, and that the B_i have total volume $< \epsilon$.

Proof of Lemma 4.4.10. We will again argue by contradiction. Suppose the lemma is false. Then for every N, there exists a C_N not a subset of L such that $\text{osc}_{C_N} f > \epsilon$. In other words,

$$\exists \text{ points } \mathbf{x}_N, \mathbf{y}_N, \mathbf{z}_N \text{ in } C_N, \text{ with } \mathbf{z}_N \notin L \text{ and } |f(\mathbf{x}_N) - f(\mathbf{y}_N)| > \epsilon. \tag{4.4.5}$$

Since $\mathbf{x}_N, \mathbf{y}_N$, and $\mathbf{z}_N$ are infinite sequences (for $N = 1, 2, \dots$), there exist convergent subsequences $\mathbf{x}_{N_i}$, $\mathbf{y}_{N_i}$, and $\mathbf{z}_{N_i}$, all converging to the same point, which we will call $\mathbf{q}$.

What do we know about $\mathbf{q}$?

- $\mathbf{q} \in \Delta$: i.e., it is a discontinuity of f – no matter how close $\mathbf{x}_{N_i}$ and $\mathbf{y}_{N_i}$ get to $\mathbf{q}$, we have $|f(\mathbf{x}_{N_i}) - f(\mathbf{y}_{N_i})| > \epsilon$. Therefore (since all the discontinuities of f are contained in the B_i), it is in some box B_i, which we'll call $B_{\mathbf{q}}$.
- $\mathbf{q} \notin L$. (The set L is open, since it is a union of open sets; see Exercise 1.5.3. Therefore its complement is closed; since no point of the sequence $\mathbf{z}_{N_i}$ is in L, its limit, $\mathbf{q}$, is not in L either.)

Since $\mathbf{q} \in B_{\mathbf{q}}$, and $\mathbf{q} \notin L$, we know that $B_{\mathbf{q}}$ is not one of the boxes with osc $> \epsilon$. But $\mathbf{x}_{N_i}$ and $\mathbf{y}_{N_i}$ are in $B_{\mathbf{q}}$ for N_i large enough, so that $\operatorname{osc}_{B_{\mathbf{q}}} f < \epsilon$ contradicts $|f(\mathbf{x}_{N_i}) - f(\mathbf{y}_{N_i})| > \epsilon$. □ Lemma 4.4.10

You may ask, how do we know they converge to the same point? Because $\mathbf{x}_{N_i}$, $\mathbf{y}_{N_i}$, and $\mathbf{z}_{N_i}$ are all in the same cube, which is shrinking to a point as $N \to \infty$.

Recall that Δ is the set of points where f is not continuous, and L is the finite union of the boxes B_i on which $\operatorname{osc}_{B_i}(f) > \epsilon$.

The boxes B_i making up L are in $\mathbb{R}^n$, so the complement of L is $\mathbb{R}^n - L$. Recall (Definition 1.5.4) that a closed set $C \subset \mathbb{R}^n$ is a set whose complement $\mathbb{R}^n - C$ is open.

The proof of Lemma 4.4.10 means that we have proved Theorem 4.4.8 in one direction: if a bounded function with bounded support is continuous except on a set of measure 0, then it is integrable.

Now we need to prove the other direction: if a function $f : \mathbb{R}^n \to \mathbb{R}$, bounded and with bounded support, is integrable, then it is continuous almost everywhere. This is easier, but the fact that we chose our dyadic cubes half-open and our boxes open introduces a little complication.

Definition 4.4.1 specifies that the boxes B_i are open. Equation 4.1.12 defining dyadic cubes shows that they are half-open: x_i is greater than or equal to one amount, but strictly less than another:

$$\frac{k_i}{2^N} \le x_i < \frac{k_i + 1}{2^N}.$$

Choose $\delta > 0$. Since f is integrable, we know (Theorem 4.3.1) that for any $\epsilon > 0$, there exists N such that the finite union of cubes

$$\{ C \in \mathcal{D}_N(\mathbb{R}^n) \mid \operatorname{osc}_C(f) > \epsilon \} \qquad 4.4.6$$

has total volume less than ϵ. So we can choose N_1 such that if $\mathcal{C}_{N_1}$ denotes the finite collection of cubes $C \in \mathcal{D}_{N_1}(\mathbb{R}^n)$ with $\operatorname{osc}_C f > \delta/4$, these cubes have total volume less than $\delta/4$.

Now let $\mathcal{C}_{N_2}$ be the finite collection of cubes with $\operatorname{osc}_C f > \delta/8$; these cubes have total volume less than $\delta/8$. Continue with $\delta/16, \dots$. Finally, consider the infinite sequence of open boxes $B_1, B_2, \dots$ obtained by listing first the interiors of the cubes in $\mathcal{C}_{N_1}$, then those of the cubes in $\mathcal{C}_{N_2}$, etc.

This almost solves our problem: the sums of the volumes of the boxes in our sequence is at most $\delta/4 + \delta/8 + \cdots = \delta/2$. The problem is that discontinuities on the boundary of dyadic cubes may go undetected by oscillation on dyadic cubes: the value of the function over one cube could be 0, and the value over an adjacent cube could be 1; in each case the oscillation over the cube would be 0, but the function would be discontinuous at points on the border between the two cubes.

This is easily dealt with: the union of all the boundaries of all dyadic cubes (over all N) has measure 0. To see this, denote by $\delta\mathcal{D}_N(\mathbb{R}^n)$ the union of the boundaries of the dyadic cubes of $\mathcal{D}_N$. Then

1. For each N, the boundary $\delta\mathcal{D}_N(\mathbb{R}^n)$ is a countable union of translates of subspaces of dimension $n-1$, hence has measure 0 by Theorem 4.4.4 and Corollary 4.4.5.

2. The set $\cup_{N=1}^{\infty} \delta\mathcal{D}_N(\mathbb{R}^n)$ has measure 0, since it is a countable union of sets of measure 0. $\square$ Theorem 4.4.8.

Corollary 4.4.11. *Let f and g be integrable functions on $\mathbb{R}^n$ such that $f \geq g$ and $\int f(\mathbf{x})|d^n\mathbf{x}| = \int g(\mathbf{x})|d^n\mathbf{x}|$. Then*

$$\{\mathbf{x} \mid f(\mathbf{x}) \neq g(\mathbf{x})\} \text{ has measure 0.}$$

Proof. The function $f - g$ is integrable, hence continuous almost everywhere. Thus if $f > g$ on a set not of measure 0, there exists $\mathbf{x}_0$ such that $f(\mathbf{x}_0) > g(\mathbf{x}_0)$ and $f - g$ is continuous at $\mathbf{x}_0$. Then there exist $\epsilon > 0$ and $r > 0$ such that $f(\mathbf{x}) - g(\mathbf{x}) > \epsilon$ when $|\mathbf{x} - \mathbf{x}_0| \leq r$. We then have

$$\int_{\mathbb{R}^n} (f-g)(\mathbf{x})|d^n\mathbf{x}| \geq \int_{B_r(\mathbf{x}_0)} (f-g)(\mathbf{x})|d^n\mathbf{x}| \geq \epsilon \ \mathrm{vol}_n(B_r(\mathbf{x}_0)) > 0, \quad 4.4.7$$

contradicting our assumption that $\int f(\mathbf{x})|d^n\mathbf{x}| = \int g(\mathbf{x})|d^n\mathbf{x}|$. $\square$

In Proposition 4.1.14 we saw that the sum of two integrable functions is integrable, and that the product of a real number and an integrable function is integrable. It is also true that if two functions $f, g : \mathbb{R}^n \to \mathbb{R}$ are both integrable, their product is integrable (although the integral of the product is in general *not* the product of the integrals).

Corollary 4.4.12: The integral of the product is usually *not* the product of the integrals. However (Proposition 4.1.16), if f and g are functions of *different* variables, then the integral of the product over the product of the domains *is* the product of the integrals.

Corollary 4.4.12 (Product of integrable functions is integrable). *If two functions $f, g : \mathbb{R}^n \to \mathbb{R}$ are both integrable, then fg is also integrable.*

Proof. If f is continuous except on a set X_f of measure 0, and g is continuous except on a set X_g of measure 0, then (Proposition 1.5.29) fg is continuous except on a subset of $X_f \cup X_g$. By Theorem 4.4.4, this subset has measure 0. Thus by Theorem 4.4.8, fg is integrable. $\square$

Corollary 4.4.13 follows from Theorem 4.4.8.

Corollary 4.4.13 (Integration is translation invariant). *If a function $f : \mathbb{R}^n \to \mathbb{R}$ is integrable, then for any $\vec{\mathbf{v}} \in \mathbb{R}^n$, the function $\mathbf{x} \mapsto f(\mathbf{x} - \vec{\mathbf{v}})$ is integrable.*

Exercises for Section 4.4

4.4.1 Use Definition 4.1.17 of n-dimensional volume to show that the same sets have measure 0 whether you define measure 0 using open boxes or closed boxes.

4.4.2 Show that $X \subset \mathbb{R}^n$ has measure 0 if and only if for any $\epsilon > 0$ there exists an infinite sequence of balls

$$B_i = \{\mathbf{x} \in \mathbb{R}^n \mid |\mathbf{x} - \mathbf{a}_i| < r_i\} \text{ with } \sum_{i=1}^{\infty} r_i^n < \epsilon \text{ such that } X \subset \cup_{i=1}^{\infty} B_i.$$

4.4.3 Prove that any compact subset of $\mathbb{R}^n$ of measure 0 has volume 0.

Hint for Exercise 4.4.3: Use the Heine Borel theorem in Appendix A3.

****4.4.4** Consider the subset $U \subset [0,1]$ that is the union of the open intervals $\left(\frac{p}{q} - \frac{C}{q^3}, \frac{p}{q} + \frac{C}{q^3}\right)$ for all rational numbers $p/q \in [0,1]$. Show that for $C > 0$ sufficiently small, U is not pavable. What would happen if the 3 were replaced by 2? (This is really hard.)

4.4.5 Show that Corollary 4.4.11 is false if you replace "measure" by "volume".

4.4.6 Show that if the boundary of $A \subset \mathbb{R}^n$ has a well-defined n-dimensional volume, that volume is 0: $\mathrm{vol}_n\, \partial A = 0$. *Hint*: Use Theorem 4.4.8.

4.4.7 Show that the set of numbers which in base 10 can be written using only the digits 1–6 has measure 0.

FIGURE 4.5.1.

The Italian mathematician Guido Fubini (1879–1943) was forced to retire when the Manifesto of Fascist Racism made anti-Semitism official policy of Italy in 1938. He joined the Institute for Advanced Study in Princeton the following year.

4.5 Fubini's theorem and iterated integrals

We now know – in principle, at least – how to determine whether a function is integrable. Assuming it is, how do we integrate it? Fubini's theorem allows us to compute multiple integrals by hand, or at least reduce them to the computation of one-dimensional integrals. It asserts that if $f : \mathbb{R}^n \to \mathbb{R}$ is integrable, then

$$\int_{\mathbb{R}^n} f(\mathbf{x})\,|d^n\mathbf{x}| = \int_{-\infty}^{\infty}\left(\dots\left(\int_{-\infty}^{\infty} f\begin{pmatrix} x_1 \\ \vdots \\ x_n \end{pmatrix} dx_1\right)\dots\right) dx_n. \qquad 4.5.1$$

That is, first we hold the variables $x_2, \dots, x_n$ constant and integrate with respect to x_1; then we integrate the resulting (no doubt complicated) function with respect to x_2, and so on. Note that the expression on the left side of equation 4.5.1 doesn't specify the order in which the variables are taken, so the iterated integral on the right can be written in any order: we could integrate first with respect to x_n, or any other variable. This is important for both theoretical and computational uses of Fubini's theorem. Note also that in one dimension, equation 4.5.1 becomes equation 4.1.30.

Remark. Equation 4.5.1 is not quite correct, because some of the functions in parentheses on the right side of equation 4.5.1 may not be integrable; this problem is discussed in Appendix A17. We state Fubini's theorem correctly at the end of this section. For now, assume that we are in the (common) situation where equation 4.5.1 works. △

In practice, the main difficulty in setting up a multiple integral as an iterated one-dimensional integral is dealing with the "boundary" of the region over which we wish to integrate the function. We tried to sweep difficulties like the fractal coastline of Britain under the rug by choosing to integrate over all of $\mathbb{R}^n$, but of course those difficulties are still there. This is where we have to come to terms with them: we have to figure out the upper and lower limits of the integrals.

If the domain of integration looks like the coastline of Britain, it is not obvious how to go about this. For domains of integration bounded by smooth curves and surfaces, formulas exist in many cases that are of interest (particularly during calculus exams), but this is still the part that gives students the most trouble. So before computing any multiple integrals, let's see how to set them up. While a multiple integral is *computed* from inside out – first with respect to the variable in the inner parentheses – we recommend *setting up* the problem from outside in, as shown in Examples 4.5.1 and 4.5.2.

By "integrate over the triangle" we mean that we imagine that the function f is defined by some formula inside the triangle, and by $f = 0$ outside the triangle.

Example 4.5.1 (Setting up multiple integrals: an easy example). Suppose we want to integrate a function $f : \mathbb{R}^2 \to \mathbb{R}$ over the triangle

$$T = \left\{ \begin{pmatrix} x \\ y \end{pmatrix} \in \mathbb{R}^2 \,\middle|\, 0 \le 2x \le y \le 2 \right\} \qquad 4.5.2$$

shown in Figure 4.5.2. This triangle is the intersection of the three regions (here, halfplanes) defined by the inequalities $0 \le x$, $2x \le y$, and $y \le 2$.

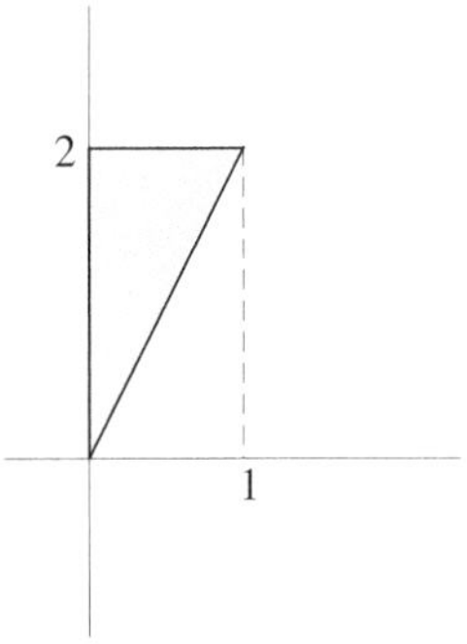

FIGURE 4.5.2.

The triangle defined by equation 4.5.2.

Say we want to integrate first with respect to y. We set up the integral as follows, temporarily omitting the limits of integration:

$$\int_{\mathbb{R}^2} f\begin{pmatrix} x \\ y \end{pmatrix} |dx\,dy| = \int \left(\int f\,dy \right) dx. \qquad 4.5.3$$

(We just write f for the function, as we don't want to complicate issues by specifying a particular function.) Starting with the outer integral – thinking first about x – we hold a pencil parallel to the y-axis and roll it over the triangle from left to right. We see that the triangle (the domain of integration) starts at $x = 0$ and ends at $x = 1$, so we write in those limits:

$$\int_0^1 \left(\int f\,dy \right) dx. \qquad 4.5.4$$

Now think about what happened as you rolled the pencil from $x = 0$ to $x = 1$: for each value of x, what values of y were in the triangle? In this simple case, for every value of x, the pencil intersects the triangle in an interval going from the hypotenuse to $y = 2$. So the upper value is $y = 2$, and the lower value is $y = 2x$, the line on which the hypotenuse lies. Thus we have

$$\int_0^1 \left(\int_{2x}^2 f\,dy \right) dx. \qquad 4.5.5$$

If we want to start by integrating f with respect to x, we write

$$\int_{\mathbb{R}^2} f\begin{pmatrix} x \\ y \end{pmatrix} |dx\,dy| = \int \left(\int f\,dx \right) dy; \qquad 4.5.6$$

starting with the outer integral, we hold our pencil parallel to the x-axis and roll it from the bottom of the triangle to the top, from $y = 0$ to $y = 2$. As we roll the pencil, we ask what are the lower and upper values of x for each value of y. The lower value is always $x = 0$, and the upper value is set by the hypotenuse, but we express it as $x = y/2$. This gives

$$\int_0^2 \left(\int_0^{\frac{y}{2}} f\, dx \right) dy. \qquad 4.5.7$$

Now suppose we are integrating over only part of the triangle, as shown in Figure 4.5.3. What limits do we put in the expression $\int (\int f\, dy)\, dx$? Try it yourself before checking the answer in the footnote.[5] △

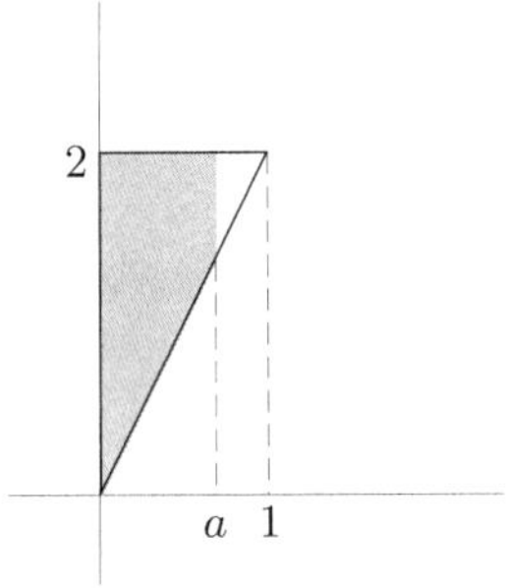

FIGURE 4.5.3.
The shaded area represents a truncated part of the triangle of Figure 4.5.2.

Exercise 4.5.3 asks you to set up the multiple integral for Example 4.5.2 when the outer integral is with respect to y. The answer will be a sum of integrals.

Example 4.5.2 (Setting up multiple integrals: a somewhat harder example). Now let's integrate an unspecified function $f : \mathbb{R}^2 \to \mathbb{R}$ over the area bordered on the top by the parabolas $y = x^2$ and $y = (x-2)^2$ and on the bottom by the straight lines $y = -x$ and $y = x - 2$, as shown in Figure 4.5.4.

Let's start again by sweeping our pencil from left to right, which corresponds to the outer integral being with respect to x. The limits for the outer integral are clearly $x = 0$ and $x = 2$, giving

$$\int_0^2 \left(\int f\, dy \right) dx. \qquad 4.5.8$$

As we sweep our pencil from left to right, we see that the lower limit for y is set by the straight line $y = -x$, and the upper limit by the parabola $y = x^2$, so we are tempted to write $\displaystyle\int_0^2 \left(\int_{-x}^{x^2} f\, dy \right) dx$, but at $x = 1$, we have a problem. The lower limit is now set by the straight line $y = x - 2$, and the upper limit by the parabola $y = (x-2)^2$. How can we express this? Try it yourself before looking at the answer in the footnote.[6] △

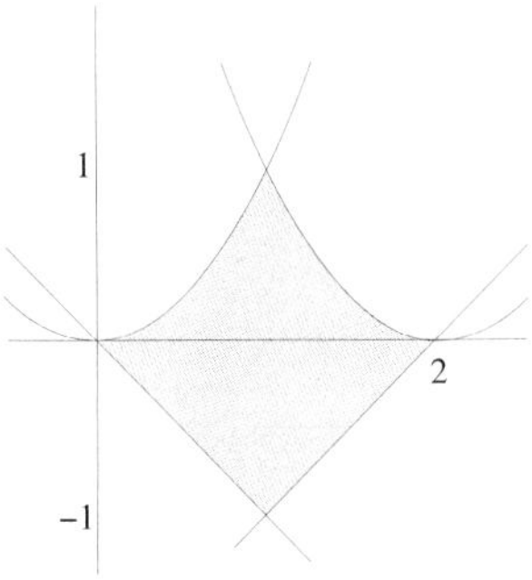

FIGURE 4.5.4.
The region of integration for Example 4.5.2.

Example 4.5.3 (Setting up a multiple integral in $\mathbb{R}^3$). Already in $\mathbb{R}^3$ this kind of visualization becomes much harder. Suppose we want to integrate a function over the pyramid P shown in Figure 4.5.5 and given by the formula

$$P = \left\{ \begin{pmatrix} x \\ y \\ z \end{pmatrix} \in \mathbb{R}^3 \mid 0 \le x;\ 0 \le y;\ 0 \le z;\ x + y + z \le 1 \right\}. \qquad 4.5.9$$

[5]When the domain of integration is the truncated triangle in Figure 4.5.3, the integral is written

$$\int_0^a \left(\int_{2x}^2 f\, dy \right) dx.$$

The other direction is harder; we will return to it in Exercise 4.5.2.

[6]We need to break up this integral into a sum of integrals:

$$\int_0^1 \left(\int_{-x}^{x^2} f\, dy \right) dx + \int_1^2 \left(\int_{x-2}^{(x-2)^2} f\, dy \right) dx.$$

Exercise 4.5.1 asks you to justify ignoring that we have counted the line $x = 1$ twice.

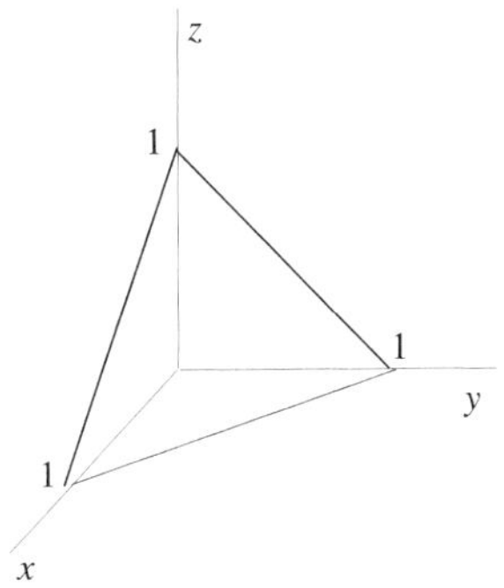

FIGURE 4.5.5.
Pyramid for Example 4.5.3.

Note that when setting up a multiple integral, the upper and lower limits for the outer integral are just numbers. Then (going from outer to inner) the limits for the next integral may depend on the variable of the outer integral. See, for instance the upper limit $y/2$ in integral 4.5.7, and the upper limit $1-z$ in integral 4.5.11. The limits for the next integral may depend on the variables of the two outer integrals; see the upper limit $1-y-z$ in formula 4.5.12.

Formula 4.5.12: When we integrate with regard to x, why is the upper value $1-y-z$, when the upper value with regard to y is $1-z$, not $1-z-x$? Since the inner integral, first to be computed, is with regard to x, we must consider all values of y and z. But when we integrate with regard to y, we have already integrated with regard to x. Clearly the value of x for which $y = 1-z-x$ is largest is $x=0$, giving upper value $y = 1-z$.

We want to figure out the limits of integration for the multiple integral

$$\int_P f\begin{pmatrix} x \\ y \\ z \end{pmatrix} |dx\,dy\,dz| = \int\left(\int\left(\int f\,dx\right)dy\right)dz. \qquad 4.5.10$$

There are six ways to apply Fubini's theorem, which in this case because of symmetries will result in the same expressions with the variables permuted. Let us vary z first. For instance, we can lift a piece of paper and see how it intersects the pyramid at various heights. Clearly the paper will only intersect the pyramid when its height is between 0 and 1. This gives $\int_0^1 (\quad)dz$, where the space needs be filled in by the double integral of f over the part of the pyramid P at height z, pictured at left in Figure 4.5.6, and again at right, this time drawn flat.

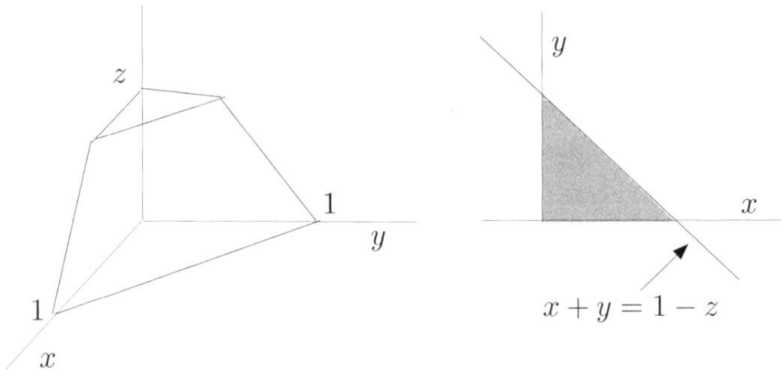

FIGURE 4.5.6. LEFT: The pyramid of Figure 4.5.5, truncated at height z. RIGHT: The plane at height z, put flat.

This time we are integrating over a triangle (which depends on z); see the right side of Figure 4.5.6. So we are in the situation of Example 4.5.1. Let's vary y first. Holding a pencil parallel to the x-axis and rolling it over the triangle from bottom to top, we see that the relevant y-values are between 0 and $1-z$ (the upper value is the largest value of y on the part of the line $y = 1-z-x$ that is in the triangle). So we write

$$\int_0^1\left(\int_0^{1-z}\left(\qquad\qquad \right)dy\right)dz, \qquad 4.5.11$$

where now the blank represents the integral over part of the horizontal line segment at height z and "depth" y. These x-values are between 0 and $1-z-y$, so finally the integral is

$$\int_0^1\left(\int_0^{1-z}\left(\int_0^{1-y-z} f\begin{pmatrix} x \\ y \\ z \end{pmatrix} dx\right)dy\right)dz. \quad \triangle \qquad 4.5.12$$

Example 4.5.4 (A less visual example). In the preceding examples, we could deduce the upper and lower limits from pictures of the region over which we want to integrate. Here is a case where the picture doesn't give the answer. Let us set up a multiple integral over the ellipse E defined by

$$E = \left\{ \begin{pmatrix} x \\ y \end{pmatrix} \mid x^2 + xy + y^2 \le 1 \right\}. \qquad 4.5.13$$

The boundary of E is defined by $y^2 + xy + x^2 - 1 = 0$, so on the boundary,

$$y = \frac{-x \pm \sqrt{x^2 - 4(x^2-1)}}{2} = \frac{-x \pm \sqrt{4-3x^2}}{2}. \qquad 4.5.14$$

Such values exist only if $4 - 3x^2 \ge 0$, i.e., $|x| \le 2\sqrt{3}/3$. So to set up the integral of a function f over E we write

Here, x and y play symmetric roles, so to integrate first with respect to x, we would just replace every y by x and vice versa.

$$\int_{-2\sqrt{3}/3}^{2\sqrt{3}/3} \left(\int_{\frac{-x-\sqrt{4-3x^2}}{2}}^{\frac{-x+\sqrt{4-3x^2}}{2}} f\begin{pmatrix} x \\ y \end{pmatrix} dy \right) dx. \quad \triangle \qquad 4.5.15$$

Now let's actually compute a few multiple integrals.

Example 4.5.5 (Computing a multiple integral). Let $f : \mathbb{R}^2 \to \mathbb{R}$ be the function $f\begin{pmatrix} x \\ y \end{pmatrix} = xy\,\mathbf{1}_S\begin{pmatrix} x \\ y \end{pmatrix}$, where S is the unit square, as shown in Figure 4.5.7. Then

$$\begin{aligned} \int_{\mathbb{R}^2} f\begin{pmatrix} x \\ y \end{pmatrix} |dx\,dy| &= \int_0^1 \left(\int_0^1 xy\,dx \right) dy \\ &= \int_0^1 \left[\frac{x^2 y}{2} \right]_{x=0}^{x=1} dy = \int_0^1 \frac{y}{2}\,dy = \frac{1}{4}. \quad \triangle \end{aligned} \qquad 4.5.16$$

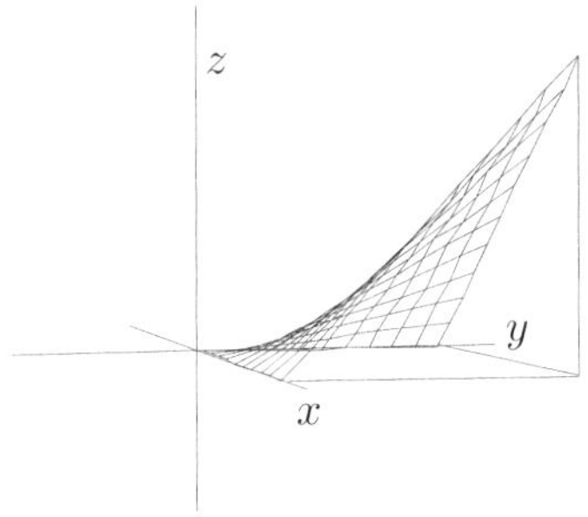

FIGURE 4.5.7.

The integral in equation 4.5.16 is 1/4: the volume under the surface defined by $f\begin{pmatrix} x \\ y \end{pmatrix} = xy$ and above the unit square is 1/4.

In Example 4.5.5 it is clear that we could have taken the integral in the opposite order and found the same result, since our function is $f\begin{pmatrix} x \\ y \end{pmatrix} = xy$. Fubini's theorem says that this is always true as long as the functions involved are integrable. This fact is useful; sometimes a multiple integral can be computed in elementary terms when written in one direction, but not in the other, as you will see in Example 4.5.6. It may also be easier to determine the limits of integration if the problem is set up in one direction rather than another, as we saw in the case of the truncated triangle shown in Figure 4.5.3.

Example 4.5.6 (Choose the easy direction). Let us integrate the function e^{-y^2} over the triangle shown in Figure 4.5.8:

$$T = \left\{ \begin{pmatrix} x \\ y \end{pmatrix} \in \mathbb{R}^2 \mid 0 \le x \le y \le 1 \right\}. \qquad 4.5.17$$

Fubini's theorem gives us two ways to write this integral as an iterated one-dimensional integral:

$$\int_0^1 \left(\int_x^1 e^{-y^2} dy \right) dx \quad \text{and} \quad \int_0^1 \left(\int_0^y e^{-y^2} dx \right) dy. \qquad 4.5.18$$

The first cannot be computed in elementary terms, since e^{-y^2} does not have an elementary antiderivative. But the second can:

$$\int_0^1 \left(\int_0^y e^{-y^2}\, dx \right) dy = \int_0^1 y e^{-y^2}\, dy = -\frac{1}{2}\left[e^{-y^2} \right]_0^1 = \frac{1}{2}\left(1 - \frac{1}{e}\right). \qquad 4.5.19$$

(The y in ye^{-y^2} makes it possible to integrate in elementary terms.) △

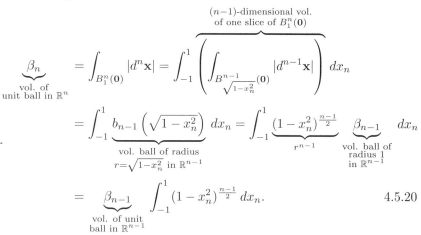

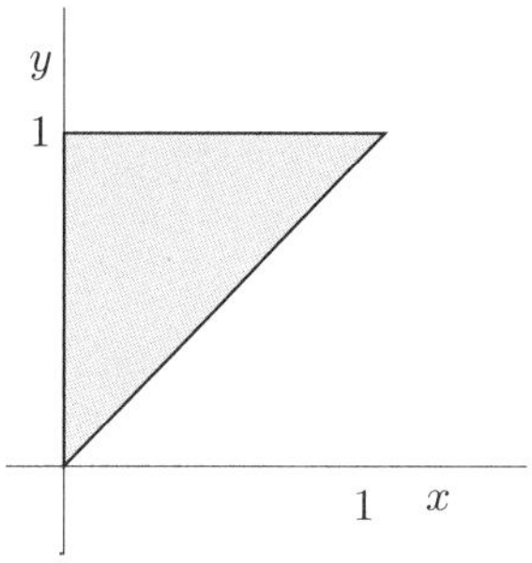

FIGURE 4.5.8.
The triangle
$$\left\{ \begin{pmatrix} x \\ y \end{pmatrix} \in \mathbb{R}^2 \mid 0 \le x \le y \le 1 \right\}$$
of Example 4.5.6.

Older textbooks contain many examples of such computational miracles. We believe it is sounder to take a serious interest in the numerical theory and go lightly over computational tricks, which do not work in any great generality in any case.

Example 4.5.7 (Volume of a ball in $\mathbb{R}^n$). Let $B_R^n(\mathbf{0})$ be the ball of radius R in $\mathbb{R}^n$, centered at $\mathbf{0}$, and let $b_n(R)$ be its volume. By Proposition 4.1.24, $b_n(R) = R^n b_n(1)$. We will denote by $\beta_n = b_n(1)$ the volume of the unit ball. By Fubini's theorem,

$$\begin{aligned}
\underbrace{\beta_n}_{\substack{\text{vol. of}\\ \text{unit ball in } \mathbb{R}^n}} &= \int_{B_1^n(\mathbf{0})} |d^n\mathbf{x}| = \int_{-1}^1 \overbrace{\left(\int_{B^{n-1}_{\sqrt{1-x_n^2}}(\mathbf{0})} |d^{n-1}\mathbf{x}| \right)}^{(n-1)\text{-dimensional vol. of one slice of } B_1^n(\mathbf{0})} dx_n \\
&= \int_{-1}^1 \underbrace{b_{n-1}\left(\sqrt{1-x_n^2}\right)}_{\substack{\text{vol. ball of radius}\\ r=\sqrt{1-x_n^2} \text{ in } \mathbb{R}^{n-1}}} dx_n = \int_{-1}^1 \underbrace{(1-x_n^2)^{\frac{n-1}{2}}}_{r^{n-1}} \underbrace{\beta_{n-1}}_{\substack{\text{vol. ball of}\\ \text{radius 1}\\ \text{in } \mathbb{R}^{n-1}}} dx_n \\
&= \underbrace{\beta_{n-1}}_{\substack{\text{vol. of unit}\\ \text{ball in } \mathbb{R}^{n-1}}} \int_{-1}^1 (1-x_n^2)^{\frac{n-1}{2}}\, dx_n. \qquad 4.5.20
\end{aligned}$$

Example 4.5.7: The ball $B_1^n(\mathbf{0})$ is the ball of radius 1 in $\mathbb{R}^n$, centered at the origin; the ball
$$B^{n-1}_{\sqrt{1-x_n^2}}(\mathbf{0})$$
is the ball of radius $\sqrt{1-x_n^2}$ in $\mathbb{R}^{n-1}$, still centered at the origin.

In the first line of equation 4.5.20 we imagine slicing the n-dimensional ball horizontally and computing the $(n-1)$-dimensional volume of each slice.

This reduces the computation of β_n to computing the integral

$$c_n = \int_{-1}^1 (1-t^2)^{\frac{n-1}{2}}\, dt. \qquad 4.5.21$$

This is a standard tricky problem from one-variable calculus: Exercise 4.5.4 asks you to show that $c_0 = \pi$ and $c_1 = 2$ and that for $n \ge 2$,

$$c_n = \frac{n-1}{n} c_{n-2}. \qquad 4.5.22$$

This allows us to make Table 4.5.9. It is easy to continue the table. (What is β_6? Check below.[7]) △

If you enjoy inductive proofs, you might try Exercise 4.5.5, which asks you to find a formula for β_i for all i. Exercise 4.5.6 asks you to show that as n increases, the volume of the n-dimensional unit ball becomes a smaller and smaller proportion of the smallest n-dimensional cube that contains it.

---

[7] $c_6 = \frac{5}{6} c_4 = \frac{15\pi}{48}$, so $\beta_6 = c_6 \beta_5 = \frac{\pi^3}{6}$.

n	$c_n = \frac{n-1}{n} c_{n-2}$	Volume of ball $\beta_n = c_n \beta_{n-1}$
0	π	
1	2	2
2	$\frac{\pi}{2}$	π
3	$\frac{4}{3}$	$\frac{4\pi}{3}$
4	$\frac{3\pi}{8}$	$\frac{\pi^2}{2}$
5	$\frac{16}{15}$	$\frac{8\pi^2}{15}$

TABLE 4.5.9. Computing the volume of the unit ball in $\mathbb{R}^1$ through $\mathbb{R}^5$.

Computing probabilities using integrals

We saw in Section 4.2 that integrals are useful in computing probabilities.

Example 4.5.8 (Using Fubini to compute probabilities). Recall (equation 4.2.12) that the expectation (expected value) of a random variable f is given by $E(f) = \int_S f(s)\mu(s)\,|ds|$, where the function μ weights outcomes according to how likely they are, and S is the sample space. If we choose two points x, y in the interval $[0,1]$, without privileging any part of that interval, what is the expected value of the function $f\begin{pmatrix} x \\ y \end{pmatrix} = (x-y)^2$?

"Without privileging any part of that interval" means that $\mu = 1$. Thus

$$\begin{aligned} E\big((x-y)^2\big) &= \int_0^1 \int_0^1 (x-y)^2 dx\,dy = \int_0^1 \left(\int_0^1 x^2 - 2xy + y^2\,dx \right) dy \\ &= \int_0^1 \left[\frac{x^3}{3} - x^2 y + y^2 x \right]_0^1 dy = \int_0^1 \frac{1}{3} - y + y^2\,dy \\ &= \left[\frac{y}{3} - \frac{y^2}{2} + \frac{y^3}{3} \right]_0^1 = \frac{1}{6}. \end{aligned} \qquad 4.5.23$$

Exercise 4.5.10 asks you to show that if two points $\mathbf{x}, \mathbf{y}$ are chosen in the unit square, without privileging any part of the square, then $E\Big(|\mathbf{x}-\mathbf{y}|^2\Big) = 1/3$.

Note that the expected value of $|x-y|$ is more complicated to compute than the expected value of $(x-y)^2$. Asked to give the expected distance between x and y, a probabilist would most likely compute $E((x-y)^2)$ and take the square root, getting $1/\sqrt{6}$, which is not the actual expected distance of $1/3$. Computing $E((x-y)^2)$ and taking the square root is analogous to computing the standard deviation; computing $E(|x-y|)$ is analogous to computing the mean absolute deviation.

Given x, y in the same interval, what is the expected value of the function $f\begin{pmatrix} x \\ y \end{pmatrix} = |x-y|$? Now we want to include positive values of $x-y$ and positive values of $y-x$. To do this, let's assume that $x > y$ (so the integral for y goes from 0 to x) and double the integral to account for the cases where $y > x$:

$$\begin{aligned} E\big(|x-y|\big) &= 2\int_0^1 \left(\int_0^x x - y\,dy \right) dx = 2\int_0^1 \left[xy - \frac{y^2}{2} \right]_0^x dx \\ &= 2\int_0^1 x^2 - \frac{x^2}{2}\,dx = 2\left[\frac{x^3}{3} - \frac{x^3}{6} \right]_0^1 = \frac{1}{3}. \quad \triangle \end{aligned} \qquad 4.5.24$$

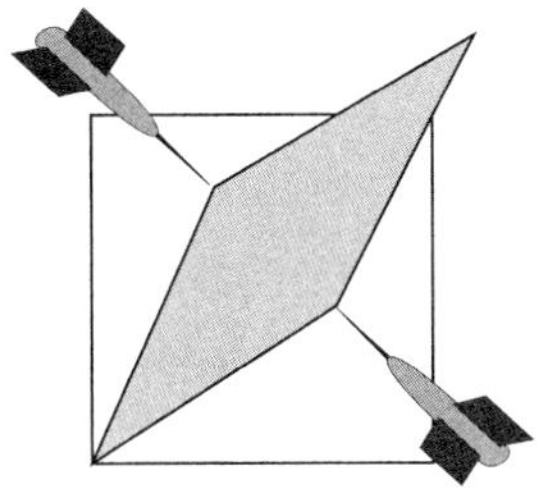

FIGURE 4.5.10.
Choosing a random parallelogram: One dart lands at $\begin{pmatrix} x_1 \\ y_1 \end{pmatrix}$, the other at $\begin{pmatrix} x_2 \\ y_2 \end{pmatrix}$. Saying that we are choosing our points at random in the square means that the density of probability for each dart is the indicator function of the square.

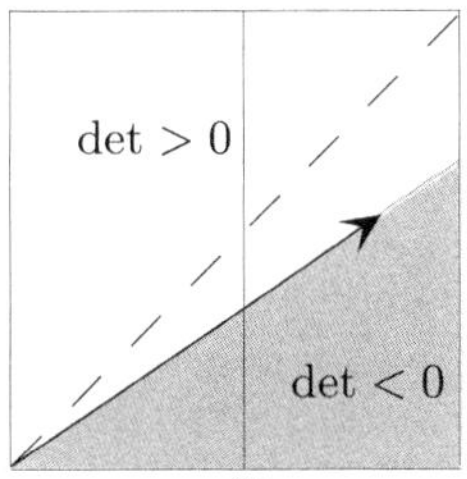

FIGURE 4.5.11.
The arrow is the first dart. If the second dart lands in the shaded area, the determinant will be negative, because the second vector is clockwise from the first (Proposition 1.4.14).

Example 4.5.9 (Computing a probability: A harder example). Choose at random two pairs of positive numbers between 0 and 1 and use those numbers as the coordinates (x_1, y_1), (x_2, y_2) of two vectors anchored at the origin, as shown in Figure 4.5.10. (You might imagine throwing darts at the unit square.) What is the expected (average) area of the parallelogram spanned by those vectors? In other words (recall Proposition 1.4.14), what is the expected value of $|\det|$? This average is

$$\int_C |\underbrace{x_1y_2 - y_1x_2}_{\det}||d^2\mathbf{x}||d^2\mathbf{y}|, \qquad 4.5.25$$

where C is the unit cube in $\mathbb{R}^4$. (Each possible parallelogram corresponds to two points in the unit square, each with two coordinates, so each point in C corresponds to one parallelogram.) Our computation will be simpler if we consider only the cases $x_1 \geq y_1$ – i.e., we assume that our first dart lands below the diagonal of the square. Since the diagonal divides the square symmetrically, the cases where the first dart lands below the diagonal and the cases where it lands above contribute the same amount to the integral. Thus we want to compute *twice* the quadruple integral

$$\int_0^1 \int_0^{x_1} \int_0^1 \int_0^1 |x_1y_2 - x_2y_1| \, dy_2 \, dx_2 \, dy_1 \, dx_1. \qquad 4.5.26$$

(Note that the integral $\int_0^{x_1}$ goes with dy_1: the innermost integral goes with the innermost integrand, and so on. The second integral $\int_0^{x_1}$ has limits of integration 0, x_1 because $0 \leq y_1 \leq x_1$.)

To get rid of the absolute values, we consider separately the case where $\det = x_1y_2 - x_2y_1$ is negative, and the case where it is positive. Note that on one side of the line $y_2 = \frac{y_1}{x_1}x_2$ (the shaded side in Figure 4.5.11), the determinant is negative, and on the other side it is positive.

Since we have assumed that the first dart lands below the diagonal, when we integrate with respect to y_2, we have two choices: if y_2 is in the shaded part, the determinant will be negative; otherwise it will be positive. So we break up the innermost integral into two parts:

$$\int_0^1 \int_0^{x_1} \int_0^1 \Bigg(\overbrace{\int_0^{(y_1x_2)/x_1} (x_2y_1 - x_1y_2) \, dy_2}^{x_1y_2 - x_2y_1 < 0, \text{ i.e., } \det<0} + \overbrace{\int_{(y_1x_2)/x_1}^1 (x_1y_2 - x_2y_1) \, dy_2}^{\det>0} \Bigg) dx_2 \, dy_1 \, dx_1. \qquad 4.5.27$$

Now we carefully compute four ordinary integrals, keeping straight what is constant and what varies. First we compute the inner integral, with respect to y_2. The first term gives

$$\left[x_2y_1y_2 - x_1\frac{y_2^2}{2} \right]_0^{y_1x_2/x_1} = \overbrace{\frac{x_2^2y_1^2}{x_1} - \frac{1}{2}x_1\frac{y_1^2x_2^2}{x_1^2}}^{\text{eval. at } y_2 = y_1x_2/x_1} - \overbrace{0}^{\text{eval. at } y_2=0} = \frac{1}{2}\frac{y_1^2x_2^2}{x_1}. \qquad 4.5.28$$

The second gives

$$\left[\frac{x_1y_2^2}{2} - x_2y_1y_2\right]^1_{y_1x_2/x_1} = \overbrace{\left(\frac{x_1}{2} - x_2y_1\right)}^{\text{eval. at } y_2=1} - \overbrace{\left(\frac{x_1y_1^2x_2^2}{2x_1^2} - \frac{x_2y_1^2x_2}{x_1}\right)}^{\text{eval. at } y_2=y_1x_2/x_1} \qquad 4.5.29$$

$$= \frac{x_1}{2} - x_2y_1 + \frac{x_2^2y_1^2}{2x_1}.$$

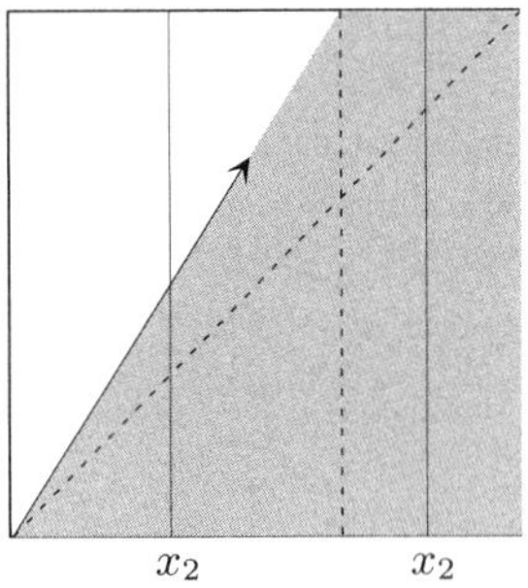

FIGURE 4.5.12.

The situation if the first dart lands above the diagonal. If the second dart lands in the shaded area, we have det < 0. For values of x_2 to the right of the vertical dotted line, y_2 goes from 0 to 1. For values of x_2 to the left of that line, we must consider separately

values of y_2 from 0 to $\dfrac{y_1x_2}{x_1}$

(the shaded region, with $-\det$) and

values of y_2 from $\dfrac{y_1x_2}{x_1}$ to 1

(the unshaded region, with $+\det$).

Continuing with the integral 4.5.27, we get

$$\int_0^1\int_0^{x_1}\int_0^1\left(\int_0^{(y_1x_2)/x_1}(x_2y_1-x_1y_2)dy_2 + \int_{(y_1x_2)/x_1}^1(x_1y_2-x_2y_1)dy_2\right)dx_2\,dy_1\,dx_1$$

$$= \int_0^1\int_0^{x_1}\underbrace{\int_0^1\left(\frac{x_1}{2} - x_2y_1 + \frac{x_2^2y_1^2}{x_1}\right)dx_2}_{\left[\frac{x_1x_2}{2} - \frac{x_2^2y_1}{2} + \frac{x_2^3y_1^2}{3x_1}\right]_{x_2=0}^{x_2=1}}\,dy_1\,dx_1$$

$$= \int_0^1\underbrace{\int_0^{x_1}\left(\frac{x_1}{2} - \frac{y_1}{2} + \frac{y_1^2}{3x_1}\right)dy_1}_{\left[\frac{x_1y_1}{2} - \frac{y_1^2}{4} + \frac{y_1^3}{9x_1}\right]_{y_1=0}^{y_1=x_1}}\,dx_1 = \int_0^1\left(\frac{x_1^2}{2} - \frac{x_1^2}{4} + \frac{x_1^2}{9}\right)dx_1$$

$$= \frac{13}{36}\int_0^1 x_1^2\,dx_1 = \frac{13}{36}\left[\frac{x_1^3}{3}\right]_0^1 = \frac{13}{108}. \qquad 4.5.30$$

So the expected area is twice 13/108, i.e., 13/54, slightly less than 1/4. (If we had not restricted the first dart to below the diagonal, we would have the situation of Figure 4.5.12, and our integral would be more complicated.[8])

What if we randomly choose three vectors in the unit cube? Then we would be integrating over a nine-dimensional cube. Computing this integral with a Riemann sum would be forbidding. Say we divide each side into 10, choosing the midpoint of each mini-cube. We turn the nine coordinates of that point into a 3×3 matrix and compute its determinant. That gives 10^9 determinants to compute, each requiring 18 multiplications and five additions.

[8]Exercise 4.5.9 asks you to compute the integral that way. The integral is then the sum of the integral for the half of the square below the diagonal (given by formula 4.5.27), and the integral for the half above the diagonal. The latter is

$$\int_0^1\int_{x_1}^1\int_0^{\frac{x_1}{y_1}}\overbrace{\left(\int_0^{\frac{y_1x_2}{x_1}}\underbrace{(x_2y_1-x_1y_2)}_{-\det}\,dy_2 + \int_{\frac{y_1x_2}{x_1}}^1\underbrace{(x_1y_2-x_2y_1)}_{+\det}\,dy_2\right)}^{\text{for } x_2 \text{ to the left of the vertical line with } x\text{-coordinate } x_1}dx_2\,dy_1\,dx_1$$

$$+ \int_0^1\int_{x_1}^1\int_{\frac{x_1}{y_1}}^1\underbrace{\left(\int_0^1(x_2y_1-x_1y_2)\,dy_2\right)}_{\substack{\text{for } x_2 \text{ to the right of the vertical line} \\ \text{with } x\text{-coordinate } x_1;\ -\det}}\,dx_2\,dy_1\,dx_1.$$

You should learn how to handle simple examples using Fubini's theorem, and you should learn some of the standard tricks that work in some more complicated situations; these will be handy on exams, particularly in physics and engineering classes.

In real life you are likely to come across nastier problems, which even a professional mathematician would have trouble solving "by hand"; most often you will want to use a computer to compute integrals for you. We discuss numerical methods of computing integrals in Section 4.6.

Go up one more dimension and the computation is really out of hand. Yet physicists routinely work with integrals in dimensions of thousands or more. The technique most often used is a sophisticated version of throwing dice, known as *Monte Carlo integration*. It is discussed in Section 4.6. △

We now give a precise statement of Fubini's theorem. A stronger version is proved in Appendix A17. Note that the left side of equation 4.5.32 is symmetric with respect to $\mathbf{x}$ and $\mathbf{y}$, so the integral on the right could be written with $|d^n\mathbf{x}|$ as the inner integrand.

Theorem 4.5.10 (Fubini's theorem). *Let $f:\mathbb{R}^n \times \mathbb{R}^m \to \mathbb{R}$ be an integrable function, and suppose that for each $\mathbf{x} \in \mathbb{R}^n$, the function $\mathbf{y} \mapsto f(\mathbf{x},\mathbf{y})$ is integrable. Then the function*

$$\mathbf{x} \mapsto \int_{\mathbb{R}^m} f(\mathbf{x},\mathbf{y})|d^m\mathbf{y}| \qquad 4.5.31$$

is integrable, and

$$\int_{\mathbb{R}^{n+m}} f(\mathbf{x},\mathbf{y})|d^n\mathbf{x}||d^m\mathbf{y}| = \int_{\mathbb{R}^n}\left(\int_{\mathbb{R}^m} f(\mathbf{x},\mathbf{y})|d^m\mathbf{y}|\right)|d^n\mathbf{x}|. \qquad 4.5.32$$

Exercises for Section 4.5

4.5.1 In Example 4.5.2, why can you ignore the fact that the line $x = 1$ is counted twice?

4.5.2 Set up the multiple integral $\int(\int f\,dx)\,dy$ for the truncated triangle shown in Figure 4.5.3.

4.5.3 a. Set up the multiple integral for Example 4.5.2, but with the outer integral with respect to y. Be careful about which square root you are using.

b. If in part a you replace $+\sqrt{y}$ by $-\sqrt{y}$ and vice versa, what would be the corresponding region of integration?

4.5.4 a. In the context of Example 4.5.7 (equation 4.5.22), show that if

$$c_n = \int_{-1}^{1}(1-t^2)^{(n-1)/2}\,dt, \quad \text{then} \quad c_n = \frac{n-1}{n}c_{n-2}, \quad \text{for } n \geq 2.$$

b. Show that $c_0 = \pi$ and $c_1 = 2$.

4.5.5 For Example 4.5.7, show that

$$\text{a.}\quad \beta_{2k} = \frac{\pi^k}{k!} \qquad \text{b.}\quad \beta_{2k+1} = \frac{\pi^k\, k!\, 2^{2k+1}}{(2k+1)!}.$$

4.5.6 a. Show that as n increases, the volume of the n-dimensional unit sphere becomes a smaller and smaller proportion of the smallest n-dimensional cube that contains it.

b. What is the first n for which the ratio of volumes is smaller than 10^{-2}?

c. What is the first n for which it is smaller than 10^{-6}?

d. For what n is the volume of the n-dimensional unit ball biggest? Justify your answer.

4.5.7 Write as an iterated integral, and in six different ways, the triple integral of xyz over the region $x, y, z \geq 0,\ x + 2y + 3z \leq 1$. You need not compute the integrals.

4.5.8 Set up the iterated integral for finding the volume of the slice of cylinder $x^2 + y^2 \leq 1$ between the planes

$$z = 0,\ z = 2,\ y = \frac{1}{2},\ y = -\frac{1}{2}.$$

4.5.9 In Example 4.5.9, compute the integral without assuming that the first dart falls below the diagonal (see the footnote after formula 4.5.27).

4.5.10 Choose two points $\mathbf{x}, \mathbf{y}$ in the unit square (square with sidelength 1, with lower left corner at the origin), so as not to privilege any part of the square. Show that $E\Big(|\mathbf{x}-\mathbf{y}|^2\Big) = 1/3$ (where E is the expected value; see Example 4.5.8).

4.5.11 a. Use Fubini's theorem to express $\displaystyle\int_0^\pi \left(\int_y^\pi \frac{\sin x}{x}\,dx\right) dy$ as an integral over a region of $\mathbb{R}^2$.

b. Write the integral as an iterated integral in the other order.

c. Compute the integral.

4.5.12 a. Represent the iterated integral $\displaystyle\int_0^a \left(\int_{x^2}^{a^2} \sqrt{y}e^{-y^2}\,dy\right) dx$ as the integral of $\sqrt{y}e^{-y^2}$ over a region of the plane. Sketch this region.

b. Use Fubini's theorem to make this integral into an iterated integral in the opposite order.

c. Evaluate the integral.

Theorem 3.3.8 on the equality of crossed partials requires only that all partial derivatives $D_i f$ be differentiable; the proof in Appendix A9 is surprisingly difficult. Exercise 4.5.13 gives an easier proof, under the stronger condition that the second partials of a function are continuous. It uses Fubini's theorem.

4.5.13 a. Show that if $U \subset \mathbb{R}^2$ is open, and $f : U \to \mathbb{R}$ is a function such that $D_2(D_1(f))$ and $D_1(D_2(f))$ exist and are continuous and

$$D_1(D_2(f))\begin{pmatrix} a \\ b \end{pmatrix} \neq D_2(D_1(f))\begin{pmatrix} a \\ b \end{pmatrix}$$

for some point $\begin{pmatrix} a \\ b \end{pmatrix}$, then there exists a square $S \subset U$ such that either

$$D_2(D_1(f)) > D_1(D_2(f)) \text{ on } S \quad \text{or} \quad D_1(D_2(f)) > D_2(D_1(f)) \text{ on } S.$$

b. Derive a contradiction by applying Fubini's theorem to the integral

$$\int_S \Big(D_2\big(D_1(f)\big) - D_1\big(D_2(f)\big)\Big)\,|dx\,dy|.$$

c. We saw in Example 3.3.9 the standard example of a function where $D_1(D_2 f)) \neq D_2(D_1(f))$:

$$f\begin{pmatrix} x \\ y \end{pmatrix} = \begin{cases} xy\frac{x^2-y^2}{x^2+y^2} & \text{if } \begin{pmatrix} x \\ y \end{pmatrix} \neq \begin{pmatrix} 0 \\ 0 \end{pmatrix} \\ 0 & \text{otherwise.} \end{cases}$$

What happens to the proof?

4.5.14 a. Set up the integral of $\sin y$ over the region

$$0 \le x \le \cos y, \; 0 \le y \le \pi/6$$

as an iterated integral, in two different ways.

b. Write $\int_1^2 \int_{y^3}^{3y^3} \frac{1}{x}\, dx\, dy$ as an integral where one first integrates with respect to y, then with respect to x.

4.5.15 Find the volume of the region

$$z \ge x^2 + y^2, \quad z \le 10 - x^2 - y^2.$$

4.5.16 Compute the integral of the function $|y - x^2|$ over the unit square defined by $0 \le x \le 1$ and $0 \le y \le 1$.

Exercise 4.5.17 is borrowed from Tiberiu Trif, "Multiple Integrals of Symmetric Functions," *American Mathematical Monthly*, Vol. 104, No. 7 (1997), pp. 605–608.

4.5.17 Recall from Exercise 4.3.4 the definitions of $P^n_{a,b} \subset Q^n_{a,b}$; use the result of that exercise to compute the following integrals.

a. Let $M_r(\mathbf{x})$ be the rth smallest of the coordinates $x_1, \dots, x_n$ of $\mathbf{x}$. Show that

$$\int_{Q^n_{0,1}} M_r(\mathbf{x})\, |d^n\mathbf{x}| = \frac{r}{n+1}.$$

b. Let $n \ge 2$ and $0 < b < 1$. Show that then

$$\int_{Q^n_{0,1}} \min\left(1, \frac{b}{x_1}, \dots, \frac{b}{x_n}\right) |d^n\mathbf{x}| = \frac{nb - b^n}{n-1}.$$

4.5.18 a. Transform the iterated integral $\int_0^1 \left(\int_y^{y^{1/3}} e^{-x^2}\, dx \right) dy$ into an integral over a subset of the plane. Sketch this subset.

b. Compute the integral.

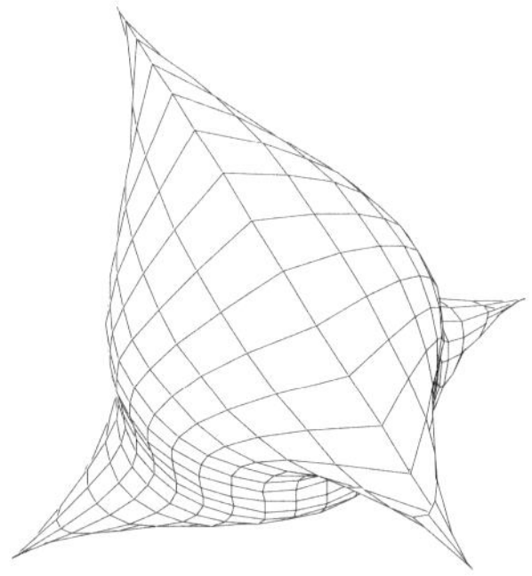

FIGURE FOR EXERCISE 4.5.19.

The region described in Exercise 4.5.19 looks like a peculiar pillow.

4.5.19 What is the volume of the region shown in the figure at left, given by

$$\frac{x^2}{(z^3-1)^2} + \frac{y^2}{(z^3+1)^2} \le 1, \quad -1 \le z \le 1?$$

4.6 NUMERICAL METHODS OF INTEGRATION

Usually Fubini's theorem does not lead to expressions that can be computed in closed form, and integrals must be computed numerically. In one dimension, there is an enormous literature on the subject. In higher dimensions, the literature is extensive but the field is not nearly so well known.

One-dimensional integrals

In first year calculus you probably heard of the trapezoidal rule and Simpson's rule for computing ordinary integrals (quite likely you've forgotten them, too). The trapezoidal rule is not of much practical interest, but Simpson's rule is probably good enough for anything you will need unless you become an engineer or physicist. In it, the function is sampled at regular intervals and different "weights" are assigned the samples.

In Simpson's rule, why do we multiply by $(b-a)/6n$? Think of the integral as the sum of the areas of n "rectangles". The width of each rectangle is $(b-a)/n$; the height should be some sort of average of the value of the function over the interval, but we have weighted each subinterval by $1+4+1=6$. Dividing by $6n$ rather than n adjusts for that weight.

Definition 4.6.1 (Simpson's rule). Let f be a function on $[a,b]$, choose an integer n, and sample f at $2n+1$ equally distributed points $x_0, x_1, \dots, x_{2n}$, where $x_0 = a$ and $x_{2n} = b$. Simpson's approximation to

$$\int_a^b f(x)\,dx \qquad \text{in } n \text{ steps is}$$

$$S^n_{[a,b]}(f) \stackrel{\text{def}}{=} \frac{b-a}{6n}\Big(f(x_0)+4f(x_1)+2f(x_2)+4f(x_3)+\cdots \\ \cdots + 4f(x_{2n-1}) + f(x_{2n})\Big).$$

For example, if $n=3$, $a=-1$, and $b=1$, we divide the interval $[-1,1]$ into six equal parts and compute

$$\frac{1}{9}\Big(f(-1)+4f(-2/3)+2f(-1/3)+4f(0)+2f(1/3)+4f(2/3)+f(1)\Big). \qquad 4.6.1$$

Why do the weights start and end with 1, and alternate between 4 and 2 for the intermediate samples? As shown in Figure 4.6.2, the pattern of weights is not $1,4,2,\dots,4,1$ but $1,4,1,1,4,1,\dots,1,4,1$: each 1 that is not the beginning point a or the endpoint b is counted twice, so it becomes 2. The weight 4 given to the midpoint is chosen so that the formula will integrate cubics exactly; see Figure 4.6.3.

FIGURE 4.6.1.
Thomas Simpson (1710–1761) worked as a weaver, then as an itinerant teacher of mathematics in London coffee houses. Another English mathematician, Charles Hutton, wrote that "it has been said that Mr Simpson frequented low company, with whom he used to guzzle porter and gin: but it must be observed that the misconduct of his family put it out of his power to keep the company of gentlemen, as well as to procure better liquor."

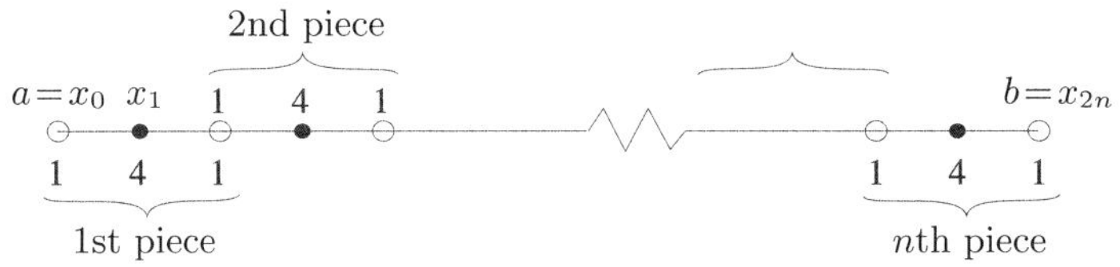

FIGURE 4.6.2. Simpson's rule breaks $[a,b]$ into n pieces (the first from $a=x_0$ to x_2, the next from x_2 to x_4, and so forth); f is evaluated at the beginning, midpoint, and endpoint of each piece, with weight 1 for the beginning and endpoint, and 4 for the midpoint. The endpoint of one interval is the beginning point of the next, so it is counted twice. The result is multiplied by $(b-a)/6n$.

We are actually breaking up the interval into n subintervals and integrating the function over each subpiece:

$$\int_a^b f(x)\,dx = \int_a^{x_2} f(x)\,dx + \int_{x_2}^{x_4} f(x)\,dx + \cdots + \int_{x_{2n-2}}^{b} f(x)\,dx. \qquad 4.6.2$$

Each subintegral is approximated by sampling the function at the beginning point and endpoint of the subpiece (with weight 1) and at the center of the subpiece (with weight 4), giving a total of 6.

Theorem 4.6.2 (Simpson's rule).

1. *If $f:[a,b] \to \mathbb{R}$ is a continuous piecewise cubic function, exactly equal to a cubic polynomial on the intervals $[x_{2i}, x_{2i+2}]$, then Simpson's rule computes the integral exactly.*
2. *If a function f is four times continuously differentiable, then there exists $c \in (a,b)$ such that*

$$S^n_{[a,b]}(f) - \int_a^b f(x)\,dx = \frac{(b-a)^5}{2880n^4} f^{(4)}(c). \tag{4.6.3}$$

By cubic polynomial we mean polynomials of degree up to and including 3: constant functions, linear functions, quadratic polynomials and cubic polynomials.

In the world of computer graphics, piecewise cubic polynomials are everywhere. When you construct a smooth curve using a drawing program, the computer is actually making a piecewise cubic curve, usually using the Bézier algorithm for cubic interpolation. Curves drawn this way are known as *cubic splines*.

When one first draws curves with such a program, it comes as a surprise how few control points are needed.

Proof. Table 4.6.3 proves part 1. A proof of part 2 is sketched in Exercise 4.6.6. Since Simpson's rule integrates piecewise cubics exactly, it is not surprising that the fourth derivative shows up in the error bound.

	Simpson's rule	Integration
Function	$1/3\,\big(f(-1) + 4(f(0)) + f(1)\big)$	$\int_{-1}^1 f(x)\,dx$
$f(x) = 1$	$1/3\,(1+4+1) = 2$	2
$f(x) = x$	0	0
$f(x) = x^2$	$1/3\,(1+0+1) = 2/3$	$\int_{-1}^1 x^2\,dx = 2/3$
$f(x) = x^3$	0	0

TABLE 4.6.3. Above we compute the integral for constant, linear, quadratic, and cubic functions, over the interval $[-1,1]$, with $n = 1$. Simpson's rule gives the same result as computing the integral directly.

Of course, you don't often encounter in real life a piecewise cubic polynomial (the exception being computer graphics). Usually, Simpson's method is used to approximate integrals, not to compute them exactly.

Simpson's rule is a *fourth-order method*; the error (if f is sufficiently differentiable) is of order h^4, where h is the step size:

$$h = (b-a)/n.$$

Thus if you increase the number of steps (pieces) by a factor of 10, the error decreases by a factor of 10,000.

Example 4.6.3 (Approximating integrals with Simpson's rule). Use Simpson's rule with $n = 100$ to approximate

$$\int_1^4 \frac{1}{x}\,dx = \ln 4 = 2\ln 2, \tag{4.6.4}$$

which is infinitely differentiable. Since $f^{(4)} = 24/x^5$, which is largest at $x = 1$, Theorem 4.6.2 asserts that the result will be correct to within

$$\frac{3^5 \cdot 24}{2880 \cdot 100^4} = 2.025 \cdot 10^{-8} \tag{4.6.5}$$

so at least seven decimals will be correct. $\triangle$

Simpson's rule is not the only way to approximate integrals by weighted averages of the values at points:

$$\int_a^b f(x)\,dx \sim \sum_i w_i f(x_i). \qquad 4.6.6$$

Some methods give equal or better precision with far fewer function evaluations. Particularly notable are the remarkably accurate *Gaussian rules*, where one takes p to be a polynomial of degree $< 2k$, and views

$$\int_a^b p(x)\,dx = \sum_{i=1}^k w_i f(x_i). \qquad 4.6.7$$

as a system of $2k$ equations (one for each of $p(x) = 1,\ x, x^2, \dots, x^{2k-1}$) in the $2k$ variables $x_1, \dots, x_k, w_1, \dots w_k$. This system has a unique solution with $a < x_1 < \cdots < x_k < b$, and moreover all the weights satisfy $w_i > 0$.

The proof is very elegant, but it requires *orthogonal polynomials*, a topic we don't develop here for lack of space. But we should note that the same proof works just as well to approximate integrals of the form

$$\int_0^\infty e^{-x} f(x)\,dx, \quad \int_{-\infty}^\infty e^{-x^2} f(x)\,dx, \quad \int_{-1}^1 \frac{f(x)\,dx}{\sqrt{1-x^2}} \qquad 4.6.8$$

which are integrals over unbounded domains, or of unbounded functions; methods like Simpson's method really do not work well for such integrals.

Exercise 4.6.4 explores Gaussian rules for computing the integral

$$\int_0^\infty e^{-x} f(x)\,dx;$$

Exercise 4.16 explores Gaussian rules for

$$\int_{-\infty}^\infty e^{-x^2} f(x)\,dx.$$

Product rules

Every one-dimensional integration rule has a higher-dimensional counterpart, called a *product rule*. If the rule in one dimension is

$$\int_a^b f(x)\,dx \approx \sum_{i=1}^k w_i f(p_i), \qquad 4.6.9$$

then the corresponding rule in n dimensions is

$$\int_{[a,b]^n} f(\mathbf{x})\,|d^n\mathbf{x}| \approx \sum_{1\le i_1,\dots,i_n\le k} w_{i_1}\dots w_{i_n} f\begin{pmatrix} p_{i_1} \\ \vdots \\ p_{i_n}\end{pmatrix}. \qquad 4.6.10$$

A weakness of Proposition 4.6.4 is that it applies to parallelograms and higher-dimensional analogues. It can be used when integrating over a region A with a less regular shape, using the indicator function (see the discussion of the indicator function in Section 4.1). When integrals are computed from the definition, using Riemann sums, then the border between A and the rest of the parallelogram does not affect the integral.

But when numerical methods are used, the fact that $\mathbf{1}_A$ is discontinuous affects the integral in serious ways. Recall, for example, that the error bound in Simpson's rule requires that the function being integrated be four times continuously differentiable.

Proposition 4.6.4 (Product rules). *If $f_1, \dots, f_n$ are functions that are integrated exactly by an integration rule, i.e.,*

$$\int_a^b f_j(x)\,dx = \sum_i w_i f_j(p_i) \quad \text{for } j = 1, \dots, n, \qquad 4.6.11$$

then the product

$$f(\mathbf{x}) \overset{\text{def}}{=} f_1(x_1) f_2(x_2) \dots f_n(x_n) \qquad 4.6.12$$

is integrated exactly by the corresponding product rule over $[a,b]^n$.

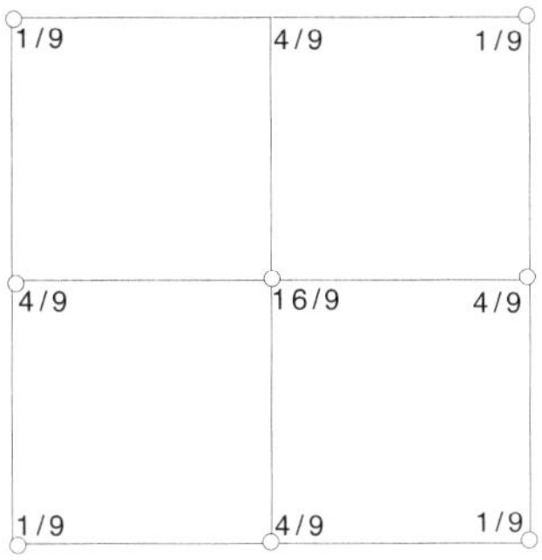

FIGURE 4.6.4.

The square of sidelength 2 divided into four subsquares, with weights given by Simpson's method.

Proof. This follows immediately from Proposition 4.1.16. Indeed,

$$\begin{aligned}\int_{[a,b]^n} f(\mathbf{x})\,|d^n\mathbf{x}| &= \left(\int_a^b f_1(x_1)\,dx_1\right)\dots\left(\int_a^b f_n(x_n)\,dx_n\right)\\ &= \left(\sum w_i f_1(p_i)\right)\dots\left(\sum w_i f_n(p_i)\right)\\ &= \sum_{i_1,\dots i_n} w_{i_1}\dots w_{i_n} f\begin{pmatrix}p_{i_1}\\ \vdots\\ p_{i_n}\end{pmatrix}. \quad \square\end{aligned} \tag{4.6.13}$$

Example 4.6.5 (Simpson's rule in two dimensions). Each weight for the two-dimensional Simpson's rule is the product of two one-dimensional Simpson weights from Definition 4.6.1. In the very simple case where we wish to integrate over a square, dividing it into only four subsquares, and sampling the function at each vertex, we have nine samples in all.

Let us do this with the square of sidelength $b-a=2$. Since $n=1$ (this square corresponds, in the one-dimensional case, to the first piece in Figure 4.6.2), the one-dimensional weights are

$$w_1 = w_3 = \frac{b-a}{6n}\cdot 1 = \frac{1}{3} \quad \text{and} \quad w_2 = \frac{b-a}{6n}\cdot 4 = \frac{4}{3}. \tag{4.6.14}$$

Equation 4.6.10 then becomes

$$\begin{aligned}\int_{[-1,1]^2} f(\mathbf{x})\,|d^2\mathbf{x}| \approx\ & \underbrace{w_1w_1}_{1/9} f\begin{pmatrix}-1\\-1\end{pmatrix} + \underbrace{w_1w_2}_{4/9} f\begin{pmatrix}-1\\0\end{pmatrix} + \underbrace{w_1w_3}_{1/9} f\begin{pmatrix}-1\\1\end{pmatrix}\\ &+ \underbrace{w_2w_1}_{4/9} f\begin{pmatrix}0\\-1\end{pmatrix} + \underbrace{w_2w_2}_{16/9} f\begin{pmatrix}0\\0\end{pmatrix} + \underbrace{w_2w_3}_{4/9} f\begin{pmatrix}0\\1\end{pmatrix}\\ &+ \underbrace{w_3w_1}_{1/9} f\begin{pmatrix}1\\-1\end{pmatrix} + \underbrace{w_3w_2}_{4/9} f\begin{pmatrix}1\\0\end{pmatrix} + \underbrace{w_3w_3}_{1/9} f\begin{pmatrix}1\\1\end{pmatrix}.\end{aligned} \tag{4.6.15}$$

This gives the weights shown in Figure 4.6.4, where we think of the square as $[-1,1]\times[-1,1]$.

Theorem 4.6.2 and Proposition 4.6.4 tell us that this two-dimensional Simpson's method will integrate exactly the polynomials

$$1,\ x,\ y,\ x^2,\ xy,\ y^2, x^3,\ x^2y,\ xy^2,\ y^3, \tag{4.6.16}$$

and many others (for instance, x^2y^3), but not x^4. They will also integrate exactly functions that are piecewise polynomials of degree at most three on each of the unit squares, using the weights shown in Figure 4.6.5. △

1	4	2	4	…	4	1
4	16	8	16	…	16	4
2	8	4	8	…	8	2
4	16	8	16	…	16	4
⋮	⋮	⋮	⋮	⋱	⋮	⋮
4	16	8	16	…	16	4
1	4	2	4	…	4	1

FIGURE 4.6.5.

Weights for approximating the integral over a square, using the two-dimensional Simpson's rule, dividing the square into $4n^2$ subsquares. Each weight is multiplied by

$$\frac{(b-a)^2}{36n^2}.$$

In Figure 4.6.4, $b-a=2$ and $n=1$, so each weight is multiplied by $1/9$.

Problems with higher-dimensional Riemann sums

Both Simpson's rule and Gaussian rules are versions of Riemann sums. Riemann sums in higher dimensions have at least two serious difficulties. One is that the fancier the method, the smoother the function to be integrated needs to be in order for the method to work according to specs. In one dimension this usually isn't serious; if there are discontinuities, you break up the interval into several intervals at the discontinuities. But in several dimensions, especially if you are trying to compute a volume by integrating an indicator function, you will only be able to maneuver around the discontinuity if you already know the answer.

The other problem has to do with the magnitude of the computation. In one dimension, there is nothing unusual in using 100 or 1000 points for Simpson's method, in order to gain the desired accuracy (which might be 10 significant digits). As the dimension goes up, this sort of thing becomes first alarmingly expensive, and then utterly impossible. In dimension 4, a Simpson approximation using 100 points to a side involves 100 000 000 function evaluations, within reason for today's computers if you are willing to wait a while; with 1000 points to a side it involves 10^{12} function evaluations, which would tie up the biggest computers for several days.

By the time you get to dimension 9, this sort of thing becomes totally unreasonable unless you decrease your desired accuracy: $100^9 = 10^{18}$ function evaluations would take more than a billion seconds (about 32 years) even on the very fastest computers, but 10^9 is within reason, and should give a couple of significant digits.

When the dimension gets higher than 10, Simpson's method and all similar methods become totally impossible, even if you are satisfied with one significant digit, just to give an order of magnitude. These situations call for the probabilistic methods described below. They quickly give a couple of significant digits (with high probability: you are never sure), but we will see that it is next to impossible to get really good accuracy (say six significant digits).

Monte Carlo methods

Suppose that we want to find the average of $|\det A|$ for all $n \times n$ matrices with all entries chosen at random in the unit interval. We computed this integral in Example 4.5.9 for $n = 2$, and found $13/54$. The thought of computing the integral exactly for 3×3 matrices is awe-inspiring, even with a computer. How about numerical integration? If we want to use Simpson's rule, even with just 10 points on the side of the cube, we will need to evaluate 10^9 determinants, each a sum of six products of three numbers. This is not out of the question with today's computers, but it's a pretty massive computation. Even then, we shouldn't expect much in the way of significant digits, since the bound for the error in Simpson's method requires the function being four times differentiable, and $|\det A|$ is not differentiable at all.

There is a much better approach, called the *Monte Carlo* method: pick points at random in the nine-dimensional cube, evaluate the absolute value of the determinant of the 3×3 matrix that you make from the coordinates of each point, and take the average. A similar method will allow you to evaluate (with some precision) integrals over domains of dimension 20 or 100 or more. Physicists use Monte Carlo methods to approximate integrals over domains of dimension Avogadro's number, $6 \cdot 10^{23}$. (That number, the number of molecules in a mole of any substance, is so large that some fancy tricks are also required, known as renormalization.)

Monte Carlo methods approximate *normalized integrals*, that is, ratios

$$\frac{\int_A f(\mathbf{x})|d^n\mathbf{x}|}{\int_A |d^n\mathbf{x}|} = \frac{\int_A f(\mathbf{x})|d^n\mathbf{x}|}{\operatorname{vol}_n(A)}. \tag{4.6.17}$$

We will think of equation 4.6.17 as the expected value of the experiment consisting of evaluating f at a point chosen at random in A:

$$E(f) = \frac{\int_A f(\mathbf{x})|d^n\mathbf{x}|}{\operatorname{vol}_n(A)}. \tag{4.6.18}$$

(The division by $\operatorname{vol}_n(A)$ makes the certain event have probability 1.) The Monte Carlo algorithm gives an approximation of $E(f)$ and $\operatorname{Var}(f)$. Since we are repeating the same experiment over and over, the *central limit theorem* from probability (Theorem 4.2.7) then allows us to estimate how accurate these approximations are.

In Definition 4.6.6, $\bar{a}$ approximates $E(f)$, the normalized integral in equation 4.6.18; $\bar{s}$ approximates the standard deviation.

Equation 4.6.17: If you need the value of the integral

$$\int_A f(\mathbf{x})|d^n\mathbf{x}|,$$

not the ratio in equation 4.6.17, you will need to compute $\operatorname{vol}_n(A)$. This can be done by the same method: put A in a box B of known volume, then pick points at random in B. The proportion of these in A provides an estimate of $\operatorname{vol}_n(A)/\operatorname{vol}_n(B)$. This is the Monte Carlo method as applied to $\int_B \mathbf{1}_A(\mathbf{x})|d^n\mathbf{x}|$.

Definition 4.6.6 (Monte Carlo method). Let $f: A \to \mathbb{R}$ be a function, and let A be a set with a probability measure. The *Monte Carlo algorithm* for computing the integral of f over A consists of

1. Choosing points $\mathbf{x}_i, i = 1, \dots, N$ in A at random with respect to the given probability measure on A
2. Evaluating $a_i = f(\mathbf{x}_i)$
3. Computing

$$\bar{a} = \frac{1}{N}\sum_{i=1}^{N} a_i \tag{4.6.19}$$

$$\bar{s}^2 = \frac{1}{N}\sum_{i=1}^{N}(a_i - \bar{a})^2 = \left(\frac{1}{N}\sum_{i=1}^{N} a_i^2\right) - \bar{a}^2 \tag{4.6.20}$$

Equation 4.6.20: Recall the two expressions for the variance (Definition 3.8.6). It is easier to compute s^2 using the second expression; the first requires keeping the values of the samples a_i in memory until the end of the computation, whereas the second only requires keeping a running tally of the b_i.

Even if we are only really interested in the expectation, we need to compute an approximation $\bar{s}$ of the standard deviation, since the central limit theorem requires the standard deviation.

The points in A referred to in Definition 4.6.6 will no doubt be chosen using a random number generator. If this is biased, the entire scheme becomes unreliable. On the other hand, off-the-shelf random number generators come with an implicit guarantee, since if you can detect a bias, you can use that information to crack most commercial encoding schemes.

Remark. The experts will tell you that formula 4.6.20 for $\bar{s}^2$ is biased, and that N should be replaced by $N-1$. It is fairly easy to see that our formula is biased: you would expect that for any N repetitions of a measurement, each measurement would be closer to the average of the measurements than to the true average. It is not at all clear that replacing N by $N-1$ gets rid of the bias (or even what it means to be unbiased); Exercise 4.6.7 spells this out and shows how to prove it. In practical applications, we don't need to know $\bar{s}$ with any great precision, and once N is large, the two formulas give similar results. △

To use a random number generator to construct a code, you can add a random sequence of bits to your message, bit by bit (with no carries, so that $1+1=0$); to decode, subtract it again. If your message (encoded as bits) is the first line below, and the second line is generated by a random number generator, the sum of the two will appear random as well, and thus indecipherable:

10 11 10 10 1111 01 01

01 01 10 10 0000 11 01

11 10 00 00 1111 10 00.

Probabilistic methods of integration are like political polls. You don't pay much for going to higher dimensions, just as you don't need to poll more people for a Presidential race than for a Senate race.

The real difficulty with Monte Carlo methods is making a good random number generator, just as in polling the real problem is making sure your sample is not biased. In the 1936 presidential election, the *Literary Digest* predicted that Alf Landon would beat Franklin D. Roosevelt, on the basis of two million mock ballots returned from a mass mailing. The mailing list was composed of people who owned cars or telephones, which during the Depression was hardly a random sampling.

Pollsters then began polling far fewer people (typically, about 10 thousand), paying more attention to getting representative samples. Still, in 1948 the *Tribune* in Chicago went to press with the headline, "Dewey Defeats Truman"; polls had unanimously predicted a crushing defeat for Truman. One problem was that some interviewers avoided low-income neighborhoods. Another was that Gallup stopped polling two weeks before the election.

The central limit theorem asserts that the probability that $\overline{a}$ is between

$$E + \frac{a\sigma}{\sqrt{N}} \quad \text{and} \quad E + \frac{b\sigma}{\sqrt{N}} \qquad 4.6.21$$

is approximately

$$\frac{1}{\sqrt{2\pi}} \int_a^b e^{-\frac{t^2}{2}}\, dt, \qquad 4.6.22$$

the area under the bell curve between a and b (see Figure 4.2.5). In principle, everything can be derived from this formula. Let us see how this allows us to see how many times the experiment needs to be repeated in order to know an integral with a certain precision and a certain confidence.

Suppose we want to compute an integral to within one part in a thousand. With Monte Carlo, we can never be *sure* of anything. But we can ask for what N (how many times we repeat the experiment) can we be 98% sure that the estimate $\overline{a}$ is correct to one part in a thousand, i.e., that

$$\left| \frac{E - \overline{a}}{E} \right| \le .001? \qquad 4.6.23$$

This requires knowing something about the bell curve: with probability 98% the result is within 2.4 standard deviations of the mean: $|E - \overline{a}| \le \frac{2.4\sigma}{\sqrt{N}}$. So we need

$$\frac{2.4\sigma}{\sqrt{N}E} \le .001, \quad \text{which gives} \quad N \ge \frac{5.56 \cdot 10^6 \cdot \sigma^2}{E^2}. \qquad 4.6.24$$

Of course when using inequality 4.6.24 we will need to replace σ by its approximation $\overline{s}$.

Example 4.6.7 (Monte Carlo). In Example 4.5.9 we computed the expected value for the determinant of a 2×2 matrix. Now let try the same for 3×3 matrices, using MATHEMATICA's Montecarlo program to approximate

$$\int_C |\det A||d^9\mathbf{x}|, \qquad 4.6.25$$

i.e., to evaluate the average absolute value of the determinant of a 3×3 matrix with entries chosen at random in $[0, 1]$.

Four runs of length 50000 gave the following estimates of the integral:

$$.13453, \ .13519, \ .13409, \ .13456. \qquad 4.6.26$$

How accurate are these guesses? We might expect the first two digits to be correct, and the third to be off by 1 or 2. What does the theory say?

The same program calculates the standard deviation of the experiment of choosing a single matrix and computing $|\det A|$ to be about .129 for the first and fourth runs in 4.6.26 and .127 for the second and third. The central limit says that the standard deviation of repeating the experiment 50000 times should be approximately $.129/\sqrt{50000} \sim .00057$. Figure 4.2.5

says that 99% of the probability should be within 2.5 standard deviations, i.e., within .0014, confirming that the third digit might be off by 1 or 2.

What if we repeat the experiment a million times or more? A single run of one million gave an estimate of .1343961 with a standard deviation of .129. Dividing this by the square root of one million gives a standard deviation of .000129; multiply this by 2.5 gives .00032; with 99% certainly we should expect the first three digits to be correct, with the fourth digit off by at most 3.

A single run of 5 million gives .134396, and again estimates the standard deviation to be .129. Dividing by the square root of 5 000 000 gives a standard deviation of .000057; multiplying by 2.5 gives .00014. (Note that when estimating how many times we need to repeat an experiment, we don't need several digits of σ; only the order of magnitude matters. If we substitute .14 for .129 we still get .00014.) △

Exercises for Section 4.6

4.6.1 a. Write out the sum given by Simpson's method with one step, for the integral $\int_Q f(\mathbf{x})|d^n\mathbf{x}|$ when Q is the unit square in $\mathbb{R}^2$ and when it is the unit cube in $\mathbb{R}^3$. (Recall that the unit square and unit cube are not centered at 0; see Figure 1.3.8.) There should be 9 and 27 terms, respectively.

b. Evaluate the sum for the unit square when $f\begin{pmatrix} x \\ y \end{pmatrix} = \dfrac{1}{1+x+y}$, and compare the approximation to the exact value of the integral.

Exercises 4.6.2–4.6.4 explore Gaussian quadrature.

Exercise 4.6.2: The pattern should be fairly clear; experiment to find initial conditions for $k = 5$. *Hint*: Entering the equations is fairly easy. The hard part is finding good initial conditions. The following work:

$k=1$	$w_1 = .7$	$x_1 = .5$
$k=2$	$w_1 = .6$	$x_1 = .3$
	$w_2 = .4$	$x_2 = .8$
$k=3$	$w_1 = .5$	$x_1 = .2$
	$w_2 = .3$	$x_2 = .7$
	$w_3 = .2$	$x_3 = .9$
$k=4$	$w_1 = .35$	$x_1 = .2$
	$w_2 = .3$	$x_2 = .5$
	$w_3 = .2$	$x_3 = .8$
	$w_4 = .1$	$x_4 = .95$

4.6.2 Find the weights and control points for the Gaussian integration scheme (see equation 4.6.7 and the margin note) by solving the system of equations 4.6.7, for $k = 2, 3, 4, 5$, where $a = -1$ and $b = 1$.

4.6.3 Find the formula relating the weights W_i and the sampling points X_i needed to compute $\int_a^b f(x)\,dx$ to the weights w_i and the points x_i appropriate for $\int_{-1}^1 f(x)\,dx$ (see equation 4.6.7).

4.6.4 a. Find the equations that must be satisfied by points $x_1 < \cdots < x_m$ and weights $w_1 < \cdots < w_m$ so that the equation

$$\int_0^\infty p(x)e^{-x}\,dx = \sum_{k=1}^m w_k p(x_k)$$

is true for all polynomials p of degree $\leq d$.

b. For what number d does this lead to as many equations as unknowns?

c. Solve the system of equations when $m = 1$.

d. Use Newton's method to solve the system for $m = 2, 3, 4$.

e. For $m = 1, \dots, 4$, approximate

$$\int_0^\infty e^{-x}\sin x\,dx \quad \text{and} \quad \int_0^\infty e^{-x}\ln x\,dx,$$

and compare the approximations with the exact values.

4.6.5 a. Show that if

$$\int_a^b f(x)\,dx = \sum_{i=1}^n c_i f(x_i) \quad \text{and} \quad \int_a^b g(x)\,dx = \sum_{i=1}^n c_i g(x_i),$$

then

$$\int_{[a,b]\times[a,b]} f(x)g(y)\,|dx\,dy| = \sum_{i=1}^n \sum_{j=1}^n c_i c_j f(x_i) g(x_j).$$

b. What is the Simpson approximation with one step of the integral

$$\int_{[0,1]\times[0,1]} x^2 y^3\,|dx\,dy|\,?$$

Exercise 4.6.6 was largely inspired by an exercise in Michael Spivak's *Calculus*.

Part b: A function f *vanishes with multiplicity* $k+1$ at a if

$$f(a) = f'(a) = \cdots = f^{(k)}(a) = 0.$$

Part c: Show that the function

$$g(t) = q(x)(f(t) - p(t)) - q(t)(f(x) - p(x))$$

vanishes $n+2$ times; recall that the $n+1$st derivative of a polynomial of degree n is zero.

***4.6.6** In this exercise we will sketch a proof of equation 4.6.3. There are many parts to the proof, and many of the intermediate steps are of independent interest.

a. Show that if the function f is continuous on $[a_0, a_n]$ and n times differentiable on (a_0, a_n), and f vanishes at the $n+1$ distinct points $a_0 < a_1 < \cdots < a_n$, then there exists $c \in (a_0, a_n)$ such that $f^{(n)}(c) = 0$.

b. Now prove the same thing if the function vanishes with *multiplicities*: show that if f vanishes with multiplicity $k_i + 1$ at a_i, and if f is $N = n + \sum_{i=0}^n k_i$ times differentiable, then there exists $c \in (a_0, a_n)$ such that $f^{(N)}(c) = 0$.

c. Let f be n times differentiable on $[a_0, a_n]$; let p be a polynomial of degree n (in fact, the unique one, by Exercise 2.5.17) such that $f(a_i) = p(a_i)$. Set $q(x) = \prod_{i=0}^n (x - a_i)$. Show that there exists $c \in (a_0, a_n)$ such that

$$f(x) - p(x) = \frac{f^{(n+1)}(c)}{(n+1)!} q(x).$$

d. Let f be four times continuously differentiable on $[a, b]$, and let p be the polynomial of degree 3 such that

$$f(a) = p(a),\ f\left(\frac{a+b}{2}\right) = p\left(\frac{a+b}{2}\right),\ f'\left(\frac{a+b}{2}\right) = p'\left(\frac{a+b}{2}\right),\ f(b) = p(b).$$

Show that

$$\int_a^b f(x)\,dx = \frac{b-a}{6}\left(f(a) + 4f\left(\frac{a+b}{2}\right) + f(b)\right) - \frac{(b-a)^5}{2880} f^{(4)}(c)$$

for some $c \in [a, b]$.

e. Prove part 2 of Simpson's rule (Theorem 4.6.2): If f is four times continuously differentiable, then there exists $c \in (a, b)$ such that

$$S^n_{[a,b]}(f) - \int_a^b f(x)\,dx = \frac{(b-a)^5}{2880 n^4} f^{(4)}(c).$$

Exercise 4.6.7: The first expression for g is like the sample variance $\bar{s}^2$ given in equation 4.6.20:

$$\bar{s}^2 = \frac{1}{N}\sum_{i=1}^N (a_i - \bar{a})^2$$

except that here we have $\frac{1}{N-1}$ not $\frac{1}{N}$. To go from the second to the third expression for g, recall (equation 3.8.14) that

$$\mathrm{Var}(f) = E\Big(\big(f - E(f)\big)^2\Big).$$

Part d: The equation

$$\mathrm{Var}_{\mathbf{T}}\left(\overline{f}_T\right) = \frac{1}{N}\mathrm{Var}_S(f)$$

is at the heart of statistics. Dividing the variance by N corresponds to dividing the standard deviation by $\sqrt{N}$; we encountered this when discussing the central limit theorem. It explains why when polling, doubling the number of people polled divides the confidence interval by $\sqrt{2}$.

4.6.7 Let $(S, \mathbf{P})$ be a finite probability space with all outcomes equally likely, and let $\mathbf{T}$ be the set of subsets of S with N elements. The space $\mathbf{T}$ is itself a probability space, as is each $T \in \mathbf{T}$; we will denote the expectations of random variables on these spaces by E_S, $E_{\mathbf{T}}$, E_T. Let $u : S \to \mathbb{R}$ be a random variable; we will say that $v : \mathbf{T} \to \mathbb{R}$ is an *unbiased estimator* of f if $E_{\mathbf{T}}(v) = E_S(u)$.

Set $\overline{f}_T = E_T(f)$; it is the average of f over the sample T. Define $g : \mathbf{T} \to \mathbb{R}$, equivalently, by

$$g(T) = \frac{1}{N-1}\sum_{t \in T}\Big(f(t) - \overline{f}_T\Big)^2 = \frac{N}{N-1}E_T\Big(\big(f - \overline{f}_T\big)^2\Big) = \frac{N}{N-1}\mathrm{Var}_T(f).$$

Saying that g is an unbiased estimator of the variance of f means that

$$\mathrm{Var}_S(f) = E_{\mathbf{T}}(g) = E_{\mathbf{T}}\frac{N}{N-1}\left(E_T(f - \overline{f}_T)^2\right) \tag{1}$$

This is what this exercise will show.

a. Show that

$$E_{\mathbf{T}}(E_T(f^2)) = E_S(f^2), \quad E_T(\overline{f}_T^2) = \overline{f}_T^2, \quad E_T(f\overline{f}_T) = \overline{f}_T^2.$$

b. Develop the square on the right of equation (1), to find

$$\frac{N}{N-1}E_{\mathbf{T}}\left(E_T(f - \overline{f}_T)^2\right) = \frac{N}{N-1}\left(E_S(f^2) - E_{\mathbf{T}}(\overline{f}_T^2)\right)$$

c. On both terms above, use $\mathrm{Var}(f) + (E(f))^2 = E(f^2)$ (equation 3.8.14) to get

$$\begin{aligned} E_{\mathbf{T}}(g) &= \frac{N}{N-1}E_{\mathbf{T}}\left(E_T(f - \overline{f}_T)^2\right) \\ &= \frac{N}{N-1}\left(\mathrm{Var}_S(f) + (E_S(f))^2 - \mathrm{Var}_{\mathbf{T}}(\overline{f}_T) - (E_{\mathbf{T}}(\overline{f}_T))^2\right) \end{aligned}$$

and show that the second and fourth terms cancel.

d. Show that $\mathrm{Var}_{\mathbf{T}}(\overline{f}_T) = \frac{1}{N}\mathrm{Var}_S(f)$, to conclude that g is an unbiased estimator of the variance of f.

4.7 Other pavings

The dyadic paving is the most rigid and restrictive we can think of, making most theorems easiest to prove. But in many settings the rigidity of the dyadic paving $\mathcal{D}_N$ is not necessary or best. Often we will want to have more "paving tiles" where the function varies rapidly, and bigger ones elsewhere, shaped to fit our domain of integration. In some situations, a particular paving is more or less imposed.

To measure the standard deviation of the income of Americans, you would want to subdivide the U.S. by census tracts, not by closely spaced latitudes and longitudes, because that is how the data are collected.

Example 4.7.1 (Measuring rainfall). Imagine that you wish to measure rainfall in South America during October 2015, in liters per square kilometer. One approach would be to use dyadic cubes (squares in this case), measuring the rainfall at the center of each cube and seeing what happens as the decomposition gets finer and finer. One problem with this approach, which we discuss in Chapter 5, is that the dyadic squares lie in a plane, and the surface of South America does not. Another is that dyadic cubes would complicate the collection of data. It would be easier to break up South America into countries and assign to each the product of its area and the rainfall that fell at a particular point in the country, perhaps its capital; you would then add these products together. To get a more accurate estimate of the integral you would use a finer decomposition, like provinces or counties. △

Here we will show that very general pavings can be used to compute integrals. Part 1 of Definition 4.7.2 says that the set of all $P \in \mathcal{P}$ completely paves $X \subset \mathbb{R}^n$, and part 2 says that two "tiles" can overlap only in a set of volume 0. Part 4 prevents tiles from being pathologically funny-shaped (there are limits to the gerrymandering allowed).

It follows from part 4 Definition 4.7.2 that each P is pavable: has a well-defined volume.

It would be hard to come up with an example where condition 4 fails. It is possible, but such an example would not look like anything you would be tempted to call a tile.

Definition 4.7.2 (A paving of $X \subset \mathbb{R}^n$). A *paving* of a subset $X \subset \mathbb{R}^n$ is a collection $\mathcal{P}$ of bounded subsets $P \subset X$ such that

1. $\cup_{P \in \mathcal{P}} P = X$
2. $\text{vol}_n(P_1 \cap P_2) = 0$ when $P_1, P_2 \in \mathcal{P}$ and $P_1 \neq P_2$.
3. Any bounded subset of X intersects only finitely many $P \in \mathcal{P}$.
4. For all $P \in \mathcal{P}$, we have $\text{vol}_n(\partial P) = 0$.

Paving the United States by counties refines the paving by states: no county lies partly in one state and partly in another. A further refinement is provided by census tracts. But this is not a nested partition, since we have no (infinite) sequence of partitions.

Definition 4.7.3 (Nested partition). A sequence $\mathcal{P}_N$ of pavings of $X \subset \mathbb{R}^n$ is called a *nested partition* of X if

1. $\mathcal{P}_{N+1}$ refines $\mathcal{P}_N$: every piece of $\mathcal{P}_{N+1}$ is contained in a piece of $\mathcal{P}_N$.
2. The pieces of $\mathcal{P}_N$ shrink to points as $N \to \infty$:
$$\lim_{N\to\infty} \sup_{P \in \mathcal{P}_N} \text{diam}(P) = 0, \qquad 4.7.1$$
where the diameter of P, denoted $\text{diam}(P)$, is the supremum of the distance between points $\mathbf{x}, \mathbf{y} \in P$.

What we called $U_N(f)$ in Section 4.1 would be called $U_{\mathcal{D}_N}(f)$ using this notation. We will often omit the subscript $\mathcal{D}_N$ (which you will recall denotes the collection of cubes C at a single level N) when referring to the dyadic decompositions, both to lighten the notation and to avoid confusion between $\mathcal{D}$ and $\mathcal{P}$, which look similar when set in small subscript type.

In contrast to the upper and lower sums of the dyadic decompositions (equation 4.1.17), where $\text{vol}_n C$ is the same for any cube C at a given resolution N, in equation 4.7.2, $\text{vol}_n P$ is not necessarily the same for all "paving tiles" $P \in \mathcal{P}_N$.

The two sides of equation 4.7.4 are equal by definition.

Theorem 4.7.4 is proved in Appendix A18.

For any bounded function f with bounded support, we can define an upper sum $U_{\mathcal{P}_N}(f)$ and a lower sum $L_{\mathcal{P}_N}(f)$ with respect to any paving:
$$U_{\mathcal{P}_N}(f) = \sum_{P \in \mathcal{P}_N} M_P(f)\,\text{vol}_n P \text{ and } L_{\mathcal{P}_N}(f) = \sum_{P \in \mathcal{P}_N} m_P(f)\,\text{vol}_n P. \qquad 4.7.2$$

Theorem 4.7.4 (Integrals using arbitrary pavings). *Let $X \subset \mathbb{R}^n$ be a bounded subset, and $\mathcal{P}_N$ a nested partition of X.*

1. *If $f : \mathbb{R}^n \to \mathbb{R}$ is integrable, then the limits*
$$\lim_{N\to\infty} U_{\mathcal{P}_N}(f\mathbf{1}_X) \quad \text{and} \quad \lim_{N\to\infty} L_{\mathcal{P}_N}(f\mathbf{1}_X) \qquad 4.7.3$$
exist and are both equal to
$$\int_X f(\mathbf{x})\,|d^n\mathbf{x}| = \int_{\mathbb{R}^n} f(\mathbf{x})\mathbf{1}_X(\mathbf{x})\,|d^n\mathbf{x}|. \qquad 4.7.4$$
2. *Conversely, if the limits in formula 4.7.3 are equal, $f\mathbf{1}_X$ is integrable and*
$$\int_{\mathbb{R}^n} f(\mathbf{x})\mathbf{1}_X(\mathbf{x})\,|d^n\mathbf{x}| \qquad 4.7.5$$
is equal to the common limit.

Exercises for Section 4.7

Exercise 4.7.1: This is a fairly obvious Riemann sum. You are allowed (and encouraged) to use all the theorems of Section 4.3. Using Riemann sums to show that limits exist is a fairly common tactic; see also Exercise 4.30.

4.7.1 a. Show that the limit $\lim_{N\to\infty} \frac{1}{N^3} \sum_{0\le n,m<N} me^{-nm/N^2}$ exists.

b. Compute the limit in part a.

4.7.2 a. Let A(R) be the number of points with integer entries in the disc given by $x^2+y^2 \le R^2$. Show that the limit $\lim_{R\to\infty} \frac{A(R)}{R^2}$ exists, and evaluate it.

b. Repeat, for the function $B(R)$ that counts how many points of the triangular grid $\left\{ n\begin{pmatrix}1\\0\end{pmatrix} + m\begin{pmatrix}1/2\\ \sqrt{3}/2\end{pmatrix} \;\middle|\; n,m\in\mathbb{Z}\right\}$ are in the disc.

4.8 Determinants

We will use determinants heavily throughout the remainder of the book: forms, to be discussed in Chapter 6, are built on the determinant.

The function D takes n vectors in $\mathbb{R}^n$ and returns a number. We will think of the determinant as a function of n vectors rather than as a function of an $n\times n$ matrix.

"Antisymmetric" and "alternating" are synonymous.

In Theorem and Definition 4.8.1 we are assuming that the entries of the matrix A are real numbers. This is not essential; they could be complex numbers, polynomials, or even functions. But it is essential that whatever the entries are, multiplication of the entries be commutative. There is no determinant function for matrices of matrices, for instance (see Exercise 4.8.4).

The *determinant* is a function of square matrices. In Section 1.4 we saw that determinants of 2×2 and 3×3 matrices have a geometric interpretation: the first gives the area of the parallelogram spanned by two vectors; the second gives the volume of the parallelepiped spanned by three vectors. In higher dimensions the determinant also has a geometric interpretation, as a *signed volume*; this is what makes the determinant important.

In order to obtain the volume interpretation most readily, we shall define the determinant by the three properties that characterize it. When proving Theorem and Definition 4.8.1 we will first prove existence, then uniqueness.

Theorem and Definition 4.8.1 (The determinant). *There exists a unique function $D\colon (\mathbb{R}^n)^n \to \mathbb{R}$ that is*

1. *Multilinear: D is linear with respect to each of its arguments.*
2. *Antisymmetric: Exchanging any two arguments changes its sign.*
3. *Normalized: $D(\mathbf{e}_1,\dots,\mathbf{e}_n)=1$.*

The determinant of an $n\times n$ matrix $A=[\vec{\mathbf{a}}_1,\dots,\vec{\mathbf{a}}_n]$ is

$$\det A = D(\vec{\mathbf{a}}_1,\dots,\vec{\mathbf{a}}_n).$$

Antisymmetry means that for all $1\le i<j\le n$, we have

$$\det[\vec{\mathbf{a}}_1,\dots,\vec{\mathbf{a}}_i,\dots,\vec{\mathbf{a}}_j,\dots,\vec{\mathbf{a}}_n] = -\det[\vec{\mathbf{a}}_1,\dots,\vec{\mathbf{a}}_j,\dots,\vec{\mathbf{a}}_i,\dots,\vec{\mathbf{a}}_n]. \quad 4.8.1$$

Let's spell out what multilinearity means: if one of the columns of A is written

$$\vec{\mathbf{a}}_i = \alpha\vec{\mathbf{u}} + \beta\vec{\mathbf{w}}, \quad 4.8.2$$

then

$$\det\left[\vec{\mathbf{a}}_1,\dots,\vec{\mathbf{a}}_{i-1},(\alpha\vec{\mathbf{u}}+\beta\vec{\mathbf{w}}),\vec{\mathbf{a}}_{i+1},\dots,\vec{\mathbf{a}}_n\right]$$
$$=\alpha\det[\vec{\mathbf{a}}_1,\dots,\vec{\mathbf{a}}_{i-1},\vec{\mathbf{u}},\vec{\mathbf{a}}_{i+1},\dots,\vec{\mathbf{a}}_n] \qquad 4.8.3$$
$$+\beta\det[\vec{\mathbf{a}}_1,\dots,\vec{\mathbf{a}}_{i-1},\vec{\mathbf{w}},\vec{\mathbf{a}}_{i+1},\dots,\vec{\mathbf{a}}_n].$$

For instance, if

$$\vec{\mathbf{u}}=\begin{bmatrix}1\\0\\1\end{bmatrix},\ \vec{\mathbf{w}}=\begin{bmatrix}2\\2\\3\end{bmatrix},\ \text{so that } -1\vec{\mathbf{u}}+2\vec{\mathbf{w}}=\begin{bmatrix}-1\\0\\-1\end{bmatrix}+\begin{bmatrix}4\\4\\6\end{bmatrix}=\begin{bmatrix}3\\4\\5\end{bmatrix},$$

then

$$\underbrace{\det\begin{bmatrix}1&3&3\\2&4&1\\0&5&1\end{bmatrix}}_{23}=\underbrace{-1\det\begin{bmatrix}1&1&3\\2&0&1\\0&1&1\end{bmatrix}}_{-1\times3=-3}\underbrace{+2\det\begin{bmatrix}1&2&3\\2&2&1\\0&3&1\end{bmatrix}}_{2\times13=26}. \qquad 4.8.4$$

One could define the determinant by a formula, but once matrices are bigger than 3×3, the formula is far too messy for hand computation – and even too time-consuming for computers, once a matrix is even moderately large. We will see (equation 4.8.20) that the determinant can be computed much more reasonably by row reduction or column reduction.

Exercise 4.8.7 asks you to show that if a matrix has a column of zeros, or if it has two identical columns, its determinant is 0.

Many variants of the function Δ_n exist, for instance development according to the last column, or according to the first row. These are all equivalent. In particular, Theorem 4.8.8 shows that it doesn't matter whether we work with columns or rows.

There is analogous formula for the development according to any column, or any row.

Proving existence

To prove existence we construct a function Δ_n that takes an $n\times n$ matrix and returns a number; we then show that it satisfies conditions 1–3 of Theorem and Definition 4.8.1.

If A is an $n\times n$ matrix with $n>1$, denote by $A_{[i,j]}$ the $(n-1)\times(n-1)$ matrix obtained from A by erasing the ith row and the jth column, as illustrated by Example 4.8.2. Define $\Delta_n:\text{Mat}\,(n,n)\to\mathbb{R}$ by

$$\Delta_1([a])\overset{\text{def}}{=}a$$

$$\Delta_n(A)\overset{\text{def}}{=}\sum_{i=1}^{n}\overbrace{(-1)^{1+i}}^{\substack{\text{tells whether}\\+\text{ or }-}}\ \overbrace{a_{i,1}\Delta_{n-1}(A_{[i,1]})}^{\substack{\text{product of } a_{i,1}\text{ and}\\ \Delta\text{ of smaller matrix}}},\ \text{for } n>1, \qquad 4.8.5$$

where $a_{i,1}$ is the entry in the ith row and first column of the matrix A, and $[a]$ is a 1×1 matrix (the number a). We will denote by Δ (without a subscript) the function of square matrices that is Δ_n on $n\times n$ matrices. This is called *development according to the first column.*

Our candidate determinant function is thus recursive: $\Delta_n(A)$ is the sum of n terms, each involving Δ_{n-1} of an $(n-1)\times(n-1)$ matrix; in turn, the Δ_{n-1} of each $(n-1)\times(n-1)$ matrix is the sum of $n-1$ terms, each involving Δ_{n-2} of an $(n–2)\times(n–2)$ matrix .... Eventually, we get down to computing (many) Δ_1 of 1×1 matrices, and Δ_1 simply returns the single entry of the matrix.

Example 4.8.2 (The function Δ_3). If

$$A=\begin{bmatrix}1&3&4\\0&1&1\\1&2&0\end{bmatrix},\ \text{then } A_{[2,1]}=\begin{bmatrix}\not1&3&4\\\not0&\not1&\not1\\\not1&2&0\end{bmatrix}=\begin{bmatrix}3&4\\2&0\end{bmatrix}, \qquad 4.8.6$$

and equation 4.8.5 corresponds to

$$\Delta_3(A) = \underbrace{1\ \Delta_2\left(\begin{bmatrix}1 & 1\\ 2 & 0\end{bmatrix}\right)}_{i=1} \underbrace{-0\ \Delta_2\left(\begin{bmatrix}3 & 4\\ 2 & 0\end{bmatrix}\right)}_{i=2} \underbrace{+1\ \Delta_2\left(\begin{bmatrix}3 & 4\\ 1 & 1\end{bmatrix}\right)}_{i=3}. \quad 4.8.7$$

The first term comes with a + sign because $i = 1$ so $1 + i = 2$ and we have $(-1)^2 = 1$; the second comes with a − sign, because $(-1)^3 = -1$, and so on.

Column operations are defined by replacing the word "row" in Definition 2.1.1 of row operations by the word "column". In particular, row reducing $A^\top$ is the same as column reducing A. Thus by Theorem 2.1.7 every matrix can be column reduced, and a square matrix in column echelon form is either the identity or has a column of zeros.

We use column operations rather than row operations in our construction because we defined the determinant as a function of the n column vectors. This convention makes the interpretation in terms of volumes simpler, and Theorem 4.8.8 shows that rows and columns work equally well.

Applying equation 4.8.5 to each of these 2×2 matrices gives

$$\begin{aligned}
\Delta_2\left(\begin{bmatrix}1 & 1\\ 2 & 0\end{bmatrix}\right) &= 1\Delta_1[0] - 2\Delta_1[1] = -2;\\
\Delta_2\left(\begin{bmatrix}3 & 4\\ 2 & 0\end{bmatrix}\right) &= 3\Delta_1[0] - 2\Delta_1[4] = -8;\\
\Delta_2\left(\begin{bmatrix}3 & 4\\ 1 & 1\end{bmatrix}\right) &= 3\Delta_1[1] - 1\Delta_1[4] = -1,
\end{aligned} \quad 4.8.8$$

so that Δ_3 of our original 3×3 matrix is $1(-2) - 0 + 1(-1) = -3$. △

Now we will verify that the function Δ satisfies properties 1, 2, and 3.

1. *Multilinearity* Let $\mathbf{b}, \mathbf{c} \in \mathbb{R}^n$, and suppose $\mathbf{a}_k = \beta\mathbf{b} + \gamma\mathbf{c}$. Set

$$\begin{aligned}
A &= [\mathbf{a}_1, \dots, \mathbf{a}_k, \dots, \mathbf{a}_n],\\
B &= [\mathbf{a}_1, \dots, \mathbf{b}, \dots, \mathbf{a}_n],\\
C &= [\mathbf{a}_1, \dots, \mathbf{c}, \dots, \mathbf{a}_n].
\end{aligned} \quad 4.8.9$$

The matrices A, B, C of equations 4.8.9 are identical except for the kth column.

The object is to show that $\Delta(A) = \beta\Delta(B) + \gamma\Delta(C)$. We need to distinguish two cases: $k = 1$ (i.e., $\mathbf{a}_k$ is the first column) and $k > 1$.

The case $k > 1$ is proved by induction. Clearly multilinearity is true for Δ_1 (if $[a]$ is a 1×1 matrix, $\Delta_1[a] = a$). We will suppose multilinearity is true for Δ_{n-1}, and prove it for Δ_n. Just write

Equation 4.8.10: The first line is equation 4.8.5; the second is the inductive assumption.

$$\begin{aligned}
\Delta_n(A) &= \sum_{i=1}^{n} (-1)^{1+i} a_{i,1} \Delta_{n-1}(A_{[i,1]})\\
&= \sum_{i=1}^{n} (-1)^{1+i} a_{i,1} \big(\beta\Delta_{n-1}(B_{[i,1]}) + \gamma\Delta_{n-1}(C_{[i,1]})\big)\\
&= \beta\sum_{i=1}^{n} (-1)^{1+i} a_{i,1} \Delta_{n-1}(B_{[i,1]}) + \gamma\sum_{i=1}^{n} (-1)^{1+i} a_{i,1}\Delta_{n-1}(C_{[i,1]})\\
&= \beta\Delta_n(B) + \gamma\Delta_n(C).
\end{aligned} \quad 4.8.10$$

This proves the case $k > 1$. For the case $k = 1$,

$$\Delta_n(A) = \sum_{i=1}^{n}(-1)^{1+i}\, a_{i,1}\, \Delta_{n-1}(A_{[i,1]}) = \sum_{i=1}^{n}(-1)^{1+i} \underbrace{(\beta\, b_{i,1} + \gamma\, c_{i,1})}_{=\, a_{i,1} \text{ by definition}} \Delta_{n-1}(A_{[i,1]})$$

In the second line of equation 4.8.11, $A_{[i,1]} = B_{[i,1]} = C_{[i,1]}$ because A, B, and C are identical except for the first column, which is erased to produce $A_{[i,1]}$, $B_{[i,1]}$, and $C_{[i,1]}$.

$$= \beta \sum_{i=1}^{n}(-1)^{1+i}\, b_{i,1}\, \Delta_{n-1} \underbrace{(A_{[i,1]})}_{=\,(B_{[i,1]})} + \gamma \sum_{i=1}^{n}(-1)^{1+i}\, c_{i,1}\, \Delta_{n-1} \underbrace{(A_{[i,1]})}_{=\,(C_{[i,1]})}$$

$$= \beta\Delta_n(B) + \gamma\Delta_n(C). \qquad 4.8.11$$

This proves multilinearity of our function Δ.

2. *Antisymmetry* We want to prove $\Delta_n(A) = -\Delta_n(\widetilde{A})$, where $\widetilde{A}$ is formed by exchanging the mth and pth columns of A.

Again, we have two cases to consider. The first, where both m and p are greater than 1, is proved by induction. It is vacuously true for $n = 2$ since $m, p > 1$ implies $n \geq 3$. We assume that $n \geq 3$, and that the function Δ_{n-1} is antisymmetric, so that $\Delta_{n-1}(A_{[i,1]}) = -\Delta_{n-1}(\widetilde{A}_{[i,1]})$ for each i. Then

$$\Delta_n(A) = \sum_{i=1}^{n}(-1)^{i+1}\, a_{i,1}\, \Delta_{n-1}(A_{[i,1]}) \overbrace{=}^{\text{by induction}} -\sum_{i=1}^{n}(-1)^{i+1}\, a_{i,1}\, \Delta_{n-1}(\widetilde{A}_{[i,1]})$$

$$= -\Delta_n(\widetilde{A}). \qquad 4.8.12$$

The case where either m or p equals 1 is trickier. Let's assume $m = 1$ and $p = 2$.[9] It will be convenient to denote by

$(a_{[i,j]})_{k,l}$ the k,l-entry of $A_{[i,j]}$

$A_{[i,j][k,l]}$ the matrix $A_{[i,j]}$ with the kth row and lth column omitted.

The numbers $(a_{[i,j]})_{k,l}$ and entries of $A_{[i,j][k,l]}$ are all entries of A, but sometimes with different indices; Figure 4.8.1 should help you figure out what corresponds to what.

Using this notation, we can go one level deeper into our recursive formula, writing

$$\Delta_n(A) = \sum_{i=1}^{n}(-1)^{i+1} a_{i,1} \Delta_{n-1}(A_{[i,1]}) = \sum_{i=1}^{n}(-1)^{i+1} a_{i,1} \left(\sum_{j=1}^{n-1}(-1)^{j+1} (a_{[i,1]})_{j,1} \Delta_{n-2}(A_{[i,1][j,1]}) \right).$$

Of course there is a similar formula for $\widetilde{A}$. We need to show that each term in the double sum for A occurs also in the double sum for $\widetilde{A}$, with opposite sign.

[9]We can restrict ourselves to $p = 2$ because if $p > 2$, we can switch the pth column with the second, then the second with the first, then the second with the pth again. By the inductive hypothesis, the first and third exchanges would each change the sign of the determinant, resulting in no net change; the only exchange that "counts" is the change of the first and second positions.

The unshaded part C on left and right in Figure 4.8.1 are the same $(n-2)\times(n-2)$ matrix, though it is called $C = A_{[i,1][j-1,1]}$ on the left and $C = A_{[j,1][i,1]}$ on the right. Moreover the coefficient

$$ab = a_{i,1}a_{j,2} = ba = \widetilde{a}_{j,1}\widetilde{a}_{i,2} \qquad 4.8.13$$

of $\Delta_{n-2}(C)$ is called $a_{i,1}(a_{[i,1]})_{j-1,1}$ on the left and $\widetilde{a}_{j,1}(\widetilde{a}_{[j,1]})_{i,1}$ on the right.

Thus the term $ab\Delta_{n-2}(C)$ comes with the sign

$$(-1)^{i+1+j-1+1} \text{ on the left and } (-1)^{j+1+i+1} \text{ on the right,} \qquad 4.8.14$$

i.e., with opposite signs.

The key to the proof is the fact (see the caption to Figure 4.8.1) that

$$A_{[i,1][j-1,1]} = \widetilde{A}_{[j,1][i,1]}.$$

This is because, after removing the ith row of A, the jth row becomes the $(j-1)$st row of $A_{[i,1]}$. This leads to the opposite signs in formula 4.8.14.

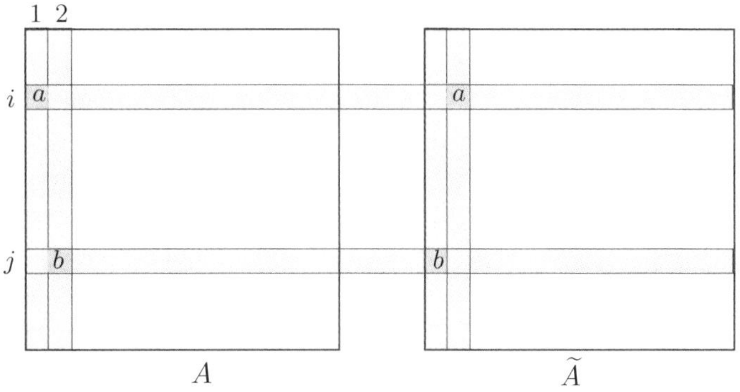

FIGURE 4.8.1. In both matrices, C is the unshaded part. LEFT: The $n \times n$ matrix A, the ith and jth row and the first two columns shaded. RIGHT: The matrix $\widetilde{A}$, identical to A except that the first two columns are exchanged. Let a, b be the shaded entries $a = a_{i,1} = \widetilde{a}_{i,2}$ and $b = a_{j,2} = \widetilde{a}_{j,1}$. Note that the unshaded parts coincide: $A_{[i,1][j-1,1]} = \widetilde{A}_{[j,1][i,1]}$.

3. *Normalization* This is much simpler. If $A = [\vec{\mathbf{e}}_1, \dots, \vec{\mathbf{e}}_n]$, then in the first column, only the first entry $a_{1,1} = 1$ is nonzero, and $A_{1,1}$ is the identity matrix one size smaller, so that Δ_{n-1} of it is 1 by induction. So

$$\Delta_n(A) = a_{1,1}\Delta_{n-1}(A_{1,1}) = 1, \qquad 4.8.15$$

and we have also proved property 3. This completes the proof of existence.

In Figure 4.8.1 we have assumed that $i < j$. For the case $i > j$ we can simply reverse the roles of A and $\widetilde{A}$.

Proving uniqueness

We have shown that Δ satisfies conditions 1–3 of Theorem and Definition 4.8.1. To prove uniqueness of the determinant, suppose Δ' is another function satisfying those conditions; we must show that $\Delta(A) = \Delta'(A)$ for any square matrix A.

First, note that if a matrix A_2 is obtained from A_1 by a column operation, then

$$\Delta'(A_2) = \mu\Delta'(A_1) \qquad 4.8.16$$

for a number μ depending on the type of column operation:

1. *A_2 is obtained by multiplying some column of A_1 by $m \neq 0$.* By multilinearity, $\Delta'(A_2) = m\Delta'(A_1)$, so $\mu = m$.
2. *A_2 is obtained by adding a multiple of the ith column of A_1 to the jth column, with $i \neq j$.* By property 1, this does not change the determinant, because

Equation 4.8.17: The second term on the right is zero, since two columns are equal (Exercise 4.8.7).

$$\begin{aligned}&\Delta'\left[\vec{\mathbf{a}}_1, \dots, \vec{\mathbf{a}}_i, \dots, (\vec{\mathbf{a}}_j + \beta\vec{\mathbf{a}}_i), \dots, \vec{\mathbf{a}}_n\right] \\ &= \Delta'\left[\vec{\mathbf{a}}_1, \dots, \vec{\mathbf{a}}_i, \dots, \vec{\mathbf{a}}_j, \dots, \vec{\mathbf{a}}_n\right] + \underbrace{\beta\,\Delta'\left[\vec{\mathbf{a}}_1, \dots, \vec{\mathbf{a}}_i, \dots, \vec{\mathbf{a}}_i, \dots, \vec{\mathbf{a}}_n\right]}_{=0 \text{ because 2 identical columns } \vec{\mathbf{a}}_i}.\end{aligned} \qquad 4.8.17$$

Thus in this case, $\mu = 1$.

3. *A_2 is obtained by exchanging two columns of A_1.* By antisymmetry, this changes the sign of the determinant, so $\mu = -1$.

Any square matrix can be column reduced until at the end, you either get the identity, or a matrix with a column of zeros.

Column reducing a matrix A to column echelon form A_p can be expressed as follows, where the μ_i corresponding to each column operation is on top of an arrow denoting the operation:

$$A \xrightarrow{\mu_1} A_1 \xrightarrow{\mu_2} A_2 \xrightarrow{\mu_3} \cdots \xrightarrow{\mu_{p-1}} A_{p-1} \xrightarrow{\mu_p} A_p. \qquad 4.8.18$$

The factors $1/\mu$ are imposed by multilinearity and antisymmetry, not by any specifics of how a function is defined, so if Δ' is any function satisfying those conditions, we must have

$$\Delta'(A_{i-1}) = \frac{1}{\mu_i}\Delta'(A_i).$$

Uniqueness also requires normalization, which allows us to assert that if A_p is in column echelon form, then $\Delta'(A_p) = \Delta(A_p)$.

Now let us see that $\Delta(A) = \Delta'(A)$.
If $A_p \neq I$, then by property 1, $\Delta(A_p) = \Delta'(A_p) = 0$ (see Exercise 4.8.7).
If $A_p = I$, then by property 3, $\Delta(A_p) = \Delta'(A_p) = 1$.
Then, working backward,

$$\begin{aligned}\Delta(A_{p-1}) &= \frac{1}{\mu_p}\Delta(A_p) = \frac{1}{\mu_p}\Delta'(A_p) = \Delta'(A_{p-1}),\\ \Delta(A_{p-2}) &= \frac{1}{\mu_p\mu_{p-1}}\Delta(A_p) = \frac{1}{\mu_p\mu_{p-1}}\Delta'(A_p) = \Delta'(A_{p-2}),\\ \dots &= \dots \\ \Delta(A) &= \frac{1}{\mu_p\mu_{p-1}\cdots\mu_1}\Delta(A_p) = \frac{1}{\mu_p\mu_{p-1}\cdots\mu_1}\Delta'(A_p) = \Delta'(A).\end{aligned} \qquad 4.8.19$$

This proves uniqueness. □

Remark. Equation 4.8.19 also provides an effective way to compute the determinant:

$$\det A = \frac{1}{\mu_p\mu_{p-1}\cdots\mu_1}\det A_p = \begin{cases}\frac{1}{\mu_p\mu_{p-1}\cdots\mu_1} & \text{if } A_p = I\\ 0 & \text{if } A_p \neq I.\end{cases} \qquad 4.8.20$$

This is a much better algorithm for computing determinants than is development according to the first column, which is *very* slow.

How slow? It takes time $T(k)$ to compute the determinant of a $k \times k$ matrix. Then to compute the determinant of a $(k+1) \times (k+1)$ matrix we will need to compute $k+1$ determinants of $k \times k$ matrices, and $k+1$ multiplications and k additions:

$$T(k+1) = (k+1)T(k) + (k+1) + k. \qquad 4.8.21$$

In particular, $T(k) > k!$. For a 15×15 matrix, this means $15! \approx 1.3 \times 10^{12}$ calls or operations; even the biggest computers of 2015 could barely do this in a second. And 15×15 is not a big matrix; engineers modeling bridges or airplanes and economists modeling a large company routinely use matrices that are more than 1000×1000. For a 40×40 matrix, the number of operations that would be needed to compute the determinant using development by the first column is *bigger than the number of seconds that have elapsed since the beginning of the universe.* In fact, bigger than the number of billionths of seconds that have elapsed. If you had set a computer computing the determinant back in the days of the dinosaurs, it would have barely begun.

But *column reduction* of an $n \times n$ matrix, using equation 4.8.20, takes about n^3 operations (see Exercise 2.2.11). Using column reduction, one can compute the determinant of a 40×40 matrix in 64 000 operations, which would take a garden-variety laptop only a fraction of a second. △

Theorems relating matrices and determinants

Theorem 4.8.3. *A matrix A is invertible if and only if* $\det A \neq 0$.

Theorem 4.8.3: Equivalently, $\det A = 0$ if and only if its columns are linearly dependent. In particular, $\det A = 0$ if there is a column or row of 0's, or if two columns or two rows are equal.

Proof. This follows immediately from the column reduction algorithm and the uniqueness proof, since along the way we showed that a square matrix has a nonzero determinant if and only if it can be column reduced to the identity. We know from Theorem 2.3.2 that a matrix is invertible if and only if it can be row reduced to the identity; the same argument applies to column reduction. □

Now we come to the key property of the determinant, for which we will see a geometric interpretation later. It was in order to prove this theorem that we defined the determinant by its properties.

Theorem 4.8.4. *If A and B are $n \times n$ matrices, then*

$$\det A \det B = \det(AB). \qquad 4.8.22$$

A definition that defines an object or operation by its properties is called an *axiomatic* definition. The proof of Theorem 4.8.4 should convince you that this can be a fruitful approach. Imagine trying to prove

$$D(A)D(B) = D(AB)$$

from the recursive definition.

Proof. The serious case is the one in which A is invertible. If A is invertible, consider the function

$$f(B) = \frac{\det(AB)}{\det A}. \qquad 4.8.23$$

As you can readily check (Exercise 4.8.8), it has properties 1, 2, and 3, which characterize the determinant function. Since the determinant is uniquely characterized by those properties, it follows that $f(B) = \det B$.

The case where A is not invertible is easy, using what we know about images and dimensions of linear transformations. If A is not invertible, $\det A = 0$ (Theorem 4.8.3), so the left side of equation 4.8.22 is zero. The right side must be zero also: since A is not invertible, $\operatorname{rank} A < n$. Since $\operatorname{img}(AB) \subset \operatorname{img} A$, then $\operatorname{rank}(AB) \leq \operatorname{rank} A < n$, so AB is not invertible either, and $\det(AB) = 0$. □

Corollary 4.8.5. *If a matrix A is invertible, then*

$$\det A^{-1} = \frac{1}{\det A}. \qquad 4.8.24$$

Proof. Just compute: $\det A \det A^{-1} = \det(AA^{-1}) = \det I = 1$. □

The next theorem says that the determinant function is basis independent.

Recall the discussion of change of basis in Section 2.7.

Theorem 4.8.6. *If P is invertible, then*

$$\det A = \det(P^{-1}AP). \qquad 4.8.25$$

$$\det \begin{bmatrix} 1 & 0 & 0 & 0 \\ 0 & 1 & 0 & 0 \\ 0 & 0 & 2 & 0 \\ 0 & 0 & 0 & 1 \end{bmatrix} = 2$$

The determinant of the type 1 elementary matrix $E_{3,2}$ is 2.

$$\det \begin{bmatrix} 1 & 0 & -3 \\ 0 & 1 & 0 \\ 0 & 0 & 1 \end{bmatrix} = 1$$

The determinant of all type 2 elementary matrices is 1.

$$\det \begin{bmatrix} 0 & 1 & 0 \\ 1 & 0 & 0 \\ 0 & 0 & 1 \end{bmatrix} = -1$$

A type 3 elementary matrix E_3 exchanges two rows so

$$\det E_3 = -\det I = -1.$$

All type 1 matrices are diagonal and all type 2 matrices are triangular, so equations 4.8.27 and 4.8.28 illustrate Theorem 4.8.9.

Proof. This follows immediately from Theorems 4.8.4 and 4.8.5. (Remember that the determinant is a number, so $\det A \det B = \det B \det A$). □

Remark 4.8.7. It follows from Theorem 4.8.6 that we can speak of the determinant of an abstract linear transformation $A: V \to V$. If $\{\mathbf{v}\}$ and $\{\mathbf{w}\}$ are two bases of V and T is the change of basis matrix, we can write A as the matrix $[A]_{\{\mathbf{v}\},\{\mathbf{v}\}}$ or as the matrix $T^{-1}[A_{\{\mathbf{w}\},\{\mathbf{w}\}}]T$; by Theorem 4.8.6 both have the same determinant. But the determinant is not defined for a linear transformation $A: V \to W$ with V different from W. △

Theorem 4.8.8 is a major result. One important consequence is that throughout this text, whenever we spoke of column operations, we could just as well have spoken of row operations.

Theorem 4.8.8. *For any $n \times n$ matrix A,*

$$\det A = \det A^{\top}. \qquad 4.8.26$$

Proof. First we will see that the theorem is true for elementary matrices. The equation $D(A_2) = \mu D(A_1)$ in the proof of uniqueness can be rewritten as $\det E = \mu \det I = \mu$, where E is an elementary matrix and I the identity matrix. So

$$\det E_1(i, m) = m \qquad 4.8.27$$

$$\det E_2(i, j, x) = 1 \qquad 4.8.28$$

$$\det E_3(i, j) = -1. \qquad 4.8.29$$

It follows that if E is any elementary matrix, $\det E = \det E^{\top}$, since

$$E_1(i, m)^{\top} = E_1(i, m), \quad E_2(i, j, x)^{\top} = E_2(j, i, x), \quad E_3(i, j)^{\top} = E_3(i, j),$$

and in all three cases the determinants are equal.

Now recall from Section 2.3 that for any matrix A there exist elementary matrices $E_1, \dots, E_k$ such that

$$\widetilde{A} = E_k \dots E_1 A. \qquad 4.8.30$$

where $\widetilde{A}$ is in row echelon form. Thus

$$\widetilde{A}^\top = A^\top E_1^\top \dots E_k^\top, \qquad 4.8.31$$

where $\widetilde{A}^\top$ is in column echelon form, and if A is square,

$$\det A = \frac{\det \widetilde{A}}{(\det E_k)\dots(\det E_1)}, \qquad \det A^\top = \frac{\det \widetilde{A}^\top}{(\det E_1^\top)\dots(\det E_k^\top)}. \qquad 4.8.32$$

There are two possibilities: $\widetilde{A} = I$ or $\widetilde{A} \neq I$. If $\widetilde{A} = I$, then $\widetilde{A}^\top = I$, and $\det \widetilde{A} = \det \widetilde{A}^\top = 1$, which gives

$$\det A = \frac{1}{(\det E_k)\dots(\det E_1)} = \frac{1}{(\det E_1^\top)\dots(\det E_k^\top)} = \det A^\top. \qquad 4.8.33$$

If $\widetilde{A} \neq I$, then (by Proposition 2.5.11) $\det A = \det A^\top = 0$. □

Equation 4.8.33: Since determinants are numbers,

$$(\det E_k)\cdots(\det E_1) = (\det E_1)\cdots(\det E_k).$$

The fact that determinants are numbers, so that multiplication of determinants is commutative, is much of the point of determinants; essentially everything having to do with matrices that does not involve non-commutativity can be done using determinants.

The determinant of a triangular matrix is easy to compute.

Theorem 4.8.9 (Determinant of triangular matrix). *If a matrix is triangular, then its determinant is the product of the entries along the diagonal.*

Recall from Definition 1.2.19 that an upper triangular matrix is a square matrix with nonzero entries only on or above the main diagonal; a lower triangular matrix is a square matrix with nonzero entries only on or below the main diagonal. The diagonal goes from top left to bottom right.

Of course a diagonal matrix is triangular, so the determinant of a diagonal matrix is also the product of its diagonal entries.

Proof. We will prove the result for upper triangular matrices; the result for lower triangular matrices then follows from Theorem 4.8.8. The proof is by induction. Theorem 4.8.9 is clearly true for a 1×1 triangular matrix (note that any 1×1 matrix is triangular). If A is triangular of size $n\times n$ with $n > 1$, the submatrix $A_{1,1}$ (A with its first row and first column removed) is also triangular, of size $(n-1) \times (n-1)$, so we may assume by induction that

$$\det A_{1,1} = a_{2,2} \dots a_{n,n}. \qquad 4.8.34$$

An alternative proof is sketched in Exercise 4.8.22.

Since $a_{1,1}$ is the only nonzero entry in the first column, development according to the first column gives

$$\det A = (-1)^2 a_{1,1} \det A_{1,1} = a_{1,1}a_{2,2}\dots a_{n,n}. \quad \square \qquad 4.8.35$$

Theorem 4.8.10: We could also put the nonzero matrix C (now $m \times n$) in the lower left corner:

$$\det \begin{bmatrix} A & \mid & 0 \\ C & \mid & B \end{bmatrix} = \det A \det B.$$

You are asked to prove Theorem 4.8.10 in Exercise 4.8.11.

Theorem 4.8.10. *If A is an $n \times n$ matrix, B an $m \times m$ matrix, and C an $n \times m$ matrix, then*

$$\det \begin{bmatrix} A & \mid & C \\ 0 & \mid & B \end{bmatrix} = \det A \det B. \qquad 4.8.36$$

Permutations and their signatures

A *permutation* of a set X is a bijective map $f: X \to X$; the word is usually used only when X is finite. The set of permutations of the set $X = \{1, 2, \dots, n\}$ is denoted Perm_n, which has $n!$ elements. It has the following properties:

A set X plus a binary operation $X \times X \to X$ satisfying these three properties is called a *group*. Thus Perm_n with the binary operation composition is a group. We encountered another group, the orthogonal group $O(n)$, in Exercise 3.2.11. The set Perm_n with the binary operation composition is a subgroup of $O(n)$, where the binary operation is matrix multiplication.

1. Composition is associative: if $\sigma_1, \sigma_2, \sigma_3 \in \text{Perm}_n$, then
$$(\sigma_1 \circ \sigma_2) \circ \sigma_3 = \sigma_1 \circ (\sigma_2 \circ \sigma_3) \qquad 4.8.37$$
(see Proposition 0.4.15).
2. There is an identity id for composition: $\sigma \circ \text{id} = \text{id} \circ \sigma = \sigma$.
3. For every $\sigma \in \text{Perm}_n$, the permutation σ^{-1} satisfies
$$\sigma \circ \sigma^{-1} = \sigma^{-1} \circ \sigma = \text{id}. \qquad 4.8.38$$

A first step is to learn how to write permutations. One way is simply to list the values of a permutation: for instance, one element $\sigma \in \text{Perm}_5$ is
$$\sigma(1) = 3,\ \sigma(2) = 4,\ \sigma(3) = 5,\ \sigma(4) = 2,\ \sigma(5) = 1. \qquad 4.8.39$$

We can write this permutation more concisely as $(1,3,5)(2,4)$. The terms $(1,3,5)$ and $(2,4)$ are called *cycles*. The cycle $(1,3,5)$ means that σ takes 1 to 3, takes 3 to 5, and takes 5 back to 1. The cycle $(2,4)$ means that σ takes 2 to 4 and 4 to 2. The identity for $(1,2,3)$ is $(1),(2),(3)$.

Cycle notation has virtues other than concision. The *order* of a permutation σ is the smallest m such that $\sigma^m = \text{id}$. (By σ^m we mean σ composed with itself m times; thus, $\sigma^3 = \sigma \circ \sigma \circ \sigma$.) When a permutation is written in cycle notation, the order is the least common multiple of the lengths of the cycles.

For instance, the cycle $(1,3,5)$ has length 3, and $(2,4)$ has length 2, so the permutation $(1,3,5)(2,4)$ has order 6, the least common multiple of 3 and 2.

You can also compute the signature as a function of the lengths of the cycles; see Exercise 4.8.10.

A *transposition* is a permutation that exchanges two symbols and leaves all others fixed; for instance, $(1)(2,4)(3)(5)$ is the transposition that exchanges 2 and 4.

Permutations come in two flavors: *even* if the signature is 1 and *odd* if the signature is -1.

Theorem and Definition 4.8.11 (Signature of permutation). *There exists a unique map*
$$\text{sgn} : \text{Perm}_n \to \{-1, 1\}$$
called the signature, such that

1. $\text{sgn}(\sigma_1 \circ \sigma_2) = \text{sgn}(\sigma_1)\,\text{sgn}(\sigma_2)$ *for all* $\sigma_1, \sigma_2 \in \text{Perm}_n$
2. $\text{sgn}(\tau) = -1$ *for all transpositions* $\tau \in \text{Perm}_n$

Let $\sigma = (1,3)(2,4)$, i.e.,
$$\sigma(1) = 3, \qquad \sigma(2) = 4$$
$$\sigma(3) = 1, \qquad \sigma(4) = 2.$$
The *permutation matrix* M_σ is then
$$M_\sigma = \begin{bmatrix} 0 & 0 & 1 & 0 \\ 0 & 0 & 0 & 1 \\ 1 & 0 & 0 & 0 \\ 0 & 1 & 0 & 0 \end{bmatrix};$$
the first column is $\vec{\mathbf{e}}_{\sigma(1)} = \vec{\mathbf{e}}_3$, the second is $\vec{\mathbf{e}}_{\sigma(2)} = \vec{\mathbf{e}}_4$, and so on.

Proof. Define for every $\sigma \in \text{Perm}_n$ the *permutation matrix* M_σ by
$$M_\sigma \vec{\mathbf{e}}_i = \vec{\mathbf{e}}_{\sigma(i)}. \qquad 4.8.40$$
Let $\sigma, \tau \in \text{Perm}_n$. Then $M_{\sigma\circ\tau} = M_\sigma M_\tau$, since
$$M_{\sigma\circ\tau}\vec{\mathbf{e}}_i = \vec{\mathbf{e}}_{\sigma(\tau(i))} = M_\sigma \vec{\mathbf{e}}_{\tau(i)} = M_\sigma M_\tau \vec{\mathbf{e}}_i, \qquad 4.8.41$$
so $M_\sigma M_\tau$ and $M_{\sigma\circ\tau}$ coincide on every basis vector. For every $\sigma \in \text{Perm}_n$ define
$$\text{sgn}(\sigma) \stackrel{\text{def}}{=} \det M_\sigma. \qquad 4.8.42$$

Property 1 of Theorem and Definition 4.8.11 follows from Theorem 4.8.4. Property 2 follows from the antisymmetry of the determinant: if τ is a transposition, then M_τ is obtained from the identity by exchanging two columns. This proves existence. Any rule satisfying properties 1 and 2 must give the same value on any permutation: every permutation can be written as a composition of transpositions, and the signature is $+1$ if you need an even number of permutations and -1 if you need an odd number. This proves uniqueness. $\square$

> Property 1:
> $$\begin{aligned}\operatorname{sgn}(\sigma_1 \circ \sigma_2) &= \det M_{\sigma_1 \circ \sigma_2} \\ &= \det(M_{\sigma_1} M_{\sigma_2}) \\ &= (\det M_{\sigma_1})(\det M_{\sigma_2}) \\ &= \operatorname{sgn}\sigma_1 \operatorname{sgn}\sigma_2.\end{aligned}$$

Remark. In the process of the proof we showed that the parity of the number of transpositions required (i.e., whether the number is even or odd) is independent of the decomposition as a product of transpositions. $\triangle$

In equation 4.8.42 we defined the signature in terms of the determinant: let us see that we can go backwards and define the determinant in terms of the signature.

Theorem 4.8.12 (Determinant in terms of permutations). *Let A be an $n \times n$ matrix with entries $a_{i,j}$. Then*

$$\det A = \sum_{\sigma \in \operatorname{Perm}_n} \operatorname{sgn}(\sigma) a_{1,\sigma(1)} \ldots a_{n,\sigma(n)}. \qquad 4.8.43$$

> Equation 4.8.43: Each term of the sum is the product of n entries of the matrix A, chosen so that there is exactly one from each row and one from each column; no two are from the same column or the same row. These products are then added together, with an appropriate sign.

Example 4.8.13 (Computing the determinant by permutations). Let $n = 3$, and let A be the matrix $A = \begin{bmatrix} 1 & 2 & 3 \\ 4 & 5 & 6 \\ 7 & 8 & 9 \end{bmatrix}$. There are six possible permutations of $n = 3$ numbers, so we have the following, where the labeling of the σ is arbitrary:

> Example 4.8.13: You are asked to confirm these signatures in Exercise 4.8.13.
>
> It would be quicker to compute the determinant using Definition 1.4.15. Theorem 4.8.12 does not provide an effective algorithm for computing determinants; for 2×2 and 3×3 matrices, which are standard in the classroom (but not anywhere else), we have explicit and manageable formulas. When they are large, column reduction (equation 4.8.20) is immeasurably faster: for a 30×30 matrix, roughly the difference between one second and the age of the universe.

$\sigma_1 = (1)(2)(3)$	$+$	$a_{1,1}a_{2,2}a_{3,3} = 1 \cdot 5 \cdot 9 = 45$
$\sigma_2 = (1,2,3)$	$+$	$a_{1,2}a_{2,3}a_{3,1} = 2 \cdot 6 \cdot 7 = 84$
$\sigma_3 = (1,3,2)$	$+$	$a_{1,3}a_{2,1}a_{3,2} = 3 \cdot 4 \cdot 8 = 96$
$\sigma_4 = (1)(2,3)$	$-$	$a_{1,1}a_{2,3}a_{3,2} = 1 \cdot 6 \cdot 8 = 48$
$\sigma_5 = (1,2)(3)$	$-$	$a_{1,2}a_{2,1}a_{3,3} = 2 \cdot 4 \cdot 9 = 72$
$\sigma_6 = (1,3)(2)$	$-$	$a_{1,3}a_{2,2}a_{3,1} = 3 \cdot 5 \cdot 7 = 105$

So $\det A = 45 + 84 + 96 - 48 - 72 - 105 = 0$. Can you see why this determinant had to be 0?[10] $\triangle$

Proof of Theorem 4.8.12. So as not to prejudice the issue, let us temporarily call the function of Theorem 4.8.12 D:

$$D(A) = \sum_{\sigma \in \operatorname{Perm}(1,\ldots,n)} \operatorname{sgn}(\sigma) a_{1,\sigma(1)} \ldots a_{n,\sigma(n)}. \qquad 4.8.44$$

---

[10] Denote by $\vec{\mathbf{a}}_1, \vec{\mathbf{a}}_2, \vec{\mathbf{a}}_3$ the columns of A. Then $2\vec{\mathbf{a}}_2 - \vec{\mathbf{a}}_1 = \vec{\mathbf{a}}_3$; the columns are linearly dependent, so the matrix is not invertible, and its determinant is 0.

We will show that D has the three properties that characterize the determinant. *Normalization* is satisfied: $D(I) = 1$, since the only $\sigma \in \text{Perm}_n$ that contributes a nonzero term to the sum is the identity permutation. Indeed, if σ is not the identity permutation, the corresponding product $\text{sgn}(\sigma)a_{1,\sigma(1)} \cdots a_{n,\sigma(n)}$ contains at least one off-diagonal element and gives 0. So the sum has only one term, corresponding to the product of the diagonal entries, which are all 1, and its sign is the signature of the identity permutation, which is $+1$.

Multilinearity is straightforward: each term $a_{1,\sigma(1)} \cdots a_{n,\sigma(n)}$ is multilinear as a function of the columns, so any linear combination of such terms is also multilinear.

To prove *antisymmetry*, let A be an $n \times n$ matrix, and let τ be the permutation of $\{1, \dots, n\}$ that exchanges i and j and leaves the other elements of $\{1, \dots, n\}$ unchanged. Note that $\text{sgn}(\tau) = \text{sgn}(\tau^{-1}) = -1$. Denote by A' the matrix formed by exchanging the ith and jth columns of A. Then equation 4.8.44, applied to the matrix A', gives

$$\begin{aligned} D(A') &= \sum_{\sigma \in \text{Perm}(1,\dots,n)} \text{sgn}(\sigma) a'_{1,\sigma(1)} \cdots a'_{n,\sigma(n)} \\ &= \sum_{\sigma \in \text{Perm}(1,\dots,n)} \text{sgn}(\sigma) a_{1,\tau\circ\sigma(1)} \cdots a_{n,\tau\circ\sigma(n)}. \end{aligned} \tag{4.8.45}$$

As σ runs through all permutations, $\sigma' = \tau \circ \sigma$ does too, so we can write

Note that $\tau^{-1} \circ \sigma' = \sigma$. We are using Theorem 4.8.4 when we write

$$\begin{aligned} \text{sgn}(\tau^{-1} \circ \sigma') &= \text{sgn}(\tau^{-1})\,\text{sgn}(\sigma') \\ &= -\text{sgn}(\sigma') \end{aligned}$$

$$\begin{aligned} D(A') &= \sum_{\sigma' \in \text{Perm}(1,\dots,n)} \underbrace{\text{sgn}(\tau^{-1} \circ \sigma')}_{\text{sgn}(\tau^{-1})\,\text{sgn}(\sigma') = -\text{sgn}(\sigma')} a_{1,\sigma'(1)} \cdots a_{n,\sigma'(n)} \\ &= -\sum_{\sigma' \in \text{Perm}(1,\dots,n)} \text{sgn}(\sigma') a_{1,\sigma'(1)} \cdots a_{n,\sigma'(n)} = -D(A). \quad \square \end{aligned} \tag{4.8.46}$$

The trace and the derivative of the determinant

The trace of

$$\begin{bmatrix} 1 & 0 & 3 \\ 1 & 2 & 1 \\ 0 & 1 & -1 \end{bmatrix}$$

is $1 + 2 + (-1) = 2$.

Definition 4.8.14 (The trace of a matrix). The *trace* of an $n \times n$ matrix A is the sum of its diagonal elements:

$$\text{tr}\, A \overset{\text{def}}{=} \sum_{i=1}^{n} a_{i,i} = a_{1,1} + a_{2,2} + \cdots + a_{n,n}. \tag{4.8.47}$$

The trace, which is obviously easy to compute (much easier than the determinant), is a linear function of A: if A, B are $n \times n$ matrices, then

$$\text{tr}(aA + bB) = a\, \text{tr}\, A + b\, \text{tr}\, B. \tag{4.8.48}$$

Theorem 4.8.15 shows that the trace and the determinant are closely related. It allows easy proofs of many properties of the trace that are not at all obvious from the definition.

Theorem 4.8.15 (Derivative of the determinant).

1. *The determinant function* $\det : \text{Mat}\,(n,n) \to \mathbb{R}$ *is differentiable.*
2. *The derivative of the determinant at the identity is given by*

$$[\mathbf{D}\det(I)]B = \operatorname{tr} B. \tag{4.8.49}$$

3. *If* $\det A \neq 0$, *then* $[\mathbf{D}\det(A)]B = \det A \operatorname{tr}(A^{-1}B)$.

Part 2 of Theorem 4.8.15 is a special case of part 3, but it is interesting in its own right. We prove it first, so we state it separately. Computing the derivative when A is not invertible is a bit trickier, and is explored in Exercise 4.8.19.

Example 4.8.16 (Derivative of the determinant of 2×2 matrices). The determinant function of 2×2 matrices can be considered as the function $\det : \mathbb{R}^4 \to \mathbb{R}$ given by

$$\det \begin{pmatrix} a \\ b \\ c \\ d \end{pmatrix} = ad - bc, \quad \text{with derivative } \left[\mathbf{D}\det \begin{bmatrix} a & b \\ c & d \end{bmatrix}\right] = [d\,, -c\,, -b\,, a].$$

At the identity this derivative is $[1\,,0\,,0\,,1]$; if we identify $B = \begin{bmatrix} \alpha & \beta \\ \gamma & \delta \end{bmatrix}$ with $\begin{pmatrix} \alpha \\ \beta \\ \gamma \\ \delta \end{pmatrix}$, we have $[1\,,0\,,0\,,1] \begin{pmatrix} \alpha \\ \beta \\ \gamma \\ \delta \end{pmatrix} = \alpha + \delta = \operatorname{tr} B$ △

Proof. 1. By Theorem 4.8.12, the determinant is a polynomial in the entries of the matrix, hence certainly differentiable. (For instance, the formula $ad - bc$ is a polynomial in the variables a, b, c, d.)

2. Since (Proposition 1.7.14) $[\mathbf{D}\det(I)]B$ is the directional derivative

$$\lim_{h \to 0} \frac{\det(I + hB) - \det I}{h}, \tag{4.8.50}$$

it is enough to evaluate that limit. Put another way, we want to find the terms that are linear in h of the expansion given by equation 4.8.43, for

$$\det(I + hB) = \det \begin{bmatrix} 1 + hb_{1,1} & hb_{1,2} & \dots & hb_{1,n} \\ hb_{2,1} & 1 + hb_{2,2} & \dots & hb_{2,n} \\ \vdots & \vdots & \ddots & \vdots \\ hb_{n,1} & hb_{n,2} & \dots & 1 + hb_{n,n} \end{bmatrix}. \tag{4.8.51}$$

Try the 2×2 case of equation 4.8.51:

$$\begin{aligned} &\det\left(I + h\begin{bmatrix} a & b \\ c & d \end{bmatrix}\right) \\ &= \det \begin{bmatrix} 1 + ha & hb \\ hc & 1 + hd \end{bmatrix} \\ &= (1 + ha)(1 + hd) - h^2bc \\ &= 1 + h(a + d) + h^2(ad - bc). \end{aligned}$$

Equation 4.8.43 shows that if a term has one factor off the diagonal, it must have at least two (as illustrated for the 2×2 case in the margin): a permutation that permutes all symbols but one to themselves must also take the last symbol to itself: it has nowhere else to go. But the off-diagonal terms all contain a factor of h, so only the term corresponding to the identity permutation contributes any linear terms in h. That term, which has signature $+1$, is

$$\begin{aligned} &(1 + hb_{1,1})(1 + hb_{2,2}) \dots (1 + hb_{n,n}) \\ &\quad = 1 + h(b_{1,1} + b_{2,2} + \dots + b_{n,n}) + \dots + h^n b_{1,1} b_{2,2} \dots b_{n,n}, \end{aligned} \tag{4.8.52}$$

and we see that the linear term is exactly $b_{1,1} + b_{2,2} + \dots + b_{n,n} = \operatorname{tr} B$.

3. Again, take directional derivatives:

$$\lim_{h\to 0}\frac{\det(A+hB)-\det A}{h} = \lim_{h\to 0}\frac{\det(A(I+hA^{-1}B))-\det A}{h}$$

$$= \lim_{h\to 0}\frac{\det A\det(I+hA^{-1}B)-\det A}{h} = \det A \lim_{h\to 0}\frac{\det(I+hA^{-1}B)-1}{h}$$

$$= \det A \lim_{h\to 0}\frac{\det(I+hA^{-1}B)-\det I}{h} = \det A\ \operatorname{tr}(A^{-1}B). \quad \square \qquad 4.8.53$$

Equation 4.8.53: The limit in the last line is the directional derivative of det at I in the direction $A^{-1}B$, which by Proposition 1.7.14 can be written

$$[\mathbf{D}\det(I)](A^{-1}B),$$

which by part 2 of the theorem is $\operatorname{tr}(A^{-1}B)$.

Corollary 4.8.17. *If P is invertible, then for any matrix A we have*

$$\operatorname{tr}(P^{-1}AP) = \operatorname{tr} A. \qquad 4.8.54$$

Equation 4.8.54 looks like equation 4.8.25 in Theorem 4.8.6, but it is not true for the same reason. Theorem 4.8.6 follows from Theorem 4.8.4:

$$\det(AB) = \det A \det B.$$

This is not true for the trace: the trace of a product is not the product of the traces.

Most proofs of Corollary 4.8.17 begin by showing that

$$\operatorname{tr} AB = \operatorname{tr} BA.$$

Exercise 4.8.5 asks you to prove this algebraically; Exercise 4.18 asks you to prove it using Corollary 4.8.17.

Proof. This uses Theorem 4.8.15 and Theorem 4.8.6 (basis independence of the determinant):

$$\begin{aligned}\operatorname{tr}(P^{-1}AP) &\overbrace{=}^{\text{Theorem 4.8.15}} [\mathbf{D}\det(I)](P^{-1}AP)\\ &= \lim_{h\to 0}\frac{\det(I+hP^{-1}AP)-\det I}{h}\\ &= \lim_{h\to 0}\frac{\det\big(P^{-1}(P+hAP)\big)-\det I}{h}\\ &= \lim_{h\to 0}\frac{\det\big(P^{-1}(I+hA)P\big)-\det I}{h}\\ &\underbrace{=}_{\text{Theorem 4.8.6}} \lim_{h\to 0}\frac{\det(I+hA)-\det I}{h} = [\mathbf{D}\det(I)]A\\ &\underbrace{=}_{\text{Theorem 4.8.15}} \operatorname{tr} A. \quad \square\end{aligned} \qquad 4.8.55$$

The determinant and eigenvalues

In Section 2.7 we gave a first treatment of eigenvalues and eigenvectors. Here we will give a different treatment; it is not better than the earlier one, and in some ways it is worse. But it is far more traditional, the formulas are very neat, and in some ways the two approaches are complementary.

Definition 4.8.18 (Characteristic polynomial). Let A be a square matrix. Then the *characteristic polynomial* of A is

$$\chi_A(t) \overset{\text{def}}{=} \det(tI - A).$$

Example 4.8.19. If $A = \begin{bmatrix} 0 & 1\\ 1 & 1\end{bmatrix}$, then

$$\chi_A(t) = \det\left(\begin{bmatrix} t & 0\\ 0 & t\end{bmatrix} - \begin{bmatrix} 0 & 1\\ 1 & 1\end{bmatrix}\right) = \det\begin{bmatrix} t & -1\\ -1 & t-1\end{bmatrix} = t^2 - t - 1. \quad \triangle$$

Clearly if A is an $n \times n$ matrix, its characteristic polynomial χ_A is a polynomial of degree n.

Theorem 4.8.20. *Let A be a square matrix. The eigenvalues of A are the roots of χ_A.*

One problem with the characteristic polynomial is that it is more or less uncomputable for moderately large matrices. Development according to the first column, or using permutations (equation 4.8.43), involves at least $n!$ computations for an $n \times n$ matrix, certainly unreasonable for $n > 20$.

Row reduction is much better, but to row reduce $tI - A$ we must do arithmetic not with numbers but with polynomials, and since row reduction involves divisions, we must compute with rational functions (ratios of polynomials).

This is much harder than arithmetic with numbers: much more time-consuming and much more numerically unstable, in the sense that round-off errors grow much faster.

In contrast, our polynomials p_i in Section 2.7 involve only row reduction of matrices of numbers.

Proof. If λ is a root of χ_A, then $\det(\lambda I - A) = 0$, so (by Theorem 4.8.3 and the dimension formula) $\ker(\lambda I - A) \neq \{\vec{0}\}$. If $\vec{v} \in \ker(\lambda I - A)$ is a nonzero vector, it is an eigenvector with eigenvalue λ since $A\vec{v} = \lambda\vec{v}$. Conversely, if $A\vec{v} = \lambda\vec{v}$ and $\vec{v} \neq \vec{0}$, then $\vec{v} \in \ker(\lambda I - A)$, so $\lambda I - A$ is not invertible, so $\det(\lambda I - A) = 0$. □

Corollary 4.8.21. *If A is a triangular matrix, the diagonal entries of A are the eigenvalues of A, each appearing as many times as its multiplicity as a root of χ_A.*

The proof is the object of Exercise 4.8.12.

Corollary 4.8.22. *If the roots of χ_A are simple, then $\mathbb{C}^n$ admits an eigenbasis for A.*

Unlike Theorem 2.7.9, Corollary 4.8.22 is not an "if and only if" statement: even if the roots of χ_A aren't simple, there may well be an eigenbasis.

Proof. If the roots are simple, there are n of them, and the corresponding eigenvectors are linearly independent by Theorem 2.7.7, providing an eigenbasis. □

How does Corollary 4.8.22 relate to our earlier treatment, more particularly to the polynomial p defined (see equations 2.7.21 and 2.7.22) by the smallest m such that $A^m\vec{w}$ is a linear combination of $\vec{w}, A\vec{w}, \dots, A^{m-1}\vec{w}$? Since the roots of p are eigenvalues of A, there is every reason to expect p to be related to χ_A, and indeed this is the case.

Theorem 4.8.23. *For any vector $\vec{w}$, the polynomial p divides χ_A.*

The proof of Theorem 4.8.23 is surprisingly difficult. The theorem is clearly true when A has an eigenbasis; then the roots of p, which are also roots of χ_A, are simple, so p divides χ_A.

The roots of p are eigenvalues, so they are also roots of χ_A. The problem is that we need to show that a root of p cannot have higher multiplicity than it does as a root of χ_A. This requires three intermediate statements, all of great interest in their own right.

Proposition 4.8.24. *If A is an $n \times n$ complex matrix, there exists an invertible matrix P such that $P^{-1}AP$ is upper triangular. Equivalently, there is a basis $\vec{v}_1, \dots, \vec{v}_n$ such that the matrix of A in that basis is upper triangular.*

Proof. The proof is by induction on n. It is obvious if $n = 1$, so suppose $n \geq 2$ and assume the result for all $(n-1) \times (n-1)$ matrices.

Find an eigenvector $\vec{\mathbf{v}}_1$ with eigenvalue λ_1 (which exists by the fundamental theorem of algebra). Choose vectors $\vec{\mathbf{w}}_2, \dots, \vec{\mathbf{w}}_n$ such that the vectors $\vec{\mathbf{v}}_1, \vec{\mathbf{w}}_2, \dots, \vec{\mathbf{w}}_n$ form a basis of $\mathbb{R}^n$. Then $T \stackrel{\text{def}}{=} [\vec{\mathbf{v}}_1, \vec{\mathbf{w}}_2, \dots, \vec{\mathbf{w}}_n]$ is invertible, and since $\vec{\mathbf{v}}_1$ is an eigenvector of A, the first column of $B \stackrel{\text{def}}{=} T^{-1}AT$ is $\lambda_1\vec{\mathbf{e}}_1$, i.e., we can write

The first column of B is

$$\begin{aligned} B\vec{\mathbf{e}}_1 &= (T^{-1}AT)\vec{\mathbf{e}}_1 = (T^{-1}A)T\vec{\mathbf{e}}_1 \\ &= T^{-1}A\vec{\mathbf{v}}_1 = T^{-1}\lambda_1\vec{\mathbf{v}}_1 \\ &= \lambda_1\vec{\mathbf{e}}_1 \quad (\text{since } T\vec{\mathbf{e}}_1 = \vec{\mathbf{v}}_1). \end{aligned}$$

$$B \stackrel{\text{def}}{=} T^{-1}AT = \left[\begin{array}{c|c} \lambda_1 & \beta \\ \hline 0 & \\ \vdots & \widetilde{B} \\ 0 & \end{array}\right], \qquad 4.8.56$$

where β is some $1 \times (n-1)$ matrix, and $\widetilde{B}$ is an $(n-1) \times (n-1)$ matrix.

By our inductive hypothesis, we can find an invertible matrix $\widetilde{Q}$ such that $\widetilde{Q}^{-1}\widetilde{B}\widetilde{Q}$ is upper triangular. Set

$$Q = \left[\begin{array}{c|c} 1 & \dots \ 0 \ \dots \\ \hline 0 & \\ \vdots & \widetilde{Q} \\ 0 & \end{array}\right] \quad \text{so that} \quad Q^{-1} = \left[\begin{array}{c|c} 1 & \dots \ 0 \ \dots \\ \hline 0 & \\ \vdots & \widetilde{Q}^{-1} \\ 0 & \end{array}\right]$$

$$\text{and} \quad Q^{-1}BQ = \left[\begin{array}{c|c} \lambda_1 & \beta\widetilde{Q} \\ \hline 0 & \\ \vdots & \widetilde{Q}^{-1}\widetilde{B}\widetilde{Q} \\ 0 & \end{array}\right]. \qquad 4.8.57$$

In particular, $Q^{-1}BQ$ is upper triangular. Set $P = TQ$, then

$$P^1AP = Q^{-1}T^{-1}ATQ = Q^{-1}BQ \qquad 4.8.58$$

is upper triangular. □

Corollary 4.8.25. *Show that if A an $n \times n$ matrix with eigenvalues $\lambda_1, \dots, \lambda_n$, then*

$$\det A = \lambda_1 \cdots \lambda_n.$$

You are asked to prove Corollary 4.8.25 in Exercise 4.8.20.

Clearly diagonalizable matrices are "nice". On the basis of general pessimism, you might expect they they are also exceptional. The opposite is true: most square matrices are diagonalizable.

Theorem 4.8.26. *All square matrices are in the closure of the diagonalizable ones: for every complex square matrix A there is a sequence of complex diagonalizable matrices A_i that converges to A.*

Proof. Suppose that $B \stackrel{\text{def}}{=} P^{-1}AP$ is upper triangular, with entries $\lambda_1, \dots, \lambda_n$ on the diagonal. Choose sequences $\lambda_{i,m}$ such that for all m the numbers $\lambda_{1,m}, \dots, \lambda_{n,m}$ are distinct, and such that $\lambda_{i,m} \to \lambda_i$ when $m \to \infty$. Let B_m be the matrix B with the λ_i replaced by $\lambda_{i,m}$. Then the sequence PB_mP^{-1} satisfies our requirements. □

Theorem 4.8.27: We cannot simply argue that $\chi_A(A)$ equals $\det(IA - A) = \det[0] = 0$. Since A is a matrix, $\chi_A(A)$ is a matrix, not a number; see equation 2.7.20. We must first compute $\chi_A(t)$, then substitute A for t. For instance, for the A in Example 4.8.19,

$$\chi_A(A) = A^2 - A - I.$$

Cayley "proved" Theorem 4.8.27 by saying that for $A = \begin{bmatrix} a & b \\ c & d \end{bmatrix}$,

$$\det(tI - A) = \det \begin{bmatrix} t-a & -b \\ -c & t-d \end{bmatrix} = t^2 - (a+d)t + ad - bc;$$

he confirmed by computation that

$$\begin{bmatrix} a & b \\ c & d \end{bmatrix}^2 - (a+d)\begin{bmatrix} a & b \\ c & d \end{bmatrix} + (ad-bc)I = [0].$$

He claimed also to have checked it for 3×3 matrices, but wrote that *I have not thought it necessary to undertake the labour of a formal proof of the theorem in the general case of a matrix of any degree.*

We would strongly disapprove of this attitude in a student.

Our next result is a famous result of linear algebra. Below, $[0]$ denotes the matrix with all entries 0.

Theorem 4.8.27 (The Cayley-Hamilton theorem). *If A is any square matrix, then $\chi_A(A) = [0]$.*

Algebraic proofs of this theorem are quite difficult, but with Theorem 4.8.26 it is easy, and its proof (with hints) is the object of Exercise 4.8.14.

Proof of Theorem 4.8.23. Now we can prove Theorem 4.8.23. Using division with remainder, write $\chi_A = qp + r$, with $\deg r < \deg p$ if $r \neq 0$. By Theorem 4.8.27, $\chi_A(A) = [0]$, so $\chi_A(A)\vec{\mathbf{w}} = \vec{\mathbf{0}}$. Thus the left side of

$$\chi_A(A)\vec{\mathbf{w}} = q(A)p(A)\vec{\mathbf{w}} + r(A)\vec{\mathbf{w}} \qquad 4.8.59$$

vanishes, and so does the first term on the right, since $p(A)\vec{\mathbf{w}} = \vec{\mathbf{0}}$; see equation 2.7.22. It follows that $r(A)\vec{\mathbf{w}} = \vec{\mathbf{0}}$. Since p is the lowest degree nonzero polynomial such that $p(A)\vec{\mathbf{w}} = \vec{\mathbf{0}}$ and $\deg r < \deg p$, this implies that $r = 0$. □

Exercises for Section 4.8

4.8.1 Compute the determinants of the matrices A, B, C in the margin, using development by the first column or development by the first row.

$$\underbrace{\begin{bmatrix} 1 & -2 & 3 & 0 \\ 4 & 0 & 1 & 2 \\ 5 & -1 & 2 & 1 \\ 3 & 2 & 1 & 0 \end{bmatrix}}_{A}, \quad \underbrace{\begin{bmatrix} 1 & 1 & 2 & 1 \\ 0 & 3 & 4 & 1 \\ 1 & 2 & 3 & 1 \\ 2 & 1 & 0 & 4 \end{bmatrix}}_{B}$$

$$\underbrace{\begin{bmatrix} 1 & 2 & 3 & 4 \\ 0 & 1 & -1 & 3 \\ 3 & 0 & 1 & 1 \\ 1 & 2 & -2 & 0 \end{bmatrix}}_{C}$$

Matrices for Exercise 4.8.1

4.8.2 a. What is the determinant of the matrix $\begin{bmatrix} b & a & 0 & 0 \\ 0 & b & a & 0 \\ 0 & 0 & b & a \\ a & 0 & 0 & b \end{bmatrix}$?

b. What is the determinant of the corresponding $n \times n$ matrix, with entries b on the diagonal and entries a on the slanted line above the diagonal and in the lower left corner?

c. For each n, for what values of a and b is the matrix in part b not invertible? *Hint*: Remember complex numbers.

4.8.3 Spell out exactly what the three conditions defining the determinant (Definition 4.8.1) mean for 2×2 matrices, and prove them.

4.8.4 Let A, B, C, D be 2×2 matrices. Consider the function Δ of 4×4 matrices of the form $M = \begin{bmatrix} [A] & [B] \\ [C] & [D] \end{bmatrix}$ given by

$$\Delta \begin{bmatrix} [A] & [B] \\ [C] & [D] \end{bmatrix} = \det A \det D - \det B \det C.$$

a. Is the function Δ multilinear with respect to the four columns of M?

b. Is it antisymmetric with respect to those columns?

c. Does it satisfy $\Delta \begin{bmatrix} [I] & [0] \\ [0] & [I] \end{bmatrix} = 1$?

4.8.5 Show by direct computation that if A, B are 2×2 matrices, then $\operatorname{tr}(AB) = \operatorname{tr}(BA)$.

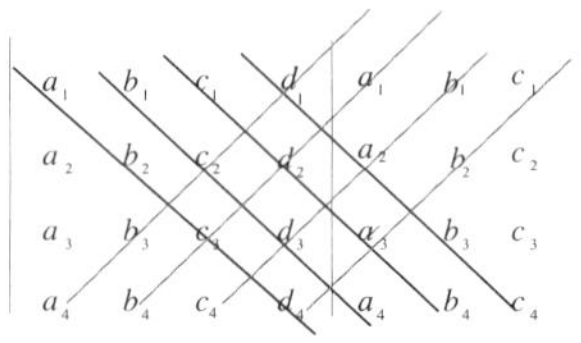

FIGURE FOR EXERCISE 4.8.6

4.8.6 In Section 1.4 (Figure 1.4.11) a shortcut was given for computing the determinant of a 3×3 matrix. The figure in the margin shows the analogous algorithm for a 4×4 matrix $\begin{bmatrix} a_1 & b_1 & c_1 & d_1 \\ a_2 & b_2 & c_2 & d_2 \\ a_3 & b_3 & c_3 & d_3 \\ a_4 & b_4 & c_4 & d_4 \end{bmatrix}$: add the products of the entries on each solid diagonal line, and subtract the products of the entries on each dotted line. Does this algorithm compute the determinant? If so, prove it. If not, what properties of the determinant does it lack?

Hint for Exercise 4.8.7, part a: Think of multiplying the column through by 2, or by -4.

4.8.7 a. Use multilinearity to show that if a square matrix has a column of zeros, its determinant must be zero.

b. Show that if two columns of a square matrix A are equal, $\det A = 0$.

4.8.8 Let A and B be $n \times n$ matrices, with A invertible. Show that the function

$$f(B) = \frac{\det(AB)}{\det A}$$

satisfies multilinearity, antisymmetry, normalization, so $f(B) = \det B$.

4.8.9 a. Find the permutation matrices of the following permutations.

i. $(1,3,2)$ ii. $(1,2,4,3)$ iii. $\tau \circ \sigma$, where $\sigma = (1)(2,3)$ and $\tau = (1,3)(2)$.

b. For part iii, confirm by matrix multiplication that $M_\tau M_\sigma = M_{\tau\circ\sigma}$.

Exercise 4.8.10: For example, the permutation

$$\sigma : 1 \mapsto 2, 2 \mapsto 1, 3 \mapsto 5, \\ 4 \mapsto 7, 5 \mapsto 4, 6 \mapsto 6, 7 \mapsto 3$$

is written in cyclic form

$$(1,2)(3,5,4,7)(6),$$

so it has signature

$$(-1)^{1+3+0} = 1.$$

4.8.10 Show that if a permutation σ is written in cycle form, its signature is

$$\operatorname{sgn}(\sigma) = (-1)^{\sum (\text{length of cycle } -1)}$$

(This is sometimes given as the definition of the signature.)

4.8.11 Prove Theorem 4.8.10: If A is an $n \times n$ matrix, B is an $m \times m$ matrix, and C is an arbitrary $n \times m$ matrix, then

$$\det \begin{bmatrix} A & | & C \\ 0 & | & B \end{bmatrix} = \det A \det B.$$

4.8.12 Prove Corollary 4.8.21.

4.8.13 Confirm that the six permutations of the numbers 1, 2, 3 have the signatures listed in Example 4.8.13.

4.8.14 Prove the Cayley-Hamilton theorem:

a. First prove it for diagonal matrices.

b. Next show that $\chi_{P^{-1}BP} = \chi_B$. Use this and part a to prove the theorem for diagonalizable matrices.

c. Finally, use Theorem 4.8.26 to prove it in general.

$$\begin{bmatrix} 2 & 1 & 0 & 1 \\ 1 & 1 & 3 & 2 \\ 2 & 0 & 2 & 1 \\ 1 & 0 & 4 & 2 \end{bmatrix}$$

Matrix for Exercise 4.8.15

4.8.15 a. Use row (or column) operations and equation 4.8.20 to compute the determinant of the matrix in the margin.

b. Compute the determinant of the same matrix, using permutations.

Exercise 4.8.16: This sort of computation is best done by computer, although we did it by hand.

4.8.16 "Prove" the Cayley-Hamilton theorem the way Cayley did: show by computation that if A is a 3×3 matrix, $\chi_A(A) = [0]$.

Note how much easier Exercise 4.8.17 is than the identical exercise using the techniques of Chapter 2 (Exercise 2.9.8).

4.8.17 Using Proposition 3.8.3, prove that for a 2×2 matrix A,

$$\|A\| = \sqrt{\frac{|A|^2 + \sqrt{|A|^4 - 4(\det A)^2}}{2}},$$

4.8.18 If $A \in \operatorname{Mat}(n,n)$ and $\vec{\mathbf{b}} \in \mathbb{R}^n$, denote by $A_i(\vec{\mathbf{b}})$ the matrix A where the ith column has been replaced by $\vec{\mathbf{b}}$.

a. If $A\vec{\mathbf{x}} = \vec{\mathbf{b}}$, show that $x_i \det A = \det A_i(\vec{\mathbf{b}})$. *Hint*: Using the rules for determinants and column operations, show that $\det A_i(\vec{\mathbf{b}}) = \det A_i(A\vec{\mathbf{x}}) = x_i \det A$.

b. Show that if A is an invertible $n \times n$ matrix of integers, then A^{-1} is a matrix of integers if and only if $\det A = \pm 1$.

Exercise 4.8.18: If $\det A \neq 0$, part a gives the formula

$$x_i = \frac{\det A_i(\vec{\mathbf{b}})}{\det A},$$

known as *Cramer's rule*. This explicit formula for solving linear equations isn't numerically effective except when n is small, but it is of considerable theoretical interest, as illustrated by part b.

Hint for Exercise 4.8.19: See Exercise 2.39.

***4.8.19** Let A be an $n \times n$ matrix.

a. Show that if $\operatorname{rank}(A) = n-1$, then $[\mathbf{D}\det(A)] : \operatorname{Mat}(n,n) \to \mathbb{R}$ is not the zero transformation.

b. Show that if $\operatorname{rank}(A) \leq n-2$, then $[\mathbf{D}\det(A)]$ is the zero transformation.

4.8.20 Let A be an $n \times n$ matrix with eigenvalues $\lambda_1, \dots, \lambda_n$ (not necessarily distinct: each appears as many times as its multiplicity as a root of the characteristic polynomial). Show that $\det A = \lambda_1 \cdots \lambda_n$.

4.8.21 Given two permutations, σ and τ, show that the transformation that associates to each its matrix (M_σ and M_τ respectively) is a *group homomorphism*: it satisfies $M_{\sigma \circ \tau} = M_\sigma M_\tau$.

4.8.22 Give an alternative proof of Theorem 4.8.9, by showing that

a. If all the diagonal entries of an upper triangle matrix are nonzero, you can use column operations (of type 2) to make the matrix diagonal, without changing the entries on the main diagonal.

b. If some entry on the main diagonal is zero, column operations can be used to get a column of zeros.

4.8.23 Let A be an orthogonal $n \times n$ matrix.

a. Show that $\det A = \pm 1$.

b. Show that all eigenvalues λ of A satisfy $|\lambda| = 1$.

c. Show that $\det A = -1$ if and only if -1 is an eigenvalue of A with odd multiplicity.

d. Show that A is a composition of reflections and rotations.

4.9 Volumes and determinants

We will see in Chapter 6 that the determinant measures "signed volume" of k-dimensional parallelograms, i.e., the volume of oriented k-parallelograms.

Theorem 4.9.1: Recall (Definition 4.1.18) that a pavable set has a well-defined volume.

In this section, we show that *in all dimensions the determinant measures volume*. This generalizes Proposition 1.4.14, which says that the area of the parallelogram spanned by $\vec{\mathbf{a}}$ and $\vec{\mathbf{b}}$ is $|\det[\vec{\mathbf{a}}, \vec{\mathbf{b}}]|$, and Proposition 1.4.21, which says that the volume of the parallelepiped spanned by $\vec{\mathbf{a}}, \vec{\mathbf{b}}, \vec{\mathbf{c}}$ is $|\det[\vec{\mathbf{a}}, \vec{\mathbf{b}}, \vec{\mathbf{c}}]|$.

Theorem 4.9.1 (Determinant scales volume). *Let $T : \mathbb{R}^n \to \mathbb{R}^n$ be a linear transformation given by the matrix $[T]$. Then for any pavable set $A \subset \mathbb{R}^n$, its image $T(A)$ is pavable, and*

$$\operatorname{vol}_n T(A) = |\det[T]| \operatorname{vol}_n A. \qquad 4.9.1$$

The absolute value of the determinant, $|\det[T]|$, scales the volume of A up or down to get the volume of $T(A)$; it measures the ratio of the volume of $T(A)$ to the volume of A. If T is linear (as in Figure 4.9.1), this ratio depends *only* on T, not on A. The set A can be a cube, or a ball, or some random potato-shaped object; if it is distorted by T its volume will always scale the same way.[11]

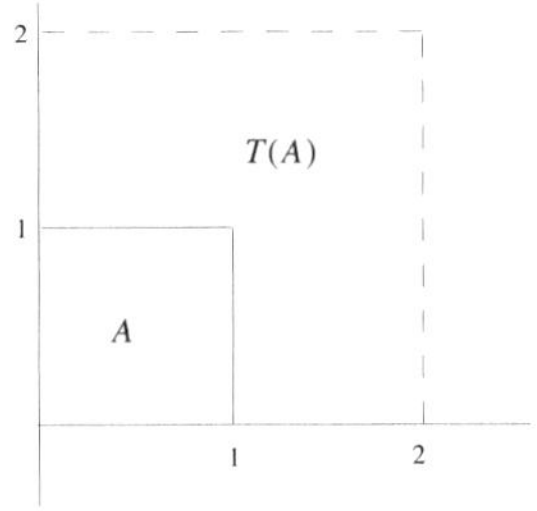

FIGURE 4.9.1.
The scaling transformation given by $\begin{bmatrix} 2 & 0 \\ 0 & 2 \end{bmatrix}$ turns the unit square with sidelength 1 into the square with sidelength 2. The area of the first is 1; the area of the second is $|\det[T]|$ times 1 (i.e., 4).

Corollary 4.9.2. *If S is an orthogonal matrix, then*

$$\text{vol}_n\, S(A) = \text{vol}_n\, A. \qquad 4.9.2$$

Proof. Since $S^\top S = I$ (margin note next to Definition 2.4.15) we have $(\det S)(\det S^\top) = (\det S)^2 = 1$, so $|\det S| = 1$. $\square$

For this section and for Chapter 5 we need to define what we mean by a k-dimensional parallelogram, also called a k-parallelogram.

In Definition 4.9.3, the business with t_i is a precise way of saying that a k-parallelogram is the object spanned by $\vec{\mathbf{v}}_1, \dots, \vec{\mathbf{v}}_k$, including its boundary and its inside.

Definition 4.9.3 (k-parallelogram in $\mathbb{R}^n$). Let $\vec{\mathbf{v}}_1, \dots, \vec{\mathbf{v}}_k$ be k vectors in $\mathbb{R}^n$. The *k-parallelogram* spanned by $\vec{\mathbf{v}}_1, \dots, \vec{\mathbf{v}}_k$ is the set of all

$$t_1\vec{\mathbf{v}}_1 + \cdots + t_k\vec{\mathbf{v}}_k \qquad 4.9.3$$

with $0 \le t_i \le 1$ for i from 1 to k. It is denoted $P(\vec{\mathbf{v}}_1, \dots, \vec{\mathbf{v}}_k)$.

Thus $P(\vec{\mathbf{v}})$ is a line segment, $P(\vec{\mathbf{v}}_1, \vec{\mathbf{v}}_2)$ is a parallelogram, $P(\vec{\mathbf{v}}_1, \vec{\mathbf{v}}_2, \vec{\mathbf{v}}_3)$ is a parallelepiped, etc.[12] Note that the order in which we take the vectors doesn't matter:

$$\begin{aligned} P(\vec{\mathbf{v}}_1, \vec{\mathbf{v}}_2) &= P(\vec{\mathbf{v}}_2, \vec{\mathbf{v}}_1), \\ P(\vec{\mathbf{v}}_1, \vec{\mathbf{v}}_2, \vec{\mathbf{v}}_3) &= P(\vec{\mathbf{v}}_1, \vec{\mathbf{v}}_3, \vec{\mathbf{v}}_2), \quad \text{and so on.} \end{aligned} \qquad 4.9.4$$

In the proof of Theorem 4.9.1 we will use a special case of the k-parallelogram: the n-dimensional unit cube. While the unit disc is traditionally centered at the origin, our unit cube has the lower left corner anchored at the origin.

Definition 4.9.4: Anchoring Q at the origin is a convenience; if we cut it from its moorings and let it float freely in n-dimensional space, it still has n-dimensional volume 1.

Definition 4.9.4 (Unit n-dimensional cube Q). The *unit n-dimensional cube Q_n* is the n-dimensional parallelogram spanned by $\vec{\mathbf{e}}_1, \dots, \vec{\mathbf{e}}_n$. When there is no ambiguity, we denote it by Q.

[11] Theorem 4.9.1 explains why the area of an ellipse with semimajor axes a and b is πab; such an ellipse is a circle with radius r distorted by the linear transformation $\begin{bmatrix} a & 0 \\ 0 & b \end{bmatrix}$. But arc length does not scale in a similar way, and there is no simple formula for the arc length of an ellipse. Indeed, *elliptic functions* are called elliptic because they come from trying to compute the arc length of the ellipse.

[12] We originally used the term "k-parallelepiped"; we dropped it when one of our daughters said "piped" made her think of a creature with 3.1415 ... legs.

Note that if we apply a linear transformation T to Q, the resulting $T(Q)$ is the n-dimensional parallelogram spanned by the columns of $[T]$. This is nothing more than the fact, illustrated in Example 1.2.6, that the ith column of a matrix $[T]$ is $[T]\vec{\mathbf{e}}_i$; if the vectors making up $[T]$ are $\vec{\mathbf{v}}_1, \dots, \vec{\mathbf{v}}_n$, this gives $\vec{\mathbf{v}}_i = [T]\vec{\mathbf{e}}_i$, and we can write

$$T(Q) = P(\vec{\mathbf{v}}_1, \dots, \vec{\mathbf{v}}_n). \qquad 4.9.5$$

This statement and Theorem 4.9.1 give the following proposition.

Notice that a 3-parallelogram in $\mathbb{R}^2$ must be squashed flat, and it can perfectly well be squashed flat in $\mathbb{R}^n$ for $n > 2$: this will happen if $\vec{\mathbf{v}}_1, \vec{\mathbf{v}}_2, \vec{\mathbf{v}}_3$ are linearly dependent.

Proposition 4.9.5 (Volume of n-parallelogram in $\mathbb{R}^n$). *Let $\vec{\mathbf{v}}_1, \dots, \vec{\mathbf{v}}_n$ be vectors in $\mathbb{R}^n$. Then*

$$\mathrm{vol}_n\, P(\vec{\mathbf{v}}_1, \dots, \vec{\mathbf{v}}_n) = \left|\det[\vec{\mathbf{v}}_1, \dots, \vec{\mathbf{v}}_n]\right|. \qquad 4.9.6$$

You may recall that in $\mathbb{R}^2$ and especially $\mathbb{R}^3$, the proof that the determinant measures volume was a lengthy computation (Propositions 1.4.14 and 1.4.21). In $\mathbb{R}^n$, such a computational proof is out of the question.

Proof of Theorem 4.9.1. If $[T]$ is not invertible, the theorem is true because both sides of equation 4.9.1 vanish. The right side vanishes because $\det[T] = 0$ when $[T]$ is not invertible (Theorem 4.8.3). The left side vanishes because if $[T]$ is not invertible, then $T(\mathbb{R}^n)$ is a subspace of $\mathbb{R}^n$ of dimension less than n, and $T(A)$ is a bounded subset of this subspace, so (by Proposition 4.3.5) it has n-dimensional volume 0.

The case where $[T]$ is invertible is much more involved. We will start by showing that the $T(C)$ for $C \in \mathcal{D}_N(\mathbb{R}^n)$ form a paving of $\mathbb{R}^n$. The first condition of Definition 4.7.2 is met because T is invertible.

Formula 4.9.7: If $\vec{\mathbf{v}} \in T(C)$, then $T^{-1}(\vec{\mathbf{v}}) \in T^{-1}T(C) = C$. If $\vec{\mathbf{v}} \in B_R(\mathbf{0})$, then $|\vec{\mathbf{v}}| \le R$, and $|T^{-1}\vec{\mathbf{v}}| \le |T^{-1}|R$, so

$$T^{-1}\vec{\mathbf{v}} \in B_{R|T^{-1}|}(\mathbf{0}).$$

Condition 2 is met because $\mathbf{x} \in T(C_1) \cap T(C_2) \implies T^{-1}(\mathbf{x}) \in C_1 \cap C_2$, but $C_1 \cap C_2 = \emptyset$ if $C_1 \ne C_2$, by the definition of the dyadic decomposition $\mathcal{D}_N$. Condition 3 is met because

$$T(C) \cap B_R(\mathbf{0}) \ne \emptyset \implies C \cap B_{R|T^{-1}|}(\mathbf{0}) \ne \emptyset; \qquad 4.9.7$$

see the note in the margin and Figure 4.9.2. Condition 4 is met because the boundary of $T(C)$ is contained in finitely many subspaces of lower dimension.

Next we need to prove the following statements:

1. The sequence of pavings $N \mapsto T\big(\mathcal{D}_N(\mathbb{R}^n)\big)$ is a nested partition.
2. If $C \in \mathcal{D}_N(\mathbb{R}^n)$, and Q is the unit n-dimensional cube (so that $\mathrm{vol}_n\, Q = 1$), then

$$\mathrm{vol}_n\, T(C) = \mathrm{vol}_n\, T(Q)\, \mathrm{vol}_n\, C. \qquad 4.9.8$$

3. If A is pavable, then its image $T(A)$ is pavable, and

$$\mathrm{vol}_n\, T(A) = \mathrm{vol}_n\, T(Q)\, \mathrm{vol}_n\, A. \qquad 4.9.9$$

4. $\mathrm{vol}_n\, T(Q) = |\det[T]|. \qquad 4.9.10$

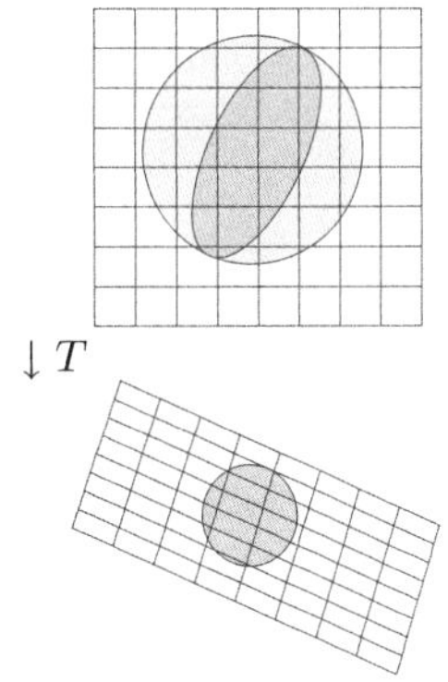

FIGURE 4.9.2.

The lower paving is $T(\mathcal{P}(\mathcal{D}_N))$. A ball of radius R intersects only finitely many pieces, since its inverse image (an ellipsoid) is contained in a ball of $R|T^{-1}|$.

We will take them in order.

Lemma 4.9.6. *The sequence of pavings $N \mapsto T\big(\mathcal{D}_N(\mathbb{R}^n)\big)$ is a nested partition.*

Proof of Lemma 4.9.6. We must check the two conditions of Definition 4.7.3 of a nested partition. The first condition is that small paving pieces must fit inside big paving pieces: if we pave $\mathbb{R}^n$ with blocks $T(C)$, then if

$$C_1 \in \mathcal{D}_{N_1}(\mathbb{R}^n),\ C_2 \in \mathcal{D}_{N_2}(\mathbb{R}^n),\ \text{and } C_1 \subset C_2, \qquad 4.9.11$$

we have

$$T(C_1) \subset T(C_2). \qquad 4.9.12$$

This is clearly met; for example, if you divide the square A of Figure 4.9.1 into four smaller squares, the image of each small square will fit inside $T(A)$.

We use the linearity of T to meet the second condition – that the pieces $T(C)$ shrink to points as $N \to \infty$. This is met: for $C \in \mathcal{D}_N(\mathbb{R}^n)$, we have

$$\operatorname{diam}(C) = \frac{\sqrt{n}}{2^N}, \quad \text{so} \quad \operatorname{diam}\big(T(C)\big) \le |T|\frac{\sqrt{n}}{2^N}. \qquad 4.9.13$$

(Recall from Definition 4.7.3 that $\operatorname{diam}(X)$ is the maximum distance between points $\mathbf{x}, \mathbf{y} \in X$. If inequality 4.9.13 isn't clear, see the footnote.[13]) So $\operatorname{diam}\big(T(C)\big) \to 0$ as $N \to \infty$. □ Lemma 4.9.6

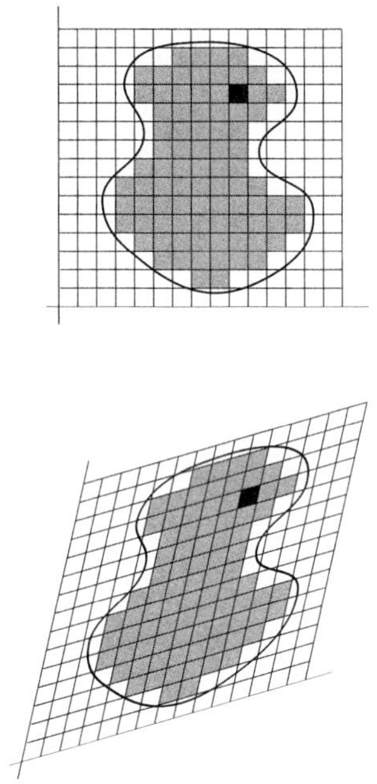

FIGURE 4.9.3.
The potato-shaped area at top is the set A; it is mapped by T to its image $T(A)$, at bottom. If C is the small black square in the top figure, $T(C)$ is the small black parallelogram in the bottom figure. The ensemble of all the $T(C)$ for C in $\mathcal{D}_N(\mathbb{R}^n)$ is denoted $T(\mathcal{D}_N)$.
The volume of $T(A)$ is the limit of the sum of the volumes of the $T(C)$, where $C \in \mathcal{D}_N(\mathbb{R}^n)$ and $C \subset A$. Each of these has the same volume: by equation 4.9.15,

$$\operatorname{vol}_n T(C) = \operatorname{vol}_n C \operatorname{vol}_n T(Q).$$

Proof of Theorem 4.9.1: second statement. Now we need to prove that if $C \in \mathcal{D}_N(\mathbb{R}^n)$, then

$$\operatorname{vol}_n T(C) = \operatorname{vol}_n T(Q) \operatorname{vol}_n C. \qquad 4.9.14.$$

We showed above that $T(Q)$ is pavable, as are all $T(C)$ for $C \in \mathcal{D}_N$. Since C is Q scaled up or down by 2^N in all directions, and $T(C)$ is $T(Q)$ scaled by the same factor, we have

$$\frac{\operatorname{vol}_n T(C)}{\operatorname{vol}_n T(Q)} = \frac{\operatorname{vol}_n C}{\operatorname{vol}_n Q} = \frac{\operatorname{vol}_n C}{1}. \qquad 4.9.15$$

Proof of Theorem 4.9.1: third statement. Now we need to prove that if A is pavable, then $T(A)$ is pavable, and

$$\operatorname{vol}_n T(A) = \operatorname{vol}_n T(Q) \operatorname{vol}_n A. \qquad 4.9.16$$

We know that A is pavable; as illustrated in Figure 4.9.3, we can compute its volume by taking the limit of the lower sum (the cubes $C \in \mathcal{D}$ that are entirely inside A) or the limit of the upper sum (the cubes either entirely inside A or straddling A).

[13]For any points $\mathbf{a}, \mathbf{b} \in C$ (which we can think of as joined by the vector $\vec{\mathbf{v}}$),

$$|T(\mathbf{a}) - T(\mathbf{b})| = |T(\mathbf{a} - \mathbf{b})| = \Big|[T]\vec{\mathbf{v}}\Big| \underbrace{\le}_{\text{Prop. 1.4.11}} |[T]||\vec{\mathbf{v}}|.$$

Thus the diameter of $T(C)$ can be at most $|[T]|$ times the length of the longest vector joining two points of C (i.e. $\sqrt{n}/2^N$, by inequality 4.1.16).

Since (by Lemma 4.9.6) $T(\mathcal{D}_N)$ is a nested partition, we can use it as a paving to measure the volume of $T(A)$, with upper and lower sums:

$$\overbrace{\sum_{T(C)\cap T(A)\neq \phi} \operatorname{vol}_n T(C)}^{\text{upper sum for } \mathbf{1}_{T(A)}} = \sum_{C\cap A\neq \phi} \overbrace{\operatorname{vol}_n(C)\operatorname{vol}_n T(Q)}^{\operatorname{vol}_n T(C) \text{ by eq. 4.9.14}} = \operatorname{vol}_n T(Q) \overbrace{\sum_{C\cap A\neq \phi} \operatorname{vol}_n C}^{\text{limit is } \operatorname{vol}_n A}; \qquad 4.9.17$$

$$\overbrace{\sum_{T(C)\subset T(A)} \operatorname{vol}_n T(C)}^{\text{lower sum for } \mathbf{1}_{T(A)}} = \sum_{C\subset A} \operatorname{vol}_n C \operatorname{vol}_n T(Q) = \operatorname{vol}_n T(Q) \overbrace{\sum_{C\subset A} \operatorname{vol}_n C}^{\text{limit is } \operatorname{vol}_n A}. \qquad 4.9.18$$

Subtracting the lower sum from the upper sum, we get

$$\underbrace{U_{T(\mathcal{D}_N)}(\mathbf{1}_{T(A)}) - L_{T(\mathcal{D}_N)}(\mathbf{1}_{T(A)})}_{\substack{\text{difference of upper and lower sums}\\ \text{with respect to nested partition } T(\mathcal{D}_N)}} = \operatorname{vol}_n T(Q) \sum_{\substack{C \text{ straddles}\\ \text{boundary of } A}} \operatorname{vol}_n C. \qquad 4.9.19$$

In equations 4.9.17 and 4.9.18, it's important to pay attention to which cubes C we are summing over:

$C \cap A \neq \phi =$ C in A or straddling A

$C \subset A =$ C entirely in A.

Subtracting the second from the first gives those C that straddle A.

Since A is pavable, the right side can be made arbitrarily small, so the upper and lower sums have a common limit, $T(A)$ is pavable, and

$$\operatorname{vol}_n T(A) = \operatorname{vol}_n T(Q) \operatorname{vol}_n A. \qquad 4.9.20$$

Proof of Theorem 4.9.1: fourth statement. This leaves part 4: why is $\operatorname{vol}_n T(Q)$ the same as $|\det[T]|$? There is no obvious relation between volumes and the immensely complicated formula for the determinant. Our strategy will be to reduce the theorem to the case where T is given by an elementary matrix, since the determinant of elementary matrices is straightforward. The following lemma is the key.

Lemma 4.9.7. *If $S, T : \mathbb{R}^n \to \mathbb{R}^n$ are linear transformations, then*

$$\operatorname{vol}_n (S \circ T)(Q) = \operatorname{vol}_n S(Q) \operatorname{vol}_n T(Q). \qquad 4.9.21$$

Proof of Lemma 4.9.7. This follows from equation 4.9.20, substituting S for T and $T(Q)$ for A:

$$\operatorname{vol}_n (S \circ T)(Q) = \operatorname{vol}_n S(T(Q)) = \operatorname{vol}_n S(Q) \operatorname{vol}_n T(Q). \quad \square \qquad 4.9.22$$

What does $E(A)$ mean when the set A is defined in geometric terms, as in Figure 4.9.4? We think of E as a transformation; applying that transformation to A means multiplying each point of A by E to obtain the corresponding point of $E(A)$. We discussed applying linear transformations to subsets in Section 1.3.

Any invertible linear transformation T, identified to its matrix, can be written as the product of elementary matrices,

$$[T] = E_k E_{k-1} \cdots E_1 \qquad 4.9.23$$

(see equation 2.3.5). So by Lemma 4.9.7 and Theorem 4.8.4, it is enough to prove part 4 for elementary matrices: i.e., to prove

$$\operatorname{vol}_n E(Q) = |\det E|. \qquad 4.9.24$$

Elementary matrices come in three kinds, as described in Definition 2.3.5. (Here we discuss them in terms of columns, not in terms of rows.)

1. If E is a type 1 elementary matrix, multiplying a column by a nonzero number m, then $\det E = m$ (equation 4.8.27), and equation

4.9.24 becomes $\operatorname{vol}_n E(Q) = |m|$. This result was proved in Proposition 4.1.20, because $E(Q)$ is then a parallelepiped all of whose sides are 1 except one side, whose length is $|m|$.

2. The case where E is type 2, adding a multiple of one column onto another, is a bit more complicated. Without loss of generality, we may assume that a multiple of the first is being added to the second.

   First let us verify it for the case $n = 2$, where E is the matrix

$$E = \begin{bmatrix} 1 & a \\ 0 & 1 \end{bmatrix}, \quad \text{with} \quad \det E = 1, \qquad 4.9.25$$

   and Q is the unit square. Then (see Figure 4.9.4), $E(Q)$ is a parallelogram with base 1 and height 1, so $\operatorname{vol}_2(E(Q)) = |\det E| = 1$.[14]

   If $n > 2$, write $\mathbb{R}^n = \mathbb{R}^2 \times \mathbb{R}^{n-2}$. Correspondingly, we can write $Q = Q_1 \times Q_2$, and $E = E_1 \times E_2$, where E_2 is the identity, as shown in Figure 4.9.5. Then by Proposition 4.1.16,

$$\operatorname{vol}_n E(Q) = \operatorname{vol}_2(E_1(Q_1)) \operatorname{vol}_{n-2}(Q_2) = 1 \cdot 1 = 1. \qquad 4.9.26$$

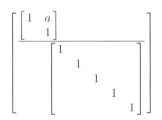

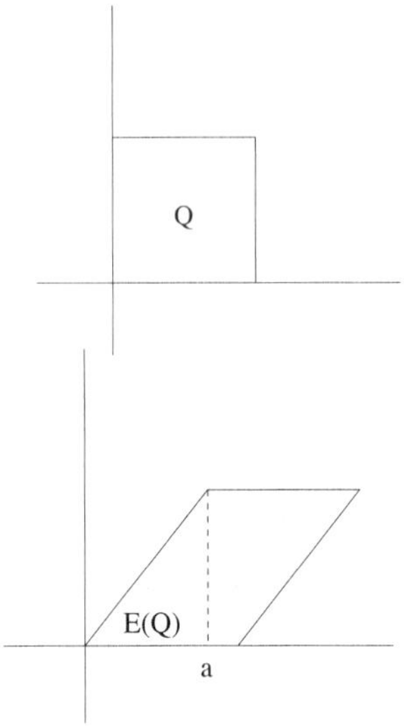

FIGURE 4.9.4.
The second type of elementary matrix, in $\mathbb{R}^2$, simply takes the unit square Q to a parallelogram with base 1 and height 1.

3. If E is type 3, then $\det E = -1$, so $|\det E| = 1$, and equation 4.9.24 becomes $\operatorname{vol}_n E(Q) = 1$. Indeed, since $E(Q)$ is just Q with vertices relabeled, its volume is 1. $\square$

Note that since $\operatorname{vol}_n T(Q) = |\det[T]|$, equation 4.9.21 can be rewritten

$$|\det[ST]| = |\det[S]|\,|\det[T]|. \qquad 4.9.27$$

Of course, this was clear from Theorem 4.8.4. But that result did not have a very transparent proof, whereas equation 4.9.21 has a clear geometric meaning. Thus this interpretation of the determinant as a volume gives a reason why Theorem 4.8.4 should be true.

$$\left[\begin{array}{c|c} \begin{bmatrix} 1 & a \\ & 1 \end{bmatrix} & \\ \hline & \begin{bmatrix} 1 & & & & \\ & 1 & & & \\ & & 1 & & \\ & & & 1 & \\ & & & & 1 \end{bmatrix} \end{array}\right]$$

FIGURE 4.9.5.
Here $n = 7$; from the 7×7 matrix E we created the 2×2 matrix

$$E_1 = \begin{bmatrix} 1 & a \\ 0 & 1 \end{bmatrix}$$

and the 5×5 identity matrix E_2.

Linear change of variables

It is always more or less equivalent to speak about volumes or to speak about integrals; translating Theorem 4.9.1 ("the determinant measures volume") into the language of integrals gives the following theorem.

---

[14]But is this a proof? Are we using our definition of volume (area in this case) using pavings, or some "geometric intuition", which is right but difficult to justify precisely? One rigorous justification uses Fubini's theorem:

$$\operatorname{vol}_2\Big(E(Q)\Big) = \int_0^1 \left(\int_{ay}^{ay+1} dx \right) dy = 1.$$

Another possibility is suggested in Exercise 4.9.2.

Theorem 4.9.8 (Linear change of variables). *Let $T : \mathbb{R}^n \to \mathbb{R}^n$ be an invertible linear transformation, and $f : \mathbb{R}^n \to \mathbb{R}$ an integrable function. Then $f \circ T$ is integrable, and*

$$\int_{\mathbb{R}^n} f(\mathbf{y})\,|d^n\mathbf{y}| = \underbrace{|\det T|}_{\substack{\text{corrects for} \\ \text{stretching by } T}} \int_{\mathbb{R}^n} \underbrace{f\big(T(\mathbf{x})\big)}_{f(\mathbf{y})}\,|d^n\mathbf{x}|. \qquad 4.9.28$$

In equation 4.9.28, $|\det T|$ corrects for the linear distortion induced by T. The $\mathbb{R}^n$ on the left is the codomain of T (and the domain of f); the $\mathbb{R}^n$ on the right is the domain of T.

Example 4.9.9 (Linear change of variables). The linear transformation given by $T = \begin{bmatrix} a & 0 \\ 0 & b \end{bmatrix}$ transforms the unit disc into an ellipse, as shown in Figure 4.9.6. The area of the ellipse is then given by

$$\text{Area of ellipse} = \int_{\text{ellipse}} |d^2\mathbf{y}| = \underbrace{\left|\det \begin{bmatrix} a & 0 \\ 0 & b \end{bmatrix}\right|}_{ab} \underbrace{\int_{\text{disc}} |d^2\mathbf{x}|}_{\pi=\text{area of unit disc}} = |ab|\pi. \qquad 4.9.29$$

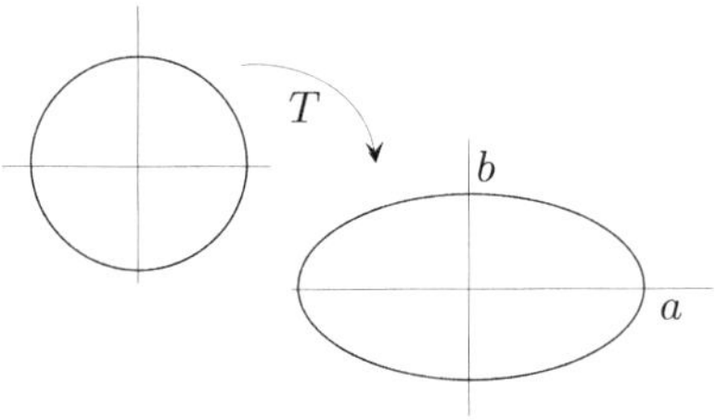

FIGURE 4.9.6.
The linear transformation
$$T = \begin{bmatrix} a & 0 \\ 0 & b \end{bmatrix}$$
takes the unit disc to the (filled-in) ellipse.

If we had integrated some function $f : \mathbb{R}^2 \to \mathbb{R}$ over the unit disc and wanted to know what the same function would give when integrated over the ellipse, we would use the formula

$$\int_{\text{ellipse}} f(\mathbf{y})|d^2\mathbf{y}| = |ab| \int_{\text{disc}} \underbrace{f\begin{pmatrix} ax_1 \\ bx_2 \end{pmatrix}}_{T(\mathbf{x})} |d^2\mathbf{x}| \qquad \triangle \quad 4.9.30$$

Proof of Theorem 4.9.8. Recall (Definition 4.1.3) that $M_C(g)$ is the supremum of $g(\mathbf{x})$ for $\mathbf{x} \in C$. Recall (Lemma 4.9.6) that $T(\mathcal{D}_N(\mathbb{R}^n))$ is a nested partition. Theorems 4.7.4 and 4.9.1 give the following:

$$\begin{aligned} \int_{\mathbb{R}^n} f\big(T(\mathbf{x})\big)|\det T||d^n\mathbf{x}| &= \lim_{N\to\infty} \sum_{C\in\mathcal{D}_N(\mathbb{R}^n)} M_C\Big((f\circ T)|\det T|\Big) \operatorname{vol}_n C \\ &\underbrace{=}_{\text{Thm. 4.9.1}} \lim_{N\to\infty} \sum_{C\in\mathcal{D}_N(\mathbb{R}^n)} M_C(f\circ T) \underbrace{\operatorname{vol}_n\big(T(C)\big)}_{=|\det T| \operatorname{vol}_n C} \qquad 4.9.31 \\ &\underbrace{=}_{\text{Thm. 4.7.4}} \lim_{N\to\infty} \sum_{P\in(T(\mathcal{D}_N(\mathbb{R}^n)))} M_P(f)\operatorname{vol}_n(P) = \int_{\mathbb{R}^n} f(\mathbf{y})|d^n\mathbf{y}|. \quad \square \end{aligned}$$

$$\begin{bmatrix} 1 & 0 & 0 & \dots & 0 \\ 2 & 2 & 0 & \dots & 0 \\ 3 & 3 & 3 & \dots & 0 \\ \vdots & \vdots & \vdots & \ddots & \vdots \\ n & n & n & \dots & n \end{bmatrix}$$

Matrix for Exercise 4.9.1.

Exercises for Section 4.9

4.9.1 Let $T : \mathbb{R}^n \to \mathbb{R}^n$ be given by the matrix in the margin, and let $A \subset \mathbb{R}^n$ be the region given by

$$|x_1| + |x_2|^2 + |x_3|^3 + \dots + |x_n|^n \le 1.$$

What is $\operatorname{vol}_n T(A)/\operatorname{vol}_n A$?

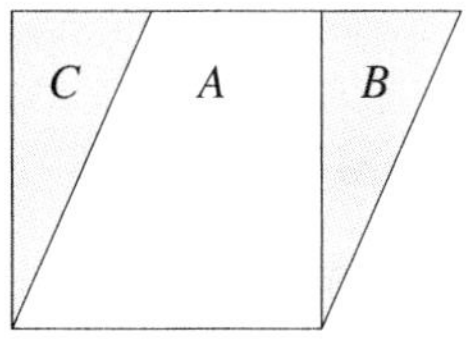

Figure for Exercise 4.9.2.

4.9.2 Use "dissection" (as suggested in the figure in the margin) to prove equation 4.9.24 when E is type 2.

Exercise 4.9.3, part a: Use Fubini. Part b: Don't use Fubini. Find a linear transformation S such that $S(T_1) = T_2$.

4.9.3 a. What is the volume of the tetrahedron T_1 with vertices

$$\begin{bmatrix}0\\0\\0\end{bmatrix}, \begin{bmatrix}1\\0\\0\end{bmatrix}, \begin{bmatrix}0\\1\\0\end{bmatrix}, \begin{bmatrix}0\\0\\1\end{bmatrix}?$$

b. What is the volume of the tetrahedron T_2 with vertices

$$\begin{bmatrix}0\\0\\0\end{bmatrix}, \begin{bmatrix}2\\1\\1\end{bmatrix}, \begin{bmatrix}-1\\3\\1\end{bmatrix}, \begin{bmatrix}-2\\-5\\2\end{bmatrix}?$$

4.9.4 What is the n-dimensional volume of the region

$$\{\, \mathbf{x} \in \mathbb{R}^n \mid x_i \geq 0 \text{ for all } i = 1, \dots, n \text{ and } x_1 + \dots + x_n \leq 1 \,\}?$$

4.9.5 Let q be a continuous function on $\mathbb{R}$, and suppose that f and g satisfy the differential equations

$$f''(x) = q(x)f(x), \qquad g''(x) = q(x)g(x).$$

Express the area $A(x)$ of the parallelogram spanned by $\begin{bmatrix}f(x)\\f'(x)\end{bmatrix}, \begin{bmatrix}g(x)\\g'(x)\end{bmatrix}$ in terms of $A(0)$. *Hint*: You may want to differentiate $A(x)$.

a. $\begin{bmatrix}1\\3\\-1\end{bmatrix}, \begin{bmatrix}2\\-1\\-1\end{bmatrix}, \begin{bmatrix}3\\6\\0\end{bmatrix}$

b. $\begin{bmatrix}1\\1\\1\\0\end{bmatrix}, \begin{bmatrix}1\\1\\0\\1\end{bmatrix}, \begin{bmatrix}1\\0\\1\\1\end{bmatrix}, \begin{bmatrix}0\\1\\1\\1\end{bmatrix}$

c. $\begin{bmatrix}4\\3\\2\\1\\0\end{bmatrix}, \begin{bmatrix}3\\2\\1\\0\\1\end{bmatrix}, \begin{bmatrix}2\\1\\0\\1\\2\end{bmatrix}, \begin{bmatrix}1\\0\\1\\2\\3\end{bmatrix}, \begin{bmatrix}0\\1\\2\\3\\4\end{bmatrix}$

Vectors for Exercise 4.9.6

4.9.6 Compute the volumes of the three k-parallelograms spanned by the vectors at left.

Exercise 4.9.7, part a: It is true, but not immediately obvious, that the first property follows from the others and is thus not actually necessary

***4.9.7** a. Show that $\widetilde{\Delta}(T) = |\det T|$ is the unique map $\operatorname{Mat}(n,n) \to \mathbb{R}$ that satisfies

1. For all $T \in \operatorname{Mat}(n,n)$, we have $\widetilde{\Delta}(T) \geq 0$.
2. The function $\widetilde{\Delta}$ is a symmetric function of the columns (i.e., switching two columns does not change the value).
3. For all $T = [\vec{\mathbf{v}}_1, \vec{\mathbf{v}}_2, \dots, \vec{\mathbf{v}}_n] \in \operatorname{Mat}(n,n)$,

$$\widetilde{\Delta}[a\vec{\mathbf{v}}_1, \vec{\mathbf{v}}_2, \dots, \vec{\mathbf{v}}_n] = |a|\widetilde{\Delta}[\vec{\mathbf{v}}_1, \vec{\mathbf{v}}_2, \dots, \vec{\mathbf{v}}_n].$$

4. For all $T = [\vec{\mathbf{v}}_1, \vec{\mathbf{v}}_2, \dots, \vec{\mathbf{v}}_n] \in \operatorname{Mat}(n,n)$,

$$\widetilde{\Delta}[\vec{\mathbf{v}}_1, \vec{\mathbf{v}}_2, \dots, \vec{\mathbf{v}}_n] = \widetilde{\Delta}[\vec{\mathbf{v}}_1 + a\vec{\mathbf{v}}_2, \vec{\mathbf{v}}_2, \dots, \vec{\mathbf{v}}_n].$$

5. $\widetilde{\Delta}(I_n) = 1$.

b. Show that $T \mapsto \operatorname{vol}_n(T(Q))$ satisfies the properties that characterize Δ.

4.10 The change of variables formula

We discussed linear changes of variables in Section 4.9. This section is devoted to nonlinear changes of variables in higher dimensions. You will no doubt have run into changes of variables in one-dimensional integrals, perhaps under the name the *substitution method*.

Example 4.10.1 (Substitution method). To compute

$$\int_0^\pi \sin x e^{\cos x}\, dx, \qquad 4.10.1$$

traditionally, one sets $u = \cos x$, so that $du = -\sin x\, dx$. Then for $x = 0$, we have $u = \cos 0 = 1$, and for $x = \pi$, we have $u = \cos \pi = -1$, so

$$\int_0^\pi \sin x e^{\cos x} dx = \int_1^{-1} -e^u du = \int_{-1}^1 e^u du = e - \frac{1}{e}. \quad \triangle \qquad 4.10.2$$

The meaning of expressions like du is explored in Chapter 6.

In this section we generalize this sort of computation to several variables. There are two parts to this: transforming the integrand, and transforming the domain of integration. In Example 4.10.1 we transformed the integrand by setting $u = \cos x$, so that $du = -\sin x\, dx$ (whatever du means), and we transformed the domain of integration by noting that $x = 0$ corresponds to $u = \cos 0 = 1$, and $x = \pi$ corresponds to $u = \cos \pi = -1$.

Both parts are harder in several variables, especially the second. In one dimension, the domain of integration is usually an interval, and it is not too hard to see how intervals correspond. Domains of integration in $\mathbb{R}^n$, even in the traditional cases of discs, sectors, balls, cylinders, etc., are quite a bit harder to handle. Much of our treatment will be concerned with making precise the "correspondences of domains" under change of variables.

Three important changes of variables

Before stating the general change of variables formula, we will explore what it says for polar coordinates in the plane, and spherical and cylindrical coordinates in space. This will help you understand the general case. In addition, many real systems (encountered, for instance, in physics courses) have a central symmetry in the plane or in space, or an axis of symmetry in space, and in all those cases, these changes of variables are the useful ones. Finally, a great many of the standard multiple integrals are computed using these changes of variables.

Polar coordinates

If you draw the (x, y)-plane in the standard way, then the polar angle θ increases as one travels counterclockwise on the unit circle.

Polar coordinates are appropriate when we have rotational symmetry in the plane.

Definition 4.10.2 (Polar coordinates map). The *polar coordinate map* P maps a point in the (r, θ)-plane to a point in the (x, y)-plane:

$$P : \begin{pmatrix} r \\ \theta \end{pmatrix} \mapsto \begin{pmatrix} x = r\cos\theta \\ y = r\sin\theta \end{pmatrix}, \qquad 4.10.3$$

where r measures distance from the origin along the spokes, and the polar angle θ measures the angle (in radians) formed by a spoke and the positive x-axis.

Thus, as shown in Figure 4.10.1, a rectangle in the domain of P becomes a curvilinear "rectangle" in the image of P.

In equation 4.10.4, the r in $r\,dr\,d\theta$ plays the role of $|\det T|$ in the linear change of variables formula (Theorem 4.9.8): it corrects for the distortion induced by the polar coordinate map P. We could put $|\det T|$ in front of the integral in the linear formula because it is a constant.

Here, we cannot put r in front of the integral; since P is nonlinear, the amount of distortion is not constant but depends on the point at which P is applied.

In equation 4.10.4, we could replace

$$\int_B f\begin{pmatrix} r\cos\theta \\ r\sin\theta \end{pmatrix}$$

by $$\int_B f\left(P\begin{pmatrix} r \\ \theta \end{pmatrix}\right),$$

the format used in Theorem 4.9.8 concerning the linear case.

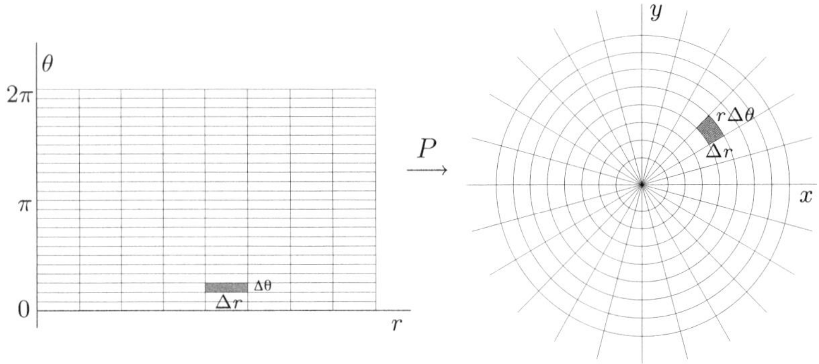

FIGURE 4.10.1. The polar coordinate map P maps the rectangle at left, with dimensions Δr and $\Delta\theta$, to the curvilinear box at right, with two straight sides of length Δr and two curved sides measuring $r\Delta\theta$ (for different values of r).

Proposition 4.10.3 (Change of variables for polar coordinates). *Suppose f is an integrable function defined on $\mathbb{R}^2$, and suppose that the polar coordinate map P maps a region $B \subset (0,\infty)\times[0,2\pi)$ of the (r,θ)-plane to a region A in the (x,y)-plane. Then*

$$\int_A f\begin{pmatrix} x \\ y \end{pmatrix} |dx\,dy| = \int_B f\begin{pmatrix} r\cos\theta \\ r\sin\theta \end{pmatrix} r\,|dr\,d\theta|. \qquad 4.10.4$$

Note that the map $P : B \to A$ is bijective (one to one and onto), since we required $\theta \in [0, 2\pi)$. This choice is essentially arbitrary; the interval $[-\pi, \pi)$ would have done just as well, as would $(0, 2\pi]$. (There is no need to worry about what happens when $\theta = 0$ or $\theta = 2\pi$, since those are sets of volume 0 and by Corollary 4.3.10 do not affect integrals.) Requiring r to be in $(0, \infty)$ sidesteps the problem that there is no well-defined polar angle at the origin; again, omitting $r = 0$ does not affect the integral.

We will postpone the discussion of where equation 4.10.4 comes from and proceed to some examples.

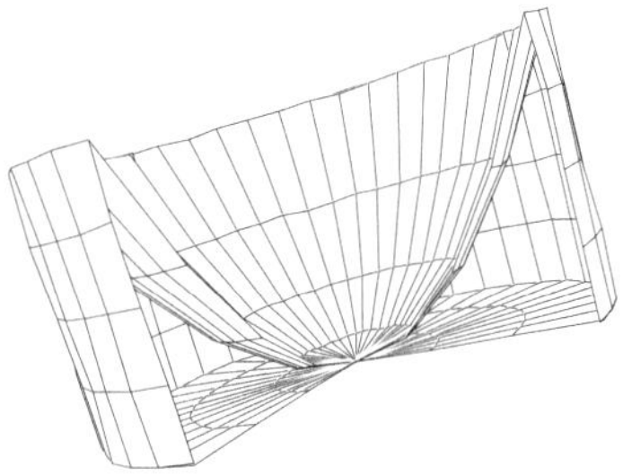

FIGURE 4.10.2.

In Example 4.10.4 the "volume under the graph" is the region inside the cylinder and outside the paraboloid. This volume was first computed by Archimedes, who invented a lot of the integral calculus in the process. No one understood what he was doing for about 2000 years.

Example 4.10.4 (Volume beneath a paraboloid of revolution). Consider the paraboloid of Figure 4.10.2, given by

$$z = f\begin{pmatrix} x \\ y \end{pmatrix} = \begin{cases} x^2 + y^2 & \text{if } x^2 + y^2 \le R^2 \\ 0 & \text{if } x^2 + y^2 > R^2. \end{cases} \qquad 4.10.5$$

Usually one would write the integral

$$\int_{\mathbb{R}^2} f\begin{pmatrix} x \\ y \end{pmatrix} |dx\,dy| \quad \text{as} \quad \int_{D_R} (x^2 + y^2)\,|dx\,dy|, \qquad 4.10.6$$

where

$$D_R = \left\{ \begin{pmatrix} x \\ y \end{pmatrix} \in \mathbb{R}^2 \;\middle|\; x^2 + y^2 \le R^2 \right\} \qquad 4.10.7$$

is the disc of radius R centered at the origin.

This integral is fairly complicated to compute using Fubini's theorem; Exercise 4.10.1 asks you to do this. Using the change of variables formula 4.10.4, it is straightforward:

$$\begin{aligned}\int_{\mathbb{R}^2} f\begin{pmatrix}x\\y\end{pmatrix}|dx\,dy| &= \int_0^{2\pi}\int_0^R f\begin{pmatrix}r\cos\theta\\r\sin\theta\end{pmatrix} r\,dr\,d\theta\\ &= \int_0^{2\pi}\int_0^R (r^2)(\cos^2\theta+\sin^2\theta)\,r\,dr\,d\theta \qquad 4.10.8\\ &= \int_0^{2\pi}\int_0^R (r^2)\,r\,dr\,d\theta = 2\pi\left[\frac{r^4}{4}\right]_0^R = \frac{\pi}{2}R^4. \quad \triangle\end{aligned}$$

Most often, polar coordinates are used when the domain of integration is a disc or a sector of a disc, but they are also useful in many cases where the equation of the boundary is well suited to polar coordinates, as in Example 4.10.5.

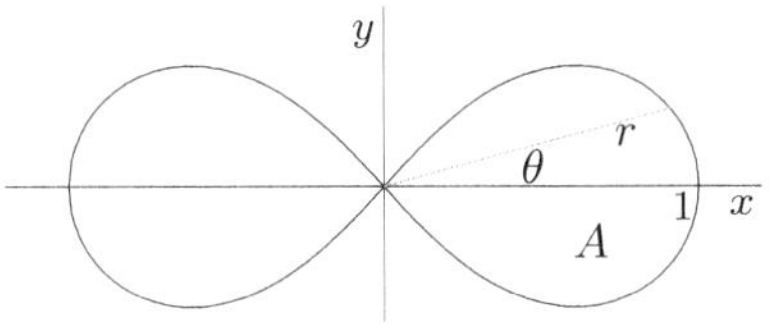

FIGURE 4.10.3.
The lemniscate of equation

$$r^2 = \cos 2\theta,$$

shown in the (x, y)-plane, using polar coordinates:

$$x = r\cos\theta,\ y = r\sin\theta.$$

(As shown in Figure 4.10.1, θ is the polar angle, and r the distance from the origin.) The region A is the right loop. Exercise 4.10.3 asks you to write the equation of the lemniscate in complex notation.

Example 4.10.5 (Area of a lemniscate). The *lemniscate* looks like a figure eight; the name comes from the Latin word for ribbon. We will compute the area of the right lobe A of the lemniscate given by the equation $r^2 = \cos 2\theta$ (i.e., the area bounded by the right loop of the curve in Figure 4.10.3).

This area can be written $\int_A dx\,dy$, which could be computed by Riemann sums, but the expressions you get applying Fubini's theorem are dismayingly complicated. Using polar coordinates simplifies the computations. The region A corresponds to the region B in the (r, θ)-plane where

$$B = \left\{ \begin{pmatrix}r\\\theta\end{pmatrix} \;\middle|\; -\frac{\pi}{4} \le \theta \le \frac{\pi}{4},\ 0 < r < \sqrt{\cos 2\theta} \right\}. \qquad 4.10.9$$

Thus in polar coordinates, the integral becomes

$$\begin{aligned}\int_{-\pi/4}^{\pi/4}\left(\int_0^{\sqrt{\cos 2\theta}} r\,dr\right)d\theta &= \int_{-\pi/4}^{\pi/4}\left[\frac{r^2}{2}\right]_0^{\sqrt{\cos 2\theta}} d\theta \qquad 4.10.10\\ &= \int_{-\pi/4}^{\pi/4}\frac{\cos 2\theta}{2}\,d\theta = \left[\frac{\sin 2\theta}{4}\right]_{-\pi/4}^{\pi/4} = \frac{1}{2}.\end{aligned}$$

(The formula for change of variables for polar coordinates, equation 4.10.4, has a function f on both sides of the equation. Since we are computing area here, the function is simply 1.) $\triangle$

Spherical coordinates

Spherical coordinates are important whenever you have a center of symmetry in $\mathbb{R}^3$. Although in mathematics the word "sphere" refers to the two-dimensional surface of a three-dimensional ball, the spherical coordinates map parametrizes space, as shown in Figure 4.10.4; it is a map from $\mathbb{R}^3$ (or a subset of $\mathbb{R}^3$) to $\mathbb{R}^3$.

Definition 4.10.6 (Spherical coordinates map). The *spherical coordinate map* S maps a point in space (e.g., a point inside the earth) known by its distance r from the center, its longitude θ, and its latitude φ, to a point in (x, y, z)-space:

$$S : \begin{pmatrix} r \\ \theta \\ \varphi \end{pmatrix} \mapsto \begin{pmatrix} x = r\cos\theta\cos\varphi \\ y = r\sin\theta\cos\varphi \\ z = r\sin\varphi \end{pmatrix}. \qquad 4.10.11$$

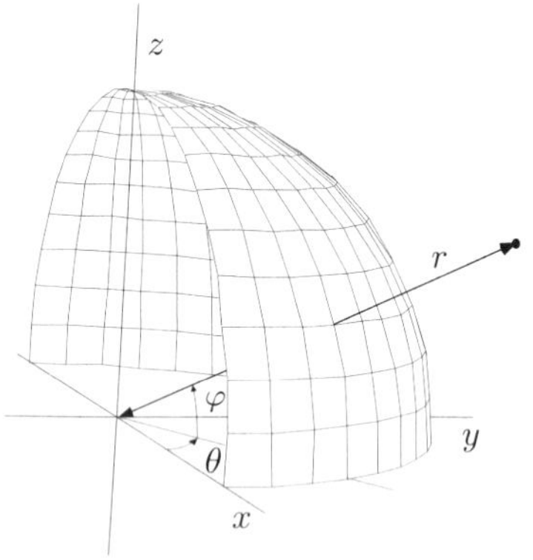

FIGURE 4.10.4.

In spherical coordinates, a point is specified by its distance r from the origin, its longitude θ, and its latitude φ; longitude and latitude are measured in radians, not in degrees.

Many authors use the angle from the North Pole rather than latitude. Mainly because most people are comfortable with the standard latitude, we prefer this form. The formulas using the North Pole are given in Exercise 5.2.3.

Proposition 4.10.7 (Change of variables for spherical coordinates). *Let f be an integrable function defined on $\mathbb{R}^3$, and let the spherical coordinate map S map a region B of (r, θ, φ)-space to a region A in (x, y, z)-space. Suppose that*

$$B \subset (0, \infty) \times [0, 2\pi) \times (-\pi/2, \pi/2).$$

Then

$$\int_A f\begin{pmatrix} x \\ y \\ z \end{pmatrix} |dx\, dy\, dz| = \int_B f\begin{pmatrix} r\cos\theta\cos\varphi \\ r\sin\theta\cos\varphi \\ r\sin\varphi \end{pmatrix} r^2\cos\varphi\, |dr\, d\theta\, d\varphi|. \qquad 4.10.12$$

The $r^2\cos\varphi$ corrects for distortion induced by the spherical coordinates map. Again, we postpone the justification for this formula.

Example 4.10.8 (Spherical coordinates). Let's integrate the function z over the upper half of the unit ball, denoted A:

$$A = \left\{ \begin{pmatrix} x \\ y \\ z \end{pmatrix} \in \mathbb{R}^3 \mid x^2 + y^2 + z^2 \le 1,\ z \ge 0 \right\}. \qquad 4.10.13$$

The region B corresponding to A under the spherical coordinates map S is

$$B = \left\{ \begin{pmatrix} r \\ \theta \\ \varphi \end{pmatrix} \in \underbrace{(0, \infty)}_{r} \times \underbrace{[0, 2\pi)}_{\theta} \times \underbrace{(-\pi/2, \pi/2)}_{\varphi} \Big| \, r \le 1,\ \varphi \ge 0 \right\}. \qquad 4.10.14$$

For the upper half of the unit ball, r goes from 0 to 1, φ from 0 to $\pi/2$ (from the Equator to the North Pole), and θ from 0 to 2π. Thus our integral $\int_A z\, |dx\, dy\, dz|$ becomes

$$\int_B \underbrace{(r\sin\varphi)}_{z}(r^2\cos\varphi)\, |dr\, d\theta\, d\varphi| = \int_0^1 \left(\int_0^{\pi/2} \left(\int_0^{2\pi} r^3 \sin\varphi\cos\varphi\, d\theta \right) d\varphi \right) dr$$

$$= 2\pi \int_0^1 r^3 \left[\frac{\sin^2\varphi}{2} \right]_0^{\pi/2} dr = 2\pi \int_0^1 \frac{r^3}{2} dr = 2\pi \left[\frac{r^4}{8} \right]_0^1 = \frac{\pi}{4}. \quad \triangle \qquad 4.10.15$$

Cylindrical coordinates

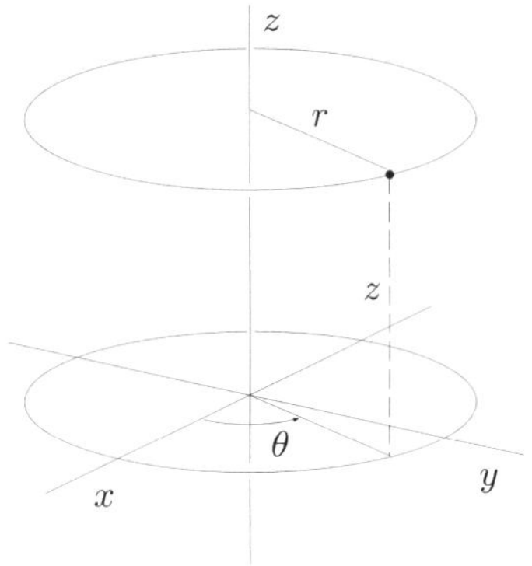

FIGURE 4.10.5.
In cylindrical coordinates, a point is specified by its distance r from the z-axis, the polar angle θ, and the z coordinate.

Cylindrical coordinates, shown in Figure 4.10.5, are important whenever you have an axis of symmetry.

Definition 4.10.9 (Cylindrical coordinates map). The *cylindrical coordinates map* C maps a point in space known by its altitude z and by the polar coordinates r, θ of its projection onto the (x, y)-plane, to a point in (x, y, z)-space:

$$C : \begin{pmatrix} r \\ \theta \\ z \end{pmatrix} \mapsto \begin{pmatrix} r\cos\theta \\ r\sin\theta \\ z \end{pmatrix}. \qquad 4.10.16$$

Proposition 4.10.10 (Change of variables for cylindrical coordinates). *Let f be an integrable function defined on $\mathbb{R}^3$, and suppose that the cylindrical coordinate map C maps a region $B \subset (0, \infty) \times [0, 2\pi) \times \mathbb{R}$ of (r, θ, z)-space to a region A in (x, y, z)-space. Then*

$$\int_A f\begin{pmatrix} x \\ y \\ z \end{pmatrix} |dx\, dy\, dz| = \int_B f\begin{pmatrix} r\cos\theta \\ r\sin\theta \\ z \end{pmatrix} r\, |dr\, d\theta\, dz|. \qquad 4.10.17$$

In equation 4.10.17, the r in $r\, dr\, d\theta\, dz$ corrects for distortion induced by the cylindrical coordinate map C. This is the same "distortion corrector" as for polar coordinates (see equation 4.10.4).

Example 4.10.11 (Integrating a function over a cone). Let us integrate $(x^2 + y^2)z$ over the region $A \subset \mathbb{R}^3$ that is the part of the cone $z^2 \geq x^2 + y^2$ where $0 \leq z \leq 1$ (see Figure 4.10.6). This corresponds under the map C to the region B in (r, θ, z)-space where $r \leq z \leq 1$. Thus our integral is

$$\int_A (x^2 + y^2)z\, |dx\, dy\, dz| = \int_B r^2 z \underbrace{(\cos^2\theta + \sin^2\theta)}_{=1} r\, |dr\, d\theta\, dz| = \int_B (r^2 z)\, r\, |dr\, d\theta\, dz|$$

$$= \int_0^{2\pi}\left(\int_0^1\left(\int_r^1 r^3 z\, dz\right) dr\right) d\theta = 2\pi \int_0^1 r^3 \left[\frac{z^2}{2}\right]_{z=r}^{z=1} dr$$

$$= 2\pi \int_0^1 r^3\left(\frac{1}{2} - \frac{r^2}{2}\right) dr = 2\pi\left[\frac{r^4}{8} - \frac{r^6}{12}\right]_0^1$$

$$= 2\pi\left(\frac{1}{8} - \frac{1}{12}\right) = \frac{\pi}{12}. \qquad 4.10.18$$

Note that it would have been unpleasant to express the flat top of the cone in spherical coordinates. △

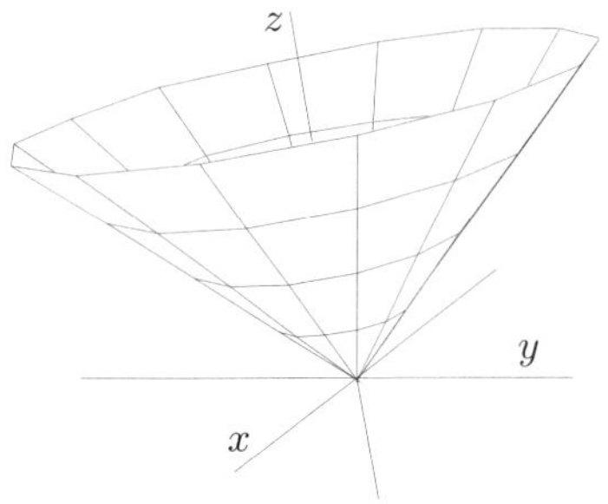

FIGURE 4.10.6.
Example 4.10.11: The region we are integrating over is the part of the cone $z^2 \geq x^2 + y^2$ where $0 \leq z \leq 1$.

General change of variables formula

We have seen how integrals transform under polar, spherical, and cylindrical changes of variables; we will now see what happens to multiple integrals under general changes of variables.

Theorem 4.10.12 (Change of variables formula). *Let X be a compact subset of $\mathbb{R}^n$ with boundary ∂X of volume 0; let $U \subset \mathbb{R}^n$ be an open set containing X. Let $\Phi : U \to \mathbb{R}^n$ be a C^1 mapping that is injective on $(X - \partial X)$ and has Lipschitz derivative, with $[\mathbf{D}\Phi(\mathbf{x})]$ invertible at every $\mathbf{x} \in (X - \partial X)$. Set $Y = \Phi(X)$.*

Then if $f : Y \to \mathbb{R}$ is integrable, $(f \circ \Phi)|\det[\mathbf{D}\Phi]|$ is integrable on X, and

$$\int_Y f(\mathbf{y})\,|d^n\mathbf{y}| = \int_X (f \circ \Phi)(\mathbf{x})\,\big|\det[\mathbf{D}\Phi(\mathbf{x})]\big|\,|d^n\mathbf{x}|. \qquad 4.10.19$$

You might think that the condition in Theorem 4.10.12 that the boundary have n-dimensional volume 0 is always met, but this isn't true; compact subsets of $\mathbb{R}^n$ exist whose boundaries do not have volume; for instance, the complement of the set U_ϵ in Example 4.4.3 is compact, but the boundary of U_ϵ does not have volume.

It is possible to define integration over such a set; see Exercise 5.2.5. But in that case integrating over the set and integrating over its closure are different.

The determinant $\det[\mathbf{D}\Phi(\mathbf{x})]$ that appears in equation 4.10.19 is often called the *Jacobian determinant* or even simply the *Jacobian*.

The theorem is proved in Appendix A19.

The map Φ is a "relaxed parametrization", where we are ignoring what happens on sets of volume 0. This is a looser use of the word "parametrize" than the one given in Section 3.1, but it is better adapted to integration problems. We discuss this further in Section 5.2.

Now let us check that the change of variables for polar and spherical coordinates are special cases of Theorem 4.10.12. (The argument for cylindrical coordinates is essentially identical to that for polar coordinates.)

When we have Lebesgue integrals, in Section 4.11, we will be able to give a cleaner version (Theorem 4.11.20) of the change of variables theorem.

The polar coordinates map P:

$$P : \begin{pmatrix} r \\ \theta \end{pmatrix} \mapsto \begin{pmatrix} x = r\cos\theta \\ y = r\sin\theta \end{pmatrix}.$$

Here we repeat equation 4.10.4 giving the polar change of variables:

$$\int_A f\begin{pmatrix} x \\ y \end{pmatrix} |dx\,dy| = \int_B f\begin{pmatrix} r\cos\theta \\ r\sin\theta \end{pmatrix} r\,|dr\,d\theta|.$$

Example 4.10.13 (Polar coordinates and the change of variables formula). It is easy to check that equation 4.10.4 for polar coordinates is a special case of equation 4.10.19, if P plays the role of Φ. The $f\begin{pmatrix} r\cos\theta \\ r\sin\theta \end{pmatrix}$ on the right side of equation 4.10.4 is $(f \circ P)(\mathbf{x})$. The "distortion corrector" r for the polar change of coordinates corresponds to $|\det[\mathbf{D}\Phi(\mathbf{x})]|$ in equation 4.10.19, since

$$\left[\mathbf{D}P\begin{pmatrix} r \\ \theta \end{pmatrix}\right] = \begin{bmatrix} \cos\theta & -r\sin\theta \\ \sin\theta & r\cos\theta \end{bmatrix}, \quad \text{giving} \quad \left|\det\left[\mathbf{D}P\begin{pmatrix} r \\ \theta \end{pmatrix}\right]\right| = r. \qquad 4.10.20$$

Checking that P meets the requirements for Φ takes a little more work. Let $f : \mathbb{R}^2 \to \mathbb{R}$ be an integrable function. Suppose that the support of f is contained in the disc of radius R. Then set

$$X = \left\{ \begin{pmatrix} r \\ \theta \end{pmatrix} \mid 0 \le r \le R,\ 0 \le \theta \le 2\pi \right\}, \qquad 4.10.21$$

and take U to be any bounded neighborhood of X, for instance, the disc centered at the origin in the (r, θ)-plane of radius $R + 2\pi$. We claim that all the requirements are satisfied: P is of class C^1 in U with Lipschitz derivative, and it is injective on $\overset{\circ}{X}$ (although not on the boundary). Moreover, $[\mathbf{D}P]$ is invertible in $\overset{\circ}{X}$, since $\det[\mathbf{D}P] = r$, which is only zero on the boundary of X. △

Recall that $\overset{\circ}{X}$ denotes the interior of X:

$$\overset{\circ}{X} = X - \partial X.$$

Example 4.10.14 (Spherical coordinates and change of variables formula). The $f\begin{pmatrix} r\cos\varphi\cos\theta \\ r\cos\varphi\sin\theta \\ r\sin\varphi \end{pmatrix}$ of the change of variables for spherical coordinates is $(f \circ \Phi)(\mathbf{x})$, where S plays the role of Φ. The spherical "distortion corrector" $r^2 \cos\varphi$ corresponds to $|\det[\mathbf{D}\Phi(\mathbf{x})]|$ in equation 4.10.19, since

The spherical coordinates map:

$$S : \begin{pmatrix} r \\ \theta \\ \varphi \end{pmatrix} = \begin{pmatrix} r\cos\varphi\cos\theta \\ r\cos\varphi\sin\theta \\ r\sin\varphi \end{pmatrix}$$

$$\left[\mathbf{D}S\begin{pmatrix} r \\ \theta \\ \varphi \end{pmatrix}\right] = \begin{bmatrix} \cos\theta\cos\varphi & -r\sin\theta\cos\varphi & -r\cos\theta\sin\varphi \\ \sin\theta\cos\varphi & r\cos\theta\cos\varphi & -r\sin\theta\sin\varphi \\ \sin\varphi & 0 & r\cos\varphi \end{bmatrix}, \qquad 4.10.22$$

so that

$$\left|\det\left[\mathbf{D}S\begin{pmatrix} r \\ \theta \\ \varphi \end{pmatrix}\right]\right| = r^2\cos\varphi. \qquad 4.10.23$$

Again, we need to check that S qualifies as a change of variables mapping Φ. If the function f to be integrated has its support in the ball of radius R around the origin, take

$$X = \left\{ \begin{pmatrix} r \\ \theta \\ \varphi \end{pmatrix} \;\middle|\; 0 \le r \le R,\ -\frac{\pi}{2} \le \varphi \le \frac{\pi}{2}, 0 \le \theta \le 2\pi \right\}, \qquad 4.10.24$$

and U any bounded open neighborhood of X. Then indeed S is C^1 on U with Lipschitz derivative; it is injective on $\overset{\circ}{X}$, and its derivative is invertible there, since $\det[\mathbf{D}S] = r^2\cos\varphi$, which vanishes only on ∂X. △

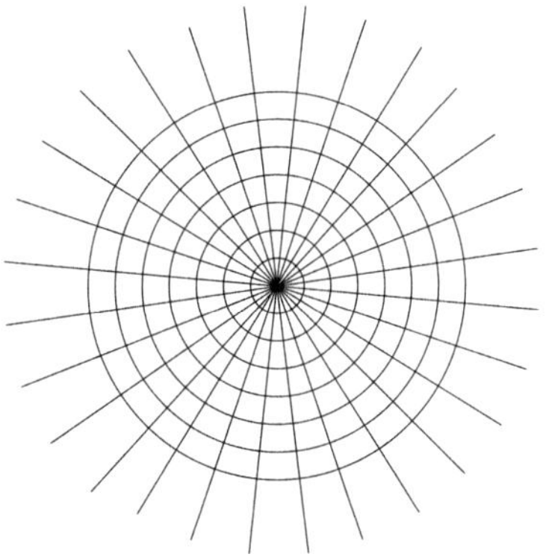

FIGURE 4.10.7.

The paving $P(\mathcal{D}_N^{\text{new}})$ of $\mathbb{R}^2$ under the polar coordinates map P of Definition 4.10.2.

The dimension of each block in the angular direction (the direction of the spokes) is $\pi/2^N$.

Remark 4.10.15. The requirement that Φ be injective (one to one) often creates great difficulties. In first year calculus, you didn't have to worry about the function being injective. This was because the integrand dx of one-dimensional calculus is actually a *form field*, integrated over an oriented domain: $\int_a^b f\,dx = -\int_b^a f\,dx$.

For instance, consider $\int_1^4 dx$. If we set $x = u^2$, so that $dx = 2u\,du$, then $x = 4$ corresponds to $u = \pm 2$, while $x = 1$ corresponds to $u = \pm 1$. If we choose $u = -2$ for the first and $u = 1$ for the second, then the change of variables is *not* injective. But this doesn't matter; the change of variable formula gives

$$3 = \int_1^4 dx = \int_1^{-2} 2u\,du = [u^2]_1^{-2} = 4 - 1 = 3; \qquad 4.10.25$$

as u travels from 1 to -2, x travels from 1 down to 0, then from 0 up to 4. In particular, it travels the segment from 1 to 0 twice, in opposite directions, and that contribution to the integral cancels. Our set-up for multiple integrals does not allow for this kind of cancellation, because it does not consider orientation. Thus our changes of variables must be injective. In Chapter 6 we will address this problem, using forms, but orientation in higher dimensions is much more difficult than in one dimension. (The best statement of the change of variables formula uses differential forms, but it is beyond the scope of this book.) △

A heuristic derivation of the change of variables formulas

It is not hard to see why the change of variables formulas are correct. In each case, the standard paving $\mathcal{D}_N$ in the new space induces a paving in the original space.

Actually, when using polar, spherical, or cylindrical coordinates, we will be better off if we use paving blocks with sidelength $\pi/2^N$ in the angular directions, rather than the $1/2^N$ of standard dyadic cubes. (Since π is irrational, dyadic fractions of radians do not fill up the circle exactly, but dyadic pieces of turns do.) We will call this paving $\mathcal{D}_N^{\text{new}}$, partly to specify these dimensions, but mainly to remember what space is being paved.

The image paving $P(\mathcal{D}_N^{\text{new}})$ of $\mathbb{R}^2$ under the polar coordinates map P is shown in Figure 4.10.7; the paving of $\mathbb{R}^3$ corresponding to spherical coordinates is shown in Figure 4.10.8.

In the case of polar, spherical, and cylindrical coordinates, the image pavings $P(\mathcal{D}_N^{\text{new}})$, $S(\mathcal{D}_N^{\text{new}})$, and $C(\mathcal{D}_N^{\text{new}})$ clearly form nested partitions. (When we make more general changes of variables Φ, we will need to impose requirements that make this true.) Thus given a change of variables mapping $\Phi : X \to Y$ we have

$$\begin{aligned}\int_Y f(\mathbf{y})|d^n\mathbf{y}| &= \lim_{N\to\infty} \sum_{C\in\mathcal{D}_N^{\text{new}}} M_{\Phi(C)}(f)\,\text{vol}_n\,\Phi(C)\\ &= \lim_{N\to\infty} \sum_{C\in\mathcal{D}_N^{\text{new}}} \underbrace{M_C(f\circ\Phi)}_{=M_{\Phi(C)}(f)} \frac{\text{vol}_n\,\Phi(C)}{\text{vol}_n\,C}\,\text{vol}_n\,C.\end{aligned} \qquad 4.10.26$$

Theorem 4.7.4 justifies the first equality. The second line (see Definitions 4.1.8 and 4.1.10) looks like the integral over X of the product of $f\circ\Phi$ and the limit of the ratio

$$\frac{\text{vol}_n\,\Phi(C)}{\text{vol}_n\,C} \qquad 4.10.27$$

as $N\to\infty$, so that C becomes small. This would give

$$\int_Y f(\mathbf{y})|d^n\mathbf{y}| \sim \int_X \left((f\circ\Phi)(\mathbf{x}) \lim_{N\to\infty} \frac{\text{vol}_n\,\Phi(C)}{\text{vol}_n\,C}\right)|d^n\mathbf{x}|. \qquad 4.10.28$$

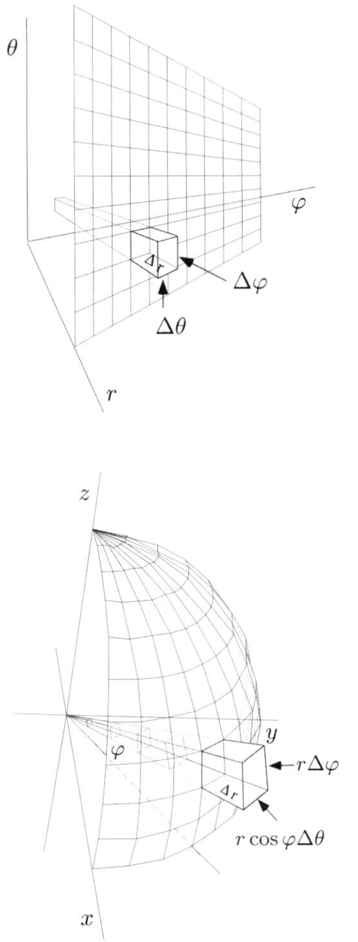

FIGURE 4.10.8.
Under the spherical coordinate map S, a box with dimensions $\Delta r, \Delta\theta$, and $\Delta\varphi$, and anchored at $\begin{pmatrix} r\\ \theta\\ \varphi\end{pmatrix}$ (top) is mapped to a curvilinear "box" with dimensions

$\Delta r, r\cos\varphi\Delta\theta,$ and $r\Delta\varphi$.

It is not obvious that the integrand is a function and therefore can be integrated; how do we know the limit exists? But by equation 4.9.1, the determinant is precisely designed to measure ratios of volumes under linear transformations. Of course, our change of variables map Φ isn't linear, but if it is differentiable, then it is almost linear on small cubes, so we would expect (see equations 4.9.8 and 4.9.10) that if $C_{\mathbf{x}}$ is the cube in $\mathcal{D}_N^{\text{new}}(\mathbb{R}^n)$ containing $\mathbf{x}$, then

$$\lim_{N\to\infty} \frac{\text{vol}_n\,\Phi(C_{\mathbf{x}})}{\text{vol}_n\,C_{\mathbf{x}}} = \left|\det[\mathbf{D}\Phi(\mathbf{x})]\right|. \qquad 4.10.29$$

So we might expect our integral $\int_A f$ to be equal to

$$\int_Y f(\mathbf{y})\,|d^n\mathbf{y}| = \int_X (f\circ\Phi)(\mathbf{x})\,\left|\det[\mathbf{D}\Phi(\mathbf{x})]\right|\,|d^n\mathbf{x}|. \qquad 4.10.30$$

We turn this argument into a rigorous proof in Appendix A19.

The hard part in applying Theorem 4.10.12 is finding an appropriate change of variables mapping Φ, when it is not immediately apparent that polar, spherical, or cylindrical mappings will do. This is similar to the difficulty of finding parametrizations as defined in Section 3.1 (but not as hard, since Φ is required to be injective, and $[\mathbf{D}\Phi]$ to be invertible, only on the interior of X, not on all of X).

Example 4.10.16 (Ratio of areas for polar coordinates). Consider the ratio of equation 4.10.29 in the case of polar coordinates, when $\Phi = P$. If a rectangle C in the (r, θ) plane, containing the point $\begin{pmatrix} r_0 \\ \theta_0 \end{pmatrix}$, has sides of length Δr and $\Delta\theta$, then the corresponding piece $P(C)$ of the (x, y) plane is approximately a rectangle with sides $r_0 \Delta\theta$ and Δr, as shown by Figure 4.10.1. Its area is approximately $r_0 \Delta r \Delta\theta$, and the ratio of areas is approximately r_0, leading to

$$\int_Y f|d^n\mathbf{y}| = \int_X (f \circ P)\, r\, |dr\, d\theta|, \qquad 4.10.31$$

where r is the ratio of the volumes of infinitesimal paving blocks. △

Example 4.10.17 (Ratio of volumes for spherical coordinates). For spherical coordinates, where $\Phi = S$, the image $S(C)$ of a box C in $\mathcal{D}_N^{\text{new}}(\mathbb{R}^3)$ with sides $\Delta r, \Delta\theta, \Delta\varphi$ is approximately a box with sides Δr, $r\Delta\varphi$, $r\cos\varphi\Delta\theta$, so the ratio of the volumes is approximately $r^2 \cos\varphi$. △

Finding other changes of variables

There is no simple recipe for finding an appropriate change of variables mapping Φ. Of course, the first thing to do is to consider whether one of the standard mappings will work. If you have an axis of symmetry, consider cylindrical coordinates; a center of symmetry in $\mathbb{R}^3$, spherical coordinates; a center of symmetry in $\mathbb{R}^2$, polar coordinates.

If an appropriate mapping isn't immediately obvious, try sketching the domain (a good graphing program helps). If the domain of integration is defined by equations, the equations may suggest a useful change of variables.

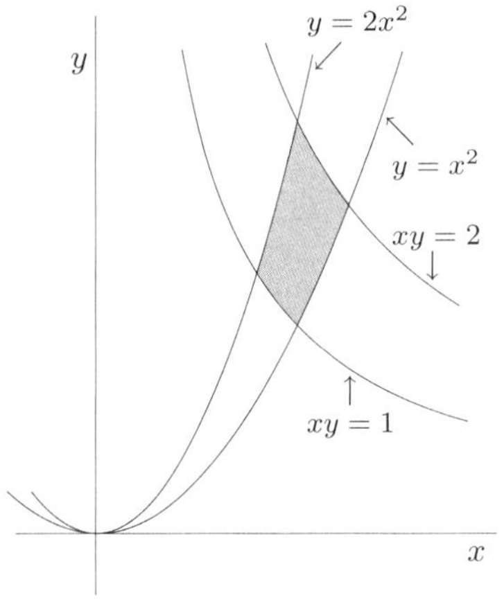

FIGURE 4.10.9.
Example 4.10.18: The shaded region is the region X defined by

$1 \le xy \le 2, \quad x^2 \le y \le 2x^2.$

Example 4.10.18 (Finding a change of variables map). Suppose you wish to find the area of the region $X \subset \mathbb{R}^2$ given by

$$1 \le xy \le 2 \quad \text{and} \quad x^2 \le y \le 2x^2. \qquad 4.10.32$$

What change of variables is appropriate? First we draw the hyperbolas given by the equalities $1 = xy$ and $xy = 2$, and the parabolas given by $x^2 = y$ and $y = 2x^2$, as shown in Figure 4.10.9. This figure suggests that setting $u = xy$ would be one good choice; the equation $x^2 \le y \le 2x^2$ suggests that setting $v = y/x^2$ would be another; the region X would then be defined by $1 \le u \le 2$ and $1 \le v \le 2$.

The next step is to express x and y in terms of u and v. This isn't always possible, but in this case we can solve $xy = u$ and $y = vx^2$ for x and y, getting $x = \sqrt[3]{u/v}$ and $y = \sqrt[3]{u^2 v}$. This gives the change of variables map

$$\Phi\begin{pmatrix} u \\ v \end{pmatrix} = \begin{pmatrix} \sqrt[3]{u/v} \\ \sqrt[3]{u^2 v} \end{pmatrix}. \qquad 4.10.33$$

(We solved explicitly for x, y in terms of u, v, so, in X, where $v \neq 0$, the mapping Φ is not only injective, it is invertible. Exercise 4.10.6 asks you to show that its derivative $[\mathbf{D}\Phi]$ is invertible in $\overset{\circ}{X}$, and to compute the area of X.) $\triangle$

The following example uses a trickier change of variables.

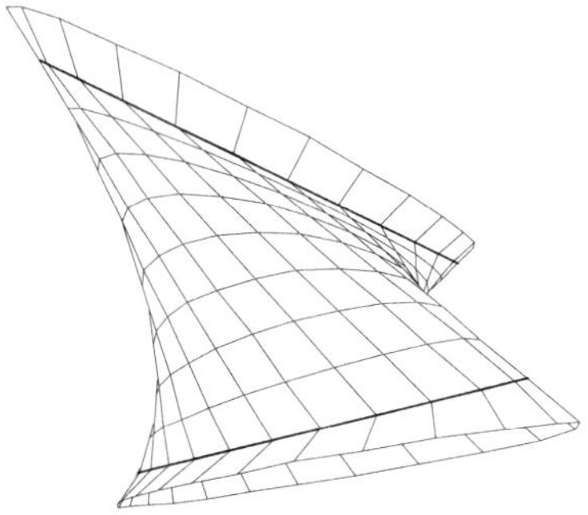

FIGURE 4.10.10.
The region T resembles a cylinder flattened at the ends. Horizontal sections of T are ellipses, which degenerate to lines when $z = \pm 1$.

Example 4.10.19 (A less standard change of variables). The region T defined by

$$\left(\frac{x}{1-z}\right)^2 + \left(\frac{y}{1+z}\right)^2 \leq 1, \quad -1 < z < 1 \qquad 4.10.34$$

looks like the curvy-sided tetrahedron pictured in Figure 4.10.10; we will compute its volume. Notice that horizontal slices of T are ellipses, so we will use "polar coordinates for the ellipse": the map

$$\begin{pmatrix} r \\ \theta \end{pmatrix} \mapsto \begin{pmatrix} ar\cos\theta \\ br\sin\theta \end{pmatrix}, \; 0 \leq r \leq 1, \; 0 \leq \theta \leq 2\pi \quad \text{parametrizes} \quad \frac{x^2}{a^2} + \frac{y^2}{b^2} \leq 1.$$

Therefore, we can set $a = 1 - z$ and $b = 1 + z$; the map

$$\gamma : [0,1] \times [0, 2\pi] \times [-1, 1] \to \mathbb{R}^3 \text{ given by } \gamma \begin{pmatrix} r \\ \theta \\ z \end{pmatrix} = \begin{pmatrix} r(1-z)\cos\theta \\ r(1+z)\sin\theta \\ z \end{pmatrix}$$

parametrizes the region T, since for each fixed $z = c$ it parametrizes the ellipse $T \cap \{z = c\}$. The determinant of $[\mathbf{D}\gamma]$ is

$$\det \begin{bmatrix} (1-z)\cos\theta & -r(1-z)\sin\theta & -r\cos\theta \\ (1+z)\sin\theta & r(1+z)\cos\theta & r\sin\theta \\ 0 & 0 & 1 \end{bmatrix} = r(1-z^2). \qquad 4.10.35$$

Thus the volume is given by the integral

Exercise 4.10.4 asks you to solve a problem of the same sort.

$$\int_0^{2\pi} \int_0^1 \int_{-1}^1 \left| r(1-z^2) \right| dz \, dr \, d\theta = \frac{4\pi}{3}. \quad \triangle \qquad 4.10.36$$

Exercises for Section 4.10

4.10.1 Using Fubini, compute $\int_{D_R} (x^2 + y^2) \, dx \, dy$ (see Example 4.10.4), where

$$D_R = \left\{ \begin{pmatrix} x \\ y \end{pmatrix} \in \mathbb{R}^2 \mid x^2 + y^2 \leq R^2 \right\}.$$

4.10.2 Derive the change of variables formula for cylindrical coordinates from the polar formula and Fubini's theorem.

4.10.3 Show that in complex notation, with $z = x + iy$, the equation of the lemniscate of Figure 4.10.3 can be written $|z^2 - \frac{1}{2}| = \frac{1}{2}$.

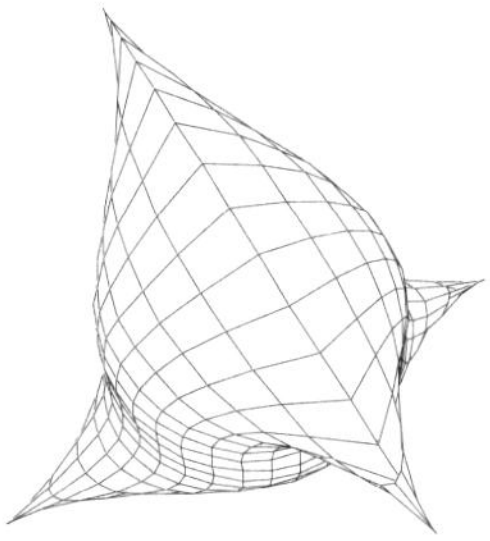

FIGURE FOR EXERCISE 4.10.4. You were asked to compute the volume of this region in Exercise 4.5.19; now you are asked to compute the volume using the change of variables formula.

4.10.4 Use the change of variables formula to compute the volume of the region $\frac{x^2}{(z^3-1)^2} + \frac{y^2}{(z^3+1)^2} \le 1$, $-1 \le z \le 1$, shown in the figure in the margin.

4.10.5 a. What is the area of the ellipse $x^2/a^2 + y^2/b^2 \le 1$? *Hint*: Use the change of variables $u = x/a$, $v = y/b$.

b. What is the volume of the ellipsoid $x^2/a^2 + y^2/b^2 + z^2/c^2 \le 1$?

4.10.6 Show that the derivative of Φ in Example 4.10.18 is invertible in $\overset{\circ}{X}$, and compute the area of X.

Exercise 4.10.7: You may want to use Theorem 3.7.15 (the spectral theorem).

***4.10.7** Let A be an $n \times n$ symmetric matrix such that $Q_A(\vec{\mathbf{x}}) = \vec{\mathbf{x}} \cdot A\vec{\mathbf{x}}$ is positive definite. What is the volume of the region $Q_A(\vec{\mathbf{x}}) \le 1$?

Exercise 4.10.8, part b: The form of $U_{a,b}$ should suggest an appropriate change of variables.

4.10.8 a. For fixed $a, b > 1$, let $U_{a,b}$ be the plane region in the first quadrant defined by the inequalities $1 \le xy \le a$, $x \le y \le bx$. Sketch $U_{2,4}$.

b. Compute $\int_{U_{a,b}} x^2y^2|dx\,dy|$.

4.10.9 Let $V = \left\{ \begin{pmatrix} x \\ y \\ z \end{pmatrix} \in \mathbb{R}^3 \;\middle|\; x > 0,\ y > 0,\ z > 0,\ \frac{x^2}{a^2} + \frac{y^2}{b^2} + \frac{z^2}{c^2} \le 1 \right\}$.

Compute $\int_V xyz\,|dx\,dy\,dz|$.

4.10.10 a. What is the analogue of spherical coordinates in four dimensions? What does the change of variables formula say in that case?

b. What is the integral of $|\mathbf{x}|$ over the ball of radius R in $\mathbb{R}^4$?

Exercise 4.10.12: First transform it into an integral over a subset of $\mathbb{R}^3$ (it will look like a quarter of a melon), then pass to spherical coordinates.

4.10.11 Justify that the volume of the ball of radius R in $\mathbb{R}^3$ is $\frac{4}{3}\pi R^3$.

4.10.12 Evaluate the iterated integral

$$\int_{-2}^{2}\int_{0}^{\sqrt{4-x^2}}\int_{0}^{\sqrt{4-x^2-y^2}} (x^2+y^2+z^2)^{3/2}\,dz\,dy\,dx.$$

4.10.13 Find the volume of the region between the cone of equation $z^2 = x^2+y^2$ and the paraboloid of equation $z = x^2 + y^2$.

Exercise 4.10.15: You may wish to use the trigonometric formulas

$$4\cos^3\theta = \cos 3\theta + 3\cos\theta$$
$$2\cos\varphi\cos\theta = \cos(\theta+\varphi) + \cos(\theta-\varphi).$$

4.10.14 Find the volume of the region $z \ge x^2+y^2$, $z \le 10 - x^2 - y^2$ (Exercise 4.5.15), but this time use cylindrical change of variables.

4.10.15 a. Sketch the curve given in polar coordinates by $r = \cos 2\theta$, $|\theta| < \pi/4$.

b. Where is the center of gravity of the region $0 < r \le \cos 2\theta$, $|\theta| < \pi/4$?

4.10.16 What is the center of gravity of the region A defined by the inequalities

$$x^2 + y^2 \le z \le 1,\ x \ge 0,\ y \ge 0?$$

4.10.17 Let $Q_a = [0,a]\times[0,a] \subset \mathbb{R}^2$ be the square of side a in the first quadrant, with two sides on the axes, and let $\Phi : \mathbb{R}^2 \to \mathbb{R}^2$ be given by

$$\Phi\begin{pmatrix} u \\ v \end{pmatrix} = \begin{pmatrix} u - v \\ e^u + e^v \end{pmatrix}.$$

Set $A = \Phi(Q_a)$.

a. Sketch A, by computing the image of each of the sides of Q_a. It might help to begin by drawing carefully the curves of equation $y = e^x + 1$ and $y = e^{-x} + 1$.

b. Show that $\Phi : Q_a \to A$ is one to one.

c. What is $\int_A y\,|dx\,dy|$?

4.10.18 What is the volume of the part of the ball of equation $x^2+y^2+z^2 \le 4$ where $z^2 \ge x^2+y^2,\ z > 0$?

4.10.19 Let $Q = [0,1] \times [0,1]$ be the unit square in $\mathbb{R}^2$, and let $\Phi : \mathbb{R}^2 \to \mathbb{R}^2$ be defined by

$$\Phi\begin{pmatrix} u \\ v \end{pmatrix} = \begin{pmatrix} u - v^2 \\ u^2 + v \end{pmatrix}. \quad \text{Set } A = \Phi(Q).$$

a. Sketch A, by computing the image of each of the sides of Q (they are all arcs of parabolas).

b. Show that $\Phi : Q \to A$ is 1–1.

c. What is $\int_A x\,|dx\,dy|$?

Change of variables for Exercise 4.10.20:

$$\Phi\begin{pmatrix} u \\ v \\ w \end{pmatrix} = \begin{pmatrix} (w^3-1)u \\ (w^3+1)v \\ w \end{pmatrix}.$$

4.10.20 Solve Exercise 4.5.19 again, using the change of variables in the margin.

4.10.21 The *moment of inertia* of a body $X \subset \mathbb{R}^3$ around an axis is the integral $\int_X (r(\mathbf{x}))^2 |d^3\mathbf{x}|$, where $r(\mathbf{x})$ is the distance from $\mathbf{x}$ to the axis.

a. Let f be a nonnegative continuous function of $x \in [a,b]$, and let B be the body obtained by rotating the region $0 \le y \le f(x),\ a \le x \le b$ around the x-axis. What is the moment of inertia of B around the x-axis?

b. What number does this give when $f(x) = \cos x, \quad a = -\dfrac{\pi}{2}, \quad b = \dfrac{\pi}{2}$?

4.11 Lebesgue integrals

This new integral of Lebesgue is proving itself a wonderful tool. I might compare it with a modern Krupp gun, so easily does it penetrate barriers which were impregnable.—Edward Van Vleck, *Bulletin of the American Mathematical Society*, vol. 23, 1916.

There are many reasons to study Lebesgue integrals. An essential one is the Fourier transform, the fundamental tool of engineering and signal processing, not to mention harmonic analysis. (The Fourier transform is discussed at the end of this section.) Lebesgue integrals are also ubiquitous in probability theory.

So far we have restricted ourselves to integrals of bounded functions with bounded support, whose upper and lower sums are equal. But often we will want to integrate functions that are not bounded or do not have bounded support. Lebesgue integration makes this possible. It also has two other advantages:

1. Lebesgue integrals exist for functions plagued with the kind of “local nonsense” that we saw in the function that is 1 at rational numbers in $[0,1]$ and 0 elsewhere (Example 4.3.3). The Lebesgue integral ignores local nonsense on sets of measure 0.
2. Lebesgue integrals are better behaved with respect to limits.

Remark. If Lebesgue integration is superior to Riemann integration, why did we put so much emphasis on Riemann integration earlier in this chapter? Riemann integrals have one great advantage over Lebesgue integrals: they can be computed using Riemann sums. Lebesgue integrals can only be computed via Riemann integrals (or perhaps by using Monte Carlo methods). Thus our approach is in keeping with our emphasis on computationally effective algorithms. △

Our approach to Lebesgue integration is very different from the standard one. The usual way of defining the Lebesgue integral

$$\int_{\mathbb{R}^n} f(\mathbf{x})|d^n\mathbf{x}|$$

is to cut up the *codomain* $\mathbb{R}$ into small intervals $I_i = [x_i, x_{i+1}]$, and to approximate the integral by

$$\sum_i x_i \mu(f^{-1}(I_i)),$$

where $\mu(A)$ is the *measure of* A, then letting the decomposition of the codomain become arbitrarily fine. Of course, this requires saying what subsets are measurable, and defining their measure. This is the main task with the standard approach, and for this reason the theory of Lebesgue integration is often called *measure theory*.

It is surprising how much more powerful the theory is when one decomposes the codomain rather than the domain. But one pays a price: it isn't at all clear how one would approximate a Lebesgue integral: figuring out what the sets $f^{-1}(I_i)$ are, never mind finding their measure, is difficult or impossible even for the simplest functions.

We take a different tack, building on the theory of Riemann integrals, and defining the integral directly by taking limits of functions that are Riemann integrable. We get measure theory at the end as a byproduct: just as Riemann integrals are used to define volume, Lebesgue integrals can be used to define measure. This is discussed in Appendix A21.

Before defining the Lebesgue integral, we will discuss the behavior of Riemann integrals with respect to limits.

Integrals and limits

The behavior of integrals under limits is often important. Here we give the best general statements about Riemann integrals and limits.

We would like to be able to say that if $k \mapsto f_k$ is a convergent sequence of functions, then, as $k \to \infty$,

$$\int \lim f_k = \lim \int f_k. \qquad 4.11.1$$

In one setting this is true and straightforward: when $k \mapsto f_k$ is a *uniformly convergent* sequence of integrable functions, all with support in the same bounded set. The key condition in Definition 4.11.1 is that given ϵ, the same K works for all $\mathbf{x}$.

Definition 4.11.1 (Uniform convergence). A sequence $k \mapsto f_k$ of functions $f_k : \mathbb{R}^n \to \mathbb{R}$ *converges uniformly* to a function f if for every $\epsilon > 0$, there exists K such that when $k \geq K$, then, for all $\mathbf{x} \in \mathbb{R}^n$, $|f_k(\mathbf{x}) - f(\mathbf{x})| < \epsilon$.

The three sequences of functions in Example 4.11.3 do *not* converge uniformly, although they do converge. Uniform convergence on all of $\mathbb{R}^n$ isn't a very common phenomenon, unless something is done to cut down the domain. For instance, suppose that $k \mapsto p_k$ is a sequence of polynomials

$$p_k(x) = a_{0,k} + a_{1,k}x + \cdots + a_{m,k}x^m \qquad 4.11.2$$

all of degree $\leq m$, and that this sequence "converges" in the "obvious" sense that for each degree i (i.e., each x^i), the sequence of coefficients $a_{i,0}, a_{i,1}, a_{i,2}, \ldots$ converges. Then $k \mapsto p_k$ does not converge uniformly on $\mathbb{R}$. But for any bounded set A, the sequence $k \mapsto p_k\mathbf{1}_A$ does converge uniformly.[15]

[15] Instead of writing $p_k\mathbf{1}_A$ we could write "p_k restricted to A".

Theorem 4.11.2 (Convergence for Riemann integrals). *Let $k \mapsto f_k$ be a sequence of integrable functions $\mathbb{R}^n \to \mathbb{R}$, all with support in a fixed ball $B \subset \mathbb{R}^n$, and converging uniformly to a function f. Then f is integrable, and*

$$\lim_{k\to\infty} \int_{\mathbb{R}^n} f_k(\mathbf{x})\,|d^n\mathbf{x}| = \int_{\mathbb{R}^n} f(\mathbf{x})\,|d^n\mathbf{x}|. \qquad 4.11.3$$

Proof. Choose $\epsilon > 0$ and K so large that $\sup_{\mathbf{x}\in\mathbb{R}^n} |f(\mathbf{x}) - f_k(\mathbf{x})| < \epsilon$ when $k > K$. Then when $k > K$, we have, for any N,

$$L_N(f) > L_N(f_k) - \epsilon \operatorname{vol}_n(B) \quad \text{and} \quad U_N(f) < U_N(f_k) + \epsilon \operatorname{vol}_n(B). \qquad 4.11.4$$

Now choose N so large that $U_N(f_k) - L_N(f_k) < \epsilon$; we get

$$U_N(f) - L_N(f) < \underbrace{U_N(f_k) - L_N(f_k)}_{<\epsilon} + 2\epsilon \operatorname{vol}_n(B), \qquad 4.11.5$$

yielding $U(f) - L(f) \le \epsilon(1 + 2\operatorname{vol}_n(B))$. □

Inequality 4.11.4: If you picture this as Riemann sums in one variable, ϵ is the difference between the height of the lower rectangles for f, and the height of the lower rectangles for f_k. The total width of all the rectangles is $\operatorname{vol}_1(B)$, since B is the support for f_k.

In many cases Theorem 4.11.2 is good enough, but it cannot deal with unbounded functions or functions with unbounded support. Example 4.11.3 shows some of the things that can go wrong.

Example 4.11.3 (Cases where the mass of an integral gets lost). Here are three sequences of functions where the limit of the integral is *not* the integral of the limit.

Exercise 4.11.13 asks you to verify that mass is indeed lost in the three sequences of functions in Example 4.11.3.

1. When f_k is defined by

$$f_k(x) = \begin{cases} 1 & \text{if } k \le x \le k+1 \\ 0 & \text{otherwise,} \end{cases} \qquad 4.11.6$$

the mass of the integral is contained in a square 1 high and 1 wide. As $k \to \infty$ this mass drifts off to infinity and gets lost:

$$\lim_{k\to\infty} \int_0^\infty f_k(x)\,dx = 1, \quad \text{but} \quad \int_0^\infty \lim_{k\to\infty} f_k(x)\,dx = \int_0^\infty 0\,dx = 0. \qquad 4.11.7$$

2. For the function

$$f_k(x) = \begin{cases} k & \text{if } 0 < x \le \frac{1}{k} \\ 0 & \text{otherwise,} \end{cases} \qquad 4.11.8$$

the mass of the integral is contained in a rectangle k high and $1/k$ wide. As $k \to \infty$, the height of the rectangle tends to ∞ and its width to 0:

$$\lim_{k\to\infty} \int_0^1 f_k(x)\,dx = 1, \text{ but } \int_0^1 \lim_{k\to\infty} f_k(x)\,dx = \int_0^1 0\,dx = 0. \qquad 4.11.9$$

3. The third example is our standard example of a function that is not integrable using the Riemann integral. Make a list $a_1, a_2, \dots$ of the rational numbers between 0 and 1. Now define

$$f_k(x) = \begin{cases} 1 & \text{if } x \in \{a_1, \dots, a_k\} \\ 0 & \text{otherwise.} \end{cases} \qquad 4.11.10$$

FIGURE 4.11.1.
Henri Lebesgue (1875–1941)

After Lebesgue's father died of tuberculosis, leaving three children, the oldest five years old, his mother cleaned houses to support them. Lebesgue later wrote, "My first good fortune was to be born to intelligent parents, then to have been sickly and extremely poor, which kept me from violent games and distractions, ... and most of all to have an extraordinary mother even for France, this country of good mothers."

When one of Lebesgue's students apprehensively arrived for her first teaching job, in which she was to replace a popular substitute teacher, she found a note from him waiting for her. *Faites-vous aimer là-bas comme partout,* he had written ("make yourself beloved there as you are everywhere"). "It was a ray of sunshine," she recalled in a note published in *Message d'un mathématicien: Henri Lebesgue, pour le centenaire de sa naissance*, Paris, A. Blanchard, 1974.

Proposition 4.11.5 is proved in Appendix A21.

Recall (Definition 4.4.2) that "except on a set of measure 0" is also written "almost everywhere".

Then we have

$$\int_0^1 f_k(x)\,dx = 0 \quad \text{for all } k, \tag{4.11.11}$$

but $\lim_{k\to\infty} f_k$ is the function that is 1 on the rationals and 0 on the irrationals between 0 and 1, and hence not integrable. △

The pitfalls of disappearing mass can be avoided by the dominated convergence theorem for Riemann integrals, Theorem 4.11.4. In practice this theorem is not as useful as one might hope, because the hypothesis that the limit is Riemann integrable is rarely satisfied unless the convergence is uniform, in which case the much easier Theorem 4.11.2 applies. But Theorem 4.11.4 is the key tool in our approach to Lebesgue integration. The proof, in Appendix A21, is quite difficult and very tricky.

Theorem 4.11.4 (Dominated convergence for Riemann integrals). *Let $f_k : \mathbb{R}^n \to \mathbb{R}$ be a sequence of R-integrable functions. Suppose there exists R such that all f_k have their support in $B_R(\mathbf{0})$ and satisfy $|f_k| \le R$. Let $f : \mathbb{R}^n \to \mathbb{R}$ be an R-integrable function such that $f(\mathbf{x}) = \lim_{k\to\infty} f_k(\mathbf{x})$ except on a set of measure 0. Then*

$$\lim_{k\to\infty} \int_{\mathbb{R}^n} f_k(\mathbf{x})\,|d^n\mathbf{x}| = \int_{\mathbb{R}^n} f(\mathbf{x})\,|d^n\mathbf{x}|. \tag{4.11.12}$$

The first two functions in Example 4.11.3 fail to meet the conditions of Theorem 4.11.4: for the first, the f_k do not have uniformly bounded support; for the second, the f_k are not uniformly bounded. The third does meet the conditions; the f_k converge a.e. to $f = 0$; both integrals in equation 4.11.12 are 0. Exercise 4.34 gives a case where this does not happen.

Defining the Lebesgue integral

The weakness of Theorem 4.11.4 is that we have to know that the limit is integrable. Usually we don't know this; most often, we need to deal with the limit of a sequence of functions, and all we know is that it is a limit. But we will now see that Theorem 4.11.4 can be used to construct the Lebesgue integral, which is much better behaved under limits.

We abbreviate "Riemann integrable" as "R-integrable" and "Lebesgue-integrable" as "L-integrable."

Proposition 4.11.5 (Convergence except on a set of measure 0). *If f_k for $k = 1, 2, \ldots$ are Riemann-integrable functions on $\mathbb{R}^n$ such that*

$$\sum_{k=1}^{\infty} \int_{\mathbb{R}^n} |f_k(\mathbf{x})||d^n\mathbf{x}| < \infty, \tag{4.11.13}$$

then the series $\sum_{k=1}^{\infty} f_k(\mathbf{x})$ converges almost everywhere.

We can now define *equal in the sense of Lebesgue*, which we denote by $\underset{L}{=}$. Recall that "a.e." means "almost everywhere", or, equivalently, "for almost all $\mathbf{x} \in \mathbb{R}^n$" or "except on a set of measure 0".

This notion of "Lebesgue equality" is fairly subtle, as sets of measure 0 can be quite complicated. For instance, if you only know a function almost everywhere, then you can never evaluate it at any point: you never know whether this is a point at which you know the function. The moral: functions that you know except on a set of measure 0 should only appear under integral signs.

Definition 4.11.6 (Lebesgue equality). Let $k \mapsto f_k$, $k \mapsto g_k$ be two sequences of R-integrable functions $\mathbb{R}^n \to \mathbb{R}$ such that

$$\sum_{k=1}^{\infty} \int_{\mathbb{R}^n} |f_k(\mathbf{x})||d^n\mathbf{x}| < \infty \quad \text{and} \quad \sum_{k=1}^{\infty} \int_{\mathbb{R}^n} |g_k(\mathbf{x})||d^n\mathbf{x}| < \infty. \qquad 4.11.14$$

We will say that

$$\sum_{k=1}^{\infty} f_k \underset{L}{=} \sum_{k=1}^{\infty} g_k \quad \text{if} \quad \sum_{k=1}^{\infty} f_k(\mathbf{x}) = \sum_{k=1}^{\infty} g_k(\mathbf{x}) \ \text{a.e.} \qquad 4.11.15$$

Lebesgue integration does not require functions to be bounded with bounded support, it ignores "local nonsense" on sets of measure 0, and it is better behaved with respect to limits.

But if you want to compute integrals, the Riemann integral is still essential. Lebesgue integrals are more or less uncomputable unless you know a function as a limit of Riemann-integrable functions in an appropriate sense – in the sense of Proposition 4.11.5, for instance.

Theorem 4.11.7. *Let $k \mapsto f_k$, $k \mapsto g_k$ be two sequences of R-integrable functions such that*

$$\sum_{k=1}^{\infty} \int_{\mathbb{R}^n} |f_k(\mathbf{x})||d^n\mathbf{x}| < \infty, \qquad \sum_{k=1}^{\infty} \int_{\mathbb{R}^n} |g_k(\mathbf{x})||d^n\mathbf{x}| < \infty, \qquad 4.11.16$$

and

$$\sum_{k=1}^{\infty} f_k \underset{L}{=} \sum_{k=1}^{\infty} g_k. \qquad 4.11.17$$

Then

$$\sum_{k=1}^{\infty} \int_{\mathbb{R}^n} f_k(\mathbf{x})|d^n\mathbf{x}| = \sum_{k=1}^{\infty} \int_{\mathbb{R}^n} g_k(\mathbf{x})|d^n\mathbf{x}|. \qquad 4.11.18$$

Theorem 4.11.7 is proved after Definition 4.11.8.

Since f_k and g_k converge to the same function f almost everywhere, the integral of a function f that is the sum of a series of R-integrable functions as in equation 4.11.13 depends only on f and not on the series. So we can now define the Lebesgue integral.

Equation 4.11.20: By Theorem 0.5.8, the series on the right is convergent, since (by part 4 of Proposition 4.1.14) it is absolutely convergent.

Note that equation 4.11.20 can be rewritten as

$$\int_{\mathbb{R}^n} \sum_{k=1}^{\infty} f_k(\mathbf{x})|d^n\mathbf{x}| = \sum_{k=1}^{\infty} \int_{\mathbb{R}^n} f_k(\mathbf{x})|d^n\mathbf{x}|,$$

exchanging the sum and the integral.

Definition 4.11.8 (Lebesgue integral). Let $k \mapsto f_k$ be a sequence of R-integrable functions such that

$$\sum_{k=1}^{\infty} \int_{\mathbb{R}^n} |f_k(\mathbf{x})||d^n\mathbf{x}| < \infty. \qquad 4.11.19$$

Then the Lebesgue integral of $f \underset{L}{=} \sum_{k=1}^{\infty} f_k$ is

$$\int_{\mathbb{R}^n} f(\mathbf{x})|d^n\mathbf{x}| \overset{\text{def}}{=} \sum_{k=1}^{\infty} \int_{\mathbb{R}^n} f_k(\mathbf{x})|d^n\mathbf{x}|. \qquad 4.11.20$$

Proof of Theorem 4.11.7. Set $h_k = f_k - g_k$, and $H_l = \sum_{k=1}^{l} h_k$. To prove equation 4.11.18 we need to show that

$$\lim_{l\to\infty}\int_{\mathbb{R}^n} H_l(\mathbf{x})\,|d^n\mathbf{x}| = 0. \qquad 4.11.21$$

The H_l form a sequence of Riemann-integrable functions converging to 0 almost everywhere (by equation 4.11.17). If in addition they all have support in $B_R(\mathbf{0})$ and $|H_l| \le R$ for all l, then H_l would meet the conditions for f_k in the dominated convergence theorem (Theorem 4.11.4), so

$$\lim_{l\to\infty}\int_{\mathbb{R}^n} H_l(\mathbf{x})\,|d^n\mathbf{x}| = 0, \quad \text{which would give} \qquad 4.11.22$$

$$\lim_{l\to\infty}\sum_{k=1}^{l}\int_{\mathbb{R}^n} h_k(\mathbf{x})\,|d^n\mathbf{x}| = 0, \qquad 4.11.23$$

proving the result.

Equation 4.11.22 uses Theorem 4.11.4: if the H_l were bounded with bounded support, we would have

$$\lim_{l\to\infty}\int_{\mathbb{R}^n} H_l(\mathbf{x})\,|d^n\mathbf{x}| = \int_{\mathbb{R}^n} 0|d^n\mathbf{x}|,$$

where the function 0 plays the role of f in Theorem 4.11.4.

To go from equation 4.11.22 to equation 4.11.23, we can switch integral and sum because the sum is finite. (Remember that H_l in equation 4.11.22 is a finite sum.)

Our strategy will be to reduce the general case, where H_l is not bounded with bounded support, to this one, by appropriately truncating the H_l: we will define the truncation $[H_l]_R$ and we will consider H_l as the sum $[H_l]_R + H_l - [H_l]_R$ and consider separately the integral of $[H_l]_R$ (see equation 4.11.27) and the integral of $H_l - [H_l]_R$ (the remainder of the proof).

Choose $\epsilon > 0$ and choose M such that

$$\sum_{k=M+1}^{\infty}\int_{\mathbb{R}^n} |h_k(\mathbf{x})||d^n\mathbf{x}| < \epsilon, \qquad 4.11.24$$

so that for $l > M$ we have

$$\begin{aligned}\int_{\mathbb{R}^n} |H_l(\mathbf{x}) - H_M(\mathbf{x})||d^n\mathbf{x}| &\le \sum_{k=M+1}^{l}\int_{\mathbb{R}^n}|h_k(\mathbf{x})||d^n\mathbf{x}| \\ &\le \sum_{k=M+1}^{\infty}\int_{\mathbb{R}^n}|h_k(\mathbf{x})||d^n\mathbf{x}| < \epsilon.\end{aligned} \qquad 4.11.25$$

By the definitions of f_k and g_k, the series $\sum_k^\infty \int |h_k|$ is convergent, so by choosing M large enough, we can make the "tail" of the series as small as we like. This is why we choose M first, then the R of the truncation.

Next choose R such that $\sup|H_M(\mathbf{x})| < R/2$ and $H_M(\mathbf{x}) = 0$ when $|\mathbf{x}| \ge R$. We define the R-truncation of a function f, denoted by $[f]_R$, by

$$[f]_R = \sup\Big(-R\mathbf{1}_{B_R(\mathbf{0})}, \inf\big(R\mathbf{1}_{B_R(\mathbf{0})}, f\big)\Big), \qquad 4.11.26$$

illustrated by Figure 4.11.2, and shown for $[H_l]_R$ in the margin. These $[H_l]_R$ form a sequence of Riemann-integrable functions all with support in $B_R(\mathbf{0})$, satisfying $|[H_l]_R| \le R$ and (by equation 4.11.17) tending to 0 almost everywhere, so, by Theorem 4.11.4,

$$\underbrace{\lim_{l\to\infty}\int_{\mathbb{R}^n} [H_l]_R(\mathbf{x})\,|d^n\mathbf{x}| = 0.}_{\text{main motor of the proof}} \qquad 4.11.27$$

The definition of $[H_l]_R$ may be easier to understand if equation 4.11.26 is rewritten as follows:

$$[H_l]_R(\mathbf{x}) = \begin{cases} 0 & \text{if } |\mathbf{x}| > R \\ R & \text{if } |\mathbf{x}| \le R, H_l(\mathbf{x}) > R \\ -R & \text{if } |\mathbf{x}| \le R, H_l(\mathbf{x}) < -R \\ H_l(\mathbf{x}) & \text{otherwise.}\end{cases}$$

The virtue of defining $[H_l]_R$ as in equation 4.11.26 is that it shows that $[H_l]_R$ is Riemann-integrable (see Corollary 4.1.15).

We have now done most of the work (the hard part was proving Theorem 4.11.4). But for $l > M$, we still need to deal with the difference

$$H_l - [H_l]_R = (H_l - H_M) - ([H_l]_R - H_M) \qquad 4.11.28$$

By inequality 4.11.25, the integral of $|H_l - H_M|$ is less than ϵ, so we only need to consider the integral of $|[H_l]_R - H_M|$. Outside $B_R(\mathbf{0})$ we have $H_M = 0$ and $[H_l]_R = 0$, so we need only consider

$$\int_{B_R(\mathbf{0})} \big|[H_l]_R - H_M(\mathbf{x})\big||d^n\mathbf{x}|. \tag{4.11.29}$$

To see that this integral is small, first find N such that $U_N(|H_l - H_M|) < \epsilon$. Then consider the union A of the cubes $C \in \mathcal{D}_N(\mathbb{R}^n)$ that intersect $B_R(\mathbf{0})$ and where $M_C(|H_l - H_M|) > R/2$. Since the upper sum $U_N(|H_l - H_M|)$ is small, A must have small volume; in fact, it is at most $2\epsilon/R$:

$$\begin{aligned}\epsilon > U_N|H_l - H_M| &= \sum_{C\in\mathcal{D}_N(\mathbb{R}^n)} M_C|H_l - H_M| \operatorname{vol}_n C \\ &\geq \sum_{\substack{C\in\mathcal{D}_N(\mathbb{R}^n)\\ C\subset A}} M_C|H_l - H_M| \operatorname{vol}_n C \geq \sum_{\substack{C\in\mathcal{D}_N(\mathbb{R}^n)\\ C\subset A}} \frac{R}{2}\operatorname{vol}_n C = \frac{R}{2}\operatorname{vol}_n A.\end{aligned} \tag{4.11.30}$$

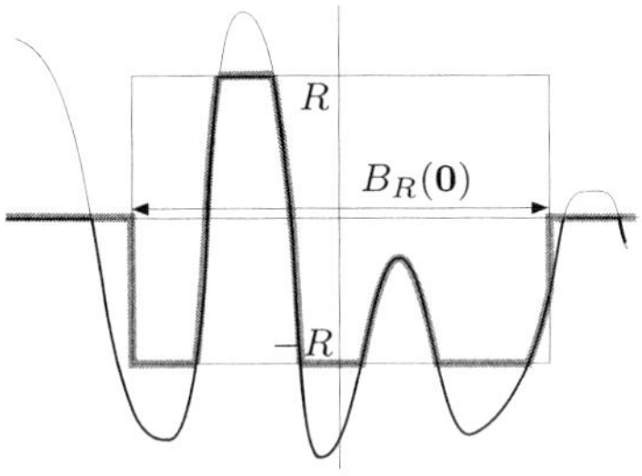

FIGURE 4.11.2.

The R-truncation of a function; see equation 4.11.26. The thin line is the graph of f; the dark line $\inf(R\mathbf{1}_{B_R(\mathbf{0})}, f)$. We take the sup of the dark line and $-R\mathbf{1}_{B_R(\mathbf{0})}$, to get the thick, light gray line representing $[f]_R$.

Let B be the union of the cubes $C \in \mathcal{D}_N(\mathbb{R}^n)$ that intersect $B_R(\mathbf{0})$ and such that $M_C(|H_l - H_M|) \leq R/2$; on these,

$$|H_l| \leq |H_l - H_M| + |H_M| \leq R/2 + R/2 = R \tag{4.11.31}$$

(remember that $\sup|H_M(\mathbf{x})| < R/2$), so on B, we have $[H_l]_R = H_l$. Thus[16]

$$\begin{aligned}&\int_{B_R(\mathbf{0})} \Big|[H_l]_R(\mathbf{x}) - H_M(\mathbf{x})\Big||d^n\mathbf{x}| \\ &= \int_A \Big|[H_l]_R(\mathbf{x}) - H_M(\mathbf{x})\Big||d^n\mathbf{x}| + \int_B \Big|[H_l]_R(\mathbf{x}) - H_M(\mathbf{x})\Big||d^n\mathbf{x}| \\ &\leq \frac{3R}{2}\operatorname{vol}_n(A) + \underbrace{\int_B \Big|H_l(\mathbf{x}) - H_M(\mathbf{x})\Big||d^n\mathbf{x}|}_{<\epsilon \text{ by inequality 4.11.25}} \leq 3\epsilon + \epsilon = 4\epsilon.\end{aligned} \tag{4.11.32}$$

In summary: For any R,

$$\lim_{l\to\infty}\int_{\mathbb{R}^n} H_l(\mathbf{x})\,|d^n\mathbf{x}| = \underbrace{\lim_{l\to\infty}\int_{\mathbb{R}^n} [H_l]_R(\mathbf{x})\,|d^n\mathbf{x}|}_{0 \text{ by equation 4.11.27}} + \underbrace{\lim_{l\to\infty}\int_{\mathbb{R}^n} (H_l(\mathbf{x}) - [H_l]_R(\mathbf{x}))\,|d^n\mathbf{x}|}_{0 \text{ as shown below}}. \tag{4.11.33}$$

For the second integral on the right, first choose $\epsilon > 0$, then M satisfying equation 4.11.24, then R as above (text immediately before equation 4.11.26). For these choices we have shown that for all $l > M$, we have

$$\int_{\mathbb{R}^n} H_l(\mathbf{x}) - [H_l]_R(\mathbf{x})\,|d^n\mathbf{x}| = \underbrace{\int_{\mathbb{R}^n} H_l(\mathbf{x}) - H_M(\mathbf{x})\,|d^n\mathbf{x}|}_{<\epsilon \text{ by ineq. 4.11.25}} - \underbrace{\int_{\mathbb{R}^n} [H_l]_R(\mathbf{x}) - H_M(\mathbf{x})\,|d^n\mathbf{x}|}_{\leq 4\epsilon \text{ by ineq. 4.11.32}}. \tag{4.11.34}$$

[16] Inequality 4.11.32: To get the $3R/2$ in the last line, remember that

$$\sup|H_M(\mathbf{x})| < R/2 \quad \text{and} \quad |[H_l]_R(\mathbf{x})| \leq R.$$

So

$$\lim_{l\to\infty}\int_{\mathbb{R}^n}(H_l(\mathbf{x})-[H_l]_R(\mathbf{x}))\,|d^n\mathbf{x}|=0. \qquad 4.11.35$$

Since (by equation 4.11.27)

$$\lim_{l\to\infty}\int_{\mathbb{R}^n}[H_l]_R(\mathbf{x}))\,|d^n\mathbf{x}|=0, \quad \text{this gives} \quad \lim_{l\to\infty}\int_{\mathbb{R}^n}H_l(\mathbf{x})\,|d^n\mathbf{x}|=0. \qquad 4.11.36$$

This completes the proof of Theorem 4.11.7. □

Some examples of Lebesgue integrals

All the integrals we computed earlier in this book are examples of Lebesgue integrals, because of the following result.

Proposition 4.11.9 justifies using the same integration symbol for Lebesgue integrals that we use for Riemann integrals.

Proposition 4.11.9. *If f is R-integrable, then it is L-integrable, and its Lebesgue integral equals its Riemann integral.*

Proof. Just take $f_1 = f$, and set $f_k = 0$ for $k = 2, 3, \dots$. Clearly $\sum f_k = f$ everywhere, and inequality 4.11.19 is satisfied:

$$\sum_{k=1}^{\infty}\int|f_k(\mathbf{x})||d^n\mathbf{x}|=\int|f_1(\mathbf{x})||d^n\mathbf{x}|<\infty, \quad \text{so} \qquad 4.11.37$$

$$\underbrace{\sum_{k=1}^{\infty}\int f_k(\mathbf{x})|d^n\mathbf{x}|}_{\text{Lebesgue integral}}=\underbrace{\int f_1(\mathbf{x})|d^n\mathbf{x}|}_{\text{Riemann integral}}<\infty. \quad \square \qquad 4.11.38$$

Example 4.11.10: In Exercise 4.11.20 you are asked to find a series of Riemann-integrable functions that converge to f exactly.

It is extraordinarily difficult to invent bounded functions with bounded support that are not Lebesgue integrable; constructing them requires the axiom of choice, hence depends on your model of set theory. See Example A21.7.

We consider the function in Example 4.11.10 a "pathological" function, not really a good example of the power of Lebesgue integration. The functions in Examples 4.11.11 and 4.11.12 are our real motivation for introducing Lebesgue integrals. Such integrals show up everywhere in physics, for example.

Exercise 4.11.19 asks you to show that if an improper integral of a function is absolutely convergent (like those of equations 4.11.40–4.11.42), then the integral exists as a Lebesgue integral.

We can also integrate all sorts of functions that we couldn't have previously thought of integrating, like the function of Example 4.3.3.

Example 4.11.10 (L-integrable function that is not R-integrable). Let f be the indicator function of the rationals, i.e.,

$$f(x)=\begin{cases}1 & \text{if } x\in\mathbb{Q}\\ 0 & \text{otherwise.}\end{cases} \qquad 4.11.39$$

This function equals 0 almost everywhere, so it is Lebesgue integrable with integral 0. △

Example 4.11.11 (An integrable function with unbounded support). In one variable, you probably studied improper integrals of functions that are not bounded or do not have bounded support: integrals like

$$\int_{-\infty}^{\infty}\frac{1}{1+x^2}\,dx=[\arctan x]_{-\infty}^{\infty}=\pi \qquad 4.11.40$$

$$\int_0^{\infty}x^n e^{-x}\,dx=n! \qquad 4.11.41$$

$$\int_0^1\frac{1}{\sqrt{x}}\,dx=[2\sqrt{x}]_0^1=2. \qquad 4.11.42$$

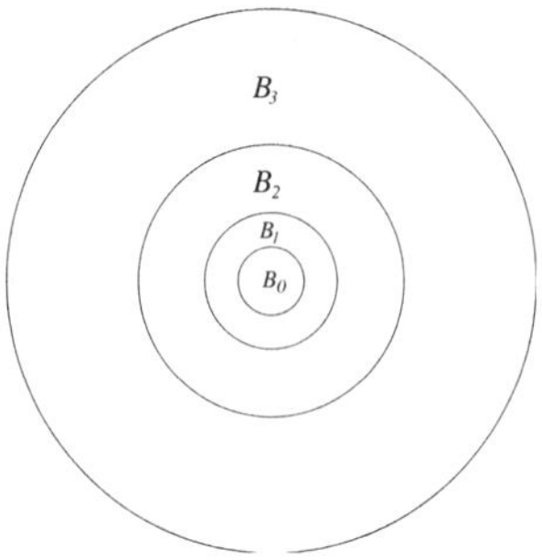

FIGURE 4.11.3.

Example 4.11.11: In $\mathbb{R}^2$, the boundaries of the B_i form a sequence of concentric circles: B_0 is the unit disc, B_1 is the circular strip between the circle of radius 1 and the circle of radius 2, and so forth.

These improper integrals have analogues in higher dimensions, like

$$\int_{\mathbb{R}^n} \frac{1}{1+|\mathbf{x}|^{n+1}}\,|d^n\mathbf{x}|. \qquad 4.11.43$$

We will show that if $m > n$, the function

$$f(\mathbf{x}) = \frac{1}{1+|\mathbf{x}|^m} \qquad 4.11.44$$

is L-integrable on $\mathbb{R}^n$. As shown in Figure 4.11.3, define

$$B_i = \left\{ \mathbf{x} \in \mathbb{R}^n \;\middle|\; 2^{i-1} < |\mathbf{x}| \le 2^i \right\}, \text{ for } i = 1, 2, \dots \qquad 4.11.45$$

and let B_0 be the unit ball. (We could set $2^{i-1} \le |\mathbf{x}| \le 2^i$; this would not affect any integrals, since the overlap would have measure 0.) Then

$$f = \sum_{i=0}^{\infty} f_i, \text{ where } f_i(\mathbf{x}) = \mathbf{1}_{B_i}(\mathbf{x})\frac{1}{1+|\mathbf{x}|^m}. \qquad 4.11.46$$

(The function f_0 is identical to f on the unit ball, and 0 elsewhere; f_1 is identical to f on B_1, and 0 elsewhere, ...)

Clearly the f_i are R-integrable: they are bounded with bounded support, and continuous except on two spheres, which certainly have measure 0. Thus the substance of this example is to show that

$$\sum_{i=1}^{\infty} \int_{\mathbb{R}^n} |f_i(\mathbf{x})|\,|d^n\mathbf{x}| < \infty. \qquad 4.11.47$$

The map $\Phi_i : \mathbf{x} \mapsto 2^{i-1}\mathbf{x}$, for $i \ge 1$, is a change of variables that takes B_1 to B_i; note that $\det[\mathbf{D}\Phi_i(\mathbf{x})] = 2^{(i-1)n}(\mathbf{x})$. The change of variables formula gives

$$\begin{aligned}
\int_{\mathbb{R}^n} |f_i(\mathbf{x})||d^n\mathbf{x}| &= \int_{B_i} f(\mathbf{x})|d^n\mathbf{x}| = \int_{B_i} \frac{1}{1+|\mathbf{x}|^m}|d^n\mathbf{x}| \\
&= \int_{B_1} (f \circ \Phi_i)(\mathbf{x})\big|\det[\mathbf{D}\Phi_i(\mathbf{x})]\big||d^n\mathbf{x}| \\
&= \int_{B_1} \frac{1}{1+|2^{i-1}\mathbf{x}|^m}\,2^{(i-1)n}|d^n\mathbf{x}| \qquad 4.11.48 \\
&\le \int_{B_1} \frac{2^{(i-1)n}}{2^{(i-1)m}|\mathbf{x}|^m}|d^n\mathbf{x}| = 2^{(i-1)(n-m)} \int_{B_1} \frac{1}{|\mathbf{x}|^m}|d^n\mathbf{x}|.
\end{aligned}$$

If $m > n$, this geometric series converges, since the sum $\sum_i 2^{(i-1)(n-m)}$ converges when $m > n$. Exercise 4.11.21 asks you to prove that if $m \le n$, then f is not L-integrable on $\mathbb{R}^n$. △

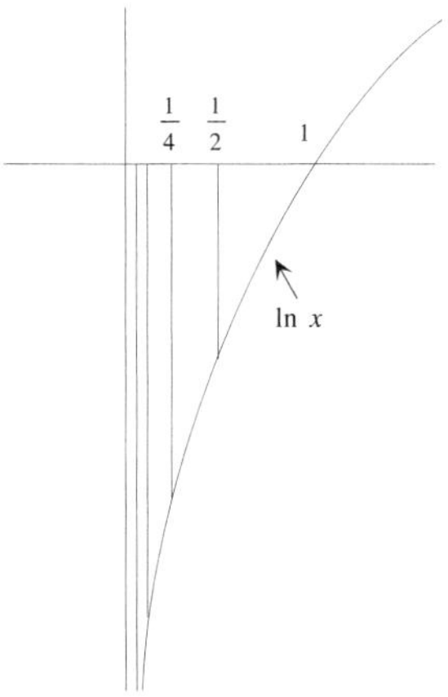

FIGURE 4.11.4.

Example 4.11.12: We consider

$$f(x) \underset{L}{=} \mathbf{1}_{[0,1]}(x)\ln x$$

as a sum of functions: the function that is $\ln x$ between $1/2$ and 1 and 0 elsewhere, plus the function that is $\ln x$ between $1/4$ and $1/2$ and 0 elsewhere, plus the function that is $\ln x$ between $1/8$ and $1/4$ and 0 elsewhere, ...

Example 4.11.12 (An unbounded, integrable function). The function $f(x) \underset{L}{=} \mathbf{1}_{[0,1]}(x)\ln x$ is L-integrable, even though it isn't bounded. As shown in Figure 4.11.4, it can be written as a sum of bounded functions:

$$f(x) \underset{L}{=} \sum_{i=0}^{\infty} f_i(x), \quad \text{where } f_i = \left(\mathbf{1}_{(2^{-(i+1)},2^{-i}]}(x)\right)\ln x. \qquad 4.11.49$$

The functions f_i are R-integrable, and the series

$$\sum_{i=0}^{\infty} \int_{1/2^{i+1}}^{1/2^i} |\ln x| \, dx \qquad 4.11.50$$

is convergent, as Exercise 4.11.15 asks you to check. △

Example 4.11.13 shows that Lebesgue integrals are not a strict generalization of improper integrals: some improper integrals, those which depend on cancellations, do not exist as Lebesgue integrals. Since there is no version of Fubini's theorem or the change of variables formula for integrals that depend on cancellations, Lebesgue integration wisely forbids them.

Exercise 4.11.16 asks you to justify the argument in Example 4.11.13.

Note that we would not want this integral to exist, since the change of variables formula fails for this sort of improper integral (see Exercise 4.11.17).

Example 4.11.13 (A function that is not Lebesgue integrable). Some improper one-dimensional integrals do *not* correspond to Lebesgue-integrable functions: integrals whose existence depends on cancellations, like

$$\int_0^{\infty} \frac{\sin x}{x} \, dx. \qquad 4.11.51$$

As you may recall from one-variable calculus, this improper integral is defined to be

$$\lim_{A \to \infty} \int_0^{A} \frac{\sin x}{x} \, dx, \qquad 4.11.52$$

and we can show that the limit exists, for instance, by saying that the series

$$\sum_{k=0}^{\infty} \int_{k\pi}^{(k+1)\pi} \frac{\sin x}{x} \, dx \qquad 4.11.53$$

is a decreasing alternating series whose terms go to 0 as $k \to \infty$. But this works only because positive and negative terms cancel: the area between the graph of $\sin x/x$ and the x axis is infinite, and there is no limit

$$\lim_{A \to \infty} \int_0^{A} \left| \frac{\sin x}{x} \right| \, dx. \qquad 4.11.54$$

The integral in 4.11.51 does not exist as a Lebesgue integral. By our definition, "f L-integrable" implies that $|f|$ is L-integrable, which is not the case here. △

Some elementary properties of the Lebesgue integral

Proposition 4.11.14 (The Lebesgue integral is linear). *If f and g are L-integrable and a, b are constants, then $af + bg$ is L-integrable and*

$$\int_{\mathbb{R}^n} (af + bg)(\mathbf{x}) |d^n\mathbf{x}| = a \int_{\mathbb{R}^n} f(\mathbf{x}) |d^n\mathbf{x}| + b \int_{\mathbb{R}^n} g(\mathbf{x}) |d^n\mathbf{x}|.$$

Proof. If $f \underset{L}{=} \sum f_k$ and $g \underset{L}{=} \sum g_k$, then $af + bg \underset{L}{=} \sum (af_k + bg_k)$. Indeed, the series $\sum (af_k + bg_k)$ will converge except on the union of the sets where the series $\sum_k |f(\mathbf{x})|$ and $\sum_k |g(\mathbf{x})|$ diverge, and the union of two sets of measure 0 is still of measure 0 (see Theorem 4.4.4). For $\mathbf{x}$ not in this set, the result follows from the corresponding statement for series of numbers. □

It is not true that the product of two L-integrable functions is necessarily L-integrable; for instance, the function $\dfrac{\mathbf{1}_{[-1,1]}(x)}{\sqrt{|x|}}$ is L-integrable, but its square is not. However, we have the following result.

Proposition 4.11.15. *If f is L-integrable on $\mathbb{R}^n$, and g is R-integrable on $\mathbb{R}^n$, then fg is L-integrable.*

Proof. Since f is L-integrable, we can set $f \underset{L}{=} \sum_k f_k$, where the functions f_k are R-integrable and

$$\sum_k \int_{\mathbb{R}^n} |f_k(\mathbf{x})|\,|d^n\mathbf{x}| < \infty. \qquad 4.11.55$$

We have $fg \underset{L}{=} \sum_k f_k g$, where $f_k g$ is R-integrable; since g is bounded,

$$\sum_k \int_{\mathbb{R}^n} |f_k(\mathbf{x})g(\mathbf{x})|\,|d^n\mathbf{x}| \le \sup_{\mathbf{x}\in\mathbb{R}^n} |g(\mathbf{x})| \sum_k \int_{\mathbb{R}^n} |f_k(\mathbf{x})|\,|d^n\mathbf{x}| < \infty. \qquad 4.11.56$$

Therefore fg is L-integrable. □

We use the symbol $\underset{L}{\le}$ to mean "$\le$ except for $\mathbf{x}$ on a set of measure 0".

Proposition 4.11.16. *If f and g are L-integrable and $f \underset{L}{\le} g$, then*

$$\int_{\mathbb{R}^n} f(\mathbf{x})|d^n\mathbf{x}| \;\le\; \int_{\mathbb{R}^n} g(\mathbf{x})|d^n\mathbf{x}|. \qquad 4.11.57$$

Proof. By Proposition 4.11.14, the statement is equivalent to saying that $0 \underset{L}{\le} g - f$ implies $0 \le \int_{\mathbb{R}^n}(g-f)(\mathbf{x})\,|d^n\mathbf{x}|$. Replace $g - f$ by f; the statement becomes $0 \underset{L}{\le} f \implies 0 \le \int_{\mathbb{R}^n} f(\mathbf{x})|d^n\mathbf{x}|$. Assume $0 \underset{L}{\le} f$, and write $f \underset{L}{=} \sum_{k=1}^{\infty} f_k$ with all f_k R-integrable and $\sum_{k=1}^{\infty} \int_{\mathbb{R}^n} |f_k|(\mathbf{x})|d^n\mathbf{x}| < \infty$. Define $F_m = \sum_{k=1}^{m} f_k$ and

$$H_0 = 0, \quad H_m = \sup\{0, F_m\} \text{ for } m > 0, \quad h_k = H_k - H_{k-1}. \qquad 4.11.58$$

We have four cases:

1. If $F_{k-1}(\mathbf{x}) \ge 0,\ F_k(\mathbf{x}) \ge 0$, then $|h_k(\mathbf{x})| = |f_k(\mathbf{x})|$
2. If $F_{k-1}(\mathbf{x}) < 0,\ F_k(\mathbf{x}) \ge 0$, then $|h_k(\mathbf{x})| = |F_k(\mathbf{x}) - 0| < |F_k(\mathbf{x}) - F_{k-1}(\mathbf{x})| = |f_k(\mathbf{x})|$
3. If $F_{k-1}(\mathbf{x}) \ge 0,\ F_k(\mathbf{x}) < 0$, then $|h_k(\mathbf{x})| = |0 - F_{k-1}(\mathbf{x})| < |F_k(\mathbf{x}) - F_{k-1}(\mathbf{x})| = |f_k(\mathbf{x})|$
4. If $F_{k-1}(\mathbf{x}) < 0,\ F_k(\mathbf{x}) < 0$, then $|h_k(\mathbf{x})| = 0 \le |f_k(\mathbf{x})|$.

In all four cases, $|h_k| \le |f_k|$, as shown in Figure 4.11.5. Moreover, h_k is an R-integrable function for every k, so by Theorem 4.11.4,

$$\sum_{k=1}^{\infty} \int_{\mathbb{R}^n} |h_k(\mathbf{x})||d^n\mathbf{x}| \le \sum_{k=1}^{\infty} \int_{\mathbb{R}^n} |f_k(\mathbf{x})||d^n\mathbf{x}| < \infty. \qquad 4.11.59$$

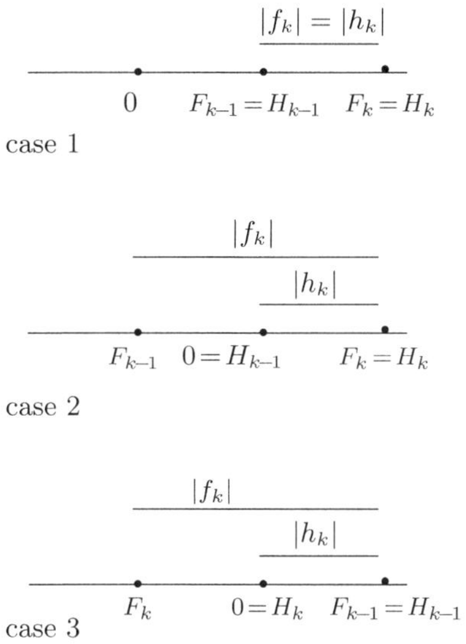

FIGURE 4.11.5.
In all three cases above, since

$$|h_k| = |H_k - H_{k-1}|$$

and

$$|f_k| = |F_k - F_{k-1}|,$$

we have

$$|h_k| \le |f_k|.$$

In case 1, we could have

$$F_k < F_{k-1},$$

since the f_k can be negative, but this would not change the result. Case 4 is obvious.

Since $H_0 = 0$ we can write H_m as

$$H_m = \underbrace{H_m - H_{m-1}}_{h_m} + \underbrace{H_{m-1} - H_{m-2}}_{h_{m-1}} + \cdots + \underbrace{H_2 - H_1}_{h_2} + \underbrace{H_1 - H_0}_{h_1}, \qquad 4.11.60$$

so $\lim_{m\to\infty} H_m(\mathbf{x}) = \sum_{k=1}^{\infty} h_k(\mathbf{x})$. Since we have assumed $0 \underset{L}{\le} f$ we have $\lim_{m\to\infty} H_m(\mathbf{x}) = f(\mathbf{x})$ a.e. Thus we have $f(\mathbf{x}) \underset{L}{=} \sum_{k=1}^{\infty} h_k(\mathbf{x})$, and

$$\int_{\mathbb{R}^n} f(\mathbf{x})|d^n\mathbf{x}| = \sum_{k=1}^{\infty} \int_{\mathbb{R}^n} h_k(\mathbf{x})|d^n\mathbf{x}| \qquad 4.11.61$$

$$= \lim_{m\to\infty} \sum_{k=1}^{m} \int_{\mathbb{R}^n} h_k(\mathbf{x})|d^n\mathbf{x}| = \lim_{m\to\infty} \int_{\mathbb{R}^n} H_m(\mathbf{x}) \ge 0. \quad \square$$

Theorems about Lebesgue integrals

The important theorems about Lebesgue integrals are Fubini's theorem, the change of variables theorem, the monotone convergence theorem, and the dominated convergence theorem. They are all easier to state for Lebesgue integrals than for Riemann integrals.

We start with Theorem 4.11.17 on series; this isn't the usual way to approach the topic, but in our treatment of Lebesgue integrals, Theorem 4.11.17 is the foundational result. (In the standard treatment of Lebesgue integration, Theorem 4.11.17 says that $L^1(\mathbb{R}^n, |d^n\mathbf{x}|)$ is complete.) Note that we are not assuming (as we do in Definition 4.11.8) that the f_k are Riemann-integrable. The theorem is proved in Appendix A21.

Theorem 4.11.17: If the f_k are Riemann-integrable, this is simply the definition of the Lebesgue integral.

Theorem 4.11.17 (A first limit theorem for Lebesgue integrals). *Let $k \mapsto f_k$ be a series of L-integrable functions such that*

$$\sum_{k=1}^{\infty} \int_{\mathbb{R}^n} |f_k(\mathbf{x})||d^n\mathbf{x}| < \infty. \qquad 4.11.62$$

Then $f(\mathbf{x}) = \sum_{k=1}^{\infty} f_k(\mathbf{x})$ exists for almost all $\mathbf{x}$, the function f is L-integrable, and

$$\int_{\mathbb{R}^n} f(\mathbf{x})|d^n\mathbf{x}| = \int_{\mathbb{R}^n} \sum_{k=1}^{\infty} f_k(\mathbf{x})|d^n\mathbf{x}| = \sum_{k=1}^{\infty} \int_{\mathbb{R}^n} f_k(\mathbf{x})|d^n\mathbf{x}|. \qquad 4.11.63$$

Of course, the monotone convergence theorem also applies to monotone decreasing sequences that are bounded below by an L-integrable function.

Theorem 4.11.18. (Monotone convergence theorem: Lebesgue integrals). *1. Let* $0 \le f_1 \underset{L}{\le} f_2 \underset{L}{\le} \cdots$ *be a sequence of L-integrable nonnegative functions. If*

$$\sup_k \int_{\mathbb{R}^n} f_k(\mathbf{x})|d^n\mathbf{x}| < \infty, \qquad 4.11.64$$

then the limit $f(\mathbf{x}) = \lim_{k\to\infty} f_k(\mathbf{x})$ *exists for almost all* $\mathbf{x}$, *the function* f *is L-integrable, and*

$$\int_{\mathbb{R}^n} f(\mathbf{x})|d^n\mathbf{x}| = \sup_k \int_{\mathbb{R}^n} f_k(\mathbf{x})|d^n\mathbf{x}|. \qquad 4.11.65$$

2. Conversely, if $f(\mathbf{x}) \stackrel{\text{def}}{=} \lim_{k\to\infty} f_k(\mathbf{x})$ *exists almost everywhere and* $\sup_k \int_{\mathbb{R}^n} f_k(\mathbf{x})|d^n\mathbf{x}| = \infty$, *then* f *is not Lebesgue integrable.*

Proof. 1. Apply Theorem 4.11.17 to the sup, rewritten as the series

$$f = f_1 + (f_2 - f_1) + \cdots + (f_k - f_{k-1}) + \cdots = g_1 + g_2 + \cdots + g_k + \cdots .$$

2. If f were L-integrable, then by Proposition 4.11.16 and $f \ge f_k$,

$$\int_{\mathbb{R}^n} f(\mathbf{x})|d^n\mathbf{x}| \ge \int_{\mathbb{R}^n} f_k(\mathbf{x})|d^n\mathbf{x}| \quad \text{for all } k.$$

Thus

$$\int_{\mathbb{R}^n} f(\mathbf{x})|d^n\mathbf{x}| \ge \sup_k \int_{\mathbb{R}^n} f_k(\mathbf{x})|d^n\mathbf{x}| = \infty.$$

So f is not integrable. □

In Theorem 4.11.19, the f_k are "dominated" by F. Theorems 4.11.19, 4.11.20, and 4.11.21 are proved in Appendix A21.

Theorem 4.11.19 (Dominated convergence theorem: Lebesgue integrals). *Let* $k \mapsto f_k$ *be a sequence of L-integrable functions that converges pointwise almost everywhere to some function* f. *Suppose there is an L-integrable function* $F : \mathbb{R}^n \to \mathbb{R}$ *such that* $|f_k(\mathbf{x})| \le F(\mathbf{x})$ *for almost all* $\mathbf{x}$. *Then* f *is L-integrable and its integral is*

$$\int_{\mathbb{R}^n} f(\mathbf{x})\,|d^n\mathbf{x}| = \lim_{k\to\infty} \int_{\mathbb{R}^n} f_k(\mathbf{x})\,|d^n\mathbf{x}|. \qquad 4.11.66$$

Theorem 4.11.20 (The change of variables formula: Lebesgue integrals). *Let* U, V *be open subsets of* $\mathbb{R}^n$, *and let* $\Phi : U \to V$ *be bijective, of class* C^1, *with inverse of class* C^1, *such that both* Φ *and* Φ^{-1} *have Lipschitz derivatives. Let* $f : V \to \mathbb{R}$ *be defined except perhaps on a set of measure* 0. *Then*

f *is L-integrable on* $V \iff (f \circ \Phi)|\det[\mathbf{D}\Phi]\,|$ *is L-integrable on* U, *and*

$$\int_V f(\mathbf{v})|d^n\mathbf{v}| = \int_U (f\circ\Phi)(\mathbf{u})\big|\det[\mathbf{D}\Phi(\mathbf{u})]\big|\,|d^n\mathbf{u}|. \qquad 4.11.67$$

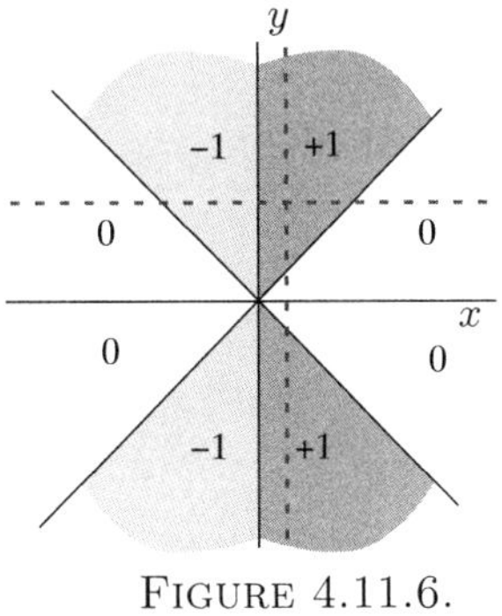

FIGURE 4.11.6.
Part 3 of Fubini's theorem: Consider the function $f:\mathbb{R}^2 \to \mathbb{R}$ that is $+1$ in the darkly shaded region, -1 in the lightly shaded region, and 0 elsewhere. For every y, the function $x \mapsto f(x,y)$ is integrable, and in fact

$$\int_{\mathbb{R}} f(x,y)\,dx = 0.$$

The function

$$y \mapsto \int_{\mathbb{R}} f(x,y)\,|dx|$$

is L-integrable on $\mathbb{R}$, and

$$\int_{\mathbb{R}} \left(\int_{\mathbb{R}} f(x,y)\,dx \right) dy = \int_{\mathbb{R}} 0\,dy = 0.$$

But f is *not* integrable on $\mathbb{R}^2$, and the iterated integral written the other way:

$$\int_{\mathbb{R}} \left(\int_{\mathbb{R}} f(x,y)\,dy \right) dx$$

is nonsense; the inner integral is identically $+\infty$ for $x > 0$ and identically $-\infty$ for $x < 0$. The problem is that $y \mapsto \int_{\mathbb{R}} |f(x,y)|\,|dx|$ (note the absolute values) is not integrable.

Theorem 4.11.21 (Fubini's theorem for the Lebesgue integral). *Let $f : \mathbb{R}^n \times \mathbb{R}^m \to \mathbb{R}$ be an L-integrable function. Then the function*

$$\mathbf{y} \mapsto \int_{\mathbb{R}^n} f(\mathbf{x},\mathbf{y})|d^n\mathbf{x}| \qquad 4.11.68$$

is defined for almost all $\mathbf{y} \in \mathbb{R}^m$ and is L-integrable on $\mathbb{R}^m$, and

$$\int_{\mathbb{R}^n\times\mathbb{R}^m} f(\mathbf{x},\mathbf{y})|d^n\mathbf{x}||d^m\mathbf{y}| = \int_{\mathbb{R}^m} \left(\int_{\mathbb{R}^n} f(\mathbf{x},\mathbf{y})|d^n\mathbf{x}| \right) |d^m\mathbf{y}|. \qquad 4.11.69$$

Conversely, if $f : \mathbb{R}^n \times \mathbb{R}^m \to \mathbb{R}$ is a function such that

1. *every point $(\mathbf{x},\mathbf{y}) \in \mathbb{R}^{n+m}$ is the center of a ball on which f is L-integrable*
2. *the function $\mathbf{x} \mapsto |f(\mathbf{x},\mathbf{y})|$ is L-integrable on $\mathbb{R}^n$ for almost all $\mathbf{y}$*
3. *the function $\mathbf{y} \mapsto \int_{\mathbb{R}^n} |f(\mathbf{x},\mathbf{y})||d^n\mathbf{x}|$ is L-integrable on $\mathbb{R}^m$*

then f is L-integrable on $\mathbb{R}^{n+m}$, and

$$\int_{\mathbb{R}^n\times\mathbb{R}^m} f(\mathbf{x},\mathbf{y})|d^n\mathbf{x}||d^m\mathbf{y}| = \int_{\mathbb{R}^m} \left(\int_{\mathbb{R}^n} f(\mathbf{x},\mathbf{y})|d^n\mathbf{x}| \right) |d^m\mathbf{y}|.$$

The absolute values in parts 2 and 3 of the converse are necessary, as shown by the function $f:\mathbb{R}^2 \to \mathbb{R}$ represented in Figure 4.11.6.

The Gaussian integral

The integral of the Gaussian bell curve is one of the most important integrals in all mathematics. The central limit theorem (Theorem 4.2.7) asserts that if you repeat the same experiment n times, independently each time, and make some measurement each time, then the probability that the average of the measurements will lie in an interval $[A, B]$ is (see formula 4.2.16)

$$\int_A^B \frac{1}{\sqrt{2\pi}\,\tau} e^{-\frac{(x-E)^2}{2\tau^2}}\,dx, \qquad 4.11.70$$

where E is the expected value of x and σ is the standard deviation, so that $\tau = \sigma/\sqrt{n}$ is the standard deviation of the repeated experiment. Since most of probability is concerned with repeating experiments, the Gaussian integral is of the greatest importance.

For the central limit theorem to make sense, the integrand of formula 4.11.70 must be a probability density, i.e., we must have

$$\int_{-\infty}^{\infty} \frac{1}{\sqrt{2\pi}\,\tau} e^{-\frac{(x-E)^2}{2\tau^2}}\,dx = 1. \qquad 4.11.71$$

Exercise 4.11.10 asks you to use a change of variables to derive equation 4.11.71 from the simpler equation

$$\int_{-\infty}^{\infty} e^{-x^2}\,dx = \sqrt{\pi}. \qquad 4.11.72$$

The function e^{-x^2} doesn't have an antiderivative that can be computed in elementary terms.[17] One way to compute it is to use Lebesgue integrals in two dimensions. Set $\int_{-\infty}^{\infty} e^{-x^2}\,dx = C$. By Fubini's theorem,

$$\int_{\mathbb{R}^2} e^{-(x^2+y^2)}\,|d^2\mathbf{x}| = \left(\int_{-\infty}^{\infty} e^{-x^2}\,dx\right)\left(\int_{-\infty}^{\infty} e^{-y^2}\,dy\right) = C^2. \qquad 4.11.73$$

We now use the change of variables formula, passing to polar coordinates:

$$\int_{\mathbb{R}^2} e^{-(x^2+y^2)}\,|d^2\mathbf{x}| = \int_0^{2\pi}\int_0^{\infty} e^{-r^2} r\,dr\,d\theta. \qquad 4.11.74$$

The r from the change of variables makes this straightforward to evaluate:

$$C^2 = \int_0^{2\pi}\int_0^{\infty} re^{-r^2}\,dr\,d\theta = 2\pi\left[-\frac{e^{-r^2}}{2}\right]_0^{\infty} = 2\pi\left(0 - \frac{-1}{2}\right) = \pi. \qquad 4.11.75$$

So $C = \sqrt{\pi}$.

Equation 4.11.73: To apply Fubini's theorem (Theorem 4.11.21), we need to check that the function

$$f : \mathbb{R}\times\mathbb{R} \to \mathbb{R},\ f\begin{pmatrix} x \\ y \end{pmatrix} = e^{-(x^2+y^2)}$$

is L-integrable over $\mathbb{R}^2$. First note that $x \mapsto e^{-x^2}$ is L-integrable over $\mathbb{R}$ (see Definition 4.11.8; note that $e^{-x^2} > 0$ and write e^{-x^2} as $\sum_{j=-\infty}^{\infty} e^{-x^2}\mathbf{1}_{[j,j+1)}$):

$$\sum_{-\infty}^{\infty}\int_j^{j+1} e^{-x^2}|dx| \le 2\sum_{j=0}^{\infty} e^{-j^2}$$

$$< 2\sum_{j=0}^{\infty}\frac{1}{2^j} = 4.$$

Next, set

$$\sum_{j,k=-\infty}^{\infty} = \sum_{j=-\infty}^{\infty}\sum_{k=-\infty}^{\infty}$$

and write

$$\sum_{j,k=-\infty}^{\infty}\int_j^{j+1}\int_k^{k+1} e^{-(x^2+y^2)}|dx\,dy|$$

$$\le 4\sum_{j=0}^{\infty}\sum_{k=0}^{\infty} e^{-(j^2+k^2)}$$

$$= 4\sum_{j=0}^{\infty} e^{-j^2}\left(\sum_{k=0}^{\infty} e^{-k^2}\right)$$

$$< 16.$$

Theorem 4.11.22 is a major result and has far-reaching consequences.

We need condition 4.11.77 so that we can apply the dominated convergence theorem in the proof.

When does the integral of a derivative equal the derivative of an integral?

Very often we need to differentiate a function that is itself an integral. This is particularly the case for Laplace transforms and Fourier transforms, as we will see. If we integrate a function with respect to one variable, and differentiate with respect to a different variable, under what circumstances does first integrating and then differentiating give the same result as first differentiating, then integrating? The dominated convergence theorem gives the following very general result.

Theorem 4.11.22 (Differentiating under the integral sign). *Let $f(t, \mathbf{x}) : \mathbb{R}^{n+1} \to \mathbb{R}$ be a function such that for each fixed t, the integral*

$$F(t) = \int_{\mathbb{R}^n} f(t,\mathbf{x})\,|d^n\mathbf{x}| \qquad 4.11.76$$

exists. Suppose that $D_t f$ exists for almost all $\mathbf{x}$. If there exists $\epsilon > 0$ and an L-integrable function g such that for all $s \ne t$,

$$|s-t| < \epsilon \implies \left|\frac{f(s,\mathbf{x}) - f(t,\mathbf{x})}{s-t}\right| \le g(\mathbf{x}), \qquad 4.11.77$$

then F is differentiable, and its derivative is

$$F'(t) = \int_{\mathbb{R}^n} D_t f(t,\mathbf{x})\,|d^n\mathbf{x}|. \qquad 4.11.78$$

[17] This is a fairly difficult result; see *Integration in Finite Terms* by R. Ritt, Columbia University Press, New York, 1948. Of course, it depends on your definition of "elementary"; an antiderivative can be computed from the *error function* $\operatorname{erf}(x) \stackrel{\text{def}}{=} \frac{2}{\sqrt{\pi}}\int_0^x e^{-t^2}\,dt$, which is a tabulated function and can be looked up in tables or typed into a calculator.

Proof. Just compute

$$F'(t) = \lim_{h\to 0} \frac{F(t+h)-F(t)}{h} = \lim_{h\to 0} \int_{\mathbb{R}^n} \frac{f(t+h,\mathbf{x})-f(t,\mathbf{x})}{h} |d^n\mathbf{x}|$$
$$= \int_{\mathbb{R}^n} \lim_{h\to 0} \frac{f(t+h,\mathbf{x})-f(t,\mathbf{x})}{h} |d^n\mathbf{x}| = \int_{\mathbb{R}^n} D_t f(t,\mathbf{x})\, |d^n\mathbf{x}|. \qquad 4.11.79$$

Moving the limit inside the integral sign is justified by the dominated convergence theorem for Lebesgue integrals (Theorem 4.11.19). □

Renowned physicist Richard Feynman (Nobel laureate 1965) taught himself to integrate using a book given to him by his high school physics teacher. It showed how to differentiate under the integral sign, a technique that (at least then) was not often taught in universities. As he wrote in *Surely You're Joking, Mr. Feynman!*, he was often able to compute integrals that others couldn't do with the standard methods they had learned. "So I got a great reputation for doing integrals, only because my box of tools was different from everybody else's."

Sometimes integration under the integral sign allows us to compute integrals that cannot be found using antiderivatives.

Example 4.11.23. Consider the function

$$F(t) = \int_0^1 f(t,x)\,dx = \int_0^1 \frac{x^t-1}{\ln x}\,dx, \quad \text{defined for } t \ge 0. \qquad 4.11.80$$

The derivative $D_t f$ exists and is

$$D_t \frac{x^t-1}{\ln x} = D_t \frac{e^{t\ln x}-1}{\ln x} = \frac{(\ln x)e^{t\ln x}}{\ln x} = x^t. \qquad 4.11.81$$

Condition 4.11.77 is satisfied, so

$$F'(t) = \int_0^1 \frac{(\ln x)x^t}{\ln x}\,dt = \int_0^1 x^t\,dt = \left[\frac{x^{t+1}}{t+1}\right]_{x=0}^{x=1} = \frac{1}{t+1}. \qquad 4.11.82$$

Thus

$$F(t) = \ln(t+1) + C, \text{ and } f(0) = C = 0, \text{ so } f(t) = \ln(t+1). \qquad \triangle$$

Example 4.11.23: The function
$$\frac{x^t-1}{\ln x}$$
has no elementary antiderivative. If it did, then we could write
$$\int_0^a \frac{x^t-1}{\ln x}dx$$
in elementary terms, and this leads to integrating e^x/x, which is known not to have an elementary antiderivative.

Applications to the Fourier and Laplace transforms

Fourier transforms and Laplace transforms are essential tools of analysis. Here we will study them only as examples of differentiation under the integral sign. We will see that these transforms turn differentiation (an analytical construction) into multiplication (algebraic). This is a large part of why these transforms are so important.

If f is an L-integrable function on $\mathbb{R}$, so is $f(x)e^{i\xi x}$ for each $\xi \in \mathbb{R}$, since

$$|f(x)e^{i\xi x}| = |f(x)|. \qquad 4.11.83$$

Definition 4.11.24 (Fourier transform). The function $\hat{f}$ defined by

$$\hat{f}(\xi) \overset{\text{def}}{=} \int_{\mathbb{R}} f(x)e^{2\pi i\xi x}\,dx. \qquad 4.11.84$$

is called the *Fourier transform* of f.

Equation 4.11.84: Note that this is an integral of a complex-valued function. So far, we have discussed only integrals of real-valued functions, but complex-valued functions introduce no new difficulties. If $f = f_1 + if_2$, with f_1 and f_2 real valued, so that they are the real and imaginary parts of f, then we define

$$\int f(x)\,dx = \int f_1(x)\,dx + i\int f_2(x)\,dx.$$

The Greek letter ξ is xi, pronounced "ksee" or "kseye". Usually people use the Roman alphabet for the variable of f, and the corresponding Greek letter for the variable of the Fourier transform: x and ξ, t and τ, z and ζ.

The derivative $D\widehat{f}$ of $\widehat{f}$ can also be written $(\widehat{f})'$.

Equation 4.11.86:

$$\left|\frac{e^{2\pi i h x}-1}{h}\right||f(x)|$$

corresponds to

$$\left|\frac{f(s,\mathbf{x})-f(t,\mathbf{x})}{s-t}\right|$$

in Theorem 4.11.22.

Saying that $x \mapsto x^p|f(x)|$ is L-integrable is saying that f decreases at infinity faster and faster as $p \to \infty$.

Remark. It would be hard to exaggerate the importance of the Fourier transform, in pure and applied mathematics. Passing from f to $\widehat{f}$ ("f hat") is one of the central constructions of mathematical analysis (and of physics, where passing from f to $\widehat{f}$ means passing from physical space to momentum space). △

According to Theorem 4.11.22, the derivative $D\widehat{f}$ of $\widehat{f}$ is

$$\begin{aligned} D\widehat{f}(\xi) &= \int_{\mathbb{R}} D_\xi\big(f(x)e^{2\pi i x\xi}\big)\,dx = \int_{\mathbb{R}} f(x)D_\xi(e^{2\pi i x\xi})\,dx \\ &= \int_{\mathbb{R}} 2\pi i\, x f(x)e^{2\pi i x\xi}\,dx = 2\pi i\,\widehat{xf}\,(\xi), \end{aligned} \qquad 4.11.85$$

provided that the difference quotients

$$\left|\frac{f(x)e^{2\pi i(\xi+h)x}-f(x)e^{2\pi i\xi x}}{h}\right| = \underbrace{|e^{2\pi i\xi x}|}_{1}\left|\frac{e^{2\pi i h x}-1}{h}\right||f(x)| \qquad 4.11.86$$

are all bounded by a single L-integrable function. Since

$$|e^{2\pi i a}-1| = 2|\sin(\pi a)| \le 2\pi|a| \quad \text{for any } a\in\mathbb{R}, \qquad 4.11.87$$

we see that

$$\left|\frac{e^{2\pi i h x}-1}{h}\right||f(x)| \le \frac{|2\pi h x||f(x)|}{|h|} = |2\pi x f(x)|, \qquad 4.11.88$$

so this is satisfied if $x \mapsto |xf(x)|$ is an L-integrable function. More generally, if $x \mapsto x^p|f(x)|$ is L-integrable, then $\widehat{f}$ is p times differentiable. To see this, set $(Mf)(x) = 2\pi i x\, f(x)$, so that $(M^pf)(x) = (2\pi i x)^p\, f(x)$. Then

$$\begin{aligned} D^p\widehat{f}(\xi) &= \int_{\mathbb{R}} D^p_\xi\left(e^{2\pi i x\xi}f(x)\right)\,dx = \int_{\mathbb{R}} (M^pf)(x)e^{2\pi i x\xi}\,dx \\ &= \widehat{M^pf}\,(\xi). \end{aligned} \qquad 4.11.89$$

Thus the faster f decreases at infinity, the smoother the Fourier transform $\widehat{f}$ is. This brings out one feature of the Fourier transform: growth conditions on the original function f get translated into smoothness conditions for the Fourier transform.

You might think that if f is L-integrable, then f must tend to 0 at infinity. This isn't true; for instance f could have spikes of height 1 and width $1/n^2$ at all the integers. △

Example 4.11.25 (The Fourier transform of the Gaussian). Let

$$f_a(x) = e^{-ax^2}, \quad \text{so that} \quad \widehat{f_a}(\xi) = \int_{\mathbb{R}} e^{-ax^2}e^{2\pi i\xi x}\,dx. \qquad 4.11.90$$

We can't compute this Fourier transform directly, but equation 4.11.85 (or Theorem 4.11.22) gives

$$\widehat{f_a}'(\xi) = \pi i\int_{-\infty}^{\infty} e^{2\pi i x\xi}(2xe^{-ax^2})\,dx. \qquad 4.11.91$$

This can be integrated by parts (justified by Exercise 4.11.8), to find

$$\widehat{f_a}'(\xi) = \pi i \left[e^{2\pi i x \xi} \frac{e^{-ax^2}}{-a} \right]_{-\infty}^{\infty} - \frac{2\pi^2 \xi}{a} \int_{-\infty}^{\infty} e^{2\pi i x \xi} e^{-ax^2}\, dx = -\frac{2\pi^2 \xi}{a} \widehat{f_a}(\xi).$$

Equation 4.11.92: By the definition of the Fourier transform,

$$\widehat{f}(0) = \int_{\mathbb{R}} f(x)\, dx.$$

It follows that $\widehat{f_a}(\xi) = Ce^{-\frac{\pi^2 \xi^2}{a}}$ for some constant C, and we can determine C by evaluating both sides at $\xi = 0$ (and making the change of variable $\sqrt{a}x = u$, so that $\sqrt{a}\, dx = du$):

$$C = \widehat{f_a}(0) = \int_{-\infty}^{\infty} e^{-ax^2}\, dx = \frac{1}{\sqrt{a}} \int_{-\infty}^{\infty} e^{-u^2}\, du \underbrace{=}_{\text{eq. 4.11.72}} \sqrt{\frac{\pi}{a}}. \qquad 4.11.92$$

For any function f, the pth moment is

$$m_p(f) = \int_{-\infty}^{\infty} x^p f(x)\, dx.$$

To what extent one can infer the behavior of a function from its moments is a big subject. When f is the function μ giving density, then $m_0(\mu)$ is the total mass (see equation 4.2.2) and the ith coordinate of the center of gravity is the ith coordinate of the 1st moment divided by the total mass (see equation 4.2.3).

Thus

$$\widehat{f_a}(\xi) = \sqrt{\frac{\pi}{a}} e^{-\frac{\pi^2 \xi^2}{a}}. \qquad 4.11.93$$

A similar computation allows us to compute the *moments* of the Gaussian: the integrals $m_p(f_a) \overset{\text{def}}{=} \int_{-\infty}^{\infty} x^p e^{-ax^2}\, dx$. For p odd, clearly $m_p = 0$. For $p \geq 2$ even, integration by parts gives

$$\begin{aligned} m_p(f_a) \overset{\text{def}}{=} \int_{-\infty}^{\infty} x^{p-1} x e^{-ax^2}\, dx &= \left[x^{p-1} \frac{e^{-ax^2}}{-2a} \right]_{-\infty}^{\infty} + \frac{p-1}{2a} \int_{-\infty}^{\infty} x^{p-2} e^{-ax^2}\, dx \\ &= \frac{p-1}{2a} m_{p-2}. \end{aligned} \qquad 4.11.94$$

In particular $m_2 = m_0/2a$. △

This replacement of the analytic operation of differentiation by the algebraic operation of multiplication is one reason why the Fourier transform is so important in the theory of differential equations, especially partial differential equations.

Example 4.11.26 is step zero of an immense field of mathematics. The zeros of the polynomial in the denominator in equation 4.11.100 (those values of ξ where the denominator vanishes) pose serious problems, and the analogues in higher dimensions are much more serious yet.

Rather than differentiating the Fourier transform, we might want to Fourier transform the derivative, which we can do if f is of class C^1 and both f and f' are L-integrable. This is best done by integration by parts:

$$\begin{aligned} \widehat{f'}(\xi) &= \int_{-\infty}^{\infty} f'(x) e^{2\pi i \xi x} dx = \lim_{A \to \infty} \int_{-A}^{A} f'(x) e^{2\pi i \xi x} dx \\ &= \lim_{A \to \infty} \left[f(x) e^{2\pi i \xi x} \right]_{x=-A}^{A} - \lim_{A \to \infty} \int_{-A}^{A} 2\pi i \xi f(x) e^{2\pi i \xi x} dx. \end{aligned} \qquad 4.11.95$$

Since f and f' are continuous and L-integrable, $\lim_{x \to \pm\infty} f(x) = 0$, so

$$\lim_{A \to \infty} \left[f(x) e^{2\pi i \xi x} \right]_{-A}^{A} = 0, \qquad 4.11.96$$

$$\widehat{f'}(\xi) = -2\pi i \xi \int_{\mathbb{R}} f(x) e^{2\pi i \xi x} dx = -2\pi i \xi \widehat{f}(\xi). \qquad 4.11.97$$

Thus the Fourier transform turns differentiation into multiplication. Example 4.11.26 shows that it turns differential operators into multiplication by polynomials, and hence solving differential equations into division by polynomials.

Example 4.11.26 (Fourier transform of a differential equation). The Fourier transform of both sides of the differential equation

$$a_p D^p f + \cdots + a_0 f = g \tag{4.11.98}$$

is

$$\underbrace{\big(a_p(-2\pi i\xi)^p + a_{p-1}(-2\pi i\xi)^{p-1} + \cdots + a_0\big)\widehat{f}}_{\text{product of } \widehat{f} \text{ and a polynomial}} = \widehat{g}, \tag{4.11.99}$$

which gives

$$\widehat{f} = \frac{\widehat{g}}{(-2\pi i\xi)^p a_p + \cdots + a_0}. \tag{4.11.100}$$

If you know how to undo the Fourier transform, you can compute f from this formula for $\widehat{f}$. △

Heisenberg's uncertainty principle

In quantum mechanics, a state of a 1-dimensional particle is described by the *wave function* $\psi : \mathbb{R} \to \mathbb{C}$ such that

$$\int_{-\infty}^{\infty} |\psi(x)|^2 \, dx = 1. \tag{4.11.101}$$

Parseval's theorem asserts that

$$\int_{-\infty}^{\infty} |\widehat{\psi}(\xi)|^2 \, dx = 1,$$

so $|\widehat{\psi}(\xi)|^2$ is a probability density; see Definition 4.2.3.

Just how ψ describes the system is fraught with philosophical difficulties, but part of the interpretation is that $|\psi|^2$ describes the "probability density of position": the probability that the particle is in $[a, b]$ is

$$\int_a^b |\psi(x)|^2 \, dx. \tag{4.11.102}$$

The "probability density of momentum" is given by $|\widehat{\psi}(\xi)|^2$. Heisenberg's uncertainty principle then says that the product of the variance of the random variable x^2 (with respect to $|\psi|^2$) and the variance of the random variable ξ^2 (with respect to $|\widehat{\psi}(\xi)|^2$) must be at least $1/(16\pi^2)$:

Inequality 4.11.103: If we want more precise knowledge about position, we pay for it by lack of knowledge of momentum, and vice versa.

$$\mathrm{Var}_{|\psi|^2}(x^2)\, \mathrm{Var}_{|\widehat{\psi}|^2}(\xi^2) \geq \frac{1}{16\pi^2}. \tag{4.11.103}$$

We will see that inequality 4.11.103 becomes an equality when ψ is the Gaussian

$$\psi_a(x) \stackrel{\text{def}}{=} \left(\frac{a}{\pi}\right)^{1/4} e^{-ax^2/2}. \tag{4.11.104}$$

Equation 4.11.104: The coefficient $\left(\frac{a}{\pi}\right)^{1/4}$ in the definition of ψ_a gives

$$\int_{\mathbb{R}} |\psi_a(x)|^2 dx = 1$$

by equation 4.11.92, so that $|\psi_a|^2$ is a probability density, as required by quantum mechanics.

In this case the probability densities for position and momentum are

$$|\psi_a(x)|^2 = \sqrt{\frac{a}{\pi}} e^{-ax^2} \quad \text{and} \quad |\widehat{\psi}_a(\xi)|^2 = 2\sqrt{\frac{\pi}{a}} e^{-4\pi^2\xi^2/a}. \tag{4.11.105}$$

As $a \to \infty$, the probability density $|\psi_a|^2$ for position becomes more and more concentrated near 0, but the probability of the momentum becomes more and more diffuse: if $a \to \infty$, then $1/a \to 0$.

To get equations 4.11.106 and 4.11.107 we first use equation 4.2.12 to compute the expectation of x and ξ, and then use

$$\operatorname{Var}(f) = E(f^2) - (E(f))^2.$$

(equation 3.8.14), noting that the term $(E(f))^2$ vanishes: $E(f) = 0$ because xe^{-ax^2} is an odd function.

Equation 4.11.108 says that the Gaussian wave function maximizes simultaneous information about position and momentum.

More precisely, the variance of the random variable x with respect to the probability density $|\psi_a|^2$ is

$$\operatorname{Var}(x) = \overbrace{\sqrt{\frac{a}{\pi}} \int_{-\infty}^{\infty} x^2 e^{-ax^2}\, dx}^{E(f^2)} - \overbrace{0}^{(E(f))^2} = \frac{1}{2a} \qquad 4.11.106$$

and the variance of the random variable ξ with respect to the probability density $|\widehat{\psi}_a|^2$ is

$$2\sqrt{\frac{\pi}{a}} \int_{-\infty}^{\infty} \xi^2 e^{-4\pi^2\xi^2/a}\, d\xi = \frac{a}{8\pi^2}, \quad \text{so} \qquad 4.11.107$$

$$\operatorname{Var}_{|\psi_a|^2}(x^2)\, \operatorname{Var}_{|\widehat{\psi}_a|^2}(\xi^2) = \frac{1}{16\pi^2}. \qquad 4.11.108$$

The Laplace transform

Definition 4.11.27 (Laplace transform). The *Laplace transform* $\mathcal{L}(f)$ of f is defined by the formula

$$\mathcal{L}(f)(s) \stackrel{\text{def}}{=} \int_0^{\infty} f(t)e^{-st} dt. \qquad 4.11.109$$

Depending on the range of values of s you are interested in, the Laplace transform $\mathcal{L}(f)$ exists for quite a broad range of functions f. For instance, if f is L-integrable, then $\mathcal{L}(f)$ is a continuous function of $s \in [0, \infty)$, and if f grows more slowly than some polynomial, then $\mathcal{L}(f)$ is defined and continuous on $(0, \infty)$.

Even if f grows as fast as e^{at}, the Laplace transform $\mathcal{L}f$ is differentiable when $s > a$.

Note that the integral is from 0 to ∞, not $-\infty$ to ∞. Again, under appropriate circumstances, we can differentiate under the integral sign:

$$D(\mathcal{L}f)(s) = \int_0^{\infty} D_s\big(f(t)e^{-st}\big)\, dt = \int_0^{\infty} -tf(t)e^{-st}\, dt = \big(\mathcal{L}(-tf)\big)(s).$$

Let us spell out the differentiability, under the hypothesis that for fixed $s > 0$, the function $t \mapsto te^{-st/2}f(t)$ is L-integrable. We need to show that the limit

$$\begin{aligned} \lim_{h\to 0} \frac{\mathcal{L}f(s+h) - \mathcal{L}f(s)}{h} &= \lim_{h\to 0} \int_0^{\infty} \frac{e^{-(s+h)t} - e^{-st}}{h} f(t)\, dt \\ &= \lim_{h\to 0} \int_0^{\infty} f(t)e^{-st}\frac{e^{-ht}-1}{h}\, dt \qquad 4.11.110 \end{aligned}$$

exists. When $0 < |h| \le s/2$, the mean value theorem gives

$$\left|\frac{e^{-ht}-1}{h}\right| \le te^{|ht|} \le te^{st/2}, \text{ so } \left|e^{-st}f(t)\frac{e^{-ht}-1}{h}\right| \le te^{-st/2}|f(t)|. \qquad 4.11.111$$

Thus, by the dominated convergence theorem (Theorem 4.11.19), we have

$$\begin{aligned} (\mathcal{L}f)'(s) &= \lim_{h\to 0} \int_0^{\infty} e^{-st}f(t)\frac{e^{-ht}-1}{h}\, dt \underbrace{=}_{\text{Thm. 4.11.19}} \int_0^{\infty} \lim_{h\to 0}\left(e^{-st}f(t)\frac{e^{-ht}-1}{h}\right) dt \\ &= -\int_0^{\infty} e^{-st}tf(t)\, dt. \qquad 4.11.112 \end{aligned}$$

Exercises for Section 4.11

4.11.1 a. Let $\mathbf{x} \in \mathbb{R}^2$. For what values of $p \in \mathbb{R}$ is $|\mathbf{x}|^p$ L-integrable over the unit disc in $\mathbb{R}^2$? (Think of polar coordinates.)

b. For those values, compute the integral.

4.11.2 a. Let $\mathbf{x} \in \mathbb{R}^3$. For what values of $p \in \mathbb{R}$ is $|\mathbf{x}|^p$ L-integrable over the unit ball in $\mathbb{R}^3$? (Think of spherical coordinates.)

b. For those values, compute the integral.

4.11.3 a. For what values of $p \in \mathbb{R}$ does $\displaystyle\int_{\mathbb{R}^2 - B_1(\mathbf{0})} |\mathbf{x}|^p \, |d^2\mathbf{x}|$ exist as a Lebesgue integral? (Think of polar coordinates.)

b. For those values, compute the integral.

4.11.4 a. For what values of $p \in \mathbb{R}$ does $\displaystyle\int_{\mathbb{R}^3 - B_1(\mathbf{0})} |\mathbf{x}|^p \, |d^3\mathbf{x}|$ exist as a Lebesgue integral? (Think of spherical coordinates.)

b. For those values, compute the integral.

4.11.5 For what values of $p \in \mathbb{R}$ does $\displaystyle\int_{\mathbb{R}^n - B_1(\mathbf{0})} |\mathbf{x}|^p \, |d^n\mathbf{x}|$ exist as a Lebesgue integral? (The answer depends on n.)

4.11.6 Set $A_m = \{0 \le y \le x^m\,,\ 0 \le x \le 1\}$, for $m \in \mathbb{R}$, $m \ge 0$. For what values of $p \in \mathbb{R}$ does $\displaystyle\int_{A_m} \frac{1}{(x^2+y^2)^p} |dx\, dy|$ exist as a Lebesgue integral?

4.11.7 Repeat Exercise 4.11.6, setting $A_m = \{0 \le y \le x^m\,,\ 1 \le x \le \infty\}$.

$$F(x) = \int_0^x f(t)\, dt,$$
$$G(x) = \int_0^x g(t)\, dt$$

Functions F and G for Exercise 4.11.8, part b.

4.11.8 a. Let f be L-integrable on $\mathbb{R}$. Show that $F(x) = \displaystyle\int_0^x f(t)\, dt$ is continuous.

b. Show that integration by parts holds for Lebesgue integrals: if f, g are L-integrable, and F and G are the functions defined in the margin, then

$$\int_a^b f(t)G(t)\, dt = F(b)G(b) - F(a)G(a) - \int_a^b F(t)g(t)\, dt.$$

c. Let $f,\ F$ be as in part a. Show that if F is L-integrable, then

$$\lim_{x \to \pm\infty} F(x) = 0.$$

d. Show that if f, F, g, G are as in part b, and in addition F is L-integrable and G is bounded, then

$$\int_{-\infty}^{\infty} f(t)G(t)\, dt = -\int_{-\infty}^{\infty} F(t)g(t)\, dt.$$

4.11.9 Compute the Fourier transform of the indicator function of $[-1, 1]$.

4.11.10 Use a change of variables and equation 4.11.72 to show that the integrand of equation 4.11.70 is a probability density.

4.11.11 Show that for all polynomials p, the Lebesgue integral

$$\int_{\mathbb{R}^n} p(\mathbf{x}) e^{-|\mathbf{x}|^2} \, |d^n\mathbf{x}| \quad \text{exists.}$$

Exercise 4.11.12: Use Theorem 4.11.20 and the changes of variables $xy = u$ and $y/x = v$.

4.11.12 For what values of a and b is $f\begin{pmatrix} x \\ y \end{pmatrix} = x^a y^b$ integrable over the region A defined by the inequalities $0 \le x \le y$ and $xy \le 1$?

4.11.13 a. Show that the three sequences of functions in Example 4.11.3 do not converge uniformly.

*b. Show that the sequence $k \mapsto p_k$ of polynomials

$$p_k(x) = a_{0,k} + a_{1,k}x + \cdots + a_{m,k}x^m \quad \text{of degree} \le m$$

does not converge uniformly on $\mathbb{R}$, unless the sequence $k \mapsto a_{i,k}$ is eventually constant for all $i > 0$ and the sequence $k \mapsto a_{0,k}$ converges.

c. Show that if the sequences $k \mapsto a_{i,k}$ converge for each $i \le m$, and A is a bounded set, then the sequence $k \mapsto p_k \mathbf{1}_A$ converges uniformly.

4.11.14 For the first two sequences of functions in Example 4.11.3, show that

$$\lim_{k\to\infty} \lim_{R\to\infty} \int_{\mathbb{R}} [f_k]_R(x)\, dx \ne \lim_{R\to\infty} \lim_{k\to\infty} \int_{\mathbb{R}} [f_k]_R(x)\, dx.$$

4.11.15 Show that the series $\displaystyle\sum_{i=0}^{\infty} \int_{1/2^{i+1}}^{1/2^i} |\ln x|\, dx$ of Example 4.11.12 converges.

4.11.16 Show that the integral $\int_0^\infty \frac{\sin x}{x}\, dx$ (formula 4.11.51) is equal to the series

$$\sum_{k=0}^{\infty} \left(\int_{k\pi}^{(k+1)\pi} \frac{\sin x}{x}\, dx \right), \text{ and that this series converges.}$$

4.11.17 Make the change of variables $u = 1/x$ in the integral $\displaystyle\int_0^\infty \frac{\sin x}{x}\, dx.$ Does the resulting integral exist as an improper integral, as described in Example 4.11.13?

Exercise 4.11.18: You will need the dominated convergence theorem (Theorem 4.11.4) to prove this.

Part b: "Except at 0" means "except at $0 \in P_k$", i.e., the zero polynomial.

4.11.18 Let P_k be the space of polynomials of degree at most k. Consider the function $F : P_k \to \mathbb{R}$ given by $p \mapsto \int_0^1 |p(x)|\, dx$.

a. Compute F when $k = 1$ and $k = 2$, i.e., evaluate the integrals $\int_0^1 |a + bx|\, dx$ and $\int_0^1 |a + bx + cx^2|\, dx$ (the second one is hard).

b. Show that F is differentiable except at 0. Compute the derivative.

*c. Show that if p has only simple roots between 0 and 1, then F is twice differentiable at p.

4.11.19 Assume f has its support on $[a, b)$, where b may be ∞. Suppose that f is L-integrable on $[a, c]$ for every c satisfying $a < c < b$. Show that if $\lim_{c\to b} \int_a^c |f(x)|\, dx$ exists, then f is L-integrable, and

$$\lim_{c\to b} \int_a^c f(x)\, dx = \int_a^b f(x)\, dx$$

4.11.20 Find a series of Riemann-integrable functions that converge exactly to the function of Example 4.11.10.

4.11.21 Prove that the function f in equation 4.11.44 is not L-integrable on $\mathbb{R}^n$ if $m \le n$.

4.12 Review exercises for Chapter 4

Exercise 4.1:

$U_N(\mathbf{1}_C) = L_N(\mathbf{1}_C)$.

4.1 Show that if $C \in \mathcal{D}(\mathbb{R}^m)$, then $\mathbf{1}_C$ is integrable.

4.2 An integrand should take a piece of the domain and return a number, in such a way that if we decompose a domain into little pieces, evaluate the integrand on the pieces, and add, the sums should have a limit as the decomposition becomes infinitely fine (and the limit should not depend on how the domain is decomposed). What will happen if we break up $[0,1]^2$ into rectangles defined by $a < x < b$ and $c < y < d$ and assign one of the numbers below to each rectangle?

a. $|ac - bd|$ b. $(ad - bc)^2$.

4.3 Let A be an $n \times n$ matrix of integers, viewed as a map $\mathbb{Z}^n \to \mathbb{Z}^n$. Which of the following are true?

1. $\ker A = 0 \implies A$ is onto.
2. A onto $\implies \ker A = 0$.
3. $\det A \neq 0 \implies \ker A = 0$.
4. $\det A \neq 0 \implies A$ is onto.

4.4 Which elementary matrices are permutation matrices? Describe the corresponding permutations.

4.5 Evaluate $\displaystyle \lim_{N\to\infty} \frac{1}{N^2} \sum_{k=1}^{N} \sum_{l=1}^{2N} e^{\frac{k+l}{N}}$.

4.6 What are the expectation, variance, and standard deviation, if they exist, of the random variable $f(x) = x$, for the following probability densities.

a. $\mu(x) = e^{-x}\mathbf{1}_{[0,\infty]}$ b. $\mu(x) = \dfrac{1}{(x+1)^2}\mathbf{1}_{[0,\infty)}$ c. $\mu(x) = \dfrac{2}{(x+1)^3}\mathbf{1}_{[0,\infty)}$

4.7 Let A and B be two disjoint bodies, with densities μ_1 and μ_2 and masses $M(A)$ and $M(B)$. Set $C = A \cup B$. Show that the center of gravity of C is

$$\overline{\mathbf{x}}(C) = \frac{M(A)\overline{\mathbf{x}}(A) + M(B)\overline{\mathbf{x}}(B)}{M(A) + M(B)}.$$

4.8 Choose r and R with $0 < r < R < \infty$.

a. Find the integral

$$\int_{A_{r,R}} \frac{e^{-(x^2+y^2+z^2)}}{\sqrt{x^2+y^2+z^2}} \quad \text{over the region } r^2 \le x^2+y^2+z^2 \le R^2.$$

b. Does this integral have a limit as $R \to \infty$? As $r \to 0$?

4.9 Let X be a subset of $\mathbb{R}^n$ such that for any $\epsilon > 0$, there exists a sequence $i \mapsto B_i$ of pavable sets satisfying $X \subset \bigcup_{i=1}^{\infty} B_i$ and $\sum_{i=1}^{\infty} \mathrm{vol}_n(B_i) < \epsilon$. Show that X has measure 0.

4.10 Give an explicit upper bound (in terms of N) for the number of cubes in $\mathcal{D}_N(\mathbb{R}^3)$ needed to cover the unit sphere $S^2 \subset \mathbb{R}^3$, such that the volume of the cubes tends to 0 as N tends to infinity.

4.11 Write each of the following double integrals as iterated integrals in two ways, and compute them:

a. The integral of $\sin(x+y)$ over the region $x^2 < y < 2$

b. The integral of $x^2 + y^2$ over the region where $1 \le |x| \le 2$ and $1 \le |y| \le 2$

4.12 Compute the integral of the function z over the region R described by the inequalities $x > 0,\ y > 0,\ z > 0,\ x + 2y + 3z < 1$.

4.13 a. If $f\begin{pmatrix} x \\ y \end{pmatrix} = a + bx + cy$, what are

$$\int_0^1 \int_0^2 f\begin{pmatrix} x \\ y \end{pmatrix} |dx\, dy| \quad \text{and} \quad \int_0^1 \int_0^2 \left(f\begin{pmatrix} x \\ y \end{pmatrix} \right)^2 |dx\, dy|?$$

b. Let f be as in part a. What is the minimum of $\int_0^1 \int_0^2 \left(f\begin{pmatrix} x \\ y \end{pmatrix} \right)^2 |dx\, dy|$ among all functions f such that

$$\int_0^1 \int_0^2 f\begin{pmatrix} x \\ y \end{pmatrix} |dx\, dy| = 1?$$

4.14 What is the z-coordinate of the center of gravity of the region

$$\frac{x^2}{(z^3-1)^2} + \frac{y^2}{(z^3+1)^2} \le 1, \quad 0 \le z \le 1?$$

4.15 Show that there exist c and u such that when f is any polynomial of degree $d \le 3$,

$$\int_{-1}^1 f(x) \frac{1}{\sqrt{1-x^2}}\, dx = c\Big(f(u) + f(-u) \Big).$$

4.16 Repeat Exercise 4.6.4, parts a–d, but this time for the weight e^{-x^2} and limits of integration $-\infty$ to ∞; i.e., find points x_i and w_i such that

$$\int_{-\infty}^{\infty} p(x) e^{-x^2}\, dx = \sum_{i=0}^{k} w_i p(x_i)$$

for all polynomials of degree $\le 2k - 1$.

e. For each of the four values of m in part d, approximate

$$\int_{-\infty}^{\infty} e^{-x^2} \cos x\, dx \quad \text{and} \quad \int_{-\infty}^{\infty} \frac{e^{-x^2}}{1+x^2}\, dx.$$

Compare the approximations with the exact values.

Exercise 4.18: Start with Corollary 4.8.17, and set

$$C = P \text{ and } D = AP^{-1}.$$

This proves the formula when C is invertible. Complete the proof by showing that if $n \mapsto C_n$ is a sequence of matrices converging to C, and $\operatorname{tr}(C_n D) = \operatorname{tr}(DC_n)$ for all n, then $\operatorname{tr}(CD) = \operatorname{tr}(DC)$.

4.17 Check part 3 of Theorem 4.8.15 when $A = \begin{bmatrix} 1 & 3 \\ 2 & 1 \end{bmatrix}$ and $B = \begin{bmatrix} a & b \\ c & d \end{bmatrix}$; that is, show that $[\mathbf{D} \det(A)]B = \det A \operatorname{tr}(A^{-1}B)$.

4.18 Show that if A and B are $n \times n$ matrices, then $\operatorname{tr}(AB) = \operatorname{tr}(BA)$.

4.19 What is the n-dimensional volume of the region

$$\{ \mathbf{x} \in \mathbb{R}^n \mid x_i \ge 0 \text{ for all } i = 1, \dots, n \text{ and } x_1 + 2x_2 + \dots + nx_n \le n \}?$$

4.20 a. Find an expression for the area of the parallelogram spanned by $\vec{\mathbf{v}}_1$ and $\vec{\mathbf{v}}_2$, in terms of $|\vec{\mathbf{v}}_1|, |\vec{\mathbf{v}}_2|$, and $|\vec{\mathbf{v}}_1 - \vec{\mathbf{v}}_2|$.

b. Prove Heron's formula: A triangle with sides of length a, b, c, has area

$$\sqrt{p(p-a)(p-b)(p-c)}\,, \quad \text{where} \quad p = \frac{a+b+c}{2}.$$

4.21 a. Sketch the curve in the plane given in polar coordinates by the equation $r = 1 + \sin\theta, \quad 0 \le \theta \le 2\pi$.

b. Find the area that the curve encloses.

4.22 A semicircle of radius R has density $\rho\begin{pmatrix} x \\ y \end{pmatrix} = m(x^2+y^2)$ proportional to the square of the distance to the center. What is its mass?

4.23 a. Let Q be the part of the unit ball $x^2+y^2+z^2 \leq 1$ where $x, y, z \geq 0$. Using spherical coordinates, set up $\int_Q (x+y+z)\,|d^3\mathbf{x}|$ as an iterated integral.

b. Compute the integral.

4.24 Let A be a square matrix. Show that A and $A^\top$ have the same eigenvalues, with the same multiplicities.

4.25 Let $A \subset \mathbb{R}^3$ be the region defined by the inequalities $x^2+y^2 \leq z \leq 1$. What is the center of gravity of A?

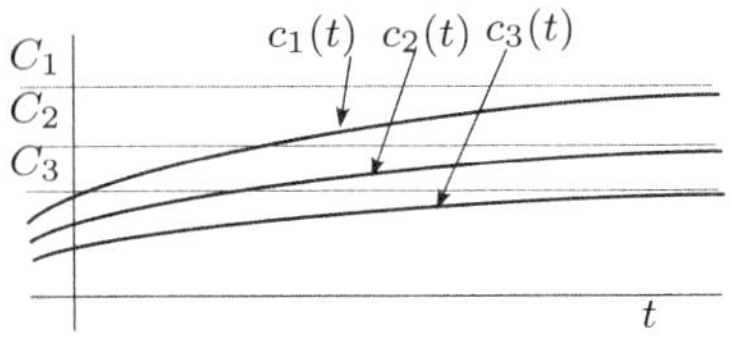

FIGURE FOR EXERCISE 4.26 The series $n \mapsto c_n(t)$ for part d. The functions $c_n(t)$ are positive monotone increasing as functions of t, and decreasing as functions of n.

4.26 In this exercise we will show that $\int_0^\infty \frac{\sin x}{x}\,dx = \frac{\pi}{2}$. This function is not Lebesgue integrable, and the integral should be understood as

$$\int_0^\infty \frac{\sin x}{x}\,dx = \lim_{a\to\infty}\int_0^a \frac{\sin x}{x}\,dx.$$

a. Show that for all $0 < a < b < \infty$,

$$\int_a^b \left(\int_0^\infty e^{-px}\sin x\,dx\right)dp = \int_0^\infty \left(\int_a^b e^{-px}\sin x\,dp\right)dx.$$

b. Use part a to show

$$\arctan b - \arctan a = \int_0^\infty \frac{(e^{-ax}-e^{-bx})\sin x}{x}\,dx.$$

c. Why does Theorem 4.11.4 not imply that

$$\lim_{a\to 0}\lim_{b\to\infty}\int_0^\infty \frac{(e^{-ax}-e^{-bx})\sin x}{x}\,dx = \int_0^\infty \frac{\sin x}{x}\,dx? \qquad (1)$$

Exercise 4.26, part d: Remember that the next omitted term is a bound for the error for each partial sum.

d. Prove that equation 1 in part c is true anyway. The following lemma is the key: if $n \mapsto c_n(t)$ is a sequence of positive monotone increasing functions of t, with $\lim_{t\to\infty} c_n(t) = C_n$, and decreasing as a function of n for each fixed t, tending to 0 (see the figure in the margin), then

$$\lim_{t\to\infty}\sum_{n=1}^\infty (-1)^n c_n(t) = \sum_{n=1}^\infty (-1)^n C_n.$$

e. Write

$$\int_0^\infty \frac{(e^{-ax}-e^{-bx})\sin x}{x}\,dx = \sum_{n=0}^\infty \int_{k\pi}^{(k+1)\pi} (-1)^k\frac{(e^{-ax}-e^{-bx})|\sin x|}{x}\,dx,$$

and use part d to prove the equation $\int_0^\infty \frac{\sin x}{x}\,dx = \frac{\pi}{2}$.

$$f(x) = \sum_{k=1}^\infty \frac{1}{2^k}\frac{1}{\sqrt{|x-a_k|}}.$$

Function f for Exercise 4.27

Exercise 4.27, part c: This depends on the order chosen.

4.27 Let $a_1, a_2, \dots$ be a list of the rationals in $[0,1]$.

a. Show that the function f given in the margin is L-integrable on $[0,1]$.

b. Show that the series converges for all x except x on a set of measure 0.

*c. Find an x for which the series converges.

4.28 a. Show that $\frac{1}{x^2}\mathbf{1}_{[1,\infty]}(x)$ is a probability density.

b. Show that for the probability density found in part a, the random variable $f(x) = x$ does not have an expectation (i.e., the expectation is infinite).

c. Show that $\frac{2}{x^3}\mathbf{1}_{[1,\infty]}(x)$ is a probability density.

d. Show that the random variable $f(x) = x$ has an expectation with respect to the probability density of part c; compute it. Show that it does not have a variance (i.e., the variance is infinite).

4.29 Let $T : \mathbb{R}^2 \to \mathbb{R}^2$ be the linear transformation given by $T(u) = au + b\bar{u}$, where we identify $\mathbb{R}^2$ with $\mathbb{C}$ in the standard way. Show that

$$\det T = |a|^2 - |b|^2 \quad \text{and} \quad \|T\| = |a| + |b|.$$

Exercise 4.30: Think of Riemann sums.

4.30 a. Find the unique integer p such that the limit

$$\lim_{n\to\infty} \frac{1}{n^p} \sum_{\substack{k,l,m \text{ integers} \\ 0\le k,\, 0\le l,\, 0\le m \\ k^2+l^2+m^2\le n^2}} \frac{klm}{k^2+l^2+m^2}$$

exists and is nonzero. For that value of p, compute the limit.

b. What happens to the limit if we replace p by p' with $p' < p$? with $p' > p$?

4.31 For $a > 1$, $b > 1$, $c > 1$, what is the volume of the part of the region

$$A = \left\{ \begin{pmatrix} x \\ y \\ z \end{pmatrix} \in \mathbb{R}^3 \mid 1 \le xyz \le a,\ 1 \le x \le b,\ xz \le y \le cxz \right\}$$

where $x > 0$, $y > 0$, $z > 0$?

a. Find a change of variables, expressing appropriate $\begin{pmatrix} u \\ v \\ w \end{pmatrix}$ in terms of $\begin{pmatrix} x \\ y \\ z \end{pmatrix}$, so that A becomes a parallelopiped in the new variables.

b. Invert the change of variables, and find its derivative and the determinant of the derivative.

c. Compute the integral.

$$\int_0^1 \left(\int_y^{y^{1/3}} \sin(x^2)\, dx \right) dy$$

Integral for Exercise 4.32.

4.32 a. Transform the iterated integral in the margin into an integral over a subset of $\mathbb{R}^2$. Sketch this subset.

b. Compute the integral.

***4.33** Find an open subset $U \subset [0,1]$ and a continuous function $f: U \to [0,1]$ whose graph $\Gamma_f \subset [0,1] \times [0,1]$ does not have volume. Show that it has 2-dimensional measure 0. *Hint*: Look at Remark 5.2.5.

4.34 Let $i \mapsto a_i$ be a list of the rationals in $[0,1]$, and set

$$f_k(x) = \begin{cases} 1 & \text{if } x \in [0,1] \text{ and } |x - a_i| < \frac{1}{10^i} \text{ for } i \le k \\ 0 & \text{otherwise.} \end{cases}$$

Show that the f_k are Riemann integrable but that they cannot be modified on a set of measure 0 to converge to a Riemann-integrable function.

4.35 Let $\mathbf{x} \in \mathbb{R}^n$. For what values of $p \in \mathbb{R}$ does $\int_{B_1(\mathbf{0})} |\mathbf{x}|^p\, |d^n\mathbf{x}|$ exist as a Lebesgue integral? (The answer depends on n.)

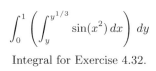

5

Volumes of manifolds

5.0 Introduction

In Chapter 4 we saw how to integrate over subsets of $\mathbb{R}^n$, first using dyadic pavings, then more general pavings. But these subsets were flat n-dimensional subsets of $\mathbb{R}^n$. What if we want to integrate over a (curvy) surface in $\mathbb{R}^3$? Many situations of obvious interest, like the area of a surface, or the total energy stored in the surface tension of a soap bubble, or the amount of fluid flowing through a pipe, are clearly some sort of surface integral. In a physics course you may have learned that the *electric flux* through a closed surface is proportional to the electric charge inside that surface.

When we say that in Chapter 4 we had "flat domains" we mean we had n-dimensional subsets of $\mathbb{R}^n$. A disc in the plane is flat, even though its boundary is a circle: we cannot bend a disc and have it remain a subset of the plane. A subset of $\mathbb{R}$ is necessarily straight; if we want a wiggly line we must allow for at least two dimensions.

In this chapter we will show how to compute the area of a surface in $\mathbb{R}^3$, or more generally, the k-dimensional volume of a k-manifold in $\mathbb{R}^n$, where $k < n$. We can't use the approach given in Section 4.1, where we saw that when integrating over a subset $A \subset \mathbb{R}^n$,

$$\int_A g(\mathbf{x})\,|d^n\mathbf{x}| = \int_{\mathbb{R}^n} g(\mathbf{x})\mathbf{1}_A(\mathbf{x})\,|d^n\mathbf{x}|. \tag{4.1.8}$$

If we try to use this equation to integrate a function in $\mathbb{R}^3$ over a surface, the integral will certainly vanish, since the surface has three-dimensional volume 0. For any k-manifold M embedded in $\mathbb{R}^n$, with $k < n$, the integral would certainly vanish, since M has n-dimensional volume 0. Instead, we need to rethink the whole process of integration.

At heart, integration is always the same:

> *Break up the domain into little pieces, assign a little number to each little piece, and finally add together all the numbers.* Then break the domain into littler pieces and repeat, taking the limit as the decomposition becomes infinitely fine. The *integrand* is the thing that assigns the number to the little piece of the domain.

There is quite a lot of leeway when choosing what kind of "little pieces" to use; choosing a decomposition of a surface into little pieces is analogous to choosing a paving, and as we saw in Section 4.7, there are many possible choices besides the dyadic paving.

In Chapter 6 we will study a different kind of integrand, which assigns numbers to little pieces of *oriented* manifolds.

The words "little piece" in this heuristic description need to be pinned down before we can do anything useful. We will break the domain into k-dimensional parallelograms, and the "little number" we attach to each little parallelogram will be its k-dimensional volume. In Section 5.1 we will see how to compute this volume.

We can only integrate over parametrized domains, and if we use the definition of parametrizations given in Chapter 3, we will not be able to parametrize even such simple objects as the circle. Section 5.2 gives a looser definition of parametrization, sufficient for integration. In Section 5.3 we compute volumes of k-manifolds; in Section 5.4 we prove three theorems relating curvature to integration, including two results that explain the meaning of Gaussian and mean curvature. Fractals and fractional dimension are discussed in Section 5.5.

5.1 Parallelograms and their volumes

We saw in Section 4.9 that the volume of a k-parallelogram in $\mathbb{R}^k$ is

$$\text{vol}_k\, P(\vec{\mathbf{v}}_1, \dots, \vec{\mathbf{v}}_k) = |\det[\vec{\mathbf{v}}_1, \dots, \vec{\mathbf{v}}_k]|. \qquad 5.1.1$$

What about a k-parallelogram in $\mathbb{R}^n$? Clearly if we draw a parallelogram on a rigid piece of cardboard, cut it out, and move it about in space, its area will not change. This area should depend only on the lengths of the vectors spanning the parallelogram and the angle between them, not on where they are placed in $\mathbb{R}^3$. It isn't obvious how to compute this volume: equation 5.1.1 can't be applied, since the determinant only exists for square matrices. A formula involving the cross product (see Proposition 1.4.20) exists for a 2-parallelogram in $\mathbb{R}^3$. How will we compute the area of a 2-parallelogram in $\mathbb{R}^4$, never mind a 3-parallelogram in $\mathbb{R}^5$?

Exercise 5.1.3 asks you to show that if $\vec{\mathbf{v}}_1, \dots, \vec{\mathbf{v}}_k$ are linearly dependent, then

$$\text{vol}_k(P(\vec{\mathbf{v}}_1, \dots, \vec{\mathbf{v}}_k)) = 0.$$

In particular, this shows that if $k > n$, then

$$\text{vol}_k(P(\vec{\mathbf{v}}_1, \dots, \vec{\mathbf{v}}_k)) = 0.$$

The following proposition is the key. It concerns k-parallelograms in $\mathbb{R}^k$, but we will be able to apply it to k-parallelograms in $\mathbb{R}^n$.

Proposition 5.1.1 (Volume of a k-parallelogram in $\mathbb{R}^k$). *Let $\vec{\mathbf{v}}_1, \dots, \vec{\mathbf{v}}_k$ be k vectors in $\mathbb{R}^k$, so that $T = [\vec{\mathbf{v}}_1, \dots, \vec{\mathbf{v}}_k]$ is a square $k \times k$ matrix. Then*

$$\text{vol}_k\, P(\vec{\mathbf{v}}_1, \dots, \vec{\mathbf{v}}_k) = \sqrt{\det(T^\top T)}. \qquad 5.1.2$$

Proof of Proposition 5.1.1: Recall that if A and B are $n \times n$ matrices, then

$$\det A \det B = \det(AB)$$
$$\det A = \det A^\top$$

(Theorems 4.8.4 and 4.8.8).

Proof. $\sqrt{\det(T^\top T)} = \sqrt{(\det T^\top)(\det T)} = \sqrt{(\det T)^2} = |\det T|$ $\square$

Example 5.1.2 (Volume of two-dimensional and three-dimensional parallelograms). When $k = 2$, we have

$$\det(T^\top T) = \det\left(\begin{bmatrix} \vec{\mathbf{v}}_1^\top \\ \vec{\mathbf{v}}_2^\top \end{bmatrix} \begin{bmatrix} \vec{\mathbf{v}}_1 & \vec{\mathbf{v}}_2 \end{bmatrix}\right) = \det \begin{bmatrix} |\vec{\mathbf{v}}_1|^2 & \vec{\mathbf{v}}_1 \cdot \vec{\mathbf{v}}_2 \\ \vec{\mathbf{v}}_2 \cdot \vec{\mathbf{v}}_1 & |\vec{\mathbf{v}}_2|^2 \end{bmatrix}$$
$$= |\vec{\mathbf{v}}_1|^2 |\vec{\mathbf{v}}_2|^2 - (\vec{\mathbf{v}}_1 \cdot \vec{\mathbf{v}}_2)^2. \qquad 5.1.3$$

Recall (Definition 1.4.6) that

$$\vec{\mathbf{x}} \cdot \vec{\mathbf{y}} = |\vec{\mathbf{x}}|\,|\vec{\mathbf{y}}| \cos\alpha,$$

where α is the angle between the vectors $\vec{\mathbf{x}}$ and $\vec{\mathbf{y}}$.

If we write $\vec{\mathbf{v}}_1 \cdot \vec{\mathbf{v}}_2 = |\vec{\mathbf{v}}_1||\vec{\mathbf{v}}_2| \cos\theta$ (where θ is the angle between $\vec{\mathbf{v}}_1$ and $\vec{\mathbf{v}}_2$), this becomes

$$\det(T^\top T) = |\vec{\mathbf{v}}_1|^2 |\vec{\mathbf{v}}_2|^2 (1 - \cos^2\theta) = |\vec{\mathbf{v}}_1|^2 |\vec{\mathbf{v}}_2|^2 \sin^2\theta. \qquad 5.1.4$$

Thus Proposition 5.1.1 asserts that the area of the 2-parallelogram spanned by $\vec{\mathbf{v}}_1$, $\vec{\mathbf{v}}_2$ is

$$\sqrt{\det(T^\top T)} = |\vec{\mathbf{v}}_1||\vec{\mathbf{v}}_2||\sin\theta|, \qquad 5.1.5$$

which agrees with the formula of height times base given in high school: if $\vec{\mathbf{v}}_2$ is the base, the height is $\vec{\mathbf{v}}_1 \sin\theta$.

The same computation in the case $k = 3$ leads to a much less familiar formula. Suppose $T = [\vec{\mathbf{v}}_1, \vec{\mathbf{v}}_2, \vec{\mathbf{v}}_3]$, and that the angle between $\vec{\mathbf{v}}_2$ and $\vec{\mathbf{v}}_3$ is θ_1, that between $\vec{\mathbf{v}}_1$ and $\vec{\mathbf{v}}_3$ is θ_2, and that between $\vec{\mathbf{v}}_1$ and $\vec{\mathbf{v}}_2$ is θ_3. Then

$$T^\top T = \begin{bmatrix} |\vec{\mathbf{v}}_1|^2 & \vec{\mathbf{v}}_1\cdot\vec{\mathbf{v}}_2 & \vec{\mathbf{v}}_1\cdot\vec{\mathbf{v}}_3 \\ \vec{\mathbf{v}}_2\cdot\vec{\mathbf{v}}_1 & |\vec{\mathbf{v}}_2|^2 & \vec{\mathbf{v}}_2\cdot\vec{\mathbf{v}}_3 \\ \vec{\mathbf{v}}_3\cdot\vec{\mathbf{v}}_1 & \vec{\mathbf{v}}_3\cdot\vec{\mathbf{v}}_2 & |\vec{\mathbf{v}}_3|^2 \end{bmatrix} \tag{5.1.6}$$

and $\det T^\top T$ is given by

$$\begin{aligned} \det T^\top T &= |\vec{\mathbf{v}}_1|^2|\vec{\mathbf{v}}_2|^2|\vec{\mathbf{v}}_3|^2 + 2(\vec{\mathbf{v}}_1\cdot\vec{\mathbf{v}}_2)(\vec{\mathbf{v}}_2\cdot\vec{\mathbf{v}}_3)(\vec{\mathbf{v}}_1\cdot\vec{\mathbf{v}}_3) \\ &\quad - |\vec{\mathbf{v}}_1|^2(\vec{\mathbf{v}}_2\cdot\vec{\mathbf{v}}_3)^2 - |\vec{\mathbf{v}}_2|^2(\vec{\mathbf{v}}_1\cdot\vec{\mathbf{v}}_3)^2 - |\vec{\mathbf{v}}_3|^2(\vec{\mathbf{v}}_1\cdot\vec{\mathbf{v}}_2)^2 \\ &= |\vec{\mathbf{v}}_1|^2|\vec{\mathbf{v}}_2|^2|\vec{\mathbf{v}}_3|^2\big(1 + 2\cos\theta_1\cos\theta_2\cos\theta_3 - (\cos^2\theta_1 + \cos^2\theta_2 + \cos^2\theta_3)\big). \end{aligned} \tag{5.1.7}$$

It follows from Proposition 5.1.1 and equation 5.1.7 that we can express the volume of a n-parallelogram in $\mathbb{R}^n$ in terms of the lengths of its vectors and the angles between them: purely geometric information.

Equation 5.1.8: To use equation 5.1.1 to compute the volume a parallelepiped spanned by three unit vectors, each making an angle of $\pi/4$ with the others, we would first have to find appropriate vectors, which would not be easy.

For instance, the volume of a parallelepiped P spanned by three unit vectors, each making an angle of $\pi/4$ with the others, is

$$\operatorname{vol}_3 P = \sqrt{1 + 2\cos^3\frac{\pi}{4} - 3\cos^2\frac{\pi}{4}} = \sqrt{\frac{\sqrt{2}-1}{2}}. \quad \triangle \tag{5.1.8}$$

Volume of a k-parallelogram in $\mathbb{R}^n$

The formula $\operatorname{vol}_k P(\vec{\mathbf{v}}_1,\dots,\vec{\mathbf{v}}_k) = \sqrt{\det(T^\top T)}$ was useful in equation 5.1.8. But what really makes Proposition 5.1.1 interesting is that the same formula can be used to compute the area of a k-parallelogram in $\mathbb{R}^n$. Note that if T is an $n \times k$ matrix with columns $\vec{\mathbf{v}}_1,\dots,\vec{\mathbf{v}}_k$, then the product $T^\top T$ is a symmetric $k \times k$ matrix whose entries are dot products of the vectors $\vec{\mathbf{v}}_i$:

$$\overbrace{\begin{bmatrix} \vdots & \vdots & \dots & \vdots \\ \vec{\mathbf{v}}_1 & \vec{\mathbf{v}}_2 & \dots & \vec{\mathbf{v}}_k \\ \vdots & \vdots & \dots & \vdots \end{bmatrix}}^{T}$$

$$\underbrace{\begin{bmatrix} \dots & \vec{\mathbf{v}}_1^\top & \dots \\ \dots & \vec{\mathbf{v}}_2^\top & \dots \\ \dots & \dots & \dots \\ \dots & \vec{\mathbf{v}}_k^\top & \dots \end{bmatrix}}_{T^\top} \underbrace{\begin{bmatrix} |\vec{\mathbf{v}}_1|^2 & \vec{\mathbf{v}}_1\cdot\vec{\mathbf{v}}_2 & \dots & \vec{\mathbf{v}}_1\cdot\vec{\mathbf{v}}_k \\ \vec{\mathbf{v}}_2\cdot\vec{\mathbf{v}}_1 & |\vec{\mathbf{v}}_2|^2 & \dots & \vec{\mathbf{v}}_2\cdot\vec{\mathbf{v}}_k \\ \vdots & \vdots & \ddots & \dots \\ \vec{\mathbf{v}}_k\cdot\vec{\mathbf{v}}_1 & \vec{\mathbf{v}}_k\cdot\vec{\mathbf{v}}_2 & \dots & |\vec{\mathbf{v}}_k|^2 \end{bmatrix}}_{T^\top T}. \tag{5.1.9}$$

There is another way to see that rotating the vectors $\vec{\mathbf{v}}_i$ by an orthogonal matrix A does not change $T^\top T$: the matrix T becomes the matrix AT, and

$$(AT)^\top AT = T^\top A^\top AT = T^\top T.$$

Exercise 5.1.6 asks you to use the singular value decomposition to give a different justification for using

$$\sqrt{\det T^\top T}$$

to define k-dimensional volume in $\mathbb{R}^n$.

Although T itself is not square, the matrix $T^\top T$ is square, and its entries can be computed from the lengths of the k vectors and the angles between them. No further information is needed. In particular, if the vectors are all rotated by the same orthogonal matrix, $T^\top T$ will be unchanged. Moreover, we do not need to know where the vectors are: at what point the parallelogram is anchored. Thus we can use $\sqrt{\det(T^\top T)}$ to define k-dimensional volume in $\mathbb{R}^n$.

Definition 5.1.3 (Volume of a k-parallelogram in $\mathbb{R}^n$). Let $T = [\vec{\mathbf{v}}_1, \dots, \vec{\mathbf{v}}_k]$ be an $n \times k$ real matrix. Then the *k-dimensional volume* of $P(\vec{\mathbf{v}}_1, \dots, \vec{\mathbf{v}}_k)$ is

$$\text{vol}_k\, P(\vec{\mathbf{v}}_1, \dots, \vec{\mathbf{v}}_k) \stackrel{\text{def}}{=} \sqrt{\det(T^\top T)}. \qquad 5.1.10$$

Exercise 5.1.5 asks you to show that $\det(T^\top T) \geq 0$, so that Definition 5.1.3 makes sense.

Example 5.1.4 (Volume of a 3-parallelogram in $\mathbb{R}^4$). Let P be the 3-parallelogram P in $\mathbb{R}^4$ spanned by $\vec{\mathbf{v}}_1 = \begin{bmatrix} 1 \\ 0 \\ 0 \\ 1 \end{bmatrix}$, $\vec{\mathbf{v}}_2 = \begin{bmatrix} 0 \\ 1 \\ 0 \\ 1 \end{bmatrix}$, $\vec{\mathbf{v}}_3 = \begin{bmatrix} 0 \\ 0 \\ 1 \\ 1 \end{bmatrix}$.
What is its 3-dimensional volume? Set $T = [\vec{\mathbf{v}}_1, \vec{\mathbf{v}}_2, \vec{\mathbf{v}}_3]$; then

$$T^\top T = \begin{bmatrix} 2 & 1 & 1 \\ 1 & 2 & 1 \\ 1 & 1 & 2 \end{bmatrix} \quad \text{and} \quad \det(T^\top T) = 4, \quad \text{so } \text{vol}_3\, P = 2. \quad \triangle$$

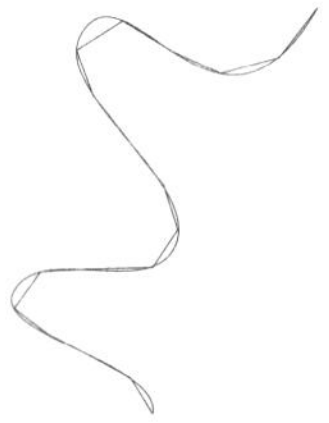

FIGURE 5.1.1.
A curve approximated by an inscribed polygon, shown already in Section 3.9.

Archimedes (287–212 BC) used this process to prove that

$$223/71 < \pi < 22/7.$$

In his famous paper *The Measurement of the Circle*, he approximated the circle by an inscribed and a circumscribed 96-sided regular polygon. That was the beginning of integral calculus.

The anchored k-parallelograms are the "little pieces" we will use when breaking up the domain.

The "little number" assigned to each piece will be its volume.

Volume of anchored k-parallelograms

To break up a domain into little k-parallelograms we will need parallelograms "anchored" at different points in the domain. We denote by $P_{\mathbf{x}}(\vec{\mathbf{v}}_1, \dots, \vec{\mathbf{v}}_k)$ a k-parallelogram in $\mathbb{R}^n$ anchored at $\mathbf{x} \in \mathbb{R}^n$. Then

$$\text{vol}_k\, P(\vec{\mathbf{v}}_1, \dots, \vec{\mathbf{v}}_k) = \text{vol}_k\, P_{\mathbf{x}}(\vec{\mathbf{v}}_1, \dots, \vec{\mathbf{v}}_k). \qquad 5.1.11$$

The need for parametrizations

Now we must address a more complex issue. The first step in integration is to "break up the domain into little pieces". In Chapter 4 we had flat domains. Now we must break up a curvy domain into flat k-parallelograms.

For a curve, this is not hard. If $C \subset \mathbb{R}^n$ is a smooth curve, the integral $\int_C |d^1\mathbf{x}|$ is the number obtained by the following process: approximate C by little line segments as in Figure 5.1.1, apply $|d^1\mathbf{x}|$ to each to get its length, and add. Then let the approximation become infinitely fine; the limit is by definition the length of C.

It is much harder to define surface area. The obvious idea of taking the limit of the area of inscribed triangles as the triangles become smaller and smaller only works if we are careful to prevent the triangles from becoming skinny as they get small, and then it isn't obvious that such inscribed polyhedra exist at all (see Exercise 5.3.14).[1] The difficulties are not insurmountable, but they are daunting.

Instead we will base our definition of surface area (and, more generally, of k-dimensional volume of k-manifolds) on a parametrization: when

[1] We speak of triangles rather than parallelograms for the same reason that you would want a three-legged stool rather than a chair if your floor were uneven. You can make all three vertices of a triangle touch a curved surface; you can't do this for the four vertices of a parallelogram.

computing the k-dimensional volume of a manifold, the first step will be to parametrize the manifold. This raises several issues. First, it forces us to relax our definition of a parametrization. Second, it requires knowing an appropriate parametrization, which can range from simple to tricky to impossible. Third, since there is more than one way to parametrize a manifold, we must make sure that the integral we compute does not depend on our choice of parametrization. We discuss the first two issues in the next section, the third in Section 5.3.

Exercises for Section 5.1

$$\begin{bmatrix}1\\0\\1\\1\end{bmatrix}, \begin{bmatrix}0\\2\\1\\1\end{bmatrix}, \begin{bmatrix}1\\1\\0\\2\end{bmatrix}$$

Vectors for Exercise 5.1.1

Hint for Exercise 5.1.3: Show that $\operatorname{rank}(T^\top T) \leq \operatorname{rank} T < k$.

Exercise 5.1.6: Satisfying

$$V(\sigma_1\vec{\mathbf{e}}_1, \dots, \sigma_k\vec{\mathbf{e}}_k) = |\sigma_1, \dots, \sigma_k|$$

means, for example, that the volume of a rectangle in $\mathbb{R}^3$ that is in the (x, y)-plane should be the product of the lengths of its sides.

5.1.1 What is vol_3 of the 3-parallelogram in $\mathbb{R}^4$ spanned by the vectors in the margin?

5.1.2 What is the volume of a parallelepiped with three sides emanating from the same vertex having lengths 1, 2, and 3, and with angles between them $\pi/3, \pi/4$, and $\pi/6$?

5.1.3 Show that if $\vec{\mathbf{v}}_1, \dots, \vec{\mathbf{v}}_k$ are linearly dependent, $\operatorname{vol}_k(P(\vec{\mathbf{v}}_1, \dots, \vec{\mathbf{v}}_k)) = 0$.

5.1.4 Show that for $\vec{\mathbf{v}}_1, \vec{\mathbf{v}}_2 \in \mathbb{R}^3$, we have $|\vec{\mathbf{v}}_1 \times \vec{\mathbf{v}}_2| = \sqrt{\det\Big([\vec{\mathbf{v}}_1, \vec{\mathbf{v}}_2]^\top [\vec{\mathbf{v}}_1, \vec{\mathbf{v}}_2]\Big)}$.

5.1.5 Let T be an $n \times k$ matrix. Show that $\det(T^\top T) \geq 0$, so that Definition 5.1.3 makes sense.

5.1.6 Use the singular value decomposition (Theorem 3.8.1) to show that there is a unique function V of k vectors in $\mathbb{R}^n$ that it is invariant under rotation and satisfies $V(\sigma_1\vec{\mathbf{e}}_1, \dots, \sigma_k\vec{\mathbf{e}}_k) = |\sigma_1, \dots, \sigma_k|$ for any numbers $\sigma_1, \dots, \sigma_k$. Show that $V = \operatorname{vol}_k$ as defined in Defintion 5.1.3.

5.2 Parametrizations

In Chapter 3 (Definition 3.1.18) we said that a parametrization of a manifold M is a C^1 mapping γ from an open subset $U \subset \mathbb{R}^k$ to M that is one to one and onto and whose derivative is one to one.

The problem with this definition is that most manifolds do not admit a parametrization. Even the circle does not; neither does the sphere, nor the torus. But our entire theory of integration over manifolds depends on parametrizations, and we cannot simply give up on most examples.

Let us examine what goes wrong for the circle. The most obvious parametrization of the circle is $\gamma : t \mapsto \begin{pmatrix}\cos t\\ \sin t\end{pmatrix}$. The problem is choosing a domain. If we choose $(0, 2\pi)$, then γ is not onto. If we choose $[0, 2\pi]$, the domain is not open, and γ is not one to one. If we choose $[0, 2\pi)$, the domain is not open. In Section 3.1 (discussion following Definition 3.1.18) we saw that the same problem arises when we parametrize the sphere by longitude and latitude.

The key point for both these examples is that *the trouble occurs on sets of volume 0*, and therefore it should not matter when we integrate. Our new definition of a parametrization will be exactly the old one, except that we allow things to go wrong on sets of volume 0.

The formulas given in Section 4.10 for changes of variables are special cases of relaxed parametrizations of open subsets of $\mathbb{R}^n$, going from a set to a set of the same dimension; formula 4.10.11 parametrized chunks of (x, y, z)-space by chunks of (r, θ, φ)-space. Here our emphasis is on parametrizing k-dimensional submanifolds in $\mathbb{R}^n$, where $k < n$, for example, parametrizing the sphere in $\mathbb{R}^3$ by pieces of the (θ, φ)-plane.

Sets of k-dimensional volume 0 in $\mathbb{R}^n$

We need to know when a subset $X \subset \mathbb{R}^n$ is negligible as far as k-dimensional integrals are concerned. Intuitively it should be fairly clear what this means: points are negligible for one-dimensional integrals or higher, points and curves are negligible for two-dimensional integrals, and so on.

It is possible to define vol_k of an arbitrary subset $X \subset \mathbb{R}^n$, and we will touch on this in Section 5.5 on fractals. That definition is quite elaborate; it is far simpler to say when such a subset has k-dimensional volume 0.

The cubes in equation 5.2.1 have sidelength $1/2^N$. We are summing over cubes that intersect X. Of course we could replace cubes by balls.

Definition 5.2.1 (k-dimensional volume 0 of a subset of $\mathbb{R}^n$).

1. A bounded subset $X \subset \mathbb{R}^n$ has k-dimensional volume 0 if

$$\lim_{N\to\infty} \sum_{\substack{C\in\mathcal{D}_N(\mathbb{R}^n)\\ C\cap X\neq\emptyset}} \Big(\underbrace{\frac{1}{2^N}}_{\text{sidelength of } C} \Big)^k = 0. \qquad 5.2.1$$

2. An arbitrary subset $X \subset \mathbb{R}^n$ has k-dimensional volume 0 if for all R, the bounded set $X \cap B_R(\mathbf{0})$ has k-dimensional volume 0.

This condition is fairly complicated to verify, but there is a criterion which applies in almost all cases encountered in practice.

Proposition 5.2.2 is proved in Appendix A20.

Proposition 5.2.2 (k-dimensional volume 0 of a manifold). *If integers m, k, n satisfy $0 \le m < k \le n$, and $M \subset \mathbb{R}^n$ is a manifold of dimension m, any closed subset $X \subset M$ has k-dimensional volume 0.*

Definition 5.2.3: Whenever we want to determine the domain of an implicit function, we will need $\gamma : (U - X) \to M$ to have locally Lipschitz derivative.

Definition 5.2.3 ("Relaxed" parametrization of a manifold). Let $M \subset \mathbb{R}^n$ be a k-dimensional manifold and let $U \subset \mathbb{R}^k$ be a subset with boundary of k-dimensional volume 0. Let $X \subset U$ be such that $U - X$ is open. Then a continuous mapping $\gamma : U \to \mathbb{R}^n$ parametrizes M if

1. $\gamma(U) \supset M$;
2. $\gamma(U - X) \subset M$;
3. $\gamma : (U - X) \to M$ is one to one, of class C^1;
4. the derivative $[D\gamma(\mathbf{u})]$ is one to one for all $\mathbf{u}$ in $U - X$;
5. X has k-dimensional volume 0, as does $\gamma(X) \cap C$ for any compact subset $C \subset M$.

In Definition 5.2.3 we require that γ be well behaved on $U - X$, not on all of U. The set X includes all the trouble spots referred to above; excluding it does not affect integrals, since it has k-dimensional volume 0. Typically, U will be closed, and X will be its boundary. But there are many cases where it is desirable to allow X to be larger; see for instance Example 5.2.4. In other cases, X may be empty.

In Example 5.2.4, $\gamma(U)$ is the union of M, the origin, and the circle of radius 1 in the plane $z = 1$, centered on the z-axis. The set $\gamma(U - X)$ is M with the line segment $x = z$, $y = 0$ removed.

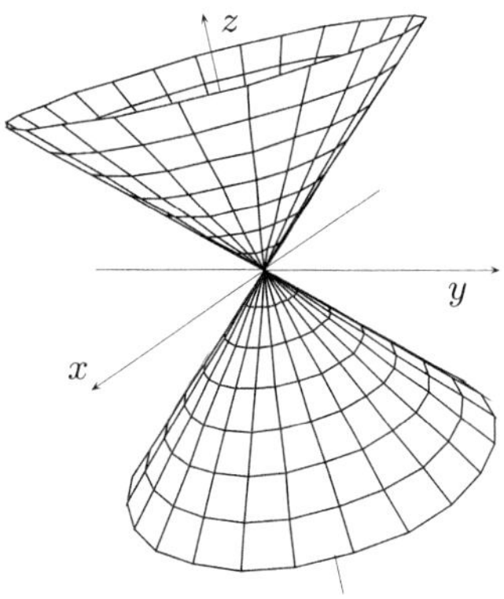

FIGURE 5.2.1.

The subset of $\mathbb{R}^3$ of equation $x^2 + y^2 = z^2$ is not a manifold at the vertex. The manifold M is the part where $0 < z < 1$.

Often condition 1 will be an equality. For example, if M is a sphere and U is a closed rectangle mapped to M by spherical coordinates, then $\gamma(U) = M$. In that case, we can take X to be the boundary of U, and $\gamma(X)$ consists of the poles and half a great circle (the international date line, for example), giving $\gamma(U - X) \subset M$ for condition 2.

From now on, the word "parametrization" will refer to Definition 5.2.3; we will call the parametrization of Definition 3.1.18 a "strict parametrization". Note that $\gamma : U - X \to M - \gamma(X)$ is a strict parametrization: we have removed the trouble spots (of k-dimensional volume 0).

Example 5.2.4 (Parametrization of a cone). The subset of $\mathbb{R}^3$ of equation $x^2 + y^2 - z^2 = 0$, shown in Figure 5.2.1, is not a manifold in the neighborhood of the vertex, which is at the origin. However, the subset

$$M = \left\{ \begin{pmatrix} x \\ y \\ z \end{pmatrix} \;\middle|\; x^2 + y^2 - z^2 = 0, \quad 0 < z < 1 \right\} \qquad 5.2.2$$

is a manifold. Let us check that the map $\gamma : [0, 1] \times [0, 2\pi] \to \mathbb{R}^3$ given by

$$\gamma : \begin{pmatrix} r \\ \theta \end{pmatrix} \mapsto \begin{pmatrix} r \cos \theta \\ r \sin \theta \\ r \end{pmatrix} \text{ parametrizes } M.$$

In the language of Definition 5.2.3 we have $U = [0, 1] \times [0, 2\pi]$. We will set $X = \partial U$, so that $U - X = (0, 1) \times (0, 2\pi)$, and X consists of the four line segments $\begin{pmatrix} 0 \\ \theta \end{pmatrix}$, $\begin{pmatrix} 1 \\ \theta \end{pmatrix}$, $\begin{pmatrix} r \\ 0 \end{pmatrix}$, and $\begin{pmatrix} r \\ 2\pi \end{pmatrix}$, for $0 \leq r \leq 1$, $0 \leq \theta \leq 2\pi$.

Condition 1 is satisfied, since $\gamma([0, 1] \times [0, 2\pi]) \supset M$. (It contains the vertex and the circle of radius 1 in the plane $z = 1$, in addition to M).

Condition 5 is satisfied: The line segments making up X have two-dimensional volume 0 by Proposition 5.2.2. Moreover, γ maps $\begin{pmatrix} 0 \\ \theta \end{pmatrix}$ to the origin; it maps $\begin{pmatrix} 1 \\ \theta \end{pmatrix}$ to the circle of radius 1 at $z = 1$; and it maps $\begin{pmatrix} r \\ 0 \end{pmatrix}$ and $\begin{pmatrix} r \\ 2\pi \end{pmatrix}$ to the line segment given by $x = z$, $y = 0$, for $0 \leq x, y \leq 1$. All have two-dimensional volume 0, by Proposition 5.2.2.

Condition 2 is satisfied, since $\gamma((0, 1) \times (0, 2\pi))$ consists of M minus the line segment $x = z$, $y = 0$. Exercise 5.2.1 asks you to check that conditions 3 and 4 are satisfied. $\triangle$

Remark 5.2.5. It might appear that the requirement in Definition 5.2.3 that the boundary of $U \subset \mathbb{R}^k$ have k-dimensional volume 0 is unnecessary; intuitively it is clear (and is justified by Proposition 5.2.2) that the boundary of a disc has two-dimensional volume 0, and the boundary of a cube has three-dimensional volume 0. But the requirement is necessary. We have run into this kind of thing before, in Example 4.4.3 and especially in Theorem 4.10.12. Let's look again at Example 4.4.3. Choose an integer $N > 2$, make a list $a_1, a_2, \ldots$ of the rationals in $[0, 1]$, and consider the set U consisting of tiny open intervals around the a_i:

$$U = \bigcup_{i=1}^{\infty} \left(a_i - \frac{1}{2^{N+i}}, \; a_i + \frac{1}{2^{N+i}} \right). \qquad 5.2.3$$

This is an open subset of $\mathbb{R}$, since it is a union of open sets, and it is dense in $[0,1]$, since it contains the rationals. You might expect that its boundary would consist of the boundaries of the segments, i.e., a countable union of points. This is not the case! By Definition 1.5.10, every point of $[0,1]-U$ is in ∂U, since every neighborhood of such a point contains points of U and points of $[0,1]-U$.

Since U is an open subset of $\mathbb{R}$, it is a manifold, and Theorem 5.2.6 says that it can be parametrized. However, since its boundary does not have 1-dimensional volume 0, Definition 5.2.3 says that the identity cannot be used to parametrize it. Exercise 5.2.5 asks you to find a parametrization.

Since some intervals will be contained in others,

$$\text{length of } U \;<\; \sum_{n=1}^{\infty} \frac{2}{2^{N+i}} = \frac{1}{2^{N-1}}. \qquad 5.2.4$$

If N is fairly large, say $N = 10$, then this length is very small. Therefore, if ∂U has length (1-dimensional volume), that "length" is the length of $[0,1]$ minus the length of U; it is not just nonzero, it is almost 1:

$$\text{length of } \partial U \;>\; 1 - \frac{1}{2^{N-1}}. \qquad 5.2.5$$

By definition, $[0,1]-U$ has length if the indicator function $\mathbf{1}_{[0,1]-U}$ is R-integrable. But the upper sums and lower sums do not converge to a common limit.

In fact, (as we saw in Example 4.4.3) ∂U does not have 1-dimensional volume, but it does have 1-dimensional *measure* and its measure satisfies inequality 5.2.5. △

Our entire theory of integrals over manifolds will be based on parametrizations. Fortunately, with our relaxed definition, all manifolds can be parametrized.

Theorem 5.2.6 (Existence of parametrizations). *All manifolds can be parametrized.*

Remark. Theorem 5.2.6 doesn't mean that parametrizations can be written explicitly: in the proof (the object of Exercise 5.2.8, not tremendously difficult), we will need to know that they can be locally parametrized, which requires the implicit function theorem: the parametrizing mappings only exist in the sense that the implicit function theorem guarantees their existence. △

A small catalog of parametrizations

Often Theorem 5.2.6 is cold comfort. Constructing a parametrization usually requires Newton's method and the implicit function theorem; such parametrizations can't be written explicitly. Fortunately, some can be written explicitly, and they turn up, disproportionately, in exam problems and in applications, especially applications where there is an underlying symmetry. Below we give examples. There are two classes: graphs and surfaces of revolution.

• **Graphs.** If $U \subset \mathbb{R}^k$ is open with boundary of k-dimensional volume 0, and $\mathbf{f} : U \to \mathbb{R}^{n-k}$ is a C^1 map, the graph of $\mathbf{f}$ is a manifold in $\mathbb{R}^n$, and

$$\mathbf{x} \mapsto \begin{pmatrix} \mathbf{x} \\ \mathbf{f}(\mathbf{x}) \end{pmatrix} \quad \text{is a parametrization.} \qquad 5.2.6$$

Example 5.2.7 (Parametrizing as a graph). The surface of equation $z = x^2 + y^2$ is parametrized by $\begin{pmatrix} x \\ y \end{pmatrix} \mapsto \begin{pmatrix} x \\ y \\ x^2+y^2 \end{pmatrix}$. Another example was given in equation 3.1.24. △

There are many cases where parametrizing as a graph still works, even though the conditions above are not satisfied: those where you can "solve" the defining equation for $n-k$ of the variables in terms of the other k.

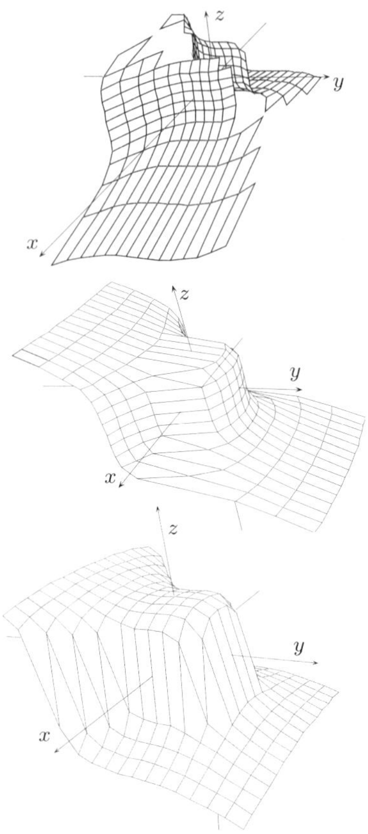

FIGURE 5.2.2.
The surface of equation

$$x^2 + y^3 + z^5 = 1$$

discussed in Example 5.2.8.
TOP: The surface seen as a graph of x as a function of y and z (i.e., parametrized by y and z). The graph consists of two pieces: the positive square root and the negative square root.
MIDDLE: Surface parametrized by x and z.
BOTTOM: Surface parametrized by x and y.

Note that the lines in the middle and bottom graphs are drawn differently: different parametrizations give different resolutions to different areas.

Example 5.2.8 (Parametrizing as a graph: a more complicated case). Consider the surface in $\mathbb{R}^3$ of equation $x^2 + y^3 + z^5 = 1$. In this case you can "solve" for x as a function of y and z:

$$x = \pm\sqrt{1 - y^3 - z^5}. \tag{5.2.7}$$

You could also solve for y or for z, as a function of the other variables, and the three approaches give different views of the surface, as shown in Figure 5.2.2. Of course, before you can call any of these a parametrization, you have to specify the domain of the function. When the equation is solved for x, the domain is the subset of the (y,z)-plane where $1 - y^3 - z^5 \geq 0$. When solving for y, remember that every real number has a unique real cube root, so the function $y = \left(1 - x^2 - z^5\right)^{1/3}$ is defined at every point, but it is not differentiable when $x^2 + z^5 = 1$, so this curve must be included in the set X of trouble points that can be ignored. △

• **Surfaces of revolution.** A *surface of revolution* can be obtained by rotating a curve, whether the curve is the graph of a function or is known as a parametrization.

1. *The case where the curve is the graph of a function.*

Suppose the curve C is the graph of a function $f(x) = y$. Let us suppose that f takes only positive values, and rotate C around the x-axis, to get the surface of revolution of equation

$$y^2 + z^2 = \big(f(x)\big)^2. \tag{5.2.8}$$

This surface can be parametrized by

$$\gamma : \begin{pmatrix} x \\ \theta \end{pmatrix} \mapsto \begin{pmatrix} x \\ f(x)\cos\theta \\ f(x)\sin\theta \end{pmatrix}, \tag{5.2.9}$$

where the angle θ measures the rotation. Why this parametrization? Since the curve is rotated around the x-axis, the x-coordinate remains unchanged. Certainly equation 5.2.8 is satisfied: $(f(x)\cos\theta)^2 + (f(x)\sin\theta)^2 = \big(f(x)\big)^2$.

Again, to be precise one must specify the domain of γ. Suppose that $f : (a,b) \to \mathbb{R}$ is defined and continuously differentiable on (a,b). Then we can choose the domain of γ to be $(a,b) \times [0, 2\pi]$, and γ is one to one, with derivative also one to one on $(a,b) \times (0, 2\pi)$.

2. *The case where the curve to be rotated is a parametrized curve.*

If C is a curve in the (x,y)-plane, parametrized by $t \mapsto \begin{pmatrix} u(t) \\ v(t) \end{pmatrix}$, the surface obtained by rotating C around the x-axis can be parametrized by

$$\begin{pmatrix} t \\ \theta \end{pmatrix} \mapsto \begin{pmatrix} u(t) \\ v(t)\cos\theta \\ v(t)\sin\theta \end{pmatrix}. \qquad 5.2.10$$

Spherical coordinates on the sphere of radius R are a special case of this construction. If C is the semicircle of radius R in the (x,z)-plane, centered at the origin, and parametrized by

$$\varphi \mapsto \begin{pmatrix} x = R\cos\varphi \\ z = R\sin\varphi \end{pmatrix}, \quad -\pi/2 \le \varphi \le \pi/2, \qquad 5.2.11$$

then the surface obtained by rotating this semicircle around the z-axis is the sphere of radius R centered at the origin in $\mathbb{R}^3$, parametrized by

$$\begin{pmatrix} \varphi \\ \theta \end{pmatrix} \mapsto \begin{pmatrix} R\cos\varphi\cos\theta \\ R\cos\varphi\sin\theta \\ R\sin\varphi \end{pmatrix}, \qquad 5.2.12$$

the parametrization of the sphere by latitude φ and longitude θ. (This does not look like formula 5.2.10 because now we are rotating around the z-axis, not the x-axis, so it is the z-coordinate that remains unchanged.

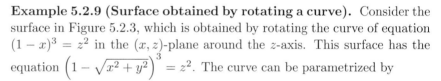

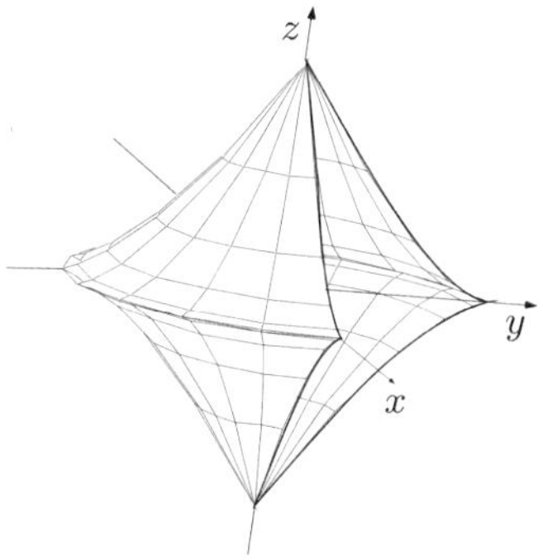

FIGURE 5.2.3.

The surface discussed in Example 5.2.9 is obtained by rotating the curve of equation

$$(1-x)^3 = z^2$$

around the z-axis. The surface drawn corresponds to the region where $|z| \le 1$, rotated only three quarters of a full turn.

If you take any equation representing a curve in the (x,z)-plane and replace x by $\sqrt{x^2+y^2}$, as we do in Example 5.2.9, you get the equation of the surface obtained by rotating the original curve around the z-axis. The surface is symmetric about the z-axis.

Example 5.2.9 (Surface obtained by rotating a curve). Consider the surface in Figure 5.2.3, which is obtained by rotating the curve of equation $(1-x)^3 = z^2$ in the (x,z)-plane around the z-axis. This surface has the equation $\left(1-\sqrt{x^2+y^2}\right)^3 = z^2$. The curve can be parametrized by

$$t \mapsto \begin{pmatrix} x = 1-t^2 \\ z = t^3 \end{pmatrix}, \qquad 5.2.13$$

(as you can check by substituting $1-t^2$ for x and t^3 for z in $(1-x)^3 = z^2$), so the surface can be parametrized by

$$\begin{pmatrix} t \\ \theta \end{pmatrix} \mapsto \begin{pmatrix} (1-t^2)\cos\theta \\ (1-t^2)\sin\theta \\ t^3 \end{pmatrix}. \qquad 5.2.14$$

Figure 5.2.3 shows the image of this parametrization for $0 \le \theta \le 3\pi/2$ and $|t| \le 1$. It can be guessed from the picture, and proved from the formula, that the points in $[-1,1] \times [0, 3\pi/2]$ where $t = \pm 1$ are trouble points (they correspond to the top and bottom "cone points"). The subset $\{0\} \times [0, 3\pi/2]$ also gives trouble; it corresponds to a "curve of cusps". △

Change of parametrization

In Section 5.3 we show that the length of a curve, the area of a surface, and the volume of a manifold are independent of the parametrization used in computing the length, area, or volume. In all three cases, the tool we use is the change of variables formula for Lebesgue integrals, Theorem 4.11.20: we set up a change of variables mapping and apply the change of variables formula to it. Here we show how to construct a change of variables map that satisfies the hypotheses of Theorem 4.11.20.

Example 5.2.10: S^2 is standard notation for the unit sphere in $\mathbb{R}^3$.

Suppose we have a k-dimensional manifold M and two parametrizations

$$\gamma_1 : \overline{U}_1 \to M \quad \text{and} \quad \gamma_2 : \overline{U}_2 \to M, \qquad 5.2.15$$

where U_1 and U_2 are subsets of $\mathbb{R}^k$. Our candidate for the change of variables mapping is $\Phi = \gamma_2^{-1} \circ \gamma_1$, i.e.,

$$\overline{U}_1 \underset{\gamma_1}{\to} M \underset{\gamma_2^{-1}}{\to} \overline{U}_2. \qquad 5.2.16$$

But serious difficulties can arise, as shown in the following example.

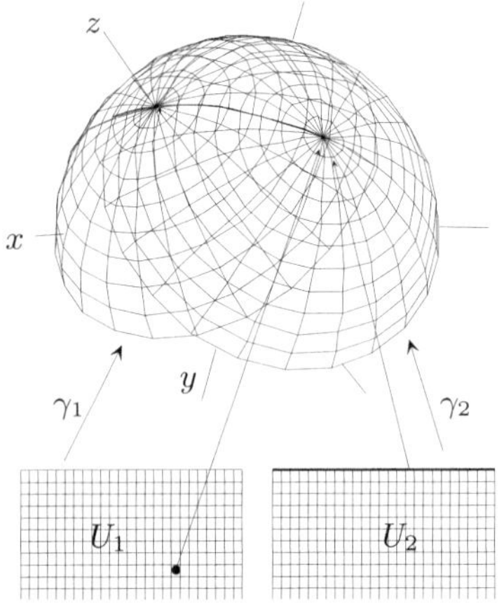

FIGURE 5.2.4.

Here X_1 consists of the entire boundary of the rectangle at left. If in our spherical coordinates we take $-\pi \le \theta \le \pi$, then the vertical parts of the boundary map to the date line; the top and bottom lines map to the poles. The dark box in the rectangle at left is a point of Y_2; it is mapped to a pole of γ_2 and then to the dark line at right.

Excluding X_1 from the domain of Φ ensures that it is injective; excluding Y_2 ensures that it is well defined. Excluding X_2 and Y_1 from the codomain ensures that when X_1 and Y_2 are removed from the domain, Φ is still surjective.

Example 5.2.10 (Problems when changing parametrizations). Let γ_1 and γ_2 be two parametrizations of S^2 by spherical coordinates, but with different poles. Call P_1, P_1' the poles for γ_1 and P_2, P_2' the poles for γ_2. Then $\gamma_2^{-1} \circ \gamma_1$ is not defined at $\gamma_1^{-1}(P_2)$ or at $\gamma_1^{-1}(P_2')$. Indeed, some single point in the domain of γ_1 maps to P_2.[2] But as shown in Figure 5.2.4, γ_2 maps a whole segment to P_2, so that $\gamma_2^{-1} \circ \gamma_1$ "maps" the point $\gamma_1^{-1}(P_2)$ to a line segment, which is nonsense. The same is true of $\gamma_1^{-1}(P_2')$. Our mapping $\gamma_2^{-1} \circ \gamma_1$ is not well defined; we can't possibly apply Theorem 4.11.20 to it in order to prove that integrals are independent of the choice of parametrization.

The only way to deal with this is to remove $\gamma_1^{-1}(\{P_2, P_2'\})$ from the domain of $\Phi = \gamma_2^{-1} \circ \gamma_1$ and hope that the removed locus has k-dimensional volume 0. In this case $k = 2$ and this is no problem: we just removed two points from the domain, and two points certainly have area 0. △

Definition 5.2.3 of a parametrization was carefully calculated to make the analogous statement true in general.

Let us set up our change of variables with a bit more precision. Let U_1 and U_2 be subsets of $\mathbb{R}^k$. Following the notation of Definition 5.2.3, denote by X_1 the negligible "trouble spots" of γ_1, and by X_2 the trouble spots of γ_2. In Example 5.2.10, X_1 and X_2 consist of the points that are mapped to the poles; in Figure 5.2.4, X_2 is the line marked in bold.

Set

$$Y_1 = (\gamma_2^{-1} \circ \gamma_1)(X_1), \quad \text{and} \quad Y_2 = (\gamma_1^{-1} \circ \gamma_2)(X_2). \qquad 5.2.17$$

In Figure 5.2.4, the dark dot in the rectangle at left is Y_2, which is mapped by γ_1 to a pole of γ_2 and then by γ_2^{-1} to the dark line at right; Y_1 is the (unmarked) dot in the right rectangle that maps to the pole of γ_1.

Now define

$$U_1^{\text{OK}} = U_1 - (X_1 \cup Y_2) \quad \text{and} \quad U_2^{\text{OK}} = U_2 - (X_2 \cup Y_1); \qquad 5.2.18$$

[2] If P_2 happens to be on the date line with respect to γ_1, two points map to P_2: in Figure 5.2.4, a point on the right boundary of the rectangle, and the corresponding point on the left boundary.

that is, we use the superscript "OK" (okay) to denote the domain or codomain of a change of variables mapping with any trouble spots of volume 0 removed.

Theorem 5.2.11 says that $\Phi : U_1^{\text{OK}} \to U_2^{\text{OK}}$ is something to which the change of variables formula applies. Recall that a diffeomorphism is a differentiable mapping with differentiable inverse.

Theorem 5.2.11. *Both* $U_1^{\text{OK}} = U_1 - (X_1 \cup Y_2)$ *and* $U_2^{\text{OK}} = U_2 - (X_2 \cup Y_1)$ *are open subsets of* $\mathbb{R}^k$ *with boundaries of* k*-dimensional volume* 0, *and*

$$\Phi \stackrel{\text{def}}{=} \gamma_2^{-1} \circ \gamma_1 : U_1^{\text{OK}} \to U_2^{\text{OK}} \qquad 5.2.19$$

is a C^1 *diffeomorphism with locally Lipschitz inverse.*

Proof. By Proposition 3.2.11 we have

$$[\mathbf{D}\Phi(\mathbf{x})] = [\mathbf{D}\gamma_2^{-1}(\gamma_1(\mathbf{x}))][\mathbf{D}\gamma_1(\mathbf{x})]; \qquad 5.2.20$$

since both $[\mathbf{D}\gamma_2^{-1}(\gamma_1(\mathbf{x}))]$ and $[\mathbf{D}\gamma_1(\mathbf{x})]$ are isomorphisms, their composition is also. So Φ is a local C^1 diffeomorphism, hence a global diffeomorphism since $\Phi : U_1^{OK} \to U_2^{OK}$ is 1–1 and onto.

The only thing left to prove is that the boundaries of U_1^{OK} and U_2^{OK} have k-dimensional volume 0. It is enough to show it for one, say, U_1^{OK}. The boundary of U_1^{OK} is contained in the union of

1. the boundary of U_1
2. X_1 (the negligible "trouble spots" of γ_1)
3. the set $Y_2 = (\gamma_1^{-1} \circ \gamma_2)(X_2)$

By part 2 of Definition 5.2.1, it is enough to show that for any $R > 0$, the part of the boundary of U_1^{OK} in the ball of radius R has volume 0; thus we may (and will) assume that both U_1 and U_2 are bounded. Choose $\epsilon > 0$, and cover $X_1 \cup \partial U_1$ by finitely many open balls $B_{r_i}(\mathbf{x}_i)$, $i = 1, \dots, p$ satisfying $\sum_{i=1}^{p} (r_i)^k < \epsilon/2$. We can do this because by hypothesis ∂U_1 and X_1 have k-dimensional volume 0. Let B be the union of the B_{r_i}.

We need to show that Y_2 also has k-dimensional volume 0. The set

$$\widetilde{Y}_2 = \overline{Y}_2 - (\overline{Y}_2 \cap B). \qquad 5.2.21$$

is closed and bounded in $\mathbb{R}^k$, so it is compact, as is $\gamma_1(\widetilde{Y}_2)$. In particular, by part 5 of Definition 5.2.3, $\gamma_2(X_2) \cap \gamma_1(\widetilde{Y}_2)$ has k-dimensional volume 0. Since $\gamma_1(\widetilde{Y}_2) \subset \gamma_2(X_2)$, it follows that $\gamma_1(\widetilde{Y}_2)$ also has k-dimensional volume 0. Now the result follows from Lemma 5.2.12.

Lemma 5.2.12. *There exists* L *such that for every* $\mathbf{y} \in \widetilde{Y}_2$ *and every* $r > 0$ *we have*

$$B_r(\gamma_1(\mathbf{y})) \subset \gamma_1(B_{Lr}(\mathbf{y})). \qquad 5.2.22$$

Let us complete the proof of Theorem 5.2.11 using the lemma. Find a cover of $\gamma_1(\widetilde{Y}_2)$ by finitely many balls $B_{r'_j}(\gamma_1(\mathbf{y}_j))$ with

$$\sum_j (r'_j)^k < \frac{\epsilon}{2L^k}. \qquad 5.2.23$$

Theorem 4.11.20, the change of variables for Lebesgue integrals, says nothing about the behavior of the change of variables map Φ on the boundary of the domain. So we can, with impunity, remove any "trouble spots" from the domain (as we do in Example 5.2.10) as long as they have volume 0.

Our earlier version of the change of variables formula using Riemann integrals (Theorem 4.10.12) requires the domain of integration X to be compact, and Φ to be defined in a neighborhood U of X. Using that theorem we would not be able to simply remove any troublesome spots (of volume 0) from U and define Φ on what is left. So we would not be able to use Theorem 4.10.12 to justify our claim that integration does not depend on the choice of parametrization.

The inverse images $\gamma_1^{-1}\big(B_{r'_j}(\gamma_1(\mathbf{y}_j))\big)$ cover $\widetilde{Y}_2$ and by Lemma 5.2.12,

$$\gamma_1^{-1}B'_{r'_j}\gamma_1(\mathbf{y}_j) \subset B_{Lr'_j}(\mathbf{y}_j); \tag{5.2.24}$$

these larger balls of radius Lr'_j therefore also cover $\widetilde{Y}_2$. Thus $\widetilde{Y}_2$ has k-dimensional volume 0:

$$\sum_j (Lr'_j)^k \le L^k \sum_j {r'_j}^k \le L^k \frac{\epsilon}{2L^k} = \frac{\epsilon}{2}; \tag{5.2.25}$$

and finally so does Y_2.

Proof of Lemma 5.2.12. By contradiction, if the lemma is false, there exist sequences $i \mapsto \mathbf{y}_i, i \mapsto \mathbf{y}'_i$ with $\mathbf{y}_i \ne y'_i$ such that

$$\lim_{i\to\infty} \frac{\gamma_1(\mathbf{y}_i) - \gamma_1(\mathbf{y}'_i)}{|\mathbf{y}_i - \mathbf{y}'_i|} = 0. \tag{5.2.26}$$

By compactness, we may assume that $i \mapsto \mathbf{y}_i$ converges to some $\mathbf{y}_0$; then since γ_1 is injective on the compact set $\tilde{Y}_2$ the sequence $i \mapsto \mathbf{y}'_i$ also converges to $\mathbf{y}_0$. Further, by the compactness of the unit sphere of $\mathbb{R}^k$ we may assume that the set of unit vectors sequence of unit vectors

$$i \mapsto \mathbf{v}_i \stackrel{\text{def}}{=} \frac{\mathbf{y}'_i - \mathbf{y}_i}{|\mathbf{y}_i - \mathbf{y}_i|} \tag{5.2.27}$$

also converges to some unit vector $\mathbf{v}_0$. Then

$$\begin{aligned}
[\mathbf{D}\gamma_1(\mathbf{y}_0)]\mathbf{v}_0 &= \lim_{i\to\infty}[\mathbf{D}\gamma_1(\mathbf{y}_i)]\mathbf{v}_0 = \lim_{i\to\infty}\lim_{j\to\infty}[\mathbf{D}\gamma_1(\mathbf{y}_i)]\frac{\mathbf{y}'_j - \mathbf{y}_j}{|\mathbf{y}_j - \mathbf{y}_j|}\\
&= \lim_{i\to\infty}\lim_{j\to\infty}\frac{1}{|\mathbf{y}'_j - \mathbf{y}_j|}[\mathbf{D}\gamma_1(\mathbf{y}_i)](\mathbf{y}'_j - \mathbf{y}_j) \qquad (5.2.28)\\
&= \lim_{i\to\infty}\lim_{j\to\infty}\frac{1}{|\mathbf{y}'_j - \mathbf{y}_j|}\Big(\gamma_1(\mathbf{y}'_j) - \gamma_1(\mathbf{y}_j) + o(|\mathbf{y}'_j - \mathbf{y}_j|)\Big) = 0
\end{aligned}$$

This contradicts the injectivity of $[\mathbf{D}\gamma_1(\mathbf{y}_0)]$. □ Lemma 5.2.12

EXERCISES FOR SECTION 5.2

5.2.1 Check that the mapping γ of Example 5.2.4 satisfies conditions 3 and 4 of Definition 5.2.3.

5.2.2 a. Show that the segment of diagonal $\left\{\begin{pmatrix} x \\ x \end{pmatrix} \in \mathbb{R}^2 \mid |x| \le 1\right\}$ does not have one-dimensional volume 0.

b. Show that the curve in $\mathbb{R}^3$ parametrized by $t \mapsto \begin{pmatrix} \cos t \\ \cos t \\ \sin t \end{pmatrix}$ has two-dimensional volume 0 but does not have one-dimensional volume 0.

5.2.3 Show that the map

$$S : \begin{pmatrix} r \\ \theta \\ \varphi \end{pmatrix} \mapsto \begin{pmatrix} r\sin\varphi\cos\theta \\ r\sin\varphi\sin\theta \\ r\cos\varphi \end{pmatrix}$$

with $0 \le r < \infty$, $0 \le \theta \le 2\pi$, and $0 \le \varphi \le \pi$, parametrizes space by the distance from the origin r, the polar angle θ, and the angle φ from the north pole.

5.2.4 Choose numbers $0 < r < R$, and consider the circle in the (x, z)-plane of radius r centered at $\begin{pmatrix} x = R \\ z = 0 \end{pmatrix}$. Let S be the surface obtained by rotating this circle around the z-axis (this is the torus shown in Figure 5.3.2).

a. Write an equation for S and check that S is a smooth surface.

b. Write a parametrization for S, paying attention to the sets U and X used.

c. Parametrize the part of S where

i. $z > 0$; ii. $x > 0$, $y > 0$; iii. $z > x + y$.

Exercise 5.2.4: Part c, iii is much harder than the others; even after finding an equation for the curve bounding the parametrizing region, you may need a computer to visualize it.

***5.2.5** Consider the open subset of $\mathbb{R}$ constructed in Example 4.4.3: list the rationals between 0 and 1, say $a_1, a_2, a_3, \dots$, and take the union

$$U = \bigcup_{i=1}^{\infty} \left(a_i - \frac{1}{2^{i+k}}\, ,\ a_i + \frac{1}{2^{i+k}} \right)$$

for some integer $k > 2$. Show that U is a one-dimensional manifold and that it can be parametrized according to Definition 5.2.3.

5.2.6 Use Definition 5.2.1 to show that a single point in any $\mathbb{R}^n$ never has 0-dimensional volume 0.[3]

5.2.7 a. Show that if $X \subset \mathbb{R}^n$ is a bounded subset of k-dimensional volume 0, then its projection onto the subspace spanned by any k standard basis vectors also has k-dimensional volume 0.

b. Show that this is not true if X is unbounded. For instance, produce an unbounded subset of $\mathbb{R}^2$ of length 0, whose projection onto the x-axis does not have length 0.

***5.2.8** Prove Theorem 5.2.6. There are many approaches, all quite fiddly. One possibility is to cover M by a sequence of cubes C_i such that $M \cap C_i$ is the graph Γ_i of a map $\mathbf{f}_i$ expressing $n - k$ variables in terms of the other k. These Γ_i will probably not have disjoint interiors: doctor up C_i to remove the part of Γ_i that is in the interior of some Γ_j, $j < i$.

Spread out the domains of these $\mathbf{f}_i$ so they are well separated. The first condition of part 5 of the theorem is then easy, and the second follows from the Heine-Borel theorem (Theorem A3.3 in Appendix A3).

[3]We did not ask what the 0-dimensional volume of a single point is. In fact, 0-dimensional volume of a set simply counts its points. For this to be a special case of Definition 5.1.3, you need to see that the determinant of the empty matrix (not the 0-matrix) is 1, and that is what the normalization condition of the determinant says: the determinant of the identity $\mathbb{R}^n \to \mathbb{R}^n$ is 1 for all n, including $n = 0$, when the matrix of the identity is the empty matrix. If you feel that this is a stretch, we sympathize.

5.3 Computing volumes of manifolds

The k-dimensional volume of a k-manifold M embedded in $\mathbb{R}^n$ is given by

$$\operatorname{vol}_k M = \int_M |d^k\mathbf{x}|, \tag{5.3.1}$$

where $|d^k\mathbf{x}|$ is the integrand that takes a k-parallelogram and returns its k-dimensional volume. Thus the length of a curve C can be written $\int_C |d^1\mathbf{x}|$, and the area of a surface S can be written $\int_S |d^2\mathbf{x}|$. Often, $|d^1\mathbf{x}|$ is written dl or ds, and called the *element of length*; $|d^2\mathbf{x}|$ is written dS or dA, and called the *element of area*; and $|d^3\mathbf{x}|$ is often written dV, and called the *element of volume*.

We feel that dl, ds, dS, dA, and dV are unfortunate, because this notation does not generalize to higher dimensions.

Heuristically, the integral in equation 5.3.1 is defined by cutting up the manifold into little anchored k-parallelograms, adding their k-dimensional volumes and taking the limits of the sums as the decomposition becomes infinitely fine.

We know how to compute $|d^k\mathbf{x}|$ of a k-parallelogram: if $T = [\vec{\mathbf{v}}_1, \dots, \vec{\mathbf{v}}_k]$,

$$|d^k\mathbf{x}|P_{\mathbf{x}}(\vec{\mathbf{v}}_1, \dots, \vec{\mathbf{v}}_k) = \operatorname{vol}_k P_{\mathbf{x}}(\vec{\mathbf{v}}_1, \dots, \vec{\mathbf{v}}_k) = \sqrt{\det(T^\top T)}. \tag{5.3.2}$$

To compute the volume of a k-manifold M, we parametrize M by a mapping γ and then compute the volume of the k-parallelograms spanned by the *partial derivatives* of γ, sum them, and take the limit as the decomposition becomes infinitely fine. This gives the following definition.

In equation 5.3.3,

$$P_{\gamma(\mathbf{u})}\Big(\overrightarrow{D_1\gamma}(\mathbf{u}), \dots, \overrightarrow{D_k\gamma}(\mathbf{u})\Big)$$

is the k-parallelogram anchored at $\gamma(\mathbf{u})$ and spanned by the partial derivatives

$$\overrightarrow{D_1\gamma}(\mathbf{u}), \dots, \overrightarrow{D_k\gamma}(\mathbf{u});$$

$|d^k\mathbf{x}|$ of this parallelogram is the volume of the parallelogram (see Proposition 5.1.1).

Note that the volume of a manifold may be infinite.

Definition 5.3.1 (Volume of manifold). Let $M \subset \mathbb{R}^n$ be a smooth k-dimensional manifold, U a pavable subset of $\mathbb{R}^k$, and $\gamma : U \to M$ a parametrization according to Definition 5.2.3. Let X be as in that definition. Then

$$\begin{aligned}\operatorname{vol}_k M &\overset{\text{def}}{=} \int_{\gamma(U-X)} |d^k\mathbf{x}| \\ &= \int_{U-X} \Big(|d^k\mathbf{x}| \big(P_{\gamma(\mathbf{u})}(\overrightarrow{D_1\gamma}(\mathbf{u}), \dots, \overrightarrow{D_k\gamma}(\mathbf{u}))\big)\Big)|d^k\mathbf{u}| \\ &= \int_{U-X} \sqrt{\det([\mathbf{D}\gamma(\mathbf{u})]^\top[\mathbf{D}\gamma(\mathbf{u})])}\ |d^k\mathbf{u}|.\end{aligned} \tag{5.3.3}$$

In Definitions 5.3.1 and 5.3.2 we integrate over $U - X$, not over U, because γ may not be differentiable on X. But X has k-dimensional volume 0, so this doesn't affect the integral.

Remark. When the manifold is a curve parametrized by $\gamma : [a, b] \to C$, equation 5.3.3 can be written

$$\int_C |d^1\mathbf{x}| = \int_{[a,b]} \sqrt{\det(\vec{\gamma}'(t) \cdot \vec{\gamma}'(t))}\ |dt| = \int_a^b |\vec{\gamma}'(t)|\, dt, \tag{5.3.4}$$

which is compatible with Definition 3.9.5 of arc length. △

Definition 5.3.1 is a special case of the following. In Definition 5.3.2 recall that $\mathbf{1}$ denotes the indicator function (Definition 4.1.1); note that $\mathbf{1}_Y(\gamma(\mathbf{u})) = \mathbf{1}_{\gamma^{-1}(Y)}(\mathbf{u})$.

Definition 5.3.2: In Chapter 6 we will study integrals of *differential forms* over manifolds. The integral of Definition 5.3.2 is sometimes referred to as the *integral of a density*. Differential forms are better, but less intuitive. The reason is that for forms there is a notion of derivative, which makes possible the generalization of the fundamental theorem of calculus. There is no such thing for densities.

Definition 5.3.2 (Integrals over manifolds, with respect to volume). Let $M \subset \mathbb{R}^n$ be a smooth k-dimensional manifold, U a pavable subset of $\mathbb{R}^k$, and $\gamma : U \to M$ a parametrization. Let X be as in Definition 5.2.3. Then $f : M \to \mathbb{R}$ is integrable over M with respect to volume if the integral on the right of equation 5.3.5 exists, and then

$$\int_M f(\mathbf{x})|d^k\mathbf{x}| \stackrel{\text{def}}{=} \int_{U-X} f(\gamma(\mathbf{u}))\sqrt{\det([\mathbf{D}\gamma(\mathbf{u})]^\top[\mathbf{D}\gamma(\mathbf{u})])}\,|d^k\mathbf{u}|. \qquad 5.3.5$$

In particular, if $Y \subset M$ is a subset such that $\mathbf{1}_{\gamma^{-1}(Y)}$ is integrable, then

$$\text{vol}_k Y \stackrel{\text{def}}{=} \int_{U-X} \mathbf{1}_Y(\gamma(\mathbf{u}))\sqrt{\det([\mathbf{D}\gamma(\mathbf{u})]^\top[\mathbf{D}\gamma(\mathbf{u})])}\,|d^k\mathbf{u}|.$$

Let us see why Definition 5.3.2 corresponds to our notion of volume (or area). To simplify the discussion, let us consider the area of a surface S parametrized by $\gamma : U \to \mathbb{R}^3$. This area should be

$$\lim_{N\to\infty} \sum_{C\in\mathcal{D}_N(\mathbb{R}^2)} \text{Area of } \gamma(C \cap U). \qquad 5.3.6$$

That is, we make a dyadic decomposition of $\mathbb{R}^2$ and see how γ maps to S the dyadic squares C that are in U or straddle it. We then sum the areas of the resulting regions $\gamma(C \cap U)$. For $C \subset U$, this is the same as $\gamma(C)$; for C straddling U, we add to the sum the area of the part of C that is in U.

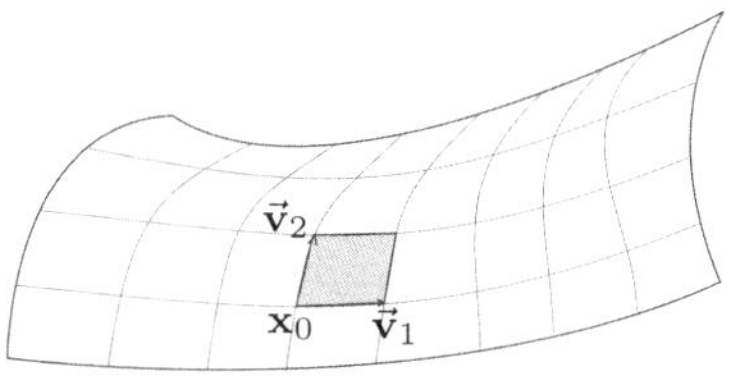

FIGURE 5.3.1.

A surface approximated by parallelograms. The point $\mathbf{x}_0$ corresponds to $\gamma(\mathbf{u})$, and the vectors $\vec{\mathbf{v}}_1$ and $\vec{\mathbf{v}}_2$ correspond to the vectors

$$\frac{1}{2^N}\overrightarrow{D_1\gamma}(\mathbf{u}) \text{ and } \frac{1}{2^N}\overrightarrow{D_2\gamma}(\mathbf{u}).$$

The sidelength of a square $C \in \mathcal{D}_N(\mathbb{R}^2)$ is $1/2^N$, so at least when $C \subset U$, the set $\gamma(C \cap U)$ is, as shown in Figure 5.3.1, approximately the parallelogram

$$P_{\gamma(\mathbf{u})}\Big(\frac{1}{2^N}\overrightarrow{D_1\gamma}(\mathbf{u}), \frac{1}{2^N}\overrightarrow{D_2\gamma}(\mathbf{u})\Big), \qquad 5.3.7$$

where $\mathbf{u}$ is the lower left corner of C. That parallelogram has area

$$\frac{1}{2^{2N}}\sqrt{\det[\mathbf{D}\gamma(\mathbf{u})]^\top[\mathbf{D}\gamma(\mathbf{u})]}. \qquad 5.3.8$$

So it seems reasonable to expect that the error we make by replacing

$$\text{Area of } \gamma(C \cap U) \quad \text{by} \quad \text{vol}_2(C)\sqrt{\det[\mathbf{D}\gamma(\mathbf{u})]^\top[\mathbf{D}\gamma(\mathbf{u})]} \qquad 5.3.9$$

in formula 5.3.6 will disappear in the limit as $N \to \infty$, and we have a Riemann sum for the integral giving surface area:

$$\underbrace{\lim_{N\to\infty} \sum_{C\in\mathcal{D}_N(\mathbb{R}^2)} \text{vol}_2\, C\sqrt{\det[\mathbf{D}\gamma(\mathbf{u})]^\top[\mathbf{D}\gamma(\mathbf{u})]}}_{\text{formula 5.3.6 with substitution given by 5.3.9}} = \underbrace{\int_U \sqrt{\det[\mathbf{D}\gamma(\mathbf{u})]^\top[\mathbf{D}\gamma(\mathbf{u})]}\,|d^2\mathbf{u}|}_{\text{area of surface by eq. 5.3.3}}. \qquad 5.3.10$$

This argument isn't entirely convincing. We can imagine the parallelograms as tiling the surface: we glue small flat tiles at the corners of a grid drawn on the surface, like using ceramic tiles to cover a curved counter. We get a better and better fit by choosing smaller and smaller tiles, but is it good enough? To be sure that Definition 5.3.2 is correct, we need to show that the integral is independent of the choice of parametrization.

Proposition 5.3.3 (Integral independent of parametrization). *Let M be a k-dimensional manifold in $\mathbb{R}^n$ and $f: M \to \mathbb{R}$ a function. If U and V are subsets of $\mathbb{R}^k$ and $\gamma_1 : U \to M$, $\gamma_2 : V \to M$ are two parametrizations of M, then*

$$\int_U f(\gamma_1(\mathbf{u}))\sqrt{\det([\mathbf{D}\gamma_1(\mathbf{u})]^\top[\mathbf{D}\gamma_1(\mathbf{u})])}\,|d^k\mathbf{u}| \qquad 5.3.11$$

exists if and only if

$$\int_V f(\gamma_2(\mathbf{v}))\sqrt{\det([\mathbf{D}\gamma_2(\mathbf{v})]^\top[\mathbf{D}\gamma_2(\mathbf{v})])}\,|d^k\mathbf{v}| \qquad 5.3.12$$

exists, and in that case, the integrals are equal.

In the one-dimensional case, this gives

$$\int_{I_1} |\vec{\gamma}_1'(t_1)|\,|dt_1| = \int_{I_2} |\vec{\gamma}_2'(t_2)|\,|dt_2|. \qquad 5.3.13$$

Integrating the length of the velocity vector, $|\vec{\gamma}'(t)|$, gives us distance along the curve. Thus Proposition 5.3.3 says that if you take the same route from New York to Boston two times, making good time the first and caught in a traffic jam the second, giving two different parametrizations by time of your route, you will go the same distance both times.

Proof. Define $\Phi = \gamma_2^{-1} \circ \gamma_1 : U^{\text{OK}} \to V^{\text{OK}}$ to be the "change of parameters" map such that $\mathbf{v} = \Phi(\mathbf{u})$. Notice that the chain rule applied to the equation

$$\gamma_1 = \gamma_2 \circ \underbrace{\gamma_2^{-1} \circ \gamma_1}_{\Phi} = \gamma_2 \circ \Phi \qquad 5.3.14$$

gives

$$[\mathbf{D}\gamma_1(\mathbf{u})] = [\mathbf{D}(\gamma_2 \circ \Phi)(\mathbf{u})] = [\mathbf{D}\gamma_2(\Phi(\mathbf{u}))][\mathbf{D}\Phi(\mathbf{u})]. \qquad 5.3.15$$

Applying the change of variables formula for Lebesgue integrals, we find

$$\int_V \overbrace{\sqrt{\det[\mathbf{D}\gamma_2(\mathbf{v})]^\top[\mathbf{D}\gamma_2(\mathbf{v})])}\, f(\gamma_2(\mathbf{v}))}^{\text{corresponds to } f(\mathbf{v}) \text{ in change of var. formula}}\ |d^k\mathbf{v}| \qquad 5.3.16$$

$$\underset{1}{=} \int_U \sqrt{\det\big([\mathbf{D}\gamma_2(\Phi(\mathbf{u}))]^\top[\mathbf{D}\gamma_2(\Phi(\mathbf{u}))]\big)}\ |\det[\mathbf{D}\Phi(\mathbf{u})]|\, f(\underbrace{\gamma_2(\Phi(\mathbf{u}))}_{\gamma_1(\mathbf{u})})\,|d^k\mathbf{u}|$$

$$\underset{2}{=} \int_U \underbrace{\sqrt{\det\big([\mathbf{D}\gamma_2(\Phi(\mathbf{u}))]^\top[\mathbf{D}\gamma_2(\Phi(\mathbf{u}))]\big)}}_{\text{square matrix}}\,\underbrace{\sqrt{\det\big([\mathbf{D}\Phi(\mathbf{u})]^\top[\mathbf{D}\Phi(\mathbf{u})]\big)}}_{\text{two square matrices}}\, f(\gamma_1(\mathbf{u}))|d^k\mathbf{u}|$$

$$\underset{3}{=} \int_U \sqrt{\det\Big([\mathbf{D}\Phi(\mathbf{u})]^\top[\mathbf{D}\gamma_2(\Phi(\mathbf{u}))]^\top[\mathbf{D}\gamma_2(\Phi(\mathbf{u}))][\mathbf{D}\Phi(\mathbf{u})]\Big)}\ f(\gamma_1(\mathbf{u}))\,|d^k\mathbf{u}|$$

Equation 5.3.16: The equality marked 1 uses the change of variables formula

$$\int_V f(\mathbf{v})|d^n\mathbf{v}| =$$

$$\int_U f\big(\Phi(\mathbf{u})\big)|\det[\mathbf{D}\Phi(\mathbf{u})]|\,|d^n\mathbf{u}|.$$

Note that we replace each $\mathbf{v}$ in the integrand in the first line by $\Phi(\mathbf{u})$. This application of the change of variables formula is justified by Theorem 5.2.11.

Equation 5.3.16, equality 2: $[\mathbf{D}\Phi(\mathbf{u})]$ is a square matrix, and for a square matrix A,

$$|\det A| = \sqrt{\det A^\top A}.$$

Equality 3: If A, B are square matrices, then

$$\det AB = \det A \det B.$$

In the third line we have

$$\sqrt{\det A}\sqrt{\det BC},$$

where A, B, C are square, so

$$\begin{aligned}\sqrt{\det A}\sqrt{\det BC} &= \sqrt{\det A \det B \det C}\\ &= \sqrt{\det B \det A \det C}\\ &= \sqrt{\det BAC}\end{aligned}$$

We continue this computation on the next page.

This continues the computation on the previous page. Equality 4 uses the chain rule. Equality 5 uses the definition $\Phi = \gamma_2^{-1} \circ \gamma_1$.

$$\int_U \sqrt{\det\Big([\mathbf{D}\Phi(\mathbf{u})]^\top [\mathbf{D}\gamma_2(\Phi(\mathbf{u}))]^\top [\mathbf{D}\gamma_2(\Phi(\mathbf{u}))][\mathbf{D}\Phi(\mathbf{u})]\Big)}\, f(\gamma_1(\mathbf{u}))\, |d^k\mathbf{u}|$$
$$\underset{4}{=} \int_U \sqrt{\det\left([\mathbf{D}(\gamma_2 \circ \Phi)(\mathbf{u})]^\top [\mathbf{D}(\gamma_2 \circ \Phi)(\mathbf{u})]\right)}\, f(\gamma_1(\mathbf{u}))\, |d^k\mathbf{u}|$$
$$\underset{5}{=} \int_U \sqrt{\det[\mathbf{D}\gamma_1(\mathbf{u})])^\top [\mathbf{D}\gamma_1(\mathbf{u})])}\, f(\gamma_1(\mathbf{u}))\, |d^k\mathbf{u}|. \quad \square \qquad 5.3.16 \text{ (cont.)}$$

Corollary 5.3.4. *Every point of a k-dimensional manifold $M \subset \mathbb{R}^n$ has a neighborhood with finite k-dimensional volume.*

Proof. This follows from Theorem 5.2.6 and Proposition 5.3.3 (and Theorem 4.3.6, which says that the integral, hence volume, exists). □

Example 5.3.5 (Two parametrizations of a halfcircle). We can parametrize the upper half of the unit circle by

$$x \mapsto \begin{pmatrix} x \\ \sqrt{1-x^2} \end{pmatrix}, -1 \le x \le 1, \quad \text{or by} \quad t \mapsto \begin{pmatrix} \cos t \\ \sin t \end{pmatrix}, 0 \le t \le \pi. \qquad 5.3.17$$

Equation 5.3.18: Here we write $\int_{[-1,1]}$ rather than $\int_{-1}^{1}$ because we are not concerned with orientation: it doesn't matter whether we go from -1 to 1, or from 1 to -1. For the same reason we write $|dx|$ not dx.

We use equation 5.3.4 to compute its length. The first parametrization gives

$$\int_{[-1,1]} \left| \left[\begin{matrix} 1 \\ \frac{-x}{\sqrt{1-x^2}} \end{matrix} \right] \right| |dx| = \int_{[-1,1]} \sqrt{1 + \frac{x^2}{1-x^2}}\, |dx| \qquad 5.3.18$$
$$= \int_{[-1,1]} \frac{1}{\sqrt{1-x^2}}\, |dx| = [\arcsin x]_{-1}^{1} = \frac{\pi}{2} - \left(-\frac{\pi}{2}\right) = \pi.$$

The second gives

$$\int_{[0,\pi]} \left| \left[\begin{matrix} -\sin t \\ \cos t \end{matrix} \right] \right| |dt| = \int_{[0,\pi]} \sqrt{\sin^2 t + \cos^2 t}\, |dt|$$
$$= \int_{[0,\pi]} |dt| = \pi. \quad \triangle \qquad 5.3.19$$

Example 5.3.6 (Length of a graph of function). The graph of a C^1 function $f : [a,b] \to \mathbb{R}$ is parametried by $x \mapsto \begin{pmatrix} x \\ f(x) \end{pmatrix}$; hence its arc length is given by the integral

Formula 5.3.20 for finding the arc length of an ellipse leads to a non-elementary "elliptic integral". The theory of such integrals and the associated elliptic functions and elliptic curves is a central part of complex analysis, number theory, and coding theory.

$$\int_{[a,b]} \left| \left[\begin{matrix} 1 \\ f'(x) \end{matrix} \right] \right| |dx| = \int_a^b \sqrt{1 + (f'(x))^2}\, dx. \qquad 5.3.20$$

Because of the square root, these integrals tend to be unpleasant or impossible to calculate in elementary terms. The following example, already pretty hard, is still one of the simplest. The length of the arc of parabola $y = ax^2$ for $0 \le x \le A$ is given by

$$\int_0^A \left| \left[\begin{matrix} 1 \\ f'(x) \end{matrix} \right] \right| |dx| = \int_0^A \left| \left[\begin{matrix} 1 \\ 2ax \end{matrix} \right] \right| |dx| = \int_0^A \sqrt{1 + 4a^2x^2}\, dx. \qquad 5.3.21$$

A table of integrals will tell you that

$$\int \sqrt{1+u^2}\,du = \frac{u}{2}\sqrt{u^2+1} + \frac{1}{2}\ln\left|u+\sqrt{1+u^2}\right|. \qquad 5.3.22$$

Setting $2ax = u$, so that $dx = \frac{du}{2a}$, gives the following length of the arc of parabola:

$$\begin{aligned}\int_0^A \sqrt{1+4a^2x^2}\,dx &= \frac{1}{4a}\left[2ax\sqrt{1+4a^2x^2} + \ln\left|2ax+\sqrt{1+4a^2x^2}\right|\right]_0^A \\ &= \frac{1}{4a}\left(2aA\sqrt{1+4a^2A^2} + \ln\left|2aA+\sqrt{1+4a^2A^2}\right|\right). \quad \triangle\end{aligned} \qquad 5.3.23$$

We can compute lengths of curves in $\mathbb{R}^n$ even when $n > 3$.

Example 5.3.7 (Length of a curve in $\mathbb{R}^4$). Let p, q be two integers, and consider the curve in $\mathbb{R}^4$ parametrized by

$$\gamma(t) = \begin{pmatrix} \cos pt \\ \sin pt \\ \cos qt \\ \sin qt \end{pmatrix}, \quad 0 \le t \le 2\pi. \qquad 5.3.24$$

The curve of Example 5.3.7 is important in mechanics, as well as in geometry. It is contained in the unit sphere $S^3 \subset \mathbb{R}^4$, and it is knotted when p and q are both greater than 1 and are relatively prime (share no common factor).

Its length is given by

$$\begin{aligned}&\int_0^{2\pi} \sqrt{(-p\sin pt)^2 + (p\cos pt)^2 + (-q\sin qt)^2 + (q\cos qt)^2}\,dt \\ &\qquad = 2\pi\sqrt{p^2+q^2}. \quad \triangle\end{aligned} \qquad 5.3.25$$

We can also measure data other than pure arc length. Suppose $I \subset \mathbb{R}$ is an interval and $\gamma : I \to C$ parametrizes a curve C. Then we can define

$$\int_C f(\mathbf{x})\,|d^1\mathbf{x}| \overset{\text{def}}{=} \int_I f(\gamma(t))\,|\vec{\gamma}'(t)|\,|dt|. \qquad 5.3.26$$

For instance, if $f(\mathbf{x})$ gives the density of a wire of variable density, the integral 5.3.26 gives the mass of the wire. Here is another example.

Example 5.3.8 (The total curvature of a closed curve). Recall that the curvature κ of a curve is defined in Definition 3.9.1. Let C be the intersection of the surfaces of equations $x = y^2 - 2$ and $x = 2 - z^2$. We will integrate κ over this curve to get the *total curvature*.

Setting $y^2 - 2 = 2 - z^2$ gives $y^2 + z^2 = 4$, which suggests setting $y = 2\cos\theta$, $z = 2\sin\theta$, and $x = 4\cos^2\theta - 2 = 2(2\cos^2\theta - 1) = 2\cos 2\theta$. That is, we parametrize C by $\gamma(\theta) = \begin{pmatrix} 2\cos 2\theta \\ 2\cos\theta \\ 2\sin\theta \end{pmatrix}$. Thus to integrate κ over C we need to compute

$$\begin{aligned}\int_C \kappa\,|d^1\mathbf{x}| &= \int_0^{2\pi} \kappa(\gamma(\theta))\,|\vec{\gamma}'(\theta)|\,d\theta = \int_0^{2\pi} \kappa(\gamma(\theta)) \left|\begin{bmatrix} -4\sin 2\theta \\ -2\sin\theta \\ 2\cos\theta \end{bmatrix}\right| d\theta \\ &= \int_0^{2\pi} \kappa(\gamma(\theta))2\sqrt{4\sin^2 2\theta + 1}\,d\theta.\end{aligned} \qquad 5.3.27$$

We use Proposition 3.9.17 (see the margin note) to compute κ:

$$\kappa\big(\gamma(t)\big) = \frac{\sqrt{16 + (16)^2\cos^2 2\theta + 8^2\sin^2 2\theta}}{2^3(\sqrt{4\sin^2 2\theta + 1})^3}. \qquad 5.3.28$$

So we have

$$\int_C \kappa\,|d^1\mathbf{x}| = \int_0^{2\pi} \frac{4\sqrt{1 + 16\cos^2 2\theta + 4\sin^2 2\theta}\,(2\sqrt{4\sin^2 2\theta + 1})}{2^3(\sqrt{4\sin^2 2\theta + 1})^3}\,d\theta$$
$$= \int_0^{2\pi} \frac{\sqrt{5 + 12\cos^2 2\theta}}{4\sin^2 2\theta + 1}\,d\theta \approx 10.107018. \quad \triangle \qquad 5.3.29$$

Proposition 3.9.17: The curvature κ of a curve parametrized by $\gamma : \mathbb{R} \to \mathbb{R}^3$ is

$$\kappa\Big(\gamma(t)\Big) = \frac{|\vec{\gamma}'(t) \times \vec{\gamma}''(t)|}{|\vec{\gamma}'(t)|^3}.$$

We used MAPLE to compute the integral in equation 5.3.29. Many graphing calculators will also compute this integral.

Example 5.3.9 (Area of a torus). Choose $R > r > 0$. We obtain the torus shown in Figure 5.3.2 by taking the circle of radius r in the (x, z)-plane that is centered at $x = R$, $z = 0$, and rotating it around the z-axis. This surface is parametrized by

$$\gamma\begin{pmatrix} u \\ v \end{pmatrix} = \begin{pmatrix} (R + r\cos u)\cos v \\ (R + r\cos u)\sin v \\ r\sin u \end{pmatrix}, \qquad 5.3.30$$

as Exercise 5.1 asks you to verify. Then

$$[\mathbf{D}\gamma(\mathbf{u})] = \begin{bmatrix} -r\sin u\cos v & -(R + r\cos u)\sin v \\ -r\sin u\sin v & (R + r\cos u)\cos v \\ r\cos u & 0 \end{bmatrix} \qquad 5.3.31$$

and the surface area of the torus is given by the integral

$$\int_{[0,2\pi]\times[0,2\pi]} \sqrt{\det[\mathbf{D}\gamma(\mathbf{u})]^\top[\mathbf{D}\gamma(\mathbf{u})]}\,|du\,dv|$$
$$= \int_{[0,2\pi]\times[0,2\pi]} \sqrt{\det\begin{bmatrix} r^2 & 0 \\ 0 & (R + r\cos u)^2 \end{bmatrix}}\,|du\,dv|$$
$$= \int_0^{2\pi}\int_0^{2\pi} r(R + r\cos u)\,du\,dv = 4\pi^2 rR. \qquad 5.3.32$$

Note that the answer has the right units: r and R have units of length, so $4\pi^2 rR$ has units length squared. $\triangle$

Equation 5.3.32: When computing

$$[\mathbf{D}\gamma(\mathbf{u})]^\top[\mathbf{D}\gamma(\mathbf{u})],$$

be sure to put the matrices in the right order (transpose to the left), and simplify the entries of the resulting matrix as much as possible before computing the determinant.

We write $|du\,dv|$ in equation 5.3.32 to avoid having to put subscripts on our variables; we could have used u_1 and u_2 rather than u and v, and then used the integrand $|d^2\mathbf{u}|$. In the final, double integral, we are integrating over an *oriented* interval, from 0 to 2π, so we write $du\,dv$ rather than $|du\,dv|$.

It is exceptional that the square root of any function can be integrated in elementary terms, as it can in Example 5.3.9. Example 5.3.10 is more typical.

Remark. Recall that in $\mathbb{R}^3$,

$$\sqrt{\det[\mathbf{D}\gamma(\mathbf{u})]^\top[\mathbf{D}\gamma(\mathbf{u})]} = \big|\overrightarrow{D_1\gamma} \times \overrightarrow{D_2\gamma}\big| \qquad 5.3.33$$

(as Exercise 5.1.4 asked you to show), so you could also compute the area in Example 5.3.9 using the cross product. $\triangle$

Example 5.3.10 (Surface area: a harder problem). What is the area of the graph of the function $x^2 + y^3$ above the unit square $Q \subset \mathbb{R}^2$? Applying formula 5.2.6, we parametrize the surface by

$$\gamma\begin{pmatrix} x \\ y \end{pmatrix} \mapsto \begin{pmatrix} x \\ y \\ x^2 + y^3 \end{pmatrix} \qquad 5.3.34$$

and use equation 5.3.33 to compute

$$\int_Q \left| \overbrace{\begin{bmatrix} 1 \\ 0 \\ 2x \end{bmatrix}}^{\overrightarrow{D_1\gamma}} \times \overbrace{\begin{bmatrix} 0 \\ 1 \\ 3y^2 \end{bmatrix}}^{\overrightarrow{D_2\gamma}} \right| |dx\, dy| = \int_0^1 \int_0^1 \sqrt{1+4x^2+9y^4}\, dx\, dy. \qquad 5.3.35$$

(On the right, we have $dx\, dy$, not $|dx\, dy|$, because we are now integrating over an oriented domain.) The integral with respect to x is one we can compute (just barely in our case, checking our result with a table of integrals). First we get

$$\int \sqrt{u^2+a^2}\, du = \frac{u\sqrt{u^2+a^2}}{2} + \frac{a^2 \ln(u+\sqrt{u^2+a^2})}{2}. \qquad 5.3.36$$

This leads to the integral

$$\begin{aligned} &\int_0^1 \int_0^1 \sqrt{1+4x^2+9y^4}\, dx\, dy \\ &= \int_0^1 \left[\frac{x\sqrt{4x^2+1+9y^4}}{2} + \frac{(1+9y^4)\ln\big(2x+\sqrt{4x^2+1+9y^4}\,\big)}{4} \right]_0^1 dy \\ &= \int_0^1 \left(\frac{\sqrt{5+9y^4}}{2} + \frac{1+9y^4}{4} \ln \frac{2+\sqrt{5+9y^4}}{\sqrt{1+9y^4}} \right) dy. \end{aligned} \qquad 5.3.37$$

It is hopeless to try to integrate this mess in elementary terms. The first term requires elliptic functions, and we don't know of any class of special functions in which the second term could be expressed. But numerically, this is no big problem; Simpson's method with 20 steps gives the approximation 1.93224957.... △

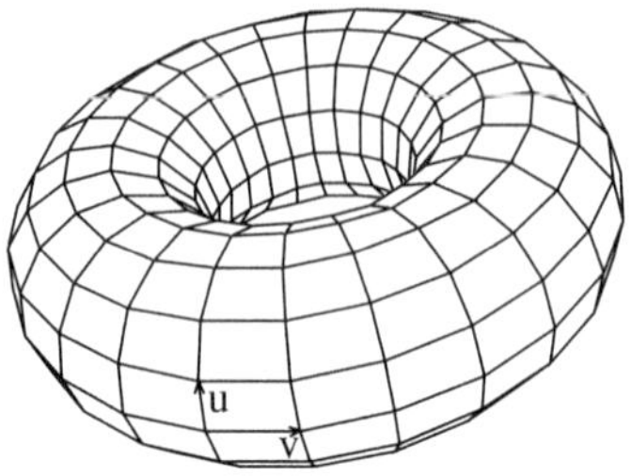

FIGURE 5.3.2.

The torus of Example 5.3.9 with the u, v coordinates drawn. You should imagine the straight lines as curved. By "torus" we mean the surface of the object. The solid object is called a "solid torus".

Even Example 5.3.10 is computationally nicer than is standard: we were able to integrate with respect to x. If we had asked for the area of the graph of $x^3 + y^4$, then we couldn't have integrated in elementary terms with respect to either variable, and would have needed the computer to evaluate the double integral.

And what's wrong with that? Integrals exist whether or not they can be computed in elementary terms, and a fear of numerical integrals is inappropriate in this age of computers. If you restrict yourself to surfaces whose areas can be computed in elementary terms, you are restricting yourself to a minute class of surfaces.

Example 5.3.11 (Area of a surface in $\mathbb{R}^4$). The subset of $\mathbb{R}^4$ given by the two equations in four unknowns

$$x_1^2 + x_2^2 = r_1^2 \quad \text{and} \quad x_3^2 + x_4^2 = r_2^2 \qquad 5.3.38$$

is a surface. It can be parametrized by

$$\gamma\begin{pmatrix} u \\ v \end{pmatrix} = \begin{pmatrix} r_1 \cos u \\ r_1 \sin u \\ r_2 \cos v \\ r_2 \sin v \end{pmatrix}, 0 \le u, v \le 2\pi, \qquad 5.3.39$$

and since

$$\overbrace{\begin{bmatrix} -r_1 \sin u & r_1 \cos u & 0 & 0 \\ 0 & 0 & -r_2 \sin v & r_2 \cos v \end{bmatrix}}^{\left[\mathbf{D}\gamma\left(\begin{pmatrix} u \\ v \end{pmatrix}\right)\right]^\top} \overbrace{\begin{bmatrix} -r_1 \sin u & 0 \\ r_1 \cos u & 0 \\ 0 & -r_2 \sin v \\ 0 & r_2 \cos v \end{bmatrix}}^{\left[\mathbf{D}\gamma\left(\begin{pmatrix} u \\ v \end{pmatrix}\right)\right]}$$

$$= \begin{bmatrix} r_1^2 & 0 \\ 0 & r_2^2 \end{bmatrix}, \qquad 5.3.40$$

the area of the surface is given by

$$\int_{[0,2\pi]\times[0,2\pi]} \sqrt{\det\left(\left[\mathbf{D}\gamma\begin{pmatrix}u\\v\end{pmatrix}\right]^\top \left[\mathbf{D}\gamma\begin{pmatrix}u\\v\end{pmatrix}\right]\right)}\,|du\,dv| \\ = \int_{[0,2\pi]\times[0,2\pi]} \sqrt{r_1^2 r_2^2}\,|du\,dv| = 4\pi^2 r_1 r_2. \quad \triangle \qquad 5.3.41$$

Another class of surfaces in $\mathbb{R}^4$, important in many applications, is described using complex variables. This leads to remarkably simpler computations than one might expect, as shown in the following example.

Example 5.3.12:

$$\left[\mathbf{D}\gamma\begin{pmatrix}r\\\theta\end{pmatrix}\right] = \begin{bmatrix}\cos\theta & -r\sin\theta\\ \sin\theta & r\cos\theta\\ 2r\cos 2\theta & -2r^2\sin 2\theta\\ 2r\sin 2\theta & 2r^2\cos 2\theta\end{bmatrix}$$

Example 5.3.12 (Area of a surface in $\mathbb{C}^2$). Consider the graph of the function $f(z) = z^2$, where z is complex, i.e., $z = x + iy$. This graph has the equation $z_2 = z_1^2$ in $\mathbb{C}^2$, or

$$x_2 = x_1^2 - y_1^2, \quad y_2 = 2x_1y_1, \quad \text{in } \mathbb{R}^4. \qquad 5.3.42$$

Let us compute the area of the part of the surface of equation $z_2 = z_1^2$ where $|z_1| \le 1$. Polar coordinates for z_1 give a nice parametrization:

$$\gamma\begin{pmatrix}r\\\theta\end{pmatrix} = \begin{pmatrix}r\cos\theta\\ r\sin\theta\\ r^2\cos 2\theta\\ r^2\sin 2\theta\end{pmatrix}, \quad 0 \le r \le 1, \quad 0 \le \theta \le 2\pi. \qquad 5.3.43$$

Again we need to compute the area of the parallelogram spanned by the two partial derivatives. Since

$$\left[\mathbf{D}\gamma\begin{pmatrix}r\\\theta\end{pmatrix}\right]^\top \left[\mathbf{D}\gamma\begin{pmatrix}r\\\theta\end{pmatrix}\right] = \begin{bmatrix}1+4r^2 & 0\\ 0 & r^2(1+4r^2)\end{bmatrix}, \qquad 5.3.44$$

the area is

$$\int_{[0,1]\times[0,2\pi]} \underbrace{\sqrt{r^2(1+4r^2)^2}}_{\substack{\text{square root}\\ \text{of a perfect square}}}\,|dr\,d\theta| = 2\pi\left[\frac{r^2}{2} + r^4\right]_0^1 = 3\pi. \qquad 5.3.45$$

Remark. Notice that the determinant in Example 5.3.12 turns out to be a perfect square; the square root that created such trouble in Example 5.3.6, which deals with the real curve of equation $y = x^2$, causes none for the complex surface with the same equation, $z_2 = z_1^2$. This "miracle" happens for *all* manifolds in $\mathbb{C}^n$ given by complex equations, such as polynomials in complex variables. Additional examples are explored in Exercises 5.3.16 and 5.3.21. Exercise 6.3.15 explains the reason behind the "miracle": complex vector spaces are naturally oriented. $\triangle$

Example 5.3.13 (Volume of a three-dimensional manifold in $\mathbb{R}^4$).
Let $U \subset \mathbb{R}^3$ be an open set, and let $f : U \to \mathbb{R}$ be a C^1 function. The graph of f is a three-dimensional manifold in $\mathbb{R}^4$, and it comes with the natural parametrization γ shown in the margin. We then have

$$\det\left(\left[\mathbf{D}\gamma\begin{pmatrix}x\\y\\z\end{pmatrix}\right]^\top \left[\mathbf{D}\gamma\begin{pmatrix}x\\y\\z\end{pmatrix}\right]\right) = \det\left(\begin{bmatrix}1 & 0 & 0 & D_1 f\\ 0 & 1 & 0 & D_2 f\\ 0 & 0 & 1 & D_3 f\end{bmatrix}\begin{bmatrix}1 & 0 & 0\\ 0 & 1 & 0\\ 0 & 0 & 1\\ D_1 f & D_2 f & D_3 f\end{bmatrix}\right)$$

$$= \det\begin{bmatrix}1+(D_1 f)^2 & (D_1 f)(D_2 f) & (D_1 f)(D_3 f)\\ (D_1 f)(D_2 f) & 1+(D_2 f)^2 & (D_2 f)(D_3 f)\\ (D_1 f)(D_3 f) & (D_2 f)(D_3 f) & 1+(D_3 f)^2\end{bmatrix} \quad 5.3.46$$

$$= 1 + (D_1 f)^2 + (D_2 f)^2 + (D_3 f)^2.$$

So the three-dimensional volume of the graph of f is

$$\int_U \sqrt{1+(D_1 f)^2+(D_2 f)^2+(D_3 f)^2}\,|d^3\mathbf{x}|. \quad 5.3.47$$

It is a challenge to find any function for which this can be integrated in elementary terms. Let us try to find the area of the graph of

$$f\begin{pmatrix}x\\y\\z\end{pmatrix} = \frac{1}{2}(x^2+y^2+z^2) \quad 5.3.48$$

above the ball of radius R centered at the origin, $B_R(\mathbf{0})$. Using spherical coordinates, we get

$$\int_{B_R(\mathbf{0})} \sqrt{1+(D_1 f)^2+(D_2 f)^2+(D_3 f)^2}\,|d^3\mathbf{x}|$$

$$= \int_{B_R(\mathbf{0})} \sqrt{1+x^2+y^2+z^2}\,|d^3\mathbf{x}|$$

$$= \int_0^{2\pi}\int_{-\pi/2}^{\pi/2}\int_0^R \sqrt{1+(r\cos\theta\cos\varphi)^2+(r\sin\theta\cos\varphi)^2+r^2\sin^2\varphi}\,r^2\cos\varphi\,dr\,d\varphi\,d\theta$$

$$= \int_0^{2\pi}\int_{-\pi/2}^{\pi/2}\int_0^R \sqrt{1+r^2}\,r^2\cos\varphi\,dr\,d\varphi\,d\theta = 4\pi\int_0^R \sqrt{1+r^2}\,r^2\,dr \quad 5.3.49$$

$$= \pi\left(R(1+R^2)^{3/2} - \frac{1}{2}\ln(R+\sqrt{1+R^2}) - \frac{1}{2}R\sqrt{1+R^2}\right). \quad \triangle$$

Example 5.3.14 (Volume of an n-dimensional sphere in $\mathbb{R}^{n+1}$). Let us compute $\mathrm{vol}_n S^n$, where $S^n \subset \mathbb{R}^{n+1}$ is the unit sphere. We could do this using some generalization of spherical coordinates, and you are asked to do so for the 3-sphere in Exercise 5.3.15. These computations become quite cumbersome, and there is an easier method. It relates $\mathrm{vol}_n S^n$ to the $(n+1)$-dimensional volume of an $(n+1)$-dimensional ball, $\mathrm{vol}_{n+1} B^{n+1}$.

How might we relate the *length* of a circle (i.e., a one-dimensional sphere, S^1) to the *area* of a disc (a two-dimensional ball, B^2)? We could fill up

---

$$\gamma\begin{pmatrix}x\\y\\z\end{pmatrix} = \begin{pmatrix}x\\y\\z\\f\begin{pmatrix}x\\y\\z\end{pmatrix}\end{pmatrix}$$

Parametrization for Example 5.3.13

Equation 5.3.49: This integral will challenge your ability with techniques of integration. You are asked in Exercise 5.3.19 to justify the last step of this computation.

the disc with concentric rings and add their areas, each approximately the length of the corresponding circle times some δr representing the spacing of the circles. The length of the circle of radius r is r times the length of the circle of radius 1. More generally, the volume of the part of B^{n+1} between r and $r + \Delta r$ should be Δr times $(\mathrm{vol}_n(S^n(r)))$. This leads to

$$\begin{aligned}\mathrm{vol}_{n+1} B^{n+1} &= \int_0^1 \mathrm{vol}_n S^n(r)\, dr \\ &= \int_0^1 r^n \mathrm{vol}_n S^n\, dr = \frac{1}{n+1} \mathrm{vol}_n(S^n),\end{aligned} \qquad 5.3.50$$

which you are asked to justify in Exercise 5.6. This allows us to add one more column to Table 4.5.9 to get Table 5.3.3.

We can compute the area of the unit disc from the length of the unit circle:

$$\mathrm{vol}_2 B^2 = \int_0^1 r \,\mathrm{vol}_1 S^1\, dr$$
$$= \int_0^1 2\pi r\, dr = \left[\frac{2\pi r^2}{2}\right]_0^1 = \pi.$$

Another way to get equation 5.3.50 is to think of the n-dimensional volume of the n-dimensional sphere of radius r as the derivative of the $(n+1)$-dimensional volume of the $(n+1)$ ball of radius r. Exercise 5.6 explores a third approach.

The 0-dimensional sphere in $\mathbb{R}$ consists of the points -1 and 1.

n	$c_n = \frac{n-1}{n} c_{n-2}$	Volume of ball $\beta_n = c_n \beta_{n-1}$	Volume of sphere $\mathrm{vol}_n S^n = (n+1)\beta_{n+1}$
0	π		2
1	2	2	2π
2	$\frac{\pi}{2}$	π	4π
3	$\frac{4}{3}$	$\frac{4\pi}{3}$	$2\pi^2$
4	$\frac{3\pi}{8}$	$\frac{\pi^2}{2}$	$\frac{8\pi^2}{3}$
5	$\frac{16}{15}$	$\frac{8\pi^2}{15}$	π^3

TABLE 5.3.3. The volume β_n of the n-dimensional unit ball $B^n \subset \mathbb{R}^n$, for $n = 1, \dots, 5$, and for the n-dimensional unit sphere in $\mathbb{R}^{n+1}$, for $n = 0, 1 \dots, 5$.

EXERCISES FOR SECTION 5.3

5.3.1 a. Let $\begin{pmatrix} r(t) \\ \theta(t) \end{pmatrix}$ be a parametrization of a curve in polar coordinates. Show that the length of the piece of curve between $t = a$ and $t = b$ is given by the integral $\int_a^b \sqrt{(r'(t))^2 + (r(t))^2(\theta'(t))^2}\, dt$.

b. Consider the spiral in polar coordinates when $r(t) = e^{-\alpha t}$ and $\theta(t) = t$, for $\alpha > 0$. What is its length between $t = 0$ and $t = b$? What is the limit of this length as $\alpha \to 0$?

c. Show that the spiral turns infinitely many times around the origin as $t \to \infty$. Does the length tend to ∞ as $b \to \infty$?

5.3.2 Use the result of Exercise 5.3.1, part a to get the length of the curve

$$\begin{pmatrix} r(t) \\ \theta(t) \end{pmatrix} = \begin{pmatrix} 1/t \\ t \end{pmatrix}$$

between $t = 1$ and $t = a$. Is the limit of the length finite as $a \to \infty$?

5.3.3 a. Suppose that $t \mapsto \begin{pmatrix} r(t) \\ \theta(t) \\ \varphi(t) \end{pmatrix}$ is a parametrization of a curve in $\mathbb{R}^3$, written in spherical coordinates. Find the formula analogous to the integral in Exercise 5.3.1, part a, for the length of the arc between $t = a$ and $t = b$.

b. What is the length of the curve parametrized by $r(t) = \cos t$, $\theta(t) = \tan t$, $\varphi(t) = t$, between $t = 0$ and $t = a$, where $0 < a < \pi/2$?

5.3.4 a. Set up an integral to compute the surface area of the unit sphere.

b. Compute the surface area (if you know the formula, as we certainly hope you do, simply giving it is not good enough).

5.3.5 a. Set up (but do not compute) the integral giving the surface area of the part of the surface of equation $z = \dfrac{x^2}{4} + \dfrac{y^2}{9}$, where $z \le a^2$.

b. What is the volume of the region $\dfrac{x^2}{4} + \dfrac{y^2}{9} \le z \le a^2$?

5.3.6 Let S be the part of the paraboloid of revolution $z = x^2 + y^2$ where $z \le 9$. Compute the integral $\displaystyle\int_S \left(x^2 + y^2 + 3z^2\right) |d^2\mathbf{x}|$.

Exercise 5.3.7: An *ellipsoid* is a solid of which every plane section is an ellipse or a circle.

5.3.7 a. Give a parametrization of the surface of the ellipsoid $\dfrac{x^2}{a^2} + \dfrac{y^2}{b^2} + \dfrac{z^2}{c^2} = 1$ analogous to spherical coordinates.

b. Set up an integral to compute the surface area of the ellipsoid.

5.3.8 What is the surface area of the part of the paraboloid of revolution $z = x^2 + y^2$ where $z \le 1$?

Exercise 5.3.9, part b: A computer and appropriate software will help.

5.3.9 a. Set up an integral to compute the integral $\int_S (x + y + z)\, |d^2\mathbf{x}|$, where S is the part of the graph of $x^3 + y^4$ above the unit circle.

b. Evaluate this integral numerically.

5.3.10 Let $X \subset \mathbb{C}^2$ be the graph of the function $w = z^k$, where $z = x + iy = re^{i\theta}$ and $w = u + iv$ are both complex variables.

Exercise 5.3.10, part a: Since $z = x + iy$, using polar coordinates $x = r\cos\theta$, $y = r\sin\theta$ for z means setting $z = r(\cos\theta + i\sin\theta)$.

a. Parametrize X in terms of the polar coordinates r, θ for z.

b. What is the area of the part of X where $|z| \le R$?

Exercise 5.3.11: Theorem 5.4.6 will show how to do this kind of problem systematically. For now, you need to do it by bare hands.

5.3.11 The *total curvature* $\mathbf{K}(S)$ of a surface $S \subset \mathbb{R}^3$ is given by integrating the Gaussian curvature (Definition 3.9.8) over the surface:

$$\mathbf{K}(S) = \int_S |K(\mathbf{x})|\,|d^2\mathbf{x}|.$$

a. What is the total curvature of the sphere $S^2_R \subset \mathbb{R}^3$ of radius R?

b. What is the total curvature of the graph of $f\begin{pmatrix} x \\ y \end{pmatrix} = x^2 - y^2$? (See Example 3.9.12.)

5.3.12 The *total curvature* of a curve C is $\int_C \kappa\,|d^1\mathbf{x}|$. Let $C \subset \mathbb{R}^3$ be the curve that is the intersection of the parabolic cylinder of equation $y = x^2 - 1$ and the parabolic cylinder of equation $z = y^2 - 1$.

a. Show that C can be parametrized by $\gamma(t) = \begin{pmatrix} t \\ t^2 - 1 \\ t^4 - 2t^2 \end{pmatrix}$.

b. Find $\kappa(\gamma(t))$ and $\tau(\gamma(t))$.

c. Find an integral for the total curvature of C between $\begin{pmatrix} 0 \\ -1 \\ 0 \end{pmatrix}$ and $\begin{pmatrix} a \\ a^2 - 1 \\ a^4 - 2a^2 \end{pmatrix}$.

d. Is the total curvature of all of C finite or infinite?

Exercise 5.3.13: Part a was discovered by Archimedes; a sculpture on his tomb with a sphere and cylinder alluded to this result. The tomb was discovered 137 years after Archimedes's death by Cicero, then questor in Sicily.

Exercise 5.3.13, part b: The circumference of the earth is 40 000 kilometers. The polar caps are the regions bounded by the Arctic and the Antarctic Circles, which are $23°27'$ from the North and South Poles respectively. The Tropics is the region between the tropic of Cancer ($23°27'$ north of the equator) and the tropic of Capricorn ($23°27'$ south of the equator).

5.3.13 a. Let S^2 be the unit sphere and let S_1 be the part of the cylinder of equation $x^2 + y^2 = 1$ with $-1 \le z \le 1$. Show that the horizontal radial projection $S_1 \to S^2$ preserves area.

b. What is the area of the polar caps on earth? The Tropics?

c. Find a formula for the area $A_R(r)$ of a disc of radius r on a sphere of radius R (the radius of the disc is measured on the sphere, not inside the ball). What is the Taylor polynomial of $A_R(r)$ to degree 4?

Exercise 5.3.14: Triangulation of the unit square means decomposing the square into triangles.

***5.3.14** Let $\mathbf{f}\begin{pmatrix} u \\ v \end{pmatrix} = \begin{pmatrix} u \\ u^2 \\ v \end{pmatrix}$ be a parametrization of a parabolic cylinder. If T is a triangle with vertices $\mathbf{a}, \mathbf{b}, \mathbf{c} \in \mathbb{R}^2$, the image triangle will be by definition the triangle in $\mathbb{R}^3$ with vertices $\mathbf{f}(\mathbf{a}), \mathbf{f}(\mathbf{b}), \mathbf{f}(\mathbf{c})$. Show that there exists a triangulation of the unit square in the (u, v)-plane such that the sum of the areas of the image triangles is arbitrarily large.

$$\gamma : \begin{pmatrix} \theta \\ \varphi \\ \psi \end{pmatrix} \mapsto \begin{pmatrix} \cos\psi\cos\varphi\cos\theta \\ \cos\psi\cos\varphi\sin\theta \\ \cos\psi\sin\varphi \\ \sin\psi \end{pmatrix}$$

Map γ for Exercise 5.3.15

5.3.15 a. Show that when φ, ψ, θ satisfy

$$-\pi/2 \le \varphi \le \pi/2, \quad -\pi/2 \le \psi \le \pi/2, \quad 0 \le \theta < 2\pi,$$

the mapping γ shown in the margin parametrizes the unit sphere S^3 in $\mathbb{R}^4$.

b. Use this parametrization to compute $\text{vol}_3(S^3)$.

5.3.16 Define $X = \left\{ \begin{pmatrix} z \\ w \end{pmatrix} \in \mathbb{C}^2 \mid w = e^z + e^{-z} \right\}$, i.e., X is the graph of the function $e^z + e^{-z}$. What is the area of the part of X where $-1 \le x \le 1$ and $-1 \le y \le 1$?

5.3.17 What is the center of gravity of a uniform wire, whose position is the parabola of equation $y = x^2$, where $0 \le x \le a$?

5.3.18 A gas has density C/r, where $r = \sqrt{x^2 + y^2 + z^2}$. If $0 < a < b$, what is the mass of the gas between the concentric spheres $r = a$ and $r = b$?

5.3.19 Justify equation 5.3.49 by computing the integral $4\pi \int_0^R \sqrt{1 + r^2}\, r^2\, dr$ in the third line.

5.3.20 Let $M_1(n, m)$ be the space of $n \times m$ matrices of rank 1. What is the three-dimensional volume of the part of $M_1(2, 2)$ made up of matrices A with $|A| \le R$, for $R > 0$?

5.3.21 What is the area of the surface in $\mathbb{C}^3$ parametrized by $\gamma(z) = \begin{pmatrix} z^p \\ z^q \\ z^r \end{pmatrix}$, for $z \in \mathbb{C}$, $|z| \le 1$?

5.4 Integration and curvature

In this section we prove three theorems relating curvature to integration: Gauss's *remarkable theorem*, which explains Gaussian curvature; Theorem 5.4.4, which explains mean curvature; and Theorem 5.4.6, which relates curvature to the Gauss map.

Gauss, writing in Latin, called this result the "remarkable theorem: If a curved surface is developed upon any other surface whatever, the measure of curvature in each point remains unchanged." Gauss's own proof is a complicated and unmotivated computation, the kind of thing that led Jacobi to liken Gauss to the *fox who erases his tracks in the sand with his tail*; see page 4.

The Latin word "egregius" (literally, *not of the common herd*) is translated as *admirable, excellent, extraordinary, distinguished, ...* . This laudatory meaning persisted in English through the 19th century (Thackeray wrote of "some one splendid and egregious"), but the word was also used in its current meaning, "remarkably bad"; in Shakespeare's *Cymbeline*, a character declares himself to be an "egregious murderer".

An *isometry* is a map that preserves lengths of all curves.

Gauss's remarkable theorem

Perhaps the most famous theorem of differential geometry is Gauss's *Theorema Egregium* (Latin for *remarkable theorem*), which asserts that the Gaussian curvature of a surface is "intrinsic": it can be computed from lengths and angles measured in the surface.

It is remarkable because the Gaussian curvature is the product of two quantities that themselves are *extrinsic*: they depend on how the surface is embedded in $\mathbb{R}^3$.

We saw in Section 3.9 that in "best coordinates", a surface $S \subset \mathbb{R}^3$ is locally near every $\mathbf{p} \in S$ the graph of a map from its tangent space $T_\mathbf{p}S$ to its normal line N_p (see Figure 3.9.6). As we saw in equation 3.9.28, the Taylor polynomial of this map starts with quadratic terms:

$$Z = f\begin{pmatrix} X \\ Y \end{pmatrix} = \frac{1}{2}(A_{2,0}X^2 + 2A_{1,1}XY + A_{0,2}Y^2) + o(X^2+Y^2). \qquad 5.4.1$$

The above is *extrinsic*: if $\varphi : S \to S'$ is an isometry between embedded surfaces, the coefficients $A_{2,0}$, $A_{1,1}$, and $A_{0,2}$ of S and the coefficients $A'_{2,0}$, $A'_{1,1}$, and $A'_{0,2}$ of S' will usually be different. Yet Gauss showed that the Gaussian curvature is intrinsic:

$$K(\mathbf{p}) = A_{2,0}A_{0,2} - A_{1,1}^2 = A'_{2,0}A'_{0,2} - {A'_{1,1}}^2. \qquad 5.4.2$$

We said in Section 3.9 that we can express the Theorema Egregium as Theorem 3.9.9, which we repeat below as Theorem 5.4.1.

Theorem 5.4.1. *Let $D_r(\mathbf{p})$ be the set of all points $\mathbf{q}$ in a surface $S \subset \mathbb{R}^3$ such that there exists a curve of length $\le r$ in S joining $\mathbf{p}$ to $\mathbf{q}$. Then*

$$\text{Area}(D_r(\mathbf{p})) = \pi r^2 - \frac{\pi K(\mathbf{p})}{12} r^4 + o(r^4). \qquad 5.4.3$$

Since the area of $D_r(\mathbf{p})$, the disc of radius r around $\mathbf{p}$, is obviously an "intrinsic" function and it determines the curvature $K(\mathbf{p})$, the Theorema Egregium follows.

The other proofs of Theorem 5.4.1 that we know all involve either the Christoffel symbol $\Gamma_{1,2}$ (hence the Levi-Civita connection associated to the Riemann metric on S inherited from the embedding), or the Jacobi second variation equation, which describes how geodesics on S spread apart.

These tools are crucial to any serious study of differential geometry, but they are beyond the scope of this book. The present proof uses instead techniques and concepts developed in Chapters 3, 4, and 5, as well as rules for "little o" and "big O" given in Appendix A11.

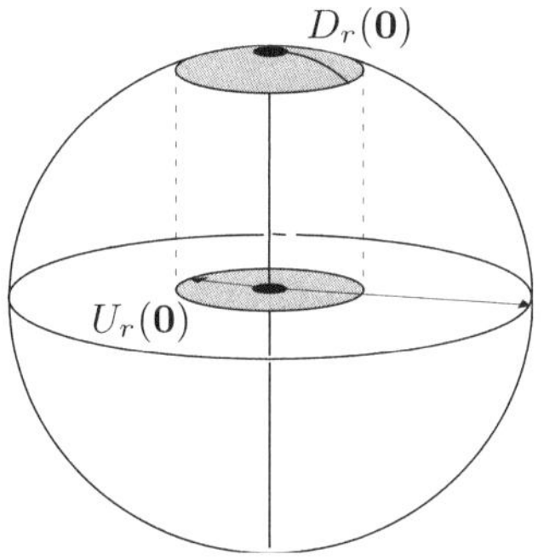

FIGURE 5.4.1.
The disc of radius r around the north pole on the unit sphere projects to the disc U_r of radius

$$\sin r = r - r^3/6 + o(r^3)$$

in the (x, y)-plane. This corresponds to equation 5.4.19 in the case $a = b = 1$.

Proof. The proof will take about four pages. Since Theorem 5.4.1 is local, we may assume that $\mathbf{p} = \mathbf{0}$ and that S is the graph of a smooth function defined near the origin of $\mathbb{R}^2$, whose Taylor polynomial starts with quadratic terms, as in equation 5.4.1. These quadratic terms are a quadratic form Q_A on $\mathbb{R}^2$, where $A = \begin{bmatrix} A_{2,0} & A_{1,1} \\ A_{1,1} & A_{0,2} \end{bmatrix}$. Since A is symmetric, by the spectral theorem (Theorem 3.7.15) there exists an orthonormal basis in which the matrix A is diagonal. To lighten notation, we replace $A_{2,0}$ by a and $A_{0,2}$ by b. Thus we may assume that our surface is the graph of a function

$$f\begin{pmatrix} x \\ y \end{pmatrix} = \frac{1}{2}(ax^2 + by^2) + o(x^2 + y^2), \qquad 5.4.4$$

so that the Gaussian curvature at the origin is $K(\mathbf{0}) = ab$. Thus to prove Theorem 5.4.1, we need to show that

$$\text{Area}(D_r(\mathbf{0})) = \pi r^2 - \frac{ab\pi}{12} r^4 + o(r^4). \qquad 5.4.5$$

We know from Definition 5.3.1 how to compute the volume of a manifold known by a parametrization. The first step is to parametrize the surface S. We will use the "radial parametrization"

$$g\begin{pmatrix} \rho \\ \theta \end{pmatrix} = \begin{pmatrix} \rho\cos\theta \\ \rho\sin\theta \\ f\begin{pmatrix} \rho\cos\theta \\ \rho\sin\theta \end{pmatrix} \end{pmatrix} = \begin{pmatrix} \rho\cos\theta \\ \rho\sin\theta \\ \frac{1}{2}\rho^2(a\cos^2\theta + b\sin^2\theta) + o(\rho^2) \end{pmatrix}. \qquad 5.4.6$$

However, to apply Definition 5.3.1, we need to find a subset $U_r \subset \mathbb{R}^2$ such that $g(U_r) = D_r(\mathbf{0})$, or, rather, that they have the same area to within terms in $o(r^4)$. Then we will be able to write

$$\text{Area}\, D_r(\mathbf{0}) = \int_{U_r} \sqrt{\det\left[\mathbf{D}g\begin{pmatrix} \rho \\ \theta \end{pmatrix}\right]^\top \left[\mathbf{D}g\begin{pmatrix} \rho \\ \theta \end{pmatrix}\right]} |d\rho\, d\theta| + o(r^4). \qquad 5.4.7$$

We will begin by finding a parametrized curve $\widetilde{\gamma}$ in S such that the distance along the curve from $\mathbf{0}$ to another point in S is sufficiently close to minimal so that our error is in $o(r^4)$. We can think of the arc $\widetilde{\gamma}([0, r])$ between $\widetilde{\gamma}(0)$ and $\widetilde{\gamma}(r)$ as the "very slightly wiggly" *lift* to S of a straight spoke (of length r) of the disc of radius r in the (ρ, θ)-plane. We will determine a set $D'_r(\mathbf{0}) \subset S$ that approximates $D_r(\mathbf{0})$ to adequate precision by determining this "almost minimal"[4] length; the desired subset $U_r \subset \mathbb{R}^2$ will be the projection of $D'_r(\mathbf{0})$ onto $\mathbb{R}^2$ (see Figure 5.4.1).

[4]We speak of curves that almost minimize distance because proving that geodesics (curves that actually minimize distances) exist in S is beyond the scope of this book.

Lengths of curves on S

How do we find a curve in S of "almost minimal" length? Suppose a smooth curve $\widetilde{\gamma}$ on S projects to $\mathbb{R}^2$ as the curve γ of equation $\theta = \alpha(\rho)$ in polar coordinates, so we can write

$$\gamma(\rho) = \begin{pmatrix} \rho\cos\alpha(\rho) \\ \rho\sin\alpha(\rho) \end{pmatrix} \quad \text{and} \quad \widetilde{\gamma}(\rho) = \begin{pmatrix} \gamma(\rho) \\ f(\gamma(\rho)) \end{pmatrix}, \qquad 5.4.8$$

where f is as in equation 5.4.4. If γ and $\widetilde{\gamma}$ are smooth, the function $\rho \mapsto \alpha(\rho)$ that specifies them has the Taylor polynomial

$$\alpha(\rho) = \theta_0 + k\rho + \frac{m}{2}\rho^2 + o(\rho^2), \qquad 5.4.9$$

where $\theta_0 = \alpha(0)$, $k = \alpha'(0)$, and $m = \alpha''(0)$. We shall see that curves that minimize the distance from $\mathbf{0}$ to a point on S must satisfy $k = 0$.

Proposition 5.4.2 is intuitive: to travel fast, you should go as straight as possible and hence have no sideways acceleration.

Note that the only difference between $\widetilde{\gamma}$ and $\widetilde{\delta}_r$ is that for the former, α is a function of ρ; for the latter, $\alpha(r)$ is constant.

Proposition 5.4.2. *Set* $\gamma(\rho) = \begin{pmatrix} \rho\cos\alpha(\rho) \\ \rho\sin\alpha(\rho) \end{pmatrix}$ *and* $\delta_r(\rho) = \begin{pmatrix} \rho\cos\alpha(r) \\ \rho\sin\alpha(r) \end{pmatrix}$, *where α is as in equation 5.4.9, and let $\widetilde{\gamma}$ and $\widetilde{\delta}_r$ be the parametrized curves*

$$\widetilde{\gamma}(\rho) = \begin{pmatrix} \gamma(\rho) \\ f(\gamma(\rho)) \end{pmatrix} \quad \text{and} \quad \widetilde{\delta}_r(\rho) = \begin{pmatrix} \delta_r(\rho) \\ f(\delta_r(\rho)) \end{pmatrix}, \qquad 5.4.10$$

If $k \neq 0$ and $r > 0$ is sufficiently small, then the arc $\widetilde{\delta}_r([0,r])$ is shorter than the arc $\widetilde{\gamma}([0,r])$.

Thus when trying to find curves that "adequately minimize" the distance between $\mathbf{0} \in S$ and a point on the boundary $\partial D_r(\mathbf{0})$, we need only consider the case $k = 0$, i.e., curves with $\alpha(\rho) = \theta_0 + m\rho^2/2 + o(\rho^2)$.

Equation 5.4.11: Definition 3.9.5 gives the formula for arc length; equation 5.3.4 shows that this definition is compatible with Definition 5.3.1 for the volume of a manifold.

In equation 5.4.12, the cross product of the first square,

$$-2\rho\cos\alpha(\rho)\sin\alpha(\rho)\alpha'(\rho)$$

cancels that of the second square.

Proof of Proposition 5.4.2. The proof consists of computing Taylor polynomials. We will compute, through terms in r^3, the Taylor polynomials of the lengths of $\widetilde{\delta}_r([0,r])$ and $\widetilde{\gamma}([0,r])$, as functions of r. The formula for arc length, the definition of $\widetilde{\gamma}$, and the chain rule tell us that

$$\begin{aligned} \text{Length}(\widetilde{\gamma}([0,r])) &= \int_0^r |\widetilde{\gamma}'(\rho)|\,d\rho = \int_0^r \left| \begin{pmatrix} \gamma'(\rho) \\ (f\circ\gamma)'(\rho) \end{pmatrix} \right| d\rho \\ &= \int_0^r \sqrt{|\gamma'(\rho)|^2 + \big|[\mathbf{D}f(\gamma(\rho))]\gamma'(\rho)\big|^2}\,d\rho. \end{aligned} \qquad 5.4.11$$

We will evaluate the two terms under the square root through terms of order ρ^2 (which integrate to give terms in r^3). For the first term we get

$$\begin{aligned} |\gamma'(\rho)|^2 &= \Big(\cos\alpha(\rho) - \rho\,\alpha'(\rho)\sin\alpha(\rho)\Big)^2 + \Big(\sin\alpha(\rho) + \rho\,\alpha'(\rho)\cos\alpha(\rho)\Big)^2 \\ &= 1 + \rho^2\big(\alpha'(\rho)\big)^2 = 1 + k^2\rho^2 + o(\rho^2). \end{aligned} \qquad 5.4.12$$

For the second term under the square root in equation 5.4.11 we get

$$\left|[\mathbf{D}f(\gamma(\rho))]\gamma'(\rho)\right|^2 = \left|\left[\mathbf{D}f\begin{pmatrix}\rho\cos\alpha(\rho)\\ \rho\sin\alpha(\rho)\end{pmatrix}\right]\gamma'(\rho)\right|^2$$

$$= \left|\begin{bmatrix} a\rho\cos\alpha(\rho)+o(\rho) & b\rho\sin\alpha(\rho)+o(\rho)\end{bmatrix}\begin{bmatrix}\cos\alpha(\rho)-\rho\sin\alpha(\rho)\alpha'(\rho)\\ \sin\alpha(\rho)+\rho\cos\alpha(\rho)\alpha'(\rho)\end{bmatrix}\right|^2$$

$$= \Big(a\rho\cos^2\alpha(\rho)+o(\rho)+b\rho\sin^2\alpha(\rho)+o(\rho)\Big)^2 \qquad 5.4.13$$

$$= \rho^2\big(a\cos^2\alpha(\rho)+b\sin^2\alpha(\rho)\big)^2 + o(\rho^2) = \rho^2\big(a\cos^2\theta_0+b\sin^2\theta_0\big)^2+o(\rho^2).$$

Equation 5.4.13: Recall the definition of f (equation 5.4.4):

$$f\begin{pmatrix}x\\y\end{pmatrix} = \frac{1}{2}(ax^2+by^2)+o(x^2+y^2).$$

The first and second equalities use Proposition A11.2. The last equality uses the Taylor polynomial for α:

$$\alpha(\rho) = \theta_0 + k\rho + \frac{m}{2}\rho^2 + o(\rho^2).$$

Putting these formulas together with $\sqrt{1+t} = 1+t/2+o(t)$ gives

$$\text{Length}(\widetilde{\gamma}([0,r]) = \int_0^r \sqrt{1+\rho^2\Big(k^2+(a\cos^2\theta_0+b\sin^2\theta_0)^2\Big)+o(\rho^2)}\,d\rho$$

$$= \int_0^r \left(1+\frac{\rho^2}{2}\Big(k^2+(a\cos^2\theta_0+b\sin^2\theta_0)^2\Big)+o(\rho^2)\right)d\rho$$

$$= r + \frac{r^3}{6}\Big(k^2+(a\cos^2\theta_0+b\sin^2\theta_0)^2\Big)+o(r^3). \qquad 5.4.14$$

The corresponding formula for the length of $\delta_r([0,r])$ is the subcase of this one where $k=0$, giving

$$\text{Length}(\delta_r([0,r]) = r + \frac{r^3}{6}\Big(a\cos^2\theta_0+b\sin^2\theta_0\Big)^2+o(r^3). \qquad 5.4.15$$

Clearly if $k\neq 0$, so that $k^2>0$, we have, for $r>0$ sufficiently small,

$$\text{Length}(\widetilde{\gamma}([0,r]) > \text{Length}(\widetilde{\delta}_r([0,r]). \qquad 5.4.16$$

□ Proposition 5.4.2

The projection of $\mathbf{D}_r(\mathbf{0})$

The proof of Proposition 5.4.2 shows that to define $D_r(\mathbf{0})$, we need only consider curves γ with Taylor polynomial $\alpha(\rho) = \theta_0+m\rho^2/2+o(\rho^2)$, which have arc length

$$\text{Length}(\gamma([0,r])) = r + \frac{r^3}{6}\Big(a\cos^2\theta_0+b\sin^2\theta_0\Big)^2+o(r^3). \qquad 5.4.17$$

Equation 5.4.17: Note that m does not appear in this equation; through terms in r^3, the length depends only on θ_0 and r.

Since the inverse function of $r\mapsto r+\frac{r^3}{6}\left(a\cos^2\theta_0+b\sin^2\theta_0\right)^2$ is

$$r\mapsto r-\frac{r^3}{6}\left(a\cos^2\theta_0+b\sin^2\theta_0\right)^2+o(r^3), \qquad 5.4.18$$

it follows (see Figure 5.4.1) that the disc $D_r(\mathbf{0})\subset S$ projects to a region $U_r\subset\mathbb{R}^2$ given in polar coordinates as

$$U_r = \left\{\begin{pmatrix}\rho\\ \theta\end{pmatrix}\;\middle|\;\rho\le r-\frac{r^3}{6}\left(a\cos^2\theta+b\sin^2\theta\right)^2+o(r^3)\right\}. \qquad 5.4.19$$

We can now evaluate, through terms in r^4, the area of $D_r(\mathbf{0})$:

$$\text{Area}\, D_r(\mathbf{0}) = \int_{U_r}\sqrt{\det\left[\mathbf{D}g\begin{pmatrix}\rho\\ \theta\end{pmatrix}\right]^\top\left[\mathbf{D}g\begin{pmatrix}\rho\\ \theta\end{pmatrix}\right]}\,|d\rho\, d\theta| + o(r^4), \qquad 5.4.20$$

where g is as in equation 5.4.6. First, we will compute the integrand.

Proposition 5.4.3. *We have*

$$\sqrt{\det\left[\mathbf{D}g\begin{pmatrix}\rho\\\theta\end{pmatrix}\right]^{\top}\left[\mathbf{D}g\begin{pmatrix}\rho\\\theta\end{pmatrix}\right]} = \rho + \frac{\rho^3}{2}(a^2\cos^2\theta + b^2\sin^2\theta) + o(\rho^3).$$

Equation 5.4.22: We go from line 2 to line 3 thanks to a handy cancellation and the fact

$$\begin{aligned} &b^2\sin^2\theta\cos^2\theta + b^2\sin^4\theta \\ &= b^2\sin^2\theta(\cos^2\theta+\sin^2\theta)\\ &= b^2\sin^2\theta \end{aligned}$$

and similarly for the terms containing $a^2\cos^2\theta$. The final equality uses equation 3.4.9 for the Taylor polynomial of $(1+x)^m$.

Proof of Proposition 5.4.3. This is a straightforward computation:

$$\left[\mathbf{D}g\begin{pmatrix}\rho\\\theta\end{pmatrix}\right] = \begin{bmatrix}\cos\theta & -\rho\sin\theta\\ \sin\theta & \rho\cos\theta\\ \rho(a\cos^2\theta + b\sin^2\theta)+o(\rho) & \rho^2\sin\theta\cos\theta(b-a)+o(\rho^2)\end{bmatrix}$$

so

$$\left[\mathbf{D}g\begin{pmatrix}\rho\\\theta\end{pmatrix}\right]^{\top}\left[\mathbf{D}g\begin{pmatrix}\rho\\\theta\end{pmatrix}\right] = \qquad 5.4.21$$

$$\begin{bmatrix} 1+\rho^2(a\cos^2\theta+b\sin^2\theta)^2+o(\rho^2) & \rho^3\cos\theta\sin\theta(a\cos^2\theta+b\sin^2\theta)(b-a)+o(\rho^3)\\ \rho^3\cos\theta\sin\theta(a\cos^2\theta+b\sin^2\theta)(b-a)+o(\rho^3) & \rho^2+\rho^4\sin^2\theta\cos^2\theta(b-a)^2+o(\rho^4)\end{bmatrix}.$$

We can then compute the determinant to within $o(\rho^3)$; in particular, we can ignore the off-diagonal terms, whose product is of order ρ^6. We get

Equation 5.4.23: To get the last equality, note that

$$\int_0^{2\pi}\sin^4\theta\,d\theta = \int_0^{2\pi}\cos^4\theta\,d\theta = \frac{3}{4}\pi$$

$$\int_0^{2\pi}\sin^2\theta\,d\theta = \int_0^{2\pi}\cos^2\theta\,d\theta = \pi,$$

so in the next-to-last line, the integral multiplying a^2+b^2 vanishes. Moreover, $\sin 2\theta = 2\cos\theta\sin\theta$, so

$$\int_0^{2\pi}\cos^2\theta\sin^2\theta\,d\theta = \frac{\pi}{4}.$$

$$\begin{aligned}&\sqrt{\det\left[\mathbf{D}g\begin{pmatrix}\rho\\\theta\end{pmatrix}\right]^{\top}\left[\mathbf{D}g\begin{pmatrix}\rho\\\theta\end{pmatrix}\right]} \qquad 5.4.22\\ &= \sqrt{\rho^2+\rho^4\Big(\sin^2\theta\cos^2\theta(b-a)^2 + (a\cos^2\theta+b\sin^2\theta)^2\Big)+o(\rho^4)}\\ &= \rho\sqrt{1+\rho^2(a^2\cos^2\theta+b^2\sin^2\theta)+o(\rho^2)}\\ &= \rho+\frac{\rho^3}{2}(a^2\cos^2\theta+b^2\sin^2\theta)+o(\rho^3) \qquad \square\end{aligned}$$

We can now compute Area $D_r(0)$ through terms of order r^4:

$$\begin{aligned}&\int_{U_r}\left(\rho+\frac{\rho^3}{2}(a^2\cos^2\theta+b^2\sin^2\theta)+o(\rho^3)\right)|d\rho\,d\theta|\\ &=\int_0^{2\pi}\int_0^{r-\frac{r^3}{6}(a\cos^2\theta+b\sin^2\theta)^2+o(r^3)}\left(\rho+\frac{\rho^3}{2}(a^2\cos^2\theta+b^2\sin^2\theta)+o(\rho^3)\right)d\rho\,d\theta\\ &=\int_0^{2\pi}\left(\frac{1}{2}\left(r-\frac{r^3}{6}(a\cos^2\theta+b\sin^2\theta)^2\right)^2+\frac{r^4}{8}(a^2\cos^2\theta+b^2\sin^2\theta)\right)d\theta+o(r^4)\\ &=\pi r^2+r^4\left(\int_0^{2\pi}a^2\left(\frac{\cos^2\theta}{8}-\frac{\cos^4\theta}{6}\right)+b^2\left(\frac{\sin^2\theta}{8}-\frac{\sin^4\theta}{6}\right)-\frac{1}{3}ab\cos^2\theta\sin^2\theta\right)d\theta+o(r^4)\\ &=\pi r^2-\frac{ab\pi}{12}r^4+o(r^4). \qquad 5.4.23\end{aligned}$$

Since the Gaussian curvature of S at $\mathbf{0}$ is $K(\mathbf{0}) = ab$, this proves Theorem 5.4.1, which proves Gauss's Theorema Egregium. $\square$

The meaning of the mean curvature

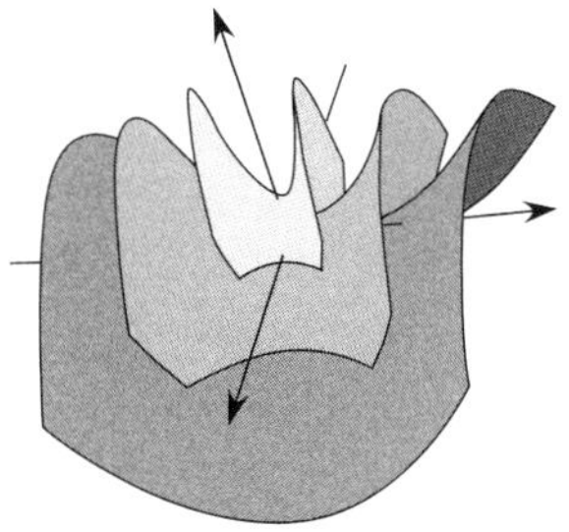

FIGURE 5.4.2.

Moving a surface along a normal vector field $\vec{\mathbf{w}}$. In this case S is the surface of equation $z = y^2 - x^2$, the vector field is $\begin{bmatrix} 2x \\ -2y \\ 1 \end{bmatrix}$ and the surfaces represented correspond to $t = -.3, 0, .3$.

Next we will justify the statement in Section 3.9 that the mean curvature measures how far a surface is from being minimal. More generally, the mean curvature vector $\vec{H}$ (Definition 3.9.7) measures how the area of a surface S varies as the surface is moved along a normal vector field. Let $\vec{\mathbf{w}}$ be a normal vector field on S. Consider (see Figure 5.4.2) the family of surfaces S_t that are the images of $\varphi_t : S \to \mathbb{R}^3$ given by

$$\varphi_t(\mathbf{x}) = \mathbf{x} + t\vec{\mathbf{w}}(\mathbf{x}). \qquad 5.4.24$$

Theorem 5.4.4. *The area of S_t is given by the formula*

$$\text{Area}\, S_t = \text{Area}\, S - 2t \int_S \vec{H}(\mathbf{x}) \cdot \vec{\mathbf{w}}(\mathbf{x}) |d^2\mathbf{x}| + o(t). \qquad 5.4.25$$

Proof. Let $U \subset \mathbb{R}^2$ be open and let $\gamma : U \to S$ be a parametrization. Rather than working directly with the vector field $\vec{\mathbf{w}}$ on S, we will work with the map $\vec{W} : U \to \mathbb{R}^3$ given by $\vec{W}(\mathbf{u}) = \vec{\mathbf{w}}(\gamma(\mathbf{u}))$; we think of $\vec{W}(\mathbf{u})$ as a vector in $\mathbb{R}^3$ anchored at $\gamma(\mathbf{u})$. Now S_t is parametrized by

$$\mathbf{u} \mapsto \varphi_t(\gamma(\mathbf{u})) = \gamma(\mathbf{u}) + t\vec{W}(\mathbf{u}), \qquad 5.4.26$$

so equation 5.3.3 says that

$$\begin{aligned}\text{Area}\, S_t &= \int_U \sqrt{\det[\mathbf{D}(\gamma + t\vec{W})(\mathbf{u})]^\top [\mathbf{D}(\gamma + t\vec{W})(\mathbf{u})]}\, |d^2\mathbf{u}| \qquad 5.4.27\\ &= \int_U \sqrt{\det\Big(\big([\mathbf{D}\gamma(\mathbf{u})]^\top + t[\mathbf{D}\vec{W}(\mathbf{u})]^\top\big)\big([\mathbf{D}\gamma(\mathbf{u})] + t[\mathbf{D}\vec{W}(\mathbf{u})]\big)\Big)}\\ &= \int_U \sqrt{\det\Big([\mathbf{D}\gamma(\mathbf{u})]^\top[\mathbf{D}\gamma(\mathbf{u})] + t\left([\mathbf{D}\gamma(\mathbf{u})]^\top[\mathbf{D}\vec{W}(\mathbf{u})] + [\mathbf{D}\vec{W}(\mathbf{u})]^\top[\mathbf{D}\gamma(\mathbf{u})]\right)\Big)}\, |d^2\mathbf{u}| + o(t).\end{aligned}$$

Moreover, using equation 5.3.5 we can rewrite equation 5.4.25 as

$$\text{Area}\, S_t = \int_U \sqrt{\det[\mathbf{D}\gamma(\mathbf{u})]^\top[\mathbf{D}\gamma(\mathbf{u})]}\Big(1 - 2t\vec{H}(\gamma(\mathbf{u})) \cdot \vec{W}(\mathbf{u})\Big)|d^2\mathbf{u}| + o(t). \qquad 5.4.28$$

In equation 5.4.28,

$$\vec{H}(\gamma(\mathbf{u})) \cdot \vec{W}(\mathbf{u})$$

plays the role of $f(\gamma(\mathbf{u}))$ in equation 5.3.5.

Proposition 5.3.3 justifies our using the parametrization $\tilde{f}$ rather than γ.

We need to show that the integrands in equations 5.4.27 and 5.4.28 are equal. First we will compute the integrand in the second line of equation 5.4.27. As we did in equation 5.4.4, we will suppose that S is the graph of

$$f\begin{pmatrix} x \\ y \end{pmatrix} = \frac{1}{2}(ax^2 + by^2) + o(x^2 + y^2), \text{ parametrized by } \tilde{f}\begin{pmatrix} x \\ y \end{pmatrix} = \begin{pmatrix} x \\ y \\ f\begin{pmatrix} x \\ y \end{pmatrix} \end{pmatrix},$$

so that γ in equation 5.4.27 is replaced by $\tilde{f}$. We will compute the integrand at the origin, so we can ignore all the terms that vanish at the origin (after computing the appropriate derivatives, of course). In these adapted coordinates, $[\mathbf{D}\gamma(\mathbf{u})]^\top[\mathbf{D}\gamma(\mathbf{u})]$ becomes

$$\left[\mathbf{D}\tilde{f}\begin{pmatrix} 0 \\ 0 \end{pmatrix}\right]^\top \left[\mathbf{D}\tilde{f}\begin{pmatrix} 0 \\ 0 \end{pmatrix}\right] = \begin{bmatrix} 1 & 0 & 0 \\ 0 & 1 & 0 \end{bmatrix}\begin{bmatrix} 1 & 0 \\ 0 & 1 \\ 0 & 0 \end{bmatrix} = \begin{bmatrix} 1 & 0 \\ 0 & 1 \end{bmatrix}. \qquad 5.4.29$$

Since $\vec{\mathbf{w}}$ is orthogonal to S, so is $\vec{W}$; thus (Proposition 1.4.20) we can write $\vec{W}$ as a multiple of the cross product of the partial derivatives of $\widetilde{f}$ by some scalar-valued function α:

$$\vec{W}\begin{pmatrix} x \\ y \end{pmatrix} = \alpha\begin{pmatrix} x \\ y \end{pmatrix}\left(\begin{bmatrix} 1 \\ 0 \\ ax + o(|x|+|y|) \end{bmatrix} \times \begin{bmatrix} 0 \\ 1 \\ by + o(|x|+|y|) \end{bmatrix}\right) = \alpha\begin{pmatrix} x \\ y \end{pmatrix}\begin{bmatrix} -ax + o(|x|+|y|) \\ -by + o(|x|+|y|) \\ 1 \end{bmatrix}.$$

This gives

$$\left[\mathbf{D}\vec{W}\begin{pmatrix} x \\ y \end{pmatrix}\right] = \left[D_1\alpha\begin{pmatrix} x \\ y \end{pmatrix}\begin{bmatrix} -ax \\ -by \\ 1 \end{bmatrix}, D_2\alpha\begin{pmatrix} x \\ y \end{pmatrix}\begin{bmatrix} -ax \\ -by \\ 1 \end{bmatrix}\right] + \alpha\begin{pmatrix} x \\ y \end{pmatrix}\begin{bmatrix} -a & 0 \\ 0 & -b \\ 0 & 0 \end{bmatrix} + o(1) \qquad 5.4.30$$

and finally

$$\left[\mathbf{D}\vec{W}\begin{pmatrix} 0 \\ 0 \end{pmatrix}\right] = \begin{bmatrix} -a\alpha\begin{pmatrix} 0 \\ 0 \end{pmatrix} & 0 \\ 0 & -b\alpha\begin{pmatrix} 0 \\ 0 \end{pmatrix} \\ D_1\alpha\begin{pmatrix} 0 \\ 0 \end{pmatrix} & D_2\alpha\begin{pmatrix} 0 \\ 0 \end{pmatrix} \end{bmatrix}. \qquad 5.4.31$$

Thus the integrand at the origin, with γ replaced by $\widetilde{f}$, is

$$\sqrt{\det\left(\begin{bmatrix} 1 & 0 \\ 0 & 1 \end{bmatrix} + t\begin{bmatrix} 1 & 0 \\ 0 & 1 \\ 0 & 0 \end{bmatrix}^{\top}\begin{bmatrix} -a\,\alpha\begin{pmatrix} 0 \\ 0 \end{pmatrix} & 0 \\ 0 & -b\,\alpha\begin{pmatrix} 0 \\ 0 \end{pmatrix} \\ D_1\alpha\begin{pmatrix} 0 \\ 0 \end{pmatrix} & D_2\alpha\begin{pmatrix} 0 \\ 0 \end{pmatrix} \end{bmatrix} + t\begin{bmatrix} -a\,\alpha\begin{pmatrix} 0 \\ 0 \end{pmatrix} & 0 \\ 0 & -b\,\alpha\begin{pmatrix} 0 \\ 0 \end{pmatrix} \\ D_1\alpha\begin{pmatrix} 0 \\ 0 \end{pmatrix} & D_2\alpha\begin{pmatrix} 0 \\ 0 \end{pmatrix} \end{bmatrix}^{\top}\begin{bmatrix} 1 & 0 \\ 0 & 1 \\ 0 & 0 \end{bmatrix}\right)}$$

$$= \sqrt{1 - 2t(a+b)\alpha\begin{pmatrix} 0 \\ 0 \end{pmatrix} + 4t^2ab\left(\alpha\begin{pmatrix} 0 \\ 0 \end{pmatrix}\right)^2} = \sqrt{1 - 2t(a+b)\alpha\begin{pmatrix} 0 \\ 0 \end{pmatrix}} \qquad 5.4.32$$

$$= 1 - t(a+b)\alpha\begin{pmatrix} 0 \\ 0 \end{pmatrix} \quad \text{(using equation 3.4.9).}$$

This completes the computation for equation 5.4.27.

Definition 3.9.7 of the mean curvature H and the mean curvature vector $\vec{H}$:

$$H(\mathbf{a}) = \frac{1}{2}(A_{2,0} + A_{0,2})$$

$$\vec{H}(\mathbf{a}) = H(\mathbf{a})\vec{\mathbf{v}}_3,$$

where $\vec{\mathbf{v}}_3$ is the chosen unit normal.

Now consider the integrand in equation 5.4.28. In our adapted coordinates, $A_{2,0} = a$, $A_{0,2} = b$, and the unit normal called $\vec{\mathbf{v}}_3$ in Definition 3.9.7 is $\begin{bmatrix} 0 \\ 0 \\ 1 \end{bmatrix}$. So the mean curvature vector $\vec{H}$ at the origin is $\begin{bmatrix} 0 \\ 0 \\ \frac{a+b}{2} \end{bmatrix}$, and the integrand for equation 5.4.28 is

$$\sqrt{\det\begin{bmatrix} 1 & 0 \\ 0 & 1 \end{bmatrix}}\left(1 - 2t\begin{bmatrix} 0 \\ 0 \\ \frac{a+b}{2} \end{bmatrix} \cdot \begin{bmatrix} 0 \\ 0 \\ \alpha\begin{pmatrix} 0 \\ 0 \end{pmatrix} \end{bmatrix}\right) = 1 - t(a+b)\,\alpha\begin{pmatrix} 0 \\ 0 \end{pmatrix}. \qquad \square$$

5.4.33

Corollary 5.4.5. *A minimal surface has mean curvature* 0.

Proof. Suppose that at some point $\mathbf{a} \in S$ we have $\vec{H}(\mathbf{a}) \neq \vec{\mathbf{0}}$. The vector field $\vec{H}$ then points on one side of the surface in some neighborhood V of

a, and we can find a little "bump" vector field $\vec{\mathbf{w}}$ that vanishes outside V, and within V points to the same side of S as $\vec{H}$; see Figure 5.4.3. In that case the integral

$$\int_S \vec{H}(\mathbf{x}) \cdot \vec{\mathbf{w}}(\mathbf{x})|d^2\mathbf{x}| \qquad 5.4.34$$

is strictly positive, so for $t > 0$ sufficiently small, $\text{Area}\, S_t < \text{Area}\, S$ (see equation 5.4.25). Thus S_t is a surface close to S, coinciding with S outside V, and with smaller area, so S does not have minimal area among nearby surfaces with the same boundary. □

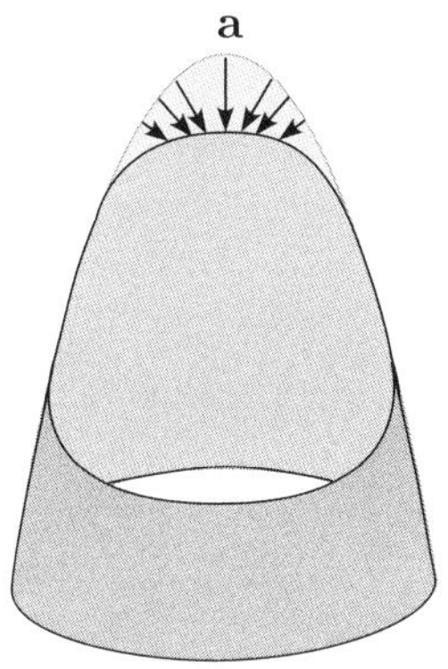

FIGURE 5.4.3.
The vector field $\vec{\mathbf{w}}$ has small support near **a** and points in the same direction as the mean curvature vector $\vec{H}(\mathbf{x})$. If we push the top down, even just locally as indicated by the vector field, the area decreases.

Gaussian curvature and the Gauss map

Let $S \subset \mathbb{R}^3$ be a surface, and $\vec{\mathbf{n}}$ a unit normal vector field. One can imagine $\vec{\mathbf{n}}(\mathbf{x})$ anchored at the point $\mathbf{x} \in S$, but if the vector $\vec{\mathbf{n}}(\mathbf{x})$ is translated so that its tail is at the origin, then its head is a point of the unit sphere S^2. This gives a map $\mathbf{n} : S \to S^2$, called the *Gauss map*, shown in Figure 5.4.4.

The Gauss map is closely related to curvature: the domain and image of the derivative $[\mathbf{Dn}(\mathbf{x})] : T_{\mathbf{x}}S \to T_{\mathbf{n}(\mathbf{x})}S^2$ coincide: both are $(\vec{\mathbf{n}}(\mathbf{x}))^{\perp}$. So (see Remark 4.8.7) it makes sense to speak of the *Jacobian determinant* $\det[\mathbf{Dn}(\mathbf{x})]$; this turns out to be Gaussian curvature $K(\mathbf{x})$.

Theorem 5.4.6 (Gaussian curvature and Gauss map). *Let $S \subset \mathbb{R}^3$ be a surface, and let $\mathbf{n} : S \to S^2$ be the Gauss map. Then*

1. $K(\mathbf{x}) = \det[\mathbf{Dn}(\mathbf{x})]$
2. *If* $\mathbf{n}$ *is injective on* S, *then*

$$\int_S |K(\mathbf{x})||d^2\mathbf{x}| = \int_{\mathbf{n}(S)} |d^2\mathbf{x}| = \textit{Area}\ \mathbf{n}(S). \qquad 5.4.35$$

If S is a small neighborhood of $\mathbf{x}$, where K can be considered as a constant, equation 5.4.35 becomes

$$|K(\mathbf{x})| \int_S |d^2\mathbf{x}| \approx \int_{\mathbf{n}(S)} |d^2\mathbf{x}|, \text{ i.e., } |K(\mathbf{x})| \approx \frac{\int_{\mathbf{n}(S)} |d^2\mathbf{x}|}{\int_S |d^2\mathbf{x}|}. \qquad 5.4.36$$

Thus the Gaussian curvature at $\mathbf{x}$ is the limit, as the neighborhood S of $\mathbf{x}$ becomes small, of the ratio of the area of $\mathbf{n}(S)$ to the area of S.

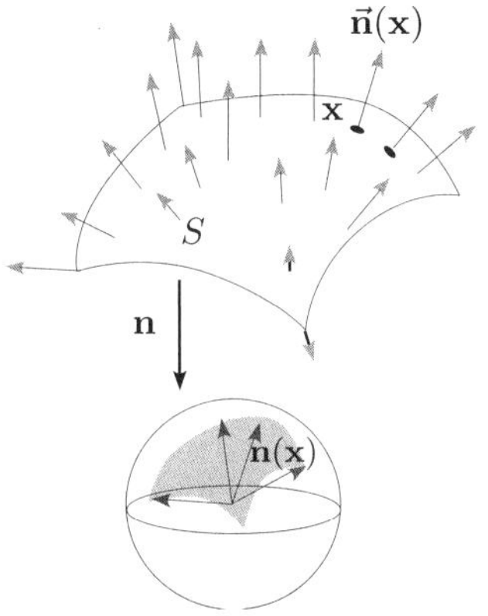

FIGURE 5.4.4.
The vector field $\vec{\mathbf{n}}$ is a unit normal vector field on a surface $S \subset \mathbb{R}^3$. The Gauss map $\mathbf{n}$ drags each unit vector $\vec{\mathbf{n}}(\mathbf{x})$ to the origin, so that its arrowhead is a point on the unit sphere S^2.

Example 5.4.7. Let $S \subset \mathbb{R}^3$ be an ellipsoid. Then $\mathbf{n}(S) = S^2$ and

$$\int_S |K(\mathbf{x})||d^2\mathbf{x}| = 4\pi, \qquad 5.4.37$$

the area of the unit sphere. More generally, this holds when S is the boundary of any bounded convex subset of $\mathbb{R}^3$ with smooth boundary. In all these cases the map $\mathbf{n} : S \to S^2$ is bijective. △

Proof. 1. This is a statement about a single point $\mathbf{x} \in S$, and is clearly independent of translations and rotations. Thus we may assume that $\mathbf{x} = \mathbf{0}$, and that locally, near $\mathbf{x}$, the surface S is the graph of a map

$$f\begin{pmatrix} X \\ Y \end{pmatrix} \stackrel{\text{def}}{=} \frac{1}{2}(aX^2 + 2bXY + cY^2) + o(|X|^2 + |Y^2|); \tag{5.4.38}$$

see equation 3.9.28 and Exercise 3.3.15. Thus

$$\gamma\begin{pmatrix} X \\ Y \end{pmatrix} = \begin{pmatrix} X \\ Y \\ f\begin{pmatrix} X \\ Y \end{pmatrix} \end{pmatrix}. \tag{5.4.39}$$

In equation 5.4.38 we could perform a rotation to eliminate the term $2bXY$; see equation 5.4.4.

is a parametrization of the surface S near $\mathbf{0}$.

The map $\mathbf{n}$ is then given by

$$(\mathbf{n} \circ \gamma)\begin{pmatrix} X \\ Y \end{pmatrix} = \underbrace{\frac{1}{\sqrt{\left(D_1 f\begin{pmatrix} X \\ Y \end{pmatrix}\right)^2 + \left(D_2 f\begin{pmatrix} X \\ Y \end{pmatrix}\right)^2 + 1}} \overbrace{\begin{bmatrix} 1 \\ 0 \\ D_1 f\begin{pmatrix} X \\ Y \end{pmatrix} \end{bmatrix}}^{\text{tangent to } S} \times \overbrace{\begin{bmatrix} 0 \\ 1 \\ D_2 f\begin{pmatrix} X \\ Y \end{pmatrix} \end{bmatrix}}^{\text{tangent to } S}}_{\text{unit vector orthogonal to surface } S} \tag{5.4.40}$$

This expression looks pretty formidable, and we need to differentiate it. But if we replace every expression by its Taylor polynomial of degree 1, it is easy, because the the denominator becomes just 1: the derivatives are linear, but they are squared, hence quadratic: they are in $o(\sqrt{x^2 + y^2})$ and can be ignored. This gives

$$\begin{aligned} \mathbf{n} \circ \gamma\begin{pmatrix} X \\ Y \end{pmatrix} &= \begin{bmatrix} 1 \\ 0 \\ aX + bY \end{bmatrix} \times \begin{bmatrix} 0 \\ 1 \\ bX + cY \end{bmatrix} + o(\sqrt{X^2 + Y^2}) \\ &= \begin{bmatrix} -(aX + bY) \\ -(bX + cY) \\ 1 \end{bmatrix} + o(\sqrt{X^2 + Y^2}), \end{aligned} \tag{5.4.41}$$

As we mentioned for the little o notation (discussion before equation 3.4.4) the first line of equation 5.4.41 is sloppily written; to be correct we should write that the difference between the left side and the cross product is in $o(\sqrt{X^2 + Y^2})$.

and the derivative $[\mathbf{D}(\mathbf{n} \circ \gamma)(\mathbf{0})]$ is given by the linear terms:

$$[\mathbf{D}(\mathbf{n} \circ \gamma)(\mathbf{0})]\begin{bmatrix} \dot{X} \\ \dot{Y} \end{bmatrix} = \begin{bmatrix} -a\dot{X} - b\dot{Y} \\ -b\dot{X} - c\dot{Y} \\ 0 \end{bmatrix}. \tag{5.4.42}$$

The third entry 0 reflects the fact that the domain and codomain of $[\mathbf{Dn}(\mathbf{x})]$ coincide, they are both $T_\mathbf{x}S$, in this case the horizontal plane, and thus the Jacobian determinant is

$$\det\begin{bmatrix} -a & -b \\ -b & -c \end{bmatrix} = ac - b^2 = K(\mathbf{0}). \tag{5.4.43}$$

Equation 5.4.43: In Chapter 3 (equation 3.9.37) we have

$$K(\mathbf{0}) = \frac{a_{2,0}a_{0,2} - a_{1,1}^2}{(1 + c^2)^2},$$

where $c = \sqrt{a_1^2 + a_2^2}$, but that was for a function whose Taylor polynomial at the origin began with linear terms $a_1 x + a_2 y$. Here the linear terms are 0.

This proves part 1.

For part 2, it is enough to prove the result locally, i.e., in a part of S parametrized as the graph of $f : U \to \mathbb{R}$ as above. If $\mathbf{n}$ is injective on $f(U)$, the map $\mathbf{n} \circ \gamma$ parametrizes $f(U) \subset S^2$. To prove part 2, we need to show that

$$\det\Big([\mathbf{D}(\mathbf{n} \circ \gamma)(\mathbf{x})]^\top [\mathbf{D}(\mathbf{n} \circ \gamma)(\mathbf{x})]\Big) = |K(\mathbf{x})|^2; \tag{5.4.44}$$

it is enough to show this at $\mathbf{x} = 0$. We computed

$$[\mathbf{D}(\mathbf{n} \circ \gamma)(\mathbf{0})] = \begin{bmatrix} -a & -b \\ -b & -c \\ 0 & 0 \end{bmatrix} \qquad 5.4.45$$

in equation 5.4.42, so

$$\det \left(\begin{bmatrix} -a & -b \\ -b & -c \\ 0 & 0 \end{bmatrix}^{\top} \begin{bmatrix} -a & -b \\ -b & -c \\ 0 & 0 \end{bmatrix} \right) = \det \begin{bmatrix} a^2+b^2 & ab+bc \\ ab+bc & b^2+c^2 \end{bmatrix} \qquad 5.4.46$$

$$= a^2b^2 + a^2c^2 + b^4 + b^2c^2 - a^2b^2 - 2ab^2c - b^2c^2 = (ac-b^2)^2. \qquad \square$$

Exercises for Section 5.4

5.4.1 a. Let X be the hyperboloid of revolution of equation $x^2+y^2-a^2z^2=1$. Show that

$$\int_X |K(\mathbf{x})||d^2\mathbf{x}| = \frac{4a\pi}{\sqrt{1+a^2}}.$$

b. What is $\int_X |K(\mathbf{x})||d^2\mathbf{x}|$ if X is the part of the hyperboloid of equation

$$x^2+y^2-a^2z^2 = -1 \quad \text{where } z > 0?$$

5.4.2 Let X be the surface of revolution obtained by rotating the circle of equation $(x-2)^2+z^2=1$ in the (x,z)-plane around the z axis. Compute $\int_X |K(\mathbf{x})||d^2\mathbf{x}|$ and $\int_X K(\mathbf{x})|d^2\mathbf{x}|$.

***5.4.3** Let X be the helicoid of equation $x\cos z + y \sin z = 0$, and $X_{a,R}$ the part of X where $0 \le z \le a$ and $x^2+y^2 \le R^2$. What is $\int_{X_{a,R}} |K(\mathbf{x})||d^2\mathbf{x}|$?

Hint for Exercise 5.4.4: Consider the "lunes" formed by two of the great circles.

***5.4.4** Let $T \subset S^2$ be the spherical triangle whose sides are arcs of great circles, and with angles α, β, γ. Show that Area $T = \alpha+\beta+\gamma-\pi$.

The three theorems of this section (Theorems 5.4.1, 5.4.4, and 5.4.6) are stated for surfaces in $\mathbb{R}^3$. Exercise 5.4.5 explores how these theorems generalize to hypersurfaces in $\mathbb{R}^{n+1}$ for any $n > 1$ (i.e., n-dimensional submanifolds of $\mathbb{R}^{n+1}$).

5.4.5 Let M be an n-dimensional smooth manifold in $\mathbb{R}^{n+1}$, $\mathbf{x}_0 \in M$ a point, and $\vec{\mathbf{n}} \in \mathbb{R}^{n+1}$ a unit normal to M at $\mathbf{x}_0$, i.e., a unit vector orthogonal to $T_{\mathbf{x}_0}M$.

a. Show that in best coordinates for M at $\mathbf{x}_0$, the manifold M has the equation

$$x_{n+1} = \frac{1}{2}(a_1x_1^2 + \cdots + a_nx_n^2) + o(x_1^2+\cdots+x_n^2)$$

for some numbers a_i, known as the *principal curvatures*.

The number $S(\mathbf{x}_0) \overset{\text{def}}{=} 2\sum_{i<j} a_ia_j$ is called the *scalar curvature* of M at $\mathbf{x}_0$.

We will call the number $K(\mathbf{x}_0) \overset{\text{def}}{=} (-1)^n a_1 \cdots a_n$ the *Gaussian curvature* of M at $\mathbf{x}_0$.[5]

b. The number $H(\mathbf{x}_0) \overset{\text{def}}{=} a_1+\cdots+a_n$ is called the *mean curvature* of M at $\mathbf{x}_0$. Show that $\vec{H}(\mathbf{x}_0) \overset{\text{def}}{=} H(\mathbf{x}_0)\vec{\mathbf{n}}$ does not depend on the choice of $\vec{\mathbf{n}}$; it is called the mean curvature vector.

[5]This terminology is consistent with that of Shlomo Sternberg in *Lectures in Differential Geometry*. M. M. Postnikov, in *Geometry VI* (*Encyclopaedia of Mathematical Sciences Vol. 91*), calls it (without the sign) the *total curvature*.

****5.4.6** Let M be as in Exercise 5.4.5. Show that the n-dimensional volume of the set $D_r(\mathbf{x})$ of points of M that can be connected to $\mathbf{x}$ by a curve *in* M of length $\le r$ has Taylor polynomial

$$\text{vol}_n(D_r(\mathbf{x})) = \Big(\text{vol}_n B_r\Big)\left(1 - \frac{r^2}{n+2}\frac{S(\mathbf{x})}{6}\right) + o(r^{n+2}),$$

where $\text{vol}_n B_r$ is the volume of the ordinary ball of radius r in $\mathbb{R}^n$ (i.e., $r^n\beta_n$ is the volume of the unit n-ball; see Table 5.3.3), and S is the scalar curvature.

5.4.7 Let M be as in Exercise 5.4.5. Let $\vec{\mathbf{w}}$ be a normal vector field on M. Consider the family of surfaces M_t that are the images of $\varphi_t : M \to \mathbb{R}^{n+1}$ given by

$$\varphi_t(\mathbf{x}) = \mathbf{x} + t\vec{\mathbf{w}}(\mathbf{x}).$$

Show that $\text{vol}_n(M_t)$ has the Taylor polynomial

$$\text{vol}_n(M_t) = \text{vol}_n M - 2t\int_S \vec{H}(\mathbf{x}) \cdot \vec{\mathbf{w}}(\mathbf{x})|d^n\mathbf{x}| + o(t).$$

5.4.8 Let M be as in Exercise 5.4.5. Let $\vec{\mathbf{n}} : M \to S^n$ be a unit normal vector field on M.

a. Show that the domain and image of $[\mathbf{D}\vec{\mathbf{n}}(\mathbf{x})]$ are both $T_{\mathbf{x}}M$, so $\det[\mathbf{D}\vec{\mathbf{n}}(\mathbf{x})]$ makes sense (see Remark 4.8.7.)

b. Show that $\det[\mathbf{D}\vec{\mathbf{n}}(\mathbf{x})] = K(\mathbf{x})$, where K is as in Exercise 5.4.5.

FIGURE 5.5.1.
Felix Hausdorff (1868–1942)
Hausdorff worked in Bonn until 1935, when he was forced to retire because he was Jewish. In 1942 he and his wife committed suicide rather than be sent to a concentration camp.

5.5 Fractals and fractional dimension

In 1919, Felix Hausdorff showed that dimensions are not limited to length, area, volume, and so on: we can also speak of *fractional dimension*. This discovery acquired much greater significance with the work of Benoit Mandelbrot, who showed that many objects in nature (the lining of the lungs, the patterns of frost on windows, the patterns formed by a film of gasoline on water) are fractals, with fractional dimension.

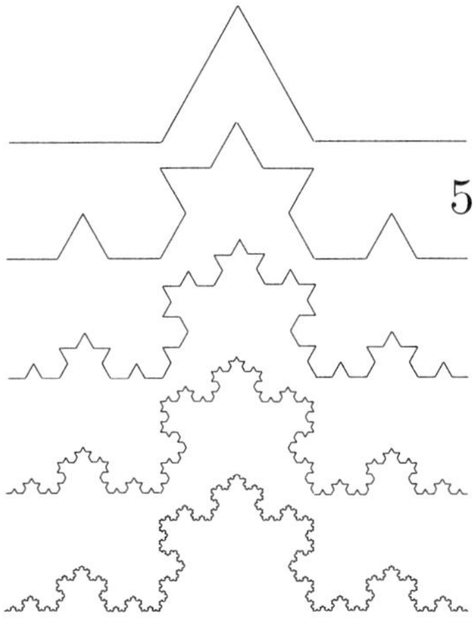

FIGURE 5.5.2.
The first five steps in constructing the Koch snowflake. Its length is infinite, but length is the wrong way to measure this fractal object. The Koch snowflake was first described by the Swedish mathematician Helge von Koch in 1904.

Example 5.5.1 (Koch snowflake). To construct the Koch snowflake curve K, start with a line segment, say $0 \le x \le 1$, $y = 0$ in $\mathbb{R}^2$. Replace its middle third by the top of an equilateral triangle, as shown in Figure 5.5.2. This gives four segments, each one-third the length of the original segment. Replace the middle third of each by the top of an equilateral triangle, and so on. What is the length of this "curve"? At resolution $N = 0$, we get length 1. At $N = 1$, when the curve consists of four segments, we get length $4 \cdot 1/3$. At $N = 2$, the length is $16 \cdot 1/9$; at the Nth, it is $(4/3)^N$. As the resolution becomes infinitely fine, the length becomes infinite!

"Length" is the wrong word to apply to the Koch snowflake, which is neither a curve nor a surface. It is a fractal, with fractional dimension: the Koch snowflake has dimension $\ln 4/\ln 3 \approx 1.26$. Let us see why this might

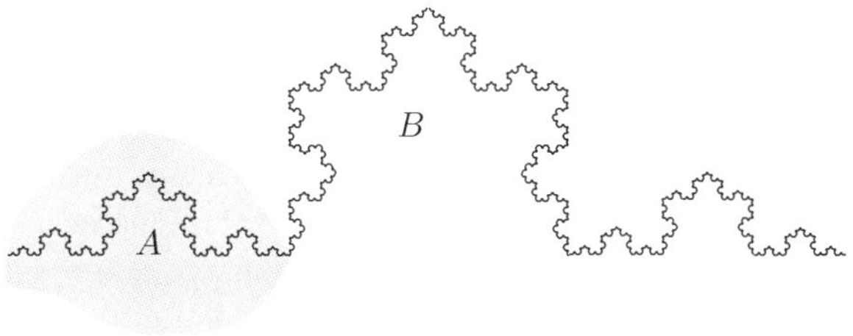

FIGURE 5.5.3. The curve B consists of four copies of A. It is also A scaled up by a factor of 3. It is the difference between these factors 3 and 4 that results in the fractional dimension of the curve.

be the case. Call A the part of the curve constructed on $[0, 1/3]$, and B the whole curve, as in Figure 5.5.3. Then B consists of four copies of A. (This is true at any level, but it is easiest to see at the first level, the top graph in Figure 5.5.2.). Therefore, in any dimension d, it should be true that $\mathrm{vol}_d(B) = 4\,\mathrm{vol}_d(A)$.

However, if you expand A by a factor of 3, you get B. (This is true in the limit, after the construction has been carried out infinitely many times.) According to the principle that area goes as the square of the length, volume goes as the cube of the length, etc., we would expect d-dimensional volume to go as the dth power of the length, which leads to

$$\mathrm{vol}_d(B) = 3^d\,\mathrm{vol}_d(A). \qquad 5.5.1$$

If you put this equation together with $\mathrm{vol}_d(B) = 4\,\mathrm{vol}_d(A)$, you will see that the only dimension in which the volume of the Koch curve can be different from 0 or ∞ is the one for which $4 = 3^d$ (i.e., $d = \ln 4/\ln 3$).

If we break up the Koch curve into the pieces built on the sides constructed at the nth level (of which there are 4^n, each of length $1/3^n$) and raise their sidelengths to the dth power, we find

$$4^n\left(\frac{1}{3}\right)^{n\ln 4/\ln 3} = 4^n e^{n\frac{\ln 4}{\ln 3}(\ln\frac{1}{3})} = 4^n e^{n\frac{\ln 4}{\ln 3}(-\ln 3)} = 4^n e^{-n\ln 4} = \frac{4^n}{4^n} = 1. \qquad 5.5.2$$

(In equation 5.5.2 we use the fact that $a^x = e^{x\ln a}$.) Although the terms have not been defined precisely, you might expect this computation to mean

$$\int_K |d\mathbf{x}^{\ln 4/\ln 3}| = 1. \quad \triangle \qquad 5.5.3$$

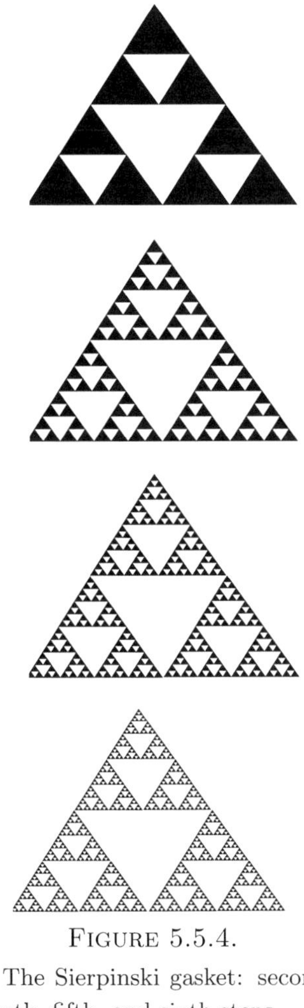

FIGURE 5.5.4.
The Sierpinski gasket: second, fourth, fifth, and sixth steps.

Example 5.5.2 (Sierpinski gasket). While the Koch snowflake looks like a thick curve, the *Sierpinski gasket* looks more like a thin surface. This is the subset of the plane obtained by taking a filled equilateral triangle of sidelength l, removing the central inscribed subtriangle, then removing the central subtriangles from the three triangles that are left, and so on, as sketched in Figure 5.5.4. We claim that this is a set of dimension $\ln 3/\ln 2$. At the nth stage of the construction, sum, over all the little pieces, the sidelength to the power p:

$$3^n\left(\frac{l}{2^n}\right)^p. \qquad 5.5.4$$

(If measuring length, $p = 1$; if measuring area, $p = 2$.) If the set had a length, the sum would converge when $p = 1$, as $n \to \infty$; in fact, the sum is infinite. If it really had an area, then the power $p = 2$ would lead to a finite limit; in fact, the sum is 0. But when $p = \ln 3/\ln 2$, the sum converges to $l^{\ln 3/\ln 2} \approx l^{1.58}$. This is the only dimension in which the Sierpinski gasket has finite, nonzero volume; in larger dimensions, the volume is 0; in smaller dimensions, it is infinite. △

Exercises for Section 5.5

5.5.1 Consider the *triadic Cantor set* C obtained by removing from $[0, 1]$ first the open middle third $(1/3, 2/3)$, then the open middle third of each of the segments left, then the open middle third of each of the segments left, etc.:

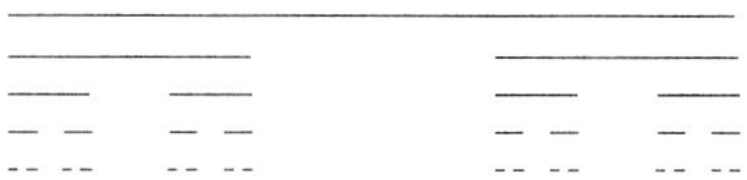

a. Show that an alternative description of C is that it is the set of points that can be written in base 3 without using the digit 1. Use this to show that C is an uncountable set.

b. Show that C is a pavable set, with one-dimensional volume 0.

c. Show that the only dimension in which C can have volume different from 0 or infinity is $\ln 2/\ln 3$.

Hint for Exercise 5.5.1, part a: the number written as

$$.02220000022202002222\ldots$$

in base 3 is in C.

5.5.2 This time let the set C be obtained from the unit interval by omitting the open middle $1/n$th, then the open middle $1/n$th of each of the remaining intervals, then the open middle $1/n$th of the remaining intervals, etc. (When n is even, this means omitting an open interval equivalent to $1/n$th of the unit interval, leaving equal amounts on both sides, and so on.)

a. Show that C is a pavable set, with one-dimensional volume 0.

b. What is the only dimension in which C can have volume different from 0 or infinity? What is this dimension when $n = 2$?

5.6 Review exercises for Chapter 5

5.1 Verify that equation 5.3.30 parametrizes the torus obtained by rotating around the z-axis the circle of radius r in the (x, z)-plane that is centered at $x = R$, $z = 0$.

5.2 Let $f : [a, b] \to \mathbb{R}$ be a smooth positive function. Find a parametrization for the surface of equation $\dfrac{x^2}{A^2} + \dfrac{y^2}{B^2} = \big(f(z)\big)^2$.

5.3 For what values of α does the spiral $\begin{pmatrix} r(t) \\ \theta(t) \end{pmatrix} = \begin{pmatrix} 1/t^\alpha \\ t \end{pmatrix}$, $\alpha > 0$ between $t = 1$ and $t = \infty$ have finite length?

5.4 Compute the area of the graph of the function $f\begin{pmatrix} x \\ y \end{pmatrix} = \frac{2}{3}(x^{3/2} + y^{3/2})$ above the region $0 \le x \le 1,\ \ 0 \le y \le 1$.

Exercise 5.5, part b: The answer should be in the form of a one-dimensional integral.

5.5 Let f be a positive C^1 function of $x \in [a, b]$.

a. Find a parametrization of the surface in $\mathbb{R}^3$ obtained by rotating the graph of f around the x-axis.

b. What is the area of this surface?

***5.6** Let

$$w_{n+1}(r) = \mathrm{vol}_{n+1}(B_r^{n+1}(\mathbf{0}))$$

be the $(n+1)$-dimensional volume of the ball of radius r in $\mathbb{R}^{n+1}$, and let $\mathrm{vol}_n(S_r^n)$ be the n-dimensional volume of the sphere of radius r in $\mathbb{R}^{n+1}$.

a. Show that $w'_{n+1}(r) = \mathrm{vol}_n(S_r^n)$.

b. Show that $\mathrm{vol}_n(S_r^n) = r^n \,\mathrm{vol}_n(S_1^n)$.

c. Derive equation 5.3.50, using $w_{n+1}(1) = \displaystyle\int_0^1 w'_{n+1}(r)\,dr$.

The *total curvature* of a surface is defined in Exercise 5.3.11.

5.7 Let H be the helicoid of equation $y\cos z = x \sin z$ (see Example 3.9.13). What is the total curvature of the part of H with $0 \le z \le a$?

5.8 For $z \in \mathbb{C}$, the function $\cos z$ is by definition $\cos z = \dfrac{e^{iz} + e^{-iz}}{2}$.

a. If $z = x + iy$, write the real and imaginary parts of $\cos z$ in terms of x, y.

b. What is the area of the part of the graph of $\cos z$ where $-\pi \le x \le \pi$ and $-1 \le y \le 1$?

5.9 Let the set C be obtained from the unit interval $[0, 1]$ by omitting the open middle 1/5th, then the open middle fifth of each remaining interval, then the open middle fifth of each remaining interval, etc.

a. Show that an alternative description of C is that it is the set of points that can be written in base 5 without using the digit 2. Use this to show that C is an uncountable set.

b. Show that C is a pavable set, with one-dimensional volume 0.

c. What is the only dimension in which C can have volume different from 0 or infinity?

Exercise 5.10 is inspired by a proposition in *Functional Analysis, Volume 1: A Gentle Introduction*, by Dzung Minh Ha (Matrix Editions, 2006).

$$\underbrace{\begin{bmatrix} 1 \\ 1 \\ 1 \\ 1 \end{bmatrix}}_{\vec{\mathbf{x}}_0}, \quad \underbrace{\begin{bmatrix} 1 \\ 2 \\ 3 \\ 0 \end{bmatrix}}_{\vec{\mathbf{x}}_1}, \quad \underbrace{\begin{bmatrix} 0 \\ 1 \\ 2 \\ 3 \end{bmatrix}}_{\vec{\mathbf{x}}_2}$$

Vectors for Exercise 5.10, part b

5.10 Let $\vec{\mathbf{x}}_0, \vec{\mathbf{x}}_1, \dots, \vec{\mathbf{x}}_k$ be vectors in $\mathbb{R}^n$, with $\vec{\mathbf{x}}_1, \dots, \vec{\mathbf{x}}_k$ linearly independent, and let $M \subset \mathbb{R}^n$ be the subspace spanned by $\vec{\mathbf{x}}_1, \dots, \vec{\mathbf{x}}_k$. Let G be the $k \times k$ matrix $G = [\vec{\mathbf{x}}_1 \dots \vec{\mathbf{x}}_k]^\top [\vec{\mathbf{x}}_1 \dots \vec{\mathbf{x}}_k]$ and let G^+ be the $(k+1) \times (k+1)$ matrix

$$G^+ = [\vec{\mathbf{x}}_0\, \vec{\mathbf{x}}_1 \dots \vec{\mathbf{x}}_k]^\top [\vec{\mathbf{x}}_0\, \vec{\mathbf{x}}_1 \dots \vec{\mathbf{x}}_k].$$

The *distance* $d(\vec{\mathbf{x}}, M)$ is by definition $d(\vec{\mathbf{x}}, M) = \inf_{\vec{\mathbf{y}} \in M}(\vec{\mathbf{x}} - \vec{\mathbf{y}})$.

a. Show that

$$\Big(d(\vec{\mathbf{x}}_0, M)\Big)^2 = \frac{\det G^+}{\det G}.$$

b. What is the distance between $\vec{\mathbf{x}}_0$ and the plane M spanned by $\vec{\mathbf{x}}_1$ and $\vec{\mathbf{x}}_2$ (as defined in the margin)?

6

Forms and vector calculus

Gradient a 1-form? How so? Hasn't one always known the gradient as a vector? Yes, indeed, but only because one was not familiar with the more appropriate 1-form concept.—C. Misner, K. S. Thorne, J. Wheeler, *Gravitation*

6.0 Introduction

What really makes calculus work is the fundamental theorem of calculus: that differentiation, having to do with speeds, and integration, having to do with areas, are somehow inverse operations.

We want to generalize the fundamental theorem of calculus to higher dimensions. Unfortunately, we cannot do so with the techniques of Chapters 4 and 5, where we integrated using $|d^n\mathbf{x}|$. The reason is that $|d^n\mathbf{x}|$ always returns a positive number; it does not concern itself with the orientation of the subset over which it is integrating, unlike the dx of one-dimensional calculus, which does:

$$\int_a^b f(x)\,dx = -\int_b^a f(x)\,dx. \tag{6.0.1}$$

The cancellations due to opposite orientations make possible the fundamental theorem of calculus. To get a fundamental theorem of calculus in higher dimensions, we need to define orientation in higher dimensions, and we need an integrand that gives one number when integrating over a domain with one orientation, and the opposite number when integrating over a domain with the opposite orientation.

It follows that orientation in higher dimensions must be defined in such a way that choosing an orientation is always a choice between one orientation and its opposite. It is fairly clear that you can orient a curve by drawing an arrow on it; orientation then means, what direction are you going along the curve, with the arrow or against it? For a surface in $\mathbb{R}^3$, an orientation is a specification of a direction in which to go through the surface, such as crossing a sphere "from the inside to the outside" or "from the outside to the inside". These two notions of orientation, for a curve and for a surface, are actually two instances of a single notion: we will provide a single definition of orientation that covers these cases and all others as well (including 0-manifolds, or points, which in other approaches to orientation are sometimes left out).

Once we have determined how to orient our objects, we must choose our *integrands*: the mathematical creature that assigns a little number to

In one-variable calculus, the standard integrand $f(x)\,dx$ takes a piece $[x_i, x_{i+1}]$ of the domain and returns the number

$$f(x_i)(x_{i+1} - x_i):$$

the area of a rectangle with height $f(x_i)$ and width $x_{i+1} - x_i$. Note that dx returns $x_{i+1} - x_i$, not $|x_{i+1} - x_i|$; this accounts for equation 6.0.1.

In Chapter 4 we studied the integrand $|d^n\mathbf{x}|$, which takes a (flat) subset $A \subset \mathbb{R}^n$ and returns its n-dimensional volume. In Chapter 5 we showed how to integrate $|d^k\mathbf{x}|$ over a (curvy) k-dimensional manifold in $\mathbb{R}^n$ to determine its k-dimensional volume. Such integrands require no mention of the orientation of the piece.

Differential forms are a special case of *tensors*. A tensor on a manifold is "anything you can build out of tangent vectors and duals of tangent vectors": a vector field is a tensor, as is a quadratic form on tangent vectors.

Although tensor calculus is a powerful tool, especially in computations, we find that speaking of tensors tends to obscure the nature of the objects under consideration.

a little piece of the domain. If we were willing to restrict ourselves to $\mathbb{R}^2$ and $\mathbb{R}^3$, we could use the techniques of vector calculus. Instead we will use *forms*, also known as *differential forms*. Forms make possible a unified treatment of differentiation and of the fundamental theorem of calculus: one operator (the *exterior derivative*) works in all dimensions, and one short, elegant statement (the generalized Stokes's theorem) generalizes the fundamental theorem of calculus to all dimensions. In contrast, vector calculus requires special formulas, operators, and theorems for each dimension where it works.

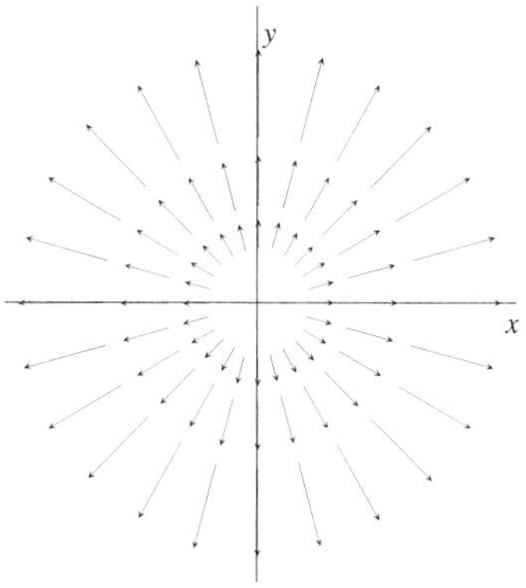

FIGURE 6.0.1. The radial vector field $\vec{F}\begin{pmatrix} x \\ y \end{pmatrix} = \begin{bmatrix} x \\ y \end{bmatrix}$.

But the language of vector calculus is used in many science courses, particularly at the undergraduate level. In addition, the functions and vector fields of vector calculus are more intuitive than forms. A vector field is an object that one can picture, as in Figure 6.0.1. Coming to terms with forms requires more effort: we can't draw you a picture of a form. A k-form is, as we shall see, something like the determinant: it takes k vectors, fiddles with them until it has a square matrix, and then takes its determinant.

The important difference between determinants and k-forms is that a k-form on $\mathbb{R}^n$ is a function of k vectors, while the determinant on $\mathbb{R}^n$ is a function of n vectors; determinants are defined only for square matrices.

For these two reasons we have devoted three sections to translating between forms and vector calculus: Section 6.5 relates forms on $\mathbb{R}^3$ to functions and vector fields, Section 6.8 shows that the exterior derivative we define using forms has three separate incarnations in the language of vector calculus: grad, curl, and div. Section 6.11 shows how Stokes's theorem, a single statement in the language of forms, becomes four more complicated statements in the language of vector calculus.

In Section 6.9 we discuss the *pullback* of form fields, which describes how integrands transform under changes of variables.

Because forms work in any dimension, they are the natural way to approach two towering subjects that are inherently four-dimensional: electromagnetism and the theory of relativity. Electromagnetism is the subject of Section 6.12. Section 6.13 introduces the *cone operator* to deal with potentials in sufficient generality to apply to electromagnetism.

Section 6.12 is an ambitious treatment of electromagnetism using forms; we will see that Maxwell's laws can be written in the elegant form

$$\mathbf{d}\mathbb{F} = 0, \quad \mathbf{d}\mathbb{M} = 4\pi\mathbb{J}.$$

We begin by introducing forms; we will then see (Section 6.2) how to integrate forms over parametrized domains (domains that come with an inherent orientation), before tackling the issue of orientation in Sections 6.3 and 6.4.

6.1 Forms on $\mathbb{R}^n$

In Section 4.8 we saw that the determinant is the unique antisymmetric and multilinear function of n vectors in $\mathbb{R}^n$ that gives 1 if evaluated on the standard basis vectors. Because of the connection between the determinant and volume described in Section 4.9, the determinant is fundamental to changes of variables in multiple integrals, as we saw in Section 4.10.

Our treatment of forms, especially the exterior derivative, was influenced by Vladimir Arnold's book *Mathematical Methods of Classical Mechanics*.

Here we will study the multilinear antisymmetric functions of k vectors in $\mathbb{R}^n$, where $k \geq 0$ may be any integer, though we will see that the only interesting case is when $k \leq n$. Again there is a close relation to volumes; these objects, called *forms* or *k-forms*, are the right integrands for integrating over oriented k-dimensional domains.

Definition 6.1.1 (k-form on $\mathbb{R}^n$). A *k-form* on $\mathbb{R}^n$ is a function φ that takes k vectors in $\mathbb{R}^n$ and returns a number $\varphi(\vec{\mathbf{v}}_1, \dots, \vec{\mathbf{v}}_k)$, such that φ is multilinear and antisymmetric as a function of the vectors.

The number k is called the *degree* of the form.

The next example is the fundamental example.

Example 6.1.2 (k-form). Let $i_1, \dots, i_k$ be any k integers between 1 and n. Then $dx_{i_1} \wedge \cdots \wedge dx_{i_k}$ is that function of k vectors $\vec{\mathbf{v}}_1, \dots, \vec{\mathbf{v}}_k$ in $\mathbb{R}^n$ that puts these vectors side by side, making the $n \times k$ matrix

$$\begin{bmatrix} v_{1,1} & \dots & v_{1,k} \\ \vdots & \dots & \vdots \\ v_{n,1} & \dots & v_{n,k} \end{bmatrix} \qquad 6.1.1$$

and selects k rows: first row i_1, then row i_2, etc., and finally row i_k, making a square $k \times k$ matrix, and finally takes its determinant. For instance,

$$\underbrace{dx_1 \wedge dx_2}_{\text{2-form}} \left(\begin{bmatrix} 1 \\ 2 \\ -1 \\ 1 \end{bmatrix}, \begin{bmatrix} 3 \\ -2 \\ 1 \\ 2 \end{bmatrix} \right) = \det \underbrace{\begin{bmatrix} 1 & 3 \\ 2 & -2 \end{bmatrix}}_{\substack{\text{1st and 2nd rows} \\ \text{of original matrix}}} = -8. \qquad 6.1.2$$

$$\underbrace{dx_1 \wedge dx_2 \wedge dx_4}_{\text{3-form}} \left(\begin{bmatrix} 1 \\ 2 \\ -1 \\ 1 \end{bmatrix}, \begin{bmatrix} 3 \\ -2 \\ 1 \\ 2 \end{bmatrix}, \begin{bmatrix} 0 \\ 1 \\ 2 \\ 1 \end{bmatrix} \right) = \det \begin{bmatrix} 1 & 3 & 0 \\ 2 & -2 & 1 \\ 1 & 2 & 1 \end{bmatrix} = -7 \qquad 6.1.3$$

$$\underbrace{dx_2 \wedge dx_1 \wedge dx_4}_{\text{3-form}} \left(\begin{bmatrix} 1 \\ 2 \\ -1 \\ 1 \end{bmatrix}, \begin{bmatrix} 3 \\ -2 \\ 1 \\ 2 \end{bmatrix}, \begin{bmatrix} 0 \\ 1 \\ 2 \\ 1 \end{bmatrix} \right) = \det \begin{bmatrix} 2 & -2 & 1 \\ 1 & 3 & 0 \\ 1 & 2 & 1 \end{bmatrix} = 7 \qquad \triangle$$

Example 6.1.3 (0-form). Definition 6.1.1 makes sense even if $k = 0$: a 0-form on $\mathbb{R}^n$ takes no vectors and returns a number. In other word, it is that number. $\triangle$

Remarks. 1. For now think of a form like $dx_1 \wedge dx_2$ or $dx_1 \wedge dx_2 \wedge dx_4$ as a single item, without worrying about the component parts. The reason for the wedge $\wedge$ will be explained at the end of this section, where we discuss the *wedge product*; we will see that the use of $\wedge$ in the wedge product is consistent with its use here. In Section 6.8 we will see that the use of d in our notation here is consistent with its use to denote the *exterior derivative*.

2. The integrand $|d^k\mathbf{x}|$ of Chapter 5 also takes k vectors in $\mathbb{R}^n$ and gives a number:

$$|d^k\mathbf{x}|(\vec{\mathbf{v}}_1, \dots, \vec{\mathbf{v}}_k) = \sqrt{\det\Big([\vec{\mathbf{v}}_1, \dots, \vec{\mathbf{v}}_k]^\top [\vec{\mathbf{v}}_1, \dots, \vec{\mathbf{v}}_k]\Big)}. \qquad 6.1.4$$

But these integrands are neither multilinear nor antisymmetric. $\triangle$

Definition 6.1.1 is actually the definition of a *constant* k-form. In this section we mainly discuss the algebra of constant forms, which we will refer to as forms. Later in this chapter we will use k-form *fields*, which have a k-form at every point; see Definition 6.1.16. There is the same relationship between constant forms and form fields as between numbers and functions.

Antisymmetry

If you exchange any two of the arguments of φ, you change the sign of φ:

$$\varphi(\vec{\mathbf{v}}_1, \dots, \vec{\mathbf{v}}_i, \dots, \vec{\mathbf{v}}_j, \dots, \vec{\mathbf{v}}_k) = -\varphi(\vec{\mathbf{v}}_1, \dots, \vec{\mathbf{v}}_j, \dots, \vec{\mathbf{v}}_i, \dots, \vec{\mathbf{v}}_k).$$

Multilinearity

If φ is a k-form and

$$\vec{\mathbf{v}}_i = a\vec{\mathbf{u}} + b\vec{\mathbf{w}},$$

then

$$\varphi\Big(\vec{\mathbf{v}}_1, \dots, (a\vec{\mathbf{u}} + b\vec{\mathbf{w}}), \dots, \vec{\mathbf{v}}_k\Big) = a\varphi(\vec{\mathbf{v}}_1, \dots, \vec{\mathbf{v}}_{i-1}, \vec{\mathbf{u}}, \vec{\mathbf{v}}_{i+1}, \dots, \vec{\mathbf{v}}_k) + b\varphi(\vec{\mathbf{v}}_1, \dots, \vec{\mathbf{v}}_{i-1}, \vec{\mathbf{w}}, \vec{\mathbf{v}}_{i+1}, \dots, \vec{\mathbf{v}}_k).$$

Equation 6.1.3: Note that to give an example of a 3-form we had to add a third vector. You cannot evaluate a 3-form on two vectors (or on four); a k-form is a function of k vectors. But you *can* evaluate a 2-form on two vectors in $\mathbb{R}^4$ (as we did in equation 6.1.2) or in $\mathbb{R}^{16}$. This is not the case for the determinant, which is a function of n vectors in $\mathbb{R}^n$.

Note there are no nonzero k-forms on $\mathbb{R}^n$ when $k > n$. If $\vec{\mathbf{v}}_1, \dots, \vec{\mathbf{v}}_k$ are vectors in $\mathbb{R}^n$ and $k > n$, then the vectors are not linearly independent, and at least one of them is a linear combination of the others, say

$$\vec{\mathbf{v}}_k = \sum_{i=1}^{k-1} a_i \vec{\mathbf{v}}_i. \qquad 6.1.5$$

Then if φ is a k-form on $\mathbb{R}^n$, evaluation on $\vec{\mathbf{v}}_1, \dots, \vec{\mathbf{v}}_k$ gives

$$\varphi(\vec{\mathbf{v}}_1, \dots, \vec{\mathbf{v}}_k) = \varphi\left(\vec{\mathbf{v}}_1, \dots, \sum_{i=1}^{k-1} a_i \vec{\mathbf{v}}_i\right) = \sum_{i=1}^{k-1} a_i \varphi(\vec{\mathbf{v}}_1, \dots, \vec{\mathbf{v}}_{k-1}, \vec{\mathbf{v}}_i). \qquad 6.1.6$$

The first term of the sum at right is $a_1\varphi(\vec{\mathbf{v}}_1, \dots, \vec{\mathbf{v}}_{k-1}, \vec{\mathbf{v}}_1)$, the second is $a_2\varphi(\vec{\mathbf{v}}_1, \vec{\mathbf{v}}_2, \dots, \vec{\mathbf{v}}_{k-1}, \vec{\mathbf{v}}_2)$, and so on; each term evaluates φ on k vectors, two of which are equal, and so (by antisymmetry) the k-form returns 0.

Geometric meaning of k-forms

Evaluating the 2-form $dx_1 \wedge dx_2$ on vectors $\vec{\mathbf{a}}, \vec{\mathbf{b}} \in \mathbb{R}^3$, we have

$$dx_1 \wedge dx_2 \left(\begin{bmatrix} a_1 \\ a_2 \\ a_3 \end{bmatrix}, \begin{bmatrix} b_1 \\ b_2 \\ b_3 \end{bmatrix} \right) = \det \begin{bmatrix} a_1 & b_1 \\ a_2 & b_2 \end{bmatrix} = a_1 b_2 - a_2 b_1. \qquad 6.1.7$$

Rather than imagining projecting $\vec{\mathbf{a}}$ and $\vec{\mathbf{b}}$ onto the plane to get the vectors of equation 6.1.8, we could imagine projecting the parallelogram spanned by $\vec{\mathbf{a}}$ and $\vec{\mathbf{b}}$ onto the plane to get the parallelogram spanned by the vectors of formula 6.1.8.

If we project $\vec{\mathbf{a}}$ and $\vec{\mathbf{b}}$ onto the (x_1, x_2)-plane, we get the vectors

$$\begin{bmatrix} a_1 \\ a_2 \end{bmatrix} \quad \text{and} \quad \begin{bmatrix} b_1 \\ b_2 \end{bmatrix}; \qquad 6.1.8$$

the determinant in equation 6.1.7 gives the *signed area of the parallelogram spanned by the vectors in equation 6.1.8.*

Thus $dx_1 \wedge dx_2$ deserves to be called the (x_1, x_2)-component of signed area. Similarly, $dx_2 \wedge dx_3$ and $dx_1 \wedge dx_3$ deserve to be called the (x_2, x_3)- and (x_1, x_3)-components of signed area.

We can now interpret equations 6.1.2 and 6.1.3 geometrically. The 2-form $dx_1 \wedge dx_2$ tells us that the (x_1, x_2)-component of signed area of the parallelogram spanned by the two vectors in equation 6.1.2 is -8. The 3-form $dx_1 \wedge dx_2 \wedge dx_4$ tells us that the (x_1, x_2, x_4)-component of signed volume of the parallelepiped spanned by the three vectors in equation 6.1.3 is -7.

Similarly, the 1-form dx gives the x-component of signed length of a vector, while dy gives its y-component:

$$dx \left(\begin{bmatrix} 2 \\ -3 \\ 1 \end{bmatrix} \right) = \det 2 = 2 \quad \text{and} \quad dy \left(\begin{bmatrix} 2 \\ -3 \\ 1 \end{bmatrix} \right) = \det(-3) = -3. \qquad 6.1.9$$

More generally (and an advantage of k-forms is that they generalize so easily to higher dimensions), we see that

$$dx_i \left(\begin{bmatrix} v_1 \\ \vdots \\ v_n \end{bmatrix} \right) = \det[v_i] = v_i \qquad 6.1.10$$

is the ith component of the signed length of $\vec{\mathbf{v}}$, and that

$$dx_{i_1} \wedge \cdots \wedge dx_{i_k}, \quad \text{evaluated on} \quad (\vec{\mathbf{v}}_1, \dots, \vec{\mathbf{v}}_k), \tag{6.1.11}$$

gives the $(x_{i_1}, \dots, x_{i_k})$-component of signed k-dimensional volume of the k-parallelogram spanned by $\vec{\mathbf{v}}_1, \dots, \vec{\mathbf{v}}_k$.

Figure 6.1.1 shows a parallelogram P spanned by two vectors, $\vec{\mathbf{v}}_1$ and $\vec{\mathbf{v}}_2$, both anchored at $\begin{pmatrix} 1 \\ 2 \\ 1.5 \end{pmatrix}$, and the projections of P onto the (x, y)-plane (P_3), the (x, z)-plane (P_2), and the (y, z)-plane (P_1). Evaluating $dx \wedge dy$ on $\vec{\mathbf{v}}_1, \vec{\mathbf{v}}_2$ gives the signed area of P_3, while $dx \wedge dz$ gives the signed area of P_2 and $dy \wedge dz$ gives the signed area of P_1. Note that although we defined a k-form as a function of k vectors, we can also think of it as a function of the k-parallelogram spanned by those k vectors.

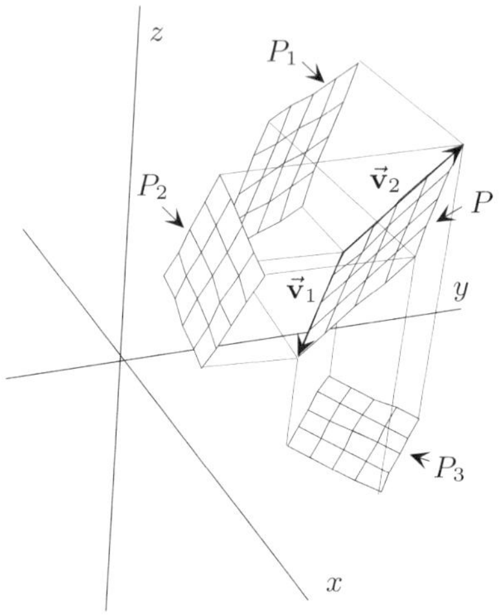

FIGURE 6.1.1.
The parallelogram P spanned by

$$\vec{\mathbf{v}}_1 = \begin{bmatrix} 1 \\ -1 \\ -0.5 \end{bmatrix}, \vec{\mathbf{v}}_2 = \begin{bmatrix} 1 \\ 0.5 \\ 1.5 \end{bmatrix}$$

and its projections onto the (x, y)-plane (P_3), the (x, z)-plane (P_2), and the (y, z)-plane (P_1). Evaluating appropriate 2-forms on $\vec{\mathbf{v}}_1, \vec{\mathbf{v}}_2$ gives the area of those projections:

$$\text{vol}_2\, P_1 = dy \wedge dz(\vec{\mathbf{v}}_1, \vec{\mathbf{v}}_2) = -5/4$$
$$\text{vol}_2\, P_2 = dx \wedge dz(\vec{\mathbf{v}}_1, \vec{\mathbf{v}}_2) = 2$$
$$\text{vol}_2\, P_3 = dx \wedge dy(\vec{\mathbf{v}}_1, \vec{\mathbf{v}}_2) = 3/2.$$

In terms of this geometric description, the statement that there are no nonzero k-forms when $k > n$ should come as no surprise: you would expect any kind of three-dimensional volume in $\mathbb{R}^2$ to be zero, and more generally any k-dimensional volume in $\mathbb{R}^n$ to be 0 when $k > n$.

Elementary forms

There is a great deal of redundancy in the expressions $dx_{i_1} \wedge \cdots \wedge dx_{i_k}$. Consider $dx_1 \wedge dx_3 \wedge dx_1$. This 3-form stacks three vectors in $\mathbb{R}^n$ side by side to make an $n \times 3$ matrix, selects the first row, then the third, then the first again, to make a 3×3 matrix and takes its determinant. This determinant is always 0. (Do you see why?[1])

But $dx_1 \wedge dx_3 \wedge dx_1$ is not the only way to write the form that takes three vectors and returns 0; both $dx_1 \wedge dx_1 \wedge dx_3$ and $dx_2 \wedge dx_3 \wedge dx_3$ do so as well. Indeed, if any two of the indices $i_1, \dots, i_k$ are equal, then $dx_{i_1} \wedge \cdots \wedge dx_{i_k} = 0$.

Next, consider $dx_1 \wedge dx_3$ and $dx_3 \wedge dx_1$. When these forms are evaluated on $\vec{\mathbf{a}} = \begin{bmatrix} a_1 \\ \vdots \\ a_n \end{bmatrix}$ and $\vec{\mathbf{b}} = \begin{bmatrix} b_1 \\ \vdots \\ b_n \end{bmatrix}$, we find

$$\begin{aligned} dx_1 \wedge dx_3(\vec{\mathbf{a}}, \vec{\mathbf{b}}) &= \det \begin{bmatrix} a_1 & b_1 \\ a_3 & b_3 \end{bmatrix} = a_1 b_3 - a_3 b_1 \\ dx_3 \wedge dx_1(\vec{\mathbf{a}}, \vec{\mathbf{b}}) &= \det \begin{bmatrix} a_3 & b_3 \\ a_1 & b_1 \end{bmatrix} = a_3 b_1 - a_1 b_3. \end{aligned} \tag{6.1.12}$$

Clearly $dx_1 \wedge dx_3 = -dx_3 \wedge dx_1$; these two 2-forms, evaluated on the same two vectors, always return opposite numbers.

[1]The determinant of a square matrix containing two identical columns is always 0, since exchanging them reverses the sign of the determinant, while keeping it the same. Since (Theorem 4.8.8), $\det A = \det A^\top$, the determinant of a matrix is also 0 if two of its rows are identical.

Recall that the signature of a permutation σ is denoted $\operatorname{sgn}(\sigma)$; see Theorem and Definition 4.8.11. Theorem 4.8.12 gives a formula for det using the signature.

Definition 6.1.4: Putting the indices in increasing order selects one particular permutation for any set of distinct integers $j_1, \dots, j_k$. For example, if $n = 4$ and $k = 3$, then $1 \le i_1 < \cdots < i_k \le n$ is the set

$$\begin{aligned}\{(i_1 = 1, i_2 = 2, i_3 = 3),\\ (i_1 = 1, i_2 = 2, i_3 = 4),\\ (i_1 = 1, i_2 = 3, i_3 = 4),\\ (i_1 = 2, i_2 = 3, i_3 = 4)\},\end{aligned}$$

which corresponds to

$$\begin{aligned}&dx_1 \wedge dx_2 \wedge dx_3,\\ &dx_1 \wedge dx_2 \wedge dx_4,\end{aligned}$$

and so on. The k-form

$$dx_2 \wedge dx_1 \wedge dx_3$$

exists, but it is not an *elementary* k-form.

More generally, if $i_1, \dots, i_k$ and $j_1, \dots, j_k$ are the same integers, just taken in a different order, so that $j_1 = i_{\sigma(1)}, j_2 = i_{\sigma(2)}, \dots, j_k = i_{\sigma(k)}$ for some permutation σ of $\{1, \dots, k\}$, then

$$dx_{j_1} \wedge \cdots \wedge dx_{j_k} = \operatorname{sgn}(\sigma) dx_{i_1} \wedge \cdots \wedge dx_{i_k}. \qquad 6.1.13$$

Indeed, $dx_{j_1} \wedge \cdots \wedge dx_{j_k}$ computes the determinant of the same matrix as $dx_{i_1} \wedge \cdots \wedge dx_{i_k}$, only with the rows permuted by σ. For instance,

$$dx_1 \wedge dx_2 \wedge dx_3 = dx_2 \wedge dx_3 \wedge dx_1 = dx_3 \wedge dx_1 \wedge dx_2. \qquad 6.1.14$$

We will find it useful to define a special class of k-forms, *elementary* k-forms, which avoid this redundancy:

Definition 6.1.4 (Elementary k-forms on $\mathbb{R}^n$). An *elementary k-form* on $\mathbb{R}^n$ is an expression of the form

$$dx_{i_1} \wedge \cdots \wedge dx_{i_k}, \qquad 6.1.15$$

where $1 \le i_1 < \cdots < i_k \le n$. Evaluated on the vectors $\vec{\mathbf{v}}_1, \dots, \vec{\mathbf{v}}_k$, it gives the determinant of the $k \times k$ matrix obtained by selecting rows $i_1, \dots, i_k$ of the matrix whose columns are the vectors $\vec{\mathbf{v}}_1, \dots, \vec{\mathbf{v}}_k$. The only elementary 0-form is the form, denoted 1, which evaluated on zero vectors returns 1.

We saw that there are no nonzero k-forms on $\mathbb{R}^n$ when $k > n$; not surprisingly, there are no elementary k-forms on $\mathbb{R}^n$ when $k > n$. (That there are no elementary k-forms when $k > n$ is an example of the "pigeonhole" principle: if more than n pigeons fit in n holes, one hole must contain at least two pigeons. Here we would need to select more than n distinct integers between 1 and n.)

All forms are linear combinations of elementary forms

We said that $dx_{i_1} \wedge \cdots \wedge dx_{i_k}$ is the fundamental example of a k-form. We will now justify this statement by showing that any k-form is a linear combination of elementary k-forms. The following definitions say that speaking of such linear combinations makes sense: we can add k-forms and multiply them by scalars in the obvious way.

Adding k-forms

$$3\,dx \wedge dy + 2\,dx \wedge dy = 5\,dx \wedge dy$$
$$dx \wedge dy + dy \wedge dx = 0$$

Multiplying k-forms by scalars

$$\begin{aligned}&5\,(dx \wedge dy + 2\,dx \wedge dz)\\ &= 5\,dx \wedge dy + 10\,dx \wedge dz\end{aligned}$$

Definition 6.1.5 (Addition of k-forms). Let φ and ψ be two k-forms. Then

$$(\varphi + \psi)(\vec{\mathbf{v}}_1, \dots, \vec{\mathbf{v}}_k) \stackrel{\text{def}}{=} \varphi(\vec{\mathbf{v}}_1, \dots, \vec{\mathbf{v}}_k) + \psi(\vec{\mathbf{v}}_1, \dots, \vec{\mathbf{v}}_k). \qquad 6.1.16$$

Definition 6.1.6 (Multiplication of k-forms by scalars). If φ is a k-form and a is a scalar, then

$$(a\varphi)(\vec{\mathbf{v}}_1, \dots, \vec{\mathbf{v}}_k) \stackrel{\text{def}}{=} a(\varphi(\vec{\mathbf{v}}_1, \dots, \vec{\mathbf{v}}_k)). \qquad 6.1.17$$

Using these definitions, the space of k-forms on $\mathbb{R}^n$ is a vector space. The elementary k-forms form a basis of this space.

Saying that the space of k-forms on $\mathbb{R}^n$ is a vector space (Definition 2.6.1) means that k-forms can be manipulated in familiar ways, for example, for scalars α, β,

$$\alpha(\beta\varphi) = (\alpha\beta)\varphi$$

(associativity for multiplication),

$$(\alpha+\beta)\varphi = \alpha\varphi + \beta\varphi$$

(distributive law for scalar addition), and

$$\alpha(\varphi+\psi) = \alpha\varphi + \alpha\psi$$

(distributive law for addition of forms).

Definition 6.1.7 ($A_c^k(\mathbb{R}^n)$). The space of k-forms on $\mathbb{R}^n$ is denoted $A_c^k(\mathbb{R}^n)$.

We use the notation A_c^k because we want to save the notation A^k for the space of *form fields* (Definition 6.1.16). The c in A_c^k stands for "constant": elements of $A_c^k(\mathbb{R}^n)$ can be thought of as constant form fields.

Theorem 6.1.8 (Elementary k-forms form a basis for $A_c^k(\mathbb{R}^n)$). *The elementary k-forms form a basis for $A_c^k(\mathbb{R}^n)$: every k-form can be uniquely written*

$$\varphi = \sum_{1\le i_1<\cdots<i_k\le n} a_{i_1\dots i_k} dx_{i_1}\wedge\cdots\wedge dx_{i_k}, \qquad 6.1.18$$

where the coefficients $a_{i_1\dots i_k}$ are given by

$$a_{i_1\dots i_k} = \varphi\left(\vec{\mathbf{e}}_{i_1},\dots,\vec{\mathbf{e}}_{i_k}\right). \qquad 6.1.19$$

When $k=n$, Theorem 6.1.8 is Theorem and Definition 4.8.1, which says that the determinant is the unique function from $(\mathbb{R}^n)^n$ to $\mathbb{R}$ that satisfies multilinearity, antisymmetry, and normalization.

Remark. In Section 1.3 we saw that the ith column of a matrix T is given by $T\vec{\mathbf{e}}_i$: once we know what a linear transformation does to the standard basis vectors, we can deduce what it does to any vector in its domain.

Equation 6.1.19 says something similar about k-forms: once we know what number a k-form on $\mathbb{R}^n$ returns when evaluated on appropriate standard basis vectors, we know what it will return when evaluated on any k vectors, since knowing the coefficients allows us to write the form in complete detail. By "appropriate" we mean all combinations of k standard basis vectors in $\mathbb{R}^n$ listed in ascending order: for a 2-form on $\mathbb{R}^2$, just $(\vec{\mathbf{e}}_1,\vec{\mathbf{e}}_2)$; for a 2-form on $\mathbb{R}^3$, $(\vec{\mathbf{e}}_1,\vec{\mathbf{e}}_2)$, $(\vec{\mathbf{e}}_1,\vec{\mathbf{e}}_3)$, and $(\vec{\mathbf{e}}_2,\vec{\mathbf{e}}_3)$. △

Example 6.1.9: The function $W_{\vec{\mathbf{v}}}(\vec{\mathbf{w}}) = \vec{\mathbf{v}}\cdot\vec{\mathbf{w}}$ is a 1-form on $\mathbb{R}^n$ because it is a function of one vector and it is linear as a function of $\vec{\mathbf{w}}$. The requirement that it be antisymmetric is automatically satisfied, since it is a function of only one vector. We will meet this function again in Definition 6.5.1, where we will see that it is the work form of a vector field.

Example 6.1.9 (Finding coefficients for a 1-form). The function defined by $W_{\vec{\mathbf{v}}}(\vec{\mathbf{w}}) = \vec{\mathbf{v}}\cdot\vec{\mathbf{w}}$, where both the fixed vector $\vec{\mathbf{v}}$ and the variable $\vec{\mathbf{w}}$ are elements of $\mathbb{R}^n$, is a 1-form on $\mathbb{R}^n$ (see the margin note). Therefore, Theorem 6.1.8 says that it can be written as a linear combination of elementary 1-forms, i.e., in the form

$$W_{\vec{\mathbf{v}}} = a_1\,dx_1 + \cdots + a_n\,dx_n. \qquad 6.1.20$$

The a_i are given by equation 6.1.19: $a_i = W_{\vec{\mathbf{v}}}(\vec{\mathbf{e}}_i) = \vec{\mathbf{v}}\cdot\vec{\mathbf{e}}_i = v_i$. Thus

$$W_{\vec{\mathbf{v}}} = v_1\,dx_1+\cdots+v_n\,dx_n \quad\text{and}\quad W_{\vec{\mathbf{v}}}(\vec{\mathbf{w}}) = v_1w_1+\cdots+v_nw_n. \quad △ \qquad 6.1.21$$

Proof of Theorem 6.1.8. First we use a 2-form φ on $\mathbb{R}^3$ to justify our statement that a k-form is completely determined by what it does to the standard basis vectors. The computation uses multilinearity in all equalities except the first; it uses antisymmetry in the last equality:

$$\varphi\Bigg(\underbrace{\begin{bmatrix} v_1 \\ v_2 \\ v_3 \end{bmatrix}}_{\vec{\mathbf{v}}}, \underbrace{\begin{bmatrix} w_1 \\ w_2 \\ w_3 \end{bmatrix}}_{\vec{\mathbf{w}}}\Bigg) \underset{1}{=} \varphi(\underbrace{v_1\vec{\mathbf{e}}_1 + v_2\vec{\mathbf{e}}_2 + v_3\vec{\mathbf{e}}_3}_{\vec{\mathbf{v}}}, \underbrace{w_1\vec{\mathbf{e}}_1 + w_2\vec{\mathbf{e}}_2 + w_3\vec{\mathbf{e}}_3}_{\vec{\mathbf{w}}})$$

$$= \varphi(v_1\vec{\mathbf{e}}_1, \underbrace{w_1\vec{\mathbf{e}}_1 + w_2\vec{\mathbf{e}}_2 + w_3\vec{\mathbf{e}}_3}_{\vec{\mathbf{w}}}) + \varphi(v_2\vec{\mathbf{e}}_2, \underbrace{w_1\vec{\mathbf{e}}_1 + w_2\vec{\mathbf{e}}_2 + w_3\vec{\mathbf{e}}_3}_{\vec{\mathbf{w}}}) + \varphi(v_3\vec{\mathbf{e}}_3, \underbrace{w_1\vec{\mathbf{e}}_1 + w_2\vec{\mathbf{e}}_2 + w_3\vec{\mathbf{e}}_3}_{\vec{\mathbf{w}}})$$

$$= \varphi(v_1\vec{\mathbf{e}}_1, w_1\vec{\mathbf{e}}_1) + \varphi(v_1\vec{\mathbf{e}}_1, w_2\vec{\mathbf{e}}_2) + \varphi(v_1\vec{\mathbf{e}}_1, w_3\vec{\mathbf{e}}_3) + \cdots \text{ (9 terms in all)} \qquad 6.1.22$$

$$\underset{2}{=} \underbrace{v_1 w_1 \varphi(\vec{\mathbf{e}}_1, \vec{\mathbf{e}}_1)}_{0} + v_1 w_2 \varphi(\vec{\mathbf{e}}_1, \vec{\mathbf{e}}_2) + v_1 w_3 \varphi(\vec{\mathbf{e}}_1, \vec{\mathbf{e}}_3) + \underbrace{v_2 w_1 \varphi(\vec{\mathbf{e}}_2, \vec{\mathbf{e}}_1)}_{-v_2 w_1 \varphi(\vec{\mathbf{e}}_1, \vec{\mathbf{e}}_2)} + \cdots \text{ (9 terms in all)}$$

$$\underset{3}{=} (v_1 w_2 - v_2 w_1)\varphi(\vec{\mathbf{e}}_1, \vec{\mathbf{e}}_2) + (v_1 w_3 - v_3 w_1)\varphi(\vec{\mathbf{e}}_1, \vec{\mathbf{e}}_3) + (v_2 w_3 - v_3 w_2)\varphi(\vec{\mathbf{e}}_2, \vec{\mathbf{e}}_3.)$$

Now consider the corresponding computation for an k-form φ on $\mathbb{R}^n$, where each equality corresponds to the equality marked with the same number in equation 6.1.22:

$$\varphi(\vec{\mathbf{v}}_1, \dots, \vec{\mathbf{v}}_k) \underset{1}{=} \varphi\left(\left(\sum_{i_1=1}^{n} v_{i_1,1}\vec{\mathbf{e}}_{i_1}\right), \left(\sum_{i_2=1}^{n} v_{i_2,2}\vec{\mathbf{e}}_{i_2}\right), \dots, \left(\sum_{i_k=1}^{n} v_{i_k,k}\vec{\mathbf{e}}_{i_k}\right)\right)$$

$$\underset{2}{=} \sum_{i_1=1}^{n}\sum_{i_2}^{n}\cdots\sum_{i_k=1}^{n} v_{i_1,1}\cdots v_{i_k,k}\varphi(\vec{\mathbf{e}}_{i_1}, \dots, \vec{\mathbf{e}}_{i_k}) \quad (n^k \text{ terms}) \qquad 6.1.23$$

$$\underset{3}{=} \sum_{1 \le i_1 < \cdots < i_k \le n} \underbrace{\left(\sum_{\sigma \in \text{Perm}_k} (\text{sgn}\,\sigma)\; v_{i_1,\sigma(1)} \cdots v_{i_k,\sigma(k)}\right.}_{\text{det of } k\times k \text{ matrix, by Theorem 4.8.12}} \left.\varphi(\vec{\mathbf{e}}_{i_1}, \dots, \vec{\mathbf{e}}_{i_k})\right).$$

Using the margin note, this becomes

$$\varphi(\vec{\mathbf{v}}_1, \dots, \vec{\mathbf{v}}_k) = \sum_{1 \le i_1 < \cdots < i_k \le n} dx_{i_1} \wedge \cdots \wedge dx_{i_k}(\vec{\mathbf{v}}_1, \dots, \vec{\mathbf{v}}_k)\underbrace{\varphi(\vec{\mathbf{e}}_{i_1}, \dots, \vec{\mathbf{e}}_{i_k})}_{a_{i_1 \dots i_k}}$$

$$= \sum_{1 \le i_1 < \cdots < i_k \le n} a_{i_1 \dots i_k} dx_{i_1} \wedge \cdots \wedge dx_{i_k}(\vec{\mathbf{v}}_1, \dots, \vec{\mathbf{v}}_k), \qquad 6.1.24$$

which completes the proof. □

Equation 6.1.22 says that for some coefficients d_1, d_2, d_3,

$$\varphi(\vec{\mathbf{v}}, \vec{\mathbf{w}}) = d_1\varphi(\vec{\mathbf{e}}_1, \vec{\mathbf{e}}_2) + d_2\varphi(\vec{\mathbf{e}}_1, \vec{\mathbf{e}}_3) + d_3\varphi(\vec{\mathbf{e}}_2, \vec{\mathbf{e}}_3);$$

φ is determined by what it does to $\vec{\mathbf{e}}_1$, $\vec{\mathbf{e}}_2$, and $\vec{\mathbf{e}}_3$. The coefficients d_i are all determinants; for instance,

$$v_1 w_2 - v_2 w_1 = \det \begin{bmatrix} v_1 & w_1 \\ v_2 & w_2 \end{bmatrix}.$$

Equation 6.1.23: The matrix referred to in the last line is the matrix formed by taking rows i_1 through i_k of the $n \times k$ matrix $[\vec{\mathbf{v}}_1, \dots, \vec{\mathbf{v}}_k]$, i.e., it is

$$dx_{i_1} \wedge \cdots \wedge dx_{i_k}(\vec{\mathbf{v}}_1, \dots, \vec{\mathbf{v}}_k).$$

Theorem 6.1.10 (Dimension of $A_c^k(\mathbb{R}^n)$). *The space $A_c^k(\mathbb{R}^n)$ has dimension equal to the binomial coefficient*

$$\binom{n}{k} = \frac{n!}{k!(n-k)!}. \qquad 6.1.25$$

Proof. Just count the elements of the basis: the elementary k-forms on $\mathbb{R}^n$. Not for nothing is the binomial coefficient called "n choose k". □

Theorem 6.1.10: Remember that $0! = 1$.

When $n = k$, the space $A_c^k(\mathbb{R}^k)$ is a line consisting of multiples of the determinant, since det is the unique function of k vectors in $\mathbb{R}^k$ that satisfies multilinearity, antisymmetry, and normalization; elements of $A_c^k(\mathbb{R}^k)$ satisfy the first two properties.

Note that

$$\binom{n}{k} = \binom{n}{n-k}, \qquad 6.1.26$$

since each time you choose k out of n things you have chosen the remaining $n-k$ things.[2] So for a given n, there are equal numbers of elementary k-forms and elementary $(n-k)$-forms: $A_c^k(\mathbb{R}^n)$ and $A_c^{n-k}(\mathbb{R}^n)$ have the same dimension. Thus, in $\mathbb{R}^3$, there is one elementary 0-form and one elementary 3-form, so the spaces $A_c^0(\mathbb{R}^3)$ and $A_c^3(\mathbb{R}^3)$ can be identified with $\mathbb{R}$.

Equation 6.1.26: The fact that for a given n, there are equal numbers of elementary k-forms and elementary $(n-k)$-forms may appear to be a coincidence, but such algebraic "coincidences" are often the outward sign of an underlying geometric symmetry. This particular coincidence is explained by *Poincaré duality*: an n-dimensional manifold has the same number of k-dimensional holes and $(n-k)$-dimensional holes. (Seeing the connection takes some advanced mathematics, including the *star operator*.)

On $\mathbb{R}^3$ there are three elementary 1-forms and three elementary 2-forms, so the spaces $A_c^1(\mathbb{R}^3)$ and $A_c^2(\mathbb{R}^3)$ can be identified with vectors in $\mathbb{R}^3$.

Example 6.1.11 (Dimension of $A_c^k(\mathbb{R}^3)$). The vector spaces $A_c^0(\mathbb{R}^3)$ and $A_c^3(\mathbb{R}^3)$ have dimension 1, and $A_c^1(\mathbb{R}^3)$ and $A_c^2(\mathbb{R}^3)$ have dimension 3, since on $\mathbb{R}^3$ we have

$$\binom{3}{0} = \frac{3!}{0!3!} = 1 \text{ elementary 0-form;} \quad \binom{3}{1} = \frac{3!}{1!2!} = 3 \text{ elementary 1-forms}$$

$$\binom{3}{2} = \frac{3!}{2!1!} = 3 \text{ elementary 2-forms;} \quad \binom{3}{3} = \frac{3!}{3!0!} = 1 \text{ elementary 3-form.}$$

The wedge product

We have used the wedge $\wedge$ to write down forms. Now we will see what it means: it denotes the *wedge product*, also known as the *exterior product*. The wedge product is a messy thing: a summation, over various shuffles of vectors, of the product of two k-forms, each given the sign $+$ or $-$ according to the rule for permutations. Figure 6.1.2 explains the use of the word "shuffle".

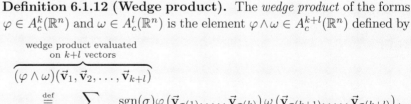

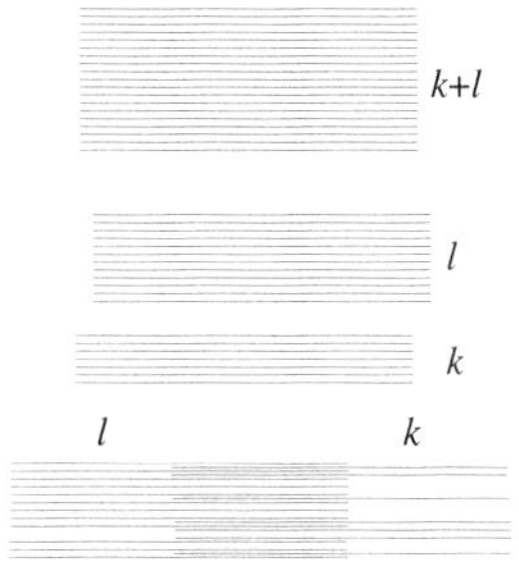

FIGURE 6.1.2.

Take a pack of $k+l$ cards, cut it to produce subpacks of k cards and l cards, and shuffle them. The order of the cards in the subpacks remains unchanged.

Definition 6.1.12 (Wedge product). The *wedge product* of the forms $\varphi \in A_c^k(\mathbb{R}^n)$ and $\omega \in A_c^l(\mathbb{R}^n)$ is the element $\varphi \wedge \omega \in A_c^{k+l}(\mathbb{R}^n)$ defined by

$$\overbrace{(\varphi \wedge \omega)(\vec{\mathbf{v}}_1, \vec{\mathbf{v}}_2, \dots, \vec{\mathbf{v}}_{k+l})}^{\text{wedge product evaluated on } k+l \text{ vectors}}$$

$$\overset{\text{def}}{=} \sum_{\substack{\text{shuffles}\\ \sigma\in\text{Perm}(k,l)}} \operatorname{sgn}(\sigma)\varphi\underbrace{\left(\vec{\mathbf{v}}_{\sigma(1)}, \dots, \vec{\mathbf{v}}_{\sigma(k)}\right)}_{k \text{ vectors}}\omega\underbrace{\left(\vec{\mathbf{v}}_{\sigma(k+1)}, \dots, \vec{\mathbf{v}}_{\sigma(k+l)}\right)}_{l \text{ vectors}}.$$

where the sum is over all permutations σ of the numbers $1, 2, \dots, k+l$ such that $\sigma(1) < \sigma(2) < \cdots < \sigma(k)$ and $\sigma(k+1) < \cdots < \sigma(k+l)$.

[2]The following computation also justifies this equality:

$$\binom{n}{n-k} = \frac{n!}{(n-k)!\big(n-(n-k)\big)!} = \frac{n!}{k!(n-k)!}.$$

On the left we have a $(k+l)$-form evaluated on $k+l$ vectors. On the right we have a complicated expression involving a k-form φ acting on k vectors and a l-form ω acting on l vectors. To understand the right side, first consider all possible permutations of the $k+l$ vectors $\vec{\mathbf{v}}_1, \vec{\mathbf{v}}_2, \dots, \vec{\mathbf{v}}_{k+l}$, dividing each permutation with a bar line $|$ so that there are k vectors to the left and l vectors to the right, since φ acts on k vectors and ω acts on l vectors. For example, if $k=2$ and $l=1$, we have six permutations:

$$\vec{\mathbf{v}}_1\vec{\mathbf{v}}_2|\vec{\mathbf{v}}_3, \quad \vec{\mathbf{v}}_1\vec{\mathbf{v}}_3|\vec{\mathbf{v}}_2, \quad \vec{\mathbf{v}}_2\vec{\mathbf{v}}_3|\vec{\mathbf{v}}_1, \quad \vec{\mathbf{v}}_2\vec{\mathbf{v}}_1|\vec{\mathbf{v}}_3, \quad \vec{\mathbf{v}}_3\vec{\mathbf{v}}_2|\vec{\mathbf{v}}_1, \quad \vec{\mathbf{v}}_3\vec{\mathbf{v}}_1|\vec{\mathbf{v}}_2. \qquad 6.1.27$$

Choose only those permutations where the indices for the k-form and the indices for the l-form are each, separately and independently, in *ascending* order (above, the first three permutations satisfy that requirement). We assign each chosen permutation its sign, according to the rule given in Theorem and Definition 4.8.11. Finally, we take the sum.

Permutation matrices are defined in equation 4.8.40. More simply, we note that the first permutation involves an even number (0) of exchanges, or *transpositions*, so the signature is positive, while the second involves an odd number (1), so the signature is negative.

Example 6.1.13 (The wedge product of two 1-forms). If φ and ω are both 1-forms, we have two permutations, $\vec{\mathbf{v}}_1|\vec{\mathbf{v}}_2$ and $\vec{\mathbf{v}}_2|\vec{\mathbf{v}}_1$, both allowable under our "ascending order" rule. The sign for the first is positive, since

$$(\vec{\mathbf{v}}_1, \vec{\mathbf{v}}_2) \to (\vec{\mathbf{v}}_1, \vec{\mathbf{v}}_2) \quad \text{corresponds to the permutation matrix} \quad \begin{bmatrix} 1 & 0 \\ 0 & 1 \end{bmatrix},$$

with determinant $+1$. The sign for the second is negative, since

$$(\vec{\mathbf{v}}_1, \vec{\mathbf{v}}_2) \to (\vec{\mathbf{v}}_2, \vec{\mathbf{v}}_1) \quad \text{corresponds to the permutation matrix} \quad \begin{bmatrix} 0 & 1 \\ 1 & 0 \end{bmatrix},$$

with determinant -1. So Definition 6.1.12 gives

$$(\varphi \wedge \omega)(\vec{\mathbf{v}}_1, \vec{\mathbf{v}}_2) = \varphi(\vec{\mathbf{v}}_1)\,\omega(\vec{\mathbf{v}}_2) - \varphi(\vec{\mathbf{v}}_2)\,\omega(\vec{\mathbf{v}}_1). \qquad 6.1.28$$

Thus the 2-form

$$dx_1 \wedge dx_2\,(\vec{\mathbf{a}}, \vec{\mathbf{b}}) = \det \begin{bmatrix} a_1 & b_1 \\ a_2 & b_2 \end{bmatrix} = a_1 b_2 - a_2 b_1, \qquad 6.1.29$$

is indeed equal to the wedge product of the 1-forms dx_1 and dx_2, which, evaluated on the same two vectors, gives

$$dx_1 \wedge dx_2(\vec{\mathbf{a}}, \vec{\mathbf{b}}) = dx_1(\vec{\mathbf{a}})\,dx_2(\vec{\mathbf{b}}) - dx_1(\vec{\mathbf{b}})\,dx_2(\vec{\mathbf{a}}) = a_1 b_2 - a_2 b_1. \qquad 6.1.30$$

So our use of the wedge in naming the elementary forms is coherent with its use to denote this special kind of multiplication. △

The wedge product of a 0-form α with a k-form ω is a k-form, $\alpha \wedge \omega = \alpha\omega$. In this case, the wedge product coincides with multiplication by numbers.

Formula 6.1.31: It takes just one transposition to switch

$$\vec{\mathbf{v}}_1\vec{\mathbf{v}}_2|\vec{\mathbf{v}}_3 \quad \text{to} \quad \vec{\mathbf{v}}_1\vec{\mathbf{v}}_3|\vec{\mathbf{v}}_2,$$

accounting for the minus sign.

Example 6.1.14 (The wedge product of a 2-form and a 1-form). If φ is a 2-form and ω is a 1-form, then (as we saw in formula 6.1.27) we have six permutations. Three have the first two indices in ascending order:

$$+(\vec{\mathbf{v}}_1\vec{\mathbf{v}}_2|\vec{\mathbf{v}}_3), \quad -(\vec{\mathbf{v}}_1\vec{\mathbf{v}}_3|\vec{\mathbf{v}}_2), \quad +(\vec{\mathbf{v}}_2\vec{\mathbf{v}}_3|\vec{\mathbf{v}}_1), \qquad 6.1.31$$

giving the wedge product

$$\varphi \wedge \omega(\vec{\mathbf{v}}_1, \vec{\mathbf{v}}_2, \vec{\mathbf{v}}_3) = \varphi(\vec{\mathbf{v}}_1, \vec{\mathbf{v}}_2)\,\omega(\vec{\mathbf{v}}_3) - \varphi(\vec{\mathbf{v}}_1, \vec{\mathbf{v}}_3)\,\omega(\vec{\mathbf{v}}_2) + \varphi(\vec{\mathbf{v}}_2, \vec{\mathbf{v}}_3)\,\omega(\vec{\mathbf{v}}_1). \qquad 6.1.32$$

Again, let's compare this with the result we get using Definition 6.1.4, setting $\varphi = dx_1 \wedge dx_2$ and $\omega = dx_3$; to avoid double indices we will rename the vectors $\vec{\mathbf{v}}_1, \vec{\mathbf{v}}_2, \vec{\mathbf{v}}_3$, calling them $\vec{\mathbf{u}}, \vec{\mathbf{v}}, \vec{\mathbf{w}}$. Using Definition 6.1.4, we get

The wedge product $\varphi \wedge \omega$ satisfies the requirements of Definition 6.1.1 for a form (multilinearity and antisymmetry). Multilinearity is not hard to see; antisymmetry is harder (as was the proof of antisymmetry for the determinant).

$$\underbrace{(dx_1 \wedge dx_2)}_{\varphi} \wedge \underbrace{dx_3}_{\omega}(\vec{\mathbf{u}}, \vec{\mathbf{v}}, \vec{\mathbf{w}}) = \det \begin{bmatrix} u_1 & v_1 & w_1 \\ u_2 & v_2 & w_2 \\ u_3 & v_3 & w_3 \end{bmatrix} \qquad 6.1.33$$

$$= u_1v_2w_3 - u_1v_3w_2 - u_2v_1w_3 + u_2v_3w_1 + u_3v_1w_2 - u_3v_2w_1.$$

If instead we use equation 6.1.32 for the wedge product, we get

$$\begin{aligned}(dx_1 \wedge dx_2) \wedge dx_3(\vec{\mathbf{u}}, \vec{\mathbf{v}}, \vec{\mathbf{w}}) &= (dx_1 \wedge dx_2)(\vec{\mathbf{u}}, \vec{\mathbf{v}})\, dx_3(\vec{\mathbf{w}}) \\ &\quad - (dx_1 \wedge dx_2)(\vec{\mathbf{u}}, \vec{\mathbf{w}})\, dx_3(\vec{\mathbf{v}}) \\ &\quad + (dx_1 \wedge dx_2)(\vec{\mathbf{v}}, \vec{\mathbf{w}})\, dx_3(\vec{\mathbf{u}}) \\ &= \det \begin{bmatrix} u_1 & v_1 \\ u_2 & v_2 \end{bmatrix} w_3 - \det \begin{bmatrix} u_1 & w_1 \\ u_2 & w_2 \end{bmatrix} v_3 + \det \begin{bmatrix} v_1 & w_1 \\ v_2 & w_2 \end{bmatrix} u_3 \qquad 6.1.34 \\ &= u_1v_2w_3 - u_2v_1w_3 - u_1v_3w_2 + u_2v_3w_1 + u_3v_1w_2 - u_3v_2w_1. \quad \triangle\end{aligned}$$

Properties of the wedge product

The wedge product behaves much like ordinary multiplication, except that we need to be careful about the sign, because of skew commutativity:

You are asked to prove Proposition 6.1.15 in Exercise 6.1.13. Part 2 is quite a bit harder than parts 1 and 3.

Part 2 justifies the omission of parentheses in the k-form

$$dx_{i_1} \wedge dx_{i_2} \wedge \cdots \wedge dx_{i_k};$$

all the ways of putting parentheses in the expression give the same result.

Proposition 6.1.15 (Properties of the wedge product). *The wedge product has the following properties:*

1. *distributivity:* $\varphi \wedge (\omega_1 + \omega_2) = \varphi \wedge \omega_1 + \varphi \wedge \omega_2$. 6.1.35
2. *associativity:* $(\varphi_1 \wedge \varphi_2) \wedge \varphi_3 = \varphi_1 \wedge (\varphi_2 \wedge \varphi_3)$. 6.1.36
3. *skew commutativity: If φ is a k-form and ω is an l-form, then* $\varphi \wedge \omega = (-1)^{kl} \omega \wedge \varphi$. 6.1.37

Note that in equation 6.1.37 the φ and ω change positions. For example, if $\varphi = dx_1 \wedge dx_2$ and $\omega = dx_3$, skew commutativity says that

$$(dx_1 \wedge dx_2) \wedge dx_3 = (-1)^2 dx_3 \wedge (dx_1 \wedge dx_2), \quad \text{that is,}$$

$$\det \begin{bmatrix} u_1 & v_1 & w_1 \\ u_2 & v_2 & w_2 \\ u_3 & v_3 & w_3 \end{bmatrix} = \det \begin{bmatrix} u_3 & v_3 & w_3 \\ u_1 & v_1 & w_1 \\ u_2 & v_2 & w_2 \end{bmatrix}, \qquad 6.1.38$$

which you can confirm either by observing that the two matrices differ by two exchanges of rows (changing the sign twice) or by carrying out the computation.

Exercise 6.1.8 asks you to verify that the wedge product of two 1-forms does not commute, and that the wedge product of a 2-form and a 1-form does commute.

It follows from part 3 of Proposition 6.1.15 that if φ is a form of even degree, then $\varphi \wedge \psi = \psi \wedge \varphi$. In particular, 2-forms commute with all forms.

Form fields

Most often, rather than integrate a k-form, we will integrate a *k-form field*, also called a "differential form".

Definition 6.1.16 (k-form field). A *k-form field* on an open subset $U \subset \mathbb{R}^n$ is a map $\varphi : U \to A_c^k(\mathbb{R}^n)$. The space of k-form fields on U is denoted $A^k(U)$.

A "— field" means that you have a "—" at every point. A "vector field" means you have attached a vector to every point. A "strawberry field" has a strawberry at every point. A "k-form field" has a form at every point: the number a form field returns depends on k vectors and on a point at which you should imagine the vectors anchored.

A 0-form field takes a "parallelogram" formed of zero vectors and anchored at $\mathbf{x}$, and returns a number; in other words, a 0-form field on U is a function on U.

To lighten terminology, we will often refer to a form field simply as a form; from now on, when we wish to speak of a form that returns the same number regardless of the point $\mathbf{x}$ at which a parallelogram is anchored, we will call it a *constant form.*

We find "differential" a mystifying word; it is almost impossible to make sense of the word "differential" as it is used in first year calculus. We know a professor who claims that he has been teaching differentials for 20 years and still doesn't know what they are.

We think of $\varphi \in A^k(U)$ as taking k-dimensional parallelograms anchored at points of U and returning numbers, according to the formula

$$\varphi\big(P_{\mathbf{x}}(\vec{\mathbf{v}}_1, \dots, \vec{\mathbf{v}}_k)\big) \overset{\text{def}}{=} \varphi(\mathbf{x})(\vec{\mathbf{v}}_1, \dots, \vec{\mathbf{v}}_k) \qquad 6.1.39$$

The function $\vec{\mathbf{v}}_1, \dots, \vec{\mathbf{v}}_k \mapsto \varphi\big(P_{\mathbf{x}}(\vec{\mathbf{v}}_1, \dots, \vec{\mathbf{v}}_k)\big)$ is multilinear and antisymmetric.

We already know how to write elements of $A^k(U)$: they are expressions of the form

$$\varphi = \sum_{1 \le i_1 < \cdots < i_k \le n} a_{i_1,\dots,i_k}(\mathbf{x})\, dx_{i_1} \wedge \cdots \wedge dx_{i_k}, \qquad 6.1.40$$

where the $a_{i_1,\dots,i_k}$ are real-valued functions of $\mathbf{x} \in U$. We ought to specify how the forms depend on $\mathbf{x}$. A k-form field φ is of class C^p if the coefficients $a_{i_1,\dots,i_k}$ are of class C^p. Usually we omit this data and assume that our form fields are as differentiable as needed, always C^0 but C^1 if we wish to differentiate φ, and sometimes C^2 (for instance, in Theorems 6.7.4 and 6.7.8).

Example 6.1.17 (A 2-form field on $\mathbb{R}^3$). The form field $\cos(xz)\, dx \wedge dy$ is a 2-form field on $\mathbb{R}^3$. Here we evaluate it twice, each time on the same vectors, but at different points:

$$\cos(xz)\, dx \wedge dy \left(P_{\begin{pmatrix}1\\2\\\pi\end{pmatrix}} \left(\begin{bmatrix}1\\0\\1\end{bmatrix}, \begin{bmatrix}2\\2\\3\end{bmatrix} \right) \right) = \big(\cos(1 \cdot \pi)\big) \det \begin{bmatrix}1 & 2\\0 & 2\end{bmatrix} = -2.$$

$$\cos(xz)\, dx \wedge dy \left(P_{\begin{pmatrix}1/2\\2\\\pi\end{pmatrix}} \left(\begin{bmatrix}1\\0\\1\end{bmatrix}, \begin{bmatrix}2\\2\\3\end{bmatrix} \right) \right) = \big(\cos(1/2 \cdot \pi)\big) \det \begin{bmatrix}1 & 2\\0 & 2\end{bmatrix} = 0.$$

Of course, if we reverse the order of the vectors in the first equation in Example 6.1.17, we get 2, not -2. The integrand $|d^k\mathbf{x}|$ is an element of volume; a k-form is an element of *signed volume.* △

Exercises for Section 6.1

6.1.1 a. List the elementary k-forms on $\mathbb{R}^3$.

b. List the elementary k-forms on $\mathbb{R}^4$.

c. How many elementary 4-forms are there on $\mathbb{R}^5$? List them.

$$\vec{\mathbf{v}}_1 = \begin{bmatrix}1\\2\\3\\4\end{bmatrix}, \vec{\mathbf{v}}_2 = \begin{bmatrix}0\\1\\-1\\1\end{bmatrix}, \vec{\mathbf{v}}_3 = \begin{bmatrix}1\\0\\3\\2\end{bmatrix}$$

Vectors for part a, Exercise 6.1.2

6.1.2 Compute the following numbers, where $\vec{\mathbf{v}}_1\vec{\mathbf{v}}_2, \vec{\mathbf{v}}_3$ are shown in the margin:

a. $dx_3 \wedge dx_2 \wedge dx_4\,(\vec{\mathbf{v}}_1, \vec{\mathbf{v}}_2, \vec{\mathbf{v}}_3)$ b. $3\,dy\left(\begin{bmatrix}1\\2\end{bmatrix}\right)$

6.1.3 Compute the following numbers:

a. $dx_1 \wedge dx_4 \left(\begin{bmatrix}1\\0\\1\\2\end{bmatrix}, \begin{bmatrix}1\\-3\\-1\\2\end{bmatrix}\right)$ b. $(dx_1 \wedge dx_2 + 2dx_2 \wedge dx_3)\left(\begin{bmatrix}1\\0\\1\end{bmatrix}, \begin{bmatrix}-2\\1\\0\end{bmatrix}\right)$

c. $dx_4 \wedge dx_2 \left(\begin{bmatrix}1\\0\\1\\2\end{bmatrix}, \begin{bmatrix}1\\-3\\-1\\2\end{bmatrix}\right)$ d. $dx_1 \wedge dx_2 \wedge dx_2 \left(\begin{bmatrix}1\\3\\1\end{bmatrix}, \begin{bmatrix}-2\\1\\4\end{bmatrix}, \begin{bmatrix}2\\2\\2\end{bmatrix}\right)$

$$\underbrace{\begin{bmatrix}1\\2\\0\end{bmatrix}}_{\vec{\mathbf{v}}_1}, \underbrace{\begin{bmatrix}2\\1\\1\end{bmatrix}}_{\vec{\mathbf{v}}_2}$$

Vectors for Exercise 6.1.4, part a

6.1.4 a. Using the integrand $|d^k\mathbf{x}|$ of Chapter 5, and the vectors $\vec{\mathbf{v}}_1, \vec{\mathbf{v}}_2$ shown in the margin, compute

i. $|d^2\mathbf{x}|(\vec{\mathbf{v}}_1, \vec{\mathbf{v}}_2)$ and $|d^2\mathbf{x}|(\vec{\mathbf{v}}_2, \vec{\mathbf{v}}_1)$.

ii. $dx \wedge dy(\vec{\mathbf{v}}_1, \vec{\mathbf{v}}_2)$ and $dx \wedge dy(\vec{\mathbf{v}}_2, \vec{\mathbf{v}}_1)$.

b. For the vectors $\vec{\mathbf{v}}_1, \vec{\mathbf{v}}_2, \vec{\mathbf{v}}_3$ shown in the margin, compute

i. $|d^3\mathbf{x}|(\vec{\mathbf{v}}_1, \vec{\mathbf{v}}_2, \vec{\mathbf{v}}_3)$ and $|d^3\mathbf{x}|(\vec{\mathbf{v}}_1, \vec{\mathbf{v}}_3, \vec{\mathbf{v}}_2)$.

ii. $dx \wedge dy \wedge dz(\vec{\mathbf{v}}_1, \vec{\mathbf{v}}_2, \vec{\mathbf{v}}_3)$ and $dx \wedge dy \wedge dz(\vec{\mathbf{v}}_1, \vec{\mathbf{v}}_3, \vec{\mathbf{v}}_2)$.

$$\underbrace{\begin{bmatrix}1\\0\\2\end{bmatrix}}_{\vec{\mathbf{v}}_1}, \underbrace{\begin{bmatrix}2\\1\\1\end{bmatrix}}_{\vec{\mathbf{v}}_2}, \underbrace{\begin{bmatrix}3\\1\\0\end{bmatrix}}_{\vec{\mathbf{v}}_3}$$

Vectors for Exercise 6.1.4, part b

6.1.5 a. What is the y-component of signed length of the vector $\begin{bmatrix}2\\-3\\4\end{bmatrix}$?

b. What is the (x_2, x_4)-component of signed volume of the parallelogram $P \subset \mathbb{R}^4$ spanned by the vectors $\vec{\mathbf{w}}_1$, $\vec{\mathbf{w}}_2$ shown in the margin?

$$\underbrace{\begin{bmatrix}2\\1\\0\\4\end{bmatrix}}_{\vec{\mathbf{w}}_1}, \underbrace{\begin{bmatrix}-2\\1\\2\\-3\end{bmatrix}}_{\vec{\mathbf{w}}_2}$$

Vectors for Exercise 6.1.5, part b

6.1.6 Which of the following expressions make sense? Evaluate those that do.

a. $dx_1 \wedge dx_2 \left(\begin{bmatrix}1\\0\\1\end{bmatrix}, \begin{bmatrix}2\\3\\1\end{bmatrix}\right)$ b. $dx_1 \wedge dx_3 \left(\begin{bmatrix}1\\1\end{bmatrix}, \begin{bmatrix}2\\3\end{bmatrix}\right)$

c. $dx_1 \wedge dx_2 \left(\begin{bmatrix}1\\1\end{bmatrix}, \begin{bmatrix}2\\3\end{bmatrix}, \begin{bmatrix}-2\\1\end{bmatrix}\right)$ d. $dx_1 \wedge dx_2 \wedge dx_4 \left(\begin{bmatrix}1\\0\\3\end{bmatrix}, \begin{bmatrix}3\\7\\2\end{bmatrix}, \begin{bmatrix}2\\0\\1\end{bmatrix}\right)$

e. $dx_1 \wedge dx_2 \wedge dx_3 \left(\begin{bmatrix}1\\1\end{bmatrix}, \begin{bmatrix}2\\3\end{bmatrix}\right)$ f. $dx_1 \wedge dx_2 \wedge dx_3 \left(\begin{bmatrix}1\\0\\3\end{bmatrix}, \begin{bmatrix}3\\7\\2\end{bmatrix}, \begin{bmatrix}2\\0\\1\end{bmatrix}\right)$

6.1.7 a. Use Theorem 6.1.10 to confirm that on $\mathbb{R}^4$ there are one elementary 0-form, four elementary 1-forms, six elementary 2-forms, four elementary 3-forms, and one elementary 4-form.

b. Use Theorem 6.1.10 to determine how many elementary 1-forms, 2-forms, 3-forms, 4-forms, and 5-forms there are on $\mathbb{R}^5$. List them.

6.1.8 Verify that the wedge product of two 1-forms does not commute, and that the wedge product of a 2-form and a 1-form does commute.

6.1.9 Show that when $k = 2$ and $n = 3$, the last line of equation 6.1.23 is identical to the last line of equation 6.1.22.

6.1.10 Let $\vec{\mathbf{a}}, \vec{\mathbf{v}}, \vec{\mathbf{w}}$ be vectors in $\mathbb{R}^3$ and let φ be the 2-form on $\mathbb{R}^3$ given by $\varphi(\vec{\mathbf{v}}, \vec{\mathbf{w}}) = \det[\vec{\mathbf{a}}, \vec{\mathbf{v}}, \vec{\mathbf{w}}]$. Write φ as a linear combination of elementary 2-forms on $\mathbb{R}^3$, in terms of the coordinates of $\vec{\mathbf{a}}$.

6.1.11 Let φ and ψ be 2-forms. Use Definition 6.1.12 to write the wedge product $\varphi \wedge \psi(\vec{\mathbf{v}}_1, \vec{\mathbf{v}}_2, \vec{\mathbf{v}}_3, \vec{\mathbf{v}}_4)$ as a combination of values of φ and ψ evaluated on appropriate vectors (as in equations 6.1.28 and 6.1.32).

6.1.12 Compute the following:

Exercise 6.1.14 amounts to proving Theorem 6.1.8 in the case where $n = k$, and not restricting oneself to 2-forms on $\mathbb{R}^2$ or 3-forms on $\mathbb{R}^3$, i.e., going through "an analogous but messier computation".

a. $(x_1 - x_4)\, dx_3 \wedge dx_2 \left(P_{\mathbf{0}} \left(\begin{bmatrix} 1 \\ 2 \\ 3 \\ 4 \end{bmatrix}, \begin{bmatrix} 0 \\ 1 \\ -1 \\ 1 \end{bmatrix} \right) \right)$ b. $e^x dy \left(P_{\begin{pmatrix} 2 \\ 1 \end{pmatrix}} \begin{pmatrix} 3 \\ 2 \end{pmatrix} \right)$

c. $x_1^2\, dx_3 \wedge dx_2 \wedge dx_1 \left(P_{\begin{pmatrix} 2 \\ 0 \\ 0 \\ 0 \end{pmatrix}} \left(\begin{bmatrix} 1 \\ 2 \\ 3 \\ 4 \end{bmatrix}, \begin{bmatrix} 0 \\ 1 \\ -1 \\ 1 \end{bmatrix}, \begin{bmatrix} 1 \\ -1 \\ -1 \\ 0 \end{bmatrix} \right) \right)$

***6.1.13** Prove Proposition 6.1.15.

6.1.14 Use Theorem and Definition 4.8.1 to show that a k-form on $\mathbb{R}^k$ can be written as a multiple of the determinant: $\omega = a \det$. Give a formula for the coefficient a.

6.1.15 Compute the following:

a. $\sin(x_4)\, dx_3 \wedge dx_2 \left(P_{\begin{pmatrix} x_1 \\ x_2 \\ x_3 \\ x_4 \end{pmatrix}} \left(\begin{bmatrix} x_2 \\ 2x_1 \\ 3 \\ 4 \end{bmatrix}, \begin{bmatrix} 0 \\ x_4 \\ -1 \\ -x_4 \end{bmatrix} \right) \right)$ b. $e^x dy \left(P_{\begin{pmatrix} x \\ y \end{pmatrix}} \begin{pmatrix} 3 \\ 2 \end{pmatrix} \right)$

c. $x_1^2 e^{x_2}\, dx_3 \wedge dx_2 \wedge dx_1 \left(P_{\begin{pmatrix} -x_1 \\ x_2 \\ -x_3 \\ x_4 \end{pmatrix}} \left(\begin{bmatrix} 1 \\ 2 \\ 3 \\ 4 \end{bmatrix}, \begin{bmatrix} 0 \\ 1 \\ -1 \\ 1 \end{bmatrix}, \begin{bmatrix} 1 \\ -1 \\ -1 \\ 0 \end{bmatrix} \right) \right)$

6.2 Integrating form fields over parametrized domains

The object of the first half of this chapter is to define *integration of forms over oriented manifolds*. Orientation is often said to be the hardest concept of vector calculus, and takes a while to define. But there is a way to sweep the difficulties under the rug, by defining our domains of integration so that their orientation is built in. In this section we adopt this approach; in the next we will look under the rug, and see what is hiding there.

Let $U \subset \mathbb{R}^k$ be a bounded open set, with boundary of k-dimensional volume 0. A C^1 mapping $\gamma : U \to \mathbb{R}^n$ defines a *domain in* $\mathbb{R}^n$ *parametrized by* U. We will denote the pair (U, γ) by $[\gamma(U)]$; we think of it as the image set $\gamma(U)$ with some extra data, namely how γ maps U to it, which includes conferring an orientation on $\gamma(U)$.[3] For now just accept the fact that $[\gamma(U)]$ is a domain over which we can integrate a k-form. As shown in Figure 6.2.1, it need not be a smooth manifold.

FIGURE 6.2.1.

You may wish to think that γ parametrizes a manifold, but for the present purposes this is irrelevant; the image of γ might be arbitrarily horrible, like the (definitely not smooth) curve shown here, which is a parametrized domain $[\gamma(U)]$, where γ is a function from $\mathbb{R}$ to $\mathbb{R}^2$.

Definition 6.2.1 (Integrating a k-form field over a parametrized domain). Let $U \subset \mathbb{R}^k$ be a bounded open set with $\mathrm{vol}_k\, \partial U = 0$. Let $V \subset \mathbb{R}^n$ be open, and let $[\gamma(U)]$ be a parametrized domain in V. Let φ be a k-form field on V.

Then the *integral of* φ over $[\gamma(U)]$ is

$$\int_{[\gamma(U)]} \varphi \overset{\text{def}}{=} \int_U \varphi\Big(P_{\gamma(\mathbf{u})}\big(\overrightarrow{D_1\gamma(\mathbf{u})}, \dots, \overrightarrow{D_k\gamma(\mathbf{u})}\big)\Big)|d^k\mathbf{u}|. \qquad 6.2.1$$

Note both the similarity with equation 5.3.3 for computing the volume of a manifold:

$$\int_M |d^k\mathbf{x}| = \int_{\gamma(U)} |d^k\mathbf{x}| = \int_U |d^k\mathbf{x}|\Big(P_{\gamma(\mathbf{u})}\big(\overrightarrow{D_1\gamma(\mathbf{u})}, \dots, \overrightarrow{D_k\gamma(\mathbf{u})}\big)\Big)|d^k\mathbf{u}|,$$

and the important difference. When we evaluate $|d^k\mathbf{x}|$ on the k-parallelogram by computing

$$|d^k\mathbf{x}|\Big(P_{\gamma(\mathbf{u})}\big(\overrightarrow{D_1\gamma(\mathbf{u})}, \dots, \overrightarrow{D_k\gamma(\mathbf{u})}\big)\Big) = \sqrt{\det([\mathbf{D}\gamma(\mathbf{u})]^\top[\mathbf{D}\gamma(\mathbf{u})])}, \qquad 6.2.2$$

the order of the vectors does not affect the result. (We saw in Chapter 5 that $\sqrt{\det(T^\top T)} = |\det T|$; exchanging columns of T changes the sign of $\det T$, but that change is erased by taking the absolute value.) In equation 6.2.1, clearly the order of the vectors (i.e., the partial derivatives) does matter; a form is an antisymmetric function.

Example 6.2.2 (Integrating a 1-form field over a parametrized curve). Consider a case where $k = 1$, $n = 2$, and $\gamma(u) = \begin{pmatrix} R\cos u \\ R\sin u \end{pmatrix}$. We will take U to be the interval $[0, a]$, for some $a > 0$. If we integrate

[3]We will see in Section 6.4 that the details of this extra data do not affect integrals, but the orientation of the parametrized domain does.

$x\,dy - y\,dx$ over $[\gamma(U)]$ using Definition 6.2.1, we find

$$\int_{[\gamma(U)]} (x\,dy - y\,dx) = \int_{[0,a]} (x\,dy - y\,dx)\left(P_{\begin{pmatrix} R\cos u \\ R\sin u \end{pmatrix}}\overbrace{\begin{bmatrix} -R\sin u \\ R\cos u \end{bmatrix}}^{\overrightarrow{D_1\gamma}(\mathbf{u})}\right)|du| \qquad 6.2.3$$

$$= \int_{[0,a]} \big(R\cos u R\cos u - (R\sin u)(-R\sin u)\big)|du| = \int_{[0,a]} R^2\,|du| = \int_0^a R^2\,du = R^2 a. \quad \triangle$$

Equation 6.2.3: Recall from equation 6.1.39 that

$$\varphi\Big(P_{\mathbf{x}}(\vec{\mathbf{v}}_1, \dots, \vec{\mathbf{v}}_k)\Big) = \varphi(\mathbf{x})(\vec{\mathbf{v}}_1, \dots, \vec{\mathbf{v}}_k)$$

In the second line of equation 6.2.3, the first $R\cos u$ is x, while the second is the number given by dy evaluated on the parallelogram. Similarly, $R\sin u$ is y, and $(-R\sin u)$ is the number given by dx evaluated on the parallelogram.

If in Example 6.2.4 we use the mapping

$$\gamma\begin{pmatrix} s \\ t \end{pmatrix} = \begin{pmatrix} s+t \\ t^2 \\ s^2 \end{pmatrix},$$

which should parametrize the same surface since s and t play symmetric roles and we integrate the same 2-form over that parametrized domain, we will get $2/3$, not $-2/3$. We have not yet defined "orientation" of a surface, but you can think that one parametrization corresponds to measuring flow through the surface in one direction; the other corresponds to measuring flow through the surface in the opposite direction.

Example 6.2.3 (An anchored k-parallelogram). An important example of a parametrized domain is $P_{\mathbf{x}}(\vec{\mathbf{v}}_1, \dots, \vec{\mathbf{v}}_k)$, parametrized by

$$\gamma\begin{pmatrix} t_1 \\ \vdots \\ t_k \end{pmatrix} = \mathbf{x} + t_1\vec{\mathbf{v}}_1 + \dots + t_k\vec{\mathbf{v}}_k, \quad 0 \le t_i \le 1. \qquad 6.2.4$$

In this case, Definition 6.2.1 gives

$$\int_{[P_{\mathbf{x}}(\vec{\mathbf{v}}_1,\dots,\vec{\mathbf{v}}_k)]} \varphi = \int_0^1 \cdots \int_0^1 \varphi\Big(P_{\gamma(\mathbf{t})}(\vec{\mathbf{v}}_1, \dots, \vec{\mathbf{v}}_k)\Big)\,dt_1 \dots dt_k. \qquad 6.2.5$$

The vectors $\vec{\mathbf{v}}_1, \dots, \vec{\mathbf{v}}_k$ are the partial derivatives of the parametrization.

Example 6.2.4 (Integrating a 2-form field over a parametrized surface in $\mathbb{R}^3$). Let us integrate $dx \wedge dy + y\,dx \wedge dz$ over the parametrized domain $[\gamma(S)]$ where

$$\gamma\begin{pmatrix} s \\ t \end{pmatrix} = \begin{pmatrix} s+t \\ s^2 \\ t^2 \end{pmatrix}, \quad S = \left\{\begin{pmatrix} s \\ t \end{pmatrix} \,\middle|\, 0 \le s \le 1,\ 0 \le t \le 1\right\}. \qquad 6.2.6$$

Applying Definition 6.2.1, we find

$$\int_{[\gamma(S)]} dx \wedge dy + y\,dx \wedge dz$$
$$= \int_0^1\int_0^1 (dx \wedge dy + y\,dx \wedge dz)\left(P_{\begin{pmatrix} s+t \\ s^2 \\ t^2 \end{pmatrix}}\left(\begin{bmatrix} 1 \\ 2s \\ 0 \end{bmatrix}, \begin{bmatrix} 1 \\ 0 \\ 2t \end{bmatrix}\right)\right) ds\,dt$$
$$= \int_0^1\int_0^1 \left(\det\begin{bmatrix} 1 & 1 \\ 2s & 0 \end{bmatrix} + s^2 \det\begin{bmatrix} 1 & 1 \\ 0 & 2t \end{bmatrix}\right) ds\,dt \qquad 6.2.7$$
$$= \int_0^1\int_0^1 \big(-2s + s^2(2t)\big)\,ds\,dt = \int_0^1 \left[-s^2 + \frac{s^3}{3}2t\right]_{s=0}^{s=1} dt$$
$$= \int_0^1 \left(-1 + \frac{2t}{3}\right) dt = \left[-t + \frac{t^2}{3}\right]_0^1 = -\frac{2}{3}. \quad \triangle$$

Why look under the rug?

If we can use Definition 6.2.1 to integrate forms over parametrized domains without once mentioning orientation, why bother with orientation when integrating over a manifold? Why can't we just parametrize our manifold and use equation 6.2.1, saving ourselves a lot of trouble? Let's look at what can go wrong. When we computed the volume of a manifold in Chapter 5 by integrating $|d^k\mathbf{x}|$ over the parametrized manifold, we saw (Proposition 5.3.3) that the choice of parametrization didn't matter. Now we must be more careful.

Example 6.2.5 (Different parametrizations give different results). Suppose you and your neighbor wish to integrate dy over the upper right quadrant of the unit circle. This can be interpreted as determining the y-component of that arc of circle; if you walk along the arc, how far will you go in the y direction?

Equation 6.2.8: When we start at $t = 0$ and go to $t = \pi/2$ we are following the "standard orientation" of $\mathbb{R}$, from negative to positive. In higher dimensions, when the partial derivatives of equation 6.2.1 are listed in the order

$$\overrightarrow{D_1}\gamma(\mathbf{u}), \dots, \overrightarrow{D_k}\gamma(\mathbf{u})$$

and not, for example, in the order

$$\overrightarrow{D_2}\gamma(\mathbf{u}), \overrightarrow{D_1}\gamma(\mathbf{u}), \dots, \overrightarrow{D_k}\gamma(\mathbf{u}),$$

we are following the "standard orientation" of $\mathbb{R}^k$.

You choose the parametrization $\gamma(t) = \begin{pmatrix} \cos t \\ \sin t \end{pmatrix}$; applying equation 6.2.1, you get

$$\int_{[0,\pi/2]} dy \left(P_{\begin{pmatrix} \cos t \\ \sin t \end{pmatrix}} \begin{bmatrix} -\sin t \\ \cos t \end{bmatrix} \right) |dt| = \int_0^{\pi/2} \cos t \, dt = \Big[\sin t\Big]_0^{\pi/2} = 1. \qquad 6.2.8$$

Your neighbor chooses the parametrization $\delta(t) = \begin{pmatrix} \sin t \\ \cos t \end{pmatrix}$ and gets

$$\int_{[0,\pi/2]} dy \left(P_{\begin{pmatrix} \sin t \\ \cos t \end{pmatrix}} \begin{bmatrix} \cos t \\ -\sin t \end{bmatrix} \right) |dt| = \int_0^{\pi/2} -\sin t \, dt = \Big[\cos t\Big]_0^{\pi/2} = -1. \qquad 6.2.9$$

Your parametrization corresponds to starting at the point $\begin{pmatrix} 1 \\ 0 \end{pmatrix}$ in the (x, y)-plane (γ evaluated at $t = 0$) and walking counterclockwise to $\begin{pmatrix} 0 \\ 1 \end{pmatrix}$ (γ evaluated at $t = \pi/2$). Your neighbor's parametrization corresponds to starting at the point $\begin{pmatrix} 0 \\ 1 \end{pmatrix}$ and walking clockwise to $\begin{pmatrix} 1 \\ 0 \end{pmatrix}$.

Getting opposite results is bad enough, but suppose we were summing the integral over the circle in the first quadrant and the integral over the second quadrant. If we used the same parametrization for both, we would get 0; if we used your parametrization for the first quadrant and your neighbor's for the second quadrant, we would get 2; if we used yours for the second quadrant and your neighbor's for the first, we would get -2. △

The point of Example 6.2.5 is that you and your neighbor were solving different problems. The direction inherent in the parametrization, counterclockwise for you, clockwise for your neighbor, was *part of the data of the domain of integration*. The question "what is the integral of dy over the first quadrant of the unit circle" is incomplete. To answer it, you must first ask, "do you mean the circle oriented clockwise or the circle oriented counterclockwise?"

Example 6.2.6 (Up the down staircase). Often, manifolds represent real physical objects that come with an inherent orientation. In one dimension, you can think of this orientation as direction on a curve. If a curve represents a one-way conveyor belt lifting grain to the top of a silo, and you parametrize the curve and integrate a 1-form over it, then to interpret your answer, you had better know whether your parametrization agrees that the direction on the curve is up, not down.

For a surface, you can think that orienting a surface means choosing a direction for a fluid flowing through the surface. For example, Gauss's law says that the flux of an electric field through the boundary of a region, from the inside of the region to the outside, is equal to the amount of charge inside the region. If we wish to compute that flux by integrating a 2-form over the parametrized boundary, we need to know how to choose a parametrization so that outward flow will count positively to the integral and inward flow will count negatively. △

The same problems exist in higher dimensions, where we will no longer be able to rely on intuitive notions of "direction on a curve" or "flow of a fluid through a surface". Thus we need to define more generally what it means for a k-manifold to be oriented, and we need to see how the choice of parametrization relates to the orientation of the manifold. We discuss orientation of manifolds in the next section; in Section 6.4, we show how to integrate forms over parametrized oriented manifolds.

Exercises for Section 6.2

Exercises 6.2.1 and 6.2.2, part b: You are strongly encouraged to use MAPLE or the equivalent.

6.2.1 Set up each of the following integrals of form fields over parametrized domains as an ordinary multiple integral, and compute it.

a. $\int_{[\gamma(I)]} x\,dy + y\,dz$, where $I = [-1, 1]$, and $\gamma(t) = \begin{pmatrix} \sin t \\ \cos t \\ t \end{pmatrix}$

b. $\int_{[\gamma(U)]} x_1\,dx_2 \wedge dx_3 + x_2\,dx_3 \wedge dx_4$, where $U = \left\{ \begin{pmatrix} u \\ v \end{pmatrix} \middle| 0 \le u, v;\ u + v \le 2 \right\}$

and $\gamma\begin{pmatrix} u \\ v \end{pmatrix} = \begin{pmatrix} uv \\ u^2 + v^2 \\ u - v \\ \ln(u + v + 1) \end{pmatrix}$

6.2.2 Repeat Exercise 6.2.1, for the following.

a. $\int_{[\gamma(U)]} x\,dy \wedge dz$, where $U = [-1, 1] \times [-1, 1]$, and $\gamma\begin{pmatrix} u \\ v \end{pmatrix} = \begin{pmatrix} u^2 \\ u + v \\ v^3 \end{pmatrix}$

b. $\int_{[\gamma(U)]} x_2\,dx_1 \wedge dx_3 \wedge dx_4$, where

$$U = \left\{ \begin{pmatrix} u \\ v \\ w \end{pmatrix} \middle| 0 \le u, v, w;\ u + v + w \le 3 \right\} \quad \text{and} \quad \gamma\begin{pmatrix} u \\ v \\ w \end{pmatrix} = \begin{pmatrix} uv \\ u^2 + w^2 \\ u - v \\ w \end{pmatrix}.$$

6.2.3 Set up each of the following integrals of form fields over parametrized domains as an ordinary multiple integral.

a. $\int_{[\gamma(U)]}(x_1 + x_4)\, dx_2 \wedge dx_3$, where $U = \left\{ \begin{pmatrix} u \\ v \end{pmatrix} \,\middle|\, |v| \le u \le 1 \right\}$

and $\gamma\begin{pmatrix} u \\ v \end{pmatrix} = \begin{pmatrix} e^u \\ e^{-v} \\ \cos u \\ \sin v \end{pmatrix}$.

b. $\int_{[\gamma(U)]} x_2 x_4\, dx_1 \wedge dx_3 \wedge dx_4$, where

$$U = \left\{ \begin{pmatrix} u \\ v \\ w \end{pmatrix} \,\middle|\, (w-1)^2 \ge u^2 + v^2,\ 0 \le w \le 1 \right\} \quad \text{and} \quad \gamma\begin{pmatrix} u \\ v \\ w \end{pmatrix} = \begin{pmatrix} u+v \\ u-v \\ w+v \\ w-v \end{pmatrix}.$$

6.2.4 Let $z_1 = x_1 + iy_1, z_2 = x_2 + iy_2$ be coordinates in $\mathbb{C}^2$. Let $S \subset \mathbb{C}$ be the square $\{z = x + iy |\, |x| \le 1, |y| \le 1\}$, and define $\gamma : S \to \mathbb{C}^2$ by

$$\gamma : z \mapsto \begin{pmatrix} e^z \\ e^{-z} \end{pmatrix}, \quad z = x + iy, |x| \le 1, |y| \le 1.$$

What is $\displaystyle\int_{[\gamma(S)]} dx_1 \wedge dy_1 + dy_1 \wedge dx_2 + dx_2 \wedge dy_2$?

6.2.5 Let $S \subset \mathbb{C}$ be the set $\{z |\, |z| < 1\}$; define $\gamma : S \to \mathbb{C}^3$ by $\gamma : z \mapsto \begin{pmatrix} z \\ z^2 \\ z^3 \end{pmatrix}$.

What is $\displaystyle\int_{[\gamma(S)]} dx_1 \wedge dy_1 \; + \; dx_2 \wedge dy_2 \; + \; dx_3 \wedge dy_3$?

6.2.6 Let $S \subset \mathbb{C}$ be the set $\{z |\, |z| < 1\}$, and let $\gamma : S \to \mathbb{C}^3$ be given by $\gamma : z \mapsto \begin{pmatrix} z^p \\ z^q \\ z^r \end{pmatrix}$. What is $\displaystyle\int_{[\gamma(S)]} dx_1 \wedge dy_1 \; + \; dx_2 \wedge dy_2 \; + \; dx_3 \wedge dy_3$?

6.2.7 Show that

$$\int_{[P_{\mathbf{x}}(h\vec{\mathbf{v}}_1, \ldots, h\vec{\mathbf{v}}_k)]} \varphi = h^k \varphi\Big(P_{\mathbf{x}}(\vec{\mathbf{v}}_1, \ldots, \vec{\mathbf{v}}_k) \Big) + o(h^k).$$

6.3 Orientation of manifolds

"The great thing in this world is not so much where we stand, as in what direction we are moving."—Oliver Wendell Holmes

We said in Section 6.0 that orientation means choosing one of two possibilities, like distinguishing left from right. We touched on what this means for a curve and a surface in Section 6.2. Here we state the definitions in complete generality. Definition 6.3.1 says that an orientation Ω of a vector space V divides bases of V into two groups: *direct* bases and *indirect* bases.

In Definition 6.3.1, sgn det is $+1$ if det > 0; it is -1 if det < 0. Since a change of basis matrix is always invertible, $\det[P_{\mathbf{v}' \to \mathbf{v}}]$ cannot be 0.

Definition 6.3.1 (Orientation of vector space). Let V be a finite-dimensional real vector space, and let $\mathcal{B}_V$ be the set of bases of V. An *orientation* of V is a map $\Omega : \mathcal{B}_V \to \{+1, -1\}$ such that if $\{\mathbf{v}\}$ and $\{\mathbf{v}'\}$ are two bases with change of basis matrix $[P_{\mathbf{v}' \to \mathbf{v}}]$, then

$$\Omega(\{\mathbf{v}'\}) = \operatorname{sgn}(\det[P_{\mathbf{v}' \to \mathbf{v}}])\, \Omega(\{\mathbf{v}\}). \qquad 6.3.1$$

A basis $\{\mathbf{w}\} \in \mathcal{B}_V$ is called *direct* if $\Omega(\{\mathbf{w}\}) = +1$; it is called *indirect* if $\Omega(\{\mathbf{w}\}) = -1$.

To orient V, it is sufficient to choose one basis of V and declare it to be direct.

If a vector space V is one dimensional, i.e., a line, our intuition says that it has two directions; an orientation of V is a choice of one of these two directions. Let us translate this in terms of bases: any nonzero vector $\mathbf{v} \in V$ is a basis of V; our direction orients V by saying that $\mathbf{v}$ is direct if it points in our chosen direction, and indirect if it points in the opposite direction.

Distinguishing the change of basis matrix from its inverse is important (and a constant source of confusion), but in the present setting, it doesn't matter: the sign of the determinant of a matrix and of its inverse coincide (Corollary 4.8.5).

If a vector space V is complex, then the change of basis matrix between two bases of V still exists, but its determinant is a complex number and doesn't have a sign, so Definition 6.3.1 does not apply. But the underlying real vector space can be oriented and in fact has a natural orientation; this is explored in Exercise 6.3.15.

Equation 6.3.3: You can think of $\mathcal{B}(M)$ as a subset of the space of $n \times (k+1)$ matrices. We define $\mathcal{B}(M)$ in order to give a meaning to the word "continuous" in Definition 6.3.3.

A choice of a basis $\{\mathbf{v}\}$ for V specifies two orientations for V,

$$\Omega(\{\mathbf{v}'\}) = \operatorname{sgn}\det[P_{\mathbf{v}' \to \mathbf{v}}] \quad \text{and} \quad \Omega(\{\mathbf{v}'\}) = -\operatorname{sgn}\det[P_{\mathbf{v}' \to \mathbf{v}}]; \qquad 6.3.2$$

the first is the unique orientation for which $\{\mathbf{v}\}$ is direct, and the second is the unique orientation for which $\{\mathbf{v}\}$ is indirect. In particular, every vector space has precisely two orientations. We denote the orientation for which $\{\mathbf{v}\}$ is direct by $\Omega^{\{\mathbf{v}\}}$ and call it the *orientation specified by* $\{\mathbf{v}\}$.

It follows from Definition 6.3.1 that two bases $\{\mathbf{v}\}$ and $\{\mathbf{w}\}$ define the same orientation if $\operatorname{sgn}\det[P_{\{\mathbf{v}\} \to \{\mathbf{w}\}}] = +1$ and opposite orientations if $\operatorname{sgn}\det[P_{\{\mathbf{v}\} \to \{\mathbf{w}\}}] = -1$.

The *standard orientation* of $\mathbb{R}^n$, denoted Ω^{st}, is defined by declaring the standard basis to be direct.

What if V is 0-dimensional? It still has two orientations. The empty set is a basis (indeed, the unique basis): it is a maximal linearly independent set of vectors. So the set of bases $\mathcal{B}_V$ is $\{\phi\}$, and there are two maps $\{\phi\} \to \{+1, -1\}$, namely $\phi \mapsto +1$ and $\phi \mapsto -1$. Since there is only one basis, equation 6.3.1 is vacuously satisfied. Thus an orientation of a 0-dimensional vector space is simply a choice of $+1$ or -1.

Example 6.3.2 (Orienting a subspace). The vector space $\mathbb{R}^n$ has a standard orientation, but subspaces do not. For instance,

$$\underbrace{\begin{bmatrix} 1 \\ -1 \\ 0 \end{bmatrix}, \begin{bmatrix} 1 \\ 0 \\ -1 \end{bmatrix}}_{\{\mathbf{v}_1\}};\quad \underbrace{\begin{bmatrix} 1 \\ 0 \\ -1 \end{bmatrix}, \begin{bmatrix} 0 \\ 1 \\ -1 \end{bmatrix}}_{\{\mathbf{v}_2\}};\quad \underbrace{\begin{bmatrix} 0 \\ 1 \\ -1 \end{bmatrix}, \begin{bmatrix} 1 \\ -1 \\ 0 \end{bmatrix}}_{\{\mathbf{v}_3\}}$$

are three bases of the subspace $V \subset \mathbb{R}^3$ of equation $x + y + z = 0$; we invite you to check that the orientations defined by the first two coincide, and that the third defines the opposite orientation.[4] Both orientations seem equally natural. △

Orienting a manifold is more involved than orienting a vector space: you must choose an orientation of the tangent space at every point, and it must vary continuously. Let $M \subset \mathbb{R}^n$ be a k-dimensional manifold. Define

$$\mathcal{B}(M) \stackrel{\text{def}}{=} \left\{ (\mathbf{x}, \vec{\mathbf{v}}_1, \ldots, \vec{\mathbf{v}}_k) \in \mathbb{R}^{n(k+1)} \right\}, \qquad 6.3.3$$

[4] $\det[P_{\{\mathbf{v}_1\} \to \{\mathbf{v}_2\}}] = \det \begin{bmatrix} 1 & 1 \\ -1 & 0 \end{bmatrix} = 1$, $\det[P_{\{\mathbf{v}_1\} \to \{\mathbf{v}_3\}}] = \det \begin{bmatrix} 0 & 1 \\ 1 & 1 \end{bmatrix} = -1$

where $\mathbf{x} \in M$ and $\vec{\mathbf{v}}_1, \dots, \vec{\mathbf{v}}_k$ is a basis of $T_\mathbf{x}M$. Let $\mathcal{B}_\mathbf{x}(M) \subset \mathcal{B}(M)$ be the subset where the first coordinate is $\mathbf{x}$, i.e., $\mathcal{B}_\mathbf{x}(M) = \{\mathbf{x}\} \times \mathcal{B}_{T_\mathbf{x}M}$.

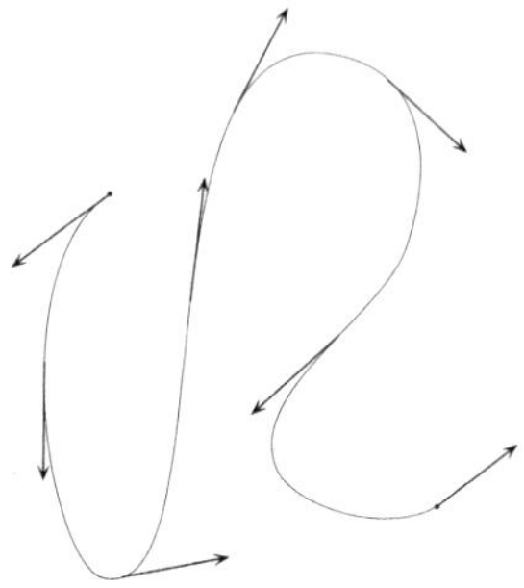

FIGURE 6.3.1.

Above we choose a tangent vector field (a field of vectors tangent to the curve) that depends continuously on $\mathbf{x}$ and never vanishes.

Definition 6.3.3 (Orientation of manifold). An *orientation* of a k-dimensional manifold $M \subset \mathbb{R}^n$ is a continuous map $\mathcal{B}(M) \to \{+1, -1\}$ whose restriction to each $\mathcal{B}_\mathbf{x}M$ is an orientation of $T_\mathbf{x}M$.

Quite often, a k-dimensional manifold M is oriented by choosing a k-form ω and defining $\Omega = \operatorname{sgn} \omega$. This works if $\omega(P_\mathbf{x}(\vec{\mathbf{v}}_1, \dots, \vec{\mathbf{v}}_k)) \neq 0$ for all $\mathbf{x} \in M$ and all bases $\vec{\mathbf{v}}_1, \dots, \vec{\mathbf{v}}_k$ of $T_\mathbf{x}M$.

Proposition 6.3.4 shows how to describe orientations of points, open subsets of $\mathbb{R}^n$, curves, and surfaces in $\mathbb{R}^3$; more general submanifolds are discussed in Proposition 6.3.9. Figures 6.3.1 and 6.3.2 illustrate the orientation of curves and surfaces.

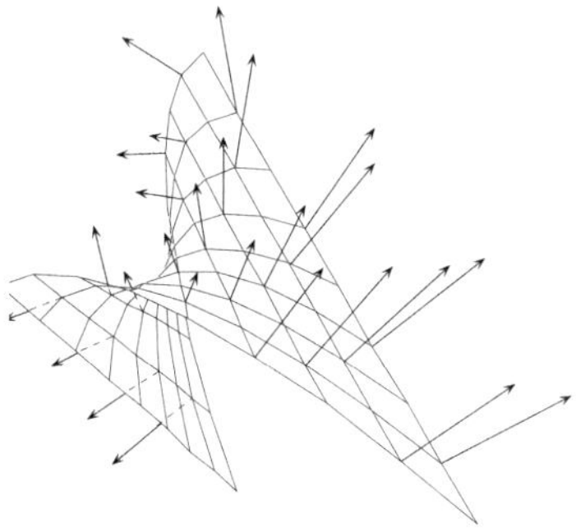

FIGURE 6.3.2.

The vectors of a transverse vector field should be thought of as anchored at points of S. The vector field specifies a side of S: the side the vector field points to, as opposed to the side it points away from. The best way to be transverse is to be *normal*, i.e., orthogonal to the surface; in practice, transverse vector fields used to orient a surface are often normal.

Proposition 6.3.4.

1. **Orienting points.** *An orientation for a 0-dimensional manifold in $\mathbb{R}^n$, i.e., a discrete set of points, is simply assigning to each point either $+1$ or -1.*
2. **Orienting open subsets of $\mathbb{R}^n$.** *An open subset $U \subset \mathbb{R}^n$ (the trivial case of an n-dimensional manifold) carries the standard orientation Ω^{st} of $\mathbb{R}^n$: at every point $\mathbf{x} \in U$ we have $T_\mathbf{x}U = \mathbb{R}^n$ (think of the vectors of $\mathbb{R}^n$ as anchored at $\mathbf{x}$) and $\Omega^{st}_\mathbf{x} = \Omega^{st}$.*
3. **Orienting a curve.** *Let $C \subset \mathbb{R}^n$ be a smooth curve. A nonvanishing tangent vector field $\vec{\mathbf{t}}$ that varies continuously with $\mathbf{x}$ defines an orientation of C by the formula*
$$\Omega^{\vec{\mathbf{t}}}_\mathbf{x}(\vec{\mathbf{v}}) \stackrel{\text{def}}{=} \operatorname{sgn}(\vec{\mathbf{t}}(\mathbf{x}) \cdot \vec{\mathbf{v}}) \qquad 6.3.4$$
for every basis $\vec{\mathbf{v}}$ of $T_\mathbf{x}C$. For $\Omega^{\vec{\mathbf{t}}}$, the basis $\vec{\mathbf{t}}(\mathbf{x})$ of $T_\mathbf{x}C$ is direct for all $\mathbf{x} \in C$.
4. **Orienting a surface in $\mathbb{R}^3$.** *Let $S \subset \mathbb{R}^3$ be a smooth surface, and let $\vec{\mathbf{n}} : S \to \mathbb{R}^3$ be a transverse vector field on S, i.e., a vector field defined at every $\mathbf{x} \in S$ and varying continuously with $\mathbf{x} \in S$, such that $\vec{\mathbf{n}}(\mathbf{x})$ does not belong to $T_\mathbf{x}S$ (and in particular is not $\vec{\mathbf{0}}$). Then we can define an orientation $\Omega^{\vec{\mathbf{n}}}$ of S by*
$$\Omega^{\vec{\mathbf{n}}}\left(\vec{\mathbf{v}}_1, \vec{\mathbf{v}}_2\right) \stackrel{\text{def}}{=} \operatorname{sgn}\left(\det[\vec{\mathbf{n}}(\mathbf{x}), \vec{\mathbf{v}}_1, \vec{\mathbf{v}}_2]\right) \qquad 6.3.5$$
for all $\mathbf{x} \in S$ and all bases $\vec{\mathbf{v}}_1, \vec{\mathbf{v}}_2$ of $T_\mathbf{x}S$. We call this the orientation of S by the transverse vector field $\vec{\mathbf{n}}$.

Proof. 1. For each point, the tangent space is a 0-dimensional vector space, and we have seen that an orientation of a 0-dimensional vector space is either $\{\phi\} \mapsto +1$ or $\{\phi\} \mapsto -1$.

2. This should be clear.

3. If $\vec{\mathbf{v}}_1$ and $\vec{\mathbf{v}}_2$ are two bases of $T_{\mathbf{x}}C$, then $\vec{\mathbf{v}}_2 = c\vec{\mathbf{v}}_1$ for some $c \neq 0$; c is the 1×1 change of basis matrix. Then

$$\operatorname{sgn}(\vec{\mathbf{t}}(\mathbf{x}) \cdot \vec{\mathbf{v}}_2) = \operatorname{sgn}(\vec{\mathbf{t}}(\mathbf{x}) \cdot c\vec{\mathbf{v}}_1) = (\operatorname{sgn} \det[c])(\operatorname{sgn}(\vec{\mathbf{t}}(\mathbf{x}) \cdot \vec{\mathbf{v}}_1). \qquad 6.3.6$$

Thus $\Omega^{\vec{\mathbf{t}}}$ satisfies equation 6.3.1. Moreover, the map $(\mathbf{x}, \vec{\mathbf{v}}) \mapsto \vec{\mathbf{t}}(\mathbf{x}) \cdot \vec{\mathbf{v}}$ is continuous on $C \times \mathbb{R}^n$ and does not vanish on the subset of $C \times \mathbb{R}^n$ where $\vec{\mathbf{v}}$ is a basis of $T_{\mathbf{x}}C$, and sgn is a continuous function of $\mathbb{R} - \{0\}$, so $\Omega^{\vec{\mathbf{t}}}$ is continuous on the set of pairs $(\mathbf{x}, \vec{\mathbf{v}})$ where $\vec{\mathbf{v}}$ is a basis of $T_{\mathbf{x}}C$.

4. The map $S \times \mathbb{R}^3 \times \mathbb{R}^3 \to \mathbb{R}$ given by

$$(\mathbf{x}, \vec{\mathbf{v}}_1, \vec{\mathbf{v}}_2) \mapsto \det[\vec{\mathbf{n}}(\mathbf{x}), \vec{\mathbf{v}_1}, \vec{\mathbf{v}_2}] \qquad 6.3.7$$

is continuous, and since $\vec{\mathbf{n}}(\mathbf{x})$ is not an element of $T_{\mathbf{x}}S$, the determinant does not vanish on $\mathcal{B}(S)$, so that

$$\Omega^{\vec{\mathbf{n}}} : (\mathbf{x}, \vec{\mathbf{v}}_1, \vec{\mathbf{v}}_2) \mapsto \operatorname{sgn}(\det[\vec{\mathbf{n}}(\mathbf{x}), \vec{\mathbf{v}_1}, \vec{\mathbf{v}_2}]). \qquad 6.3.8$$

is also continuous.

If $\vec{\mathbf{v}}_1, \vec{\mathbf{v}}_2$ and $\vec{\mathbf{w}}_1, \vec{\mathbf{w}}_2$ are two bases of $T_{\mathbf{x}}S$ with change of basis matrix $P = \begin{bmatrix} p_{1,1} & p_{1,2} \\ p_{2,1} & p_{2,2} \end{bmatrix}$, then the 3×3 change of basis matrix for the bases $\vec{\mathbf{n}}(\mathbf{x}), \vec{\mathbf{v}}_1, \vec{\mathbf{v}}_2$ and $\vec{\mathbf{n}}(\mathbf{x}), \vec{\mathbf{w}}_1, \vec{\mathbf{w}}_2$ is $\widetilde{P} = \begin{bmatrix} 1 & 0 & 0 \\ 0 & p_{1,1} & p_{1,2} \\ 0 & p_{2,1} & p_{2,2} \end{bmatrix}$, and since $\det \widetilde{P} = \det P$, we see that

$$\begin{aligned} \Omega^{\vec{\mathbf{n}}}(\vec{\mathbf{w}}_1, \vec{\mathbf{w}_2}) &= \operatorname{sgn}(\det[\vec{\mathbf{n}}(\mathbf{x}), \vec{\mathbf{w}}_1, \vec{\mathbf{w}_2}]) = \operatorname{sgn}(\det \widetilde{P}) \operatorname{sgn}(\det[\vec{\mathbf{n}}(\mathbf{x}), \vec{\mathbf{v}}_1, \vec{\mathbf{v}_2}]) \\ &= \operatorname{sgn}(\det P)\Omega^{\vec{\mathbf{n}}}(\vec{\mathbf{v}}_1, \vec{\mathbf{v}_2}). \end{aligned} \qquad 6.3.9$$

So $\Omega^{\vec{\mathbf{n}}}$ satisfies equation 6.3.1. □

Example 6.3.5 (A nonorientable surface). The Möbius strip shown in Figure 6.3.4 is not orientable; it is not possible to choose a transverse vector field $\vec{\mathbf{n}}$ varying continuously with $\mathbf{x}$. If you imagine yourself walking along the surface of a Möbius strip, planting a forest of normal vectors, one at each point, all pointing "up" (in the direction of your head), then when you get back to where you started there will be vectors anchored arbitrarily close to each other, pointing in opposite directions. Note that although a form cannot be integrated over a Möbius strip, the Möbius strip does have a well-defined area, obtained by integrating $|d^2\mathbf{x}|$ over it. △

Example 6.3.6 (Orienting a circle centered at 0). For every $R > 0$, the vector field

$$\vec{\mathbf{t}}\begin{pmatrix} x \\ y \end{pmatrix} = \begin{bmatrix} -y \\ x \end{bmatrix} \qquad 6.3.10$$

is a nonvanishing vector field tangent to the circle of equation $x^2 + y^2 = R^2$, defining the *counterclockwise* orientation. △

Have you ever tried to teach a child the difference between right and left? Usually one asks, what hand do you draw with? For this to work, of course, the child must have a dominant hand, and the teacher must know which one it is. Thus although orientation of space by the right hand is "standard", it may not appear particularly "natural". Imagine that you are in radio communication with an alien civilization living on a planet shrouded by clouds, so that you cannot tell them to look at anything. Is there any way you could communicate what is meant by "right" and "left"? Could you give them instructions to build a right-handed corkscrew, as opposed to its mirror image? If you think about it, you will probably come to the conclusion that there is no way to communicate this information.

Physicists tell us this is not true: when a neutron decays into a proton, electron, and neutrino, the proton and electron do not move in opposite directions, and the changing electric field of the proton makes the electron move in a right-handed corkscrew. The experiment of watching this decay could be communicated.

However, the direction is opposite when an antineutron decays, so if the aliens lived in a world of antimatter, we would have communicated the wrong handedness. According to astronomers, there are no galaxies in our universe made up of antimatter, and the question "why not?" is a central puzzle of cosmology.

FIGURE 6.3.3.

The Möbius strip is named after the German mathematician and astronomer August Möbius (1790–1868), who discovered it in 1858; it had been found four years earlier by Johann Listing.

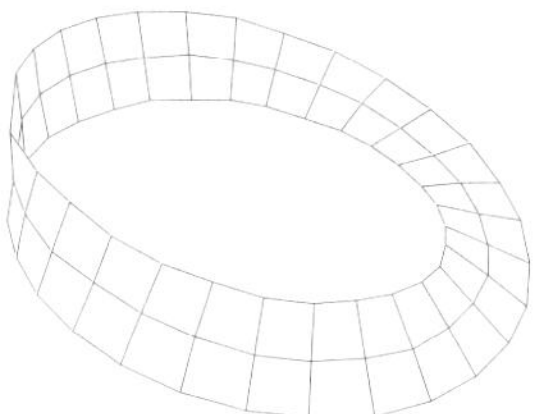

FIGURE 6.3.4.

Make a big Möbius strip out of paper. Give one child a yellow crayon, another a blue crayon, and start them coloring on opposite sides of the strip. Compare this figure with Figure 6.3.2.

Example 6.3.8: The gradient $\vec{\nabla} f(\mathbf{x})$ was defined in equation 1.7.22 as

$$\vec{\nabla} f(\mathbf{x}) \stackrel{\text{def}}{=} [\mathbf{D}f(\mathbf{x})]^\top$$

Usually we prefer to work with the line matrix $[\mathbf{D}f(\mathbf{x})]$ rather than the vector $\vec{\nabla} f(\mathbf{x})$, but this is an exception. If a manifold M is defined by an equation $f(\mathbf{x}) = 0$ (i.e., M has dimension one less than the ambient space), then at every $\mathbf{x} \in M$ the vector $\vec{\nabla} f(\mathbf{x})$ is orthogonal to $T_\mathbf{x}M$.

Example 6.3.7 (Orienting a surface in $\mathbb{R}^3$). Let $S \subset \mathbb{R}^3$ be the surface of equation $ax^2 + by^2 + cz^2 = R^2$, where we assume that $abc \neq 0$ (i.e., a, b, and c are all nonzero) and $R > 0$. Then the radial vector field

$$\vec{\mathbf{n}}\begin{pmatrix} x \\ y \\ z \end{pmatrix} = \begin{bmatrix} x \\ y \\ z \end{bmatrix} \qquad 6.3.11$$

defines an orientation of S. Indeed, if $f\begin{pmatrix} x \\ y \\ z \end{pmatrix} = ax^2 + by^2 + cz^2 - R^2$, then

$$\left[\mathbf{D}f\begin{pmatrix} x \\ y \\ z \end{pmatrix}\right]\overbrace{\begin{bmatrix} x \\ y \\ z \end{bmatrix}}^{\vec{\mathbf{n}}} = [2ax,\ 2by,\ 2cz]\begin{bmatrix} x \\ y \\ z \end{bmatrix} = 2R^2 \neq 0, \qquad 6.3.12$$

so $\vec{\mathbf{n}}$ is not in $\ker[\mathbf{D}f(\mathbf{x})]$, hence (by Theorem 3.2.4) not in the tangent space to S. Hence it is transverse to S. It goes from the side of S where $ax^2 + by^2 + cz^2 < R^2$ to the side where $ax^2 + by^2 + cz^2 > R^2$. △

Example 6.3.7 is a bit misleading: it seems to suggest that in order to orient a surface you need to guess a vector field. In reality, when a surface $S \subset \mathbb{R}^3$ is given by an equation $f(\mathbf{x}) = 0$, and moreover $[\mathbf{D}f(\mathbf{x})] \neq [0]$ at all points of S, then the gradient $\vec{\nabla} f(\mathbf{x}) = [\mathbf{D}f(\mathbf{x})]^\top$ is a transverse vector field (in fact, normal) on S, and hence always defines an orientation.

Example 6.3.8. Consider the surface S defined by $f(\mathbf{x}) = \sin(x + yz) = 0$. We saw in Example 3.1.13 that the vector field

$$\vec{\nabla} f(\mathbf{x}) \stackrel{\text{def}}{=} \left[\mathbf{D}f\begin{pmatrix} x \\ y \\ z \end{pmatrix}\right]^\top = \begin{bmatrix} \cos(x+yz) \\ z\cos(x+yz) \\ y\cos(x+yz) \end{bmatrix} \qquad 6.3.13$$

never vanishes at points of S. Better yet, it is orthogonal to S since, by Theorem 3.2.4, $\vec{\mathbf{v}} \in \ker[\mathbf{D}f(\mathbf{x})]$ implies $\vec{\mathbf{v}} \in T_\mathbf{x}S$. Thus, for $\vec{\mathbf{v}} \in T_\mathbf{x}S$,

$$0 = [\mathbf{D}f(\mathbf{x})]\vec{\mathbf{v}} = \vec{\nabla} f(\mathbf{x}) \cdot \vec{\mathbf{v}}. \qquad 6.3.14$$

So the vector $\vec{\mathbf{x}} = \vec{\nabla} f(\mathbf{x})$ defines an orientation of S. △

This construction can be considerably generalized.

Proposition 6.3.9 (Orienting manifolds given by equations). *Let $U \subset \mathbb{R}^n$ be open, and let $\mathbf{f}: U \to \mathbb{R}^{n-k}$ be a map of class C^1 such that $[\mathbf{Df}(\mathbf{x})]$ is surjective at all $\mathbf{x} \in M \stackrel{\text{def}}{=} \mathbf{f}^{-1}(\mathbf{0})$. Then the map $\Omega_\mathbf{x}: \mathcal{B}(T_\mathbf{x}M) \to \{+1, -1\}$ given by*

$$\Omega(\vec{\mathbf{v}}_1, \dots, \vec{\mathbf{v}}_k) \stackrel{\text{def}}{=} \operatorname{sgn}\det[\vec{\nabla} f_1(\mathbf{x}), \dots, \vec{\nabla} f_{n-k}(\mathbf{x}), \vec{\mathbf{v}}_1, \dots, \vec{\mathbf{v}}_k] \qquad 6.3.15$$

is an orientation of M.

Proposition 6.3.9 is perhaps a bit misleading, tempting the reader to think that all surfaces are orientable. Remember that manifolds can always

FIGURE 6.3.5.
Johann Listing (1808–1882)
Listing was the first to describe the Möbius strip; he also coined the words "topology" and "micron". At age 13, he helped support his family with money he earned drawing and doing calligraphy. He became a physicist as well as mathematician, and worked with Gauss on experiments concerning terrestrial magnetism.

be given *locally* by equations (one equation for a surface in $\mathbb{R}^3$), but it may be impossible to define a k-dimensional manifold in $\mathbb{R}^n$ by $n-k$ equations with linearly independent derivatives. In particular, it will be impossible if the manifold is nonorientable. See Example 6.4.9.

Orientation of connected and unconnected manifolds

Recall from the discussion in Example 3.1.8 (margin note below Figure 3.1.9) that a manifold M is *connected* if given any two points $\mathbf{u}_1, \mathbf{u}_2 \in M$, there exists a path in M connecting them: a continuous map $\delta : [a, b] \to M$ such that $\delta(a) = \mathbf{u}_1$ and $\delta(b) = \mathbf{u}_2$.

Proposition 6.3.10 (Orientation of connected, orientable manifold). *If M is a connected manifold, then either M is not orientable, or it has two orientations. If M is orientable, then specifying an orientation of $T_{\mathbf{x}}M$ at one point defines the orientation at every point.*

Note that if an orientable manifold is not connected, knowing its orientation at one point does not tell you its orientation everywhere.

Proof. If M is orientable and $\Omega : \mathcal{B}(M) \to \{+1, -1\}$ is an orientation of M, then $-\Omega$ is also an orientation, so an orientable manifold has at least two orientations.

We must show that if M is connected and Ω', Ω'' are two orientations of M such that $\Omega'_{\mathbf{x}_0} = \Omega''_{\mathbf{x}_0}$ for one point $\mathbf{x}_0 \in M$, then $\Omega'_{\mathbf{x}} = \Omega''_{\mathbf{x}}$ for all $\mathbf{x} \in M$. Choose a path $\gamma : [a, b] \to M$ with $\gamma(a) = \mathbf{x}_0$ and $\gamma(b) = \mathbf{x}$. Then for all t there exists $s(t) \in \{+1, -1\}$ such that $\Omega'_{\gamma(t)} = s(t)\Omega''_{\gamma(t)}$. Moreover, the function s is continuous, and a continuous function on $[a, b]$ that takes only the values -1 and $+1$ must be constant, by the intermediate value theorem. Since $\Omega'_{\mathbf{x}_0} = \Omega''_{\mathbf{x}_0}$, we have $s(a) = +1$, so $s(b) = +1$. $\square$

EXERCISES FOR SECTION 6.3

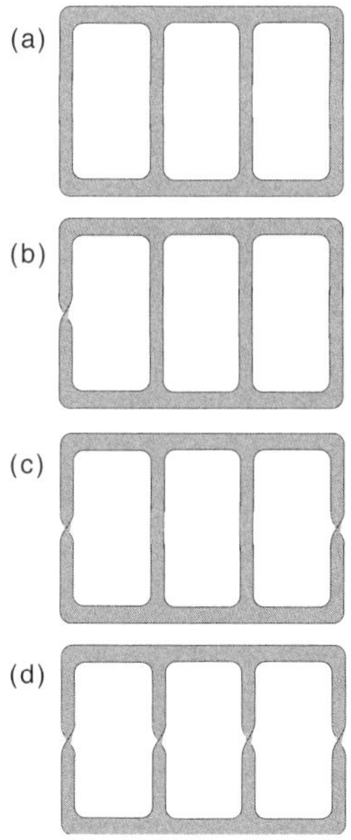

Surfaces for Exercise 6.3.2

6.3.1 Is the constant vector field $\begin{bmatrix} 1 \\ 1 \end{bmatrix}$ a tangent vector field defining an orientation of the line of equation $x + y = 0$? How about the line of equation $x - y = 0$?

6.3.2 Which of the surfaces in the margin are orientable?

6.3.3 Does any constant vector field define an orientation of the unit sphere in $\mathbb{R}^3$?

6.3.4 Find a vector field that orients the curve given by $x + x^2 + y^2 = 2$.

6.3.5 Which of the vector fields

$$\begin{bmatrix} 1 \\ 1 \\ 1 \end{bmatrix}, \begin{bmatrix} -1 \\ 1 \\ 1 \end{bmatrix}, \begin{bmatrix} -1 \\ -1 \\ 1 \end{bmatrix}, \begin{bmatrix} -1 \\ -1 \\ -1 \end{bmatrix}$$

define an orientation of the plane $P \subset \mathbb{R}^3$ of equation $x+y+z=0$, and among these, which pairs define the same orientation?

6.3.6 Find a vector field that orients the surface $S \subset \mathbb{R}^3$ given by $x^2+y^3+z=1$.

6.3.7 Let V be the plane of equation $x+2y-z=0$. Show that the bases

$$\vec{\mathbf{v}}_1 = \begin{bmatrix}1\\0\\1\end{bmatrix}, \vec{\mathbf{v}}_2 = \begin{bmatrix}0\\1\\2\end{bmatrix} \quad \text{and} \quad \vec{\mathbf{w}}_1 = \begin{bmatrix}2\\-3\\-4\end{bmatrix}, \vec{\mathbf{w}}_2 = \begin{bmatrix}1\\2\\5\end{bmatrix}$$

give the same orientation.

6.3.8 Let P be the plane of equation $x+y+z=0$.

a. Of the three bases

$$\begin{bmatrix}1\\0\\-1\end{bmatrix}, \begin{bmatrix}0\\1\\-1\end{bmatrix}, \quad \begin{bmatrix}-1\\0\\1\end{bmatrix}, \begin{bmatrix}-1\\1\\0\end{bmatrix}, \quad \begin{bmatrix}1\\-1\\0\end{bmatrix}, \begin{bmatrix}0\\-1\\1\end{bmatrix}$$

which gives a different orientation than the other two?

b. Find a normal vector to P that gives the same orientation as that basis.

6.3.9 a. Let $C \subset \mathbb{R}^2$ be the circle of equation $(x-1)^2+y^2=4$. Find the unit tangent vector field $\vec{T}$ describing the orientation "increasing polar angle". (A "unit vector field" is a vector field in which all the vectors have length 1.)

b. Explain carefully why the phrase "increasing polar angle" does not describe an orientation of the circle of equation $(x-2)^2+y^2=1$.

6.3.10 Let $M \subset \mathbb{R}^2$ be a manifold consisting of two circles of radius 1/2, one centered at $\begin{pmatrix}-1\\0\end{pmatrix}$, the other at $\begin{pmatrix}1\\0\end{pmatrix}$. How many orientations does M have? Describe them.

6.3.11 Let $S \subset \mathbb{R}^4$ be the locus given by the equations $x_1^2 - x_2^2 = x_3$ and $2x_1x_2 = x_4$.

a. Show that S is a surface.

b. Find a basis for the tangent space to S at the origin that is direct for the orientation given by Proposition 6.3.9.

6.3.12 Consider the manifold $M \subset \mathbb{R}^4$ of equation $x_1^2+x_2^2+x_3^2-x_4=0$. Find a basis for the tangent space to M at the point $\begin{pmatrix}1\\0\\0\\1\end{pmatrix}$ that is direct for the orientation given by Proposition 6.3.9.

6.3.13 a. Find two continuous maps $[0,1] \to \mathbb{R}^3$, denoted $\mathbf{v}$ and $\mathbf{w}$, such that

$$\mathbf{v}(0) = \mathbf{e}_1, \quad \mathbf{v}(1) = \vec{\mathbf{e}}_2, \quad \mathbf{w}(0) = \vec{\mathbf{e}}_2, \quad \mathbf{w}(1) = \vec{\mathbf{e}}_1,$$

and $\mathbf{v}(t)$ and $\mathbf{w}(t)$ are linearly independent for all t.

b. Show that there are no such mappings $[0,1] \to \mathbb{R}^2$.

c. Show that given any two linearly independent vectors $\vec{\mathbf{u}}_1, \vec{\mathbf{u}}_2$ in $\mathbb{R}^n$, $n>2$, there exist maps $\mathbf{v}, \mathbf{w} : [0,1] \to \mathbb{R}^n$ such that

$$\mathbf{v}(0) = \vec{\mathbf{u}}_1, \quad \mathbf{v}(1) = \vec{\mathbf{u}}_2, \quad \mathbf{w}(0) = \vec{\mathbf{u}}_2, \quad \mathbf{w}(1) = \vec{\mathbf{u}}_1,$$

and for each t, $\mathbf{v}(t)$ and $\mathbf{w}(t)$ are linearly independent.

Exercise 6.3.13: Part a shows that there is no way to choose an orientation for all planes in $\mathbb{R}^3$ that depends continuously on the plane. If there were such a standard orientation, then given any two linearly independent vectors $\vec{\mathbf{v}}, \vec{\mathbf{w}}$ in $\mathbb{R}^3$, it would be possible to choose one: the plane that they span would have an orientation, and either $(\vec{\mathbf{v}}, \vec{\mathbf{w}})$ in that order, or $(\vec{\mathbf{w}}, \vec{\mathbf{v}})$ in that order would be a direct basis. We could then choose the first vector of the direct basis.

Part b shows that in $\mathbb{R}^2$, this does work. Either $\det[\vec{\mathbf{v}}, \vec{\mathbf{w}}] > 0$ or $\det[\vec{\mathbf{w}}, \vec{\mathbf{v}}] > 0$, and we can choose the first column of whichever matrix has positive determinant.

Part c shows that it does not work in $\mathbb{R}^n$, $n>2$. As in $\mathbb{R}^3$, we can exchange the vectors by a continuous motion, keeping them linearly independent at all times.

6.3.14 Where is the vector field $\vec{\mathbf{e}}_1$ transverse to the unit sphere in $\mathbb{R}^3$? Where does the orientation given by this vector field to the unit sphere coincide with the orientation given by the outward-pointing normal?

Exercise 6.3.15 shows how to orient an n-dimensional complex vector space E: given a complex basis $\vec{\mathbf{v}}_1, \dots, \vec{\mathbf{v}}_n$, we can orient E by declaring the real basis

$$\vec{\mathbf{v}}_1, i\vec{\mathbf{v}}_1, \dots, \vec{\mathbf{v}}_n, i\vec{\mathbf{v}}_n$$

to be direct. We will say that this orientation is standard. The exercise shows that this orientation does not depend on the choice of complex basis of E.

6.3.15 Let E be a complex vector space of dimension n. A basis of E is a set of vectors $\mathbf{v}_1, \dots, \mathbf{v}_n$ such that every vector of E can be uniquely written $c_1\mathbf{v}_1 + \dots + c_n\mathbf{v}_n$, where the coefficients $c_1, \dots, c_n$ are complex numbers.

a. Show that the $2n$ vectors $\mathbf{v}_1, i\mathbf{v}_1, \dots, \mathbf{v}_n, i\mathbf{v}_n$ are a basis of E viewed as a real vector space, i.e., every vector $\mathbf{x} \in E$ can be written uniquely as

$$\mathbf{x} = a_1\mathbf{v}_1 + b_1(i\mathbf{v}_1) + \dots + a_n\mathbf{v}_n + b_n(i\mathbf{v}_n),$$

with $a_1, \dots, a_n, b_1, \dots, b_n \in \mathbb{R}$. Denote this basis by $\{\widetilde{\mathbf{v}}\}$.

b. Let $\mathbf{w}_1, \dots, \mathbf{w}_n$ be another basis of E, with complex change of basis matrix $C = [P_{\mathbf{w}\to\mathbf{v}}]$; see Proposition and Definition 2.6.17. Denote the basis $\mathbf{w}_1, i\mathbf{w}_1, \dots, \mathbf{w}_n, i\mathbf{w}_n$ by $\{\widetilde{\mathbf{w}}\}$. Write the change of basis matrix $\widetilde{C} = [P_{\widetilde{\mathbf{w}}\to\widetilde{\mathbf{v}}}]$ in terms of the real and imaginary parts of the $c_{j,i}$. *Hint*: Try $n = 1$ first, then $n = 2$; the pattern should become clear.

**c. Show that $\det \widetilde{C} = |\det C|^2$. Again, do $n = 1$ and $n = 2$ first.

d. Let Ω be the orientation of E such that $\mathbf{v}_1, i\mathbf{v}_1, \dots, \mathbf{v}_n, i\mathbf{v}_n$ is a direct basis. Show that then $\mathbf{w}_1, i\mathbf{w}_1, \dots, \mathbf{w}_n, i\mathbf{w}_n$ is also a direct basis.

6.4 Integrating forms over oriented manifolds

The object of this section is to define the integral of a k-form over an oriented k-dimensional manifold. We saw in Chapter 5 that when we integrate $|d^k\mathbf{x}|$ over a manifold, the integral does not depend on the choice of parametrization; this enabled us to define the integral of $|d^k\mathbf{x}|$ over a manifold. In Example 6.2.5 we saw that the choice of parametrization does matter when we integrate a k-form over an oriented manifold. To define the integral of a k-form over an oriented manifold, we must stipulate that the parametrization *preserve* the orientation of the manifold. Are we going down the down staircase (the parametrization preserves the orientation) or up the down staircase (the parametrization reverses the orientation)?

Definition 6.4.1: For a linear transformation to preserve orientation, the domain and codomain must be oriented, and must have the same dimension. Of course,

$$T(\vec{\mathbf{e}}_1), \dots, T(\vec{\mathbf{e}}_k)$$

are the columns of T.

Definition 6.4.1 (Orientation-preserving linear transformation). Let V be a k-dimensional vector space oriented by $\Omega : \mathcal{B}(V) \to \{+1, -1\}$. A linear transformation $T : \mathbb{R}^k \to V$ is *orientation preserving* if

$$\Omega\big(T(\vec{\mathbf{e}}_1), \dots, T(\vec{\mathbf{e}}_k)\big) = +1. \qquad 6.4.1$$

It is *orientation reversing* if

$$\Omega\big(T(\vec{\mathbf{e}}_1), \dots, T(\vec{\mathbf{e}}_k)\big) = -1. \qquad 6.4.2$$

Note that if T in Definition 6.4.1 is not invertible, then it neither preserves nor reverses orientation, since in that case $T(\vec{\mathbf{e}}_1), \dots, T(\vec{\mathbf{e}}_k)$ is not a basis of V.

Once more, faced with a nonlinear problem, we linearize it: a parametrization of a manifold preserves orientation if its derivative preserves orientation.

Definition 6.4.2: Since

$$(U - X) \subset \mathbb{R}^k$$

is open, it is necessarily k-dimensional. (We haven't defined dimension of a set; what we mean is that $U - X$ is a k-dimensional manifold.) Think of a disc in $\mathbb{R}^3$ and a ball in $\mathbb{R}^3$; the ball can be open but the disc cannot. The "fence" of a disc in $\mathbb{R}^3$ is the entire disc, and it is impossible (Definition 1.5.2) to surround each point of the disc with a ball contained entirely in the disc.

We could write the left side of equation 6.4.3 as

$$\Omega\Big([\mathbf{D}\gamma(\mathbf{u})]\vec{\mathbf{e}}_1, \dots, [\mathbf{D}\gamma(\mathbf{u})]\vec{\mathbf{e}}_k\Big).$$

Definition 6.4.2 (Orientation-preserving parametrization of a manifold). Let $M \subset \mathbb{R}^m$ be a k-dimensional manifold oriented by Ω, and let $U \subset \mathbb{R}^k$ be a subset with boundary of k-dimensional volume 0. Let $\gamma : U \to \mathbb{R}^m$ parametrize M as described in Definition 5.2.3, so that the set X of "trouble spots" satisfies all conditions of that definition; in particular, X has k-dimensional volume 0. Then γ is *orientation preserving* if for all $\mathbf{u} \in (U - X)$, we have

$$\Omega(\overrightarrow{D_1\gamma}(\mathbf{u}), \dots, \overrightarrow{D_k\gamma}(\mathbf{u})) = +1. \qquad 6.4.3$$

Note that the vectors in equation 6.4.3 are linearly independent by part 4 of Definition 5.2.3, so, by Proposition 3.2.7, they form a basis of $T_{\gamma(\mathbf{u})}M$.

Example 6.4.3 (Parametrizations of unit circle). Let C be the circle of equation $x^2+y^2 = R^2, R > 0$ oriented by the vector field $\vec{\mathbf{t}}\begin{pmatrix} x \\ y \end{pmatrix} = \begin{bmatrix} -y \\ x \end{bmatrix}$, as in Example 6.3.6. This is the *counterclockwise orientation.*

The parametrization $\gamma(t) = \begin{pmatrix} R\cos t \\ R\sin t \end{pmatrix}$ preserves that orientation since

$$\vec{\mathbf{t}}(\gamma(t)) \cdot \vec{\gamma}'(t) = \begin{bmatrix} -R\sin t \\ R\cos t \end{bmatrix} \cdot \begin{bmatrix} -R\sin t \\ R\cos t \end{bmatrix} = R^2 > 0. \qquad 6.4.4$$

As t increases, $\gamma(t)$ travels around the circle counterclockwise.

The parametrization $\gamma_1(t) = \begin{pmatrix} R\sin t \\ R\cos t \end{pmatrix}$ reverses the orientation since

$$\vec{\mathbf{t}}(\gamma_1(t)) \cdot \vec{\gamma}_1'(t) = \begin{bmatrix} -R\cos t \\ R\sin t \end{bmatrix} \cdot \begin{bmatrix} R\cos t \\ -R\sin t \end{bmatrix} = -R^2 < 0. \qquad 6.4.5$$

As t increases, $\gamma_1(t)$ travels around the circle clockwise. $\triangle$

Recall from Proposition 6.3.4 that a transverse vector field $\vec{\mathbf{n}}$ orients a surface $S \subset \mathbb{R}^3$ by the formula

$$\Omega_{\mathbf{x}}(\vec{\mathbf{v}}_1, \vec{\mathbf{v}}_2) = \operatorname{sgn}\det[\vec{\mathbf{n}}(\mathbf{x}), \vec{\mathbf{v}}_1, \vec{\mathbf{v}}_2]. \qquad 6.4.6$$

If γ is a parametrization of S, and if we set $\vec{\mathbf{n}} = \overrightarrow{D_1\gamma} \times \overrightarrow{D_2\gamma}$, then γ is orientation preserving, since

$$\det[\overrightarrow{D_1\gamma} \times \overrightarrow{D_2\gamma}, \overrightarrow{D_1\gamma}, \overrightarrow{D_2\gamma}] > 0. \qquad 6.4.7$$

This is a restatement of the fact (Proposition 1.4.20) that if $\vec{\mathbf{v}}$ and $\vec{\mathbf{w}}$ are linearly independent vectors in $\mathbb{R}^3$, then $\det[\vec{\mathbf{v}} \times \vec{\mathbf{w}}, \vec{\mathbf{v}}, \vec{\mathbf{w}}] > 0$. $\triangle$

Example 6.4.4 (Orientation-preserving parametrization of surface in $\mathbb{C}^3$). Consider the surface $S \subset \mathbb{C}^3$ parametrized by $z \mapsto \begin{pmatrix} z \\ z^2 \\ z^3 \end{pmatrix}$, $|z| < 1$.

The tangent space to S at $\begin{pmatrix} z \\ z^2 \\ z^3 \end{pmatrix}$ is the complex line consisting of the complex multiples of $\begin{pmatrix} 1 \\ 2z \\ 3z^2 \end{pmatrix}$; in particular, it carries a standard orientation for which the real basis $v_1 = \begin{pmatrix} 1 \\ 2z \\ 3z^2 \end{pmatrix}$, $v_2 = \begin{pmatrix} i \\ 2iz \\ 3iz^2 \end{pmatrix}$ is direct; see Exercise 6.3.15. Think of S as a surface in $\mathbb{R}^6$, parametrized by the function γ in the margin.

$$\gamma : \begin{pmatrix} x \\ y \end{pmatrix} \mapsto \begin{pmatrix} x_1 &=& x \\ y_1 &=& y \\ x_2 &=& x^2 - y^2 \\ y_2 &=& 2xy \\ x_3 &=& x^3 - 3xy^2 \\ y_3 &=& 3x^2y - y^3 \end{pmatrix}$$

Parametrization for Example 6.4.4: The first two entries correspond to the real and imaginary parts of z, the third entry to the real part of z^2, the fourth entry to the imaginary part of z^2, and so on.

Does this parametrization preserve the standard orientation? Using Definition 6.3.1, we must compute the determinant of the change of matrix basis between the basis

$$\overrightarrow{D_1\gamma}\begin{pmatrix} x \\ y \end{pmatrix} = \begin{pmatrix} 1 \\ 0 \\ 2x \\ 2y \\ 3x^2 - 3y^2 \\ 6xy \end{pmatrix}, \quad \overrightarrow{D_2\gamma}\begin{pmatrix} x \\ y \end{pmatrix} = \begin{pmatrix} 0 \\ 1 \\ -2y \\ 2x \\ -6xy \\ 3x^2 - 3y^2 \end{pmatrix} \qquad 6.4.8$$

and the basis (v_1, v_2). If we write v_1 and v_2 in real and imaginary parts, we get the vectors in equation 6.4.8, so the change of basis matrix is the identity and γ is orientation preserving. △

Example 6.4.5: It is important to realize that a parametrization may be neither orientation preserving nor orientation reversing: it may preserve orientation on one part of the manifold and reverse it on another.

Example 6.4.5 (A parametrization that neither preserves nor reverses orientation). Consider the unit sphere parametrized by

$$\gamma\begin{pmatrix} \varphi \\ \theta \end{pmatrix} = \begin{pmatrix} \cos\varphi\cos\theta \\ \cos\varphi\sin\theta \\ \sin\varphi \end{pmatrix}, \quad \text{for } 0 \le \theta \le \pi,\ -\pi \le \varphi \le \pi, \qquad 6.4.9$$

and oriented by the outward-pointing normal, $\vec{\mathbf{n}} = \begin{bmatrix} x \\ y \\ z \end{bmatrix}$. We have

$$\det\left[\vec{\mathbf{n}}\left(\gamma\begin{pmatrix} \varphi \\ \theta \end{pmatrix}\right), \overrightarrow{D_1\gamma}\begin{pmatrix} \varphi \\ \theta \end{pmatrix}, \overrightarrow{D_2\gamma}\begin{pmatrix} \varphi \\ \theta \end{pmatrix}\right] = -\cos\varphi. \qquad 6.4.10$$

For $-\pi \le \varphi < -\pi/2$, we have $-\cos\varphi > 0$; for $-\pi/2 < \varphi < \pi/2$, we have $-\cos\varphi < 0$; for $\pi/2 < \varphi \le \pi$, we have $-\cos\varphi > 0$; the sign of the determinant changes, so the parametrization keeps switching between reversing and preserving orientation.

What goes wrong here? The mapping γ takes the entire line $\varphi = -\pi/2$ in the (φ, θ)-plane to the single point $\begin{pmatrix} 0 \\ 0 \\ -1 \end{pmatrix}$; it takes the line $\varphi = \pi/2$ to the point $\begin{pmatrix} 0 \\ 0 \\ 1 \end{pmatrix}$; for those values, γ is not one to one. It is only a strict parametrization on the (φ, θ)-plane *minus* those two lines. But that domain is not connected; it consists of three separate pieces. △

We will use the following two propositions when we show that the manifold of Example 6.4.9 is not orientable.

Proposition 6.4.6 (Checking orientation at a single point). *Let M be an oriented manifold, and let $\gamma : U \to M$ be a parametrization of an open subset of M, with $U - X$ connected, where X is as in Definition 5.2.3. Then if γ preserves orientation at a single point of U, it preserves orientation at every point of U.*

Proposition 6.4.6: With the relaxed definition of parametrization (Definition 5.2.3), U might be connected, but $U - X$ might not. In that case checking at a single point will only give "orientation preserving" in the set of points that can be connected to that point in $U - X$.

The proof of Proposition 6.4.6 is similar to the proof of Proposition 6.3.10 and is left as Exercise 6.4.8.

Example 6.4.7 (Checking parametrization of an oriented torus). Let M be the torus of Example 5.3.9, obtained by choosing $R > r > 0$ and taking the circle of radius r in the (x, z)-plane that is centered at $x = R$, $z = 0$, and rotating it around the z-axis, as shown in Figure 6.4.1. (This is a two-dimensional object, not the solid torus.) Assume it is oriented by the outward-pointing normal. Does the parametrization

$$\gamma \begin{pmatrix} u \\ v \end{pmatrix} = \begin{pmatrix} (R + r\cos u)\cos v \\ (R + r\cos u)\sin v \\ r\sin u \end{pmatrix} \qquad 6.4.11$$

preserve that orientation?

Since

$$\overrightarrow{D_1\gamma} = \begin{bmatrix} -r\sin u\cos v \\ -r\sin u\sin v \\ r\cos u \end{bmatrix}, \overrightarrow{D_2\gamma} = \begin{bmatrix} -(R + r\cos u)\sin v \\ (R + r\cos u)\cos v \\ 0 \end{bmatrix}, \qquad 6.4.12$$

we can take as our normal vector field $\vec{\mathbf{n}}$ the vector field

$$\overrightarrow{D_1\gamma} \times \overrightarrow{D_2\gamma} = -r(R + r\cos u) \begin{bmatrix} \cos u\cos v \\ \cos u\sin v \\ \sin u \end{bmatrix}. \qquad 6.4.13$$

Does this normal vector field point in or out? At $\gamma\begin{pmatrix} 0 \\ 0 \end{pmatrix}$, we have

$$\vec{\mathbf{n}}\left(\gamma\begin{pmatrix} 0 \\ 0 \end{pmatrix}\right) = -r(r + R)\begin{bmatrix} 1 \\ 0 \\ 0 \end{bmatrix} \qquad 6.4.14$$

which points in, so γ does not preserve the orientation given by the outward-pointing vector. △

Example 6.4.7: By Proposition 6.4.6, it is sufficient to check orientation at a single point, i.e., to evaluate $\vec{\mathbf{n}}$ at one point on the torus. We chose the point $\gamma\begin{pmatrix} 0 \\ 0 \end{pmatrix}$, which simplifies the computation.

Proposition 6.4.8 (Orientation-preserving parametrizations). *Let $M \subset \mathbb{R}^n$ be a k-dimensional oriented manifold. Let U_1, U_2 be subsets of $\mathbb{R}^k$, and let $\gamma_1 : U_1 \to \mathbb{R}^n$ and $\gamma_2 : U_2 \to \mathbb{R}^n$ be two parametrizations of M. Then γ_1 and γ_2 are either both orientation preserving or both orientation reversing if and only if for all $\mathbf{u}_1 \in U_1^{\text{OK}}$ and all $\mathbf{u}_2 \in U_2^{\text{OK}}$ with $\gamma_1(\mathbf{u}_1) = \gamma_2(\mathbf{u}_2)$ we have*

$$\det[\mathbf{D}(\gamma_2^{-1} \circ \gamma_1)(\mathbf{u}_1)] > 0. \qquad 6.4.15$$

Proposition 6.4.8: The mapping $\gamma_2^{-1} \circ \gamma_1$ is the "change of parameters" mapping. Recall from Section 5.2 that the superscript "OK" (okay) denotes the domain or codomain of a change of variables mapping from which we have removed any trouble spots, assumed to be of volume 0.

Proof. Set $\mathbf{x} = \gamma_1(\mathbf{u}_1) = \gamma_2(\mathbf{u}_2)$. Then (by Proposition 3.2.7), the columns of the $n \times k$ matrix $[\mathbf{D}\gamma_1(\mathbf{u}_1)]$ form a basis for $T_\mathbf{x}M$, as do the columns of $[\mathbf{D}\gamma_2(\mathbf{u}_2)]$. Write

$$\begin{aligned}[\mathbf{D}\gamma_1(\mathbf{u}_1)] &= [\mathbf{D}(\gamma_2 \circ \gamma_2^{-1} \circ \gamma_1)(\mathbf{u}_1)] \\ &= [\mathbf{D}\gamma_2(\underbrace{\gamma_2^{-1} \circ \gamma_1)(\mathbf{u}_1)}_{\gamma_2(\mathbf{u}_2)}][\mathbf{D}(\gamma_2^{-1} \circ \gamma_1)(\mathbf{u}_1)] \\ &= [\mathbf{D}\gamma_2(\mathbf{u}_2)][\mathbf{D}(\gamma_2^{-1} \circ \gamma_1)(\mathbf{u}_1)]\end{aligned} \qquad 6.4.16$$

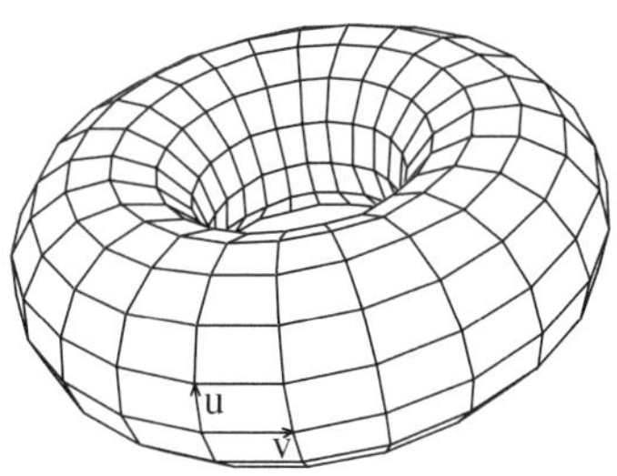

FIGURE 6.4.1.

Example 6.4.7: The torus with the u and v coordinates drawn. You should imagine the straight lines as curved. By "torus" we mean the surface of the object. The solid object is called a solid torus.

This is equation 2.6.22, as rewritten in matrix multiplication form in the margin note next to that equation: $[\mathbf{D}\gamma_1(\mathbf{u}_1)]$ and $[\mathbf{D}\gamma_2(\mathbf{u}_2)]$ are two bases, and $[\mathbf{D}(\gamma_2^{-1} \circ \gamma_1)(\mathbf{u}_1)]$ is the change of basis matrix. By Definition 6.3.1,

$$\begin{aligned}&\Omega(\overrightarrow{D_1\gamma_1}(\mathbf{u}), \dots, \overrightarrow{D_k\gamma_1}(\mathbf{u})) \\ &\quad = \operatorname{sgn}\big(\det[\mathbf{D}(\gamma_2^{-1} \circ \gamma_1)(\mathbf{u}_1)]\big)\, \Omega(\overrightarrow{D_1\gamma_2}(\mathbf{u}), \dots, \overrightarrow{D_k\gamma_2}(\mathbf{u})).\end{aligned} \qquad 6.4.17$$

It then follows from Definition 6.4.2 that if $\det[\mathbf{D}(\gamma_2^{-1} \circ \gamma_1)(\mathbf{u}_1)] > 0$, then if γ_1 is orientation preserving, γ_2 will also preserve orientation, and if γ_1 is orientation reversing, γ_2 will also reverse orientation. Conversely, if γ_1 and γ_2 both preserve orientation, or both reverse orientation, then we must have $\det[\mathbf{D}(\gamma_2^{-1} \circ \gamma_1)(\mathbf{u}_1)] > 0$. □

A nonorientable manifold

We saw that the Möbius strip can't be oriented, but there is something unsatisfactory about a proof that requires the evidence of one's eyes. Mathematics should not be limited to objects (however curious) brought to class for show and tell. Now we can give an example of a manifold that we can prove is not orientable, without benefit of a drawing or scissors and glue.

Example 6.4.9 (A nonorientable manifold). Consider the subset $X \subset \text{Mat}\,(2,3)$ made of all 2×3 matrices of rank 1. This is a manifold of dimension 4 in Mat (2, 3), as Exercise 6.4.2 asks you to show; we will see that it is nonorientable. Indeed, suppose that X is orientable. Let

$$\gamma_1, \gamma_2 : \left(\mathbb{R}^2 - \left\{\begin{pmatrix}0\\0\end{pmatrix}\right\}\right) \times \mathbb{R}^2 \to X \qquad 6.4.18$$

Exercise 3.8 spells out some of the details used without proof in Example 6.4.9.

be given by

$$\gamma_1 : \begin{pmatrix}a_1\\b_1\\c_1\\d_1\end{pmatrix} \mapsto \begin{bmatrix}a_1 & c_1a_1 & d_1a_1\\ b_1 & c_1b_1 & d_1b_1\end{bmatrix}, \quad \gamma_2 : \begin{pmatrix}a_2\\b_2\\c_2\\d_2\end{pmatrix} \mapsto \begin{bmatrix}c_2a_2 & a_2 & d_2a_2\\ c_2b_2 & b_2 & d_2b_2\end{bmatrix}. \qquad 6.4.19$$

Both maps are strict parametrizations of their images, and since the domain of each is connected, by Proposition 6.4.6 they are both either orientation preserving on the whole domain or orientation reversing on the whole domain. In particular, $\det\left[\mathbf{D}(\gamma_2^{-1} \circ \gamma_1)\right]$ must be either everywhere positive or everywhere negative where the composition is defined. In this case, it is easy to compute $\gamma_2^{-1} \circ \gamma_1$, by setting the corresponding entries of the image matrices to be equal; the equation

$$\gamma_1 \begin{pmatrix} a_1 \\ b_1 \\ c_1 \\ d_1 \end{pmatrix} = \gamma_2 \begin{pmatrix} a_2 \\ b_2 \\ c_2 \\ d_2 \end{pmatrix} \quad \text{leads to} \quad \begin{aligned} a_2 &= c_1 a_1 \\ b_2 &= c_1 b_1 \\ c_2 &= \frac{a_1}{a_2} = \frac{1}{c_1} \\ d_2 &= \frac{d_1 a_1}{a_2} = \frac{d_1}{c_1} \end{aligned}. \qquad 6.4.20$$

It follows from formula 6.4.19 that the first two variables of γ_1 and γ_2 aren't simultaneously 0: for γ_1, we never have $a_1 = b_1 = 0$ and for γ_2, we never have

$$a_2 = b_2 = 0.$$

Thus the first column of the first matrix in equation 6.4.19 is never $\vec{\mathbf{0}}$, and the second column of the second matrix is never $\vec{\mathbf{0}}$.

Thus we see that $\gamma_2^{-1} \circ \gamma_1$ takes the region in $\mathbb{R}^4$ where $\begin{pmatrix} a_1 \\ b_1 \end{pmatrix} \neq \begin{pmatrix} 0 \\ 0 \end{pmatrix}$ and $c_1 \neq 0$ to the region where $\begin{pmatrix} a_2 \\ b_2 \end{pmatrix} \neq \begin{pmatrix} 0 \\ 0 \end{pmatrix}$ and $c_2 \neq 0$.

If we compose γ_1 and γ_2 on the left in equation 6.4.20 by γ_2^{-1}, we get

$$(\gamma_2^{-1} \circ \gamma_1) \begin{pmatrix} a_1 \\ b_1 \\ c_1 \\ d_1 \end{pmatrix} = \begin{pmatrix} c_1 a_1 \\ c_1 b_1 \\ 1/c_1 \\ d_1/c_1 \end{pmatrix},$$

which allows us to compute the derivative of the composition.

Now we compute

$$\det \left[\mathbf{D}(\gamma_2^{-1} \circ \gamma_1)\right] = \det \begin{bmatrix} c_1 & 0 & a_1 & 0 \\ 0 & c_1 & b_1 & 0 \\ 0 & 0 & -\frac{1}{c_1^2} & 0 \\ 0 & 0 & -\frac{d_1}{c_1^2} & \frac{1}{c_1} \end{bmatrix} = -\frac{1}{c_1}. \qquad 6.4.21$$

This determinant is negative when $c_1 > 0$ and positive when $c_1 < 0$, contradicting the hypothesis that X is orientable. △

The locus X is defined by the equation

$$\mathbf{f} \begin{pmatrix} x_1 \\ x_2 \\ x_3 \\ y_1 \\ y_2 \\ y_3 \end{pmatrix} = \begin{pmatrix} x_1 y_2 - x_2 y_1 \\ x_1 y_3 - x_3 y_1 \\ x_2 y_3 - x_3 y_2 \end{pmatrix}.$$

Thus it does not meet the requirements of Proposition 6.3.9 concerning orientation of a manifold known by equations; the codomain of $\mathbf{f}$ is of the wrong dimension. By Theorem 3.1.10, the codomain should be of dimension $n - k = 2$, not 3.

Why doesn't Example 6.4.9 contradict Proposition 6.3.9, which shows how to orient a manifold defined by equations? The locus X is the set of matrices $\begin{bmatrix} x_1 & x_2 & x_3 \\ y_1 & y_2 & y_3 \end{bmatrix}$ defined by the equations

$$x_1 y_2 - x_2 y_1 = 0, \; x_1 y_3 - x_3 y_1 = 0, \; x_2 y_3 - x_3 y_2 = 0, \qquad 6.4.22$$

i.e., the equation $\mathbf{f}(\mathbf{x}) = \mathbf{0}$, where $\mathbf{f}$ is the function defined in the margin. The codomain of this function has the wrong dimension; to define a four-dimensional manifold in $\mathbb{R}^6$, we would expect to need two equations, not three. Near any point of X, two of the three equations do define X. But you can't choose the same two at every point. For example, the first equation says that columns 1 and 2 are linearly dependent, while the second says that columns 1 and 3 are linearly dependent. So columns 1, 2, and 3 are linearly dependent and that the matrix is of rank one, *unless* $x_1 = y_1 = 0$, in which case it may well have rank 2: the matrix $\begin{bmatrix} 0 & x_2 & x_3 \\ 0 & y_2 & y_3 \end{bmatrix}$ will satisfy the first two equations of 6.4.22, but will usually not satisfy the third. So the first two equations in 6.4.22 define X except in a neighborhood of those points in X where $x_1 = y_1 = 0$.

Thus Proposition 6.3.9 does not guarantee that X is orientable, and the fact that X is not orientable proves that it cannot be defined by two equations with linearly independent derivatives.

Integrating forms over oriented manifolds

We will now see that the integral of a k-form over an oriented manifold is independent of the choice of parametrization, as long as the parametrization preserves orientation. This will allow us to define integrals of forms over oriented manifolds.

Proof of Theorem 6.4.10: We found this argument by working through the proof of the statement that integrals of manifolds with respect to $|d^k\mathbf{x}|$ are independent of parametrization (Proposition 5.3.3), and noting the differences. You may find it instructive to compare the two arguments. Superficially, the equations may seem very different, but note the similarities. The first line of equation 5.3.16 corresponds to the right side of the first line of equation 6.4.25; in both we have

$$\int_{U_2} (--)|d^k\mathbf{u}_2|.$$

The second lines of both have

$$\int_{U_1} (--)\ |\det[\mathbf{D}\Phi(\mathbf{u}_1)]|.$$

(In Proposition 5.3.3 we used U rather than U_1 and V rather than U_2.)

Theorem 6.4.10 (Integral independent of orientation-preserving parametrizations). *Let $M \subset \mathbb{R}^n$ be a k-dimensional oriented manifold, let U_1, U_2 be open subsets of $\mathbb{R}^k$, and let $\gamma_1 : U_1 \to \mathbb{R}^n$ and $\gamma_2 : U_2 \to \mathbb{R}^n$ be two orientation-preserving parametrizations of M. Then for any k-form φ defined on a neighborhood of M,*

$$\int_{[\gamma_1(U_1)]} \varphi = \int_{[\gamma_2(U_2)]} \varphi. \qquad 6.4.23$$

If γ_1, γ_2 are both orientation reversing, the integrals are also equal; if one map preserves orientation and the other reverses it, they are opposite:

$$\int_{[\gamma_1(U_1)]} \varphi = -\int_{[\gamma_2(U_2)]} \varphi. \qquad 6.4.24$$

But if $\det[\mathbf{D}(\gamma_2^{-1} \circ \gamma_1)]$ is positive in some regions of U_1 and negative in others, then the integrals *are probably unrelated*, as we saw already in Example 6.2.5.

Proof of Theorem 6.4.10. We will prove this by trying to prove the (false) statement that the integrals are equal for any parametrizations (whether orientation preserving or reversing) and discovering where we go wrong.

Define the "change of parameters" map $\Phi = \gamma_2^{-1} \circ \gamma_1 : U_1^{\text{OK}} \to U_2^{\text{OK}}$. Then Definition 6.2.1 and the change of variables formula give

$$\begin{aligned} \int_{[\gamma_2(U_2)]} \varphi &= \int_{U_2} \varphi\Big(P_{\gamma_2(\mathbf{u}_2)}\big(\overrightarrow{D_1\gamma_2(\mathbf{u}_2)}, \dots, \overrightarrow{D_k\gamma_2(\mathbf{u}_2)}\big)\Big)|d^k\mathbf{u}_2| \\ &= \int_{U_1} \varphi\Big(P_{\gamma_2\circ\Phi(\mathbf{u}_1)}\big(\overrightarrow{D_1\gamma_2(\Phi(\mathbf{u}_1))}, \dots, \overrightarrow{D_k\gamma_2(\Phi(\mathbf{u}_1))}\big)\Big)\Big|\det[\mathbf{D}\Phi(\mathbf{u}_1)]\Big||d^k\mathbf{u}_1|. \end{aligned} \qquad 6.4.25$$

We want to express everything in terms of γ_1. There is no trouble with the point $(\gamma_2 \circ \Phi)(\mathbf{u}_1) = \gamma_1(\mathbf{u}_1)$ where the parallelogram is anchored, but the vectors that span it are more troublesome and require the following lemma.

Lemma 6.4.11. *If $\vec{\mathbf{w}}_1, \dots, \vec{\mathbf{w}}_k$ are any k vectors in $\mathbb{R}^k$, then*

$$\begin{aligned} &\varphi\Big(P_{\gamma_2(\mathbf{u}_2)}\big([\mathbf{D}\gamma_2(\mathbf{u}_2)]\vec{\mathbf{w}}_1, \dots, [\mathbf{D}\gamma_2(\mathbf{u}_2)]\vec{\mathbf{w}}_k\big)\Big) \\ &\quad = \varphi\Big(P_{\gamma_2(\mathbf{u}_2)}\big(\overrightarrow{D_1\gamma_2(\mathbf{u}_2)}, \dots, \overrightarrow{D_k\gamma_2(\mathbf{u}_2)}\big)\Big)\ \det[\vec{\mathbf{w}}_1, \dots, \vec{\mathbf{w}}_k]. \end{aligned} \qquad 6.4.26$$

Proof of Lemma 6.4.11. Think of the first line of equation 6.4.26 as a function F of $\vec{\mathbf{w}}_1, \dots, \vec{\mathbf{w}}_k$:

$$\underbrace{F(\vec{\mathbf{w}}_1, \dots, \vec{\mathbf{w}}_k)}_{a(\mathbf{u}_2)\det(\vec{\mathbf{w}}_1,\dots,\vec{\mathbf{w}}_k)} = \varphi\Big(P_{\gamma_2(\mathbf{u}_2)}\big([\mathbf{D}\gamma_2(\mathbf{u}_2)]\vec{\mathbf{w}}_1, \dots, [\mathbf{D}\gamma_2(\mathbf{u}_2)]\vec{\mathbf{w}}_k\big)\Big). \qquad 6.4.27$$

This function F is a multilinear, antisymmetric function of k vectors in $\mathbb{R}^k$, so it is a multiple of the determinant: $F = a(\mathbf{u}_2)\det$. By Theorem 6.1.8, the coefficient $a(\mathbf{u}_2)$ is found by evaluating F on the standard basis vectors:

Equation 6.4.28: The ith column of a matrix T is $T\vec{\mathbf{e}}_i$, so $[\mathbf{D}\gamma_2(\mathbf{u}_2)]\vec{\mathbf{e}}_i = \overrightarrow{D_i\gamma_2(\mathbf{u}_2)}$.

$$\begin{aligned} a(\mathbf{u}_2) = F(\vec{\mathbf{e}}_1, \dots, \vec{\mathbf{e}}_k) &= \varphi\Big(P_{\gamma_2(\mathbf{u}_2)}\big([\mathbf{D}\gamma_2(\mathbf{u}_2)]\vec{\mathbf{e}}_1, \dots, [\mathbf{D}\gamma_2(\mathbf{v})]\vec{\mathbf{e}}_k\big)\Big) \qquad 6.4.28\\ &= \varphi\Big(P_{\gamma_2(\mathbf{v})}\big(\overrightarrow{D_1\gamma_2(\mathbf{u}_2)}, \dots, \overrightarrow{D_k\gamma_2(\mathbf{u}_2)}\big)\Big). \qquad \square \end{aligned}$$

Equation 6.4.29: We use the chain rule to go from the second to the third line. In applying the chain rule, remember that

$$\gamma_1 = \gamma_2 \circ \gamma_2^{-1} \circ \gamma_1 = \gamma_2 \circ \Phi.$$

and that $[\mathbf{D}\gamma_2(\Phi(\mathbf{u}))](\overrightarrow{D_i\Phi(\mathbf{u})})$ is the ith column of the matrix

$$\overbrace{[\mathbf{D}(\gamma_2 \circ \Phi)(\mathbf{u})]}^{[\mathbf{D}\gamma_1(\mathbf{u})]} = [\mathbf{D}\gamma_2(\Phi(\mathbf{u}))][\mathbf{D}\Phi(\mathbf{u})]$$

since the ith column of AB is $A\vec{\mathbf{b}}_i$.

Now we write the function being integrated on the second line of equation 6.4.25, but take $\det[\mathbf{D}\Phi(\mathbf{u}_1)]$ out of absolute value signs. We use Lemma 6.4.11 ("backwards") to go from line 1 to line 2 in equation 6.4.29; $\mathbf{u}_2$ in that lemma corresponds to $\Phi(\mathbf{u}_1)$; the determinant $\det[\vec{\mathbf{w}}_1, \dots, \vec{\mathbf{w}}_k]$ corresponds to $\det[\mathbf{D}\Phi(\mathbf{u}_1)] = \det[D_1\Phi(\mathbf{u}_1), \dots, D_k\Phi(\mathbf{u}_1)]$:

$$\begin{aligned} &\varphi\Big(P_{\gamma_2\circ\Phi(\mathbf{u}_1)}\big(\overrightarrow{D_1\gamma_2(\Phi(\mathbf{u}_1))}, \dots, \overrightarrow{D_k\gamma_2(\Phi(\mathbf{u}_1))}\big)\Big)\det \overbrace{[\mathbf{D}\Phi(\mathbf{u}_1)]}^{\overrightarrow{D_1\Phi(\mathbf{u}_1)},\dots,\overrightarrow{D_k\Phi(\mathbf{u}_1)}}\\ &= \varphi\Big(P_{\gamma_2\circ\Phi(\mathbf{u}_1)}\big([\mathbf{D}\gamma_2(\Phi(\mathbf{u}_1))](\overrightarrow{D_1\Phi(\mathbf{u}_1)}), \dots, [\mathbf{D}\gamma_2(\Phi(\mathbf{u}_1))](\overrightarrow{D_k\Phi(\mathbf{u}_1)})\big)\Big)\\ &= \varphi\Big(P_{\gamma_1(\mathbf{u}_1)}\big(\overrightarrow{D_1\gamma_1(\mathbf{u}_1)}, \dots, \overrightarrow{D_k\gamma_1(\mathbf{u}_1)}\big)\Big). \qquad 6.4.29 \end{aligned}$$

The key point is that going from the first to the second line of equation 6.4.25 introduces $|\det[\mathbf{D}\Phi(\mathbf{u}_1)]|$, while going from the last line to the first line of equation 6.4.29 introduces $\det[\mathbf{D}\Phi(\mathbf{u}_1)]$, with no absolute value signs. Except for the absolute values, the integrands are the same. Therefore, the integral we get using γ_1 and the integral we get using γ_2 will be the same if $\det[\mathbf{D}\Phi] = \det[\mathbf{D}(\gamma_2^{-1} \circ \gamma_1)] > 0$ for all $\mathbf{u}_1 \in U^{\text{OK}}$, which (by Proposition 6.4.8) will be true if both parametrizations are orientation preserving. $\square$

Now we can define the integral of a form field over an oriented manifold.

Definition 6.4.12 (Integral of a form field over an oriented manifold). Let $M \subset \mathbb{R}^n$ be a k-dimensional oriented manifold, φ a k-form field on a neighborhood of M, and $\gamma : U \to M$ any orientation-preserving parametrization of M. Then

$$\int_M \varphi \stackrel{\text{def}}{=} \int_{[\gamma(U)]} \varphi = \int_U \varphi\Big(P_{\gamma(\mathbf{u})}\big(\overrightarrow{D_1\gamma(\mathbf{u})}, \dots, \overrightarrow{D_k\gamma(\mathbf{u})}\big)\Big)|d^k\mathbf{u}|. \qquad 6.4.30$$

Example 6.4.13 (Integrating a 2-form over an oriented surface). What is the integral of the 2-form $\omega = y\,dy \wedge dz + x\,dx \wedge dz + z\,dx \wedge dy$

Example 6.4.13: This is an example of the first class of parametrizations listed in Section 5.2, parametrizations as graphs; see formula 5.2.6.

through the piece P of the plane defined by $x+y+z=1$ where $x, y, z \geq 0$, and oriented by $\vec{\mathbf{n}} = \begin{bmatrix} 1 \\ 1 \\ 1 \end{bmatrix}$? This surface is the graph of $z = 1 - x - y$, so

$$\gamma \begin{pmatrix} x \\ y \end{pmatrix} = \begin{pmatrix} x \\ y \\ 1-x-y \end{pmatrix} \qquad 6.4.31$$

is a parametrization, if x and y are in the triangle $T \subset \mathbb{R}^2$ given by $x, y \geq 0$, $x+y \leq 1$. Moreover, this parametrization preserves orientation by $\vec{\mathbf{n}}$, since

$$\overrightarrow{D_1\gamma} \times \overrightarrow{D_2\gamma} = \begin{bmatrix} 1 \\ 0 \\ -1 \end{bmatrix} \times \begin{bmatrix} 0 \\ 1 \\ -1 \end{bmatrix} = \begin{bmatrix} 1 \\ 1 \\ 1 \end{bmatrix}. \qquad 6.4.32$$

(As in Example 6.4.7, we use the cross product to check that γ preserves orientation; see also inequality 6.4.7.)

Using Definition 6.4.12, we see that the integral is

$$\begin{aligned}
\int_P \omega &= \int_T \Big(y\,dy \wedge dz + x\,dx \wedge dz + z\,dx \wedge dy\Big)\Bigg(P_{\begin{pmatrix} x \\ y \\ 1-x-y \end{pmatrix}}\Big(\underbrace{\begin{bmatrix} 1 \\ 0 \\ -1 \end{bmatrix}}_{\overrightarrow{D_1\gamma}}, \underbrace{\begin{bmatrix} 0 \\ 1 \\ -1 \end{bmatrix}}_{\overrightarrow{D_2\gamma}}\Big)\Bigg)\,|dx\,dy| \\
&= \int_T \left(y \det \begin{bmatrix} 0 & 1 \\ -1 & -1 \end{bmatrix} + x \det \begin{bmatrix} 1 & 0 \\ -1 & -1 \end{bmatrix} + (1-x-y) \det \begin{bmatrix} 1 & 0 \\ 0 & 1 \end{bmatrix}\right) |dx\,dy| \\
&= \int_T \Big(y - x + 1 - x - y\Big)\,|dx\,dy| \\
&= \int_T (1-2x)\,|dx\,dy| = \int_0^1 \left(\int_0^{1-y} (1-2x)\,dx\right) dy \qquad 6.4.33 \\
&= \int_0^1 \Big[x - x^2\Big]_0^{1-y}\,dy = \int_0^1 (y - y^2)\,dy = \left[\frac{y^2}{2} - \frac{y^3}{3}\right]_0^1 = \frac{1}{6}.
\end{aligned}$$

Note that the formula for integrating a 2-form over a surface in $\mathbb{R}^3$ enables us to transform an integral over a surface in $\mathbb{R}^3$ into an integral over a piece of $\mathbb{R}^2$, as studied in Chapter 4. △

Example 6.4.14 (Integrating a 2-form field over a parametrized surface in $\mathbb{C}^3$). Consider again the oriented surface $S \subset \mathbb{C}^3$ which we saw in Example 6.4.4: the surface parametrized by $z \mapsto \begin{pmatrix} z \\ z^2 \\ z^3 \end{pmatrix}$, $|z| < 1$, and oriented as in that example. What is

$$\int_S dx_1 \wedge dy_1 + dx_2 \wedge dy_2 + dx_3 \wedge dy_3\,? \qquad 6.4.34$$

Example 6.4.14: To parametrize S using polar coordinates (equation 6.4.35) we use

$$z = r\cos\theta + ir\sin\theta$$

(equation 0.7.10); the first two entries correspond to the real and imaginary parts of z, the third and fourth to squaring the real and imaginary parts, and so on. Remember De Moivre's formula

$$z^n = r^n(\cos n\theta + i\sin n\theta).$$

This parametrization preserves orientation as long as $r > 0$, since

$$dx_1 \wedge dy_1\left(\overrightarrow{D_1\delta}, \overrightarrow{D_2\delta}\right) = \det\begin{bmatrix}\cos\theta & -r\sin\theta\\ \sin\theta & r\cos\theta\end{bmatrix} = r.$$

(We know from equation 6.4.8 that $dx_1 \wedge dy_1(D_1\gamma, D_2\gamma) = 1$, so a basis of the tangent space will be direct if $dx_1 \wedge dy_1$ evaluated on the basis vectors returns a positive number.)

In equations 6.4.39, both integrals are over the *same* oriented point, $\mathbf{x} = +2$. We use curly brackets to avoid confusion between integrating over the point $+2$ with negative orientation, and integrating over the point -2.

We can parametrize the surface using polar coordinates:[5]

$$\delta : \begin{pmatrix} r\\ \theta\end{pmatrix} \mapsto \begin{pmatrix} r\cos\theta\\ r\sin\theta\\ r^2\cos 2\theta\\ r^2\sin 2\theta\\ r^3\cos 3\theta\\ r^3\sin 3\theta\end{pmatrix}, \quad \text{for } 0 \le \theta \le 2\pi,\ 0 \le r < 1. \tag{6.4.35}$$

(Do you see why we set $r < 1$?[6]) Then

$$(dx_1\wedge dy_1 + dx_2\wedge dy_2 + dx_3\wedge dy_3)\left(P_{\delta\begin{pmatrix} r\\ \theta\end{pmatrix}}\left(\overrightarrow{D_1\delta}\begin{pmatrix} r\\ \theta\end{pmatrix}, \overrightarrow{D_2\delta}\begin{pmatrix} r\\ \theta\end{pmatrix}\right)\right) \tag{6.4.36}$$

$$= \det\begin{bmatrix}\cos\theta & -r\sin\theta\\ \sin\theta & r\cos\theta\end{bmatrix} + \det\begin{bmatrix}2r\cos 2\theta & -2r^2\sin 2\theta\\ 2r\sin 2\theta & 2r^2\cos 2\theta\end{bmatrix} + \det\begin{bmatrix}3r^2\cos 3\theta & -3r^3\sin 3\theta\\ 3r^2\sin 3\theta & 3r^3\cos 3\theta\end{bmatrix}$$

$$= r + 4r^3 + 9r^5.$$

So we find the following for our integral:

$$2\pi\int_0^1 (r + 4r^3 + 9r^5)\,dr = 6\pi. \quad \triangle \tag{6.4.37}$$

Example 6.4.15 (Integrating a 0-form over an oriented point). Let $\mathbf{x}$ be an oriented point and f a function (i.e., a 0-form field) defined in some neighborhood of $\mathbf{x}$. Then

$$\int_{+\mathbf{x}} f = +f(\mathbf{x}) \quad \text{and} \quad \int_{-\mathbf{x}} f = -f(\mathbf{x}). \tag{6.4.38}$$

Thus

$$\int_{+\{+2\}} x^2 = 4 \quad \text{and} \quad \int_{-\{+2\}} x^2 = -4. \quad \triangle \tag{6.4.39}$$

Exercises for Section 6.4

6.4.1 If the cone M of equation $f\begin{pmatrix} x\\ y\\ z\end{pmatrix} = x^2 + y^2 - z^2 = 0$ (Example 5.2.4) is oriented by $\vec{\nabla} f$, does the parametrization $\gamma : \begin{pmatrix} r\\ \theta\end{pmatrix} \mapsto \begin{pmatrix} r\cos\theta\\ r\sin\theta\\ r\end{pmatrix}$ preserve orientation?

6.4.2 Confirm (Example 6.4.9) that the subset $X \subset \text{Mat}\,(2,3)$ made of all 2×3 matrices of rank 1 is a manifold of dimension 4 in $\text{Mat}\,(2,3)$.

[5] It would be considerably harder to compute the integral using the parametrization of Example 6.4.4.

[6] Recall that $r = |z|$, which was stipulated in Example 6.4.4 to be less than 1.

6.4.3 In Example 6.4.5 we saw that the map γ (equation 6.4.9) parametrizing the unit sphere oriented by the outward normal was neither orientation preserving nor orientation reversing. Can you make that mapping either orientation preserving for the entire sphere, or orientation reversing for the entire sphere, by choosing a different range of values of θ and φ?

Exercises 6.4.4, 6.4.5, 6.4.6: Recall (discussion after Definition 6.3.3) that a k-dimensional manifold M can oriented by choosing a k-form ω and defining

$$\Omega = \operatorname{sgn} \omega,$$

so long as $\omega(P_{\mathbf{x}}(\vec{\mathbf{v}}_1, \dots, \vec{\mathbf{v}}_k) \neq 0$ for all $\mathbf{x} \in M$ and for all bases $\vec{\mathbf{v}}_1, \dots, \vec{\mathbf{v}}_k$ of $T_{\mathbf{x}}M$.

6.4.4 What is the integral $\int_S x_3\, dx_1 \wedge dx_2 \wedge dx_4$, where S is the part of the three-dimensional manifold of equation

$$x_4 = x_1 x_2 x_3 \quad \text{where } 0 \le x_1, x_2, x_3 \le 1,$$

oriented by $\Omega = \operatorname{sgn} dx_1 \wedge dx_2 \wedge dx_3$? *Hint*: This surface is a graph, so it is easy to parametrize.

6.4.5 Let $z_1 = x_1 + iy_1$, $z_2 = x_2 + iy_2$ be coordinates in $\mathbb{C}^2$. Compute the integral of $dx_1 \wedge dy_1 + dy_1 \wedge dx_2$ over the part of the locus of equation $z_2 = z_1^k$ where $|z_1| < 1$, oriented by $\Omega = \operatorname{sgn} dx_1 \wedge dy_1$.

6.4.6 Let $z_1 = x_1 + iy_1$, $z_2 = x_2 + iy_2$ be coordinates in $\mathbb{C}^2 = \mathbb{R}^4$. Integrate the 2-form $dx_1 \wedge dy_1 + dx_2 \wedge dy_2$ over the part of the surface X of equation $z_2 = e^{z_1}$ where $|\operatorname{Re} z_1| \le a$, $|\operatorname{Im} z_1| \le b$, oriented by $\Omega = \operatorname{sgn} dx_1 \wedge dy_1$. *Hint*: Use Euler's formula (equation 1.5.55).

Part b of Exercise 6.4.7: Use the same method as in Example 6.4.9; this time you can find an orientation of $M_1(3,3)$ such that parametrizations φ_1, φ_2, and φ_3 of parts of $M_1(3,3)$ are all orientation preserving.

6.4.7 Let $M_1(n,m)$ be the space of $n \times m$ matrices of rank 1.

a. Show that $M_1(2,2)$ is orientable. *Hint*: $M_1(2,2)$ is three dimensional in $\mathbb{R}^4$, so it can be oriented by choosing a normal vector field.

*b. Show that $M_1(3,3)$ is orientable.

6.4.8 Prove Proposition 6.4.6.

6.5 Forms in the language of vector calculus

FIGURE 6.5.1.

One of the foremost American scientists of the nineteenth century, the theoretical physicist and chemist Josiah Willard Gibbs was also one of the founders of vector calculus, producing printed notes for the use of his own students in 1881 and 1884.

The real difficulty with forms is acquiring a mental image of what they are. What "is" $dx_1 \wedge dx_2 + dx_3 \wedge dx_4$? We saw in Section 6.1 that it is the function that takes two vectors in $\mathbb{R}^4$, projects them first onto the (x_1, x_2)-plane and takes the signed area of the resulting parallelogram, then projects them onto the (x_3, x_4)-plane, takes the signed area of that parallelogram, and finally adds the two signed areas. But that description is convoluted, and although it isn't hard to use it in computations, it hardly expresses understanding.

Acquiring an intuitive understanding of what sort of information a k-form encodes really is difficult. In some sense, a first course in electromagnetism is largely a matter of understanding what sort of beast the electromagnetic field is, namely a 2-form field on $\mathbb{R}^4$. However, in $\mathbb{R}^3$, it really is possible to visualize all forms and form fields, because they can be described in terms of functions and vector fields. Each of the four kinds of forms on $\mathbb{R}^3$ (0-forms, 1-forms, 2-forms, and 3-forms) has its own personality.

0-forms and 0-form fields

In any $\mathbb{R}^n$, a constant 0-form is simply a number, and a 0-form field is simply a function. If $U \subset \mathbb{R}^n$ is open and $f : U \to \mathbb{R}$ is a function, then the rule $f(P_{\mathbf{x}}) = f(\mathbf{x})$ makes f into a 0-form field.

1-forms and the work of a vector field

Every 1-form is the *work form* of a vector field.

Definition 6.5.1 (Work form). The *work form* $W_{\vec{F}}$ of a vector field $\vec{F} = \begin{bmatrix} F_1 \\ \vdots \\ F_n \end{bmatrix}$ is the 1-form field defined by

$$W_{\vec{F}}\big(P_{\mathbf{x}}(\vec{\mathbf{v}})\big) \overset{\text{def}}{=} \vec{F}(\mathbf{x}) \cdot \vec{\mathbf{v}}. \qquad 6.5.1$$

In coordinates, this gives $W_{\vec{F}} = F_1 dx_1 + \cdots + F_n dx_n$. For example, the 1-form $y\,dx - x\,dy$ is the work form of $\vec{F} = \begin{bmatrix} y \\ -x \end{bmatrix}$:

$$W_{\vec{F}}\big(P_{\mathbf{x}}(\vec{\mathbf{v}})\big) = \begin{bmatrix} y \\ -x \end{bmatrix} \cdot \begin{bmatrix} v_1 \\ v_2 \end{bmatrix} = yv_1 - xv_2 = (y\,dx - x\,dy)\big(P_{\mathbf{x}}(\vec{\mathbf{v}})\big). \qquad 6.5.2$$

We already saw the constant 1-form $W_{\vec{\mathbf{v}}}$ in Example 6.1.9. We can now think of that 1-form as the work form $W_{\vec{F}}$ of the constant vector field $\vec{F}(\mathbf{x}) = \vec{\mathbf{v}}$.

What have we gained by relating 1-forms to vector fields? Mainly that it is easy to visualize $W_{\vec{F}}$ and to understand what it measures. If $\vec{F}$ is a force field, its work form associates to a little line segment the work that the force field does along the line segment. Recall (Corollary 1.4.4) that the dot product $\vec{\mathbf{x}} \cdot \vec{\mathbf{y}}$ is the product of $|\vec{\mathbf{x}}|$ and the signed length of the projection of $\vec{\mathbf{y}}$ onto the line spanned by $\vec{\mathbf{x}}$, positive if the projection points in the direction of $\vec{\mathbf{x}}$, negative if it points in the opposite direction.

Thus (see Figure 6.5.2) the 1-form $x\,dx + y\,dy$ associated to the radial vector field $\begin{bmatrix} x \\ y \end{bmatrix}$ does very little work over the path consisting of little line segments approximating the unit circle: evaluated on each little line segment it will return a number close to 0. It does a lot of work over the path from the origin to the point $\begin{pmatrix} 10 \\ 10 \end{pmatrix}$; over each little line segment making up that path, it will return a positive number. It does much less work over the path of the same length from $\begin{pmatrix} 0 \\ -5 \end{pmatrix}$ to $\begin{pmatrix} 10 \\ 5 \end{pmatrix}$; over the little segments in the fourth quadrant, the work will be small.

For a simple example from physics, consider the vector field $\vec{F}$ representing the force field of gravity. This is the constant vector field $\begin{bmatrix} 0 \\ 0 \\ -gm \end{bmatrix}$ associated to the work form $-gm\,dz$, where g is the Earth's gravitational

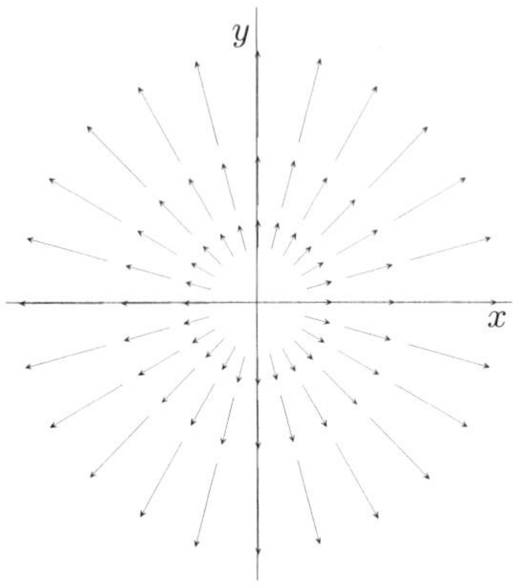

FIGURE 6.5.2.

The 1-form field $x\,dx + y\,dy$ is the work form field of the radial vector field

$$\vec{F}\begin{pmatrix} x \\ y \end{pmatrix} = \begin{bmatrix} x \\ y \end{bmatrix}$$

shown above. Thus $x\,dx + y\,dy$ measures the "work" of this vector field on a little vector $\vec{\mathbf{v}}$ anchored at $\begin{pmatrix} x \\ y \end{pmatrix}$. If $\vec{\mathbf{v}}$ is aligned with the vector field and points in the same direction, the work will be large. If it is aligned with the vector field and points in the opposite direction, the work will be large but negative; if it is close to orthogonal to the vector field, the work will be very small.

For this image to be meaningful, you should picture $\vec{\mathbf{v}}$ as short; the work of $\vec{F}$ on a vector is measured *at the point at which the vector is anchored*. The work of $\vec{F}$ on $\begin{bmatrix} 10 \\ 10 \end{bmatrix}$ (i.e., in the direction of that vector) at $\begin{pmatrix} 1 \\ -1 \end{pmatrix}$ is 0, although if you drew $\begin{bmatrix} 10 \\ 10 \end{bmatrix}$ it would spill over onto the first quadrant, where the work would be positive.

constant and m is mass. In the absence of friction, it requires no work to push a wagon of mass m horizontally from $\mathbf{a}$ to $\mathbf{b}$; the vector $\overrightarrow{\mathbf{b}-\mathbf{a}}$ and the constant vector field representing gravity are orthogonal to each other:

$$-gm\,dz\left(\begin{bmatrix} b_1-a_1\\ b_2-a_2\\ 0\end{bmatrix}\right)=\begin{bmatrix}0\\0\\-gm\end{bmatrix}\cdot\begin{bmatrix} b_1-a_1\\ b_2-a_2\\ 0\end{bmatrix}=0. \qquad 6.5.3$$

In equation 6.5.3, g represents the acceleration of gravity at the surface of the earth, and m is mass; $-gm$ is weight, a force; it is negative because it points down.

But if the wagon rolls down an inclined plane from $\mathbf{a}$ to $\mathbf{b}$, the force field of gravity does "work" on the wagon equal to the dot product of gravity and the displacement vector of the wagon:

$$-gm\,dz\left(\begin{bmatrix} b_1-a_1\\ b_2-a_2\\ b_3-a_3\end{bmatrix}\right)=\begin{bmatrix}0\\0\\-gm\end{bmatrix}\cdot\begin{bmatrix} b_1-a_1\\ b_2-a_2\\ b_3-a_3\end{bmatrix}=-gm(b_3-a_3), \qquad 6.5.4$$

which is positive, since b_3-a_3 is negative. If you want to push the wagon back up the inclined plane, you will need to furnish the work, and the force field of gravity will do negative work.

Remark. A force field like the gravitation field has "dimensions" energy per length. It's the "per length" that tells us that the integrand to associate to a force field should be something to be integrated over curves. Since direction makes a difference – it takes work to push a wagon uphill but the wagon rolls down by itself – the appropriate integrand is a 1-form field, which is integrated over *oriented* curves. $\triangle$

Physicists use the word "dimension" to describe what is being measured, as opposed to the specific units used. A force field might be measured in joules/meter or in ergs/centimeter; both correspond to "dimensions" energy/length.

2-forms and the flux of a vector field

If $\vec{F}$ is a vector field on an open subset $U\subset\mathbb{R}^3$, then we can associate to it a 2-form field on U called its *flux form* $\Phi_{\vec{F}}$.

Definition 6.5.2 (Flux form). The *flux form* $\Phi_{\vec{F}}$ is the 2-form field

$$\Phi_{\vec{F}}\big(P_{\mathbf{x}}(\vec{\mathbf{v}},\vec{\mathbf{w}})\big)\stackrel{\text{def}}{=}\det[\vec{F}(\mathbf{x}),\ \vec{\mathbf{v}},\ \vec{\mathbf{w}}]. \qquad 6.5.5$$

In coordinates, this is $\Phi_{\vec{F}}=F_1\,dy\wedge dz-F_2\,dx\wedge dz+F_3\,dx\wedge dy$:

$$(F_1\,dy\wedge dz-F_2\,dx\wedge dz+F_3\,dx\wedge dy)\;P_{\mathbf{x}}\left(\begin{bmatrix}v_1\\v_2\\v_3\end{bmatrix},\begin{bmatrix}w_1\\w_2\\w_3\end{bmatrix}\right) \qquad 6.5.6$$
$$=F_1(\mathbf{x})(v_2w_3-v_3w_2)-F_2(\mathbf{x})(v_1w_3-v_3w_1)+F_3(\mathbf{x})(v_1w_2-v_2w_1)$$
$$=\det[\vec{F}(\mathbf{x}),\ \vec{\mathbf{v}},\ \vec{\mathbf{w}}].$$

Equation 6.5.6: It may be easier to remember the coordinate definition of $\Phi_{\vec{F}}$ if it is written

$$\Phi_{\vec{F}}=$$
$$F_1dy\wedge dz+F_2dz\wedge dx+F_3dx\wedge dy$$

(changing the order and sign of the middle term). Then (setting $x=1, y=2, z=3$) you can think that the first term goes (1,2,3), the second (2,3,1), and the third (3,1,2). Thus $\Phi_{\begin{pmatrix}x\\y\\z\end{pmatrix}}$ is the 2-form

$$x\,dy\wedge dz+y\,dz\wedge dx+z\,dx\wedge dy.$$

In this form, it is clear, again from Theorem 6.1.8, that every 2-form field on $\mathbb{R}^3$ is the flux form of a vector field; it is just a question of using the coefficients of the elementary 2-forms to make a vector field.

Once more, what we have gained is an ability to visualize. The flux form of a vector field associates to a parallelogram the flow of the vector field through the parallelogram, since

$$\det[\vec{F}(\mathbf{x}),\ \vec{\mathbf{v}},\ \vec{\mathbf{w}}]=\vec{F}(\mathbf{x})\cdot(\vec{\mathbf{v}}\times\vec{\mathbf{w}}) \qquad 6.5.7$$

gives the signed volume of the 3-parallelogram spanned by the three vectors. Thus if $\vec{F}$ is the velocity vector field of a fluid, the integral of its flux form over a surface measures the amount of fluid flowing through the surface in unit time, as suggested by Figure 6.5.3.

Indeed, the fluid that flows through the parallelogram $P_{\mathbf{x}}(\vec{\mathbf{v}}, \vec{\mathbf{w}})$ in unit time will fill the parallelepiped $P_{\mathbf{x}}(\vec{F}(\mathbf{x}), \vec{\mathbf{v}}, \vec{\mathbf{w}})$: the particle which at time 0 was at the corner $\mathbf{x}$ is now at $\mathbf{x} + \vec{F}(\mathbf{x})$. The sign of the flux is positive

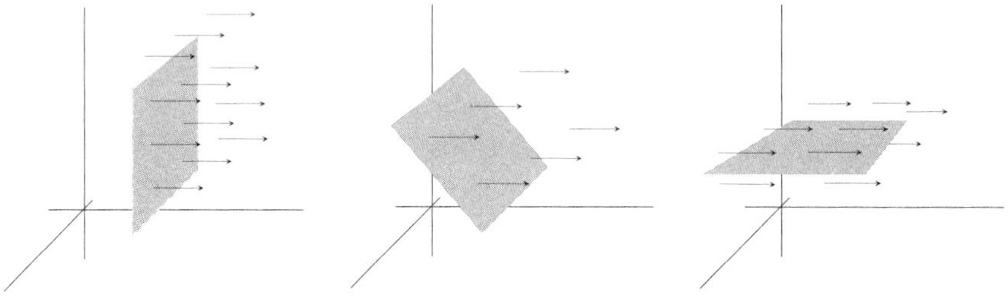

FIGURE 6.5.3. The flow of $\vec{F}$ through a surface depends on the angle between $\vec{F}$ and the surface. LEFT: $\vec{F}$ is orthogonal to the surface, providing maximum flow. MIDDLE: $\vec{F}$ is not orthogonal to the surface, allowing less flow. RIGHT: $\vec{F}$ is parallel to the surface; the flow through the surface is 0.

if $\vec{F}$ is on the same side of the parallelogram as $\vec{\mathbf{v}} \times \vec{\mathbf{w}}$, and negative if it is on the other side.

Flow is maximum when $\vec{F}$ is orthogonal to the surface: the volume of the parallelepiped spanned by $\vec{F}, \vec{\mathbf{v}}, \vec{\mathbf{w}}$ is greatest when the angle θ between the vectors $\vec{F}$ and $\vec{\mathbf{v}} \times \vec{\mathbf{w}}$ is 0, since

$$\vec{\mathbf{x}} \cdot \vec{\mathbf{y}} = |\vec{\mathbf{x}}||\vec{\mathbf{y}}| \cos \theta. \qquad 6.5.8$$

Note that when we identify k-forms on $\mathbb{R}^3$ to work forms, flux forms, and mass forms, we use the geometry of $\mathbb{R}^3$: the dot product for the work form, and the determinant for the flux form and mass form. Defining k-forms on any $\mathbb{R}^n$ requires no geometry.

Nothing flows through the surface when $\vec{F}$ is parallel to the surface. This corresponds to $\vec{F} \cdot (\vec{\mathbf{v}} \times \vec{\mathbf{w}}) = 0$ ($\vec{F}$ is perpendicular to $\vec{\mathbf{v}} \times \vec{\mathbf{w}}$ and therefore parallel to the parallelogram spanned by $\vec{\mathbf{v}}$ and $\vec{\mathbf{w}}$). In this case the parallelepiped spanned by $\vec{F}, \vec{\mathbf{v}}, \vec{\mathbf{w}}$ is flat.

Remark. If a vector field represents the flow of a fluid, what dimensions will it have? The vector field measures how much fluid flows in unit time through a unit surface perpendicular to the direction of flow, so the dimensions should be mass/(length^2). The length squared in the denominator tips us off that the appropriate integrand to associate to this vector field is a 2-form, or at least an integrand to be integrated over a surface.[7] △

Example 6.5.3 (Flux form of vector field). The 2-form

$$\Phi = y\, dy \wedge dz + x\, dx \wedge dz - z\, dx \wedge dy \qquad 6.5.9$$

[7]You might go one step further and say it is a 3-form on spacetime: the result of integrating it over a surface in space and an interval in time is the total mass flowing through that region of spacetime. In general, any $(n-1)$-form field in $\mathbb{R}^n$ can be considered a flux form field.

is the flux form of the vector field $\vec{F}\begin{pmatrix} x \\ y \\ z \end{pmatrix} = \begin{bmatrix} y \\ -x \\ -z \end{bmatrix}$ shown in Figure 6.5.4. In the (x, y)-plane, this vector field is simply rotation clockwise around the origin. As you would expect, Φ returns 0 when evaluated on any parallelogram lying in the (x, y)-plane. For example,

$$y\,dy\wedge dz + x\,dx\wedge dz - z\,dx\wedge dy\left(P_{\begin{pmatrix} 1 \\ 1 \\ 0 \end{pmatrix}}\left(\begin{bmatrix} 1 \\ 0 \\ 0 \end{bmatrix}, \begin{bmatrix} 0 \\ 1 \\ 0 \end{bmatrix}\right)\right) = 1\cdot 0 + 1\cdot 0 - 0\cdot 1 = 0.$$

But Φ evaluated on the same parallelogram anchored at $\mathbf{e}_3$ gives -1. Indeed, you would expect a negative number: Figure 6.5.4 shows that the z-component of the vectors is negative where $z > 0$ and positive where $z < 0$. $\triangle$

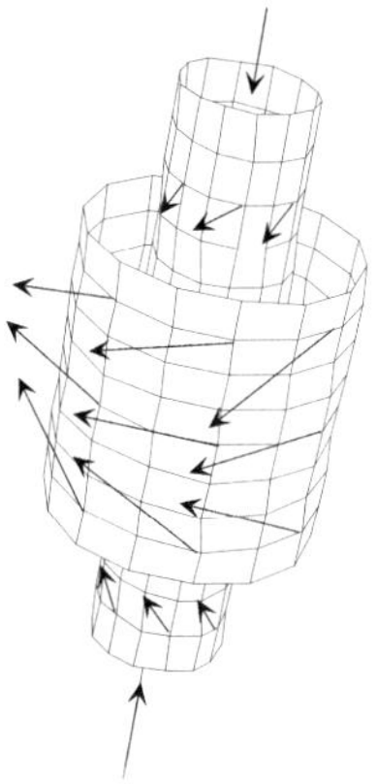

FIGURE 6.5.4.
The 2-form

$$y\,dy \wedge dz + x\,dx \wedge dz - z\,dx \wedge dy$$

is the flux form of the vector field

$$\vec{F}\begin{pmatrix} x \\ y \\ z \end{pmatrix} = \begin{bmatrix} y \\ -x \\ -z \end{bmatrix}.$$

Vectors are tangent to concentric cylinders around the z-axis; the z-component of the vectors is negative where $z > 0$ and positive where $z < 0$.

3-forms and the mass form of a function

Any 3-form on an open subset of $\mathbb{R}^3$ is the 3-form $dx \wedge dy \wedge dz$ multiplied by a function: we call this 3-form the *mass form of f*, and denote it by M_f: $M_f = f\,dx \wedge dy \wedge dz$. Note that the constant 3-form $dx \wedge dy \wedge dz$ is another name for the determinant:

$$dx \wedge dy \wedge dz(\vec{\mathbf{v}}_1, \vec{\mathbf{v}}_2, \vec{\mathbf{v}}_3) = \det[\vec{\mathbf{v}}_1, \vec{\mathbf{v}}_2, \vec{\mathbf{v}}_3].$$

Definition 6.5.4 (Mass form of a function). Let U be a subset of $\mathbb{R}^3$ and $f : U \to \mathbb{R}$ a function. The *mass form* M_f is the 3-form defined by

$$\underbrace{M_f}_{\substack{\text{mass}\\ \text{form}\\ \text{of } f}}\left(P_{\mathbf{x}}(\vec{\mathbf{v}}_1, \vec{\mathbf{v}}_2, \vec{\mathbf{v}}_3)\right) \stackrel{\text{def}}{=} f(\mathbf{x})\,\underbrace{\det\left[\vec{\mathbf{v}}_1, \vec{\mathbf{v}}_2, \vec{\mathbf{v}}_3\right]}_{\text{signed volume of } P}. \qquad 6.5.10$$

The mass form of a function is the natural thing to consider if the dimensions of the function are *something/cubic length*, such as ordinary density (which might be measured using units kg/m^3) or charge density (which might be measured using units coulombs/m^3).

Summary: work, flux, and mass forms on $\mathbb{R}^3$

Let f be a function on $\mathbb{R}^3$ and $\vec{F} = \begin{bmatrix} F_1 \\ F_2 \\ F_3 \end{bmatrix}$ be a vector field. Then

$$W_{\vec{F}} = F_1\,dx + F_2\,dy + F_3\,dz \qquad 6.5.11$$

$$\Phi_{\vec{F}} = F_1\,dy \wedge dz - F_2\,dx \wedge dz + F_3\,dx \wedge dy \qquad 6.5.12$$

$$M_f = f\,dx \wedge dy \wedge dz. \qquad 6.5.13$$

Integrating work, flux, and mass forms over oriented manifolds

Now let us translate Definition 6.4.12 (integrating a k-form field over an oriented manifold) into the language of vector calculus.

- **Integrating work forms.**

Let C be an oriented curve and $\gamma : [a,b] \to C$ an orientation-preserving parametrization. Then Definition 6.4.12 gives

$$\int_C W_{\vec{F}} = \int_{[a,b]} W_{\vec{F}}\Big(P_{\gamma(u)}\big(\underbrace{\overrightarrow{D_1\gamma(u)}}_{\vec{\gamma}'(u)}\big)\Big)|du| = \int_a^b \vec{F}\big(\gamma(u)\big)\cdot\vec{\gamma}'(u)\,du. \qquad 6.5.14$$

Definition 6.5.5 (Work). The *work* of a vector field $\vec{F}$ along an oriented curve C is $\displaystyle\int_C W_{\vec{F}}$.

Equation 6.5.15: Recall from equation 6.3.4 that a curve C can be oriented by

$$\operatorname{sgn}(\vec{\mathbf{t}}(\mathbf{x})\cdot\vec{\mathbf{v}}),$$

where $\vec{\mathbf{t}}$ is a nonvanishing tangent vector field that varies continuously with $\mathbf{x}$ and $\vec{\mathbf{v}}$ is a basis of $T_{\mathbf{x}}C$. By Definition 6.4.2, a map γ parametrizing the curve preserves this orientation if $\Omega(\vec{\gamma}') = +1$, i.e.,

$$\operatorname{sgn}(\vec{\mathbf{t}}(t)\cdot\vec{\gamma}'(t)) = +1.$$

Example 6.5.6 (Integrating a work form over a helix). What is the work of $\vec{F}\begin{pmatrix}x\\y\\z\end{pmatrix} = \begin{bmatrix}y\\-x\\0\end{bmatrix}$ over the helix oriented by the tangent vector field $\vec{\mathbf{t}} = \begin{bmatrix}-\sin t\\ \cos t\\ 1\end{bmatrix}$, and parametrized by $\gamma(t) = \begin{pmatrix}\cos t\\ \sin t\\ t\end{pmatrix}$, for $0 < t < 4\pi$?

The parametrization preserves orientation: since

$$\underbrace{\begin{bmatrix}-\sin t\\ \cos t\\ 1\end{bmatrix}}_{\vec{\mathbf{t}}(t)}\cdot\underbrace{\begin{bmatrix}-\sin t\\ \cos t\\ 1\end{bmatrix}}_{\vec{\gamma}'(t)} = 2, \quad \text{we have } \Omega(\vec{\gamma}') = +1. \qquad 6.5.15$$

So by equation 6.5.14 the work of the vector field $\vec{F}$ over the helix is

Equation 6.5.16: Recall that

$$W_{\vec{F}}\big(P_{\mathbf{x}}(\vec{\mathbf{v}})\big) = \vec{F}(\mathbf{x})\cdot\vec{\mathbf{v}}.$$

$$\int_0^{4\pi} \overbrace{\begin{bmatrix}\sin t\\ -\cos t\\ 0\end{bmatrix}}^{\vec{F}(\gamma(t))}\cdot\overbrace{\begin{bmatrix}-\sin t\\ \cos t\\ 1\end{bmatrix}}^{\gamma'(t)}\,dt = \int_0^{4\pi}(-\sin^2 t - \cos^2 t)\,dt = -4\pi. \quad \triangle \qquad 6.5.16$$

The terms "work" and "flux" in Definitions 6.5.5 and 6.5.7 are standard terms in physics.

- **Integrating flux forms.**

Let S be an oriented surface and let $\gamma : U \to S$ be an orientation-preserving parametrization. Then Definition 6.4.12 gives

Equation 6.5.17: Recall that

$$\Phi_{\vec{F}}\big(P_{\mathbf{x}}(\vec{\mathbf{v}},\vec{\mathbf{w}})\big) = \det[\vec{F}(\mathbf{x}),\vec{\mathbf{v}},\vec{\mathbf{w}}].$$

$$\begin{aligned}\int_S \Phi_{\vec{F}} &= \int_U \Phi_{\vec{F}}\Big(P_{\gamma(\mathbf{u})}\big(\overrightarrow{D_1\gamma(\mathbf{u})},\overrightarrow{D_2\gamma(\mathbf{u})}\big)\Big)|d^2\mathbf{u}|\\ &= \int_U \det\big[\vec{F}\big(\gamma(\mathbf{u})\big),\overrightarrow{D_1\gamma(\mathbf{u})},\overrightarrow{D_2\gamma(\mathbf{u})}\big]\,|d^2\mathbf{u}|.\end{aligned} \qquad 6.5.17$$

Definition 6.5.7 (Flux). The *flux* of a vector field $\vec{F}$ over an oriented surface S is $\int_S \Phi_{\vec{F}}$.

Example 6.5.8 (Integrating a flux form). The flux of the vector field

$$\vec{F}\begin{pmatrix} x \\ y \\ z \end{pmatrix} = \begin{bmatrix} x \\ y^2 \\ z \end{bmatrix} \qquad 6.5.18$$

through the parametrized domain $\begin{pmatrix} u \\ v \end{pmatrix} \mapsto \begin{pmatrix} u^2 \\ uv \\ v^2 \end{pmatrix}$, $0 \le u, v \le 1$ is

$$\int_0^1 \int_0^1 \det \begin{bmatrix} u^2 & 2u & 0 \\ u^2v^2 & v & u \\ v^2 & 0 & 2v \end{bmatrix} du\, dv = \int_0^1 \int_0^1 (2u^2v^2 - 4u^3v^3 + 2u^2v^2)\, du\, dv$$

$$= \int_0^1 \left[\frac{4}{3}u^3v^2 - u^4v^3\right]_{u=0}^1 dv = \int_0^1 \left(\frac{4}{3}v^2 - v^3\right) dv = \frac{7}{36}. \quad \triangle \qquad 6.5.19$$

Recall that

$$M_f\Big(P_{\mathbf{x}}(\mathbf{v}_1, \mathbf{v}_2, \mathbf{v}_3)\Big) = f(\mathbf{x}) \det[\vec{\mathbf{v}}_1, \vec{\mathbf{v}}_2, \vec{\mathbf{v}}_3].$$

In coordinates, M_f is written

$$f(\mathbf{x})\, dx \wedge dy \wedge dz.$$

For symmetry's sake, we could define the mass of a function f over a region $U \subset \mathbb{R}^3$ as $\int_U M_f$, but in practice, this language isn't used; one speaks of the integral of $f\, dx \wedge dy \wedge dz$. (People often speak of the integral of f, but that is a misuse of language.)

- **Integrating a mass form M_f over an oriented region of $\mathbb{R}^3$.**

Let $U, V \subset \mathbb{R}^3$ be open sets, and let $\gamma : U \to V$ be a C^1 map preserving the standard orientation. If $f : V \to \mathbb{R}$ is a function, then

$$\int_V M_f = \int_U M_f\left(P_{\gamma(\mathbf{u})}(\overrightarrow{D_1\gamma}(\mathbf{u}), \overrightarrow{D_2\gamma}(\mathbf{u}), \overrightarrow{D_3\gamma}(\mathbf{u}))\right) |d^3\mathbf{u}|$$
$$= \int_U f(\gamma(\mathbf{u})) \det[\mathbf{D}\gamma(\mathbf{u})]\, |d^3\mathbf{u}|. \qquad 6.5.20$$

If $V = U$ and $\gamma(\mathbf{x}) = \mathbf{x}$ is the identity, equation 6.5.20 becomes

$$\int_{[\gamma(U)]} M_f = \int_U f(\mathbf{u})\, |d^3\mathbf{u}|; \qquad 6.5.21$$

the integral of M_f is simply what we called the integral of f in Section 4.2. If f is the density of some object, then this integral measures its mass. $\triangle$

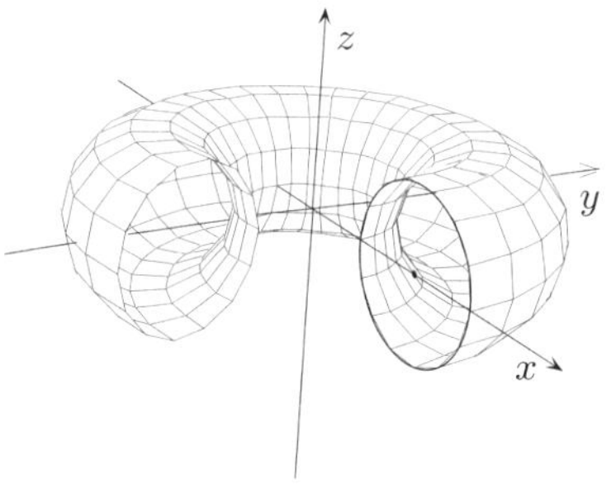

FIGURE 6.5.5.
This torus was discussed in Example 5.3.9. This time we are interested in the solid torus.

Example 6.5.9 (Integrating a mass form). Let f be the function

$$f\begin{pmatrix} x \\ y \\ z \end{pmatrix} = x^2 + y^2. \qquad 6.5.22$$

For $r < R$, let $T_{r,R}$ be the torus obtained by rotating the circle of radius r centered at $\begin{pmatrix} R \\ 0 \end{pmatrix}$ in the (x, z)-plane around the z-axis; see Figure 6.5.5.

Suppose we wish to compute the integral of M_f over the region bounded by $T_{r,R}$ (i.e., the inside of the torus). Since we are integrating over an open subset of $\mathbb{R}^3$, this is (except for the question of orientation) a problem that could be solved with the techniques of Chapter 4: either the "identity parametrization" as in equation 6.5.21, or a change of variables, as in Section 4.10.

Assume that the solid torus has the standard orientation of $\mathbb{R}^3$. Clearly the identity parametrization preserves that orientation: the partial derivatives of the identity are the standard basis vectors, and $\det[\vec{\mathbf{e}}_1, \vec{\mathbf{e}}_2, \vec{\mathbf{e}}_3] = 1$. But the identity parametrization would lead to quite a clumsy integral.

Equation 6.5.23: When computing integrals by hand, it's a good idea to choose a parametrization that reflects the symmetries of the problem. Here the torus is symmetrical around the z-axis; equation 6.5.23 reflects that symmetry. See Example 5.2.9.

The following parametrization, with $0 \le u \le r$ and $0 \le v, w \le 2\pi$, is better adapted:

$$\gamma \begin{pmatrix} u \\ v \\ w \end{pmatrix} = \begin{pmatrix} (R + u\cos v)\cos w \\ (R + u\cos v)\sin w \\ u \sin v \end{pmatrix}. \qquad 6.5.23$$

Exercise 6.5.21 asks you to determine whether γ is orientation preserving and to compute the integral of M_f over the inside of the torus. △

Although we are used to thinking of the flux form as a function of two vectors, this is true only in $\mathbb{R}^3$. In $\mathbb{R}^n$, it is a function of $n-1$ vectors. Those vectors, plus the vector $\vec{F}(\mathbf{x})$, can be used to form the $n \times n$ matrix

$$\left[\vec{F}(\mathbf{x}), \vec{\mathbf{v}}_1, \dots, \vec{\mathbf{v}}_{n-1}\right],$$

which has a determinant. Geometrically, this makes sense. For example, in $\mathbb{R}^3$, flux flows through a surface, so the flux form is evaluated on a parallelogram spanned by two vectors, but in $\mathbb{R}^2$, flux flows through a curve, so the flux form is evaluated on one vector.

Notice that a vector field in $\mathbb{R}^2$ defines two different 1-forms, the work and the flux. Exercise 6.12 asks you to show that they are related by the rule

$$W_{\vec{F}}(\vec{\mathbf{v}}) = \Phi_{\vec{F}}\left(\begin{bmatrix} 0 & -1 \\ 1 & 0 \end{bmatrix} \vec{\mathbf{v}}\right).$$

Work, flux, and mass forms on $\mathbb{R}^n$

In all dimensions n,

1. 0-form fields are functions.
2. Every 1-form field is the work form of a vector field.
3. Every $(n-1)$-form field is the flux form of a vector field.
4. Every n-form field is the mass form of a function.

We've already seen this for 0-forms and 1-forms. In $\mathbb{R}^3$, the flux form is, of course, a 2-form. Its definition can be generalized:

Definition 6.5.10 (Flux form on $\mathbb{R}^n$). If $\vec{F}$ is a vector field on $U \subset \mathbb{R}^n$, and $\vec{\mathbf{v}}_1, \dots, \vec{\mathbf{v}}_{n-1}$ are vectors in $\mathbb{R}^n$, then the *flux form* $\Phi_{\vec{F}}$ is the $(n-1)$-form defined by

$$\Phi_{\vec{F}} P_{\mathbf{x}}(\vec{\mathbf{v}}_1, \dots, \vec{\mathbf{v}}_{n-1}) \stackrel{\text{def}}{=} \det\left[\vec{F}(\mathbf{x}), \vec{\mathbf{v}}_1, \dots, \vec{\mathbf{v}}_{n-1}\right]. \qquad 6.5.24$$

In coordinates, this becomes (by the development of the determinant by the first column)

$$\begin{aligned} \Phi_{\vec{F}} &= \sum_{i=1}^{n} (-1)^{i-1} F_i dx_1 \wedge \dots \wedge \widehat{dx_i} \wedge \dots \wedge dx_n \\ &= F_1 dx_2 \wedge \dots \wedge dx_n - F_2 dx_1 \wedge dx_3 \wedge \dots \wedge dx_n + \cdots \\ &\quad \cdots + (-1)^{n-1} F_n dx_1 \wedge dx_2 \wedge \dots \wedge dx_{n-1}, \end{aligned} \qquad 6.5.25$$

where the term under the hat is omitted.

For instance, the flux of the radial vector field $\vec{F}\begin{pmatrix} x_1 \\ \vdots \\ x_n \end{pmatrix} = \begin{bmatrix} x_1 \\ \vdots \\ x_n \end{bmatrix}$ is

$$\Phi_{\vec{F}} = (x_1 dx_2 \wedge \cdots \wedge dx_n) - (x_2 dx_1 \wedge dx_3 \wedge \cdots \wedge dx_n) + \cdots$$
$$\cdots + (-1)^{n-1} x_n dx_1 \wedge \cdots \wedge dx_{n-1}. \qquad 6.5.26$$

Equation 6.5.26: In the first term, where $i = 1$, we omit dx_1; in the second term, where $i = 2$, we omit dx_2, and so on. This corresponds to omitting the ith row when computing the determinant; see Example 4.8.2.

In any dimension n, an n-form is a multiple of the determinant, so every n-form is the mass form of a function:

Definition 6.5.11 (Mass form on $\mathbb{R}^n$). Let U be a subset of $\mathbb{R}^n$. The *mass form* M_f of a function $f : U \to \mathbb{R}$ is given by

$$M_f \overset{\text{def}}{=} f dx_1 \wedge \cdots \wedge dx_n. \qquad 6.5.27$$

The correspondences between forms, functions, and vectors explain why vector calculus works in $\mathbb{R}^3$, and why it doesn't work in higher dimensions. Table 6.5.6 shows that when k is anything other than 0, 1, $n-1$, or n, a k-form on $\mathbb{R}^n$ cannot be interpreted in terms of functions or vector fields.

In dimensions higher than $\mathbb{R}^3$, some forms cannot be expressed in terms of vector fields and functions: in particular, 2-forms on $\mathbb{R}^4$, which are of great interest in physics, since the electromagnetic field is such a 2-form on spacetime. The language of vector calculus is not suited to describing integrands over surfaces in higher dimensions, while the language of forms is.

Forms	Vector Calculus	
	$\mathbb{R}^3$	$\mathbb{R}^n$, $n > 3$
0-Form	Function	Function
1-Form	Vector field (via work form)	Vector field
$(n-2)$-Form	Same as 1-form	No equivalent
$(n-1)$-Form	Vector field (via flux form)	Vector field
n-Form	Function (via mass form)	Function

TABLE 6.5.6. In all dimensions, 0-forms, 1-forms, $(n-1)$-forms, and n-forms can be identified to vector fields or functions. Other form fields have no equivalent in vector calculus.

It is also possible to relate the wedge product of forms to the products of vector fields.

Proposition 6.5.12. *For any two vector fields $\vec{F}$ and $\vec{G}$ on $\mathbb{R}^3$, we have*

$$\Phi_{\vec{F}\times\vec{G}} = W_{\vec{F}} \wedge W_{\vec{G}} \quad \text{and} \quad M_{\vec{F}\cdot\vec{G}} = W_{\vec{F}} \wedge \Phi_{\vec{G}} = W_{\vec{G}} \wedge \Phi_{\vec{F}}.$$

The proof is left as Exercises 6.5.4 and 6.5.5. It is possible to give conceptual proofs, but it is easier to just compute both sides and check.

Example 6.5.13 (Electromagnetic field). The electromagnetic field, a six-component object, cannot be represented either as a function (a one-component object) or as a vector field (in $\mathbb{R}^4$, a four-component object).

The standard way to deal with the problem is to choose coordinates x, y, z, t, in particular choosing a specific space-like subspace and a specific time-like subspace, quite likely those of your laboratory. Experiments indicate the following force law: there are two time-dependent vector fields, $\vec{\mathbf{E}}$ (the electric field) and $\vec{\mathbf{B}}$ (the magnetic field), with the property that a charge q at $(\mathbf{x}, t)$ and with velocity $\vec{\mathbf{v}}$ (in the laboratory coordinates) is subject to the force

$$q\left(\vec{\mathbf{E}}(\mathbf{x},t) + \left(\frac{\vec{\mathbf{v}}}{c} \times \vec{\mathbf{B}}(\mathbf{x},t)\right)\right), \qquad 6.5.28$$

Equation 6.5.28: We are using the *cgs* system of units (centimeter, gram, second). In this system the unit of charge is the *statcoulomb*, which is designed to remove constants from Coulomb's law. (The SI system, or *système international d'unités*, based on kilograms and seconds, uses a different unit of charge, the *coulomb*, which results in constants μ_0 and ϵ_0 that clutter up the equations).

We could go one step further and use the mathematicians' privilege of choosing units arbitrarily, setting $c = 1$, but that offends intuition.

where c represents the speed of light. But $\vec{\mathbf{E}}$ and $\vec{\mathbf{B}}$ are not really vector fields. A true vector field "keeps its identity" when you change coordinates. In particular, if a vector field is $\vec{\mathbf{0}}$ in one coordinate system, it will be $\vec{\mathbf{0}}$ in every coordinate system. This is not true of the electric and magnetic fields. If in one coordinate system the charge is at rest and the electric field is $\vec{\mathbf{0}}$, then the particle will not be accelerated in those coordinates. In another system moving at constant velocity with respect to the first (on a train rolling through the laboratory, for instance) it will still not be accelerated. But it now feels a force from the magnetic field, which must be compensated for by an electric field, which cannot now be zero.

Is there something natural that the electric field and the magnetic field together represent? The answer is yes: there is a 2-form field on $\mathbb{R}^4$, namely

$$\mathbb{F} = E_x dx \wedge c\,dt + E_y dy \wedge c\,dt + E_z dz \wedge c\,dt \; + B_x dy \wedge dz + B_y dz \wedge dx + B_z dx \wedge dy,$$

$$\text{i.e.,} \qquad \mathbb{F} = W_{\vec{\mathbf{E}}} \wedge c\,dt \; + \; \Phi_{\vec{\mathbf{B}}}. \qquad 6.5.29$$

In expression 6.5.29, the work and the flux are with respect to the spatial variables only.

In the cgs system, $\vec{\mathbf{E}}$ and $\vec{\mathbf{B}}$ have the same dimensions, force per charge. The c (speed of light) is then necessary so that

$$W_{\vec{\mathbf{E}}} \wedge c\,dt \text{ and } \Phi_{\vec{\mathbf{B}}}$$

have the same dimensions:

$$\frac{\text{force}}{\text{charge} \times \text{length}^2}.$$

This 2-form (really a 2-form field) is called the *Faraday* by the distinguished physicists Charles Misner, Kip Thorne, and J. Archibald Wheeler (in their book *Gravitation*, the bible of general relativity). It is a natural object, the same in every frame. Thus form fields are really the natural language in which to write Maxwell's equations. We explore this further in Section 6.12. △

Exercises for Section 6.5

6.5.1 Which of the following expressions are identical?

a. $F_1\,dx + F_2\,dy + F_3\,dz$	b. $W_{\vec{F}}\left(P_{\mathbf{x}}(\vec{\mathbf{v}})\right)$	c. $\Phi_{\vec{F}}\left(P_{\mathbf{x}}(\vec{\mathbf{v}}, \vec{\mathbf{w}})\right)$
d. $F_1\,dy \wedge dz - F_2\,dx \wedge dz + F_3\,dx \wedge dy$	e. $\det[\vec{F}(\mathbf{x}), \vec{\mathbf{v}}, \vec{\mathbf{w}}]$	f. $\vec{F}(\mathbf{x}) \cdot (\vec{\mathbf{v}} \times \vec{\mathbf{w}})$
g. $f\,dx \wedge dy \wedge dz$	h. $\Phi_{\vec{F}}$	i. $\vec{F}(\mathbf{x}) \cdot \vec{\mathbf{v}}$
j. work form of $\vec{F}$	k. flux form of $\vec{F}$	l. $W_{\vec{F}}$

$$\vec{F} = \begin{bmatrix} x^2 \\ xy \\ -z \end{bmatrix}, \quad \vec{G} = \begin{bmatrix} x^2 \\ xy \\ x \end{bmatrix}$$

Vector fields for Exercise 6.5.2.

6.5.2 a. Write in coordinate form the work form field and the flux form field of the vector fields $\vec{F}$ and $\vec{G}$ shown in the margin.

b. For what vector field $\vec{F}$ is each of the following 1-form fields on $\mathbb{R}^3$ the work form field $W_{\vec{F}}$?

i. $xy\,dx - y^2\,dz$ ii. $y\,dx + 2\,dy - 3x\,dz$

c. For what vector field $\vec{F}$ is each of the following 2-form fields on $\mathbb{R}^3$ the flux form field $\Phi_{\vec{F}}$?

i. $2z^4\,dx \wedge dy + 3y\,dy \wedge dz - x^2 z\,dx \wedge dz$

ii. $yz\,dx \wedge dz - xy^2 z\,dy \wedge dz$

6.5.3 Some of the following expressions do not make sense. Correct them so that they do.

a. $\Phi_{\vec{F}}\Big(P_{\mathbf{x}}(\vec{\mathbf{v}}_1, \vec{\mathbf{v}}_2, \vec{\mathbf{v}}_3)\Big)$ b. W_f c. $M_f\Big(P_{\mathbf{x}}(\vec{\mathbf{v}}, \vec{\mathbf{w}})\Big)$

d. $(\vec{\mathbf{v}}_1 \cdot \vec{\mathbf{v}}_2) \times \vec{\mathbf{v}}_3$ e. $\Phi_{\vec{F}}$ f. $\Phi_{\vec{F}} = F_1\,dx \wedge dy - F_2\,dy \wedge dz + F_3\,dz \wedge dx$

g. $W_{\vec{F}}\Big(P_{\mathbf{x}}(\vec{\mathbf{v}}_1, \vec{\mathbf{v}}_2)\Big)$ h. $M_{\vec{F}}$ i. $W_{\vec{F}} = F_1\,dx + F_2\,dy + F_3\,dz$

In Exercises 6.5.4 and 6.5.5, $\vec{F}$ and $\vec{G}$ are vector fields on a subset of $\mathbb{R}^3$.

6.5.4 Show that $\Phi_{\vec{F}\times\vec{G}} = W_{\vec{F}} \wedge W_{\vec{G}}$.

6.5.5 Show that $M_{\vec{F}\cdot\vec{G}} = W_{\vec{F}} \wedge \Phi_{\vec{G}} = W_{\vec{G}} \wedge \Phi_{\vec{F}}$.

6.5.6 What is the work form field $W_{\vec{F}}\Big(P_{\mathbf{a}}(\vec{\mathbf{u}})\Big)$ of the vector field

$$\vec{F}\begin{pmatrix} x \\ y \\ z \end{pmatrix} = \begin{bmatrix} x^2 y \\ x - y \\ -z \end{bmatrix}, \text{ at } \mathbf{a} = \begin{pmatrix} 0 \\ 1 \\ 2 \end{pmatrix}, \text{ evaluated on the vector } \vec{\mathbf{u}} = \begin{bmatrix} 1 \\ -1 \\ 1 \end{bmatrix}?$$

6.5.7 For the following 1-forms on $\mathbb{R}^2$, write down the corresponding vector field. Sketch the vector field. Describe a path over which the work of the 1-form would be 0. Describe a path over which the work would be large.

a. dy b. $x\,dx$ c. $x\,dx - y\,dy$

Exercise 6.5.8: Part c is considerably more difficult than parts a and b. Some knowledge of differential equations helps. In any case, do Exercise 1.5.10 first.

6.5.8 Repeat Exercise 6.5.7 for the following 1-forms on $\mathbb{R}^2$:

a. $y\,dx + x\,dy$ b. $-y\,dx + x\,dy$ c. $(x - y)\,dx + (x + y)\,dy$

6.5.9 a. Construct an oriented parallelogram anchored at $\begin{pmatrix} 1 \\ 1 \\ 0 \end{pmatrix}$ to which the 2-form $\Phi = y\,dy \wedge dz + x\,dx \wedge dz - z\,dx \wedge dy$ of Example 6.5.3 will assign a positive number.

b. At what point $\mathbf{x}$ might you anchor $P_{\mathbf{x}}(\vec{\mathbf{e}}_1, \vec{\mathbf{e}}_2)$ if you wanted Φ evaluated on the parallelogram to return a positive number? A negative number?

6.5.10 What is the flux form field $\Phi_{\vec{F}}$ of the vector field $\vec{F}\begin{pmatrix} x \\ y \\ z \end{pmatrix} = \begin{bmatrix} -x \\ y^2 \\ xy \end{bmatrix}$,

evaluated on $P\left(\begin{bmatrix} 1 \\ 0 \\ 1 \end{bmatrix}, \begin{bmatrix} 0 \\ 1 \\ 0 \end{bmatrix}\right)$ at the point $\mathbf{x} = \begin{pmatrix} 1 \\ 2 \\ -1 \end{pmatrix}$?

6.5.11 Evaluate the work of each the following vector fields $\vec{F}$ on the given 1-parallelograms:

a. $\vec{F} = \begin{bmatrix} x \\ y \end{bmatrix}$ on $P_{\begin{pmatrix} 1 \\ 1 \end{pmatrix}}\begin{bmatrix} 2 \\ 3 \end{bmatrix}$ b. $\vec{F} = \begin{bmatrix} x^2 \\ \sin xy \end{bmatrix}$ on $P_{\begin{pmatrix} -1 \\ -\pi \end{pmatrix}}\begin{bmatrix} e \\ \pi \end{bmatrix}$

c. $\vec{F} = \begin{bmatrix} y \\ x \\ z \end{bmatrix}$ on $P_{\begin{pmatrix} 1 \\ 0 \\ 1 \end{pmatrix}} \begin{bmatrix} 2 \\ 3 \\ -1 \end{bmatrix}$ d. $\vec{F} = \begin{bmatrix} \sin y \\ \cos(x+z) \\ e^x \end{bmatrix}$ on $P_{\begin{pmatrix} 0 \\ 1 \\ -1 \end{pmatrix}} \begin{bmatrix} 0 \\ 1 \\ 0 \end{bmatrix}$

6.5.12 What is the mass form of the function $f\begin{pmatrix} x \\ y \\ z \end{pmatrix} = xy + z^2$, evaluated at the point $\mathbf{x} = \begin{pmatrix} 1 \\ 2 \\ 1 \end{pmatrix}$ on the vectors $\begin{bmatrix} 1 \\ 0 \\ 1 \end{bmatrix}, \begin{bmatrix} 2 \\ 1 \\ 1 \end{bmatrix}$, and $\begin{bmatrix} 0 \\ 1 \\ 1 \end{bmatrix}$?

6.5.13 Verify that $\det[\vec{F}(\mathbf{x}), \vec{\mathbf{v}}_1, \dots, \vec{\mathbf{v}}_{n-1}]$ is an $(n-1)$-form field, so that Definition 6.5.10 of the flux form on $\mathbb{R}^n$ makes sense.

6.5.14 Rewrite the computation of Example 6.5.6, using the language of forms.

$$\underbrace{\begin{bmatrix} 0 \\ 1 \\ 1 \end{bmatrix}}_{\vec{\mathbf{v}}_1}, \underbrace{\begin{bmatrix} 1 \\ 1 \\ 0 \end{bmatrix}}_{\vec{\mathbf{v}}_2}, \underbrace{\begin{bmatrix} -1 \\ 1 \\ 1 \end{bmatrix}}_{\vec{\mathbf{v}}_3}$$

Vectors for Exercise 6.5.15

6.5.15 Given $\vec{F}\begin{pmatrix} x \\ y \\ z \end{pmatrix} = \begin{bmatrix} y^2 \\ x+z \\ xz \end{bmatrix}$, $f\begin{pmatrix} x \\ y \\ z \end{pmatrix} = xz + zy$, the point $\mathbf{x} = \begin{pmatrix} 1 \\ 1 \\ -1 \end{pmatrix}$, and the vectors $\vec{\mathbf{v}}_1, \vec{\mathbf{v}}_2, \vec{\mathbf{v}}_3$ in the margin, what is

a. the work form $W_{\vec{F}}\big(P_{\mathbf{x}}(\vec{\mathbf{v}}_1)\big)$? b. the flux form $\Phi_{\vec{F}}\big(P_{\mathbf{x}}(\vec{\mathbf{v}}_1, \vec{\mathbf{v}}_2)\big)$?

c. the mass form $M_f\big(P_{\mathbf{x}}(\vec{\mathbf{v}}_1, \vec{\mathbf{v}}_2, \vec{\mathbf{v}}_3)\big)$?

6.5.16 Evaluate the flux of the following vector fields on the 2-parallelograms:

a. $\vec{F} = \begin{bmatrix} y \\ x \\ z \end{bmatrix}$ on $P_{\begin{pmatrix} 1 \\ 0 \\ 1 \end{pmatrix}}\left(\begin{bmatrix} 1 \\ 1 \\ 0 \end{bmatrix}, \begin{bmatrix} 0 \\ 0 \\ -1 \end{bmatrix}\right)$ b. $\vec{F} = \begin{bmatrix} \sin y \\ \cos(x+z) \\ e^x \end{bmatrix}$ on $P_{\begin{pmatrix} 0 \\ 1 \\ -1 \end{pmatrix}}\left(\begin{bmatrix} 0 \\ 1 \\ 0 \end{bmatrix}, \begin{bmatrix} 1 \\ 2 \\ 0 \end{bmatrix}\right)$

6.5.17 Let R be the rectangle with vertices $\begin{pmatrix} 0 \\ 0 \end{pmatrix}, \begin{pmatrix} 0 \\ a \end{pmatrix}, \begin{pmatrix} b \\ a \end{pmatrix}, \begin{pmatrix} b \\ 0 \end{pmatrix}$, with $a, b > 0$, and oriented so that these vertices appear in that order. Find the work of the vector field $\vec{F}\begin{pmatrix} x \\ y \end{pmatrix} = \begin{bmatrix} xy \\ ye^x \end{bmatrix}$ around the boundary of R.

Exercise 6.5.18: We could also describe this arc of helix as "oriented by increasing t".

6.5.18 Find the work of $\vec{F}\begin{pmatrix} x \\ y \\ z \end{pmatrix} = \begin{bmatrix} x^2 \\ y^2 \\ z^2 \end{bmatrix}$ over the arc of helix parametrized by $\gamma : t \mapsto \begin{pmatrix} \cos t \\ \sin t \\ at \end{pmatrix}$, for $0 \le t \le \alpha$, and oriented so that γ is orientation preserving.

Exercise 6.5.19: The answer should be some function of a and R. Part of the problem is finding parametrizations of S that preserve orientation.

6.5.19 Find the flux of the vector field $\vec{F}\begin{pmatrix} x \\ y \\ z \end{pmatrix} = r^a \begin{bmatrix} x \\ y \\ z \end{bmatrix}$, where a is a number and $r = \sqrt{x^2+y^2+z^2}$, through the surface S, where S is the sphere of radius R oriented by the outward-pointing normal.

6.5.20 What is the flux of the vector field $\vec{F}\begin{pmatrix} x \\ y \\ z \end{pmatrix} = \begin{bmatrix} x \\ -y \\ xy \end{bmatrix}$ through the surface $z = \sqrt{x^2+y^2}$, $x^2+y^2 \le 1$, oriented by the outward normal?

6.5.21 a. In Example 6.5.9, does γ preserve orientation?

b. Compute the integral.

6.6 Boundary orientation

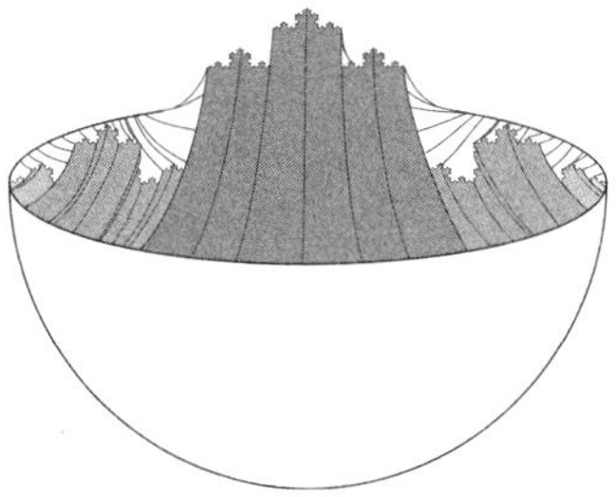

FIGURE 6.6.1.

To form this piece-with-boundary, "bridge" the cut edge of an orange with a Koch snowflake (see Figure 5.5.2). Now dip it into soapy water, so that a film of soap bubble connects the rim of the cut orange and the snowflake. We think of this figure as a model for a breaking wave: the complexity of the Koch curve is quite like that of water in a cresting wave. A breaking wave is just the kind of thing to which a physicist would want to apply Stokes theorem: the balance between surface tension (a surface integral), and gravity and momentum (a volume integral) is the key to the cohesion of the water. The crashing of the wave is poorly understood; we speculate that it corresponds to the loss of the conditions to be a piece-with-boundary: the crest acquires positive area but no longer carries surface tension, breaking the balance.

The standard definition of a manifold-with-boundary leads to an unpalatable choice: either forgo using Stokes's theorem in the numerous important cases where the pieces do not have smooth boundaries, or use it knowing that there is a very real possibility that the theorem does not apply, and that one's results are invalid.

Stokes's theorem, the generalization of the fundamental theorem of calculus, is about comparing integrals over pieces of manifolds and integrals over the boundaries of those pieces. Here we will define a "piece-with-boundary" of a manifold. We will see that if the manifold is oriented, then the boundary of such a piece carries a natural orientation, called the *boundary orientation*.

For Stokes's theorem to be true, the boundary of a piece-with-boundary can't have just any shape. In many treatments the boundaries are restricted to being smooth: if the manifold is three-dimensional, the boundary is required to be a smooth surface; if the manifold is 2 dimensional, the boundary is required to be a smooth curve. Such pieces are called *manifolds-with-boundary*. (They are not manifolds; manifolds don't have boundaries.)

Requiring pieces to have smooth boundaries is an unnecessary and burdensome condition, excluding many standard examples of vector calculus. It excludes k-parallelograms, whose boundaries are not smooth at the corners, and it excludes the region bounded by a cone, which is not smooth at the vertex or along the boundary of the base. The typical solution is to fudge and apply Stokes's theorem anyway, on the correct grounds that the nonsmooth part of the boundary is too small to affect the integral.

Requiring boundaries to be smooth also excludes many real-life examples. In material science, meteorology, biology, physics, etc., one is often interested in the flow of a substance through a boundary – the movement of ions through the wall of a cell, for example. Stokes's theorem is a powerful tool for studying such flow. But the boundaries are rarely smooth; think of capillaries and alveoli, or the boundary of a cloud. Figures 6.6.1 and 6.6.2 show other instances.

FIGURE 6.6.2. Suture line of Cladiscites, an ammonite that lived in the upper Triassic period. Ammonites, extinct for some 65 million years, are related to the octopus. An ammonite lived in the last of a series of air-filled chambers making up its shell, which resembles that of today's chambered nautilus. The suture line marks the juncture of the shell wall and the wall separating the chambers. To study the metabolism of such a creature, one might wish to integrate over one chamber and its boundary. Such a boundary, which includes the suture line, is obviously not smooth. (Picture courtesy of Jean Guex, University of Lausanne.)

In these cases it is less obvious that the nonsmooth part of the boundary does not affect the integral. Can Stokes's theorem be used anyway? It can if the conditions in our definition of piece-with-boundary are met. Thus we prefer to be precise about allowable boundaries. We will begin by defining the boundary of a subset of a manifold; then we will define smooth points of the boundary; finally we will rquire that the nonsmooth boundary of k-dimensional manifold have $(k-1)$-dimensional volume 0.

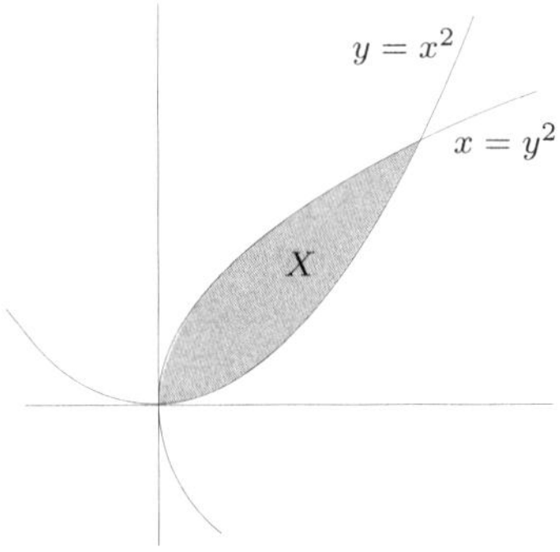

FIGURE 6.6.3.
The shaded region satisfying $y \geq x^2$ and $x \geq y^2$ is a subset X of the manifold $\mathbb{R}^2$.

Definition 6.6.1 (Boundary of a subset of a manifold). Let $M \subset \mathbb{R}^n$ be a k-dimensional manifold, and $X \subset M$ a subset. The *boundary of X in M*, written $\partial_M X$, is the set of points $\mathbf{x} \in M$ such that every neighborhood of $\mathbf{x}$ contains points of X and points of $M - X$.

Note that unless M is an open subset of $\mathbb{R}^n$ (a "fat" piece of $\mathbb{R}^n$), the boundary of X in M is not the boundary of X in $\mathbb{R}^n$ as defined in Definition 1.5.10.

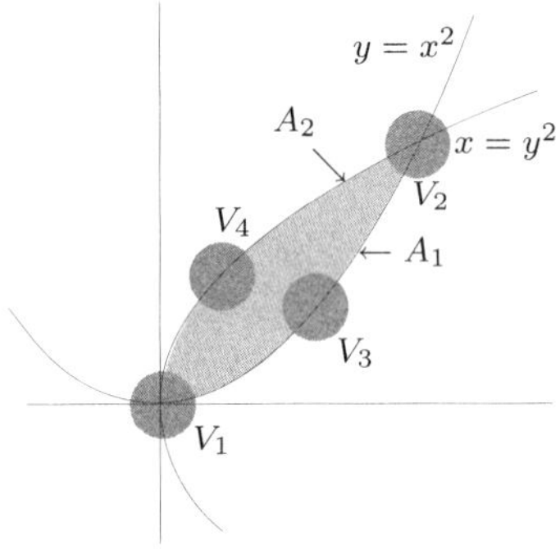

FIGURE 6.6.4.
Figure 6.6.3, with more detail. Points of the boundary $\partial_{\mathbb{R}^2} X$ that are in the balls V_3 and V_4 are "smooth points" of the boundary.

Smooth boundary

By Theorem 3.1.10, a manifold M is defined by *equalities*: every $\mathbf{x} \in M$ has a neighborhood $U \subset \mathbb{R}^n$ such that $M \cap U$ is the locus of equation $\mathbf{F} = \mathbf{0}$ for some C^1 function $\mathbf{F} : U \to \mathbb{R}^{n-k}$ with $[\mathbf{DF}(\mathbf{x})]$ surjective. A piece $X \subset M$ is defined by *inequalities* saying how to carve the piece out of the manifold. For example, the piece $X \subset \mathbb{R}^2$ shown in Figure 6.6.3 is set off from the rest of $\mathbb{R}^2$ by the inequalities $y \geq x^2$ and $x \geq y^2$. The boundary of the piece X consists of points where at least one inequality is an equality: points $\mathbf{x} \in X$ such that either $g_1(\mathbf{x}) = 0$ or $g_2(\mathbf{x}) = 0$ where $g_1(\mathbf{x}) = y - x^2$ and $g_2(\mathbf{x}) = x - y^2$.

Definition 6.6.2 (Smooth point of boundary, smooth boundary). Let $M \subset \mathbb{R}^n$ be a k-dimensional manifold and $X \subset M$ a subset. A point $\mathbf{x} \in \partial_M X$ is a *smooth point* of the boundary of X if there exist a neighborhood $V \subset \mathbb{R}^n$ of $\mathbf{x}$ and a single C^1 function $g : V \cap M \to \mathbb{R}$ with

$$g(\mathbf{x}) = 0, \quad X \cap V = \{g \geq 0\}, \quad [\mathbf{D}g(\mathbf{x})] : T_{\mathbf{x}} M \to \mathbb{R} \text{ onto}.$$

The set of smooth points of the boundary of X is the *smooth boundary* of X, denoted $\partial_M^s X$.

Definition 6.6.2: The formula $g \geq 0$ defines X in $V \cap M$; the equality $g(\mathbf{x}) = 0$ says that $\mathbf{x}$ is a point of the boundary of X. The derivative being onto says that $\mathbf{x}$ is a smooth point of the boundary.

If there are several functions g_i defining X, usually the smooth part of the boundary will be the locus where exactly one is an equality and the remainder are strict inequalities. In Figure 6.6.4, the origin is not a smooth point: $V_1 \cap X$ cannot be defined by a single g with onto derivative.

Note that we could just as well have required $X \cap V = \{g \leq 0\}$.

The *nonsmooth boundary* of X is that part of the boundary $\partial_M X$ that is not smooth.

Proposition 6.6.3. *The smooth boundary $\partial_M^s X$ is a $(k-1)$-dimensional manifold.*

Proof. This follows from Proposition 3.2.10. $\square$

The proof of Proposition 6.6.4 is left as Exercise 6.6.11.

Proposition 6.6.4. *Let U be an open subset of $\mathbb{R}^n$, and let $M \subset U$ be the k-domensional manifold defined by $\mathbf{f} = \mathbf{0}$, where $\mathbf{f}: U \to \mathbb{R}^{n-k}$ is a C^1 function with $[\mathbf{Df}(\mathbf{x})]$ onto at all $\mathbf{x} \in M$. Let $g: U \to \mathbb{R}$ be of class C^1. Then*

$$[\mathbf{D}g(\mathbf{x})]: T_{\mathbf{x}}M \to \mathbb{R} \text{ onto} \iff \left[\mathbf{D}\begin{pmatrix}\mathbf{f}\\ g\end{pmatrix}(\mathbf{x})\right]: \mathbb{R}^n \to \mathbb{R}^{n-k+1} \text{ onto.}$$

Corner points

Corner points are closely related to smooth boundary points. Remember (see Theorem 2.7.10) that the inequality $\vec{\mathbf{v}} \geq \vec{\mathbf{w}}$ between vectors means that $v_i \geq w_i$ for each coordinate of $\vec{\mathbf{v}}$ and $\vec{\mathbf{w}}$.

Definition 6.6.5 (Corner point). Let $M \subset \mathbb{R}^n$ be a k-dimensional manifold and $X \subset M$ a subset. A point $\mathbf{x} \in X$ is a *corner point of codimension m* if there exist a neighborhood $V \subset \mathbb{R}^n$ of $\mathbf{x}$ and a C^1 function $\mathbf{g}: V \cap M \to \mathbb{R}^m$ with

$$X \cap V = \{\mathbf{g} \geq \mathbf{0}\}, \quad \mathbf{g}(\mathbf{x}) = \mathbf{0}, \quad [\mathbf{Dg}(\mathbf{x})] \text{ onto.}$$

Every point of the boundary of a cube is a corner point: points of faces have codimension 1 (they are the smooth boundary), points of edges codimension 2, and vertices codimension 3.

Interior points are corner points of codimension 0.

It may seem counterintuitive to call a smooth boundary point a corner point; where is the corner? But if you set $m = 1$ in Definition 6.6.5 you get the definition of a smooth boundary point. Allowing smooth points to be corner points simplifies the definition of piece-with-corners (Definition 6.6.9).

Note that a corner point of codimension 1 is a smooth boundary point. Note also that different points of X will almost always have different functions $\mathbf{g}$ with different codomains with different m's.

Corner points of codimension $m \geq 2$ are nonsmooth.

Example 6.6.6 (Smooth points and corner points). The region X satisfying $y \geq x^2$ and $x \geq y^2$ is a subset of the manifold $\mathbb{R}^2$; it is shown as the shaded area in Figure 6.6.4. Its boundary consists of

- the arc A_1 of the parabola $y = x^2$ where $0 \leq x \leq 1$, and
- the arc A_2 of the parabola $x = y^2$ where $0 \leq y \leq 1$.

The parts of A_1 where $0 < x < 1$ and the part of A_2 where $0 < y < 1$ (for instance, $A_1 \cap V_3$ and $A_2 \cap V_4$) are smooth. For a point $\begin{pmatrix}x\\y\end{pmatrix} \in A_1$ we can take $g\begin{pmatrix}x\\y\end{pmatrix} = y - x^2$, since $\left[\mathbf{D}g\begin{pmatrix}x\\y\end{pmatrix}\right] = [-2x\,, 1]$ is onto $\mathbb{R}$. For a boundary point in A_2, we can take $g\begin{pmatrix}x\\y\end{pmatrix} = x - y^2$, with onto derivative $[1\,, -2y]$.

The points $\begin{pmatrix}0\\0\end{pmatrix}$ and $\begin{pmatrix}1\\1\end{pmatrix}$ are in the boundary $\partial_M X$, but they are not smooth points: $V_1 \cap X$ cannot be defined by a single g with onto derivative; nor can $V_2 \cap X$. They are, however, corner points of codimension 2. Indeed, at both points we can take

$$\mathbf{g}\begin{pmatrix}x\\y\end{pmatrix} = \begin{pmatrix}x - y^2\\ y - x^2\end{pmatrix}, \quad \text{with onto derivatives}$$

$$\left[\mathbf{Dg}\begin{pmatrix}0\\0\end{pmatrix}\right] = \begin{bmatrix}1 & 0\\0 & 1\end{bmatrix} \quad \text{and} \quad \left[\mathbf{Dg}\begin{pmatrix}1\\1\end{pmatrix}\right] = \begin{bmatrix}1 & -2\\-2 & 1\end{bmatrix}. \qquad \triangle$$

Example 6.6.7 (A subset with no smooth boundary). The subset pictured in Figure 6.6.5 is bounded by three copies of the Koch snowflake of Example 5.5.1, rotated by 120° and 240°, and translated so they fit end to end. This region has no smooth boundary: as we saw in Example 5.5.1, the length of any little piece of the boundary is always infinite, but Corollary 5.3.4 says that every point of a k-dimensional manifold has a neighborhood with finite k-dimensional volume, so by Proposition 6.6.3, the region has no smooth boundary. $\triangle$

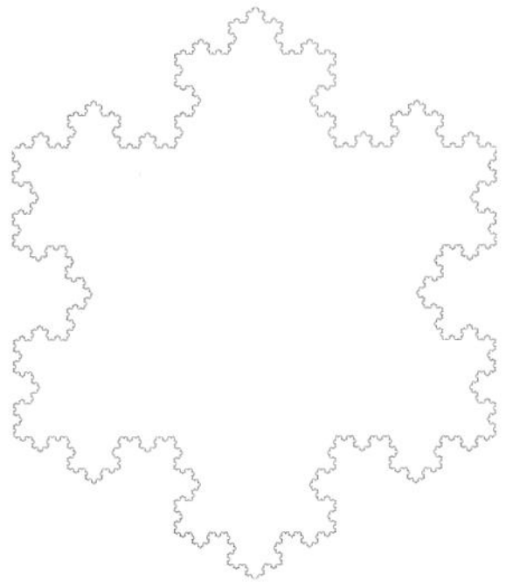

FIGURE 6.6.5.
The region pictured above is often called the Koch snowflake. It is a bounded closed subset of $\mathbb{R}^2$, but has no smooth boundary.

Definition 6.6.8 (Piece-with-boundary of a manifold). A *piece-with-boundary* of a k-dimensional manifold M is a compact subset $X \subset M$ such that

1. The set of nonsmooth points in $\partial_M X$ has $(k-1)$-dimensional volume 0
2. The smooth boundary has finite $(k-1)$-dimensional volume:

$$\mathrm{vol}_{k-1}(\partial^s_M X) < \infty \tag{6.6.1}$$

Remark. The nonsmooth points in part 1 are analogous to the "trouble spots" that we excluded from our domains of integration in Chapter 5. As long as they have $(k-1)$-dimensional volume 0, they will not affect integrals of $(k-1)$-forms over the boundary. The condition of part 2 is necessary for the proof of Stoke's theorem; see equation 6.10.75. $\triangle$

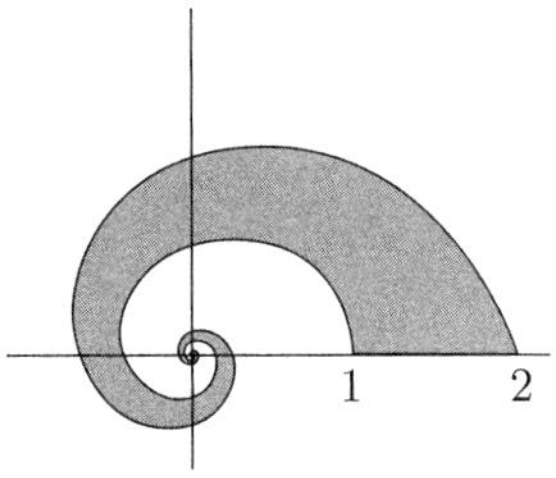

FIGURE 6.6.6.
This shaded region is a piece-with-boundary: the nonsmooth boundary consists of three points on the x-axis, with length 0, and the smooth boundary has finite length, since it spirals quickly to the nonsmooth point at the origin.

In contrast, the region shown in Figure 6.6.7 is not a piece-with-boundary.

Definition 6.6.9 (Piece-with-corners). Let X be a compact subset of a k-dimensional manifold $M \subset \mathbb{R}^n$. If every point of the boundary $\partial_M X$ is a corner point, then X is a *piece-with-corners*.

Exercise 6.6.9 asks you to show that a piece-with-corners is a piece-with-boundary.

Example 6.6.10 (Piece-with-boundary). We saw in Example 6.6.6 that the shaded region shown in Figures 6.6.3 and 6.6.4 (the region set off from the rest of $\mathbb{R}^2$ by the inequalities $y \geq x^2$ and $x \geq y^2$) is a piece-with-corners. It follows from Exercise 6.6.9 that it is a piece-with-boundary. $\triangle$

Example 6.6.11. The region in Figure 6.6.6 is a piece-with-boundary but not a piece-with-corners. It is bounded by the two spirals $r = e^{-\theta}$ and $r = 2e^{-\theta}$ for $\theta \geq 0$, and the segment $1 \leq x \leq 2$, $y = 0$. Except for the points $\begin{pmatrix}0\\0\end{pmatrix}, \begin{pmatrix}1\\0\end{pmatrix}, \begin{pmatrix}2\\0\end{pmatrix}$, the boundary is obvioiusly smooth, and the smooth boundary has finite length; this is not obvious from the figure, but is justified in Exercise 5.3.1, part c. The points $\begin{pmatrix}1\\0\end{pmatrix}, \begin{pmatrix}2\\0\end{pmatrix}$ are corner points, but the origin is not: we cannot straighten out the spirals by a C^1 map $\mathbf{g} : \mathbb{R}^2 \to \mathbb{R}^2$ with surjective derivative at the origin. $\triangle$

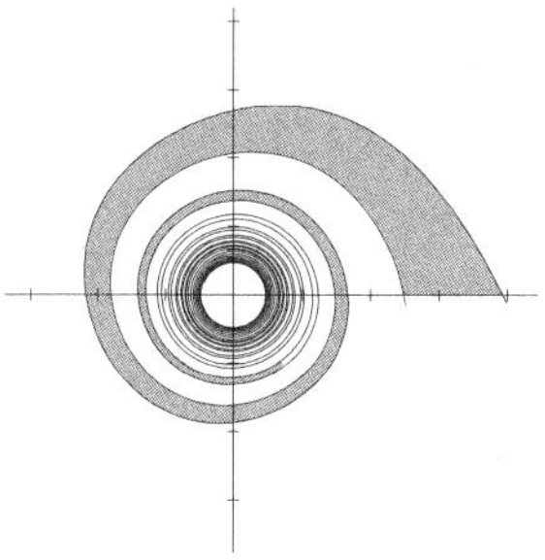

FIGURE 6.6.7.

Example 6.6.12: The shaded region spirals inward. If the spirals bounding the shaded region are chosen as in Exercise 5.3.2, they have infinite length near the origin. The nonsmooth boundary of the shaded region consists of a single point, but the smooth boundary has infinite length, so condition 2 of Definition 6.6.8 is not satisfied.

Example 6.6.12 (Region that is not a piece-with-boundary). The compact subset of $\mathbb{R}^2$ shown in Figure 6.6.7 is bounded by the two spirals parametrized by $r = 1/\theta$ and $r = 2/\theta$, for $\theta \geq 1/(2\pi)$. Again the nonsmooth boundary is just three points, so condition 1 of Definition 6.6.8, is satisfied. But it does not satisfy condition 2, since the 1-dimensional volume (length) of the smooth boundary is infinite (see Exercise 5.3.2). So it is not a piece-with-boundary. $\triangle$

Example 6.6.13 (Another locus not a piece-with-boundary). Let $M \subset \mathbb{R}^2$ be the x-axis and let $X \subset M$ be the locus defined by $g \leq 0$, for

$$g(x) = \begin{cases} x^3 \sin \frac{1}{x} & \text{if } x \neq 0 \\ 0 & \text{if } x = 0. \end{cases} \qquad 6.6.2$$

This function g is C^1, as required by Definition 6.6.2; this can be seen by the same computation as in Example 1.9.4, replacing the x^2 by x^3.

As shown in Figure 6.6.8, the locus $X \subset M = \mathbb{R}$ consists of disjoint intervals $[x_{2i-1}, x_{2i}]$, $i = 1, 2, \dots$ of the x-axis. Away from 0, each such interval has a civilized boundary: the endpoints x_j of the interval. These points are smooth points of the boundary, since M is smooth and $g'(x_j) \neq 0$.

The only nonsmooth point of the boundary is the point $\begin{pmatrix}x\\y\end{pmatrix} = \begin{pmatrix}0\\0\end{pmatrix}$; it is not smooth because $g'(0) = 0$, so g' is not onto $\mathbb{R}$.[8] The second condition of Definition 6.6.8 is not met because, as Exercise 5.2.6 asked you to show, the 0-dimensional volume of a single point is 1. Indeed, a subset of a curve cannot be a piece-with-boundary if it has any nonsmooth boundary. $\triangle$

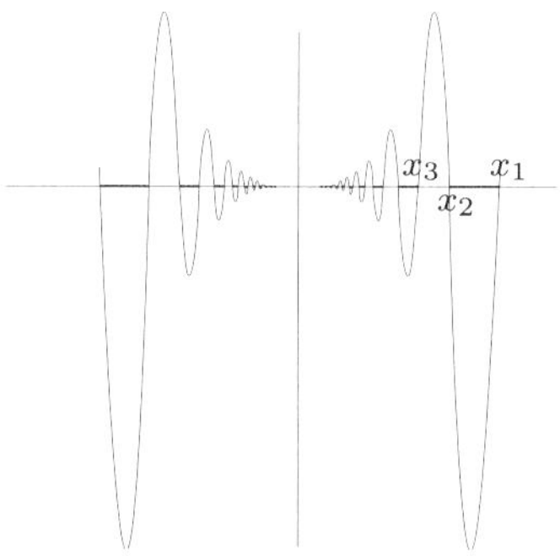

FIGURE 6.6.8.

Near 0, the one-dimensional locus of Example 6.6.13 consists of infinitely many arbitrarily small intervals on the x-axis. This locus is not a piece-with-boundary of $\mathbb{R}$.

Remark. Our definition of a piece-with-boundary is designed to make Stokes's theorem true. Stokes's theorem is the higher dimensional form of the fundamental theorem of calculus. In the case of our set X, the fundamental theorem of calculus would say

$$\int_X h'(x)\, dx = \sum_{i=1}^{\infty} \big(h(x_{2i-1}) - h(x_{2i})\big). \qquad 6.6.3$$

To prove such a statement, we would need to think about the convergence of the series on the right. This is possible, but the series is not absolutely convergent so the issue is a bit delicate. It would become much more delicate in higher dimensions. It is to avoid such complications that we put the second condition in our definition of a piece-with-boundary. $\triangle$

[8]Actually, showing that g' is not onto is not good enough; we need to show that there is no C^1 function defining X whose derivative is onto. If there were such a function, then the boundary of X near 0 would be a manifold, by Theorem 3.1.10, but it isn't; it isn't the graph of a map from one point into $\mathbb{R}$.

Example 6.6.14. In constrast to the locus X of Example 6.6.13, the two-dimensional locus consisting of the area between the graph and the x-axis in Figure 6.6.8 is a piece-with-boundary of $\mathbb{R}^2$. The boundary is the x-axis and the graph of g; the points where they intersect are nonsmooth points. Both conditions of Definition 6.6.8 are met: here $k-1=1$ and the nonsmooth points have 1-dimensional volume (i.e., length) 0, and for every $\epsilon > 0$ we can find an open subset $U \subset \mathbb{R}^2$ that covers all the nonsmooth points and such that the length of $\partial^s_{\mathbb{R}^2} Y \cap U$ is less than ϵ. (Remember that U need not be connected.) This locus is not a piece-with-corners because the origin is not a corner point. △

Example 6.6.15 (A k-parallelogram seen as a piece-with-boundary of a manifold). Since k-parallelograms are the crucial construction in our theory of integration, we need to see that a k-parallelogram $P_{\mathbf{x}}(\vec{\mathbf{v}}_1, \dots, \vec{\mathbf{v}}_k)$ is a piece-with-corners of a k-dimensional submanifold of $\mathbb{R}^n$.

We define a locus by a set of equations, usually but not always of the form $\mathbf{F}(\mathbf{x}) = \mathbf{0}$; see Theorem 3.1.10 and Example 3.2.5. We define a piece of a manifold by inequalities saying where the piece begins and ends; see Figure 6.6.3. So we begin by finding the relevant equations and inequalities.

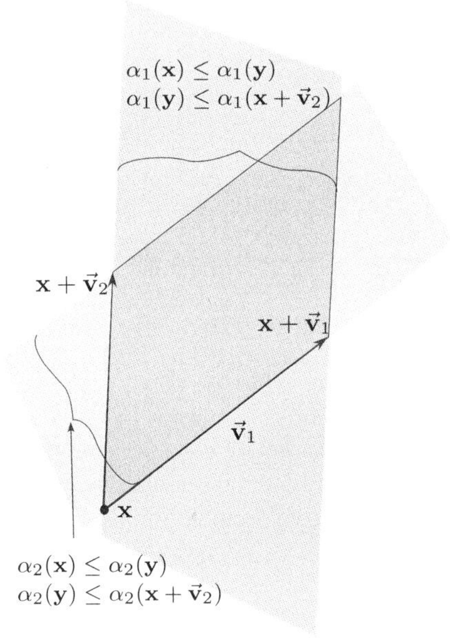

FIGURE 6.6.9.
The pale gray band is defined by the two inequalities

$$\alpha_1(\mathbf{x}) \le \alpha_1(\mathbf{y}) \le \alpha_1(\mathbf{x}+\vec{\mathbf{v}}_2).$$

The medium gray band is defined by

$$\alpha_2(\mathbf{x}) \le \alpha_2(\mathbf{y}) \le \alpha_2(\mathbf{x}+\vec{\mathbf{v}}_2).$$

The intersection of the two is the parallelogram $P_{\mathbf{x}}(\vec{\mathbf{v}}_1, \vec{\mathbf{v}}_2)$.

Denote by $V \subset \mathbb{R}^n$ the subspace spanned by $\vec{\mathbf{v}}_1, \dots, \vec{\mathbf{v}}_k$. Choose a linear transformation $A: \mathbb{R}^n \to \mathbb{R}^{n-k}$ (i.e., an $(n-k) \times n$ matrix) such that $\ker A = V$. Let $\mathbf{F}: \mathbb{R}^n \to \mathbb{R}^{n-k}$ be the mapping $\mathbf{F}(\mathbf{y}) = A\mathbf{y} - A\mathbf{x}$. Then $P_{\mathbf{x}}(\vec{\mathbf{v}}_1, \dots, \vec{\mathbf{v}}_k)$ is a subset of the k-dimensional manifold M defined by $\mathbf{F}(\mathbf{y}) = \mathbf{0}$. The manifold M is an affine subspace of $\mathbb{R}^n$.

We can further define linear functions $\alpha_i : M \to \mathbb{R}$ such that

$$\ker \alpha_i = \operatorname{Span}(\vec{\mathbf{v}}_1, \dots, \widehat{\vec{\mathbf{v}}_i}, \dots, \vec{\mathbf{v}}_k), \qquad 6.6.4$$

where the hat on $\vec{\mathbf{v}}_i$ indicates that $\vec{\mathbf{v}}_i$ is omitted. Since the vectors $\vec{\mathbf{v}}_1, \dots, \vec{\mathbf{v}}_k$ are linearly independent and

$$\vec{\mathbf{v}}_i \notin \operatorname{Span}(\vec{\mathbf{v}}_1, \dots, \widehat{\vec{\mathbf{v}}_i}, \dots, \vec{\mathbf{v}}_k), \qquad 6.6.5$$

we have $\alpha_i(\vec{\mathbf{v}}_i) \ne 0$. Changing the sign of α_i if necessary, we may assume that $\alpha_i(\vec{\mathbf{v}}_i) > 0$. Then, as shown in Figure 6.6.9, $P_{\mathbf{x}}(\vec{\mathbf{v}}_1, \dots, \vec{\mathbf{v}}_k)$ is defined in M by the $2k$ inequalities

$$\alpha_i(\mathbf{x}) \le \alpha_i(\mathbf{y}) \le \alpha_i(\mathbf{x}+\vec{\mathbf{v}}_i). \qquad 6.6.6$$

The boundary $\partial_M P_{\mathbf{x}}(\vec{\mathbf{v}}_1, \dots, \vec{\mathbf{v}}_k)$ is the set where $\mathbf{F} = \mathbf{0}$ (because the k-parallelogram is in M) and at least one of the $2k$ inequalities in formula 6.6.6 is an equality. The *smooth part* of the boundary is the locus where exactly one inequality in 6.6.6 is an equality and the others are all strict inequalities. In this case the function g of Definition 6.6.2 is

$$g(\mathbf{y}) = \alpha_i(\mathbf{y}) - \alpha_i(\mathbf{x}) \quad \text{or} \quad g(\mathbf{y}) = \alpha_i(\mathbf{x}+\vec{\mathbf{v}}_i) - \alpha_i(\mathbf{y}), \qquad 6.6.7$$

depending on whether the inequality realized as an equality is on the left or the right of equation 6.6.6.

The *nonsmooth part* of the boundary is where at least two of the $2k$ inequalities (necessarily coming form $i_1 \ne i_2$) are satisfied as equalities.

Condition 1 of Definition 6.6.8 says that for $P_{\mathbf{x}}(\vec{\mathbf{v}}_1, \dots, \vec{\mathbf{v}}_k)$ to be a piece-with-boundary, its nonsmooth boundary must have $(k-1)$-dimensional volume 0. This follows from Proposition 5.2.2; indeed this set is a subset of finitely many affine subspaces of dimension $k-2$.

Condition 2 is also satisfied: if $T_i = [\vec{\mathbf{v}}_1, \dots, \widehat{\vec{\mathbf{v}}_i}, \dots, \vec{\mathbf{v}}_k]$, then the $(k-1)$-dimensional volume of the smooth boundary of $P_{\mathbf{x}}(\vec{\mathbf{v}}_1, \dots, \vec{\mathbf{v}}_k)$ is

$$2\sum_{i=1}^{k} \sqrt{\det(T_i^\top T_i)}; \qquad 6.6.8$$

for each i, there are two faces (in dimension 3, a top and bottom, left and right, front and back). In particular, the $(k-1)$-dimensional volume of the smooth boundary is finite.

Exercise 6.6.10 asks you to show the slightly stronger statement: that when $\mathbf{v}_1, \dots, \mathbf{v}_k$ are linearly independent, $P_{\mathbf{x}}(\vec{\mathbf{v}}_1, \dots, \vec{\mathbf{v}}_k)$ is a piece-with-corners. $\triangle$

Theorem 6.6.16. *If $X \subset M$ is a k-dimensional piece-with-boundary, then X has finite k-dimensional volume.*

Proof. By Corollary 5.3.4, every point has a neighborhood with finite k-dimensional volume. Since a piece-with-boundary is compact, it can be covered by finitely many such neighborhoods. $\square$

Proposition 6.6.17 (Products of pieces-with-boundary). *Let $M \subset \mathbb{R}^m$ and $P \subset \mathbb{R}^p$ be oriented manifolds, M of dimension k and P of dimension l. If $X \subset M$ and $Y \subset P$ are pieces-with-boundary, then $M \times P \subset \mathbb{R}^{m+p}$ is a $(k+l)$-dimensional oriented manifold, and $X \times Y$ is a piece-with-boundary of $M \times P$.*

Proposition 6.6.17 is proved in Appendix A20.

Translating and rotating pieces-with-boundary

We saw in Section 3.1 that a smooth manifold that is translated, rotated, or "linearly distorted" in the sense of Corollary 3.1.17 remains a smooth manifold. The following proposition shows that the same is true of a piece-with-boundary of a manifold.

Proposition 6.6.18 (Piece-with-boundary independent of coordinates). *Let $\mathbf{g} : \mathbb{R}^n \to \mathbb{R}^n$ be a mapping of the form*

$$\mathbf{g}(\mathbf{x}) = A\mathbf{x} + \mathbf{c}, \qquad 6.6.9$$

where A is an invertible matrix. If $M \subset \mathbb{R}^n$ is a smooth k-dimensional manifold, and $X \subset M$ is a piece-with-boundary, then $\mathbf{g}(X)$ is a piece-with-boundary of $\mathbf{g}(M)$.

Proof. Let V_N be the union of the cubes of $\mathcal{D}_N(\mathbb{R}^n)$ that contain nonsmooth points of $\partial_M X$; suppose that V_N is made up of b_N cubes. Then $\mathbf{g}(V_N)$ contains all the nonsmooth points of $\partial_{\mathbf{g}(M)}\mathbf{g}(X)$, since by Corollary 3.1.17, the image of the smooth boundary is smooth. But for each cube C of V_N, the image $\mathbf{g}(C)$ intersects at most $(2(\sqrt{n}+1))^n$ cubes of $\mathcal{D}_N(\mathbb{R}^n)$; see Exercise 6.6.7, part a. In particular, there are at most

$$(2(\sqrt{n}+1))^n b_N \tag{6.6.10}$$

cubes of $\mathcal{D}_N(\mathbb{R}^n)$ that contain nonsmooth points of $\partial_{\mathbf{g}(M)}\mathbf{g}(X)$.

Thus the hypothesis

$$\underbrace{\lim_{N\to\infty} \frac{b_N}{2^{(k-1)N}} = 0}_{\text{Def. 6.6.8, condition 1, for } X} \qquad \text{implies} \qquad \underbrace{\lim_{N\to\infty} \frac{(2(\sqrt{n}+1))^n b_N}{2^{(k-1)N}} = 0}_{\text{Def. 6.6.8, condition 1, for } \mathbf{g}(X)}. \tag{6.6.11}$$

Thus $\mathbf{g}(X)$ meets the first condition for a piece-with-boundary.

We will first check the second condition when A is orthogonal, i.e., when $A^\top A = I$. By hypothesis, the smooth boundary of X has finite $(k-1)$-dimensional volume, so, by Definition 5.3.1, there exist a bounded open set $W \subset \mathbb{R}^{k-1}$ with boundary of volume 0 and a parametrization $\gamma : W \to \partial^s_M(X)$ such that

$$\mathrm{vol}_{k-1}\, \partial^s_M X = \int_W \sqrt{\det[\mathbf{D}\gamma(\mathbf{w})]^\top[\mathbf{D}\gamma(\mathbf{w})]}\,|d^{k-1}\mathbf{w}| < \infty. \tag{6.6.12}$$

Moreover, $\mathbf{g}\circ\gamma$ parametrizes the smooth boundary of $\mathbf{g}(X)$, and

$$\begin{aligned} \mathrm{vol}_{k-1}\partial^s_{\mathbf{g}(M)}\mathbf{g}(X) &= \int_W \sqrt{\det[\mathbf{D}(\mathbf{g}\circ\gamma)(\mathbf{w})]^\top[\mathbf{D}(\mathbf{g}\circ\gamma)(\mathbf{w})]}\,|d^{k-1}\mathbf{w}| \\ &= \int_W \sqrt{\det[\mathbf{D}\gamma(\mathbf{w})]^\top[\mathbf{D}\mathbf{g}(\gamma(\mathbf{w}))]^\top[\mathbf{D}\mathbf{g}(\gamma(\mathbf{w}))][\mathbf{D}\gamma(\mathbf{w})]}\,|d^{k-1}\mathbf{w}| \\ &\underset{1}{=} \int_W \sqrt{\det[\mathbf{D}\gamma(\mathbf{w})]^\top A^\top A[\mathbf{D}\gamma(\mathbf{w})]}\,|d^{k-1}\mathbf{w}| \\ &\underset{2}{=} \int_W \sqrt{\det[\mathbf{D}\gamma(\mathbf{w})]^\top[\mathbf{D}\gamma(\mathbf{w})]}\,|d^{k-1}\mathbf{w}|. \end{aligned} \tag{6.6.13}$$

Thus the area of the smooth boundary is unchanged by rotations and translations, so it is still finite, so condition 2 is met. This proves Proposition 6.6.18 when A is orthogonal. Proving the proposition when A is just invertible, not necessarily orthogonal, requires two statements, which you are asked to prove in Exercise 6.6.7:

1. The image $\mathbf{g}(C)$ of a cube $C \in V_N$ intersects at most $\big((|A|+1)\sqrt{n}+2\big)^n$ cubes of $\mathcal{D}_N(\mathbb{R}^n)$.

2. If A is an $n\times n$ matrix, then there exists a number M such that for all $n \times k$ matrices B,

$$\det\big((AB)^\top AB\big) \le M \det B^\top B, \tag{6.6.14}$$

giving a bound rather than an equality in the equation corresponding to equation 6.6.13. □

Proof of Proposition 6.6.18: Recall that an $n\times n$ orthogonal matrix is a matrix whose columns form an orthonormal basis of $\mathbb{R}^n$, and that a square matrix A is orthogonal if and only if $A^\top A = I$. All orthogonal matrices have determinant $+1$ or -1:

$$1 = \det I = \det(A^\top A) = (\det A)^2.$$

Those with determinant $+1$ give rotations, those with determinant -1 give reflections.

Equation 6.6.11: By Definition 6.6.8, a subset must meet two conditions in order to be a piece-with-boundary. The first is that the set of nonsmooth points in its boundary must have $(k-1)$-dimensional volume 0. Since X is a piece-with-boundary, we know this is true for X, and it follows that it is also true for $\mathbf{g}(X)$.

The equality marked 1 in equation 6.6.13: Since

$$\mathbf{g}(\mathbf{x}) = A\mathbf{x} + \mathbf{c},$$

we have

$$[\mathbf{D}(\mathbf{g}(\gamma))] = A.$$

The equality marked 2: Going from line 3 to line 4: Since A is orthogonal, $A^\top A = I$.

Orientation of a piece-with-boundary

Let $M \subset \mathbb{R}^n$ be a smooth submanifold, and $X \subset M$ a piece-with-boundary of M. An *orientation of* X is simply an orientation of M, or at least an orientation of a subset $U \cap M$, where $U \subset \mathbb{R}^n$ is an open subset containing the piece-with-boundary X.[9]

Example 6.6.19. (Orienting a parallelogram). Let $\vec{\mathbf{v}}_1, \dots, \vec{\mathbf{v}}_k \in \mathbb{R}^n$ be linearly independent vectors spanning a subspace V; let $\mathbf{x} \in \mathbb{R}^n$ be a point. We saw in Example 6.6.15 that $P_{\mathbf{x}}(\vec{\mathbf{v}}_1, \dots, \vec{\mathbf{v}}_k)$ is a piece-with-boundary of the manifold M given by $A\mathbf{y} - A\mathbf{x} = \mathbf{0}$, where $A : \mathbb{R}^n \to \mathbb{R}^{n-k}$ is a linear transformation such that $\ker A = V$. The *standard orientation of* $P_{\mathbf{x}}(\vec{\mathbf{v}}_1, \dots, \vec{\mathbf{v}}_k)$ is the orientation such that $\vec{\mathbf{v}}_1, \dots, \vec{\mathbf{v}}_k$ in that order form a direct basis of $T_{\mathbf{y}}M$ for every $\mathbf{y} \in M$; note that $T_{\mathbf{y}}M = V$ for every $\mathbf{y} \in M$. When referring to $P_{\mathbf{x}}(\vec{\mathbf{v}}_1, \dots, \vec{\mathbf{v}}_k)$ as an oriented piece-with-boundary, we mean "with the standard orientation". We will denote by $-P_{\mathbf{x}}(\vec{\mathbf{v}}_1, \dots, \vec{\mathbf{v}}_k)$ the parallelogram with the opposite orientation. $\triangle$

Boundary orientation

Definition 6.6.20: The inequality $[\mathbf{D}g(\mathbf{x})]\vec{\mathbf{v}} > 0$ means that g is increasing in the direction $\vec{\mathbf{v}}$, so $\vec{\mathbf{v}}$ points into the region where $g > 0$. Recall from Definition 6.6.2 that X was by definition the region $g \geq 0$. If we had chosen the opposite convention, that would change the signs of the outward-pointing and inward-pointing vectors.

If P is a piece-with-boundary of a surface M, then $T_{\mathbf{x}}M$ is a plane and $T_{\mathbf{x}}\partial_M^s P$ is a line in that plane. If P is a piece-with-boundary of a three-dimensional manifold $M \subset \mathbb{R}^n$, then $T_{\mathbf{x}}M$ is a three-dimensional vector space and $T_{\mathbf{x}}\partial_M^s P$ is a plane.

We need orientation of domains and their boundaries so that we can integrate and differentiate forms, but orientation is important for other reasons. Homology theory, one of the big branches of algebraic topology, is an enormous abstraction of the constructions in our discussion of orientation.

To orient boundaries, we need the following definition. It is illustrated by Figures 6.6.10 and 6.6.11.

Definition 6.6.20 (Outward-pointing and inward-pointing vectors). Let $M \subset \mathbb{R}^n$ be a manifold, $X \subset M$ a piece-with-boundary, $\mathbf{x}$ a smooth point in $\partial_M X$, and g the function referred to in Definition 6.6.2. Let $\vec{\mathbf{v}}$ be tangent to M at $\mathbf{x}$. Then $\vec{\mathbf{v}}$

$$\textit{points outward from } X \text{ if } [\mathbf{D}g(\mathbf{x})]\vec{\mathbf{v}} < 0, \quad \text{and}$$
$$\textit{points inward to } X \text{ if } [\mathbf{D}g(\mathbf{x})]\vec{\mathbf{v}} > 0.$$

A vector that points out of a piece-with-boundary is an *outward-pointing vector*; a vector that points in is an *inward-pointing vector*.

Why do the inequalities $[\mathbf{D}g(\mathbf{x})]\vec{\mathbf{v}} < 0$ and $[\mathbf{D}g(\mathbf{x})]\vec{\mathbf{v}} > 0$ define outward-pointing and inward-pointing vectors? The tangent space at $\mathbf{x}$ to the smooth boundary consists of vectors $\vec{\mathbf{v}}$ anchored at $\mathbf{x}$ that are tangent to both M and $\partial_M^s X$: $[\mathbf{Df}(\mathbf{x})]\vec{\mathbf{v}} = \mathbf{0}$ and $[\mathbf{D}g(\mathbf{x})]\vec{\mathbf{v}} = 0$. (If M is $\mathbb{R}^n$, or an open subset of $\mathbb{R}^n$, then the tangent space $T_{\mathbf{x}}M$ is all of $\mathbb{R}^n$, or, rather, vectors anchored at $\mathbf{x}$ and pointing in every possible direction.)

This tangent space separates $T_{\mathbf{x}}M$ into two sets of vectors: those that point into the piece-with-boundary and those that point away from it. If $\vec{\mathbf{v}}$ points from a point where g is 0 to the region where g is negative (away from the piece), the directional derivative $[\mathbf{D}g(\mathbf{x})]\vec{\mathbf{v}}$ is negative. If $\vec{\mathbf{v}}$ points

[9] To orient a piece-with-boundary of M, you do not need to orient all of M, just a part of M big enough to contain the piece, so it may be possible, for instance, to orient a piece-with-boundary of a Möbius band.

from a point where g is 0 to the region where g is positive (into the piece), the directional derivative $[\mathbf{D}g(\mathbf{x})]\vec{\mathbf{v}}$ is positive.

Definition 6.6.21 (Oriented boundary of piece-with-boundary of an oriented manifold). Let M be a k-dimensional manifold oriented by Ω, and P a piece-with-boundary of M. Let $\mathbf{x}$ be a point of the smooth boundary $\partial_M^s P$ and let $\vec{\mathbf{v}}_{\text{out}} \in T_{\mathbf{x}}M$ be an outward-pointing vector. Then the function $\Omega^\partial : \mathcal{B}(T_{\mathbf{x}}\partial P) \to \{+1, -1\}$ given by

$$\overbrace{\underbrace{\Omega^\partial_{\mathbf{x}}(\vec{\mathbf{v}}_1, \dots, \vec{\mathbf{v}}_{k-1})}_{\text{basis of } T_{\mathbf{x}}\partial_M^s P}}^{\text{orienting smooth boundary}} \stackrel{\text{def}}{=} \overbrace{\Omega_{\mathbf{x}}\underbrace{(\vec{\mathbf{v}}_{\text{out}}, \vec{\mathbf{v}}_1, \dots, \vec{\mathbf{v}}_{k-1})}_{\text{basis of } T_{\mathbf{x}}M}}^{\text{orienting manifold}} \qquad 6.6.15$$

defines an orientation on the smooth boundary $\partial_M^s P$.

When reading Definition 6.6.21, note that the boundary orientation of a piece-with-boundary depends on the orientation of the manifold; if the orientation of the manifold is changed, the boundary orientation will change also.

Example 6.6.22 (Oriented boundary of a piece-with-boundary of an oriented curve). Let C be a curve oriented by Ω, P a piece-with-boundary of C, and $\mathbf{x}$ a smooth point in its boundary. Then

$$\Omega^\partial_{\mathbf{x}}(\phi) = \Omega_{\mathbf{x}}(\vec{\mathbf{v}}_{\text{out}}), \qquad 6.6.16$$

where ϕ is the unique basis of the trivial vector space $T_{\mathbf{x}}\{\mathbf{x}\}$; here $\{\mathbf{x}\}$ is the 0-dimensional manifold consisting of $\mathbf{x}$.

If the outward-pointing vector $\vec{\mathbf{v}}_{\text{out}}$ is a direct basis of $T_{\mathbf{x}}C$ (pointing in the same direction as the tangent vector field orienting the curve), then $\Omega^\partial_{\mathbf{x}} = +1$; if it is an indirect basis, then $\Omega^\partial_{\mathbf{x}} = -1$.

Thus at the endpoint of the piece-with-boundary, $\Omega^\partial = +1$, and at the beginning point $\Omega^\partial = -1$; see Figure 6.6.10. △

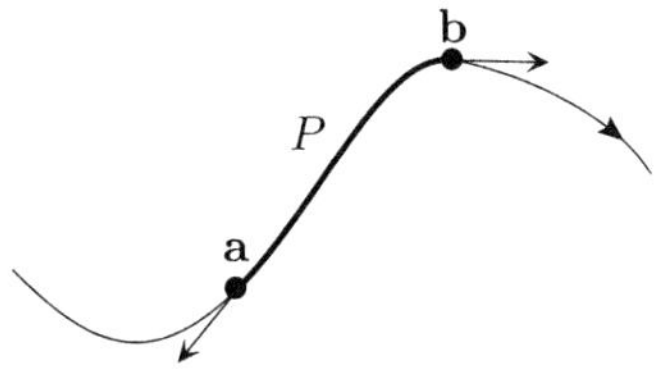

FIGURE 6.6.10. Example 6.6.22: An oriented curve with a piece-with-boundary P marked in bold. The boundary of P consists of the points $\mathbf{a}$ and $\mathbf{b}$. The vector pointing out of P at $\mathbf{b}$ points in the same direction as the curve, so $\Omega^\partial_{\mathbf{b}} = +1$; the vector pointing out of P at $\mathbf{a}$ points in the opposite direction as the curve, so $\Omega^\partial_{\mathbf{a}} = -1$. Thus the oriented boundary of P is $+\mathbf{b} - \mathbf{a}$.

Example 6.6.23 (Oriented boundary of a piece-with-boundary of $\mathbb{R}^2$). Let a smooth curve C be the smooth boundary of a piece-with-boundary $S \subset \mathbb{R}^2$. Give $\mathbb{R}^2$ the standard orientation $\Omega = \operatorname{sgn}\det$. Then at a point $\mathbf{x} \in C$, the boundary C is oriented by

$$\Omega^\partial(\vec{\mathbf{v}}) = \operatorname{sgn}\det(\vec{\mathbf{v}}_{\text{out}}, \vec{\mathbf{v}}), \quad \text{where } \vec{\mathbf{v}} \in T_{\mathbf{x}}C. \qquad 6.6.17$$

To relate this to our notion of "right" and "left", assume that we draw the standard basis vectors in the plane in the standard way, with $\vec{\mathbf{e}}_2$ counterclockwise (to the left) from $\vec{\mathbf{e}}_1$. Then

$$\det(\vec{\mathbf{v}}_{\text{out}}, \vec{\mathbf{v}}) > 0 \qquad 6.6.18$$

corresponds to the vector $\vec{\mathbf{v}}_{\text{out}}$ being on your right when you stand on the curve and look in the direction of $\vec{\mathbf{v}}$ (see part 2 of Proposition 1.4.14). Then S is on your left, as shown in Figure 6.6.11. △

Example 6.6.24 (Oriented boundary of a piece-with-boundary of an oriented surface in $\mathbb{R}^3$). Let $P \subset S$ be a piece-with-boundary of a surface S oriented by a normal vector field $\vec{\mathbf{n}}$; it is shown as the shaded region in Figure 6.6.12. In this case Definition 6.6.21 says that at $\mathbf{x}$, the curve $\partial_S P$ is oriented by

$$\Omega^{\partial}_{\mathbf{x}}(\vec{\mathbf{v}}) = \Omega^{\vec{\mathbf{n}}}_{\mathbf{x}}(\vec{\mathbf{v}}_{\text{out}}, \vec{\mathbf{v}}), \quad \text{where } \vec{\mathbf{v}} \in T_{\mathbf{x}}\partial_S P. \qquad 6.6.19$$

By Proposition 6.3.4,

$$\Omega^{\vec{\mathbf{n}}}_{\mathbf{x}}(\vec{\mathbf{v}}_1, \vec{\mathbf{v}}_2) = \operatorname{sgn}\det[\vec{\mathbf{n}}(\mathbf{x}), \vec{\mathbf{v}}_1, \vec{\mathbf{v}}_2], \qquad 6.6.20$$

where $\vec{\mathbf{v}}_1, \vec{\mathbf{v}}_2$ form a basis of $T_{\mathbf{x}}S$. This gives

$$\Omega^{\partial}_{\mathbf{x}}(\vec{\mathbf{v}}) = \Omega^{\vec{\mathbf{n}}}_{\mathbf{x}}(\vec{\mathbf{v}}_{\text{out}}, \vec{\mathbf{v}}) = \operatorname{sgn}\det\big(\vec{\mathbf{n}}(\mathbf{x}), \vec{\mathbf{v}}_{\text{out}}, \vec{\mathbf{v}}\big). \qquad 6.6.21$$

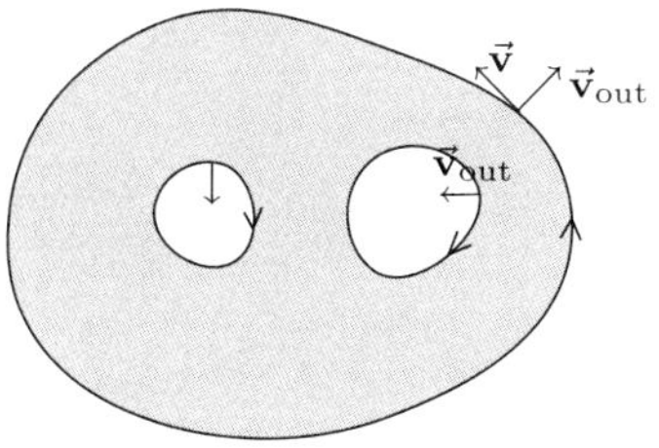

FIGURE 6.6.11.
The oriented boundary of the shaded region of $\mathbb{R}^2$ consists of three curves. If you walk along them following the arrows, the outward-pointing vector is to your right, and the region to your left.

The basis $\vec{\mathbf{v}}$ of $T_{\mathbf{x}}\partial_S P$ is direct for the boundary orientation if the sgn det in equation 6.6.21 is $+1$.

Again, to relate this to our notion of "right" and "left", draw $\vec{\mathbf{e}}_1, \vec{\mathbf{e}}_2, \vec{\mathbf{e}}_3$ the standard way. By Proposition 1.4.21, the sgn det of equation 6.6.20 will be $+1$ if $\vec{\mathbf{n}}(\mathbf{x}), \vec{\mathbf{v}}_{\text{out}}, \vec{\mathbf{v}}$ (in that order) satisfy the right-hand rule. If you walk on the boundary of P with your head pointing in the direction of $\vec{\mathbf{n}}$, then you are walking in the "direct" direction for the boundary orientation if P is to your left.

For example, consider the United States as a piece-with-boundary of the surface of the earth, oriented by the outward-pointing normal. At Boston, stand at the land's edge and face the ocean (in the direction of the outward-pointing vector). Stick out your left arm; it is pointing in the direction of the boundary orientation: north. (But at Los Angeles, the boundary orientation points south.) At a point of the boundary with Canada, the boundary orientation *of the United States* points west, but at the same point, the boundary orientation *of Canada* points east. $\triangle$

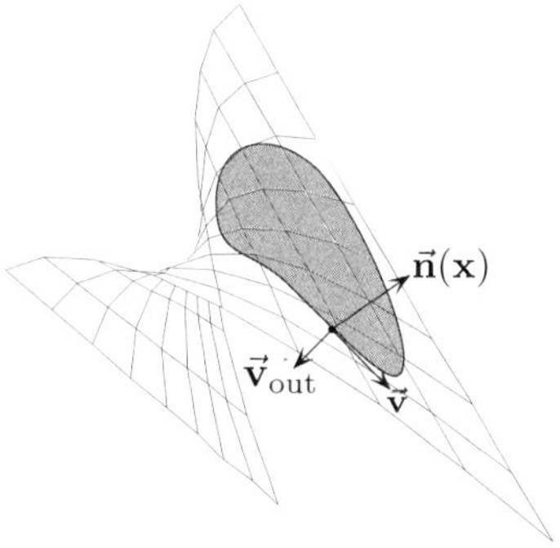

FIGURE 6.6.12.
The shaded area is the piece-with-boundary P of the surface S. The vector $\vec{\mathbf{v}}_{\text{out}}$ is tangent to S at a point in the boundary of S and points out of P. The vector $\vec{\mathbf{v}}$ is tangent to the boundary. The vectors $\vec{\mathbf{n}}(\mathbf{x}), \vec{\mathbf{v}}_{\text{out}}, \vec{\mathbf{v}}$ satisfy the right-hand rule, so (by Proposition 1.4.21) $\Omega^{\partial}_{\mathbf{x}}(\vec{\mathbf{v}}) = +1$ since

$$\operatorname{sgn}\det[\vec{\mathbf{n}}(\mathbf{x}), \vec{\mathbf{v}}_{\text{out}}, \vec{\mathbf{v}}] = +1.$$

So $\vec{\mathbf{v}}$ is direct for the boundary orientation.

Example 6.6.25 (Oriented boundary of a piece-with-boundary of $\mathbb{R}^3$). Suppose U is a piece-with-boundary of $\mathbb{R}^3$ with the standard orientation by det. Then by Definition 6.6.21 its boundary S is oriented by

$$\Omega^{\partial}(\vec{\mathbf{v}}_1, \vec{\mathbf{v}}_2) = \operatorname{sgn}\det(\vec{\mathbf{v}}_{\text{out}}, \vec{\mathbf{v}}_1, \vec{\mathbf{v}}_2). \qquad 6.6.22$$

Thus, if a surface S bounds a piece $U \subset \mathbb{R}^3$, with $\mathbb{R}^3$ given the standard orientation, the orientation of S by the outward-pointing normal is the boundary orientation of U.

A basis $\vec{\mathbf{u}}, \vec{\mathbf{v}}$ for $T_{\mathbf{x}}S$ is direct for that boundary orientation if $\vec{\mathbf{u}} \times \vec{\mathbf{v}}$ points out, since $\det[\vec{\mathbf{u}} \times \vec{\mathbf{v}}, \vec{\mathbf{u}}, \vec{\mathbf{v}}] > 0$. (Recall that by Proposition 1.4.20, if $\vec{\mathbf{u}}$ and $\vec{\mathbf{v}}$ are not collinear, which they certainly aren't, since they form a basis, then $\det[\vec{\mathbf{u}}, \vec{\mathbf{v}}, \vec{\mathbf{u}} \times \vec{\mathbf{v}}] > 0$. It takes two transpositions to go from $\vec{\mathbf{u}}, \vec{\mathbf{v}}, \vec{\mathbf{u}} \times \vec{\mathbf{v}}$ to $\vec{\mathbf{u}} \times \vec{\mathbf{v}}, \vec{\mathbf{u}}, \vec{\mathbf{v}}$, so if $\det[\vec{\mathbf{u}}, \vec{\mathbf{v}}, \vec{\mathbf{u}} \times \vec{\mathbf{v}}] > 0$, then $\det[\vec{\mathbf{u}} \times \vec{\mathbf{v}}, \vec{\mathbf{u}}, \vec{\mathbf{v}}] > 0$.) $\triangle$

The oriented boundary of an oriented k-parallelogram

We saw in Example 6.6.15 that when the vectors $\vec{\mathbf{v}}_1, \dots, \vec{\mathbf{v}}_k$ are linearly independent, an oriented k-parallelogram $P_{\mathbf{x}}(\vec{\mathbf{v}}_1, \dots, \vec{\mathbf{v}}_k)$ is a piece-with-boundary of an affine subspace M parallel to $\operatorname{Span}(\vec{\mathbf{v}}_1, \dots, \vec{\mathbf{v}}_k)$. Since

Span $(\vec{\mathbf{v}}_1, \dots, \vec{\mathbf{v}}_k)$ is oriented by the order of the vectors, a k-parallelogram is an *oriented* piece-with-boundary. As such its boundary carries an orientation.

Proposition 6.6.26 (Oriented boundary of oriented k-parallelogram). *Let the k-parallelogram $P_{\mathbf{x}}(\vec{\mathbf{v}}_1, \dots, \vec{\mathbf{v}}_k)$ have the standard orientation (see Example 6.6.19). Then its oriented boundary is given by the following, where a hat over a term indicates that it is omitted:*

$$\partial P_{\mathbf{x}}(\vec{\mathbf{v}}_1, \dots, \vec{\mathbf{v}}_k) = \sum_{i=1}^{k} (-1)^{i-1} \Big(P_{\mathbf{x}+\vec{\mathbf{v}}_i}(\vec{\mathbf{v}}_1, \dots, \widehat{\vec{\mathbf{v}}_i}, \dots, \vec{\mathbf{v}}_k) - P_{\mathbf{x}}(\vec{\mathbf{v}}_1, \dots, \widehat{\vec{\mathbf{v}}_i}, \dots, \vec{\mathbf{v}}_k) \Big). \qquad 6.6.23$$

Proposition 6.6.26 is proved following Example 6.6.29.

What does the sum in the expression 6.6.23 mean? We can't add and subtract parallelograms! The sum really means the following: to integrate a $(k-1)$-form φ over $\partial P_{\mathbf{x}}(\vec{\mathbf{v}}_1, \dots, \vec{\mathbf{v}}_k)$, we integrate φ over the summands, which are $(k-1)$-parallelograms, to which we give the standard orientation multiplied by the indicated sign, and add the results. Alternately, we integrate over the summand parallelograms with their standard orientations, and combine the resulting numbers with the indicated signs.

Why does each term omit a vector? The smooth boundary of an object always has one dimension less than the object itself. The boundary of a k-dimensional parallelogram is made up of $(k-1)$-parallelograms, so omitting a vector gives the right number of vectors.

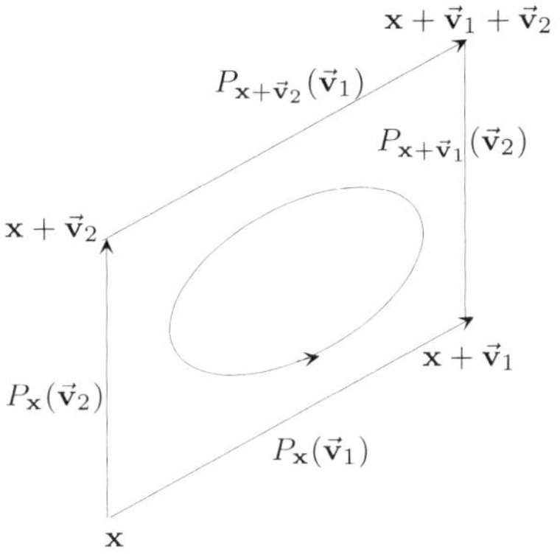

FIGURE 6.6.13.

If you go around the boundary of the parallelogram counterclockwise, starting at $\mathbf{x}$, you will find the sum in equation 6.6.25. The last two edges of that sum are negative because you are traveling against the direction of the vectors in question.

The important consequence of preceding each term by $(-1)^{i-1}$ is that the *oriented boundary of the oriented boundary* is 0. For a cube, the boundary of the boundary consists of the edges of each face. Each edge appears twice, once with a + sign and once with a − sign, so that the two cancel.

Remark. A 0-dimensional manifold is oriented by the choice of sign (see Proposition 6.3.4). Thus an oriented 0-parallelogram $P_{\mathbf{x}}$ is either $+P_{\mathbf{x}}$ or $-P_{\mathbf{x}}$. Since $P_{\mathbf{x}}$ is itself a manifold, its boundary is empty, which is what Proposition 6.6.26 says when $k = 0$. $\triangle$

Example 6.6.27 (The boundary of an oriented 1-parallelogram). The boundary of $P_{\mathbf{x}}(\vec{\mathbf{v}})$ is

$$\partial P_{\mathbf{x}}(\vec{\mathbf{v}}) = P_{\mathbf{x}+\vec{\mathbf{v}}} - P_{\mathbf{x}}, \qquad 6.6.24$$

where $-P_{\mathbf{x}}$ is the point $\mathbf{x}$ with a minus sign and $P_{\mathbf{x}+\vec{\mathbf{v}}}$ is the point $\mathbf{x} + \vec{\mathbf{v}}$ with a plus sign. (Did you expect the right side of equation 6.6.24 to be $P_{\mathbf{x}+\vec{\mathbf{v}}}(\vec{\mathbf{v}}) - P_{\mathbf{x}}(\vec{\mathbf{v}})$? Remember that $\vec{\mathbf{v}}$ is the $\vec{\mathbf{v}}_i$ that is being omitted.) So the boundary of an oriented line segment is its *end minus its beginning*.

Example 6.6.28 (The boundary of an oriented 2-parallelogram). As shown by Figure 6.6.13, the boundary of an oriented parallelogram is

$$\underbrace{\partial P_{\mathbf{x}}(\vec{\mathbf{v}}_1, \vec{\mathbf{v}}_2)}_{\text{boundary}} = \underbrace{P_{\mathbf{x}}(\vec{\mathbf{v}}_1)}_{\text{1st side}} + \underbrace{P_{\mathbf{x}+\vec{\mathbf{v}}_1}(\vec{\mathbf{v}}_2)}_{\text{2nd side}} - \underbrace{P_{\mathbf{x}+\vec{\mathbf{v}}_2}(\vec{\mathbf{v}}_1)}_{\text{3rd side}} - \underbrace{P_{\mathbf{x}}(\vec{\mathbf{v}}_2)}_{\text{4th side}}. \quad \triangle \qquad 6.6.25$$

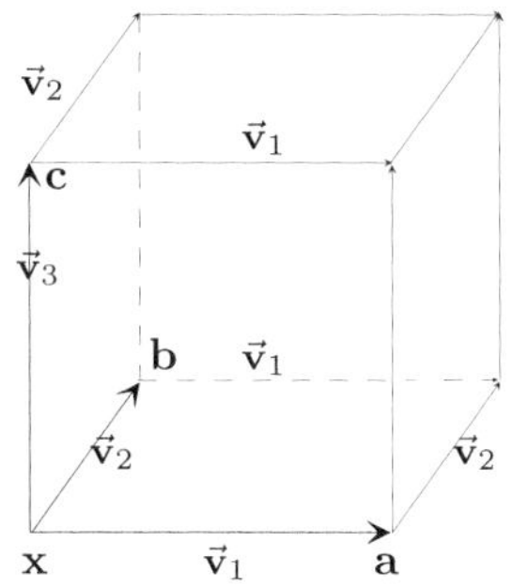

FIGURE 6.6.14 The cube spanned by the vectors $\vec{\mathbf{v}}_1, \vec{\mathbf{v}}_2, \vec{\mathbf{v}}_3$, anchored at $\mathbf{x}$. To lighten notation set $\mathbf{a} = \mathbf{x} + \vec{\mathbf{v}}_1$, $\mathbf{b} = \mathbf{x} + \vec{\mathbf{v}}_2$, and $\mathbf{c} = \mathbf{x} + \vec{\mathbf{v}}_3$. The original three vectors are drawn in dark lines; the translates in lighter or dotted lines.

Example 6.6.29 (Boundary of a cube). For the faces of a cube shown in Figure 6.6.14 we have

$$\begin{aligned}
&(i=1 \quad \text{so} \quad (-1)^{i-1} = \;\;\, 1); \qquad +\Big(\underbrace{P_{\mathbf{x}+\vec{\mathbf{v}}_1}(\vec{\mathbf{v}}_2,\vec{\mathbf{v}}_3)}_{\text{right side}} - \underbrace{P_{\mathbf{x}}(\vec{\mathbf{v}}_2,\vec{\mathbf{v}}_3)}_{\text{left side}}\Big)\\
&(i=2 \quad \text{so} \quad (-1)^{i-1} = -1); \qquad -\Big(\underbrace{P_{\mathbf{x}+\vec{\mathbf{v}}_2}(\vec{\mathbf{v}}_1,\vec{\mathbf{v}}_3)}_{\text{back}} - \underbrace{P_{\mathbf{x}}(\vec{\mathbf{v}}_1,\vec{\mathbf{v}}_3)}_{\text{front}}\Big) \qquad 6.6.26\\
&(i=3 \quad \text{so} \quad (-1)^{i-1} = \;\;\, 1); \qquad +\Big(\underbrace{P_{\mathbf{x}+\vec{\mathbf{v}}_3}(\vec{\mathbf{v}}_1,\vec{\mathbf{v}}_2)}_{\text{top}} - \underbrace{P_{\mathbf{x}}(\vec{\mathbf{v}}_1,\vec{\mathbf{v}}_2)}_{\text{bottom}}\Big). \quad \triangle
\end{aligned}$$

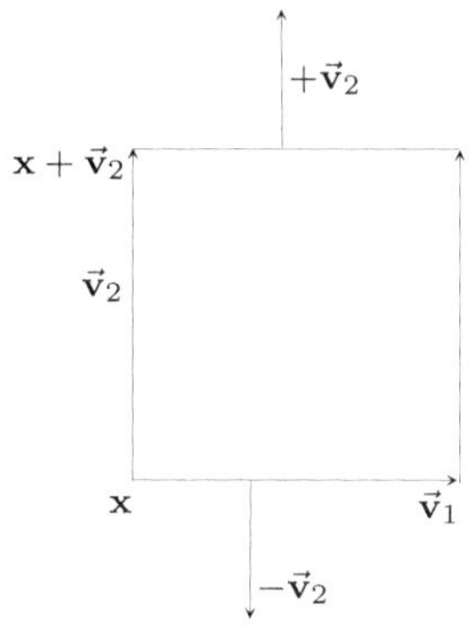

FIGURE 6.6.15. The vector $\vec{\mathbf{v}}_1$ anchored at $\mathbf{x}$ is identical to $P_{\mathbf{x}}(\vec{\mathbf{v}}_1, \widehat{\vec{\mathbf{v}}_2})$. On this segment of the boundary of the parallelogram $P_{\mathbf{x}}(\vec{\mathbf{v}}_1, \vec{\mathbf{v}}_2)$, the outward-pointing vector is $-\vec{\mathbf{v}}_2$. The top edge of the parallelogram is $P_{\mathbf{x}+\vec{\mathbf{v}}_2}(\vec{\mathbf{v}}_1, \widehat{\vec{\mathbf{v}}_2})$; on it, the outward-pointing vector is $+\vec{\mathbf{v}}_2$. (We have shortened the outward-pointing and inward-pointing vectors.)

How many "faces" make up the boundary of a 4-parallelogram? What is each face? How would you describe the boundary following the format used for the cube in Figure 6.6.14? Check your answer.[10]

Proof of Proposition 6.6.26. The boundary of $P_{\mathbf{x}}(\vec{\mathbf{v}}_1, \dots, \vec{\mathbf{v}}_k)$ is composed of its $2k$ faces (four for a parallelogram, six for a cube ...), each of the form

$$P_{\mathbf{x}+\vec{\mathbf{v}}_i}(\vec{\mathbf{v}}_1, \dots, \widehat{\vec{\mathbf{v}}_i}, \dots, \vec{\mathbf{v}}_k), \quad \text{or} \quad P_{\mathbf{x}}(\vec{\mathbf{v}}_1, \dots, \widehat{\vec{\mathbf{v}}_i}, \dots, \vec{\mathbf{v}}_k), \qquad 6.6.27$$

where a hat over a term indicates that it is being omitted. We need to show that the orientation of this boundary is consistent with Definition 6.6.21 of the oriented boundary of a piece-with-boundary.

Let M be the affine k-dimensional subspace containing $P_{\mathbf{x}}(\vec{\mathbf{v}}_1, \dots, \vec{\mathbf{v}}_k)$, oriented so that $\vec{\mathbf{v}}_1, \dots, \vec{\mathbf{v}}_k$ is a direct basis of $T_{\mathbf{x}}M$.

At a point of $P_{\mathbf{x}+\vec{\mathbf{v}}_i}(\vec{\mathbf{v}}_1, \dots, \widehat{\vec{\mathbf{v}}_i}, \dots, \vec{\mathbf{v}}_k)$, the vector $\vec{\mathbf{v}}_i$ is outward pointing, whereas at a point of $P_{\mathbf{x}}(\vec{\mathbf{v}}_1, \dots, \widehat{\vec{\mathbf{v}}_i}, \dots, \vec{\mathbf{v}}_k)$, the vector $-\vec{\mathbf{v}}_i$ is outward pointing. This is illustrated by Figure 6.6.15; Exercise 6.6.4 justifies it rigorously.

---

[10] A 4-parallelogram spanned by the vectors $\vec{\mathbf{v}}_1, \vec{\mathbf{v}}_2, \vec{\mathbf{v}}_3, \vec{\mathbf{v}}_4$, anchored at $\mathbf{x}$, is denoted $P_{\mathbf{x}}(\vec{\mathbf{v}}_1, \vec{\mathbf{v}}_2, \vec{\mathbf{v}}_3, \vec{\mathbf{v}}_4)$. It has eight "faces," each a 3-parallelogram. They are

$$\begin{aligned}
&+P_{\mathbf{x}}(\vec{\mathbf{v}}_1,\vec{\mathbf{v}}_2,\vec{\mathbf{v}}_3), \quad -P_{\mathbf{x}}(\vec{\mathbf{v}}_2,\vec{\mathbf{v}}_3,\vec{\mathbf{v}}_4), \quad +P_{\mathbf{x}}(\vec{\mathbf{v}}_1,\vec{\mathbf{v}}_3,\vec{\mathbf{v}}_4), \quad -P_{\mathbf{x}}(\vec{\mathbf{v}}_1,\vec{\mathbf{v}}_2,\vec{\mathbf{v}}_4),\\
&+P_{\mathbf{x}+\vec{\mathbf{v}}_1}(\vec{\mathbf{v}}_2,\vec{\mathbf{v}}_3,\vec{\mathbf{v}}_4), \qquad -P_{\mathbf{x}+\vec{\mathbf{v}}_2}(\vec{\mathbf{v}}_1,\vec{\mathbf{v}}_3,\vec{\mathbf{v}}_4),\\
&+P_{\mathbf{x}+\vec{\mathbf{v}}_3}(\vec{\mathbf{v}}_1,\vec{\mathbf{v}}_2,\vec{\mathbf{v}}_4), \qquad -P_{\mathbf{x}+\vec{\mathbf{v}}_4}(\vec{\mathbf{v}}_1,\vec{\mathbf{v}}_2,\vec{\mathbf{v}}_3).
\end{aligned}$$

Thus the standard orientation of $P_{\mathbf{x}+\vec{\mathbf{v}}_i}(\vec{\mathbf{v}}_1,\dots,\widehat{\vec{\mathbf{v}}_i},\dots,\vec{\mathbf{v}}_k)$ is consistent with the boundary orientation of $P_{\mathbf{x}}(\vec{\mathbf{v}}_1,\dots,\vec{\mathbf{v}}_i,\dots,\vec{\mathbf{v}}_k)$ precisely if

$$\Omega(\vec{\mathbf{v}}_i,\vec{\mathbf{v}}_1,\dots,\widehat{\vec{\mathbf{v}}_i},\dots,\vec{\mathbf{v}}_k) = +1, \qquad 6.6.28$$

i.e., precisely if the permutation σ_i on k symbols which consists of taking the ith element and putting it in first position is a positive permutation. But the signature of σ_i is $(-1)^{i-1}$, because you can obtain σ_i by switching the ith symbol first with the $(i-1)$th, then the $(i-2)$th, etc., and finally the first, doing $i-1$ transpositions. This explains why $P_{\mathbf{x}+\vec{\mathbf{v}}_i}(\vec{\mathbf{v}}_1,\dots,\widehat{\vec{\mathbf{v}}_i},\dots,\vec{\mathbf{v}}_k)$ occurs with sign $(-1)^{i-1}$.

A similar argument holds for $P_{\mathbf{x}}(\vec{\mathbf{v}}_1,\dots,\widehat{\vec{\mathbf{v}}_i},\dots,\vec{\mathbf{v}}_k)$. The orientation of this parallelogram is compatible with the boundary orientation precisely if $\Omega(-\vec{\mathbf{v}}_i,\vec{\mathbf{v}}_1,\dots,\widehat{\vec{\mathbf{v}}_i},\dots,\vec{\mathbf{v}}_k) = +1$, which occurs if the permutation σ_i is odd. This explains why $P_{\mathbf{x}}(\vec{\mathbf{v}}_1,\dots,\widehat{\vec{\mathbf{v}}_i},\dots,\vec{\mathbf{v}}_k)$ occurs with sign $(-1)^i$ in the sum in equation 6.6.23. □

Exercises for Section 6.6

Exercise 6.6.3: We already saw $\vec{\nabla}$ ("nabla") in equation 1.7.22. It denotes the transpose of the derivative, called the gradient:

$$\vec{\nabla} f(\mathbf{x}) = [\mathbf{D}f(\mathbf{x})]^\top.$$

Note that $\vec{\nabla} f(\mathbf{x}_0)$ is orthogonal at $\mathbf{x}_0$ to the manifold X of equation $f(\mathbf{x}) = 0$: since

$$T_{\mathbf{x}_0}X = \ker[\mathbf{D}f(\mathbf{x}_0)],$$

if $\vec{\mathbf{v}} \in T_{\mathbf{x}_0}X$, then

$$\vec{\nabla} f(\mathbf{x}_0)\cdot\vec{\mathbf{v}} = [\mathbf{D}f(\mathbf{x}_0)]\vec{\mathbf{v}} = 0.$$

This is discussed at greater length in Section 6.8, particularly the subsection "geometric interpretation of the gradient".

6.6.1 Verify that the following inequalities describe a piece-with-boundary of $\mathbb{R}^3$. (You may wish to use a computer to help you visualize the loci.)

a. $$\begin{aligned} xyz &\le 1\\ x^2+y^2+z^2 &\le 4 \end{aligned}$$

b. $$\begin{aligned} xyz &\le 1\\ x^2+y^2+z^2 &\le 4\\ x+y+z &\ge 0 \end{aligned}$$

What parts of the boundary are smooth?

6.6.2 Verify that the following inequalities describe a piece-with-boundary of $\mathbb{R}^2$. Sketch the loci.

a. $x^2 \ge y^2$ b. $xy \ge 0$

c. $xy \ge 1$ and $x+y \le 4$ d. $y \le 1$ and $y \ge x^2$

6.6.3 a. Let $X \subset \mathbb{R}^n$ be a manifold of the form $X = f^{-1}(0)$ where $f:\mathbb{R}^n \to \mathbb{R}$ is a C^1 function and $[\mathbf{D}f(\mathbf{x})] \ne 0$ for all $\mathbf{x} \in X$. Let $\vec{\mathbf{v}}_1,\dots,\vec{\mathbf{v}}_{n-1}$ be a basis for $T_{\mathbf{x}}(X)$. Show that

$$\Omega(\vec{\mathbf{v}}_1,\dots,\vec{\mathbf{v}}_{n-1}) = \operatorname{sgn}\det\left(\vec{\nabla} f(\mathbf{x}),\vec{\mathbf{v}}_1,\dots,\vec{\mathbf{v}}_{n-1}\right)$$

defines an orientation of X (see the margin for the definition of $\vec{\nabla}$).

b. How does this definition relate to the definition of the boundary orientation?

6.6.4 Show that at a point of $P_{\mathbf{x}}(\vec{\mathbf{v}}_1,\dots,\widehat{\vec{\mathbf{v}}_i},\dots,\vec{\mathbf{v}}_k)$, the vector $-\vec{\mathbf{v}}_i$ is outward pointing, and at a point of $P_{\mathbf{x}+\vec{\mathbf{v}}_i}(\vec{\mathbf{v}}_1,\dots,\widehat{\vec{\mathbf{v}}_i},\dots,\vec{\mathbf{v}}_k)$, the vector $\vec{\mathbf{v}}_i$ is outward pointing. (This is not difficult, but requires keeping a lot of notation straight. You will want to use the linear functions α_i from Example 6.6.15, and you will also need Definitions 6.6.2 and 6.6.20.)

6.6.5 Consider the region $X = P \cap B \subset \mathbb{R}^3$, where P is the plane of equation $x+y+z=0$ and B is the ball $x^2+y^2+z^2 \leq 1$. Orient P by the normal $\vec{N} = \begin{bmatrix} 1 \\ 1 \\ 1 \end{bmatrix}$ and orient the sphere $x^2+y^2+z^2=1$ by the outward-pointing normal.

a. Which of $\operatorname{sgn} dx \wedge dy$, $\operatorname{sgn} dx \wedge dz$, $\operatorname{sgn} dy \wedge dz$ give the same orientation of P as $\vec{N}$?

b. Show that X is a piece-with-boundary of P and that the mapping shown in the margin, for $0 \leq t \leq 2\pi$, is a parametrization of ∂X.

$$t \mapsto \begin{pmatrix} \frac{\cos t}{\sqrt{2}} - \frac{\sin t}{\sqrt{6}} \\ -\frac{\cos t}{\sqrt{2}} - \frac{\sin t}{\sqrt{6}} \\ 2\frac{\sin t}{\sqrt{6}} \end{pmatrix}$$

Map for Exercise 6.6.5, part b

c. Is the parametrization in part b compatible with the boundary orientation of ∂X?

d. Do any of $\operatorname{sgn} dx$, $\operatorname{sgn} dy$, $\operatorname{sgn} dz$ define the orientation of ∂X at every point?

e. Do any of

$$\operatorname{sgn} x\,dy - y\,dx, \quad \operatorname{sgn} x\,dz - z\,dx, \quad \operatorname{sgn} y\,dz - z\,dy$$

define the orientation of ∂X at every point?

6.6.6 Consider the curve C of equation $x^2+y^2=1$, oriented by the tangent vector $\begin{bmatrix} 0 \\ 1 \end{bmatrix}$ at the point $\begin{pmatrix} 1 \\ 0 \end{pmatrix}$.

a. Show that the subset X where $x \geq 0$ is a piece-with-boundary of C. What is its oriented boundary?

b. Show that the subset Y where $|x| \leq 1/2$ is a piece-with-boundary of C. What is its oriented boundary?

c. Is the subset Z where $x > 0$ a piece-with-boundary of C? If so, what is its boundary?

Exercise 6.6.7: Part c is a lot trickier than it may look. The key is to apply the singular value decomposition to B. The solution also uses Corollary 4.8.25 and Exercise 3.2.11.

6.6.7 a. Justify the statement (proof of Proposition 6.6.18) that when A is orthogonal, then for each C of V_N, the image $\mathbf{g}(C)$ intersects at most $(2(\sqrt{n}+1))^n$ cubes of $\mathcal{D}_N(\mathbb{R}^n)$.

b. If $C \in \mathcal{D}_N(\mathbb{R}^n)$, and $\mathbf{g}(\mathbf{x}) = A\mathbf{x}+\mathbf{b}$ is an affine mapping, show that $\mathbf{g}(C)$ intersects at most $\left((|A|+1)\sqrt{n}+2\right)^n$ cubes of $\mathcal{D}_N(\mathbb{R}^n)$.

c. Show that if A is an $n \times n$ matrix, there exists a number M such that for all $n \times k$ matrices B, we have

$$\det\left((AB)^\top AB\right) \leq M \det B^\top B.$$

Exercise 6.6.8: When finding a basis for the tangent space, you will need to think separately about the cases $x_1 \neq 0$, $x_2 \neq 0$, and $x_3 \neq 0$.

6.6.8 a. Let $M \subset \mathbb{R}^4$ be a manifold defined by the equation $x_4 = x_1^2+x_2^2+x_3^2$ and oriented by $\operatorname{sgn} dx_1 \wedge dx_2 \wedge dx_3$. Consider the subset $X \subset M$ where $x_4 \leq 1$. Show that it is a piece-with-boundary.

b. Let $\mathbf{x}$ be a point of ∂X. Find a basis for the tangent space $T_\mathbf{x}\partial X$ that is direct for the boundary orientation.

Hint for Exercise 6.6.9: Look at Proposition 6.10.7 and Heine-Borel (Theorem A3.3).

***6.6.9** Show that a piece-with-corners is a piece-with-boundary.

6.6.10 Show that when $\mathbf{v}_1, \dots, \mathbf{v}_k$ are linearly independent, $P_\mathbf{x}(\vec{\mathbf{v}}_1, \dots, \vec{\mathbf{v}}_k)$ is a piece-with-corners.

6.6.11 Prove Proposition 6.6.4.

6.7 The exterior derivative

Now we come to the construction that gives the theory of forms its power, making possible a fundamental theorem of calculus in higher dimensions. The derivative for forms, the *exterior derivative*, generalizes the derivative of ordinary functions. What is the ordinary derivative? You know that

$$f'(x) = \lim_{h \to 0} \frac{1}{h}\big(f(x+h) - f(x)\big), \qquad 6.7.1$$

but we will reinterpret this formula as

$$f'(x) = \lim_{h \to 0} \frac{1}{h} \int_{\partial P_x(h)} f. \qquad 6.7.2$$

This formula uses different notation to describe the same operation. Instead of saying that we are evaluating f at the two points $x+h$ and x, we say that we are integrating the 0-form f over the oriented boundary of the oriented segment $[x, x+h] = P_x(h)$. This boundary consists of its endpoint $+P_{x+h}$ and its beginning point $-P_x$. Integrating the 0-form f over these two oriented points means evaluating f on those points (Example 6.4.15). So equations 6.7.1 and 6.7.2 say exactly the same thing.

It may seem absurd to take equation 6.7.1, which everyone understands perfectly well, and turn it into equation 6.7.2, which looks more complicated. But the language generalizes nicely to forms.

In order to have a fundamental theorem of calculus for differential forms, we need derivatives as well as integrals. We have integrals: forms were designed to be integrands. Here we show that forms also have derivatives.

We will see in Section 6.12 that two of Maxwell's equations say that a certain 2-form on $\mathbb{R}^4$ has exterior derivative zero; a course in electromagnetism might spend six months trying to really understand what this means.

Defining the exterior derivative

The exterior derivative $\mathbf{d}$ is an operator that takes a k-form φ and gives a $(k+1)$-form, $\mathbf{d}\varphi$. Since a $(k+1)$-form takes an oriented $(k+1)$-dimensional parallelogram and gives a number, to define the exterior derivative of a k-form φ, we must say what number $\mathbf{d}\varphi$ returns when evaluated on an oriented $(k+1)$-parallelogram. We will work quite hard to see what the exterior derivative gives in particular cases, and to see how to compute it.

Equation 6.7.3: We are integrating φ over the boundary, just as in equation 6.7.2 we are integrating f over the boundary.

Even if the limit in equation 6.7.3 exists, it isn't obvious that $\mathbf{d}\varphi$ is a $(k+1)$-form, i.e., a multilinear and alternating function of $(k+1)$ vectors. Both are the object of part 1 of Theorem 6.7.4, a major result.

Definition 6.7.1 (Exterior derivative). Let $U \subset \mathbb{R}^n$ be an open subset. The exterior derivative $\mathbf{d}: A^k(U) \to A^{k+1}(U)$ is defined by the formula

$$\underbrace{\mathbf{d}\varphi}_{(k+1)-\text{form}} \overbrace{\big(P_{\mathbf{x}}(\vec{\mathbf{v}}_1, \dots, \vec{\mathbf{v}}_{k+1})\big)}^{(k+1)\text{-parallelogram}} \stackrel{\text{def}}{=} \lim_{h \to 0} \frac{1}{h^{k+1}} \overbrace{\int_{\underbrace{\partial P_{\mathbf{x}}(h\vec{\mathbf{v}}_1, \dots, h\vec{\mathbf{v}}_{k+1})}_{\text{boundary of } k+1\text{-parallelogram, smaller and smaller as } h \to 0}} \varphi}^{\text{integrating } \varphi \text{ over boundary}}. \qquad 6.7.3$$

Remark 6.7.2. The integral on the right side of equation 6.7.3 makes sense: since $P_{\mathbf{x}}(h\vec{\mathbf{v}}_1, \dots, h\vec{\mathbf{v}}_{k+1})$ is a $(k+1)$-dimensional parallelogram, its boundary consists of k-dimensional faces, so the boundary is something over which we can integrate the k-form φ.

But it isn't obvious that the limit exists. What is the problem? The faces making up the boundary of $P_{\mathbf{x}}(\vec{\mathbf{v}}_1, \dots, \vec{\mathbf{v}}_{k+1})$ are k-dimensional, so multiplying $\vec{\mathbf{v}}_1, \dots, \vec{\mathbf{v}}_{k+1}$ by h to get $P_{\mathbf{x}}(h\vec{\mathbf{v}}_1, \dots, h\vec{\mathbf{v}}_{k+1})$ should scale the integral over each face roughly by h^k. Thus the integral over each face contributes something of the form $h^k(a_0 + a_1 h + o(h))$ with $a_0 \neq 0$ in general; dividing by h^{k+1} gives a term a_0/h, which blows up as $h \to 0$.

When we write that "the integral over each face contributes something of the form

$$h^k(a_0 + a_1 h + o(h))$$

with $a_0 \neq 0$ in general" we are thinking of the Taylor polynomial of the integral as a function of h.

The only way that the limit can exist is if the terms a_0/h from different faces cancel. And they do: opposite faces carry opposite orientations, and if one face contributes $a_0/h + \cdots$ the other contributes $-a_0/h + \cdots$. (The terms indicated by dots are different for the different faces. These differences contribute to the exterior derivative.)

This is why we have put so much emphasis on orientation. Without a theory of integration over oriented domains we could have no derivative, hence no fundamental theorem of calculus in higher dimensions. △

Example 6.7.3 (Exterior derivative of function). When φ is a 0-form field (i.e., a function), the exterior derivative is just the derivative:

$$\mathbf{d}f\big(P_{\mathbf{x}}(\vec{\mathbf{v}})\big) = \lim_{h\to 0} \frac{1}{h}\int_{\partial P_{\mathbf{x}}(h\vec{\mathbf{v}})} f = \lim_{h\to 0} \frac{f(\mathbf{x}+h\vec{\mathbf{v}}) - f(\mathbf{x})}{h} = [\mathbf{D}f(\mathbf{x})]\vec{\mathbf{v}}. \qquad 6.7.4$$

The last equality is Proposition 1.7.14. Thus the exterior derivative generalizes the derivative as studied in Chapter 1 to k-forms for $k \geq 1$. △

Theorem 6.7.4, part 3: Recall that a constant form returns the same number regardless of the point at which it is evaluated.

Part 4: It may seem odd that the matrix $[\mathbf{D}f] = [D_1 f \dots D_k f]$ is written as a sum, but note that

$$\mathbf{d}f(\vec{\mathbf{v}}) = [D_1 f\, dx_1 + \cdots + D_n f\, dx_n]\begin{bmatrix} v_1 \\ \vdots \\ v_n \end{bmatrix} = \sum_{i=1}^{n} D_i f dx_i \begin{bmatrix} v_1 \\ \vdots \\ v_n \end{bmatrix} = D_1 f v_1 + \cdots + D_n f v_n = [\mathbf{D}f]\vec{\mathbf{v}}.$$

Since the dx_i are elementary 1-forms on $\mathbb{R}^n$, writing

$$\mathbf{d}f = \sum_{i=1}^{n} (D_i f)\, dx_i$$

is analogous to writing

$$\vec{\mathbf{v}} = \sum_{i=1}^{n} v_i \vec{\mathbf{e}}_i.$$

In part 5, note that the form $dx_{i_1} \wedge \cdots \wedge dx_{i_k}$ is an elementary form. The general case is given by Theorem 6.7.9.

Theorem 6.7.4 (Computing the exterior derivative). *Let*

$$\varphi = \sum_{1 \leq i_1 < \cdots < i_k \leq n} a_{i_1,\dots,i_k} dx_{i_1} \wedge \cdots \wedge dx_{i_k} \qquad 6.7.5$$

be a k-form of class C^2 on an open subset $U \subset \mathbb{R}^n$.

1. *The limit in equation 6.7.3 exists and defines a $(k+1)$-form.*
2. *The exterior derivative is linear over $\mathbb{R}$: if φ and ψ are k-forms on $U \subset \mathbb{R}^n$, and a and b are numbers (not functions), then*
$$\mathbf{d}(a\,\varphi + b\,\psi) = a\,\mathbf{d}\varphi + b\,\mathbf{d}\psi. \qquad 6.7.6$$
$$\text{In particular,}\quad \mathbf{d}(\varphi + \psi) = \mathbf{d}\varphi + \mathbf{d}\psi. \qquad 6.7.7$$
3. *The exterior derivative of a constant form is 0.*
4. *The exterior derivative of the 0-form f is given by the formula*
$$\mathbf{d}f = [\mathbf{D}f] = \sum_{i=1}^{n} (D_i f)\, dx_i. \qquad 6.7.8$$
5. *If $f : U \to \mathbb{R}$ is a C^2 function, then*
$$\mathbf{d}\left(f\, dx_{i_1} \wedge \cdots \wedge dx_{i_k}\right) = \mathbf{d}f \wedge dx_{i_1} \wedge \cdots \wedge dx_{i_k}. \qquad 6.7.9$$

Part 4 of Theorem 6.7.4 is proved in Example 6.7.3. The remainder of the proof is in Appendix A22.

These rules allow you to compute the exterior derivative of any k-form:

Equation 6.7.10: The first line just says that $\mathbf{d}\varphi = \mathbf{d}\varphi$. The second line says that the exterior derivative of the sum is the sum of the exterior derivatives. For example,

$$\mathbf{d}\Big(f(dx \wedge dy) + g(dy \wedge dz)\Big) =$$
$$\mathbf{d}\Big(f(dx \wedge dy)\Big) + \mathbf{d}\Big(g(dy \wedge dz)\Big).$$

The third line says that

$$\mathbf{d}\Big(f(dx \wedge dy)\Big) = \mathbf{d}f \wedge dx \wedge dy.$$

$$\begin{aligned}\mathbf{d}\varphi &= \mathbf{d}\overbrace{\sum_{1\le i_1<\cdots<i_k\le n} a_{i_1,\dots,i_k}\,dx_{i_1}\wedge\cdots\wedge dx_{i_k}}^{\text{writing } \varphi \text{ in full}} \\ &\underbrace{\underbrace{=}_{(2)} \sum_{1\le i_1<\cdots<i_k\le n} \mathbf{d}\big((a_{i_1,\dots,i_k})(dx_{i_1}\wedge\cdots\wedge dx_{i_k})\big)}_{\text{exterior derivative of sum equals sum of exterior derivatives}} \\ &\underbrace{\underbrace{=}_{(5)} \sum_{1\le i_1<\cdots<i_k\le n} (\mathbf{d}\underbrace{a_{i_1,\dots,i_k}}_{f})\wedge dx_{i_1}\wedge\cdots\wedge dx_{i_k}.}_{\text{problem reduced to computing ext. deriv. of function}}\end{aligned} \qquad 6.7.10$$

Going from the first to the second line reduces the computation to computing exterior derivatives of elementary forms; going from the second to the third line reduces it to computing exterior derivatives of functions. In applying part 5 we think of the coefficients $a_{i_1,\dots,i_k}$ as the function f.

We compute the exterior derivative of $f = a_{i_1,\dots,i_k}$ from part 4:

$$\mathbf{d}a_{i_1,\dots,i_k} = \sum_{j=1}^{n} D_j a_{i_1,\dots,i_k}\,dx_j. \qquad 6.7.11$$

For example, if f is a function in the three variables x, y, and z, then

$$\mathbf{d}f = D_1 f\,dx + D_2 f\,dy + D_3 f\,dz, \qquad 6.7.12$$

so

$$\begin{aligned}\mathbf{d}f\wedge dx\wedge dy &= \Big(D_1 f\,dx + D_2 f\,dy + D_3 f\,dz\Big)\wedge dx\wedge dy \\ &= D_1 f\,\underbrace{dx\wedge dx\wedge dy}_{0} + D_2 f\,\underbrace{dy\wedge dx\wedge dy}_{0} + D_3 f\,dz\wedge dx\wedge dy \\ &= D_3 f\,dz\wedge dx\wedge dy = D_3 f\,dx\wedge dy\wedge dz.\end{aligned} \qquad 6.7.13$$

Exchanging two terms changes the sign of the wedge product, so exchanging two identical terms changes the sign while leaving it unchanged. Thus a wedge product with two identical terms must be 0. With a bit of practice you will learn to ignore such wedge products.

You will usually want to put the dx_i in ascending order, which may change the sign, as in the third line of equation 6.7.14.

Going from the first to the second line of equation 6.7.13 uses the distributive property of the wedge product (property 6.1.15). In the second line, $dx \wedge dx \wedge dy$ is 0 because it contains two dx; similarly, $dy \wedge dx \wedge dy$ is 0 because it contains two dy.

Example 6.7.5 (Computing the exterior derivative of an elementary 2-form on $\mathbb{R}^4$). The exterior derivative of $x_2x_3(dx_2 \wedge dx_4)$ is

$$\begin{aligned}\mathbf{d}(x_2x_3)\wedge dx_2\wedge dx_4 &= \underbrace{\Big(\overbrace{D_1(x_2x_3)}^{0}\,dx_1 + D_2(x_2x_3)\,dx_2 + D_3(x_2x_3)\,dx_3 + \overbrace{D_4(x_2x_3)}^{0}\,dx_4\Big)}_{\mathbf{d}(x_2x_3)}\wedge dx_2\wedge dx_4 \\ &= (x_3\,dx_2 + x_2\,dx_3)\wedge dx_2\wedge dx_4 = (x_3\,dx_2\wedge dx_2\wedge dx_4) + (x_2\,dx_3\wedge dx_2\wedge dx_4) \\ &= \underbrace{x_2\,(dx_3\wedge dx_2\wedge dx_4)}_{dx\text{'s out of order}} = \underbrace{-x_2\,(dx_2\wedge dx_3\wedge dx_4)}_{\text{sign changes as order is corrected}}. \quad \triangle\end{aligned} \qquad 6.7.14$$

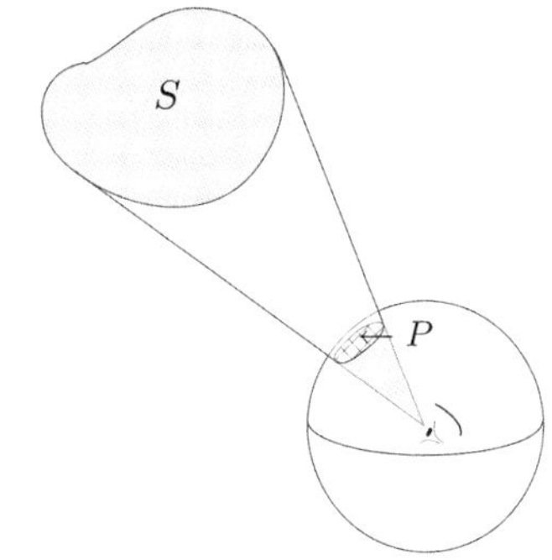

FIGURE 6.7.1.
Example 6.7.7: The origin is your eye; the "solid angle" with which you see the surface S is the cone. The intersection of the cone and the sphere of radius 1 around your eye is the region P. You are asked in Exercise 6.7.12 to show that the integral of $\Phi_{\vec{F}_3}$ over S and the integral over P are equal.

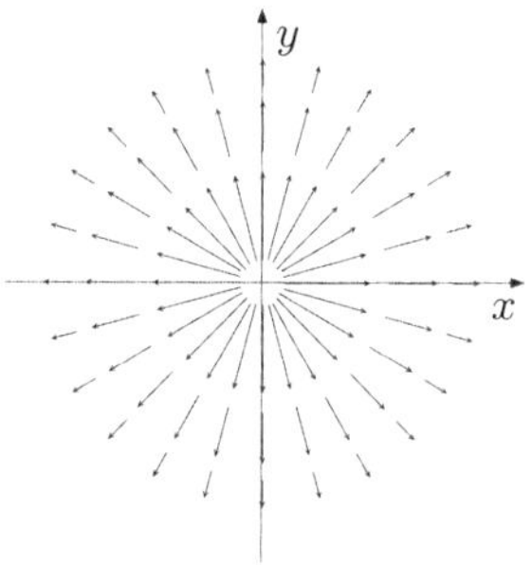

FIGURE 6.7.2.
The vector field $\vec{F}_2$. The vectors are scaled by $\frac{1}{|\mathbf{x}|}$, so they get smaller the further they are from the origin, and blow up near the origin. This scaling means that if $\vec{F}_2$ represents the flow of an incompressible fluid, then just as much flows through the unit circle as flows through a larger circle.

What is the exterior derivative of the 2-form $x_1x_3^2\,dx_1 \wedge dx_2$ on $\mathbb{R}^3$? Check your answer in the footnote.[11]

Example 6.7.6 (Computing the exterior derivative of a 2-form). Compute the exterior derivative of the 2-form on $\mathbb{R}^4$,

$$\psi = x_1x_2\,dx_2 \wedge dx_4 - x_2^2\,dx_3 \wedge dx_4, \qquad 6.7.15$$

which is the sum of two elementary 2-forms. We have

$$\begin{aligned} \mathbf{d}\psi &= \mathbf{d}(x_1x_2\,dx_2 \wedge dx_4) - \mathbf{d}(x_2^2\,dx_3 \wedge dx_4) \\ &= \Big(D_1(x_1x_2)\,dx_1 + D_2(x_1x_2)\,dx_2 + D_3(x_1x_2)\,dx_3 + D_4(x_1x_2)dx_4\Big) \wedge dx_2 \wedge dx_4 \\ &\quad - \Big(D_1(x_2^2)\,dx_1 + D_2(x_2^2)\,dx_2 + D_3(x_2^2)\,dx_3 + D_4(x_2^2)\,dx_4\Big) \wedge dx_3 \wedge dx_4 \\ &= (x_2\,dx_1 + x_1\,dx_2) \wedge dx_2 \wedge dx_4 - (2x_2\,dx_2 \wedge dx_3 \wedge dx_4) \\ &= x_2\,dx_1 \wedge dx_2 \wedge dx_4 + \underbrace{x_1\,dx_2 \wedge dx_2 \wedge dx_4}_{=0} - 2x_2\,dx_2 \wedge dx_3 \wedge dx_4 \qquad 6.7.16 \\ &= x_2\,dx_1 \wedge dx_2 \wedge dx_4 - 2x_2\,dx_2 \wedge dx_3 \wedge dx_4. \quad \triangle \end{aligned}$$

Example 6.7.7 (Element of angle). The vector fields

$$\vec{F}_2\begin{pmatrix} x \\ y \end{pmatrix} = \frac{1}{x^2+y^2}\begin{bmatrix} x \\ y \end{bmatrix} \quad \text{and} \quad \vec{F}_3\begin{pmatrix} x \\ y \\ z \end{pmatrix} = \frac{1}{(x^2+y^2+z^2)^{3/2}}\begin{bmatrix} x \\ y \\ z \end{bmatrix} \qquad 6.7.17$$

satisfy $\mathbf{d}\Phi_{\vec{F}_2} = 0$ and $\mathbf{d}\Phi_{\vec{F}_3} = 0$, as you are asked to confirm in Exercise 6.7.3. The forms $\Phi_{\vec{F}_2}$ and $\Phi_{\vec{F}_3}$ can be called the *element of polar angle* and the *element of solid angle*; the latter is depicted in Figure 6.7.1.

We will now find analogues in any dimension. Using again a hat to denote a term that is omitted, our candidate is the $(n-1)$-form on $\mathbb{R}^n$

$$\omega_n = \frac{1}{(x_1^2+\cdots+x_n^2)^{n/2}}\sum_{i=1}^{n}(-1)^{i-1}x_i\,dx_1 \wedge \cdots \wedge \widehat{dx_i} \wedge \cdots \wedge dx_n, \qquad 6.7.18$$

which (by equation 6.5.25) can be thought of as the flux of the vector field

$$\vec{F}_n(\vec{\mathbf{x}}) = \frac{1}{(x_1^2+\cdots+x_n^2)^{n/2}}\begin{bmatrix} x_1 \\ \vdots \\ x_n \end{bmatrix} = \frac{\vec{\mathbf{x}}}{|\vec{\mathbf{x}}|^n}. \qquad 6.7.19$$

The scaling $\frac{\mathbf{x}}{|\mathbf{x}|^n}$ means that the vectors get smaller the further they are from the origin (as shown for $n = 2$ in Figure 6.7.2). It is chosen so that the flux outward through a series of ever-larger spheres all centered at the

---

11

$$\begin{aligned} \mathbf{d}(x_1x_3^2\,dx_1 \wedge dx_2) &= \mathbf{d}(x_1x_3^2) \wedge dx_1 \wedge dx_2 \\ &= D_1(x_1x_3^2)\,dx_1 \wedge dx_1 \wedge dx_2 + D_2(x_1x_3^2)\,dx_2 \wedge dx_1 \wedge dx_2 + D_3(x_1x_3^2)\,dx_3 \wedge dx_1 \wedge dx_2 \\ &= 2x_1x_3\,dx_3 \wedge dx_1 \wedge dx_2 = 2x_1x_3\,dx_1 \wedge dx_2 \wedge dx_3. \end{aligned}$$

origin remains constant, by Stokes's theorem; in fact, the flux is equal to $\text{vol}_{(n-1)} S^{n-1}$ (see Exercise 6.31).

Remark. The flux remaining constant (i.e., $\mathbf{d}\Phi_{\vec{F}_3} = 0$) is a conservation law. Note that the vector field $\vec{F}_3$ on $\mathbb{R}^3$, the space in which we live, scales as the inverse square of the distance from the origin; this is easier to see if we write equation 6.7.19 as

$$\vec{F}_3(\vec{\mathbf{x}}) = \frac{1}{|\vec{\mathbf{x}}|^2}\frac{\vec{\mathbf{x}}}{|\vec{\mathbf{x}}|}, \qquad 6.7.20$$

where $\vec{\mathbf{x}}/|\vec{\mathbf{x}}|$ is the unit vector in the direction of $\vec{\mathbf{x}}$. All fundamental force fields of physics are conservative, and they are all variants of $\vec{F}_3$. This largely explains the appearance of inverse square laws throughout physics; we will see examples in Section 6.12 on electromagnetism. If we lived in a space of dimension n then (in order to have conservation laws) we would have inverse $(n-1)$-power laws. △

If the spheres are oriented as boundaries of balls (with the standard orientation), then the flux of this vector field through each sphere is positive.

The following shows that the exterior derivative of this flux form is 0:

$$\begin{aligned}
\mathbf{d}\omega_n = \mathbf{d}\Phi_{\vec{F}_n} &= \mathbf{d}\left(\frac{1}{(x_1^2+\cdots+x_n^2)^{n/2}}\sum_{i=1}^n(-1)^{i-1}x_i\,dx_1\wedge\cdots\wedge\widehat{dx_i}\wedge\cdots\wedge dx_n\right)\\
&= \sum_{i=1}^n(-1)^{i-1}D_i\frac{x_i}{(x_1^2+\cdots+x_n^2)^{n/2}}dx_i\wedge dx_1\wedge\cdots\wedge\widehat{dx_i}\wedge\cdots\wedge dx_n\\
&= \sum_{i=1}^n D_i\left(\frac{x_i}{(x_1^2+\cdots+x_n^2)^{n/2}}\right)dx_1\wedge\cdots\wedge dx_n\\
&= \sum_{i=1}^n\left(\frac{(x_1^2+\cdots+x_n^2)^{n/2}-nx_i^2(x_1^2+\cdots+x_n^2)^{n/2-1}}{(x_1^2+\cdots+x_n^2)^n}\right)dx_1\wedge\cdots\wedge dx_n\\
&= \sum_{i=1}^n\left(\frac{x_1^2+\cdots+x_n^2-nx_i^2}{(x_1^2+\cdots+x_n^2)^{n/2+1}}\right)dx_1\wedge\cdots\wedge dx_n = 0. \qquad 6.7.21
\end{aligned}$$

Equation 6.7.21: In going from line 1 to line 2 we compute only the partial derivative with respect to x_i, omitting the partial derivatives with respect to the variables that appear among the dx_j to the right, since they would result in something of the form $dx_j \wedge dx_j$, which is 0.

Going from line 2 to 3, we move dx_i to its proper position, which also gets rid of the $(-1)^{i-1}$. The move requires $i-1$ transpositions. If $i-1$ is odd, this changes the sign to negative, but $(-1)^{i-1} = -1$. If $i-1$ is even, moving dx_i leaves the sign unchanged, and $(-1)^{i-1} = 1$.

Going from line 3 to 4 is merely calculating the partial derivative, and going from 4 to 5 involves factoring out $(x_1^2+\cdots+x_n^2)^{n/2-1}$ from the numerator and canceling with the same factor in the denominator.

We get the last equality because the sum of the numerators cancel. For instance, when $n = 2$ we have $x_1^2 + x_2^2 - 2x_1^2 + x_1^2 + x_2^2 - 2x_2^2 = 0$. △

Taking the exterior derivative twice

The exterior derivative of a k-form is a $(k+1)$-form; the exterior derivative of that $(k+1)$-form is a $(k+2)$-form. One remarkable property of the exterior derivative is that if you take it twice, you always get 0. (To be precise, we must specify that φ be twice continuously differentiable.)

Theorem 6.7.8. *For any k-form φ of class C^2 on an open subset $U \subset \mathbb{R}^n$,*

$$\mathbf{d}(\mathbf{d}\varphi) = 0. \qquad 6.7.22$$

Proof. This can just be computed. Let us see it first for 0-forms:

$$\mathbf{dd}f = \mathbf{d}\Big(\sum_{i=1}^{n} D_i f\, dx_i\Big) = \sum_{i=1}^{n} \mathbf{d}\big(D_i f\, dx_i\big)$$
$$= \sum_{i=1}^{n} \mathbf{d}D_i f \wedge dx_i = \sum_{i=1}^{n}\sum_{j=1}^{n} D_j D_i f\, dx_j \wedge dx_i = 0. \qquad 6.7.23$$

Equation 6.7.23: In the double sum, the terms corresponding to $i = j$ vanish, since they are followed by $dx_i \wedge dx_i$. If $i \neq j$, the terms $D_j D_i f\, dx_j \wedge dx_i$ and $D_i D_j f\, dx_i \wedge dx_j$ cancel, since the crossed partials are equal, and $dx_j \wedge dx_i = -dx_i \wedge dx_j$.

(The second equality in the second line is part 4 of Theorem 6.7.4. Here $D_i f$ plays the role of f in part 4, giving $\mathbf{d}(D_i f) = \sum_{j=1}^{n} D_j D_i f\, dx_j$.)

If $k > 0$, it is enough to make the following computation:

$$\mathbf{d}\big(\mathbf{d}(f\, dx_{i_1} \wedge \dots \wedge dx_{i_k})\big) = \mathbf{d}(\mathbf{d}f \wedge dx_{i_1} \wedge \dots \wedge dx_{i_k})$$
$$= \underbrace{(\mathbf{dd}f)}_{=0} \wedge dx_{i_1} \wedge \dots \wedge dx_{i_k} = 0. \quad \square \qquad 6.7.24$$

There is also a conceptual proof of Theorem 6.7.8. Suppose φ is a k-form and we want to evaluate $\mathbf{d}(\mathbf{d}\varphi)$ on $k+2$ vectors. We get $\mathbf{d}(\mathbf{d}\varphi)$ by integrating $\mathbf{d}\varphi$ over the boundary of the oriented $(k+2)$-parallelogram spanned by the vectors (i.e., by integrating φ over the boundary of the boundary). But what is the boundary of the boundary? It is empty! One way of saying this is that each face of the $(k+2)$-parallelogram is a $(k+1)$-dimensional parallelogram, and each edge of the $(k+1)$-parallelogram is also the edge of another $(k+1)$-parallelogram, but with opposite orientation, as Figure 6.7.3 suggests for $k = 1$, and as Exercise 6.7.10 asks you to prove.

The boundary of the boundary of a 2-parallelogram is its four vertices: the sum of the boundaries of its four edges. When the parallelogram is oriented, each vertex is counted twice, with opposite orientation. For instance, the vertex $P_{\mathbf{x}+\vec{\mathbf{v}}_1}$ shows up twice, once as the end of $P_{\mathbf{x}}(\vec{\mathbf{v}}_1)$, and once as the beginning of $P_{\mathbf{x}+\vec{\mathbf{v}}_1}(\vec{\mathbf{v}}_2)$.

Exterior derivative of a wedge product

The following result is one more basic building block in the theory of the exterior derivative. It says that the exterior derivative of a wedge product satisfies an analogue of Leibniz's rule for differentiating products. There is a sign that comes in to complicate matters.

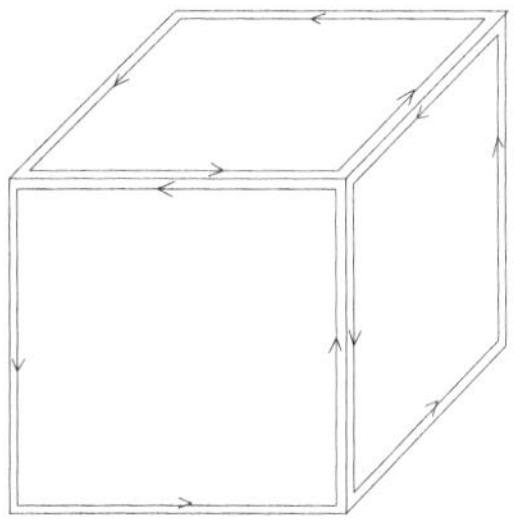

FIGURE 6.7.3.
Each edge of the cube is an edge of two faces of the cube and is taken twice, with opposite orientations.

Theorem 6.7.9 (Exterior derivative of wedge product). *If φ is a k-form and ψ is an l-form, then*

$$\mathbf{d}(\varphi \wedge \psi) = \mathbf{d}\varphi \wedge \psi + (-1)^k \varphi \wedge \mathbf{d}\psi. \qquad 6.7.25$$

Exercise 6.7.11 asks you to prove Theorem 6.7.9. It is an application of Theorem 6.7.4. Part 5 of Theorem 6.7.4 is a special case of Theorem 6.7.9, the case where φ is a function and ψ is elementary.

Exercises for Section 6.7

6.7.1 What is the exterior derivative of

a. $x_1 x_3\, dx_3 \wedge dx_4$ on $\mathbb{R}^4$? b. $\cos xy\, dx \wedge dy$ on $\mathbb{R}^3$?

6.7.2 What is the exterior derivative of

a. $\sin(xyz)\, dx$ on $\mathbb{R}^3$?

b. $x_1 x_3\, dx_2 \wedge dx_4$ on $\mathbb{R}^4$?

c. $\displaystyle\sum_{i=1}^{n} x_i^2\, dx_1 \wedge \cdots \wedge \widehat{dx_i} \wedge \cdots \wedge dx_n$ on $\mathbb{R}^n$?

6.7.3 In Example 6.7.7, confirm that a. $\mathbf{d}\Phi_{\vec{F}_2} = 0$ b. $\mathbf{d}\Phi_{\vec{F}_3} = 0$

6.7.4 Let φ be the 2-form on $\mathbb{R}^4$ given by $\varphi = x_1^2 x_3\, dx_2 \wedge dx_3 + x_1 x_3 dx_1 \wedge dx_4$. Compute $\mathbf{d}\varphi$.

6.7.5 a. Compute from the definition the exterior derivative of $z^2\ dx \wedge dy$.

b. Compute the same derivative using the formulas, stating clearly at each stage what property you are using.

6.7.6 Let f be a function from $\mathbb{R}^3$ to $\mathbb{R}$. Compute the exterior derivatives

a. $\mathbf{d}(f\, dx \wedge dz)$ b. $\mathbf{d}(f\, dy \wedge dz)$

6.7.7 a. Let $\varphi = x_2^2\, dx_3$. Compute from the definition $\mathbf{d}\varphi\Big(P_{-\mathbf{e}_2}(\vec{\mathbf{e}}_2, \vec{\mathbf{e}}_3)\Big)$.

b. Compute $\mathbf{d}\varphi$ from the formulas. Use your result to check the computation in part a.

6.7.8 a. Let $\varphi = x_1 x_3\, dx_2 \wedge dx_4$. Compute from the definition the number

$$\mathbf{d}\varphi\Big(P_{\mathbf{e}_1}(\vec{\mathbf{e}}_2, \vec{\mathbf{e}}_3, \vec{\mathbf{e}}_4)\Big).$$

b. What is $\mathbf{d}\varphi$? Use your result to check the computation in part a.

***6.7.9** Find all the 1-forms $\omega = p(y,z)\, dx + q(x,z)\, dy$ such that

$$\mathbf{d}\omega = x\, dy \wedge dz + y\, dx \wedge dz.$$

6.7.10 Show that each face of a $(k+2)$-parallelogram is a $(k+1)$-dimensional parallelogram, and that each face of the $(k+1)$-parallelogram is also the face of another $(k+1)$-parallelogram, but with opposite orientation.

6.7.11 Prove Theorem 6.7.9 concerning the exterior derivative of the wedge product:

a. Show it for 0-forms:

$$\mathbf{d}(fg) = f\,\mathbf{d}g + g\,\mathbf{d}f.$$

b. Show that it is enough to prove the theorem when

$$\varphi = a(\mathbf{x})\, dx_{i_1} \wedge \cdots \wedge dx_{i_k}$$
$$\psi = b(\mathbf{x})\, dx_{j_1} \wedge \cdots \wedge dx_{j_l}$$

c. Prove the theorem when φ and ψ are as in part b, using that

$$\varphi \wedge \psi = a(\mathbf{x}) b(\mathbf{x})\, dx_{i_1} \wedge \cdots \wedge dx_{i_k} \wedge dx_{j_1} \wedge \cdots \wedge dx_{j_l}.$$

Exercise 6.7.12: This problem will be easier using Stokes's theorem and $\mathbf{d}\Phi_{\vec{F}_3} = 0$; see Exercise 6.10.10.

6.7.12 Prove the statement in the caption of Figure 6.7.1, in the special case where the surface S projects injectively to the subset P of the unit sphere. That is, prove that the integral of the "element of solid angle" $\Phi_{\vec{F}_3}$ over such a surface S is the same as its integral over the corresponding P.

6.8 Grad, curl, div, and all that

The operators *grad*, *div*, and *curl* are the workhorses of vector calculus, and are essential in many physics and engineering courses. They are three different incarnations of the exterior derivative. Thus the exterior derivative provides a unifying concept for understanding grad, curl, and div, which otherwise might seem to be unrelated. In turn, this allows us to give physical meaning to the rather abstract notion of exterior derivative.

Grad, curl, and div are *differential operators*; they are all some combination of partial derivatives. We already saw grad (the gradient) in Chapter 1; see equation 1.7.22. The formulas for the gradient and the divergence work in any $\mathbb{R}^n$, but there is no obvious generalization of the curl, other than the exterior derivative.

Definition 6.8.1: We denote by $\vec{\nabla}$ ("nabla") the operator

$$\vec{\nabla} = \begin{bmatrix} D_1 \\ D_2 \\ D_3 \end{bmatrix}.$$

Some authors call $\vec{\nabla}$ "del". We use $\vec{\nabla}$ to make it easier to remember the formulas, which we can summarize:

$$\begin{aligned} \operatorname{grad} f &= \vec{\nabla} f \\ \operatorname{curl} \vec{F} &= \vec{\nabla} \times \vec{F} \\ \operatorname{div} \vec{F} &= \vec{\nabla} \cdot \vec{F}. \end{aligned}$$

Mnemonic: Both curl and cross product start with "c"; both divergence and dot product start with "d".

Definition 6.8.1 (Grad, curl, and div). Let $U \subset \mathbb{R}^3$ be an open set, $f : U \to \mathbb{R}$ a C^1 function, and $\vec{F}$ a C^1 vector field on U. The *gradient* of a function, the *curl* of a vector field, and the *divergence* of a vector field are given by the following formulas:

$$\operatorname{grad} f \stackrel{\text{def}}{=} \begin{bmatrix} D_1 f \\ D_2 f \\ D_3 f \end{bmatrix} = \vec{\nabla} f$$

$$\operatorname{curl} \vec{F} = \operatorname{curl} \begin{bmatrix} F_1 \\ F_2 \\ F_3 \end{bmatrix} \stackrel{\text{def}}{=} \vec{\nabla} \times \vec{F} = \begin{bmatrix} D_1 \\ D_2 \\ D_3 \end{bmatrix} \times \begin{bmatrix} F_1 \\ F_2 \\ F_3 \end{bmatrix} = \underbrace{\begin{bmatrix} D_2F_3 - D_3F_2 \\ D_3F_1 - D_1F_3 \\ D_1F_2 - D_2F_1 \end{bmatrix}}_{\text{cross product of } \vec{\nabla} \text{ and } \vec{F}}$$

$$\operatorname{div} \vec{F} = \operatorname{div} \begin{bmatrix} F_1 \\ F_2 \\ F_3 \end{bmatrix} \stackrel{\text{def}}{=} \vec{\nabla} \cdot \vec{F} = \begin{bmatrix} D_1 \\ D_2 \\ D_3 \end{bmatrix} \cdot \begin{bmatrix} F_1 \\ F_2 \\ F_3 \end{bmatrix} = \underbrace{D_1F_1 + D_2F_2 + D_3F_3}_{\text{dot product of } \vec{\nabla} \text{ and } \vec{F}}.$$

Note that both the gradient of a function and the curl of a vector field are vector fields, while the divergence of a vector field is a function.

Example 6.8.2 (Curl and div). Let $\vec{F}\begin{pmatrix} x \\ y \\ z \end{pmatrix} = \begin{bmatrix} x+y \\ x^2yz \\ yz \end{bmatrix}$. Then

$$\operatorname{curl} \vec{F} = \begin{bmatrix} D_2(yz) - D_3(x^2yz) \\ D_3(x+y) - D_1(yz) \\ D_1(x^2yz) - D_2(x+y) \end{bmatrix} = \begin{bmatrix} z - x^2y \\ 0 \\ 2xyz - 1 \end{bmatrix}. \qquad 6.8.1$$

$$\operatorname{div} \vec{F} = D_1(x+y) + D_2(x^2yz) + D_3(yz) = 1 + x^2z + y. \quad \triangle \qquad 6.8.2$$

When φ is a k-form, $\mathbf{d}\varphi$ is a $(k+1)$-form, so since f is a 0-form, $\mathbf{d}f$ should be a 1-form, which it is. Since $W_{\vec{F}}$ is a 1-form, $\mathbf{d}W_{\vec{F}}$ should be a 2-form, which it is, ...

The following theorem says that in $\mathbb{R}^3$, the rather abstract notion of a k-form takes on a concrete embodiment: every 1-form is the work form of a vector field, every 2-form is the flux form of a vector field, and every 3-form is the mass form of a function. Note that this embodiment requires the geometry of $\mathbb{R}^3$: the dot product and the cross product. So grad, curl, and div are dependent on the geometry of $\mathbb{R}^3$, while the exterior derivative is not; it is "intrinsic" or "natural". This is discussed further in Section 6.9.

Theorem 6.8.3 (Exterior derivative of form fields on $\mathbb{R}^3$). *Let f be a function on $\mathbb{R}^3$ and let $\vec{F}$ be a vector field. Then*

1. $\mathbf{d}f = W_{\vec{\nabla} f} = W_{\text{grad } f}$ ($\mathbf{d}f$ *is the work form field of* $\text{grad } f$)
2. $\mathbf{d}W_{\vec{F}} = \Phi_{\vec{\nabla} \times \vec{F}} = \Phi_{\text{curl } \vec{F}}$ ($\mathbf{d}W_{\vec{F}}$ *is the flux form field of* $\text{curl } \vec{F}$)
3. $\mathbf{d}\Phi_{\vec{F}} = M_{\vec{\nabla} \cdot \vec{F}} = M_{\text{div } \vec{F}}$ ($\mathbf{d}\Phi_{\vec{F}}$ *is the mass form field of* $\text{div } \vec{F}$)

Part 1 of Theorem 6.8.3 says

$$\mathbf{d}f\Big(P_{\mathbf{x}}(\vec{\mathbf{v}})\Big) = \vec{\nabla} f(\mathbf{x}) \cdot \vec{\mathbf{v}}.$$

Part 2 says

$$\mathbf{d}W_{\vec{F}}\Big(P_{\mathbf{x}}(\vec{\mathbf{v}}_1, \vec{\mathbf{v}}_2)\Big) = \det[(\vec{\nabla} \times \vec{F})(\mathbf{x}), \vec{\mathbf{v}}_1, \vec{\mathbf{v}}_2].$$

Part 3 says

$$\mathbf{d}\Phi_{\vec{F}}\Big(P_{\mathbf{x}}(\vec{\mathbf{v}}_1, \vec{\mathbf{v}}_2, \vec{\mathbf{v}}_3)\Big) = (\vec{\nabla} \cdot \vec{F})(\mathbf{x}) \det[\vec{\mathbf{v}}_1, \vec{\mathbf{v}}_2, \vec{\mathbf{v}}_3].$$

This is summarized by Table 6.8.1.

Vector calculus in $\mathbb{R}^3$		Form fields in $\mathbb{R}^3$
functions	$=$	0-forms
$\downarrow$ gradient		$\downarrow \mathbf{d}$
vector fields	work form W $\longrightarrow$	1-forms
$\downarrow$ curl		$\downarrow \mathbf{d}$
vector fields	flux form Φ $\longrightarrow$	2-forms
$\downarrow$ div		$\downarrow \mathbf{d}$
functions	mass form M $\longrightarrow$	3-forms

TABLE 6.8.1. In $\mathbb{R}^3$, 0-form fields and 3-form fields can be identified with functions, and 1-form fields and 2-form fields can be identified with vector fields. The diagram commutes; if you start anywhere on the left, and go down and right, or right and down, you will get the same result. (This is exactly the content of Theorem 6.8.3.) For instance, if you start with a function, take its gradient to get a vector field, and then take the work form of that vector field, that work form is the 1-form one gets by taking the exterior derivative of f: $\mathbf{d}f = W_{\text{grad } f}$.

Proof of Theorem 6.8.3. Exercise 6.8.4 asks you to prove part 3. The following proves parts 1 and 2:

$$\mathbf{d}f = D_1 f\, dx + D_2 f\, dy + D_3 f\, dz = W_{\begin{bmatrix} D_1 f \\ D_2 f \\ D_3 f \end{bmatrix}} = W_{\vec{\nabla} f} \qquad 6.8.3$$

$$\begin{aligned} \mathbf{d}W_{\vec{F}} &= \mathbf{d}(F_1\, dx + F_2\, dy + F_3\, dz) = \mathbf{d}F_1 \wedge dx + \mathbf{d}F_2 \wedge dy + \mathbf{d}F_3 \wedge dz \\ &= (D_1 F_1\, dx + D_2 F_1\, dy + D_3 F_1\, dz) \wedge dx \\ &\quad + (D_1 F_2\, dx + D_2 F_2\, dy + D_3 F_2\, dz) \wedge dy \\ &\quad + (D_1 F_3\, dx + D_2 F_3\, dy + D_3 F_3\, dz) \wedge dz \qquad 6.8.4 \\ &= (D_1 F_2 - D_2 F_1)\, dx \wedge dy + (D_1 F_3 - D_3 F_1)\, dx \wedge dz \\ &\quad + (D_2 F_3 - D_3 F_2)\, dy \wedge dz \\ &= \Phi_{\begin{bmatrix} D_2 F_3 - D_3 F_2 \\ D_3 F_1 - D_1 F_3 \\ D_1 F_2 - D_2 F_1 \end{bmatrix}} = \Phi_{\vec{\nabla} \times \vec{F}}. \qquad \square \end{aligned}$$

Remark 6.8.4. What does Table 6.8.1 tell us about compositions of the differential operators grad, curl, and div? We know from Theorem 6.7.8 that $\mathbf{d}(\mathbf{d}\varphi) = 0$, so $\operatorname{curl}\operatorname{grad} f = \vec{\mathbf{0}}$, since the flux form of $\operatorname{curl}\operatorname{grad} f$ is $\mathbf{dd}f = 0$. Thus if $\operatorname{curl}\vec{F}$ is not the zero vector, there is no function f such that $\vec{F} = \operatorname{grad} f$.

Similarly, it follows from Table 6.8.1 that $\operatorname{div}\operatorname{curl}\vec{F} = 0$.

What about div grad, grad div, curl curl, which are also some versions of $\mathbf{d}^2$? These differential operators make sense: grad f is a vector field, and div takes vector fields to functions, so $\operatorname{div}\operatorname{grad} f$ takes functions to functions. Similarly, grad div and curl curl take vector fields to vector fields. But this requires identifying 1-forms to 2-forms (both identified with vector fields) and identifying 0-forms and 3-forms (both identified with functions).

These operators cannot be written as $\mathbf{d}(\mathbf{d}\varphi)$, since $\mathbf{d}$ always takes a k-form to a $(k+1)$-form. Consider div grad: grad f corresponds to $\mathbf{d}$ taking 0-forms to 1-forms, and div corresponds to $\mathbf{d}$ taking 2-forms to 3-forms. Using the identification of 1-forms and 2-forms to vector fields, div grad makes sense, but it cannot be written as $\mathbf{dd}$.

Div grad is the most important differential operator in existence: the *Laplacian*. In $\mathbb{R}^3$ the Laplacian Δf is

$$\Delta f \overset{\text{def}}{=} (D_1^2 + D_2^2 + D_3^2)(f) = \begin{bmatrix} D_1 \\ D_2 \\ D_3 \end{bmatrix} \cdot \begin{bmatrix} D_1 f \\ D_2 f \\ D_3 f \end{bmatrix} = \operatorname{div}\operatorname{grad} f. \qquad 6.8.5$$

It measures to what extent the graph of a function is "tight". It shows up in electromagnetism, relativity, elasticity, complex analysis ... .

The operators grad div $\vec{F}$ and curl curl $\vec{F}$ are related to the Laplacian: the first minus the second gives the Laplacian $\vec{\Delta}$ acting on each coordinate of a vector field:

$$\vec{\Delta}\vec{F} = \operatorname{grad}\operatorname{div}\vec{F} - \operatorname{curl}\operatorname{curl}\vec{F} = \begin{bmatrix} \Delta F_1 \\ \Delta F_2 \\ \Delta F_3 \end{bmatrix}. \quad \triangle \qquad 6.8.6$$

Remark 6.8.4: We explore this issue further in Section 6.13, where we see that if $\vec{F}$ is a vector field *on a convex subset* of $\mathbb{R}^3$, then $\operatorname{curl}\vec{F} = \vec{\mathbf{0}}$ is also a sufficient condition for $\vec{F}$ to be a gradient.

We can also see that

$$\operatorname{div}\operatorname{curl}\vec{F} = 0$$

by computing

$$\operatorname{div}\operatorname{curl}\vec{F} = \vec{\nabla}\cdot(\vec{\nabla}\times\vec{F}) = 0.$$

(This uses Proposition 1.4.20. There the result is proved for vectors but the computation is purely algebraic and works for the entries D_i of $\vec{\nabla}$.)

Equation 6.8.5: Since the Laplacian in $\mathbb{R}^3$ is the dot product $\vec{\nabla}\cdot\vec{\nabla}$, it is sometimes denoted $\vec{\nabla}^2$.

Geometric interpretation of the gradient

The gradient of a function is gotten by putting the entries of the line matrix $[\mathbf{D}f(\mathbf{x})]$ in a column: $\operatorname{grad} f(\mathbf{x}) = [\mathbf{D}f(\mathbf{x})]^\top$. In particular,

$$(\operatorname{grad} f(\mathbf{x}))\cdot\vec{\mathbf{v}} = [\mathbf{D}f(\mathbf{x})]\vec{\mathbf{v}}; \qquad 6.8.7$$

the dot product of $\vec{\mathbf{v}}$ with the gradient is the directional derivative in the direction $\vec{\mathbf{v}}$.

If θ is the angle between $\operatorname{grad} f(\mathbf{x})$ and $\vec{\mathbf{v}}$, then, by Proposition 1.4.3,

$$\operatorname{grad} f(\mathbf{x})\cdot\vec{\mathbf{v}} = |\operatorname{grad} f(\mathbf{x})|\,|\vec{\mathbf{v}}|\cos\theta, \qquad 6.8.8$$

which becomes $|\operatorname{grad} f(\mathbf{x})|\cos\theta$ if $\vec{\mathbf{v}}$ is constrained to have length 1. This is maximal when $\theta = 0$, giving $[\mathbf{D}f(\mathbf{x})]\vec{\mathbf{v}} = \operatorname{grad} f(\mathbf{x})\cdot\vec{\mathbf{v}} = |\operatorname{grad} f(\mathbf{x})|$. So grad $f(\mathbf{x})$ *points in the direction in which* f *increases the fastest at* $\mathbf{x}$*; its length equals the rate of increase of* f *in that direction, measured in units of the codomain per unit of the domain.*

Remark. Some people find it easier to think of the gradient, which is a vector, and thus an element of $\mathbb{R}^n$, than to think of the derivative, which is a line matrix, and thus a linear *function* $\mathbb{R}^n \to \mathbb{R}$. They also find it easier to think that the gradient is orthogonal to the curve (or surface, or higher-dimensional manifold) of equation $f(\mathbf{x}) - c = 0$ than to think that ker $[\mathbf{D}f(\mathbf{x})]$ is the tangent space to the curve (or surface or manifold). Since the derivative is the transpose of the gradient and vice versa, it may not seem to make any difference which perspective one chooses.

But the derivative has an advantage that the gradient lacks: it needs no extra geometric structure on $\mathbb{R}^n$, whereas, as equation 6.8.7 makes clear, the gradient requires the dot product. Often there is no natural dot product available. Thus the derivative of a function is the natural thing to consider.

Note the words "more appropriate 1-form concept" in the quotation at the beginning of this chapter.

But in physics, gradients of functions really matter: gradients of potential energy functions are *force fields*, and we really want to think of force fields as vectors. For example, the gravitational force field is the vector $-gm\,\vec{\mathbf{e}}_3$, which we saw in equation 6.5.3; this is the gradient of the height function (or rather, minus the gradient of the height function).

Force fields are *conservative* exactly when they are gradients of functions, called potentials, discussed in Section 6.13. However, the potential is not observable, and discovering whether it exists from examining the force field is a big chapter in mathematical physics. △

By "conservative" we mean that the integral on a closed path (i.e., the total energy expended) is zero. The gravity force field is conservative, but any force field involving friction is not: the potential energy you lose going down a hill on a bicycle is never quite recouped when you roll up the other side.

In physics, a vector field is very often a force field. If a vector field is the gradient of a potential energy function, it is then possible to analyze the behavior of the system in terms of conservation of energy. This is usually very enlightening, as well as immensely simpler than solving the equations of motion.

Geometric interpretation of the curl

The peculiar mixture of partials that go into the curl seems impenetrable. We aim to justify a description in terms of the *curl probe*. Key to the description is that rotational flow – for instance, the motion of a spinning top or whirlpool or tornado or ocean eddy – is described mathematically by an *angular velocity vector*. This is a vector in $\mathbb{R}^3$, aligned with the axis of rotation, and with length representing angular velocity (how many rotations per unit time). By convention, the vector is oriented to point in the direction in which a screw moves when turned in the direction of rotation.

The angular velocity vector is different from a "velocity vector" pointing in the direction of motion, with length representing speed. Things actually move in the direction of the velocity vector, but nothing moves in the direction of the angular velocity vector. A spinning skater has an angular velocity vector pointing up or down – and if we know that vector we can infer her motion – but she is neither rising to the ceiling nor sinking down through the ice.

The *curl probe* embodies this representation of rotational flow.

In French the curl is known as the *rotationnel*, and originally in English it was called the *rotation* of the vector field. It was to avoid the abbreviation "rot" that the word "curl" was substituted.

The curl probe. Consider a paddle attached to a long handle that is free to rotate, but in the presence of friction, so that its angular speed (not acceleration) is proportional to the torque exerted by the paddles. We will orient the handle of the probe away from the paddle. Imagine sticking the paddle into a rotating fluid or gas at a point $\mathbf{x}$, with the handle pointing in a particular direction; see Figure 6.8.2. If the fluid or gas is in constant motion with velocity vector field $\vec{F}$, then

> *the speed at which the handle rotates counterclockwise is proportional to the component of the curl $(\vec{\nabla} \times \vec{F})(\mathbf{x})$ of the vector field at $\mathbf{x}$ in the direction of the handle of the probe.*

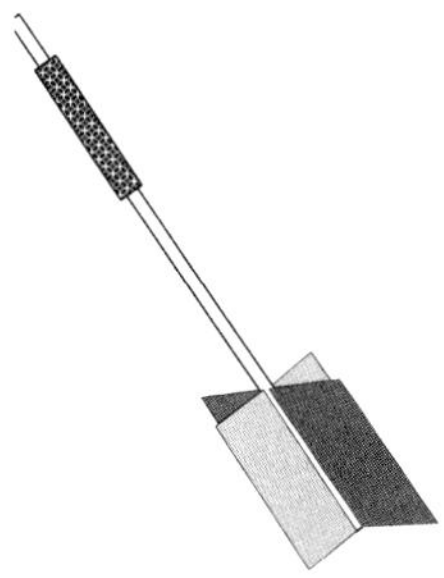

FIGURE 6.8.2.

Curl probe: Put the paddle wheels at some spot of the fluid; the speed at which the paddle rotates counterclockwise will be proportional to the component of the curl in the direction of the axis of the probe, i.e., towards its handle. (That speed can be negative.)

Thus if we first align the curl probe with the x-axis, the number it returns is the first component of the curl: $D_2F_3 - D_3F_2$. If we next align it with the y-axis, we get the second component of the curl: $D_3F_1 - D_1F_3$, and so on. If we align the curl probe with an arbitrary unit vector $\mathbf{a}$, then

$$\underbrace{\begin{bmatrix} D_2F_3 - D_3F_2 \\ D_3F_1 - D_1F_3 \\ D_1F_2 - D_2F_1 \end{bmatrix}}_{\text{curl } \vec{F}} \cdot \underbrace{\begin{bmatrix} a_1 \\ a_2 \\ a_3 \end{bmatrix}}_{\mathbf{a}} \qquad 6.8.9$$

is the component of $\text{curl}\,\vec{F}$ in the direction $\mathbf{a}$; see Corollary 1.4.4.

Why should this be the case? Using Definition 6.5.2 of the flux form, Theorem 6.8.3, and Definition 6.7.1 of the exterior derivative, we see that

$$\begin{aligned} \det[\text{curl}\,\vec{F}(\mathbf{x}), \vec{\mathbf{v}}_1, \vec{\mathbf{v}}_2] &= \underbrace{\Phi_{\text{curl}\,\vec{F}}}_{\mathbf{d}W_{\vec{F}}}\big(P_{\mathbf{x}}(\vec{\mathbf{v}}_1, \vec{\mathbf{v}}_2)\big) \\ &= \lim_{h \to 0} \frac{1}{h^2} \int_{\partial P_{\mathbf{x}}(h\vec{\mathbf{v}}_1, h\vec{\mathbf{v}}_2)} W_{\vec{F}} \end{aligned} \qquad 6.8.10$$

measures the work of $\vec{F}$ over the oriented boundary of the parallelogram spanned by $\vec{\mathbf{v}}_1$ and $\vec{\mathbf{v}}_2$. If $\vec{\mathbf{v}}_1$ and $\vec{\mathbf{v}}_2$ are unit vectors orthogonal to the axis of the probe and to each other, this work is approximately proportional to the torque to which the probe will be subjected.

The exterior derivative of a 1-form $W_{\vec{F}}$ evaluated on a parallelogram acts like the curl probe positioned perpendicular to the parallelogram: both $\mathbf{d}W_{\vec{F}}$ and the curl probe return a number that measures the extent to which the vector field wraps around the boundary of the parallelogram.

FIGURE 6.8.3.

During an earthquake, a fault line in front of the building opens up and the sidewalk is split, the part leading to the door shifting south on the fault line, and the remainder shifting north. This creates rotational forces that attempt to move the building counterclockwise, twisting it.

Moral: In areas prone to earthquakes, building codes must take rotational forces into account.

Remark. We only understood the curl *as an exterior derivative;* we never actually came to terms with the partial derivatives involved. This is another advantage of the differential form interpretation of vector calculus over the standard one. △

Example 6.8.5 (Curl and earthquakes). Suppose a building is located near a fault line. In Figure 6.8.3, we will suppose that the fault line is roughly north-south, which we think of as being on the y-axis, with north positive; the front of the building (the side with the door) is parallel to the fault line. During an earthquake, the lot on which the house is placed moves south, except for the part on the other side of the fault line, which moves north. What happens to the building? The vector field describing the motion can be written approximately $\begin{bmatrix} 0 \\ -1/x \\ 0 \end{bmatrix}$: there is motion only in the direction of the y-axis (along the fault), and that motion is negative and falls off with distance from the y-axis. Since

$$\text{curl} \begin{bmatrix} 0 \\ -1/x \\ 0 \end{bmatrix} = \begin{bmatrix} D_1 \\ D_2 \\ D_3 \end{bmatrix} \times \begin{bmatrix} 0 \\ -1/x \\ 0 \end{bmatrix} = \begin{bmatrix} 0 \\ 0 \\ 1/x^2 \end{bmatrix}, \qquad 6.8.11$$

the building is subject to rotational force (torque), and (since the curl points up) that force tries to rotate the building counterclockwise. The roof resists. △

Geometric interpretation of the divergence

The divergence is easier to interpret than the curl. If φ is a 2-form on $\mathbb{R}^3$, then $\mathbf{d}\varphi$ evaluated on a little three-dimensional box (with its standard orientation) returns approximately the volume of the box times the integral of φ over the box's boundary. If $\varphi = \Phi_{\vec{F}}$ is the flux of a vector field, then

$$\text{div}\,\vec{F}\,dx \wedge dy \wedge dz = \mathbf{d}\Phi_{\vec{F}} \qquad 6.8.12$$

says that the divergence of $\vec{F}$ at a point $\mathbf{x}$ is proportional to the *flux* of $\vec{F}$ through the boundary of a small box around $\mathbf{x}$, i.e., the *net flow out* of the box. In particular, if the fluid is incompressible, so that the density cannot change, the divergence of its velocity vector field is 0: exactly as much must flow in as out. Thus, *the divergence of a vector field measures the change in density of a fluid whose motion is described by the vector field.*

Exercises for Section 6.8

6.8.1 Let $U \subset \mathbb{R}^3$ be open, $\mathbf{x} \in U$ a point, $\vec{\mathbf{u}}, \vec{\mathbf{v}}, \vec{\mathbf{w}}$ elements of $\mathbb{R}^3$, f a function on U, and $\vec{F}$ a vector field on U.

a. Which of the following are numbers? vectors? functions? vector fields?

i. $\text{grad}\, f$ ii. $\text{curl}\, \vec{F}$ iii. $dx \wedge dy(\vec{\mathbf{v}}, \vec{\mathbf{w}})$

iv. $\vec{\mathbf{u}} \cdot (\vec{\mathbf{v}} \times \vec{\mathbf{w}})$ v. $\text{grad}\, f(\mathbf{x}) \cdot \vec{\mathbf{v}}$ vi. $\text{div}\, \vec{F}$

b. Which of the following expressions are identical?

$\text{div}\, \vec{F}$	$\vec{\nabla} \times \vec{F}$	$\vec{\nabla} \cdot \vec{F}$	$D_1 f\, dx_1 + D_2 f\, dx_2 + D_3 f\, dx_3$
$\text{curl}\, \vec{F}$	$\vec{\nabla} f$	$\mathbf{d}\Phi_{\vec{F}}$	$M_{\vec{\nabla} \cdot \vec{F}}$
$M_{\text{div}\, \vec{F}}$	$\mathbf{d}f$	$\Phi_{\text{curl}\, \vec{F}}$	$\mathbf{d}W_{\vec{F}}$
$W_{\vec{\nabla} f}$	$\Phi_{\vec{\nabla} \times \vec{F}}$	$W_{\text{grad}\, f}$	$\text{grad}\, f$

c. Some of the expressions below do not make sense. Correct them so that they do.

i. $\text{grad}\, \vec{F}$ ii. $\text{curl}\, f$ iii. $\Phi_{\vec{F}}(\vec{\mathbf{v}}_1, \vec{\mathbf{v}}_2, \vec{\mathbf{v}}_3)$

iv. W_f v. $\text{div}\, \vec{F}$ vi. $W_{\vec{F}}$

6.8.2 Compute the gradients of the following functions:

a. $f\begin{pmatrix} x \\ y \end{pmatrix} = x$ b. $f\begin{pmatrix} x \\ y \end{pmatrix} = x^2 + y^2$ c. $f\begin{pmatrix} x \\ y \end{pmatrix} = \sin(x+y)$

d. $f\begin{pmatrix} x \\ y \\ z \end{pmatrix} = xyz$ e. $f\begin{pmatrix} x \\ y \\ z \end{pmatrix} = \dfrac{xyz}{x^2+y^2+z^2}$

6.8.3 What is the grad of the function $f = x^2y + z$? What are the curl and div of the vector field $\vec{F} = \begin{bmatrix} -y \\ x \\ xz \end{bmatrix}$?

6.8.4 Prove part 3 of Theorem 6.8.3.

6.8.5 Compute the gradients of the following functions:

a. $f\begin{pmatrix} x \\ y \end{pmatrix} = y^2$ b. $f\begin{pmatrix} x \\ y \end{pmatrix} = x^2 - y^2$

c. $f\begin{pmatrix} x \\ y \end{pmatrix} = \ln(x^2+y^2)$ d. $f\begin{pmatrix} x \\ y \\ z \end{pmatrix} = \ln|x+y+z|$

6.8.6 Show that $\mathbf{d}f = W_{\text{grad}\, f}$ when $f\begin{pmatrix} x \\ y \\ z \end{pmatrix} = xyz$ by computing both from the definitions and evaluating on a vector $\vec{\mathbf{v}} = \begin{pmatrix} a \\ b \\ c \end{pmatrix}$.

6.8.7 Let $\varphi = xy\,dx + z\,dy + yz\,dz$ be a 1-form on $\mathbb{R}^3$. For what vector field $\vec{F}$ can φ be written $dW_{\vec{F}}$? Show the equivalence of $\mathbf{d}W_{\vec{F}}$ and $\Phi_{\vec{\nabla}\times\vec{F}}$ by computing both from the definitions.

6.8.8 a. For what vector field $\vec{F}$ is the 2-form $xy\,dx\wedge dy + x\,dy\wedge dz + xy\,dx\wedge dz$ on $\mathbb{R}^3$ the flux form field $\Phi_{\vec{F}}$?

b. Compute the exterior derivative of $xy\,dx\wedge dy + x\,dy\wedge dz + xy\,dx\wedge dz$ using Theorem 6.7.4. Show that it is the same as the mass form field of $\operatorname{div}\vec{F}$.

6.8.9 a. Let $\vec{F} = \begin{bmatrix} F_1 \\ F_2 \end{bmatrix}$ be a vector field in the plane. Show that if $\vec{F}$ is the gradient of a C^2 function, then $D_2F_1 = D_1F_2$.

b. Show that this is not necessarily true if f is twice differentiable but the second derivatives are not continuous.

6.8.10 a. What is $\mathbf{d}W_{\begin{bmatrix} 0 \\ 0 \\ x \end{bmatrix}}$? What is $\mathbf{d}W_{\begin{bmatrix} 0 \\ 0 \\ x \end{bmatrix}}(P_{\begin{bmatrix} 0 \\ 0 \\ 0 \end{bmatrix}}(\vec{\mathbf{e}}_1, \vec{\mathbf{e}}_3))$?

b. Compute $\mathbf{d}W_{\begin{bmatrix} 0 \\ 0 \\ x \end{bmatrix}}\left(P_{\begin{bmatrix} 0 \\ 0 \\ 0 \end{bmatrix}}(\vec{\mathbf{e}}_1, \vec{\mathbf{e}}_3)\right)$ directly from the definition.

6.8.11 What is the exterior derivative of a. $W_{\begin{bmatrix} x \\ y \\ z \end{bmatrix}}$? b. $\Phi_{\begin{bmatrix} x \\ y \\ z \end{bmatrix}}$?

6.8.12 a. What is the divergence of $\vec{F}\begin{pmatrix} x \\ y \\ z \end{pmatrix} = \begin{bmatrix} x^2 \\ y^2 \\ yz \end{bmatrix}$?

b. Use part a to compute $\mathbf{d}\Phi_{\vec{F}} P_{\begin{pmatrix} 1 \\ 1 \\ 2 \end{pmatrix}}(\vec{\mathbf{e}}_1, \vec{\mathbf{e}}_2, \vec{\mathbf{e}}_3)$.

c. Compute it again, directly from the definition of the exterior derivative.

$$\begin{bmatrix} x^2y \\ -2yz \\ x^3y^2 \end{bmatrix}, \quad \begin{bmatrix} \sin xz \\ \cos yz \\ xyz \end{bmatrix}$$

Vector fields for Exercise 6.8.13.

6.8.13 a. Compute the divergence and curl of the vector fields in the margin.

b. Compute them again, directly from the definition of the exterior derivative.

6.8.14 Show that $\operatorname{grad}\operatorname{div}\vec{\mathbf{F}} - \operatorname{curl}\operatorname{curl}\vec{\mathbf{F}} = \begin{bmatrix} \Delta F_1 \\ \Delta F_2 \\ \Delta F_3 \end{bmatrix}$.

6.9 The pullback

To prove Stokes's theorem, we will need the *pullback* of form fields. The pullback describes how integrands transform under changes of variables; it was used for some 200 years before being properly defined. When you write "let $x = f(u)$, so that $dx = f'(u)\,du$", you are computing a pullback, $f^*\,dx = f'(u)\,du$. This kind of computation, and more doubtful ones in several variables, were used throughout the 18th and 19th centuries. It has been used implicitly throughout Chapter 6, and indeed underlies the change of variables formula for integrals both in elementary calculus and as developed in Section 4.10. Elie Cartan developed the concept of differential forms and the exterior derivative around 1900, largely to make these computations precise. The central result of this section is Theorem 6.9.8.

FIGURE 6.9.1.
Elie Cartan (1869–1951) formalized the theory of differential forms. Other names associated with the generalized Stokes's theorem include Henri Poincaré, Vito Volterra, and Luitzen Brouwer.

Cartan's son Henri (who served as president of the jury for John Hubbard's PhD) was a renowned mathematician; he died August 13, 2008, age 104. Another son, a physicist, was arrested by the Germans in 1942 and executed 15 months later.

The pullback by a linear transformation

We will begin by the simplest case, pullbacks of forms by linear transformations.

Definition 6.9.1 (Pullback by a linear transformation). Let V, W be vector spaces, $T : V \to W$ a linear transformation, and φ a constant k-form on W. Then the *pullback* by T is the mapping $T^* : A_c^k(W) \to A_c^k(V)$ defined by

$$T^*\varphi(\vec{\mathbf{v}}_1, \dots, \vec{\mathbf{v}}_k) \stackrel{\text{def}}{=} \varphi\big(T(\vec{\mathbf{v}}_1), \dots, T(\vec{\mathbf{v}}_k)\big), \qquad 6.9.1$$

and the k-form $T^*\varphi$ is called the *pullback of φ by T*.

The pullback $T^*\varphi$ of φ, acting on k vectors $\vec{\mathbf{v}}_1, \dots, \vec{\mathbf{v}}_k$ in the domain of T, gives the same result as φ, acting on the vectors $T(\vec{\mathbf{v}}_1), \dots, T(\vec{\mathbf{v}}_k)$ in the codomain.

Note that both forms must have the same degree: they both act on the same number of vectors. But the domain V and codomain W can be of different dimensions.

The pullback $T^*\varphi$ of φ is pronounced "T upper star phi".

It follows immediately from Definition 6.9.1 that $T^* : A_c^k(W) \to A_c^k(V)$ is linear, as you are asked to show in Exercise 6.9.1. The following proposition and the linearity of T^* give a cumbersome but straightforward way to compute the pullback of any form by a linear transformation $T : \mathbb{R}^n \to \mathbb{R}^m$.

Proposition 6.9.2 (Computing the pullback by a linear transformation). *Let $T : \mathbb{R}^n \to \mathbb{R}^m$ be a linear transformation. Denote by $x_1, \dots, x_n$ the coordinates in $\mathbb{R}^n$ and by $y_1, \dots, y_m$ the coordinates in $\mathbb{R}^m$. Then*

$$T^*(dy_{i_1} \wedge \cdots \wedge dy_{i_k}) = \sum_{1 \le j_1 < \cdots < j_k \le n} b_{j_1,\dots,j_k} dx_{j_1} \wedge \cdots \wedge dx_{j_k}, \qquad 6.9.2$$

where $b_{j_1,\dots,j_k}$ is the number obtained by taking the matrix of T; selecting its rows $i_1, \dots, i_k$, in that order; selecting its columns $j_1, \dots, j_k$; and taking the determinant of the resulting matrix.

A square submatrix of a matrix is called a *minor*. Determinants of minors occur in many settings. The real meaning of this construction is given by Proposition 6.9.2.

Example 6.9.3 (Computing the pullback). Let $T : \mathbb{R}^4 \to \mathbb{R}^3$ be the linear transformation given by the matrix $[T] = \begin{bmatrix} 1 & 0 & 0 & 1 \\ 0 & 1 & 0 & 1 \\ 0 & 0 & 1 & 1 \end{bmatrix}$. Then

$$\begin{aligned} T^*\big(dy_2 \wedge dy_3\big) &= b_{1,2}\, dx_1 \wedge dx_2 \;+\; b_{1,3}\, dx_1 \wedge dx_3 \;+\; b_{1,4}\, dx_1 \wedge dx_4 \\ &\quad + b_{2,3}\, dx_2 \wedge dx_3 \;+\; b_{2,4}\, dx_2 \wedge dx_4 \;+\; b_{3,4}\, dx_3 \wedge dx_4, \end{aligned} \qquad 6.9.3$$

Since we are computing the pullback $T^* dy_2 \wedge dy_3$, we take the second and third rows of T, and then select out columns 1 and 2 for $b_{1,2}$, columns 1 and 3 for $b_{1,3}$, and so on.

where

$$b_{1,2} = \det \begin{bmatrix} 0 & 1 \\ 0 & 0 \end{bmatrix} = 0, \quad b_{1,3} = \det \begin{bmatrix} 0 & 0 \\ 0 & 1 \end{bmatrix} = 0, \quad b_{1,4} = \det \begin{bmatrix} 0 & 1 \\ 0 & 1 \end{bmatrix} = 0,$$

$$b_{2,3} = \det \begin{bmatrix} 1 & 0 \\ 0 & 1 \end{bmatrix} = 1, \quad b_{2,4} = \det \begin{bmatrix} 1 & 1 \\ 0 & 1 \end{bmatrix} = 1, \quad b_{3,4} = \det \begin{bmatrix} 0 & 1 \\ 1 & 1 \end{bmatrix} = -1.$$

So

$$T^* dy_2 \wedge dy_3 = dx_2 \wedge dx_3 + dx_2 \wedge dx_4 - dx_3 \wedge dx_4. \quad \triangle \qquad 6.9.4$$

Proof. Since any k-form on $\mathbb{R}^n$ is of the form

$$\sum_{1 \le j_1 < \cdots < j_k \le n} b_{j_1,\dots,j_k} dx_{j_1} \wedge \cdots \wedge dx_{j_k}, \qquad 6.9.5$$

the only problem is to compute the coefficients. This is analogous to equation 6.1.24 in the proof of Theorem 6.1.8:

$$\begin{aligned} b_{j_1,\dots,j_k} &= (T^* dy_{i_1} \wedge \cdots \wedge dy_{i_k})(\vec{\mathbf{e}}_{j_1}, \dots, \vec{\mathbf{e}}_{j_k}) \\ &= (dy_{i_1} \wedge \cdots \wedge dy_{i_k})\big(T(\vec{\mathbf{e}}_{j_1}), \dots, T(\vec{\mathbf{e}}_{j_k})\big). \end{aligned} \qquad 6.9.6$$

This is what we needed: $dy_{i_1} \wedge \cdots \wedge dy_{i_k}$ selects the corresponding rows from the matrix $[(T(\vec{\mathbf{e}}_{j_1}), \dots, T(\vec{\mathbf{e}}_{j_k})]$. This is precisely the matrix made up of the columns $j_1, \dots, j_k$ of $[T]$. $\square$

Pullback of a k-form field by a C^1 mapping

If $X \subset \mathbb{R}^n$, $Y \subset \mathbb{R}^m$ are open subsets and $\mathbf{f} : X \to Y$ is a C^1 mapping, then we can use $\mathbf{f}$ to pull k-form fields on Y back to k-form fields on X.

Definition 6.9.4 (Pullback by a C^1 mapping). If φ is a k-form field on Y, and $\mathbf{f} : X \to Y$ is a C^1 mapping, then $\mathbf{f}^* : A^k(Y) \to A^k(X)$ is defined by

$$(\mathbf{f}^*\varphi)\Big(P_{\mathbf{x}}(\vec{\mathbf{v}}_1, \dots, \vec{\mathbf{v}}_k)\Big) \overset{\text{def}}{=} \varphi\Big(P_{\mathbf{f}(\mathbf{x})}\big([\mathbf{Df}(\mathbf{x})]\vec{\mathbf{v}}_1, \dots, [\mathbf{Df}(\mathbf{x})]\vec{\mathbf{v}}_k\big)\Big). \qquad 6.9.7$$

Definition 6.9.4 says that $\mathbf{f}^*$ is a linear transformation from $A^k(Y)$ to $A^k(X)$:

$$\mathbf{f}^*(\varphi_1 + \varphi_2) = \mathbf{f}^*(\varphi_1) + \mathbf{f}^*(\varphi_2)$$
$$\mathbf{f}^*(a\varphi) = a\mathbf{f}^*(\varphi).$$

If pullbacks seem unnatural to you, note that when φ is a 0-form (i.e., a function g), the pullback is another way of expressing compositions:

$$\mathbf{f}^* g = g \circ \mathbf{f},$$

since

$$\mathbf{f}^* g(P_{\mathbf{x}}) = g(P_{\mathbf{f}(\mathbf{x})}) = g(\mathbf{f}(\mathbf{x})) = g \circ \mathbf{f}(P_{\mathbf{x}}).$$

If $k = n$, so that $\mathbf{f}(X)$ can be viewed as a parametrized domain, then Definition 6.2.1 of the integral over a parametrized domain can be written

$$\int_{\mathbf{f}(X)} \varphi = \int_X \mathbf{f}^*\varphi. \qquad 6.9.8$$

Thus we have been using pullbacks throughout Chapter 6.

Example 6.9.5 (Pullback by C^1 mapping). Define $\mathbf{f} : \mathbb{R}^2 \to \mathbb{R}^3$ by

$$\mathbf{f}\begin{pmatrix} x_1 \\ x_2 \end{pmatrix} = \begin{pmatrix} x_1^2 \\ x_1x_2 \\ x_2^2 \end{pmatrix}. \qquad 6.9.9$$

We will compute $\mathbf{f}^*(y_2\, dy_1 \wedge dy_3)$. Certainly

$$\mathbf{f}^*(y_2\, dy_1 \wedge dy_3) = b\; dx_1 \wedge dx_2 \qquad 6.9.10$$

for some function $b : \mathbb{R}^2 \to \mathbb{R}$. The object is to compute that function:

$$\begin{aligned} b\begin{pmatrix} x_1 \\ x_2 \end{pmatrix} &= b\begin{pmatrix} x_1 \\ x_2 \end{pmatrix} dx_1 \wedge dx_2\left(\begin{bmatrix} 1 \\ 0 \end{bmatrix}, \begin{bmatrix} 0 \\ 1 \end{bmatrix}\right) = \mathbf{f}^*(y_2\, dy_1 \wedge dy_3)\left(P_{\begin{pmatrix} x_1 \\ x_2 \end{pmatrix}}\left(\begin{bmatrix} 1 \\ 0 \end{bmatrix}, \begin{bmatrix} 0 \\ 1 \end{bmatrix}\right)\right) \\ &= (y_2\, dy_1 \wedge dy_3)\left(P_{\begin{pmatrix} x_1^2 \\ x_1x_2 \\ x_2^2 \end{pmatrix}}\left(\begin{bmatrix} 2x_1 \\ x_2 \\ 0 \end{bmatrix}, \begin{bmatrix} 0 \\ x_1 \\ 2x_2 \end{bmatrix}\right)\right) \\ &= x_1x_2 \det\begin{bmatrix} 2x_1 & 0 \\ 0 & 2x_2 \end{bmatrix} = 4x_1^2x_2^2. \end{aligned} \qquad 6.9.11$$

So we have $\mathbf{f}^*(y_2\, dy_1 \wedge dy_3) = 4x_1^2x_2^2\; dx_1 \wedge dx_2$. △

Pullbacks and compositions

To prove Stokes's theorem, we will need to know how pullbacks behave under composition. If S and T are linear transformations, then

$$\begin{aligned} (S \circ T)^*\varphi(\vec{\mathbf{v}}_1, \dots, \vec{\mathbf{v}}_k) &= \varphi\Big((S \circ T)(\vec{\mathbf{v}}_1), \dots, (S \circ T)(\vec{\mathbf{v}}_k)\Big) \\ &= S^*\varphi\big(T(\vec{\mathbf{v}}_1), \dots, T(\vec{\mathbf{v}}_k)\big) = T^*S^*\varphi(\vec{\mathbf{v}}_1, \dots, \vec{\mathbf{v}}_k). \end{aligned} \qquad 6.9.12$$

Thus $(S \circ T)^* = T^* S^*$. The same formula holds for pullbacks of form fields by C^1 mappings.

Proposition 6.9.6 (Pullbacks by compositions). *If $X \subset \mathbb{R}^n$, $Y \subset \mathbb{R}^m$ and $Z \subset \mathbb{R}^p$ are open, $\mathbf{f} : X \to Y$ and $\mathbf{g} : Y \to Z$ are C^1 mappings, and φ is a k-form on Z, then*

$$(\mathbf{g} \circ \mathbf{f})^* \varphi = \mathbf{f}^* \mathbf{g}^* \varphi. \qquad 6.9.13$$

The first, third, and fourth equalities in equation 6.9.14 are the definition of the pullback for $\mathbf{g} \circ \mathbf{f}$, $\mathbf{g}$, and $\mathbf{f}$ respectively; the second equality is the chain rule.

Proof. This follows from the chain rule:

$$(\mathbf{g} \circ \mathbf{f})^* \varphi(P_{\mathbf{x}}(\vec{\mathbf{v}}_1, \dots, \vec{\mathbf{v}}_k)) = \varphi\Big(P_{(\mathbf{g}(\mathbf{f}(\mathbf{x}))}\big([\mathbf{D}(\mathbf{g} \circ \mathbf{f})(\mathbf{x})]\vec{\mathbf{v}}_1, \dots, [\mathbf{D}(\mathbf{g} \circ \mathbf{f})(\mathbf{x})]\vec{\mathbf{v}}_k\big)\Big)$$

$$= \varphi\Big(P_{(\mathbf{g}(\mathbf{f}(\mathbf{x}))}\big([\mathbf{Dg}(\mathbf{f}(\mathbf{x}))][\mathbf{Df}(\mathbf{x})]\vec{\mathbf{v}}_1, \dots, [\mathbf{Dg}(\mathbf{f}(\mathbf{x}))][\mathbf{Df}(\mathbf{x})]\vec{\mathbf{v}}_k\big)\Big) \qquad 6.9.14$$

$$= \mathbf{g}^* \varphi\Big(P_{\mathbf{f}(\mathbf{x})}\big([\mathbf{Df}(\mathbf{x})]\vec{\mathbf{v}}_1, \dots, [\mathbf{Df}(\mathbf{x})]\vec{\mathbf{v}}_k\big)\Big) = \mathbf{f}^* \mathbf{g}^* \varphi(P_{\mathbf{x}}(\vec{\mathbf{v}}_1, \dots, \vec{\mathbf{v}}_k)). \quad \square$$

The wedge product and exterior derivative are intrinsic

Proposition 6.9.7 and Theorem 6.9.8 say that the wedge product and the exterior derivative are independent of coordinates.

In Example 6.9.5 we worked from the definition. But using Proposition 6.9.7 we could compute the pullback more easily:

$$\begin{aligned} &\mathbf{f}^*(y_2\, dy_1 \wedge dy_3) \\ &= (x_1 x_2)\big(\mathbf{d}(x_1^2) \wedge \mathbf{d}(x_2^2)\big) \\ &= (x_1 x_2)(2x_1 dx_1) \wedge (2x_2 dx_2) \\ &= 4x_1^2 x_2^2\, dx_1 \wedge dx_2. \end{aligned}$$

We denote these permutations by Perm(k, l), as we did in Definition 6.1.12.

Proposition 6.9.7 (Pullback and wedge products). *Let $X \subset \mathbb{R}^n$ and $Y \subset \mathbb{R}^m$ be open, $\mathbf{f} : X \to Y$ a C^1 mapping, and φ and ψ respectively a k-form and an l-form on Y. Then*

$$\mathbf{f}^* \varphi \wedge \mathbf{f}^* \psi = \mathbf{f}^*(\varphi \wedge \psi). \qquad 6.9.15$$

Proof. This is one of those proofs where you write down the definitions and follow your nose. Let us spell it out when $\mathbf{f} = T$ is linear; we leave the general case as Exercise 6.9.2. Recall that the wedge product is a certain sum over all permutations σ of $\{1, \dots, k + l\}$ such that

$$\sigma(1) < \cdots < \sigma(k) \quad \text{and} \quad \sigma(k+1) < \cdots < \sigma(k+l). \qquad 6.9.16$$

We find

$$\begin{aligned} &T^*(\varphi \wedge \psi)(\vec{\mathbf{v}}_1, \dots, \vec{\mathbf{v}}_{k+l}) = (\varphi \wedge \psi)\big(T(\vec{\mathbf{v}}_1), \dots, T(\vec{\mathbf{v}}_{k+l})\big) \\ &= \sum_{\sigma \in \text{Perm}(k,l)} \text{sgn}(\sigma) \varphi\Big(T(\vec{\mathbf{v}}_{\sigma(1)}), \dots, T(\vec{\mathbf{v}}_{\sigma(k)})\Big) \psi\Big(T(\vec{\mathbf{v}}_{\sigma(k+1)}), \dots, T(\vec{\mathbf{v}}_{\sigma(k+l)})\Big) \\ &= \sum_{\sigma \in \text{Perm}(k,l)} \text{sgn}(\sigma) T^* \varphi(\vec{\mathbf{v}}_{\sigma(1)}, \dots, \vec{\mathbf{v}}_{\sigma(k)})\; T^* \psi(\vec{\mathbf{v}}_{\sigma(k+1)}, \dots, \vec{\mathbf{v}}_{\sigma(k+l)}) \\ &= (T^* \varphi \wedge T^* \psi)(\vec{\mathbf{v}}_1, \dots, \vec{\mathbf{v}}_{k+l}). \quad \square \end{aligned} \qquad 6.9.17$$

The next theorem has the innocent appearance $\mathbf{df}^* = \mathbf{f}^* \mathbf{d}$. But this formula says something quite deep. To define the exterior derivative, we used parallelograms $P_{\mathbf{x}}(\vec{\mathbf{v}}_1, \dots, \vec{\mathbf{v}}_k)$. To do this, we had to know how to draw straight lines from one point to another; we were using the linear (straight) structure of a vector space. (We used $\mathbb{R}^n$, but any vector space would have done.) Theorem 6.9.8 says that "curved parallelograms" (little bits of manifolds) would have worked as well. Thus the exterior derivative is not restricted to forms defined on vector spaces.

In this book we have discussed forms on vector spaces, but differential forms can also be defined on manifolds embedded in $\mathbb{R}^n$ and on abstract manifolds. Theorem 6.9.8 says that an exterior derivative exists for such forms.

By "intrinsic" we mean "inherent: independent of some external conditions or circumstances". The pullback of a form by a C^1 mapping is a C^1 change of variables. Equation 6.9.18 says that when a form is pulled back by a C^1 mapping, its exterior derivative remains the same, translated appropriately into the new variables.

Elie Cartan proved that the exterior derivative is the only differential operator (other than multiples of the identity) that can be defined on a manifold without specifying any extra structure.

In particular, the gradient requires extra structure: the inner product.

Theorem 6.9.8 (Exterior derivative is intrinsic). *Let $X \subset \mathbb{R}^n$, $Y \subset \mathbb{R}^m$ be open sets, and let $\mathbf{f} : X \to Y$ be a C^1 mapping. If φ is a k-form field on Y, then*

$$\mathbf{d}\mathbf{f}^*\varphi = \mathbf{f}^*\mathbf{d}\varphi. \tag{6.9.18}$$

Proof. We will prove this theorem by induction on k. The case $k = 0$, where $\varphi = g$ is a function, is an application of the chain rule:

$$\begin{aligned}\mathbf{f}^*\mathbf{d}g\big(P_{\mathbf{x}}(\vec{\mathbf{v}})\big) &= \mathbf{d}g\Big(P_{\mathbf{f}(\mathbf{x})}\big([\mathbf{Df}(\mathbf{x})]\vec{\mathbf{v}}\big)\Big) = [\mathbf{D}g(\mathbf{f}(\mathbf{x}))][\mathbf{Df}(\mathbf{x})]\vec{\mathbf{v}} \\ &= [\mathbf{D}(g\circ\mathbf{f})(\mathbf{x})]\vec{\mathbf{v}} = \mathbf{d}(g\circ\mathbf{f})\big(P_{\mathbf{x}}(\vec{\mathbf{v}})\big) = \mathbf{d}(\mathbf{f}^*g)\big(P_{\mathbf{x}}(\vec{\mathbf{v}})\big).\end{aligned} \tag{6.9.19}$$

If $k > 0$, it is enough to prove the result when we can write $\varphi = \psi \wedge dx_i$, where ψ is a $(k-1)$-form. Then

$$\begin{aligned}\mathbf{f}^*\mathbf{d}(\psi\wedge dx_i) &\overbrace{=}^{\text{Theorem 6.7.9}} \mathbf{f}^*\Big(\mathbf{d}\psi\wedge dx_i + \overbrace{(-1)^{k-1}\psi\wedge\mathbf{d}dx_i}^{0\text{ since }\mathbf{d}dx_1=0}\Big) \\ &\underbrace{=}_{\text{Prop. 6.9.7}} \mathbf{f}^*(\mathbf{d}\psi)\wedge\mathbf{f}^*dx_i \underbrace{=}_{\text{inductive hypothesis}} \mathbf{d}(\mathbf{f}^*\psi)\wedge\mathbf{f}^*dx_i,\end{aligned} \tag{6.9.20}$$

whereas

$$\begin{aligned}\mathbf{d}\mathbf{f}^*(\psi\wedge dx_i) = \mathbf{d}\big(\mathbf{f}^*\psi\wedge\mathbf{f}^*dx_i\big) &\overbrace{=}^{\text{Theorem 6.7.9}} \big(\mathbf{d}(\mathbf{f}^*\psi)\big)\wedge\mathbf{f}^*dx_i + (-1)^{k-1}\mathbf{f}^*\psi\wedge\mathbf{d}(\mathbf{f}^*dx_i) \\ &= \big(\mathbf{d}(\mathbf{f}^*\psi)\big)\wedge\mathbf{f}^*dx_i + (-1)^{k-1}\mathbf{f}^*\psi\wedge\mathbf{d}\mathbf{d}\mathbf{f}^*x_i \\ &= \big(\mathbf{d}(\mathbf{f}^*\psi)\big)\wedge\mathbf{f}^*dx_i. \quad \square\end{aligned} \tag{6.9.21}$$

Equation 6.9.21: The $\mathbf{d}(\mathbf{f}^*dx_i)$ in the first line becomes $\mathbf{ddf}^*x_i$ in the second line. This substitution is allowed by induction (it is the case $k = 0$) because x_i is a function. In fact $\mathbf{f}^*x_i = f_i$, the ith component of $\mathbf{f}$. Of course, $\mathbf{ddf}^*x_i = 0$ since it is the exterior derivative taken twice.

Remark. This proof of Theorem 6.9.8 is thoroughly unsatisfactory; it does not explain why the result is true. It is quite possible to give a conceptual proof, but that proof is as hard as (and largely a repetition of) the proof of Theorem 6.7.4. The proof of Theorem 6.7.4 was quite difficult, and the present proof really builds on the work we did there. △

Exercises for Section 6.9

6.9.1 a. Show that if $T : V \to W$ is a linear transformation, the pullback $T^* : A_c^k(W) \to A_c^k(V)$ is linear.

b. Show that the pullback by a C^1 mapping is linear.

6.9.2 In the text we proved Proposition 6.9.7 in the special case where the mapping $\mathbf{f}$ is linear. Prove the general statement, where $\mathbf{f}$ is only assumed to be of class C^1.

6.9.3 Let $U \subset \mathbb{R}^n$ be open, $f: U \to \mathbb{R}^m$ of class C^1, and ξ a vector field on $\mathbb{R}^m$.

a. Show that $(f^* W_\xi)(\mathbf{x}) = W_{[\mathbf{D}f(\mathbf{x})]^\top \xi(\mathbf{x})}$.

b. Let $m = n$. Show that if $[\mathbf{D}f(\mathbf{x})]$ is invertible, then

$$(f^*\Phi_\xi)(\mathbf{x}) = \det[\mathbf{D}f(\mathbf{x})]\Phi_{[\mathbf{D}f(\mathbf{x})]^{-1}\xi(\mathbf{x})}.$$

c. Let $m = n$, let A be a square matrix, and let $A_{[j,i]}$ be the matrix obtained from A by erasing the jth row and the ith column. Let $\operatorname{adj}(A)$ be the matrix whose (i,j)th entry is $(\operatorname{adj}(A))_{i,j} = (-1)^{i+j} \det A_{[j,i]}$. Show that

$$A(\operatorname{adj}(A)) = (\det A)I \quad \text{and} \quad f^*\Phi_\xi(\mathbf{x}) = \Phi_{\operatorname{adj}([\mathbf{D}f(\mathbf{x})])\xi(\mathbf{x})}$$

Exercise 6.9.3: The matrix $\operatorname{adj}(A)$ of part c is the *adjoint matrix* of A. The equation in part b is unsatisfactory: it does not say how to represent $(f^*\Phi_\xi)(\mathbf{x})$ as the flux of a vector field when the $n \times n$ matrix $[\mathbf{D}f(\mathbf{x})]$ is not invertible. Part c deals with this situation.

6.10 The generalized Stokes's theorem

We worked hard to define the exterior derivative and to define orientation of manifolds and of boundaries. Now we are going to reap some rewards for our labor: a higher-dimensional analogue of the fundamental theorem of calculus, Stokes's theorem. It covers in one statement the four integral theorems of vector calculus, which are explored in Section 6.11.

Recall the fundamental theorem of calculus:

Theorem 6.10.1 (Fundamental theorem of calculus). *If f is a C^1 function on a neighborhood of $[a,b]$, then*

$$\int_a^b f'(t)\,dt = f(b) - f(a). \qquad 6.10.1$$

Restate this as

$$\int_{[a,b]} \mathbf{d}f = \int_{\partial[a,b]} f, \qquad 6.10.2$$

i.e., the integral of $\mathbf{d}f$ over the oriented interval $[a,b]$ is equal to the integral of f over the oriented boundary $+b - a$ of the interval. This is the case $k = n = 1$ of Theorem 6.10.2.

Special cases of the generalized Stokes's theorem are discussed in Section 6.11.

Theorem 6.10.2 is probably the best tool mathematicians have for deducing global properties from local properties. It is a wonderful theorem. It is often called the generalized Stokes's theorem, to distinguish it from the special case (surfaces in $\mathbb{R}^3$) also known as Stokes's theorem.

Theorem 6.10.2 (Generalized Stokes's theorem). *Let X be a piece-with-boundary of a k-dimensional oriented smooth manifold M in $\mathbb{R}^n$. Give the boundary ∂X of X the boundary orientation, and let φ be a $(k-1)$-form of class C^2 defined on an open set containing X. Then*

$$\int_{\partial X} \varphi = \int_X \mathbf{d}\varphi. \qquad 6.10.3$$

Theorem 6.10.2: A lot is hidden in the equation

$$\int_{\partial X} \varphi = \int_X \mathbf{d}\varphi.$$

First, we can only integrate forms over manifolds, and the boundary of X is not a manifold, so despite its elegance, the equation does not make sense. What we mean is

$$\int_{\partial^s_M X} \varphi = \int_X \mathbf{d}\varphi;$$

by Proposition 6.6.3 and Definition 6.6.21, the smooth boundary is an oriented manifold. Since by the definition of a piece-with-boundary the set of nonsmooth points in the boundary has $(k-1)$-dimensional volume 0, we can set

$$\int_{\partial X} \varphi \overset{\text{def}}{=} \int_{\partial^s_M X} \varphi.$$

Second, the smooth boundary will usually not be compact, so how do we know the integral is defined (that it doesn't diverge)? This is why we require the second condition of a piece-with-boundary: that its smooth boundary have finite $(k-1)$-dimensional volume.

Recall (Definition 6.6.8) that a piece-with-boundary is compact. Without that hypothesis, Stokes's theorem would be false: the proof uses Heine-Borel (Theorem A3.3), which applies to compact subsets of $\mathbb{R}^n$.

For the proof of Stokes's theorem, the manifold M must be at least of class C^2; in practice, it will always be of class C^∞.

This beautiful, short statement is the main result of the theory of forms. Note that the dimensions in equation 6.10.3 make sense: if X is k-dimensional, ∂X is $(k-1)$-dimensional, and if φ is a $(k-1)$-form, $\mathbf{d}\varphi$ is a k-form, so $\mathbf{d}\varphi$ can be integrated over X, and φ can be integrated over ∂X.

You apply Stokes's theorem every time you use antiderivatives to compute an integral: to compute the integral of the 1-form $f\,dx$ over the oriented line segment $[a,b]$, you begin by finding a function g such that $\mathbf{d}g = f\,dx$, and then say

$$\int_a^b f\,dx = \int_{[a,b]} \mathbf{d}g = \int_{\partial[a,b]} g = g(b) - g(a). \qquad 6.10.4$$

This isn't quite the way Stokes's theorem is usually used in higher dimensions, where "looking for antiderivatives" has a different flavor.

Example 6.10.3 (Integrating over the boundary of a square). Let S be the square described by the inequalities $|x|, |y| \le 1$, with the standard orientation. To compute the integral $\int_C x\,dy - y\,dx$, where C is the boundary of S, with the boundary orientation, one possibility is to parametrize the four sides of the square (being careful to get the orientations right), then to integrate $x\,dy - y\,dx$ over all four sides and add. Another possibility is to apply Stokes's theorem:

$$\int_C x\,dy - y\,dx = \int_S \mathbf{d}(x\,dy - y\,dx) = \int_S 2\,dx \wedge dy = \int_S 2|dx\,dy| = 8. \qquad 6.10.5$$

(The square S has sidelength 2, so its area is 4.) △

What is the integral over C of $x\,dy + y\,dx$? Check below.[12]

Example 6.10.4 (Integrating over the boundary of a cube). Let us integrate the 2-form

$$\varphi = (x - y^2 + z^3)\,(dy \wedge dz + dx \wedge dz + dx \wedge dy) \qquad 6.10.6$$

over the boundary of the cube C_a given by $0 \le x, y, z \le a$.

It is quite possible to do this directly, parametrizing all six faces of the cube, but Stokes's theorem simplifies things substantially.

Computing the exterior derivative of φ gives

$$\begin{aligned}\mathbf{d}\varphi &= dx \wedge dy \wedge dz - 2y\,dy \wedge dx \wedge dz + 3z^2\,dz \wedge dx \wedge dy \\ &= (1 + 2y + 3z^2)\,dx \wedge dy \wedge dz,\end{aligned} \qquad 6.10.7$$

[12] $\mathbf{d}(x\,dy + y\,dx) = dx \wedge dy + dy \wedge dx = 0$, so the integral is 0.

so we have

$$\int_{\partial C_a} \varphi = \int_{C_a} (1+2y+3z^2)\, dx \wedge dy \wedge dz$$
$$= \int_0^a \int_0^a \int_0^a (1+2y+3z^2)\, dx\, dy\, dz \qquad 6.10.8$$
$$= a^2\big([x]_0^a + [y^2]_0^a + [z^3]_0^a\big) = a^2\big(a+a^2+a^3\big). \quad \triangle$$

Example 6.10.5 (Stokes's theorem: a harder example). Now let's try something similar but harder, integrating

$$\varphi = (x_1 - x_2^2 + x_3^3 - \cdots \pm x_n^n)\left(\sum_{i=1}^n dx_1 \wedge \cdots \wedge \widehat{dx_i} \wedge \cdots \wedge dx_n\right) \qquad 6.10.9$$

over the boundary of the n-dimensional cube C_a given by $0 \le x_j \le a$, for $j = 1, \dots, n$, with the standard orientation.

This time, the idea of computing the integral directly is pretty awesome: parametrizing all $2n$ faces of the cube, etc. Doing it using Stokes's theorem is also pretty awesome, but much more manageable. We know how to compute $\mathbf{d}\varphi$, and it comes out to

$$\mathbf{d}\varphi = \underbrace{(1 + 2x_2 + 3x_3^2 + \cdots + nx_n^{n-1})}_{\sum_{j=1}^n jx_j^{j-1}}\, dx_1 \wedge \cdots \wedge dx_n, \qquad 6.10.10$$

The integral of $jx_j^{j-1}\, dx_1 \wedge \cdots \wedge dx_n$ over C_a is

$$\int_0^a \cdots \int_0^a jx_j^{j-1} dx_1 \dots dx_n = a^{j+n-1}, \qquad 6.10.11$$

so the whole integral is $\sum_{j=1}^n a^{j+n-1} = a^n(1 + a + \cdots + a^{n-1})$. $\triangle$

Example 6.10.5: Computing this exterior derivative is less daunting if you are alert for terms that can be discarded. Denote

$$(x_1 - x_2^2 + x_3^3 - \cdots \pm x_n^n)$$

by f. Then

$$D_1 f = dx_1,$$
$$D_2 f = -2x_2\, dx_2,$$
$$D_3 f = 3x_3^2\, dx_3$$

and so on, ending with

$$\pm nx_n^{n-1}\, dx_n.$$

For $D_1 f$, the only term of

$$\sum_{i=1}^n dx_1 \wedge \cdots \wedge \widehat{dx_i} \wedge \cdots \wedge dx_n$$

that survives is that in which $i = 1$, giving

$$dx_1 \wedge dx_2 \wedge \cdots \wedge dx_n.$$

For $D_2 f$, the only term of the sum that survives is

$$dx_1 \wedge dx_3 \wedge \cdots \wedge dx_n,$$

giving

$$-2x_2\, dx_2 \wedge dx_1 \wedge dx_3 \wedge \cdots \wedge dx_n;$$

when the order is corrected this gives

$$2x_2\, dx_1 \wedge dx_2 \wedge \cdots \wedge dx_n.$$

In the end, all the terms are followed simply by $dx_1 \wedge \cdots \wedge dx_n$, and any minus signs have become plus signs.

These examples bring out one unpleasant feature of Stokes's theorem: it relates the integral of a $(k-1)$-form to the integral of a k-form only if the former is integrated over a boundary. It is often possible to skirt this difficulty, as in the next example.

Example 6.10.6 (Integrating over faces of a cube). Let S be an open box without its top: the union of the faces of the cube C given by $-1 \le x, y, z \le 1$ except the top face, oriented by the outward-pointing normal. What is $\int_S \Phi_{\vec{F}}$, where $\vec{F} = \begin{bmatrix} x \\ y \\ z \end{bmatrix}$? Stokes's theorem says that the integral of $\Phi_{\vec{F}}$ over the entire boundary ∂C is the integral over C of $\mathbf{d}\Phi_{\vec{F}} = M_{\operatorname{div} \vec{F}} = M_3 = 3\, dx \wedge dy \wedge dz$, so

$$\int_{\partial C} \Phi_{\vec{F}} = \int_C \mathbf{d}\Phi_{\vec{F}} = \int_C 3\, dx \wedge dy \wedge dz = 3 \int_C dx \wedge dy \wedge dz = 24. \qquad 6.10.12$$

Now we must subtract from that the integral over the top. We parametrize the top of the cube using the obvious parametrization[13] $\gamma\begin{pmatrix} s \\ t \end{pmatrix} = \begin{pmatrix} s \\ t \\ 1 \end{pmatrix}$, which preserves orientation. This gives

The matrix in equation 6.10.13 is

$$\left[\vec{F}\left(\gamma\begin{pmatrix} s \\ t \end{pmatrix}\right), \overrightarrow{D_1\gamma}\begin{pmatrix} s \\ t \end{pmatrix}, \overrightarrow{D_2\gamma}\begin{pmatrix} s \\ t \end{pmatrix}\right]$$

(see equation 6.5.17).

$$\int_{\text{top}} \Phi_{\vec{F}} = \int_{-1}^{1}\int_{-1}^{1} \det \begin{bmatrix} s & 1 & 0 \\ t & 0 & 1 \\ 1 & 0 & 0 \end{bmatrix} |ds\, dt| = 4. \qquad 6.10.13$$

So the whole integral is $24 - 4 = 20$. Of course it would be easier to argue that all faces must contribute the same amount to the flux, so the top must contribute $24/6 = 4$. △

Before proving the generalized Stokes's theorem, we want to sketch two proofs of the fundamental theorem of calculus in one dimension. You probably have seen the first; it is the second that generalizes to higher dimensions.

Two proofs of the fundamental theorem of calculus

First proof

Let $F(x) = \int_a^x f(t)\, dt$. We will show that

$$F'(x) = f(x), \qquad 6.10.14$$

as Figure 6.10.1 suggests. Indeed,

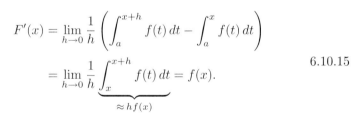

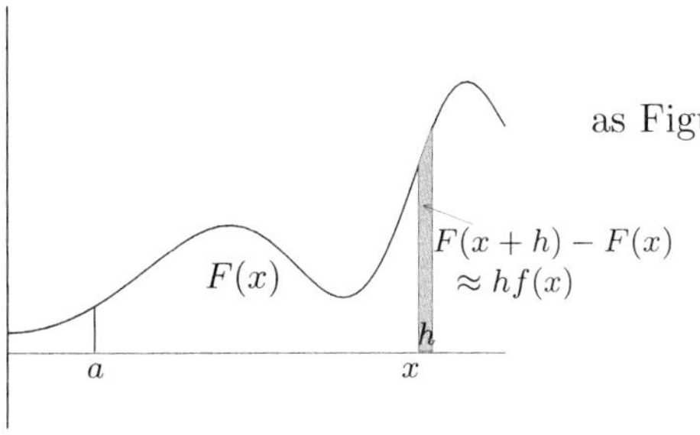

FIGURE 6.10.1.
Computing the derivative of F.

$$\begin{aligned} F'(x) &= \lim_{h\to 0} \frac{1}{h}\left(\int_a^{x+h} f(t)\, dt - \int_a^x f(t)\, dt\right) \\ &= \lim_{h\to 0} \frac{1}{h} \underbrace{\int_x^{x+h} f(t)\, dt}_{\approx hf(x)} = f(x). \end{aligned} \qquad 6.10.15$$

The last integral is approximately $hf(x)$; the error disappears in the limit.

Now consider the function

$$f(x) - \underbrace{\int_a^x f'(t)\, dt.}_{\text{with deriv. } f'(x)} \qquad 6.10.16$$

The argument above shows that its derivative is zero, so it is constant; evaluating the function at $x = a$, we see that the constant is $f(a)$. Thus

$$f(b) - \int_a^b f'(t)\, dt = f(a). \quad \square \qquad 6.10.17$$

[13]This parametrization is "obvious" because x and y parametrize the top of the cube, and at the top, $z = 1$.

Second proof

Here the appropriate drawing is the Riemann sum shown in Figure 6.10.2. By the definition of the integral,

$$\int_a^b f(x)\,dx \approx \sum_i f(x_i)(x_{i+1}-x_i), \qquad 6.10.18$$

where $x_0 < x_1 < \cdots < x_m$ decompose $[a,b]$ into m little pieces, with $a = x_0$ and $b = x_m$. By Taylor's theorem,

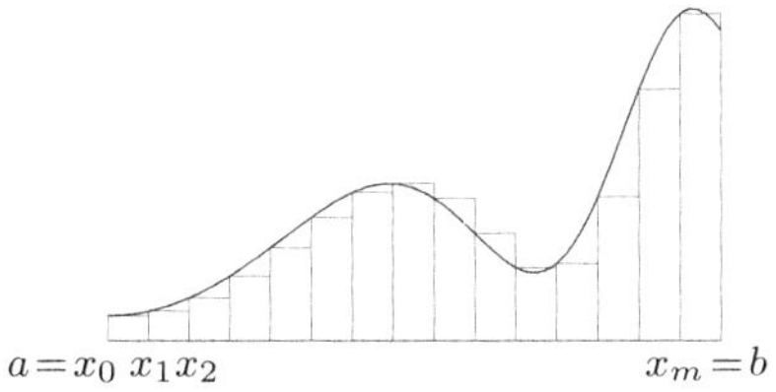

FIGURE 6.10.2.

A Riemann sum as an approximation to the integral in equation 6.10.18.

$$f(x_{i+1}) \approx f(x_i) + f'(x_i)(x_{i+1}-x_i). \qquad 6.10.19$$

These two statements together give

$$\int_a^b f'(x)\,dx \approx \sum_{i=0}^{m-1} f'(x_i)(x_{i+1}-x_i) \approx \sum_{i=0}^{m-1}\big(f(x_{i+1})-f(x_i)\big). \qquad 6.10.20$$

In the last sum, all the x_i other than $a = x_0$ and $b = x_m$ cancel, giving

$$\sum_{i=0}^{m-1}\big(f(x_{i+1})-f(x_i)\big) = f(x_m) - f(x_0) = f(b) - f(a). \qquad 6.10.21$$

Let us show that the two $\approx$ of equation 6.10.20 become equalities in the limit as $m \to \infty$. The first $\approx$ in equation 6.10.20 is simply the convergence of Riemann sums to the integral: since the derivative f' is continuous on $[a,b]$, Theorem 4.3.6 applies.

Here we are working under the assumption that f is of class C^1. Our arguments are nonconstructive, since they use Theorem 4.3.6, which itself uses the Bolzano-Weierstrass theorem (Theorem 1.6.3). If f is of class C^2, then we can bound the errors.

For the second $\approx$, note that by the mean value theorem (Theorem 1.6.13), there exists $c_i \in [x_i, x_{i+1}]$ such that

$$f(x_{i+1}) - f(x_i) = f'(c_i)(x_{i+1}-x_i). \qquad 6.10.22$$

Since f' is continuous on the compact set $[a,b]$, it is uniformly continuous (see the proof of Theorem 4.3.6), and so for all $\epsilon > 0$, there exists M such that when $m > M$ we have $|f'(c_i) - f'(x_i)| < \epsilon$. So for $m > M$ we have

$$\begin{aligned}&\left|\sum_{i=0}^{m-1}\big(f(x_{i+1})-f(x_i)-f'(x_i)(x_{i+1}-x_i)\big)\right| \\ &\qquad \le \sum_{i=0}^{m-1}|f'(c_i)-f'(x_i)||x_{i+1}-x_i| < \epsilon|b-a|.\end{aligned} \qquad 6.10.23$$

If f is assumed to be of class C^2, the justification is much easier and more precise: by Corollary 1.9.2,

$$\begin{aligned}\left|\int_{x_i}^{x_{i+1}} f'(x)\,dx - f'(x_i)(x_{i+1}-x_i)\right| &= \left|\int_{x_i}^{x_{i+1}} f'(x) - f'(x_i)\,dx\right| \\ &\le \sup|f''||x_{i+1}-x_i|^2,\end{aligned} \qquad 6.10.24$$

and

$$\begin{aligned}\Big|f(x_{i+1}) - f(x_i) - f'(x_i)(x_{i+1}-x_i)\Big| \\ = \Big|\Big(f(x_{i+1}) - f(x_i) - f'(x_i)(x_{i+1}-x_i)\Big) \\ - \Big(f(x_{i+1}) - f(x_i) - f'(c_i)(x_{i+1}-x_i)\Big)\Big| \\ = |(f'(x_i) - f'(c_i)||x_{i+1}-x_i| \le \sup|f''||x_{i+1}-x_i|^2.\end{aligned} \qquad 6.10.25$$

Recall that

$$|x_{i+1} - x_i| = \frac{|b-a|}{m}.$$

So all in all, we made $2m$ errors, each of which is $\le \sup|f''|\frac{|b-a|^2}{m^2}$. Adding them up over all the pieces leaves an m in the denominator and a constant in the numerator, so the error tends to 0 as the decomposition becomes finer and finer. □

Stokes's theorem: An informal proof

Suppose you decompose X into little pieces that are approximated by oriented k-parallelograms P_i:

$$P_i = P_{\mathbf{x}_i}(\vec{\mathbf{v}}_{1,i}, \vec{\mathbf{v}}_{2,i}, \dots, \vec{\mathbf{v}}_{k,i}). \qquad 6.10.26$$

Then

$$\int_X \mathbf{d}\varphi \approx \sum_i \mathbf{d}\varphi(P_i) \approx \sum_i \int_{\partial P_i} \varphi \approx \int_{\partial X} \varphi. \qquad 6.10.27$$

The first $\approx$ is the definition of the integral; it becomes an equality in the limit as the decomposition becomes infinitely fine. The second $\approx$ comes from our definition of the exterior derivative. For the third $\approx$, when we add over all the P_i, all the internal boundaries cancel, leaving $\int_{\partial X} \varphi$.

As in the case of Riemann sums, we need to understand the errors that are signaled by our $\approx$ signs. If our parallelograms P_i have sidelength ϵ, then there are on the order of ϵ^{-k} such parallelograms (precisely that many if X is a cube of volume 1). The errors in the first and second replacements are of order ϵ^{k+1}. For the first, it is our definition of the integral, and the error becomes small as the decomposition becomes infinitely fine. For the second, from the definition of the exterior derivative,

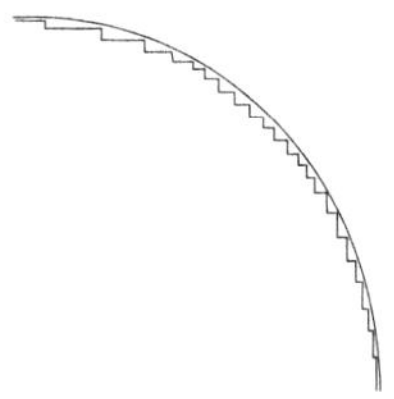

FIGURE 6.10.3.
The staircase is very close to the curve, but its length is not close to the length of the curve; the curve doesn't fit well with a dyadic decomposition. Here the informal proof of Stokes's theorem is not enough.

$$\mathbf{d}\varphi(P_i) = \int_{\partial P_i} \varphi + \text{ terms of order at least } k+1. \qquad 6.10.28$$

So the total error is at least of order $\epsilon^{-k}(\epsilon^{k+1}) = \epsilon$, and indeed the errors disappear in the limit. □

We find this informal argument convincing, but it is not quite rigorous. The problem is that the boundary of X does not necessarily fit well with the boundaries of the little cubes, as illustrated by Figure 6.10.3.

Stokes's theorem: A rigorous proof

Recall from Section 6.6 why we state and prove Stokes's theorem for pieces-with-boundary of manifolds: the typical approach, which requires boundaries to be smooth, excludes many model and real-life examples. Even restricting Stokes's theorem to pieces-with-corners means excluding the cone, which is a serious obstacle to understanding electromagnetism and the nature of light: the main tool for solving Maxwell's equations is the Liénard–Wiechert formula, which consists of an integral over the light cone.

Proving Stokes's theorem rigorously and in full generality is really difficult and uses just about everything in the book. There are three parts to the proof:

Recall that we defined a piece-with-corners in Definition 6.6.9.

1. We prove it for a piece-with-corners Y (this is quite lengthy).
2. We show that we can trim our domain of integration X to turn it into a piece-with-corners X_ϵ, with all nonsmooth points of the boundary removed (this is hard, but the hard part has been relegated to Appendix A23).
3. We show that

$$\lim_{\epsilon \to 0} \int_{\partial X_\epsilon} \varphi = \int_{\partial X} \varphi \quad \text{and} \quad \lim_{\epsilon \to 0} \int_{X_\epsilon} \mathbf{d}\varphi = \int_X \mathbf{d}\varphi. \qquad 6.10.29$$

Part 1: Proving Stokes's theorem for pieces-with-corners

Let $Y \subset M$ be a piece-with-corners. Proposition 6.10.7 says that a diffeomorphi sm will locally straighten its boundary. A straightened boundary will make life easier, since then we can use our notion of integration based on (straight) dyadic cubes.

Proposition 6.10.7. *Let $M \subset \mathbb{R}^n$ be a k-dimensional manifold, and let $Y \subset M$ be a piece-with-corners. Then every $\mathbf{x} \in Y$ is the center of a ball U in $\mathbb{R}^n$ such that there exists a diffeomorphism $\mathbf{F}: U \to \mathbb{R}^n$ satisfying*

$$\mathbf{F}(U \cap M) = \mathbf{F}(U) \cap \mathbb{R}^k \qquad 6.10.30$$

$$\mathbf{F}(U \cap Y) = \mathbf{F}(U) \cap \mathbb{R}^k \cap Z, \qquad 6.10.31$$

where Z is a region $x_1 \geq 0, \dots, x_j \geq 0$ for some $j \leq k$.

Proposition 6.10.7: The idea is to use the functions defining M and Y within M as coordinate functions of a new space; this has the result of straightening out M and ∂Y, since these become sets where appropriate coordinate functions vanish.

Such a region Z will be referred to as a *quadrant*.

Proof. The proof of Proposition 6.10.7 is illustrated by Figure 6.10.4. At any $\mathbf{x} \in Y$, Definition 6.6.5 of a corner point and Definition 3.1.10 defining a manifold known by equations give us a neighborhood $V \subset \mathbb{R}^n$ of $\mathbf{x}$, and a collection of C^1 functions $V \to \mathbb{R}$ with linearly independent derivatives: functions $f_1, \dots, f_{n-k}$ (the coordinates of the function $\mathbf{f}: \mathbb{R}^n \to \mathbb{R}^{n-k}$ defining $M \cap V$) and functions $\widetilde{g}_1, \dots, \widetilde{g}_m$ (extensions of the coordinates $g_i, \dots, g_m$ of the map $\mathbf{g}$ of Definition 6.6.5). The number m may be 0; this will happen if $\mathbf{x}$ is in the interior of Y.

The ball B in Figure 6.10.4 refers to the trimming we will do in Proposition 6.10.10.

In Figure 6.10.4, we have three instances of $\mathbf{x}$ and three instances of U, corresponding to the three maps $\mathbf{F}_1$, $\mathbf{F}_2$, and $\mathbf{F}_3$ (instances of $\mathbf{F}$). To keep from overloading the picture, we do not label everything.

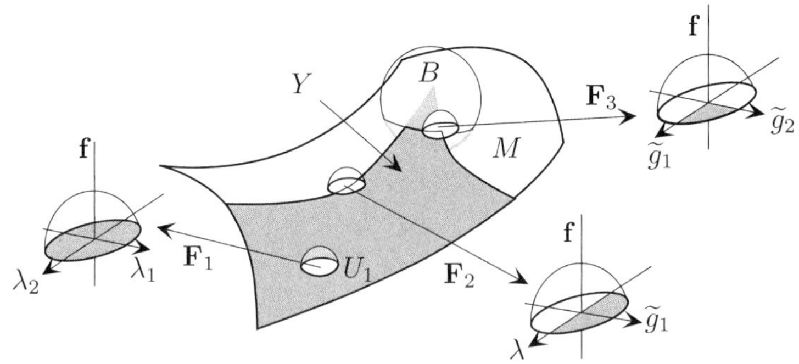

FIGURE 6.10.4. Here, $n = 3$ and $k = 2$. The darkly shaded region is Y, the piece-with-corners. At all points in the manifold M, we have $\mathbf{f} = \mathbf{0}$, so $\mathbf{F}$ goes from a neighborhood $V \subset \mathbb{R}^3$ of $\mathbf{x} \in Y$ (more generally, $V \subset \mathbb{R}^n$) to a subset of $\mathbb{R}^3$, taking M to a neighborhood of $\mathbf{0}$ in the (x, y)-plane (in the general case, to a neighborhood of the origin in a subspace of dimension k, where the other $n - k$ coordinates vanish). Thus $\mathbf{F}$ flattens M and straightens out its boundary. Here we have three instances of the function $\mathbf{F}$,

$$\mathbf{F}_1 = \begin{pmatrix} \lambda_1 \\ \lambda_2 \\ \mathbf{f} \end{pmatrix}, \quad \mathbf{F}_2 = \begin{pmatrix} \tilde{g}_1 \\ \lambda \\ \mathbf{f} \end{pmatrix} \quad \mathbf{F}_3 = \begin{pmatrix} \tilde{g}_1 \\ \tilde{g}_2 \\ \mathbf{f} \end{pmatrix}.$$

Since the derivative of $\mathbf{F}$ at $\mathbf{x}$ is an isomorphism (for each instance of $\mathbf{F}$), the map $\mathbf{F}$ is a diffeomorphism on some subset $U \subset V$.

In each case we extend the collection of the derivatives of the $n - k$ functions f_i, $i = 1, \dots, n - k$ and the m functions $\tilde{g}_j$, $j = 1, \dots, m$ to a maximal collection of n linearly independent linear functions, by adding $k - m$ linearly independent linear functions $\lambda_1, \dots, \lambda_{k-m} : \mathbb{R}^n \to \mathbb{R}$. These n functions define a map $\mathbf{F} : V \to \mathbb{R}^n$ whose derivative at $\mathbf{x}$ is invertible. We can apply the inverse function theorem to say that locally, $\mathbf{F}$ is invertible, so locally it is a diffeomorphism on some neighborhood $U \subset V$ of $\mathbf{x}$. Equations 6.10.30 and 6.10.31 then follow, as shown in Figure 6.10.4. □

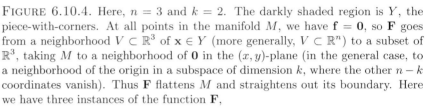

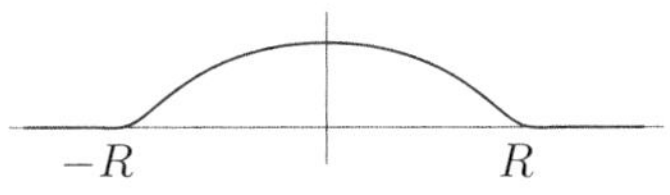

FIGURE 6.10.5.

Graph of the bump function β_R of equation 6.10.32. As $|\mathbf{x}| < R$ tends to R, the quantity $R^2 - |\mathbf{x}|^2$ becomes very small, so the number $1/(R^2 - |\mathbf{x}|^2)$ becomes large and positive and $-1/(R^2 - |\mathbf{x}|^2)$ gets closer and closer to $-\infty$. Then

$$e^{-1/(R^2 - |\mathbf{x}|^2)}$$

becomes *very small*, fitting seamlessly with 0 for $|\mathbf{x}| \geq R$.

• Partitioning the domain of integration

Since Y is compact, the Heine-Borel theorem (Theorem A3.3) says that we can cover it by *finitely* many (overlapping) $U_1, \dots, U_N$ satisfying the properties specified in Proposition 6.10.7; suppose that they are balls of radius $R_1, \dots, R_N$ centered at $\mathbf{x}_1, \dots, \mathbf{x}_N$.

Let $\mathbf{F}_i : U_i \to \mathbb{R}^n$ be the functions defined in Proposition 6.10.7 (one for each $\mathbf{x}_i$). Set $W_i = \mathbf{F}_i(U_i)$ and let $\mathbf{h}_i : W_i \cap (\mathbb{R}^k \times \{\mathbf{0}\}) \to U_i$ be the restriction of $\mathbf{F}_i^{-1}$. It is a parametrization of a neighborhood of $M \cap U_i$, and maps some quadrant $Z \subset \mathbb{R}^k$ to $Y \cap U_i$.

We will now write our form φ (the k-form of the generalized Stokes's theorem) as a sum of forms φ_i, each with support in only one piece U_i. Rather than hacking Y apart, we will use a softer technique: functions (the α_i of equation 6.10.35) that fade out one parametrization as we bring another in. This is done as follows.

Define $\beta_R : \mathbb{R}^n \to \mathbb{R}$ to be the "bump" function

$$\beta_R(\mathbf{x}) = \begin{cases} e^{-1/(R^2-|\mathbf{x}|^2)} & \text{if } |\mathbf{x}|^2 \le R^2 \\ 0 & \text{if } |\mathbf{x}|^2 > R^2, \end{cases} \qquad 6.10.32$$

shown in Figure 6.10.5. Exercise 6.10.7 asks you to show that β_R is of class C^∞ on all of $\mathbb{R}^n$. Let $\beta_i : \mathbb{R}^n \to \mathbb{R}$ be the C^∞ function

$$\beta_i(\mathbf{x}) = \beta_{R_i}(\mathbf{x} - \mathbf{x}_i), \qquad 6.10.33$$

with support in the ball V_i around $\mathbf{x}_i$. Set

$$\beta(\mathbf{x}) = \sum_{i=1}^{N} \beta_i(\mathbf{x}), \qquad 6.10.34$$

where N is the number of U_i given to us by Heine-Borel; this is a finite set of overlapping bump functions (corresponding to overlapping U_i), so that we have $\beta(\mathbf{x}) > 0$ on a neighborhood of Y. Then the functions

$$\alpha_i(\mathbf{x}) = \frac{\beta_i(\mathbf{x})}{\beta(\mathbf{x})} \qquad 6.10.35$$

are C^∞ on some neighborhood of $\mathbf{x} \in \mathbb{R}^n$. (These α_i are the "fade-in" functions referred to above.) Clearly $\sum_{i=1}^{N} \alpha_i(\mathbf{x}) = 1$ in a neighborhood of Y, so that if we set $\varphi_i = \alpha_i \varphi$, we can write the $(k-1)$-form φ of Stokes's theorem as

$$\varphi = \sum_{i=1}^{N} \alpha_i \, \varphi = \sum_{i=1}^{N} \varphi_i. \qquad 6.10.36$$

Equation 6.10.36: The sum

$$\sum_{i=1}^{N} \alpha_i = 1$$

is called a *partition of unity* because it breaks up 1 into a sum of functions.

The functions α_i making up a partition of unity have the interesting property that they can have small support, which makes it possible to piece together global functions, forms, etc., from local ones. We used to think that they were a purely theoretical tool, but we were wrong. They are important in signal processing: the "windows" used in image processing are a partition of unity.

Hence, by Theorem 6.7.4,

$$\mathbf{d}\varphi = \sum_{i=1}^{N} \mathbf{d}\varphi_i. \qquad 6.10.37$$

• Proving Stokes's theorem for quadrants

We used Proposition 6.10.7 to turn the piece-with-corners Y into a collection of quadrants of $\mathbb{R}^k$. We now prove Stokes's theorem for such quadrants.

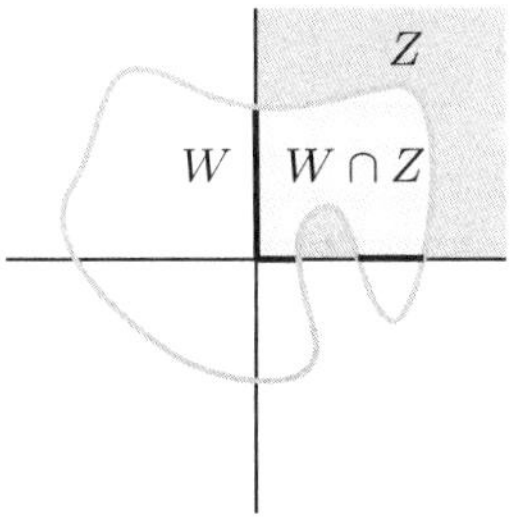

FIGURE 6.10.6.

Above, W is the entire odd-shaped region and Z is the shaded quadrant. Since $\varphi = 0$ outside W, when we integrate φ over the boundary of $W \cap Z$, the only part of the boundary that matters are the straight lines marked in bold.

Proposition 6.10.8 (Stokes's theorem for quadrants). *Let $Z \subset \mathbb{R}^k$ be a closed quadrant; let W be a bounded open subset of $\mathbb{R}^k$, oriented by* sgn det *on $\mathbb{R}^k$. Give $\partial(W \cap Z)$ the boundary orientation. Let φ be a $(k-1)$–form on $\mathbb{R}^k$ of class C^2 with compact support in W. Then*

$$\int_{\partial Z} \varphi = \int_{W \cap Z} \mathbf{d}\varphi. \qquad 6.10.38$$

Note that φ vanishes on the boundary of W; this is why on the left side of equation 6.10.38 we can integrate over ∂Z rather than $\partial(W \cap Z)$. As shown in Figure 6.10.6, the only part of $\partial(W \cap Z)$ that matters is the boundary of Z, which is straight, and fits well with the boundaries of dyadic cubes.

Proof. Choose $\epsilon > 0$. Take the dyadic decomposition $\mathcal{D}_N(\mathbb{R}^k)$, where the sidelength of the cubes is $h = 2^{-N}$. By taking N sufficiently large, we can guarantee that the difference between the integral of $\mathbf{d}\varphi$ over $W \cap Z$ and the Riemann sum in inequality 6.10.39 is less than $\epsilon/2$:

$$\left|\int_{W\cap Z} \mathbf{d}\varphi - \sum_{C\in\mathcal{D}_N,\ C\subset Z} \mathbf{d}\varphi(C)\right| < \frac{\epsilon}{2}. \qquad 6.10.39$$

Now we evaluate the total difference between the exterior derivative of φ over the cubes C and the integral of φ over the boundaries of the C:

$$\left|\sum_{C\in\mathcal{D}_N,\ C\subset Z} \mathbf{d}\varphi(C) - \sum_{C\in\mathcal{D}_N,\ C\subset Z} \int_{\partial C} \varphi\right|. \qquad 6.10.40$$

The number of cubes $C \in \mathcal{D}_N$ that intersect the support of φ is at most $L2^{kN}$ for some constant L. To determine the bound for the error for each cube, we use the fact (from inequality A22.11 in the proof of Theorem 6.7.4)[14] that there exist a constant K and $\delta > 0$ such that when $|h| < \delta$,

$$\left|\mathbf{d}\varphi\big(P_{\mathbf{x}}(h\vec{\mathbf{e}}_1, \dots, h\vec{\mathbf{e}}_k)\big) - \int_{\partial P_{\mathbf{x}}(h\vec{\mathbf{e}}_1,\dots,h\vec{\mathbf{e}}_k)} \varphi\right| \le Kh^{k+1}. \qquad 6.10.41$$

Inequality 6.10.41 follows from Taylor's theorem with remainder.

Thus if we replace the k-parallelograms of inequality 6.10.41 by dyadic cubes, with $h = 2^{-N}$, we see that the bound for the error for one cube is

$$K2^{-N(k+1)}. \qquad 6.10.42$$

This gives

$$\left|\sum_{C\in\mathcal{D}_N,\ C\subset Z} \mathbf{d}\varphi(C) - \sum_{C\in\mathcal{D}_N,\ C\subset Z} \int_{\partial C} \varphi\right| \le \underbrace{L2^{kN}}_{\text{No. of cubes}} \ \underbrace{K2^{-N(k+1)}}_{\substack{\text{bound for}\\ \text{each cube}}} = LK2^{-N}. \qquad 6.10.43$$

This can also be made $< \epsilon/2$ by taking N sufficiently large – i.e., by taking

$$N \ge \frac{\ln(2LK) - \ln\epsilon}{\ln 2}. \qquad 6.10.44$$

Putting these inequalities together, we get

$$\overbrace{\left|\int_{W\cap Z} \mathbf{d}\varphi - \sum_{C\in\mathcal{D}_N,\ C\subset Z} \mathbf{d}\varphi(C)\right|}^{\le\epsilon/2 \text{ by inequality 6.10.39}} + \overbrace{\left|\sum_{C\in\mathcal{D}_N,\ C\subset Z} \mathbf{d}\varphi(C) - \sum_{C\in\mathcal{D}_N,\ C\subset Z} \int_{\partial C} \varphi\right|}^{\le\epsilon/2} \le \epsilon, \qquad 6.10.45$$

$$\text{i.e.,} \quad \left|\int_{W\cap Z} \mathbf{d}\varphi - \sum_{C\in\mathcal{D}_N,\ C\subset Z} \int_{\partial C} \varphi\right| \le \epsilon. \qquad 6.10.46$$

All the internal boundaries in the sum

$$\sum_{C\in\mathcal{D}_N,\ C\subset Z} \int_{\partial C} \varphi \qquad 6.10.47$$

[14] There we have h^{k+2} not h^{k+1}, since there φ is a k-form; here it is a $(k-1)$-form.

cancel, since each appears twice with opposite orientations. So we have

$$\sum_{C\in\mathcal{D}_N,\ C\subset Z}\int_{\partial C}\varphi=\int_{\partial W\cap Z}\varphi, \quad \text{which gives} \tag{6.10.48}$$

$$\left|\int_{W\cap Z}\mathbf{d}\varphi-\int_{\partial(W\cap Z)}\varphi\right|\le\epsilon \tag{6.10.49}$$

in the limit as $N\to\infty$. Since ϵ is arbitrary, the proposition follows. □

Since φ vanishes outside W,

$$\int_{\partial Z}\varphi=\int_{\partial(W\cap Z)}\varphi,$$

justifying equation 6.10.48.

- **Proof of Stokes's theorem for pieces-with-corners**

We now use Stokes's theorem for quadrants to prove Stokes's theorem for a piece-with-corners Y.

Theorem 6.10.9 (Stokes's theorem for pieces-with-corners). *Let $M\subset\mathbb{R}^n$ be a k-dimensional manifold, and let $Y\subset M$ be a piece-with-corners. Let φ be a $(k-1)$-form defined on a neighborhood of Y. Then*

$$\int_{\partial_M Y}\varphi=\int_{\partial_M^s Y}\varphi=\int_Y\mathbf{d}\varphi. \tag{6.10.50}$$

Proof. It is enough to prove $\int_Y\mathbf{d}\varphi_i=\int_{\partial_M Y}\varphi_i$, since then

$$\int_Y\mathbf{d}\varphi\underbrace{=}_{\text{eq. 6.10.37}}\int_Y\sum_i^N\mathbf{d}\varphi_i=\sum_i^N\int_Y\mathbf{d}\varphi_i$$
$$=\sum_i^N\int_{\partial_M Y}\varphi_i=\int_{\partial_M Y}\sum_i^N\varphi_i=\int_{\partial_M Y}\varphi. \tag{6.10.51}$$

To prove $\int_Y\mathbf{d}\varphi_i=\int_{\partial_M Y}\varphi_i$, we pull each φ_i back to $\mathbb{R}^k$ using the pullback $\mathbf{h}_i^*(\varphi_i)$. Below, the U_i are in $\mathbb{R}^n$ and the W_i are in $\mathbb{R}^k$:

$$\int_Y\mathbf{d}\varphi_i\underset{1}{=}\int_{Y\cap U_i}\mathbf{d}\varphi_i\underset{2}{=}\int_{W_i\cap Z_i}\mathbf{h}_i^*(\mathbf{d}\varphi_i)\underset{3}{=}\int_{W_i\cap Z_i}\mathbf{d}(\mathbf{h}_i^*\varphi_i)$$
$$\underset{4}{=}\int_{W_i\cap\partial Z_i}\mathbf{h}_i^*(\varphi_i)\underset{5}{=}\int_{\partial_M Y\cap U_i}\varphi_i\underset{6}{=}\int_{\partial_M Y}\varphi_i. \tag{6.10.52}$$

The great mathematician Alexander Grothendieck (1928–2014) claimed that one has achieved a good proof when the definitions and intermediate steps are such that the proof appears natural. This is a lofty goal, but we think that the sequence of equalities in equation 6.10.52 qualifies.

Equalities 1 and 6: The support of φ_i is in U_i.

Equalities 2 and 5: $\mathbf{h}_i$ parametrizes $Y\cap U_i$.

Equality 3 is the first crucial step, using $\mathbf{d}\mathbf{h}_i^*=\mathbf{h}_i^*\mathbf{d}$ (Theorem 6.9.8).

Equality 4 (also a crucial step) is Proposition 6.10.8, Stokes's theorem for quadrants. We have satisfied the conditions of that proposition: each $W_i\cap Z_i$ satisfies the conditions for $W\cap Z$ in that proposition, and using all the $W_i\cap Z_i$ we can account for all of Y.

Part 2. Trimming X to make a piece-with-corners X_ϵ

Let $X\subset M$ be a piece-with-boundary. We now trim X to make a piece-with-corners X_ϵ to which we can apply Stokes's theorem. The trimmed-off pieces are the interiors of balls B_i. (You may imagine that we are scooping out little balls with an ice cream scoop.) By Definitions 6.6.8 and 5.2.1, when we scoop out the balls B_i, we are removing all the bad (nonsmooth points) of the boundary of X: Definition 6.6.8 says that all points in the nonsmooth boundary of X are contained in a set with $(k-1)$-dimensional volume 0; by Definition 5.2.1, this means that for any $\epsilon>0$, we can cover

Definition 5.2.1 was stated in terms of cubes rather than balls, but as we mentioned in Chapter 5, we could have replaced cubes by balls.

the nonsmooth boundary of X with finitely many balls B_i, $i = 1, \dots, p$ centered at $\mathbf{x}_i \in X$ and with radii r_i small enough that

$$\sum_{i=1}^{p} r_i^{k-1} < \epsilon. \qquad 6.10.53$$

Proposition A6.10.10 says that if we position these balls B_i correctly, we can turn X into a piece-with-corners, as illustrated by Figure 6.10.7.

As usual, we denote by $B_r(\mathbf{x})$ be the open ball of radius r around $\mathbf{x}$.

Proposition 6.10.10: Removing the balls $B_{r_i}(\mathbf{x}_i)$ means taking the intersection of X with the complement of all the balls. Saying that X_ϵ is a piece-with-corners is saying that the balls are in "general position": various exceptional accidents do not occur, such as balls being tangent to each other or tangent to M or tangent to the boundary of X.

Proposition 6.10.10 (Trimming X to make a piece-with-corners). *For all $\epsilon > 0$, there exist points $\mathbf{x}_1, \dots, \mathbf{x}_p \in X$ and $r_1 > 0, \dots, r_p > 0$, such that*

$$\sum_{i=1}^{p} r_i^{k-1} < \epsilon \quad \text{and} \quad X_\epsilon \stackrel{\text{def}}{=} X - \bigcup_{i=1}^{p} B_{r_i}(\mathbf{x}_i) \qquad 6.10.54$$

is a piece-with-corners.

Although Proposition 6.10.10 is almost obvious, its proof is the hardest part of the proof of the generalized Stokes's theorem; we have relegated it to Appendix A23.

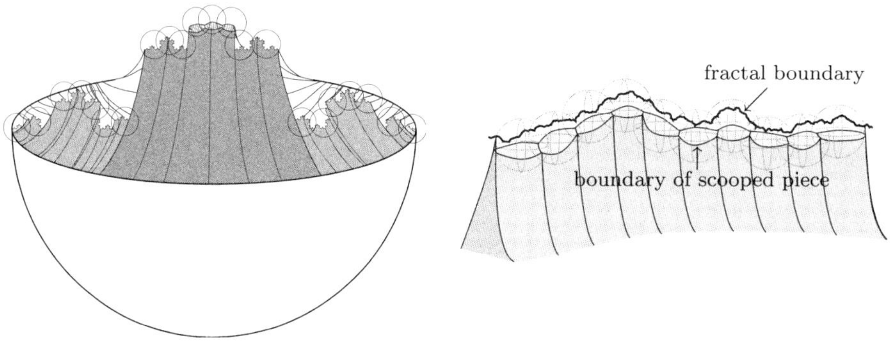

FIGURE 6.10.7. LEFT: We saw this piece-with-complicated-boundary in Figure 6.6.1; its boundary includes a Koch snowflake, with all its fractal complexity. "Scooping out" balls B_i as described in Proposition 6.10.10 turns the piece-with-complicated-boundary into a piece-with-corners (above, the top three scoops have been removed). RIGHT: A closeup of the fractal boundary and the boundary of the scooped piece.

Part 3. Completing the proof of Stokes's theorem

We now know that we can trim a piece-with-boundary $X \subset M$ to make a piece-with-corners X_ϵ to which Stokes's theorem applies:

$$\int_{\partial_M X_\epsilon} \varphi = \int_{X_\epsilon} \mathbf{d}\varphi, \quad \text{so} \quad \lim_{\epsilon \to 0} \int_{\partial_M X_\epsilon} \varphi = \lim_{\epsilon \to 0} \int_{X_\epsilon} \mathbf{d}\varphi. \qquad 6.10.55$$

We have to show that

$$\lim_{\epsilon\to 0}\int_{\partial_M X_\epsilon}\varphi = \int_{\partial_M X}\varphi \quad\text{and}\quad \lim_{\epsilon\to 0}\int_{X_\epsilon}\mathbf{d}\varphi = \int_X \mathbf{d}\varphi \qquad 6.10.56$$

To do this we need to show that a limit of integrals is an integral of limits; our only tool is the dominated convergence theorem. That theorem requires that we bound the functions to be integrated by an integrable function. Proposition 6.10.11 is the main tool.

Proposition 6.10.11. *Let*

$$\psi = \sum_{1\le i_1<\cdots<i_k\le n} a_{i_1,\dots,i_k}\,dx_{i_1}\wedge\cdots\wedge dx_{i_k} \qquad 6.10.57$$

be a k-form on an open set $U\subset\mathbb{R}^n$, and let $M\subset U$ be a k-dimensional oriented manifold. Then

$$\left|\int_M \psi\right| \le \sqrt{\sum_{1\le i_1<\cdots<i_k\le n}\Big(\sup_{\mathbf{x}\in U}|a_{i_1,\dots,i_k}(\mathbf{x})|\Big)^2}\;\mathrm{vol}_k(M). \qquad 6.10.58$$

Proof. We need the following lemma from linear algebra; you are asked to prove it in Exercise 6.10.9, which is not all that easy.

Lemma 6.10.12. *Let $A=[\vec{\mathbf{a}}_1,\dots,\vec{\mathbf{a}}_k]$ be an $n\times k$ matrix. Then*

$$\det(A^\top A) = \sum_{1\le i_1<\cdots<i_k\le n}\Big(dx_{i_1}\wedge\cdots\wedge dx_{i_k}(\vec{\mathbf{a}}_1,\dots,\vec{\mathbf{a}}_k)\Big)^2. \qquad 6.10.59$$

Equation 6.10.60, going from line 2 to line 3: The second term of the second line is the right side of equation 6.10.59, replacing $\vec{\mathbf{a}}_i$ by $\vec{\mathbf{v}}_i$. By Definition 5.1.3,

$$\mathrm{vol}_k P(\vec{\mathbf{v}}_1,\dots,\vec{\mathbf{v}}_k) = \sqrt{\det(A^\top A)}.$$

It follows from Schwarz's inequality and Lemma 6.10.12 that

$$\begin{aligned}
\Big|\psi(P_{\mathbf{x}}(\vec{\mathbf{v}}_1,\dots,\vec{\mathbf{v}}_k))\Big|^2 &\underbrace{=}_{\text{def. of }\psi} \left|\sum_{1\le i_1<\cdots<i_k\le n} a_{i_1,\dots,i_k}(\mathbf{x})\,dx_{i_1}\wedge\cdots\wedge dx_{i_k}(\vec{\mathbf{v}}_1,\dots,\vec{\mathbf{v}}_k)\right|^2\\
&\underbrace{\le}_{\text{Schwarz}} \left(\sum_{1\le i_1<\cdots<i_k\le n}|a_{i_1,\dots,i_k}(\mathbf{x})|^2\right)\left(\sum_{1\le i_1<\cdots<i_k\le n}\Big(dx_{i_1}\wedge\cdots\wedge dx_{i_k}(\vec{\mathbf{v}}_1,\dots,\vec{\mathbf{v}}_k)\Big)^2\right)\\
&\underbrace{=}_{\substack{\text{Lemma 6.10.12}\\ \text{(see margin)}}} \left(\sum_{1\le i_1<\cdots<i_k\le n}|a_{i_1,\dots,i_k}(\mathbf{x})|^2\right)\underbrace{\Big(\mathrm{vol}_k\big(P_{\mathbf{x}}(\vec{\mathbf{v}}_1,\dots,\vec{\mathbf{v}}_k)\big)\Big)^2}_{\det(A^\top A)}. \qquad 6.10.60
\end{aligned}$$

Choose a parametrization $\gamma: W\to M$ by a subset $W\subset\mathbb{R}^k$. Inequality 6.10.60 gives

$$\begin{aligned}
&\Big|\psi(P_{\gamma(\mathbf{w})}(\overrightarrow{D_1\gamma}(\mathbf{w}),\dots,\overrightarrow{D_k\gamma}(\mathbf{w}))\Big| \qquad 6.10.61\\
&\le\left(\sqrt{\sum_{1\le i_1<\cdots<i_k\le n}|a_{i_1,\dots,i_k}(\gamma(\mathbf{w}))|^2}\right)\mathrm{vol}_k(P_{\gamma(\mathbf{w})}(\overrightarrow{D_1\gamma}(\mathbf{w}),\dots,\overrightarrow{D_k\gamma}(\mathbf{w}))
\end{aligned}$$

Both sides are functions of $\mathbf{w}$, and

$$\begin{aligned}\left|\int_M \psi\right| &= \left|\int_W \psi(P_{\gamma(\mathbf{w})}(\overrightarrow{D_1\gamma}(\mathbf{w}),\dots,\overrightarrow{D_k\gamma}(\mathbf{w}))|d^k\mathbf{w}|\right| \\ &\le \int_W \left|\psi(P_{\gamma(\mathbf{w})}(\overrightarrow{D_1\gamma}(\mathbf{w}),\dots,\overrightarrow{D_k\gamma}(\mathbf{w}))\right||d^k\mathbf{w}|,\end{aligned} \qquad 6.10.62$$

which is then at most the integral of the right side. But

$$\operatorname{vol}_k M = \int_W \operatorname{vol}_k\Big(P_{\gamma(\mathbf{w})}(\overrightarrow{D_1\gamma}(\mathbf{w}),\dots,\overrightarrow{D_k\gamma}(\mathbf{w}))\Big)|d^k\mathbf{w}| \qquad 6.10.63$$

and we can bound $\sqrt{\sum_{1\le i_1<\dots<i_k\le n}|a_{i_1,\dots,i_k}(\gamma(\mathbf{w}))|^2}$ by its sup, proving inequality 6.10.58. □ Proposition 6.10.11

Proposition 6.10.13 says that when we trim X to make X_ϵ, the boundaries of the scoops have volumes behaving roughly as if M were flat.

Proposition 6.10.13. *Let $M\subset\mathbb{R}^n$ be a k-dimensional manifold of class C^2, and $P\subset M$ a compact subset. Then for any $\epsilon>0$, there exists $\rho>0$ such that for $r\le\rho$ and any ball $B_r(\mathbf{p})\subset\mathbb{R}^n$ centered at a point $\mathbf{p}\in P$ and of radius r, we have*

$$\begin{aligned}\operatorname{vol}_{k-1}\big(\partial B_r(\mathbf{p})\cap M\big) &\le (1+\epsilon)\operatorname{vol}_{k-1}\big(\partial B_r(\mathbf{p})\cap T_\mathbf{p}M\big)\\ &= (1+\epsilon)Kr^{k-1},\end{aligned} \qquad 6.10.64$$

where K is the volume of the $(k-1)$-dimensional unit sphere.

You are asked to prove Proposition 6.10.13 in Exercise 6.10.11.

Stokes's theorem then follows from the series of equalities below. Let X be a piece-with-boundary of a k-dimensional oriented smooth manifold M in $\mathbb{R}^n$. Then

$$\int_{\partial_M X}\varphi \overset{1}{=} \int_{\partial^s_M X}\varphi \overset{2}{=} \lim_{\epsilon\to 0}\int_{\partial^s_M X_\epsilon}\varphi \underbrace{\overset{3}{=}}_{\text{Thm. 6.10.9}} \lim_{\epsilon\to 0}\int_{X_\epsilon}\mathbf{d}\varphi \overset{4}{=} \int_X \mathbf{d}\varphi. \qquad 6.10.65$$

The first equality is a definition: integrals over boundaries are by definition integrals over the smooth boundary. Equalities 2 and 4 are applications of the dominated convergence theorem, which requires interpretation in terms of parametrizations.

The constant C in inequality 6.10.67 is the square root in Proposition 6.10.11, as applied to $\psi=\mathbf{d}\varphi$. The sup that occurs is finite, since X is compact and φ is of class C^2, so $\mathbf{d}\varphi$ is of class C^1, so its coefficients are continuous.

In inequality 6.10.68, the equality is Definition 5.3.1 and the inequality is Theorem 6.6.16.

Start with equality 4, which is a bit simpler. Let W be a subset of $\mathbb{R}^k$, and let $\gamma: W\to M$ be a parametrization of an open set of M containing X. Set

$$Y_\epsilon=\gamma^{-1}(X_\epsilon) \text{ and } Y=\gamma^{-1}(X), \text{ so } \lim_{\epsilon\to 0}\mathbf{1}_{Y_\epsilon}=\mathbf{1}_Y. \qquad 6.10.66$$

Since φ is defined on an open set containing X, by inequality 6.10.60 there is a constant C such that

$$|\mathbf{d}\varphi(P_\mathbf{x}(\vec{\mathbf{v}}_1,\dots,\vec{\mathbf{v}}_k))| \le C\underbrace{|d^k\mathbf{x}|(P_\mathbf{x}(\vec{\mathbf{v}}_1,\dots,\vec{\mathbf{v}}_k))}_{\operatorname{vol}_k(P_\mathbf{x}(\vec{\mathbf{v}}_1,\dots,\vec{\mathbf{v}}_k))} \qquad 6.10.67$$

for all $\mathbf{x}\in X$, and any vectors $\vec{\mathbf{v}}_1,\dots,\vec{\mathbf{v}}_k\in\mathbb{R}^n$. Moreover,

$$\int_Y\Big(|d^k\mathbf{x}|(P_{\gamma(\mathbf{w})}(\overrightarrow{D_1\gamma}(\mathbf{w}),\dots,\overrightarrow{D_k\gamma}(\mathbf{w}))\Big)|d^k\mathbf{w}| = \int_X |d^k\mathbf{x}| < \infty. \qquad 6.10.68$$

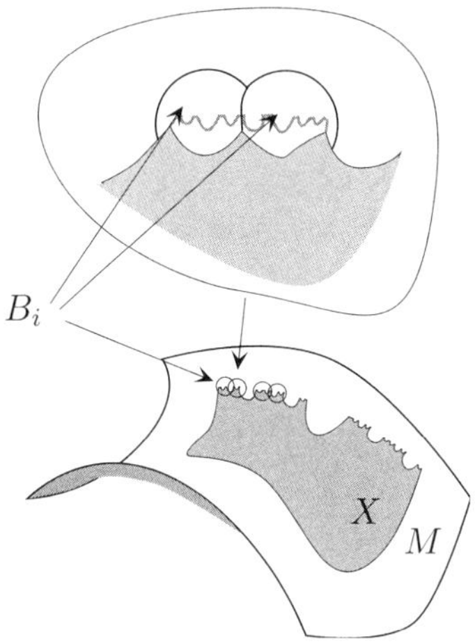

FIGURE 6.10.8.

This figure illustrates how we use Proposition 6.10.13 to show that the part $\partial_2 X_\epsilon$ of the smooth boundary can be made arbitrarily small, so that we can prove that

$$\int_{\partial^s_M X} \varphi = \lim_{\epsilon \to 0} \int_{\partial^s_M X_\epsilon} \varphi.$$

BOTTOM: A 2-dimensional piece-with-boundary, with nonsmooth boundary of 1-dimensional volume (length) 0.

TOP: Blow-up of balls covering the nonsmooth boundary.

The nonsmooth boundary is covered by balls B_i with radii r_i. Here, $k = 2$, so saying that the nonsmooth boundary has $(k-1)$-dimensional volume means that we can make $\sum r_i = \sum r_i^{k-1}$ arbitrarily small.

The curves $\partial B_i \cap M$ are almost but not quite plane curves; the intersections of the balls with planes through the center have lengths $2\pi r_i$, so the total length is arbitrarily small. The proposition says that the total length of

$$\bigcup_i (\partial B_i \cap M)$$

is at most $2\pi(1+\epsilon)\sum r_i$ as soon as the balls are sufficiently small.

Thus the functions

$$(\mathbf{1}_{Y_\epsilon})\mathbf{d}\varphi\Big(P_{\gamma(\mathbf{w})}(\overrightarrow{D_1\gamma}(\mathbf{w}), \dots, \overrightarrow{D_k\gamma}(\mathbf{w}))\Big) \qquad 6.10.69$$

converge pointwise to

$$(\mathbf{1}_{Y})\mathbf{d}\varphi\Big(P_{\gamma(\mathbf{w})}(\overrightarrow{D_1\gamma}(\mathbf{w}), \dots, \overrightarrow{D_k\gamma}(\mathbf{w}))\Big), \qquad 6.10.70$$

and they are all bounded in absolute value by the same integrable function:

$$\begin{aligned}(\mathbf{1}_{Y_\epsilon})\,&\Big|\mathbf{d}\varphi(P_{\gamma(\mathbf{w})}(\overrightarrow{D_1\gamma}(\mathbf{w}), \dots, \overrightarrow{D_k\gamma}(\mathbf{w}))\Big)\Big| \\ &\le C\mathbf{1}_Y\Big(|d^k\mathbf{x}|(P_{\gamma(\mathbf{w})}(\overrightarrow{D_1\gamma}(\mathbf{w}), \dots, \overrightarrow{D_k\gamma}(\mathbf{w}))\Big),\end{aligned} \qquad 6.10.71$$

so the dominated convergence theorem (Theorem 4.11.19) says that

$$\lim_{\epsilon\to 0}\int_{X_\epsilon}\mathbf{d}\varphi = \int_X \mathbf{d}\varphi. \qquad 6.10.72$$

The proof of equality 2 in equation 6.10.65 is a bit more complicated. The smooth boundary $\partial^s_M X_\epsilon$ has two parts:

$$\partial_1 X_\epsilon \overset{\text{def}}{=} X_\epsilon \cap (\partial^s_M X) \quad \text{and} \quad \partial_2 X_\epsilon = X_\epsilon \cap \bigcup_i^N \partial B_i \qquad 6.10.73$$

• For $\partial_1 X_\epsilon$, the same proof as above works. Let $\delta : W \to \partial^s_M X$ be a parametrization, where W is an appropriate subset of $\mathbb{R}^{k-1}$, and set $W_\epsilon = \delta^{-1}(X_\epsilon)$. Then

$$\begin{aligned}&\Big|\varphi\Big(P_{\delta(\mathbf{w})}(\overrightarrow{D_1\delta}(\mathbf{w}), \dots, \overrightarrow{D_{k-1}\delta}(\mathbf{w}))\Big)\Big| \\ &\qquad\le C|d^{k-1}\mathbf{x}|\Big(P_{\delta(\mathbf{w})}(\overrightarrow{D_1\delta}(\mathbf{w}), \dots, \overrightarrow{D_{k-1}\delta}(\mathbf{w})\Big),\end{aligned} \qquad 6.10.74$$

where C is the sup of the square root in Proposition 6.10.11, as applied to $\psi = \varphi$. The integral

$$\int_W |d^{k-1}\mathbf{w}| \underbrace{=}_{\text{Def. 5.3.1}} \int_W \underbrace{|d^{k-1}\mathbf{x}|\Big(P_{\delta(\mathbf{w})}(\overrightarrow{D_1\delta}(\mathbf{w}), \dots, \overrightarrow{D_{k-1}\delta}(\mathbf{w})\Big)}_{\text{this function of } \mathbf{w} \text{ is the dominating function}} |d^{k-1}\mathbf{w}| \qquad 6.10.75$$

is finite since $\partial^s_M X$ has finite $(k-1)$-volume (condition 2 of Definition 6.6.8), and we can use the "dominating function" above to prove that

$$\lim_{\epsilon\to 0}\int_{W_\epsilon}|d^{k-1}\mathbf{w}| = \int_W |d^{k-1}\mathbf{w}|. \qquad 6.10.76$$

• The other part, $\partial_2 X_\epsilon \subset \partial^s_M X_\epsilon$, is the part of the boundary contained in the boundary of the balls B_i. Let K be the volume of the $(k-1)$-dimensional unit sphere (see Example 5.3.14). As illustrated by Figure 6.10.8, Proposition 6.10.13 says that for r sufficiently small, if $B_r(\mathbf{x})$ is a ball centered at $\mathbf{x} \in X$ then

$$\begin{aligned}\mathrm{vol}_{k-1}\big(\partial B_r(\mathbf{x}) \cap M\big) &\le (1+\epsilon)\,\mathrm{vol}_{k-1}\big(\partial B_r(\mathbf{x}) \cap T_\mathbf{x}M\big) \\ &= (1+\epsilon)Kr^{k-1}.\end{aligned} \qquad 6.10.77$$

Thus the condition $\sum_i r_i^{k-1} < \epsilon$ from Proposition 6.10.10, together with inequality 6.10.74 bounding φ in terms of volume, shows that

$$\lim_{\epsilon \to 0} \int_{\partial_2 X_\epsilon} \varphi = 0. \qquad 6.10.78$$

In the third edition, we ended the proof of Kantorovich's theorem with a simple □ and no summary. A reader objected: "a neat and intricate proof like this one should end with a trumpet blast, not just stop." We believe the proof of Stokes's theorem also deserves a "trumpet blast".

This completes the proof of Stokes's theorem in full generality, applied to pieces-with-boundary as defined in Definition 6.6.8. 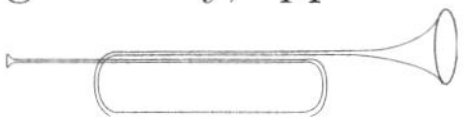

Exercises for Section 6.10

6.10.1 Let U be a compact piece-with-boundary of $\mathbb{R}^3$. Show that

$$\text{vol}_3\, U = \int_{\partial U} \frac{1}{3}\left(z\, dx \wedge dy + y\, dz \wedge dx + x\, dy \wedge dz\right).$$

Exercise 6.10.2: The volume of a full cone is $\frac{1}{3}$·height·area of base.

6.10.2 Let C be that part of the cone of equation $z = a - \sqrt{x^2 + y^2}$ where $z \geq 0$, oriented by the upward-pointing normal. What is the integral

$$\int_C x\, dy \wedge dz + y\, dz \wedge dx + z\, dx \wedge dy\,?$$

6.10.3 Compute the integral of $x_1\, dx_2 \wedge dx_3 \wedge dx_4$ over the part of the 3-dimensional manifold of equation $x_1 + x_2 + x_3 + x_4 = a$ where $x_1, x_2, x_3, x_4 \geq 0$, oriented so that the projection to (x_1, x_2, x_3)-coordinate 3-space is orientation preserving.

6.10.4 a. Show that the 2-form on $\mathbb{R}^3 - \{\mathbf{0}\}$ given by

$$\varphi = \frac{x\, dy \wedge dz + y\, dz \wedge dx + z\, dx \wedge dy}{(x^2 + y^2 + z^2)^{3/2}} \quad \text{satisfies } \mathbf{d}\varphi = 0.$$

b. Compute $\int_S \varphi$, where S is the sphere of radius $R \neq \sqrt{3}$ centered at $\begin{bmatrix} 1 \\ 1 \\ 1 \end{bmatrix}$, oriented by the outward-pointing normal. (The result depends on R, of course.)

6.10.5 Let S be the part of the ellipsoid $\frac{x^2}{a^2} + \frac{y^2}{b^2} + \frac{z^2}{c^2} = 1$ where x, y, z are nonnegative; orient S by the outward-pointing normal. What is the integral of

$$\omega = x\, dy \wedge dz + y\, dz \wedge dx + z\, dx \wedge dy$$

over S? (You may use Stokes's theorem or parametrize the surface.)

6.10.6 a. Parametrize the surface in 4-space given by the equations

$$x_1^2 + x_2^2 = a^2 \quad \text{and} \quad x_3^2 + x_4^2 = b^2, \quad \text{for } a, b > 0.$$

b. Show that

$$\operatorname{sgn}(x_1\, dx_2 - x_2\, dx_1) \wedge (x_3\, dx_4 - x_4\, dx_3)$$

is an orientation of the surface. Does your parametrization preserve or reverse orientation?

c. Integrate the 2-form $x_1 x_4\, dx_2 \wedge dx_3$ over this surface.

d. Compute $\mathbf{d}(x_1 x_4\, dx_2 \wedge dx_3)$.

e. Represent the surface as the boundary of a 3-dimensional manifold in $\mathbb{R}^4$, and verify that Stokes's theorem is true in this case.

6.10.7 Show that the bump function β_R is of class C^∞ on all of $\mathbb{R}^n$.

***6.10.8** a. Sketch in the (x, y)-plane the curve of polar equation $r^2 = 2\cos 2\theta$.

b. Orient this curve by increasing θ and integrate over the curve the 1-form

$$\frac{-y\,dx + (x+1)\,dy}{(x+1)^2 + y^2} + \frac{-(y+1)\,dx + x\,dy}{x^2 + (y+1)^2}.$$

Exercise 6.10.10: That is, prove that the integral of the "element of solid angle" $\Phi_{\vec{F}_3}$ over such a surface S is the same as its integral over the corresponding P. You were asked to do this in Exercise 6.7.12, but it is easier with Stokes.

6.10.9 Prove Lemma 6.10.12. One possible proof goes as follows.

a. Show that the mapping $\operatorname{Mat}(n,k) \to \mathbb{R}$ given by $B \mapsto \det A^\top B$ is multilinear and alternating as a function of the columns of B. By Theorem 6.1.8, part 1, it is of the form

$$\sum_{1 \le i_1 < \cdots < i_k \le 1} c_{i_1,\dots i_k}\, dx_{i_1} \wedge \cdots \wedge dx_{i_k}.$$

b. Determine the coefficients $c_{i_1,\dots i_k}$, using Theorem 6.1.8, part 2.

c. Evaluate the mapping when $B = A$.

6.10.10 Use Stokes's theorem to prove the statement in the caption of Figure 6.7.1, in the special case where the surface S projects injectively to the unit sphere.

Exercise 6.10.11: Note that in equation (1), $\left|\begin{pmatrix}\mathbf{u}\\ \mathbf{f}(\mathbf{u})\end{pmatrix}\right|$ is a number, so $\frac{\mathbf{u}}{|\mathbf{u}|}\left|\begin{pmatrix}\mathbf{u}\\ \mathbf{f}(\mathbf{u})\end{pmatrix}\right|$ is an element of $\mathbb{R}^k$. Since $\mathbf{y} - \mathbf{f}(\mathbf{u})$ is an element of $\mathbb{R}^{n-k}$, an element of the codomain of F is indeed in $\mathbb{R}^n$.

6.10.11 Prove Proposition 6.10.13. One approach goes as follows: rotate and translate M so that $\mathbf{p}$ is at the origin of $\mathbb{R}^n$ and $T_\mathbf{p}M = \mathbb{R}^k \times \{\mathbf{0}\}$, so that near the origin M is locally the graph of a C^2 map $\mathbf{f}: U \to \mathbb{R}^{n-k}$ with U a neighborhood of $\mathbf{0}$ in $\mathbb{R}^k$, and $[\mathbf{Df}(\mathbf{0})] = [0]$, so the Taylor polynomial of $\mathbf{f}$ starts with quadratic terms.

Define $F: U \times \mathbb{R}^{n-k} \to \mathbb{R}^n$ by

$$F\begin{pmatrix}\mathbf{u}\\ \mathbf{y}\end{pmatrix} = \begin{pmatrix}\frac{\mathbf{u}}{|\mathbf{u}|}\left|\begin{pmatrix}\mathbf{u}\\ \mathbf{f}(\mathbf{u})\end{pmatrix}\right| \\ \mathbf{y} - \mathbf{f}(\mathbf{u})\end{pmatrix}. \tag{1}$$

a. Show that F is of class C^2 and that $\left[\mathbf{D}F\begin{pmatrix}\mathbf{0}\\ \mathbf{0}\end{pmatrix}\right]$ is the identity.

b. Show that $F(M \cap \partial B_r(\mathbf{0}) = T_\mathbf{0}M \cap \partial B_r(\mathbf{0})$.

c. Show that a map close to the identity with derivative close to the identity modifies volumes arbitrarily little.

6.11 The integral theorems of vector calculus

In this section we discuss the four special cases of the generalized Stokes's theorem that make sense in $\mathbb{R}^2$ and $\mathbb{R}^3$: the fundamental theorem for line integrals, Green's theorem, Stokes's theorem for surfaces in $\mathbb{R}^3$, and the divergence theorem (also known as Gauss's theorem). These theorems don't say anything that is not contained in the generalized Stokes's theorem, but they are used everywhere in electromagnetism, fluid mechanics, and many other fields. Thus these theorems should all become personal friends, or at least acquaintances.

We could call Theorem 6.11.1 the fundamental theorem for integrals over curves; "line integrals" is more traditional. It is also called the independence of path theorem. Using a parametrization, it can easily be reduced to the ordinary fundamental theorem of calculus, Theorem 6.10.1, which it is if $n = 1$.

Theorem 6.11.1 (Fundamental theorem for line integrals). *Let C be an oriented curve in $\mathbb{R}^n$, with oriented boundary $(P_{\mathbf{b}} - P_{\mathbf{a}})$, and let f be a function defined on a neighborhood of C. Then*

$$\int_C \mathbf{d}f = f(\mathbf{b}) - f(\mathbf{a}). \tag{6.11.1}$$

Green's theorem is the special case of Stokes's theorem for surface integrals when the surface is flat. (Yes, we do need both *bounded*'s in Theorem 6.11.2. The exterior of the unit disc is bounded by the unit circle but it is not bounded.)

George Green was the son of a baker and himself became a miller. He left school at age nine and it is not clear how he learned the mathematics needed to publish in 1828 *An Essay on the Application of Mathematical Analysis to the Theories of Electricity and Magnetism*, which has been described as "one of the most important mathematical works of all time" (J. J. O'Connor and E. F. Robertson, MacTutor History of Mathematics web site). At age 40 he became an undergraduate at Cambridge University.

Theorem 6.11.2 (Green's theorem). *Let S be a bounded region of $\mathbb{R}^2$, bounded by a curve C (or several curves C_i), carrying the boundary orientation as described in Definition 6.6.21. Let $\vec{F}$ be a vector field defined on a neighborhood of S. Then*

$$\int_S \mathbf{d}W_{\vec{F}} = \int_C W_{\vec{F}} \quad \text{or} \quad \int_S \mathbf{d}W_{\vec{F}} = \sum_i \int_{C_i} W_{\vec{F}}. \tag{6.11.2}$$

Suppose $\vec{F} = \begin{pmatrix} f \\ g \end{pmatrix}$. Then Green's theorem is traditionally written

$$\int_S (D_1 g - D_2 f)\, dx\, dy = \int_C f\, dx + g\, dy. \tag{6.11.3}$$

To see that the two versions are the same, write

$$W_{\vec{F}} = f\begin{pmatrix} x \\ y \end{pmatrix} dx + g\begin{pmatrix} x \\ y \end{pmatrix} dy \tag{6.11.4}$$

and use Theorem 6.7.4 to compute its exterior derivative:

$$\begin{aligned} \mathbf{d}W_{\vec{F}} &= \mathbf{d}(f\,dx + g\,dy) = \mathbf{d}f \wedge dx + \mathbf{d}g \wedge dy \\ &= (D_1 f\, dx + D_2 f\, dy) \wedge dx + (D_1 g\, dx + D_2 g\, dy) \wedge dy \\ &= D_2 f\, dy \wedge dx + D_1 g\, dx \wedge dy = (D_1 g - D_2 f)\, dx \wedge dy. \end{aligned} \tag{6.11.5}$$

When we say that you should learn the special cases of Stokes's theorem, we don't mean that you should memorize their classical forms or use them in your computations. But you should know their names, and what

$$\int_{\partial X} \varphi = \int_X \mathbf{d}\varphi$$

becomes in each case. What kind of manifold is X? What kind of object is its boundary? What kind of forms are φ and $\mathbf{d}\varphi$?

Example 6.11.3 (Green's theorem). What is the integral

$$\int_{\partial U} 2xy\, dy + x^2\, dx, \tag{6.11.6}$$

where U is the part of the disc of radius R centered at the origin where $y \geq 0$, with the standard orientation? This corresponds to Green's theorem, with $f\begin{pmatrix} x \\ y \end{pmatrix} = x^2$ and $g\begin{pmatrix} x \\ y \end{pmatrix} = 2xy$, so that $D_1 g = 2y$ and $D_2 f = 0$. Using polar coordinates (see Proposition 4.10.3) we have

$$\int_{\partial U} 2xy\,dy + x^2\,dx = \int_U (D_1 g - D_2 f)\,dx\,dy = \int_U 2y\,dx\,dy \qquad 6.11.7$$
$$= \int_0^\pi \int_0^R (2r\sin\theta)\,r\,dr\,d\theta = \frac{2R^3}{3}\int_0^\pi \sin\theta\,d\theta = \frac{4R^3}{3}.$$

In the language of forms we would write

$$\int_{\partial U} 2xy\,dy + x^2\,dx = \int_U \mathbf{d}(2xy\,dy + x^2\,dx) = \int_U 2y\,dx \wedge dy, \qquad 6.11.8$$

and then continue as above.

What happens if we integrate over the boundary of the entire disc?[15]

Theorem 6.11.4 (Stokes's theorem for surfaces in $\mathbb{R}^3$). *Let S be an oriented surface in $\mathbb{R}^3$, bounded by a curve C that is given the boundary orientation. Let φ be a 1-form field defined on a neighborhood of S. Then*

$$\int_S \mathbf{d}\varphi = \int_C \varphi. \qquad 6.11.9$$

FIGURE 6.11.1.

George Stokes asked what we now know as Stokes's theorem for surfaces in $\mathbb{R}^3$ on an examination in Cambridge in 1854. Traditionally, the theorem is simply called "Stokes's theorem".

Equation 6.11.11: In the traditional version, $|d^2\mathbf{x}|$ is written dS and $|d^1\mathbf{x}|$ is written dl or ds. Sometimes $\vec{N}(\mathbf{x})\ |d^2\mathbf{x}|$ is written $\overrightarrow{dS}$.

Note that the curve C in Theorem 6.11.4 may well consist of several pieces C_i.

Again, let's translate this into classical notation. Suppose C is the union of disjoint simple closed curves C_i. Without loss of generality, we can write $\varphi = W_{\vec{F}}$, so that Theorem 6.11.4 becomes

$$\int_S \mathbf{d}W_{\vec{F}} = \int_S \Phi_{\operatorname{curl}\vec{F}} = \sum_i \int_{C_i} W_{\vec{F}}. \qquad 6.11.10$$

Now let $\vec{N}$ be the normal unit vector field defining the orientation of S, and $\vec{T}$ the unit tangent vector field defining the orientation of the C_i. This gives the classical version of Stokes's theorem:

$$\iint_S (\operatorname{curl}\vec{F}(\mathbf{x}))\cdot\vec{N}(\mathbf{x})\ |d^2\mathbf{x}| = \sum_i \int_{C_i} \vec{F}(\mathbf{x})\cdot\vec{T}(\mathbf{x})\,|d^1\mathbf{x}|. \qquad 6.11.11$$

Let us check that equations 6.11.10 and 6.11.11 are equivalent, starting with the left sides. In equation 6.11.11, the $\vec{N}|d^2\mathbf{x}|$ on the left takes a parallelogram $P_{\mathbf{x}}(\vec{\mathbf{v}},\vec{\mathbf{w}})$ and returns the vector

$$\vec{N}(\mathbf{x})|\vec{\mathbf{v}}\times\vec{\mathbf{w}}|, \qquad 6.11.12$$

since the integrand $|d^2\mathbf{x}|$ is the element of area; given a parallelogram, it returns its area (the length of the cross product of its sides). When integrating over S, we evaluate the integrand only on parallelograms $P_{\mathbf{x}}(\vec{\mathbf{v}},\vec{\mathbf{w}})$ that are tangent to S at $\mathbf{x}$, with compatible orientation. Since $\vec{\mathbf{v}}$, $\vec{\mathbf{w}}$ are

[15] It is 0, by symmetry: the integral of $2y$ over the top semidisc cancels the integral over the bottom semidisc.

tangent to S, the cross product $\vec{\mathbf{v}} \times \vec{\mathbf{w}}$ is perpendicular to S. So $\vec{\mathbf{v}} \times \vec{\mathbf{w}}$ is a positive multiple of $\vec{N}(\mathbf{x})$. Since $\vec{N}(\mathbf{x})$ has length 1,

$$\vec{\mathbf{v}} \times \vec{\mathbf{w}} = |\vec{\mathbf{v}} \times \vec{\mathbf{w}}| \vec{N}(\mathbf{x}). \qquad 6.11.13$$

So the left sides of equations 6.11.10 and 6.11.11 are equal, since

$$\begin{aligned} \left(\operatorname{curl} \vec{F}(\mathbf{x})\right) \cdot \left(\vec{N}(\mathbf{x})\right) |d^2\mathbf{x}|(\vec{\mathbf{v}}, \vec{\mathbf{w}}) &= \left(\operatorname{curl} \vec{F}(\mathbf{x})\right) \cdot (\vec{\mathbf{v}} \times \vec{\mathbf{w}}) \\ &= \det[\operatorname{curl} \vec{F}(\mathbf{x}), \vec{\mathbf{v}}, \vec{\mathbf{w}}] \\ &= \Phi_{\operatorname{curl} \vec{F}}(\vec{\mathbf{v}}, \vec{\mathbf{w}}). \end{aligned} \qquad 6.11.14$$

i.e., the flux of $\operatorname{curl} \vec{F}$ acting on $\vec{\mathbf{v}}$ and $\vec{\mathbf{w}}$.

Now let's compare the right sides. Set $\vec{F} = \begin{bmatrix} F_1 \\ F_2 \\ F_3 \end{bmatrix}$. On the right side of equation 6.11.10, the integrand is

$$W_{\vec{F}} = F_1\, dx + F_2\, dy + F_3\, dz; \qquad 6.11.15$$

given a vector $\vec{\mathbf{v}}$, it returns the number $F_1 v_1 + F_2 v_2 + F_3 v_3$.

On the right side of equation 6.11.11, $\vec{T}(\mathbf{x})\, |d^1\mathbf{x}|$ is a complicated way of expressing the identity. Since the element of arc length $|d^1\mathbf{x}|$ takes a vector and returns its length, $\vec{T}(\mathbf{x})\, |d^1\mathbf{x}|$ takes a vector and returns $\vec{T}(\mathbf{x})$ times its length. Since $\vec{T}(\mathbf{x})$ has length 1, the result is a vector with length $|\vec{\mathbf{v}}|$, tangent to the curve. When integrating, we are only going to evaluate the integrand on vectors tangent to the curve and pointing in the direction of $\vec{T}$, so this process just takes such a vector and returns precisely the same vector. So $\vec{F}(\mathbf{x}) \cdot \vec{T}(\mathbf{x})\, |d^1\mathbf{x}|$ takes a vector $\vec{\mathbf{v}}$ and returns the number

$$\big(\vec{F}(\mathbf{x}) \cdot \underbrace{\vec{T}(\mathbf{x})\, |d^1\mathbf{x}|\big)(\vec{\mathbf{v}})}_{\vec{\mathbf{v}}} = \begin{bmatrix} F_1 \\ F_2 \\ F_3 \end{bmatrix} \cdot \begin{bmatrix} v_1 \\ v_2 \\ v_3 \end{bmatrix} = F_1 v_1 + F_2 v_2 + F_3 v_3 = W_{\vec{F}}(\vec{\mathbf{v}}). \qquad 6.11.16$$

Example 6.11.5: Exercise 6.3.9 shows that for appropriate curves, orienting by decreasing polar angle means that the curve is oriented clockwise.

The cylinder is like an empty paper towel roll. The surface S is obtained by cutting the top of the roll in some irregular fashion corresponding to its intersection with the surface $z = \sin xy + 2$; its boundary consists of the two curves: $\partial S = C + C_1$.

Example 6.11.5 (Stokes's theorem). Let C be the intersection of the cylinder of equation $x^2 + y^2 = 1$ with the surface of equation $z = \sin xy + 2$. Orient C so that the polar angle decreases along C. What is the work over C of the vector field

$$\vec{\mathbf{F}} \begin{pmatrix} x \\ y \\ z \end{pmatrix} = \begin{bmatrix} y^3 \\ x \\ z \end{bmatrix} ? \qquad 6.11.17$$

It's not so obvious how to visualize C, much less integrate over it. Stokes's theorem says there is an easier approach: compute the integral over the subsurface S consisting of the cylinder $x^2 + y^2 = 1$ bounded at the top by C and at the bottom by the unit circle C_1 in the (x, y)-plane, oriented counterclockwise.

By Stokes's theorem, $\int_C \varphi + \int_{C_1} \varphi = \int_S \mathbf{d}\varphi$, so rather than integrate over the irregular curve C, we will integrate over S and then subtract the integral over C_1. First we integrate over S:

Since C is oriented clockwise, and C_1 counterclockwise, $C + C_1$ form the oriented boundary of S. If you walk on S clockwise along C, with your head pointing away from the z-axis, the surface is to your left; if you do the same along C_1, counterclockwise, the surface is still to your left.

FIGURE 6.11.2.
Mikhail Ostrogradski (1801–1862)

There is much contention as to who should get credit for the integral theorems of vector calculus. The divergence theorem was stated by Lagrange in 1762 and rediscovered by Gauss in 1813; according to some accounts, Ostrogradski gave the first proof (the St. Petersburg Academy of Sciences, 1828).

George Green published his paper privately in 1828; Lord Kelvin rediscovered Green's result in 1846. George Stokes (1819–1903) proved Stokes's theorem.

$$\int_C W_{\vec{F}} + \int_{C_1} W_{\vec{F}} = \int_S \Phi_{\text{curl}\,\vec{F}} = \int_S \Phi_{\begin{bmatrix} 0 \\ 0 \\ 1-3y^2 \end{bmatrix}} = 0. \qquad 6.11.18$$

The integral is 0 because the vector field $\begin{bmatrix} 0 \\ 0 \\ 1-3y^2 \end{bmatrix}$ is vertical and has no flow through the vertical cylinder. Parametrize C_1 in the obvious way:

$$\gamma(t) = \begin{bmatrix} \cos t \\ \sin t \\ 0 \end{bmatrix}. \qquad 6.11.19$$

This parametrization is compatible with the counterclockwise orientation of C_1 (see Examples 6.3.6 and 6.4.3). Using equation 6.5.14, compute

$$\begin{aligned}\int_{C_1} W_{\vec{F}} &= \int_0^{2\pi} \overbrace{\begin{bmatrix} (\sin t)^3 \\ \cos t \\ 0 \end{bmatrix}}^{\vec{F}(\gamma(t))} \cdot \overbrace{\begin{bmatrix} -\sin t \\ \cos t \\ 0 \end{bmatrix}}^{\gamma'(t)} dt \\ &= \int_0^{2\pi} -(\sin t)^4 + (\cos t)^2 \, dt = -\frac{3}{4}\pi + \pi = \frac{\pi}{4}.\end{aligned} \qquad 6.11.20$$

So the work over C is

$$\int_C W_{\vec{F}} = \int_S \Phi_{\text{curl}\,\vec{F}} - \int_{C_1} W_{\vec{F}} = -\frac{\pi}{4}. \qquad 6.11.21$$

What if both curves were oriented clockwise? counterclockwise?[16] △

The divergence theorem

The divergence theorem is also known as *Gauss's theorem*.

---

[16]Denote these curves by C^+ and C_1^+, and denote by C^- and C_1^- the curves oriented counterclockwise. Then (leaving out the integrands to simplify notation) we would have

$$\int_{C^+} + \Big(-\int_{C_1^+}\Big) = \int_S, \quad \text{but} \quad -\int_{C_1^+} = \int_{C_1^-},$$

so $\int_{C^+} W_{\vec{F}}$ remains unchanged. If both were oriented counterclockwise, so that C did not have the boundary orientation of S, we would have

$$-\int_{C^-} + \int_{C_1^-} = \int_S = 0 \quad \text{and} \quad \int_{C^-} W_{\vec{F}} = \int_{C_1^-} W_{\vec{F}} = \frac{\pi}{4}.$$

Theorem 6.11.6 (The divergence theorem). *Let X be a bounded domain in $\mathbb{R}^3$ with the standard orientation of space, and let its boundary ∂X be a union of surfaces S_i, each oriented by the outward normal. Let φ be a 2-form field defined on a neighborhood of X. Then*

$$\int_X \mathbf{d}\varphi = \sum_i \int_{S_i} \varphi. \tag{6.11.22}$$

FIGURE 6.11.3.
Born in 287 BC, Archimedes was killed in 212 BC during the capture of Syracuse by the Romans. He was about 22 years old when he discovered "Archimedes' principle" (Example 6.11.8).

"It is not possible to find in all geometry more difficult and intricate questions, or more simple and lucid explanations," wrote Plutarch several centuries later.

Again, let's make this look a bit more classical. Write $\varphi = \Phi_{\vec{F}}$, so that $\mathbf{d}\varphi = \mathbf{d}\Phi_{\vec{F}} = M_{\text{div}\,\vec{F}}$, and let $\vec{N}$ be the unit outward-pointing vector field on the S_i; then equation 6.11.22 can be rewritten

$$\int_X M_{\text{div}\,\vec{F}} = \iiint_X \text{div}\,\vec{F}\,dx\,dy\,dz = \sum_i \iint_{S_i} \vec{F}\cdot\vec{N}\,|d^2\mathbf{x}|, \tag{6.11.23}$$

the classical version of the divergence theorem. Why is this right? When we discussed Stokes's theorem, we saw that $\vec{F}\cdot\vec{N}$, evaluated on a parallelogram tangent to the surface, is the same thing as the flux of $\vec{F}$ evaluated on the same parallelogram. So indeed equation 6.11.23 is the same as

$$\int_X \mathbf{d}\Phi_{\vec{F}} = \int_X M_{\text{div}\,\vec{F}} = \sum_i \int_{S_i} \Phi_{\vec{F}}. \tag{6.11.24}$$

Remark. We think equations 6.11.11 and 6.11.23 are a good reason to avoid the classical notation. They bring in $\vec{N}$, which usually involves dividing by the square root of the length; this is both messy and unnecessary, since the $|d^2\mathbf{x}|$ term cancels with the denominator. More seriously, the classical notation hides the resemblance of the special Stokes's theorem and the divergence theorem to the general one, Theorem 6.10.2. But the classical notation has a geometric immediacy that speaks to people who are used to it. $\triangle$

Example 6.11.7: The unit cube is the cube spanned by $\vec{\mathbf{e}}_1$, $\vec{\mathbf{e}}_2$, and $\vec{\mathbf{e}}_3$; it is not centered at the origin.

Equation 6.11.25: Our general formula for integrating a mass form over a parametrized piece of $\mathbb{R}^3$ is (equation 6.5.20)

$$\int_{\gamma(U)} M_f = \int_U f\big(\gamma(\mathbf{u})\big)\det[\mathbf{D}\gamma(\mathbf{u})]\,|d^3\mathbf{u}|.$$

Here we parametrize by the identity $\gamma(\mathbf{x}) = \mathbf{x}$ and use the special case (equation 6.5.21)

$$\int_{\gamma(U)} M_f = \int_U f(\mathbf{u})\,|d^3\mathbf{u}|.$$

Example 6.11.7 (Divergence theorem). Let Q be the unit cube. What is the flux of the vector field $\begin{bmatrix} x^2y \\ -2yz \\ x^3y^2 \end{bmatrix}$ through the boundary of Q if Q carries the standard orientation of $\mathbb{R}^3$ and the boundary has the boundary orientation?

Since $\mathbf{d}\Phi_{\vec{F}} = M_{\text{div}\,\vec{F}}$, the divergence theorem asserts that

$$\int_{\partial Q} \Phi_{\begin{bmatrix} x^2y \\ -2yz \\ x^3y^2 \end{bmatrix}} = \int_Q M_{\text{div}\begin{bmatrix} x^2y \\ -2yz \\ x^3y^2 \end{bmatrix}} = \int_Q (2xy - 2z)\,|d^3\mathbf{x}|. \tag{6.11.25}$$

This can readily be computed by Fubini's theorem:

$$\int_0^1\int_0^1\int_0^1 (2xy - 2z)\,dx\,dy\,dz = \frac{1}{2} - 1 = -\frac{1}{2}. \quad \triangle \tag{6.11.26}$$

Example 6.11.8 (The principle of Archimedes). Archimedes (Figure 6.11.3) is said to have been asked by Hiero, king of Syracuse, to determine whether a crown he had ordered constructed was really made of gold. Archimedes discovered that by weighing the crown suspended in water, he could determine whether or not it was counterfeit. According to legend, he made the discovery in the bath, and proceeded to run naked through the streets, crying "Eureka" ("*I found it*").

The principle he claimed is the following: *A body immersed in a fluid receives a buoyant force equal to the weight of the displaced fluid.*

We do not understand how he came to this conclusion; the derivation we will give uses mathematics that was certainly not available to Archimedes.

The force the fluid exerts on the immersed body is due to the pressure p. Suppose that the body is X, with boundary ∂X made up of little oriented parallelograms P_i. The fluid exerts a force approximately

$$p(\mathbf{x}_i)\,\text{Area}\,(P_i)\vec{\mathbf{n}}_i, \qquad 6.11.27$$

where $\vec{\mathbf{n}}$ is an inward-pointing unit vector perpendicular to P_i and $\mathbf{x}_i$ is a point of P_i; this becomes a better and better approximation as P_i becomes small, so that the pressure on it becomes approximately constant. The total force exerted by the fluid is the sum of the forces exerted on all the little pieces of the boundary.

Thus the force is naturally a surface integral, really an integral of a 2-form field, since the orientation of ∂X matters. But we can't think of it as a single 2-form field: the force has three components, and we have to think of each one as a 2-form field. In fact, the force is

$$\begin{bmatrix} \int_{\partial X} p\Phi_{\vec{\mathbf{e}}_1} \\ \int_{\partial X} p\Phi_{\vec{\mathbf{e}}_2} \\ \int_{\partial X} p\Phi_{\vec{\mathbf{e}}_3} \end{bmatrix}, \qquad 6.11.28$$

since

$$\begin{bmatrix} p(\mathbf{x})\Phi_{\vec{\mathbf{e}}_1} \\ p(\mathbf{x})\Phi_{\vec{\mathbf{e}}_2} \\ p(\mathbf{x})\Phi_{\vec{\mathbf{e}}_3} \end{bmatrix} (P_{\mathbf{x}}(\vec{\mathbf{v}}_1, \vec{\mathbf{v}}_2)) = p(\mathbf{x}) \begin{bmatrix} \det[\vec{\mathbf{e}}_1, \vec{\mathbf{v}}_1, \vec{\mathbf{v}}_2] \\ \det[\vec{\mathbf{e}}_2, \vec{\mathbf{v}}_1, \vec{\mathbf{v}}_2] \\ \det[\vec{\mathbf{e}}_3, \vec{\mathbf{v}}_1, \vec{\mathbf{v}}_2] \end{bmatrix} \qquad 6.11.29$$

$$= p(\mathbf{x})(\vec{\mathbf{v}}_1 \times \vec{\mathbf{v}}_2) \underbrace{=}_{\text{Prop. 1.4.20}} p(\mathbf{x}) \underbrace{\text{Area}\,\big(P_{\mathbf{x}}(\vec{\mathbf{v}}_1, \vec{\mathbf{v}}_2)\big)\,\vec{\mathbf{n}}}_{\substack{\text{vector pointing} \\ \text{in direction of } \vec{\mathbf{n}}}}.$$

In an incompressible fluid on the surface of the earth, the pressure is of the form $p(\mathbf{x}) = -\mu g z$, where μ is the density and g the gravitational constant. The divergence theorem says that if ∂X is oriented by the outward-pointing normal, then

$$\underbrace{\begin{bmatrix} \int_{\partial X} \mu g z \Phi_{\vec{\mathbf{e}}_1} \\ \int_{\partial X} \mu g z \Phi_{\vec{\mathbf{e}}_2} \\ \int_{\partial X} \mu g z \Phi_{\vec{\mathbf{e}}_3} \end{bmatrix}}_{\text{total force exerted by liquid}} = \begin{bmatrix} \int_X M_{\vec{\nabla}\cdot(\mu g z \vec{\mathbf{e}}_1)} \\ \int_X M_{\vec{\nabla}\cdot(\mu g z \vec{\mathbf{e}}_2)} \\ \int_X M_{\vec{\nabla}\cdot(\mu g z \vec{\mathbf{e}}_3)} \end{bmatrix}. \qquad 6.11.30$$

A crown made partly of gold weighs less than one made of pure gold of the same volume. The solution Archimedes found was to measure the water displaced by the crown and by an equal weight of gold.

According to Marcus Vitruvius Pollio, writing in the first century BC, Hiero gave a contractor a precise amount of gold to make a crown to be placed in a temple to honor the gods. The crown was "an exquisitely finished piece of handiwork . . . . But afterwards a charge was made that gold had been abstracted and an equivalent weight of silver had been added . . . ." Applying his principle, Archimedes "detected the mixing of silver with the gold, and made the theft of the contractor perfectly clear" (translation from the Latin by Morris Hicky Morgan in *Vitruvius: The Ten Books on Architecture*, Harvard University Press, Cambridge, 1914).

How did Archimedes find his result without the divergence theorem? He may have thought of an object as made up of little cubes, perhaps separated by little sheets of water. Then the force exerted on the object is the sum of the forces exerted on all the little cubes. Archimedes' law is easy to see for one cube of side s, where the vertical component of top of the cube is z, which is a negative number ($z = 0$ is the surface of the water).

The lateral forces obviously cancel, the force on the top is vertical, of magnitude $s^2 g\mu z$, and the force on the bottom is also vertical, of magnitude $-s^2 g\mu(z-s)$, so the total force is $s^3\mu g$, which is precisely the weight of a cube of the fluid of side s.

If a body is made of lots of little cubes separated by sheets of water, all the forces on the interior walls cancel, so it doesn't matter whether the sheets of water are there or not, and the total force on the body is buoyant, of magnitude equal to the weight of the displaced fluid. Note how similar this ad hoc argument is to the proof of Stokes's theorem.

Exercise 6.11.3: Use cylindrical coordinates.

Exercise 6.11.5 is a "shaggy dog" exercise, with lots of irrelevant detail.

The divergences are

$$\vec{\nabla} \cdot (\mu g z \vec{\mathbf{e}}_1) = \vec{\nabla} \cdot (\mu g z \vec{\mathbf{e}}_2) = 0 \quad \text{and} \quad \vec{\nabla} \cdot (\mu g z \vec{\mathbf{e}}_3) = \mu g. \qquad 6.11.31$$

Thus the total force is $\begin{bmatrix} 0 \\ 0 \\ \int_X M_{\mu g} \end{bmatrix}$, and the third component is the weight of the displaced fluid; the force is oriented upwards. This proves the Archimedes principle. $\triangle$

Exercises for Section 6.11

6.11.1 Let S be the torus obtained by rotating around the z-axis the circle of equation $(x-2)^2 + z^2 = 1$. Orient S by the outward-pointing normal. Compute $\int_S \Phi_{\vec{F}}$, where $\vec{F} = \begin{bmatrix} x + \cos(yz) \\ y + e^{x+z} \\ z - x^2 y^2 \end{bmatrix}$.

6.11.2 Suppose $U \subset \mathbb{R}^3$ is open, $\vec{F}$ is a C^1 vector field on U, and $\mathbf{a}$ is a point of U. Let $S_r(\mathbf{a})$ be the sphere of radius r centered at $\mathbf{a}$, oriented by the outward-pointing normal. Compute $\lim_{r\to 0} \frac{1}{r^3} \int_{S_r(\mathbf{a})} \Phi_{\vec{F}}$.

6.11.3 a. Let X be a bounded region in the (x, z)-plane where $x > 0$, and call Z_α the part of $\mathbb{R}^3$ swept out by rotating X around the z-axis by an angle α. Find a formula for the volume of Z_α, in terms of an integral over X.

b. Let X be the circle of radius 1 in the (x, z)-plane, centered at the point $x = 2, z = 0$. What is the volume of the torus obtained by rotating it around the z-axis by a full circle?

c. What is the flux of the vector field $\begin{bmatrix} x \\ y \\ z \end{bmatrix}$ through the part of the boundary of this torus where $y \geq 0$, oriented by the normal pointing out of the torus?

6.11.4 Let $\vec{F}$ be the vector field $\vec{F} = \vec{\nabla}\left(\frac{\sin xyz}{xy}\right)$. What is the work of $\vec{F}$ along the parametrized curve $\gamma(t) = \begin{pmatrix} t\cos \pi t \\ t \\ t \end{pmatrix}$, for $0 \leq t \leq 1$, oriented so that γ preserves orientation?

6.11.5 What is the integral of $W_{\begin{pmatrix} -y/(x^2+y^2) \\ x/(x^2+y^2) \end{pmatrix}}$ around the boundary of the 11-sided regular polygon inscribed in the unit circle, with a vertex at $\begin{pmatrix} 1 \\ 0 \end{pmatrix}$, oriented as the boundary of the polygon?

6.11.6 Let C be the intersection of the cylinder of equation $x^2 + y^2 = 1$ with the surface of equation $z = \sin xy + 2$. Orient C so that the polar angle decreases along C. What is the work of the vector field $\begin{bmatrix} 2y \\ x \\ 3z^2 \end{bmatrix}$ over C?

6.11.7 Let C be a closed curve in the plane. Show that the two vector fields $\begin{bmatrix} y \\ 0 \end{bmatrix}$ and $\begin{bmatrix} 0 \\ x \end{bmatrix}$ do opposite work around C.

6.11.8 Let C be a closed curve in the plane. Show that $\begin{bmatrix} xy^2 \\ 0 \end{bmatrix}$ and $\begin{bmatrix} 0 \\ -x^2y \end{bmatrix}$ do equal work around C.

6.11.9 Suppose $U \subset \mathbb{R}^3$ is open, $\vec{F}$ is a vector field on U, $\mathbf{a}$ is a point of U, and $\vec{\mathbf{v}} \neq \mathbf{0}$ is a vector in $\mathbb{R}^3$. Let U_R be the disc of radius R in the plane of equation $(\mathbf{x} - \mathbf{a}) \cdot \vec{\mathbf{v}} = 0$, centered at $\mathbf{a}$, oriented by the normal vector field $\vec{\mathbf{v}}$, and let ∂U_R be its boundary, with the boundary orientation. Compute $\lim_{R \to 0} \frac{1}{R^2} \int_{\partial U_R} W_{\vec{F}}$.

Exercise 6.11.10: The first step is to find a closed curve of which C is a piece.

6.11.10 Compute the integral $\int_C W_{\vec{F}}$, where $\vec{F}\begin{pmatrix} x \\ y \end{pmatrix} = \begin{bmatrix} xy \\ \frac{\cos y}{y+1} + x \end{bmatrix}$, and C is the upper halfcircle $x^2 + y^2 = 1$, $y \geq 0$, oriented clockwise.

6.11.11 Find the flux of the vector field $\begin{bmatrix} x^2 \\ y^2 \\ z^2 \end{bmatrix}$ through the surface of the unit sphere, oriented by the outward-pointing normal.

Exercise 6.11.12: Think of integrating $x\,dy$ around the triangle.

6.11.12 Compute the area of the triangle with vertices $\begin{bmatrix} a_1 \\ b_1 \end{bmatrix}, \begin{bmatrix} a_2 \\ b_2 \end{bmatrix}, \begin{bmatrix} a_3 \\ b_3 \end{bmatrix}$.

6.11.13 What is the work of the vector field $\begin{bmatrix} -3y \\ 3x \\ 1 \end{bmatrix}$ around the circle given by $x^2 + y^2 = 1$, $z = 3$, oriented by the tangent vector $\begin{bmatrix} 0 \\ -1 \\ 0 \end{bmatrix}$ at $\begin{pmatrix} 1 \\ 0 \\ 3 \end{pmatrix}$?

$$\vec{F}\begin{pmatrix} x \\ y \\ z \end{pmatrix} = \begin{bmatrix} x + yz \\ y + xz \\ z + xy \end{bmatrix}$$

Vector field for Exercise 6.11.14

6.11.14 What is the flux of the vector field $\vec{F}$ in the margin through the boundary of the region in the first octant $x, y, z \geq 0$, where $z \leq 4$ and $x^2 + y^2 \leq 4$, oriented by the outward-pointing normal?

Part a of Exercise 6.11.15 is an easy special case of Exercise 6.3.15.

6.11.15 a. Let $L \subset \mathbb{C}^2$ be a one-dimensional complex subspace. Show that L, viewed as a real plane in $\mathbb{R}^4$, has a unique orientation specified as follows: if $\vec{\mathbf{v}} \in L$ is a nonzero vector, then $\vec{\mathbf{v}}, i\vec{\mathbf{v}}$ is a direct basis of L.

*b. Let $z_1 = x_1 + iy_1$, $z_2 = x_2 + iy_2$ be coordinates in $\mathbb{C}^2 = \mathbb{R}^4$. Choose two relatively prime positive integers p and q. Integrate the 2-form $dx_1 \wedge dy_1 + dx_2 \wedge dy_2$ over the part of the surface $X_{p,q}$ of equation

$$z_1^p + z_2^q = 0 \quad \text{where} \quad |z_1| \leq R_1, |z_2| \leq R_2,$$

oriented by the requirement that if $\vec{\mathbf{v}} \neq \mathbf{0}$ is an element of the tangent space $T_{\mathbf{x}}X$, then $\vec{\mathbf{v}}, i\vec{\mathbf{v}}$ is a direct basis of $T_{\mathbf{x}}X$ viewed as a real plane.

6.12 Electromagnetism

In the introduction to this chapter, we said that forms are the natural way to approach two subjects of immense importance in understanding the physical world: electromagnetism and the theory of relativity. Indeed, this is one reason we wrote this chapter on differential forms.

The theory of relativity is beyond the scope of this book, but in this section we hope to convince you that forms are the natural language for electromagnetism. We also find the history of electromagnetism a wonderful illustration of the interaction between physics and mathematics: curl, grad, and div were largely devised by James Clerk Maxwell in order to understand the relationship between electricity and magnetism explored by Michael Faraday in his experiments.

History is complicated, and the history of electromagnetism in the nineteenth century (actually 1785–1905) is no exception. The scientists involved showed amazing insight, but they also took wrong turns and explored many dead ends. Even assuming such a thing is possible, we do not have the ability or knowledge to give a genuine history, which would in any case take up far more space than is available here. But we do want to explore the meaning of Maxwell's equations.

We also want to discuss one of the grand surprises of history, where a quantity measured in the laboratory, using batteries, magnets, pith balls, and the like turned out to be equal to the speed of light. This realization led to the unification of electromagnetism and optics, which existed then (Newton wrote an entire book on optics) but as unrelated fields. To say that this development was important is the rankest of understatements. In particular, nothing involving computers, radio, television, or telephones can be imagined without Maxwell's unification of electromagnetism and optics.

In the cgs system, charge is measured in *statcoulombs*: two objects a centimeter apart, each with a charge of one statcoulomb, repel each other with a force of one *dyne*, the force needed to accelerate a mass of one gram by one centimeter per second squared. In cgs, current is measured in statcoulombs per second.

The cgs system is better suited to insects than to people: a beetle might weigh a tenth of a gram, and a dyne is the amount of force one might associate with an ant. The MKS (meter, kilogram, second) system, now more commonly called SI for *système international d'unités*, is more adapted to people: people are reasonably measured in kilograms, and 100 amp is a standard amount of current for running a household (the ampere is the unit of current in MKS).

The units of length in the two systems differ by factor of a hundred, and the units of weight by a factor of a thousand, but the units of charge differ by approximately 10^{10}: an amp is a coulomb per second, and a coulomb is approximately 3.34×10^{10} statcoulomb. The statcoulomb is an appropriate measure for electrostatics, the ampere for current. Two objects a meter apart, each with a charge of one coulomb, would generate a force of about 10 billion newtons – enough to accelerate, in one second, a mass of 10 billion kilograms from a standstill to a speed of one meter per second (roughly, a slow walk).

Maxwell's equations and the exterior derivative

We saw in Example 6.5.13 that in the cgs (centimeter, gram, second) system, and described in the standard "vector calculus" language, the electromagnetic field consists of two time-dependent vector fields[17], the electric field $\vec{\mathbf{E}}$ and the magnetic field $\vec{\mathbf{B}}$, which exert on a particle of charge q at $\mathbf{x}$ and moving with velocity $\vec{\mathbf{v}}$ the force

$$\vec{F} = q\left(\vec{\mathbf{E}}(\mathbf{x},t) + \left(\frac{\vec{\mathbf{v}}}{c} \times \vec{\mathbf{B}}(\mathbf{x},t)\right)\right), \qquad 6.12.1$$

where c is the speed of light.

This force $\vec{F}$ is called the *Lorentz force*, and equation 6.12.1 is called the *Lorentz force law*. Although it may not be obvious, all everyday events, like leaning on a table, making a telephone call, or turning on the the stove, are essentially completely explained by the Lorentz force, which is observed at every instant. It is hard to think of anything that is not an illustration of the Lorentz force law.

[17]By time-dependent we mean that each entry of the vector field depends on the three spatial variables and time, but when we take curl and div of these vector fields, the partial derivatives involved are only with respect to the spatial variables. Of course the partial derivatives with respect to time can't be ignored; they show up explicitly in Maxwell's equations.

FIGURE 6.12.1.
The Scottish physicist James Clerk Maxwell (1831–1879). In 1931, Albert Einstein described Maxwell's influence on physicists' perception of reality as "the most profound and the most fruitful that physics has experienced since the time of Newton."

The electromagnetic field is subject to the following four laws, collectively known as *Maxwell's equations*:

$$-\frac{1}{c}D_t\vec{\mathbf{B}} = \operatorname{curl}\vec{\mathbf{E}} \qquad \text{(Faraday's law)} \qquad 6.12.2$$

$$\operatorname{div}\vec{\mathbf{B}} = 0 \qquad \text{(no magnetic charge)} \qquad 6.12.3$$

$$\frac{1}{c}D_t\vec{\mathbf{E}} = \operatorname{curl}\vec{\mathbf{B}} - \frac{4\pi}{c}\vec{\mathbf{j}} \qquad \text{(Ampère's law, corrected)} \qquad 6.12.4$$

$$\operatorname{div}\vec{\mathbf{E}} = 4\pi\rho \qquad \text{(Gauss's law,)} \qquad 6.12.5$$

where ρ is the charge density and $\vec{\mathbf{j}}$ is the current density.

In traditional notation, equations 6.12.2 and 6.12.4 are written

$$-\frac{1}{c}\frac{\partial\vec{\mathbf{B}}}{\partial t} = \operatorname{curl}\vec{\mathbf{E}} \quad \text{and} \quad \frac{1}{c}\frac{\partial\vec{\mathbf{E}}}{\partial t} = \operatorname{curl}\vec{\mathbf{B}} - \frac{4\pi}{c}\vec{\mathbf{j}} \qquad 6.12.6$$

Above, these equations are written in the cgs system (see the margin note page 670). In the cgs system, the current density $\vec{\mathbf{j}}$ is measured in statcoulombs per square centimeter per second, and the charge density ρ is measured in statcoulombs per centimeter cubed:

$$\underbrace{\frac{\text{statC}}{(\text{cm})^2 s}}_{\text{measure of current density}} \qquad \underbrace{\frac{\text{statC}}{(\text{cm})^3}.}_{\text{measure of charge density}} \qquad 6.12.7$$

We will now reinterpret these equations in terms of differential forms; we will see that exterior derivatives are a very natural way to write them.

The vector fields $\vec{\mathbf{E}}$ and $\vec{\mathbf{B}}$ are functions of x, y, z, and t, but the operators div and curl take derivatives only with respect to the spatial variables. Maxwell's equations as written in equations 6.12.2–6.12.5 treat space and time in different ways, and make it difficult to work in spacetime. For instance, it is difficult to pass from one frame of reference to another with respect to which it is moving. By contrast, the formulation 6.12.15 of Maxwell's laws in terms of differential forms treats space and time on an equal footing.

Define the two 2-forms

$$\begin{aligned} \mathbb{F} &= W_{\vec{\mathbf{E}}} \wedge c\,dt + \Phi_{\vec{\mathbf{B}}} \\ \mathbb{M} &= W_{\vec{\mathbf{B}}} \wedge c\,dt - \Phi_{\vec{\mathbf{E}}}, \end{aligned} \qquad 6.12.8$$

known as the *Faraday 2-form* and the *Maxwell 2-form*.

(In equations 6.12.8, the work and the flux are with respect to the spatial variables only.) As usual, we denote by Φ the flux form: $\Phi_{\vec{\mathbf{B}}}$ is the flux form of $\vec{\mathbf{B}}$. We already saw the Faraday 2-form in formula 6.5.29. As Exercise 6.12.1 asks you to confirm,

$$\mathbf{d}\mathbb{F} = \Phi_{\vec{\nabla}\times\vec{\mathbf{E}}} \wedge c\,dt + M_{\vec{\nabla}\cdot\vec{\mathbf{B}}} + \Phi_{\frac{1}{c}\frac{\partial\vec{\mathbf{B}}}{\partial t}} \wedge c\,dt. \qquad 6.12.9$$

It follows that the first two of Maxwell's equations,

$$\operatorname{curl}\vec{\mathbf{E}} + \frac{1}{c}D_t\vec{\mathbf{B}} = \begin{bmatrix} 0 \\ 0 \\ 0 \end{bmatrix} = \vec{\mathbf{0}} \quad \text{and} \quad \operatorname{div}\vec{\mathbf{B}} = 0, \qquad 6.12.10$$

are equivalent to $\mathbf{d}\mathbb{F} = 0$:

$$\mathbf{d}\mathbb{F} = \underbrace{\Phi_{\vec{\nabla}\times\vec{\mathbf{E}}} \wedge c\,dt + \Phi_{\frac{1}{c}D_t\vec{\mathbf{B}}} \wedge c\,dt}_{\Phi_{\vec{0}}=0} + \underbrace{M_{\operatorname{div}\vec{\mathbf{B}}}}_{0} = 0. \qquad 6.12.11$$

Let us define the "charge-current" 3-form

$$\mathbb{J} = \frac{1}{c}\Phi_{\vec{\mathbf{j}}} \wedge c\,dt - M_\rho\,; \tag{6.12.12}$$

we will see that the equation $\mathbf{d}\mathbb{M} = 4\pi\mathbb{J}$ is equivalent to the other two of Maxwell's equations. Compute

$$\begin{aligned}\mathbf{d}\mathbb{M} = \mathbf{d}\big(W_{\vec{\mathbf{B}}} \wedge c\,dt - \Phi_{\vec{\mathbf{E}}}\big) &= \Phi_{\vec{\nabla}\times\vec{\mathbf{B}}} \wedge c\,dt - M_{\vec{\nabla}\cdot\vec{\mathbf{E}}} - \Phi_{\frac{1}{c}D_t\vec{\mathbf{E}}} \wedge c\,dt\\ &= 4\pi\big(\Phi_{\frac{\vec{\mathbf{j}}}{c}} \wedge c\,dt - M_\rho\big).\end{aligned} \tag{6.12.13}$$

If we break up this equation into components involving only "space-like" terms and components involving two space-like terms and dt, we get

$$\operatorname{div}\vec{\mathbf{E}} = 4\pi\rho \quad \text{and} \quad \frac{1}{c}D_t\vec{\mathbf{E}} = \operatorname{curl}\vec{\mathbf{B}} - 4\pi\frac{\vec{\mathbf{j}}}{c}, \tag{6.12.14}$$

the other two of Maxwell's equations.

Thus we can summarize Maxwell's equations, elegantly and concisely, as

$$\mathbf{d}\mathbb{F} = 0, \quad \text{and} \quad \mathbf{d}\mathbb{M} = 4\pi\mathbb{J}. \tag{6.12.15}$$

Equations 6.12.15: One consequence of $\mathbf{d}\mathbb{M} = 4\pi\mathbb{J}$ is that

$$\mathbf{d}\mathbb{J} = \frac{1}{4\pi}\mathbf{d}\mathbf{d}\mathbb{M} = 0.$$

Using equation 6.12.12, we can express this in terms of $\vec{\mathbf{j}}$ and ρ: computing

$$\begin{aligned}0 = \mathbf{d}\mathbb{J} = \mathbf{d}\left(\frac{1}{c}\Phi_{\vec{\mathbf{j}}} \wedge c\,dt - M_\rho\right) &= \frac{1}{c}M_{\operatorname{div}\vec{\mathbf{j}}} \wedge c\,dt\\ &\quad - d\rho\,dx \wedge dy \wedge dz\\ &= \frac{1}{c}(\operatorname{div}\vec{\mathbf{j}})\,dx \wedge dy \wedge dz \wedge c\,dt\\ &\quad - D_t\rho\,dt \wedge dx \wedge dy \wedge dz\\ &= \operatorname{div}\vec{\mathbf{j}}\,dx \wedge dy \wedge dz \wedge dt\\ &\quad + D_t\rho\,dx \wedge dy \wedge dz \wedge dt\end{aligned}$$

gives

$$D_t\rho + \operatorname{div}\vec{\mathbf{j}} = 0.$$

This is called the *equation of continuity*, and intuitively corresponds to conservation of charge.

What does this tell us? It should be clear from equation 6.12.1 describing the Lorentz force that our intuition about space and time is fundamentally flawed: force should not depend on velocity, unless there is a background reference frame with respect to which velocity is being measured. This happens for instance in hydrodynamics: motion in a fluid depends on the velocity with respect to the ambient fluid. Early attempts to understand Maxwell's equations were based on just this idea: they postulated an ambient "fluid", the *ether*, with respect to which velocity is measured. But all attempts to find the ether failed, most notably in the Michelson-Morley experiment, which showed that the speed of light is the same in all directions, and hence that the earth has no motion with respect to the ether.

The only successful resolution of these difficulties is the special theory of relativity, which says that four-dimensional "spacetime" is the stage for physical reality, but that splitting spacetime into space and time is always unnatural. Within this context, the 2-forms $\mathbb{F}$ and $\mathbb{M}$ are natural, but the vector fields $\vec{\mathbf{E}}$ and $\vec{\mathbf{B}}$ are not; they depend on such an unnatural splitting.

The crucial issue is the relation between $\mathbb{F}$ and $\mathbb{M}$. This is a fairly elaborate issue.

Coulomb's experiment

The first quantitative experiment in electricity was apparently performed by Charles Augustin Coulomb in 1785.[18] He had invented the torsion balance, which allowed him to measure accurately very small forces. Using his balance, he justified that electric charges of the same sign repel with a

[18]A translation of his report to the Académie des Sciences can be found at http://ppp.unipv.it/Coulomb/Pages/eF3opAI.html.

force proportional to the product of the charges and inversely proportional to the square of the distance separating them:

Coulomb's law is like Newton's law of gravitation: the gravitational attraction between two point masses is proportional to the product of the masses and inversely proportional to the square of the distance separating them.

$$F = k_1 \frac{q_1 q_2}{r^2} \qquad \text{(Coulomb's law)}, \tag{6.12.16}$$

where q_1 and q_2 are the charges, r is the distance separating them, and k_1 is the ratio of proportionality. Until you have chosen a unit of charge, it doesn't make sense to ask what the value of k_1 is, but in whatever units you might use, it will have dimensions

$$\frac{\text{force} \times \text{length}^2}{\text{charge}^2}. \tag{6.12.17}$$

Ampère's experiment

The practical as well as theoretical importance of Oersted's discovery of the connection between electricity and magnetism cannot be exaggerated. For instance, nothing involving electric motors can be imagined without it. Work on electromagnetism was continued notably by Michael Faraday.

Skip ahead to 1820. The Danish physicist Hans Christian Oersted has just announced that an electric current can deflect a magnetized needle like a compass. Within a week, André Marie Ampère shows that two parallel wires carrying electric currents attract (or repel if the currents are in opposite directions), as shown in Figure 6.12.2. The wires attract or repel each other by a force per unit length that is proportional to the product of the currents and inversely proportional to the distance separating the wires:

$$\frac{F}{L} = k_2 \frac{2 I_1 I_2}{d}, \tag{6.12.18}$$

where F/L is force per length, I_1 and I_2 are the currents in the two wires, and d is the distance between them. Until you have chosen units, it doesn't make sense to speak of the numerical value of k_2, but in whatever units you use, it will have dimensions

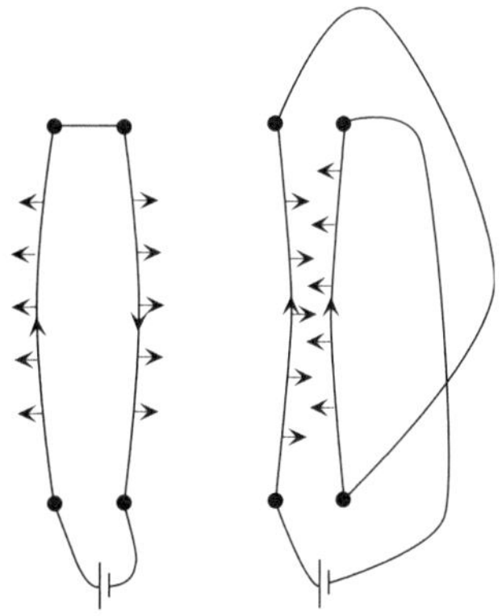

FIGURE 6.12.2.

LEFT: Ampère found that two parallel wires carrying opposite electric currents repel. RIGHT: If the current goes the same direction in both wires, the wires attract.

$$\frac{\text{force} \times \text{length} \times \text{time}^2}{\text{charge}^2 \times \text{length}}. \tag{6.12.19}$$

Notice that the ratio k_1/k_2 has dimensions

$$\frac{\text{force} \times \text{length}^2}{\text{charge}^2} \frac{\text{charge}^2 \times \text{length}}{\text{force} \times \text{length} \times \text{time}^2} = \frac{\text{length}^2}{\text{time}^2}; \tag{6.12.20}$$

independent of any choice of a unit of charge, the ratio will be a velocity squared.

Let us now run the numbers, making our measurements in SI units. The unit of charge in that system is the coulomb; it is chosen so that $k_2 = 10^{-7}$. If you measure the repulsion of two charges of 1 coulomb one meter apart, you find a force of $\sim 8.99 \times 10^9$ newtons. This leads to $k_1 \sim 8.99 \times 10^9$.

The great surprise hidden in these values for k_1 and k_2 is that if you compute the ratio k_1/k_2, you land on a figure very close to the square of the speed of light:

$$\frac{k_1}{k_2} \approx 8.99 \times 10^{16} \frac{m^2}{s^2} \approx c^2. \tag{6.12.21}$$

This appearance of the speed of light in experiments involving electrical charges and currents wasn't apparently noticed until 1862. By then, Faraday and Maxwell had developed the theory of electric and magnetic fields, and one consequence of Maxwell's equations was that these fields should propagate as waves with velocity $\sqrt{k_1/k_2}$. The significance of this quantity was then clear. As Maxwell wrote:

> *We can scarcely avoid the inference that light consists in the traverse undulations of the same medium which is the cause of electric and magnetic phenomena.*

This quotation is from "On the Physical Lines of Force," Part III, *Philosophical Magazine and Journal of Science*, Fourth Series, Jan. 1862.

Faraday's experiment

There was a flurry of work after Oersted's and Ampère's experiments. The really crucial experiment was performed by Faraday in 1831: *Faraday's law*, which he deduced from this experiment, explains electric motors and generators, and lies behind all the changes in human civilization arising from the use of electricity. In some sense you perform this experiment many times every day, when you ring a doorbell or start a car, for example. The engineering that goes into these devices cloaks the underlying laws, so instead we will describe something that is clearly "physics", and which is quite close to Faraday's initial setup. It is also quite easy to perform with measurable results if you have access to a fairly hefty magnet and coil.

The experiment is described in Figure 6.12.3. Suppose that you place a bar magnet upright on a table. Connect the coil to a galvanometer (the simplest multi-tester will do). First move the coil fairly fast up and down around the magnet: you will see the indicator of the galvanometer move back and forth. Next, do almost the same thing: put the coil on the table and jiggle the magnet in and out of the coil. You will see the same thing, of course: the indicator of the galvanometer will move back and forth.

Why "of course"? The two experiments look almost identical; if you allow for relativity, they are identical. But one experiment is fairly easy to understand from the Lorentz force, the other not.

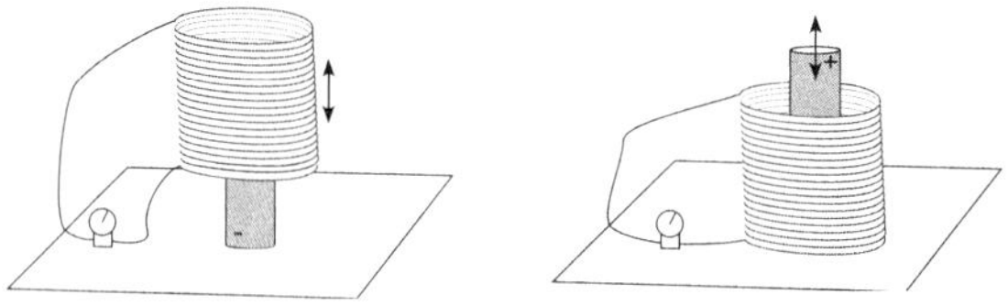

FIGURE 6.12.3. You can put the magnet on the table and jiggle the coil up and down around it (left), or you can put the coil on the table and jiggle the magnet in and out of the coil (right). The result is the same: an alternating current in the coil.

Due to Oersted's experiment, Faraday knew how to build quite a sensitive galvanometer. With this setup it isn't too hard to produce enough current to light a flashlight bulb.

When you move the coil, the electrons in the wire are being moved up and down, with a velocity something like $(\cos \omega t)\vec{\mathbf{e}}_3$. The magnetic field is (mainly) radial, something like

$$\vec{\mathbf{B}}\begin{pmatrix} x \\ y \\ z \end{pmatrix} = \begin{bmatrix} a \\ a \\ 0 \end{bmatrix}, \tag{6.12.22}$$

where a depends on the strength of the magnet and the relative radii of the magnet and of the coil. Thus by the Lorentz force law (equation 6.12.1) the electrons should "feel" a force

$$\begin{bmatrix} 0 \\ 0 \\ \cos \omega t \end{bmatrix} \times \begin{bmatrix} a \\ a \\ 0 \end{bmatrix} = \begin{bmatrix} -a \cos \omega t \\ a \cos \omega t \\ 0 \end{bmatrix}, \tag{6.12.23}$$

i.e., a force in the direction of the wire. So the electrons move along the wire, creating the current that the galvanometer measures.

Now consider moving the magnet. Where is the moving charge? There aren't any obvious such charges, and by the Lorentz force law, an electric field must be pushing the charges; this electric field must be created by the changing magnetic field, since that is the only thing that is changing. Thus if the magnet is pulled away from the wire, the charges must be pushed by an electric field created by the changing magnetic field.

After a number of experiments, Faraday concluded that the current in the wire is proportional to the rate of change of the magnetic field through the loop formed by the wire. This is a variant of Faraday's law.

In the remainder of this section, we have three objectives:

- to interpret Coulomb's, Ampère's, and Faraday's experiments in terms of Maxwell's laws, conveying some of the physical meaning of those laws
- to show how Maxwell was led to his conclusion about light on the basis of a correction to the work of Ampère and Faraday
- to explain the difficulties involved in the appearance of velocity in the Lorentz force, and how this led to Einstein's special theory of relativity

Coulomb's experiment and the electric field

Now we will see that Coulomb's result leads to Gauss's law $\operatorname{div} \vec{\mathbf{E}} = 4\pi\rho$, one of Maxwell's equations. We will use SI units, in which Gauss's law becomes $\rho(\mathbf{x}) = \epsilon_0 \operatorname{div} \vec{\mathbf{E}}(\mathbf{x})$. One interpretation of Coulomb's result is that a (stationary) charge q placed at $\mathbf{y} \in \mathbb{R}^3$ creates electric and magnetic fields

$$\vec{\mathbf{E}}(\mathbf{x}) = \frac{q}{4\pi\epsilon_0} \frac{\overrightarrow{\mathbf{x}-\mathbf{y}}}{|\mathbf{x}-\mathbf{y}|^3}, \qquad \vec{\mathbf{B}}(\mathbf{x}) = \vec{\mathbf{0}}, \tag{6.12.24}$$

where ϵ_0 (the *permettivity of free space*) is a proportionality constant and doesn't have a value until units of charge and electric field are chosen. Assume that a *charge density* ρ is C^1 with compact support. This charge density ρ creates an electric field

$$\vec{\mathbf{E}}(\mathbf{x}) = \frac{1}{4\pi\epsilon_0} \int_{\mathbb{R}^3} \rho(\mathbf{y}) \frac{\overrightarrow{\mathbf{x}-\mathbf{y}}}{|\mathbf{x}-\mathbf{y}|^3} |d^3\mathbf{y}|. \tag{6.12.25}$$

Equation 6.12.27: Equality 1 is true because the flow of $\vec{F}_3$ through $\partial B_r(\mathbf{x})$ is unchanged as r decreases; it is always 4π. See Figure 6.7.1. The integral to the right of equality 1 is the average of ρ over the sphere of radius r around $\mathbf{x}$. In equality 2 we make the change of variables $\mathbf{y} = \mathbf{x}+\mathbf{u}$; note that $\vec{F}_3(-\mathbf{u}) = -\vec{F}_3(\mathbf{u})$; this accounts for the change of sign.

Equality 3 uses Stokes; we integrate only over $\partial B_r(\mathbf{0})$ because there is no charge on the "outer" boundary component of $A_{R,r}$; see Figure 6.12.4. The sign changes in line 2 because the boundary of $B_r(\mathbf{0})$ is oriented by the outward-pointing normal, which points in the opposite direction of the outward-pointing normal defining the orientation of the inner component of $\partial A_{R,r}$; again see Figure 6.12.4.

Equality 4: Since ρ has support in $B_R(\mathbf{x})$, we can go from the compact domain of integration $A_{R,r}$ in line 2 to $\mathbb{R}^3 - B_r(\mathbf{0})$ in line 3. We also go from the divergence with respect to $\mathbf{u}$ to the divergence with respect to $\mathbf{x}$, which we can do because when we apply Leibniz's rule to compute $\operatorname{div}_\mathbf{u}$, the divergence of $\vec{F}_3$ is 0. Switching from $\operatorname{div}_\mathbf{u}$ to $\operatorname{div}_\mathbf{x}$ is important because $\vec{F}_3(\mathbf{u}) = \frac{\mathbf{u}}{|\mathbf{u}|^3}$ blows up near the origin, so we can't differentiate with respect to $\mathbf{u}$ there.

In equality 5 we take the limit as $r \to 0$ and differentiate under the integral sign. In equality 6 we change variables back and use equation 6.12.25.

Exercise 6.12.11 asks you to use the 1-dimensional analogue of equation 6.12.27 to understand the signs and the passage from $\operatorname{div}_\mathbf{u}$ to $\operatorname{div}_\mathbf{x}$.

Note that the electric field given by equation 6.12.24 obeys an inverse square law: the $\overrightarrow{\mathbf{x}-\mathbf{y}}$ in the numerator cancels one of the powers in the denominator, and in addition gives the direction of the field. It can be expressed in terms of the vector field $\vec{F}_3$ of Example 6.7.7:

$$\frac{\overrightarrow{\mathbf{x}-\mathbf{y}}}{|\mathbf{x}-\mathbf{y}|^3} = \vec{F}_3(\mathbf{x}-\mathbf{y}). \tag{6.12.26}$$

Recall that the divergence of $\vec{F}_3$ (i.e., the exterior derivative of the flux) is 0 away from $\mathbf{0}$.

Set $A_{R,r} = \{\, \mathbf{u} \mid r \le |\mathbf{u}| \le R \,\}$, where R is chosen so large that all the charge is contained in the ball $B_R(\mathbf{x})$ around every point $\mathbf{x}$ in the support of ρ (the set $A_{R,r}$ is used in the integral in equation 6.12.27, line 2). Compute the following, where in line 2, $\operatorname{div}_\mathbf{u}$ is divergence with respect to $\mathbf{u}$ and in line 3, $\operatorname{div}_\mathbf{x}$ is divergence with respect to $\mathbf{x}$:

$$\begin{aligned}\frac{\rho(\mathbf{x})}{\epsilon_0} &\overset{1}{=} -\lim_{r\to 0}\frac{1}{4\pi\epsilon_0}\int_{\partial B_r(\mathbf{x})}\rho(\mathbf{y})\Phi_{\vec{F}_3(\mathbf{x}-\mathbf{y})} \overset{2}{=} +\lim_{r\to 0}\frac{1}{4\pi\epsilon_0}\int_{\partial B_r(\mathbf{0})}\rho(\mathbf{x}+\mathbf{u})\Phi_{\vec{F}_3(\mathbf{u})}\\ &\overset{3}{=} -\lim_{r\to 0}\frac{1}{4\pi\epsilon_0}\int_{A_{R,r}}\operatorname{div}_\mathbf{u}\left(\rho(\mathbf{x}+\vec{\mathbf{u}})\vec{F}_3(\mathbf{u})\right)|d^3\mathbf{u}|\\ &\overset{4}{=} -\lim_{r\to 0}\frac{1}{4\pi\epsilon_0}\int_{\mathbb{R}^3-B_r(\mathbf{0})}\operatorname{div}_\mathbf{x}\left(\rho(\mathbf{x}+\vec{\mathbf{u}})\vec{F}_3(\mathbf{u})\right)|d^3\mathbf{u}|\\ &\overset{5}{=} -\operatorname{div}_\mathbf{x}\left(\frac{1}{4\pi\epsilon_0}\int_{\mathbb{R}^3}\rho(\mathbf{x}+\vec{\mathbf{u}})\vec{F}_3(\mathbf{u})\right)|d^3\mathbf{u}| \overset{6}{=} \operatorname{div}\vec{\mathbf{E}}(\mathbf{x})\end{aligned} \tag{6.12.27}$$

This is the SI version of Gauss's law.

Remark 6.12.1. This rather delicate computation "justifies" the assertion

$$\operatorname{div}\vec{F}_n(\mathbf{x}) = \operatorname{vol}_{n-1} S^{n-1}\delta(\mathbf{x}),$$

where $\delta(\mathbf{x})$ is the unit mass at the origin, also called the *Dirac delta function*. Indeed, if we could differentiate (if we could "divergence")

$$\vec{\mathbf{E}}(\mathbf{x}) = \frac{1}{4\pi\epsilon_0}\int_{\mathbb{R}^3}\rho(\mathbf{y})\vec{F}_3(\mathbf{x}-\mathbf{y})|d^3\mathbf{y}| \tag{6.12.28}$$

under the integral sign, we would get (remember $\operatorname{vol}_2(S^2) = 4\pi$)

$$\operatorname{div}\vec{\mathbf{E}}(\mathbf{x}) = \frac{1}{4\pi\epsilon_0}\int_{\mathbb{R}^3}\rho(\mathbf{y})(4\pi)\delta(\mathbf{x}-\mathbf{y}) = \frac{1}{\epsilon_0}\rho(\mathbf{x}). \tag{6.12.29}$$

precisely what the computation 6.12.27 proves. The theory of distributions allows us to define *distributional derivatives*, and asserts that $\vec{F}_n$ does have a *distributional divergence* that is a point mass. In fact, the computation above is the proof of this fact. △

Let us check that this description of the electric field really corresponds to Coulomb's law

$$F = k_1\frac{q_1q_2}{r^2}, \tag{6.12.30}$$

which we first saw as equation 6.12.16. In Coulomb's experiment there are two charges q_1 and q_2 a distance d apart; let us assume that the first is at the origin and the other at $d\vec{\mathbf{e}}_1$. Then, by equation 6.12.24, the first creates an electric field which at the second is

$$\frac{q_1}{4\pi\epsilon_0}\frac{1}{d^2}\vec{\mathbf{e}}_1 \qquad 6.12.31$$

and by the Lorentz force law (normalized to have no constants), the force on the second charge is

$$\vec{F} = q_2\big(\vec{\mathbf{E}} + (\vec{\mathbf{v}} \times \vec{\mathbf{B}})\big) = q_2(\vec{\mathbf{E}}) = \frac{q_1 q_2}{4\pi\epsilon_0}\frac{1}{d^2}\vec{\mathbf{e}}_1. \qquad 6.12.32$$

In particular, $1/(4\pi\epsilon_0)$ is what we called k_1 earlier (see equation 6.12.21).

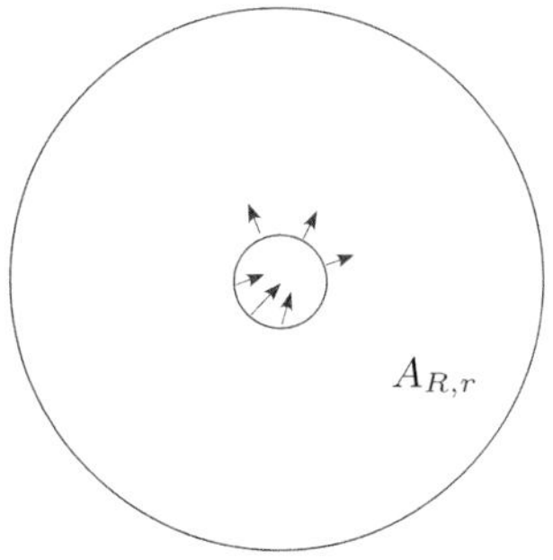

FIGURE 6.12.4.

This figure illustrates equation 6.12.27. The shaded region is $A_{R,r}$; the small white disc is $B_r(\mathbf{0})$. When integrating over $\partial A_{R,r}$ we ignore the outer boundary component, on which there is no charge. In line 1 of equation 6.12.27, $\partial B_r(\mathbf{0})$ has the boundary orientation of $B_r(\mathbf{0})$ (arrows pointing out of $B_r(\mathbf{0})$ into $A_{R,r}$); when we consider $\partial B_r(\mathbf{0})$ as one component of the boundary of $A_{R,r}$, it is oriented by arrows that point out of $A_{R,r}$. This explains the change of sign from line 1 to line 2.

The Biot-Savart law and Maxwell's laws

Ampère's experiment in 1820 showed that a current in a wire creates a magnetic field around the wire. Jean-Baptiste Biot and Felix Savart, as well as Ampère himself, went about studying this magnetic field quantitatively. They discovered that a steady current j in a wire Γ of any shape gives rise to a magnetic field

$$\vec{\mathbf{B}}(\mathbf{x}) = \frac{\mu_0}{4\pi}\int_\Gamma \vec{\mathbf{j}}(\mathbf{y}) \times \frac{\overrightarrow{(\mathbf{x}-\mathbf{y})}}{|\mathbf{x}-\mathbf{y}|^3}|d^1\mathbf{y}|, \qquad \text{(Biot-Savart law)} \quad 6.12.33$$

illustrated by Figure 6.12.5.

The Biot-Savart law is a (rather elaborate) experimental observation. It is not a fundamental law of physics, and in fact is only true for steady currents, although the law $\operatorname{div}\vec{\mathbf{B}} = 0$ derived from it is universally true (it doesn't involve any time derivatives). A correction term is needed for varying currents (in fact, Maxwell's correction term $\epsilon_0\mu_0 D_t\vec{\mathbf{E}}$; see equation 6.12.62).

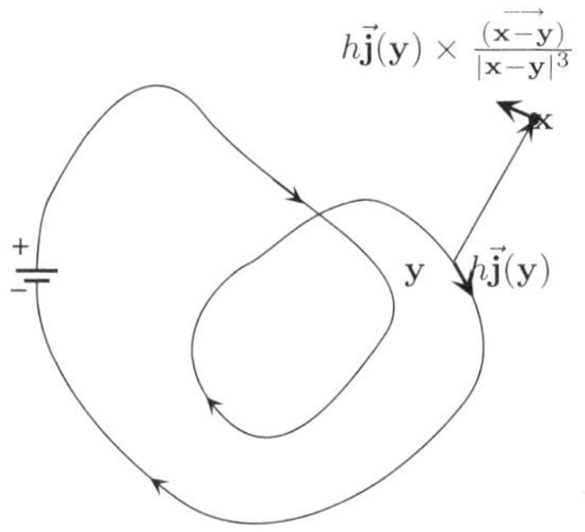

FIGURE 6.12.5. The cross product

$$h\vec{\mathbf{j}}(\mathbf{y}) \times \frac{\overrightarrow{\mathbf{x}-\mathbf{y}}}{|\mathbf{x}-\mathbf{y}|^3}$$

is perpendicular to the plane spanned by $\overrightarrow{\mathbf{x}-\mathbf{y}}$ and the small segment of wire of length h anchored at $\mathbf{y}$. The Biot-Savart law says the magnetic field is approximated by the sum of such contributions as the wire is approximated by shorter and shorter pieces. Thus the Biot-Savart law says that integrating an appropriate multiple of this cross product over the wire gives the magnetic field.

In this expression, μ_0 is a constant of proportionality (known as the *permeability of free space*, or the *magnetic constant*), and the vector $\vec{\mathbf{j}}(\mathbf{y})$

represents the current: it has magnitude j and is tangent to the wire at $\mathbf{y} \in \Gamma$, pointing in the direction of the current.

Let us assume that $\mathbf{x}$ is outside the wire; then two of Maxwell's equations are obtained by computing the divergence and the curl of the magnetic field $\vec{\mathbf{B}}$ as given by the Biot-Savart law. We will justify this statement for $\operatorname{div} \vec{\mathbf{B}} = 0$. Recall (Proposition 6.5.12) that if $\vec{F}$, $\vec{G}$ are vector fields, then $\Phi_{\vec{F}\times\vec{G}} = W_{\vec{F}} \wedge W_{\vec{G}}$. This plus Theorem 6.7.9 gives

$$\mathbf{d}\Phi_{\vec{F}\times\vec{G}} = \mathbf{d}W_{\vec{F}} \wedge W_{\vec{G}} - W_{\vec{F}} \wedge \mathbf{d}W_{\vec{G}}, \qquad 6.12.34$$

which can be rewritten as

$$M_{\operatorname{div}(\vec{F}\times\vec{G})} \overbrace{=}^{\text{Thm. 6.8.3}} \Phi_{\operatorname{curl}\vec{F}} \wedge W_{\vec{G}} - W_{\vec{F}} \wedge \Phi_{\operatorname{curl}\vec{G}}. \qquad 6.12.35$$

Now set $\vec{F} = \vec{\mathbf{j}}(\mathbf{y})$ and $\vec{G} = \frac{1}{|\mathbf{x}-\mathbf{y}|^3}(\overrightarrow{\mathbf{x}-\mathbf{y}})$. Since the expression for $\vec{\mathbf{B}}$ in equation 6.12.33 is an integral depending on the parameter $\mathbf{x}$, we will consider the curl with respect only to $\mathbf{x}$, i.e., we set $\vec{\nabla}_{\mathbf{x}} = \begin{bmatrix} D_{x_1} \\ D_{x_2} \\ D_{x_3} \end{bmatrix}$, where the partial derivatives are with respect to $\mathbf{x}$. Then

$$\operatorname{curl}\vec{F} = \operatorname{curl}\vec{\mathbf{j}}(\mathbf{y}) = \vec{\nabla}_{\mathbf{x}} \times \vec{\mathbf{j}}(\mathbf{y}) = \vec{\mathbf{0}}. \qquad 6.12.36$$

We also have $\operatorname{curl}\vec{G} = \vec{\mathbf{0}}$, since $\operatorname{curl}\operatorname{grad} = \vec{\mathbf{0}}$ (see Remark 6.8.4), and since (as you are asked to show in Exercise 6.12.2)

$$\frac{\overrightarrow{\mathbf{x}-\mathbf{y}}}{|\mathbf{x}-\mathbf{y}|^3} = -\operatorname{grad}\frac{1}{|\mathbf{x}-\mathbf{y}|}, \qquad 6.12.37$$

where the partial derivatives in the gradient are computed only with respect to $\mathbf{x}$. Thus (by equation 6.12.35) $M_{\operatorname{div}\vec{F}\times\vec{G}} = 0$, so

$$\operatorname{div}\vec{F}\times\vec{G} = \operatorname{div}\left(\vec{\mathbf{j}}(\mathbf{y}) \times \frac{\overrightarrow{\mathbf{x}-\mathbf{y}}}{|\mathbf{x}-\mathbf{y}|^3}\right) = 0. \qquad 6.12.38$$

Note that when computing the divergence (and the curl) of $\vec{\mathbf{B}}$, it is natural to differentiate under the integral sign, as in Theorem 4.11.22. Thus

$$\begin{aligned}\operatorname{div}\vec{\mathbf{B}} &= \operatorname{div}\frac{\mu_0}{4\pi}\int_\Gamma \left(\vec{\mathbf{j}}(\mathbf{y}) \times \frac{\overrightarrow{\mathbf{x}-\mathbf{y}}}{|\mathbf{x}-\mathbf{y}|^3}\right)|d^1\mathbf{y}| \\ &= \frac{\mu_0}{4\pi}\int_\Gamma \underbrace{\operatorname{div}\left(\vec{\mathbf{j}}(\mathbf{y}) \times \frac{\overrightarrow{\mathbf{x}-\mathbf{y}}}{|\mathbf{x}-\mathbf{y}|^3}\right)}_{0 \text{ by eq. 6.12.38}}|d^1\mathbf{y}| = 0.\end{aligned} \qquad 6.12.39$$

Equation 6.12.39: Theorem 4.11.22 says that under appropriate conditions,

$$D_x\int_{\mathbb{R}^n} f(x,\mathbf{y})\,|d^n\mathbf{y}| = \int_{\mathbb{R}^n} D_x f(x,\mathbf{y})\,|d^n\mathbf{y}|,$$

If we denote the integrand on the right side of equation 6.12.33 by A, this amounts to saying that we can "curl" and "divergence" under the integral sign, computing (equation 6.12.39)

$$\operatorname{div}\vec{\mathbf{B}} = \frac{\mu_0}{4\pi}\int_\Gamma \operatorname{div} A = 0$$

and (equation 6.12.44)

$$\operatorname{curl}\vec{\mathbf{B}} = \frac{\mu_0}{4\pi}\int_\Gamma \operatorname{curl} A = \mu_0\vec{\mathbf{j}}.$$

The derivation $\operatorname{div}\vec{\mathbf{B}} = 0$ above works everywhere, including inside the wire, if we are careful to think that we are in $\mathbb{R}^3$, with the wire a 3-dimensional object, and current distributed smoothly. More precisely:

If $\vec{\mathbf{j}}$ is a C^1 vector field with compact support on $\mathbb{R}^3$, and

$$\vec{\mathbf{B}}(\mathbf{x}) = \frac{\mu_0}{4\pi}\int_{\mathbb{R}^3} \vec{\mathbf{j}}(\mathbf{y}) \times \frac{\overrightarrow{\mathbf{x}-\mathbf{y}}}{|\mathbf{x}-\mathbf{y}|^3}|d^3\mathbf{y}| \qquad 6.12.40$$

(i.e., the Biot-Savart law 6.12.33 with the wire Γ replaced by $\mathbb{R}^3$ and $|d^1\mathbf{y}|$ by $|d^3\mathbf{y}|$), then $\operatorname{div}\vec{\mathbf{B}} = 0$.

First we must check that $\vec{\mathbf{B}}$ is defined everywhere, in particular when $\mathbf{x}$ is in the support of $\vec{\mathbf{j}}$, where the vector field $\overrightarrow{(\mathbf{x}-\mathbf{y})}/|\mathbf{x}-\mathbf{y}|^3$ blows up. Exercise 6.12.4 asks you to check that

$$\int_{B_1(\mathbf{0})} \frac{1}{|\mathbf{u}|^2} |d^3\mathbf{u}| = 4\pi < \infty, \qquad 6.12.41$$

and derive from this that $\vec{\mathbf{B}}$ is well defined.

That doesn't mean we can blithely differentiate under the integral sign: the condition given in inequality 4.11.77 of Theorem 4.11.22 is not met. But if we make a preliminary change of variables by writing $\overrightarrow{\mathbf{x}-\mathbf{y}} = \vec{\mathbf{u}}$, so that $\vec{\mathbf{y}} = \overrightarrow{\mathbf{x}-\mathbf{u}}$, the formula for $\vec{\mathbf{B}}$ becomes

$$\vec{\mathbf{B}}(\mathbf{x}) = \frac{\mu_0}{4\pi} \int_{\mathbb{R}^3} \vec{\mathbf{j}}(\mathbf{x}-\mathbf{u}) \times \frac{\vec{\mathbf{u}}}{|\mathbf{u}|^3} |d^3\mathbf{u}|, \qquad 6.12.42$$

and this time the conditions are met; you are asked in Exercise 6.12.5 to check that for any fixed $\mathbf{x}_1$ and for $0 \le |\mathbf{x}_2 - \mathbf{x}_1| \le 1$, there exists R such that

$$\frac{1}{|\mathbf{x}_1 - \mathbf{x}_2|} \left| \left(\vec{\mathbf{j}}(\mathbf{x}_1 - \mathbf{u}) \times \frac{\vec{\mathbf{u}}}{|\mathbf{u}|^3} \right) - \left(\vec{\mathbf{j}}(\mathbf{x}_2 - \mathbf{u}) \times \frac{\vec{\mathbf{u}}}{|\mathbf{u}|^3} \right) \right| \le \sup_{\mathbb{R}^3} |[\mathbf{D}\vec{\mathbf{j}}]| \frac{1}{|\mathbf{u}|^2} \mathbf{1}_{B_R(\mathbf{0})}(\mathbf{u}), \qquad 6.12.43$$

and that the expression on the right is integrable over $\mathbb{R}^3$. The proof then continues as above, where we were concerned with the magnetic field outside the wire.

The curl is harder to compute, but under the assumption of *steady currents*, i.e., when $\operatorname{div}\vec{\mathbf{j}} = 0$, the Biot-Savart law simplifies to give

$$\operatorname{curl}\vec{\mathbf{B}} = \mu_0 \vec{\mathbf{j}}; \qquad 6.12.44$$

this is Ampère's law before Maxwell's correction.

Remark. The description of the Biot-Savart law in terms of div and curl is due to Maxwell. Faraday was uncomfortable with math[19]; his analysis of experiments was done with drawings, with curves representing "lines of force" swirling around, the density of lines corresponding to the intensity of whatever field he was studying. Maxwell invented the divergence and curl largely to give a mathematical form to Faraday's descriptions. △

[19] In 1857, age 63, Faraday wrote Maxwell, then 26, "when a mathematician investigating physical actions ... has arrived at his conclusions, may they not be expressed in common language as fully, clearly, and definitely as in mathematical formulae? If so, would it not be a great boon to such as I to express them so? – translating them out of their hieroglyphics .... I think it must be so, because I have always found that you could convey to me a perfectly clear idea of your conclusions ... " (letter quoted in L. Campbell and W. Garnett, *The Life of James Clerk Maxwell*, Macmillan and Co., London, 1884).

Ampère's experiment and the magnetic field

Now we want to explain Ampère's observation about forces between parallel wires, one with current I_1, the other with current I_2:

Recall that $\frac{F}{L}$ represents force per length.

$$\frac{F}{L} = k_2 \frac{2I_1I_2}{d} \qquad 6.12.45$$

in terms of the magnetic field as described by the Biot-Savart law. (We first saw equation 6.12.45 as equation 6.12.18.) In particular, we will see how the μ_0 of the magnetic field (which we saw in equation 6.12.33) relates to Ampère's k_2. First let us see what the Biot-Savart law gives for a steady current I in an infinite straight wire, which we will place on the x-axis, so that for $\mathbf{x} = \begin{pmatrix} x \\ y \\ z \end{pmatrix}$, we have

$$\begin{aligned} \vec{\mathbf{B}}(\mathbf{x}) &= \frac{\mu_0}{4\pi} \int_{-\infty}^{\infty} \frac{1}{((x-s)^2 + y^2 + z^2)^{3/2}} \left(\begin{bmatrix} I \\ 0 \\ 0 \end{bmatrix} \times \begin{bmatrix} x-s \\ y \\ z \end{bmatrix} \right) ds \\ &= \frac{\mu_0}{2\pi} \frac{I}{y^2+z^2} \begin{bmatrix} 0 \\ -z \\ y \end{bmatrix}. \end{aligned} \qquad 6.12.46$$

Equation 6.12.46: Exercise 6.12.3 asks you to confirm this computation.

Two parallel wires distance d apart carry currents I_1 and I_2. Suppose both wires are in the (x, y)-plane, and the current is in the positive x-direction. Place the first wire on the x-axis, with the second wire parallel to it, on the line $y = d$. Then if $\mathbf{w} = \begin{pmatrix} w_1 \\ w_2 \\ w_3 \end{pmatrix}$ is a point on the second wire, we have $w_2 = d$ and $w_3 = 0$, so the magnetic field due to the first wire is

$$\vec{\mathbf{B}}(\mathbf{w}) = \frac{\mu_0}{2\pi} \frac{I}{w_2^2} \begin{bmatrix} 0 \\ 0 \\ w_2 \end{bmatrix} = \frac{\mu_0}{2\pi d} I_1 \, \vec{\mathbf{e}}_3. \qquad 6.12.47$$

Equation 6.12.48: Note that $\vec{\mathbf{E}}$ and $\vec{\mathbf{v}} \times \vec{\mathbf{B}}$ must have the same units for the equation to make sense, so $\vec{\mathbf{E}}$ and $\vec{\mathbf{B}}$ have different units. In the cgs system (equation 6.12.1):

$$\vec{F} = q\left(\vec{\mathbf{E}} + \left(\frac{\vec{\mathbf{v}}}{c} \times \vec{\mathbf{B}}\right)\right),$$

$\vec{\mathbf{E}}$ and $\vec{\mathbf{B}}$ have the same units.

Even after choosing a unit of charge, the proportionality constant μ_0 has no particular value until a unit is chosen for the magnetic field. In SI units, this is done by setting the Lorentz force to be

$$\vec{F} = q\Big(\vec{\mathbf{E}} + (\vec{\mathbf{v}} \times \vec{\mathbf{B}})\Big), \qquad 6.12.48$$

(where q is charge and $\vec{\mathbf{v}}$ is velocity) with no constants; this sets the dimensions of the electric field to be $\frac{\text{force}}{\text{charge}}$ and the dimensions of the magnetic field to be $\frac{\text{force}}{\text{charge} \times \text{velocity}}$.

In our system of two wires we have $\vec{\mathbf{E}} = \vec{\mathbf{0}}$, since the wires are electrically neutral[20]. So the force of the magnetic field acting on the second wire, and measured per unit length, is

$$\begin{aligned}\vec{F} &= q(\vec{\mathbf{v}} \times \vec{\mathbf{B}}) = q\vec{\mathbf{v}} \times \vec{\mathbf{B}} = q\vec{\mathbf{v}} \times \frac{\mu_0}{2\pi d}\, I_1\, \vec{\mathbf{e}}_3 \\ &= I_2\vec{\mathbf{e}}_1 \times \frac{\mu_0}{2\pi d}\, I_1\, \vec{\mathbf{e}}_3 = \frac{\mu_0}{4\pi}\frac{2I_1I_2}{d}\, \vec{\mathbf{e}}_1 \times \vec{\mathbf{e}}_3 \text{ newtons} \\ &= -\frac{\mu_0}{4\pi}\frac{2I_1I_2}{d}\, \vec{\mathbf{e}}_2 \text{ newtons.}\end{aligned} \qquad 6.12.49$$

Equation 6.12.49: To go from line 1 to line 2, note that $q\vec{\mathbf{v}}$ has dimensions charge times distance over time, i.e., current times distance, and that the current by definition is moving in the direction of $\vec{\mathbf{e}}_1$. So $q\vec{\mathbf{v}} = I_2\vec{\mathbf{e}}_1$.

The minus sign in the last line indicates that the force is attracting between the wires, and the number $\mu_0/4\pi$ is precisely the coefficient called k_2 in equation 6.12.18.

Recall that $k_1 = 1/(4\pi\epsilon_0)$. Thus the observation $k_1/k_2 \approx c^2$ (equation 6.12.21) becomes

$$\frac{1}{\epsilon_0\mu_0} \approx c^2. \qquad 6.12.50$$

The choice of the unit of charge (and of length, time, and mass), together with the Lorentz force law, impose the units of the electric and magnetic field. The unit of charge can be chosen so as to make either ϵ_0 (equation 6.12.24) or μ_0 (equation 6.12.33) come out to whatever you like, but not both. The other is then a quantity that needs to be measured in the laboratory. Alternatively, one can measure the speed of light; in SI, the permeability μ_0 is defined to be $4\pi/10^7$, and then equation 6.12.50 says that the permittivity ϵ_0 is $1/(\mu_0 c^2)$.

Faraday's experiment and Maxwell's laws

Now we will relate to Maxwell's laws Faraday's observation that the current in the wire is proportional to the rate of change of the magnetic field through the loop formed by the wire. Figure 6.12.6 shows a setup similar to Figure 6.12.3, but better suited to quantitative analysis. We will assume that the magnetic field is constant: $\vec{\mathbf{B}} = B\vec{\mathbf{e}}_3$, and that the distance separating the strands of the wires is w. If the wire is moved right at speed v, then the force on the electrons on the crosspiece of the wire is $\frac{v}{c}Bw$ per unit charge, by the Lorentz force law.

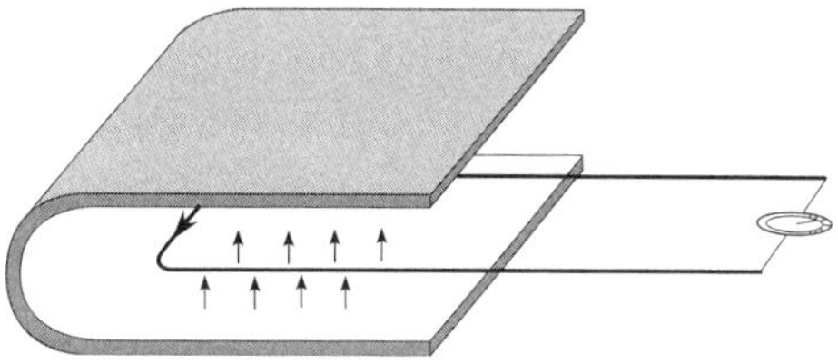

FIGURE 6.12.6. A loop of wire is inserted in the magnetic field due to a permanent U-shaped magnet. Outside the magnet the wire is connected to an ammeter. If the wire is pulled out of the magnet, the charges in the piece of wire transverse to the motion are "pushed" by the Lorentz force $\frac{\vec{\mathbf{v}}}{c} \times \vec{\mathbf{B}}$, in the direction indicated by the arrow.

[20]The protons occupy fixed positions in the wire and the electrons slosh around, but in any macroscopic length of the wire, there are *very nearly* the same number of electrons as protons. Any local predominance of electrons over protons or vice-versa would result in immense electrostatic forces, and these forces push the system back to electric neutrality.

Suppose instead that the magnet is moved right at speed v. The force per unit charge on the electrons in the cross piece must be the same $\frac{v}{c}Bw$, hence the integral of the electric field around the circuit should be

$$\frac{v}{c}Bw = \int_C W_{\vec{\mathbf{E}}} \qquad 6.12.51$$

Note that in both scenarios, we have

$$vBw = -D_t \int_S \Phi_{\vec{\mathbf{B}}}, \qquad 6.12.52$$

where S is the piece of plane bounded by the circuit C.

Thus we are led to the equation

$$-\frac{1}{c}D_t \int_S \Phi_{\vec{\mathbf{B}}} = \int_C W_{\vec{\mathbf{E}}}. \qquad 6.12.53$$

Stokes's theorem now asserts that

$$\int_C W_{\vec{\mathbf{E}}} = \int_S \Phi_{\vec{\nabla}\times\vec{\mathbf{E}}}. \qquad 6.12.54$$

Differentiating under the integral sign leads to

$$-\frac{1}{c}\int_S D_t \Phi_{\vec{\mathbf{B}}} = \int_S \Phi_{\vec{\nabla}\times\vec{\mathbf{E}}}, \qquad 6.12.55$$

and clearly this is a consequence of Faraday's law

$$-\frac{1}{c}D_t\vec{\mathbf{B}} = \vec{\nabla}\times\vec{\mathbf{E}}. \qquad 6.12.56$$

Faraday's version of Maxwell's laws and Maxwell's correction

In the 1830s and 1840s, Faraday performed an enormous number of experiments, which ended up, after Maxwell had appropriately interpreted them, as the following four equations, written in SI units:

$$\begin{aligned} \operatorname{div}\vec{\mathbf{E}} &= \frac{1}{\epsilon_0}\rho & \qquad \operatorname{curl}\vec{\mathbf{E}} &= -D_t\vec{\mathbf{B}} \\ \operatorname{div}\vec{\mathbf{B}} &= 0 & \qquad \operatorname{curl}\vec{\mathbf{B}} &= \quad \mu_0\vec{\mathbf{j}} \end{aligned} \qquad 6.12.57$$

He also had the *equation of continuity*

$$D_t\rho + \operatorname{div}\vec{\mathbf{j}} = 0, \qquad 6.12.58$$

expressing conservation of charge, and he had the Lorentz force law.

Maxwell soon noticed that these equations are mathematically inconsistent. Recall that for any vector field $\vec{F}$, we have $\operatorname{div}\operatorname{curl}\vec{F} = 0$, which is equivalent to $\mathbf{dd}W_{\vec{F}} = 0$. This comes out right for

$$\operatorname{div}\operatorname{curl}\vec{\mathbf{E}} = -\operatorname{div} D_t\vec{\mathbf{B}} = -D_t \operatorname{div}\vec{\mathbf{B}} = 0, \qquad 6.12.59$$

but

$$\operatorname{div}\operatorname{curl}\vec{\mathbf{B}} = \mu_0 \operatorname{div}\vec{\mathbf{j}}, \qquad 6.12.60$$

and there is no reason to expect the right side to be 0. Recall that the divergence measures compressibility. Water is incompressible: for any bounded region, in any water flow, at every instant as much water leaves it as enters it. This is definitely not true of electric current: electrons can accumulate, on a capacitor plate for instance, and if you take a region surrounding the plate, as you charge the plate more electrons enter the region than leave. It is in fact no surprise that there is trouble involving $\operatorname{div}\vec{\mathbf{j}}$ since we needed to assume steady currents, i.e., $\operatorname{div}\vec{\mathbf{j}} = 0$, in the derivation of $\operatorname{curl}\vec{\mathbf{B}} = \mu_0\vec{\mathbf{j}}$ (see equation 6.12.44).

In traditional notation,

$$\epsilon_0\mu_0 D_t\vec{\mathbf{E}} \quad \text{is} \quad \epsilon_0\mu_0\frac{\partial\vec{\mathbf{E}}}{\partial t}.$$

Recall (equation 6.12.50) that

$$\frac{1}{\epsilon_0\mu_0} \approx c^2,$$

so $\epsilon_0\mu_0$ is very small. Therefore Maxwell's correction term was not detectable in the kinds of table-top experiments Faraday performed.

Maxwell cast around for a term to add to make the equations consistent, and he observed that adding $\epsilon_0\mu_0 D_t\vec{\mathbf{E}}$ to $\mu_0\vec{\mathbf{j}}$ to get

$$\operatorname{curl}\vec{\mathbf{B}} = \mu_0\vec{\mathbf{j}} + \epsilon_0\mu_0 D_t\vec{\mathbf{E}} \quad \text{(Ampère's law, corrected, in SI)} \qquad 6.12.61$$

solves the problem:

$$\begin{aligned}\operatorname{div}\operatorname{curl}\vec{\mathbf{B}} &= \mu_0\operatorname{div}\vec{\mathbf{j}} + \epsilon_0\mu_0\operatorname{div}D_t\vec{\mathbf{E}} = \mu_0\operatorname{div}\vec{\mathbf{j}} + \epsilon_0\mu_0 D_t\operatorname{div}\vec{\mathbf{E}}\\ &= \mu_0\big(\underbrace{\operatorname{div}\vec{\mathbf{j}} + D_t\rho}_{0\text{ by eq. }6.12.58}\big) = 0,\end{aligned} \qquad 6.12.62$$

where the first equality on the second line comes from $\operatorname{div}\vec{\mathbf{E}} = \frac{1}{\epsilon_0}\rho$ (equation 6.12.57).

The wave equation and electromagnetic radiation

Why did Maxwell conclude from the observation that $\epsilon_0\mu_0 \approx 1/c^2$ that light is an electromagnetic wave? (See the quotation page 674.) To understand the connection, you need to know something about the wave equation.

In acoustics, one studies the wave equation with a equal to the speed of sound. In hydrodynamics, a is the speed of sound in water, and so forth. Each medium has its characteristic speed.

In its simplest form, the wave equation for a function $f(t,x)$ of one temporal variable and one spatial variable is

$$D_t^2 f = a^2 D_x^2 f, \quad \text{or, in classical notation,} \quad \frac{\partial^2 f}{\partial t^2} = a^2\frac{\partial^2 f}{\partial x^2}, \qquad 6.12.63$$

where a is a constant whose value depends on what is being described. This equation describes all sorts of things: a vibrating string, sound, etc. It has been known since Euler and Bernoulli, and had been studied extensively by many mathematicians long before Maxwell: Euler, Bernoulli, Lagrange, Laplace, d'Alembert, Fourier, Poisson, and others.

The one thing you need to know about the wave equation is that if g is any function of one variable, then $f(t,x) = g(x - at)$ is a solution of the wave equation. (This uses the chain rule; see the margin note for equation 6.12.66.) It is easy to visualize this function: its graph is the graph of g, moving to the right at speed a. Actually, $f(t,x) = g(x + at)$ is just as good a solution; then the graph of g moves to the left, but *always at speed a*.

The wave equation in $\mathbb{R}^3$ is written

$$D_t^2 f = a^2\Delta f, \quad \text{or, in classical notation,} \quad \frac{\partial^2 f}{\partial t^2} = a^2\Delta f, \qquad 6.12.64$$

where Δ is the Laplacian

$$\Delta f = \operatorname{div}\operatorname{grad} f = D_1^2 f + D_2^2 f + D_3^2 f. \qquad 6.12.65$$

Again, the only thing you need to know about this equation is that if $g : \mathbb{R} \to \mathbb{R}$ is any function and $\vec{\mathbf{v}} \in \mathbb{R}^3$ is any unit vector, then

$$f(t, \mathbf{x}) = g(\vec{\mathbf{x}} \cdot \vec{\mathbf{v}} - at) \qquad 6.12.66$$

is a solution of the wave equation, as you are asked to show in Exercise 6.12.6; its graph is the graph of g, moving in the direction $\vec{\mathbf{v}}$ at speed a.

Equation 6.12.66: To see that $f(t,x) = g(x - at)$ is a solution to the wave equation 6.12.63, write $f = g \circ h$, where $h : (t,x) \mapsto x - at$; $g : u \mapsto g(u)$, then use the chain rule to compute

$$\begin{aligned}[\mathbf{D}(g \circ h)(t,x)] &= [\mathbf{D}g(h(t,x))][\mathbf{D}h(t,x)] \\ &= g'(h(t,x))[-a, 1].\end{aligned}$$

Then the first partial derivatives of f are

$$\begin{aligned} D_t f(t,x) &= -ag'(x - at) \\ D_x f(t,x) &= g'(x - at) \end{aligned},$$

and the second partials are

$$D_t^2 f(t,x) = a^2 g''(x - at)$$

$$D_x^2 f(t,x) = g''(x - at).$$

giving

$$D_t^2 f = a^2 D_x^2 f$$

This is still not quite what we need: we need the wave equation for vector fields like $\vec{\mathbf{E}}$ and $\vec{\mathbf{B}}$. The following formula (which you were asked to justify in Exercise 6.8.14) provides the key:

$$\vec{\Delta}\vec{\mathbf{F}} \stackrel{\text{def}}{=} \operatorname{grad} \operatorname{div} \vec{\mathbf{F}} - \operatorname{curl} \operatorname{curl} \vec{\mathbf{F}} = \begin{bmatrix} \Delta F_1 \\ \Delta F_2 \\ \Delta F_3 \end{bmatrix}. \qquad 6.12.67$$

Thus if a vector field $\vec{\mathbf{F}}$ satisfies the equation

$$D_t^2 \vec{\mathbf{F}} = a^2 \left(\operatorname{grad} \operatorname{div} \vec{\mathbf{F}} - \operatorname{curl} \operatorname{curl} \vec{\mathbf{F}} \right), \qquad 6.12.68$$

each of its three components will satisfy the wave equation, and have solutions that are waves moving at speed a.

In the absence of charges or currents, Maxwell's equations 6.12.57 (with Ampère's law in the corrected version 6.12.61 – the correction is crucial here) are written

$$\begin{aligned} \operatorname{div} \vec{\mathbf{E}} &= 0 & \operatorname{curl} \vec{\mathbf{E}} &= -D_t \vec{\mathbf{B}} \\ \operatorname{div} \vec{\mathbf{B}} &= 0 & \operatorname{curl} \vec{\mathbf{B}} &= \epsilon_0 \mu_0 D_t \vec{\mathbf{E}}. \end{aligned} \qquad 6.12.69$$

Equation 6.12.69: In traditional notation, the equations for $\operatorname{curl} \vec{\mathbf{E}}$ and $\operatorname{curl} \vec{\mathbf{B}}$ are written

$$\operatorname{curl} \vec{\mathbf{E}} = -\frac{\partial \vec{\mathbf{B}}}{\partial t}$$

$$\operatorname{curl} \vec{\mathbf{B}} = \epsilon_0 \mu_0 \frac{\partial \vec{\mathbf{E}}}{\partial t}$$

We have

$$\operatorname{grad} \operatorname{div} \vec{\mathbf{E}} = 0$$

because $\operatorname{div} \vec{\mathbf{E}} = 0$, not because Theorem 6.7.8 says that

$$\mathbf{dd}\varphi = 0;$$

as discussed in Remark 6.8.4, that theorem doesn't apply to grad div.

If we differentiate the last equation with respect to time, we find

$$\epsilon_0 \mu_0 D_t^2 \vec{\mathbf{E}} = \operatorname{curl} D_t \vec{\mathbf{B}} = -\operatorname{curl} \operatorname{curl} \vec{\mathbf{E}}. \qquad 6.12.70$$

Clearly $\operatorname{grad} \operatorname{div} \vec{\mathbf{E}} = 0$, so altogether we find

$$\epsilon_0 \mu_0 D_t^2 \vec{\mathbf{E}} = \left(\operatorname{grad} \operatorname{div} \vec{\mathbf{E}} - \operatorname{curl} \operatorname{curl} \vec{\mathbf{E}} \right). \qquad 6.12.71$$

A similar computation leads to

$$\epsilon_0 \mu_0 D_t^2 \vec{\mathbf{B}} = \left(\operatorname{grad} \operatorname{div} \vec{\mathbf{B}} - \operatorname{curl} \operatorname{curl} \vec{\mathbf{B}} \right). \qquad 6.12.72$$

Thus, by equations 6.12.21, 6.12.50, and 6.12.68, both the electric field and the magnetic field obey the wave equation with speed of propagation

$$1/\sqrt{\epsilon_0 \mu_0} = c. \qquad 6.12.73$$

In other words, Maxwell's equations predict waves moving at the speed of light (within the margin of error of contemporary measurements of the speed of light). You can really see why Maxwell guessed that light was electromagnetic radiation.

The ether, the Michelson-Morley experiment, and relativity

Note, however, that Maxwell did not say that light was electromagnetic radiation, he said that "light consists in the traverse undulations of the

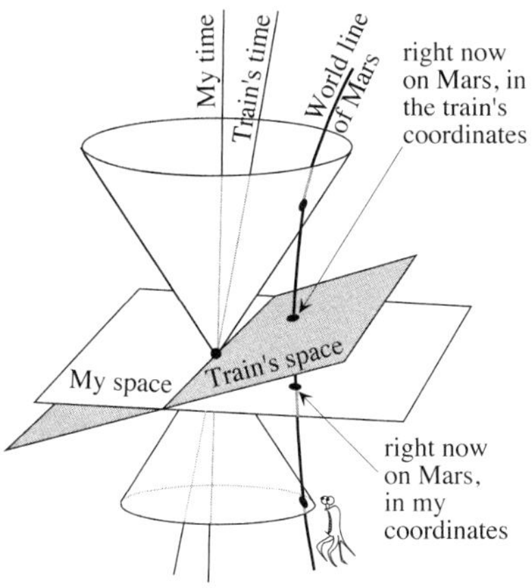

FIGURE 6.12.7.
The light cone

$$x^2 + y^2 + z^2 - c^2t^2 = 0$$

at a particular point in spacetime. If you are at the vertex, the cone below represents everything you can see: you can see the moon at a particular place and time, and Mars at a different place and different time, since it takes light longer to arrive from Mars than from the moon. The upper (forward) cone represents every point in spacetime to which you can send a light message, or, conversely, every point in spacetime that can see you.

The shaded plane represents a different frame of reference, for instance, someone on a moving train. The light cone is the same, regardless of the frame of reference. This is counterintuitive. If you drew the forward cone representing points of spacetime you could hit by throwing a ball, then the "thrown ball" cone would be different if you were standing still or on on a moving train.

(The creature on the "world line" of Mars is inspired by the Martian in Bill Watterson's *The Calvin and Hobbes Tenth Anniversary Book.*)

same medium which is the cause of electric and magnetic phenomena." Just as the wave equation in acoustics has as its characteristic speed the speed of sound in air, Maxwell very sensibly thought that his equation described the vibrations of some medium, which he called the *ether*. What could be more natural than to wonder what the speed of the earth might be with respect to the ether? For about 30 years after Maxwell's publication, the hunt was on to "find the ether".

The final nail was driven into the coffin of this program by the Michelson-Morley experiment in 1887. This experiment showed that the speed of light is constant, the same whether the light is going in the direction of the earth around the sun or against it. Many attempts were made to come to terms with this, in particular by Lorentz and Poincaré. But a truly coherent description was only given in the 1905 paper by Einstein in which he created the special theory of relativity.

This requires a complete recasting of one's intuition about space and time. It might seem clear what the expression "right now, on Mars" means, but according to Einstein, it doesn't mean anything: the notion of simultaneity for distant events is an illusion. Spacetime is not "space, evolving through time". Instead, spacetime is a 4-dimensional space whose points are called *events*, and the only physically meaningful structure on this space is given by the light cones, as illustrated by Figure 6.12.7. A light-ray can go from one event to a second event if one event lies on the light cone emanating from the second event.

In Figure 6.12.7, the "world line" of Mars represents the movement of Mars through spacetime. At the particular place and time represented by the vertex of the light cone, you can only see Mars at one point of that world line, the point where Mars's world line intersects the backward light cone, and the t of that point is not "right now" in your coordinates. Nor is it the "right now" of an observer at the same point and time as you, but in a moving train. Since "right now" should mean the same thing to you and the train observer, we are forced to conclude that "right now on Mars" is meaningless.

This spacetime is called *Minkowski space*, which is $\mathbb{R}^4$ with coordinates x, y, z, t, endowed with the quadratic form $x^2 + y^2 + z^2 - c^2t^2$, called the *Lorentz pseudometric*. If one set of coordinates differs from another by an affine transformation preserving the pseudometric, the laws of physics in the two sets of coordinates should be the same. Such transformations are called *Lorentz transformations*.

For instance, you might think that if you were in a train moving due east (along the x-axis) at constant speed t, then your coordinates, compared to someone with x, y, z, t coordinates outside the train, could be written

$$x' = x - vt, \quad y' = y, \quad z' = z, \quad t' = t. \qquad 6.12.74$$

(To keep the formulas simple, we allow for motion only in the x-direction.) But this is not true, because the system of equations 6.12.74 is not a Lorentz transformation. The transformation that does preserve light cones is

$$x' = \frac{x - vt}{\sqrt{1 - v^2/c^2}}, \quad y' = y, \quad z' = z, \quad t' = \frac{t - vx/c^2}{\sqrt{1 - v^2/c^2}}, \qquad 6.12.75$$

as you are asked to check in Exercise 6.12.7. Note that the two equations are almost identical when v is much smaller than c, as it almost always is in ordinary life.

Remark. In equation 6.12.75, the term

$$\frac{1}{\sqrt{1-v^2/c^2}} \tag{6.12.76}$$

is called the *relativistic correction factor* or the *Lorentz factor*. It shows up any time you want to change coordinates while respecting the Lorentz pseudometric. It is often denoted by γ.

It follows from equations 6.12.75 that if one twin brother returns to Earth after a voyage by spaceship, he will be younger than the twin who stayed home. Since the location of the clock on the spaceship (relative to the spaceship) remains unchanged, we have

$$x' = \frac{x - vt}{\sqrt{1-v^2/c^2}} = 0, \tag{6.12.77}$$

i.e., $x = vt$. Substituting this value for x in the equation for t' in 6.12.75 gives $t' = t\sqrt{1-v^2/c^2}$: if the twins meet again, the traveler's clock will be slow compared to clocks on Earth. △

This slowing down of time is not something observable only in imaginary experiments. It was confirmed in 1971 by physicists Hafele and Keating when they flew around the world in commercial airlines with four atomic clocks.

By endowing spacetime with geometry, the Lorentz pseudometric allows us to relate $\mathbb{M}$ and $\mathbb{F}$: we can derive $\mathbb{M}$ from $\mathbb{F}$, or $\mathbb{F}$ from $\mathbb{M}$, using

$$\mathbb{M} = *\mathbb{F}, \tag{6.12.78}$$

where the "star operator" $*$ represents a kind of perpendicularity. Maxwell's equations, the Lorentz pseudometric, and equation 6.12.78 give a complete description of special relativity; incorporating gravitation into this description of physical reality gives the general theory of relativity. The difference is the same as the difference between $\mathbb{R}^n$ and a manifold: *locally, special relativity describes general relativity, just as locally, the tangent space to a manifold describes the manifold.* A complete description of physics – the much-desired "unified field theory" or "theory of everything" – would require incorporating also the strong and weak nuclear forces described by quantum mechanics.

Recall that Maxwell's laws describing electromagnetism can be written as

$$\mathbf{d}\mathbb{F} = 0 \quad \text{and} \quad \mathbf{d}\mathbb{M} = 4\pi\mathbb{J},$$

where

$$\mathbb{F} = W_{\vec{\mathbf{E}}} \wedge c\,dt + \Phi_{\vec{\mathbf{B}}}$$
$$\mathbb{M} = W_{\vec{\mathbf{B}}} \wedge c\,dt - \Phi_{\vec{\mathbf{E}}}.$$

Writing Maxwell's laws using curl and div requires the geometry of $\mathbb{R}^3$; writing them using $\mathbb{F}$ and $\mathbb{M}$ requires no geometry. The Lorentz pseudometric provides the geometry of spacetime.

Exercises for Section 6.12

6.12.1 Justify equation 6.12.9 by computing $\mathbf{d}\mathbb{F}$.

6.12.2 Justify equation 6.12.37.

6.12.3 Justify the second equality in equation 6.12.46.

6.12.4 Show that equation 6.12.41 is correct, and that it follows that $\vec{\mathbf{B}}$ is well defined.

Exercise 6.12.5: We are assuming that $\vec{\mathbf{j}}$ is a vector field with compact support on $\mathbb{R}^3$.

Exercise 6.12.6, part a: $h\mathbf{v}$ is the function $\mathbf{x} \mapsto h(\mathbf{x})\vec{\mathbf{v}}$.

Exercise 6.12.8 describes *transverse waves*: waves that oscillate orthogonally to the direction of propagation. Here, the oscillations of $\vec{\mathbf{E}}$ are in the y-direction and the oscillations of $\vec{\mathbf{B}}$ are in the z-direction, whereas the wave propagates in the x-direction.

Waves oscillating in the direction of propagation are called *longitudinal waves*. Exercise 6.12.9 says that there are no longitudinal electromagnetic waves. Sound, by contrast, is mainly carried by longitudinal pressure waves in air.

We found some very nice animations of longitudinal and transverse waves on a web site created by Daniel A. Russell:
http://www.acs.psu.edu/drussell/Demos/waves/wavemotion.html

6.12.5 Show that equation 6.12.43 is correct and that the function on the right is integrable over $\mathbb{R}^3$.

6.12.6 a. Show that if $\vec{\mathbf{v}} \in \mathbb{R}^n$ is a vector and $h: \mathbb{R}^n \to \mathbb{R}$ is a function,

$$(\operatorname{div}(h\vec{\mathbf{v}})(\mathbf{x}) = (\overrightarrow{\operatorname{grad}}\, h(\mathbf{x})) \cdot \vec{\mathbf{v}}.$$

b. Show that if $g: \mathbb{R} \to \mathbb{R}$ is any function and $\vec{\mathbf{v}} \in \mathbb{R}^3$ is any unit vector, then $f(t, \mathbf{x}) = g(\vec{\mathbf{x}} \cdot \vec{\mathbf{v}} - at)$ is a solution of the wave equation

$$\frac{\partial^2 f}{\partial t^2} = a^2 \Delta f.$$

6.12.7 Show that the transformation in equation 6.12.75 is indeed a Lorentz transformation, i.e., that

$$|\mathbf{x}'|^2 - c^2(t')^2 = |\mathbf{x}|^2 - c^2t^2.$$

6.12.8 Show that in the absence of charges or currents, if g is any function of one variable, then

$$\vec{\mathbf{E}} = \begin{bmatrix} 0 \\ g(x-ct) \\ 0 \end{bmatrix}, \qquad \vec{\mathbf{B}} = \begin{bmatrix} 0 \\ 0 \\ g(x-ct) \end{bmatrix}$$

is a solution of Maxwell's equations.

6.12.9 a. In the absence of charges or currents, show that if g is a function of one variable such that for some $a \in \mathbb{R}$, there exists $\vec{\mathbf{B}}$ such that

$$\vec{\mathbf{E}} = \begin{bmatrix} g(x-at) \\ 0 \\ 0 \end{bmatrix} \quad \text{and} \quad \vec{\mathbf{B}}$$

are a solution to Maxwell's equations, then g is constant.

b. In the absence of charges or currents, is

$$\vec{\mathbf{E}} = \begin{bmatrix} g(y-ct) \\ 0 \\ 0 \end{bmatrix}, \quad \vec{\mathbf{B}} = \begin{bmatrix} 0 \\ 0 \\ g(y-ct) \end{bmatrix} \quad \text{a solution?}$$

6.12.10 Compute in terms of $\vec{\mathbf{E}}$ and $\vec{\mathbf{B}}$ the 4-forms $\mathbb{F} \wedge \mathbb{F}$, $\mathbb{F} \wedge \mathbb{M}$, $\mathbb{M} \wedge \mathbb{M}$.

6.12.11 a. Using Definition 6.5.10, show that in $\mathbb{R}$ (i.e., $\mathbb{R}^n$ for $n = 1$), the flux form of a vector field is a 0-form (a function). Show that $\vec{F}_1(x) = \vec{x}/|x|$ (see Example 6.7.7), and that its flux form $\Phi_{\vec{F}_1}$ is the function $\operatorname{sgn}(x)$, which is $+1$ if $x > 0$ and -1 if $x < 0$.

b. Let $f: \mathbb{R} \to \mathbb{R}$ be C^1 with compact support. Show the 1-dimensional analogue of equation 6.12.27: if

$$g(x) = \frac{1}{2}\int_{\mathbb{R}} f(y)\operatorname{sgn}(x-y)\,dy, \quad \text{then } g'(x) = f(x).$$

Go through the sequence of equalities of 6.12.27, write the 1-dimensional analogue and verify that it is true.

6.13 Potentials

A question that comes up constantly in physics is whether a force field is *conservative*. The gravitational force field is conservative: if you hop on your bicycle on a hill at 100 meters elevation, zoom down to sea level, then go up another hill until you reach your starting elevation, the total work against gravity is zero, whatever your actual path. Friction (with air, with the road, with your brakes) is not conservative, which is why you get tired during such a ride.

A similar question, just as important for physics, is whether the force fields of electromagnetism are "conservative", whatever "conservative" means in this context.

An important question that constantly comes up in geometry is, when does a space have "holes"? Again, it isn't obvious what this means: the example to keep in mind is the hole in a doughnut: a torus has a "hole".

We will see in this section that these two questions are closely related. They are both variants of the question "when can a k-form φ be written $\mathbf{d}\psi$ for some $(k-1)$-form ψ?"

Conservative vector fields and their potentials

It is much easier to deal with functions than to deal with vector fields, but when describing work or flux, the real actors are vector fields. In the special case where a vector field $\vec{F}$ is conservative, we can think of $\vec{F}$ in terms of a function.

Altitude is the potential of the gravity force field. Think of how much easier it is to measure the difference of altitude between the top and bottom of a hill than to measure the work it takes to get from bottom to top along some winding path.

Note that altitude, like all potentials, is only defined up to an additive constant; you have to choose a base level (such as sea level, or the center of the earth, or the altitude of your base camp in the Himalayas). Adding a constant does not change the difference of potential.

A function f is a *potential* for a vector field $\vec{F}$ if $\operatorname{grad} f = \vec{F}$; the potential at a point x is then $f(x)$. The following justifies calling a vector field *conservative* if it is the gradient of a potential.

Let $U \subset \mathbb{R}^3$ be open. If a C^1 function f on U is a potential for $\vec{F}$, and $\gamma : [a, b] \to U$ is a C^1 path in U, then by Theorem 6.11.1,

$$\int_{[\gamma]} W_{\vec{F}} = f(\gamma(b)) - f(\gamma(a)). \qquad 6.13.1$$

The work of the force field $\operatorname{grad} f$ on the path γ is the difference of the potential at the endpoints.

Thus if γ is a *closed* C^1 path (i.e., $\gamma(a) = \gamma(b)$), then

$$\int_{[\gamma]} W_{\vec{F}} = f(\gamma(b)) - f(\gamma(a)) = 0, \qquad 6.13.2$$

which justifies calling a vector field conservative if it is the gradient of a potential.

Suppose $n = 3$. Since $\operatorname{curl} \operatorname{grad} f = \vec{0}$ for any C^1 function f, a necessary condition for a vector field $\vec{F}$ to be the gradient of a potential is that $\operatorname{curl} \vec{F} = \vec{0}$. But is it sufficient? Example 6.13.1 says the answer is no.

Example 6.13.1 ($\operatorname{curl} \vec{F} = \vec{0}$ **not sufficient).** Consider an important example from physics: the magnetic field due to a constant current in an infinite straight wire. We computed in equation 6.12.46 the magnetic field due to a constant current I in a wire along the x-axis, going in the positive x direction, and found

$$\vec{\mathbf{B}}\begin{pmatrix} x \\ y \\ z \end{pmatrix} = \frac{\mu_0 I}{2\pi} \frac{1}{y^2 + z^2} \begin{bmatrix} 0 \\ -z \\ y \end{bmatrix}. \qquad 6.13.3$$

We encountered this vector field (up to a constant multiple) in Example 6.7.7. You were asked there to show that $\mathbf{d}W_{\vec{\mathbf{B}}} = \Phi_{\operatorname{curl}\vec{\mathbf{B}}} = 0$; here we do it in the language of vector calculus:

$$\operatorname{curl}\vec{\mathbf{B}} = \frac{\mu_0 I}{2\pi}\begin{bmatrix} D_2\dfrac{y}{y^2+z^2} - D_3\dfrac{-z}{y^2+z^2} \\ 0 \\ 0 \end{bmatrix}. \qquad 6.13.4$$

The first entry gives

$$\frac{(y^2+z^2)-2y^2}{(y^2+z^2)^2} + \frac{(y^2+z^2)-2z^2}{(y^2+z^2)^2} = 0. \qquad 6.13.5$$

It is essential to realize that $\vec{\mathbf{B}}$ is not defined on all of $\mathbb{R}^3$: it is defined on $\mathbb{R}^3$ with the x-axis removed. The unit circle in the (y,z)-plane surrounds the "hole" in the domain of $\vec{\mathbf{B}}$.

Recall (equation 6.5.14) that the formula for integrating a work form over an oriented curve is

$$\int_C W_{\vec{F}} = \int_a^b \vec{F}\big(\gamma(t)\big)\cdot\vec{\gamma}'(t)\,dt.$$

Recall that the unit circle is often denoted S^1.

Yet the vector field $\vec{\mathbf{B}}$ cannot be written as $\operatorname{grad} f$ for any function f from $(\mathbb{R}^3 - z\text{-axis})$ to $\mathbb{R}$. Indeed, using the standard parametrization

$$\gamma(t) = \begin{pmatrix} 0 \\ \cos t \\ \sin t \end{pmatrix} \qquad 6.13.6$$

of the unit circle in the (y,z)-plane, oriented as the boundary of the unit disc, we find that the work of $\vec{\mathbf{B}}$ around the circle is

$$\int_{S^1} W_{\vec{\mathbf{B}}} = \frac{\mu_0 I}{4\pi}\int_0^{2\pi} \overbrace{\underbrace{\frac{1}{\cos^2 t+\sin^2 t}\begin{bmatrix}0\\-\sin t\\ \cos t\end{bmatrix}}_{\vec{F}(\gamma(t))}\cdot\underbrace{\begin{bmatrix}0\\-\sin t\\ \cos t\end{bmatrix}}_{\vec{\gamma}'(t)}}^{1}\,dt = \frac{\mu_0 I}{2}. \qquad 6.13.7$$

This cannot occur for the work of a conservative vector field: we started at one point and returned to the same point, so if the vector field were conservative, the work would be zero. △

Potentials and forms

In Example 6.13.1 we showed that there exists a vector field $\vec{F}$ on an open subset $U \subset \mathbb{R}^3$ such that $\operatorname{curl}\vec{F} = \vec{\mathbf{0}}$, but no function f on U is a potential for $\vec{F}$, i.e., no f on U satisfies $\vec{F} = \operatorname{grad} f$. Restated in terms of forms, this says that the 1-form $W_{\vec{F}}$ satisfies $\mathbf{d}W_{\vec{F}} = 0$, but there is no function f on U such that $\mathbf{d}f = W_{\vec{F}}$. In this language, all the above generalizes to k-forms on open subsets of $\mathbb{R}^n$.

If ψ is a $(k-1)$-form on U, we say that ψ is a *potential* for $\mathbf{d}\psi$; for a k-form $\varphi \in A^k(U)$ to admit a potential means that we can write φ as $\mathbf{d}\psi$ for some $\psi \in A^{k-1}(U)$. Recall that $\mathbf{dd}\varphi = 0$ for all forms of all degrees on all open sets. Thus, a necessary condition for $\varphi \in A^k(U)$ to admit a potential ψ is that $\mathbf{d}\varphi = \mathbf{dd}\psi = 0$. As we saw in Example 6.13.1, this is not sufficient.

Again we need a different (stronger) criterion to check that $\varphi \in A^k(U)$ cannot be written $\mathbf{d}\psi$, analogous to the condition that the work of the gradient of a potential around a closed loop must be 0.

Proposition 6.13.2. *If $U \subset \mathbb{R}^n$ is open and $\varphi \in A^k(U)$, $\psi \in A^{k-1}(U)$ satisfy $\varphi = \mathbf{d}\psi$, then for every compact oriented k-manifold $M \subset U$ we have $\int_M \varphi = 0$.*

By Definition 6.6.1, the boundary of M in M is always empty. This does not mean that we can apply Proposition 6.13.2 to any manifold; for M to be a piece-with-boundary of itself, it must be compact.

Proof. The boundary of M is empty. Thus

$$\int_M \varphi = \int_M \mathbf{d}\psi = \int_{\partial M} \psi = 0. \quad \square \qquad 6.13.8$$

Example 6.13.3. The form $\omega_n \in A^{n-1}(\mathbb{R}^n - \{\mathbf{0}\})$ was given in equation 6.7.18 as

$$\omega_n = \frac{1}{(x_1^2 + \cdots + x_n^2)^{n/2}} \sum_{i=1}^{n} (-1)^{i-1} x_i \, dx_1 \wedge \cdots \wedge \widehat{dx_i} \wedge \cdots \wedge dx_n. \qquad 6.13.9$$

There we showed that $\mathbf{d}\omega_n = 0$, but by Exercise 6.31, $\int_{S^{n-1}} \omega_n > 0$. Thus ω_n is an $(n-1)$-form with $\mathbf{d}\omega_n = 0$, which cannot be written $\mathbf{d}\psi$. $\triangle$

Forms of different types detect holes of different kinds; saying what this means is the object of *de Rham cohomology*, which is beyond (but not far beyond) the scope of this book.

Note that the forms in Examples 6.13.1 and 6.13.3 are defined on domains with "holes", and moreover, holes of different types. The hole in the case of the magnetic field is the z-axis; the hole for ω_3 is just the origin of $\mathbb{R}^3$. You might imagine catching the z-axis with a lasso, but it would require a butterfly net to catch the origin.

Star-shaped domains

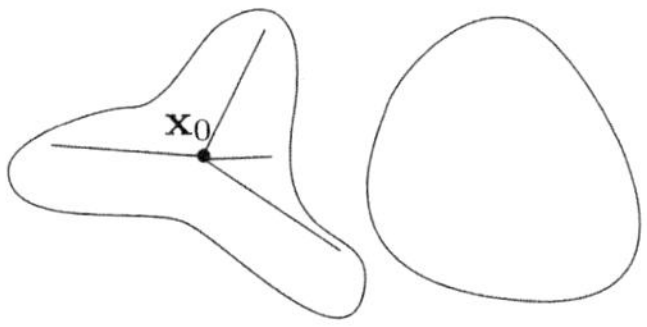

FIGURE 6.13.1.

LEFT: A star-shaped domain; every point can be seen from the black point $\mathbf{x}_0$. RIGHT: A convex domain; every point can be seen from every point.

The forms ω_n of Example 6.13.3 show that $\mathbf{d}\varphi = 0$ does not imply the existence of ψ such that $\mathbf{d}\psi = \varphi$. We now show that in *domains without holes*, $\mathbf{d}\varphi = 0$ does imply the existence of ψ with $\mathbf{d}\psi = \varphi$. The very best way of being without holes is to be convex. The second best way is to be *star-shaped*, illustrated by Figure 6.13.1, left. The proof of the Poincaré lemma is no harder in that setting than for convex domains, and the added generality is sometimes useful.

Definition 6.13.4 (Convex domain, star-shaped domain). A domain $U \subset \mathbb{R}^n$ is *convex* if for any two points $\mathbf{x}, \mathbf{y} \in U$, the straight line segment $[\mathbf{x}, \mathbf{y}]$ joining $\mathbf{x}$ to $\mathbf{y}$ lies entirely in U.

A domain $U \subset \mathbb{R}^n$ is *star-shaped with respect to* $\mathbf{x}_0 \in U$ if for all $\mathbf{x} \in U$, the straight line segment $[\mathbf{x}_0, \mathbf{x}]$ joining $\mathbf{x}_0$ to $\mathbf{x}$ lies entirely in U. It is *star-shaped* if it is star-shaped with respect to one of its points.

Poincaré's lemma has innumerable applications, for example in electromagnetism and gravitation.

Poincaré called this result a lemma, but it is a major theorem. This is not the only case where an author underestimated the importance of a result.

Theorem 6.13.5 (Poincaré's lemma). *Let $U \subset \mathbb{R}^n$ be open and star-shaped. Then $\varphi \in A^k(U)$ can be written $\varphi = \mathbf{d}\psi$ for some $(k-1)$-form $\psi \in A^{k-1}(U)$ if and only if $\mathbf{d}\varphi = 0$.*

In particular, Theorem 6.13.5 is true if U is open and convex.

The case relevant to mechanics in $\mathbb{R}^3$ is an immediate consequence.

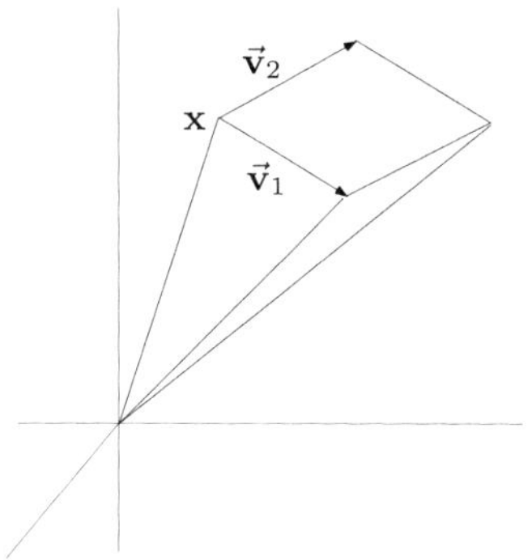

FIGURE 6.13.2.

The cone over the 2-parallelogram $P_{\mathbf{x}}(\vec{\mathbf{v}}_1, \vec{\mathbf{v}}_2)$ is obtained by joining every point of $P_{\mathbf{x}}(\vec{\mathbf{v}}_1, \vec{\mathbf{v}}_2)$ to the origin. It is 3-dimensional: a full ice cream cone, not an empty one. The *cone over the boundary* of the parallelogram is 2-dimensional, consisting of the four triangles joining the origin to the edges of the parallelogram. The *boundary of the cone* includes the original parallelogram, so it has five faces.

Corollary 6.13.6. *If $U \subset \mathbb{R}^3$ is convex (or, more generally, star-shaped) and $\vec{F}$ is a C^1 vector field on U, then $\vec{F}$ is the gradient of a C^1 function f on U if and only if* $\operatorname{curl} \vec{F} = \vec{0}$.

Proof of Theorem 6.13.5. That $\mathbf{d}\varphi = 0$ is a necessary condition is Theorem 6.7.8: if $\varphi = \mathbf{d}\psi$, then $\mathbf{d}\varphi = \mathbf{dd}\psi = 0$.

The converse is the delicate direction. We will prove it as the slightly stronger statement Theorem 6.13.12. This requires some new vocabulary: the *cone over a k-parallelogram* and the *cone operator*. These constructions are analogous to the notions of boundary and the exterior derivative.

Cones and their boundaries

The *cone* over a parallelogram is illustrated by Figures 6.13.2 and 6.13.3.

Definition 6.13.7 (Cone over a k-parallelogram). If $P_{\mathbf{x}}(\vec{\mathbf{v}}_1, \dots, \vec{\mathbf{v}}_k)$ is an oriented k-parallelogram in $\mathbb{R}^n$, then the *cone* over the parallelogram, denoted $CP_{\mathbf{x}}(\vec{\mathbf{v}}_1, \dots, \vec{\mathbf{v}}_k)$, is the parametrized domain

$$\gamma : \mathbf{t} \mapsto t_0(\mathbf{x} + t_1\vec{\mathbf{v}}_1 + \dots + t_k\vec{\mathbf{v}}_k), \qquad 6.13.10$$

with all t_i satisfying $0 \le t_i \le 1$.

(We label the variables $t_0, \dots, t_k$ rather than $t_1, \dots, t_{k+1}$ because t_0 plays a distinguished role.)

This is "dual" to the construction of the boundary. The boundary of a parallelogram is one dimension less than the parallelogram, while the cone is one dimension more: the boundary of a 2-dimensional parallelogram is its 1-dimensional edges, while its cone is 3-dimensional.

The boundary of the cone over $P_{\mathbf{x}}(\vec{\mathbf{v}}_1, \dots, \vec{\mathbf{v}}_k)$ consists of $P_{\mathbf{x}}(\vec{\mathbf{v}}_1, \dots, \vec{\mathbf{v}}_k)$ together with the union of the cones over the faces of $\partial P_{\mathbf{x}}(\vec{\mathbf{v}}_1, \dots, \vec{\mathbf{v}}_k)$:

$$|\partial CP_{\mathbf{x}}(\vec{\mathbf{v}}_1, \dots, \vec{\mathbf{v}}_k)| = |P_{\mathbf{x}}(\vec{\mathbf{v}}_1, \dots, \vec{\mathbf{v}}_k)| + |C\partial P_{\mathbf{x}}(\vec{\mathbf{v}}_1, \dots, \vec{\mathbf{v}}_k)|, \qquad 6.13.11$$

where the absolute values mean that we are not considering boundary orientation.

The cone is naturally oriented by its parametrization. The following lemma tells how the boundary is oriented.

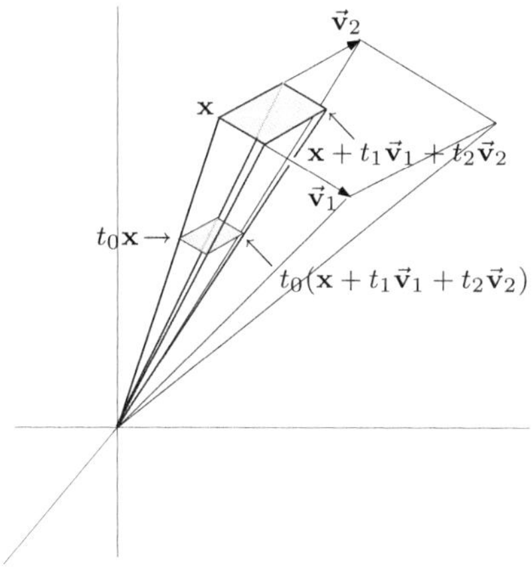

FIGURE 6.13.3.

Any point in the cone can be identified using the parametrization described in formula 6.13.10.

Proposition 6.13.8 (Boundary orientation of a cone). *The oriented boundary of a cone over a k-parallelogram is given by*

$$\partial CP_{\mathbf{x}}(\vec{\mathbf{v}}_1, \dots, \vec{\mathbf{v}}_k) = +\underbrace{P_{\mathbf{x}}(\vec{\mathbf{v}}_1, \dots, \vec{\mathbf{v}}_k)}_{\substack{\text{base of cone} \\ \text{over } P_{\mathbf{x}}(\vec{\mathbf{v}}_1, \dots, \vec{\mathbf{v}}_k)}} - \underbrace{C\partial P_{\mathbf{x}}(\vec{\mathbf{v}}_1, \dots, \vec{\mathbf{v}}_k)}_{\substack{\text{cone over boundary of} \\ P_{\mathbf{x}}(\vec{\mathbf{v}}_1, \dots, \vec{\mathbf{v}}_k)}}. \qquad 6.13.12$$

Thus equation 6.13.11 becomes

$$\partial CP = P - C\partial P; \qquad 6.13.13$$

∂CP consists of P taken with its orientation and $C\partial P$ taken against its orientation.

The boundary of a 2-parallelogram is its four edges; the cone over each edge is the triangle that connects the edge to the origin. So the cone over the boundary of a 2-parallelogram consists of four triangles, while the boundary of the cone over a 2-parallelogram consists of those four triangles plus the parallelogram itself; see Figure 6.13.2.

Remark. The oriented boundary of the oriented boundary of a cone is 0, as it should be. In the case of a 2-parallelogram, $\partial\partial CP_{\mathbf{x}}(\vec{\mathbf{v}}_1, \vec{\mathbf{v}}_2)$ consists of the parallelogram's four edges, plus the four vectors connecting the vertices of the parallelogram to the origin. But each edge of the parallelogram is also an edge of a triangle, and each vector connecting the parallelogram to the origin is an edge of two triangles. Equation 6.13.12 deals with this by counting each part of $\partial\partial C$ twice, with opposite orientations. △

Since $\mathbf{dd}\varphi = 0$, in order for Stokes's theorem to be true the oriented boundary of an oriented boundary must be 0:

$$\int_{\partial\partial A} \varphi = \int_{\partial A} \mathbf{d}\varphi = \int_A \mathbf{dd}\varphi = 0.$$

Proof of Proposition 6.13.8. The cone over an oriented k-parallelogram carries the orientation given by the parametrization γ defined in Definition 6.13.7. Denote that orientation by Ω. Then (Definition 6.4.2) the partial derivatives of γ,

$$\underbrace{\mathbf{x} + t_1\vec{\mathbf{v}}_1 + \cdots + t_k\vec{\mathbf{v}}_k}_{\overrightarrow{D_0\gamma}},\ \underbrace{t_0\vec{\mathbf{v}}_1}_{\overrightarrow{D_1\gamma}},\ \dots,\ \underbrace{t_0\vec{\mathbf{v}}_k}_{\overrightarrow{D_k\gamma}}, \qquad 6.13.14$$

form a direct basis for $CP_{\mathbf{x}}(\vec{\mathbf{v}}_1, \dots, \vec{\mathbf{v}}_k)$.

The change of basis matrix from the basis of partial derivatives to the basis $\mathbf{x}, \vec{\mathbf{v}}_1, \dots, \vec{\mathbf{v}}_k$ is the lower-triangular, $(k+1)\times(k+1)$ matrix

$$\begin{bmatrix} 1 & 0 & 0 & \dots & 0 \\ t_1 & t_0 & 0 & \dots & 0 \\ t_2 & 0 & t_0 & \dots & 0 \\ \vdots & \vdots & \vdots & \ddots & \vdots \\ t_k & 0 & 0 & \dots & t_0 \end{bmatrix}, \qquad 6.13.15$$

with 0's everywhere except the diagonal and the first column. Since (Theorem 4.8.9) the determinant of this matrix is $t_0^k > 0$ when $t_0 \neq 0$, the basis $\vec{\mathbf{x}}, \vec{\mathbf{v}}_1, \dots, \vec{\mathbf{v}}_k$ is direct for $CP_{\mathbf{x}}(\vec{\mathbf{v}}_1, \dots, \vec{\mathbf{v}}_n)$. Thus $\Omega(\vec{\mathbf{x}}, \vec{\mathbf{v}}_1, \dots, \vec{\mathbf{v}}_k) = +1$.

When we say that $\vec{\mathbf{x}}$ points out of the cone at any point of $P_{\mathbf{x}}(\vec{\mathbf{v}}_1, \dots, \vec{\mathbf{v}}_k)$, we are treating $\vec{\mathbf{x}}$ as a vector that can be moved about.

That part of the boundary of a cone over a base B that corresponds to B contributes positively to the boundary orientation.

Now we can check the + sign in equation 6.13.12 for the k-parallelogram $P_{\mathbf{x}}(\vec{\mathbf{v}}_1, \dots, \vec{\mathbf{v}}_k)$. The vector $\vec{\mathbf{x}}$ points out of the cone at any point of $P_{\mathbf{x}}(\vec{\mathbf{v}}_1, \dots, \vec{\mathbf{v}}_k)$, so $\vec{\mathbf{x}}$ can play the role of $\vec{\mathbf{v}}_{\text{out}}$ in equation 6.6.15 and

$$\underbrace{\Omega^{\partial}(\vec{\mathbf{v}}_1, \dots, \vec{\mathbf{v}}_k)}_{\substack{\text{orientation of } P_{\mathbf{x}} \\ \text{as boundary of } CP_{\mathbf{x}}}} = \underbrace{\Omega(\vec{\mathbf{x}}, \vec{\mathbf{v}}_1, \dots, \vec{\mathbf{v}}_k)}_{\text{orientation of the cone } CP_{\mathbf{x}}} = +1. \qquad 6.13.16$$

Thus the given orientation of $P_{\mathbf{x}}(\vec{\mathbf{v}}_1, \dots, \vec{\mathbf{v}}_k)$ coincides with its orientation as part of the boundary of $CP_{\mathbf{x}}(\vec{\mathbf{v}}_1, \dots, \vec{\mathbf{v}}_n)$, justifying the + sign in equation 6.13.12.

To summarize: If P is any oriented parallelogram, it contributes positively to the parametrized domain ∂CP. In particular, if we take any face σ of ∂P (with its orientation as part of ∂P), then σ contributes positively to $\partial C\sigma$ (where $C\sigma$ is the cone over the oriented parallelogram σ). So $|\sigma|$

In Definition 6.13.9, the hypothesis that U is open and star-shaped guarantees that

$$CP_{\mathbf{x}}(h\vec{\mathbf{v}}_1, \dots, h\vec{\mathbf{v}}_{k-1})$$

is contained in U for all $\mathbf{x} \in U$ and all h with $|h|$ sufficiently small, so that the integral in equation 6.13.18 makes sense for such h. We require U to be star-shaped with respect to $\mathbf{0}$ since our cones are anchored at the origin.

Compare the exterior derivative and the cone operator: The exterior derivative integrates over the boundary and takes a limit; the cone operator integrates over the cone and takes a limit.

Compare formula 6.13.20 to equation 6.2.1 in Definition 6.2.1:

$$\int_{[\gamma(U)]} \varphi = \int_U \varphi(P_{\gamma(\mathbf{u})}(\overrightarrow{D_1\gamma}(\mathbf{u}), .., \overrightarrow{D_k\gamma}(\mathbf{u})))|d^k\mathbf{u}|.$$

To evaluate $\varphi = f\, dx_{i_1} \wedge \dots \wedge dx_{i_k}$ on the parallelogram

$$P_{\gamma(\mathbf{t})}(\overrightarrow{D_1\gamma}(\mathbf{t}), \dots, \overrightarrow{D_k\gamma}(\mathbf{t}))$$

we evaluate f at $\gamma(\mathbf{t})$ and we evaluate $dx_{i_1} \wedge \dots \wedge dx_{i_k}$ on the partial derivatives of γ at $\mathbf{t}$.

contributes $+2\sigma$ to $\partial P + \partial C \partial P$, and 0σ to $\partial P - \partial C \partial P$. Since the oriented boundary of an oriented boundary must be 0, then σ must contribute 0 to $\partial\partial CP$, so faced with the choice between

$$\partial\partial CP = +\partial P + \partial C \partial P \quad \text{and} \quad \partial\partial CP = +\partial P - \partial C \partial P, \qquad 6.13.17$$

we must choose the second. □

The cone operator

Just as the boundary ∂ allowed us to associate the $(k+1)$-form $\mathbf{d}\varphi$ to a k-form φ, the cone C allows us to associate a $(k-1)$-form $\mathbf{c}\varphi$ to a k-form:

Definition 6.13.9 (Cone operator). Let $U \subset \mathbb{R}^n$ be open and star-shaped with respect to $\mathbf{0}$. The *cone operator* $\mathbf{c}\colon A^k(U) \to A^{k-1}(U)$ is defined by the formula

$$\mathbf{c}\varphi(P_{\mathbf{x}}(\vec{\mathbf{v}}_1, \dots, \vec{\mathbf{v}}_{k-1})) \overset{\text{def}}{=} \lim_{h \to 0} \frac{1}{h^{k-1}} \int_{CP_{\mathbf{x}}(h\vec{\mathbf{v}}_1, \dots, h\vec{\mathbf{v}}_{k-1})} \varphi. \qquad 6.13.18$$

We will see in Theorem 6.13.12 that this form $\mathbf{c}\varphi$ is the solution to our problems; it is the $(k-1)$-form ψ of Theorem 6.13.5. If $\mathbf{d}\varphi = 0$, then $\mathbf{d}(\mathbf{c}\varphi) = \varphi$. But first we need to show that the limit in equation 6.13.18 exists.

Lemma 6.13.10. *The limit given in equation 6.13.18 exists, and defines a $(k-1)$-form on U.*

Proof. It is enough to prove the result when

$$\varphi = f\, dx_{i_1} \wedge \dots \wedge dx_{i_k}, \text{ where } f \text{ is a } C^1 \text{ function.} \qquad 6.13.19$$

Since the cone $CP_{\mathbf{x}}(\vec{\mathbf{v}}_1, \dots, \vec{\mathbf{v}}_{k-1})$ comes parametrized by γ (see Definition 6.13.7), we can write

$$\begin{aligned}
\mathbf{c}\varphi\big(P_{\mathbf{x}}(\vec{\mathbf{v}}_1, \dots, \vec{\mathbf{v}}_{k-1})\big) &= \lim_{h\to 0} \frac{1}{h^{k-1}} \int_{CP_{\mathbf{x}}(h\vec{\mathbf{v}}_1, \dots, h\vec{\mathbf{v}}_{k-1})} \overbrace{f\, dx_{i_1} \wedge \dots \wedge dx_{i_k}}^{\varphi} \\
&= \lim_{h\to 0} \frac{1}{h^{k-1}} \int_{[0,1]\times[0,h]\times\cdots\times[0,h]} \varphi(P_{\gamma(\mathbf{t})}(\overrightarrow{D_1\gamma}, \dots, \overrightarrow{D_k\gamma}))|d\mathbf{t}^k| \qquad 6.13.20 \\
&= \lim_{h\to 0} \frac{1}{h^{k-1}} \int_0^1 \int_0^h \cdots \int_0^h \overbrace{f\big(t_0(\mathbf{x} + t_1\vec{\mathbf{v}}_1 + \dots + t_{k-1}\vec{\mathbf{v}}_{k-1})\big)}^{f(\gamma(\mathbf{t}))} \\
&\quad \underbrace{dx_{i_1} \wedge \ldots \wedge dx_{i_k}}_{\text{elem. } k\text{-form}} \underbrace{\Big(\mathbf{x} + t_1\vec{\mathbf{v}}_1 + \dots + t_{k-1}\vec{\mathbf{v}}_{k-1},\ t_0\vec{\mathbf{v}}_1, \dots, t_0\vec{\mathbf{v}}_{k-1}\Big)}_{k \text{ vectors in } \mathbb{R}^n\text{: } \overrightarrow{D_0\gamma}\overrightarrow{D_1\gamma}, \dots, \overrightarrow{D_{k-1}\gamma}} |dt_{k-1} \dots dt_1\, dt_0|.
\end{aligned}$$

This looks frightful, but it is just a function (on the third line) times a k-form evaluated on k vectors (on the fourth line), and in any case the integral simplifies remarkably. First, by multilinearity and antisymmetry,

$$\begin{aligned} dx_{i_1} \wedge \cdots \wedge dx_{i_k}[\mathbf{x} + t_1\vec{\mathbf{v}}_1 + \cdots + t_{k-1}\vec{\mathbf{v}}_{k-1},\ t_0\vec{\mathbf{v}}_1, \dots, t_0\vec{\mathbf{v}}_{k-1}] \\ = t_0^{k-1} dx_{i_1} \wedge \cdots \wedge dx_{i_k}[\mathbf{x}, \vec{\mathbf{v}}_1, \dots, \vec{\mathbf{v}}_{k-1}]. \end{aligned} \qquad 6.13.21$$

(You are asked to justify equation 6.13.21 in Exercise 6.13.6.) Next, observe that (by Corollary 1.9.2) there exists a constant K such that

$$\Big|f\big(\underbrace{t_0(\mathbf{x} + t_1\vec{\mathbf{v}}_1 + \cdots + t_{k-1}\vec{\mathbf{v}}_{k-1})}_{\mathbf{b}}\big) - f\big(\underbrace{t_0\mathbf{x}}_{\mathbf{a}}\big)\Big| \le hK. \qquad 6.13.22$$

Corollary 1.9.2:

$$|f(\mathbf{b}) - f(\mathbf{a})| \le \left(\sup_{\mathbf{c}\in[\mathbf{a},\mathbf{b}]} \Big|[\mathbf{D}f(\mathbf{c})]\Big|\right)|\overrightarrow{\mathbf{b}-\mathbf{a}}|.$$

Here we have (remember $|t_0| \le 1$ and that $0 \le t_1, \dots, t_{k-1} \le h$)

$$\begin{aligned} |\mathbf{b}-\mathbf{a}| &= |t_0(t_1 v_1 + \cdots + t_{k-1}v_{k-1}| \\ &\le |t_0||t_1 v_1 + \cdots + t_{k-1}v_{k-1}| \\ &\le |t_1 v_1 + \cdots + t_{k-1}v_{k-1}| \\ &\le h(|v_1| + \cdots + |v_{k-1}|). \end{aligned}$$

The sum $(|v_1| + \cdots + |v_{k-1}|)$ contributes to K in equation 6.13.22, as does the sup of the derivative.

Since

$$\frac{1}{h^{k-1}} \int_0^h \cdots \int_0^h hK\ dt_{k-1} \dots dt_1 = hK \qquad 6.13.23$$

will disappear in the limit, we can replace $f\big(t_0(\mathbf{x} + t_1\vec{\mathbf{v}}_1 + \cdots + t_{k-1}\vec{\mathbf{v}}_{k-1})\big)$ by $f(t_0\mathbf{x})$. After these replacements, the variables $t_i, \dots, t_{k-1}$ no longer appear in the integrand, so the integrals from 0 to h each contribute h, giving h^{k-1}, which cancels with the h^{k-1} in the denominator. Thus we can write

$$\begin{aligned} \mathbf{c}\varphi\big(P_{\mathbf{x}}(\vec{\mathbf{v}}_1, \dots, \vec{\mathbf{v}}_{k-1})\big) &= \lim_{h\to 0} \frac{1}{h^{k-1}} \int_{CP_{\mathbf{x}}(h\vec{\mathbf{v}}_1, \dots, h\vec{\mathbf{v}}_{k-1})} \varphi \\ &= \int_0^1 f(t_0\mathbf{x}) t_0^{k-1}\Big(dx_{i_1} \wedge \cdots \wedge dx_{i_k}\big(P_{t_0\mathbf{x}}(\mathbf{x}, \vec{\mathbf{v}}_1, \dots, \vec{\mathbf{v}}_{k-1})\big)\Big)\, dt_0. \end{aligned} \qquad 6.13.24$$

Some texts simply define $\mathbf{c}\varphi$ using equation 6.13.24, giving no geometrical interpretation in terms of cones. We find the geometrical interpretation helpful. It also makes the proof of the Poincaré lemma much more transparent.

Thus the limit exists, and is certainly multilinear and antisymmetric as a function of the $\vec{\mathbf{v}}$'s. □

Example 6.13.11 (Computing the cone of a form). We will compute $\mathbf{c}\varphi$ when $\varphi = x_3\, dx_1 \wedge dx_3$. Since φ is a 2-form on $\mathbb{R}^3$, $\mathbf{c}\varphi$ is a 1-form on $\mathbb{R}^3$ and can be written

$$\mathbf{c}\varphi = f\, dx_1 + g\, dx_2 + h\, dx_3. \qquad 6.13.25$$

To find the coefficients f, g, and h, we evaluate $\mathbf{c}\varphi$ on the standard basis vectors (see Theorem 6.1.8). To get f, we evaluate $\mathbf{c}\varphi$ on $P_{\mathbf{x}}(\vec{\mathbf{e}}_1)$:

Equation 6.13.26, first line: The $t_0 x_3$ is the third coordinate of $t_0\mathbf{x}$; it corresponds to $f(t_0\mathbf{x})$ in equation 6.13.24. The second t_0 corresponds to t_0^{k-1} in equation 6.13.24.

$$\begin{aligned} f(\mathbf{x}) = \mathbf{c}(\underbrace{x_3\, dx_1 \wedge dx_3}_{\varphi})\left(P_{\mathbf{x}}\begin{pmatrix}1\\0\\0\end{pmatrix}\right) &\underbrace{=}_{\substack{\text{eq.}\\ 6.13.24}} \int_0^1 \underbrace{t_0 x_3}_{\substack{\text{see}\\ \text{margin}}}\ t_0\, dx_1 \wedge dx_3\left(P_{t_0\mathbf{x}}\left(\begin{bmatrix}x_1\\x_2\\x_3\end{bmatrix}, \begin{bmatrix}1\\0\\0\end{bmatrix}\right)\right) dt_0 \\ &= \int_0^1 -x_3^2 t_0^2\, dt_0 = -\frac{x_3^2}{3}. \end{aligned} \qquad 6.13.26$$

Exercise 6.13.10 asks you to show that

$$g(\mathbf{x}) = 0 \quad \text{and} \quad h(\mathbf{x}) = \frac{x_1 x_3}{3}. \qquad 6.13.27$$

Thus

$$\mathbf{c}(x_3\, dx_1 \wedge dx_3) = -\frac{x_3^2}{3}\, dx_1 + \frac{x_1 x_3}{3}\, dx_3. \quad \triangle \qquad 6.13.28$$

Statement and proof of a strong form of Poincaré's lemma

We could state Theorem 6.13.12 for a subset U that is star-shaped with respect to any point $\mathbf{x}_0$, but then we would have to anchor our cones at $\mathbf{x}_0$, which would complicate notation.

The idea behind the proof of Theorem 6.13.12 is that the boundary of the cone is the cone over the boundary plus the original face. When $k = 2$, the cone over the boundary of the parallelogram consists of four triangles, while the boundary of the cone consists of those four triangles plus the original parallelogram; see Figure 6.13.2. As shown in equation 6.13.30, $\mathbf{d}(\mathbf{c}\varphi)$ is $\mathbf{c}\varphi$ integrated over the boundary of the parallelogram P; this equals φ integrated over the cone over the boundary of P. In equation 6.13.29, the plus sign may seem confusing; you might expect $\mathbf{c}(\mathbf{d}\varphi)$ to be the sum of $\mathbf{d}(\mathbf{c}\varphi)$ and φ, but the orientation of the various domains of integration makes the arithmetic come out right.

Theorem 6.13.12 (Poincaré's lemma: a variant). *Let $U \subset \mathbb{R}^n$ be open and star-shaped with respect to $\mathbf{0}$. Let φ be a k-form on U. Then*

$$\varphi = \mathbf{d}(\mathbf{c}\varphi) + \mathbf{c}(\mathbf{d}\varphi). \qquad 6.13.29$$

In particular, if $\mathbf{d}\varphi = 0$, then $\varphi = \mathbf{d}(\mathbf{c}\varphi)$.

Proof. Written without all the proper detail, we have

$$\mathbf{d}(\mathbf{c}\varphi)(P) \overset{1}{=} (\lim) \int_{\partial P_h} \mathbf{c}\varphi \overset{2}{=} (\lim) \int_{C\partial P_h} \varphi \overset{3}{=} (\lim) \int_{P_h} \varphi - (\lim) \int_{\partial C P_h} \varphi$$
$$\overset{4}{=} (\lim) \int_{P_h} \varphi - (\lim) \int_{C P_h} \mathbf{d}\varphi = \varphi(P) - \mathbf{c}(\mathbf{d}\varphi)(P). \qquad 6.13.30$$

Equality (1) is the definition of the exterior derivative, (2) is the definition of the cone operator, (3) is equation 6.13.13, and (4) is Stokes's theorem. □

Exercise 6.13.9 asks you to rewrite the proof of Theorem 6.13.12 with the proper detail and to justify the last equality.

Example 6.13.13 (Poincaré's lemma). Let $\varphi = x_3\, dx_1 \wedge dx_3$, as in Example 6.13.11. Note that $\mathbf{d}(x_3\, dx_1 \wedge dx_3) = 0$. Therefore, by Theorem 6.13.12, we should have

$$\underbrace{\mathbf{d}(\mathbf{c}\, x_3\, dx_1 \wedge dx_3)}_{\mathbf{d}\mathbf{c}\varphi} = \underbrace{x_3\, dx_1 \wedge dx_3}_{\varphi}. \qquad 6.13.31$$

Indeed,

$$\mathbf{d}(\mathbf{c}x_3\, dx_1 \wedge dx_3) \overset{\text{eq. 6.13.28}}{=} \mathbf{d}\left(-\frac{x_3^2}{3}\, dx_1 + \frac{x_1 x_3}{3}\, dx_3\right) \qquad 6.13.32$$
$$= -\frac{2x_3}{3}\, dx_3 \wedge dx_1 + \frac{x_3}{3}\, dx_1 \wedge dx_3 = x_3\, dx_1 \wedge dx_3. \quad \triangle$$

Potentials and electromagnetism

Equation 6.13.33: A force field $\vec{F}$ is conservative if $\mathbf{d}W_{\vec{F}} = 0$. In some sense $\mathbf{d}\mathbb{F} = 0$ says that electromagnetism is conservative.

Spacetime is of course convex.

Recall that one way of stating Maxwell's equations is (equation 6.12.15)

$$\mathbf{d}\mathbb{F} = 0, \qquad \mathbf{d}\mathbb{M} = 4\pi\mathbb{J}. \qquad 6.13.33$$

From the first of these equations and the Poincaré lemma (and the fact that $\mathbb{R}^4$ is convex), it follows that there is a 1-form $\mathbb{A}$ on spacetime such that $\mathbf{d}\mathbb{A} = \mathbb{F}$, i.e., $\mathbb{A}$ is a potential for $\mathbb{F}$: one possible choice of $\mathbb{A}$ is $\mathbf{c}\mathbb{F}$.

In any splitting of spacetime into space and time, any 1-form on spacetime can be written

$$\mathbb{A} = \frac{1}{c} W_{\vec{A}} - V c\, dt, \qquad 6.13.34$$

where $W_{\vec{A}}$ is the work of some vector field $\vec{A}$ (called the *vector potential*) and V is a function (called the *scalar potential*). Then, since

$$\mathbf{d}\mathbb{A} = \underbrace{\frac{1}{c}\Phi_{\vec{\nabla}\times\vec{A}} - \frac{1}{c^2}W_{D_t\vec{A}} \wedge c\,dt}_{\mathbf{d}\left(\frac{1}{c}W_{\vec{A}}\right)} - W_{\vec{\nabla}V} \wedge c\,dt = \mathbb{F}, \qquad 6.13.35$$

by $\mathbb{F} = W_{\vec{\mathbf{E}}} \wedge c\,dt + \Phi_{\vec{\mathbf{B}}}$ (equation 6.12.8) we have

$$\vec{\mathbf{E}} = -(\vec{\nabla}V + \frac{1}{c^2}D_t\vec{A}) \quad \text{and} \quad \vec{\mathbf{B}} = \frac{1}{c}\vec{\nabla}\times\vec{A}, \qquad 6.13.36$$

so that $\mathbb{M} \stackrel{\text{def}}{=} W_{\vec{\mathbf{B}}} \wedge cdt - \Phi_{\vec{\mathbf{E}}}$ (the second equation in 6.12.8) becomes

$$\mathbb{M} = W_{\frac{1}{c}\vec{\nabla}\times\vec{A}} \wedge c\,dt + \Phi_{(\vec{\nabla}V + \frac{1}{c^2}D_t\vec{A})}. \qquad 6.13.37$$

This writes $\mathbb{M}$ in terms of $\mathbb{A}$, i.e., in terms of $\vec{A}$ and V. Since $\mathbf{d}\mathbb{F} = 0$ has been built into our equations via $\mathbf{d}\mathbb{A} = \mathbb{F}$, the equation $\mathbf{d}\mathbb{M} = 4\pi\mathbb{J}$ encodes all of electromagnetism, becoming

$$\begin{aligned}\mathbf{d}\mathbb{M} &= \Phi_{\left(\frac{1}{c}\vec{\nabla}\times(\vec{\nabla}\times\vec{A})\right) + \frac{1}{c^3}(D_t^2\vec{A} + \frac{1}{c}\vec{\nabla}D_tV)} \wedge c\,dt + M_{\vec{\nabla}\cdot(\frac{1}{c^2}D_t\vec{A} + \vec{\nabla}V)} \\ &= 4\pi\mathbb{J} = 4\pi\left(\frac{1}{c}\Phi_{\vec{\mathbf{j}}} \wedge c\,dt - M_\rho\right).\end{aligned} \qquad 6.13.38$$

(The second line uses equations 6.12.12 and 6.12.13.)

Written in components, changing sign and multiplying through by c for the first equation below, this becomes

$$\underbrace{\vec{\nabla}\times(\vec{\nabla}\times\vec{A})}_{\text{curl curl }\vec{A}} + \left(\frac{1}{c^2}D_t^2\vec{A} + \vec{\nabla}D_tV\right) = 4\pi\vec{\mathbf{j}} \qquad 6.13.39$$

$$\vec{\nabla}\cdot\left(\frac{1}{c^2}D_t\vec{A} + \vec{\nabla}V\right) = \underbrace{\vec{\nabla}\cdot\vec{\nabla}V}_{\text{div grad }V} + \vec{\nabla}\left(\frac{1}{c^2}D_t\vec{A}\right) = -4\pi\rho. \qquad 6.13.40$$

A first way to improve these equations is to recall (equation 6.12.67) that the Laplacian $\vec{\Delta}\vec{A}$ of a vector field $\vec{A}$ is $\vec{\Delta}\vec{A} = \text{grad div }\vec{A} - \text{curl curl }\vec{A}$, and the Laplacian ΔV of a function V is $\Delta V = \text{div grad }V$. Using these relations, we can rewrite equations 6.13.39 and 6.13.40 as

$$(\vec{\Delta}\vec{A} - \frac{1}{c^2}D_t^2\vec{A}) - \vec{\nabla}(\vec{\nabla}\cdot\vec{A} + D_tV) = -4\pi\vec{\mathbf{j}} \qquad 6.13.41$$

$$\Delta V + \frac{1}{c^2}\vec{\nabla}\cdot D_t\vec{A} = -4\pi\rho. \qquad 6.13.42$$

This system of equations, although still scary, is much better than Maxwell's equation: it is an equation for four unknown functions instead of six. But they can be simplified further. Note (see the comment in the margin) that $\mathbb{A}$ is only defined up to addition of $\mathbf{d}f$ for some function f on spacetime. It turns out that we can choose f so that after adding $\frac{1}{c}\mathbf{d}f$ to our potential $\mathbb{A}$, the new potential will satisfy the *Lorenz gauge condition*

$$\vec{\nabla}\cdot\vec{A} + D_tV = 0, \quad \text{i.e.,} \quad \vec{\nabla}\cdot\vec{A} = -D_tV, \qquad 6.13.43$$

In equation 6.13.35, the derivative $\mathbf{d}\left(\frac{1}{c}W_{\vec{A}}\right)$ is split into the derivative with respect to space variables and the derivative with respect to time variables.

The splitting of $\mathbb{A}$ into scalar potential and vector potential depends on the splitting of spacetime into space and time. In addition, if $\mathbf{d}\mathbb{A} = \mathbb{F}$, then for any function f we have $\mathbf{d}(\mathbb{A} + \mathbf{d}f) = \mathbb{F}$.

Thus the potential $\mathbb{A}$ is only defined up to adding $\mathbf{d}f$ for an arbitrary function f. It seems that this should mean that $\mathbb{A}$ is a mathematical construct without physical reality. This is not true: Hermann Weyl, in 1928, found that $\mathbb{A}$ can be understood as a "connection in a 1-dimensional bundle"; just as gravity is the curvature of spacetime in general relativity, the curvature of the connection $\mathbb{A}$ is the electromagnetic field. Adding $\mathbf{d}f$ means "using different coordinates in the bundle". Physicists call this "working in a different gauge".

This idea turned out to be immensely important: in 1954 Yang and Mills interpreted the strong nuclear force, which keeps atomic nuclei together, as the curvature of a connection in a 2-dimensional bundle. Soon thereafter, gauge theory took over particle physics completely; today particle physics is gauge theory.

In particular particle physicists are now differential geometers, and the interaction between these branches of mathematics and physics has been immensely profitable to both fields. Simon Donaldson, Michael Freedman, and Edward Witten each received the Fields Medal for work in gauge field theory.

which, differentiated with respect to t, gives

$$\vec{\nabla} \cdot D_t \vec{A} = D_t \vec{\nabla} \cdot \vec{A} = -D_t^2 V. \tag{6.13.44}$$

Using this identity, our system of equations decouples into

$$\vec{\Delta} \vec{A} - \frac{1}{c^2} D_t^2 \vec{A} = -4\pi \vec{\mathbf{j}} \tag{6.13.45}$$

$$\Delta V - \frac{1}{c^2} D_t^2 V = -4\pi \rho. \tag{6.13.46}$$

Equation 6.13.45 depends only on $\vec{A}$, not on V. It is really three equations for three scalar-valued functions (see equation 6.8.6). Equation 6.13.46 defines a single scalar-valued function that depends only on V. So Maxwell's complicated equations are now reduced to four separate scalar wave equations.

Equation 6.13.48 is just equation 6.13.43, where $\vec{A}$ has been replaced by $\vec{A} + \vec{\nabla} f$ and V has been replaced by $V - \frac{1}{c^2} D_t f$.

The equation $\Delta f - \frac{1}{c^2} D_t^2 f = g$, where g is known and f is unknown, is called the *inhomogeneous wave equation*; it has solutions for very general inhomogeneities (i.e., right side of the equation).

The above derivation depends on our assumption that we can choose f such that the potential $\mathbb{A} + \frac{1}{c}\mathbf{d}f$ satisfies the Lorenz gauge condition. Let us justify that assumption.

Equation 6.13.34 tells us how to split the 1-form $\mathbb{A}$ into scalar and vector potentials: $\mathbb{A} = \frac{1}{c} W_{\vec{A}} - V c\, dt$. This gives

$$\mathbb{A} + \frac{1}{c}\mathbf{d}f = \frac{1}{c} W_{\vec{A} + \vec{\nabla} f} - (V - \frac{1}{c^2} D_t f) \wedge c dt. \tag{6.13.47}$$

The new vector potential $\vec{A} + \vec{\nabla} f$ and the new scalar potential $V - \frac{1}{c^2} D_t f$ will satisfy the Lorenz gauge condition if

$$\vec{\nabla} \cdot (\vec{A} + \vec{\nabla} f) + D_t \left(V - \frac{1}{c^2} D_t f \right) = 0. \tag{6.13.48}$$

Note that we can rewrite the Lorenz gauge condition as

$$\Delta f - \frac{1}{c^2} D_t^2 f = -(\vec{\nabla} \cdot \vec{A} + D_t V), \tag{6.13.49}$$

which is an *inhomogeneous wave equation*, which can be solved for f. In fact, there are explicit formulas for its solution.[21] Thus it is possible to find potentials that satisfy the Lorenz gauge for any electromagnetic field.

The Lorenz gauge condition is named after Danish physicist Ludwig Lorenz (1829–1891). It is sometimes mistakenly attributed to the Dutch physicist H. Lorentz, the Lorentz of the "Lorentz force".

Moreover, equations 6.13.45 and 6.13.46 are themselves a system of four inhomogeneous wave equations, one for the scalar potential and one for each component of the vector potential $\vec{A}$. Solving these for V and $\vec{A}$ allow us to find the potentials, hence also the fields, for any distribution of charge and current.

Exercises for Section 6.13

6.13.1 Which of the vector fields of Exercise 1.1.6 are gradients of functions?

---

[21] Stating and checking these formulas would take us too far afield. The interested reader will find a very readable account in D. Griffiths, *Introduction to Electrodynamics*, Chapter 9.

6.13.2 a. Is there a function f on $\mathbb{R}^3$ such that

i. $\mathbf{d}f = \cos(x+yz)\,dx + y\cos(x+yz)\,dy + z\cos(x+yz)\,dz$?

ii. $\mathbf{d}f = \cos(x+yz)\,dx + z\cos(x+yz)\,dy + y\cos(x+yz)\,dz$?

b. Find the function when it exists.

6.13.3 A charge of c coulombs per meter on a vertical wire $x = a$, $y = b$ creates an electric potential $V\begin{pmatrix} x \\ y \\ z \end{pmatrix} = c\ln\Big((x-a)^2 + (y-b)^2\Big)$. Several such wires produce a potential that is the sum of the potentials due to the individual wires.

a. What is the electric field due to a single wire going through the point $\begin{pmatrix} 0 \\ 0 \end{pmatrix}$, with charge per length $c = 1\,\text{coul/m}$?

Exercise 6.13.3: "coul" stands for "coulomb" and "m" for "meter".

b. Sketch the electric field due to two wires, both charged with 1 coul/m, one going through the point $\begin{pmatrix} 1 \\ 0 \end{pmatrix}$ and the other through $\begin{pmatrix} -1 \\ 0 \end{pmatrix}$.

c. Repeat for the first wire charged with 1 coul/m, the other with −1 coul/m.

6.13.4 a. Is $\begin{bmatrix} \dfrac{x}{x^2+y^2} \\ \dfrac{y}{x^2+y^2} \end{bmatrix}$ the gradient of a function on $\mathbb{R}^2 - \{\mathbf{0}\}$?

b. Is the vector field $\begin{bmatrix} x \\ y \\ z \end{bmatrix}$ on $\mathbb{R}^3$ the curl of another vector field?

6.13.5 a. Let $\varphi = x\,dx + y\,dy$. Compute $\mathbf{c}(\varphi)$ and check that $\mathbf{dc}\varphi = \varphi$.

b. Let $\varphi = x\,dy - y\,dx$. Compute $\mathbf{c}(\varphi)$ and check that $\mathbf{dc}\varphi + \mathbf{cd}\varphi = \varphi$.

6.13.6 Use the properties of multilinearity and antisymmetry to justify equation 6.13.21.

6.13.7 a. Find the Faraday 2-form $\mathbb{F}$ of the electromagnetic field defined by the $\vec{\mathbf{E}}$ and $\vec{\mathbf{B}}$ of Exercise 6.12.8.

b. Using the cone operator, find a potential $\mathbb{A}$ for this $\mathbb{F}$.

c. Check that $\mathbf{d}\mathbb{A} = \mathbb{F}$.

****6.13.8** a. Show that a 1-form φ on $\mathbb{R}^2 - \{\mathbf{0}\}$ can be written $\mathbf{d}f$ if and only if $\mathbf{d}\varphi = 0$ and $\int_{S^1}\varphi = 0$, where S^1 is the unit circle, oriented counterclockwise.

b. Show that a 1-form φ on $\mathbb{R}^2 - \left\{\begin{pmatrix} -1 \\ 0 \end{pmatrix}, \begin{pmatrix} 1 \\ 0 \end{pmatrix}\right\}$ can be written $\mathbf{d}f$ if and only if $\mathbf{d}\varphi = 0$ and both $\int_{S_1}\varphi = 0$, $\int_{S_2}\varphi = 0$, where S_1 is the circle of radius 1 centered at $\begin{pmatrix} -1 \\ 0 \end{pmatrix}$ and S_2 is the circle of radius 1 centered at $\begin{pmatrix} 1 \\ 0 \end{pmatrix}$, both oriented counterclockwise.

Exercise 6.13.9, part a: By "proper detail and notation" we mean replace (lim) with the correct description of each limit and replace P_h by the correct notation for the appropriate parallelograms.

6.13.9 a. Rewrite equation 6.13.30 with the proper detail and notation.

b. Carefully justify each equality in equation 6.13.30.

6.13.10 Confirm the computations for g and h in equations 6.13.27.

6.14 Review exercises for Chapter 6

In Exercise 6.1, B is an $n \times n$ matrix and φ and ψ are both 1-forms on $\mathbb{R}^3$; $\vec{\mathbf{v}}$ and $\vec{\mathbf{w}}$ are vectors in $\mathbb{R}^3$; f is integrable.

6.1 Which of the following are numbers? Identify those that are not.

a. $\vec{\mathbf{v}} \cdot \vec{\mathbf{w}}$	b. $dx_1 \wedge dx_2(\vec{\mathbf{v}}, \vec{\mathbf{w}})$	c. $\vec{\mathbf{v}} \times \vec{\mathbf{w}}$	d. $\det B$
e. $\operatorname{rank} B$	f. $\operatorname{tr} B$	g. $\dim A^k(\mathbb{R}^n)$	h. $\lvert\vec{\mathbf{v}}\rvert$
i. $A^k(\mathbb{R}^k)$	j. $\varphi \wedge \psi(\vec{\mathbf{v}}, \vec{\mathbf{w}})$	k. $\int_{\mathbb{R}} f(x)\,dx$	l. $\operatorname{sgn}(\sigma)$

6.2 Let $\vec{F}$ be a vector field in $\mathbb{R}^n$, let $\vec{\mathbf{v}}_1, \dots, \vec{\mathbf{v}}_{n-1}$ be vectors in $\mathbb{R}^n$, and let φ be the $(n-1)$-form on $\mathbb{R}^n$ given by

$$\varphi(\vec{\mathbf{v}}_1, \dots, \vec{\mathbf{v}}_{n-1}) = \det[\vec{F}(\mathbf{x}), \vec{\mathbf{v}}_1, \dots, \vec{\mathbf{v}}_{n-1}].$$

Write φ as a linear combination of elementary $(n-1)$-forms on $\mathbb{R}^n$, in terms of the coordinates of $\vec{F}$.

6.3 Use Definition 6.1.12 to write the wedge product $\varphi \wedge \psi(\vec{\mathbf{v}}_1, \vec{\mathbf{v}}_2, \vec{\mathbf{v}}_3, \vec{\mathbf{v}}_4)$, where φ is a 1-form and ψ is a 3-form, as a combination of values of φ and ψ evaluated on appropriate vectors (as in equations 6.1.28 and 6.1.32).

6.4 Set up each of the following integrals of form fields over parametrized domains as an ordinary multiple integral.

a. $\int_{[\gamma(I)]} y^2 dy + x^2 dz$, where $I = [0, a]$ and $\gamma(t) = \begin{pmatrix} t^3 \\ t^2+1 \\ t^2-1 \end{pmatrix}$

b. $\int_{[\gamma(U)]} \sin y^2\, dx \wedge dz$, where $U = [0,a] \times [0,b]$, and $\gamma\begin{pmatrix} u \\ v \end{pmatrix} = \begin{pmatrix} u^2 - v \\ uv \\ v^4 \end{pmatrix}$

6.5 Find a 1-form field on $\mathbb{R}^2$ whose sign orients the circle of radius 1 centered at $\begin{pmatrix} 3 \\ 0 \end{pmatrix}$ in the clockwise direction.

6.6 Find an orientation Ω for the surface $S \subset \mathbb{R}^3$ given by

$$x^2 + y^3 + z^4 = 1.$$

6.7 Consider the manifold $S^3 \subset \mathbb{R}^4$ of equation $x_1^2 + x_2^2 + x_3^2 + x_4^2 = 1$.

a. Show that $\operatorname{sgn} dx_1 \wedge dx_2 \wedge dx_3$ is not an orientation of S^3.

b. Show that $\Omega_{\mathbf{x}}(\vec{\mathbf{v}}_1, \vec{\mathbf{v}}_2, \vec{\mathbf{v}}_3) = \operatorname{sgn} \det[\mathbf{x}, \vec{\mathbf{v}}_1, \vec{\mathbf{v}}_2, \vec{\mathbf{v}}_3]$ is an orientation.

6.8 a. Show that the locus $M \subset \mathbb{R}^4$ given by $x_1^2 + x_2^2 + x_1 x_4^2 = 1$ is a smooth manifold.

b. Find a 3-form field whose sign orients M.

6.9 In Example 6.4.3 we saw that $\gamma_1(\theta) = \begin{pmatrix} \cos\theta \\ \sin\theta \end{pmatrix}$ and $\gamma_2(\theta) = \begin{pmatrix} \sin\theta \\ \cos\theta \end{pmatrix}$ give opposite orientations. Confirm (Proposition 6.4.8) that $\det[\mathbf{D}(\gamma_2^{-1} \circ \gamma_1)] < 0$.

6.10 Let $z_1 = x_1 + iy_1$, $z_2 = x_2 + iy_2$ be coordinates in $\mathbb{C}^2$. Compute the integral of $dx_1 \wedge dy_1 + dx_2 \wedge dy_2$ over the part of the locus of equation $z_2 = z_1^k$ where $|z_1| < 1$, oriented by $\operatorname{sgn} dx_1 \wedge dy_1$.

6.11 For the following 1-forms, write down the corresponding vector field. Sketch the vector field. Describe a path over which the work of the 1-form would be small. Describe a path over which the work would be large.

a. $(x^2 + y^2)\,dz$, on $\mathbb{R}^3$ b. $y\,dx - x\,dy - z\,dz$, on $\mathbb{R}^3$

6.12 a. In $\mathbb{R}^2$, a vector field defines two 1-forms: the work and the flux. Show that they are related by formula

$$W_{\vec{F}}(\vec{\mathbf{v}}) = \Phi_{\vec{F}}\left(\begin{bmatrix} 0 & -1 \\ 1 & 0 \end{bmatrix}\vec{\mathbf{v}}\right).$$

b. What does the transformation $\begin{bmatrix} 0 & -1 \\ 1 & 0 \end{bmatrix}$ correspond to geometrically?

$$\vec{F}\begin{pmatrix} x \\ y \\ z \end{pmatrix} = \begin{bmatrix} y \\ -z \\ yz \end{bmatrix}$$

Vector field for Exercise 6.13.

c. Can you explain why the work and the flux on $\mathbb{R}^2$ are related by the formula in part a?

6.13 Find the flux of the vector field $\vec{F}$ shown in the margin, through S, where S is the part of the cone $z = \sqrt{x^2+y^2}$ where $x, y \geq 0$, $x^2+y^2 \leq R$, and S is oriented by the inward-pointing normal (i.e., the flux measures flow into the cone).

6.14 Let S be the part of the surface of equation $z = \sin xy + 2$ where

$$x^2 + y^2 \leq 1 \quad \text{and} \quad x \geq 0,$$

oriented by the upward-pointing normal; let $\vec{F} = \begin{bmatrix} 0 \\ 0 \\ x+y \end{bmatrix}$. What is the flux of $\vec{F}$ through S?

6.15 Consider the manifold $S^3 \subset \mathbb{R}^4$ of equation $x_1^2+x_2^2+x_3^2+x_4^2 = 1$, oriented by $\Omega_{\mathbf{x}}(\vec{\mathbf{v}}_1, \vec{\mathbf{v}}_2, \vec{\mathbf{v}}_3) = \operatorname{sgn}\det[\mathbf{x}, \vec{\mathbf{v}}_1, \vec{\mathbf{v}}_2, \vec{\mathbf{v}}_3]$. Let X be the subset of S^3 where $x_4 \leq 0$.

a. Show that X is a piece-with-boundary of S^3.

b. Find a basis for the tangent space $T_{\mathbf{x}}(\partial X)$ at the point $\mathbf{x} = \begin{pmatrix} 1 \\ 0 \\ 0 \\ 0 \end{pmatrix} \in \partial X$ which is direct for the boundary orientation.

6.16 a. Compute the derivative of $xy\, dz$ from the definition.

b. Compute the same derivative using the formulas given in Theorem 6.7.4, stating clearly at each stage what property you are using.

6.17 a. Let $\varphi = xyz\, dy$. Compute from the definition the number

$$\mathbf{d}\varphi\left(P_{\begin{pmatrix} 1 \\ 2 \\ 3 \end{pmatrix}}(\vec{\mathbf{e}}_2, \vec{\mathbf{e}}_3)\right).$$

b. What is $\mathbf{d}\varphi$? Use your result to check the computation in part a.

6.18 Let $\vec{\mathbf{r}} = \begin{bmatrix} x_1 \\ \vdots \\ x_n \end{bmatrix}$ be the radial vector field in $\mathbb{R}^n$.

Exercise 6.18 gives another way to derive equation 5.3.50.

a. Show that $\mathbf{d}(\Phi_{\vec{\mathbf{r}}}) = n(dx_1 \wedge \cdots \wedge dx_n)$.

b. Let $B_1^n(\mathbf{0})$ and S^{n-1} be the unit ball and the unit sphere in $\mathbb{R}^n$, the ball with the standard orientation and the sphere with the boundary orientation. Use Stokes's theorem to prove

$$\operatorname{vol}_n\left(B_1^n(\mathbf{0})\right) = \frac{1}{n}\operatorname{vol}_{n-1}(S^{n-1}).$$

6.19 Using Theorem 6.8.3, prove the equations

$$\operatorname{curl}(\operatorname{grad} f) = \vec{\mathbf{0}} \quad \text{and} \quad \operatorname{div}(\operatorname{curl} \vec{F}) = 0$$

for any function f and any vector field $\vec{F}$ (of class at least C^2).

6.20 a. For what vector field $\vec{F}$ is the 1-form on $\mathbb{R}^3$

$$x^2\,dx + y^2 z\,dy + xy\,dz \quad \text{the work form field } W_{\vec{F}}?$$

b. Compute the exterior derivative of $x^2\,dx + y^2 z\,dy + xy\,dz$ using Theorem 6.7.4. Show that it is the same as $\Phi_{\vec{\nabla}\times\vec{F}}$.

6.21 a. There is an exponent m such that

$$\vec{\nabla}\cdot(x^2+y^2+z^2)^m \begin{bmatrix} x \\ y \\ z \end{bmatrix} = 0; \quad \text{find it.}$$

Exercise 6.21, part b: The subscript on Φ may be hard to read. It is $r^{2m}\vec{r}$.

*b. More generally, there is an exponent m (depending on n) such that the $(n-1)$-form $\Phi_{r^{2m}\vec{r}}$ has exterior derivative 0, when $\vec{r}$ is the vector field $\begin{bmatrix} x_1 \\ \vdots \\ x_n \end{bmatrix}$, and $r = |\vec{r}|$. Can you find it? (Start with $n=1$ and $n=2$.)

6.22 a. Find the unique polynomial p such that $p(1)=1$ and such that if

$$\omega = x\,dy\wedge dz - 2zp(y)\,dx\wedge dy + yp(y)\,dz\wedge dx,$$

then $\mathbf{d}\omega = dx\wedge dy\wedge dz$.

b. For this polynomial p, find the integral $\int_S \omega$, where S is that part of the sphere $x^2+y^2+z^2=1$ where $z \geq \sqrt{2}/2$, oriented by the outward-pointing normal.

6.23 a. Compute the exterior derivative of the 2-form

$$\varphi = \frac{1}{(x^2+y^2+z^2)^{3/2}}\Big(x\,dy\wedge dz + y\,dz\wedge dx + z\,dx\wedge dy\Big).$$

b. Compute the integral of φ over the unit sphere $x^2+y^2+z^2=1$, oriented by the outward-pointing normal.

c. Compute the integral of φ over the boundary of the cube of side 4, centered at the origin, and oriented by the outward-pointing normal.

d. Can φ be written $\mathbf{d}\psi$ for some 1-form ψ on $\mathbb{R}^3 - \{\mathbf{0}\}$?

6.24 Let S be the surface of equation $z = 9 - y^2$, oriented by the upward-pointing normal.

a. Sketch the piece $X \subset S$ where $x \geq 0$, $z \geq 0$, and $y \geq x$, indicating carefully the boundary orientation.

b. Give a parametrization of X, being careful about the domain of the parametrizing map and whether it is orientation preserving.

c. Find the work of the vector field $\begin{bmatrix} 0 \\ xz \\ 0 \end{bmatrix}$ around the boundary of X.

6.25 Let $U \subset \mathbb{R}^3$ be a subset bounded by a surface S, to which we will give the boundary orientation. What relation is there between the volume of U and the flux $\int_S \Phi_{\begin{bmatrix} x \\ y \\ z \end{bmatrix}}$?

6.26 Let $\vec{F}$ be the vector field $\vec{F}\begin{pmatrix} x \\ y \\ z \end{pmatrix} = \begin{bmatrix} F_1(x,y) \\ F_2(x,y) \\ 0 \end{bmatrix}$, where F_1 and F_2 are defined on all of $\mathbb{R}^2$. Suppose $D_2F_1 = D_1F_2$. Show that there exists a function $f : \mathbb{R}^3 \to \mathbb{R}$ such that $\vec{F} = \vec{\nabla} f$.

6.27 Show that the electromagnetic field of a charge q moving in the direction of the x-axis at constant speed v is

$$\vec{\mathbf{E}} = \frac{q\gamma}{4\pi\Big((\gamma x - \gamma vt)^2 + y^2 + z^2\Big)^{3/2}} \begin{bmatrix} x - vt \\ y \\ z \end{bmatrix}$$

$$\vec{\mathbf{B}} = \frac{v}{c}\,\frac{q\gamma}{4\pi\Big((\gamma x - \gamma vt)^2 + y^2 + z^2\Big)^{3/2}} \begin{bmatrix} 0 \\ -z \\ y \end{bmatrix}, \quad \text{where } \gamma = \frac{1}{\sqrt{1 - v^2/c^2}}.$$

Exercise 6.27: Equation 6.12.75 relates the coordinates x, y, z, t of a frame of reference moving at speed v in the direction of the x-axis with respect to a frame with coordinates x', y', z', t'. Equation 6.12.24 gives the electromagnetic field of a charge at rest.

It is often said that the magnetic field is a relativistic side effect of the electric field of moving charges. Exercise 6.27 illustrates this: with the particle at rest the magnetic field is $\vec{\mathbf{0}}$, but a moving charge will deflect a magnetic needle. Note the v/c in the magnetic field of the moving charge. At ordinary (human) speeds, the magnetic field will be extremely small.

6.28 Find a 1-form φ such that $\mathbf{d}\varphi = y\,dz \wedge dx - x\,dy \wedge dz$.

6.29 Let $U \subset \mathbb{R}^3$ be a 3-dimensional piece-with-boundary.

a. What does Stokes's theorem say about $\int_U \mathbf{d}\mathbb{F}$?

b. What does Stokes's theorem say about $\int_U \mathbf{d}\mathbb{M}$?

6.30 Let $S \subset \mathbb{R}^3$ be a smooth oriented surface, and $X \subset S$ a 2-dimensional piece-with-boundary. Let $I = [t_0, t_1]$ be a time interval.

a. Show that $V = S \times I$ is a 3-dimensional piece-with-boundary of $\mathbb{R}^4$.

b. What does Stokes's theorem say about $\int_V \mathbf{d}\mathbb{F}$? Show that if we divide the integral by $t_1 - t_0$ and let t_1 tend to t_0, we find Faraday's law.

c. What does Stokes's theorem say about $\int_V \mathbf{d}\mathbb{M}$? Show that if we divide the integral by $t_1 - t_0$ and let t_1 tend to t_0, we find Ampère's law.

Exercise 6.30: The surface S might be very complicated, for instance, the boundary of a cloud. Understanding the charge on the boundary of a cloud is essential for understanding lightning and thunderstorms.

6.31 Let $\vec{F}_n = \frac{\vec{\mathbf{x}}}{|\vec{\mathbf{x}}|^n}$ be the vector field defined in Example 6.7.7, and let S_R^{n-1} be the sphere $|\vec{\mathbf{x}}| = R$, oriented by the outward-pointing normal. Show that $\int_{S_R^{n-1}} \Phi_{\vec{F}_n}$ does not depend on R, and that it equals $\mathrm{vol}_{n-1}(S_1^{n-1})$.

6.32 Compute the integral $\int_S \Phi_{\vec{F}}$, where $\vec{F}\begin{pmatrix} x \\ y \\ z \end{pmatrix} = \begin{bmatrix} -x^2yz \\ y \\ (z^2-1)xy \end{bmatrix}$, and S is the part of the parabolic cylinder of equation $y = 9 - x^2$ where $y \geq 0$ and $0 \leq z \leq 1$, oriented by the transverse vector field $\vec{\mathbf{e}}_2$.

6.33 Let V, W, be two finite-dimensional real vector spaces, oriented by

$$\Omega^V : \mathcal{B}(V) \to \{\pm 1\} \quad \text{and} \quad \Omega^W : \mathcal{B}(W) \to \{\pm 1\}.$$

Show that

$$\Omega^{V\times W}(\{v\},\{w\}) \stackrel{\text{def}}{=} \Omega^V(\{v\})\,\Omega^W(\{w\})$$

orients $V \times W$.

****6.34** Let $\vec{\mathbf{r}} = \begin{bmatrix} x \\ y \\ z \end{bmatrix}$ and $r = |\vec{\mathbf{r}}| = \sqrt{x^2 + y^2 + z^2}$. Show that

$$\mathbb{F} = qW_{\vec{\mathbf{r}}/r^3} \wedge c\,dt$$

is an electromagnetic field. What are the corresponding charges and currents?

Exercise 6.34 asks you to show that $\mathbf{d}\mathbb{F} = 0$ and to compute $\mathbf{d}\mathbb{M}$. In both cases, these exterior derivatives are "in the sense of distributions"; see Remark 6.12.1 (page 676).

Remarks. 1. Exercise 6.34 isn't a straightforward computation. For $\mathbb{F}$ to be an electromagnetic field, we must have $\mathbf{d}\mathbb{F} = 0$ everywhere, and it isn't clear what this means when $r = 0$. In Example 6.13.1 we saw that trouble may be hiding at points where a form is not defined. We deal with this as follows.

A 3-dimensional piece-with-boundary $X \subset \mathbb{R}^4$ will be called *r-adapted* if locally near points where $r = 0$ it represents t as a function of x, y, and z; in that case we write $X_\epsilon = X - \{r < \epsilon\}$ and we define

$$\partial_{inn} X_\epsilon \subset \partial X_\epsilon$$

to be the subset where $r = \epsilon$, with the boundary orientation. If φ is a 2-form on $\mathbb{R}^4 - \{r = 0\}$ and $X \subset \mathbb{R}^4$ is an r-adapted oriented 3-dimensional piece-with-boundary, we define

$$\int_X \mathbf{d}\varphi \overset{\text{def}}{=} \lim_{\epsilon \to 0} \left(\int_{X_\epsilon} \mathbf{d}\varphi - \int_{\partial_{inn} X_\epsilon} \varphi \right).$$

If φ is well defined where $r = 0$, then, by Stokes's theorem,

$$\int_X \mathbf{d}\varphi = \int_{X_\epsilon} \mathbf{d}\varphi + \int_{X - X_\epsilon} \mathbf{d}\varphi = \int_{X_\epsilon} \mathbf{d}\varphi - \int_{\partial_{inn} X_\epsilon} \varphi,$$

so the formula is correct in that case, and otherwise the boundary term captures whatever is hiding on $\{r = 0\}$.

2. Note that

$$\mathbf{d}\left(\frac{1}{r}\right) = -W_{\vec{\mathbf{r}}/r^3}.$$

This contrasts with Example 6.13.1. There we had

$$\mathbf{d} \arctan \frac{y}{x} = \frac{x\,dy - y\,dx}{x^2 + y^2}.$$

But $\arctan \frac{y}{x}$ *is not* a well-defined function on $\mathbb{R}^2 - \{\mathbf{0}\}$, whereas $1/r$ is a well-defined function on $\mathbb{R}^3 - \{\mathbf{0}\}$.

Forms were defined to be integrands, so it is quite reasonable to define $\mathbf{d}\mathbb{F}$ via its integrals over oriented pieces-with-boundary. It is also quite reasonable to restrict to r-adapted pieces: pieces-with-boundary for which, locally near points where $r = 0$, the piece X is the graph of a map expressing t as a function of $\begin{pmatrix} x \\ y \\ z \end{pmatrix}$. Pieces that are not r-adapted intersect the locus $r = 0$ in exceptional ways, and it may be difficult to say what share of whatever nastiness is hiding there is carried by X. Such pieces are exceptional; by budging them arbitrarily little we can make them r-adapted. △

Appendix: Analysis

A0 Introduction

This appendix is intended for students using this book for a class in analysis, and for the occasional student in a beginning course who has mastered the statement of the theorem and wishes to delve further.

In addition to proofs of statements not proved in the main text, it includes a justification of arithmetic (Appendix A1), a discussion of cubic and quartic equations (Appendix A2), the Heine-Borel theorem (Appendix A3), a definition of "big O" (Appendix A11), Stirling's formula (Appendix A16), and a definition of Lebesgue measure and a discussion of what sets are measurable (Appendix A21).

A1 Arithmetic of real numbers

Because you learned to add, subtract, divide, and multiply in elementary school, the algorithms used may seem obvious. But understanding how computers simulate real numbers is not nearly as routine as you might imagine. A real number involves an infinite amount of information, and computers cannot handle such things: they compute with finite decimals. This inevitably involves rounding off, and writing arithmetic subroutines that minimize round-off errors is a whole art in itself. In particular, computer addition and multiplication are not commutative or associative. Anyone who really wants to understand numerical problems has to take a serious interest in "computer arithmetic".

Most equivalence classes consist of a single expression, but the equivalence class 0 has two:

$$+\dots 00.00\dots \text{ and } -\dots 00.00\dots$$

as do the finite decimals. For instance, the equivalence class 1 consists of

$$+\dots 00.99\dots \text{ and } +\dots 01.00\dots.$$

It is harder than one might think to define arithmetic for the reals – addition, multiplication, subtraction, and division – and to show that the usual rules of arithmetic hold. Addition and multiplication as taught in elementary school always start at the right, and for reals there is no right.

Recall that in Section 0.5 we defined the reals as "the set of infinite decimals". For rigor's sake we will now spell out exactly what this means; to avoid making special conventions, we will write our infinite decimals with leading 0's.

Definition A1.1 (Real numbers). The set of real numbers is the set of equivalence classes of expressions

$$\pm\dots 000a_n a_{n-1}\dots a_0 \underset{\uparrow}{.} a_{-1}a_{-2}\dots, \tag{A1.1}$$

where all a_i are in $\{0,1,2,3,4,5,6,7,8,9\}$, and (as indicated by the arrow) a decimal point separates a_0 and a_{-1}. Two such expressions

$$a = \pm\dots 0a_n a_{n-1}\dots a_0.a_{-1}\dots \text{ and } b = \pm\dots 0b_m b_{m-1}\dots b_0.b_{-1}\dots$$

are equivalent if and only if any of the following conditions is met:

1. They are equal.
2. All a_i and all b_i are 0, and the signs are opposite (this equivalence class is called 0).
3. a and b have the same sign; there exists k such that $a_k \neq 9$ and $a_{k-1} = a_{k-2} = \dots = 9$, and

$$b_j = a_j \text{ for } j > k, \quad b_k = a_k + 1,\ b_{k-1} = b_{k-2} = \dots = 0.$$

Definition A1.2 (k-truncation). The *k-truncation* of a real number $a = \dots 000a_n a_{n-1} \cdots a_0 . a_{-1}a_{-2}\dots$ is the finite decimal

$$[a]_k \stackrel{\text{def}}{=} \dots a_n \dots a_k 000 \dots \qquad A1.2$$

Equation AA1.2 contains a decimal point but we don't know where to put it. It is among the a_i or among the 0's, depending on whether k is negative or positive.

For instance, if $a = 21.3578$, then $[a]_{-2} = 21.35$.

It is tempting to say that if you take two reals, truncate (cut) them further and further to the right and add them (or multiply them, or subtract them, etc.) and look only at the digits to the left of any fixed position, the digits we see will not be affected by where the truncation takes place, once it is well beyond where we are looking. The problem with this is that it isn't quite true.

Example A1.3 (Addition). Consider adding the following two numbers:

$$\begin{matrix} .222222\dots 222\dots \\ .777777\dots 778\dots. \end{matrix} \qquad A1.3$$

If we truncate before the position of the 8, the sum of the truncated numbers will be $.9999\dots 9$; if we truncate after the 8, it will be $1.0000\dots 0$. So there cannot be any rule which says, "the 100th digit will stay the same if you truncate after the Nth digit, however large N is." The "carry" can come from arbitrarily far to the right. △

Definition A1.5: Since we don't yet have a notion of subtraction in $\mathbb{R}$, we can't write $|x-y| < \epsilon$ for $x, y \in \mathbb{R}$, much less

$$\sum (x_i - y_i)^2 < \epsilon^2,$$

which involves addition and multiplication besides. Our definition of k-close uses only subtraction of finite decimals.

It is possible to define the arithmetic of real numbers in terms of digits, but it is quite involved. Even showing that addition is associative involves at least six different cases. None is hard, but keeping straight what you are doing is quite delicate. Exercise A1.6 should give you enough of a taste of this approach. We will use a different approach, based on Definition A1.5, which says what it means for two points to be "k-close".

Let us denote by $\mathbb{D}$ the set of finite decimals.

For instance, if $a = 1.23000013$ and $b = 1.22999903$, then a and b are not 7-close, since

$$[a]_{-7} - [b]_{-7} = 11 \times 10^{-7} > 10^{-7}$$

but they are 6-close, since

$$[a]_{-6} - [b]_{-6} = 10^{-6}.$$

The notion of k-close is the correct way of saying that two numbers agree to k digits after the decimal point. It takes into account the convention by which a number ending in all 9's is equal to the rounded up number ending in all 0's.

The numbers .9998 and 1.0001 are 3-close (but not 4-close).

Definition A1.4 (Finite decimal continuity). A map $f : \mathbb{D}^n \to \mathbb{D}$ is called *finite decimal continuous* (or $\mathbb{D}$-continuous) if for all integers N and k, there exists an integer l such that if $(x_1, \dots, x_n)$ and $(y_1, \dots, y_n)$ are two elements of $\mathbb{D}^n$ with all $|x_i|, |y_i| < N$, and if $|x_i - y_i| < 10^{-l}$ for all $i = 1, \dots, n$, then

$$|f(x_1, \dots, x_n) - f(y_1, \dots, y_n)| < 10^{-k}. \qquad A1.4$$

Exercise A1.2 asks you to show that the functions $A(x,y) = x+y$, $M(x,y) = xy$, $S(x,y) = x-y$, and $\text{Assoc}(x,y,z) = (x+y)+z$ are $\mathbb{D}$-continuous and that $1/x$ is not.

To see why Definition A1.4 is the right definition, we need to say what it means for two points $\mathbf{x}, \mathbf{y} \in \mathbb{R}^n$ to be close.

Definition A1.5 (k-close). Two points $\mathbf{x}, \mathbf{y} \in \mathbb{R}^n$ are *k-close* if $\big|[x_i]_{-k} - [y_i]_{-k}\big| \le 10^{-k}$ for each $i = 1, \dots, n$.

Notice that if two numbers are k-close for all k, then they are equal (see Exercise A1.1).

If $f : \mathbb{D}^n \to \mathbb{D}$ is $\mathbb{D}$-continuous, then define $\widetilde{f} : \mathbb{R}^n \to \mathbb{R}$ by the formula

$$\widetilde{f}(\mathbf{x}) = \sup_k \inf_{l \le -k} f([x_1]_l, \dots, [x_n]_l). \qquad A1.5$$

Proposition A1.6. *The function $\widetilde{f} : \mathbb{R}^n \to \mathbb{R}$ is the unique function that coincides with f on $\mathbb{D}^n$ and satisfies the continuity condition that for all $k \in \mathbb{N}$ and all $N \in \mathbb{N}$, there exists $l \in \mathbb{N}$ such that when $\mathbf{x}, \mathbf{y} \in \mathbb{R}^n$ are l-close and all coordinates x_i of $\mathbf{x}$ satisfy $|x_i| < N$, then $\widetilde{f}(\mathbf{x})$ and $\widetilde{f}(\mathbf{y})$ are k-close.*

The functions $\widetilde{A}$ and $\widetilde{M}$ satisfy the conditions of Proposition A1.6; thus they apply to the real numbers, while A and M without tildes apply to finite decimals.

The proof is the object of Exercise A1.4. Now setting up arithmetic for the reals is plain sailing: we can define addition and multiplication of reals by setting

$$x + y = \widetilde{A}(x, y) \quad \text{and} \quad xy = \widetilde{M}(x, y), \qquad A1.6$$

where $A(x, y) = x + y$ and $M(x, y) = xy$. It isn't harder to show that the basic laws of arithmetic hold:

$x + y = y + x$	Addition is commutative.
$(x + y) + z = x + (y + z)$	Addition is associative.
$x + (-x) = 0$	Existence of additive inverse.
$xy = yx$	Multiplication is commutative.
$(xy)z = x(yz)$	Multiplication is associative.
$x(y + z) = xy + xz$	Multiplication is distributive over addition.

These are all proved the same way. Let us prove the last. Consider the function $\mathbb{D}^3 \to \mathbb{D}$ given by

$$F(x, y, z) = \overbrace{M\big(x, A(y, z)\big)}^{x(y+z)} - \overbrace{A\big(M(x, y), M(x, z)\big)}^{xy+xz}. \qquad A1.7$$

We leave it to you to check that F is $\mathbb{D}$-continuous, and that

$$\widetilde{F}(x, y, z) = \widetilde{M}\big(x, \widetilde{A}(y, z)\big) - \widetilde{A}\big(\widetilde{M}(x, y), \widetilde{M}(x, z)\big). \qquad A1.8$$

But F is identically 0 on $\mathbb{D}^3$, and the identically 0 function on $\mathbb{R}^3$ coincides with 0 on $\mathbb{D}^3$ and satisfies the continuity condition of Proposition A1.6, so $\widetilde{F}$ vanishes identically by the uniqueness part of Proposition A1.6.

It is one of the basic irritants of elementary school math that division is not defined in the world of finite decimals.

This sets up almost all of arithmetic; the missing piece is division. Exercise A1.3 asks you to define division in the reals.

Exercises for Section A1

A1.1 Show that if two numbers are k-close for all k, then they are equal.

Asterisks (*) denote difficult exercises. Two stars indicate a particularly challenging exercise.

FIGURE A1.1.

Giuseppe Peano (1858–1932)

Peano was the son of Italian farmers. He discovered Peano curves in 1890. He was noted for his rigor and his ability to disprove theorems by other mathematicians by finding exceptions. He also proposed a universal language with words from English, French, German, and Latin but no grammar.

Exercise A1.5: Peano curves give onto, continuous mappings from $\mathbb{R} \to \mathbb{R}^2$. Analogues of Peano curves can be constructed from $\mathbb{R}$ to the infinite-dimensional vector space $\mathcal{C}[0,1]$!

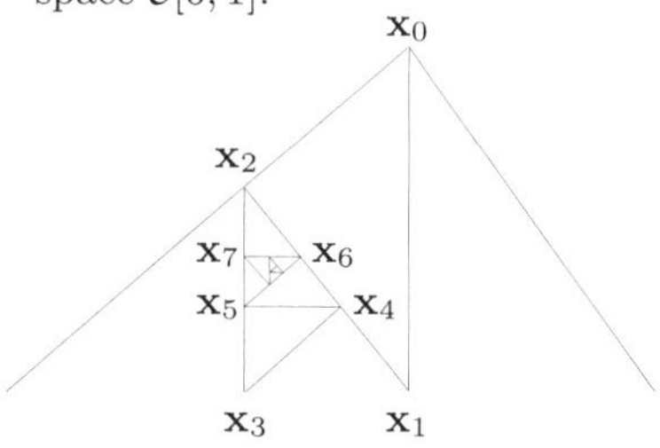

FIGURE FOR EXERCISE A1.5

This sequence corresponds to the string of digits

00100010010....

***A1.2** Show that the functions $A(x,y) = x+y$, $M(x,y) = xy$, $S(x,y) = x-y$, and $\text{Assoc}(x,y,z) = (x+y)+z$ are $\mathbb{D}$-continuous, and that $1/x$ is not. Notice that for A and S, the l of Definition A1.4 does not depend on N, but that for M, l does depend on N.

***A1.3** Define division of reals, using the following steps.

a. Show that the algorithm of long division of a positive finite decimal a by a positive finite decimal b defines a repeating decimal a/b, and that $b(a/b) = a$.

b. Show that the function $\text{inv}(x)$ defined for $x > 0$ by the formula

$$\text{inv}(x) = \inf_k \frac{1}{[x]_k}$$

satisfies $x\,\text{inv}(x) = 1$ for all $x > 0$.

c. Define the inverse for any $x \neq 0$, and show that $x\,\text{inv}(x) = 1$ for all $x \neq 0$.

****A1.4** Prove Proposition A1.6. This can be broken into the following steps.

a. Show that $\sup_k \inf_{l \geq k} f([x_1]_l, \dots, [x_n]_l)$ is well defined (i.e., that the sets of numbers involved are bounded). Looking at the function S from Exercise A1.2, explain why both the sup and the inf are there.

b. Show that the function $\widetilde{f}$ has the required continuity properties.

c. Show the uniqueness.

****A1.5** In this exercise we will construct a continuous map $\gamma : [0,1] \to \mathbb{R}^2$, the image of which is a (full) triangle T; such a mapping is called a *Peano curve*. We will write our numbers in $[0,1]$ in base 2, so such a number might be something like .0011101000011..., and we will use the table below.

digit / position	0	1
even	left	right
odd	right	left

Take a right triangle T. We will associate to a string $\underline{s} = s_1, s_2, \dots$ of digits 0 and 1 a sequence of points $\mathbf{x}_0, \mathbf{x}_1, \mathbf{x}_2, \dots$ of T by starting at the point $\mathbf{x}_0$, dropping the perpendicular to the opposite side, landing at $\mathbf{x}_1$, and turning left or right according to the digit s_1, as interpreted by the bottom line of the table, since this digit is the first digit (and therefore in an odd position): on 0 turn right and on 1 turn left. (The figure corresponds to the number 00100010010..., so at $\mathbf{x}_1$ we turn right.)

Now drop the perpendicular to the opposite side, landing at $\mathbf{x}_2$, and turn right or left according to the digit s_2, as interpreted by the top line of the table, etc.

This construction is illustrated in Figure A1.5.

a. Show that for any string of digits $(\underline{s})$, the sequence $n \mapsto \mathbf{x}_n(\underline{s})$ converges.

b. Suppose $t \in [0,1]$ can be written in base 2 in two different ways (one ending in 0's and the other in 1's), and call $(\underline{s})$, $(\underline{s}')$ the two strings of digits. Show that

$$\lim_{n\to\infty} \mathbf{x}_n(\underline{s}) = \lim_{n\to\infty} \mathbf{x}_n(\underline{s}').$$

Hint: Construct the sequences associated to .1000... and .0111....

This allows us to define $\gamma(t) = \lim_{n\to\infty} \mathbf{x}_n(\underline{s})$.

c. Show that γ is continuous.

d. Show that every point in T is in the image of γ. What is the maximum number of distinct numbers $t_1, \dots, t_k$ such that $\gamma(t_1) = \cdots = \gamma(t_k)$? *Hint*: Choose a point in T, and draw a path of the sort above that leads to it.

****A1.6** a. Let x and y be two strictly positive reals. Show that $x + y$ is well defined by showing that for any k, the digit in the kth position of $[x]_N + [y]_N$ is the same for all sufficiently large N. Note that N cannot depend just on k; it must also depend on x and y.

*b. Now drop the hypothesis that the numbers are positive, and try to define addition. You will find that this is quite a bit harder than part a.

*c. Show that addition is commutative. Again, this is a lot easier when the numbers are positive.

**d. Show that addition is associative: $x + (y + z) = (x + y) + z$. This is much harder and requires separate consideration of the cases where each of x, y, and z is positive and negative.

Exercise A1.6, part d is perhaps the hardest exercise in the book. It is also rather dull, mainly because it requires a very careful analysis of lots and lots of different cases.

A2 Cubic and quartic equations

We will show that a cubic equation can be solved using formulas analogous to the formula

$$\frac{-b \pm \sqrt{b^2 - 4ac}}{2a} \qquad \text{A2.1}$$

for the quadratic equation $ax^2 + bx + c = 0$.

Let us start with two examples; the explanation of the tricks will follow.

Example A2.1 (Solving a cubic equation). Let us solve the equation $x^3 + x + 1 = 0$. First substitute $x = u - 1/3u$, to get

$$\left(u - \frac{1}{3u}\right)^3 + \left(u - \frac{1}{3u}\right) + 1 = u^3 - u + \frac{1}{3u} - \frac{1}{27u^3} + u - \frac{1}{3u} + 1 = 0. \qquad \text{A2.2}$$

After simplification and multiplication by u^3, this becomes

$$u^6 + u^3 - \frac{1}{27} = 0. \qquad \text{A2.3}$$

This is a quadratic equation for u^3, which can be solved by formula A2.1, to yield

$$u^3 = \frac{1}{2}\left(-1 \pm \sqrt{\frac{31}{27}}\right) \approx 0.0358, -1.0358. \qquad \text{A2.4}$$

Both values of u^3 have real cube roots: $u_1 \approx 0.3295$ and $u_2 \approx -1.0118$. This allows us to find $x = u - 1/3u$:

$$x = u_1 - \frac{1}{3u_1} = u_2 - \frac{1}{3u_2} \approx -0.6823. \quad \triangle \qquad \text{A2.5}$$

Here we see something bizarre: in Example A2.1, the polynomial has only one real root and we can find it using only real numbers, but here in Example A2.2 there are three real roots, and we can't find any of them using only real numbers.

We will see that when Cardano's formula is used, if a real polynomial has one real root, we can always find it using only real numbers, but if it has three real roots, we never can find any of them using real numbers. However, it is possible to find such real roots using real numbers, if one applies trigonometry instead of Cardano's formula; this is explored in Exercise A2.6.

Example A2.2 (A cubic equation with three distinct real roots). Let us solve the equation $x^3 - 3x + 1 = 0$. The right substitution to make in this case (one that turns the cubic equation into a quadratic equation for u^3) is $x = u + 1/u$, which leads to

$$\left(u + \frac{1}{u}\right)^3 - 3\left(u + \frac{1}{u}\right) + 1 = 0. \qquad A2.6$$

After multiplying out, canceling, and multiplying by u^3, this gives the quadratic equation $u^6 + u^3 + 1 = 0$ with solutions

$$\frac{-1 \pm i\sqrt{3}}{2} = \cos\frac{2\pi}{3} \pm i \sin\frac{2\pi}{3}. \qquad A2.7$$

Denote by v_1 the solution with positive imaginary part. The cube roots of v_1 are

$$\cos\frac{2\pi}{9} + i\sin\frac{2\pi}{9}, \quad \cos\frac{8\pi}{9} + i\sin\frac{8\pi}{9}, \quad \cos\frac{14\pi}{9} + i\sin\frac{14\pi}{9}. \qquad A2.8$$

In all three cases,[1] we have $1/u = \overline{u}$, so that $u + 1/u = 2\mathrm{Re}\, u$, leading to the three roots

$$x_1 = 2\cos\frac{2\pi}{9} \approx 1.532088, \quad x_2 = 2\cos\frac{8\pi}{9} \approx -1.879385,$$
$$x_3 = 2\cos\frac{14\pi}{9} \approx 0.347296. \quad \triangle \qquad A2.9$$

Derivation of Cardano's formula for cubic equations

The substitutions $x = u - 1/3u$ and $x = u + 1/u$ in Examples A2.1 and A2.2 were special cases.

Eliminating the term in x^2 means changing the roots so that their sum is 0: if the roots of a cubic polynomial are a_1, a_2, a_3, then we can write the polynomial as

$$\begin{aligned} p &= (x - a_1)(x - a_2)(x - a_3) \\ &= x^3 - (a_1 + a_2 + a_3)x^2 \\ &\quad + (a_1a_2 + a_1a_3 + a_2a_3)x \\ &\quad - a_1a_2a_3. \end{aligned}$$

Thus eliminating the term in x^2 means that $a_1 + a_2 + a_3 = 0$. We will use this to prove Proposition A2.4.

If we start with the equation $x^3 + ax^2 + bx + c = 0$, we can eliminate the term in x^2 by setting $x = y - a/3$. The equation then becomes

$$y^3 + py + q = 0, \quad \text{where } p = b - \frac{a^2}{3} \text{ and } q = c - \frac{ab}{3} + \frac{2a^3}{27}. \qquad A2.10$$

Now set $y = u - \frac{p}{3u}$; the equation $y^3 + py + q = 0$ then becomes

$$u^3 + q - \frac{p^3}{27u^3} = 0, \quad \text{i.e.,} \quad u^6 + qu^3 - \frac{p^3}{27} = 0, \qquad A2.11$$

which is a quadratic equation for u^3. If we solve this quadratic equation and substitute the value of u in $y = u - \frac{p}{3u}$ we find Cardano's formula

$$y = \left(\frac{-q \pm \sqrt{q^2 + \frac{4p^3}{27}}}{2}\right)^{1/3} - \frac{p}{3}\left(\frac{-q \pm \sqrt{q^2 + \frac{4p^3}{27}}}{2}\right)^{-1/3}. \qquad A2.12$$

[1] For any complex number $u \in \mathbb{C}$, we have $\frac{1}{u} = \frac{1}{u} \cdot \frac{\bar{u}}{\bar{u}} = \frac{\bar{u}}{|u|^2}$, so that $1/u = \bar{u}$ if and only if $|u| = 1$. The complex numbers in formula A2.8 all have absolute value 1, since $|u| = |\cos\theta + i\sin\theta| = \sqrt{\cos^2\theta + \sin^2\theta} = 1$, so in each case we have $u + 1/u = u + \bar{u} = 2\mathrm{Re}\, u$.

Let v_1, v_2 be the two solutions of the quadratic equation $v^2+qv-\frac{p^3}{27}=0$, and for $i=1,2$ let $u_{i,1},u_{i,2},u_{i,3}$ be the three cubic roots of v_i. We now have apparently six roots for the equation $x^3+px+q=0$: the numbers

$$y_{i,j}=u_{i,j}-\frac{p}{3u_{i,j}}, \qquad i=1,2;\ j=1,2,3. \qquad \text{A2.13}$$

Exercise A2.2 asks you to show that $-p/(3u_{1,j})$ is a cubic root of v_2, and that we can renumber the cube roots of v_2 so that $-p/(3u_{1,j})=u_{2,j}$. This gives $y_{1,j}=y_{2,j}$ for $j=1,2,3$, explaining why the apparently six roots are really only three.

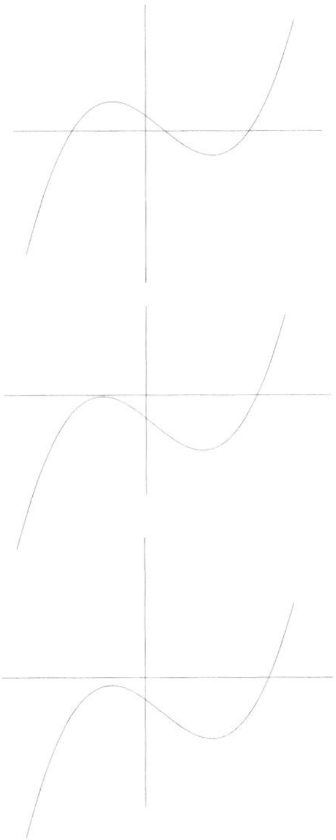

FIGURE A2.1.

The graphs of three cubic polynomials. The polynomial at the top has three roots. As it is varied, the two roots to the left coalesce to give a double root, as shown by the middle figure. If the polynomial is varied a bit further, the double root vanishes (actually becoming a pair of complex conjugate roots).

The discriminant of a cubic equation

Definition A2.3 (Discriminant of cubic equation). The *discriminant* of the cubic polynomial x^3+px+q is the number $\Delta \stackrel{\text{def}}{=} 27q^2+4p^3$.

Proposition A2.4 (Discriminants and double roots). *The discriminant Δ vanishes if and only if $x^3+px+q=0$ has a double root.*

Proof. If there is a double root, then the roots are necessarily $\{a,a,-2a\}$ for some number a, since the sum of the roots is 0. Multiply out

$$(x-a)^2(x+2a)=x^3-3a^2x+2a^3, \quad \text{so } p=-3a^2 \text{ and } q=2a^3, \qquad \text{A2.14}$$

and indeed $4p^3+27q^2=-4\cdot 27a^6+4\cdot 27a^6=0$.

Now we need to show that if the discriminant is 0, the polynomial has a double root. Set $\Delta=0$, and call α the square root of $-p/3$ such that $2\alpha^3=q$; such a square root exists since $4\alpha^6=4(-p/3)^3=-4p^3/27=q^2$. Since

$$(x-\alpha)^2(x+2\alpha)=x^3+x(-4\alpha^2+\alpha^2)+2\alpha^3=x^3+px+q, \qquad \text{A2.15}$$

we see that α is a double root of our cubic polynomial. □

Cardano's formula for real polynomials

Suppose p,q are real. Figure A2.1 should explain why equations with double roots are the boundary between equations with one real root and equations with three real roots.

Proposition A2.5 (Real roots of a real cubic polynomial). *The real cubic polynomial x^3+px+q has one real root if $27q^2+4p^3>0$ and three real roots if $27q^2+4p^3<0$.*

Proof. If the polynomial has three real roots, then it has a local positive maximum at $\sqrt{-p/3}$ and a local negative minimum at $-\sqrt{-p/3}$. In particular, p must be negative. Thus we must have

$$\left(\left(\sqrt{-\frac{p}{3}}\right)^3+p\sqrt{-\frac{p}{3}}+q\right)\left(\left(-\sqrt{-\frac{p}{3}}\right)^3-p\sqrt{-\frac{p}{3}}+q\right)<0. \qquad \text{A2.16}$$

After a bit of computation, this becomes the result we want:

$$q^2 + \frac{4p^3}{27} < 0. \quad \square \qquad A2.17$$

Faced with this dilemma, the Italians of the sixteenth century, and their successors until about 1800, held their noses and computed with complex numbers. The name "imaginary" expresses what they thought of such numbers.

Thus if a real cubic polynomial has three real roots and you want to find them by Cardano's formula, you must use complex numbers, even though both the problem and the result involve only reals.

Exercises for Section A2

A2.1 a. In Exercise 0.7.8 you were asked to solve $x^3 - x^2 - x = 2$ by first finding the real root by trial and error. Find the real root now following the procedure outlined under "derivation of Cardano's formula".

b. In part a you will have turned the equation $x^3 - x^2 - x = 2$ into an equation of the form $y^3 + py + q = 0$. What is the discriminant of this cubic equation? What does it say about the existence of a double root?

c. In part a you should have landed on two different ways of computing the real root. What sentence in the text asserts that this does not contradict your conclusion from part b? What exercise justifies that assertion?

d. Compute the two imaginary roots of $x^3 - x^2 - x = 2$, using Proposition 0.7.7. What other technique could you use to compute these roots?

Hint for Exercise A2.2: See the subsection on the derivation of Cardano's formulas, in particular equation A2.13.

A2.2 Show that $-p/(3u_{1,j})$ is a cubic root of v_2, and that we can renumber the cube roots of v_2 so that $-p/(3u_{1,j}) = u_{2,j}$.

A2.3 Confirm (equation A2.10) that if you set $x = y - a/3$, the equation

$$x^3 + ax^2 + bx + c = 0$$

becomes $y^3 + py + q = 0$, where $p = b - \frac{a^2}{3}$ and $q = c - \frac{ab}{3} + \frac{2a^3}{27}$.

A2.4 Show that the following cubics have exactly one real root, and find it.
a. $x^3 - 18x + 35 = 0$ b. $x^3 + 3x^2 + x + 2 = 0$

Exercise A2.6 gives an alternative to Cardano's formula that applies to cubics with three real roots. In part a, use de Moivre's formula:

$$\cos n\theta + i \sin n\theta = (\cos\theta + i\sin\theta)^n.$$

A2.5 Show that the polynomial $x^3 - 7x + 6$ has three real roots. Find them.

A2.6 There is a way of finding the roots of real cubics with three real roots, using only real numbers and a bit of trigonometry:

a. Prove the formula $4\cos^3\theta - 3\cos\theta - \cos 3\theta = 0$.

b. Set $x = y/a$ in the equation $x^3 + px + q = 0$, and show that there is a value of a for which the equation becomes $4y^3 - 3y - q_1 = 0$; find the value of a and q_1.

c. Show that there exists an angle θ such that $\cos(3\theta) = q_1$ precisely when $27q^2 + 4p^3 < 0$ (i.e., precisely when the original polynomial has three real roots).

d. Find a formula (involving arccos) for all three roots of a real cubic polynomial with three real roots.

Exercise A2.7 sketches how to deal with quartic equations. It uses results from Section 3.1.

****A2.7** In this exercise, we will find formulas for the solution of fourth degree equations, known as *quartics*. Let $w^4 + aw^3 + bw^2 + cw + d$ be a quartic polynomial.
a. Show that if we set $w = x - a/4$, the quartic equation becomes

$$x^4 + px^2 + qx + r = 0.$$

Compute p, q, and r in terms of a, b, c, d.

b. Now set $y = x^2 + p/2$, and show that solving the quartic is equivalent to finding the intersections of the parabolas Γ_1 and Γ_2 of equation

$$x^2 - y + p/2 = 0 \quad \text{and} \quad y^2 + qx + r - \frac{p^2}{4} = 0$$

respectively, pictured in Figure A2.7.1. The parabolas Γ_1 and Γ_2 intersect (usually) in four points, and the curves of equation

$$f_m\begin{pmatrix} x \\ y \end{pmatrix} = y^2 + qx + r - \frac{p^2}{4} + m(x^2 - y + p/2) = 0 \qquad (1)$$

are exactly the curves given by quadratic equations which pass through those four points; some of these curves are shown in Figure A2.7.3.

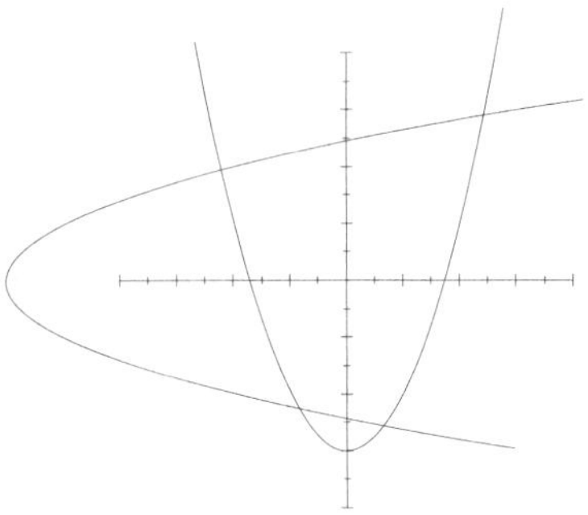

FIGURE A2.7.1. The two parabolas of equation (1); their axes are respectively the y-axis and the x-axis.

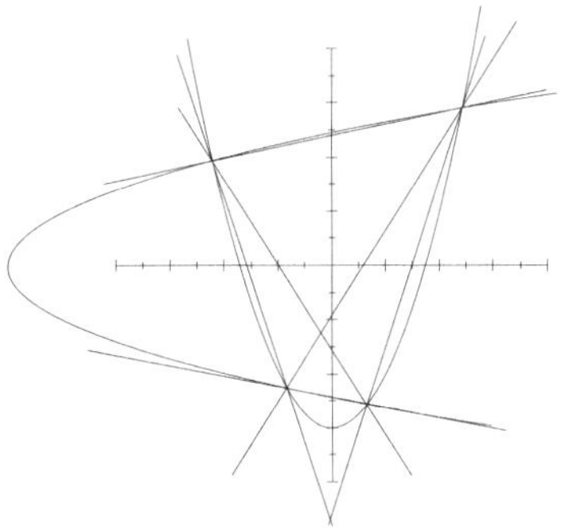

FIGURE A2.7.2. The three pairs of lines that go through the intersections of the two parabolas.

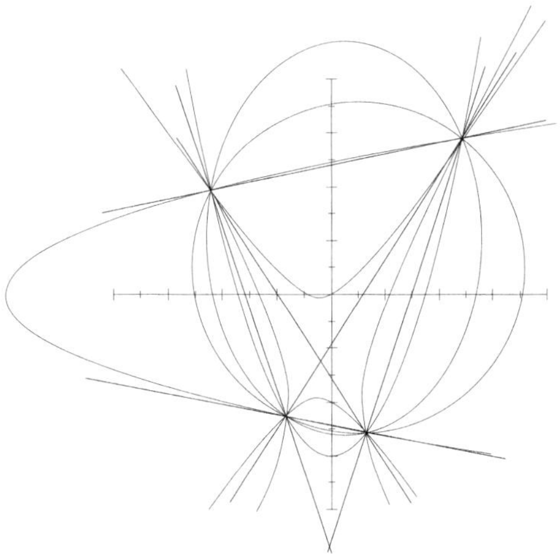

FIGURE A2.7.3. The curves $f_m\begin{pmatrix} x \\ y \end{pmatrix} = y^2 + qx + r - \frac{p^2}{4} + m(x^2 - y + p/2) = 0$ for seven different values of m.

c. What can you say about the curve given by equation (1) when $m = 1$? When m is negative? When m is positive?

d. The assertion in part b is not quite correct. There is one curve that passes through those four points and is given by a quadratic equation, that is missing from the family given by equation (1). Find it.

e. The next step is the really clever part of the solution. Among these curves, there are three, shown in Figure A2.7.2, that consist of a pair of lines. Each such "degenerate" curve consists of a pair of diagonals of the quadrilateral formed by the intersection points of the parabolas. Since there are three of these, we may hope that the corresponding values of m are solutions of a cubic equation, and this is the case. Using the fact that a pair of lines is not a smooth curve near the point where they intersect, show that the numbers m for which the equation $f_m = 0$ defines a pair of lines, and the coordinates x, y of the point where they intersect, are the solutions of the system of three equations in three unknowns,

$$\begin{aligned} y^2 + qx + r - \frac{p^2}{4} + m(x^2 - y + p/2) &= 0 \\ 2y - m &= 0 \\ q + 2mx &= 0. \end{aligned}$$

f. Expressing x and y in terms of m using the last two equations, show that m satisfies the equation

$$m^3 - 2pm^2 + (p^2 - 4r)m + q^2 = 0,$$

called the *resolvent cubic* of the original quartic equation. Let m_1, m_2, and m_3 be the roots of the equation, and let $\begin{pmatrix} x_1 \\ y_1 \end{pmatrix}, \begin{pmatrix} x_2 \\ y_2 \end{pmatrix}, \begin{pmatrix} x_3 \\ y_3 \end{pmatrix}$ be the corresponding points of intersection of the diagonals. This doesn't quite give the equations of the lines forming the two diagonals. The next part gives a way of finding them.

g. Let $\begin{pmatrix} x_1 \\ y_1 \end{pmatrix}$ be one of the points of intersection, as before, and consider the line l_k through the point $\begin{pmatrix} x_1 \\ y_1 \end{pmatrix}$ with slope k, of equation

$$y - y_1 = k(x - x_1).$$

Show that the values of k for which l_k is a diagonal are also the values for which the restrictions to l_k of the two quadratic functions $y^2 + qx + r - \frac{p^2}{4}$ and $x^2 - y - p/2$ are proportional. Show that this gives the equations

You can ignore the third root of the resolvent cubic or use it to check your answers.

$$\frac{1}{k^2} = \frac{-k}{2k(-kx_1 + y_1) + q} \quad \text{and} \quad \frac{1}{k^2} = \frac{kx_1 - y_1 + p/2}{(kx_1 - y_1)^2 - p^2/4 + r},$$

which can be reduced to the single quadratic equation

$$k^2(x_1^2 - y_1 + p/2) = y_1^2 + qx_1 - p^2/4 + r. \tag{2}$$

Now the full solution is at hand. Compute (m_1, x_1, y_1) and (m_2, x_2, y_2). Then for each of these compute the slopes $k_{i,1}$ and $k_{i,2} = -k_{i,1}$ from equation (2). You now have four lines, two through A and two through B. Intersect them in pairs to find the four intersections of the parabolas.

h. Solve the quartic equations $x^4 - 4x^2 + x + 1 = 0$ and $x^4 + 4x^3 + x - 1 = 0$.

Exercise A2.8: The parabola of equation $y = x^2$ is obviously closed, so its complement is open. But it is one thing to say that around every point off the parabola there exists a disc that does not intersect the parabola, and another thing to find the radius of such a disc.

A2.8 Find a formula for the radius r of the largest disc centered at $\begin{pmatrix} 2 \\ 3 \end{pmatrix}$ that doesn't intersect the parabola of equation $y = x^2$, using the following steps:

a. Find the distance squared $\left| \begin{pmatrix} 2 \\ 3 \end{pmatrix} - \begin{pmatrix} x \\ x^2 \end{pmatrix} \right|^2$ as a fourth-degree polynomial in x.

b. Find the zeros of the derivative by the method of Exercise A2.6.

c. Find r.

A3 Two results in topology: Nested compact sets and Heine-Borel

In this section we give two more properties of compact subsets of $\mathbb{R}^n$, which we will need for proofs in Appendices A20, A21, and A23 and for the proof of Stokes's theorem in Section 6.10.

In Theorem A3.1, note that the hypothesis that the X_k are compact is essential. For instance, the intervals $(0, 1/n)$ form a decreasing intersection of nonempty sets, but their intersection is empty; the sequence of unbounded intervals $[k,\infty)$ is a decreasing sequence of nonempty closed subsets, but its intersection is also empty.

FIGURE A3.1.
Eduard Heine (1821–1881)
The history of the Heine-Borel theorem is complicated. Dirichlet, Weierstrass, and Pincherle could also claim some credit.

Theorem A3.1 (Decreasing intersection of nested compact sets). *Let $k \mapsto X_k \subset \mathbb{R}^n$ be a sequence of nonempty compact sets such that $X_1 \supset X_2 \supset \dots$. Then*

$$\bigcap_{k=1}^{\infty} X_k \neq \emptyset. \qquad \text{A3.1}$$

Proof. For each k, choose $\mathbf{x}_k \in X_k$ (using the hypothesis that $X_k \neq \emptyset$). Since this is in particular a sequence in X_1, choose a convergent subsequence $i \mapsto \mathbf{x}_{k_i}$. Since for every m the sequence $\mathbf{x}_m, \mathbf{x}_{m+1}, \dots$ is contained in X_m, the limit of this sequence is a point of the intersection $\cap_{m=1}^{\infty} X_m$, and so is the limit, since each X_m is closed. □

Definition A3.2 (Open cover). A collection $\mathcal{U}$ of open subsets of $\mathbb{R}^n$ *covers* $X \subset \mathbb{R}^n$ if $X \subset \bigcup_{U \in \mathcal{U}} U$. Such a collection is called an *open cover* of X.

The next statement constitutes the definition of "compact" in general topology; all other properties of compact sets can be derived from it. It will not play such a central role for us, but we will need it to prove the general Stokes's theorem.

FIGURE A3.2.
Emile Borel (1871–1956) was minister of the French Navy from 1925 to 1940; after World War II he received a medal for his work in the Resistance (*International Mathematical Congresses: An Illustrated History 1893–1986*, D. Albers, G. Alexanderson, C. Reid. Springer-Verlag, 1986).

Theorem A3.3 (Heine-Borel theorem). *If $X \subset \mathbb{R}^n$ is compact and $\mathcal{U}$ is an open cover of X, then $\mathcal{U}$ contains a finite subcover: there exist finitely many open sets $U_1, \dots, U_m \in \mathcal{U}$ such that*

$$X \subset \bigcup_{i=1}^{m} U_i. \qquad \text{A3.2}$$

Proof. This is very similar to the proof of Theorem 1.6.3. We argue by contradiction: suppose we need infinitely many U_i to cover X.

The set X is contained in a box $-10^N \le x_i < 10^N$ for some N. Decompose this box into finitely many closed boxes of side 1 in the obvious way. If each box is covered by finitely many U_i, then all of X is also, so at least one box requires infinitely many of the U_i to cover it; denote that box by B_0.

Now cut up B_0 into 10^n closed boxes of side $1/10$ (in the plane, 100 boxes; in $\mathbb{R}^3$, 1,000 boxes). At least one of these smaller boxes must again require infinitely many of the U_i to cover it. Call such a box B_1, and keep going: cut up B_1 into 10^n boxes of side $1/10^2$; again, at least one of these boxes must require infinitely many U_i to cover it; call one such box B_2, etc.

The boxes B_i form a decreasing sequence of compact sets, so there exists a point $\mathbf{x} \in \cap B_i$. This point is in X, so it is in one of the U_i. That U_i contains the ball of radius r around $\mathbf{x}$ for some $r > 0$, and hence around all the boxes B_j for j sufficiently large (to be precise, as soon as $\sqrt{n}/10^j < r$).

This is a contradiction. □

A4 Proof of the chain rule

Proving the chain rule is surprisingly difficult. Even in one dimension, if you try to prove it using the definitions

$$f'(x) = \lim_{h \to 0} \frac{f(x+h) - f(x)}{h}$$

and

$$g'(y) = \lim_{k \to 0} \frac{g(y+k) - g(y)}{k},$$

treating $g(y+k) - g(y)$ as an h in the first equation, you will not succeed, since $g(y+k) - g(y)$ could be 0. Instead you have to define remainders and show that they are small.

The proof of the chain rule would be substantially simplified by using the big O and little o notation discussed in Definition 3.4.1 and Appendix A.11: much of the work is deriving, in special cases, the rules proved in Appendix A.11. See Exercise A11.4.

Equation A4.6: In the first line, we are evaluating $\mathbf{f}$ at $\mathbf{g}(\mathbf{a}+\vec{\mathbf{h}})$, plugging in the value for $\mathbf{g}(\mathbf{a}+\vec{\mathbf{h}})$ given by the right side of equation A4.4. We then see that

$$[\mathbf{Dg}(\mathbf{a})]\vec{\mathbf{h}} + \mathbf{r}(\vec{\mathbf{h}})$$

plays the role of $\vec{\mathbf{k}}$ in the left side of equation A4.5. In the second line we plug this value for $\vec{\mathbf{k}}$ into the right side of equation A4.6.

To go from the second to the third line of equation A4.6, we use the linearity of $[\mathbf{Df}(\mathbf{g}(\mathbf{a}))]$:

$$[\mathbf{Df}(\mathbf{g}(\mathbf{a}))]\Big([\mathbf{Dg}(\mathbf{a})]\vec{\mathbf{h}} + \mathbf{r}(\vec{\mathbf{h}})\Big)$$
$$= [\mathbf{Df}(\mathbf{g}(\mathbf{a}))][\mathbf{Dg}(\mathbf{a})]\vec{\mathbf{h}}$$
$$+ [\mathbf{Df}(\mathbf{g}(\mathbf{a}))]\mathbf{r}(\vec{\mathbf{h}}).$$

Theorem 1.8.3 (Chain rule). *Let $U \subset \mathbb{R}^n, V \subset \mathbb{R}^m$ be open sets, let $\mathbf{g}: U \to V$ and $\mathbf{f}: V \to \mathbb{R}^p$ be mappings, and let $\mathbf{a}$ be a point of U. If $\mathbf{g}$ is differentiable at $\mathbf{a}$ and $\mathbf{f}$ is differentiable at $\mathbf{g}(\mathbf{a})$, then the composition $\mathbf{f} \circ \mathbf{g}$ is differentiable at $\mathbf{a}$, and its derivative is given by*

$$[\mathbf{D}(\mathbf{f} \circ \mathbf{g})(\mathbf{a})] = [\mathbf{Df}(\mathbf{g}(\mathbf{a}))] \circ [\mathbf{Dg}(\mathbf{a})].$$

Proof. We will define two "remainder" functions, $\mathbf{r}$ and $\mathbf{s}$, the function $\mathbf{r}$ giving the difference between the increment to $\mathbf{g}$ and its linear approximation at $\mathbf{a}$, the function $\mathbf{s}$ giving the difference between the increment to $\mathbf{f}$ and its linear approximation at $\mathbf{g}(\mathbf{a})$:

$$\underbrace{\mathbf{g}(\mathbf{a}+\vec{\mathbf{h}}) - \mathbf{g}(\mathbf{a})}_{\text{increment to function}} - \underbrace{[\mathbf{Dg}(\mathbf{a})]\vec{\mathbf{h}}}_{\text{linear approx.}} = \mathbf{r}(\vec{\mathbf{h}}) \qquad A4.1$$

$$\underbrace{\mathbf{f}\big(\mathbf{g}(\mathbf{a})+\vec{\mathbf{k}}\big) - \mathbf{f}\big(\mathbf{g}(\mathbf{a})\big)}_{\text{increment to } \mathbf{f}} - \underbrace{[\mathbf{Df}(\mathbf{g}(\mathbf{a}))]\vec{\mathbf{k}}}_{\text{linear approx.}} = \mathbf{s}(\vec{\mathbf{k}}). \qquad A4.2$$

The hypotheses that $\mathbf{g}$ is differentiable at $\mathbf{a}$ and that $\mathbf{f}$ is differentiable at $\mathbf{g}(\mathbf{a})$ say exactly that

$$\lim_{\vec{\mathbf{h}} \to \vec{\mathbf{0}}} \frac{\mathbf{r}(\vec{\mathbf{h}})}{|\vec{\mathbf{h}}|} = \vec{\mathbf{0}} \quad \text{and} \quad \lim_{\vec{\mathbf{k}} \to \vec{\mathbf{0}}} \frac{\mathbf{s}(\vec{\mathbf{k}})}{|\vec{\mathbf{k}}|} = \vec{\mathbf{0}}. \qquad A4.3$$

Now we rewrite equations A4.1 and A4.2 in more convenient form:

$$\mathbf{g}(\mathbf{a}+\vec{\mathbf{h}}) = \mathbf{g}(\mathbf{a}) + [\mathbf{Dg}(\mathbf{a})]\vec{\mathbf{h}} + \mathbf{r}(\vec{\mathbf{h}}) \qquad A4.4$$

$$\mathbf{f}\big(\mathbf{g}(\mathbf{a})+\vec{\mathbf{k}}\big) = \mathbf{f}\big(\mathbf{g}(\mathbf{a})\big) + [\mathbf{Df}(\mathbf{g}(\mathbf{a}))]\vec{\mathbf{k}} + \mathbf{s}(\vec{\mathbf{k}}), \qquad A4.5$$

and then write

$$\mathbf{f}\big(\mathbf{g}(\mathbf{a}+\vec{\mathbf{h}})\big) = \mathbf{f}\Big(\overbrace{\mathbf{g}(\mathbf{a}) + \underbrace{[\mathbf{Dg}(\mathbf{a})]\vec{\mathbf{h}} + \mathbf{r}(\vec{\mathbf{h}})}_{\text{treat as } \vec{\mathbf{k}}, \text{ left side eq. A4.5}}}^{\text{from equation A4.4}}\Big) \qquad A4.6$$

$$= \mathbf{f}(\mathbf{g}(\mathbf{a})) + [\mathbf{Df}(\mathbf{g}(\mathbf{a}))]\underbrace{\Big([\mathbf{Dg}(\mathbf{a})]\vec{\mathbf{h}} + \mathbf{r}(\vec{\mathbf{h}})\Big)}_{\vec{\mathbf{k}}} + \mathbf{s}\underbrace{\Big([\mathbf{Dg}(\mathbf{a})]\vec{\mathbf{h}} + \mathbf{r}(\vec{\mathbf{h}})\Big)}_{\vec{\mathbf{k}}}$$

$$= \mathbf{f}\big(\mathbf{g}(\mathbf{a})\big) + [\mathbf{Df}(\mathbf{g}(\mathbf{a}))]\big([\mathbf{Dg}(\mathbf{a})]\vec{\mathbf{h}}\big) + \underbrace{[\mathbf{Df}(\mathbf{g}(\mathbf{a}))]\big(\mathbf{r}(\vec{\mathbf{h}})\big) + \mathbf{s}\big([\mathbf{Dg}(\mathbf{a})]\vec{\mathbf{h}} + \mathbf{r}(\vec{\mathbf{h}})\big)}_{\text{remainder}}.$$

We can subtract $\mathbf{f}\big(\mathbf{g}(\mathbf{a})\big)$ from both sides of equation A4.6, to get

$$\underbrace{\mathbf{f}\big(\mathbf{g}(\mathbf{a}+\vec{\mathbf{h}})\big) - \mathbf{f}\big(\mathbf{g}(\mathbf{a})\big)}_{\text{increment to composition}} = \underbrace{\overbrace{[\mathbf{Df}(\mathbf{g}(\mathbf{a}))]}^{\substack{\text{linear approx.}\\ \text{to } \Delta\mathbf{f} \text{ at } \mathbf{g}(\mathbf{a})}} \overbrace{\big([\mathbf{Dg}(\mathbf{a})]\ \vec{\mathbf{h}}\big)}^{\substack{\text{linear approx.}\\ \text{to } \Delta\mathbf{g} \text{ at } \mathbf{a}}}}_{\text{composition of linear approximations}} + \text{remainder}. \qquad A4.7$$

So to prove the chain rule, we need to prove (see Proposition and Definition 1.7.9) that

$$\lim_{\vec{\mathbf{h}}\to\vec{\mathbf{0}}} \frac{1}{|\vec{\mathbf{h}}|}\Big(\mathbf{f}\big(\mathbf{g}(\mathbf{a}+\vec{\mathbf{h}})\big) - \mathbf{f}\big(\mathbf{g}(\mathbf{a})\big) - [\mathbf{Df}(\mathbf{g}(\mathbf{a}))][\mathbf{Dg}(\mathbf{a})]\Big) = \vec{\mathbf{0}}, \text{ i.e., that}$$

$$\lim_{\vec{\mathbf{h}}\to\vec{\mathbf{0}}} \frac{1}{|\vec{\mathbf{h}}|} \overbrace{[\mathbf{Df}(\mathbf{g}(\mathbf{a}))]\big(\mathbf{r}(\vec{\mathbf{h}})\big) + \mathbf{s}\big([\mathbf{Dg}(\mathbf{a})]\vec{\mathbf{h}} + \mathbf{r}(\vec{\mathbf{h}})\big)}^{\text{remainder, eq. A4.6}} = \vec{\mathbf{0}}. \qquad A4.8$$

Let us look separately at the two terms in the limit in equation A4.8. The first is straightforward. Since (Proposition 1.4.11)

$$\big|[\mathbf{Df}(\mathbf{g}(\mathbf{a}))]\mathbf{r}(\vec{\mathbf{h}})\big| \le \big|[\mathbf{Df}(\mathbf{g}(\mathbf{a}))]\big|\,|\mathbf{r}(\vec{\mathbf{h}})|, \quad \text{we have} \qquad A4.9$$

$$\lim_{\vec{\mathbf{h}}\to\vec{\mathbf{0}}} \frac{\big|[\mathbf{Df}(\mathbf{g}(\mathbf{a}))]\mathbf{r}(\vec{\mathbf{h}})\big|}{|\vec{\mathbf{h}}|} \le \Big|[\mathbf{Df}(\mathbf{g}(\mathbf{a}))]\Big| \underbrace{\lim_{\vec{\mathbf{h}}\to\vec{\mathbf{0}}} \frac{|\mathbf{r}(\vec{\mathbf{h}})|}{|\vec{\mathbf{h}}|}}_{=0 \text{ by eq. A4.3}} = 0. \qquad A4.10$$

The second term is harder. We want to show that

$$\lim_{\vec{\mathbf{h}}\to\vec{\mathbf{0}}} \frac{\mathbf{s}\big([\mathbf{Dg}(\mathbf{a})]\vec{\mathbf{h}} + \mathbf{r}(\vec{\mathbf{h}})\big)}{|\vec{\mathbf{h}}|} = \vec{\mathbf{0}}. \qquad A4.11$$

First note that there exists $\delta > 0$ such that $|\mathbf{r}(\vec{\mathbf{h}})| \le |\vec{\mathbf{h}}|$ when $|\vec{\mathbf{h}}| < \delta$ (by equation A4.3). Thus, when $|\vec{\mathbf{h}}| < \delta$, we have

$$\big|[\mathbf{Dg}(\mathbf{a})]\vec{\mathbf{h}} + \underbrace{\mathbf{r}(\vec{\mathbf{h}})}_{\le|\vec{\mathbf{h}}|}\big| \le \big|[\mathbf{Dg}(\mathbf{a})]\vec{\mathbf{h}}\big| + |\vec{\mathbf{h}}| \le \big(|[\mathbf{Dg}(\mathbf{a})]| + 1\big)|\vec{\mathbf{h}}|. \qquad A4.12$$

We specify "$|\vec{\mathbf{k}}|$ sufficiently small" by $|\vec{\mathbf{k}}| \le \delta'$.

For $|\vec{\mathbf{k}}|$ sufficiently small,

$$\frac{|\mathbf{s}(\vec{\mathbf{k}})|}{|\vec{\mathbf{k}}|} \le \epsilon, \quad \text{i.e.,} \quad |\mathbf{s}(\vec{\mathbf{k}})| \le \epsilon|\vec{\mathbf{k}}|;$$

otherwise the limit as $\vec{\mathbf{k}} \to \vec{\mathbf{0}}$ would not be $\vec{\mathbf{0}}$, which would contradict equation A4.3.

Equation A4.3 also tells us that for any $\epsilon > 0$, there exists $0 < \delta' < \delta$ such that when $|\vec{\mathbf{k}}| \le \delta'$, then $|\mathbf{s}(\vec{\mathbf{k}})| \le \epsilon|\vec{\mathbf{k}}|$ (see the margin note). When

$$|\vec{\mathbf{h}}| \le \frac{\delta'}{|[\mathbf{Dg}(\mathbf{a})])| + 1}; \quad \text{i.e.,} \quad \big(|[\mathbf{Dg}(\mathbf{a})]| + 1\big)|\vec{\mathbf{h}}| \le \delta', \qquad A4.13$$

then equation A4.12 gives

$$|[\mathbf{Dg}(\mathbf{a})]\vec{\mathbf{h}} + \mathbf{r}(\vec{\mathbf{h}})| \le \delta', \qquad A4.14$$

so we can substitute the expression $|[\mathbf{Dg}(\mathbf{a})]\vec{\mathbf{h}}+\mathbf{r}(\vec{\mathbf{h}})|$ for $|\vec{\mathbf{k}}|$ in the equation $|\mathbf{s}(\vec{\mathbf{k}})| \le \epsilon|\vec{\mathbf{k}}|$, which is true when $|\vec{\mathbf{k}}| < \delta'$. This gives

$$\underbrace{\big|\mathbf{s}\big([\mathbf{Dg}(\mathbf{a})]\vec{\mathbf{h}} + \mathbf{r}(\vec{\mathbf{h}})\big)\big|}_{|\mathbf{s}(\vec{\mathbf{k}})|} \le \underbrace{\epsilon\big|[\mathbf{Dg}(\mathbf{a})]\vec{\mathbf{h}} + \mathbf{r}(\vec{\mathbf{h}})\big|}_{\epsilon|\vec{\mathbf{k}}|} \underbrace{\le}_{\text{ineq. A4.12}} \epsilon\big(|[\mathbf{Dg}(\mathbf{a})]| + 1\big)\,|\vec{\mathbf{h}}|. \qquad A4.15$$

Dividing by $|\vec{\mathbf{h}}|$ gives

$$\frac{\big|\mathbf{s}\big([\mathbf{Dg}(\mathbf{a})]\vec{\mathbf{h}} + \mathbf{r}(\vec{\mathbf{h}})\big)\big|}{|\vec{\mathbf{h}}|} \le \epsilon\big(|[\mathbf{Dg}(\mathbf{a})]| + 1\big), \qquad A4.16$$

and since this is true for every $\epsilon > 0$, equation A4.11 is correct. □

A5 Proof of Kantorovich's theorem

Theorem 2.8.13 (Kantorovich's theorem). *Let $\mathbf{a}_0$ be a point in $\mathbb{R}^n$, U an open neighborhood of $\mathbf{a}_0$ in $\mathbb{R}^n$, and $\mathbf{f} : U \to \mathbb{R}^n$ a differentiable mapping, with its derivative $[\mathbf{Df}(\mathbf{a}_0)]$ invertible. Define*

$$\vec{\mathbf{h}}_0 \stackrel{\text{def}}{=} -[\mathbf{Df}(\mathbf{a}_0)]^{-1}\mathbf{f}(\mathbf{a}_0), \quad \mathbf{a}_1 \stackrel{\text{def}}{=} \mathbf{a}_0 + \vec{\mathbf{h}}_0, \quad U_1 \stackrel{\text{def}}{=} B_{|\vec{\mathbf{h}}_0|}(\mathbf{a}_1). \qquad A5.1$$

If $\overline{U_1} \subset U$ and the derivative $[\mathbf{Df}(\mathbf{x})]$ satisfies the Lipschitz condition

$$\big|[\mathbf{Df}(\mathbf{u}_1)] - [\mathbf{Df}(\mathbf{u}_2)]\big| \le M|\mathbf{u}_1 - \mathbf{u}_2| \quad \textit{for all points } \mathbf{u}_1, \mathbf{u}_2 \in \overline{U_1}, \qquad A5.2$$

and if the inequality

$$\big|\mathbf{f}(\mathbf{a}_0)\big|\,\big|[\mathbf{Df}(\mathbf{a}_0)]^{-1}\big|^2 M \le \frac{1}{2} \qquad A5.3$$

is satisfied, the equation $\mathbf{f}(\mathbf{x}) = \mathbf{0}$ has a unique solution in the closed ball $\overline{U_1}$, and Newton's method with initial guess $\mathbf{a}_0$ converges to it.

Facts 1, 2, and 3 guarantee that the hypotheses about $\mathbf{a}_0$ of our theorem are also true of $\mathbf{a}_1$. We need statement 1 in order to define $\vec{\mathbf{h}}_1$, $\mathbf{a}_2$, and U_2. Statement 2 guarantees that $U_2 \subset U_1$, hence $[\mathbf{Df}(\mathbf{x})]$ satisfies the same Lipschitz condition on U_2 as on U_1. Statement 3 is needed to show that inequality A5.3 is satisfied at $\mathbf{a}_1$. (Remember that the ratio M has not changed.)

Proof. The proof is fairly involved, so we will first outline our approach. We prove existence by showing the following four facts:

1. $[\mathbf{Df}(\mathbf{a}_1)]$ is invertible, allowing us to define $\vec{\mathbf{h}}_1 = -[\mathbf{Df}(\mathbf{a}_1)]^{-1}\mathbf{f}(\mathbf{a}_1)$
2. $|\vec{\mathbf{h}}_1| \le \dfrac{|\vec{\mathbf{h}}_0|}{2}$ A5.4
3. $\big|\mathbf{f}(\mathbf{a}_1)\big|\,\big|[\mathbf{Df}(\mathbf{a}_1)]^{-1}\big|^2 \le \big|\mathbf{f}(\mathbf{a}_0)\big|\,\big|[\mathbf{Df}(\mathbf{a}_0)]^{-1}\big|^2$ A5.5
4. $|\mathbf{f}(\mathbf{a}_1)| \le \dfrac{M}{2}|\vec{\mathbf{h}}_0|^2$ A5.6

If statements 1, 2, and 3 are true, we can define sequences $\vec{\mathbf{h}}_i, \mathbf{a}_{i+1}, U_{i+1}$:

$$\begin{aligned} \vec{\mathbf{h}}_i &= -[\mathbf{Df}(\mathbf{a}_i)]^{-1}\mathbf{f}(\mathbf{a}_i) \\ \mathbf{a}_{i+1} &= \mathbf{a}_i + \vec{\mathbf{h}}_i \\ U_{i+1} &= \left\{ \mathbf{x} \;\middle|\; |\mathbf{x} - \mathbf{a}_{i+1}| \le |\vec{\mathbf{h}}_i| \right\} \end{aligned} \qquad A5.7$$

and at each stage all the hypotheses of Theorem 2.8.13 are true. Statement 2, together with Proposition 1.5.35, proves that the sequence $i \mapsto \mathbf{a}_i$ converges; call the limit $\mathbf{a}$. Statement 4 then says that $\mathbf{a}$ satisfies $\mathbf{f}(\mathbf{a}) = \mathbf{0}$. Indeed, by part 2,

$$|\vec{\mathbf{h}}_i| \le \frac{|\vec{\mathbf{h}}_0|}{2^i} \qquad A5.8$$

so by part 4,

$$|\mathbf{f}(\mathbf{a}_i)| \le \frac{M}{2}|\vec{\mathbf{h}}_{i-1}|^2 \le \frac{M}{2^i}|\vec{\mathbf{h}}_0|^2, \qquad A5.9$$

and in the limit as $i \to \infty$, we have $|\mathbf{f}(\mathbf{a})| = 0$.

First we need to prove Proposition A5.1 and Lemma A5.3.

Proposition A5.1 is a variant of Taylor's theorem with remainder; see Theorems A12.1 and A12.5.

Proposition A5.1. *Let $U \subset \mathbb{R}^n$ be an open ball, $V \subset \mathbb{R}^n$ a neighborhood of $\overline{U}$, and $\mathbf{f} : V \to \mathbb{R}^m$ a differentiable mapping whose derivative is Lipschitz with Lipschitz ratio M. Then for $\mathbf{x}, \mathbf{y} \in \overline{U}$,*

$$\Big|\underbrace{\mathbf{f}(\mathbf{y}) - \mathbf{f}(\mathbf{x})}_{\text{increment to } \mathbf{f}} - \underbrace{[\mathbf{Df}(\mathbf{x})](\mathbf{y} - \mathbf{x})}_{\substack{\text{linear approx.} \\ \text{of increment to } \mathbf{f}}}\Big| \le \frac{M}{2}|\mathbf{y} - \mathbf{x}|^2. \qquad A5.10$$

Before giving the proof, let us see why this statement is reasonable. The term $[\mathbf{Df}(\mathbf{x})](\mathbf{y} - \mathbf{x})$ is the linear approximation of the increment to the function in terms of the increment $\mathbf{y} - \mathbf{x}$ to the variable. You would expect the error term to be of second degree: some multiple of $|\mathbf{y} - \mathbf{x}|^2$, which gets very small as $\mathbf{y} \to \mathbf{x}$. That is what Proposition A5.1 says, and it identifies the Lipschitz ratio M as the main ingredient of the coefficient of $|\mathbf{y} - \mathbf{x}|^2$.

The coefficient $M/2$ is the smallest coefficient that works for *all* functions $\mathbf{f} : U \to \mathbb{R}^m$, although for most functions, the inequality holds with a smaller coefficient. We know it is the smallest possible coefficient because equality is achieved for $f(x) = x^2$:

$$[\mathbf{D}f(x)] = f'(x) = 2x, \quad \text{so} \quad |[\mathbf{D}f(x)] - [\mathbf{D}f(y)]| = 2|x - y|, \qquad A5.11$$

and the best Lipschitz ratio is $M = 2$:

$$|y^2 - x^2 - 2x(y - x)| = (y - x)^2 = \frac{2}{2}(y - x)^2 = \frac{M}{2}(y - x)^2. \qquad A5.12$$

In Example A5.2 we have no guarantee that the difference between the increment to f and its approximation by $f'(x)(y-x)$ behaves like $(y - x)^2$.

Example A5.2 (A derivative that is not Lipschitz). If the derivative of $\mathbf{f}$ is *not* Lipschitz, then usually there exists no C such that

$$|\mathbf{f}(\mathbf{y}) - \mathbf{f}(\mathbf{x}) - [\mathbf{Df}(\mathbf{x})](\mathbf{y} - \mathbf{x})| \le C|\mathbf{y} - \mathbf{x}|^2. \qquad A5.13$$

Let $f(x) = x^{4/3}$, so $[\mathbf{D}f(x)] = f'(x) = \frac{4}{3}x^{1/3}$. In particular, $f'(0) = 0$, so

$$|f(y) - f(0) - f'(0)(y - 0)| = |f(y)| = y^{4/3}. \qquad A5.14$$

But $y^{4/3}$ is not $\le C|y|^2$ for any C, since $y^{4/3}/y^2 = 1/y^{2/3} \to \infty$ as $y \to 0$. $\triangle$

Proof of Proposition A5.1. Choose $\mathbf{x}, \mathbf{y} \in \overline{U}$, set $\vec{\mathbf{h}} = \mathbf{y} - \mathbf{x}$, and consider the function $\mathbf{g}(t) = \mathbf{f}(\mathbf{x} + t\vec{\mathbf{h}})$. Each coordinate of $\mathbf{g}$ is a differentiable function of the single variable t in some neighborhood of $[0, 1]$, whose derivative is continuous on $[0, 1]$, so by the fundamental theorem of calculus

$$\overbrace{\mathbf{f}(\mathbf{x} + \vec{\mathbf{h}})}^{\mathbf{g}(1)} - \overbrace{\mathbf{f}(\mathbf{x})}^{\mathbf{g}(0)} = \mathbf{g}(1) - \mathbf{g}(0) = \int_0^1 \mathbf{g}'(t)\, dt. \qquad A5.15$$

Writing $\mathbf{g}$ as the composition $\mathbf{f} \circ \mathbf{s}$ for $\mathbf{s}(t) = \mathbf{x} + t\vec{\mathbf{h}}$ and using the chain rule, we see that

$$\mathbf{g}'(t) = [\mathbf{Df}(\mathbf{x} + t\vec{\mathbf{h}})]\vec{\mathbf{h}}, \qquad A5.16$$

which we will write as

$$\mathbf{g}'(t) = [\mathbf{Df}(\mathbf{x})]\vec{\mathbf{h}} + \Big([\mathbf{Df}(\mathbf{x} + t\vec{\mathbf{h}})]\vec{\mathbf{h}} - [\mathbf{Df}(\mathbf{x})]\vec{\mathbf{h}}\Big). \qquad A5.17$$

This leads to

$$\mathbf{f}(\mathbf{x}+\vec{\mathbf{h}})-\mathbf{f}(\mathbf{x}) = \int_0^1 [\mathbf{Df}(\mathbf{x})]\vec{\mathbf{h}}\,dt + \int_0^1 \Big([\mathbf{Df}(\mathbf{x}+t\vec{\mathbf{h}})]\vec{\mathbf{h}} - [\mathbf{Df}(\mathbf{x})]\vec{\mathbf{h}}\Big)\,dt. \qquad A5.18$$

To go from the first to the second line of inequality A5.19, the Lipschitz condition on the derivative of **f** says that

$$\Big|[\mathbf{Df}(\mathbf{x} + t\vec{\mathbf{h}})] - [\mathbf{Df}(\mathbf{x})]\Big| \le M|\mathbf{x} + t\vec{\mathbf{h}} - \mathbf{x}| = Mt|\vec{\mathbf{h}}|,$$

where we replace $\mathbf{x}$ by $\mathbf{x} + t\vec{\mathbf{h}}$ and $\mathbf{y}$ by $\mathbf{x}$.

The first term on the right is the integral from 0 to 1 of a constant, so it is simply that constant, and we can rewrite equation A5.18 as

$$\big|\mathbf{f}(\mathbf{x} + \vec{\mathbf{h}}) - \mathbf{f}(\mathbf{x}) - [\mathbf{Df}(\mathbf{x})]\vec{\mathbf{h}}\big| = \Big|\int_0^1 \Big([\mathbf{Df}(\mathbf{x} + t\vec{\mathbf{h}})]\vec{\mathbf{h}} - [\mathbf{Df}(\mathbf{x})]\vec{\mathbf{h}}\Big)\,dt\Big|$$

$$\le \int_0^1 Mt|\vec{\mathbf{h}}||\vec{\mathbf{h}}|\,dt = \frac{M}{2}|\vec{\mathbf{h}}|^2. \qquad A5.19$$

□ Proposition A5.1

Proving Lemma A5.3 is the hardest part of proving Kantorovich's theorem.

Lemma A5.3. *The matrix* $[\mathbf{Df}(\mathbf{a}_1)]$ *is invertible, and*

$$\big|[\mathbf{Df}(\mathbf{a}_1)]^{-1}\big| \le 2\big|[\mathbf{Df}(\mathbf{a}_0)]^{-1}\big|. \qquad A5.20$$

Proof. We have required (inequality A5.2) that the derivative matrix not vary too fast, so it is reasonable to hope that

$$[\mathbf{Df}(\mathbf{a}_0)]^{-1}[\mathbf{Df}(\mathbf{a}_1)] \qquad A5.21$$

is not too far from the identity. Indeed, set

$$A = I-\big([\mathbf{Df}(\mathbf{a}_0)]^{-1}[\mathbf{Df}(\mathbf{a}_1)]\big) = \underbrace{[\mathbf{Df}(\mathbf{a}_0)]^{-1}[\mathbf{Df}(\mathbf{a}_0)]}_{I} - \big([\mathbf{Df}(\mathbf{a}_0)]^{-1}[\mathbf{Df}(\mathbf{a}_1)]\big)$$

$$= [\mathbf{Df}(\mathbf{a}_0)]^{-1}\Big([\mathbf{Df}(\mathbf{a}_0)] - [\mathbf{Df}(\mathbf{a}_1)]\Big). \qquad A5.22$$

By inequality A5.2, $\big|[\mathbf{Df}(\mathbf{a}_0)]-[\mathbf{Df}(\mathbf{a}_1)]\big| \le M|\mathbf{a}_0-\mathbf{a}_1|$, and by definition (equation A5.1), $|\vec{\mathbf{h}}_0| = |\mathbf{a}_1 - \mathbf{a}_0|$. So

$$|A| \le \big|[\mathbf{Df}(\mathbf{a}_0)]^{-1}\big|\;|\vec{\mathbf{h}}_0|M. \qquad A5.23$$

By definition (equation A5.1 again),

$$\vec{\mathbf{h}}_0 = -[\mathbf{Df}(\mathbf{a}_0)]^{-1}\mathbf{f}(\mathbf{a}_0), \qquad \text{so} \qquad |\vec{\mathbf{h}}_0| \le \big|[\mathbf{Df}(\mathbf{a}_0)]^{-1}\big||\mathbf{f}(\mathbf{a}_0)| \qquad A5.24$$

(Proposition 1.4.11, once more). This gives us

$$|A| \le \underbrace{\big|[\mathbf{Df}(\mathbf{a}_0)]^{-1}\big|\,\big|[\mathbf{Df}(\mathbf{a}_0)]^{-1}\big|\,|\mathbf{f}(\mathbf{a}_0)|M}_{\text{left side of inequality A5.3}}. \qquad A5.25$$

Inequality A5.3 now guarantees

$$|A| \le \frac{1}{2}, \qquad A5.26$$

□ Lemma A5.3

Now we can show that $[\mathbf{Df}(\mathbf{a}_1)]$ is invertible: by the definition of A we have $I-A = [\mathbf{Df}(\mathbf{a}_0)]^{-1}[\mathbf{Df}(\mathbf{a}_1)]$; since $|A| \le 1/2$, we can apply Proposition

1.5.38 to say that $I - A$ is invertible, so by Proposition 1.2.15, $[\mathbf{Df}(\mathbf{a}_1)]$ is invertible and

$$B \stackrel{\text{def}}{=} (I-A)^{-1} = [\mathbf{Df}(\mathbf{a}_1)]^{-1}[\mathbf{Df}(\mathbf{a}_0)]. \qquad A5.27$$

Then

$$\begin{aligned}[\mathbf{Df}(\mathbf{a}_1)]^{-1} &= B[\mathbf{Df}(\mathbf{a}_0)]^{-1} = \overbrace{(I + A + A^2 + \cdots)}^{B \text{ by Prop. 1.5.38)}}[\mathbf{Df}(\mathbf{a}_0)]^{-1} \\ &= [\mathbf{Df}(\mathbf{a}_0)]^{-1} + A[\mathbf{Df}(\mathbf{a}_0)]^{-1} + \cdots . \end{aligned} \qquad A5.28$$

Hence (by the triangle inequality and Proposition 1.4.11),

$$\begin{aligned}\left|[\mathbf{Df}(\mathbf{a}_1)]^{-1}\right| &\le \left|[\mathbf{Df}(\mathbf{a}_0)]^{-1}\right| + |A|\left|[\mathbf{Df}(\mathbf{a}_0)]^{-1}\right| + \cdots \\ &= \left|[\mathbf{Df}(\mathbf{a}_0)]^{-1}\right|\left(1 + |A| + |A|^2 + \cdots\right) \\ &\le \left|[\mathbf{Df}(\mathbf{a}_0)]^{-1}\right| \underbrace{\left(1 + 1/2 + 1/4 + \cdots\right)}_{\text{since } |A| \le 1/2, \text{ ineq. A5.26}} = 2\left|[\mathbf{Df}(\mathbf{a}_0)]^{-1}\right|. \end{aligned} \qquad A5.29$$

□ Lemma A5.3

Inequality A5.29: Note that we use the number 1, not the identity matrix: $1 + |A| + |A|^2 + \cdots$, not $I + A + A^2 + \cdots$. This is crucial because since $|A| \le 1/2$, we have

$$|A|^0 + |A|^1 + |A|^2 + \cdots \le 2,$$

since we have a geometric series (see Example 0.5.6).

When we first wrote this proof, adapting a proof using the norm of a matrix rather than the length, we factored before using the triangle inequality and ended up with $|I| + |A| + |A^2| + \cdots$. This was disastrous: $|I| = \sqrt{n}$, *not* 1. The discovery that this could be fixed by factoring *after* using the triangle inequality was most welcome.

So far we have proved statement 1; this enables us to define the next step of Newton's method:

$$\vec{\mathbf{h}}_1 = -[\mathbf{Df}(\mathbf{a}_1)]^{-1}\mathbf{f}(\mathbf{a}_1), \quad \mathbf{a}_2 = \mathbf{a}_1 + \vec{\mathbf{h}}_1, \quad U_2 = \left\{ \mathbf{x} \;\middle|\; |\mathbf{x} - \mathbf{a}_2| \le |\vec{\mathbf{h}}_1| \right\}. \qquad A5.30$$

Now we will prove statement 4, which we will call Lemma A5.4:

Lemma A5.4. *We have the inequality* $|\mathbf{f}(\mathbf{a}_1)| \le \frac{M}{2}|\vec{\mathbf{h}}_0|^2$.

Proof. Since $\vec{\mathbf{h}}_0 = \mathbf{a}_1 - \mathbf{a}_0$, Proposition A5.1 says that

$$\left|\mathbf{f}(\mathbf{a}_1) \overbrace{- \mathbf{f}(\mathbf{a}_0) - [\mathbf{Df}(\mathbf{a}_0)]\vec{\mathbf{h}}_0}^{\vec{0}}\right| \le \frac{M}{2}|\vec{\mathbf{h}}_0|^2, \qquad A5.31$$

but a miracle happens during the computation: the second and third terms on the left cancel, since

$$-[\mathbf{Df}(\mathbf{a}_0)]\vec{\mathbf{h}}_0 = [\mathbf{Df}(\mathbf{a}_0)] \overbrace{[\mathbf{Df}(\mathbf{a}_0)]^{-1}\mathbf{f}(\mathbf{a}_0)}^{-\vec{\mathbf{h}}_0 \text{ by def. (eq. A5.1)}} = \mathbf{f}(\mathbf{a}_0). \qquad A5.32$$

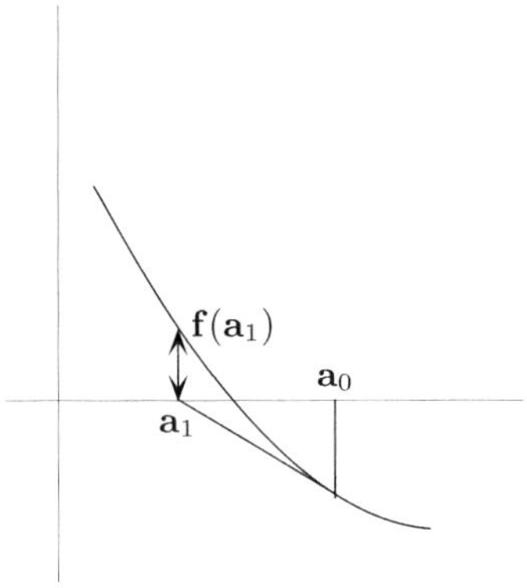

FIGURE A5.1.

The terms that cancel are exactly the value of the linearization to $\mathbf{f}$ at $\mathbf{a}_0$, evaluated at $\mathbf{a}_1$.

(Figure A5.1 explains why the cancellation occurs.) So we get

$$|\mathbf{f}(\mathbf{a}_1)| \le \frac{M}{2}|\vec{\mathbf{h}}_0|^2 \text{ as required.} \quad \square \quad \text{Lemma A5.4} \qquad A5.33$$

Proof of Kantorovich theorem, continued. Now we just string together the inequalities. We have proved statements 1 and 4. To prove statement 2, i.e., $|\vec{\mathbf{h}}_1| \le |\vec{\mathbf{h}}_0|/2$, we consider

$$|\vec{\mathbf{h}}_1| \le \left|\mathbf{f}(\mathbf{a}_1)\right|\left|[\mathbf{Df}(\mathbf{a}_1)]^{-1}\right| \le \frac{M|\vec{\mathbf{h}}_0|^2}{2} 2\left|[\mathbf{Df}(\mathbf{a}_0)]^{-1}\right|. \qquad A5.34$$

The first inequality in inequality A5.34 uses the definition of $\vec{\mathbf{h}}_1$ (equation A5.30) and Proposition 1.4.11. The second inequality uses Lemmas A5.3 and A5.4.

Note that the middle term of inequality A5.34 has $\mathbf{a}_1$, while the right term has $\mathbf{a}_0$.

Cancel the 2's and write $|\vec{\mathbf{h}}_0|^2$ as two factors:

$$|\vec{\mathbf{h}}_1| \le |\vec{\mathbf{h}}_0| M \left|[\mathbf{Df}(\mathbf{a}_0)]^{-1}\right| |\vec{\mathbf{h}}_0|. \qquad A5.35$$

Replace the second $|\vec{\mathbf{h}}_0|$, using $\vec{\mathbf{h}}_0 = -[\mathbf{Df}(\mathbf{a}_0)]^{-1}\mathbf{f}(\mathbf{a}_0)$, to get

$$|\vec{\mathbf{h}}_1| \le |\vec{\mathbf{h}}_0|\Big(\underbrace{M|[\mathbf{Df}(\mathbf{a}_0)]^{-1}|\,|\mathbf{f}(\mathbf{a}_0)|\,|[\mathbf{Df}(\mathbf{a}_0)]^{-1}|}_{\le 1/2 \text{ by inequality A5.3}}\Big) \le \frac{|\vec{\mathbf{h}}_0|}{2}. \qquad A5.36$$

Now we will prove part 3, i.e.,

$$|\mathbf{f}(\mathbf{a}_1)|\,\big|[\mathbf{Df}(\mathbf{a}_1)]^{-1}\big|^2 \le |\mathbf{f}(\mathbf{a}_0)|\,\big|[\mathbf{Df}(\mathbf{a}_0)]^{-1}\big|^2. \qquad A5.37$$

Using Lemma A5.4 to get a bound for $|\mathbf{f}(\mathbf{a}_1)|$, and inequality A5.20 to get a bound for $|[\mathbf{Df}(\mathbf{a}_1)]^{-1}|$, we write

$$\begin{aligned}|\mathbf{f}(\mathbf{a}_1)|\,|[\mathbf{Df}(\mathbf{a}_1)]^{-1}|^2 &\le \frac{M|\mathbf{h}_0|^2}{2}\left(4|[\mathbf{Df}(\mathbf{a}_0)]^{-1}|^2\right)\\ &\le 2|[\mathbf{Df}(\mathbf{a}_0)]^{-1}|^2 M\,\overbrace{\left(|[\mathbf{Df}(\mathbf{a}_0)]^{-1}||\mathbf{f}(\mathbf{a}_0)|\right)^2}^{\ge|\vec{\mathbf{h}}_0|^2} \qquad A5.38\\ &\le |[\mathbf{Df}(\mathbf{a}_0)]^{-1}|^2\;|\mathbf{f}(\mathbf{a}_0)|\,2\underbrace{|\mathbf{f}(\mathbf{a}_0)||[\mathbf{Df}(\mathbf{a}_0)]^{-1}|^2 M}_{\text{at most } 1/2 \text{ by A5.3}}\\ &\le |[\mathbf{Df}(\mathbf{a}_0)]^{-1}|^2|\mathbf{f}(\mathbf{a}_0)|.\end{aligned}$$

In the third edition, we ended the proof of existence with a simple □ and no summary. A reader objected: "a neat and intricate proof like this one should end with a trumpet blast, not just stop."

We have now proved statements 1–4. This proves that a solution in the closed ball $\overline{U_1}$ exists, and that Newton's method with initial guess $\mathbf{a}_0$ converges to it. Next we prove that the solution is unique.

Proof of uniqueness. To prove that the solution in $\overline{U}_1$ is unique, we will prove that if $\mathbf{y} \in \overline{U}_1$ and $\mathbf{f}(\mathbf{y}) = \mathbf{0}$, then

$$|\mathbf{y} - \mathbf{a}_{i+1}| \le \frac{1}{2}|\mathbf{y} - \mathbf{a}_i|. \qquad A5.39$$

This will prove that $\mathbf{y} = \lim \mathbf{a}_i$, and thus that $\lim \mathbf{a}_i$ is the unique solution of $\mathbf{f}(\mathbf{x}) = \mathbf{0}$ in $\overline{U}_1$. First, set

$$\mathbf{f}(\mathbf{y}) = \mathbf{f}(\mathbf{a}_i) + [\mathbf{Df}(\mathbf{a}_i)](\mathbf{y} - \mathbf{a}_i) + \vec{\mathbf{r}}_i, \qquad A5.40$$

where $\vec{\mathbf{r}}_i$ is the remainder necessary for the equality to be true. Then since $\mathbf{f}(\mathbf{y}) = \mathbf{0}$,

$$\mathbf{y} - \mathbf{a}_i = \overbrace{-[\mathbf{Df}(\mathbf{a}_i)]^{-1}\mathbf{f}(\mathbf{a}_i)}^{+\vec{\mathbf{h}}_i \text{ by eq. A5.7}} - [\mathbf{Df}(\mathbf{a}_i)]^{-1}\vec{\mathbf{r}}_i, \qquad A5.41$$

which we can rewrite as

$$\mathbf{y} - \overbrace{(\mathbf{a}_i + \vec{\mathbf{h}}_i)}^{\mathbf{a}_{i+1}} = \mathbf{y} - \mathbf{a}_{i+1} = -[\mathbf{Df}(\mathbf{a}_i)]^{-1}\vec{\mathbf{r}}_i. \qquad A5.42$$

Now we return to equation A5.40. By Proposition A5.1, since $\mathbf{y} \in \overline{U}_1$ and $\mathbf{a}_i \in U_1 \subset \overline{U}_1$, we have

$$|\vec{\mathbf{r}}_i| = \big|\underbrace{\mathbf{f}(\mathbf{y}) - \mathbf{f}(\mathbf{a}_i)}_{\text{increment to } \mathbf{f}} - \underbrace{[\mathbf{Df}(\mathbf{a}_i)](\mathbf{y} - \mathbf{a}_i)}_{\substack{\text{linear approx.}\\ \text{of increment to } \mathbf{f}}}\big| \le \frac{M}{2}|\mathbf{y} - \mathbf{a}_i|^2. \qquad A5.43$$

Equation A5.42 and inequality A5.43 then give

$$|\mathbf{y}-\mathbf{a}_{i+1}| \le \left|[\mathbf{Df}(\mathbf{a}_i)]^{-1}\right| \frac{M}{2}|\mathbf{y}-\mathbf{a}_i|^2. \qquad \text{A5.44}$$

We will use this result to prove inequality A5.39 by induction. Note that

$$\begin{aligned}|\mathbf{y}-\mathbf{a}_1| &\underbrace{\le}_{\text{eq. A5.44}} \left|[\mathbf{Df}(\mathbf{a}_0)]^{-1}\right| \frac{M}{2}|\mathbf{y}-\mathbf{a}_0|^2 \underbrace{\le}_{\text{see margin}} \left|[\mathbf{Df}(\mathbf{a}_0)]^{-1}\right| \frac{M}{2}|2\vec{\mathbf{h}}_0||\mathbf{y}-\mathbf{a}_0|\\ &\le \left|[\mathbf{Df}(\mathbf{a}_0)]^{-1}\right| M \underbrace{|\mathbf{f}(\mathbf{a}_0)|\left|[\mathbf{Df}(\mathbf{a}_0)]^{-1}\right|}_{\ge|\vec{\mathbf{h}}_0| \text{ by eq. A5.1}} |\mathbf{y}-\mathbf{a}_0|\\ &\le \frac{1}{2}|\mathbf{y}-\mathbf{a}_0|.\end{aligned} \qquad \text{A5.45}$$

Inequality A5.45: For the second inequality in the first line, note that since $\mathbf{y} \in \overline{U_1}$, it is in a ball with radius $|\vec{\mathbf{h}}_0|$ and center $\mathbf{a}_1$, and thus can be at most $|2\vec{\mathbf{h}}_0|$ away from $\mathbf{a}_0$. So we can replace one of the $|\mathbf{y}-\mathbf{a}_0|$ by $|2\vec{\mathbf{h}}_0|$.

Going from line 2 to line 3 uses inequality A5.3.

Now assume by induction that

$$|\mathbf{y}-\mathbf{a}_j| \le \frac{1}{2}|\mathbf{y}-\mathbf{a}_{j-1}| \quad \text{for all } j \le i, \qquad \text{A5.46}$$

and rewrite inequality A5.44, dividing each side by $|\mathbf{y}-\mathbf{a}_i|$:

$$\begin{aligned}\frac{|\mathbf{y}-\mathbf{a}_{i+1}|}{|\mathbf{y}-\mathbf{a}_i|} &\le \left|[\mathbf{Df}(\mathbf{a}_i)]^{-1}\right| \frac{M}{2}|\mathbf{y}-\mathbf{a}_i| \le 2\left|[\mathbf{Df}(\mathbf{a}_{i-1})]^{-1}\right| \frac{M}{2}\frac{|\mathbf{y}-\mathbf{a}_{i-1}|}{2}\\ &\le \dots\\ &\le \frac{1}{2}M|\mathbf{y}-\mathbf{a}_0|\left|[\mathbf{Df}(\mathbf{a}_0)]^{-1}\right| \qquad \text{A5.47}\\ &\le \frac{1}{2}M|2\vec{\mathbf{h}}_0|\left|[\mathbf{Df}(\mathbf{a}_0)]^{-1}\right| \le |\mathbf{f}(\mathbf{a}_0)|\left|[\mathbf{Df}(\mathbf{a}_0)]^{-1}\right|^2 M\\ &\le \frac{1}{2}. \qquad \text{(We again use inequality A5.3.)}\end{aligned}$$

Inequality A5.47: The inductive hypothesis is used to replace the $|\mathbf{y}-\mathbf{a}_i|$ in the middle of the first line by $\frac{1}{2}|\mathbf{y}-\mathbf{a}_{i-1}|$ at the end of that line. Lemma A5.3 justifies replacing $|[\mathbf{Df}(\mathbf{a}_i)]^{-1}|$ in the middle of the first line by $2|[\mathbf{Df}(\mathbf{a}_{i-1})]^{-1}|$ at the end of the line.

As in inequality A5.45, we can replace $|\mathbf{y}-\mathbf{a}_0|$ in the third line by $|2\vec{\mathbf{h}}_0|$ Then we use

$$\vec{\mathbf{h}}_0 = -[\mathbf{Df}(\mathbf{a}_0)]^{-1}\mathbf{f}(\mathbf{a}_0)$$

(equation A.5.1).

Thus $|\mathbf{y}-\mathbf{a}_{i+1}| \le \frac{1}{2}|\mathbf{y}-\mathbf{a}_i|$. This proves that $\mathbf{y} = \lim \mathbf{a}_i$, and that $\lim \mathbf{a}_i$ is the unique solution of $\mathbf{f}(\mathbf{x}) = \mathbf{0}$ in $\overline{U_1}$. □

The same argument proves Proposition 2.8.14. Existence is no problem: since U_1 is a subset of U_0, a solution in U_1 is a solution in U_0. For uniqueness, the key point is that the $|\mathbf{y}-\mathbf{a}_0| \le 2|\vec{\mathbf{h}}_0|$ used in inequality A5.45 is still true if $\mathbf{y}$ is chosen in U_0 (see Figure 2.8.7, right).

Exercise for Section A.5

We thank Robert Terrell for Exercise A5.1.

When $\alpha = 1$, the two conditions are the same.

The function

$$f(x) = x^{4/3}$$

does not have a Lipschitz derivative but does satisfy condition 1 with $\alpha = 1/3$. (We are most interested in the case when $\mathbf{u}_1 - \mathbf{u}_2$ is small, so if $\alpha < 1$, the condition is weaker.)

The following exercise shows that Kantorovich's theorem still works if the Lipschitz condition for the derivative is replaced by a weaker notion of continuity.

****A5.1** Define U, $\mathbf{a}_0$, $\mathbf{f}$, $\mathbf{a}_1$, and U_1, as in Theorem 2.8.13. Suppose $U_1 \subset U$, and that for some α with $0 < \alpha \le 1$, the derivative $[\mathbf{D}\vec{\mathbf{f}}(\mathbf{x})]$ satisfies

$$\left|[\mathbf{D}\vec{\mathbf{f}}(\mathbf{u}_1)] - [\mathbf{D}\vec{\mathbf{f}}(\mathbf{u}_2)]\right| \le M|\mathbf{u}_1 - \mathbf{u}_2|^\alpha \quad \text{for all } \mathbf{u}_1, \mathbf{u}_2 \in U_1. \qquad (1)$$

Show that if the inequality

$$\left|[\mathbf{D}\vec{\mathbf{f}}(\mathbf{a}_0)]^{-1}\right|^{\alpha+1}|\vec{\mathbf{f}}(\mathbf{a}_0)|^\alpha M \le k \quad \text{is satisfied, with} \quad \left|\frac{k}{(\alpha+1)(1-k)}\right|^\alpha \le 1-k,$$

then the equation $\vec{\mathbf{f}}(\mathbf{x}) = \vec{\mathbf{0}}$ has a unique solution in $\overline{U_1}$, and Newton's method starting at $\mathbf{a}_0$ converges to it.

Hint: You will need to modify each step of the proof of Theorem 2.8.13. Inequality A5.10 can be replaced by

$$\left|\vec{\mathbf{f}}(\mathbf{u}+\vec{\mathbf{h}}) - \vec{\mathbf{f}}(\mathbf{u}) - [\mathbf{D}\vec{\mathbf{f}}(\mathbf{u})]\vec{\mathbf{h}}\right| \le \frac{M}{\alpha+1}|\vec{\mathbf{h}}|^{\alpha+1}.$$

Since we base the inverse and implicit function theorems on the Kantorovich theorem, it follows from Exercise A5.1 that the inverse and implicit function theorems also work with this weaker condition. Of course, the domains of the inverse and implicit functions have to be adapted.

Inequality A5.20 becomes

$$|[\mathbf{D}\vec{\mathbf{f}}(\mathbf{a}_1)]^{-1}| \le \frac{1}{1-k}|[\mathbf{D}\vec{\mathbf{f}}(\mathbf{a}_0)]^{-1}|$$

(replace inequality A5.26 by $|I - [\mathbf{D}\vec{\mathbf{f}}(\mathbf{a}_0)]^{-1}[\mathbf{D}\vec{\mathbf{f}}(\mathbf{a}_1)]| \le k$). Lemma A5.4 becomes

$$|\vec{\mathbf{f}}(\mathbf{a}_1)| \le \frac{M}{\alpha+1}|\vec{\mathbf{h}}_0|^{\alpha+1}, \quad \text{which leads to replacing inequality A5.36 by}$$

$$|\vec{\mathbf{h}}_1| \le \frac{k}{(1-k)(\alpha+1)}|\vec{\mathbf{h}}_0|.$$

The proof now ends as in inequalities A5.37 and A5.38.

A6 Proof of Lemma 2.9.5 (superconvergence)

Here we prove Lemma 2.9.5, which was used in proving that Newton's method superconverges. Recall that

$$c = \frac{1-k}{1-2k}|[\mathbf{Df}(\mathbf{a}_0)]^{-1}|\frac{M}{2}.$$

Conditions of Theorem 2.9.4:

$$\left|[\mathbf{D}\vec{\mathbf{f}}(\mathbf{u}_1)] - [\mathbf{D}\vec{\mathbf{f}}(\mathbf{u}_2)]\right| \le M|\mathbf{u}_1 - \mathbf{u}_2|$$

$$\left|\vec{\mathbf{f}}(\mathbf{a}_0)\right|\left|[\mathbf{D}\vec{\mathbf{f}}(\mathbf{a}_0)]^{-1}\right|^2 M = k < \frac{1}{2}.$$

Lemma 2.9.5. *If the conditions of Theorem 2.9.4 are satisfied, then*

$$|\vec{\mathbf{h}}_{i+1}| \le c|\vec{\mathbf{h}}_i|^2 \quad \text{for all } i. \qquad A6.1$$

Proof. Look back at Lemma A5.4 (rewritten for $\mathbf{a}_i$):

$$|\mathbf{f}(\mathbf{a}_i)| \le \frac{M}{2}|\vec{\mathbf{h}}_{i-1}|^2. \qquad A6.2$$

The definition $\vec{\mathbf{h}}_i = -[\mathbf{Df}(\mathbf{a}_i)]^{-1}\mathbf{f}(\mathbf{a}_i)$ and inequality A6.2 give

$$|\vec{\mathbf{h}}_i| \le |[\mathbf{Df}(\mathbf{a}_i)]^{-1}|\,|\mathbf{f}(\mathbf{a}_i)| \le \frac{M}{2}|[\mathbf{Df}(\mathbf{a}_i)]^{-1}||\vec{\mathbf{h}}_{i-1}|^2. \qquad A6.3$$

If we have such a bound, superconvergence will occur sooner or later.

This is an equation almost of the form $|\vec{\mathbf{h}}_i| \le c|\vec{\mathbf{h}}_{i-1}|^2$; the difference is that the coefficient $\frac{M}{2}|[\mathbf{Df}(\mathbf{a}_i)]^{-1}|$ is not a constant but depends on $\mathbf{a}_i$. So the $\vec{\mathbf{h}}_i$ will superconverge if we can find a bound on $|[\mathbf{Df}(\mathbf{a}_i)]|^{-1}$ valid for all i. (The term $M/2$ is not a problem because it is a constant.) We cannot find such a bound if the derivative $[\mathbf{Df}(\mathbf{a})]$ is not invertible at the limit point $\mathbf{a}$. (We saw this in one dimension in Example 2.9.1, where $f'(1) = 0$.) In such a case $|[\mathbf{Df}(\mathbf{a}_i)]^{-1}| \to \infty$ as $\mathbf{a}_i \to \mathbf{a}$. But Lemma A6.1 says that if the product of the Kantorovich inequality is strictly less than 1/2, we have such a bound.

Lemma A6.1 (A bound on $|[\mathbf{Df}(\mathbf{a}_n)]|^{-1}$). *If*

$$\left|\mathbf{f}(\mathbf{a}_0)\right|\left|[\mathbf{Df}(\mathbf{a}_0)]^{-1}\right|^2 M = k, \qquad \textit{where} \quad k < 1/2, \tag{A6.4}$$

then all $[\mathbf{Df}(\mathbf{a}_n)]^{-1}$ *exist and satisfy*

$$\left|[\mathbf{Df}(\mathbf{a}_n)]^{-1}\right| \le \left|[\mathbf{Df}(\mathbf{a}_0)]^{-1}\right|\frac{1-k}{1-2k}. \tag{A6.5}$$

You may find it helpful to refer to the proof of Lemma A5.3, as we are more concise here. Note that while Lemma A5.3 compares the derivative at $\mathbf{a}_1$ to the derivative at $\mathbf{a}_0$, here we compare the derivative at $\mathbf{a}_n$ to the derivative at $\mathbf{a}_0$.

Proof of Lemma A6.1. The proof of this lemma is a rerun of the proof of Lemma A5.3. We will use inequality A5.36 in the form $|\vec{\mathbf{h}}_1| \le k|\vec{\mathbf{h}}_0|$, which by induction gives $|\vec{\mathbf{h}}_i| \le k|\vec{\mathbf{h}}_{i-1}|$, so that

$$|\mathbf{a}_n - \mathbf{a}_0| = \left|\sum_{i=0}^{n-1}\vec{\mathbf{h}}_i\right| \underbrace{\le}_{\text{triangle inequality}} \sum_{i=0}^{n-1}|\vec{\mathbf{h}}_i| \le |\vec{\mathbf{h}}_0|\overbrace{\left(1+k+\cdots+k^{n-1}\right)}^{\le\sum_{n=0}^{\infty}k^n=1/(1-k),\ \text{by eq. 0.5.4}} \le \frac{|\vec{\mathbf{h}}_0|}{1-k}. \tag{A6.6}$$

The A_n in equation A6.7 corresponds to A in equation A5.22; $\mathbf{a}_n$ in equation A6.7 corresponds to $\mathbf{a}_1$ in equation A5.22.

Inequality A6.8: The second inequality uses inequality A6.6. The third uses the inequality

$$|\vec{\mathbf{h}}_0| \le |[\mathbf{Df}(\mathbf{a}_0)]^{-1}||\mathbf{f}(\mathbf{a}_0)|;$$

see inequality A5.24. The last equality uses the hypothesis of the lemma, equation A6.4.

Next write

$$A_n = I - [\mathbf{Df}(\mathbf{a}_0)]^{-1}[\mathbf{Df}(\mathbf{a}_n)] = [\mathbf{Df}(\mathbf{a}_0)]^{-1}\overbrace{\left([\mathbf{Df}(\mathbf{a}_0)] - [\mathbf{Df}(\mathbf{a}_n)]\right)}^{\le M|\mathbf{a}_0-\mathbf{a}_n|\ \text{by Lipschitz condition}}, \tag{A6.7}$$

so that

$$\begin{aligned}|A_n| &\le |[\mathbf{Df}(\mathbf{a}_0)]^{-1}|\,M\,|\mathbf{a}_0 - \mathbf{a}_n| \;\le\; \left|[\mathbf{Df}(\mathbf{a}_0)]^{-1}\right|M\frac{|\vec{\mathbf{h}}_0|}{1-k}\\ &\le \frac{\left|[\mathbf{Df}(\mathbf{a}_0)]^{-1}\right|^2 M|\mathbf{f}(\mathbf{a}_0)|}{1-k} = \frac{k}{1-k}.\end{aligned} \tag{A6.8}$$

We are assuming $k < 1/2$, so $I - A_n$ is invertible (by Proposition 1.5.38), and the same argument that led to inequality A5.29 here gives

$$\begin{aligned}\left|[\mathbf{Df}(\mathbf{a}_n)]^{-1}\right| &\le \left|[\mathbf{Df}(\mathbf{a}_0)]^{-1}\right|\underbrace{\left(1+|A_n|+|A_n|^2+\cdots\right)}_{=\frac{1}{1-|A_n|}}\\ &\le \frac{1-k}{1-2k}\left|[\mathbf{Df}(\mathbf{a}_0)]^{-1}\right|. \qquad \square\end{aligned} \tag{A6.9}$$

A7 Proof of differentiability of the inverse function

In Section 2.10 we proved the existence of an inverse function $\mathbf{g}$; more precisely, we proved that there exists a function $\mathbf{g}: V \to W_0$ such that $\mathbf{f}\circ\mathbf{g}$ is the identity on V. It follows immediately that $\mathbf{g}\circ\mathbf{f}$ is the identity on $\mathbf{g}(V)$. We want more. We want $\mathbf{f}\circ\mathbf{g}$ to be the identity on a neighborhood of $\mathbf{x}_0$: this will be true if $\mathbf{g}(V)$ contains a neighborhood of $\mathbf{x}_0$, as is guaranteed by part 2 of the conclusion of Theorem 2.10.7. We also want $\mathbf{g}$ to be continuous and continuously differentiable. Finally, we want $\mathbf{g}$ to be the unique local inverse of $\mathbf{f}$ defined in V with these properties. We address these issues here, after restating the theorem.

Theorem 2.10.7 (The inverse function theorem). *Let $W \subset \mathbb{R}^m$ be an open neighborhood of $\mathbf{x}_0$, and let $\mathbf{f} : W \to \mathbb{R}^m$ be a continuously differentiable function. Set $\mathbf{y}_0 = \mathbf{f}(\mathbf{x}_0)$.*

If the derivative $[\mathbf{Df}(\mathbf{x}_0)]$ is invertible, then $\mathbf{f}$ is invertible in some neighborhood of $\mathbf{y}_0$, and the inverse is differentiable.

To quantify this statement, we will specify the radius R of a ball V centered at $\mathbf{y}_0$, in which the inverse function is defined. First simplify notation by setting $L = [\mathbf{Df}(\mathbf{x}_0)]$. Now find $R > 0$ satisfying the following conditions:

1. *The ball W_0 of radius $2R|L^{-1}|$ centered at $\mathbf{x}_0$ is contained in W.*
2. *In W_0, the derivative of $\mathbf{f}$ satisfies the Lipschitz condition*

$$|[\mathbf{Df}(\mathbf{u})] - [\mathbf{Df}(\mathbf{v})]| \le \frac{1}{2R|L^{-1}|^2}|\mathbf{u} - \mathbf{v}|. \qquad A7.1$$

Set $V = B_R(\mathbf{y}_0)$. Then

1. *There exists a unique continuously differentiable mapping $\mathbf{g} : V \to W_0$ such that*

$$\mathbf{g}(\mathbf{y}_0) = \mathbf{x}_0 \quad \text{and} \quad \mathbf{f}\big(\mathbf{g}(\mathbf{y})\big) = \mathbf{y} \text{ for all } \mathbf{y} \in V.$$

   By the chain rule, the derivative of $\mathbf{g}$ is $[\mathbf{Dg}(\mathbf{y})] = [\mathbf{Df}(\mathbf{g}(\mathbf{y}))]^{-1}$.
2. *The image of $\mathbf{g}$ contains the ball of radius R_1 around $\mathbf{x}_0$, where*

$$R_1 = R\left|L^{-1}\right|^2\left(\sqrt{|L|^2 + \frac{2}{|L^{-1}|^2}} - |L|\right). \qquad A7.2$$

We begin by proving that $\mathbf{f}$ is injective on W_0, which will prove that $\mathbf{g}$ is unique. To show that $\mathbf{f}$ is injective on W_0, we will show that the function

$$\mathbf{F}(\mathbf{z}) \stackrel{\text{def}}{=} \frac{1}{2R|L^{-1}|}L^{-1}\Big(\mathbf{f}\Big(\mathbf{x}_0 + 2R|L^{-1}|\mathbf{z}\Big) - \mathbf{y}_0\Big) \qquad A7.3$$

satisfies the hypotheses of Lemma A7.1.

The function $\mathbf{F}$ is $\mathbf{f}$ appropriately translated and linearly distorted so that it is defined in the unit ball of $\mathbb{R}^n$, with $\mathbf{F}(\mathbf{0}) = \mathbf{0}$ and $[\mathbf{DF}(\mathbf{0})] = I$. Let us check that the Lipschitz condition on $\mathbf{f}$ given by inequality A7.1 translates into the simpler Lipschitz condition of Lemma A7.1:

Inequality A7.4: Taking the derivative of $\mathbf{F}$ with respect of $\mathbf{z}$ brings out a factor of $2R|L^{-1}|$, which cancels with the denominator.

$$\begin{aligned}|[\mathbf{DF}(\mathbf{z}_1)] - [\mathbf{DF}(\mathbf{z}_2)]| &= \Big|L^{-1}[\mathbf{Df}(\overbrace{\mathbf{x}_0 + 2R|L^{-1}|\mathbf{z}_1}^{\mathbf{u}})] - L^{-1}[\mathbf{Df}(\overbrace{\mathbf{x}_0 + 2R|L^{-1}|\mathbf{z}_2}^{\mathbf{v}})]\Big| \\ &\le |L^{-1}|\frac{1}{2R|L^{-1}|^2}\Big|\overbrace{2R|L^{-1}|\mathbf{z}_1 - 2R|L^{-1}|\mathbf{z}_2}^{\mathbf{u}-\mathbf{v}}\Big| = |\mathbf{z}_1 - \mathbf{z}_2|.\end{aligned} \qquad A7.4$$

Lemma A7.1. *Let B be the open unit ball of $\mathbb{R}^n$, and let $\mathbf{F} : B \to \mathbb{R}^n$ be a C^1 mapping such that*

$$\mathbf{F}(\mathbf{0}) = \mathbf{0}, \quad [\mathbf{DF}(\mathbf{0})] = I, \quad \big|[\mathbf{DF}(\mathbf{x})] - [\mathbf{DF}(\mathbf{y})]\big| \le |\mathbf{x} - \mathbf{y}| \qquad A7.5$$

for all $\mathbf{x}, \mathbf{y} \in B$. Then $\mathbf{F}$ is injective.

Proof. Using Corollary 1.9.2, we can write

$$\begin{aligned}|\mathbf{F}(\mathbf{x})-\mathbf{F}(\mathbf{y})| &= \Big|(\mathbf{x}-\mathbf{y})+\big(\mathbf{F}(\mathbf{x})-\mathbf{x}\big)-\big(\mathbf{F}(\mathbf{y})-\mathbf{y}\big)\Big| \\ &= \Big|(\mathbf{x}-\mathbf{y})+\big(\mathbf{F}-I\big)(\mathbf{x})-\big(\mathbf{F}-I\big)(\mathbf{y})\Big| \\ &\geq |\mathbf{x}-\mathbf{y}| - \sup_{\mathbf{z}\in[\mathbf{x},\mathbf{y}]}\Big|[\mathbf{DF}(\mathbf{z})]-I\Big|\,|\mathbf{x}-\mathbf{y}| \\ &= |\mathbf{x}-\mathbf{y}| - \sup_{\mathbf{z}\in[\mathbf{x},\mathbf{y}]}\Big|[\mathbf{DF}(\mathbf{z})]-[\mathbf{DF}(\mathbf{0})]\Big|\,|\mathbf{x}-\mathbf{y}| \\ &\geq |\mathbf{x}-\mathbf{y}| - \sup_{\mathbf{z}\in[\mathbf{x},\mathbf{y}]}|\mathbf{z}-\mathbf{0}|\,|\mathbf{x}-\mathbf{y}| \\ &= |\mathbf{x}-\mathbf{y}|\big(1-\sup(|\mathbf{x}|,|\mathbf{y}|)\big).\end{aligned} \qquad A7.6$$

Thus $\mathbf{F}(\mathbf{x})=\mathbf{F}(\mathbf{y})$ implies $\mathbf{x}=\mathbf{y}$. □

Inequality A7.6: Going from the 2nd to the 3rd line uses the mean value theorem and the triangle inequality. Going from the 3rd to 4th line uses $[\mathbf{DF}(\mathbf{0})]=I$. To go from the next-to-last line to the last line, we use that

$$\sup_{\mathbf{z}\in[\mathbf{x},\mathbf{y}]}\Big|\mathbf{z}\Big| = \sup(|\mathbf{x}|,|\mathbf{y}|),$$

since the point of a line segment farthest from the origin is always one of the endpoints.

It follows that $\mathbf{f}$ is injective. Therefore, $\mathbf{g}$ is the unique map $V\to W_0$ such that $\mathbf{f}\circ\mathbf{g}$ is the identity on V. If $\mathbf{g}_1$ is such a map, then $\mathbf{g}_1(\mathbf{y})$ is an inverse image of $\mathbf{y}$ under $\mathbf{f}$ that is an element of W_0, and there is at most one (hence exactly one) such inverse image, namely $\mathbf{g}(\mathbf{y})$.

Proving that g is continuous on V.

Inequality A7.6 gives us more: it gives us a Lipschitz ratio for $\mathbf{g}$ on the ball $B_{R'}(\mathbf{y}_0)=\{|\mathbf{y}-\mathbf{y}_0|\leq R'\}$ for any $R'<R$. Note that if the inverse function theorem holds for some R, it also holds for R' (with a smaller V and a smaller W_0); in particular if $|\mathbf{y}-\mathbf{y}_0|\leq R'$, then $|\mathbf{g}(\mathbf{y})-\mathbf{x}_0|\leq 2R'|L^{-1}|$.

So choose $\mathbf{y}_1,\mathbf{y}_2\in B_{R'}(\mathbf{y}_0)$ and set $\mathbf{x}_1=\mathbf{g}(\mathbf{y}_1)$, $\mathbf{x}_2=\mathbf{g}(\mathbf{y}_2)$. Define $\mathbf{z}_i$ by $\mathbf{x}_i=\mathbf{x}_0+2R|L^{-1}|\mathbf{z}_i$ (see the margin). Then

$$\begin{aligned}|\mathbf{y}_1-\mathbf{y}_2| = |\mathbf{f}(\mathbf{x}_1)-\mathbf{f}(\mathbf{x}_2)| &= \Big|2R|L^{-1}|L\mathbf{F}(\mathbf{z}_1)-2R|L^{-1}|L\mathbf{F}(\mathbf{z}_2)\Big| \\ &= 2R|L^{-1}|\left|L\left(\mathbf{F}\left(\frac{\mathbf{x}_1-\mathbf{x}_0}{2R|L^{-1}|}\right)-\mathbf{F}\left(\frac{\mathbf{x}_2-\mathbf{x}_0}{2R|L^{-1}|}\right)\right)\right| \\ &\geq 2R\left|\mathbf{F}\left(\frac{\mathbf{x}_1-\mathbf{x}_0}{2R|L^{-1}|}\right)-\mathbf{F}\left(\frac{\mathbf{x}_2-\mathbf{x}_0}{2R|L^{-1}|}\right)\right| \\ &\underbrace{\geq}_{\text{ineq. A7.6}} 2R\frac{|\mathbf{x}_1-\mathbf{x}_2|}{2R|L^{-1}|}\left(1-\frac{\sup(|\mathbf{x}_1-\mathbf{x}_0|,|\mathbf{x}_2-\mathbf{x}_0|)}{2R|L^{-1}|}\right) \\ &= \frac{1}{|L^{-1}|}\left(1-\frac{\sup(|\mathbf{x}_1-\mathbf{x}_0|,|\mathbf{x}_2-\mathbf{x}_0|)}{2R|L^{-1}|}\right)|\mathbf{x}_1-\mathbf{x}_2| \\ &\geq \frac{R-R'}{|L^{-1}|R}|\mathbf{x}_1-\mathbf{x}_2|.\end{aligned} \qquad A7.7$$

Thus if $\mathbf{y}_1,\mathbf{y}_2$ are in the ball $B_{R'}(\mathbf{y}_0)$,

$$|\mathbf{g}(\mathbf{y}_1)|-\mathbf{g}(\mathbf{y}_2)| = |\mathbf{x}_1-\mathbf{x}_2| \leq \frac{|L^{-1}|R}{R-R'}|\mathbf{y}_1-\mathbf{y}_2|. \qquad A7.8$$

Inequality A7.7: It follows from equation A7.3 that

$$2R|L^{-1}|L\mathbf{F}(\mathbf{z}_i) = \mathbf{f}\Big(\mathbf{x}_0+2R|L^{-1}|\mathbf{z}_i\Big)-\mathbf{y}_0;$$

set

$$\mathbf{x}_i=\mathbf{x}_0+2R|L^{-1}|\mathbf{z}_i,$$

so

$$\mathbf{f}(\mathbf{x}_i)=2R|L^{-1}|\,L\mathbf{F}(z_i)+\mathbf{y}_0.$$

Note also that

$$\mathbf{z}_i=\frac{\mathbf{x}_i-\mathbf{x}_0}{2R|L^{-1}|}.$$

To go from the 2nd to 3rd line of inequality A7.7 we use the following inequality: if $L:\mathbb{R}^n\to\mathbb{R}^n$ is an invertible linear map, then

$$|L^{-1}||L\mathbf{v}|\geq|L^{-1}L\mathbf{v}|=|\mathbf{v}|.$$

In the second line we have $|L^{-1}|$ times the length of (L times something complicated); in the third line this is replaced by the length of "something complicated".

Changing the base point

First we will see that in order to prove that $\mathbf{g}$ is continuously differentiable on its domain, it is sufficient to show that it is continuously differentiable at the "base point" $\mathbf{y}_0$. Note (see Remark 2.10.8) that to show the existence of $R > 0$ and of $\mathbf{g}$, the only hypotheses about $\mathbf{f}$ we used were that $\mathbf{f}(\mathbf{x}_0) = \mathbf{y}_0$, that $[\mathbf{Df}(\mathbf{x}_0)]$ is invertible, and that $\mathbf{x} \mapsto [\mathbf{Df}(\mathbf{x})]$ is Lipschitz in a neighborhood of $\mathbf{x}_0$. For any points $\mathbf{y}_0' \in V$ and $\mathbf{x}_0' = \mathbf{g}(\mathbf{y}_0')$, the same hypotheses are true.

Write $\mathbf{g} = \mathbf{g}_{\mathbf{x}_0,\mathbf{y}_0}$. For any $\mathbf{y}_0' \in V$ and $\mathbf{x}_0' \stackrel{\text{def}}{=} \mathbf{g}(\mathbf{y}_0')$, there is an analogous map $\mathbf{g}_{\mathbf{x}_0',\mathbf{y}_0'}$. This map also specifies an inverse image of $\mathbf{y}$ under $\mathbf{f}$, and since $\mathbf{f}$ is injective on W_0, it must be the same inverse image in some neighborhood of $\mathbf{y}_0'$. Thus if we prove that $\mathbf{g} = \mathbf{g}_{\mathbf{x}_0,\mathbf{y}_0}$ is differentiable at $\mathbf{y}_0$, the same proof will show that $\mathbf{g}_{\mathbf{x}_0',\mathbf{y}_0'}$ is differentiable at $\mathbf{y}_0'$, hence $\mathbf{g}$ is also differentiable at $\mathbf{y}_0'$, since it coincides with $\mathbf{g}_{\mathbf{x}_0',\mathbf{y}_0'}$ in a neighborhood of $\mathbf{y}_0'$.

Let us see that $\mathbf{g}(V)$ is open. From the argument above, if $\mathbf{g}(V)$ contains a neighborhood of $\mathbf{x}_0$, then it will contain a neighborhood of $\mathbf{x}_0' \in \mathbf{g}(V)$. By the injectivity of $\mathbf{f}$ on W_0, the set $\mathbf{g}(V)$ does contain a neighborhood of $\mathbf{x}_0$: if $\mathbf{x}$ is sufficiently close to $\mathbf{x}_0$ then $\mathbf{f}(\mathbf{x})$ is in V so we can consider the element $\mathbf{x}_1 \stackrel{\text{def}}{=} \mathbf{g}(\mathbf{f}(\mathbf{x}))$ of W_0, and

$$\mathbf{f}(\mathbf{x}_1) = \mathbf{f}(\mathbf{g}(\mathbf{f}(\mathbf{x})) = \mathbf{f}(\mathbf{x}), \qquad A7.9$$

which by injectivity of $\mathbf{f}$ implies $\mathbf{x} = \mathbf{x}_1 \in \mathbf{g}(V)$.

Proving that $\mathbf{g}$ is differentiable at $\mathbf{y}_0$

Here we show that $\mathbf{g}$ is differentiable at $\mathbf{y}_0$, with derivative $[\mathbf{Dg}(\mathbf{y}_0)] = L^{-1}$, i.e., that

$$\lim_{\vec{\mathbf{k}}\to\vec{\mathbf{0}}} \frac{(\mathbf{g}(\mathbf{y}_0 + \vec{\mathbf{k}}) - \mathbf{g}(\mathbf{y}_0)) - L^{-1}\vec{\mathbf{k}}}{|\vec{\mathbf{k}}|} = \vec{\mathbf{0}}. \qquad A7.10$$

When $|\mathbf{y}_0 + \vec{\mathbf{k}}| \in V$, define $\vec{\mathbf{r}}(\vec{\mathbf{k}})$ to be the increment to $\mathbf{x}_0$ that under $\mathbf{f}$ gives the increment $\vec{\mathbf{k}}$ to $\mathbf{y}_0$:

$$\mathbf{f}\big(\mathbf{x}_0 + \vec{\mathbf{r}}(\vec{\mathbf{k}})\big) = \mathbf{y}_0 + \vec{\mathbf{k}}, \qquad A7.11$$

It follows from equation A7.11 that if $\vec{\mathbf{k}}$ is nonzero and small enough that $\mathbf{y}_0 + \vec{\mathbf{k}}$ is in the image of $\mathbf{f}$, then $\vec{\mathbf{r}}(\vec{\mathbf{k}}) \neq \vec{\mathbf{0}}$.

which implies that

$$\mathbf{g}(\mathbf{y}_0 + \vec{\mathbf{k}}) = \mathbf{x}_0 + \vec{\mathbf{r}}(\vec{\mathbf{k}}). \qquad A7.12$$

Substitute the right side of equation A7.12 for $\mathbf{g}(\mathbf{y}_0 + \vec{\mathbf{k}})$ in the left side of equation A7.10, remembering that $\mathbf{g}(\mathbf{y}_0) = \mathbf{x}_0$. This gives

$$\lim_{\vec{\mathbf{k}}\to\vec{\mathbf{0}}} \frac{\mathbf{x}_0 + \vec{\mathbf{r}}(\vec{\mathbf{k}}) - \mathbf{x}_0 - L^{-1}\vec{\mathbf{k}}}{|\vec{\mathbf{k}}|} = \lim_{\vec{\mathbf{k}}\to\vec{\mathbf{0}}} \frac{\vec{\mathbf{r}}(\vec{\mathbf{k}}) - L^{-1}\vec{\mathbf{k}}}{|\vec{\mathbf{k}}|}\frac{|\vec{\mathbf{r}}(\vec{\mathbf{k}})|}{|\vec{\mathbf{r}}(\vec{\mathbf{k}})|}$$

$$= \lim_{\vec{\mathbf{k}}\to\vec{\mathbf{0}}} \frac{L^{-1}\Big(L\vec{\mathbf{r}}(\vec{\mathbf{k}}) - \overbrace{\big(\mathbf{f}(\mathbf{x}_0 + \vec{\mathbf{r}}(\vec{\mathbf{k}})) - \mathbf{f}(\mathbf{x}_0)\big)}^{\vec{\mathbf{k}} \text{ by equation A7.11}}\Big)}{|\vec{\mathbf{r}}(\vec{\mathbf{k}})|}\frac{|\vec{\mathbf{r}}(\vec{\mathbf{k}})|}{|\vec{\mathbf{k}}|}. \qquad A7.13$$

Since $\mathbf{f}$ is differentiable at $\mathbf{x}_0$ with derivative L, the term

$$\frac{L\vec{\mathbf{r}}(\vec{\mathbf{k}}) - \mathbf{f}(\mathbf{x}_0 + \vec{\mathbf{r}}(\vec{\mathbf{k}})) + \mathbf{f}(\mathbf{x}_0)}{|\vec{\mathbf{r}}(\vec{\mathbf{k}})|} \qquad A7.14$$

has limit $\vec{\mathbf{0}}$ as $\vec{\mathbf{r}}(\vec{\mathbf{k}}) \to \vec{\mathbf{0}}$. The differentiability of $\mathbf{g}$ at $\mathbf{y}_0$ (equation A7.10) will follow from part 6 of Theorem 1.5.26 if we show that $\vec{\mathbf{r}}(\vec{\mathbf{k}}) \to \vec{\mathbf{0}}$ when $\vec{\mathbf{k}} \to \vec{\mathbf{0}}$, fast enough that $|\vec{\mathbf{r}}(\vec{\mathbf{k}})|/|\vec{\mathbf{k}}|$ remains bounded. Equation A7.12 tells us that

Here, $|\vec{\mathbf{r}}(\vec{\mathbf{k}})|/|\vec{\mathbf{k}}|$ plays the role of h in part 6 of Theorem 1.5.26, while L^{-1} applied to the term in formula A7.14 plays the role of $\mathbf{f}$.

$$|\vec{\mathbf{r}}(\vec{\mathbf{k}})| = |\mathbf{g}(\mathbf{y}_0 + \vec{\mathbf{k}}) - \mathbf{g}(\mathbf{y}_0)| \qquad A7.15$$

and since $\mathbf{g}$ is continuous, $|\vec{\mathbf{r}}(\vec{\mathbf{k}})|$ does tend to 0 with $|\vec{\mathbf{k}}|$. The proof that $\mathbf{g}$ is continuous (equation A7.8) shows more. Let $\mathbf{x} = \mathbf{g}(\mathbf{y}_0 + \vec{\mathbf{k}})$; then as $\vec{\mathbf{k}}$ tends to $\vec{\mathbf{0}}$, the length $|\mathbf{x} - \mathbf{x}_0| = |\mathbf{g}(\mathbf{y}_0 + \vec{\mathbf{k}}) - \mathbf{g}(\mathbf{y}_0)|$ also tends to 0 since $\mathbf{g}$ is continuous. In particular $|\mathbf{x} - \mathbf{x}_0| < R|L^{-1}|$ when $|\vec{\mathbf{k}}|$ is sufficiently small. Inequality A7.8, with $R' = R/2$, then says that for these values of $\vec{\mathbf{k}}$ we have

$$\frac{|\vec{\mathbf{r}}(\vec{\mathbf{k}})|}{|\vec{\mathbf{k}}|} = \frac{|\mathbf{g}(\mathbf{y}_0 + \vec{\mathbf{k}}) - \mathbf{g}(\mathbf{y}_0)|}{|\vec{\mathbf{k}}|} \le \frac{R|L^{-1}|}{R - R/2}\,\frac{|\mathbf{y}_0 + \vec{\mathbf{k}} - \mathbf{y}_0|}{|\vec{\mathbf{k}}|} = 2|L^{-1}| \qquad A7.16$$

so $|\vec{\mathbf{r}}(\vec{\mathbf{k}})|/|\vec{\mathbf{k}}|$ remains bounded as $\vec{\mathbf{k}} \to \vec{\mathbf{0}}$.

Proving that g is continuously differentiable on V. This follows immediately from $[\mathbf{Dg}(\mathbf{y})] = [\mathbf{Df}(\mathbf{g}(\mathbf{y}))]^{-1}$, derived from the chain rule.

Proving equation A7.2 Suppose $|\mathbf{x}-\mathbf{x}_0| < R_1$. Then, by Corollary 1.9.2,

$$|\mathbf{f}(\mathbf{x})-\mathbf{f}(\mathbf{x}_0)| \le |\mathbf{x}-\mathbf{x}_0| \sup_{|\mathbf{z}-\mathbf{x}_0|<R_1} \big|[\mathbf{Df}(\mathbf{z})]\big| < R_1 \sup_{|\mathbf{z}-\mathbf{x}_0|<R_1} \big|[\mathbf{Df}(\mathbf{z})]\big|. \qquad A7.17$$

So if $R_1 \sup_{|\mathbf{z}-\mathbf{x}_0|<R_1} \big|[\mathbf{Df}(\mathbf{z})]\big| < R$, then $\mathbf{f}(\mathbf{x})$ is in V. We find a bound for $|[\mathbf{Df}(\mathbf{z})]|$, when $|\mathbf{z} - \mathbf{x}_0| < R_1$:

$$|[\mathbf{Df}(\mathbf{z})] - [\mathbf{Df}(\mathbf{x}_0)]| = |[\mathbf{Df}(\mathbf{z})] - L| \underbrace{\le}_{\text{ineq. A7.1}} \frac{1}{2R|L^{-1}|^2}\,|\mathbf{z} - \mathbf{x}_0| < \frac{R_1}{2R|L^{-1}|^2}$$

so

$$|[\mathbf{Df}(\mathbf{z})]| \le |L| + \frac{R_1}{2R|L^{-1}|^2}, \text{ i.e., } \sup_{|\mathbf{z}-\mathbf{x}_0|<R_1} \big|[\mathbf{Df}(\mathbf{z})]\big| \le |L| + \frac{R_1}{2R|L^{-1}|^2}. \qquad A7.18$$

Substituting this bound for $R_1 \sup_{|\mathbf{z}-\mathbf{x}_0|<R_1} \big|[\mathbf{Df}(\mathbf{z})]\big|$ in inequality A7.17, we see that if

Remember that R is the radius of V, the domain of $\mathbf{g}$.

$$R \ge \left(|L| + \frac{R_1}{2R|L^{-1}|^2}\right) R_1, \qquad A7.19$$

then $|\mathbf{x} - \mathbf{x}_0| < R_1$ implies $\mathbf{f}(\mathbf{x}) \in V$. Then $\mathbf{g}(\mathbf{f}(\mathbf{x}))$ is an inverse image of $\mathbf{f}(\mathbf{x})$ in W_0, but since $\mathbf{f}$ is injective on W_0 there only is one, so $\mathbf{g}(\mathbf{f}(\mathbf{x})) = \mathbf{x}$ and $\mathbf{x} \in \mathbf{g}(V)$. We leave it to you to check that the largest R_1 satisfying inequality A7.19 is

$$R_1 = R|L^{-1}|^2 \left(-|L| + \sqrt{|L|^2 + \frac{2}{|L^{-1}|^2}}\right). \qquad A7.20$$

A8 Proof of the implicit function theorem

Theorem 2.10.14 (The implicit function theorem). *Let W be an open neighborhood of $\mathbf{c} \in \mathbb{R}^n$, and let $\mathbf{F} : W \to \mathbb{R}^{n-k}$ be a differentiable function, with $\mathbf{F}(\mathbf{c}) = \mathbf{0}$ and $[\mathbf{DF}(\mathbf{c})]$ onto.*

Order the variables in the domain so that the matrix consisting of the first $n-k$ columns of $[\mathbf{DF}(\mathbf{c})]$ row reduces to the identity. Set $\mathbf{c} \overset{\text{def}}{=} \begin{pmatrix}\mathbf{a}\\ \mathbf{b}\end{pmatrix}$, where the entries of $\mathbf{a}$ correspond to the $n-k$ pivotal unknowns, and the entries of $\mathbf{b}$ correspond to the k nonpivotal (active) unknowns. Then there exists a unique continuously differentiable mapping $\mathbf{g}$ from a neighborhood of $\mathbf{b}$ to a neighborhood of $\mathbf{a}$ such that $\mathbf{F}\begin{pmatrix}\mathbf{x}\\ \mathbf{y}\end{pmatrix} = \mathbf{0}$ expresses the first $n-k$ variables as $\mathbf{g}$ applied to the last k variables: $\mathbf{x} = \mathbf{g}(\mathbf{y})$.

To specify the domain of $\mathbf{g}$, let L be the $n \times n$ matrix

$$L = \begin{bmatrix} [D_1\mathbf{F}(\mathbf{c}), \dots, D_{n-k}\mathbf{F}(\mathbf{c})] & [D_{n-k+1}\mathbf{F}(\mathbf{c}), \dots, D_n\mathbf{F}(\mathbf{c})] \\ [0] & I_k \end{bmatrix}. \qquad \text{A8.1}$$

Then L is invertible. Now find a number $R > 0$ satisfying the following hypotheses:

1. *The ball W_0 with radius $2R|L^{-1}|$ centered at $\mathbf{c}$ is contained in W.*
2. *For $\mathbf{u}$, $\mathbf{v}$ in W_0 the derivative satisfies the Lipschitz condition*

$$\Big|[\mathbf{DF}(\mathbf{u})] - [\mathbf{DF}(\mathbf{v})]\Big| \le \frac{1}{2R|L^{-1}|^2}|\mathbf{u} - \mathbf{v}|.$$

Then there exists a unique continuously differentiable mapping

$$\mathbf{g} : B_R(\mathbf{b}) \to B_{2R|L^{-1}|}(\mathbf{a})$$

such that

$$\mathbf{g}(\mathbf{b}) = \mathbf{a} \quad \text{and} \quad \mathbf{F}\begin{pmatrix}\mathbf{g}(\mathbf{y})\\ \mathbf{y}\end{pmatrix} = \mathbf{0} \quad \text{for all } \mathbf{y} \in B_R(\mathbf{b}).$$

By the chain rule, the derivative of this implicit function $\mathbf{g}$ at $\mathbf{b}$ is

$$[\mathbf{Dg}(\mathbf{b})] = -\underbrace{[D_1\mathbf{F}(\mathbf{c}), \dots, D_{n-k}\mathbf{F}(\mathbf{c})]}_{\substack{\text{partial deriv. for the}\\ n-k \text{ pivotal variables}}}{}^{-1}\underbrace{[D_{n-k+1}\mathbf{F}(\mathbf{c}), \dots, D_n\mathbf{F}(\mathbf{c})]}_{\substack{\text{partial deriv. for the}\\ k \text{ nonpivotal variables}}}. \qquad \text{A8.2}$$

It would be possible, and in some sense more natural, to prove the theorem directly, using the Kantorovich theorem. But this approach will avoid our having to go through all the work of proving that the implicit function is continuously differentiable.

Proof. The inverse function theorem is obviously a special case of the implicit function theorem: the special case where $\mathbf{F}\begin{pmatrix}\mathbf{x}\\ \mathbf{y}\end{pmatrix} = \mathbf{f}(\mathbf{x}) - \mathbf{y}$; i.e., we can separate out the $\mathbf{y}$ from $\mathbf{F}\begin{pmatrix}\mathbf{x}\\ \mathbf{y}\end{pmatrix}$. There is a sneaky way of making the implicit function theorem into a special case of the inverse function theorem. We will create a new function $\widetilde{\mathbf{F}}$ to which we can apply the inverse function theorem. Then we will show how the inverse of $\widetilde{\mathbf{F}}$ will give us our implicit function $\mathbf{g}$.

When we add a tilde to $\mathbf{F}$, creating the function $\widetilde{\mathbf{F}}$ of equation A8.3, we use $\mathbf{F}\begin{pmatrix}\mathbf{x}\\ \mathbf{y}\end{pmatrix}$ as the first $n-k$ coordinates of $\widetilde{\mathbf{F}}$ and stick on $\mathbf{y}$ (k coordinates) at the bottom; $\mathbf{y}$ just goes along for the ride. We do this to fix the dimensions:

$$\widetilde{\mathbf{F}} : \mathbb{R}^n \to \mathbb{R}^n$$

can have an inverse function, while $\mathbf{F}$ can't.

Equation A8.4: The derivative

$$L = [\mathbf{D}\widetilde{\mathbf{F}}(\mathbf{c})] = \begin{bmatrix} [\mathbf{DF}(\mathbf{c})] \\ [0] \mid I \end{bmatrix}$$

is an $n \times n$ matrix; the entry

$$[\mathbf{DF}(\mathbf{c})] = [D_1\mathbf{F}(\mathbf{c}) \dots D_n\mathbf{F}(\mathbf{c})]$$

is a matrix $n-k$ tall and n wide; the $[0]$ matrix is k high and $n-k$ wide; the identity matrix is $k \times k$.

Consider the function $\widetilde{\mathbf{F}} : W \to \mathbb{R}^{n-k} \times \mathbb{R}^k$, defined by

$$\widetilde{\mathbf{F}}\begin{pmatrix}\mathbf{x}\\ \mathbf{y}\end{pmatrix} = \begin{pmatrix}\mathbf{F}\begin{pmatrix}\mathbf{x}\\ \mathbf{y}\end{pmatrix}\\ \mathbf{y}\end{pmatrix}, \qquad A8.3$$

where $\mathbf{x}$ represents $n-k$ variables, which we have put as the first variables, and $\mathbf{y}$ the remaining k variables. Whereas $\mathbf{F}$ goes from $W \subset \mathbb{R}^n$ to the lower-dimensional space $\mathbb{R}^{n-k}$, and thus has no hope of having an inverse, the domain and codomain of $\widetilde{\mathbf{F}}$ have the same dimension: n, as illustrated by Figure A8.1.

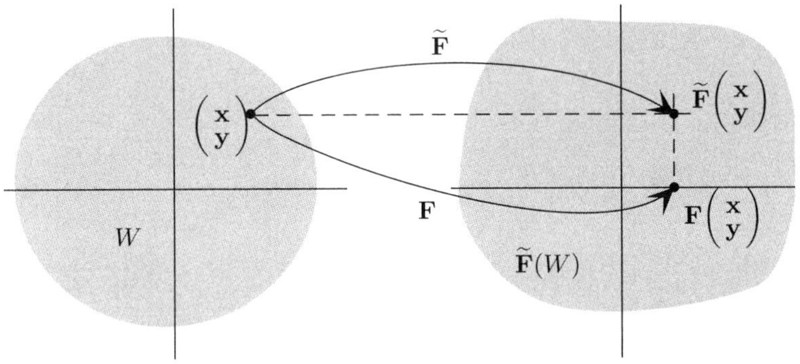

FIGURE A8.1. The mapping $\widetilde{\mathbf{F}}$ is designed to add dimensions to the image of $\mathbf{F}$ so that the image has the same dimension as the domain. Here, $n = 2$ and $k = n - k = 1$; $\mathbf{F}$ maps $\begin{pmatrix}\mathbf{x}\\ \mathbf{y}\end{pmatrix}$ to a point in $\mathbb{R}$, while $\widetilde{\mathbf{F}}$ maps it to a point in $\mathbb{R}^2$.

Now we will find an inverse of $\widetilde{\mathbf{F}}$, and we will show that the first coordinates of that inverse are precisely the implicit function $\mathbf{g}$.

The derivative of $\widetilde{\mathbf{F}}$ at $\mathbf{c}$ is the matrix L of equation A8.1:

$$[\mathbf{D}\widetilde{\mathbf{F}}(\mathbf{c})] = \begin{bmatrix} [D_1\mathbf{F}(\mathbf{c}), \dots, D_{n-k}\mathbf{F}(\mathbf{c})] & [D_{n-k+1}\mathbf{F}(\mathbf{c}), \dots, D_n\mathbf{F}(\mathbf{c})] \\ [0] & I \end{bmatrix}. \qquad A8.4$$

It follows from Exercise 2.5 that $[\mathbf{D}\widetilde{\mathbf{F}}(\mathbf{c})]$ is invertible precisely when $[D_1\mathbf{F}(\mathbf{c}), \dots, D_{n-k}\mathbf{F}(\mathbf{c})]$ is invertible, i.e., the hypothesis of the inverse function theorem is met.

Note that if the derivative of $\mathbf{F}$ is Lipschitz on $W_0 \subset W$, as described in conditions 1 and 2 of the implicit function theorem, then the derivative of $\widetilde{\mathbf{F}}$ is Lipschitz on $W_0 \subset W$, as described in conditions 1 and 2 of the inverse function theorem. There is something to check about condition 2, since the derivative of $\widetilde{\mathbf{F}}$ is not quite the derivative of $\mathbf{F}$. But this is not a problem: since the derivative of $\widetilde{\mathbf{F}}$ is $[\mathbf{D}\widetilde{\mathbf{F}}(\mathbf{u})] = \begin{bmatrix} [\mathbf{DF}(\mathbf{u})] \\ [0] \mid I \end{bmatrix}$, when we compute $|[\mathbf{D}\widetilde{\mathbf{F}}(\mathbf{u})] - [\mathbf{D}\widetilde{\mathbf{F}}(\mathbf{v})]|$, the identity matrices cancel, giving

$$\Big|[\mathbf{D}\widetilde{\mathbf{F}}(\mathbf{u})] - [\mathbf{D}\widetilde{\mathbf{F}}(\mathbf{v})]\Big| = \Big|[\mathbf{DF}(\mathbf{u})] - [\mathbf{DF}(\mathbf{v})]\Big|. \qquad A8.5$$

Thus $\widetilde{\mathbf{F}}$ has a unique inverse $\widetilde{\mathbf{G}} : B_R\begin{pmatrix}\mathbf{0}\\ \mathbf{b}\end{pmatrix} \to W_0$; when $|\mathbf{y} - \mathbf{b}| < R$,

$$\widetilde{\mathbf{F}} \circ \widetilde{\mathbf{G}}\begin{pmatrix}\mathbf{x}\\ \mathbf{y}\end{pmatrix} = \begin{pmatrix}\mathbf{x}\\ \mathbf{y}\end{pmatrix}. \qquad A8.6$$

Now define $\mathbf{g}$ by $\widetilde{\mathbf{G}}\begin{pmatrix}\mathbf{0}\\ \mathbf{y}\end{pmatrix} = \begin{pmatrix}\mathbf{g}(\mathbf{y})\\ \mathbf{y}\end{pmatrix}$. Then

$$\begin{pmatrix}\mathbf{0}\\ \mathbf{y}\end{pmatrix} = \widetilde{\mathbf{F}} \circ \widetilde{\mathbf{G}}\begin{pmatrix}\mathbf{0}\\ \mathbf{y}\end{pmatrix} = \widetilde{\mathbf{F}}\begin{pmatrix}\mathbf{g}(\mathbf{y})\\ \mathbf{y}\end{pmatrix} = \begin{pmatrix}\mathbf{F}\begin{pmatrix}\mathbf{g}(\mathbf{y})\\ \mathbf{y}\end{pmatrix}\\ \mathbf{y}\end{pmatrix}. \qquad A8.7$$

We denote by $B_R\begin{pmatrix}\mathbf{0}\\ \mathbf{b}\end{pmatrix}$ the ball of radius R centered at $\begin{pmatrix}\mathbf{0}\\ \mathbf{b}\end{pmatrix}$.

While $\widetilde{\mathbf{G}}$ is defined on all of $B_R\begin{pmatrix}\mathbf{0}\\ \mathbf{b}\end{pmatrix}$, we will only be interested in points $\widetilde{\mathbf{G}}\begin{pmatrix}\mathbf{0}\\ \mathbf{y}\end{pmatrix}$.

Thus

$$\mathbf{F}\begin{pmatrix}\mathbf{g}(\mathbf{y})\\ \mathbf{y}\end{pmatrix} = \mathbf{0}; \qquad A8.8$$

$\mathbf{g}$ is the required "implicit function": $\mathbf{F}\begin{pmatrix}\mathbf{x}\\ \mathbf{y}\end{pmatrix} = \mathbf{0}$ implicitly defines $\mathbf{x}$ in terms of $\mathbf{y}$, and $\mathbf{g}$ makes this relationship explicit.

Remark. Equation A8.8 says that a point $\begin{pmatrix}\mathbf{g}(\mathbf{y})\\ \mathbf{y}\end{pmatrix} \in B_R\begin{pmatrix}\mathbf{0}\\ \mathbf{b}\end{pmatrix}$, i.e., a point of the graph of $\mathbf{g}$, is a solution to $\mathbf{F}\begin{pmatrix}\mathbf{x}\\ \mathbf{y}\end{pmatrix} = \mathbf{0}$. We can also show that in an appropriate region, the set of equation $\mathbf{F}\begin{pmatrix}\mathbf{x}\\ \mathbf{y}\end{pmatrix} = \mathbf{0}$ is contained in the graph of $\mathbf{g}$. If $\mathbf{F}\begin{pmatrix}\mathbf{x}\\ \mathbf{y}\end{pmatrix} = \mathbf{0}$, then

Exercise A8.1 asks you to show that the implicit function found this way is unique.

$$\begin{pmatrix}\mathbf{x}\\ \mathbf{y}\end{pmatrix} = \widetilde{\mathbf{G}} \circ \widetilde{\mathbf{F}}\begin{pmatrix}\mathbf{x}\\ \mathbf{y}\end{pmatrix} = \widetilde{\mathbf{G}}\begin{pmatrix}\mathbf{F}\begin{pmatrix}\mathbf{x}\\ \mathbf{y}\end{pmatrix}\\ \mathbf{y}\end{pmatrix} = \widetilde{\mathbf{G}}\begin{pmatrix}\mathbf{0}\\ \mathbf{y}\end{pmatrix} = \begin{pmatrix}\mathbf{g}(\mathbf{y})\\ \mathbf{y}\end{pmatrix}, \qquad A8.9$$

so $\mathbf{x} = \mathbf{g}(\mathbf{y})$, proving that in $\widetilde{\mathbf{G}}\left(B_R\begin{pmatrix}\mathbf{0}\\ \mathbf{b}\end{pmatrix}\right)$, the set of equation $\mathbf{F}\begin{pmatrix}\mathbf{x}\\ \mathbf{y}\end{pmatrix} = \mathbf{0}$ is contained in the graph of $\mathbf{g}$. Thus the graph of $\mathbf{g}$ and solutions to $\mathbf{F}\begin{pmatrix}\mathbf{x}\\ \mathbf{y}\end{pmatrix} = \mathbf{0}$ coincide in $\widetilde{\mathbf{G}}\left(B_R\begin{pmatrix}\mathbf{0}\\ \mathbf{b}\end{pmatrix}\right)$. $\triangle$

In equation A8.10 $[\mathbf{Dg}(\mathbf{b})]$ is an $(n-k) \times k$ matrix, I is $k \times k$, and $[0]$ is the $(n-k)\times k$ zero matrix. In this equation we are using the fact that $\mathbf{g}$ is differentiable; otherwise we could not apply the chain rule.

Now we need to prove equation A2.10.29. Define $\tilde{\mathbf{g}}$ by $\tilde{\mathbf{g}}(\mathbf{y}) = \begin{pmatrix}\mathbf{g}(\mathbf{y})\\ \mathbf{y}\end{pmatrix}$. By equation A8.8, the function $\mathbf{F} \circ \tilde{\mathbf{g}}$ is identically $\mathbf{0}$, so its derivative is $[0]$, which gives (by the chain rule),

$$[\mathbf{D}(\mathbf{F} \circ \tilde{\mathbf{g}})(\mathbf{b})] = \left[\mathbf{DF}\begin{pmatrix}\mathbf{g}(\mathbf{b})\\ \mathbf{b}\end{pmatrix}\right][\mathbf{D}\tilde{\mathbf{g}}(\mathbf{b})] = [0], \quad \text{i.e.,} \qquad A8.10$$

$$\begin{bmatrix} | & & | & | & & | \\ D_1\mathbf{F}(\mathbf{c}) & \dots & D_{n-k}\mathbf{F}(\mathbf{c}), & D_{n-k+1}\mathbf{F}(\mathbf{c}), & \dots, & D_n\mathbf{F}(\mathbf{c}) \\ | & & | & | & & | \end{bmatrix} \begin{bmatrix}\mathbf{Dg}(\mathbf{b})\\ I\end{bmatrix} = [0]. \qquad A8.11$$

If A denotes the first $n-k$ columns of $[\mathbf{DF}(\mathbf{c})]$ and B the last k columns, we have

$$A[\mathbf{Dg}(\mathbf{b})] + B = [0], \text{ so } A[\mathbf{Dg}(\mathbf{b})] = -B, \text{ so } [\mathbf{Dg}(\mathbf{b})] = -A^{-1}B. \qquad A8.12$$

Substituting back, this is exactly what we wanted to prove:

$$[\mathbf{Dg}(\mathbf{b})] = \underbrace{-[D_1\mathbf{F}(\mathbf{c}), \dots, D_{n-k}\mathbf{F}(\mathbf{c})]^{-1}}_{-A^{-1}} \underbrace{[D_{n-k+1}\mathbf{F}(\mathbf{c}), \dots, D_n\mathbf{F}(\mathbf{c})]}_{B}. \quad \square$$

Exercise for Section A.8

A8.1 Using the notation of Theorem 2.10.14, show that the implicit function found by setting $\mathbf{g}(\mathbf{y}) = \mathbf{G}\begin{pmatrix}\mathbf{0}\\ \mathbf{y}\end{pmatrix}$ is the *unique* continuous function defined on $B_R(\mathbf{b})$ satisfying

$$\mathbf{F}\begin{pmatrix}\mathbf{g}(\mathbf{y})\\ \mathbf{y}\end{pmatrix} = \mathbf{0} \quad \text{and} \quad \mathbf{g}(\mathbf{b}) = \mathbf{a}.$$

A9 Proving the Equality of Crossed Partials

Theorem 3.3.8 (Equality of crossed partials). *Let U be an open subset of $\mathbb{R}^n$ and $f : U \to \mathbb{R}$ a function such that all partial derivatives $D_i f$ are themselves differentiable at $\mathbf{a} \in U$. Then for every pair of variables x_i, x_j, the crossed partials are equal:*

$$D_j(D_i f)(\mathbf{a}) = D_i(D_j f)(\mathbf{a}).$$

Proof. We will show that

$$D_j(D_i f)(\mathbf{a}) = \lim_{t \to 0} \frac{1}{t^2}\Big(f(\mathbf{a} + t\vec{\mathbf{e}}_i + t\vec{\mathbf{e}}_j) - f(\mathbf{a} + t\vec{\mathbf{e}}_i) - f(\mathbf{a} + t\vec{\mathbf{e}}_j) + f(\mathbf{a})\Big).$$

Since the expression on the right is symmetric with respect to i and j, this proves the theorem. Write

$$g(s) = f(\mathbf{a} + s\vec{\mathbf{e}}_i + t\vec{\mathbf{e}}_j) - f(\mathbf{a} + s\vec{\mathbf{e}}_i), \quad \text{so that} \qquad A9.1$$

$$f(\mathbf{a} + t\vec{\mathbf{e}}_i + t\vec{\mathbf{e}}_j) - f(\mathbf{a} + t\vec{\mathbf{e}}_i) - f(\mathbf{a} + t\vec{\mathbf{e}}_j) + f(\mathbf{a}) = g(t) - g(0). \qquad A9.2$$

Since f is differentiable, g is also, so by the mean value theorem there exists c_t between 0 and t such that

We can't write $c_t \in (0, t)$ because t may be negative.

$$f(\mathbf{a} + t\vec{\mathbf{e}}_i + t\vec{\mathbf{e}}_j) - f(\mathbf{a} + t\vec{\mathbf{e}}_i) - f(\mathbf{a} + t\vec{\mathbf{e}}_j) + f(\mathbf{a}) = tg'(c_t). \qquad A9.3$$

Note that differentiating g in equation A9.1 gives

$$g'(c_t) = D_i f(\mathbf{a} + c_t\vec{\mathbf{e}}_i + t\vec{\mathbf{e}}_j) - D_i f(\mathbf{a} + c_t\vec{\mathbf{e}}_i). \qquad A9.4$$

Since $D_i f$ is differentiable at $\mathbf{a}$, we can write

$$0 = \lim_{\sqrt{c_t^2+t^2}\to 0} \frac{1}{\sqrt{c_t^2+t^2}} \Big(D_i f(\mathbf{a} + c_t \mathbf{e}_i + t\mathbf{e}_j) - D_i f(\mathbf{a}) - [\mathbf{D}(D_i f)(\mathbf{a})](c_t \vec{\mathbf{e}}_i + t\vec{\mathbf{e}}_j) \Big)$$
$$= \lim_{|t|\to 0} \frac{1}{|t|} \Big(D_i f(\mathbf{a} + c_t \mathbf{e}_i + t\mathbf{e}_j) - D_i f(\mathbf{a}) - c_t (D_i^2 f)(\mathbf{a}) - t D_j (D_i f)(\mathbf{a}) \Big) \qquad \text{A9.5}$$

and

In equation A9.5 we are treating $c_t \vec{\mathbf{e}}_i + t\vec{\mathbf{e}}_j$ as the increment $\vec{\mathbf{h}}$ in Proposition and Definition 1.7.9.

$$0 = \lim_{c_t\to 0} \frac{1}{|c_t|} \Big(D_i f(\mathbf{a} + c_t \mathbf{e}_i) - D_i f(\mathbf{a}) - [\mathbf{D} D_i f(\mathbf{a})](c_t \vec{\mathbf{e}}_i) \Big)$$
$$= \lim_{t\to 0} \frac{1}{|t|} \Big(D_i f(\mathbf{a} + c_t \mathbf{e}_i) - D_i f(\mathbf{a}) - c_t (D_i^2 f(\mathbf{a}) \Big). \qquad \text{A9.6}$$

Passing from the first line to the second in both equations, we are using $|c_t| \le |t|$, and the fact that $D_i f$ is differentiable at $\mathbf{a}$, so its derivative $[\mathbf{D} D_i f(\mathbf{a})]$ is a linear transformation given by the Jacobian matrix (a line matrix in this case) with entries $D_k(D_i f)(\mathbf{a}), k = 1, \dots, n$. So we have

Equation A9.7: Going from line 1 to line 2 uses equation A9.3. Exercise A9.1 asks you to justify the final equality.

$$\lim_{t\to 0} \frac{1}{t^2} \Big(f(\mathbf{a} + t\mathbf{e}_i + t\mathbf{e}_j) - f(\mathbf{a} + t\mathbf{e}_i) - f(\mathbf{a} + t\mathbf{e}_j) + f(\mathbf{a}) \Big)$$
$$= \lim_{t\to 0} \frac{g'(c_t)}{t} = D_j(D_i f)(\mathbf{a}). \qquad \text{A9.7}$$

The expression in the first line is symmetric with respect to i and j. Thus $D_j(D_i f)(\mathbf{a}) = D_i(D_j f)(\mathbf{a})$. □

Exercises for Section A.9

A9.1 Justify the final equality in equation A9.7.

A9.2 Compute $\lim_{x\to 0} \lim_{y\to 0} \frac{y}{x+y}$ and $\lim_{y\to 0} \lim_{x\to 0} \frac{y}{x+y}$.

A10 Functions with many vanishing partial derivatives

Proposition 3.3.17 (Size of a function with many vanishing partial derivatives). *Let U be an open subset of $\mathbb{R}^n$ and let $g : U \to \mathbb{R}$ be a C^k function. If at $\mathbf{a} \in U$ all partial derivatives of g up to order k vanish (including the 0th partial derivative, $g(\mathbf{a})$), then*

$$\lim_{\vec{\mathbf{h}}\to\vec{\mathbf{0}}} \frac{g(\mathbf{a}+\vec{\mathbf{h}})}{|\vec{\mathbf{h}}|^k} = 0.$$

We could start the induction at $k=0$, and Theorem 3.3.17 would contain Theorem 1.9.8 as a special case; you are asked to show this in Exercise A10.1.

Proof. The proof is by induction on k, starting with $k=1$. The case $k=1$ follows from Theorem 1.9.8: if g vanishes at $\mathbf{a}$, and its first partials are continuous, g is differentiable at $\mathbf{a}$, and its derivative is given by the Jacobian matrix. So if the first partials vanish at $\mathbf{a}$, the derivative is 0:

$$0 = \underbrace{\lim_{\vec{\mathbf{h}}\to\vec{\mathbf{0}}} \frac{g(\mathbf{a}+\vec{\mathbf{h}}) - \overbrace{g(\mathbf{a})}^{0} - \overbrace{[\mathbf{D}g(\mathbf{a})]}^{[0]}\vec{\mathbf{h}}}{|\vec{\mathbf{h}}|}}_{=0 \text{ since } g \text{ is differentiable}} = \lim_{\vec{\mathbf{h}}\to\vec{\mathbf{0}}} \frac{g(\mathbf{a}+\vec{\mathbf{h}})}{|\vec{\mathbf{h}}|}. \qquad A10.1$$

This proves the case $k=1$. Now set

Equation A10.2: There is no a_0, so the point $\mathbf{c}_0(\vec{\mathbf{h}})$ starts with the $(i+1)$st term:

$$\mathbf{c}_0(\vec{\mathbf{h}}) = \begin{pmatrix} a_1+h_1 \\ \vdots \\ a_n+h_n \end{pmatrix}.$$

$$\mathbf{c}_i(\vec{\mathbf{h}}) = \begin{pmatrix} a_1 \\ \vdots \\ a_i \\ a_{i+1}+h_{i+1} \\ \vdots \\ a_n+h_n \end{pmatrix}, \quad \text{so } \mathbf{c}_0(\vec{\mathbf{h}}) = \mathbf{a}+\vec{\mathbf{h}} \text{ and } \mathbf{c}_n(\vec{\mathbf{h}}) = \mathbf{a}. \qquad A10.2$$

This gives

$$\begin{aligned} g(\mathbf{a}+\vec{\mathbf{h}}) - g(\mathbf{a}) &= g(\mathbf{c}_0(\vec{\mathbf{h}})) - g(\mathbf{c}_n(\vec{\mathbf{h}})) \\ &= \Big(g(\mathbf{c}_0(\vec{\mathbf{h}}) - g(\mathbf{c}_1(\vec{\mathbf{h}})\Big) + \Big(g(\mathbf{c}_1(\vec{\mathbf{h}}) - g(\mathbf{c}_2(\vec{\mathbf{h}})\Big) + \cdots + \Big(g(\mathbf{c}_{n-1}(\vec{\mathbf{h}}) - g(\mathbf{c}_n(\vec{\mathbf{h}})\Big) \end{aligned} \qquad A10.3$$

By the mean value theorem there exists a point

If we define $k:\mathbb{R}\to\mathbb{R}$ by

$$k(x) = g\begin{pmatrix} a_1 \\ \vdots \\ a_{i-1} \\ x \\ a_{i+1}+h_{i+1} \\ \vdots \\ a_n+h_n \end{pmatrix},$$

then

$$\mathbf{g}(\mathbf{c}_{i-1}(\vec{\mathbf{h}})) - g(\mathbf{c}_i(\vec{\mathbf{h}})) = h_i D_i g(\vec{\mathbf{b}}_i)$$

is the standard one variable mean value theorem:

$$k(a_i+h_i) - k(a_i) = h_i k'(b_i).$$

$$\mathbf{b}_i = \begin{pmatrix} a_1 \\ \vdots \\ a_{i-1} \\ b_i \\ a_{i+1}+h_{i+1} \\ \vdots \\ a_n+h_n \end{pmatrix} \text{ such that } \quad g(\mathbf{c}_{i-1}(\vec{\mathbf{h}})) - g(\mathbf{c}_i(\vec{\mathbf{h}})) = h_i D_i g(\mathbf{b}_i). \qquad A10.4$$

Thus equation A10.3 can be rewritten

$$g(\mathbf{a}+\vec{\mathbf{h}}) - g(\mathbf{a}) = \sum_{i=1}^{n} h_i D_i g(\mathbf{b}_i) \qquad A10.5$$

Now we can restate our problem; we want to prove that

$$\lim_{\vec{\mathbf{h}}\to\vec{\mathbf{0}}} \frac{g(\mathbf{a}+\vec{\mathbf{h}})}{|\vec{\mathbf{h}}|^k} = \lim_{\vec{\mathbf{h}}\to\vec{\mathbf{0}}} \frac{g(\mathbf{a}+\vec{\mathbf{h}})}{|\vec{\mathbf{h}}||\vec{\mathbf{h}}|^{k-1}} = \sum_{i=1}^{n} \lim_{\vec{\mathbf{h}}\to\vec{\mathbf{0}}} \frac{h_i}{|\vec{\mathbf{h}}|} \frac{D_i g(\mathbf{b}_i)}{|\vec{\mathbf{h}}|^{k-1}} = 0. \qquad A10.6$$

Since $|h_i|/|\vec{\mathbf{h}}| \le 1$, this comes down to proving that

$$\lim_{\vec{\mathbf{h}}\to\vec{\mathbf{0}}} \frac{D_i g(\mathbf{b}_i)}{|\vec{\mathbf{h}}|^{k-1}} = 0. \qquad A10.7$$

Set $\mathbf{b}_i = \mathbf{a} + \vec{\mathbf{c}}_i$; i.e., $\vec{\mathbf{c}}_i$ is the increment to $\mathbf{a}$ that produces $\mathbf{b}_i$. If we substitute this value for $\mathbf{b}_i$ in equation A10.7, we now need to prove

$$\lim_{\vec{\mathbf{h}} \to \mathbf{0}} \frac{D_i g(\mathbf{a} + \vec{\mathbf{c}}_i)}{|\vec{\mathbf{h}}|^{k-1}} = 0. \qquad A10.8$$

By definition, all partial derivatives of g to order k exist, are continuous on U and vanish at $\mathbf{a}$. By induction we may assume that Proposition 3.3.17 is true for $D_i g$, so that

In equation A10.9 (left) we are substituting $D_i g$ for g and $\vec{\mathbf{c}}_i$ for $\vec{\mathbf{h}}$ in Proposition 3.3.17. You may object that in the denominator we now have $k-1$ instead of k. But the proposition is true when g is a C^k function, and if g is a C^k function, then $D_i g$ is a C^{k-1} function.

$$\lim_{\vec{\mathbf{c}}_i \to \vec{\mathbf{0}}} \frac{D_i g(\mathbf{a} + \vec{\mathbf{c}}_i)}{|\vec{\mathbf{c}}_i|^{k-1}} = 0, \quad \text{so} \quad \lim_{\vec{\mathbf{h}} \to \vec{\mathbf{0}}} \frac{D_i g(\mathbf{a} + \vec{\mathbf{c}}_i)}{|\vec{\mathbf{h}}|^{k-1}} = 0 \qquad A10.9$$

This tells us that for any ϵ, there exists δ such that if $|\vec{\mathbf{h}}| < \delta$, then

$$\frac{D_i g(\mathbf{a} + \vec{\mathbf{h}})}{|\vec{\mathbf{h}}|^{k-1}} < \epsilon. \qquad A10.10$$

Why can we switch $\vec{\mathbf{c}}_i$ to $\vec{\mathbf{h}}$ in the denominator in equations A10.9? We know that $|\vec{\mathbf{c}}_i| \le |\vec{\mathbf{h}}|$, since c_i is between 0 and h_i and

$$\vec{\mathbf{c}}_i = \begin{bmatrix} 0 \\ 0 \\ \vdots \\ c_i \\ h_{i+1} \\ \vdots \\ h_n \end{bmatrix}.$$

So the left equation in A10.9 a stronger statement than the right equation.

If $|\vec{\mathbf{h}}| < \delta$, then $|\vec{\mathbf{c}}_i| < \delta$. And putting the bigger number $|\vec{\mathbf{h}}|^{k-1}$ in the denominator just makes that quantity smaller. So we're done:

$$\lim_{\vec{\mathbf{h}} \to \mathbf{0}} \frac{g(\mathbf{a} + \vec{\mathbf{h}})}{|\vec{\mathbf{h}}|^k} = \sum_{i=1}^{n} \lim_{\vec{\mathbf{h}} \to \vec{\mathbf{0}}} \frac{h_i}{|\vec{\mathbf{h}}|} \frac{D_i g(\mathbf{a} + \vec{\mathbf{c}}_i)}{|\vec{\mathbf{h}}|^{k-1}} = 0. \quad \square \qquad A10.11$$

Exercise for Section A.10

A10.1 In the proof of Proposition 3.3.17, show that you could start the induction at $k = 0$ and that then the proposition contains Theorem 1.9.8 as a special case.

A11 Proving rules for Taylor polynomials

Notation with big O "significantly simplifies calculations because it allows us to be sloppy – but in a satisfactorily controlled way."—Donald Knuth, Stanford University (*Notices of the AMS*, Vol. 45, No. 6, p. 688).

Propositions 3.4.3 and 3.4.4 on computing Taylor polynomials follow from some rules for doing arithmetic with o and O ("little o" and "big O").

Definition A11.1 (Big O). Let $U \subset \mathbb{R}^n$ be a neighborhood of $\mathbf{0}$ and $h: U \to \mathbb{R}$ a function satisfying $h(\mathbf{x}) > 0$ if $\mathbf{x} \ne \mathbf{0}$. Then a function $f: U \to \mathbb{R}$ is in $O(h)$ if there exist $\delta > 0$ and a constant C such that $|f(\mathbf{x})| \le Ch(\mathbf{x})$ when $0 < |\mathbf{x}| < \delta$; this should be read "$f$ is at most of order $h(\mathbf{x})$".

Recall from Definition 3.4.1 that f is in $o(h)$ if

$$\lim_{\mathbf{x}\to\mathbf{0}} \frac{f(\mathbf{x})}{h(\mathbf{x})} = 0.$$

Alternatively, we can say that f is in $o(h)$ if for all $\epsilon > 0$, there exists $\delta > 0$ such that

$$|\mathbf{x}| < \delta \implies |f(\mathbf{x})| \le \epsilon h(\mathbf{x}).$$

Clearly this is a stronger requirement than the requirement for O that there exists a constant C such that

$$|\mathbf{x}| < \delta \implies f(\mathbf{x}) \le Ch(\mathbf{x}).$$

Proposition A11.2: If $k < l$, then $|\mathbf{x}|^k$ is bigger than $|\mathbf{x}|^l$ for $\mathbf{x}$ near $\mathbf{0}$: compare for instance $|1/8|^2$ and $|1/8|^3$. For $\mathbf{x}$ near $\mathbf{0}$, think of $O(|\mathbf{x}|^k)$ as the weight of an elephant and $O(|\mathbf{x}|^l)$ as the weight of a mouse. Their combined weight is still, roughly, the weight of the elephant. This image is somewhat shaky mathematically but it makes the point that Big O and little o are intended to distinguish what is predominant from what is not.

In inequality A11.3, note that the terms to the left and right of the second inequality are identical except that the $C_2|\mathbf{x}|^l$ on the left becomes $C_2|\mathbf{x}|^k$ on the right.

All these proofs are essentially identical; they are exercises in fine shades of meaning.

Note that

$$f \in o(h) \implies f \in O(h). \qquad A11.1$$

Below, to lighten the notation, we write $O(|\mathbf{x}|^k) + O(|\mathbf{x}|^l) = O(|\mathbf{x}|^k)$ to mean that if $f \in O(|\mathbf{x}|^k)$ and $g \in O(|\mathbf{x}|^l)$, then $f + g \in O(|\mathbf{x}|^k)$; we use similar notation for products and compositions.

Proposition A11.2 (Addition and multiplication rules for o and O). *Suppose that $0 \le k \le l$ are two integers. Then*

1. $O(|\mathbf{x}|^k) + O(|\mathbf{x}|^l) = O(|\mathbf{x}|^k)$
2. $o(|\mathbf{x}|^k) + o(|\mathbf{x}|^l) = o(|\mathbf{x}|^k)$ — *formulas for addition*
3. $o(|\mathbf{x}|^k) + O(|\mathbf{x}|^l) = o(|\mathbf{x}|^k)$ *if $k < l$*

4. $O(|\mathbf{x}|^k)\, O(|\mathbf{x}|^l) = O(|\mathbf{x}|^{k+l})$ — *formulas for multiplication*
5. $O(|\mathbf{x}|^k)\, o(|\mathbf{x}|^l) = o(|\mathbf{x}|^{k+l})$

Proof. The formulas for addition and multiplication are more or less obvious; half the work is figuring out exactly what they mean. Suppose $U \subset \mathbb{R}^n$ is a neighborhood of $\mathbf{0}$, and $f, g : U \to \mathbb{R}$ are functions.

Addition formulas. For the first of the addition formulas, the hypothesis is that there exist $\delta > 0$ and constants C_1, C_2 such that when $0 < |\mathbf{x}| < \delta$,

$$|f(\mathbf{x})| \le C_1|\mathbf{x}|^k \quad \text{and} \quad |g(\mathbf{x})| \le C_2|\mathbf{x}|^l. \qquad A11.2$$

If $\delta_1 = \inf\{\delta, 1\}$, $C = C_1 + C_2$, and $|\mathbf{x}| < \delta_1$, then

$$f(\mathbf{x}) + g(\mathbf{x}) \le C_1|\mathbf{x}|^k + C_2|\mathbf{x}|^l \le C_1|\mathbf{x}|^k + C_2|\mathbf{x}|^k = C|\mathbf{x}|^k. \qquad A11.3$$

For the second, the hypothesis is that

$$\lim_{|\mathbf{x}|\to 0} \frac{f(\mathbf{x})}{|\mathbf{x}|^k} = 0 \quad \text{and} \quad \lim_{|\mathbf{x}|\to 0} \frac{g(\mathbf{x})}{|\mathbf{x}|^l} = 0. \qquad A11.4$$

Since $l \ge k$, we have $\lim_{|\mathbf{x}|\to 0} \dfrac{g(\mathbf{x})}{|\mathbf{x}|^k} = 0$ also, so

$$\lim_{|\mathbf{x}|\to 0} \frac{f(\mathbf{x}) + g(\mathbf{x})}{|\mathbf{x}|^k} = 0. \qquad A11.5$$

The third follows from the second, since $g \in O(|\mathbf{x}|^l)$ implies that $g \in o(|\mathbf{x}|^k)$ when $l > k$. (Can you justify that statement?[2])

Multiplication formulas. The multiplication formulas are similar; suppose functions f and g as above. For the first multiplication formula (part 4 of Proposition A11.2), by hypothesis there exist $\delta > 0$ and constants C_1 and C_2 such that when $|\mathbf{x}| < \delta$,

$$|f(\mathbf{x})| \le C_1|\mathbf{x}|^k, \quad |g(\mathbf{x})| \le C_2|\mathbf{x}|^l. \qquad A11.6$$

[2]Set $l = 3$ and $k = 2$. Then in an appropriate neighborhood, we have $g(\mathbf{x}) \le C|\mathbf{x}|^3 = C|\mathbf{x}||\mathbf{x}|^2$; by taking $|\mathbf{x}|$ sufficiently small, we can make $C|\mathbf{x}| < \epsilon$.

Then $f(\mathbf{x})g(\mathbf{x}) \le C_1C_2|\mathbf{x}|^{k+l}$.

For the second (part 5 of Proposition A11.2), the hypothesis is the same for f, and for g we know that for every ϵ, there exists η such that if $|\mathbf{x}| < \eta$, then $|g(\mathbf{x})| \le \epsilon|\mathbf{x}|^l$. When $|\mathbf{x}| < \eta$,

The Greek letter η is called "eta".

$$|f(\mathbf{x})g(\mathbf{x})| \le C_1\epsilon|\mathbf{x}|^{k+l}, \quad \text{so} \quad \lim_{|\mathbf{x}|\to 0} \frac{|f(\mathbf{x})g(\mathbf{x})|}{|\mathbf{x}|^{k+l}} = 0. \qquad \square \qquad A11.7$$

Composition rules for o and O

To speak of Taylor polynomials of compositions, we need to be sure that the compositions are defined. Let U be a neighborhood of $\mathbf{0}$ in $\mathbb{R}^n$, and let V be a neighborhood of 0 in $\mathbb{R}$. We will write Taylor polynomials for compositions $g \circ f$, where $f : U - \{\mathbf{0}\} \to \mathbb{R}$ and $g : V \to \mathbb{R}$:

$$\begin{array}{ccccc} U - \{\mathbf{0}\} & \xrightarrow{f} & \mathbb{R} & & \\ & & \cup & & \\ & & V & \xrightarrow{g} & \mathbb{R}. \end{array} \qquad A11.8$$

Proposition A11.3: f goes from a subset of $\mathbb{R}^n$ to $V \subset \mathbb{R}$, while g goes from V to $\mathbb{R}$, so $g \circ f$ goes from a subset of $\mathbb{R}^n$ to $\mathbb{R}$. Since the domain of g is a subset of $\mathbb{R}$, the variable for the first term is x, not $\mathbf{x}$.

To prove statements 1 and 3, the requirement that $l > 0$ is essential. When $l = 0$, saying that

$$f \in O(|x|^l) = O(1)$$

is just saying that f is bounded in a neighborhood of 0; that does not guarantee that its values can be the input for g, or be in the region where we know anything about g.

In statement 2, saying $f \in o(1)$ precisely says that for all ϵ, there exists δ such that when $\mathbf{x} < \delta$, then $f(\mathbf{x}) \le \epsilon$; i.e.

$$\lim_{\mathbf{x}\to\mathbf{0}} f(\mathbf{x}) = 0.$$

So the values of f are in the domain of g for $|\mathbf{x}|$ sufficiently small.

We must insist that g be defined at 0, since no reasonable condition will prevent 0 from being a value of f. In particular, when we require that g be in $O(x^k)$, we need to specify $k \ge 0$. Moreover, $f(\mathbf{x})$ must be in V when $|\mathbf{x}|$ is sufficiently small; so if $f \in O(|\mathbf{x}|^l)$ we must have $l > 0$, and if $f \in o(|\mathbf{x}|^l)$ we must have $l \ge 0$. This explains the restrictions on the exponents in Proposition A11.3.

Proposition A11.3 (Composition rules for o and O). *Let U be a neighborhood of $\mathbf{0}$ in $\mathbb{R}^n$, and $V \subset \mathbb{R}$ a neighborhood of 0; let $f : U - \{\mathbf{0}\} \to \mathbb{R}$ and $g : V \to \mathbb{R}$ be functions. We will assume throughout that $k \ge 0$.*

1. *If $g \in O(|x|^k)$ and $f \in O(|\mathbf{x}|^l)$ with $l > 0$, then $g \circ f \in O(|\mathbf{x}|^{kl})$.*
2. *If $g \in O(|x|^k)$ and $f \in o(|\mathbf{x}|^l)$ with $l \ge 0$, then $g \circ f \in o(|\mathbf{x}|^{kl})$.*
3. *If $g \in o(|x|^k)$ and $f \in O(|\mathbf{x}|^l)$ with $l > 0$, then $g \circ f \in o(|\mathbf{x}|^{kl})$.*

Proof. 1. The hypothesis is that there exist $\delta_1 > 0$, $\delta_2 > 0$ and constants C_1 and C_2 such that when $|x| < \delta_1$ and $|\mathbf{x}| < \delta_2$,

$$|g(x)| \le C_1|x|^k, \quad |f(\mathbf{x})| \le C_2|\mathbf{x}|^l. \qquad A11.9$$

Since $l > 0$, $f(\mathbf{x})$ is small when $|\mathbf{x}|$ is small, so the composition $g(f(\mathbf{x}))$ is defined for $|\mathbf{x}|$ sufficiently small: i.e., we may suppose that $\eta > 0$ is chosen so that $\eta < \delta_2$, and that $|f(\mathbf{x})| < \delta_1$ when $|\mathbf{x}| < \eta$. Then

$$\big|g(f(\mathbf{x}))\big| \le C_1|f(\mathbf{x})|^k \le C_1\left(C_2|\mathbf{x}|^l\right)^k = C_1C_2^k|\mathbf{x}|^{kl}. \qquad A11.10$$

2. We know as before that there exist C and $\delta_1 > 0$ such that when $|x| < \delta_1$, we have $|g(x)| < C|x|^k$. Choose $\epsilon > 0$; for f we know that there

exists $\delta_2 > 0$ such that $|f(\mathbf{x})| \le \epsilon |\mathbf{x}|^l$ when $|\mathbf{x}| < \delta_2$. Taking δ_2 smaller if necessary, we may also suppose $\epsilon|\delta_2|^l < \delta_1$. Then when $|\mathbf{x}| < \delta_2$, we have

$$\big|g\big(f(\mathbf{x})\big)\big| \le C|f(\mathbf{x})|^k \le C\left(\epsilon|\mathbf{x}|^l\right)^k = \underbrace{C\epsilon^k}_{\text{arbitrarily small}} |\mathbf{x}|^{kl}. \qquad A11.11$$

Thus $g \circ f$ is in $o(|\mathbf{x}|^{kl}|$.

3. Our hypothesis $g \in o(|x|^k)$ asserts that for any $\epsilon > 0$ there exists $\delta_1 > 0$ such that $|g(x)| < \epsilon |x|^k$ when $0 \le |x| < \delta_1$.

Now our hypothesis on f says that there exist C and $\delta_2 > 0$ such that $|f(\mathbf{x})| < C|\mathbf{x}|^l$ when $|\mathbf{x}| < \delta_2$; taking δ_2 smaller if necessary, we may further assume that $C|\delta_2|^l < \delta_1$. Then if $|\mathbf{x}| < \delta_2$,

This is where we use the fact that $l > 0$. If $l = 0$, then making δ_2 small would not make $C|\delta_2|^l$ small, since $C|\delta_2|^0 = C$.

$$\big|g\big(f(\mathbf{x})\big)\big| \le \epsilon |f(\mathbf{x})|^k \le \epsilon \big|C|\mathbf{x}|^l\big|^k = \epsilon C^k |\mathbf{x}|^{lk}. \quad \square \qquad A11.12$$

Proving Propositions 3.4.3 and 3.4.4

Now we are ready now to use Propositions A11.2 and A11.3 to prove Propositions 3.4.3 and 3.4.4, which we repeat below.

Proposition 3.4.3 (Sums and products of Taylor polynomials). *Let $U \subset \mathbb{R}^n$ be open, and let $f, g : U \to \mathbb{R}$ be C^k functions. Then $f + g$ and fg are also of class C^k, and*

1. *The Taylor polynomial of the sum is the sum of the Taylor polynomials:*

$$P^k_{f+g,\mathbf{a}}(\mathbf{a}+\vec{\mathbf{h}}) = P^k_{f,\mathbf{a}}(\mathbf{a}+\vec{\mathbf{h}}) + P^k_{g,\mathbf{a}}(\mathbf{a}+\vec{\mathbf{h}})$$

2. *The Taylor polynomial of the product fg is obtained by taking the product*

$$P^k_{f,\mathbf{a}}(\mathbf{a}+\vec{\mathbf{h}}) \cdot P^k_{g,\mathbf{a}}(\mathbf{a}+\vec{\mathbf{h}})$$

and discarding the terms of degree $> k$

Proposition 3.4.4 (Chain rule for Taylor polynomials). *Let $U \subset \mathbb{R}^n$ and $V \subset \mathbb{R}$ be open, and $g : U \to V$, $f : V \to \mathbb{R}$ be of class C^k. Then $f \circ g : U \to \mathbb{R}$ is of class C^k, and if $g(\mathbf{a}) = b$, then the Taylor polynomial $P^k_{f\circ g,\mathbf{a}}(\mathbf{a}+\vec{\mathbf{h}})$ is obtained by considering the polynomial*

$$P^k_{f,b}\big(P^k_{g,\mathbf{a}}(\mathbf{a}+\vec{\mathbf{h}})\big)$$

and discarding the terms of degree $> k$.

Each proposition has two parts: one asserts that sums, products, and compositions of C^k functions are of class C^k; the other tells how to compute their Taylor polynomials.

The first part is proved by induction on k, using the second part. The rules for computing Taylor polynomials say that the $(k-1)$-partial derivatives of a sum, product, or composition are complicated sums of products and compositions of derivatives, of order at most $k-1$, of the given C^k functions. As such, they are themselves continuously differentiable, by Theorems 1.8.1 and 1.8.3. So the sums, products, and compositions are of class C^k.

Sums and products of Taylor polynomials. The case of sums follows immediately from part 2 of Proposition A11.2. For products, suppose

$$f(\mathbf{x}) = p_k(\mathbf{x}) + r_k(\mathbf{x}) \quad \text{and} \quad g(\mathbf{x}) = q_k(\mathbf{x}) + s_k(\mathbf{x}), \qquad \textit{A11.13}$$

with $r_k, s_k \in o(|\mathbf{x}|^k)$. Multiply

$$f(\mathbf{x})g(\mathbf{x}) = \big(p_k(\mathbf{x}) + r_k(\mathbf{x})\big)\big(q_k(\mathbf{x}) + s_k(\mathbf{x})\big) = P_k(\mathbf{x}) + R_k(\mathbf{x}), \qquad \textit{A11.14}$$

where $P_k(\mathbf{x})$ is obtained by multiplying $p_k(\mathbf{x})q_k(\mathbf{x})$ and keeping the terms of degree between 1 and k. The remainder $R_k(\mathbf{x})$ contains the higher-degree terms of the product $p_k(\mathbf{x})q_k(\mathbf{x})$, which of course are in $o(|\mathbf{x}|^k)$. It also contains the products $r_k(\mathbf{x})s_k(\mathbf{x}), r_k(\mathbf{x})q_k(\mathbf{x})$, and $p_k(\mathbf{x})s_k(\mathbf{x})$, which are of the following forms:

$$\begin{aligned} O(1)s_k(\mathbf{x}) &\in o(|\mathbf{x}|^k);\\ r_k(\mathbf{x})O(1) &\in o(|\mathbf{x}|^k); \qquad \textit{A11.15}\\ r_k(\mathbf{x})s_k(\mathbf{x}) &\in o(|\mathbf{x}|^{2k}). \end{aligned}$$

Compositions of Taylor polynomials. Let us write $g(\mathbf{a}+\vec{\mathbf{h}})$ as

$$g(\mathbf{a}+\vec{\mathbf{h}}) = \underbrace{b}_{\substack{\text{constant}\\ \text{term}}} + \underbrace{Q^k_{g,\mathbf{a}}(\vec{\mathbf{h}})}_{\substack{\text{polynomial terms}\\ 1\le \text{ degree } \le k}} + \underbrace{r^k_{g,\mathbf{a}}(\vec{\mathbf{h}})}_{\text{remainder}}, \qquad \textit{A11.16}$$

separating out the constant term, of degree 0; the sum of the polynomial terms of degree between 1 and k, called $Q^k_{g,\mathbf{a}}(\vec{\mathbf{h}})$; and the remainder $r^k_{g,\mathbf{a}}(\vec{\mathbf{h}})$ (terms of degree $> k$). This gives

$$|Q^k_{g,\mathbf{a}}(\vec{\mathbf{h}})| \in O(|\vec{\mathbf{h}}|) \quad \text{and} \quad r^k_{g,\mathbf{a}}(\vec{\mathbf{h}}) \in o(|\vec{\mathbf{h}}|^k). \qquad \textit{A11.17}$$

Then

$$(f\circ g)(\mathbf{a}+\vec{\mathbf{h}}) = P^k_{f,b}\Big(b+Q^k_{g,\mathbf{a}}(\vec{\mathbf{h}})+r^k_{g,\mathbf{a}}(\vec{\mathbf{h}})\Big)+r^k_{f,b}\Big(b+Q^k_{g,\mathbf{a}}(\vec{\mathbf{h}})+r^k_{g,\mathbf{a}}(\vec{\mathbf{h}})\Big). \qquad \textit{A11.18}$$

Among the terms on the right side of equation A11.18, there are the terms of $P^k_{f,b}(b + Q^k_{g,\mathbf{a}}(\vec{\mathbf{h}}))$ of degree at most k in $\vec{\mathbf{h}}$; we must show that all the others are in $o(|\vec{\mathbf{h}}|^k)$. We have

$$r^k_{f,b}\Big(b + Q^k_{g,\mathbf{a}}(\vec{\mathbf{h}}) + r^k_{g,\mathbf{a}}(\vec{\mathbf{h}})\Big) \in o(|\vec{\mathbf{h}}|^k) \qquad \textit{A11.19}$$

by the computation in equation A11.20, where we use equation 3.4.2 in the first line, equation A11.17 in the second line, equation A11.1 and part 1 of Proposition A11.2 for the first equality in the third line, and part 3 of Proposition A11.3 (applied to the composition $o \circ O$) for the last equality:

$$r^k_{f,b}\big(\underbrace{b + Q^k_{g,\mathbf{a}}(\vec{\mathbf{h}}) + r^k_{g,\mathbf{a}}(\vec{\mathbf{h}})}_{\vec{\mathbf{h}} \text{ in eq. 3.4.2}}\big) \in o\Big(\big|Q^k_{g,\mathbf{a}}(\vec{\mathbf{h}}) + r^k_{g,\mathbf{a}}(\vec{\mathbf{h}})\big|^k\Big)$$

$$\subset o\left(\Big|O(|\vec{\mathbf{h}}|) + o(|\vec{\mathbf{h}}|^k)\Big|^k\right) \qquad A11.20$$

$$= o\Big(\big|O(|\vec{\mathbf{h}}|)\big|^k\Big) = o(|\vec{\mathbf{h}}|^k).$$

Formula A11.21: f is a function of a single variable, so when we apply Definition 3.3.13, we are in the trivial case where the multi-index I is the single index m.

The other terms are of the form

$$\frac{1}{m!} D_m f(b) \Big(b + Q^k_{g,\mathbf{a}}(\vec{\mathbf{h}}) + r^k_{g,\mathbf{a}}(\vec{\mathbf{h}})\Big)^m. \qquad A11.21$$

If we multiply out the power, we find some terms of degree at most k in the coordinates h_i of $\vec{\mathbf{h}}$, and no factors $r^k_{g,\mathbf{a}}(\vec{\mathbf{h}})$. These terms are precisely the terms we are keeping in our candidate Taylor polynomial for the composition. Then there are terms of degree greater than k in the h_i, with no factors $r^k_{g,\mathbf{a}}(\vec{\mathbf{h}})$, which are evidently in $o(|\vec{\mathbf{h}}|^k)$. Finally, there are terms that contain at least one factor $r^k_{g,\mathbf{a}}(\vec{\mathbf{h}})$. These last are in $O(1)o(|\vec{\mathbf{h}}|^k) = o(|\vec{\mathbf{h}}|^k)$. □

Exercises for Section A.11

A11.1 True or false? (Explain your reasoning). For the function $|x|^{3/2}$,

a. $|x|^{3/2} \in o\big(|x|\big)$ b. $|x|^{3/2} \in O\big(|x|\big)$

c. $|x|^{3/2} \in o\big(|x|^2\big)$ d. $|x|^{3/2} \in O\big(|x|^2\big)$

A11.2 Repeat Exercise A11.1 for the function $\dfrac{x}{\ln |x|}$.

A11.3 Repeat Exercise A11.1 for the function $x \ln |x|$.

A11.4 a. Show that Proposition 3.4.4 (the chain rule for Taylor polynomials) contains the chain rule (Theorem 1.8.3) as a special case, as long as the mappings are continuously differentiable.

b. Go back to Appendix A.4 and show how o and O notation can be used to shorten the proof of the chain rule.

A12 Taylor's theorem with remainder

Let U be an open subset of $\mathbb{R}$ and let $f : U \to \mathbb{R}$ be k times continuously differentiable on U. It is all very well to claim (Theorem 3.3.16) that

$$\lim_{\vec{\mathbf{h}} \to \vec{\mathbf{0}}} \frac{f(\mathbf{a} + \vec{\mathbf{h}}) - P^k_{f,\mathbf{a}}(\mathbf{a} + \vec{\mathbf{h}})}{|\vec{\mathbf{h}}|^k} = 0; \qquad A12.1$$

that doesn't tell you how small the difference $f(\mathbf{a}+\vec{\mathbf{h}}) - P^k_{f,\mathbf{a}}(\mathbf{a}+\vec{\mathbf{h}})$ is for any particular $\vec{\mathbf{h}} \neq \vec{\mathbf{0}}$. Taylor's theorem with remainder gives such a bound, in the form of a multiple of $|\vec{\mathbf{h}}|^{k+1}$. To get such a result we must require bit more about the function f; we will assume that f is of class C^{k+1}.

Recall Taylor's theorem with remainder in one dimension:

In Landau's notation, equation A12.1 says that if f is of class C^{k+1} near $\mathbf{a}$, then not only is

$$f(\mathbf{a}+\vec{\mathbf{h}}) - P^k_{f,\mathbf{a}}(\mathbf{a}+\vec{\mathbf{h}})$$

in $o(|\vec{\mathbf{h}}|^k)$; it is in fact in $O(|\vec{\mathbf{h}}|^{k+1})$. Theorem A12.7 gives a formula for the constant implicit in the O.

Theorem A12.1 (Taylor's theorem with remainder in one dimension). *If $f : (a-R, a+R) \to \mathbb{R}$ is $(k+1)$-times continuously differentiable and $|h| < R$, then*

$$f(a+h) = \overbrace{f(a) + f'(a)h + \cdots + \frac{1}{k!}f^{(k)}(a)h^k}^{\substack{P^k_{f,a}(a+h) \\ \text{(Taylor polynomial of } f \text{ at } a\text{, of degree } k)}} + \underbrace{\frac{1}{k!}\int_0^h (h-t)^k f^{(k+1)}(a+t)\,dt}_{\text{remainder}}. \qquad \text{A12.2}$$

When $k = 0$, equation A12.2 is the fundamental theorem of calculus:

$$f(a+h) = f(a) + \underbrace{\int_0^h f'(a+t)\,dt}_{\text{remainder}}.$$

Proof. The standard proof is by repeated integration by parts; you are asked to do that in Exercise A12.4. Here is a slicker and less natural proof. First, rewrite equation A12.2, setting $h = x - a$ and $s = a + t$:

$$f(x) = f(a) + f'(a)(x-a) + \cdots + \frac{1}{k!}f^{(k)}(a)(x-a)^k + \frac{1}{k!}\int_a^x (x-s)^k f^{(k+1)}(s)\,ds. \qquad \text{A12.3}$$

Equation A12.3: Since

$$h = x - a, \quad s = a + t,$$

as t goes from 0 to h, the variable s goes from a to x.

Since a is the point at which we are approximating the function, and $x = a + h$ is the point at which we are evaluating how good the approximation is, it is natural to think of holding a constant and varying x. This proof is "unnatural" because we are holding x constant and varying a.

To show that the right side of equation A12.3 is constant, we show that its derivative is 0.

The two sides are equal when $a = x$: all the terms on the right vanish except the first, giving $f(x) = f(x)$. If we can show that as a varies and x stays fixed, the right side stays constant, then we will know that the two sides are always equal. So we compute the derivative of the right side:

$$\overbrace{f'(a) + \Big(-f'(a)}^{=0} + \overbrace{(x-a)f''(a)\Big) + \Big(-(x-a)f''(a)}^{=0} + \frac{(x-a)^2 f'''(a)}{2!}\Big) + \cdots$$

$$\cdots + \Big(-\frac{(x-a)^{k-1}f^{(k)}(a)}{(k-1)!} + \overbrace{\frac{(x-a)^k f^{(k+1)}(a)}{k!}\Big) \underbrace{- \frac{(x-a)^k f^{(k+1)}(a)}{k!}}_{\text{derivative of the remainder}}}^{=0},$$

where the last term is the derivative of the integral, computed by the fundamental theorem of calculus. A careful look shows that everything drops out. □

Evaluating the remainder: in one dimension

To use Taylor's theorem with remainder, you must find a bound for the remainder. It is *not* useful to compute the integral; if you do this, by repeated integration by parts, you get exactly $f(a+h)$ minus the Taylor polynomial; you are right back to equation A12.2 where you started.

Theorem A12.2. *There exists c between a and $a+h$ such that*

$$f(a+h) = P^k_{f,a}(a+h) + \frac{f^{(k+1)}(c)}{(k+1)!}h^{k+1}. \qquad \text{A12.4}$$

Proof of Theorem A12.2: The statement that there exists c with $G(c) = p$ requires G to be continuous, since it uses the intermediate value theorem. That is why we need f to be of class C^{k+1}.

Proof. Suppose F, G are functions on $[a, b]$, with G continuous and $F > 0$ integrable; set $m = \inf_{[a,b]} G$ and $M = \sup_{[a,b]} G$. Then $mF \le FG \le MF$, so there exists p with $m \le p \le M$ and $c \in [a, b]$ with $G(c) = p$ such that

$$\int_a^b F(t)G(t)\,dt = p\int_a^b F(t)\,dt = G(c)\int_a^b F(t)\,dt. \qquad \text{A12.5}$$

Applying this to the remainder in equation A12.2, with $a = 0$, $b = h$ and $F(t) = (h-t)^k$, $G(t) = f^{(k+1)}(a+t)$ leads to the existence of $t_0 \in [0, h]$ such that, setting $c = a + t_0$, we have

$$\underbrace{\frac{1}{k!}\int_0^h (h-t)^k f^{(k+1)}(a+t)\,dt}_{\text{remainder, eq. A12.2}} = f^{(k+1)}(c)\frac{1}{k!}\int_0^h (h-t)^k dt = \frac{h^{k+1}}{(k+1)!}f^{(k+1)}(c). \qquad \square$$

Corollary A12.3. *If $|f^{(k+1)}(a+t)| \le C$ for t between 0 and h, then*

$$|f(a+h) - P^k_{f,a}(a+h)| \le \frac{C}{(k+1)!}h^{k+1}. \qquad \text{A12.6}$$

A calculator that computes to eight places can store formula A12.7 and spit it out when you evaluate sines; even hand calculation isn't out of the question. This is how the original trigonometric tables were computed.

Computing large factorials is quicker if you know that $6! = 720$.

It isn't often that high derivatives of functions can be so easily bounded; usually using Taylor's theorem with remainder is much messier.

Example A12.4 (Finding a bound for the remainder in one dimension). A standard example is to compute $\sin\theta$ to eight decimals when $|\theta| \le \pi/6$. Since the successive derivatives of $\sin\theta$ are all sines and cosines, they are all bounded by 1, so the remainder after taking k terms of the Taylor polynomial is at most

$$\frac{1}{(k+1)!}\left(\frac{\pi}{6}\right)^{k+1} \quad \text{for } |\theta| \le \pi/6. \qquad \text{A12.7}$$

By trial and error, we discover that $k = 8$ is good enough; we have $1/9! = 3.2002048 \times 10^{-6}$ and $(\pi/6)^9 \approx 2.76349 \times 10^{-3}$; the error is then at most 8.8438×10^{-9}. Thus we can be sure that when $|\theta| \le \pi/6$,

$$\sin\theta = \theta - \frac{\theta^3}{3!} + \frac{\theta^5}{5!} - \frac{\theta^7}{7!} \qquad \text{A12.8}$$

to eight decimals, i.e., $|\sin\theta - (\theta - \frac{\theta^3}{3!} + \frac{\theta^5}{5!} - \frac{\theta^7}{7!})| \le 10^{-8}$ △

Theorem A12.5 (Taylor's theorem with remainder in higher dimensions). *Let $U \subset \mathbb{R}^n$ be open, and let $f : U \to \mathbb{R}$ be a function of class C^{k+1}. Suppose that the interval $[\mathbf{a}, \mathbf{a} + \vec{\mathbf{h}}]$ is contained in U. Then there exists $\mathbf{c} \in [\mathbf{a}, \mathbf{a} + \vec{\mathbf{h}}]$ such that*

$$f(\mathbf{a} + \vec{\mathbf{h}}) = P^k_{f,\mathbf{a}}(\mathbf{a} + \vec{\mathbf{h}}) + \sum_{I \in \mathcal{I}_n^{k+1}} \frac{1}{I!} D_I f(\mathbf{c})\,\vec{\mathbf{h}}^I. \qquad \text{A12.9}$$

Proof. Define $\varphi : \mathbb{R} \to \mathbb{R}^n$ by $\varphi(t) = \mathbf{a} + t\vec{\mathbf{h}}$, and let $g : \mathbb{R} \to \mathbb{R}$ be the function $g = f \circ \varphi$. Theorem A12.2 applied to g when $h = 1$ and $a = 0$ says that there exists c with $0 < c < 1$ such that

Equation A12.10: Recall that when writing out the Taylor polynomial $P^k_{g,a}(a+h)$, the function g and its derivatives are evaluated at a, not at $a+h$. This accounts for the $g(0), \dots, g^{(k)}(0)$.

$$g(1) = \underbrace{g(0) + \cdots + \frac{g^{(k)}(0)}{k!}}_{\text{Taylor polynomial}} + \underbrace{\frac{1}{(k+1)!} g^{(k+1)}(c)}_{\text{remainder}}. \qquad A12.10$$

We need to show that the terms of equation A12.10 are the same as the corresponding terms of equation A12.9. It is obvious that the left sides are equal; by definition, $g(1) = f(\mathbf{a} + \vec{\mathbf{h}})$. That the Taylor polynomials on the right sides are the same follows from the chain rule for Taylor polynomials (Proposition 3.4.4):

Equation A12.11:

$$(t\vec{\mathbf{h}})^I = t^{\deg I}\vec{\mathbf{h}}^I = t^m\vec{\mathbf{h}}^I.$$

For instance, for $I = (2,3,7)$ with $\deg I = 12$,

$$(th_1)^2(th_2)^3(th_3)^7 = t^{12}h_1^2h_2^3h_3^7.$$

$$\begin{aligned} P^k_{g,0}(t) = P^k_{f,\mathbf{a}}\big(P^k_{\varphi,0}(t)\big) &= \sum_{m=0}^{k} \sum_{I \in \mathcal{I}^m_n} \frac{1}{I!} D_I f(\mathbf{a})(t\vec{\mathbf{h}})^I \\ &= \sum_{m=0}^{k} \Big(\sum_{I \in \mathcal{I}^m_n} \frac{1}{I!} D_I f(\mathbf{a})(\vec{\mathbf{h}})^I \Big) t^m. \end{aligned} \qquad A12.11$$

This shows that

$$P^k_{f,\mathbf{a}}(0+1) = g(0) + \cdots + \frac{g^{(k)}(0)}{k!} = P^k_{f,\mathbf{a}}(\mathbf{a} + \vec{\mathbf{h}}). \qquad A12.12$$

To see that the remainders are the same, set $\mathbf{c} = \varphi(c)$. Again the chain rule for Taylor polynomials gives

$$\begin{aligned} P^{k+1}_{g,c}(t) = P^{k+1}_{f,\mathbf{c}}\big(P^{k+1}_{\varphi,c}(t)\big) &= \sum_{m=0}^{k+1} \sum_{I \in \mathcal{I}^m_n} \frac{1}{I!} D_I f(\mathbf{c})(t\vec{\mathbf{h}})^I \\ &= \sum_{m=0}^{k+1} \Big(\sum_{I \in \mathcal{I}^m_n} \frac{1}{I!} D_I f(\mathbf{c})(\vec{\mathbf{h}})^I \Big) t^m. \end{aligned} \qquad A12.13$$

Looking at the terms of degree $k+1$ on both sides gives the desired result:

$$\underbrace{\frac{1}{(k+1)!} g^{(k+1)}(c)}_{\text{remainder, eq. A12.10}} = \underbrace{\sum_{I \in \mathcal{I}^{k+1}_n} \frac{1}{I!} D_I f(\mathbf{c})(\vec{\mathbf{h}})^I}_{\text{remainder, eq. A12.9}}. \quad \square \qquad A12.14$$

There are several ways to turn this into a bound on the remainder; they yield somewhat different results. We will use the following lemma. We call it the *polynomial formula* because it generalizes the binomial formula to polynomials. It is rather nice in its own right and shows how multi-index notation can simplify complicated formulas.

Lemma A12.6 (Polynomial formula).

$$\sum_{I \in \mathcal{I}^k_n} \frac{1}{I!} \vec{\mathbf{h}}^I = \frac{1}{k!}(h_1 + \cdots + h_n)^k \qquad A12.15$$

Proof. The proof is by induction on n. When $n = 1$, the lemma simply asserts that $h^k = h^k$. Suppose the formula is true for n, and let us prove it for $n+1$. In this case, $\vec{\mathbf{h}} = \begin{bmatrix} h_1 \\ \vdots \\ h_{n+1} \end{bmatrix}$; we denote $\begin{bmatrix} h_1 \\ \vdots \\ h_n \end{bmatrix}$ by $\vec{\mathbf{h}}'$. Compute:

Equation A12.16: The last step is the binomial formula

$$(a+b)^k = \sum_{m=0}^{k} \frac{k!}{m!(k-m)!} a^m b^{k-m}.$$

Equation 3.4.9 is a somewhat more elaborate version.

$$\begin{aligned}\sum_{I \in \mathcal{I}^k_{n+1}} \frac{1}{I!} \vec{\mathbf{h}}^I &= \sum_{m=0}^{k} \sum_{J \in \mathcal{I}^m_n} \frac{1}{J!} (\vec{\mathbf{h}}')^J \frac{1}{(k-m)!} h_{n+1}^{k-m} \\ &= \sum_{m=0}^{k} \underbrace{\frac{1}{m!}(h_1 + \cdots + h_n)^m}_{\text{by induction on } n} \frac{1}{(k-m)!} h_{n+1}^{k-m} \\ &= \frac{1}{k!} \sum_{m=0}^{k} \frac{k!}{(k-m)!m!} (h_1 + \cdots + h_n)^m h_{n+1}^{k-m} \\ &= \frac{1}{k!}(h_1 + \cdots + h_n + h_{n+1})^k. \quad \square\end{aligned} \qquad A12.16$$

This, together with Theorem A12.5, gives the following result.

Theorem A12.7 (An explicit bound for the Taylor remainder). *Let $U \subset \mathbb{R}^n$ be open, and let $f : U \to \mathbb{R}$ be a function of class C^{k+1}. Suppose that the interval $[\mathbf{a}, \mathbf{a} + \vec{\mathbf{h}}]$ is contained in U. If*

$$\sup_{I \in \mathcal{I}^{k+1}_n} \sup_{\mathbf{c} \in [\mathbf{a}, \mathbf{a}+\vec{\mathbf{h}}]} |D_I f(\mathbf{c})| \le C, \quad \textit{then} \qquad A12.17$$

$$\left| f(\mathbf{a} + \vec{\mathbf{h}}) - P^k_{f,\mathbf{a}}(\mathbf{a} + \vec{\mathbf{h}}) \right| \le C \left(\sum_{i=1}^{n} |h_i| \right)^{k+1}. \qquad A12.18$$

Exercises for Section A12

A12.1 a. Let $f(x) = e^x$, so that $f(1) = e$. Use Corollary A12.3 to show that

$$e = \sum_{i=0}^{k} \frac{1}{i!} + r_{k+1}, \quad \text{where} \quad |r_{k+1}| \le \frac{3}{(k+1)!}.$$

b. If $e = a/b$ for some integers a and b, deduce from part a that

$$|k!a - bm| \le \frac{3b}{k+1}, \quad \text{where } m \text{ is the integer } \frac{k!}{0!} + \frac{k!}{1!} + \frac{k!}{2!} + \cdots + \frac{k!}{k!}.$$

Conclude that if k is large enough, then $k!a - bm$ is an integer that is arbitrarily small, and therefore 0.

c. Observe that k does not divide m evenly, since it divides every summand but the last one. Since k may be freely chosen, provided only that it is sufficiently large, take k to be a prime number larger than b. Then k divides the left side of $k!a = bm$ but does not divide m. What conclusion do you reach?

A12.2 Let $f\begin{pmatrix} x \\ y \end{pmatrix} = e^{\sin(x+y^2)}$. Use MAPLE, MATHEMATICA, or similar software.

a. Calculate the Taylor polynomial $P^k_{f,\mathbf{a}}$ of degree $k = 1, 2$, and 4 at $\mathbf{a} = \begin{pmatrix} 1 \\ 1 \end{pmatrix}$.

b. Estimate the maximum error $|P^k_{f,\mathbf{a}} - f|$ on the region $|x-1| < .5$ and $|y-1| < .5$, for $k = 1$ and $k = 2$.

c. Estimate the maximum error in the region $|x-1| < .25$ and $|y-1| < .25$, for $k = 1$ and $k = 2$.

A12.3 a. Write the integral form of the remainder when $\sin(xy)$ is approximated by its Taylor polynomial of degree 2 at the origin.

b. Give an upper bound for the remainder when $x^2 + y^2 \le 1/4$.

A12.4 Prove equation A12.2 by induction, by first checking that when $k = 0$, it is the fundamental theorem of calculus, and using integration by parts to prove

$$\frac{1}{k!}\int_0^h (h-t)^k g^{(k+1)}(a+t)\,dt = \frac{1}{(k+1)!}g^{(k+1)}(a)h^{k+1} + \frac{1}{(k+1)!}\int_0^h (h-t)^{k+1} g^{(k+2)}(a+t)\,dt.$$

A12.5 Let f be the function $f\begin{pmatrix} x \\ y \end{pmatrix} = \operatorname{sgn}(y)\sqrt{\dfrac{-x+\sqrt{x^2+y^2}}{2}}$ where $\operatorname{sgn}(y)$ is the sign of y (i.e., $+1$ when $y > 0$, 0 when $y = 0$, and -1 when $y < 0$).

Exercise A12.5: Part a is almost obvious except when

$$y = 0,\ x > 0,$$

where y changes sign. It may help to show that this mapping can be written $(r, \theta) \mapsto \sqrt{r}\sin(\theta/2)$ in polar coordinates.

a. Show that f is continuously differentiable on the complement of the half-line $y = 0$, $x \le 0$.

b. Show that if $\mathbf{a} = \begin{pmatrix} -1 \\ -\epsilon \end{pmatrix}$ and $\vec{\mathbf{h}} = \begin{bmatrix} 0 \\ 2\epsilon \end{bmatrix}$, then although both $\mathbf{a}$ and $\mathbf{a} + \vec{\mathbf{h}}$ are in the domain of definition of f, Taylor's theorem with remainder (Theorem A12.5) is not true.

c. What part of the statement is violated? Where does the proof fail?

A13 PROVING THEOREM 3.5.3 (COMPLETING SQUARES)

Below we repeat Theorem 3.5.3.

Theorem 3.5.3 (Quadratic forms as sums of squares).
1. *For any quadratic form $Q : \mathbb{R}^n \to \mathbb{R}$, there exist $m = k + l$ linearly independent linear functions $\alpha_1, \dots, \alpha_m : \mathbb{R}^n \to \mathbb{R}$ such that*

$$Q(\mathbf{x}) = \big(\alpha_1(\mathbf{x})\big)^2 + \dots + \big(\alpha_k(\mathbf{x})\big)^2 - \big(\alpha_{k+1}(\mathbf{x})\big)^2 - \dots - \big(\alpha_{k+l}(\mathbf{x})\big)^2.$$

2. *The number k of plus signs and the number l of minus signs depends only on Q and not on the specific linear functions chosen.*

Proof. Part 2 is proved in Section 3.5. To prove part 1, we will argue by induction on the number of variables x_i appearing in Q.

Let $Q : \mathbb{R}^n \to \mathbb{R}$ be a quadratic form. Clearly, if only one variable x_i appears, then $Q(\vec{\mathbf{x}}) = \pm a x_i^2$ with $a > 0$, so $Q(\vec{\mathbf{x}}) = \pm(\sqrt{a}\, x_i)^2$, and the theorem is true. So suppose it is true for all quadratic forms in which at most $p-1$ variables appear, and suppose p variables appear in the expression of Q. Let x_i be such a variable; then either a term $\pm a x_i^2$ appears with $a > 0$, or it doesn't.

This is just formalizing the completion of squares.

a. If a term $\pm a x_i^2$ appears with $a > 0$, we can then write

$$\begin{aligned} Q(\vec{\mathbf{x}}) &= \pm\Big(a x_i^2 + \beta(\vec{\mathbf{x}}) x_i + \frac{\big(\beta(\vec{\mathbf{x}})\big)^2}{4a}\Big) + Q_1(\vec{\mathbf{x}}) \\ &= \pm\underbrace{\Big(\sqrt{a}\, x_i + \frac{\beta(\vec{\mathbf{x}})}{2\sqrt{a}}\Big)^2}_{\alpha_0(\mathbf{x})} + Q_1(\vec{\mathbf{x}}) = \pm\alpha_0(\mathbf{x}) + Q_1(\vec{\mathbf{x}}), \end{aligned} \qquad A13.1$$

where α_0 and β are linear functions, and Q_1 a quadratic function (both β and Q_1 are functions of the $p-1$ variables appearing in Q other than x_i). By induction, we can write

$$Q_1(\vec{\mathbf{x}}) = \pm\big(\alpha_1(\vec{\mathbf{x}})\big)^2 \pm \cdots \pm \big(\alpha_q(\vec{\mathbf{x}})\big)^2 \qquad A13.2$$

for some linearly independent linear functions α_j, for $j = 1, \dots, q$, of the $p-1$ variables appearing in Q other than x_i.

We must check that the linear functions $\alpha_0, \alpha_1, \dots, \alpha_q$ are linearly independent. Suppose $c_0\alpha_0 + \cdots + c_q\alpha_q = 0$; then

$$(c_0\alpha_0 + \cdots + c_q\alpha_q)(\vec{\mathbf{x}}) = 0 \quad \text{for every } \vec{\mathbf{x}}, \qquad A13.3$$

Recall that β is a function of the variables other than x_i; thus when those variables are 0, for instance, when $\vec{\mathbf{x}} = \vec{\mathbf{e}}_i$, then so is $\beta(\vec{\mathbf{x}})$, as are $\alpha_1(\vec{\mathbf{x}}), \dots, \alpha_q(\vec{\mathbf{x}})$.

in particular, when $\vec{\mathbf{x}} = \vec{\mathbf{e}}_i$ (when $x_i = 1$ and all the other variables are 0).

This leads to $c_0\sqrt{a} = 0$, so $c_0 = 0$, so equation A13.3 and the linear independence of $\alpha_1, \dots, \alpha_q$ imply that $c_1 = \cdots = c_q = 0$.

b. If no term $\pm a x_i^2$ appears, then there must be a term of the form $\pm a x_i x_j$ with $a > 0$. Make the substitution $x_j = x_i + u$; we can now write

$$\begin{aligned} Q(\vec{\mathbf{x}}) &= a x_i^2 + \beta(\vec{\mathbf{x}}, u) x_i + \frac{\big(\beta(\vec{\mathbf{x}}, u)\big)^2}{4a} + Q_1(\vec{\mathbf{x}}, u) \\ &= \pm\left(\sqrt{a}\, x_i + \frac{\beta(\vec{\mathbf{x}}, u)}{2\sqrt{a}}\right)^2 + Q_1(\vec{\mathbf{x}}, u), \end{aligned} \qquad A13.4$$

where β and Q_1 are functions (linear and quadratic respectively) of u and of the variables that appear in Q other than x_i and x_j. Now argue exactly as above; the only subtle point is that in order to prove $c_0 = 0$ you need to set $u = 0$ (i.e., to set $x_i = x_j = 1$). □

A14 Classifying constrained critical points

In this appendix we justify using the augmented Hessian matrix to compute the signature of constrained critical points.

Theorem 3.7.13 (Computing the signature of a constrained critical point). *Let $U \subset \mathbb{R}^{n+m}$ be open, $f: U \to \mathbb{R}$ a C^2 function, and $\mathbf{F}: U \to \mathbb{R}^m$ a C^2 map defining a manifold $Z \stackrel{\text{def}}{=} \mathbf{F}^{-1}(\mathbf{0})$, with $[\mathbf{DF}(\mathbf{z})]$ onto at every point $\mathbf{z} \in Z$. Let $\mathbf{z}_0 \in Z$ be a critical point of f restricted to Z of signature (p, q), and let (p_1, q_1) be the signature of the quadratic form defined by the augmented Hessian matrix H. Then*

$$(p, q) = (p_1 - m, q_1 - m).$$

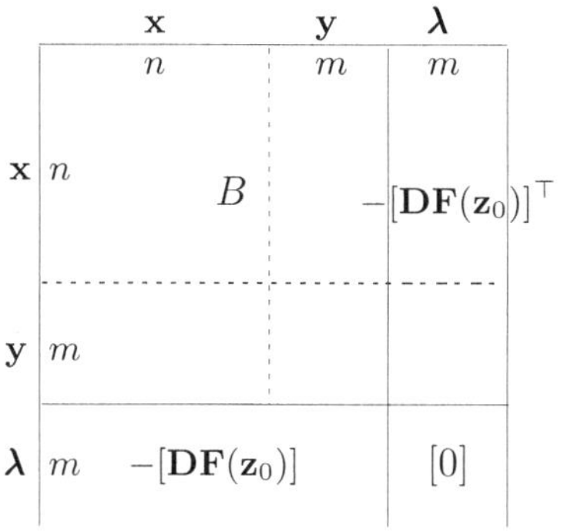

FIGURE A14.1.

The augmented Hessian matrix is the Hessian matrix (matrix of second derivatives) of the Lagrangian function defined in equation A.14.1. To get the matrix at bottom left, first compute the derivative of $L_{f,\mathbf{F}}$ with respect to $\boldsymbol{\lambda}$, which is

$$D_{\boldsymbol{\lambda}} L_{f,\mathbf{F}} \begin{pmatrix} \mathbf{z} \\ \boldsymbol{\lambda} \end{pmatrix} = -D_{\boldsymbol{\lambda}} \boldsymbol{\lambda} \mathbf{F}(\mathbf{z}) = -\begin{bmatrix} F_1(\mathbf{z}) & \cdots & F_m(\mathbf{z}) \end{bmatrix};$$

then take derivatives with respect to $z_1, \dots, z_{n+m}$. The 0 matrix at bottom right represents the second derivatives with respect to $\boldsymbol{\lambda}$: $D_{\lambda_i} D_{\lambda_j} L_{f,\mathbf{F}}$.

Recall that the matrix denoted B in Figure A14.1 is the matrix defined in equation 3.7.42, with entries

$$B_{i,j} = D_i D_j f - \sum_{k=1}^{m} \lambda_k D_i D_j F_k.$$

with all second partials evaluated at $\mathbf{z}_0$.

This proof was influenced by a paper by Catherine Hassell and Elmer Rees.

Proof. Denote by $\boldsymbol{\lambda}$ the $1 \times m$ row matrix $\boldsymbol{\lambda} = [\lambda_1 \ \cdots \ \lambda_m]$. Note that the augmented Hessian matrix (see Figure A14.1) is the Hessian matrix (matrix of second derivatives) of the *Lagrangian function*

$$L_{f,\mathbf{F}}: \mathbb{R}^{n+2m} \to \mathbb{R} \quad \text{given by} \quad L_{f,\mathbf{F}} \begin{pmatrix} \mathbf{z} \\ \boldsymbol{\lambda}^\top \end{pmatrix} = f(\mathbf{z}) - \boldsymbol{\lambda}\mathbf{F}(\mathbf{z}). \qquad \text{A14.1}$$

Since $\mathbf{z}_0$ is a critical point of $f|_Z$, it follows (by the Lagrange multiplier theorem, Theorem 3.7.5) that there exists $\boldsymbol{\lambda}_0 \in \text{Mat}(1, m)$ with

$$[\mathbf{D}f(\mathbf{z}_0)] = \boldsymbol{\lambda}_0 [\mathbf{DF}(\mathbf{z}_0)], \qquad \text{A14.2}$$

so $[\mathbf{D}L_{f,\mathbf{F}}]$ vanishes at the point $\begin{pmatrix} \mathbf{z}_0 \\ \boldsymbol{\lambda}_0^\top \end{pmatrix}$. (Note that both $[\mathbf{D}f(\mathbf{z}_0)]$ and $\boldsymbol{\lambda}_0[\mathbf{DF}(\mathbf{z}_0)]$ are row matrices $n+m$ wide.) Thus when $\mathbf{z}_0$ is a critical point of f *constrained to* Z, then $\begin{pmatrix} \mathbf{z}_0 \\ \boldsymbol{\lambda}_0^\top \end{pmatrix}$ is an *unconstrained* critical point of the Lagrangian function $L_{f,\mathbf{F}}$. Since $[\mathbf{DF}(\mathbf{z}_0)]$ is onto, we may relabel the variables so that $\mathbf{z} = \begin{pmatrix} \mathbf{x} \\ \mathbf{y} \end{pmatrix}$, and

$$\begin{bmatrix} \overrightarrow{D_{n+1}}\mathbf{F}(\mathbf{z}_0) & \dots & \overrightarrow{D_{n+m}}\mathbf{F}(\mathbf{z}_0) \end{bmatrix} \quad \text{is invertible.} \qquad \text{A14.3}$$

Perform the change of variables in $\mathbb{R}^{n+m}$ that consists of replacing the y_i by the F_i, i.e., consider the map

$$\Psi: \mathbb{R}^n \times \mathbb{R}^m \to \mathbb{R}^n \times \mathbb{R}^m \quad \text{given by} \quad \Psi\begin{pmatrix} \mathbf{x} \\ \mathbf{y} \end{pmatrix} = \begin{pmatrix} \mathbf{x} \\ \mathbf{F}\begin{pmatrix} \mathbf{x} \\ \mathbf{y} \end{pmatrix} \end{pmatrix}. \qquad \text{A14.4}$$

The derivative of Ψ at $\mathbf{z}_0$ is

$$[\mathbf{D}\Psi(\mathbf{z}_0)] = \begin{bmatrix} I & [0] \\ [D_1\mathbf{F}(\mathbf{z}_0) \ \dots \ D_n\mathbf{F}(\mathbf{z}_0)] & [D_{n+1}\mathbf{F}(\mathbf{z}_0) \ \dots \ D_{n+m}\mathbf{F}(\mathbf{z}_0)] \end{bmatrix}, \qquad \text{A14.5}$$

which is invertible (by a variant of Exercise 2.5, or, more easily, by Theorem 4.8.9), since $[D_{n+1}\mathbf{F}(\mathbf{z}_0) \ \dots \ D_{n+m}\mathbf{F}(\mathbf{z}_0)]$ is invertible.

Write $\mathbf{z}_0$ as $\begin{pmatrix} \mathbf{x}_0 \\ \mathbf{y}_0 \end{pmatrix}$. By the inverse function theorem, there are neighborhoods

$$W \subset U \subset \mathbb{R}^{n+m} \text{ of } \begin{pmatrix} \mathbf{x}_0 \\ \mathbf{y}_0 \end{pmatrix} \quad \text{and} \quad \widetilde{W} \subset \mathbb{R}^{n+m} \text{ of } \begin{pmatrix} \mathbf{x}_0 \\ \mathbf{0} \end{pmatrix} \qquad \text{A14.6}$$

such that $\Psi: W \to \widetilde{W}$ is a C^2 invertible function with C^2 inverse. Let $\Phi: \widetilde{W} \to W$ be its inverse, and denote by $\begin{pmatrix} \mathbf{x} \\ \mathbf{u} \end{pmatrix}$ the variables of $\widetilde{W}$. The point of Φ is to replace something "curvy" by something "flat": the constraint on $f\circ\Phi$ is the flat set $\mathbf{u}=\mathbf{0}$, whereas the original constraint on f, defined by $\mathbf{F}\begin{pmatrix}\mathbf{x}\\ \mathbf{y}\end{pmatrix}=\mathbf{0}$, is curvy; see Figure A14.2. Moreover, $\gamma(\mathbf{x}) \stackrel{\text{def}}{=} \Phi\begin{pmatrix}\mathbf{x}\\ \mathbf{0}\end{pmatrix}$ is a parametrization of the manifold Z, with $\gamma(\mathbf{x}_0)=\mathbf{z}_0$. Define

$$\widetilde{f} \stackrel{\text{def}}{=} f\circ\Phi, \qquad \widetilde{\mathbf{F}} \stackrel{\text{def}}{=} \mathbf{F}\circ\Phi, \tag{A14.7}$$

so that[3]

$$\widetilde{f}\begin{pmatrix}\mathbf{x}\\ \mathbf{0}\end{pmatrix} = f(\gamma(\mathbf{x})), \qquad \widetilde{\mathbf{F}}\begin{pmatrix}\mathbf{x}\\ \mathbf{u}\end{pmatrix}=\mathbf{u}. \tag{A14.8}$$

Then $L_{\widetilde{f},\widetilde{\mathbf{F}}}\begin{pmatrix}\mathbf{x}\\ \mathbf{u}\\ \boldsymbol{\lambda}^\top\end{pmatrix} = \widetilde{f}\begin{pmatrix}\mathbf{x}\\ \mathbf{u}\end{pmatrix} - \boldsymbol{\lambda}\mathbf{u}$. In particular,

$$L_{\widetilde{f},\widetilde{\mathbf{F}}}\begin{pmatrix}\mathbf{x}\\ \mathbf{0}\\ \mathbf{0}\end{pmatrix} = \widetilde{f}\begin{pmatrix}\mathbf{x}\\ \mathbf{0}\end{pmatrix} \underbrace{=}_{\text{eq. A14.8}} f\circ\gamma(\mathbf{x}) \quad \text{and} \tag{A14.9}$$

$$\begin{aligned} L_{\widetilde{f},\widetilde{\mathbf{F}}}\begin{pmatrix}\mathbf{x}\\ \mathbf{u}\\ \boldsymbol{\lambda}^\top\end{pmatrix} &= \widetilde{f}\begin{pmatrix}\mathbf{x}\\ \mathbf{u}\end{pmatrix} - \boldsymbol{\lambda}\mathbf{u} = f\left(\Phi\begin{pmatrix}\mathbf{x}\\ \mathbf{u}\end{pmatrix}\right) - \boldsymbol{\lambda}\mathbf{F}\left(\Phi\begin{pmatrix}\mathbf{x}\\ \mathbf{u}\end{pmatrix}\right)\\ &= L_{f,\mathbf{F}}\begin{pmatrix}\Phi\begin{pmatrix}\mathbf{x}\\ \mathbf{u}\end{pmatrix}\\ \boldsymbol{\lambda}^\top\end{pmatrix}. \end{aligned} \tag{A14.10}$$

(Equation A14.10 uses equations A14.7 and A14.8.) Equation A14.9 tells us that the signature (p,q) of f constrained to X, i.e, the signature of the critical point $\mathbf{x}_0$ of $f\circ\gamma$, is the signature of the quadratic form given by the matrix of second derivatives of $L_{\widetilde{f},\widetilde{\mathbf{F}}}$ *with respect to just the* $\mathbf{x}$*-variables.* Equation A14.10 tells us that $L_{f,\mathbf{F}}$ and $L_{\widetilde{f},\widetilde{\mathbf{F}}}$ are the "same" functions, related by a change of variables. The signature is invariant under changes of variables (see Proposition 3.5.12), so the signature of the critical point $\begin{pmatrix}\mathbf{x}_0\\ \mathbf{0}\\ \boldsymbol{\lambda}_0^\top\end{pmatrix}$ of $L_{\widetilde{f},\widetilde{\mathbf{F}}}$ is the same as the signature of the critical point $\begin{pmatrix}\mathbf{x}_0\\ \mathbf{y}_0\\ \boldsymbol{\lambda}_0^\top\end{pmatrix}$ of $L_{f,\mathbf{F}}$. Thus we will be done if we can prove Proposition A14.1.

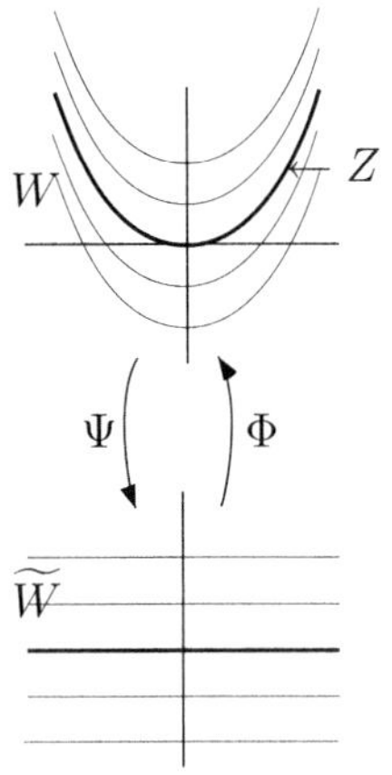

FIGURE A14.2.
Let Z be the dark parabola above:

$$Z=\left\{\begin{pmatrix}x\\ y\end{pmatrix} \mid y=x^2\right\},$$

so that Z is defined by $F=0$, where $F\begin{pmatrix}x\\ y\end{pmatrix}=x^2-y$. Let $\Psi:\mathbb{R}^2\to\mathbb{R}^2$ be defined by

$$\Psi\begin{pmatrix}x\\ y\end{pmatrix}=\begin{pmatrix}x\\ x^2-y\end{pmatrix}.$$

In this case we can skip the inverse function theorem: it is easy to calculate $\Phi\begin{pmatrix}x\\ u\end{pmatrix}=\begin{pmatrix}x\\ x^2-u\end{pmatrix}$, and to see that it maps the straight line $u=0$ to the parabola $y=x^2$.

[3] Second equation in A14.8: Define φ by setting $\Phi\begin{pmatrix}\mathbf{x}\\ \mathbf{u}\end{pmatrix}=\begin{pmatrix}\mathbf{x}\\ \varphi\begin{pmatrix}\mathbf{x}\\ \mathbf{u}\end{pmatrix}\end{pmatrix}$. Then

$$\begin{pmatrix}\mathbf{x}\\ \mathbf{u}\end{pmatrix} = \Psi\left(\Phi\begin{pmatrix}\mathbf{x}\\ \mathbf{u}\end{pmatrix}\right) = \Psi\begin{pmatrix}\mathbf{x}\\ \varphi\begin{pmatrix}\mathbf{x}\\ \mathbf{u}\end{pmatrix}\end{pmatrix} = \begin{pmatrix}\mathbf{x}\\ \mathbf{F}\begin{pmatrix}\mathbf{x}\\ \varphi\begin{pmatrix}\mathbf{x}\\ \mathbf{u}\end{pmatrix}\end{pmatrix}\end{pmatrix}.$$ So

$$\mathbf{u} = \mathbf{F}\begin{pmatrix}\mathbf{x}\\ \varphi\begin{pmatrix}\mathbf{x}\\ \mathbf{u}\end{pmatrix}\end{pmatrix} = \mathbf{F}\circ\Phi\begin{pmatrix}\mathbf{x}\\ \mathbf{u}\end{pmatrix} = \widetilde{\mathbf{F}}\begin{pmatrix}\mathbf{x}\\ \mathbf{u}\end{pmatrix}.$$

Proposition A14.1.

1. *The* $(n+2m)\times(n+2m)$ *Hessian matrix of* $L_{\widetilde{f},\widetilde{\mathbf{F}}}$ *at* $\begin{pmatrix}\mathbf{x}_0\\ \mathbf{0}\\ \boldsymbol{\lambda}_0^\top\end{pmatrix}$ *is*

$$\begin{bmatrix}[D_{\mathbf{x}}D_{\mathbf{x}}\widetilde{L}] & [D_{\mathbf{x}}D_{\mathbf{u}}\widetilde{L}] & [D_{\mathbf{x}}D_{\boldsymbol{\lambda}}\widetilde{L}]\\ [D_{\mathbf{u}}D_{\mathbf{x}}\widetilde{L}] & [D_{\mathbf{u}}D_{\mathbf{u}}\widetilde{L}] & [D_{\mathbf{u}}D_{\boldsymbol{\lambda}}\widetilde{L}]\\ [D_{\boldsymbol{\lambda}}D_{\mathbf{x}}\widetilde{L}] & [D_{\boldsymbol{\lambda}}D_{\mathbf{u}}\widetilde{L}] & [D_{\boldsymbol{\lambda}}D_{\boldsymbol{\lambda}}\widetilde{L}]\end{bmatrix} = \begin{bmatrix}[D_{x_i}D_{x_j}\widetilde{f}] & 0 & 0\\ 0 & [D_{u_k}D_{u_l}\widetilde{f}] & -I\\ 0 & -I & 0\end{bmatrix},$$

where $\widetilde{L} \stackrel{\text{def}}{=} L_{\widetilde{f},\widetilde{\mathbf{F}}} : \mathbb{R}^{n+2m} \to \mathbb{R}$.

2. *The corresponding quadratic form has signature* $(p+m, q+m)$.

Proof of Proposition A14.1: Recall from equation A14.10 that

$$L_{\widetilde{f},\widetilde{\mathbf{F}}}\begin{pmatrix}\mathbf{x}\\ \mathbf{u}\\ \boldsymbol{\lambda}^\top\end{pmatrix} = \widetilde{f}\begin{pmatrix}\mathbf{x}\\ \mathbf{u}\end{pmatrix} - \boldsymbol{\lambda}\mathbf{u}.$$

Proof. 1. The Hessian matrix in Proposition A14.1 has nine submatrices; by symmetry, only six need to be examined:

$[D_{\mathbf{x}}D_{\mathbf{x}}\widetilde{L}]$ We have $[D_{\mathbf{x}}D_{\mathbf{x}}\widetilde{L}] = [D_{x_i}D_{x_j}\widetilde{f}]$, since there are no $\mathbf{x}$-terms in $\boldsymbol{\lambda}\mathbf{u}$.

$[D_{\mathbf{u}}D_{\mathbf{u}}\widetilde{L}]$ The $\mathbf{u}$ variables only appear to degree 1 in $\boldsymbol{\lambda}\mathbf{u}$, so the second derivatives with respect to $\mathbf{u}$ vanish, giving $[D_{\mathbf{u}}D_{\mathbf{u}}\widetilde{L}] = [D_{u_k}D_{u_l}\widetilde{f}]$

$[D_{\mathbf{u}}D_{\boldsymbol{\lambda}}\widetilde{L}]$ The term $\widetilde{f}$ does not contribute, since it contains no $\boldsymbol{\lambda}$ terms; the second derivatives of $-\boldsymbol{\lambda}\mathbf{u}$ form $-I$, where I is the $m\times m$ identity matrix.

$[D_{\boldsymbol{\lambda}}D_{\boldsymbol{\lambda}}\widetilde{L}]$ The second derivatives of $L_{\widetilde{f}}$ with respect to the $\boldsymbol{\lambda}$ variables vanish.

$[D_{\mathbf{x}}D_{\boldsymbol{\lambda}}\widetilde{L}]$ The derivatives first with respect to $\boldsymbol{\lambda}$-variables and then with respect to the $\mathbf{x}$-variables vanish.

$[D_{\mathbf{u}}D_{\mathbf{x}}\widetilde{L}]$ These derivatives vanish because $\mathbf{x} \mapsto \widetilde{f}\begin{pmatrix}\mathbf{x}\\ \mathbf{0}\end{pmatrix}$ has a critical point at $\begin{pmatrix}\mathbf{x}_0\\ \mathbf{0}\end{pmatrix}$, so its Taylor polynomial of degree 2 at $\begin{pmatrix}\mathbf{x}_0\\ \mathbf{0}\end{pmatrix}$ has terms not involving $\mathbf{x}$-variables that will disappear when differentiated with respect to $\mathbf{x}$-variables and *no linear terms* with respect to $\mathbf{x}-\mathbf{x}_0$. Thus after differentiating once with respect to the $\mathbf{x}$-variables there will be a linear factor in $\mathbf{x}-\mathbf{x}_0$ left, which will still be there after differentiating with respect to the $\mathbf{u}$ variables, and which will make the second derivative vanish when evaluated at $\begin{pmatrix}\mathbf{x}_0\\ \mathbf{0}\end{pmatrix}$.

2. Part 1 tells us that in the quadratic form in $n+2m$ variables given by the Hessian matrix in Proposition A14.1, there are no cross terms between the first n variables and the last $2m$. We have seen that the quadratic form corresponding to the first n variables is (p,q), so if the quadratic form with respect to the last $2m$ variables, corresponding to the symmetric matrix

$$\begin{bmatrix}[D_{u_k}D_{u_l}\widetilde{f}] & -I\\ -I & 0\end{bmatrix}, \qquad \text{A14.11}$$

has signature (p',q'), then the signature of the whole matrix is $(p+p', q+q')$. Thus we are left with the following lemma.

Lemma A14.2. *If A is any $m \times m$ matrix, then the signature of the quadratic form on $\mathbb{R}^{2m}$ with matrix* $\begin{bmatrix} A & -I \\ -I & 0 \end{bmatrix}$ *is (m, m).*

Proof of Lemma A14.2. In this case it is easier to think of symmetric matrices than quadratic forms. Consider the quadratic forms Q_t corresponding to the matrices

$$\begin{bmatrix} tA & -I \\ -I & 0 \end{bmatrix}. \qquad A14.12$$

These matrices are all invertible, so their eigenvalues never vanish, and in particular they have the same signs for $t = 0$ and $t = 1$. So it is enough to show that the quadratic form Q_0 has signature (m, m). Denote by $\begin{pmatrix} \mathbf{v} \\ \mathbf{w} \end{pmatrix}$ the elements of $\mathbb{R}^{2m}$. Then Q_0 is the quadratic form

$$-2\vec{\mathbf{v}} \cdot \vec{\mathbf{w}} = \frac{1}{2}|\vec{\mathbf{v}} - \vec{\mathbf{w}}|^2 - \frac{1}{2}|\vec{\mathbf{v}} + \vec{\mathbf{w}}|^2, \qquad A14.13$$

so its signature is indeed (m, m). □ Lemma A14.2

Equation A14.13: The signature is (m, m) not $(1, 1)$ because $\vec{\mathbf{v}}$ and $\vec{\mathbf{w}}$ are in $\mathbb{R}^m$ so $|\vec{\mathbf{v}} - \vec{\mathbf{w}}|^2$ is

$$(v_1 - w_1)^2 + \cdots + (v_m - w_m)^2.$$

and $-|\vec{\mathbf{v}} + \vec{\mathbf{w}}|^2$ is

$$-(v_1 + w_1)^2 - \cdots - (v_m + w_m)^2.$$

The $2m$ linear functions $v_i - w_i$ and $v_i + w_i$ are linearly independent.

This proves Proposition A14.1, which proves Theorem 3.7.13.

Remark. There are many ways to see that the matrices A14.12 are invertible. For instance, by exchanging the last and first m columns, the matrices become $\begin{bmatrix} -I & tA \\ 0 & -I \end{bmatrix}$; by Theorem 4.8.10, the determinant of this matrix is 1; the determinant of the matrix A14.12 is $(-1)^m$. △

Exercise for Section A14

A14.1 a. Work through the proof of Proposition A14.1 in the case where X is the circle of equation $x^2 + y^2 = 1$ and $f\begin{pmatrix} x \\ y \end{pmatrix} = y$, for the critical points $\begin{pmatrix} 0 \\ 1 \end{pmatrix}$ and $\begin{pmatrix} 0 \\ -1 \end{pmatrix}$: i.e., identify $\mathbf{F}$, $\widetilde{\mathbf{F}}$, $\tilde{f}$, $\mathbf{g}$, $\widetilde{\mathbf{g}}$, Ψ, Φ, $L_{f,\mathbf{F}}$, and $L_{\tilde{f},\widetilde{\mathbf{F}}}$ and check that Proposition A14.1 is true.

b. Repeat part a when $f\begin{pmatrix} x \\ y \end{pmatrix} = y - ax^2$, where a is a parameter, for all the critical points.

A15 Geometry of curves and surfaces: proofs

Here we prove three results of Section 3.9: Proposition 3.9.10 on curvature of surfaces and Propositions 3.9.15 and 3.9.16 on the Frenet frame. In Proposition 3.9.10, note that if $c = 0$, then the surface S is already in best coordinates, since then $a_1 = 0$ and $a_2 = 0$ and the tangent plane to the surface at the origin is spanned by two of the standard basis vectors.

Proposition 3.9.10 (Putting a surface into "best" coordinates). *Let $U \subset \mathbb{R}^2$ be open, $f : U \to \mathbb{R}$ a C^2 function, and S the graph of f. Let the Taylor polynomial of f at the origin be*

$$z = f\begin{pmatrix} x \\ y \end{pmatrix} = a_1 x + a_2 y + \frac{1}{2}\left(a_{2,0}x^2 + 2a_{1,1}xy + a_{0,2}y^2\right) + \cdots. \qquad A15.1$$

Set $c \stackrel{\text{def}}{=} \sqrt{a_1^2 + a_2^2}$. If $c \neq 0$, then with respect to the orthonormal basis

$$\vec{\mathbf{v}}_1 \stackrel{\text{def}}{=} \begin{bmatrix} -\frac{a_2}{c} \\ \frac{a_1}{c} \\ 0 \end{bmatrix}, \quad \vec{\mathbf{v}}_2 \stackrel{\text{def}}{=} \begin{bmatrix} \frac{a_1}{c\sqrt{1+c^2}} \\ \frac{a_2}{c\sqrt{1+c^2}} \\ \frac{c}{\sqrt{1+c^2}} \end{bmatrix}, \quad \vec{\mathbf{v}}_3 \stackrel{\text{def}}{=} \begin{bmatrix} \frac{a_1}{\sqrt{1+c^2}} \\ \frac{a_2}{\sqrt{1+c^2}} \\ \frac{-1}{\sqrt{1+c^2}} \end{bmatrix},$$

S is the graph of Z as a function

$$F\begin{pmatrix} X \\ Y \end{pmatrix} = \frac{1}{2}\left(A_{2,0}X^2 + 2A_{1,1}XY + A_{0,2}Y^2\right) + \cdots$$

which starts with quadratic terms, where

$$A_{2,0} \stackrel{\text{def}}{=} \frac{-1}{c^2\sqrt{1+c^2}}\left(a_{2,0}a_2^2 - 2a_{1,1}a_1a_2 + a_{0,2}a_1^2\right)$$

$$A_{1,1} \stackrel{\text{def}}{=} \frac{1}{c^2(1+c^2)}\left(a_1a_2(a_{2,0} - a_{0,2}) + a_{1,1}(a_2^2 - a_1^2)\right)$$

$$A_{0,2} \stackrel{\text{def}}{=} \frac{-1}{c^2(1+c^2)^{3/2}}\left(a_{2,0}a_1^2 + 2a_{1,1}a_1a_2 + a_{0,2}a_2^2\right).$$

Equations A15.2: To express x, y, z in the X, Y, Z coordinates, we compute

$$\begin{bmatrix} x \\ y \\ z \end{bmatrix} = [\vec{\mathbf{v}}_1\, \vec{\mathbf{v}}_2\, \vec{\mathbf{v}}_3] \begin{bmatrix} X \\ Y \\ Z \end{bmatrix};$$

the matrix $[\vec{\mathbf{v}}_1\, \vec{\mathbf{v}}_2\, \vec{\mathbf{v}}_3]$ is the change of basis matrix $[P_{\vec{\mathbf{v}}\to\vec{\mathbf{e}}}]$.

Equation A15.3 expresses implicitly Z in terms of X and Y. Theorem 3.4.7 tells us how to find its Taylor polynomial. That is what we are doing in the remainder of the proof.

Proof. In the X, Y, Z coordinates,

$$x = -\frac{a_2}{c}X + \frac{a_1}{c\sqrt{1+c^2}}Y + \frac{a_1}{\sqrt{1+c^2}}Z \qquad y = \frac{a_1}{c}X + \frac{a_2}{c\sqrt{1+c^2}}Y + \frac{a_2}{\sqrt{1+c^2}}Z$$

$$z = \frac{c}{\sqrt{1+c^2}}Y - \frac{1}{\sqrt{1+c^2}}Z. \qquad A15.2$$

Thus equation A15.1 for S becomes

$$\begin{aligned}
\overbrace{\frac{c}{\sqrt{1+c^2}}Y - \frac{1}{\sqrt{1+c^2}}Z}^{z \text{ from equation A15.2}} &= a_1\Bigg(\overbrace{-\frac{a_2}{c}X + \frac{a_1}{c\sqrt{1+c^2}}Y + \frac{a_1}{\sqrt{1+c^2}}Z}^{x \text{ from equation A15.2}}\Bigg) + a_2\Bigg(\overbrace{\frac{a_1}{c}X + \frac{a_2}{c\sqrt{1+c^2}}Y + \frac{a_2}{\sqrt{1+c^2}}Z}^{y \text{ from equation A15.2}}\Bigg) \\
&+ \frac{a_{2,0}}{2}\left(-\frac{a_2}{c}X + \frac{a_1}{c\sqrt{1+c^2}}Y + \frac{a_1}{\sqrt{1+c^2}}Z\right)^2 \\
&+ \frac{2a_{1,1}}{2}\left(-\frac{a_2}{c}X + \frac{a_1}{c\sqrt{1+c^2}}Y + \frac{a_1}{\sqrt{1+c^2}}Z\right)\left(\frac{a_1}{c}X + \frac{a_2}{c\sqrt{1+c^2}}Y + \frac{a_2}{\sqrt{1+c^2}}Z\right) \\
&+ \frac{a_{0,2}}{2}\left(\frac{a_1}{c}X + \frac{a_2}{c\sqrt{1+c^2}}Y + \frac{a_2}{\sqrt{1+c^2}}Z\right)^2 + \cdots \qquad A15.3
\end{aligned}$$

All the linear terms in X and Y cancel, showing that this is an adapted system. The only remaining linear term is $-\sqrt{1+c^2}Z$ and the coefficient of Z is not 0, so $D_3F \neq 0$, and the implicit function theorem applies. Thus in these coordinates, equation A15.3 expresses Z as a function of X and Y that starts with quadratic terms.

To compute $A_{2,0}$, $A_{1,1}$, and $A_{0,2}$, we need to multiply out the right side of equation A15.3. The linear terms in X and Y cancel. Since we are interested only in terms up to degree 2, we can ignore the terms in Z in the quadratic terms on the right, which contribute terms of degree at least 3 (since Z, as a function of X and Y, starts with quadratic terms). So we can rewrite equation A15.3 as

$$-\sqrt{1+c^2}\,Z = \frac{1}{2}\Bigg(a_{2,0}\left(-\frac{a_2}{c}X + \frac{a_1}{c\sqrt{1+c^2}}Y\right)^2 + 2a_{1,1}\left(-\frac{a_2}{c}X + \frac{a_1}{c\sqrt{1+c^2}}Y\right)\left(\frac{a_1}{c}X + \frac{a_2}{c\sqrt{1+c^2}}Y\right) + a_{0,2}\left(\frac{a_1}{c}X + \frac{a_2}{c\sqrt{1+c^2}}Y\right)^2\Bigg) + \cdots. \qquad \text{A15.4}$$

If we multiply out, collect terms, and divide by $-\sqrt{1+c^2}$, this becomes

$$Z = \frac{1}{2}\Bigg(\overbrace{-\frac{1}{c^2\sqrt{1+c^2}}\left(a_{2,0}a_2^2 - 2a_{1,1}a_1a_2 + a_{0,2}a_1^2\right)}^{A_{2,0}}X^2 + 2\frac{1}{c^2(1+c^2)}\Big(a_1a_2(a_{2,0}-a_{0,2}) + a_{1,1}(a_2^2 - a_1^2)\Big)XY - \frac{1}{c^2(1+c^2)^{3/2}}\left(a_{2,0}a_1^2 + 2a_{1,1}a_1a_2 + a_{0,2}a_2^2\right)Y^2\Bigg) + \cdots. \qquad \text{A15.5}$$

This concludes the proof of Proposition 3.9.10. □

In equation A15.3 we were glad to see that the linear terms in X and Y cancel, showing that we had indeed chosen adapted coordinates. For the linear terms in Y, remember that $c = \sqrt{a_1^2 + a_2^2}$. So the linear terms on the right are

$$\frac{a_1^2 Y}{c\sqrt{1+c^2}} + \frac{a_2^2 Y}{c\sqrt{1+c^2}} = \frac{c^2 Y}{c\sqrt{1+c^2}}$$

and on the left we have

$$\frac{cY}{\sqrt{1+c^2}}.$$

Equation A15.4: Since the expression of Z in terms of X and Y starts with quadratic terms, a term that is linear in Z is actually quadratic in X and Y.

We get $-\sqrt{1+c^2}\,Z$ on the left side of equation A15.4 by collecting all the linear terms in Z from equation A15.3:

$$-\frac{1}{\sqrt{1+c^2}} - \frac{a_1^2}{\sqrt{1+c^2}} - \frac{a_2^2}{\sqrt{1+c^2}};$$

remember that $c = \sqrt{a_1^2 + a_2^2}$.

Equation A15.5 says

$$Z = \frac{1}{2}(A_{2,0}X^2 + 2A_{1,1}XY + A_{0,2}Y^2) + \cdots.$$

Proposition 3.9.15 (Frenet frame). *The coordinates with respect to the Frenet frame at 0,*

$$\vec{\mathbf{t}}(0),\ \vec{\mathbf{n}}(0),\ \vec{\mathbf{b}}(0),$$

are adapted to the curve C at $\mathbf{a} = \delta(0)$.

Proposition 3.9.16 (Frenet frame related to curvature and torsion). *The Frenet frame satisfies the following equations, where κ is the curvature of the curve at $\mathbf{a} = \delta(0)$ and τ is its torsion:*

$$\begin{aligned}
\vec{\mathbf{t}}'(0) &= \qquad\quad \kappa\vec{\mathbf{n}}(0)\\
\vec{\mathbf{n}}'(0) &= -\kappa\vec{\mathbf{t}}(0) \qquad\qquad + \tau\vec{\mathbf{b}}(0)\\
\vec{\mathbf{b}}'(0) &= \qquad\quad -\tau\vec{\mathbf{n}}(0).
\end{aligned}$$

Proof of Propositions 3.9.15 and 3.9.16. We may assume that the curve C is written in its adapted coordinates, i.e., as in equation 3.9.56, which we repeat here:

When equation A15.6 first appeared (as equation 3.9.56) we used dots (...) to denote the terms that can be ignored. Here we denote these terms by $o(X^3)$.

We can divide by $\sqrt{a_2^2+b_2^2}$ because we are assuming that

$$\sqrt{a_2^2+b_2^2} = A_2 \neq 0;$$

otherwise the torsion is not defined. See the margin note next to Definition 3.9.14.

$$Y = \frac{1}{2}\sqrt{a_2^2+b_2^2}X^2 + \frac{a_2a_3+b_2b_3}{6\sqrt{a_2^2+b_2^2}}X^3 = \frac{A_2}{2}X^2 + \frac{A_3}{6}X^3 + o(X^3)$$
$$Z = \frac{-b_2a_3+a_2b_3}{6\sqrt{a_2^2+b_2^2}}X^3 + \cdots = \frac{B_3}{6}X^3 + o(X^3). \qquad \text{A15.6}$$

This means that we know (locally) the parametrization as a graph

$$\delta : X \mapsto \begin{pmatrix} X \\ \frac{A_2}{2}X^2 + \frac{A_3}{6}X^3 + o(X^3) \\ \frac{B_3}{6}X^3 + o(X^3) \end{pmatrix}, \qquad \text{A15.7}$$

whose derivative at X is

$$\delta'(X) = \begin{bmatrix} 1 \\ A_2X + \frac{A_3X^2}{2} + \cdots \\ \frac{B_3X^2}{2} + \cdots \end{bmatrix}. \qquad \text{A15.8}$$

Equation 3.9.25:

$$s(t) = \int_{t_0}^{t} |\vec{\gamma}'(u)|\, du.$$

Parametrizing C by arc length means calculating X as a function of arc length s, or rather calculating the Taylor polynomial of $X(s)$ to degree 3. Equation 3.9.25 (repeated at left) tells us how to compute $s(X)$, the arc length corresponding to X; we will need to invert this to find $X(s)$, the X-coordinate of the point a distance s along the curve.

Lemma A15.1. 1. *The function*

$$s(X) = \int_0^X \underbrace{\sqrt{1 + \left(A_2t + \frac{A_3}{2}t^2\right)^2 + \left(\frac{B_3}{2}t^2\right)^2 + o(t^2)}}_{\text{length of } \delta'(t)}\, dt \qquad \text{A15.9}$$

has the Taylor polynomial

$$s(X) = X + \frac{1}{6}A_2^2X^3 + o(X^3). \qquad \text{A15.10}$$

2. *The inverse function $X(s)$ has the Taylor polynomial*

$$X(s) = s - \frac{1}{6}A_2^2s^3 + o(s^3) \qquad \textit{to degree 3.} \qquad \text{A15.11}$$

Proof of Lemma A15.1. 1. Using the binomial formula (equation 3.4.9), we have, to degree 2,

$$\sqrt{1 + \left(A_2t + \frac{A_3}{2}t^2\right)^2 + \left(\frac{B_3}{2}t^2\right)^2 + o(t^2)} = 1 + \frac{1}{2}A_2^2t^2 + o(t^2) \qquad \text{A15.12}$$

and integrating this gives

$$s(X) = \int_0^X \left(1 + \frac{1}{2}A_2^2t^2 + o(t^2)\right) dt = X + \frac{1}{6}A_2^2X^3 + o(X^3) \qquad \text{A15.13}$$

to degree 3. This proves part 1.

2. The inverse function $X(s)$ has a Taylor polynomial; write it as $X(s) = \alpha s + \beta s^2 + \gamma s^3 + o(s^3)$, and use the equation $s\big(X(s)\big) = s$ and

equation A15.13 to write

$$\begin{aligned} s(X(s)) &= X(s) + \frac{1}{6}A_2^2\big(X(s)\big)^3 + o(s^3) \\ &= \big(\alpha s + \beta s^2 + \gamma s^3 + o(s^3)\big) + \frac{1}{6}A_2^2\big(\alpha s + \beta s^2 + \gamma s^3 + o(s^3)\big)^3 + o(s^3) \\ &= s. \end{aligned} \qquad A15.14$$

Develop the cube and identify the coefficients of like powers to find

$$\alpha = 1, \qquad \beta = 0, \qquad \gamma = -\frac{A_2^2}{6}, \qquad A15.15$$

proving part 2. □ Lemma A15.1

Proof of Propositions 3.9.15 and 3.9.16, continued. Replacing X in equation A15.7 by the value of $X(s)$ given in equation A15.11, we see that up to degree 3, the parametrization of our curve by arc length is given by

The X-component starts with s, the Y-component starts with s^2, and the Z-component starts with s^3; thus the X, Y, Z coordinate system is adapted.

$$\begin{aligned} X(s) &= s - \frac{1}{6}A_2^2 s^3 + o(s^3) \\ Y(s) &= \frac{1}{2}A_2\left(s - \frac{1}{6}A_2^2 s^3\right)^2 + \frac{1}{6}A_3 s^3 + o(s^3) \;=\; \frac{1}{2}A_2 s^2 + \frac{1}{6}A_3 s^3 + o(s^3) \\ Z(s) &= \frac{1}{6}B_3 s^3 + o(s^3). \end{aligned} \qquad A15.16$$

This proves Proposition 3.9.15. Differentiating these functions gives us the velocity vector

$$\vec{\mathbf{t}}(s) = \begin{bmatrix} 1 - \dfrac{A_2^2}{2}s^2 + o(s^2) \\ A_2 s + \dfrac{1}{2}A_3 s^2 + o(s^2) \\ \dfrac{B_3}{2}s^2 + o(s^2) \end{bmatrix} \text{ to degree 2, hence } \vec{\mathbf{t}}(0) = \begin{bmatrix} 1 \\ 0 \\ 0 \end{bmatrix}. \qquad A15.17$$

Now we want to compute $\vec{\mathbf{n}}(s)$. We have

$$\vec{\mathbf{n}}(s) \underbrace{=}_{\text{eq. 3.9.57}} \frac{\vec{\mathbf{t}}'(s)}{|\vec{\mathbf{t}}'(s)|} = \frac{1}{|\vec{\mathbf{t}}'(s)|}\begin{bmatrix} -A_2^2 s + o(s) \\ A_2 + A_3 s + o(s) \\ B_3 s + o(s) \end{bmatrix}. \qquad A15.18$$

We need to evaluate $|\vec{\mathbf{t}}'(s)|$:

$$\begin{aligned} |\vec{\mathbf{t}}'(s)| &= \sqrt{A_2^4 s^2 + A_2^2 + A_3^2 s^2 + 2A_2A_3 s + B_3^2 s^2 + o(s^2)} \\ &= \sqrt{A_2^2 + 2A_2A_3 s + o(s)}. \end{aligned} \qquad A15.19$$

Therefore,

Remember that $A_2 \neq 0$.

$$\begin{aligned} \frac{1}{|\vec{\mathbf{t}}'(s)|} &= \big(A_2^2 + 2A_2A_3 s + o(s)\big)^{-1/2} = \left(A_2^2\left(1 + \frac{2A_3}{A_2}s\right) + o(s)\right)^{-1/2} \\ &= \frac{1}{A_2}\left(1 + \frac{2A_3 s}{A_2}\right)^{-1/2} + o(s). \end{aligned} \qquad A15.20$$

Again using the binomial formula (equation 3.4.9),

$$\frac{1}{|\vec{\mathbf{t}}'(s)|} = \frac{1}{A_2}\left(1 - \frac{1}{2}\left(\frac{2A_3}{A_2}s\right) + o(s)\right) = \frac{1}{A_2} - \frac{A_3}{A_2^2}s + o(s). \qquad A15.21$$

Substituting this value for $1/|\vec{\mathbf{t}}'(s)|$ in equation A15.18 gives

$$\vec{\mathbf{n}}(s) = \begin{bmatrix} \left(\frac{1}{A_2} - \frac{A_3}{A_2^2}s\right)(-A_2^2 s) + o(s) \\ \left(\frac{1}{A_2} - \frac{A_3}{A_2^2}s\right)(A_2 + A_3 s) + o(s) \\ \left(\frac{1}{A_2} - \frac{A_3}{A_2^2}s\right)(B_3 s) + o(s) \end{bmatrix} = \begin{bmatrix} -A_2 s + o(s) \\ 1 + o(s) \\ \frac{B_3}{A_2}s + o(s) \end{bmatrix}. \qquad A15.22$$

Hence $\vec{\mathbf{n}}(0) = \begin{bmatrix} 0 \\ 1 \\ 0 \end{bmatrix}$, and (recall $\vec{\mathbf{t}}(0)$ from equation A15.17)

$$\vec{\mathbf{b}}(0) \underbrace{=}_{\text{eq. 3.9.57}} \vec{\mathbf{t}}(0) \times \vec{\mathbf{n}}(0) = \begin{bmatrix} 1 \\ 0 \\ 0 \end{bmatrix} \times \begin{bmatrix} 0 \\ 1 \\ 0 \end{bmatrix} = \begin{bmatrix} 0 \\ 0 \\ 1 \end{bmatrix}. \qquad A15.23$$

By Definition 3.9.14 of κ and τ,

$$\vec{\mathbf{n}}'(0) \underbrace{=}_{\text{eq. A15.22}} \begin{bmatrix} -A_2 \\ 0 \\ B_3/A_2 \end{bmatrix} = -\kappa\vec{\mathbf{t}}(0) + \tau\vec{\mathbf{b}}(0), \quad \text{and} \quad \vec{\mathbf{t}}'(0) = \kappa\vec{\mathbf{n}}(0). \qquad A15.24$$

All that remains is to prove that $\vec{\mathbf{b}}'(0) = -\tau\vec{\mathbf{n}}(0)$, i.e., $\vec{\mathbf{b}}'(0) = -\frac{B_3}{A_2}\begin{bmatrix} 0 \\ 1 \\ 0 \end{bmatrix}$.

Ignoring terms of degree > 1,

$$\vec{\mathbf{b}}(s) \approx \underbrace{\begin{bmatrix} 1 \\ A_2 s \\ 0 \end{bmatrix} \times \begin{bmatrix} -A_2 s \\ 1 \\ \frac{B_3}{A_2}s \end{bmatrix}}_{\vec{\mathbf{t}}(s)\times\vec{\mathbf{n}}(s)} = \begin{bmatrix} 0 \\ -\frac{B_3}{A_2}s \\ 1 \end{bmatrix}, \text{ so } \vec{\mathbf{b}}'(0) = \begin{bmatrix} 0 \\ -\frac{B_3}{A_2} \\ 0 \end{bmatrix} = -\tau\vec{\mathbf{n}}(0).$$

□ Proposition 3.9.16

Exercise for Section A15

A15.1 Using the notation and the computations in the proof of Proposition 3.9.10, show that the mean curvature is given by the formula

$$H(\mathbf{0}) = \frac{1}{2(1+c^2)^{3/2}}\Big(a_{2,0}(1+a_2^2) - 2a_1 a_2 a_{1,1} + a_{0,2}(1+a_1^2)\Big),$$

which we already saw as equation 3.9.38.

A16 STIRLING'S FORMULA AND THE CENTRAL LIMIT THEOREM

To see why the central limit theorem is true, we need to understand how the factorial $n!$ behaves as n becomes large. How big is 100!? How many digits does it have? *Stirling's formula* gives a very useful approximation.

Proposition A16.1 (Stirling's formula). *The number $n!$ is approximately*

$$n! \approx \sqrt{2\pi}\left(\frac{n}{e}\right)^n \sqrt{n}, \qquad \text{A16.1}$$

in the sense that the ratio of the two sides tends to 1 as n tends to ∞.

Stirling's formula was actually due to the English mathematician Abraham de Moivre (1667–1754), not the Scottish mathematician James Stirling (1692–1770), but de Moivre credited Stirling with improving it.

For instance,

$$\sqrt{2\pi}\,(100/e)^{100}\sqrt{100} \approx 9.3248 \cdot 10^{157} \text{ and } 100! \approx 9.3326 \cdot 10^{157}, \qquad \text{A16.2}$$

for a ratio of about 1.0008.

Proof. Define the number R_n by the formula

$$\ln n! = \underbrace{\ln 1 + \ln 2 + \cdots + \ln n}_{\text{midpoint Riemann sum}} = \int_{1/2}^{n+1/2} \ln x\, dx + R_n \qquad \text{A16.3}$$

(as illustrated by Figure A16.1, the left side is a midpoint Riemann sum, where each subrectangle has width 1). This formula is justified by the following computation, which shows that the R_n form a convergent sequence:

$$|R_n - R_{n-1}| = \left|\ln n - \int_{n-1/2}^{n+1/2} \ln x\, dx\right| = \left|\int_{-1/2}^{1/2} \ln\left(1+\frac{t}{n}\right) dt\right| \qquad \text{A16.4}$$

$$= \left|\int_{-1/2}^{1/2}\left(\ln\left(1+\frac{t}{n}\right) - \frac{t}{n}\right) dt\right| \le \left|-2\int_{-1/2}^{1/2}\left(\frac{t}{n}\right)^2 dt\right| = \frac{1}{6n^2},$$

Equation A16.4: The second equality comes from setting $x = n+t$ and writing

$$x = n + t = n\left(1+\frac{t}{n}\right),$$

so

$$\ln x = \ln\Big(n(1+t/n)\Big) = \ln n + \ln(1+t/n)$$

and

$$\int_{-1/2}^{1/2} \ln n\, dt = \ln n.$$

To get the next equality we subtract

$$\int_{-1/2}^{1/2} \frac{t}{n}\, dt = 0.$$

The inequality is Taylor's theorem with remainder:

$$\ln(1+h) = h + \frac{1}{2}\left(-\frac{1}{(1+c)^2}\right)h^2$$

for some c with $|c| < |h|$; in our case, $h = t/n$ with $t \in [-1/2, 1/2]$ and $c = -1/2$ is the worst value. (See Theorem A12.2, applied to $f(a+h) = \ln(1+\frac{t}{n})$ and $k = 1$.)

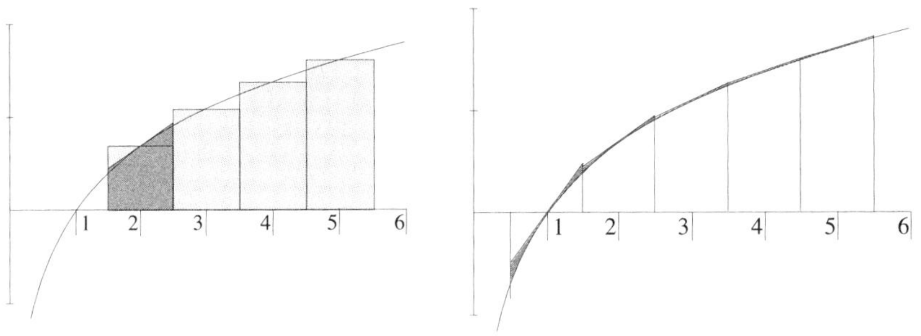

FIGURE A16.1. LEFT: The sum $\ln 1 + \ln 2 + \cdots + \ln n$ is a midpoint Riemann sum for the integral $\int_{1/2}^{n+1/2} \ln x\, dx$. The kth rectangle has the same area as the trapezoid whose top edge is tangent to the graph of $\ln x$ at $\ln n$, as illustrated when $k = 2$. RIGHT: The difference between the areas of the trapezoids and the area under the graph of the logarithm is the shaded region. It has finite total area, as shown in equation A16.3.

so the series formed by the $R_n - R_{n-1}$ is convergent, and the sequence converges to some limit R. Thus we can rewrite equation A16.3 as follows:

$$\ln n! = \left(\int_{1/2}^{n+1/2} \ln x \, dx\right) + R + \epsilon_1(n)$$

$$= \left[x \ln x - x\right]_{1/2}^{n+1/2} + R + \epsilon_1(n) \qquad A16.5$$

$$= \left(\left(n + \frac{1}{2}\right)\ln\left(n + \frac{1}{2}\right) - \left(n + \frac{1}{2}\right)\right) - \left(\frac{1}{2}\ln\frac{1}{2} - \frac{1}{2}\right) + R + \epsilon_1(n),$$

where $\epsilon_1(n)$ tends to 0 as n tends to ∞. Now notice that

$$\left(n + \frac{1}{2}\right)\ln\left(n + \frac{1}{2}\right) = \left(n + \frac{1}{2}\right)\ln n + \frac{1}{2} + \epsilon_2(n), \qquad A16.6$$

where $\epsilon_2(n)$ includes all the terms that tend to 0 as $n \to \infty$. Putting all this together, we see that there is a constant

$$c = R - \left(\frac{1}{2}\ln\frac{1}{2} - \frac{1}{2}\right) \quad \text{such that} \qquad A16.7$$

$$\ln n! = n \ln n + \frac{1}{2}\ln n - n + c + \overbrace{\epsilon(n)}^{\epsilon_1(n) + \epsilon_2(n)}, \qquad A16.8$$

where $\epsilon(n) \to 0$ as $n \to \infty$. Exponentiating this and setting $C = e^c$ gives

$$n! = e^c n^n e^{-n} \sqrt{n} \underbrace{e^{\epsilon(n)}}_{\to 1 \text{ as } n \to \infty} = C n^n e^{-n} \sqrt{n} \underbrace{e^{\epsilon(n)}}_{\to 1 \text{ as } n \to \infty}. \qquad A16.9$$

Except for the determination of C, this is exactly Stirling's formula. There isn't any obvious reason why it should be possible to evaluate C exactly, but we will see in equation A16.21 that $C = \sqrt{2\pi}$. □

We now prove the following version of the central limit theorem:

Theorem A16.2 (Central limit theorem). *If a fair coin is tossed $2n$ times, the probability that the number of heads is between $n + a\sqrt{n}$ and $n + b\sqrt{n}$ tends to*

$$\frac{1}{\sqrt{\pi}} \int_a^b e^{-t^2} dt \quad \text{as } n \text{ tends to } \infty. \qquad A16.10$$

Proof. The probability of having between $n + a\sqrt{n}$ and $n + b\sqrt{n}$ heads is

$$\frac{1}{2^{2n}} \sum_{k=a\sqrt{n}}^{b\sqrt{n}} \binom{2n}{n+k} = \frac{1}{2^{2n}} \sum_{k=a\sqrt{n}}^{b\sqrt{n}} \frac{(2n)!}{(n+k)!(n-k)!}. \qquad A16.11$$

The idea is to use Stirling's formula to rewrite the sum on the right, cancel everything we can, and see that what is left is a Riemann sum for the integral in formula A16.10 (more precisely, $1/\sqrt{\pi}$ times that Riemann sum).

Equation A16.6 comes from:

$$\ln\left(n + \frac{1}{2}\right) = \ln\left(n\left(1 + \frac{1}{2n}\right)\right)$$

$$= \ln n + \ln\left(1 + \frac{1}{2n}\right)$$

$$= \ln n + \frac{1}{2n} + O\left(\frac{1}{n^2}\right).$$

(The last equality uses equation 3.4.8 for the Taylor polynomial of $\ln$.)

The epsilons $\epsilon_1(n)$ and $\epsilon_2(n)$ are unrelated, but both go to 0 as $n \to \infty$, as does

$$\epsilon(n) = \epsilon_1(n) + \epsilon_2(n).$$

Exercise A16.3 asks you to justify that the left side of equation A16.11 is the desired probability.

We use the version of Stirling's formula given in equation A16.9, using C rather than $\sqrt{2\pi}$, since we have not yet proved that $C = \sqrt{2\pi}$.

Let us begin by writing $k = t\sqrt{n}$, so that the sum is over those values of t between a and b such that $t\sqrt{n}$ is an integer; we will denote this set by $T_{[a,b]}$. These points are regularly spaced, $1/\sqrt{n}$ apart, between a and b, and hence are good candidates for the points at which to evaluate a function when forming a Riemann sum. With this notation, our sum becomes

$$\frac{1}{2^{2n}} \sum_{t\in T_{[a,b]}} \frac{(2n)!}{(n+t\sqrt{n})!(n-t\sqrt{n})!} \qquad A16.12$$

$$\approx \frac{1}{2^{2n}} \sum_{t\in T_{[a,b]}} \frac{C(2n)^{2n}e^{-2n}\sqrt{2n}}{\left(C(n+t\sqrt{n})^{(n+t\sqrt{n})}e^{-(n+t\sqrt{n})}\sqrt{n+t\sqrt{n}}\right)\left(C(n-t\sqrt{n})^{(n-t\sqrt{n})}e^{-(n-t\sqrt{n})}\sqrt{n-t\sqrt{n}}\right)}.$$

Now for some of the cancellations: $(2n)^{2n} = 2^{2n}n^{2n}$, and the powers of 2 cancel with the fraction in front of the sum. All the exponential terms cancel, since $e^{-(n+t\sqrt{n})}e^{-(n-t\sqrt{n})} = e^{-2n}$. Also, one power of C cancels. This leaves

$$\cdots = \frac{1}{C} \sum_{t\in T_{[a,b]}} \frac{n^{2n}\sqrt{2n}}{\sqrt{n^2-t^2n}\;(n+t\sqrt{n})^{(n+t\sqrt{n})}(n-t\sqrt{n})^{(n-t\sqrt{n})}}. \qquad A16.13$$

Next, write $(n+t\sqrt{n})^{(n+t\sqrt{n})} = n^{(n+t\sqrt{n})}(1+t/\sqrt{n})^{(n+t\sqrt{n})}$, and similarly for the term in $n - t\sqrt{n}$. Note that the powers of n cancel with the n^{2n} in the numerator, to find

$$\cdots = \frac{1}{C} \sum_{t\in T_{[a,b]}} \underbrace{\sqrt{\frac{2n}{n^2-t^2n}}}_{\text{base}} \underbrace{\frac{1}{(1+t/\sqrt{n})^{(n+t\sqrt{n})}(1-t/\sqrt{n})^{(n-t\sqrt{n})}}}_{\text{height of rectangles for Riemann sum}}. \qquad A16.14$$

We denote by Δt the spacing of the points t (i.e., $1/\sqrt{n}$).

As $n \to \infty$, the term under the square root converges to $\sqrt{2/n} = \sqrt{2}\Delta t$, so it is the length of the base of the rectangles we need for our Riemann sum. For the other term, remember that

$$\lim_{x\to\infty}\left(1+\frac{a}{x}\right)^x = e^a. \qquad A16.15$$

We use equation A16.15 repeatedly in the following calculation:

$$\frac{1}{(1+t/\sqrt{n})^{(n+t\sqrt{n})}(1-t/\sqrt{n})^{(n-t\sqrt{n})}}$$
$$= \frac{1}{(1+t/\sqrt{n})^n(1+t/\sqrt{n})^{t\sqrt{n}}(1-t/\sqrt{n})^n(1-t/\sqrt{n})^{-t\sqrt{n}}} \qquad A16.16$$
$$= \frac{1}{(1-t^2/n)^n}\frac{(1-\frac{t^2}{t\sqrt{n}})^{t\sqrt{n}}}{(1+\frac{t^2}{t\sqrt{n}})^{t\sqrt{n}}} \to \frac{1}{e^{-t^2}}\frac{e^{-t^2}}{e^{t^2}} = e^{-t^2}.$$

Third line of equation A16.16: The denominator of the first term tends to e^{-t^2} as $n \to \infty$, by equation A16.15. By the same equation, the numerator of the second term tends to e^{-t^2}, and the denominator of the second term tends to e^{t^2}.

Putting this together, we see that

$$\frac{1}{C} \sum_{t\in T_{[a,b]}} \sqrt{\frac{2n}{n^2-t^2n}} \frac{1}{(1+t/\sqrt{n})^{(n+t\sqrt{n})}(1-t/\sqrt{n})^{(n-t\sqrt{n})}} \qquad A16.17$$

from equation A16.14 converges to

$$\frac{\sqrt{2}}{C}\frac{1}{\sqrt{n}}\sum_{t\in T_{[a,b]}} e^{-t^2}, \tag{A16.18}$$

which is the desired Riemann sum. Thus as $n \to \infty$,

$$\frac{1}{2^{2n}}\sum_{k=a\sqrt{n}}^{b\sqrt{n}}\binom{2n}{n+k} \to \frac{\sqrt{2}}{C}\int_a^b e^{-t^2}\,dt. \tag{A16.19}$$

We finally need to invoke a fact justified in Section 4.11 (equation 4.11.72):

$$\int_{-\infty}^{\infty} e^{-t^2}\,dt = \sqrt{\pi}. \tag{A16.20}$$

When we set $a = -\infty$ and $b = +\infty$ in the integral in equation A16.19, we must have

Exercise A16.1 gives another way to derive that $C = \sqrt{2\pi}$.

$$\frac{\sqrt{2}}{C}\sqrt{\pi} = \frac{\sqrt{2}}{C}\int_{-\infty}^{\infty} e^{-t^2}\,dt = 1, \tag{A16.21}$$

since the probability of a certain event is 1. So $C = \sqrt{2\pi}$, and finally

$$\underbrace{\frac{1}{2^{2n}}\sum_{k=a\sqrt{n}}^{b\sqrt{n}}\binom{2n}{n+k}}_{\substack{\text{prob.of having between}\\ n+a\sqrt{n}\text{ and }n+b\sqrt{n}\text{ heads}}} \quad \text{converges to} \quad \frac{1}{\sqrt{\pi}}\int_a^b e^{-t^2}\,dt. \quad \square \tag{A16.22}$$

Exercises for Section A16

A16.1 Show that if there is a constant C such that $n! = C\sqrt{n}\left(\frac{n}{e}\right)^n(1+o(1))$, as is proved in Proposition A16.1, then $C = \sqrt{2\pi}$. The argument is fairly elementary, but not at all obvious. Let $c_n = \int_0^\pi \sin^n x\,dx$.

Exercise A16.1 sketches another way to find the constant in Stirling's formula. Hint for part b: Write

$$\sin^n x = \sin x \sin^{n-1} x$$

and integrate by parts.

a. Show that $c_n < c_{n-1}$ for all $n = 1, 2, \dots$.

b. Show that for $n \geq 2$, we have $c_n = \frac{n-1}{n}c_{n-2}$.

c. Show that $c_0 = \pi$ and $c_1 = 2$, and use this and part b to show that

$$c_{2n} = \frac{2n-1}{2n}\cdot\frac{2n-3}{2n-2}\cdots\frac{1}{2}\pi = \frac{(2n)!\pi}{2^{2n}(n!)^2}$$

$$c_{2n+1} = \frac{2n}{2n+1}\cdot\frac{2n-2}{2n-1}\cdots\frac{2}{3}2 = \frac{2^{2n}(n!)^2 2}{(2n+1)!}.$$

d. Use Stirling's formula with constant C to show that

$$c_{2n} = \frac{1}{C}\sqrt{\frac{2}{n}}\,\pi\Big(1+o(1)\Big) \quad \text{and} \quad c_{2n+1} = \frac{C}{\sqrt{2n+1}}\Big(1+o(1)\Big).$$

Now use part a to show that $C^2 \leq 2\pi + o(1)$ and $C^2 \geq 2\pi + o(1)$.

A16.2 Show that the version of the central limit theorem given here (Theorem A16.2) is a special case of the version given in Chapter 4 (Theorem 4.2.7).

A16.3 Justify that the left side of equation A16.11 is the desired probability.

A17 Proving Fubini's theorem

We will prove a stronger theorem (Theorem A17.2) than the version given in Theorem 4.5.10: the assumption "that for each $\mathbf{x} \in \mathbb{R}^n$, the function $\mathbf{y} \mapsto f(\mathbf{x}, \mathbf{y})$ is integrable" is not really necessary. But we need to be careful; it is not quite true that just because f is integrable, the function $\mathbf{y} \mapsto f(\mathbf{x}, \mathbf{y})$ is integrable, and we can't simply remove that hypothesis. The following example illustrates the difficulty.

By "rough statement" we mean equation 4.5.1:

$$\int_{\mathbb{R}^n} f\,|d^n\mathbf{x}| = \int_{-\infty}^{\infty}\left(\dots\left(\int_{-\infty}^{\infty} f\begin{bmatrix} x_1 \\ \vdots \\ x_n \end{bmatrix} dx_1\right)\dots\right)dx_n.$$

Example A17.1 (A case where the rough statement of Fubini's theorem does not work). Consider the function $f:\mathbb{R}^2 \to \mathbb{R}$ that equals 0 outside the unit square, and 1 both inside the square and on its boundary, *except* for the part of boundary where $x = 1$. On that boundary, $f = 1$ when y is rational, and $f = 0$ when y is irrational:

$$f\begin{pmatrix} x \\ y \end{pmatrix} = \begin{cases} 1 & \text{if } 0 \le x < 1 \text{ and } 0 \le y \le 1 \\ 1 & \text{if } x = 1 \text{ and } y \text{ is rational} \\ 0 & \text{otherwise.} \end{cases} \qquad A17.1$$

Following the procedure of Section 4.5, we write the double integral

$$\iint_{\mathbb{R}^2} f\begin{pmatrix} x \\ y \end{pmatrix} dx\,dy = \int_0^1 \left(\int_0^1 f\begin{pmatrix} x \\ y \end{pmatrix} dy\right) dx. \qquad A17.2$$

However, the inner integral $F(x) = \int_0^1 f\begin{pmatrix} x \\ y \end{pmatrix} dy$ does not make sense when $x = 1$. Our function f is integrable on $\mathbb{R}^2$, but $f\begin{pmatrix} 1 \\ y \end{pmatrix}$ is not an integrable function of y. $\triangle$

The function F could be undefined on a much more complicated set than a single point, but this set will necessarily have volume 0, so it does not affect the integral $\int_{\mathbb{R}} F(x)\,dx$.

Fortunately, this is not a problem. Since a point has one-dimensional volume 0, hence measure 0, you could define $F(1)$ to be anything you want, without affecting the integral $\int_0^1 F(x)\,dx$. This always happens: if $f : \mathbb{R}^{n+m} \to \mathbb{R}$ is integrable, then $\mathbf{y} \mapsto f(\mathbf{x}, \mathbf{y})$ is always integrable except for a set of $\mathbf{x}$ of measure 0, which doesn't matter. We deal with this problem by using upper integrals and lower integrals for the inner integral.

For example, if we have an integrable function $f\begin{pmatrix} x_1 \\ x_2 \\ y \end{pmatrix}$, we can think of it as a function on $\mathbb{R}^2 \times \mathbb{R}$, where we consider x_1 and x_2 as the horizontal variables and y as the vertical variable.

Suppose $\mathbf{x} \in \mathbb{R}^n$ denotes the first n variables of the domain of a function $f : \mathbb{R}^{n+m} \to \mathbb{R}$, and $\mathbf{y} \in \mathbb{R}^m$ denotes the last m variables. We will think of the $\mathbf{x}$ variables as "horizontal" and the $\mathbf{y}$ variables as "vertical". We denote by $f_{\mathbf{x}}$ the restriction of f to the vertical subset where the horizontal coordinate is fixed to be $\mathbf{x}$. We denote by $f^{\mathbf{y}}$ the restriction of the function to the horizontal subset where the vertical coordinate is fixed at $\mathbf{y}$. Then $f(\mathbf{x}, \mathbf{y})$ can also be written as $f^{\mathbf{y}}(\mathbf{x})$ or as $f_{\mathbf{x}}(\mathbf{y})$:

$$f_{\mathbf{x}}(\mathbf{y}) = f^{\mathbf{y}}(\mathbf{x}) = f(\mathbf{x}, \mathbf{y}). \qquad A17.3$$

As we saw in Example A17.1, it is not true that if f is integrable, then $f_{\mathbf{x}}$ and $f^{\mathbf{y}}$ are also integrable for every $\mathbf{x}$ and $\mathbf{y}$. But the following *is* true:

Theorem A17.2 (Fubini's theorem). *Let $f : \mathbb{R}^n \times \mathbb{R}^m \to \mathbb{R}$ be an integrable function. Then the functions $U(f_{\mathbf{x}})$, $L(f_{\mathbf{x}})$, $U(f^{\mathbf{y}})$, $L(f^{\mathbf{y}})$ are all integrable, and*

$$\int_{\mathbb{R}^n} U(f_{\mathbf{x}})\,|d^n\mathbf{x}| = \int_{\mathbb{R}^n} L(f_{\mathbf{x}})\,|d^n\mathbf{x}| = \int_{\mathbb{R}^m} U(f^{\mathbf{y}})\,|d^m\mathbf{y}| = \int_{\mathbb{R}^m} L(f^{\mathbf{y}})\,|d^m\mathbf{y}| = \int_{\mathbb{R}^n\times\mathbb{R}^m} f\,|d^n\mathbf{x}|\,|d^m\mathbf{y}|. \qquad \text{A17.4}$$

Equation A17.4 says that adding the upper sums for all columns gives the same result as adding lower sums for all columns, as the decomposition becomes infinitely fine. The same is true for upper and lower sums for all rows. All these sums tend to the integral.

Corollary A17.3. *The set of $\mathbf{x}$ such that $U(f_{\mathbf{x}}) \neq L(f_{\mathbf{x}})$ and the set of $\mathbf{y}$ such that $U(f^{\mathbf{y}}) \neq L(f^{\mathbf{y}})$ both have measure 0. Thus the set of $\mathbf{x}$ such that $f_{\mathbf{x}}$ is not integrable has n-dimensional measure 0, and the set of $\mathbf{y}$ where $f^{\mathbf{y}}$ is not integrable has m-dimensional measure 0.*

Proof of Corollary A17.3. If these measures were not 0, the first and third equalities of equation A17.4 would not be true. (See Corollary 4.4.11.) □

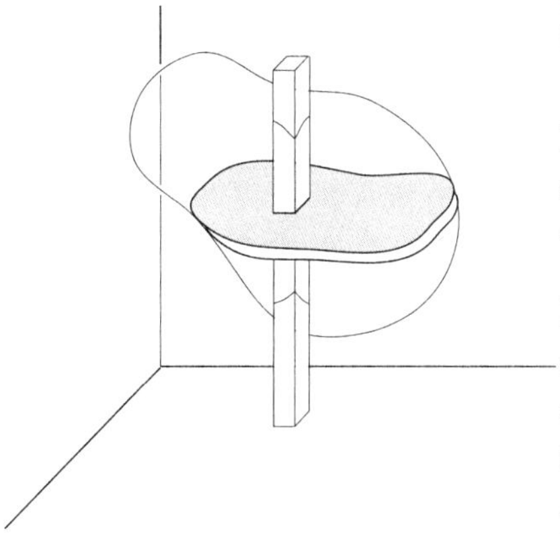

FIGURE A17.1.

Here we imagine that x and y are horizontal variables and z is vertical. Fixing a value of the horizontal variable picks out a French fry; choosing a value of z picks out a flat potato chip. Integrating over all the French fries and adding them up gives the same result as integrating over all the potato chips and adding them.

Proof of Theorem A17.2. Consider a double integral over some bounded domain in $\mathbb{R}^2$. For every N, we sum over all the squares of some dyadic decomposition of the plane. The squares can be taken in any order, since only finitely many contribute a nonzero term (because the domain is bounded). Adding together the entries of each column and then adding the totals is like integrating $f_{\mathbf{x}}$; adding together the entries of each row and then adding the totals together is like integrating $f^{\mathbf{y}}$, as illustrated in Table A17.2.

1		5			$1+5=$ 6
2		6			$2+6=$ 8
3		7		gives the same result as	$3+7=$ 10
+4		+8			$4+8=$ 12
10	+	26	= 36		36

TABLE A17.2. LEFT: Like integrating $f_{\mathbf{x}}$. RIGHT: Like integrating $f^{\mathbf{y}}$.

Putting this in practice requires a little attention to limits. The inequality that makes things work is that for any $N' \geq N$, we have

$$U_N(f) \geq U_N\big(U_{N'}(f_{\mathbf{x}})\big) \qquad \text{A17.5}$$

(see Proposition 4.1.9 and note that we have f on the left and $f_{\mathbf{x}}$ on the right). The expression $U_N(U_{N'}(f_{\mathbf{x}}))$ may seem strange, but $U_{N'}(f_{\mathbf{x}})$ is just a function of $\mathbf{x}$, bounded with bounded support, so we can take Nth upper sums of it. Indeed,

$$U_N(f) = \sum_{C \in \mathcal{D}_N(\mathbb{R}^n \times \mathbb{R}^m)} M_C(f) \operatorname{vol}_{n+m} C$$

$$\geq \sum_{C_1 \in \mathcal{D}_N(\mathbb{R}^n)} \sum_{C_2 \in \mathcal{D}_{N'}(\mathbb{R}^m)} M_{C_1 \times C_2}(f) \operatorname{vol}_n C_1 \operatorname{vol}_m C_2 \qquad \text{A17.6}$$

$$= \underbrace{\sum_{C_1 \in \mathcal{D}_N(\mathbb{R}^n)} M_{C_1} \Bigg(\underbrace{\sum_{C_2 \in \mathcal{D}_{N'}(\mathbb{R}^m)} M_{C_2}(f_{\mathbf{x}}) \operatorname{vol}_m C_2}_{U_{N'}(f_{\mathbf{x}})} \Bigg) \operatorname{vol}_n C_1}_{U_N U_{N'}(f_x)}.$$

Inequality A17.6: The first line is the definition of an upper sum. To go from the first to the second line, note that the decomposition of $\mathbb{R}^n \times \mathbb{R}^m$ into $C_1 \times C_2$ with $C_1 \in \mathcal{D}_N(\mathbb{R}^n)$ and $C_2 \in \mathcal{D}_{N'}(\mathbb{R}^m)$ is finer than $\mathcal{D}_N(\mathbb{R}^{n+m})$.

The third line: for each C_1 in $\mathcal{D}_N(\mathbb{R}^n)$ we choose a point $\mathbf{x} \in C_1$, and for each $C_2 \in \mathcal{D}_{N'}(\mathbb{R}^m)$ we find the $\mathbf{y} \in C_2$ such that $f(\mathbf{x}, \mathbf{y})$ is maximal, and add these maxima. These maxima are restricted to all have the same $\mathbf{x}$-coordinate, so they are at most $M_{C_1 \times C_2} f$, and even if we now maximize over all $\mathbf{x} \in C_1$, we will still find less than if we had added the maxima independently; equality will occur only if all the maxima are above each other (i.e., all have the same $\mathbf{x}$-coordinate).

An analogous argument about lower sums gives, for $N' \geq N$,

$$U_N(f) \geq U_N\big(U_{N'}(f_{\mathbf{x}})\big) \geq L_N\big(L_{N'}(f_{\mathbf{x}})\big) \geq L_N(f). \qquad \text{A17.7}$$

Since f is integrable, we can make $U_N(f)$ and $L_N(f)$ arbitrarily close, by choosing N sufficiently large; we will squeeze the two ends of equation A17.7 together, squeezing everything inside in the process.

The limits as $N' \to \infty$ of $U_{N'}(f_{\mathbf{x}})$ and $L_{N'}(f_{\mathbf{x}})$ are, by Definition 4.1.12, the upper and lower integrals $U(f_{\mathbf{x}})$ and $L(f_{\mathbf{x}})$, so we can rewrite inequality A17.7:

$$U_N(f) \geq U_N\big(U(f_{\mathbf{x}})\big) \geq L_N\big(L(f_{\mathbf{x}})\big) \geq L_N(f). \qquad \text{A17.8}$$

Given a function f, we have $U(f) \geq L(f)$; in addition, if $f \geq g$, then $U_N(f) \geq U_N(g)$. So $U_N\big(L(f_{\mathbf{x}})\big)$ and $L_N\big(U(f_{\mathbf{x}})\big)$ are between the inner values of inequality A17.8:

$$\begin{aligned} U_N\big(U(f_{\mathbf{x}})\big) &\geq U_N\big(L(f_{\mathbf{x}})\big) \geq L_N\big(L(f_{\mathbf{x}})\big) \\ U_N\big(U(f_{\mathbf{x}})\big) &\geq L_N\big(U(f_{\mathbf{x}})\big) \geq L_N\big(L(f_{\mathbf{x}})\big). \end{aligned} \qquad \text{A17.9}$$

We don't know which is bigger, $U_N(L(f_{\mathbf{x}}))$ or $L_N(U(f_{\mathbf{x}}))$, but that doesn't matter. We know they are between the first and last terms of inequality A17.8, which themselves have a common limit as $N \to \infty$.

So $U_N\big(L(f_{\mathbf{x}})\big)$ and $L_N\big(L(f_{\mathbf{x}})\big)$ have a common limit, as do $U_N\big(U(f_{\mathbf{x}})\big)$ and $L_N\big(U(f_{\mathbf{x}})\big)$, showing that $L(f_{\mathbf{x}})$ and $U(f_{\mathbf{x}})$ are integrable, and their integrals are equal, since both are equal to $\int_{\mathbb{R}^n \times \mathbb{R}^m} f$.

The argument about the functions $f^{\mathbf{y}}$ is similar. □ Theorem A17.2

A18 Justifying the use of other pavings

In this section we prove Theorem 4.7.4, which justifies using pavings other than dyadic cubes when integrating.

Theorem 4.7.4 (Integrals using arbitrary pavings). *Let $X \subset \mathbb{R}^n$ be a bounded subset, and $\mathcal{P}_N$ a nested partition of X.*

If $f : \mathbb{R}^n \to \mathbb{R}$ is integrable, then the limits

$$\lim_{N\to\infty} U_{\mathcal{P}_N}(f\mathbf{1}_X) \quad \text{and} \quad \lim_{N\to\infty} L_{\mathcal{P}_N}(f\mathbf{1}_X). \qquad A18.1$$

exist and are both equal to

$$\int_X f(\mathbf{x})\,|d^n\mathbf{x}| = \int_{\mathbb{R}^n} f(\mathbf{x})\mathbf{1}_X(\mathbf{x})\,|d^n\mathbf{x}|.$$

2. Conversely, if the limits in A18.1 are equal, $f\mathbf{1}_X$ is integrable and

$$\int_{\mathbb{R}^n} f(\mathbf{x})\mathbf{1}_X(\mathbf{x})\,|d^n\mathbf{x}|$$

is equal to the common limit.

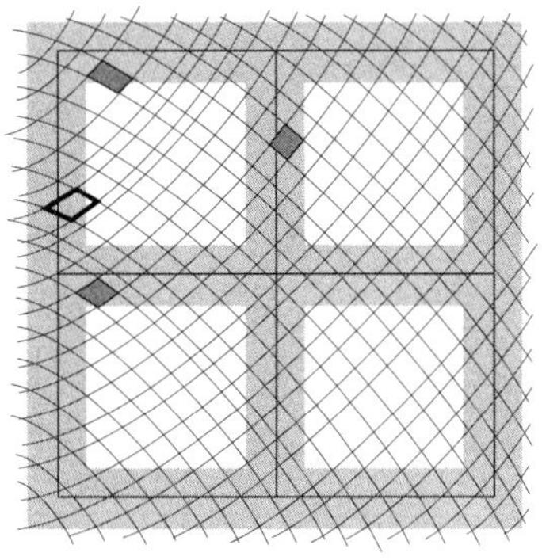

FIGURE A18.1.

This picture illustrates the choice of N, N', and N''. The dyadic decomposition $\mathcal{D}_N$ is represented by the four large squares outlined in black. The set B (pale grey) is the thickening of $\partial\mathcal{D}_N$ made up of the cubes of $\mathcal{D}_{N'}$ that touch $\partial\mathcal{D}_N$. The tiles of $\mathcal{P}_{N''}$ are supposed to be so small that they are either completely contained in B or completely contained in some $C \in \mathcal{D}_N$. Here we didn't quite take N'' big enough: the darkly shaded tiles are okay, but the tile with emphasized boundary isn't.

Proof. 1. The boundary ∂X has measure 0 since for any N it is contained in the union of the boundaries of the pieces of $\mathcal{P}_N$, and this is a finite union of sets of volume 0. The indicator function $\mathbf{1}_X$ is integrable (why?[4]). By replacing f by $\mathbf{1}_X f$, we may suppose that the support of f is in X. We need to prove that for any ϵ, we can find M such that

$$U_{\mathcal{P}_M}(f) - L_{\mathcal{P}_M}(f) < \epsilon. \qquad A18.2$$

Since we know that the analogous statement for dyadic pavings is true, the idea is to use "other pavings" small enough so that each paving piece P either is entirely inside a dyadic cube, or (if it touches or intersects a boundary between dyadic cubes) contributes a negligible amount to the upper and lower sums. The proof is illustrated by Figure A18.1.

First, using the fact that f is integrable, find N such that the difference between upper and lower sums of dyadic decompositions is less than $\epsilon/2$:

Replacing f by $\mathbf{1}_X f$ uses the fact that the product of two R-integrable functions is integrable. This is proved in Corollary 4.4.12.

$$U_N(f) - L_N(f) < \frac{\epsilon}{2}. \qquad A18.3$$

Next, find $N' > N$ such that if B is the union of the cubes $C \in \mathcal{D}_{N'}$ whose closures intersect $\partial\mathcal{D}_N$, then

Why the 8 in the denominator in inequality A18.4? Because it will give us the result we want; the ends justify the means.

$$\mathrm{vol}_n\, B \le \frac{\epsilon}{8 \sup |f|}. \qquad A18.4$$

Now, find N'' such that every $P \in \mathcal{P}_{N''}$ either is entirely contained in B or is entirely contained in some $C \in \mathcal{D}_N$, or both. We claim that this N'' works, in the sense that

We know this N'' exists because the diameters of the tiles go to 0.

$$U_{\mathcal{P}_{N''}}(f) - L_{\mathcal{P}_{N''}}(f) < \epsilon, \qquad A18.5$$

but it takes a bit of doing to prove it.

[4]Theorem 4.3.9: A function $f : \mathbb{R}^n \to \mathbb{R}$, bounded with bounded support, is integrable if it is continuous except on a set of volume 0.

Every $\mathbf{x}$ is contained in some dyadic cube C. Let $C_N(\mathbf{x})$ be the cube at level N that contains $\mathbf{x}$. Define the function $\overline{f}$ that assigns to each $\mathbf{x}$ the maximum of f over its cube:

$$\overline{f}(\mathbf{x}) = M_{C_N(\mathbf{x})}(f). \qquad A18.6$$

Every $\mathbf{x}$ is also in some paving tile P. If $\mathbf{x}$ is contained in a single tile at level L, let $P_L(\mathbf{x})$ be that tile. Define $\overline{g}$ to be the function that assigns to each $\mathbf{x}$ the maximum of f over its paving tile $P_{N''}(\mathbf{x})$ if $P_{N''}(\mathbf{x})$ is entirely within a dyadic cube at level N and $\mathbf{x}$ is contained in no other tile, and minus $\sup|f|$ otherwise:

A point $\mathbf{x}$ may be in more than one paving tile at a given level, but, by Definition 4.7.2 and Corollary 4.3.10, such points don't affect integrals.

$$\overline{g}(\mathbf{x}) = \begin{cases} M_{P_{N''}(\mathbf{x})}(f) & \text{if } P_{N''}(\mathbf{x}) \cap \partial\mathcal{D}_N = \phi \text{ and } \mathbf{x} \text{ is contained} \\ & \quad \text{in a single tile of } \mathcal{P}_{N''} \\ -\sup|f| & \text{otherwise.} \end{cases} \qquad A18.7$$

Then $\overline{g} \le \overline{f}$; hence, by part 3 of Proposition 4.1.14,

$$\int_{\mathbb{R}^n} \overline{g}|d^n\mathbf{x}| \le \int_{\mathbb{R}^n} \overline{f}|d^n\mathbf{x}| = U_N(f). \qquad A18.8$$

Now we compute the upper sum $U_{\mathcal{P}_{N''}}(f)$:

$$\begin{aligned} U_{\mathcal{P}_{N''}}(f) &= \sum_{P\in\mathcal{P}_{N''}} M_P(f)\,\text{vol}_n P \qquad A18.9 \\ &= \underbrace{\sum_{\substack{P\in\mathcal{P}_{N''},\\ P\cap\partial\mathcal{D}_N=\phi}} M_P(f)\,\text{vol}_n P}_{\substack{\text{contribution from } P\\ \text{entirely in dyadic cubes}}} + \underbrace{\sum_{\substack{P\in\mathcal{P}_{N''},\\ P\cap\partial\mathcal{D}_N\neq\phi}} \big(M_P(f)\,\text{vol}_n P.}_{\substack{\text{contribution from } P \text{ that intersect}\\ \text{the boundary of dyadic cubes}}} \end{aligned}$$

The sum of indicator functions is the constant function 1 except on a set of volume 0.

We want a statement that relates integrals computed using dyadic cubes and paving tiles. Since $\sum_{P\in\mathcal{P}} \mathbf{1}_P = 1$ except on a set of volume 0,

$$\begin{aligned} \int_{\mathbb{R}^n} \overline{g}(\mathbf{x})|d^n\mathbf{x}| &= \sum_{P\in\mathcal{P}_{N''}} \int_{\mathbb{R}^n} \overline{g}(\mathbf{x})\mathbf{1}_P(\mathbf{x})|d^n\mathbf{x}| \qquad A18.10 \\ &= \sum_{\substack{P\in\mathcal{P}_{N''},\\ P\cap\partial\mathcal{D}_N=\phi}} M_P(f)\,\text{vol}_n P + \sum_{\substack{P\in\mathcal{P}_{N''},\\ P\cap\partial\mathcal{D}_N\neq\phi}} (-\sup|f|)\,\text{vol}_n P. \end{aligned}$$

Note that we can write the last term in equation A18.9 as

$$\begin{aligned} \sum_{\substack{P\in\mathcal{P}_{N''},\\ P\cap\partial\mathcal{D}_N\neq\phi}} \big(M_P(f)\,\text{vol}_n P &= \sum_{\substack{P\in\mathcal{P}_{N''},\\ P\cap\partial\mathcal{D}_N\neq\phi}} \big(M_P(f) \overbrace{-\sup|f| + \sup|f|}^{\text{cancels out}}\big)\,\text{vol}_n P \\ &= \sum_{\substack{P\in\mathcal{P}_{N''},\\ P\cap\partial\mathcal{D}_N\neq\phi}} \big(-\sup|f|\big)\,\text{vol}_n P + \sum_{\substack{P\in\mathcal{P}_{N''},\\ P\cap\partial\mathcal{D}_N\neq\phi}} \big(M_P(f) + \sup|f|\big)\,\text{vol}_n P. \end{aligned}$$

So we can rewrite equation A18.9 as

Equation A18.11: Since $M_P(f)$ is the least upper bound over P while $\sup|f|$ is the least upper bound over $\mathbb{R}^n$, we have

$$M_P(f) + \sup|f| \le 2\sup|f|.$$

$$U_{\mathcal{P}_{N''}}(f) = \int_{\mathbb{R}^n} \overline{g}|d^n\mathbf{x}| + \sum_{\substack{P \in \mathcal{P}_{N''}, \\ P \cap \partial\mathcal{D}_N \neq \phi}} \overbrace{\big(M_P(f) + \sup|f|\big)}^{\le 2\sup|f| \text{ (see note in margin)}} \operatorname{vol}_n P. \qquad A18.11$$

Using inequality A18.4 to give an upper bound on the volume of the paving pieces P that intersect the boundary, we get

$$\left| U_{\mathcal{P}_{N''}}(f) - \int_{\mathbb{R}^n} \overline{g}|d^n\mathbf{x}| \right| \le 2\sup|f| \operatorname{vol}_n B \le 2\sup|f| \frac{\epsilon}{8\sup|f|} = \frac{\epsilon}{4}. \qquad A18.12$$

Substituting $U_N(f)$ for the integral of $\overline{g}$ (justified by inequality A18.8) gives

$$\left| U_{\mathcal{P}_{N''}}(f) - \overbrace{U_N(f)}^{\ge \int_{\mathbb{R}^n} \overline{g}|d^n\mathbf{x}|} \right| \le \frac{\epsilon}{4}, \quad \text{so} \quad U_{\mathcal{P}_{N''}}(f) \le U_N(f) + \frac{\epsilon}{4}. \qquad A18.13$$

An exactly analogous argument leads to

$$L_{\mathcal{P}_{N''}}(f) \ge L_N(f) - \frac{\epsilon}{4}, \quad \text{i.e.,} \quad -L_{\mathcal{P}_{N''}}(f) \le -L_N(f) + \frac{\epsilon}{4}. \qquad A18.14$$

Adding these together gives

$$U_{\mathcal{P}_{N''}}(f) - L_{\mathcal{P}_{N''}}(f) \le \underbrace{U_N(f) - L_N(f)}_{<\epsilon/2 \text{ by ineq. A18.3}} + \frac{\epsilon}{2} < \epsilon. \qquad A18.15$$

2. The same argument works if we exchange $\mathcal{D}_N$ and $\mathcal{P}_N$. □

A19 The change of variables formula: a rigorous proof

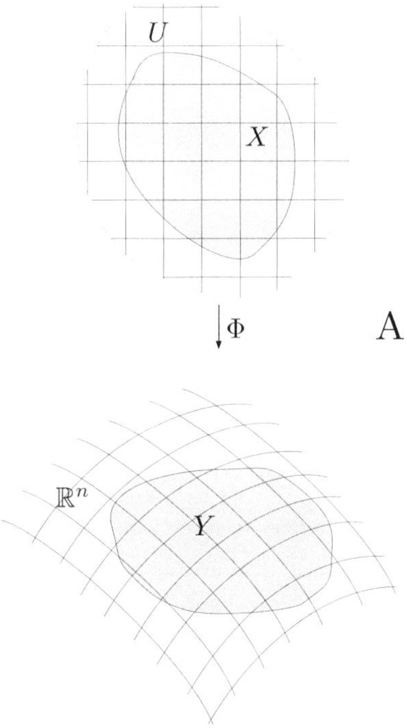

FIGURE A19.1.

The C^1 mapping Φ maps X to Y. In the proof of the change of variables formula we will use the fact that Φ is defined on U, not just on X.

Theorem 4.10.12 (Change of variables formula). *Let X be a compact subset of $\mathbb{R}^n$ with boundary ∂X of volume 0; let $U \subset \mathbb{R}^n$ be an open set containing X. Let $\Phi : U \to \mathbb{R}^n$ be a C^1 mapping that is injective on $(X - \partial X)$ and has Lipschitz derivative, with $[\mathbf{D}\Phi(\mathbf{x})]$ invertible at every $\mathbf{x} \in (X - \partial X)$. Set $Y = \Phi(X)$. If $f : Y \to \mathbb{R}$ is integrable, $(f \circ \Phi)|\det[\mathbf{D}\Phi]|$ is integrable on X, and*

$$\int_Y f(\mathbf{y})\,|d^n\mathbf{y}| = \int_X (f \circ \Phi)(\mathbf{x})\,\big|\det[\mathbf{D}\Phi(\mathbf{x})]\big|\,|d^n\mathbf{x}|.$$

Proof. The proof is a (lengthy) matter of dotting the *i*'s of the sketch in Section 4.10. As shown in Figure A19.1, we use the dyadic decomposition of X, and the image decomposition for Y, whose paving blocks are the $\Phi(C \cap X)$, $C \in \mathcal{D}_N(\mathbb{R}^n)$. We call this image partition $\Phi(\mathcal{D}_N(X))$. The outline of the proof is as follows, where $\mathbf{x}_C$ denotes the point in C where Φ

is evaluated:

$$\begin{aligned}\int_Y f|d^n\mathbf{y}| &\approx \sum_{C\in\mathcal{D}_N(X)} \overbrace{M_{\Phi(C)}f\Big(\text{vol}_n\,\Phi(C)\Big)}^{\substack{\sup f \text{ over curvy cube}\\ \text{times vol. of curvy cube}}}\\ &\approx \sum_{C\in\mathcal{D}_N(X)} M_C(f\circ\Phi)\Big(\text{vol}_n\,C|\det[\mathbf{D}\Phi(\mathbf{x}_C)]|\Big)\\ &\approx \int_X (f\circ\Phi)\mathbf{x}|\det[\mathbf{D}\Phi(\mathbf{x})]|)\,|d^n\mathbf{x}|.\end{aligned} \qquad A19.1$$

Equation A19.1: The second line is a Riemann sum; $\mathbf{x}_C$ is the point in C where Φ is evaluated: midpoint, lower left corner, or some other choice. The $\approx$ become equalities in the limit.

1. To justify the first $\approx$, we need to show that $\Phi(\mathcal{D}_N(X))$, the image decomposition of Y, is a nested partition.
2. To justify the second (this is the hard part) we need to show that as $N\to\infty$, the volume of a curvy cube of the image decomposition equals the volume of a cube of the original dyadic decomposition times $|\det[\mathbf{D}\Phi(\mathbf{x}_C)]|$.
3. The third $\approx$ is the definition of the integral as the limit of a Riemann sum.

We need Proposition A19.1 (of interest in its own right) to prove part 1.

Proposition A19.1 (Volume of image by a C^1 map). *Let $Z\subset\mathbb{R}^n$ be a compact pavable subset of $\mathbb{R}^n$, let $U\subset\mathbb{R}^n$ be an open set containing Z, and let $\Phi: U\to\mathbb{R}^n$ be a C^1 mapping with bounded derivative. Set $K=\sup_{\mathbf{x}\in U}|[\mathbf{D}\Phi(\mathbf{x})]|$. Then*

$$\text{vol}_n\,\Phi(Z)\ \le\ (K\sqrt{n})^n\,\text{vol}_n\,Z. \qquad A19.2$$

In particular, if $\text{vol}_n Z=0$, then $\text{vol}_n\,\Phi(Z)=0$.

Equation A19.3: The subset A consists of the closure of all the C either entirely within Z or straddling the boundary of Z. Recall that $\overline{C}$ denotes the closure of C.

A cube $C\in\mathcal{D}_N(\mathbb{R}^n)$ has sidelength $1/2^N$, and the distance between a point $\mathbf{x}$ in C and its center $\mathbf{x}_0$ is

$$|\mathbf{x}-\mathbf{x}_0|\le\frac{\sqrt{n}}{2\cdot 2^N}$$

(see Exercise 4.1.8). So $\Phi(C)$ is contained in the box C' centered at $\Phi(\mathbf{z}_C)$ with sidelength $K\sqrt{n}/2^N$.

Proof. Choose $\epsilon>0$ and $N\ge 0$ so large that if we set

$$A\overset{\text{def}}{=}\bigcup_{\substack{C\in\mathcal{D}_N(\mathbb{R}^n),\\ C\cap Z\neq\phi}}\overline{C},\quad\text{then}\quad A\subset U\quad\text{and}\quad \text{vol}_n\,A\le\text{vol}_n\,Z+\epsilon. \qquad A19.3$$

Let $\mathbf{z}_C$ be the center of one of the cubes C above. Then by Corollary 1.9.2, when $\mathbf{z}\in C$ we have

$$|\Phi(\mathbf{z}_C)-\Phi(\mathbf{z})|\le K|\mathbf{z}_C-\mathbf{z}|\le K\frac{\sqrt{n}}{2\cdot 2^N}. \qquad A19.4$$

(The distance between any two points in the image is at most K times the distance between the corresponding points of the domain.) Therefore, $\Phi(C)$ is contained in the box C' centered at $\Phi(\mathbf{z}_C)$ with sidelength $K\sqrt{n}/2^N$.

Finally,

$$\Phi(Z)\subset\bigcup_{\substack{C\in\mathcal{D}_N(\mathbb{R}^n),\\ C\cap Z\neq\phi}} C',\qquad\text{so} \qquad A19.5$$

Inequality A19.6: To go from line 1 to line 2 we simultaneously multiply and divide by

$$\text{vol}_n C = \left(\frac{1}{2^N}\right)^n.$$

$$\text{vol}_n \Phi(Z) \le \sum_{\substack{C \in \mathcal{D}_N(\mathbb{R}^n),\\ C\cap Z \ne \emptyset}} \text{vol}_n C' \le \sum_{\substack{C \in \mathcal{D}_N(\mathbb{R}^n),\\ C\cap Z \ne \emptyset}} \left(\frac{K\sqrt{n}}{2^N}\right)^n$$

$$= \underbrace{\frac{\left(\frac{K\sqrt{n}}{2^N}\right)^n}{\left(\frac{1}{2^N}\right)^n}}_{\substack{\text{ratio vol}_n C' \\ \text{to vol}_n C}} \sum_{\substack{C \in \mathcal{D}_N(\mathbb{R}^n),\\ C\cap Z \ne \emptyset}} \text{vol}_n C = (K\sqrt{n})^n \,\text{vol}_n A$$

$$\le (K\sqrt{n})^n(\text{vol}_n Z + \epsilon). \quad \square \qquad \text{A19.6}$$

Corollary A19.2. *For each N, the partition $\Phi(\mathcal{D}_N(X))$ is a paving of Y, and the sequence $N \mapsto \Phi(\mathcal{D}_N(X))$ is a nested partition of Y.*

Proof. Conditions 1 and 3 of Definition 4.7.2 of a paving are obvious; conditions 2 and 4 follow from the last sentence of Proposition A19.1. To show that the sequence is a nested partition (Definition 4.7.3), we need to check that the pieces are nested and that the diameters tend to 0 as N tends to infinity. The first is clear: if $C_1 \subset C_2$, then $\Phi(C_1) \subset \Phi(C_2)$. The second is inequality A19.4. $\square$

Our next proposition contains the real substance of the change of variables theorem. It says exactly why we can replace the volume of the little curvy parallelogram $\Phi(C)$ by its approximate volume $|\det[\mathbf{D}\Phi(\mathbf{x})]| \,\text{vol}_n C$: for a change of variables map Φ such that $\Phi(\mathbf{0}) = \mathbf{0}$, the image $\Phi(C)$ of a cube C centered at $\mathbf{0}$ is arbitrarily close to the image of C by the derivative of Φ at $\mathbf{0}$, as shown in Figures A19.3 and A19.4.

Proposition A19.3. *Let U, V be open subsets of $\mathbb{R}^n$ with $\mathbf{0} \in U$ and $\mathbf{0} \in V$. Let $\Phi : U \to V$ be a bijective differentiable mapping with $\Phi(\mathbf{0}) = \mathbf{0}$ such that $[\mathbf{D}\Phi]$ is Lipschitz and $\Phi^{-1} : V \to U$ is also differentiable with Lipschitz derivative. Let M be a Lipschitz ratio for both $[\mathbf{D}\Phi]$ and $[\mathbf{D}\Phi^{-1}]$. Then*

1. *For any $\epsilon > 0$, there exists $\delta > 0$ such that if C is a cube centered at $\mathbf{0}$ of sidelength $< 2\delta$, then*

$$(1-\epsilon)\,[\mathbf{D}\Phi(\mathbf{0})]C \subset \underbrace{\Phi(C)}_{\substack{\text{squeezed between}\\ \text{right and left sides}\\ \text{as } \epsilon\to 0}} \subset (1+\epsilon)\,[\mathbf{D}\Phi(\mathbf{0})]C. \qquad \text{A19.7}$$

2. *We can choose δ to depend only on ϵ, $\big|[\mathbf{D}\Phi(\mathbf{0})]\big|$, $\big|[\mathbf{D}\Phi(\mathbf{0})]^{-1}\big|$, and the Lipschitz ratio M, but no other information about Φ.*

Whenever we compare balls and cubes in $\mathbb{R}^n$, a pesky $\sqrt{n}$ complicates the formulas. We will need to deal with this several times in the proof of Proposition A19.3; the following lemma isolates what we need.

Lemma A19.4. *Choose $0 < a < b$, and let $C_a, C_b \subset \mathbb{R}^n$ be the cubes centered at the origin of sidelength $2a$ and $2b$ respectively. Then the ball B of radius $\frac{(b-a)|\mathbf{x}|}{a\sqrt{n}}$ around any point $\mathbf{x} \in C_a$ is contained in C_b.*

Proof. If $\mathbf{x} \in C_a$, then $|\mathbf{x}| < a\sqrt{n}$, so the radius of B is less than $b-a$. See Figure A19.2. □

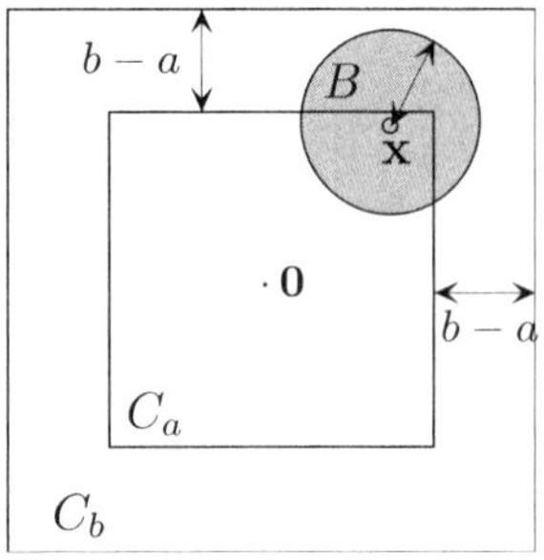

FIGURE A19.2.
This illustrates Lemma A19.4. Since the ball B centered at a point in C_a has radius

$$\frac{(b-a)|\mathbf{x}|}{a\sqrt{n}} < b-a,$$

B is contained in C_b.

Proof of Proposition A19.3. The right and the left inclusions of equation A19.7 require slightly different treatments. Both follow from Proposition A5.1. Remember that the largest n-dimensional cube contained in a ball of radius r has sidelength $2r/\sqrt{n}$.

The right inclusion, illustrated by Figure A19.3, is obtained by finding δ such that if the sidelength of C is at most 2δ, and $\mathbf{x} \in C$, then

$$\left|[\mathbf{D}\Phi(\mathbf{0})]^{-1}\Big(\Phi(\mathbf{x}) - [\mathbf{D}\Phi(\mathbf{0})](\mathbf{x})\Big)\right| \le \frac{\epsilon|\mathbf{x}|}{\sqrt{n}}. \qquad A19.8$$

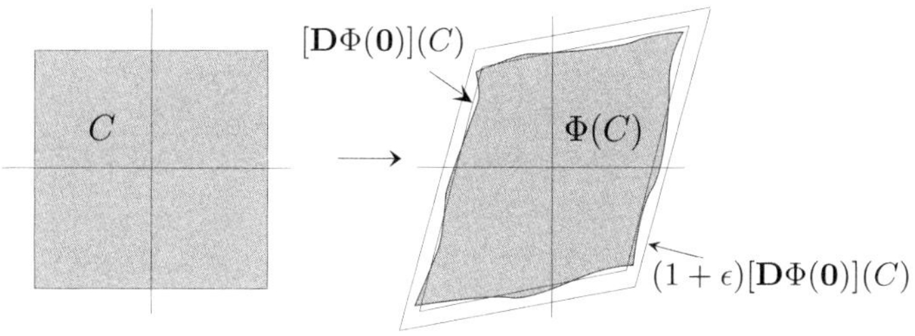

FIGURE A19.3. The cube C is mapped to $\Phi(C)$, which is almost $[\mathbf{D}\Phi(\mathbf{0})](C)$, and definitely inside $(1+\epsilon)[\mathbf{D}\Phi(\mathbf{0})](C)$. As $\epsilon \to 0$, the image $\Phi(C)$ becomes more and more exactly the parallelepiped $[\mathbf{D}\Phi(\mathbf{0})]C$.

Why does inequality A19.8 prove the right inclusion? We want to know that if $\mathbf{x} \in C$, then

$\Phi(\mathbf{x}) \in (1+\epsilon)[\mathbf{D}\Phi(\mathbf{0})](C)$, i.e.,

$$[\mathbf{D}\Phi(\mathbf{0})]^{-1}\Phi(\mathbf{x}) \in (1+\epsilon)C.$$

Since

$$[\mathbf{D}\Phi(\mathbf{0})]^{-1}\Phi(\mathbf{x}) = \mathbf{x} + [\mathbf{D}\Phi(\mathbf{0})]^{-1}\Big(\Phi(\mathbf{x}) - [\mathbf{D}\Phi(\mathbf{0})](\mathbf{x})\Big),$$

$[\mathbf{D}\Phi(\mathbf{0})]^{-1}\Phi(\mathbf{x})$ is distance

$$\left|[\mathbf{D}\Phi(\mathbf{0})]^{-1}\Big(\Phi(\mathbf{x}) - [\mathbf{D}\Phi(\mathbf{0})](\mathbf{x})\Big)\right|$$

from $\mathbf{x}$. But the ball of radius $\epsilon|\mathbf{x}|/\sqrt{n}$ around any point $\mathbf{x} \in C$ is completely contained in $(1+\epsilon)C$, by Lemma A19.4 (setting $a = 1$ and $b - a = \epsilon$).

By Proposition A5.1 (see the margin),

$$\left|[\mathbf{D}\Phi(\mathbf{0})]^{-1}\Big(\Phi(\mathbf{x}) - \underbrace{\Phi(\mathbf{0})}_{\mathbf{0}\text{ by def.}} - [\mathbf{D}\Phi(\mathbf{0})](\mathbf{x})\Big)\right| \le \left|[\mathbf{D}\Phi(\mathbf{0})]^{-1}\right|\frac{M|\mathbf{x}|^2}{2}, \qquad A19.9$$

so it is enough to require that when $\mathbf{x} \in C$,

$$\frac{\left|[\mathbf{D}\Phi(\mathbf{0})]^{-1}\right|M|\mathbf{x}|^2}{2} \le \frac{\epsilon|\mathbf{x}|}{\sqrt{n}}, \quad \text{i.e.,} \quad |\mathbf{x}| \le \frac{2\epsilon}{M\sqrt{n}\left|[\mathbf{D}\Phi(\mathbf{0})]^{-1}\right|}. \qquad A19.10$$

Since $\mathbf{x} \in C$ and C has sidelength at most 2δ, we have $|\mathbf{x}| \le \delta\sqrt{n}$, so the right inclusion will be satisfied if

$$\delta = \frac{2\epsilon}{Mn\left|[\mathbf{D}\Phi(\mathbf{0})]^{-1}\right|}. \qquad A19.11$$

By Proposition A5.1, if the derivative of $\mathbf{f}$ is Lipschitz with Lipschitz ratio M, then

$$\Big|\mathbf{f}(\mathbf{x}) - \mathbf{f}(\mathbf{0}) - [\mathbf{D}\mathbf{f}(\mathbf{0})]\mathbf{x}\Big| \le \frac{M}{2}|\mathbf{x}|^2.$$

In inequality A19.9, Φ plays the role of $\mathbf{f}$.

For the left inclusion, illustrated by Figure A19.4, we need to find δ such that when C has sidelength $\le 2\delta$ and $\mathbf{x} \in (1-\epsilon)C$, then

$$(1-\epsilon)\,[\mathbf{D}\Phi(\mathbf{0})]C \subset \Phi(C), \quad \text{i.e.,} \quad \Phi^{-1}[\mathbf{D}\Phi(\mathbf{0})]\mathbf{x} \in C \qquad A19.12$$

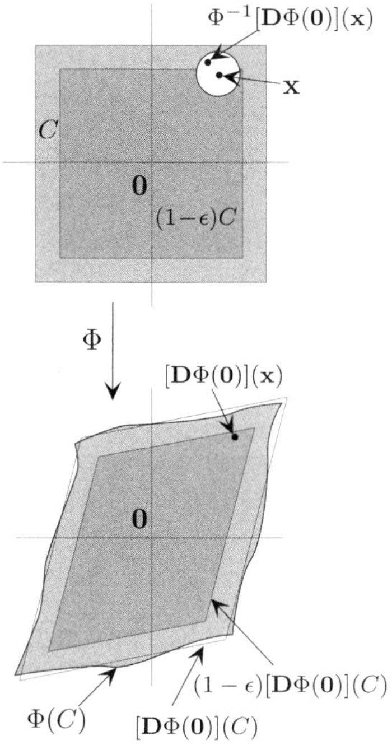

FIGURE A19.4.

We will prove the left inclusion

$$(1-\epsilon)[\mathbf{D}\Phi(\mathbf{0})]C \subset \Phi(C)$$

of Proposition A19.3 by showing that for any $\mathbf{x} \in (1-\epsilon)C$ the point $\Phi^{-1}([\mathbf{D}\Phi(\mathbf{0})]\mathbf{x})$ belongs to the ball of radius

$$\frac{\epsilon}{\sqrt{n}(1-\epsilon)}|\mathbf{x}|$$

centered at $\mathbf{x}$. Lemma A19.4 guarantees that this ball is contained in C, so that

$$[\mathbf{D}\Phi(\mathbf{0})]\mathbf{x} = \Phi(\Phi^{-1}[\mathbf{D}\Phi(\mathbf{0})]\mathbf{x}) \in \Phi(C).$$

Substituting δ for b and $(1-\epsilon)\delta$ for a in Lemma A19.4, this is equivalent to showing that we can find δ such that if $\mathbf{x} \in (1-\epsilon)C$, then

$$\left|\Phi^{-1}([\mathbf{D}\Phi(\mathbf{0})]\mathbf{x}-\mathbf{x}\right| \le \frac{\epsilon}{\sqrt{n}(1-\epsilon)}|\mathbf{x}|. \qquad A19.13$$

(Note that since the derivative $[\mathbf{D}\Phi]$ approximates Φ, we should expect $\Phi^{-1}([\mathbf{D}\Phi(\mathbf{0})]$ to be close to the identity, so the left side of the inequality should be in $o(|\mathbf{x}|)$.) To find such a δ, we use Proposition A5.1 (third line of inequality A19.14). Set $\mathbf{y} = [\mathbf{D}\Phi(\mathbf{0})]\mathbf{x}$. This time, Φ^{-1} plays the role of $\mathbf{f}$ in Proposition A5.1. We find

$$\begin{aligned}
\left|\Phi^{-1}\big([\mathbf{D}\Phi(\mathbf{0})]\mathbf{x}\big) - \mathbf{x}\right| &= |\Phi^{-1}\mathbf{y} - [\mathbf{D}\Phi^{-1}(\mathbf{0})][\mathbf{D}\Phi(\mathbf{0})]\mathbf{x}| \\
&= |\Phi^{-1}(\mathbf{0}+\mathbf{y}) - \underbrace{\Phi^{-1}(\mathbf{0})}_{\mathbf{0}\text{ by def.}} - [\mathbf{D}\Phi^{-1}(\mathbf{0})]\mathbf{y}| \\
&\overbrace{\le}^{\text{Prop. A5.1}} \frac{M}{2}|\mathbf{y}|^2 = \frac{M}{2}\Big|[\mathbf{D}\Phi(\mathbf{0})]\mathbf{x}\Big|^2 \\
&\le \frac{M}{2}\Big|[\mathbf{D}\Phi(\mathbf{0})]\Big|^2 \big|\mathbf{x}\big|^2.
\end{aligned} \qquad A19.14$$

Thus inequality A19.13 will be satisfied if

$$\frac{M}{2}|[\mathbf{D}\Phi(\mathbf{0})]|^2|\mathbf{x}|^2 \le \frac{\epsilon}{\sqrt{n}(1-\epsilon)}|\mathbf{x}|, \quad \text{i.e.,} \qquad A19.15$$

$$|\mathbf{x}| \le \frac{2\epsilon}{M(1-\epsilon)\sqrt{n}|[\mathbf{D}\Phi(\mathbf{0})]|^2}. \qquad A19.16$$

Remember that $\mathbf{x} \in (1-\epsilon)C$, so $|\mathbf{x}| < (1-\epsilon)\delta\sqrt{n}$, so the left inclusion is satisfied if we take

$$\delta = \frac{2\epsilon}{(1-\epsilon)^2 Mn\big|[\mathbf{D}\Phi(\mathbf{0})]\big|^2}. \qquad A19.17$$

Choose the smaller of this δ and the δ in equation A19.11.

2. This is clear from equations A19.11 and A19.17. □ Proposition A19.3

The integral is defined in terms of upper and lower sums, and we must now translate Proposition A19.3 into that language.

Proposition A19.5. *Let U and V be bounded subsets in $\mathbb{R}^n$ and let $\Phi : U \to V$ be a bijective differentiable mapping with Lipschitz derivative such that $\Phi^{-1} : V \to U$ is also differentiable with Lipschitz derivative. Then for any $\eta > 0$, there exists N such that if $C \in \mathcal{D}_N(\mathbb{R}^n)$ and $C \subset U$,*

$$\begin{aligned}
(1-\eta)\, M_C\big(|\det[\mathbf{D}\Phi]|\big)\ \mathrm{vol}_n C &\le \mathrm{vol}_n \Phi(C) \\
&\le (1+\eta)\, m_C\big(|\det[\mathbf{D}\Phi]|\big)\ \mathrm{vol}_n C.
\end{aligned} \qquad A19.18$$

Proof of Proposition A19.5. Choose $\eta > 0$, and find $\epsilon > 0$ so that

$$(1+\epsilon)^{n+1} < 1+\eta \quad \text{and} \quad (1-\epsilon)^{n+1} > 1-\eta. \qquad A19.19$$

For this ϵ, find N_1 such that Proposition A19.3 is true for every cube $C \in \mathcal{D}_{N_1}(\mathbb{R}^n)$ such that $C \subset U$. For each cube C, let $\mathbf{z}_C$ be its center. Then (as discussed in the margin) if $N \geq N_1$ and $C \in \mathcal{D}_N(\mathbb{R}^n)$, it follows from Proposition A19.3 and Theorem 4.9.1 that

$$\operatorname{vol}_n \Phi(C) < (1+\epsilon)^n \big|\det[\mathbf{D}\Phi(\mathbf{z}_C)]\big| \operatorname{vol}_n C. \qquad A19.20$$

Inequality A 19.20: For an appropriate N_1, the right in inclusion of Proposition A19.3 tells us that

$$\operatorname{vol}_n \Phi(C) \leq \operatorname{vol}_n\Big((1+\epsilon)[\mathbf{D}\Phi(\mathbf{z}_C)]\, C\Big).$$

Theorem 4.9.1 says that if T is linear,

$$\operatorname{vol}_n T(A) = |\det[T]| \operatorname{vol}_n A;$$

replacing T by $[\mathbf{D}\Phi(\mathbf{z}_C)]$ and A by C gives

$$\operatorname{vol}_n [\mathbf{D}\Phi(\mathbf{z}_C)](C) = |\det[\mathbf{D}\Phi(\mathbf{z}_C)]| \operatorname{vol}_n C.$$

So for an appropriate N_1, we have

$$\operatorname{vol}_n \Phi(C) \leq \operatorname{vol}_n\Big((1+\epsilon)[\mathbf{D}\Phi(\mathbf{z}_C)]\, C\Big) = (1+\epsilon)^n |\det[\mathbf{D}\Phi(\mathbf{z}_C)]| \operatorname{vol}_n C;$$

the equality is Theorem 4.9.1.

Next find N_2 such that for every cube $C \in \mathcal{D}_{N_2}(\mathbb{R}^n)$ with $C \subset U$, we have

$$\frac{M_C|\det[\mathbf{D}\Phi]|}{m_C|\det[\mathbf{D}\Phi]|} < 1+\epsilon \quad \text{and} \quad \frac{m_C|\det[\mathbf{D}\Phi]|}{M_C|\det[\mathbf{D}\Phi]|} > 1-\epsilon. \qquad A19.21$$

Actually the second inequality follows from the first, since $1/(1+\epsilon) > 1-\epsilon$.

In inequalities A19.21,

$$M_C|\det[\mathbf{D}\Phi]|$$

is not M_C times $[\mathbf{D}\Phi]|$; it is the least upper bound of $|\det[\mathbf{D}\Phi]|$. Similarly, $m_C|\det[\mathbf{D}\Phi]|$ is the greatest lower bound of $|\det[\mathbf{D}\Phi]|$.

Remark. How do we know there exists N_2 such that inequalities A19.21 are satisfied, i.e., that the ratios are close to 1? Note that $[\mathbf{D}\Phi]$ Lipschitz implies that $\det[\mathbf{D}\Phi]$ is Lipschitz and bounded, since det is a polynomial function of the entries, and the same for $[\mathbf{D}\Phi^{-1}]$. By the chain rule,

$$[\mathbf{D}\Phi^{-1}(\Phi(\mathbf{x}))][\mathbf{D}\Phi(\mathbf{x})] = I, \qquad A19.22$$

and by $\det AB = \det A \det B$, we have $\det[\mathbf{D}\Phi^{-1}(\Phi(\mathbf{x}))]\det[\mathbf{D}\Phi(\mathbf{x})] = 1$. Since $\det[\mathbf{D}\Phi]$ is Lipschitz, $\exists M$ such that for $\mathbf{x}, \mathbf{y} \in C$, $C \in \mathcal{D}_{N_2}(\mathbb{R}^n)$,

$$\Big|\det[\mathbf{D}\Phi(\mathbf{x})] - \det[\mathbf{D}\Phi(\mathbf{y})]\Big| \leq M|\mathbf{x}-\mathbf{y}|, \text{ i.e.} \qquad A19.23$$

$$\left|\frac{\det[\mathbf{D}\Phi(\mathbf{x})]}{\det[\mathbf{D}\Phi(\mathbf{y})]} - 1\right| \leq \frac{M|\mathbf{x}-\mathbf{y}|}{|\det[\mathbf{D}\Phi(\mathbf{y})]|} = M|\mathbf{x}-\mathbf{y}||\det[\mathbf{D}\Phi^{-1}(\Phi(\mathbf{y})]|;$$

since $[\mathbf{D}\Phi^{-1}]$ is bounded, the right side can be made arbitrarily small by taking $\mathbf{x}$ and $\mathbf{y}$ close enough together, i.e., by taking N_2 large. △

If N is the larger of N_1 and N_2, then for $C \in \mathcal{D}_N(\mathbb{R}^n)$, the first inequality in A19.21 gives

$$\Big|\det[\mathbf{D}\Phi(\mathbf{z}_C)]\Big| \leq M_C\Big|\det[\mathbf{D}\Phi]\Big| \underbrace{<}_{\text{ineq. A19.21}} (1+\epsilon)m_C|\det[\mathbf{D}\Phi]|, \qquad A19.24$$

which together with inequality A19.20 gives

$$\operatorname{vol}_n \Phi(C) < (1+\epsilon)^{n+1} m_C\big|\det[\mathbf{D}\Phi]\big| \operatorname{vol}_n C. \qquad A19.25$$

An exactly similar argument leads to

$$\operatorname{vol}_n \Phi(C) > (1-\epsilon)^{n+1} M_C\big|\det[\mathbf{D}\Phi]\big| \operatorname{vol}_n C. \quad \square \qquad A19.26$$

Completing the proof of the change of variables formula

We may assume that $f \geq 0$, and we may extend it to be 0 on $\Phi(U) - Y$ (recall that $\Phi(X) = Y$). Choose $\eta > 0$, and choose N_1 sufficiently large so that two conditions are met: first, the union of the cubes $C \in \mathcal{D}_{N_1}(X)$ whose closures intersect the boundary of X have total volume $< \eta$, and second, the closures of these C are contained in U.

We will denote by Z the union of the closure of these cubes; thus, we want N_1 to be large enough so that

We can think of Z as a thickening of ∂X, the boundary of X.

$$\mathrm{vol}_n Z < \eta \quad \text{and} \quad Z \subset U. \qquad A19.27$$

Then (since X is compact), $X \cup Z$ is a compact subset of U. Denote by M the Lipschitz ratio of $[\mathbf{D}\Phi]$, and set

$$K = \sup_{\mathbf{x}\in X\cup Z} |\det[\mathbf{D}\Phi(\mathbf{x})]| \quad \text{and} \quad L = \sup_{\mathbf{y}\in\Phi(U)} |f(\mathbf{y})|. \qquad A19.28$$

We know that K is well defined because $|\det[\mathbf{D}\Phi(\mathbf{x})]|$ is continuous on the compact set $X \cup Z$.

Lemma A19.6. *The closure of $X - Z$ is compact and contains no point of ∂X.*

Proof. Since X is bounded, $X - Z$ is bounded, so its closure is closed and bounded. For the second statement, every point of ∂X has a neighborhood contained in Z, and so cannot be in the closure of $X - Z$. □

Of course we need the first part of Lemma A19.6 to apply Theorem 1.6.9. We need the second part to know that $[\mathbf{D}\Phi]^{-1}$ exists; recall that Theorem 4.10.12 specifies that $[\mathbf{D}\Phi(\mathbf{x})]$ is invertible at every $\mathbf{x} \in (X - \partial X)$.

It follows from Lemma A19.6 and Theorem 1.6.9 that $[\mathbf{D}\Phi]^{-1}$ is bounded on $X - Z$, say by $\widetilde{K}$, and it is also Lipschitz:

Inequality A19.29: Since $\widetilde{K}$ is a bound for $|[\mathbf{D}\Phi]^{-1}|$, it is also a bound for $|[\mathbf{D}\Phi(\mathbf{y})]^{-1}|$. This accounts for the $\widetilde{K}^2$ in the second line.

$$\begin{aligned} |[\mathbf{D}\Phi(\mathbf{x})]^{-1} - [\mathbf{D}\Phi(\mathbf{y})]^{-1}| &= \Big|[\mathbf{D}\Phi(\mathbf{x})]^{-1} \overbrace{\Big([\mathbf{D}\Phi(\mathbf{y})] - [\mathbf{D}\Phi(\mathbf{x})]\Big)}^{\le M|\mathbf{y}-\mathbf{x}|} [\mathbf{D}\Phi(\mathbf{y})]^{-1}\Big| \\ &\le (\widetilde{K})^2 M|\mathbf{x} - \mathbf{y}|. \end{aligned} \qquad A19.29$$

Thus we can apply Proposition A19.5, choosing $N_2 > N_1$ so that inequality A19.18 is true for all cubes in $\mathcal{D}_{N_2}$ contained in $X - Z$. By Definition 4.1.8 we have

$$U_{N_2}\big((f\circ\Phi)|\det[\mathbf{D}\Phi]|\big) = \sum_{C\in\mathcal{D}_{N_2}(\mathbb{R}^n)} M_C\big((f\circ\Phi)|\det[\mathbf{D}\Phi]|\big)\,\mathrm{vol}_n C.$$

Inequality A19.30: To go from line 1 to line 2, remember that if f and g are positive,

$$M_C(fg) \le M_C f M_C g.$$

Remember that

$$\mathrm{vol}_n Z < \eta$$

(inequality A19.27) and that

$$|\det[\mathbf{D}\Phi]| \le K$$

(equation A19.28).

We will consider separately the cubes of $\mathcal{D}_{N_2}$ in Z, which we will call *boundary cubes*, and the others, which we will call *interior cubes*. For the boundary cubes, we have (recall that K and L are defined in equation A19.28)

$$\begin{aligned} &\sum_{\text{boundary cubes } C} M_C\Big((f\circ\Phi)|\det[\mathbf{D}\Phi]|\Big)\,\mathrm{vol}_n C \\ \le &\sum_{\text{boundary cubes } C} \underbrace{M_C(f\circ\Phi)}_{\le L}\,\underbrace{M_C|\det[\mathbf{D}\Phi]|}_{\le K}\,\mathrm{vol}_n C \\ \le &\, LK\,\mathrm{vol}_n Z \;<\; LK\eta. \end{aligned} \qquad A19.30$$

For the interior cubes, we have

$$\sum_{\text{interior cubes } C} M_C\Big((f\circ\Phi)|\det[\mathbf{D}\Phi]|\Big)\operatorname{vol}_n C$$

$$\le \sum_{\text{int. cubes } C} \underbrace{M_C(f\circ\Phi)}_{M_{\Phi(C)}(f)} \;\underbrace{M_C|\det[\mathbf{D}\Phi]|\operatorname{vol}_n C}_{\le\frac{\operatorname{vol}_n\Phi(C)}{1-\eta},\text{ by ineq. A19.18}} \qquad A19.31$$

$$\le \frac{1}{1-\eta}\sum_{\text{int. cubes } C} M_{\Phi(C)}(f)\operatorname{vol}_n\Phi(C) = \frac{1}{1-\eta}U_{\Phi(\mathcal{D}_{N_2}(\mathbb{R}^n))}(f).$$

Putting these results together gives

$$U_{N_2}\big((f\circ\Phi)|\det[\mathbf{D}\Phi]|\big) \le \frac{1}{1-\eta}U_{\Phi(\mathcal{D}_{N_2}(\mathbb{R}^n))}(f) + LK\,\eta. \qquad A19.32$$

A similar argument about lower sums, using N_3, leads to

$$L_{N_3}\big((f\circ\Phi)|\det[\mathbf{D}\Phi]|\big) \ \ge\ \frac{1}{1+\eta}L_{\Phi(\mathcal{D}_{N_3}(\mathbb{R}^n))}(f) - LK\eta. \qquad A19.33$$

Denoting by N the larger of N_2 and N_3, we get

$$\frac{1}{1+\eta}L_{\Phi(\mathcal{D}_N(\mathbb{R}^n))}(f) - LK\,\eta \underbrace{\le}_{\text{ineq. A19.33}} L_N\big((f\circ\Phi)|\det[\mathbf{D}\Phi]|\big) \qquad A19.34$$

$$\le U_N\big((f\circ\Phi)|\det[\mathbf{D}\Phi]|\big) \underbrace{\le}_{\text{ineq. A19.32}} \frac{1}{1-\eta}U_{\Phi(\mathcal{D}_N(\mathbb{R}^n))}(f) + LK\eta.$$

Since f is integrable and $\Phi(\mathcal{D}_N(\mathbb{R}^n))$ is a nested partition (by Corollary A19.2), the lower and upper sums in the outer terms both converge as $N\to\infty$ to

$$\int_Y f(\mathbf{y})|d^n\mathbf{y}|. \qquad A19.35$$

Since $\eta > 0$ is arbitrary, this shows that the inner terms have a common limit. Thus $(f\circ\Phi)|\det[\mathbf{D}\Phi]|$ is integrable, and

$$\int_X (f\circ\Phi)(\mathbf{x})|\det[\mathbf{D}\Phi(\mathbf{x})]||d^n\mathbf{x}| = \int_Y f(\mathbf{y})|d^n\mathbf{y}|. \quad \square \qquad A19.36$$

A20 Volume 0 and related results

Here we prove three statements concerning volume and volume 0: Proposition 4.3.4, Proposition 5.2.2, and Proposition 6.6.17.

Proposition 4.3.4 (Bounded part of graph has volume 0). *Let $f:\mathbb{R}^n\to\mathbb{R}$ be an integrable function with graph $\Gamma(f)$, and let $C_0\subset\mathbb{R}^n$ be any dyadic cube. Then*

$$\operatorname{vol}_{n+1}\big(\underbrace{\Gamma(f)\cap(C_0\times\mathbb{R})}_{\text{bounded part of graph}}\big) = 0.$$

Proof. The proof is not so very hard, but we have two types of dyadic cubes that we need to keep straight: the $(n+1)$-dimensional cubes that intersect the graph of the function, and the n-dimensional cubes over which the function itself is evaluated.

Figure A20.1 illustrates the proof. In this case we have squares that intersect the graph, and intervals over which the function is evaluated. In keeping with that figure, let us denote the cubes in $\mathbb{R}^{n+1}$ by S (for squares) and the cubes in $\mathbb{R}^n$ by I (for intervals).

We need to show that the total volume of the cubes $S \in \mathcal{D}_N(\mathbb{R}^{n+1})$ that intersect $\Gamma(f) \cap (C_0 \times \mathbb{R})$ is small when N is large. Let us choose ϵ, and N satisfying the requirement of equation 4.3.1 for that ϵ: we decompose C_0 into n-dimensional I-cubes small enough so that the total n-dimensional volume of the I-cubes over which $\text{osc}(f) > \epsilon$ is less than ϵ.

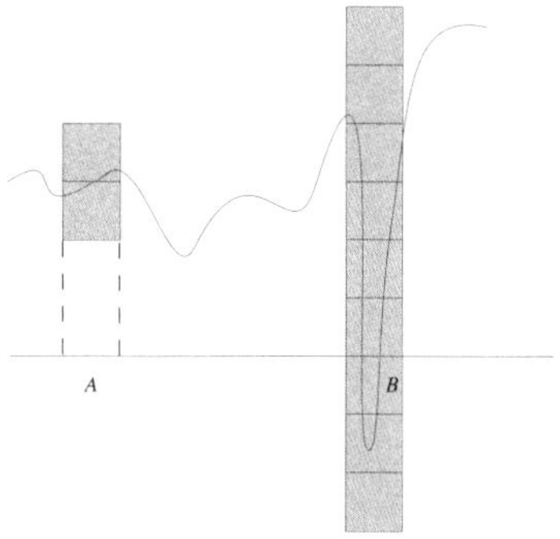

FIGURE A20.1.

The graph of a function from $\mathbb{R} \to \mathbb{R}$. Here the x-axis plays the role of $\mathbb{R}^n$ in the theorem, and the (x,y)-plane plays the role of $\mathbb{R}^{n+1}$. Over the interval A, the function has osc $< \epsilon$; over the interval B, it has osc $> \epsilon$. Above A, we keep the two cubes that intersect the graph; above B, we keep the entire tower, including the basement.

Now we count the $(n+1)$-dimensional S-cubes that intersect the graph. There are two kinds: those whose projection on $\mathbb{R}^n$ are I-cubes with $\text{osc}(f) > \epsilon$, and the others.

For those with large oscillation, think of each n-dimensional I-cube over which $\text{osc}(f) > \epsilon$ as the ground floor of a tower of $(n+1)$-dimensional S-cubes that is at most $\sup|f|$ high and goes down (into the basement) at most $-\sup|f|$. To be sure we have enough, we add an extra S-cube at top and bottom. Each tower then contains $2(\sup|f|+1)\cdot 2^N$ such cubes. (We multiply by 2^N because that is the inverse of the height of an S-cube. At $N=0$, the height of a cube is 1; at $N=1$, the height is $1/2$, so we need twice as many cubes to make the same height tower.) You will see from Figure A20.1 that we are counting more squares than we actually need.

How many such towers of cubes will we need? We chose N large enough so that the total n-dimensional volume of all I-cubes with osc $> \epsilon$ is less than ϵ. The volume of an I-cube is 2^{-nN}, so there are $2^{nN}\epsilon$ cubes I for which we need towers. So the number of S-cubes we need to cover the region of large oscillation is the product

$$\underbrace{\epsilon 2^{nN}}_{\substack{\text{no. of } I\text{-cubes} \\ \text{with osc}>\epsilon}} \quad \underbrace{2(\sup|f|+1)2^N}_{\substack{\text{no. of } S\text{-cubes above} \\ \text{one } I\text{-cube with osc}(f)>\epsilon}}. \qquad A20.1$$

In formula A20.2 we are counting more cubes than necessary: we are using the entire n-dimensional volume of C_0, rather than subtracting the parts over which $\text{osc}(f) > \epsilon$.

For the second sort (small oscillation), for each I-cube we require at most $2^N\epsilon + 2$ S-cubes, giving the product

$$\underbrace{2^{nN}\,\text{vol}_n(C_0)}_{\substack{\text{no. of } I\text{-cubes} \\ \text{to cover } C_0}} \quad \underbrace{(2^N\epsilon+2)}_{\substack{\text{no. of } S\text{-cubes above} \\ \text{one } I\text{-cube with osc}(f)<\epsilon}}. \qquad A20.2$$

Adding these numbers (A20.1 and A20.2), we find that the bounded part of the graph is covered by

$$2^{(n+1)N}\left(2\epsilon(\sup|f|+1) + \left(\epsilon + \frac{2}{2^N}\right)\text{vol}_n(C_0)\right) \qquad A20.3$$

S-cubes. This is an enormous number, but each cube has $(n+1)$-dimensional volume $1/2^{(n+1)N}$, so the total volume is

$$2\epsilon(\sup|f|+1) + \left(\epsilon + \frac{2}{2^N}\right)\text{vol}_n(C_0), \qquad A20.4$$

which can be made arbitrarily small. □

Proposition 4.3.5 is a special case of Proposition 5.2.2, restated below.

Proposition 5.2.2 (k-dimensional volume of a manifold). *If integers m, k, n satisfy $0 \le m < k \le n$, and $M \subset \mathbb{R}^n$ is a manifold of dimension m, any closed subset $X \subset M$ has k-dimensional volume* 0.

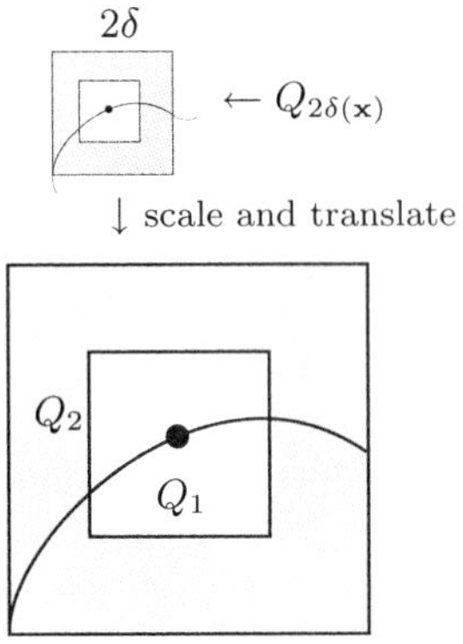

FIGURE A20.2.
TOP: Our C^1 map $\mathbf{f}$ is defined on the bottom edge of the larger box $Q_{2\delta(\mathbf{x})}$ (of sidelength 2δ), not just on the (bottom edge of the) smaller box $Q_{\delta(\mathbf{x})}$ inside it. We require $\mathbf{f}$ to be defined on the larger box to be sure that its derivative is bounded on the smaller box. This is a standard trick. BOTTOM: We translate and scale $Q_{\delta(\mathbf{x})}$ to turn it into the unit cube $Q = Q_1 \times Q_2$. Our map $\widetilde{\mathbf{f}}$ extends to the larger box Q_1'.

Proof. By Definition 5.2.1, it is enough to show that for any $R > 0$ the set $X \cap \overline{B}_R(\mathbf{0})$ has k-dimensional volume 0; such an intersection is closed and bounded, hence compact. Denote by $Q_r(\mathbf{x}) \subset \mathbb{R}^n$ the open box of sidelength r centered at $\mathbf{x}$, i.e., the set $\{\, \mathbf{y} \in \mathbb{R}^n \mid |y_i - x_i| < r/2 \,\}$. For every $\mathbf{x} \in X$ there exists $\delta(\mathbf{x}) > 0$ such that $Q_{2\delta(\mathbf{x})}(\mathbf{x}) \cap M$ is the graph of a C^1 map $\mathbf{f}$ representing $n-m$ variables in terms of the other m. The smaller boxes $Q_{\delta(\mathbf{x})}(\mathbf{x})$ form an open cover of X, so by Theorem A3.3 (Heine-Borel), there are finitely many points $\mathbf{x}_i \in X,\ i = 1, \dots, K$ such that

$$X \cap \overline{B}_R(\mathbf{0}) \subset \bigcup_{i=1}^{K} Q_{\delta(\mathbf{x}_i)}(\mathbf{x}_i). \qquad \text{A20.5}$$

It is then enough to prove Proposition 5.2.2 for one such box $Q_{\delta(\mathbf{x}_i)}$.

By translating and scaling, we may assume that this box is the unit cube $Q = Q_1 \times Q_2 \subset \mathbb{R}^n$, where Q_1 is the unit cube in $\mathbb{R}^m$ and Q_2 is the unit cube in $\mathbb{R}^{n-m}$. By reordering the variables if necessary, we may assume that $M \cap Q$ is the graph of a C^1 mapping $\widetilde{\mathbf{f}} : Q_1 \to Q_2$ representing $x_{m+1}, \dots, x_n$ in terms of $x_1, \dots, x_m$.

Moreover, $\widetilde{\mathbf{f}}$ extends to a concentric cube Q_1' with sidelength 2. Thus $\mathbf{y} \mapsto [\mathbf{D}\widetilde{\mathbf{f}}(\mathbf{y})]$ is continuous on the closed cube $\overline{Q_1}$, and $|[\mathbf{D}\widetilde{\mathbf{f}}(\mathbf{y})]|$ is bounded on Q_1: by Theorem 1.6.9 there exists L such that for all $\mathbf{y} \in Q_1$ we have $|[\mathbf{D}\widetilde{\mathbf{f}}(\mathbf{y})]| \le L$; see Figure A20.2.

It follows that for any N, and for any cube $C_1 \in \mathcal{D}_N(\mathbb{R}^m)$ with $C_1 \subset \overline{Q_1}$, the set of cubes $C = C_1 \times C_2 \in \mathcal{D}_N(\mathbb{R}^n)$ that intersect M has cardinality at most $(L\sqrt{m}/2)^{n-m}$. Indeed, if $\mathbf{y}_1$ is the center of C_1 and $\mathbf{y}_2$ is some other point in C_1, then

$$|\mathbf{y}_1 - \mathbf{y}_2| \le \frac{\sqrt{m}}{2}\frac{1}{2^N}, \quad \text{so (Cor. 1.9.2)} \quad |\mathbf{f}(\mathbf{y}_1) - \mathbf{f}(\mathbf{y}_2)| \le \frac{L\sqrt{m}}{2^N}. \qquad \text{A20.6}$$

Within that distance of $\mathbf{f}(\mathbf{y}_1)$, there are at most

$$\left(\frac{L\sqrt{m}}{2}\right)^{n-m} \quad \text{cubes of} \quad \mathcal{D}_N(\mathbb{R}^{n-m}). \qquad \text{A20.7}$$

Thus the total number of cubes in $\mathcal{D}_N(\mathbb{R}^n)$ needed to cover $M \cap Q$ is at most $2^{mN}(L\sqrt{m}/2)^{n-m}$, with k-dimensional volume

$$\left(2^{mN}\left(\frac{L\sqrt{m}}{2}\right)^{n-m}\right)2^{-kN} = 2^{-(k-m)N}\left(\frac{L\sqrt{m}}{2}\right)^{n-m}. \qquad \text{A20.8}$$

Since $k > m$, this tends to 0 as N tends to ∞. □

Proposition 6.6.17 (Products of pieces-with-boundary). *Let $M \subset \mathbb{R}^m$ and $P \subset \mathbb{R}^p$ be oriented manifolds, M of dimension k and P of dimension l. If $X \subset M$ and $Y \subset P$ are pieces-with-boundary, then $M \times P \subset \mathbb{R}^{m+p}$ is a $(k+l)$-dimensional oriented manifold and $X \times Y$ is a piece-with-boundary of $M \times P$.*

Proof. The first statement is the object of Exercises 3.30 and 6.33. We will show that $X \times Y$ is a piece-with-boundary. Clearly $X \times Y$ is compact, since it is closed and bounded. It should also be clear that

$$\partial_{M\times P}(X \times Y) = \Big((\partial_M X) \times Y\Big) \cup \Big(X \times (\partial_P Y)\Big). \qquad A20.9$$

Below we write the matrix multiplication

$$\left[\mathbf{D}\varphi\begin{pmatrix}\mathbf{x}\\ \mathbf{y}\end{pmatrix}\right]^\top \left[\mathbf{D}\varphi\begin{pmatrix}\mathbf{x}\\ \mathbf{y}\end{pmatrix}\right]$$

from equation A20.12:

$$\overbrace{\begin{bmatrix}\mathbf{D}\gamma & [0]\\ [0] & \mathbf{D}\delta\end{bmatrix}}^{\left[\mathbf{D}\varphi\begin{pmatrix}\mathbf{x}\\ \mathbf{y}\end{pmatrix}\right]}$$

$$\underbrace{\begin{bmatrix}\mathbf{D}\gamma^\top & [0]\\ [0] & \mathbf{D}\delta^\top\end{bmatrix}}_{\left[\mathbf{D}\varphi\begin{pmatrix}\mathbf{x}\\ \mathbf{y}\end{pmatrix}\right]^\top} \begin{bmatrix}\mathbf{D}\gamma^\top\mathbf{D}_\gamma & [0]\\ [0] & \mathbf{D}\delta^\top\mathbf{D}\delta\end{bmatrix}$$

By Theorem 4.8.10, the det on the left of equation A20.12 equals the product of det's on the right.

Moreover, the points of

$$(\partial^s_M X \times \overset{\circ}{Y}) \cup (\overset{\circ}{X} \times \partial^s_P Y) \subset \partial_{M\times P}(X \times Y) \qquad A20.10$$

are smooth points of the boundary, defined by the functions

$$\widetilde{g}\begin{pmatrix}\mathbf{x}\\ \mathbf{y}\end{pmatrix} = g(\mathbf{x}) \quad \text{and} \quad \widetilde{h}\begin{pmatrix}\mathbf{x}\\ \mathbf{y}\end{pmatrix} = h(\mathbf{y}) \qquad A20.11$$

where g and h define X in M near $\mathbf{x}$ and Y in P near $\mathbf{y}$ respectively. The derivatives of $\widetilde{g}$ and $\widetilde{h}$ are onto, since they have the same codomain as g and h, and these have derivatives that are onto at $\mathbf{x}$ and $\mathbf{y}$ respectively.

Let us see that the smooth boundary of $X \times Y$ has finite $(k+l-1)$-volume. We will show it for $(\partial_M X) \times Y$; the argument for $X \times \partial_P Y$ is identical. Let $\gamma : U \to \mathbb{R}^m$ parametrize $\partial^s_M X$ (with $U \subset \mathbb{R}^{k-1}$) and let $\delta : V \to \mathbb{R}^p$ parametrize Y (with $V \subset \mathbb{R}^l$). By hypothesis, the integrals

$$\underbrace{\int_U \sqrt{\det([\mathbf{D}\gamma(\mathbf{x})]^\top[\mathbf{D}\gamma(\mathbf{x})]}|d^{k-1}\mathbf{x}|}_{\operatorname{vol}_{k-1}\partial^s_M X<\infty} \quad \text{and} \quad \underbrace{\int_V \sqrt{\det([\mathbf{D}\delta(\mathbf{y})]^\top[\mathbf{D}\delta(\mathbf{y})]}|d^l\mathbf{y}|}_{\operatorname{vol}_l Y<\infty}$$

are both finite. The map $\varphi \overset{\text{def}}{=} \begin{pmatrix}\gamma\\ \delta\end{pmatrix} : U \times V \to \mathbb{R}^m \times \mathbb{R}^p$ parametrizes $(\partial^s_M X) \times Y$, and

$$\det\left[\mathbf{D}\varphi\begin{pmatrix}\mathbf{x}\\ \mathbf{y}\end{pmatrix}\right]^\top \left[\mathbf{D}\varphi\begin{pmatrix}\mathbf{x}\\ \mathbf{y}\end{pmatrix}\right] = \Big(\det[\mathbf{D}\gamma(\mathbf{x})]^\top[\mathbf{D}\gamma(\mathbf{x})]\Big)\Big(\det[\mathbf{D}\delta(\mathbf{y})]^\top[\mathbf{D}\delta(\mathbf{y})]\Big) \qquad A20.12$$

since the derivative of φ is an $(m+p)\times(k-1+l)$-matrix with the $m \times (k-1)$-matrix $[\mathbf{D}\gamma(\mathbf{x})]$ in the upper left corner, and the $p \times l$ matrix $[\mathbf{D}\delta(\mathbf{y})]$ in the bottom right, with all other entries 0 (see the margin). Then we have

$$\int_{U\times V} \sqrt{\det\left[\mathbf{D}\varphi\begin{pmatrix}\mathbf{x}\\ \mathbf{y}\end{pmatrix}\right]^\top \left[\mathbf{D}\varphi\begin{pmatrix}\mathbf{x}\\ \mathbf{y}\end{pmatrix}\right]}|d^{k-1}\mathbf{x}\, d^l\mathbf{y}| = \int_U \sqrt{\det[\mathbf{D}\gamma(\mathbf{x})]^\top[\mathbf{D}\gamma(\mathbf{x})]}|d^{k-1}\mathbf{x}| \int_V \sqrt{\det[\mathbf{D}\delta(\mathbf{y})]^\top[\mathbf{D}\delta(\mathbf{y})]}|d^l\mathbf{y}|.$$

Use the superscript ns to denote nonsmooth. The description in equation A20.10 of the smooth boundary of $X \times Y$ tells us that the nonsmooth boundary of $X \times Y$ is

$$\Big((\partial^{ns}_M X) \times Y\Big) \cup \Big(X \times (\partial^{ns}_P Y)\Big) \cup \Big(\partial_M X \times \partial_P Y\Big), \qquad A20.13$$

and we must show that these have $(k+l-1)$-volume 0. The second term is just like the first. For the first, note that the cubes of $\mathcal{D}_N(\mathbb{R}^{m+p})$ that intersect $(\partial_M^{ns} X) \times Y$ are the cubes $C' \times C'' \in \mathcal{D}_N(\mathbb{R}^m) \times \mathcal{D}_N(\mathbb{R}^p)$ such that $C' \cap \partial_M^{ns} X \neq \emptyset$ and $C'' \cap Y \neq \emptyset$. We have

$$\lim_{N\to\infty} \sum_{\substack{C' \in \mathcal{D}_N(\mathbb{R}^m) \\ C' \cap \partial_M^{ns} X \neq \emptyset}} \left(\frac{1}{2^N}\right)^{k-1} = 0, \qquad A20.14$$

where $1/2^N$ is the sidelength of the cubes. We need to see that the sequence

$$N \mapsto \sum_{\substack{C'' \in \mathcal{D}_N(\mathbb{R}^p) \\ C'' \cap Y \neq \emptyset}} \left(\frac{1}{2^N}\right)^{l} \qquad A20.15$$

remains bounded as $N \to \infty$. Since Y is compact, by Heine-Borel we can cover Y by finitely many open sets in which P is the graph of a C^1 function expressing $p-l$ variables in terms of the other l; it is enough to show that the sequence A20.15 remains bounded for such a graph $f : W \to \mathbb{R}^{p-l}$. Further we may assume that all the partial derivatives of f are bounded on W, and it then follows that there exists a constant K such that for any cube $C'' \in \mathcal{D}_N(\mathbb{R}^l)$ with $C'' \subset W$, the graph of f restricted to C'' intersects at most K' cubes of $\mathcal{D}_N(\mathbb{R}^m)$. This shows that the sequence A20.15 is bounded: there are at most $K''2^{Nl}$ cubes of $\mathcal{D}_N(\mathbb{R}^l)$ in W for some constant K'', which is essentially the l-dimensional volume of W, so it requires at most $K'K''2^{Nl}$ cubes of $\mathcal{D}_N(\mathbb{R}^{m+p})$ to cover the graph of f, and the product of this by the sidelength raised to the lth power, i.e., 2^{-Nl}, is bounded by $K'K''$. This uses part 4 of Theorem 1.5.16.

Finally, we need to show that $\partial_M X \times \partial_N Y$ (the third term in A20.13) has $(k+l-1)$-volume 0. This is similar: we can find constants L' and L'' such that for any N, the boundary $\partial_M X$ can be covered by at most $L'2^{N(k-1)}$ cubes of $\mathcal{D}_N(\mathbb{R}^m)$ and the boundary $\partial_P Y$ can be covered by at most $L''2^{N(l-1)}$ cubes of $\mathcal{D}_N(\mathbb{R}^p)$. Thus $\partial_M X \times \partial_P Y$ can be covered by at most $L'L''2^{N(k+l-2)}$ cubes of $\mathcal{D}_N(\mathbb{R}^{m+p})$, and raising the sidelength to the $(k+l-1)$th power leaves $L'L''2^{-N}$, which goes to 0 with N. □

A21 Lebesgue measure and proofs for Lebesgue integrals

Theorem 4.11.4: Many mathematicians would say one should put this result off until Lebesgue measure is available, where it is easy and natural. With our approach, the theorem is harder, but setting up the Lebesgue integral is immensely easier.

The function doing the "dominating" is $R\mathbf{1}_{B_R(\mathbf{0})}$.

This section contains proofs of various statements in Section 4.11, as well as a brief discussion of Lebesgue measure.

Theorem 4.11.4 (Dominated convergence for Riemann integrals). *Let $f_k : \mathbb{R}^n \to \mathbb{R}$ be a sequence of R-integrable functions. Suppose there exists R such that all f_k have their support in $B_R(\mathbf{0})$ and satisfy $|f_k| \le R$. Let $f : \mathbb{R}^n \to \mathbb{R}$ be an R-integrable function such that the set of $\mathbf{x}$ where $\lim_{k\to\infty} f_k(\mathbf{x}) \neq f(\mathbf{x})$ has measure 0. Then*

$$\lim_{k\to\infty} \int_{\mathbb{R}^n} f_k(\mathbf{x})\,|d^n\mathbf{x}| = \int_{\mathbb{R}^n} f(\mathbf{x})\,|d^n\mathbf{x}|.$$

The dominated convergence theorem for Riemann integrals was proved by the Italian mathematician Cesare Arzela in 1885. Our proof follows the proof of a related result due to W. F. Eberlein, *Communications on Pure and Applied Mathematics*, 10 (1957), pp. 357–360. The trick of translating the problem into an inner product space so that the parallelogram law could be used (in equation A21.20) is due to Marcel Riesz.

To prove this theorem, we begin by proving an innocent-looking result about exchanging limits and integrals. Much of the difficulty is concentrated in this proposition, which could be used as the basis of the entire theory. Recall (Definition 4.1.18) that a set is pavable if it has a well-defined volume – i.e., if its indicator function is integrable.

Proposition A21.1 (Monotone convergence for Riemann integrals). *Let $f_k : \mathbb{R}^n \to \mathbb{R}$ be a sequence of R-integrable functions, all with support in the unit cube $Q \subset \mathbb{R}^n$ and satisfying $1 \geq f_1 \geq f_2 \geq \cdots \geq 0$. Let $B \subset Q$ be a pavable subset with $\mathrm{vol}_n(B) = 0$, and suppose that*

$$\lim_{k\to\infty} f_k(\mathbf{x}) = 0 \quad \text{if } \mathbf{x} \in Q,\ \mathbf{x} \notin B. \qquad \text{A21.1}$$

Then

$$\lim_{k\to\infty} \int_{\mathbb{R}^n} f_k(\mathbf{x})\,|d^n\mathbf{x}| = \int_{\mathbb{R}^n} \lim_{k\to\infty} f_k(\mathbf{x})\,|d^n\mathbf{x}| = 0. \qquad \text{A21.2}$$

In inequality A21.3,

$$\int_{A_k} f_k |d^n\mathbf{x}|$$

is $\leq K$ because $f_k \leq 1$ and we have assumed $\mathrm{vol}_n A_k < K$. The other term,

$$\int_{Q-A_k} f_k |d^n\mathbf{x}|,$$

is $\leq K$ because $f_k \leq K$ on $Q - A_k$ and

$$\mathrm{vol}_n(Q - A_k) \leq \mathrm{vol}_n Q = 1.$$

Proof. The sequence $k \mapsto \int_{\mathbb{R}^n} f_k(\mathbf{x})\,|d^n\mathbf{x}|$ is nonincreasing and nonnegative, so it has a limit, which we call $2K$. We will suppose that $K > 0$ and derive a contradiction: we will find a set of positive volume of points $\mathbf{x} \notin B$ such that $\lim_{k\to\infty} f_k(\mathbf{x}) \geq K$, which contradicts equation A21.1.

Let $A_k \subset Q$ be the set $A_k = \{\, \mathbf{x} \in Q \mid f_k(\mathbf{x}) \geq K \,\}$, so that since the sequence f_k is nonincreasing, the sets A_k are nested: $A_1 \supset A_2 \supset \cdots$.

It is tempting to try to show that $\mathrm{vol}_n \cap_k (A_k) \neq 0$, which seems obvious, since the A_k are nested, and $\mathrm{vol}_n A_k \geq K$ for each k. Indeed, if $\mathrm{vol}_n A_k < K$ for some k, then

$$2K \leq \int_{\mathbb{R}^n} f_k |d^n\mathbf{x}| = \int_{A_k} f_k |d^n\mathbf{x}| + \int_{Q-A_k} f_k |d^n\mathbf{x}| < K + K, \qquad \text{A21.3}$$

i.e., $2K < 2K$. Thus the intersection should have volume at least K, and since B has volume 0, there should be points in the intersection $\cap_k(A_k)$ that are not in B, i.e., points $\mathbf{x} \notin B$ such that $\lim_{k\to\infty} f_k(\mathbf{x}) \geq K$.

The problem with this argument is that A_k might fail to be pavable (see Exercise A21.1), so we cannot blithely speak of its volume. Even if the A_k are pavable, their intersection might not be pavable (see Exercise A21.2). In this particular case this is just an irritant, not a fatal flaw; we need to doctor the A_k a bit. We can replace $\mathrm{vol}_n(A_k)$ by the *lower volume* $\underline{\mathrm{vol}}_n(A_k) \stackrel{\text{def}}{=} L(\mathbf{1}_{A_k})$, i.e., the sum of the volumes of all dyadic cubes (of all sizes) contained in a maximal collection of disjoint dyadic cubes contained in A_k. (Recall from Definition 4.1.10 that we denote by $L(f)$ the lower integral of f: $L(f) = \lim_{N\to\infty} L_N(f)$.)

Inequality A21.4: The equality marked 1 is justified by Corollary 4.1.15, which says that

$$\sup(f_k(\mathbf{x}), K) \text{ and } \inf(f_k(\mathbf{x}), K)$$

are integrable. For the inequality marked 2, see the margin note on the next page.

Even this lower volume is at least K, since for any numbers a and b, we have $a = \inf(a, b) + \sup(a, b) - b$:

$$\begin{aligned} 2K &\leq \int_Q f_k(\mathbf{x})|d^n\mathbf{x}| = \int_Q \inf(f_k(\mathbf{x}), K)|d^n\mathbf{x}| + \int_Q \sup(f_k(\mathbf{x}), K)|d^n\mathbf{x}| - K \\ &\leq \int_Q \sup(f_k(\mathbf{x}), K)|d^n\mathbf{x}| \underset{1}{=} L(\sup(f_k(\mathbf{x}), K)) \underset{2}{\leq} K + \underline{\mathrm{vol}}_n(A_k). \end{aligned} \qquad \text{A21.4}$$

So $\underline{\text{vol}}_n(A_k) \geq K$. Now we want to see that if we remove all of B from the A_k, what is left still has positive volume. To find points in the intersection of the A_k that are not in B, we adjust our A_k. First, choose a number N such that the union of all the dyadic cubes in $\mathcal{D}_N(\mathbb{R}^n)$ whose closures intersect B has total volume $< K/3$. Let B' be the union of all these cubes, so that $\text{vol}_n B' < K/3$, and let $A'_k = A_k - B'$. Note that the A'_k are still nested, and $\underline{\text{vol}}_n(A'_k) \geq 2K/3$. Next choose ϵ so small that $\epsilon/(1-\epsilon) < K/3$, and for each k let $A''_k \subset A'_k$ be a finite union of closed dyadic cubes, such that $\underline{\text{vol}}_n(A'_k - A''_k) < \epsilon^k$.

Since the difference $(A'_k - A''_k)$ is very small, the A''_k almost fill up the A'_k, so their volume is also nonnegligible.

Unfortunately, now the A''_k are no longer nested, so define

$$A'''_k = A''_1 \cap A''_2 \cap \cdots \cap A''_k. \qquad A21.5$$

We need to show that the A'''_k are nonempty; this is true, since

$$\begin{aligned}\underbrace{\text{vol}_n A'''_k}_{\text{vol}_n(A''_1\cap A''_2\cap\cdots\cap A''_k)} &\geq \underbrace{\underline{\text{vol}}_n A'_k}_{\underline{\text{vol}}_n(A'_1\cap A'_2\cap\cdots\cap A'_k)} - \left(\epsilon + \epsilon^2 + \cdots + \epsilon^k\right)\\ &> \frac{2K}{3} - \frac{\epsilon}{1-\epsilon}\\ &> \frac{K}{3}. \qquad A21.6\end{aligned}$$

Now the punch line: The A'''_k form a decreasing intersection of compact sets, so their intersection is nonempty (see Theorem A3.1). Let $\mathbf{x} \in \cap_k A'''_k$, then all $f_k(\mathbf{x}) \geq K$, but $\mathbf{x} \notin B$, contradicting equation A21.1. □

Inequality A21.4 on page 777: The inequality marked 2 isn't obvious. It is enough to show that

$$L_N(\sup(f_k(\mathbf{x}), K)) \leq K + L_N(\mathbf{1}_{A_k})$$

for any N. Take any cube

$$C \in \mathcal{D}_N(\mathbb{R}^n).$$

Either $m_C(f_k) \leq K$, in which case,

$$m_C(f_k)\,\text{vol}_n C \leq K\,\text{vol}_n C,$$

or $m_C(f_k) > K$. In the latter case, since $f_k \leq 1$,

$$m_C(f_k)\,\text{vol}_n C \leq \text{vol}_n C.$$

The first case contributes at most $K\,\text{vol}_n Q = K$ to the lower integral. The second contributes at most $L_N(\mathbf{1}_{A_k}) = \underline{\text{vol}}_n(A_k)$, since any cube for which $m_C(f_k) > K$ is entirely within A_k, and thus contributes to the lower sum. This is why the possible nonpavability of A_k is just an irritant. For typical nonpavable sets, like the rationals or the irrationals, the lower volume is 0. Here there definitely are whole dyadic cubes completely contained in A_k.

Inequality A21.6: The inequality

$$\epsilon + \epsilon^2 + \cdots + \epsilon^k < \frac{\epsilon}{1-\epsilon}.$$

follows from equation 0.5.4 with $a = r = \epsilon$

Corollary A21.2. *Let $k \mapsto h_k$ be a sequence of R-integrable nonnegative functions on the unit cube $Q \subset \mathbb{R}^n$, and let h be an R-integrable function on Q, satisfying $0 \leq h(\mathbf{x}) \leq 1$. Let $B \subset Q$ be a pavable set of volume 0. If $\sum_{k=1}^{\infty} h_k(\mathbf{x}) \geq h(\mathbf{x})$ when $\mathbf{x} \notin B$, then*

$$\sum_{k=1}^{\infty} \int_Q h_k(\mathbf{x})|d^n\mathbf{x}| \geq \int_Q h(\mathbf{x})|d^n\mathbf{x}|. \qquad A21.7$$

Proof. Set $g_k = \sum_{i=1}^{k} h_i$, which is a nondecreasing sequence of nonnegative integrable functions, and $\widetilde{g}_k = \inf(g_k, h)$, which is still a nondecreasing sequence of nonnegative integrable functions. Finally, set $f_k = h - \widetilde{g}_k$; these functions f_k satisfy the hypotheses of Proposition A21.1. So

$$\begin{aligned}0 \underbrace{=}_{\text{eq. A21.2}} \lim_{k\to\infty} \int f_k|d^n\mathbf{x}| &= \int h|d^n\mathbf{x}| - \lim_{k\to\infty} \int \widetilde{g}_k|d^n\mathbf{x}|\\ &\geq \int h|d^n\mathbf{x}| - \lim_{k\to\infty} \int g_k|d^n\mathbf{x}| = \int h|d^n\mathbf{x}| - \lim_{k\to\infty} \sum_{i=1}^{k} \int h_i|d^n\mathbf{x}|\\ &= \int h|d^n\mathbf{x}| - \sum_{i=1}^{\infty} \int h_i|d^n\mathbf{x}|. \quad \square \qquad A21.8\end{aligned}$$

In the second line of inequality A21.8 we can exchange sum and integral because the sum is finite.

Simplifications to the dominated convergence theorem

Let us simplify Theorem 4.11.4. First, it is enough prove the statement when $f = 0$. Indeed, we can replace the f_k by $f_k - f$, which converge to 0. So we need to prove that if all f_k are uniformly bounded with the same bounded support and converge to 0 except on a set X of measure 0, then their integrals also converge to 0.

Note that the function h of equation A21.9 is not related to the h in Corollary A21.2. It is a deformation of the constant function 1. It "dips" for $\mathbf{x} \in X_\epsilon$; it dips only slightly because it is continuous.

Property 2 involves two cases. If $\mathbf{x} \in X_\epsilon$, then $h(\mathbf{x}) < 1$ and

$$h^k(\mathbf{x}) \to 0 \text{ as } k \to \infty.$$

If $\mathbf{x} \notin X_\epsilon$, then $h(\mathbf{x}) = 1$ but $f_k(\mathbf{x}) \to 0$ as $k \to \infty$, since we are assuming here that $f = 0$.

Property 3: If $\mathbf{x} \notin X_\epsilon$, then $h(\mathbf{x}) = 1$, so the difference on the left side is 0. If $\mathbf{x} \in X_\epsilon$, then $0 \le h(\mathbf{x}) < 1$, so

$$\begin{aligned} |f_k - g_k| &= |f_k - h^k f_k| \\ &\le |1 - h^k||f_k| \le |f_k| \le R, \end{aligned}$$

and we are integrating over X_ϵ, which has volume $\le \epsilon$.

The next step is to show that if the statement is true when the f_k converge everywhere, then it is true when they converge almost everywhere. Choose $\epsilon > 0$, and cover X by a union $X_\epsilon \stackrel{\text{def}}{=} \cup_i B_i$ of open boxes with $\sum_i \text{vol}_n B_i \le \epsilon$. Now find a continuous function $h : \mathbb{R}^n \to [0,1]$ such that[5]

$$h(\mathbf{x}) = 1 \text{ when } \mathbf{x} \notin X_\epsilon \text{ and } 0 \le h(\mathbf{x}) < 1 \text{ for } \mathbf{x} \in X_\epsilon. \qquad A21.9$$

Now consider the functions $g_k(\mathbf{x}) = (h(\mathbf{x}))^k f_k(\mathbf{x})$. These functions are Riemann integrable, since each g_k is the product of a continuous function and a Riemann integrable function: it is continuous where f_k is continuous, so except on a set of measure 0, and it is bounded with bounded support. The g_k also have the following properties:

1. All g_k have support in $B_R(\mathbf{0})$ and they satisfy $|g_k(\mathbf{x})| \le R$ for all $\mathbf{x} \in B_R(\mathbf{0})$, where R is as in Theorem 4.11.4.
2. $\lim_{k\to\infty} g_k(\mathbf{x}) = 0$ for every $\mathbf{x} \in \mathbb{R}^n$, as explained in the margin.
3. $\left| \int_{\mathbb{R}^n} g_k(\mathbf{x})|d^n\mathbf{x}| - \int_{\mathbb{R}^n} f_k(\mathbf{x})|d^n\mathbf{x}| \right| \le R\epsilon.$

Thus if we can show that $\lim_{k\to\infty} \int_{\mathbb{R}^n} g_k(\mathbf{x})|d^n\mathbf{x}| = 0$, we will have shown

$$\lim_{k\to\infty} \left| \int_{\mathbb{R}^n} f_k(\mathbf{x})|d^n\mathbf{x}| \right| < R\epsilon, \qquad A21.10$$

and as this is true for any ϵ, this yields the result.

Finally, write $g_k = g_k^+ - g_k^-$; it is clearly enough to prove the result for the positive and the negative parts separately. Taking these simplifications into account, it is enough to prove Proposition A21.3.

Since all f_k have their support in $B_R(\mathbf{0})$ and satisfy $|f_k| \le R$, by scaling, it is easy to see that we may assume that the f_k are bounded by 1 and have their support in the unit cube.

Proposition A21.3 (Simplified dominated convergence theorem). *Suppose $k \mapsto f_k$ is a sequence of R-integrable functions all satisfying $0 \le f_k \le 1$, and all with support in the unit cube Q. If for all $\mathbf{x} \in Q$ we have $\lim_{k\to\infty} f_k(\mathbf{x}) = 0$, then*

$$\lim_{k\to\infty} \int_{\mathbb{R}^n} f_k(\mathbf{x})|d^n\mathbf{x}| = \int_{\mathbb{R}^n} \lim_{k\to\infty} f_k(\mathbf{x})|d^n\mathbf{x}| = 0. \qquad A21.11$$

---

[5]For instance, if we denote by $d(\mathbf{x}, Y)$ the distance between $\mathbf{x}$ and the nearest point in Y: $d(\mathbf{x}, Y) \stackrel{\text{def}}{=} \inf_{\mathbf{y}\in Y} |\mathbf{x} - \mathbf{y}|$, then one might set

$$h(\mathbf{x}) = \Big(1 - d(\mathbf{x}, \mathbb{R}^n - X_\epsilon)\Big)^+.$$

(The $^+$ refers to the positive part of a function; see Definition 4.1.15.) Exercise A21.3 asks you to show this function h satisfies the required properties.

Proof. By passing to a subsequence, we may assume (Theorem 1.6.3) that

$$\lim_{k\to\infty}\int_{\mathbb{R}^n} f_k(\mathbf{x})|d^n\mathbf{x}| = M; \qquad A21.12$$

we will assume that $M > 0$ and derive a contradiction. Consider the following set K_p of linear combinations of the functions f_m:

$$K_p = \left\{ \sum_{m=p}^{\infty} a_m f_m \;\middle|\; a_m \geq 0,\ \text{finitely many terms nonzero}, \sum_{m=p}^{\infty} a_m = 1 \right\}. \qquad A21.13$$

We will need three properties of functions $k \in K_p$:

1. For any $k_p \in K_p$ and for any $\mathbf{x} \in Q$, we have $0 \leq k_p(\mathbf{x}) \leq 1$, since k_p is a weighted average of the f_k, and $0 \leq f_k \leq 1$.
2. For any point $\mathbf{x} \in Q$, and any sequence $p \mapsto k_p$ with $k_p \in K_p$, we have $\lim_{p\to\infty} k_p(\mathbf{x}) = 0$. Indeed, for any $\epsilon > 0$ and any fixed $\mathbf{x} \in Q$, we can find N such that all $f_m(\mathbf{x})$ satisfy $0 \leq f_m(\mathbf{x}) \leq \epsilon$ when $m \geq N$, so that when $p \geq N$ and $k_p \in K_p$, we have

$$k_p(\mathbf{x}) = \sum_{m=p}^{\infty} a_m f_m(\mathbf{x}) \leq \sum_{m=p}^{\infty} (a_m \epsilon) = \epsilon. \qquad A21.14$$

3. For any sequence $p \mapsto k_p$ with $k_p \in K_p$,

$$\lim_{p\to\infty}\int_Q k_p(\mathbf{x})|d^n\mathbf{x}| = M. \qquad A21.15$$

To see that property 3 is true, choose $\epsilon > 0$ and choose N so large that $|\int_Q f_m |d^n\mathbf{x}| - M| < \epsilon$ when $m \geq N$. Then, when $p > N$,

$$\begin{aligned}\left|\int_Q k_p(\mathbf{x})|d^n\mathbf{x}| - M\right| &= \left|\Big(\sum_{m=p}^{\infty} a_m \underbrace{\int_Q f_m(\mathbf{x})|d^n\mathbf{x}|\Big) - M}_{<\epsilon}\right| \\ &< \sum_{m=p}^{\infty}(a_m\epsilon) = \epsilon.\end{aligned} \qquad A21.16$$

Now we consider integrals of squares (this was Marcel Riesz's contribution). As p increases, the sets K_p are nested: $K_1 \supset K_2 \supset \cdots$, so if we set

$$d_p = \inf_{k_p \in K_p} \int_Q k_p^2(\mathbf{x})|d^n\mathbf{x}|, \qquad A21.17$$

From here on, k_p is a specific element of K_p. The function k_p is a substitute for the infimum of elements of K_p. This inf exists as a function but will usually not be an element of K_p; it may include infinitely many nonzero a_m (see equation A21.13), so will usually not be Riemann integrable.

the infima are nondecreasing. Thus the d_p form a nondecreasing sequence, bounded by 1 (by property 1 above), hence convergent. Choose a specific $k_p \in K_p$ such that

$$\int_Q k_p^2\,|d^n\mathbf{x}| < d_p + 1/p \qquad A21.18$$

Lemma A21.4. *For all $\epsilon > 0$, there exists N such that when $p, q \geq N$,*

$$\int_Q (k_p - k_q)^2|d^n\mathbf{x}| < \epsilon. \qquad A21.19$$

Proof of Lemma A21.4. Algebra says that

$$\int_Q \left(\frac{1}{2}(k_p - k_q)\right)^2 |d^n\mathbf{x}| + \int_Q \left(\frac{1}{2}(k_p + k_q)\right)^2 |d^n\mathbf{x}| = \frac{1}{2}\int_Q k_p^2 |d^n\mathbf{x}| + \frac{1}{2}\int_Q k_q^2 |d^n\mathbf{x}|. \qquad A21.20$$

Equation A21.20 is simple algebra, but it can also be thought of as an application of the parallelogram law.

But when $p, q \geq N$, then $\frac{1}{2}(k_p + k_q)$ is in K_N, so $\int_Q (\frac{1}{2}(k_p+k_q))^2 |d^n\mathbf{x}| \geq d_N$. This and inequality A21.18 give

$$\int_Q \left(\frac{1}{2}(k_p - k_q)\right)^2 |d^n\mathbf{x}| \leq \frac{1}{2}\left(d_p + \frac{1}{p}\right) + \frac{1}{2}\left(d_q + \frac{1}{q}\right) - d_N. \qquad A21.21$$

As $N \to \infty$, this can be made arbitrarily small: the $1/p$ and $1/q$ go to 0, and since the d_N converge, $d_p/2 + d_q/2$ cancels the d_N in the limit. □

Actually, the d_N are identically 0; they are a nondecreasing sequence of nonnegative numbers and by inequality A21.26 they converge to 0.

Using this lemma, we can choose a further subsequence h_q of the k_p such that

$$\sum_{q=1}^{\infty} \left(\int_Q (h_q - h_{q+1})^2(\mathbf{x})\, |d^n\mathbf{x}| \right)^{1/2} \qquad A21.22$$

For example, we can take

$$h_i = k_{N_i},$$

where N_i is chosen using Lemma A21.4, for $\epsilon = 1/2^i$.

converges. Notice that

$$h_q(\mathbf{x}) = (h_q - h_{q+1})(\mathbf{x}) + (h_{q+1} - h_{q+2})(\mathbf{x}) + \cdots, \qquad A21.23$$

since

$$h_q(\mathbf{x}) - \sum_{i=q}^{m} (h_i - h_{i+1})(\mathbf{x}) = h_{m+1}(\mathbf{x}), \qquad A21.24$$

which tends to 0 when $m \to \infty$ by inequality A21.14. In particular,

$$|h_q| \leq \sum_{m=q}^{\infty} |h_{m+1} - h_m|, \qquad A21.25$$

so we can apply Corollary A21.2 to get the first inequality below:

$$\int_Q h_q |d^n\mathbf{x}| \leq \sum_{m=q}^{\infty} \int_Q |h_m - h_{m+1}| |d^n\mathbf{x}| \leq \sum_{m=q}^{\infty} \left| \int_Q (h_m - h_{m+1})^2 |d^n\mathbf{x}| \right|^{1/2}. \qquad A21.26$$

Inequality A21.26: The second inequality follows from Schwarz's lemma for integrals (see Exercise A21.4). Write

$$\left(\int_Q |h_m - h_{m+1}| \cdot 1 |d^n\mathbf{x}|\right)^2 \leq \left(\int_Q |h_m - h_{m+1}|^2 |d^n\mathbf{x}|\right) \cdot \left(\int_Q 1^2 |d^n\mathbf{x}|\right) = \left(\int_Q |h_m - h_{m+1}|^2 |d^n\mathbf{x}|\right).$$

The sum on the right can be made arbitrarily small by taking q sufficiently large. This contradicts inequality A21.16 and the assumption $M > 0$. This proves Proposition A21.3, hence also Theorem 4.11.4. □

Proposition 4.11.5 (Convergence except on a set of measure 0). *If f_k for $k = 1, 2, \dots$ are Riemann-integrable functions on $\mathbb{R}^n$ such that*

$$\sum_{k=1}^{\infty} \int_{\mathbb{R}^n} |f_k(\mathbf{x})||d^n\mathbf{x}| < \infty,$$

then the series $\sum_{k=1}^{\infty} f_k(\mathbf{x})$ converges almost everywhere.

Proof of Proposition 4.11.5: We want to show that X (the set of points $\mathbf{x}$ where

$$\sum_{k=1}^{\infty} \left| f_k(\mathbf{x}) \right|$$

diverges) has measure 0. The strategy of this proof is that we construct a union of Y_i that covers X and whose total volume is at most ϵ.

We thank John Milnor for his contributions to this proof.

Proof. Set

$$A \stackrel{\text{def}}{=} \sum_{k=1}^{\infty} \int_{\mathbb{R}^n} |f_k(\mathbf{x})||d^n\mathbf{x}| \qquad A21.27$$

We will take X to be the set of $\mathbf{x} \in \mathbb{R}^n$ such that $\sum_{k=1}^{\infty} |f_k(\mathbf{x})|$ diverges. By Theorem 0.5.8, $\sum_{k=1}^{\infty} f_k(\mathbf{x})$ converges for all $\mathbf{x} \notin X$, since the series is absolutely convergent; the hard part is to show that X has measure 0.

Remark. This may seem a triviality: by Proposition 4.1.14, for any m,

$$\begin{aligned} \int_{\mathbb{R}^n} \left(\sum_{k=1}^{m} |f_k(\mathbf{x})| \right) |d^n\mathbf{x}| &= \sum_{k=1}^{m} \int_{\mathbb{R}^n} |f_k(\mathbf{x})||d^n\mathbf{x}| \\ &\le \sum_{k=1}^{\infty} \int_{\mathbb{R}^n} |f_k(\mathbf{x})||d^n\mathbf{x}| = A, \end{aligned} \qquad A21.28$$

so the set of $\mathbf{x} \in \mathbb{R}^n$ where $\sum_{k=1}^{m} |f_k(\mathbf{x})| \ge A/\epsilon$ cannot have volume $> \epsilon$. As m increases, we should get an increasing sequence of sets, all of volume $\le \epsilon$, whose union should still have arbitrarily small volume. Unfortunately, this "obvious" argument doesn't work: the sets of $\mathbf{x}$ where $\sum_{k=1}^{m} |f_k(\mathbf{x})| \ge A/\epsilon$ may fail to have a well-defined volume; see Exercise A21.1.

The only case where it does work is when $A = 0$. If $A = 0$, the function f_k cannot be continuous at $\mathbf{x}$ if $f_k(\mathbf{x}) \ne 0$. Since (Theorem 4.4.8) f_k is continuous almost everywhere, it must vanish almost everywhere. So $\sum_{k=1}^{\infty} f_k$ must vanish except on a countable union of sets of measure 0, and such a union has measure 0, by Theorem 4.4.4. $\triangle$

Note that p_0 may be a large number, and p_1 may be a much bigger number yet (not just 1 more than p_0). The definition of h_m says that the sum of the first p_0 terms $|f_k|$ accounts for 15/16 of the integral A:

$$\begin{aligned} &\int_{\mathbb{R}^n} h_0(\mathbf{x})\,|d^n\mathbf{x}| \\ &= \int_{\mathbb{R}^n} \sum_{k=1}^{p_0} |f_k||d^n\mathbf{x}| \\ &= \sum_{k=1}^{p_0} \int_{\mathbb{R}^n} |f_k||d^n\mathbf{x}| \\ &\ge \frac{15}{16} \sum_{k=1}^{\infty} \int_{\mathbb{R}^n} |f_k(\mathbf{x})||d^n\mathbf{x}| \\ &= \frac{15}{16} A. \end{aligned}$$

The sum of the first p_1 terms accounts for 63/64 of the integral A, the first p_2 terms account for 255/256 of A, and so on. The object of all this is to replace the f_k by the $g_k = h_k - h_{k-1}$; these g_k get small at a geometric rate.

If we had a definition of measure (as opposed to just measure 0), it would be simple to prove the statement when $A > 0$. Without such a definition, we will need to be considerably more elaborate. First, let us choose a sequence of integers $m \mapsto p_m$, $m = 0, 1, 2, \dots$ such that if we set

$$h_m = \sum_{k=1}^{p_m} |f_k|, \quad \text{then} \quad \int_{\mathbb{R}^n} h_m(\mathbf{x})\,|d^n\mathbf{x}| \ge A\left(1 - \frac{1}{4^{m+2}}\right); \qquad A21.29$$

for each $\mathbf{x}$, divergence of the series $\sum_{k=1}^{\infty} |f_k(\mathbf{x})|$ is the same as divergence of the sequence $m \mapsto h_m(\mathbf{x})$. Now set $g_m = h_m - h_{m-1}$, for $m = 1, 2, \dots$. Since

$$h_m = h_0 + (h_1 - h_0) + \cdots + (h_m - h_{m+1}) = h_0 + g_1 + \cdots + g_m, \qquad A21.30$$

divergence of $m \mapsto h_m(\mathbf{x})$ is the same as divergence of the series $\sum_{m=1}^{\infty} g_m(\mathbf{x})$. By inequality A21.29, the integrals of the h_m "eat" almost all of A, so the integrals of the g_m are tiny:

As m increases, the integrals of the h_m are an increasing sequence of positive numbers converging to A. The g_m are ≥ 0, so since the integral $\int_{\mathbb{R}^n} g_m(\mathbf{x})|d^n\mathbf{x}|$ is small, the g_m can only be big on a tiny set; there can be no cancellations accounting for the small integral.

$$\begin{aligned}\int_{\mathbb{R}^n} g_m(\mathbf{x})|d^n\mathbf{x}| &= \left(A - \int_{\mathbb{R}^n} h_{m-1}(\mathbf{x})|d^n\mathbf{x}|\right) - \overbrace{\left(A - \int_{\mathbb{R}^n} h_m(\mathbf{x})|d^n\mathbf{x}|\right)}^{>0}\\ &\leq A - \int_{\mathbb{R}^n} h_{m-1}(\mathbf{x})|d^n\mathbf{x}| \qquad A21.31\\ &\leq \frac{A}{4^{m+1}}.\end{aligned}$$

Choose $\epsilon > 0$. For $m = 1, 2, \dots$, let Y_m be the union of the cubes $C \in \mathcal{D}_{Nm}(\mathbb{R}^n)$ such that $M_C(g_m) > \frac{A}{2^m\epsilon}$; see Figure A21.1. We will find $N_1 < N_2 < N_3, \dots$ such that

$$\text{if } M_C(g_m) > \frac{A}{2^m\epsilon}, \text{ then } \operatorname{vol}_n(Y_m) \leq \frac{\epsilon}{2^m}. \qquad A21.32$$

Let N_m be so large that

$$U_{N_m}(g_m) - L_{N_m}(g_m) \leq \frac{A}{4^{m+1}}. \qquad A21.33$$

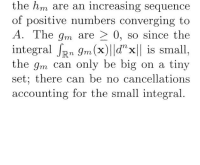

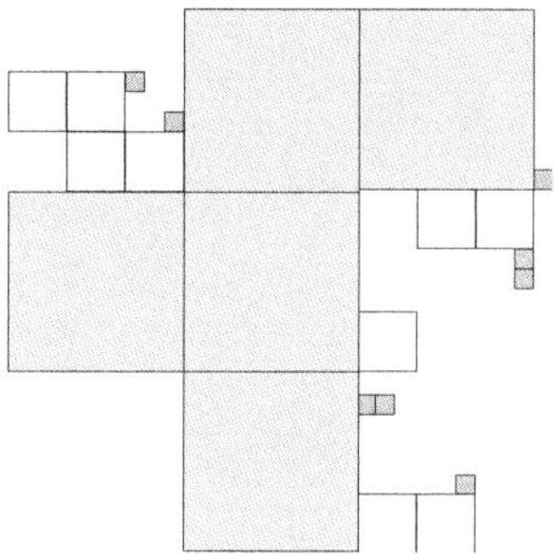

FIGURE A21.1.

The large shaded cubes, in $\mathcal{D}_{N_1}$, make up Y_1; by inequality A21.35, their combined volume is $\leq \epsilon/2$. The small white cubes, in $\mathcal{D}_{N_2}$, make up Y_2, with total volume $\leq \epsilon/4$. The tiny shaded cubes, in $\mathcal{D}_{N_3}$, make up Y_3, with total volume $\leq \epsilon/8$.

Then

$$\begin{aligned}\operatorname{vol}_n(Y_m)\frac{A}{2^m\epsilon} &\leq \sum_{\substack{C \subset Y_m\\ C \in \mathcal{D}_{N_m}(\mathbb{R}^n)}} M_C(g_m)\operatorname{vol}_n(C) \leq U_{N_m}(g_m) \qquad A21.34\\ &\leq \int_{\mathbb{R}^n} g_m(\mathbf{x})|d^n\mathbf{x}| + \frac{A}{4^{m+1}} \leq \frac{2A}{4^{m+1}} < \frac{A}{4^m}.\end{aligned}$$

This gives

$$\operatorname{vol}_n(Y_m) \leq \frac{A}{4^m}\frac{2^m\epsilon}{A} = \frac{\epsilon}{2^m}. \qquad A21.35$$

Thus if $Y = \bigcup_m Y_m$, then Y is a countable union of cubes of total volume $\epsilon = \frac{\epsilon}{2} + \frac{\epsilon}{2^2} + \frac{\epsilon}{2^3} + \cdots$. But if $\mathbf{x} \notin Y$, then by the definition of Y

$$g_m(\mathbf{x}) \leq \frac{A}{2^m\epsilon} \quad \text{for all } m = 1, 2, \dots, \qquad A21.36$$

and

$$\begin{aligned}\sum_{i=1}^{\infty}|f_i(\mathbf{x})| &= \lim_{k\to\infty} h_k(\mathbf{x}) = h_0(\mathbf{x}) + \sum_{i=1}^{\infty} g_i(\mathbf{x}) \qquad A21.37\\ &\leq h_0(\mathbf{x}) + \sum_{m=1}^{\infty}\frac{A}{2^m\epsilon} = h_0(\mathbf{x}) + \frac{A}{\epsilon}.\end{aligned}$$

In particular for $\mathbf{x} \notin Y$ the series $\sum_{i=1}^{\infty}|f_i(\mathbf{x})|$ converges. So the set X of $\mathbf{x} \in \mathbb{R}^n$ such that $\sum_{k=1}^{\infty}|f_k(\mathbf{x})|$ diverges is entirely contained in Y, which is a countable union of cubes of total volume at most ϵ. Thus $\sum_{k=1}^{\infty}|f_k(\mathbf{x})|$ converges almost everywhere. □

Proofs of major theorems involving Lebesgue integrals

Theorem 4.11.17 (A first limit theorem for Lebesgue integrals). *Let $k \mapsto f_k$ be a series of L-integrable functions such that*

$$\sum_{k=1}^{\infty} \int_{\mathbb{R}^n} |f_k(\mathbf{x})||d^n\mathbf{x}| < \infty.$$

Then $f(\mathbf{x}) = \sum_{k=1}^{\infty} f_k(\mathbf{x})$ exists for almost all $\mathbf{x}$, the function f is L-integrable, and

$$\int_{\mathbb{R}^n} f(\mathbf{x})|d^n\mathbf{x}| = \int_{\mathbb{R}^n} \sum_{k=1}^{\infty} f_k(\mathbf{x})|d^n\mathbf{x}| = \sum_{k=1}^{\infty} \int_{\mathbb{R}^n} f_k(\mathbf{x})|d^n\mathbf{x}|.$$

Proof. The idea is to write $f_k \underset{L}{=} \sum_{i=1}^{\infty} f_{k,i}$, where the $f_{k,i}$ are Riemann integrable with

$$\sum_{i=1}^{\infty} \int_{\mathbb{R}^n} |f_{k,i}(\mathbf{x})||d^n\mathbf{x}| < \infty, \qquad A21.38$$

and write f as a double series: $f \underset{L}{=} \sum_{k=1}^{\infty} \sum_{i=1}^{\infty} f_{k,i}$.

The difficulty with this scheme is that the series

$$\sum_{k=1}^{\infty}\sum_{i=1}^{\infty} \int_{\mathbb{R}^n} |f_{k,i}(\mathbf{x})||d^n\mathbf{x}| \qquad A21.39$$

might fail to converge and therefore f may not be L-integrable. The equality $f = \sum_{k=1}^{\infty} \sum_{i=1}^{\infty} f_{k,i}$ could be due to cancellations, which aren't available in A21.39, where the $|f_{k,i}|$ are all positive.

This is easily remedied: by Theorem 4.11.7, we are not limited to one particular way of representing the f_k as a sum of a series. We can define another by choosing for each k a number $m(k)$ such that

$$\sum_{i=m(k)+1}^{\infty} \int_{\mathbb{R}^n} |f_{k,i}(\mathbf{x})||d^n\mathbf{x}| < \frac{1}{2^k}. \qquad A21.40$$

Then set

$$g_{k,1} = \sum_{j=1}^{m(k)} f_{k,j} \quad \text{and} \quad g_{k,j} = f_{k,m(k)+j-1}, \text{ for } 2 \le j < \infty, \qquad A21.41$$

so that $f = \sum_{k,j=1}^{\infty} g_{k,j}$. The series

$$\sum_{j=1}^{\infty} g_{k,j} = \sum_{j=1}^{m(k)} f_{k,j} + \overbrace{\sum_{j=2}^{\infty} f_{k,m(k)+j-1}}^{f_{k,m(k)+1}+f_{k,m(k)+2}+\cdots} \qquad A21.42$$

Proof of Theorem 4.11.17: Strictly speaking, we haven't quite shown that f is L-integrable, since inequality A21.43 involves a double sum, and the definition of integrability only allows for a single sum. However, the double sum can be turned into a single sum, for instance using the functions $h_l = \sum_{j=1}^{l} g_{j,l-j+1}$ such that the series

$$\sum_{l=1}^{\infty} |h_l(\mathbf{x})||d^n\mathbf{x}|$$

converges. This is the definition of L-integrability, and by definition

$$\int_{\mathbb{R}^n} f(\mathbf{x})|d^n\mathbf{x}| = \int_{\mathbb{R}^n} \sum_{l=1}^{\infty} h_l(\mathbf{x})|d^n\mathbf{x}| = \int_{\mathbb{R}^n} \sum_{k=1}^{\infty} f_k(\mathbf{x})|d^n\mathbf{x}|,$$

where since the series are absolutely convergent, the sum doesn't depend on the order of summation.

converges to f_k a.e., just as the series $\sum_i f_{k,i}$ does, but much faster, since the first term of $\sum_j g_{k,j}$ gobbles up most of the sum.

Now write (using inequality A21.44 to go from the first to second lines)

$$\begin{aligned}\sum_{k=1}^{\infty}\sum_{j=1}^{\infty}\int_{\mathbb{R}^n}|g_{k,j}(\mathbf{x})||d^n\mathbf{x}| &= \overbrace{\sum_{k=1}^{\infty}\int_{\mathbb{R}^n}|g_{k,1}(\mathbf{x})||d^n\mathbf{x}|}^{\text{big term}} + \overbrace{\sum_{k=1}^{\infty}\sum_{j=2}^{\infty}\int_{\mathbb{R}^n}|g_{k,j}(\mathbf{x})||d^n\mathbf{x}|}^{\text{little term}} \\ &\le \sum_{k=1}^{\infty}\left(\int_{\mathbb{R}^n}|f_k(\mathbf{x})||d^n\mathbf{x}| + \frac{1}{2^k}\right) + \sum_{k=1}^{\infty}\frac{1}{2^k} \\ &\le \sum_{k=1}^{\infty}\int_{\mathbb{R}^n}|f_k(\mathbf{x})||d^n\mathbf{x}| + 2 < \infty. \qquad A21.43\end{aligned}$$

Inequality A21.44 uses the following, where we use abbreviated notation, set $j = m(k)+1$, and omit $\mathbf{x}$ and $|d^n\mathbf{x}|$ where they don't fit:

$$\begin{aligned}&\int \overbrace{\left|\sum_{i=1}^{\infty} f_{k,i} - \sum_{i=j}^{\infty} f_{k,i}\right|}^{f \text{ in Prop. 4.11.16}} |d^n\mathbf{x}| \\ &\underset{(1)}{\le} \int \overbrace{\left(\left|\sum_{i}^{\infty} f_{k,i}\right| + \left|\sum_{i=j}^{\infty} f_{k,i}\right|\right)}^{g \text{ in Prop. 4.11.16}} \\ &\underset{(2)}{=} \int\left|\sum_{i}^{\infty} f_{k,i}\right| + \int\left|\sum_{i=j}^{\infty} f_{k,i}\right| \\ &\underset{(3)}{\le} \int\left|\sum_{i}^{\infty} f_{k,i}\right| + \int\sum_{i=j}^{\infty}|f_{k,i}| \\ &\underset{(4)}{=} \int\left|\sum_{i}^{\infty} f_{k,i}\right| + \sum_{i=j}^{\infty}\int|f_{k,i}|\end{aligned}$$

Inequality 1 uses Proposition 4.11.16; equality 2 uses Proposition 4.11.14; inequality 3 again uses Proposition 4.11.16; equality 4 is Definition 4.11.8 of the Lebesgue integral.

In the second line in inequality A21.43, the first $1/2^k$ allows for the possibility that $\sum_{i=m(k)+1}^{\infty}\int f_{k,i}(\mathbf{x})$ might be negative, as shown below, where we use inequality 21.40 to go from the second to third lines:

$$\begin{aligned}\int_{\mathbb{R}^n}|\overbrace{g_{k,1}(\mathbf{x})}^{\sum_{j=1}^{m(k)} f_{k,j}}||d^n\mathbf{x}| &= \int_{\mathbb{R}^n}\left|\sum_{i=1}^{\infty} f_{k,i}(\mathbf{x}) - \sum_{i=m(k)+1}^{\infty} f_{k,i}(\mathbf{x})\right||d^n\mathbf{x}| \\ &\le \int_{\mathbb{R}^n}\left|\sum_{i=1}^{\infty} f_{k,i}(\mathbf{x})\right||d^n\mathbf{x}| + \sum_{i=m(k)+1}^{\infty}\int_{\mathbb{R}^n}|f_{k,i}(\mathbf{x})|\,|d^n\mathbf{x}| \\ &\le \int_{\mathbb{R}^n}|f_k(\mathbf{x})|\,|d^n\mathbf{x}| + \frac{1}{2^k}. \qquad A21.44\end{aligned}$$

It follows from inequality 21.43 that the double series $\sum_{k,j} g_{k,j}(\mathbf{x})$ converges except on the set of $\mathbf{x}$ for which the series $\sum_{k,j}|g_{k,j}(\mathbf{x})|$ diverges, which has measure 0. Thus f can be represented as the sum of a series of R-integrable functions, such that the integrals of their absolute values converge. □

Many famous mathematicians (Banach, Riesz, Landau, Hausdorff) proved versions of the dominated convergence theorem. The main contribution is certainly Lebesgue's.

Theorem 4.11.19 (Dominated convergence for Lebesgue integrals). *Let $k \mapsto f_k$ be a sequence of L-integrable functions that converges pointwise to some function f almost everywhere. Suppose there is an L-integrable function $F : \mathbb{R}^n \to \mathbb{R}$ such that $|f_k(\mathbf{x})| \le F(\mathbf{x})$ for almost all $\mathbf{x}$. Then f is L-integrable and its integral is*

$$\int_{\mathbb{R}^n} f(\mathbf{x})\,|d^n\mathbf{x}| = \lim_{k\to\infty}\int_{\mathbb{R}^n} f_k(\mathbf{x})\,|d^n\mathbf{x}|.$$

Proof. We will make some preliminary simplifications. First, the numbers $A_k \overset{\text{def}}{=} \int_{\mathbb{R}^n} f_k(\mathbf{x})|d^n\mathbf{x}|$ are a bounded sequence of numbers, so by the

Bolzano-Weierstrass theorem (Theorem 1.6.3) we can pass to a subsequence and assume that they converge to a limit A: $\int_{\mathbb{R}^n} f_k(\mathbf{x})|d^n\mathbf{x}| \to A$. Thus we need to show that f is L-integrable and that

$$\int_{\mathbb{R}^n} f(\mathbf{x})|d^n\mathbf{x}| = A. \qquad A21.45$$

To prove equation A21.45, it is enough to prove that $\int_{\mathbb{R}^n} f(\mathbf{x})|d^n\mathbf{x}| \geq A$. Indeed, we can then also show that, $\int_{\mathbb{R}^n} f(\mathbf{x})|d^n\mathbf{x}| \leq A$ by applying the same argument to the functions $-f_k$, which converge pointwise to $-f$ almost everywhere.

Note that $F(\mathbf{x}) \geq |f_k(\mathbf{x})| \geq 0$ almost everywhere.

Finally, we may assume that all $f_k \geq 0$. Indeed, the functions $\widetilde{f}_k = F + f_k$ satisfy $|\widetilde{f}_k| = |F + f_k| \leq 2F$ but have the virtue that $\widetilde{f}_k \geq 0$. If we can prove the theorem for the $\widetilde{f}_k$, by subtracting F we recover the theorem for the f_k.

So we will assume $f_k \geq 0$ and $\int_{\mathbb{R}^n} f_k(\mathbf{x})|d^n\mathbf{x}| \to A$, and will prove

$$\int_{\mathbb{R}^n} f(\mathbf{x})|d^n\mathbf{x}| \geq A. \qquad A21.46$$

Although each g_i is a limit of a monotone increasing sequence, the g_i themselves are a monotone decreasing sequence, since the functions defining g_i and g_{i+1} are the same except for f_i, missing from the definition of g_{i+1}.

Consider the functions

$$\begin{aligned} g_1 &= \lim\,(f_1,\ \sup(f_1, f_2),\ \sup(f_1, f_2, f_3), \dots) \\ g_2 &= \lim\,(f_2,\ \sup(f_2, f_3),\ \sup(f_2, f_3, f_4), \dots) \\ g_3 &= \lim\,(f_3,\ \sup(f_3, f_4),\ \sup(f_3, f_4, f_5), \dots) \\ \dots &= \lim\,(\qquad \dots \quad \dots \qquad). \end{aligned} \qquad A21.47$$

Inequality A21.48: We get the last inequality because F is L-integrable.

Each g_i is a limit of a monotone increasing sequence of L-integrable functions, since $0 \leq f_i \leq \sup(f_i, f_{i+1}) \leq \sup(f_i, f_{i+1}, f_{i+2}) \leq \dots$. These functions satisfy the hypotheses of the monotone convergence theorem (Theorem 4.11.18): by hypothesis, $|f_k(\mathbf{x})| \leq F(\mathbf{x})$ for almost all $\mathbf{x}$, so

$$\int_{\mathbb{R}^n} f_i(\mathbf{x})|d^n\mathbf{x}| \leq \int_{\mathbb{R}^n} \sup(f_i, \dots, f_{i+j})(\mathbf{x})\,|d^n\mathbf{x}| \ \leq\ \int_{\mathbb{R}^n} F(\mathbf{x})|d^n\mathbf{x}| < \infty. \qquad A21.48$$

So by the monotone convergence theorem, each g_i is L-integrable, and

$$\begin{aligned} \int_{\mathbb{R}^n} f_i(\mathbf{x})|d^n\mathbf{x}| &\leq \sup_{j\to\infty} \int_{\mathbb{R}^n} \sup(f_i, \dots, f_{i+j})(\mathbf{x})|d^n\mathbf{x}| \\ &\underbrace{=}_{\text{Thm. 4.11.18}} \int_{\mathbb{R}^n} g_i(\mathbf{x})|d^n\mathbf{x}| < \infty. \end{aligned} \qquad A21.49$$

Now we can apply the monotone convergence theorem again, this time to the monotone decreasing sequence $i \mapsto g_i$, to say that the limit of the g_i is L-integrable, and

$$\int_{\mathbb{R}^n} \lim_{i\to\infty} g_i(\mathbf{x})|d^n\mathbf{x}| = \lim_{i\to\infty} \int_{\mathbb{R}^n} g_i(\mathbf{x})|d^n\mathbf{x}| \geq \lim_{i\to\infty} \int_{\mathbb{R}^n} f_i(\mathbf{x})|d^n\mathbf{x}| = A. \qquad A21.50$$

But $\lim_{i\to\infty} g_i(\mathbf{x}) = f(\mathbf{x})$ for every $\mathbf{x}$ where the $f_k(\mathbf{x})$ converge, i.e, almost everywhere, so f is L-integrable, and $\int_{\mathbb{R}^n} f(\mathbf{x})|d^n\mathbf{x}| \geq A$. $\square$

Theorem 4.11.21 (Fubini's theorem for the Lebesgue integral). *Let $f : \mathbb{R}^n \times \mathbb{R}^m \to \mathbb{R}$ be an L-integrable function. Then the function*

$$\mathbf{y} \mapsto \int_{\mathbb{R}^n} f(\mathbf{x}, \mathbf{y}) |d^n\mathbf{x}|$$

is defined for almost all $\mathbf{y}$ *and is L-integrable on* $\mathbb{R}^m$, *and in that case*

$$\int_{\mathbb{R}^n \times \mathbb{R}^m} f(\mathbf{x}, \mathbf{y}) |d^n\mathbf{x}||d^m\mathbf{y}| = \int_{\mathbb{R}^m} \left(\int_{\mathbb{R}^n} f(\mathbf{x}, \mathbf{y}) |d^n\mathbf{x}| \right) |d^m\mathbf{y}|. \qquad A21.51$$

Conversely, if $f : \mathbb{R}^n \times \mathbb{R}^m \to \mathbb{R}$ *is a function defined except on a set of measure 0 and such that*

1. *every point* $(\mathbf{x}, \mathbf{y}) \in \mathbb{R}^{n+m}$ *is the center of a ball on which* f *is L-integrable*
2. *the function* $\mathbf{x} \mapsto |f(\mathbf{x}, \mathbf{y})|$ *is L-integrable on* $\mathbb{R}^n$ *for almost all* $\mathbf{y}$
3. *the function* $\mathbf{y} \mapsto \int_{\mathbb{R}^n} |f(\mathbf{x}, \mathbf{y})||d^n\mathbf{x}|$ *is L-integrable on* $\mathbb{R}^m$

then f *is L-integrable on* $\mathbb{R}^{n+m}$, *and*

$$\int_{\mathbb{R}^n \times \mathbb{R}^m} f(\mathbf{x}, \mathbf{y}) |d^n\mathbf{x}||d^m\mathbf{y}| = \int_{\mathbb{R}^m} \left(\int_{\mathbb{R}^n} f(\mathbf{x}, \mathbf{y}) |d^n\mathbf{x}| \right) |d^m\mathbf{y}|.$$

Note that in equation A21.51, the integrals on the right could be written in the opposite order:

$$\int_{\mathbb{R}^n \times \mathbb{R}^m} f(\mathbf{x}, \mathbf{y}) |d^n\mathbf{x}||d^m\mathbf{y}| = \int_{\mathbb{R}^n} \left(\int_{\mathbb{R}^m} f(\mathbf{x}, \mathbf{y}) |d^m\mathbf{y}| \right) |d^n\mathbf{x}|. \qquad A21.52$$

Proof. Assume first that f is L-integrable. Then $f \underset{L}{=} \sum_k \widetilde{f}_k$, where the $\widetilde{f}_k$ are R-integrable on $\mathbb{R}^{n+m}$, with

$$\sum_k \int_{\mathbb{R}^n \times \mathbb{R}^m} |\widetilde{f}_k(\mathbf{x}, \mathbf{y})||d^n\mathbf{x}||d^m\mathbf{y}| < \infty. \qquad A21.53$$

To avoid the minor but unpleasant difficulties with upper and lower sums in Theorem A17.2, we replace $\widetilde{f}_k$ by f_k, where

$$f_k(\mathbf{x}, \mathbf{y}) = \begin{cases} 0 & \text{if } U((\widetilde{f}_k)_\mathbf{x}) \neq L((\widetilde{f}_k)_\mathbf{x}) \text{ or } U((\widetilde{f}_k)^\mathbf{y}) \neq L((\widetilde{f}_k)^\mathbf{y}) \\ \widetilde{f}(\mathbf{x}, \mathbf{y}) & \text{otherwise.} \end{cases}$$

Any set is the union of its points, so a union of sets of measure 0 is usually not of measure 0. But (Theorem 4.4.4) a *countable* union of sets of measure 0 has measure 0, and (since a set of volume 0 has measure 0) a countable union of sets of volume 0 has measure 0. Lebesgue integration would not make sense without the notions of countable and uncountable sets due to Cantor.

Then $f \underset{L}{=} \sum_{k=1}^{\infty} f_k$. Each f_k is still R-integrable, since we have modified $\widetilde{f}$ only on a set of volume 0 (see Corollary A17.3), so Fubini's theorem for R-integrable functions applies to the f_k. Moreover,

$$\sum_k \int_{\mathbb{R}^n \times \mathbb{R}^m} |f_k(\mathbf{x}, \mathbf{y})||d^n\mathbf{x}||d^m\mathbf{y}| < \infty, \qquad A21.54$$

since a countable union of sets of volume 0 (one for each k) has measure 0.

Thus

$$\begin{aligned}\int_{\mathbb{R}^n\times\mathbb{R}^m} f(\mathbf{x},\mathbf{y})|d^n\mathbf{x}||d^m\mathbf{y}| &\underset{1}{=} \sum_k \int_{\mathbb{R}^n\times\mathbb{R}^m} f_k(\mathbf{x},\mathbf{y})|d^n\mathbf{x}||d^m\mathbf{y}|\\ &\underset{2}{=} \sum_k \int_{\mathbb{R}^m}\left(\int_{\mathbb{R}^n} f_k(\mathbf{x},\mathbf{y})|d^n\mathbf{x}|\right)|d^m\mathbf{y}|\\ &\underset{3}{=} \int_{\mathbb{R}^m}\left(\sum_k \int_{\mathbb{R}^n} f_k(\mathbf{x},\mathbf{y})|d^n\mathbf{x}|\right)|d^m\mathbf{y}|\\ &\underset{4}{=} \int_{\mathbb{R}^m}\left(\int_{\mathbb{R}^n} \sum_k f_k(\mathbf{x},\mathbf{y})|d^n\mathbf{x}|\right)|d^m\mathbf{y}|\\ &= \int_{\mathbb{R}^m}\left(\int_{\mathbb{R}^n} f(\mathbf{x},\mathbf{y})|d^n\mathbf{x}|\right)|d^m\mathbf{y}|.\end{aligned} \qquad A21.55$$

Equation A21.55: The first equality is the definition of the Lebesgue integral; the second is Fubini's theorem for functions that are R-integrable.

The exchanges of sums and integrals (equalities 3 and 4) are justified by Theorem 4.11.17. We can apply that theorem to get equality 3 because

$$\begin{aligned}\sum_k \int_{\mathbb{R}^m}\left|\int_{\mathbb{R}^n} f_k(\mathbf{x},\mathbf{y})|d^n\mathbf{x}|\right||d^m\mathbf{y}| &\le \sum_k \int_{\mathbb{R}^m}\left(\int_{\mathbb{R}^n} |f_k(\mathbf{x},\mathbf{y})||d^n\mathbf{x}|\right)|d^m\mathbf{y}|\\ &= \sum_k \int_{\mathbb{R}^n\times\mathbb{R}^m} |f_k(\mathbf{x},\mathbf{y})||d^n\mathbf{x}||d^m\mathbf{y}|\\ &< \infty,\end{aligned} \qquad A21.56$$

where we have used the triangle inequality (part 4 of Proposition 4.1.14) for the first inequality, Fubini's theorem for the R-integrable $|f_k|$ for the equality, and the definition of the L-integrability of f for the final inequality.

To apply Theorem 4.11.17 to get equality 4, we need to know that

$$\sum_k \int_{\mathbb{R}^n} |f_k(\mathbf{x},\mathbf{y})||d^n\mathbf{x}| < \infty \text{ for almost all } \mathbf{y}. \qquad A21.57$$

We see this as follows: the sequence of functions

$$g_k(\mathbf{y}) = \sum_{i=1}^{k} \int_{\mathbb{R}^n} |f_i(\mathbf{x},\mathbf{y})||d^n\mathbf{x}| \qquad A21.58$$

is increasing, so by the monotone convergence theorem 4.11.18, if

$$\sup_k \int_{\mathbb{R}^m} g_k(\mathbf{y})|d^m\mathbf{y}| < \infty, \qquad A21.59$$

then pointwise we have $\sup_{k\to\infty} g_k(\mathbf{y}) < \infty$ for almost every $\mathbf{y}$; this is precisely what we need. We have already done all the work to bound the integral in equation A21.56: simply replace $f_k(\mathbf{x},\mathbf{y})$ on the left with $|f_k(\mathbf{x},\mathbf{y})|$.

This proves the first half of Theorem 4.11.21.

For the converse, note that all we have to prove is that f is L-integrable; the last equation then follows from the first part. First let us see that the function $f\mathbf{1}_C$ is L-integrable for each cube $C \in \mathcal{D}_0(\mathbb{R}^{n+m})$.

Indeed, the closure $\overline{C}$ is compact, hence, by Theorem A3.3 (Heine-Borel) it is covered by finitely many of the balls given by hypothesis 1, say $B_1, \dots, B_k$. We can then write

$$f\mathbf{1}_C = \Big(\underbrace{f\mathbf{1}_{B_1} + f(\mathbf{1}_{B_2} - \mathbf{1}_{B_1\cap B_2}) + \cdots + f(\mathbf{1}_{B_k} - \mathbf{1}_{(B_1\cup\cdots\cup B_{k-1})\cap B_k})}_{\text{finite sum of L-integrable functions}}\Big)\mathbf{1}_C. \qquad A21.60$$

Each function $\mathbf{1}_{B_k}$ is Riemann integrable, and by assumption 1, f is L-integrable on each B_k, so, by Proposition 4.11.15, in the parentheses in equation A21.60 we have a finite sum of L-integrable functions; this sum is L-integrable by Proposition 4.11.14.

Thus $f\mathbf{1}_C$ is the product of an L-integrable function and an R-integrable function, hence (again by Proposition 4.11.15) L-integrable. So by the first part of Theorem 4.11.21, we can apply Fubini to each $f\mathbf{1}_C$, to get the first equality below:

$$\begin{aligned}
&\sum_{C\in\mathcal{D}_0(\mathbb{R}^{n+m})} \int_{\mathbb{R}^{n+m}} |f(\mathbf{x},\mathbf{y})|\, \mathbf{1}_C(\mathbf{x},\mathbf{y})|d^n\mathbf{x}||d^m\mathbf{y}| \\
&\underset{1}{=} \sum_{C\in\mathcal{D}_0(\mathbb{R}^{n+m})} \int_{\mathbb{R}^m}\left(\int_{\mathbb{R}^n} |f(\mathbf{x},\mathbf{y})|\, \mathbf{1}_C(\mathbf{x},\mathbf{y})|d^n\mathbf{x}|\right)|d^m\mathbf{y}| \\
&\underset{2}{=} \sum_{C_2\in\mathcal{D}_0(\mathbb{R}^m)} \int_{\mathbb{R}^m} \mathbf{1}_{C_2}(\mathbf{y})\Big(\overbrace{\sum_{C_1\in\mathcal{D}_0(\mathbb{R}^n)} \int_{\mathbb{R}^n} \mathbf{1}_{C_1}(\mathbf{x})\, |f(\mathbf{x},\mathbf{y})||d^n\mathbf{x}|}^{<\infty \text{ by hypoth. 2, so can apply Thm. 4.11.17}}\Big)|d^m\mathbf{y}| \\
&\underset{3}{=} \sum_{C_2\in\mathcal{D}_0(\mathbb{R}^m)} \int_{\mathbb{R}^m} \mathbf{1}_{C_2}(\mathbf{y})\Big(\underbrace{\int_{\mathbb{R}^n} |f(\mathbf{x},\mathbf{y})||d^n\mathbf{x}|}_{<\infty \text{ by hypothesis 3}}\Big)|d^m\mathbf{y}| \qquad A21.61 \\
&\underset{4}{=} \int_{\mathbb{R}^m}\left(\int_{\mathbb{R}^n} |f(\mathbf{x},\mathbf{y})||d^n\mathbf{x}|\right)|d^m\mathbf{y}| \\
&< \infty.
\end{aligned}$$

Inequality A21.61:
C is a cube in $\mathcal{D}_0(\mathbb{R}^{n+m})$
C_1 is a cube in $\mathcal{D}_0(\mathbb{R}^n)$
C_2 is a cube in $\mathcal{D}_0(\mathbb{R}^m)$.

Note that in line 1 of inequality A21.61 we write $|f(\mathbf{x},\mathbf{y})|$, not $f(\mathbf{x},\mathbf{y})$. This will allow us, when we get to line 3, to apply Theorem 4.11.17.

Equality 3 of inequality A21.61 uses Theorem 4.11.17:

$$\begin{aligned}
&\sum_{C_1}\int_{\mathbb{R}^n} \mathbf{1}_{C_1}(\mathbf{x})\, |f(\mathbf{x},\mathbf{y})||d^n\mathbf{x}| \\
&= \int_{\mathbb{R}^n}\sum_{C_1} \mathbf{1}_{C_1}(\mathbf{x})\, |f(\mathbf{x},\mathbf{y})||d^n\mathbf{x}| \\
&= \int_{\mathbb{R}^n} |f(\mathbf{x},\mathbf{y})|\, |d^n\mathbf{x}|.
\end{aligned}$$

Equality 4 is similar: use Theorem 4.11.17 to switch the sum and first integral in line 4, and note that

$$\int_{\mathbb{R}^m}\sum_{C_2\in\mathcal{D}_0(\mathbb{R}^m)} \mathbf{1}_{C_2}(\mathbf{y})g|d^m\mathbf{y}| = \int_{\mathbb{R}^m} g|d^m\mathbf{y}|.$$

(To save space, we have replaced the inner integral by g.)

The equality marked 2 in inequality A21.61 is the identity

$$\mathbf{1}_C(\mathbf{x},\mathbf{y}) = \mathbf{1}_{C_2}(\mathbf{y})\mathbf{1}_{C_1}(\mathbf{x}), \qquad A21.62$$

and the fact that $\mathbf{1}_{C_2}(\mathbf{y})$ is a constant and can be factored out as far as the inner integral is concerned. The third and fourth are applications of Theorem 4.11.17, as noted in the margin. We get the $< \infty$ at the end by hypothesis 3: the function $\mathbf{y} \mapsto \int_{\mathbb{R}^n} |f(\mathbf{x},\mathbf{y})||d^n\mathbf{x}|$ is integrable, so its integral is finite.

It then follows from Theorem 4.11.17 that f itself (without the absolute values) is L-integrable: that the first line in inequality A21.61 is $< \infty$ is the condition needed to apply that theorem. $\square$

Theorem 4.11.20 is true if Φ is of class C^1 but does not have Lipschitz derivative, but our version of the implicit function theorem doesn't allow us to prove it.

Theorem 4.11.20 (The change of variables formula for Lebesgue integrals). *Let U, V be open subsets of $\mathbb{R}^n$, and let $\Phi : U \to V$ be bijective, of class C^1, with inverse of class C^1, such that both Φ and Φ^{-1} have Lipschitz derivatives. Let $f : V \to \mathbb{R}$ be defined except perhaps on a set of measure 0. Then*

$$f \text{ is L-integrable on } V \iff (f \circ \Phi)|\det[\mathbf{D}\Phi]| \text{ is L-integrable on } U.$$

In that case,

$$\int_V f(\mathbf{v})|d^n\mathbf{v}| = \int_U (f \circ \Phi)(\mathbf{u})\big|\det[\mathbf{D}\Phi(\mathbf{u})]\big|\,|d^n\mathbf{u}|. \qquad A21.63$$

Proof. First we will show that if f is L-integrable, then $|\det[\mathbf{D}\Phi]|(f \circ \Phi)$ is L-integrable and equation A21.63 is correct. Take all the cubes of $\mathcal{D}_1(\mathbb{R}^n)$ with closures completely contained in V, then all the cubes of $\mathcal{D}_2(\mathbb{R}^n)$ with closures completely contained in V and not contained in the previous cubes, etc. We will find a countable collection of cubes $C_1, C_2, \dots$ completely covering V.

Since we are assuming that f is L-integrable, we can write $f \underset{L}{=} \sum_{k=1}^{\infty} f_k$, where the f_k are R-integrable

$$\sum_{k=1}^{\infty} \int_{\mathbb{R}^n} |f_k(\mathbf{x})||d^n\mathbf{x}| < \infty. \qquad A21.64$$

If we set $f_{k,i} = f_k \mathbf{1}_{C_i}$, we can further write

$$f \underset{L}{=} \sum_{k=1}^{\infty} f_k = \sum_{k=1}^{\infty}\sum_{k=1}^{\infty} f_i \mathbf{1}_{C_i} = \sum_{k,i=1}^{\infty} f_{k,i}, \qquad A21.65$$

since the C_i are disjoint and cover V. By the monotone convergence theorem, we then have

$$\sum_{k,i=1}^{\infty} \int_{\mathbb{R}^n} |f_{k,i}(\mathbf{x})||d^n\mathbf{x}| = \sum_{k=1}^{\infty} \int_{\mathbb{R}^n} |f_k(\mathbf{x})||d^n\mathbf{x}| \underbrace{<}_{\text{ineq. A21.64}} \infty, \qquad A21.66$$

so that, by equations A21.65 and A21.66, Theorem 4.11.7, and the definition of the Lebesgue integral, we have

$$\int_V f(\mathbf{v})|d^n\mathbf{v}| = \sum_{k,i=1}^{\infty} \int_V f_{k,i}(\mathbf{v})|d^n\mathbf{v}|. \qquad A21.67$$

We can apply Theorem 4.10.12 (the change variables formula for Riemann integrals) to each $f_{k,i}$: they have support in the closure of a cube, which is compact, with a boundary of volume 0, and Φ is defined and of class C^1 on a neighborhood of the cube (namely V). Thus

$$\int_V f(\mathbf{v})|d^n\mathbf{v}| = \sum_{k,i=1}^{\infty} \int_V f_{k,i}(\mathbf{v})|d^n\mathbf{v}|$$
$$\underbrace{=}_{\text{Thm. 4.10.12}} \sum_{k,i=1}^{\infty} \int_U (f_{k,i}\circ\Phi)(\mathbf{u})\big|\det[\mathbf{D}\Phi(\mathbf{u})]\big||d^n\mathbf{u}| \qquad \text{A21.68}$$
$$\underbrace{=}_{\text{Thm. 4.11.17}} \int_U \sum_{k,i=1}^{\infty} (f_{k,i}\circ\Phi)(\mathbf{u})\big|\det[\mathbf{D}\Phi(\mathbf{u})]\big||d^n\mathbf{u}|$$
$$= \int_U (f\circ\Phi)(\mathbf{u})\big|\det[\mathbf{D}\Phi(\mathbf{u})]\big||d^n\mathbf{u}|.$$

To prove ($\Longleftarrow$), exchange the roles of U and V and use the chain rule. □

Equation A21.68: When we use Theorem 4.11.17 to exchange sum and integral (going from line 2 to line 3), the hypothesis of Theorem 4.11.17 is met because if we replace $(f_{k,i}\circ\Phi)(\mathbf{u})$ in line 2 by $|(f_{k,i}\circ\Phi)(\mathbf{u})|$, then, term by term,

$$\int_U |(f_{k,i}\circ\Phi)(\mathbf{u})||\det\Phi(\mathbf{u})||d^n\mathbf{u}|$$

equals

$$\int_{\mathbb{R}^n} |f_{k,i}(\mathbf{x})||d^n\mathbf{x}|.$$

Now apply inequality A21.66.

Lebesgue measure

The first thing we did with Riemann integrals was to define volume. We now use Lebesgue integrals to define "volume in the sense of Lebesgue", known as *Lebesgue measure*.

Definition A21.5 (Lebesgue measure). A bounded set $X \subset \mathbb{R}^n$ is *measurable* if its indicator function is L-integrable; in that case its *Lebesgue measure* $\mu(X)$ is the integral of the indicator function:

$$\mu(X) \stackrel{\text{def}}{=} \int_{\mathbb{R}^n} \mathbf{1}_X(\mathbf{x})|d^n\mathbf{x}|. \qquad \text{A21.69}$$

A general subset $X \subset \mathbb{R}^n$ is *measurable* if for all R the intersection $X \cap B_R(\mathbf{0})$ is measurable, and its measure (which may be infinite) is

$$\mu(X) \stackrel{\text{def}}{=} \sup_R \mu\Big(X \cap B_R(\mathbf{0})\Big). \qquad \text{A21.70}$$

For instance, the rationals in $[0,1]$ are not pavable (Example 4.3.3) but they have a well-defined measure (Example 4.11.10).

It is impossible to *construct* a non-measurable set; showing that one exists requires the axiom of choice. See Example A21.7.

If a set has volume, then, by Proposition 4.11.9, its measure coincides with its volume. But many sets that are not pavable are measurable.

Theorem A21.6 (Lebesgue measurable sets).

1. *The complement in $\mathbb{R}^n$ of a measurable set is measurable.*
2. *Any countable union or intersection of measurable sets is measurable.*
3. *If $X_1, X_2, \dots$ is a sequence of disjoint measurable sets, then*

$$\mu(\cup_i X_i) = \sum_i \mu(X_i). \qquad \text{A21.71}$$

4. *All open subsets and all closed subsets of $\mathbb{R}^n$ are measurable.*
5. *Any subset of a set of measure 0 is measurable, with measure 0.*

In the standard treatment of measure theory, parts 1 and 2 of Theorem A21.6 say that the Lebesgue measurable sets form a σ-algebra. Part 5 says that the σ-algebra is complete. Part 3 says that Lebesgue measure is countably additive. Part 4 says that it contains the Borel σ-algebra.

There are many measures other than Lebesgue measure; some contain the Borel σ-algebra and some do not.

Proof. 1. By definition, $X \subset \mathbb{R}^n$ is measurable if for any R, the function $\mathbf{1}_X \mathbf{1}_{B_R(\mathbf{0})}$ is L-integrable. But then

$$(1 - \mathbf{1}_X)\mathbf{1}_{B_R(\mathbf{0})} = \mathbf{1}_{B_R(\mathbf{0})} - \mathbf{1}_X \mathbf{1}_{B_R(\mathbf{0})} \qquad \text{A21.72}$$

is also L-integrable.

2. This follows from the monotone convergence theorem. Let $i \mapsto X_i$ be a sequence of measurable sets. The functions

$$\mathbf{1}_{X_1}\mathbf{1}_{B_R(\mathbf{0})} \leq (\mathbf{1}_{X_1 \cup X_2})\mathbf{1}_{B_R(\mathbf{0})} \leq (\mathbf{1}_{X_1 \cup X_2 \cup X_3})\mathbf{1}_{B_R(\mathbf{0})} \leq \cdots . \qquad \text{A21.73}$$

form a monotone increasing sequence of L-integrable functions, all bounded by the integrable function $\mathbf{1}_{B_R(\mathbf{0})}$. Thus the limit, which is $\mathbf{1}_{\cup X_i}\mathbf{1}_{B_R(\mathbf{0})}$, is L-integrable, showing that the union is measurable. The case of intersections is shown using the fact that the complement of $\cap X_i$ is the union of the complements of the X_i.

3. If the X_i are disjoint, $X = \cup_i X_i$, and $\sum_{i=1}^\infty \mu(X_i) < \infty$, we can apply Theorem 4.11.17 to $\mathbf{1}_X = \sum_{k=1}^\infty \mathbf{1}_{X_i}$ to get

$$\int_{\mathbb{R}^n} \mathbf{1}_X(\mathbf{x})|d^n\mathbf{x}| = \sum_{k=1}^\infty \int_{\mathbb{R}^n} \mathbf{1}_{X_i}(\mathbf{x})|d^n\mathbf{x}|, \quad \text{i.e.,} \quad \mu(X) = \sum_{i=1}^\infty \mu(X_i). \qquad \text{A21.74}$$

If $\sum_{i=1}^\infty \mu(X_i)$ is divergent, then for any A there exists R such that

$$\sum_{i=1}^\infty \mu(X_i \cap B_R(\mathbf{0})) > A, \qquad \text{A21.75}$$

and then $\mu(X \cap B_R(\mathbf{0})) > A$ also, so $\mu(X)$ is also infinite.

4. Any open set is a union of open balls with rational centers and rational radii, and there are only a countable number of such balls, so by part 2, the union is measurable. The closed sets are the complements of the open sets, hence measurable, by part 1.

5. By Definition 4.4.1 of measure 0, any subset X of a set of measure 0 still has measure 0, so $\mathbf{1}_X \underset{L}{=} 0$. $\square$

Do there exist *any* nonmeasurable sets? If you allow yourself the highly nonconstructive axiom of choice, then such things definitely exist.

Example A21.7 (A nonmeasurable set). Let us divide the numbers $x \in [0,1]$ into classes, where two numbers x and y belong to the same class if and only if $x - y$ is rational. Now take one element from each class, and call X the set of these samples.

This set X is not measurable (and thus its indicator function is not L-integrable, although it is bounded with bounded support). If $a_1, a_2, \ldots$ is a listing of the rationals between -1 and 1, then

$$[0,1] \subset \bigcup_n (X + a_n) \subset [-1,2]. \qquad \text{A21.76}$$

By the left inclusion, X cannot have measure 0. But if X has measure $\delta > 0$, then since all the sets in the union are translates of each other, they

---

The indicator function of the complement of X (i.e., of $\mathbb{R}^n - X$), is $1 - \mathbf{1}_X$.

A "rational center" is a center whose coordinates are all rational numbers.

A fairly detailed treatment of the axiom of choice can be found in *Naive Set Theory* by Paul R. Halmos.

Logicians tells us that the axiom of choice is a consequence of the continuum hypothesis (see Section 0.6). Gödel proved that both are consistent with the other axioms of set theory, and Cohen proved that both are independent of the other axioms of set theory.

Example A21.7: For example, the rational numbers form a class; the rational numbers plus π form another, as do the rational numbers plus $\sqrt{2}$.

Formula A21.76: The set
$$X + a_n$$
is the set of all $x + a_n$ for all $x \in X$.

all have the same measure δ, and since the sets in the union are disjoint, that says that $[-1, 2]$ has infinite measure, whereas of course its length is 3. So X cannot be measurable. $\triangle$

The Banach-Tarski paradox asserts that you can cut the unit ball in $\mathbb{R}^3$ into five pieces, move them around by rigid motions, and reassemble them to make two balls exactly like the first one. Some of the pieces must be nonmeasurable. For further reading: *The Banach-Tarski Paradox* by Stan Wagon, Cambridge University Press, 1993.

Remark. Above we wrote, "now take one element from each class." The axiom of choice says that we can do this: given any collection of nonempty sets, there exists a set formed of one element of each set.

But this is totally nonconstructive. In the case above, it is a *theorem* that no rule to do this can be written down. Further, the axiom of choice has been proved independent of the other axioms of set theory. So you are free to accept it or reject it, and seeing Example A21.7 above (and others more horrible, for instance, the *Banach-Tarski paradox*) you might be tempted to reject it, especially since you can then adopt the axiom that all sets are measurable.

The problem is that you would then lose a lot of modern mathematics, including just about all of functional analysis. A keystone of functional analysis is the *Hahn-Banach theorem*, which is known to be equivalent to the axiom of choice. $\triangle$

Exercises for Section A21

A21.1 Show that there exists a continuous function $f : \mathbb{R} \to \mathbb{R}$, bounded with bounded support (and in particular integrable), such that the set

$$\{ x \in \mathbb{R} \mid f(x) \geq 0 \}$$

is not pavable. For instance, follow the following steps.

a. Show that if $X \subset \mathbb{R}^n$ is any nonempty subset, then the function

$$f_X : \mathbb{R}^n \to \mathbb{R} \quad \text{given by} \quad f_X(\mathbf{x}) = \inf_{\mathbf{y} \in X} |\mathbf{x} - \mathbf{y}|$$

is continuous. Show that $f_X(\mathbf{x}) = 0$ if and only if $\mathbf{x} \in \overline{X}$.

b. Take any nonpavable closed subset $X \subset [0, 1]$, such as the complement of the set U_ϵ that is constructed in Example 4.4.3, and set $X' = X \cup \{0, 1\}$. Set

$$f(x) = -\mathbf{1}_{[0,1]}(x) f_{X'}(x).$$

Show that this function f satisfies our requirements.

A21.2 Make a list $a_1, a_2, \ldots$ of the rationals in $[0, 1]$. Consider the function f_k such that

$f_k(x) = 0$ if $x \notin [0, 1]$, or if $x \in \{a_1, \ldots, a_k\}$;

$f_k(x) = 1$ if $x \in [0, 1]$ and $x \notin \{a_1, \ldots, a_k\}$.

Show that all the f_k are R-integrable and that $f(x) = \lim_{k \to \infty} f_k(x)$ exists for every x, but that f is not R-integrable (but is L-integrable).

A21.3 Show that the h defined by $h(\mathbf{x}) = \Big(1 - d(\mathbf{x}, \mathbb{R}^n - X_\epsilon)\Big)^+$ satisfies $|h(\mathbf{x}) - h(\mathbf{y})| \leq |\mathbf{x} - \mathbf{y}|$ and hence is continuous. Show also that $h(\mathbf{x}) = 1$ when $\mathbf{x} \notin X_\epsilon$, and $0 \leq h(\mathbf{x}) < 1$ when $\mathbf{x} \in X_\epsilon$.

A21.4 Show that if f and g are any L-integrable functions on $\mathbb{R}^n$, then

It is possible for either the right side of inequality 1 to be infinite, or for both sides to be infinite.

$$\left(\int_{\mathbb{R}^n} f(\mathbf{x})g(\mathbf{x})|d^n\mathbf{x}|\right)^2 \leq \left(\int_{\mathbb{R}^n} (f(\mathbf{x}))^2|d^n\mathbf{x}|\right)\left(\int_{\mathbb{R}^n} (g(\mathbf{x}))^2|d^n\mathbf{x}|\right). \qquad (1)$$

Hint: Follow the proof of Schwarz's inequality (Theorem 1.4.5). Consider the quadratic polynomial

$$\int_{\mathbb{R}^n} ((f+tg)(\mathbf{x}))^2|d^n\mathbf{x}| = \int_{\mathbb{R}^n} (f(\mathbf{x}))^2|d^n\mathbf{x}| + 2t\int_{\mathbb{R}^n} f(\mathbf{x})g(\mathbf{x})|d^n\mathbf{x}| + t^2\int_{\mathbb{R}^n} (g(\mathbf{x}))^2|d^n\mathbf{x}|.$$

Since the polynomial is ≥ 0, its discriminant is nonpositive.

A21.5 Justify the ($\Longleftarrow$) part of Theorem 4.11.20.

A22 Computing the exterior derivative

In this appendix we prove Theorem 6.7.4, which says how to compute the exterior derivative of a k-form.

Theorem 6.7.4 (Computing the exterior derivative of a k-form). *Let*

$$\varphi = \sum_{1\leq i_1<\cdots<i_k\leq n} a_{i_1,\dots,i_k}\, dx_{i_1}\wedge\cdots\wedge dx_{i_k}$$

be a k-form of class C^2 on an open subset $U\subset\mathbb{R}^n$.

1. *The limit in equation 6.7.3 exists and defines a $(k+1)$-form.*
2. *The exterior derivative is linear over $\mathbb{R}$: if φ and ψ are k-forms on $U\subset\mathbb{R}^n$, and a and b are numbers (not functions), then*
$$\mathbf{d}(a\,\varphi + b\,\psi) = a\,\mathbf{d}\varphi + b\,\mathbf{d}\psi.$$
In particular, $\mathbf{d}(\varphi+\psi) = \mathbf{d}\varphi + \mathbf{d}\psi.$
3. *The exterior derivative of a constant form is 0.*
4. *The exterior derivative of the 0-form f is given by the formula*
$$\mathbf{d}f = [\mathbf{D}f] = \sum_{i=1}^{n}(D_i f)\, dx_i.$$
5. *If $f: U\to\mathbb{R}$ is a C^2 function, then*
$$\mathbf{d}\,(f\, dx_{i_1}\wedge\cdots\wedge dx_{i_k}) = \mathbf{d}f\wedge dx_{i_1}\wedge\cdots\wedge dx_{i_k}.$$

Proof. Part 4 is proved in Example 6.7.3. We will first prove part 5, then part 1. Parts 2 and 3 will follow immediately.

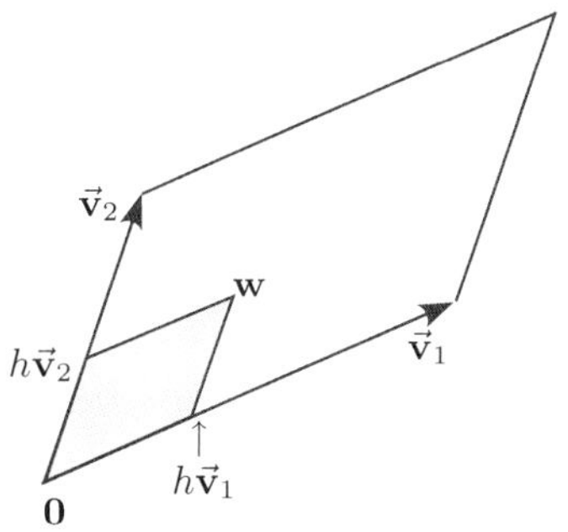

FIGURE A22.1.

Note that in the sums $\gamma_{1,i}$ and $\gamma_{0,i}$, the term $t_i\vec{\mathbf{v}}_i$ is omitted. Here $k+1=2$. The segment $[h\vec{\mathbf{v}}_1, \mathbf{w}]$ is parametrized by

$$\gamma_{1,1}(\mathbf{t}) = h\vec{\mathbf{v}}_1 + t_1\vec{\mathbf{v}}_2;$$

the segment $[h\vec{\mathbf{v}}_2, \mathbf{w}]$ is parametrized by

$$\gamma_{1,2}(\mathbf{t}) = h\vec{\mathbf{v}}_2 + t_1\vec{\mathbf{v}}_1.$$

The segment $[\mathbf{0}, h\vec{\mathbf{v}}_1]$ is parametrized by

$$\gamma_{0,2}(\mathbf{t}) = t_1\vec{\mathbf{v}}_1;$$

the segment $[\mathbf{0}, h\vec{\mathbf{v}}_2]$ is parametrized by

$$\gamma_{0,1}(\mathbf{t}) = t_2\vec{\mathbf{v}}_2.$$

It is enough to prove part 5 at the origin, which simplifies the notation; this amounts to translating φ. The idea is to write $f = T^0(f)+T^1(f)+R(f)$ as a Taylor polynomial with remainder at the origin, where

the constant term is $\quad T^0(f)(\vec{\mathbf{x}}) = f(\mathbf{0})$,
the linear term is $\quad T^1(f)(\vec{\mathbf{x}}) = D_1f(\mathbf{0})x_1 + \dots + D_nf(\mathbf{0})x_n = [\mathbf{D}f(\mathbf{0})]\vec{\mathbf{x}}$,
the remainder satisfies $\quad |R(f)(\vec{\mathbf{x}})| \le C|\vec{\mathbf{x}}|^2$ for some constant C.

We will then see that only the linear terms contribute to the limit.

Since φ is a k-form, the exterior derivative $\mathbf{d}\varphi$ should be a $(k+1)$-form. Thus we need to evaluate it on $k+1$ vectors and check that it is multilinear and alternating. This involves integrating φ over the boundary (i.e., the faces) of $P_{\mathbf{0}}(h\vec{\mathbf{v}}_1, \dots, h\vec{\mathbf{v}}_{k+1})$. As illustrated by Figure A22.1, we can parametrize those faces by the $2(k+1)$ mappings

$$\gamma_{1,i}\begin{pmatrix} t_1 \\ \vdots \\ t_k \end{pmatrix} = \gamma_{1,i}(\mathbf{t}) = h\vec{\mathbf{v}}_i + t_1\vec{\mathbf{v}}_1 + \dots + t_{i-1}\vec{\mathbf{v}}_{i-1} + t_i\vec{\mathbf{v}}_{i+1} + \dots + t_k\vec{\mathbf{v}}_{k+1},$$

$$\gamma_{0,i}\begin{pmatrix} t_1 \\ \vdots \\ t_k \end{pmatrix} = \gamma_{0,i}(\mathbf{t}) = \qquad t_1\vec{\mathbf{v}}_1 + \dots + t_{i-1}\vec{\mathbf{v}}_{i-1} + t_i\vec{\mathbf{v}}_{i+1} + \dots + t_k\vec{\mathbf{v}}_{k+1},$$

for i from 1 to $k+1$, and where $0 \le t_j \le h$ for each $j = 1, \dots, k$. (Note that on the right, $\vec{\mathbf{v}}_i$ is omitted.) We will denote by Q_h the domain of this parametrization. Notice that $\gamma_{1,i}$ and $\gamma_{0,i}$ have the same partial derivatives, the k vectors $\vec{\mathbf{v}}_1, \dots, \vec{\mathbf{v}}_{k+1}$, *excluding* the vector $\vec{\mathbf{v}}_i$; we will write the integrals over the faces parametrized by $\gamma_{1,i}$ and $\gamma_{0,i}$ under the same integral sign. So we can write the exterior derivative as the limit as $h \to 0$ of the following sum, where the hat on $\vec{\mathbf{v}}_i$ indicates that $\vec{\mathbf{v}}_i$ is omitted:

$$\sum_{i=1}^{k+1}(-1)^{i-1}\frac{1}{h^{k+1}}\int_{Q_h}\underbrace{\Big(f\big(\gamma_{1,i}(\mathbf{t})\big) - f\big(\gamma_{0,i}(\mathbf{t})\big)\Big)}_{\text{coefficient (function of }\mathbf{t})}\underbrace{dx_{i_1}\wedge\dots\wedge dx_{i_k}}_{k\text{-form}}\underbrace{\overbrace{(\vec{\mathbf{v}}_1, \dots, \widehat{\vec{\mathbf{v}}_i}, \dots, \vec{\mathbf{v}}_{k+1})}^{\text{partial derivatives of }\gamma_{1,i}\text{ and }\gamma_{0,i}}}_{k\text{ vectors}}|d^k\mathbf{t}|, \tag{A22.1}$$

where each term

$$\int_{Q_h}\Big(f\big(\gamma_{1,i}(\mathbf{t})\big) - f\big(\gamma_{0,i}(\mathbf{t})\big)\Big)\,dx_{i_1}\wedge\dots\wedge dx_{i_k}(\vec{\mathbf{v}}_1, \dots, \widehat{\vec{\mathbf{v}}_i}, \dots, \vec{\mathbf{v}}_{k+1})|d^k\mathbf{t}| \tag{A22.2}$$

is the sum of three terms, of which the second is the only one that counts (most of the work is in proving that the third one doesn't count):

There are $k+1$ mappings $\gamma_{1,i}(\mathbf{t})$, one for each i from 1 to $k+1$: for each mapping, a different $\vec{\mathbf{v}}_i$ is omitted. The same is true for $\gamma_{0,i}(\mathbf{t})$.

constant term: $\displaystyle\int_{Q_h}\overbrace{\Big(T^0(f)\big(\gamma_{1,i}(\mathbf{t})\big) - T^0(f)\big(\gamma_{0,i}(\mathbf{t})\big)\Big)}^{\text{coefficient of }k\text{-form}}\overbrace{dx_{i_1}\wedge\dots\wedge dx_{i_k}}^{k\text{-form}}(\vec{\mathbf{v}}_1, \dots, \widehat{\vec{\mathbf{v}}_i}, \dots, \vec{\mathbf{v}}_{k+1})|d^k\mathbf{t}| +$

linear term: $\displaystyle\int_{Q_h}\Big(T^1(f)\big(\gamma_{1,i}(\mathbf{t})\big) - T^1(f)\big(\gamma_{0,i}(\mathbf{t})\big)\Big)dx_{i_1}\wedge\dots\wedge dx_{i_k}(\vec{\mathbf{v}}_1, \dots, \widehat{\vec{\mathbf{v}}_i}, \dots, \vec{\mathbf{v}}_{k+1})|d^k\mathbf{t}| +$

remainder: $\displaystyle\int_{Q_h}\Big(R(f)\big(\gamma_{1,i}(\mathbf{t})\big) - R(f)\big(\gamma_{0,i}(\mathbf{t})\big)\Big)dx_{i_1}\wedge\dots\wedge dx_{i_k}(\vec{\mathbf{v}}_1, \dots, \widehat{\vec{\mathbf{v}}_i}, \dots, \vec{\mathbf{v}}_{k+1})|d^k\mathbf{t}|.$

The constant term cancels, since

$$\underbrace{T^0(f)(\text{anything})}_{\text{constant}} - \underbrace{T^0(f)(\text{anything})}_{\text{same constant}} = 0. \qquad A22.3$$

In equation A22.4 the derivatives are evaluated at $\mathbf{0}$ because that is where the Taylor polynomial is being computed. The second equality in equation A22.4 comes from linearity.

For the linear term, note that

$$\begin{aligned} T^1(f)\big(\gamma_{1,i}(\mathbf{t})\big) - T^1(f)\big(\gamma_{0,i}(\mathbf{t})\big) &= [\mathbf{D}f(\mathbf{0})]\big(\overbrace{h\vec{\mathbf{v}}_i + \gamma_{0,i}(\mathbf{t})}^{\gamma_{1,i}(\mathbf{t})}\big) - [\mathbf{D}f(\mathbf{0})]\big(\gamma_{0,i}(\mathbf{t})\big) \\ &= h[\mathbf{D}f(\mathbf{0})]\vec{\mathbf{v}}_i. \end{aligned} \qquad A22.4$$

In the next-to-last line of equation A22.5, one h in the numerator comes from the $h[\mathbf{D}f(\mathbf{0})]\vec{\mathbf{v}}_i$ in equation A22.4. The other h^k come from the fact that we are integrating over Q_h, a cube of sidelength h in $\mathbb{R}^k$.

This is a constant with respect to $\mathbf{t}$, so the sum for the linear terms is

$$\begin{aligned} &\sum_{i=1}^{k+1}(-1)^{i-1}\frac{1}{h^{k+1}}\int_{Q_h}\Big(T^1(f)\big(\gamma_{1,i}(\mathbf{t})\big) - T^1(f)\big(\gamma_{0,i}(\mathbf{t})\big)\Big)dx_{i_1}\wedge\cdots\wedge dx_{i_k} \\ &\qquad\qquad (\vec{\mathbf{v}}_1,\dots,\widehat{\vec{\mathbf{v}}_i},\dots,\vec{\mathbf{v}}_{k+1})\,|d^k\mathbf{t}| \\ &\underbrace{=}_{\text{eq. A22.4}} \sum_{i=1}^{k+1}(-1)^{i-1}\frac{h^{k+1}}{h^{k+1}}\big([\mathbf{D}f(\mathbf{0})]\vec{\mathbf{v}}_i\big)dx_{i_1}\wedge\cdots\wedge dx_{i_k}(\vec{\mathbf{v}}_1,\dots,\widehat{\vec{\mathbf{v}}_i},\dots,\vec{\mathbf{v}}_{k+1}) \\ &= (\mathbf{d}f\wedge dx_{i_1}\wedge\cdots\wedge dx_{i_k})\big(P_{\mathbf{0}}(\vec{\mathbf{v}}_1,\dots,\vec{\mathbf{v}}_{k+1})\big) \end{aligned} \qquad A22.5$$

by Definition 6.1.12 of the wedge product, in this case the wedge product of the 1-form $\mathbf{d}f$ with a k-form.

Definition 6.1.12: Since $\mathbf{d}f$ is a 1-form, there are $k+1$ shuffles σ: ways to write

$$\vec{\mathbf{v}}_1,\dots,\vec{\mathbf{v}}_{k+1} \text{ as}$$

$$\vec{\mathbf{v}}_i|\vec{\mathbf{v}}_1,\dots,\widehat{\vec{\mathbf{v}}_i},\dots,\vec{\mathbf{v}}_{k+1}$$

with the vectors to the right of the bar in ascending order. It is easy to check that $\operatorname{sgn}(\sigma) = (-1)^{i-1}$. For instance, the "identity shuffle" where $i = 1$ gives signature $+1$, while the shuffle that exchanges $\vec{\mathbf{v}}_1$ and $\vec{\mathbf{v}}_2$ has signature -1.

The last equality in equation A22.5 explains one reason why we defined the oriented boundary as we did, each part of the boundary being given the sign $(-1)^{i-1}$: this makes it compatible with the wedge product.

We now show that the remainder is in $O(h^2)$. We see this as follows: by the triangle inequality and the fact that all t_i satisfy $|t_i| \le h$, we see that

$$|\gamma_{0,i}| \le |h|\sum_{i=1}^{k+1}|\vec{\mathbf{v}}_i|, \quad \text{and} \quad |\gamma_{1,i}| \le |h|\sum_{i=1}^{k+1}|\vec{\mathbf{v}}_i| \qquad A22.6$$

Now apply Taylor's theorem A12.7 to the Taylor polynomial of degree 1:

$$\Big|f(\mathbf{a}+\vec{\mathbf{h}}) - P^1_{f,\mathbf{a}}(\mathbf{a}+\vec{\mathbf{h}})\Big| \le C\left(\sum_{i=1}^{n}|h_i|\right)^2, \qquad A22.7$$

where

$$C = \sup_{I\in\mathcal{I}^2_n}\ \sup_{\mathbf{c}\in[\mathbf{a},\mathbf{a}+\vec{\mathbf{h}}]}|D_I f(\mathbf{c})|. \qquad A22.8$$

This gives

$$|R(f)\big(\gamma_{0,i}(\mathbf{t})\big)| \le Kh^2 \quad \text{and} \quad |R(f)\big(\gamma_{1,i}(\mathbf{t})\big)| \le Kh^2, \qquad A22.9$$

where

$$K = C\left(\sum_{i=1}^{k+1}|\vec{\mathbf{v}}_i|\right)^2. \qquad A22.10$$

Now we see that the remainder disappears in the limit, using

$$\Big|R(f)\big(\gamma_{1,i}(\mathbf{t})\big) - R(f)\big(\gamma_{0,i}(\mathbf{t})\big)\Big| \le \Big|R(f)\big(\gamma_{1,i}(\mathbf{t}\big)\Big| + \Big|R(f)\big(\gamma_{0,i}(\mathbf{t})\big)\Big| \le 2h^2K.$$

Inserting this into the integral in formula A22.1 leads to

$$\left|\int_{Q_h} \underbrace{\Big(R(f)\big(\gamma_{1,i}(\mathbf{t})\big) - R(f)\big(\gamma_{0,i}(\mathbf{t})\big)\Big)}_{\le 2Kh^2} dx_{i_1}\wedge\ldots\wedge dx_{i_k}(\vec{\mathbf{v}}_1,\ldots,\widehat{\vec{\mathbf{v}}}_i,\ldots,\vec{\mathbf{v}}_{k+1})\,|d^k\mathbf{t}|\right|$$

$$\le \int_{Q_h}\Big(2h^2K\left|dx_{i_1}\wedge\cdots\wedge dx_{i_k}(\vec{\mathbf{v}}_1,\ldots,\widehat{\vec{\mathbf{v}}}_i,\ldots,\vec{\mathbf{v}}_{k+1})\right|\,|d^k\mathbf{t}|\Big) \qquad A22.11$$

$$\le h^{k+2}K(\sup_j|\vec{\mathbf{v}}_j|)^k,$$

which disappears in the limit after dividing by h^{k+1}. This proves part 5.

Equation A22.12: In particular, the limit in the second line exists because the limit in the third line exists, by part 5.

The following proves part 1:

$$\mathbf{d}\Big(\sum a_{i_1\ldots i_k}dx_{i_1}\wedge\cdots\wedge dx_{i_k}\Big)P_{\mathbf{x}}(\vec{\mathbf{v}}_1,\ldots,\vec{\mathbf{v}}_{k+1})$$

$$= \lim_{h\to 0}\frac{1}{h^{k+1}}\int_{\partial P_{\mathbf{x}}(\vec{\mathbf{v}}_1,\ldots,\vec{\mathbf{v}}_{k+1})}\Big(\sum a_{i_1\ldots i_k}dx_{i_1}\wedge\cdots\wedge dx_{i_k}\Big) \qquad A22.12$$

$$= \sum_{1\le i_1<\cdots<i_k\le n}\lim_{h\to 0}\left(\frac{1}{h^{k+1}}\int_{\partial P_{\mathbf{x}}(\vec{\mathbf{v}}_1,\ldots,\vec{\mathbf{v}}_{k+1})} a_{i_1\ldots i_k}dx_{i_1}\wedge\cdots\wedge dx_{i_k}\right)$$

$$\overbrace{=}^{\text{part 5}} \sum_{1\le i_1<\cdots<i_k\le n}(\mathbf{d}a_{i_1\ldots i_k}\wedge dx_{i_1}\wedge\cdots\wedge dx_{i_k})\Big(P_{\mathbf{x}}(\vec{\mathbf{v}}_1,\ldots,\vec{\mathbf{v}}_{k+1})\Big).$$

Part 2 is now clear, and part 3 follows immediately from parts 5 and 1. □

Exercise for Section A22

A22.1 Show that the last equality of equation A22.5 is "by the definition of the wedge product".

A23 Proof of Proposition 6.10.10 (for Stokes's theorem)

Here we prove Proposition 6.10.10, which was used in the proof of Stokes's theorem in Section 6.10.

Proposition 6.10.10 (Trimming X to make X_ϵ). *Let $B_r(\mathbf{x})$ be the open ball of radius r around $\mathbf{x}$. For all $\epsilon > 0$, there exist points $\mathbf{x}_1,\ldots,\mathbf{x}_p \in X$ and $r_1 > 0,\ldots,r_p > 0$, such that $\sum_{i=1}^p r_i^{k-1} < \epsilon$ and*

$$X_\epsilon \overset{\text{def}}{=} X - \bigcup_{i=1}^p B_{r_i}(\mathbf{x}_i) \quad \text{is a piece-with-corners.}$$

Proof. Proposition 6.10.10 will follow easily from the modified Whitney transversality theorem, Theorem A23.3.

Transversality corresponds to "general position"; it is the usual situation, not something exceptional. A line and a plane in $\mathbb{R}^3$ are transversal if the line doesn't lie in the plane; two planes are transversal if they do not coincide. Two 1-dimensional subspaces of $\mathbb{R}^3$ (lines through the origin) are never transversal.

Two curves in $\mathbb{R}^2$ are transversal if at every point of intersection the tangent spaces are distinct (i.e., at a point of intersection the curves are not tangent to each other). Two curves in $\mathbb{R}^3$ are traversal only if they don't intersect. (At a point of intersection, the two tangent spaces could be distinct, but two lines don't span space.)

FIGURE A23.1.

Hassler Whitney (1907–1989) was on the faculty at Harvard before moving to the Institute for Advanced Study at Princeton. His many contributions included the first usable definition of a manifold (1936). He was described by mathematician Stanislaw Ulam as "friendly, but rather taciturn ... with wry humour, shyness but self-assurance, a probity which shines through, and a certain genius for persistent and deep follow-through in mathematics."

Definition A23.1 (Transversality).

1. Two subspaces of $\mathbb{R}^n$ are *transversal* if together they span $\mathbb{R}^n$.
2. Two manifolds M and N are *transversal* if at every point where they intersect, their tangent spaces are transversal.

More generally, if M and N are manifolds and $X \subset M$ is a subset, we will say that M is transversal to N on X if for all points of $\mathbf{x} \in X \cap N$, the tangent spaces $T_{\mathbf{x}}M$ and $T_{\mathbf{x}}N$ are transversal.

Let G be the group of rigid motions of $\mathbb{R}^n$ (maps of the form $\mathbf{x} \mapsto A\mathbf{x}+\vec{\mathbf{b}}$, where A is an $n \times n$ orthogonal matrix and $\vec{\mathbf{b}} \in \mathbb{R}^n$). Exercise A23.1 asks you to prove that G is a manifold in $\mathbb{R}^n \times \operatorname{Mat}(n,n)$ of dimension

$$\dim G = n + \frac{n(n-1)}{2} = \frac{n(n+1)}{2}. \qquad A23.1$$

Transversality is an *open condition*: if two transversal manifolds are modified slightly, they are still transversal. But we have to be careful how we state this.

Proposition A23.2 (Stability of transversality). *If $M \subset \mathbb{R}^n$ is a manifold, $X \subset M$ is compact, and $V \subset \mathbb{R}^n$ is a subspace that is transversal to M on X, then for all $g \in G$ sufficiently close to the identity, M is transversal to $g(V)$ on X.*

Proof. By contradiction, suppose that $i \mapsto g_i$ is a sequence tending to the identity such that M is not transversal to $g_i(V)$ on X. Then there exists a sequence $i \mapsto \mathbf{x}_i$ in X and unit vectors $\vec{\mathbf{w}}_i \in \mathbb{R}^n$ such that $\vec{\mathbf{w}}_i$ is orthogonal to both $g_i(V)$ and to $T_{\mathbf{x}_i}M$. Since both X and the unit sphere are compact, by choosing a subsequence we may suppose that the $\mathbf{x}_i$ converge to some $\mathbf{x} \in X$ and that the $\vec{\mathbf{w}}_i$ converge to some unit vector $\vec{\mathbf{w}} \in \mathbb{R}^n$. Since the g_i converge to the identity, the subspaces $g_i(V)$ converge to V. At the limit, $\vec{\mathbf{w}}$ is a unit vector orthogonal to both V and $T_{\mathbf{x}}M$, contradicting the hypothesis that V and M are transversal on X. Thus no such sequence can exist. □

Theorem A23.3 says that by a small rigid motion, we can make M transversal to a fixed sphere.

Theorem A23.3 (Modified Whitney transversality theorem). *Let $S \subset \mathbb{R}^n$ be a sphere, M a manifold, and $X \subset M$ a compact subset. Then there exists a rigid motion $g \in G$ arbitrarily close to the identity such that M is transversal to $g(S)$ on X.*

Proof. First, in Lemma A23.4, we state the result for manifolds that are graphs (not just locally graphs); we will see that the general case follows.

As usual, we denote by $\Gamma(\mathbf{f})$ the graph of a map $\mathbf{f}$. The case where we will apply this is where $\mathbb{R}^n = \mathbb{R}^k \times \mathbb{R}^{n-k}$, where $U \subset \mathbb{R}^k$ is a subset and $\mathbf{f}: U \to \mathbb{R}^{n-k}$ is a mapping. In that case the graph is

$$\Gamma(\mathbf{f}) = \left\{ \begin{pmatrix} \mathbf{x} \\ \mathbf{f}(\mathbf{x}) \end{pmatrix} \mid \mathbf{x} \in U \right\}. \qquad A23.2$$

Proof of Theorem A23.3: The space $E_{\mathbf{x}}$ can be chosen to be generated by k standard basis vectors, so $E_{\mathbf{x}}^{\perp}$ is generated by the other standard basis vectors.

Lemma A23.4. *Let $S \subset \mathbb{R}^n$ be a sphere. If $U \subset \mathbb{R}^k$ is open and $\mathbf{f}: U \to \mathbb{R}^{n-k}$ is of class C^1, then there exists $g \in G$ arbitrarily close to the identity such that the graph $\Gamma(\mathbf{f})$ is transversal to $g(S)$.*

We prove Lemma A23.4 after using it to prove Theorem A23.3.

Proof of Theorem A23.3 from Lemma A23.4. By Definition 3.1.2 of a manifold, for every point of $\mathbf{x} \in M$, there exists a subspace $E_{\mathbf{x}} \subset \mathbb{R}^n$, its orthogonal complement $E_{\mathbf{x}}^{\perp}$, a ball $B_{\mathbf{x}} \subset E_{\mathbf{x}}$ of some radius $r_{\mathbf{x}} > 0$, and a C^1 map $\mathbf{f}_{\mathbf{x}}: B_{\mathbf{x}} \to E_{\mathbf{x}}^{\perp}$ such that

$$M = \cup_{\mathbf{x} \in M} \Gamma(\mathbf{f}_{\mathbf{x}}) \qquad A23.3$$

For all ϵ with $0 < \epsilon < 1$, let $\Gamma^{\epsilon}(\mathbf{f}_{\mathbf{x}})$ be the graph of the restriction of $\mathbf{f}_{\mathbf{x}}$ to the concentric ball $B_{\mathbf{x}}^{\epsilon}$ of radius $r_{\mathbf{x}}(1-\epsilon)$. These balls form an open cover of M, so, since X is compact, finitely many cover X. Thus there exist $\rho > 0$ and $\mathbf{x}_1, \dots, \mathbf{x}_p$ such that the $\Gamma^{\rho}(\mathbf{f}_{\mathbf{x}_i})$ cover X, and the closures $\overline{\Gamma}^{\rho}(\mathbf{f}_{\mathbf{x}_i})$ are compact.

By Lemma A23.4, we can find $g_1 \in G$ such that $\Gamma(\mathbf{f}_{\mathbf{x}_1})$ is transversal to $g_1(S)$. Then by Proposition A23.2, for all h sufficiently close to the identity, $\Gamma(\mathbf{f}_{\mathbf{x}_1})$ is transversal to $h \circ g_1(S)$ on the part $\overline{\Gamma}^{\rho}(\mathbf{f}_{\mathbf{x}_1})$, since this part is compact. Denote by h_1 such an h with $\Gamma(\mathbf{f}_{\mathbf{x}_2})$ transversal to $h_1(g_1(S))$.

Set $g_2 = h_1 \circ g_1$. For all $h \in G$ sufficiently close to the identity, $h \circ g_2(S)$ is transversal to

$$\Gamma(\mathbf{f}_{\mathbf{x}_1}) \cup \Gamma(\mathbf{f}_{\mathbf{x}_2}) \quad \text{on} \quad \overline{\Gamma}^{\rho}(\mathbf{f}_{\mathbf{x}_1}) \cup \overline{\Gamma}^{\rho}(\mathbf{f}_{\mathbf{x}_2}). \qquad A23.4$$

We can continue this way, choosing h_2, defining $g_3 = h_2 \circ g_2$, choosing h_3, and so on, so that

$$\Gamma(\mathbf{f}_{\mathbf{x}_i}) \text{ is transversal to } g_i(S), \text{ and } \bigcup_{j=1}^{i-1} \overline{\Gamma}^{\rho}(\mathbf{f}_{\mathbf{x}_j}) \text{ is transversal to } g_i(S).$$

When we get to $i = p$, we are done. □

Proof of Lemma A23.4. Let $\mathbf{p} \in \mathbb{R}^n$ be the center of the sphere S of radius R. Define $\Phi : \mathbb{R}^n \times U \times \mathbb{R}^n \times G \to \mathbb{R}^{1+k+n+n}$ given by

$$\Phi : (\vec{\mathbf{w}}, \mathbf{x}, \mathbf{y}, g) \mapsto \begin{bmatrix} |\vec{\mathbf{w}}|^2 - 1 \\ \begin{bmatrix} I_k & [\mathbf{Df}(\mathbf{x})]^\top \end{bmatrix} \vec{\mathbf{w}} \\ R\vec{\mathbf{w}} - [\mathbf{D}g](\mathbf{y} - \mathbf{p}) \\ \begin{pmatrix} \mathbf{x} \\ \mathbf{f}(\mathbf{x}) \end{pmatrix} - g(\mathbf{y}) \end{bmatrix} \qquad A23.5$$

Equation A23.5: The first entry on the right is 1 high. The second is k high: I_k is the $k \times k$ identity matrix and $[\mathbf{Df}(\mathbf{x})]^\top$ is a $k \times (n-k)$ matrix, so $[I_k \; [\mathbf{Df}(\mathbf{x})]^\top]$ is a $k \times n$ matrix; multiplying it by $\vec{\mathbf{w}}$ gives a column vector k high.

The third and fourth entries are both n high. In the third entry, $[\mathbf{D}(g)]$ is an orthogonal $n \times n$ matrix. We don't need to specify where this derivative is evaluated because g is a rigid motion, a composition of a rotation and a translation, written

$$g(\mathbf{x}) = A\mathbf{x} - \vec{\mathbf{b}},$$

where A is an orthogonal matrix; as such the derivative is constant, everywhere the rotation matrix A.

Equation A23.5: If

$$\Phi(\vec{\mathbf{w}}, \mathbf{x}, \mathbf{y}, g) = \vec{\mathbf{0}},$$

then

$$\begin{aligned} |\mathbf{w}|^2 &= 1 \\ [I_k \; [\mathbf{Df}(\mathbf{x})]^\top]\, \vec{\mathbf{w}} &= \vec{\mathbf{0}} \\ [\mathbf{D}g](\mathbf{y} - \mathbf{p}) &= R\vec{\mathbf{w}} \\ \begin{pmatrix} \mathbf{x} \\ \mathbf{f}(\mathbf{x}) \end{pmatrix} &= g(\mathbf{y}) \end{aligned}$$

The map Φ is illustrated by Figure A23.2.

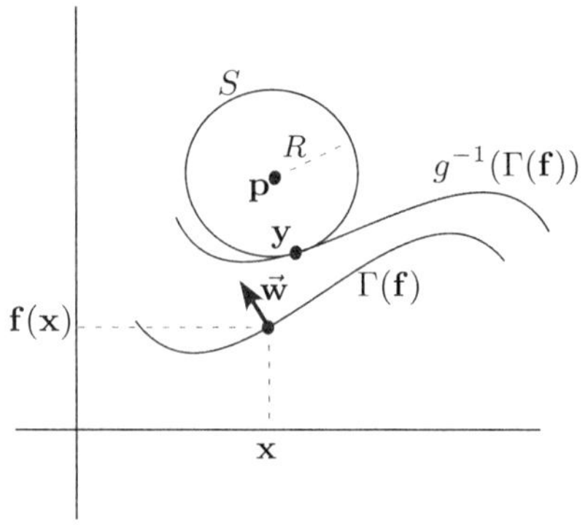

FIGURE A23.2. The moved graph $g^{-1}(\Gamma(\mathbf{f}))$ fails to be transversal to the sphere S (here a circle) if there is a point $\mathbf{y} = g^{-1}\begin{pmatrix}\mathbf{x}\\ \mathbf{f}(\mathbf{x})\end{pmatrix}$ and a unit vector $\vec{\mathbf{w}}$ orthogonal to $\Gamma(\mathbf{f})$ at $\begin{pmatrix}\mathbf{x}\\ \mathbf{f}(\mathbf{x})\end{pmatrix}$ such that $R\vec{\mathbf{w}} = [\mathbf{D}g](\mathbf{y}-\mathbf{p})$.

We will show that

$$\Gamma(\mathbf{f}),\ g(S) \text{ not transversal} \iff \exists\ \vec{\mathbf{w}}, \mathbf{x}, \mathbf{y} \text{ with } (\vec{\mathbf{w}}, \mathbf{x}, \mathbf{y}, g) \in \Phi^{-1}(\mathbf{0}).$$

Moreover, we will show (Theorem A23.6) that the set $\Phi^{-1}(\mathbf{0})$ is "small" (in the sense that it cannot contain any open subset of G), so near any point of G there exists g such that $\Gamma(\mathbf{f})$ and $g(S)$ are transversal.

Let us examine what an element of $\Phi^{-1}(\mathbf{0})$ is. The point $\begin{pmatrix}\mathbf{x}\\ \mathbf{f}(\mathbf{x})\end{pmatrix} = g(\mathbf{y})$ is a point of $\Gamma(\mathbf{f}) \cap g(S)$. By Definition 3.2.1, the tangent space to the manifold $\Gamma(\mathbf{f})$ at $\begin{pmatrix}\mathbf{x}\\ \mathbf{f}(\mathbf{x})\end{pmatrix}$ is the graph of $[\mathbf{Df}(\mathbf{x})]$; the vectors in the graph of $[\mathbf{Df}(\mathbf{x})]$ are those of the form

$$\begin{bmatrix} I_k \\ [\mathbf{Df}(\mathbf{x})] \end{bmatrix} \vec{\mathbf{v}} \quad \text{for some } \vec{\mathbf{v}} \in \mathbb{R}^k. \qquad A23.6$$

The equation $[I_k\ [\mathbf{Df}(\mathbf{x})]^\top]\vec{\mathbf{w}} = \mathbf{0}$ says that the dot product of $\vec{\mathbf{w}}$ with each column of $\begin{bmatrix} I_k \\ [\mathbf{Df}(\mathbf{x})] \end{bmatrix}$ is 0, i.e.,

$$\begin{bmatrix} I_k \\ [\mathbf{Df}(\mathbf{x})] \end{bmatrix} \vec{\mathbf{e}}_i \cdot \vec{\mathbf{w}} = 0. \qquad A23.7$$

So $\vec{\mathbf{w}}$ is orthogonal to the tangent space of $\Gamma(\mathbf{f})$ at $\begin{pmatrix}\mathbf{x}\\ \mathbf{f}(\mathbf{x})\end{pmatrix}$.

Since $R\vec{\mathbf{w}} = [\mathbf{D}g](\mathbf{y}-\mathbf{p})$, the vector $\vec{\mathbf{w}}$ (which has length 1 by the first entry of equation A23.5) is a multiple of the radius vector of the sphere from the center $g(\mathbf{p})$ to $g(\mathbf{y})$, hence orthogonal to the sphere $g(S)$ at $g(\mathbf{y})$, hence orthogonal to $T_{g(\mathbf{y})}g(S)$.

Thus $\vec{\mathbf{w}}$ is not in the span of $T_{g(\mathbf{y})}\Gamma(\mathbf{f}) \cup T_{g(\mathbf{y})}g(S)$, so those tangent spaces are not transversal. Hence, at a point $(\vec{\mathbf{w}}, \mathbf{x}, \mathbf{y}, g) \in \Phi^{-1}(\mathbf{0})$, the manifolds $\Gamma(\mathbf{f})$ and $g(S)$ are not transversal.

Since $\dim T_{g(\mathbf{y})}g(S) = n-1$, if $T_{g(\mathbf{y})}\Gamma(\mathbf{f})$ is not transversal to $T_{g(\mathbf{y})}g(S)$, then

$$T_{g(\mathbf{y})}\Gamma(\mathbf{f}) \subset T_{g(\mathbf{y})}g(S).$$

Indeed, if $\vec{\mathbf{v}} \notin T_{g(\mathbf{y})}g(S)$, then $\vec{\mathbf{v}}$ together with $T_{g(\mathbf{y})}g(S)$ span $\mathbb{R}^n$.

For the converse, suppose that $g(S)$ and $\Gamma(\mathbf{f})$ are not transversal at some point $g(\mathbf{y}) = \begin{pmatrix} \mathbf{x} \\ \mathbf{f}(\mathbf{x}) \end{pmatrix} \in \Gamma(\mathbf{f}) \cap g(S)$. Then $g(\mathbf{y}-\mathbf{p})$ is orthogonal to $T_{g(\mathbf{y})}g(S)$ and to $T_{g(\mathbf{y})}\Gamma(\mathbf{f})$. Take $\vec{\mathbf{w}} = g(\mathbf{y}-\mathbf{p})/R$. It should be clear that $\Phi(\vec{\mathbf{w}}, \mathbf{x}, \mathbf{y}, g) = \vec{\mathbf{0}}$. □

The object now is to show that $\Phi^{-1}(\mathbf{0})$ is small. We will show that $\Phi^{-1}(\mathbf{0})$ is a manifold of dimension $< \dim G$.

Lemma A23.5. 1. *The derivative of Φ is surjective at every point of $\Phi^{-1}(\mathbf{0})$.*

2. *The manifold $\Phi^{-1}(\mathbf{0})$ has dimension* $\dim G - 1$.

Proof of Lemma A23.5. 1. The Jacobian matrix $[\mathbf{D}\Phi]$ has the form shown in Figure A23.3. You are asked in Exercise A23.2 to check that it is enough to show that each of the shaded matrices is surjective, and that the entries above them vanish. The entry in the column labeled $\mathbf{y}$ is $[\mathbf{D}g]$, which is an orthogonal matrix, hence certainly invertible, and the entry in the column labeled "translations" is simply the identity.

	$\vec{\mathbf{w}}$	$\mathbf{x}$	$\mathbf{y}$	translations	rotations	
$\lvert\vec{\mathbf{w}}\rvert^2 - 1$		0	0	0	0	1
$\vec{\mathbf{w}} \cdot (\vec{\mathbf{e}}_i + \vec{D}_i\mathbf{f}(\mathbf{x}))$		*	0	0	0	k
$R\vec{\mathbf{w}} - [\mathbf{D}g](\mathbf{y}-\mathbf{p})$	R Id	0		0	$[\mathbf{D}g]$	n
$\begin{pmatrix} \mathbf{x} \\ \mathbf{f}(\mathbf{x}) \end{pmatrix} - g(\mathbf{y})$	0	*	*		*	n
	n	k	n	n	$n(n-1)/2$	

FIGURE A23.3. The columns represent the variables of the domain, and the rows represent the coordinate functions. The numbers at right and at bottom give the dimensions of the submatrices. We have divided G into translations (dimension n) and rotations (dimension $n(n-1)/2$).

The shaded entry (actually, the two matrices) in the first column (labeled $\vec{\mathbf{w}}$) requires a bit more care. That matrix is

$$\begin{bmatrix} 2w_1 & \dots & 2w_k & 2w_{k+1} & \dots & 2w_n \\ 1 & \dots & 0 & * & \dots & * \\ \vdots & \ddots & \vdots & \vdots & & \vdots \\ 0 & \dots & 1 & * & \dots & * \end{bmatrix}. \qquad A23.8$$

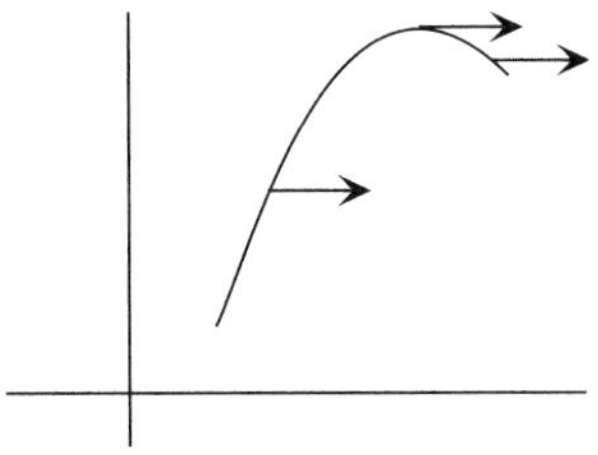

FIGURE A23.4.

A horizontal vector (one whose coordinates $w_{k+1}, \dots, w_n$ in the codomain $\mathbb{R}^{n-k}$ all vanish) can never be orthogonal to the graph of $\mathbf{f}$.

It should be clear that this matrix is surjective if and only if at least one of $w_{k+1}, \dots, w_n$ does not vanish. But if they all vanish, then $\vec{\mathbf{w}}$ cannot be orthogonal to the graph of $\mathbf{f}$ at any point, so this $\vec{\mathbf{w}}$ cannot be one entry of an element of $\Phi^{-1}(\mathbf{0})$, as illustrated in Figure A23.4. This proves part 1.

2. Since $[\mathbf{D}\Phi]$ is surjective at every point of $\Phi^{-1}(\mathbf{0})$, we can apply Theorem 3.1.10 to say that $\Phi^{-1}(\mathbf{0})$ is a manifold of dimension

$$\dim(\text{domain } \Phi) - \dim(\text{codomain } \Phi) = (n+k+n+\dim G) - (1+k+n+n),$$

i.e., of dimension $\dim G - 1$. □ Lemma A23.5

The next result is an easy case of Sard's theorem. The statement may seem obvious, but remember Peano curves, where the image of a one-dimensional line becomes a full triangle. Theorem A23.6 says that Peano curves cannot be C^1.

Arthur Sard (1909–1980) got his PhD at Harvard University in 1936, then joined the faculty at Queen's College (part of the City University of New York.) During World War 2 he worked with the Applied Mathematics Panel, especially on control for machine guns on bombers.

Theorem A23.6 (An easy case of Sard's theorem). *Let M be a manifold of dimension p, let N be a manifold of dimension q, let Y be a compact subset of M, and let $f : M \to N$ be a C^1 mapping. If $p < q$, then $f(Y)$ has q-dimensional volume 0.*

Proof of Theorem A23.6. You are asked in Exercise A23.3 to show that it is enough to prove this result if $U \subset \mathbb{R}^p$ is open, $\mathbf{g}: U \to \mathbb{R}^{n-p}$ for some n is a C^1 mapping, and Y is a compact part of the graph $\Gamma(\mathbf{g})$.

The original Sard's theorem puts no restrictions on the dimensions of M and N, but requires f to be more differentiable than just class C^1. It says that the set of critical values of f then has measure 0. See J. Milnor, *Topology from the Differentiable Viewpoint*, Princeton University Press, 1997 (our candidate for the best mathematics book ever written).

Next, you are asked in Exercise A23.4 to show that there exists C_1 such that for any $r > 0$ (however small), Y can be covered by $\leq C_1/r^p$ balls of radius r.

Let $B \subset \mathbb{R}^n$ be a ball of radius r centered at $\mathbf{x} \in \Gamma(\mathbf{g})$. Then $f(B)$ is contained in the ball centered at $f(\mathbf{x})$ of radius $|[\mathbf{D}f(\mathbf{x})]|r$. Thus if we set

$$C_2 = \sup_Y |[\mathbf{D}f(\mathbf{x})]|, \qquad A23.9$$

we see that $f(Y)$ is covered by $C_1 r^{-p}$ balls of radius $C_2 r$. By Definition 5.2.1, $f(Y)$ has q-dimensional volume 0 if

$$\left(\frac{C_1}{r^p}\right)(C_2 r)^q \qquad A23.10$$

can be made arbitrarily small, and if $q > p$ this tends to 0 as $r \to 0$. □

In particular, the part of $\Phi^{-1}(\mathbf{0})$ where the point of non-transversality lies in $\Gamma^\rho(\mathbf{f})$ is compact, so its projection to G has zero volume in dimension $n(n+1)/2$, i.e., the dimension of G (see equation A23.1 and Exercise

A23.1). In particular, it is compact and contains no open set of G, and there exist rigid motions $g \in G$ arbitrarily close to the identity that have a neighborhood not in the projection of $\Phi^{-1}(\mathbf{0})$. A look at the statement of Theorem A23.3 shows that such a g solves our problem. This completes the proof of Theorem A23.3. □

Proof of Proposition 6.10.10. Using Heine-Borel, cover $\partial^{ns}X$ by finitely many balls $B_i \overset{\text{def}}{=} B_{r_i}(\mathbf{x}_i)$, $i = 1, \dots, p$. Note that if we move these balls sufficiently little and modify their radii sufficiently little, the new balls will still cover $\partial^{ns}X$.

By Theorem A23.3, we can move $\mathbf{x}_1$ to $\mathbf{x}_1'$ by an arbitrarily small rigid motion, and modify r_1 arbitrarily little, so that if we write $B_1' = B_{r_1'}(\mathbf{x}_1')$, then $\partial B_i'$ is transversal to $\partial^s X$, and $B_1', B_2, \dots, B_p$ still cover $\partial^{ns}X$.

Now let B_2' be a similar small modification of B_2, so that ∂B_2 is transversal to $\partial^s X$ and to $\partial B_1'$, and $B'1, B'2, B_3, \dots, B_p$ still cover $\partial^{ns}X$.

Continue this way; when you get to B_p, you will be done. □

Exercises for Section A23

Exercise A23.1: Note that
$$\frac{n(n-1)}{2}$$
is the dimension of the space of antisymmetric $n \times n$ matrices. It is also the dimension of the orthogonal group $O(n)$, since the space of antisymmetric $n \times n$ matrices is the tangent space to the orthogonal group at the origin. See Exercise 3.2.11.

A23.1 Prove that the group G of rigid motions of $\mathbb{R}^n$ is a manifold in $\mathbb{R}^n \times \text{Mat}\,(n,n)$ of dimension

$$\dim G = n + \frac{n(n-1)}{2} = \frac{n(n+1)}{2}.$$

A23.2 Suppose that a matrix A contains surjective rectangular submatrices with no column intersecting more than one submatrix, and every row intersecting at least one submatrix. Suppose that there are no nonzero entries above the surjective submatrices. Show that A is surjective.

A23.3 Show that to prove Theorem A23.6, it is enough to prove it if $U \subset \mathbb{R}^p$ is open, $\mathbf{g}: U \to \mathbb{R}^{n-p}$ for some n is a C^1 mapping, and Y is a compact part of the graph $\Gamma(\mathbf{g})$. *Hint*: This follows almost immediately from Heine-Borel (Theorem A3.3).

A23.4 Show that there exists C_1 such that for any $r > 0$ (however small), the subset Y of Theorem A23.6 can be covered by $\leq C_1/r^p$ balls of radius r. *Hint*: Use Heine-Borel.

Bibliography

The following books were referred to in the text, are natural continuations of topics addressed in the text, or influenced our treatment of some topic. The list is certainly not exhaustive.

Analysis

Dzung Minh HA, *Functional Analysis, Volume 1: A Gentle Introduction*, Matrix Editions, 2006.

Walter RUDIN, *Real and Complex Analysis*, third edition, McGraw-Hill Higher Education, 1986.

Calculus and Forms

Henri CARTAN, *Differential Forms*, Hermann, Paris; H. Mifflin Co., Boston, 1970.

Jean DIEUDONNÉ, *Infinitesimal Calculus*, Houghton Mifflin Co., Boston, 1971.

Stanley GROSSMAN, *Multivariable Calculus, Linear Algebra and Differential Equations*, Harcourt Brace College Publishers, Fort Worth, 1995.

Lynn LOOMIS and Shlomo STERNBERG, *Advanced Calculus*, Addison-Wesley, Reading, MA, 1968.

Michael SPIVAK, *Calculus*, second edition, Publish or Perish, Inc., Wilmington, DE, 1980 (one-variable calculus); *Calculus on Manifolds*, W. A. Benjamin, Inc., NY, 1965; available from Publish or Perish, Inc., Wilmington, DE.

Differential Equations

What it Means to Understand a Differential Equation, The College Mathematics Journal, Vol. 25, Nov. 5 (Nov. 1994), 372-384.

John HUBBARD and Beverly WEST, *Differential Equations, A Dynamical Systems Approach, Part I*, Texts in Applied Mathematics No. 5, Springer-Verlag, N.Y., 1991.

Differential Geometry

Manfredo P. DO CARMO, *Differential Geometry of Curves and Surfaces*, Prentice-Hall, Inc., 1976.

John MILNOR, *Morse Theory*, Princeton University Press, 1963. (Lemma 2.2, p. 6, is referred to in section 3.6.)

Frank MORGAN, *Riemannian Geometry, A Beginner's Guide*, second edition, AK Peters, Ltd., Wellesley, MA, 1998.

Fourier Analysis

T. W. KÖRNER, *Fourier Analysis*, Cambridge University Press, 1990.

Barbara Burke HUBBARD, *The World According to Wavelets*, A K Peters, 1998.

Fractals and Chaos

John HUBBARD, *The Beauty and Complexity of the Mandelbrot Set*, Science TV, N.Y., Aug. 1989.

Benoit B. MANDELBROT, *Fractal Geometry of Nature*, W. H. Freeman & Co., 1988.

History

John STILLWELL, *Mathematics and Its History*, Springer-Verlag, New York, 1989.

Linear Algebra

Bernard KOLMAN, *Elementary Linear Algebra*, Prentice Hall, 1999.

Gilbert STRANG, *Introduction to Linear Algebra*, Wellesley Cambridge Press, 1993.

Set Theory

Paul R. HALMOS, *Naive Set Theory*, Springer Verlag, 1987.

Topology

Colin C. ADAMS, *Knot Book: An Elementary Introduction to the Mathematical Theory of Knots*, W. H. Freeman & Co., 2001, (paperback edition).

John MILNOR, *Topology from the Differentiable Viewpoint*, Princeton University Press, 1997.

Jeffrey R. WEEKS, *Shape of Space: How to Visualize Surfaces and Three-Dimensional Manifolds*, Marcel Dekker, 1985.

Paul B. YALE, *Geometry and Symmetry* (Dover Books on Advanced Mathematics), Dover Publications, 1988.

Photo Credits

Preface

p. vii Joseph Fourier, provided by Pour la Science (8 rue Férou, 75006 Paris), which published the same picture on page 33 of *Poincaré: Philosophe et Mathématicien*, N° 4 in the series "Les Génies de la Science."

p. xi Jean Dieudonné, from the Internet.

Chapter 0

p. 4 Nathaniel Bowditch, from the Internet.

p. 6 Euclid, Pour la Science (p. 50 of *Poincaré: Philosophe et Mathématicien*).

p. 8 Sketch of Russell by Roger Hayward, Pour la Science (p. 22 of *Poincaré: Philosophe et Mathématicien*).

p. 22 Georg Cantor, Pour la Science (p. 22 of *Poincaré: Philosophe et Mathématicien*).

p. 22 Joseph Liouville, Pour la Science (p. 67 of *Poincaré: Philosophe et Mathématicien*).

p. 23 Charles Hermite, *Acta Mathematica*. Table Generale des Tomes 1-35 (1913). By due permission of Institut Mittag-Leffler.

p. 24 Paul Cohen, Stanford University News Service.

p. 24 Kurt Gödel, from the Internet. This picture appears in *International Mathematical Congresses. An illustrated history 1893–1986*, by Donald J. Albers, G. L. Alexanderson, and Constance Reid (Rev. ed. Springer-Verlag, N. Y., 1986). One of the authors wrote that the picture comes from the Seely G. Mudd Manuscript Library; an archivist there was unable to trace it.

p. 26 Girolamo Cardano, from the Internet. First published in 1554 in *De subtilitate libri XXI* by Girolamo Cardano.

p. 28 Abraham de Moivre, from the Internet. The Royal Society of London holds rights to a black and white negative of a portrait by Joseph Highmore (1736). A copy of the picture in this text was sent to the society, which did not respond as to whether the picture in this text is the one to which the society holds rights.

p. 29 Niccolo Fontana Tartaglia, from the Internet.

Chapter 1

p. 46 Arthur Cayley, from the Internet. Originally published in *Harper's Weekly*, v. 28, 1884 Aug. 23. Also available from the Library of Congress (reproduction number LC-USZ62-120356).

p. 58 Salvatore Pincherle, from the Internet. Attempts to trace this picture through the American Mathematical Society and Italian Mathematical Union were unsuccessful.

p. 70 Hermann Schwarz, from the Internet.

p. 71 Viktor Bunyakovsky, from the Internet.

p. 87 Karl Weierstrass, Pour la Science (p. 29 *Poincaré: Philosophe et Mathématicien*).

p. 112 Niels Abel, Pour la Science (p. 43 of *Poincaré: Philosophe et Mathématicien*).

p. 113 Jean d'Alembert, from the Internet, probably originally from lithograph in Mme. Delpech, *Iconographie Française*. (Library of Congress LC-USZ62 10165 or LC-USZ62 26176.)

p. 114 Evariste Galois, from the Internet. This is a photograph of a heliogravure made by Dujardin, 1897, from a drawing by Alfred Galois. It appears in the frontispiece of *Ouevres mathématiques* of Galois.

p. 125 Carl Jacobi, Pour la Science (p. 65 of *Poincaré: Philosophe et Mathématicien*).

p. 138 Gottfried Leibniz, from the Internet.

Chapter 2

p. 180 Leonhard Euler, Pour la Science (p. 42 of *Poincaré: Philosophe et Mathématicien*).

p. 198 Carl Gauss, Pour la Science (p. 51 of *Poincaré: Philosophe et Mathématicien*).

p. 232 Isaac Newton, Pour la Science (p. 39 of *Poincaré: Philosophe et Mathématicien*).

p. 234 Augustin Cauchy, Pour la Science (p. 43 of *Poincaré: Philosophe et Mathématicien*).

p. 236 Rudolf Lipschitz, from the Internet.

p. 243 Leonid Kantorovich, from the Internet.

Chapter 3

p. 284 Felix Klein, Pour la Science (p. 44 of *Poincaré: Philosophe et Mathématicien*).

p. 315 Taylor, from the Internet. This picture appeared in 1793 in Taylor's *Contemplio Philosophica* and was reprinted in *Brook Taylor: der Mathematiker und Philosoph*, Wurzburg, K. 1937.) The Royal Society of London holds reproduction rights to a photograph of a portrait by an unknown artist. A copy of the picture in this text was sent to the society, which did not respond as to whether the picture in this text is the one to which it holds rights.

p. 325 Edmund Landau, from the Internet. This picture appears, "courtesy of Mrs. Elizabeth Brody," in *Edmund Landau Collected Works*, vol. 1, Thales Verlag, probably 1985, GmbH D-4300 Essen. Attempts to contact the publisher were unsuccessful.

p. 353 Joseph Lagrange, Pour la Science (p. 33 of *Poincaré: Philosophe et Mathématicien*).

p. 369 Andrei Kolmogorov, from the Internet. Believed to be the photograph taken by Jürgen Moser that appeared in *Foundations of Mechanics* by Ralph H. Abraham.

p. 376 *Principal Component Analysis in the Eigenface Technique for Facial Recognition*, Kevin Huang, Trinity College Digital Repository, April 1, 2012. Reprinted with permission.

Chapter 4

p. 401 Lord Kelvin, Pour la Science (p. 64 of *Poincaré: Philosophe et Mathématicien*).

p. 402 Bernhard Riemann, Pour la Science (p. 46 of *Poincaré: Philosophe et Mathématicien*).

p. 419 Georges Buffon, from the Internet. The Royal Society of London has a negative of an engraving by Robert Hart from an original picture by Drouais, reproduction rights unknown.

p. 433 Ludwig Boltzmann, Pour la Science (p. 89 of *Poincaré: Philosophe et Mathématicien*).

p. 438 Guido Fubini, from the Internet. According to an archivist at the Institute for Advanced Studies (Princeton), this photograph appeared in volume one Fubini's collected works (*Opere Scelte*), published in 1957 by Edizioni Cremonese in Rome. According to Edizioni Cremonese, the copyright belongs to the Italian Mathematical Union.

p. 450 Thomas Simpson, from the Internet.

p. 501 Henri Lebesgue, from the Internet.

Chapter 5

p. 560 Felix Hausdorff, from the Internet.

Chapter 6

p. 586 August Möbius, from the Internet.

p. 587 Johann Listing, from the Internet.

p. 599 Josiah Willard Gibbs, from the Internet.

p. 611 fossil suture line, from Jean Guex, Institute of Geology, University of Lausanne, who reports that it is in the public domain.

p. 640 Elie Cartan, from the Internet.

p. 663 George Stokes, from the Internet. The picture was apparently taken in 1857; it appears in *Memoir and Scientific Correspondence of the Late Sir George Gabriel Stokes*, vol. 1, Cambridge: At the University Press, 1907.

p. 665 Mikhail Ostrogradski, from the Internet.

p. 666 Archimedes, from the Internet.

p. 671 James Clerk Maxwell, from the Internet.

Appendix

p. 707 Guiseppe Peano, Pour la Science (p. 19 of *Poincaré: Philosophe et Mathématicien*), from the University of Turin.

p. 714 Eduard Heine, from the Internet.

p. 714 Emile Borel, from the Internet. This picture appeared in *An Illustrated History 1893–1986*, D. J. Albers, G. L. Alexanderson, C. Reid, p. 15, 1986, © Springer-Verlag); it is reproduced with permission from Springer and the authors.

p. 776 Hassler Whitney, from the Internet. This picture was taken by Don Albers and is reproduced with his permission.

INDEX

Bold page numbers indicate a page where a term is defined, formally or informally.
Page numbers in italics indicate that the term is used in a theorem, proposition, lemma, or corollary.